T0295397

Climate Change and Land

An IPCC Special Report on climate change, desertification, land degradation, sustainable land management, food security, and greenhouse gas fluxes in terrestrial ecosystems

Edited by

Valérie Masson-Delmotte
Co-Chair Working Group I

Panmao Zhai
Co-Chair Working Group I

Hans-Otto Pörtner
Co-Chair Working Group II

Debra Roberts
Co-Chair Working Group II

Jim Skea
Co-Chair Working Group III

Eduardo Calvo Buendía
Co-Chair TFI

Priyadarshi R. Shukla
Co-Chair Working Group III

Raphael Slade
Head of TSU (Science)

Sarah Connors
Senior Science Officer

Renée van Diemen
Scientist

Marion Ferrat
Head of Communications

Eamon Haughey
Senior Scientist

Sigourney Luz
Communications Manager

Suvadip Neogi
Scientist

Minal Pathak
Senior Scientist

Jan Petzold
Science Officer

Joana Portugal Pereira
Senior Scientist

Purvi Vyas
Science Officer

Elizabeth Huntley
Head of TSU (Operations)

Katie Kissick
Head of TSU (Operations)

Malek Belkacemi
IT/Web Manager

Juliette Malley
Senior Administrator

CAMBRIDGE
UNIVERSITY PRESS

University Printing House, Cambridge CB2 8BS, United Kingdom

One Liberty Plaza, 20th Floor, New York, NY 10006, USA

477 Williamstown Road, Port Melbourne, VIC 3207, Australia

314–321, 3rd Floor, Plot 3, Splendor Forum, Jasola District Centre, New Delhi – 110025, India

103 Penang Road, #05-06/07, Visioncrest Commercial, Singapore 238467

Cambridge University Press is part of the University of Cambridge.

It furthers the University's mission by disseminating knowledge in the pursuit of education,
learning, and research at the highest international levels of excellence.

www.cambridge.org
Information on this title: www.cambridge.org/9781009158015
DOI: 10.1017/9781009157988

First published 2022

Printed in the United Kingdom by TJ Books Limited, Padstow Cornwall

A catalogue record for this publication is available from the British Library.

ISBN 978-1-009-15801-5 Paperback

Cambridge University Press has no responsibility for the persistence or accuracy of URLs for external or third-party internet websites
referred to in this publication and does not guarantee that any content on such websites is, or will remain, accurate or appropriate.

Use the following reference to cite the entire volume:
IPCC, 2019: *Climate Change and Land: an IPCC special report on climate change, desertification, land degradation, sustainable land management, food security, and greenhouse gas fluxes in terrestrial ecosystems* [P.R. Shukla, J. Skea, E. Calvo Buendia, V. Masson-Delmotte, H.-O. Pörtner, D. C. Roberts, P. Zhai, R. Slade, S. Connors, R. van Diemen, M. Ferrat, E. Haughey, S. Luz, S. Neogi, M. Pathak, J. Petzold, J. Portugal Pereira, P. Vyas, E. Huntley, K. Kissick, M. Belkacemi, J. Malley, (eds.)]. Cambridge University Press, Cambridge, UK and New York, NY, USA, 896 pp. https://doi.org/10.1017/9781009157988.

Electronic copies of this report are available from the IPCC website www.ipcc.ch

Front cover photograph: Agricultural landscape between Ankara and Hattusha, Anatolia, Turkey (40°00′N–33°35′E)
www.yannarthusbertrand.org | www.goodplanet.org. © Yann Arthus-Bertrand

Contents

Front Matter

Foreword V

Preface VII

SPM

Summary for Policymakers 3

TS

Technical Summary 37

Chapters

Chapter 1   Framing and context 77

Chapter 2   Land–climate interactions 131

Chapter 3   Desertification 249

Chapter 4   Land degradation 345

Chapter 5   Food security 437

Chapter 6   Interlinkages between desertification, land degradation, food security and greenhouse gas fluxes: Synergies, trade-offs and integrated response options 551

Chapter 7   Risk management and decision-making in relation to sustainable development 673

Annexes

Annex I   Glossary 803

Annex II   Acronyms 831

Annex III   Contributors to the IPCC Special Report on Climate Change and Land 839

Annex IV   Reviewers of the IPCC Special Report on Climate Change and Land 847

Index 865

Foreword
and Preface

Foreword

Climate Change and Land: an IPCC special report on climate change, desertification, land degradation, sustainable land management, food security, and greenhouse gas fluxes in terrestrial ecosystems, also known as the Special Report on Climate Change and Land (SRCCL), is the second Special Report to be produced in the Intergovernmental Panel on Climate Change's (IPCC) Sixth Assessment Cycle (AR6). It has been jointly produced by IPCC Working Groups I, II and III in association with the Task Force on National Greenhouse Gas Inventories.

Over two years in the making, this report highlights the multiple interactions between climate change and land. It assesses the dynamics of the land-climate system, and the economic and social dimensions of addressing the challenges of land degradation, desertification and food security in a changing climate. It also assesses the options for governance and decision-making across multiple scales. This report is interdisciplinary in nature and brings together an unprecedented number of experts from varying fields of research. Their expertise ranges from agricultural systems and rural livelihoods to nutrition and forestry. Over 52 different countries from all regions of the world were represented in the chapter teams, and, for the first time in an IPCC report, a majority of authors – 53% – were from developing countries. This reflects the important role that developing countries play in climate change research and decision-making, particularly in the context of land.

The IPCC provides policymakers with regular scientific assessments on climate change, its implications and risks, as well as adaptation and mitigation options. Since it was established jointly in 1988 by the World Meteorological Organization (WMO) and the United Nations Environment Programme (UNEP), the IPCC has produced a series of Assessment Reports, Special Reports, Technical Papers and Methodological Reports which have informed international negotiations and actions to tackle climate change.

The participation and collaboration of hundreds of experts worldwide underpins the success of IPCC reports. It is their knowledge, enthusiasm and dedication, as well as their willingness to work across disciplines, which gives IPCC reports their impact and policy relevance. We would like to express our gratitude to all the Coordinating Lead Authors, Lead Authors, Contributing Authors, Review Editors, Chapter Scientists and Expert and Government Reviewers who devoted their time and effort to make the Special Report on Climate Change and Land possible. We would also like to thank the members of the IPCC Bureau, especially members of the SRCCL Steering Committee, for their scientific leadership and support. Last, but by no means least, we would like to thank the staff of the Working Group I, II and III Technical Support Units and the IPCC Secretariat for their unwavering commitment to the development of this IPCC Special Report.

This report would not have been possible without governments supporting their scientists' participation in this process, contributing to the IPCC, hosting meetings and facilitating the essential participation of authors and experts from developing countries. We would like to share our appreciation to the government of Ireland for hosting the Scoping Meeting and to the governments of Norway, New Zealand, Ireland and Colombia for hosting Lead Author Meetings. Our thanks also to the governments of France, Germany, the United Kingdom and Japan for funding the Technical Support Units of Working Groups I, II and III, and the Task Force on National Greenhouse Gas Inventories, respectively. We also acknowledge the government of Norway's generous support for communications and outreach activities, and the support of the Irish Environmental Protection Agency for an additional post in the Working Group III Technical Support Unit.

We especially wish to thank the IPCC Chair, Hoesung Lee, for his overall leadership, the IPCC Vice-Chairs Youba Sokona, Thelma Krug and Ko Barrett for their guidance and deep knowledge of the IPCC, and the Co-Chairs of Working Groups I, II and III Valérie Masson-Delmotte, Panmao Zhai, Hans-Otto Pörtner, Debra Roberts, Jim Skea and Priyadarshi Shukla, as well as Eduardo Calvo Buendía, Co-Chair of the TFI for their tireless leadership throughout the process.

We are also grateful for the very professional work of the IPCC Secretariat and WMO LCP Department in facilitating the work and numerous meetings.

Petteri Taalas
Secretary-General
World Meteorological Organization

Inger Andersen
Executive Director
United Nations Environment Programme

Preface

This IPCC special report on climate change, desertification, land degradation, sustainable land management, food security, and greenhouse gas fluxes in terrestrial ecosystems, also known as the Special Report on Climate Change and Land (SRCCL), is the second Special Report to be produced in the Intergovernmental Panel on Climate Change's (IPCC) Sixth Assessment Cycle (AR6). The report was jointly prepared by Working Groups I, II and III in association with the Task Force on National Greenhouse Gas Inventories (TFI). The Working Group III Technical Support Unit was responsible for logistical and technical support for the preparation of this Special Report. This Special Report builds upon the IPCC's Fifth Assessment Report (AR5) in 2013–2014 and on relevant research subsequently published in the scientific, technical and socio-economic literature. It was prepared following IPCC principles and procedures. This Special Report is the second of three cross-Working Group Special Reports to be published in the AR6, accompanying the three main Working Group Reports, the Synthesis Report and a Refinement to the 2006 IPCC Guidelines for National Greenhouse Gas Inventories.

Scope of the Report

Previous IPCC reports made reference to land and its role in the climate system. Threats to agriculture, forestry and other ecosystems, but also the role of land and forest management in climate change, have been documented since the IPCC Second Assessment Report, especially so in the Special Report on Land Use, Land-Use Change and Forestry. The IPCC Special Report on Extreme Events discussed sustainable land management, including land use planning and ecosystem management and restoration, among the potential low-regret measures that provide benefits under current climate and a range of future climate change scenarios. The IPCC SRCCL responds to proposals for Special Reports from governments and observer organisations provided at the start of the IPCC AR6. It addresses greenhouse gas (GHG) fluxes in terrestrial ecosystems and sustainable land management in relation to climate adaptation and mitigation, desertification, land degradation and food security. The report sits alongside other IPCC reports, including the Special Report on Global Warming of 1.5°C, the Special Report on Oceans and Cryosphere in a Changing Climate (SROCC), and related reports from other UN Bodies. It was produced giving careful attention to these other assessments, with the aim of achieving coherence and complementarity, as well as providing an updated assessment of the current state of knowledge. The Special Report is an assessment of the relevant state of knowledge, based on the scientific and technical literature available and accepted for publication up to 7 April 2019, totalling over 7,000 publications.

Structure of the Report

This report consists of a short Summary for Policymakers, a Technical Summary, seven Chapters, and Annexes, as well as online chapter Supplementary Material.

Chapter 1 provides a synopsis of the main issues addressed in the report, which are explored in more detail in Chapters 2–7. It also introduces important concepts and definitions and highlights discrepancies with previous reports that arise from different objectives.

Chapter 2 focuses on the natural system and dynamics, assessing recent progress towards understanding the impacts of climate change on land, and the feedbacks arising from biogeochemical and biophysical exchange fluxes.

Chapter 3 examines how the world's dryland populations are uniquely vulnerable to desertification and climate change, but also have significant knowledge in adapting to climate variability and addressing desertification.

Chapter 4 assesses the urgency of tackling land degradation across all land ecosystems. Despite accelerating trends of land degradation, reversing these trends is attainable through restoration efforts and proper implementation of sustainable land management, which is expected to improve resilience to climate change, mitigate climate change and ensure food security for generations to come.

Chapter 5 focuses on food security, with an assessment of the risks and opportunities that climate change presents to food systems, considering how mitigation and adaptation can contribute to both human and planetary health.

Chapter 6 focuses on the response options within the land system that deal with trade-offs and increase benefits in an integrated way in support of the Sustainable Development Goals.

Finally, Chapter 7 highlights these aspects further, by assessing the opportunities, decision making and policy responses to risks in the climate-land-human system.

The Process

The IPCC SRCCL was prepared in accordance with the principles and procedures established by the IPCC and represents the combined efforts of leading experts in the field of climate change. A scoping meeting for the SRCCL was held in Dublin, Ireland, in 2017, and the final outline was approved by the Panel at its 45th Session in March 2017 in Guadalajara, Mexico. Governments and IPCC observer organisations nominated 640 experts for the author team. The team of 15 Coordinating Lead Authors and 71 Lead Authors plus 21 Review Editors were selected by Working Groups I, II and III Bureaux, in collaboration with the Task Force on National Greenhouse Gas Inventories. In addition, 96 Contributing Authors were invited by chapter teams to provide technical information in the form of text, graphs or data for assessment. Report drafts prepared by the authors were subject to two rounds of formal review and revision followed by a final round of government comments on the Summary for Policymakers. The enthusiastic participation of the scientific community and governments to the review process resulted in more than 28,000 written review comments, submitted by 596 individual expert reviewers and 42 governments.

The Review Editors for the chapters monitored the review process to ensure that all substantive review comments received appropriate consideration. The Summary for Policymakers (SPM) was approved line-by-line at the joint meeting of Working Groups I, II and III; the SPM and the underlying chapters were then accepted at the 50th Session of the IPCC, 2–6 August 2019 in Geneva, Switzerland.

Acknowledgements

The Special Report on Climate Change and Land broke new ground for IPCC. It was the first IPCC report to be produced by all three Working Groups in collaboration with the Task Force on National Greenhouse Gas Inventories (TFI), and it was the first IPCC report with more authors from developing countries than authors from developed countries. It was marked by an inspiring degree of collaboration and interdisciplinarity, reflecting the wide scope of the mandate given to authors by the Panel. It brought together authors not only from the IPCC's traditional scientific communities, but also those from sister UN organisations including the Intergovernmental Science-Policy Platform on Biodiversity and Ecosystem Services (IPBES), the Science-Policy Interface of the UN Convention to Combat Desertification (UNCCD) and the Food and Agriculture Organization of the UN (FAO).

We must pay tribute to the 107 Coordinating Lead Authors, Lead Authors and Review Editors, from 52 countries, who were responsible for the report. They gave countless hours of their time, on a voluntary basis, and attended four Lead Author meetings in widely scattered parts of the globe. The constructive interplay between the authors, who draft the report, and the Review Editors, who provide assurance that all comments are responded to, greatly helped the process. Throughout, all demonstrated scientific rigour while at the same time maintaining good humour and a spirit of true collaboration. They did so against a very tight timetable which allowed no scope for slippage. They were supported by input from 96 Contributing Authors.

We would like to acknowledge especially the support of the Chapter Scientists who took time out from their emerging careers to support the production of the report. We thank Yuping Bai, Aliyu Barau, Erik Contreras, Abdoul Aziz Diouf, Baldur Janz, Frances Manning, Dorothy Nampanzira, Chuck Chuan Ng, Helen Paulos, Xiyan Xu and Thobekile Zikhali. We very much hope that the experience will help them in their future careers and that their vital role will be suitably recognised.

The production of the report was guided by a Steering Committee drawn from across the IPCC Bureau. We would like to thank our colleagues who served on this committee including: the Co-Chairs of Working Groups and the TFI: Priyadarshi Shukla, Jim Skea, Valérie Masson-Delmotte, Panmao Zhai, Hans-Otto Pörtner, Debra Roberts, Eduardo Calvo Buendía; Working Group Vice-Chairs: Mark Howden, Nagmeldin Mahmoud, Ramón Pichs-Madruga, Andy Reisinger, Noureddine Yassaa; and Youba Sokona, Vice-Chair of IPCC. Youba Sokona acted as champion for the report and his wise council was valued by all. Further support came from IPCC Bureau members: Edvin Aldrian, Fatima Driouech, Gregory Flato, Jan Fuglestvedt, Muhammad Tariq and Carolina Vera (Working Group I); Andreas Fischlin, Carlos Méndez, Joy Jacqueline Pereira, Roberto A. Sánchez-Rodríguez, Sergey Semenov, Pius Yanda and Taha M. Zatari (Working Group II); and Amjad Abdulla, Carlo Carraro, Diriba Korecha Dadi and Diana Ürge-Vorsatz (Working Group III).

Several governments and other bodies hosted and supported the scoping meeting, the four Lead Author meetings, and the final IPCC Plenary. These were: the Government of Norway and the Norwegian Environment Agency, the Government of New Zealand and the University of Canterbury, the Government of Ireland and the Environmental Protection Agency, the Government of Colombia and the International Centre for Tropical Agriculture (CIAT), the Government of Switzerland and the World Meteorological Organization.

The staff of the IPCC Secretariat based in Geneva provided a wide range of support for which we would like to thank Abdalah Mokssit, Secretary of the IPCC, and his colleagues: Kerstin Stendahl, Jonathan Lynn, Sophie Schlingemann, Jesbin Baidya, Laura Biagioni, Annie Courtin, Oksana Ekzarkho, Judith Ewa, Joelle Fernandez, Andrea Papucides Bach, Nina Peeva, Mxolisi Shongwe, and Werani Zabula. Thanks are due to Elhousseine Gouaini who served as the conference officer for the 50th Session of the IPCC.

A number of individuals provided support for the visual elements of the report and its communication. We would single out Jordan Harold of the University of East Anglia, Susan Escott of Escott Hunt Ltd, Angela Morelli and Tom Gabriel Johansen of Info Design Lab, and Polly Jackson, Ian Blenkinsop, Autumn Forecast, Francesca Romano and Alice Woodward of Soapbox Communications Ltd.

The report was managed by the Technical Support Unit of IPCC Working Group III which has the generous financial support of the UK Engineering and Physical Sciences Research Council (EPSRC) and the UK Government through its Department of Business, Energy and Industrial Strategy (BEIS). In addition, the Irish Environmental Protection Agency provided support for two secondees to the WG III Technical Support Unit, while the Norwegian Environment Agency enabled an expanded set of communication activities. Without the support of all these bodies this report would not have been possible.

Our particular appreciation goes to the Working Group Technical Support Units whose tireless dedication, professionalism and enthusiasm led the production of this Special Report. This Report could not have been prepared without the commitment of members of the Working Group III Technical Support Unit, all new to the IPCC, who rose to the unprecedented Sixth Assessment Report challenge and were pivotal in all aspects of the preparation of the Report: Raphael Slade, Lizzie Huntley, Katie Kissick, Malek Belkacemi, Renée van Diemen, Marion Ferrat, Eamon Haughey, Bhushan Kankal, Géninha Lisboa, Sigourney Luz, Juliette Malley, Suvadip Neogi, Minal Pathak, Joana Portugal Pereira and Purvi Vyas. Our warmest thanks go to the collegial and collaborative support provided by Sarah Connors, Melissa Gomis, Robin Matthews, Wilfran Moufouma-Okia, Clotilde Péan, Roz Pidcock, Anna Pirani, Tim Waterfield and Baiquan Zhou from the WG I Technical Support Unit, and Jan Petzold, Bard Rama, Maike Nicolai, Elvira Poloczanska, Melinda Tignor and Nora Weyer from the WG II Technical Support Unit.

And a final deep thanks to family and friends who indirectly supported the work by tolerating the periods authors spent away from home, the long hours and their absorption in the process of producing this report.

SIGNED

Valérie Masson-Delmotte
Co-Chair Working Group I

Panmao Zhai
Co-Chair Working Group I

Hans-Otto Pörtner
Co-Chair Working Group II

Debra Roberts
Co-Chair Working Group II

Jim Skea
Co-Chair Working Group III

Eduardo Calvo Buendía
Co-Chair TFI

Priyadarshi R. Shukla
Co-Chair Working Group III

Summary for Policymakers

Summary
for Policymakers

Drafting Authors:
Almut Arneth (Germany), Humberto Barbosa (Brazil), Tim Benton (United Kingdom), Katherine Calvin (The United States of America), Eduardo Calvo (Peru), Sarah Connors (United Kingdom), Annette Cowie (Australia), Edouard Davin (France/Switzerland), Fatima Denton (The Gambia), Renée van Diemen (The Netherlands/United Kingdom), Fatima Driouech (Morocco), Aziz Elbehri (Morocco), Jason Evans (Australia), Marion Ferrat (France), Jordan Harold (United Kingdom), Eamon Haughey (Ireland), Mario Herrero (Australia/Costa Rica), Joanna House (United Kingdom), Mark Howden (Australia), Margot Hurlbert (Canada), Gensuo Jia (China), Tom Gabriel Johansen (Norway), Jagdish Krishnaswamy (India), Werner Kurz (Canada), Christopher Lennard (South Africa), Soojeong Myeong (Republic of Korea), Nagmeldin Mahmoud (Sudan), Valérie Masson-Delmotte (France), Cheikh Mbow (Senegal), Pamela McElwee (The United States of America), Alisher Mirzabaev (Germany/Uzbekistan), Angela Morelli (Norway/Italy), Wilfran Moufouma-Okia (France), Dalila Nedjraoui (Algeria), Suvadip Neogi (India), Johnson Nkem (Cameroon), Nathalie De Noblet-Ducoudré (France), Lennart Olsson (Sweden), Minal Pathak (India), Jan Petzold (Germany), Ramón Pichs-Madruga (Cuba), Elvira Poloczanska (United Kingdom/Australia), Alexander Popp (Germany), Hans-Otto Pörtner (Germany), Joana Portugal Pereira (United Kingdom), Prajal Pradhan (Nepal/Germany), Andy Reisinger (New Zealand), Debra C. Roberts (South Africa), Cynthia Rosenzweig (The United States of America), Mark Rounsevell (United Kingdom/Germany), Elena Shevliakova (The United States of America), Priyadarshi R. Shukla (India), Jim Skea (United Kingdom), Raphael Slade (United Kingdom), Pete Smith (United Kingdom), Youba Sokona (Mali), Denis Jean Sonwa (Cameroon), Jean-Francois Soussana (France), Francesco Tubiello (The United States of America/Italy), Louis Verchot (The United States of America/Colombia), Koko Warner (The United States of America/Germany), Nora M. Weyer (Germany), Jianguo Wu (China), Noureddine Yassaa (Algeria), Panmao Zhai (China), Zinta Zommers (Latvia).

This Summary for Policymakers should be cited as:
IPCC, 2019: Summary for Policymakers. In: *Climate Change and Land: an IPCC special report on climate change, desertification, land degradation, sustainable land management, food security, and greenhouse gas fluxes in terrestrial ecosystems* [P.R. Shukla, J. Skea, E. Calvo Buendia, V. Masson-Delmotte, H.- O. Pörtner, D. C. Roberts, P. Zhai, R. Slade, S. Connors, R. van Diemen, M. Ferrat, E. Haughey, S. Luz, S. Neogi, M. Pathak, J. Petzold, J. Portugal Pereira, P. Vyas, E. Huntley, K. Kissick, M. Belkacemi, J. Malley, (eds.)]. https://doi.org/10.1017/9781009157988.001

Acknowledgements

The Special Report on Climate Change and Land broke new ground for IPCC. It was the first IPCC report to be produced by all three Working Groups in collaboration with the Task Force on National Greenhouse Gas Inventories (TFI), and it was the first IPCC report with more authors from developing countries than authors from developed countries. It was marked by an inspiring degree of collaboration and interdisciplinarity, reflecting the wide scope of the mandate given to authors by the Panel. It brought together authors not only from the IPCC's traditional scientific communities, but also those from sister UN organisations including the Intergovernmental Science-Policy Platform on Biodiversity and Ecosystem Services (IPBES), the Science-Policy Interface of the UN Convention to Combat Desertification (UNCCD) and the Food and Agriculture Organization of the UN (FAO).

We must pay tribute to the 107 Coordinating Lead Authors, Lead Authors and Review Editors, from 52 countries, who were responsible for the report. They gave countless hours of their time, on a voluntary basis, and attended four Lead Author meetings in widely scattered parts of the globe. The constructive interplay between the authors, who draft the report, and the Review Editors, who provide assurance that all comments are responded to, greatly helped the process. Throughout, all demonstrated scientific rigour while at the same time maintaining good humour and a spirit of true collaboration. They did so against a very tight timetable which allowed no scope for slippage. They were supported by input from 96 Contributing Authors.

We would like to acknowledge especially the support of the Chapter Scientists who took time out from their emerging careers to support the production of the report. We thank Yuping Bai, Aliyu Barau, Erik Contreras, Abdoul Aziz Diouf, Baldur Janz, Frances Manning, Dorothy Nampanzira, Chuck Chuan Ng, Helen Paulos, Xiyan Xu and Thobekile Zikhali. We very much hope that the experience will help them in their future careers and that their vital role will be suitably recognised.

The production of the report was guided by a Steering Committee drawn from across the IPCC Bureau. We would like to thank our colleagues who served on this committee including: the Co-Chairs of Working Groups and the TFI: Priyadarshi Shukla, Jim Skea, Valérie Masson-Delmotte, Panmao Zhai, Hans-Otto Pörtner, Debra Roberts, Eduardo Calvo Buendía; Working Group Vice-Chairs: Mark Howden, Nagmeldin Mahmoud, Ramón Pichs-Madruga, Andy Reisinger, Noureddine Yassaa; and Youba Sokona, Vice-Chair of IPCC. Youba Sokona acted as champion for the report and his wise council was valued by all. Further support came from IPCC Bureau members: Edvin Aldrian, Fatima Driouech, Gregory Flato, Jan Fuglestvedt, Muhammad Tariq and Carolina Vera (Working Group I); Andreas Fischlin, Carlos Méndez, Joy Jacqueline Pereira, Roberto A. Sánchez-Rodríguez, Sergey Semenov, Pius Yanda and Taha M. Zatari (Working Group II); and Amjad Abdulla, Carlo Carraro, Diriba Korecha Dadi and Diana Ürge-Vorsatz (Working Group III).

Several governments and other bodies hosted and supported the scoping meeting, the four Lead Author meetings, and the final IPCC Plenary. These were: the Government of Norway and the Norwegian Environment Agency, the Government of New Zealand and the University of Canterbury, the Government of Ireland and the Environmental Protection Agency, the Government of Colombia and the International Centre for Tropical Agriculture (CIAT), the Government of Switzerland and the World Meteorological Organization.

The staff of the IPCC Secretariat based in Geneva provided a wide range of support for which we would like to thank Abdalah Mokssit, Secretary of the IPCC, and his colleagues: Kerstin Stendahl, Jonathan Lynn, Sophie Schlingemann, Jesbin Baidya, Laura Biagioni, Annie Courtin, Oksana Ekzarkho, Judith Ewa, Joelle Fernandez, Andrea Papucides Bach, Nina Peeva, Mxolisi Shongwe, and Werani Zabula. Thanks are due to Elhousseine Gouaini who served as the conference officer for the 50th Session of the IPCC.

A number of individuals provided support for the visual elements of the report and its communication. We would single out Jordan Harold of the University of East Anglia, Susan Escott of Escott Hunt Ltd, Angela Morelli and Tom Gabriel Johansen of Info Design Lab, and Polly Jackson, Ian Blenkinsop, Autumn Forecast, Francesca Romano and Alice Woodward of Soapbox Communications Ltd.

The report was managed by the Technical Support Unit of IPCC Working Group III which has the generous financial support of the UK Engineering and Physical Sciences Research Council (EPSRC) and the UK Government through its Department of Business, Energy and Industrial Strategy (BEIS). In addition, the Irish Environmental Protection Agency provided support for two secondees to the WG III Technical Support Unit, while the Norwegian Environment Agency enabled an expanded set of communication activities. Without the support of all these bodies this report would not have been possible.

Our particular appreciation goes to the Working Group Technical Support Units whose tireless dedication, professionalism and enthusiasm led the production of this Special Report. This Report could not have been prepared without the commitment of members of the Working Group III Technical Support Unit, all new to the IPCC, who rose to the unprecedented Sixth Assessment Report challenge and were pivotal in all aspects of the preparation of the Report: Raphael Slade, Lizzie Huntley, Katie Kissick, Malek Belkacemi, Renée van Diemen, Marion Ferrat, Eamon Haughey, Bhushan Kankal, Géninha Lisboa, Sigourney Luz, Juliette Malley, Suvadip Neogi, Minal Pathak, Joana Portugal Pereira and Purvi Vyas. Our warmest thanks go to the collegial and collaborative support provided by Sarah Connors, Melissa Gomis, Robin Matthews, Wilfran Moufouma-Okia, Clotilde Péan, Roz Pidcock, Anna Pirani, Tim Waterfield and Baiquan Zhou from the WG I Technical Support Unit, and Jan Petzold, Bard Rama, Maike Nicolai, Elvira Poloczanska, Melinda Tignor and Nora Weyer from the WG II Technical Support Unit.

And a final deep thanks to family and friends who indirectly supported the work by tolerating the periods authors spent away from home, the long hours and their absorption in the process of producing this report.

SIGNED

Valérie Masson-Delmotte
Co-Chair Working Group I

Panmao Zhai
Co-Chair Working Group I

Hans-Otto Pörtner
Co-Chair Working Group II

Debra Roberts
Co-Chair Working Group II

Jim Skea
Co-Chair Working Group III

Eduardo Calvo Buendía
Co-Chair TFI

Priyadarshi R. Shukla
Co-Chair Working Group III

Introduction

This Special Report on Climate Change and Land[1] responds to the Panel decision in 2016 to prepare three Special Reports[2] during the Sixth Assessment cycle, taking account of proposals from governments and observer organisations.[3] This report addresses greenhouse gas (GHG) fluxes in land-based ecosystems, land use and sustainable land management[4] in relation to climate change adaptation and mitigation, desertification[5], land degradation[6] and food security[7]. This report follows the publication of other recent reports, including the IPCC *Special Report on Global Warming of 1.5°C* (SR15), the thematic assessment of the Intergovernmental Science-Policy Platform on Biodiversity and Ecosystem Services (IPBES) on *Land Degradation and Restoration*, the IPBES *Global Assessment Report on Biodiversity and Ecosystem Services*, and the *Global Land Outlook* of the UN Convention to Combat Desertification (UNCCD). This report provides an updated assessment of the current state of knowledge[8] while striving for coherence and complementarity with other recent reports.

This Summary for Policymakers (SPM) is structured in four parts: *A) People, land and climate in a warming world; B) Adaptation and mitigation response options; C) Enabling response options; and, D) Action in the near-term.*

Confidence in key findings is indicated using the IPCC calibrated language; the underlying scientific basis of each key finding is indicated by references to the main report.[9]

---

[1] The terrestrial portion of the biosphere that comprises the natural resources (soil, near-surface air, vegetation and other biota, and water), the ecological processes, topography, and human settlements and infrastructure that operate within that system.

[2] The three Special reports are: *Global Warming of 1.5°C: an IPCC special report on the impacts of global warming of 1.5°C above pre-industrial levels and related global greenhouse gas emission pathways, in the context of strengthening the global response to the threat of climate change, sustainable development, and efforts to eradicate poverty; Climate Change and Land: an IPCC special report on climate change, desertification, land degradation, sustainable land management, food security, and greenhouse gas fluxes in terrestrial ecosystems; The Ocean and Cryosphere in a Changing Climate.*

[3] Related proposals were: climate change and desertification; desertification with regional aspects; land degradation – an assessment of the interlinkages and integrated strategies for mitigation and adaptation; agriculture, forestry and other land use; food and agriculture; and food security and climate change.

[4] Sustainable land management is defined in this report as 'the stewardship and use of land resources, including soils, water, animals and plants, to meet changing human needs, while simultaneously ensuring the long-term productive potential of these resources and the maintenance of their environmental functions'.

[5] Desertification is defined in this report as 'land degradation in arid, semi-arid, and dry sub-humid areas resulting from many factors, including climatic variations and human activities'.

[6] Land degradation is defined in this report as 'a negative trend in land condition, caused by direct or indirect human induced processes, including anthropogenic climate change, expressed as long-term reduction and as loss of at least one of the following: biological productivity; ecological integrity; or value to humans'.

[7] Food security is defined in this report as 'a situation that exists when all people, at all times, have physical, social, and economic access to sufficient, safe and nutritious food that meets their dietary needs and food preferences for an active and healthy life'.

[8] The assessment covers literature accepted for publication by 7th April 2019.

[9] Each finding is grounded in an evaluation of underlying evidence and agreement. A level of confidence is expressed using five qualifiers: very low, low, medium, high and very high, and typeset in italics, for example, *medium confidence*. The following terms have been used to indicate the assessed likelihood of an outcome or a result: virtually certain 99–100% probability, very likely 90–100%, likely 66–100%, about as likely as not 33–66%, unlikely 0–33%, very unlikely 0–10%, exceptionally unlikely 0–1%. Additional terms (extremely likely 95–100%, more likely than not >50–100%, more unlikely than likely 0–<50%, extremely unlikely 0–5%) may also be used when appropriate. Assessed likelihood is typeset in italics, for example, very likely. This is consistent with IPCC AR5.

A. People, land and climate in a warming world

A.1 **Land provides the principal basis for human livelihoods and well-being including the supply of food, freshwater and multiple other ecosystem services, as well as biodiversity. Human use directly affects more than 70% (*likely* 69–76%) of the global, ice-free land surface (*high confidence*). Land also plays an important role in the climate system. (Figure SPM.1) {1.1, 1.2, 2.3, 2.4}**

A.1.1   People currently use one quarter to one third of land's potential net primary production[10] for food, feed, fibre, timber and energy. Land provides the basis for many other ecosystem functions and services,[11] including cultural and regulating services, that are essential for humanity (*high confidence*). In one economic approach, the world's terrestrial ecosystem services have been valued on an annual basis to be approximately equivalent to the annual global Gross Domestic Product[12] (*medium confidence*). (Figure SPM.1) {1.1, 1.2, 3.2, 4.1, 5.1, 5.5}

A.1.2   Land is both a source and a sink of GHGs and plays a key role in the exchange of energy, water and aerosols between the land surface and atmosphere. Land ecosystems and biodiversity are vulnerable to ongoing climate change, and weather and climate extremes, to different extents. Sustainable land management can contribute to reducing the negative impacts of multiple stressors, including climate change, on ecosystems and societies (*high confidence*). (Figure SPM.1) {1.1, 1.2, 3.2, 4.1, 5.1, 5.5}

A.1.3   Data available since 1961[13] show that global population growth and changes in per capita consumption of food, feed, fibre, timber and energy have caused unprecedented rates of land and freshwater use (*very high confidence*) with agriculture currently accounting for ca. 70% of global fresh-water use (*medium confidence*). Expansion of areas under agriculture and forestry, including commercial production, and enhanced agriculture and forestry productivity have supported consumption and food availability for a growing population (*high confidence*). With large regional variation, these changes have contributed to increasing net GHG emissions (*very high confidence*), loss of natural ecosystems (e.g., forests, savannahs, natural grasslands and wetlands) and declining biodiversity (*high confidence*). (Figure SPM.1) {1.1, 1.3, 5.1, 5.5}

A.1.4   Data available since 1961 shows the per capita supply of vegetable oils and meat has more than doubled and the supply of food calories per capita has increased by about one third (*high confidence*). Currently, 25–30% of total food produced is lost or wasted (*medium confidence*). These factors are associated with additional GHG emissions (*high confidence*). Changes in consumption patterns have contributed to about two billion adults now being overweight or obese (*high confidence*). An estimated 821 million people are still undernourished (*high confidence*). (Figure SPM.1) {1.1, 1.3, 5.1, 5.5}

A.1.5   About a quarter of the Earth's ice-free land area is subject to human-induced degradation (*medium confidence*). Soil erosion from agricultural fields is estimated to be currently 10 to 20 times (no tillage) to more than 100 times (conventional tillage) higher than the soil formation rate (*medium confidence*). Climate change exacerbates land degradation, particularly in low-lying coastal areas, river deltas, drylands and in permafrost areas (*high confidence*). Over the period 1961–2013, the annual area of drylands in drought has increased, on average by slightly more than 1% per year, with large inter-annual variability. In 2015, about 500 (380-620) million people lived within areas which experienced desertification between the 1980s and 2000s. The highest numbers of people affected are in South and East Asia, the circum Sahara region including North Africa, and the Middle East including the Arabian Peninsula (*low confidence*). Other dryland regions have also experienced desertification. People living in already degraded or desertified areas are increasingly negatively affected by climate change (*high confidence*). (Figure SPM.1) {1.1, 1.2, 3.1, 3.2, 4.1, 4.2, 4.3}

---

[10]   Land's potential net primary production (NPP) is defined in this report as 'the amount of carbon accumulated through photosynthesis minus the amount lost by plant respiration over a specified time period that would prevail in the absence of land use'.

[11]   In its conceptual framework, IPBES uses 'nature's contribution to people' in which it includes ecosystem goods and services.

[12]   I.e., estimated at $75 trillion for 2011, based on US dollars for 2007.

[13]   This statement is based on the most comprehensive data from national statistics available within FAOSTAT, which starts in 1961. This does not imply that the changes started in 1961. Land use changes have been taking place from well before the pre-industrial period to the present.

Land use and observed climate change

A. Observed temperature change relative to 1850-1900

Since the pre-industrial period (1850-1900) the observed mean land surface air temperature has risen considerably more than the global mean surface (land and ocean) temperature (GMST).

CHANGE in TEMPERATURE rel. to 1850-1900 (°C)

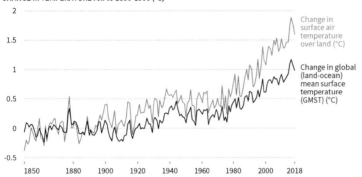

Change in surface air temperature over land (°C)

Change in global (land-ocean) mean surface temperature (GMST) (°C)

B. GHG emissions

An estimated 23% of total anthropogenic greenhouse gas emissions (2007-2016) derive from Agriculture, Forestry and Other Land Use (AFOLU).

CHANGE in EMISSIONS since 1961
1. Net CO₂ emissions from FOLU (GtCO₂ yr⁻¹)
2. CH₄ emissions from Agriculture (GtCO₂eq yr⁻¹)
3. N₂O emissions from Agriculture (GtCO₂eq yr⁻¹)

GtCO₂eq yr⁻¹

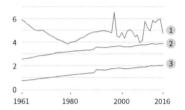

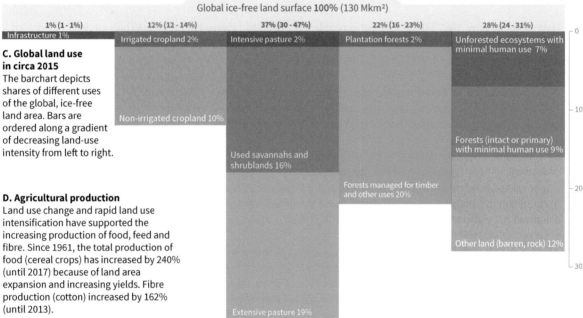

Global ice-free land surface 100% (130 Mkm²)

| 1% (1 - 1%) | 12% (12 - 14%) | 37% (30 - 47%) | 22% (16 - 23%) | 28% (24 - 31%) |

Infrastructure 1%
Irrigated cropland 2%
Intensive pasture 2%
Plantation forests 2%
Unforested ecosystems with minimal human use 7%
Non-irrigated cropland 10%
Used savannahs and shrublands 16%
Forests (intact or primary) with minimal human use 9%
Forests managed for timber and other uses 20%
Other land (barren, rock) 12%
Extensive pasture 19%

C. Global land use in circa 2015

The barchart depicts shares of different uses of the global, ice-free land area. Bars are ordered along a gradient of decreasing land-use intensity from left to right.

D. Agricultural production

Land use change and rapid land use intensification have supported the increasing production of food, feed and fibre. Since 1961, the total production of food (cereal crops) has increased by 240% (until 2017) because of land area expansion and increasing yields. Fibre production (cotton) increased by 162% (until 2013).

CHANGE in % rel. to 1961
1. Inorganic N fertiliser use
2. Cereal yields
3. Irrigation water volume
4. Total number of ruminant livestock

%

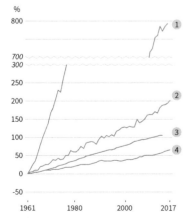

E. Food demand

Increases in production are linked to consumption changes.

CHANGE in % rel. to 1961 and 1975
1. Population
2. Prevalence of overweight + obese
3. Total calories per capita
4. Prevalence of underweight

%

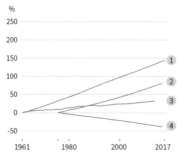

F. Desertification and land degradation

Land-use change, land-use intensification and climate change have contributed to desertification and land degradation.

CHANGE in % rel. to 1961 and 1970
1. Population in areas experiencing desertification
2. Dryland areas in drought annually
3. Inland wetland extent

%

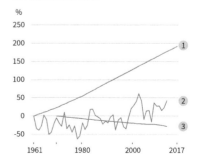

Figure SPM.1: Land use and observed climate change | A representation of the land use and observed climate change covered in this assessment report. Panels A-F show the status and trends in selected land use and climate variables that represent many of the core topics covered in this report. The annual time series in B and D-F are based on the most comprehensive, available data from national statistics, in most cases from FAOSTAT which starts in 1961. Y-axes in panels D-F are expressed relative to the starting year of the time series (rebased to zero). Data sources and notes: **A:** The warming curves are averages of four datasets {2.1, Figure 2.2, Table 2.1} **B:** N_2O and CH_4 from agriculture are from FAOSTAT; Net CO_2 emissions from FOLU using the mean of two bookkeeping models (including emissions from peatland fires since 1997). All values expressed in units of CO_2-eq are based on AR5 100-year Global Warming Potential values without climate-carbon feedbacks (N_2O=265; CH_4=28). (Table SPM.1) {1.1, 2.3} **C:** Depicts shares of different uses of the global, ice-free land area for approximately the year 2015, ordered along a gradient of decreasing land-use intensity from left to right. Each bar represents a broad land cover category; the numbers on top are the total percentage of the ice-free area covered, with uncertainty ranges in brackets. Intensive pasture is defined as having a livestock density greater than 100 animals/km². The area of 'forest managed for timber and other uses' was calculated as total forest area minus 'primary/intact' forest area. {1.2, Table 1.1, Figure 1.3} **D:** Note that fertiliser use is shown on a split axis. The large percentage change in fertiliser use reflects the low level of use in 1961 and relates to both increasing fertiliser input per area as well as the expansion of fertilised cropland and grassland to increase food production. {1.1, Figure 1.3} **E:** Overweight population is defined as having a body mass index (BMI) > 25 kg m⁻²; underweight is defined as BMI < 18.5 kg m⁻². {5.1, 5.2} **F:** Dryland areas were estimated using TerraClimate precipitation and potential evapotranspiration (1980-2015) to identify areas where the Aridity Index is below 0.65. Population data are from the HYDE3.2 database. Areas in drought are based on the 12-month accumulation Global Precipitation Climatology Centre Drought Index. The inland wetland extent (including peatlands) is based on aggregated data from more than 2000 time series that report changes in local wetland area over time. {3.1, 4.2, 4.6}

A.2   **Since the pre-industrial period, the land surface air temperature has risen nearly twice as much as the global average temperature (*high confidence*). Climate change, including increases in frequency and intensity of extremes, has adversely impacted food security and terrestrial ecosystems as well as contributed to desertification and land degradation in many regions (*high confidence*). {2.2, 3.2, 4.2, 4.3, 4.4, 5.1, 5.2, Executive Summary Chapter 7, 7.2}**

A.2.1   Since the pre-industrial period (1850-1900) the observed mean land surface air temperature has risen considerably more than the global mean surface (land and ocean) temperature (GMST) (*high confidence*). From 1850-1900 to 2006-2015 mean land surface air temperature has increased by 1.53°C (*very likely* range from 1.38°C to 1.68°C) while GMST increased by 0.87°C (*likely* range from 0.75°C to 0.99°C). (Figure SPM.1) {2.2.1}

A.2.2   Warming has resulted in an increased frequency, intensity and duration of heat-related events, including heatwaves[14] in most land regions (*high confidence*). Frequency and intensity of droughts has increased in some regions (including the Mediterranean, west Asia, many parts of South America, much of Africa, and north-eastern Asia) (*medium confidence*) and there has been an increase in the intensity of heavy precipitation events at a global scale (*medium confidence*). {2.2.5, 4.2.3, 5.2}

A.2.3   Satellite observations[15] have shown vegetation greening[16] over the last three decades in parts of Asia, Europe, South America, central North America, and southeast Australia. Causes of greening include combinations of an extended growing season, nitrogen deposition, Carbon Dioxide (CO_2) fertilisation[17], and land management (*high confidence*). Vegetation browning[18] has been observed in some regions including northern Eurasia, parts of North America, Central Asia and the Congo Basin, largely as a result of water stress (*medium confidence*). Globally, vegetation greening has occurred over a larger area than vegetation browning (*high confidence*). {2.2.3, Box 2.3, 2.2.4, 3.2.1, 3.2.2, 4.3.1, 4.3.2, 4.6.2, 5.2.2}

A.2.4   The frequency and intensity of dust storms have increased over the last few decades due to land use and land cover changes and climate-related factors in many dryland areas resulting in increasing negative impacts on human health, in regions such as the Arabian Peninsula and broader Middle East, Central Asia (*high confidence*).[19] {2.4.1, 3.4.2}

A.2.5   In some dryland areas, increased land surface air temperature and evapotranspiration and decreased precipitation amount, in interaction with climate variability and human activities, have contributed to desertification. These areas include Sub-Saharan Africa, parts of East and Central Asia, and Australia. (*medium confidence*) {2.2, 3.2.2, 4.4.1}

---

14   A heatwave is defined in this report as 'a period of abnormally hot weather'. Heatwaves and warm spells have various and, in some cases, overlapping definitions.

15   The interpretation of satellite observations can be affected by insufficient ground validation and sensor calibration. In addition their spatial resolution can make it difficult to resolve small-scale changes.

16   Vegetation greening is defined in this report as 'an increase in photosynthetically active plant biomass which is inferred from satellite observations'.

17   CO_2 fertilisation is defined in this report as 'the enhancement of plant growth as a result of increased atmospheric carbon dioxide (CO_2) concentration'. The magnitude of CO_2 fertilisation depends on nutrients and water availability.

18   Vegetation browning is defined in this report as 'a decrease in photosynthetically active plant biomass which is inferred from satellite observations'.

19   Evidence relative to such trends in dust storms and health impacts in other regions is limited in the literature assessed in this report.

A.2.6   Global warming has led to shifts of climate zones in many world regions, including expansion of arid climate zones and contraction of polar climate zones (*high confidence*). As a consequence, many plant and animal species have experienced changes in their ranges, abundances, and shifts in their seasonal activities (*high confidence*). {2.2, 3.2.2, 4.4.1}

A.2.7   Climate change can exacerbate land degradation processes (*high confidence*) including through increases in rainfall intensity, flooding, drought frequency and severity, heat stress, dry spells, wind, sea-level rise and wave action, and permafrost thaw with outcomes being modulated by land management. Ongoing coastal erosion is intensifying and impinging on more regions with sea-level rise adding to land use pressure in some regions (*medium confidence*). {4.2.1, 4.2.2, 4.2.3, 4.4.1, 4.4.2, 4.9.6, Table 4.1, 7.2.1, 7.2.2}

A.2.8   Climate change has already affected food security due to warming, changing precipitation patterns, and greater frequency of some extreme events (*high confidence*). Studies that separate out climate change from other factors affecting crop yields have shown that yields of some crops (e.g., maize and wheat) in many lower-latitude regions have been affected negatively by observed climate changes, while in many higher-latitude regions, yields of some crops (e.g., maize, wheat, and sugar beets) have been affected positively over recent decades (*high confidence*). Climate change has resulted in lower animal growth rates and productivity in pastoral systems in Africa (*high confidence*). There is robust evidence that agricultural pests and diseases have already responded to climate change resulting in both increases and decreases of infestations (*high confidence*). Based on indigenous and local knowledge, climate change is affecting food security in drylands, particularly those in Africa, and high mountain regions of Asia and South America.[20] {5.2.1, 5.2.2, 7.2.2}

A.3   Agriculture, Forestry and Other Land Use (AFOLU) activities accounted for around 13% of CO$_2$, 44% of methane (CH$_4$), and 81% of nitrous oxide (N$_2$O) emissions from human activities globally during 2007-2016, representing 23% (12.0 ± 2.9 GtCO$_2$eq yr^{-1}) of total net anthropogenic emissions of GHGs (*medium confidence*).[21] The natural response of land to human-induced environmental change caused a net sink of around 11.2 GtCO$_2$ yr^{-1} during 2007–2016 (equivalent to 29% of total CO2 emissions) (*medium confidence*); the persistence of the sink is uncertain due to climate change (*high confidence*). If emissions associated with pre- and post-production activities in the global food system[22] are included, the emissions are estimated to be 21–37% of total net anthropogenic GHG emissions (*medium confidence*). {2.3, Table 2.2, 5.4}

A.3.1   Land is simultaneously a source and a sink of CO$_2$ due to both anthropogenic and natural drivers, making it hard to separate anthropogenic from natural fluxes (*very high confidence*). Global models estimate net CO$_2$ emissions of 5.2 ± 2.6 GtCO$_2$ yr^{-1} (*likely* range) from land use and land-use change during 2007–2016. These net emissions are mostly due to deforestation, partly offset by afforestation/reforestation, and emissions and removals by other land use activities (*very high confidence*).[23] There is no clear trend in annual emissions since 1990 (*medium confidence*). (Figure SPM.1, Table SPM.1) {1.1, 2.3, Table 2.2, Table 2.3}

A.3.2   The natural response of land to human-induced environmental changes such as increasing atmospheric CO$_2$ concentration, nitrogen deposition, and climate change, resulted in global net removals of 11.2 ± 2.6 GtCO$_2$ yr^{-1} (*likely* range) during 2007–2016. The sum of the net removals due to this response and the AFOLU net emissions gives a total net land-atmosphere flux that removed 6.0 ± 3.7 GtCO$_2$ yr^{-1} during 2007–2016 (*likely* range). Future net increases in CO$_2$ emissions from vegetation and soils due to climate change are projected to counteract increased removals due to CO$_2$ fertilisation and longer growing seasons (*high confidence*). The balance between these processes is a key source of uncertainty for determining the future of the land carbon sink. Projected thawing of permafrost is expected to increase the loss of soil carbon (*high confidence*). During the 21st century, vegetation growth in those areas may compensate in part for this loss (*low confidence*). (Table SPM.1) {Box 2.3, 2.3.1, 2.5.3, 2.7, Table 2.3}

---

[20]   The assessment covered literature whose methodologies included interviews and surveys with indigenous peoples and local communities.

[21]   This assessment only includes CO$_2$, CH$_4$ and N$_2$O.

[22]   Global food system in this report is defined as 'all the elements (environment, people, inputs, processes, infrastructures, institutions, etc.) and activities that relate to the production, processing, distribution, preparation and consumption of food, and the output of these activities, including socioeconomic and environmental outcomes at the global level'. These emissions data are not directly comparable to the national inventories prepared according to the 2006 IPCC Guidelines for National Greenhouse Gas Inventories.

[23]   The net anthropogenic flux of CO$_2$ from 'bookkeeping' or 'carbon accounting' models is composed of two opposing gross fluxes: gross emissions (about 20 GtCO$_2$ yr^{-1}) are from deforestation, cultivation of soils, and oxidation of wood products; gross removals (about 14 GtCO$_2$ yr^{-1}) are largely from forest growth following wood harvest and agricultural abandonment (*medium confidence*).

A.3.3   Global models and national GHG inventories use different methods to estimate anthropogenic CO_2 emissions and removals for the land sector. Both produce estimates that are in close agreement for land-use change involving forest (e.g., deforestation, afforestation), and differ for managed forest. Global models consider as managed forest those lands that were subject to harvest whereas, consistent with IPCC guidelines, national GHG inventories define managed forest more broadly. On this larger area, inventories can also consider the natural response of land to human-induced environmental changes as anthropogenic, while the global model approach (Table SPM.1) treats this response as part of the non-anthropogenic sink. For illustration, from 2005 to 2014, the sum of the national GHG inventories net emission estimates is 0.1 ± 1.0 GtCO$_2$ yr^{-1}, while the mean of two global bookkeeping models is 5.2 ± 2.6 GtCO$_2$ yr^{-1} (*likely* range). Consideration of differences in methods can enhance understanding of land sector net emission estimates and their applications. {2.4.1, 2.7.3, Fig 2.5, Box 2.2}

Net anthropogenic emissions due to Agriculture, Forestry, and other Land Use (AFOLU) (Panel 1) and global food systems (average for 2007–2016)[1] (Panel 2). Positive values represent emissions; negative values represent removals.

Panel 1: Contribution of AFOLU

Gas	Units	Net anthropogenic emissions due to Agriculture, Forestry, and Other Land Use (AFOLU)			Direct Anthropogenic				Natural response of land to human-induced environmental change[7]	Net land – atmosphere flux from all lands
		FOLU	Agriculture	Total	Non-AFOLU anthropogenic GHG emissions[6]	Total net anthropogenic emissions (AFOLU + non-AFOLU) by gas	AFOLU as a % of total net anthropogenic emissions, by gas		
		A	B	C = A + B	D	E = C + D	F = (C/E) ×100	G	A + G
CO_2[2]	$GtCO_2\,yr^{-1}$	5.2 ± 2.6	No data[11]	5.2 ± 2.6	33.9 ± 1.8	39.1 ± 3.2	13%	-11.2 ± 2.6	-6.0 ± 3.7
CH_4[3,8]	$MtCH_4\,yr^{-1}$	19.2 ± 5.8	142 ± 42	161 ± 43	201 ± 101	362 ± 109	44%		
	$GtCO_2eq\,yr^{-1}$	0.5 ± 0.2	4.0 ± 1.2	4.5 ± 1.2	5.6 ± 2.8	10.1 ± 3.1			
N_2O[3,8]	$MtN_2O\,yr^{-1}$	0.3 ± 0.1	8.3 ± 2.5	8.7 ± 2.5	2.0 ± 1.0	10.6 ± 2.7	81%		
	$GtCO_2eq\,yr^{-1}$	0.09 ± 0.03	2.2 ± 0.7	2.3 ± 0.7	0.5 ± 0.3	2.8 ± 0.7			
Total (GHG)	$GtCO_2eq\,yr^{-1}$	5.8 ± 2.6	6.2 ± 1.4	12.0 ± 2.9	40.0 ± 3.4	52.0 ± 4.5	23%		

Panel 2: Contribution of global food system

Gas	Units	Land-use change	Agriculture	Non-AFOLU[5] other sectors pre- to post-production	Total global food system emissions
CO_2 Land-use change[4]	$GtCO_2\,yr^{-1}$	4.9 ± 2.5			
CH_4 Agriculture[3,8,9]	$GtCO_2eq\,yr^{-1}$		4.0 ± 1.2		
N_2O Agriculture[3,8,9]	$GtCO_2eq\,yr^{-1}$		2.2 ± 0.7		
CO_2 other sectors[5]	$GtCO_2\,yr^{-1}$			2.6 – 5.2	
Total[10]	$GtCO_2eq\,yr^{-1}$	4.9 ± 2.5	6.2 ± 1.4	2.6 – 5.2	10.8 – 19.1

Table SPM.1 | Data sources and notes:

[1] Estimates are only given until 2016 as this is the latest date when data are available for all gases.

[2] Net anthropogenic flux of CO_2 due to land cover change such as deforestation and afforestation, and land management including wood harvest and regrowth, as well as peatland burning, based on two bookkeeping models as used in the Global Carbon Budget and for AR5. Agricultural soil carbon stock change under the same land use is not considered in these models. {2.3.1.2.1, Table 2.2, Box 2.2}

[3] Estimates show the mean and assessed uncertainty of two databases, FAOSTAT and USEPA. 2012 {2.3, Table 2.2}

[4] Based on FAOSTAT. Categories included in this value are 'net forest conversion' (net deforestation), drainage of organic soils (cropland and grassland), biomass burning (humid tropical forests, other forests, organic soils). It excludes 'forest land' (forest management plus net forest expansion), which is primarily a sink due to afforestation. Note: Total FOLU emissions from FAOSTAT are 2.8 (±1.4) $GtCO_2$ yr^{-1} for the period 2007–2016. {Table 2.2, Table 5.4}

[5] CO_2 emissions induced by activities not included in the AFOLU sector, mainly from energy (e.g., grain drying), transport (e.g., international trade), and industry (e.g., synthesis of inorganic fertilisers) part of food systems, including agricultural production activities (e.g., heating in greenhouses), pre-production (e.g., manufacturing of farm inputs) and post-production (e.g., agri-food processing) activities. This estimate is land based and hence excludes emissions from fisheries. It includes emissions from fibre and other non-food agricultural products since these are not separated from food use in databases. The CO_2 emissions related to the food system in sectors other than AFOLU are 6--13% of total anthropogenic CO_2 emissions. These emissions are typically low in smallholder subsistence farming. When added to AFOLU emissions, the estimated share of food systems in global anthropogenic emissions is 21--37%. {5.4.5, Table 5.4}

[6] Total non-AFOLU emissions were calculated as the sum of total CO_2eq emissions values for energy, industrial sources, waste and other emissions with data from the Global Carbon Project for CO_2, including international aviation and shipping and from the PRIMAP database for CH_4 and N_2O averaged over 2007–2014 only as that was the period for which data were available. {2.3, Table 2.2}.

[7] The natural response of land to human-induced environmental changes is the response of vegetation and soils to environmental changes such as increasing atmospheric CO_2 concentration, nitrogen deposition, and climate change. The estimate shown represents the average from Dynamic Global Vegetation Models {2.3.1.2, Box 2.2, Table 2.3}

[8] All values expressed in units of CO_2eq are based on AR5 100-year Global Warming Potential (GWP) values without climate-carbon feedbacks (N_2O = 265; CH_4 = 28). Note that the GWP has been used across fossil fuel and biogenic sources of methane. If a higher GWP for fossil fuel CH_4 (30 per AR5) were used, then total anthropogenic CH_4 emissions expressed in CO_2eq would be 2% greater.

[9] This estimate is land based and hence excludes emissions from fisheries and emissions from aquaculture (except emissions from feed produced on land and used in aquaculture), and also includes non-food use (e.g. fibre and bioenergy) since these are not separated from food use in databases. It excludes non-CO_2 emissions associated with land use change (FOLU category) since these are from fires in forests and peatlands.

[10] Emissions associated with food loss and waste are included implicitly, since emissions from the food system are related to food produced, including food consumed for nutrition and to food loss and waste. The latter is estimated at 8–10% of total anthropogenic emissions in CO_2eq. {5.5.2.5}

[11] No global data are available for agricultural CO_2 emissions.

A.3.4   Global AFOLU emissions of methane in the period 2007–2016 were 161 ± 43 $MtCH_4$ yr^{-1} (4.5 ± 1.2 $GtCO_2$eq yr^{-1}) (*medium confidence*). The globally averaged atmospheric concentration of CH_4 shows a steady increase between the mid-1980s and early 1990s, slower growth thereafter until 1999, a period of no growth between 1999–2006, followed by a resumption of growth in 2007 (*high confidence*). Biogenic sources make up a larger proportion of emissions than they did before 2000 (*high confidence*). Ruminants and the expansion of rice cultivation are important contributors to the rising concentration (*high confidence*). (Figure SPM.1) {Table 2.2, 2.3.2, 5.4.2, 5.4.3}

A.3.5   Anthropogenic AFOLU N_2O emissions are rising, and were 8.7 ± 2.5 MtN_2O yr^{-1} (2.3 ± 0.7 $GtCO_2$eq yr^{-1}) during the period 2007-2016. Anthropogenic N_2O emissions {Figure SPM.1, Table SPM.1} from soils are primarily due to nitrogen application including inefficiencies (over-application or poorly synchronised with crop demand timings) (*high confidence*). Cropland soils emitted around 3 MtN_2O yr^{-1} (around 795 $MtCO_2$ eq yr^{-1}) during the period 2007–2016 (*medium confidence*). There has been a major growth in emissions from managed pastures due to increased manure deposition (*medium confidence*). Livestock on managed pastures and rangelands accounted for more than one half of total anthropogenic N_2O emissions from agriculture in 2014 (*medium confidence*). {Table 2.1, 2.3.3, 5.4.2, 5.4.3}

A.3.6   Total net GHG emissions from AFOLU emissions represent 12.0 ± 2.9 $GtCO_2$eq yr^{-1} during 2007–2016. This represents 23% of total net anthropogenic emissions {Table SPM.1}.[24] Other approaches, such as global food system, include agricultural emissions and land use change (i.e., deforestation and peatland degradation), as well as outside farm gate emissions from energy, transport and industry sectors for food production. Emissions within farm gate and from agricultural land expansion contributing to the global food system represent 16–27% of total anthropogenic emissions (*medium confidence*). Emissions outside the farm gate represent 5–10% of total anthropogenic emissions (*medium confidence*). Given the diversity of food systems, there are large regional differences in the contributions from different components of the food system (*very high confidence*). Emissions from agricultural production are projected to increase (*high confidence*), driven by population and income growth and changes in consumption patterns (*medium confidence*). {5.5, Table 5.4}

---

[24]   This assessment only includes CO_2, CH_4 and N_2O.

SPM

A.4 Changes in land conditions,[25] either from land-use or climate change, affect global and regional climate (*high confidence*). At the regional scale, changing land conditions can reduce or accentuate warming and affect the intensity, frequency and duration of extreme events. The magnitude and direction of these changes vary with location and season (*high confidence*). {Executive Summary Chapter 2, 2.3, 2.4, 2.5, 3.3}

A.4.1 Since the pre-industrial period, changes in land cover due to human activities have led to both a net release of CO_2 contributing to global warming (*high confidence*), and an increase in global land albedo[26] causing surface cooling (*medium confidence*). Over the historical period, the resulting net effect on globally averaged surface temperature is estimated to be small (*medium confidence*). {2.4, 2.6.1, 2.6.2}

A.4.2 The likelihood, intensity and duration of many extreme events can be significantly modified by changes in land conditions, including heat related events such as heatwaves (*high confidence*) and heavy precipitation events (*medium confidence*). Changes in land conditions can affect temperature and rainfall in regions as far as hundreds of kilometres away (*high confidence*). {2.5.1, 2.5.2, 2.5.4, 3.3, Cross-Chapter Box 4 in Chapter 2}

A.4.3 Climate change is projected to alter land conditions with feedbacks on regional climate. In those boreal regions where the treeline migrates northward and/or the growing season lengthens, winter warming will be enhanced due to decreased snow cover and albedo while warming will be reduced during the growing season because of increased evapotranspiration (*high confidence*). In those tropical areas where increased rainfall is projected, increased vegetation growth will reduce regional warming (*medium confidence*). Drier soil conditions resulting from climate change can increase the severity of heat waves, while wetter soil conditions have the opposite effect (*high confidence*). {2.5.2, 2.5.3}

A.4.4 Desertification amplifies global warming through the release of CO_2 linked with the decrease in vegetation cover (*high confidence*). This decrease in vegetation cover tends to increase local albedo, leading to surface cooling (*high confidence*). {3.3}

A.4.5 Changes in forest cover, for example from afforestation, reforestation and deforestation, directly affect regional surface temperature through exchanges of water and energy (*high confidence*).[27] Where forest cover increases in tropical regions cooling results from enhanced evapotranspiration (*high confidence*). Increased evapotranspiration can result in cooler days during the growing season (*high confidence*) and can reduce the amplitude of heat related events (*medium confidence*). In regions with seasonal snow cover, such as boreal and some temperate regions, increased tree and shrub cover also has a wintertime warming influence due to reduced surface albedo (*high confidence*).[28] {2.3, 2.4.3, 2.5.1, 2.5.2, 2.5.4}

A.4.6 Both global warming and urbanisation can enhance warming in cities and their surroundings (heat island effect), especially during heat related events, including heat waves (*high confidence*). Night-time temperatures are more affected by this effect than daytime temperatures (*high confidence*). Increased urbanisation can also intensify extreme rainfall events over the city or downwind of urban areas (*medium confidence*). {2.5.1, 2.5.2, 2.5.3, 4.9.1, Cross-Chapter Box 4 in Chapter 2}

---

[25] Land conditions encompass changes in land cover (e.g., deforestation, afforestation, urbanisation), in land use (e.g., irrigation), and in land state (e.g., degree of wetness, degree of greening, amount of snow, amount of permafrost).

[26] Land with high albedo reflects more incoming solar radiation than land with low albedo.

[27] The literature indicates that forest cover changes can also affect climate through changes in emissions of reactive gases and aerosols. {2.4, 2.5}

[28] Emerging literature shows that boreal forest-related aerosols may counteract at least partly the warming effect of surface albedo. {2.4.3}

Box SPM. 1 | Shared Socio-economic Pathways (SSPs)

In this report the implications of future socio-economic development on climate change mitigation, adaptation and land-use are explored using shared socio-economic pathways (SSPs). The SSPs span a range of challenges to climate change mitigation and adaptation.

- SSP1 includes a peak and decline in population (~7 billion in 2100), high income and reduced inequalities, effective land-use regulation, less resource intensive consumption, including food produced in low-GHG emission systems and lower food waste, free trade and environmentally-friendly technologies and lifestyles. Relative to other pathways, SSP1 has low challenges to mitigation and low challenges to adaptation (i.e., high adaptive capacity)

- SSP2 includes medium population growth (~9 billion in 2100), medium income, technological progress, production and consumption patterns are a continuation of past trends, and only a gradual reduction in inequality occurs. Relative to other pathways, SSP2 has medium challenges to mitigation and medium challenges to adaptation (i.e., medium adaptive capacity).

- SSP3 includes high population growth (~13 billion in 2100), low income and continued inequalities, material-intensive consumption and production, barriers to trade, and slow rates of technological change. Relative to other pathways, SSP3 has high challenges to mitigation and high challenges to adaptation (i.e., low adaptive capacity).

- SSP4 includes medium population growth (~9 billion in 2100), medium income, but significant inequality within and across regions. Relative to other pathways, SSP4 has low challenges to mitigation, but high challenges to adaptation (i.e., low adaptive capacity).

- SSP5 includes a peak and decline in population (~7 billion in 2100), high income, reduced inequalities, and free trade. This pathway includes resource-intensive production, consumption and lifestyles. Relative to other pathways, SSP5 has high challenges to mitigation, but low challenges to adaptation (i.e., high adaptive capacity).

- The SSPs can be combined with Representative Concentration Pathways (RCPs) which imply different levels of mitigation, with implications for adaptation. Therefore, SSPs can be consistent with different levels of global mean surface temperature rise as projected by different SSP-RCP combinations. However, some SSP-RCP combinations are not possible; for instance RCP2.6 and lower levels of future global mean surface temperature rise (e.g., 1.5°C) are not possible in SSP3 in modelled pathways. {1.2.2, 6.1.4, Cross-Chapter Box 1 in Chapter 1, Cross-Chapter Box 9 in Chapter 6}

A. Risks to humans and ecosystems from changes in land-based processes as a result of climate change

Increases in global mean surface temperature (GMST), relative to pre-industrial levels, affect processes involved in **desertification** (water scarcity), **land degradation** (soil erosion, vegetation loss, wildfire, permafrost thaw) and **food security** (crop yield and food supply instabilities). Changes in these processes drive risks to food systems, livelihoods, infrastructure, the value of land, and human and ecosystem health. Changes in one process (e.g. wildfire or water scarcity) may result in compound risks. Risks are location-specific and differ by region.

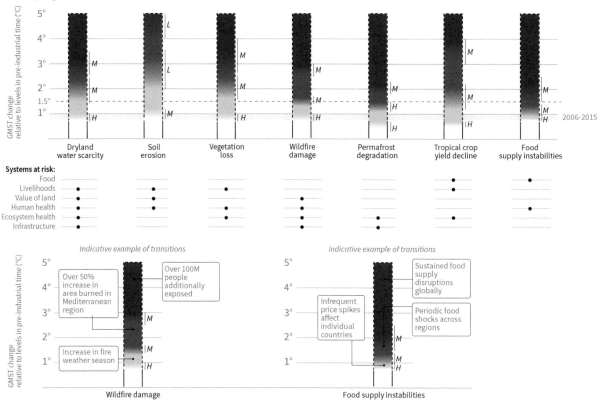

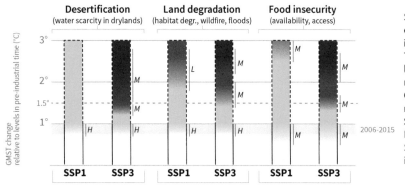

B. Different socioeconomic pathways affect levels of climate related risks

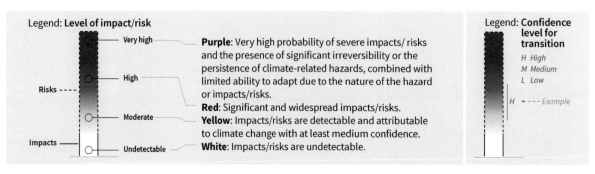

Socio-economic choices can reduce or exacerbate climate related risks as well as influence the rate of temperature increase. The **SSP1** pathway illustrates a world with low population growth, high income and reduced inequalities, food produced in low GHG emission systems, effective land use regulation and high adaptive capacity. The **SSP3** pathway has the opposite trends. Risks are lower in SSP1 compared with SSP3 given the same level of GMST increase.

Legend: Level of impact/risk

Very high

High

Risks ----

Moderate

Impacts ——

Undetectable

Purple: Very high probability of severe impacts/ risks and the presence of significant irreversibility or the persistence of climate-related hazards, combined with limited ability to adapt due to the nature of the hazard or impacts/risks.
Red: Significant and widespread impacts/risks.
Yellow: Impacts/risks are detectable and attributable to climate change with at least medium confidence.
White: Impacts/risks are undetectable.

Legend: Confidence level for transition

H High
M Medium
L Low

H ◄--- Example

Figure SPM.2: Risks to land-related human systems and ecosystems from global climate change, socio-economic development and mitigation choices in terrestrial ecosystems. | As in previous IPCC reports the literature was used to make expert judgements to assess the levels of global warming at which levels of risk are undetectable, moderate, high or very high, as described further in Chapter 7 and other parts of the underlying report. The Figure indicates assessed risks at approximate warming levels which may be influenced by a variety of factors, including adaptation responses. The assessment considers adaptive capacity consistent with the SSP pathways as described below. **Panel A**: Risks to selected elements of the land system as a function of global mean surface temperature {2.1, Box 2.1, 3.5, 3.7.1.1, 4.4.1.1, 4.4.1.2, 4.4.1.3, 5.2.2, 5.2.3, 5.2.4, 5.2.5, 7.2, 7.3, Table SM7.1}. Links to broader systems are illustrative and not intended to be comprehensive. Risk levels are estimated assuming medium exposure and vulnerability driven by moderate trends in socioeconomic conditions broadly consistent with an SSP2 pathway. {Table SM7.4} **Panel B**: Risks associated with desertification, land degradation and food security due to climate change and patterns of socio-economic development. Increasing risks associated with desertification include population exposed and vulnerable to water scarcity in drylands. Risks related to land degradation include increased habitat degradation, population exposed to wildfire and floods and costs of floods. Risks to food security include availability and access to food, including population at risk of hunger, food price increases and increases in disability adjusted life years attributable due to childhood underweight. Risks are assessed for two contrasted socio-economic pathways (SSP1 and SSP3 {Box SPM.1}) excluding the effects of targeted mitigation policies. {3.5, 4.2.1.2, 5.2.2, 5.2.3, 5.2.4, 5.2.5, 6.1.4, 7.2, Table SM7.5} Risks are not indicated beyond 3°C because SSP1 does not exceed this level of temperature change. **All panels:** As part of the assessment, literature was compiled and data extracted into a summary table. A formal expert elicitation protocol (based on modified-Delphi technique and the Sheffield Elicitation Framework), was followed to identify risk transition thresholds. This included a multi-round elicitation process with two rounds of independent anonymous threshold judgement, and a final consensus discussion. Further information on methods and underlying literature can be found in Chapter 7 Supplementary Material.

A.5   Climate change creates additional stresses on land, exacerbating existing risks to livelihoods, biodiversity, human and ecosystem health, infrastructure, and food systems (*high confidence*). Increasing impacts on land are projected under all future GHG emission scenarios (*high confidence*). Some regions will face higher risks, while some regions will face risks previously not anticipated (*high confidence*). Cascading risks with impacts on multiple systems and sectors also vary across regions (*high confidence*). (Figure SPM.2) {2.2, 3.5, 4.2, 4.4, 4.7, 5.1, 5.2, 5.8, 6.1, 7.2, 7.3, Cross-Chapter Box 9 in Chapter 6}

A.5.1   With increasing warming, the frequency, intensity and duration of heat related events including heatwaves are projected to continue to increase through the 21st century (*high confidence*). The frequency and intensity of droughts are projected to increase particularly in the Mediterranean region and southern Africa (*medium confidence*). The frequency and intensity of extreme rainfall events are projected to increase in many regions (*high confidence*). {2.2.5, 3.5.1, 4.2.3, 5.2}

A.5.2   With increasing warming, climate zones are projected to further shift poleward in the middle and high latitudes (*high confidence*). In high-latitude regions, warming is projected to increase disturbance in boreal forests, including drought, wildfire, and pest outbreaks (*high confidence*). In tropical regions, under medium and high GHG emissions scenarios, warming is projected to result in the emergence of unprecedented[29] climatic conditions by the mid to late 21st century (*medium confidence*). {2.2.4, 2.2.5, 2.5.3, 4.3.2}

A.5.3   Current levels of global warming are associated with moderate risks from increased dryland water scarcity, soil erosion, vegetation loss, wildfire damage, permafrost thawing, coastal degradation and tropical crop yield decline (*high confidence*). Risks, including cascading risks, are projected to become increasingly severe with increasing temperatures. At around 1.5°C of global warming the risks from dryland water scarcity, wildfire damage, permafrost degradation and food supply instabilities are projected to be high (*medium confidence*). At around 2°C of global warming the risk from permafrost degradation and food supply instabilities are projected to be very high (*medium confidence*). Additionally, at around 3°C of global warming risk from vegetation loss, wildfire damage, and dryland water scarcity are also projected to be very high (*medium confidence*). Risks from droughts, water stress, heat related events such as heatwaves and habitat degradation simultaneously increase between 1.5°C and 3°C warming (*low confidence*). (Figure SPM.2) {7.2.2, Cross-Chapter Box 9 in Chapter 6, Chapter 7 Supplementary Material}

A.5.4   The stability of food supply[30] is projected to decrease as the magnitude and frequency of extreme weather events that disrupt food chains increases (*high confidence*). Increased atmospheric CO_2 levels can also lower the nutritional quality of crops (*high confidence*). In SSP2, global crop and economic models project a median increase of 7.6% (range of 1–23%) in cereal prices in 2050 due to climate change (RCP6.0), leading to higher food prices and increased risk of food insecurity and hunger (*medium*

---

[29]   Unprecedented climatic conditions are defined in this report as 'not having occurred anywhere during the 20th century'. They are characterised by high temperature with strong seasonality and shifts in precipitation. In the literature assessed, the effect of climatic variables other than temperature and precipitation were not considered.

[30]   The supply of food is defined in this report as 'encompassing availability and access (including price)'. Food supply instability refers to variability that influences food security through reducing access.

confidence). The most vulnerable people will be more severely affected (*high confidence*). {5.2.3, 5.2.4, 5.2.5, 5.8.1, 7.2.2.2, 7.3.1}

A.5.5   In drylands, climate change and desertification are projected to cause reductions in crop and livestock productivity (*high confidence*), modify the plant species mix and reduce biodiversity (*medium confidence*). Under SSP2, the dryland population vulnerable to water stress, drought intensity and habitat degradation is projected to reach 178 million people by 2050 at 1.5°C warming, increasing to 220 million people at 2°C warming, and 277 million people at 3°C warming (*low confidence*). {3.5.1, 3.5.2, 3.7.3}

A.5.6   Asia and Africa[31] are projected to have the highest number of people vulnerable to increased desertification. North America, South America, Mediterranean, southern Africa and central Asia may be increasingly affected by wildfire. The tropics and subtropics are projected to be most vulnerable to crop yield decline. Land degradation resulting from the combination of sea-level rise and more intense cyclones is projected to jeopardise lives and livelihoods in cyclone prone areas (*very high confidence*). Within populations, women, the young, elderly and poor are most at risk (*high confidence*). {3.5.1, 3.5.2, 4.4, Table 4.1, 5.2.2, 7.2.2, Cross-Chapter Box 3 in Chapter 2}

A.5.7   Changes in climate can amplify environmentally induced migration both within countries and across borders (*medium confidence*), reflecting multiple drivers of mobility and available adaptation measures (*high confidence*). Extreme weather and climate or slow-onset events may lead to increased displacement, disrupted food chains, threatened livelihoods (*high confidence*), and contribute to exacerbated stresses for conflict (*medium confidence*). {3.4.2, 4.7.3, 5.2.3, 5.2.4, 5.2.5, 5.8.2, 7.2.2, 7.3.1}

A.5.8   Unsustainable land management has led to negative economic impacts (*high confidence*). Climate change is projected to exacerbate these negative economic impacts (*high confidence*). {4.3.1, 4.4.1, 4.7, 4.8.5, 4.8.6, 4.9.6, 4.9.7, 4.9.8, 5.2, 5.8.1, 7.3.4, 7.6.1, Cross-Chapter Box 10 in Chapter 7}

A.6   The level of risk posed by climate change depends both on the level of warming and on how population, consumption, production, technological development, and land management patterns evolve (*high confidence*). Pathways with higher demand for food, feed, and water, more resource-intensive consumption and production, and more limited technological improvements in agriculture yields result in higher risks from water scarcity in drylands, land degradation, and food insecurity (*high confidence*). (Figure SPM.2b) {5.1.4, 5.2.3, 6.1.4, 7.2, Cross-Chapter Box 9 in Chapter 6}

A.6.1   Projected increases in population and income, combined with changes in consumption patterns, result in increased demand for food, feed, and water in 2050 in all SSPs (*high confidence*). These changes, combined with land management practices, have implications for land-use change, food insecurity, water scarcity, terrestrial GHG emissions, carbon sequestration potential, and biodiversity (*high confidence*). Development pathways in which incomes increase and the demand for land conversion is reduced, either through reduced agricultural demand or improved productivity, can lead to reductions in food insecurity (*high confidence*). All assessed future socio-economic pathways result in increases in water demand and water scarcity (*high confidence*). SSPs with greater cropland expansion result in larger declines in biodiversity (*high confidence*). {6.1.4}

A.6.2   Risks related to water scarcity in drylands are lower in pathways with low population growth, less increase in water demand, and high adaptive capacity, as in SSP1. In these scenarios the risk from water scarcity in drylands is moderate even at global warming of 3°C (*low confidence*). By contrast, risks related to water scarcity in drylands are greater for pathways with high population growth, high vulnerability, higher water demand, and low adaptive capacity, such as SSP3. In SSP3 the transition from moderate to high risk occurs between 1.2°C and 1.5°C (*medium confidence*). (Figure SPM.2b, Box SPM.1) {7.2}

A.6.3   Risks related to climate change driven land degradation are higher in pathways with a higher population, increased land-use change, low adaptive capacity and other barriers to adaptation (e.g., SSP3). These scenarios result in more people exposed to ecosystem degradation, fire, and coastal flooding (*medium confidence*). For land degradation, the projected transition from moderate to high risk occurs for global warming between 1.8°C and 2.8°C in SSP1 (*low confidence*) and between 1.4°C and 2°C in SSP3 (*medium confidence*). The projected transition from high to very high risk occurs between 2.2°C and 2.8°C for SSP3 (*medium confidence*). (Figure SPM.2b) {4.4, 7.2}

---

[31]   West Africa has a high number of people vulnerable to increased desertification and yield decline. North Africa is vulnerable to water scarcity.

A.6.4   Risks related to food security are greater in pathways with lower income, increased food demand, increased food prices resulting from competition for land, more limited trade, and other challenges to adaptation (e.g., SSP3) (*high confidence*). For food security, the transition from moderate to high risk occurs for global warming between 2.5°C and 3.5°C in SSP1 (*medium confidence*) and between 1.3°C and 1.7°C in SSP3 (*medium confidence*). The transition from high to very high risk occurs between 2°C and 2.7°C for SSP3 (*medium confidence*). (Figure SPM.2b) {7.2}

A.6.5   Urban expansion is projected to lead to conversion of cropland leading to losses in food production (*high confidence*). This can result in additional risks to the food system. Strategies for reducing these impacts can include urban and peri-urban food production and management of urban expansion, as well as urban green infrastructure that can reduce climate risks in cities[32] (*high confidence*). (Figure SPM.3) {4.9.1, 5.5, 5.6, 6.3, 6.4, 7.5.6}

---

[32]   The land systems considered in this report do not include urban ecosystem dynamics in detail. Urban areas, urban expansion, and other urban processes and their relation to land-related processes are extensive, dynamic, and complex. Several issues addressed in this report such as population, growth, incomes, food production and consumption, food security, and diets have close relationships with these urban processes. Urban areas are also the setting of many processes related to land-use change dynamics, including loss of ecosystem functions and services, that can lead to increased disaster risk. Some specific urban issues are assessed in this report.

B. Adaptation and mitigation response options

B.1 **Many land-related responses that contribute to climate change adaptation and mitigation can also combat desertification and land degradation and enhance food security. The potential for land-related responses and the relative emphasis on adaptation and mitigation is context specific, including the adaptive capacities of communities and regions. While land-related response options can make important contributions to adaptation and mitigation, there are some barriers to adaptation and limits to their contribution to global mitigation. (*very high confidence*) (Figure SPM.3) {2.6, 4.8, 5.6, 6.1, 6.3, 6.4}**

B.1.1 Some land-related actions are already being taken that contribute to climate change adaptation, mitigation and sustainable development. The response options were assessed across adaptation, mitigation, combating desertification and land degradation, food security and sustainable development, and a select set of options deliver across all of these challenges. These options include, but are not limited to, sustainable food production, improved and sustainable forest management, soil organic carbon management, ecosystem conservation and land restoration, reduced deforestation and degradation, and reduced food loss and waste (*high confidence*). These response options require integration of biophysical, socioeconomic and other enabling factors. {6.3, 6.4.5, 7.5.6, Cross-Chapter Box 10 in Chapter 7}

B.1.2 While some response options have immediate impacts, others take decades to deliver measurable results. Examples of response options with immediate impacts include the conservation of high-carbon ecosystems such as peatlands, wetlands, rangelands, mangroves and forests. Examples that provide multiple ecosystem services and functions, but take more time to deliver, include afforestation and reforestation as well as the restoration of high-carbon ecosystems, agroforestry, and the reclamation of degraded soils (*high confidence*). {6.4.5, 7.5.6, Cross-Chapter Box 10 in Chapter 7}

B.1.3 The successful implementation of response options depends on consideration of local environmental and socio-economic conditions. Some options such as soil carbon management are potentially applicable across a broad range of land use types, whereas the efficacy of land management practices relating to organic soils, peatlands and wetlands, and those linked to freshwater resources, depends on specific agro-ecological conditions (*high confidence*). Given the site-specific nature of climate change impacts on food system components and wide variations in agroecosystems, adaptation and mitigation options and their barriers are linked to environmental and cultural context at regional and local levels (*high confidence*). Achieving land degradation neutrality depends on the integration of multiple responses across local, regional and national scales and across multiple sectors including agriculture, pasture, forest and water (*high confidence*). {4.8, 6.2, 6.3, 6.4.4, 7.5.6}

B.1.4 Land-based options that deliver carbon sequestration in soil or vegetation, such as afforestation, reforestation, agroforestry, soil carbon management on mineral soils, or carbon storage in harvested wood products, do not continue to sequester carbon indefinitely (*high confidence*). Peatlands, however, can continue to sequester carbon for centuries (*high confidence*). When vegetation matures or when vegetation and soil carbon reservoirs reach saturation, the annual removal of CO_2 from the atmosphere declines towards zero, while carbon stocks can be maintained (*high confidence*). However, accumulated carbon in vegetation and soils is at risk from future loss (or sink reversal) triggered by disturbances such as flood, drought, fire, or pest outbreaks, or future poor management (*high confidence*). {6.4.1}

B.2 **Most of the response options assessed contribute positively to sustainable development and other societal goals (*high confidence*). Many response options can be applied without competing for land and have the potential to provide multiple co-benefits (*high confidence*). A further set of response options has the potential to reduce demand for land, thereby enhancing the potential for other response options to deliver across each of climate change adaptation and mitigation, combating desertification and land degradation, and enhancing food security (*high confidence*). (Figure SPM.3) {4.8, 6.2, 6.3.6, 6.4.3}**

B.2.1 A number of land management options, such as improved management of cropland and grazing lands, improved and sustainable forest management, and increased soil organic carbon content, do not require land use change and do not create demand for more land conversion (*high confidence*). Further, a number of response options such as increased food productivity, dietary choices and food losses, and waste reduction, can reduce demand for land conversion, thereby potentially freeing land and creating opportunities for enhanced implementation of other response options (*high confidence*). Response options that reduce competition for land are possible and are applicable at different scales, from farm to regional (*high confidence*). (Figure SPM.3) {4.8, 6.3.6, 6.4}

B.2.2   A wide range of adaptation and mitigation responses, e.g., preserving and restoring natural ecosystems such as peatland, coastal lands and forests, biodiversity conservation, reducing competition for land, fire management, soil management, and most risk management options (e.g., use of local seeds, disaster risk management, risk sharing instruments) have the potential to make positive contributions to sustainable development, enhancement of ecosystem functions and services and other societal goals (*medium confidence*). Ecosystem-based adaptation can, in some contexts, promote nature conservation while alleviating poverty and can even provide co-benefits by removing GHGs and protecting livelihoods (e.g., mangroves) (*medium confidence*). {6.4.3, 7.4.6.2}

B.2.3   Most of the land management-based response options that do not increase competition for land, and almost all options based on value chain management (e.g., dietary choices, reduced post-harvest losses, reduced food waste) and risk management, can contribute to eradicating poverty and eliminating hunger while promoting good health and wellbeing, clean water and sanitation, climate action, and life on land (*medium confidence*). {6.4.3}

B.3   Although most response options can be applied without competing for available land, some can increase demand for land conversion (*high confidence*). At the deployment scale of several GtCO$_2$ yr^{-1}, this increased demand for land conversion could lead to adverse side effects for adaptation, desertification, land degradation and food security (*high confidence*). If applied on a limited share of total land and integrated into sustainably managed landscapes, there will be fewer adverse side-effects and some positive co-benefits can be realised (*high confidence*). (Figure SPM.3) {4.5, 6.2, 6.4, Cross-Chapter Box 7 in Chapter 6}

B.3.1   If applied at scales necessary to remove CO$_2$ from the atmosphere at the level of several GtCO$_2$ yr^{-1}, afforestation, reforestation and the use of land to provide feedstock for bioenergy with or without carbon capture and storage, or for biochar, could greatly increase demand for land conversion (*high confidence*). Integration into sustainably managed landscapes at appropriate scale can ameliorate adverse impacts (*medium confidence*). Reduced grassland conversion to croplands, restoration and reduced conversion of peatlands, and restoration and reduced conversion of coastal wetlands affect smaller land areas globally, and the impacts on land use change of these options are smaller or more variable (*high confidence*). (Figure SPM.3) {Cross-Chapter Box 7 in Chapter 6, 6.4}

B.3.2   While land can make a valuable contribution to climate change mitigation, there are limits to the deployment of land-based mitigation measures such as bioenergy crops or afforestation. Widespread use at the scale of several millions of km^2 globally could increase risks for desertification, land degradation, food security and sustainable development (*medium confidence*). Applied on a limited share of total land, land-based mitigation measures that displace other land uses have fewer adverse side-effects and can have positive co-benefits for adaptation, desertification, land degradation or food security. (*high confidence*) (Figure SPM.3) {4.2, 4.5, 6.4; Cross-Chapter Box 7 in Chapter 6}

B.3.3   The production and use of biomass for bioenergy can have co-benefits, adverse side-effects, and risks for land degradation, food insecurity, GHG emissions and other environmental and sustainable development goals (*high confidence*). These impacts are context specific and depend on the scale of deployment, initial land use, land type, bioenergy feedstock, initial carbon stocks, climatic region and management regime, and other land-demanding response options can have a similar range of consequences (*high confidence*). The use of residues and organic waste as bioenergy feedstock can mitigate land use change pressures associated with bioenergy deployment, but residues are limited and the removal of residues that would otherwise be left on the soil could lead to soil degradation (*high confidence*). (Figure SPM.3) {2.6.1.5, Cross-Chapter Box 7 in Chapter 6}

B.3.4   For projected socioeconomic pathways with low population, effective land-use regulation, food produced in low-GHG emission systems and lower food loss and waste (SSP1), the transition from low to moderate risk to food security, land degradation and water scarcity in dry lands occur between 1 and 4 million km^2 of bioenergy or bioenergy with carbon capture and storage (BECCS) (*medium confidence*). By contrast, in pathways with high population, low income and slow rates of technological change (SSP3), the transition from low to moderate risk occurs between 0.1 and 1 million km^2 (*medium confidence*). (Box SPM.1) {6.4, Table SM7.6, Cross Chapter Box 7 in Chapter 6}

B.4 **Many activities for combating desertification can contribute to climate change adaptation with mitigation co-benefits, as well as to halting biodiversity loss with sustainable development co-benefits to society (*high confidence*). Avoiding, reducing and reversing desertification would enhance soil fertility, increase carbon storage in soils and biomass, while benefitting agricultural productivity and food security (*high confidence*). Preventing desertification is preferable to attempting to restore degraded land due to the potential for residual risks and maladaptive outcomes (*high confidence*). {3.6.1, 3.6.2, 3.6.3, 3.6.4, 3.7.1, 3.7.2}**

B.4.1 Solutions that help adapt to and mitigate climate change while contributing to combating desertification are site and regionally specific and include *inter alia*: water harvesting and micro-irrigation, restoring degraded lands using drought-resilient ecologically appropriate plants, agroforestry, and other agroecological and ecosystem-based adaptation practices (*high confidence*). {3.3, 3.6.1, 3.7.2, 3.7.5, 5.2, 5.6}

B.4.2 Reducing dust and sand storms and sand dune movement can lessen the negative effects of wind erosion and improve air quality and health (*high confidence*). Depending on water availability and soil conditions, afforestation, tree planting and ecosystem restoration programs, which aim for the creation of windbreaks in the form of 'green walls' and 'green dams' using native and other climate resilient tree species with low water needs, can reduce sand storms, avert wind erosion, and contribute to carbon sinks, while improving micro-climates, soil nutrients and water retention (*high confidence*). {3.3, 3.6.1, 3.7.2, 3.7.5}

B.4.3 Measures to combat desertification can promote soil carbon sequestration (*high confidence*). Natural vegetation restoration and tree planting on degraded land enriches, in the long term, carbon in the topsoil and subsoil (*medium confidence*). Modelled rates of carbon sequestration following the adoption of conservation agriculture practices in drylands depend on local conditions (*medium confidence*). If soil carbon is lost, it may take a prolonged period of time for carbon stocks to recover. {3.1.4, 3.3, 3.6.1, 3.6.3, 3.7.1, 3.7.2}

B.4.4 Eradicating poverty and ensuring food security can benefit from applying measures promoting land degradation neutrality (including avoiding, reducing and reversing land degradation) in rangelands, croplands and forests, which contribute to combating desertification, while mitigating and adapting to climate change within the framework of sustainable development. Such measures include avoiding deforestation and locally suitable practices including management of rangeland and forest fires (*high confidence*). {3.4.2, 3.6.1, 3.6.2, 3.6.3, 4.8.5}

B.4.5 Currently there is a lack of knowledge of adaptation limits and potential maladaptation to combined effects of climate change and desertification. In the absence of new or enhanced adaptation options, the potential for residual risks and maladaptive outcomes is high (*high confidence*). Even when solutions are available, social, economic and institutional constraints could pose barriers to their implementation (*medium confidence*). Some adaptation options can become maladaptive due to their environmental impacts, such as irrigation causing soil salinisation or over extraction leading to ground-water depletion (*medium confidence*). Extreme forms of desertification can lead to the complete loss of land productivity, limiting adaptation options or reaching the limits to adaptation (*high confidence*). {Executive Summary Chapter 3, 3.6.4, 3.7.5, 7.4.9}

B.4.6 Developing, enabling and promoting access to cleaner energy sources and technologies can contribute to adaptation and mitigating climate change and combating desertification and forest degradation through decreasing the use of traditional biomass for energy while increasing the diversity of energy supply (*medium confidence*). This can have socioeconomic and health benefits, especially for women and children. (*high confidence*). The efficiency of wind and solar energy infrastructures is recognised; the efficiency can be affected in some regions by dust and sand storms (*high confidence*). {3.5.3, 3.5.4, 4.4.4, 7.5.2, Cross-Chapter Box 12 in Chapter 7}

B.5 Sustainable land management,[33] including sustainable forest management,[34] can prevent and reduce land degradation, maintain land productivity, and sometimes reverse the adverse impacts of climate change on land degradation (*very high confidence*). It can also contribute to mitigation and adaptation (*high confidence*). Reducing and reversing land degradation, at scales from individual farms to entire watersheds, can provide cost effective, immediate, and long-term benefits to communities and support several Sustainable Development Goals (SDGs) with co-benefits for adaptation (*very high confidence*) and mitigation (*high confidence*). Even with implementation of sustainable land management, limits to adaptation can be exceeded in some situations (*medium confidence*). {1.3.2, 4.1.5, 4.8, 7.5.6, Table 4.2}

B.5.1 Land degradation in agriculture systems can be addressed through sustainable land management, with an ecological and socioeconomic focus, with co-benefits for climate change adaptation. Management options that reduce vulnerability to soil erosion and nutrient loss include growing green manure crops and cover crops, crop residue retention, reduced/zero tillage, and maintenance of ground cover through improved grazing management (*very high confidence*). {4.8}

B.5.2 The following options also have mitigation co-benefits. Farming systems such as agroforestry, perennial pasture phases and use of perennial grains, can substantially reduce erosion and nutrient leaching while building soil carbon (*high confidence*). The global sequestration potential of cover crops would be about 0.44 ± 0.11 $GtCO_2$ yr^{-1} if applied to 25% of global cropland (*high confidence*). The application of certain biochars can sequester carbon (*high confidence*), and improve soil conditions in some soil types/climates (*medium confidence*). {4.8.1.1, 4.8.1.3, 4.9.2, 4.9.5, 5.5.1, 5.5.4, Cross-Chapter Box 6 in Chapter 5}

B.5.3 Reducing deforestation and forest degradation lowers GHG emissions (*high confidence*), with an estimated technical mitigation potential of 0.4–5.8 $GtCO_2$ yr^{-1}. By providing long-term livelihoods for communities, sustainable forest management can reduce the extent of forest conversion to non-forest uses (e.g., cropland or settlements) (*high confidence*). Sustainable forest management aimed at providing timber, fibre, biomass, non-timber resources and other ecosystem functions and services, can lower GHG emissions and can contribute to adaptation (*high confidence*). {2.6.1.2, 4.1.5, 4.3.2, 4.5.3, 4.8.1.3, 4.8.3, 4.8.4}

B.5.4 Sustainable forest management can maintain or enhance forest carbon stocks, and can maintain forest carbon sinks, including by transferring carbon to wood products, thus addressing the issue of sink saturation (*high confidence*). Where wood carbon is transferred to harvested wood products, these can store carbon over the long-term and can substitute for emissions-intensive materials reducing emissions in other sectors (*high confidence*). Where biomass is used for energy, e.g., as a mitigation strategy, the carbon is released back into the atmosphere more quickly (*high confidence*). (Figure SPM.3) {2.6.1, 2.7, 4.1.5, 4.8.4, 6.4.1, Cross-Chapter Box 7 in Chapter 6}

B.5.5 Climate change can lead to land degradation, even with the implementation of measures intended to avoid, reduce or reverse land degradation (*high confidence*). Such limits to adaptation are dynamic, site-specific and are determined through the interaction of biophysical changes with social and institutional conditions (*very high confidence*). In some situations, exceeding the limits of adaptation can trigger escalating losses or result in undesirable transformational changes (*medium confidence*) such as forced migration (*low confidence*), conflicts (*low confidence*) or poverty (*medium confidence*). Examples of climate change induced land degradation that may exceed limits to adaptation include coastal erosion exacerbated by sea level rise where land disappears (*high confidence*), thawing of permafrost affecting infrastructure and livelihoods (*medium confidence*), and extreme soil erosion causing loss of productive capacity (*medium confidence*). {4.7, 4.8.5, 4.8.6, 4.9.6, 4.9.7, 4.9.8}

B.6 Response options throughout the food system, from production to consumption, including food loss and waste, can be deployed and scaled up to advance adaptation and mitigation (*high confidence*). The total technical mitigation potential from crop and livestock activities, and agroforestry is estimated as 2.3 – 9.6 $GtCO_2eq$ yr^{-1} by 2050 (*medium confidence*). The total technical mitigation potential of dietary changes is estimated as 0.7 – 8 $GtCO_2eq$ yr^{-1} by 2050 (*medium confidence*). {5.3, 5.5, 5.6}

---

[33] Sustainable land management is defined in this report as 'the stewardship and use of land resources, including soils, water, animals and plants, to meet changing human needs, while simultaneously ensuring the long-term productive potential of these resources and the maintenance of their environmental functions'. Examples of options include, *inter alia*, agroecology (including agroforestry), conservation agriculture and forestry practices, crop and forest species diversity, appropriate crop and forest rotations, organic farming, integrated pest management, the conservation of pollinators, rain water harvesting, range and pasture management, and precision agriculture systems.

[34] Sustainable forest management is defined in this report as 'the stewardship and use of forests and forest lands in a way, and at a rate, that maintains their biodiversity, productivity, regeneration capacity, vitality, and their potential to fulfil now and in the future, relevant ecological, economic and social functions at local, national and global levels and that does not cause damage to other ecosystems'.

B.6.1   Practices that contribute to climate change adaptation and mitigation in cropland include increasing soil organic matter, erosion control, improved fertiliser management, improved crop management, for example paddy rice management, and use of varieties and genetic improvements for heat and drought tolerance. For livestock, options include better grazing land management, improved manure management, higher-quality feed, and use of breeds and genetic improvement. Different farming and pastoral systems can achieve reductions in the emissions intensity of livestock products. Depending on the farming and pastoral systems and level of development, reductions in the emissions intensity of livestock products may lead to absolute reductions in GHG emissions (*medium confidence*). Many livestock related options can enhance the adaptive capacity of rural communities, in particular, of smallholders and pastoralists. Significant synergies exist between adaptation and mitigation, for example through sustainable land management approaches (*high confidence*). {4.8, 5.3.3, 5.5.1, 5.6}

B.6.2   Diversification in the food system (e.g., implementation of integrated production systems, broad-based genetic resources, and diets) can reduce risks from climate change (*medium confidence*). Balanced diets, featuring plant-based foods, such as those based on coarse grains, legumes, fruits and vegetables, nuts and seeds, and animal-sourced food produced in resilient, sustainable and low-GHG emission systems, present major opportunities for adaptation and mitigation while generating significant co-benefits in terms of human health (*high confidence*). By 2050, dietary changes could free several million km^2 (*medium confidence*) of land and provide a technical mitigation potential of 0.7 to 8.0 GtCO$_2$eq yr^{-1}, relative to business as usual projections (*high confidence*). Transitions towards low-GHG emission diets may be influenced by local production practices, technical and financial barriers and associated livelihoods and cultural habits (*high confidence*). {5.3, 5.5.2, 5.5, 5.6}

B.6.3   Reduction of food loss and waste can lower GHG emissions and contribute to adaptation through reduction in the land area needed for food production (*medium confidence*). During 2010-2016, global food loss and waste contributed 8 –10% of total anthropogenic GHG emissions (*medium confidence*). Currently, 25 –30% of total food produced is lost or wasted (*medium confidence*). Technical options such as improved harvesting techniques, on-farm storage, infrastructure, transport, packaging, retail and education can reduce food loss and waste across the supply chain. Causes of food loss and waste differ substantially between developed and developing countries, as well as between regions (*medium confidence*). By 2050, reduced food loss and waste can free several million km^2 of land (*low confidence*). {5.5.2, 6.3.6}

B.7   Future land use depends, in part, on the desired climate outcome and the portfolio of response options deployed (*high confidence*). All assessed modelled pathways that limit warming to 1.5°C or well below 2°C require land-based mitigation and land-use change, with most including different combinations of reforestation, afforestation, reduced deforestation, and bioenergy (*high confidence*). A small number of modelled pathways achieve 1.5°C with reduced land conversion (*high confidence*) and thus reduced consequences for desertification, land degradation, and food security (*medium confidence*). (Figure SPM.4) {2.6, 6.4, 7.4, 7.6, Cross-Chapter Box 9 in Chapter 6}

B.7.1   Modelled pathways limiting global warming to 1.5°C[35] include more land-based mitigation than higher warming level pathways (*high confidence*), but the impacts of climate change on land systems in these pathways are less severe (*medium confidence*). (Figure SPM.2, Figure SPM.4) {2.6, 6.4, 7.4, Cross-Chapter Box 9 in Chapter 6}

B.7.2   Modelled pathways limiting global warming to 1.5°C and 2°C project a 2 million km^2 reduction to a 12 million km^2 increase in forest area in 2050 relative to 2010 (*medium confidence*). 3°C pathways project lower forest areas, ranging from a 4 million km^2 reduction to a 6 million km^2 increase (*medium confidence*). (Figure SPM.3, Figure SPM.4) {2.5, 6.3, 7.3, 7.5, Cross-Chapter Box 9 in Chapter 6}

B.7.3   The land area needed for bioenergy in modelled pathways varies significantly depending on the socio-economic pathway, the warming level, and the feedstock and production system used (*high confidence*). Modelled pathways limiting global warming to 1.5°C use up to 7 million km^2 for bioenergy in 2050; bioenergy land area is smaller in 2°C (0.4 to 5 million km^2) and 3°C pathways (0.1 to 3 million km^2) (*medium confidence*). Pathways with large levels of land conversion may imply adverse side-effects impacting water scarcity, biodiversity, land degradation, desertification, and food security, if not adequately and carefully managed, whereas best practice implementation at appropriate scales can have co-benefits, such as management of dryland salinity, enhanced biocontrol and biodiversity and enhancing soil carbon sequestration (*high confidence*). (Figure SPM.3) {2.6, 6.1, 6.4, 7.2, Cross-Chapter Box 7 in Chapter 6}

---

[35]   In this report references to pathways limiting global warming to a particular level are based on a 66% probability of staying below that temperature level in 2100 using the MAGICC model.

B.7.4   Most mitigation pathways include substantial deployment of bioenergy technologies. A small number of modelled pathways limit warming to 1.5°C with reduced dependence on bioenergy and BECCS (land area below <1 million km^2 in 2050) and other carbon dioxide removal (CDR) options (*high confidence*). These pathways have even more reliance on rapid and far-reaching transitions in energy, land, urban systems and infrastructure, and on behavioural and lifestyle changes compared to other 1.5°C pathways. {2.6.2, 5.5.1, 6.4, Cross-Chapter Box 7 in Chapter 6}

B.7.5   These modelled pathways do not consider the effects of climate change on land or CO_2 fertilisation. In addition, these pathways include only a subset of the response options assessed in this report (*high confidence*); the inclusion of additional response options in models could reduce the projected need for bioenergy or CDR that increases the demand for land. {6.4.4, Cross-Chapter Box 9 in Chapter 6}

Potential global contribution of response options to mitigation, adaptation, combating desertification and land degradation, and enhancing food security

Panel A shows response options that can be implemented without or with limited competition for land, including some that have the potential to reduce the demand for land. Co-benefits and adverse side effects are shown quantitatively based on the high end of the range of potentials assessed. Magnitudes of contributions are categorised using thresholds for positive or negative impacts. Letters within the cells indicate confidence in the magnitude of the impact relative to the thresholds used (see legend). Confidence in the direction of change is generally higher.

Response options based on land management

	Response option	Mitigation	Adaptation	Desertification	Land Degradation	Food Security	Cost
Agriculture	Increased food productivity	L	M	L	M	H	—
	Agro-forestry	M	M	M	M	L	●
	Improved cropland management	M	L	L	L	L	●●
	Improved livestock management	M	L	L	L	L	●●●
	Agricultural diversification	L	L	L	M	L	●
	Improved grazing land management	M	L	L	L	L	—
	Integrated water management	L	L	L	L	L	●●
	Reduced grassland conversion to cropland	L	—	L	L	L	●
Forests	Forest management	M	L	L	L	L	●●
	Reduced deforestation and forest degradation	H	L	L	L	L	●●
Soils	Increased soil organic carbon content	H	L	M	M	L	●●
	Reduced soil erosion	←→ L	L	M	M	L	●●
	Reduced soil salinization		L	L	M	L	●●
	Reduced soil compaction		L		L	L	●
Other ecosystems	Fire management	M	M	M	M	L	●
	Reduced landslides and natural hazards	L	L	L	L	L	—
	Reduced pollution including acidification	←→ M	M	L	L	L	—
	Restoration & reduced conversion of coastal wetlands	M	L	M	M	←→ L	—
	Restoration & reduced conversion of peatlands	M	—	na	M	L	●

Response options based on value chain management

	Response option	Mitigation	Adaptation	Desertification	Land Degradation	Food Security	Cost
Demand	Reduced post-harvest losses	H	M	L	L	H	—
	Dietary change	H	—	L	H	H	—
	Reduced food waste (consumer or retailer)	H	—	L	M	M	—
Supply	Sustainable sourcing	—	L		L	L	—
	Improved food processing and retailing	L	L			L	—
	Improved energy use in food systems	L	L			L	—

Response options based on risk management

	Response option	Mitigation	Adaptation	Desertification	Land Degradation	Food Security	Cost
Risk	Livelihood diversification	—	L		L	L	—
	Management of urban sprawl	—	L	L	M	L	—
	Risk sharing instruments	←→ L	L		←→ L	L	●●

Options shown are those for which data are available to assess global potential for three or more land challenges.
The magnitudes are assessed independently for each option and are not additive.

Key for criteria used to define magnitude of impact of each integrated response option

		Mitigation Gt CO₂-eq yr⁻¹	Adaptation Million people	Desertification Million km²	Land Degradation Million km²	Food Security Million people
Positive	Large	More than 3	Positive for more than 25	Positive for more than 3	Positive for more than 3	Positive for more than 100
	Moderate	0.3 to 3	1 to 25	0.5 to 3	0.5 to 3	1 to 100
	Small	Less than 0.3	Less than 1	Less than 0.5	Less than 0.5	Less than 1
	Negligible	No effect	No effect	No effect	No effect	No effect
Negative	Small	Less than -0.3	Less than 1	Less than 0.5	Less than 0.5	Less than 1
	Moderate	-0.3 to -3	1 to 25	0.5 to 3	0.5 to 3	1 to 100
	Large	More than -3	Negative for more than 25	Negative for more than 3	Negative for more than 3	Negative for more than 100

←→ **Variable:** Can be positive or negative   ——— no data   na not applicable

Confidence level
Indicates confidence in the estimate of magnitude category.

H High confidence
M Medium confidence
L Low confidence

Cost range
See technical caption for cost ranges in US$ tCO₂e⁻¹ or US$ ha⁻¹.

●●● High cost
●● Medium cost
● Low cost
— no data

Potential global contribution of response options to mitigation, adaptation, combating desertification and land degradation, and enhancing food security

Panel B shows response options that rely on additional land-use change and could have implications across three or more land challenges under different implementation contexts. For each option, the first row (high level implementation) shows a quantitative assessment (as in Panel A) of implications for global implementation at scales delivering CO_2 removals of more than 3 $GtCO_2 \, yr^{-1}$ using the magnitude thresholds shown in Panel A. The red hatched cells indicate an increasing pressure but unquantified impact. For each option, the second row (best practice implementation) shows qualitative estimates of impact if implemented using best practices in appropriately managed landscape systems that allow for efficient and sustainable resource use and supported by appropriate governance mechanisms. In these qualitative assessments, green indicates a positive impact, grey indicates a neutral interaction.

Bioenergy and BECCS

Mitigation	Adaptation	Desertification	Land degradation	Food security	Cost
H	L	(increasing pressure)	(increasing pressure)	L	● / ●●●

High level: Impacts on adaptation, desertification, land degradation and food security are maximum potential impacts, assuming carbon dioxide removal by BECCS at a scale of 11.3 $GtCO_2 \, yr^{-1}$ in 2050, and noting that bioenergy without CCS can also achieve emissions reductions of up to several $GtCO_2 \, yr^{-1}$ when it is a low carbon energy source {2.6.1; 6.3.1}. Studies linking bioenergy to food security estimate an increase in the population at risk of hunger to up to 150 million people at this level of implementation {6.3.5}. The red hatched cells for desertification and land degradation indicate that while up to 15 million km^2 of additional land is required in 2100 in 2°C scenarios which will increase pressure for desertification and land degradation, the actual area affected by this additional pressure is not easily quantified {6.3.3; 6.3.4}.

Mitigation	Adaptation	Desertification	Land degradation	Food security

Best practice: The sign and magnitude of the effects of bioenergy and BECCS depends on the scale of deployment, the type of bioenergy feedstock, which other response options are included, and where bioenergy is grown (including prior land use and indirect land use change emissions). For example, limiting bioenergy production to marginal lands or abandoned cropland would have negligible effects on biodiversity, food security, and potentially co-benefits for land degradation; however, the benefits for mitigation could also be smaller. {Table 6.58}

Reforestation and forest restoration

Mitigation	Adaptation	Desertification	Land degradation	Food security	Cost
M	M	M	M	M	●●

High level: Impacts on adaptation, desertification, land degradation and food security are maximum potential impacts assuming implementation of reforestation and forest restoration (partly overlapping with afforestation) at a scale of 10.1 $GtCO_2 \, yr^{-1}$ removal {6.3.1}. Large-scale afforestation could cause increases in food prices of 80% by 2050, and more general mitigation measures in the AFOLU sector can translate into a rise in undernourishment of 80–300 million people; the impact of reforestation is lower {6.3.5}.

Mitigation	Adaptation	Desertification	Land degradation	Food security

Best practice: There are co-benefits of reforestation and forest restoration in previously forested areas, assuming small scale deployment using native species and involving local stakeholders to provide a safety net for food security. Examples of sustainable implementation include, but are not limited to, reducing illegal logging and halting illegal forest loss in protected areas, reforesting and restoring forests in degraded and desertified lands {Box6.1C; Table 6.6}.

Afforestation

Mitigation	Adaptation	Desertification	Land degradation	Food security	Cost
M	M	M	L	M	●●

High level: Impacts on adaptation, desertification, land degradation and food security are maximum potential impacts assuming implementation of afforestation (partly overlapping with reforestation and forest restoration) at a scale of 8.9 $GtCO_2 \, yr^{-1}$ removal {6.3.1}. Large-scale afforestation could cause increases in food prices of 80% by 2050, and more general mitigation measures in the AFOLU sector can translate into a rise in undernourishment of 80–300 million people {6.3.5}.

Mitigation	Adaptation	Desertification	Land degradation	Food security

Best practice: Afforestation is used to prevent desertification and to tackle land degradation. Forested land also offers benefits in terms of food supply, especially when forest is established on degraded land, mangroves, and other land that cannot be used for agriculture. For example, food from forests represents a safety-net during times of food and income insecurity {6.3.5}.

Biochar addition to soil

Mitigation	Adaptation	Desertification	Land degradation	Food security	Cost
M			L	L	●●●

High level: Impacts on adaptation, desertification, land degradation and food security are maximum potential impacts assuming implementation of biochar at a scale of 6.6 $GtCO_2 \, yr^{-1}$ removal {6.3.1}. Dedicated biomass crops required for feedstock production could occupy 0.4–2.6 Mkm^2 of land, equivalent to around 20% of the global cropland area, which could potentially have a large effect on food security for up to 100 million people {6.3.5}.

Mitigation	Adaptation	Desertification	Land degradation	Food security

Best practice: When applied to land, biochar could provide moderate benefits for food security by improving yields by 25% in the tropics, but with more limited impacts in temperate regions, or through improved water holding capacity and nutrient use efficiency. Abandoned cropland could be used to supply biomass for biochar, thus avoiding competition with food production; 5-9 Mkm^2 of land is estimated to be available for biomass production without compromising food security and biodiversity, considering marginal and degraded land and land released by pasture intensification {6.3.5}.

27

Figure SPM.3: Potential global contribution of response options to mitigation, adaptation, combating desertification and land degradation, and enhancing food security. | This Figure is based on an aggregation of information from studies with a wide variety of assumptions about how response options are implemented and the contexts in which they occur. Response options implemented differently at local to global scales could lead to different outcomes. **Magnitude of potential:** For panel A, magnitudes are for the technical potential of response options globally. For each land challenge, magnitudes are set relative to a marker level as follows. For mitigation, potentials are set relative to the approximate potentials for the response options with the largest individual impacts ($\sim$3 GtCO$_2$-eq yr^{-1}). The threshold for the 'large' magnitude category is set at this level. For adaptation, magnitudes are set relative to the 100 million lives estimated to be affected by climate change and a carbon-based economy between 2010 and 2030. The threshold for the 'large' magnitude category represents 25% of this total. For desertification and land degradation, magnitudes are set relative to the lower end of current estimates of degraded land, 10–60 million km^2. The threshold for the 'large' magnitude category represents 30% of the lower estimate. For food security, magnitudes are set relative to the approximately 800 million people who are currently undernourished. The threshold for the 'large' magnitude category represents 12.5% of this total. For panel B, for the first row (high level implementation) for each response option, the magnitude and thresholds are as defined for panel A. In the second row (best practice implementation) for each response option, the qualitative assessments that are green denote potential positive impacts, and those shown in grey indicate neutral interactions. Increased food production is assumed to be achieved through sustainable intensification rather than through injudicious application of additional external inputs such as agrochemicals. **Levels of confidence:** Confidence in the magnitude category (high, medium or low) into which each option falls for mitigation, adaptation, combating desertification and land degradation, and enhancing food security. *High confidence* means that there is a high level of agreement and evidence in the literature to support the categorisation as high, medium or low magnitude. *Low confidence* denotes that the categorisation of magnitude is based on few studies. *Medium confidence* reflects medium evidence and agreement in the magnitude of response. **Cost ranges:** Cost estimates are based on aggregation of often regional studies and vary in the components of costs that are included. In panel B, cost estimates are not provided for best practice implementation. One coin indicates low cost (<USD10 tCO$_2$-eq^{-1} or <USD20 ha^{-1}), two coins indicate medium cost (USD10-USD100 tCO$_2$-eq^{-1} or USD20 –USD200 ha^{-1}), and three coins indicate high cost (>USD100 tCO$_2$-eq^{-1} or USD200 ha^{-1}). Thresholds in USD ha^{-1} are chosen to be comparable, but precise conversions will depend on the response option. **Supporting evidence:** Supporting evidence for the magnitude of the quantitative potential for land management-based response options can be found as follows: for mitigation Table's 6.13 to 6.20, with further evidence in Section 2.7.1; for adaptation Table's 6.21 to 6.28; for combating desertification Table's 6.29 to 6.36, with further evidence in Chapter 3; for combating degradation tables 6.37 to 6.44, with further evidence in Chapter 4; for enhancing food security Table's 6.45 to 6.52, with further evidence in Chapter 5. Other synergies and trade-offs not shown here are discussed in Chapter 6. Additional supporting evidence for the qualitative assessments in the second row for each option in panel B can be found in the Table's 6.6, 6.55, 6.56 and 6.58, Section 6.3.5.1.3, and Box 6.1c.

C.   Enabling response options

C.1   Appropriate design of policies, institutions and governance systems at all scales can contribute to land-related adaptation and mitigation while facilitating the pursuit of climate-adaptive development pathways (*high confidence*). Mutually supportive climate and land policies have the potential to save resources, amplify social resilience, support ecological restoration, and foster engagement and collaboration between multiple stakeholders (*high confidence*). (Figure SPM.1, Figure SPM.2, Figure SPM.3) {3.6.2, 3.6.3, 4.8, 4.9.4, 5.7, 6.3, 6.4, 7.2.2, 7.3, 7.4, 7.4.7, 7.4.8, 7.5, 7.5.5, 7.5.6, 7.6.6, Cross-Chapter Box 10 in Chapter 7}

C.1.1   Land-use zoning, spatial planning, integrated landscape planning, regulations, incentives (such as payment for ecosystem services), and voluntary or persuasive instruments (such as environmental farm planning, standards and certification for sustainable production, use of scientific, local and indigenous knowledge and collective action), can achieve positive adaptation and mitigation outcomes (*medium confidence*). They can also contribute revenue and provide incentive to rehabilitate degraded lands and adapt to and mitigate climate change in certain contexts (*medium confidence*). Policies promoting the target of land degradation neutrality can also support food security, human wellbeing and climate change adaptation and mitigation (*high confidence*). (Figure SPM.2) {3.4.2, 4.1.6, 4.7, 4.8.5, 5.1.2, 5.7.3, 7.3, 7.4.6, 7.4.7, 7.5}

C.1.2   Insecure land tenure affects the ability of people, communities and organisations to make changes to land that can advance adaptation and mitigation (*medium confidence*). Limited recognition of customary access to land and ownership of land can result in increased vulnerability and decreased adaptive capacity (*medium confidence*). Land policies (including recognition of customary tenure, community mapping, redistribution, decentralisation, co-management, regulation of rental markets) can provide both security and flexibility response to climate change (*medium confidence*). {3.6.1, 3.6.2, 5.3, 7.2.4, 7.6.4, Cross-Chapter Box 6 in Chapter 5}

C.1.3   Achieving land degradation neutrality will involve a balance of measures that avoid and reduce land degradation, through adoption of sustainable land management, and measures to reverse degradation through rehabilitation and restoration of degraded land. Many interventions to achieve land degradation neutrality commonly also deliver climate change adaptation and mitigation benefits. The pursuit of land degradation neutrality provides impetus to address land degradation and climate change simultaneously (*high confidence*). {4.5.3, 4.8.5, 4.8.7, 7.4.5}

C.1.4   Due to the complexity of challenges and the diversity of actors involved in addressing land challenges, a mix of policies, rather than single policy approaches, can deliver improved results in addressing the complex challenges of sustainable land management and climate change (*high confidence*). Policy mixes can strongly reduce the vulnerability and exposure of human and natural systems to climate change (*high confidence*). Elements of such policy mixes may include weather and health insurance, social protection and adaptive safety nets, contingent finance and reserve funds, universal access to early warning systems combined with effective contingency plans (*high confidence*). (Figure SPM.4) {1.2, 4.8, 4.9.2, 5.3.2, 5.6, 5.6.6, 5.7.2, 7.3.2, 7.4, 7.4.2, 7.4.6, 7.4.7, 7.4.8, 7.5.5, 7.5.6, 7.6.4}

C.2   Policies that operate across the food system, including those that reduce food loss and waste and influence dietary choices, enable more sustainable land-use management, enhanced food security and low emissions trajectories (*high confidence*). Such policies can contribute to climate change adaptation and mitigation, reduce land degradation, desertification and poverty as well as improve public health (*high confidence*). The adoption of sustainable land management and poverty eradication can be enabled by improving access to markets, securing land tenure, factoring environmental costs into food, making payments for ecosystem services, and enhancing local and community collective action (*high confidence*). {1.1.2, 1.2.1, 3.6.3, 4.7.1, 4.7.2, 4.8, 5.5, 6.4, 7.4.6, 7.6.5}

C.2.1   Policies that enable and incentivise sustainable land management for climate change adaptation and mitigation include improved access to markets for inputs, outputs and financial services, empowering women and indigenous peoples, enhancing local and community collective action, reforming subsidies and promoting an enabling trade system (*high confidence*). Land restoration and rehabilitation efforts can be more effective when policies support local management of natural resources, while strengthening cooperation between actors and institutions, including at the international level. {3.6.3, 4.1.6, 4.5.4, 4.8.2, 4.8.4, 5.7, 7.2, 7.3}

C.2.2   Reflecting the environmental costs of land-degrading agricultural practices can incentivise more sustainable land management (*high confidence*). Barriers to the reflection of environmental costs arise from technical difficulties in estimating these costs and those embodied in foods. {3.6.3, 5.5.1, 5.5.2, 5.6.6, 5.7, 7.4.4, Cross-Chapter Box 10 in Chapter 7}

C.2.3   Adaptation and enhanced resilience to extreme events impacting food systems can be facilitated by comprehensive risk management, including risk sharing and transfer mechanisms (*high confidence*). Agricultural diversification, expansion of market access, and preparation for increasing supply chain disruption can support the scaling up of adaptation in food systems (*high confidence*). {5.3.2, 5.3.3, 5.3.5}

C.2.4   Public health policies to improve nutrition, such as increasing the diversity of food sources in public procurement, health insurance, financial incentives, and awareness-raising campaigns, can potentially influence food demand, reduce healthcare costs, contribute to lower GHG emissions and enhance adaptive capacity (*high confidence*). Influencing demand for food, through promoting diets based on public health guidelines, can enable more sustainable land management and contribute to achieving multiple SDGs (*high confidence*). {3.4.2, 4.7.2, 5.1, 5.7, 6.3, 6.4}

C.3   Acknowledging co-benefits and trade-offs when designing land and food policies can overcome barriers to implementation (*medium confidence*). Strengthened multi-level, hybrid and cross-sectoral governance, as well as policies developed and adopted in an iterative, coherent, adaptive and flexible manner can maximise co-benefits and minimise trade-offs, given that land management decisions are made from farm level to national scales, and both climate and land policies often range across multiple sectors, departments and agencies (*high confidence*). (Figure SPM.3) {4.8.5, 4.9, 5.6, 6.4, 7.3, 7.4.6, 7.4.8, 7.4.9, 7.5.6, 7.6.2}

C.3.1   Addressing desertification, land degradation, and food security in an integrated, coordinated and coherent manner can assist climate resilient development and provides numerous potential co-benefits (*high confidence*). {3.7.5, 4.8, 5.6, 5.7, 6.4, 7.2.2, 7.3.1, 7.3.4, 7.4.7, 7.4.8, 7.5.6, 7.5.5}

C.3.2   Technological, biophysical, socio-economic, financial and cultural barriers can limit the adoption of many land-based response options, as can uncertainty about benefits (*high confidence*). Many sustainable land management practices are not widely adopted due to insecure land tenure, lack of access to resources and agricultural advisory services, insufficient and unequal private and public incentives, and lack of knowledge and practical experience (*high confidence*). Public discourse, carefully designed policy interventions, incorporating social learning and market changes can together help reduce barriers to implementation (*medium confidence*). {3.6.1, 3.6.2, 5.3.5, 5.5.2, 5.6, 6.2, 6.4, 7.4, 7.5, 7.6}

C.3.3   The land and food sectors face particular challenges of institutional fragmentation and often suffer from a lack of engagement between stakeholders at different scales and narrowly focused policy objectives (*medium confidence*). Coordination with other sectors, such as public health, transportation, environment, water, energy and infrastructure, can increase co-benefits, such as risk reduction and improved health (*medium confidence*). {5.6.3, 5.7, 6.2, 6.4.4, 7.1, 7.3, 7.4.8, 7.6.2, 7.6.3}

C.3.4   Some response options and policies may result in trade-offs, including social impacts, ecosystem functions and services damage, water depletion, or high costs, that cannot be well-managed, even with institutional best practices (*medium confidence*). Addressing such trade-offs helps avoid maladaptation (*medium confidence*). Anticipation and evaluation of potential trade-offs and knowledge gaps supports evidence-based policymaking to weigh the costs and benefits of specific responses for different stakeholders (*medium confidence*). Successful management of trade-offs often includes maximising stakeholder input with structured feedback processes, particularly in community-based models, use of innovative fora like facilitated dialogues or spatially explicit mapping, and iterative adaptive management that allows for continuous readjustments in policy as new evidence comes to light (*medium confidence*). {5.3.5, 6.4.2, 6.4.4, 6.4.5, 7.5.6, Cross-Chapter Box 9 in Chapter 7}

C.4   The effectiveness of decision-making and governance is enhanced by the involvement of local stakeholders (particularly those most vulnerable to climate change including indigenous peoples and local communities, women, and the poor and marginalised) in the selection, evaluation, implementation and monitoring of policy instruments for land-based climate change adaptation and mitigation (*high confidence*). Integration across sectors and scales increases the chance of maximising co-benefits and minimising trade-offs (*medium confidence*). {1.4, 3.1, 3.6, 3.7, 4.8, 4.9, 5.1.3, Box 5.1, 7.4, 7.6}

C.4.1   Successful implementation of sustainable land management practices requires accounting for local environmental and socio-economic conditions (*very high confidence*). Sustainable land management in the context of climate change is typically advanced by involving all relevant stakeholders in identifying land-use pressures and impacts (such as biodiversity decline, soil loss, over-extraction of groundwater, habitat loss, land-use change in agriculture, food production and forestry) as well as preventing, reducing and restoring degraded land (*medium confidence*). {1.4.1, 4.1.6, 4.8.7, 5.2.5, 7.2.4, 7.6.2, 7.6.4}

C.4.2   Inclusiveness in the measurement, reporting and verification of the performance of policy instruments can support sustainable land management (*medium confidence*). Involving stakeholders in the selection of indicators, collection of climate data, land modelling and land-use planning, mediates and facilitates integrated landscape planning and choice of policy (*medium confidence*). {3.7.5, 5.7.4, 7.4.1, 7.4.4, 7.5.3, 7.5.4, 7.5.5, 7.6.4, 7.6.6}

C.4.3   Agricultural practices that include indigenous and local knowledge can contribute to overcoming the combined challenges of climate change, food security, biodiversity conservation, and combating desertification and land degradation (*high confidence*). Coordinated action across a range of actors including businesses, producers, consumers, land managers and policymakers in partnership with indigenous peoples and local communities enable conditions for the adoption of response options (*high confidence*) {3.1.3, 3.6.1, 3.6.2, 4.8.2, 5.5.1, 5.6.4, 5.7.1, 5.7.4, 6.2, 7.3, 7.4.6, 7.6.4}

C.4.4   Empowering women can bring synergies and co-benefits to household food security and sustainable land management (*high confidence*). Due to women's disproportionate vulnerability to climate change impacts, their inclusion in land management and tenure is constrained. Policies that can address land rights and barriers to women's participation in sustainable land management include financial transfers to women under the auspices of anti-poverty programmes, spending on health, education, training and capacity building for women, subsidised credit and program dissemination through existing women's community-based organisations (*medium confidence*). {1.4.1, 4.8.2, 5.1.3, Cross-Chapter Box 11 in Chapter 7}

A. Pathways linking socioeconomic development, mitigation responses and land

Socioeconomic development and land management influence the evolution of the land system including the relative amount of land allocated to CROPLAND, PASTURE, BIOENERGY CROPLAND, FOREST, and NATURAL LAND. The lines show the median across Integrated Assessment Models (IAMs) for three alternative shared socioeconomic pathways (**SSP1**, **SSP2** and **SSP5** at **RCP1.9**); shaded areas show the range across models. Note that pathways illustrate the effects of climate change mitigation but not those of climate change impacts or adaptation.

A. Sustainability-focused (SSP1)
Sustainability in land management, agricultural intensification, production and consumption patterns result in reduced need for agricultural land, despite increases in per capita food consumption. This land can instead be used for reforestation, afforestation, and bioenergy.

B. Middle of the road (SSP2)
Societal as well as technological development follows historical patterns. Increased demand for land mitigation options such as bioenergy, reduced deforestation or afforestation decreases availability of agricultural land for food, feed and fibre.

C. Resource intensive (SSP5)
Resource-intensive production and consumption patterns, results in high baseline emissions. Mitigation focuses on technological solutions including substantial bioenergy and BECCS . Intensification and competing land uses contribute to declines in agricultural land.

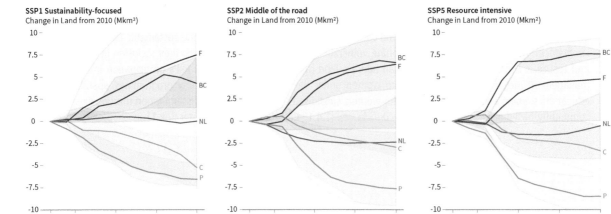

B. Land use and land cover change in the SSPs

	Quantitative indicators for the SSPs	Count of models included*	Change in Natural Land from 2010 Mkm²	Change in Bioenergy Cropland from 2010 Mkm²	Change in Cropland from 2010 Mkm²	Change in Forest from 2010 Mkm²	Change in Pasture from 2010 Mkm²
SSP1	RCP1.9 in 2050	5/5	0.5 (-4.9, 1)	2.1 (0.9, 5)	-1.2 (-4.6, -0.3)	3.4 (-0.1, 9.4)	-4.1 (-5.6, -2.5)
	↳ 2100		0 (-7.3, 7.1)	4.3 (1.5, 7.2)	-5.2 (-7.6, -1.8)	7.5 (0.4, 15.8)	-6.5 (-12.2, -4.8)
	RCP2.6 in 2050	5/5	-0.9 (-2.2, 1.5)	1.3 (0.4, 1.9)	-1 (-4.7, 1)	2.6 (-0.1, 8.4)	-3 (-4, -2.4)
	↳ 2100		0.2 (-3.5, 1.1)	5.1 (1.6, 6.3)	-3.2 (-7.7, -1.8)	6.6 (-0.1, 10.5)	-5.5 (-9.9, -4.2)
	RCP4.5 in 2050	5/5	0.5 (-1, 1.7)	0.8 (0.5, 1.3)	0.1 (-3.2, 1.5)	0.6 (-0.7, 4.2)	-2.4 (-3.3, -0.9)
	↳ 2100		1.8 (-1.7, 6)	1.9 (1.4, 3.7)	-2.3 (-6.4, -1.6)	3.9 (0.2, 8.8)	-4.6 (-7.3, -2.7)
	Baseline in 2050	5/5	0.3 (-1.1, 1.8)	0.5 (0.2, 1.4)	0.2 (-1.6, 1.9)	-0.1 (-0.8, 1.1)	-1.5 (-2.9, -0.2)
	↳ 2100		3.3 (-0.3, 5.9)	1.8 (1.4, 2.4)	-1.5 (-5.7, -0.9)	0.9 (0.3, 3)	-2.1 (-7, 0)
SSP2	RCP1.9 in 2050	4/5	-2.2 (-7, 0.6)	4.5 (2.1, 7)	-1.2 (-2, 0.3)	3.4 (-0.9, 7)	-4.8 (-6.2, -0.4)
	↳ 2100		-2.3 (-9.6, 2.7)	6.6 (3.6, 11)	-2.9 (-4, 0.1)	6.4 (-0.8, 9.5)	-7.6 (-11.7, -1.3)
	RCP2.6 in 2050	5/5	-3.2 (-4.2, 0.1)	2.2 (1.7, 4.7)	0.6 (-1.9, 1.9)	1.6 (-0.9, 4.2)	-1.4 (-3.7, 0.4)
	↳ 2100		-5.2 (-7.2, 0.5)	6.9 (2.3, 10.8)	-1.4 (-4, 0.8)	5.6 (-0.9, 5.9)	-7.2 (-8, 0.5)
	RCP4.5 in 2050	5/5	-2.2 (-2.2, 0.7)	1.5 (0.1, 2.1)	1.2 (-0.9, 2.7)	-0.9 (-2.5, 2.9)	-0.1 (-2.5, 1.6)
	↳ 2100		-3.4 (-4.7, 1.5)	4.1 (0.4, 6.3)	0.7 (-2.6, 3.1)	-0.5 (-3.1, 5.9)	-2.8 (-5.3, 1.9)
	Baseline in 2050	5/5	-1.5 (-2.6, -0.2)	0.7 (0, 1.5)	1.3 (1, 2.7)	-1.3 (-2.5, -0.4)	-0.1 (-1.2, 1.6)
	↳ 2100		-2.1 (-5.9, 0.3)	1.2 (0.1, 2.4)	1.9 (0.8, 2.8)	-1.3 (-2.7, -0.2)	-0.2 (-1.9, 2.1)
SSP3	RCP1.9 in 2050	Infeasible in all assessed models	-	-	-	-	-
	↳ 2100		-	-	-	-	-
	RCP2.6 in 2050	Infeasible in all assessed models	-	-	-	-	-
	↳ 2100		-	-	-	-	-
	RCP4.5 in 2050	3/3	-3.4 (-4.4, -2)	1.3 (1.3, 2)	2.3 (1.2, 3)	-2.4 (-4, -1)	2.1 (-0.1, 3.8)
	↳ 2100		-6.2 (-6.8, -5.4)	4.6 (1.5, 7.1)	3.4 (1.9, 4.5)	-3.1 (-5.5, -0.3)	2 (-2.5, 4.4)
	Baseline in 2050	4/4	-3 (-4.6, -1.7)	1 (0.2, 1.5)	2.5 (1.5, 3)	-2.5 (-4, -1.5)	2.4 (0.6, 3.8)
	↳ 2100		-5 (-7.1, -4.2)	1.1 (0.9, 2.5)	5.1 (3.8, 6.1)	-5.3 (-6, -2.6)	3.4 (0.9, 6.4)
SSP4	RCP1.9 in 2050	Infeasible in all assessed models**	-	-	-	-	-
	↳ 2100		-	-	-	-	-
	RCP2.6 in 2050	3/3	-4.5 (-6, -2.1)	3.3 (1.5, 4.5)	0.5 (-0.1, 0.9)	0.7 (-0.3, 2.2)	-0.6 (-0.7, 0.1)
	↳ 2100		-5.8 (-10.2, -4.7)	2.5 (2.3, 15.2)	-0.8 (-0.8, 1.8)	1.4 (-1.7, 4.1)	-1.2 (-2.5, -0.2)
	RCP4.5 in 2050	3/3	-2.7 (-4.4, -0.4)	1.7 (1, 1.9)	1.1 (-0.1, 1.7)	-1.8 (-2.3, 2.1)	0.8 (-0.5, 1.5)
	↳ 2100		-2.8 (-7.8, -2)	2.7 (2.3, 4.7)	1.1 (0.2, 1.2)	-0.7 (-2.6, 1)	1.4 (-1, 1.8)
	Baseline in 2050	3/3	-2.8 (-2.9, -0.2)	1.1 (0.7, 2)	1.1 (0.7, 1.8)	-1.8 (-2.3, -1)	1.5 (-0.5, 2.1)
	↳ 2100		-2.4 (-5, -1)	1.7 (1.4, 2.6)	1.2 (1.2, 1.9)	-2.4 (-2.5, -2)	1.3 (-1, 4.4)
SSP5	RCP1.9 in 2050	2/4	-1.5 (-3.9, 0.9)	6.7 (6.2, 7.2)	-1.9 (-3.5, -0.4)	3.1 (-0.1, 6.3)	-6.4 (-7.7, -5.1)
	↳ 2100		-0.5 (-4.2, 3.2)	7.6 (7.2, 8)	-3.4 (-6.2, -0.5)	4.7 (0.1, 9.4)	-8.5 (-10.7, -6.2)
	RCP2.6 in 2050	4/4	-3.4 (-6.9, 0.3)	4.8 (3.8, 5.1)	-2.1 (-4, 1)	3.9 (-0.1, 6.7)	-4.4 (-5, 0.2)
	↳ 2100		-4.3 (-8.4, 0.5)	9.1 (7.7, 9.2)	-3.3 (-6.5, -0.5)	3.9 (-0.1, 9.3)	-6.3 (-9.1, -1.4)
	RCP4.5 in 2050	4/4	-2.5 (-3.7, 0.2)	1.7 (0.6, 2.9)	0.6 (-3.3, 1.9)	-0.1 (-1.7, 6)	-1.2 (-2.6, 2.3)
	↳ 2100		-4.1 (-4.6, 0.7)	4.8 (2, 8)	-1 (-5.5, 1)	-0.2 (-1.4, 9.1)	-3 (-5.2, 2.1)
	Baseline in 2050	4/4	-0.6 (-3.8, 0.4)	0.8 (0, 2.1)	1.5 (-0.7, 3.3)	-1.9 (-3.4, 0.5)	-0.1 (-1.5, 2.9)
	↳ 2100		-0.2 (-2.4, 1.8)	1 (0.2, 2.3)	1 (-2, 2.5)	-2.1 (-3.4, 1.1)	-0.4 (-2.4, 2.8)

* Count of models included / Count of models attempted. One model did not provide land data and is excluded from all entries.

** One model could reach RCP1.9 with SSP4, but did not provide land data

Figure SPM.4: Pathways linking socioeconomic development, mitigation responses and land | Future scenarios provide a framework for understanding the implications of mitigation and socioeconomics on land. The Shared Socioeconomic Pathways (SSPs) span a range of different socioeconomic assumptions (Box SPM.1). They are combined with Representative Concentration Pathways (RCPs)[36] which imply different levels of mitigation. The changes in cropland, pasture, bioenergy cropland, forest, and natural land from 2010 are shown. For this Figure, Cropland includes all land in food, feed, and fodder crops, as well as other arable land (cultivated area). This category includes first generation non-forest bioenergy crops (e.g., corn for ethanol, sugar cane for ethanol, soybeans for biodiesel), but excludes second generation bioenergy crops. Pasture includes categories of pasture land, not only high-quality rangeland, and is based on FAO definition of 'permanent meadows and pastures'. Bioenergy cropland includes land dedicated to second generation energy crops (e.g., switchgrass, miscanthus, fast-growing wood species). Forest includes managed and unmanaged forest. Natural land includes other grassland, savannah, and shrubland. **Panel A:** This panel shows integrated assessment model (IAM)[37] results for SSP1, SSP2 and SSP5 at RCP1.9.[38] For each pathway, the shaded areas show the range across all IAMs; the line indicates the median across models. For RCP1.9, SSP1, SSP2 and SSP5 results are from five, four and two IAMs respectively. **Panel B:** Land use and land cover change are indicated for various SSP-RCP combinations, showing multi-model median and range (min, max). (Box SPM.1) {1.3.2, 2.7.2, 6.1, 6.4.4, 7.4.2, 7.4.4, 7.4.5, 7.4.6, 7.4.7, 7.4.8, 7.5.3, 7.5.6, Cross-Chapter Box 1 in Chapter 1, Cross-Chapter Box 9 in Chapter 6}

---

[36]   Representative Concentration Pathways (RCPs) are scenarios that include timeseries of emissions and concentrations of the full suite of greenhouse gases (GHGs) and aerosols and chemically active gases, as well as land use/land cover.

[37]   Integrated Assessment Models (IAMs) integrate knowledge from two or more domains into a single framework. In this figure, IAMs are used to assess linkages between economic, social and technological development and the evolution of the climate system.

[38]   The RCP1.9 pathways assessed in this report have a 66% chance of limiting warming to 1.5°C in 2100, but some of these pathways overshoot 1.5°C of warming during the 21st century by >0.1°C.

D. Action in the near-term

D.1   Actions can be taken in the near-term, based on existing knowledge, to address desertification, land degradation and food security while supporting longer-term responses that enable adaptation and mitigation to climate change. These include actions to build individual and institutional capacity, accelerate knowledge transfer, enhance technology transfer and deployment, enable financial mechanisms, implement early warning systems, undertake risk management and address gaps in implementation and upscaling (*high confidence*). {3.6.1, 3.6.2, 3.7.2, 4.8, 5.3.3, 5.5, 5.6.4, 5.7, 6.2, 6.4, 7.3, 7.4, 7.6, Cross-Chapter Box 10 in Chapter 7}

D.1.1   Near-term capacity-building, technology transfer and deployment, and enabling financial mechanisms can strengthen adaptation and mitigation in the land sector. Knowledge and technology transfer can help enhance the sustainable use of natural resources for food security under a changing climate (*medium confidence*). Raising awareness, capacity building and education about sustainable land management practices, agricultural extension and advisory services, and expansion of access to agricultural services to producers and land users can effectively address land degradation (*medium confidence*). {3.1, 5.7.4, 7.2, 7.3.4, 7.5.4}

D.1.2   Measuring and monitoring land use change including land degradation and desertification is supported by the expanded use of new information and communication technologies (cell phone based applications, cloud-based services, ground sensors, drone imagery), use of climate services, and remotely sensed land and climate information on land resources (*medium confidence*). Early warning systems for extreme weather and climate events are critical for protecting lives and property and enhancing disaster risk reduction and management (*high confidence*). Seasonal forecasts and early warning systems are critical for food security (famine) and biodiversity monitoring including pests and diseases and adaptive climate risk management (*high confidence*). There are high returns on investments in human and institutional capacities. These investments include access to observation and early warning systems, and other services derived from in-situ hydro-meteorological and remote sensing-based monitoring systems and data, field observation, inventory and survey, and expanded use of digital technologies (*high confidence*). {1.2, 3.6.2, 4.2.2, 4.2.4, 5.3.1, 5.3.6, 6.4, 7.3.4, 7.4.3, 7.5.4, 7.5.5, 7.6.4, Cross-Chapter Box 5 in Chapter 3}

D.1.3   Framing land management in terms of risk management, specific to land, can play an important role in adaptation through landscape approaches, biological control of outbreaks of pests and diseases, and improving risk sharing and transfer mechanisms (*high confidence*). Providing information on climate-related risk can improve the capacity of land managers and enable timely decision making (*high confidence*). {5.3.2, 5.3.5, 5.6.2, 5.6.3 5.6.5, 5.7.1, 5.7.2, 7.2.4, Cross-Chapter Box 6 in Chapter 5}

D.1.4   Sustainable land management can be improved by increasing the availability and accessibility of data and information relating to the effectiveness, co-benefits and risks of emerging response options and increasing the efficiency of land use (*high confidence*). Some response options (e.g., improved soil carbon management) have been implemented only at small-scale demonstration facilities and knowledge, financial, and institutional gaps and challenges exist with upscaling and the widespread deployment of these options (*medium confidence*). {4.8, 5.5.1, 5.5.2, 5.6.1, 5.6.5, 5.7.5, 6.2, 6.4}

D.2   Near-term action to address climate change adaptation and mitigation, desertification, land degradation and food security can bring social, ecological, economic and development co-benefits (*high confidence*). Co-benefits can contribute to poverty eradication and more resilient livelihoods for those who are vulnerable (*high confidence*). {3.4.2, 5.7, 7.5}

D.2.1   Near-term actions to promote sustainable land management will help reduce land and food-related vulnerabilities, and can create more resilient livelihoods, reduce land degradation and desertification, and loss of biodiversity (*high confidence*). There are synergies between sustainable land management, poverty eradication efforts, access to market, non-market mechanisms and the elimination of low-productivity practices. Maximising these synergies can lead to adaptation, mitigation, and development co-benefits through preserving ecosystem functions and services (*medium confidence*). {3.4.2, 3.6.3, Table 4.2, 4.7, 4.9, 4.10, 5.6, 5.7, 7.3, 7.4, 7.5, 7.6, Cross-Chapter Box 12 in Chapter 7}

D.2.2   Investments in land restoration can result in global benefits and in drylands can have benefit-cost ratios of between three and six in terms of the estimated economic value of restored ecosystem services (*medium confidence*). Many sustainable land management technologies and practices are profitable within three to ten years (*medium confidence*). While they can require upfront investment, actions to ensure sustainable land management can improve crop yields and the economic value of pasture. Land restoration and rehabilitation measures improve livelihood systems and provide both short-term positive economic returns and longer-term benefits in terms of climate change adaptation and mitigation, biodiversity and enhanced ecosystem functions and services (*high confidence*). {3.6.1, 3.6.3, 4.8.1, 7.2.4, 7.2.3, 7.3.1, 7.4.6, Cross-Chapter Box 10 in Chapter 7}

D.2.3   Upfront investments in sustainable land management practices and technologies can range from about USD20 ha^{-1} to USD5000 ha^{-1}, with a median estimated to be around USD500 ha^{-1}. Government support and improved access to credit can help overcome barriers to adoption, especially those faced by poor smallholder farmers (*high confidence*). Near-term change to balanced diets (SPM B6.2.) can reduce the pressure on land and provide significant health co-benefits through improving nutrition (*medium confidence*). {3.6.3, 4.8, 5.3, 5.5, 5.6, 5.7, 6.4, 7.4.7, 7.5.5, Cross-Chapter Box 9 in Chapter 6}

D.3   **Rapid reductions in anthropogenic GHG emissions across all sectors following ambitious mitigation pathways reduce negative impacts of climate change on land ecosystems and food systems (*medium confidence*). Delaying climate mitigation and adaptation responses across sectors would lead to increasingly negative impacts on land and reduce the prospect of sustainable development (*medium confidence*). (Box SPM.1, Figure SPM.2) {2.5, 2.7, 5.2, 6.2, 6.4, 7.2, 7.3.1, 7.4.7, 7.4.8, 7.5.6, Cross-Chapter Box 9 in Chapter 6, Cross-Chapter Box 10 in Chapter 7}**

D.3.1   Delayed action across sectors leads to an increasing need for widespread deployment of land-based adaptation and mitigation options and can result in a decreasing potential for the array of these options in most regions of the world and limit their current and future effectiveness (*high confidence*). Acting now may avert or reduce risks and losses, and generate benefits to society (*medium confidence*). Prompt action on climate mitigation and adaptation aligned with sustainable land management and sustainable development depending on the region could reduce the risk to millions of people from climate extremes, desertification, land degradation and food and livelihood insecurity (*high confidence*). {1.3.5, 3.4.2, 3.5.2, 4.1.6, 4.7.1, 4.7.2, 5.2.3, 5.3.1, 6.3, 6.5, 7.3.1}

D.3.2   In future scenarios, deferral of GHG emissions reductions implies trade-offs leading to significantly higher costs and risks associated with rising temperatures (*medium confidence*). The potential for some response options, such as increasing soil organic carbon, decreases as climate change intensifies, as soils have reduced capacity to act as sinks for carbon sequestration at higher temperatures (*high confidence*). Delays in avoiding or reducing land degradation and promoting positive ecosystem restoration risk long-term impacts including rapid declines in productivity of agriculture and rangelands, permafrost degradation and difficulties in peatland rewetting (*medium confidence*). {1.3.1, 3.6.2, 4.8, 4.9, 4.9.1, 5.5.2, 6.3, 6.4, 7.2, 7.3; Cross-Chapter Box 10 in Chapter 7}

D.3.3   Deferral of GHG emissions reductions from all sectors implies trade-offs including irreversible loss in land ecosystem functions and services required for food, health, habitable settlements and production, leading to increasingly significant economic impacts on many countries in many regions of the world (*high confidence*). Delaying action as is assumed in high emissions scenarios could result in some irreversible impacts on some ecosystems, which in the longer-term has the potential to lead to substantial additional GHG emissions from ecosystems that would accelerate global warming (*medium confidence*). {1.3.1, 2.5.3, 2.7, 3.6.2, 4.9, 4.10.1, 5.4.2.4, 6.3, 6.4, 7.2, 7.3, Cross-Chapter Box 9 in Chapter 6, Cross-Chapter Box 10 in Chapter 7}

Technical
Summary

TS

Technical Summary

Editors:

Priyadarshi R. Shukla (India), Jim Skea (United Kingdom), Raphael Slade (United Kingdom) Renée van Diemen (The Netherlands/United Kingdom), Eamon Haughey (Ireland), Juliette Malley (United Kingdom), Minal Pathak (India), Joana Portugal Pereira (United Kingdom)

Drafting Authors:

Fahmuddin Agus (Indonesia), Almut Arneth (Germany), Paulo Artaxo (Brazil), Humberto Barbosa (Brazil), Luis G. Barioni (Brazil), Tim G. Benton (United Kingdom), Suruchi Bhadwal (India), Katherine Calvin (The United States of America), Eduardo Calvo (Peru), Donovan Campbell (Jamaica), Francesco Cherubini (Italy), Sarah Connors (France/United Kingdom), Annette Cowie (Australia), Edouard Davin (France/Switzerland), Kenel Delusca (Haiti), Fatima Denton (The Gambia), Aziz Elbehri (Morocco), Karlheinz Erb (Italy), Jason Evans (Australia), Dulce Flores-Renteria (Mexico), Felipe Garcia-Oliva (Mexico), Giacomo Grassi (Italy/European Union), Kathleen Hermans (Germany), Mario Herrero (Australia/Costa Rica), Richard Houghton (The United States of America), Joanna House (United Kingdom), Mark Howden (Australia), Margot Hurlbert (Canada), Ismail Abdel Galil Hussein (Egypt), Muhammad Mohsin Iqbal (Pakistan), Gensuo Jia (China), Esteban Jobbagy (Argentina), Francis X. Johnson (Sweden), Joyce Kimutai (Kenya), Kaoru Kitajima (Japan), Tony Knowles (South Africa), Vladimir Korotkov (The Russian Federation), Murukesan V. Krishnapillai (Micronesia/India), Jagdish Krishnaswamy (India), Werner Kurz (Canada), Anh Le Hoang (Viet Nam), Christopher Lennard (South Africa), Diqiang Li (China), Emma Liwenga (The United Republic of Tanzania), Shuaib Lwasa (Uganda), Nagmeldin Mahmoud (Sudan), Valérie Masson-Delmotte (France), Cheikh Mbow (Senegal), Pamela McElwee (The United States of America), Carlos Fernando Mena (Ecuador), Francisco Meza (Chile), Alisher Mirzabaev (Germany/Uzbekistan), John Morton (United Kingdom), Wilfran Moufouma-Okia (France), Soojeong Myeong (The Republic of Korea), Dalila Nedjraoui (Algeria), Johnson Nkem (Cameroon), Ephraim Nkonya (The United Republic of Tanzania), Nathalie De Noblet-Ducoudré (France), Lennart Olsson (Sweden), Balgis Osman Elasha (Côte d'Ivoire), Jan Petzold (Germany), Ramón Pichs-Madruga (Cuba), Elvira Poloczanska (United Kingdom), Alexander Popp (Germany), Hans-Otto Pörtner (Germany), Prajal Pradhan (Germany/Nepal), Mohammad Rahimi (Iran), Andy Reisinger (New Zealand), Marta G. Rivera-Ferre (Spain), Debra C. Roberts (South Africa), Cynthia Rosenzweig

(The United States of America), Mark Rounsevell (United Kingdom), Nobuko Saigusa (Japan), Tek Sapkota (Canada/Nepal), Elena Shevliakova (The United States of America), Andrey Sirin (The Russian Federation), Pete Smith (United Kingdom), Youba Sokona (Mali), Denis Jean Sonwa (Cameroon), Jean-Francois Soussana (France), Adrian Spence (Jamaica), Lindsay Stringer (United Kingdom), Raman Sukumar (India), Miguel Angel Taboada (Argentina), Fasil Tena (Ethiopia), Francesco N. Tubiello (The United States of America/Italy), Murat Türkeş (Turkey), Riccardo Valentini (Italy), Ranses José Vázquez Montenegro (Cuba), Louis Verchot (Colombia/The United States of America), David Viner (United Kingdom), Koko Warner (The United States of America), Mark Weltz (The United States of America), Nora M. Weyer (Germany), Anita Wreford (New Zealand), Jianguo Wu (China), Yinlong Xu (China), Noureddine Yassaa (Algeria), Sumaya Zakieldeen (Sudan), Panmao Zhai (China), Zinta Zommers (Latvia)

Chapter Scientists:
Yuping Bai (China), Aliyu Salisu Barau (Nigeria), Abdoul Aziz Diouf (Senegal), Baldur Janz (Germany), Frances Manning (United Kingdom), Erik Mencos Contreras (The United States of America/Mexico), Dorothy Nampanzira (Uganda), Chuck Chuan Ng (Malaysia), Helen Berga Paulos (Ethiopia), Xiyan Xu (China), Thobekile Zikhali (Zimbabwe)

This Technical Summary should be cited as:
P.R. Shukla, J. Skea, R. Slade, R. van Diemen, E. Haughey, J. Malley, M. Pathak, J. Portugal Pereira (eds.) Technical Summary, 2019. In: *Climate Change and Land: an IPCC special report on climate change, desertification, land degradation, sustainable land management, food security, and greenhouse gas fluxes in terrestrial ecosystems* [P.R. Shukla, J. Skea, E. Calvo Buendia, V. Masson-Delmotte, H.-O. Pörtner, D. C. Roberts, P. Zhai, R. Slade, S. Connors, R. van Diemen, M. Ferrat, E. Haughey, S. Luz, S. Neogi, M. Pathak, J. Petzold, J. Portugal Pereira, P. Vyas, E. Huntley, K. Kissick, M, Belkacemi, J. Malley, (eds.)]. https://doi.org/10.1017/9781009157988.002

Table of Contents

TS.0   Introduction 40

TS.1   Framing and context 40

TS.2   Land–climate interactions 44

TS.3   Desertification 50

TS.4   Land degradation 53

TS.5   Food security 56

TS.6   Interlinkages between desertification, land
        degradation, food security and greenhouse
        gas fluxes 61

TS.7   Risk management and decision making
        in relation to sustainable development 67

TS

TS.0   Introduction

This Technical Summary to the IPCC Special Report on Climate Change and Land (SRCCL)[1] comprises a compilation of the chapter executive summaries illustrated with figures from the report. It follows the structure of the SRCCL (Figure TS.1) and is presented in seven parts. TS.1 (Chapter 1) provides a synopsis of the main issues addressed in the Special Report, introducing key concepts and definitions and highlighting where the report builds on previous publications. TS.2 (Chapter 2) focuses on the dynamics of the land–climate system (Figure TS.2). It assesses recent progress towards understanding the impacts of climate change on land, and the feedbacks land has on climate and which arise from altered biogeochemical and biophysical fluxes between the atmosphere and the land surface. TS.3 (Chapter 3) examines how the world's dryland populations are uniquely vulnerable to desertification and climate change, but also have significant knowledge in adapting to climate variability and addressing desertification. TS.4 (Chapter 4) assesses the urgency of tackling land degradation across all land ecosystems. Despite accelerating trends of land degradation, reversing these trends is attainable through restoration efforts and improved land management, which is expected to improve resilience to climate change, mitigate climate change, and ensure food security for generations to come. TS.5 (Chapter 5) focuses on food security, with an assessment of the risks and opportunities that climate change presents to food systems. It considers how mitigation and adaptation can contribute to both human and planetary health. TS.6 (Chapter 6) introduces options for responding to the challenges of desertification, land degradation and food security and evaluates the trade-offs for sustainable land management, climate adaptation and mitigation, and the sustainable development goals. TS.7 (Chapter 7) further assesses decision making and policy responses to risks in the climate-land-human system.

TS.1   Framing and context

Land, including its water bodies, provides the basis for human livelihoods and well-being through primary productivity, the supply of food, freshwater, and multiple other ecosystem services (*high confidence*). Neither our individual or societal identities, nor the world's economy would exist without the multiple resources, services and livelihood systems provided by land ecosystems and biodiversity. The annual value of the world's total terrestrial ecosystem services has been estimated at 75 trillion USD in 2011, approximately equivalent to the annual global Gross Domestic Product (based on USD2007 values) (*medium confidence*). Land and its biodiversity also represent essential, intangible benefits to humans, such as cognitive and spiritual enrichment, sense of belonging and aesthetic and recreational values. Valuing ecosystem services with monetary methods often overlooks these intangible services that shape societies, cultures and quality of life and the intrinsic value of biodiversity. The Earth's land area is finite. Using land resources sustainably is fundamental for human well-being (*high confidence*). {1.1.1}

The current geographic spread of the use of land, the large appropriation of multiple ecosystem services and the loss of biodiversity are unprecedented in human history (*high confidence*). By 2015, about three-quarters of the global ice-free land surface was affected by human use. Humans appropriate one-quarter to one-third of global terrestrial potential net primary production (*high confidence*). Croplands cover 12–14% of the global ice-free surface. Since 1961, the supply of global per capita food calories increased by about one-third, with the consumption of vegetable oils and meat more than doubling. At the same time, the use of inorganic nitrogen fertiliser increased by nearly ninefold, and the use of irrigation water roughly doubled (*high confidence*). Human use, at varying intensities, affects about 60–85% of forests and 70–90% of other natural ecosystems (e.g., savannahs, natural grasslands) (*high confidence*). Land use caused global biodiversity to decrease by around 11–14% (*medium confidence*). (Figure TS.2). {1.1.2}

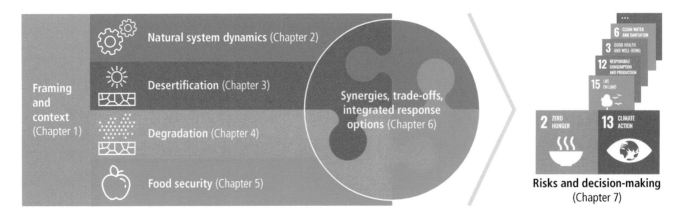

Figure TS.1 | Overview of the IPCC Special Report on Climate Change and Land (SRCCL).

---

[1]   The full title of the report is the *IPCC special report on climate change, desertification, land degradation, sustainable land management, food security, and greenhouse gas fluxes in terrestrial ecosystems*

Land use and observed climate change

A. Observed temperature change relative to 1850–1900
Since the pre-industrial period (1850–1900) the observed mean land surface air temperature has risen considerably more than the global mean surface (land and ocean) temperature (GMST).

CHANGE in TEMPERATURE rel. to 1850–1900 (°C)

Change in surface air temperature over land (°C)

Change in global (land-ocean) mean surface temperature (GMST) (°C)

B. GHG emissions
An estimated 23% of total anthropogenic greenhouse gas emissions (2007–2016) derive from Agriculture, Forestry and Other Land Use (AFOLU).

CHANGE in EMISSIONS since 1961
1 Net CO_2 emissions from FOLU ($GtCO_2$ yr^{-1})
2 CH_4 emissions from Agriculture ($GtCO_2eq$ yr^{-1})
3 N_2O emissions from Agriculture ($GtCO_2eq$ yr^{-1})

$GtCO_2eq$ yr^{-1}

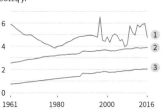

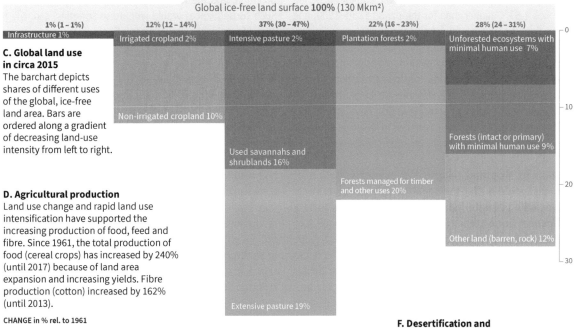

Global ice-free land surface **100%** (130 Mkm²)

| 1% (1 – 1%) | 12% (12 – 14%) | 37% (30 – 47%) | 22% (16 – 23%) | 28% (24 – 31%) |

Infrastructure 1%
Irrigated cropland 2%
Intensive pasture 2%
Plantation forests 2%
Unforested ecosystems with minimal human use 7%
Non-irrigated cropland 10%
Used savannahs and shrublands 16%
Forests (intact or primary) with minimal human use 9%
Forests managed for timber and other uses 20%
Other land (barren, rock) 12%
Extensive pasture 19%

C. Global land use in circa 2015
The barchart depicts shares of different uses of the global, ice-free land area. Bars are ordered along a gradient of decreasing land-use intensity from left to right.

D. Agricultural production
Land use change and rapid land use intensification have supported the increasing production of food, feed and fibre. Since 1961, the total production of food (cereal crops) has increased by 240% (until 2017) because of land area expansion and increasing yields. Fibre production (cotton) increased by 162% (until 2013).

CHANGE in % rel. to 1961
1 Inorganic N fertiliser use
2 Cereal yields
3 Irrigation water volume
4 Total number of ruminant livestock

%

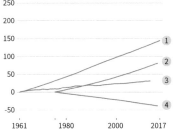

E. Food demand
Increases in production are linked to consumption changes.

CHANGE in % rel. to 1961 and 1975
1 Population
2 Prevalence of overweight + obese
3 Total calories per capita
4 Prevalence of underweight

%

F. Desertification and land degradation
Land-use change, land-use intensification and climate change have contributed to desertification and land degradation.

CHANGE in % rel. to 1961 and 1970
1 Population in areas experiencing desertification
2 Dryland areas in drought annually
3 Inland wetland extent

%

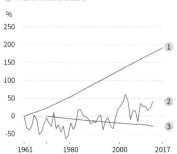

Figure TS.2 | Land use and observed climate change: A representation of the principal land challenges and land–climate system processes covered in this assessment report.

Figure TS.2 (continued): Panels A-F show the status and trends in selected land use and climate variables that represent many of the core topics covered in this report. The annual time series in B and D–F are based on the most comprehensive, available data from national statistics, in most cases from FAOSTAT which starts in 1961. Y-axes in panels D–F are expressed relative to the starting year of the time series (rebased to zero). Data sources and notes: **A:** The warming curves are averages of four datasets {2.1; Figure 2.2; Table 2.1} **B:** N_2O and CH_4 from agriculture are from FAOSTAT; Net CO_2 emissions from FOLU using the mean of two bookkeeping models (including emissions from peatland fires since 1997). All values expressed in units of CO_2-eq are based on AR5 100-year Global Warming Potential values without climate-carbon feedbacks (N_2O = 265; CH_4 = 28). {see Table SPM.1, 1.1, 2.3} **C:** Depicts shares of different uses of the global, ice-free land area for approximately the year 2015, ordered along a gradient of decreasing land-use intensity from left to right. Each bar represents a broad land cover category; the numbers on top are the total % of the ice-free area covered, with uncertainty ranges in brackets. Intensive pasture is defined as having a livestock density greater than 100 animals/km². The area of 'forest managed for timber and other uses' was calculated as total forest area minus 'primary/intact' forest area. {1.2, Table 1.1, Figure 1.3} **D:** Note that fertiliser use is shown on a split axis. The large percentage change in fertiliser use reflects the low level of use in 1961 and relates to both increasing fertiliser input per area as well as the expansion of fertilised cropland and grassland to increase food production. {1.1, Figure 1.3} **E:** Overweight population is defined as having a body mass index (BMI) >25 kg m^{-2}; underweight is defined as BMI <18.5 kg m^{-2}. {5.1, 5.2} **F:** Dryland areas were estimated using TerraClimate precipitation and potential evapotranspiration (1980–2015) to identify areas where the Aridity Index is below 0.65. Population data are from the HYDE3.2 database. Areas in drought are based on the 12-month accumulation Global Precipitation Climatology Centre Drought Index. The inland wetland extent (including peatlands) is based on aggregated data from more than 2000 time series that report changes in local wetland area over time. {3.1, 4.2, 4.6}

Warming over land has occurred at a faster rate than the global mean and this has had observable impacts on the land system (*high confidence*). The average temperature over land for the period 2006–2015 was 1.53°C higher than for the period 1850–1900, and 0.66°C larger than the equivalent global mean temperature change. These warmer temperatures (with changing precipitation patterns) have altered the start and end of growing seasons, contributed to regional crop yield reductions, reduced freshwater availability, and put biodiversity under further stress and increased tree mortality (*high confidence*). Increasing levels of atmospheric CO_2, have contributed to observed increases in plant growth as well as to increases in woody plant cover in grasslands and savannahs (*medium confidence*). {1.1.2}

Urgent action to stop and reverse the over-exploitation of land resources would buffer the negative impacts of multiple pressures, including climate change, on ecosystems and society (*high confidence*). Socio-economic drivers of land use change such as technological development, population growth and increasing per capita demand for multiple ecosystem services are projected to continue into the future (*high confidence*). These and other drivers can amplify existing environmental and societal challenges, such as the conversion of natural ecosystems into managed land, rapid urbanisation, pollution from the intensification of land management and equitable access to land resources (*high confidence*). Climate change will add to these challenges through direct, negative impacts on ecosystems and the services they provide (*high confidence*). Acting immediately and simultaneously on these multiple drivers would enhance food, fibre and water security, alleviate desertification, and reverse land degradation, without compromising the non-material or regulating benefits from land (*high confidence*). {1.1.2, 1.2.1, 1.3.2–1.3.6, Cross-Chapter Box 1 in Chapter 1}

Rapid reductions in anthropogenic greenhouse gas (GHG) emissions that restrict warming to "well-below" 2°C would greatly reduce the negative impacts of climate change on land ecosystems (*high confidence*). In the absence of rapid emissions reductions, reliance on large-scale, land-based, climate change mitigation is projected to increase, which would aggravate existing pressures on land (*high confidence*). Climate change mitigation efforts that require large land areas (e.g., bioenergy and afforestation/reforestation) are projected to compete with existing uses of land (*high confidence*). The competition for land could increase food prices and lead to further intensification (e.g., fertiliser and water use) with implications for water and air pollution, and the further loss of biodiversity (*medium confidence*). Such consequences would jeopardise societies' capacity to achieve many Sustainable Development Goals (SDG) that depend on land (*high confidence*). {1.3.1, Cross-Chapter Box 2 in Chapter 1}

Nonetheless, there are many land-related climate change mitigation options that do not increase the competition for land (*high confidence*). Many of these options have co-benefits for climate change adaptation (*medium confidence*). Land use contributes about one-quarter of global greenhouse gas emissions, notably CO_2 emissions from deforestation, CH_4 emissions from rice and ruminant livestock and N_2O emissions from fertiliser use (*high confidence*). Land ecosystems also take up large amounts of carbon (*high confidence*). Many land management options exist to both reduce the magnitude of emissions and enhance carbon uptake. These options enhance crop productivity, soil nutrient status, microclimate or biodiversity, and thus, support adaptation to climate change (*high confidence*). In addition, changes in consumer behaviour, such as reducing the over-consumption of food and energy would benefit the reduction of GHG emissions from land (*high confidence*). The barriers to the implementation of mitigation and adaptation options include skills deficit, financial and institutional barriers, absence of incentives, access to relevant technologies, consumer awareness and the limited spatial scale at which the success of these practices and methods have been demonstrated. {1.2.1, 1.3.2, 1.3.3, 1.3.4, 1.3.5, 1.3.6}

Sustainable food supply and food consumption, based on nutritionally balanced and diverse diets, would enhance food security under climate and socio-economic changes (*high confidence*). Improving food access, utilisation, quality and safety to enhance nutrition, and promoting globally equitable diets compatible with lower emissions have demonstrable positive impacts on land use and food security (*high confidence*). Food security is also negatively affected by food loss and waste (estimated as 25–30% of total food produced) (*medium confidence*). Barriers to improved food security include economic drivers (prices, availability and stability of supply) and traditional, social and cultural norms around food eating practices. Climate change is expected to increase variability in food production and prices globally (*high confidence*), but the trade in food commodities can buffer these effects. Trade can provide embodied

TS

flows of water, land and nutrients (*medium confidence*). Food trade can also have negative environmental impacts by displacing the effects of overconsumption (*medium confidence*). Future food systems and trade patterns will be shaped as much by policies as by economics (*medium confidence*). {1.2.1, 1.3.3}

A gender-inclusive approach offers opportunities to enhance the sustainable management of land (*medium confidence*). Women play a significant role in agriculture and rural economies globally. In many world regions, laws, cultural restrictions, patriarchy and social structures such as discriminatory customary laws and norms reduce women's capacity in supporting the sustainable use of land resources (*medium confidence*). Therefore, acknowledging women's land rights and bringing women's land management knowledge into land-related decision-making would support the alleviation of land degradation, and facilitate the take-up of integrated adaptation and mitigation measures (*medium confidence*). {1.4.1, 1.4.2}

Regional and country specific contexts affect the capacity to respond to climate change and its impacts, through adaptation and mitigation (*high confidence*). There is large variability in the availability and use of land resources between regions, countries and land management systems. In addition, differences in socio-economic conditions, such as wealth, degree of industrialisation, institutions and governance, affect the capacity to respond to climate change, food insecurity, land degradation and desertification. The capacity to respond is also strongly affected by local land ownership. Hence, climate change will affect regions and communities differently (*high confidence*). {1.3, 1.4}

Cross-scale, cross-sectoral and inclusive governance can enable coordinated policy that supports effective adaptation and mitigation (*high confidence*). There is a lack of coordination across governance levels, for example, local, national, transboundary and international, in addressing climate change and sustainable land management challenges. Policy design and formulation is often strongly sectoral, which poses further barriers when integrating international decisions into relevant (sub)national policies. A portfolio of policy instruments that are inclusive of the diversity of governance actors would enable responses to complex land and climate challenges (*high confidence*). Inclusive governance that considers women's and indigenous people's rights to access and use land enhances the equitable sharing of land resources, fosters food security and increases the existing knowledge about land use, which can increase opportunities for adaptation and mitigation (*medium confidence*). {1.3.5, 1.4.1, 1.4.2, 1.4.3}

Scenarios and models are important tools to explore the trade-offs and co-benefits of land management decisions under uncertain futures (*high confidence*). Participatory, co-creation processes with stakeholders can facilitate the use of scenarios in designing future sustainable development strategies (*medium confidence*). In addition to qualitative approaches, models are critical in quantifying scenarios, but uncertainties in models arise from, for example, differences in baseline datasets, land cover classes and modelling paradigms (*medium confidence*). Current scenario approaches are limited in quantifying time-dependent policy and management decisions that can lead from today to desirable futures or visions. Advances in scenario analysis and modelling are needed to better account for full environmental costs and non-monetary values as part of human decision-making processes. {1.2.2, Cross-Chapter Box 1 in Chapter 1}

TS.2   Land–climate interactions

Implications of climate change, variability and extremes for land systems

It is certain that globally averaged land surface air temperature (LSAT) has risen faster than the global mean surface temperature (i.e., combined LSAT and sea surface temperature) from the preindustrial period (1850–1900) to the present day (1999–2018). According to the single longest and most extensive dataset, from 1850–1900 to 2006–2015 mean land surface air temperature has increased by 1.53°C (*very likely range* from 1.38°C to 1.68°C) while global mean surface temperature has increased by 0.87°C (*likely range* from 0.75°C to 0.99°C). For the 1881–2018 period, when four independently produced datasets exist, the LSAT increase was 1.41°C (1.31–1.51°C), where the range represents the spread in the datasets' median estimates. Analyses of paleo records, historical observations, model simulations and underlying physical principles are all in agreement that LSATs are increasing at a higher rate than SST as a result of differences in evaporation, land–climate feedbacks and changes in the aerosol forcing over land (*very high confidence*). For the 2000–2016 period, the land-to-ocean warming ratio (about 1.6) is in close agreement between different observational records and the CMIP5 climate model simulations (the *likely range* of 1.54–1.81). {2.2.1}

Anthropogenic warming has resulted in shifts of climate zones, primarily as an increase in dry climates and decrease of polar climates (*high confidence*). Ongoing warming is projected to result in new, hot climates in tropical regions and to shift climate zones poleward in the mid- to high latitude and upward in regions of higher elevation (*high confidence*). Ecosystems in these regions will become increasingly exposed to temperature and rainfall extremes beyond the climate regimes they are currently adapted to (*high confidence*), which can alter their structure, composition and functioning. Additionally, high-latitude warming is projected to accelerate permafrost thawing and increase disturbance in boreal forests through abiotic (e.g., drought, fire) and biotic (e.g., pests, disease) agents (*high confidence*). {2.2.1, 2.2.2, 2.5.3}

Globally, greening trends (trends of increased photosynthetic activity in vegetation) have increased over the last 2–3 decades by 22–33%, particularly over China, India, many parts of Europe, central North America, southeast Brazil and southeast Australia (*high confidence*). This results from a combination of direct (i.e., land use and management, forest conservation and expansion) and indirect factors (i.e., CO_2 fertilisation, extended growing season, global warming, nitrogen deposition, increase of diffuse radiation) linked to human activities (*high confidence*). Browning trends (trends of decreasing photosynthetic activity) are projected in many regions where increases in drought and heatwaves are projected in a warmer climate. There is *low confidence* in the projections of global greening and browning trends. {2.2.4, Cross-Chapter Box 4 in Chapter 2}

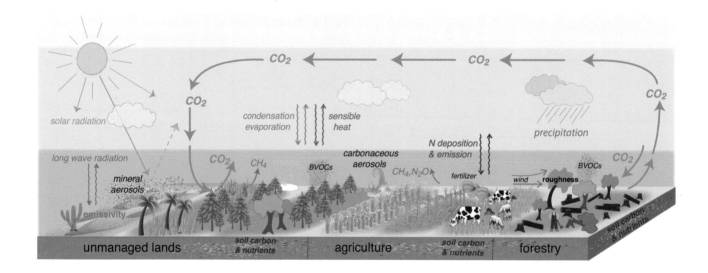

Figure TS.3 | The structure and functioning of managed and unmanaged ecosystems that affect local, regional and global climate. Land surface characteristics such as albedo and emissivity determine the amount of solar and long-wave radiation absorbed by land and reflected or emitted to the atmosphere. Surface roughness influences turbulent exchanges of momentum, energy, water and biogeochemical tracers. Land ecosystems modulate the atmospheric composition through emissions and removals of many GHGs and precursors of SLCFs, including biogenic volatile organic compounds (BVOCs) and mineral dust. Atmospheric aerosols formed from these precursors affect regional climate by altering the amounts of precipitation and radiation reaching land surfaces through their role in clouds physics.

The frequency and intensity of some extreme weather and climate events have increased as a consequence of global warming and will continue to increase under medium and high emission scenarios (*high confidence*). Recent heat-related events, for example, heatwaves, have been made more frequent or intense due to anthropogenic GHG emissions in most land regions and the frequency and intensity of drought has increased in Amazonia, north-eastern Brazil, the Mediterranean, Patagonia, most of Africa and north-eastern China (*medium confidence*). Heatwaves are projected to increase in frequency, intensity and duration in most parts of the world (*high confidence*) and drought frequency and intensity is projected to increase in some regions that are already drought prone, predominantly in the Mediterranean, central Europe, the southern Amazon and southern Africa (*medium confidence*). These changes will impact ecosystems, food security and land processes including GHG fluxes (*high confidence*). {2.2.5}

Climate change is playing an increasing role in determining wildfire regimes alongside human activity (*medium confidence*), with future climate variability expected to enhance the risk and severity of wildfires in many biomes such as tropical rainforests (*high confidence*). Fire weather seasons have lengthened globally between 1979 and 2013 (*low confidence*). Global land area burned has declined in recent decades, mainly due to less burning in grasslands and savannahs (*high confidence*). While drought remains the dominant driver of fire emissions, there has recently been increased fire activity in some tropical and temperate regions during normal to wetter than average years due to warmer temperatures that increase vegetation flammability (*medium confidence*). The boreal zone is also experiencing larger and more frequent fires, and this may increase under a warmer climate (*medium confidence*). {Cross-Chapter Box 4 in Chapter 2}

Terrestrial greenhouse gas fluxes on unmanaged and managed lands

Agriculture, forestry and other land use (AFOLU) is a significant net source of GHG emissions (*high confidence*), contributing to about 23% of anthropogenic emissions of carbon dioxide (CO_2), methane (CH_4) and nitrous oxide (N_2O) combined as CO_2 equivalents in 2007–2016 (*medium confidence*). AFOLU results in both emissions and removals of CO_2, CH_4 and N_2O to and from the atmosphere (*high confidence*). These fluxes are affected simultaneously by natural and human drivers, making it difficult to separate natural from anthropogenic fluxes (*very high confidence*). (Figure TS.3) {2.3}

The total net land-atmosphere flux of CO_2 on both managed and unmanaged lands *very likely* provided a global net removal from 2007 to 2016 according to models (-6.0 ± 3.7 $GtCO_2$ yr^{-1}, *likely range*). This net removal is comprised of two major components: (i) modelled net anthropogenic emissions from AFOLU are 5.2 ± 2.6 $GtCO_2$ yr^{-1} (*likely range*) driven by land cover change, including deforestation and afforestation/reforestation, and wood harvesting (accounting for about 13% of total net anthropogenic emissions of CO_2) (*medium confidence*), and (ii) modelled net removals due to non-anthropogenic processes are 11.2 ± 2.6 $GtCO_2$ yr^{-1} (*likely*

range) on managed and unmanaged lands, driven by environmental changes such as increasing CO_2, nitrogen deposition and changes in climate (accounting for a removal of 29% of the CO_2 emitted from all anthropogenic activities (fossil fuel, industry and AFOLU) (*medium confidence*). {2.3.1}

Global models and national GHG inventories use different methods to estimate anthropogenic CO_2 emissions and removals for the land sector. Consideration of differences in methods can enhance understanding of land sector net emission such as under the Paris Agreement's global stocktake (*medium confidence*). Both models and inventories produce estimates that are in close agreement for land-use change involving forest (e.g., deforestation, afforestation), and differ for managed forest. Global models consider as managed forest those lands that were subject to harvest whereas, consistent with IPCC guidelines, national GHG inventories define managed forest more broadly. On this larger area, inventories can also consider the natural response of land to human-induced environmental changes as anthropogenic, while the global model approach treats this response as part of the non-anthropogenic sink. For illustration, from 2005 to 2014, the sum of the national GHG inventories net emission estimates is 0.1 ± 1.0 $GtCO_2$ yr^{-1}, while the mean of two global bookkeeping models is 5.1 ± 2.6 $GtCO_2$yr^{-1} (*likely range*). {Table SPM.1}

The gross emissions from AFOLU (one-third of total global emissions) are more indicative of mitigation potential of reduced deforestation than the global net emissions (13% of total global emissions), which include compensating deforestation and afforestation fluxes (*high confidence*). The net flux of CO_2 from AFOLU is composed of two opposing gross fluxes: (i) gross emissions (20 $GtCO_2$ yr^{-1}) from deforestation, cultivation of soils and oxidation of wood products, and (ii) gross removals (–14 $GtCO_2$ yr^{-1}), largely from forest growth following wood harvest and agricultural abandonment (*medium confidence*). (Figure TS.4) {2.3.1}

Land is a net source of CH_4, accounting for 44% of anthropogenic CH_4 emissions for the 2006–2017 period (*medium confidence*). The pause in the rise of atmospheric CH_4 concentrations between 2000 and 2006 and the subsequent renewed increase appear to be partially associated with land use and land use change. The recent depletion trend of the 13C isotope in the atmosphere indicates that higher biogenic sources explain part of the current CH_4 increase and that biogenic sources make up a larger proportion of the source mix than they did before 2000 (*high confidence*). In agreement with the findings of AR5, tropical wetlands and peatlands continue to be important drivers of inter-annual variability and current CH_4 concentration increases (*medium evidence, high agreement*). Ruminants and the expansion of rice cultivation are also important contributors to the current trend (*medium evidence, high agreement*). There is significant and ongoing accumulation of CH_4 in the atmosphere (*very high confidence*). {2.3.2}

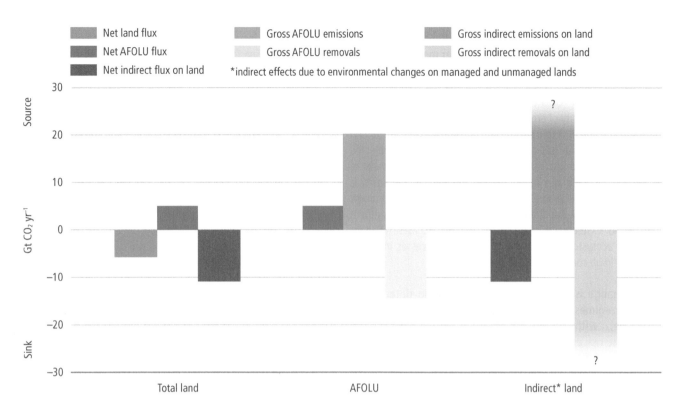

Figure TS.4 | Net and gross fluxes of CO$_2$ from land (annual averages for 2008–2017). Left: The total net flux of CO$_2$ between land and atmosphere (grey) is shown with its two component fluxes, (i) net AFOLU emissions (blue), and (ii) the net land sink (brown), due to indirect environmental effects and natural effects on managed and unmanaged lands. Middle: The gross emissions and removals contributing to the net AFOLU flux. Right: The gross emissions and removals contributing to the land sink.

AFOLU is the main anthropogenic source of N$_2$O primarily due to nitrogen application to soils (*high confidence*). In croplands, the main driver of N$_2$O emissions is a lack of synchronisation between crop nitrogen demand and soil nitrogen supply, with approximately 50% of the nitrogen applied to agricultural land not taken up by the crop. Cropland soils emit over 3 MtN$_2$O-N yr^{-1} (*medium confidence*). Because the response of N$_2$O emissions to fertiliser application rates is non-linear, in regions of the world where low nitrogen application rates dominate, such as sub-Saharan Africa and parts of Eastern Europe, increases in nitrogen fertiliser use would generate relatively small increases in agricultural N$_2$O emissions. Decreases in application rates in regions where application rates are high and exceed crop demand for parts of the growing season will have very large effects on emissions reductions (*medium evidence, high agreement*). {2.3.3}

While managed pastures make up only one-quarter of grazing lands, they contributed more than three-quarters of N$_2$O emissions from grazing lands between 1961 and 2014 with rapid recent increases of nitrogen inputs resulting in disproportionate growth in emissions from these lands (*medium confidence*). Grazing lands (pastures and rangelands) are responsible for more than one-third of total anthropogenic N$_2$O emissions or more than one-half of agricultural emissions (*high confidence*). Emissions are largely from North America, Europe, East Asia, and South Asia, but hotspots are shifting from Europe to southern Asia (*medium confidence*). {2.3.3}

Increased emissions from vegetation and soils due to climate change in the future are expected to counteract potential sinks due to CO$_2$ fertilisation (*low confidence*). Responses of vegetation and soil organic carbon (SOC) to rising atmospheric CO$_2$ concentration and climate change are not well constrained by observations (*medium confidence*). Nutrient (e.g., nitrogen, phosphorus) availability can limit future plant growth and carbon storage under rising CO$_2$ (*high confidence*). However, new evidence suggests that ecosystem adaptation through plant-microbe symbioses could alleviate some nitrogen limitation (*medium evidence, high agreement*). Warming of soils and increased litter inputs will accelerate carbon losses through microbial respiration (*high confidence*). Thawing of high latitude/altitude permafrost will increase rates of SOC loss and change the balance between CO$_2$ and CH$_4$ emissions (*medium confidence*). The balance between increased respiration in warmer climates and carbon uptake from enhanced plant growth is a key uncertainty for the size of the future land carbon sink (*medium confidence*). {2.3.1, 2.7.2, Box 2.3}

Biophysical and biogeochemical land forcing and feedbacks to the climate system

Changes in land conditions from human use or climate change in turn affect regional and global climate (*high confidence*). On the global scale, this is driven by changes in emissions or removals of CO$_2$, CH$_4$ and N$_2$O by land (biogeochemical effects) and by changes in the surface albedo (*very high confidence*). Any local land changes that redistribute energy and water vapour between the land and the atmosphere influence regional climate (biophysical effects; *high confidence*). However, there is *no confidence* in whether such biophysical effects influence global climate. {2.1, 2.3, 2.5.1, 2.5.2}

Changes in land conditions modulate the likelihood, intensity and duration of many extreme events including heatwaves (*high confidence*) and heavy precipitation events (*medium confidence*). Dry soil conditions favour or strengthen summer heatwave conditions through reduced evapotranspiration and increased sensible heat. By contrast wet soil conditions, for example from irrigation or crop management practices that maintain a cover crop all year round, can dampen extreme warm events through increased evapotranspiration and reduced sensible heat. Droughts can be intensified by poor land management. Urbanisation increases extreme rainfall events over or downwind of cities (*medium confidence*). {2.5.1, 2.5.2, 2.5.3}

Historical changes in anthropogenic land cover have resulted in a mean annual global warming of surface air from biogeochemical effects (*very high confidence*), dampened by a cooling from biophysical effects (*medium confidence*). Biogeochemical warming results from increased emissions of GHGs by land, with model-based estimates of +0.20 ± 0.05°C (global climate models) and +0.24 ± 0.12°C – dynamic global vegetation models (DGVMs) as well as an observation-based estimate of +0.25 ± 0.10°C. A net biophysical cooling of −0.10 ± 0.14°C has been derived from global climate models in response to the increased surface albedo and decreased turbulent heat fluxes, but it is smaller than the warming effect from land-based emissions. However, when both biogeochemical and biophysical effects are accounted for within the same global climate model, the models do not agree on the sign of the net change in mean annual surface air temperature. {2.3, 2.5.1, Box 2.1}

The future projected changes in anthropogenic land cover that have been examined for AR5 would result in a biogeochemical warming and a biophysical cooling whose magnitudes depend on the scenario (*high confidence*). Biogeochemical warming has been projected for RCP8.5 by both global climate models (+0.20 ± 0.15°C) and DGVMs (+0.28 ± 0.11°C) (*high confidence*). A global biophysical cooling of 0.10 ± 0.14°C is estimated from global climate models and is projected to dampen the land-based warming (*low confidence*). For RCP4.5, the biogeochemical warming estimated from global climate models (+0.12 ± 0.17°C) is stronger than the warming estimated by DGVMs (+0.01 ± 0.04°C) but based on limited evidence, as is the biophysical cooling (−0.10 ± 0.21°C). {2.5.2}

Regional climate change can be dampened or enhanced by changes in local land cover and land use (*high confidence*) but this depends on the location and the season (*high confidence*). In boreal regions, for example, where projected climate change will migrate the treeline northward, increase the growing season length and thaw permafrost, regional winter warming will be enhanced by decreased surface albedo and snow, whereas warming will be dampened during the growing season due to larger evapotranspiration (*high confidence*). In the tropics, wherever climate

change will increase rainfall, vegetation growth and associated increase in evapotranspiration will result in a dampening effect on regional warming (*medium confidence*). {2.5.2, 2.5.3}

According to model-based studies, changes in local land cover or available water from irrigation will affect climate in regions as far as few hundreds of kilometres downwind (*high confidence*). The local redistribution of water and energy following the changes on land affect the horizontal and vertical gradients of temperature, pressure and moisture, thus altering regional winds and consequently moisture and temperature advection and convection and subsequently, precipitation. {2.5.2, 2.5.4, Cross-Chapter Box 4 in Chapter 2}

Future increases in both climate change and urbanisation will enhance warming in cities and their surroundings (urban heat island), especially during heatwaves (*high confidence*). Urban and peri-urban agriculture, and more generally urban greening, can contribute to mitigation (*medium confidence*) as well as to adaptation (*high confidence*), with co-benefits for food security and reduced soil-water-air pollution. {Cross-Chapter Box 4 in Chapter 2}

Regional climate is strongly affected by natural land aerosols (*medium confidence*) (e.g., mineral dust, black, brown and organic carbon), but there is *low confidence* in historical trends, inter-annual and decadal variability and future changes. Forest cover affects climate through emissions of biogenic volatile organic compounds (BVOC) and aerosols (*low confidence*). The decrease in the emissions of BVOC resulting from the historical conversion of forests to cropland has resulted in a positive radiative forcing through direct and indirect aerosol effects, a negative radiative forcing through the reduction in the atmospheric lifetime of methane and it has contributed to increased ozone concentrations in different regions (*low confidence*). {2.4, 2.5}

Consequences for the climate system of land-based adaptation and mitigation options, including carbon dioxide removal (negative emissions)

About one-quarter of the 2030 mitigation pledged by countries in their initial Nationally Determined Contributions (NDCs) under the Paris Agreement is expected to come from land-based mitigation options (*medium confidence*). Most of the NDCs submitted by countries include land-based mitigation, although many lack details. Several refer explicitly to reduced deforestation and forest sinks, while a few include soil carbon sequestration, agricultural management and bioenergy. Full implementation of NDCs (submitted by February 2016) is expected to result in net removals of 0.4–1.3 GtCO$_2$ y^{-1} in 2030 compared to the net flux in 2010, where the range represents low to high mitigation ambition in pledges, not uncertainty in estimates (*medium confidence*). {2.6.3}

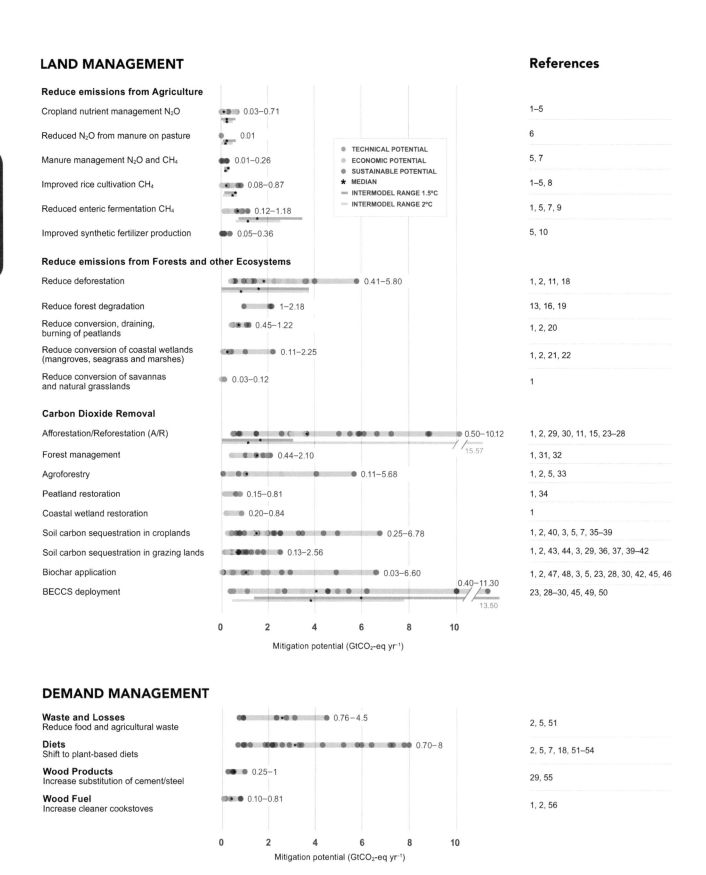

Figure TS.5 | Mitigation potential of response options in 2020–2050, measured in GtCO2-eq yr⁻¹, adapted from Roe et al. (2017).

Figure TS.5 (continued): Mitigation potentials reflect the full range of low to high estimates from studies published after 2010, differentiated according to technical (possible with current technologies), economic (possible given economic constraints) and sustainable potential (technical or economic potential constrained by sustainability considerations). Medians are calculated across all potentials in categories with more than four data points. We only include references that explicitly provide mitigation potential estimates in CO_2-eq yr^{-1} (or a similar derivative) by 2050. Not all options for land management potentials are additive, as some may compete for land. Estimates reflect a range of methodologies (including definitions, global warming potentials and time horizons) that may not be directly comparable or additive. Results from IAMs are shown to compare with single option 'bottom-up' estimates, in available categories from the 2°C and 1.5°C scenarios in the SSP Database (version 2.0). The models reflect land management changes, yet in some instances, can also reflect demand-side effects from carbon prices, so may not be defined exclusively as 'supply-side'.

Several mitigation response options have technical potential for >3 GtCO$_2$-eq yr^{-1} by 2050 through reduced emissions and Carbon Dioxide Removal (CDR) (*high confidence*), some of which compete for land and other resources, while others may reduce the demand for land (*high confidence*). Estimates of the technical potential of individual response options are not necessarily additive. The largest potential for reducing AFOLU emissions are through reduced deforestation and forest degradation (0.4–5.8 GtCO$_2$-eq yr^{-1}) (*high confidence*), a shift towards plant-based diets (0.7–8.0 GtCO$_2$-eq yr^{-1}) (*high confidence*) and reduced food and agricultural waste (0.8–4.5 CO$_2$-eq y$^{(-1)}$) (*high confidence*). Agriculture measures combined could mitigate 0.3–3.4 GtCO$_2$-eq yr^{-1} (*medium confidence*). The options with largest potential for CDR are afforestation/reforestation (0.5–10.1 CO$_2$-eq yr^{-1}) (*medium confidence*), soil carbon sequestration in croplands and grasslands (0.4–8.6 CO$_2$-eq yr^{-1}) (*high confidence*) and Bioenergy with Carbon Capture and Storage (BECCS) (0.4–11.3 CO$_2$-eq yr^{-1}) (*medium confidence*). While some estimates include sustainability and cost considerations, most do not include socio-economic barriers, the impacts of future climate change or non-GHG climate forcings. {2.6.1}

Response options intended to mitigate global warming will also affect the climate locally and regionally through biophysical effects (*high confidence*). Expansion of forest area, for example, typically removes CO_2 from the atmosphere and thus dampens global warming (biogeochemical effect, *high confidence*), but the biophysical effects can dampen or enhance regional warming depending on location, season and time of day. During the growing season, afforestation generally brings cooler days from increased evapotranspiration, and warmer nights (*high confidence*). During the dormant season, forests are warmer than any other land cover, especially in snow-covered areas where forest cover reduces albedo (*high confidence*). At the global level, the temperature effects of boreal afforestation/reforestation run counter to GHG effects, while in the tropics they enhance GHG effects. In addition, trees locally dampen the amplitude of heat extremes (*medium confidence*). {2.5.2, 2.5.4, 2.7, Cross-Chapter Box 4 in Chapter 2}

Mitigation response options related to land use are a key element of most modelled scenarios that provide strong mitigation, alongside emissions reduction in other sectors (*high confidence*). More stringent climate targets rely more heavily on land-based mitigation options, in particular, CDR (*high confidence*). Across a range of scenarios in 2100, CDR is delivered by both afforestation (median values of −1.3, −1.7 and −2.4 GtCO$_2$yr^{-1} for scenarios RCP4.5, RCP2.6 and RCP1.9 respectively) and BECCS (−6.5, −11 and −14.9 GtCO$_2$ yr^{-1} respectively). Emissions of

CH$_4$ and N$_2$O are reduced through improved agricultural and livestock management as well as dietary shifts away from emission-intensive livestock products by 133.2, 108.4 and 73.5 MtCH$_4$ yr^{-1}; and 7.4, 6.1 and 4.5 MtN$_2$O yr^{-1} for the same set of scenarios in 2100 (*high confidence*). High levels of bioenergy crop production can result in increased N$_2$O emissions due to fertiliser use. The Integrated Assessment Models that produce these scenarios mostly neglect the biophysical effects of land-use on global and regional warming. {2.5, 2.6.2}

Large-scale implementation of mitigation response options that limit warming to 1.5 or 2°C would require conversion of large areas of land for afforestation/reforestation and bioenergy crops, which could lead to short-term carbon losses (*high confidence*). The change of global forest area in mitigation pathways ranges from about −0.2 to +7.2 Mkm^2 between 2010 and 2100 (median values across a range of models and scenarios: RCP4.5, RCP2.6, RCP1.9), and the land demand for bioenergy crops ranges from about 3.2 to 6.6 Mkm^2 in 2100 (*high confidence*). Large-scale land-based CDR is associated with multiple feasibility and sustainability constraints. In high carbon lands such as forests and peatlands, the carbon benefits of land protection are greater in the short-term than converting land to bioenergy crops for BECCS, which can take several harvest cycles to 'pay-back' the carbon emitted during conversion (carbon-debt), from decades to over a century (*medium confidence*). (Figure TS.5) {2.6.2, Chapters 6, 7}

It is possible to achieve climate change targets with low need for land-demanding CDR such as BECCS, but such scenarios rely more on rapidly reduced emissions or CDR from forests, agriculture and other sectors. Terrestrial CDR has the technical potential to balance emissions that are difficult to eliminate with current technologies (including food production). Scenarios that achieve climate change targets with less need for terrestrial CDR rely on agricultural demand-side changes (diet change, waste reduction), and changes in agricultural production such as agricultural intensification. Such pathways that minimise land use for bioenergy and BECCS are characterised by rapid and early reduction of GHG emissions in all sectors, as well as earlier CDR in through afforestation. In contrast, delayed mitigation action would increase reliance on land-based CDR (*high confidence*). {2.6.2}

TS.3 Desertification

Desertification is land degradation in arid, semi-arid, and dry sub-humid areas, collectively known as drylands, resulting from many factors, including human activities and climatic variations. The range and intensity of desertification have increased in some dryland areas over the past several decades (*high confidence*). Drylands currently cover about 46.2% (±0.8%) of the global land area and are home to 3 billion people. The multiplicity and complexity of the processes of desertification make its quantification difficult. Desertification hotspots, as identified by a decline in vegetation productivity between the 1980s and 2000s, extended to about 9.2% of drylands (±0.5%), affecting about 500 (±120) million people in 2015. The highest numbers of people affected are in South and East Asia, the circum Sahara region including North Africa and the Middle East including the Arabian Peninsula (*low confidence*). Other dryland regions have also experienced desertification. Desertification has already reduced agricultural productivity and incomes (*high confidence*) and contributed to the loss of biodiversity in some dryland regions (*medium confidence*). In many dryland areas, spread of invasive plants has led to losses in ecosystem services (*high confidence*), while over-extraction is leading to groundwater depletion (*high confidence*). Unsustainable land management, particularly when coupled with droughts, has contributed to higher dust-storm activity, reducing human well-being in drylands and beyond (*high confidence*). Dust storms were associated with global cardiopulmonary mortality of about 402,000 people in 2005. Higher intensity of sand storms and sand dune movements are causing disruption and damage to transportation and solar and wind energy harvesting infrastructures (*high confidence*). (Figure TS.6) {3.1.1, 3.1.4, 3.2.1, 3.3.1, 3.4.1, 3.4.2, 3.4.2, 3.7.3, 3.7.4}

Attribution of desertification to climate variability and change, and to human activities, varies in space and time (*high confidence*). Climate variability and anthropogenic climate change, particularly through increases in both land surface air temperature and evapotranspiration, and decreases in precipitation, are likely to have played a role, in interaction with human activities, in causing desertification in some dryland areas. The major human drivers of desertification interacting with climate change are expansion of croplands, unsustainable land management practices and increased pressure on land from population and income growth. Poverty is limiting both capacities to adapt to climate change and availability of financial resources to invest in sustainable land management (SLM) (*high confidence*). {3.1.4, 3.2.2, 3.4.2}

Climate change will exacerbate several desertification processes (*medium confidence*). Although CO_2 fertilisation effect is enhancing vegetation productivity in drylands (*high confidence*), decreases in water availability have a larger effect than CO_2 fertilisation in many dryland areas. There is *high confidence* that aridity will increase in some places, but no evidence for a projected global trend in dryland aridity (*medium confidence*). The area at risk of salinisation is projected to increase in the future (*limited evidence, high agreement*). Future climate change is projected to increase the potential for water driven soil erosion in many dryland areas (*medium*

confidence*), leading to soil organic carbon decline in some dryland areas. {3.1.1, 3.2.2, 3.5.1, 3.5.2, 3.7.1, 3.7.3}

Risks from desertification are projected to increase due to climate change (*high confidence*). Under shared socio-economic pathway SSP2 ('Middle of the Road') at 1.5°C, 2°C and 3°C of global warming, the number of dryland population exposed (vulnerable) to various impacts related to water, energy and land sectors (e.g. water stress, drought intensity, habitat degradation) is projected to reach 951 (178) million, 1152 (220) million and 1285 (277) million, respectively. While at global warming of 2°C, under SSP1 ('Sustainability'), the exposed (vulnerable) dryland population is 974 (35) million, and under SSP3 ('Fragmented World') it is 1267 (522) million. Around half of the vulnerable population is in South Asia, followed by Central Asia, West Africa and East Asia. {2.2, 3.1.1, 3.2.2, 3.5.1, 3.5.2, 7.2.2}

Desertification and climate change, both individually and in combination, will reduce the provision of dryland ecosystem services and lower ecosystem health, including losses in biodiversity (*high confidence*). Desertification and changing climate are projected to cause reductions in crop and livestock productivity (*high confidence*), modify the composition of plant species and reduce biological diversity across drylands (*medium confidence*). Rising CO_2 levels will favour more rapid expansion of some invasive plant species in some regions. A reduction in the quality and quantity of resources available to herbivores can have knock-on consequences for predators, which can potentially lead to disruptive ecological cascades (*limited evidence, low agreement*). Projected increases in temperature and the severity of drought events across some dryland areas can increase chances of wildfire occurrence (*medium confidence*). {3.1.4, 3.4.1, 3.5.2, 3.7.3}

Increasing human pressures on land, combined with climate change, will reduce the resilience of dryland populations and constrain their adaptive capacities (*medium confidence*). The combination of pressures coming from climate variability, anthropogenic climate change and desertification will contribute to poverty, food insecurity, and increased disease burden (*high confidence*), as well as potentially to conflicts (*low confidence*). Although strong impacts of climate change on migration in dryland areas are disputed (*medium evidence, low agreement*), in some places, desertification under changing climate can provide an added incentive to migrate (*medium confidence*). Women will be impacted more than men by environmental degradation, particularly in those areas with higher dependence on agricultural livelihoods (*medium evidence, high agreement*). {3.4.2, 3.6.2}

Desertification exacerbates climate change through several mechanisms such as changes in vegetation cover, sand and dust aerosols and greenhouse gas fluxes (*high confidence*). The extent of areas in which dryness (rather than temperature) controls CO_2 exchange has increased by 6% between 1948 and 2012, and is projected to increase by at least another 8% by 2050 if the expansion continues at the same rate. In these areas, net carbon uptake is about 27% lower than in other areas (*low confidence*). Desertification also tends to increase

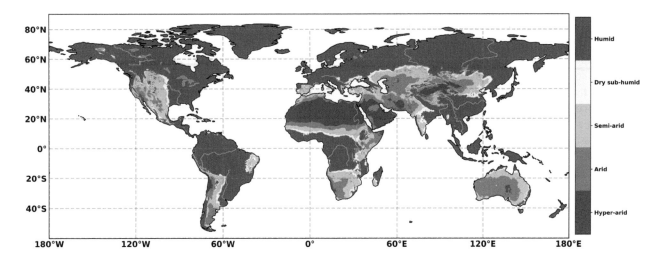

Figure TS.6 | Geographical distribution of drylands, delimited based on the aridity index (AI). The classification of AI is: Humid AI > 0.65, Dry sub-humid 0.50 < AI ≤ 0.65, Semi-arid 0.20 < AI ≤ 0.50, Arid 0.05 < AI ≤ 0.20, Hyper-arid AI < 0.05. Data: TerraClimate precipitation and potential evapotranspiration (1980–2015) (Abatzoglou et al. 2018).

albedo, decreasing the energy available at the surface and associated surface temperatures, producing a negative feedback on climate change (*high confidence*). Through its effect on vegetation and soils, desertification changes the absorption and release of associated greenhouse gases (GHGs). Vegetation loss and drying of surface cover due to desertification increases the frequency of dust storms (*high confidence*). Arid ecosystems could be an important global carbon sink, depending on soil water availability (*medium evidence, high agreement*). {3.3.3, 3.4.1, 3.5.2}

Site and regionally-specific technological solutions, based both on new scientific innovations and indigenous and local knowledge (ILK), are available to avoid, reduce and reverse desertification, simultaneously contributing to climate change mitigation and adaptation (*high confidence*). SLM practices in drylands increase agricultural productivity and contribute to climate change adaptation with mitigation co-benefits (*high confidence*). Integrated crop, soil and water management measures can be employed to reduce soil degradation and increase the resilience of agricultural production systems to the impacts of climate change (*high confidence*). These measures include crop diversification and adoption of drought-resilient econogically appropriate plants, reduced tillage, adoption of improved irrigation techniques (e.g. drip irrigation) and moisture conservation methods (e.g. rainwater harvesting using indigenous and local practices), and maintaining vegetation and mulch cover. Conservation agriculture increases the capacity of agricultural households to adapt to climate change (*high confidence*) and can lead to increases in soil organic carbon over time, with quantitative estimates of the rates of carbon sequestration in drylands following changes in agricultural practices ranging between 0.04 and 0.4 t ha⁻¹ (*medium confidence*). Rangeland management systems based on sustainable grazing and re-vegetation increase rangeland productivity and the flow of ecosystem services (*high confidence*). The combined use of salt-tolerant crops, improved irrigation practices, chemical remediation measures and appropriate mulch and compost is effective in reducing the impact of secondary salinisation (*medium confidence*). Application of sand dune stabilisation techniques contributes to reducing sand and dust storms (*high confidence*). Agroforestry practices and shelterbelts help reduce soil erosion and sequester carbon. Afforestation programmes aimed at creating windbreaks in the form of 'green walls' and 'green dams' can help stabilise and reduce dust storms, avert wind erosion, and serve as carbon sinks, particularly when done with locally adapted native and other climate resilient tree species (*high confidence*). {3.4.2, 3.6.1, 3.7.2}

Investments into SLM, land restoration and rehabilitation in dryland areas have positive economic returns (*high confidence*). Each USD invested into land restoration can have social returns of about 3–6 USD over a 30-year period. Most SLM practices can become financially profitable within 3 to 10 years (*medium evidence, high agreement*). Despite their benefits in addressing desertification, mitigating and adapting to climate change, and increasing food and economic security, many SLM practices are not widely adopted due to insecure land tenure, lack of access to credit and agricultural advisory services, and insufficient incentives for private land-users (*robust evidence, high agreement*). {3.6.3}

Indigenous and local knowledge often contributes to enhancing resilience against climate change and combating desertification (*medium confidence*). Dryland populations have developed traditional agroecological practices which are well adapted to resource-sparse dryland environments. However, there is *robust evidence* documenting losses of traditional agroecological knowledge. Traditional agroecological practices are also increasingly unable to cope with growing demand for food. Combined use of ILK and new SLM technologies can contribute to raising the resilience to the challenges of climate change and desertification (*high confidence*). {3.1.3, 3.6.1, 3.6.2}

Policy frameworks promoting the adoption of SLM solutions contribute to addressing desertification as well as mitigating and adapting to climate change, with co-benefits for poverty eradication and food security among dryland populations (*high confidence*). Implementation of Land Degradation Neutrality (LDN) policies allows populations to avoid, reduce and reverse desertification, thus contributing to climate change adaptation with mitigation co-benefits (*high confidence*). Strengthening land tenure security is a major factor contributing to the adoption of soil conservation measures in croplands (*high confidence*). On-farm and off-farm livelihood diversification strategies increase the resilience of rural households against desertification and extreme weather events, such as droughts (*high confidence*). Strengthening collective action is important for addressing causes and impacts of desertification, and for adapting to climate change (*medium confidence*). A greater emphasis on understanding gender-specific differences over land use and land management practices can help make land restoration projects more successful (*medium confidence*). Improved access to markets raises agricultural profitability and motivates investment into climate change adaptation and SLM (*medium confidence*). Payments for ecosystem services give additional incentives to land users to adopt SLM practices (*medium confidence*). Expanding access to rural advisory services increases the knowledge on SLM and facilitates their wider adoption (*medium confidence*). Developing, enabling and promoting access to cleaner energy sources and technologies can contribute to reducing desertification and mitigating climate change through decreasing the use of fuelwood and crop residues for energy (*medium confidence*). Policy responses to droughts based on proactive drought preparedness and drought risk mitigation are more efficient in limiting drought-caused damages than reactive drought relief efforts (*high confidence*). {3.4.2, 3.6.2, 3.6.3, Cross-Chapter Box 5 in Chapter 3}

The knowledge on limits of adaptation to the combined effects of climate change and desertification is insufficient. However, the potential for residual risks and maladaptive outcomes is high (*high confidence*). Empirical evidence on the limits to adaptation in dryland areas is limited. Potential limits to adaptation include losses of land productivity due to irreversible forms of desertification. Residual risks can emerge from the inability of SLM measures to fully compensate for yield losses due to climate change impacts. They also arise from foregone reductions in ecosystem services due to soil fertility loss even when applying SLM measures could revert land to initial productivity after some time. Some activities favouring agricultural intensification in dryland areas can become maladaptive due to their negative impacts on the environment (*medium confidence*) Even when solutions are available, social, economic and institutional constraints could pose barriers to their implementation (*medium confidence*) {3.6.4}.

Improving capacities, providing higher access to climate services, including local-level early warning systems, and expanding the use of remote sensing technologies are high-return investments for enabling effective adaptation and mitigation responses that help address desertification (*high confidence*). Reliable and timely climate services, relevant to desertification, can aid the development of appropriate adaptation and mitigation options reducing, the impact of desertification on human and natural systems (*high confidence*), with quantitative estimates showing that every USD invested in strengthening hydro-meteorological and early warning services in developing countries can yield between 4 and 35 USD (*low confidence*). Knowledge and flow of knowledge on desertification is currently fragmented. Improved knowledge and data exchange and sharing will increase the effectiveness of efforts to achieve LDN (*high confidence*). Expanded use of remotely sensed information for data collection helps in measuring progress towards achieving LDN (*low evidence, high agreement*). {3.2.1, 3.6.2, 3.6.3, Cross-Chapter Box 5 in Chapter 3}

TS.4   Land degradation

Land degradation affects people and ecosystems throughout the planet and is both affected by climate change and contributes to it. In this report, land degradation is defined as a *negative trend in land condition, caused by direct or indirect human-induced processes including anthropogenic climate change, expressed as long-term reduction or loss of at least one of the following: biological productivity, ecological integrity, or value to humans.* Forest degradation is land degradation that occurs in forest land. Deforestation is the conversion of forest to non-forest land and can result in land degradation. {4.1.3}

Land degradation adversely affects people's livelihoods (*very high confidence*) and occurs over a quarter of the Earth's ice-free land area (*medium confidence*). The majority of the 1.3 to 3.2 billion affected people (*low confidence*) are living in poverty in developing countries (*medium confidence*). Land-use changes and unsustainable land management are direct human causes of land degradation (*very high confidence*), with agriculture being a dominant sector driving degradation (*very high confidence*). Soil loss from conventionally tilled land exceeds the rate of soil formation by >2 orders of magnitude (*medium confidence*). Land degradation affects humans in multiple ways, interacting with social, political, cultural and economic aspects, including markets, technology, inequality and demographic change (*very high confidence*). Land degradation impacts extend beyond the land surface itself, affecting marine and freshwater systems, as well as people and ecosystems far away from the local sites of degradation (*very high confidence*). {4.1.6, 4.2.1, 4.2.3, 4.3, 4.6.1, 4.7, Table 4.1}

Climate change exacerbates the rate and magnitude of several ongoing land degradation processes and introduces new degradation patterns (*high confidence*). Human-induced global warming has already caused observed changes in two drivers of land degradation: increased frequency, intensity and/or amount of heavy precipitation (*medium confidence*); and increased heat stress (*high confidence*). In some areas sea level rise has exacerbated coastal erosion (*medium confidence*). Global warming beyond present day will further exacerbate ongoing land degradation processes through increasing floods (*medium confidence*), drought frequency and severity (*medium confidence*), intensified cyclones (*medium confidence*), and sea level rise (*very high confidence*), with outcomes being modulated by land management (*very high confidence*). Permafrost thawing due to warming (*high confidence*), and coastal erosion due to sea level rise and impacts of changing storm paths (*low confidence*), are examples of land degradation affecting places where it has not typically been a problem. Erosion of coastal areas because of sea level rise will increase worldwide (*high confidence*). In cyclone prone areas, the combination of sea level rise and more intense cyclones will cause land degradation with serious consequences for people and livelihoods (*very high confidence*). {4.2.1, 4.2.2, 4.2.3, 4.4.1, 4.4.2, 4.9.6, Table 4.1}

Land degradation and climate change, both individually and in combination, have profound implications for natural resource-based livelihood systems and societal groups (*high

confidence). The number of people whose livelihood depends on degraded lands has been estimated to be about 1.5 billion worldwide (*very low confidence*). People in degraded areas who directly depend on natural resources for subsistence, food security and income, including women and youth with limited adaptation options, are especially vulnerable to land degradation and climate change (*high confidence*). Land degradation reduces land productivity and increases the workload of managing the land, affecting women disproportionally in some regions. Land degradation and climate change act as threat multipliers for already precarious livelihoods (*very high confidence*), leaving them highly sensitive to extreme climatic events, with consequences such as poverty and food insecurity (*high confidence*) and, in some cases, migration, conflict and loss of cultural heritage (*low confidence*). Changes in vegetation cover and distribution due to climate change increase the risk of land degradation in some areas (*medium confidence*). Climate change will have detrimental effects on livelihoods, habitats and infrastructure through increased rates of land degradation (*high confidence*) and from new degradation patterns (*low evidence, high agreement*). {4.1.6, 4.2.1, 4.7}

Land degradation is a driver of climate change through emission of greenhouse gases (GHGs) and reduced rates of carbon uptake (*very high confidence*). Since 1990, globally the forest area has decreased by 3% (*low confidence*) with net decreases in the tropics and net increases outside the tropics (*high confidence*). Lower carbon density in re-growing forests compared, to carbon stocks before deforestation, results in net emissions from land-use change (*very high confidence*). Forest management that reduces carbon stocks of forest land also leads to emissions, but global estimates of these emissions are uncertain. Cropland soils have lost 20–60% of their organic carbon content prior to cultivation, and soils under conventional agriculture continue to be a source of GHGs (*medium confidence*). Of the land degradation processes, deforestation, increasing wildfires, degradation of peat soils, and permafrost thawing contribute most to climate change through the release of GHGs and the reduction in land carbon sinks following deforestation (*high confidence*). Agricultural practices also emit non-CO_2 GHGs from soils and these emissions are exacerbated by climate change (*medium confidence*). Conversion of primary to managed forests, illegal logging and unsustainable forest management result in GHG emissions (*very high confidence*) and can have additional physical effects on the regional climate including those arising from albedo shifts (*medium confidence*). These interactions call for more integrative climate impact assessments. {4.2.2, 4.3, 4.5.4, 4.6}

Large-scale implementation of dedicated biomass production for bioenergy increases competition for land with potentially serious consequences for food security and land degradation (*high confidence*). Increasing the extent and intensity of biomass production, for example, through fertiliser additions, irrigation or monoculture energy plantations, can result in local land degradation. Poorly implemented intensification of land management contributes to land degradation (e.g., salinisation from irrigation) and disrupted livelihoods (*high confidence*). In areas where afforestation and reforestation occur on previously degraded lands, opportunities exist to restore and rehabilitate lands with potentially significant co-benefits (*high confidence*) that depend on whether restoration involves natural or plantation forests. The total area of degraded lands has been estimated at 10–60 Mkm² (*very low confidence*). The extent of degraded and marginal lands suitable for dedicated biomass production is highly uncertain and cannot be established without due consideration of current land use and land tenure. Increasing the area of dedicated energy crops can lead to land degradation elsewhere through indirect land-use change (*medium confidence*). Impacts of energy crops can be reduced through strategic integration with agricultural and forestry systems (*high confidence*) but the total quantity of biomass that can be produced through synergistic production systems is unknown. {4.1.6, 4.4.2, 4.5, 4.7.1, 4.8.1, 4.8.3, 4.8.4, 4.9.3}

Reducing unsustainable use of traditional biomass reduces land degradation and emissions of CO₂ while providing social and economic co-benefits (*very high confidence*). Traditional biomass in the form of fuelwood, charcoal and agricultural residues remains a primary source of energy for more than one-third of the global population, leading to unsustainable use of biomass resources and forest degradation and contributing around 2% of global GHG emissions (*low confidence*). Enhanced forest protection, improved forest and agricultural management, fuel-switching and adoption of efficient cooking and heating appliances can promote more sustainable biomass use and reduce land degradation, with co-benefits of reduced GHG emissions, improved human health, and reduced workload especially for women and youth (*very high confidence*). {4.1.6, 4.5.4}

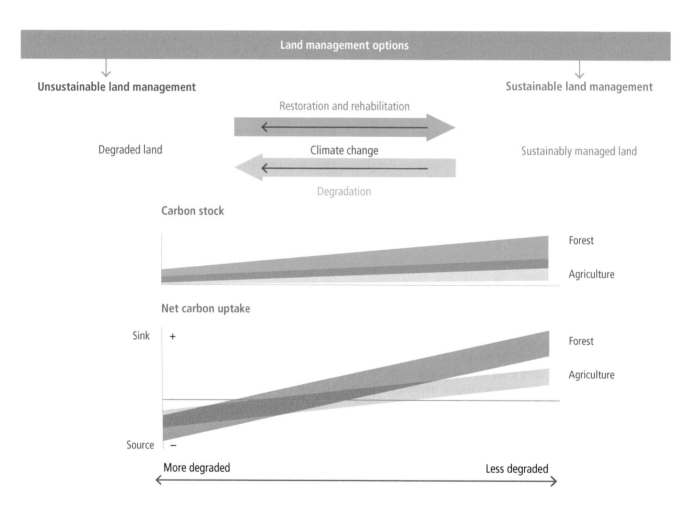

Figure TS.7 | Conceptual figure illustrating that climate change impacts interact with land management to determine sustainable or degraded outcome. Climate change can exacerbate many degradation processes (Table 4.1) and introduce novel ones (e.g., permafrost thawing or biome shifts), hence management needs to respond to climate impacts in order to avoid, reduce or reverse degradation. The types and intensity of human land-use and climate change impacts on lands affect their carbon stocks and their ability to operate as carbon sinks. In managed agricultural lands, degradation typically results in reductions of soil organic carbon stocks, which also adversely affects land productivity and carbon sinks. In forest land, reduction in biomass carbon stocks alone is not necessarily an indication of a reduction in carbon sinks. Sustainably managed forest landscapes can have a lower biomass carbon density but the younger forests can have a higher growth rate, and therefore contribute stronger carbon sinks, than older forests. Ranges of carbon sinks in forest and agricultural lands are overlapping. In some cases, climate change impacts may result in increased productivity and carbon stocks, at least in the short term.

Issue/ syndrome	Impact on climate change	Human driver	Climate driver	Land management options	References
Erosion of agricultural soils	Emission: CO_2, N_2O			Increase soil organic matter, no-till, perennial crops, erosion control, agroforestry, dietary change	3.1.4, 3.4.1, 3.5.2, 3.7.1, 4.8.1, 4.8.5, 4.9.2, 4.9.5
Deforestation	Emission of CO_2			Forest protection, sustainable forest management and dietary change	4.1.5, 4.5, 4.8.3, 4.8.4, 4.9.3
Forest degradation	Emission of CO_2 Reduced carbon sink			Forest protection, sustainable forest management	4.1.5, 4.5, 4.8.3, 4.8.4, 4.9.3
Overgrazing	Emission: CO_2, CH_4 Increasing albedo			Controlled grazing, rangeland management	3.1.4.2, 3.4.1, 3.6.1, 3.7.1, 4.8.1.4
Firewood and charcoal production	Emission: CO_2, CH_4 Increasing albedo			Clean cooking (health co-benefits, particularly for women and children)	3.6.3, 4.5.4, 4.8.3, 4.8.4
Increasing fire frequency and intensity	Emission: CO_2, CH_4, N_2O Emission: aerosols, increasing albedo			Fuel management, fire management	3.1.4, 3.6.1, 4.1.5, 4.8.3, Cross-Chapter Box 3 in Chp 2
Degradation of tropical peat soils	Emission: CO_2, CH_4			Peatland restoration, erosion control, regulating the use of peat soils	4.9.4
Thawing of permafrost	Emission: CO_2, CH_4			Relocation of settlement and infrastructure	4.8.5.1
Coastal erosion	Emission: CO_2, CH_4			Wetland and coastal restoration, mangrove conservation, long-term land-use planning	4.9.6, 4.9.7, 4.9.8
Sand and dust storms, wind erosion	Emission: aerosols			Vegetation management, afforestation, windbreaks	3.3.1, 3.4.1, 3.6.1, 3.7.1, 3.7.2
Bush encroachment	Capturing: CO_2, Decreasing albedo			Grazing land management, fire management	3.6.1.3, 3.7.3.2

Human driver	Climate driver
Grazing pressure	Warming trend
Agriculture practice	Extreme temperature
Expansion of agriculture	Drying trend
Forest clearing	Extreme rainfall
Wood fuel	Shifting rains
	Intensifying cyclones
	Sea level rise

Figure TS.8 | **Interaction of human and climate drivers can exacerbate desertification and land degradation. Figure shows key desertification and land degradation issues, how they impact climate change, and the key drivers, with potential solutions.** Climate change exacerbates the rate and magnitude of several ongoing land degradation and desertification processes. Human drivers of land degradation and desertification include expanding agriculture, agricultural practices and forest management. In turn, land degradation and desertification are also drivers of climate change through GHG emissions, reduced rates of carbon uptake, and reduced capacity of ecosystems to act as carbon sinks into the future. Impacts on climate change are either warming (in red) or cooling (in blue).

Land degradation can be avoided, reduced or reversed by implementing sustainable land management, restoration and rehabilitation practices that simultaneously provide many co-benefits, including adaptation to and mitigation of climate change (*high confidence*). Sustainable land management involves a comprehensive array of technologies and enabling conditions, which have proven to address land degradation at multiple landscape scales, from local farms (*very high confidence*) to entire watersheds (*medium confidence*). Sustainable forest management can prevent deforestation, maintain and enhance carbon sinks and can contribute towards GHG emissions-reduction goals. Sustainable forest management generates socio-economic benefits, and provides fibre, timber and biomass to meet society's growing needs. While sustainable forest management sustains high carbon sinks, the conversion from primary forests to sustainably managed forests can result in carbon emission during the transition and loss of biodiversity (*high confidence*). Conversely, in areas of degraded forests, sustainable forest management can increase carbon stocks and biodiversity (*medium confidence*). Carbon storage in long-lived wood products and reductions of emissions from use of wood products to substitute for emissions-intensive materials also contribute to mitigation objectives. (Figure TS.8) {4.8, 4.9, Table 4.2}

Lack of action to address land degradation will increase emissions and reduce carbon sinks and is inconsistent with the emissions reductions required to limit global warming to 1.5°C or 2°C. (*high confidence*). Better management of soils can offset 5–20% of current global anthropogenic GHG emissions (*medium confidence*). Measures to avoid, reduce and reverse land degradation are available but economic, political, institutional, legal and socio-cultural barriers, including lack of access to resources and knowledge, restrict their uptake (*very high confidence*). Proven measures that facilitate implementation of practices that avoid, reduce, or reverse land degradation include tenure reform, tax incentives, payments for ecosystem services, participatory integrated land-use planning, farmer networks and rural advisory services. Delayed action increases the costs of addressing land degradation, and can lead to irreversible biophysical and human outcomes (*high confidence*). Early actions can generate both site-specific and immediate benefits to communities affected by land degradation, and contribute to long-term global benefits through climate change mitigation (*high confidence*). (Figure TS.7) {4.1.5, 4.1.6, 4.7.1, 4.8, Table 4.2}

Even with adequate implementation of measures to avoid, reduce and reverse land degradation, there will be residual degradation in some situations (*high confidence*). Limits to adaptation are dynamic, site specific and determined through the interaction of biophysical changes with social and institutional conditions. Exceeding the limits of adaptation will trigger escalating losses or result in undesirable changes, such as forced migration, conflicts, or poverty. Examples of potential limits to adaptation due to climate-change-induced land degradation are coastal erosion (where land disappears, collapsing infrastructure and livelihoods due to thawing of permafrost), and extreme forms of soil erosion. {4.7, 4.8.5, 4.8.6, 4.9.6, 4.9.7, 4.9.8}

Land degradation is a serious and widespread problem, yet key uncertainties remain concerning its extent, severity, and linkages to climate change (*very high confidence*). Despite the difficulties of objectively measuring the extent and severity of land degradation, given its complex and value-based characteristics, land degradation represents – along with climate change – one of the biggest and most urgent challenges for humanity (*very high confidence*). The current global extent, severity and rates of land degradation are not well quantified. There is no single method by which land degradation can be measured objectively and consistently over large areas because it is such a complex and value-laden concept (*very high confidence*). However, many existing scientific and locally based approaches, including the use of ILK, can assess different aspects of land degradation or provide proxies. Remote sensing, corroborated by other data, can generate geographically explicit and globally consistent data that can be used as proxies over relevant time scales (several decades). Few studies have specifically addressed the impacts of proposed land-based negative emission technologies on land degradation. Much research has tried to understand how livelihoods and ecosystems are affected by a particular stressor – for example, drought, heat stress, or waterlogging. Important knowledge gaps remain in understanding how plants, habitats and ecosystems are affected by the cumulative and interacting impacts of several stressors, including potential new stressors resulting from large-scale implementation of negative emission technologies. {4.10}

TS.5 Food security

The current food system (production, transport, processing, packaging, storage, retail, consumption, loss and waste) feeds the great majority of world population and supports the livelihoods of over 1 billion people. Since 1961, food supply per capita has increased more than 30%, accompanied by greater use of nitrogen fertilisers (increase of about 800%) and water resources for irrigation (increase of more than 100%). However, an estimated 821 million people are currently undernourished, 151 million children under five are stunted, 613 million women and girls aged 15 to 49 suffer from iron deficiency, and 2 billion adults are overweight or obese. The food system is under pressure from non-climate stressors (e.g., population and income growth, demand for animal-sourced products), and from climate change. These climate and non-climate stresses are impacting the four pillars of food security (availability, access, utilisation, and stability). (Figure TS.9) {5.1.1, 5.1.2}

Observed climate change is already affecting food security through increasing temperatures, changing precipitation patterns, and greater frequency of some extreme events (*high confidence*). Studies that separate out climate change from other factors affecting crop yields have shown that yields of some crops (e.g., maize and wheat) in many lower-latitude regions have been affected negatively by observed climate changes, while in many higher-latitude regions, yields of some crops (e.g., maize, wheat, and sugar beets) have been affected positively over recent decades. Warming compounded by drying has caused large negative effects on yields in parts of the Mediterranean. Based on ILK, climate change is affecting food security in drylands, particularly those in Africa, and high mountain regions of Asia and South America. (Figure TS.10) {5.2.2}

Food security will be increasingly affected by projected future climate change (*high confidence*). Across SSPs 1, 2, and 3, global crop and economic models projected a 1–29% cereal price increase in 2050 due to climate change (RCP 6.0), which would impact consumers globally through higher food prices; regional effects will vary (*high confidence*). Low-income consumers are particularly at risk, with models projecting increases of 1–183 million additional people at risk of hunger across the SSPs compared to a no climate change scenario (*high confidence*). While increased CO_2 is projected to be beneficial for crop productivity at lower temperature increases, it is projected to lower nutritional quality (*high confidence*) (e.g., wheat grown at 546–586 ppm CO_2 has 5.9–12.7% less protein, 3.7–6.5% less zinc, and 5.2–7.5% less iron). Distributions of pests and diseases will change, affecting production negatively in many regions (*high confidence*). Given increasing extreme events and interconnectedness, risks of food system disruptions are growing (*high confidence*). {5.2.3, 5.2.4}

Vulnerability of pastoral systems to climate change is very high (*high confidence*). Pastoralism is practiced in more than 75% of countries by between 200 and 500 million people, including nomadic communities, transhumant herders, and agropastoralists. Impacts in pastoral systems in Africa include lower pasture and animal productivity, damaged reproductive function, and biodiversity loss. Pastoral system vulnerability is exacerbated by non-climate factors

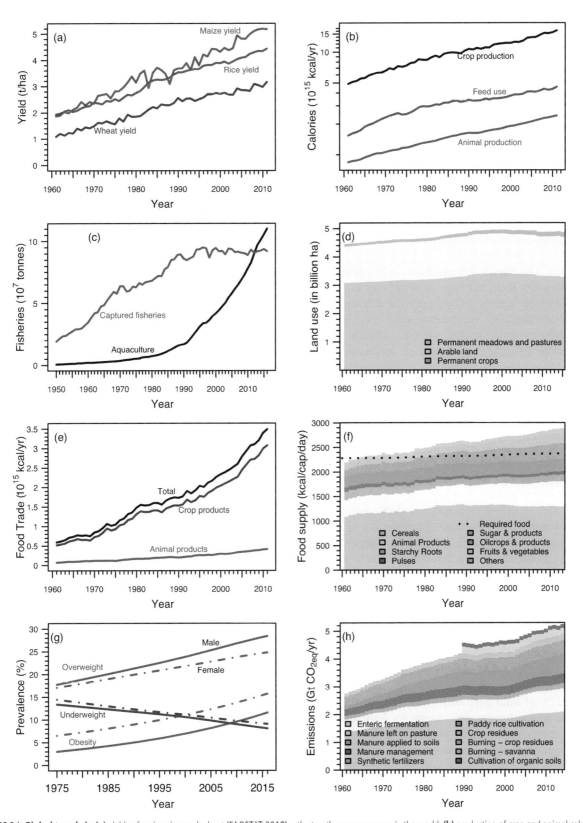

Figure TS.9 | Global trends in (a) yields of maize, rice, and wheat (FAOSTAT 2018) – the top three crops grown in the world; **(b)** production of crop and animal calories and use of crop calories as livestock feed (FAOSTAT 2018); **(c)** production from marine and aquaculture fisheries (FishStat 2019); **(d)** land used for agriculture (FAOSTAT 2018); **(e)** food trade in calories (FAOSTAT 2018); **(f)** food supply and required food (i.e., based on human energy requirements for medium physical activities) from 1961–2012 (FAOSTAT 2018; Hiç et al. 2016); **(g)** prevalence of overweight, obesity and underweight from 1975–2015 (Abarca-Gómez et al. 2017); and **(h)** GHG emissions for the agriculture sector, excluding land-use change (FAOSTAT 2018). For figures (b) and (e), data provided in mass units were converted into calories using nutritive factors (FAO 2001b). Data on emissions due to burning of savanna and cultivation of organic soils is provided only after 1990 (FAOSTAT 2018).

(land tenure, sedentarisation, changes in traditional institutions, invasive species, lack of markets, and conflicts). {5.2.2}

Fruit and vegetable production, a key component of healthy diets, is also vulnerable to climate change (*medium evidence, high agreement*). Declines in yields and crop suitability are projected under higher temperatures, especially in tropical and semi-tropical regions. Heat stress reduces fruit set and speeds up development of annual vegetables, resulting in yield losses, impaired product quality, and increasing food loss and waste. Longer growing seasons enable a greater number of plantings to be cultivated and can contribute to greater annual yields. However, some fruits and vegetables need a period of cold accumulation to produce a viable harvest, and warmer winters may constitute a risk. {5.2.2}

Food security and climate change have strong gender and equity dimensions (*high confidence*). Worldwide, women play a key role in food security, although regional differences exist. Climate change impacts vary among diverse social groups depending on age, ethnicity, gender, wealth, and class. Climate extremes have immediate and long-term impacts on livelihoods of poor and vulnerable communities, contributing to greater risks of food insecurity that can be a stress multiplier for internal and external migration (*medium confidence*). Empowering women and rights-based approaches to decision-making can create synergies among household food security, adaptation, and mitigation. {5.2.6, 5.6.4}

Many practices can be optimised and scaled up to advance adaptation throughout the food system (*high confidence*). Supply-side options include increased soil organic matter and erosion control, improved cropland, livestock, grazing land management, and genetic improvements for tolerance to heat and drought. Diversification in the food system (e.g., implementation of integrated production systems, broad-based genetic resources, and heterogeneous diets) is a key strategy to reduce risks (*medium confidence*). Demand-side adaptation, such as adoption of healthy and sustainable diets, in conjunction with reduction in food loss and waste, can contribute to adaptation through reduction in additional land area needed for food production and associated food system vulnerabilities. ILK can contribute to enhancing food system resilience (*high confidence*). {5.3, 5.6.3 Cross-Chapter Box 6 in Chapter 5}.

About 21–37% of total greenhouse gas (GHG) emissions are attributable to the food system. These are from agriculture and land use, storage, transport, packaging, processing, retail, and consumption (*medium confidence*). This estimate includes emissions of 9–14% from crop and livestock activities within the farm gate and 5–14% from land use and land-use change including deforestation and peatland degradation (*high confidence*); 5–10% is from supply chain activities (*medium confidence*). This estimate includes GHG emissions from food loss and waste. Within the food system, during the period 2007–2016, the major sources of emissions from the supply side were agricultural production, with crop and livestock activities within the farm gate generating respectively 142 ± 42 TgCH$_4$ yr^{-1} (*high confidence*) and 8.0 ± 2.5 TgN$_2$O yr^{-1} (high confidence), and CO$_2$ emissions linked to relevant land-use change dynamics such as deforestation and peatland degradation, generating 4.9 ± 2.5 GtCO$_2$ yr^{-1}. Using 100-year GWP values (no

climate feedback) from the IPCC AR5, this implies that total GHG emissions from agriculture were 6.2 ± 1.4 GtCO$_2$-eq yr^{-1}, increasing to 11.1 ± 2.9 GtCO$_2$-eq yr^{-1} including relevant land use. Without intervention, these are likely to increase by about 30–40% by 2050, due to increasing demand based on population and income growth and dietary change (*high confidence*). {5.4}

Supply-side practices can contribute to climate change mitigation by reducing crop and livestock emissions, sequestering carbon in soils and biomass, and by decreasing emissions intensity within sustainable production systems (*high confidence*). Total technical mitigation potential from crop and livestock activities and agroforestry is estimated as 2.3–9.6 GtCO$_2$-eq yr^{-1} by 2050 (*medium confidence*). Options with large potential for GHG mitigation in cropping systems include soil carbon sequestration (at decreasing rates over time), reductions in N$_2$O emissions from fertilisers, reductions in CH$_4$ emissions from paddy rice, and bridging of yield gaps. Options with large potential for mitigation in livestock systems include better grazing land management, with increased net primary production and soil carbon stocks, improved manure management, and higher-quality feed. Reductions in GHG emissions intensity (emissions per unit product) from livestock can support reductions in absolute emissions, provided appropriate governance to limit total production is implemented at the same time (*medium confidence*). {5.5.1}

Consumption of healthy and sustainable diets presents major opportunities for reducing GHG emissions from food systems and improving health outcomes (*high confidence*). Examples of healthy and sustainable diets are high in coarse grains, pulses, fruits and vegetables, and nuts and seeds; low in energy-intensive animal-sourced and discretionary foods (such as sugary beverages); and with a carbohydrate threshold. Total technical mitigation potential of dietary changes is estimated as 0.7–8.0 GtCO$_2$-eq yr^{-1} by 2050 (*medium confidence*). This estimate includes reductions in emissions from livestock and soil carbon sequestration on spared land, but co-benefits with health are not taken into account. Mitigation potential of dietary change may be higher, but achievement of this potential at broad scales depends on consumer choices and dietary preferences that are guided by social, cultural, environmental, and traditional factors, as well as income growth. Meat analogues such as imitation meat (from plant products), cultured meat, and insects may help in the transition to more healthy and sustainable diets, although their carbon footprints and acceptability are uncertain. {5.5.2, 5.6.5}

Reduction of food loss and waste could lower GHG emissions and improve food security (*medium confidence*). Combined food loss and waste amount to 25–30% of total food produced (*medium confidence*). During 2010–2016, global food loss and waste equalled 8–10% of total anthropogenic GHG emissions (*medium confidence*); and cost about 1 trillion USD2012 per year (*low confidence*). Technical options for reduction of food loss and waste include improved harvesting techniques, on-farm storage, infrastructure, and packaging. Causes of food loss (e.g., lack of refrigeration) and waste (e.g., behaviour) differ substantially in developed and developing countries, as well as across regions (*robust evidence, medium agreement*). {5.5.2}

TS

GGCMs with explicit N stress

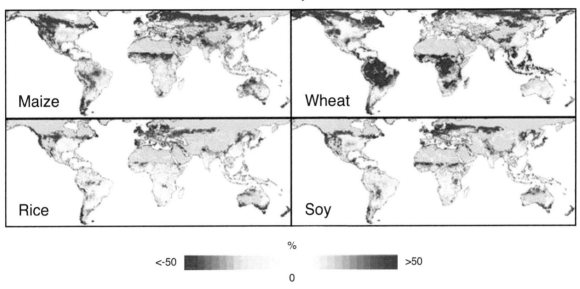

%
<-50 [gradient] >50
0

Figure TS.10 | AgMIP median yield changes (%) for RCP8.5 (2070–2099 in comparison to 1980–2010 baseline) with CO_2 effects and explicit nitrogen stress over five GCMs **x** four Global Gridded Crop Models (GGCMs) for rainfed maize, wheat, rice, and soy (20 ensemble members from EPIC, GEPIC, pDSSAT, and PEGASUS; except for rice which has 15). Grey areas indicate historical areas with little to no yield capacity. All models use a 0.5° grid, but there are differences in grid cells simulated to represent agricultural land. While some models simulated all land areas, others simulated only potential suitable cropland area according to evolving climatic conditions. Others utilised historical harvested areas in 2000 according to various data sources (Rosenzweig et al. 2014).

Agriculture and the food system are key to global climate change responses. Combining supply-side actions such as efficient production, transport, and processing with demand-side interventions such as modification of food choices, and reduction of food loss and waste, reduces GHG emissions and enhances food system resilience (*high confidence*). Such combined measures can enable the implementation of large-scale land-based adaptation and mitigation strategies without threatening food security from increased competition for land for food production and higher food prices. Without combined food system measures in farm management, supply chains, and demand, adverse effects would include increased numbers of malnourished people and impacts on smallholder farmers (*medium evidence, high agreement*). Just transitions are needed to address these effects. (Figure TS.11) {5.5, 5.6, 5.7}

For adaptation and mitigation throughout the food system, enabling conditions need to be created through policies, markets, institutions, and governance (*high confidence*). For adaptation, resilience to increasing extreme events can be accomplished through risk sharing and transfer mechanisms such as insurance markets and index-based weather insurance (*high confidence*). Public health policies to improve nutrition – such as school procurement, health insurance incentives, and awareness-raising campaigns – can potentially change demand, reduce healthcare costs, and contribute to lower GHG emissions (*limited evidence, high agreement*). Without inclusion of comprehensive food system responses in broader climate change policies, the mitigation and adaptation potentials assessed in Chapter 5 will not be realised and food security will be jeopardised (*high confidence*). {5.7.5}

TS

Food system response options

Mitigation and adaptation potential

- ▇ Very high
- ▨ High
- ▨ Limited
- ⁄⁄⁄⁄ None

Response options	Mitigation	Adaptation
Improved crop management		
Increased soil organic matter content	Very high	High
Change in crop variety	Limited	Very high
Improved water management	Limited	Very high
Adjustment of planting dates	None	Very high
Precision fertiliser management	High	High
Integrated pest management	None	High
Counter season crop production	None	High
Biochar application	High	Very high
Agroforestry	High	High
Changing monoculture to crop diversification	Limited	Very high
Changes in cropping area, land rehabilitation (enclosures, afforestation) perennial farming	High	High
Tillage and crop establishment	High	High
Residue management	High	High
Crop–livestock systems	High	High
Improved livestock managment		
Silvopastural system	Very high	Very high
New livestock breed	Limited	Limited
Livestock fattening	Limited	Limited
Shifting to small ruminants or drought-resistant livestock or fish farming	Limited	High
Feed and fodder banks	High	High
Methane inhibitors	Very high	None
Thermal stress control	High	Very high
Seasonal feed supplementation	High	High
Improved animal health and parasites control	High	Very high
Climate services		
Early warning systems	None	Very high
Planning and prediction at seasonal to intra-seasonal climate risk	None	Very high
Crop and livestock insurance	None	Very high
Improved supply chain		
Food storage infrastructures	Limited	Very high
Shortening supply chains	High	Limited
Improved food transport and distribution	High	Very high
Improved efficiency and sustainability of food processing, retail and agrifood industries	High	Very high
Improved energy efficiencies of agriculture	High	Limited
Reduce food loss	Very high	Limited
Urban and peri-urban agriculture	Limited	Limited
Bioeconomy (e.g. energy from waste)	None	Limited
Demand management		
Dietary changes	Very high	Very high
Reduce food waste	Very high	Very high
Packaging reductions	Limited	Limited
New ways of selling (e.g. direct sales)	Limited	Limited
Transparency of food chains and external costs	High	High

Figure TS.11 | Response options related to food system and their potential impacts on mitigation and adaptation. Many response options offer significant potential for both mitigation and adaptation.

TS.6 Interlinkages between desertification, land degradation, food security and GHG fluxes: Synergies, trade-offs and integrated response options

The land challenges, in the context of this report, are climate change mitigation, adaptation, desertification, land degradation, and food security. The chapter also discusses implications for Nature's Contributions to People (NCP), including biodiversity and water, and sustainable development, by assessing intersections with the Sustainable Development Goals (SDGs). The chapter assesses response options that could be used to address these challenges. These response options were derived from the previous chapters and fall into three broad categories: land management, value chain, and risk management.

The land challenges faced today vary across regions; climate change will increase challenges in the future, while socio-economic development could either increase or decrease challenges (*high confidence*). Increases in biophysical impacts from climate change can worsen desertification, land degradation, and food insecurity (*high confidence*). Additional pressures from socio-economic development could further exacerbate these challenges; however, the effects are scenario dependent. Scenarios with increases in income and reduced pressures on land can lead to reductions in food insecurity; however, all assessed scenarios result in increases in water demand and water scarcity (*medium confidence*). {6.1}

The applicability and efficacy of response options are region and context specific; while many value chain and risk management options are potentially broadly applicable, many land management options are applicable on less than 50% of the ice-free land surface (*high confidence*). Response options are limited by land type, bioclimatic region, or local food system context (*high confidence*). Some response options produce adverse side effects only in certain regions or contexts; for example, response options that use freshwater may have no adverse side effects in regions where water is plentiful, but large adverse side effects in regions where water is scarce (*high confidence*). Response options with biophysical climate effects (e.g., afforestation, reforestation) may have different effects on local climate, depending on where they are implemented (*medium confidence*). Regions with more challenges have fewer response options available for implementation (*medium confidence*). {6.1, 6.2, 6.3, 6.4}

Nine options deliver medium-to-large benefits for all five land challenges (*high confidence*). The options with medium-to-large benefits for all challenges are increased food productivity, improved cropland management, improved grazing land management, improved livestock management, agroforestry, forest management, increased soil organic carbon content, fire management and reduced post-harvest losses. A further two options, dietary change and reduced food waste, have no global estimates for adaptation but have medium-to-large benefits for all other challenges (*high confidence*). {6.3, 6.4}

Five options have large mitigation potential (>3 GtCO$_2$e yr^{-1}) without adverse impacts on the other challenges (*high confidence*). These are: increased food productivity; reduced deforestation and forest degradation; increased soil organic carbon content; fire management; and reduced post-harvest losses. Two further options with large mitigation potential, dietary change and reduced food waste, have no global estimates for adaptation but show no negative impacts across the other challenges. Five options: improved cropland management; improved grazing land managements; agroforestry; integrated water management; and forest management, have moderate mitigation potential, with no adverse impacts on the other challenges (*high confidence*). {6.3.6}

Sixteen response options have large adaptation potential (more than 25 million people benefit), without adverse side effects on other land challenges (*high confidence*). These are increased food productivity, improved cropland management, agroforestry, agricultural diversification, forest management, increased soil organic carbon content, reduced landslides and natural hazards, restoration and reduced conversion of coastal wetlands, reduced post-harvest losses, sustainable sourcing, management of supply chains, improved food processing and retailing, improved energy use in food systems, livelihood diversification, use of local seeds, and disaster risk management (*high confidence*). Some options (such as enhanced urban food systems or management of urban sprawl) may not provide large global benefits but may have significant positive local effects without adverse effects (*high confidence*). (Figure TS.13) {6.3, 6.4}

Seventeen of 40 options deliver co-benefits or no adverse side effects for the full range of NCPs and SDGs; only three options (afforestation, BECCS), and some types of risk sharing instruments, such as insurance) have potentially adverse side effects for five or more NCPs or SDGs (*medium confidence*). The 17 options with co-benefits and no adverse side effects include most agriculture- and soil-based land management options, many ecosystem-based land management options, forest management, reduced post-harvest losses, sustainable sourcing, improved energy use in food systems, and livelihood diversification (*medium confidence*). Some of the synergies between response options and SDGs include positive poverty eradication impacts from activities like improved water management or improved management of supply chains. Examples of synergies between response options and NCPs include positive impacts on habitat maintenance from activities like invasive species management and agricultural diversification. However, many of these synergies are not automatic, and are dependent on well-implemented activities requiring institutional and enabling conditions for success. {6.4}

Most response options can be applied without competing for available land; however, seven options result in competition for land (*medium confidence*). A large number of response options do not require dedicated land, including several land management options, all value chain options, and all risk management options. Four options could greatly increase competition for land if applied at scale: afforestation, reforestation, and land used to provide feedstock for BECCS or biochar, with three further options: reduced grassland

61

conversion to croplands, restoration and reduced conversion of peatlands and restoration, and reduced conversion of coastal wetlands having smaller or variable impacts on competition for land. Other options such as reduced deforestation and forest degradation, restrict land conversion for other options and uses. Expansion of the current area of managed land into natural ecosystems could have negative consequences for other land challenges, lead to the loss of biodiversity, and adversely affect a range of NCPs (*high confidence*). {6.3.6, 6.4}

Some options, such as bioenergy and BECCS, are scale dependent. The climate change mitigation potential for bioenergy and BECCS is large (up to 11 $GtCO_2$ yr^{-1}); however, the effects of bioenergy production on land degradation, food insecurity, water scarcity, GHG emissions, and other environmental goals are scale- and context-specific (*high confidence*). These effects depend on the scale of deployment, initial land use, land type, bioenergy feedstock, initial carbon stocks, climatic region and management regime (*high confidence*). Large areas of monoculture bioenergy crops that displace other land uses can result in land competition, with adverse effects for food production, food consumption, and thus food security, as well as adverse effects for land degradation, biodiversity, and water scarcity (*medium confidence*). However, integration of bioenergy into sustainably managed agricultural landscapes can ameliorate these challenges (*medium confidence*). {6.2, 6.3, 6.4, Cross-Chapter Box 7 in Chapter 6}

Response options are interlinked; some options (e.g., land sparing and sustainable land management options) can enhance the co-benefits or increase the potential for other options (*medium confidence*). Some response options can be more effective when applied together (*medium confidence*); for example, dietary change and waste reduction expand the potential to apply other options by freeing as much as 5.8 Mkm^2 (0.8–2.4 Mkm^2 for dietary change; about 2 Mkm^2 for reduced post-harvest losses, and 1.4 Mkm^2 for reduced food waste) of land (*low confidence*). Integrated water management and increased soil organic carbon can increase food productivity in some circumstances. {6.4}

Other response options (e.g., options that require land) may conflict; as a result, the potentials for response options are not all additive, and a total potential from the land is currently unknown (*high confidence*). Combining some sets of options (e.g., those that compete for land) may mean that maximum potentials cannot be realised, for example, reforestation, afforestation, and bioenergy and BECCS, all compete for the same finite land resource so the combined potential is much lower than the sum of potentials of each individual option, calculated in the absence of alternative uses of the land (*high confidence*). Given the interlinkages among response options and that mitigation potentials for individual options assume that they are applied to all suitable land, the total mitigation potential is much lower than the sum of the mitigation potential of the individual response options (*high confidence*). (Figure TS.12) {6.4}

The feasibility of response options, including those with multiple co-benefits, is limited due to economic, technological, institutional, socio-cultural, environmental and geophysical barriers (*high confidence*). A number of response options (e.g., most agriculture-based land management options, forest management, reforestation and restoration) have already been implemented widely to date (*high confidence*). There is *robust evidence* that many other response options can deliver co-benefits across the range of land challenges, yet these are not being implemented. This limited application is evidence that multiple barriers to implementation of response options exist (*high confidence*). {6.3, 6.4}

Coordinated action is required across a range of actors, including business, producers, consumers, land managers, indigenous peoples and local communities and policymakers to create enabling conditions for adoption of response options (*high confidence*). The response options assessed face a variety of barriers to implementation (economic, technological, institutional, socio-cultural, environmental and geophysical) that require action across multiple actors to overcome (*high confidence*). There are a variety of response options available at different scales that could form portfolios of measures applied by different stakeholders – from farm to international scales. For example, agricultural diversification and use of local seeds by smallholders can be particularly useful poverty eradication and biodiversity conservation measures, but are only successful when higher scales, such as national and international markets and supply chains, also value these goods in trade regimes, and consumers see the benefits of purchasing these goods. However, the land and food sectors face particular challenges of institutional fragmentation, and often suffer from a lack of engagement between stakeholders at different scales (*medium confidence*). {6.3, 6.4}

Delayed action will result in an increased need for response to land challenges and a decreased potential for land-based response options due to climate change and other pressures (*high confidence*). For example, failure to mitigate climate change will increase requirements for adaptation and may reduce the efficacy of future land-based mitigation options (*high confidence*). The potential for some land management options decreases as climate change increases; for example, climate alters the sink capacity for soil and vegetation carbon sequestration, reducing the potential for increased soil organic carbon (*high confidence*). Other options (e.g., reduced deforestation and forest degradation) prevent further detrimental effects to the land surface; delaying these options could lead to increased deforestation, conversion, or degradation, serving as increased sources of GHGs and having concomitant negative impacts on NCPs (*medium confidence*). Carbon dioxide removal (CDR) options – such as reforestation, afforestation, bioenergy and BECCS – are used to compensate for unavoidable emissions in other sectors; delayed action will result in larger and more rapid deployment later (*high confidence*). Some response options will not be possible if action is delayed too long; for example, peatland restoration might not be possible after certain thresholds of degradation have been exceeded, meaning that peatlands could not be restored in certain locations (*medium confidence*) {6.2, 6.3, 6.4}.

Early action, however, has challenges including technological readiness, upscaling, and institutional barriers (*high confidence*). Some of the response options have technological barriers that may limit their wide-scale application in the near term (*high confidence*). Some response options, for example, BECCS, have only been implemented at small-scale demonstration facilities; challenges exist with upscaling these options to the levels discussed in Chapter 6 (*medium confidence*). Economic and institutional barriers, including governance, financial incentives and financial resources, limit the near-term adoption of many response options, and 'policy lags', by which implementation is delayed by the slowness of the policy implementation cycle, are significant across many options (*medium confidence*). Even some actions that initially seemed like 'easy wins' have been challenging to implement, with stalled policies for reducing emissions from deforestation and forest degradation and fostering conservation (REDD+) providing clear examples of how response options need sufficient funding, institutional support, local buy-in, and clear metrics for success, among other necessary enabling conditions. {6.2, 6.4}

Some response options reduce the consequences of land challenges, but do not address underlying drivers (*high confidence*). For example, management of urban sprawl can help reduce the environmental impact of urban systems; however, such management does not address the socio-economic and demographic changes driving the expansion of urban areas. By failing to address the underlying drivers, there is a potential for the challenge to re-emerge in the future (*high confidence*). {6.4}

Many response options have been practised in many regions for many years; however, there is limited knowledge of the efficacy and broader implications of other response options (*high confidence*). For the response options with a large evidence base and ample experience, further implementation and upscaling would carry little risk of adverse side effects (*high confidence*). However, for other options, the risks are larger as the knowledge gaps are greater; for example, uncertainty in the economic and social aspects of many land response options hampers the ability to predict their effects (*medium confidence*). Furthermore, Integrated Assessment Models, like those used to develop the pathways in the IPCC Special Report on Global Warming of 1.5°C (SR15), omit many of these response options and do not assess implications for all land challenges (*high confidence*). {6.4}

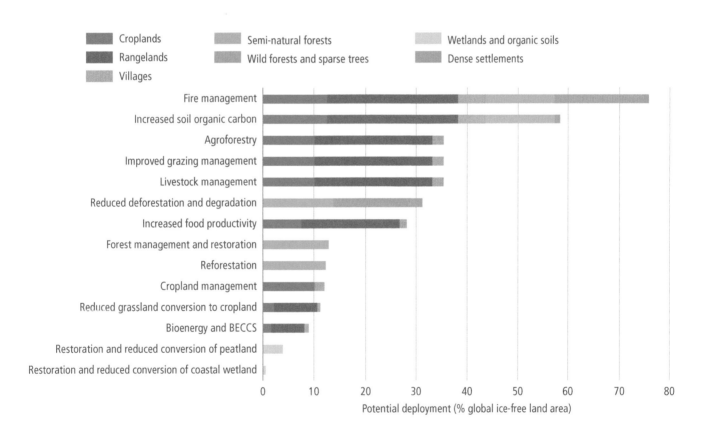

Figure TS.12 | Potential deployment area of land management responses (see Table 6.1) across land-use types (or anthromes, see Section 6.3), when selecting responses having only co-benefits for local challenges and for climate change mitigation and no large adverse side effects on global food security. See Figure 6.2 for the criteria used to map challenges considered (desertification, land degradation, climate change adaptation, chronic undernourishment, biodiversity, groundwater stress and water quality). No response option was identified for barren lands.

Potential global contribution of response options to mitigation, adaptation, combating desertification and land degradation, and enhancing food security

Panel A shows response options that can be implemented without or with limited competition for land, including some that have the potential to reduce the demand for land. Co-benefits and adverse side effects are shown quantitatively based on the high end of the range of potentials assessed. Magnitudes of contributions are categorised using thresholds for positive or negative impacts. Letters within the cells indicate confidence in the magnitude of the impact relative to the thresholds used (see legend). Confidence in the direction of change is generally higher.

Response options based on land management

	Response option	Mitigation	Adaptation	Desertification	Land Degradation	Food Security	Cost
Agriculture	Increased food productivity	L	M	L	M	H	
Agriculture	Agro-forestry	M	M	M	M	L	●
Agriculture	Improved cropland management	M	L	L	L	L	●●
Agriculture	Improved livestock management	M	L	L	L	L	●●●
Agriculture	Agricultural diversification	L	L	L	M	L	
Agriculture	Improved grazing land management	M	L	L	L	L	
Agriculture	Integrated water management	L	L	L	L	L	●●
Agriculture	Reduced grassland conversion to cropland	L		L	L -	L	●
Forests	Forest management	M	L	L	L	L	●●
Forests	Reduced deforestation and forest degradation	H	L	L	L	L	●●
Soils	Increased soil organic carbon content	H	L	M	L	L	●●
Soils	Reduced soil erosion	←→ L	L	M	L	L	●●
Soils	Reduced soil salinization		L	L	L	L	●●
Soils	Reduced soil compaction		L		L	L	●
Other ecosystems	Fire management	M	M	M	M	L	●
Other ecosystems	Reduced landslides and natural hazards	L	L	L	L	L	
Other ecosystems	Reduced pollution including acidification	←→ M	M	L	L	L	
Other ecosystems	Restoration & reduced conversion of coastal wetlands	M	L	M	M	←→ L	
Other ecosystems	Restoration & reduced conversion of peatlands	M		na	M -	L	●

Response options based on value chain management

	Response option	Mitigation	Adaptation	Desertification	Land Degradation	Food Security	Cost
Demand	Reduced post-harvest losses	H	M	L	L	H	
Demand	Dietary change	H		L	H	H	
Demand	Reduced food waste (consumer or retailer)	H		L	M	M	
Supply	Sustainable sourcing		L		L	L	
Supply	Improved food processing and retailing	L	L			L	
Supply	Improved energy use in food systems	L	L			L	

Response options based on risk management

	Response option	Mitigation	Adaptation	Desertification	Land Degradation	Food Security	Cost
Risk	Livelihood diversification		L		L	L	
Risk	Management of urban sprawl		L	L	M	L	
Risk	Risk sharing instruments	←→ L	L		←→ L	L	●●

Options shown are those for which data are available to assess global potential for three or more land challenges.
The magnitudes are assessed independently for each option and are not additive.

Key for criteria used to define magnitude of impact of each integrated response option

		Mitigation Gt CO₂-eq yr⁻¹	Adaptation Million people	Desertification Million km²	Land Degradation Million km²	Food Security Million people
Positive	Large	More than 3	Positive for more than 25	Positive for more than 3	Positive for more than 3	Positive for more than 100
Positive	Moderate	0.3 to 3	1 to 25	0.5 to 3	0.5 to 3	1 to 100
Positive	Small	Less than 0.3	Less than 1	Less than 0.5	Less than 0.5	Less than 1
	Negligible	No effect	No effect	No effect	No effect	No effect
Negative	Small	Less than -0.3	Less than 1	Less than 0.5	Less than 0.5	Less than 1
Negative	Moderate	-0.3 to -3	1 to 25	0.5 to 3	0.5 to 3	1 to 100
Negative	Large	More than -3	Negative for more than 25	Negative for more than 3	Negative for more than 3	Negative for more than 100

←→ **Variable:** Can be positive or negative   ——— no data   na not applicable

Confidence level
Indicates confidence in the estimate of magnitude category.

H High confidence
M Medium confidence
L Low confidence

Cost range
See technical caption for cost ranges in US$ tCO₂e⁻¹ or US$ ha⁻¹.

●●● High cost
●● Medium cost
● Low cost
——— no data

Figure TS.13 | Potential global contribution of response options to mitigation, adaptation, combating desertification and land degradation, and enhancing food security (Panel A).

Potential global contribution of response options to mitigation, adaptation, combating desertification and land degradation, and enhancing food security

Panel B shows response options that rely on additional land-use change and could have implications across three or more land challenges under different implementation contexts. For each option, the first row (high level implementation) shows a quantitative assessment (as in Panel A) of implications for global implementation at scales delivering CO_2 removals of more than 3 $GtCO_2$ yr^{-1} using the magnitude thresholds shown in Panel A. The red hatched cells indicate an increasing pressure but unquantified impact. For each option, the second row (best practice implementation) shows qualitative estimates of impact if implemented using best practices in appropriately managed landscape systems that allow for efficient and sustainable resource use and supported by appropriate governance mechanisms. In these qualitative assessments, green indicates a positive impact, grey indicates a neutral interaction.

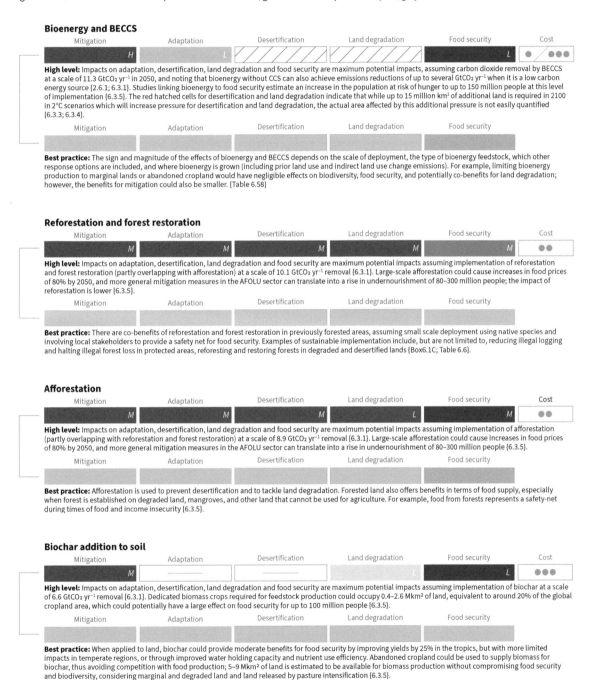

Bioenergy and BECCS

High level: Impacts on adaptation, desertification, land degradation and food security are maximum potential impacts, assuming carbon dioxide removal by BECCS at a scale of 11.3 $GtCO_2$ yr^{-1} in 2050, and noting that bioenergy without CCS can also achieve emissions reductions of up to several $GtCO_2$ yr^{-1} when it is a low carbon energy source {2.6.1; 6.3.1}. Studies linking bioenergy to food security estimate an increase in the population at risk of hunger to up to 150 million people at this level of implementation {6.3.5}. The red hatched cells for desertification and land degradation indicate that while up to 15 million km^2 of additional land is required in 2100 in 2°C scenarios which will increase pressure for desertification and land degradation, the actual area affected by this additional pressure is not easily quantified {6.3.3; 6.3.4}.

Best practice: The sign and magnitude of the effects of bioenergy and BECCS depends on the scale of deployment, the type of bioenergy feedstock, which other response options are included, and where bioenergy is grown (including prior land use and indirect land use change emissions). For example, limiting bioenergy production to marginal lands or abandoned cropland would have negligible effects on biodiversity, food security, and potentially co-benefits for land degradation; however, the benefits for mitigation could also be smaller. {Table 6.58}

Reforestation and forest restoration

High level: Impacts on adaptation, desertification, land degradation and food security are maximum potential impacts assuming implementation of reforestation and forest restoration (partly overlapping with afforestation) at a scale of 10.1 $GtCO_2$ yr^{-1} removal {6.3.1}. Large-scale afforestation could cause increases in food prices of 80% by 2050, and more general mitigation measures in the AFOLU sector can translate into a rise in undernourishment of 80–300 million people; the impact of reforestation is lower {6.3.5}.

Best practice: There are co-benefits of reforestation and forest restoration in previously forested areas, assuming small scale deployment using native species and involving local stakeholders to provide a safety net for food security. Examples of sustainable implementation include, but are not limited to, reducing illegal logging and halting illegal forest loss in protected areas, reforesting and restoring forests in degraded and desertified lands {Box6.1C; Table 6.6}.

Afforestation

High level: Impacts on adaptation, desertification, land degradation and food security are maximum potential impacts assuming implementation of afforestation (partly overlapping with reforestation and forest restoration) at a scale of 8.9 $GtCO_2$ yr^{-1} removal {6.3.1}. Large-scale afforestation could cause increases in food prices of 80% by 2050, and more general mitigation measures in the AFOLU sector can translate into a rise in undernourishment of 80–300 million people {6.3.5}.

Best practice: Afforestation is used to prevent desertification and to tackle land degradation. Forested land also offers benefits in terms of food supply, especially when forest is established on degraded land, mangroves, and other land that cannot be used for agriculture. For example, food from forests represents a safety-net during times of food and income insecurity {6.3.5}.

Biochar addition to soil

High level: Impacts on adaptation, desertification, land degradation and food security are maximum potential impacts assuming implementation of biochar at a scale of 6.6 $GtCO_2$ yr^{-1} removal {6.3.1}. Dedicated biomass crops required for feedstock production could occupy 0.4–2.6 Mkm^2 of land, equivalent to around 20% of the global cropland area, which could potentially have a large effect on food security for up to 100 million people {6.3.5}.

Best practice: When applied to land, biochar could provide moderate benefits for food security by improving yields by 25% in the tropics, but with more limited impacts in temperate regions, or through improved water holding capacity and nutrient use efficiency. Abandoned cropland could be used to supply biomass for biochar, thus avoiding competition with food production; 5–9 Mkm^2 of land is estimated to be available for biomass production without compromising food security and biodiversity, considering marginal and degraded land and land released by pasture intensification {6.3.5}.

Figure TS.13 | Potential global contribution of response options to mitigation, adaptation, combating desertification and land degradation, and enhancing food security (Panel B).

Figure TS.13 (continued): This Figure is based on an aggregation of information from studies with a wide variety of assumptions about how response options are implemented and the contexts in which they occur. Response options implemented differently at local to global scales could lead to different outcomes. **Magnitude of potential:** For panel A, magnitudes are for the technical potential of response options globally. For each land challenge, magnitudes are set relative to a marker level as follows. For mitigation, potentials are set relative to the approximate potentials for the response options with the largest individual impacts ($\sim$3 GtCO$_2$-eq yr^{-1}). The threshold for the 'large' magnitude category is set at this level. For adaptation, magnitudes are set relative to the 100 million lives estimated to be affected by climate change and a carbon-based economy between 2010 and 2030. The threshold for the 'large' magnitude category represents 25% of this total. For desertification and land degradation, magnitudes are set relative to the lower end of current estimates of degraded land, 10–60 million km^2. The threshold for the 'large' magnitude category represents 30% of the lower estimate. For food security, magnitudes are set relative to the approximately 800 million people who are currently undernourished. The threshold for the 'large' magnitude category represents 12.5% of this total. For panel B, for the first row (high level implementation) for each response option, the magnitude and thresholds are as defined for panel A. In the second row (best practice implementation) for each response option, the qualitative assessments that are green denote potential positive impacts, and those shown in grey indicate neutral interactions. Increased food production is assumed to be achieved through sustainable intensification rather than through injudicious application of additional external inputs such as agrochemicals. **Levels of confidence:** Confidence in the magnitude category (high, medium or low) into which each option falls for mitigation, adaptation, combating desertification and land degradation, and enhancing food security. *High confidence* means that there is a high level of agreement and evidence in the literature to support the categorisation as high, medium or low magnitude. *Low confidence* denotes that the categorisation of magnitude is based on few studies. *Medium confidence* reflects medium evidence and agreement in the magnitude of response. **Cost ranges:** Cost estimates are based on aggregation of often regional studies and vary in the components of costs that are included. In panel B, cost estimates are not provided for best practice implementation. One coin indicates low cost (<USD10 tCO$_2$-eq^{-1} or <USD20 ha^{-1}), two coins indicate medium cost (USD10–USD100 tCO$_2$-eq^{-1} or USD20–USD200 ha^{-1}), and three coins indicate high cost (>USD100 tCO$_2$-eq^{-1} or USD200 ha^{-1}). Thresholds in USD ha^{-1} are chosen to be comparable, but precise conversions will depend on the response option. **Supporting evidence:** Supporting evidence for the magnitude of the quantitative potential for land management-based response options can be found as follows: for mitigation Tables 6.13 to 6.20, with further evidence in Section 2.7.1; for adaptation Tables 6.21 to 6.28; for combating desertification Tables 6.29 to 6.36, with further evidence in Chapter 3; for combating degradation tables 6.37 to 6.44, with further evidence in Chapter 4; for enhancing food security Table's 6.45 to 6.52, with further evidence in Chapter 5. Other synergies and trade-offs not shown here are discussed in Chapter 6. Additional supporting evidence for the qualitative assessments in the second row for each option in panel B can be found in the Table's 6.6, 6.55, 6.56 and 6.58, Section 6.3.5.1.3, and Box 6.1c.

TS.7 Risk management and decision making in relation to sustainable development

Increases in global mean surface temperature are projected to result in continued permafrost degradation and coastal degradation (*high confidence*), increased wildfire, decreased crop yields in low latitudes, decreased food stability, decreased water availability, vegetation loss (*medium confidence*), decreased access to food and increased soil erosion (*low confidence*). There is *high agreement* and *high evidence* that increases in global mean temperature will result in continued increase in global vegetation loss, coastal degradation, as well as decreased crop yields in low latitudes, decreased food stability, decreased access to food and nutrition, and *medium confidence* in continued permafrost degradation and water scarcity in drylands. Impacts are already observed across all components (*high confidence*). Some processes may experience irreversible impacts at lower levels of warming than others. There are high risks from permafrost degradation, and wildfire, coastal degradation, stability of food systems at 1.5°C while high risks from soil erosion, vegetation loss and changes in nutrition only occur at higher temperature thresholds due to increased possibility for adaptation (*medium confidence*). {7.2.2.1, 7.2.2.2, 7.2.2.3; 7.2.2.4; 7.2.2.5; 7.2.2.6; 7.2.2.7; Figure 7.1}

These changes result in compound risks to food systems, human and ecosystem health, livelihoods, the viability of infrastructure, and the value of land (*high confidence*). The experience and dynamics of risk change over time as a result of both human and natural processes (*high confidence*). There is *high confidence* that climate and land changes pose increased risks at certain periods of life (i.e. to the very young and ageing populations) as well as sustained risk to those living in poverty. Response options may also increase risks. For example, domestic efforts to insulate populations from food price spikes associated with climatic stressors in the mid-2000s inadequately prevented food insecurity and poverty, and worsened poverty globally. (Figure TS.14) {7.2.1, 7.2.2, 7.3, Table 7.1}

There is significant regional heterogeneity in risks: tropical regions, including Sub-Saharan Africa, Southeast Asia and Central and South America are particularly vulnerable to decreases in crop yield (*high confidence*). Yield of crops in higher latitudes may initially benefit from warming as well as from higher carbon dioxide (CO_2) concentrations. But temperate zones, including the Mediterranean, North Africa, the Gobi desert, Korea and western United States are susceptible to disruptions from increased drought frequency and intensity, dust storms and fires (*high confidence*). {7.2.2}

Risks related to land degradation, desertification and food security increase with temperature and can reverse development gains in some socio-economic development pathways (*high confidence*). SSP1 reduces the vulnerability and exposure of human and natural systems and thus limits risks resulting from desertification, land degradation and food insecurity compared to SSP3 (*high confidence*). SSP1

is characterized by low population growth, reduced inequalities, land-use regulation, low meat consumption, increased trade and few barriers to adaptation or mitigation. SSP3 has the opposite characteristics. Under SSP1, only a small fraction of the dryland population (around 3% at 3°C for the year 2050) will be exposed and vulnerable to water stress. However under SSP3, around 20% of dryland populations (for the year 2050) will be exposed and vulnerable to water stress by 1.5°C and 24% by 3°C. Similarly under SSP1, at 1.5°C, 2 million people are expected to be exposed and vulnerable to crop yield change. Over 20 million are exposed and vulnerable to crop yield change in SSP3, increasing to 854 million people at 3°C (*low confidence*). Livelihoods deteriorate as a result of these impacts, livelihood migration is accelerated, and strife and conflict is worsened (*medium confidence*). {Cross-Chapter Box 9 in Chapter 6, 7.2.2, 7.3.2, Table 7.1, Figure 7.2}

Land-based adaptation and mitigation responses pose risks associated with the effectiveness and potential adverse side-effects of measures chosen (*medium confidence*). Adverse side-effects on food security, ecosystem services and water security increase with the scale of BECCS deployment. In a SSP1 future, bioenergy and BECCS deployment up to 4 million km^2 is compatible with sustainability constraints, whereas risks are already high in a SSP3 future for this scale of deployment. {7.2.3}

There is *high confidence* that policies addressing vicious cycles of poverty, land degradation and greenhouse gas (GHG) emissions implemented in a holistic manner can achieve climate-resilient sustainable development. Choice and implementation of policy instruments determine future climate and land pathways (*medium confidence*). Sustainable development pathways (described in SSP1) supported by effective regulation of land use to reduce environmental trade-offs, reduced reliance on traditional biomass, low growth in consumption and limited meat diets, moderate international trade with connected regional markets, and effective GHG mitigation instruments can result in lower food prices, fewer people affected by floods and other climatic disruptions, and increases in forested land (*high agreement, limited evidence*) (SSP1). A policy pathway with limited regulation of land use, low technology development, resource intensive consumption, constrained trade, and ineffective GHG mitigation instruments can result in food price increases, and significant loss of forest (*high agreement, limited evidence*) (SSP3). {3.7.5, 7.2.2, 7.3.4, 7.5.5, 7.5.6, Table 7.1, Cross-Chapter Box 9 in Chapter 6, Cross-Chapter Box 12 in Chapter 7}

Delaying deep mitigation in other sectors and shifting the burden to the land sector, increases the risk associated with adverse effects on food security and ecosystem services (*high confidence*). The consequences are an increased pressure on land with higher risk of mitigation failure and of temperature overshoot and a transfer of the burden of mitigation and unabated climate change to future generations. Prioritising early decarbonisation with minimal reliance on CDR decreases the risk of mitigation failure (*high confidence*). {2.5, 6.2, 6.4, 7.2.1, 7.2.2, 7.2.3, 7.5.6, 7.5.7, Cross-Chapter Box 9 in Chapter 6}

Trade-offs can occur between using land for climate mitigation or Sustainable Development Goal (SDG) 7 (affordable clean energy) with biodiversity, food, groundwater and riverine ecosystem services (*medium confidence*). There is *medium confidence* that trade-offs currently do not figure into climate policies and decision making. Small hydro power installations (especially in clusters) can impact downstream river ecological connectivity for fish (*high agreement, medium evidence*). Large scale solar farms and wind turbine installations can impact endangered species and disrupt habitat connectivity (*medium agreement, medium evidence*). Conversion of rivers for transportation can disrupt fisheries and endangered species (through dredging and traffic) (*medium agreement, low evidence*). {7.5.6}

The full mitigation potential assessed in this report will only be realised if agricultural emissions are included in mainstream climate policy (*high agreement, high evidence*). Carbon markets are theoretically more cost-effective than taxation but challenging to implement in the land-sector (*high confidence*) Carbon pricing (through carbon markets or carbon taxes) has the potential to be an effective mechanism to reduce GHG emissions, although it remains relatively untested in agriculture and food systems. Equity considerations can be balanced by a mix of both market and non-market mechanisms (*medium evidence, medium agreement*). Emissions leakage could be reduced by multi-lateral action (*high agreement, medium evidence*). {7.4.6, 7.5.5, 7.5.6, Cross Chapter Box 9 in Chapter 6}

A suite of coherent climate and land policies advances the goal of the Paris Agreement and the land-related SDG targets on poverty, hunger, health, sustainable cities and communities, responsible consumption and production, and life on land. There is *high confidence* that acting early will avert or minimise risks, reduce losses and generate returns on investment. The economic costs of action on sustainable land management (SLM), mitigation, and adaptation are less than the consequences of inaction for humans and ecosystems (*medium confidence*). Policy portfolios that make ecological restoration more attractive, people more resilient – expanding financial inclusion, flexible carbon credits, disaster risk and health insurance, social protection and adaptive safety nets, contingent finance and reserve funds, and universal access to early warning systems – could save 100 billion USD a year, if implemented globally. {7.3.1, 7.4.7, 7.4.8, 7.5.6, Cross-Chapter Box 10 in Chapter 7}

Coordination of policy instruments across scales, levels, and sectors advances co-benefits, manages land and climate risks, advances food security, and addresses equity concerns (*medium confidence*). Flood resilience policies are mutually reinforcing and include flood zone mapping, financial incentives to move, and building restrictions, and insurance. Sustainability certification, technology transfer, land-use standards and secure land tenure schemes, integrated with early action and preparedness, advance response options. SLM improves with investment in agricultural research, environmental farm practices, agri-environmental payments, financial support for sustainable agricultural water infrastructure (including dugouts), agriculture emission trading, and elimination

of agricultural subsidies (*medium confidence*). Drought resilience policies (including drought preparedness planning, early warning and monitoring, improving water use efficiency), synergistically improve agricultural producer livelihoods and foster SLM. (Figure TS.15) {3.7.5, Cross-Chapter Box 5 in Chapter 3, 7.4.3, 7.4.6, 7.5.6, 7.4.8, 7.5.6, 7.6.3}

Technology transfer in land use sectors offers new opportunities for adaptation, mitigation, international cooperation, R&D collaboration, and local engagement (*medium confidence*). International cooperation to modernise the traditional biomass sector will free up both land and labour for more productive uses. Technology transfer can assist the measurement and accounting of emission reductions by developing countries. {7.4.4, 7.4.6, Cross-Chapter Box 12 in Chapter 7}

Measuring progress towards goals is important in decision-making and adaptive governance to create common understanding and advance policy effectiveness (*high agreement, medium evidence*). Measurable indicators, selected with the participation of people and supporting data collection, are useful for climate policy development and decision-making. Indicators include the SDGs, nationally determined contributions (NDCs), land degradation neutrality (LDN) core indicators, carbon stock measurement, measurement and monitoring for REDD+, metrics for measuring biodiversity and ecosystem services, and governance capacity. {7.5.5, 7.5.7, 7.6.4, 7.6.6}

The complex spatial, cultural and temporal dynamics of risk and uncertainty in relation to land and climate interactions and food security, require a flexible, adaptive, iterative approach to assessing risks, revising decisions and policy instruments (*high confidence*). Adaptive, iterative decision-making moves beyond standard economic appraisal techniques to new methods such as dynamic adaptation pathways with risks identified by trigger points through indicators. Scenarios can provide valuable information at all planning stages in relation to land, climate and food; adaptive management addresses uncertainty in scenario planning with pathway choices made and reassessed to respond to new information and data as it becomes available. {3.7.5, 7.4.4, 7.5.2, 7.5.3, 7.5.4, 7.5.7, 7.6.1, 7.6.3}

ILK can play a key role in understanding climate processes and impacts, adaptation to climate change, SLM across different ecosystems, and enhancement of food security (*high confidence*). ILK is context-specific, collective, informally transmitted, and multi-functional, and can encompass factual information about the environment and guidance on management of resources and related rights and social behaviour. ILK can be used in decision-making at various scales and levels, and exchange of experiences with adaptation and mitigation that include ILK is both a requirement and an entry strategy for participatory climate communication and action. Opportunities exist for integration of ILK with scientific knowledge. {7.4.1, 7.4.5, 7.4.6, 7.6.4, Cross-Chapter Box 13 in Chapter 7}

A. Risks to humans and ecosystems from changes in land-based processes as a result of climate change

Increases in global mean surface temperature (GMST), relative to pre-industrial levels, affect processes involved in **desertification** (water scarcity), **land degradation** (soil erosion, vegetation loss, wildfire, permafrost thaw) and **food security** (crop yield and food supply instabilities). Changes in these processes drive risks to food systems, livelihoods, infrastructure, the value of land, and human and ecosystem health. Changes in one process (e.g. wildfire or water scarcity) may result in compound risks. Risks are location-specific and differ by region.

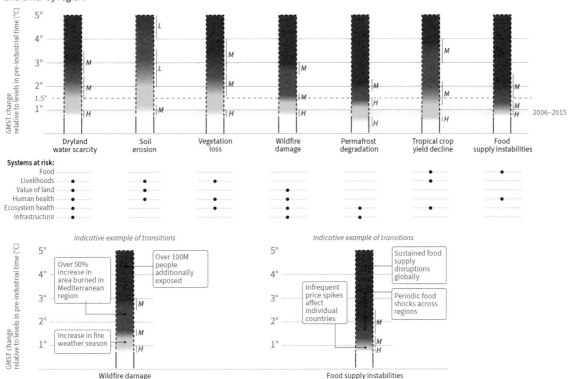

B. Different socioeconomic pathways affect levels of climate related risks

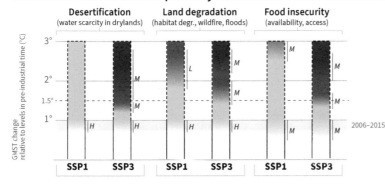

Socio-economic choices can reduce or exacerbate climate related risks as well as influence the rate of temperature increase. The **SSP1** pathway illustrates a world with low population growth, high income and reduced inequalities, food produced in low GHG emission systems, effective land use regulation and high adaptive capacity. The **SSP3** pathway has the opposite trends. Risks are lower in SSP1 compared with SSP3 given the same level of GMST increase.

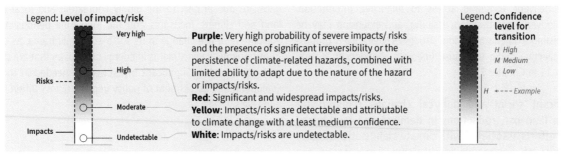

Legend: Level of impact/risk

Purple: Very high probability of severe impacts/ risks and the presence of significant irreversibility or the persistence of climate-related hazards, combined with limited ability to adapt due to the nature of the hazard or impacts/risks.
Red: Significant and widespread impacts/risks.
Yellow: Impacts/risks are detectable and attributable to climate change with at least medium confidence.
White: Impacts/risks are undetectable.

Legend: Confidence level for transition
H High
M Medium
L Low

Figure TS.14 | Risks to land-related human systems and ecosystems from global climate change, socio-economic development and mitigation choices.

Figure TS.14 (continued): As in previous IPCC reports the literature was used to make expert judgements to assess the levels of global warming at which levels of risk are undetectable, moderate, high or very high, as described further in Chapter 7 and other parts of the underlying report. The figure indicates assessed risks at approximate warming levels which may be influenced by a variety of factors, including adaptation responses. The assessment considers adaptive capacity consistent with the SSP pathways as described below. **Panel A:** Risks to selected elements of the land system as a function of global mean surface temperature {2.1; Box 2.1; 3.5; 3.7.1.1; 4.4.1.1; 4.4.1.2; 4.4.1.3; 5.2.2; 5.2.3; 5.2.4; 5.2.5; 7.2;7.3, Table SM7.1}. Links to broader systems are illustrative and not intended to be comprehensive. Risk levels are estimated assuming medium exposure and vulnerability driven by moderate trends in socioeconomic conditions broadly consistent with an SSP2 pathway. {Table SM7.4}. **Panel B:** Risks associated with desertification, land degradation and food security due to climate change and patterns of socio-economic development. Increasing risks associated with desertification include population exposed and vulnerable to water scarcity in drylands. Risks related to land degradation include increased habitat degradation, population exposed to wildfire and floods and costs of floods. Risks to food security include availability and access to food, including population at risk of hunger, food price increases and increases in disability adjusted life years attributable due to childhood underweight. Risks are assessed for two contrasted socio-economic pathways (SSP1 and SSP3 {SPM Box 1}) excluding the effects of targeted mitigation policies {3.5; 4.2.1.2; 5.2.2; 5.2.3; 5.2.4; 5.2.5; 6.1.4; 7.2, Table SM7.5}. Risks are not indicated beyond 3°C because SSP1 does not exceed this level of temperature change. All panels: As part of the assessment, literature was compiled and data extracted into a summary table. A formal expert elicitation protocol (based on modified-Delphi technique and the Sheffield Elicitation Framework), was followed to identify risk transition thresholds. This included a multi-round elicitation process with two rounds of independent anonymous threshold judgement, and a final consensus discussion. Further information on methods and underlying literature can be found in Chapter 7 Supplementary Material.

Participation of people in land and climate decision making and policy formation allows for transparent effective solutions and the implementation of response options that advance synergies, reduce trade-offs in sustainable land management (*high confidence*), and overcomes barriers to adaptation and mitigation (*high confidence*). Improvements to sustainable land management are achieved by: (1) engaging people in citizen science by mediating and facilitating landscape conservation planning, policy choice, and early warning systems (*medium confidence*); (2) involving people in identifying problems (including species decline, habitat loss, land use change in agriculture, food production and forestry), selection of indicators, collection of climate data, land modelling, agricultural innovation opportunities. When social learning is combined with collective action, transformative change can occur addressing tenure issues and changing land use practices (*medium confidence*). Meaningful participation overcomes barriers by opening up policy and science surrounding climate and land decisions to inclusive discussion that promotes alternatives. {3.8.5, 7.5.1, 7.5.9; 7.6.1, 7.6.4, 7.6.5, 7.6.7, 7.7.4, 7.7.6}

Empowering women can bolster synergies among household food security and sustainable land management (*high confidence*). This can be achieved with policy instruments that account for gender differences. The overwhelming presence of women in many land-based activities including agriculture provides opportunities to mainstream gender policies, overcome gender barriers, enhance gender equality, and increase sustainable land management and food security (*high confidence*). Policies that address barriers include gender qualifying criteria and gender appropriate delivery, including access to financing, information, technology, government transfers, training, and extension may be built into existing women's programs, structures (civil society groups) including collective micro enterprise (*medium confidence*). {Cross-Chapter Box 11 in Chapter 7}

The significant social and political changes required for sustainable land use, reductions in demand and land-based mitigation efforts associated with climate stabilisation require a wide range of governance mechanisms. The expansion and diversification of land use and biomass systems and markets requires hybrid governance: public-private partnerships, transnational, polycentric, and state governance to insure opportunities are maximised, trade-offs are managed equitably, and negative impacts are minimised (*medium confidence*). {7.5.6, 7.7.2, 7.7.3, Cross-Chapter Box 7 in Chapter 6}

Land tenure systems have implications for both adaptation and mitigation, which need to be understood within specific socio-economic and legal contexts, and may themselves be impacted by climate change and climate action (*limited evidence, high agreement*). Land policy (in a diversity of forms beyond focus on freehold title) can provide routes to land security and facilitate or constrain climate action, across cropping, rangeland, forest, fresh-water ecosystems and other systems. Large-scale land acquisitions are an important context for the relations between tenure security and climate change, but their scale, nature and implications are imperfectly understood. There is *medium confidence* that land titling and recognition programs, particularly those that authorise and respect indigenous and communal tenure, can lead to improved management of forests, including for carbon storage. Strong public coordination (government and public administration) can integrate land policy with national policies on adaptation and reduce sensitivities to climate change. {7.7.2; 7.7.3; 7.7.4, 7.7.5}

Significant gaps in knowledge exist when it comes to understanding the effectiveness of policy instruments and institutions related to land use management, forestry, agriculture and bioenergy. Interdisciplinary research is needed on the impacts of policies and measures in land sectors. Knowledge gaps are due in part to the highly contextual and local nature of land and climate measures and the long time periods needed to evaluate land use change in its socio-economic frame, as compared to technological investments in energy or industry that are somewhat more comparable. Significant investment is needed in monitoring, evaluation and assessment of policy impacts across different sectors and levels. {7.8}

Table TS.1 | Selection of Policies/Programmes/Instruments that support response options.

Category	Intergrated Response Option	Policy instrument supporting response option
Land management in agriculture	Increased food productivity	Investment in agricultural research for crop and livestock improvement, agricultural technology transfer, inland capture fisheries and aquaculture {7.4.7} agricultural policy reform and trade liberalisation
	Improved cropland, grazing and livestock management	Environmental farm programs/agri-environment schemes, water efficiency requirements and water transfer {3.8.5}, extension services
	Agroforestry	Payment for ecosystem services (ES) {7.4.6}
	Agricultural diversification	Elimination of agriculture subsidies {5.7.1}, environmental farm programs, agri-environmental payments {7.5.6}, rural development programmes
	Reduced grassland conversion to cropland	Elimination of agriculture subsidies, remove insurance incentives, ecological restoration {7.4.6}
	Integrated water management	Integrated governance {7.6.2}, multi-level instruments {7.4.1}
Land management in forests	Forest management, reduced deforestation and degradation, reforestation and forest restoration, afforestation	REDD+, forest conservation regulations, payments for ES, recognition of forest rights and land tenure {7.4.6}, adaptive management of forests {7.5.4}, land-use moratoriums, reforestation programmes and investment {4.9.1}
Land management of soils	Increased soil organic carbon content, reduced soil erosion, reduced soil salinisation, reduced soil compaction, biochar addition to soil	Land degradation neutrality (LDN) {7.4.5}, drought plans, flood plans, flood zone mapping {7.4.3}, technology transfer {7.4.4}, land-use zoning {7.4.6}, ecological service mapping and stakeholder-based quantification {7.5.3}, environmental farm programmes/agri-environment schemes, water-efficiency requirements and water transfer {3.7.5}
Land management in all other ecosystems	Fire management	Fire suppression, prescribed fire management, mechanical treatments {7.4.3}
	Reduced landslides and natural hazards	Land-use zoning {7.4.6}
	Reduced pollution – acidification	Environmental regulations, climate mitigation (carbon pricing) {7.4.4}
	Management of invasive species/ encroachment	Invasive species regulations, trade regulations {5.7.2, 7.4.6}
	Restoration and reduced conversion of coastal wetlands	Flood zone mapping {7.4.3}, land-use zoning {7.4.6}
	Restoration and reduced conversion of peatlands	Payment for ES {7.4.6; 7.5.3}, standards and certification programmes {7.4.6}, land-use moratoriums
	Biodiversity conservation	Conservation regulations, protected areas policies
Carbon dioxide removal (CDR) land management	Enhanced weathering of minerals	No data
	Bioenergy and bioenergy with carbon capture and storage (BECCS)	Standards and certification for sustainability of biomass and land use {7.4.6}
Demand management	Dietary change	Awareness campaigns/education, changing food choices through nudges, synergies with health insurance and policy {5.7.2}
	Reduced post-harvest losses Reduced food waste (consumer or retailer), material substitution	Agricultural business risk programmes {7.4.8}; regulations to reduce and taxes on food waste, improved shelf life, circularising the economy to produce substitute goods, carbon pricing, sugar/fat taxes {5.7.2}
Supply management	Sustainable sourcing	Food labelling, innovation to switch to food with lower environmental footprint, public procurement policies {5.7.2}, standards and certification programmes {7.4.6}
	Management of supply chains	Liberalised international trade {5.7.2}, food purchasing and storage policies of governments, standards and certification programmes {7.4.6}, regulations on speculation in food systems
	Enhanced urban food systems	Buy local policies; land-use zoning to encourage urban agriculture, nature-based solutions and green infrastructure in cities; incentives for technologies like vertical farming
	Improved food processing and retailing, improved energy use in food systems	Agriculture emission trading {7.4.4}; investment in R&D for new technologies; certification
Risk management	Management of urban sprawl	Land-use zoning {7.4.6}
	Livelihood diversification	Climate-smart agriculture policies, adaptation policies, extension services {7.5.6}
	Disaster risk management	Disaster risk reduction {7.5.4; 7.4.3}, adaptation planning
	Risk-sharing instruments	Insurance, iterative risk management, CAT bonds, risk layering, contingency funds {7.4.3}, agriculture business risk portfolios {7.4.8}

A. Pathways linking socioeconomic development, mitigation responses and land

Socioeconomic development and land management influence the evolution of the land system including the relative amount of land allocated to CROPLAND, PASTURE, BIOENERGY CROPLAND, FOREST, and NATURAL LAND. The lines show the median across Integrated Assessment Models (IAMs) for three alternative shared socioeconomic pathways (**SSP1**, **SSP2** and **SSP5** at **RCP1.9**); shaded areas show the range across models. Note that pathways illustrate the effects of climate change mitigation but not those of climate change impacts or adaptation.

A. Sustainability-focused (SSP1)
Sustainability in land management, agricultural intensification, production and consumption patterns result in reduced need for agricultural land, despite increases in per capita food consumption. This land can instead be used for reforestation, afforestation, and bioenergy.

B. Middle of the road (SSP2)
Societal as well as technological development follows historical patterns. Increased demand for land mitigation options such as bioenergy, reduced deforestation or afforestation decreases availability of agricultural land for food, feed and fibre.

C. Resource intensive (SSP5)
Resource-intensive production and consumption patterns, results in high baseline emissions. Mitigation focuses on technological solutions including substantial bioenergy and BECCS . Intensification and competing land uses contribute to declines in agricultural land.

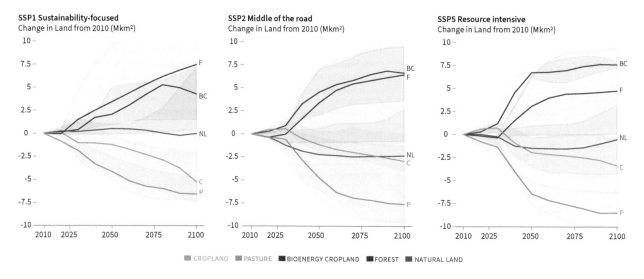

Figure TS.15 | Pathways linking socioeconomic development, mitigation responses and land (Panel A).

B. Land use and land cover change in the SSPs

Quantitative indicators for the SSPs		Count of models included*	Change in Natural Land from 2010 Mkm²	Change in Bioenergy Cropland from 2010 Mkm²	Change in Cropland from 2010 Mkm²	Change in Forest from 2010 Mkm²	Change in Pasture from 2010 Mkm²
SSP1	RCP1.9 in 2050	5/5	0.5 (-4.9, 1)	2.1 (0.9, 5)	-1.2 (-4.6, -0.3)	3.4 (-0.1, 9.4)	-4.1 (-5.6, -2.5)
	↳ 2100		0 (-7.3, 7.1)	4.3 (1.5, 7.2)	-5.2 (-7.6, -1.8)	7.5 (0.4, 15.8)	-6.5 (-12.2, -4.8)
	RCP2.6 in 2050	5/5	-0.9 (-2.2, 1.5)	1.3 (0.4, 1.9)	-1 (-4.7, 1)	2.6 (-0.1, 8.4)	-3 (-4, -2.4)
	↳ 2100		0.2 (-3.5, 1.1)	5.1 (1.6, 6.3)	-3.2 (-7.7, -1.8)	6.6 (-0.1, 10.5)	-5.5 (-9.9, -4.2)
	RCP4.5 in 2050	5/5	0.5 (-1, 1.7)	0.8 (0.5, 1.3)	0.1 (-3.2, 1.5)	0.6 (-0.7, 4.2)	-2.4 (-3.3, -0.9)
	↳ 2100		1.8 (-1.7, 6)	1.9 (1.4, 3.7)	-2.3 (-6.4, -1.6)	3.9 (0.2, 8.8)	-4.6 (-7.3, -2.7)
	Baseline in 2050	5/5	0.3 (-1.1, 1.8)	0.5 (0.2, 1.4)	0.2 (-1.6, 1.9)	-0.1 (-0.8, 1.1)	-1.5 (-2.9, -0.2)
	↳ 2100		3.3 (-0.3, 5.9)	1.8 (1.4, 2.4)	-1.5 (-5.7, -0.9)	0.9 (0.3, 3)	-2.1 (-7, 0)
SSP2	RCP1.9 in 2050	4/5	-2.2 (-7, 0.6)	4.5 (2.1, 7)	-1.2 (-2, 0.3)	3.4 (-0.9, 7)	-4.8 (-6.2, -0.4)
	↳ 2100		-2.3 (-9.6, 2.7)	6.6 (3.6, 11)	-2.9 (-4, 0.1)	6.4 (-0.8, 9.5)	-7.6 (-11.7, -1.3)
	RCP2.6 in 2050	5/5	-3.2 (-4.2, 0.1)	2.2 (1.7, 4.7)	0.6 (-1.9, 1.9)	1.6 (-0.9, 4.2)	-1.4 (-3.7, 0.4)
	↳ 2100		-5.2 (-7.2, 0.5)	6.9 (2.3, 10.8)	-1.4 (-4, 0.8)	5.6 (-0.9, 5.9)	-7.2 (-8, 0.5)
	RCP4.5 in 2050	5/5	-2.2 (-2.2, 0.7)	1.5 (0.1, 2.1)	1.2 (-0.9, 2.7)	-0.9 (-2.5, 2.9)	-0.1 (-2.5, 1.6)
	↳ 2100		-3.4 (-4.7, 1.5)	4.1 (0.4, 6.3)	0.7 (-2.6, 3.1)	-0.5 (-3.1, 5.9)	-2.8 (-5.3, 1.9)
	Baseline in 2050	5/5	-1.5 (-2.6, -0.2)	0.7 (0, 1.5)	1.3 (1, 2.7)	-1.3 (-2.5, -0.4)	-0.1 (-1.2, 1.6)
	↳ 2100		-2.1 (-5.9, 0.3)	1.2 (0.1, 2.4)	1.9 (0.8, 2.8)	-1.3 (-2.7, -0.2)	-0.2 (-1.9, 2.1)
SSP3	RCP1.9 in 2050		Infeasible in all assessed models	-	-	-	-
	↳ 2100			-	-	-	-
	RCP2.6 in 2050		Infeasible in all assessed models	-	-	-	-
	↳ 2100			-	-	-	-
	RCP4.5 in 2050	3/3	-3.4 (-4.4, -2)	1.3 (1.3, 2)	2.3 (1.2, 3)	-2.4 (-4, -1)	2.1 (-0.1, 3.8)
	↳ 2100		-6.2 (-6.8, -5.4)	4.6 (1.5, 7.1)	3.4 (1.9, 4.5)	-3.1 (-5.5, -0.3)	2 (-2.5, 4.4)
	Baseline in 2050	4/4	-3 (-4.6, -1.7)	1 (0.2, 1.5)	2.5 (1.5, 3)	-2.5 (-4, -1.5)	2.4 (0.6, 3.8)
	↳ 2100		-5 (-7.1, -4.2)	1.1 (0.9, 2.5)	5.1 (3.8, 6.1)	-5.3 (-6, -2.6)	3.4 (0.9, 6.4)
SSP4	RCP1.9 in 2050		Infeasible in all assessed models**	-	-	-	-
	↳ 2100			-	-	-	-
	RCP2.6 in 2050	3/3	-4.5 (-6, -2.1)	3.3 (1.5, 4.5)	0.5 (-0.1, 0.9)	0.7 (-0.3, 2.2)	-0.6 (-0.7, 0.1)
	↳ 2100		-5.8 (-10.2, -4.7)	2.5 (2.3, 15.2)	-0.8 (-0.8, 1.8)	1.4 (-1.7, 4.1)	-1.2 (-2.5, -0.2)
	RCP4.5 in 2050	3/3	-2.7 (-4.4, -0.4)	1.7 (1, 1.9)	1.1 (-0.1, 1.7)	-1.8 (-2.3, 2.1)	0.8 (-0.5, 1.5)
	↳ 2100		-2.8 (-7.8, -2)	2.7 (2.3, 4.7)	1.1 (0.2, 1.2)	-0.7 (-2.6, 1)	1.4 (-1, 1.8)
	Baseline in 2050	3/3	-2.8 (-2.9, -0.2)	1.1 (0.7, 2)	1.1 (0.7, 1.8)	-1.8 (-2.3, -1)	1.5 (-0.5, 2.1)
	↳ 2100		-2.4 (-5, -1)	1.7 (1.4, 2.6)	1.2 (1.2, 1.9)	-2.4 (-2.5, -2)	1.3 (-1, 4.4)
SSP5	RCP1.9 in 2050	2/4	-1.5 (-3.9, 0.9)	6.7 (6.2, 7.2)	-1.9 (-3.5, -0.4)	3.1 (-0.1, 6.3)	-6.4 (-7.7, -5.1)
	↳ 2100		-0.5 (-4.2, 3.2)	7.6 (7.2, 8)	-3.4 (-6.2, -0.5)	4.7 (0.1, 9.4)	-8.5 (-10.7, -6.2)
	RCP2.6 in 2050	4/4	-3.4 (-6.9, 0.3)	4.8 (3.8, 5.1)	-2.1 (-4, 1)	3.9 (-0.1, 6.7)	-4.4 (-5, 0.2)
	↳ 2100		-4.3 (-8.4, 0.5)	9.1 (7.7, 9.2)	-3.3 (-6.5, -0.5)	3.9 (-0.1, 9.3)	-6.3 (-9.1, -1.4)
	RCP4.5 in 2050	4/4	-2.5 (-3.7, 0.2)	1.7 (0.6, 2.9)	0.6 (-3.3, 1.9)	-0.1 (-1.7, 6)	-1.2 (-2.6, 2.3)
	↳ 2100		-4.1 (-4.6, 0.7)	4.8 (2, 8)	-1 (-5.5, 1)	-0.2 (-1.4, 9.1)	-3 (-5.2, 2.1)
	Baseline in 2050	4/4	-0.6 (-3.8, 0.4)	0.8 (0, 2.1)	1.5 (-0.7, 3.3)	-1.9 (-3.4, 0.5)	-0.1 (-1.5, 2.9)
	↳ 2100		-0.2 (-2.4, 1.8)	1 (0.2, 2.3)	1 (-2, 2.5)	-2.1 (-3.4, 1.1)	-0.4 (-2.4, 2.8)

* Count of models included / Count of models attempted. One model did not provide land data and is excluded from all entries.

** One model could reach RCP1.9 with SSP4, but did not provide land data.

Figure TS.15 | Pathways linking socioeconomic development, mitigation responses and land (Panel B).

Figure TS.15 (continued): Future scenarios provide a framework for understanding the implications of mitigation and socioeconomics on land. The SSPs span a range of different socioeconomic assumptions (Box SPM.1). They are combined with Representative Concentration Pathways (RCPs)[2] which imply different levels of mitigation. The changes in cropland, pasture, bioenergy cropland, forest, and natural land from 2010 are shown. For this Figure, Cropland includes all land in food, feed, and fodder crops, as well as other arable land (cultivated area). This category includes first generation non-forest bioenergy crops (e.g., corn for ethanol, sugar cane for ethanol, soybeans for biodiesel), but excludes second generation bioenergy crops. Pasture includes categories of pasture land, not only high-quality rangeland, and is based on FAO definition of 'permanent meadows and pastures'. Bioenergy cropland includes land dedicated to second generation energy crops (e.g., switchgrass, miscanthus, fast-growing wood species). Forest includes managed and unmanaged forest. Natural land includes other grassland, savannah, and shrubland. **Panel A:** This panel shows integrated assessment model (IAM)[3] results for SSP1, SSP2 and SSP5 at RCP1.9.[4] For each pathway, the shaded areas show the range across all IAMs; the line indicates the median across models. For RCP1.9, SSP1, SSP2 and SSP5 results are from five, four and two IAMs respectively. **Panel B:** Land use and land cover change are indicated for various SSP-RCP combinations, showing multi-model median and range (min, max). (Box SPM.1) {1.3.2, 2.7.2, 6.1, 6.4.4, 7.4.2, 7.4.4, 7.4.5, 7.4.6, 7.4.7, 7.4.8, 7.5.3, 7.5.6, Cross-Chapter Box 1 in Chapter 1, Cross-Chapter Box 9 in Chapter 6}

---

2 Representative Concentration Pathways (RCPs) are scenarios that include timeseries of emissions and concentrations of the full suite of GHGs and aerosols and chemically active gases, as well as land use/land cover.

3 Integrated Assessment Models (IAMs) integrate knowledge from two or more domains into a single framework. In this figure, IAMs are used to assess linkages between economic, social and technological development and the evolution of the climate system.

4 The RCP1.9 pathways assessed in this report have a 66% chance of limiting warming to 1.5°C in 2100, but some of these pathways overshoot 1.5°C of warming during the 21st century by >0.1°C.

Chapters

1

Framing and context

Coordinating Lead Authors:
Almut Arneth (Germany), Fatima Denton (The Gambia)

Lead Authors:
Fahmuddin Agus (Indonesia), Aziz Elbehri (Morocco), Karlheinz Erb (Italy), Balgis Osman Elasha (Côte d'Ivoire), Mohammad Rahimi (Iran), Mark Rounsevell (United Kingdom), Adrian Spence (Jamaica), Riccardo Valentini (Italy)

Contributing Authors:
Peter Alexander (United Kingdom), Yuping Bai (China), Ana Bastos (Portugal/Germany), Niels Debonne (The Netherlands), Jan Fuglestvedt (Norway), Rafaela Hillerbrand (Germany), Baldur Janz (Germany), Thomas Kastner (Austria), Ylva Longva (United Kingdom), Patrick Meyfroidt (Belgium), Michael O'Sullivan (United Kingdom)

Review Editors:
Edvin Aldrian (Indonesia), Bruce McCarl (The United States of America), María José Sanz Sánchez (Spain)

Chapter Scientists:
Yuping Bai (China), Baldur Janz (Germany)

This chapter should be cited as:
Arneth, A., F. Denton, F. Agus, A. Elbehri, K. Erb, B. Osman Elasha, M. Rahimi, M. Rounsevell, A. Spence, R. Valentini, 2019: Framing and Context. In: *Climate Change and Land: an IPCC special report on climate change, desertification, land degradation, sustainable land management, food security, and greenhouse gas fluxes in terrestrial ecosystems* [P.R. Shukla, J. Skea, E. Calvo Buendia, V. Masson-Delmotte, H.-O. Pörtner, D.C. Roberts, P. Zhai, R. Slade, S. Connors, R. van Diemen, M. Ferrat, E. Haughey, S. Luz, S. Neogi, M. Pathak, J. Petzold, J. Portugal Pereira, P. Vyas, E. Huntley, K. Kissick, M. Belkacemi, J. Malley, (eds.)]. https://doi.org/10.1017/9781009157988.003

1

Table of contents

Executive summary 79

1.1 Introduction and scope of the report 81

    1.1.1 Objectives and scope of the assessment 81

    Box 1.1 | Land in previous IPCC
    and other relevant reports 83

    1.1.2 Status and dynamics of
    the (global) land system 84

1.2 Key challenges related to land use change 88

    1.2.1 Land system change, land degradation,
    desertification and food security 88

    1.2.2 Progress in dealing with uncertainties
    in assessing land processes
    in the climate system 91

    Cross-Chapter Box 1 | Scenarios and other
    methods to characterise the future of land 93

1.3 Response options to the key challenges 96

    1.3.1 Targeted decarbonisation relying
    on large land-area need 97

    Cross-Chapter Box 2 | Implications of large-scale
    conversion from non-forest to forest land 98

    1.3.2 Land management 100

    1.3.3 Value chain management 100

    1.3.4 Risk management 102

    1.3.5 Economics of land-based mitigation
    pathways: Costs versus benefits of
    early action under uncertainty 102

    1.3.6 Adaptation measures and scope
    for co-benefits with mitigation 102

1.4 Enabling the response 103

    1.4.1 Governance to enable the response 103

    1.4.2 Gender agency as a critical factor in climate
    and land sustainability outcomes 104

    1.4.3 Policy instruments 105

1.5 The interdisciplinary nature of the SRCCL 106

Frequently Asked Questions 107

    FAQ 1.1: What are the approaches to study
    the interactions between land and climate? 107

    FAQ 1.2: How region-specific are the impacts
    of different land-based adaptation
    and mitigation options? 107

    FAQ 1.3: What is the difference between
    desertification and land degradation?
    And where are they happening? 107

References 108

Appendix 125

References to Appendix 128

Executive summary

Land, including its water bodies, provides the basis for human livelihoods and well-being through primary productivity, the supply of food, freshwater, and multiple other ecosystem services (*high confidence*). Neither our individual or societal identities, nor the world's economy would exist without the multiple resources, services and livelihood systems provided by land ecosystems and biodiversity. The annual value of the world's total terrestrial ecosystem services has been estimated at 75 trillion USD in 2011, approximately equivalent to the annual global Gross Domestic Product (based on USD2007 values) (*medium confidence*). Land and its biodiversity also represent essential, intangible benefits to humans, such as cognitive and spiritual enrichment, sense of belonging and aesthetic and recreational values. Valuing ecosystem services with monetary methods often overlooks these intangible services that shape societies, cultures and quality of life and the intrinsic value of biodiversity. The Earth's land area is finite. Using land resources sustainably is fundamental for human well-being (*high confidence*). {1.1.1}

The current geographic spread of the use of land, the large appropriation of multiple ecosystem services and the loss of biodiversity are unprecedented in human history (*high confidence*). By 2015, about three-quarters of the global ice-free land surface was affected by human use. Humans appropriate one-quarter to one-third of global terrestrial potential net primary production (*high confidence*). Croplands cover 12–14% of the global ice-free surface. Since 1961, the supply of global per capita food calories increased by about one-third, with the consumption of vegetable oils and meat more than doubling. At the same time, the use of inorganic nitrogen fertiliser increased by nearly ninefold, and the use of irrigation water roughly doubled (*high confidence*). Human use, at varying intensities, affects about 60–85% of forests and 70–90% of other natural ecosystems (e.g., savannahs, natural grasslands) (*high confidence*). Land use caused global biodiversity to decrease by around 11–14% (*medium confidence*). {1.1.2}

Warming over land has occurred at a faster rate than the global mean and this has had observable impacts on the land system (*high confidence*). The average temperature over land for the period 2006–2015 was 1.53°C higher than for the period 1850–1900, and 0.66°C larger than the equivalent global mean temperature change. These warmer temperatures (with changing precipitation patterns) have altered the start and end of growing seasons, contributed to regional crop yield reductions, reduced freshwater availability, and put biodiversity under further stress and increased tree mortality (*high confidence*). Increasing levels of atmospheric CO_2, have contributed to observed increases in plant growth as well as to increases in woody plant cover in grasslands and savannahs (*medium confidence*). {1.1.2}

Urgent action to stop and reverse the over-exploitation of land resources would buffer the negative impacts of multiple pressures, including climate change, on ecosystems and society (*high confidence*). Socio-economic drivers of land-use change such as technological development, population growth and increasing per capita demand for multiple ecosystem services are projected to continue into the future (*high confidence*). These and other drivers can amplify existing environmental and societal challenges, such as the conversion of natural ecosystems into managed land, rapid urbanisation, pollution from the intensification of land management and equitable access to land resources (*high confidence*). Climate change will add to these challenges through direct, negative impacts on ecosystems and the services they provide (*high confidence*). Acting immediately and simultaneously on these multiple drivers would enhance food, fibre and water security, alleviate desertification, and reverse land degradation, without compromising the non-material or regulating benefits from land (*high confidence*). {1.1.2, 1.2.1, 1.3.2–1.3.6, Cross-Chapter Box 1 in Chapter 1}

Rapid reductions in anthropogenic greenhouse gas (GHG) emissions that restrict warming to *"well-below"* 2°C would greatly reduce the negative impacts of climate change on land ecosystems (*high confidence*). In the absence of rapid emissions reductions, reliance on large-scale, land-based, climate change mitigation is projected to increase, which would aggravate existing pressures on land (*high confidence*). Climate change mitigation efforts that require large land areas (e.g., bioenergy and afforestation/reforestation) are projected to compete with existing uses of land (*high confidence*). The competition for land could increase food prices and lead to further intensification (e.g., fertiliser and water use) with implications for water and air pollution, and the further loss of biodiversity (*medium confidence*). Such consequences would jeopardise societies' capacity to achieve many Sustainable Development Goals (SDGs) that depend on land (*high confidence*). {1.3.1, Cross-Chapter Box 2 in Chapter 1}

Nonetheless, there are many land-related climate change mitigation options that do not increase the competition for land (*high confidence*). Many of these options have co-benefits for climate change adaptation (*medium confidence*). Land use contributes about one-quarter of global greenhouse gas emissions, notably CO_2 emissions from deforestation, CH_4 emissions from rice and ruminant livestock and N_2O emissions from fertiliser use (*high confidence*). Land ecosystems also take up large amounts of carbon (*high confidence*). Many land management options exist to both reduce the magnitude of emissions and enhance carbon uptake. These options enhance crop productivity, soil nutrient status, microclimate or biodiversity, and thus, support adaptation to climate change (*high confidence*). In addition, changes in consumer behaviour, such as reducing the over-consumption of food and energy would benefit the reduction of GHG emissions from land (*high confidence*). The barriers to the implementation of mitigation and adaptation options include skills deficit, financial and institutional barriers, absence of incentives, access to relevant technologies, consumer awareness and the limited spatial scale at which the success of these practices and methods have been demonstrated. {1.2.1, 1.3.2, 1.3.3, 1.3.4, 1.3.5, 1.3.6}

Sustainable food supply and food consumption, based on nutritionally balanced and diverse diets, would enhance food security under climate and socio-economic changes (*high confidence*). Improving food access, utilisation, quality and safety to enhance nutrition, and promoting globally equitable diets compatible with lower emissions have demonstrable positive impacts on land use and food security (*high confidence*). Food security is also negatively affected by food loss and waste (estimated as 25–30% of total food produced) (*medium confidence*). Barriers to improved food security include economic drivers (prices, availability and stability of supply) and traditional, social and cultural norms around food eating practices. Climate change is expected to increase variability in food production and prices globally (*high confidence*), but the trade in food commodities can buffer these effects. Trade can provide embodied flows of water, land and nutrients (*medium confidence*). Food trade can also have negative environmental impacts by displacing the effects of overconsumption (*medium confidence*). Future food systems and trade patterns will be shaped as much by policies as by economics (*medium confidence*). {1.2.1, 1.3.3}

A gender-inclusive approach offers opportunities to enhance the sustainable management of land (*medium confidence*). Women play a significant role in agriculture and rural economies globally. In many world regions, laws, cultural restrictions, patriarchy and social structures such as discriminatory customary laws and norms reduce women's capacity in supporting the sustainable use of land resources (*medium confidence*). Therefore, acknowledging women's land rights and bringing women's land management knowledge into land-related decision-making would support the alleviation of land degradation, and facilitate the take-up of integrated adaptation and mitigation measures (*medium confidence*). {1.4.1, 1.4.2}

Regional and country specific contexts affect the capacity to respond to climate change and its impacts, through adaptation and mitigation (*high confidence*). There is large variability in the availability and use of land resources between regions, countries and land management systems. In addition, differences in socio-economic conditions, such as wealth, degree of industrialisation, institutions and governance, affect the capacity to respond to climate change, food insecurity, land degradation and desertification. The capacity to respond is also strongly affected by local land ownership. Hence, climate change will affect regions and communities differently (*high confidence*). {1.3, 1.4}

Cross-scale, cross-sectoral and inclusive governance can enable coordinated policy that supports effective adaptation and mitigation (*high confidence*). There is a lack of coordination across governance levels, for example, local, national, transboundary and international, in addressing climate change and sustainable land management challenges. Policy design and formulation is often strongly sectoral, which poses further barriers when integrating international decisions into relevant (sub)national policies. A portfolio of policy instruments that are inclusive of the diversity of governance actors would enable responses to complex land and climate challenges (*high confidence*). Inclusive governance that considers women's and indigenous people's rights to access and use land enhances the equitable sharing of land resources, fosters food security and increases the existing knowledge about land use, which can increase opportunities for adaptation and mitigation (*medium confidence*). {1.3.5, 1.4.1, 1.4.2, 1.4.3}

Scenarios and models are important tools to explore the trade-offs and co-benefits of land management decisions under uncertain futures (*high confidence*). Participatory, co-creation processes with stakeholders can facilitate the use of scenarios in designing future sustainable development strategies (*medium confidence*). In addition to qualitative approaches, models are critical in quantifying scenarios, but uncertainties in models arise from, for example, differences in baseline datasets, land cover classes and modelling paradigms (*medium confidence*). Current scenario approaches are limited in quantifying time-dependent policy and management decisions that can lead from today to desirable futures or visions. Advances in scenario analysis and modelling are needed to better account for full environmental costs and non-monetary values as part of human decision-making processes. {1.2.2, Cross-Chapter Box 1 in Chapter 1}

1.1 Introduction and scope of the report

1.1.1 Objectives and scope of the assessment

Land, including its water bodies, provides the basis for our livelihoods through basic processes such as net primary production that fundamentally sustain the supply of food, bioenergy and freshwater, and the delivery of multiple other ecosystem services and biodiversity (Hoekstra and Wiedmann 2014; Mace et al. 2012; Newbold et al. 2015; Runting et al. 2017; Isbell et al. 2017) (Cross-Chapter Box 8 in Chapter 6). The annual value of the world's total terrestrial ecosystem services has been estimated to be about 75 trillion USD in 2011, approximately equivalent to the annual global Gross Domestic Product (based on USD2007 values) (Costanza et al. 2014; IMF 2018). Land also supports non-material ecosystem services such as cognitive and spiritual enrichment and aesthetic values (Hernández-Morcillo et al. 2013; Fish et al. 2016), intangible services that shape societies, cultures and human well-being. Exposure of people living in cities to (semi-)natural environments has been found to decrease mortality, cardiovascular disease and depression (Rook 2013; Terraube et al. 2017). Non-material and regulating ecosystem services have been found to decline globally and rapidly, often at the expense of increasing material services (Fischer et al. 2018; IPBES 2018a). Climate change will exacerbate diminishing land and freshwater resources, increase biodiversity loss, and will intensify societal vulnerabilities, especially in regions where economies are highly dependent on natural resources. Enhancing food security and reducing malnutrition, whilst also halting and reversing desertification and land degradation, are fundamental societal challenges that are increasingly aggravated by the need to both adapt to and mitigate climate change impacts without compromising the non-material benefits of land (Kongsager et al. 2016; FAO et al. 2018).

Annual emissions of GHGs and other climate forcers continue to increase unabatedly. *Confidence* is *very high* that the window of opportunity, the period when significant change can be made, for limiting climate change within tolerable boundaries is rapidly narrowing (Schaeffer et al. 2015; Bertram et al. 2015; Riahi et al. 2015; Millar et al. 2017; Rogelj et al. 2018a). The Paris Agreement formulates the goal of limiting global warming this century to well below 2°C above pre-industrial levels, for which rapid actions are required across the energy, transport, infrastructure and agricultural sectors, while factoring in the need for these sectors to accommodate a growing human population (Wynes and Nicholas 2017; Le Quere et al. 2018). Conversion of natural land, and land management, are significant net contributors to GHG emissions and climate change, but land ecosystems are also a GHG sink (Smith et al. 2014; Tubiello et al. 2015; Le Quere et al. 2018; Ciais et al. 2013a). It is not surprising, therefore, that land plays a prominent role in many of the Nationally Determined Contributions (NDCs) of the parties to the Paris Agreement (Rogelj et al. 2018a,b; Grassi et al. 2017; Forsell et al. 2016), and land-measures will be part of the NDC review by 2023.

A range of different climate change mitigation and adaptation options on land exist, which differ in terms of their environmental and societal implications (Meyfroidt 2018; Bonsch et al. 2016; Crist et al. 2017; Humpenoder et al. 2014; Harvey and Pilgrim 2011; Mouratiadou et al. 2016; Zhang et al. 2015; Sanz-Sanchez et al. 2017; Pereira et al. 2010; Griscom et al. 2017; Rogelj et al. 2018a) (Chapters 4–6). The Special Report on climate change, desertification, land degradation, sustainable land management, food security, and GHG fluxes in terrestrial ecosystems (SRCCL) synthesises the current state of scientific knowledge on the issues specified in the report's title (Figure 1.1 and Figure 1.2). This knowledge is assessed in the context of the Paris Agreement, but many of the SRCCL issues concern other international conventions such as the United Nations Convention on Biodiversity (UNCBD), the UN Convention to Combat Desertification (UNCCD), the UN Sendai Framework for Disaster Risk Reduction (UNISDR) and the UN Agenda 2030 and its Sustainable Development Goals (SDGs). The SRCCL is the first report in which land is the central focus since the IPCC Special Report on land use, land-use change and forestry (Watson et al. 2000) (Box 1.1). The main objectives of the SRCCL are to:

1. Assess the current state of the scientific knowledge on the impacts of socio-economic drivers and their interactions with climate change on land, including degradation, desertification and food security;
2. Evaluate the feasibility of different land-based response options to GHG mitigation, and assess the potential synergies and trade-offs with ecosystem services and sustainable development;
3. Examine adaptation options under a changing climate to tackle land degradation and desertification and to build resilient food systems, as well as evaluating the synergies and trade-offs between mitigation and adaptation;
4. Delineate the policy, governance and other enabling conditions to support climate mitigation, land ecosystem resilience and food security in the context of risks, uncertainties and remaining knowledge gaps.

Land use and observed climate change

A. Observed temperature change relative to 1850–1900

Since the pre-industrial period (1850-1900) the observed mean land surface air temperature has risen considerably more than the global mean surface (land and ocean) temperature (GMST).

CHANGE in TEMPERATURE rel. to 1850-1900 (°C)

Change in surface air temperature over land (°C)

Change in global (land-ocean) mean surface temperature (GMST) (°C)

B. GHG emissions

An estimated 23% of total anthropogenic greenhouse gas emissions (2007–2016) derive from Agriculture, Forestry and Other Land Use (AFOLU).

CHANGE in EMISSIONS since 1961
1. Net CO_2 emissions from FOLU (GtCO₂ yr⁻¹)
2. CH_4 emissions from Agriculture (GtCO₂eq yr⁻¹)
3. N_2O emissions from Agriculture (GtCO₂eq yr⁻¹)

$GtCO_2eq\ yr^{-1}$

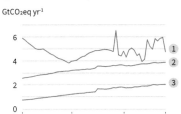

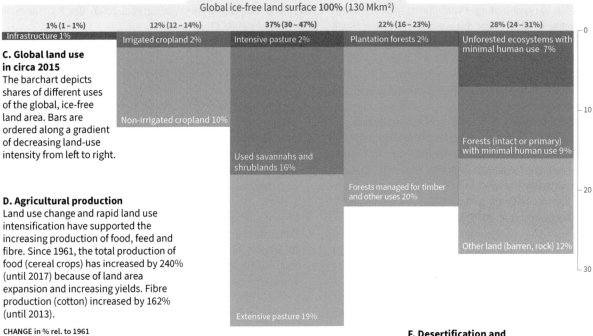

Global ice-free land surface **100%** (130 Mkm²)

| 1% (1 – 1%) | 12% (12 – 14%) | 37% (30 – 47%) | 22% (16 – 23%) | 28% (24 – 31%) |

Infrastructure 1%

Irrigated cropland 2%

Intensive pasture 2%

Plantation forests 2%

Unforested ecosystems with minimal human use 7%

C. Global land use in circa 2015

The barchart depicts shares of different uses of the global, ice-free land area. Bars are ordered along a gradient of decreasing land-use intensity from left to right.

Non-irrigated cropland 10%

Used savannahs and shrublands 16%

Forests (intact or primary) with minimal human use 9%

Forests managed for timber and other uses 20%

D. Agricultural production

Land use change and rapid land use intensification have supported the increasing production of food, feed and fibre. Since 1961, the total production of food (cereal crops) has increased by 240% (until 2017) because of land area expansion and increasing yields. Fibre production (cotton) increased by 162% (until 2013).

Other land (barren, rock) 12%

Extensive pasture 19%

CHANGE in % rel. to 1961
1. Inorganic N fertiliser use
2. Cereal yields
3. Irrigation water volume
4. Total number of ruminant livestock

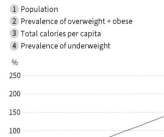

E. Food demand

Increases in production are linked to consumption changes.

CHANGE in % rel. to 1961 and 1975
1. Population
2. Prevalence of overweight + obese
3. Total calories per capita
4. Prevalence of underweight

F. Desertification and land degradation

Land-use change, land-use intensification and climate change have contributed to desertification and land degradation.

CHANGE in % rel. to 1961 and 1970
1. Population in areas experiencing desertification
2. Dryland areas in drought annually
3. Inland wetland extent

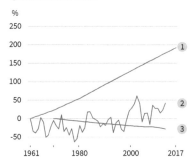

Figure 1.1 | A representation of the principal land challenges and land-climate system processes covered in this assessment report.

Figure 1.1 (continued): A. The warming curves are averages of four datasets (Section 2.1, Figure 2.2 and Table 2.1). **B.** N_2O and CH_4 from agriculture are from FAOSTAT; Net land-use change emissions of CO_2 from forestry and other land use (including emissions from peatland fires since 1997) are from the annual Global Carbon Budget, using the mean of two bookkeeping models. All values expressed in units of CO_2-eq are based on AR5 100-year Global Warming Potential values without climate-carbon feedbacks (N_2O = 265; CH_4 = 28) (Table SPM.1 and Section 2.3). **C.** Depicts shares of different uses of the global, ice-free land area for approximately the year 2015, ordered along a gradient of decreasing land-use intensity from left to right. Each bar represents a broad land cover category; the numbers on top are the total percentage of the ice-free area covered, with uncertainty ranges in brackets. Intensive pasture is defined as having a livestock density greater than 100 animals/km^2. The area of 'forest managed for timber and other uses' was calculated as total forest area minus 'primary/intact' forest area. (Section 1.2, Table 1.1, Figure 1.3). **D.** Note that fertiliser use is shown on a split axis (source: International Fertiliser Industry Association, www.ifastat.org/databases). The large percentage change in fertiliser use reflects the low level of use in 1961 and relates to both increasing fertiliser input per area as well as the expansion of fertilised cropland and grassland to increase food production (1.1, Figure 1.3). **E.** Overweight population is defined as having a body mass index (BMI) >25 kg m^{-2} (source: Abarca-Gómez et al. 2017); underweight is defined as BMI <18.5 kg m^{-2}. (Population density, source: *United Nations, Department of Economic and Social Affairs 2017*) (Sections 5.1 and 5.2). **F.** Dryland areas were estimated using TerraClimate precipitation and potential evapotranspiration (1980–2015) (Abatzoglou et al. 2018) to identify areas where the Aridity Index is below 0.65. Areas experiencing human caused desertification, after accounting for precipitation variability and CO_2 fertilisation, are identified in Le et al. 2016. Population data for these areas were extracted from the gridded historical population database HYDE3.2 (Goldewijk et al. 2017). Areas in drought are based on the 12-month accumulation Global Precipitation Climatology Centre Drought Index (Ziese et al. 2014). The area in drought was calculated for each month (Drought Index below −1), and the mean over the year was used to calculate the percentage of drylands in drought that year. The inland wetland extent (including peatlands) is based on aggregated data from more than 2000 time series that report changes in local wetland area over time (Dixon et al. 2016; Darrah et al. 2019) (Sections 3.1, 4.2 and 4.6).

Box 1.1 | Land in previous IPCC and other relevant reports

Previous IPCC reports have made reference to land and its role in the climate system. Threats to agriculture, forestry and other ecosystems, but also the role of land and forest management in climate change, have been documented since the IPCC Second Assessment Report, especially so in the Special Report on land use, land-use change and forestry (Watson et al. 2000). The IPCC Special Report on extreme events (SREX) discussed sustainable land management, including land-use planning, and ecosystem management and restoration among the potential low-regret measures that provide benefits under current climate and a range of future, climate change scenarios. Low-regret measures are defined in the report as those with the potential to offer benefits now and lay the foundation for tackling future, projected change. Compared to previous IPCC reports, the SRCCL offers a more integrated analysis of the land system as it embraces multiple direct and indirect drivers of natural resource management (related to food, water and energy securities), which have not previously been addressed to a similar depth (Field et al. 2014a; Edenhofer et al. 2014).

The recent IPCC Special Report on Global Warming of 1.5°C (SR15) targeted specifically the Paris Agreement, without exploring the possibility of future global warming trajectories above 2°C (IPCC 2018). Limiting global warming to 1.5°C compared to 2°C is projected to lower the impacts on terrestrial, freshwater and coastal ecosystems and to retain more of their services for people. In many scenarios proposed in this report, large-scale land use features as a mitigation measure. In the reports of the Food and Agriculture Organization (FAO), land degradation is discussed in relation to ecosystem goods and services, principally from a food security perspective (FAO and ITPS 2015). The UNCCD report (2014) discusses land degradation through the prism of desertification. It devotes due attention to how land management can contribute to reversing the negative impacts of desertification and land degradation. The IPBES assessments (2018a, b, c, d, e) focus on biodiversity drivers, including a focus on land degradation and desertification, with poverty as a limiting factor. The reports draw attention to a world in peril in which resource scarcity conspires with drivers of biophysical and social vulnerability to derail the attainment of sustainable development goals. As discussed in Chapter 4 of the SRCCL, different definitions of degradation have been applied in the IPBES degradation assessment (IPBES 2018b), which potentially can lead to different conclusions for restoration and ecosystem management.

The SRCCL complements and adds to previous assessments, whilst keeping the IPCC-specific 'climate perspective'. It includes a focussed assessment of risks arising from maladaptation and land-based mitigation (i.e. not only restricted to direct risks from climate change impacts) and the co-benefits and trade-offs with sustainable development objectives. As the SRCCL cuts across different policy sectors it provides the opportunity to address a number of challenges in an integrative way at the same time, and it progresses beyond other IPCC reports in having a much more comprehensive perspective on land.

The SRCCL identifies and assesses land-related challenges and response options in an integrative way, aiming to be policy relevant across sectors. Chapter 1 provides a synopsis of the main issues addressed in this report, which are explored in more detail in Chapters 2–7. Chapter 1 also introduces important concepts and definitions and highlights discrepancies with previous reports that arise from different objectives (a full set of definitions is provided in the Glossary). Chapter 2 focuses on the natural system dynamics, assessing recent progress towards understanding the impacts of climate change on land, and the feedbacks arising from altered biogeochemical and biophysical exchange fluxes (Figure 1.2 | Overview over the SRCCL.1.2).

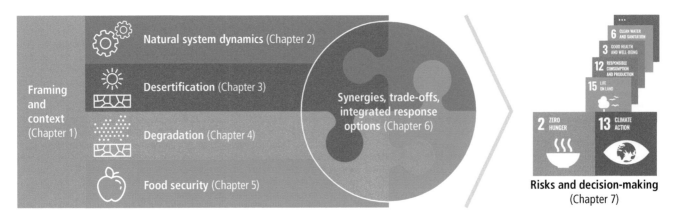

Figure 1.2 | Overview over the SRCCL.

Chapter 3 examines how the world's dryland populations are uniquely vulnerable to desertification and climate change, but also have significant knowledge in adapting to climate variability and addressing desertification. Chapter 4 assesses the urgency of tackling land degradation across all land ecosystems. Despite accelerating trends of land degradation, reversing these trends is attainable through restoration efforts and proper implementation of sustainable land management (SLM), which is expected to improve resilience to climate change, mitigate climate change, and ensure food security for generations to come. Food security is the focus of Chapter 5, with an assessment of the risks and opportunities that climate change presents to food systems, considering how mitigation and adaptation can contribute to both human and planetary health.

Chapter 6 focuses on the response options within the land system that deal with trade-offs and increase benefits in an integrated way in support of the SDGs. Chapter 7 highlights these aspects further, by assessing the opportunities, decision-making and policy responses to risks in the climate-land-human system.

1.1.2   Status and dynamics of the (global) land system

1.1.2.1   Land ecosystems and climate change

Land ecosystems play a key role in the climate system, due to their large carbon pools and carbon exchange fluxes with the atmosphere (Ciais et al. 2013b). Land use, the total of arrangements, activities and inputs applied to a parcel of land (such as agriculture, grazing, timber extraction, conservation or city dwelling; see Glossary), and land management (sum of land-use practices that take place within broader land-use categories; see Glossary) considerably alter terrestrial ecosystems and play a key role in the global climate system. An estimated one-quarter of total anthropogenic GHG emissions arise mainly from deforestation, ruminant livestock and fertiliser application (Smith et al. 2014; Tubiello et al. 2015; Le Quere et al. 2018; Ciais et al. 2013a), and especially methane (CH_4) and nitrous oxide (N_2O) emissions from agriculture have been rapidly increasing over the last decades (Hoesly et al. 2018; Tian et al. 2019) (Figure 1.1 and Sections 2.3.2–2.3.3).

Globally, land also serves as a large CO_2 sink, which was estimated for the period 2008–2017 to be nearly 30% of total anthropogenic emissions (Le Quere et al. 2015; Canadell and Schulze 2014; Ciais et al. 2013a; Zhu et al. 2016) (Section 2.3.1). This sink has been attributed to increasing atmospheric CO_2 concentration, a prolonged growing season in cool environments, or forest regrowth (Le Quéré et al. 2013; Pugh et al. 2019; Le Quéré et al. 2018; Ciais et al. 2013a; Zhu et al. 2016). Whether or not this sink will persist into the future is one of the largest uncertainties in carbon cycle and climate modelling (Ciais et al. 2013a; Bloom et al. 2016; Friend et al. 2014; Le Quere et al. 2018). In addition, changes in vegetation cover caused by land use (such as conversion of forest to cropland or grassland, and vice versa) can result in regional cooling or warming through altered energy and momentum transfer between ecosystems and the atmosphere. Regional impacts can be substantial, but whether the effect leads to warming or cooling depends on the local context (Lee et al. 2011; Zhang et al. 2014; Alkama and Cescatti 2016) (Section 2.6). Due to the current magnitude of GHG emissions and CO_2 carbon dioxide removal in land ecosystems, there is *high confidence* that GHG reduction measures in agriculture, livestock management and forestry would have substantial climate change mitigation potential, with co-benefits for biodiversity and ecosystem services (Smith and Gregory 2013; Smith et al. 2014; Griscom et al. 2017) (Sections 2.6 and 6.3).

The mean temperature over land for the period 2006–2015 was 1.53°C higher than for the period 1850–1900, and 0.66°C larger than the equivalent global mean temperature change (Section 2.2). Climate change affects land ecosystems in various ways (Section 7.2). Growing seasons and natural biome boundaries shift in response to warming or changes in precipitation (Gonzalez et al. 2010; Wärlind et al. 2014; Davies-Barnard et al. 2015; Nakamura et al. 2017). Atmospheric CO_2 increases have been attributed to underlie, at least partially, observed woody plant cover increase in grasslands and savannahs (Donohue et al. 2013). Climate change-induced shifts in habitats, together with warmer temperatures, cause pressure on plants and animals (Pimm et al. 2014; Urban et al. 2016). National cereal crop losses of nearly 10% have been estimated for the period 1964–2007 as a consequence of heat and drought weather extremes (Deryng et al. 2014; Lesk et al. 2016). Climate change is expected to reduce yields in areas that are already under heat and water stress (Schlenker and Lobell 2010; Lobell et al. 2011, 2012; Challinor et al. 2014) (Section 5.2.2). At the same time, warmer temperatures can increase productivity in cooler regions (Moore and Lobell 2015) and might open opportunities for crop area expansion, but any overall benefits might be counterbalanced by reduced suitability in warmer regions (Pugh et al. 2016; Di Paola et al. 2018). Increasing atmospheric CO_2 is expected to increase productivity and water use efficiency in crops and in forests (Muller et al. 2015; Nakamura et al. 2017; Kimball 2016). The increasing number of extreme weather events linked to climate change is also expected to result in forest losses; heat waves and droughts foster wildfires (Seidl et al. 2017; Fasullo et al. 2018) (Cross-Chapter Box 3 in Chapter 2). Episodes of observed enhanced tree mortality across many world regions have been attributed to heat and drought stress (Allen et al. 2010; Anderegg et al. 2012), whilst weather extremes also impact local

infrastructure and hence transportation and trade in land-related goods (Schweikert et al. 2014; Chappin and van der Lei 2014). Thus, adaptation is a key challenge to reduce adverse impacts on land systems (Section 1.3.6).

1.1.2.2 Current patterns of land use and land cover

Around three-quarters of the global ice-free land, and most of the highly productive land area, are by now under some form of land use (Erb et al. 2016a; Luyssaert et al. 2014; Venter et al. 2016) (Table 1.1). One-third of used land is associated with changed land cover. Grazing land is the single largest land-use category, followed by used forestland and cropland. The total land area used to raise livestock is notable: it includes all grazing land and an estimated additional

Table 1.1 | Extent of global land use and management around the year 2015.

	Best guess	Range		Range	Type	Reference
	[million km²]		[% of total]			
Total	130.4		100%			
USED LAND	**92.6**	**90.0–99.3**	**71%**	**69–76%**		
Infrastructure (settlements, mining, etc.)	**1.4**	**1.2–1.9**	**1%**		**LCC**	1,2,3,4,5,6
Cropland	**15.9**	**15.9–18.8**	**12%**	**12–14%**		1,7
Irrigated cropland	3.1		2%		LCC	8
Non-irrigated cropland	12.8	12.8–15.7	10%		LCC	8
Grazing land	**48.0**	**38.8–61.9**	**37%**	**30–47%**		
Permanent pastures	**27.1**	**22.8–32.8**	**21%**	**17–25%**		5,7,8
Intensive permanent pastures[a]	2.6		2%		LCC	8,9
Extensive permanent pastures, on potential forest sites[b]	8.7		7%		LCC	9
Extensive permanent pastures, on natural grasslands[b]	15.8	11.5–21.6	12%	9–16%	LM	
Non-forested, used land, multiple uses[c]	**20.1**	**6.1–39.1**	**16%**	**5–30%**	**LM**	
Used forests[d]	**28.1**	**20.3–30.5**	**22%**	**16–23%**		10,11,12
Planted forests	2.9		2%		LCC	12
Managed for timber and other uses	25.2	17.4–27.6	20%	13–21%	LM	12
UNUSED LAND	**37.0**	**31.1–40.4**	**28%**	**24–31%**		5,11,13
Unused, unforested ecosystems, including grasslands and wetlands	9.4	5.9–10.4	7%	5–8%		1,13
Unused forests (intact or primary forests)	12.0	11.7–12.0	9%			11,12
Other land (barren wilderness, rocks, etc.)	15.6	13.5 18.0	12%	10–14%		4,5,13,14
Land-cover conversions (sum of LCC)	31.5	31.3–34.9	24%	24–27%		
Land-use occurring within natural land-cover types (sum of LM)	61.1	55.1–68.0	47%	42–52%		

[a] >100 animals/km².

[b] <100 animals/km², residual category within permanent pastures.

[c] Calculated as residual category. Contains land not classified as forests or cropland, such as savannah and tundra used as rangelands, with extensive uses like seasonal, rough grazing, hunting, fuelwood collection outside forests, wild products harvesting, etc.

[d] Used forest calculated as total forest minus unused forests.

Note: This table is based on data and approaches described in Lambin and Meyfroidt (2011, 2014); Luyssaert et al. (2014); Erb et al. (2016a), and references below. The target year for data is 2015, but proportions of some subcategories are from 2000 (the year with the most reconciled datasets available) and their relative extent was applied to some broad land-use categories for 2015. Sources: Settlements (1) Luyssaert et al. 2014; (2) Lambin and Meyfroidt 2014; (3) Global Human Settlements dataset, https://ghsl. jrc.ec.europa.eu/. Total infrastructure including transportation (4) Erb et al. 2007; (5) Stadler et al. 2018; mining (6) Cherlet et al. 2018; (7) FAOSTAT 2018; (8) proportions from Erb et al. 2016a; (9) Ramankutty et al. 2008 extrapolated from 2000–2010 trend for permanent pastures from (7); (9) Erb et al. 2017; (10) Schepaschenko et al. 2015; (11) Potapov et al. 2017; (12) FAO 2015a; (13) Venter et al. 2016; (14) Ellis et al. 2010.

one-fifth of cropland for feed production (Foley et al. 2011). Globally, 60–85% of the total forested area is used, at different levels of intensity, but information on management practices globally is scarce (Erb et al. 2016a). Large areas of unused (primary) forests remain only in the tropics and northern boreal zones (Luyssaert et al. 2014; Birdsey and Pan 2015; Morales-Hidalgo et al. 2015; Potapov et al. 2017; Erb et al. 2017), while 73–89% of other, non-forested natural ecosystems (natural grasslands, savannahs, etc.) are used. Large uncertainties relate to the extent of forest (32.0–42.5 million km^2) and grazing land (39–62 million km^2), due to discrepancies in definitions and observation methods (Luyssaert et al. 2014; Erb et al. 2017; Putz and Redford 2010; Schepaschenko et al. 2015; Birdsey and Pan 2015; FAO 2015a; Chazdon et al. 2016a; FAO 2018a). Infrastructure areas (including settlements, transportation and mining), while being almost negligible in terms of extent, represent particularly pervasive land-use activities, with far-reaching ecological, social and economic implications (Cherlet et al. 2018; Laurance et al. 2014).

The large imprint of humans on the land surface has led to the definition of anthromes, i.e. large-scale ecological patterns created by the sustained interactions between social and ecological drivers. The dynamics of these 'anthropogenic biomes' are key for land-use impacts as well as for the design of integrated response options (Ellis and Ramankutty 2008; Ellis et al. 2010; Cherlet et al. 2018; Ellis et al. 2010) (Chapter 6).

The intensity of land use varies hugely within and among different land-use types and regions. Averaged globally, around 10% of the ice-free land surface was estimated to be intensively managed (such as tree plantations, high livestock density grazing, large agricultural inputs), two-thirds moderately and the remainder at low intensities (Erb et al. 2016a). Practically all cropland is fertilised, with large regional variations. Irrigation is responsible for 70% of ground- or surface-water withdrawals by humans (Wisser et al. 2008; Chaturvedi et al. 2015; Siebert et al. 2015; FAOSTAT 2018). Humans appropriate one-quarter to one-third of the total potential net primary production (NPP), i.e. the NPP that would prevail in the absence of land use (estimated at about 60 GtC yr^{-1}; Bajželj et al. 2014; Haberl et al. 2014), about equally through biomass harvest and changes in NPP due to land management. The current total of agricultural (cropland and grazing) biomass harvest is estimated at about 6 GtC yr^{-1}, around 50–60% of this is consumed by livestock. Forestry harvest for timber and wood fuel amounts to about 1 GtC yr^{-1} (Alexander et al. 2017; Bodirsky and Müller 2014; Lassaletta et al. 2014, 2016; Mottet et al. 2017; Haberl et al. 2014; Smith et al. 2014; Bais et al. 2015; Bajželj et al. 2014) (Cross-Chapter Box 7 in Chapter 6).

1.1.2.3   Past and ongoing trends

Globally, cropland area changed by +15% and the area of permanent pastures by +8% since the early 1960s (FAOSTAT 2018), with strong regional differences (Figure 1.3). In contrast, cropland production since 1961 increased by about 3.5 times, the production of animal products by 2.5 times, and forestry by 1.5 times; in parallel with strong yield (production per unit area) increases (FAOSTAT 2018) (Figure 1.3). Per capita calorie supply increased by 17% since 1970 (Kastner et al. 2012), and diet composition changed markedly,

tightly associated with economic development and lifestyle: since the early 1960s, per capita dairy product consumption increased by a factor of 1.2, and meat and vegetable oil consumption more than doubled (FAO 2017, 2018b; Tilman and Clark 2014; Marques et al. 2019). Population and livestock production represent key drivers of the global expansion of cropland for food production, only partly compensated by yield increases at the global level (Alexander et al. 2015). A number of studies have reported reduced growth rates or stagnation in yields in some regions in the last decades (*medium evidence, high agreement*; Lin and Huybers 2012; Ray et al. 2012; Elbehri, Aziz, Joshua Elliott 2015) (Section 5.2.2).

The past increases in agricultural production have been associated with strong increases in agricultural inputs (Foley et al. 2011; Siebert et al. 2015; Lassaletta et al. 2016) (Figures 1.1 and 1.3). Irrigation area doubled, total nitrogen fertiliser use increased by 800% (FAOSTAT 2018; IFASTAT 2018) since the early 1960s. Biomass trade volumes grew by a factor of nine (in tonnes dry matter yr^{-1}) in this period, which is much stronger than production (FAOSTAT 2018), resulting in a growing spatial disconnect between regions of production and consumption (Friis et al. 2016; Friis and Nielsen 2017; Schröter et al. 2018; Liu et al. 2013; Krausmann and Langthaler 2019). Urban and other infrastructure areas expanded by a factor of two since 1960 (Krausmann et al. 2013), resulting in disproportionally large losses of highly fertile cropland (Seto and Reenberg 2014; Martellozzo et al. 2015; Bren d'Amour et al. 2016; Seto and Ramankutty 2016; van Vliet et al. 2017). World regions show distinct patterns of change (Figure 1.3).

While most pastureland expansion replaced natural grasslands, cropland expansion replaced mainly forests (Ramankutty et al. 2018; Ordway et al. 2017; Richards and Friess 2016). Noteworthy large conversions occurred in tropical dry woodlands and savannahs, for example, in the Brazilian Cerrado (Lehmann and Parr 2016; Strassburg et al. 2017), the South American Caatinga and Chaco regions (Parr et al. 2014; Lehmann and Parr 2016) or African savannahs (Ryan et al. 2016). More than half of the original 4.3–12.6 million km^2 global wetlands (Erb et al. 2016a; Davidson 2014; Dixon et al. 2016) have been drained; since 1970 the wetland extent index, developed by aggregating data field-site time series that report changes in local inland wetland area, indicates a decline of more than 30% (Darrah et al. 2019) (Figure 1.1 and Section 4.2.1). Likewise, one-third of the estimated global area that in a non-used state would be covered in forests (Erb et al. 2017) has been converted to agriculture.

Global forest area declined by 3% since 1990 (about –5% since 1960) and continues to do so (FAO 2015a; Keenan et al. 2015; MacDicken et al. 2015; FAO 1963; Figure 1.1), but uncertainties are large. *Low agreement* relates to the concomitant trend of global tree cover. Some remote-sensing based assessments show global net-losses of forest or tree cover (Li et al. 2016; Nowosad et al. 2018; Hansen et al. 2013); others indicate a net gain (Song et al. 2018). Tree-cover gains would be in line with observed and modelled increases in photosynthetic active tissues ('greening'; Chen et al. 2019; Zhu et al. 2016; Zhao et al. 2018; de Jong et al. 2013; Pugh et al. 2019; De Kauwe et al. 2016; Kolby Smith et al. 2015) (Box 2.3 in Chapter 2), but *confidence* remains *low* whether gross forest or tree-cover gains are

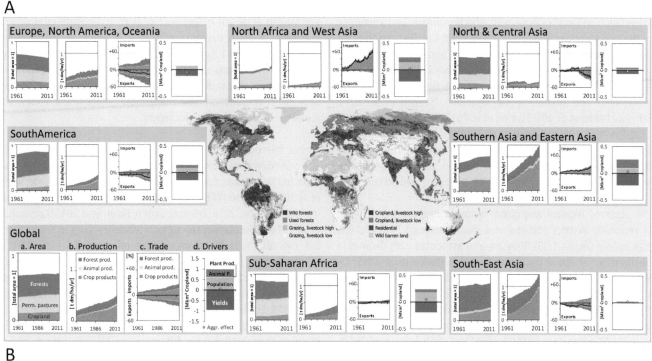

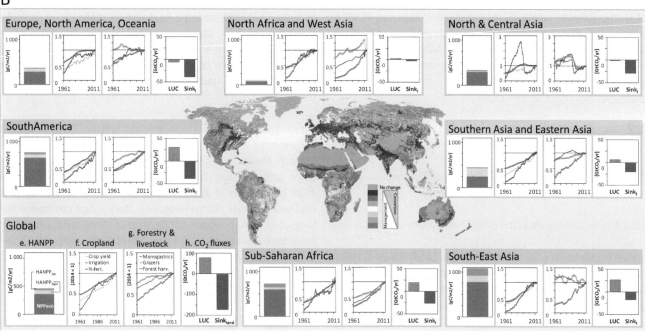

Figure 1.3 | Status and trends in the global land system: A. Trends in area, production and trade, and drivers of change. The map shows the global pattern of land systems (combination of maps Nachtergaele (2008); Ellis et al. (2010); Potapov et al. (2017); FAO's Animal Production and Health Division (2018); livestock low/high relates to low or high livestock density, respectively). The inlay figures show, for the globe and seven world regions, from left to right: (a) Cropland, permanent pastures and forest (used and unused) areas, standardised to total land area, (b) production in dry matter per year per total land area, (c) trade in dry matter in percent of total domestic production, all for 1961 to 2014 (data from FAOSTAT (2018) and FAO (1963) for forest area 1961). (d) drivers of cropland for food production between 1994 and 2011 (Alexander et al. 2015). See panel "global" for legend. "Plant Produc., Animal P.": changes in consumption of plant-based products and animal-products, respectively. **B.** Selected land-use pressures and impacts. The map shows the ratio between impacts on biomass stocks of land-cover conversions and of land management (changes that occur with land-cover types; only changes larger than 30 gC m^{-2} displayed; Erb et al. 2017), compared to the biomass stocks of the potential vegetation (vegetation that would prevail in the absence of land use, but with current climate). The inlay figures show, from left to right (e) the global Human Appropriation of Net Primary production (HANPP) in the year 2005, in gC m^{-2} yr^{-1} (Krausmann et al. 2013). The sum of the three components represents the NPP of the potential vegetation and consist of: (i) NPP$_{eco}$, i.e. the amount of NPP remaining in ecosystem after harvest, (ii) HANPP$_{harv}$, i.e. NPP harvested or killed during harvest, and (iii) HANPP$_{luc}$, i.e. NPP foregone due to land-use change. The sum of NPP$_{eco}$ and HANPP$_{harv}$ is the NPP of the actual vegetation (Haberl et al. 2014; Krausmann et al. 2013). The two central inlay figures show changes in land-use intensity, standardised to 2014, related to (f) cropland (yields, fertilisation, irrigated area) and (g) forestry harvest per forest area, and grazers and monogastric livestock density per agricultural area (FAOSTAT 2018). (h) Cumulative CO$_2$ fluxes between land and the atmosphere between 2000 and 2014. LUC: annual CO$_2$ land use flux due to changes in land cover and forest management; Sink$_{land}$: the annual CO$_2$ land sink caused mainly by the indirect anthropogenic effects of environmental change (e.g. climate change and the fertilising effects of rising CO$_2$ and N concentrations), excluding impacts of land-use change (Le Quéré et al. 2018) (Section 2.3).

as large, or larger, than losses. This uncertainty, together with poor information on forest management, affects estimates and attribution of the land carbon sink (Sections 2.3, 4.3 and 4.6). Discrepancies are caused by different classification schemes and applied thresholds (e.g., minimum tree height and tree-cover thresholds used to define a forest), the divergence of forest and tree cover, and differences in methods and spatiotemporal resolution (Keenan et al. 2015; Schepaschenko et al. 2015; Bastin et al. 2017; Sloan and Sayer 2015; Chazdon et al. 2016a; Achard et al. 2014). However, there is *robust evidence* and *high agreement* that a net loss of forest and tree cover prevails in the tropics and a net gain, mainly of secondary, semi-natural and planted forests, in the temperate and boreal zones.

The observed regional and global historical land-use trends result in regionally distinct patterns of C fluxes between land and the atmosphere (Figure 1.3B). They are also associated with declines in biodiversity, far above background rates (Ceballos et al. 2015; De Vos et al. 2015; Pimm et al. 2014; Newbold et al. 2015; Maxwell et al. 2016; Marques et al. 2019). Biodiversity losses from past global land-use change have been estimated to be about 8–14%, depending on the biodiversity indicator applied (Newbold et al. 2015; Wilting et al. 2017; Gossner et al. 2016; Newbold et al. 2018; Paillet et al. 2010). In future, climate warming has been projected to accelerate losses of species diversity rapidly (Settele et al. 2014; Urban et al. 2016; Scholes et al. 2018; Fischer et al. 2018; Hoegh-Guldberg et al. 2018). The concomitance of land-use and climate change pressures render ecosystem restoration a key challenge (Anderson-Teixeira 2018; Yang et al. 2019) (Sections 4.8 and 4.9).

1.2   Key challenges related to land use change

1.2.1   Land system change, land degradation, desertification and food security

1.2.1.1   Future trends in the global land system

Human population is projected to increase to nearly 9.8 (± 1) billion people by 2050 and 11.2 billion by 2100 (United Nations 2018). More people, a growing global middle class (Crist et al. 2017), economic growth, and continued urbanisation (Jiang and O'Neill 2017) increase the pressures on expanding crop and pasture area and intensifying land management. Changes in diets, efficiency and technology could reduce these pressures (Billen et al. 2015; Popp et al. 2016; Muller et al. 2017; Alexander et al. 2015; Springmann et al. 2018; Myers et al. 2017; Erb et al. 2016c; FAO 2018b) (Sections 5.3 and 6.2.2).

Given the large uncertainties underlying the many drivers of land use, as well as their complex relation to climate change and other biophysical constraints, future trends in the global land system are explored in scenarios and models that seek to span across these uncertainties (Cross-Chapter Box 1 in Chapter 1). Generally, these scenarios indicate a continued increase in global food demand, owing to population growth and increasing wealth. The associated land area needs are a key uncertainty, a function of the interplay between production, consumption, yields, and production efficiency (in particular for livestock and waste) (FAO 2018b; van Vuuren et al. 2017; Springmann et al. 2018; Riahi et al. 2017; Prestele et al. 2016; Ramankutty et al. 2018; Erb et al. 2016b; Popp et al. 2016) (Section 1.3 and Cross-Chapter Box 1 in Chapter 1). Many factors, such as climate change, local contexts, education, human and social capital, policy-making, economic framework conditions, energy availability, degradation, and many more, affect this interplay, as discussed in all chapters of this report.

Global telecouplings in the land system, the distal connections and multidirectional flows between regions and land systems, are expected to increase, due to urbanisation (Seto et al. 2012; van Vliet et al. 2017; Jiang and O'Neill 2017; Friis et al. 2016), and international trade (Konar et al. 2016; Erb et al. 2016b; Billen et al. 2015; Lassaletta et al. 2016). Telecoupling can support efficiency gains in production, but can also lead to complex cause–effect chains and indirect effects such as land competition or leakage (displacement of the environmental impacts; see Glossary), with governance challenges (Baldos and Hertel 2015; Kastner et al. 2014; Liu et al. 2013; Wood et al. 2018; Schröter et al. 2018; Lapola et al. 2010; Jadin et al. 2016; Erb et al. 2016b; Billen et al. 2015; Chaudhary and Kastner 2016; Marques et al. 2019; Seto and Ramankutty 2016) (Section 1.2.1.5). Furthermore, urban growth is anticipated to occur at the expense of fertile (crop)land, posing a food security challenge, in particular in regions of high population density and agrarian-dominated economies, with limited capacity to compensate for these losses (Seto et al. 2012; Güneralp et al. 2013; Aronson et al. 2014; Martellozzo et al. 2015; Bren d'Amour et al. 2016; Seto and Ramankutty 2016; van Vliet et al. 2017).

Future climate change and increasing atmospheric CO_2 concentration are expected to accentuate existing challenges by, for example, shifting biomes or affecting crop yields (Baldos and Hertel 2015; Schlenker and Lobell 2010; Lipper et al. 2014; Challinor et al. 2014; Myers et al. 2017) (Section 5.2.2), as well as through land-based climate change mitigation. There is *high confidence* that large-scale implementation of bioenergy or afforestation can further exacerbate existing challenges (Smith et al. 2016) (Section 1.3.1 and Cross-Chapter Box 7 in Chapter 6).

1.2.1.2   Land degradation

As discussed in Chapter 4, the concept of land degradation, including its definition, has been used in different ways in different communities and in previous assessments (such as the IPBES Land Degradation and Restoration Assessment). In the SRCCL, land degradation is defined as a *negative trend in land condition, caused by direct or indirect human-induced processes including anthropogenic climate change, expressed as long-term reduction or loss of at least one of the following: biological productivity, ecological integrity or value to humans*. This definition applies to forest and non-forest land (Chapter 4 and Glossary).

Land degradation is a critical issue for ecosystems around the world due to the loss of actual or potential productivity or utility (Ravi et al. 2010; Mirzabaev et al. 2015; FAO and ITPS 2015; Cerretelli et al. 2018). Land degradation is driven to a large degree by unsustainable agriculture and forestry, socio-economic pressures, such as rapid urbanisation and population growth, and unsustainable production practices in combination with climatic factors (Field et al. 2014b; Lal 2009; Beinroth et al. 1994; Abu Hammad and Tumeizi 2012; Ferreira et al. 2018; Franco and Giannini 2005; Abahussain et al. 2002).

Global estimates of the total degraded area vary from less than 10 million km^2 to over 60 million km^2, with additionally large disagreement regarding the spatial distribution (Gibbs and Salmon 2015) (Section 4.3). The annual increase in the degraded land area has been estimated as 50,000–100,000 million km^2 yr^{-1} (Stavi and Lal 2015), and the loss of total ecosystem services equivalent to about 10% of the world's GDP in the year 2010 (Sutton et al. 2016). Although land degradation is a common risk across the globe, poor countries remain most vulnerable to its impacts. Soil degradation is of particular concern, due to the long period necessary to restore soils (Lal 2009; Stockmann et al. 2013; Lal 2015), as well as the rapid degradation of primary forests through fragmentation (Haddad et al. 2015). Among the most vulnerable ecosystems to degradation are high-carbon-stock wetlands (including peatlands). Drainage of natural wetlands for use in agriculture leads to high CO_2 emissions and degradation (*high confidence*) (Strack 2008; Limpens et al. 2008; Aich et al. 2014; Murdiyarso et al. 2015; Kauffman et al. 2016; Dohong et al. 2017; Arifanti et al. 2018; Evans et al. 2019). Land degradation is an important factor contributing to uncertainties in the mitigation potential of land-based ecosystems (Smith et al. 2014). Furthermore, degradation that reduces forest (and agricultural) biomass and soil organic carbon leads to higher rates of runoff (*high confidence*) (Molina et al. 2007; Valentin et al. 2008; Mateos et al. 2017; Noordwijk et al. 2017) and hence to increasing flood risk (*low confidence*) (Bradshaw et al. 2007; Laurance 2007; van Dijk et al. 2009).

1.2.1.3 Desertification

The SRCCL adopts the definition of the UNCCD of desertification, being land degradation in arid, semi-arid and dry sub-humid areas (drylands) (Glossary and Section 3.1.1). Desertification results from various factors, including climate variations and human activities, and is not limited to irreversible forms of land degradation (Tal 2010; Bai et al. 2008). A critical challenge in the assessment of desertification is to identify a 'non-desertified' reference state (Bestelmeyer et al. 2015). While climatic trends and variability can change the intensity of desertification processes, some authors exclude climate effects, arguing that desertification is a purely human-induced process of land degradation with different levels of severity and consequences (Sivakumar 2007).

As a consequence of varying definitions and different methodologies, the area of desertification varies widely (D'Odorico et al. 2013; Bestelmeyer et al. 2015; and references therein). Arid regions of the world cover up to about 46% of the total terrestrial surface (about 60 million km^2) (Pravalie 2016; Koutroulis 2019). Around 3 billion people reside in dryland regions (D'Odorico et al. 2013; Maestre et al. 2016) (Section 3.1.1). In 2015, about 500 (360–620) million people lived within areas which experienced desertification between 1980s and 2000s (Figure 1.1 and Section 3.1.1). The combination of

low rainfall with frequently infertile soils renders these regions, and the people who rely on them, vulnerable to both climate change, and unsustainable land management (*high confidence*). In spite of the national, regional and international efforts to combat desertification, it remains one of the major environmental problems (Abahussain et al. 2002; Cherlet et al. 2018).

1.2.1.4 Food security, food systems and linkages to land-based ecosystems

The High Level Panel of Experts of the Committee on Food Security define the food system as to "gather all the elements (environment, people, inputs, processes, infrastructures, institutions, etc.) and activities that relate to the production, processing, distribution, preparation and consumption of food, and the output of these activities, including socio-economic and environmental outcomes" (HLPE 2017). Likewise, food security has been defined as "a situation that exists when all people, at all times, have physical, social and economic access to sufficient, safe and nutritious food that meets their dietary needs and food preferences for an active and healthy life" (FAO 2017). By this definition, food security is characterised by food availability, economic and physical access to food, food utilisation and food stability over time. Food and nutrition security is one of the key outcomes of the food system (FAO 2018b; Figure 1.4).

After a prolonged decline, world hunger appears to be on the rise again, with the number of undernourished people having increased to an estimated 821 million in 2017, up from 804 million in 2016 and 784 million in 2015, although still below the 900 million reported in 2000 (FAO et al. 2018) (Section 5.1.2). Of the total undernourished in 2018, for example, 256.5 million lived in Africa, and 515.1 million in Asia (excluding Japan). The same FAO report also states that child undernourishment continues to decline, but levels of overweight populations and obesity are increasing. The total number of overweight children in 2017 was 38–40 million worldwide, and globally up to around two billion adults are by now overweight (Section 5.1.2). FAO also estimated that close to 2000 million people suffer from micronutrient malnutrition (FAO 2018b).

Food insecurity most notably occurs in situations of conflict, and conflict combined with droughts or floods (Cafiero et al. 2018; Smith et al. 2017). The close parallel between food insecurity prevalence and poverty means that tackling development priorities would enhance sustainable land use options for climate mitigation.

Climate change affects the food system as changes in trends and variability in rainfall and temperature variability impact crop and livestock productivity and total production (Osborne and Wheeler 2013; Tigchelaar et al. 2018; Iizumi and Ramankutty 2015), the nutritional quality of food (Loladze 2014; Myers et al. 2014; Ziska et al. 2016; Medek et al. 2017), water supply (Nkhonjera 2017), and incidence of pests and diseases (Curtis et al. 2018). These factors also impact on human health, increasing morbidity and affecting human ability to process ingested food (Franchini and Mannucci 2015; Wu et al. 2016; Raiten and Aimone 2017). At the same time, the food system generates negative externalities (the environmental effects of production and consumption) in the form of GHG emissions

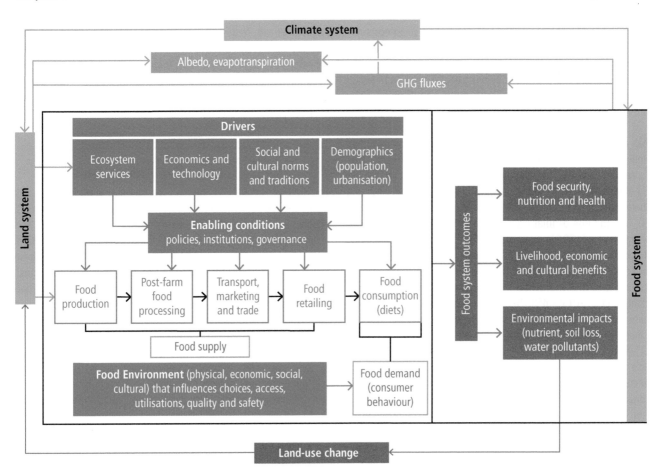

Figure 1.4 | Food system (and its relations to land and climate): The food system is conceptualised through supply (production, processing, marketing and retailing) and demand (consumption and diets) that are shaped by physical, economic, social and cultural determinants influencing choices, access, utilisation, quality, safety and waste. Food system drivers (ecosystem services, economics and technology, social and cultural norms and traditions, and demographics) combine with the enabling conditions (policies, institutions and governance) to affect food system outcomes including food security, nutrition and health, livelihoods, economic and cultural benefits as well as environmental outcomes or side-effects (nutrient and soil loss, water use and quality, GHG emissions and other pollutants). Climate and climate change have direct impacts on the food system (productivity, variability, nutritional quality) while the latter contributes to local climate (albedo, evapotranspiration) and global warming (GHGs). The land system (function, structures, and processes) affects the food system directly (food production) and indirectly (ecosystem services) while food demand and supply processes affect land (land-use change) and land-related processes (e.g., land degradation, desertification) (Chapter 5).

(Sections 1.1.2 and 2.3), pollution (van Noordwijk and Brussaard 2014; Thyberg and Tonjes 2016; Borsato et al. 2018; Kibler et al. 2018), water quality (Malone et al. 2014; Norse and Ju 2015), and ecosystem services loss (Schipper et al. 2014; Eeraerts et al. 2017) with direct and indirect impacts on climate change and reduced resilience to climate variability. As food systems are assessed in relation to their contribution to global warming and/or to land degradation (e.g., livestock systems) it is critical to evaluate their contribution to food security and livelihoods and to consider alternatives, especially for developing countries where food insecurity is prevalent (Röös et al. 2017; Salmon et al. 2018).

1.2.1.5   Challenges arising from land governance

Land-use change has both positive and negative effects: it can lead to economic growth, but it can become a source of tension and social unrest leading to elite capture, and competition (Haberl 2015). Competition for land plays out continuously among different use types (cropland, pastureland, forests, urban spaces, and conservation and protected lands) and between different users within the same land-use category (subsistence vs commercial farmers) (Dell'Angelo

et al. 2017b). Competition is mediated through economic and market forces (expressed through land rental and purchases, as well as trade and investments). In the context of such transactions, power relations often disfavour disadvantaged groups such as small-scale farmers, indigenous communities or women (Doss et al. 2015; Ravnborg et al. 2016). These drivers are influenced to a large degree by policies, institutions and governance structures. Land governance determines not only who can access the land, but also the role of land ownership (legal, formal, customary or collective) which influences land use, land-use change and the resulting land competition (Moroni 2018).

Globally, there is competition for land because it is a finite resource and because most of the highly productive land is already exploited by humans (Lambin and Meyfroidt 2011; Lambin 2012; Venter et al. 2016). Driven by growing population, urbanisation, demand for food and energy, as well as land degradation, competition for land is expected to accentuate land scarcity in the future (Tilman et al. 2011; Foley et al. 2011; Lambin 2012; Popp et al. 2016) (*robust evidence, high agreement*). Climate change influences land use both directly and indirectly, as climate policies can also a play a role in increasing land competition via forest conservation policies, afforestation, or energy crop production (Section 1.3.1), with the potential for implications for food security (Hussein et al. 2013) and local land-ownership.

An example of large-scale change in land ownership is the much-debated large-scale land acquisition (LSLA) by investors which peaked in 2008 during the food price crisis, the financial crisis, and has also been linked to the search for biofuel investments (Dell'Angelo et al. 2017a). Since 2000, almost 50 million hectares of land have been acquired, and there are no signs of stagnation in the foreseeable future (Land Matrix 2018). The LSLA phenomenon, which largely targets agriculture, is widespread, including Sub-Saharan Africa, Southeast Asia, Eastern Europe and Latin America (Rulli et al. 2012; Nolte et al. 2016; Constantin et al. 2017). LSLAs are promoted by investors and host governments on economic grounds (infrastructure, employment, market development) (Deininger et al. 2011), but their social and environmental impacts can be negative and significant (Dell'Angelo et al. 2017a).

Much of the criticism of LSLA focuses on its social impacts, especially the threat to local communities' land rights (especially indigenous people and women) (Anseeuw et al. 2011) and displaced communities creating secondary land expansion (Messerli et al. 2014; Davis et al. 2015). The promises that LSLAs would develop efficient agriculture on non-forested, unused land (Deininger et al. 2011) has so far not been fulfilled. However, LSLA is not the only outcome of weak land governance structures (Wang et al. 2016): other forms of inequitable or irregular land acquisition can also be home-grown, pitting one community against a more vulnerable group (Xu 2018) or land capture by urban elites (McDonnell 2017). As demands on land are increasing, building governance capacity and securing land tenure becomes essential to attain sustainable land use, which has the potential to mitigate climate change, promote food security, and potentially reduce risks of climate-induced migration and associated risks of conflicts (Section 7.6).

1.2.2 Progress in dealing with uncertainties in assessing land processes in the climate system

1.2.2.1 Concepts related to risk, uncertainty and confidence

In context of the SRCCL, risk refers to the potential for the adverse consequences for human or (land-based) ecological systems, arising from climate change or responses to climate change. Risk related to climate change impacts integrates across the hazard itself, the time of exposure and the vulnerability of the system; the assessment of all three of these components, their interactions and outcomes, is uncertain (see Glossary for expanded definition, and Section 7.1.2). For instance, a risk to human society is the continued loss of productive land which might arise from climate change, mismanagement, or a combination of both factors. However, risk can also arise from the potential for adverse consequences from responses to climate change, such as widespread deployment of bioenergy which is intended to reduce GHG emissions and thus limit climate change, but can present its own risks to food security (Chapters 5–7).

Demonstrating with some statistical certainty that the climate or the land system affected by climate or land use has changed (detection),

and evaluating the relative contributions of multiple causal factors to that change (with a formal assessment of confidence (attribution); see Glossary) remain challenging aspects in both observations and models (Rosenzweig and Neofotis 2013; Gillett et al. 2016; Lean 2018). Uncertainties arising for example, from missing or imprecise data, ambiguous terminology, incomplete process representation in models, or human decision-making contribute to these challenges, and some examples are provided in this subsection. In order to reflect various sources of uncertainties in the state of scientific understanding, IPCC assessment reports provide estimates of confidence (Mastrandrea et al. 2011). This confidence language is also used in the SRCCL (Figure 1.5).

1.2.2.2 Nature and scope of uncertainties related to land use

Identification and communication of uncertainties is crucial to support decision making towards sustainable land management. Providing a robust, and comprehensive understanding of uncertainties in observations, models and scenarios is a fundamental first step in the IPCC confidence framework (see above). This will remain a challenge in future, but some important progress has been made over recent years.

Uncertainties in observations

The detection of changes in vegetation cover and structural properties underpins the assessment of land-use change, degradation and desertification. It is continuously improving by enhanced Earth observation capacity (Hansen et al. 2013; He et al. 2018; Ardö et al. 2018; Spennemann et al. 2018) (see also Table SM.1.1 in Supplementary Material). Likewise, the picture of how soil organic carbon, and GHG and water fluxes, respond to land-use change and land management continues to improve through advances in methodologies and sensors (Kostyanovsky et al. 2018; Brümmer et al. 2017; Iwata et al. 2017; Valayamkunnath et al. 2018). In both cases, the relative shortness of the record, data gaps, data treatment algorithms and – for remote sensing – differences in the definitions of major vegetation-cover classes limit the detection of trends (Alexander et al. 2016a; Chen et al. 2014; Yu et al. 2014; Lacaze et al. 2015; Song 2018; Peterson et al. 2017). In many developing countries, the cost of satellite remote sensing remains a challenge, although technological advances are starting to overcome this problem (Santilli et al. 2018), while ground-based observations networks are often not available.

Integration of multiple data sources in model and data assimilation schemes reduces uncertainties (Li et al. 2017; Clark et al. 2017; Lees et al. 2018), which might be important for the advancement of early warning systems. Early warning systems are a key feature of short-term (i.e. seasonal) decision-support systems and are becoming increasingly important for sustainable land management and food security (Shtienberg 2013; Jarroudi et al. 2015) (Sections 6.2.3 and 7.4.3). Early warning systems can help to optimise fertiliser and water use, aid disease suppression, and/or increase the economic benefit by enabling strategic farming decisions on when and what to plant (Caffi et al. 2012; Watmuff et al. 2013; Jarroudi et al. 2015; Chipanshi et al. 2015). Their suitability depends on the capability of the methods to accurately predict crop or pest developments, which in turn depends on expert agricultural knowledge, and the accuracy of

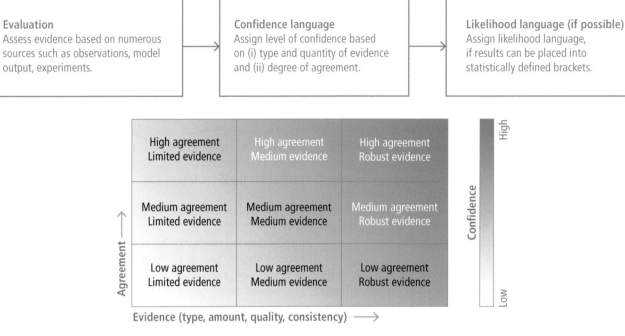

Evaluation	Confidence language	Likelihood language (if possible)
Assess evidence based on numerous sources such as observations, model output, experiments.	Assign level of confidence based on (i) type and quantity of evidence and (ii) degree of agreement.	Assign likelihood language, if results can be placed into statistically defined brackets.

High agreement Limited evidence	High agreement Medium evidence	High agreement Robust evidence	High
Medium agreement Limited evidence	Medium agreement Medium evidence	Medium agreement Robust evidence	Confidence
Low agreement Limited evidence	Low agreement Medium evidence	Low agreement Robust evidence	Low

Agreement ⟶

Evidence (type, amount, quality, consistency) ⟶

Figure 1.5 | Use of confidence language.

the weather data used to run phenological models (Caffi et al. 2012; Shtienberg 2013).

Uncertainties in models

Model intercomparison is a widely used approach to quantify some sources of uncertainty in climate change, land-use change and ecosystem modelling, often associated with the calculation of model-ensemble medians or means (see e.g., Sections 2.2 and 5.2). Even models of broadly similar structure differ in their projected outcome for the same input, as seen for instance in the spread in climate change projections from Earth System Models (ESMs) to similar future anthropogenic GHG emissions (Parker 2013; Stocker et al. 2013a). These uncertainties arise, for instance, from different parameter values, different processes represented in models, or how these processes are mathematically described. If the outputs of ESM simulations are used as input to impact models, these uncertainties can propagate to projected impacts (Ahlstrom et al. 2013).

Thus, the increased quantification of model performance in benchmarking exercises (the repeated confrontation of models with observations to establish a track-record of model developments and performance) is an important development to support the design and the interpretation of the outcomes of model ensemble studies (Randerson et al. 2009; Luo et al. 2012; Kelley et al. 2013). Since observational datasets in themselves are uncertain, benchmarking benefits from transparent information on the observations that are used, and the inclusion of multiple, regularly updated data sources (Luo et al. 2012; Kelley et al. 2013). Improved benchmarking approaches and the associated scoring of models may support weighted model means contingent on model performance. This could be an important step forward when calculating ensemble means across a range of models (Buisson et al. 2009; Parker 2013; Prestele et al. 2016).

Uncertainties arising from unknown futures

Large differences exist in projections of future land-cover change, both between and within scenario projections (Fuchs et al. 2015; Eitelberg et al. 2016; Popp et al. 2016; Krause et al. 2017; Alexander et al. 2016a). These differences reflect the uncertainties associated with baseline data, thematic classifications, different model structures and model parameter estimation (Alexander et al. 2017a; Prestele et al. 2016; Cross-Chapter Box 1 in Chapter 1). Likewise, projections of future land-use change are also highly uncertain, reflecting – among other factors – the absence of important crop, pasture and management processes in Integrated Assessment Models (Rose 2014) (Cross-Chapter Box 1 in Chapter 1) and in models of the terrestrial carbon cycle (Arneth et al. 2017). These processes have been shown to have large impacts on carbon stock changes (Arneth et al. 2017). Common scenario frameworks are used to capture the range of future uncertainties in scenarios. The most commonly used recent framework in climate change studies is based on the Representative Concentration Pathways (RCPs) and the Shared Socio-economic Pathways (SSPs) (Popp et al. 2016; Riahi et al. 2017). The RCPs prescribe levels of radiative forcing (W m^{-2}) arising from different atmospheric concentrations of GHGs that lead to different levels of climate change. For example, RCP2.6 (2.6 W m^{-2}) is projected to lead to global mean temperature changes of about 0.9°C–2.3°C, and RCP8.5 (8.5 W m^{-2}) to global mean temperature changes of about 3.2°C–5.4°C (van Vuuren et al. 2014).

The SSPs describe alternative trajectories of future socio-economic development with a focus on challenges to climate mitigation and challenges to climate adaptation (O'Neill et al. 2014). SSP1 represents a sustainable and cooperative society with a low-carbon economy and high capacity to adapt to climate change. SSP3 has social inequality that entrenches reliance on fossil fuels and limits adaptive capacity. SSP4 has large differences in income within and across world regions; it facilitates low-carbon economies in places, but limits adaptive capacity everywhere. SSP5 is a technologically advanced world with a strong economy that is heavily dependent on fossil fuels, but with high adaptive capacity. SSP2 is an intermediate case between SSP1 and SSP3 (O'Neill et al. 2014). The SSPs are commonly used with models to project future land-use change (Cross-Chapter Box 1 in Chapter 1).

The SSPs map onto the RCPs through shared assumptions. For example, a higher level of climate change (RCP8.5) is associated with higher challenges for climate change mitigation (SSP5). Not all SSPs are, however, associated with all RCPs. For example, an SSP5 world is committed to high fossil fuel use, associated GHG emissions, and this is not easily commensurate with lower levels of climate change (e.g., RCP2.6). Engstrom et al. (2016) took this approach further by ascribing levels of probability that associate an SSP with an RCP, contingent on the SSP scenario assumptions (Cross-Chapter Box 1 in Chapter 1).

Cross-Chapter Box 1 | Scenarios and other methods to characterise the future of land

Mark Rounsevell (United Kingdom/Germany), Almut Arneth (Germany), Katherine Calvin (The United States of America), Edouard Davin (France/Switzerland), Jan Fuglestvedt (Norway), Joanna House (United Kingdom), Alexander Popp (Germany), Joana Portugal Pereira (United Kingdom), Prajal Pradhan (Nepal/Germany), Jim Skea (United Kingdom), David Viner (United Kingdom).

About this box
The land-climate system is complex and future changes are uncertain, but methods exist (collectively known as *futures analysis*) to help decision-makers in navigating through this uncertainty. Futures analysis comprises a number of different and widely used methods, such as scenario analysis (Rounsevell and Metzger 2010), envisioning or target setting (Kok et al. 2018), pathways analysis (IPBES 2016; IPCC 2018),[1] and conditional probabilistic futures (Vuuren et al. 2018; Engstrom et al. 2016; Henry et al. 2018) (Table 1 in this Cross-Chapter Box). Scenarios and other methods to characterise the future can support a discourse with decision-makers about the sustainable development options that are available to them. All chapters of this assessment draw conclusions from futures analysis and so, the purpose of this box is to outline the principal methods used, their application domains, their uncertainties and their limitations.

Exploratory scenario analysis
Many exploratory scenarios are reported in climate and land system studies on climate change (Dokken 2014), such as related to land-based, climate change mitigation via reforestation/afforestation, avoided deforestation or bioenergy (Kraxner et al. 2013; Humpenoder et al. 2014; Krause et al. 2017) and climate change impacts and adaptation (Warszawski et al. 2014). There are global-scale scenarios of food security (Foley et al. 2011; Pradhan et al. 2013, 2014), but fewer scenarios of desertification, land degradation and restoration (Wolff et al. 2018). Exploratory scenarios combine qualitative 'storylines' or descriptive narratives of the underlying causes (or drivers) of change (Nakicenovic and Swart 2000; Rounsevell and Metzger 2010; O'Neill et al. 2014) with quantitative projections from computer models. Different types of models are used for this purpose based on very different modelling paradigms, baseline data and underlying assumptions (Alexander et al. 2016a; Prestele et al. 2016). Figure 1 in this Cross-Chapter Box below outlines how a combination of models can quantify these components as well as the interactions between them.

Exploratory scenarios often show that socio-economic drivers have a larger effect on land-use change than climate drivers (Harrison et al. 2014, 2016). Of these, technological development is critical in affecting the production potential (yields) of food and bioenergy and the feed conversion efficiency of livestock (Rounsevell et al. 2006; Wise et al. 2014; Kreidenweis et al. 2018), as well as the area of land needed for food production (Foley et al. 2011; Weindl et al. 2017; Kreidenweis et al. 2018). Trends in consumption, for example, diets or waste reduction, are also fundamental in affecting land-use change (Pradhan et al. 2013; Alexander et al. 2016b; Weindl et al. 2017; Alexander et al. 2017; Vuuren et al. 2018; Bajželj et al. 2014). Scenarios of land-based mitigation through large-scale bioenergy production and afforestation often lead to negative trade-offs with food security (food prices), water resources and biodiversity (Cross-Chapter Box 7 in Chapter 6).

Many exploratory scenarios are based on common frameworks such as the Shared Socio-economic Pathways (SSPs) (Popp et al. 2016; Riahi et al. 2017; Doelman et al. 2018)) (Section 1.2). However, other methods are used. *Stylised scenarios* prescribe assumptions about climate and land-use change solutions, for example, dietary change, food waste reduction and afforestation areas

---

[1] Different communities have a different understanding of the concept of pathways (IPCC 2018). Here, we refer to pathways as a description of the time-dependent actions required to move from today's world to a set of future visions (IPCC 2018). However, the term pathways is commonly used in the climate change literature as a synonym for projections or trajectories (e.g., shared socio-economic pathways).

Cross-Chapter Box 1 (continued)

(Pradhan et al. 2013, 2014; Kreidenweis et al. 2016; Rogelj et al. 2018b; Seneviratne et al. 2018; Vuuren et al. 2018). These scenarios provide useful thought experiments, but the feasibility of achieving the stylised assumptions is often unknown. *Shock scenarios* explore the consequences of low probability, high-impact events such as pandemic diseases, cyber-attacks and failures in food supply chains (Challinor et al. 2018), often in food security studies. Because of the diversity of exploratory scenarios, attempts have been made to categorise them into 'archetypes' based on the similarity between their assumptions in order to facilitate communication (IPBES 2018a).

Conditional probabilistic futures explore the consequences of model parameter uncertainty in which these uncertainties are conditional on scenario assumptions (Neill 2004). Only a few studies have applied the conditional probabilistic approach to land-use futures (Brown et al. 2014; Engstrom et al. 2016; Henry et al. 2018). By accounting for uncertainties in key drivers these studies show large ranges in land-use change, for example, global cropland areas of 893–2380 Mha by the end of the 21st century (Engstrom et al. 2016). They also find that land-use targets may not be achieved, even across a wide range of scenario parameter settings, because of trade-offs arising from the competition for land (Henry et al. 2018; Heck et al. 2018). Accounting for uncertainties across scenario assumptions can lead to convergent outcomes for land-use change, which implies that certain outcomes are more robust across a wide range of uncertain scenario assumptions (Brown et al. 2014).

In addition to global scale scenario studies, sub-national studies demonstrate that regional climate change impacts on the land system are highly variable geographically because of differences in the spatial patterns of both climate and socio-economic change (Harrison et al. 2014). Moreover, the capacity to adapt to these impacts is strongly dependent on the regional, socio-economic context and coping capacity (Dunford et al. 2014); processes that are difficult to capture in global scale scenarios. Regional scenarios are often co-created with stakeholders through participatory approaches (Kok et al. 2014), which are powerful in reflecting diverse worldviews and stakeholder values. Stakeholder participatory methods provide additional richness and context to storylines, as well as providing salience and legitimacy for local stakeholders (Kok et al. 2014).

Cross-Chapter Box 1, Table 1 | Description of the principal methods used in land and climate futures analysis.

Futures method	Description and subtypes	Application domain	Time horizon	Examples in this assessment
Exploratory scenarios. Trajectories of change in system components from the present to contrasting, alternative futures based on plausible and internally consistent assumptions about the underlying drivers of change	*Long-term projections* quantified with models	Climate system, land system and other components of the environment (e.g., biodiversity, ecosystem functioning, water resources and quality), for example the SSPs	10–100 years	2.3, 2.6.2, 5.2.3, 6.1.4, 6.4.4, 7.2
	Business-as-usual scenarios (including 'outlooks')	A continuation into the future of current trends in key drivers to explore the consequences of these in the near term	5–10 years, 20–30 years for outlooks	1.2.1, 2.6.2, 5.3.4, 6.1.4
	Policy and planning scenarios (including business planning)	*Ex ante* analysis of the consequences of alternative policies or decisions based on known policy options or already implemented policy and planning measures	5–30 years	2.6.3, 5.5.2, 5.6.2, 6.4.4
	Stylised scenarios (with single and multiple options)	Afforestation/reforestation areas, bioenergy areas, protected areas for conservation, consumption patterns (e.g., diets, food waste)	10–100 years	2.6.1, 5.5.1, 5.5.2, 5.6.1, 5.6.2, 6.4.4, 7.2
	Shock scenarios (high impact single events)	Food supply chain collapses, cyberattacks, pandemic diseases (humans, crops and livestock)	Near-term events (up to 10 years) leading to long-term impacts (10–100 years)	5.8.1
	Conditional probabilistic futures ascribe probabilities to uncertain drivers that are conditional on scenario assumptions	Where some knowledge is known about driver uncertainties, for example, population, economic growth, land-use change	10–100 years	1.2
Normative scenarios. Desired futures or outcomes that are aspirational and how to achieve them	*Visions, goal-seeking or target-seeking scenarios*	Environmental quality, societal development, human well-being, the Representative Concentration Pathways (RCPs,) 1.5°C scenarios	5–10 years to 10–100 years	2.6.2, 6.4.4, 7.2, 5.5.2
	Pathways as alternative sets of choices, actions or behaviours that lead to a future vision (goal or target)	Socio-economic systems, governance and policy actions	5–10 years to 10–100 years	5.5.2, 6.4.4, 7.2

Cross-Chapter Box 1 (continued)

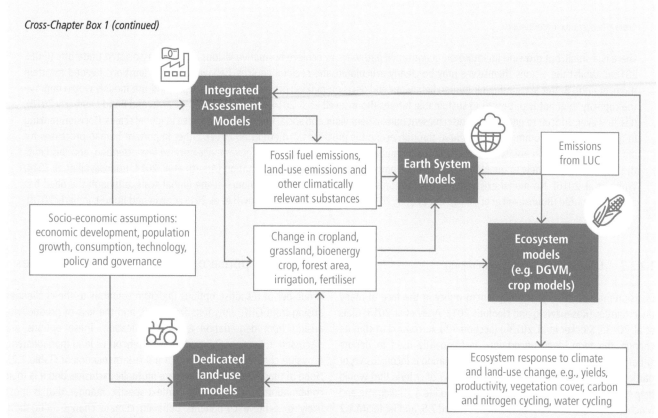

Cross-Chapter Box 1, Figure 1 | Interactions between land and climate system components and models in scenario analysis. The blue text describes selected model inputs and outputs.

Normative scenarios: visions and pathways analysis

Normative scenarios reflect a desired or target-seeking future. Pathways analysis is important in moving beyond the '*what if?*' perspective of exploratory scenarios to evaluate how normative futures might be achieved in practice, recognising that multiple pathways may achieve the same future vision. Pathways analysis focuses on consumption and behavioural changes through transitions and transformative solutions (IPBES 2018a). Pathways analysis is highly relevant in support of policy, since it outlines sets of time-dependent actions and decisions to achieve future targets, especially with respect to sustainable development goals, as well as highlighting trade-offs and co-benefits (IPBES 2018a). Multiple, alternative pathways have been shown to exist that mitigate trade-offs whilst achieving the priorities for future sustainable development outlined by governments and societal actors. Of these alternatives, the most promising focus on long-term societal transformations through education, awareness raising, knowledge sharing and participatory decision-making (IPBES 2018a).

What are the limitations of land-use scenarios?

Applying a common scenario framework (e.g., RCPs/SSPs) supports the comparison and integration of climate- and land-system scenarios, but a 'climate-centric' perspective can limit the capacity of these scenarios to account for a wider range of land-relevant drivers (Rosa et al. 2017). For example, in climate mitigation scenarios it is important to assess the impact of mitigation actions on the broader environment such as biodiversity, ecosystem functioning, air quality, food security, desertification/degradation and water cycles (Rosa et al. 2017). This implies the need for a more encompassing and flexible approach to creating scenarios that considers other environmental aspects, not only as a part of impact assessment, but also during the process of creating the scenarios themselves.

A limited number of models can quantify global scale, land-use change scenarios, and there is large variance in the outcomes of these models (Alexander et al. 2016a; Prestele et al. 2016). In some cases, there is greater variability between the models themselves than between the scenarios that they are quantifying, and these differences vary geographically (Prestele et al. 2016). These differences arise from variations in baseline datasets, thematic classes and modelling paradigms (Alexander et al. 2016a; Popp et al. 2016; Prestele et al. 2016). Model evaluation is critical in establishing confidence in the outcomes of modelled futures (Ahlstrom et al. 2012; Kelley et al. 2013). Some, but not all, land-use models are evaluated against observational data and model evaluation is rarely reported. Hence, there is a need for more transparency in land-use modelling, especially in evaluation and testing, as well as making model code available with complete sets of scenario outputs (e.g., Dietrich et al. 2018).

Cross-Chapter Box 1 (continued)

There is a small, but growing literature on quantitative pathways to achieve normative visions and their associated trade-offs (IPBES 2018a). Whilst the visions themselves may be clearly articulated, the societal choices, behaviours and transitions needed to attain them, are not. Better accounting for human behaviour and decision-making processes in global scale land-use models would improve the capacity to quantify pathways to sustainable futures (Rounsevell et al. 2014; Arneth et al. 2014; Calvin and Bond-Lamberty 2018). It is, however, difficult to understand and represent human behaviour and social interaction processes at global scales. Decision-making in global models is commonly represented through economic processes (Arneth et al. 2014). Other important human processes for land systems including equity, fairness, land tenure and the role of institutions and governance, receive less attention, and this limits the use of global models to quantify transformative pathways, adaptation and mitigation (Arneth et al. 2014; Rounsevell et al. 2014; Wang et al. 2016). No model exists at present to represent complex human behaviours at the global scale, although the need has been highlighted (Rounsevell et al. 2014; Arneth et al. 2014; Robinson et al. 2017; Brown et al. 2017; Calvin and Bond-Lamberty 2018).

1.2.2.3   Uncertainties in decision-making

Decision-makers develop and implement policy in the face of many uncertainties (Rosenzweig and Neofotis 2013; Anav et al. 2013; Ciais et al. 2013a; Stocker et al. 2013b) (Section 7.5). In context of climate change, the term 'deep uncertainty' is frequently used to denote situations in which either the analysis of a situation is inconclusive, or parties to a decision cannot agree on a number of criteria that would help to rank model results in terms of likelihood (e.g., Hallegatte and Mach 2016; Maier et al. 2016) (Sections 7.1 and 7.5, and Table SM.1.2 in Supplementary Material). However, existing uncertainty does not support societal and political inaction.

The many ways of dealing with uncertainty in decision-making can be summarised by two decision approaches: (economic) cost-benefit analysis, and the precautionary approach. A typical variant of cost-benefit analysis is the minimisation of negative consequences. This approach needs reliable probability estimates (Gleckler et al. 2016; Parker 2013) and tends to focus on the short term. The precautionary approach does not take account of probability estimates (cf. Raffensperger and Tickner 1999), but instead focuses on avoiding the worst outcome (Gardiner 2006).

Between these two extremes, various decision approaches seek to address uncertainties in a more reflective manner that avoids the limitations of cost-benefit analysis and the precautionary approach. Climate-informed decision analysis combines various approaches to explore options and the vulnerabilities and sensitivities of certain decisions. Such an approach includes stakeholder involvement (e.g., elicitation methods), and can be combined with, for example, analysis of climate or land-use change modelling (Hallegatte and Rentschler 2015; Luedeling and Shepherd 2016).

Flexibility is facilitated by political decisions that are not set in stone and can change over time (Walker et al. 2013; Hallegatte and Rentschler 2015). Generally, within the research community that investigates deep uncertainty, a paradigm is emerging that requires the development of a strategic vision of the long – or mid-term future, while committing to short-term actions and establishing a framework to guide future actions, including revisions and flexible adjustment of decisions (Haasnoot 2013) (Section 7.5).

1.3   Response options to the key challenges

A number of response options underpin solutions to the challenges arising from GHG emissions from land, and the loss of productivity arising from degradation and desertification. These options are discussed in Sections 2.5 and 6.2 and rely on (i) land management, (ii) value chain management, and (iii) risk management (Table 1.2). None of these response options are mutually exclusive, and it is their combination in a regionally, context-specific manner that is most likely to achieve co-benefits between climate change mitigation, adaptation and other environmental challenges in a cost-effective way (Griscom et al. 2017; Kok et al. 2018). Sustainable solutions affecting both demand and supply are expected to yield most co-benefits if these rely not only on the carbon footprint, but are extended to other vital ecosystems such as water, nutrients and biodiversity footprints (van Noordwijk and Brussaard 2014; Cremasch 2016). As an entry point to the discussion in Chapter 6, we introduce here a selected number of examples that cut across climate change mitigation, food security, desertification, and degradation issues, including potential trade-offs and co-benefits.

Table 1.2 | Broad categorisation of response options into three main classes and eight sub-classes. For illustration, the table includes examples of individual response options. A complete list and description is provided in Chapter 6.

Response options based on land management	
in agriculture	Improved management of: cropland, grazing land, livestock; agro-forestry; avoidance of conversion of grassland to cropland; integrated water management
in forests	Improved management of forests and forest restoration; reduced deforestation and degradation; afforestation
of soils	Increased soil organic carbon content; reduced soil erosion; reduced soil salinisation
across all/other ecosystems	Reduced landslides and natural hazards; reduced pollution including acidification; biodiversity conservation; restoration and reduced conversion of peatlands
specifically for CO_2 removal	Enhanced weathering of minerals; bioenergy and BECCS
Response options based on value chain management	
through demand management	Dietary change; reduced post-harvest losses; reduced food waste
through supply management	Sustainable sourcing; improved energy use in food systems; improved food processing and retailing
Response options based on risk management	
Risk management	Risk-sharing instruments; use of local seeds; disaster risk management

1.3.1   Targeted decarbonisation relying on large land-area need

Most global future scenarios that aim to achieve global warming of 2°C or well below rely on bioenergy (BE; BECCS, with carbon capture and storage; Cross-Chapter Box 7 in Chapter 6) or afforestation and reforestation (de Coninck et al. 2018; Rogelj et al. 2018b,a; Anderson and Peters 2016; Popp et al. 2016; Smith et al. 2016) (Cross-Chapter Box 2 in Chapter 1). In addition to the very large area requirements projected for 2050 or 2100, several other aspects of these scenarios have also been criticised. For instance, they simulate very rapid technological and societal uptake rates for the land-related mitigation measures, when compared with historical observations (Turner et al. 2018; Brown et al. 2019; Vaughan and Gough 2016). Furthermore, *confidence* in the projected bioenergy or BECCS net carbon uptake potential is *low*, because of many diverging assumptions. This includes assumptions about bioenergy crop yields, the possibly large energy demand for CCS, which diminishes the net-GHG-saving of bioenergy systems, or the incomplete accounting for ecosystem processes and of the cumulative carbon-loss arising from natural vegetation clearance for bioenergy crops or bioenergy forests and subsequent harvest regimes (Anderson and Peters 2016; Bentsen 2017; Searchinger et al. 2017; Bayer et al. 2017; Fuchs et al. 2017; Pingoud et al. 2018; Schlesinger 2018). Bioenergy provision under politically unstable conditions may also be a problem (Erb et al. 2012; Searle and Malins 2015).

Large-scale bioenergy plantations and forests may compete for the same land area (Harper et al. 2018). Both potentially have adverse side effects on biodiversity and ecosystem services, as well as socio-economic trade-offs such as higher food prices due to land-area competition (Shi et al. 2013; Bárcena et al. 2014; Fernandez-Martinez et al. 2014; Searchinger et al. 2015; Bonsch et al. 2016; Creutzig et al. 2015; Kreidenweis et al. 2016; Santangeli et al. 2016; Williamson 2016; Graham et al. 2017; Krause et al. 2017; Hasegawa et al. 2018; Humpenoeder et al. 2018). Although forest-based mitigation could have co-benefits for biodiversity and many ecosystem services, this depends on the type of forest planted and the vegetation cover it replaces (Popp et al. 2014; Searchinger et al. 2015) (Cross-Chapter Box 2 in Chapter 1).

There is *high confidence* that scenarios with large land requirements for climate change mitigation may not achieve SDGs, such as no poverty, zero hunger and life on land, if competition for land and the need for agricultural intensification are greatly enhanced (Creutzig et al. 2016; Dooley and Kartha 2018; Hasegawa et al. 2015; Hof et al. 2018; Roy et al. 2018; Santangeli et al. 2016; Boysen et al. 2017; Henry et al. 2018; Kreidenweis et al. 2016; UN 2015). This does not mean that smaller-scale land-based climate mitigation could not have positive outcomes for then achieving these goals (e.g., Sections 6.2, and 4.5, Cross-Chapter Box 7 in Chapter 6).

Cross-Chapter Box 2 | Implications of large-scale conversion from non-forest to forest land

Baldur Janz (Germany), Almut Arneth (Germany), Francesco Cherubini (Norway/Italy), Edouard Davin (Switzerland/France), Aziz Elbehri (Morocco), Kaoru Kitajima (Japan), Werner Kurz (Canada).

Efforts to increase forest area

While deforestation continues in many world regions, especially in the tropics, large expansion of mostly managed forest area has taken place in some countries. In the IPCC context, reforestation (conversion to forest of land that previously contained forests but has been converted to some other use) is distinguished from afforestation (conversion to forest of land that historically has not contained forests; see Glossary). Past expansion of managed forest area occurred in many world-regions for a variety of reasons, from meeting needs for wood fuel or timber (Vadell et al. 2016; Joshi et al. 2011; Zaloumis and Bond 2015; Payn et al. 2015; Shoyama 2008; Miyamoto et al. 2011) to restoration-driven efforts, with the aim of enhancing ecological function (Filoso et al. 2017; Salvati and Carlucci 2014; Ogle et al. 2018; Crouzeilles et al. 2016; FAO 2016) (Sections 3.7 and 4.9).

In many regions, net forest area increase includes deforestation (often of native forests) alongside increasing forest area (often managed forest, but also more natural forest restoration efforts) (Heilmayr et al. 2016; Scheidel and Work 2018; Hua et al. 2018; Crouzeilles et al. 2016; Chazdon et al. 2016b). China and India have seen the largest net forest area increase, aiming to alleviate soil erosion, desertification and overgrazing (Ahrends et al. 2017; Cao et al. 2016; Deng et al. 2015; Chen et al. 2019) (Sections 3.7 and 4.9) but uncertainties in exact forest area changes remain large, mostly due to differences in methodology and forest classification (FAO 2015a; Song et al. 2018; Hansen et al. 2013; MacDicken et al. 2015).

What are the implications for ecosystems?

1. Implications for biogeochemical and biophysical processes

There is *robust evidence* and *medium agreement* that whilst forest area expansion increases ecosystem carbon storage, the magnitude of the increased stock depends on the type and length of former land use, forest type planted, and climatic regions (Bárcena et al. 2014; Poeplau et al. 2011; Shi et al. 2013; Li et al. 2012) (Section 4.3). While reforestation of former croplands increases net ecosystem carbon storage (Bernal et al. 2018; Lamb 2018), afforestation on native grassland results in reduction of soil carbon stocks, which can reduce or negate the net carbon benefits which are dominated by increases in biomass, dead wood and litter carbon pools (Veldman et al. 2015, 2017).

Forest vs non-forest lands differ in land surface reflectiveness of shortwave radiation and evapotranspiration (Anderson et al. 2011; Perugini et al. 2017) (Section 2.4). Evapotranspiration from forests during the growing season regionally cools the land surface and enhances cloud cover that reduces shortwave radiation reaching the land, an impact that is especially pronounced in the tropics. However, dark evergreen conifer-dominated forests have low surface reflectance, and tend to cause warming of the near-surface atmosphere compared to non-forest land, especially when snow cover is present such as in boreal regions (Duveiller et al. 2018; Alkama and Cescatti 2016; Perugini et al. 2017) (*medium evidence, high agreement*).

2. Implications for water balance

Evapotranspiration by forests reduces surface runoff and erosion of soil and nutrients (Salvati et al. 2014). Planting of fast-growing species in semi-arid regions or replacing natural grasslands with forest plantations can divert soil water resources to evapotranspiration from groundwater recharge (Silveira et al. 2016; Zheng et al. 2016; Cao et al. 2016). Multiple cases are reported from China where afforestation programs, some with irrigation, without having been tailored to local precipitation conditions, resulted in water shortages and tree mortality (Cao et al. 2016; Yang et al. 2014; Li et al. 2014; Feng et al. 2016). Water shortages may create long-term water conflicts (Zheng et al. 2016). However, reforestation (in particular for restoration) is also associated with improved water filtration, groundwater recharge (Ellison et al. 2017) and can reduce risk of soil erosion, flooding, and associated disasters (Lee et al. 2018) (Section 4.9).

3. Implications for biodiversity

Impacts of forest area expansion on biodiversity depend mostly on the vegetation cover that is replaced: afforestation on natural non-tree-dominated ecosystems can have negative impacts on biodiversity (Abreu et al. 2017; Griffith et al. 2017; Veldman et al. 2015; Parr et al. 2014; Wilson et al. 2017; Hua et al. 2016; see also IPCC 1.5° report (2018)). Reforestation with monocultures of fast-growing, non-native trees has little benefit to biodiversity (Shimamoto et al. 2018; Hua et al. 2016). There are also concerns regarding some commonly used plantation species (e.g., Acacia and Pinus species) to become invasive (Padmanaba and Corlett 2014; Cunningham et al. 2015b).

1

Reforestation with mixes of native species, especially in areas that retain fragments of native forest, can support ecosystem services and biodiversity recovery, with positive social and environmental co-benefits (Cunningham et al. 2015a; Dendy et al. 2015; Chaudhary and Kastner 2016; Huang et al. 2018; Locatelli et al. 2015b) (Section 4.5). Even though species diversity in re-growing forests is typically lower than in primary forests, planting native or mixed species can have positive effects on biodiversity (Brockerhoff et al. 2013; Pawson et al. 2013; Thompson et al. 2014). Reforestation has been shown to improve links among existing remnant forest patches, increasing species movement, and fostering gene flow between otherwise isolated populations (Gilbert-Norton et al. 2010; Barlow et al. 2007; Lindenmayer and Hobbs 2004).

4. Implications for other ecosystem services and societies
Forest area expansion could benefit recreation and health, preservation of cultural heritage and local values and knowledge, livelihood support (via reduced resource conflicts, restoration of local resources). These social benefits could be most successfully achieved if local communities' concerns are considered (Le et al. 2012). However, these co-benefits have rarely been assessed due to a lack of suitable frameworks and evaluation tools (Baral et al. 2016).

Industrial forest management can be in conflict with the needs of forest-dependent people and community-based forest management over access to natural resources (Gerber 2011; Baral et al. 2016) and/or loss of customary rights over land use (Malkamäki et al. 2018; Cotula et al. 2014). A common result is out-migration from rural areas and diminishing local uses of ecosystems (Gerber 2011). Policies promoting large-scale tree plantations gain traction if these are reappraised in view of potential co-benefits with several ecosystem services and local societies (Bull et al. 2006; Le et al. 2012).

Scenarios of forest area expansion for land-based climate change mitigation
Conversion of non-forest to forest land has been discussed as a relatively cost-effective climate change mitigation option when compared to options in the energy and transport sectors (*medium evidence, medium agreement*) (de Coninck et al. 2018; Griscom et al. 2017; Fuss et al. 2018), and can have co-benefits with adaptation.

Sequestration of CO_2 from the atmosphere through forest area expansion has become a fundamental part of stringent climate change mitigation scenarios (Rogelj et al. 2018a; Fuss et al. 2018) (e.g., Sections 2.5, 4.5 and 6.2). The estimated mitigation potential ranges from about 0.5 to 10 $GtCO_2$ yr^{-1} (*robust evidence, medium agreement*), and depends on assumptions regarding available land and forest carbon uptake potential (Houghton 2013; Houghton and Nassikas 2017; Griscom et al. 2017; Lenton 2014; Fuss et al. 2018; Smith 2016) (Section 2.5.1). In climate change mitigation scenarios, typically, no differentiation is made between reforestation and afforestation despite different overall environmental impacts between these two measures. Likewise, biodiversity conservation, impacts on water balances, other ecosystem services, or land-ownership – as constraints when simulating forest area expansion (Cross-Chapter Box 1 in Chapter 1) – tend not to be included as constraints when simulating forest area expansion.

Projected forest area increases, relative to today's forest area, range from approximately 25% in 2050 and increase to nearly 50% by 2100 (Rogelj et al. 2018a; Kreidenweis et al. 2016; Humpenoder et al. 2014). Potential adverse side-effects of such large-scale measures, especially for low-income countries, could be increasing food prices from the increased competition for land (Kreidenweis et al. 2016; Hasegawa et al. 2015, 2018; Boysen et al. 2017) (Section 5.5). Forests also emit large amounts of biogenic volatile compounds that under some conditions contribute to the formation of atmospherically short-lived climate forcing compounds, which are also detrimental to health (Ashworth et al. 2013; Harrison et al. 2013). Recent analyses argued for an upper limit of about 5 million km^2 of land globally available for climate change mitigation through reforestation, mostly in the tropics (Houghton 2013) – with potential regional co-benefits.

Since forest growth competes for land with bioenergy crops (Harper et al. 2018) (Cross-Chapter Box 7 in Chapter 6), global area estimates need to be assessed in light of alternative mitigation measures at a given location. In all forest-based mitigation efforts, the sequestration potential will eventually saturate unless the area keeps expanding, or harvested wood is either used for long-term storage products or for carbon capture and storage (Fuss et al. 2018; Houghton et al. 2015) (Section 2.5.1). Considerable uncertainty in forest carbon uptake estimates is further introduced by potential forest losses from fire or pest outbreaks (Allen et al. 2010; Anderegg et al. 2015) (Cross-Chapter Box 3 in Chapter 2). And like all land-based mitigation measures, benefits may be diminshed by land-use displacement, and through trade of land-based products, especially in poor countries that experience forest loss (e.g., Africa) (Bhojvaid et al. 2016; Jadin et al. 2016).

Cross-Chapter Box 2 (continued)

Conclusion

Reforestation is a mitigation measure with potential co-benefits for conservation and adaptation, including biodiversity habitat, air and water filtration, flood control, enhanced soil fertility and reversal of land degradation. Potential adverse side-effects of forest area expansion depend largely on the state of the land it displaces as well as tree species selections. Active governance and planning contribute to maximising co-benefits while minimising adverse side-effects (Laestadius et al. 2011; Dinerstein et al. 2015; Veldman et al. 2017) (Section 4.8 and Chapter 7). At large spatial scales, forest expansion is expected to lead to increased competition for land, with potentially undesirable impacts on food prices, biodiversity, non-forest ecosystems and water availability (Bryan and Crossman 2013; Boysen et al. 2017; Kreidenweis et al. 2016; Egginton et al. 2014; Cao et al. 2016; Locatelli et al. 2015a; Smith et al. 2013).

1.3.2   Land management

1.3.2.1   Agricultural, forest and soil management

Sustainable land management (SLM) describes "the stewardship and use of land resources, including soils, water, animals and plants, to meet changing human needs while simultaneously assuring the long-term productive potential of these resources and the maintenance of their environmental functions" (Alemu 2016; Altieri and Nicholls 2017) (e.g., Section 4.1.5), and includes ecological, technological and governance aspects.

The choice of SLM strategy is a function of regional context and land-use types, with *high agreement* on (a combination of) choices such as agroecology (including agroforestry), conservation agriculture and forestry practices, crop and forest species diversity, appropriate crop and forest rotations, organic farming, integrated pest management, the preservation and protection of pollination services, rainwater harvesting, range and pasture management, and precision agriculture systems (Stockmann et al. 2013; Ebert, 2014; Schulte et al. 2014; Zhang et al. 2015; Sunil and Pandravada 2015; Poeplau and Don 2015; Agus et al. 2015; Keenan 2015; MacDicken et al. 2015; Abberton et al. 2016). Conservation agriculture and forestry uses management practices with minimal soil disturbance such as no tillage or minimum tillage, permanent soil cover with mulch, combined with rotations to ensure a permanent soil surface, or rapid regeneration of forest following harvest (Hobbs et al. 2008; Friedrich et al. 2012). Vegetation and soils in forests and woodland ecosystems play a crucial role in regulating critical ecosystem processes, therefore reduced deforestation together with sustainable forest management are integral to SLM (FAO 2015b) (Section 4.8). In some circumstances, increased demand for forest products can also lead to increased management of carbon storage in forests (Favero and Mendelsohn 2014). Precision agriculture is characterised by a "management system that is information and technology based, is site specific and uses one or more of the following sources of data: soils, crops, nutrients, pests, moisture, or yield, for optimum profitability, sustainability, and protection of the environment" (USDA 2007) (Cross-Chapter Box 6 in Chapter 5). The management of protected areas that reduce deforestation also plays an important role in climate change mitigation and adaptation while delivering numerous ecosystem services and sustainable development benefits

(Bebber and Butt 2017). Similarly, when managed in an integrated and sustainable way, peatlands are also known to provide numerous ecosystem services, as well as socio-economic and mitigation and adaptation benefits (Ziadat et al. 2018).

Biochar is an organic compound used as soil amendment and is believed to be potentially an important global resource for mitigation. Enhancing the carbon content of soil and/or use of biochar (Chapter 4) have become increasingly important as a climate change mitigation option with possibly large co-benefits for other ecosystem services. Enhancing soil carbon storage and the addition of biochar can be practiced with limited competition for land, provided no productivity/ yield loss and abundant unused biomass, but evidence is limited and impacts of large scale application of biochar on the full GHG balance of soils, or human health are yet to be explored (Gurwick et al. 2013; Lorenz and Lal 2014; Smith 2016).

1.3.3   Value chain management

1.3.3.1   Supply management

Food losses from harvest to retailer. Approximately one-third of losses and waste in the food system occurs between crop production and food consumption, increasing substantially if losses in livestock production and overeating are included (Gustavsson et al. 2011; Alexander et al. 2017). This includes on-farm losses, farm to retailer losses, as well retailer and consumer losses (Section 1.3.3.2).

Post-harvest food loss – on farm and from farm to retailer – is a widespread problem, especially in developing countries (Xue et al. 2017), but are challenging to quantify. For instance, averaged for eastern and southern Africa an estimated 10–17% of annual grain production is lost (Zorya et al. 2011). Across 84 countries and different time periods, annual median losses in the supply chain before retailing were estimated at about 28 kg per capita for cereals or about 12 kg per capita for eggs and dairy products (Xue et al. 2017). For the year 2013, losses prior to the reaching retailers were estimated at 20% (dry weight) of the production amount (22% wet weight) (Gustavsson et al. 2011; Alexander et al. 2017). While losses of food cannot be realistically reduced to zero, advancing harvesting technologies (Bradford et al. 2018; Affognon et al. 2015), storage capacity (Chegere 2018) and efficient transportation could all contribute to reducing these losses with co-benefits for food availability, the land area needed for food production and related GHG emissions.

Stability of food supply, transport and distribution. Increased climate variability enhances fluctuations in world food supply and price variability (Warren 2014; Challinor et al. 2015; Elbehri et al. 2017). 'Food price shocks' need to be understood regarding their transmission across sectors and borders and impacts on poor and food insecure populations, including urban poor subject to food deserts and inadequate food accessibility (Widener et al. 2017; Lehmann et al. 2013; Le 2016; FAO 2015b). Trade can play an important stabilising role in food supply, especially for regions with agro-ecological limits to production, including water scarce regions, as well as regions that experience short-term production variability due to climate, conflicts or other economic shocks (Gilmont 2015; Marchand et al. 2016). Food trade can either increase or reduce the overall environmental impacts of agriculture (Kastner et al. 2014). Embedded in trade are virtual transfers of water, land area, productivity, ecosystem services, biodiversity, or nutrients (Marques et al. 2019; Wiedmann and Lenzen 2018; Chaudhary and Kastner 2016) with either positive or negative implications (Chen et al. 2018; Yu et al. 2013). Detrimental consequences in countries in which trade dependency may accentuate the risk of food shortages from foreign production shocks could be reduced by increasing domestic reserves or importing food from a diversity of suppliers (Gilmont 2015; Marchand et al. 2016).

Climate mitigation policies could create new trade opportunities (e.g., biomass) (Favero and Massetti 2014) or alter existing trade patterns. The transportation GHG footprints of supply chains may be causing a differentiation between short and long supply chains (Schmidt et al. 2017) that may be influenced by both economics and policy measures (Section 5.4). In the absence of sustainable practices and when the ecological footprint is not valued through the market system, trade can also exacerbate resource exploitation and environmental leakages, thus weakening trade mitigation contributions (Dalin and Rodríguez-Iturbe 2016; Mosnier et al. 2014; Elbehri et al. 2017). Ensuring stable food supply while pursuing climate mitigation and adaptation will benefit from evolving trade rules and policies that allow internalisation of the cost of carbon (and costs of other vital resources such as water, nutrients). Likewise, future climate change mitigation policies would gain from measures designed to internalise the environmental costs of resources and the benefits of ecosystem services (Elbehri et al. 2017; Brown et al. 2007).

1.3.3.2   Demand management

Dietary change. Demand-side solutions to climate mitigation are an essential complement to supply-side, technology and productivity driven solutions (*high confidence*) (Creutzig et al. 2016; Bajželj et al. 2014; Erb et al. 2016b; Creutzig et al. 2018) (Sections 5.5.1 and 5.5.2). The environmental impacts of the animal-rich 'western diets' are being examined critically in the scientific literature (Hallström et al. 2015; Alexander et al. 2016h; Alexander et al. 2015; Tilman and Clark 2014; Aleksandrowicz et al. 2016; Poore and Nemecek 2018)

(Section 5.4.6). For example, if the average diet of each country were consumed globally, the agricultural land area needed to supply these diets would vary 14-fold, due to country differences in ruminant protein and calorific intake (–55% to +178% compared to existing cropland areas). Given the important role enteric fermentation plays in methane (CH_4) emissions, a number of studies have examined the implications of lower animal-protein diets (Swain et al. 2018; Röös et al. 2017; Rao et al. 2018). Reduction of animal protein intake has been estimated to reduce global green water (from precipitation) use by 11% and blue water (from rivers, lakes, groundwater) use by 6% (Jalava et al. 2014). By avoiding meat from producers with above-median GHG emissions and halving animal-product intake, consumption change could free-up 21 million km^2 of agricultural land and reduce GHG emissions by nearly 5 $GtCO_2$-eq yr^{-1} or up to 10.4 $GtCO_2$-eq yr^{-1} when vegetation carbon uptake is considered on the previously agricultural land (Poore and Nemecek 2018, 2019).

Diets can be location and community specific, are rooted in culture and traditions while responding to changing lifestyles driven for instance by urbanisation and changing income. Changing dietary and consumption habits would require a combination of non-price (government procurement, regulations, education and awareness raising) and price incentives (Juhl and Jensen 2014) to induce consumer behavioural change with potential synergies between climate, health and equity (addressing growing global nutrition imbalances that emerge as undernutrition, malnutrition, and obesity) (FAO 2018b).

Reduced waste and losses in the food demand system. Global averaged per capita food waste and loss (FWL) have increased by 44% between 1961 and 2011 (Porter et al. 2016) and are now around 25–30% of global food produced (Kummu et al. 2012; Alexander et al. 2017). Food waste occurs at all stages of the food supply chain from the household to the marketplace (Parfitt et al. 2010) and is found to be larger at household than at supply chain levels. A meta-analysis of 55 studies showed that the highest share of food waste was at the consumer stage (43.9% of total) with waste increasing with per capita GDP for high-income countries until a plateaux at about 100 kg cap^{-1} yr^{-1} (around 16% of food consumption) above about 70,000 USD cap^{-1} (van der Werf and Gilliland 2017; Xue et al. 2017). Food loss from supply chains tends to be more prevalent in less developed countries where inadequate technologies, limited infrastructure, and imperfect markets combine to raise the share of the food production lost before use.

There are several causes behind food waste including economics (cheap food), food policies (subsidies) as well as individual behaviour (Schanes et al. 2018). Household level food waste arises from overeating or overbuying (Thyberg and Tonjes 2016). Globally, overconsumption was found to waste 9–10% of food bought (Alexander et al. 2017).

Solutions to FWL thus need to address technical and economic aspects. Such solutions would benefit from more accurate data on the loss-source, loss-magnitude and causes along the food supply chain. In the long run, internalising the cost of food waste into the product price would more likely induce a shift in consumer behaviour towards less waste and more nutritious, or alternative, food intake (FAO 2018b). Reducing FWL would bring a range of benefits for health, reducing pressures on land, water and nutrients, lowering emissions and safeguarding food security. Reducing food waste by 50% would generate net emissions reductions in the range of 20 to 30% of total food-sourced GHGs (Bajželj et al. 2014). SDG 12 ("Ensure sustainable consumption and production patterns") calls for per capita global food waste to be reduced by one-half at the retail and consumer level, and reducing food losses along production and supply chains by 2030.

1.3.4   Risk management

Risk management refers to plans, actions, strategies or policies to reduce the likelihood and/or magnitude of adverse potential consequences, based on assessed or perceived risks. Insurance and early warning systems are examples of risk management, but risk can also be reduced (or resilience enhanced) through a broad set of options ranging from seed sovereignty, livelihood diversification, to reducing land loss through urban sprawl. Early warning systems support farmer decision-making on management strategies (Section 1.2) and are a good example of an adaptation measure with mitigation co-benefits such as reducing carbon losses (Section 1.3.6). Primarily designed to avoid yield losses, early warning systems also support fire management strategies in forest ecosystems, which prevents financial as well as carbon losses (de Groot et al. 2015). Given that over recent decades on average around 10% of cereal production was lost through extreme weather events (Lesk et al. 2016), where available and affordable, insurance can buffer farmers and foresters against the financial losses incurred through such weather and other (fire, pests) extremes (Falco et al. 2014) (Sections 7.2 and 7.4). Decisions to take up insurance are influenced by a range of factors such as the removal of subsidies or targeted education (Falco et al. 2014). Enhancing access and affordability of insurance in low-income countries is a specific objective of the UNFCCC (Linnerooth-Bayer and Mechler 2006). A global mitigation co-benefit of insurance schemes may also include incentives for future risk reduction (Surminski and Oramas-Dorta 2014).

1.3.5   Economics of land-based mitigation pathways: Costs versus benefits of early action under uncertainty

The overarching societal costs associated with GHG emissions and the potential implications of mitigation activities can be measured by various metrics (cost-benefit analysis, cost effectiveness analysis) at different scales (project, technology, sector or the economy) (IPCC 2018) (Section 1.4). The social cost of carbon (SCC) measures the total net damages of an extra metric tonne of CO_2 emissions due to the associated climate change (Nordhaus 2014; Pizer et al. 2014). Both negative and positive impacts are monetised and discounted to arrive at the net value of consumption loss. As the SCC depends on discount rate assumptions and value judgements (e.g., relative weight given to current vs future generations), it is not a straightforward policy tool to compare alternative options. At the sectoral level, marginal abatement cost curves (MACCs) are widely used for the assessment of costs related to GHG emissions reduction. MACCs measure the cost of reducing one more GHG unit and are either expert-based or model-derived and offer a range of approaches and assumptions on discount rates or available abatement technologies (Kesicki 2013). In land-based sectors, Gillingham and Stock (2018) reported short-term static abatement costs for afforestation of between 1 and 10 USD2017 per tCO_2, soil management at 57 and livestock management at 71 USD2017 per tCO_2. MACCs are more reliable when used to rank alternative options compared to a baseline (or business as usual) rather than offering absolute numerical measures (Huang et al. 2016). The economics of land-based mitigation options encompass also the "costs of inaction" that arise either from the economic damages due to continued accumulation of GHGs in the atmosphere and from the diminution in value of ecosystem services or the cost of their restoration where feasible (Rodriguez-Labajos 2013; Ricke et al. 2018). Overall, it remains challenging to estimate the costs of alternative mitigation options owing to the context – and scale-specific interplay between multiple drivers (technological, economic, and socio-cultural) and enabling policies and institutions (IPCC 2018) (Section 1.4).

The costs associated with mitigation (both project-linked such as capital costs or land rental rates, or sometimes social costs) generally increase with stringent mitigation targets and over time. Sources of uncertainty include the future availability, cost and performance of technologies (Rosen and Guenther 2015; Chen et al. 2016) or lags in decision-making, which have been demonstrated by the uptake of land use and land utilisation policies (Alexander et al. 2013; Hull et al. 2015; Brown et al. 2018b). There is growing evidence of significant mitigation gains through conservation, restoration and improved land management practices (Griscom et al. 2017; Kindermann et al. 2008; Golub et al. 2013; Favero et al. 2017) (Chapters 4 and 6), but the mitigation cost efficiency can vary according to region and specific ecosystem (Albanito et al. 2016). Recent model developments that treat process-based, human–environment interactions have recognised feedbacks that reinforce or dampen the original stimulus for land-use change (Robinson et al. 2017; Walters and Scholes 2017). For instance, land mitigation interventions that rely on large-scale, land-use change (e.g., afforestation) would need to account for the rebound effect (which dampens initial impacts due to feedbacks) in which raising land prices also raises the cost of land-based mitigation (Vivanco et al. 2016). Although there are few direct estimates, indirect assessments strongly point to much higher costs if action is delayed or limited in scope (*medium confidence*). Quicker response options are also needed to avoid loss of high-carbon ecosystems and other vital ecosystem services that provide multiple services that are difficult to replace (peatlands, wetlands, mangroves, forests) (Yirdaw et al. 2017; Pedrozo-Acuña et al. 2015). Delayed action would raise relative costs in the future or could make response options less feasible (*medium confidence*) (Goldstein et al. 2019; Butler et al. 2014).

1.3.6   Adaptation measures and scope for co-benefits with mitigation

Adaptation and mitigation have generally been treated as two separate discourses, both in policy and practice, with mitigation addressing cause and adaptation dealing with the consequences of climate change (Hennessey et al. 2017). While adaptation (e.g., reducing flood risks) and mitigation (e.g., reducing non-CO_2 emissions from agriculture) may have different objectives and operate at different scales, they can also generate joint outcomes (Locatelli et al. 2015b) with adaptation generating mitigation co-benefits. Seeking to integrate strategies for achieving adaptation and mitigation goals is attractive in order to reduce competition for limited resources and trade-offs (Lobell et al. 2013; Berry et al. 2015; Kongsager and Corbera 2015). Moreover, determinants that can foster adaptation and mitigation practices are similar. These tend to include available technology and resources, and credible information for policymakers to act on (Yohe 2001).

Four sets of mitigation–adaptation interrelationships can be distinguished: (i) mitigation actions that can result in adaptation benefits; (ii) adaptation actions that have mitigation benefits; (iii) processes that have implications for both adaptation and mitigation; and (iv) strategies and policy processes that seek to promote an integrated set of responses for both adaptation and mitigation (Klein et al. 2007). A high level of adaptive capacity is a key ingredient to developing successful mitigation policy. Implementing mitigation action can result in increasing resilience especially if it is able to reduce risks. Yet, mitigation and adaptation objectives, scale of implementation, sector and even metrics to identify impacts tend to differ (Ayers and Huq 2009), and institutional setting, often does not enable an environment where synergies are sought (Kongsager et al. 2016). Trade-offs between adaptation and mitigation exist as well and need to be understood (and avoided) to establish win-win situations (Porter et al. 2014; Kongsager et al. 2016).

Forestry and agriculture offer a wide range of lessons for the integration of adaptation and mitigation actions given the vulnerability of forest ecosystems or cropland to climate variability and change (Keenan 2015; Gaba et al. 2015) (Sections 5.6 and 4.8). Increasing adaptive capacity in forested areas has the potential to prevent deforestation and forest degradation (Locatelli et al. 2011). Reforestation projects, if well managed, can increase community economic opportunities that encourage conservation (Nelson and de Jong 2003), build capacity through training of farmers and installation of multifunctional plantations with income generation (Reyer et al. 2009), strengthen local institutions (Locatelli et al. 2015a) and increase cash-flow to local forest stakeholders from foreign donors (West 2016). A forest plantation that sequesters carbon for mitigation can also reduce water availability to downstream populations and heighten their vulnerability to drought. Inversely, not recognising mitigation in adaptation projects may yield adaptation measures that increase greenhouse gas emissions, a prime example of 'maladaptation'. Analogously, 'mal-mitigation' would result in reducing GHG emissions, but increasing vulnerability (Barnett and O'Neill 2010; Porter et al. 2014). For instance, the cost of pursuing large-scale adaptation and mitigation projects has been associated with higher failure risks, onerous transactions costs and the complexity of managing big projects (Swart and Raes 2007).

Adaptation encompasses both biophysical and socio-economic vulnerability and underlying causes (informational, capacity,

financial, institutional, and technological; Huq et al. 2014) and it is increasingly linked to resilience and to broader development goals (Huq et al. 2014). Adaptation measures can increase performance of mitigation projects under climate change and legitimise mitigation measures through the more immediately felt effects of adaptation (Locatelli et al. 2011; Campbell et al. 2014; Locatelli et al. 2015b). Effective climate policy integration in the land sector is expected to gain from (i) internal policy coherence between adaptation and mitigation objectives, (ii) external climate coherence between climate change and development objectives, (iii) policy integration that favours vertical governance structures to foster effective mainstreaming of climate change into sectoral policies, and (iv) horizontal policy integration through overarching governance structures to enable cross-sectoral coordination (Sections 1.4 and 7.4).

1.4   Enabling the response

Climate change and sustainable development are challenges to society that require action at local, national, transboundary and global scales. Different time-perspectives are also important in decision-making, ranging from immediate actions to long-term planning and investment. Acknowledging the systemic link between food production and consumption, and land-resources more broadly is expected to enhance the success of actions (Bazilian et al. 2011; Hussey and Pittock 2012). Because of the complexity of challenges and the diversity of actors involved in addressing these challenges, decision-making would benefit from a portfolio of policy instruments. Decision-making would also be facilitated by overcoming barriers such as inadequate education and funding mechanisms, as well as integrating international decisions into all relevant (sub)national sectoral policies (Section 7.4).

'Nexus thinking' emerged as an alternative to the sector-specific governance of natural resource use to achieve global securities of water (D'Odorico et al. 2018), food and energy (Hoff 2011; Allan et al. 2015), and also to address biodiversity concerns (Fischer et al. 2017). Yet, there is no agreed definition of "nexus" nor a uniform framework to approach the concept, which may be land-focused (Howells et al. 2013), water-focused (Hoff 2011) or food-centred (Ringler and Lawford 2013; Biggs et al. 2015). Significant barriers remain to establish nexus approaches as part of a wider repertoire of responses to global environmental change, including challenges to cross-disciplinary collaboration, complexity, political economy and the incompatibility of current institutional structures (Hayley et al. 2015; Wichelns 2017) (Sections 7.5.6 and 7.6.2).

1.4.1   Governance to enable the response

Governance includes the processes, structures, rules and traditions applied by formal and informal actors including governments, markets, organisations, and their interactions with people. Land governance actors include those affecting policies and markets, and those directly changing land use (Hersperger et al. 2010). The former includes governments and administrative entities, large companies investing in land, non-governmental institutions and international institutions. It also includes UN agencies that are working at the interface between climate change and land management, such as the FAO and the World Food Programme that have *inter alia* worked on advancing knowledge to support food security through the improvement of techniques and strategies for more resilient farm systems. Farmers and foresters directly act on land (actors in proximate causes) (Hersperger et al. 2010) (Chapter 7).

Policy design and formulation has often been strongly sectoral. For example, agricultural policy might be concerned with food security, but have little concern for environmental protection or human health. As food, energy and water security and the conservation of biodiversity rank highly on the Agenda 2030 for Sustainable Development, the promotion of synergies between and across sectoral policies is important (IPBES 2018a). This can also reduce the risks of anthropogenic climate forcing through mitigation, and bring greater collaboration between scientists, policymakers, the private sector and land managers in adapting to climate change (FAO 2015a). Polycentric governance (Section 7.6) has emerged as an appropriate way of handling resource management problems, in which the decision-making centres take account of one another in competitive and cooperative relationships and have recourse to conflict resolution mechanisms (Carlisle and Gruby 2017). Polycentric governance is also multi-scale and allows the interaction between actors at different levels (local, regional, national and global) in managing common pool resources such as forests or aquifers.

Implementation of systemic, nexus approaches has been achieved through socio-ecological systems (SES) frameworks that emerged from studies of how institutions affect human incentives, actions and outcomes (Ostrom and Cox 2010). Recognition of the importance of SES laid the basis for alternative formulations to tackle the sustainable management of land resources focusing specifically on institutional and governance outcomes (Lebel et al. 2006; Bodin 2017). The SES approach also addresses the multiple scales in which the social and ecological dimensions interact (Veldkamp et al. 2011; Myers et al. 2016; Azizi et al. 2017) (Section 6.1).

Adaptation or resilience pathways within the SES frameworks require several attributes, including indigenous and local knowledge (ILK) and trust building for deliberative decision-making and effective collective action, polycentric and multi-layered institutions and responsible authorities that pursue just distributions of benefits to enhance the adaptive capacity of vulnerable groups and communities (Lebel et al. 2006). The nature, source and mode of knowledge generation are critical to ensure that sustainable solutions are community-owned and fully integrated within the local context (Mistry and Berardi 2016; Schneider and Buser 2018). Integrating ILK with scientific information is a prerequisite for such community-owned solutions (Cross-Chapter Box 13 in Chapter 7). ILK is context-specific, transmitted orally or through imitation and demonstration, adaptive to changing environments, and collectivised through a shared social memory (Mistry and Berardi 2016). ILK is also holistic since indigenous people do not seek solutions aimed at adapting to climate change alone, but instead look for solutions to increase their resilience to a wide range of shocks and stresses (Mistry and Berardi 2016). ILK can be deployed in the practice of

climate governance, especially at the local level where actions are informed by the principles of decentralisation and autonomy (Chanza and de Wit 2016). ILK need not be viewed as needing confirmation or disapproval by formal science, but rather it can complement scientific knowledge (Klein et al. 2014).

The capacity to apply individual policy instruments and policy mixes is influenced by governance modes. These modes include hierarchical governance that is centralised and imposes policy through top-down measures, decentralised governance in which public policy is devolved to regional or local government, public-private partnerships that aim for mutual benefits for the public and private sectors and self or private governance that involves decisions beyond the realms of the public sector (IPBES 2018a). These governance modes provide both constraints and opportunities for key actors that impact the effectiveness, efficiency and equity of policy implementation.

1.4.2 Gender agency as a critical factor in climate and land sustainability outcomes

Environmental resource management is not gender neutral. Gender is an essential variable in shaping ecological processes and change, building better prospects for livelihoods and sustainable development (Resurrección 2013) (Cross-Chapter Box 11 in Chapter 7). Entrenched legal and social structures and power relations constitute additional stressors that render women's experience of natural resources disproportionately negative when compared to men. Socio-economic drivers and entrenched gender inequalities affect land-based management (Agarwal 2010). The intersections between climate change, gender and climate adaptation takes place at multiple scales: household, national and international, and adaptive capacities are shaped through power and knowledge.

Germaine to the gender inequities is the unequal access to land-based resources. Women play a significant role in agriculture (Boserup 1989; Darity 1980) and rural economies globally (FAO 2011), but are well below their share of labour in agriculture globally (FAO 2011). In 59% of 161 surveyed countries, customary, traditional and religious practices hinder women's land rights (OECD 2014). Moreover, women typically shoulder disproportionate responsibility for unpaid domestic work including care-giving activities (Beuchelt and Badstue 2013) and the provision of water and firewood (UNEP 2016). Exposure to violence restricts, in large regions, their mobility for capacity-building activities and productive work outside the home (Day et al. 2005; UNEP 2016). Large-scale development projects can erode rights, and lead to over-exploitation of natural resources. Hence, there are cases where reforms related to land-based management, instead of enhancing food security, have tended to increase the vulnerability of both women and men and reduce their ability to adapt to climate change (Pham et al. 2016). Access to, and control over, land and land-based resources is essential in taking concrete action on land-based mitigation, and inadequate access can affect women's rights and participation in land governance and management of productive assets.

Timely information, such as from early warning systems, is critical in managing risks, disasters, and land degradation, and in enabling land-based adaptation. Gender, household resources and social status, are all determinants that influence the adoption of land-based strategies (Theriault et al. 2017). Climate change is not a lone driver in the marginalisation of women; their ability to respond swiftly to its impacts will depend on other socio-economic drivers that may help or hinder action towards adaptive governance. Empowering women and removing gender-based inequities constitutes a mechanism for greater participation in the adoption of sustainable practices of land management (Mello and Schmink 2017). Improving women's access to land (Arora-Jonsson 2014) and other resources (water) and means of economic livelihoods (such as credit and finance) are the prerequisites to enable women to participate in governance and decision-making structures (Namubiru-Mwaura 2014). Still, women are not a homogenous group, and distinctions through elements of ethnicity, class, age and social status, require a more nuanced approach and not a uniform treatment through vulnerability lenses only. An intersectional approach that accounts for various social identifiers under different situations of power (Rao 2017) is considered suitable to integrate gender into climate change research and helps to recognise overlapping and interdependent systems of power (Djoudi et al. 2016; Kaijser and Kronsell 2014; Moosa and Tuana 2014; Thompson-Hall et al. 2016).

1.4.3   Policy instruments

Policy instruments enable governance actors to respond to environmental and societal challenges through policy action. Examples of the range of policy instruments available to public policymakers are discussed below based on four categories of instruments: (i) legal and regulatory instruments, (ii) rights-based instruments and customary norms, (iii) economic and financial instruments, and (iv) social and cultural instruments.

1.4.3.1   Legal and regulatory instruments

Legal and regulatory instruments deal with all aspects of intervention by public policy organisations to correct market failures, expand market reach, or intervene in socially relevant areas with inexistent markets. Such instruments can include legislation to limit the impacts of intensive land management, for example, protecting areas that are susceptible to nitrate pollution or soil erosion. Such instruments can also set standards or threshold values, for example, mandated water quality limits, organic production standards, or geographically defined regional food products. Legal and regulatory instruments may also define liability rules, for example, where environmental standards are not met, as well as establishing long-term agreements for land resource protection with land owners and land users.

1.4.3.2   Economic and financial instruments

Economic (such as taxes, subsidies) and financial (weather-index insurance) instruments deal with the many ways in which public policy organisations can intervene in markets. A number of instruments are available to support climate mitigation actions including public provision, environmental regulations, creating property rights and markets (Sterner 2003). Market-based policies such as carbon

taxes, fuel taxes, cap and trade systems or green payments have been promoted (mostly in industrial economies) to encourage markets and businesses to contribute to climate mitigation, but their effectiveness to date has not always matched expectations (Grolleau et al. 2016) (Section 7.4.4). Market-based instruments in ecosystem services generate both positive (incentives for conservation), but also negative environmental impacts, and also push food prices up or increase price instability (Gómez-Baggethun and Muradian 2015; Farley and Voinov 2016). Footprint labels can be an effective means of shifting consumer behaviour. However, private labels focusing on a single metric (e.g., carbon) may give misleading signals if they target a portion of the life cycle (e.g., transport) (Appleton 2009) or ignore other ecological indicators (water, nutrients, biodiversity) (van Noordwijk and Brussaard 2014).

Effective and durable, market-led responses for climate mitigation depend on business models that internalise the cost of emissions into economic calculations. Such 'business transformation' would itself require integrated policies and strategies that aim to account for emissions in economic activities (Biagini and Miller 2013; Weitzman 2014; Eidelwein et al. 2018). International initiatives such as REDD+ and agricultural commodity roundtables (beef, soybeans, palm oil, sugar) are expanding the scope of private sector participation in climate mitigation (Nepstad et al. 2013), but their impacts have not always been effective (Denis et al. 2014). Payments for environmental services (PES) defined as "voluntary transactions between service users and service providers that are conditional on agreed rules of natural resource management for generating offsite services" (Wunder 2015) have not been widely adopted and have not yet been demonstrated to deliver as effectively as originally hoped (Börner et al. 2017) (Sections 7.4 and 7.5). PES in forestry were shown to be effective only when coupled with appropriate regulatory measures (Alix-Garcia and Wolff 2014). Better designed and expanded PES schemes would encourage integrated soil–water–nutrient management packages (Stavi et al. 2016), services for pollinator protection (Nicole 2015), water use governance under scarcity, and engage both public and private actors (Loch et al. 2013). Effective PES also requires better economic metrics to account for human-directed losses in terrestrial ecosystems and to food potential, and to address market failures or externalities unaccounted for in market valuation of ecosystem services.

Resilient strategies for climate adaptation can rely on the construction of markets through social networks as in the case of livestock systems (Denis et al. 2014) or when market signals encourage adaptation through land markets or supply chain incentives for sustainable land management practices (Anderson et al. 2018). Adequate policy (through regulations, investments in research and development or support to social capabilities) can support private initiatives for effective solutions to restore degraded lands (Reed and Stringer 2015), or mitigate against risk and to avoid shifting risks to the public (Biagini and Miller 2013). Governments, private business, and community groups could also partner to develop sustainable production codes (Chartres and Noble 2015), and in co-managing land-based resources (Baker and Chapin 2018), while public-private partnerships can be effective mechanisms in deploying infrastructure to cope with climatic events (floods) and for climate-indexed

1

insurance (Kunreuther 2015). Private initiatives that depend on trade for climate adaptation and mitigation require reliable trading systems that do not impede climate mitigation objectives (Elbehri et al. 2015; Mathews 2017).

1.4.3.3 Rights-based instruments and customary norms

Rights-based instruments and customary norms deal with the equitable and fair management of land resources for all people (IPBES 2018a). These instruments emphasise the rights in particular of indigenous peoples and local communities, including for example, recognition of the rights embedded in the access to, and use of, common land. Common land includes situations without legal ownership (e.g., hunter-gathering communities in South America or Africa, and bushmeat), where the legal ownership is distinct from usage rights (Mediterranean transhumance grazing systems), or mixed ownership-common grazing systems (e.g., crofting in Scotland). A lack of formal (legal) ownership has often led to the loss of access rights to land, where these rights were also not formally enshrined in law, which especially effects indigenous communities, for example, deforestation in the Amazon basin. Overcoming the constraints associated with common-pool resources (forestry, fisheries, water) are often of economic and institutional nature (Hinkel et al. 2014) and require tackling the absence or poor functioning of institutions and the structural constraints that they engender through access and control levers using policies and markets and other mechanisms (Schut et al. 2016). Other examples of rights-based instruments include the protection of heritage sites, sacred sites and peace parks (IPBES 2018a). Rights-based instruments and customary norms are consistent with the aims of international and national human rights, and the critical issue of liability in the climate change problem.

1.4.3.4 Social and cultural norms

Social and cultural instruments are concerned with the communication of knowledge about conscious consumption patterns and resource-effective ways of life through awareness raising, education and communication of the quality and the provenance of land-based products. Examples of the latter include consumption choices aided by ecolabelling (Section 1.4.3.2) and certification. Cultural indicators (such as social capital, cooperation, gender equity, women's knowledge, socio-ecological mobility) contribute to the resilience of social-ecological systems (Sterling et al. 2017). Indigenous communities (such as the Inuit and Tsleil Waututh Nation in Canada) that continue to maintain traditional foods exhibit greater dietary quality and adequacy (Sheehy et al. 2015). Social and cultural instruments also include approaches to self-regulation and voluntary agreements, especially with respect to environmental management and land resource use. This is becoming especially irrelevant for the increasingly important domain of corporate social responsibility (Halkos and Skouloudis 2016).

1.5 The interdisciplinary nature of the SRCCL

Assessing the land system in view of the multiple challenges that are covered by the SRCCL requires a broad, inter-disciplinary perspective. Methods, core concepts and definitions are used differently in different sectors, geographic regions, and across academic communities addressing land systems, and these concepts and approaches to research are also undergoing a change in their interpretation through time. These differences reflect varying perspectives, in nuances or emphasis, on land as components of the climate and socio-economic systems. Because of its inter-disciplinary nature, the SRCCL can take advantage of these varying perspectives and the diverse methods that accompany them. That way, the report aims to support decision-makers across sectors and world regions in the interpretation of its main findings and support the implementation of solutions.

FAQ 1.1 | What are the approaches to study the interactions between land and climate?

Climate change shapes the way land is able to support supply of food and water for humans. At the same time the land surface interacts with the overlying atmosphere, thus human modifications of land use, land cover and urbanisation affect global, regional and local climate. The complexity of the land–climate interactions requires multiple study approaches embracing different spatial and temporal scales. Observations of land atmospheric exchanges, such as of carbon, water, nutrients and energy can be carried out at leaf level and soil with gas exchange systems, or at canopy scale by means of micrometeorological techniques (i.e. eddy covariance). At regional scale, atmospheric measurements by tall towers, aircraft and satellites can be combined with atmospheric transport models to obtain spatial explicit maps of relevant greenhouse gases fluxes. At longer temporal scale (>10 years) other approaches are more effective, such as tree-ring chronologies, satellite records, population and vegetation dynamics and isotopic studies. Models are important to bring information from measurement together and to extend the knowledge in space and time, including the exploration of scenarios of future climate–land interactions.

FAQ 1.2 | How region-specific are the impacts of different land-based adaptation and mitigation options?

Land-based adaptation and mitigation options are closely related to region-specific features for several reasons. Climate change has a definite regional pattern with some regions already suffering from enhanced climate extremes and others being impacted little, or even benefiting. From this point of view increasing confidence in regional climate change scenarios is becoming a critical step forward towards the implementation of adaptation and mitigation options. Biophysical and socio-economic impacts of climate change depend on the exposures of natural ecosystems and economic sectors, which are again specific to a region, reflecting regional sensitivities due to governance. The overall responses in terms of adaptation or mitigation capacities to avoid and reduce vulnerabilities and enhance adaptive capacity, depend on institutional arrangements, socio-economic conditions, and implementation of policies, many of them having definite regional features. However global drivers, such as agricultural demand, food prices, changing dietary habits associated with rapid social transformations (i.e. urban vs rural, meat-eating vs vegetarian) may interfere with region-specific policies for mitigation and adaptation options and need to be addressed at the global level.

FAQ 1.3 | What is the difference between desertification and land degradation? And where are they happening?

The difference between land degradation and desertification is geographic. Land degradation is a general term used to describe a negative trend in land condition caused by direct or indirect human-induced processes (including anthropogenic climate change). Degradation can be identified by the long-term reduction or loss in biological productivity, ecological integrity or value to humans. Desertification is land degradation when it occurs in arid, semi-arid, and dry sub-humid areas, which are also called drylands. Contrary to some perceptions, desertification is not the same as the expansion of deserts. Desertification is also not limited to irreversible forms of land degradation.

References

Abahussain, A.A., A.S. Abdu, W.K. Al-Zubari, N.A. El-Deen and M. Abdul-Raheem, 2002: Desertification in the Arab region: Analysis of current status and trends. *J. Arid Environ.*, **51**, 521–545, doi:10.1006/jare.2002.0975.

Abarca-Gómez, L. et al., 2017: Worldwide trends in body-mass index, underweight, overweight and obesity from 1975 to 2016: A pooled analysis of 2416 population-based measurement studies in 128.9 million children, adolescents and adults. *Lancet*, **390**, 2627–2642, doi:10.1016/S0140-6736(17)32129-3.

Abatzoglou, J.T., S.Z. Dobrowski, S.A. Parks and K.C. Hegewisch, 2018: TerraClimate, a high-resolution global dataset of monthly climate and climatic water balance from 1958–2015. *Sci. Data*, **5**, 170191, doi:10.1038/sdata.2017.191.

Abberton, M., 2016: Global agricultural intensification during climate change: a role for genomics. *Plant Biotechnol. J.*, **14**, 1095–1098, doi:10.1111/pbi.12467.

Abreu, R.C.R. et al., 2017: The biodiversity cost of carbon sequestration in tropical savanna. *Sci. Adv.*, **3**, e1701284, doi:10.1126/sciadv.1701284.

Abu Hammad, A. and A. Tumeizi, 2012: Land degradation: Socioeconomic and environmental causes and consequences in the eastern Mediterranean. *L. Degrad. Dev.*, **23**, 216–226, doi:10.1002/ldr.1069.

Achard, F. et al., 2014: Determination of tropical deforestation rates and related carbon losses from 1990 to 2010. *Glob. Chang. Biol.*, **20**, 2540–2554, doi:10.1111/GCB.12605.

Affognon, H. et al., 2015: Unpacking postharvest losses in Sub-Saharan Africa: A meta-analysis. *World Dev.*, **66**, 49–68, doi:10.1016/J.WORLDDEV.2014.08.002.

Agarwal, B., 2010: *Gender and Green Governance*. Oxford University Press, Oxford, UK.

Agus, F., H. Husnain and R.D. Yustika, 2015: Improving agricultural resilience to climate change through soil management. *J. Penelit. dan Pengemb. Pertan.*, **34**, 147–158. doi:10.21082/jp3.v34n4.2015. pp. 147–158.

Ahlstrom, A., G. Schurgers, A. Arneth and B. Smith, 2012: Robustness and uncertainty in terrestrial ecosystem carbon response to CMIP5 climate change projections. *Environ. Res. Lett.*, **7**, doi:04400810.1088/1748-9326/7/4/044008.

Ahlstrom, A., B. Smith, J. Lindstrom, M. Rummukainen and C.B. Uvo, 2013: GCM characteristics explain the majority of uncertainty in projected 21st century terrestrial ecosystem carbon balance. *Biogeosciences*, **10**, 1517–1528, doi:10.5194/bg-10-1517-2013.

Ahrends, A. et al., 2017: China's fight to halt tree cover loss. *Proc. R. Soc. B Biol. Sci.*, **284**, 1–10, doi:10.1098/rspb.2016.2559.

Aich, S., S.M.L. Ewe, B. Gu, T.W. Dreschel, 2014: An evaluation of peat loss from an Everglades tree island, Florida, USA. *Mires Peat*, **14**, 1–15.

Albanito, F. et al., 2016: Carbon implications of converting cropland to bioenergy crops or forest for climate mitigation: A global assessment. *GCB Bioenergy*, **8**, 81–95, doi:10.1111/gcbb.12242.

Aleksandrowicz, L., R. Green, E.J.M. Joy, P. Smith, and A. Haines, 2016: The impacts of dietary change on greenhouse gas emissions, land use, water use and health: A systematic review. *PLoS One*, **11**, e0165797, doi:10.1371/journal.pone.0165797.

Alemu, M.M., 2016: Sustainable land management. *J. Environ. Prot.*, **7**, 502–506, doi:10.4236/jep.2016.74045.

Alexander, P., D. Moran, M.D.A. Rounsevell, and P. Smith, 2013: Modelling the perennial energy crop market: The role of spatial diffusion. *J.R. Soc. Interface*, **10**, in press, doi:10.1098/rsif.2013.0656.

Alexander, P. et al., 2015: Drivers for global agricultural land use change: The nexus of diet, population, yield and bioenergy. *Glob. Environ. Chang.*, doi:10.1016/j.gloenvcha.2015.08.011.

Alexander, P. et al., 2016a: Assessing uncertainties in land cover projections. *Glob. Chang. Biol.*, doi:10.1111/gcb.13447.

Alexander, P., C. Brown, A. Arneth, J. Finnigan, and M.D.A. Rounsevell, 2016b: Human appropriation of land for food: The role of diet. *Glob. Environ. Chang. Policy Dimens.*, **41**, 88–98, doi:10.1016/j.gloenvcha.2016.09.005.

Alexander, P. et al., 2017: Losses, inefficiencies and waste in the global food syste. *Agric. Syst.*, **153**, 190–200, doi:10.1016/j.agsy.2017.01.014.

Alexander, P. et al., 2018: Adaptation of global land use and management intensity to changes in climate and atmospheric carbon dioxide. *Glob. Chang. Biol.*, doi:10.1111/gcb.14110.

Alix-Garcia, J. and H. Wolff, 2014: Payment for ecosystem services from forests. *Annu. Rev. Resour. Econ.*, **6**, 361–380, doi:10.1146/annurev-resource-100913-012524.

Alkama, R. and A. Cescatti, 2016: Biophysical climate impacts of recent changes in global forest cover. *Science*, **351**, 600–604, doi:10.1126/science.aac8083.

Allan, T., M. Keulertz, and E. Woertz, 2015: The water–food–energy nexus: An introduction to nexus concepts and some conceptual and operational problems (vol 31, pg 301, 2015). *Int. J. Water Resour. Dev.*, **31**, 800, doi:10.1080/07900627.2015.1060725.

Allen, C.D. et al., 2010: A global overview of drought and heat-induced tree mortality reveals emerging climate change risks for forests. *For. Ecol. Manage.*, **259**, 660–684, doi:10.1016/j.foreco.2009.09.001.

Altieri, M.A., and C.I. Nicholls, 2017: The adaptation and mitigation potential of traditional agriculture in a changing climate. *Clim. Change*, **140**, 33–45, doi:10.1007/s10584-013-0909-y.

Anav, A. et al., 2013: Evaluating the land and ocean components of the global carbon cycle in the CMIP5 Earth system models. *J. Clim.*, **26**, 6801–6843, doi:10.1175/jcli-d-12-00417.1.

Anderegg, W.R.L., J.M. Kane, and L.D.L. Anderegg, 2012: Consequences of widespread tree mortality triggered by drought and temperature stress. *Nat. Clim. Chang.*, **3**, 30.

Anderegg, W.R.L. et al., 2015: Tree mortality from drought, insects, and their interactions in a changing climate. *New Phytol.*, **208**, 674–683, doi:10.1111/nph.13477.

Anderson-Teixeira, K.J., 2018: Prioritizing biodiversity and carbon. *Nat. Clim. Chang.*, **8**, 667–668, doi:10.1038/s41558-018-0242-6.

Anderson, K. and G.P. Peters, 2016: The trouble with negative emissions. *Science*, **354**, 182–183, doi:10.1126/science.aah4567.

Anderson, R.G. et al., 2011: Biophysical considerations in forestry for climate protection. *Front. Ecol. Environ.*, **9**, 174–182, doi:10.1890/090179.

Anderson, S.E. et al., 2018: *The Critical Role of Markets in Climate Change Adaptation*. National Bureau of Economic Research.

Anseeuw, W., L.A. Wily, L. Cotula, and M. Taylor, 2011: *Land Rights and the Rush for Land: Findings of the Global Commercial Pressures on Land Research Project*. International Land Coalition, Rome, Italy, 72 pp.

Appleton, A.E., 2009: Private climate change standards and labelling schemes under the WTO agreement on technical barriers to trade. In: *International trade regulation and the mitigation of climate change: World Trade Forum*. Cambridge University Press, Cambridge, United Kingdom, pp. 131–152.

Ardö, J., T. Tagesson, S. Jamali, and A. Khatir, 2018: MODIS EVI-based net primary production in the Sahel 2000–2014. *Int. J. Appl. Earth Obs. Geoinf.*, **65**, 35–45, doi:10.1016/j.jag.2017.10.002.

Arifanti, V.B., J.B. Kauffman, D. Hadriyanto, D. Murdiyarso, and R. Diana, 2018: Carbon dynamics and land use carbon footprints in mangrove-converted aquaculture: The case of the Mahakam Delta, Indonesia. *For. Ecol. Manage.*, **432**, 17–29, doi:10.1016/j.foreco.2018.08.047.

Arneth, A., C. Brown, and M.D.A. Rounsevell, 2014: Global models of human decision-making for land-based mitigation and adaptation assessment. *Nat. Clim. Chang.*, **4**, 550–557, doi:10.1038/nclimate2250.

Arneth, A. et al., 2017: Historical carbon dioxide emissions caused by land-use changes are possibly larger than assumed. *Nat. Geosci.*, **10**, 79, doi:10.1038/ngeo2882.

Aronson, M.F.J. et al., 2014: A global analysis of the impacts of urbanization on bird and plant diversity reveals key anthropogenic drivers. *Proc. R. Soc. B Biol. Sci.*, **281**, 20133330–20133330, doi:10.1098/rspb.2013.3330.

Arora-Jonsson, S., 2014: Forty years of gender research and environmental policy: Where do we stand? *Womens. Stud. Int. Forum*, **47**, 295–308, doi:10.1016/J.WSIF.2014.02.009.

Ashworth, K., O. Wild, and C.N. Hewitt, 2013: Impacts of biofuel cultivation on mortality and crop yields. *Nat. Clim. Chang.*, **3**, 492–496, doi:10.1038/nclimate1788.

Ayers, J.M. and S. Huq, 2009: The value of linking mitigation and adaptation: A case study of Bangladesh. *Environ. Manage.*, **43**, 753–764, doi:10.1007/s00267-008-9223-2.

Azizi, A., A. Ghorbani, B. Malekmohammadi, and H.R. Jafari, 2017: Government management and overexploitation of groundwater resources: absence of local community initiatives in Ardabil plain-Iran. *J. Environ. Plan. Manag.*, **60**, 1785–1808, doi:10.1080/09640568.2016.1257975.

Bai, Z.G., D.L. Dent, L. Olsson, and M.E. Schaepman, 2008: Proxy global assessment of land degradation. *Soil Use Manag.*, **24**, 223–234, doi:10.1111/j.1475-2743.2008.00169.x.

Bais, A.L.S., C. Lauk, T. Kastner, and K. Erb, 2015: Global patterns and trends of wood harvest and use between 1990 and 2010. *Ecol. Econ.*, **119**, 326–337, doi:10.1016/j.ecolecon.2015.09.011.

Bajželj, B. et al., 2014: Importance of food-demand management for climate mitigation. *Nat. Clim. Chang.*, **4**, 924, doi:10.1038/nclimate2353.

Baker, S. and F.S. Chapin III, 2018: Going beyond "it depends:" the role of context in shaping participation in natural resource management. *Ecol. Soc.*, **23**, doi:10.5751/ES-09868-230120.

Baldos, U.L.C. and T.W. Hertel, 2015: The role of international trade in managing food security risks from climate change. *Food Secur.*, **7**, 275–290, doi:10.1007/s12571-015-0435-z.

Baral, H., M.R. Guariguata, and R.J. Keenan, 2016: A proposed framework for assessing ecosystem goods and services from planted forests. *Ecosyst. Serv.*, **22**, 260–268, doi:10.1016/j.ecoser.2016.10.002.

Bárcena, T.G. et al., 2014: Soil carbon stock change following afforestation in Northern Europe: a meta-analysis. *Glob. Chang. Biol.*, **20**, 2393–2405, doi:10.1111/gcb.12576.

Barlow, J. et al., 2007: Quantifying the biodiversity value of tropical primary, secondary, and plantation forests. *Proc. Natl. Acad. Sci. U.S.A.*, **104**, 18555–18560, doi:10.1073/pnas.0703333104.

Barnett, J., and S. O'Neill, 2010: Maladaptation. *Glob. Environ. Chang.*, **2**, 211–213, doi:10.1016/j.gloenvcha.2009.11.004.

Bastin, J.-F. et al., 2017: The extent of forest in dryland biomes. *Science*, **356**, 635–638, doi:10.1126/science.aam6527.

Bayer, A.D. et al., 2017: Uncertainties in the land-use flux resulting from land-use change reconstructions and gross land transitions. *Earth Syst. Dyn.*, **8**, 91–111, doi:10.5194/esd-8-91-2017.

Bazilian, M. et al., 2011: Considering the energy, water and food nexus: Towards an integrated modelling approach. *Energy Policy*, **39**, 7896–7906, doi:10.1016/J.ENPOL.2011.09.039.

Bebber, D.P. and N. Butt, 2017: Tropical protected areas reduced deforestation carbon emissions by one third from 2000–2012. *Sci. Rep.*, **7**, doi:1400510.1038/s41598-017-14467-w.

Beinroth, F.H., H. Eswaran, P.F. Reich and E. Van Den Berg, 1994: Land related stresses. In: *Stressed Ecosystems and Sustainable Agriculture* [Virmani, S.M., J.C. Katyal, H. Eswaran and I.P. Abrol, (eds.)]. Oxford and IBH, New Delhi, India.

Bentsen, N.S., 2017: Carbon debt and payback time – Lost in the forest? *Renew. Sustain. Energy Rev.*, **73**, 1211–1217, doi:10.1016/j.rser.2017.02.004.

Bernal, B., L.T. Murray, and T.R.H. Pearson, 2018: Global carbon dioxide removal rates from forest landscape restoration activities. *Carbon Balance Manag.*, **13**, doi:10.1186/s13021-018-0110-8.

Berry, P.M. et al., 2015: Cross-sectoral interactions of adaptation and mitigation measures. *Clim. Change*, **128**, 381–393, doi:10.1007/s10584-014-1214-0.

Bertram, C. et al., 2015: Carbon lock-in through capital stock inertia associated with weak near-term climate policies. *Technol. Forecast. Soc. Change*, **90**, 62–72, doi:10.1016/j.techfore.2013.10.001.

Bestelmeyer, B.T. et al., 2015: Desertification, land use and the transformation of global drylands. *Front. Ecol. Environ.*, **13**, 28–36, doi:10.1890/140162.

Beuchelt, T.D. and L. Badstue, 2013: Gender, nutrition – and climate-smart food production: Opportunities and trade-offs. *Food Secur.*, **5**, 709–721, doi:10.1007/s12571-013-0290-8.

Bhojvaid, P.P., M.P. Singh, S.R. Reddy, and J. Ashraf, 2016: Forest transition curve of India and related policies, acts and other major factors. *Trop. Ecol.*, **57**, 133–141.

Biagini, B. and A. Miller, 2013: Engaging the private sector in adaptation to climate change in developing countries: importance, status and challenges. *Clim. Dev.*, **5**, 242–252, doi:10.1080/17565529.2013.821053.

Biggs, E.M. et al., 2015: Sustainable development and the water–energy–food nexus: A perspective on livelihoods. *Environ. Sci. Policy*, **54**, 389–397, doi:10.1016/J.ENVSCI.2015.08.002.

Billen, G., L. Lassaletta, and J. Garnier, 2015: A vast range of opportunities for feeding the world in 2050: Trade-off between diet, N contamination and international trade. *Environ. Res. Lett.*, **10**, doi:10.1088/1748-9326/10/2/025001.

Birdsey, R. and Y. Pan, 2015: Trends in management of the world's forests and impacts on carbon stocks. *For. Ecol. Manage.*, **355**, 83–90, doi:10.1016/j.foreco.2015.04.031.

Bloom, A.A., J.-F. Exbrayat, I.R. van der Velde, L. Feng, and M. Williams, 2016: The decadal state of the terrestrial carbon cycle: Global retrievals of terrestrial carbon allocation, pools, and residence times. *Proc. Natl. Acad. Sci.*, **113**, 1285–1290, doi:10.1073/pnas.1515160113.

Bodin, Ö., 2017: Collaborative environmental governance: Achieving collective action in social-ecological systems. *Science*, **357**, eaan1114, doi:10.1126/science.aan1114.

Bodirsky, B.L. and C. Müller, 2014: Robust relationship between yields and nitrogen inputs indicates three ways to reduce nitrogen pollution. *Environ. Res. Lett.*, doi:10.1088/1748-9326/9/11/111005.

Bonsch, M. et al., 2016: Trade-offs between land and water requirements for large-scale bioenergy production. *GCB Bioenergy*, **8**, 11–24, doi:10.1111/gcbb.12226.

Börner, J. et al., 2017: The effectiveness of payments for environmental services. *World Dev.*, **96**, 359–374, doi:10.1016/J.WORLDDEV.2017.03.020.

Borsato, E., P. Tarolli, and F. Marinello, 2018: Sustainable patterns of main agricultural products combining different footprint parameters. *J. Clean. Prod.*, **179**, 357–367, doi:10.1016/J.JCLEPRO.2018.01.044.

Boserup, E., 1989: Population, the status of women, and rural development. *Popul. Dev. Rev.*, **15**, 45–60, doi:10.2307/2807921.

Boysen, L.R., W. Lucht, and D. Gerten, 2017: Trade-offs for food production, nature conservation and climate limit the terrestrial carbon dioxide removal potential. *Glob. Chang. Biol.*, **23**, 4303–4317, doi:10.1111/gcb.13745.

Bradford, K.J. et al., 2018: The dry chain: Reducing postharvest losses and improving food safety in humid climates. *Trends Food Sci. Technol.*, **71**, 84–93, doi:10.1016/J.TIFS.2017.11.002.

Bradshaw, C.J.A., N.S. Sodhi, K.S.-H. Peh, and B.W. Brook, 2007: Global evidence that deforestation amplifies flood risk and severity in the developing world. *Glob. Chang. Biol.*, **13**, 2379–2395, doi:10.1111/j.1365-2486.2007.01446.x.

Bren d'Amour, C. et al., 2016: Future urban land expansion and implications for global croplands. *Proc. Natl. Acad. Sci.*, **114**, 201606036, doi:10.1073/pnas.1606036114.

Brockerhoff, E.G., H. Jactel, J.A. Parrotta, and S.F.B. Ferraz, 2013: Role of eucalypt and other planted forests in biodiversity conservation and the provision of biodiversity-related ecosystem services. *For. Ecol. Manage.*, **301**, 43–50, doi:10.1016/j.foreco.2012.09.018.

Brown, C. et al., 2014: Analysing uncertainties in climate change impact assessment across sectors and scenarios. *Clim. Change*, **128**, 293–306, doi:10.1007/s10584-014-1133-0.

Brown, C., P. Alexander, S. Holzhauer, and M.D.A. Rounsevell, 2017: Behavioral models of climate change adaptation and mitigation in land-based sectors. *Wiley Interdiscip. Rev. Clim. Chang.*, **8**, e448, doi:10.1002/wcc.448.

Brown, C., P. Alexander, A. Arneth, I. Holman, and M. Rounsevell, 2019: Achievement of Paris climate goals unlikely due to time lags in the land system. *Nat. Clim. Chang.*, **9**, 203–208, doi:10.1038/s41558-019-0400-5.

Brown, C., P. Alexander, and M. Rounsevell, 2018: Empirical evidence for the diffusion of knowledge in land use change. *J. Land Use Sci.*, **13(3)**, 269–283, doi:10.1080/1747423X.2018.1515995.

Brümmer, C. et al., 2017: Gas chromatography vs. quantum cascade laser-based N2O flux measurements using a novel chamber design. *Biogeosciences*, **14**, 1365–1381, doi:10.5194/bg-14-1365-2017.

Bryan, B.A. and N.D. Crossman, 2013: Impact of multiple interacting financial incentives on land use change and the supply of ecosystem services. *Ecosyst. Serv.*, **4**, 60–72, doi:10.1016/j.ecoser.2013.03.004.

Buisson, L., W. Thuiller, N. Casajus, S. Lek, and G. Grenouillet, 2009: Uncertainty in ensemble forecasting of species distribution. *Glob. Chang. Biol.*, **16**, 1145–1157, doi:10.1111/j.1365-2486.2009.02000.x.

Bull, G.Q. et al., 2006: Industrial forest plantation subsidies: Impacts and implications. *For. Policy Econ.*, **9**, 13–31, doi:10.1016/j.forpol.2005.01.004.

Butler, M.P., P.M. Reed, K. Fisher-Vanden, K. Keller, and T. Wagener, 2014: Inaction and climate stabilization uncertainties lead to severe economic risks. *Clim. Change*, **127**, 463–474, doi:10.1007/s10584-014-1283-0.

Caffi, T., S.E. Legler, V. Rossi, and R. Bugiani, 2012: Evaluation of a warning system for early-season control of grapevine powdery mildew. *Plant Dis.*, **96**, 104–110, doi:10.1094/PDIS-06-11-0484.

Cafiero, C., S. Viviani, and M. Nord, 2018: Food security measurement in a global context: The food insecurity experience scale. *Measurement*, **116**, 146–152, doi:10.1016/J.MEASUREMENT.2017.10.065.

Calvin, K. and B. Bond-Lamberty, 2018: Integrated human-earth system modeling – State of the science and future directions. *Environ. Res. Lett.*, **13**, doi:10.1088/1748-9326/aac642.

Campbell, B.M., P. Thornton, R. Zougmoré, P. van Asten, and L. Lipper, 2014: Sustainable intensification: What is its role in climate smart agriculture? *Curr. Opin. Environ. Sustain.*, **8**, 39–43, doi:10.1016/j.cosust.2014.07.002.

Canadell, J.G., and E.D. Schulze, 2014: Global potential of biospheric carbon management for climate mitigation. *Nat. Commun.*, **5**, doi:528210.1038/ncomms6282.

Cao, S., J. Zhang, L. Chen, and T. Zhao, 2016: Ecosystem water imbalances created during ecological restoration by afforestation in China, and lessons for other developing countries. *J. Environ. Manage.*, **183**, 843–849, doi:10.1016/j.jenvman.2016.07.096.

Carlisle, K., and R.L. Gruby, 2017: Polycentric systems of governance: A theoretical model for the commons. *Policy Stud. J.*, doi:10.1111/psj.12212.

Ceballos, G. et al., 2015: Accelerated modern human-induced species losses: Entering the sixth mass extinction. *Sci. Adv.*, doi:10.1126/sciadv.1400253.

Cerretelli, S. et al., 2018: Spatial assessment of land degradation through key ecosystem services: The role of globally available data. *Sci. Total Environ.*, **628–629**, 539–555, doi:10.1016/J.SCITOTENV.2018.02.085.

Challinor, A.J. et al., 2014: A meta-analysis of crop yield under climate change and adaptation. *Nat. Clim. Chang.*, **4**, 287–291, doi:10.1038/nclimate2153.

Challinor, A.J., B. Parkes, and J. Ramirez-Villegas, 2015: Crop yield response to climate change varies with cropping intensity. *Glob. Chang. Biol.*, **21**, 1679–1688, doi:10.1111/gcb.12808.

Challinor, A.J. et al., 2018: Transmission of climate risks across sectors and borders Subject Areas.

Chanza, N. and A. de Wit, 2016: Enhancing climate governance through indigenous knowledge: Case in sustainability science. *S. Afr. J. Sci.*, **112**, 1–7, doi:10.17159/sajs.2016/20140286.

Chappin, E.J.L. and T. van der Lei, 2014: Adaptation of interconnected infrastructures to climate change: A socio-technical systems perspective. *Util. Policy*, **31**, 10–17, doi: 10.1016/j.jup.2014.07.003.

Chartres, C.J. and A. Noble, 2015: Sustainable intensification: Overcoming land and water constraints on food production. *Food Secur.*, **7**, 235–245, doi:10.1007/s12571-015-0425-1.

Chaturvedi, V. et al., 2015: Climate mitigation policy implications for global irrigation water demand. *Mitig. Adapt. Strateg. Glob. Chang.*, **20**, 389–407, doi:10.1007/s11027-013-9497-4.

Chaudhary, A. and T. Kastner, 2016: Land use biodiversity impacts embodied in international food trade. *Glob. Environ. Chang.*, **38**, 195–204, doi:10.1016/J.GLOENVCHA.2016.03.013.

Chazdon, R.L. et al., 2016a: When is a forest a forest? Forest concepts and definitions in the era of forest and landscape restoration. *Ambio*, doi:10.1007/s13280-016-0772-y.

Chazdon, R.L. et al., 2016b: Carbon sequestration potential of second-growth forest regeneration in the Latin American tropics. *Sci. Adv.*, **2**, e1501639–e1501639, doi:10.1126/sciadv.1501639.

Chegere, M.J., 2018: Post-harvest losses reduction by small-scale maize farmers: The role of handling practices. *Food Policy*, **77**, 103–115, doi:10.1016/J.FOODPOL.2018.05.001.

Chen, B. et al., 2018: Global land-water nexus: Agricultural land and freshwater use embodied in worldwide supply chains. *Sci. Total Environ.*, **613–614**, 931–943, doi:10.1016/J.SCITOTENV.2017.09.138.

Chen, C. et al., 2019: China and India lead in greening of the world through land-use management. *Nat. Sustain.*, **2**, 122–129, doi:10.1038/s41893-019-0220-7.

Chen, J. et al., 2014: Global land cover mapping at 30 m resolution: A POK-based operational approach. *ISPRS J. Photogramm. Remote Sens.*, **103**, 7–27, doi:10.1016/j.isprsjprs.2014.09.002.

Chen, Y.-H., M. Babiker, S. Paltsev, and J. Reilly, 2016: *Costs of Climate Mitigation Policies*. MIT Joint Program on the Science and Policy of Global Change, Massachusetts Institute of Technology, Cambridge, MA, USA, 22 pp.

Cherlet, M. et al., (eds.), 2018: *World Atlas of Desertification: Rethinking Land Degradation and Sustainable Land Management* (3rd edition). Publication Office of the European Union, Luxembourg, 247 pp.

Chipanshi, A. et al., 2015: Evaluation of the Integrated Canadian Crop Yield Forecaster (ICCYF) model for in-season prediction of crop yield across the Canadian agricultural landscape. *Agric. For. Meteorol.*, **206**, 137–150, doi:10.1016/J.AGRFORMET.2015.03.007.

Clark, D.A. et al., 2017: Reviews and syntheses: Field data to benchmark the carbon cycle models for tropical forests. *Biogeosciences*, **14**, 4663–4690, doi:10.5194/bg-14-4663-2017.

de Coninck, H., A. Revi, M. Babiker, P. Bertoldi, M. Buckeridge, A. Cartwright, W. Dong, J. Ford, S. Fuss, J.-C. Hourcade, D. Ley, R. Mechler, P. Newman, A. Revokatova, S. Schultz, L. Steg, and T. Sugiyama, 2018: Strengthening and implementing the global response. In: Global Warming of 1.5°C: An IPCC special report on the impacts of global warming of 1.5°C above pre-industrial levels and related global greenhouse gas emission pathways, in the context of strengthening the global response to the threat of climate change [V. Masson-Delmotte, P. Zhai, H.-O. Pörtner, D. Roberts, J. Skea, P.R. Shukla, A. Pirani, W. Moufouma-Okia, C. Péan, R. Pidcock, S. Connors, J.B.R. Matthews, Y. Chen, X. Zhou, M.I. Gomis, E. Lonnoy, T. Maycock, M. Tignor, and T. Waterfield (eds.)]. In press.

Constantin, C., C. Luminița, and A.J. Vasile, 2017: Land grabbing: A review of extent and possible consequences in Romania. *Land use policy*, **62**, 143–150, doi:10.1016/j.landusepol.2017.01.001.

Costanza, R. et al., 2014: Changes in the global value of ecosystem services. *Glob. Environ. Chang.*, **26**, 152–158, doi:10.1016/j.gloenvcha.2014.04.002.

Cotula, L. et al., 2014: Testing claims about large land deals in Africa: Findings from a multi-country study. *J. Dev. Stud.*, **50**, 903–925, doi:10.1080/0022 0388.2014.901501.

Coyle, D.R. et al., 2017: Soil fauna responses to natural disturbances, invasive species and global climate change: Current state of the science and a call to action. *Soil Biol. Biochem.*, **110**, 116–133, doi:10.1016/J. SOILBIO.2017.03.008.

Cremasch, G.D., 2016: Sustainability Metrics for Agri-food Supply Chains. PhD Thesis, Wageningen School of Social Sciences (WASS), Wageningen, Netherlands, doi:10.18174/380247.

Creutzig, F. et al., 2015: Bioenergy and climate change mitigation: an assessment. *Glob. Chang. Biol. Bioenergy*, **7**, 916–944, doi:10.1111/ gcbb.12205.

Creutzig, F. et al., 2016: Beyond technology: Demand-side solutions for climate change mitigation. *Annu. Rev. Environ. Resour.*, **41**, 173–198, doi:10.1146/ annurev-environ-110615-085428.

Creutzig, F. et al., 2018: Towards demand-side solutions for mitigating climate change. *Nat. Clim. Chang.*, **8**, 260–263, doi:10.1038/s41558-018-0121-1.

Crist, E., C. Mora, and R. Engelman, 2017: The interaction of human population, food production, and biodiversity protection. *Science*, **356**, 260–264, doi:10.1126/science.aal2011.

Crouzeilles, R. et al., 2016: A global meta-analysis on the ecological drivers of forest restoration success. *Nat. Commun.*, **7**, 1–8, doi:1166610.1038/ ncomms11666.

Cunningham, S.C. et al., 2015a: Reforestation with native mixed-species plantings in a temperate continental climate effectively sequesters and stabilizes carbon within decades. *Glob. Chang. Biol.*, **21**, 1552–1566, doi:10.1111/gcb.12746.

Cunningham, S.C. et al., 2015b: Balancing the environmental benefits of reforestation in agricultural regions. *Perspect. Plant Ecol. Evol. Syst.*, **17**, 301–317, doi:10.1016/J.PPEES.2015.06.001.

Curtis, P.G., C.M. Slay, N.L. Harris, A. Tyukavina, and M.C. Hansen, 2018: Classifying drivers of global forest loss. *Science*, **361**, 1108–1111, doi:10.1126/science.aau3445.

D'Odorico, P., A. Bhattachan, K.F. Davis, S. Ravi, and C.W. Runyan, 2013: Global desertification: Drivers and feedbacks. *Adv. Water Resour.*, **51**, 326–344, doi:10.1016/j.advwatres.2012.01.013.

D'Odorico, P. et al., 2018: The Global Food-Energy-Water Nexus. *Rev. Geophys.*, **56**, 456–531, doi:10.1029/2017RG000591.

Daliakopoulos, I.N. et al., 2016: The threat of soil salinity: A European scale review. *Sci. Total Environ.*, **573**, 727–739, doi:10.1016/J. SCITOTENV.2016.08.177.

Dalin, C. and I. Rodríguez-Iturbe, 2016: Environmental impacts of food trade via resource use and greenhouse gas emissions. *Environ. Res. Lett.*, **11**, 035012, doi:10.1088/1748-9326/11/3/035012.

Darity, W.A., 1980: The Boserup theory of agricultural growth: A model for anthropological economics. *J. Dev. Econ.*, **7**, 137–157, doi:10.1016/0304-3878(80)90001-2.

Darrah, S.E. et al., 2019: Improvements to the Wetland Extent Trends (WET) index as a tool for monitoring natural and human-made wetlands. *Ecol. Indic.*, **99**, 294–298, doi:10.1016/j.ecolind.2018.12.032.

Davidson, N.C., 2014: How much wetland has the world lost? Long-term and recent trends in global wetland area. *Mar. Freshw. Res.*, **65**, 934–941, doi:10.1071/MF14173.

Davies-Barnard, T., P.J. Valdes, J.S. Singarayer, A.J. Wiltshire and C.D. Jones, 2015: Quantifying the relative importance of land cover change from climate and land-use in the representative concentration pathway. *Global Biogeochem. Cycles*, 842–853, doi:10.1002/2014GB004949.

Davis, K.F., K. Yu, M.C. Rulli, L. Pichdara and P. D'Odorico, 2015: Accelerated deforestation driven by large-scale land acquisitions in Cambodia. *Nat. Geosci.*, **8**, 772–775, doi:10.1038/ngeo2540.

Day, T., K. McKenna, and A. Bowlus, 2005: *The Economic Costs of Violence Against Women: An Evaluation of the Literature. Expert brief compiled in preparation for the Secretary-General's in-depth study on all forms of violence against women.* NY: United Nations. Barzman, New York City, pp. 1–66.

Deininger, K. et al., 2011: *Rising Global Interest in Farmland: Can it Yield Sustainable and Equitable Benefits?* 1st ed. The World Bank, Washington D.C., 164 pp. doi:10.1596/978-0-8213-8591-3.

Dell'Angelo, J., P. D'Odorico, and M.C. Rulli, 2017a: Threats to sustainable development posed by land and water grabbing. *Curr. Opin. Environ. Sustain.*, **26–27**, 120–128, doi:10.1016/j.cosust.2017.07.007.

Dell'Angelo, J., P. D'Odorico, M.C. Rulli, and P. Marchand, 2017b: The tragedy of the grabbed commons: Coercion and dispossession in the global land rush. *World Dev.*, **92**, 1–12, doi:10.1016/J.WORLDDEV.2016.11.005.

Dendy, J., S. Cordell, C.P. Giardina, B. Hwang, E. Polloi and K. Rengulbai, 2015: The role of remnant forest patches for habitat restoration in degraded areas of Palau. *Restor. Ecol.*, **23**, 872–881, doi:10.1111/rec.12268.

Deng, L., Z. Shangguan, and S. Sweeney, 2015: "Grain for Green" driven land use change and carbon sequestration on the Loess Plateau, China. *Sci. Rep.*, **4**, 7039, doi:10.1038/srep07039.

Denis, G. et al., 2014: Global changes, livestock and vulnerability: The social construction of markets as an adaptive strategy. *Geogr. J.*, **182**, 153–164, doi:10.1111/geoj.12115.

Dietrich, J.P. et al., 2018: MAgPIE 4 – A modular open source framework for modeling global land-systems. *Geosci. Model Dev.* **12(4)** 1299–1317, doi:10.5194/gmd-2018-295.

van Dijk, A.I.J.M. et al., 2009: Forest – flood relation still tenuous – comment on 'Global evidence that deforestation amplifies flood risk and severity in the developing world' by C.J.A. Bradshaw, N.S. Sodi, K.S.-H. Peh and B.W. Brook. *Glob. Chang. Biol.*, **15**, 110–115, doi:10.1111/j.1365-2486.2008.01708.x.

Dinerstein, E. et al., 2015: Guiding agricultural expansion to spare tropical forests. *Conserv. Lett.*, **8**, 262–271, doi:10.1111/conl.12149.

Dixon, M.J.R. et al., 2016: Tracking global change in ecosystem area: The Wetland Extent Trends index. *Biol. Conserv.*, doi:10.1016/j. biocon.2015.10.023.

Djoudi, H. et al., 2016: Beyond dichotomies: Gender and intersecting inequalities in climate change studies. *Ambio*, **45**, 248–262, doi:10.1007/ s13280-016-0825-2.

Doelman, J.C. et al., 2018: Exploring SSP land-use dynamics using the IMAGE model: Regional and gridded scenarios of land-use change and land-based climate change mitigation. *Glob. Environ. Chang.*, **48**, 119–135, doi:10.1016/j.gloenvcha.2017.11.014.

Dohong, A., A.A. Aziz, and P. Dargusch, 2017: A review of the drivers of tropical peatland degradation in South-East Asia. *Land use policy*, **69**, 349–360, doi:10.1016/j.landusepol.2017.09.035.

IPCC, 2014: Climate Change 2014: Impacts, Adaptation and Vulnerability. Part A: Global and Sectoral Aspects. Contribution of Working Group II to the Fifth Assessment Report of the Intergovernmental Panel on Climate Change [Field, C.B., V.R. Barros, D.J. Dokken, K.J. Mach, M.D. Mastrandrea, T.E. Bilir, M. Chatterjee, K.L. Ebi, Y.O. Estrada, R.C. Genova, B. Girma, E.S. Kissel, A.N. Levy, S. MacCracken, P.R. Mastrandrea, and L.L. White (eds.)]. Cambridge University Press, Cambridge, United Kingdom and New York, NY, USA, 1132 pp.

Donohue, R.J., M.L. Roderick, T.R. McVicar, and G.D. Farquhar, 2013: Impact of CO_2 fertilization on maximum foliage cover across the globe's warm, arid environments. *Geophys. Res. Lett.*, **40**, 3031–3035, doi:10.1002/grl.50563.

Dooley, K. and S. Kartha, 2018: Land-based negative emissions: risks for climate mitigation and impacts on sustainable development. *Int. Environ. Agreements Polit. Law Econ.*, **18**, 79–98, doi:10.1007/s10784-017-9382-9.

Doss, C., C. Kovarik, A. Peterman, A. Quisumbing, and M. van den Bold, 2015: Gender inequalities in ownership and control of land in Africa: Myth and reality. *Agric. Econ.*, **46**, 403–434, doi:10.1111/agec.12171.

Dunford, R., P.A. Harrison, J. Jäger, M.D.A. Rounsevell, and R. Tinch, 2014: Exploring climate change vulnerability across sectors and scenarios using indicators of impacts and coping capacity. *Clim. Change*, **128**, 339–354, doi:10.1007/s10584-014-1162-8.

Duveiller, G., J. Hooker, and A. Cescatti, 2018: The mark of vegetation change on Earth's surface energy balance. *Nat. Commun.*, **9**, 679, doi:10.1038/s41467-017-02810-8.

Ebert, A.W., 2014: Potential of underutilized traditional vegetables and legume crops to contribute to food and nutritional security, income and more sustainable production systems. *Sustainability*, **6**, 319–335, doi:10.3390/su6010319.

Eeraerts, M., I. Meeus, S. Van Den Berge, and G. Smagghe, 2017: Landscapes with high intensive fruit cultivation reduce wild pollinator services to sweet cherry. *Agric. Ecosyst. Environ.*, **239**, 342–348, doi:10.1016/J.AGEE.2017.01.031.

Egginton, P., F. Beall, and J. Buttle, 2014: Reforestation – Climate change and water resource implications. *For. Chron.*, **90**, 516–524, doi:10.5558/tfc2014-102.

Eidelwein, F., D.C. Collatto, L.H. Rodrigues, D.P. Lacerda, and F.S. Piran, 2018: Internalization of environmental externalities: Development of a method for elaborating the statement of economic and environmental results. *J. Clean. Prod.*, **170**, 1316–1327, doi:10.1016/J.JCLEPRO.2017.09.208.

Eitelberg, D.A., J. van Vliet, J.C. Doelman, E. Stehfest, and P.H. Verburg, 2016: Demand for biodiversity protection and carbon storage as drivers of global land change scenarios. *Glob. Environ. Chang.*, **40**, 101–111, doi:10.1016/j.gloenvcha.2016.06.014.

Elbehri, A., J. Elliott, and T. Wheeler, 2015: Climate change, food security and trade: An overview of global assessments and policy insights. In: *Climate Change and Food Systems: Global assessments and implications for food security and trade* [Elbehri, A. (ed.)]. FAO, Rome, Italy, pp. 1–27.

Elbehri, A. et al., 2017: FAO-IPCC Expert Meeting on Climate Change, Land Use and Food Security: Final Meeting Report; January 23–25, 2017, FAO and IPCC, Food and Agriculture Organization of the United Nations, Rome, Italy, pp. 1–27.

Ellis, E.C. and N. Ramankutty, 2008: Putting people in the map: Anthropogenic biomes of the world. *Front. Ecol. Environ.*, **6**, 439–447, doi:10.1890/070062.

Ellis, E.C., K.K. Goldewijk, S. Siebert, D. Lightman, and N. Ramankutty, 2010: Anthropogenic transformation of the biomes, 1700 to 2000. *Glob. Ecol. Biogeogr.*, doi:10.1111/j.1466-8238.2010.00540.x.

Ellison, D. et al., 2017: Trees, forests and water: Cool insights for a hot world. *Glob. Environ. Chang.*, **43**, 51–61, doi:10.1016/j.gloenvcha.2017.01.002.

Engstrom, K. et al., 2016: Assessing uncertainties in global cropland futures using a conditional probabilistic modelling framework. *Earth Syst. Dyn.*, **7**, 893–915, doi:10.5194/esd-7-893-2016.

Erb, K.-H. et al., 2007: A comprehensive global 5 min resolution land-use data set for the year 2000 consistent with national census data. *J. Land Use Sci.*, **2**, 191–224, doi:10.1080/17474230701622981.

Erb, K.-H. et al., 2016a: Land management: Data availability and process understanding for global change studies. *Glob. Chang. Biol.*, **23**, 512–533, doi:10.1111/gcb.13443.

Erb, K.-H. et al., 2016b: Exploring the biophysical option space for feeding the world without deforestation. *Nat. Commun.*, **7**.

Erb, K.-H., H. Haberl, and C. Plutzar, 2012: Dependency of global primary bioenergy crop potentials in 2050 on food systems, yields, biodiversity conservation and political stability. *Energy Policy*, **47**, 260–269, doi:10.1016/j.enpol.2012.04.066.

Erb, K.-H. et al., 2016c: Biomass turnover time in terrestrial ecosystems halved by land use. *Nat. Geosci.*, **9**, 674–678, doi:10.1038/ngeo2782.

Erb, K.-H. et al., 2017: Unexpectedly large impact of forest management and grazing on global vegetation biomass. *Nature*, **553**, 73–76, doi:10.1038/nature25138.

Evans, C.D. et al., 2019: Rates and spatial variability of peat subsidence in Acacia plantation and forest landscapes in Sumatra, Indonesia. *Geoderma*, **338**, 410–421, doi:10.1016/j.geoderma.2018.12.028.

Falco, S. Di, F. Adinolfi, M. Bozzola, and F. Capitanio, 2014: Crop Insurance as a Strategy for Adapting to Climate Change. *J. Agric. Econ.*, **65**, 485–504, doi:10.1111/1477-9552.12053.

FAO, 1963: World Forest Inventory 1963. Food and Agriculture Organization of the United Nations, Rome, 113 pp.

FAO, 2011: The State of Food and Agriculture: Women in agriculture – Closing the gender gap for development. Food and Agriculture Organization of the United Nations, Rome, Italy.

FAO, 2015a: Global Forest Resources Assessments 2015. Food and Agriculture Organization of the United Nations, Rome.

FAO, 2015b: Learning Tool on Nationally Appropriate Mitigation Actions (NAMAs) in the agriculture, forestry and other land use (AFOLU) sector. Food and Agriculture Organization of the United Nations, Rome, Italy, 162 pp.

FAO, 2016: State of the World's Forests 2016. Forests and agriculture: Land-use challenges and opportunities. Food and Agriculture Organization of the United Nations, Rome, Italy.

FAO, 2017: The Future of Food and Agriculture: Trends and Challenges. Food and Agriculture Organization of the United Nations, Rome, Italy.

FAO, 2018a: The State of the World's Forests 2018 – Forest Pathways to Sustainable Development. Food and Agriculture Organization of the United Nations, Rome, Italy, 139 pp.

FAO, 2018b: The Future of Food and Agriculture: Alternative Pathways to 2050. Food and Agricultural Organization of the United Nations, Rome, Italy, 228 pp.

FAO, IFAD, UNICEF, WFP and WHO, 2018: The State of Food Security and Nutrition in the World 2018. Building climate resilience for food security and nutrition. Food and Agriculture Organization of the United Nations, Rome, Italy.

FAO and ITPS, 2015: Status of the World's Soil Resources (SWSR) – Main Report. Food and Agriculture Organization of the United Nations, Rome, Italy.

FAOSTAT, 2018: Statistical Databases. http://faostat.fao.org.

Farley, J. and A. Voinov, 2016: Economics, socio-ecological resilience and ecosystem services. *J. Environ. Manage.*, **183**, 389–398, doi:10.1016/J.JENVMAN.2016.07.065.

Fasullo, J.T., B.L. Otto-Bliesner and S. Stevenson, 2018: ENSO's changing influence on temperature, precipitation, and wildfire in a warming climate. *Geophys. Res. Lett.*, **0**, doi:10.1029/2018GL079022.

Favero, A. and E. Massetti, 2014: Trade of woody biomass for electricity generation under climate mitigation policy. *Resour. Energy Econ.*, **36**, 166–190, doi:10.1016/J.RESENEECO.2013.11.00.

Favero, A. and R. Mendelsohn, 2014: Using markets for woody biomass energy to sequester carbon in forests. *J. Assoc. Environ. Resour. Econ.*, **1**, 75–95, doi:10.1086/676033.

Favero, A., R. Mendelsohn, and B. Sohngen, 2017: Using forests for climate mitigation: Sequester carbon or produce woody biomass? *Clim. Change*, **144**, 195–206, doi:10.1007/s10584-017-2034-9.

Feng, X. et al., 2016: Revegetation in China's Loess Plateau is approaching sustainable water resource limits. *Nat. Clim. Chang.*, **6**, 1019–1022, doi:10.1038/nclimate3092.

Fernandez-Martinez, M. et al., 2014: Nutrient availability as the key regulator of global forest carbon balance. *Nat. Clim. Chang.*, **4**, 471–476, doi:10.1038/nclimate2177.

Ferreira, C.S.S., R.P.D. Walsh and A.J.D. Ferreira, 2018: Degradation in urban areas. *Curr. Opin. Environ. Sci. Heal.*, **5**, 19–25, doi:10.1016/j.coesh.2018.04.001.

IPCC, 2014: Summary for policymakers. In: *Climate Change 2014: Impacts, Adaptation, and Vulnerability. Contribution of Working Group II to the Fifth Assessment Report of the Intergovernmental Panel on Climate Change* [Field, C.B., V.R. Barros, D.J. Dokken, K.J. Mach, M.D. Mastrandrea, T.E. Bilir, M. Chatterjee, K.L. Ebi, Y.O. Estrada, R.C. Genova, B. Girma, E.S. Kissel, A.N. Levy, S. MacCracken, P.R. Mastrandrea and L.L.White (eds.)]. Cambridge University Press, Cambridge, United Kingdom and New York, NY, USA, 1–32 pp.

Field, C.B., V.R. Barros, K.J. Mach, M.D. Mastrandrea, M. van Aalst, W.N. Adger, D.J. Arent, J. Barnett, R. Betts, T.E. Bilir, J. Birkmann, J. Carmin, D.D. Chadee, A.J. Challinor, M. Chatterjee, W. Cramer, D.J. Davidson, Y.O. Estrada, J.-P. Gattuso, Y. Hijioka, O. Hoegh-Guldberg, H.Q. Huang, G.E. Insarov, R.N. Jones, R.S. Kovats, P. Romero-Lankao, J.N. Larsen, I.J. Losada, J.A. Marengo, R.F. McLean, L.O. Mearns, R. Mechler, J.F. Morton, I. Niang, T. Oki, J.M. Olwoch, M. Opondo, E.S. Poloczanska, H.-O. Pörtner, M.H. Redsteer, A. Reisinger, A. Revi, D.N. Schmidt, M.R. Shaw, W. Solecki, D.A. Stone, J.M.R. Stone, K.M. Strzepek, A.G. Suarez, P. Tschakert, R. Valentini, S. Vicuña, A. Villamizar, K.E. Vincent, R. Warren, L.L. White, T.J. Wilbanks, P.P. Wong and G.W. Yohe., 2014b: Technical Summary. In: Climate Change 2014: Impacts, Adaptation and Vulnerability. Part A: Global and Sectoral Aspects. Contribution of Working Group II to the Fifth Assessment Report of the Intergovernmental Panel on Climate Change [Field, C.B., V.R. Barros, D.J. Dokken, K.J. Mach, M.D. Mastrandrea, T.E. Bilir, M. Chatterjee, K.L. Ebi, Y.O. Estrada, R.C. Genova, B. Girma, E.S. Kissel, A.N. Levy, S. MacCracken, P.R. Mastrandrea and L.L.White (eds.)]. Cambridge University Press, Cambridge, United Kingdom and New York, NY, USA, 35–94 pp.

Filoso, S., M.O. Bezerra, K.C.B. Weiss and M.A. Palmer, 2017: Impacts of forest restoration on water yield: A systematic review. *PLoS One*, **12**, e0183210, doi:10.1371/journal.pone.0183210.

Fischer, J. et al., 2017: Reframing the food–biodiversity challenge. *Trends Ecol. Evol.*, **32**, 335–345, doi:10.1016/j.tree.2017.02.009.

Fischer, M. et al., 2018: *IPBES: Summary for Policymakers of the Regional Assessment Report on Biodiversity and Ecosystem Services for Europe and Central Asia of the Intergovernmental Science-Policy Platform on Biodiversity and Ecosystem Services*. Bonn, Germany, 48 pp.

Fish, R., A. Church, and M. Winter, 2016: Conceptualising cultural ecosystem services: A novel framework for research and critical engagement. *Ecosyst. Serv.*, **21**, 208–217, doi:10.1016/j.ecoser.2016.09.002.

Foley, J.A. et al., 2011: Solutions for a cultivated planet. *Nature*, **478**, 337–342, doi:10.1038/nature10452.

Font Vivanco, D., R. Kemp, and E. van der Voet, 2016: How to deal with the rebound effect? A policy-oriented approach. *Energy Policy*, **94**, 114–125, doi:10.1016/J.ENPOL.2016.03.054.

Forsell, N., O. Turkovska, M. Gusti, M. Obersteiner, M. Elzen and P. Havlík, 2016: *Assessing the INDCs' land use, land use change and forest emission projections. Carbon Balance Manage.*, **11**, 1–17, doi:10.1186/s13021-016-0068-3.

Franchini, M. and P.M. Mannucci, 2015: Impact on human health of climate changes. *Eur. J. Intern. Med.*, **26**, 1–5, doi:10.1016/j.ejim.2014.12.008.

Franco, A. and N. Giannini, 2005: Perspectives for the use of biomass as fuel in combined cycle power plants. *Int. J. Therm. Sci.*, **44**, 163–177, doi:10.1016/J.IJTHERMALSCI.2004.07.005.

Friedrich, T., R. Derpsch, and A. Kassam, 2012: Overview of the global spread of conservation agriculture. *F. Actions Sci. Reports*, 1–7, doi:10.1201/9781315365800-4.

Friend, A.D. et al., 2014: Carbon residence time dominates uncertainty in terrestrial vegetation responses to future climate and atmospheric CO_2. *Proc. Natl. Acad. Sci.*, **111**, 3280–3285, doi:10.1073/pnas.1222477110.

Friis, C. and J.Ø. Nielsen, 2017: Land-use change in a telecoupled world: The relevance and applicability of the telecoupling framework in the case of banana plantation expansion in Laos. *Ecol. Soc.*, doi:10.5751/ES-09480-220430.

Friis, C. et al., 2016: From teleconnection to telecoupling: Taking stock of an emerging framework in land system science. *J. Land Use Sci.*, doi:10.1080/1747423X.2015.1096423.

Fuchs, R., M. Herold, P.H. Verburg, J.G.P.W. Clevers, and J. Eberle, 2015: Gross changes in reconstructions of historic land cover/use for Europe between 1900 and 2010. *Glob. Chang. Biol.*, **21**, 299–313, doi:10.1111/gcb.12714.

Fuchs, R., R. Prestele, and P.H. Verburg, 2017: A global assessment of gross and net land change dynamics for current conditions and future scenarios. *Earth Syst. Dyn. Discuss.*, 1–29, doi:10.5194/esd-2017-121.

Fuss, S. et al., 2018: Negative emissions—Part 2: Costs, potentials and side effects. *Environ. Res. Lett.*, **13**, 063002, doi:10.1088/1748-9326/aabf9f.

Gaba, S. et al., 2015: Multiple cropping systems as drivers for providing multiple ecosystem services: From concepts to design. *Agron. Sustain. Dev.*, **35**, 607–623, doi:10.1007/s13593-014-0272-z.

Gardiner, S.M., 2006: A Core Precautionary Principle. *J. Polit. Philos.*, **14**, 33–60, doi:10.1111/j.1467-9760.2006.00237.x.

Gerber, J.F., 2011: Conflicts over industrial tree plantations in the South: Who, how and why? *Glob. Environ. Chang.*, **21**, 165–176, doi:10.1016/j.gloenvcha.2010.09.005.

Gibbs, H.K. and J.M. Salmon, 2015: Mapping the world's degraded lands. *Appl. Geogr.*, **57**, 12–21, doi:10.1016/j.apgeog.2014.11.024.

Gilbert, M. et al., 2018: Global distribution data for cattle, buffaloes, horses, sheep, goats, pigs, chickens and ducks in 2010. *Scientific Data*, **5**, doi:10.1038/sdata.2018.227.

Gilbert-Norton, L., R. Wilson, J.R. Stevens and K.H. Beard, 2010: A meta-analytic review of corridor effectiveness. *Conserv. Biol.*, **24**, 660–668, doi:10.1111/j.1523-1739.2010.01450.x.

Gillett, N.P. et al., 2016: The detection and attribution model intercomparison project (DAMIP v1.0) contribution to CMIP6. *Geosci. Model Dev.*, **9**, 3685–3697, doi:10.5194/gmd-9-3685-2016.

Gillingham, K. and J.H. Stock, 2018: The cost of reducing greenhouse gas emissions. *J. Econ. Perspect.*, **32**, 53–72, doi:10.1257/jep.32.4.53.

Gilmont, M., 2015: Water resource decoupling in the MENA through food trade as a mechanism for circumventing national water scarcity. *Food Secur.*, **7**, 1113–1131, doi:10.1007/s12571-015-0513-2.

Gleckler P.J., et al., 2016: A more powerful reality test for climate models. *Eos (Washington. DC).*, **97**, doi:10.1029/2016EO051663.

Goldewijk, K.K., A. Beusen, J. Doelman, and E. Stehfest, 2017: Anthropogenic land use estimates for the Holocene – HYDE 3.2. *Earth Syst. Sci. Data*, **9**, 927–953, doi:10.5194/essd-9-927-2017.

Goldstein, A., W.R. Turner, J. Gladstone, and D.G. Hole, 2019: The private sector's climate change risk and adaptation blind spots. *Nat. Clim. Chang.*, **9**, 18–25, doi:10.1038/s41558-018-0340-5.

Golub, A.A., et al., 2013: Global climate policy impacts on livestock, land use, livelihoods and food security. *Proc. Natl. Acad. Sci. U.S.A.*, **110**, 20894–20899, doi:10.1073/pnas.1108772109.

Gómez-Baggethun, E. and R. Muradian, 2015: In markets we trust? Setting the boundaries of market-based instruments in ecosystem services governance. *Ecol. Econ.*, **117**, 217–224, doi:10.1016/J.ECOLECON.2015.03.016.

Gonzalez, P., R.P. Neilson, J.M. Lenihan, and R.J. Drapek, 2010: Global patterns in the vulnerability of ecosystems to vegetation shifts due to climate change. *Glob. Ecol. Biogeogr.*, **19**, 755–768, doi:10.1111/j.1466-8238.2010.00558.x.

Gossner, M.M. et al., 2016: Land-use intensification causes multitrophic homogenization of grassland communities. *Nature*, **540**, 266–269, doi:10.1038/nature20575.

Graham, C.T., et al., 2017: Implications of afforestation for bird communities: the importance of preceding land-use type. *Biodivers. Conserv.*, **26**, 3051–3071, doi:10.1007/s10531-015-0987-4.

Grassi, G., et al., 2017: The key role of forests in meeting climate targets requires science for credible mitigation. *Nat. Clim. Chang.*, **7**, 220–226, doi:10.1038/nclimate3227.

Griffith, D.M. et al., 2017: Comment on "The extent of forest in dryland biomes". *Science*, **358**, eaao1309, doi:10.1126/science.aao1309.

Griscom, B.W. et al., 2017: Natural climate solutions. *Proc. Natl. Acad. Sci. USA*, **114**, 11645–11650, doi:10.1073/pnas.1710465114.

Grolleau, G., L. Ibanez, N. Mzoughi, and M. Teisl, 2016: Helping eco-labels to fulfil their promises. *Clim. Policy*, **16**, 792–802, doi:10.1080/14693062.2015.1033675.

de Groot, W.J., B.M. Wotton, and M.D. Flannigan, 2015: Chapter 11 – Wildland Fire Danger Rating and Early Warning Systems. In:*Wildfire Hazards, Risks and Disasters* [Shroder, J.F. and D. Paton (eds.)], Elsevier, Oxford, pp. 207–228, doi:10.1016/B978-0-12-410434-1.00011-7.

Güneralp, B., K.C. Seto, B. Gueneralp, and K.C. Seto, 2013: Futures of global urban expansion: Uncertainties and implications for biodiversity conservation. *Environ. Res. Lett.*, **8**, doi:10.1088/1748-9326/8/1/014025.

Gurwick, N.P., L.A. Moore, C. Kelly, and P. Elias, 2013: A systematic review of biochar research, with a focus on its stability in situ and its promise as a climate mitigation strategy. *PLoS One*, **8**, doi:10.1371/journal.pone.0075932.

Gustavsson, J., C. Cederberg, U. Sonesson, R. van Otterdijk and A. Meybeck, 2011: *Global Food Losses and Food Waste – Extent, Causes and Prevention*. Study conducted for the International Congress, Swedish Institute for Food and Biotechnology (SIK), Gothenburg, Sweden.

Haasnoot, M., 2013: Dynamic adaptive policy pathways: A method for crafting robust decisions for a deeply uncertain world. *Glob. Environ. Chang.*, **23**, 485–498, doi:10.1016/j.gloenvcha.2012.12.006.

Haberl, H., 2015: *Competition for land: A sociometabolic perspective*. Elsevier, **119**, 424–431, doi:10.1016/j.ecolecon.2014.10.002.

Haberl, H., K.-H. Erb and F. Krausmann, 2014: Human appropriation of net primary production: Patterns, trends and planetary boundaries. *Annu. Rev. Environ. Resour.*, **39**, 363–391, doi:10.1146/annurev-environ-121912-094620.

Haddad, N.M. et al., 2015: Habitat fragmentation and its lasting impact on Earth's ecosystems. *Sci. Adv.*, **1**, doi:10.1126/sciadv.1500052.

Halkos, G. and A. Skouloudis, 2016: *Cultural dimensions and corporate social responsibility: A cross-country analysis*. MPRA Paper 6922, University Library of Munich, Germany.

Hallegatte, S. and J. Rentschler, 2015: Risk management for development-assessing obstacles and prioritizing action. *Risk Anal.*, **35**, 193–210, doi:10.1111/risa.12269.

Hallegatte, S. and K.J. Mach, 2016: Make climate-change assessments more relevant. *Nature*, **534**, 613–615, doi:10.1038/534613a.

Hallström, E., A. Carlsson-Kanyama and P. Börjesson, 2015: Environmental impact of dietary change: A systematic review. *J. Clean. Prod.*, **91**, 1–11, doi:10.1016/J.JCLEPRO.2014.12.008.

Hansen, M.C. et al., 2013: High-resolution global maps of 21st-century forest cover change. *Science*, **342**, 850–853, doi:10.1126/science.1244693.

Harper, A.B. et al., 2018: Land-use emissions play a critical role in land-based mitigation for Paris climate targets. *Nat. Commun.*, **9**, doi:10.1038/s41467-018-05340-z.

Harrison, P.A., R. Dunford, C. Savin, M.D.A. Rounsevell, I.P. Holman, A.S. Kebede and B. Stuch, 2014: Cross-sectoral impacts of climate change and socio-economic change for multiple, European land – and water-based sectors. *Clim. Change*, **128**, 279–292, doi:10.1007/s10584-014-1239-4.

Harrison, P.A., R.W. Dunford, I.P. Holman, and M.D.A. Rounsevell, 2016: Climate change impact modelling needs to include cross-sectoral interactions. *Nat. Clim. Chang.*, **6**, 885–890, doi:10.1038/nclimate3039.

Harrison, S.P. et al., 2013: Volatile isoprenoid emissions from plastid to planet. *New Phytol.*, **197**, 49–57, doi:10.1111/nph.12021.

Harvey, M. and S. Pilgrim, 2011: The new competition for land: Food, energy and climate change. *Food Policy*, **36**, S40–S51, doi:10.1016/J.FOODPOL.2010.11.009.

Hasegawa, T. et al., 2015: Consequence of climate mitigation on the risk of hunger. *Environ. Sci. Technol.*, **49**, 7245–7253, doi:10.1021/es5051748.

Hasegawa, T. et al., 2018: Risk of increased food insecurity under stringent global climate change mitigation policy. *Nat. Clim. Chang.*, **8**, 699–703, doi:10.1038/s41558-018-0230-x.

Hayley, L., C. Declan, B. Michael, and R. Judith, 2015: Tracing the water–energy–food nexus: Description, theory and practice. *Geogr. Compass*, **9**, 445–460, doi:10.1111/gec3.12222.

He, T. et al., 2018: Evaluating land surface albedo estimation from Landsat MSS, TM, ETM + and OLI data based on the unified direct estimation approach. *Remote Sens. Environ.*, **204**, 181–196, doi:10.1016/j.rse.2017.10.031.

Heck, V., D. Gerten, W. Lucht and A. Popp, 2018: Biomass-based negative emissions difficult to reconcile with planetary boundaries. *Nat. Clim. Chang.*, **8**, 151–155, doi:10.1038/s41558-017-0064-y.

Heilmayr, R., C. Echeverría, R. Fuentes, and E.F. Lambin, 2016: A plantation-dominated forest transition in Chile. *Appl. Geogr.*, **75**, 71–82, doi:10.1016/j.apgeog.2016.07.014.

Hennessey, R., J. Pittman, A. Morand, and A. Douglas, 2017: Co-benefits of integrating climate change adaptation and mitigation in the Canadian energy sector. *Energy Policy*, **111**, 214–221, doi:10.1016/J.ENPOL.2017.09.025.

Henry, R.C. et al., 2018: Food supply and bioenergy production within the global cropland planetary boundary. *PLoS One*, **13**, e0194695–e0194695, doi:10.1371/journal.pone.0194695.

Hernández-Morcillo, M., T. Plieninger and C. Bieling, 2013: An empirical review of cultural ecosystem service indicators. *Ecol. Indic.*, **29**, 434–444, doi:10.1016/j.ecolind.2013.01.013.

Hersperger, A.M., M.-P. Gennaio, P.H. Verburg and M. Bürgi, 2010: Linking land change with driving forces and actors: Four conceptual models. *Ecol. Soc.*, **15**, doi:10.5751/ES-03562-150401.

Hinkel, J., P.W.G. Bots and M. Schlüter, 2014: Enhancing the Ostrom social-ecological system framework through formalization. *Ecol. Soc.*, **19(3)**, doi:10.5751/ES-06475-190351.

HLPE, 2017: *Nutrition and Food Systems. A report by the High Level Panel of Experts on Food Security and Nutrition of the Committee on World Food Security*. High Level Panel of Experts on Food Security and Nutrition of the Committee on World Food Security, Rome, Italy, 151 pp.

Hobbs, P.R., K. Sayre, and R. Gupta, 2008: The role of conservation agriculture in sustainable agriculture. *Philos. Trans. R. Soc. B Biol. Sci.*, **363**, 543–555, doi:10.1098/rstb.2007.2169.

Hoegh-Guldberg, O., D. Jacob, M. Taylor, M. Bindi, S. Brown, I. Camilloni, A. Diedhiou, R. Djalante, K.L. Ebi, F. Engelbrecht, J. Guiot, Y. Hijioka, S. Mehrotra, A. Payne, S.I. Seneviratne, A. Thomas, R. Warren, and G. Zhou, 2018: Impacts of 1.5°C Global Warming on Natural and Human Systems. In: Global warming of 1.5°C. An IPCC Special Report on the impacts of global warming of 1.5°C above pre-industrial levels and related global greenhouse gas emission pathways, in the context of strengthening the global response to the threat of climate change [Masson-Delmotte, V.P. Zhai, H.-O. Pörtner, D. Roberts, J. Skea, P.R. Shukla, A. Pirani, W. Moufouma-Okia, C. Péan, R. Pidcock, S. Connors, J.B.R. Matthews, Y. Chen, X. Zhou, M.I. Gomis, E. Lonnoy, T. Maycock, M. Tignor and T. Waterfield (eds.)]. In press.

Hoekstra, A.Y. and T.O. Wiedmann, 2014: Humanity's unsustainable environmental footprint. *Science*, **344**, 1114–1117, doi:10.1126/science.1248365.

Hoesly, R.M. et al., 2018: Historical (1750–2014) anthropogenic emissions of reactive gases and aerosols from the Community Emissions Data System (CEDS). *Geosci. Model Dev.*, **11**, 369–408, doi:10.5194/gmd-11-369-2018.

Hof, C., et al., 2018: Bioenergy cropland expansion may offset positive effects of climate change mitigation for global vertebrate diversity. *Proc. Natl. Acad. Sci.*, **115**, 13294–13299, doi:10.1073/pnas.1807745115.

Hoff, H., 2011: Bonn 2011 Conference: The Water, Energy and Food Security Nexus – Solutions for the Green Economy. Stockholm, 1–52 pp.

Houghton, R.A., 2013: The emissions of carbon from deforestation and degradation in the tropics: Past trends and future potential. *Carbon Manag.*, **4**, 539–546, doi:10.4155/cmt.13.41.

Houghton, R.A. and A.A. Nassikas, 2017: Global and regional fluxes of carbon from land use and land cover change 1850–2015. *Global Biogeochem. Cycles*, **31**, 456–472, doi:10.1002/2016GB005546.

Houghton, R.A., B. Byers, and A.A. Nassikas, 2015: A role for tropical forests in stabilizing atmospheric CO_2. *Nat. Clim. Chang.*, **5**, 1022–1023, doi:10.1038/nclimate2869.

Howells, M. et al., 2013: Integrated analysis of climate change, land-use, energy and water strategies. *Nat. Clim. Chang.*, **3**, 621–626. doi:10.1038/nclimate1789.

Hua, F. et al., 2016: Opportunities for biodiversity gains under the world's largest reforestation programme. *Nat. Commun.*, **7**, 1–11, doi:10.1038/ncomms12717.

Hua, F. et al., 2018: Tree plantations displacing native forests: The nature and drivers of apparent forest recovery on former croplands in Southwestern China from 2000 to 2015. *Biol. Conserv.*, **222**, 113–124, doi:10.1016/j.biocon.2018.03.034.

Huang, S.K., L. Kuo, and K.-L. Chou, 2016: The applicability of marginal abatement cost approach: A comprehensive review. *J. Clean. Prod.*, **127**, 59–71, doi:10.1016/J.JCLEPRO.2016.04.013.

Huang, Y. et al., 2018: Impacts of species richness on productivity in a large-scale subtropical forest experiment. *Science*, **362**, 80–83, doi:10.1126/science.aat6405.

Hull, V., M.-N. Tuanniu and J. Liu, 2015: Synthesis of human-nature feedbacks. *Ecol. Soc.*, **20**(3), doi.org/10.5751/ES-07404-200317.

Humpenoder, F. et al., 2014: Investigating afforestation and bioenergy CCS as climate change mitigation strategies. *Environ. Res. Lett.*, **9**, 064029, doi:10.1088/1748-9326/9/6/064029.

Humpenoeder, F. et al., 2018: Large-scale bioenergy production: How to resolve sustainability trade-offs? *Environ. Res. Lett.*, **13**, doi:10.1088/1748-9326/aa9e3b.

Noble, I.R., S. Huq, Y.A. Anokhin, J. Carmin, D. Goudou, F.P. Lansigan, B. Osman-Elasha and A. Villamizar, 2014: Adaptation needs and options. In: Climate Change 2014: Impacts, Adaptation and Vulnerability. Part A: Global and Sectoral Aspects. Contribution of Working Group II to the Fifth Assessment Report of the Intergovernmental Panel on Climate Change [Field, C.B., V.R. Barros, D.J. Dokken, K.J. Mach, M.D. Mastrandrea, T.E. Bilir, M. Chatterjee, K.L. Ebi, Y.O. Estrada, R.C. Genova, B. Girma, E.S. Kissel, A.N. Levy, S. MacCracken, P.R. Mastrandrea and L.L. White (eds.)]. Cambridge University Press, Cambridge, United Kingdom and New York, NY, USA, pp. 833–868.

Hussein, Z., T. Hertel and A. Golub, 2013: Climate change mitigation policies and poverty in developing countries. *Environ. Res. Lett.*, **8**, 035009, [PAGE CITATION?] doi:10.1088/1748-9326/8/3/035009.

Hussey, K. and J. Pittock, 2012: The energy–water nexus: Managing the links between energy and water for a sustainable future. *Ecol. Soc.*, **17**, doi:10.5751/ES-04641-170131.

IFASTAT, 2018: Statistical Databases. www.ifastat.org/.

Iizumi, T. and N. Ramankutty, 2015: How do weather and climate influence cropping area and intensity? *Glob. Food Sec.*, **4**, 46–50, doi:10.1016/j.gfs.2014.11.003.

IMF, 2018: *World Economic Outlook*. World Economic Outlook Database, International Monetary Fund, Washington D.C., USA.

IPBES, 2016: The Methodological Assessment Report on Scenarios and Models of Biodiversity and Ecosystem Services [S. Ferrier et al., (eds.)]. Secretariat of the Intergovernmental Science-Policy Platform on Biodiversity and Ecosystem Services, Bonn, Germany, 348 pp.

IPBES, 2018a: *The Regional Assessment Report on Biodiversity and Ecosystem services from Europe and Central Asia Biodiversity* [Rounsevell, M., Fischer, M., Torre-Marin Rando, A. and Mader, A. (eds.)]. Secretariat of the Intergovernmental Science-Policy Platform on Biodiversity and Ecosystem Services, Bonn, Germany, 892 pp.

IPBES, 2018b: *The IPBES Assessment Report on Land Degradation and Restoration* [Montanarella, L., Scholes, R. and Brainich, A. (eds.)]. Secretariat of the Intergovernmental Science-Policy Platform on Biodiversity and Ecosystem Services, Bonn, Germany, 744 pp.

IPBES, 2018c: *The Regional Assessment Report on Biodiversity and Ecosystem Services for Africa* [Archer, E. Dziba, L., Mulongoy, K.J., Maoela, M.A. and Walters, M. (eds.)]. Secretariat of the Intergovernmental Science-Policy Platform on Biodiversity and Ecosystem Services, Bonn, Germany, 492 pp.

IPBES, 2018d: *The IPBES Regional Assessment Report on Biodiversity and Ecosystem Services for the Americas* [Rice, J., Seixas, C.S., Zaccagnini, M.E., Bedoya-Gaitán, M. and Valderrama N. (eds.)]. Secretariat of the Intergovernmental Science-Policy Platform on Biodiversity and Ecosystem Services, Bonn, Germany, 656 pp.

IPBES, 2018e: *The IPBES Regional Assessment Report on Biodiversity and Ecosystem Services for Asia and the Pacific* [Karki, M., Senaratna Sellamuttu, S., Okayasu, S. and Suzuki, W. (eds.)]. Secretariat of the Intergovernmental Science-Policy Platform on Biodiversity and Ecosystem Services, Bonn, Germany, 612 pp.

IPCC, 2000: Special Report on Emissions Scenarios. Nature Publishing Group [Nakićenović, N. and R. Swart (eds.)]. Cambridge University Press, Cambridge, United Kingdom and New York, NY, USA, 612 pp.

IPCC, 2000: Land Use, Land-Use Change and Forestry: A special report of the Intergovernmental Panel on Climate Change [Watson, R.T., I.R. Noble, B. Bolin, N.H. Ravindranath, D.J. Verardo and D.J. Dokken (eds.).]. Cambridge University Press Cambridge, United Kingdom, pp 375.

IPCC, 2013: Summary for policymakers. In: Climate Change 2013: The Physical Science Basis. Contribution of Working Group I to the Fifth Assessment Report of the Intergovernmental Panel on Climate Change [Stocker, T.F., D. Qin, G.-K. Plattner, M. Tignor, S.K. Allen, J. Boschung, A. Nauels, Y. Xia, V. Bex and P.M. Midgley (eds.)], Cambridge University Press, Cambridge, United Kingdom and New York, NY, USA, 1–29.

IPCC, 2014: Summary for policymakers. In: Climate Change 2014: Impacts, Adaptation, and Vulnerability. Contribution of Working Group II to the Fifth Assessment Report of the Intergovernmental Panel on Climate Change [Field, C.B., V.R. Barros, D.J. Dokken, K.J. Mach, M.D. Mastrandrea, T.E. Bilir, M. Chatterjee, K.L. Ebi, Y.O. Estrada, R.C. Genova, B. Girma, E.S. Kissel, A.N. Levy, S. MacCracken, P.R. Mastrandrea and L.L. White (eds.)]. Cambridge University Press, Cambridge, United Kingdom and New York, NY, USA, 1–32 pp.

IPCC, 2018: Global Warming of 1.5°C: An IPCC special report on the impacts of global warming of 1.5°C above pre-industrial levels and related global greenhouse gas emission pathways, in the context of strengthening the global response to the threat of climate change, sustainable development and efforts to eradicate poverty [Masson-Delmotte, V., P. Zhai, H.-O. Pörtner, D. Roberts, J. Skea, P.R. Shukla, A. Pirani, W. Moufouma-Okia, C. Péan, R. Pidcock, S. Connors, J.B.R. Matthews, Y. Chen, X. Zhou, M.I. Gomis, E. Lonnoy, T. Maycock, M. Tignor and T. Waterfield (eds.)]. In press, 1552 pp.

Isbell, F. et al., 2017: Linking the influence and dependence of people on biodiversity across scales. *Nature*, **546**, 65–72, doi:10.1038/nature22899.

Iwata, Y., I. Miyamoto, K. Kameyama and M. Nishiya, 2017: Effect of sensor installation on the accurate measurement of soil water content. *Eur. J. Soil Sci.*, **68**, 817–828, doi:10.1111/ejss.12493.

Jadin, I., P. Meyfroidt and E.F. Lambin, 2016: International trade and land use intensification and spatial reorganization explain Costa Rica's forest transition. *Environ. Res. Lett.*, **11**, 035005, doi:10.1088/1748-9326/11/3/035005.

Jalava, M., M. Kummu, M. Porkka, S. Siebert, and O. Varis, 2014: Diet change—a solution to reduce water use? *Environ. Res. Lett.*, **9**, 74016.

EL Jarroudi, M. et al., 2015: Economics of a decision-support system for managing the main fungal diseases of winter wheat in the Grand-Duchy of Luxembourg. *F. Crop. Res.*, **172**, 32–41, doi:10.1016/J.FCR.2014.11.012.

Jiang, L. and B.C. O'Neill, 2017: Global urbanization projections for the shared socioeconomic pathways. *Glob. Environ. Chang.*, **42**, 193–199, doi:10.1016/J.GLOENVCHA.2015.03.008.

1

de Jong, R., M.E. Schaepman, R. Furrer, S. de Bruin, and P.H. Verburg, 2013: Spatial relationship between climatologies and changes in global vegetation activity. *Glob. Chang. Biol.*, **19**, 1953–1964, doi:10.1111/gcb.12193.

Joshi, A.K., P. Pant, P. Kumar, A. Giriraj, and P.K. Joshi, 2011: National forest policy in India: Critique of targets and implementation. *Small-scale For.*, **10**, 83–96, doi:10.1007/s11842-010-9133-z.

Juhl, H.J. and M.B. Jensen, 2014: Relative price changes as a tool to stimulate more healthy food choices – A Danish household panel study. *Food Policy*, **46**, 178–182, doi:10.1016/J.FOODPOL.2014.03.008.

Kaijser, A. and A. Kronsell, 2014: Climate change through the lens of intersectionality. *Env. Polit.*, **23**, 417–433, doi:10.1080/09644016.2013.835203.

Kanter, D.R. et al., 2016: Evaluating agricultural trade-offs in the age of sustainable development. *AGSY*, **163**, 73–88, doi:10.1016/j.agsy.2016.09.010.

Kastner, T., M.J.I. Rivas, W. Koch, and S. Nonhebel, 2012: Global changes in diets and the consequences for land requirements for food. *Proc. Natl. Acad. Sci.*, doi:10.1073/pnas.1117054109.

Kastner, T., K.H. Erb, and H. Haberl, 2014: Rapid growth in agricultural trade: Effects on global area efficiency and the role of management. *Environ. Res. Lett.*, **9**, doi:10.1088/1748-9326/9/3/034015.

Kauffman, J.B., H. Hernandez Trejo, M. del Carmen Jesus Garcia, C. Heider, and W.M. Contreras, 2016: Carbon stocks of mangroves and losses arising from their conversion to cattle pastures in the Pantanos de Centla, Mexico. *Wetl. Ecol. Manag.*, **24**, 203–216, doi:10.1007/s11273-015-9453-z.

De Kauwe, M.G., T.F. Keenan, B.E. Medlyn, I.C. Prentice, and C. Terrer, 2016: Satellite-based estimates underestimate the effect of CO_2 fertilization on net primary productivity. *Nat. Clim. Chang.*, **6**, pages 892–893, doi:10.1038/nclimate3105.

Keenan, R.J., 2015: Climate change impacts and adaptation in forest management: A review. *Ann. For. Sci.*, **72**, 145–167, doi:10.1007/s13595-014-0446-5.

Keenan, R.J. et al., 2015: Dynamics of global forest area: Results from the FAO Global Forest Resources Assessment 2015. *For. Ecol. Manage.*, **352**, 9–20, doi:10.1016/j.foreco.2015.06.014.

Kelley, D.I. et al., 2013: A comprehensive benchmarking system for evaluating global vegetation models. *Biogeosciences*, **10**, 3313–3340, doi:10.5194/bg-10-3313-2013.

Kesicki, F., 2013: What are the key drivers of MAC curves? A partial-equilibrium modelling approach for the UK. *Energy Policy*, **58**, 142–151, doi:10.1016/J.ENPOL.2013.02.043.

Kibler, K.M., D. Reinhart, C. Hawkins, A.M. Motlagh, and J. Wright, 2018: Food waste and the food-energy-water nexus: A review of food waste management alternatives. *Waste Manag.*, **74**, 52–62, doi:10.1016/J.WASMAN.2018.01.014.

Kimball, B.A., 2016: Crop responses to elevated CO_2 and interactions with H2O, N, and temperature. *Curr. Opin. Plant Biol.*, **31**, 36–43, doi:10.1016/j.pbi.2016.03.006.

Kindermann, G., I. McCallum, S. Fritz, and M. Obersteiner, 2008: A global forest growing stock, biomass and carbon map based on FAO statistics. *Silva Fenn.*, **42**, doi:10.14214/sf.244.

Klein, J.A. et al., 2014: Unexpected climate impacts on the Tibetan Plateau: Local and scientific knowledge in findings of delayed summer. *Glob. Environ. Chang.*, **28**, 141–152, doi:10.1016/J.GLOENVCHA.2014.03.007.

Kok, K. et al., 2014: European participatory scenario development: Strengthening the link between stories and models. *Clim. Change*, **128**, 187–200, doi:10.1007/s10584-014-1143-y.

Kok, M.T.J. et al., 2018: Pathways for agriculture and forestry to contribute to terrestrial biodiversity conservation: A global scenario-study. *Biol. Conserv.*, **221**, 137–150, doi:10.1016/j.biocon.2018.03.003.

Kolby Smith, W. et al., 2015: Large divergence of satellite and earth system model estimates of global terrestrial CO_2 fertilization. *Nat. Clim. Chang.*, **6**, 306–310, doi:10.1038/nclimate2879.

Konar, M., J.J. Reimer, Z. Hussein, and N. Hanasaki, 2016: The water footprint of staple crop trade under climate and policy scenarios. *Environ. Res. Lett.*, **11**, 035006, doi:10.1088/1748-9326/11/3/035006.

Kongsager, R. and E. Corbera, 2015: Linking mitigation and adaptation in carbon forestry projects: Evidence from Belize. *World Dev.*, **76**, 132–146, doi:10.1016/J.WORLDDEV.2015.07.003.

Kongsager, R., B. Locatelli, and F. Chazarin, 2016: Addressing climate change mitigation and adaptation together: A global assessment of agriculture and forestry projects. *Environ. Manage.*, **57**, 271–282, doi:10.1007/s00267-015-0605-y.

Kostyanovsky, K.I., D.R. Huggins, C.O. Stockle, S. Waldo, and B. Lamb, 2018: Developing a flow through chamber system for automated measurements of soil N2O and CO_2 emissions. *Meas. J. Int. Meas. Confed.*, **113**, 172–180, doi:10.1016/j.measurement.2017.05.040.

Koutroulis, A.G., 2019: Dryland changes under different levels of global warming. *Sci. Total Environ.*, **655**, 482–511, doi:10.1016/J.SCITOTENV.2018.11.215.

Krause, A. et al., 2017: Global consequences of afforestation and bioenergy cultivation on ecosystem service indicators. *Biogeosciences*, 4829–4850, doi:10.5194/bg-2017-160.

Krausmann, F. and E. Langthaler, 2019: Food regimes and their trade links: A socio-ecological perspective. *Ecol. Econ.*, **160**, 87–95, doi:10.1016/J.ECOLECON.2019.02.011.

Krausmann, F. et al., 2013: Global human appropriation of net primary production doubled in the 20th century. *Proc. Natl. Acad. Sci. U.S.A.*, **110**, 10324–10329, doi:10.1073/pnas.1211349110.

Kraxner, F. et al., 2013: Global bioenergy scenarios – Future forest development, land-use implications and trade-offs. *Biomass and Bioenergy*, **57**, 86–96, doi:10.1016/j.biombioe.2013.02.003.

Kreidenweis, U. et al., 2016: Afforestation to mitigate climate change: impacts on food prices under consideration of albedo effects. *Environ. Res. Lett.*, **11**, 1–12, doi:10.1088/1748-9326/11/8/085001.

Kreidenweis, U. et al., 2018: Pasture intensification is insufficient to relieve pressure on conservation priority areas in open agricultural markets. *Glob. Chang. Biol.*, **24**, 3199–3213, doi:10.1111/gcb.14272.

Kummu, M. et al., 2012: Lost food, wasted resources: Global food supply chain losses and their impacts on freshwater, cropland and fertiliser use. *Sci. Total Environ.*, **438**, 477–489, doi:10.1016/J.SCITOTENV.2012.08.092.

Kunreuther, H., 2015: The role of insurance in reducing losses from extreme events: The need for public–private partnerships. *Geneva Pap. Risk Insur. Issues Pract.*, **40**, 741–762, doi:10.1057/gpp.2015.14.

Lacaze, R. et al., 2015: Operational 333m biophysical products of the copernicus global land service for agriculture monitoring. *Int. Arch. Photogramm. Remote Sens. Spat. Inf. Sci. – ISPRS Arch.*, **40**, 53–56, doi:10.5194/isprsarchives-XL-7-W3-53-2015.

Laestadius, L. et al., 2011: Mapping opportunities for forest landscape restoration. *Unasylva*, **62**, 47–48.

Lal, R., 2009: Soils and world food security. *Soil and Tillage Research*, **102**, 1–4, doi:10.1016/j.still.2008.08.001.

Lal, R., 2015: Restoring soil quality to mitigate soil degradation. *Sustainability*, **7**, 5875, doi:10.3390/su7055875.

Lamb, D., 2018: Undertaking large-scale forest restoration to generate ecosystem services. *Restor. Ecol.*, **26**, 657–666, doi:10.1111/rec.12706.

Lambin, E.F., 2012: Global land availability: Malthus versus Ricardo. *Global Food Security*, **1**, 83–87, doi:10.1016/j.gfs.2012.11.002.

Lambin, E.F. and P. Meyfroidt, 2011: Global land use change, economic globalization and the looming land scarcity. *Proc Natl Acad Sci U S A*, **108**, 3465–3472, doi:10.1073/pnas.1100480108.

Lambin, E.F. and P Patrick Meyfroidt, 2014: Trends in gobal land-use competition. In *Rethinking Global Land Use in an Urban Era*, Vol. 14,

[Seto, K.C. and A. Reenberg (eds.)]. The MIT Press, Cambridge, Massachusetts, pp. 11–22.

Land Matrix, 2018: Land Matrix Global Observatory. www.landmatrix.org.

Lapola, D.M. et al., 2010: Indirect land-use changes can overcome carbon savings from biofuels in Brazil. *Proc. Natl. Acad. Sci. U.S.A.*, **107**, 3388–3393, doi:10.1073/pnas.0907318107.

Lassaletta, L., G. Billen, B. Grizzetti, J. Anglade, and J. Garnier, 2014: 50 year trends in nitrogen use efficiency of world cropping systems: The relationship between yield and nitrogen input to cropland. *Environ. Res. Lett.*, doi:10.1088/1748-9326/9/10/105011.

Lassaletta, L. et al., 2016: Nitrogen use in the global food system: Past trends and future trajectories of agronomic performance, pollution, trade and dietary demand. *Environ. Res. Lett.*, **11**, 095007, doi:10.1088/1748-9326/11/9/095007.

Laurance, W.F., 2007: Forests and floods. *Nature*, **449**, 409–410, doi:10.1038/449409a.

Laurance, W.F., J. Sayer and K.G. Cassman, 2014: Agricultural expansion and its impacts on tropical nature. *Trends Ecol. Evol.*, **29**, 107–116, doi:10.1016/J.TREE.2013.12.001.

Le, H.D., C. Smith, J. Herbohn and S. Harrison, 2012: More than just trees: Assessing reforestation success in tropical developing countries. *J. Rural Stud.*, **28**, 5–19, doi:10.1016/j.jrurstud.2011.07.006.

Le, Q.B., E. Nkonya and A. Mirzabaev, 2016: Biomass productivity-based mapping of global land degradation hotspots. In: *Economics of Land Degradation and Improvement – A Global Assessment for Sustainable Development* [Nkonya, E., A. Mirzabaev and J. Von Braun, (eds.)]. Springer International Publishing, Cham, Switzerland, pp. 55–84, doi:10.1007/978-3-319-19168-3_4.

Le, T.T.H. Trang, 2016: Effects of climate change on rice yield and rice market in Vietnam. *J. Agric. Appl. Econ.*, **48**, 366–382, doi:10.1017/aae.2016.21.

Lean, J.L., 2018: Observation-based detection and attribution of 21st century climate change. *Wiley Interdiscip. Rev. Chang.*, **9**, doi:0.1002/wcc.511.

Lebel, L. et al., 2006: Governance and the capacity to manage resilience in regional social-ecological systems. *Ecol. Soc.*, **11**, 19, doi:10.5751/ES-01606-110119.

Lee, J. et al., 2018: Economic viability of the national-scale forestation program: The case of success in the Republic of Korea. *Ecosyst. Serv.*, **29**, 40–46, doi:10.1016/j.ecoser.2017.11.001.

Lee, X. et al., 2011: Observed increase in local cooling effect of deforestation at higher latitudes. *Nature*, **479**, 384–387, doi:10.1038/nature10588.

Lees, K.J., T. Quaife, R.R.E. Artz, M. Khomik, and J.M. Clark, 2018: Potential for using remote sensing to estimate carbon fluxes across northern peatlands – A review. *Sci. Total Environ.*, **615**, 857–874, doi:10.1016/j.scitotenv.2017.09.103.

Lehmann, C.E.R. and C.L. Parr, 2016: Tropical grassy biomes: Linking ecology, human use and conservation. *Philos. Trans. R. Soc. B-Biological Sci.*, **371**, 20160329, doi:20160329 10.1098/rstb.2016.0329.

Lehmann, N., S. Briner, and R. Finger, 2013: The impact of climate and price risks on agricultural land use and crop management decisions. *Land use policy*, **35**, 119–130, doi:10.1016/J.LANDUSEPOL.2013.05.008.

Lempert, R., N. Nakicenovic, D. Sarewitz, and M. Schlesinger, 2004: Characterizing climate-change uncertainties for decision-makers: An editorial essay. *Clim. Change*, **65**, 1–9, doi:10.1023/B:CLIM.0000037561.75281.b3.

Lenton, T.M., 2014: The global potential for carbon dioxide removal. In: *Geoengineering of the Climate System* [Hester, R.E. and R.M. Harrison, (eds.)]. Royal Society of Chemistry, pp. 52–79.

Lesk, C., P. Rowhani, and N. Ramankutty, 2016: Influence of extreme weather disasters on global crop production. *Nature*, **529**, 84, doi:10.1038/nature16467.

Li, D., S. Niu, and Y. Luo, 2012: Global patterns of the dynamics of soil carbon and nitrogen stocks following afforestation: A meta-analysis. *New Phytol.*, **195**, 172–181, doi:10.1111/j.1469-8137.2012.04150.x.

Li, S., M. Xu and B. Sun, 2014: Long-term hydrological response to reforestation in a large watershed in southeastern China. *Hydrol. Process.*, **28**, 5573–5582, doi:10.1002/hyp.10018.

Li, W. et al., 2016: Major forest changes and land cover transitions based on plant functional types derived from the ESA CCI Land Cover product. *Int. J. Appl. Earth Obs. Geoinf.*, **47**, 30–39, doi:10.1016/J.JAG.2015.12.006.

Li, W. et al., 2017: Land-use and land-cover change carbon emissions between 1901 and 2012 constrained by biomass observations. *Biogeosciences*, **145194**, 5053–5067, doi:10.5194/bg-14-5053-2017.

Limpens, J. et al., 2008: Peatlands and the carbon cycle: from local processes to global implications – a synthesis. *Biogeosciences*, **5**, 1475–1491, doi:10.5194/bg-5-1475-2008.

Lin, M. and P. Huybers, 2012: Reckoning wheat yield trends. *Environ. Res. Lett.*, **7**, 24016, doi:10.1088/1748-9326/7/2/024016.

Lindenmayer, D.B. and R.J. Hobbs, 2004: Fauna conservation in Australian plantation forests – a review. *Biol. Conserv.*, **119**, 151–168, doi:10.1016/J.BIOCON.2003.10.028.

Linnerooth-Bayer, J. and R. Mechler, 2006: Insurance for assisting adaptation to climate change in developing countries: A proposed strategy. *Clim. Policy*, **6**, 621–636, doi:10.1080/14693062.2006.9685628.

Lipper, L. et al., 2014: Climate-smart agriculture for food security. *Nat. Clim. Chang.*, **4**, 1068–1072, doi:10.1038/nclimate2437.

Liu, J. et al., 2013: Framing Sustainability in a Telecoupled World. *Ecol. Soc.*, **2**, doi:10.5751/ES-05873-180226.

Lobell, D.B., W. Schlenker, and J. Costa-Roberts, 2011: Climate trends and global crop production since 1980. *Science*, **333**, 616–620, doi:10.1126/science.1204531.

Lobell, D.B., A. Sibley, and J. Ivan Ortiz-Monasterio, 2012: Extreme heat effects on wheat senescence in India. *Nat. Clim. Chang.*, **2**, 186–189, doi:10.1038/nclimate1356.

Lobell, D.B., C.B. Uris Lantz, and T.W. Hertel, 2013: Climate adaptation as mitigation: The case of agricultural investments. *Environ. Res. Lett.*, **8**, 15012, doi:10.1088/1748-9326/8/1/015012.

Locatelli, B., V. Evans, A. Wardell, A. Andrade, and R. Vignola, 2011: Forests and climate change in Latin America: Linking adaptation and mitigation. *Forests*, **2**, doi:10.3390/f2010431.

Locatelli, B. et al., 2015a: Tropical reforestation and climate change: Beyond carbon. *Restor. Ecol.*, **23**, 337–343, doi:10.1111/rec.12209.

Locatelli, B., C. Pavageau, E. Pramova, and M. Di Gregorio, 2015b: Integrating climate change mitigation and adaptation in agriculture and forestry: Opportunities and trade-offs. *Wiley Interdiscip. Rev. Clim. Chang.*, **6**, 585-598, doi:10.1002/wcc.357.

Loch, A. et al., 2013: *The Role of Water Markets in Climate Change Adaptation.* National Climate Change Adaptation Research Facility, Gold Coast, Australia, 125 pp.

Loladze, I., 2014: Hidden shift of the ionome of plants exposed to elevated CO_2 depletes minerals at the base of human nutrition. *Elife*, **3**, e02245, doi:10.7554/eLife.02245.

Lorenz, K. and R. Lal, 2014: Biochar application to soil for climate change mitigation by soil organic carbon sequestration. *J. Plant Nutr. Soil Sci.*, **177**, 651–670, doi:10.1002/jpln.201400058.

Luedeling, E. and E. Shepherd, 2016: Decision-Focused Agricultural Research. *Solutions*, **7**, 46–54.

Luo, Y.Q. et al., 2012: A framework of benchmarking land models. *Biogeosciences*, **10**, 3857–3874, doi:10.5194/bgd-9-1899-2012.

Luyssaert, S. et al., 2014: Land management and land-cover change have impacts of similar magnitude on surface temperature. *Nat. Clim. Chang.*, **4**, 389–393, doi:10.1038/nclimate2196.

MacDicken, K.G., 2015: Global forest resources assessment 2015: What, why and how? *For. Ecol. Manage.*, **352**, 3–8, doi:10.1016/j.foreco.2015.02.006.

MacDicken, K.G. et al., 2015: Global progress toward sustainable forest management. *For. Ecol. Manage.*, **352**, 47–56, doi:10.1016/j.foreco.2015.02.005.

Mace, G.M., K. Norris, and A.H. Fitter, 2012: Biodiversity and ecosystem services: A multilayered relationship. *Trends Ecol. Evol.*, **27**, 19–25, doi:10.1016/j.tree.2011.08.006.

Maestre, F.T. et al., 2012: Plant species richness and ecosystem multifunctionality in global drylands. *Science*, **335**, 214–218, doi:10.1126/science.1215442.

Maestre F.T. et al., 2016: Structure and functioning of dryland ecosystems in a changing world. *Annual Review of Ecology, Evolution and Systematics*, **47**, 215–237, doi:10.1146/annurev-ecolsys-121415-032311.

Maier, H.R. et al., 2016: An uncertain future, deep uncertainty, scenarios, robustness and adaptation: How do they fit together? *Environ. Model. Softw.*, **81**, 154–164, doi:10.1016/j.envsoft.2016.03.014.

Malkamäki, A. et al., 2018: A systematic review of the socio-economic impacts of large-scale tree plantations, worldwide. *Glob. Environ. Chang.*, **53**, 90–103, doi:10.1016/j.gloenvcha.2018.09.001.

Malone, R.W. et al., 2014: Cover crops in the upper midwestern United States: Simulated effect on nitrate leaching with artificial drainage. *J. Soil Water Conserv.*, **69**, 292–305, doi:10.2489/jswc.69.4.292.

Marchand, P. et al., 2016: Reserves and trade jointly determine exposure to food supply shocks. *Environ. Res. Lett.*, **11**, 095009, doi:10.1088/1748-9326/11/9/095009.

Marques, A. et al., 2019: Increasing impacts of land use on biodiversity and carbon sequestration driven by population and economic growth. *Nat. Ecol. Evol.*, **3**, 628–637, doi:10.1038/s41559-019-0824-3.

Martellozzo, F. et al., 2015: Urbanization and the loss of prime farmland: A case study in the Calgary–Edmonton corridor of Alberta. *Reg. Environ. Chang.*, **15**, 881–893, doi:10.1007/s10113-014-0658-0.

Martin-Guay, M.O., A. Paquette, J. Dupras, and D. Rivest, 2018: The new Green Revolution: Sustainable intensification of agriculture by intercropping. *Sci. Total Environ.*, **615**, 767–772, doi:10.1016/j.scitotenv.2017.10.024.

Mastrandrea, M.D. et al., 2011: The IPCC AR5 guidance note on consistent treatment of uncertainties: A common approach across the working groups. *Clim. Change*, **108**, 675, doi:10.1007/s10584-011-0178-6.

Mateos, E., J.M. Edeso, and L. Ormaetxea, 2017: Soil erosion and forests biomass as energy resource in the basin of the Oka River in Biscay. *Forests*, **8**, 1–20, doi:10.3390/f8070258.

Mathews, J.A., 2017: Global trade and promotion of cleantech industry: A post-Paris agenda. *Clim. Policy*, **17**, 102–110, doi:10.1080/14693062.2016.1215286.

Maxwell, S.L., R.A. Fuller, T.M. Brooks, and J.E.M. Watson, 2016: Biodiversity: The ravages of guns, nets and bulldozers. *Nature*, **536**, 143–145, doi:10.1038/536143a.

McDonnell, S., 2017: Urban land grabbing by political elites: Exploring the political economy of land and the challenges of regulation. In: *Kastom, property and ideology: Land transformations in Melanesia* [McDonnell, S., M.G. Allen, C. Filer (Eds.)]. Australian National University Press, Canberra, Australia, pp. 283–304.

Medek, Danielle E., Joel Schwartz, S.S.M., 2017: Estimated effects of future atmospheric CO_2 concentrations on protein intake and the risk of protein deficiency by country and region. *Env. Heal. Perspect*, **125**, 087002. doi:10.1289/EHP41.

Mello, D. and M. Schmink, 2017: Amazon entrepreneurs: Women's economic empowerment and the potential for more sustainable land use practices. *Womens. Stud. Int. Forum*, **65**, 28–36, doi:10.1016/J.WSIF.2016.11.008.

Messerli, P., M. Giger, M.B. Dwyer, T. Breu and S. Eckert, 2014: The geography of large-scale land acquisitions: Analysing socio-ecological patterns of target contexts in the global South. *Appl. Geogr.*, **53**, 449–459, doi:10.1016/j.apgeog.2014.07.005.

Meyfroidt, P., 2018: Trade-offs between environment and livelihoods: Bridging the global land use and food security discussions. *Glob. Food Sec.*, **16**, 9–16, doi:10.1016/J.GFS.2017.08.001.

Millar, R.J. et al., 2017: Emission budgets and pathways consistent with limiting warming to 1.5°C. *Nat. Geosci.*, **10**, 741–747, doi:10.1038/NGEO3031.

Mirzabaev, A., E. Nkonya and J. von Braun, 2015: *Economics of sustainable land management. Elsevier*, **15**, 9–19, doi:10.1016/j.cosust.2015.07.004.

Mistry, J. and A. Berardi, 2016: Bridging indigenous and scientific knowledge. *Science*, **352**, 1274–1275, doi:10.1126/science.aaf1160.

Miyamoto, A., M. Sano, H. Tanaka and K. Niiyama, 2011: Changes in forest resource utilization and forest landscapes in the southern Abukuma Mountains, Japan during the twentieth century. *J. For. Res.*, **16**, 87–97, doi:10.1007/s10310-010-0213-x.

Molina, A., G. Govers, V. Vanacker, and J. Poesen, 2007: Runoff generation in a degraded Andean ecosystem: Interaction of vegetation cover and land use. *Catena*, **71**, 357–370, doi:10.1016/j.catena.2007.04.002.

Moore, F.C. and D.B. Lobell, 2015: The fingerprint of climate trends on European crop yields. *Proc. Natl. Acad. Sci.*, **112**, 2670–2675, doi:10.1073/pnas.1409606112.

Moosa, C.S. and N. Tuana, 2014: Mapping a research agenda concerning gender and climate change: A review of the literature. *Hypatia*, **29**, 677–694, doi:10.1111/hypa.12085.

Morales-Hidalgo, D., S.N. Oswalt, and E. Somanathan, 2015: Status and trends in global primary forest, protected areas, and areas designated for conservation of biodiversity from the Global Forest Resources Assessment 2015. *For. Ecol. Manage.*, **352**, 68–77, doi:10.1016/j.foreco.2015.06.011.

Moroni, S., 2018: Property as a human right and property as a special title. Rediscussing private ownership of land. *Land use policy*, **70**, 273–280, doi:10.1016/J.LANDUSEPOL.2017.10.037.

Mosnier, A. et al., 2014: Global food markets, trade and the cost of climate change adaptation. *Food Secur.*, **6**, 29–44, doi:10.1007/s12571-013-0319-z.

Mottet, A. et al., 2017: Livestock: On our plates or eating at our table? A new analysis of the feed/food debate. *Glob. Food Sec.*, **14**, 1–8, doi:10.1016/J.GFS.2017.01.001.

Mouratiadou, I. et al., 2016: The impact of climate change mitigation on water demand for energy and food: An integrated analysis based on the Shared Socioeconomic Pathways. *Environ. Sci. Policy*, **64**, 48–58, doi:10.1016/J.ENVSCI.2016.06.007.

Muller, A. et al., 2017: Strategies for feeding the world more sustainably with organic agriculture. *Nat. Commun.*, **8**, doi:10.1038/s41467-017-01410-w.

Muller, C. et al., 2015: Implications of climate mitigation for future agricultural production. *Environ. Res. Lett.*, **10**, doi:12500410.1088/1748-9326/10/12/125004.

Murdiyarso, D. et al., 2015: The potential of Indonesian mangrove forests for global climate change mitigation. *Nat. Clim. Chang.*, **5**, 1089–1092, doi:10.1038/NCLIMATE2734.

Myers, S.S., Zanobetti, A., Kloog, I., Huybers, P., Leakey, A.D., Bloom, A.J., 2014: Increasing CO_2 threatens human nutrition. *Nature*, **510**, 139, doi:10.1038/nature13179.

Myers, S.S. et al., 2017: Climate change and global food systems: Potential impacts on food security and undernutrition. *Annu. Rev. Public Health*, **38**, 259–277, doi:10.1146/annurev-publhealth-031816-044356.

Nachtergaele, F., 2008: *Mapping Land Use Systems at Global and Regional Scales for Land Degradation Assessment Analysis Version 1.0*, Food and Agriculture Organization of the United Nations, Rome, Italy, 77 pp.

Nakamura, A. et al., 2017: Forests and their canopies: Achievements and horizons in canopy science, *Trends Ecol. Evol.*, **32**, 438–451, doi:10.1016/j.tree.2017.02.020.

Namubiru-Mwaura, E., 2014: *Land tenure and gender: Approaches and challenges for strengthening rural women's land rights*. Women's Voice, Agency, & Participation Research Series No. 06, World Bank, Washington, DC, USA, 32 pp.

Nelson, K.C. and B.H.J. de Jong, 2003: Making global initiatives local realities: Carbon mitigation projects in Chiapas, Mexico. *Glob. Environ. Chang.*, **13**, 19–30, doi: 10.1016/S0959-3780(02)00088-2.

Nepstad, D.C., W. Boyd, C.M. Stickler, T. Bezerra, and A.A. Azevedo, 2013: Responding to climate change and the global land crisis: REDD+, market transformation and low-emissions rural development. *Philos. Trans. R. Soc. Lond. B. Biol. Sci.*, **368**, 20120167, doi:10.1098/rstb.2012.0167.

Newbold, T. et al., 2015: Global effects of land use on local terrestrial biodiversity. *Nature*, **520**, 45–50, doi:10.1038/nature14324.

Newbold, T., D.P. Tittensor, M.B.J. Harfoot, J.P.W. Scharlemann, and D.W. Purves, 2018: Non-linear changes in modelled terrestrial ecosystems subjected to perturbations. *bioRxiv*, doi:10.1101/439059.

Nicole, W., 2015: Pollinator power: Nutrition security benefits of an ecosystem service. *Environ. Health Perspect.*, **123**, A210–A215, doi:10.1289/ehp.123-A210.

Nkhonjera, G.K., 2017: Understanding the impact of climate change on the dwindling water resources of South Africa, focusing mainly on Olifants River basin: A review. *Environ. Sci. Policy*, **71**, 19–29, doi:10.1016/J.ENVSCI.2017.02.004.

Nolte, K., W. Chamberlain and M. Giger, 2016: *International Land Deals for Agriculture: Fresh insights from the Land Matrix: Analytical Report II*. Centre for Development and Environment, University of Bern; Centre de coopération internationale en recherche agronomique pour le développement; German Institute of Global and Area Studies; University of Pretoria; Bern Open Publishing, Bern, Montpellier, Hamburg, Pretoria, 1–56 pp.

van Noordwijk, M. and L. Brussaard, 2014: Minimizing the ecological footprint of food: Closing yield and efficiency gaps simultaneously? *Curr. Opin. Environ. Sustain.*, **8**, 62–70, doi:10.1016/J.COSUST.2014.08.008.

Noordwijk, M. Van, L. Tanika and B. Lusiana, 2017: Flood risk reduction and flow buffering as ecosystem services – Part 2: Land use and rainfall intensity effects in Southeast Asia. *Hydrol. Earth Syst. Sci.*, 2341–2360, doi:10.5194/hess-21-2341-2017.

Nordhaus, W., 2014: Estimates of the social cost of carbon: Concepts and results from the DICE-2013R model and alternative approaches. *J. Assoc. Environ. Resour. Econ.*, **1**, 273–312, doi:10.1086/676035.

Norse, D. and X. Ju, 2015: Environmental costs of China's food security. *Agric. Ecosyst. Environ.*, **209**, 5–14, doi:10.1016/J.AGEE.2015.02.014.

Nowosad, J., T.F. Stepinski and P. Netzel, 2018: Global assessment and mapping of changes in mesoscale landscapes: 1992–2015. *Int. J. Appl. Earth Obs. Geoinf.*, **78**, 332–40, doi:10.1016/j.jag.2018.09.013.

O'Neill, B.C. et al., 2014: A new scenario framework for climate change research: The concept of shared socioeconomic pathways. *Clim. Change*, **122**, 387–400, doi:10.1007/s10584-013-0905-2.

OECD, 2014: *Social Institutions and Gender Index (SIGI)*. OECD Development Centre's Social Cohesion Unit, Paris, France, www.oecd.org/dev/development-gender/BrochureSIGI2015-web.pdf.

Ogle, S.M. et al., 2018: Delineating managed land for reporting national greenhouse gas emissions and removals to the United Nations framework convention on climate change. *Carbon Balance Manag.*, **13**, doi:10.1186/s13021-018-0095-3.

O'Neill, B. C, 2004: Conditional Probabilistic Population Projections: An Application to Climate Change. *International Statistical Institute*, **72**(2), 167–184.

Ordway, E.M., G.P. Asner and E.F. Lambin, 2017: Deforestation risk due to commodity crop expansion in sub-Saharan Africa. *Environ. Res. Lett.*, **12**, 044015, doi:10.1088/1748-9326/aa6509.

Osborne, T.M. and T.R. Wheeler, 2013: Evidence for a climate signal in trends of global crop yield variability over the past 50 years. *Environ. Res. Lett.*, **8**, 024001, doi:10.1088/1748-9326/8/2/024001.

Ostrom, E. and M. Cox, 2010: Moving beyond panaceas: A multi-tiered diagnostic approach for social-ecological analysis. *Environ. Conserv.*, **37**, 451–463, doi:10.1017/S0376892910000834.

Padmanaba, M. and R.T. Corlett, 2014: Minimizing risks of invasive alien plant species in tropical production forest management. *Forests*, **5**, 1982–1998, doi:10.3390/f5081982.

Paillet, Y. et al., 2010: Biodiversity differences between managed and unmanaged forests: Meta-analysis of species richness in Europe, *Conserv. Biol.*, **24**(1), 101–112, doi:10.1111/j.1523-1739.2009.01399.x.

Di Paola, A. et al., 2018: The expansion of wheat thermal suitability of Russia in response to climate change. *Land use policy*, **78**, 70–77, doi:10.1016/J.LANDUSEPOL.2018.06.035.

Parfitt, J., M. Barthel, and S. Macnaughton, 2010: Food waste within food supply chains: quantification and potential for change to 2050. *Philos. Trans. R. Soc. Lond. B. Biol. Sci.*, **365**, 3065–3081, doi:10.1098/rstb.2010.0126.

Parker, W.S., 2013: Ensemble modeling, uncertainty and robust predictions. *Wiley Interdiscip. Rev. Chang.*, **4**, 213–223, doi:10.1002/wcc.220.

Parr, C.L., C.E.R.R. Lehmann, W.J. Bond, W.A. Hoffmann A.N. Andersen, 2014: Tropical grassy biomes: misunderstood, neglected and under threat. *Trends Ecol. Evol.*, **29**, 205–213, doi:10.1016/j.tree.2014.02.004.

Pawson, S.M. et al., 2013: Plantation forests, climate change and biodiversity. *Biodivers. Conserv.*, **22**, 1203–1227, doi:10.1007/s10531-013-0458-8.

Payn, T. et al., 2015: Changes in planted forests and future global implications. *For. Ecol. Manage.*, **352**, 57–67, doi:10.1016/J.FORECO.2015.06.021.

Pedrozo-Acuña, A., R. Damania, M.A. Laverde-Barajas, and D. Mira-Salama, 2015: Assessing the consequences of sea-level rise in the coastal zone of Quintana Roo, México: The costs of inaction. *J. Coast. Conserv.*, **19**, 227–240, doi:10.1007/s11852-015-0383-y.

Pereira, H.M. et al., 2010: Scenarios for global biodiversity in the 21st century. *Science*, **330**, 1496–1501, doi:10.1126/science.1196624.

Perugini, L. et al., 2017: Biophysical effects on temperature and precipitation due to land cover change. *Environ. Res. Lett.*, **12**, 1–21, doi:10.1088/1748-9326/aa6b3f.

Peterson, E.E., S.A. Cunningham, M. Thomas, S. Collings, G.D. Bonnett and B. Harch, 2017: An assessment framework for measuring agroecosystem health. *Ecol. Indic.*, **79**, 265–275, doi:10.1016/j.ecolind.2017.04.002.

Pham, P., P. Doneys, and D.L. Doane, 2016: Changing livelihoods, gender roles and gender hierarchies: The impact of climate, regulatory and socio-economic changes on women and men in a Co Tu community in Vietnam. *Women's Stud. Int. Forum*, **54**, 48–56, doi:10.1016/J.WSIF.2015.10.001.

Pimm, S.L. et al., 2014: The biodiversity of species and their rates of extinction, distribution, and protection. *Science*, **344**, 1246752–1246752, doi:10.1126/science.1246752.

Pingoud, K., T. Ekholm, R. Sievänen, S. Huuskonen, and J. Hynynen, 2018: Trade-offs between forest carbon stocks and harvests in a steady state – A multi-criteria analysis. *J. Environ. Manage.*, **210**, 96–103, doi:10.1016/J.JENVMAN.2017.12.076.

Pizer, W. et al., 2014: Using and improving the social cost of carbon. *Science*, **346**, 1189–1190, doi:10.1126/science.1259774.

Poeplau, C. and A. Don, 2015: Carbon sequestration in agricultural soils via cultivation of cover crops – A meta-analysis. *Agric. Ecosyst. Environ.*, **200**, 33–41, doi:10.1016/J.AGEE.2014.10.024.

Poeplau, C. et al., 2011: Temporal dynamics of soil organic carbon after land-use change in the temperate zone – carbon response functions as a model approach. *Glob. Chang. Biol.*, **17**, 2415–2427, doi:10.1111/j.1365-2486.2011.02408.x.

Poore, J. and T. Nemecek, 2018: Reducing food's environmental impacts through producers and consumers. *Science*, **360**, 987–992, doi:10.1126/science.aaq0216.

Popp, A. et al., 2014: Land-use protection for climate change mitigation. *Nat. Clim. Chang.*, **4**, 1095–1098, doi:10.1038/nclimate2444.

Popp, A. et al., 2016: Land-use futures in the shared socio-economic pathways. *Glob. Environ. Chang.*, **42**, doi:10.1016/j.gloenvcha.2016.10.002.

Porter, J.R., L. Xie, A.J. Challinor, K. Cochrane, S.M. Howden, M.M. Iqbal, D.B. Lobell, and M.I. Travasso, 2014: Food security and food production systems. In: Climate Change 2014: Impacts, Adaptation, and Vulnerability. Part A: Global and Sectoral Aspects. Contribution of Working Group II to the Fifth Assessment Report of the Intergovernmental Panel on Climate

Change [Field, C.B., V.R. Barros, D.J. Dokken, K.J. Mach, M.D. Mastrandrea, T.E. Bilir, M. Chatterjee, K.L. Ebi, Y.O. Estrada, R.C. Genova, B. Girma, E.S. Kissel, A.N. Levy, S. MacCracken, P.R. Mastrandrea, and L.L. White (eds.)]. Cambridge University Press, Cambridge, United Kingdom, pp. 485–533, doi:10.1017/CBO9781107415379.

Porter, S.D., D.S. Reay, P. Higgins and E. Bomberg, 2016: A half-century of production-phase greenhouse gas emissions from food loss and waste in the global food supply chain. *Sci. Total Environ.*, **571**, 721–729, doi:10.1016/J.SCITOTENV.2016.07.041.

Potapov, P. et al., 2017: The last frontiers of wilderness: Tracking loss of intact forest landscapes from 2000 to 2013. *Sci. Adv.*, **3**, e1600821, doi:10.1126/sciadv.1600821.

Pradhan, P., M.K.B. Lüdeke, D.E. Reusser, and J.P. Kropp, 2013: Embodied crop calories in animal products. *Environ. Res. Lett.*, **8**, doi:10.1088/1748-9326/8/4/044044.

Pradhan, P., M.K.B. Lüdeke, D.E. Reusser, and J.P. Kropp, 2014: Food Self-Sufficiency across Scales: How Local Can We Go? **15**, 9779, doi:10.1021/es5005939.

Pravalie, R., 2016: Drylands extent and environmental issues. A global approach. *Earth-Science Rev.*, **161**, 259–278, doi:10.1016/j.earscirev.2016.08.003.

Prestele, R. et al., 2016: Hotspots of uncertainty in land-use and land-cover change projections: A global-scale model comparison. *Glob. Chang. Biol.*, **22**, 3967–3983, doi:10.1111/gcb.13337.

Pugh, T.A.M. et al., 2016: Climate analogues suggest limited potential for intensification of production on current croplands under climate change. *Nat. Commun.*, **7**, doi:1260810.1038/ncomms12608.

Pugh, T.A.M. et al., 2019: Role of forest regrowth in global carbon sink dynamics. *Proc. Natl. Acad. Sci.*, 201810512, doi:10.1073/pnas.1810512116.

Putz, F.E. and K.H. Redford, 2010: The importance of defining "Forest": Tropical forest degradation, deforestation, long-term phase shifts and further transitions. *Biotropica*, doi:10.1111/j.1744-7429.2009.00567.x.

Le Quéré, C. et al., 2015: Global Carbon Budget 2015. *Earth Syst. Sci. Data*, **7**, 349–396, doi:10.5194/essd-7-349-2015.

Le Quéré, C. et al., 2018: Global Carbon Budget 2017. *Earth Syst. Sci. Data*, **10**, 405–448, doi:10.5194/essd-10-405-2018.

Le Quéré, C. et al., 2013: The global carbon budget 1959–2011. *Earth Syst. Sci. Data*, **5**, 165–185, doi:10.5194/essd-5-165-2013.

Le Quéré, C. et al., 2018: Global Carbon Budget 2018. *Earth Syst. Sci. Data Discuss.*, 1–3, doi:10.5194/essd-2018-120.

Raffensperger, C. and J.A. Tickner, 1999: Introduction: To Foresee and Forestall. In: *Protecting Public Health & The Environment: Implementing The Precautionary Principle*, [Raffensperger, C. and J.A. Tickner (eds.)]. Island Press, Washington, DC, USA, pp. 1–11.

Raiten, D.J. and A.M. Aimone, 2017: The intersection of climate/environment, food, nutrition and health: Crisis and opportunity. *Curr. Opin. Biotechnol.*, **44**, 52–62, doi:10.1016/J.COPBIO.2016.10.006.

Ramankutty, N., A.T. Evan, C. Monfreda and J.A. Foley, 2008: Farming the planet: 1. Geographic distribution of global agricultural lands in the year 2000. *Global Biogeochem. Cycles*, **22**, GB1003, doi:10.1029/2007GB002952.

Ramankutty, N. et al., 2018: Trends in global agricultural land use: Implications for environmental health and food security. *Annu. Rev. Plant Biol.*, **69**, 789–815, doi:10.1146/annurev-arplant-042817-040256.

Randerson, J.T. et al., 2009: Systematic assessment of terrestrial biogeochemistry in coupled climate-carbon models. *Glob. Chang. Biol.*, **15**, 2462–2484, doi:10.1111/j.1365-2486.2009.01912.x.

Rao, N., 2017: Assets, agency and legitimacy: Towards a relational understanding of gender equality policy and practice. *World Dev.*, **95**, 43–54, doi:10.1016/J.WORLDDEV.2017.02.018.

Rao, Y. et al., 2018: Integrating ecosystem services value for sustainable land-use management in semi-arid region. *J. Clean. Prod.*, **186**, 662–672, doi:10.1016/J.JCLEPRO.2018.03.119.

Ravi, S., D.D. Breshears, T.E. Huxman, and P. D'Odorico, 2010: Land degradation in drylands: Interactions among hydrologic–aeolian erosion

and vegetation dynamics. *Geomorphology*, **116**, 236–245, doi:10.1016/j.geomorph.2009.11.023.

Ravnborg, H.M., R. Spichiger, R.B. Broegaard, and R.H. Pedersen, 2016: Land governance, gender equality and development: Past achievements and remaining challenges. *J. Int. Dev.*, **28**, 412–427, doi:10.1002/jid.3215.

Ray, D.K., N. Ramankutty, N.D. Mueller, P.C. West and J.A. Foley, 2012: Recent patterns of crop yield growth and stagnation. *Nat. Commun.*, **3**, doi:10.1038/ncomms2296.

Reed, M. and L.C. Stringer, 2015: *Climate change and desertification: Anticipating, assessing & adapting to future change in drylands*. Impulse Report for the 3rd UNCCD Scientific Conference, Agropolis International, Montpellier, France, 1–140 pp.

Resurrección, B.P., 2013: Persistent women and environment linkages in climate change and sustainable development agendas. *Womens. Stud. Int. Forum*, **40**, 33–43, doi:10.1016/J.WSIF.2013.03.011.

Reyer, C., M. Guericke, and P.L. Ibisch, 2009: Climate change mitigation via afforestation, reforestation and deforestation avoidance: and what about adaptation to environmental change? *New For.*, **38**, 15–34, doi:10.1007/s11056-008-9129-0.

Riahi, K. et al., 2015: Locked into Copenhagen pledges – implications of short-term emission targets for the cost and feasibility of long-term climate goals. *Technol. Forecast. Soc. Change*, **90**, 8–23, doi:10.1016/j.techfore.2013.09.016.

Riahi, K. et al., 2017: The shared socioeconomic pathways and their energy, land use and greenhouse gas emissions implications: An overview. *Glob. Environ. Chang.*, **42**, 153–168, doi:10.1016/j.gloenvcha.2016.05.009.

Richards, D.R. and D.A. Friess, 2016: Rates and drivers of mangrove deforestation in Southeast Asia, 2000-2012. *Proc. Natl. Acad. Sci. U.S.A.*, **113**, 344–349, doi:10.1073/pnas, 1510272113.

Ricke, K., L. Drouet, K. Caldeira, and M. Tavoni, 2018: Country-level social cost of carbon. *Nat. Clim. Chang.*, **8**, 895–900, doi:10.1038/s41558-018-0282-y.

Ringler, C. and R. Lawford, 2013: The nexus across water, energy, land and food (WELF): Potential for improved resource use efficiency? *Curr. Opin. Environ. Sustain.*, **5**, 617–624, doi:10.1016/J.COSUST.2013.11.002.

Robinson, D.A. et al., 2017: Modelling feedbacks between human and natural processes in the land system. *Earth Syst. Dyn. Discuss.*, doi:10.5194/esd-2017-68.

Myers, R., A.J.P. Sanders, A.M. Larson, R.D. Prasti, A. Ravikumar, 2016: *Analyzing multilevel governance in Indonesia: Lessons for REDD+ from the study of landuse change in Central and West Kalimantan*, CIFOR Working Paper no. 202, Center for International Forestry Research (CIFOR), Bogor, Indonesia, 69 pp.

Rodriguez-Labajos, B., 2013: Climate change, ecosystem services and costs of action and inaction: Scoping the interface. *Wiley Interdiscip. Rev. Chang.*, **4**, 555–573, doi:10.1002/wcc.247.

Rogelj, J.D. Shindell, K. Jiang, S. Fifita, P. Forster, V. Ginzburg, C. Handa, H. Kheshgi, S. Kobayashi, E. Kriegler, L. Mundaca, R. Séférian and M.V.Vilariño, 2018a: Mitigation Pathways Compatible with 1.5°C in the Context of Sustainable Development. In: Global Warming of 1.5°C. An IPCC Special Report on the impacts of global warming of 1.5°C above pre-industrial levels and related global greenhouse gas emission pathways, in the context of strengthening the global response to the threat of climate change, sustainable development and efforts to eradicate poverty [Masson-Delmotte, V., P. Zhai, H.-O. Pörtner, D. Roberts, J. Skea, P.R. Shukla, A. Pirani, W. Moufouma-Okia, C. Péan, R. Pidcock, S. Connors, J.B.R. Matthews, Y. Chen, X. Zhou, M.I. Gomis, E. Lonnoy, T. Maycock, M. Tignor, and T. Waterfield (eds.)]. In press, pp. 93–174.

Rogelj, J. et al., 2018b: Scenarios towards limiting global mean temperature increase below 1.5 degrees C. *Nat. Clim. Chang.*, **8**, 325–332, doi:10.1038/s41558-018-0091-3.

Rook, G.A., 2013: Regulation of the immune system by biodiversity from the natural environment: An ecosystem service essential to health. *Proc. Natl. Acad. Sci.*, **110**, 18360–18367, doi:10.1073/pnas.1313731110.

Röös, E. et al., 2017: Greedy or needy? Land use and climate impacts of food in 2050 under different livestock futures. *Glob. Environ. Chang.*, **47**, 1–12, doi:10.1016/J.GLOENVCHA.2017.09.001.

Rosa, I.M.D.I.M.D. et al., 2017: Multiscale scenarios for nature futures. *Nat. Ecol. Evol.*, **1**, 1416–1419, doi:10.1038/s41559-017-0273-9.

Rose, S.K., 2014: Integrated assessment modeling of climate change adaptation in forestry and pasture land use: A review. *Energy Econ.*, **46**, 548–554, doi:10.1016/J.ENECO.2014.09.018.

Rosen, R.A. and E. Guenther, 2015: The economics of mitigating climate change: What can we know? *Technol. Forecast. Soc. Change*, **91**, 93–106, doi:10.1016/J.TECHFORE.2014.01.013.

Rosenzweig, C. and P. Neofotis, 2013: Detection and attribution of anthropogenic climate change impacts. *Wiley Interdiscip. Rev. Chang.*, **4**, 121–150, doi:10.1002/wcc.209.

Rosenzweig, C. et al., 2014: Assessing agricultural risks of climate change in the 21st century in a global gridded crop model intercomparison. *Proc. Natl. Acad. Sci. U.S.A.*, **111**, 3268–3273, doi:10.1073/pnas.1222463110.

Rounsevell, M.D.A. and M.J. Metzger, 2010: Developing qualitative scenario storylines for environmental change assessment. *Wiley Interdiscip. Rev. Clim. Chang.*, **1**, 606–619, doi:10.1002/wcc.63.

Rounsevell, M.D.A. et al., 2006: A coherent set of future land use change scenarios for Europe. *Agric. Ecosyst. Environ.*, **114**, 57–68, doi:10.1016/j.agee.2005.11.027.

Rounsevell, M.D.A. et al., 2014: Towards decision-based global land use models for improved understanding of the Earth system. *Earth Syst. Dyn.*, **5**, 117–137, doi:10.5194/esd-5-117-2014.

Roy, J., P. Tschakert, H. Waisman, S. Abdul Halim, P. Antwi-Agyei, P. Dasgupta, B. Hayward, M. Kanninen, D. Liverman, C. Okereke, P.F. Pinho, K. Riahi, and A.G. Suarez Rodriguez, 2018: Sustainable Development, Poverty Eradication and Reducing Inequalities. In: Global Warming of 1.5°C. An IPCC Special Report on the impacts of global warming of 1.5°C above pre-industrial levels and related global greenhouse gas emission pathways, in the context of strengthening the global response to the threat of climate change, sustainable development and efforts to eradicate poverty [Masson-Delmotte, V., P. Zhai, H.-O. Pörtner, D. Roberts, J. Skea, P.R. Shukla, A. Pirani, W. Moufouma-Okia, C. Péan, R. Pidcock, S. Connors, J.B.R. Matthews, Y. Chen, X. Zhou, M.I. Gomis, E. Lonnoy, T. Maycock, M. Tignor, and T. Waterfield (eds.)]. In press, pp. 445–538.

Rulli, M.C., A. Saviori, and P. D'Odorico, 2012: Global land and water grabbing. *Pnas*, **110**, 892–897, doi:10.1073/pnas.1213163110/-/DCSupplemental.

Runting, R.K. et al., 2017: Incorporating climate change into ecosystem service assessments and decisions: a review. *Glob. Chang. Biol.*, **23**, 28–41, doi:10.1111/gcb.13457.

Ryan, C.M. et al., 2016: Ecosystem services from southern African woodlands and their future under global change. *Philos. Trans. R. Soc. B-Biological Sci.*, **371**, doi:2015031210.1098/rstb.2015.0312.

Salmon, G. et al., 2018: Trade-offs in livestock development at farm level: Different actors with different objectives. *Glob. Food Sec.*, doi:10.1016/J.GFS.2018.04.002.

Salvati, L. and M. Carlucci, 2014: Zero net land degradation in Italy: The role of socioeconomic and agro-forest factors. *J. Environ. Manage.*, **145**, 299–306, doi:10.1016/J.JENVMAN.2014.07.006.

Salvati, L., A. Sabbi, D. Smiraglia, and M. Zitti, 2014: Does forest expansion mitigate the risk of desertification? Exploring soil degradation and land-use changes in a Mediterranean country. *Int. For. Rev.*, **16**, 485–496, doi:10.1505/146554814813484149.

Santangeli, A. et al., 2016: Global change synergies and trade-offs between renewable energy and biodiversity. *Glob. Chang. Biol. Bioenergy*, **8**, doi:10.1111/gcbb.12299.

Santilli, G., C. Vendittozzi, C. Cappelletti, S. Battistini, and P. Gessini, 2018: CubeSat constellations for disaster management in remote areas. *Acta Astronaut.*, **145**, 11–17, doi:10.1016/j.actaastro.2017.12.050.

Sanz-Sanchez, M.-J. et al., 2017: *Sustainable Land Management Contribution to Successful Land-based Climate Change Adaptation and Mitigation*. A Report of the Science-Policy Interface, United Nations Convention to Combat Desertification (UNCCD), Bonn, Germany, 170 pp.

Schaeffer, M. et al.., 2015: Mid – and long-term climate projections for fragmented and delayed-action scenarios. *Technol. Forecast. Soc. Change*, **90**, 257–268, doi:10.1016/j.techfore.2013.09.013.

Schanes, K., K. Dobernig and B. Gözet, 2018: Food waste matters – A systematic review of household food waste practices and their policy implications. *J. Clean. Prod.*, **182**, 978–991, doi:10.1016/J.JCLEPRO.2018.02.030.

Schauberger, B. et al., 2017: Consistent negative response of US crops to high temperatures in observations and crop models. *Nat. Commun.*, **8**, doi:10.1038/ncomms13931.

Scheidel, A. and C. Work, 2018: Forest plantations and climate change discourses: New powers of 'green' grabbing in Cambodia. *Land use policy*, **77**, 9–18, doi:10.1016/j.landusepol.2018.04.057.

Schepaschenko, D. et al., 2015: Development of a global hybrid forest mask through the synergy of remote sensing, crowdsourcing and FAO statistics. *Remote Sens. Environ.*, **162**, 208–220, doi:10.1016/j.rse.2015.02.011.

Schipper, L.A., R.L. Parfitt, S. Fraser, R.A. Littler, W.T. Baisden and C. Ross, 2014: Soil order and grazing management effects on changes in soil C and N in New Zealand pastures. *Agric. Ecosyst. Environ.*, **184**, 67–75, doi:10.1016/J.AGEE.2013.11.012.

Schlenker, W. and D.B. Lobell, 2010: Robust negative impacts of climate change on African agriculture. *Environ. Res. Lett.*, **5**, 14010, doi:10.1186/s13021-018-0095-3.

Schlesinger, W.H., 2018: Are wood pellets a green fuel? *Science*, **359**, 1328–1329, doi:10.1126/science.aat2305.

Schmidt, C.G., K. Foerstl, and B. Schaltenbrand, 2017: The supply chain position paradox: Green practices and firm performance. *J. Supply Chain Manag.*, **53**, 3–25, doi:10.1111/jscm.12113.

Schneider, F. and T. Buser, 2018: Promising degrees of stakeholder interaction in research for sustainable development. *Sustain. Sci.*, **13**, 129–142, doi:10.1007/s11625-017-0507-4.

Scholes, R. et al., 2018: *IPBES: Summary for policymakers of the thematic assessment report on land degradation and restoration of the Intergovernmental Science-Policy Platform on Biodiversity and Ecosystem Services*. IPBES secretariat, Bonn, Germany, 44 pp.

Schröter, M. et al., 2018: Interregional flows of ecosystem services: Concepts, typology and four cases. *Ecosyst. Serv.*, doi:10.1016/j.ecoser.2018.02.003.

Schulte, R.P.O. et al., 2014: Functional land management: A framework for managing soil-based ecosystem services for the sustainable intensification of agriculture. *Environ. Sci. Policy*, **38**, 45–58, doi:10.1016/J.ENVSCI.2013.10.002.

Schut, M. et al., 2016: Sustainable intensification of agricultural systems in the Central African Highlands: The need for institutional innovation. *Agric. Syst.*, **145**, 165–176, doi:10.1016/J.AGSY.2016.03.005.

Schweikert, A., P. Chinowsky, X. Espinet, and M. Tarbert, 2014: Climate change and infrastructure impacts: Comparing the impact on roads in ten countries through 2100. *Procedia Eng.*, **78**, 306–316, doi:10.1016/j.proeng.2014.07.072.

Searchinger, T.D. et al., 2015: High carbon and biodiversity costs from converting Africa's wet savannahs to cropland. *Nat. Clim. Chang.*, **5**, 481–486, doi:10.1038/nclimate2584.

Searchinger, T.D., T. Beringer, and A. Strong, 2017: Does the world have low-carbon bioenergy potential from the dedicated use of land? *Energy Policy*, **110**, 434–446, doi:10.1016/j.enpol.2017.08.016.

Searle, S. and C. Malins, 2015: A reassessment of global bioenergy potential in 2050. *GCB Bioenergy*, **7**, 328–336, doi:10.1111/gcbb.12141.

Seidl, R. et al., 2017: Forest disturbances under climate change. *Nat. Clim. Chang.*, doi:10.1038/nclimate3303.

Seneviratne, S.I. et al., 2018: Climate extremes, land-climate feedbacks and land-use forcing at 1.5 degrees C. *Philos. Trans. R. Soc. a-Mathematical Phys. Eng. Sci.*, **376**, doi:2016045010.1098/rsta.2016.0450.

Seto, K.C. and A. Reenberg (eds.), 2014: *Rethinking Global Land Use in an Urban Era*. The MIT Press, Cambridge, Massachusetts, USA, 408 pp.

Seto, K.C. and N. Ramankutty, 2016: Hidden linkages between urbanization and food systems. *Science*, **352**, 943–945, doi:10.1126/science.aaf7439.

Seto, K.C., B. Guneralp and L.R. Hutyra, 2012: Global forecasts of urban expansion to 2030 and direct impacts on biodiversity and carbon pools. *Proc. Natl. Acad. Sci.*, **109**, 16083–16088, doi:10.1073/pnas.1211658109.

Settele, J., R. Scholes, R. Betts, S. Bunn, P. Leadley, D. Nepstad, J.T. Overpeck, and M.A. Taboada, 2014: Terrestrial and inland water systems. In: Climate Change 2014: Impacts, Adaptation, and Vulnerability. Part A: Global and Sectoral Aspects. Contribution of Working Group II to the Fifth Assessment Report of the Intergovernmental Panel on Climate Change [Field, C.B., V.R. Barros, D.J. Dokken, K.J. Mach, M.D. Mastrandrea, T.E. Bilir, M. Chatterjee, K.L. Ebi, Y.O. Estrada, R.C. Genova, B. Girma, E.S. Kissel, A.N. Levy, S. MacCracken, P.R. Mastrandrea, and L.L.White (eds.)]. Cambridge University Press, Cambridge, United Kingdom and New York, NY, USA, pp. 271–359.

Sheehy, T., F. Kolahdooz, C. Roache, and S. Sharma, 2015: Traditional food consumption is associated with better diet quality and adequacy among Inuit adults in Nunavut, Canada. *Int. J. Food Sci. Nutr.*, **66**, 445–451, doi: 10.3109/09637486.2015.1035232.

Shi, S., W. Zhang, P. Zhang, Y. Yu, and and F. Ding, 2013: A synthesis of change in deep soil organic carbon stores with afforestation of agricultural soils. *For. Ecol. Manage.*, **296**, 53–63, doi:10.1016/j.foreco.2013.01.026.

Shimamoto, C.Y., A.A. Padial, C.M. Da Rosa, and M.C.M.M. Marques, 2018: Restoration of ecosystem services in tropical forests: A global meta-analysis. *PLoS One*, **13**, 1–16, doi:10.1371/journal.pone.0208523.

Shoyama, K., 2008: Reforestation of abandoned pasture on Hokkaido, northern Japan: Effect of plantations on the recovery of conifer-broadleaved mixed forest. *Landsc. Ecol. Eng.*, **4**, 11–23, doi:10.1007/s11355-008-0034-7.

Shtienberg, D., 2013: Will decision-support systems be widely used for the management of plant diseases? *Annu. Rev. Phytopathol.*, **51**, 1–16, doi:10.1146/annurev-phyto-082712-102244.

Siebert, S., M. Kummu, M. Porkka, P. Döll, N. Ramankutty and B.R. Scanlon, 2015: A global data set of the extent of irrigated land from 1900 to 2005. *Hydrol. Earth Syst. Sci.*, doi:10.5194/hess-19-1521-2015.

Silveira, L., P. Gamazo, J. Alonso and L. Martínez, 2016: Effects of afforestation on groundwater recharge and water budgets in the western region of Uruguay. *Hydrol. Process.*, **30**, 3596–3608, doi:10.1002/hyp.10952.

Sivakumar, M.V.K., 2007: Interactions between climate and desertification. *Agric. For. Meteorol.* **142**, 143–155, doi:10.1016/j.agrformet.2006.03.025.

Sloan, S. and J.A. Sayer, 2015: Forest Resources Assessment of 2015 shows positive global trends but forest loss and degradation persist in poor tropical countries. *For. Ecol. Manage.*, **352**, 134–145, doi:10.1016/j.foreco.2015.06.013.

Smith, M.D., M.P. Rabbitt and A. Coleman – Jensen, 2017: Who are the world's food insecure? New evidence from the food and agriculture organization's food insecurity experience scale. *World Dev.*, **93**, 402–412, doi:10.1016/J.WORLDDEV.2017.01.006.

Smith, P., 2016: Soil carbon sequestration and biochar as negative emission technologies. *Glob. Chang. Biol.*, **22**, 1315–1324, doi:10.1111/gcb.13178.

Smith, P. and P.J. Gregory, 2013: Climate change and sustainable food production. *Proceedings of the Nutrition Society*, Vol. 72 of, 21–28, doi:10.1017/S0029665112002832.

Smith P., M. Bustamante, H. Ahammad, H. Clark, H. Dong, E.A. Elsiddig, H. Haberl, R. Harper, J. House, M. Jafari, O. Masera, C. Mbow, N.H. Ravindranath, C.W. Rice, C. Robledo Abad, A. Romanovskaya, F. Sperling, and F. Tubiello, 2014: Agriculture, Forestry and Other Land Use (AFOLU). In: Climate Change 2014: Mitigation of Climate Change. Contribution of Working Group III to the Fifth Assessment Report of the Intergovernmental Panel on Climate Change [Edenhofer, O., R. Pichs-Madruga, Y. Sokona, E. Farahani, S. Kadner, K. Seyboth, A. Adler, I. Baum, S. Brunner, P. Eickemeier, B. Kriemann, J. Savolainen, S. Schlömer, C. von Stechow, T. Zwickel and J.C. Minx (eds.)]. Cambridge University Press, Cambridge, United Kingdom and New York, NY, USA.

Smith, P. et al., 2016: Biophysical and economic limits to negative CO_2 emissions. *Nat. Clim. Chang.*, **6**, 42–50, doi:DOI: 10.1038/NCLIMATE2870.

Song, X.-P., 2018: Global estimates of ecosystem service value and change: Taking into account uncertainties in satellite-based land cover data. *Ecol. Econ.*, **143**, 227–235, doi:10.1016/j.ecolecon.2017.07.019.

Song, X.-P. et al., 2018: Global land change from 1982 to 2016. *Nature*, **560**, 639–643, doi:10.1038/s41586-018-0411-9.

Spennemann, P.C. et al., 2018: Land-atmosphere interaction patterns in southeastern South America using satellite products and climate models. *Int. J. Appl. Earth Obs. Geoinf.*, **64**, 96–103, doi:10.1016/j.jag.2017.08.016.

Springmann, M. et al., 2018: Options for keeping the food system within environmental limits. *Nature*, **562**, 1, doi:10.1038/s41586-018-0594-0.

Ssmith, P. et al., 2013: How much land-based greenhouse gas mitigation can be achieved without compromising food security and environmental goals? *Glob. Chang. Biol.*, **19**, 2285–2302, doi:10.1111/gcb.12160.

Stadler, K. et al., 2018: EXIOBASE 3: Developing a time series of detailed environmentally extended multi-regional input-output tables. *Journal of Industrial Ecology*, **22**, 502–515, doi:10.1111/jiec.12715.

Stavi, I. and R. Lal, 2015: Achieving zero net land degradation: Challenges and opportunities. *J. Arid Environ.*, **112**, 44–51, doi:10.1016/j.jaridenv.2014.01.016.

Stavi, I., G. Bel and E. Zaady, 2016: Soil functions and ecosystem services in conventional, conservation and integrated agricultural systems. A review. *Agron. Sustain. Dev.*, **36**, 32, doi:10.1007/s13593-016-0368-8.

Sterling, E. et al., 2017: Culturally grounded indicators of resilience in social-ecological systems. *Environ. Soc.*, **8**, 63–95, doi:10.3167/ares.2017.080104.

Sterner, T. and Coria, J. (eds.), 2003: *Policy Instruments for Environmental and Natural Resource Management*. Resources for the Future Press, Washington, DC, USA, 504 pp.

Stocker, T.F. et al., 2013b: Technical Summary. In: Climate Change 2013: The Physical Science Basis. Contribution of Working Group I to the Fifth Assessment Report of the Intergovernmental Panel on Climate Change [Stocker, T.F., D. Qin, G.-K. Plattner, M. Tignor, S.K. Allen, J. Boschung, A. Nauels, Y. Xia, V. Bex and P.M. Midgley (eds.)]. Cambridge University Press, Cambridge, United Kingdom and New York, NY, USA, 33–115 pp.

Stockmann, U. et al., 2013: The knowns, known unknowns and unknowns of sequestration of soil organic carbon. *Agric. Ecosyst. Environ.*, **164**, 80–99, doi:10.1016/J.AGEE.2012.10.001.

Strack, M., 2008: *Peatland and Climate Change*. International Peat Society and Saarijärven Offset Oy, Jyväskylä, Finland, 223 pp.

Strassburg, B.B.N. et al., 2017: Moment of truth for the Cerrado hotspot. *Nat. Ecol. Evol.*, **1**, 0099, doi:10.1038/s41559-017-0099.

Sunil, N. and S.R. Pandravada, 2015: Alien Crop Resources and Underutilized Species for Food and Nutritional Security of India. In: *Plant Biology and Biotechnology*, Springer India, New Delhi, pp. 757–775.

Surminski, S. and D. Oramas-Dorta, 2014: Flood insurance schemes and climate adaptation in developing countries. *Int. J. Disaster Risk Reduct.*, **7**, 154–164, doi: 10.1016/j.ijdrr.2013.10.005.

Sutton, P.C., S.J. Anderson, R. Costanza, and I. Kubiszewski, 2016: The ecological economics of land degradation: Impacts on ecosystem service values. *Ecol. Econ.*, **129**, 182–192, doi:10.1016/j.ecolecon.2016.06.016.

Swain, M., L. Blomqvist, J. McNamara, and W.J. Ripple, 2018: Reducing the environmental impact of global diets. *Sci. Total Environ.*, **610–611**, 1207–1209, doi:10.1016/J.SCITOTENV.2017.08.125.

Swart, R.O.B. and F. Raes, 2007: Making integration of adaptation and mitigation work: mainstreaming into sustainable development policies? *Clim. Policy*, **7**, 288–303, doi:10.1080/14693062.2007.9685657.

Tal, A., 2010: Desertification. In: *The Turning Points of Environmental History* [Uekoetter, F. (ed.)]. University of Pittsburgh Press, Pittsburgh, Pennsylvania, USA, pp. 146–161.

Taylor, A., J. Downing, B. Hassan, F. Denton and T.E. Downing, 2007: Inter-relationships between adaptation and mitigation. *Climate Change 2007: Impacts, Adaptation and Vulnerability. Contribution of Working Group II to the Fourth Assessment Report of the Intergovernmental Panel on Climate Change* [M.L. Parry, O.F. Canziani, J.P. Palutikof, P.J. van der Linden, and C.E. Hanson, (eds.)]. Cambridge University Press, Cambridge, UK, 745–777 pp.

Terraube, J., A. Fernandez-Llamazares, and M. Cabeza, 2017: The role of protected areas in supporting human health: A call to broaden the assessment of conservation outcomes. *Curr. Opin. Environ. Sustain.*, **25**, 50–58, doi.org/10.1016/j.cosust.2017.08.005.

Theriault, V., M. Smale, and H. Haider, 2017: How does gender affect sustainable intensification of cereal production in the West African Sahel? Evidence from Burkina Faso. *World Dev.*, **92**, 177–191, doi:10.1016/J.WORLDDEV.2016.12.003.

Thompson-Hall, M., E.R. Carr, and U. Pascual, 2016: Enhancing and expanding intersectional research for climate change adaptation in agrarian settings. *Ambio*, **45**, 373–382, doi:10.1007/s13280-016-0827-0.

Thompson, I.D. et al., 2014: Biodiversity and ecosystem services: Lessons from nature to improve management of planted forests for REDD-plus. *Biodivers. Conserv.*, **23**, 2613–2635, doi:10.1007/s10531-014-0736-0.

Thyberg, K.L. and D.J. Tonjes, 2016: Drivers of food waste and their implications for sustainable policy development. *Resour. Conserv. Recycl.*, **106**, 110–123, doi:10.1016/J.RESCONREC.2015.11.016.

Tian, H. et al., 2019: Global soil nitrous oxide emissions since the preindustrial era estimated by an ensemble of terrestrial biosphere models: Magnitude, attribution and uncertainty. *Glob. Chang. Biol.*, **25**, 640–659, doi:10.1111/gcb.14514.

Tigchelaar, M., D.S. Battisti, R.L. Naylor and D.K. Ray, 2018: Future warming increases probability of globally synchronized maize production shocks. *Proc. Natl. Acad. Sci.*, **115**, 6644–6649, doi:10.1073/pnas.1718031115.

Tilman, D. and M. Clark, 2014: Global diets link environmental sustainability and human health. *Nature*, **515**, 518–522, doi:10.1038/nature13959.

Tilman, D., C. Balzer, J. Hill, and B.L. Befort, 2011: Global food demand and the sustainable intensification of agriculture. *Proc. Natl. Acad. Sci.*, **108**, 20260–20264, doi:10.1073/pnas.1116437108.

Tom Veldkamp, Nico Polman, Stijn Reinhard, M.S., 2011: From scaling to governance of the land system: Bridging ecological and economic perspectives. *Ecol. Soc.*, **16**, 1, doi: 10.5751/ES-03691-160101.

Tubiello, F.N. et al., 2015: The contribution of agriculture, forestry and other land use activities to global warming, 1990–2012. *Glob. Chang. Biol.*, **21**, 2655–2660, doi:10.1111/gcb.12865.

Turner, P.A., C.B. Field, D.B. Lobell, D.L. Sanchez, and K.J. Mach, 2018: Unprecedented rates of land-use transformation in modelled climate change mitigation pathways. *Nature* Sust., **1**, 240–245, doi: 10.1038/s41893-018-0063-7.

UNCCD, 2014: Desertification: The Invisible Frontline, Secretariat of the United Nations Convention to Combat Desertification, United Nations Convention to Combat Desertification, Bonn, Germany.

UNEP, 2016: Global Gender and Environment Outlook, UN Environment, Nairobi, Kenya, 222 pp.

United Nations, 2015: Transforming Our World: The 2030 Agenda for Sustainable Development. United Nations, New York, NY, USA, 41 pp.

United Nations, 2018: *2018 Revision of World Urbanization Prospects*. www.un.org/development/desa/publications/2018-revision-of-world-urbanization-prospects.html.

United Nations Department of Economic and Social Affairs, 2017: World Population Prospects: The 2017 Revision, DVD Edition.

Urban, M.C. et al., 2016: Improving the forecast for biodiversity under climate change. *Science*, **353**, aad8466, doi:10.1126/science.aad8466.

USDA, 2007: *Precision Agriculture: NRCS Support for Emerging Technologies*. Agronomy Technical Note No. 1, Soil Quality National Technology Development Team, East National Technology Support Center, Natural Resources Conservation Service, Greensboro, North Carolina, USA, 9 pp.

Vadell, E., S. De-Miguel, and J. Pemán, 2016: Large-scale reforestation and afforestation policy in Spain: A historical review of its underlying ecological, socioeconomic and political dynamics. *Land use policy*, **55**, 37–48, doi:10.1016/J.LANDUSEPOL.2016.03.017.

Valayamkunnath, P., V. Sridhar, W. Zhao, and R.G. Allen, 2018: Intercomparison of surface energy fluxes, soil moisture, and evapotranspiration from eddy covariance, large-aperture scintillometer, and modeling across three ecosystems in a semiarid climate. *Agric. For. Meteorol.*, **248**, 22–47, doi:10.1016/j.agrformet.2017.08.025.

Valentin, C. et al., 2008: Agriculture, ecosystems and environment runoff and sediment losses from 27 upland catchments in Southeast Asia: Impact of rapid land use changes and conservation practices. *Agric. Ecosyst. Environ.*, **128**, 225–238, doi:10.1016/j.agee.2008.06.004.

Vaughan, N.E. and C. Gough, 2016: Expert assessment concludes negative emissions scenarios may not deliver. *Environ. Res. Lett.*, **11**, 95003, doi:10.1088/1748-9326/11/9/095003.

Veldman, J.W. et al., 2015: Where tree planting and forest expansion are bad for biodiversity and ecosystem services. *Bioscience*, **65**, 1011–1018, doi:10.1093/biosci/biv118.

Veldman, J.W., F.A.O. Silveira, F.D. Fleischman, N.L. Ascarrunz and G. Durigan, 2017: Grassy biomes: An inconvenient reality for large-scale forest restoration? A comment on the essay by Chazdon and Laestadius. *Am. J. Bot.*, **104**, 649–651, doi:10.3732/ajb.1600427.

Venter, O. et al., 2016: Sixteen years of change in the global terrestrial human footprint and implications for biodiversity conservation. *Nat. Commun.*, **7**, doi:10.1038/ncomms12558.

van Vliet, J., D.A. Eitelberg, and P.H. Verburg, 2017: A global analysis of land take in cropland areas and production displacement from urbanization. *Glob. Environ. Chang.*, **43**, 107–115, doi:10.1016/j.gloenvcha.2017.02.001.

De Vos, J.M., L.N. Joppa, J.L. Gittleman, P.R. Stephens, and S.L. Pimm, 2015: Estimating the normal background rate of species extinction. *Conserv. Biol.*, **29**, 452–462, doi:10.1111/cobi.12380.

van Vuuren, D.P. and T.R. Carter, 2014: Climate and socio-economic scenarios for climate change research and assessment: reconciling the new with the old. *Clim. Change*, **122**, 415–429, doi:10.1007/s10584-013-0974-2.

van Vuuren, D.P. et al., 2017: Energy, land-use and greenhouse gas emissions trajectories under a green growth paradigm. *Glob. Environ. Chang.*, **42**, 237–250, doi:10.1016/J.GLOENVCHA.2016.05.008.

Vuuren, D.P. Van et al., 2018: The need for negative emission technologies. *Nat. Clim. Chang.*, **8**, 391–397, doi:10.1038/s41558-018-0119-8.

Walker, W.E., M. Haasnoot and J.H. Kwakkel, 2013: Adapt or perish: A review of planning approaches for adaptation under deep uncertainty. *Sustainability*, **5**, 955–979, doi:10.3390/su5030955.

Walters, M. and R.J. Scholes (eds.), 2017: *The GEO handbook on biodiversity observation networks*. Springer International Publishing, Cham, Switzerland, 326 pp. doi:10.1007/978-3-319-27288-7.

Wang, X. et al., 2016: Taking account of governance: Implications for land-use dynamics, food prices and trade patterns. *Ecol. Econ.*, **122**, 12–24, doi:10.1016/j.ecolecon.2015.11.018.

Wärlind, D. et al., 2014: Nitrogen feedbacks increase future terrestrial ecosystem carbon uptake in an individual-based dynamic vegetation model. *Biogeosciences*, **11**, 6131–6146, doi:10.5194/bg-11-6131-2014.

Warren, D.D. and D.C. and N.R. and J.P. and R., 2014: Global crop yield response to extreme heat stress under multiple climate change futures. *Environ. Res. Lett.*, **9**, 34011.

Warszawski, L., K. Frieler, V. Huber, F. Piontek, O. Serdeczny and J. Schewe, 2014: The inter-sectoral impact model intercomparison project (ISI–MIP): Project framework. *Proc. Natl. Acad. Sci.*, **111**, 3228–3232, doi:10.1073/pnas.1312330110.

Watmuff, G., D.J. Reuter and S.D. Speirs, 2013: Methodologies for assembling and interrogating N, P, K, and S soil test calibrations for Australian cereal, oilseed and pulse crops. *Crop Pasture Sci.*, **64**, 424, doi:10.1071/CP12424.

Weindl, I. et al., 2017: Livestock and human use of land: Productivity trends and dietary choices as drivers of future land and carbon dynamics. *Glob. Planet. Change*, **159**, 1–10, doi:10.1016/j.gloplacha.2017.10.002.

Weitzman, M.L., 2014: Can negotiating a uniform carbon price help to internalize the global warming externality? *J. Assoc. Environ. Resour. Econ.*, **1**, 29–49, doi:10.3386/w19644.

van der Werf, P. and J.A. Gilliland, 2017: A systematic review of food losses and food waste generation in developed countries. *Proc. Inst. Civ. Eng. – Waste Resour. Manag.*, **170**, 66–77, doi:10.1680/jwarm.16.00026.

West, T.A.P., 2016: Indigenous community benefits from a de-centralized approach to REDD+ in Brazil. *Clim. Policy*, **16**, 924–939, doi:10.1080/14693062.2015.1058238.

Wichelns, D., 2017: The water-energy-food nexus: Is the increasing attention warranted, from either a research or policy perspective? *Environ. Sci. Policy*, **69**, 113–123, doi:10.1016/J.ENVSCI.2016.12.018.

Widener, M.J., L. Minaker, S. Farber, J. Allen, B. Vitali, P.C. Coleman and B. Cook, 2017: How do changes in the daily food and transportation environments affect grocery store accessibility? *Appl. Geogr.*, **83**, 46–62, doi:10.1016/J.APGEOG.2017.03.018.

Wiedmann, T. and M. Lenzen, 2018: Environmental and social footprints of international trade. *Nat. Geosci.*, **11**, 314–321, doi:10.1038/s41561-018-0113-9.

Williamson, P., 2016: Emissions reduction: Scrutinize CO$_2$ removal methods. *Nature*, **530**, 153–155, doi:10.1038/530153a.

Wilson, S.J., J. Schelhas, R. Grau, A.S. Nanni and S. Sloan, 2017: Forest ecosystem-service transitions: The ecological dimensions of the forest transition. *Ecol. Soc.*, **22**, doi:10.5751/es-09615-220438.

Wilting, H.C., A.M. Schipper, M. Bakkenes, J.R. Meijer and M.A.J. Huijbregts, 2017: Quantifying biodiversity losses due to human consumption: A global-scale footprint analysis. *Environ. Sci. Technol.*, **51**, 3298–3306, doi:10.1021/acs.est.6b05296.

Wise, R.M., I. Fazey, M.S. Smith, S.E. Park, H.C. Eakin, E.R.M.A. Van Garderen and B. Campbell, 2014: Reconceptualising adaptation to climate change as part of pathways of change and response. *Glob. Environ. Chang.*, **28**, 325–336, doi:10.1016/j.gloenvcha.2013.12.002.

Wisser, D., S. Frolking, E.M. Douglas, B.M. Fekete, C.J. Vörösmarty and A.H. Schumann, 2008: Global irrigation water demand: Variability and uncertainties arising from agricultural and climate data sets. *Geophys. Res. Lett.*, doi:10.1029/2008GL035296.

Wolff, S., E.A. Schrammeijer, C. Schulp and P.H. Verburg, 2018: Meeting global land restoration and protection targets: What would the world look like in 2050? *Glob. Environ. Chang.*, **52**, 259–272, doi:10.1016/j.gloenvcha.2018.08.002.

Wood, S.A., M.R. Smith, J. Fanzo, R. Remans and R.S. DeFries, 2018: Trade and the equitability of global food nutrient distribution. *Nat. Sustain.*, **1**, 34–37, doi:10.1038/s41893-017-0008-6.

Wu, X., Y. Lu, S. Zhou, L. Chen and B. Xu, 2016: Impact of climate change on human infectious diseases: Empirical evidence and human adaptation. *Environ. Int.*, **86**, 14–23, doi:10.1016/J.ENVINT.2015.09.007.

Wunder, S., 2015: Revisiting the concept of payments for environmental services. *Ecol. Econ.*, **117**, 234–243, doi:10.1016/J.ECOLECON.2014.08.016.

Wynes, S. and K.A. Nicholas, 2017: The climate mitigation gap: Education and government recommendations miss the most effective individual actions. *Environ. Res. Lett.*, **12**, 74024, doi:10.1088/1748-9326/aa7541.

Xu, Y., 2018: Political economy of land grabbing inside China involving foreign investors. *Third World Q.*, **39**(11), 2069–2084, doi:10.1080/01436597.2018.1447372.

Xue, L. et al., 2017: Missing food, missing data? A critical review of global food losses and food waste data. *Environ. Sci. Technol.*, **51**, 6618–6633, doi:10.1021/acs.est.7b00401.

Yang, L., L. Chen, W. Wei, Y. Yu and H. Zhang, 2014: Comparison of deep soil moisture in two re-vegetation watersheds in semi-arid regions. *J. Hydrol.*, **513**, 314–321, doi:10.1016/j.jhydrol.2014.03.049.

Yang, Y., D. Tilman, G. Furey and C. Lehman, 2019: Soil carbon sequestration accelerated by restoration of grassland biodiversity. *Nat. Commun.*, **10**, 718, doi:10.1038/s41467-019-08636-w.

Yirdaw, E., M. Tigabu and A. Monge, 2017: Rehabilitation of degraded dryland ecosystems – Review. *Silva Fenn.*, **51**, doi:10.14214/sf.1673.

Yohe, G.W., 2001: Mitigative capacity – The mirror image of adaptive capacity on the emissions side. *Clim. Change*, **49**, 247–262, doi:10.1023/A:1010677916703.

Yu, L. et al., 2014: Meta-discoveries from a synthesis of satellite-based land-cover mapping research. *Int. J. Remote Sens.*, **35**, 4573–4588, doi:10.1080/01431161.2014.930206.

Yu, Y., K. Feng and K. Hubacek, 2013: Tele-connecting local consumption to global land use. *Glob. Environ. Chang.*, **23**, 1178–1186, doi:10.1016/J.GLOENVCHA.2013.04.006.

Zaloumis, N.P. and W.J. Bond, 2015: Reforestation of afforestation? The attributes of old growth grasslands in South Africa. *Philos. Trans. R. Soc. B*, **371**, 1–9, doi:10.1098/rstb.2015.0310.

Zhang, M. et al., 2014: Response of surface air temperature to small-scale land clearing across latitudes. *Environ. Res. Lett.*, **9**, 34002, doi:10.1088/1748-9326/9/3/034002.

Zhang, X. et al., 2015: Managing nitrogen for sustainable development. *Nature*, **528**, 51–59, doi:10.1038/nature15743.

Zhao, L., A. Dai and B. Dong, 2018: Changes in global vegetation activity and its driving factors during 1982–2013. *Agric. For. Meteorol.*, doi:10.1016/j.agrformet.2017.11.013.

Zheng, H., Y. Wang, Y. Chen and T. Zhao, 2016: Effects of large-scale afforestation project on the ecosystem water balance in humid areas: An example for southern China. *Ecol. Eng.*, **89**, 103–108, doi:10.1016/j.ecoleng.2016.01.013.

Zhu, Z. et al., 2016: Greening of the Earth and its drivers. *Nat. Clim. Chang.*, **6**, 791–795, doi:10.1038/nclimate3004.

Ziadat, F., S. Bunning, S. Corsi, and R. Vargas, 2018: Sustainable soil and land management for climate smart agriculture. In: *Climate Smart Agriculture Sourcebook* [Ziadat, F., S. Bunning, S. Corsi and R. Vargas (eds.)]. Food and Agriculture Organization of the United Nations, Rome, Italy, pp 1–33.

Ziese, M. et al., 2014: The GPCC Drought Index – A new, combined and gridded global drought index. *Earth Syst. Sci. Data*, **6**, 285–295, doi:10.5194/essd-6-285-2014.

Ziska, L.H. et al., 2016: Rising atmospheric CO$_2$ is reducing the protein concentration of a floral pollen source essential for North American bees. *Proceedings. Biol. Sci.*, **283**, 20160414, doi:10.1098/rspb.2016.0414.

Zorya, S. et al., 2011: *Missing food: The Case of Postharvest Grain Losses in Sub-Saharan Africa*. The International Bank for Reconstruction and Development/The World Bank Report No. 60371-AFR, Washington, DC, USA, 96 pp.

Appendix

Table Appendix 1.1 | Observations related to variables indicative of land management (LM), and their uncertainties.

LM-related process	Observations methodology	Scale of observations (space and time)	Uncertainties[2]	Pros and cons	Select literature
GHG emissions	Micrometeorological fluxes (CO_2) Micrometeorological fluxes (CH_4) Micrometeorological fluxes (N_2O)	1–10 ha 0.5 hr – >10 y	5–15% 10–40% 20–50%	**Pros** – Larger footprints – Continuous monitoring – Less disturbance on monitored system – Detailed protocols **Cons** – Limitations by fetch and turbulence scale – Not all trace gases	Richardson et al. 2006; Luyssaert et al. 2007; Foken and Napo 2008; Mauder et al. 2013; Peltola et al. 2014; Wang et al. 2015; Rannik et al. 2015; Campioli et al. 2016; Rannik et al. 2016; Wang et al. 2017a; Brown and Wagner-Riddle 2017; Desjardins et al. 2018
	Soil chambers (CO_2) Soil chambers (CH_4) Soil chambers (N_2O)	0.01–1 ha 0.5 hr – 1 y	5–15% 5–25% 53–100%[3]	**Pros** – Relatively inexpensive – Possibility of manipulation experiments – Large range of trace gases **Cons** – Smaller footprint – Complicated upscaling – Static pressure interference	Vargas and Allen 2008; Lavoie et al. 2015; Barton et al. 2015; Dossa et al. 2015; Ogle et al. 2016; Pirk et al. 2016; Morin et al. 2017; Lammirato et al. 2018
	Atmospheric inversions (CO_2) Atmospheric inversions (CH_4)	Regional 1 – >10 y	50% 3–8%	**Pros** – Integration on large scale – Attribution detection (with 14C) – Rigorously derived uncertainty **Cons** – Not suited at farm scale – Large high-precision observation network required	Wang et al. 2017b Pison et al. 2018
Carbon balance	Soil carbon point measurements	0.01–1 ha >5 y	5–20%	**Pros** – Easy protocol – Well established analytics **Cons** – Need high number of samples for upscaling – Detection limit is high	Chiti et al. 2018; Castaldi et al. 2018; Chen et al. 2018; Deng et al. 2018
	Biomass measurements	0.01–1 ha 1–5 y	2–8%	**Pros** – Well established allometric equations – High accuracy at plot level **Cons** – Difficult to scale up – Labour intensive	Pelletier et al. 2012; Henry et al. 2015; Vanguelova et al. 2016; Djomo et al. 2016; Forrester et al. 2017; Xu et al. 2017 Marziliano et al. 2017; Clark et al. 2017; Disney et al. 2018; Urbazaev et al. 2018; Paul et al. 2018

---

[2]   Uncertainty here is defined as the coefficient of variation CV. In the case of micrometeorological fluxes they refer to random errors and CV of daily average.

[3]   >100 for fluxes less than 5 gN_2O-N ha^{-1} d^{-1}.

LM-related process	Observations methodology	Scale of observations (space and time)	Uncertainties[2]	Pros and cons	Select literature
Water balance	Soil moisture (IoT sensors, Cosmic rays, Thermo-optical sensing etc.)	0.01 ha – regional 0.5 hr – <1 y	3–5% vol	**Pros** – New technology – Big data analytics – Relatively inexpensive **Cons** – Scaling problems	Yu et al. 2013; Zhang and Zhou 2016; Iwata et al. 2017; McJannet et al. 2017; Karthikeyan et al. 2017; Iwata et al. 2017; Cao et al. 2018; Amaral et al. 2018; Moradizadeh and Saradjian 2018; Strati et al. 2018
	Evapotranspiration	0.01 ha – regional 0.5 hr – >10 y	10–20%	**Pros** – Well established methods – Easy integration in models and DSS **Cons** – Partition of fluxes need additional measurements	Zhang et al. 2017; Papadimitriou et al. 2017; Kaushal et al. 2017; Valayamkunnath et al. 2018; Valayamkunnath et al. 2018; Tie et al. 2018; Wang et al. 2018
Soil erosion	Sediment transport	1 ha – regional 1d – >10y	21–34%	**Pros** – Long history of methods – Integrative tools **Cons** – Validation is lacking – Labour intensive	Efthimiou 2018; García-Barrón et al. 2018; Fiener et al. 2018
Land cover	Satellite	0.01 ha – regional 1 d – >10 y	16–100%	**Pros** – Increasing platforms available – Consolidated algorithms **Cons** – Need validation – Lack of common land-use definitions	Olofsson et al. 2014; Liu et al. 2018; Yang et al. 2018

Table Appendix 1.2 | Possible uncertainties decision-making faces (following Hansson and Hadorn 2016).

Type	Knowledge gaps	Understanding the uncertainties
Uncertainty of consequences	Do the model(s) adequately represent the target system? What are the numerical values of input parameters, boundary conditions, or initial conditions? What are all potential events that we would take into account if we were aware of them? Will future events relevant for our decisions, including expected impacts from these decisions, in fact take place?	Ensemble approaches; downscaling Benchmarking, sensitivity analyses Scenario approaches
Moral uncertainty	How to (ethically) evaluate the decisions? What values to base the decision on (often unreliable ranking of values not doing justice to the range of values at stake, see Sen 1992), including choice of discount rate, risk attitude (risk aversion, risk neutral, …). Which ethical principles? (i.e. utilitarian, deontic, virtue, or other?).	Possibly scenario analysis; Identification of lock-in effects and path-dependency (e.g., Kinsley et al. 2016)
Uncertainty of demarcation	What are the options that we can actually choose between? (not fully known because 'decision costs' may be high, certain options are not 'seen' as they are outside current ideologies). How can the mass of decisions be divided into individual decisions? e.g., how this influences international negotiations and the question who does what and when (cp. Hammond et al. 1999).	Possibly scenario analysis
Uncertainty of consequences and uncertainty of demarcation	What effects does a decision have when combined with the decisions of others? (e.g., other countries may follow the inspiring example in climate reduction of country X, or they may use it solely in their own economic interest).	Games
Uncertainty of demarcation and moral uncertainty	How would we decide in the future? (Spohn 1977; Rabinowicz 2002).	

References to Appendix

Amaral, A.M. et al., 2018: Uncertainty of weight measuring systems applied to weighing lysimeters. *Comput. Electron. Agric.*, **145**, 208–216, doi:10.1016/j.compag.2017.12.033.

Barton, L., B. Wolf, D. Rowlings, C. Scheer, R. Kiese, P. Grace, K. Stefanova and K. Butterbach-Bahl, 2015: Sampling frequency affects estimates of annual nitrous oxide fluxes. *Sci. Rep.*, **5**, 1–9, doi:10.1038/srep15912.

Brown, S.E. and C. Wagner-Riddle, 2017: Assessment of random errors in multi-plot nitrous oxide flux gradient measurements. *Agric. For. Meteorol.*, **242**, 10–20, doi:10.1016/j.agrformet.2017.04.005.

Campioli, M. et al., 2016: Article evaluating the convergence between eddy-covariance and biometric methods for assessing carbon budgets of forests. *Nat. Commun.*, **7**, doi:10.1038/ncomms13717.

Cao, D.-F., B. Shi, G.-Q. Wei, S.-E. Chen and H.-H. Zhu, 2018: An improved distributed sensing method for monitoring soil moisture profile using heated carbon fibers. *Meas. J. Int. Meas. Confed.*, **123**, doi:10.1016/j.measurement.2018.03.052.

Castaldi, F. et al., 2018: Estimation of soil organic carbon in arable soil in Belgium and Luxembourg with the LUCAS topsoil database. *Eur. J. Soil Sci.*, doi:10.1111/ejss.12553.

Chen, S. et al., 2018: Fine resolution map of top – and subsoil carbon sequestration potential in France. *Sci. Total Environ.*, **630**, 389–400, doi:10.1016/J.SCITOTENV.2018.02.209.

Chiti, T. et al., 2018: Soil organic carbon pool's contribution to climate change mitigation on marginal land of a Mediterranean montane area in Italy. *J. Environ. Manage.*, **218**, 593–601, doi:10.1016/j.jenvman.2018.04.093.

Clark, D.A. et al., 2017: Reviews and syntheses: Field data to benchmark the carbon cycle models for tropical forests. *Biogeosciences*, **14**, 4663–4690, doi:10.5194/bg-14-4663-2017.

Deng, X. et al., 2018: Baseline map of organic carbon stock in farmland topsoil in East China. *Agric. Ecosyst. Environ.*, **254**, 213–223, doi:10.1016/J.AGEE.2017.11.022.

Desjardins, R.L. et al., 2018: The challenge of reconciling bottom-up agricultural methane emissions inventories with top-down measurements. *Agric. For. Meteorol.*, **248**, 48–59, doi:10.1016/j.agrformet.2017.09.003.

Disney, M.I. et al., 2018: Weighing trees with lasers: Advances, challenges and opportunities. *Interface Focus*, **8**, 20170048, doi:10.1098/rsfs.2017.0048.

Djomo, A.N. et al., 2016: Tree allometry for estimation of carbon stocks in African tropical forests. *Forestry*, **89**, 446–455, doi:10.1093/forestry/cpw025.

Dossa, G.G.O. et al., 2015: Correct calculation of CO_2 efflux using a closed-chamber linked to a non-dispersive infrared gas analyzer. *Methods Ecol. Evol.*, **6**, 1435–1442, doi:10.1111/2041-210X.12451.

Efthimiou, N., 2018: The importance of soil data availability on erosion modeling. *CATENA*, **165**, 551–566, doi:10.1016/J.CATENA.2018.03.002.

Fiener, P. et al., 2018: Uncertainties in assessing tillage erosion – How appropriate are our measuring techniques? *Geomorphology*, **304**, 214–225, doi:10.1016/J.GEOMORPH.2017.12.031.

Foken, T. and C.J. Napo, 2008: *Micrometeorology*. Springer International Publishing, Cham, Switzerland.

Forrester, D.I. et al., 2017: Generalized biomass and leaf area allometric equations for European tree species incorporating stand structure, tree age and climate. *For. Ecol. Manage.*, **396**, 160–175, doi:10.1016/j.foreco.2017.04.011.

García-Barrón, L., J. Morales, and A. Sousa, 2018: A new methodology for estimating rainfall aggressiveness risk based on daily rainfall records for multi-decennial periods. *Sci. Total Environ.*, **615**, 564–571, doi:10.1016/j.scitotenv.2017.09.305.

Hammond, J.S., R.L. Keeney, and H.R., 1999: Smart choices: A practical guide to making better life decisions. *Broadway Books*, New York, NY, USA.

Hansson, S.O. and G.H. Hadorn, 2016: Introducing the Argumentative Turn in Policy Analysis. *The Argumentative Turn in Policy Analysis*, Logic, Argumentation & Reasoning, Springer International Publishing, Cham, Switzerland, 11–35, doi:10.1007/978-3-319-30549-3.

Henry, M. et al., 2015: Recommendations for the use of tree models to estimate national forest biomass and assess their uncertainty. *Ann. For. Sci.*, **72**, 769–777, doi:10.1007/s13595-015-0465-x.

Iwata, Y., T. Miyamoto, K. Kameyama, and M. Nishiya, 2017: Effect of sensor installation on the accurate measurement of soil water content. *Eur. J. Soil Sci.*, **68**, 817–828, doi:10.1111/ejss.12493.

Karthikeyan, L., M. Pan, N. Wanders, D.N. Kumar, and E.F. Wood, 2017: Four decades of microwave satellite soil moisture observations: Part 1. A review of retrieval algorithms. *Adv. Water Resour.*, **109**, 106–120, doi:10.1016/j.advwatres.2017.09.006.

Kaushal, S.S., A.J. Gold and P.M. Mayer, 2017: Land use, climate, and water resources-global stages of interaction. *Water (Switzerland)*, **9**, 815, doi:10.3390/w9100815.

Lammirato, C., U. Lebender, J. Tierling, and J. Lammel, 2018: Analysis of uncertainty for N 2 O fluxes measured with the closed-chamber method under field conditions: Calculation method, detection limit and spatial variability. *J. Plant Nutr. Soil Sci.*, **181**, 78–89, doi:10.1002/jpln.201600499.

Lavoie, M., C.L. Phillips, and D. Risk, 2015: A practical approach for uncertainty quantification of high-frequency soil respiration using Forced Diffusion chambers. *J. Geophys. Res. Biogeosciences*, **120**, 128–146, doi:10.1002/2014JG002773.

Liu, X. et al., 2018: Comparison of country-level cropland areas between ESA-CCI land cover maps and FAOSTAT data. *Int. J. Remote Sens.*, doi:10.1080/01431161.2018.1465613.

Luyssaert, S. et al., 2007: CO_2 balance of boreal, temperate and tropical forests derived from a global database. *Glob. Chang. Biol.*, **13**, 2509–2537, doi:10.1111/j.1365-2486.2007.01439.x.

Marziliano, P., G. Menguzzato and V. Coletta, 2017: Evaluating Carbon Stock Changes in Forest and Related Uncertainty. *Sustainability*, **9**, 1702, doi:10.3390/su9101702.

Mauder, M. et al., 2013: A strategy for quality and uncertainty assessment of long-term eddy-covariance measurements. *Agric. For. Meteorol.*, **169**, 122–135, doi:10.1016/j.agrformet.2012.09.006.

McJannet, D., A. Hawdon, B. Baker, L. Renzullo, and R. Searle, 2017: Multiscale soil moisture estimates using static and roving cosmic-ray soil moisture sensors. *Hydrol. Earth Syst. Sci. Discuss.*, 1–28, doi:10.5194/hess-2017-358.

Moradizadeh, M. and M.R. Saradjian, 2018: Estimation of improved resolution soil moisture in vegetated areas using passive AMSR-E data. *J. Earth Syst. Sci.*, **127**, 24, doi:10.1007/s12040-018-0925-4.

Morin, T.H. et al., 2017: Combining eddy-covariance and chamber measurements to determine the methane budget from a small, heterogeneous urban floodplain wetland park. *Agric. For. Meteorol.*, **237238**, 160–170, doi:10.1016/j.agrformet.2017.01.022.

Ogle, K., E. Ryan, F.A. Dijkstra E. Pendall, 2016: Quantifying and reducing uncertainties in estimated soil CO_2 fluxes with hierarchical data-model integration. *J. Geophys. Res. Biogeosciences*, **121**, 2935–2948, doi:10.1002/2016JG003385.

Olofsson, P. et al., 2014: Good practices for estimating area and assessing accuracy of land change. *Remote Sens. Environ.*, **148**, 42–57, doi:10.1016/j.rse.2014.02.015.

Papadimitriou, L.V., A.G. Koutroulis, M.G. Grillakis and I.K. Tsanis, 2017: The effect of GCM biases on global runoff simulations of a land surface model. *Hydrol. Earth Syst. Sci.*, **21**, 4379–4401, doi:10.5194/hess-21-4379-2017.

Paul, K.I. et al., 2018: Using measured stocks of biomass and litter carbon to constrain modelled estimates of sequestration of soil organic carbon under contrasting mixed-species environmental plantings. *Sci. Total Environ.*, **615**, 348–359, doi:10.1016/j.scitotenv.2017.09.263.

Pelletier, J., K.R. Kirby, and C. Potvin, 2012: Significance of carbon stock uncertainties on emission reductions from deforestation and forest degradation in developing countries. *For. Policy Econ.*, **24**, 3–11, doi:10.1016/j.forpol.2010.05.005.

Peltola, O. et al., 2014: Evaluating the performance of commonly used gas analysers for methane eddy covariance flux measurements: the InGOS inter-comparison field experiment. *Biogeosciences*, **11**, 3163–3186, doi:10.5194/bg-11-3163-2014, www.biogeosciences.net/11/3163/2014/.

Pirk, N. et al., 2016: Calculations of automatic chamber flux measurements of methane and carbon dioxide using short time series of concentrations. *Biogeosciences*, **13**, 903–912, doi:10.5194/bg-13-903-2016, www.biogeosciences.net/13/903/2016/.

Pison, I. et al., 2018: How a European network may help with estimating methane emissions on the French national scale. *Atmos. Chem. Phys*, **185194**, 3779–3798, doi:10.5194/acp-18-3779-2018.

Rabinowicz, W., 2002: Does practical deliberation crowd out self-prediction? *Erkenntnis*, **57**, 91–122, doi:10.1023/A:1020106622032.

Rannik, Ü., O. Peltola, and I. Mammarella, 2016: Random uncertainties of flux measurements by the eddy covariance technique. *Atmos. Meas. Tech*, **9**, 5163–5181, doi:10.5194/amt-9-5163-2016, www.atmos-meas-tech.net/9/5163/2016/.

Spohn, W., 1977: "Where Luce and Krantz do really generalize Savage's decision model." *Erkenntnis*, **11**, 113–134.

Strati, V. et al., 2018: Modelling Soil Water Content in a Tomato Field: Proximal Gamma Ray Spectroscopy and Soil–Crop System Models. *Agriculture*, **8**, 60, doi:10.3390/agriculture8040060.

Tie, Q., H. Hu, F. Tian, and N.M. Holbrook, 2018: Comparing different methods for determining forest evapotranspiration and its components at multiple temporal scales. *Sci. Total Environ.*, **633**, 12–29, doi:10.1016/j.scitotenv.2018.03.082.

Urbazaev, M., C. Thiel, F. Cremer, R. Dubayah, M. Migliavacca, M. Reichstein and C. Schmullius, 2018: Estimation of forest aboveground biomass and uncertainties by integration of field measurements, airborne LiDAR, and SAR and optical satellite data in Mexico. *Carbon Balance Manag.*, **13**, doi:10.1186/s13021-018-0093-5.

Valayamkunnath, P., V. Sridhar, W. Zhao, and R.G. Allen, 2018: Intercomparison of surface energy fluxes, soil moisture, and evapotranspiration from eddy covariance, large-aperture scintillometer, and modeling across three ecosystems in a semiarid climate. *Agric. For. Meteorol.*, **248**, 22–47, doi:10.1016/j.agrformet.2017.08.025.

Vanguelova, E.I. et al., 2016: Sources of errors and uncertainties in the assessment of forest soil carbon stocks at different scales—review and recommendations. *Environ. Monit. Assess.*, **188**, doi:10.1007/s10661-016-5608-5.

Vargas, R. and M.F. Allen, 2008: Environmental controls and the influence of vegetation type, fine roots and rhizomorphs on diel and seasonal variation in soil respiration. *New Phytol.*, **179**, 460–471, doi:10.1111/j.1469-8137.2008.02481.x.

Wang, E. et al., 2018: Making sense of cosmic-ray soil moisture measurements and eddy covariance data with regard to crop water use and field water balance. *Agric. Water Manag.*, **204**, 271–280, doi:10.1016/J.AGWAT.2018.04.017.

Wang, X., C. Wang, and B. Bond-Lamberty, 2017a: Quantifying and reducing the differences in forest CO_2 – fluxes estimated by eddy covariance, biometric and chamber methods: A global synthesis. doi:10.1016/j.agrformet.2017.07.023.

Wang, Y. et al., 2017b: Estimation of observation errors for large-scale atmospheric inversion of CO_2 emissions from fossil fuel combustion. *Tellus B Chem. Phys. Meteorol.*, **69**, 1325723, doi:10.1080/16000889.2017.1325723.

Xu, L. et al., 2017: Spatial distribution of carbon stored in forests of the Democratic Republic of Congo. *Sci. Rep.*, **7**, 1–12, doi:10.1038/s41598-017-15050-z.

Yang, L. et al., 2018: Spatio-temporal analysis and uncertainty of fractional vegetation cover change over northern China during 2001–2012 based on multiple vegetation data sets. *Remote Sens.*, **10**, 549, doi:10.3390/rs10040549.

Zhang, D. and G. Zhou, 2016: Estimation of Soil moisture from optical and thermal remote sensing: A review. *Sensors*, **16**, 1308, doi:10.3390/s16081308.

Zhang, Y. et al., 2017: Global variation of transpiration and soil evaporation and the role of their major climate drivers. *J. Geophys. Res.*, **122**, 6868–6881, doi:10.1002/2017JD027025.

1

2

Land–climate interactions

Coordinating Lead Authors:
Gensuo Jia (China), Elena Shevliakova (The United States of America)

Lead Authors:
Paulo Artaxo (Brazil), Nathalie De Noblet-Ducoudré (France), Richard Houghton (The United States of America), Joanna House (United Kingdom), Kaoru Kitajima (Japan), Christopher Lennard (South Africa), Alexander Popp (Germany), Andrey Sirin (The Russian Federation), Raman Sukumar (India), Louis Verchot (Colombia/The United States of America)

Contributing Authors:
William Anderegg (The United States of America), Edward Armstrong (United Kingdom), Ana Bastos (Portugal/Germany), Terje Koren Bernsten (Norway), Peng Cai (China), Katherine Calvin (The United States of America), Francesco Cherubini (Italy), Sarah Connors (France/United Kingdom), Annette Cowie (Australia), Edouard Davin (Switzerland/France), Cecile De Klein (New Zealand), Giacomo Grassi (Italy/European Union), Rafiq Hamdi (Belgium), Florian Humpenöder (Germany), David Kanter (The United States of America), Gerhard Krinner (France), Sonali McDermid (India/The United States of America), Devaraju Narayanappa (India/France), Josep Peñuelas (Spain), Prajal Pradhan (Nepal), Benjamin Quesada (Colombia), Stephanie Roe (The Philippines/The United States of America), Robert A. Rohde (The United States of America), Martijn Slot (Panama), Rolf Sommer (Germany), Moa Sporre (Norway), Benjamin Sulman (The United States of America), Alasdair Sykes (United Kingdom), Phil Williamson (United Kingdom), Yuyu Zhou (China)

Review Editors:
Pierre Bernier (Canada), Jhan Carlo Espinoza (Peru), Sergey Semenov (The Russian Federation)

Chapter Scientist:
Xiyan Xu (China)

This chapter should be cited as:
Jia, G., E. Shevliakova, P. Artaxo, N. De Noblet-Ducoudré, R. Houghton, J. House, K. Kitajima, C. Lennard, A. Popp, A. Sirin, R. Sukumar, L. Verchot, 2019: Land–climate interactions. In: *Climate Change and Land: an IPCC special report on climate change, desertification, land degradation, sustainable land management, food security, and greenhouse gas fluxes in terrestrial ecosystems* [P.R. Shukla, J. Skea, E. Calvo Buendia, V. Masson-Delmotte, H.-O. Pörtner, D.C. Roberts, P. Zhai, R. Slade, S. Connors, R. van Diemen, M. Ferrat, E. Haughey, S. Luz, S. Neogi, M. Pathak, J. Petzold, J. Portugal Pereira, P. Vyas, E. Huntley, K. Kissick, M, Belkacemi, J. Malley, (eds.)]. https://doi.org/10.1017/9781009157988.004

Table of contents

Executive summary 133

2.1 Introduction: Land–climate interactions 137

  2.1.1 Recap of previous IPCC and other
         relevant reports as baselines 137

  2.1.2 Introduction to the chapter structure 138

  Box 2.1: Processes underlying
  land–climate interactions 139

2.2 The effect of climate variability
     and change on land 140

  2.2.1 Overview of climate impacts on land 140

  2.2.2 Climate-driven changes in aridity 142

  2.2.3 The influence of climate
         change on food security 142

  2.2.4 Climate-driven changes
         in terrestrial ecosystems 143

  2.2.5 Climate extremes and their impact
         on land functioning 144

  Cross-Chapter Box 3 | Fire and climate change 148

2.3 Greenhouse gas fluxes between
     land and atmosphere 151

  2.3.1 Carbon dioxide 152

  2.3.2 Methane 157

  2.3.3 Nitrous oxide 160

  Box 2.2: Methodologies for estimating national
  to global scale anthropogenic land carbon fluxes 163

  Box 2.3: CO_2 fertilisation and enhanced
  terrestrial uptake of carbon 165

2.4 Emissions and impacts of short-lived
     climate forcers (SLCF) from land 166

  2.4.1 Mineral dust 166

  2.4.2 Carbonaceous aerosols 167

  2.4.3 Biogenic volatile organic compounds 169

2.5 Land impacts on climate and weather through
     biophysical and GHG effects 171

  2.5.1 Impacts of historical and future
         anthropogenic land cover changes 171

  2.5.2 Impacts of specific land use changes 176

  2.5.3 Amplifying/dampening climate changes
         via land responses 182

  2.5.4 Non-local and downwind effects resulting
         from changes in land cover 184

  Cross-Chapter Box 4 | Climate change
  and urbanisation 186

2.6 Climate consequences of response options 188

  2.6.1 Climate impacts of individual response options 189

  2.6.2 Integrated pathways for
         climate change mitigation 195

  2.6.3 The contribution of response options
         to the Paris Agreement 199

2.7 Plant and soil processes underlying
     land–climate interactions 201

  2.7.1 Temperature responses of plant
         and ecosystem production 201

  2.7.2 Water transport through
         soil-plant-atmosphere continuum
         and drought mortality 202

  2.7.3 Soil microbial effects on soil nutrient dynamics
         and plant responses to elevated CO_2 202

  2.7.4 Vertical distribution of soil organic carbon 203

  2.7.5 Soil carbon responses to warming
         and changes in soil moisture 203

  2.7.6 Soil carbon responses to changes
         in organicmatter inputs by plants 204

Frequently Asked Questions 205

  FAQ 2.1: How does climate change affect
  land use and land cover? 205

  FAQ 2.2: How do the land and land use
  contribute to climate change? 205

  FAQ 2.3: How does climate change affect
  water resources? 205

References 206

Appendix 243

Executive summary

Land and climate interact in complex ways through changes in forcing and multiple biophysical and biogeochemical feedbacks across different spatial and temporal scales. This chapter assesses climate impacts on land and land impacts on climate, the human contributions to these changes, as well as land-based adaptation and mitigation response options to combat projected climate changes.

Implications of climate change, variability and extremes for land systems

It is certain that globally averaged land surface air temperature (LSAT) has risen faster than the global mean surface temperature (i.e., combined LSAT and sea surface temperature) from the preindustrial period (1850–1900) to the present day (1999–2018). According to the single longest and most extensive dataset, from 1850–1900 to 2006–2015 mean land surface air temperature has increased by 1.53°C (*very likely* range from 1.38°C to 1.68°C) while global mean surface temperature has increased by 0.87°C (*likely range* from 0.75°C to 0.99°C). For the 1880–2018 period, when four independently produced datasets exist, the LSAT increase was 1.41°C (1.31–1.51°C), where the range represents the spread in the datasets' median estimates. Analyses of paleo records, historical observations, model simulations and underlying physical principles are all in agreement that LSATs are increasing at a higher rate than SST as a result of differences in evaporation, land–climate feedbacks and changes in the aerosol forcing over land (*very high confidence*). For the 2000–2016 period, the land-to-ocean warming ratio (about 1.6) is in close agreement between different observational records and the CMIP5 climate model simulations (the *likely* range of 1.54–1.81). {2.2.1}

Anthropogenic warming has resulted in shifts of climate zones, primarily as an increase in dry climates and decrease of polar climates (*high confidence*). Ongoing warming is projected to result in new, hot climates in tropical regions and to shift climate zones poleward in the mid- to high latitudes and upward in regions of higher elevation (*high confidence*). Ecosystems in these regions will become increasingly exposed to temperature and rainfall extremes beyonwd the climate regimes they are currently adapted to (*high confidence*), which can alter their structure, composition and functioning. Additionally, high-latitude warming is projected to accelerate permafrost thawing and increase disturbance in boreal forests through abiotic (e.g., drought, fire) and biotic (e.g., pests, disease) agents (*high confidence*). {2.2.1, 2.2.2, 2.5.3}

Globally, greening trends (trends of increased photosynthetic activity in vegetation) have increased over the last 2–3 decades by 22–33%, particularly over China, India, many parts of Europe, central North America, southeast Brazil and southeast Australia (*high confidence*). This results from a combination of direct (i.e., land use and management, forest conservation and expansion) and indirect factors (i.e., CO_2 fertilisation, extended growing season, global warming, nitrogen deposition, increase of diffuse radiation) linked to human activities (*high confidence*). Browning trends (trends of decreasing photosynthetic activity) are projected in many regions where increases in drought and heatwaves are projected in a warmer climate. There is *low confidence* in the projections of global greening and browning trends. {2.2.4, Cross-Chapter Box 4 in this chapter}

The frequency and intensity of some extreme weather and climate events have increased as a consequence of global warming and will continue to increase under medium and high emission scenarios (*high confidence*). Recent heat-related events, for example, heatwaves, have been made more frequent or intense due to anthropogenic greenhouse gas (GHG) emissions in most land regions and the frequency and intensity of drought has increased in Amazonia, north-eastern Brazil, the Mediterranean, Patagonia, most of Africa and north-eastern China (*medium confidence*). Heatwaves are projected to increase in frequency, intensity and duration in most parts of the world (*high confidence*) and drought frequency and intensity is projected to increase in some regions that are already drought prone, predominantly in the Mediterranean, central Europe, the southern Amazon and southern Africa (*medium confidence*). These changes will impact ecosystems, food security and land processes including GHG fluxes (*high confidence*). {2.2.5}

Climate change is playing an increasing role in determining wildfire regimes alongside human activity (*medium confidence*), with future climate variability expected to enhance the risk and severity of wildfires in many biomes such as tropical rainforests (*high confidence*). Fire weather seasons have lengthened globally between 1979 and 2013 (*low confidence*). Global land area burned has declined in recent decades, mainly due to less burning in grasslands and savannahs (*high confidence*). While drought remains the dominant driver of fire emissions, there has recently been increased fire activity in some tropical and temperate regions during normal to wetter than average years due to warmer temperatures that increase vegetation flammability (*medium confidence*). The boreal zone is also experiencing larger and more frequent fires, and this may increase under a warmer climate (*medium confidence*). {Cross-Chapter Box 4 in this chapter}

Terrestrial greenhouse gas fluxes on unmanaged and managed lands

Agriculture, forestry and other land use (AFOLU) is a significant net source of GHG emissions (*high confidence*), contributing to about 23% of anthropogenic emissions of carbon dioxide (CO_2), methane (CH_4) and nitrous oxide (N_2O) combined as CO_2 equivalents in 2007–2016 (*medium confidence*). AFOLU results in both emissions and removals of CO_2, CH_4 and N_2O to and from the atmosphere (*high confidence*). These fluxes are affected simultaneously by natural and human drivers, making it difficult to separate natural from anthropogenic fluxes (*very high confidence*). {2.3}

The total net land-atmosphere flux of CO_2 on both managed and unmanaged lands *very likely* provided a global net removal from 2007 to 2016 according to models (-6.0 ± 3.7 $GtCO_2$ yr^{-1},

likely range). This net removal is comprised of two major components: (i) modelled net anthropogenic emissions from AFOLU are 5.2 ± 2.6 GtCO$_2$ yr^{-1} (*likely range*) driven by land cover change, including deforestation and afforestation/reforestation, and wood harvesting (accounting for about 13% of total net anthropogenic emissions of CO$_2$) (*medium confidence*), and (ii) modelled net removals due to non-anthropogenic processes are 11.2 ± 2.6 GtCO$_2$ yr^{-1} (*likely range)* on managed and unmanaged lands, driven by environmental changes such as increasing CO$_2$, nitrogen deposition and changes in climate (accounting for a removal of 29% of the CO$_2$ emitted from all anthropogenic activities (fossil fuel, industry and AFOLU) (*medium confidence*). {2.3.1}

Global models and national GHG inventories use different methods to estimate anthropogenic CO$_2$ emissions and removals for the land sector. Consideration of differences in methods can enhance understanding of land sector net emission such as under the Paris Agreement's global stocktake (*medium confidence*). Both models and inventories produce estimates that are in close agreement for land-use change involving forest (e.g., deforestation, afforestation), and differ for managed forest. Global models consider as managed forest those lands that were subject to harvest whereas, consistent with IPCC guidelines, national GHG inventories define managed forest more broadly. On this larger area, inventories can also consider the natural response of land to human-induced environmental changes as anthropogenic, while the global model approach {Table SPM.1} treats this response as part of the non-anthropogenic sink. For illustration, from 2005 to 2014, the sum of the national GHG inventories net emission estimates is 0.1 ± 1.0 GtCO$_2$ yr^{-1}, while the mean of two global bookkeeping models is 5.1 ± 2.6 GtCO$_2$ yr^{-1} (*likely range*).

The gross emissions from AFOLU (one-third of total global emissions) are more indicative of mitigation potential of reduced deforestation than the global net emissions (13% of total global emissions), which include compensating deforestation and afforestation fluxes (*high confidence*). The net flux of CO$_2$ from AFOLU is composed of two opposing gross fluxes: (i) gross emissions (20 GtCO$_2$ yr^{-1}) from deforestation, cultivation of soils and oxidation of wood products, and (ii) gross removals (–14 GtCO$_2$ yr^{-1}), largely from forest growth following wood harvest and agricultural abandonment (*medium confidence*). {2.3.1}

Land is a net source of CH$_4$, accounting for 44% of anthropogenic CH$_4$ emissions for the 2006–2017 period (*medium confidence*). The pause in the rise of atmospheric CH$_4$ concentrations between 2000 and 2006 and the subsequent renewed increase appear to be partially associated with land use and land use change. The recent depletion trend of the ^{13}C isotope in the atmosphere indicates that higher biogenic sources explain part of the current CH$_4$ increase and that biogenic sources make up a larger proportion of the source mix than they did before 2000 (*high confidence*). In agreement with the findings of AR5, tropical wetlands and peatlands continue to be important drivers of inter-annual variability and current CH$_4$ concentration increases (*medium evidence, high agreement*). Ruminants and the expansion of rice cultivation are also important contributors to the current trend (*medium evidence, high agreement*).

There is significant and ongoing accumulation of CH$_4$ in the atmosphere (*very high confidence*). {2.3.2}

AFOLU is the main anthropogenic source of N$_2$O primarily due to nitrogen application to soils (*high confidence*). In croplands, the main driver of N$_2$O emissions is a lack of synchronisation between crop nitrogen demand and soil nitrogen supply, with approximately 50% of the nitrogen applied to agricultural land not taken up by the crop. Cropland soils emit over 3 MtN$_2$O-N yr^{-1} (*medium confidence*). Because the response of N$_2$O emissions to fertiliser application rates is non-linear, in regions of the world where low nitrogen application rates dominate, such as sub-Saharan Africa and parts of Eastern Europe, increases in nitrogen fertiliser use would generate relatively small increases in agricultural N$_2$O emissions. Decreases in application rates in regions where application rates are high and exceed crop demand for parts of the growing season will have very large effects on emissions reductions (*medium evidence, high agreement*). {2.3.3}

While managed pastures make up only one-quarter of grazing lands, they contributed more than three-quarters of N$_2$O emissions from grazing lands between 1961 and 2014 with rapid recent increases of nitrogen inputs resulting in disproportionate growth in emissions from these lands (*medium confidence*). Grazing lands (pastures and rangelands) are responsible for more than one-third of total anthropogenic N$_2$O emissions or more than one-half of agricultural emissions (*high confidence*). Emissions are largely from North America, Europe, East Asia, and South Asia, but hotspots are shifting from Europe to southern Asia (*medium confidence*). {2.3.3}

Increased emissions from vegetation and soils due to climate change in the future are expected to counteract potential sinks due to CO$_2$ fertilisation (*low confidence*). Responses of vegetation and soil organic carbon (SOC) to rising atmospheric CO$_2$ concentration and climate change are not well constrained by observations (*medium confidence*). Nutrient (e.g., nitrogen, phosphorus) availability can limit future plant growth and carbon storage under rising CO$_2$ (*high confidence*). However, new evidence suggests that ecosystem adaptation through plant-microbe symbioses could alleviate some nitrogen limitation (*medium evidence, high agreement*). Warming of soils and increased litter inputs will accelerate carbon losses through microbial respiration (*high confidence*). Thawing of high latitude/altitude permafrost will increase rates of SOC loss and change the balance between CO$_2$ and CH$_4$ emissions (*medium confidence*). The balance between increased respiration in warmer climates and carbon uptake from enhanced plant growth is a key uncertainty for the size of the future land carbon sink (*medium confidence*). {2.3.1, 2.7.2, Box 2.3}

Biophysical and biogeochemical land forcing and feedbacks to the climate system

Changes in land conditions from human use or climate change in turn affect regional and global climate (*high confidence*). On the global scale, this is driven by changes in emissions or removals of CO$_2$, CH$_4$ and N$_2$O by land (biogeochemical effects) and by changes in the surface albedo (*very high confidence*). Any local land changes that redistribute energy and water vapour between the land and the atmosphere influence regional climate (biophysical effects; *high confidence*). However, there is *no confidence* in whether such biophysical effects influence global climate. {2.1, 2.3, 2.5.1, 2.5.2}

Changes in land conditions modulate the likelihood, intensity and duration of many extreme events including heatwaves (*high confidence*) and heavy precipitation events (*medium confidence*). Dry soil conditions favour or strengthen summer heatwave conditions through reduced evapotranspiration and increased sensible heat. By contrast wet soil conditions, for example from irrigation or crop management practices that maintain a cover crop all year round, can dampen extreme warm events through increased evapotranspiration and reduced sensible heat. Droughts can be intensified by poor land management. Urbanisation increases extreme rainfall events over or downwind of cities (*medium confidence*). {2.5.1, 2.5.2, 2.5.3}

Historical changes in anthropogenic land cover have resulted in a mean annual global warming of surface air from biogeochemical effects (*very high confidence*), dampened by a cooling from biophysical effects (*medium confidence*). Biogeochemical warming results from increased emissions of GHGs by land, with model-based estimates of +0.20 ± 0.05°C (global climate models) and +0.24 ± 0.12°C – dynamic global vegetation models (DGVMs) as well as an observation-based estimate of +0.25 ± 0.10°C. A net biophysical cooling of −0.10 ± 0.14°C has been derived from global climate models in response to the increased surface albedo and decreased turbulent heat fluxes, but it is smaller than the warming effect from land-based emissions. However, when both biogeochemical and biophysical effects are accounted for within the same global climate model, the models do not agree on the sign of the net change in mean annual surface air temperature. {2.3, 2.5.1, Box 2.1}

The future projected changes in anthropogenic land cover that have been examined for AR5 would result in a biogeochemical warming and a biophysical cooling whose magnitudes depend on the scenario (*high confidence*). Biogeochemical warming has been projected for RCP8.5 by both global climate models (+0.20 ± 0.15°C) and DGVMs (+0.28 ± 0.11°C) (*high confidence*). A global biophysical cooling of 0.10 ± 0.14°C is estimated from global climate models and is projected to dampen the land-based warming (*low confidence*). For RCP4.5, the biogeochemical warming estimated from global climate models (+0.12 ± 0.17°C) is stronger than the warming estimated by DGVMs (+0.01 ± 0.04°C) but based on *limited evidence*, as is the biophysical cooling (−0.10 ± 0.21°C). {2.5.2}

Regional climate change can be dampened or enhanced by changes in local land cover and land use (*high confidence*) but this depends on the location and the season (*high confidence*). In boreal regions, for example, where projected climate change will migrate the treeline northward, increase the growing season length and thaw permafrost, regional winter warming will be enhanced by decreased surface albedo and snow, whereas warming will be dampened during the growing season due to larger

evapotranspiration (*high confidence*). In the tropics, wherever climate change will increase rainfall, vegetation growth and associated increase in evapotranspiration will result in a dampening effect on regional warming (*medium confidence*). {2.5.2, 2.5.3}

According to model-based studies, changes in local land cover or available water from irrigation will affect climate in regions as far as few hundreds of kilometres downwind (*high confidence*). The local redistribution of water and energy following the changes on land affect the horizontal and vertical gradients of temperature, pressure and moisture, thus altering regional winds and consequently moisture and temperature advection and convection and subsequently, precipitation. {2.5.2, 2.5.4, Cross-Chapter Box 4}

Future increases in both climate change and urbanisation will enhance warming in cities and their surroundings (urban heat island), especially during heatwaves (*high confidence*). Urban and peri-urban agriculture, and more generally urban greening, can contribute to mitigation (*medium confidence*) as well as to adaptation (*high confidence*), with co-benefits for food security and reduced soil-water-air pollution. {Cross-Chapter Box 4}

Regional climate is strongly affected by natural land aerosols (*medium confidence*) (e.g., mineral dust, black, brown and organic carbon), but there is *low confidence* in historical trends, inter-annual and decadal variability and future changes. Forest cover affects climate through emissions of biogenic volatile organic compounds (BVOC) and aerosols (*low confidence*). The decrease in the emissions of BVOC resulting from the historical conversion of forests to cropland has resulted in a positive radiative forcing through direct and indirect aerosol effects, a negative radiative forcing through the reduction in the atmospheric lifetime of methane and it has contributed to increased ozone concentrations in different regions (*low confidence*). {2.4, 2.5}

Consequences for the climate system of land-based adaptation and mitigation options, including carbon dioxide removal (negative emissions)

About one-quarter of the 2030 mitigation pledged by countries in their initial nationally determined contributions (NDCs) under the Paris Agreement is expected to come from land-based mitigation options (*medium confidence*). Most of the NDCs submitted by countries include land-based mitigation, although many lack details. Several refer explicitly to reduced deforestation and forest sinks, while a few include soil carbon sequestration, agricultural management and bioenergy. Full implementation of NDCs (submitted by February 2016) is expected to result in net removals of 0.4–1.3 $GtCO_2$ y^{-1} in 2030 compared to the net flux in 2010, where the range represents low to high mitigation ambition in pledges, not uncertainty in estimates (*medium confidence*). {2.6.3}

Several mitigation response options have technical potential for >3 $GtCO_2$-eq yr^{-1} by 2050 through reduced emissions and Carbon Dioxide Removal (CDR) (*high confidence*), some of which compete for land and other resources, while others may

reduce the demand for land (*high confidence*). Estimates of the technical potential of individual response options are not necessarily additive. The largest potential for reducing AFOLU emissions are through reduced deforestation and forest degradation (0.4–5.8 $GtCO_2$-eq yr^{-1}) (*high confidence*), a shift towards plant-based diets (0.7–8.0 $GtCO_2$-eq yr^{-1}) (*high confidence*) and reduced food and agricultural waste (0.8–4.5 $GtCO_2$-eq yr^{-1}) (*high confidence*). Agriculture measures combined could mitigate 0.3–3.4 $GtCO_2$-eq yr^{-1} (*medium confidence*). The options with largest potential for CDR are afforestation/reforestation (0.5–10.1 $GtCO_2$-eq yr^{-1}) (*medium confidence*), soil carbon sequestration in croplands and grasslands (0.4–8.6 $GtCO_2$-eq yr^{-1}) (*high confidence*) and Bioenergy with Carbon Capture and Storage (BECCS) (0.4–11.3 $GtCO_2$-eq yr^{-1}) (*medium confidence*). While some estimates include sustainability and cost considerations, most do not include socio-economic barriers, the impacts of future climate change or non-GHG climate forcings. {2.6.1}

Response options intended to mitigate global warming will also affect the climate locally and regionally through biophysical effects (*high confidence*). Expansion of forest area, for example, typically removes CO_2 from the atmosphere and thus dampens global warming (biogeochemical effect, *high confidence*), but the biophysical effects can dampen or enhance regional warming depending on location, season and time of day. During the growing season, afforestation generally brings cooler days from increased evapotranspiration, and warmer nights (*high confidence*). During the dormant season, forests are warmer than any other land cover, especially in snow-covered areas where forest cover reduces albedo (*high confidence*). At the global level, the temperature effects of boreal afforestation/reforestation run counter to GHG effects, while in the tropics they enhance GHG effects. In addition, trees locally dampen the amplitude of heat extremes (*medium confidence*). {2.5.2, 2.5.4, 2.7, Cross-Chapter Box 4}

Mitigation response options related to land use are a key element of most modelled scenarios that provide strong mitigation, alongside emissions reduction in other sectors (*high confidence*). More stringent climate targets rely more heavily on land-based mitigation options, in particular, CDR (*high confidence*). Across a range of scenarios in 2100, CDR is delivered by both afforestation (median values of –1.3, –1.7 and –2.4 $GtCO_2$ yr^{-1} for scenarios RCP4.5, RCP2.6 and RCP1.9 respectively) and bioenergy with carbon capture and storage (BECCS) (–6.5, –11 and –14.9 $GtCO_2$ yr^{-1} respectively). Emissions of CH_4 and N_2O are reduced through improved agricultural and livestock management as well as dietary shifts away from emission-intensive livestock products by 133.2, 108.4 and 73.5 $MtCH_4$ yr^{-1}; and 7.4, 6.1 and 4.5 MtN_2O yr^{-1} for the same set of scenarios in 2100 (*high confidence*). High levels of bioenergy crop production can result in increased N_2O emissions due to fertiliser use. The Integrated Assessment Models that produce these scenarios mostly neglect the biophysical effects of land-use on global and regional warming. {2.5, 2.6.2}

Large-scale implementation of mitigation response options that limit warming to 1.5°C or 2°C would require conversion of large areas of land for afforestation/reforestation and bioenergy crops, which could lead to short-term carbon losses (*high confidence*). The change of global forest area in mitigation pathways ranges from about –0.2 to +7.2 Mkm^2 between 2010 and 2100 (median values across a range of models and scenarios: RCP4.5, RCP2.6, RCP1.9), and the land demand for bioenergy crops ranges from about 3.2 to 6.6 Mkm^2 in 2100 (*high confidence*). Large-scale land-based CDR is associated with multiple feasibility and sustainability constraints (Chapters 6 and 7). In high carbon lands such as forests and peatlands, the carbon benefits of land protection are greater in the short-term than converting land to bioenergy crops for BECCS, which can take several harvest cycles to 'pay-back' the carbon emitted during conversion (carbon-debt), from decades to over a century (*medium confidence*). {2.6.2, Chapters 6, 7}

It is possible to achieve climate change targets with low need for land-demanding CDR such as BECCS, but such scenarios rely more on rapidly reduced emissions or CDR from forests, agriculture and other sectors. Terrestrial CDR has the technical potential to balance emissions that are difficult to eliminate with current technologies (including food production). Scenarios that achieve climate change targets with less need for terrestrial CDR rely on agricultural demand-side changes (diet change, waste reduction), and changes in agricultural production such as agricultural intensification. Such pathways that minimise land use for bioenergy and BECCS are characterised by rapid and early reduction of GHG emissions in all sectors, as well as earlier CDR in through afforestation. In contrast, delayed mitigation action would increase reliance on land-based CDR (*high confidence*). {2.6.2}

2.1 Introduction: Land–climate interactions

This chapter assesses the literature on two-way interactions between climate and land, with focus on scientific findings published since AR5 and some aspects of the land–climate interactions that were not assessed in previous IPCC reports. Previous IPCC assessments recognised that climate affects land cover and land surface processes, which in turn affect climate. However, previous assessments mostly focused on the contribution of land to global climate change via its role in emitting and absorbing greenhouse gases (GHGs) and short-lived climate forcers (SLCFs), or via implications of changes in surface reflective properties (i.e., albedo) for solar radiation absorbed by the surface. This chapter examines scientific advances in understanding the interactive changes of climate and land, including impacts of climate change, variability and extremes on managed and unmanaged lands. It assesses climate forcing of land changes from direct (e.g., land use change and land management) and indirect (e.g., increasing atmospheric CO_2 concentration and nitrogen deposition) effects at local, regional and global scales.

2.1.1 Recap of previous IPCC and other relevant reports as baselines

The evidence that land cover matters for the climate system have long been known, especially from early paleoclimate modelling studies and impacts of human-induced deforestation at the margin of deserts (de Noblet et al. 1996; Kageyama et al. 2004). The understanding of how land use activities impact climate has been put forward by the pioneering work of Charney (1975) who examined the role of overgrazing-induced desertification on the Sahelian climate.

Since then there have been many modelling studies that reported impacts of idealised or simplified land cover changes on weather patterns (e.g., Pielke et al. 2011). The number of studies dealing with such issues has increased significantly over the past 10 years, with more studies that address realistic past or projected land changes. However, very few studies have addressed the impacts of land cover changes on climate as very few land surface models embedded within climate models (whether global or regional), include a representation of land management. Observation-based evidence of land-induced climate impacts emerged even more recently (e.g., Alkama and Cescatti 2016; Bright et al. 2017; Lee et al. 2011; Li et al. 2015; Duveiller et al. 2018; Forzieri et al. 2017) and the literature is therefore limited.

In previous IPCC reports, the interactions between climate change and land were covered separately by three working groups. AR5 WGI assessed the role of land use change in radiative forcing, land-based GHGs source and sink, and water cycle changes that focused on changes of evapotranspiration, snow and ice, runoff and humidity. AR5 WGII examined impacts of climate change on land, including terrestrial and freshwater ecosystems, managed ecosystems, and cities and settlements. AR5 WGIII assessed land-based climate change mitigation goals and pathways related to the agriculture, forestry and other land use (AFOLU). Here, this chapter assesses land–climate interactions from all three working groups. It also

builds on previous special reports such as the Special Report on Global Warming of 1.5°C (SR15). It links to the IPCC Guidelines on National Greenhouse Gas Inventories in the land sector. Importantly, this chapter assesses knowledge that has never been reported in any of those previous reports. Finally, the chapter also tries to reconcile the possible inconsistencies across the various IPCC reports.

Land-based water cycle changes

AR5 reported an increase in global evapotranspiration from the early 1980s to 2000s, but a constraint on further increases from low soil moisture availability. Rising CO_2 concentration limits stomatal opening and thus also reduces transpiration, a component of evapotranspiration. Increasing aerosol levels, declining surface wind speeds and declining levels of solar radiation reaching the ground are additional regional causes of the decrease in evapotranspiration.

Land area precipitation change

Averaged over the mid-latitude land areas of the northern hemisphere, precipitation has increased since 1901 (*medium confidence* before 1951 and *high confidence* thereafter). For other latitudes, area-averaged long-term positive or negative trends have *low confidence*. There are *likely* more land regions where the number of heavy precipitation events has increased than where it has decreased. Extreme precipitation events over most of the mid-latitude land masses and over wet tropical regions will very likely become more intense and more frequent (IPCC 2013a).

Land-based GHGs

AR5 reported that annual net CO_2 emissions from anthropogenic land use change were 0.9 [0.1–1.7] GtC yr^{-1} on average during 2002–2011 (*medium confidence*). From 1750–2011, CO_2 emissions from fossil fuel combustion have released an estimated 375 [345–405] GtC to the atmosphere, while deforestation and other land use change have released an estimated 180 [100–260] GtC. Of these cumulative anthropogenic CO_2 emissions, 240 [230–250] GtC have accumulated in the atmosphere, 155 [125–185] GtC have been taken up by the ocean and 160 [70–250] GtC have accumulated in terrestrial ecosystems (i.e., the cumulative residual land sink) (Ciais et al. 2013a). Updated assessment and knowledge gaps are covered in Section 2.3.

Future terrestrial carbon source/sink

AR5 projected with *high confidence* that tropical ecosystems will uptake less carbon and with *medium confidence* that at high latitudes, land carbon sink will increase in a warmer climate. Thawing permafrost in the high latitudes is potentially a large carbon source in warmer climate conditions, however the magnitude of CO_2 and CH_4 emissions due to permafrost thawing is still uncertain. The SR15 further indicates that constraining warming to 1.5°C would prevent the melting of an estimated permafrost area of 2 million km^2 over the next centuries compared to 2°C. Updates to these assessments are found in Section 2.3.

Land use change altered albedo

AR5 stated with *high confidence* that anthropogenic land use change has increased the land surface albedo, which has led to a RF of -0.15 ± 0.10 W m^{-2}. However, it also underlined that the sources of the large spread across independent estimates were caused by differences in assumptions for the albedo of natural and managed surfaces and for the fraction of land use change before 1750. Generally, our understanding of albedo changes from land use change has been enhanced from AR4 to AR5, with a narrower range of estimates and a higher confidence level. The radiative forcing from changes in albedo induced by land use changes was estimated in AR5 at -0.15 W m^{-2} (-0.25 to about -0.05), with *medium confidence* in AR5 (Myhre et al. 2013). This was an improvement over AR4 in which it was estimated at -0.2 W m^{-2} (-0.4 to about 0), with *low to medium confidence* (Forster et al. 2007). Section 2.5 shows that albedo is not the only source of biophysical land-based climate forcing to be considered.

Hydrological feedback to climate

Land use changes also affect surface temperatures through non-radiative processes, and particularly through the hydrological cycle. These processes are less well known and are difficult to quantify but tend to offset the impact of albedo changes. As a consequence, there is low agreement on the sign of the net change in global mean temperature as a result of land use change (Hartmann et al. 2013a). An updated assessment on these points is covered in Sections 2.5 and 2.2.

Climate-related extremes on land

AR5 reported that impacts from recent climate-related extremes reveal significant vulnerability and exposure of some ecosystems to current climate variability. Impacts of such climate-related extremes include alteration of ecosystems, disruption of food production and water supply, damage to infrastructure and settlements, morbidity and mortality, and consequences for mental health and human well-being (Burkett et al. 2014). The SR15 further indicates that limiting global warming to 1.5°C limits the risks of increases in heavy precipitation events in several regions (*high confidence*). In urban areas, climate change is projected to increase risks for people, assets, economies and ecosystems (*very high confidence*). These risks are amplified for those lacking essential infrastructure and services or living in exposed areas. An updated assessment and a knowledge gap for this chapter are covered in Section 2.2 and Cross-Chapter Box 4.

Land-based climate change adaptation and mitigation

AR5 reported that adaptation and mitigation choices in the near-term will affect the risks related to climate change throughout the 21st century (Burkett et al. 2014). AFOLU are responsible for about 10–12 GtCO$_2$eq yr^{-1} anthropogenic greenhouse gas emissions, mainly from deforestation and agricultural production. Global CO$_2$ emissions from forestry and other land use have declined since AR4, largely due to increased afforestation. The SR15 further indicates that afforestation and bioenergy with carbon capture and storage

(BECCS) are important land-based carbon dioxide removal (CDR) options. It also states that land use and land-use change emerge as a critical feature of virtually all mitigation pathways that seek to limit global warming to 1.5°C. The Climate Change 2014 Synthesis Report concluded that co-benefits and adverse side effects of mitigation could affect achievement of other objectives, such as those related to human health, food security, biodiversity, local environmental quality, energy access, livelihoods and equitable sustainable development. Updated assessment and knowledge gaps are covered in Section 2.6 and Chapter 7.

Overall, sustainable land management is largely constrained by climate change and extremes, but also puts bounds on the capacity of land to effectively adapt to climate change and mitigate its impacts. Scientific knowledge has advanced on how to optimise our adaptation and mitigation efforts while coordinating sustainable land management across sectors and stakeholders. Details are assessed in subsequent sections.

2.1.2   Introduction to the chapter structure

This chapter assesses the consequences of changes in land cover and functioning, resulting from both land use and climate change, to global and regional climates. The chapter starts with an assessment of the historical and projected responses of land processes to climate change and extremes (Section 2.2). Subsequently, the chapter assesses historical and future changes in terrestrial GHG fluxes (Section 2.3) as well as non-GHG fluxes and precursors of SLCFs (Section 2.4). Section 2.5 focuses on how historical and future changes in land use and land cover influence climate change/variability through biophysical and biogeochemical forcing and feedbacks, how specific land management affects climate, and how, in turn, climate-induced land changes feed back to climate. Section 2.6 assesses the consequences of land-based adaptation and mitigation options for the climate system in GHG and non-GHG exchanges. Sections 2.3 and 2.6 address implications of the Paris Agreement for land–climate interactions, and the scientific evidence base for ongoing negotiations around the Paris rulebook, the global stocktake and credibility in measuring, reporting and verifying the climate impacts of anthropogenic activities on land. This chapter also examines how land use and management practices may affect climate change through biophysical feedbacks and radiative forcing (Section 2.5), and assesses policy-relevant projected land use changes and sustainable land management for mitigation and adaptation (Section 2.6). Finally, the chapter concludes with a brief assessment of advances in the understanding of the ecological and biogeochemical processes underlying land–climate interactions (Section 2.7).

The chapter includes three chapter boxes providing general overview of (i) processes underlying land–climate interactions (Box 2.1), (ii) methodological approaches for estimating anthropogenic land carbon fluxes from national to global scales (Box 2.2), and (iii) CO$_2$ fertilisation and enhanced terrestrial uptake of carbon (Box 2.3). In addition, this chapter includes two cross-chapter boxes on climate change and fire (Cross-Chapter Box 3), and on urbanisation and climate change (Cross-Chapter Box 4).

In summary, the chapter assesses scientific understanding related to (i) how a changing climate affects terrestrial ecosystems, including those on managed lands, (ii) how land affects climate through biophysical and biogeochemical feedbacks, and (iii) how land use or cover change and land management play an important and complex role in the climate system. This chapter also pays special attention to advances in understanding cross-scale interactions, emerging issues, heterogeneity and teleconnections.

Box 2.1 | Processes underlying land–climate interactions

Land continuously interacts with the atmosphere through exchanges of, for instance, GHGs (e.g., CO_2, CH_4, N_2O), water, energy or precursors of short lived-climate forcers (e.g., biogenic volatile organic compounds, dust, black carbon). The terrestrial biosphere also interacts with oceans through processes such as the influx of freshwater, nutrients, carbon and particles. These interactions affect where and when rain falls and thus irrigation needs for crops, frequency and intensity of heatwaves, and air quality. They are modified by global and regional climate change, decadal, inter-annual and seasonal climatic variations, and weather extremes, as well as human actions on land (e.g., crop and forest management, afforestation and deforestation). This in turn affects atmospheric composition, surface temperature, hydrological cycle and thus local, regional and global climate. This box introduces some of the fundamental land processes governing biophysical and biogeochemical effects and feedbacks to the climate (Box 2.1, Figure 1).

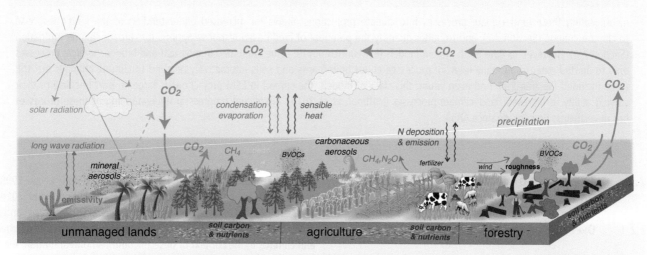

Box 2.1, Figure 1 | The structure and functioning of managed and unmanaged ecosystems that affect local, regional and global climate. Land surface characteristics such as albedo and emissivity determine the amount of solar and long-wave radiation absorbed by land and reflected or emitted to the atmosphere. Surface roughness influences turbulent exchanges of momentum, energy, water and biogeochemical tracers. Land ecosystems modulate the atmospheric composition through emissions and removals of many GHGs and precursors of SLCFs, including biogenic volatile organic compounds (BVOCs) and mineral dust. Atmospheric aerosols formed from these precursors affect regional climate by altering the amounts of precipitation and radiation reaching land surfaces through their role in clouds physics.

'Biophysical interactions' are exchanges of water and energy between the land and the atmosphere (Section 2.5). Land warms up from absorbing solar and long-wave radiation; it cools down through transfers of sensible heat (via conduction and convection) and latent heat (energy associated with water evapotranspiration) to the atmosphere and through long-wave radiation emission from the land surface (Box 2.1, Figure 1). These interactions between the land and the atmosphere depend on land surface characteristics, including reflectivity of shortwave radiation (albedo), emissivity of long wave radiation by vegetation and soils, surface roughness and soil water access by vegetation, which depends on both soil characteristics and amounts of roots. Over seasonal, inter-annual and decadal timescales, these characteristics vary among different land cover and land-use types and are affected by both natural processes and land management (Anderson et al. 2011). A dense vegetation with high leaf area index, like forests, may absorb more energy than nearby herbaceous vegetation partly due to differences in surface albedo (especially when snow is on the ground). However, denser vegetation also sends more energy back to the atmosphere in the form of evapotranspiration (Bonan, 2008; Burakowski et al., 2018; Ellison et al., 2017) (Section 2.5.2) and this contributes to changes in atmospheric water vapour content, and subsequently to changes in rainfall.

Particularly in extra-tropical regions, these characteristics exhibit strong seasonal patterns with the development and senescence of the vegetation (e.g., leaf colour change and drop). For example, in deciduous forests, seasonal growth increases albedo by 20–50% from the spring minima to growing season maxima, followed by rapid decrease during leaf fall, whereas in grasslands, spring greening causes albedo decreases and only increases with vegetation browning (Hollinger et al. 2010). The seasonal patterns of sensible and latent heat fluxes are also driven by the cycle of leaf development and senescence in temperate deciduous forests: sensible heat fluxes peak in spring and autumn and latent heat fluxes peak in mid-summer (Moore et al. 1996; Richardson et al. 2013).

Box 2.1 (continued)

Exchanges of GHGs between the land and the atmosphere are referred to as 'biogeochemical interactions' (Section 2.3), which are driven mainly by the balance between photosynthesis and respiration by plants, and by the decomposition of soil organic matter by microbes. The conversion of atmospheric carbon dioxide into organic compounds by plant photosynthesis, known as terrestrial net primary productivity, is the source of plant growth, food for human and other organisms, and soil organic carbon. Due to strong seasonal patterns of growth, northern hemisphere terrestrial ecosystems are largely responsible for the seasonal variations in global atmospheric CO_2 concentrations. In addition to CO_2, soils emit methane (CH_4) and nitrous oxide (N_2O) (Section 2.3). Soil temperature and moisture strongly affect microbial activities and resulting fluxes of these three GHGs.

Much like fossil fuel emissions, GHG emissions from anthropogenic land cover change and land management are 'forcers' on the climate system. Other land-based changes to climate are described as 'feedbacks' to the climate system – a process by which climate change influences some property of land, which in turn diminishes (negative feedback) or amplifies (positive feedback) climate change. Examples of feedbacks include the changes in the strength of land carbon sinks or sources, soil moisture and plant phenology (Section 2.5.3).

Incorporating these land–climate processes into climate projections allows for increased understanding of the land's response to climate change (Section 2.2), and to better quantify the potential of land-based response options for climate change mitigation (Section 2.6). However, to date Earth system models (ESMs) incorporate some combined biophysical and biogeochemical processes only to limited extent and many relevant processes about how plants and soils interactively respond to climate changes are still to be included (Section 2.7). And even within this class of models, the spread in ESM projections is large, in part because of their varying ability to represent land–climate processes (Hoffman et al. 2014). Significant progress in understanding of these processes has nevertheless been made since AR5.

2.2   The effect of climate variability and change on land

2.2.1   Overview of climate impacts on land

2.2.1.1   Climate drivers of land form and function

Energy is redistributed from the warm equator to the colder poles through large-scale atmospheric and oceanic processes driving the Earth's weather and climate (Oort and Peixóto 1983; Carissimo et al. 1985; Yang et al. 2015a). Subsequently, a number of global climate zones have been classified ranging from large-scale primary climate zones (tropical, sub-tropical, temperate, sub-polar, polar) to much higher-resolution, regional climate zones (e.g., the Köppen-Geiger classification, Kottek et al. 2006). Biomes are adapted to regional climates (Figure 2.1) and may shift as climate, land surface characteristics (e.g., geomorphology, hydrology), CO_2 fertilisation and fire interact. These biomes and the processes therein are subject to modes of natural variability in the ocean-atmosphere system that result in regionally wetter/dryer or hotter/cooler periods having temporal scales from weeks to months (e.g., Southern Annular Mode), months to seasons (e.g., Madden-Julian Oscillation), years (e.g., El Niño Southern Oscillation) and decades (e.g., Pacific Decadal Oscillation). Furthermore, climate and weather extremes (such as drought, heatwaves, very heavy rainfall, strong winds), whose frequency, intensity and duration are often a function of large-scale modes of variability, impact ecosystems at various space and timescales.

It is *very likely* that changes to natural climate variability as a result of global warming has and will continue to impact terrestrial ecosystems

with subsequent impacts on land processes (Hulme et al. 1999; Parmesan and Yohe 2003; Di Lorenzo et al. 2008; Kløve et al. 2014; Berg et al. 2015; Lemordant et al. 2016; Pecl et al. 2017). This chapter assesses climate variability and change, particularly extreme weather and climate, in the context of desertification, land degradation, food security and terrestrial ecosystems more generally. This section does specifically assess the impacts of climate variability and climate change on desertification, land degradation and food security as these impacts are assessed respectively in Chapters 3, 4 and 5. This chapter begins with an assessment of observed warming on land.

2.2.1.2   Changes in global land surface air temperature

Based on analysis of several global and regional land surface air temperature (LSAT) datasets, AR5 concluded that the global LSAT had increased over the instrumental period of record, with the warming rate approximately double that reported over the oceans since 1979 and that 'it is certain that globally averaged LSAT has risen since the late 19th century and that this warming has been particularly marked since the 1970s'. Warming found in the global land datasets is also in a broad agreement with station observations (Hartmann et al. 2013a).

Since AR5, LSAT datasets have been improved and extended. The National Center for Environmental Information, which is a part of the US National Oceanic and Atmospheric Administration (NOAA), developed a new, fourth version of the Global Historical Climatology Network monthly dataset (GHCNm, v4). The dataset provides an expanded set of station temperature records with more than 25,000 total monthly temperature stations compared to 7200 in versions v2 and v3 (Menne et al. 2018). Goddard Institute for Space Studies, which is a part of

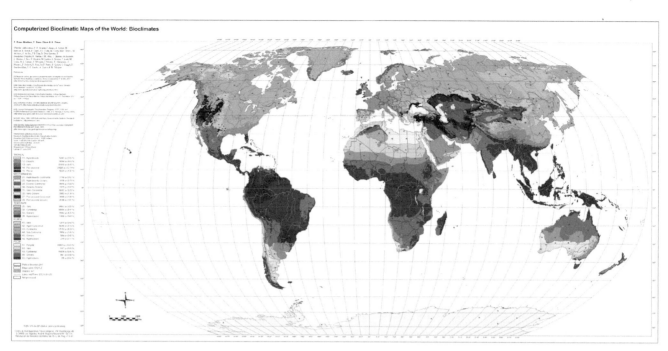

Figure 2.1 | Worldwide Bioclimatic Classification System, 1996–2018. Source: Rivas-Martinez et al. (2011). Online at www.globalbioclimatics.org.

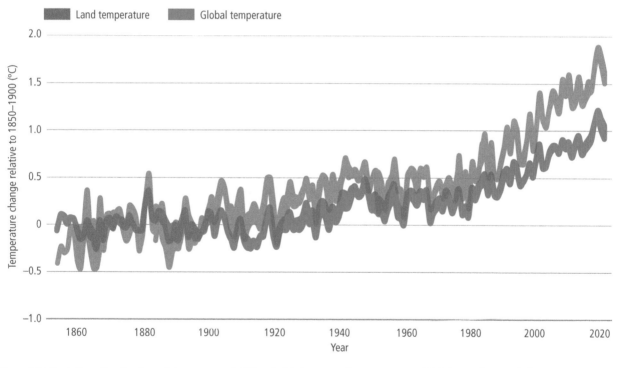

Figure 2.2 | Evolution of land surface air temperature (LSAT) and global mean surface temperature (GMST) over the period of instrumental observations. The brown line shows annual mean LSAT in the BEST, CRUTEM4.6, GHCNmv4 and GISTEMP datasets, expressed as departures from global average LSAT in 1850–1900, with the brown line thickness indicating inter-dataset range. The blue line shows annual mean GMST in the HadCRUT4, NOAAGlobal Temp, GISTEMP and Cowtan&Way datasets (monthly values of which were reported in the Special Report on Global Warming of 1.5°C; Allen et al. 2018).

the US National Aeronautics and Space Administration, (NASA/GISS), provides estimate of land and ocean temperature anomalies (GISTEMP). The GISTEMP land temperature anomalies are based upon primarily NOAA/GHCN version 3 dataset (Lawrimore et al. 2011) and account for urban effects through nightlight adjustments (Hansen et al. 2010). The Climatic Research Unit (CRU) of the University of East Anglia, UK (CRUTEM) dataset, now version CRUTEM4.6,

incorporates additional stations (Jones et al. 2012). Finally, the Berkeley Earth Surface Temperature (BEST) dataset provides LSAT from 1750 to present based on almost 46,000 time series and has the longest temporal coverage of the four datasets (Rohde et al. 2013). This dataset was derived with methods distinct from those used for development of the NOAA GHCNm, NASA/GISS GISTEMP and the University of East Anglia CRUTEM datasets.

Table 2.1 | Increases in land surface air temperature (LSAT) from preindustrial period and the late 19th century to present day.

Time period	Dataset of LSAT increase (°C)			
	BEST	CRUTEM4.6	GHCNm, v4	GISTEMP
From 1850–1900 to 2006–2015	1.53 1.38–1.68 (95% confidence)	1.32*		
From 1850–1900 to 1999–2018	1.52 1.39–1.66 (95% confidence)	1.31	NA	NA
From 1881–1900 to 1999–2018	1.51 1.40–1.63 (95% confidence)	1.31	1.37	1.45

* CRUTEM4.6 LSAT increase is computed from 1856–1900 average.

According to the available observations in the four datasets, the globally averaged LSAT increased by 1.44°C from the preindustrial period (1850–1900) to the present day (1999–2018). The warming from the late 19th century (1881–1900) to the present day (1999–2018) was 1.41°C (1.31°C–1.51°C) (Table 2.1). The 1.31°C–1.51°C range represents the spread in median estimates from the four available land datasets and does not reflect uncertainty in data coverage or methods used. Based on the BEST dataset (the longest dataset with the most extensive land coverage) the total observed increase in LSAT between the average of the 1850–1900 period and the 2006–2015 period was 1.53°C, (1.38–1.68°C; 95% confidence), while the GMST increase for the same period was 0.87°C (0.75–0.99°C; 90% confidence) (IPCC, 2018: Summary for policymakers, Allen et al. 2018).

The extended and improved land datasets reaffirmed the AR5 conclusion that it is certain that globally averaged LSAT has risen since the preindustrial period and that this warming has been particularly marked since the 1970s (Figure 2.2).

Recent analyses of LSAT and sea surface temperature (SST) observations, as well as analyses of climate model simulations, have refined our understanding of underlying mechanisms responsible for a faster rate of warming over land than over oceans. Analyses of paleo records, historical observations, model simulations and underlying physical principles are all in agreement that the land is warming faster than the oceans as a result of differences in evaporation, land–climate feedbacks (Section 2.5) and changes in the aerosol forcing over land (*very high confidence*) (Braconnot et al. 2012; Joshi et al. 2013; Sejas et al. 2014; Byrne and O'Gorman 2013, 2015; Wallace and Joshi 2018; Allen et al. 2019). There is also *high confidence* that difference in land and ocean heat capacity is not the primary reason for faster land than ocean warming. For the recent period, the land-to-ocean warming ratio is in close agreement between different observational records (about 1.6) and the CMIP5 climate model simulations (the *likely* range of 1.54°C to 1.81°C). Earlier studies analysing slab ocean models (models in which it is assumed that the deep ocean has equilibrated) produced a higher land temperature increases than sea surface temperature (Manabe et al. 1991; Sutton et al. 2007).

It is certain that globally averaged LSAT has risen faster than GMST from the preindustrial period (1850–1900) to the present day (1999–2018). This is because the warming rate of the land compared to the ocean is substantially higher over the historical period (by approximately 60%) and because the Earth's surface is approximately one-third land and two-thirds ocean. This enhanced land warming impacts land processes with implications for desertification (Section 2.2.2 and Chapter 3), food security (Section 2.2.3 and Chapter 5), terrestrial ecosystems (Section 2.2.4), and GHG and non-GHG fluxes between the land and climate (Sections 2.3 and 2.4). Future changes in land characteristics through adaptation and mitigation processes and associated land–climate feedbacks can dampen warming in some regions and enhance warming in others (Section 2.5).

2.2.2 Climate-driven changes in aridity

Desertification is defined and discussed at length in Chapter 3 and is a function of both human activity and climate variability and change. There are uncertainties in distinguishing between historical climate-caused aridification and desertification and future projections of aridity as different measurement methods of aridity do not agree on historical or projected changes (Sections 3.1.1 and 3.2.1). However, warming trends over drylands are twice the global average (Lickley and Solomon 2018) and some temperate drylands are projected to convert to subtropical drylands as a result of an increased drought frequency causing reduced soil moisture availability in the growing season (Engelbrecht et al. 2015; Schlaepfer et al. 2017). We therefore assess with *medium confidence* that a warming climate will result in regional increases in the spatial extent of drylands under mid- and high emission scenarios and that these regions will warm faster than the global average warming rate.

2.2.3 The influence of climate change on food security

Food security and the various components thereof are addressed in depth in Chapter 5. Climate variables relevant to food security and food systems are predominantly temperature and precipitation-related, but also include integrated metrics that combine these and other variables (e.g., solar radiation, wind, humidity) and extreme weather and climate events including storm surge (Section 5.2.1). The impact of climate change through changes in these variables is projected to negatively impact all aspects of food security (food availability, access, utilisation and stability), leading to complex impacts on global food security (*high confidence*) (Chapter 5, Table 5.1).

Climate change will have regionally distributed impacts, even under aggressive mitigation scenarios (Howden et al. 2007; Rosenzweig et al. 2013; Challinor et al. 2014; Parry et al. 2005; Lobell and Tebaldi 2014; Wheeler and Von Braun 2013). For example, in the northern hemisphere the northward expansion of warmer temperatures in the middle and higher latitudes will lengthen the growing season (Gregory and Marshall 2012; Yang et al. 2015b) which may benefit crop productivity (Parry et al. 2004; Rosenzweig et al., 2014; Deryng et al. 2016). However, continued rising temperatures are expected to impact global wheat yields by about 4–6% reductions for every degree of temperature rise (Liu et al. 2016a; Asseng et al. 2015) and across both mid- and low latitude regions, rising temperatures are also expected to be a constraining factor for maize productivity by the end of the century (Bassu et al. 2014; Zhao et al. 2017). Although there has been a general reduction in frost occurrence during winter and spring, and a lengthening of the frost free season in response to growing concentrations of GHGs (Fischer and Knutti 2014; Wypych et al. 2017), there are regions where the frost season length has increased, for example, in southern Australia (Crimp et al. 2016). Despite the general reduced frost season length, late spring frosts may increase risk of damage to warming induced precocious vegetation growth and flowering (Meier et al. 2018). Observed and projected warmer minimum temperatures have, and will continue to, reduce the number of winter chill units required by temperate fruit and nut trees (Luedeling 2012). Crop yields are impacted negatively by increases of seasonal rainfall variability in the tropics, sub-tropics, water-limited and high elevation environments, while drought severity and growing season temperatures also have a negative impact on crop yield (Nelson et al. 2009; Schlenker and Lobell 2010; Müller et al. 2017; Parry et al. 2004; Wheeler and Von Braun 2013; Challinor et al. 2014).

Changes in extreme weather and climate (Section 2.2.5) have negative impacts on food security through regional reductions of crop yields. A recent study shows that 18–43% of the explained yield variance of four crops (maize, soybeans, rice and spring wheat) is attributable to extremes of temperature and rainfall, depending on the crop type (Vogel et al. 2019). Climate shocks, particularly severe drought impact low-income small-holder producers disproportionately (Vermeulen et al. 2012; Rivera Ferre 2014). Extremes also compromise critical food supply chain infrastructure, making transport of and access to harvested food more difficult (Brown et al. 2015; Fanzo et al. 2018). There is *high confidence* that the impacts of enhanced climate extremes, together with non-climate factors such as nutrient limitation, soil health and competitive plant species, generally outweighs the regionally positive impacts of warming (Lobell et al. 2011; Leakey et al. 2012; Porter et al. 2014; Gray et al. 2016; Pugh et al. 2016; Wheeler and Von Braun 2013; Beer 2018).

2.2.4   Climate-driven changes in terrestrial ecosystems

Previously, the IPCC AR5 reported high confidence that the Earth's biota composition and ecosystem processes have been strongly affected by past changes in global climate and that the magnitudes of projected changes for the 21st century under high warming scenarios (for example, RCP8.5) are higher than those under historic climate change (Settele et al. 2014). There is *high confidence* that as a

result of climate changes over recent decades many plant and animal species have experienced range size and location changes, altered abundances and shifts in seasonal activities (Urban 2015; Ernakovich et al. 2014; Elsen and Tingley 2015; Hatfield and Prueger 2015; Savage and Vellend 2015; Yin et al. 2016; Pecl et al. 2017; Gonsamo et al. 2017; Fadrique et al. 2018; Laurance et al. 2018). There is *high confidence* that climate zones have already shifted in many parts of the world, primarily as an increase of dry, arid climates accompanied by a decrease of polar climates (Chan and Wu 2015; Chen and Chen 2013; Spinoni et al. 2015b). Regional climate zones shifts have been observed over the Asian monsoon region (Son and Bae 2015), Europe (Jylhä et al. 2010), China (Yin et al. 2019), Pakistan (Adnan et al. 2017), the Alps (Rubel et al. 2017) and north-eastern Brazil, southern Argentina, the Sahel, Zambia and Zimbabwe, the Mediterranean area, Alaska, Canada and north-eastern Russia (Spinoni et al. 2015b).

There is *high confidence* that bioclimates zones will further shift as the climate warms (Williams et al. 2007; Rubel and Kottek 2010; Garcia et al. 2016; Mahony et al. 2017; Law et al. 2018). There is also *high confidence* that novel, unprecedented climates (climate conditions with no analogue in the observational record) will emerge, particularly in the tropics (Williams and Jackson 2007; Colwell et al. 2008a; Mora et al. 2013, 2014; Hawkins et al. 2014; Mahony et al. 2017; Maule et al. 2017). It is *very likely* that terrestrial ecosystems and land processes will be exposed to disturbances beyond the range of current natural variability as a result of global warming, even under low- to medium-range warming scenarios, and that these disturbances will alter the structure, composition and functioning of the system (Settele et al. 2014; Gauthier et al. 2015; Seddon et al. 2016).

In a warming climate, many species will be unable to track their climate niche as it moves, especially those in extensive flat landscapes with low dispersal capacity and in the tropics whose thermal optimum is already near current temperature (Diffenbaugh and Field 2013; Warszawski et al. 2013). Range expansion in higher latitudes and elevations as a result of warming often, but not exclusively, occurs in abandoned lands (Harsch et al. 2009; Landhäusser et al. 2010; Gottfried et al. 2012; Boisvert-Marsh et al. 2014; Bryn and Potthoff 2018; Rumpf et al. 2018; Buitenwerf et al. 2018; Steinbauer et al. 2018). This expansion typically favours thermophilic species at the expense of cold adapted species as the climate becomes suitable for lower latitude/altitude species (Rumpf et al. 2018). In temperate drylands, however, range expansion can be countered by intense and frequent drought conditions which result in accelerated rates of taxonomic change and spatial heterogeneity in an ecotone (Tietjen et al. 2017).

Since the advent of satellite observation platforms, a global increase in vegetation photosynthetic activity (i.e., greening) as evidenced through remotely sensed indices such as leaf area index (LAI) and normalised difference vegetation index (NDVI). Three satellite-based leaf area index records (GIMMS3g, GLASS and GLOMAP) imply increased growing season LAI (greening) over 25–50% and browning over less than 4% of the global vegetated area, resulting in greening trend of $0.068 \pm 0.045\,m^2\,m^{-2}\,yr^{-1}$ over 1982–2009 (Zhu et al. 2016). Greening has been observed in southern Amazonia, southern Australia, the Sahel and central Africa, India, eastern China and the northern extratropical latitudes (Myneni et al. 1997; de Jong et al.

2012; Los 2013; Piao et al. 2015; Mao et al. 2016; Zhu et al. 2016; Carlson et al. 2017; Forzieri et al. 2017; Pan et al. 2018; Chen et al. 2019). Greening has been attributed to direct factors, namely human land use management and indirect factors such as CO_2 fertilisation, climate change, and nitrogen deposition (Donohue et al. 2013; Keenan et al. 2016; Zhu et al. 2016). Indirect factors have been used to explain most greening trends primarily through CO_2 fertilisation in the tropics and through an extended growing season and increased growing season temperatures as a result of climate change in the high latitudes (Fensholt et al. 2012; Zhu et al. 2016). The extension of the growing season in high latitudes has occurred together with an earlier spring greenup (the time at which plants begin to produce leaves in northern mid- and high-latitude ecosystems) (Goetz et al. 2015; Xu et al. 2016a, 2018) with subsequent earlier spring carbon uptake (2.3 days per decade) and gross primary productivity (GPP) (Pulliainen et al. 2017). The role of direct factors of greening are being increasingly investigated and a recent study has attributed over a third of observed global greening between 2000 and 2017 to direct factors, namely afforestation and croplands, in China and India (Chen et al. 2019).

It should be noted that measured greening is a product of satellite-derived radiance data and, as such, does not provide information on ecosystem health indicators such as species composition and richness, homeostasis, absence of disease, vigour, system resilience and the different components of ecosystems (Jørgensen et al. 2016). For example, a regional greening attributable to croplands expansion or intensification might occur at the expense of ecosystem biodiversity.

Within the global greening trend are also detected regional decreases in vegetation photosynthetic activity (i.e., browning) in northern Eurasia, the southwestern USA, boreal forests in North America, inner Asia and the Congo Basin, largely as a result of intensified drought stress. Since the late 1990s rates and extents of browning have exceeded those of greening in some regions, the collective result of which has been a slowdown of the global greening rate (de Jong et al. 2012; Pan et al. 2018). Within these long-term trends, inter-annual variability of regional greening and browning is attributable to regional climate variability, responses to extremes such as drought, disease and insect infestation and large-scale tele-connective controls such as ENSO and the Atlantic Multi-decadal Organization (Verbyla 2008; Revadekar et al. 2012; Epstein et al. 2018; Zhao et al. 2018).

Projected increases in drought conditions in many regions suggest long-term global vegetation greening trends are at risk of reversal to browning in a warmer climate (de Jong et al. 2012; Pan et al. 2018; Pausas and Millán 2018). On the other hand, in higher latitudes vegetation productivity is projected to increase as a result of higher atmospheric CO_2 concentrations and longer growing periods as a result of warming (Ito et al. 2016) (Section 2.3 and Box 2.3). Additionally, climate-driven transitions of ecosystems, particularly range changes, can take years to decades for the equilibrium state to be realised and the rates of these 'committed ecosystem changes' (Jones et al. 2009) vary between low and high latitudes (Jones et al. 2010). Furthermore, as direct factors are poorly integrated into Earth

systems models (ESMs) uncertainties in projected trends of greening and browning are further compounded (Buitenwerf et al. 2018; Chen et al. 2019). Therefore, there is *low confidence* in the projection of global greening and browning trends.

Increased atmospheric CO_2 concentrations have both direct and indirect effects on terrestrial ecosystems (Sections 2.2.2 and 2.2.3, and Box 2.3). The direct effect is primarily through increased vegetation photosynthetic activity as described above. Indirect effects include decreased evapotranspiration that may offset the projected impact of drought in some water-stressed plants through improved water use efficiency in temperate regions, suggesting that some rain-fed cropping systems and grasslands will benefit from elevated atmospheric CO_2 concentrations (Roy et al. 2016; Milly and Dunne 2016; Swann et al. 2016; Chang et al. 2017; Zhu et al. 2017). In tropical regions, increased flowering activity is associated primarily with increasing atmospheric CO_2, suggesting that a long-term increase in flowering activity may persist in some vegetation, particularly mid-story trees and tropical shrubs, and may enhance reproduction levels until limited by nutrient availability or climate factors such as drought frequency, rising temperatures, and reduced insolation (Pau et al. 2018).

2.2.5 Climate extremes and their impact on land functioning

Extreme weather events are generally defined as the upper or lower statistical tails of the observed range of values of climate variables or climate indicators (e.g., temperature/rainfall or drought/aridity indices respectively). Previous IPCC reports have reported with *high confidence* on the increase of many types of observed extreme temperature events (Seneviratne et al. 2012; Hartmann et al. 2013b; Hoegh-Guldberg et al. 2018). However, as a result of observational constraints, increases in precipitation extremes are less confident, except in observations-rich regions with dense, long-lived station networks, such as Europe and North America, where there have been likely increases in the frequency or intensity of heavy rainfall.

Extreme events occur across a wide range of time and space scales (Figure 2.3) and may include individual, relatively short-lived weather events (e.g., extreme thunderstorms storms) or a combination or accumulation of non-extreme events (Colwell et al. 2008b; Handmer et al. 2012), for example, moderate rainfall in a saturated catchment having the flood peak at mean high tide (Leonard et al. 2014). Combinatory processes leading to a significant impact are referred to as a compound event and are a function of the nature and number of physical climate and land variables, biological agents such as pests and disease, the range of spatial and temporal scales, the strength of dependence between processes and the perspective of the stakeholder who defines the impact (Leonard et al. 2014; Millar and Stephenson 2015). Currently, there is *low confidence* in the impact of compound events on land as the multi-disciplinary approaches needed to address the problem are few (Zscheischler et al. 2018) and the rarity of compound extreme climatic events renders the analysis of impacts difficult.

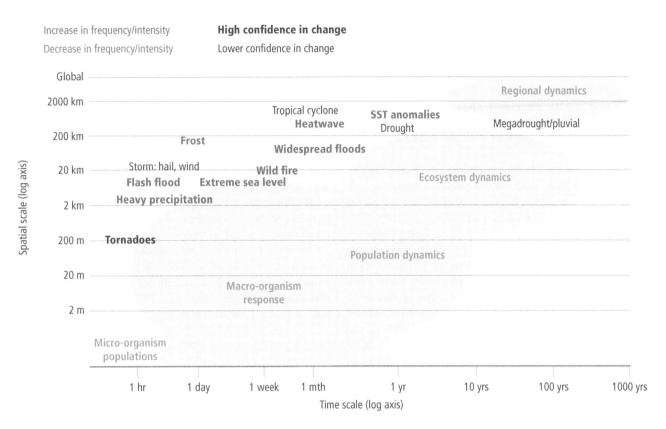

Figure 2.3 | **Spatial and temporal scales of typical extreme weather and climate events and the biological systems they impact (shaded grey).** Individuals, populations and ecosystems within these space-time ranges respond to relevant climate stressors. Orange (blue) labels indicate an increase (decrease) in the frequency or intensity of the event, with bold font reflecting confidence in the change. Non-bold black labels indicate low confidence in observed changes in frequency or intensity of these events. Each event type indicated in the figure is likely to affect biological systems at all temporal and spatial scales located to the left and below the specific event position in the figure. From Ummenhofer and Meehl (2017).

2.2.5.1 Changes in extreme temperatures, heatwaves and drought

It is *very likely* that most land areas have experienced a decrease in the number of cold days and nights, and an increase in the number of warm days and unusually hot nights (Orlowsky and Seneviratne 2012; Seneviratne et al. 2012; Mishra et al. 2015; Ye et al. 2018). Although there is no consensus definition of heatwaves, as some heatwave indices have relative thresholds and others absolute thresholds, trends between indices of the same type show that recent heat-related events have been made more frequent or more intense due to anthropogenic GHG emissions in most land regions (Lewis and Karoly 2013; Smith et al. 2013b; Scherer and Diffenbaugh 2014; Fischer and Knutti 2015; Ceccherini et al. 2016; King et al. 2016; Bador et al. 2016; Stott et al. 2016; King 2017; Hoegh-Guldberg et al. 2018). Globally, 50–80% of the land fraction is projected to experience significantly more intense hot extremes than historically recorded (Fischer and Knutti 2014; Diffenbaugh et al. 2015; Seneviratne et al. 2016). There is *high confidence* that heatwaves will increase in frequency, intensity and duration into the 21st century (Russo et al. 2016; Ceccherini et al. 2017; Herrera-Estrada and Sheffield 2017) and under high emission scenarios, heatwaves by the end of the century may become extremely long (more than 60 consecutive days) and frequent (once every two years) in Europe, North America, South America, Africa, Indonesia, the Middle East, South and Southeast Asia and Australia (Rusticucci 2012; Cowan et al. 2014; Russo et al. 2014;

Scherer and Diffenbaugh 2014; Pal and Eltahir 2016; Rusticucci et al. 2016; Schär 2016; Teng et al. 2016; Dosio 2017; Mora et al. 2017; Dosio et al. 2018; Lehner et al. 2018; Lhotka et al. 2018; Lopez et al. 2018; Tabari and Willems 2018). Furthermore, unusual heatwave conditions today will occur regularly by 2040 under the RCP 8.5 scenario (Russo et al. 2016). The intensity of heat events may be modulated by land cover and soil characteristics (Miralles et al. 2014; Lemordant et al. 2016; Ramarao et al. 2016). Where temperature increase results in decreased soil moisture, latent heat flux is reduced while sensible heat fluxes are increased, allowing surface air temperature to rise further. However, this feedback may be diminished if the land surface is irrigated through enhanced evapotranspiration (Mueller et al. 2015; Siebert et al. 2017) (Section 2.5.2.2).

Drought (IPCC 2013c), including megadroughts of the last century, for example, the Dustbowl drought (Hegerl et al. 2018) (Chapter 5), is a normal component of climate variability (Hoerling et al. 2010; Dai 2011) and may be seasonal, multi-year (Van Dijk et al. 2013) or multi-decadal (Hulme 2001) with increasing degrees of impact on regional activities. This inter-annual variability is controlled particularity through remote sea surface temperature (SST) forcings, such as the Inter-decadal Pacific Oscillation (IPO) and the Atlantic Multi-decadal Oscillation (AMO), El Niño/Southern Oscillation (ENSO) and Indian Ocean Dipole (IOD), that cause drought as a result of reduced rainfall (Kelley et al. 2015; Dai 2011; Hoell et al. 2017; Espinoza et al. 2018). In some cases however, large scale SST modes do not fully explain the severity of drought some recent event attribution studies have identified a climate change fingerprint in several regional droughts, for example, the western Amazon (Erfanian et al. 2017), southern Africa (Funk et al. 2018; Yuan et al. 2018), southern Europe and the Mediterranean including North Africa (Kelley et al. 2015; Wilcox et al. 2018), parts of North America (Williams et al. 2015; Mote et al. 2016), Russia (Otto et al. 2012), India (Ramarao et al. 2015) and Australia (Lewis and Karoly 2013).

Long-term global trends in drought are difficult to determine because of this natural variability, potential deficiencies in drought indices (especially in how evapotranspiration is treated) and the quality and availability of precipitation data (Sheffield et al. 2012; Dai 2013; Trenberth et al. 2014; Nicholls and Seneviratne 2015; Mukherjee et al. 2018). However, regional trends in frequency and intensity of drought are evident in several parts of the world, particularly in low latitude land areas, such as the Mediterranean, North Africa and the Middle East (Vicente-Serrano et al. 2014; Spinoni et al. 2015a; Dai and Zhao 2017; Páscoa et al. 2017), many regions of sub-Saharan Africa (Masih et al. 2014; Dai and Zhao 2017), central China (Wang et al. 2017e), the southern Amazon (Fu et al. 2013; Espinoza et al. 2018), India (Ramarao et al. 2016), east and south Asia, parts of North America and eastern Australia (Dai and Zhao 2017). A recent analysis of 4500 meteorological droughts globally found increased drought frequency over the East Coast of the USA, Amazonia and north-eastern Brazil, Patagonia, the Mediterranean region, most of Africa and north-eastern China with decreased drought frequency over northern Argentina, Uruguay and northern Europe (Spinoni et al. 2019). The study also found drought intensity has become more severe over north-western USA, parts of Patagonia and southern Chile, the Sahel, the Congo River basin, southern Europe, north-eastern China, and south-eastern Australia, whereas the eastern USA, south-eastern Brazil, northern Europe, and central-northern Australia experienced less severe droughts. In addition to the IPCC SR15 assessment of medium confidence in increased drying over the Mediterranean region (Hoegh-Guldberg et al. 2018), it is further assessed with *medium confidence* that frequency and intensity of droughts in Amazonia, north-eastern Brazil, Patagonia, most of Africa, and north-eastern China has increased.

There is *low confidence* in how large-scale modes of variability will respond to a warming climate (Deser et al. 2012; Liu 2012; Christensen et al. 2013; Hegerl et al. 2015; Newman et al. 2016). Although, there is evidence for an increased frequency of extreme ENSO events, such as the 1997/98 El Niño and 1988/89 La Niña (Cai et al. 2014a, 2015) and extreme positive phases of the IOD (Christensen et al. 2013; Cai et al. 2014b). However, the assessment by the SR15 was retained on an increased regional drought risk (*medium confidence)*, specifically over the Mediterranean and South Africa at both 1.5°C and 2°C warming levels compared to present day, with drought risk at 2°C being significantly higher than at 1.5°C (Hoegh-Guldberg et al. 2018).

2.2.5.2   Impacts of heat extremes and drought on land

There is *high confidence* that heat extremes such as unusually hot nights, extremely high daytime temperatures, heatwaves and drought are damaging to crop production (Chapter 5). Extreme heat

events impact a wide variety of tree functions including reduced photosynthesis, increased photooxidative stress, leaves abscise, a decreased growth rate of remaining leaves and decreased growth of the whole tree (Teskey et al. 2015). Although trees are more resilient to heat stress than grasslands (Teuling et al. 2010), it has been observed that different types of forest (e.g., needleleaf vs broadleaf) respond differently to drought and heatwaves (Babst et al. 2012). For example, in the Turkish Anatolian forests net primary productivity (NPP) generally decreased during drought and heatwave events between 2000 and 2010 but in a few other regions, NPP of needle leaf forests increased (Erşahin et al. 2016). However, forests may become less resilient to heat stress in future due to the long recovery period required to replace lost biomass and the projected increased frequency of heat and drought events (Frank et al. 2015a; McDowell and Allen 2015; Johnstone et al. 2016; Stevens-Rumann et al. 2018). Additionally, widespread regional tree mortality may be triggered directly by drought and heat stress (including warm winters) and exacerbated by insect outbreak and fire (Neuvonen et al. 1999; Breshears et al. 2005; Berg et al. 2006; Soja et al. 2007; Kurz et al. 2008; Allen et al. 2010).

Gross primary production (GPP) and soil respiration form the first and second largest carbon fluxes from terrestrial ecosystems to the atmosphere in the global carbon cycle (Beer et al. 2010; Bond-Lamberty and Thomson 2010). Heat extremes impact the carbon cycle through altering these and change ecosystem-atmosphere CO_2 fluxes and the ecosystem carbon balance. Compound heat and drought events result in a stronger carbon sink reduction compared to single-factor extremes as GPP is strongly reduced and ecosystem respiration less so (Reichstein et al. 2013; Von Buttlar et al. 2018). In forest biomes, however, GPP may increase temporarily as a result of increased insolation and photosynthetic activity as was seen during the 2015–2016 ENSO related drought over Amazonia (Zhu et al. 2018). Longer extreme events (heatwave or drought or both) result in a greater reduction in carbon sequestration and may also reverse long-term carbon sinks (Ciais et al. 2005; Phillips et al. 2009; Wolf et al. 2016b; Ummenhofer and Meehl 2017; Von Buttlar et al. 2018; Reichstein et al. 2013). Furthermore, extreme heat events may impact the carbon cycle beyond the lifetime of the event. These lagged effects can slow down or accelerate the carbon cycle: it will slow down if reduced vegetation productivity and/or widespread mortality after an extreme drought are not compensated by regeneration, or speed up if productive tree and shrub seedlings cause rapid regrowth after windthrow or fire (Frank et al. 2015a). Although some ecosystems may demonstrate resilience to a single heat climate stressor like drought (e.g., forests), compound effects of, for example, deforestation, fire and drought, potentially can result in changes to regional precipitation patterns and river discharge, losses of carbon storage and a transition to a disturbance-dominated regime (Davidson et al. 2012). Additionally, adaptation to seasonal drought may be overwhelmed by multi-year drought and their legacy effects (Brando et al. 2008; da Costa et al. 2010).

Under medium- and high-emission scenarios, global warming will exacerbate heat stress, thereby amplifying deficits in soil moisture and runoff despite uncertain precipitation changes (Ficklin and Novick 2017; Berg and Sheffield 2018; Cook et al. 2018; Dai et al. 2018; Engelbrecht et al. 2015; Ramarao et al. 2015; Grillakis 2019). This will increase the rate of drying causing drought to set in quicker, become more intense and widespread, last longer and could result in an increased global aridity (Dai 2011; Prudhomme et al. 2014).

The projected changes in the frequency and intensity of extreme temperatures and drought is expected to result in decreased carbon sequestration by ecosystems and degradation of ecosystems health and loss of resilience (Trumbore et al. 2015). Also affected are many aspects of land functioning and type including agricultural productivity (Lesk et al. 2016), hydrology (Mosley 2015; Van Loon and Laaha 2015), vegetation productivity and distribution (Xu et al. 2011; Zhou et al. 2014), carbon fluxes and stocks, and other biogeochemical cycles (Frank et al. 2015b; Doughty et al. 2015; Schlesinger et al. 2016). Carbon stocks are particularly vulnerable to extreme events due to their large carbon pools and fluxes, potentially large lagged impacts and long recovery times to regain lost stocks (Frank et al. 2015a) (Section 2.2).

2.2.5.3 Changes in heavy precipitation

A large number of extreme rainfall events have been documented over the past decades (Coumou and Rahmstorf 2012; Seneviratne et al. 2012; Trenberth 2012; Westra et al. 2013; Espinoza et al. 2014; Guhathakurta et al. 2017; Taylor et al. 2017; Thompson et al. 2017; Zilli et al. 2017). The observed shift in the trend distribution of precipitation extremes is more distinct than for annual mean precipitation and the global land fraction experiencing more intense precipitation events is larger than expected from internal variability (Fischer and Knutti 2014; Espinoza et al. 2018; Fischer et al. 2013). As a result of global warming, the number of record-breaking rainfall events globally has increased significantly by 12% during the period 1981–2010 compared to those expected due to natural multi-decadal climate variability (Lehmann et al. 2015). The IPCC SR15 reports robust increases in observed precipitation extremes for annual maximum 1-day precipitation (RX1day) and consecutive 5-day precipitation (RX5day) (Hoegh-Guldberg et al. 2018; Schleussner et al. 2017). A number of extreme rainfall events have been attributed to human influence (Min et al. 2011; Pall et al. 2011; Sippel and Otto 2014; Trenberth et al. 2015; Krishnan et al. 2016) and the largest fraction of anthropogenic influence is evident in the most rare and extreme events (Fischer and Knutti 2014).

A warming climate is expected to intensify the hydrological cycle as a warmer climate facilitates more water vapour in the atmosphere, as approximated by the Clausius-Clapeyron (C-C) relationship, with subsequent effects on regional extreme precipitation events (Christensen and Christensen 2003; Pall et al. 2007; Berg et al. 2013; Wu et al. 2013; Guhathakurta et al. 2017; Thompson et al. 2017; Taylor et al. 2017; Zilli et al. 2017; Manola et al. 2018). Furthermore, changes to the dynamics of the atmosphere amplify or weaken future precipitation extremes at the regional scale (O'Gorman 2015; Pfahl et al. 2017). Continued anthropogenic warming is very likely to increase the frequency and intensity of extreme rainfall in many regions of the globe (Seneviratne et al. 2012; Mohan and Rajeevan 2017; Prein et al. 2017; Stott et al. 2016) although many general circulation models (GCMs) underestimate observed increased trends in heavy precipitation suggesting a substantially stronger intensification

of future heavy rainfall than the multi-model mean (Borodina et al. 2017; Min et al. 2011). Furthermore, the response of extreme convective precipitation to warming remains uncertain because GCMs and regional climate models (RCMs) are unable to explicitly simulate sub-grid scale processes such as convection, the hydrological cycle and surface fluxes and have to rely on parameterisation schemes for this (Crétat et al. 2012; Rossow et al. 2013; Wehner 2013; Kooperman et al. 2014; O'Gorman 2015; Larsen et al. 2016; Chawla et al. 2018; Kooperman et al. 2018; Maher et al. 2018; Rowell and Chadwick 2018). High-resolution RCMs that explicitly resolve convection have a better representation of extreme precipitation but are dependent on the GCM to capture the large scale environment in which the extreme event may occur (Ban et al. 2015; Prein et al. 2015; Kendon et al. 2017). Inter-annual variability of precipitation extremes in the convective tropics are not well captured by global models (Allan and Liu 2018).

There is *low confidence* in the detection of long-term observed and projected seasonal and daily trends of extreme snowfall. The narrow rain–snow transition temperature range at which extreme snowfall can occur is relatively insensitive to climate warming and subsequent large interdecadal variability (Kunkel et al. 2013; O'Gorman 2014, 2015).

2.2.5.4 Impacts of precipitation extremes on different land cover types

More intense rainfall leads to water redistribution between surface and ground water in catchments as water storage in the soil decreases (green water) and runoff and reservoir inflow increases (blue water) (Liu and Yang 2010; Eekhout et al. 2018). This results in increased surface flooding and soil erosion, increased plant water stress and reduced water security, which in terms of agriculture means an increased dependency on irrigation and reservoir storage (Nainggolan et al. 2012; Favis-Mortlock and Mullen 2011; García-Ruiz et al. 2011; Li and Fang 2016; Chagas and Chaffe 2018). As there is high confidence of a positive correlation between global warming and future flood risk, land cover and processes are likely to be negatively impacted, particularly near rivers and in floodplains (Kundzewicz et al. 2014; Alfieri et al. 2016; Winsemius et al. 2016; Arnell and Gosling 2016; Alfieri et al. 2017; Wobus et al. 2017).

In agricultural systems, heavy precipitation and inundation can delay planting, increase soil compaction and cause crop losses through anoxia and root diseases (Posthumus et al. 2009). In tropical regions, flooding associated with tropical cyclones can lead to crop failure from both rainfall and storm surge. In some cases, flooding can affect yield more than drought, particularly in tropical regions (e.g., India) and in some mid/high latitude regions such as China and central and northern Europe (Zampieri et al. 2017). Waterlogging of croplands and soil erosion also negatively affect farm operations and block important transport routes (Vogel and Meyer 2018; Kundzewicz and Germany 2012). Flooding can be beneficial in drylands if the floodwaters infiltrate and recharge alluvial aquifers along ephemeral river pathways, extending water availability into dry seasons and drought years, and supporting riparian systems and human communities (Kundzewicz and Germany 2012; Guan et al. 2015). Globally, the impact of rainfall extremes on agriculture

2

is less than that of temperature extremes and drought, although in some regions and for some crops, extreme precipitation explains a greater component of yield variability, for example, of maize in the Midwestern USA and southern Africa (Ray et al. 2015; Lesk et al. 2016; Vogel et al. 2019).

Although many soils on floodplains regularly suffer from inundation, the increases in the magnitude of flood events mean that new areas with no recent history of flooding are now becoming severely affected (Yellen et al. 2014). Surface flooding and associated soil saturation often results in decreased soil quality through nutrient loss, reduced plant productivity, stimulated microbial growth and microbial community composition, negatively impacted soil redox and increased GHG emissions (Bossio and Scow 1998; Niu et al. 2014; Barnes et al. 2018; Sánchez-Rodríguez et al. 2019). The impact of flooding on soil quality is influenced by management systems that may mitigate or exacerbate the impact. Although soils tend to recover quickly after floodwater removal, the impact of repeated extreme flood events over longer timescales on soil quality and function is unclear (Sánchez-Rodríguez et al. 2017).

Flooding in ecosystems may be detrimental through erosion or permanent habitat loss, or beneficial, as a flood pulse brings nutrients to downstream regions (Kundzewicz et al. 2014). Riparian forests can be damaged through flooding; however, increased flooding may also be of benefit to forests where upstream water demand has lowered stream flow, but this is difficult to assess and the effect of flooding on forests is not well studied (Kramer et al. 2008; Pawson et al. 2013). Forests may mitigate flooding, however flood mitigation potential is limited by soil saturation and rainfall intensity (Pilaš et al. 2011; Ellison et al. 2017). Some grassland species under heavy rainfall and soil saturated conditions responded negatively with decreased reproductive biomass and germination rates (Gellesch et al. 2017), however overall productivity in grasslands remains constant in response to heavy rainfall (Grant et al. 2014).

Extreme rainfall alters responses of soil CO_2 fluxes and CO_2 uptake by plants within ecosystems, and therefore result in changes in ecosystem carbon cycling (Fay et al. 2008; Frank et al. 2015a). Extreme rainfall and flooding limits oxygen in soil which may suppress the activities of soil microbes and plant roots and lower soil respiration, therefore lowering carbon cycling (Knapp et al. 2008; Rich and Watt 2013; Philben et al. 2015). However, the impact of extreme rainfall on carbon fluxes in different biomes differs. For example, extreme rainfall in mesic biomes reduces soil CO_2 flux to the atmosphere and GPP whereas in xeric biomes the opposite is true, largely as a result of increased soil water availability (Knapp and Smith 2001; Heisler and Knapp 2008; Heisler-White et al. 2009; Zeppel et al. 2014; Xu and Wang 2016; Liu et al. 2017b; Connor and Hawkes 2018).

As shown above GHG fluxes between the land and atmosphere are affected by climate. The next section assesses these fluxes in greater detail and the potential for land as a carbon sink.

Cross-Chapter Box 3 | Fire and climate change

Raman Sukumar (India), Almut Arneth (Germany), Werner Kurz (Canada), Andrey Sirin (Russian Federation), Louis Verchot (Colombia/The United States of America)

Fires have been a natural part of Earth's geological past and its biological evolution since at least the late Silurian, about 400 million years ago (Scott 2000). Presently, roughly 3% of the Earth's land surface burns annually which affects both energy and matter exchanges between the land and atmosphere (Stanne et al. 2009). Climate is a major determinant of fire regimes through its control of fire weather, as well as through its interaction with vegetation productivity (fuel availability) and structure (fuel distribution and flammability) (Archibald et al. 2013) at the global (Krawchuk and Moritz 2011), regional (Pausas and Paula 2012) and local (Mondal and Sukumar 2016) landscape scales. Presently, humans are the main cause of fire ignition with lightning playing a lesser role globally (Bowman et al. 2017; Harris et al. 2016), although the latter factor has been predominantly responsible for large fires in regions such as the North American boreal forests (Veraverbeke et al. 2017). Humans also influence fires by actively extinguishing them, reducing spread and managing fuels.

Historical trends and drivers in land area burnt
While precipitation has been the major influence on fire regimes before the Holocene, human activities have become the dominant drivers since then (Bowman et al. 2011). There was less biomass burning during the 20th century than at any time during the past two millennia as inferred from charcoal sedimentary records (Doerr and Santín 2016), though there has been an increase in the most recent decades (Marlon et al. 2016). Trends in land area burnt have varied regionally (Giglio et al. 2013). Northern hemisphere Africa has experienced a fire decrease of 1.7 Mha yr^{-1} (–1.4% yr^{-1}) since 2000, while southern hemisphere Africa saw an increase of 2.3 Mha yr^{-1} (+1.8% yr^{-1}) during the same period. Southeast Asia witnessed a small increase of 0.2 Mha yr^{-1} (+2.5% yr^{-1}) since 1997, while Australia experienced a sharp decrease of about 5.5 Mha yr^{-1} (–10.7% yr^{-1}) during 2001–2011, followed by an upsurge in 2011 that exceeded the annual area burned in the previous 14 years. A recent analysis using the Global Fire Emissions Database v.4 (GFED4s) that includes small fires concluded that the net reduction in land area burnt globally during 1998–2015 was –24.3 ± 8.8% (–1.35 ± 0.49% yr^{-1}) (Andela et al. 2017). However, from the point of fire emissions it is important to consider the land cover types which have experienced changes in area burned; in this instance, most of the declines have come from grasslands, savannas and other non-forest land cover types (Andela et al. 2017). Significant increases in forest area burned (with higher fuel consumption per unit area) have been recorded in western and boreal

Cross-Chapter Box 3 (continued)

North America (Abatzoglou and Williams 2016; Ansmann et al. 2018) and in boreal Siberia (Ponomarev et al. 2016) in recent times. The 2017 and 2018 fires in British Columbia, Canada, were the largest ever recorded since the 1950s with 1.2 Mha and 1.4 Mha of forest burnt, respectively (Hanes et al. 2018) and smoke from these fires reaching the stratosphere over central Europe (Ansmann et al. 2018).

Climate variability and extreme climatic events such as severe drought, especially those associated with the El Niño Southern Oscillation (ENSO), play a major role in fire upsurges, as in equatorial Asia (Huijnen et al. 2016). Fire emissions in tropical forests increased by 133% on average during and following six El Niño years compared to six La Niña years during 1997–2016, due to reductions in precipitation and terrestrial water storage (Chen et al. 2017). The expansion of agriculture and deforestation in the humid tropics has also made these regions more vulnerable to drought-driven fires (Davidson et al. 2012; Brando et al. 2014). Even when deforestation rates were overall declining, as in the Brazilian Amazon during 2003–2015, the incidence of fire increased by 36% during the drought of 2015 (Aragão et al. 2018).

GHG emissions from fires

Emissions from wildfires and biomass burning are a significant source of GHGs (CO_2, CH_4, N_2O), carbon monoxide (CO), carbonaceous aerosols, and an array of other gases including non-methane volatile organic compounds (NMVOC) (Akagi et al. 2011; Van Der Werf et al. 2010). GFED4s has updated fire-related carbon emission estimates biome-wise (regionally and globally), using higher resolution input data gridded at 0.25°, a new burned area dataset with small fires, improved fire emission factors (Akagi et al. 2011; Urbanski 2014) and better fire severity characterisation of boreal forests (van der Werf et al. 2017). The estimates for the period 1997–2016 are 2.2 GtC yr^{-1}, being highest in the 1997 El Nino (3.0 GtC yr^{-1}) and lowest in 2013 (1.8 GtC yr^{-1}). Furthermore, fire emissions during 1997–2016 were dominated by savanna (65.3%), followed by tropical forest (15.1%), boreal forest (7.4%), temperate forest (2.3%), peatland (3.7%) and agricultural waste burning (6.3%) (van der Werf et al. 2017).

Fires not only transfer carbon from land to the atmosphere but also between different terrestrial pools: from live to dead biomass to soil, including partially charred biomass, charcoal and soot constituting 0.12–0.39 GtC yr^{-1} or 0.2–0.6% of annual terrestrial NPP (Doerr and Santín 2016). Carbon from the atmosphere is sequestered back into regrowing vegetation at rates specific to the type of vegetation and other environmental variables (Loehman et al. 2014). Fire emissions are thus not necessarily a net source of carbon into the atmosphere, as post-fire recovery of vegetation can sequester a roughly equivalent amount back into biomass over a time period of one to a few years (in grasslands and agricultural lands) to decades (in forests) (Landry and Matthews 2016). Fires from deforestation (for land use change) and on peatlands (which store more carbon than terrestrial vegetation) obviously are a net source of carbon from the land to the atmosphere (Turetsky et al. 2014); these types of fires were estimated to emit 0.4 GtC yr^{-1} in recent decades (van der Werf et al. 2017). Peatland fires dominated by smouldering combustion under low temperatures and high moisture conditions can burn for long periods (Turetsky et al. 2014).

Fires, land degradation/desertification and land-atmosphere exchanges

Flammable ecosystems are generally adapted to their specific fire regimes (Bond et al. 2005). A fire regime shift alters vegetation and soil properties in complex ways, both in the short- and the long-term, with consequences for carbon stock changes, albedo, fire-atmosphere-vegetation feedbacks and the ultimate biological capacity of the burnt land (Bond et al. 2004; Bremer and Ham 1999; MacDermott et al. 2016; Tepley et al. 2018; Moody et al. 2013; Veraverbeke et al. 2012) A fire-driven shift in vegetation from a forested state to an alternative stable state such as a grassland (Fletcher et al. 2014; Moritz 2015) with much less carbon stock is a distinct possibility. Fires cause soil erosion through action of wind and water (Moody et al. 2013) thus resulting in land degradation (Chapter 4) and eventually desertification (Chapter 3). Fires also affect carbon exchange between land and atmosphere through ozone (which retards photosynthesis) and aerosol (which slightly increases diffuse radiation) emissions. The net effect from fire on global GPP during 2002–2011 is estimated to be −0.86 ± 0.74 GtC yr^{-1} (Yue and Unger 2018).

Fires under future climate change

Temperature increase and precipitation decline would be the major driver of fire regimes under future climates as evapotranspiration increases and soil moisture decreases (Pechony and Shindell 2010; Aldersley et al. 2011; Abatzoglou and Williams 2016; Fernandes et al. 2017). The risk of wildfires in future could be expected to change, increasing significantly in North America, South America, central Asia, southern Europe, southern Africa and Australia (Liu et al. 2010). There is emerging evidence that recent regional surges in wildland fires are being driven by changing weather extremes, thereby signalling geographical shifts in fire proneness (Jolly et al. 2015). Fire weather season has already lengthened by 18.7% globally between 1979 and 2013, with statistically significant increases across 25.3% but decreases only across 10.7% of Earth's land surface covered with vegetation. Even sharper changes have been observed during the second half of this period (Jolly et al. 2015). Correspondingly, the global area experiencing long fire weather seasons (defined as experiencing a fire weather season greater than one standard deviation (SD)

Cross-Chapter Box 3 (continued)

from the mean global value) has increased by 3.1% per annum or 108.1% during 1979–2013. Fire frequencies under 2050 conditions are projected to increase by approximately 27% globally, relative to the 2000 levels, with changes in future fire meteorology playing the most important role in enhancing global wildfires, followed by land cover changes, lightning activities and land use, while changes in population density exhibit the opposite effects (Huang et al. 2014).

However, climate is only one driver of a complex set of environmental, ecological and human factors in influencing fire regimes (Bowman et al. 2011). While these factors lead to complex projections of future burnt area and fire emissions (Knorr et al. 2016a, b), human exposure to wildland fires could still increase due to population expansion into areas already under high risk of fires (Knorr et al. 2016a, b). There are still major challenges in projecting future fire regimes and how climate, vegetation and socio/economic factors will interact (Hantson et al. 2016; Harris et al. 2016). There is also need for integrating various fire management strategies, such as fuel-reduction treatments in natural and planted forests, with other environmental and societal considerations to achieve the goals of carbon emissions reductions, maintain water quality, biodiversity conservation and human safety (Moritz et al. 2014; Gharun et al. 2017).

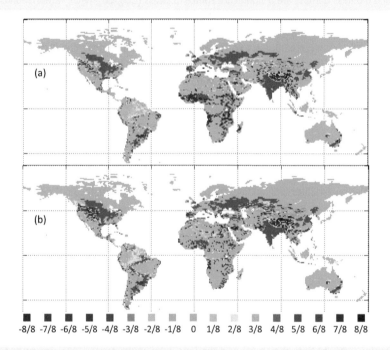

-8/8 -7/8 -6/8 -5/8 -4/8 -3/8 -2/8 -1/8   0   1/8 2/8 3/8 4/8 5/8 6/8 7/8 8/8

Cross-Chapter Box 3, Figure 1 | The probability of low-fire regions becoming fire prone (positive values), or of fire-prone areas changing to a low-fire state (negative values) between 1971–2000 and 2071–2100 based on eight-Earth system model (ESM) ensembles, two Shared Socio-economic Pathways (SSPs) and two Representative Concentration Pathways (RCPs). Light grey: areas where at least one ensemble simulation predicts a positive and one a negative change (lack of agreement). Dark grey: area with >50% past or future cropland. Fire-prone areas are defined as having a fire frequency of >0.01 yr⁻¹, **(a)** RCP4.5 emissions with SSP3 demographics, and **(b)** RCP8.5 emissions with SSP5 demographics (Knorr et al. 2016a).

In summary, climate change is playing an increasing role in determining wildfire regimes alongside human activity (*medium confidence*), with future climate variability expected to enhance the risk and severity of wildfires in many biomes, such as tropical rainforests (*high confidence*). Fire weather seasons have lengthened globally between 1979 and 2013 (*low confidence*). Global land area burned has declined in recent decades, mainly due to less burning in grasslands and savannas (*high confidence*). While drought remains the dominant driver of fire emissions, there has recently been increased fire activity in some tropical and temperate regions during normal to wetter-than-average years due to warmer temperatures that increase vegetation flammability (*medium confidence*). The boreal zone is also experiencing larger and more frequent fires, and this may increase under a warmer climate (*medium confidence*).

2.3 Greenhouse gas fluxes between land and atmosphere

Land is simultaneously a source and sink for several GHGs. Moreover, both natural and anthropogenic processes determine fluxes of GHGs, making it difficult to separate 'anthropogenic' and 'non-anthropogenic' emissions and removals. A meeting report by the IPCC (2010) divided the processes responsible for fluxes from land into three categories: (i) the *direct effects* of anthropogenic activity due to changing land cover and land management, (ii) the *indirect effects* of anthropogenic environmental change, such as climate change, carbon dioxide (CO_2) fertilisation, nitrogen deposition, and (iii) *natural* climate variability and natural disturbances (e.g., wildfires, windrow, disease). The meeting report (IPCC 2010) noted that it was impossible with any direct observation to separate direct anthropogenic effects from non-anthropogenic (indirect and natural) effects in the land sector.

As a result, different approaches and methods for estimating the anthropogenic fluxes have been developed by different communities to suit their individual purposes, tools and data availability.

The major GHGs exchanged between land and the atmosphere discussed in this chapter are CO_2 (Section 2.3.1), methane (CH_4) (Section 2.3.2) and nitrous oxide (N_2O) (Section 2.3.3). We estimate the total emissions from AFOLU to be responsible for approximately 23% of global anthropogenic GHG emissions over the period 2007–2016 (Smith et al. 2013a; Ciais et al. 2013a) (Table 2.2). The estimate is similar to that reported in AR5 (*high confidence*), with slightly more than half these emissions coming as non-CO_2 GHGs from agriculture. Emissions from AFOLU have remained relatively constant since AR4, although their relative contribution to anthropogenic emissions has decreased due to increases in emissions from the energy sector.

Table 2.2 | Net anthropogenic emissions due to Agriculture, Forestry, and other Land Use (AFOLU) and non-AFOLU (average for 2007–2016).[1] Positive value represents emissions; negative value represents removals.

Gas	Units	Net anthropogenic emissions due to Agriculture, Forestry, and Other Land Use (AFOLU)			Non-AFOLU anthropogenic GHG emissions[4]	Total net anthropogenic emissions (AFOLU + non-AFOLU) by gas	AFOLU as a % of total net anthropogenic emissions, by gas	Natural response of land to human-induced environmental change[5]	Net land–atmosphere flux from all lands
		FOLU	Agriculture	Total					
		A	B	C = A + B	D	E = C + D	F = (C/E) × 100	G	A + G
CO_2[2]	$GtCO_2$ yr^{-1}	5.2 ± 2.6	No data	5.2 ± 2.6	33.9 ± 1.8	39.1 ± 3.2	13%	−11.2 ± 2.6	−6.0 ± 3.7
CH_4[3,6]	$MtCH_4$ yr^{-1}	19.2 ± 5.8	142 ± 43	161 ± 43	201 ± 101	362 ± 109			
	$GtCO_2$-eq yr^{-1}	0.5 ± 0.2	4.0 ± 1.2	4.5 ± 1.2	5.6 ± 2.8	10.1 ± 3.1	44%		
N_2O[3,6]	MtN_2O yr^{-1}	0.3 ± 0.1	8.3 ± 2.5	8.7 ± 2.5	2.0 ± 1.0	10.6 ± 2.7			
	$GtCO_2$-eq yr^{-1}	0.09 ± 0.03	2.2 ± 0.7	2.3 ± 0.7	0.5 ± 0.3	2.8 ± 0.7	81%		
Total (GHG)	$GtCO_2$-eq yr^{-1}	5.8 ± 2.6	6.2 ± 1.4	12.0 ± 2.9	40.0 ± 3.4	52.0 ± 4.5	23%		

[1] Estimates are only given until 2016 as this is the latest date when data are available for all gases.

[2] Net anthropogenic flux of CO_2 due to land cover change such as deforestation and afforestation, and land management including wood harvest and regrowth, as well as peatland burning, based on two bookkeeping models as used in the Global Carbon Budget and for AR5. Agricultural soil carbon stock change under the same land use is not considered in these models.

[3] Estimates show the mean and assessed uncertainty of two databases, FAOSTAT and USEPA 2012.

[4] Total non-AFOLU emissions were calculated as the sum of total CO_2-eq emissions values for energy, industrial sources, waste and other emissions with data from the Global Carbon Project for CO_2, including international aviation and shipping and from the PRIMAP database for CH_4 and N_2O averaged over 2007–2014 only as that was the period for which data were available.

[5] The natural response of land to human-induced environmental changes is the response of vegetation and soils to environmental changes such as increasing atmospheric CO_2 concentration, nitrogen deposition, and climate change. The estimate shown represents the average from Dynamic Global Vegetation Models.

[6] All values expressed in units of CO_2-eq are based on AR5 100-year Global Warming Potential (GWP) values without climate-carbon feedbacks (N_2O = 265; CH_4 = 28). Note that the GWP has been used across fossil fuel and biogenic sources of methane. If a higher GWP for fossil fuel CH_4 (30 per AR5), then total anthropogenic CH_4 emissions expressed in CO_2-eq would be 2% greater.

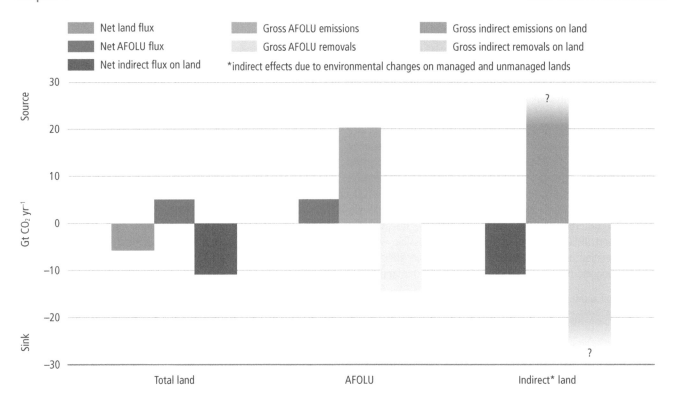

Figure 2.4 | Net and gross fluxes of CO₂ from land (annual averages for 2008–2017). Left: The total net flux of CO₂ between land and atmosphere (grey) is shown with its two component fluxes, (i) net AFOLU emissions (blue), and (ii) the net land sink (brown), due to indirect environmental effects and natural effects on managed and unmanaged lands. Middle: The gross emissions and removals contributing to the net AFOLU flux. Right: The gross emissions and removals contributing to the land sink.

2.3.1   Carbon dioxide

This section is divided into four sub-sections (Figure 2.4): (i) the total net flux of CO₂ between land and atmosphere, (ii) the contributions of AFOLU fluxes and the non-AFOLU land sink to that total net CO₂ flux, (iii) the gross emissions and removals comprising the net AFOLU flux, and (iv) the gross emissions and removals comprising the land sink. Emissions to the atmosphere are positive; removals from the atmosphere are negative.

2.3.1.1   The total net flux of CO₂ between land and atmosphere

The net effects of all anthropogenic and non-anthropogenic processes on managed and unmanaged land result in a net removal of CO₂ from the atmosphere (*high confidence*). This total net land-atmosphere removal (defined here as *the total net land flux*) is estimated to have averaged 6.0 ± 2.0 GtCO₂ yr⁻¹ (*likely range*) from 2007–2016 (Table 2.3). The estimate is determined from summing the AFOLU and non-AFOLU fluxes due to transient climate change, CO₂ fertilisation and nitrogen deposition calculated by models in the global carbon budget (Le Quéré et al. 2018), and is consistent with inverse modelling techniques based on atmospheric CO₂ concentrations and air transport (range: 5.1–8.8 GtCO₂ yr⁻¹) (Peylin et al. 2013; Van Der Laan-Luijkx et al. 2017; Saeki and Patra 2017; Le Quéré et al. 2018) (see Box 2.2 for methods). A recent inverse analysis, considering carbon transport in rivers and oceans, found a net flux of CO₂ for land within this range, but a lower source from southern lands and a lower sink in northern lands (Resplandy et al. 2018).

The net removal of CO₂ by land has generally increased over the last 60 years in proportion to total emissions of CO₂ (*high confidence*). Although land has been a net sink for CO₂ since around the middle of last century, it was a net source to the atmosphere before that time, primarily as a result of emissions from AFOLU (Le Quéré et al. 2018).

2.3.1.2   Separation of the total net land flux into AFOLU fluxes and the land sink

The total net flux of carbon between land and the atmosphere can be divided into fluxes due to direct human activities (i.e., AFOLU) and fluxes due to indirect anthropogenic and natural effects (i.e., the land sink) (Table 2.3). These two components are less certain than their sums, the total net flux of CO₂ between atmosphere and land. The land sink, estimated with DGVMs, is least certain (Figure 2.5).

Fluxes attributed to AFOLU

The modelled AFOLU flux was a net emission of 5.2 ± 2.6 GtCO₂ yr⁻¹ (*likely range*) for 2007–2016, approximately 13% of total anthropogenic CO₂ emissions (Le Quéré et al. 2018) (Table 2.3). This net flux was due to direct anthropogenic activities, predominately tropical deforestation, but also afforestation/reforestation, and fluxes due to forest management (e.g., wood harvest) and other types of land management, as well as peatland drainage and burning. The AFOLU flux is the mean of two estimates from bookkeeping models (Hansis et al. 2015; Houghton and Nassikas 2017), and this estimated

Table 2.3 | Perturbation of the global carbon cycle caused by anthropogenic activities (GtCO₂ yr⁻¹). Source: Le Quéré et al. (2018).

	CO₂ flux (GtCO₂ y⁻¹), 10-year mean					
	1960–1969	1970–1979	1980–1989	1990–1999	2000–2009	2008–2017
Emissions						
Fossil CO₂ emissions	11.4 ± 0.7	17.2 ± 0.7	19.8 ± 1.1	23.1 ± 1.1	28.6 ± 1.5	34. ± 1.8
AFOLU net emissions	5.5 ± 2.6	4.4 ± 2.6	4.4 ± 2.6	5.1 ± 2.6	4.8 ± 2.6	5.5 ± 2.6
Partitioning						
Growth in atmosphere	6.2 ± 0.3	10.3 ± 0.3	12.5 ± 0.07	11.4 ± 0.07	14.7 ± 0.07	17.2 ± 0.07
Ocean sink	3.7 ± 1.8	4.8 ± 1.8	6.2 ± 1.8	7.3 ± 1.8	7.7 ± 1.8	8.8 ± 1.8
Land sink (non-AFOLU)	4.4 ± 1.8	7.7 ± 1.5	6.6 ± 2.2	8.8 ± 1.8	9.9 ± 2.6	11.7 ± 2.6
Budget imbalance	2.2	−1.1	−1.1	0.7	0.7	1.8
Total net land flux (AFOLU – land sink)	+1.1 ± 3.2	−3.3 ± 3.0	−2.2 ± 3.4	−3.7 ± 2.2	−5.1 ± 3.2	−6.2 ± 3.7

mean is consistent with the mean obtained from an assemblage of DGVMs (Le Quéré et al. 2018) (Box 2.2 and Figure 2.5), although not all individual DGMVs include the same types of land use. Net CO₂ emissions from AFOLU have been relatively constant since 1900. AFOLU emissions were the dominant anthropogenic emissions until around the middle of the last century when fossil fuel emissions became dominant (Le Quéré et al. 2018). AFOLU activities have resulted in emissions of CO₂ over recent decades (*robust evidence, high agreement*) although there is a wide range of estimates from different methods and approaches (Smith et al. 2014; Houghton et al. 2012; Gasser and Ciais 2013; Pongratz et al. 2014; Tubiello et al. 2015; Grassi et al. 2018) (Box 2.2, Figure 2.5 and Figure 2.7).

DGVMs and one bookkeeping model (Hansis et al. 2015) used spatially explicit, harmonised land-use change data (LUH2) (Hurtt et al. 2017) based on HYDE 3.2. The HYDE data, in turn, are based on changes in the areas of croplands and pastures. In contrast, the Houghton bookkeeping approach (Houghton and Nassikas 2017) used primarily changes in forest area from the FAO Forest Resource Assessment (FAO 2015) and FAOSTAT to determine changes in land use. To the extent that forests are cleared for land uses other than crops and pastures, estimates from Houghton and Nassikas (2017, 2018) are higher than estimates from DGMVs. In addition, both bookkeeping models (Hansis et al. 2015; Houghton and Nassikas 2017) included estimates of carbon emissions in Southeast Asia from peat burning from GFED4s (Randerson et al. 2015) and from peat drainage (Hooijer et al. 2010).

Satellite-based estimates of CO₂ emissions from losses of tropical forests during 2000–2010 corroborate the modelled emissions but are quite variable; 4.8 GtCO₂ yr⁻¹ (Tyukavina et al. 2015), 3.0 GtCO₂ yr⁻¹ (Harris et al. 2015), 3.2 GtCO₂ yr⁻¹ (Achard et al. 2014) and 1.6 GtCO₂ yr⁻¹ (Baccini et al. 2017). Differences in estimates can be explained to a large extent by the different approaches used. For example, the analysis by Tyukavina et al. (2015) led to a higher estimate because they used a finer spatial resolution. Three of the estimates considered losses in forest area and ignored degradation and regrowth of forests. Baccini et al. (2017) in contrast, included both losses and gains in forest area and losses and gains of carbon within forests (i.e., forest degradation and growth). The four remote sensing studies cited above also reported committed emissions; in essence, all of the carbon lost from deforestation was assumed to

be released to the atmosphere in the year of deforestation. In reality, only some of the carbon in trees is released immediately to the atmosphere at the time of deforestation. The unburned portion is transferred to woody debris and wood products. Both bookkeeping models and DGVMs account for the delayed emissions in growth and decomposition. Finally, the satellite-based estimates do not include changes in soil carbon.

In addition to differences in land-cover data sets between models and satellites, there are many other methodological reasons for differences (Houghton et al. 2012; Gasser and Ciais 2013; Pongratz et al. 2014; Tubiello et al. 2015) (Box 2.2). There are different definitions of land-cover type, including forest (e.g., FAO uses a tree cover threshold for forests of 10%, Tyukavina et al. (2017) used 25%), different estimates of biomass and soil carbon density (MgC ha⁻¹), different approaches to tracking emissions through time (legacy effects) and different types of activity included (e.g., forest harvest, peatland drainage and fires). Most DGVMS only recently (since AR5) included forest management processes, such as tree harvesting and land clearing for shifting cultivation, leading to larger estimates of CO₂ emissions than when these processes are not considered (Arneth et al. 2017; Erb et al. 2018). Grazing management has likewise been found to have large effects (Sanderman et al. 2017), and is not included in most DGVMs (Pugh et al. 2015; Pongratz et al., 2018).

Nationally reported greenhouse gas inventories versus global model estimates

There are large differences globally between estimates of net anthropogenic land-atmosphere fluxes of CO₂ from national GHGIs and from global models, and the same is true in many regions (Figure 2.5). Fluxes reported to the UNFCCC through country GHGIs were noted as about 4.3 GtCO₂ yr⁻¹ lower (Grassi et al. 2018) than estimates from the bookkeeping model (Houghton et al. 2012) used in the carbon budget for AR5 (Ciais et al. 2013a). The anthropogenic emissions of CO₂ from AFOLU reported in countries' GHG inventories were 0.1 ± 1.0 GtCO₂ yr⁻¹ globally during 2005–2014 (Grassi et al. 2018) much lower than emission estimates from the two global bookkeeping models of 5.1 ± 2.6 GtCO₂ yr⁻¹ (*likely range*) over the same time period (Le Quéré et al. 2018). Transparency and comparability in estimates can support measuring, reporting and verifying GHG fluxes under the UNFCCC, and also the global

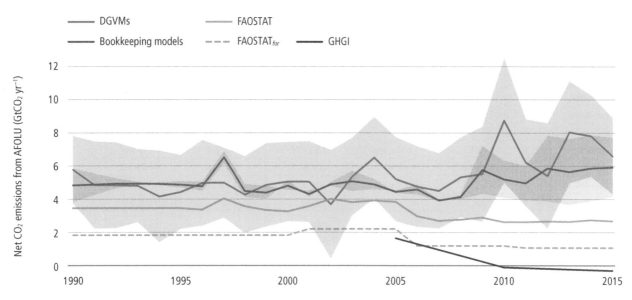

Figure 2.5 | Global net CO$_2$ emissions due to AFOLU from different approaches (in GtCO$_2$ yr^{-1}). Brown line: the mean and individual estimates (brown shading) from two bookkeeping models (Houghton and Nassikas 2017; Hansis et al. 2015). Blue line: the mean from DGVMs run with the same driving data with the pale blue shading showing the ±1 standard deviation range. Yellow line: data downloaded from FAOSTAT website (Tubiello et al. 2013); the dashed line is primarily forest-related emissions, while the solid yellow line also includes emissions from peat fires and peat draining. Orange line: Greenhouse Gas Inventories (GHGI) based on country reports to UNFCCC (Grassi et al. 2018), data are shown only from 2005 because reporting in many developing countries became more consistent/reliable after this date. For more details on methods see Box 2.2.

a) Effects of various factors on the forest CO$_2$ fluxes and where they occur

Direct-human induced effects
– Land use change
– Harvest and other management

Occur on managed land

Indirect-human induced effects
– Climate change induced change in T°, precipitation, length of growing season
– Atmospheric CO$_2$ fertilisation and N deposition, impact of air pollution
– Changes in natural disturbances regime

Natural effects
– Natural interannual variability
– Natural disturbances

Occur on managed and unmanaged land

b) Conceptual differences in defining the anthropogenic land CO$_2$ flux

Figure 2.6 | Summary of the main conceptual differences between GHG Inventories and global models in considering what is the 'anthropogenic land CO$_2$ flux'. Adapted from Grassi et al. (2018), effects of key processes on the land flux as defined by IPCC (2010) including where these effects occur (in managed and/or unmanaged lands) and how these effects are captured in (a) bookkeeping models that do not explicitly model the effects of environmental change (although some is implicitly captured in data on carbon densities and growth and decay rates), (b) DGVMs that include the effects of environmental change on all lands, and run the models with and without land use change to diagnose 'land use change'. The 'land sink' is then conceptually assumed to be a natural response of land to the anthropogenic perturbation of environmental change, DGVMs include the effects of inter-annual climate variability, and some include fires but no other natural disturbances, and (c) GHG Inventories reported by countries to the UNFCCC that report all fluxes in areas the countries define as 'managed land' but do not report unmanaged land. This is the CO$_2$ flux due to Land Use Land Use Change and Forestry (LULUCF) which is a part of the overall AFOLU flux. The area of land considered as managed in the inventories is greater than that considered as subject to direct management activities (harvest and regrowth) in the models.

stocktake, which will assess globally the progress towards achieving the long-term goals of the Paris Agreement. These differences can be reconciled largely by taking account of the different approaches to defining 'anthropogenic' in terms of different areas of land and treatment of indirect environmental change (Grassi et al. 2018).

To date there has been one study that quantitatively reconciles the global model estimates with GHGIs (Grassi et al. 2018). The separation of anthropogenic from non-anthropogenic effects is impossible with direct observation (IPCC 2010). The different approaches of models and GHGIs to estimating anthropogenic emissions and removals are shown in (Figure 2.6). The difficulty is that *indirect* effects of environmental changes (e.g., climate change and rising atmospheric CO_2) affect both manged and unmanaged lands, and some approaches treat these as anthropogenic while others do not. Bookkeeping models (e.g., Houghton and Nassikas 2017) attempt to estimate the fluxes of CO_2 driven by direct anthropogenic effects alone. DGVMs model the *indirect* environmental effects of climate and CO_2. If the indirect effects happen on land experiencing anthropogenic land cover change or management (harvest and regrowth), DGVMs treat this as anthropogenic. Country GHGIs separately report fluxes due to land conversion (e.g., forests to croplands) and fluxes due to land management (e.g., forest land remaining forest land). The 'managed land proxy' is used as a pragmatic approach to estimate anthropogenic fluxes on managed lands, whereby countries define the areas they consider managed and include all of the emissions and removals that occur on those lands. Emissions and removals are caused simultaneously by direct, indirect and natural drivers and are captured in the reporting, which often relies on inventories.

Grassi et al. (2018) demonstrated that estimates of CO_2 emissions from global models and from nationally reported GHGIs were similar for deforestation and afforestation, but different for managed forests. Countries generally reported larger areas of managed forests than the models and the carbon removals by these managed forests were also larger. The flux due to indirect effects on managed lands was quantified using post-processing of results from DGVMs, looking at the *indirect* effects of CO_2 and climate change on secondary forest areas. The derived DGVM *indirect* managed forest flux was found to account for most of the difference between the bookkeeping models and the inventories.

Regional differences

Figure 2.7 shows regional differences in emissions due to AFOLU. Recent increases in deforestation rates in some tropical countries have been partially balanced by increases in forest area in India, China, the USA and Europe (FAO-FRA 2015). The trend in emissions from AFOLU since the 1990s is *uncertain* because some data suggest a declining rate of deforestation (FAO-FRA 2015), while data from satellites suggest an increasing rate (Kim 2014; Hansen et al. 2012). The disagreement results in part from differences in the definition of forest and approaches to estimating deforestation. The FAO defines deforestation as the conversion of forest to another land use (FAO-FRA 2015), while the measurement of forest loss by satellite may include wood harvests (forests remaining forests) and natural disturbances that are not directly caused by anthropogenic activity (e.g., forest mortality from droughts and fires). Trends in anthropogenic and natural disturbances

may be in opposite directions. For example, recent drought-induced fires in the Amazon have increased the emissions from wildfires at the same time that emissions from anthropogenic deforestation have declined (Aragão et al. 2018). Furthermore, there have been advances since AR5 in estimating the GHG effects of different types of forest management (e.g., Valade et al. 2017). Overall, there is *robust evidence* and *high agreement* for a net loss of forest area and tree cover in the tropics and a net gain, mainly of secondary forests and sustainably managed forests, in the temperate and boreal zones (Chapter 1).

Processes responsible for the land sink

Just over half of total net anthropogenic CO_2 emissions (AFOLU and fossil fuels) were taken up by oceanic and land sinks (*robust evidence, high agreement*) (Table 2.3). The land sink was referred to in AR5 as the 'residual terrestrial flux', as it was not estimated directly, but calculated by difference from the other directly estimated fluxes in the budget (Table 2.3). In the 2018 budget (Le Quéré et al. 2018), the land sink term was instead estimated directly by DGVMs, leaving a budget imbalance of 2.2 $GtCO_2$ yr^{-1} (sources overestimated or sinks underestimated). The budget imbalance may result from variations in oceanic uptake or from uncertainties in fossil fuel or AFOLU emissions, as well as from land processes not included in DGVMs.

The land sink is thought to be driven largely by the indirect effects of environmental change (e.g., climate change, increased atmospheric CO_2 concentration, nitrogen deposition) on unmanaged and managed lands (*robust evidence, high agreement*). The land sink has generally increased since 1900 and was a net sink of 11.7 ± 3.7 $GtCO_2$ yr^{-1} during the period 2008–2017 (Table 2.3), absorbing 29% of global anthropogenic emissions of CO_2. The land sink has slowed the rise in global land-surface air temperature by 0.09 ± 0.02°C since 1982 (*medium confidence*) (Zeng et al. 2017).

The rate of CO_2 removal by land accelerated from −0.026 ± 0.24 $GtCO_2$ yr^{-1} during the warming period (1982–1998) to −0.436 ± 0.260 $GtCO_2$ yr^{-1} during the warming hiatus (1998–2012). One explanation is that respiration rates were lower during the warming hiatus (Ballantyne et al. 2017). However, the lower rate of growth in atmospheric CO_2 during the warming hiatus may have resulted, not from lower rates of respiration, but from declining emissions from AFOLU (lower rates of tropical deforestation and increased forest growth in northern mid-latitudes) (Piao et al. 2018). Changes in the growth rate of atmospheric CO_2, by themselves, do not identify the processes responsible and the cause of the variation is uncertain.

While year-to-year variability in the indirect land sink is high in response to climate variability, DGVM fluxes are influenced far more on decadal timescales by CO_2 fertilisation. A DGVM intercomparison (Sitch et al. 2015) for 1990–2009 found that CO_2 fertilisation alone contributed a mean global removal of −10.54 ± 3.68 $GtCO_2$ yr^{-1} (trend −0.444 ± 0.202 $GtCO_2$ yr^{-1}). Data from forest inventories around the world corroborate the modelled land sink (Pan et al. 2011). The geographic distribution of the non-AFOLU land sink is less certain. While it seems to be distributed globally, its distribution between the tropics and non-tropics is estimated to be between 1:1 (Pan et al. 2011) and 1:2 (Houghton et al. 2018).

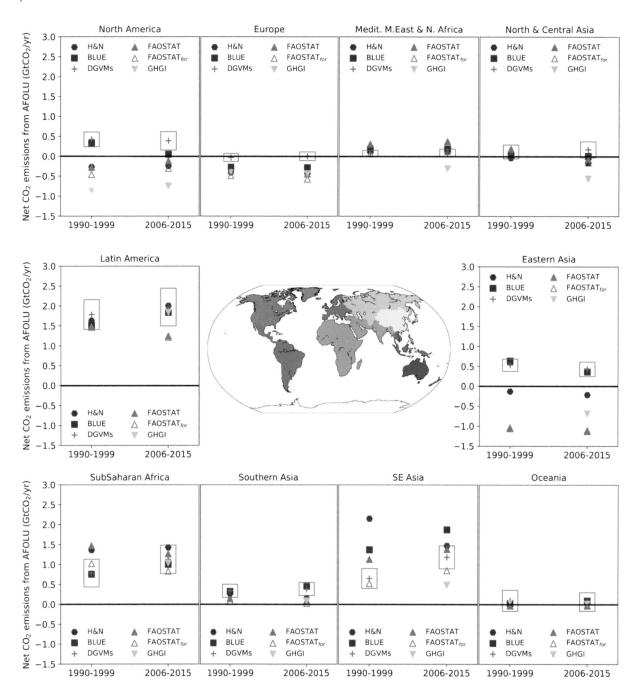

Figure 2.7 | Regional trends in net anthropogenic land-atmosphere CO$_2$ flux from a range of different approaches (in GtCO$_2$ yr^{-1}). Red symbols: bookkeeping models (hexagon: Houghton and Nassikas 2017; square: Hansis et al. 2015). Blue cross: the mean from DGMVs with the box showing the 1 standard deviation range. Green triangles: downloaded from FAOSTAT website; the open triangle is primarily forest-related emissions, while the closed triangle includes emission from peat fires and peat drainage. Yellow inverted triangle: GHGI LULUCF flux based on country reports to UNFCCC (Grassi et al. 2018). Data for developing countries are only shown for 2006–2015 because reporting in many developing countries became more consistent/reliable after 2005. For more details on methods see Box 2.2.

As described in Box 2.3, rising CO$_2$ concentrations have a fertilising effect on land, while climate has mixed effects; for example, rising temperature increases respiration rates and may enhance or reduce photosynthesis depending on location and season, while longer growing seasons might allow for higher carbon uptake. However, these processes are not included in DGVMs, which may account for at least some of the land sink. For example, a decline in the global area burned by fires each year (Andela et al. 2017) accounts for an estimated net sink (and/or reduced emissions) of 0.5 GtCO$_2$ yr^{-1} (*limited evidence, medium agreement*) (Arora and Melton 2018).

Boreal forests represent an exception to this decline (Kelly et al. 2013). The reduction in burning not only reduces emissions, but also allows more growth of recovering forests. There is also an estimated net carbon sink of about the same magnitude (0.5 GtCO$_2$ yr^{-1}) as a result of soil erosion from agricultural lands and redeposition in anaerobic environments where respiration is reduced (*limited evidence, low agreement*) (Wang et al. 2017d). A recent study attributes an increase in land carbon to a longer-term (1860–2005) aerosol-induced cooling (Zhang et al. 2019). Recent evidence also suggests that DGVMs and ESMs underestimate the effects of drought on CO_2 emissions (Humphrey et al. 2018; Green et al. 2019; Kolus et al. 2019).

2.3.1.3 Gross emissions and removals contributing to AFOLU emissions

The modelled AFOLU flux of 5.5 ± 3.7 $GtCO_2$ yr^{-1} over the period 2008–2017 represents a net value. It consists of both gross emissions of CO_2 from deforestation, forest degradation and the oxidation of wood products, as well as gross removals of CO_2 in forests and soils recovering from harvests and agricultural abandonment (Figure 2.4). The uncertainty of these gross fluxes is high because few studies report gross fluxes from AFOLU. Houghton and Nassikas (2017) estimated gross emissions to be as high as 20.2 $GtCO_2$ yr^{-1} (*limited evidence, low agreement*) (Figure 2.4), and even this may be an underestimate because the land-use change data used from FAOSTAT (Tubiello et al. 2013) is itself a net of all changes within a country.

Gross emissions and removals of CO_2 result from rotational uses of land, such as wood harvest and shifting cultivation, including regrowth. These gross fluxes are more informative for assessing the timing and potential for mitigation than estimates of net fluxes, because the gross fluxes include a more complete accounting of individual activities. Gross emissions from rotational land use in the tropics are approximately 37% of total CO_2 emissions, rather than 14%, as suggested by net AFOLU emissions (Houghton and Nassikas 2018). Further, if the forest is replanted or allowed to regrow, gross removals of nearly the same magnitude would be expected to continue for decades.

2.3.1.4 Gross emissions and removals contributing to the non-anthropogenic land sink

The *net* land sink averaged 11.2 ± 2.6 $GtCO_2$ yr^{-1} (*likely range*) over 2007–2016 (Table 2.3.2), but its gross components have not been estimated at the global level. There are many studies that suggest increasing emissions of carbon are due to indirect environmental effects and natural disturbance, for example, temperature-induced increases in respiration rates (Bond-Lamberty et al. 2018), increased tree mortality (Brienen et al. 2015; Berdanier and Clark 2016; McDowell et al. 2018) and thawing permafrost (Schuur et al. 2015). The global carbon budget indicates that land and ocean sinks have *increased* over the last six decades in proportion to total CO_2 emissions (Le Quéré et al. 2018) (*robust evidence, high agreement*). That means that any emissions must have been balanced by even larger removals (likely driven by CO_2 fertilisation, climate change, nitrogen deposition, erosion and redeposition of soil carbon, a reduction in areas burned, aerosol-induced cooling and changes in natural disturbances) (Box 2.3).

Climate change is expected to impact terrestrial biogeochemical cycles via an array of complex feedback mechanisms that will act to either enhance or decrease future CO_2 emissions from land. Because the gross emissions and removals from environmental changes are not constrained at present, the balance of future positive and negative feedbacks remains uncertain. Estimates from climate models in Coupled Model Intercomparison Project 5 (CMIP5) exhibit large

differences for different carbon and nitrogen cycle feedbacks and how they change in a warming climate (Anav et al. 2013; Friedlingstein et al. 2006; Friedlingstein, et al. 2014). The differences are in large part due to the uncertainty regarding how primary productivity and soil respiration will respond to environmental changes, with many of the models not even agreeing on the sign of change. Furthermore, many models do not include a nitrogen cycle, which may limit the CO_2 fertilisation effect in the future (Box 2.3). There is an increasing amount of observational data available and methods to constrain models (e.g., Cox et al. 2013; Prentice, et al., 2015) which can reduce uncertainty.

2.3.1.5 Potential impact of mitigation on atmospheric CO_2 concentrations

If CO_2 concentrations decline in the future as a result of low emissions and large negative emissions, the global land and ocean sinks are expected to weaken (or even reverse). The oceans are expected to release CO_2 back to the atmosphere when the concentration declines (Ciais et al. 2013a; Jones et al. 2016). This means that to maintain atmospheric CO_2 and temperature at low levels, both the excess CO_2 from the atmosphere and the CO_2 progressively outgassed from the ocean and land sinks will need to be removed. This outgassing from the land and ocean sinks is called the 'rebound effect' of the global carbon cycle (Ciais et al. 2013a). It will reduce the effectiveness of negative emissions and increase the deployment level needed to achieve a climate stabilisation target (Jackson et al. 2017; Jones et al. 2016) (*limited evidence, high agreement*).

2.3.2 Methane

2.3.2.1 Atmospheric trends

In 2017, the globally averaged atmospheric concentration of CH_4 was 1850 ± 1 ppbv (Figure 2.8A). Systematic measurements of atmospheric CH_4 concentrations began in the mid-1980s and trends show a steady increase between the mid-1980s and early-1990s, slower growth thereafter until 1999, a period of no growth between 1999 and 2006, followed by a resumption of growth in 2007. The growth rates show very high inter-annual variability with a negative trend from the beginning of the measurement period until about 2006, followed by a rapid recovery and continued high inter-annual variability through 2017 (Figure 2.8B). The growth rate has been higher over the past 4 years (*high confidence*) (Nisbet et al. 2019). The trend in $\delta^{13}C$-CH_4 prior to 2000 with less depleted ratios indicated that the increase in atmospheric concentrations was due to thermogenic (fossil) CH_4 emissions; the reversal of this trend after 2007 indicates a shift to biogenic sources (Figure 2.8C).

Understanding the underlying causes of temporal variation in atmospheric CH_4 concentrations is an active area of research. Several studies concluded that inter-annual variability of CH_4 growth was driven by variations in natural emissions from wetlands (Rice et al. 2016; Bousquet et al. 2006; Bousquet et al. 2011). These modelling efforts concluded that tropical wetlands were responsible for between 50 and 100% of the inter-annual fluctuations and the renewed growth in atmospheric concentrations after 2007. However, results were inconsistent for the magnitude and geographic distribution of the wetland sources between the models. Pison et al. (2013) used two atmospheric inversion models and the ORCHIDEE model and found greater uncertainty in the role of wetlands in inter-annual variability between 1990 and 2009 and during the 1999–2006 pause. Poulter et al. (2017) used several biogeochemical models and inventory-based wetland area data to show that wetland CH$_4$ emissions increases in the boreal zone have been offset by decreases in the tropics, and concluded that wetlands have not contributed significantly to renewed atmospheric CH$_4$ growth.

The models cited above assumed that atmospheric hydroxyl radical (OH) sink over the period analysed did not vary. OH reacts with CH$_4$ as the first step toward oxidation to CO$_2$. In global CH$_4$ budgets,

the atmospheric OH sink has been difficult to quantify because its short lifetime (about 1 second) and its distribution is controlled by precursor species that have non-linear interactions (Taraborrelli et al., 2012; Prather et al., 2017). Understanding of the atmospheric OH sink has evolved recently. The development of credible time series of methyl chloroform (MCF: CH3CCl3) observations offered a way to understand temporal dynamics of OH abundance and applying this to global budgets further weakened the argument for the role of wetlands in determining temporal trends since 1990. Several authors used the MCF approach and concluded that changes in the atmospheric OH sink explained a large portion of the suppression in global CH$_4$ concentrations relative to the pre-1999 trend (Turner et al. 2017; Rigby et al. 2013; McNorton et al. 2016). These studies could not reject the null hypothesis that OH has remained constant in recent decades and they did not suggest a mechanism

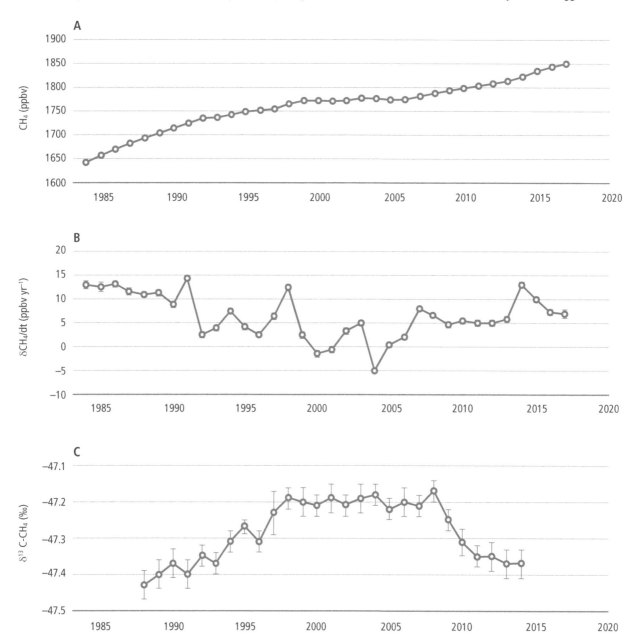

Figure 2.8 | Globally averaged atmospheric CH$_4$ mixing ratios (Frame A) and instantaneous rates of change (Frame B) and C isotope/variation (Frame C). Data sources: NOAA/ESRL (www.esrl.noaa.gov/gmd/ccgg/trends_ch4); Dlugokencky et al. (1994) and Schaefer et al. (2016).

for the inferred OH concentration changes (Nisbet et al. 2019). Nicely et al. (2018) used a mechanistic approach and demonstrated that variation in atmospheric OH was much lower than what MCF studies claimed that positive trends in OH due to the effects of water vapour, nitrogen oxides (NOx), tropospheric ozone and expansion of the tropical Hadley cells offsets the decrease in OH that is expected from increasing atmospheric CH_4 concentrations.

The depletion of $\delta^{13}C_{atm}$ beginning in 2009 could be due to changes in several sources. Decreased fire emissions combined with increased tropical wetland emissions compared to earlier years could explain the $\delta^{13}C$ perturbations to atmospheric CH_4 sources (Worden et al. 2017; Schaefer et al. 2016). However, because tropical wetland emissions are higher in the southern hemisphere, and the remote sensing observations show that CH_4 emissions increases are largely in the north tropics (Bergamaschi et al. 2013; Melton et al. 2013; Houweling et al. 2014), an increased wetland source does not fit well with the southern hemisphere $\delta^{13}C$ observations. New evidence shows that tropical wetland CH_4 emissions are significantly underestimated, perhaps by a factor of 2, because estimates do not account for release by tree stems (Pangala et al. 2017). Several authors have concluded that agriculture is a more probable source of increased emissions, particularly from rice and livestock in the tropics, which is consistent with inventory data (Wolf et al. 2017; Patra et al. 2016; Schaefer et al. 2016).

The importance of fugitive emissions in the global atmospheric accumulation rate is growing (*medium evidence, high agreement*). The increased production of natural gas in the US from the mid 2000s is of particular interest because it coincides with renewed atmospheric CH_4 growth (Rice et al. 2016; Hausmann et al. 2015). Reconciling increased fugitive emissions with increased isotopic depletion of atmospheric CH_4 indicates that there are *likely* multiple changes in emissions and sinks that affect atmospheric accumulation (*medium confidence*).

With respect to atmospheric CH_4 growth rates, we conclude that there is significant and ongoing accumulation of CH_4 in the atmosphere (*very high confidence*). The reason for the pause in growth rates and subsequent renewed growth is at least partially associated with land use and land use change. Evidence that variation in the atmospheric OH sink plays a role in the year-to-year variation of the CH_4 is accumulating, but results are contradictory (*medium evidence, low agreement*) and refining this evidence is constrained by lack of long-term isotopic measurements at remote sites, particularly in the tropics. Fugitive emissions *likely* contribute to the renewed growth after 2006 (*medium evidence, high agreement*). Additionally, the recent depletion trend of ^{13}C isotope in the atmosphere indicates that growth in biogenic sources explains part of the current growth and that biogenic sources make up a larger proportion of the source mix compared to the period before 1997 (*robust evidence, high agreement*). In agreement with the findings of AR5, we conclude that wetlands are important drivers of inter-annual variability and current growth rates (*medium evidence, high agreement*). Ruminants and the expansion of rice cultivation are also important contributors to the current growth trend (*medium evidence, high agreement*).

2.3.2.2 Land use effects

Agricultural emissions are predominantly from enteric fermentation and rice, with manure management and waste burning contributing small amounts (Figure 2.9). Since 2000, livestock production has been responsible for 33% of total global emissions and 66% of agricultural emissions (EDGAR 4.3.2 database, May 2018; USEPA 2012; Tubiello et al. 2014; Janssens-Maenhout et al. 2017b). Asia has the largest livestock emissions (37%) and emissions in the region have been growing by around 2% per year over the same period. North America is responsible for 26% and emissions are stable; Europe is responsible for around 8% of emissions, and these are decreasing slightly (<1% per year). Africa is responsible for 14%, but emissions are growing fastest in this region at around 2.5% y^{-1}. In Latin America and the Caribbean, livestock emissions are decreasing at around 1.6% per year and the region makes up 16% of emissions. Rice emissions are responsible for about 24% of agricultural emissions and 89% of these are from Asia. Rice emissions are increasing by 0.9% per year in that region. These trends are predicted to continue through 2030 (USEPA 2013).

Upland soils are a net sink of atmospheric CH_4, but soils both produce and consume the gas. On the global scale, climatic zone, soil texture and land cover have an important effect on CH_4 uptake in upland soils (Tate 2015; Yu et al. 2017; Dutaur and Verchot 2007). Boreal soils take up less than temperate or tropical soils, coarse textured soils take up more CH_4 than medium and fine textured soils, and forests take up more than other ecosystems. Low levels of nitrogen fertilisation or atmospheric deposition can affect the soil microbial community and stimulate soil CH_4 uptake in nitrogen-limited soils, while higher fertilisation rates decrease uptake (Edwards et al. 2005; Zhuang et al., 2013). Globally, nitrogen fertilisation on agricultural lands may have suppressed CH_4 oxidation by as much as 26 Tg between 1998 and 2004 (*low confidence, low agreement*) (Zhuang et al., 2013). The effect of nitrogen additions is cumulative and repeated fertilisation events have progressively greater suppression effects (*robust evidence, high agreement*) (Tate 2015). Other factors such as higher temperatures, increased atmospheric concentrations and changes in rainfall patterns stimulate soil CH_4 consumption in unfertilised ecosystems. Several studies (Yu et al. 2017; Xu et al. 2016; Curry 2009) have shown that globally, uptake has been increasing during the second half of the 20th century and is expected to continue to increase by as much as 1 Tg in the 21st century, particularly in forests and grasslands (*medium evidence, high agreement*).

Northern peatlands (40–70°N) are a significant source of atmospheric CH_4, emitting about 48 TgCH_4, or about 10% of the total emissions to the atmosphere (Zhuang et al. 2006; Wuebbles and Hayhoe 2002). CH_4 emissions from natural northern peatlands are highly variable, with the highest rate from fens (*medium evidence, high agreement*). Peatland management and restoration alters the exchange of CH_4 with the atmosphere (*medium evidence, high agreement*). Management of peat soils typically converts them from CH_4 sources to sinks (Augustin et al. 2011; Strack and Waddington 2008; Abdalla et al. 2016) (*robust evidence, high agreement*). While restoration decreases CO_2 emissions (Section 4.9.4), CH_4 emissions often increase relative to the

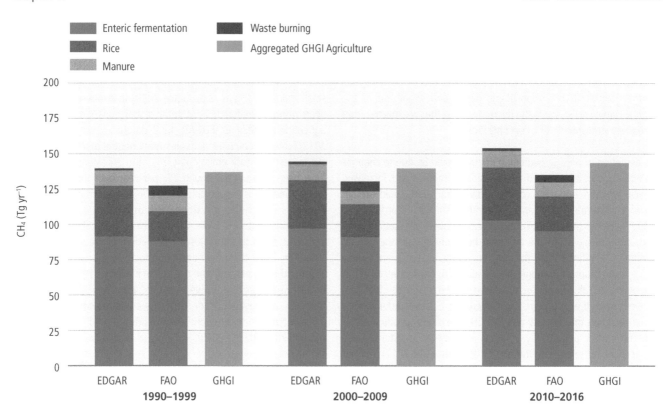

Figure 2.9 | Average agricultural CH₄ emissions estimates from 1990. Sub-sectorial agricultural emissions are based on the Emissions Database for Global Atmospheric Research (EDGAR v4.3.2; Janssens-Maenhout et al. 2017a); FAOSTAT (Tubiello et al. 2013); and National GHGI data (Grassi et al. 2018). GHGI data are aggregate values for the sector. Note that EDGAR data are complete only through 2012; the data in the right-hand panel represent the three years 2010–2012 and are presented for comparison.

drained conditions (*robust evidence, high agreement*) (Osterloh et al. 2018; Christen et al. 2016; Koskinen et al. 2016; Tuittila et al. 2000; Vanselow-Algan et al. 2015; Abdalla et al. 2016). Drained peatlands are usually considered to be negligible methane sources, but they emit CH₄ under wet weather conditions and from drainage ditches (Drösler et al. 2013; Sirin et al. 2012). While ditches cover only a small percentage of the drained area, emissions can be sufficiently high that drained peatlands emit comparable CH₄ as undrained ones (*medium evidence, medium agreement*) (Sirin et al. 2012; Wilson et al. 2016).

Because of the large uncertainty in the tropical peatland area, estimates of the global flux are highly uncertain. A meta-analysis of the effect of conversion of primary forest to rice production showed that emissions increased by a factor of four (*limited evidence, high agreement*) (Hergoualc'h and Verchot, 2012). For land uses that required drainage, emissions decreased by a factor of three (*limited evidence, high agreement*). There are no representative measurements of emissions from drainage ditches in tropical peatlands.

2.3.3 Nitrous oxide

2.3.3.1 Atmospheric trends

The atmospheric abundance of N₂O has increased since 1750, from a pre-industrial concentration of 270 ppbv to 330 ppbv in 2017 (*high agreement, robust evidence*) (US National Oceanographic and Atmospheric Agency, Earth Systems Research Laboratory)

(Figure 2.10). The rate of increase has also increased, from approximately 0.15 ppbv yr⁻¹ 100 years ago, to 0.85 ppbv yr⁻¹ over the period 2001–2015 (Wells et al. 2018). Atmospheric N₂O isotopic composition (¹⁴/¹⁵N) was relatively constant during the pre-industrial period (Prokopiou et al. 2018) and shows a decrease in the δ¹⁵N as the N₂O mixing ratio in the atmosphere has increased between 1940 and 2005. This recent decrease indicates that terrestrial sources are the primary driver of increasing trends and marine sources contribute around 25% (Snider et al. 2015). Microbial denitrification and nitrification processes are responsible for more than 80% of total global N₂O emissions, which includes natural soils, agriculture and oceans, with the remainder coming from non-biological sources such as biomass burning and fossil-fuel combustion (Fowler et al. 2015). The isotopic trend also indicates a shift from denitrification to nitrification as the primary source of N₂O as a result of the use of synthetic nitrogen fertiliser (*high evidence, high agreement*) (Park et al. 2012; Toyoda et al. 2013; Snider et al. 2015; Prokopiou et al. 2018).

The three independent sources of N₂O emissions estimates from agriculture at global, regional and national levels are: USEPA, EDGAR and FAOSTAT (USEPA 2013; Tubiello et al. 2015; Janssens-Maenhout et al. 2017a). EDGAR and FAOSTAT have temporal resolution beyond 2005 and these databases compare well with national inventory data (Figure 2.10). USEPA has historical estimates through 2005 and projections thereafter. The independent data use IPCC methods, with Tier 1 emission factors and national reporting of activity data. Tier 2 approaches are also available based on top-down and bottom-up approaches. Recent estimates using inversion modelling and process

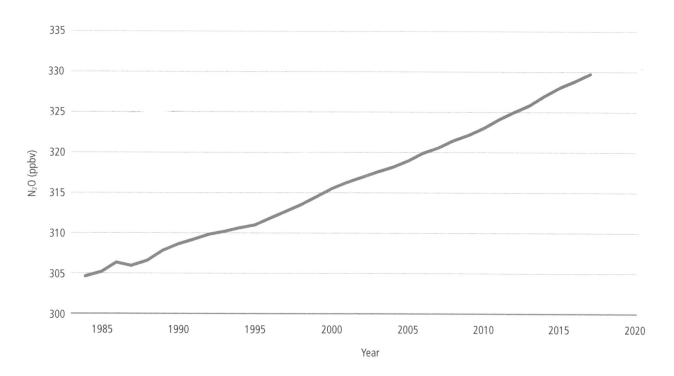

Figure 2.10 | **Globally averaged atmospheric N₂O mixing ratios since 1984.** Data source: NOAA/ESRL Global Monitoring Division (www.esrl.noaa.gov/gmd/hats/combined/N₂O.html).

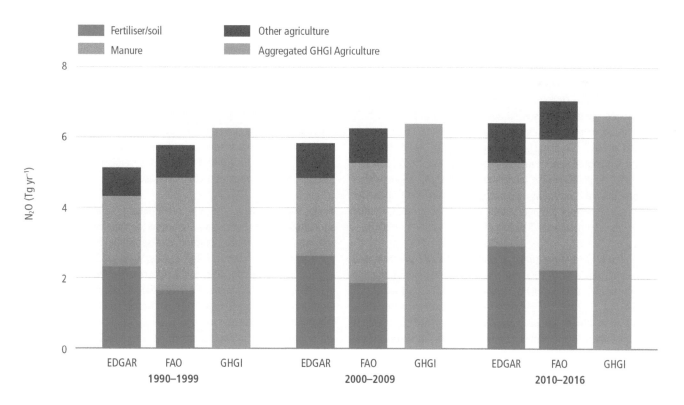

Figure 2.11 | **Average agricultural N₂O emissions estimates from 1990.** Sub-sectorial agricultural emissions are based on the Emissions Database for Global Atmospheric Research (EDGAR v4.3.2; Janssens-Maenhout et al. 2017a); FAOSTAT (Tubiello et al. 2013) and National GHGI data (Grassi et al. 2018). GHGI data are aggregate values for the sector. Note that EDGAR data are complete only through 2012; the EDGAR data in the right-hand panel represent the three years 2010–2012 and are presented for comparison. FAO data for the "other agriculture" category includes emissions from crop residues, cultivated organic soil, and burning of crop residues.

models estimate total annual global N_2O emissions of 16.1–18.7 (bottom-up) and 15.9–17.7 TgN (top-down), demonstrating relatively close agreement (Thompson et al. 2014). Agriculture is the largest source and has increased with extensification and intensification. Recent modelling estimates of terrestrial sources show a higher emissions range that is slightly more constrained than what was reported in AR5: approximately 9 (7–11) TgN_2O-N yr^{-1} (Saikawa et al. 2014; Tian et al. 2016) compared to 6.6 (3.3–9.0) TgN_2O-N yr^{-1} (Ciais et al. 2013a). Estimates of marine N_2O emissions are between 2.5 and 4.6 TgN_2O-N yr^{-1} (Buitenhuis et al., 2018; Saikawa et al., 2014).

To conclude, N_2O is continuing to accumulate in the atmosphere at an increasingly higher rate (*very high confidence*), driven primarily by increases in manure production and synthetic nitrogen fertiliser use from the mid-20th century onwards (*high confidence*). Findings since AR5 have constrained regional and global estimates of annual N_2O emissions and improved our understanding of the spatio-temporal dynamics of N_2O emissions, including soil rewetting and freeze-thaw cycles which are important determinants of total annual emission fluxes in some regions (*medium confidence*).

2.3.3.2   Land use effects

Agriculture is responsible for approximately two-thirds of N_2O emissions (*robust evidence*, *high agreement*) (Janssens-Maenhout et al. 2017b). Total emissions from this sector are the sum of direct and indirect emissions. Direct emissions from soils are the result of mineral fertiliser and manure application, manure management, deposition of crop residues, cultivation of organic soils and inorganic nitrogen inputs through biological nitrogen fixation. Indirect emissions come from increased warming, enrichment of downstream water bodies from runoff, and downwind nitrogen deposition on soils. The main driver of N_2O emissions in croplands is a lack of synchronisation between crop nitrogen demand and soil nitrogen supply, with approximately 50% of nitrogen applied to agricultural land not taken up by the crop (Zhang et al. 2017). Cropland soils emit over 3 TgN_2O-N yr^{-1} (*medium evidence*, *high agreement*) (Janssens-Maenhout et al. 2017b; Saikawa et al. 2014). Regional inverse modelling studies show larger tropical emissions than the inventory approaches and they show increases in N_2O emissions from the agricultural sector in South Asia, Central America, and South America (Saikawa et al. 2014; Wells et al. 2018).

Emissions of N_2O from pasturelands and rangelands have increased by as much as 80% since 1960 due to increased manure production and deposition (*robust evidence*, *high agreement*) (de Klein et al. 2014; Tian et al. 2018; Chadwick et al. 2018; Dangal et al. 2019; Cardenas et al. 2019). Studies consistently report that pasturelands and rangelands are responsible for around half of the total agricultural N_2O emissions (Davidson 2009; Oenema et al. 2014; Dangal et al. 2019). An analysis by Dangal et al. (2019) shows that, while managed pastures make up around one-quarter of the global grazing lands, they contribute 86% of the net global N_2O emissions from grasslands and that more than half of these emissions are related to direct deposition of livestock excreta on soils.

Many studies calculate N_2O emissions from a linear relationship between nitrogen application rates and N_2O emissions. New studies

are increasingly finding nonlinear relationships, which means that N_2O emissions per hectare are lower than the Tier 1 EFs (IPCC 2003) at low nitrogen application rates, and higher at high nitrogen application rates (*robust evidence*, *high agreement*) (Shcherbak et al. 2014; van Lent et al. 2015; Satria 2017). This not only has implications for how agricultural N_2O emissions are estimated in national and regional inventories, which now often use a linear relationship between nitrogen applied and N_2O emissions, it also means that in regions of the world where low nitrogen application rates dominate, increases in nitrogen fertiliser use would generate relatively small increases in agricultural N_2O emissions. Decreases in application rates in regions where application rates are high and exceed crop demand for parts of the growing season are likely to have very large effects on emissions reductions (*medium evidence*, *high agreement*).

Deforestation and other forms of land-use change alter soil N_2O emissions. Typically, N_2O emissions increase following conversion of native forests and grasslands to pastures or croplands (McDaniel et al. 2019; van Lent et al. 2015). This increase lasts from a few years to a decade or more, but there is a trend toward decreased N_2O emissions with time following land use change and, ultimately, lower N_2O emissions than had been occurring under native vegetation, in the absence of fertilisation (*medium evidence*, *high agreement*) (Meurer et al. 2016; van Lent et al. 2015) (Figure 2.12). Conversion of native vegetation to fertilised systems typically leads to increased N_2O emissions over time, with the rate of emission often being a function of nitrogen fertilisation rates, however, this response can be moderated by soil characteristics and water availability (*medium evidence*, *high agreement*) (van Lent et al. 2015; Meurer et al. 2016). Restoration of agroecosystems to natural vegetation, over the period of one to two decades does not lead to recovery of N_2O emissions to the levels of the original vegetation (McDaniel et al. 2019). To conclude, findings since AR5 increasingly highlight the limits of linear N_2O emission factors, particularly from field to regional scales, with emissions rising nonlinearly at high nitrogen application rates (*high confidence*). Emissions from unfertilised systems often increase and then decline over time with typically lower emissions than was the case under native vegetation (*high confidence*).

While soil emissions are the predominant source of N_2O in agriculture, other sources are important (or their importance is only just emerging). Biomass burning is responsible for approximately 0.7 TgN_2O-N yr^{-1} (0.5–1.7 TgN_2O-N yr^{-1}) or 11% of total gross anthropogenic emissions due to the release of N_2O from the oxidation of organic nitrogen in biomass (UNEP 2013). This source includes crop residue burning, forest fires, household cook stoves and prescribed savannah, pasture and cropland burning. Aquaculture is currently not accounted for in most assessments or compilations. While it is currently responsible for less than 0.1 TgN_2O-N yr^{-1}, it is one of the fastest growing sources of anthropogenic N_2O emissions (Williams and Crutzen 2010; Bouwman et al. 2013) (*limited evidence*, *high agreement*). Finally, increased nitrogen deposition from terrestrial sources is leading to greater indirect N_2O emissions, particularly since 1980 (*moderate evidence*, *high agreement*) (Tian et al. 2018, 2016). In marine systems, deposition is estimated to have increased the oceanic N_2O source by 0.2 TgN_2O-N yr^{-1} or 3% of total gross anthropogenic emissions (Suntharalingam et al. 2012).

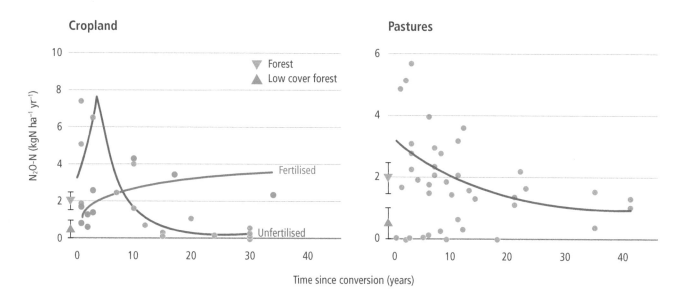

Figure 2.12 | **Effect of time since conversion on N$_2$O fluxes in unfertilised (orange circles) and fertilised (blue circles) tropical croplands (left frame) and in unfertilised tropical pastures (right frame).** Average N$_2$O flux and 95% confidence intervals are given for upland forests (orange inverted triangle) and low canopy forests (blue inverted triangle), for comparison. The solid lines represent the trends for unfertilised and fertilised cases. Data source: van Lent et al. (2015).

Box 2.2 | Methodologies for estimating national to global scale anthropogenic land carbon fluxes

Bookkeeping/accounting models calculate changes in biomass and soils that result from changes in land activity using data on biomass density and rates of growth/decomposition, typically from ground-based inventory data collection (field measurements of carbon in trees and soils) (Houghton et al. 2012; Hansis et al. 2015; Houghton and Nassikas 2017). The approach includes only those changes directly caused by major categories of land-use change and management. The models do not explicitly include the indirect effects to changing environmental conditions, although some effects are implicit in the biomass, growth rates and decay rates used. Thus, the models may overestimate past fluxes. The bookkeeping models include fluxes from peatland burning based on GFED estimates (Randerson et al. 2015).

DGVMs simulate ecological processes, such as photosynthesis, respiration, allocation, growth, decomposition etc., driven by environmental conditions (climate variability, climate change, CO$_2$, nitrogen concentrations). Models vary with respect to the processes included, with many since AR5 now including forest management, fire, nitrogen and other management (Sitch et al. 2005; Le Quéré et al. 2018). Models are forced with increasing atmospheric CO$_2$ and changing climate, and run with and without 'land use change' (land cover and forest harvest) to differentiate the anthropogenic effects from the indirect effects of climate and CO$_2$: the 'land sink'. Thus, indirect effects are explicitly included. This approach also includes a 'lost atmospheric sink capacity', or the carbon uptake due to environmental effects on forests that does not happen once the forests are removed (Pongratz et al. 2010).

Integrated assessment models (IAMs) use storylines to construct alternative future scenarios of GHG emissions and atmospheric concentrations within a global socio-economic framework, including projections of AFOLU based on assumptions of, for example, crop yields, population growth and bioenergy use (Cross-Chapter Box 1 and Chapter 1). Some models include simplified DGVMs, which may include climate and CO$_2$ effects, while others use AFOLU emissions from other sources.

ESMs couple DGVMs, surface hydrology, and energy exchange models with atmosphere, ocean, and sea ice models, enabling exploration of feedbacks between climate change and the carbon cycle (e.g., warming effects increase soil and plant respiration and lead to higher atsmpheric CO$_2$ concentrations, which in turn promote plant growth) (Friedlingstein et al. 2014). They sometimes include numerical experiments with and without land-use change to diagnose the anthropogenic AFOLU flux (Lawrence et al. 2016).

Satellite data can be used as a proxy for plant activity (e.g., greenness) and to map land cover, vegetation fires and biomass density. Algorithms, models and independent data are used to calculate fluxes of CO$_2$ from satellite data, although calculating the net carbon flux is difficult because of the lack of information on the respiratory flux. Some active satellite sensors (LiDAR) are able to measure three-dimensional structure in woody vegetation, which is closely related to biomass density (Zarin et al. 2016; Baccini et al. 2012; Saatchi et al. 2011). Together with land-cover change data, these estimates of biomass density can be used to provide

Box 2.2 (continued)

observational-based estimates of fluxes due to changes in forest area (e.g., Tyukavina et al. (2015), Harris et al. (2015) and Baccini et al. (2012) or degradation (Baccini et al. 2017)). Satellite estimates of biomass vary considerably (Mitchard et al. 2013; Saatchi et al. 2015; Avitabile et al. 2016): data are available only for recent decades, methods generally assume that all losses of carbon are immediately released to the atmosphere and changes in soil carbon are generally ignored. The approach implicitly includes indirect and natural disturbance effects as well as direct anthropogenic effects.

Atmospheric inversions use observations of atmospheric concentrations with a model of atmospheric transport, based on data for wind speed and direction, to calculate implied emissions (Gatti et al. 2014; Liu et al. 2017a; van der Laan-Luijkx et al. 2017). Since AR5, there has been an increase in availability of concentration data from flux tower networks and satellites, enabling better global coverage at finer spatial scales and some national estimates (e.g., in the UK inverse techniques are used together with national GHG inventories). A combination of concentrations of different gases and isotopes enables the separation of fossil, ocean and land fluxes. However, inversions give only the net flux of CO_2 from land; they cannot separate natural and anthropogenic fluxes.

Micrometeorological flux measurements data on CO_2 concentrations and air movements recorded on instrumented towers enable the calculation of CO_2 flux at the ecosystem scale. Global and regional Flux Networks (FluxNet (global), AsiaFlux, Ameriflux (North America), ICOS (EU), NEON (USA), and others) contribute to a global flux database, which is used to verify the results of modelling, inventory and remote sensing studies.

FAOSTAT has produced country level estimates of GHG emissions (Tubiello et al. 2013) from agriculture (1961–2016) and land use (1990–2016) using a globally consistent methodological approach based largely on IPCC Tier 1 methods of the 2006 IPCC Guidelines (FAO 2015). FAO emissions estimates were used as one of the three database inputs into the AR5 WGIII AFOLU chapter. Non-CO_2 emissions from agriculture are estimated directly from national statistics of activity data reported by countries to FAO. CO_2 emissions from land use and land-use change are computed mostly at Tier 1, albeit at fine geospatial scales to capture effects from peatland degradation and biomass fires (Rossi et al. 2016). Emissions from forest land and deforestation are based on the IPCC carbon stock change method, thus constituting a Tier 3 estimate relying on country statistics of carbon stocks and forest area collected through the FAO FRA. The carbon flux is estimated assuming instantaneous emissions in the year of forest area loss and changes in carbon stocks within extant forests, but does not distinguish 'managed' and 'unmanaged' forest areas, albeit it treats separately emissions from primary, secondary and planted forest (Federici et al. 2015).

Country Reporting of GHG Inventories (GHGIs): All parties to the UNFCCC are required to report national GHGIs of anthropogenic emissions and removals. Reporting requirements are differentiated between developed and developing countries. Because of the difficulty of separating direct anthropogenic fluxes from indirect or natural fluxes, the IPCC (2003) adopted the 'managed land' concept as a proxy to facilitate GHGI reporting. All GHG fluxes on 'managed land' are defined as anthropogenic, with each country applying their own definition of 'managed land' (i.e., 'where human interventions and practices have been applied to perform production, ecological or social functions' (IPCC 2006)). Fluxes may be determined on the basis of changes in carbon stocks (e.g., from forest inventories) or by activity data (e.g., area of land cover change management activity multiplied by emission factors or with modelled fluxes). Depending on the specific methods used, GHGIs include all direct anthropogenic effects and may include the indirect anthropogenic effects of environmental change (generally sinks) and natural effects (Section 2.3.1.2). GHG fluxes from 'unmanaged land' are not reported in GHGIs because they are assumed to be non-anthropogenic. The reported estimates may then be filtered through agreed 'accounting rules' (i.e., what countries actually count towards their mitigation targets (Cowie et al. 2007; Lee and Sanz 2017). The accounting aims to better quantify the additional mitigation actions by, for example, factoring out the impact of natural disturbances and forest age-related dynamics (Canadell et al. 2007; Grassi et al. 2018).

Box 2.3 | CO$_2$ fertilisation and enhanced terrestrial uptake of carbon

All DGVMs and ESMs represent the CO$_2$ fertilisation effect (Le Quéré et al. 2017; Hoffman et al. 2014). There is *high confidence* that elevated CO$_2$ results in increased short-term CO$_2$ uptake per unit leaf area (Swann et al. 2016; Field et al. 1995; Donohue et al. 2013), however, whether this increased CO$_2$ uptake at the leaf level translates into increased growth for the whole plant differs among plant species and environments, because growth is constrained by whole-plant resource allocation and nutrient limitation (e.g., nitrogen, phosphorus, potassium and soil water and light limitations (Körner 2006; Peñuelas et al. 2017; Friend et al. 2014a)). Interactions between plants and soil microbes further modulate the degree of nutrient limitation on CO$_2$ fertilisation (Terrer et al. 2017).

At the ecosystems level, enhanced CO$_2$ uptake at decadal or longer timescales depends on changes in plant community composition and ecosystem respiration, as well disturbance and natural plant mortality (De Kauwe et al., 2016; Farrior et al., 2015; Keenan et al., 2017; Sulman et al, 2019). The results of free-air carbon dioxide enrichment (FACE) experiments over two decades are highly variable because of these factors (Norby et al. 2010; Körner 2015; Feng et al. 2015; Paschalis et al. 2017; Terrer et al. 2017; Du et al. 2019). Under higher atmospheric CO$_2$ concentrations, the ratio of CO$_2$ uptake to water loss (water use efficiency (WUE)), increases and enhances drought tolerance of plants (*high confidence*) (Berry et al., 2010; Ainsworth and Rogers 2007).

Long-term CO$_2$ and water vapour flux measurements show that WUE in temperate and boreal forests of the northern hemisphere has increased more than predicted by photosynthetic theory and models over the past two decades (*high confidence*) (Keenan et al. 2013; Laguë and Swann 2016). New theories have emerged on how CO$_2$ uptake by trees is related to water loss and to the risk of damaging xylem (water conducting tissues) in the trunk and branches (Wolf et al. 2016a; Anderegg et al. 2018a). Tree ring studies of stable carbon and oxygen isotopes also detected increased WUE in recent decades (Battipaglia et al. 2013; Silva and Anand 2013; van der Sleen et al. 2014). Yet, tree ring studies often fail to show acceleration of tree growth rates in support of CO$_2$ fertilisation, even when they show increased WUE (van der Sleen et al. 2014). The International Tree Ring Data Bank (ITRDB) indicated that only about 20% of the sites in the database showed increasing trends in tree growth that cannot be explained by climate variability, nitrogen deposition, elevation or latitude. Thus there is *limited evidence (low agreement)* among observations of enhanced tree growth due to CO$_2$ fertilisation of forests during the 20th century (Gedalof and Berg 2010).

In grasslands, although it is possible for CO$_2$ fertilisation to alleviate the impacts of drought and heat stress on net carbon uptake (Roy et al. 2016), there is *low confidence* about its projected magnitude. Because of its effect on water use efficiency, CO$_2$ fertilisation is expected to be pronounced in semi-arid habitats; and because of different metabolic pathways, C3 plants are expected to be more sensitive to elevated CO$_2$ concentrations than C4 grasses (Donohue et al. 2013; Morgan et al. 2011; Derner et al. 2003). Neither of these expectations was observed over a 12-year study of elevated CO$_2$ in a grassland system: enhanced growth was not observed during dry summers and growth of C4 grasses was unexpectedly stimulated, while growth of C3 grasses was not (Reich et al. 2014, 2018).

There is *medium confidence* that CO$_2$ fertilisation effects have increased water use efficiency in crops and thus reduced agricultural water use per unit of crop produced (Deryng et al. 2016; Nazemi and Wheater 2015; Elliott et al. 2014). This effect could lead to near-term continued greening of agricultural areas. However, current assessments of these effects are based on limited observations, mostly from the temperate zone (Deryng et al. 2016).

One line of evidence for CO$_2$ fertilisation is the increasing land sink ('the residual land sink' in AR5) over the last 50 years as the atmospheric CO$_2$ concentration has increased (Los 2013; Sitch et al. 2015; Campbell et al. 2017; Keenan and Riley 2018). A combined analysis of atmospheric inverse analyses, ecosystem models and forest inventory data concluded that 60% of the recent terrestrial carbon sink can be directly attributed to increasing atmospheric CO$_2$ (Schimel et al. 2015). A global analysis using a 'reconstructed vegetation index' (RVI) for the period 1901–2006 from MODIS satellite-derived normalised vegetation difference index (NDVI) showed that CO$_2$ fertilisation contributed at least 40% of the observed increase in the land carbon sink (Los 2013). Without CO$_2$ fertilisation, ESMs are unable to simulate the increasing land sink and the observed atmospheric CO$_2$ concentration growth rate since the middle of the 20th century (Shevliakova et al. 2013). There are other mechanisms that could explain enhanced land carbon uptake such as increased regional forest and shrub cover (Chen et al. 2019) (Cross-Chapter Box 2 and Chapter 1), and, at higher latitudes, increasing temperatures and longer growing seasons (Zhu et al. 2016).

In summary, there is *low confidence* about the magnitude of the CO$_2$ effect and other factors that may explain at least a portion of the land sink (e.g., nitrogen deposition, increased growing season, reduced burning, erosion and re-deposition or organic sediments, aerosol-induced cooling). Increases in atmospheric CO$_2$ result in increased water use efficiency and increase leaf-level photosynthesis (*high confidence*). The extent to which CO$_2$ fertilisation results in plant- or ecosystem-level carbon accumulation is highly variable and affected by other environmental constraints (*high confidence*). Even in ecosystems where CO$_2$ fertilisation has been detected in recent decades, those effects are found to weaken as a result of physiological acclimation, soil nutrient limitation and other constraints on growth (Friend et al., 2014; Körner, 2006; Peñuelas et al., 2017).

2.4   Emissions and impacts of short-lived climate forcers (SLCF) from land

While the rising atmospheric concentration of GHGs is the largest driver of anthropogenic changes in climate, the levels of short-lived climate forcers (SLCF) can significantly modulate regional climate by altering radiation exchanges and hydrological cycle and impact ecosystems (*high confidence*) (Boucher et al. 2013; Rogelj et al. 2014; Kok et al. 2018). This section assesses the current state of knowledge with respect to past and future emissions of the three major SLCFs and their precursors: mineral dust, carbonaceous aerosols (black carbon (BC) and organic carbon (OC)) and BVOCs. This section also reports on implications of changes in their emissions for climate. Aerosols particles with diameters between about 0.010 μm to about 20 μm are recognised as SLCFs, a term that refers to their short atmospheric lifetime (a few days). BVOCs are important precursors of ozone and OC, both important climate forcing agents with short atmospheric lifetimes.

While the AR5 did not assess land aerosols emissions in depth, their findings stated that although progress in quantifying regional emissions of anthropogenic and natural land aerosols has been made, considerable uncertainty still remains about their historical trends, their inter-annual and decadal variability and about any changes in the future (Calvo et al. 2013; Klimont et al. 2017). Some new and improved understanding of processes controlling emissions and atmospheric processing has been developed since AR5, for example, a better understanding of the climatic role of BC as well as the understanding of the role of BVOCs in formation of secondary organic aerosols (SOA).

Depending on the chemical composition and size, aerosols can absorb or scatter sunlight and thus directly affect the amount of absorbed and scattered radiation (Fuzzi et al. 2015; Nousiainen 2011; de Sá et al. 2019) Aerosols affect cloud formation and development, and thus can also influence precipitation patterns and amounts (Suni et al. 2015). In addition, deposition of aerosols – especially BC – on snow and ice surfaces can reduce albedo and increase warming as a self-reinforcing feedback. Aerosols deposition also changes biogeochemical cycling in critical terrestrial ecosystems, with deposition of nutrients such as nitrogen and phosphorus (Andreae et al. 2002). Primary land aerosols are emitted directly into the atmosphere due to natural or anthropogenic processes and include mineral aerosols (or dust), volcanic dust, soot from combustion, organic aerosols from industry, vehicles or biomass burning, bioaerosols from forested regions and others. SOAs are particulates that are formed in the atmosphere by the gas-to-particles conversion processes from gaseous precursors, such as BVOCs, and account for a large fraction of fine mode (particles less than 2.5μm) aerosol mass (Hodzic et al. 2016; Manish et al. 2017). Land use change can affect the climate through changed emissions of SLCFs such as aerosols, ozone precursors and methane.

Aerosols from air pollution will decline in the coming years as a means for improving urban and regional air, but their removal will lead to additional warming (Boucher et al. 2013), with important regional variability, and partially offsetting projected mitigation effects for

two to three decades in 1.5°C consistent pathways (*high confidence*) (IPCC 2018). It is important to emphasise that changes in emissions can either be due to external forcing or through a feedback in the climate system (Box 2.1). For instance, enhanced dust emissions due to reduced vegetation could be a forcing if overgrazing is the cause of larger dust emission, or a feedback if dryer climate is the cause. This distinction is important in terms of mitigation measures to be implemented.

2.4.1   Mineral dust

One of the most abundant atmospheric aerosols emitted into the atmosphere is mineral dust, a 'natural' aerosol that is produced by wind strong enough to initiate the emissions process of sandblasting. Mineral dust is preferentially emitted from dry and unvegetated soils in topographic depressions where deep layers of alluvium have been accumulated (Prospero et al. 2002). Dust is also emitted from disturbed soils by human activities, with a 25% contribution to global emissions based on a satellite-based estimate (Ginoux et al. 2012).

Dust is then transported over long distances across continents and oceans. The dust cycle, which consists of mineral dust emission, transport, deposition and stabilisation, has multiple interactions with many climate processes and biogeochemical cycles.

2.4.1.1   Mineral dust as a short-lived climate forcer from land

Depending on the dust mineralogy, mixing state and size, dust particles can absorb or scatter shortwave and longwave radiation. Dust particles serve as cloud condensation nuclei and ice nuclei. They can influence the microphysical properties of clouds, their lifetime and precipitation rate (Kok et al. 2018). New and improved understanding of processes controlling emissions and transport of dust, its regional patterns and variability, as well as its chemical composition, has been developed since AR5.

While satellites remain the primary source of information to locate dust sources and atmospheric burden, in-situ data remains critical to constrain optical and mineralogical properties of the dust (Di Biagio et al. 2017; Rocha-Lima et al. 2018). Dust particles are composed of minerals, including iron oxides which strongly absorb shortwave radiation and provide nutrients for marine ecosystems. Another mineral such as feldspar is an efficient ice nuclei (Harrison et al. 2016). Dust mineralogy varies depending on the native soils, so global databases were developed to characterise the mineralogical composition of soils for use in weather and climate models (Journet et al. 2014; Perlwitz et al. 2015). New field campaigns, as well as new analyses of observations from prior campaigns, have produced insights into the role of dust in western Africa in climate system, such as long-ranged transport of dust across the Atlantic (Groß et al. 2015) and the characterisation of aerosol particles and their ability to act as ice and cloud condensation nuclei (Price et al. 2018). Size distribution at emission is another key parameter controlling dust interactions with radiation. Most models now use the parametrisation of Kok (2011) based on the theory of brittle material. It was shown that most models underestimate the size of the global dust cycle (Kok 2011).

Characterisation of spatial and temporal distribution of dust emissions is essential for weather prediction and climate projections (*high confidence*). Although there is a growing confidence in characterising the seasonality and peak of dust emissions (i.e., spring–summer (Wang et al. 2015)) and how the meteorological and soil conditions control dust sources, an understanding of long-term future dust dynamics, inter-annual dust variability and how they will affect future climate still requires substantial work. Dust is also important at high latitude, where it has an impact on snow-covered surface albedo and weather (Bullard et al. 2016).

2.4.1.2   Effects of past climate change on dust emissions and feedbacks

A limited number of model-based studies found that dust emissions have increased significantly since the late 19th century: by 25% from the preindustrial period to the present day (e.g., from 729 Tg yr^{-1} to 912 Tg yr^{-1}) with about 50% of the increase driven by climate change and about 40% driven by land use cover change, such as conversion of natural land to agriculture (*low confidence*) (Stanelle et al. 2014). These changes resulted in a clear sky radiative forcing at the top of the atmosphere of -0.14 Wm^{-2} (Stanelle et al. 2014). The authors found that, in North Africa, most dust is of natural origin, with a recent 15% increase in dust emissions attributed to climate change. In North America two-thirds of dust emissions take place on agricultural lands and both climate change and land-use change jointly drive the increase; between the pre-industrial period and the present day, the overall effect of changes in dust was -0.14 W m^{-2} cooling of clear sky net radiative forcing on top of the atmosphere, with -0.05 W m^{-2} from land use and -0.083 W m^{-2} from changes in climate.

The comparison of observations for vertically integrated mass of atmospheric dust per unit area (i.e., dust mass path (DMP)) obtained from the remotely sensed data and the DMP from CMIP5 models reveal that the model-simulate range of DMP was much lower than the estimates (Evan et al. 2014). ESMs typically do not reproduce inter-annual and longer timescales variability seen in observations (Evan et al. 2016). Analyses of the CMIP5 models (Evan 2018; Evan et al. 2014) reveal that all climate models systematically underestimate dust emissions, the amount of dust in the atmosphere and its inter-annual variability (*medium confidence*).

One commonly suggested reason for the lack of dust variability in climate models is the models' inability to simulate the effects of land surface changes on dust emission (Stanelle et al. 2014). Models that account for changes in land surface show more agreement with the satellite observations both in terms of aerosol optical depth and DMP (Kok et al. 2014). New prognostic dust emissions models are now able to account for both changes in surface winds and vegetation characteristics (e.g., leaf area index and stem area index) and soil water, ice and snow cover (Evans et al. 2016). As a result, new modelling studies (e.g., Evans et al. 2016) indicate that, in regions where soil and vegetation respond strongly to ENSO events, such as in Australia, inclusion of dynamic vegetation characteristics into dust emission parameterisations improves comparisons between the modelled and observed relationship with long-term climate variability (e.g., ENSO) and dust levels (Evans et al. 2016). Thus, there has been

progress in incorporating the effects of vegetation, soil moisture, surface wind and vegetation on dust emission source functions, but the number of studies demonstrating such improvement remains small (*limited evidence, medium agreement*).

2.4.1.3   Future changes of dust emissions

There is no agreement about the direction of future changes in dust emissions. Atmospheric dust loading is projected to increase over the southern edge of the Sahara in association with surface wind and precipitation changes (Pu and Ginoux, 2018), while Evan et al. (2016) project a decline in African dust emissions. Dust optical depth (DOD) is also projected to increase over the central Arabian peninsula in all seasons, and to decrease over northern China from March-April-May to September-October-November (Pu and Ginoux 2018). Climate models project rising drought risks over the south-western and central US in the 21st century. The projected drier regions largely overlay the major dust sources in the US. However, whether dust activity in the US will increase in the future is not clear, due to the large uncertainty in dust modelling (Pu and Ginoux 2017). Future trends of dust emissions will depend on changes in precipitation patterns and atmospheric circulation (*limited evidence, high agreement*). However, implication of changes in human activities, including mitigation (e.g., bioenergy production) and adaption (e.g., irrigation) are not characterised in the current literature.

2.4.2   Carbonaceous aerosols

Carbonaceous aerosols are one of the most abundant components of aerosol particles in continental areas of the atmosphere and a key land–atmosphere component (Contini et al. 2018). They can make up to 60–80% of PM2.5 (particulate matter with size less than 2.5 µm) in urban and remote atmospheres (Tsigaridis et al. 2014; Kulmala et al. 2011). It comprises an organic fraction (OC) and a refractory light-absorbing component, generally referred to as elemental carbon (EC), from which BC is the optically active absorption component of EC (Gilardoni et al. 2011; Bond et al. 2013).

2.4.2.1   Carbonaceous aerosol precursors of short-lived climate forcers from land

OC is a major component of aerosol mass concentration, and it originates from different anthropogenic (combustion processes) and natural (natural biogenic emissions) sources (Robinson et al. 2007). A large fraction of OC in the atmosphere has a secondary origin, as it can be formed in the atmosphere through condensation to the aerosol phase of low vapour pressure gaseous compounds emitted as primary pollutants or formed in the atmosphere. This component is SOA (Hodzic et al. 2016). A third component of the optically active aerosols is the so-called brown carbon (BrC), an organic material that shows enhanced solar radiation absorption at short wavelengths (Wang et al. 2016b; Laskin et al. 2015; Liu et al. 2016a; Bond et al. 2013; Saturno et al. 2018).

OC and EC have distinctly different optical properties, with OC being important for the scattering properties of aerosols and EC central for the absorption component (Rizzo et al. 2013; Tsigaridis et al. 2014; Fuzzi et al. 2015). While OC is reflective and scatters solar radiation, it has a cooling effect on climate. On the other side, BC and BrC absorb solar radiation and they have a warming effect in the climate system (Bond et al. 2013).

OC is also characterised by a high solubility with a high fraction of water-soluble organic compounds (WSOC) and it is one of the main drivers of the oxidative potential of atmospheric particles. This makes particles loaded with oxidised OC an efficient cloud condensation nuclei (CCN) in most of the conditions (Pöhlker et al. 2016; Thalman et al. 2017; Schmale et al. 2018).

Biomass burning is a major global source of carbonaceous aerosols (Bowman et al. 2011; Harrison et al. 2010; Reddington et al. 2016; Artaxo et al. 2013). As knowledge of past fire dynamics improved through new satellite observations, new fire proxies' datasets (Marlon et al. 2013; van Marle et al. 2017a), process-based models (Hantson et al. 2016) and a new historic biomass burning emissions dataset starting in 1750 have been developed (van Marle et al. 2017b) (Cross-Chapter Box 3 in this chapter). Revised versions of OC biomass burning emissions (van Marle et al. 2017b) show, in general, reduced trends compared to the emissions derived by Lamarque et al. (2010) for CMIP5. CMIP6 global emissions pathways (Gidden et al. 2018; Hoesly et al. 2018) estimate global BC emissions in 2015 at 9.8 MtBC yr^{-1}, while global OC emissions are 35 MtOC yr^{-1}.

Land use change is critically important for carbonaceous aerosols, since biomass-burning emissions consist mostly of organic aerosol, and the undisturbed forest is also a large source of organic aerosols (Artaxo et al. 2013). Additionally, urban aerosols are also mostly carbonaceous because of the source composition (traffic, combustion, industry, etc.) (Fuzzi et al. 2015). Burning of fossil fuels, biomass-burning emissions and SOA from natural BVOC emissions are the main global sources of carbonaceous aerosols. Any change in each of these components directly influence the radiative forcing (Contini et al. 2018; Boucher et al. 2013; Bond et al. 2013).

One important component of carbonaceous aerosols is the primary biological aerosol particles (PBAP), also called bioaerosols, that correspond to a significant fraction of aerosols in forested areas (Fröhlich-Nowoisky et al. 2016; Pöschl and Shiraiwa 2015). They are emitted directly by the vegetation as part of the biological processes (Huffman et al. 2012). Airborne bacteria, fungal spores, pollen, archaea, algae and other bioparticles are essential for the reproduction and spread of organisms across various terrestrial ecosystems. They can serve as nuclei for cloud droplets, ice crystals and precipitation, thus influencing the hydrological cycle and climate (Whitehead et al. 2016; Scott et al. 2015; Pöschl et al. 2010).

2.4.2.2   Effects of past climate change on carbonaceous aerosols emissions and feedbacks

Annual global emission estimates of BC range from 7.2–7.5 Tg yr^{-1} (using bottom-up inventories) (Bond et al. 2013; Klimont et al. 2017) up to 17.8 ± 5.6 Tg yr^{-1} (using a fully coupled climate-aerosol-

urban model constrained by aerosol measurements) (Cohen and Wang 2014), with considerably higher BC emissions for Eastern Europe, southern East Asia, and Southeast Asia, mostly due to higher anthropogenic BC emissions estimates. A significant source of BC, the net trend in global burned area from 2000–2012 was a modest decrease of 4.3 Mha yr^{-1} (–1.2% yr^{-1}).

Carbonaceous aerosols are important in urban areas as well as pristine continental regions, since they can be responsible for 50–85% of PM2.5 (Contini et al. 2018; Klimont et al. 2017). In boreal and tropical forests, carbonaceous aerosols originate from BVOC oxidation (Section 2.4.3). The largest global source of BC aerosols is open burning of forests, savannah and agricultural lands with emissions of about 2700 Gg yr^{-1} in the year 2000 (Bond et al. 2013).

ESMs most likely underestimate globally averaged EC emissions (Bond et al. 2013; Cohen and Wang 2014), although recent emission inventories have included an upwards adjustment in these numbers (Hoesly et al. 2018). Vertical EC profiles have also been shown to be poorly constrained (Samset et al. 2014), with a general tendency of too much EC at high altitudes. Models differ strongly in the magnitude and importance of the coating-enhancement of ambient EC absorption (Boucher et al. 2016; Gustafsson and Ramanathan 2016) in their estimated lifetime of these particles, as well as in dry and wet removal efficiency (*limited evidence, medium agreement*) (Mahmood et al. 2016).

The equilibrium in emissions and concentrations between the scattering properties of organic aerosol versus the absorption component of BC is a key ingredient in the future climatic projections of aerosol effects (*limited evidence, high agreement*). The uncertainties in net climate forcing from BC-rich sources are substantial, largely due to lack of knowledge about cloud interactions with both BC and co-emitted OC. A strong positive forcing of about 1.1 wm^{-2} was calculated by Bond et al. (2013), but this forcing is balanced by a negative forcing of –1.45 wm^{-2}, and shows clearly a need to work on the co-emission issue for carbonaceous aerosols. The forcing will also depend on the aerosol-cloud interactions, where carbonaceous aerosol can be coated and change their CCN capability. It is difficult to estimate the changes in any of these components in a future climate, but this will strongly influence the radiative forcing (*high confidence*) (Contini et al. 2018; Boucher et al. 2013; Bond et al. 2013).

De Coninck et al. (2018) reported studies estimating a lower global temperature effect from BC mitigation (e.g., Samset et al. 2014; Boucher et al. 2016), although commonly used models do not capture properly observed effects of BC and co-emissions on climate (e.g., Bond et al. 2013). Regionally, the warming effects can be substantially larger, for example, in the Arctic (Sand et al. 2015) and high mountain regions near industrialised areas or areas with heavy biomass-burning impacts (*high confidence*) (Ming et al. 2013).

2.4.2.3   Future changes of carbonaceous aerosol emissions

Due to the short atmospheric lifetime of carbonaceous aerosols in the atmosphere, of the order of a few days, most studies dealing with the future concentration levels have a regional character (Cholakian et al. 2018; Fiore et al. 2012). The studies agree that the uncertainties in changes in emissions of aerosols and their precursors are generally higher than those connected to climate change itself. Confidence in future changes in carbonaceous aerosol concentration projections is limited by the reliability of natural and anthropogenic emissions (including wildfires, largely caused by human activity) of primary aerosol as well as that of the precursors. The Aerosol Chemistry Model Intercomparison Project (AerChemMIP) is endorsed by the Coupled-Model Intercomparison Project 6 (CMIP6) and is designed to quantify the climate impacts of aerosols and chemically reactive gases (Lamarque et al. 2013). These simulations calculated future responses to SLCF emissions for the RCP scenarios in terms of concentration changes and radiative forcing. Carbonaceous aerosol emissions are expected to increase in the near future due to possible increases in open biomass-burning emissions (from forest, savannah and agricultural fires), and increase in SOA from oxidation of BVOCs (*medium confidence*) (Tsigaridis et al. 2014; van Marle et al. 2017b; Giglio et al. 2013).

More robust knowledge has been produced since the conclusions reported in AR5 (Boucher et al. 2013) and all lines of evidence now agree on a small effect on carbonaceous aerosol global burden due to climate change (*medium confidence*). The regional effects, however, are predicted to be much higher (Westervelt et al. 2015). With respect to possible changes in the chemical composition of PM as a result of future climate change, only a few sparse data are available in the literature and the results are, as yet, inconclusive. The co-benefits of reducing aerosol emissions due to air quality issues will play an important role in future carbonaceous aerosol emissions (*high confidence*) (Gonçalves et al. 2018; Shindell et al. 2017).

2.4.3 Biogenic volatile organic compounds

BVOCs are emitted in large amounts by forests (Guenther et al. 2012). They include isoprene, terpenes, alkanes, alkenes, alcohols, esters, carbonyls and acids (Peñuelas and Staudt 2010; Guenther et al. 1995, 2012). Their emissions represent a carbon loss to the ecosystem, which can be up to 10% of the carbon fixed by photosynthesis under stressful conditions (Bracho-Nunez et al. 2011). The global average emission for vegetated surfaces is 0.7g C m^{-2} yr^{-1} but can exceed 100 g C m^{-2} yr^{-1} in some tropical ecosystems (Peñuelas and Llusià 2003).

2.4.3.1 BVOC precursors of short-lived climate forcers from land

BVOCs are rapidly oxidised in the atmosphere to form less volatile compounds that can condense and form SOA. In boreal and tropical forests, carbonaceous aerosols originate from BVOC oxidation, of which isoprene and terpenes are the most important precursors (Claeys et al. 2004; Hu et al. 2015; De Sá et al. 2017; de Sá et al. 2018; Liu et al. 2016b). See the following sub-section for more detail.

BVOCs are the most important precursors of SOA. The transformation process of BVOCs affects the aerosol size distribution both by

contributing to new particle formation and to the growth of larger pre-existing particles. SOA affects the scattering of radiation by the particles themselves (direct aerosol effect), but also changes the amount of CCN and the lifetime and optical properties of clouds (indirect aerosol effect).

High amounts of SOA are observed over forest areas, in particular in boreal and tropical regions where they have been found to mostly originate from BVOC emissions (Manish et al. 2017). In particular, isoprene epoxydiol-derived SOA (IEPOX-SOA) is being identified in recent studies in North America and Amazonian forest as a major component in the oxidation of isoprene (Allan et al. 2014; Schulz et al. 2018; De Sá et al. 2017). In tropical regions, BVOCs can be convected up to the upper atmosphere, where their volatility is reduced and where they become SOA. In some cases those particles are transported back to the lower atmosphere (Schulz et al. 2018; Wang et al. 2016a; Andreae et al. 2018). In the upper troposphere in the Amazon, SOA are important CCN and are responsible for the vigorous hydrological cycle (Pöhlker et al. 2018). This strong link between BVOC emissions by plants and the hydrological cycle has been discussed in a number of studies (Fuentes et al. 2000; Schmale et al. 2018; Pöhlker et al. 2018, 2016).

Changing BVOC emissions also affect the oxidant concentrations in the atmosphere. Their impact on the concentration of ozone depends on the NOx concentrations. In polluted regions, high BVOC emissions lead to increased production of ozone, followed by the formation of more OH and a reduction in the methane lifetime. In more pristine regions (NOx-limited), increasing BVOC emissions instead lead to decreasing OH and ozone concentrations, resulting in a longer methane lifetime. The net effect of BVOCs then can change over time if NOx emissions are changing.

BVOCs' possible climate effects have received little attention because it was thought that their short lifetime would preclude them from having any significant direct influence on climate (Unger 2014a; Sporre et al. 2019). Higher temperatures and increased CO_2 concentrations are (separately) expected to increase the emissions of BVOCs (Jardine et al. 2011, 2015; Fuentes et al. 2016). This has been proposed to initiate negative climate feedback mechanisms through increased formation of SOA (Arneth et al. 2010; Kulmala 2004; Unger et al. 2017). More SOA can make clouds more reflective, which can provide a cooling effect. Furthermore, the increase in SOA formation has also been proposed to lead to increased aerosol scattering, resulting in an increase in diffuse radiation. This could boost GPP and further increase BVOC emissions (Kulmala et al. 2014; Cirino et al. 2014; Sena et al. 2016; Schafer et al. 2002; Ometto et al. 2005; Oliveira et al. 2007). This important feedback is starting to emerge (Sporre et al. 2019; Kulmala 2004; Arneth et al. 2017). However, there is evidence that this influence might be significant at different spatial scales, from local to global, through aerosol formation and through direct and indirect greenhouse effects (*limited evidence, medium agreement*). Most tropical forest BVOCs are primarily emitted from tree foliage, but soil microbes can also be a major source of some compounds including sesquiterpenes (Bourtsoukidis et al. 2018).

2.4.3.2 Historical changes of BVOCs and contribution to climate change

Climate warming over the past 30 years, together with the longer growing season experienced in boreal and temperate environments, have increased BVOC global emissions since the preindustrial times (*limited evidence, medium agreement*) (Peñuelas 2009; Sanderson et al. 2003; Pacifico et al. 2012). This was opposed by lower BVOC emissions caused by the historical conversion of natural vegetation and forests to cropland (*limited evidence, medium agreement*) (Unger 2013, 2014a; Fu and Liao 2014). The consequences of historical anthropogenic land cover change were a decrease in the global formation of SOA (–13%) (Scott et al. 2017) and tropospheric burden (–13%) (Heald and Geddes 2016). This has resulted in a positive radiative forcing (and thus warming) from 1850–2000 of 0.017 W m^{-2} (Heald and Geddes 2016), 0.025 W m^{-2} (Scott et al. 2017) and 0.09 W m^{-2} (Unger 2014b) through the direct aerosol effect. In present-day conditions, global SOA production from all sources spans between 13 and 121 Tg yr^{-1} (Tsigaridis et al. 2014). The indirect aerosol effect (change in cloud condensation nuclei), resulting from land use induced changes in BVOC emissions, adds an additional positive radiative forcing of 0.008 W m^{-2} (Scott et al. 2017). More studies with different model setups are needed to fully assess this indirect aerosol effect associated with land use change from the preindustrial to present. CMIP6 global emissions pathways (Hoesly et al. 2018; Gidden et al. 2018) estimates global VOCs emissions in 2015 at 230 MtVOC yr^{-1}. They also estimated that, from 2000–2015, emissions were up from 200–230 MtVOC yr^{-1}.

There is (*limited evidence, medium agreement*) that historical changes in BVOC emissions have also impacted on tropospheric ozone. At most surface locations where land use has changed, the NOx concentrations are sufficiently high for the decrease in BVOC emissions to lead to decreasing ozone concentrations (Scott et al. 2017). However, in more pristine regions (with low NOx concentrations), the imposed conversion to agriculture has increased ozone through decreased BVOC emissions and their subsequent decrease in OH (Scott et al. 2017; Heald and Geddes 2016). In parallel, the enhanced soil NOx emissions from agricultural land can increase the ozone concentrations in NOx limited regions (Heald and Geddes 2016).

Another impact of the historical decrease in BVOC emissions is the reduction in the atmospheric lifetime of methane (*limited evidence, medium agreement*), which results in a negative radiative forcing that ranges from –0.007 W m^{-2} (Scott et al. 2017) to –0.07 W m^{-2} (Unger 2014b). However, knowledge of the degree that BVOC emissions impact on oxidant concentrations, in particular OH (and thus methane concentrations), is still limited and therefore these numbers are very uncertain (Heald and Spracklen 2015; Scott et al. 2017). The effect of land use change on BVOC emissions are highly heterogeneous (Rosenkranz et al. 2015) and though the global values of forcing described above are small, the local or regional values can be higher, and even of opposite sign, than the global values.

2.4.3.3 Future changes of BVOCs

Studies suggest that increasing temperature will change BVOC emissions through change in species composition and rate of BVOC production. A further 2°C–3°C rise in the mean global temperature could increase BVOC global emissions by an additional 30–45% (Peñuelas and Llusià 2003). In two modelling studies, the impact on climate from rising BVOC emissions was found to become even larger with decreasing anthropogenic aerosol emissions (Kulmala et al. 2013; Sporre et al. 2019). A negative feedback on temperature, arising from the BVOC-induced increase in the first indirect aerosol effect, has been estimated by two studies to be in the order of –0.01 W m^{-2} K (Scott et al. 2018b; Paasonen et al. 2013). Enhanced aerosol scattering from increasing BVOC emissions has been estimated to contribute to a global gain in BVOC emissions of 7% (Rap et al. 2018). In a warming planet, BVOC emissions are expected to increase but magnitude of this increase is unknown and will depend on future land use change, in addition to climate (*limited evidence, medium agreement*).

There is a very limited number of studies investigating the climate impacts of BVOCs using future land use scenarios (Ashworth et al. 2012; Pacifico et al. 2012). Scott et al. (2018a) found that a future deforestation according to the land use scenario in RCP8.5 leads to a 4% decrease in BVOC emissions at the end of the century. This resulted in a direct aerosol forcing of +0.006 W m^{-2} (decreased reflection by particles in the atmosphere) and a first indirect aerosol forcing of –0.001 W m^{-2} (change in the amount of CCN). Studies not including future land use scenarios but investigating the climate feedbacks leading to increasing future BVOC emissions, have found a direct aerosol effect of –0.06 W m^{-2} (Sporre et al. 2019) and an indirect aerosol effect of –0.45 W m^{-2} (Makkonen et al. 2012; Sporre et al. 2019). The stronger aerosol effects from the feedback compared to the land use are, at least partly, explained by a much larger change in the BVOC emissions.

A positive climate feedback could happen in a future scenario with increasing BVOC emissions, where higher ozone and methane concentrations could lead to an enhanced warming which could further increase BVOC emissions (Arneth et al. 2010). This possible feedback is mediated by NOx levels. One recent study including dynamic vegetation, land use change, CO$_2$ and climate change found no increase and even a slight decrease in global BVOC emissions at the end of the century (Hantson et al. 2017). There is a lack of understanding concerning the processes governing the BVOC emissions, the oxidation processes in the atmosphere, the role of the BVOC oxidation products in new particle formation and particle growth, as well as general uncertainties in aerosol–cloud interactions. There is a need for continued research into these processes, but the current knowledge indicates that changing BVOC emissions need to be taken into consideration when assessing the future climate and how land use will affect it. In summary, the magnitude and sign of net effect of BVOC emissions on the radiation budget and surface temperature is highly uncertain.

2.5 Land impacts on climate and weather through biophysical and GHG effects

The focus of this section is summarised in Figure 2.13. We report on what we know regarding the influence land has on climate via biophysical and biogeochemical exchanges. Biogeochemical effects herein only refer to changes in net emissions of CO_2 from land. The influence of land on atmospheric composition is discussed in Section 2.3.

All sections discuss impacts of land on global and regional climate, and climate extremes, whenever the information is available. Section 2.5.1 presents effects of historical and future land use scenarios, Section 2.5.2 is devoted to impacts of specific anthropogenic land uses such as forestation, deforestation, irrigation, crop and forest management, Section 2.5.3 focuses on how climate-driven land changes feedback on climate, and Section 2.5.4 puts forward the theory that land use changes in one region can affect another region.

2.5.1 Impacts of historical and future anthropogenic land cover changes

The studies reported below focus essentially on modelling experiments, as there is no direct observation of how historical land use changes have affected the atmospheric dynamics and physics at the global and regional scales. Moreover, the climate modelling experiments only assess the impacts of anthropogenic land cover changes (e.g., deforestation, urbanisation) and neglect the effects of changes in land management (e.g., irrigation, use of fertilisers, choice of species varieties among managed forests or crops). Because of this restricted accounting for land use changes, we will use the term 'land cover changes' in Sections 2.5.1.1 and 2.5.1.2.

Each section starts by describing changes at the global scale and regional scale, and ends with what we know about the impacts of those scenarios on extreme weather events, whenever the information is available.

2.5.1.1 Impacts of global historical land cover changes on climate

At the global level

The contribution of anthropogenic land cover changes to the net global warming throughout the 20th century has been derived from few model-based estimates that account simultaneously for biogeochemical and biophysical effects of land on climate (Table 2.4). The simulated net change in mean global annual surface air temperature, averaged over all the simulations, is a small warming of 0.078 ± 0.093°C, ranging from small cooling simulated by two models (−0.05°C and −0.02°C respectively in Brovkin et al. (2004) and Simmons and Matthews (2016), to larger warming simulated by three models (>+0.14°C; Shevliakova et al. 2013; Pongratz et al. 2010; Matthews et al. 2004). When starting from the Holocene period, He et al. (2014) estimated an even larger net warming effect of anthropogenic land cover changes (+0.72°C).

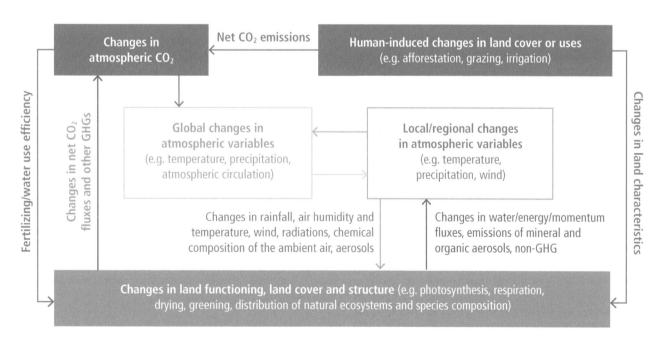

Figure 2.13 | Global, local and regional climate changes are the focus of this section. They are examined through changes in climate states (e.g., changes in air temperature and humidity, rainfall, radiation) as well as through changes in atmospheric dynamics (e.g., circulation patterns). Changes in land that influence climate are either climate- or human-driven. Dark-blue arrows and boxes refer to what we consider imposed changes (forcings). Dark-grey arrows and boxes refer to responses of land to forcings (blue boxes and blue-outline box) and feedbacks on those initial forcings. Pale-grey and pale-blue arrows and boxes refer respectively to global and local/regional climate changes and their subsequent changes on land.

Table 2.4 | Change in mean global annual surface air temperature resulting from anthropogenic land cover change over the historical period. This historical period varies from one simulation to another (middle column).

Reference of the study	Time period	Mean global annual change in surface air temperature (°C)
Simmons and Matthews (2016)	1750–2000	−0.02
Shevliakova et al. (2013)	1861–2005	+0.17
Pongratz et al. (2010)	1900–2000	+0.14
Matthews et al. (2004)	1700–2000	+0.15
Brovkin et al. (2004)	1850–2000	−0.05
Mean ± standard deviation		0.078 ± 0.093

This net small warming signal results from the competing effects of biophysical cooling (*medium confidence*) and biogeochemical warming (*very high confidence*) (Figure 2.14[1]). The global biophysical cooling alone has been estimated by a larger range of climate models and is −0.10 ± 0.14°C; it ranges from −0.57°C to +0.06°C (e.g., Zhang et al. 2013a; Hua and Chen 2013; Jones et al. 2013b; Simmons and Matthews 2016) (Table A2.1). This cooling is essentially dominated by increases in surface albedo: historical land cover changes have generally led to a dominant brightening of land as discussed in AR5 (Myhre et al. 2013). Reduced incoming longwave radiation at the surface from reduced evapotranspiration and thus less water vapour in the atmosphere has also been reported as a potential contributor to this cooling (Claussen et al. 2001). The cooling is, however, dampened by decreases in turbulent fluxes, leading to decreased loss of heat and water vapour from the land through convective processes. Those non-radiative processes are well-known to often oppose the albedo-induced surface temperature changes (e.g., Davin and de Noblet-Ducoudre (2010), Boisier et al. (2012)).

Historical land cover changes have contributed to the increase in atmospheric CO_2 content (Section 2.3) and thus to global warming (biogeochemical effect, *very high confidence*). The global mean biogeochemical warming has been calculated from observation-based estimates (+0.25 ± 0.10°C) (e.g., Li et al. (2017a), Avitabile et al. (2016), Carvalhais et al. (2014), Le Quéré et al. (2015)), or estimated from DGVMs (+0.24 ± 0.12°C) (Peng et al. 2017; Arneth et al. 2017; Pugh et al. 2015; Hansis et al. 2015) and global climate models (+0.20 ± 0.05°C) (Pongratz et al. 2010; Brovkin et al. 2004; Matthews et al. 2004; Simmons and Matthews 2016).

The magnitude of these simulated biogeochemical effects may, however, be underestimated as they do not account for a number of processes such as land management, nitrogen/phosphorus cycles, changes in the emissions of CH_4, N_2O and non-GHG emissions from land (Ward et al. 2014; Arneth et al. 2017; Cleveland et al. 2015; Pongratz et al. 2018). Two studies have accounted for those compounds and found a global net positive radiative forcing in response to historical anthropogenic land cover changes, indicating a net surface warming (Mahowald et al. 2017; Ward et al. 2014). However, first the estimated biophysical radiative forcing in those studies only accounts for changes in albedo and not for changes in turbulent fluxes. Secondly, the combined estimates also depend on other several key modelling estimates such as climate sensitivity, CO_2 fertilisation caused by land use emissions, possible synergistic effects, validity of radiative forcing concept for land forcing. The comparison with the other above-mentioned modelling studies is thus difficult.

In addition, most of those estimates do not account for the evolution of natural vegetation in unmanaged areas, while observations and numerical studies have reported a greening of the land in boreal regions resulting from both extended growing season and poleward migration of tree lines (Lloyd et al. 2003; Lucht et al. 1995; Section 2.2). This greening enhances global warming via a reduction of surface albedo (winter darkening of the land through the snow-albedo feedbacks; e.g., Forzieri et al. 2017). At the same time, cooling occurs due to increased evapotranspiration during the growing season, along with enhanced photosynthesis, in essence, increased CO_2 sink (Qian et al. 2010). When feedbacks from the poleward migration of treeline are accounted for, together with the biophysical effects of historical anthropogenic land cover change, the biophysical annual cooling (about −0.20°C to −0.22°C on land, −0.06°C globally) is significantly dampened by the warming (about +0.13°C) resulting from the movements of natural vegetation (Strengers et al. 2010). Accounting simultaneously for both anthropogenic and natural land cover changes reduces the cooling impacts of historical land cover change in this specific study.

At the regional level

The global and annual estimates reported above mask out very contrasted regional and seasonal differences. Biogeochemical effects of anthropogenic land cover change on temperature follow the spatial patterns of GHG-driven climate change with stronger warming over land than ocean, and stronger warming in northern high latitudes than in the tropics and equatorial regions (Arctic amplification). Biophysical effects on the contrary are much stronger where land cover has been modified than in their surroundings (see Section 2.5.4 for a discussion on non-local effects). Very contrasted regional temperature changes can thus result, depending on whether biophysical processes dampen or exacerbate biogeochemical impacts.

Figure 2.15 compares, for seven climate models, the biophysical effects of historical anthropogenic land cover change in North America and Eurasia (essentially cooling) to the regional warming resulting from the increased atmospheric CO_2 content since pre-industrial times (De Noblet-Ducoudré et al. 2012; comparing

---

[1]   The detailed list of all values used to construct this figure is provided in Table A2.1 in the Appendix at the end of the chapter.

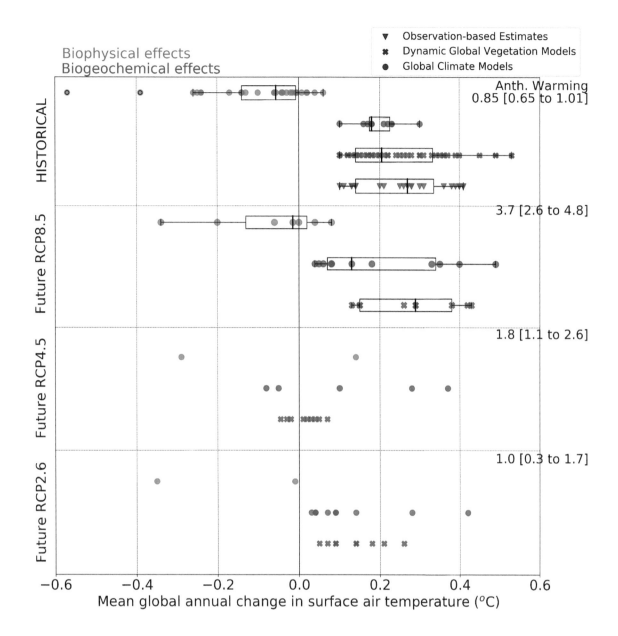

Figure 2.14 | Changes in mean global annual surface air temperature (°C) in response to historical and future anthropogenic land cover changes as estimated from a range of studies. See Table A2.1 in the Appendix for detailed information. Temperature changes resulting from biophysical processes (e.g., changes in physical land surface characteristics such as albedo, evapotranspiration and roughness length) are illustrated using blue symbols and temperature changes resulting from biogeochemical processes (e.g., changes in atmospheric CO_2 composition) use red symbols. Future changes are shown for three distinct scenarios: RCP8.5, RCP4.5 and RCP2.6. The markers 'filled circle', 'filled cross' and 'filled triangle down' represent estimates from global climate models, DGVMs and observations respectively. When the number of estimates is sufficiently large, box plots are overlaid; they show the ensemble minimum, first quartile (25th percentile), median, third quartile (75th percentile), and the ensemble maximum. Scatter points beyond the box plot are the outliers. Details about how temperature change is estimated from DGVMs and observations is provided in the Appendix. Numbers on the right-hand side give the mean and the range of simulated mean global annual warming from various climate models.

1973–2002 to 1871–1900). It shows a dominant biophysical cooling effect of changes in land cover, at all seasons, as large as the regional footprint of anthropogenic global warming. Averaged over all agricultural areas of the world (Pongratz et al. 2010) reported a 20th century biophysical cooling of –0.10°C, and Strengers et al. (2010) reported a land induced cooling as large as –1.5°C in western Russia and eastern China between 1871 and 2007. There is thus *medium confidence* that anthropogenic land cover change has dampened warming in many regions of the world over the historical period.

Very few studies have explored the effects of historical land cover changes on seasonal climate. There is, however, evidence that the seasonal magnitude and sign of those effects at the regional level are strongly related to soil-moisture/evapotranspiration and snow regimes, particularly in temperate and boreal latitudes (Teuling et al. 2010; Pitman and de Noblet-Ducoudré 2012; Alkama and Cescatti 2016). Quesada et al. (2017a) showed that atmospheric circulation changes can be significantly strengthened in winter for tropical and temperate regions. However, the lack of studies underlines the need for a more systematic assessment of seasonal, regional and other-than-mean-temperature metrics in the future.

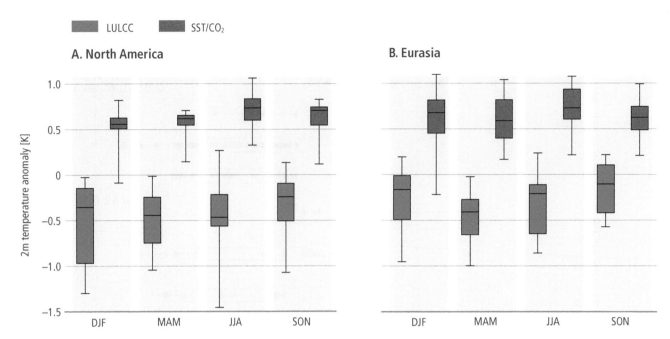

Figure 2.15 | Simulated changes in mean surface air temperature (°C) between the pre-industrial period (1870–1900) and the present-day (1972–2002) for all seasons and for (A) North America and (B) Eurasia. Source: De Noblet-Ducoudré et al. (2012). Brown boxes are the changes simulated in response to increased atmospheric GHG content between both time periods and subsequent changes in sea-surface temperature and sea-ice extent (SST/CO_2). The CO_2 changes accounted for include emissions from all sources, including land use. Blue boxes are the changes simulated in response to the biophysical effects of historical land cover changes. The box-and-whisker plots have been drawn using results from seven climate models and ensembles of 10 simulations per model and time period. The bottom and top of each grey box are the 25th and 75th percentiles, and the horizontal line within each box is the 50th percentile (the median). The whiskers (straight lines) indicate the ensemble maximum and minimum values. Seasons are respectively December-January-February (DJF), March-April-May (MAM), June-July-August (JJA) and September-October-November (SON). North America and Eurasia are extended regions where land-use changes are the largest between the two time periods considered (their contours can be found in Figure 1 of De Noblet-Ducoudré et al. (2012).

Effects on extremes

The effect of historical deforestation on extreme temperature trends is intertwined with the effect of other climate forcings, thus making it difficult to quantify based on observations. Based on results from four climate models, the impact of historical anthropogenic land cover change on temperature and precipitation extremes was found to be locally as important as changes arising from increases in atmospheric CO_2 and sea-surface temperatures, but with a lack of model agreement on the sign of changes (Pitman et al. 2012). In some regions, the impact of land cover change masks or amplifies the effect of increased CO_2 on extremes (Avila et al. 2012; Christidis et al. 2013). Using an observational constraint for the local biophysical effect of land cover change applied to a set of CMIP5 climate models, Lejeune et al. (2018) found that historical deforestation increased extreme hot temperatures in northern mid-latitudes. The results also indicate a stronger impact on the warmest temperatures compared to mean temperatures. Findell et al. (2017) reached similar conclusions, although using only a single climate model. Importantly, the climate models involved in these three studies did not consider the effect of management changes, which have been shown to be important, as discussed in Section 2.5.2.

Based on the studies discussed above, there is *limited evidence* but *high agreement* that land cover change affects local temperature extremes more than mean values. Observational studies assessing the role of land cover on temperature extremes are still very limited (Zaitchik et al. 2006; Renaud and Rebetez 2008), but suggest that

trees dampen seasonal and diurnal temperature variations at all latitudes, and even more so in temperate regions compared to short vegetation (Chen et al. 2018; Duveiller et al. 2018; Li et al. 2015a; Lee et al. 2011). Furthermore, trees also locally dampen the amplitude of heat extremes (Renaud and Rebetez 2008; Zaitchik et al. 2006) although this result depends on the forest type, coniferous trees providing less cooling effect than broadleaf trees (Renaud et al. 2011; Renaud and Rebetez 2008).

2.5.1.2 Impacts of future global land cover changes on climate

At the global level

The most extreme CMIP5 emissions scenario, RCP8.5, has received the most attention in the literature with respect to how projected future anthropogenic land use land cover changes (Hurtt et al. 2011) will affect the highest levels of global warming.

Seven model-based studies have examined both the biophysical and biogeochemical effects of anthropogenic changes in land cover, as projected in RCP8.5, on future climate change (Simmons and Matthews 2016; Davies-Barnard et al. 2014; Boysen et al. 2014) (Table 2.5). They all agree on a biogeochemical warming, ranging from +0.04°C to +0.35°C, in response to land cover change. Two models predict an additional biophysical warming, while the others agree on a biophysical cooling that dampens (or overrules) the biogeochemical warming. Using a wider range of global

Table 2.5 | **Change in mean global annual surface air temperature resulting from anthropogenic land cover changes projected for the future, according to three different scenarios: RCP8.5, RCP4.5 and RCP2.6.** Temperature changes resulting from biophysical and biogeochemical effects of land cover change are examined.

Reference of the study	Time period	Mean global annual change in surface air temperature (°C) Biophysical/biogeochemical		
		RCP2.6	RCP4.5	RCP8.5
Simmons and Matthews (2016)	2000–2100	−0.35/+0.42	−0.29/+0.37	−0.34/+0.35
Davies-Barnard et al. (2014)	2005–2100	−0.01/+0.04	+0.14/−0.08	−0.015/+0.04
Boysen et al. (2014)	2005–2100			+0.04/+0.08 0/+0.05 +0.08/+0.06 −0.20/+0.13 −0.06/+0.33

climate models, the biogeochemical warming (*high confidence*) is +0.20 ± 0.15°C whereas it is +0.28 ± 0.11°C when estimated from DGVMs (Pugh et al. 2015; Stocker et al. 2014). This biogeochemical warming is compensated for by a biophysical cooling (*medium confidence*) of −0.10 ± 0.14°C (Quesada et al. 2017a; Davies-Barnard et al. 2015; Boysen et al. 2014). The estimates of temperature changes resulting from anthropogenic land cover changes alone remain very small compared to the projected mean warming of +3.7°C by the end of the 21st century (ranging from 2.6°C–4.8°C depending on the model and compared to 1986–2005; Figure 2.14).

Two other projected land cover change scenarios have been examined (RCP4.5 and RCP2.6; Table 2.5; Figure 2.14) but only one climate modelling experiment has been carried out for each, to estimate the biophysical impacts on climate of those changes (Davies-Barnard et al. 2015). For RCP2.6, ESMs and DGVMs agree on a systematic biogeochemical warming resulting from the imposed land cover changes, ranging from +0.03 to +0.28°C (Brovkin et al. 2013), which is significant compared to the projected mean climate warming of +1°C by the end of the 21st century (ranging from 0.3°C–1.7°C depending on the models, compared to 1986–2005). A very small amount of biophysical cooling is expected from the one estimate. For RCP4.5, biophysical warming is expected from only one estimate, and results from a projected large forestation in the temperate and high latitudes. There is no agreement on the sign of the biogeochemical effect: there are as many studies predicting cooling as warming, whichever the method to compute those effects (ESMs or DGVMs).

Previous scenarios – Special Report on Emission Scenarios (SRES) results of climate studies using those scenarios were reported in AR4 – displayed larger land use changes than the more recent ones (RCP, AR5). There is *low confidence* from some of those previous scenarios (SRES A2 and B1) of a small warming effect (+0.2 to +0.3°C) of anthropogenic land cover change on mean global climate, this being dominated by the release of CO_2 in the atmosphere from land conversions (Sitch et al. 2005). This additional warming remains quite small when compared to the one resulting from the combined anthropogenic influences (+1.7°C for SRES B1 and +2.7°C for SRES A2). A global biophysical cooling of −0.14°C is estimated in response to the extreme land cover change projected in SRES A2, a value that far exceeds the impacts of historical land use changes (−0.05°C) calculated using the same climate model (Davin et al. 2007). The authors derived a biophysical climatic sensitivity to land use change of about −0.3°C W.m^{-2} for their

model, whereas a warming of about 1°C W.m^{-2} is obtained in response to changes in atmospheric CO_2 concentration.

Those studies generally do not report on changes in atmospheric variables other than surface air temperature, thereby limiting our ability to assess the effects of anthropogenic land cover changes on regional climate (Sitch et al. 2005). However, small reductions reported in rainfall via changes in biophysical properties of the land, following the massive tropical deforestation in SRES A2 (+0.5 and +0.25 mm day^{-1} respectively in the Amazon and Central Africa). They also report opposite changes – that is, increased rainfall of about 0.25 mm day^{-1} across the entire tropics and subtropics, triggered by biogeochemical effects of this same deforestation.

At the regional level

In regions that will undergo land cover changes, dampening of the future anthropogenic warming can be as large as −26% while enhancement is always smaller than 9% within RCP8.5 by the end of the 21st century (Boysen et al. 2014). Voldoire (2006) shows that, by 2050, and following the SRES B2 scenario, the contribution of land cover changes to the total temperature change can be as large as 15% in many boreal regions, and as large as 40% in south-western tropical Africa. Feddema et al. (2005) simulate large decreases in the diurnal temperature range in the future (2050 and 2100 in SRES B1 and A2) following tropical deforestation in both scenarios. In the Amazon, for example, the diurnal temperature range is lowered by 2.5°C due to increases in minimum temperature, while little change is obtained for the maximum value.

There is thus *medium evidence* that future anthropogenic land cover change will have a significant effect on regional temperature via biophysical effects in many regions of the world. There is, however, *no agreement* on whether warming will be dampened or enhanced, and there is *no agreement* on the sign of the contribution across regions.

There are very few studies that go beyond analysing the changes in mean surface air temperature. Some studies attempted to look at global changes in rainfall and found no significant influence of future land cover changes (Brovkin et al. 2013; Sitch et al. 2005; Feddema et al. 2005). Quesada et al. (2017a, b) however carried out a systematic multi-model analysis of the response of a number of atmospheric, radiative and hydrological variables (e.g., rainfall, sea level pressure,

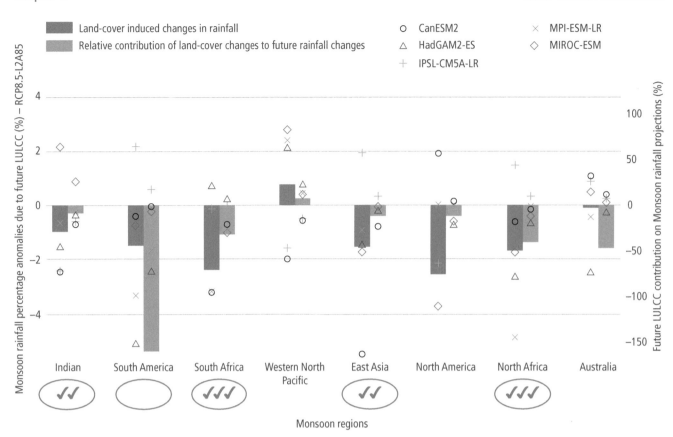

Figure 2.16 | Changes in monsoon rainfall in RCP8.5 scenario resulting from projected changes in anthropogenic land cover, in eight monsoon regions (%, blue bars). Differences are calculated between the end of the 21st century (2071–2100) and the end of the 20th century (1976–2005), and the percent change is calculated with reference to 1976–2005. Grey bars refer to the relative contribution of land-cover changes (in %) to future rainfall projections: it is the ratio between the change in rainfall responding to land cover changes and the one responding to all anthropogenic changes (Quesada et al. 2017b). Negative values mean that changes in land cover have an opposite effect (dampening) on rainfall compared to the effects of all anthropogenic changes. Monsoon regions have been defined following Yim et al. (2014). The changes have been simulated by five climate models (Brovkin et al. 2013). Results are shown for December-January-February for southern hemisphere regions, and for June-July-August for northern hemisphere regions. Statistical significance is given by blue tick marks and circles: one, two and three blue tick marks are displayed for the regions where at least 80% of the climate models have regional changes significant at the 66th, 75th and 80th confidence level, respectively; blue circles are added when the regional values are also significant at 90th confidence level. Note that future land cover change impacts on South American monsoon are neither significant nor robust among models, along with very small future projected changes in South American monsoon rainfall.

geopotential height, wind speed, soil-moisture, turbulent heat fluxes, shortwave and longwave radiation, cloudiness) to RCP8.5 land cover scenario. In particular, they found a significant reduction of rainfall in six out of eight monsoon regions studied (Figure 2.16) of about 1.9–3% (which means more than –0.5 mm day^{-1} in some areas) in response to future anthropogenic land cover changes. Including those changes in global climate models reduces the projected increase in rainfall by about 9–41% in those same regions, when all anthropogenic forcings are accounted for (30% in the global monsoon region as defined by Wang and Ding (2008)). In addition, they found a shortening of the monsoon season of one to four days. They conclude that the projected future increase in monsoon rains may be overestimated by those models that do not yet include biophysical effects of land cover changes. Overall, the regional hydrological cycle was found to be substantially reduced and wind speed significantly strengthened in response to regional deforestation within the tropics, with magnitude comparable to projected changes with all forcings (Quesada et al. 2017b).

Effects on extremes

Results from a set of climate models have shown that the impact of future anthropogenic land cover change on extreme temperatures can be of similar magnitude as the changes arising from half a degree global mean annual surface temperature change (Hirsch et al. 2018). However, this study also found a lack of agreement between models with respect to the magnitude and sign of changes, thus making land cover change a factor of uncertainty in future climate projections.

2.5.2 Impacts of specific land use changes

2.5.2.1 Impacts of deforestation and forestation

Deforestation or forestation,[2] wherever it occurs, triggers simultaneously warming and cooling of the surface and of the atmosphere via changes in its various characteristics (Pitman 2003; Strengers et al. 2010; Bonan 2008). Following deforestation, warming results from (i) the release of

---

[2] The term 'forestation' is used herein as this chapter does not distinguish between afforestation and reforestation. In model-based studies, simulations with and without trees are compared; in observation-based estimates, sites with and without trees are compared.

CO$_2$ and other GHGs in the atmosphere (biogeochemical impact) and subsequent increase in incoming infrared radiation at surface (greenhouse effect), (ii) a decrease in the total loss of energy through turbulent fluxes (latent and sensible heat fluxes) resulting from reduced surface roughness, (iii) an increased incoming solar radiation following reduced cloudiness that often (but not always) accompanies the decreased total evapotranspiration. Cooling occurs in response to (iv) increased surface albedo that reduces the amount of absorbed solar radiation, (v) reduced incoming infrared radiation triggered by the decreased evapotranspiration and subsequent decrease in atmospheric water vapour. Points ii–v are referred to as biophysical effects. Deforestation and forestation also alter rainfall and winds (horizontal as well as vertical, as will be further discussed below).

The literature that discusses the effects of forestation on climate is more limited than for deforestation, but they reveal a similar climatic response with opposite sign, as further discussed below. For each latitudinal band (tropical, temperate and boreal) we look at how very large-scale deforestation or forestation impacts on the global mean climate, followed by an examination of the large-scale changes in the specific latitudinal band, and finally more regionally focused analysis. Large-scale idealised deforestation or forestation experiments are often carried out with global or regional climate models as they allow us to understand and measure how sensitive climate is to very large changes

in land cover (similar to the instant doubling of CO$_2$ in climate models to calculate the climatic sensitivity to GHGs). Details of the model-based studies discussed below can be found in Table A2.2 in the Appendix.

Global and regional impacts of deforestation/forestation in tropical regions

A pan-tropical deforestation would lead to the net release of CO$_2$ from land, and thus to mean global annual warming, with model-based estimates of biogeochemical effects ranging from +0.19 to +1.06°C, with a mean value of +0.53 ± 0.32°C (Ganopolski et al. 2001; Snyder et al. 2004; Devaraju et al. 2015a; Longobardi et al. 2016; Perugini et al. 2017). There is, however, *no agreement* between models on the magnitude and sign of the biophysical effect of such changes at the global scale (the range spans from –0.5°C to +0.7°C with a mean value of +0.1 ± 0.27°C) (e.g., Devaraju et al. (2015b), Snyder (2010), Longobardi et al. (2016a)) (Figure 2.17). This is the result of many compensation effects in action: increased surface albedo following deforestation, decreased atmospheric water vapour content due to less tropical evapotranspiration, and decreased loss of energy from tropical land in the form of latent and sensible heat fluxes.

There is, however, *high confidence* that such large land cover change would lead to a mean biophysical warming when averaged over the

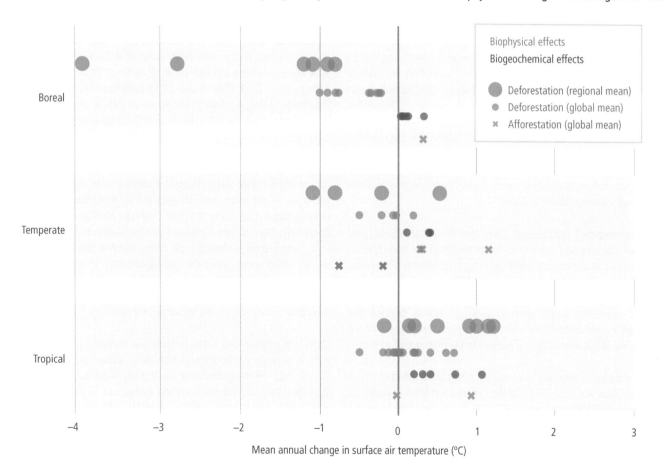

Figure 2.17 | Changes in mean annual surface air temperature (°C) in response to idealised large-scale deforestation (circles) or forestation (crosses). Estimated from a range of studies (see Table A2.2 in the Appendix for detailed information and references to the studies). Temperature changes resulting from biophysical processes (e.g., changes in physical land surface characteristics such as albedo, evapotranspiration, and roughness length) are illustrated using blue symbols and temperature changes resulting from biogeochemical processes (e.g., changes in atmospheric CO$_2$ composition) use orange symbols. Small blue and orange circles and crosses are model-based estimates of changes in temperature averaged globally. Large circles are estimates averaged only over the latitudinal band where deforestation is imposed.

deforested land. A mean warming of +0.61 ± 0.48°C is found over the entire tropics. On the other hand, biophysical regional cooling and global warming is expected from forestation (Wang et al. 2014b; Bathiany et al. 2010).

Large-scale deforestation (whether pan-tropical or imposed at the sub-continent level, e.g., the Amazon) results in significant mean rainfall decrease (Lawrence and Vandecar 2015; Lejeune et al. 2015; Perugini et al. 2017). In their review, Perugini et al. (2017) reported an average simulated decrease of −288 ± 75 mm yr^{-1} (95% confidence interval). Inversely large-scale forestation increases tropical rainfall by 41 ± 21 mm yr^{-1}. The magnitude of the change in precipitation strongly depends on the type of land cover conversion. For instance, conversion of tropical forest to bare soil causes larger reductions in regional precipitation than conversion to pasture (respectively −470 ± 60 mm yr^{-1} and −220 ± 100 mm yr^{-1}). Biogeochemical effects in response to pan-tropical deforestation, particularly CO_2 release, are generally not taken into account in those studies, but could intensify the hydrological cycle and thus precipitation (Kendra Gotangco Castillo and Gurney 2013).

Specific model-based deforestation studies have been carried out for Africa (Hagos et al. 2014; Boone et al. 2016; Xue et al. 2016; Nogherotto et al. 2013; Hartley et al. 2016; Klein et al. 2017; Abiodun et al. 2012), southern America (Butt et al. 2011; Wu et al. 2017; Spracklen and Garcia-Carreras 2015; Lejeune et al. 2015) and Southeast Asia (Ma et al. 2013b; Werth and Avissar 2005; Mabuchi et al. 2005; Tölle et al. 2017). All found decreases in evapotranspiration following deforestation (*high agreement*), resulting in surface warming, despite the competing effect from increased surface albedo (*high agreement*). Changes in thermal gradients between deforested and adjacent regions, between land and ocean, affect horizontal surface winds (*high agreement*) and thus modify the areas where rain falls, as discussed in Section 2.5.4. An increase in the land-sea thermal contrast has been found in many studies as surface friction is reduced by deforestation, thus increasing the monsoon flow in Africa and South America (Wu et al. 2017).

Observation-based estimates all agree that deforestation increases local land-surface and ambient air temperatures in the tropics, while forestation has the reverse effect (*very high confidence*) (Prevedello et al. 2019; Schultz et al. 2017; Li et al. 2015b; Alkama and Cescatti 2016). There is *very high confidence* that forests are cooler than any shorter vegetation (crops, grasses, bare soil) during daytime due to larger transpiration rates, and there is *high confidence* that the amplitude of the diurnal cycle is smaller in the presence of forests.

Large-scale forestation scenarios of West Africa (Abiodun et al. 2012), eastern China (Ma et al. 2013a) or the Saharan and Australian deserts (Ornstein et al. 2009; Kemena et al. 2017) all concluded that regional surface cooling is simulated wherever trees are grown (−2.5°C in the Sahel, −1°C in the savanna area of West Africa, up to −8°C in the western Sahara and −1.21°C over land in eastern China) while cooling of the ambient air is smaller (−0.16°C). In the case of savanna forestation, this decrease entirely compensates the GHG-induced future warming (+1°C following the SRES A1B scenario). West African countries thus have the potential to reduce, or even totally cancel in

some places, the GHG-induced warming in the deforested regions (Abiodun et al. 2012). However, this is compensated by enhanced warming in adjacent countries (non-local effect).

Global and regional impacts of deforestation/forestation in temperate regions

As for the tropics, model-based experiments show that large-scale temperate deforestation would induce a small mean global annual warming through the net release of CO_2 into the atmosphere (ranging from +0.10 to +0.40°C with a mean value of +0.20 ± 0.13°C) (Figure 2.17), whereas there is less agreement on the sign of the mean global annual temperature change resulting from biophysical processes: estimates range from −0.5°C to +0.18°C with a mean value of −0.13 ± 0.22°C. There is also *very low agreement* on the mean annual temperature change in the temperate zone (−0.4 ± 0.62°C; Phillips et al. 2007; Snyder et al. 2004; Longobardi et al. 2016a; Devaraju et al. 2015a, 2018). There is *medium agreement* on a global and latitudinal biophysical warming in response to forestation (Laguë and Swann 2016; Swann et al. 2012; Gibbard et al. 2005; Wang et al. 2014b) (Figure 2.17), but this is based on a smaller number of studies.

The lack of agreement at the annual scale among the climate models is, however, masking *rising agreement* regarding seasonal impacts of deforestation at those latitudes. There is *high agreement* that temperate deforestation leads to summer warming and winter cooling (Bright et al. 2017; Zhao and Jackson 2014; Gálos et al. 2011, 2013; Wickham et al. 2013; Ahlswede and Thomas 2017; Anderson-Teixeira et al. 2012; Anderson et al. 2011; Chen et al. 2012; Strandberg and Kjellström 2018). The winter cooling is driven by the increased surface albedo, amplified by the snow-albedo feedback. In some models, and when deforestation is simulated for very large areas, the cooling is further amplified by high latitude changes in sea-ice and snow extent (polar amplification). Summer warming occurs because the latent and sensible heat fluxes that take energy out of the surface diminish with the smaller roughness length and lower evapotranspiration efficiency of low vegetation, as compared to tree canopies (Davin and de Noblet-Ducoudre 2010; Anav et al. 2010). Conversely, there is *high agreement* that forestation in North America or in Europe cools surface climate during summer time, especially in regions where water availability can support large evapotranspiration rates. In temperate regions with water deficits, the simulated change in evapotranspiration following forestation will be insignificant, while the decreased surface albedo will favour surface warming.

Observation-based estimates confirm the existence of a seasonal pattern of response to deforestation, with colder winters any time there is snow on the ground and in any place where soils are brighter than the trees, and warmer summers (Schultz et al. 2017; Wickham et al. 2014; Juang et al. 2007; Tang et al. 2018; Peng et al. 2014; Zhang et al. 2014b; Prevedello et al. 2019; Li et al. 2015b; Alkama and Cescatti 2016). In contrast, forestation induces cooler summers wherever trees have access to sufficient soil moisture to transpire. The magnitude of the cooling depends on the wetness of the area of concern (Wickham et al. 2013) as well as on the original and targeted species and varieties implicated in the vegetation conversion (Peng et al. 2014; Juang et al. 2007).

There is also *high confidence* from observation-based estimates that mean annual daytime temperatures are warmer following deforestation, while night-time temperatures are cooler (Schultz et al. 2017; Wickham et al. 2014; Juang et al. 2007; Tang et al. 2018; Prevedello et al. 2019; Peng et al. 2014; Zhang et al. 2014b; Li et al. 2015b; Alkama and Cescatti 2016). Deforestation then increases the amplitude of diurnal temperature variations while forestation reduces it (*high confidence*). Two main reasons have been put forward to explain why nights are warmer in forested areas: their larger capacity to store heat and the existence of a nocturnal temperature inversion bringing warmer air from the higher atmospheric levels down to the surface.

In addition to those seasonal and diurnal fluctuations, Lejeune et al. (2018) found systematic warming of the hottest summer days following historical deforestation in the northern mid-latitudes, and this echoes Strandberg and Kjellström (2018) who argue that the August 2003 and July 2010 heatwaves could have been largely mitigated if Europe had been largely forested.

In a combined modelling of large-scale forestation of western Europe and climate change scenario (SRES A2), Gálos et al. (2013) found relatively small dampening potential of additional forest on ambient air temperature at the end of the 21st century when compared to the beginning (the cooling resulting from land cover changes is –0.5°C whereas the GHG-induced warming exceeds 2.5°C). Influence on rainfall was, however, much larger and significant. Projected annual rainfall decreases following warming were cancelled in Germany and significantly reduced in both France and Ukraine through forestation. In addition, forestation decreased the number of warming-induced dry days but increased the number of extreme precipitation events.

The net impact of forestation, combining both biophysical and biogeochemical effects, has been tested in the warmer world predicted by RCP 8.5 scenario (Sonntag et al. 2016, 2018). The cooling effect from the addition of 8 Mkm^2 of forests following the land use RCP 4.5 scenario was too small (–0.27°C annually) to dampen the RCP 8.5 warming. However, it reached about –1°C in some temperate regions and –2.5°C in boreal ones. This is accompanied by a reduction in the number of extremely warm days.

Global and regional impacts of deforestation/forestation in boreal regions

Consistent with what we have previously discussed for temperate and tropical regions, large-scale boreal deforestation induces a biogeochemical warming of +0.11 ± 0.09°C (Figure 2.17). But contrary to those other latitudinal bands, the biophysical effect is a consistent cooling across all models (–0.55 ± 0.29°C when averaged globally). It is also significantly larger than the biogeochemical warming (e.g., Dass et al. (2013), Longobardi et al. (2016a), Devaraju et al. (2015a), Bathiany et al. (2010), Devaraju et al. (2018)) and is driven by the increased albedo, enhanced by the snow-albedo feedback as well as by an increase in sea-ice extent in the Arctic. Over boreal lands, the cooling is as large as –1.8 ± 1.2°C. However, this means that annual cooling masks a seasonal contrast, as discussed in Strandberg and Kjellström (2018) and Gao et al. (2014): during summer time, following the removal of forest, the decreased

evapotranspiration results in a significant summer warming that outweighs the effect of an increased albedo effect.

The same observation-based estimates (as discussed in the previous subsection) show similar patterns for the temperate latitudes: seasonal and daily contrasts. Schultz et al. (2017), however, found that mean annual night-time changes are as large as daytime ones in those regions (mean annual nocturnal cooling of –1.4 ± 0.10°C, balanced by mean annual daytime warming of 1.4 ± 0.04°C). This contrasts with both temperate and tropical regions where daytime changes are always larger than the night-time ones.

Arora and Montenegro (2011) combined large-scale forestation and climate change scenario (SRES A2): forestation of either 50% or 100% of the total agricultural area was gradually prescribed between years 2011 and 2060 everywhere. In addition, boreal, temperate and tropical forestation have been tested separately. Both biophysical and biogeochemical effects were accounted for. The net simulated impact of forestation was a cooling varying from –0.04°C to –0.45°C, depending on the location and magnitude of the additional forest cover. It was, however, quite marginal compared to the large global warming resulting from anthropogenic GHG emissions (+3°C at the end of the 21st century). In their experiment, forestation in boreal regions led to biophysical warming and biogeochemical cooling that compensated each other, whereas forestation in the tropics led to both biophysical and biogeochemical cooling. The authors concluded that tropical forestation is three times more effective at cooling down climate than boreal or temperate forestation.

Conclusion

In conclusion, planting trees will always result in capturing more atmospheric CO_2, and thus will mean annual cooling of the globe (*very high confidence*). At the regional level, however, the magnitude and sign of the local temperature change depends on (i) where forestation occurs, (ii) its magnitude, (iii) the level of warming under which the land cover change is applied, and (iv) the land conversion type. This is because the background climatic conditions (e.g., precipitation and snow regimes, mean annual temperature) within which the land cover changes occur vary across regions (Pitman et al. 2011; Montenegro et al. 2009; Juang et al. 2007; Wickham et al. 2014; Hagos et al. 2014; Voldoire 2006; Feddema et al. 2005; Strandberg and Kjellström 2018). In addition, there is *high confidence* that estimates of the influence of any land cover or land use change on surface temperature from the sole consideration of the albedo and the CO_2 effects is incorrect as changes in turbulent fluxes (i.e., latent and sensible heat fluxes) are large contributors to local temperature change (Bright et al. 2017).

There is *high confidence* that, in boreal and temperate latitudes, the presence of forest cools temperature in warmer locations and seasons (provided that the soil is not dry), whereas it warms temperature in colder locations and seasons (provided the soil is brighter than the trees or covered with snow). In the humid tropics, forestation increases evapotranspiration year-round and thus decreases temperature (*high confidence*). In tropical areas with a strong seasonality of rainfall, forestation will also increase evapotranspiration year-round, unless the soil becomes too dry. In all regions there is *medium confidence* that the diurnal temperature range decreases with increasing forest cover, with potentially reduced extreme values of temperature.

Although there is not enough literature yet that rigorously compares both biophysical and biogeochemical effects of realistic scenarios of forestation, there is *high confidence* that, at the local scale (that is where the forest change occurs), biophysical effects on surface temperature are far more important than the effects resulting from the changes in emitted CO_2.

What is lacking in the literature today is an estimate of the impacts that natural disturbances in forests will have on local climates and on the build-up of atmospheric CO_2 (O'Halloran et al. 2012), illustrated with many examples that changes in albedo following disturbances can result in radiative forcing changes opposite to, and as large as, the ones resulting from the associated changes in the net release of CO_2 by land. The resulting climate effects depend on the duration of the perturbation and of the following recovery of vegetation.

2.5.2.2   Impacts of changes in land management

There have been little changes in net cropland area over the past 50 years (at the global scale) compared to continuous changes in land management (Erb et al. 2017). Similarly, in Europe, change in forest management has resulted in a very significant anthropogenic land change. Management affects water, energy and GHG fluxes exchanged between the land and the atmosphere, and thus affects temperature and rainfall, sometimes to the same extent as changes in land cover do (as discussed in Luyssaert et al. (2014)).

The effects of irrigation, which is a practice that has been substantially studied, including one attempt to manage solar radiation via increases in cropland albedo (geoengineering the land) are assessed, along with a discussion of recent findings on the effects of forest management on local climate, although there is not enough literature yet on this topic to carry out a thorough assessment. The effects of urbanisation on climate are assessed in a specific cross-chapter box within this chapter (Cross-Chapter Box 4 in this chapter).

There are a number of other practices that exist whose importance for climate mitigation has been examined (some are reported in Section 2.6 and Chapter 6). There is, however, not enough literature available for assessing their biophysical effect on climate. Few papers are generally found per agricultural practice, for example, Jeong et al. (2014b) for double cropping, Bagley et al. (2017) for the timing of the growing season and Erb et al. (2017) for a review of 10 management practices.

Similarly, there are very few studies that have examined how choosing species varieties and harvesting strategies in forest management impacts on climate through biophysical effects, and how those effects compare to the consequences of the chosen strategies on the net CO_2 sink of the managed forest. The modelling studies highlight the existence of competing effects, for example, between the capacity of certain species to store more carbon than others (thus inducing cooling) while at the same time reducing the total evapotranspiration

loss and absorbing more solar radiation via lower albedo (thus inducing warming) (Naudts et al. 2016a; Luyssaert et al. 2018).

Irrigation

There is substantial literature on the effects of irrigation on local, regional and global climate as this is a major land management issue. There is *very high confidence* that irrigation increases total evapotranspiration, increases the total amount of water vapour in the atmosphere and decreases mean surface daytime temperature within the irrigated area and during the time of irrigation (Bonfils and Lobell 2007; Alter et al. 2015; Chen and Jeong 2018; Christy et al. 2006; Im and Eltahir 2014; Im et al. 2014; Mueller et al. 2015). Decreases in maximum daytime temperature can locally be as large as −3°C to −8°C (Cook et al. 2015; Han and Yang 2013; Huber et al. 2014; Alter et al. 2015; Im et al. 2014). Estimates of the contribution of irrigation to past historical trends in ambient air temperature vary between −0.07°C and −0.014°C/decade in northern China (Han and Yang 2013; Chen and Jeong 2018) while being quite larger in California, USA (−0.14°C to −0.25°C/decade) (Bonfils and Lobell 2007). Surface cooling results from increased energy being taken up from the land via larger evapotranspiration rates. In addition, there is growing evidence from modelling studies that such cooling can locally mitigate the effect of heatwaves (Thiery et al. 2017; Mueller et al. 2015).

There is *no agreement* on changes in night-time temperatures, as discussed in Chen and Jeong (2018) who summarised the findings from observations in many regions of the world (India, China, North America and eastern Africa) (Figure 2.18). Where night-time warming is found (Chen and Jeong 2018; Christy et al. 2006), two explanations are put forward, (i) an increase in incoming longwave radiation in response to increased atmospheric water vapour content (greenhouse effect), and (ii) an increased storage of heat in the soil during daytime. Because of the larger heat capacity of moister soil, heat is then released to the atmosphere at night.

There is *robust evidence* from modelling studies that implementing irrigation enhances rainfall, although there is *very low confidence* on where this increase occurs. When irrigation occurs in Sahelian Africa during the monsoon period, rainfall is decreased over irrigated areas (*high agreement*), increased in the southwest if the crops are located in western Africa (Alter et al. 2015) and increased in the east/northeast when crops are located further east in Sudan (Im and Eltahir 2014; Im et al. 2014) The cooler irrigated surfaces in the Sahel, because of their greater evapotranspiration, inhibit convection and create an anomalous descending motion over crops that suppresses rainfall but influences the circulation of monsoon winds. Irrigation in India occurs prior to the start of the monsoon season and the resulting land cooling decreases the land-sea temperature contrast. This can delay the onset of the Indian monsoon and decrease its intensity (Niyogi et al. 2010; Guimberteau et al. 2012). Results from a modelling study by De Vrese et al. (2016) suggest that part of the excess rainfall triggered by Indian irrigation falls westward, in the horn of Africa. The theory behind those local and downwind changes in rainfall support the findings from the models, but we do not yet have sufficient literature to robustly assess the magnitude and exact location of the expected changes driven by irrigation.

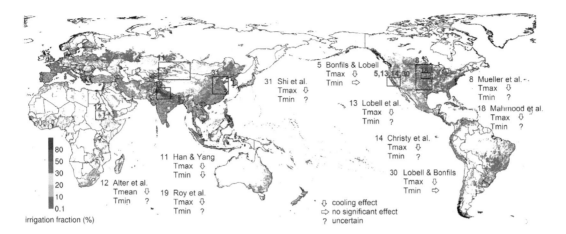

Figure 2.18 | Global map of areas equipped for irrigation (colours), expressed as a percentage of total area, or irrigation fraction. Source: Siebert et al. (2013). Numbered boxes show regions where irrigation causes cooling (down arrow) of surface mean (*Tmean*), maximum (*Tmax*) or minimum (*Tmin*) temperature, or else no significant effect (right arrow) or where the effect is uncertain (question mark), based on observational studies as reviewed in Chen and Jeong (2018). *Tmax* refers to the warmest daily temperature while *Tmin* to the coldest one, which generally occurs at night (Alter et al. 2015; Han and Yang 2013; Roy et al. 2007; Shi et al. 2013; Bonfils and Lobell 2007; Lobell et al. 2008; Lobell and Bonfils 2008; Christy et al. 2006; Mahmood et al. 2006; Mueller et al. 2015).

Cropland albedo

Various methods have been proposed to increase surface albedo in cropland and thus reduce local surface temperature (*high confidence*): choose 'brighter' crop varieties (Ridgwell et al. 2009; Crook et al. 2015; Hirsch et al. 2017; Singarayer et al. 2009; Singarayer and Davies-Barnard 2012), abandon tillage (Lobell et al. 2006; Davin et al. 2014), include cover crops into rotation in areas where soils are darker than vegetation (Carrer et al. 2018; Kaye and Quemada 2017) or use greenhouses (as in Campra et al. (2008)). See Seneviratne et al. (2018) for a review.

Whatever the solution chosen, the induced reduction in absorbed solar radiation cools the land – more specifically during the hottest summer days (*low confidence*) (Davin et al. 2014; Wilhelm et al. 2015;

Figure 2.19). Changes in temperature are essentially local and seasonal (limited to crop growth season) or sub-seasonal (when resulting from inclusion of cover crop or tillage suppression). Such management action on incoming solar radiation thus holds the potential to counteract warming in cultivated areas during crop growing season.

Introducing cover crops into a rotation can also have a warming effect in areas where vegetation has a darker albedo than soil, or in winter during snow periods if the cover crops or their residues are tall enough to overtop the snow cover (Kaye and Quemada 2017; Lombardozzi et al. 2018). In addition, evapotranspiration greater than that of bare soil during this transitional period reduces soil temperature (Ceschia et al. 2017). Such management strategy can have another substantial mitigation effect as it allows carbon to be stored in the soil and to reduce both direct and indirect N_2O

A   Albedo effect – southern Europe

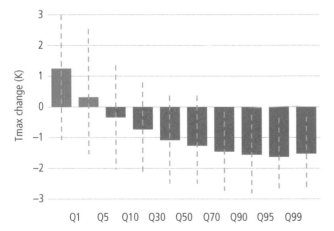

B   Albedo effect – northern Europe

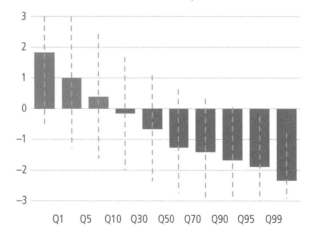

Figure 2.19 | Change in summer (July–August) daily maximum temperature (°C) resulting from increased surface albedo in unploughed versus ploughed land, in (A) southern, and (B) northern Europe, during the period 1986–2009. Changes are simulated for different quantiles of the daily maximum temperature distribution, where Q1 represents the coolest 1% and Q99 the warmest 1% of summer days. Only grid cells with more than 60% of their area in cropland are included. The dashed bars represent the standard deviation calculated across all days and grid points. Southern Europe refers to Europe below 45°N, and northern Europe refers to Europe above 45°N (Davin et al., 2014).

emissions (Basche et al. 2014; Kaye and Quemada 2017), in particular if fertilisation of the subsequent crop is reduced (Constantin et al. 2010, 2011). The use of cover crops thus substantially improves the GHG budget of croplands (Kaye and Quemada 2017; Tribouillois et al. 2018). More discussion on the role of management practices for mitigation can be found in Section 2.6 and Chapter 6.

Only a handful of modelling studies have looked at effects other than changes in atmospheric temperature in response to increased cropland albedo. Seneviratne et al. (2018) have found significant changes in rainfall following an idealised increase in cropland albedo, especially within the Asian monsoon regions. The benefits of cooler temperature on production, resulting from increased albedo, is cancelled out by decreases in rainfall that are harmful for crop productivity. The rarity of a concomitant evaluation of albedo management impact on crop productivity prevents us from providing a robust assessment of this practice in terms of both climate mitigation and food security.

2.5.3   Amplifying/dampening climate changes via land responses

Section 2.1 and Box 2.1 illustrate the various ways through which land can affect the atmosphere and thereby climate and weather. Section 2.2 illustrates the many impacts that climate changes have on the functioning of land ecosystems. Section 2.3 discusses the effects that future climatic conditions have on the capacity of the land to absorb anthropogenic CO_2, which then controls the sign

of the feedback to the initial global warming. Sections 2.5.1 and 2.5.2 show the effects of changes in anthropogenic land cover or land management on climate variables or processes. Therefore, land has the potential to dampen or amplify the GHG-induced global climate warming, or can be used as a tool to mitigate regional climatic consequences of global warming such as extreme weather events, in addition to increasing the capacity of land to absorb CO_2 (Figure 2.20).

Land-to-climate feedbacks are difficult to assess with global or regional climate models, as both types of models generally omit a large number of processes. Among these are (i) the response of vegetation to climate change in terms of growth, productivity, and geographical distribution, (ii) the dynamics of major disturbances such as fires, (iii) the nutrients dynamics, and (iv) the dynamics and effects of short-lived chemical tracers such as biogenic volatile organic compounds (Section 2.4). Therefore, only those processes that are fully accounted for in climate models are considered here.

2.5.3.1   Effects of changes in land cover and productivity resulting from global warming

In boreal regions, the combined northward migration of the treeline and increased growing season length in response to increased temperatures in those regions (Section 2.2) will have positive feedbacks both on global and regional annual warming (*high confidence*) (Garnaud and Sushama 2015; Jeong et al. 2014a; O'ishi and Abe-Ouchi 2009; Port et al. 2012; Strengers et al. 2010). The warming resulting from the decreased surface albedo remains

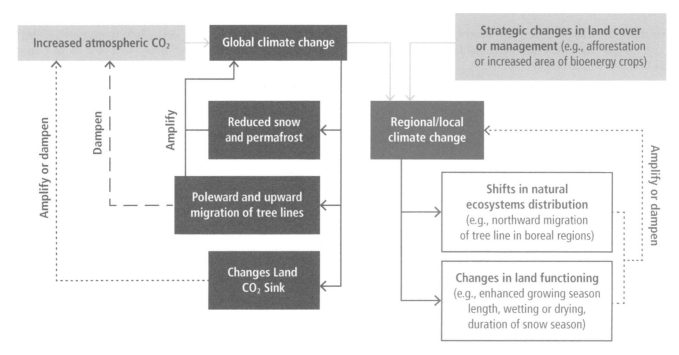

Figure 2.20 | Schematics of the various ways land has been shown in the literature to either amplify or dampen the initial GHG-induced climatic change. Brown arrows and boxes represent the global scale and blue arrows and boxes represent the regional/local level. Grey arrows and boxes refer to what we consider herein as imposed changes – that is, the initial atmospheric GHG content as well as anthropogenic land cover change and land management. Dampening feedbacks are represented with dashed lines, amplifying ones with solid lines and feedbacks where the direction may be variable are represented using dotted lines. The feedbacks initiated by changes in snow and permafrost areas in boreal regions are discussed in Section 2.5.3.2, the ones initiated by changes in ecosystem distribution are discussed in Sections 2.5.3.1, 2.5.1 and 2.5.2, and the feedbacks related to changes in the land functioning are discussed in Sections 2.5.3.3 and 2.5.1, as well as in Sections 2.3 and 2.5 (for changes in net CO_2 fluxes). References supporting this figure can be found in each of those sections.

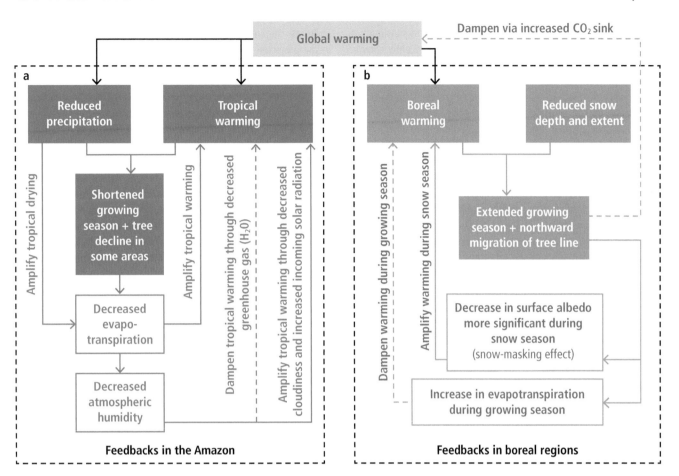

Figure 2.21 | Schematic illustration of the processes through which the effects of global warming in (a) the Amazon (blue arrows and boxes), and (b) boreal regions (grey arrows and boxes) feedback on the regional climate change. In boreal regions, the sign of the feedbacks depends on the season, although annually global warming is further enhanced in those regions. Dashed lines illustrate negative feedbacks, while solid lines indicate positive feedbacks. References supporting this figure can be found in the text.

the dominant signal in all modelling studies at the annual timescale and during the snow season, while cooling is obtained during the growing season (Section 2.5.2.1 and Figure 2.21, right panel).

In the tropics, climate change will cause both greening and browning (Section 2.2). Where global warming provokes a decrease in rainfall, the induced decrease in biomass production leads to increased local warming (*high confidence*) (Port et al. 2012; Wu et al. 2016; Yu et al. 2016). The reverse is true where warming generates increases in rainfall and thus greening. As an example, Port et al. (2012) simulated decreases in tree cover and shortened growing season in the Amazon, despite the CO_2 fertilisation effects, in response to both future tropical warming and reduced precipitation (Figure 2.21, left panel). This browning of the land decreases both evapotranspiration and atmospheric humidity. The warming driven by the drop in evapotranspiration is enhanced via a decrease in cloudiness, increasing solar radiation, and is dampened by reduced water vapour greenhouse radiation.

There is *very low confidence* on how feedbacks affect rainfall in the tropics where vegetation changes may occur, as the sign of the change in precipitation depends on where the greening occurs and on the season (as discussed in Section 2.5.2). There is, however, *high confidence* that increased vegetation growth in the southern Sahel increases African monsoon rains (Yu et al. 2016; Port et al. 2012;

Wu et al. 2016). Confidence on the direction of such feedbacks is also based on a significant number of paleoclimate studies that analysed how vegetation dynamics helped maintain a northward position of the African monsoon during the Holocene time period (9–6 kyr BP) (de Noblet-Ducoudré et al. 2000; Rachmayani et al. 2015).

2.5.3.2 Feedbacks to climate from high-latitude land-surface changes

In high latitudes, snow albedo and permafrost carbon feedbacks are the most well-known and most important surface-related climate feedbacks because of their large-scale impacts.

In response to ongoing and projected decrease in seasonal snow cover (Derksen and Brown 2012; Brutel-Vuilmet et al. 2013) warming is, and will continue to be, enhanced in boreal regions (*high confidence*) (Brutel-Vuilmet et al. 2013; Perket et al. 2014; Thackeray and Fletcher 2015; Mudryk et al. 2017). One reason for this is the large reflectivity (albedo) the snow exerts on shortwave radiative forcing: the all-sky global land snow shortwave radiative effect is evaluated to be around -2.5 ± 0.5 W m^{-2} (Flanner et al. 2011; Singh et al. 2015). In the southern hemisphere, perennial snow on the Antarctic is the dominant contribution, while in the northern hemisphere, this is essentially attributable to seasonal snow, with a smaller contribution from snow on glaciated areas. Another reason is the sensitivity of snow cover to temperature: Mudryk et al. (2017) recently showed that, in the high latitudes, climate models tend to correctly represent this sensitivity, while in mid-latitude and alpine regions, the simulated snow cover sensitivity to temperature variations tends to be biased low. In total, the global snow albedo feedback is about 0.1 W m^{-2} K^{-1}, which amounts to about 7% of the strength of the globally dominant water vapour feedback (e.g., Thackeray and Fletcher (2015). While climate models do represent this feedback, a persistent spread in the modelled feedback strength has been noticed (Qu and Hall 2014) and, on average, the simulated snow albedo feedback strength tends to be somewhat weaker than in reality (*medium confidence*) (Flanner et al. 2011; Thackeray and Fletcher 2015). Various reasons for the spread and biases of the simulated snow albedo feedback have been identified, notably inadequate representations of vegetation masking snow in forested areas (Loranty et al. 2014; Wang et al. 2016c; Thackeray and Fletcher 2015).

The second most important potential feedback from land to climate relates to permafrost decay. There is *high confidence* that, following permafrost decay from a warming climate, the resulting emissions of CO$_2$ and/or CH$_4$ (caused by the decomposition of organic matter in previously frozen soil) will produce additional GHG-induced warming. There is, however, substantial uncertainty on the magnitude of this feedback, although recent years have seen large progress in its quantification. Lack of agreement results from several critical factors that carry large uncertainties. The most important are (i) the size of the permafrost carbon pool, (ii) its decomposability, (iii) the magnitude, timing and pathway of future high-latitude climate change, and (iv) the correct identification and model representation of the processes at play (Schuur et al. 2015). The most recent comprehensive estimates establish a total soil organic carbon storage in permafrost of about 1500 ± 200 PgC (Hugelius et al. 2014, 2013; Olefeldt et al. 2016), which is about 300 Pg C lower than previous estimates (*low confidence*). Important progress has been made in recent years at incorporating permafrost-related processes in complex ESMs (e.g., McGuire et al. (2018)), but representations of some critical processes such as thermokarst formation are still in their infancy (Schuur et al. 2015). Recent model-based estimates of future permafrost carbon release (Koven et al. 2015; McGuire et al. 2018) have converged on an important insight. Their results suggest that substantial net carbon release of the coupled vegetation-permafrost system will probably not occur before about 2100 because carbon uptake by increased vegetation growth will initially compensate for GHG releases from permafrost (*limited evidence, high agreement*).

2.5.3.3　Feedbacks related to changes in soil moisture resulting from global warming

There is *medium evidence* but *high agreement* that soil moisture conditions influence the frequency and magnitude of extremes such as drought and heatwaves. Observational evidence indicates that dry soil moisture conditions favour heatwaves, in particular in regions where evapotranspiration is limited by moisture availability (Mueller and Seneviratne 2012; Quesada et al. 2012; Miralles et al. 2018; Geirinhas et al. 2018; Miralles et al. 2014; Chiang et al. 2018; Dong and Crow 2019; Hirschi et al. 2014).

In future climate projections, soil moisture plays an important role in the projected amplification of extreme heatwaves and drought in many regions of the world (*medium confidence*) (Seneviratne et al. 2013; Vogel et al. 2017; Donat et al. 2018; Miralles et al. 2018). In addition, the areas where soil moisture affects heat extremes will not be located exactly where they are today. Changes in rainfall, temperature, and thus in evapotranspiration, will induce changes in soil moisture and therefore where temperature and latent heat flux will be negatively coupled (Seneviratne et al. 2006; Fischer et al. 2012). Quantitative estimates of the actual role of soil moisture feedbacks are, however, very uncertain due to the *low confidence* in projected soil moisture changes (IPCC 2013a), to weaknesses in the representation of soil moisture–atmosphere interactions in climate models (Sippel et al. 2017; Ukkola et al. 2018; Donat et al. 2018; Miralles et al. 2018) and to methodological uncertainties associated with the soil moisture prescription framework commonly used to disentangle the effect of soil moisture on changes in temperature extremes (Hauser et al. 2017).

Where soil moisture is predicted to decrease in response to climate change in the subtropics and temperate latitudes, this drying could be enhanced by the existence of soil moisture feedbacks (*low confidence*) (Berg et al. 2016). The initial decrease in precipitation and increase in potential evapotranspiration and latent heat flux, in response to global climate change, leads to decreased soil moisture at those latitudes and can potentially amplify both. Such a feature is consistent with evidence that, in a warmer climate, land and atmosphere will be more strongly coupled via both the water and energy cycles (Dirmeyer et al. 2014; Guo et al. 2006). This increased sensitivity of atmospheric response to land perturbations implies that changes in land uses and cover are expected to have more impact on climate in the future than they do today.

Beyond temperature, it has been suggested that soil moisture feedbacks influence precipitation occurrence and intensity. But the importance, and even the sign of this feedback, is still largely uncertain and debated (Tuttle and Salvucci 2016; Yang et al. 2018; Froidevaux et al. 2014; Guillod et al. 2015).

2.5.4　Non-local and downwind effects resulting from changes in land cover

Changes in land cover or land management do not just have local consequences but also affect adjacent or more remote areas. Those non-local impacts may occur in three different ways.

1. Any action on land that affects photosynthesis and respiration has an impact on the atmospheric CO$_2$ content as this GHG is well mixed in the atmosphere. In turn, this change affects the downwelling longwave radiation everywhere on the planet and contributes to global climate change. This is more thoroughly discussed in Section 2.6 where various land-based mitigation solutions are examined. Local land use changes thus have the potential to affect the global climate via changes in atmospheric CO$_2$.
2. Any change in land cover or land management may impact on local surface air temperature and moisture, and thus sea-level pressure. Thermal, moisture and surface pressure gradients

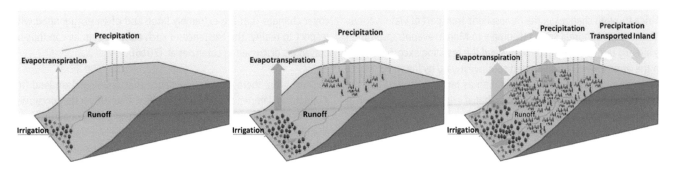

Figure 2.22 | Schematic illustration of how combined forestation and irrigation can influence downwind precipitation on mountainous areas, favour vegetation growth and feed back to the forested area via increased runoff. Showing Los Angeles, California (Layton and Ellison 2016). Areas of forest plantations and irrigation are located on the left panel, whereas consequent downwind effects and feedbacks are illustrated in the middle and right panels.

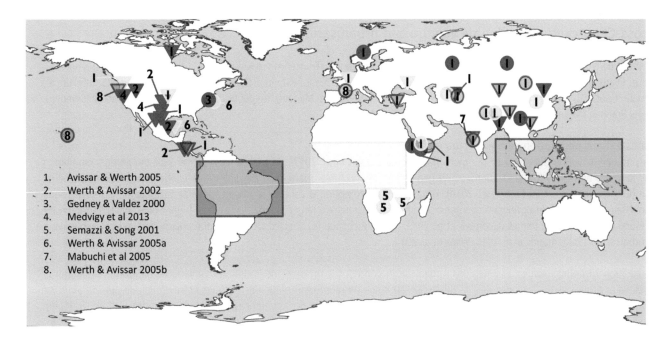

1. Avissar & Werth 2005
2. Werth & Avissar 2002
3. Gedney & Valdez 2000
4. Medvigy et al 2013
5. Semazzi & Song 2001
6. Werth & Avissar 2005a
7. Mabuchi et al 2005
8. Werth & Avissar 2005b

Figure 2.23 | Extra-tropical effects on precipitation due to deforestation in each of the three major tropical regions. Increasing (circles) and decreasing (triangles) precipitation result from complete deforestation of either Amazonia (red), Africa (yellow) or Southeast Asia (blue) as reviewed by Lawrence and Vandecar (2015). Boxes indicate the area where tropical forest was removed in each region. Numbers refer to the study the data were derived from (Avissar and Werth 2005; Gedney and Valdes 2000; Semazzi and Song 2001; Werth 2002; Mabuchi et al. 2005; Werth 2005).

between the area of change and neighbouring areas are then modified and affect the amount of heat, water vapour and pollutants flowing out (downwind) of the area (e.g., Ma et al. (2013b), McLeod et al. (2017), Abiodun et al. (2012), Keys (2012)). Forests, for example, provide water vapour to the atmosphere which supports terrestrial precipitation downwind (Ellison et al. 2017; Layton and Ellison 2016; Spracklen et al. 2012, 2018). Within a few days, water vapour can travel several hundred kilometres before being condensed into rain and potentially being transpired again (Makarieva et al. 2009). This cascading moisture recycling (succession of evapotranspiration, water vapour transport and condensation-rainfall) has been observed in South America (Spracklen et al. 2018; Zemp et al. 2014; Staal et al. 2018; Spracklen et al. 2012). Deforestation can thus potentially decrease rainfall downwind, while combining 'small-scale' forestation and irrigation, which in the semi-arid region is susceptible to boost the precipitation-recycling mechanism with

better vegetation growth downwind (Ellison et al. 2017; Layton and Ellison, 2016) (Figure 2.22).

3. Many studies using global climate models have reported that the climatic changes resulting from changes in land are not limited to the lower part of the atmosphere, but can reach the upper levels via changes in large-scale ascent (convection) or descent (subsidence) of air. This coupling to the upper atmosphere triggers perturbations in large-scale atmospheric transport (of heat, energy and water) and subsequent changes in temperature and rainfall in regions located quite far away from the original perturbation (Laguë and Swann 2016; Feddema et al. 2005, Badger and Dirmeyer 2016; Garcia 2016; Stark 2015; Devaraju 2018; Quesada et al. 2017a) (Figure 2.23).

De Vrese et al. (2016) for example, using a global climate model, found that irrigation in India could affect regions as remote as eastern

Africa through changes in the atmospheric transport of water vapour. At the onset of boreal spring (February to March) evapotranspiration is already large over irrigated crops and the resulting excess moisture in the atmosphere is transported southwestward by low-level winds. This results in increases in precipitation as large as 1mm d⁻¹ in the Horn of Africa. Such a finding implies that, if irrigation is to decrease in India, rainfall can decrease in eastern Africa where the consequences of drought are already disastrous.

Changes in sea-surface temperature have also been simulated in response to large-scale vegetation changes (Cowling et al. 2009; Davin and de Noblet-Ducoudre 2010; Wang et al. 2014b, Notaro Liu 2007). Most of those modelling studies have been carried out with land

cover changes that are extremely large and often exaggerated with respect to reality. The existence of such teleconnections can thus be biased, as discussed in Lorenz et al. (2016).

In conclusion, there is *high confidence* that any action on land (for example, to dampen global warming effects), wherever they occur, will not only have effects on local climate but also generate atmospheric changes in neighbouring regions, and potentially as far as hundreds of kilometres downwind. More remote teleconnections, thousands of kilometres away from the initial perturbation, are impossible to observe and have only been reported by modelling studies using extreme land cover changes. There is *very low confidence* that detectable changes due to such long-range processes can occur.

Cross-Chapter Box 4 | Climate change and urbanisation

Nathalie de Noblet-Ducoudré (France), Peng Cai (China), Sarah Connors (France/United Kingdom), Martin Dallimer (United Kingdom), Jason Evans (Australia), Rafiq Hamdi (Belgium), Gensuo Jia (China), Kaoru Kitajima (Japan), Christopher Lennard (South Africa), Shuaib Lwasa (Uganda), Carlos Fernando Mena (Ecuador), Soojeong Myeong (Republic of Korea), Lennart Olsson (Sweden), Prajal Pradhan (Nepal/Germany), Lindsay Stringer (United Kingdom)

Cities extent, population, and expected growth
Despite only covering 0.4–0.9% of the global land surface (Esch et al. 2017; Zhou et al. 2015), over half the world's population live in towns and cities (United Nations, 2017) generating around three-quarters of the global total carbon emissions from energy use (Creutzig et al. 2015b; Seto et al. 2014). Urban food consumption is a large source of these anthropogenic GHG emissions (Goldstein et al. 2017). In developed countries, per capita emissions are larger in small cities than bigger ones, while the opposite is found in developing countries (Gudipudi et al. 2019). Climate change is expected to increase the energy demand of people living in urban areas (Santamouris et al. 2015; Wenz et al. 2017).

In addition to being a driver of emissions, urbanisation contributes to forest degradation, converts neighbouring agricultural, forested or otherwise undeveloped land to urban use, altering natural or semi-natural ecosystems both within and outside of urban areas (Du and Huang 2017). It has been identified as a major driver of land degradation, as illustrated in Chapters 3, 4 and 5. Highly productive lands are experiencing the highest rate of conversion to urbanised landscapes (Nizeyimana et al. 2001; Pandey et al. 2018), affecting food security. Loss of agricultural land and increased pollution and waste are some of key challenges arising from urbanisation and urban growth (Chen 2007). The proportion of urban population is predicted to reach about 70% by the middle of the century (United Nations 2017) with growth especially taking place in the developing world (Angel et al. 2011; Dahiya 2012). Urban sprawl is projected to consume 1.8–2.4% and 5% of the current cultivated land by 2030 and 2050, respectively (Pradhan et al. 2014; Bren d'Amour et al. 2016), driven by both general population increase and immigration from rural areas (Adger et al. 2015; Seto et al. 2011; Geddes et al. 2012). New city dwellers in developing countries will require land for housing to be converted from non-urban to urban land (Barbero-Sierra et al. 2013), indicating future degradation. These growing urban areas will experience direct and indirect climate change impacts, such as sea level rise and storm surges (Boettle et al. 2016; Revi et al. 2014), increasing soil salinity and landslides from precipitation extremes. Furthermore, poorly planned urbanisation can increase people's risk to climate hazards as informal settlements and poorly built infrastructure are often the most exposed to hazards from fire, flooding and landslides (Adger et al. 2015; Geddes et al. 2012; Revi et al. 2014). Currently, avoiding land degradation and maintaining/enhancing ecosystem services are rarely considered in planning processes (Kuang et al. 2017).

Climate change, urban heat island and threats specific to urban populations
Cities alter the local atmospheric conditions as well as those of the surrounding areas (Wang et al. 2016b; Zhong et al. 2017). There is *high confidence* that urbanisation increases mean annual surface air temperature in cities and in their surroundings, with increases ranging from 0.19–2.60°C (Torres-Valcárcel et al. 2015; Li et al. 2018a; Doan et al. 2016) (Cross-Chapter Box 4; Figure 1). This phenomenon is referred to as the urban heat island (UHI) effect (Oke et al. 2017; Bader et al. 2018). The magnitude and diurnal amplitude of the UHI varies from one city to another and depends on the local background climate (Wienert and Kuttler 2005; Zhao et al. 2014; Ward et al. 2016). There is nevertheless *high confidence* that urbanisation affects night-time temperatures more substantially than daytime ones (Argüeso et al. 2014; Alghamdi and Moore 2015; Alizadeh-Choobari et al. 2016; Fujibe, 2009; Hausfather et al. 2013; Liao et al. 2017; Sachindra et al. 2016; Camilloni and Barrucand 2012; Wang et al. 2017a; Hamdi, 2010;

Cross-Chapter Box 4 (continued)

Arsiso et al. 2018; Elagib 2011; Lokoshchenko 2017; Robaa 2013). In addition, there is *high confidence* that the UHI effect makes heatwaves more intense in cities by 1.22–4°C, particularly at night (Li and Bou-Zeid 2013; Li et al. 2017b; Hamdi et al. 2016; Founda and Santamouris 2017; Wang et al. 2017a). As there is a well-established relationship between extremely high temperatures and morbidity, mortality (Watts et al. 2015) and labour productivity (Costa et al. 2016), an expected increase in extreme heat events with future climate change will worsen the conditions in cities.

Cross-Chapter Box 4, Figure 1 | Change in annual mean surface air temperature resulting from urbanisation (°C). The colour and size of the circles refer to the magnitude of the change. (This map has been compiled using the following studies: Kim et al. (2016), Sun et al. (2016), Chen et al. (2016a), Founda et al. (2015), Rafael et al. (2017), Hinkel and Nelson (2007), Chrysanthou et al. (2014), Dou et al. (2014), Zhou et al. (2016), (2017), Polydoros et al. (2018), Li et al. (2018a), Bader et al. (2018), Alizadeh-Choobari et al. (2016), Fujibe (2009), Lokoshchenko (2017), Torres-Valcárcel et al. (2015), Doan et al. (2016), Elagib (2011), Liao et al. (2017)).

Individual city case studies show that precipitation mean and extremes are increased over and downwind of urban areas, especially in the afternoon and early evening when convective rise of the atmosphere is the strongest (*medium confidence*). The case studies covered: different inland and coastal US cities (Haberlie et al. 2014; McLeod et al. 2017; Ganeshan and Murtugudde 2015), Dutch coastal cities (Daniels et al. 2016), Hamburg (Schlünzen et al. 2010), Shanghai (Liang and Ding 2017), Beijing (Dou et al. 2014), and Jakarta and Kuala Lumpur (Lorenz et al. 2016). Increased aerosol concentrations, however, can interrupt the precipitation formation process and thereby reduce heavy rainfall (Daniels et al. 2016; Zhong et al. 2017). Urban areas also experience altered water cycle in other aspects: the evaporative demand for plants in cities are increased by as much as 10% (Zipper et al. 2017), while the high proportion of paving in cities means that surface runoff of water is high (Hamdi et al. 2011; Pataki et al. 2011). In addition, water retention is lower in degraded, sealed soils beneath urban surfaces compared to intact soils. Increased surface water runoff, especially when and where the rainfall intensity is likely to intensify (IPCC 2013a), leads to a greater likelihood of flooding in urban areas without implementation of adaptation measures (Shade and Kremer 2019; Wang et al. 2013; EPA 2015).

Urbanisation alters the stock size of soil organic carbon (SOC) and its stability. The conversion of vegetated land to urban land results in a loss of carbon stored in plants, while stresses associated with the urban environment (e.g., heat, limited water availability, pollution) reduce plant growth and survival in cities (Xu et al. 2016b). Overall, carbon densities or stocks decrease from natural land areas to the urban core along the rural-urban gradient (Tao et al. 2015; Zhang et al. 2015). For example, the Seoul Forest Park, an urban park, shows a tenfold difference in SOC stocks across its land cover types (Bae and Ryu 2015). In Changchun in Northeast China, however, SOC density is higher in recreational forests within urban areas compared to a production forest (Zhang et al. 2015).

Urban air pollution as an environmental risk increases with climate change. Increased air temperatures can lead to reduced air quality by enhancing the formation of photochemical oxidants and increasing the concentration of air pollutants such as ozone, with corresponding threats to human health (Sharma et al. 2013). The occurrence of bronchial asthma and allergic respiratory diseases is increasing worldwide, and urban residents are experiencing poor air quality conditions more frequently than rural residents

Cross-Chapter Box 4 (continued)

(D'Amato et al. 2010). Excess morbidity and mortality related to extremely poor air quality are found in many cities worldwide (Harlan and Ruddell 2011). Some emissions that lead to reduced air quality are also contributors to climate change (Shindell et al. 2018; de Coninck et al. 2018).

Urban response options for climate change, desertification, land degradation and food security

Urban green infrastructure (UGI) has been proposed as a solution to mitigate climate change directly through carbon sequestration (Davies et al. 2011; Edmondson et al. 2014). However, compared to overall carbon emissions from cities, its mitigation effects are likely to be small (*medium confidence*). UGI nevertheless has an important role in adapting cities to climate change (Demuzere et al. 2014; Sussams et al. 2015; Elmqvist et al. 2016; Gill et al. 2007; Revi et al. 2014). Adaptation through UGIs is achieved through, for example, (i) reduction in air temperature (Cavan et al. 2014; Di Leo et al. 2016; Feyisa et al. 2014; Zölch et al. 2016; Li et al. 2019) which can help improve human health and comfort (e.g., Brown and Nicholls 2015; Klemm et al. 2015), (ii) reduction in the energy demands of buildings through the use of green roofs and walls (e.g., Coma et al. 2017), and (iii) reduction in surface water runoff and flood risk (Zeleňáková et al. 2017). Given that UGI necessarily involves the retention and management of non-sealed surfaces, co-benefits for land degradation will also be apparent (*limited evidence, high agreement*) (Murata and Kawai 2018; Scalenghe and Marsan 2009).

Urban agriculture is one aspect of UGI that has the potential to both meet some of the food needs of cities and reduce land degradation pressures in rural areas (*low confidence*) (e.g., Wilhelm and Smith (2018)). Urban agriculture has many forms, such as backyard gardening, allotments, plants on rooftops or balconies, urban-fringe/peri-urban agriculture, hydroponics, aquaponics, livestock grazing in open spaces and vertical farming (Gerster-Bentaya 2013) (Section 5.6.5).

Consuming locally produced food and enhancing the efficiency of food processing and transportation can minimise food losses, contribute to food security and, in some circumstances, reduce GHG emissions (Brodt et al. 2013; Michalský and Hooda 2015; Tobarra et al. 2018) (Section 5.5.2.3). Furthermore, urban agriculture has the potential to counteract the separation of urban populations from food production. This separation is one driver of the transition towards more homogeneous, high-protein diets, which are associated with increased GHG emissions (Goldstein et al. 2017; Moragues-Faus and Marceau 2018; Magarini and Calori 2015). Barriers to the uptake of urban agriculture as a climate change mitigation option include the need for efficient distribution systems to ensure lowered carbon emissions (Newman et al. 2012) and the concern that urban agriculture may harbour pathogenic diseases, or that its products be contaminated by soil or air pollution (Hamilton et al. 2014; Ercilla-Montserrat et al. 2018).

In summary

Climate change is already affecting the health and energy demand of large numbers of people living in urban areas (*high confidence*) (Section 2.2). Future changes to both climate and urbanisation will enhance warming in cities and their surroundings, especially during heatwaves (*high confidence*). Urban and peri-urban agriculture and, more generally, the implementation of urban green infrastructure, can contribute to climate change mitigation (*medium confidence*) as well as to adaptation (*high confidence*), including co-benefits for food security and reduced soil-water-air pollution.

2.6   Climate consequences of response options

Response options can affect climate mitigation and adaptation simultaneously, therefore this Special Report on Climate Change and Land (SRCCL) discusses land-based response options in an integrated way (Chapter 1). In this chapter, we assess response options that that have an effect on climate. A description of the full set of response options across the SRCCL can be found in Chapter 6, including the interplay between mitigation, adaptation, desertification, land degradation, food security and other co-benefits and trade-offs. Response options specific to desertification, degradation and food security are described in more detail in Chapters 3, 4 and 5.

Some response options lead to land use change and can compete with other land uses, including other response options, while others may free-up land that can be used for further mitigation/adaptation by reducing demand for land or products (e.g., agricultural intensification, diet shifts and reduction of waste) (*high confidence*).

Some response options result in a net removal of GHGs from the atmosphere and storage in living or dead organic material, or in geological stores (IPCC SR15). Such options are frequently referred to in the literature as CO_2 removal (CDR), greenhouse gas removal (GGR) or negative emissions technologies (NETs). CDR options are assessed alongside emissions reduction options. Although they have a land footprint, solar and wind farms are not are not assessed here as they affect GHG flux in the energy industrial sectors with minimal effect in the land sector, but the impact of solar farms on agricultural land competition is dealt with in Chapter 7.

A number of different types of scenario approach exist for estimating climate contribution of land-based response options (Cross-Chapter Box 1 and Chapter 1). Mitigation potentials have been estimated for single and sometimes multiple response options using stylised 'bottom-up' scenarios. Response options are not mutually exclusive (e.g., management of soil carbon and cropland management). Different options interact with each other; they may have additive effects or compete with each other for land or other resources and thus these potentials cannot necessarily be added up. The interplay between different land-based mitigation options, as well as with mitigation options in other sectors (such as energy or transport), in contributing to specific mitigation pathways has been assessed using IAMs (Section 2.7.2). These include interactions with wider socioeconomic conditions (Cross-Chapter Box 1 and Chapter 1) and other sustainability goals (Chapter 6).

2.6.1   Climate impacts of individual response options

Since AR5, there have been many new estimates of the climate impacts of single or multiple response options, summarised in Figure 2.24 and discussed in sub-sections below. Recently published syntheses of mitigation potential of land-based response options (e.g., Hawken (2017a), Smith et al. (2016b), Griscom et al. (2017), Minx et al. (2018), Fuss et al. (2018b), Nemet et al. (2018)) are also included in Figure 2.24. The wide range in mitigation estimates reflects differences in methodologies that may not be directly comparable, and estimates cannot be necessarily be added if they were calculated independently as they may be competing for land and other resources.

Some studies assess a 'technical mitigation potential' – the amount possible with current technologies. Some include resource constraints (e.g., limits to yields, limits to natural forest conversion) to assess a 'sustainable potential'. Some assess an 'economic potential' mitigation at different carbon prices. Few include social and political constraints (e.g., behaviour change, enabling conditions) (Chapter 7), the biophysical climate effects (Section 2.5) or the impacts of future climate change (Section 2.3). Carbon stored in biomass and soils may be at risk of future climate change (Section 2.2), natural disturbances such as wildfire (Cross-Chapter Box 3 in this chapter) and future changes in land use or management changes that result in a net loss of carbon (Gren and Aklilu 2016).

2.6.1.1   Land management in agriculture

Reducing non-CO_2 emissions from agriculture through cropland nutrient management, enteric fermentation, manure management, rice cultivation and fertiliser production has a total mitigation potential of 0.30–3.38 $GtCO_2$-eq yr^{-1} (*medium confidence*) (combined sub-category measures in Figure 2.24, details below) with a further 0.25–6.78 $GtCO_2$-eq yr^{-1} from soil carbon management (Section 2.6.1.3). Other literature that looks at broader categories finds mitigation potential of 1.4–2.3 $GtCO_2$-eq yr^{-1} from improved cropland management (Smith et al. 2008, 2014; Pradhan et al., 2013); 1.4–1.8 $GtCO_2$-eq yr^{-1} from improved grazing land management (Conant et al. 2017; Herrero et al. 2016; Smith et al. 2008, 2014) and 0.2–2.4 $GtCO_2$-eq yr^{-1} from improved livestock management

(Smith et al. 2008, 2014; Herrero et al. 2016, FAO 2007). Detailed discussions of the mitigation potential of agricultural response options and their co-benefits are provided in Chapter 5 and Sections 5.5 and 5.6.

The three main measures to reduce enteric fermentation include improved animal diets (higher quality, more digestible livestock feed), supplements and additives (reduce methane by changing the microbiology of the rumen), and animal management and breeding (improve husbandry practices and genetics). Applying these measures can mitigate 0.12–1.18 $GtCO_2$-eq yr^{-1} (*medium confidence*) (Hristov et al. 2013; Dickie et al. 2014; Herrero et al. 2016; Griscom et al. 2017). However, these measures may have limitations such as need of crop-based feed (Pradhan et al. 2013) and associated ecological costs, toxicity and animal welfare issues related to food additives (Llonch et al. 2017). Measures to manage manure include anaerobic digestion for energy use, composting as a nutrient source, reducing storage time and changing livestock diets, and have a potential of 0.01–0.26 $GtCO_2$-eq yr^{-1} (Herrero et al. 2016; Dickie et al. 2014).

On croplands, there is a mitigation potential of 0.03–0.71 $GtCO_2$-eq yr^{-1} for cropland nutrient management (fertiliser application) (*medium confidence*) (Griscom et al. 2017; Hawken 2017; Paustian et al. 2016; Dickie et al. 2014; Beach et al. 2015). Reducing emissions from rice production through improved water management (periodic draining of flooded fields to reduce methane emissions from anaerobic decomposition) and straw residue management (applying in dry conditions instead of on flooded fields and avoiding burning to reduce methane and N_2O emissions) has the potential to mitigate up to 60% of emissions (Hussain et al. 2015), or 0.08–0.87 $GtCO_2$-eq yr^{-1} (*medium confidence*) (Griscom et al. 2017; Hawken 2017; Paustian et al. 2016; Hussain et al. 2015; Dickie et al. 2014; Beach et al. 2015). Furthermore, sustainable intensification through the integration of crop and livestock systems can increase productivity, decrease emission intensity and act as a climate adaptation option (Section 5.5.1.4).

Agroforestry is a land management system that combines woody biomass (e.g., trees or shrubs) with crops and/or livestock). The mitigation potential from agroforestry ranges between 0.08–5.7 $GtCO_2$ yr^{-1}, (*medium confidence*) (Griscom et al. 2017; Dickie et al. 2014; Zomer et al. 2016; Hawken 2017). The high estimate is from an optimum scenario combing four agroforestry solutions (silvopasture, tree intercropping, multistrata agroforestry and tropical staple trees) of Hawken (2017a). Zomer et al. (2016) reported that the trees in agroforestry landscapes had increased carbon stock by 7.33 $GtCO_2$ between 2000 and 2010, or 0.7 $GtCO_2$ yr^{-1} (Section 5.5.1.3).

2.6.1.2   Land management in forests

The mitigation potential for reducing and/or halting deforestation and degradation ranges from 0.4–5.8 $GtCO_2$ yr^{-1} (*high confidence*) (Griscom et al. 2017; Hawken 2017; Busch and Engelmann 2017; Baccini et al. 2017; Zarin et al. 2016; Federici et al. 2015; Carter et al. 2015; Houghton et al. 2015; Smith et al. 2013a; Houghton and Nassikas 2018). The higher figure represents a complete halting of land use conversion in forests and peatlands (i.e., assuming recent

LAND MANAGEMENT

Figure 2.24 | Mitigation potential of response options in 2020–2050, measured in GtCO$_2$-eq yr^{-1}, adapted from Roe et al. (2017). Mitigation potentials reflect the full range of low to high estimates from studies published after 2010, differentiated according to technical (possible with current technologies), economic (possible given economic constraints) and sustainable potential (technical or economic potential constrained by sustainability considerations). Medians are calculated across all potentials in categories with more than four data points. We only include references that explicitly provide mitigation potential estimates in CO$_2$-eq yr^{-1} (or a similar derivative) by 2050. Not all options for land management potentials are additive, as some may compete for land. Estimates reflect a range of methodologies (including definitions, global warming potentials and time horizons) that may not be directly comparable or additive. Results from IAMs are shown to compare with single option 'bottom-up' estimates, in available categories from the 2°C and 1.5°C scenarios in the SSP Database (version 2.0). The models reflect land management changes, yet in some instances, can also reflect demand-side effects from carbon prices, so may not be defined exclusively as 'supply-side'. References: 1) Griscom et al. (2017), 2) Hawken (2017), 3) Paustian et al. (2016), 4) Beach et al. (2016), 5) Dickie et al. (2014), 6) Herrero et al. (2013), 7) Herrero et al. (2016), 8) Hussain et al. (2015), 9) Hristov, et al. (2013), 10) Zhang et al. (2013), 11) Houghton and Nassikas (2018), 12) Busch and Engelmann (2017), 13) Baccini et al. (2017), 14) Zarin et al. (2016), 15) Houghton, et al. (2015), 16) Federici et al. (2015), 17) Carter et al. (2015), 18) Smith et al. (2013), 19) Pearson et al. (2017), 20) Hooijer et al. (2010), 21) Howard (2017), 22) Pendleton et al. (2012), 23) Fuss et al. (2018), 24) Dooley and Kartha (2018), 25) Kreidenweis et al. (2016), 26) Yan et al. (2017), 27) Sonntag et al. (2016), 28) Lenton (2014), 29) McLaren (2012), 30) Lenton (2010), 31) Sasaki et al. (2016), 32) Sasaki et al. (2012), 33) Zomer et al. (2016), 34) Couwenberg et al. (2010), 35) Conant et al. (2017), 36) Sanderman et al. (2017), 37) Frank et al. (2017), 38) Henderson et al. (2015), 39) Sommer and Bossio (2014), 40. Lal (2010), 41. Zomer et al. (2017), 42. Smith et al. (2016), 43) Poeplau and Don (2015), 44) Powlson et al. (2014), 45. Powell and Lenton (2012), 46) Woolf et al. (2010), 47) Roberts et al. (2010), 48. Pratt and Moran (2010), 49. Turner et al. (2018), 50) Koornneef et al. (2012), 51) Bajželj et al. (2014), 52) Springmann et al. (2016), 53) Tilman and Clark (2014), 54) Hedenus et al. (2014), 55) Miner (2010), 56) Bailis et al. (2015).

rates of carbon loss are saved each year). Separate estimates of degradation only range from 1.0–2.18 $GtCO_2$ yr^{-1}. Reduced deforestation and forest degradation include conservation of existing carbon pools in vegetation and soil through protection in reserves, controlling disturbances such as fire and pest outbreaks, and changing management practices. Differences in estimates stem from varying land cover definitions, the time periods assessed and the carbon pools included (most higher estimates include belowground, dead wood, litter, soil and peat carbon). When deforestation and degradation are halted, it may take many decades to fully recover the biomass initially present in native ecosystems (Meli et al. 2017) (Section 4.8.3).

Afforestation/reforestation (A/R) and forest restoration can increase carbon sequestration in both vegetation and soils by 0.5–10.1 $GtCO_2$ yr^{-1} (*medium confidence*) (Fuss et al. 2018; Griscom et al. 2017; Hawken 2017; Kreidenweis et al. 2016; Li et al. 2016; Huang et al. 2017; Sonntag et al. 2016; Lenton 2014; McLaren 2012; Lenton 2010; Erb et al. 2018; Dooley and Kartha 2018; Yan et al. 2017; Houghton et al. 2015; Houghton and Nassikas 2018). Afforestation is the conversion to forest of land that historically has not contained forests. Reforestation is the conversion to forest of land that has previously contained forests but that has been converted to some other use. Forest restoration refers to practices aimed at regaining ecological integrity in a deforested or degraded forest landscape. The lower estimate represents the lowest range from an ESM (Yan et al. 2017) and of sustainable global negative emissions potential (Fuss et al. 2018), and the higher estimate reforests all areas where forests are the native cover type, constrained by food security and biodiversity considerations (Griscom et al. 2017). It takes time for full carbon removal to be achieved as the forest grows. Removal occurs at faster rates in young- to medium-aged forests and declines thereafter such that older forest stands have smaller carbon removals but larger stocks, with net uptake of carbon slowing as forests reach maturity (Yao et al. 2018; Poorter et al. 2016; Tang et al. 2014). The land intensity of afforestation and reforestation has been estimated at 0.0029 km^2 tC^{-1} yr^{-1} (Smith et al. 2016a). Boysen et al. (2017) estimated that to sequester about 100 GtC by 2100 would require 13 Mkm^2 of abandoned cropland and pastures (Section 4.8.3).

Forest management has the potential to mitigate 0.4–2.1 $GtCO_2$-eq yr^{-1} (*medium confidence*) (Sasaki et al. 2016; Griscom et al. 2017; Sasaki et al. 2012). Forest management can alter productivity, turnover rates, harvest rates carbon in soil and carbon in wood products (Erb et al. 2017; Campioli et al. 2015; Birdsey and Pan 2015; Erb et al. 2016; Noormets et al. 2015; Wäldchen et al. 2013; Malhi et al. 2015; Quesada et al. 2018; Nabuurs et al. 2017; Bosello et al. 2009) (Section 4.8.4). Fertilisation may enhance productivity but would increase N_2O emissions. Preserving and enhancing carbon stocks in forests has immediate climate benefits but the sink can saturate and is vulnerable to future climate change (Seidl et al. 2017). Wood can be harvested and used for bioenergy substituting for fossil fuels (with or without carbon capture and storage) (Section 2.6.1.5), for long-lived products such as timber (see below), to be buried as biochar (Section 2.6.1.1) or for use in the wider bioeconomy, enabling areas of land to be used continuously for mitigation. This leads to initial carbon loss and lower carbon stocks but with each harvest cycle, the

carbon loss (debt) can be paid back and after a parity time, result in net savings (Laganière et al. 2017; Bernier and Paré 2013; Mitchell et al. 2012; Haberl et al. 2012; Haberl 2013; Ter-Mikaelian et al. 2015; Macintosh et al. 2015). The trade-off between maximising forest carbon stocks and maximising substitution is highly dependent on the counterfactual assumption (no-use vs extrapolation of current management), initial forest conditions and site-specific contexts (such as regrowth rates and the displacement factors and efficiency of substitution), and relative differences in emissions released during extraction, transport and processing of the biomass- or fossil-based resources, as well as assumptions about emission associated with the product or energy source that is substituted (Grassi et al. 2018b; Nabuurs et al. 2017; Pingoud et al. 2018; Smyth et al. 2017a; Luyssaert et al. 2018; Valade et al. 2017; York 2012; Ter-Mikaelian et al. 2014; Naudts et al. 2016b; Mitchell et al. 2012; Haberl et al. 2012; Macintosh et al. 2015; Laganière et al. 2017; Haberl 2013). This leads to uncertainty about optimum mitigation strategies in managed forests, while high carbon ecosystems such as primary forests would have large initial carbon losses and long pay-back times, and thus protection of stocks would be more optimal (Lemprière et al. 2013; Kurz et al. 2016; Keith et al. 2014) (Section 4.8.4).

Global mitigation potential from increasing the demand of wood products to replace construction materials range from 0.25–1 $GtCO_2$-eq yr^{-1} (*medium confidence*) (McLaren 2012; Miner 2010), the uncertainty is determined in part by consideration of the factors described above, and is sensitive to the displacement factor, or the substitution benefit in CO_2, when wood is used instead of another material, which may vary in the future as other sectors reduce emissions (and may also vary due to market factors) (Sathre and O'Connor 2010; Nabuurs et al. 2018; Iordan et al. 2018; Braun et al. 2016; Gustavsson et al. 2017; Peñaloza et al. 2018; Soimakallio et al. 2016; Grassi et al. 2018b). Using harvested carbon in long-lived products (e.g., for construction) can represent a store that can sometimes be from decades to over a century, while the wood can also substitute for intensive building materials, avoiding emissions from the production of concrete and steel (Sathre and O'Connor 2010; Smyth et al. 2017b; Nabuurs et al. 2007; Lemprière et al. 2013). The harvest of carbon and storage in products affects the net carbon balance of the forest sector, with the aim of sustainable forest management strategies being to optimise carbon stocks and use harvested products to generate sustained mitigation benefits (Nabuurs et al. 2007).

Biophysical effects of forest response options are variable depending on the location and scale of activity (Section 2.6). Reduced deforestation or afforestation in the tropics contributes to climate mitigation through both biogeochemical and biophysical effects. It also maintains rainfall recycling to some extent. In contrast, in higher latitude boreal areas, observational and modelling studies show that afforestation and reforestation lead to local and global warming effects, particularly in snow covered regions in the winter as the albedo is lower for forests than bare snow (Bathiany et al. 2010; Dass et al. 2013; Devaraju et al. 2018; Ganopolski et al. 2001; Snyder et al. 2004; West et al. 2011; Arora and Montenegro 2011) (Section 2.6). Management, for example, thinning practices in forestry, could increase the albedo in regions where albedo decreases with age. The length of rotation cycles in forestry affects tree height and thus roughness, and through the removal of leaf mass harvest reduces evapotranspiration (Erb et al. 2017), which could lead to increased fire susceptibility in the tropics. In temperate and boreal sites, biophysical forest management effects on surface temperature were shown to be of similar magnitude than changes in land cover (Luyssaert et al. 2014). These biophysical effects could be of a magnitude to overcompensate biogeochemical effects, for example, the sink strength of regrowing forests after past depletions (Luyssaert et al. 2018; Naudts et al. 2016b), but many parameters and assumptions on counterfactual influence the account (Anderson et al. 2011; Li et al. 2015b; Bright et al. 2015).

Forest cover also affects climate through reactive gases and aerosols, with *limited evidence* and *medium agreement* that the decrease in the emissions of BVOC resulting from the historical conversion of forests to cropland has resulted in a positive radiative forcing through direct and indirect aerosol effects. A negative radiative forcing through reduction in the atmospheric lifetime of CH_4 has increased and decreased ozone concentrations in different regions (Section 2.4).

2.6.1.3   Land management of soils

The global mitigation potential for increasing soil organic matter stocks in mineral soils is estimated to be in the range of 0.4–8.64 $GtCO_2$ yr^{-1} (*high confidence*), though the full literature range is wider with high uncertainty related to some practices (Fuss et al. 2018; Sommer and Bossio 2014; Lal 2010; Lal et al. 2004; Conant et al. 2017; Dickie et al. 2014; Frank et al. 2017a; Griscom et al. 2017; Herrero et al. 2015, 2016; McLaren 2012; Paustian et al. 2016; Poeplau and Don 2015; Powlson et al. 2014; Smith et al. 2016c; Zomer et al. 2017). Some studies have separate potentials for soil carbon sequestration in croplands (0.25–6.78 $GtCO_2$ yr^{-1}) (Griscom et al. 2017; Hawken 2017; Frank et al. 2017a; Paustian et al. 2016; Herrero et al. 2016; Henderson et al. 2015; Dickie et al. 2014; Conant et al. 2017; Lal 2010) and soil carbon sequestration in grazing lands (0.13–2.56 $GtCO_2$ yr^{-1}) (Griscom et al. 2017; Hawken 2017; Frank et al. 2017a; Paustian et al. 2016; Powlson et al. 2014; McLaren 2012; Zomer et al. 2017; Smith et al. 2015; Sommer and Bossio 2014; Lal 2010). The potential for soil carbon sequestration and storage varies considerably depending on prior and current land management approaches, soil type, resource availability, environmental conditions, microbial composition and nutrient availability among other factors (Hassink and Whitmore 1997; Smith and Dukes 2013; Palm et al. 2014; Lal 2013; Six et al. 2002; Feng et al. 2013). Soils are a finite carbon sink and sequestration rates may decline to negligible levels over as little as a couple of decades as soils reach carbon saturation (West et al. 2004; Smith and Dukes 2013). The sink is at risk of reversibility, in particular due to increased soil respiration under higher temperatures (Section 2.3).

Land management practices to increase carbon interact with agricultural and fire management practices (Cross-chapter Box 3 and Chapter 5) and include improved rotations with deeper rooting cultivars, addition of organic materials and agroforestry (Lal 2011; Smith et al. 2008; Lorenz and Pitman 2014; Lal 2013; Vermeulen et al.

2012; de Rouw et al. 2010). Adoption of green manure cover crops, while increasing cropping frequency or diversity, helps sequester SOC (Poeplau and Don 2015; Mazzoncini et al. 2011; Luo et al. 2010). Studies of the long-term SOC sequestration potential of conservation agriculture (i.e., the simultaneous adoption of minimum tillage, (cover) crop residue retention and associated soil surface coverage, and crop rotations) include results that are both positive (Powlson et al. 2016; Zhang et al. 2014) and inconclusive (Cheesman et al. 2016; Palm et al. 2014; Govaerts et al. 2009).

The efficacy of reduced and zero-till practices is highly context-specific; many studies demonstrate increased carbon storage (e.g., Paustian et al. (2000), Six et al. (2004), van Kessel et al. (2013)), while others show the opposite effect (Sisti et al. 2004; Álvaro-Fuentes et al. 2008; Christopher et al. 2009). On the other hand, deep ploughing can contribute to SOC sequestration by burying soil organic matter in the subsoil where it decomposes slowly (Alcántara et al. 2016). Meta-analyses (Haddaway et al. 2017; Luo et al. 2010; Meurer et al. 2018) also show a mix of positive and negative responses, and the lack of robust comparisons of soils on an equivalent mass basis continues to be a problem for credible estimates (Wendt and Hauser 2013; Powlson et al. 2011; Powlson et al. 2014).

Soil carbon management interacts with N_2O (Paustian et al. 2016). For example, Li et al. (2005) estimate that the management strategies required to increase carbon sequestration (reduced tillage, crop residue and manure recycling) would increase N_2O emissions significantly, offsetting 75–310% of the carbon sequestered in terms of CO_2 equivalence, while other practices such as cover crops can reduce N_2O emissions (Kaye and Quemada 2017).

The management of soil erosion could avoid a net emissions of 1.36–3.67 $GtCO_2$ yr^{-1} and create a sink of 0.44–3.67 $GtCO_2$ yr^{-1} (*low confidence*) (Jacinthe and Lal 2001; Lal et al. 2004; Stallard 1998; Smith et al. 2001; Van Oost et al. 2007). The overall impact of erosion control on mitigation is context-specific and uncertain at the global level and the final fate of eroded material is still debated (Hoffmann et al., 2013).

Biochar is produced by thermal decomposition of biomass in the absence of oxygen (pyrolysis) into a stable, long-lived product like charcoal that is relatively resistant to decomposition (Lehmann et al. 2015) and which can stabilise organic matter when added to soil (Weng et al. 2017). Although charcoal has been used traditionally by many cultures as a soil amendment, 'modern biochar', produced in facilities that control emissions, is not widely used. The range of global potential of biochar is 0.03–6.6 $GtCO_2$-eq yr^{-1} by 2050, including energy substitution, with 0.03–4.9 $GtCO_2$ yr^{-1} for CDR only (*medium confidence*) (Griscom et al. 2017; Hawken 2017; Paustian et al. 2016; Fuss et al. 2018; Lenton 2014, 2010; Powell and Lenton 2012; Woolf et al. 2010; Pratt and Moran 2010; Smith 2016; Roberts et al. 2010). An analysis in which biomass supply constraints were applied to protect against food insecurity, loss of habitat and land degradation, estimated *technical potential* abatement of 3.7–6.6 $GtCO_2$-eq yr^{-1} (including 2.6–4.6 $GtCO_2$ yr^{-1} carbon stabilisation) (Woolf et al. 2010). Fuss et al. (2018) propose a range of 0.5–2 $GtCO_2$-eq yr^{-1} as the *sustainable potential* for negative emissions through biochar. Griscom et al. (2017) suggest a potential of 1.0 $GtCO_2$ yr^{-1} based on available residues. Biochar can provide additional climate change mitigation benefits by decreasing N_2O emissions from soil and reducing nitrogen fertiliser requirements in agricultural soils (Borchard et al. 2019). Application of biochar to cultivated soils can darken the surface and reduce its mitigation potential via decreases in surface albedo, but the magnitude of this effect depends on soil moisture content, biochar application method and type of land use (*low confidence*) (Verheijen et al. 2013; Bozzi et al. 2015) (Section 4.9.5).

2.6.1.4   Land management in other ecosystems

Protection and restoration of wetlands, peatlands and coastal habitats reduces net carbon loss (primarily from sediment/soils) and provides continued or enhanced natural CO_2 removal (Section 4.9.4). Reducing annual emissions from peatland conversion, draining and burning could mitigate 0.45–1.22 $GtCO_2$-eq yr^{-1} up to 2050 (*medium confidence*) (Hooijer et al. 2010; Griscom et al. 2017; Hawken 2017) and peatland restoration 0.15–0.81 (*low confidence*) (Couwenberg et al. 2010; Griscom et al. 2017). The upper end from Griscom et al. (2017) represents a maximum sustainable potential (accounting for biodiversity and food security safeguards) for rewetting and biomass enhancement. Wetland drainage and rewetting was included as a flux category under the second commitment period of the Kyoto Protocol, with significant management knowledge gained over the last decade (IPCC 2013b). However, there are high uncertainties as to carbon storage and flux rates, in particular the balance between CH_4 sources and CO_2 sinks (Spencer et al. 2016). Peatlands are sensitive to climate change which may increase carbon uptake by vegetation and carbon emissions due to respiration, with the balance being regionally dependent (*high confidence*). There is *low confidence* about the future peatland sink globally. Some peatlands have been found to be resilient to climate change (Minayeva and Sirin 2012), but the combination of land use change and climate change may make them vulnerable to fire (Sirin et al. 2011). While models show mixed results for the future sink (Spahni et al. 2013; Chaudhary et al. 2017; Ise et al. 2008), a study that used extensive historical data sets to project change under future warming scenarios found that the current global peatland sink could increase slightly until 2100 and decline thereafter (Gallego-Sala et al. 2018).

Reducing the conversion of coastal wetlands (mangroves, seagrass and marshes) could reduce emissions by 0.11–2.25 $GtCO_2$-eq yr^{-1} by 2050 (*medium confidence*) (Pendleton et al. 2012; Griscom et al. 2017; Howard et al. 2017; Hawken 2017). Mangrove restoration can mitigate the release of 0.07 $GtCO_2$ yr^{-1} through rewetting (Crooks et al. 2011) and take up 0.02–0.84 $GtCO_2$ yr^{-1} from biomass and soil enhancement (*medium confidence*) (Griscom et al. 2017). The ongoing benefits provided by mangroves as a natural carbon sink can be nationally-important for small island developing states (SIDS) and other countries with extensive coastlines, based on estimates of high carbon sequestration rates per unit area (McLeod et al. 2011; Duarte et al. 2013; Duarte 2017; Taillardat et al. 2018). There is only *medium confidence* in the effectiveness of enhanced carbon uptake using mangroves, due to the many uncertainties regarding the response of mangroves to future climate change (Jennerjahn

et al. 2017), dynamic changes in distributions (Kelleway et al. 2017) and other local-scale factors affecting long-term sequestration and climatic benefits (e.g., methane release) (Dutta et al. 2017). The climate mitigation potential of coastal vegetated habitats (mangrove forests, tidal marshes and seagrasses) is considered in Chapter 5 of the IPCC Special Report on the Ocean, Cryosphere and Climate Change (SROCC), in a wider 'blue carbon' context.

2.6.1.5   Bioenergy and bioenergy with carbon capture and storage

An introduction and overview of bioenergy and bioenergy with carbon capture and storage (BECCS) can be found in Cross-Chapter Boxes 7 and 12, and Chapters 6 and 7. CCS technologies are discussed in SR15. The discussion below refers to modern bioenergy only (e.g., liquid biofuels for transport and the use of solid biofuels in combined heat and power plants).

The mitigation potential of bioenergy coupled with CCS (i.e., BECCS), is estimated to be between 0.4 and 11.3 $GtCO_2$ yr^{-1} (*medium confidence*) based on studies that directly estimate mitigation for BECCS (not bioenergy) in units of CO_2 (not EJ) (McLaren 2012; Lenton 2014; Fuss et al. 2018; Turner et al. 2018b; Lenton 2010; Koornneef et al. 2012; Powell and Lenton 2012). SR15 reported a potential of 1–85 $GtCO_2$ yr^{-1} which they noted could be narrowed to a range of 0.5–5 $GtCO_2$ yr^{-1} when taking account of sustainability aims (Fuss et al. 2018). The upper end of the SR15 range is considered as a theoretical potential. Previously, the IPCC Special Report on Renewable Energy Sources concluded the technical potential of biomass supply for energy (without BECCS) could reach 100–300 EJ yr^{-1} by 2050, which would be 2–15 $GtCO_2$ yr^{-1} (using conversion factors 1 EJ = 0.02–0.05 $GtCO_2$ yr^{-1} emission reduction, SR15). A range of recent studies including sustainability or economic constraints estimate that 50–244 EJ (1–12 $GtCO_2$ yr^{-1} using the conversion factors above) of bioenergy could be produced on 0.1–13 Mkm^2 of land (Fuss et al. 2018; Chan and Wu 2015; Schueler et al. 2016; Wu et al. 2013; Searle and Malins 2015; Wu et al. 2019; Heck et al. 2018; Fritz et al. 2013).

There is *high confidence* that the most important factors determining future biomass supply for energy are land availability and land productivity (Berndes et al. 2013; Creutzig et al. 2015a; Woods et al. 2015; Daioglou et al. 2019). Estimates of marginal/degraded lands currently considered available for bioenergy range from 3.2–14.0 Mkm^2, depending on the adopted sustainability criteria, land class definitions, soil conditions, land mapping method and environmental and economic considerations (Campbell et al. 2008; Cai et al. 2011; Lewis and Kelly 2014).

Bioenergy production systems can lead to net emissions in the short term that can be 'paid-back' over time, with multiple harvest cycles and fossil fuel substitution, unlike fossil carbon emissions (Campbell et al. 2008; Cai et al. 2011; Lewis and Kelly 2014; De Oliveira Bordonal et al. 2015). Stabilising bioenergy crops in previous high carbon forestland or peatland results in high emissions of carbon that may take from decades to more than a century to be re-paid in terms of net CO_2 emission savings from replacing fossil fuels, depending on previous forest carbon stock, bioenergy yields and displacement efficiency (Elshout et al. 2015; Harper et al. 2018; Daioglou et al. 2017). In the case of bioenergy from managed forests, the magnitude and timing of the net mitigation benefits is controversial as it varies with differences due to local climate conditions, forest management practice, fossil fuel displacement efficiency and methodological approaches (Hudiburg et al. 2011; Berndes et al. 2013; Guest et al. 2013; Lamers and Junginger 2013; Cherubini et al. 2016; Cintas et al. 2017; Laurance et al. 2018; Valade et al. 2018; Baker et al. 2019). Suitable bioenergy crops can be integrated in agricultural landscapes to reverse ecosystem carbon depletion (Creutzig et al. 2015a; Robertson et al. 2017; Vaughan et al. 2018; Daioglou et al. 2017). Cultivation of short rotation woody crops and perennial grasses on degraded land or cropland previously used for annual crops typically accumulate carbon in soils due to their deep root systems (Don et al. 2012; Robertson et al. 2017). The use of residues and organic waste as bioenergy feedstock can mitigate land use change pressures associated with bioenergy deployment, but residues are limited and the removal of residues that would otherwise be left on the soil could lead soil degradation (Chum et al. 2011; Liska et al. 2014; Monforti et al. 2015; Zhao et al. 2015; Daioglou et al. 2016).

The steps required to cultivate, harvest, transport, process and use biomass for energy generate emissions of GHGs and other climate pollutants (Chum et al. 2011; Creutzig et al. 2015b; Staples et al. 2017; Daioglou et al. 2019). Life-cycle GHG emissions of modern bioenergy alternatives are usually lower than those for fossil fuels (*robust evidence, medium agreement*) (Chum et al. 2011; Creutzig et al. 2015b). The magnitude of these emissions largely depends on location (e.g., soil quality, climate), prior land use, feedstock used (e.g., residues, dedicated crops, algae), land use practice (e.g., soil management, fertiliser use), biomass transport (e.g., distances and transport modes) and the bioenergy conversion pathway and product (e.g., wood pellets, ethanol). Use of conventional food and feed crops as a feedstock generally provides the highest bioenergy yields per hectare, but also causes more GHG emissions per unit energy compared to agriculture residues, biomass from managed forests and lignocellulosic crops such as short-rotation coppice and perennial grasses (Chum et al. 2011; Gerbrandt et al. 2016) due to the application of fertilisers and other inputs (Oates et al. 2016; Rowe et al. 2016; Lai et al. 2017; Robertson et al. 2017).

Bioenergy from dedicated crops are in some cases held responsible for GHG emissions resulting from indirect land use change (iLUC), that is the bioenergy activity may lead to displacement of agricultural or forest activities into other locations, driven by market-mediated effects. Other mitigation options may also cause iLUC. At a global level of analysis, indirect effects are not relevant because all land-use emissions are direct. iLUC emissions are potentially more significant for crop-based feedstocks such as corn, wheat and soybean, than for advanced biofuels from lignocellulosic materials (Chum et al. 2011; Wicke et al. 2012; Valin et al. 2015; Ahlgren and Di Lucia 2014). Estimates of emissions from iLUC are inherently uncertain, widely debated in the scientific community and are highly dependent on modelling assumptions, such as supply/demand elasticities, productivity estimates, incorporation or exclusion of emission credits for coproducts and scale of biofuel

deployment (Rajagopal and Plevin 2013; Finkbeiner 2014; Kim et al. 2014; Zilberman 2017). In some cases, iLUC effects are estimated to result in emission reductions. For example, market-mediated effects of bioenergy in North America showed potential for increased carbon stocks by inducing conversion of pasture or marginal land to forestland (Cintas et al. 2017; Duden et al. 2017; Dale et al. 2017; Baker et al. 2019). There is a wide range of variability in iLUC values for different types of biofuels, from −75–55 gCO_2 MJ^{-1} (Ahlgren and Di Lucia 2014; Valin et al. 2015; Plevin et al. 2015; Taheripour and Tyner 2013; Bento and Klotz 2014). There is low confidence in attribution of emissions from iLUC to bioenergy.

Bioenergy deployment can have large biophysical effects on regional climate, with the direction and magnitude of the impact depending on the type of bioenergy crop, previous land use and seasonality (*limited evidence, medium agreement*). A study of two alternative future bioenergy scenarios using 15 Mkm^2 of intensively used managed land or conversion of natural areas showed a nearly neutral effect on surface temperature at global levels (considering biophysical effects and CO_2 and N_2O fluxes from land but not substitution effects), although there were significant seasonal and regional differences (Kicklighter et al. 2013). Modelling studies on biofuels in the US found the switch from annual crops to perennial bioenergy plantations like Miscanthus could lead to regional cooling due to increases in evapotranspiration and albedo (Georgescu et al. 2011; Harding et al. 2016), with perennial bioenergy crop expansion over suitable abandoned and degraded farmlands causing near-surface cooling up to 5°C during the growing season (Wang et al. 2017b). Similarly, growing sugarcane on existing cropland in Brazil cools down the local surface during daytime conditions up to −1°C, but warmer conditions occur if sugar cane is deployed at the expense of natural vegetation (Brazilian Cerrado) (Loarie et al. 2011). In general, bioenergy crops (as for all crops) induce a cooling of ambient air during the growing season, but after harvest the decrease in evapotranspiration can induce warming (Harding et al. 2016; Georgescu et al. 2013; Wang et al. 2017b). Bioenergy crops were found to cause increased isoprene emissions in a scenario where 0.69 Mkm^2 of oil palm for biodiesel in the tropics and 0.92 Mkm^2 of short rotation coppice (SRC) in the mid-latitudes were planted, but effects on global climate were negligible (Ashworth et al. 2012).

2.6.1.6   Enhanced weathering

Weathering is the natural process of rock decomposition via chemical and physical processes in which CO_2 is removed from the atmosphere and converted to bicarbonates and/or carbonates (IPCC 2005). Formation of calcium carbonates in the soil provides a permanent sink for mineralised organic carbon (Manning 2008; Beerling et al. 2018). Mineral weathering can be enhanced through grinding up rock material to increase the surface area, and distributing it over land to provide carbon removals of 0.5–4.0 GtCO_2 yr^{-1} (*medium confidence*) (Beerling et al. 2018; Lenton 2010; Smith et al. 2016a; Taylor et al. 2016). While the geochemical potential is quite large, agreement on the technical potential is low due to a variety of unknown parameters and limits, such as rates of mineral extraction, grinding, delivery and challenges with scaling and deployment.

2.6.1.7   Demand management in the food sector
(diet change, waste reduction)

Demand-side management has the potential for climate change mitigation via reducing emissions from production, switching to consumption of less emission intensive commodities and making land available for CO_2 removal (Section 5.5.2). Reducing food losses and waste increases the overall efficiency of food value chains (with less land and inputs needed) along the entire supply chain and has the potential to mitigate 0.8–4.5 $GtCO_2$-eq yr^{-1} (*high confidence*) (Bajželj et al. 2014; Dickie et al 2014; Hawken 2017; Hiç et al. 2016) (Section 5.5.2.5).

Shifting to diets that are lower in emissions-intensive foods like beef delivers a mitigation potential of 0.7–8.0 $GtCO_2$-eq yr^{-1} (*high confidence*) (Bajželj et al. 2014; Dickie et al. 2014; Herrero et al. 2016; Hawken 2017; Springmann et al. 2016; Tilman and Clark 2014; Hedenus et al. 2014; Stehfest et al. 2009) with most of the higher end estimates (>6 $GtCO_2$-eq yr^{-1}) based on veganism, vegetarianism or very low ruminant meat consumption (Section 5.5.2). In addition to direct mitigation gains, decreasing meat consumption, primarily of ruminants, and reducing wastes further reduces water use, soil degradation, pressure on forests and land used for feed potentially freeing up land for mitigation (Tilman and Clark 2014) (Chapters 5 and 6). Additionally, consumption of locally produced food, shortening the supply chain, can in some cases minimise food loss, contribute to food security and reduce GHG emissions associated with energy consumption and food loss (Section 5.5.2.6).

2.6.2   Integrated pathways for climate
change mitigation

Land-based response options have the potential to interact, resulting in additive effects (e.g., climate co-benefits) or negating each other (e.g., through competition for land). They also interact with mitigation options in other sectors (such as energy or transport) and thus they need to be assessed collectively under different climate mitigation targets and in combination with other sustainability goals (Popp et al. 2017; Obersteiner et al. 2016; Humpenöder et al. 2018). IAMs with distinctive land-use modules are the basis for the assessment of mitigation pathways as they combine insights from various disciplines in a single framework and cover the largest sources of anthropogenic GHG emissions from different sectors (see also SR15 Chapter 2 and Technical Annex for more details). IAMs consider a limited, but expanding, portfolio of land-based mitigation options. Furthermore, the inclusion and detail of a specific mitigation measure differs across IAMs and studies (see also SR15 and Chapter 6). For example, the IAM scenarios based on the shared socio-economic pathways (SSPs) (Riahi et al. 2017) (Cross-Chapter Box 1 and Chapter 1) include possible trends in agriculture and land use for five different socioeconomic futures, but cover a limited set of land-based mitigation options: dietary changes, higher efficiency in food processing (especially in livestock production systems), reduction of food waste, increasing agricultural productivity, methane reductions in rice paddies, livestock and grazing management for reduced methane emissions from enteric

fermentation, manure management, improvement of N-efficiency, 1st generation biofuels, reduced deforestation, afforestation, 2nd generation bioenergy crops and BECCS (Popp et al. 2017). However, many 'natural climate solutions' (Griscom et al. 2017), such as forest management, rangeland management, soil carbon management or wetland management, are not included in most of these scenarios. In addition, most IAMs neglect the biophysical effects of land-use such as changes in albedo or evapotranspiration with few exceptions (Kreidenweis et al. 2016).

Mitigation pathways, based on IAMs, are typically designed to find the least cost pathway to achieve a pre-defined climate target (Riahi et al. 2017). Such cost-optimal mitigation pathways, especially in RCP2.6 (broadly a 2°C target) and 1.9 scenarios (broadly a 1.5°C target), project GHG emissions to peak early in the 21st century, strict GHG emission reduction afterwards and, depending on the climate target, net CDR from the atmosphere in the second half of the century (Chapter 2 of SR15; Tavoni et al. 2015; Riahi et al. 2017). In most of these pathways, land use is of great importance because of its mitigation potential as discussed in Section 2.7.1: these pathways are based on the assumptions that (i) large-scale afforestation and reforestation removes substantial amounts of CO_2 from the atmosphere, (ii) biomass grown on cropland or from forestry residues can be used for energy generation or BECCS substituting fossil fuel emissions and generating CDR, and (iii) non-CO_2 emissions from agricultural production can be reduced, even under improved agricultural management (Popp et al. 2017; Rogelj et al. 2018a; Van Vuuren et al. 2018, Frank et al. 2018).

From the IAM scenarios available to this assessment, a set of feasible mitigation pathways has been identified which is illustrative of the range of possible consequences on land use and GHG emissions (presented in this chapter) and sustainable development (Chapter 6). Thus, the IAM scenarios selected here vary due to underlying socio-economic and policy assumptions, the mitigation options considered, long-term climate goals, the level of inclusion of other sustainability goals (such as land and water restrictions for biodiversity conservation or food production) and the models by which they are generated.

In the baseline case without climate change mitigation, global CO_2 emissions from land-use change decrease over time in most scenarios due to agricultural intensification and decreases in demand for agricultural commodities – some even turning negative by the end of the century due to abandonment of agricultural land and associated carbon uptake through vegetation regrowth. Median global CO_2 emissions from land-use change across 5 SSPs and 5 IAMs decrease throughout the 21st century: 3, 1.9 and –0.7 $GtCO_2$ yr^{-1} in 2030, 2050 and 2100 respectively (Figure 2.25). In contrast, CH_4 and N_2O emissions from agricultural production remain rather constant throughout the 21st century (CH_4: 214, 231.7 and 209.1 $MtCH_4$ yr^{-1} in 2030, 2050 and 2100 respectively; N_2O: 9.1, 10.1 and 10.3 MtN_2O yr^{-1} in 2030, 2050 and 2100 respectively).

In the mitigation cases (RCP4.5, RCP2.6 and RCP1.9), most of the scenarios indicate strong reductions in CO_2 emissions due to (i) reduced deforestation and (ii) carbon uptake due to afforestation. However, CO_2 emissions from land use can occur in some mitigation scenarios as a result of weak land-use change regulation (Fujimori et al. 2017; Calvin et al. 2017) or displacement effects into pasture land caused by high bioenergy production combined with forest protection only (Popp et al. 2014). The level of CO_2 removal globally (median value across SSPs and IAMs) increases with the stringency of the climate target (RCP4.5, RCP2.6 and RCP1.9) for both afforestation (–1.3, –1.7 and –2.4 $GtCO_2$ yr^{-1} in 2100) and BECCS (–6.5, –11 and –14.9 $GtCO_2$ yr^{-1} in 2100) (Cross-Chapter Box 7 and Chapter 6). In the mitigation cases (RCP4.5, RCP2.6 and RCP1.9), CH_4 and N_2O emissions are remarkably lower compared to the baseline case (CH_4: 133.2, 108.4 and 73.5 $MtCH_4$ yr^{-1} in 2100; N_2O: 7.4, 6.1 and 4.5 MtN_2O yr^{-1} in 2100; see previous paragraph for CH_4 and N_2O emissions in the baseline case). The reductions in the mitigation cases are mainly due to improved agricultural management such as improved nitrogen fertiliser management, improved water management in rice production, improved manure management (by, for example, covering of storages or adoption

of biogas plants), better herd management and better quality of livestock through breeding and improved feeding practices. In addition, dietary shifts away from emission-intensive livestock products also lead to decreased CH_4 and N_2O emissions especially in RCP2.6 and RCP1.9 scenarios. However, high levels of bioenergy production can result in increased N_2O emissions due to nitrogen fertilisation of dedicated bioenergy crops.

Such high levels of CO_2 removal through mitigation options that require land conversion (BECCS and afforestation) shape the land system dramatically (Figure 2.26). Across the different RCPs, SSPs and IAMs, median change of global forest area throughout the 21st century ranges from about –0.2 to +7.2 Mkm^2 between 2010 and 2100, and agricultural land used for 2nd generation bioenergy crop production ranges from about 3.2–6.6 Mkm^2 in 2100 (Popp et al. 2017; Rogelj et al. 2018). Land requirements for bioenergy and afforestation for a RCP1.9 scenario are higher than for a RCP2.6 scenario and

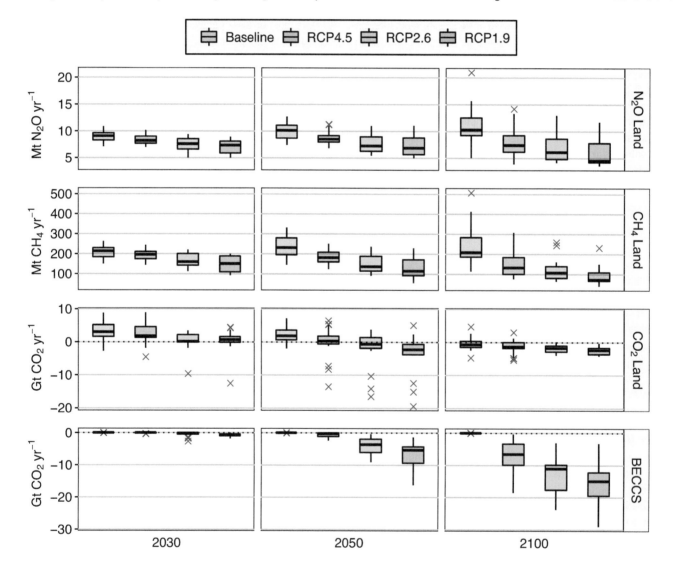

Figure 2.25 | Land-based global GHG emissions and removals in 2030, 2050 and 2100 for baseline, RCP4.5, RCP2.6 and RCP1.9 based on the SSP. Source: Popp et al. (2017), Rogelj et al. (2018), Riahi et al. (2017). Data is from an update of the IAMC Scenario Explorer developed for the SR15 (Huppmann et al. 2018; Rogelj et al. 2018). Boxplots (Tukey style) show median (horizontal line), interquartile range (IQR box) and the range of values within 1.5 × IQR at either end of the box (vertical lines) across 5 SSPs and across 5 IAMs. Outliers (red crosses) are values greater than 1.5 × IQR at either end of the box. The categories CO_2 Land, CH_4 Land and N_2O Land include GHG emissions from land-use change and agricultural land use (including emissions related to bioenergy production). In addition, the category CO_2 Land includes negative emissions due to afforestation. BECCS reflects the CO_2 emissions captured from bioenergy use and stored in geological deposits.

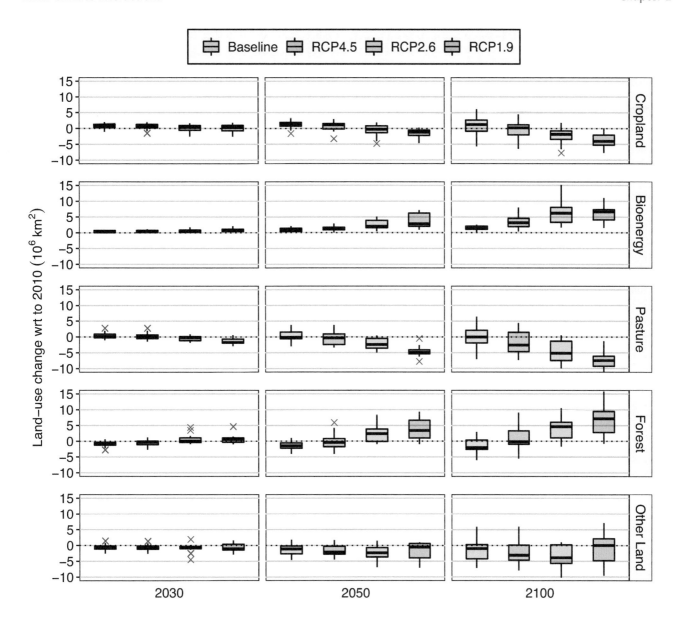

Figure 2.26 | Global change of major land cover types by 2030, 2050 and 2100 relative to 2010 for baseline, RCP4.5, RCP2.6 and RCP1.9 based on the SSP. Source: Popp et al. (2017), Rogelj et al. (2018), Riahi et al. (2017). Data is from an update of the IAMC Scenario Explorer developed for the SR15 (Huppmann et al. 2018; Rogelj et al. 2018). Boxplots (Tukey style) show median (horizontal line), interquartile range IQR (box) and the range of values within 1.5 × IQR at either end of the box (vertical lines) across 5 SSPs and across 5 IAMs. Outliers (red crosses) are values greater than 1.5 × IQR at either end of the box. In 2010, total land cover at global scale was estimated 15–16 Mkm² for cropland, 0–0.14 Mkm² for bioenergy, 30–35 Mkm² for pasture and 37–42 Mkm² for forest, across the IAMs that reported SSP pathways (Popp et al. 2017).

especially a RCP4.5 mitigation scenario. As a consequence of the expansion of mainly land-demanding mitigation options, global pasture land is reduced in most mitigation scenarios much more strongly than compared to baseline scenarios (median reduction of 0, 2.6, 5.1 and 7.5 Mkm² between 2010 and 2100 in baseline, RCP4.5, RCP2.6 and RCP1.9 respectively). In addition, cropland for food and feed production decreases with the stringency of the climate target (+1.2, +0.2, −1.8 and −4 Mkm² in 2100 compared to 2010 in baseline, RCP4.5, RCP2.6 and RCP1.9 respectively). These reductions in agricultural land for food and feed production are facilitated by agricultural intensification on agricultural land and in livestock production systems (Popp et al. 2017), but also by changes in consumption patterns (Fujimori et al. 2017; Frank et al. 2017b).

The pace of projected land-use change over the coming decades in ambitious mitigation scenarios goes well beyond historical changes in some instances (Turner et al. (2018b), see also SR15). This raises issues for societal acceptance, and distinct policy and governance for avoiding negative consequences for other sustainability goals will be required (Humpenöder et al. 2018; Obersteiner et al. 2016; Calvin et al. 2014) (Chapters 6 and 7).

Different mitigation strategies can achieve the net emissions reductions that would be required to follow a pathway that limits global warming to 2°C or 1.5°C, with very different consequences on the land system.

Figure 2.27 shows six alternative pathways (archetypes) for achieving ambitious climate targets (RCP2.6 and RCP1.9), highlighting land-based strategies and GHG emissions. All pathways are assessed by different models but are all based on the SSP2 (Riahi et al. 2017), with all based on an RCP 1.9 mitigation pathway expect for Pathway 1, which is RCP2.6. All scenarios show land-based negative emissions, but the amount varies across pathways, as do the relative contributions of different land-based CDR options, such as afforestation/reforestation and BECCS.

Pathway 1 RCP2.6 'Portfolio' (Fricko et al. 2017) shows a strong near-term decrease of CO_2 emissions from land-use change, mainly due to reduced deforestation, as well as slightly decreasing N_2O and CH_4 emissions after 2050 from agricultural production due to improved agricultural management and dietary shifts away from emissions-intensive livestock products. However, in contrast to CO_2 emissions, which turn net-negative around 2050 due to afforestation/reforestation, CH_4 and N_2O emissions persist throughout the century due to difficulties of eliminating these residual emissions based on existing agricultural management methods (Stevanović et al. 2017;

Frank et al. 2017b). In addition to abating land related GHG emissions as well as increasing the terrestrial sink, this example also shows the importance of the land sector in providing biomass for BECCS and hence CDR in the energy sector. In this scenario, annual BECCS-based CDR is about three times higher than afforestation-based CDR in 2100 (–11.4 and –3.8 $GtCO_2$ yr^{-1} respectively). Cumulative CDR throughout the century amounts to –395 $GtCO_2$ for BECCS and –73 $GtCO_2$ for afforestation. Based on these GHG dynamics, the land sector turns GHG emission neutral in 2100. However, accounting also for BECCS-based CDR taking place in the energy sector, but with biomass provided by the land sector, turns the land sector GHG emission neutral already in 2060, and significantly net-negative by the end of the century.

Pathway 2 RCP1.9 'Increased Ambition' (Rogelj et al. 2018) has dynamics of land-based GHG emissions and removals that are very similar to those in Pathway 1 (RCP2.6) but all GHG emission reductions as well as afforestation/reforestation and BECCS-based CDR start earlier in time at a higher rate of deployment. Cumulative CDR throughout the century amounts to –466 $GtCO_2$ for BECCS and –117 $GtCO_2$ for afforestation.

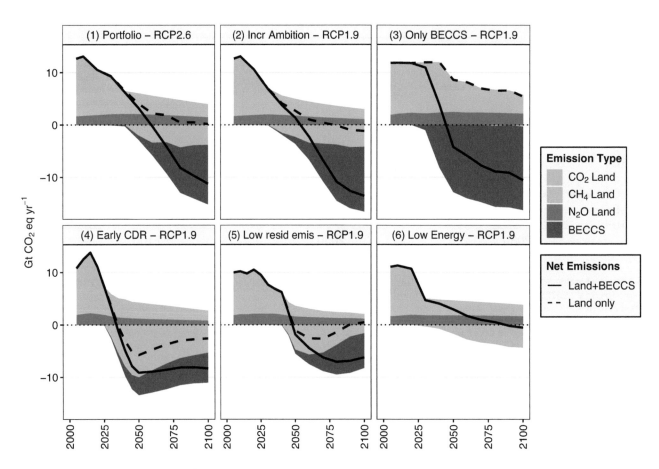

Figure 2.27 | Evolution and breakdown of global land-based GHG emissions and removals under six alternative mitigation pathways. This figure illustrates the differences in timing and magnitude of land-based mitigation approaches including afforestation and BECCS. All pathways are based on different IAM realisations of SSP2. Pathway 1 is based on RCP 2.6, while all other pathways are based on RCP 1.9. Pathway 1: MESSAGE-GLOBIOM (Fricko et al. 2017); Pathway 2: MESSAGE-GLOBIOM (Rogelj et al. 2018); Pathway 3: REMIND-MAgPIE (Kriegler et al. 2017); Pathway 4: REMIND-MAgPIE (Bertram et al. 2018); Pathway 5: IMAGE (van Vuuren et al. 2018); Pathway 6: MESSAGE-GLOBIOM (Grubler et al. 2018). Data is from an update of the IAMC Scenario Explorer developed for the SR15 (Rogelj et al. 2018). The categories CO_2 Land, CH_4 Land and N_2O Land include GHG emissions from land-use change and agricultural land use (including emissions related to bioenergy production). In addition, the category CO_2 Land includes negative emissions due to afforestation. BECCS reflects the CO_2 emissions captured from bioenergy use and stored in geological deposits. Solid lines show the net effect of all land based GHG emissions and removals (CO_2 Land, CH_4 Land, N_2O Land and BECCS), while dashed lines show the net effect excluding BECCS. CH_4 and N_2O emissions are converted to CO_2-eq using GWP factors of 28 and 265 respectively.

Pathway 3 RCP 1.9 'Only BECCS', in contrast to Pathway 2, includes only BECCS-based CDR (Kriegler et al. 2017). As a consequence, CO_2 emissions are persistent much longer, predominantly from indirect land-use change due to large-scale bioenergy cropland expansion into non-protected natural areas (Popp et al. 2017; Calvin et al. 2014). While annual BECCS CDR rates in 2100 are similar to Pathways 1 and 2 (–15.9 $GtCO_2$ yr^{-1}), cumulative BECCS-based CDR throughout the century is much larger (–944 $GtCO_2$).

Pathway 4 RCP1.9 'Early CDR' (Bertram et al. 2018) indicates that a significant reduction in the later century in the BECCS-related CDR as well as CDR in general can be achieved with earlier and mainly terrestrial CDR, starting in 2030. In this scenario, terrestrial CDR is based on afforestation but could also be supported by soil organic carbon sequestration (Paustian et al. 2016) or other natural climate solutions, such as rangeland or forest management (Griscom et al. 2017). This scenario highlights the importance of the timing for CDR-based mitigation pathways (Obersteiner et al. 2016). As a result of near-term and mainly terrestrial CDR deployment, cumulative BECCS-based CDR throughout the century is limited to –300 $GtCO_2$, while cumulative afforestation-based CDR amounts to –428 $GtCO_2$.

In Pathway 5 RCP1.9 'Low residual emissions' (van Vuuren et al. 2018), land-based mitigation is driven by stringent enforcement of measures and technologies to reduce end-of-pipe non-CO_2 emissions and by introduction of in-vitro (cultured) meat, reducing residual N_2O and CH_4 emissions from agricultural production. In consequence, much lower amounts of CDR from afforestation and BECCS are needed with much later entry points to compensate for residual emissions. Cumulative CDR throughout the century amounts to –252 $GtCO_2$ for BECCS and –128 $GtCO_2$ for afforestation. Therefore, total cumulative land-based CDR in Pathway 5 is substantially lower compared to Pathways 2–4 (380 $GtCO_2$).

Finally, Pathway 6 RCP1.9 'Low Energy' (Grubler et al. 2018), equivalent to Pathway LED in SR15, indicates the importance of other sectoral GHG emission reductions for the land sector. In this example, rapid and early reductions in energy demand and associated drops in energy-related CO_2 emissions limit overshoot and decrease the requirements for negative emissions technologies, especially for land-demanding CDR, such as biomass production for BECCS and afforestation. While BECCS is not used at all in Pathway 6, cumulative CDR throughout the century for afforestation amounts to –124 $GtCO_2$.

Besides their consequences on mitigation pathways and land consequences, those archetypes can also affect multiple other sustainable development goals that provide both challenges and opportunities for climate action (Chapter 6).

2.6.3   The contribution of response options to the Paris Agreement

The previous sections indicated how land-based response options have the potential to contribute to the Paris Agreement, not only though reducing anthropogenic emissions but also for providing anthropogenic sinks that can contribute to "…a balance between anthropogenic emissions by sources and removals by sinks of greenhouse gases in the second half of this century…" (Paris Agreement, Article 4). The balance applies globally, and relates only to GHGs, not aerosols (Section 2.4) or biophysical effects (Section 2.5).

The Paris Agreement includes an enhanced transparency framework to track countries' progress towards achieving their individual targets (i.e., nationally determined contributions (NDCs)), and a global stocktake (every five years starting in 2023), to assess the countries' collective progress towards the long-term goals of the Paris Agreement. The importance of robust and transparent definitions and methods (including the approach to separating anthropogenic from natural fluxes) (Fuglestvedt et al. 2018), and the needs for reconciling country GHG inventories and models (Grassi et al. 2018a), was highlighted in Section 2.3 in relation to estimating emissions. Issues around estimating mitigation is also key to transparency and credibility and is part of the Paris Rulebook.

The land sector is expected to deliver up to 25% of GHG mitigation pledged by countries by 2025–2030 in their NDCs, based on early assessments of 'Intended' NDCs submitted ahead of the Paris Agreement and updates immediately after (*low confidence*) (Grassi et al. 2017; Forsell et al. 2016). While most NDCs submitted to date include commitments related to the land sector, they vary with how much information is given and the type of target, with more ambitious targets for developing countries often being 'conditional' on support and climate finance. Some do not specify the role of AFOLU but include it implicitly as part of economy-wide pledges (e.g., reducing total emission or emission intensity), a few mention multi-sectoral mitigation targets which include AFOLU in a fairly unspecified manner. Many NDCs include specific AFOLU response options, with most focused on the role of forests. A few included soil carbon sequestration or agricultural mitigation and a few explicitly mentioned bioenergy (e.g., Cambodia, Indonesia and Malaysia), but this could be implicitly included with reduced emissions in energy sectors through fuel substitution (see Cross-Chapter Box 7 and Chapter 6 for discussion on cross sector flux reporting). The countries indicating AFOLU mitigation most prominently were Brazil and Indonesia, followed by other countries focusing either on avoiding carbon emissions (e.g., Ethiopia, Gabon, Mexico, DRC, Guyana and Madagascar) or on promoting the sink through large afforestation programmes (e.g., China, India) (Grassi et al. 2017).

Figure 2.28 shows the CO_2 mitigation potential of NDCs compared to historical fluxes from LULUCF.[3] It shows future fluxes based on current policies in place and on country-stated Business As Usual (BAU) activities (these are different from current policies as many countries are already implementing polices that they do not include as part of their historical business-as-usual baseline) (Grassi et al. 2017). Under implementation of unconditional pledges, the net LULUCF flux in 2030 has been estimated to be a sink of –0.41 ± 0.68 $GtCO_2$ yr^{-1}, which increases to –1.14 ± 0.48 $GtCO_2$ yr^{-1} in 2030 with conditional activities. This compares to net LULUCF in 2010 calculated from

---

[3] CO_2 fluxes due to land use, land-use change and forestry, in essence, not including the part of AFOLU fluxes that are from agriculture.

the GHG Inventories of 0.01 ± 0.86 $GtCO_2$ yr^{-1} (Grassi et al. 2017). Forsell et al. (2016) similarly find a reduction in 2030 compared to 2010 of 0.5 $GtCO_2$ yr^{-1} (range: 0.2–0.8) by 2020 and 0.9 $GtCO_2$ yr^{-1} (range: 0.5–1.3) by 2030 for unconditional and conditional cases.

The approach of countries to calculating the LULUCF contribution towards the NDC varies, with implications for comparability and transparency. For example, by following the different approaches used to include LULUCF in country NDCs, Grassi et al. (2017) found a three-fold difference in estimated mitigation: 1.2–1.9 $GtCO_2$-eq yr^{-1} when 2030 expected emissions are compared to 2005 emissions, 0.7–1.4 $GtCO_2$-eq yr^{-1} when 2030 emissions are compared to reference scenarios based on current policies or 2.3–3.0 $GtCO_2$-eq yr^{-1} when compared to BAU, and 3.0–3.8 $GtCO_2$-eq yr^{-1} when based on using each countries' approach to calculation stated in the NDC

(i.e., when based on a mix of country approaches, using either past years or BAU projections as reference).

In exploring the effectiveness of the NDCs, SR15 concluded "[e]stimates of global average temperature increase are 2.9°–3.4°C above preindustrial levels with a greater than 66% probability by 2100" (Roberts et al. 2006; Rogelj et al. 2016), under a full implementation of unconditional NDCs and a continuation of climate action similar to that of the NDCs. In order to achieve either the 1.5°C or 2°C pathways, this shortfall would imply the need for submission (and achievement) of more ambitious NDCs, and plan for a more rapid transformation of their national energy, industry, transport and land use sectors (Peters and Geden 2017; Millar et al. 2017; Rogelj et al. 2016).

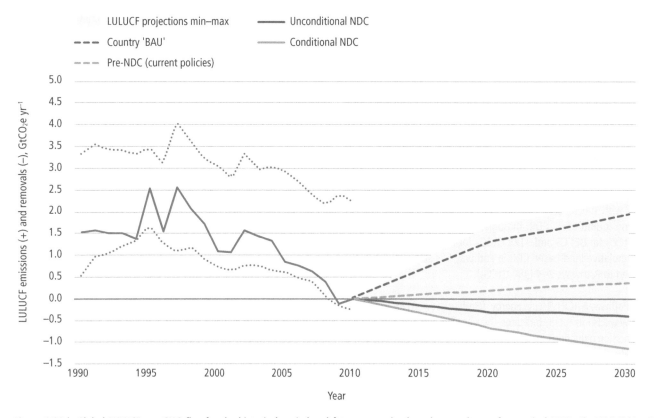

Figure 2.28 | Global LULUCF net GHG flux for the historical period and future scenarios based on analyses of countries' NDCs. The LULUCF historical data (blue solid line) reflect the following countries' documents (in order of priority): (i) data submitted to UNFCCC (NDCs[4], 2015 GHG Inventories[5] and recent National Communications[6,7]), (ii) other official countries' documents, (iii) FAO-based datasets (i.e., FAO-FRA for forest (Tian et al. 2015)) as elaborated by (Federici et al. 2015), and (iv) FAOSTAT for non-forest land use emissions (FAO 2015). The four future scenarios reflect official countries' information, mostly intended NDCs or updated NDCs available at the time of the analysis (Feb 2016), complemented by Biennial Update Reports[8] and National Communications, and show (i) the BAU scenario as defined by the country, (ii) the trend based on pre-NDC levels of activity (current policies in place in countries), and (iii) the unconditional NDC and conditional NDC scenarios. The shaded area indicates the full range of countries' available projections (min-max), expressing the available countries' information on uncertainties beyond the specific scenarios shown. The range of historical country datasets (dotted lines) reflects differences between alternative selections of country sources, in essence, GHG inventories for developed countries complemented by FAO-based datasets (upper range) or by data in National Communications (lower range) for developing countries.

---

4   UNFCCC. INDCs as communicated by Parties, www4.unfccc.int/submissions/indc/Submission%20Pages/submissions.aspx (UNFCCC, 2015).

5   UNFCCC. Greenhouse Gas Inventories, unfccc.int/national_reports/annex_i_ghg_inventories/national_inventories_submissions/items/8812.php (UNFCCC, 2015).

6   UNFCCC. National Communications Non-Annex 1, unfccc.int/nationalreports/non-annexinatcom/submittednatcom/items/653.php (UNFCCC, 2015).

7   UNFCCC. National Communications Annex 1, unfccc.int/nationalreports/annexinatcom/submittednatcom/items/7742.php (UNFCCC, 2015).

8   UNFCCC. Biennial Update Reports, unfccc.int/national_reports/non-annex_i_natcom/reporting_on_climate_change/items/8722.php (UNFCCC, 2015).

Response options relying on the use of land could provide around a third of the additional mitigation needed in the near term (2030) to close the gap between current policy trajectories based on NDCs and what is required to achieve a 2°C (>66% chance) or 1.5°C (50–66% chance) pathway according to the UNEP Emissions Gap Report (Roberts et al. 2006). The report estimates annual reduction potentials in 2030 from agriculture at 3.0 (2.3–3.7) $GtCO_2$-eq yr^{-1}, a combination of 'uncertain measures' (biochar, peat-related emission reductions and demand-side management) at 3.7 (2.6–4.8) $GtCO_2$-eq yr^{-1}; forests at 5.3 (4.1–6.5) $GtCO_2$-eq yr^{-1}, bioenergy at 0.9 $GtCO_2$-eq yr^{-1} and BECCS at 0.3 (0.2–0.4) $GtCO_2$-eq yr^{-1} (UNEP 2017) (Table 4.1). These response options account for 35% of potential reduction (or 32% without bioenergy and BECCS) out of a total (all sector) potential of 38 (35–41) $GtCO_2$-eq yr^{-1}. The potentials estimated in the UNEP Emissions Gap Report are based on the technical potential of individual response options from literature including that presented in Section 2.1. CDR related to land use, while not a substitute for strong action in the energy sector, has the technical potential to balance unavoidable emissions that are difficult to eliminate with current technologies (*high confidence*), with early action avoiding deeper and more rapid action later (*very high confidence*) (Strefler et al. 2018; Elmar et al. 2018; SR15).

2.7 Plant and soil processes underlying land–climate interactions

Projecting future complex interactions between land and climate require ESMs. A growing number of studies suggested that many processes important for interactions between land and climate were missing in the CMIP5-class ESMs and that the DGVMs used tended to elevate CO_2 emission and removals (*high confidence*) (Busch and Sage 2017; Rogers et al. 2017; Anderegg et al. 2016; Tjoelker 2018; Sulman et al. 2014; Wieder et al. 2018; Davidson et al. 2006a).

Ecosystem complexity stemming from the diversity of plants, animals and microbes, as well as their biological responses to gradual climate changes (e.g., adaptive migration) and disturbance events (e.g., extreme weather events, fire, pest outbreaks) (Section 2.2), are of potential importance. Of these processes, this section focuses on plant and soil processes as recent empirical work, including those explained in the following subsections, offers potential for improved model projections under warmer and CO_2-rich futures.

The magnitude of future uptake and release of CO_2 and other GHGs by vegetation are among the greatest uncertainties (Ciais et al. 2013b). One reason for this uncertainty stems from the lack of understanding of the mechanisms responsible for plant responses to increasing temperatures. The short- and long-term projections of gross photosynthesis responses to changes in temperature, CO_2 and nutrient availability vary greatly among the models (Busch and Sage 2017; Rogers et al. 2017). Net CO_2 exchange requires estimation of autotrophic respiration, which is another source of uncertainty in ESM projections (Malhi et al. 2011). The importance of plant acclimation of photosynthesis and respiration in understanding vegetation response to climate change is now widely recognised (*high confidence*) (Rogers et al., 2017; Tan et al., 2017; Tjoelker, 2018;

Vanderwel et al., 2015) (Section 2.7.1). Acclimation is broadly defined as the biochemical, physiological, morphological or developmental adjustments within the lifetime of organisms that result in improved performance under the new condition. Acclimation often operates over a time span of days to weeks, and can mitigate the negative effects of climate change on organismal growth and ecosystem functions (Tjoelker 2018).

Soil carbon and microbial processes, which interact with plant responses to climate, represent another large source of uncertainty in model projections (*medium confidence*) (Sections 2.7.2, 2.7.3 and 2.7.4). Given the wide range of uncertainty associated with SOC size estimates, CMIP5 models use a wide range of starting SOC stocks from 510–3040 GtC (Todd-Brown et al. 2013). Soil microbial respiration is estimated to release 40–70 GtC annually from the soil to the atmosphere globally (Hawkes et al. 2017). Projections of changes in global SOC stocks during the 21st century by CMIP5 models also ranged widely, from a loss of 37 Gt to a gain of 146 Gt, with differences largely explained by initial SOC stocks, differing carbon input rates and different decomposition rates and temperature sensitivities (Todd-Brown et al. 2013). With respect to land–climate interactions, the key processes affecting SOC stocks are warming (which is expected to accelerate SOC losses through microbial respiration) and acceleration of plant growth (which increases inputs of carbon to soils). However, complex mechanisms underlying SOC responses to moisture regimes, carbon addition, and warming drive considerable uncertainty in projections of future changes in SOC stocks (Sulman et al. 2014; Singh et al. 2010; Wieder et al. 2018).

2.7.1 Temperature responses of plant and ecosystem production

Climate-change responses of net ecosystem production cannot be modelled by simple instantaneous response functions because of thermal acclimation responses of plants and soil microbes, as well as delayed responses arising from interactions between plants and the soil (*high confidence*) (Slot et al. 2014; Rogers et al. 2017; Tan et al. 2017; Tjoelker 2018). Photosynthesis and respiration of component plant species exhibit different functional shapes among species (Slot et al. 2014), and carbon balance at the stand level is influenced by respiration of ecosystem biomass other than plants. Large uncertainty remains for thermal responses of bacteria and other soil organisms (Section 2.7.5). Bayesian statistical estimates of global photosynthesis and total ecosystem respirations suggest that they exhibit different responses to thermal anomalies during the last 35 years (Li et al. 2018b).

Thermal responses of plant respiration, which consumes approximately one half of GPP, have not been appropriately incorporated in most ESMs (Davidson et al., 2006; Tjoelker, 2018). Assumptions associated with respiration have been a major source of uncertainty for ESMs at the time of AR5. In most existing models, a simple assumption that respiration doubles with each 10°C increase of temperature (i.e., Q10 = 2) is adopted, ignoring acclimation. Even a small error stemming from this assumption can strongly influence estimated net carbon balance at large spatial scales of ecosystems and biomes over the time period of multiple decades (Smith and Dukes 2013; Smith et al. 2016b). In order to estimate more appropriate thermal response curves of respiration, a global database including data from 899 plant species has been compiled (Atkin et al. 2015), and respiration data from 231 plants species across seven biomes have been analysed (Heskel et al. 2016). These empirical data on thermal responses of respiration demonstrate a globally convergent pattern (Huntingford et al. 2017). According to a sensitivity analysis of a relatively small number of ESMs, a newly derived function of instantaneous responses of plant respiration to temperature (instead of a traditional exponential function of Q10 = 2) makes a significant difference in estimated autotrophic respiration especially in cold biomes (Heskel et al. 2016).

Acclimation results in reduced sensitivity of plant respiration with rising temperature, in essence, down regulation of warming-related increase in respiratory carbon emission (*high confidence*) (Atkin et al. 2015; Slot and Kitajima 2015; Tjoelker 2018). For example, experimental data from a tropical forest canopy show that temperature acclimation ameliorates the negative effects of rising temperature to leaf and plant carbon balance (Slot et al. 2014). Analysis of CO_2 flux data to quantify optimal temperature of net primary production of tropical forests also suggest acclimation potential for many tropical forests (Tan et al. 2017). Comparisons of models with and without thermal acclimation of respiration show that acclimation can halve the increase of plant respiration with projected temperature increase by the end of 21st century (Vanderwel et al. 2015).

It is typical that acclimation response to warming results in increases of the optimum temperature for photosynthesis and growth (Slot and Winter 2017; Yamori et al. 2014; Rogers et al. 2017). Although such shift is a result of a complex interactions of biochemical, respiratory and stomatal regulation (Lloyd and Farquhar 2008), it can be approximated by a simple algorithm to address acclimation (Kattge et al. 2007). Mercado et al. (2018), using this approach, found that inclusion of biogeographical variation in photosynthetic temperature response was critically important for estimating future land surface carbon uptake. In the tropics, CO_2 fertilisation effect (Box 2.3) is suggested to be more important for observed increases in carbon sink strength than increased leaf area index or a longer growing season (Zhu et al. 2016). Acclimation responses of photosynthesis and growth to simultaneous changes of temperature and CO_2, as well as stress responses above the optimal temperature for photosynthesis, remain a major knowledge gap in modelling responses of plant productivity under future climate change (Rogers et al. 2017).

2.7.2 Water transport through soil-plant-atmosphere continuum and drought mortality

How climate change, especially changes of precipitation patterns, influence water transport through the soil-plant-atmosphere continuum, is a key element in projecting the future of water vapour flux from land and cooling via latent heat flux (*high confidence*) (Sellers et al. 1996; Bonan 2008; Brodribb 2009; Choat et al. 2012; Sperry and Love 2015; Novick et al. 2016; Sulman et al. 2016). Even without changes in leaf area per unit area of land, when plants

close stomata in response to water shortage, dry atmosphere or soil moisture deficit, the stand-level fluxes of water (and associated latent heat flux) decrease (Seneviratne et al. 2018). Closing stomata enhances drought survival at the cost of reduced photosynthetic production, while not closing stomata avoids loss of photosynthetic production at the cost of increased drought mortality (Sperry and Love 2015). Hence, species-specific responses to drought, in terms of whether they close stomata or not, have short- and long-term consequences (Anderegg et al. 2018a; Buotte et al. 2019). Increased drought-induced mortality of forest trees, often exacerbated by insect outbreak and fire (e.g., Breshears et al. (2005), Kurz et al. (2008), Allen et al. (2010)) (Section 2.2.4), have long-term impact on hydrological interactions between land and atmosphere (Anderegg et al. 2018b).

New models linking plant water transport with canopy gas exchange and energy fluxes are expected to improve projections of climate change impacts on forests and land-atmosphere interactions (*medium confidence*) (Bohrer et al., 2005; Anderegg et al., 2016; Sperry and Love, 2015; Wolf et al., 2016). Yet, there is much uncertainty in the ability of current vegetation and land surface models to adequately capture tree mortality and the response of forests to climate extremes like drought (Rogers et al. 2017; Hartmann et al. 2018). Most vegetation models use climate stress envelopes or vegetation carbon balance estimations to project climate-driven mortality and loss of forests (McDowell et al. 2011); these may not adequately project biome shifts and impacts of disturbance in future climates. For example, a suite of vegetation models was compared to a field drought experiment in the Amazon on mature rainforest trees and all models performed poorly in projecting the timing and magnitude of biomass loss due to drought (Powell et al. 2013). More recently, the loss of water transport due to embolism (disruption of xylem water continuity) (Sperry and Love 2015), rather than carbon starvation (Rowland et al. 2015), is receiving attention as a key physiological process relevant for drought-induced tree mortality (Hartmann et al. 2018). A key challenge to modelling efforts is to consider differences among plant species and vegetation types in their drought responses. One approach is to classify plant species to 'functional types' that exhibit similar responses to environmental variations (Anderegg et al. 2016). Certain traits of species, such as tree height, is shown to be predictive of growth decline and mortality in response to drought (Xu et al. 2016a). Similarly, tree rooting depth is positively related to mortality, contrary to expectation, during prolonged droughts in tropical dry forest (Chitra-Tarak et al. 2017).

2.7.3 Soil microbial effects on soil nutrient dynamics and plant responses to elevated CO_2

Soil microbial processes influencing nutrient and carbon dynamics represent a large source of uncertainty in projecting land–climate interactions. For example, ESMs incorporating nitrogen and phosphorus limitations (but without considering the effects of mycorrhizae and rhizosphere priming) indicate that the simulated future carbon-uptake on land is reduced significantly when both nitrogen and phosphorus are limited as compared to only carbon-stimulation, by 63% (of 197 Pg C) under RCP2.6 and by 67% (of 425 Pg C) under RCP8.5 (Zhang et al. 2013c). Mineral nutrient limitation progressively reduces the CO_2 fertilisation effects on plant growth and productivity over time (*robust evidence, medium agreement*) (Norby et al. 2010; Sardans et al. 2012; Reich and Hobbie 2013; Feng et al. 2015; Terrer et al. 2017). The rates at which nutrient limitation develops differ among studies and sites. A recent meta-analysis shows that experimental CO_2 enrichment generally results in lower nitrogen and phosphorus concentrations in plant tissues (Du et al. 2019), and isotopic analysis also suggest a global trend of decreases in leaf nutrient concentration (Craine et al. 2018; Jonard et al. 2015). However, reduced responses to elevated CO_2 (eCO_2) may not be a simple function of nitrogen dilution per se, as they result from complex interactions of ecosystem factors that influence nitrogen acquisition by plants (Liang et al. 2016; Rutting 2017; Du et al. 2019).

Increasing numbers of case studies suggest that soil microbial processes, such as nitrogen mineralisation rates and symbiosis with plants, influence nutrient limitation on eCO_2 effects on plant growth (*medium confidence*) (Drake et al. 2011; Zak et al. 2011; Hungate et al. 2013; Talhelm et al. 2014; Du et al. 2019). Rhizosphere priming effects (i.e., release of organic matters by roots to stimulate microbial activities) and mycorrhizal associations are proposed to explain why some sites are becoming nitrogen limited after a few years and others are sustaining growth through accelerated nitrogen uptake (*limited evidence, medium agreement*) (Phillips et al. 2011; Terrer et al. 2017).

Model assessments that including rhizosphere priming effects and ectomycorrhizal symbiosis suggest that soil organic matter (SOM) cycling is accelerated through microbial symbiosis (*medium confidence*) (Elbert et al. 2012; Sulman et al. 2017; Orwin et al. 2011; Baskaran et al. 2017). Uncertainty exists in differences among ectomycorrhizal fungal species in their ability to decompose SOM (Pellitier and Zak 2018) and the capacity of ecosystems to sustain long-term growth with these positive symbiotic feedbacks is still under debate (Terrer et al. 2017). ESMs include only biological nitrogen cycles, even though a recent study suggests that bedrock weathering can be a significant source of nitrogen to plants (Houlton et al. 2018). In contrast, rock weathering is widely considered to be key for phosphorus availability, and tropical forests with highly weathered soils are considered to be limited by phosphorus availability rather than nitrogen availability (Reed et al. 2015). Yet evidence from phosphorus fertilisation experiments is lacking (Schulte-Uebbing and de Vries 2018) and phosphorus limitation of tropical tree growth may be strongly species-specific (Ellsworth et al. 2017; Turner et al. 2018a). Limitation by availability of soil nutrients other than nitrogen and phosphorus has not been studied in the context of land–climate interactions, except potassium as a potentially limiting factor for terrestrial plant productivity in interaction with nitrogen, phosphorus and hydrology (Sardans and Peñuelas 2015; Zhao et al. 2017; Wright et al. 2018).

Anthropogenic alteration of global and regional nitrogen and phosphorus cycles, largely through use of chemical fertilisers and pollution, has major implications for future ecosystem attributes, including carbon storage, in natural and managed ecosystems (*high confidence*) (Peñuelas et al. 2013, 2017; Wang et al. 2017c; Schulte-Uebbing and de Vries 2018, Yuan et al. 2018). During 1997–2013, the contribution of nitrogen deposition to the global carbon sink

has been estimated at 0.27 ± 0.13 GtC yr^{-1}, and the contribution of phosphorus deposition as 0.054 ± 0.10 GtC yr^{-1}; these constitute about 9% and 2% of the total land carbon sink, respectively (Wang et al. 2017c). Anthropogenic deposition of nitrogen enhances carbon sequestration by vegetation (Schulte-Uebbing and de Vries 2018), but this effect of nitrogen deposition on carbon sequestration may be offset by increased emission of GHGs such as N_2O and CH_4 (Liu and Greaver 2009). Furthermore, nitrogen deposition may lead to imbalance of nitrogen vs phosphorus availability (Peñuelas et al. 2013), soil microbial activity and SOM decomposition (Janssens et al. 2010) and reduced ecosystem stability (Chen et al. 2016b).

2.7.4 Vertical distribution of soil organic carbon

It has long been recognised that dynamics of soil organic carbon (SOC) represent a large source of uncertainties on biogeochemical interactions of land with atmosphere and climate as detailed below. Since AR5, there have been new understandings on SOC size, as well as on the microbial processes that influence SOM dynamics under climate change and LULCC. Three existing databases (SoilGrids, the Harmonized World Soil Data Base and Northern Circumpolar Soil Database) substantially differ in the estimated size of global SOC stock down to 1 m depth, varying between 2500 Pg to 3400 Pg with differences among databases largely attributable to carbon stored in permafrost (Joosten 2015; Köchy et al. 2015; Tifafi et al. 2018). These values are four to eight times larger than the carbon stock associated with the terrestrial vegetation (Bond-Lamberty et al. 2018). New estimates since AR5 show that much larger areas in the Amazon and Congo basins are peatlands (Gumbricht et al. 2017; Dargie et al. 2019).

Deep soil layers can contain much more carbon than previously assumed (*limited evidence, medium agreement*) (e.g., González-Jaramillo et al. (2016)). Based on radiocarbon measurements, deep SOC can be very old, with residence times up to several thousand years (Rumpel and Kögel-Knabner 2011) or even several tens of thousands of years (Okuno and Nakamura 2003). Dynamics associated with such deeply buried carbon remain poorly studied and ignored by the models, and are not addressed in most of the studies assessed in this subsection. Deep soil carbon is thought to be stabilised by mineral interactions, but recent experiments suggest that CO_2 release from deep soils can also be increased by warming, with a 4°C warming enhancing annual soil respiration by 34–37% (Hicks Pries et al. 2017), or with the addition of fresh carbon (Fontaine et al. 2007). While erosion is not typically modelled as a carbon flux in ESMs, erosion and burial of carbon-containing sediments is likely a significant carbon transfer from land to ocean (*medium confidence*) (Berhe et al. 2007; Asefaw et al. 2008; Wang et al. 2017e).

2.7.5 Soil carbon responses to warming and changes in soil moisture

Annually, 119 GtC is estimated to be emitted from the terrestrial ecosystem to the atmosphere, of which about 50% is attributed to soil microbial respiration (Auffret et al. 2016; Shao et al. 2013). It is yet not possible to make mechanistic and quantitative projections about how multiple environmental factors influence soil microbial respiration (Davidson et al. 2006a; Dungait et al. 2012). Soil warming experiments show significant variability in temperature and moisture responses across biomes and climates; Crowther et al. (2016) found that warming-induced SOC loss is greater in regions with high initial carbon stocks, while an analysis of an expanded version of the same dataset did not support this conclusion (Gestel et al. 2018). Studies of SOC responses to warming over time have also shown complex responses. In a multi-decadal warming experiment, Melillo et al. (2017) found that soil respiration response to warming went through multiple phases of increasing and decreasing strength, which were related to changes in microbial communities and available substrates over time. Conant et al. (2011) and Knorr et al. (2005) suggested that transient decomposition responses to warming could be explained by depletion of labile substrates, but that long-term SOC losses could be amplified by high temperature sensitivity of slowly decomposing SOC components. Overall, long-term SOC responses to warming remain uncertain (Davidson et al. 2006a; Dungait et al. 2012; Nishina et al. 2014; Tian et al. 2015).

It is widely known that soil moisture plays an important role in SOM decomposition by influencing microbial processes (e.g., Monard et al. (2012), Moyano et al. (2013), Yan et al. (2018)), as confirmed by a recent global meta-analysis (*high confidence*) (Hawkes et al. 2017). A likely mechanism is that increased soil moisture lowers carbon mineralisation rates under anaerobic conditions, resulting in enhanced carbon stocks, but experimental analyses have shown that this effect may last for only 3–4 weeks after which iron reduction can actually accelerate the loss of previously protected OC by facilitating microbial access (Huang and Hall 2017).

Experimental studies of responses of microbial respiration to warming have found variable results (Luo et al. 2001; Bradford et al. 2008; Zhou et al. 2011; Carey et al. 2016; Teramoto et al. 2016). No acclimation was observed in carbon-rich calcareous temperate forest soils (Schindlbacher et al. 2015) and arctic soils (Hartley et al. 2008), and a variety of ecosystems from the Arctic to the Amazon indicated that microbes appear to enhance the temperature sensitivity of soil respiration in Arctic and boreal soils, thereby releasing even more carbon than currently projected (Karhu et al. 2014). In tropical forests, phosphorus limitation of microbial processes is a key factor influencing soil respiration (Camenzind et al. 2018). Temperature responses of symbiotic mycorrhizae differ widely among host plant species, without a clear pattern that may allow generalisation across plant species and vegetation types (Fahey et al. 2016).

Some new insights have been obtained since AR5 from investigations of improved mechanistic understanding of factors that regulate temperature responses of soil microbial respiration. Carbon use efficiency and soil nitrogen dynamics have large influence on SOC responses to warming (*high confidence*) (Allison et al. 2010; Frey et al. 2013; Wieder, William R., Bonan, Gordon B., Allison 2013; García-Palacios et al. 2015). More complex community interactions including competitive and trophic interactions could drive unexpected responses to SOC cycling to changes in temperature, moisture and carbon inputs (Crowther et al. 2015; Buchkowski et al. 2017).

Competition for nitrogen among bacteria and fungi could also suppress decomposition (Averill et al. 2014). Overall, the roles of soil microbial community and trophic dynamics in global SOC cycling remain very uncertain.

2.7.6   Soil carbon responses to changes in organic matter inputs by plants

While current ESM structures mean that increasing carbon inputs to soils drive corresponding increases in SOC stocks, long-term carbon addition experiments have found contradictory SOC responses. Some litter addition experiments have observed increased SOC accumulation (Lajtha et al. 2014b; Liu et al. 2009), while others suggest insignificant SOC responses (Lajtha et al. 2014a; van Groenigen et al. 2014). Microbial dynamics are believed to have an important role in driving complex responses to carbon additions. The addition of fresh organic material can accelerate microbial growth and SOM decomposition via priming effects (Kuzyakov et al. 2014; Cheng et al. 2017). SOM cycling is dominated by 'hot spots' including the rhizosphere as well as areas surrounding fresh detritus (*medium evidence, high agreement*) (Finzi et al. 2015; Kuzyakov and Blagodatskaya 2015). This complicates projections of SOC responses to increasing plant productivity as increasing carbon inputs could promote higher SOC storage, but these fresh carbon inputs could also deplete SOC stocks by promoting faster decomposition (Hopkins et al. 2014; Guenet et al. 2018; Sulman et al. 2014). A meta-analysis by van Groenigen et al. (2014) suggested that elevated CO_2 accelerated SOC turnover rates across several biomes. These effects could be especially important in high-latitude regions where soils have high organic matter content and plant productivity is increasing (Hartley et al. 2012), but have also been observed in the tropics (Sayer et al. 2011).

Along with biological decomposition, another source of uncertainty in projecting responses of SOC to climate change is stabilisation via interactions with mineral particles (*high confidence*) (Kögel-Knabner et al. 2008; Kleber et al. 2011; Marschner et al. 2008; Schmidt 2011). Historically, conceptual models of SOC cycling have centred on the role of chemical recalcitrance: the hypothesis that long-lived components of SOC are formed from organic compounds that are inherently resistant to decomposition. Under the emerging new paradigm, stable SOC is primarily formed by the bonding of microbially-processed organic material to mineral particles, which limits the accessibility of organic material to microbial decomposers (Lützow et al. 2006; Keiluweit et al. 2015; Kallenbach et al. 2016; Kleber et al. 2011; Hopkins et al. 2014). SOC in soil aggregates can be protected from microbial decomposition by being trapped in soil pores too small for microbes to access (Blanco-Canqui and Lal 2004; Six et al. 2004) or by oxygen limitation (Keiluweit et al. 2016). Some new models are integrating these mineral protection processes into SOC cycling projections (Wang et al. 2017a; Sulman et al. 2014; Riley et al. 2014; Wieder et al. 2015), although the sensitivity of mineral-associated organic matter to changes in temperature, moisture, fire (Box 2.1) and carbon inputs is highly uncertain. Improved quantitative understanding of soil ecosystem processes will be critically important for projection of future land–climate feedback interactions.

Frequently Asked Questions

FAQ 2.1 | How does climate change affect land use and land cover?

Contemporary land cover and land use is adapted to current climate variability within particular temperature and/or rainfall ranges (referred to as climate envelopes). Anthropogenic GHG emissions impact land through changes in the weather and climate and also through modifications in atmospheric composition through increased GHGs, especially CO_2. A warming climate alters the current regional climate variability and results in a shift of regional climate envelopes poleward and to higher elevations. The shift of warmer climate envelopes into high latitude areas has potential benefits for agriculture here through extended growing seasons, warmer seasonal temperatures and increased atmospheric CO_2 concentrations which enhance photosynthetic activity. However, this warming will also lead to enhanced snowmelt and reduced albedo, permafrost melting and the further release of CH_4 and CO_2 into the atmosphere as the permafrost begins to decompose. Concurrent with these climate envelope shifts will be the emergence of new, hot climates in the tropics and increases in the frequency, intensity and duration of extreme events (e.g., heatwaves, very heavy rainfall, drought). These emergent hot climates will negatively affect land use (through changes in crop productivity, irrigation needs and management practices) and land cover through loss of vegetation productivity in many parts of the world, and would overwhelm any benefits to land use and land cover derived from increased atmospheric CO_2 concentrations.

FAQ 2.2 | How do the land and land use contribute to climate change?

Any changes to the land and how it is used can effect exchanges of water, energy, GHGs (e.g., CO_2, CH_4, N_2O), non-GHGs (e.g., BVOCs) and aerosols (mineral, e.g., dust, or carbonaceous, e.g., BC) between the land and the atmosphere. Land and land use change therefore alter the state (e.g., chemical composition and air quality, temperature and humidity) and the dynamics (e.g., strength of horizontal and vertical winds) of the atmosphere, which, in turn, can dampen or amplify local climate change. Land-induced changes in energy, moisture and wind can affect neighbouring, and sometimes more distant, areas. For example, deforestation in Brazil warms the surface, in addition to global warming, and enhances convection which increases the relative temperature difference between the land and the ocean, boosting moisture advection from the ocean and thus rainfall further inland. Vegetation absorbs CO_2 to use for growth and maintenance. Forests contain more carbon in their biomass and soils than croplands and so a conversion of forest to cropland, for example, results in emissions of CO_2 to the atmosphere, thereby enhancing the GHG-induced global warming. Terrestrial ecosystems are both sources and sinks of chemical compounds such as nitrogen and ozone. BVOCs contribute to forming tropospheric ozone and secondary aerosols, which respectively effect surface warming and cloud formation. Semi-arid and arid regions release dust, as do cropland areas after harvest. Increasing the amount of aerosols in the atmosphere impacts temperature in both positive and negative ways depending on the particle size, altitude and nature (carbonaceous or mineral, for example). Although global warming will impact the functioning and state of the land (FAQ 2.1), this is not a one-way interaction as changes in land and land use can also affect climate and thus modulate climate change. Understanding this two-way interaction can help improve adaptation and mitigation strategies, as well as manage landscapes.

FAQ 2.3 | How does climate change affect water resources?

Renewable freshwater resources are essential for the survival of terrestrial and aquatic ecosystems and for human use in agriculture, industry and in domestic contexts. As increased water vapour concentrations are expected in a warmer atmosphere, climate change will alter the hydrological cycle and therefore regional freshwater resources. In general, wet regions are projected to get wetter and dry regions drier, although there are regional exceptions to this. The consequent impacts vary regionally; where rainfall is projected to be lower in the future (many arid subtropical regions and those with a Mediterranean climate), a reduction of water resources is expected. Here increased temperatures and decreased rainfall will reduce surface and groundwater resources, increase plant evapotranspiration and increase evaporation rates from open water (rivers, lakes, wetlands) and water supply infrastructure (canals, reservoirs). In regions where rainfall is projected to be higher in the future (many high latitude regions and the wet tropics), an increase in water resources can be expected to benefit terrestrial and freshwater ecosystems, agriculture and domestic use, however, these benefits may be limited due to increased temperatures. An increase in extreme rainfall events is also expected which will lead to increases in surface runoff, regional flooding and nutrient removal as well as a reduction in soil water and groundwater recharge in many places. Anthropogenic land use change may amplify or moderate the climate change effect on water resources, therefore informed land management strategies need to be developed. A warming climate will exacerbate the existing pressures on renewable freshwater resources in water-stressed regions of the Earth and result in increased competition for water between human and natural systems.

2

References

Abatzoglou, J.T. and A.P. Williams, 2016: Impact of anthropogenic climate change on wildfire across western US forests. *Proc. Natl. Acad. Sci.*, **113**, 11770–11775, doi:10.1073/pnas.1607171113.

Abdalla, M. et al., 2016: Emissions of methane from northern peatlands: A review of management impacts and implications for future management options. *Ecol. Evol.*, **6**, 7080–7102, doi:10.1002/ece3.2469.

Abiodun, B.J., Z.D. Adeyewa, P.G. Oguntunde, A.T. Salami, and V.O. Ajayi, 2012: Modeling the impacts of reforestation on future climate in West Africa. *Theor. Appl. Climatol.*, **110**, 77–96, doi:10.1007/s00704-012-0614-1.

Achard, F. et al., 2014: Determination of tropical deforestation rates and related carbon losses from 1990 to 2010. *Glob. Chang. Biol.*, **20**, 2540–2554, doi:10.1111/gcb.12605.

Adger, W.N. et al., 2015: Focus on environmental risks and migration: Causes and consequences. *Environ. Res. Lett.*, **10**, doi:10.1088/1748-9326/10/6/060201.

Adnan, S., K. Ullah, S. Gao, A.H. Khosa, and Z. Wang, 2017: Shifting of agro-climatic zones, their drought vulnerability, and precipitation and temperature trends in Pakistan. *Int. J. Climatol.*, **37**, 529–543, doi:10.1002/joc.5019.

Ahlgren, S. and L. Di Lucia, 2014: Indirect land use changes of biofuel production – A review of modelling efforts and policy developments in the European Union. *Biotechnol. Biofuels*, doi:10.1186/1754-6834-7-35.

Ahlswede, B.J. and R.Q. Thomas, 2017: Community earth system model simulations reveal the relative importance of afforestation and forest management to surface temperature in eastern North America. *Forests*, **8**, 1–10, doi:10.3390/f8120499.

Ainsworth, E.A., and A. Rogers, 2007: The response of photosynthesis and stomatal conductance to rising [CO$_2$]: Mechanisms and environmental interactions. *Plant, Cell Environ.*, **30**, 258–270, doi:10.1111/j.1365-3040.2007.01641.x.

Akagi, S.K. et al., 2011: Emission factors for open and domestic biomass burning for use in atmospheric models. *Atmos. Chem. Phys.*, **11**, 4039–4072, doi:10.5194/acp-11-4039-2011.

Alcántara, V., A. Don, R. Well, and R. Nieder, 2016: Deep ploughing increases agricultural soil organic matter stocks. *Glob. Chang. Biol.*, doi:10.1111/gcb.13289.

Aldersley, A., S.J. Murray, and S.E. Cornell, 2011: Global and regional analysis of climate and human drivers of wildfire. *Sci. Total Environ.*, **409**, 3472–3481, doi:10.1016/j.scitotenv.2011.05.032.

Alfieri, L., L. Feyen, and G. Di Baldassarre, 2016: Increasing flood risk under climate change: A pan-European assessment of the benefits of four adaptation strategies. *Clim. Change*, **136**, 507–521, doi:10.1007/s10584-016-1641-1.

Alfieri, L. et al., 2017: Global projections of river flood risk in a warmer world. *Earth's Futur.*, **5**, 171–182, doi:10.1002/2016EF000485.

Alghamdi, A.S., and T.W. Moore, 2015: Detecting temporal changes in Riyadh's urban heat island. *Pap. Appl. Geogr.*, **1**, 312–325, doi:10.1080/23754931.2015.1084525.

Alizadeh-Choobari, O., P. Ghafarian, and P. Adibi, 2016: Inter-annual variations and trends of the urban warming in Tehran. *Atmos. Res.*, **170**, 176–185, doi:10.1016/j.atmosres.2015.12.001.

Alkama, R., and A. Cescatti, 2016: Biophysical climate impacts of recent changes in global forest cover. *Science*, **351**, 600–604, doi:10.1126/science.aac8083.

Allan, J.D. et al., 2014: Airborne observations of IEPOX-derived isoprene SOA in the Amazon during SAMBBA. *Atmos. Chem. Phys.*, **14**, 11393–11407, doi:10.5194/acp-14-11393-2014.

Allan, R.P., and C. Liu, 2018: Evaluating large-scale variability and change in tropical rainfall and its extremes. In: *Tropical Extremes: Natural Variability and Trends*, [Venugopal, V., J. Sukhatme, R. Murtugudde, and R. Roca, (eds.)]. Elsevier, pp. 139–163.

Allen, C.D. et al., 2010: A global overview of drought and heat-induced tree mortality reveals emerging climate change risks for forests. *For. Ecol. Manage.*, **259**, 660–684, doi:10.1016/j.foreco.2009.09.001.

Allen, M., O.P. Dube, W. Solecki, F. Aragón–Durand, W. Cramer, S. Humphreys, M. Kainuma, J. Kala, N. Mahowald, Y. Mulugetta, R. Perez, M. Wairiu, and K. Zickfeld, 2018: Framing and Context. In: Global Warming of 1.5°C. An IPCC Special Report on the impacts of global warming of 1.5°C above pre-industrial levels and related global greenhouse gas emission pathways, in the context of strengthening the global response to the threat of climate change, sustainable development, and efforts to eradicate poverty [Masson-Delmotte, V., P. Zhai, H.-O. Pörtner, D. Roberts, J. Skea, P.R. Shukla, A. Pirani, W. Moufouma-Okia, C. Péan, R. Pidcock, S. Connors, J.B.R. Matthews, Y. Chen, X. Zhou, M.I. Gomis, E. Lonnoy, T. Maycock, M. Tignor, and T. Waterfield (eds.)]. In press.

Allen, R.J., T. Hassan, C.A. Randles, and H. Su, 2019: Enhanced land–sea warming contrast elevates aerosol pollution in a warmer world. *Nat. Clim. Chang.*, **9**, 300-305, doi:10.1038/s41558-019-0401-4.

Allison, S.D., M.D. Wallenstein, and M.A. Bradford, 2010: Soil-carbon response to warming dependent on microbial physiology. *Nat. Geosci.*, **3**, 336–340, doi:10.1038/ngeo846.

Alter, R.E., E.S. Im, and E.A.B. Eltahir, 2015: Rainfall consistently enhanced around the Gezira Scheme in East Africa due to irrigation. *Nat. Geosci.*, **8**, 763–767, doi:10.1038/ngeo2514.

Álvaro-Fuentes, J., M. V López Sánchez, C. Cantero-Martínez, and J.L. Arrúe Ugarte, 2008: Tillage effects on soil organic carbon fractions in Mediterranean dryland agroecosystems. *Soil Sci. Soc. Am. J.*, **72**, 541–547, doi:10.2136/sssaj2007.0164.

Anav, A., P.M. Ruti, V. Artale, and R. Valentini, 2010: Modelling the effects of land-cover changes on surface climate in the Mediterranean region. *Clim. Res.*, **41**, 91–104, doi:10.3354/cr00841.

Anav, A. et al., 2013: Evaluating the land and ocean components of the Global Carbon Cycle in the CMIP5 Earth System Models. *J. Clim.*, **26**, 6801–6843, doi:10.1175/JCLI-D-12-00417.1.

Andela, N. et al., 2017: A human-driven decline in global burned area. *Science*, **356**, 1356–1362, doi:10.1126/science.aal4108.

Anderegg, W.R.L. et al., 2016: Meta-analysis reveals that hydraulic traits explain cross-species patterns of drought-induced tree mortality across the globe. *Proc. Natl. Acad. Sci.*, **113**, 5024–5029, doi:10.1073/pnas.1525678113.

Anderegg, W.R.L. et al., 2018a: Woody plants optimise stomatal behaviour relative to hydraulic risk. *Ecol. Lett.*, **0**, doi:10.1111/ele.12962.

Anderegg, W.R.L. et al., 2018b: Hydraulic diversity of forests regulates ecosystem resilience during drought. *Nature*, **561**, 538–541, doi:10.1038/s41586-018-0539-7.

Anderson-Teixeira, K.J. et al., 2012: Climate-regulation services of natural and agricultural ecoregions of the Americas. *Nat. Clim. Chang.*, **2**, 177–181, doi:10.1038/nclimate1346.

Anderson, R.G. et al., 2011: Biophysical considerations in forestry for climate protection. *Front. Ecol. Environ.*, **9**, 174–182, doi:10.1890/090179.

Andreae et al., 2002: Biogeochemical cycling of carbon, water, energy, trace gases, and aerosols in Amazonia: The LBA-EUSTACH experiments. *J. Geophys. Res. D Atmos.*, **107**, 1998–2000, doi:10.1029/2001JD000524.

Andreae, M.O. et al., 2018: Aerosol characteristics and particle production in the upper troposphere over the Amazon Basin. *Atmos. Chem. Phys.*, **18**, 921–961, doi:10.5194/acp-18-921-2018.

Angel, S., J. Parent, D.L. Civco, A. Blei, and D. Potere, 2011: The dimensions of global urban expansion: Estimates and projections for all countries, 2000–2050. *Prog. Plann.*, **75**, 53–107, doi.org/10.1016/j.progress.2011.04.001.

Ansmann, A. et al., 2018: Extreme levels of Canadian wildfire smoke in the stratosphere over central Europe on 21–22 August 2017. *Atmos. Chem. Phys.*, **18**, 11831–11845, doi:10.5194/acp-18-11831-2018.

Aragão, L.E.O.C. et al., 2018: 21st Century drought-related fires counteract the decline of Amazon deforestation carbon emissions. *Nat. Commun.*, **9**, 536, doi:10.1038/s41467-017-02771-y.

Archibald, S., C.E.R. Lehmann, J.L. Gómez-Dans, and R.A. Bradstock, 2013: Defining pyromes and global syndromes of fire regimes. *Proc. Natl. Acad. Sci.*, **110**, 6442–6447, doi:10.1073/pnas.1211466110.

Argüeso, D., J.P. Evans, L. Fita, and K.J. Bormann, 2014: Temperature response to future urbanization and climate change. *Clim. Dyn.*, **42**, 2183–2199, doi:10.1007/s00382-013-1789-6.

Arnell, N.W., and S.N. Gosling, 2016: The impacts of climate change on river flood risk at the global scale. *Clim. Change*, **134**, 387–401, doi:10.1007/s10584-014-1084-5.

Arneth, A. et al., 2010: Terrestrial biogeochemical feedbacks in the climate system. *Nat. Geosci.*, **3**, 525–532, doi:10.1038/ngeo905.

Arneth, A. et al., 2017: Historical carbon dioxide emissions caused by land-use changes are possibly larger than assumed. *Nat. Geosci.*, **10**, 79–84. doi:10.1038/ngeo2882.

Arora, V.K. and G.J. Boer, 2010: Uncertainties in the 20th century carbon budget associated with land use change. *Glob. Chang. Biol.*, doi:10.1111/j.1365-2486.2010.02202.x.

Arora, V.K. and A. Montenegro, 2011: Small temperature benefits provided by realistic afforestation efforts. *Nat. Geosci.*, **4**, 514–518, doi:10.1038/ngeo1182.

Arora, V.K., and J.R. Melton, 2018: Reduction in global area burned and wildfire emissions since 1930s enhances carbon uptake by land. *Nat. Commun.*, doi:10.1038/s41467-018-03838-0.

Arsiso, B.K., G. Mengistu Tsidu, G.H. Stoffberg, and T. Tadesse, 2018: Influence of urbanization-driven land use/cover change on climate: The case of Addis Ababa, Ethiopia. *Phys. Chem. Earth*, doi:10.1016/j.pce.2018.02.009.

Artaxo, P. et al., 2013: Atmospheric aerosols in Amazonia and land use change: From natural biogenic to biomass burning conditions. *Faraday Discuss.*, **165**, 203–235, doi:10.1039/c3fd00052d.

Asefaw, B.A., J.W. Harden, M.S. Torn, and J. Harte, 2008: Linking soil organic matter dynamics and erosion-induced terrestrial carbon sequestration at different landform positions. *J. Geophys. Res. Biogeosciences*, **113**, 1–12, doi:10.1029/2008JG000751. https://doi.org/10.1029/2008JG000751.

Ashworth, K., G. Folberth, C.N. Hewitt, and O. Wild, 2012: Impacts of near-future cultivation of biofuel feedstocks on atmospheric composition and local air quality. *Atmos. Chem. Phys.*, **12**, 919–939, doi:10.5194/acp-12-919-2012.

Asseng, S. et al., 2015: Rising temperatures reduce global wheat production. *Nat. Clim. Chang.*, **5**, 143–147, doi:10.1038/nclimate2470.

Atkin, O.K. et al., 2015: Global variability in leaf respiration in relation to climate, plant functional types and leaf traits. *New Phytol.*, **206**, 614–636, doi:10.1111/nph.13253.

Auffret, M.D. et al., 2016: The role of microbial community composition in controlling soil respiration responses to temperature. *PLoS One*, **11**, e0165448, doi:10.1371/journal.pone.0165448.

Augustin, J., J. Couwenberg, and M. Minke, 2011: Peatlands and greenhouse gases (Sec. 3.1). In: *Carbon Credits From Peatland Rewetting: Climate – Biodiversity – Land Use* [Parish, T.F., A.Sirin, D.Charman, H. Joosten, Minayeva and L. Silvius, M. and Stringer, (eds.)]. Global Environment Centre and Wetlands International, Kuala Lumpur, Malaysia and Wageningen, Netherlands, pp. 13–19.

Averill, C., B.L. Turner, and A.C. Finzi, 2014: Mycorrhiza-mediated competition between plants and decomposers drives soil carbon storage. *Nature*, **505**, 543–545, doi:10.1038/nature12901.

Avila, F.B., A.J. Pitman, M.G. Donat, L.V. Alexander, and G. Abramowitz, 2012: Climate model simulated changes in temperature extremes due to land cover change. *J. Geophys. Res. Atmos.*, **117**, 1–19, doi:10.1029/2011JD016382.

Avissar, R. and D. Werth, 2005: Global hydroclimatological teleconnections resulting from tropical deforestation. *J. Hydrometeorol.*, **6**, 134–145, doi:10.1175/JHM406.1.

Avitabile, V., M. Herold, G.B.M. Heuvelink, A.G. P, de B. Jong, and V.G. Laurin, 2016: An integrated pan-tropical biomass map using multiple reference datasets. *Glob. Chang. Biol.*, **22**, 1406–1420.

Babst, F. et al., 2012: 500 years of regional forest growth variability and links to climatic extreme events in Europe. *Environ. Res. Lett.*, **7**, 45705, doi:10.1088/1748-9326/7/4/045705.

Baccini, A. et al., 2012: Estimated carbon dioxide emissions from tropical deforestation improved by carbon-density maps. *Nat. Clim. Chang.*, **2**, 182–185, doi:10.1038/nclimate1354.

Baccini, A. et al., 2017: Tropical forests are a net carbon source based on aboveground measurements of gain and loss. *Science*, **358**, 230–234, doi:10.1126/science.aam5962.

Bader, D.A. et al., 2018: Urban climate science. In: *Climate Change and Cities: Second Assessment Report of the Urban Climate Change Research Network* [Rosenweigh, C. et al. (eds.)]. Urban Climate Change Research Network, Cambridge University Press, Cambridge, UK, pp. 27–60.

Bador, M., L. Terray, and J. Boé, 2016: Emergence of human influence on summer record-breaking temperatures over Europe. *Geophys. Res. Lett.*, **43**, 404–412, doi:10.1002/2015GL066560. http://doi.wiley.com/10.1002/2015GL066560 (Accessed October 28, 2018).

Bae, J., and Y. Ryu, 2015: Land use and land cover changes explain spatial and temporal variations of the soil organic carbon stocks in a constructed urban park. *Landsc. Urban Plan.*, **136**, 57–67, doi:10.1016/j.landurbplan.2014.11.015.

Bagley, J.E. et al., 2017: The influence of land cover on surface energy partitioning and evaporative fraction regimes in the U.S. Southern Great Plains. *J. Geophys. Res. Atmos.*, **122**, 5793–5807, doi:10.1002/2017JD026740.

Bajželj, B. and K. Richards, 2014: The positive feedback loop between the impacts of climate change and agricultural expansion and relocation. *Land*, **3**, 898–916, doi:10.3390/land3030898. http://www.mdpi.com/2073-445X/3/3/898/.

Baker, J.S., C.M. Wade, B.L. Sohngen, S. Ohrel, and A.A. Fawcett, 2019: Potential complementarity between forest carbon sequestration incentives and biomass energy expansion. *Energy Policy*, doi:10.1016/j.enpol.2018.10.009.

Bala, G., K. et al., 2007: Combined climate and carbon-cycle effects of large-scale deforestation. *Proc. Natl. Acad. Sci.*, **104**, 6550–6555, doi:10.1073/pnas.0608998104.

Ballantyne, A. et al., 2017: Accelerating net terrestrial carbon uptake during the warming hiatus due to reduced respiration. *Nat. Clim. Chang.*, **7**, 148–152, doi:10.1038/nclimate3204.

Ban, N., J. Schmidli, and C. Schär, 2015: Heavy precipitation in a changing climate: Does short-term summer precipitation increase faster? *Geophys. Res. Lett.*, **42**, 1165–1172, doi:10.1002/2014GL062588.

Barbero-Sierra, C., M.J. Marques, and M. Ruíz-Pérez, 2013: The case of urban sprawl in Spain as an active and irreversible driving force for desertification. *J. Arid Environ.*, **90**, 95–102, doi:10.1016/j.jaridenv.2012.10.014.

Barnes, C.J., C.J. van der Gast, N.P. Mcnamara, R. Rowe, and G.D. Bending, 2018: Extreme rainfall affects assembly of the root-associated fungal community. *New Phytol.*, doi:10.1111/nph.14990.

Baron, R.E., W.D. Montgomery, and S.D. Tuladhar, 2018: *An Analysis of Black Carbon Mitigation as a Response to Climate Change*. Copenhagen Consensus Center, 31 pp. Retrieved from http://www.jstor.org/stable/resrep16323.1.

Basche, A.D., F.E. Miguez, T.C. Kaspar, and M.J. Castellano, 2014: Do cover crops increase or decrease nitrous oxide emissions? A meta-analysis. *J. Soil Water Conserv.*, **69**, 471–482, doi:10.2489/jswc.69.6.471.

Baskaran, P., R. et al., 2017: Modelling the influence of ectomycorrhizal decomposition on plant nutrition and soil carbon sequestration in boreal forest ecosystems. *New Phytol.*, **213**, 1452–1465, doi:10.1111/nph.14213.

Bassu, S. et al., 2014: How do various maize crop models vary in their responses to climate change factors? *Glob. Chang. Biol.*, **20**, 2301–2320, doi:10.1111/gcb.12520.

Bathiany, S., M. Claussen, V. Brovkin, T. Raddatz, and V. Gayler, 2010a: Combined biogeophysical and biogeochemical effects of large-scale forest cover changes in the MPI earth system model. *Biogeosciences*, **7**, 1383–1399, doi:10.5194/bg-7-1383-2010.

Battipaglia, G. et al., 2013: Elevated CO_2 increases tree-level intrinsic water use efficiency: Insights from carbon and oxygen isotope analyses in tree rings across three forest FACE sites. *New Phytol.*, **197**, 544–554, doi:10.1111/nph.12044.

Beach, R.H. et al., 2015: Global mitigation potential and costs of reducing agricultural non-CO_2 greenhouse gas emissions through 2030. *J. Integr. Environ. Sci.*, **12**, 87–105, doi:10.1080/1943815X.2015.1110183.

Beer, C. et al., 2010: Terrestrial gross carbon dioxide uptake: global distribution and covariation with climate. *Science*, **329**, 834–838, doi:10.1126/science.1184984.

Beer, T., 2018: The impact of extreme weather events on food security. In: *Climate Change, Extreme Events and Disaster Risk Reduction* [S. Mal, R. Singh, and C. Huggel (eds.)]. Sustainable Development Goals Series. Springer, Cham, Switzerland, pp. 121–133.

Beerling, D.J. et al., 2018: Farming with crops and rocks to address global climate, food and soil security. *Nat. Plants*, **4**, 138–147, doi:10.1038/s41477-018-0108-y.

Bento, A.M. and R. Klotz, 2014: Climate policy decisions require policy-based lifecycle analysis. *Environ. Sci. Technol.*, **48**, 5379–5387, doi:10.1021/es405164g.

Berdanier, A.B. and J.S. Clark, 2016: Multiyear drought-induced morbidity preceding tree death in southeastern U.S. forests. *Ecol. Appl.*, **26**, 17–23, doi:10.1890/15-0274.

Berg, A. and J. Sheffield, 2018: Climate change and drought: The soil moisture perspective. *Curr. Clim. Chang. Reports*, **4**, 180–191, doi:10.1007/s40641-018-0095-0.

Berg, A. et al., 2015: Interannual coupling between summertime surface temperature and precipitation over land: Processes and implications for climate change. *J. Clim.*, **28**, 1308–1328, doi:10.1175/JCLI-D-14-00324.1.

Berg, A. et al., 2016: Land-atmosphere feedbacks amplify aridity increase over land under global warming. *Nat. Clim. Chang.*, **6**, 869–874, doi:10.1038/nclimate3029.

Berg, E.E., J. David Henry, C.L. Fastie, A.D. De Volder, and S.M. Matsuoka, 2006: Spruce beetle outbreaks on the Kenai Peninsula, Alaska, and Kluane National Park and Reserve, Yukon Territory: Relationship to summer temperatures and regional differences in disturbance regimes. *For. Ecol. Manage.*, **227**, 219–232, doi:10.1016/j.foreco.2006.02.038.

Berg, P., C. Moseley, and J.O. Haerter, 2013: Strong increase in convective precipitation in response to higher temperatures. *Nat. Geosci.*, **6**, 181–185, doi:10.1038/ngeo1731.

Bergamaschi, P. et al., 2013: Atmospheric CH_4 in the first decade of the 21st century: Inverse modeling analysis using SCIAMACHY satellite retrievals and NOAA surface measurements. *J. Geophys. Res. Atmos.*, **118**, 7350–7369, doi:10.1002/jgrd.50480.

Berhe, A.A., J. Harte, J.W. Harden, and M.S. Torn, 2007: The significance of the erosion-induced terrestrial carbon sink. *Bioscience*, **57**, 337–346, doi:10.1641/B570408.

Berndes, G., S. Ahlgren, P. Börjesson, and A.L. Cowie, 2013: Bioenergy and land use change—state of the art. *Wiley Interdiscip. Rev. Energy Environ.*, **2**, 282–303, doi:10.1002/wene.41.

Bernier, P. and D. Paré, 2013: Using ecosystem CO_2 measurements to estimate the timing and magnitude of greenhouse gas mitigation potential of forest bioenergy. *GCB Bioenergy*, **5**, 67–72, doi:10.1111/j.1757-1707.2012.01197.x.

Berry, J.A., D.J. Beerling, and P.J. Franks, 2010: Stomata: Key players in the earth system, past and present. *Curr. Opin. Plant Biol.*, **13**, 233–240, doi:10.1016/j.pbi.2010.04.013.

Bertram, C. et al., 2018: Targeted policies can compensate most of the increased sustainability risks in 1.5°C mitigation scenarios. *Environ. Res. Lett.*, **13**, 64038, doi:10.1088/1748-9326/aac3ec.

Betts, R.A., P.D. Falloon, K.K. Goldewijk, and N. Ramankutty, 2007: Biogeophysical effects of land use on climate: Model simulations of radiative forcing and large-scale temperature change. *Agric. For. Meteorol.*, **142**, 216–233, doi:10.1016/j.agrformet.2006.08.021.

Betts, R.A., 2001: Biogeophysical impacts of land use on present-day climate: Near-surface temperature change and radiative forcing. *Atmos. Sci. Lett.*, **2**, 1–13, doi:10.1006/asle.2001.0023.

Di Biagio, C. et al., 2017: Global scale variability of the mineral dust long-wave refractive index: A new dataset of in situ measurements for climate modeling and remote sensing. *Atmos. Chem. Phys.*, **17**, 1901–1929, doi:10.5194/acp-17-1901-2017.

Birdsey, R. and Y. Pan, 2015: Trends in management of the world's forests and impacts on carbon stocks. *For. Ecol. Manag.*, 355, 83–89, doi:10.1016/j.foreco.2015.04.031.

Blanco-Canqui, H. and R. Lal, 2004: Mechanisms of carbon sequestration in soil aggregates. *Crit. Rev. Plant Sci.*, **23**, 481–504, doi:10.1080/07352680490886842.

Boettle, M., D. Rybski, and J.P. Kropp, 2016: Quantifying the effect of sea level rise and flood defence – A point process perspective on coastal flood damage. *Nat. Hazards Earth Syst. Sci.*, **16**, 559–576, doi:10.5194/nhess-16-559-2016.

Bohrer, G. et al., 2005: Finite element tree crown hydrodynamics model (FETCH) using porous media flow within branching elements: A new representation of tree hydrodynamics. *Water Resour. Res.*, **41**, doi:10.1029/2005wr004181.

Boisier, J.P. et al., 2012: Attributing the impacts of land-cover changes in temperate regions on surface temperature and heat fluxes to specific causes: Results from the first LUCID set of simulations. *J. Geophys. Res. Atmos.*, **117**, 1–16, doi:10.1029/2011JD017106.

Boisvert-Marsh, L., C. Périé, and S. de Blois, 2014: Shifting with climate? Evidence for recent changes in tree species distribution at high latitudes. *Ecosphere*, **5** (7), 1–33, doi:10.1890/ES14-00111.1.

Bonan, G.B., 2008: Forests and climate change: Forcings, feedbacks, and the climate benefits of forests. *Science*, **320**, 1444–1449, doi:10.1126/science.1155121.

Bond-Lamberty, B. and A. Thomson, 2010: Temperature-associated increases in the global soil respiration record. *Nature*, **464**, 579–582, doi:10.1038/nature08930.

Bond-Lamberty, B., V.L. Bailey, M. Chen, C.M. Gough, and R. Vargas, 2018: Globally rising soil heterotrophic respiration over recent decades. *Nature*, **560**, 80–83, doi:10.1038/s41586-018-0358-x.

Bond, T.C. et al., 2013: Bounding the role of black carbon in the climate system: A scientific assessment. *J. Geophys. Res. Atmos.*, **118**, 5380–5552, doi:10.1002/jgrd.50171.

Bond, W.J., F.I. Woodward, G.F. Midgley, 2004: The global distribution of ecosystems in a world without fire. *New Phytol.*, **165**, 525–538, doi:10.1111/j.1469-8137.2004.01252.x.

Bonfils, C. and D. Lobell, 2007: Empirical evidence for a recent slowdown in irrigation-induced cooling. *Proc. Natl. Acad. Sci.*, **104**, 13582–13587, doi:10.1073/pnas.0700144104.

Boone, A.A. et al., 2016: The regional impact of Land-Use Land-cover Change (LULCC) over West Africa from an ensemble of global climate models under the auspices of the WAMME2 project. *Clim. Dyn.*, **47**, 3547–3573, doi:10.1007/s00382-016-3252-y.

Borchard, N. et al., 2019: Biochar, soil and land-use interactions that reduce nitrate leaching and N_2O emissions: A meta-analysis. *Sci. Total Environ.*, doi:10.1016/j.scitotenv.2018.10.060.

Borodina, A., E.M. Fischer, and R. Knutti, 2017: Models are likely to underestimate increase in heavy rainfall in the extratropical regions with high rainfall intensity. *Geophys. Res. Lett.*, **44**, 7401–7409, doi:10.1002/2017GL074530.

Bosello, F., C. Carraro, and E. De Cian, 2009: *An Analysis of Adaptation as a Response to Climate Change*. Copenhagen Consensus Center, Frederiksberg, Denmark, 29 pp.

Bossio, D.A. and K.M. Scow, 1998: Impacts of carbon and flooding on soil microbial communities: Phospholipid fatty acid profiles and substrate utilization patterns. *Microb. Ecol.*, **35**, 265–278, doi:10.1007/s002489900082.

Boucher, O., D. Randall, P. Artaxo, C. Bretherton, G. Feingold, P. Forster, V.-M. Kerminen, Y. Kondo, H. Liao, U. Lohmann, P. Rasch, S.K. Satheesh, S. Sherwood, B. Stevens and X.Y. Zhang, 2013: Clouds and aerosols. In: Climate Change 2013: The Physical Science Basis. Contribution of Working Group I to the Fifth Assessment Report of the Intergovernmental Panel on Climate Change [Stocker, T.F., D. Qin, G.-K. Plattner, M. Tignor, S.K. Allen, J. Boschung, A. Nauels, Y. Xia, V. Bex and P.M. Midgley (eds.)]. Cambridge University Press, Cambridge, United Kingdom and New York, NY, USA, pp. 571–658.

Boucher, O. et al., 2016: Jury is still out on the radiative forcing by black carbon. *Proc. Natl. Acad. Sci.*, **113**, E5092–E5093, doi:10.1073/pnas.1607005113.

Bourtsoukidis, E. et al., 2018: Strong sesquiterpene emissions from Amazonian soils. *Nat. Commun.*, **9**, doi:10.1038/s41467-018-04658-y.

Bousquet, P. et al., 2006: Contribution of anthropogenic and natural sources to atmospheric methane variability. *Nature*, **443**, 439–443, doi:10.1038/nature05132.

Bousquet, P. et al., 2011: Source attribution of the changes in atmospheric methane for 2006–2008. *Atmos. Chem. Phys.*, **11**, 3689–3700, doi:10.5194/acp-11-3689-2011.

Bouwman, A.F. et al., 2013: Global trends and uncertainties in terrestrial denitrification and N_2O emissions. *Philos. Trans. R. Soc. B Biol. Sci.*, **368**, 20130112–20130112, doi:10.1098/rstb.2013.0112.

Bowman, D.M.J.S. et al., 2011: The human dimension of fire regimes on Earth. *J. Biogeogr.*, **38**, 2223–2236, doi:10.1111/j.1365-2699.2011.02595.x.

Bowman, D.M.J.S. et al., 2017: Human exposure and sensitivity to globally extreme wildfire events. *Nat. Ecol. Evol.*, **1**, 0058, doi: 10.1038/s41559-016-0058.

Boysen, L.R. et al., 2014: Global and regional effects of land-use change on climate in 21st century simulations with interactive carbon cycle. *Earth Syst. Dyn. Discuss.*, **5**, 443–472, doi:10.5194/esdd-5-443-2014.

Boysen, L.R., W. Lucht, and D. Gerten, 2017: Trade-offs for food production, nature conservation and climate limit the terrestrial carbon dioxide removal potential. *Glob. Chang. Biol.*, **23**, 4303–4317, doi:10.1111/gcb.13745.

Bozzi, E., L. Genesio, P. Toscano, M. Pieri, and F. Miglietta, 2015: Mimicking biochar-albedo feedback in complex Mediterranean agricultural landscapes. *Environ. Res. Lett.*, **10**, doi:10.1088/1748-9326/10/8/084014.

Bracho-Nunez, A., S. Welter, M. Staudt, and J. Kesselmeier, 2011: Plant-specific volatile organic compound emission rates from young and mature leaves of Mediterranean vegetation. *J. Geophys. Res. Atmos.*, **116**, 1–13, doi:10.1029/2010JD015521.

Braconnot, P. et al., 2012: Evaluation of climate models using palaeoclimatic data. *Nat. Clim. Chang.*, **2**, 417–424, doi:10.1038/nclimate1456.

Bradford, M.A. et al., 2008: Thermal adaptation of soil microbial respiration to elevated temperature. *Ecol. Lett.*, **11**, 1316–1327, doi:10.1111/j.1461-0248.2008.01251.x.

Brando, P.M. et al., 2008: Drought effects on litterfall, wood production and belowground carbon cycling in an Amazon forest: results of a throughfall reduction experiment. *Philos. Trans. R. Soc. B Biol. Sci.*, **363**, 1839–1848, doi:10.1098/rstb.2007.0031.

Brando, P.M. et al., 2014: Abrupt increases in Amazonian tree mortality due to drought-fire interactions. *Proc. Natl. Acad. Sci.*, **111**, 6347–6352, doi:10.1073/pnas.1305499111.

Braun, M. et al., 2016: A holistic assessment of greenhouse gas dynamics from forests to the effects of wood products use in Austria. *Carbon Manag.*, **7**, 271–283, doi:10.1080/17583004.2016.1230990.

Bremer, D.J., and J.M. Ham, 1999: Effect of spring burning on the surface energy balance in a tallgrass prairie. *Agric. For. Meteorol.*, **97**, 43–54, doi:10.1016/S0168-1923(99)00034-9.

Bren d'Amour, C. et al., 2016: Future urban land expansion and implications for global croplands. *Proc. Natl. Acad. Sci.*, **114**, 8939–8944, doi:10.1073/pnas.1606036114.

Breshears, D.D. et al., 2005: Regional vegetation die-off in response to global-change-type drought. *Proc. Natl. Acad. Sci.*, **102**, 15144–15148, doi:10.1073/pnas.0505734102.

Brienen, R.J.W. et al., 2015: Long-term decline of the Amazon carbon sink. *Nature*, **519**, 344–348, doi:10.1038/nature14283.

Bright, R.M., K. Zhao, R.B. Jackson, and F. Cherubini, 2015: Quantifying surface albedo and other direct biogeophysical climate forcings of forestry activities. *Glob. Chang. Biol.*, **21**, 3246–3266, doi:10.1111/gcb.12951.

Bright, R. et al., 2017: Local temperature response to land cover and management change driven by non-radiative processes. *Nat. Clim. Chang.*, **7**, 296–302, doi:10.1038/nclimate3250.

Brodribb, T.J., 2009: Xylem hydraulic physiology: The functional backbone of terrestrial plant productivity. *Plant Sci.*, **177**, 245–251, doi:10.1016/j.plantsci.2009.06.001.

Brodt, S., K.J. Kramer, A. Kendall, and G. Feenstra, 2013: Comparing environmental impacts of regional and national-scale food supply chains: A case study of processed tomatoes. *Food Policy*, **42**, 106–114, doi:10.1016/j.foodpol.2013.07.004.

Brovkin, V. et al., 2004: Role of land cover changes for atmospheric CO_2 increase and climate change during the last 150 years. *Glob. Chang. Biol.*, **10**, 1253–1266, doi:10.1111/j.1365-2486.2004.00812.x.

Brovkin, V. et al., 2006: Biogeophysical effects of historical land cover changes simulated by six Earth system models of intermediate complexity. *Clim. Dyn.*, **26**, 587–600, doi:10.1007/s00382-005-0092-6.

Brovkin, V. et al., 2013: Effect of anthropogenic land-use and land-cover changes on climate and land carbon storage in CMIP5 projections for the twenty-first century. *J. Clim.*, **26**, 6859–6881, doi:10.1175/JCLI-D-12-00623.1.

Brovkin, V. et al., 2015: Cooling biogeophysical effect of large-scale tropical deforestation in three Earth System models. In: *EGU General Assembly Conference Abstracts*, Vol. 17 of EGU General Assembly Conference Abstracts, 8903.

Brown, M.E. et al., 2015: *Climate Change, Global Food Security, and the U.S. Food System*. US Global Change Research Program, Washington DC, USA, 146 pp. *146 pp.* doi:10.7930/J0862DC7.

Brown, S., and R.J. Nicholls, 2015: Subsidence and human influences in mega deltas: The case of the Ganges–Brahmaputra–Meghna. *Sci. Total Environ.*, **527–528**, 362–374, doi:https://doi.org/10.1016/j.scitotenv.2015.04.124.

Brutel-Vuilmet, C., M. Ménégoz, and G. Krinner, 2013: An analysis of present and future seasonal Northern Hemisphere land snow cover simulated by CMIP5 coupled climate models. *Cryosphere*, **7**, 67–80, doi:10.5194/tc-7-67-2013.

Bryn, A. and K. Potthoff, 2018: Elevational treeline and forest line dynamics in Norwegian mountain areas – a review. *Landsc. Ecol.*, **33**, 1225–1245, doi:10.1007/s10980-018-0670-8.

Buchkowski, R.W., M.A. Bradford, A.S. Grandy, O.J. Schmitz, and W.R. Wieder, 2017: Applying population and community ecology theory to advance understanding of belowground biogeochemistry. *Ecol. Lett.*, **20**, 231–245, doi:10.1111/ele.12712.

Buitenhuis, E.T., P. Suntharalingam, and C. Le Quéré, 2018: Constraints on global oceanic emissions of N_2O from observations and models. *Biogeosciences*, 15, 2161–2175, doi:10.5194/bg-15-2161-2018.

Buitenwerf, R., B. Sandel, S. Normand, A. Mimet, and J.-C. Svenning, 2018: Land surface greening suggests vigorous woody regrowth throughout European semi-natural vegetation. *Glob. Chang. Biol.*, **24**, 5789–5801, doi:10.1111/gcb.14451.

Bullard, J.E. et al., 2016: High latitude dust in the Earth system. *Rev. Geophys.*, **54**, 447–485, doi:10.1002/2016RG000518.

Buotte, P.C. et al., 2019: Near-future forest vulnerability to drought and fire varies across the western United States. *Glob. Chang. Biol.*, **25**, 290–303, doi:10.1111/gcb.14490.

Burakowski, E. et al., 2018: The role of surface roughness, albedo, and Bowen ratio on ecosystem energy balance in the Eastern United States. *Agric. For. Meteorol.*, **249**, 367–376, doi:10.1016/j.agrformet.2017.11.030.

Burkett, V.R., A.G. Suarez, M. Bindi, C. Conde, R. Mukerji, M.J. Prather, A.L. St. Clair, and G.W. Yohe, 2014: Point of departure. Impacts, Adaptation and Vulnerability. Part A: Global and Sectoral Aspects. Contribution of Working Group II to the Fifth Assessment Report of the Intergovernmental Panel on Climate Change [Field, C.B., V.R. Barros, D.J. Dokken, K.J. Mach, M.D. Mastrandrea, T.E. Bilir, M. Chatterjee, K.L. Ebi, Y.O. Estrada, R.C. Genova, B. Girma, E.S. Kissel, A.N. Levy, S. MacCracken, P.R. Mastrandrea, and L.L.White (eds.)]. Cambridge University Press, Cambridge, United Kingdom and New York, NY, USA, pp. 169–194.

Busch, F.A., and R.F. Sage, 2017: The sensitivity of photosynthesis to O2 and CO₂ concentration identifies strong Rubisco control above the thermal optimum. *New Phytol.*, **213**, 1036–1051, doi:10.1111/nph.14258.

Busch, J., and J. Engelmann, 2017: Cost-effectiveness of reducing emissions from tropical deforestation, 2016–2050. *Environ. Res. Lett.*, **13**, 15001, doi:10.1088/1748-9326/aa907c.

Butt, N., P.A. De Oliveira, and M.H. Costa, 2011: Evidence that deforestation affects the onset of the rainy season in Rondonia, Brazil. *J. Geophys. Res. Atmos.*, **116**, 2–9, doi:10.1029/2010JD015174.

Von Buttlar, J. et al., 2018: Impacts of droughts and extreme-temperature events on gross primary production and ecosystem respiration: A systematic assessment across ecosystems and climate zones. *Biogeosciences*, **15**, 1293–1318, doi:10.5194/bg-15-1293-2018.

Byrne, M.P. and P.A. O'Gorman, 2013: Land-ocean warming contrast over a wide range of climates: Convective quasi-equilibrium theory and idealized simulations. *J. Clim.*, **26**, 4000–4016, doi:10.1175/JCLI-D-12-00262.1.

Byrne, M.P., and P.A. O'Gorman, 2015: The response of precipitation minus evapotranspiration to climate warming: Why the "wet-get-wetter, dry-get-drier" scaling does not hold over land. *J. Clim.*, **28**, 8078–8092, doi:10.1175/JCLI-D-15-0369.1.

Cai, W. et al., 2014a: Increasing frequency of extreme El Niño events due to greenhouse warming. *Nat. Clim. Chang.*, **4**, 111–116, doi:10.1038/nclimate2100.

Cai, W. et al., 2014b: Increased frequency of extreme Indian ocean dipole events due to greenhouse warming. *Nature*, **510**, 254–258, doi:10.1038/nature13327.

Cai, W. et al., 2015: Increased frequency of extreme La Niña events under greenhouse warming. *Nat. Clim. Chang.*, **5**, 132–137, doi:10.1038/nclimate2492.

Cai, X., X. Zhang, and D. Wang, 2011: Land availability for biofuel production. *Environ. Sci. Technol.*, **45**, 334–339, doi:10.1021/es103338e.

Calvin, K. et al., 2014: Trade-offs of different land and bioenergy policies on the path to achieving climate targets. *Clim. Change*, **123**, 691–704, doi:10.1007/s10584-013-0897-y.

Calvin, K. et al., 2017: The SSP4: A world of deepening inequality. *Glob. Environ. Chang.*, **42**, 284–296, doi:10.1016/J.GLOENVCHA.2016.06.010.

Calvo, A.I. et al., 2013: Research on aerosol sources and chemical composition: Past, current and emerging issues. *Atmos. Res.*, **120–121**, 1–28, doi:10.1016/j.atmosres.2012.09.021.

Camenzind, T., S. Hättenschwiler, K.K. Treseder, A. Lehmann, and M.C. Rillig, 2018: Nutrient limitation of soil microbial processes in tropical forests. *Ecol. Monogr.*, **88**, 4–21, doi:10.1002/ecm.1279.

Camilloni, I., and M. Barrucand, 2012: Temporal variability of the Buenos Aires, Argentina, urban heat island. *Theor. Appl. Climatol.*, **107**, 47–58, doi:10.1007/s00704-011-0459-z.

Campbell, J.E., D.B. Lobell, R.C. Genova, and C.B. Field, 2008: The global potential of bioenergy on abandoned agriculture lands. *Environ. Sci. Technol.*, **42**, 5791–5794, doi:10.1021/es800052w.

Campbell, J.E. et al., 2017: Large historical growth in global terrestrial gross primary production. *Nature*, **544**, 84–87, doi:10.1038/nature22030.

Campioli, M. et al., 2015: Biomass production effciency controlled by management in temperate and boreal ecosystems. *Nat. Geosci.*, **8**, 843–846, doi: 10.1038/ngeo2553.

Campra, P., M. Garcia, Y. Canton, and A. Palacios-Orueta, 2008: Surface temperature cooling trends and negative radiative forcing due to land use change toward greenhouse farming in southeastern Spain. *J. Geophys. Res. Atmos.*, **113**, 1–10, doi:10.1029/2008JD009912.

Canadell, J.G. et al., 2007: Contributions to accelerating atmospheric CO₂ growth from economic activity, carbon intensity, and efficiency of natural sinks. *Proc. Natl. Acad. Sci.*, **104**, 18866–18870, doi:10.1073/pnas.0702737104.

Cardenas, L.M. et al., 2019: Nitrogen use efficiency and nitrous oxide emissions from five UK fertilised grasslands. *Sci. Total Environ.*, **661**, 696–710, doi:10.1016/j.scitotenv.2019.01.082.

Carey, J.C. et al., 2016: Temperature response of soil respiration largely unaltered with experimental warming. *Proc. Natl. Acad. Sci.*, **113**, 13797–13802, doi:10.1073/pnas.1605365113.

Carissimo, B.C., A.H. Oort, and T.H. Vonder Haar, 1985: Estimating the meridional energy transports in the atmosphere and ocean. *J. Phys. Oceanogr.*, **15**, 82–91, doi:10.1175/1520-0485(1985)015<0082:ETMETI>2.0.CO;2.

Carlson, B.Z. et al., 2017: Observed long-term greening of alpine vegetation – A case study in the French Alps. *Environ. Res. Lett.*, **12**, 114006, doi:10.1088/1748-9326/aa84bd.

Carrer, D., G. Pique, M. Ferlicoq, X. Ceamanos, and E. Ceschia, 2018: What is the potential of cropland albedo management in the fight against global warming? A case study based on the use of cover crops. *Environ. Res. Lett.*, **13**, 44030, doi:10.1088/1748-9326/aab650.

Carter, S., M. et al., 2015: Mitigation of agricultural emissions in the tropics: Comparing forest land-sparing options at the national level. *Biogeosciences*, **12**, 4809–4825, doi:10.5194/bg-12-4809-2015.

Carvalhais, N. et al., 2014: Global covariation of carbon turnover times with climate in terrestrial ecosystems. *Nature*, **514**, 213–217, doi:10.1038/nature13731.

Cavan, G. et al., 2014: Urban morphological determinants of temperature regulating ecosystem services in two African cities. *Ecol. Indic.*, **42**, 43–57, doi:10.1016/j.ecolind.2014.01.025.

Ceccherini, G., S. Russo, I. Ameztoy, C. Patricia Romero, and C. Carmona-Moreno, 2016: Magnitude and frequency of heat and cold waves in recent decades: The case of South America. *Nat. Hazards Earth Syst. Sci.*, **16**, 821–831, doi:10.5194/nhess-16-821-2016.

Ceccherini, G., S. Russo, I. Ameztoy, A. Francesco Marchese, and C. Carmona-Moreno, 2017: Heat waves in Africa 1981-2015, observations and reanalysis. *Nat. Hazards Earth Syst. Sci.*, **17**, 115–125, doi:10.5194/nhess-17-115-2017.

Ceschia, E. et al., 2017: Potentiel d'atténuation des changements climatiques par les couverts intermédiaires. *Innov. Agron.*, **62**, 43–58, doi:10.15454/1.517402718167511E12.

Chadwick, D.R. et al., 2018: The contribution of cattle urine and dung to nitrous oxide emissions: Quantification of country specific emission factors and implications for national inventories. *Sci. Total Environ.*, **635**, 607–617, doi:10.1016/j.scitotenv.2018.04.152.

Chagas, V.B.P. and P.L.B. Chaffe, 2018: The role of land cover in the propagation of rainfall into streamflow trends. *Water Resour. Res.*, **54**, 5986–6004, doi:10.1029/2018WR022947.

Challinor, A.J. et al., 2014: A meta-analysis of crop yield under climate change and adaptation. *Nat. Clim. Chang.*, **4**, 287–291, doi:10.1038/nclimate2153.

Chan, D. and Q. Wu, 2015: Significant anthropogenic-induced changes of climate classes since 1950. *Sci. Rep.*, **5**, 13487, doi:10.1038/srep13487.

Chang, J. et al., 2017: Future productivity and phenology changes in European grasslands for different warming levels: Implications for grassland management and carbon balance. *Carbon Balance Manag.*, **12**, 11, doi:10.1186/s13021-017-0079-8.

Charney, J.G., 1975: Dynamics of deserts and drought in the Sahel. *Q.J.R. Meteorol. Soc.*, **101**, 193–202, doi:10.1002/qj.49710142802.

Chase, T.N., R.A. Pielke, T.G.F. Kittel, R.R. Nemani, and S.W. Running, 2000: Simulated impacts of historical land cover changes on global climate in northern winter. *Clim. Dyn.*, doi:10.1007/s003820050007.

Chase, T.N. et al., 2001: Relative climatic effects of landcover change and elevated carbon dioxide combined with aerosols: A comparison of model results and observations. *J. Geophys. Res.*, **106**, 31685, doi:10.1029/2000JD000129.

Chaudhary, N., P.A. Miller, and B. Smith, 2017: Modelling Holocene peatland dynamics with an individual-based dynamic vegetation model. *Biogeosciences*, doi:10.5194/bg-14-2571-2017.

Chawla, I., K.K. Osuri, P.P. Mujumdar, and D. Niyogi, 2018: Assessment of the Weather Research and Forecasting (WRF) model for simulation of extreme rainfall events in the upper Ganga Basin. *Hydrol. Earth Syst. Sci.*, **22**, 1095–1117, doi:10.5194/hess-22-1095-2018.

Cheesman, S., C. Thierfelder, N.S. Eash, G.T. Kassie, and E. Frossard, 2016: Soil carbon stocks in conservation agriculture systems of Southern Africa. *Soil Tillage Res.*, **156**, 99–109, doi:10.1016/j.still.2015.09.018.

Chen, B. et al., 2016a: Exploring the possible effect of anthropogenic heat release due to global energy consumption upon global climate: a climate model study. *Int. J. Climatol.*, **36**, 4790–4796, doi:10.1002/joc.4669.

Chen, C. et al., 2019: China and India lead in greening of the world through land-use management. *Nat. Sustain.*, **2**, 122–129, doi:10.1038/s41893-019-0220-7.

Chen, D. and H.W. Chen, 2013: Using the Köppen classification to quantify climate variation and change: An example for 1901–2010. *Environ. Dev.*, **6**, 69–79, doi:10.1016/j.envdev.2013.03.007.

Chen, G.S., M. Notaro, Z. Liu, and Y. Liu, 2012: Simulated local and remote biophysical effects of afforestation over the Southeast United States in boreal summer. *J. Clim.*, **25**, 4511–4522, doi:10.1175/JCLI-D-11-00317.1.

Chen, J., 2007: Rapid urbanization in China: A real challenge to soil protection and food security. *CATENA*, **69**, 1–15, doi:https://doi.org/10.1016/j.catena.2006.04.019.

Chen, L., P.A. Dirmeyer, Z. Guo, and N.M. Schultz, 2018: Pairing FLUXNET sites to validate model representations of land-use/land-cover change. *Hydrol. Earth Syst. Sci.*, **22**, 111–125, doi:10.5194/hess-22-111-2018.

Chen, W., Y. Zhang, X. Mai, and Y. Shen, 2016b: Multiple mechanisms contributed to the reduced stability of Inner Mongolia grassland ecosystem following nitrogen enrichment. *Plant Soil*, **409**, 283–296, doi:10.1007/s11104-016-2967-1.

Chen, X. and S.J. Jeong, 2018: Irrigation enhances local warming with greater nocturnal warming effects than daytime cooling effects. *Environ. Res. Lett.*, **13**, 24005, doi:10.1088/1748-9326/aa9dea.

Chen, Y. et al., 2017: A pan-tropical cascade of fire driven by El Niño/Southern Oscillation. *Nat. Clim. Chang.*, **7**, 906–911, doi:10.1038/s41558-017-0014-8.

Cheng, L. et al., 2017: Warming enhances old organic carbon decomposition through altering functional microbial communities. *Isme J.*, **11**, 1825. https://doi.org/10.1038/ismej.2017.48.

Cherubini, F. et al., 2016: Global spatially explicit CO_2 emission metrics for forest bioenergy. *Sci. Rep.*, **6**, 20186, doi:10.1038/srep20186 http://dx.doi.org/10.1038/srep20186.

Chiang, F., O. Mazdiyasni, and A. AghaKouchak, 2018: Amplified warming of droughts in southern United States in observations and model simulations. *Sci. Adv.*, **4**, eaat2380, doi:10.1126/sciadv.aat2380.

Chitra-Tarak, R. et al., 2017: The roots of the drought: Hydrology and water uptake strategies mediate forest-wide demographic response to precipitation. *J. Ecol.*, **106**, 1495–1507, doi:10.1111/1365-2745.12925.

Choat, B. et al., 2012: Global convergence in the vulnerability of forests to drought. *Nature*, **491**, 752–755, doi:10.1038/nature11688.

Cholakian, A., A. Colette, G. Ciarelli, I. Coll, and M. Beekmann, 2018: Future climatic drivers and their effect on PM10 components in Europe and the Mediterranean Sea. *Atmos. Chem. Phys. Discuss.*, 1–35, doi:10.5194/acp-2018-868.

Christen, A. et al., 2016: Summertime greenhouse gas fluxes from an urban bog undergoing restoration through rewetting. *Mires Peat*, **17**, 1–24, doi:10.19189/MaP.2015.OMB.207.

Christensen, J.H., and O.B. Christensen, 2003: Severe summertime flooding in Europe. *Nature*, **421**, 805–806, doi:10.1038/421805a.

Christensen, J.H. et al., 2013: Climate phenomena and their relevance for future regional climate change. Climate Change 2013: The Physical Science Basis. Contribution of Working Group I to the Fifth Assessment Report of the Intergovernmental Panel on Climate Change [Stocker, T.F., D. Qin, G.-K. Plattner, M. Tignor, S.K. Allen, J. Boschung, A. Nauels, Y. Xia, V. Bex and P.M. Midgley (eds.)], Cambridge University Press, Cambridge, United Kingdom and New York, NY, USA, pp. 1217–1308.

Christidis, N., P.A. Stott, G.C. Hegerl, and R.A. Betts, 2013: The role of land use change in the recent warming of daily extreme temperatures. *Geophys. Res. Lett.*, **40**, 589–594, doi:10.1002/grl.50159.

Christopher, S.F., R. Lal, and U. Mishra, 2009: Regional study of no-till effects on carbon sequestration in the Midwestern United States. *Soil Sci. Soc. Am. J.*, **73**, 207–216, doi:10.2136/sssaj2007.0336.

Christy, J.R., W.B. Norris, K. Redmond, and K.P. Gallo, 2006: Methodology and results of calculating central California surface temperature trends: Evidence of human-induced climate change? *J. Clim.*, **19**, 548–563, doi:10.1175/JCLI3627.1.

Chrysanthou, A., G. Van Der Schrier, E.J.M. Van Den Besselaar, A.M.G. Klein Tank, and T. Brandsma, 2014: The effects of urbanization on the rise of the European temperature since 1960. *Geophys. Res. Lett.*, **41**, 7716–7722, doi:10.1002/2014GL061154.

Chum, H., A. Faaij, J. Moreira, G. Berndes, P. Dhamija, H. Dong, B. Gabrielle, A. Goss Eng, W. Lucht, M. Mapako, O. Masera Cerutti, T. McIntyre, T. Minowa, and K. Pingoud, 2011: Bioenergy. In: IPCC Special Report on Renewable Energy Sources and Climate Change Mitigation, (Edenhofer, O., R. Pichs-Madruga, Y. Sokona, K. Seyboth, P. Matschoss, S. Kadner, T. Zwickel, P. Eickemeier, G. Hansen, S. Schlömer, C. von Stechow (eds.)]. Cambridge University Press, Cambridge, United Kingdom and New York, NY, USA, United Kingdom and New York, NY, USA, pp. 209–332.

Ciais, P. et al., 2005: Europe-wide reduction in primary productivity caused by the heat and drought in 2003. *Nature*, **437**, 529–533, doi:10.1038/nature03972.

Ciais, P. Ciais, P., C. Sabine, G. Bala, L. Bopp, V. Brovkin, J. Canadell, A. Chhabra, R. DeFries, J. Galloway, M. Heimann, C. Jones, C. Le Quéré, R.B. Myneni, S. Piao, and P. Thornton, 2013a: Carbon and other biogeochemical cycles. In: Climate Change 2013: The Physical Science Basis. Contribution of Working Group I to the Fifth Assessment Report of the Intergovernmental Panel on Climate Change [Stocker, T.F., D. Qin, G.-K. Plattner, M. Tignor, S.K. Allen, J. Boschung, A. Nauels, Y. Xia, V. Bex and P.M. Midgley (eds.)]. Cambridge University Press, Cambridge, United Kingdom and New York, NY, USA, pp. 465–570.

Cintas, O. et al., 2017: Carbon balances of bioenergy systems using biomass from forests managed with long rotations: Bridging the gap between stand and landscape assessments. *GCB Bioenergy*, **9**, 1238–1251, doi:10.1111/gcbb.12425.

Cirino, G.G., R.A.F. Souza, D.K. Adams, and P. Artaxo, 2014: The effect of atmospheric aerosol particles and clouds on net ecosystem exchange in the Amazon. *Atmos. Chem. Phys.*, **14**, 6523–6543, doi:10.5194/acp-14-6523-2014.

Claeys, M. et al., 2004: Formation of secondary organic aerosols from isoprene and its gas-phase oxidation products through reaction with hydrogen peroxide. *Atmos. Environ.*, **38**, 4093–4098, doi:10.1016/j.atmosenv.2004.06.001.

2

Claussen, M. et al., 2001: Biophysical versus biogeochemical feedbacks of large-scale land cover change. *Geophys. Res. Lett.*, **28**, 1011–1014, doi:10.1029/2000GL012471.

Cleveland, C. et al., 2015: Future productivity and carbon storage limited by terrestrial nutrient availability Future productivity and carbon storage limited by terrestrial nutrient availability. *Nat. Geosci.*, **8**, 441, doi:10.1038/NGEO2413.

Cohen, J.B. and C. Wang, 2014: Estimating global black carbon emissions using a top-down Kalman Filter approach. *J. Geophys. Res. Atmos.*, **119**, 307–323, doi:10.1002/2013JD019912.

Colwell, R.K., G. Brehm, C.L. Cardelús, A.C. Gilman, and J.T. Longino, 2008a: Global warming, elevational range shifts, and lowland biotic attrition in the wet tropics. *Science*, **322**, 258–261, doi:10.1126/science.1162547.

Colwell, R.K., G. Brehm, C.L. Cardelús, A.C. Gilman, and J.T. Longino, 2008b: Global warming, elevational range shifts, and lowland biotic attrition in the wet tropics. *Science*, **322**, 258–261, doi:10.1126/science.1162547.

Coma, J. et al., 2017: Vertical greenery systems for energy savings in buildings: A comparative study between green walls and green facades. *Build. Environ.*, **111**, 228–237, doi:10.1016/j.buildenv.2016.11.014.

Conant, R.T. et al., 2011: Temperature and soil organic matter decomposition rates - synthesis of current knowledge and a way forward. *Glob. Chang. Biol.*, **17**, 3392–3404, doi:10.1111/j.1365-2486.2011.02496.x.

Conant, R.T., C.E.P. Cerri, B.B. Osborne, and K. Paustian, 2017: Grassland management impacts on soil carbon stocks: A new synthesis. *Ecol. Appl.*, doi:10.1002/eap.1473.

de Coninck, H., A. Revi, M. Babiker, P. Bertoldi, M. Buckeridge, A. Cartwright, W. Dong, J. Ford, S. Fuss, J.-C. Hourcade, D. Ley, R. Mechler, P. Newman, A. Revokatova, S. Schultz, L. Steg, and T. Sugiyama, 2018, 2018: Strengthening and implementing the global response. In: Global Warming of 1.5°C: An IPCC special report on the impacts of global warming of 1.5°C above pre-industrial levels and related global greenhouse gas emission pathways, in the context of strengthening the global response to the threat of climate change [V. Masson-Delmotte, P. Zhai, H.-O. Pörtner, D. Roberts, J. Skea, P.R. Shukla, A. Pirani, W. Moufouma-Okia, C. Péan, R. Pidcock, S. Connors, J.B.R. Matthews, Y. Chen, X. Zhou, M.I. Gomis, E. Lonnoy, T. Maycock, M. Tignor, and T. Waterfield (eds.)]. In press.

Connor, E.W. and C.V. Hawkes, 2018: Effects of extreme changes in precipitation on the physiology of C4 grasses. *Oecologia*, **188**, 355–365, doi:10.1007/s00442-018-4212-5.

Constantin, J. et al., 2010: Effects of catch crops, no till and reduced nitrogen fertilization on nitrogen leaching and balance in three long-term experiments. *Agric. Ecosyst. Environ.*, **135**, 268–278, doi:10.1016/j.agee.2009.10.005.

Constantin, J. et al., 2011: Cumulative effects of catch crops on nitrogen uptake, leaching and net mineralization. *Plant Soil*, **341**, 137–154, doi:10.1007/s11104-010-0630-9.

Contini, D., R. Vecchi, and M. Viana, 2018: Carbonaceous aerosols in the atmosphere. *Atmosphere*, **9**, 181, doi:10.3390/atmos9050181.

Cook, B.I., T.R. Ault, and J.E. Smerdon, 2015: Unprecedented 21st century drought risk in the American Southwest and Central Plains. *Sci. Adv.*, **1**, e1400082–e1400082, doi:10.1126/sciadv.1400082.

Cook, B.I., J.S. Mankin, and K.J. Anchukaitis, 2018: Climate change and drought: From past to future. *Curr. Clim. Chang. Reports*, **4**, 164–179, doi:10.1007/s40641-018-0093-2.

da Costa, A.C.L. et al., 2010: Effect of 7 yr of experimental drought on vegetation dynamics and biomass storage of an eastern Amazonian rainforest. *New Phytol.*, **187**, 579–591, doi:10.1111/j.1469-8137.2010.03309.x.

Costa, H., G. Floater, H. Hooyberghs, S. Verbeke, K. De Ridder, 2016: *Climate change , heat stress and labour productivity: A cost methodology for city economies*. Working Paper No. 278, Centre for Climate Change Economics and Policy. and Working Paper No. 248, Grantham Research Institute on Climate Change and the Environment, London, UK, 15 pp.

Coumou, D. and S. Rahmstorf, 2012: A decade of weather extremes. *Nat. Clim. Chang.*, **2**, 491–496, doi:10.1038/nclimate1452.

Couwenberg, J., R. Dommain, and H. Joosten, 2010: Greenhouse gas fluxes from tropical peatlands in south-east Asia. *Glob. Chang. Biol.*, **16**, 1715–1732, doi:10.1111/j.1365-2486.2009.02016.x.

Cowan, T. et al., 2014: More frequent, longer, and hotter heat waves for Australia in the twenty-first century. *J. Clim.*, **27**, 5851–5871, doi:10.1175/JCLI-D-14-00092.1.

Cowie, A.L., M.U.F. Kirschbaum, and M. Ward, 2007: Options for including all lands in a future greenhouse gas accounting framework. *Env. Sci Policy*, **10**, 306–321.

Cowling, S.A., C.D. Jones, and P.M. Cox, 2009: Greening the terrestrial biosphere: Simulated feedbacks on atmospheric heat and energy circulation. *Clim. Dyn.*, **32**, 287–299, doi:10.1007/s00382-008-0481-8.

Cox, P.M. et al., 2013: Sensitivity of tropical carbon to climate change constrained by carbon dioxide variability. *Nature*, **494**, 341–344, doi:10.1038/nature11882.

Craine, J.M. et al., 2018: Isotopic evidence for oligotrophication of terrestrial ecosystems. *Nat. Ecol. Evol.*, **2**, 1735–1744, doi:10.1038/s41559-018-0694-0.

Crétat, J., B. Pohl, Y. Richard, and P. Drobinski, 2012: Uncertainties in simulating regional climate of Southern Africa: Sensitivity to physical parameterizations using WRF. *Clim. Dyn.*, **38**, 613–634, doi:10.1007/s00382-011-1055-8.

Creutzig, F. et al., 2015a: Bioenergy and climate change mitigation: An assessment. *GCB Bioenergy*, **7**, 916–944, doi:10.1111/gcbb.12205.

Creutzig, F., G. Baiocchi, R. Bierkandt, P.-P. Pichler, and K.C. Seto, 2015b: Global typology of urban energy use and potentials for an urbanization mitigation wedge. *Proc. Natl. Acad. Sci.*, **112**, 6283–6288, doi:10.1073/pnas.1315545112.

Crimp, S.J. et al., 2016: Recent seasonal and long-term changes in southern Australian frost occurrence. *Clim. Change*, **139**, 115–128, doi:10.1007/s10584-016-1763-5.

Crook, J.A., L.S. Jackson, S.M. Osprey, and P.M. Forster, 2015: A comparison of temperature and precipitation responses to different earth radiation management geoengineering schemes. *J. Geophys. Res.*, **120**, 9352–9373, doi:10.1002/2015JD023269.

Crooks, S., D. Herr, J. Tamelander, D. Laffoley, and J. Vandever, 2011: *Mitigating Climate Change through Restoration and Management of Coastal Wetlands and Near-shore Marine Ecosystems: Challenges and Opportunities*. Environment department papers, no.121. Marine Ecosystem Series. World Bank, Washington, DC, USA. 69 pp.

Crowther, T.W. et al., 2015: Biotic interactions mediate soil microbial feedbacks to climate change. *Proc. Natl. Acad. Sci.*, **112**, 7033–7038, doi:10.1073/pnas.1502956112.

Crowther, T.W. et al., 2016: Quantifying global soil carbon losses in response to warming. *Nature*, **540**, 104–108, doi:10.1038/nature20150.

Curry, C.L., 2009: The consumption of atmospheric methane by soil in a simulated future climate. *Biogeosciences*, **6**, 2355–2367, doi:10.5194/bgd-6-6077-2009.

D'Amato, G., L. Cecchi, M. D'Amato, and G. Liccardi, 2010: Urban air pollution and climate change as environmental risk factors of respiratory allergy: An update. *J. Investig. Allergol. Clin. Immunol.*, **20**, 95–102.

Dahiya, B., 2012: Cities in Asia, 2012: Demographics, economics, poverty, environment and governance. *Cities*, **29**, S44–S61, doi:10.1016/j.cities.2012.06.013. doi:10.1016/j.cities.2012.06.013.

Dai, A., 2011: Drought under global warming: A review. *Wiley Interdiscip. Rev. Clim. Chang.*, **2**, 45–65, doi:10.1002/wcc.81.

Dai, A., 2013: Increasing drought under global warming in observations and models. *Nat. Clim. Chang.*, **3**, 52–58, doi:10.1038/nclimate1633.

Dai, A., and T. Zhao, 2017: Uncertainties in historical changes and future projections of drought. Part I: Estimates of historical drought changes. *Clim. Change*, **144**, 519–533, doi:10.1007/s10584-016-1705-2.

Dai, A., T. Zhao, and J. Chen, 2018: Climate change and drought: A precipitation and evaporation perspective. *Curr. Clim. Chang. Reports*, **4**, 301–312, doi:10.1007/s40641-018-0101-6.

Daioglou, V., E. Stehfest, B. Wicke, A. Faaij, and D.P. van Vuuren, 2016: Projections of the availability and cost of residues from agriculture and forestry. *GCB Bioenergy*, doi:10.1111/gcbb.12285.

Daioglou, V. et al., 2017: Greenhouse gas emission curves for advanced biofuel supply chains. *Nat. Clim. Chang.*, doi:10.1038/s41558-017-0006-8.

Daioglou, V., J.C. Doelman, B. Wicke, A. Faaij, and D.P. van Vuuren, 2019: Integrated assessment of biomass supply and demand in climate change mitigation scenarios. *Glob. Environ. Chang.*, doi:10.1016/j.gloenvcha.2018.11.012.

Dale, V.H., E. Parish, K.L. Kline, and E. Tobin, 2017: How is wood-based pellet production affecting forest conditions in the southeastern United States? *For. Ecol. Manage.*, doi:10.1016/j.foreco.2017.03.022.

Dangal, S.R.S. et al., 2019: Global nitrous oxide emissions from pasturelands and rangelands: Magnitude, spatiotemporal patterns, and attribution. *Global Biogeochem. Cycles*, **33**, 200–222, doi:10.1029/2018GB006091.

Daniels, E.E., G. Lenderink, R.W.A. Hutjes, and A.A.M. Holtslag, 2016: Observed urban effects on precipitation along the Dutch west coast. *Int. J. Climatol.*, **36**, 2111–2119, doi:10.1002/joc.4458.

Dargie, G.C. et al., 2019: Congo Basin peatlands: Threats and conservation priorities. *Mitig. Adapt. Strateg. Glob. Chang.*, **24**, 669–686, doi:10.1007/s11027-017-9774-8.

Dass, P., C. Müller, V. Brovkin, and W. Cramer, 2013: Can bioenergy cropping compensate high carbon emissions from large-scale deforestation of high latitudes? *Earth Syst. Dyn.*, **4**, 409–424, doi:10.5194/esd-4-409-2013.

Davidson, E.A., 2009: The contribution of manure and fertilizer nitrogen to atmospheric nitrous oxide since 1860. *Nat. Geosci.*, **2**, 659–662, doi:10.1038/ngeo608.

Davidson, E.A., I.A. Janssens, E.A. Davidson, and I.A. Janssens, 2006a: Temperature sensitivity of soil carbon decomposition and feedbacks to climate change. *Nature*, **440**, 165–173, doi:10.1038/nature04514.

Davidson, E.A., I.A. Janssens, and Y.Q. Luo, 2006b: On the variability of respiration in terrestrial ecosystems: Moving beyond Q(10). *Glob. Chang. Biol.*, **12**, 154–164, doi:10.1111/j.1365-2486.2005.01065.x.

Davidson, E.A. et al., 2012: The Amazon basin in transition. *Nature*, **481**, 321–328, doi:10.1038/nature10717.

Davies-Barnard, T., P.J. Valdes, J.S. Singarayer, F.M. Pacifico, and C.D. Jones, 2014: Full effects of land use change in the representative concentration pathways. *Environ. Res. Lett.*, **9**, doi:10.1088/1748-9326/9/11/114014.

Davies-Barnard, T., P.J. Valdes, J.S. Singarayer, A.J. Wiltshire, and C.D. Jones, 2015: Quantifying the relative importance of land cover change from climate and land use in the representative concentration pathways. *Global Biogeochem. Cycles*, **29**, 842–853, doi:10.1002/2014GB004949.

Davies, Z.G., J.L. Edmondson, A. Heinemeyer, J.R. Leake, and K.J. Gaston, 2011: Mapping an urban ecosystem service: Quantifying above-ground carbon storage at a city-wide scale. *J. Appl. Ecol.*, **48**, 1125–1134, doi:10.1111/j.1365-2664.2011.02021.x.

Davin, E.L., and N. de Noblet-Ducoudre, 2010: Climatic impact of global-scale deforestation: Radiative versus nonradiative processes. *J. Clim.*, **23**, 97–112, doi:10.1175/2009JCLI3102.1.

Davin, E.L., N. de Noblet-Ducoudré, and P. Friedlingstein, 2007: Impact of land cover change on surface climate: Relevance of the radiative forcing concept. *Geophys. Res. Lett.*, **34**, n/a-n/a, doi:10.1029/2007GL029678.

Davin, E.L., S.I. Seneviratne, P. Ciais, A. Olioso, and T. Wang, 2014: Preferential cooling of hot extremes from cropland albedo management. *Proc. Natl. Acad. Sci.*, **111**, 9757–9761, doi:10.1073/pnas.1317323111.

Demuzere, M. et al., 2014: Mitigating and adapting to climate change: Multifunctional and multi-scale assessment of green urban infrastructure. *J. Environ. Manage.*, **146**, 107–115, doi:10.1016/j.jenvman.2014.07.025.

Derksen, C., and R. Brown, 2012: Spring snow cover extent reductions in the 2008-2012 period exceeding climate model projections. *Geophys. Res. Lett.*, **39**, doi:10.1029/2012GL053387. doi:10.1029/2012GL053387.

Derner, J.D. et al., 2003: Above- and below-ground responses of C3-C4 species mixtures to elevated CO_2 and soil water availability. *Glob. Chang. Biol.*, **9**, 452–460, doi:10.1046/j.1365-2486.2003.00579.x.

Deryng, D. et al., 2016b: Regional disparities in the beneficial effects of rising CO_2 concentrations on crop water productivity. *Nat. Clim. Chang.*, **6**, 786–790, doi:10.1038/nclimate2995.

Deser, C., A. Phillips, V. Bourdette, and H. Teng, 2012: 1. Uncertainty in climate change projections: The role of internal variability. *Clim. Dyn.*, **38**, 527–546, doi:10.1007/s00382-010-0977-x.

Devaraju, N., G. Bala, and A. Modak, 2015a: Effects of large-scale deforestation on precipitation in the monsoon regions: Remote versus local effects. *Proc. Natl. Acad. Sci.*, **112**, 3257–3262, doi:10.1073/pnas.1423439112.

Devaraju, N., G. Bala, and R. Nemani, 2015b: Modelling the influence of land-use changes on biophysical and biochemical interactions at regional and global scales. *Plant, Cell Environ.*, 38, 1931–1946, doi:10.1111/pce.12488.

Devaraju, N., G. Bala, K. Caldeira, and R. Nemani, 2016: A model based investigation of the relative importance of CO_2-fertilization, climate warming, nitrogen deposition and land use change on the global terrestrial carbon uptake in the historical period. *Clim. Dyn.*, **47**, 173–190, doi:10.1007/s00382-015-2830-8.

Devaraju, N., N. de Noblet-Ducoudré, B. Quesada, and G. Bala, 2018: Quantifying the relative importance of direct and indirect biophysical effects of deforestation on surface temperature and teleconnections. *J. Clim.*, **31**, 3811–3829, doi:10.1175/JCLI-D-17-0563.1.

Dickie, A., C. Streck, S. Roe, M. Zurek, F. Haupt, and A. Dolginow, 2014: *Strategies for Mitigating Climate Change in Agriculture:Recommentations for Philanthropy-Executive Summary*. Climate Focus and California Environmental Associates, and the Climate and Land Use Alliance. 87 pp.

Diffenbaugh, N.S., and C.B. Field, 2013: Changes in ecologically critical terrestrial climate conditions. *Science*, **341**, 486–492, doi:10.1126/science.1237123.

Diffenbaugh, N.S., D.L. Swain, and D. Touma, 2015: Anthropogenic warming has increased drought risk in California. *Proc. Natl. Acad. Sci.*, **112**, 3931–3936, doi:10.1073/pnas.1422385112.

Van Dijk, A.I.J.M. et al., 2013: The Millennium Drought in southeast Australia (2001–2009): Natural and human causes and implications for water resources, ecosystems, economy, and society. *Water Resour. Res.*, **49**, 1040–1057, doi:10.1002/wrcr.20123.

Dirmeyer, P.A., Z. Wang, M.J. Mbuh, and H.E.Norton, 2014: Intensified land surface control on boundary layer growth in a changing climate. *Geophys. Res. Lett.*, **45**, 1290–1294, doi:10.1002/2013GL058826.Received.

Dlugokencky, E.J., L.P. Steele, P.M. Lang, and K.A. Masarie, 1994: The growth rate and distribution of atmospheric methane. *J. Geophys. Res.*, **99**, 17021, doi:10.1029/94JD01245.

Doan, Q. Van, H. Kusaka, and Q.B. Ho, 2016: Impact of future urbanization on temperature and thermal comfort index in a developing tropical city: Ho Chi Minh City. *Urban Clim.*, **17**, 20–31, doi:10.1016/j.uclim.2016.04.003.

Doerr, S.H., and C. Santín, 2016: Global trends in wildfire and its impacts: Perceptions versus realities in a changing world. *Philos. Trans. R. Soc. B Biol. Sci.*, **371**, 20150345, doi:10.1098/rstb.2015.0345.

Don, A. et al., 2012: Land-use change to bioenergy production in Europe: implications for the greenhouse gas balance and soil carbon. *GCB Bioenergy*, **4**, 372–391, doi:10.1111/j.1757-1707.2011.01116.x.

Donat, M.G., A.J. Pitman, and O. Angélil, 2018: Understanding and reducing future uncertainty in midlatitude daily heat extremes via land surface feedback constraints. *Geophys. Res. Lett.*, **45**, 10,627-10,636, doi:10.1029/2018GL079128.

Dong, J., and W.T. Crow, 2019. L-band remote-sensing increases sampled levels of global soil moisture-air temperature coupling strength. *Remote Sens. Environ.*, **220**, 51–58, doi:10.1016/j.rse.2018.10.024.

2

Donohue, R.J., M.L. Roderick, T.R. McVicar, and G.D. Farquhar, 2013: Impact of CO_2 fertilization on maximum foliage cover across the globe's warm, arid environments. *Geophys. Res. Lett.*, **40**, 3031–3035, doi:10.1002/grl.50563.

Dooley, K., and S. Kartha, 2018: Land-based negative emissions: Risks for climate mitigation and impacts on sustainable development. *Int. Environ. Agreements Polit. Law Econ.*, doi:10.1007/s10784-017-9382-9.

Dosio, A., 2017: Projection of temperature and heat waves for Africa with an ensemble of CORDEX Regional Climate Models. *Clim. Dyn.*, **49**, 493–519, doi:10.1007/s00382-016-3355-5.

Dosio, A., L. Mentaschi, E.M. Fischer, and K. Wyser, 2018: Extreme heat waves under 1.5°C and 2°C global warming. *Environ. Res. Lett.*, **13**, 54006, doi:10.1088/1748-9326/aab827.

Dou, J., Y. Wang, R. Bornstein, and S. Miao, 2014: Observed spatial characteristics of Beijing urban climate impacts on summer thunderstorms. *J. Appl. Meteorol. Climatol.*, **54**, 94–105, doi:10.1175/JAMC-D-13-0355.1.

Doughty, C.E. et al., 2015: Drought impact on forest carbon dynamics and fluxes in Amazonia. *Nature*, **519**, 78–82, doi:10.1038/nature14213.

Drake, J.E. et al., 2011: Increases in the flux of carbon belowground stimulate nitrogen uptake and sustain the long-term enhancement of forest productivity under elevated CO_2. *Ecol. Lett.*, **14**, 349–357, doi:10.1111/j.1461-0248.2011.01593.x.

Drösler, M., L.V. Verchot, A. Freibauer, G. Pan, A. Freibauer, C.D. Evans, R.A. Bourboniere, J.P.Alm, S. Page, F.Agus, S. Sabiham, C.Wang, N. Srivastava, L. Borheau-Chavez, J. Couwenberg, K. Hergoualc'h, A. Hooijer, J. Jauhiainen, K. Minkkinen, N. French, T. Strand, A. Sirin, R. Mickler, K. Tansey, N. Larkin, 2014. Drained Inland Organic Soils, 2013. In: Task Force on National Greenhouse Gas Inventories of the IPCC, [Hiraishi T., T. Krug, K. Tanabe, N. Srivastava, B. Jamsranjav, M. Fukuda, and T. Troxler (eds.)]. Supplement to the 2006 IPCC Guidelines: Wetlands. Hayama, Japan: Institute for Global Environmental Strategies (IGES) on behalf of the Intergovernmental Panel on Climate Change (IPCC).

Du, C.J., X.D. Wang, M.Y. Zhang, J. Jing, and Y.H. Gao, 2019: Effects of elevated CO_2 on plant C-N-P stoichiometry in terrestrial ecosystems: A meta-analysis. *Sci. Total Environ.*, **650**, 697–708, doi:10.1016/j.scitotenv.2018.09.051.

Du, X., and Z. Huang, 2017: Ecological and environmental effects of land use change in rapid urbanization: The case of Hangzhou, China. *Ecol. Indic.*, **81**, 243–251, doi:10.1016/j.ecolind.2017.05.040.

Duarte, C.M., 2017: Reviews and syntheses: Hidden forests, the role of vegetated coastal habitats in the ocean carbon budget. *Biogeosciences*, **14**, 301–310, doi:10.5194/bg-14-301-2017.

Duarte, C.M., I.J. Losada, I.E. Hendriks, I. Mazarrasa, and N. Marba, 2013: The role of coastal plant communities for climate change mitigation and adaptation. *Nat. Clim. Chang.*, **3**, 961–968, doi:10.1038/nclimate1970.

Duden, A.S. et al., 2017: Modeling the impacts of wood pellet demand on forest dynamics in southeastern United States. *Biofuels, Bioprod. Biorefining*, doi:10.1002/bbb.1803.

Dungait, J.A.J., D.W. Hopkins, A.S. Gregory, and A.P. Whitmore, 2012: Soil organic matter turnover is governed by accessibility not recalcitrance. *Glob. Chang. Biol.*, **18**, 1781–1796, doi:10.1111/j.1365-2486.2012.02665.x.

Dutaur, L. and L.V. Verchot, 2007: A global inventory of the soil CH_4 sink. *Global Biogeochem. Cycles*, **21**, 1–9, doi:10.1029/2006GB002734.

Dutta, M.K., T.S. Bianchi, and S.K. Mukhopadhyay, 2017: Mangrove Methane Biogeochemistry in the Indian Sundarbans: A Proposed Budget. *Front. Mar. Sci.*, **4**, 187, doi:10.3389/fmars.2017.00187.

Duveiller, G., J. Hooker, and A. Cescatti, 2018: The mark of vegetation change on Earth's surface energy balance. *Nat. Commun.*, **9**, 679, doi:10.1038/s41467-017-02810-8.

Edmondson, J.L., Z.G. Davies, K.J. Gaston, and J.R. Leake, 2014: Urban cultivation in allotments maintains soil qualities adversely affected by conventional agriculture. *J. Appl. Ecol.*, **51**, 880–889, doi:10.1111/1365-2664.12254.

Edwards, M.E., L.B. Brubaker, A. V Lozhkin, and M. Patricia, 2005: Structurally novel biomes: A response to past warming in Beringia. Ecology, **86**, 1696–1703, doi:10.1890/03-0787.

Eekhout, J.P.C., J.E. Hunink, W. Terink, and J. de Vente, 2018: Why increased extreme precipitation under climate change negatively affects water security. *Hydrol. Earth Syst. Sci. Discuss.*, 1–16, doi:10.5194/hess-2018-161.

Elagib, N.A., 2011: Evolution of urban heat island in Khartoum. *Int. J. Climatol.*, **31**, 1377–1388, doi:10.1002/joc.2159.

Elbert, W. et al., 2012: Contribution of cryptogamic covers to the global cycles of carbon and nitrogen. *Nat. Geosci.*, **5**, 459–462, doi:10.1038/ngeo1486.

Elliott, J. et al., 2014: Constraints and potentials of future irrigation water availability on agricultural production under climate change. *Proc. Natl. Acad. Sci.*, **111**, 3239–3244, doi:10.1073/pnas.1222474110.

Ellison, D. et al., 2017: Trees, forests and water: Cool insights for a hot world. *Glob. Environ. Chang.*, **43**, 51–61, doi:10.1016/j.gloenvcha.2017.01.002.

Ellsworth, D.S. et al., 2017: Elevated CO_2 does not increase eucalypt forest productivity on a low-phosphorus soil. *Nat. Clim. Chang.*, **7**, 279–282, doi:10.1038/nclimate3235.

Elmar, K. et al., 2018: Pathways limiting warming to 1.5°C: A tale of turning around in no time? *Philos. Trans. R. Soc. A Math. Phys. Eng. Sci.*, **376**, 20160457, doi:10.1098/rsta.2016.0457.

Elmqvist, T. et al., 2016: Ecosystem Services Provided by Urban Green Infrastructure, In *Routledge Handbook of Ecosystem Services* [Potschin, M., R. Haines-Young, R. Fish, and R.K. Turner (eds.)]. 1st ed. Routledge, Oxford, UK, 630 pp.

Elsen, P.R., and M.W. Tingley, 2015: Global mountain topography and the fate of montane species under climate change. *Nat. Clim. Chang.*, **5**, 772–776, doi:10.1038/nclimate2656.

Elshout, P.M.F. et al., 2015: Greenhouse-gas payback times for crop-based biofuels. *Nat. Clim. Chang.*, **5**, 604–610, doi:10.1038/nclimate2642.

Engelbrecht, F. et al., 2015: Projections of rapidly rising surface temperatures over Africa under low mitigation. *Environ. Res. Lett.*, **10**, 85004, doi:10.1088/1748-9326/10/8/085004.

EPA, 2015: *Stormwater Management in Response to Climate Change Impacts: Lessons from the Chesapeake Bay and Great Lakes Regions (Final Report)*. EPA/600/R-15/087F, 2015. U.S. Environmental Protection Agency, Washington, DC, USA, 80 pp.

Epstein, H. et al., 2018: *Tundra Greeness*. In: NOAA Arctic Report Card 2018, 46–52 pp https://www.arctic.noaa.gov/Report-Card.

Erb, K.-H. et al., 2016: Biomass turnover time in terrestrial ecosystems halved by land use.Nat. Geosci., **9**, 674–678, doi:10.1038/NGE02782.

Erb, K.-H. et al., 2017: Land management: data availability and process understanding for global change studies. *Glob. Chang. Biol.*, **23**, 512–533, doi:10.1111/gcb.13443.

Erb, K.-H. et al., 2018: Unexpectedly large impact of forest management and grazing on global vegetation biomass. *Nature*, **553**, 73–76, doi:10.1038/nature25138.

Ercilla-Montserrat, M., P. Muñoz, J.I. Montero, X. Gabarrell, and J. Rieradevall, 2018: A study on air quality and heavy metals content of urban food produced in a Mediterranean city (Barcelona). *J. Clean. Prod.*, **195**, 385–395, doi:https://doi.org/10.1016/j.jclepro.2018.05.183.

Erfanian, A., G. Wang, and L. Fomenko, 2017: Unprecedented drought over tropical South America in 2016: Significantly under-predicted by tropical SST. *Sci. Rep.*, **7**, 5811, doi:10.1038/s41598-017-05373-2.

Ernakovich, J.G. et al., 2014: Predicted responses of arctic and alpine ecosystems to altered seasonality under climate change. *Glob. Chang. Biol.*, **20**, 3256–3269, doi:10.1111/gcb.12568.

Erşahin, S., B.C. Bilgili, Ü. Dikmen, and I. Ercanli, 2016: Net primary productivity of Anatolian forests in relation to climate, 2000–2010. *For. Sci.*, **62**, 698–709, doi:10.5849/forsci.15-171.

Esch, T. et al., 2017: Breaking new ground in mapping human settlements from space – The Global Urban Footprint. *ISPRS J. Photogramm. Remote Sens.*, **134**, 30–42, doi:10.1016/j.isprsjprs.2017.10.012.

Espinoza, J.C. et al., 2014: The extreme 2014 flood in south-western Amazon basin: The role of tropical-subtropical South Atlantic SST gradient. *Environ. Res. Lett.*, **9**, 124007, doi:10.1088/1748-9326/9/12/124007.

Espinoza, J.C., J. Ronchail, J.A. Marengo, and H. Segura, 2018: Contrasting North–South changes in Amazon wet-day and dry-day frequency and related atmospheric features (1981–2017). *Climate Dynamics*, **52**, 5413–5430, doi:10.1007/s00382-018-4462-2.

Evan, A.T., 2018: Surface winds and dust biases in climate models. *Geophys. Res. Lett.*, **45**, 1079–1085, doi:10.1002/2017GL076353.

Evan, A.T., C. Flamant, S. Fiedler, and O. Doherty, 2014: An analysis of aeolian dust in climate models. *Geophys. Res. Lett.*, **41**, 5996–6001, doi:10.1002/2014GL060545.

Evan, A.T., C. Flamant, M. Gaetani, and F. Guichard, 2016: The past, present and future of African dust. *Nature*, **531**, 493–495, doi:10.1038/nature17149.

Evans, S., P. Ginoux, S. Malyshev, and E. Shevliakova, 2016: Climate-vegetation interaction and amplification of Australian dust variability. *Geophys. Res. Lett.*, **43**, 11,823-11,830, doi:10.1002/2016GL071016.

Fadrique, B. et al., 2018: Widespread but heterogeneous responses of Andean forests to climate change. *Nature*, **564**, 207–212, doi:10.1038/s41586-018-0715-9.

Fahey, C., K. Winter, M. Slot, and K. Kitajima, 2016: Influence of arbuscular mycorrhizal colonization on whole-plant respiration and thermal acclimation of tropical tree seedlings. *Ecol. Evol.*, **6**, 859–870, doi:10.1002/ece3.1952.

Fanzo, J., C. Davis, R. McLaren, and J. Choufani, 2018: The effect of climate change across food systems: Implications for nutrition outcomes. *Glob. Food Sec.*, **18**, 12–19, doi:10.1016/j.gfs.2018.06.001.

FAO, 2015: FAOSTAT. Land use emissions database. http://faostat.fao.org/site/705/default.aspx.

FAO-FRA, 2015: *The Global Forest Resources Assessment 2015*. Food and Agriculture Organization of the United Nations, Rome, Italy, 244 pp.

Farrior, C.E., I. Rodriguez-Iturbe, R. Dybzinski, S.A. Levin, and S.W. Pacala, 2015: Decreased water limitation under elevated CO_2 amplifies potential for forest carbon sinks. *Proc. Natl. Acad. Sci.*, **112**, 7213–7218, doi:10.1073/pnas.1506262112.

Favis-Mortlock, D., and D. Mullen, 2011: Soil erosion by water under future climate change. In: *Soil Hydrology, Land Use and Agriculture: Measurement and Modelling* [Shukla, M.K. (ed.)]. CABI, Wallingford, UK, 384–414 pp.

Fay, P.A., D.M. Kaufman, J.B. Nippert, J.D. Carlisle, and C.W. Harper, 2008: Changes in grassland ecosystem function due to extreme rainfall events: Implications for responses to climate change. *Glob. Chang. Biol.*, **14**, 1600–1608, doi:10.1111/j.1365-2486.2008.01605.x.

Feddema, J.J. et al., 2005: The importance of land-cover change in simulating future climates. *Science*, **310**, 1674–1678, doi:10.1126/science.1118160.

Federici, S., F.N. Tubiello, M. Salvatore, H. Jacobs, and J. Schmidhuber, 2015: New estimates of CO_2 forest emissions and removals: 1990–2015. *For. Ecol. Manage.*, **352**, 89–98, doi:10.1016/j.foreco.2015.04.022.

Feng, W., A.F. Plante, and J. Six, 2013: Improving estimates of maximal organic carbon stabilization by fine soil particles. *Biogeochemistry*, **112**, 81–93, doi:10.1007/s10533-011-9679-7.

Feng, Z.Z. et al., 2015: Constraints to nitrogen acquisition of terrestrial plants under elevated CO_2. *Glob. Chang. Biol.*, **21**, 3152–3168, doi:10.1111/gcb.12938.

Fensholt, R. et al., 2012: Greenness in semi-arid areas across the globe 1981–2007 – an Earth Observing Satellite based analysis of trends and drivers. *Remote Sens. Environ.*, **121**, 144–158, doi:10.1016/j.rse.2012.01.017.

Fernandes, K. et al., 2017: Heightened fire probability in Indonesia in non-drought conditions: The effect of increasing temperatures. *Environ. Res. Lett.*, **12**, doi:10.1088/1748-9326/aa6884.

Feyisa, G.L., K. Dons, and H. Meilby, 2014: Efficiency of parks in mitigating urban heat island effect: An example from Addis Ababa. *Landsc. Urban Plan.*, **123**, 87–95, doi:10.1016/j.landurbplan.2013.12.008.

Ficklin, D.L., and K.A. Novick, 2017: Historic and projected changes in vapor pressure deficit suggest a continental-scale drying of the United States atmosphere. *J. Geophys. Res.*, **122**, 2061–2079, doi:10.1002/2016JD025855.

Field, C.B., R.B. Jackson, and H.A. Mooney, 1995: Stomatal responses to increased CO_2: implications from the plant to the global scale. *Plant. Cell Environ.*, **18**, 1214–1225, doi:10.1111/j.1365-3040.1995.tb00630.x.

Findell, K.L., E. Shevliakova, P.C.D. MIlly, and R.J. Stouffer, 2007: Modeled impact of anthropogenic land cover change on climate. *J. Clim.*, **20**, 3621–3634, doi:10.1175/JCLI4185.1.

Findell, K.L., A.J. Pitman, M.H. England, and P.J. Pegion, 2009: Regional and global impacts of land cover change and sea surface temperature anomalies. *J. Clim.*, **22**, 3248–3269, doi:10.1175/2008JCLI2580.1.

Findell, K.L. et al., 2017: The impact of anthropogenic land use and land cover change on regional climate extremes. *Nat. Commun.*, **8**, 989, doi:10.1038/s41467-017-01038-w.

Finkbeiner, M., 2014: Indirect land use change – Help beyond the hype? *Biomass and Bioenergy*, **62**, 218–221, doi:10.1016/j.biombioe.2014.01.024.

Finzi, A.C. et al., 2015: Rhizosphere processes are quantitatively important components of terrestrial carbon and nutrient cycles. *Glob. Chang. Biol.*, **21**, 2082–2094, doi:10.1111/gcb.12816.

Fiore, A.M. et al., 2012: Global air quality and climate. *Chem. Soc. Rev.*, **41**, 6663, doi:10.1039/c2cs35095e.

Fischer, E.M. and R. Knutti, 2014: Detection of spatially aggregated changes in temperature and precipitation extremes. *Geophys. Res. Lett.*, **41**, 547–554, doi:10.1002/2013GL058499.

Fischer, E.M. and R. Knutti, 2015: Anthropogenic contribution to global occurrence of heavy-precipitation and high-temperature extremes. *Nat. Clim. Chang.*, **5**, 560–564, doi:10.1038/nclimate2617.

Fischer, E.M., J. Rajczak, and C. Schär, 2012: Changes in European summer temperature variability revisited. *Geophys. Res. Lett.*, **39**, doi:10.1029/2012GL052730.

Fischer, E.M., U. Beyerle, and R. Knutti, 2013: Robust spatially aggregated projections of climate extremes. *Nat. Clim. Chang.*, **3**, 1033–1038, doi:10.1038/nclimate2051.

Flanner, M.G., K.M. Shell, M. Barlage, D.K. Perovich, and M.A. Tschudi, 2011: Radiative forcing and albedo feedback from the Northern Hemisphere cryosphere between 1979 and 2008. *Nat. Geosci.*, **4**, 151–155, doi:10.1038/ngeo1062.

Fletcher, M.-S., S.W. Wood, and S.G. Haberle, 2014: A fire-driven shift from forest to non-forest: evidence for alternative stable states? *Ecology*, **95**, 2504–2513, doi:10.1890/12-1766.1.

Fontaine, S., P. Barre, N. Bdioui, B. Mary, and C. Rumpel, 2007: Stability of organic carbon in deep soil layers. *Nature*, **450**, 277–281. doi:10.1038/nature06275.

Forsell, N. et al., 2016: Assessing the INDCs' land use, land use change, and forest emission projections. *Carbon Balance Manag.*, **11**, 26, doi:10.1186/s13021-016-0068-3.

Forzieri, G., R. Alkama, D.G. Miralles, and A. Cescatti, 2017: Satellites reveal contrasting responses of regional climate to the widespread greening of Earth. *Science*, **356**, 1180–1184, doi:10.1126/science.aal1727.

Founda, D., and M. Santamouris, 2017: Synergies between Urban Heat Island and Heat Waves in Athens (Greece), during an extremely hot summer (2012). *Sci. Rep.*, **7**, 1–11, doi:10.1038/s41598-017-11407-6.

Founda, D., F. Pierros, M. Petrakis, and C. Zerefos, 2015: Interdecadal variations and trends of the Urban Heat Island in Athens (Greece) and its response to heat waves. *Atmos. Res.*, **161–162**, 1–13, doi:10.1016/j.atmosres.2015.03.016.

Fowler, D. et al., 2015: Effects of global change during the 21st century on the nitrogen cycle. *Atmos. Chem. Phys.*, **15**, 13849–13893, doi:10.5194/acp-15-13849-2015.

Frank, D. et al., 2015a: Effects of climate extremes on the terrestrial carbon cycle: Concepts, processes and potential future impacts. *Glob. Chang. Biol.*, **21**, 2861–2880, doi:10.1111/gcb.12916.

Frank, D. et al., 2015b: Effects of climate extremes on the terrestrial carbon cycle: Concepts, processes and potential future impacts. *Glob. Chang. Biol.*, **21**, 2861–2880, doi:10.1111/gcb.12916.

Frank, S. et al., 2017a: Reducing greenhouse gas emissions in agriculture without compromising food security? *Environ. Res. Lett.*, **12**, 105004, doi:10.1088/1748-9326/aa8c83.

Frank, S. et al., 2017b: Reducing greenhouse gas emissions in agriculture without compromising food security? *Environ. Res. Lett.*, **12**, 105004, doi:10.1088/1748-9326/aa8c83.

Frey, S.D., J. Lee, J.M. Melillo, and J. Six, 2013: The temperature response of soil microbial efficiency and its feedback to climate. *Nat. Clim. Chang.*, **3**, 395, doi:10.1038/nclimate1796.

Fricko, O. et al., 2017: The marker quantification of the Shared Socioeconomic Pathway 2: A middle-of-the-road scenario for the 21st century. *Glob. Environ. Chang.*, **42**, 251–267, doi:10.1016/J.GLOENVCHA.2016.06.004.

Friedlingstein, P. et al., 2014: Uncertainties in CMIP5 climate projections due to carbon cycle feedbacks. *J. Clim.*, doi:10.1175/JCLI-D-12-00579.1.

Friedlingstein, P. et al., 2006: Climate-carbon cycle feedback analysis: results from the C4MIP model intercomparison. *J. Clim.*, **19**, 3337–3353, doi:10.1175/JCLI3800.1.

Friend, A.D. et al., 2014a: Carbon residence time dominates uncertainty in terrestrial vegetation responses to future climate and atmospheric CO_2. *Proc. Natl. Acad. Sci. U.S.A.*, **111**, 3280–3285, doi:10.1073/pnas.1222477110.

Friend, A.D. et al., 2014b: Carbon residence time dominates uncertainty in terrestrial vegetation responses to future climate and atmospheric CO_2. *Proc. Natl. Acad. Sci.*, **111**, 3280–3285, doi:10.1073/pnas.1222477110.

Fritz, S. et al., 2013: Downgrading recent estimates of land available for biofuel production. *Environ. Sci. Technol.*, doi:10.1021/es303141h.

Fröhlich-Nowoisky, J. et al., 2016: Bioaerosols in the Earth system: Climate, health, and ecosystem interactions. *Atmos. Res.*, **182**, 346–376, doi:10.1016/j.atmosres.2016.07.018.

Froidevaux, P. et al., 2014: Influence of the background wind on the local soil moisture–precipitation feedback. *J. Atmos. Sci.*, **71**, 782–799, doi:10.1175/JAS-D-13-0180.1.

Fu, R. et al., 2013: Increased dry-season length over southern Amazonia in recent decades and its implication for future climate projection. *Proc. Natl. Acad. Sci.*, **110**, 18110–18115, doi:10.1073/pnas.1302584110.

Fu, Y., and H. Liao, 2014: Impacts of land use and land cover changes on biogenic emissions of volatile organic compounds in China from the late 1980s to the mid-2000s: implications for tropospheric ozone and secondary organic aerosol. *Tellus B Chem. Phys. Meteorol.*, **66**, 24987, doi:10.3402/tellusb.v66.24987.

Fuentes, J.D. et al., 2000: Biogenic hydrocarbons in the atmospheric boundary layer: A review. *Bull. Am. Meteorol. Soc.*, **81**, 1537–1575, doi:10.1175/1520-0477(2000)081<1537:BHITAB>2.3.CO;2.

Fuentes, J.D. et al., 2016: Linking meteorology, turbulence, and air chemistry in the amazon rain forest. *Bull. Am. Meteorol. Soc.*, **97**, 2329–2342, doi:10.1175/BAMS-D-15-00152.1.

Fuglestvedt, J. et al., 2018: Implications of possible interpretations of "greenhouse gas balance" in the Paris Agreement. *Philos. Trans. R. Soc. A Math. Phys. Eng. Sci.*, **376**, 20160445, doi:10.1098/rsta.2016.0445.

Fujibe, F., 2009: Detection of urban warming in recent temperature trends in Japan. *Int. J. Climatol.*, **29**, 1811–1822, doi:10.1002/joc.1822.

Fujimori, S. et al., 2017: SSP3: AIM implementation of Shared Socioeconomic Pathways. *Glob. Environ. Chang.*, **42**, 268–283, doi:10.1016/J.GLOENVCHA.2016.06.009.

Funk, C. et al., 2018: Anthropogenic enhancement of moderate-to-strong El Niño events likely contributed to drought and poor harvests in southern Africa during 2016. *Bull. Am. Meteorol. Soc.*, **99**, S91–S96, doi:10.1175/BAMS-D-17-0112.1.

Fuss, S. et al., 2018: Negative emissions – Part 2: Costs, potentials and side effects. *Environ. Res. Lett.*, **submitted**, 2–4, doi:10.1088/1748-9326/aabf9f.

Fuzzi, S. et al., 2015a: Particulate matter, air quality and climate: Lessons learned and future needs. *Atmos. Chem. Phys.*, **15**, 8217–8299, doi:10.5194/acp-15-8217-2015.

Gallego-Sala, A. V et al., 2018: Latitudinal limits to the predicted increase of the peatland carbon sink with warming. *Nat. Clim. Chang.*, **8**, 907–913, doi:10.1038/s41558-018-0271-1.

Gálos, B., C. Mátyás, and D. Jacob, 2011: Regional characteristics of climate change altering effects of afforestation. *Environ. Res. Lett.*, **6**, 44010, doi:10.1088/1748-9326/6/4/044010.

Gálos, B. et al., 2013: Case study for the assessment of the biogeophysical effects of a potential afforestation in Europe. *Carbon Balance Manag.*, **8**, 3, doi:10.1186/1750-0680-8-3.

Ganeshan, M. and R. Murtugudde, 2015: Nocturnal propagating thunderstorms may favor urban "hot-spots": A model-based study over Minneapolis. *Urban Clim.*, **14**, 606–621, doi:10.1016/j.uclim.2015.10.005.

Ganopolski, A. et al., 2001: Climber-2: A climate system model of intermediate complexity. Part II: Model sensitivity. *Clim. Dyn.*, **17**, 735–751, doi:10.1007/s003820000144.

Gao, Y. et al., 2014: Biogeophysical impacts of peatland forestation on regional climate changes in Finland. *Biogeosciences*, **11**, 7251–7267, doi:10.5194/bg-11-7251-2014.

García-Palacios, P. et al., 2015: Are there links between responses of soil microbes and ecosystem functioning to elevated CO_2, N deposition and warming? A global perspective. *Glob. Chang. Biol.*, **21**, 1590–1600, doi:10.1111/gcb.12788.

García-Ruiz, J.M., I.I. López-Moreno, S.M. Vicente-Serrano, T. Lasanta-Martínez, and S. Beguería, 2011: Mediterranean water resources in a global change scenario. *Earth-Science Rev.*, **105**, 121–139, doi:10.1016/j.earscirev.2011.01.006.

Garcia, R.A., M. Cabeza, R. Altwegg, and M.B. Araújo, 2016: Do projections from bioclimatic envelope models and climate change metrics match? *Glob. Ecol. Biogeogr.*, **25**, 65–74, doi:10.1111/geb.12386.

Garnaud, C. and L. Sushama, 2015: Biosphere-climate interactions in a changing climate over North America. *J. Geophys. Res. Atmos.*, **120**, 1091–1108, doi:10.1002/2014JD022055.

Gasser, T., and P. Ciais, 2013: A theoretical framework for the net land-to-atmosphere CO_2 flux and its implications in the definition of "emissions from land-use change." *Earth Syst. Dyn.*, **4**, 171–186, doi:10.5194/esd-4-171-2013.

Gatti, L.V. et al., 2014: Drought sensitivity of Amazonian carbon balance revealed by atmospheric measurements. *Nature*, **506**, 76–80, doi:10.1038/nature12957.

Gauthier, S., P. Bernier, T. Kuuluvainen, A.Z. Shvidenko, and D.G. Schepaschenko, 2015: Boreal forest health and global change. *Science*, **349**, 819–822, doi:10.1126/science.aaa9092.

Gedalof, Z. and A.A. Berg, 2010: Tree ring evidence for limited direct CO_2 fertilization of forests over the 20th century. *Global Biogeochem. Cycles*, **24**, n/a-n/a, doi:10.1029/2009GB003699.

Geddes, A., W.N. Adger, N.W. Arnell, R. Black, and D.S.G. Thomas, 2012: Migration, environmental change, and the "challenges of governance." *Environ. Plan. C Gov. Policy*, **30**, 951–967, doi:10.1068/c3006ed.

Gedney, N. and P.J. Valdes, 2000: The effect of Amazonian deforestation on the northern hemisphere circulation and climate. *Geophys. Res. Lett.*, **27**, 3053, doi:10.1029/2000GL011794.

Geirinhas, J.L., R.M. Trigo, R. Libonati, C.A.S. Coelho, and A.C. Palmeira, 2018: Climatic and synoptic characterization of heat waves in Brazil. *Int. J. Climatol.*, **38**, 1760–1776, doi:10.1002/joc.5294.

Gellesch, E., M.A.S. Arfin Khan, J. Kreyling, A. Jentsch, and C. Beierkuhnlein, 2017: Grassland experiments under climatic extremes: Reproductive fitness versus biomass. *Environ. Exp. Bot.*, **144**, 68–75, doi:10.1016/j.envexpbot.2017.10.007.

Georgescu, M., D.B. Lobell, and C.B. Field, 2011: Direct climate effects of perennial bioenergy crops in the United States. *Proc. Natl. Acad. Sci.*, **108**, 4307–4312, doi:10.1073/pnas.1008779108.

Georgescu, M., D.B. Lobell, C.B. Field, and A. Mahalov, 2013: Simulated hydroclimatic impacts of projected Brazilian sugarcane expansion. *Geophys. Res. Lett.*, **40**, 972–977, doi:10.1002/grl.50206.

Gerbrandt, K. et al., 2016: Life cycle assessment of lignocellulosic ethanol: a review of key factors and methods affecting calculated GHG emissions and energy use. *Curr. Opin. Biotechnol.*, **38**, 63–70, doi:10.1016/j.copbio.2015.12.021.

Gerster-Bentaya, M., 2013: Nutrition-sensitive urban agriculture. *Food Secur.*, **5**, 723–737, doi:10.1007/s12571-013-0295-3.

Gestel, N. Van et al., 2018: Brief communications arising predicting soil carbon loss with warming. *Nat. Publ. Gr.*, **554**, E4–E5, doi:10.1038/nature25745.

Gharun, M., M. Possell, T.L. Bell, and M.A. Adams, 2017: Optimisation of fuel reduction burning regimes for carbon, water and vegetation outcomes. *J. Environ. Manage.*, **203**, 157–170, doi:10.1016/j.jenvman.2017.07.056.

Gibbard, S., K. Caldeira, G. Bala, T.J. Phillips, and M. Wickett, 2005: Climate effects of global land cover change. *Geophys. Res. Lett.*, **32**, 1–4, doi:10.1029/2005GL024550.

Gidden, M.J. et al., 2018: Global emissions pathways under different socioeconomic scenarios for use in CMIP6: a dataset of harmonized emissions trajectories through the end of the century. *Geosci. Model Dev. Discuss.*, 1–42, doi:10.5194/gmd-2018-266.

Giglio, L., J.T. Randerson, and G.R. Van Der Werf, 2013: Analysis of daily, monthly, and annual burned area using the fourth-generation global fire emissions database (GFED4). *J. Geophys. Res. Biogeosciences*, **118**, 317–328, doi:10.1002/jgrg.20042.

Gill, S.E., J.F. Handley, A.R. Ennos, and S. Pauleit, 2007: Adapting cities for climate change: The role of the green infrastructure. *Built Environ.*, **33**, 115–133, doi:10.2148/benv.33.1.115.

Gillett, N.P., V.K. Arora, D. Matthews, and M.R. Allen, 2013: Constraining the ratio of global warming to cumulative CO2 emissions using CMIP5 simulations. *J. Clim.*, **26**, 6844–6858, doi:10.1175/JCLI-D-12-00476.1.

Ginoux, P., J.M. Prospero, T.E. Gill, N.C. Hsu, and M. Zhao, 2012: Global-scale attribution of anthropogenic and natural dust sources and their emission rates based on MODIS Deep Blue aerosol products. *Rev. Geophys.*, **50**, doi:10.1029/2012RG000388.

Goetz, S.J. et al., 2015: Measurement and monitoring needs, capabilities and potential for addressing reduced emissions from deforestation and forest degradation under measurement and monitoring needs, capabilities and potential for addressing reduced emissions from deforestation. *Environ. Res. Lett.*, **10**, 123001, doi:10.1088/1748-9326/10/12/123001.

Goldstein, B., M. Birkved, J. Fernández, and M. Hauschild, 2017: Surveying the environmental footprint of urban food consumption. *J. Ind. Ecol.*, **21**, 151–165, doi:10.1111/jiec.12384.

Gonçalves, K. dos S. et al., 2018: Development of non-linear models predicting daily fine particle concentrations using aerosol optical depth retrievals and ground-based measurements at a municipality in the Brazilian Amazon region. *Atmos. Environ.*, **184**, 156–165, doi:10.1016/j.atmosenv.2018.03.057.

Gonsamo, A., J.M. Chen, Y.W. Ooi, C.J.M., and O.Y.W., 2017: Peak season plant activity shift towards spring is reflected by increasing carbon uptake by extratropical ecosystems. *Glob. Chang. Biol.*, **24**, 2117–2128, doi:10.1111/gcb.14001.

González-Jaramillo, V. et al., 2016: Assessment of deforestation during the last decades in Ecuador using NOAA-AVHRR satellite data. *Erdkunde*, **70**, 217–235, doi:10.3112/erdkunde.2016.03.02.

Gottfried, M. et al., 2012: Continent-wide response of mountain vegetation to climate change. *Nat. Clim. Chang.*, **2**, 111–115, doi:10.1038/nclimate1329.

Govaerts, B. et al., 2009: Conservation agriculture and soil carbon sequestration: Between myth and farmer reality. *CRC. Crit. Rev. Plant Sci.*, **28**, 97–122, doi:10.1080/07352680902776358.

Grant, K., J. Kreyling, H. Heilmeier, C. Beierkuhnlein, and A. Jentsch, 2014: Extreme weather events and plant–plant interactions: Shifts between competition and facilitation among grassland species in the face of drought and heavy rainfall. *Ecol. Res.*, **29**, 991–1001, doi:10.1007/s11284-014-1187-5.

Grassi, G., M.G.J. den Elzen, A.F. Hof, R. Pilli, and S. Federici, 2012: The role of the land use, land use change and forestry sector in achieving Annex I reduction pledges. *Clim. Change*, **115**, 873–881, doi:10.1007/s10584-012-0584-4.

Grassi, G. et al., 2017: The key role of forests in meeting climate targets requires science for credible mitigation. *Nat. Clim. Chang.*, **7**, 220–226, doi:10.1038/nclimate3227.

Grassi, G. et al., 2018a: Reconciling global model estimates and country reporting of anthropogenic forest CO2 sinks. *Nat. Clim. Chang.*, 1–35.

Grassi, G., R. Pilli, J. House, S. Federici, and W.A. Kurz, 2018b: Science-based approach for credible accounting of mitigation in managed forests. *Carbon Balance Manag.*, doi:10.1186/s13021-018-0096-2.

Gray, S.B. et al., 2016: Intensifying drought eliminates the expected benefits of elevated carbon dioxide for soybean. *Nat. Plants*, **2**, 16132, doi:10.1038/nplants.2016.132.

Green, J.K. et al., 2019: Large influence of soil moisture on long-term terrestrial carbon uptake. *Nature*, **565**, 476–479, doi:10.1038/s41586-018-0848-x.

Gregory, P.J., and B. Marshall, 2012: Attribution of climate change: A methodology to estimate the potential contribution to increases in potato yield in Scotland since 1960. *Glob. Chang. Biol.*, **18**, 1372–1388, doi:10.1111/j.1365-2486.2011.02601.x.

Gren, I.-M., and A.Z. Aklilu, 2016: Policy design for forest carbon sequestration: A review of the literature. *For. Policy Econ.*, **70**, 128–136, doi:10.1016/j.forpol.2016.06.008.

Grillakis, M.G., 2019: Increase in severe and extreme soil moisture droughts for Europe under climate change. *Sci. Total Environ.*, **660**, 1245–1255, doi:10.1016/j.scitotenv.2019.01.001.

Griscom, B.W. et al., 2017a: Natural climate solutions. *Proc. Natl. Acad. Sci.*, **114**, 11645–11650, doi:10.1073/pnas.1710465114.

van Groenigen, K.J., X. Qi, C.W. Osenberg, Y. Luo, and B.A. Hungate, 2014: Faster decomposition under increased atmospheric CO2 limits soil carbon storage. *Science*, **344**, 508–509, doi:10.1126/science.1249534.

Groß, S. et al., 2015: Optical properties of long-range transported Saharan dust over Barbados as measured by dual-wavelength depolarization Raman lidar measurements. *Atmos. Chem. Phys.*, **15**, 11067–11080, doi:10.5194/acp-15-11067-2015.

Grubler, A. et al., 2018: A low energy demand scenario for meeting the 1.5 °C target and sustainable development goals without negative emission technologies. *Nat. Energy*, **3**, 515–527, doi:10.1038/s41560-018-0172-6.

Guan, K., B. Sultan, M. Biasutti, C. Baron, and D.B. Lobell, 2015: What aspects of future rainfall changes matter for crop yields in West Africa? *Geophys. Res. Lett.*, **42**, 8001–8010, doi:10.1002/2015GL063877.

Gudipudi, R. et al., 2019: The efficient, the intensive, and the productive: Insights from urban Kaya scaling. *Appl. Energy*, **236**, 155–162, doi:10.1016/j.apenergy.2018.11.054.

Guenet, B., et al., 2018: Impact of priming on global soil carbon stocks. *Glob. Chang. Biol.*, **24**, 1873–1883, doi:10.1111/gcb.14069.

Guenther, A. et al., 1995: A global model of natural volatile organic compound emissions. *J. Geophys. Res.*, **100**, 8873–8892, doi:10.1029/94JD02950.

Guenther, A.B. et al., 2012: The model of emissions of gases and aerosols from nature version 2.1 (MEGAN2.1): An extended and updated framework for modeling biogenic emissions. *Geosci. Model Dev.*, **5**, 1471–1492, doi:10.5194/gmd-5-1471-2012.

Guest, G., F. Cherubini, and A.H. Strømman, 2013: The role of forest residues in the accounting for the global warming potential of bioenergy. *GCB Bioenergy*, **5**, 459–466, doi:10.1111/gcbb.12014.

Guhathakurta, P., D.S. Pai, and M.N. Rajeevan, 2017: Variability and trends of extreme rainfall and rainstorms. In: *Observed Climate Variability and Change over the Indian Region* [Rajeevan, M. and S. Nayak (eds.)]. Springer, Singapore, pp. 37–49.

Guillod, B.P., B. Orlowsky, D.G. Miralles, A.J. Teuling, and S.I. Seneviratne, 2015: Reconciling spatial and temporal soil moisture effects on afternoon rainfall. *Nat. Commun.*, **6**, 6443, doi:10.1038/ncomms7443.

Guimberteau, M., K. Laval, A. Perrier, and J. Polcher, 2012: Global effect of irrigation and its impact on the onset of the Indian summer monsoon. *Clim. Dyn.*, **39**, 1329–1348, doi:10.1007/s00382-011-1252-5.

Gumbricht, T. et al., 2017: An expert system model for mapping tropical wetlands and peatlands reveals South America as the largest contributor. *Glob. Chang. Biol.*, **23**, 3581–3599, doi:10.1111/gcb.13689.

Guo, Z. et al., 2006: GLACE: The global land–atmosphere coupling experiment. Part II: Analysis. *J. Hydrometeorol.*, **7**, 611–625, doi:10.1175/JHM511.1.

Gustafsson, Ö., and V. Ramanathan, 2016: Convergence on climate warming by black carbon aerosols. *Proc. Natl. Acad. Sci.*, **113**, 4243–4245, doi:10.1073/pnas.1603570113.

Gustavsson, L. et al., 2017: Climate change effects of forestry and substitution of carbon-intensive materials and fossil fuels. *Renew. Sustain. Energy Rev.*, doi:10.1016/j.rser.2016.09.056.

Gütschow, J. et al., 2016: The PRIMAP-hist national historical emissions time series. *Earth Syst. Sci. Data*, **8**, 571–603, doi:10.5194/essd-8-571-2016.

Haberl, H., 2013: Net land-atmosphere flows of biogenic carbon related to bioenergy: Towards an understanding of systemic feedbacks. *GCB Bioenergy*, *5*, *351*–357, doi:10.1111/gcbb.12071.

Haberl, H. et al., 2012: Correcting a fundamental error in greenhouse gas accounting related to bioenergy. *Energy Policy*, **45**, 18–23, doi:10.1016/j.enpol.2012.02.051.

Haberlie, A.M., W.S. Ashley, T.J. Pingel, 2014: The effect of urbanisation on the climatology of thunderstorm initiation. *Q.J.R. Meteorol. Soc.*, **141**, 663–675, doi:10.1002/qj.2499.

Haddaway, N.R. et al., 2017: How does tillage intensity affect soil organic carbon? A systematic review. *Environ. Evid.*, **6**, 30, doi:10.1186/s13750-017-0108-9.

Hagos, S. et al., 2014: Assessment of uncertainties in the response of the African monsoon precipitation to land use change simulated by a regional model. *Clim. Dyn.*, **43**, 2765–2775, doi:10.1007/s00382-014-2092-x.

Hamdi, R., 2010: Estimating urban heat island effects on the temperature series of Uccle (Brussels, Belgium) using remote sensing data and a land surface scheme. *Remote Sens.*, **2**, 2773–2784, doi:10.3390/rs2122773.

Hamdi, R., P. Termonia, and P. Baguis, 2011: Effects of urbanization and climate change on surface runoff of the Brussels Capital Region: A case study using an urban soil-vegetation-atmosphere-transfer model. *Int. J. Climatol.*, **31**, 1959–1974, doi:10.1002/joc.2207.

Hamdi, R. et al., 2016: Evolution of urban heat wave intensity for the Brussels Capital Region in the ARPEGE-Climat A1B scenario. *Urban Clim.*, **17**, 176–195, doi:10.1016/j.uclim.2016.08.001.

Hamilton, A.J. et al., 2014: Give peas a chance? Urban agriculture in developing countries. A review. *Agron. Sustain. Dev.*, **34**, 45–73, doi:10.1007/s13593-013-0155-8.

Han, S., and Z. Yang, 2013: Cooling effect of agricultural irrigation over Xinjiang, Northwest China from 1959 to 2006. *Environ. Res. Lett.*, **8**, doi:10.1088/1748-9326/8/2/024039.

Handmer, J., Y. Honda, Z.W. Kundzewicz, N. Arnell, G. Benito, J. Hatfield, I.F. Mohamed, P. Peduzzi, S. Wu, B. Sherstyukov, K. Takahashi, and Z. Yan, 2012: Changes in Impacts of Climate Extremes: Human Systems and Ecosystems. In: Managing the Risks of Extreme Events and Disasters to Advance Climate Change Adaptation. A Special Report of Working Groups I and II of the Intergovernmental Panel on Climate Change (IPCC) [Field, C.B., V. Barros, T.F. Stocker, D. Qin, D.J. Dokken, K.L. Ebi, M.D. Mastrandrea, K.J. Mach, G.-K. Plattner, S.K. Allen, M. Tignor, and P.M. Midgley (eds.)]. Cambridge University Press, Cambridge, UK, and New York, NY, USA, pp. 231–290.

Hanes, C.C. et al., 2018: Fire-regime changes in Canada over the last half century. *Can. J. For. Res.*, **49**, 256–269, doi:10.1139/cjfr-2018-0293.

Hansen, J. et al., 2005: Efficacy of climate forcings. *J. Geophys. Res. D Atmos.*, **110**, 1–45, doi:10.1029/2005JD005776.

Hansen, J., R. Ruedy, M. Sato, and K. Lo, 2010: Global surface temperature change. *Rev. Geophys.*, **48**, RG4004, doi:10.1029/2010RG000345.

Hansen, J.E. et al., 1998: Climate forcings in the Industrial era. *Proc. Natl. Acad. Sci.*, **95**, 12753–12758, doi:10.1073/pnas.95.22.12753.

Hansen, M.C. et al., 2012: Carbon emissions from land use and land-cover change. *Biogeosciences*, **9**, 5125–5142, doi:10.5194/bg-9-5125-2012.

Hansis, E., S.J. Davis, and J. Pongratz, 2015: Relevance of methodological choices for accounting of land use change carbon fluxes. *Global Biogeochem. Cycles*, **29**, 1230–1246, doi:10.1002/2014GB004997.

Hantson, S. et al., 2016: The status and challenge of global fire modelling. *Biogeosciences*, **13**, 3359–3375, doi:10.5194/bg-13-3359-2016.

Hantson, S et al., 2017: Global isoprene and monoterpene emissions under changing climate, vegetation, CO_2 and land use. *Atmos. Environ.*, **155**, 35–45, doi:10.1016/j.atmosenv.2017.02.010.

Harding, K.J., T.E. Twine, A. VanLoocke, J.E. Bagley, and J. Hill, 2016: Impacts of second-generation biofuel feedstock production in the central U.S. on the hydrologic cycle and global warming mitigation potential. *Geophys. Res. Lett.*, **43**, 10,773–10,781, doi:10.1002/2016gl069981.

Harlan, S.L., and D.M. Ruddell, 2011: Climate change and health in cities: impacts of heat and air pollution and potential co-benefits from mitigation and adaptation. *Curr. Opin. Environ. Sustain.*, **3**, 126–134, doi:10.1016/j.cosust.2011.01.001.

Harper, A.B. et al., 2018: Land-use emissions play a critical role in land-based mitigation for Paris climate targets. *Nature Communications* doi:10.1038/s41467-018-05340-z.

Harris, R.M.B., T.A. Remenyi, G.J. Williamson, N.L. Bindoff, and D.M.J.S. Bowman, 2016: Climate–vegetation–fire interactions and feedbacks: Trivial detail or major barrier to projecting the future of the Earth system? *Wiley Interdiscip. Rev. Clim. Chang.*, **7**, 910–931, doi:10.1002/wcc.428.

Harris, Z.M., R. Spake, and G. Taylor, 2015: Land use change to bioenergy: A meta-analysis of soil carbon and GHG emissions. *Biomass and Bioenergy*, **82**, 27–39, doi:10.1016/j.biombioe.2015.05.008.

Harrison, A.D. et al., 2016: Not all feldspars are equal: a survey of ice nucleating properties across the feldspar group of minerals. *Atmos. Chem. Phys.*, **16**, 10927–10940, doi:10.5194/acp-16-10927-2016.

Harrison, S.P., J.R. Marlon, and P.J. Bartlein, 2010: Fire in the Earth system. *Chang. Clim. Earth Syst. Soc.*, **324**, 21–48, doi:10.1007/978-90-481-8716-4_3.

Harsch, M.A., P.E. Hulme, M.S. McGlone, and R.P. Duncan, 2009: Are treelines advancing? A global meta-analysis of treeline response to climate warming. *Ecol. Lett.*, **12**, 1040–1049, doi:10.1111/j.1461-0248.2009.01355.x.

Hartley, A.J., D.J. Parker, L. Garcia-Carreras, and S. Webster, 2016: Simulation of vegetation feedbacks on local and regional scale precipitation in West Africa. *Agric. For. Meteorol.*, **222**, 59–70, doi:10.1016/j.agrformet.2016.03.001.

Hartley, I.P., D.W. Hopkins, M.H. Garnett, M. Sommerkorn, and P.A. Wookey, 2008: Soil microbial respiration in arctic soil does not acclimate to temperature. *Ecol. Lett.*, **11**, 1092–1100, doi:10.1111/j.1461-0248.2008.01223.x.

Hartley, I.P. et al., 2012: A potential loss of carbon associated with greater plant growth in the European Arctic. *Nat. Clim. Chang.*, **2**, 875–879, doi:10.1038/nclimate1575.

Hartmann, D.L. Hartmann, D.L., A.M.G. Klein Tank, M. Rusticucci, L.V. Alexander, S. Brönnimann, Y. Charabi, F.J. Dentener, E.J. Dlugokencky, D.R. Easterling, A. Kaplan, B.J. Soden, P.W. Thorne, M. Wild and P.M. Zhai, 2013a: Observations: Atmosphere and Surface. In: Climate Change 2013: The Physical Science Basis. Contribution of Working Group I to the Fifth Assessment Report of the Intergovernmental Panel on Climate Change [Stocker, T.F., D. Qin, G.-K. Plattner, M. Tignor, S.K. Allen, J. Boschung, A. Nauels, Y. Xia, V. Bex and P.M. Midgley (eds.)]. Cambridge University Press, Cambridge, United Kingdom and New York, NY, USA, pp. 159–254.

Hartmann, H. et al., 2018: Research frontiers for improving our understanding of drought-induced tree and forest mortality. *New Phytol.*, **218**, 15–28, doi:10.1111/nph.15048.

Hassink, J. and A.P. Whitmore, 1997: A model of the physical protection of organic matter in soils. *Soil Sci. Soc. Am. J.*, **61**, 131, doi:10.2136/sssaj1997.03615995006100010020x.

Hatfield, J.L. and J.H. Prueger, 2015: Temperature extremes: Effect on plant growth and development. *Weather Clim. Extrem.*, **10**, 4–10, doi:10.1016/J.WACE.2015.08.001.

Hauser, M., R. Orth, and S.I. Seneviratne, 2017: Investigating soil moisture–climate interactions with prescribed soil moisture experiments: an assessment with the Community Earth System Model (version 1.2). *Geosci. Model Dev.*, **10**, 1665–1677, doi:10.5194/gmd-10-1665-2017.

Hausfather, Z. et al., 2013: Quantifying the effect of urbanization on U.S. historical climatology network temperature records. *J. Geophys. Res. Atmos.*, **118**, 481–494, doi:10.1029/2012JD018509.

Hausmann, P., R. Sussmann, and D. Smale, 2015: Contribution of oil and natural gas production to renewed increase of atmospheric methane (2007–2014): Top-down estimate from ethane and methane column observations. *Atmos. Chem. Phys. Discuss.*, **15**, 35991–36028, doi:10.5194/acpd-15-35991-2015.

Hawken, P. (ed.), 2017: *Drawdown: The Most Comprehensive Plan Ever Proposed to Reverse Global Warming*. Penguin Books, 240 pp.

Hawkes, C.V., B.G. Waring, J.D. Rocca, and S.N. Kivlin, 2017: Historical climate controls soil respiration responses to current soil moisture. *Proc. Natl. Acad. Sci.*, **114**, 6322–6327, doi:10.1073/pnas.1620811114.

Hawkins, E. et al., 2014: Uncertainties in the timing of unprecedented climates. *Nature*, **511**, E3–E5, doi:10.1038/nature13523.

He, F. et al., 2014: Simulating global and local surface temperature changes due to Holocene anthropogenic land cover change. *Geophys. Res. Lett.*, **41**, 623–631, doi:10.1002/2013GL058085.

Heald, C.L., and D.V. Spracklen, 2015: Land use change impacts on air quality and climate. *Chem. Rev.*, **115**, 4476–4496, doi:10.1021/cr500446g.

Heald, C.L. and J.A. Geddes, 2016: The impact of historical land use change from 1850 to 2000 on secondary particulate matter and ozone. *Atmos. Chem. Phys.*, **16**, 14997–15010, doi:10.5194/acp-16-14997-2016.

Heck, V., D. Gerten, W. Lucht, and A. Popp, 2018: Biomass-based negative emissions difficult to reconcile with planetary boundaries. *Nat. Clim. Chang.*, doi:10.1038/s41558-017-0064-y.

Hedenus, F., S. Wirsenius, and D.J.A. Johansson, 2014: The importance of reduced meat and dairy consumption for meeting stringent climate change targets. *Clim. Change*, **124**, 79–91, doi:10.1007/s10584-014-1104-5.

Hegerl, G.C. et al., 2015: Challenges in quantifying changes in the global water cycle. *Bull. Am. Meteorol. Soc.*, **96**, 1097–1115, doi:10.1175/BAMS-D-13-00212.1.

Hegerl, G.C., S. Brönnimann, A. Schurer, and T. Cowan, 2018: The early 20th century warming: Anomalies, causes, and consequences. *Wiley Interdiscip. Rev. Clim. Chang.*, **9**, e522, doi:10.1002/wcc.522.

Heisler-White, J.L., J.M. Blair, E.F. Kelly, K. Harmoney, and A.K. Knapp, 2009: Contingent productivity responses to more extreme rainfall regimes across a grassland biome. *Glob. Chang. Biol.*, **15**, 2894–2904, doi:10.1111/j.1365-2486.2009.01961.x.

Heisler, J.L., and A.K. Knapp, 2008: Temporal coherence of aboveground net primary productivity in mesic grasslands. *Ecography (Cop.).*, **31**, 408–416, doi:10.1111/j.0906-7590.2008.05351.x.

Henderson-Sellers, A., and V. Gornitz, 1984: Possible climatic impacts of land cover transformations, with particular emphasis on tropical deforestation. *Clim. Change*, **6**, 231–257, doi:10.1007/BF00142475.

Henderson, B.B. et al., 2015: Greenhouse gas mitigation potential of the world's grazing lands: Modeling soil carbon and nitrogen fluxes of mitigation practices. *Agric. Ecosyst. Environ.*, **207**, 91–100, doi:10.1016/j.agee.2015.03.029.

Hergoualc'h, K.A., and L.V. Verchot, 2012: Changes in soil CH_4 fluxes from the conversion of tropical peat swamp forests: a meta-analysis. *J. Integr. Environ. Sci.*, **9**, 93–101, doi:10.1080/1943815X.2012.679282.

Herrera-Estrada, J.E., and J. Sheffield, 2017: Uncertainties in future projections of summer droughts and heat waves over the contiguous United States. *J. Clim.*, **30**, 6225–6246, doi:10.1175/JCLI-D-16-0491.1.

Herrero, M. et al., 2015: Livestock and the environment: What have we learned in the past decade? *Annu. Rev. Environ. Resour.*, **40**, 177–202, doi:10.1146/annurev-environ-031113-093503.

Herrero, M. et al., 2016: Greenhouse gas mitigation potentials in the livestock sector. *Nat. Clim. Chang.*, **6**, 452–461, doi:10.1038/nclimate2925.

Heskel, M.A. et al., 2016: Convergence in the temperature response of leaf respiration across biomes and plant functional types. *Proc. Natl. Acad. Sci.*, **113**, 3832–3837, doi:10.1073/pnas.1520282113.

Hiç, C., P. Pradhan, D. Rybski, and J.P. Kropp, 2016: Food surplus and its climate burdens. *Environ. Sci. Technol.*, doi:10.1021/acs.est.5b05088.

Hicks Pries, C.E., C. Castanha, R.C. Porras, and M.S. Torn, 2017: The whole-soil carbon flux in response to warming. *Science*, **355**, 1420 LP-1423.

Hinkel, K.M., and F.E. Nelson, 2007: Anthropogenic heat island at Barrow, Alaska, during winter: 2001–2005. *J. Geophys. Res. Atmos.*, **112**, 2001–2005, doi:10.1029/2006JD007837.

Hirsch, A.L., M. Wilhelm, E.L. Davin, W. Thiery, and S.I. Seneviratne, 2017: Can climate-effective land management reduce regional warming? *J. Geophys. Res.*, **122**, 2269–2288, doi:10.1002/2016JD026125.

Hirsch, A.L. et al., 2018: Biogeophysical impacts of land-use change on climate extremes in low-emission scenarios: Results from HAPPI-Land. *Earth's Futur.*, **6**, 396–409, doi:10.1002/2017EF000744.

Hirschi, M., B. Mueller, W. Dorigo, and S.I. Seneviratne, 2014: Using remotely sensed soil moisture for land-atmosphere coupling diagnostics: The role of surface vs. root-zone soil moisture variability. *Remote Sens. Environ.*, **154**, 246–252, doi:10.1016/j.rse.2014.08.030.

Hodzic, A., P.S. Kasibhatla, D.S. Jo, C.D. Cappa, J.L. Jimenez, S. Madronich, and R.J. Park, 2016: Rethinking the global secondary organic aerosol (SOA) budget: Stronger production, faster removal, shorter lifetime. *Atmos. Chem. Phys.*, **16**, 7917–7941, doi:10.5194/acp-16-7917-2016.

Hoegh-Guldberg, O., D. Jacob, M. Taylor, M. Bindi, S. Brown, I. Camilloni, A. Diedhiou, R. Djalante, K.L. Ebi, F. Engelbrecht, J.Guiot, Y. Hijioka, S. Mehrotra, A. Payne, S.I. Seneviratne, A. Thomas, R. Warren, and G. Zhou, 2018: Impacts of 1.5°C of Global Warming on Natural and Human Systems. In: Global Warming of 1.5°C. An IPCC Special Report on the impacts of global warming of 1.5°C above pre-industrial levels and related global greenhouse gas emission pathways, in the context of strengthening the global response to the threat of climate change, sustainable development, and efforts to eradicate poverty [Masson-Delmotte, V., P. Zhai, H.-O. Pörtner, D. Roberts, J. Skea, P.R. Shukla, A. Pirani, W. Moufouma-Okia, C. Péan, R. Pidcock, S. Connors, J.B.R. Matthews, Y. Chen, X. Zhou, M.I.Gomis, E. Lonnoy, T.Maycock, M.Tignor, and T. Waterfield (eds.)]. In press, pp. 175–311.

Hoell, A., M. Hoerling, J. Eischeid, X.W. Quan, and B. Liebmann, 2017: Reconciling theories for human and natural attribution of recent East Africa drying. *J. Clim.*, **30**, 1939–1957, doi:10.1175/JCLI-D 16-0558.1.

Hoerling, M., J. Eischeid, and J. Perlwitz, 2010: Regional precipitation trends: Distinguishing natural variability from anthropogenic forcing. *J. Clim.*, **23**, 2131–2145, doi:10.1175/2009JCLI3420.1.

Hoesly, R.M. et al., 2018: Historical (1750–2014) anthropogenic emissions of reactive gases and aerosols from the Community Emissions Data System (CEDS). *Geosci. Model Dev.*, **11**, 369–408, doi:10.5194/gmd-11-369-2018.

Hoffman, F.M. et al., 2014: Causes and implications of persistent atmospheric carbon dioxide biases in Earth System Models. *J. Geophys. Res. Biogeosciences*, **119**, 141–162, doi:10.1002/2013JG002381.

Hoffmann, T., et al., 2013: Humans and the missing C-sink: erosion and burial of soil carbon through time. *Earth Surf. Dyn.*, **1**, 45–52, doi:10.5194/esurf-1-45-2013.

Hollinger, D.Y. et al., 2010: Albedo estimates for land surface models and support for a new paradigm based on foliage nitrogen concentration. *Glob. Chang. Biol.*, **16**, 696–710, doi:10.1111/j.1365-2486.2009.02028.x.

2

Hooijer, A. et al., 2010: Current and future CO_2 emissions from drained peatlands in Southeast Asia. *Biogeosciences*, **7**, 1505–1514, doi:10.5194/bg-7-1505-2010.

Hopkins, F.M. et al., 2014: Increased belowground carbon inputs and warming promote loss ofsoil organic carbon through complementary microbial responses. *Soil Biol. Biochem.*, **76**, 57–69, doi:10.1016/j.soilbio.2014.04.028.

Houghton, R.A. and A.A. Nassikas, 2017: Global and regional fluxes of carbon from land use and land cover change 1850–2015. *Global Biogeochem. Cycles*, **31**, 456–472, doi:10.1002/2016GB005546.

Houghton, R.A. and A.A. Nassikas, 2018: Negative emissions from stopping deforestation and forest degradation, globally. *Glob. Chang. Biol.*, **24**, 350–359, doi:10.1111/gcb.13876.

Houghton, R.A. et al., 2012: Carbon emissions from land use and land-cover change. *Biogeosciences*, **9**, 5125–5142, doi:10.5194/bg-9-5125-2012.

Houghton, R.A., B. Byers, and A.A. Nassikas, 2015: A role for tropical forests in stabilizing atmospheric CO_2. *Nat. Clim. Chang.*, doi:10.1038/nclimate2869.

Houghton, R.A., A. Baccini, and W.S. Walker, 2018: Where is the residual terrestrial carbon sink? *Glob. Chang. Biol.*, **24**, 3277–3279, doi:10.1111/gcb.14313.

Houlton, B.Z., S.L. Morford, and R.A. Dahlgren, 2018: Convergent evidence for widespread rock nitrogen sources in Earth's surface environment. *Science*, **360**, 58–62, doi:10.1126/science.aan4399.

Houweling, S. et al., 2014: A multi-year methane inversion using SCIAMACHY, accounting for systematic errors using TCCON measurements. *Atmos. Chem. Phys.*, **14**, 3991–4012, doi:10.5194/acp-14-3991-2014.

Howard, J. et al., 2017: Clarifying the role of coastal and marine systems in climate mitigation. *Front. Ecol. Environ.*, **15**, 42–50, doi:10.1002/fee.1451.

Howden, S.M. et al., 2007: Adapting agriculture to climate change. *Proc. Natl. Acad. Sci.*, **104**, 19691–19696, doi:10.1073/pnas.0701890104.

Hristov, A.N. et al., 2013: *Mitigation of Greenhouse Gas Emissions in Livestock Production: A Review of Technical Options for Non-CO_2 Emissions* [Gerber, P.J., B. Henderson, and H.P.S. Makkar (eds.)]. FAO Animal Production and Health Paper No. 177. Food and Agriculture Organization of the United Nations, Rome, Italy, 226 pp.

Hu, W.W. et al., 2015: Characterization of a real-time tracer for isoprene epoxydiols-derived secondary organic aerosol (IEPOX-SOA) from aerosol mass spectrometer measurements. *Atmos. Chem. Phys.*, **15**, 11807–11833, doi:10.5194/acp-15-11807-2015.

Hua, W., H. Chen, S. Sun, and L. Zhou, 2015: Assessing climatic impacts of future land use and land cover change projected with the CanESM2 model. *Int. J. Climatol.*, doi:10.1002/joc.4240.

Hua, W.J. and H.S. Chen, 2013: Recognition of climatic effects of land use/land cover change under global warming. *Chinese Sci. Bull.*, doi:10.1007/s11434-013-5902-3.

Huang, J. et al., 2017: Dryland climate change: Recent progress and challenges. *Rev. Geophys.*, **55**, 719–778, doi:10.1002/2016RG000550.

Huang, W., and S.J. Hall, 2017: Elevated moisture stimulates carbon loss from mineral soils by releasing protected organic matter. *Nat. Commun.*, **8**, 1774, doi:10.1038/s41467-017-01998-z.

Huang, Y., S. Wu, and J.O. Kaplan, 2014: Sensitivity of global wildfire occurrences to various factors in thecontext of global change. *Atmos. Environ.*, **121**, 86–92, doi:10.1016/j.atmosenv.2015.06.002.

Huber, D., D. Mechem, and N. Brunsell, 2014: The effects of Great Plains irrigation on the surface energy balance, regional circulation, and precipitation. *Climate*, **2**, 103–128, doi:10.3390/cli2020103.

Hudiburg, T.W., B.E. Law, C. Wirth, and S. Luyssaert, 2011: Regional carbon dioxide implications of forest bioenergy production. *Nat. Clim. Chang.*, **1**, 419–423, doi:10.1038/nclimate1264.

Huffman, J.A. et al., 2012: Size distributions and temporal variations of biological aerosol particles in the Amazon rainforest characterized by microscopy and real-time UV-APS fluorescence techniques during AMAZE-08. *Atmos. Chem. Phys.*, **12**, 11997–12019, doi:10.5194/acp-12-11997-2012.

Hugelius, G. et al., 2013: The northern circumpolar soil carbon database: Spatially distributed datasets of soil coverage and soil carbon storage in the northern permafrost regions. *Earth Syst. Sci. Data*, **5**, 3–13, doi:10.5194/essd-5-3-2013.

Hugelius, G. et al., 2014: Estimated stocks of circumpolar permafrost carbon with quantified uncertainty ranges and identified data gaps. *Biogeosciences*, **11**, 6573–6593, doi:10.5194/bg-11-6573-2014.

Huijnen, V. et al., 2016: Fire carbon emissions over maritime southeast Asia in 2015 largest since 1997. *Sci. Rep.*, **6**, 26886, doi:10.1038/srep26886.

Hulme, M., 2001: Climatic perspectives on Sahelian desiccation: 1973–1998. *Glob. Environ. Chang.*, **11**, 19–29, doi:10.1016/S0959-3780(00)00042-X.

Hulme, M. et al., 1999: Relative impacts of human-induced climate change and natural climate variability. *Nature*, **397**, 688–691, doi:10.1038/17789.

Humpenöder, F. et al., 2018: Large-scale bioenergy production: How to resolve sustainability trade-offs? *Environ. Res. Lett.*, **13**, 24011, doi:10.1088/1748-9326/aa9e3b.

Humphrey, V. et al., 2018: Sensitivity of atmospheric CO_2 growth rate to observed changes in terrestrial water storage. *Nature*, **560**, 628–631, doi:10.1038/s41586-018-0424-4. http://www.nature.com/articles/s41586-018-0424-4.

Hungate, B.A. et al., 2013: Cumulative response of ecosystem carbon and nitrogen stocks to chronic CO_2 exposure in a subtropical oak woodland. *New Phytol.*, **200**, 753–766, doi:10.1111/nph.12333.

Huntingford, C. et al., 2017: Implications of improved representations of plant respiration in a changing climate. *Nat. Commun.*, **8**, 1602, doi:10.1038/s41467-017-01774-z.

Huppmann, D., J. Rogelj, E. Kriegler, V. Krey, and K. Riahi, 2018: A new scenario resource for integrated 1.5°C research. *Nat. Clim. Chang.*, **8**, 1027–1030, doi:10.1038/s41558-018-0317-4.

Hurtt, G.C. et al., 2011: Harmonization of land-use scenarios for the period 1500–2100: 600 years of global gridded annual land-use transitions, wood harvest, and resulting secondary lands. *Clim. Change*, **109**, 117–161, doi:10.1007/s10584-011-0153-2.

Hussain, S. et al., 2015: Rice management interventions to mitigate greenhouse gas emissions: A review. *Environ. Sci. Pollut. Res.*, **22**, 3342–3360, doi:10.1007/s11356-014-3760-4.

Im, E.S. and E.A.B. Eltahir, 2014: Enhancement of rainfall and runoff upstream from irrigation location in a climate model of West Africa. *Water Resour. Res.*, **50**, 8651–8674, doi:10.1002/2014WR015592.

Im, E.S, M.P. Marcella, and E.A.B. Eltahir, 2014: Impact of potential large-scale irrigation on the West African monsoon and its dependence on location of irrigated area. *J. Clim.*, **27**, 994–1009, doi:10.1175/JCLI-D-13-00290.1.

Iordan, C.M., X. Hu, A. Arvesen, P. Kauppi, and F. Cherubini, 2018: Contribution of forest wood products to negative emissions: Historical comparative analysis from 1960 to 2015 in Norway, Sweden and Finland. *Carbon Balance Manag.*, **13**, 12, doi:10.1186/s13021-018-0101-9.

IPCC, 2003: Good Practice Guidance for Land Use, Land-Use Change and Forestry. [Penman, J., M. Gytarsky, T. Hiraishi, T. Krug, D. Kruger, R. Pipatti, L. Buendia, K. Miwa, T. Ngara, K. Tanabe and F. Wagner (eds.)]. Institute for Global Environmental Strategies (IGES) for the IPCC, Hayama, Kanagawa, Japan, 590 pp.

IPCC, 2005: Special Report on Carbon Dioxide Capture and Storage. Prepared by Working Group III of the Intergovernmental Panel on Climate Change. [Metz, B., O. Davidson, H.C. de Coninck, M. Loos, and L.A. Meyer, (eds.)]. Cambridge University Press, Cambridge, United Kingdom and New York, NY, USA, 443 pp.

IPCC, 2006: 2006 IPCC Guidelines for National Greenhouse Gas Inventories – A Primer. [Eggleston H.S., K. Miwa, N. Srivastava, and K. Tanabe (eds.)]. Institute for Global Environmental Strategies (IGES) for the Intergovernmental Panel on Climate Change. IGES, Japan, 20 pp.

2

IPCC, 2010a: Revisiting the Use of Managed Land as a Proxy for Estimating National Anthropogenic Emissions and Removals [Eggleston H.S., N. Srivastava N., K. Tanabe, and J. Baasansuren (eds.)]. Meeting Report, 5–7 May, 2009, INPE, São José dos Campos, Brazil. IGES, Japan 2010, 56 pp.

IPCC, 2013a: Summary for Policymakers. In: Climate Change 2013: The Physical Science Basis. Contribution of Working Group I to the Fifth Assessment Report of the Intergovernmental Panel on Climate Change [Stocker, T.F., D. Qin, G.-K. Plattner, M. Tignor, S.K. Allen, J. Boschung, A. Nauels, Y. Xia, V. Bex and P.M. Midgley (eds.)], Cambridge University Press, Cambridge, United Kingdom and New York, NY, USA, pp. 1–29.

IPCC, 2013b: Climate Change 2013: The Physical Science Basis. Contribution of Working Group I to the Fifth Assessment Report of the Intergovernmental Panel on Climate Change [Stocker, T.F., D. Qin, G.-K. Plattner, M. Tignor, S.K. Allen, J. Boschung, A. Nauels, Y. Xia, V. Bex and P.M. Midgley (eds.)]. Cambridge University Press, Cambridge, United Kingdom and New York, NY, USA, 1535 pp.

IPCC, 2013c: Annex III: Glossary [Planton, S. (ed.)]. In: Climate Change 2013: The Physical Science Basis. Contribution of Working Group I to the Fifth Assessment Report of the Intergovernmental Panel on Climate Change [Stocker, T.F., D. Qin, G.-K. Plattner, M. Tignor, S.K. Allen, J. Boschung, A. Nauels, Y. Xia, V. Bex and P.M. Midgley (eds.)]. Cambridge University Press, Cambridge, United Kingdom and New York, NY, USA, pp. 1447–1466.

IPCC, 2018: Global Warming of 1.5°C: An IPCC special report on the impacts of global warming of 1.5°C above pre-industrial levels and related global greenhouse gas emission pathways, in the context of strengthening the global response to the threat of climate change [V. Masson-Delmotte, P. Zhai, H.-O. Pörtner, D. Roberts, J. Skea, P.R. Shukla, A. Pirani, W. Moufouma-Okia, C. Péan, R. Pidcock, S. Connors, J.B.R. Matthews, Y. Chen, X. Zhou, M.I. Gomis, E. Lonnoy, T. Maycock, M. Tignor, and T. Waterfield (eds.)]. In press, 1552 pp.

Ise, T., A.L. Dunn, S.C. Wofsy, and P.R. Moorcroft, 2008: High sensitivity of peat decomposition to climate change through water-table feedback. *Nat. Geosci.*, **1**, 763–766, doi:10.1038/ngeo331.

Ito, A., K. Nishina, and H.M. Noda, 2016: Impacts of future climate change on the carbon budget of northern high-latitude terrestrial ecosystems: An analysis using ISI-MIP data. *Polar Sci.*, **10**, 346–355, doi:10.1016/j.polar.2015.11.002.

Jacinthe, P.A. and R. Lal, 2001: A mass balance approach to assess carbon dioxide evolution during erosional events. *L. Degrad. Dev.*, **12**, 329–339, doi:10.1002/ldr.454.

Jackson, R.B. et al., 2017: The ecology of soil carbon: Pools, vulnerabilities, and biotic and abiotic controls. *Annu. Rev. Ecol. Evol. Syst.*, **48**, 419–445, doi:10.1146/annurev-ecolsys-112414-054234.

Jacobson, M., H. Hansson, K. Noone, and R. Charlson, 2000: Organic atmospheric aerosols: Review and state of the science. *Rev. Geophys*, **38**, 267–294, doi:10.1029/1998RG000045.

Janssens-Maenhout, G. et al., 2017a: EDGAR v4.3.2 global atlas of the three major greenhouse gas emissions for the period 1970–2012. *Earth Syst. Sci. Data Discuss.*, **2017**, 1–55, doi:10.5194/essd-2017-79.

Janssens-Maenhout, G. et al., 2017b: *Fossil CO_2 & GHG Emissions of All World Countries*. JRC Science for Policy Report, EUR 28766 EN. Publications Office of the European Union, Luxembourg, doi:10.2760/709792, 239 pp.

Janssens, I.A. et al., 2010: Reduction of forest soil respiration in response to nitrogen deposition. *Nat. Geosci.*, **3**, 315–322, doi:10.1038/ngeo844.

Jardine, K. et al., 2011: Within-canopy sesquiterpene ozonolysis in Amazonia. *J. Geophys. Res. Atmos.*, **116**, 1–10, doi:10.1029/2011JD016243.

Jardine, K. et al., 2015: Green leaf volatile emissions during high temperature and drought stress in a Central Amazon rainforest. *Plants*, **4**, 678–690, doi:10.3390/plants4030678.

Jennerjahn, T.C. et al., 2017: Mangrove ecosystems under climate change. In: *Mangrove Ecosystems: A Global Biogeographic Perspective* [Rivera Monroy, V.H., S.Y. Lee, E. Kristensen, and R.R. Twilley (eds.)]. Springer International Publishing, Cham, Switzerland, pp. 211–244.

Jeong, J.H. et al., 2014a: Intensified Arctic warming under greenhouse warming by vegetation-atmosphere-sea ice interaction. *Environ. Res. Lett.*, **9**, doi:10.1088/1748-9326/9/9/094007.

Jeong, S.J. et al., 2014b: Effects of double cropping on summer climate of the North China Plain and neighbouring regions. *Nat. Clim. Chang.*, **4**, 615–619, doi:10.1038/nclimate2266.

Jiang, L., 2014: Internal consistency of demographic assumptions in the shared socioeconomic pathways. *Popul. Environ.*, **35**, 261–285, doi:10.1007/s11111-014-0206-3.

Johnstone, J.F. et al., 2016: Changing disturbance regimes, ecological memory, and forest resilience. *Front. Ecol. Environ.*, **14**, 369–378, doi:10.1002/fee.1311.

Jolly, W.M. et al., 2015: Climate-induced variations in global wildfire danger from 1979 to 2013. *Nat. Commun.*, **6**, 7537, doi:10.1038/ncomms8537.

Jonard, M. et al., 2015: Tree mineral nutrition is deteriorating in Europe. *Glob. Chang. Biol.*, **21**, 418–430, doi:10.1111/gcb.12657.

Jones, A.D. et al., 2013a: Greenhouse gas policy influences climate via direct effects of land-use change. *J. Clim.*, **26**, 3657–3670, doi:10.1175/JCLI-D-12-00377.1.

Jones, A., W.D. Collins, and M.S. Torn, 2013b: On the additivity of radiative forcing between land use change and greenhouse gases. *Geophys. Res. Lett.*, doi:10.1002/grl.50754.

Jones, C., J. Lowe, S. Liddicoat, and R. Betts, 2009: Committed terrestrial ecosystem changes due to climate change. *Nat. Geosci.*, **2**, 484–487, doi:10.1038/ngeo555.

Jones, C., S. Liddicoat, and J. Lowe, 2010: Role of terrestrial ecosystems in determining CO_2 stabilization and recovery behaviour. *Tellus, Ser. B Chem. Phys. Meteorol.*, **62**, 682–699, doi:10.1111/j.1600-0889.2010.00490.x.

Jones, C.D. et al., 2016: Simulating the Earth system response to negative emissions. *Environ. Res. Lett.*, **11**, 95012, doi:10.1088/1748-9326/11/9/095012.

Jones, P.D. et al., 2012: Hemispheric and large-scale land-surface air temperature variations: An extensive revision and an update to 2010. *J. Geophys. Res. Atmos.*, **117**, doi:10.1029/2011JD017139.

de Jong, R., J. Verbesselt, M.E. Schaepman, and S. de Bruin, 2012: Trend changes in global greening and browning: Contribution of short-term trends to longer-term change. *Glob. Chang. Biol.*, **18**, 642–655, doi:10.1111/j.1365-2486.2011.02578.x.

Joosten, H., 2015: Current soil carbon loss and land degradation globally-where are the hotspots and why there? In: *Soil Carbon: Science, Management and Policy for Multiple Benefits* [Banwart, S.A., E. Noellemeyer, and E. Milne (eds.)]. Wallingford, UK, pp. 224–234.

Jørgensen, S.E., L. Xu, J. Marques, and F. Salas, 2016: Application of indicators for the assessment of ecosystem health. In: *Handbook of Ecological Indicators for Assessment of Ecosystem Health* [Jørgensen, S.E., L. Xu, and R. Costanza (eds.)]. CRC Press, Boca Raton, Florida Taylor & Francis Group, pp. 22–89.

Joshi, M.M., F.H. Lambert, and M.J. Webb, 2013: An explanation for the difference between twentieth and twenty-first century land-sea warming ratio in climate models. *Clim. Dyn.*, **41**, 1853–1869, doi:10.1007/s00382-013-1664-5.

Journet, E., Y. Balkanski, and S.P. Harrison, 2014: A new data set of soil mineralogy for dust-cycle modeling. *Atmos. Chem. Phys.*, **14**, 3801–3816, doi:10.5194/acp-14-3801-2014.

Juang, J.Y., G. Katul, M. Siqueira, P. Stoy, and K. Novick, 2007: Separating the effects of albedo from eco-physiological changes on surface temperature along a successional chronosequence in the southeastern United States. *Geophys. Res. Lett.*, **34**, 1–5, doi:10.1029/2007GL031296.

Jylhä, K. et al., 2010: Observed and projected future shifts of climatic zones in Europe and their use to visualize climate change information. *Weather. Clim. Soc.*, **2**, 148–167, doi:10.1175/2010wcas1010.1.

2

221

Kageyama, M., S. Charbit, C. Ritz, M. Khodri, and G. Ramstein, 2004: Quantifying ice-sheet feedbacks during the last glacial inception. *Geophys. Res. Lett.*, **31**, 1–4, doi:10.1029/2004GL021339.

Kallenbach, C.M., S.D. Frey, and A.S. Grandy, 2016: Direct evidence for microbial-derived soil organic matter formation and its ecophysiological controls. *Nat. Commun.*, **7**, 13630, doi:10.1038/ncomms13630.

Karhu, K. et al., 2014: Temperature sensitivity of soil respiration rates enhanced by microbial community response. *Nature*, **513**, 81–84, doi:10.1038/nature13604.

Kattge, J. et al., 2007: Temperature acclimation in a biochemical model of photosynthesis: A reanalysis of data from 36 species. *Plant, Cell Environ.*, **30**, 1176–1190, doi:10.1111/j.1365-3040.2007.01690.x.

De Kauwe, M.G., T.F. Keenan, B.E. Medlyn, I.C. Prentice, and C. Terrer, 2016: Satellite based estimates underestimate the effect of CO_2 fertilization on net primary productivity. *Nat. Clim. Chang.*, **6**, 892–893, doi:10.1038/nclimate3105.

Kaye, J.P., and M. Quemada, 2017: Using cover crops to mitigate and adapt to climate change. A review. *Agron. Sustain. Dev.*, **37**, 4, doi:10.1007/s13593-016-0410-x.

Keenan, T.F., and W.J. Riley, 2018: Greening of the land surface in the world's cold regions consistent with recent warming. *Nat. Clim. Chang.*, **8**, 825–828, doi:10.1038/s41558-018-0258-y.

Keenan, T.F. et al., 2013: Increase in forest water-use efficiency as atmospheric carbon dioxide concentrations rise. *Nature*, **499**, 324–327, doi:10.1038/nature12291.

Keenan, T.F. et al., 2016: Recent pause in the growth rate of atmospheric CO_2 due to enhanced terrestrial carbon uptake. *Nat. Commun.*, **7**, doi:10.1038/ncomms13428.

Keenan, T.F. et al., 2017: Corrigendum: Recent pause in the growth rate of atmospheric CO_2 due to enhanced terrestrial carbon uptake. *Nat. Commun.*, **8**, 16137, doi:10.1038/ncomms16137.

Keiluweit, M. et al., 2015: Mineral protection of soil carbon counteracted by root exudates. *Nat. Clim. Chang.*, **5**, 588–595, doi:10.1038/nclimate2580.

Keiluweit, M., P.S. Nico, M. Kleber, and S. Fendorf, 2016: Are oxygen limitations under recognized regulators of organic carbon turnover in upland soils? *Biogeochemistry*, **127**, 157–171, doi:10.1007/s10533-015-0180-6.

Keith, H. et al., 2014: Managing temperate forests for carbon storage: impacts of logging versus forest protection on carbon stocks. *Ecosphere*, **5**, art75, doi:10.1890/ES14-00051.1.

Kelleway, J.J. et al., 2017: Review of the ecosystem service implications of mangrove encroachment into salt marshes. *Glob. Chang. Biol.*, **23**, 3967–3983, doi:10.1111/gcb.13727.

Kelley, C.P., S. Mohtadi, M.A. Cane, R. Seager, and Y. Kushnir, 2015: Climate change in the Fertile Crescent and implications of the recent Syrian drought. *Proc. Natl. Acad. Sci.*, **112**, 3241–3246, doi:10.1073/pnas.1421533112.

Kelly, R. et al., 2013: Recent burning of boreal forests exceeds fire regime limits of the past 10,000 years. *Proc. Natl. Acad. Sci.*, **110**, 13055–13060, doi:10.1073/pnas.1305069110.

Kemena, T.P., K. Matthes, T. Martin, S. Wahl, and A. Oschlies, 2017: Atmospheric feedbacks in North Africa from an irrigated, afforested Sahara. *Clim. Dyn.*, **0**, 1–21, doi:10.1007/s00382-017-3890-8.

Kendon, E.J. et al., 2017: Do convection-permitting regional climate models improve projections of future precipitation change? *Bull. Am. Meteorol. Soc.*, **98**, 79–93, doi:10.1175/BAMS-D-15-0004.1.

Kendra Gotangco Castillo, C., and K.R. Gurney, 2013: A sensitivity analysis of surface biophysical, carbon, and climate impacts of tropical deforestation rates in CCSM4-CNDV. *J. Clim.*, **26**, 805–821, doi:10.1175/JCLI-D-11-00382.1.

van Kessel, C. et al., 2013: Climate, duration, and N placement determine N_2O emissions in reduced tillage systems: A meta-analysis. *Glob. Chang. Biol.*, **19**, 33–44, doi:10.1111/j.1365-2486.2012.02779.x.

Kicklighter, D. et al., 2013: Climate impacts of a large-scale biofuels expansion. *Geophys. Res. Lett.*, **40**, 1624–1630, doi:10.1002/grl.50352.

Kim, D., 2014: The effect of land-use change on the net exchange rates of greenhouse gases: A meta-analytical approach. doi:10.5194/bgd-11-1053-2014.

Kim, H., Y.K. Kim, S.K. Song, and H.W. Lee, 2016: Impact of future urban growth on regional climate changes in the Seoul Metropolitan Area, Korea. *Sci. Total Environ.*, **571**, 355–363, doi:10.1016/j.scitotenv.2016.05.046.

Kim, S. et al., 2014: Indirect land use change and biofuels: Mathematical analysis reveals a fundamental flaw in the regulatory approach. *Biomass and Bioenergy*, **71**, 408–412, doi:10.1016/j.biombioe.2014.09.015.

King, A.D., 2017: Attributing changing rates of temperature record breaking to anthropogenic influences. *Earth's Futur.*, **5**, 1156–1168, doi:10.1002/2017EF000611.

King, A.D. et al., 2016: Emergence of heat extremes attributable to anthropogenic influences. *Geophys. Res. Lett.*, **43**, 3438–3443, doi:10.1002/2015GL067448.

Kleber, M. et al., 2011: Old and stable soil organic matter is not necessarily chemically recalcitrant: Implications for modeling concepts and temperature sensitivity. *Glob. Chang. Biol.*, **17**, 1097–1107, doi:10.1111/j.1365-2486.2010.02278.x.

Klein, C. et al., 2017: Feedback of observed interannual vegetation change: A regional climate model analysis for the West African monsoon. *Clim. Dyn.*, **48**, 2837–2858, doi:10.1007/s00382-016-3237-x.

de Klein, C.A.M., M.A. Shepherd, and T.J. van der Weerden, 2014: Nitrous oxide emissions from grazed grasslands: Interactions between the N cycle and climate change – A New Zealand case study. *Curr. Opin. Environ. Sustain.*, **9–10**, 131–139, doi:10.1016/j.cosust.2014.09.016.

Klemm, W., B.G. Heusinkveld, S. Lenzholzer, M.H. Jacobs, and B. Van Hove, 2015: Psychological and physical impact of urban green spaces on outdoor thermal comfort during summertime in The Netherlands. *Build. Environ.*, **83**, 120–128, doi:10.1016/j.buildenv.2014.05.013.

Klimont, Z. et al., 2017: Global anthropogenic emissions of particulate matter including black carbon. *Atmos. Chem. Phys.*, **17**, 8681–8723, doi:10.5194/acp-17-8681-2017.

Kløve, B. et al., 2014: Climate change impacts on groundwater and dependent ecosystems. *J. Hydrol.*, **518**, 250–266, doi:10.1016/j.jhydrol.2013.06.037.

Knapp, A.K., and M.D. Smith, 2001: Variation among biomes in temporal dynamics of aboveground primary production. *Science*, **291**, 481–484, doi:10.1126/science.291.5503.481.

Knapp, A.K. et al., 2008: Consequences of more extreme precipitation regimes for terrestrial ecosystems. *Bioscience*, **58**, 811–821, doi:10.1641/B580908.

Knorr, W., I.C. Prentice, J.I. House, and E.A. Holland, 2005: Long-term sensitivity of soil carbon turnover to warming. *Nature*, **433**, 298–301, doi:10.1038/nature03226.

Knorr, W., A. Arneth, and L. Jiang, 2016a: Demographic controls of future global fire risk. *Nat. Clim. Chang.*, **6**, 781–785, doi:10.1038/nclimate2999.

Knorr, W., L. Jiang, and A. Arneth, 2016b: Climate, CO_2 and human population impacts on global wildfire emissions. *Biogeosciences*, **13**, 267–282, doi:10.5194/bg-13-267-2016.

Köchy, M., R. Hiederer, and A. Freibauer, 2015: Global distribution of soil organic carbon – Part 1: Masses and frequency distributions of SOC stocks for the tropics, permafrost regions, wetlands, and the world. *SOIL*, **1**, 351–365, doi:10.5194/soil-1-351-2015.

Kögel-Knabner, I. et al., 2008: Organo-mineral associations in temperate soils: Integrating biology, mineralogy, and organic matter chemistry. *J. Plant Nutr. Soil Sci.*, **171**, 61–82, doi:10.1002/jpln.200700048.

Kok, J.F., 2011: A scaling theory for the size distribution of emitted dust aerosols suggests climate models underestimate the size of the global dust cycle. *Proc. Natl. Acad. Sci.*, **108**, 1016–1021, doi:10.1073/pnas.1014798108.

Kok, J.F., S. Albani, N.M. Mahowald, and D.S. Ward, 2014: An improved dust emission model - Part 2: Evaluation in the Community Earth System Model, with implications for the use of dust source functions. *Atmos. Chem. Phys.*, **14**, 13043–13061, doi:10.5194/acp-14-13043-2014.

Kok, J.F., D.S. Ward, N.M. Mahowald, and A.T. Evan, 2018: Global and regional importance of the direct dust-climate feedback. *Nat. Commun.*, **9**, 241, doi:10.1038/s41467-017-02620-y.

Kolus, H.R. et al., 2019: Land carbon models underestimate the severity and duration of drought's impact on plant productivity. *Sci. Rep.*, **9**, 2758, doi:10.1038/s41598-019-39373-1.

Kooperman, G.J., M.S. Pritchard, and R.C.J. Somerville, 2014: The response of US summer rainfall to quadrupled CO_2 climate change in conventional and superparameterized versions of the NCAR community atmosphere model. *J. Adv. Model. Earth Syst.*, **6**, 859–882, doi:10.1002/2014MS000306.

Kooperman, G.J., M.S. Pritchard, T.A. O'Brien, and B.W. Timmermans, 2018: Rainfall from resolved rather than parameterized processes better represents the present-day and climate change response of moderate rates in the Community Atmosphere Model. *J. Adv. Model. Earth Syst.*, **10**, 971–988, doi:10.1002/2017MS001188.

Koornneef, J. et al., 2012: Global potential for biomass and carbon dioxide capture, transport and storage up to 2050. *Int. J. Greenh. Gas Control*, doi:10.1016/j.ijggc.2012.07.027.

Körner, C., 2006: Plant CO_2 responses: An issue of definition, time and resource supply. *New Phytol.*, **172**, 393–411, doi:10.1111/j.1469-8137.2006.01886.x.

Körner, C., 2015: Paradigm shift in plant growth control. *Curr. Opin. Plant Biol.*, **25**, 107–114, doi:10.1016/j.pbi.2015.05.003.

Koskinen, M., L. Maanavilja, M. Nieminen, K. Minkkinen, and E.-S. Tuittila, 2016: High methane emissions from restored Norway spruce swamps in southern Finland over one growing season. *Mires Peat*, **17**, 1–13, doi:10.19189/MaP.2015.OMB.202.

Kottek, M., J. Grieser, C. Beck, B. Rudolf, and F. Rubel, 2006: World map of the Köppen-Geiger climate classification updated. *Meteorol. Zeitschrift*, **15**, 259–263, doi:10.1127/0941-2948/2006/0130.

Koven, C.D. et al., 2015: A simplified, data-constrained approach to estimate the permafrost carbon–climate feedback. *Philos. Trans. R. Soc. A Math. Phys. Eng. Sci.*, **373**, 20140423, doi:10.1098/rsta.2014.0423.

Kramer, K., S.J. Vreugdenhil, and D.C. van der Werf, 2008: Effects of flooding on the recruitment, damage and mortality of riparian tree species: A field and simulation study on the Rhine floodplain. *For. Ecol. Manage.*, **255**, 3893–3903, doi:10.1016/j.foreco.2008.03.044.

Krawchuk, M.A., and M.A. Moritz, 2011: Constraints on global fire activity vary across a resource gradient. *Ecology*, **92**, 121–132, doi:10.1890/09-1843.1.

Kreidenweis, U. et al., 2016: Afforestation to mitigate climate change: impacts on food prices under consideration of albedo effects. *Environ. Res. Lett.*, **11**, 85001, doi:10.1088/1748-9326/11/8/085001.

Kriegler, E. et al., 2017: Fossil-fueled development (SSP5): An energy and resource intensive scenario for the 21st century. *Glob. Environ. Chang.*, **42**, 297–315, doi:10.1016/J.GLOENVCHA.2016.05.015.

Krishnan, R. et al., 2016: Deciphering the desiccation trend of the South Asian monsoon hydroclimate in a warming world. *Clim. Dyn.*, **47**, 1007–1027, doi:10.1007/s00382-015-2886-5.

Kuang, W. et al., 2017: An EcoCity model for regulating urban land cover structure and thermal environment: Taking Beijing as an example. *Sci. China Earth Sci.*, **60**, 1098–1109, doi:10.1007/s11430-016-9032-9.

Kulmala, M., 2004: A new feedback mechanism linking forests, aerosols, and climate. *Atmos. Chem. Phys.*, **4**, 557–562, doi:10.5194/acp-4-557-2004.

Kulmala, M. et al., 2011: General overview: European Integrated project on Aerosol Cloud Climate and Air Quality interactions (EUCAARI) – Integrating aerosol research from nano to global scales. *Atmos. Chem. Phys.*, **11**, 13061–130143, doi:10.5194/acp-11-13061-2011.

Kulmala, M. et al., 2013: Climate feedbacks linking the increasing atmospheric CO_2 concentration, BVOC emissions, aerosols and clouds in forest ecosystems. In: *Biology, Controls and Models of Tree Volatile Organic Compound Emissions* [Niinemets, Ü. and R. Monson (eds.)]. Springer, Dordrecht, Netherlands, pp. 489–508.

Kulmala, M. et al., 2014: CO_2-induced terrestrial climate feedback mechanism: From carbon sink to aerosol source and back. *Boreal Environ. Res.*, **19**, 122–131. Kundzewicz, Z.W. et al., 2014: Flood risk and climate change: Global and regional perspectives. *Hydrol. Sci. J.*, **59**, 1–28, doi:10.1080/02626667.2013.857411.

Kunkel, K.E. et al., 2013: Monitoring and understanding trends in extreme storms: State of knowledge. *Bull. Am. Meteorol. Soc.*, **94**, 499–514, doi:10.1175/BAMS-D-11-00262.1.

Kurz, W.A. et al., 2008: Mountain pine beetle and forest carbon feedback to climate change. *Nature*, **452**, 987–990, doi:10.1038/nature06777.

Kurz, W.A., C. Smyth, and T. Lemprière, 2016: Climate change mitigation through forest sector activities: principles, potential and priorities. *Unasylva*, **67**, 61–67.

Kuzyakov, Y., and E. Blagodatskaya, 2015: Microbial hotspots and hot moments in soil: Concept & review. *Soil Biol. Biochem.*, **83**, 184–199, doi:10.1016/j.soilbio.2015.01.025.

Kuzyakov, Y., I. Bogomolova, and B. Glaser, 2014: Biochar stability in soil: Decomposition during eight years and transformation as assessed by compound-specific14C analysis. *Soil Biol. Biochem.*, **70**, 229–236, doi:10.1016/j.soilbio.2013.12.021.

Kvalevåg, M.M., G. Myhre, G. Bonan, and S. Levis, 2010: Anthropogenic land cover changes in a GCM with surface albedo changes based on MODIS data. *Int. J. Climatol.*, doi:10.1002/joc.2012.

van der Laan-Luijkx, I.T. et al., 2017: The CarbonTracker Data Assimilation Shell (CTDAS) v1.0: Implementation and global carbon balance 2001–2015. *Geosci. Model Dev.*, **10**, 2785–2800, doi:10.5194/gmd-10-2785-2017.

Laganière, J., D. Paré, E. Thiffault, and P.Y. Bernier, 2017: Range and uncertainties in estimating delays in greenhouse gas mitigation potential of forest bioenergy sourced from Canadian forests. *GCB Bioenergy*, **9**, 358–369, doi:10.1111/gcbb.12327.

Laguë, M.M. and A.L.S. Swann, 2016: Progressive midlatitude afforestation: Impacts on clouds, global energy transport, and precipitation. *J. Clim.*, **29**, 5561–5573, doi:10.1175/jcli-d-15-0748.1.

Lai, L. et al., 2017: Soil nitrogen dynamics in switchgrass seeded to a marginal cropland in South Dakota. *GCB Bioenergy*, n/a-n/a, doi:10.1111/gcbb.12475.

Lajtha, K., R.D. Bowden, and K. Nadelhoffer, 2014a: Litter and root manipulations provide insights into soil organic matter dynamics and stability. *Soil Sci. Soc. Am. J.*, **78**, S261, doi:10.2136/sssaj2013.08.0370nafsc.

Lajtha, K. et al., 2014b: Changes to particulate versus mineral-associated soil carbon after 50 years of litter manipulation in forest and prairie experimental ecosystems. *Biogeochemistry*, **119**, 341–360, doi:10.1007/s10533-014-9970-5.

Lal, R., 2010: Managing soils and ecosystems for mitigating anthropogenic carbon emissions and advancing global food security. *Bioscience*, doi:10.1525/bio.2010.60.9.8.

Lal, R., 2011: Sequestering carbon in soils of agro-ecosystems. *Food Policy*, **36**, S33–S39, doi:10.1016/j.foodpol.2010.12.001.

Lal, R., 2013: Soil carbon management and climate change. *Carbon Manag.*, **4**, 439–462, doi:10.4155/cmt.13.31.

Lal, R., M. Griffin, J. Apt, L. Lave, and M.G. Morgan, 2004: Managing soil carbon. *Science*, **304**, 393, doi: 10.1126/science.1093079.

Lamarque, J.-F. et al., 2010: Historical (1850–2000) gridded anthropogenic and biomass burning emissions of reactive gases and aerosols: methodology and application. *Atmos. Chem. Phys.*, **10**, 7017–7039, doi:10.5194/acp-10-7017-2010.

Lamarque, J.F. et al., 2013: The Atmospheric Chemistry and Climate Model Intercomparison Project (ACCMIP): Overview and description of models, simulations and climate diagnostics. *Geosci. Model Dev.*, **6**, 179–206, doi:10.5194/gmd-6-179-2013.

Lamers, P., and M. Junginger, 2013: The "debt" is in the detail: A synthesis of recent temporal forest carbon analyses on woody biomass for energy. *Biofuels, Bioprod. Biorefining*, **7**, 373–385, doi:10.1002/bbb.1407.

Landhäusser, S.M., D. Deshaies, and V.J. Lieffers, 2010: Disturbance facilitates rapid range expansion of aspen into higher elevations of the Rocky Mountains under a warming climate. *J. Biogeogr.*, **37**, 68–76, doi:10.1111/j.1365-2699.2009.02182.x.

Landry, J.S., and H.D. Matthews, 2016: Non-deforestation fire vs. fossil fuel combustion: The source of CO_2 emissions affects the global carbon cycle and climate responses. *Biogeosciences*, **13**, 2137–2149, doi:10.5194/bg-13-2137-2016.

Larsen, M.A.D., J.H. Christensen, M. Drews, M.B. Butts, and J.C. Refsgaard, 2016: Local control on precipitation in a fully coupled climate-hydrology model. *Sci. Rep.*, **6**, 22927, doi:10.1038/srep22927.

Laskin, A., J. Laskin, and S.A. Nizkorodov, 2015: Chemistry of atmospheric brown carbon. *Chem. Rev.*, **115**, 4335–4382, doi:10.1021/cr5006167.

Laurance, S. et al., 2018: Compositional response of Amazon forests to climate change. *Glob. Chang. Biol.*, **25**, 39–56, doi:10.1111/gcb.14413.

Law, D.J. et al., 2018: Bioclimatic envelopes for individual demographic events driven by extremes: Plant mortality from drought and warming. *Int. J. Plant Sci.*, **180**, 53–62, doi:10.1086/700702.

Lawrence, D. and K. Vandecar, 2015: Effects of tropical deforestation on climate and agriculture. *Nat. Clim. Chang.*, **5**, 27–36, doi:10.1038/nclimate2430.

Lawrence, D.M. et al., 2016: The Land Use Model Intercomparison Project (LUMIP) contribution to CMIP6: Rationale and experimental design. *Geosci. Model Dev.*, **9**, 2973–2998, doi:10.5194/gmd-9-2973-2016.

Lawrence, P.J. et al., 2012: Simulating the biogeochemical and biogeophysical impacts of transient land cover change and wood harvest in the Community Climate System Model (CCSM4) from 1850 to 2100. *J. Clim.*, doi:10.1175/JCLI-D-11-00256.1.

Lawrence, P.J., D.M. Lawrence, and G.C. Hurtt, 2018: Attributing the carbon cycle impacts of CMIP5 historical and future land use and land cover change in the Community Earth System Model (CESM1). *J. Geophys. Res. Biogeosciences*, **123**, 1732–1755, doi:10.1029/2017JG004348.

Lawrimore, J.H. et al. 2011: An overview of the Global Historical Climatology Network monthly mean temperature data set, version 3. *J. Geophys. Res. Atmos.*, **116**, doi:10.1029/2011JD016187.

Layton, K. and D. Ellison, 2016: Induced precipitation recycling (IPR): A proposed concept for increasing precipitation through natural vegetation feedback mechanisms. *Ecol. Eng.*, **91**, 553–565, doi:10.1016/j.ecoleng.2016.02.031.

Leakey, A.D.B., K.A. Bishop, and E.A. Ainsworth, 2012: A multi-biome gap in understanding of crop and ecosystem responses to elevated CO_2. *Curr. Opin. Plant Biol.*, **15**, 228–236, doi:10.1016/j.pbi.2012.01.009.

Lee, D. and Sanz, M.J., 2017: *UNFCCC Accounting for Forests: What's In and What's Out of NDCs and REDD+*. Climate and Land Use Alliance, 16 pp.

Lee, X. et al., 2011: Observed increase in local cooling effect of deforestation at higher latitudes. *Nature*, **479**, 384–387, doi:10.1038/nature10588.

Lehmann, J., D. Coumou, and K. Frieler, 2015: Increased record-breaking precipitation events under global warming. *Clim. Change*, **132**, 501–515, doi:10.1007/s10584-015-1434-y.

Lehner, F., C. Deser, and B.M. Sanderson, 2018: Future risk of record-breaking summer temperatures and its mitigation. *Clim. Change*, **146**, 363–375, doi:10.1007/s10584-016-1616-2.

Lejeune, Q., E.L. Davin, B.P. Guillod, and S.I. Seneviratne, 2015: Influence of Amazonian deforestation on the future evolution of regional surface fluxes, circulation, surface temperature and precipitation. *Clim. Dyn.*, **44**, 2769–2786, doi:10.1007/s00382-014-2203-8.

Lejeune, Q., E.L. Davin, L. Gudmundsson, and S.I. Seneviratne, 2018: Historical deforestation increased the risk of heat extremes in northern mid-latitudes 2. *Nat. Clim. Chang.*, **8**, 1–16, doi:10.1038/s41558-018-0131-z.

Lemordant, L., P. Gentine, M. Stefanon, P. Drobinski, and S. Fatichi, 2016: Modification of land-atmosphere interactions by CO_2 effects: Implications for summer dryness and heat wave amplitude. *Geophys. Res. Lett.*, **43**, 10,240-10,248, doi:10.1002/2016GL069896.

Lemprière, T.C. et al., 2013: Canadian boreal forests and climate change mitigation. *Environ. Rev.*, **21**, 293–321, doi:10.1139/er-2013-0039.

van Lent, J., K. Hergoualc'h, and L.V. Verchot, 2015: Reviews and syntheses: Soil N_2O and NO emissions from land use and land-use change in the tropics and subtropics: A meta-analysis. *Biogeosciences*, **12**, 7299–7313, doi:10.5194/bg-12-7299-2015.

Lenton, T.M., 2010: The potential for land-based biological CO_2 removal to lower future atmospheric CO_2 concentration. *Carbon Manag.*, doi:10.4155/cmt.10.12.

Lenton, T.M., 2014: The Global Potential for Carbon Dioxide Removal. *Geoengineering of the Climate System* [Harrison, R.M. and R.E. Hester (eds.)]. Royal Society of Chemistry, Cambridge, U.K., 52–79 pp. doi:10.1039/9781782621225-00052.

Di Leo, N., F.J. Escobedo, and M. Dubbeling, 2016: The role of urban green infrastructure in mitigating land surface temperature in Bobo-Dioulasso, Burkina Faso. *Environ. Dev. Sustain.*, **18**, 373–392, doi:10.1007/s10668-015-9653-y.

Leonard, M. et al., 2014: A compound event framework for understanding extreme impacts. *Wiley Interdiscip. Rev. Clim. Chang.*, **5**, 113–128, doi:10.1002/wcc.252.

Lesk, C., P. Rowhani, and N. Ramankutty, 2016: Influence of extreme weather disasters on global crop production. *Nature*, **529**, 84. doi:10.1038/nature16467.

Lewis, S.C., and D.J. Karoly, 2013: Anthropogenic contributions to Australia's record summer temperatures of 2013. *Geophys. Res. Lett.*, **40**, 3708–3709, doi:10.1002/grl.50673.

Lewis, S.M., and M. Kelly, 2014: Mapping the potential for biofuel production on marginal lands: Differences in definitions, data and models across scales. *Isprs Int. J. Geo-Information*, **3**, 430–459, doi:10.3390/ijgi3020430.

Lhotka, O., J. Kyselý, and A. Farda, 2018: Climate change scenarios of heat waves in Central Europe and their uncertainties. *Theor. Appl. Climatol.*, **131**, 1043–1054, doi:10.1007/s00704-016-2031-3.

Li, C., S. Frolking, and K. Butterbach-Bahl, 2005: Carbon sequestration in arable soils is likely to increase nitrous oxide emissions, offsetting reductions in climate radiative forcing. *Clim. Change*, **72**, 321–338, doi:10.1007/s10584-005-6791-5.

Li, D., and E. Bou-Zeid, 2013: Synergistic interactions between urban heat islands and heat waves: The impact in cities is larger than the sum of its parts. *J. Appl. Meteorol. Climatol.*, **52**, 2051–2064, doi:10.1175/JAMC-D-13-02.1.

Li, D. et al., 2015a: Contrasting responses of urban and rural surface energy budgets to heat waves explain synergies between urban heat islands and heat waves. *Environ. Res. Lett.*, **10**, 54009, doi:10.1088/1748-9326/10/5/054009.

Li, D. et al., 2019: Urban heat island: Aerodynamics or imperviousness? *Sci. Adv.*, **5**, doi:10.1126/sciadv.aau4299.

Li, H. et al., 2018a: A new method to quantify surface urban heat island intensity. *Sci. Total Environ.*, **624**, 262–272, doi:10.1016/j.scitotenv.2017.11.360.

Li, W. et al., 2017a: Land-use and land-cover change carbon emissions between 1901 and 2012 constrained by biomass observations. *Biogeosciences*, **14**, 5053–5067, doi:10.5194/bg-14-5053-2017.

Li, W. et al., 2018b: Recent changes in global photosynthesis and terrestrial ecosystem respiration constrained from multiple observations. *Geophys. Res. Lett.*, **45**, 1058–1068, doi:10.1002/2017GL076622.

Li, X., Y. Zhou, G.R. Asrar, M. Imhoff, and X. Li, 2017b: The surface urban heat island response to urban expansion: A panel analysis for the conterminous United States. *Sci. Total Environ.*, **605–606**, 426–435, doi:10.1016/j.scitotenv.2017.06.229.

Li, Y., M. Zhao, S. Motesharrei, Q. Mu, E. Kalnay, and S. Li, 2015b: Local cooling and warming effects of forests based on satellite observations. *Nat. Commun.*, **6**, 1–8, doi:10.1038/ncomms7603.

Li, Y. et al., 2016: The role of spatial scale and background climate in the latitudinal temperature response to deforestation. *Earth Syst. Dyn.*, **7**, 167–181, doi:10.5194/esd-7-167-2016.

Li, Z. and H. Fang, 2016: Impacts of climate change on water erosion: A review. *Earth-Science Rev.*, **163**, 94–117, doi:10.1016/j.earscirev.2016.10.004.

Liang, J., X. Qi, L. Souza, and Y. Luo, 2016: Processes regulating progressive nitrogen limitation under elevated carbon dioxide: A meta-analysis. *Biogeosciences*, **13**, 2689–2699, doi:10.5194/bg-13-2689-2016.

Liang, P. and Y. Ding, 2017: The long-term variation of extreme heavy precipitation and its link to urbanization effects in Shanghai during 1916–2014. *Adv. Atmos. Sci.*, **34**, 321–334, doi:10.1007/s00376-016-6120-0.

Liao, W., D. Wang, X. Liu, G. Wang, and J. Zhang, 2017: Estimated influence of urbanization on surface warming in Eastern China using time-varying land use data. *Int. J. Climatol.*, **37**, 3197–3208, doi:10.1002/joc.4908.

Lickley, M., and S. Solomon, 2018: Drivers, timing and some impacts of global aridity change. *Environ. Res. Lett.*, **13**, 104010, doi:10.1088/1748-9326/aae013.

Liska, A.J. et al., 2014: Biofuels from crop residue can reduce soil carbon and increase CO_2 emissions. *Nat. Clim. Chang.*, **4**, 398–401, doi:10.1038/nclimate2187.

Liu, B. et al., 2016a: Similar estimates of temperature impacts on global wheat yield by three independent methods. *Nat. Clim. Chang.*, **6**, 1130–1136, doi:10.1038/nclimate3115.

Liu, C., C.E. Chung, F. Zhang, and Y. Yin, 2016b: The colors of biomass burning aerosols in the atmosphere. *Sci. Rep.*, **6**, doi:10.1038/srep28267.

Liu, J., and H. Yang, 2010: Spatially explicit assessment of global consumptive water uses in cropland: Green and blue water. *J. Hydrol.*, **384**, 187–197, doi:10.1016/j.jhydrol.2009.11.024.

Liu, J. et al., 2017a: Contrasting carbon cycle responses of the tropical continents to the 2015–2016 El Niño. *Science*, **358**, eaam5690, doi:10.1126/science.aam5690.

Liu, L. and T.L. Greaver, 2009: A review of nitrogen enrichment effects on three biogenic GHGs: The CO_2 sink may be largely offset by stimulated N_2O and CH_4 emission. *Ecol. Lett.*, **12**, 1103–1117, doi:10.1111/j.1461-0248.2009.01351.x.

Liu, L. et al.., 2009: Enhanced litter input rather than changes in litter chemistry drive soil carbon and nitrogen cycles under elevated CO_2: A microcosm study. *Glob. Chang. Biol.*, **15**, 441–453, doi:10.1111/j.1365-2486.2008.01747.x. https://doi.org/10.1111/j.1365-2486.2008.01747.x.

Liu, W.J. et al., 2017b: Repackaging precipitation into fewer, larger storms reduces ecosystem exchanges of CO_2 and H2O in a semiarid steppe. *Agric. For. Meteorol.*, **247**, 356–364, doi:10.1016/j.agrformet.2017.08.029.

Liu, Y., J. Stanturf, and S. Goodrick, 2010: Trends in global wildfire potential in a changing climate. *For. Ecol. Manage.*, **259**, 685–697, doi:10.1016/j.foreco.2009.09.002.

Liu, Y., Y. Li, S. Li, and S. Motesharrei, 2015: Spatial and temporal patterns of global NDVI trends: Correlations with climate and human factors. *Remote Sens.*, **7**, 13233–13250, doi:10.3390/rs71013233.

Liu, Y. et al., 2016c: Isoprene photochemistry over the Amazon rainforest. *Proc. Natl. Acad. Sci.*, **113**, 6125–6130, doi:10.1073/pnas.1524136113.

Liu, Z., 2012: Dynamics of interdecadal climate variability: A historical perspective. *J. Clim.*, **25**, 1963–1995, doi:10.1175/2011JCLI3980.1.

Llonch, P., M.J. Haskell, R.J. Dewhurst, and S.P. Turner, 2017: Current available strategies to mitigate greenhouse gas emissions in livestock systems: An animal welfare perspective. *Animal*, doi:10.1017/S1751731116001440.

Lloyd, A.H., T.S. Rupp, C.L. Fastie, and A.M. Starfield, 2003: Patterns and dynamics of treeline advance on the Seward Peninsula, Alaska. **108**, D2, 8161, doi:10.1029/2001JD000852.

Lloyd, J., and G.D. Farquhar, 2008: Effects of rising temperatures and [CO_2] on the physiology of tropical forest trees. *Philos. Trans. R. Soc. B Biol. Sci.*, **363**, 1811–1817, doi:10.1098/rstb.2007.0032.

Loarie, S.R., D.B. Lobell, G.P. Asner, Q. Mu, and C.B. Field, 2011: Direct impacts on local climate of sugar-cane expansion in Brazil. *Nat. Clim. Chang.*, **1**, 105–109, doi:10.1038/nclimate1067.

Lobell, D.B., and C. Bonfils, 2008: The effect of irrigation on regional temperatures: A spatial and temporal analysis of trends in California, 1934–2002. *J. Clim.*, **21**, 2063–2071, doi:10.1175/2007JCLI1755.1.

Lobell, D.B., and C. Tebaldi, 2014: Getting caught with our plants down: The risks of a global crop yield slowdown from climate trends in the next two decades. *Environ. Res. Lett.*, **9**, 74003, doi:10.1088/1748-9326/9/7/074003.

Lobell, D.B., G. Bala, and P.B. Duffy, 2006: Biogeophysical impacts of cropland management changes on climate. *Geophys. Res. Lett.*, **33**, 4–7, doi:10.1029/2005GL025492.

Lobell, D.B., C.J. Bonfils, L.M. Kueppers, and M.A. Snyder, 2008: Irrigation cooling effect on temperature and heat index extremes. *Geophys. Res. Lett.*, **35**, 1–5, doi:10.1029/2008GL034145.

Lobell, D.B., W. Schlenker, and J. Costa-Roberts, 2011: Climate trends and global crop production since 1980. *Science*, **333**, 616–620, doi:10.1126/science.1204531.

Loehman, R.A., E. Reinhardt, and K.L. Riley, 2014: Wildland fire emissions, carbon, and climate: Seeing the forest and the trees – A cross-scale assessment of wildfire and carbon dynamics in fire-prone, forested ecosystems. *For. Ecol. Manage.*, **317**, 9–19, doi:10.1016/j.foreco.2013.04.014.

Lokoshchenko, M.A., 2017: Urban heat island and urban dry island in Moscow and their centennial changes. *J. Appl. Meteorol. Climatol.*, **56**, 2729–2745, doi:10.1175/JAMC-D-16-0383.1.

Lombardozzi, D.L. et al., 2018: Cover Crops May Cause Winter Warming in Snow-Covered Regions. *Geophys. Res. Lett.*, **45**, 9889–9897, doi:10.1029/2018GL079000.

Longobardi, P., A. Montenegro, H. Beltrami, and M. Eby, 2016a: Deforestation induced climate change: Effects of spatial scale. *PLoS One*, doi:10.1371/journal.pone.0153357.

Longobardi, P., A. Montenegro, H. Beltrami, and M. Eby, 2016b: Deforestation induced climate change: Effects of spatial scale. *PLoS One*, **11**, doi:10.1371/journal.pone.0153357.

Van Loon, A.F., and G. Laaha, 2015: Hydrological drought severity explained by climate and catchment characteristics. *J. Hydrol.*, **526**, 3–14, doi:10.1016/j.jhydrol.2014.10.059.

Lopez, H. et al., 2018: Early emergence of anthropogenically forced heat waves in the western United States and Great Lakes. *Nat. Clim. Chang.*, **8**, 414–420, doi:10.1038/s41558-018-0116-y.

Loranty, M.M., L.T. Berner, S.J. Goetz, Y. Jin, and J.T. Randerson, 2014: Vegetation controls on northern high latitude snow-albedo feedback: Observations and CMIP5 model simulations. *Glob. Chang. Biol.*, **20**, 594–606, doi:10.1111/gcb.12391.

Lorenz, R., and A.J. Pitman, 2014: Effect of land-atmosphere coupling strength on impacts from Amazonian deforestation. *Geophys. Res. Lett.*, **41**, 5987–5995, doi:10.1002/2014GL061017.

Lorenz, R. et al., 2016: Influence of land-atmosphere feedbacks on temperature and precipitation extremes in the GLACE-CMIP5 ensemble. *J. Geophys. Res.*, **121**, 607–623, doi:10.1002/2015JD024053.

Di Lorenzo, E. et al., 2008: North Pacific Gyre Oscillation links ocean climate and ecosystem change. *Geophys. Res. Lett.*, **35**, L08607, doi:10.1029/2007GL032838.

Los, S.O., 2013: Analysis of trends in fused AVHRR and MODIS NDVI data for 1982-2006: Indication for a CO_2 fertilization effect in global vegetation. *Global Biogeochem. Cycles*, **27**, 318–330, doi:10.1002/gbc.20027.

Lucht, W. et al., 1995: Climatic control of the high-latitude vegetation greening trend and Pinatubo effect. *Science*, **296**, 1687–1689, doi:10.1126/science.1071828.

Luedeling, E., 2012: Climate change impacts on winter chill for temperate fruit and nut production: A review. *Sci. Hortic. (Amsterdam).*, **144**, 218–229, doi:10.1016/j.scienta.2012.07.011.

Luo, Y., S. Wan, D. Hui, and L.L. Wallace, 2001: Acclimatization of soil respiration to warming in a tall grass prairie. *Nature*, **413**, 622–625, doi:10.1038/35098065.

Luo, Z., E. Wang, and O.J. Sun, 2010: Can no-tillage stimulate carbon sequestration in agricultural soils? A meta-analysis of paired experiments. *Agric. Ecosyst. Environ.*, **139**, 224–231, doi:10.1016/j.agee.2010.08.006.

Lützow, M.V. et al., 2006: Stabilization of organic matter in temperate soils: Mechanisms and their relevance under different soil conditions – A review. *Eur. J. Soil Sci.*, **57**, 426–445, doi:10.1111/j.1365-2389.2006.00809.x.

Luyssaert, S. et al., 2014: Land management and land-cover change have impacts of similar magnitude on surface temperature. *Nat. Clim. Chang.*, **4**, 389–393, doi:10.1038/nclimate2196.

Luyssaert, S. et al., 2018: Trade-offs in using European forests to meet climate objectives. *Nature*, **562**, 259–262, doi:10.1038/s41586-018-0577-1.

Ma, D., M. Notaro, Z. Liu, G. Chen, and Y. Liu, 2013a: Simulated impacts of afforestation in East China monsoon region as modulated by ocean variability. *Clim. Dyn.*, **41**, 2439–2450, doi:10.1007/s00382-012-1592-9.

Ma, E., A. Liu, X. Li, F. Wu, and J. Zhan, 2013b: Impacts of vegetation change on the regional surface climate: A scenario-based analysis of afforestation in Jiangxi Province, China. *Adv. Meteorol.*, **2013**, doi:10.1155/2013/796163.

Mabuchi, K., Y. Sato, and H. Kida, 2005: Climatic impact of vegetation change in the Asian tropical region. Part I: Case of the Northern Hemisphere summer. *J. Clim.*, **18**, 410–428, doi:10.1175/JCLI-3273.1.

MacDermott, H.J., R.J. Fensham, Q. Hua, and D.M.J.S. Bowman, 2016: Vegetation, fire and soil feedbacks of dynamic boundaries between rainforest, savanna and grassland. *Austral Ecol.*, **42**, 154–164, doi:10.1111/aec.12415.

MacDougall, A.H., N.C. Swart, and R. Knutti, 2016: The uncertainty in the transient climate response to cumulative CO_2 emissions arising from the uncertainty in physical climate parameters. *J. Clim.*, doi:10.1175/jcli-d-16-0205.1.

Macintosh, A., H. Keith, and D. Lindenmayer, 2015: Rethinking forest carbon assessments to account for policy institutions. *Nat. Clim. Chang.*, **5**, 946.

Magarini, A. and A. Calori, 2015: *Food and the Cities. Food Policies for Sustainable Cities*. 1st ed. Edizioni Ambiente, Milán.

Maher, P., G.K. Vallis, S.C. Sherwood, M.J. Webb, and P.G. Sansom, 2018: The impact of parameterized convection on climatological precipitation in atmospheric global climate models. *Geophys. Res. Lett.*, **45**, 3728–3736, doi:10.1002/2017GL076826.

Mahmood, R. et al., 2006: Impacts of irrigation on 20th century temperature in the northern Great Plains. *Glob. Planet. Change*, **54**, 1–18, doi:10.1016/j.gloplacha.2005.10.004.

Mahmood, R. et al., 2016: Seasonality of global and Arctic black carbon processes in the Arctic Monitoring and Assessment Programme models. *J. Geophys. Res. Atmos.*, **121**, 7100–7116, doi:10.1002/2016JD024849.

Mahony, C.R., A.J. Cannon, T. Wang, and S.N. Aitken, 2017: A closer look at novel climates: new methods and insights at continental to landscape scales. *Glob. Chang. Biol.*, **23**, 3934–3955, doi:10.1111/gcb.13645.

Mahowald, N.M., D.S. Ward, S.C. Doney, P.G. Hess, and J.T. Randerson, 2017: Are the impacts of land use on warming underestimated in climate policy? *Environ. Res. Lett.*, **12**, 94016, doi:10.1088/1748-9326/aa836d.

Makarieva, A.M., V.G. Gorshkov, and B.L. Li, 2009: Precipitation on land versus distance from the ocean: Evidence for a forest pump of atmospheric moisture. *Ecol. Complex.*, **6**, 302–307, doi:10.1016/j.ecocom.2008.11.004.

Makkonen, R. et al., 2012: Air pollution control and decreasing new particle formation lead to strong climate warming. *Atmos. Chem. Phys.*, **12**, 1515–1524, doi:10.5194/acp-12-1515-2012.

Malhi, Y., C. Doughty, and D. Galbraith, 2011: The allocation of ecosystem net primary productivity in tropical forests. *Philos. Trans. R. Soc. B Biol. Sci.*, **366**, 3225–3245, doi:10.1098/rstb.2011.0062.

Malhi, Y. et al., 2015: The linkages between photosynthesis, productivity, growth and biomass in lowland Amazonian forests. *Glob. Chang. Biol.*, **21**, 2283–2295, doi:10.1111/gcb.12859.

Manabe, S., R.J. Stouffer, M.J. Spelman, and K. Bryan, 1991: Transient responses of a coupled ocean–atmosphere model to gradual changes of atmospheric CO_2. Part I: Annual mean response. *J. Clim.*, **4**, 785–818, doi:10.1175/1520-0442(1991)004<0785:TROACO>2.0.CO;2.

Manish, S. et al., 2017: Recent advances in understanding secondary organic aerosol: Implications for global climate forcing. *Rev. Geophys.*, **55**, 509–559, doi:10.1002/2016RG000540.

Manning, D.A.C., 2008: Biological enhancement of soil carbonate precipitation: Passive removal of atmospheric CO_2, *Mineral. Mag.*, **72**, 639–649, doi:10.1180/minmag.2008.072.2.639.

Manola, I., B. Van Den Hurk, H. De Moel, and J.C.J.H. Aerts, 2018: Future extreme precipitation intensities based on a historic event. *Hydrol. Earth Syst. Sci.*, **22**, 3777–3788, doi:10.5194/hess-22-3777-2018.

Mao, J. et al., 2016: Human-induced greening of the northern extratropical land surface. *Nat. Clim. Chang.*, **6**, 959–963, doi:10.1038/nclimate3056.

van Marle, M.J.E. et al., 2017a: Fire and deforestation dynamics in Amazonia (1973–2014). *Global Biogeochem. Cycles*, **31**, 24–38, doi:10.1002/2016GB005445.

van Marle, M.J.E. et al., 2017b: Historic global biomass burning emissions for CMIP6 (BB4CMIP) based on merging satellite observations with proxies and fire models (1750–2015). *Geosci. Model Dev.*, **10**, 3329–3357, doi:10.5194/gmd-10-3329-2017.

Marlon, J.R. et al., 2013: Global biomass burning: A synthesis and review of Holocene paleofire records and their controls. *Quat. Sci. Rev.*, **65**, 5–25, doi:10.1016/j.quascirev.2012.11.029.

Marlon, J.R. et al., 2016: Reconstructions of biomass burning from sediment-charcoal records to improve data–model comparisons. *Biogeosciences*, **13**, 3225–3244, doi:10.5194/bg-13-3225-2016.

Marschner, B. et al., 2008: How relevant is recalcitrance for the stabilization of organic matter in soils? *J. Plant Nutr. Soil Sci.*, **171**, 91–110, doi:10.1002/jpln.200700049. doi:10.1002/jpln.200700049.

Masih, I., S. Maskey, F.E.F. Mussá, and P. Trambauer, 2014: A review of droughts on the African continent: a geospatial and long-term perspective. *Hydrol. Earth Syst. Sci.*, **18**, 3635–3649, doi:10.5194/hess-18-3635-2014.

Matthews, H.D., N.P. Gillett, P.A. Stott, and K. Zickfeld, 2009: The proportionality of global warming to cumulative carbon emissions. *Nature*, **459**, 829. doi:10.1038/nature08047.

Matthews, H.D.D., A.J.J. Weaver, K.J.J. Meissner, N.P.P. Gillett, and M. Eby, 2004: Natural and anthropogenic climate change: incorporating historical land cover change, vegetation dynamics and the global carbon cycle. *Clim. Dyn.*, **22**, 461–479, doi:10.1007/s00382-004-0392-2.

Maule, C.F., T. Mendlik, and O.B. Christensen, 2017: The effect of the pathway to a two degrees warmer world on the regional temperature change of Europe. *Clim. Serv.*, **7**, 3–11, doi:10.1016/j.cliser.2016.07.002.

Mazzoncini, M., T.B. Sapkota, P. Bàrberi, D. Antichi, and R. Risaliti, 2011: Long-term effect of tillage, nitrogen fertilization and cover crops on soil organic carbon and total nitrogen content. *Soil Tillage Res.*, **114**, 165–174, doi:10.1016/j.still.2011.05.001.

McDaniel, M.D., D. Saha, M.G. Dumont, M. Hernández, and M.A. Adams, 2019: The effect of land-use change on soil CH_4 and N_2O Fluxes: A global meta-analysis. *Ecosystems*, doi:10.1007/s10021-019-00347-z.

McDowell, N. et al., 2018: Drivers and mechanisms of tree mortality in moist tropical forests. *New Phytol.*, **219**, 851–869, doi:10.1111/nph.15027.

McDowell, N.G., and C.D. Allen, 2015: Darcy's law predicts widespread forest mortality under climate warming. *Nat. Clim. Chang.*, **5**, 669–672, doi:10.1038/nclimate2641.

McDowell, N.G. et al., 2011: The interdependence of mechanisms underlying climate-driven vegetation mortality. *Trends Ecol. Evol.*, **26**, 523–532, doi:10.1016/j.tree.2011.06.003.

McGuire, A.D. et al., 2018: Dependence of the evolution of carbon dynamics in the northern permafrost region on the trajectory of climate change. *Proc. Natl. Acad. Sci.*, **115**, 201719903, doi:10.1073/pnas.1719903115.

McLaren, D., 2012: A comparative global assessment of potential negative emissions technologies. *Process Saf. Environ. Prot.*, **90**, 489–500, doi:10.1016/j.psep.2012.10.005.

McLeod, E. et al., 2011: A blueprint for blue carbon: Toward an improved understanding of the role of vegetated coastal habitats in sequestering CO_2. *Front. Ecol. Environ.*, **9**, 552–560, doi:10.1890/110004.

McLeod, J., M. Shepherd, and C.E. Konrad, 2017: Spatio-temporal rainfall patterns around Atlanta, Georgia and possible relationships to urban land cover. *Urban Clim.*, **21**, 27–42, doi:10.1016/j.uclim.2017.03.004.

McNorton, J. et al., 2016: Role of OH variability in the stalling of the global atmospheric CH_4 growth rate from 1999 to 2006. *Atmos. Chem. Phys.*, **16**, 7943–7956, doi:10.5194/acp-16-7943-2016.

Meier, M., J. Fuhrer, and A. Holzkämper, 2018: Changing risk of spring frost damage in grapevines due to climate change? A case study in the Swiss Rhone Valley. *Int. J. Biometeorol.*, **62**, 991–1002, doi:10.1007/s00484-018-1501-y.

Meli, P. et al., 2017: A global review of past land use, climate, and active vs. passive restoration effects on forest recovery. *PLoS One*, doi:10.1371/journal.pone.0171368.

Melillo, J.M. et al., 2017: Long-term pattern and magnitude of soil carbon feedback to the climate system in a warming world. *Science*, **358**, 101–105, doi:10.1126/science.aan2874.

Melton, J.R. et al., 2013: Present state of global wetland extent and wetland methane modelling: conclusions from a model inter-comparison project (WETCHIMP). *Biogeosciences*, **10**, 753–788, doi:10.5194/bg-10-753-2013.

Menne, M.J., C.N. Williams, B.E. Gleason, J.J. Rennie, and J.H. Lawrimore, 2018: The global historical climatology network monthly temperature dataset, Version 4. *J. Clim.*, **31**, 9835–9854, doi:10.1175/JCLI-D-18-0094.1.

Mercado, L.M. et al., 2018: Large sensitivity in land carbon storage due to geographical and temporal variation in the thermal response of photosynthetic capacity. *New Phytol.*, **218**, 1462–1477, doi:10.1111/nph.15100.

Meurer, K.H.E. et al., 2016: Direct nitrous oxide (N_2O) fluxes from soils under different land use in Brazil—a critical review. *Environ. Res. Lett.*, **11**, 23001, doi:10.1088/1748-9326/11/2/023001.

Meurer, K.H.E., N.R. Haddaway, M.A. Bolinder, and T. Kätterer, 2018: Tillage intensity affects total SOC stocks in boreo-temperate regions only in the topsoil—A systematic review using an ESM approach. *Earth-Science Rev.*, **177**, 613–622, doi:10.1016/j.earscirev.2017.12.015.

Michalský, M., and P.S. Hooda, 2015: Greenhouse gas emissions of imported and locally produced fruit and vegetable commodities: A quantitative assessment. *Environ. Sci. Policy*, **48**, 32–43, doi:10.1016/j.envsci.2014.12.018.

Millar, C.I., and N.L. Stephenson, 2015: Temperate forest health in an era of emerging megadisturbance. *Science*, **349**, 823–826, doi:10.1126/science.aaa9933.

Millar, R.J. et al., 2017: Emission budgets and pathways consistent with limiting warming to 1.5°c. *Nat. Geosci.*, **10**, 741–747, doi:10.1038/NGEO3031.

Milly, P.C.D., and K.A. Dunne, 2016: Potential evapotranspiration and continental drying. *Nat. Clim. Chang.*, **6**, 946–949, doi:10.1038/nclimate3046.

Min, S.K., X. Zhang, F.W. Zwiers, and G.C. Hegerl, 2011: Human contribution to more-intense precipitation extremes. *Nature*, **470**, 378–381, doi:10.1038/nature09763.

Minayeva, T.Y. and A.A. Sirin, 2012: Peatland biodiversity and climate change. *Biol. Bull. Rev.*, **2**, 164–175, doi:10.1134/s207908641202003x.

Miner, R., 2010: *Impact of the Global Forest Industry on Atmospheric Greenhouse Gases*. Food and Agriculture Organization of the United Nations, Rome, Italy, 71 pp.

Ming, J., C. Xiao, Z. Du, and X. Yang, 2013: An overview of black carbon deposition in High Asia glaciers and its impacts on radiation balance. *Adv. Water Resour.*, **55**, 80–87, doi:10.1016/j.advwatres.2012.05.015.

Minx, J.C. et al., 2018: Negative emissions – Part 1: Research landscape and synthesis. *Environ. Res. Lett.*, doi:10.1088/1748-9326/aabf9b.

Miralles, D.G., A.J. Teuling, C.C. Van Heerwaarden, and J.V.G. De Arellano, 2014: Mega-heatwave temperatures due to combined soil desiccation and atmospheric heat accumulation. *Nat. Geosci.*, **7**, 345–349, doi:10.1038/ngeo2141.

Miralles, D.G., P. Gentine, S.I. Seneviratne, and A.J. Teuling, 2018: Land-atmospheric feedbacks during droughts and heatwaves: state of the science and current challenges. *Ann. N.Y. Acad. Sci.*, **1436**, 19–35, doi:10.1111/nyas.13912.

Mishra, V., A.R. Ganguly, B. Nijssen, and D.P. Lettenmaier, 2015: Changes in observed climate extremes in global urban areas. *Environ. Res. Lett.*, **10**, 24005, doi:10.1088/1748-9326/10/2/024005.

Mitchard, E.T.A. et al., 2013: Uncertainty in the spatial distribution of tropical forest biomass: A comparison of pan-tropical maps. *Carbon Balance Manag.*, **8**, 10, doi:10.1186/1750-0680-8-10.

Mitchell, S.R., M.E. Harmon, and K.E.B. O'Connell, 2012: Carbon debt and carbon sequestration parity in forest bioenergy production. *Glob. Chang. Biol. Bioenergy*, **4**, 818–827, doi:10.1111/j.1757-1707.2012.01173.x.

Mohan, T.S., and M. Rajeevan, 2017: Past and future trends of hydroclimatic intensity over the Indian monsoon region. *J. Geophys. Res. Atmos.*, **122**, 896–909, doi:10.1002/2016JD025301.

Monard, C., C. Mchergui, N. Nunan, F. Martin-Laurent, and L. Vieublé-Gonod, 2012: Impact of soil matric potential on the fine-scale spatial distribution and activity of specific microbial degrader communities. *FEMS Microbiol. Ecol.*, **81**, 673–683, doi:10.1111/j.1574-6941.2012.01398.x.

Mondal, N., and R. Sukumar, 2016: Fires in seasonally dry tropical forest: Testing the varying constraints hypothesis across a regional rainfall gradient. *PLoS One*, **11**, e0159691.

Monforti, F. et al., 2015: Optimal energy use of agricultural crop residues preserving soil organic carbon stocks in Europe. *Renew. Sustain. Energy Rev.*, **44**, 519–529, doi:10.1016/j.rser.2014.12.033.

Montenegro, A. et al., 2009: The net carbon drawdown of small scale afforestation from satellite observations. *Glob. Planet. Change*, **69**, 195–204, doi:10.1016/j.gloplacha.2009.08.005.

Moody, J.A., R.A. Shakesby, P.R. Robichaud, S.H. Cannon, and D.A. Martin, 2013: Current research issues related to post-wildfire runoff and erosion processes. *Earth-Science Rev.*, **122**, 10–37, doi:10.1016/j.earscirev.2013.03.004.

Moore, K.E. et al., 1996: Seasonal variation in radiative and turbulent exchange at a deciduous forest in Central Massachusetts. *J. Appl. Meterology*, **35**, 122–134, doi:10.1175/1520-0450(1996)035<0122:SVIRAT>2.0.CO;2.

Mora, C. et al., 2013: The projected timing of climate departure from recent variability. *Nature*, **502**, 183–187, doi:10.1038/nature12540.

Mora, C. et al., 2014: Mora et al. reply. *Nature*, **511**, E5–E6, doi:10.1038/nature13524.

Mora, C. et al., 2017: Global risk of deadly heat. *Nat. Clim. Chang.*, **7**, 501–506, doi:10.1038/nclimate3322.

Moragues-Faus, A., and A. Marceau, 2018: Measuring progress in sustainable food cities: An indicators toolbox for action. *Sustain.*, **11**, 1–17, doi:10.3390/su11010045.

Morgan, J.A. et al., 2011: C4 grasses prosper as carbon dioxide eliminates desiccation in warmed semi-arid grassland. *Nature*, **476**, 202–205, doi:10.1038/nature10274.

Moritz, E.B. and D.D.A. and M.A., 2015: A minimal model of fire-vegetation feedbacks and disturbance stochasticity generates alternative stable states in grassland–shrubland–woodland systems. *Environ. Res. Lett.*, **10**, 034018, doi:10.1088/1748-9326/10/3/034018.

Moritz, M.A. et al., 2014: Learning to coexist with wildfire. *Nature*, **515**, 58. doi:10.1038/nature13946.

Mosley, L.M., 2015: Drought impacts on the water quality of freshwater systems; review and integration. *Earth-Science Rev.*, **140**, 203–214, doi:10.1016/j.earscirev.2014.11.010.

Mote, P.W. et al., 2016: Perspectives on the causes of exceptionally low 2015 snowpack in the western United States. *Geophys. Res. Lett.*, **43**, 10,980-10,988, doi:10.1002/2016GL069965.

Moyano, F.E., S. Manzoni, and C. Chenu, 2013: Responses of soil heterotrophic respiration to moisture availability: An exploration of processes and models. *Soil Biol. Biochem.*, **59**, 72–85, doi:10.1016/j.soilbio.2013.01.002.

Mudryk, L.R., P.J. Kushner, C. Derksen, and C. Thackeray, 2017: Snow cover response to temperature in observational and climate model ensembles. *Geophys. Res. Lett.*, **44**, 919–926, doi:10.1002/2016GL071789.

Mueller, B., and S.I. Seneviratne, 2012: Hot days induced by precipitation deficits at the global scale. *Proc. Natl. Acad. Sci.*, **109**, 12398–12403, doi:10.1073/pnas.1204330109.

Mueller, N.D., 2015: Cooling of US Midwest summer temperature extremes from cropland intensification. *Nat. Clim. Chang.*, **6**, 317–322, doi:10.1038/nclimate2825.

Mukherjee, S., A. Mishra, and K.E. Trenberth, 2018: Climate change and drought: A perspective on drought indices. *Curr. Clim. Chang. Reports*, **4**, 145–163, doi:10.1007/s40641-018-0098-x.

Müller, C. et al., 2017: An AgMIP framework for improved agricultural representation in integrated assessment models. *Environ. Res. Lett.*, **12**, 125003, doi:10.1088/1748-9326/aa8da6.

Murata, T. and N. Kawai, 2018: Degradation of the urban ecosystem function due to soil sealing: Involvement in the heat island phenomenon and hydrologic cycle in the Tokyo Metropolitan Area. *Soil Sci. Plant Nutr.*, **64**, 145–155, doi:10.1080/00380768.2018.1439342.

Myhre, G., D. Shindell, F.-M. Bréon, W. Collins, J. Fuglestvedt, J. Huang, D. Koch, J.-F. Lamarque, D. Lee, B. Mendoza, T. Nakajima, A. Robock, G. Stephens, T. Takemura and H. Zhang, 2013: Anthropogenic and Natural Radiative Forcing. Climate Change 2013: The Physical Science Basis. Contribution of Working Group I to the Fifth Assessment Report of the Intergovernmental Panel on Climate Change [Stocker, T.F., D. Qin, G.-K. Plattner, M. Tignor, S.K. Allen, J. Boschung, A. Nauels, Y. Xia, V. Bex and P.M. Midgley (eds.)]. Cambridge University Press, Cambridge, United Kingdom and New York, NY, USA, pp. 659–740.

Myneni, R.B., C.D. Keeling, C.J. Tucker, G. Asrar, and R.R. Nemani, 1997: Increased plant growth in the northern high latitudes from 1981 to 1991. *Nature*, **386**, 698–702, doi:10.1038/386698a0.

Nabuurs, G.-J., E.J.M.M. Arets, and M.-J. Schelhaas, 2018: Understanding the implications of the EU-LULUCF regulation for the wood supply from EU forests to the EU. *Carbon Balance Manag.*, **13**, 18, doi:10.1186/s13021-018-0107-3.

Nabuurs, G.J., O. Masera, K. Andrasko, P. Benitez-Ponce, R. Boer, M. Dutschke, E. Elsiddig, J. Ford-Robertson, P. Frumhoff, T. Karjalainen, O. Krankina, W.A. Kurz, M. Matsumoto, W. Oyhantcabal, N.H. Ravindranath, M.J. Sanz Sanchez, X. Zhang, 2007: In Climate Change 2007: Mitigation. Contribution of Working Group III to the Fourth Assessment Report of the Intergovernmental Panel on Climate Change [Metz, B., O.R. Davidson, P.R. Bosch, R. Dave, L.A. Meyer (eds)]. Cambridge University Press, Cambridge, United Kingdom and New York, NY, USA, pp. 541–584.

Nabuurs, G.J., E.J.M.M. Arets, and M.J. Schelhaas, 2017: European forests show no carbon debt, only a long parity effect. *For. Policy Econ.*, **75**, 120–125, doi:10.1016/j.forpol.2016.10.009.

Nainggolan, D. et al., 2012: Afforestation, agricultural abandonment and intensification: Competing trajectories in semi-arid Mediterranean agro-ecosystems. *Agric. Ecosyst. Environ.*, **159**, 90–104, doi:10.1016/j.agee.2012.06.023.

Naudts, K. et al., 2016a: Mitigate climate warming. *Science*, **351**, 597–601, doi:10.1126/science.aac9976.

Naudts, K. et al., 2016b: Europe's forest management did not mitigate climate warming. *Science*, **351**, 597 LP-600, doi:10.1126/science.aad7270.

Nazemi, A. and H.S. Wheater, 2015: On inclusion of water resource management in Earth system models – Part 1: Problem definition and representation of water demand. *Hydrol. Earth Syst. Sci.*, **19**, 33–61, doi:10.5194/hess-19-33-2015.

Nelson, G.C., et al., 2009: *Climate Change: Impact on Agriculture and Costs of Adaptation*. Food Policy Report. International Food Policy Research Institute (IFPRI), Washington, DC, USA, doi:10.2499/0896295354.

Nemet, G.F. et al., 2018: Negative emissions – Part 3: Innovation and upscaling. *Environ. Res. Lett.*, doi:10.1088/1748-9326/aabff4.

Neuvonen, S., P. Niemela, and T. Virtanen, 1999: Climatic change and insect outbreaks in boreal forests: The role of winter temperatures. *Ecological Bulletins.*, 47, 63–67, www.jstor.org/stable/20113228.

Newman, L., C. Ling, and K. Peters, 2012: Between field and table: environmental impl ications of local food distribution. *Int. J. Sustain. Soc.*, **5**, 11, doi:10.1504/ijssoc.2013.050532.

Newman, M. et al., 2016: The Pacific Decadal Oscillation, revisited. *J. Clim.*, **29**, 4399–4427, doi:10.1175/JCLI-D-15-0508.1.

Nicely, J.M. et al., 2018: Changes in global tropospheric OH expected as a result of climate change over the last several decades. *J. Geophys. Res. Atmos.*, **123**, 10,774–10,795, doi:10.1029/2018JD028388.

Nicholls, N., and S.I. Seneviratne, 2015: Comparing IPCC assessments: How do the AR4 and SREX assessments of changes in extremes differ? *Clim. Change*, **133**, 7–21, doi:10.1007/s10584-013-0818-0.

Nisbet, E.G. et al., 2019: Very strong atmospheric methane growth in the four years 2014–2017: Implications for the Paris Agreement. *Global Biogeochem. Cycles*, **33**, 318–342, doi:10.1029/2018GB006009.

Nishina, K. et al., 2014: Quantifying uncertainties in soil carbon responses to changes in global mean temperature and precipitation. *Earth Syst. Dyn.*, **5**, 197–209, doi:10.5194/esd-5-197-2014.

Niu, S. et al., 2014: Plant growth and mortality under climatic extremes: An overview. *Environ. Exp. Bot.*, **98**, 13–19, doi:10.1016/j.envexpbot.2013.10.004.

Niyogi, D., C. Kishtawal, S. Tripathi, and R.S. Govindaraju, 2010: Observational evidence that agricultural intensification and land use change may be reducing the Indian summer monsoon rainfall. *Water Resour. Res.*, **46**, 1–17, doi:10.1029/2008WR007082.

Nizeyimana, E.L. et al., 2001: Assessing the impact of land conversion to urban use on soils with different productivity levels in the USA. *Soil Sci. Soc. Am. J.*, **65**, 391–402, doi:10.2136/sssaj2001.652391x.

de Noblet-Ducoudré, N., M. Claussen, and C. Prentice, 2000: Mid-Holocene greening of the Sahara: First results of the GAIM 6000 year BP experiment with two asynchronously coupled atmosphere/biome models. *Clim. Dyn.*, **16**, 643–659, doi:10.1007/s003820000074.

De Noblet-Ducoudré, N. et al., 2012: Determining robust impacts of land-use-induced land cover changes on surface climate over North America and Eurasia: Results from the first set of LUCID experiments. *J. Clim.*, **25**, 3261–3281, doi:10.1175/JCLI-D-11-00338.1.

de Noblet, N.I. et al., 1996: Possible role of atmosphere-biosphere interactions in triggering the last glaciation. *Geophys. Res. Lett.*, **23**, 3191–3194, doi:10.1029/96GL03004.

Nogherotto, R., E. Coppola, F. Giorgi, and L. Mariotti, 2013: Impact of Congo Basin deforestation on the African monsoon. *Atmos. Sci. Lett.*, **14**, 45–51, doi:10.1002/asl2.416.

Noormets, A. et al., 2015: Effects of forest management on productivity and carbon sequestration: A review and hypothesis. *For. Ecol. Manage.*, **355**, 124–140, doi:10.1016/j.foreco.2015.05.019.

Norby, R.J., J.M. Warren, C.M. Iversen, B.E. Medlyn, and R.E. McMurtrie, 2010: CO_2 enhancement of forest productivity constrained by limited nitrogen availability. *Proc. Natl. Acad. Sci.*, **107**, 19368–19373, doi:10.1073/pnas.1006463107.

Nousiainen, T., 2011: Optical modeling of mineral dust particles: A review. *J. Quant. Spectrosc. Radiat. Transf.*, 110, 14–16, doi:10.1016/j.jqsrt.2009.02.002.

2

Novick, K.A. et al., 2016: The increasing importance of atmospheric demand for ecosystem water and carbon fluxes. *Nat. Clim. Chang.*, **6**, 1023–1027, doi:10.1038/nclimate3114.

O'Gorman, P.A., 2014: Contrasting responses of mean and extreme snowfall to climate change. *Nature*, **512**, 416–418, doi:10.1038/nature13625.

O'Gorman, P.A., 2015: Precipitation extremes under climate change. *Curr. Clim. Chang. Reports*, **1**, 49–59, doi:10.1007/s40641-015-0009-3.

O'Halloran, T.L. et al., 2012: Radiative forcing of natural forest disturbances. *Glob. Chang. Biol.*, **18**, 555–565, doi:10.1111/j.1365-2486.2011.02577.x.

O'ishi, R., and A. Abe-Ouchi, 2009: Influence of dynamic vegetation on climate change arising from increasing CO_2. *Clim. Dyn.*, **33**, 645–663, doi:10.1007/s00382-009-0611-y.

Oates, L.G. et al., 2016: Nitrous oxide emissions during establishment of eight alternative cellulosic bioenergy cropping systems in the North Central United States. *GCB Bioenergy*, **8**, 539–549, doi:10.1111/gcbb.12268.

Obersteiner, M. et al., 2016: Assessing the land resource–food price nexus of the Sustainable Development Goals. *Sci. Adv.*, **2**, e1501499–e1501499, doi:10.1126/sciadv.1501499.

Oenema, O. et al., 2014: Reducing nitrous oxide emissions from the global food system. *Curr. Opin. Environ. Sustain.*, **9–10**, 55–64, doi:10.1016/j.cosust.2014.08.003.

Oke, T.R., G. Mills, A. Christen, and J.A. Voogt, 2017: *Urban Climates*. Cambridge University Press, Cambridge, 526 pp.

Okuno, M., and T. Nakamura, 2003: Radiocarbon dating of tephra layers: recent progress in Japan. *Quat. Int.*, **105**, 49–56, doi:10.1016/S1040-6182(02)00150-7.

Olefeldt, D. et al., 2016: Circumpolar distribution and carbon storage of thermokarst landscapes. *Nat. Commun.*, **7**, 1–11, doi:10.1038/ncomms13043.

Oliveira, P.H.F. et al., 2007: The effects of biomass burning aerosols and clouds on the CO_2 flux in Amazonia. *Tellus, Ser. B Chem. Phys. Meteorol.*, **59**, 338–349, doi:10.1111/j.1600-0889.2007.00270.x.

De Oliveira Bordonal, R. et al., 2015: Greenhouse gas balance from cultivation and direct land use change of recently established sugarcane (*Saccharum officinarum*) plantation in south-central Brazil. *Renew. Sustain. Energy Rev.*, **52**, 547–556, doi:10.1016/j.rser.2015.07.137.

Ometto, J.P.H.B., A.D. Nobre, H.R. Rocha, P. Artaxo, and L.A. Martinelli, 2005: Amazonia and the modern carbon cycle: Lessons learned. *Oecologia*, **143**, 483–500, doi:10.1007/s00442-005-0034-3.

Oort, A.H., and J.P. Peixóto, 1983: Global angular momentum and energy balance requirements from observations. *Adv. Geophys.*, **25**, 355–490, doi:10.1016/S0065-2687(08)60177-6.

Van Oost, K. et al., 2007: The impact of agricultural soil erosion on the global carbon cycle. *Science*, **318**, 626–629, doi:10.1126/science.1145724.

Orlowsky, B., and S.I. Seneviratne, 2012: Global changes in extreme events: Regional and seasonal dimension. *Clim. Change*, **110**, 669–696, doi:10.1007/s10584-011-0122-9.

Ornstein, L., I. Aleinov, and D. Rind, 2009: Irrigated afforestation of the Sahara and Australian Outback to end global warming. *Clim. Change*, **97**, 409–437, doi:10.1007/s10584-009-9626-y.

Orwin, K.H., M.U.F. Kirschbaum, M.G. St John, and I.A. Dickie, 2011: Organic nutrient uptake by mycorrhizal fungi enhances ecosystem carbon storage: A model-based assessment. *Ecol. Lett.*, **14**, 493–502, doi:10.1111/j.1461-0248.2011.01611.x.

Osterloh, K., N. Tauchnitz, O. Spott, J. Hepp, S. Bernsdorf, and R. Meissner, 2018: Changes of methane and nitrous oxide emissions in a transition bog in central Germany (German National Park Harz Mountains) after rewetting. *Wetl. Ecol. Manag.*, **26**, 87–102, doi:10.1007/s11273-017-9555-x.

Otto, F.E.L., N. Massey, G.J. Van Oldenborgh, R.G. Jones, and M.R. Allen, 2012: Reconciling two approaches to attribution of the 2010 Russian heat wave. *Geophys. Res. Lett.*, **39**, n/a-n/a, doi:10.1029/2011GL050422.

Paasonen, P. et al., 2013: Warming-induced increase in aerosol number concentration likely to moderate climate change. *Nat. Geosci.*, **6**, 438–442, doi:10.1038/ngeo1800.

Pacifico, F., G.A. Folberth, C.D. Jones, S.P. Harrison, and W.J. Collins, 2012: Sensitivity of biogenic isoprene emissions to past, present, and future environmental conditions and implications for atmospheric chemistry. *J. Geophys. Res. Atmos.*, **117**, D22302, doi:10.1029/2012JD018276.

Pal, J.S. and E.A.B. Eltahir, 2016: Future temperature in southwest Asia projected to exceed a threshold for human adaptability. *Nat. Clim. Chang.*, **6**, 197–200, doi:10.1038/nclimate2833.

Pall, P., M.R. Allen, and D.A. Stone, 2007: Testing the Clausius-Clapeyron constraint on changes in extreme precipitation under CO_2 warming. *Clim. Dyn.*, **28**, 351–363, doi:10.1007/s00382-006-0180-2.

Pall, P. et al., 2011: Anthropogenic greenhouse gas contribution to flood risk in England and Wales in autumn 2000. *Nature*, **470**, 382–385, doi:10.1038/nature09762.

Palm, C., H. Blanco-Canqui, F. DeClerck, L. Gatere, and P. Grace, 2014: Conservation agriculture and ecosystem services: An overview. *Agric. Ecosyst. Environ.*, **187**, 87–105, doi:10.1016/j.agee.2013.10.010.

Pan, N., X. Feng, B. Fu, S. Wang, F. Ji, and S. Pan, 2018: Increasing global vegetation browning hidden in overall vegetation greening: Insights from time-varying trends. *Remote Sens. Environ.*, **214**, 59–72, doi:10.1016/j.rse.2018.05.018.

Pan, Y. et al., 2011: A large and persistent carbon sink in the world's forests. *Science*, **333**, 988–993, doi:10.1126/science.1201609.

Pandey, B., Q. Zhang, and K.C. Seto, 2018: Time series analysis of satellite data to characterize multiple land use transitions: A case study of urban growth and agricultural land loss in India. *J. Land Use Sci.*, **13**, 221–237, doi:10.1080/1747423X.2018.1533042.

Pangala, S.R. et al., 2017: Large emissions from floodplain trees close the Amazon methane budget. *Nature*, **552**, 230–234, doi:10.1038/nature24639.

Park, S. et al., 2012: Trends and seasonal cycles in the isotopic composition of nitrous oxide since 1940. *Nat. Geosci.*, **5**, 261–265, doi:10.1038/ngeo1421.

Parmesan, C. and G. Yohe, 2003: A globally coherent fingerprint of climate change impacts across natural systems. *Nature*, **421**, 37–42, doi:10.1038/nature01286.

Parry, M., C. Rosenzweig, and M. Livermore, 2005: Climate change, global food supply and risk of hunger. *Philos. Trans. R. Soc. B Biol. Sci.*, **360**, 2125–2138, doi:10.1098/rstb.2005.1751.

Parry, M.L., C. Rosenzweig, A. Iglesias, M. Livermore, and G. Fischer, 2004: Effects of climate change on global food production under SRES emissions and socio-economic scenarios. *Glob. Environ. Chang.*, **14**, 53–67, doi:10.1016/j.gloenvcha.2003.10.008.

Paschalis, A., G.G. Katul, S. Fatichi, S. Palmroth, and D. Way, 2017: On the variability of the ecosystem response to elevated atmospheric CO_2 across spatial and temporal scales at the Duke Forest FACE experiment. *Agric. For. Meteorol.*, **232**, 367–383, doi:10.1016/j.agrformet.2016.09.003.

Páscoa, P., C.M. Gouveia, A. Russo, and R.M. Trigo, 2017: Drought trends in the Iberian Peninsula over the last 112 years. *Adv. Meteorol.*, **2017**, 1–13, doi:10.1155/2017/4653126.

Pataki, D.E. et al., 2011: Socio-ecohydrology and the urban water challenge. *Ecohydrology*, **4**, 341–347, doi:10.1002/eco.209.

Patra, P.K. et al., 2016: Regional methane emission estimation based on observed atmospheric concentrations (2002–2012). *J. Meteorol. Soc. Japan. Ser. II*, **94**, 91–113, doi:10.2151/jmsj.2016-006.

Pau, S., D.K. Okamoto, O. Calderón, and S.J. Wright, 2018: Long-term increases in tropical flowering activity across growth forms in response to rising CO_2 and climate change. *Glob. Chang. Biol.*, **24**, 2105–2116, doi:10.1111/gcb.14004.

Pausas, J.G., and S. Paula, 2012: Fuel shapes the fire-climate relationship: Evidence from Mediterranean ecosystems. *Glob. Ecol. Biogeogr.*, **21**, 1074–1082, doi:10.1111/j.1466-8238.2012.00769.x.

Pausas, J.G., and M.M. Millán, 2018: greening and browning in a climate change hotspot: The Mediterranean Basin. *Bioscience*, **69**, 143–151, doi:10.1093/biosci/biy157.

2

Paustian, K., J. Six, E.T. Elliott, and H.W. Hunt, 2000: Management options for reducing CO₂ emissions from agricultural soils. *Biogeochemistry*, **48**, 147–163, doi:10.1023/A:1006271331703.

Paustian, K. et al., 2016: Climate-smart soils. *Nature*, **532**, 49–57, doi:10.1038/nature17174.

Pawson, S.M. et al., 2013: Plantation forests, climate change and biodiversity. *Biodivers. Conserv.*, **22**, 1203–1227, doi:10.1007/s10531-013-0458-8.

Pechony, O. and D.T. Shindell, 2010: Driving forces of global wildfires over the past millennium and the forthcoming century. *Proc. Natl. Acad. Sci.*, **107**, 19167–19170, doi:10.1073/pnas.1003669107.

Pecl, G.T. et al., 2017: Biodiversity redistribution under climate change: Impacts on ecosystems and human well-being. *Science*, **355**, eaai9214, doi:10.1126/science.aai9214.

Pellitier, P.T. and D.R. Zak, 2018: Ectomycorrhizal fungi and the enzymatic liberation of nitrogen from soil organic matter: Why evolutionary history matters. *New Phytol.*, **217**, 68–73, doi:10.1111/nph.14598.

Peñaloza, D., M. Erlandsson, J. Berlin, M. Wålinder, and A. Falk, 2018: Future scenarios for climate mitigation of new construction in Sweden: Effects of different technological pathways. *J. Clean. Prod.*, **187**, 1025–1035, doi:10.1016/j.jclepro.2018.03.285.

Pendleton, L. et al., 2012: Estimating global "blue carbon" emissions from conversion and degradation of vegetated coastal ecosystems. *PLoS One*, **7**, e43542, doi:10.1371/journal.pone.0043542.

Peng, S. et al., 2017: Sensitivity of land use change emission estimates to historical land use and land cover mapping. *Global Biogeochem. Cycles*, **31**, 626–643, doi:10.1002/2015GB005360.

Peng, S.-S. et al., 2014: Afforestation in China cools local land surface temperature. *Proc. Natl. Acad. Sci.*, **111**, 2915–2919, doi:10.1073/pnas.1315126111.

Peñuelas, J., 2009: Phenology feedbacks on climate change. *Science*, **324**, 887–888, doi:10.1126/science.1173004.

Peñuelas, J. and J. Llusià, 2003: BVOCs: Plant defense against climate warming? *Trends Plant Sci.*, **8**, 105–109, doi:10.1016/S1360-1385(03)00008-6.

Peñuelas, J. and M. Staudt, 2010: BVOCs and global change. *Trends Plant Sci.*, **15**, 133–144, doi:10.1016/j.tplants.2009.12.005.

Peñuelas, J. et al., 2013: Human-induced nitrogen-phosphorus imbalances alter natural and managed ecosystems across the globe. *Nat. Commun.*, **4**, doi:10.1038/ncomms3934.

Peñuelas, J. et al., 2017: Shifting from a fertilization-dominated to a warming-dominated period. *Nat. Ecol. Evol.*, **1**, 1438–1445, doi:10.1038/s41559-017-0274-8.

Perket, J., M.G. Flanner, and J.E. Kay, 2014: Diagnosing shortwave cryosphere radiative effect and its 21st century evolution in CESM. *J. Geophys. Res.*, **119**, 1356–1362, doi:10.1002/2013JD021139.

Perlwitz, J.P., C. Pérez García-Pando, and R.L. Miller, 2015: Predicting the mineral composition of dust aerosols – Part 2: Model evaluation and identification of key processes with observations. *Atmos. Chem. Phys.*, **15**, 11629–11652, doi:10.5194/acp-15-11629-2015.

Perugini, L. et al., 2017: Biophysical effects on temperature and precipitation due to land cover change. *Environ. Res. Lett.*, **12**, 053002, doi:10.1088/1748-9326/aa6b3f.

Peters, G.P. and O. Geden, 2017: Catalysing a political shift from low to negative carbon. *Nat. Clim. Chang.*, **7**, 619–621, doi:10.1038/nclimate3369.

Peylin, P. et al., 2013: Global atmospheric carbon budget: results from an ensemble of atmospheric CO₂ inversions. *Biogeosciences*, **10**, 6699–6720, doi:10.5194/bg-10-6699-2013.

Pfahl, S., P.A. O'Gorman, and E.M. Fischer, 2017: Understanding the regional pattern of projected future changes in extreme precipitation. *Nat. Clim. Chang.*, **7**, 423–427, doi:10.1038/nclimate3287.

Philben, M. et al., 2015: Temperature, oxygen, and vegetation controls on decomposition in a James Bay peatland. *Global Biogeochem. Cycles*, **29**, 729–743, doi:10.1002/2014GB004989.

Phillips, O.L. et al., 2009: Drought sensitivity of the Amazon Rainforest. *Science*, **323**, 1344–1347, doi:10.1126/science.1164033.

Phillips, R.P. et al., 2011: Enhanced root exudation induces microbial feedbacks to N cycling in a pine forest under long-term CO₂ fumigation. *Ecol. Lett.*, **14**, 187–194, doi:10.1111/j.1461-0248.2010.01570.x.

Phillips, T.J. et al., 2007: Combined climate and carbon-cycle effects of large-scale deforestation. *Proc. Natl. Acad. Sci.*, **104**, 6550–6555, doi:10.1073/pnas.0608998104.

Piao, S. et al., 2015: Detection and attribution of vegetation greening trend in China over the last 30 years. *Glob. Chang. Biol.*, **21**, 1601–1609, doi:10.1111/gcb.12795.

Piao, S. et al., 2018: Lower land-use emissions responsible for increased net land carbon sink during the slow warming period. *Nat. Geosci.*, **11**, 739–743, doi:10.1038/s41561-018-0204-7.

Pielke, R.A. et al., 2011: Land use/land cover changes and climate: Modeling analysis and observational evidence. *Wiley Interdiscip. Rev. Clim. Chang.*, **2**, 828–850, doi:10.1002/wcc.144.

Pilaš, I., K.-H. Feger, U. Vilhar, and A. Wahren, 2011: Multidimensionality of scales and approaches for forest–water interactions. In: *Forest Management and the Water Cycle* [Bredemeier, M., S. Cohen, D.L. Goldbold, E. Lode, V. Pichler, and P. Schleppi (eds.)]. Springer, Heidelberg, Germany, pp. 351–380.

Pingoud, K., T. Ekholm, R. Sievänen, S. Huuskonen, and J. Hynynen, 2018: Trade-offs between forest carbon stocks and harvests in a steady state – A multi-criteria analysis. *J. Environ. Manage.*, doi:10.1016/j.jenvman.2017.12.076.

Pison, I., B. Ringeval, P. Bousquet, C. Prigent, and F. Papa, 2013: Stable atmospheric methane in the 2000s: Key-role of emissions from natural wetlands. *Atmos. Chem. Phys.*, **13**, 11609–11623, doi:10.5194/acp-13-11609-2013.

Pitman, A., and N. de Noblet-Ducoudré, 2012: Human effects on climate through land-use-induced land-cover change. *Futur. World's Clim.*, 77–95, doi:10.1016/B978-0-12-386917-3.00004-X.

Pitman, A.J., 2003: The evolution of, and revolution in, land surface schemes designed for climate models. *Int. J. Climatol.*, **23**, 479–510, doi:10.1002/joc.893.

Pitman, A.J. et al., 2011: Importance of background climate in determining impact of land-cover change on regional climate. *Nat. Clim. Chang.*, **1**, 472–475, doi:10.1038/nclimate1294.

Pitman, A.J. et al., 2012: Effects of land cover change on temperature and rainfall extremes in multi-model ensemble simulations. *Earth Syst. Dyn.*, **3**, 213–231, doi:10.5194/esd-3-213-2012.

Plevin, R.J., J. Beckman, A.A. Golub, J. Witcover, and M. O'Hare, 2015: Carbon accounting and economic model uncertainty of emissions from biofuels-induced land use change. *Environ. Sci. Technol.*, doi:10.1021/es505481d.

Poeplau, C. and A. Don, 2015: Carbon sequestration in agricultural soils via cultivation of cover crops – A meta-analysis. *Agric. Ecosyst. Environ.*, **200**, 33–41, doi:10.1016/j.agee.2014.10.024.

Pöhlker, M.L. et al., 2016: Long-term observations of cloud condensation nuclei in the Amazon rain forest – Part 1: Aerosol size distribution, hygroscopicity, and new model parametrizations for CCN prediction. *Atmos. Chem. Phys.*, **16**, 15709–15740, doi:10.5194/acp-16-15709-2016.

Pöhlker, M.L. et al., 2018: Long-term observations of cloud condensation nuclei over the Amazon rain forest – Part 2: Variability and characteristics of biomass burning, long-range transport, and pristine rain forest aerosols. *Atmos. Chem. Phys.*, **18**, 10289–10331, doi:10.5194/acp-18-10289-2018.

Polydoros, A., T. Mavrakou, and C. Cartalis, 2018: Quantifying the trends in land surface temperature and surface urban heat island intensity in Mediterranean cities in view of smart urbanization. *Urban Sci.*, **2**, 16, doi:10.3390/urbansci2010016.

Pongratz, J., C.H. Reick, T. Raddatz, and M. Claussen, 2010: Biogeophysical versus biogeochemical climate response to historical anthropogenic land cover change. *Geophys. Res. Lett.*, **37**, 1–5, doi:10.1029/2010GL043010.

Pongratz, J., C.H. Reick, R.A. Houghton, and J.I. House, 2014: Terminology as a key uncertainty in net land use and land cover change carbon flux estimates. *Earth Syst. Dyn.*, **5**, 177–195, doi:10.5194/esd-5-177-2014.

Pongratz, J. et al., 2018: *Models Meet Data: Challenges and Opportunities in Implementing Land Management in Earth System Models.* **24**, 1470–11487, doi: 10.1111/gcb.13988.

Ponomarev, I.E., I.V. Kharuk, and J.K. Ranson, 2016: Wildfires dynamics in Siberian larch forests. *For.*, **7**, 125, doi:10.3390/f7060125.

Poorter, L. et al., 2016: Biomass resilience of neotropical secondary forests. *Nature*, 530, 211–214, doi:10.1038/nature16512.

Popp, A. et al., 2014: Land-use transition for bioenergy and climate stabilization: Model comparison of drivers, impacts and interactions with other land use based mitigation options. *Clim. Change*, **123**, 495–509, doi:10.1007/s10584-013-0926-x.

Popp, A. et al., 2017: Land-use futures in the shared socio-economic pathways. *Glob. Environ. Chang.*, **42**, 331–345, doi:10.1016/J.GLOENVCHA.2016.10.002.

Port, U., V. Brovkin, and M. Claussen, 2012: The influence of vegetation dynamics on anthropogenic climate change. *Earth Syst. Dyn.*, **3**, 233–243, doi:10.5194/esd-3-233-2012.

Porter, J.R., L. Xie, A.J. Challinor, K. Cochrane, S.M. Howden, M.M. Iqbal, D.B. Lobell, and M.I. Travasso, 2014: Food Security and Food Production Systems. Climate Change 2014: Impacts, Adaptation and Vulnerability. Part A: Global and Sectoral Aspects. Contribution of Working Group II to the Fifth Assessment Report of the Intergovernmental Panel on Climate Change [Field, C.B., V.R. Barros, D.J. Dokken, K.J. Mach, M.D. Mastrandrea, T.E. Bilir, M. Chatterjee, K.L. Ebi, Y.O. Estrada, R.C. Genova, B. Girma, E.S. Kissel, A.N. Levy, S. MacCracken, P.R. Mastrandrea, and L.L. White (eds.)]. Cambridge University Press, Cambridge, United Kingdom and New York, NY, USA, pp. 485–533.

Porter, S.D., D.S. Reay, P. Higgins, and E. Bomberg, 2016: A half-century of production-phase greenhouse gas emissions from food loss & waste in the global food supply chain. *Sci. Total Environ.*, **571**, 721–729, doi:10.1016/j.scitotenv.2016.07.041.

Pöschl, U., and M. Shiraiwa, 2015: Multiphase chemistry at the atmosphere–biosphere interface influencing climate and public health in the Anthropocene. *Chem. Rev.*, **115**, 4440–4475, doi:10.1021/cr500487s.

Pöschl, U. et al., 2010: Rainforest aerosols as biogenic nuclei of clouds and precipitation in the Amazon. *Science*, **329**, 1513–1516, doi:10.1126/science.1191056.

Posthumus, H. et al., 2009: Impacts of the summer 2007 floods on agriculture in England. *J. Flood Risk Manag.*, **2**, 182–189, doi:10.1111/j.1753-318X.2009.01031.x.

Potter, G.L., H.W. Ellsaesser, M.C. MacCracken, and J.S. Ellis, 1981: Albedo change by man: Test of climatic effects. *Nature*, **291**, 47–49, doi:10.1038/291047a0.

Potter, G.L., H.W. Ellsaesser, M.C. MacCracken, and F.M. Luther, 1975: Possible climatic impact of tropical deforestation. *Nature*, **258**, 697–698, doi:10.1038/258697a0.

Poulter, B. et al., 2017: Global wetland contribution to 2000–2012 atmospheric methane growth rate dynamics. *Environ. Res. Lett.*, **12**, doi:10.1088/1748-9326/aa8391.

Powell, T.L. et al., 2013: Confronting model predictions of carbon fluxes with measurements of Amazon forests subjected to experimental drought. *New Phytol.*, **200**, 350–365, doi:10.1111/nph.12390.

Powell, T.W.R. and T.M. Lenton, 2012: Future carbon dioxide removal via biomass energy constrained by agricultural efficiency and dietary trends. *Energy Environ. Sci.*, 5, 8116-8133, doi:10.1039/c2ee21592f.

Powlson, D.S., A.P. Whitmore, and K.W.T. Goulding, 2011: Soil carbon sequestration to mitigate climate change: A critical re-examination to identify the true and the false. *Eur. J. Soil Sci.*, **62**, 42–55, doi:10.1111/j.1365-2389.2010.01342.x.

Powlson, D.S. et al., 2014: Limited potential of no-till agriculture for climate change mitigation. *Nat. Clim. Chang.*, **4**, 678–683, doi:10.1038/nclimate2292.

Powlson, D.S., C.M. Stirling, C. Thierfelder, R.P. White, and M.L. Jat, 2016: Does conservation agriculture deliver climate change mitigation through soil carbon sequestration in tropical agro-ecosystems? *Agric. Ecosyst. Environ.*, **220**, 164–174, doi:10.1016/j.agee.2016.01.005.

Pradhan, P., D.E. Reusser, and J.P. Kropp, 2013: Embodied greenhouse gas emissions in diets. *PLoS One*, **8**, e62228, doi:10.1371/journal.pone.0062228.

Pradhan, P., M.K.B. Lüdeke, D.E. Reusser, and J.P. Kropp, 2014: Food self-sufficiency across scales: How local can we go? *Environ. Sci. Technol.*, **48**, 9463–9470, doi:10.1021/es5005939.

Prather, M.J. et al., 2017: Global atmospheric chemistry – Which air matters. *Atmos. Chem. Phys.*, **17**, 9081–9102, doi:10.5194/acp-17-9081-2017.

Pratt, K., and D. Moran, 2010: Evaluating the cost-effectiveness of global biochar mitigation potential. *Biomass and Bioenergy*, **34**, 1149–1158, doi:10.1016/j.biombioe.2010.03.004.

Prein, A.F. et al., 2015: A review on regional convection-permitting climate modeling: Demonstrations, prospects, and challenges. *Rev. Geophys.*, **53**, 323–361, doi:10.1002/2014RG000475.

Prein, A.F. et al., 2017: The future intensification of hourly precipitation extremes. *Nat. Clim. Chang.*, **7**, 48–52, doi:10.1038/nclimate3168.

Prentice, I.C., Liang, X., Medlyn, B.E. and Wang, Y.P., 2015: Reliable, robust and realistic: The three R's of next-generation land-surface modelling. *Atmos. Chem. Phys.*, **15**, 5987–6005, doi:10.5194/acp-15-5987-2015.

Prevedello, J.A., G.R. Winck, M.M. Weber, E. Nichols, and B. Sinervo, 2019: Impacts of forestation and deforestation on local temperature across the globe. *PLoS One*, **14**, 1–18, doi:10.1371/journal.pone.0213368.

Price, H.C. et al., 2018: Atmospheric ice-nucleating particles in the dusty tropical Atlantic. *J. Geophys. Res. Atmos.*, **123**, 2175–2193, doi:10.1002/2017JD027560.

Prokopiou, M. et al., 2018: Changes in the isotopic signature of atmospheric nitrous oxide and its global average source during the last three millennia. *J. Geophys. Res. Atmos.*, **123**, 10,757–10,773, doi:10.1029/2018JD029008.

Prospero, J.M., P. Ginoux, O. Torres, S.E. Nicholson, and T.E. Gill, 2002: Environmental characterization of global sources of atmospheric soil dust identified with the Nimbus 7 Total Ozone Mapping Spectrometer (TOMS) absorbing aerosol product. *Rev. Geophys.*, **40**, 2–31, doi:10.1029/2000RG000095.

Prudhomme, C. et al., 2014: Hydrological droughts in the 21st century, hotspots and uncertainties from a global multimodel ensemble experiment. *Proc. Natl. Acad. Sci.*, **111**, 3262–3267, doi:10.1073/pnas.1222473110.

Pu, B. and P. Ginoux, 2017: Projection of American dustiness in the late 21st century due to climate change. *Sci. Rep.*, **7**, 5553, doi:10.1038/s41598-017-05431-9.

Pu, B. and P. Ginoux, 2018: How reliable are CMIP5 models in simulating dust optical depth? *Atmos. Chem. Phys.*, **18**, 12491–12510, doi:10.5194/acp-18-12491-2018.

Pugh, T.A.M. et al., 2015: Simulated carbon emissions from land-use change are substantially enhanced by accounting for agricultural management. *Environ. Res. Lett.*, **10**, 124008, doi:10.1088/1748-9326/10/12/124008.

Pugh, T.A.M. et al., 2016: Climate analogues suggest limited potential for intensification of production on current croplands under climate change. *Nat. Commun.*, **7**, 1–8, doi:10.1038/ncomms12608.

Pulliainen, J. et al., 2017: Early snowmelt significantly enhances boreal springtime carbon uptake. *Proc. Natl. Acad. Sci.*, **114**, 201707889, doi:10.1073/pnas.1707889114.

Qian, H., R. Joseph, and N. Zeng, 2010: Enhanced terrestrial carbon uptake in the northern high latitudes in the 21st century from the Coupled Carbon Cycle Climate Model Intercomparison Project model projections. *Glob. Chang. Biol.*, **16**, 641–656, doi:10.1111/j.1365-2486.2009.01989.x.

2

Qu, X. and A. Hall, 2014: On the persistent spread in snow-albedo feedback. *Clim. Dyn.*, **42**, 69–81, doi:10.1007/s00382-013-1774-0.

Le Quéré, C. et al., 2018: Global Carbon Budget 2018. *Earth Syst. Sci. Data*, **10**, 2141–2194, doi:10.5194/essd-10-2141-2018.

Le Quéré, C. et al., 2015: Global Carbon Budget 2015. *Earth Syst. Sci. Data*, **7**, 349–396, doi:10.5194/essd-7-349-2015.

Le Quéré, C. et al., 2017: Global Carbon Budget 2017. *Earth Syst. Sci. Data Discuss.*, **10**, 1–79, doi:10.5194/essd-2017-123.

Quesada, B., R. Vautard, P. Yiou, M. Hirschi, and S.I. Seneviratne, 2012: Asymmetric European summer heat predictability from wet and dry southern winters and springs. *Nat. Clim. Chang.*, **2**, 736–741, doi:10.1038/nclimate1536.

Quesada, B., A. Arneth, and N. de Noblet-Ducoudré, 2017a: Atmospheric, radiative, and hydrologic effects of future land use and land cover changes: A global and multimodel climate picture. *J. Geophys. Res.*, **122**, 5113–5131, doi:10.1002/2016JD025448.

Quesada, B., N. Devaraju, N. de Noblet-Ducoudré, and A. Arneth, 2017b: Reduction of monsoon rainfall in response to past and future land use and land cover changes. *Geophys. Res. Lett.*, **44**, 1041–1050, doi:10.1002/2016GL070663.

Quesada, B., A. Arneth, E. Robertson, and N. de Noblet-Ducoudré, 2018: Potential strong contribution of future anthropogenic land-use and land-cover change to the terrestrial carbon cycle. *Environ. Res. Lett.*, **13**, 64023, doi:10.1088/1748-9326/aac4c3.

Rachmayani, R., M. Prange, and M. Schulz, 2015: North African vegetation-precipitation feedback in early and mid-Holocene climate simulations with CCSM3-DGVM. *Clim. Past*, **11**, 175–185, doi:10.5194/cp-11-175-2015.

Rafael, S. et al., 2017: Quantification and mapping of urban fluxes under climate change: Application of WRF-SUEWS model to Greater Porto area (Portugal). *Environ. Res.*, **155**, 321–334, doi:10.1016/j.envres.2017.02.033.

Rajagopal, D. and R.J. Plevin, 2013: Implications of market-mediated emissions and uncertainty for biofuel policies. *Energy Policy*, **56**, 75–82, doi:10.1016/j.enpol.2012.09.076.

Ramarao, M.V.S., R. Krishnan, J. Sanjay, and T.P. Sabin, 2015: Understanding land surface response to changing South Asian monsoon in a warming climate. *Earth Syst. Dyn.*, **6**, 569–582, doi:10.5194/esd-6-569-2015.

Ramarao, M.V.S., J. Sanjay, and R. Krishnan, 2016: Modulation of summer monsoon sub-seasonal surface air temperature over India by soil moisture-temperature coupling. *Mausam*, **67**, 53–66.

Randerson, J., G.R. van der Werf, L. Giglio, G.J. Collagtz, and P.S. Kasibhatla, 2015: Global Fire Emissions Data Base, Version 4. Oak Ridge, Tennessee, USA. Retrieved from: www.globalfiredata.org/index.html.

Rap, A. et al., 2018: Enhanced global primary production by biogenic aerosol via diffuse radiation fertilization. *Nat. Geosci.*, 11, 640–644, doi:10.1038/s41561-018-0208-3.

Ray, D.K., J.S. Gerber, G.K. Macdonald, and P.C. West, 2015: Climate variation explains a third of global crop yield variability. *Nat. Commun.*, **6**, 5989, doi:10.1038/ncomms6989.

Reddington, C.L. et al., 2016: Analysis of particulate emissions from tropical biomass burning using a global aerosol model and long-term surface observations. *Atmos. Chem. Phys.*, **16**, 11083–11106, doi:10.5194/acp-16-11083-2016.

Reed, S.C., X. Yang, and P.E. Thornton, 2015: Incorporating phosphorus cycling into global modeling efforts: A worthwhile, tractable endeavor. *New Phytol.*, **208**, 324–329, doi:10.1111/nph.13521.

Reich, P.B. and S.E. Hobbie, 2013: Decade-long soil nitrogen constraint on the CO_2 fertilization of plant biomass. *Nat. Clim. Chang.*, **3**, 278–282, doi:10.1038/nclimate1694.

Reich, P.B., S.E. Hobbie, and T.D. Lee, 2014: Plant growth enhancement by elevated CO_2 eliminated by joint water and nitrogen limitation. *Nat. Geosci.*, **7**, 920–924, doi:10.1038/ngeo2284.

Reich, P.B., S.E. Hobbie, T.D. Lee, and M.A. Pastore, 2018: Unexpected reversal of C3 versus C4 grass response to elevated CO_2 during a 20-year field experiment. *Science*, **360**, 317–320, doi:10.1126/science.aas9313.

Reichstein, M. et al., 2013: Climate extremes and the carbon cycle. *Nature*, **500**, 287–295, doi:10.1038/nature12350.

Renaud, V., and M. Rebetez, 2008: Comparison between open site and below canopy climatic conditions in Switzerland during the exceptionally hot summer of 2003. *Agric. For. Meteorol.*, **149**, 873–880, doi:10.1016/j.agrformet.2008.11.006.

Renaud, V., J.L. Innes, M. Dobbertin, and M. Rebetez, 2011: Comparison between open-site and below-canopy climatic conditions in Switzerland for different types of forests over 10 years (1998–2007). *Theor. Appl. Climatol.*, **105**, 119–127, doi:10.1007/s00704-010-0361-0.

Resplandy, L. et al., 2018: Revision of global carbon fluxes based on a reassessment of oceanic and riverine carbon transport. *Nat. Geosci.*, **11**, 504–509, doi:10.1038/s41561-018-0151-3.

Revadekar, J.V., Y.K. Tiwari, and K.R. Kumar, 2012: Impact of climate variability on NDVI over the Indian region during 1981–2010. *Int. J. Remote Sens.*, **33**, 7132–7150, doi:10.1080/01431161.2012.697642.

Revi, A., D.E. Satterthwaite, F. Aragón-Durand, J. Corfee-Morlot, R.B.R. Kiunsi, M. Pelling, D.C. Roberts, and W. Solecki, 2014: Urban Areas. Climate Change 2014: Impacts, Adaptation and Vulnerability. Part A: Global and Sectoral Aspects. Contribution of Working Group II to the Fifth Assessment Report of the Intergovernmental Panel on Climate Change [Field, C.B., V.R. Barros, D.J. Dokken, K.J. Mach, M.D. Mastrandrea, T.E. Bilir, M. Chatterjee, K.L. Ebi, Y.O. Estrada, R.C. Genova, B. Girma, E.S. Kissel, A.N. Levy, S. MacCracken, P.R. Mastrandrea, and L.L.White (eds.)]. Cambridge University Press, Cambridge, United Kingdom and New York, NY, USA, pp. 535–612.

Riahi, K. et al., 2017: The Shared Socioeconomic Pathways and their energy, land use, and greenhouse gas emissions implications: An overview. *Glob. Environ. Chang.*, **42**, 153–168, doi:10.1016/j.gloenvcha.2016.05.009.

Rice, A.L. et al., 2016: Atmospheric methane isotopic record favors fossil sources flat in 1980s and 1990s with recent increase. *Proc. Natl. Acad. Sci.*, **113**, 10791–10796, doi:10.1073/pnas.1522923113.

Rich, S.M. and M. Watt, 2013: Soil conditions and cereal root system architecture: Review and considerations for linking Darwin and Weaver. *J. Exp. Bot.*, **64**, 1193–1208, doi:10.1093/jxb/ert043.

Richardson, A. et al., 2013: Climate change, phenology, and phenological controlof vegetatif feedbacks to the climate system. *Agric. For, Meteorol*, **169**, 156–157, doi:10.1016/j.agrformet.2012.09.012.

Ridgwell, A., J.S. Singarayer, A.M. Hetherington, and P.J. Valdes, 2009: Tackling regional climate change by leaf albedo bio-geoengineering. *Curr. Biol.*, **19**, 146–150, doi:10.1016/j.cub.2008.12.025.

Rigby, M. et al., 2013: Re-evaluation of the lifetimes of the major CFCs and Biogeosciences CH3CCl3 using atmospheric trends. *Atmos. Chem. Phys.*, **13**, 2691–2702, doi:10.5194/acp-13-2691-2013.

Riley, W., J. Tang, and W.J. Riley, 2014: Weaker soil carbon – Climate feedbacks resulting from microbial and abiotic interactions. *Nat. Clim. Chang.*, **5**, 56, doi:10.1038/nclimate2438.

Rivas-Martinez, S., S. Rivas-Saenz, and A. Marino, 2011: Worldwide bioclimatic classification system. *Glob. Geobot.*, **1**, 1–638, doi:10.5616/gg110001.

Rivera Ferre, M.G., 2014: Impacts of climate change on food availability: Distribution and exchange of food. In: *Global Environmental Change* [Freedman, B. (ed.)]. Springer Dordrecht, Netherlands, pp. 701–707.

Rizzo, L.V. et al., 2013: Long term measurements of aerosol optical properties at a primary forest site in Amazonia. *Atmos. Chem. Phys.*, **13**, 2391–2413, doi:10.5194/acp-13-2391-2013.

Robaa, S.M., 2013: Some aspects of the urban climates of Greater Cairo Region, Egypt. *Int. J. Climatol.*, **33**, 3206–3216, doi:10.1002/joc.3661.

Roberts, K.G., B.A. Gloy, S. Joseph, N.R. Scott, and J. Lehmann, 2010: Life cycle assessment of biochar systems: Estimating the energetic, economic, and climate change potential. *Environ. Sci. Technol.*, **44**, 827–833, doi:10.1021/es902266r.

Roberts, P.D., G.B. Stewart, and A.S. Pullin, 2006: *Are review articles a reliable source of evidence to support conservation and environmental management? A comparison with medicine.* Biol. Conserv., **132**, 409–423, doi:10.1016/j.biocon.2006.04.034.

Robertson, G.P. et al., 2017: Cellulosic biofuel contributions to a sustainable energy future: Choices and outcomes. *Science*, **356**, eaal2324, doi:10.1126/science.aal2324.

Robinson, A.L. et al., 2007: Rethinking organic aerosols: Semivolatile emissions and photochemical aging. *Science*, **315**, 1259–1262, doi:10.1126/science.1133061.

Rocha-Lima, A. et al., 2018: A detailed characterization of the Saharan dust collected during the Fennec campaign in 2011: In situ ground-based and laboratory measurements. *Atmos. Chem. Phys.*, **18**, 1023–1043, doi:10.5194/acp-18-1023-2018.

Roe, S., S. Weiner, M. Obersteiner, and S. Frank, 2017: *How improved land use can contribute to the 1.5C goal of the Paris Agreement.* Climate Focus and the International Institute for Applied Systems Analysis, 38 pp.

Rogelj, J. et al., 2014: Disentangling the effects of CO_2 and short-lived climate forcer mitigation. *Proc. Natl. Acad. Sci.*, **111**, 16325–16330, doi:10.1073/pnas.1415631111.

Rogelj, J. et al., 2016: Paris Agreement climate proposals need a boost to keep warming well below 2°C. *Nature*, **534**, 631. doi:10.1038/nature18307.

Rogelj, J. et al., 2018: Scenarios towards limiting global mean temperature increase below 1.5°c. *Nat. Clim. Chang.*, **8**, 325–332, doi:10.1038/s41558-018-0091-3.

Rogers, A. et al., 2017: A roadmap for improving the representation of photosynthesis in Earth system models. *New Phytol.*, **213**, 22–42, doi:10.1111/nph.14283.

Rohde, R. et al., 2013: Berkeley earth temperature averaging process. *Geoinformatics Geostatistics An Overv.*, **1**, 1–13, doi:10.4172/2327-4581.1000103.

Rosenkranz, M., T.A.M. Pugh, J.-P. Sghnitzler, and A. Arneth, 2015: Effect of land-use change and management on biogenic volatile organic compound emissions – selecting climate-smart cultivars. *Plant. Cell Environ.*, **38**, 1896–1912, doi:10.1111/pce.12453.

Rosenzweig, C. et al., 2013: The Agricultural Model Intercomparison and Improvement Project (AgMIP): Protocols and pilot studies. *Agric. For. Meteorol.*, **170**, 166–182, doi:10.1016/j.agrformet.2012.09.011.

Rossi, S., F.N. Tubiello, P. Prosperi, M. Salvatore, H. Jacobs, R. Biancalani, J.I. House, and L. Boschetti, 2016: FAOSTAT estimates of greenhouse gas emissions from biomass and peat fires. *Clim. Change*, **135**, 699–711, doi:10.1007/s10584-015-1584-y.

Rossow, W.B., A. Mekonnen, C. Pearl, and W. Goncalves, 2013: Tropical precipitation extremes. *J. Clim.*, **26**, 1457–1466, doi:10.1175/JCLI-D-11-00725.1. http://journals.ametsoc.org/doi/abs/10.1175/JCLI-D-11-00725.1 (Accessed October 30, 2018).

de Rouw, A. et al., 2010: Possibilities of carbon and nitrogen sequestration under conventional tillage and no-till cover crop farming (Mekong valley, Laos). *Agric. Ecosyst. Environ.*, doi:10.1016/j.agee.2009.12.013.

Rowe, R.L. et al., 2016: Initial soil C and land-use history determine soil C sequestration under perennial bioenergy crops. *GCB Bioenergy*, **8**, 1046–1060, doi:10.1111/gcbb.12311.

Rowell, D.P. and R. Chadwick, 2018: Causes of the uncertainty in projections of tropical terrestrial rainfall change: East Africa. *J. Clim.*, **31**, 5977–5995, doi:10.1175/JCLI-D-17-0830.1.

Rowland, L. et al., 2015: Death from drought in tropical forests is triggered by hydraulics not carbon starvation. *Nature*, **528**, 119. doi:10.1038/nature15539.

Roy, J. et al., 2016: Elevated CO_2 maintains grassland net carbon uptake under a future heat and drought extreme. *Proc. Natl. Acad. Sci.*, **113**, 6224–6229, doi:10.1073/pnas.1524527113.

Roy, S. et al., 2007: Impacts of the agricultural Green Revolution-induced land use changes on air temperatures in India. *J. Geophys. Res. Atmos.*, **112**, 1–13, doi:10.1029/2007JD008834.

Rubel, F. and M. Kottek, 2010: Observed and projected climate shifts 1901–2100 depicted by world maps of the Köppen-Geiger climate classification. *Meteorol. Zeitschrift*, **19**, 135–141, doi:10.1127/0941-2948/2010/0430.

Rubel, F., K. Brugger, K. Haslinger, and I. Auer, 2017: The climate of the European Alps: Shift of very high resolution Köppen-Geiger climate zones 1800–2100. *Meteorol. Zeitschrift*, **26**, 115–125, doi:10.1127/metz/2016/0816.

Rumpel, C., and I. Kögel-Knabner, 2011: Deep soil organic matter – A key but poorly understood component of terrestrial C cycle. *Plant Soil*, **338**, 143–158, doi:10.1007/s11104-010-0391-5.

Rumpf, S.B. et al., 2018: Range dynamics of mountain plants decrease with elevation. *Proc. Natl. Acad. Sci.*, **115**, 1848–1853, doi:10.1073/pnas.1713936115.

Russo, S. et al., 2014: Magnitude of extreme heat waves in present climate and their projection in a warming world. *J. Geophys. Res. Atmos.*, **119**, 12500–12512, doi:10.1002/2014JD022098.

Russo, S., A.F. Marchese, J. Sillmann, and G. Imme, 2016: When will unusual heat waves become normal in a warming Africa? *Environ. Res. Lett.*, **11**, 54016, doi:10.1088/1748-9326/11/5/054016.

Rusticucci, M., 2012: Observed and simulated variability of extreme temperature events over South America. *Atmos. Res.*, **106**, 1–17, doi:10.1016/j.atmosres.2011.11.001.

Rusticucci, M., J. Kyselý, G. Almeira, and O. Lhotka, 2016: Long-term variability of heat waves in Argentina and recurrence probability of the severe 2008 heat wave in Buenos Aires. *Theor. Appl. Climatol.*, **124**, 679–689, doi:10.1007/s00704-015-1445-7.

Rutting, T., 2017: Nitrogen mineralization, not N-2 fixation, alleviates progressive nitrogen limitation – Comment on "Processes regulating progressive nitrogen limitation under elevated carbon dioxide: a meta-analysis" by Liang et al. (2016). *Biogeosciences*, **14**, 751–754, doi:10.5194/bg-14-751-2017.

de Sá, S.S. et al., 2018: Urban influence on the concentration and composition of submicron particulate matter in central Amazonia. *Atmos. Chem. Phys. Discuss.*, doi:10.5194/acp-2018-172.

de Sá, S.S. et al., 2019: Contributions of biomass-burning, urban, and biogenic emissions to the concentrations and light-absorbing properties of particulate matter in central Amazonia during the dry season. *Atmos. Chem. Phys. Discuss.*, 1–77, doi:10.5194/acp-2018-1309.

de Sá, S.S. et al., 2017: Influence of urban pollution on the production of organic particulate matter from isoprene epoxydiols in central Amazonia. *Atmos. Chem. Phys.*, **17**, 6611–6629, doi:10.5194/acp-17-6611-2017.

Saatchi, S. et al., 2015: Seeing the forest beyond the trees. *Glob. Ecol. Biogeogr.*, **24**, 606–610, doi:10.1111/geb.12256.

Saatchi, S.S. et al., 2011: Benchmark map of forest carbon stocks in tropical regions across three continents. *Proc. Natl. Acad. Sci.*, **108**, 9899–9904, doi:10.1073/pnas.1019576108.

Sachindra, D.A., A.W.M.M. Ng, S. Muthukumaran, and B.J.C.C. Perera, 2016: Impact of climate change on urban heat island effect and extreme temperatures: A case-study. *Q.J.R. Meteorol. Soc.*, **142**, 172–186, doi:10.1002/qj.2642.

Saeki, T., and P.K. Patra, 2017: Implications of overestimated anthropogenic CO_2 emissions on East Asian and global land CO_2 flux inversion. *Geosci. Lett.*, **4**, 9, doi:10.1186/s40562-017-0074-7.

Sagan, C., O.B. Toon, and J.B. Pollack, 1979: Anthropogenic albedo changes and the Earth's climate. *Science*, **206**, 1363–1368. http://www.jstor.org/stable/1748990.

Saikawa, E. et al., 2014: Global and regional emissions estimates for N_2O. *Atmos. Chem. Phys.*, **14**, 4617–4641, doi:10.5194/acp-14-4617-2014.

Samset, B.H. et al., 2014: Modelled black carbon radiative forcing and atmospheric lifetime in AeroCom Phase II constrained by aircraft observations. *Atmos. Chem. Phys.*, **14**, 12465–12477, doi:10.5194/acp-14-12465-2014.

Sánchez-Rodríguez, A.R., P.W. Hill, D.R. Chadwick, and D.L. Jones, 2017: Crop residues exacerbate the negative effects of extreme flooding on soil quality. *Biol. Fertil. Soils*, **53**, 751–765, doi:10.1007/s00374-017-1214-0.

Sánchez-Rodríguez, A.R., C. Nie, P.W. Hill, D.R. Chadwick, and D.L. Jones, 2019: Extreme flood events at higher temperatures exacerbate the loss of soil functionality and trace gas emissions in grassland. *Soil Biol. Biochem.*, **130**, 227–236, doi:10.1016/j.soilbio.2018.12.021.

Sand, M. et al., 2015: Response of Arctic temperature to changes in emissions of short-lived climate forcers. *Nat. Clim. Chang.*, **6**, 286. doi:10.1038/nclimate2880.

Sanderman, J., T. Hengl, and G.J. Fiske, 2017: Soil carbon debt of 12,000 years of human land use. *Proc. Natl. Acad. Sci.*, **114**, 9575–9580, doi:10.1073/pnas.1706103114.

Sanderson, M.G., C.D. Jones, W.J. Collins, C.E. Johnson, and R.G. Derwent, 2003: Effect of climate change on isoprene emissions and surface ozone levels. *Geophys. Res. Lett.*, **30**, doi:10.1029/2003GL017642.

Santamouris, M., C. Cartalis, A. Synnefa, and D. Kolokotsa, 2015: On the impact of urban heat island and global warming on the power demand and electricity consumption of buildings – A review. *Energy Build.*, **98**, 119–124, doi:10.1016/j.enbuild.2014.09.052.

Sardans, J. and J. Peñuelas, 2015: Potassium: A neglected nutrient in global change. *Glob. Ecol. Biogeogr.*, **24**, 261–275, doi:10.1111/geb.12259.

Sardans, J., A. Rivas-Ubach, and J. Peñuelas, 2012: The C:N:P stoichiometry of organisms and ecosystems in a changing world: A review and perspectives. *Perspect. Plant Ecol. Evol. Syst.*, **14**, 33–47, doi:10.1016/j.ppees.2011.08.002.

Sasaki, N. et al., 2016: Sustainable management of tropical forests can reduce carbon emissions and stabilize timber production. *Front. Environ. Sci.*, **4**, 50, doi:10.3389/fenvs.2016.00050.

Sathre, R. and J. O'Connor, 2010: Meta-analysis of greenhouse gas displacement factors of wood product substitution. *Environ. Sci. Policy*, **13**, 104–114, doi:10.1016/j.envsci.2009.12.005.

Satria, 2017: Substantial N_2O emissions from peat decomposition and N fertilization in an oil palm plantation exacerbated by hotspots. *Environ. Res. Lett*, **12**, doi:10.1088/1748-9326/aa80f1.

Saturno, J. et al., 2018: Black and brown carbon over central Amazonia: Long-term aerosol measurements at the ATTO site. *Atmos. Chem. Phys.*, **18**, 12817–12843, doi:10.5194/acp-18-12817-2018.

Savage, J. and M. Velend, 2015: Elevational shifts, biotic homogenization and time lags in vegetation change during 40 years of climate warming. *Ecography (Cop.).*, **38**, 546–555, doi:10.1111/ecog.01131.

Sayer, E.J., M.S. Heard, H.K. Grant, T.R. Marthews, and E.V.J. Tanner, 2011: Soil carbon release enhanced by increased tropical forest litterfall. *Nat. Clim. Chang.*, **1**, 304–307, doi:10.1038/nclimate1190.

Scalenghe, R., and F.A. Marsan, 2009: The anthropogenic sealing of soils in urban areas. *Landsc. Urban Plan.*, **90**, 1–10, doi:10.1016/j.landurbplan.2008.10.011.

Schaefer, H. et al., 2016: A 21st-century shift from fossil-fuel to biogenic methane emissions indicated by 13CH₄. *Science*, **352**, 80–84, doi:10.1126/science.aad2705.

Schafer, J.S. et al., 2002: Observed reductions of total solar irradiance by biomass-burning aerosols in the Brazilian Amazon and Zambian Savanna. *Geophys. Res. Lett.*, **29**, 4-1-4–4, doi:10.1029/2001GL014309.

Schär, C., 2016: Climate extremes: The worst heat waves to come. *Nat. Clim. Chang.*, **6**, 128–129, doi:10.1038/nclimate2864.

Scherer, M. and N.S. Diffenbaugh, 2014: Transient twenty-first century changes in daily-scale temperature extremes in the United States. *Clim. Dyn.*, **42**, 1383–1404, doi:10.1007/s00382-013-1829-2.

Schimel, D., B.B. Stephens, and J.B. Fisher, 2015: Effect of increasing CO_2 on the terrestrial carbon cycle. *Proc. Natl. Acad. Sci.*, **112**, 436–441, doi:10.1073/pnas.1407302112.

Schindlbacher, A., J. Schnecker, M. Takriti, W. Borken, and W. Wanek, 2015: Microbial physiology and soil CO_2 efflux after 9 years of soil warming in a temperate forest – no indications for thermal adaptations. *Glob. Chang. Biol.*, **21**, 4265–4277, doi:10.1111/gcb.12996.

Schlaepfer, D.R. et al., 2017: Climate change reduces extent of temperate drylands and intensifies drought in deep soils. *Nat. Commun.*, **8**, 14196, doi:10.1038/ncomms14196.

Schlenker, W., and D.B. Lobell, 2010: Robust negative impacts of climate change on African agriculture. *Environ. Res. Lett.*, **5**, 14010, doi:10.1088/1748-9326/5/1/014010.

Schlesinger, W.H. et al., 2016: Forest biogeochemistry in response to drought. *Glob. Chang. Biol.*, **22**, 2318–2328, doi:10.1111/gcb.13105.

Schleussner, C.-F., P. Pfleiderer, and E.M. Fischer, 2017: In the observational record half a degree matters. *Nat. Clim. Chang.*, **7**, 460–462, doi:10.1038/nclimate3320.

Schlünzen, K.H., P. Hoffmann, G. Rosenhagen, and W. Riecke, 2010: Long-term changes and regional differences in temperature and precipitation in the metropolitan area of Hamburg. *Int. J. Climatol.*, **30**, 1121–1136, doi:10.1002/joc.1968.

Schmale, J. et al., 2018: Long-term cloud condensation nuclei number concentration, particle number size distribution and chemical composition measurements at regionally representative observatories. *Atmos. Chem. Phys.*, **185194**, 2853–2881, doi:10.5194/acp-18-2853-2018.

Schmidt, M. et al., 2011: Persistance of soil organic matter as an ecosystem property. *Nature*, **478**, 49–56, doi:10.1038/nature10386.

Schueler, V., S. Fuss, J.C. Steckel, U. Weddige, and T. Beringer, 2016: Productivity ranges of sustainable biomass potentials from non-agricultural land. *Environ. Res. Lett.*, **11**, 074026, doi:10.1088/1748-9326/11/7/074026.

Schulte-Uebbing, L. and W. de Vries, 2018: Global-scale impacts of nitrogen deposition on tree carbon sequestration in tropical, temperate, and boreal forests: A meta-analysis. *Glob. Chang. Biol.*, **24**, e416–e431, doi:10.1111/gcb.13862.

Schultz, N.M., P.J. Lawrence, and X. Lee, 2017: Global satellite data highlights the diurnal asymmetry of the surface temperature response to deforestation. *J. Geophys. Res. Biogeosciences*, **122**, 903–917, doi:10.1002/2016JG003653.

Schulz, C. et al., 2018: Aircraft-based observations of isoprene epoxydiol-derived secondary organic aerosol (IEPOX-SOA) in the tropical upper troposphere over the Amazon region. *Atmos. Chem. Phys. Discuss.*, **3**, 1–32, doi:10.5194/acp-2018-232.

Schuur, E.A.G. et al., 2015: Climate change and the permafrost carbon feedback. *Nature*, **520**, 171–179, doi:10.1038/nature14338.

Scott, A., 2000: The Pre-Quaternary history of fire. *Palaeogeogr. Palaeoclimatol. Palaeoecol.*, **164**, 281–329, doi:10.1016/S0031-0182(00)00192-9.

Scott, C.E. et al., 2015: Impact of gas-to-particle partitioning approaches on the simulated radiative effects of biogenic secondary organic aerosol. *Atmos. Chem. Phys.*, **15**, 12989–13001, doi:10.5194/acp-15-12989-2015.

Scott, C.E. et al., 2017: Impact on short-lived climate forcers (SLCFs) from a realistic land-use change scenario via changes in biogenic emissions. *Faraday Discuss.*, **200**, 101–120, doi:10.1039/c7fd00028f.

Scott, C.E. et al., 2018a: Impact on short-lived climate forcers increases projected warming due to deforestation. *Nat. Commun.*, **9**, 157, doi:10.1038/s41467-017-02412-4.

Scott, C.E. et al., 2018b: Substantial large-scale feedbacks between natural aerosols and climate. *Nat. Geosci.*, **11**, 44–48, doi:10.1038/s41561-017-0020-5.

Searle, S. and C. Malins, 2015: A reassessment of global bioenergy potential in 2050. *GCB Bioenergy*, **7**, 328–336, doi:10.1111/gcbb.12141.

Seddon, A.W.R., M. Macias-Fauria, P.R. Long, D. Benz, and K.J. Willis, 2016: Sensitivity of global terrestrial ecosystems to climate variability. *Nature*, **531**, 229–232, doi:10.1038/nature16986.

Seidl, R. et al., 2017: Forest disturbances under climate change. *Nat. Clim. Chang.*, **7**, 395–402, doi:10.1038/nclimate3303.

Sejas, S.A., O.S. Albert, M. Cai, and Y. Deng, 2014: Feedback attribution of the land-sea warming contrast in a global warming simulation of the NCAR CCSM4. *Environ. Res. Lett.*, **9**, 124005, doi:10.1088/1748-9326/9/12/124005.

Sellers, P.J. et al., 1996: A revised land surface parameterization (SiB2) for atmospheric GCMs. Part I: Model formulation. *J. Clim.*, **9**, 676–705, doi:10.1175/1520-0442(1996)009<0676:ARLSPF>2.0.CO;2.

Semazzi, F.H.M. and Y. Song, 2001: A GCM study of climate change induced by deforestation in Africa. *Clim. Res.*, **17**, 169–182, doi:10.3354/cr017169.

Sena, E.T., A. McComiskey, and G. Feingold, 2016: A long-term study of aerosol-cloud interactions and their radiative effect at the Southern Great Plains using ground-based measurements. *Atmos. Chem. Phys.*, **16**, 11301–11318, doi:10.5194/acp-16-11301-2016.

Seneviratne, S.I., D. Lüthi, M. Litschi, and C. Schär, 2006: Land-atmosphere coupling and climate change in Europe. *Nature*, **443**, 205–209, doi:10.1038/nature05095.

Seneviratne, S.I., N. Nicholls, D. Easterling, C.M. Goodess, S. Kanae, J. Kossin, Y. Luo, J. Marengo, K. McInnes, M. Rahimi, M. Reichstein, A. Sorteberg, C. Vera, and X. Zhang, 2012: Changes in Climate Extremes and Their Impacts on the Natural Physical Environment. In: Managing the Risks of Extreme Events and Disasters to Advance Climate Change Adaptation. A Special Report of Working Groups I and II of the Intergovernmental Panel on Climate Change (IPCC) [Field, C.B., V. Barros, T.F. Stocker, D. Qin, D.J. Dokken, K.L. Ebi, M.D. Mastrandrea, K.J. Mach, G.-K. Plattner, S.K. Allen, M. Tignor, and P.M. Midgley (eds.)]. Cambridge University Press, Cambridge, UK, and New York, NY, USA, pp. 109–230.

Seneviratne, S.I. et al., 2013: Impact of soil moisture-climate feedbacks on CMIP5 projections: First results from the GLACE-CMIP5 experiment. *Geophys. Res. Lett.*, **40**, 5212–5217, doi:10.1002/grl.50956.

Seneviratne, S.I., M.G. Donat, A.J. Pitman, R. Knutti, and R.L. Wilby, 2016: Allowable CO_2 emissions based on regional and impact-related climate targets. *Nature*, **529**, 477–483, doi:10.1038/nature16542.

Seneviratne, S.I. et al., 2018: Land radiative management as contributor to regional-scale climate adaptation and mitigation. *Nat. Geosci.*, **11**, 88–96, doi:10.1038/s41561-017-0057-5.

Seto, K.C., M. Fragkias, B. Güneralp, and M.K. Reilly, 2011: A meta-analysis of global urban land expansion. *PLoS One*, **6**, e23777, doi:10.1371/journal.pone.0023777.

Seto K.C., S. Dhakal, A. Bigio, H. Blanco, G.C. Delgado, D. Dewar, L. Huang, A. Inaba, A. Kansal, S. Lwasa, J.E. McMahon, D.B. Müller, J. Murakami, H. Nagendra, and A. Ramaswami, 2014: Human Settlements, Infrastructure and Spatial Planning. In: Climate Change 2014: Mitigation of Climate Change. Contribution of Working Group III to the Fifth Assessment Report of the Intergovernmental Panel on Climate Change [Edenhofer, O., R. Pichs-Madruga, Y. Sokona, E. Farahani, S. Kadner, K. Seyboth, A. Adler, I. Baum, S. Brunner, P. Eickemeier, B. Kriemann, J. Savolainen, S. Schlömer, C. von Stechow, T. Zwickel and J.C. Minx (eds.)]. Cambridge University Press, Cambridge, United Kingdom and New York, NY, USA, pp. 923–1000.

Settele, J., R. Scholes, R. Betts, S. Bunn, P. Leadley, D. Nepstad, J.T. Overpeck, and M.A. Taboada 2014: Terrestrial and Inland Water Systems. In: Climate Change 2014: Impacts, Adaptation and Vulnerability. Part A: Global and Sectoral Aspects. Contribution of Working Group II to the Fifth Assessment Report of the Intergovernmental Panel on Climate Change [Field, C.B., V.R. Barros, D.J. Dokken, K.J. Mach, M.D. Mastrandrea, T.E. Bilir, M. Chatterjee, K.L. Ebi, Y.O. Estrada, R.C. Genova, B. Girma, E.S. Kissel, A.N. Levy, S. MacCracken, P.R. Mastrandrea, and L.L.White (eds.)]. Cambridge University Press, Cambridge, United Kingdom and New York, NY, USA, pp. 271–360.

Shade, C. and P. Kremer, 2019: Predicting land use changes in philadelphia following green infrastructure policies. *Land*, **8**, 28, doi:10.3390/land8020028.

Shao, P. et al., 2013: Soil microbial respiration from observations and Earth System Models. *Environ. Res. Lett.*, **8**, 034034, doi:10.1088/1748-9326/8/3/034034.

Sharma, S.B., S. Jain, P. Khirwadkar, and S. Kulkarni, 2013: The effects of air pollution on the environment and human health. *Indian Journal of Research in Pharmacy and Biotechnol*, **1**(3), 2320–3471.

Shcherbak, I., N. Millar, and G.P. Robertson, 2014: Global metaanalysis of the nonlinear response of soil nitrous oxide (N_2O) emissions to fertilizer nitrogen. *Proc. Natl. Acad. Sci.*, **111**, 9199–9204, doi:10.1073/pnas.1322434111.

Sheffield, J., E.F. Wood, and M.L. Roderick, 2012: Little change in global drought over the past 60 years. *Nature*, **491**, 435–438, doi:10.1038/nature11575.

Shevliakova, E. et al. 2013: Historical warming reduced due to enhanced land carbon uptake. *Proc. Natl. Acad. Sci.*, **110**, 16730–16735, doi:10.1073/pnas.1314047110.

Shi, W., F. Tao, and J. Liu, 2013: Regional temperature change over the Huang-Huai-Hai Plain of China: The roles of irrigation versus urbanization. *Int. J. Climatol.*, **34**, 1181–1195, doi:10.1002/joc.3755.

Shindell, D. et al., 2017: A climate policy pathway for near- and long-term benefits. *Science*, **356**, 493–494, doi:10.1126/science.aak9521.

Shindell, D., G. Faluvegi, K. Seltzer, and C. Shindell, 2018: Quantified, localized health benefits of accelerated carbon dioxide emissions reductions. *Nat. Clim. Chang.*, **8**, 291–295, doi:10.1038/s41558-018-0108-y.

Siebert, S., V. Henrich, K. Frenken, and J. Burke, 2013: Global map of irrigation areas Version 5. Food and Agriculture Organization of the United Nations, Rome, Italy. http://www.fao.org/nr/water/aquastat/irrigationmap/index10.stm.

Siebert, S., H. Webber, G. Zhao, and F. Ewert, 2017: Heat stress is overestimated in climate impact studies for irrigated agriculture. *Environ. Res. Lett.*, **12**, 54023, doi:10.1088/1748-9326/aa702f.

Silva, L.C.R. and M. Anand, 2013: Probing for the influence of atmospheric CO_2 and climate change on forest ecosystems across biomes. *Glob. Ecol. Biogeogr.*, **22**, 83–92, doi:10.1111/j.1466-8238.2012.00783.x.

Simmons, C.T. and H.D. Matthews, 2016: Assessing the implications of human land-use change for the transient climate response to cumulative carbon emissions. *Environ. Res. Lett.*, **11**, doi:10.1088/1748-9326/11/3/035001.

Singarayer, J.S. and T. Davies-Barnard, 2012: Regional climate change mitigation with crops: Context and assessment. *Philos. Trans. R. Soc. A Math. Phys. Eng. Sci.*, **370**, 4301–4316, doi:10.1098/rsta.2012.0010.

Singarayer, J.S., A. Ridgwell, and P. Irvine, 2009: Assessing the benefits of crop albedo bio-geoengineering. *Environ. Res. Lett.*, **4**, doi:10.1088/1748-9326/4/4/045110.

Singh, B.K., R.D. Bardgett, P. Smith, and D.S. Reay, 2010: Microorganisms and climate change: Terrestrial feedbacks and mitigation options. *Nat. Rev. Microbiol.*, **8**, 779–790, doi:10.1038/nrmicro2439.

Singh, D., M.G. Flanner, and J. Perket, 2015: The global land shortwave cryosphere radiative effect during the MODIS era. *Cryosphere*, **9**, 2057–2070, doi:10.5194/tc-9-2057-2015.

Sippel, S. and F.E.L. Otto, 2014: Beyond climatological extremes – Assessing how the odds of hydrometeorological extreme events in South-East Europe change in a warming climate. *Clim. Change*, **125**, 381–398, doi:10.1007/s10584-014-1153-9.

Sippel, S. et al., 2017: Refining multi-model projections of temperature extremes by evaluation against land–atmosphere coupling diagnostics. *Earth Syst. Dyn.*, **8**, 387–403, doi:10.5194/esd-8-387-2017.

Sirin, A.A., G. Suvorov, M. Chistotin, and M. Glagolev, 2012: Values of methane emission from drainage channels. *Doosigik*, **2**, 1–10.

Sirin, A., T. Minayeva, A. Vozbrannaya, and S. Bartalev, 2011: How to avoid peat fires? *Sci. Russ.*, N2, 13-21.

Sisti, C.P.J. et al., 2004: Change in carbon and nitrogen stocks in soil under 13 years of conventional or zero tillage in southern Brazil. *Soil Tillage Res.*, **76**, 39–58, doi:10.1016/j.still.2003.08.007.

Sitch, S. et al. 2005: Impacts of future land cover changes on atmospheric CO_2 and climate. *Global Biogeochem. Cycles*, **19**, 1–15, doi:10.1029/2004GB002311.

Sitch, S. et al., 2015: Recent trends and drivers of regional sources and sinks of carbon dioxide. *Biogeosciences*, **12**, 653–679, doi:10.5194/bg-12-653-2015.

Six, J., R.T. Conant, E.A. Paul, and K. Paustian, 2002: Stabilization mechanisms of soil organic matter: Implications for C-saturation of soils. *Plant Soil*, **241**, 155–176, doi:10.1023/A:1016125726789.

Six, J. et al., 2004: The potential to mitigate global warming with no-tillage management is only realised when practised in the long term. *Glob. Chang. Biol.*, **10**, 155–160, doi:10.1111/j.1529-8817.2003.00730.x.

van der Sleen, P. et al., 2014: No growth stimulation of tropical trees by 150 years of CO_2 fertilization but water-use efficiency increased. *Nat. Geosci.*, **8**, 24–28, doi:10.1038/ngeo2313.

Slot, M., and K. Kitajima, 2015: General patterns of acclimation of leaf respiration to elevated temperatures across biomes and plant types. *Oecologia*, **177**, 885–900, doi:10.1007/s00442-014-3159-4.

Slot, M., and K. Winter, 2017: Photosynthetic acclimation to warming in tropical forest tree Seedlings. *J. Exp. Bot.*, **68**, 2275–2284, doi:10.1093/jxb/erx071.

Slot, M. et al, 2014: Thermal acclimation of leaf respiration of tropical trees and lianas: Response to experimental canopy warming, and consequences for tropical forest carbon balance. *Glob. Chang. Biol.*, **20**, 2915–2926, doi:10.1111/gcb.12563.

Smith, M.C., J.S. Singarayer, P.J. Valdes, J.O. Kaplan, and N.P. Branch, 2016a: The biogeophysical climatic impacts of anthropogenic land use change during the Holocene. *Clim. Past*, **12**, 923–941, doi:10.5194/cp-12-923-2016.

Smith, N.G., and J.S. Dukes, 2013: Plant respiration and photosynthesis in global-scale models: Incorporating acclimation to temperature and CO_2. *Glob. Chang. Biol.*, **19**, 45–63, doi:10.1111/j.1365-2486.2012.02797.x.

Smith, N.G., S.L. Malyshev, E. Shevliakova, J. Kattge, and J.S. Dukes, 2016b: Foliar temperature acclimation reduces simulated carbon sensitivity to climate. *Nat. Clim. Chang.*, **6**, 407–411, doi:10.1038/nclimate2878.

Smith, P., 2016: Soil carbon sequestration and biochar as negative emission technologies. *Glob. Chang. Biol.*, **22**, 1315–1324, doi:10.1111/gcb.13178.

Smith, P. et al., 2008: Greenhouse gas mitigation in agriculture. *Philos. Trans. R. Soc. B Biol. Sci.*, **363**, 789–813, doi:10.1098/rstb.2007.2184.

Smith, P. et al., 2013a: How much land-based greenhouse gas mitigation can be achieved without compromising food security and environmental goals? *Glob. Chang. Biol.*, **19**, 2285–2302, doi:10.1111/gcb.12160.

Smith, P. et al., 2015: Global change pressures on soils from land use and management. *Glob. Chang. Biol.*, **22**, 1008–1028, doi:10.1111/gcb.13068.

Smith, P. et al., 2016c: Biophysical and economic limits to negative CO_2 emissions. *Nat. Clim. Chang.*, **6**, 42–50, doi:10.1038/nclimate2870.

Smith, S. V, W.H. Renwick, R.W. Buddemeier, and C.J. Crossland, 2001: Budgets of soil erosion and deposition for sediments and sedimentary organic carbon across the conterminous United States. *Global Biogeochem. Cycles*, **15**, 697–707, doi:10.1029/2000GB001341.

Smith, T.T., B.F. Zaitchik, and J.M. Gohlke, 2013b: Heat waves in the United States: Definitions, patterns and trends. *Clim. Change*, **118**, 811–825, doi:10.1007/s10584-012-0659-2.

Smyth, C., W.A. Kurz, G. Rampley, T.C. Lemprière, and O. Schwab, 2017a: Climate change mitigation potential of local use of harvest residues for bioenergy in Canada. *GCB Bioenergy*, **9**, 817–832, doi:10.1111/gcbb.12387.

Smyth, C., G. Rampley, T.C. Lemprière, O. Schwab, and W.A. Kurz, 2017b: Estimating product and energy substitution benefits in national-scale mitigation analyses for Canada. *GCB Bioenergy*, **9**, 1071–1084, doi:10.1111/gcbb.12389.

Snider, D.M., J.J. Venkiteswaran, S.L. Schiff, and J. Spoelstra, 2015: From the ground up: Global nitrous oxide sources are constrained by stable isotope values. *PLoS One*, **10**, e0118954, doi:10.1371/journal.pone.0118954.

Snyder, P.K., 2010: The influence of tropical deforestation on the Northern Hemisphere climate by atmospheric teleconnections. *Earth Interact.*, doi:10.1175/2010EI280.1.

Snyder, P.K., C. Delire, and J.A. Foley, 2004: Evaluating the influence of different vegetation biomes on the global climate. *Clim. Dyn.*, **23**, 279–302, doi:10.1007/s00382-004-0430-0.

Soimakallio, S., L. Saikku, L. Valsta, and K. Pingoud, 2016: Climate change mitigation challenge for wood utilization – The case of Finland. *Environ. Sci. Technol.*, **50**, 5127–5134, doi:10.1021/acs.est.6b00122.

Soja, A.J. et al., 2007: Climate-induced boreal forest change: Predictions versus current observations. *Glob. Planet. Change*, **56**, 274–296, doi:10.1016/j.gloplacha.2006.07.028.

Sommer, R., and D. Bossio, 2014: Dynamics and climate change mitigation potential of soil organic carbon sequestration. *J. Environ. Manage.*, **144**, 83–87, doi:10.1016/j.jenvman.2014.05.017.

Son, K.H. and D.H. Bae, 2015: Drought analysis according to shifting of climate zones to arid climate zone over Asia monsoon region. *J. Hydrol.*, **529**, 1021–1029, doi:10.1016/j.jhydrol.2015.09.010.

Sonntag, S., J. Pongratz, C.H. Reick, and H. Schmidt, 2016: Reforestation in a high-CO_2 world—Higher mitigation potential than expected, lower adaptation potential than hoped for. *Geophys. Res. Lett.*, **43**, 6546–6553, doi:10.1002/2016GL068824.

Sonntag, S. et al., 2018: Quantifying and comparing effects of climate engineering methods on the Earth system. *Earth's Futur.*, doi:10.1002/eft2.285.

Spahni, R., F. Joos, B.D. Stocker, M. Steinacher, and Z.C. Yu, 2013: Transient simulations of the carbon and nitrogen dynamics in northern peatlands: From the Last Glacial Maximum to the 21st century. *Clim. Past*, **9**, 1287–1308, doi:10.5194/cp-9-1287-2013.

Spencer, T. et al., 2016: Global coastal wetland change under sea-level rise and related stresses: The DIVA Wetland Change Model. *Glob. Planet. Change*, **139**, 15–30, doi:10.1016/j.gloplacha.2015.12.018.

Sperry, J.S., and D.M. Love, 2015: What plant hydraulics can tell us about responses to climate-change droughts. *New Phytol.*, **207**, 14–27, doi:10.1111/nph.13354.

Spinoni, J., G. Naumann, J. Vogt, and P. Barbosa, 2015a: European drought climatologies and trends based on a multi-indicator approach. *Glob. Planet. Change*, **127**, 50–57, doi:10.1016/J.GLOPLACHA.2015.01.012.

Spinoni, J., J. Vogt, G. Naumann, H. Carrao, and P. Barbosa, 2015b: Towards identifying areas at climatological risk of desertification using the Köppen-Geiger classification and FAO aridity index. *Int. J. Climatol.*, **35**, 2210–2222, doi:10.1002/joc.4124.

Spinoni, J. et al., 2019: A new global database of meteorological drought events from 1951 to 2016. *J. Hydrol. Reg. Stud.*, **22**, 100593, doi:10.1016/j.ejrh.2019.100593.

Sporre, M.K., S.M. Blichner, I.H.H. Karset, R. Makkonen, and T.K. Berntsen, 2019: BVOC-aerosol-climate feedbacks investigated using NorESM. *Atmos. Chem. Phys.*, **19**, 4763–4782, doi:10.5194/acp-19-4763-2019.

Spracklen, D. V, J.C.A. Baker, L. Garcia-Carreras, and J.H. Marsham, 2018: The effects of tropical vegetation on rainfall. *Annu. Rev. Environ. Resour.*, **43**, 193–218, doi:10.1146/annurev-environ-102017-030136.

Spracklen, D.V., and L. Garcia-Carreras, 2015: The impact of Amazonian deforestation on Amazon Basin rainfall. *Geophys. Res. Lett.*, **42**, 9546–9552, doi:10.1002/2015GL066063.

Spracklen, D.V., S.R. Arnold, and C.M. Taylor, 2012: Observations of increased tropical rainfall preceded by air passage over forests. *Nature*, **489**, 282–285, doi:10.1038/nature11390.

Springmann, M., H.C.J. Godfray, M. Rayner, and P. Scarborough, 2016: Analysis and valuation of the health and climate change cobenefits of dietary change. *Proc. Natl. Acad. Sci.*, **113**, 4146–4151, doi:10.1073/pnas.1523119113.

Staal, A. et al., 2018: Forest-rainfall cascades buffer against drought across the Amazon. *Nat. Clim. Chang.*, **8**, 539-543, doi:10.1038/s41558-018-0177-y.

Stallard, R.F., 1998: Terrestrial sedimentation and the carbon cycle: Coupling weathering and erosion to carbon burial. *Global Biogeochem. Cycles*, **12**, 231–257, doi:10.1029/98GB00741.

Stanelle, T., I. Bey, T. Raddatz, C. Reick, and I. Tegen, 2014: Anthropogenically induced changes in twentieth century mineral dust burden and the associated impact on radiative forcing. *J. Geophys. Res.*, **119**, 13526–13546, doi:10.1002/2014JD022062.

Stanne, T.M., L.L.E. Sjögren, S. Koussevitzky, and A.K. Clarke, 2009: Identification of new protein substrates for the chloroplast ATP-dependent Clp protease supports its constitutive role in *Arabidopsis*. *Biochem. J.*, **417**, 257–269, doi:10.1042/BJ20081146.

Staples, M.D., R. Malina, and S.R.H. Barrett, 2017: The limits of bioenergy for mitigating global life-cycle greenhouse gas emissions from fossil fuels. *Nat. Energy*, **2**, 16202, doi:10.1038/nenergy.2016.202.

Steinbauer, M.J. et al., 2018: Accelerated increase in plant species richness on mountain summits is linked to warming. *Nature*, **556**, 231–234, doi:10.1038/s41586-018-0005-6.

Stevanović, M. et al., 2017: Mitigation strategies for greenhouse gas emissions from agriculture and land-use change: Consequences for food prices. *Environ. Sci. Technol.*, **51**, 365–374, doi:10.1021/acs.est.6b04291.

Stevens-Rumann, C.S. et al., 2018: Evidence for declining forest resilience to wildfires under climate change. *Ecol. Lett.*, **21**, 243–252, doi:10.1111/ele.12889.

Stocker, B.D., F. Feissli, K.M. Strassmann, R. Spahni, and F. Joos, 2014: Past and future carbon fluxes from land use change, shifting cultivation and wood harvest. *Tellus, Ser. B Chem. Phys. Meteorol.*, **66**, 23188, doi:10.3402/tellusb.v66.23188.

Stott, P.A. et al., 2016: Attribution of extreme weather and climate-related events. *Wiley Interdiscip. Rev. Clim. Chang.*, **7**, 23–41, doi:10.1002/wcc.380.

Strack, M., and J.M. Waddington, 2008: Spatiotemporal variability in peatland subsurface methane dynamics. *J. Geophys. Res. Biogeosciences*, **113**, doi:10.1029/2007JG000472.

Strandberg, G., and E. Kjellström, 2018: Climate impacts from afforestation and deforestation in Europe. *Earth Interact.*, **23**, 1–27, doi:10.1175/ei-d-17-0033.1.

Strefler, J. et al., 2018: Between Scylla and Charybdis: Delayed mitigation narrows the passage between large-scale CDR and high costs. *Environ. Res. Lett.*, **13**, 44015, doi:10.1088/1748-9326/aab2ba.

Strengers, B.J. et al., 2010: Assessing 20th century climate-vegetation feedbacks of land-use change and natural vegetation dynamics in a fully coupled vegetation-climate model. *Int. J. Climatol.*, **30**, 2055–2065, doi:10.1002/joc.2132.

Sulman, B.N. et al., 2019: Diverse mycorrhizal associations enhance terrestrial C storage in a global model. *Global Biogeochem. Cycles*, **33**, 501–523, doi:10.1029/2018GB005973.

Sulman, B.N., R.P. Phillips, A.C. Oishi, E. Shevliakova, and S.W. Pacala, 2014: Microbe-driven turnover offsets mineral-mediated storage of soil carbon under elevated CO_2. *Nat. Clim. Chang.*, **4**, 1099–1102, doi:10.1038/nclimate2436.

Sulman, B.N., D.T. Roman, K. Yi, L.X. Wang, R.P. Phillips, and K.A. Novick, 2016: High atmospheric demand for water can limit forest carbon uptake and transpiration as severely as dry soil. *Geophys. Res. Lett.*, **43**, 9686–9695, doi:10.1002/2016gl069416.

Sulman, B.N. et al., 2017: Feedbacks between plant N demand and rhizosphere priming depend on type of mycorrhizal association. *Ecol. Lett.*, **20**, 1043–1053, doi:10.1111/ele.12802.

Sun, Y., X. Zhang, G. Ren, F.W. Zwiers, and T. Hu, 2016: Contribution of urbanization to warming in China. *Nat. Clim. Chang.*, **6**, 706–709, doi:10.1038/nclimate2956.

Suni, T. et al., 2015: The significance of land-atmosphere interactions in the Earth system – ILEAPS achievements and perspectives. *Anthropocene*, **12**, 69–84, doi:10.1016/j.ancene.2015.12.001.

Suntharalingam, P. et al., 2012: Quantifying the impact of anthropogenic nitrogen deposition on oceanic nitrous oxide. *Geophys. Res. Lett.*, **39**, 1–7, doi:10.1029/2011GL050778.

Sussams, L.W., W.R. Sheate, and R.P. Eales, 2015: Green infrastructure as a climate change adaptation policy intervention: Muddying the waters or clearing a path to a more secure future? *J. Environ. Manage.*, **147**, 184–193, doi:10.1016/j.jenvman.2014.09.003.

Sutton, R.T., B. Dong, and J.M. Gregory, 2007: Land/sea warming ratio in response to climate change: IPCC AR4 model results and comparison with observations. *Geophys. Res. Lett.*, **34**, doi:10.1029/2006GL028164.

Swann, A.L.S., I.Y. Fung, and J.C.H. Chiang, 2012: Mid-latitude afforestation shifts general circulation and tropical precipitation. *Proc. Natl. Acad. Sci.*, **109**, 712–716, doi:10.1073/pnas.1116706108.

Swann, A.L.S., F.M. Hoffman, C.D. Koven, and J.T. Randerson, 2016: Plant responses to increasing CO_2 reduce estimates of climate impacts on drought severity. *Proc. Natl. Acad. Sci.*, **113**, 10019–10024, doi:10.1073/pnas.1604581113.

Tabari, H. and P. Willems, 2018: More prolonged droughts by the end of the century in the Middle East. *Environ. Res. Lett.*, **13**, 104005, doi:10.1088/1748-9326/aae09c.

Taheripour, F., and W.E. Tyner, 2013: Induced land use emissions due to first and second generation biofuels and uncertainty in land use emission factors. *Econ. Res. Int.*, **2013**, 315787, doi:10.1155/2013/315787.

Taillardat, P., D.A. Friess, and M. Lupascu, 2018: Mangrove blue carbon strategies for climate change mitigation are most effective at the national scale. *Biol. Lett.*, **14**, 20180251, doi:10.1098/rsbl.2018.0251.

Talhelm, A.F. et al., 2014: Elevated carbon dioxide and ozone alter productivity and ecosystem carbon content in northern temperate forests. *Glob. Chang. Biol.*, **20**, 2492–2504, doi:10.1111/gcb.12564.

Tan, Z.H. et al., 2017: Optimum air temperature for tropical forest photosynthesis: Mechanisms involved and implications for climate warming. *Environ. Res. Lett.*, **12**, doi:10.1088/1748-9326/aa6f97.

Tang, B., X. Zhao, and W. Zhao, 2018: Local effects of forests on temperatures across Europe. *Remote Sens.*, **10**, 1–24, doi:10.3390/rs10040529.

Tang, J., S. Luyssaert, A.D. Richardson, W. Kutsch, and I.A. Janssens, 2014: Steeper declines in forest photosynthesis than respiration explain age-driven decreases in forest growth. *Proc. Natl. Acad. Sci.*, **111**, 8856–8860, doi:10.1073/pnas.1320761111.

Tao, Y. et al., 2015: Variation in ecosystem services across an urbanization gradient: A study of terrestrial carbon stocks from Changzhou, China. *Ecol. Modell.*, **318**, 210–216, doi:10.1016/j.ecolmodel.2015.04.027.

Taraborrelli, D. et al., 2012: Hydroxyl radical buffered by isoprene oxidation over tropical forests. *Nat. Geosci.*, **5**, 190–193, doi:10.1038/ngeo1405.

Tate, K.R., 2015: Soil methane oxidation and land-use change – From process to mitigation. *Soil Biol. Biochem.*, **80**, 260–272, doi:10.1016/j.soilbio.2014.10.010.

Tavoni, M. et al., 2015: Post-2020 climate agreements in the major economies assessed in the light of global models. *Nat. Clim. Chang.*, **5**, 119–126, doi:10.1038/nclimate2475.

Taylor, C.M. et al., 2017: Frequency of extreme Sahelian storms tripled since 1982 in satellite observations. *Nature*, **544**, 475–478, doi:10.1038/nature22069.

Taylor, L.L. et al., 2016: Enhanced weathering strategies for stabilizing climate and averting ocean acidification. *Nat. Clim. Chang.*, **6**, 402–406, doi:10.1038/nclimate2882.

Teng, H., G. Branstator, G.A. Meehl, and W.M. Washington, 2016: Projected intensification of subseasonal temperature variability and heat waves in the Great Plains. *Geophys. Res. Lett.*, **43**, 2165–2173, doi:10.1002/2015GL067574.

Tepley, A.J. et al., 2018: Influences of fire–vegetation feedbacks and post-fire recovery rates on forest landscape vulnerability to altered fire regimes. *J. Ecol.*, **106**, 1925–1940, doi:10.1111/1365-2745.12950.

Ter-Mikaelian, M.T., S.J. Colombo, and J. Chen, 2014: The burning question: Does forest bioenergy reduce carbon emissions? A review of common misconceptions about forest carbon accounting. *J. For.*, **113**, 57–68, doi:10.5849/jof.14-016.

Ter-Mikaelian, M.T. et al., 2015: Carbon debt repayment or carbon sequestration parity? Lessons from a forest bioenergy case study in Ontario, Canada. *GCB Bioenergy*, **7**, 704–716, doi:10.1111/gcbb.12198.

Teramoto, M., N. Liang, M. Takagi, J. Zeng, and J. Grace, 2016: Sustained acceleration of soil carbon decomposition observed in a 6-year warming experiment in a warm-temperate forest in southern Japan. *Sci. Rep.*, **6**, 35563. doi:10.1038/srep35563.

Terrer, C. et al., 2017: Ecosystem responses to elevated CO_2 governed by plant-soil interactions and the cost of nitrogen acquisition. *New Phytol.*, **2**, 507–522, doi:10.1111/nph.14872.

Teskey, R. et al., 2015: Responses of tree species to heat waves and extreme heat events. *Plant Cell Environ.*, **38**, 1699–1712, doi:10.1111/pce.12417.

Teuling, A.J. et al., 2010: Contrasting response of European forest and grassland energy exchange to heatwaves. *Nat. Geosci.*, **3**, 722–727, doi:10.1038/ngeo950.

Thackeray, C.W. and C.G. Fletcher, 2015: Snow albedo feedback: Current knowledge, importance, outstanding issues and future directions. *Prog. Phys. Geogr.*, **40**, 392–408, doi:10.1177/0309133315620999.

Thalman, R. et al., 2017: CCN activity and organic hygroscopicity of aerosols downwind of an urban region in central Amazonia: Seasonal and diel variations and impact of anthropogenic emissions. *Atmos. Chem. Phys.*, **17**, 11779–11801, doi:10.5194/acp-17-11779-2017.

Tharammal, T., G. Bala, D. Narayanappa, and R. Nemani, 2018: Potential roles of CO_2 fertilization, nitrogen deposition, climate change, and land use and land cover change on the global terrestrial carbon uptake in the twenty-first century. *Climate Dynamics*. **52**, 4393–4406, doi:10.1007/s00382-018-4388-8.

Thiery, W. et al., 2017: Present-day irrigation mitigates heat extremes. *J. Geophys. Res.*, **122**, 1403–1422, doi:10.1002/2016JD025740.

Thompson, R.L. et al., 2014: TransCom N_2O model inter-comparison – Part 2: Atmospheric inversion estimates of N_2O emissions. *Atmos. Chem. Phys.*, **14**, 6177–6194, doi:10.5194/acp-14-6177-2014.

Thompson, V. et al., 2017: High risk of unprecedented UK rainfall in the current climate. *Nat. Commun.*, **8**, 107, doi:10.1038/s41467-017-00275-3.

Tian, H. et al., 2015: Global patterns and controls of soil organic carbon dynamics as simulated by multiple terrestrial biosphere models: Current status and future directions. *Global Biogeochem. Cycles*, **29**, 775–792, doi:10.1002/2014GB005021.

Tian, H. et al., 2016: The terrestrial biosphere as a net source of greenhouse gases to the atmosphere. *Nature*, **531**, 225–228, doi:10.1038/nature16946.

Tian, H. et al., 2018: The Global N_2O Model Intercomparison Project. *Bull. Am. Meteorol. Soc.*, **99**, 1231–1251, doi:10.1175/BAMS-D-17-0212.1.

Tietjen, B. et al., 2017: Climate change-induced vegetation shifts lead to more ecological droughts despite projected rainfall increases in many global temperate drylands. *Glob. Chang. Biol.*, **23**, 2743–2754, doi:10.1111/gcb.13598.

Tifafi, M., B. Guenet, and C. Hatté, 2018: Large differences in global and regional total soil carbon stock estimates based on SoilGrids, HWSD, and NCSCD: Intercomparison and evaluation based on field data from USA, England, Wales, and France. *Global Biogeochem. Cycles*, **32**, 42–56, doi:10.1002/2017GB005678.

Tilman, D., and M. Clark, 2014: Global diets link environmental sustainability and human health. *Nature*, **515**, 518–522, doi:10.1038/nature13958.

Tjoelker, M.G., 2018: The role of thermal acclimation of plant respiration under climate warming: Putting the brakes on a runaway train? *Plant Cell Environ.*, **41**, 501–503, doi:10.1111/pce.13126.

Tobarra, M.A., L.A. López, M.A. Cadarso, N. Gómez, and I. Cazcarro, 2018: Is seasonal households' consumption good for the nexus carbon/water footprint? The Spanish fruits and vegetables case. *Environ. Sci. Technol.*, **52**, 12066–12077, doi:10.1021/acs.est.8b00221.

Todd-Brown, K.E.O. et al., 2013: Causes of variation in soil carbon simulations from CMIP5 Earth system models and comparison with observations. *Biogeosciences*, **10**, 1717–1736, doi:10.5194/bg-10-1717-2013.

Tölle, M.H., S. Engler, and H.J. Panitz, 2017: Impact of abrupt land cover changes by tropical deforestation on Southeast Asian climate and agriculture. *J. Clim.*, **30**, 2587–2600, doi:10.1175/JCLI-D-16-0131.1.

Torres-Valcárcel, Á.R., J. Harbor, A.L. Torres-Valcárcel, and C.J. González-Avilés, 2015: Historical differences in temperature between urban and non-urban areas in Puerto Rico. *Int. J. Climatol.*, **35**, 1648–1661, doi:10.1002/joc.4083.

Toyoda, S. et al., 2013: Decadal time series of tropospheric abundance of N_2O isotopomers and isotopologues in the Northern Hemisphere obtained by the long-term observation at Hateruma Island, Japan. *J. Geophys. Res. Atmos.*, **118**, 3369–3381, doi:10.1002/jgrd.50221.

Trenberth, K.E., 2012: Framing the way to relate climate extremes to climate change. *Clim. Change*, **115**, 283–290, doi:10.1007/s10584-012-0441-5.

Trenberth, K.E. et al., 2014: Global warming and changes in drought. *Nat. Clim. Chang.*, **4**, 17–22, doi:10.1038/nclimate2067.

Trenberth, K.E., J.T. Fasullo, and T.G. Shepherd, 2015: Attribution of climate extreme events. *Nat. Clim. Chang.*, **5**, 725–730, doi:10.1038/nclimate2657.

Tribouillois, H., J. Constantin, and E. Justes, 2018: Cover crops mitigate direct greenhouse gases balance but reduce drainage under climate change scenarios in temperate climate with dry summers. *Glob. Chang. Biol.*, **24**, 2513–2529, doi:10.1111/gcb.14091.

Trumbore, S., P. Brando, and H. Hartmann, 2015: Forest health and global change. *Science*, **349**, 814–818, doi:10.1126/science.aac6759.

Tsigaridis, K. et al., 2014: The AeroCom evaluation and intercomparison of organic aerosol in global models. *Atmos. Chem. Phys.*, **14**, 10845–10895, doi:10.5194/acp-14-10845-2014.

Tubiello, F.N. et al., 2013: The FAOSTAT database of greenhouse gas emissions from agriculture. *Environ. Res. Lett.*, **8**, 15009, doi:10.1088/1748-9326/8/1/015009.

Tubiello, F.N. et al., 2014: *Agriculture, Forestry and Other Land Use Emissions by Sources and Removals by Sinks: 1990-2011 Analysis*. FAO Statistics Division Working Paper Series, ESS/14-02, 4–89 pp.

Tubiello, F.N. et al., 2015: The contribution of agriculture, forestry and other land use activities to global warming, 1990–2012. *Glob. Chang. Biol.*, **21**, 2655–2660, doi:10.1111/gcb.12865.

Tuittila, E.-S. et al., 2000: Methane dynamics of a restored cut-away peatland. *Glob. Chang. Biol.*, **6**, 569–581, doi:10.1046/j.1365-2486.2000.00341.x.

Turetsky, M.R. et al., 2014: Global vulnerability of peatlands to fire and carbon loss. *Nat. Geosci.*, **8**, 11. doi:10.1038/ngeo2325.

Turner, A.J., C. Frankenberg, P.O. Wennberg, and D.J. Jacob, 2017: Ambiguity in the causes for decadal trends in atmospheric methane and hydroxyl. *Proc. Natl. Acad. Sci.*, **114**, 5367–5372, doi:10.1073/pnas.1616020114.

Turner, B.L., T. Brenes-Arguedas, and R. Condit, 2018a: Pervasive phosphorus limitation of tree species but not communities in tropical forests. *Nature*, **555**, 367–370, doi:10.1038/nature25789.

Turner, P.A., C.B. Field, D.B. Lobell, D.L. Sanchez, and K.J. Mach, 2018b: Unprecedented rates of land-use transformation in modelled climate change mitigation pathways. *Nat. Sustain.*, **1**, 240–245, doi:10.1038/s41893-018-0063-7.

Tuttle, S. and G. Salvucci, 2016: Atmospheric science: Empirical evidence of contrasting soil moisture-precipitation feedbacks across the United States. *Science*, **352**, 825–828, doi:10.1126/science.aaa7185.

Tyukavina, A. et al., 2015: Aboveground carbon loss in natural and managed tropical forests from 2000 to 2012. *Environ. Res. Lett.*, **10**, 74002–74002, doi:10.1088/1748-9326/10/7/074002.

Tyukavina, A. et al., 2017: Types and rates of forest disturbance in Brazilian Legal Amazon, 2000–2013. *Sci. Adv.*, **3**, e1601047, doi:10.1126/sciadv.1601047.

Ukkola, A.M., A.J. Pitman, M.G. Donat, M.G. De Kauwe, and O. Angélil, 2018: Evaluating the contribution of land-atmosphere coupling to heat extremes in CMIP5 Models. *Geophys. Res. Lett.*, **45**, 9003–9012, doi:10.1029/2018GL079102.

Ummenhofer, C.C. and G.A. Meehl, 2017: Extreme weather and climate events with ecological relevance: A review. *Philos. Trans. R. Soc. B Biol. Sci.*, **372**, 20160135, doi:10.1098/rstb.2016.0135.

UNEP, 2013: Drawing Down N_2O to Protect Climate and the Ozone Layer: A UNEP Synthesis Report.. United Nations Environment Programme (UNEP), Nairobi, Kenya, 76 pp, www.unep.org/publications/ebooks/UNEPN2Oreport/.

UNEP, 2017: The Emissions Gap Report 2017: A UN Environment Synthesis Report. United Nations Environment Programme (UNEP), Nairobi, Kenya, 116 pp.

Unger, N., 2013: Isoprene emission variability through the twentieth century. *J. Geophys. Res. Atmos.*, **118**, 13, 606–613, doi:10.1002/2013JD020978.

Unger, N., 2014a: On the role of plant volatiles in anthropogenic global climate change. *Geophys. Res. Lett.*, **41**, 8563–8569, doi:10.1002/2014GL061616.

Unger, N., 2014b: Human land-use-driven reduction of forest volatiles cools global climate. *Nat. Clim. Chang.*, **4**, 907–910, doi:10.1038/nclimate2347.

Unger, N., X. Yue, and K.L. Harper, 2017: Aerosol climate change effects on land ecosystem services. *Faraday Discuss.*, **200**, 121–142, doi:10.1039/c7fd00033b.

United Nations, Department of Economic and Social Affairs, Population Division, 2017: *World Population Prospects: The 2017 Revision, Key Findings and Advance Tables*. Working paper ESA/P/WP/248, 53 pp.

Urban, M.C., 2015: Accelerating extinction risk from climate change. *Science*, **348**, 571–573, doi:10.1126/science.aaa4984.

Urbanski, S.P., 2014: Wildland fire emissions, carbon, and climate: Emission factors. *For. Ecol. Manage.*, **317**, 51–60, doi:10.1016/j.foreco.2013.05.045.

USEPA, 2012: *Global Anthropogenic Non-CO_2 Greenhouse Gas Emissions: 1990–2030*. Office of Atmospheric Programs, Climate Change Division, U.S. Environmental Protection Agency, Washington, DC, USA, 176 pp.

USEPA, 2013: *Global Mitigation of Non-CO_2 Greenhouse Gases: 2010–2030*. EPA-430-R-13-011, United States Environmental Protection Agency Office of Atmospheric Programs (6207J), Washington, DC, USA, 410 pp.

Valade, A., C. Magand, S. Luyssaert, and V. Bellassen, 2017: Sustaining the sequestration efficiency of the European forest sector. *For. Ecol. Manage.*, **405**, 44–55, doi:10.1016/j.foreco.2017.09.009.

Valade, A. et al., 2018: Carbon costs and benefits of France's biomass energy production targets. *Carbon Balance Manag.*, **13**, 26, doi:10.1186/s13021-018-0113-5.

Valin, H., et al., 2015: *The Land Use Change Impact of Biofuels Consumed in the EU: Quantification of Area and Greenhouse Gas Impacts*. ECOFYS Netherlands B.V., Utrecht, Netherlands, 261 pp.

Vanderwel, M.C. et al., 2015: Global convergence in leaf respiration from estimates of thermal acclimation across time and space. *New Phytol.*, **207**, 1026–1037, doi:10.1111/nph.13417.

Vanselow-Algan, M. et al., 2015: High methane emissions dominated annual greenhouse gas balances 30 years after bog rewetting. *Biogeosciences*, **12**, 4361–4371, doi:10.5194/bg-12-4361-2015.

Vaughan, N.E. et al., 2018: Evaluating the use of biomass energy with carbon capture and storage in low emission scenarios. *Environ. Res. Lett.*, **13**, 44014, doi:10.1088/1748-9326/aaaa02.

Veraverbeke, S. et al., 2012: Assessing post-fire vegetation recovery using red-near infrared vegetation indices: Accounting for background and vegetation variability. *ISPRS J. Photogramm. Remote Sens.*, **68**, 28–39, doi:10.1016/j.isprsjprs.2011.12.007.

Veraverbeke, S. et al., 2017: Lightning as a major driver of recent large fire years in North American boreal forests. *Nat. Clim. Chang.*, **7**, 529, doi:10.1038/nclimate3329.

Verbyla, D., 2008: The greening and browning of Alaska based on 1982–2003 satellite data. *Glob. Ecol. Biogeogr.*, **17**, 547–555, doi:10.1111/j.1466-8238.2008.00396.x.

Verheijen, F.G.A. et al., 2013: Reductions in soil surface albedo as a function of biochar application rate: Implications for global radiative forcing. *Environ. Res. Lett.*, **8**, 044008, doi:10.1088/1748-9326/8/4/044008.

Vermeulen, S.J., B.M. Campbell, and J.S.I. Ingram, 2012a: Climate change and food systems. *Annu. Rev. Environ. Resour.*, **37**, 195–222, doi:10.1146/annurev-environ-020411-130608.

Veselovskii, I. et al., 2016: Retrieval of optical and physical properties of African dust from multiwavelength Raman lidar measurements during the SHADOW campaign in Senegal. *Atmos. Chem. Phys.*, **16**, 7013–7028, doi:10.5194/acp-16-7013-2016.

Vicente-Serrano, S.M. et al., 2014: Evidence of increasing drought severity caused by temperature rise in southern Europe. *Environ. Res. Lett.*, **9**, 44001, doi:10.1088/1748-9326/9/4/044001.

Vogel, E. and R. Meyer, 2018: Climate change, climate extremes, and global food production—adaptation in the agricultural sector. *Resilience*, 31–49, doi:10.1016/B978-0-12-811891-7.00003-7.

Vogel, E. et al., 2019: The effects of climate extremes on global agricultural yields. *Environ. Res. Lett.*, doi:10.1088/1748-9326/ab154b.

Vogel, M.M. et al., 2017: Regional amplification of projected changes in extreme temperatures strongly controlled by soil moisture-temperature feedbacks. *Geophys. Res. Lett.*, **44**, 1511–1519, doi:10.1002/2016GL071235.

Voldoire, A., 2006: Quantifying the impact of future land-use changes against increases in GHG concentrations. *Geophys. Res. Lett.*, **33**, 2–5, doi:10.1029/2005GL024354.

De Vrese, P., S. Hagemann, and M. Claussen, 2016: Asian irrigation, African rain: Remote impacts of irrigation. *Geophys. Res. Lett.*, **43**, 3737–3745, doi:10.1002/2016GL068146.

van Vuuren, D.P. et al., 2018: Alternative pathways to the 1.5°C target reduce the need for negative emission technologies. *Nat. Clim. Chang.*, **8**, 1–7, doi:10.1038/s41558-018-0119-8.

Wäldchen, J., E.-D. Schulze, I. Schöning, M. Schrumpf, and C. Sierra, 2013: The influence of changes in forest management over the past 200 years on present soil organic carbon stocks. *For. Ecol. Manage.*, **289**, 243–254, doi:10.1016/j.foreco.2012.10.014.

Wallace, C.J. and M. Joshi, 2018: Comparison of land-ocean warming ratios in updated observed records and CMIP5 climate models. *Environ. Res. Lett.*, **13**, doi:10.1088/1748-9326/aae46f.

Wang, B., and Q. Ding, 2008: Global monsoon: Dominant mode of annual variation in the tropics. *Dyn. Atmos. Ocean.*, **44**, 165–183, doi:10.1016/j.dynatmoce.2007.05.002.

Wang, G.Q. et al., 2013: Simulating the impact of climate change on runoff in a typical river catchment of the Loess Plateau, China. *J. Hydrometeorol.*, **14**, 1553–1561, doi:10.1175/JHM-D-12-081.1.

Wang, J. et al., 2016a: Amazon boundary layer aerosol concentration sustained by vertical transport during rainfall. *Nature*, **539**, 416–419, doi:10.1038/nature19819.

Wang, J., B. Huang, D. Fu, P.M. Atkinson, and X. Zhang, 2016b: Response of urban heat island to future urban expansion over the Beijing–Tianjin–Hebei metropolitan area. *Appl. Geogr.*, **70**, 26–36, doi:10.1016/j.apgeog.2016.02.010.

Wang, J., Z. Yan, X.W. Quan, and J. Feng, 2017a: Urban warming in the 2013 summer heat wave in eastern China. *Clim. Dyn.*, **48**, 3015–3033, doi:10.1007/s00382-016-3248-7.

Wang, M. et al., 2017b: On the long-term hydroclimatic sustainability of perennial bioenergy crop expansion over the United States. *J. Clim.*, **30**, 2535–2557, doi:10.1175/jcli-d-16-0610.1.

Wang, R. et al., 2014a: Trend in global black carbon emissions from 1960 to 2007. *Environ. Sci. Technol.*, **48**, 6780–6787, doi:10.1021/es5021422.

Wang, R. et al., 2017c: Global forest carbon uptake due to nitrogen and phosphorus deposition from 1850 to 2100. *Glob. Chang. Biol.*, **23**, 4854–4872, doi:10.1111/gcb.13766.

Wang, W., A.T. Evan, C. Flamant, and C. Lavaysse, 2015: On the decadal scale correlation between African dust and Sahel rainfall: The role of Saharan heat low-forced winds. *Sci. Adv.*, **1**, p.e1500646, doi:10.1126/sciadv.1500646.

Wang, W. et al., 2016c: Evaluation of air-soil temperature relationships simulated by land surface models during winter across the permafrost region. *Cryosphere*, **10**, 1721–1737, doi:10.5194/tc-10-1721-2016.

Wang, X. et al., 2016d: Deriving brown carbon from multiwavelength absorption measurements: Method and application to AERONET and Aethalometer observations. *Atmos. Chem. Phys.*, **16**, 12733–12752, doi:10.5194/acp-16-12733-2016.

Wang, Y., X. Yan, and Z. Wang, 2014b: The biogeophysical effects of extreme afforestation in modeling future climate. *Theor. Appl. Climatol.*, **118**, 511–521, doi:10.1007/s00704-013-1085-8.

Wang, Z. et al., 2017d: Human-induced erosion has offset one-third of carbon emissions from land cover change. *Nat. Clim. Chang.*, **7**, 345–349, doi:10.1038/nclimate3263.

Wang, Z. et al., 2017e: Does drought in China show a significant decreasing trend from 1961 to 2009? *Sci. Total Environ.*, **579**, 314–324, doi:10.1016/j.scitotenv.2016.11.098.

Ward, D.S., N.M. Mahowald, and S. Kloster, 2014: Potential climate forcing of land use and land cover change. *Atmos. Chem. Phys.*, **14**, 12701–12724, doi:10.5194/acp-14-12701-2014.

Ward, K., S. Lauf, B. Kleinschmit, and W. Endlicher, 2016: Heat waves and urban heat islands in Europe: A review of relevant drivers. *Sci. Total Environ.*, **569–570**, 527–539, doi:10.1016/j.scitotenv.2016.06.119.

Warszawski, L. et al., 2013: A multi-model analysis of risk of ecosystem shifts under climate change. *Environ. Res. Lett.*, **8**, 44018, doi:10.1088/1748-9326/8/4/044018.

Watts, N., W.N. Adger, and P. Agnolucci, 2015: Health and climate change: Policy responses to protect public health. *Environnement, Risques et Sante*, **14**, 466–468, doi:10.1016/S0140-6736(15)60854-6.

Wehner, M.F., 2013: Very extreme seasonal precipitation in the NARCCAP ensemble: Model performance and projections. *Clim. Dyn.*, **40**, 59–80, doi:10.1007/s00382-012-1393-1.

Wells, K.C. et al., 2018: Top-down constraints on global N_2O emissions at optimal resolution: Application of a new dimension reduction technique. *Atmos. Chem. Phys.*, **18**, 735–756, doi:10.5194/acp-18-735-2018.

Wendt, J.W. and S. Hauser, 2013: An equivalent soil mass procedure for monitoring soil organic carbon in multiple soil layers. *Eur. J. Soil Sci.*, **64**, 58–65, doi:10.1111/ejss.12002.

Weng, Z. et al., 2017: Biochar built soil carbon over a decade by stabilizing rhizodeposits. *Nat. Clim. Chang.*, **7**, 371–376, doi:10.1038/nclimate3276.

Wenz, L., A. Levermann and M. Auffhammer, 2017: North–south polarization of European electricity consumption under future warming. *Proc. Natl. Acad. Sci.*, **114**, E7910–E7918, doi:10.1073/pnas.1704339114.

van der Werf, G.R. et al., 2017: Global fire emissions estimates during 1997–2015. *Earth Syst. Sci. Data Discuss.*, 1–43, doi:10.5194/essd-2016-62.

Van Der Werf, G.R. et al., 2010: Global fire emissions and the contribution of deforestation, savanna, forest, agricultural, and peat fires (1997–2009). *Atmos. Chem. Phys.*, **10**, 11707–11735, doi:10.5194/acp-10-11707-2010.

Werth, D., 2002: The local and global effects of Amazon deforestation. *J. Geophys. Res.*, **107**, 8087, doi:10.1029/2001jd000717.

Werth, D., 2005: The local and global effects of African deforestation. *Geophys. Res. Lett.*, **32**, 1–4, doi:10.1029/2005GL002969.

Werth, D. and R. Avissar, 2005: The local and global effects of Southeast Asian deforestation. *Geophys. Res. Lett.*, **32**, 1–4, doi:10.1029/2005GL022970.

West, P.C., G.T. Narisma, C.C. Barford, C.J. Kucharik, and J.A. Foley, 2011: An alternative approach for quantifying climate regulation by ecosystems. *Front. Ecol. Environ.*, **9**, 126–133, doi:10.1890/090015.

West, T.O. et al., 2004: Carbon management response curves: Estimates of temporal soil carbon dynamics. *Environ. Manage.*, **33**, 507–518, doi:10.1007/s00267-003-9108-3.

Westervelt, D.M., L.W. Horowitz, V. Naik, J.C. Golaz, and D.L. Mauzerall, 2015: Radiative forcing and climate response to projected 21st century aerosol decreases. *Atmos. Chem. Phys.*, **15**, 12681–12703, doi:10.5194/acp-15-12681-2015.

Westra, S., L.V. Alexander, and F.W. Zwiers, 2013: Global increasing trends in annual maximum daily precipitation. *J. Clim.*, **26**, 3904–3918, doi:10.1175/JCLI-D-12-00502.1.

Wheeler, T. and J. Von Braun, 2013: Climate change impacts on global food security. *Science*, **341**, 508–513, doi:10.1126/science.1239402.

Whitehead, J.D. et al., 2016: Biogenic cloud nuclei in the central Amazon during the transition from wet to dry season. *Atmos. Chem. Phys.*, **16**, 9727–9743, doi:10.5194/acp-16-9727-2016.

Wicke, B., P. Verweij, H. van Meijl, D.P. van Vuuren, and A.P.C. Faaij, 2012: Indirect land use change: Review of existing models and strategies for mitigation. *Biofuels*, **3**, 87–100, doi:10.4155/bfs.11.154.

Wickham, J., T.G. Wade, and K.H. Riitters, 2014: An isoline separating relatively warm from relatively cool wintertime forest surface temperatures for the southeastern United States. *Glob. Planet. Change*, **120**, 46–53, doi:10.1016/j.gloplacha.2014.05.012.

Wickham, J.D., T.G. Wade, and K.H. Riitters, 2013: Empirical analysis of the influence of forest extent on annual and seasonal surface temperatures for the continental United States. *Glob. Ecol. Biogeogr.*, **22**, 620–629, doi:10.1111/geb.12013.

Wieder, W.R., G.B. Bonan, and S.D. Allison, 2013: Global Soil Carbon Projections are Improved by Modeling Microbial Processes. *Nat. Clim. Chang.*, **3**, 909. doi:/10.1038/nclimate1951.

Wieder, W.R., C.C. Cleveland, W.K. Smith, and K. Todd-Brown, 2015: Future productivity and carbon storage limited by terrestrial nutrient availability. *Nat. Geosci.*, **8**, 441–444, doi:10.1038/NGEO2413.

Wieder, W.R. et al., 2018: Carbon cycle confidence and uncertainty: Exploring variation among soil biogeochemical models. *Glob. Chang. Biol.*, **24**, 1563–1579, doi:10.1111/gcb.13979.

Wienert, U., and W. Kuttler, 2005: The dependence of the urban heat island intensity on latitude – A statistical approach. *Meteorol. Zeitschrift*, **14**, 677–686, doi:10.1127/0941-2948/2005/0069.

Wilcox, L.J. et al., 2018: Multiple perspectives on the attribution of the extreme European summer of 2012 to climate change. *Clim. Dyn.*, **50**, 3537–3555, doi:10.1007/s00382-017-3822-7.

Wilhelm, J.A. and R.G. Smith, 2018: Ecosystem services and land sparing potential of urban and peri-urban agriculture: A review. *Renew. Agric. Food Syst.*, **33**, 481–494, doi:10.1017/S1742170517000205.

Wilhelm, M., E.L. Davin, and S.I. Seneviratne, 2015: Climate engineering of vegetated land for hot extremes mitigation: An Earth system model sensitivity study. *J. Geophy. Res. Atmos.*, **120**, 2612–2623, doi:10.1002/2014JD022293.

Williams, A.P. et al., 2015: Contribution of anthropogenic warming to California drought during 2012-2014. *Geophys. Res. Lett.*, **42**, 6819–6828, doi:10.1002/2015GL064924.

Williams, J. and P.J. Crutzen, 2010: Nitrous oxide from aquaculture. *Nat. Geosci.*, **3**, 143, doi:10.1038/ngeo804.

Williams, J.W. and S.T. Jackson, 2007: Novel climates, no-analog communities, and ecological surprises. *Front. Ecol. Environ.*, **5**, 475–482, doi:10.1890/070037.

Williams, J.W., S.T. Jackson, and J.E. Kutzbach, 2007: Projected distributions of novel and disappearing climates by 2100 AD. *Proc. Natl. Acad. Sci.*, **104**, 5738–5742, doi:10.1073/pnas.0606292104.

Wilson, D. et al., 2016: Multiyear greenhouse gas balances at a rewetted temperate peatland. *Glob. Chang. Biol.*, **22**, 4080–4095, doi:10.1111/gcb.13325.

Winsemius, H.C. et al., 2016: Global drivers of future river flood risk. *Nat. Clim. Chang.*, **6**, 381–385, doi:10.1038/nclimate2893.

Wobus, C. et al., 2017: Climate change impacts on flood risk and asset damages within mapped 100-year floodplains of the contiguous United States. *Nat. Hazards Earth Syst. Sci.*, **17**, 2199–2211, doi:10.5194/nhess-17-2199-2017.

Wolf, A., W.R.L. Anderegg, and S.W. Pacala, 2016a: Optimal stomatal behavior with competition for water and risk of hydraulic impairment. *Proc. Natl. Acad. Sci.*, **113**, E7222–E7230, doi:10.1073/pnas.1615144113.

Wolf, J., G.R. Asrar, and T.O. West, 2017: Revised methane emissions factors and spatially distributed annual carbon fluxes for global livestock. *Carbon Balance Manag.*, **12**, 16, doi:10.1186/s13021-017-0084-y.

Wolf, S. et al., 2016b: Warm spring reduced carbon cycle impact of the 2012 US summer drought. *Proc. Natl. Acad. Sci.*, **113**, 5880–5885, doi:10.1073/pnas.1519620113.

Woods, J., L. M, B. M, K.L. Kline, and A. Faaij, 2015: Land and bioenergy. In: *Bioenergy & Sustainability: Bridging the Gaps*, [Souza, G.M., R.L. Victoria, C.A. Joly, L.M. Verdade(eds.)], 258–295pp, São Paulo.

Woolf, D., J.E. Amonette, F.A. Street-Perrott, J. Lehmann, and S. Joseph, 2010: Sustainable biochar to mitigate global climate change. *Nat. Commun.*, **1**, doi:10.1038/ncomms1053.

Worden, J.R., et al., 2017: Reduced biomass burning emissions reconcile conflicting estimates of the post-2006 atmospheric methane budget. *Nat. Commun.*, **8**, 1–12, doi:10.1038/s41467-017-02246-0.

Wright, S.J. et al., 2018: Plant responses to fertilization experiments in lowland, species-rich, tropical forests. *Ecology*, **99**, 1129–1138, doi:10.1002/ecy.2193.

Wu, M. et al., 2016: Vegetation-climate feedbacks modulate rainfall patterns in Africa under future climate change. *Earth Syst. Dyn.*, **7**, 627–647, doi:10.5194/esd-7-627-2016.

Wu, M. et al., 2017: Impacts of land use on climate and ecosystem productivity over the Amazon and the South American continent. *Environ. Res. Lett.*, **12**, 054016, doi:10.1088/1748-9326/aa6fd6.

Wu, P., N. Christidis, and P. Stott, 2013: Anthropogenic impact on Earth's hydrological cycle. *Nat. Clim. Chang.*, **3**, 807–810, doi:10.1038/nclimate1932.

Wu, W. et al., 2019: Global advanced bioenergy potential under environmental protection policies and societal transformation measures. *GCB Bioenergy*, doi:10.1111/gcbb.12614.

Wuebbles, D. and K. Hayhoe, 2002: Atmospheric methane and global change. *Earth-Science Rev.*, **57**, 177–210, doi:10.1016/S0012-8252(01)00062-9.

Wypych, A., Z. Ustrnul, A. Sulikowska, F.M. Chmielewski, and B. Bochenek, 2017: Spatial and temporal variability of the frost-free season in Central Europe and its circulation background. *Int. J. Climatol.*, **37**, 3340–3352, doi:10.1002/joc.4920.

Xu, C., H. Liu, A.P. Williams, Y. Yin, and X. Wu, 2016a: Trends toward an earlier peak of the growing season in Northern Hemisphere mid-latitudes. *Glob. Chang. Biol.*, **22**, 2852–2860, doi:10.1111/gcb.13224.

Xu, H.-jie, and X.-P. Wang, 2016: Effects of altered precipitation regimes on plant productivity in the arid region of northern China. *Ecol. Inform.*, **31**, 137–146, doi:10.1016/j.ecoinf.2015.12.003.

Xu, L. et al., 2011: Widespread decline in greenness of Amazonian vegetation due to the 2010 drought. *Geophys. Res. Lett.*, **38**, L07402, doi:10.1029/2011GL046824.

Xu, Q., R. Yang, Y.-X.X. Dong, Y.-X.X. Liu, and L.-R.R. Qiu, 2016b: The influence of rapid urbanization and land use changes on terrestrial carbon sources/sinks in Guangzhou, China. *Ecol. Indic.*, **70**, 304–316, doi:10.1016/j.ecolind.2016.05.052.

Xu, X. et al., 2016c: Four decades of modeling methane cycling in terrestrial ecosystems. *Biogeosciences*, **13**, 3735–3755, doi:10.5194/bg-13-3735-2016.

Xu, X., W.J. Riley, C.D. Koven, and G. Jia, 2018: Observed and simulated sensitivities of spring greenup to preseason climate in northern temperate and boreal regions. *J. Geophys. Res. Biogeosciences*, **123**, 60–78, doi:10.1002/2017JG004117.

Xue, Y. et al., 2016: West African monsoon decadal variability and surface-related forcings: Second West African Monsoon Modeling and Evaluation Project Experiment (WAMME II). *Clim. Dyn.*, **47**, 3517–3545, doi:10.1007/s00382-016-3224-2.

Yamori, W., K. Hikosaka, and D.A. Way, 2014: Temperature response of photosynthesis in C3, C4, and CAM plants: Temperature acclimation and temperature adaptation. *Photosynth. Res.*, **119**, 101–117, doi:10.1007/s11120-013-9874-6.

Yan, M., J. Liu, and Z. Wang, 2017: Global climate responses to land use and land cover changes over the past two millennia. *Atmosphere*, **8**, 64, doi:10.3390/atmos8040064.

Yan, Z. et al., 2018: A moisture function of soil heterotrophic respiration that incorporates microscale processes. *Nat. Commun.*, **9**, 2562, doi:10.1038/s41467-018-04971-6.

Yang, H., Q. Li, K. Wang, Y. Sun, and D. Sun, 2015a: Decomposing the meridional heat transport in the climate system. *Clim. Dyn.*, **44**, 2751–2768, doi:10.1007/s00382-014-2380-5.

Yang, L., G. Sun, L. Zhi, and J. Zhao, 2018: Negative soil moisture-precipitation feedback in dry and wet regions. *Sci. Rep.*, **8**, 4026, doi:10.1038/s41598-018-22394-7.

Yang, X. et al., 2015b: Potential benefits of climate change for crop productivity in China. *Agric. For. Meteorol.*, **208**, 76–84, doi:10.1016/j.agrformet.2015.04.024.

Yao, Y., S. Piao, and T. Wang, 2018: Future biomass carbon sequestration capacity of Chinese forests. *Sci. Bull.*, **63**, 1108–1117, doi:10.1016/j.scib.2018.07.015.

Ye, J.S. et al., 2018: Which temperature and precipitation extremes best explain the variation of warm versus cold years and wet versus dry years? *J. Clim.*, **31**, 45–59, doi:10.1175/JCLI-D-17-0377.1.

Yellen, B. et al., 2014: Source, conveyance and fate of suspended sediments following Hurricane Irene. New England, USA. *Geomorphology*, **226**, 124–134, doi:10.1016/j.geomorph.2014.07.028.

Yim, S.Y., B. Wang, J. Liu, and Z. Wu, 2014: A comparison of regional monsoon variability using monsoon indices. *Clim. Dyn.*, **43**, 1423–1437, doi:10.1007/s00382-013-1956-9.

Yin, Y., Q. Tang, L. Wang, and X. Liu, 2016: Risk and contributing factors of ecosystem shifts over naturally vegetated land under climate change in China. *Sci. Rep.*, **6**, 1–11, doi:10.1038/srep20905.

Yin, Y., D. Ma, and S. Wu, 2019: Enlargement of the semi-arid region in China from 1961 to 2010. *Clim. Dyn.*, **52**, 509–521, doi:10.1007/s00382-018-4139-x.

York, R., 2012: Do alternative energy sources displace fossil fuels? *Nat. Clim. Chang.*, doi:10.1038/nclimate1451.

Yu, L., Y. Huang, W. Zhang, T. Li, and W. Sun, 2017: Methane uptake in global forest and grassland soils from 1981 to 2010. *Sci. Total Environ.*, **607–608**, 1163–1172, doi:10.1016/j.scitotenv.2017.07.082.

Yu, M., G. Wang, and H. Chen, 2016: Quantifying the impacts of land surface schemes and dynamic vegetation on the model dependency of projected changes in surface energy and water budgets. *J. Adv. Model. Earth Syst.*, **8**, 370–386, doi:10.1002/2015MS000492.

Yuan, X., L. Wang, and E. Wood, 2018: Anthropogenic intensification of southern African flash droughts as exemplified by the 2015/16 season. *Bull. Am. Meteorol. Soc.*, **99**, S86–S90, doi:10.1175/BAMS-D-17-0077.1.

Yue, X. and N. Unger, 2018: Fire air pollution reduces global terrestrial productivity. *Nat. Commun.*, **9**, 5413, doi:10.1038/s41467-018-07921-4. https://doi.org/10.1038/s41467-018-07921-4.

Zaitchik, B.F., A.K. Macalady, L.R. Bonneau, and R.B. Smith, 2006: Europe's 2003 heat wave: A satellite view of impacts and land–atmosphere feedbacks. *Int. J. Climatol.*, **26**, 743–769, doi:10.1002/joc.1280.

Zak, D.R., K.S. Pregitzer, M.E. Kubiske, and A.J. Burton, 2011: Forest productivity under elevated CO_2 and O_3: Positive feedbacks to soil N cycling sustain decade-long net primary productivity enhancement by CO_2. *Ecol. Lett.*, **14**, 1220–1226, doi:10.1111/j.1461-0248.2011.01692.x.

Zampieri, M., A. Ceglar, F. Dentener, and A. Toreti, 2017: Wheat yield loss attributable to heat waves, drought and water excess at the global, national and subnational scales. *Environ. Res. Lett.*, **12**, 64008, doi:10.1088/1748-9326/aa723b.

Zarin, D.J. et al., 2016: Can carbon emissions from tropical deforestation drop by 50% in 5 years? *Glob. Chang. Biol.*, **22**, 1336–1347, doi:10.1111/gcb.13153.

Zeleňáková, M., D.C. Diaconu, and K. Haarstad, 2017: Urban water retention measures. *Procedia Eng.*, **190**, 419–426, doi:10.1016/j.proeng.2017.05.358.

Zemp, D.C. et al., 2014: On the importance of cascading moisture recycling in South America. 13337–13359, doi:10.5194/acp-14-13337-2014.

Zeng, Z. et al., 2017: Climate mitigation from vegetation biophysical feedbacks during the past three decades. *Nat. Clim. Chang.*, **7**, 432–436, doi:10.1038/nclimate3299.

Zeppel, M.J.B., J.V. Wilks, and J.D. Lewis, 2014: Impacts of extreme precipitation and seasonal changes in precipitation on plants. *Biogeosciences*, **11**, 3083–3093, doi:10.5194/bg-11-3083-2014.

Zhang, D. et al., 2015: Effects of forest type and urbanization on carbon storage of urban forests in Changchun, Northeast China. *Chinese Geogr. Sci.*, **25**, 147–158, doi:10.1007/s11769-015-0743-4.

Zhang, G. et al., 2017: Extensive and drastically different alpine lake changes on Asia's high plateaus during the past four decades. *Geophys. Res. Lett.*, **44**, 252–260, doi:10.1002/2016GL072033.

Zhang, H.L., R. Lal, X. Zhao, J.F. Xue, and F. Chen, 2014a: Opportunities and challenges of soil carbon sequestration by conservation agriculture in China. *Advances in Agronomy*, **124**, 1–36, doi:10.1016/B978-0-12-800138-7.00001-2.

Zhang, M. et al., 2014b: Response of surface air temperature to small-scale land clearing across latitudes. *Environ. Res. Lett.*, **9**, 034002, doi:10.1088/1748-9326/9/3/034002.

Zhang, Q., A.J. Pitman, Y.P. Wang, Y.J. Dai, and P.J. Lawrence, 2013a: The impact of nitrogen and phosphorous limitation on the estimated terrestrial carbon balance and warming of land use change over the last 156 yr. *Earth Syst. Dyn.*, **4**, 333–345, doi:10.5194/esd-4-333-2013.

Zhang, W.-f. et al., 2013b: New technologies reduce greenhouse gas emissions from nitrogenous fertilizer in China. *Proc. Natl. Acad. Sci.*, **110**, 8375–8380, doi:10.1073/pnas.1210447110.

Zhang, W.-f. et al., 2013c: New technologies reduce greenhouse gas emissions from nitrogenous fertilizer in China. *Proc. Natl. Acad. Sci.*, **110**, 8375–8380, doi:10.1073/pnas.1210447110.

Zhang, Y. et al., 2019: increased global land carbon sink due to aerosol-induced cooling. *Global Biogeochem. Cycles*, **33**, 439–457, doi:10.1029/2018GB006051.

Zhao, C. et al., 2017: Temperature increase reduces global yields of major crops in four independent estimates. *Proc. Natl. Acad. Sci.*, **114**(35), 9326–9331, doi:10.1073/pnas.1701762114.

Zhao, G. et al., 2015: Sustainable limits to crop residue harvest for bioenergy: Maintaining soil carbon in Australia's agricultural lands. *GCB Bioenergy*, 7, 479–487, doi:10.1111/gcbb.12145.

Zhao, K. and R.B. Jackson, 2014: Biophysical forcings of land-use changes from potential forestry activities in North America. *Ecol. Monogr.*, **84**, 329–353, doi:10.1890/12-1705.1.

Zhao, L., X. Lee, R.B. Smith, and K. Oleson, 2014: Strong contributions of local background climate to urban heat islands. *Nature*, **511**, 216–219, doi:10.1038/nature13462.

Zhao, L., A. Dai, and B. Dong, 2018: Changes in global vegetation activity and its driving factors during 1982–2013. *Agric. For. Meteorol.*, **249**, 198–209, doi:10.1016/j.agrformet.2017.11.013.

Zhao, M. and A.J. Pitman, 2002: The impact of land cover change and increasing carbon dioxide on the extreme and frequency of maximum temperature and convective precipitation. *Geophys. Res. Lett.*, **29**, 1078, doi:10.1029/2001gl013476.

Zhong, S. et al., 2017: Urbanization-induced urban heat island and aerosol effects on climate extremes in the Yangtze River Delta region of China. *Atmos. Chem. Phys.*, **17**, 5439–5457, doi:10.5194/acp-17-5439-2017.

Zhou, B., D. Rybski, and J.P. Kropp, 2017: The role of city size and urban form in the surface urban heat island. *Sci. Rep.*, **7**, 4791, doi:10.1038/s41598-017-04242-2.

Zhou, D. et al., 2016: Spatiotemporal trends of urban heat island effect along the urban development intensity gradient in China. *Sci. Total Environ.*, **544**, 617–626, doi:10.1016/j.scitotenv.2015.11.168.

Zhou, J. et al., 2011: Microbial mediation of carbon-cycling feedbacks to climate warming. *Nat. clinate Chang.*, **2**, 106–110. doi:10.1038/nclimate1331.

Zhou, L. et al., 2014: Widespread decline of Congo rainforest greenness in the past decade. *Nature*, **508**, 86–90, doi:10.1038/nature13265.

Zhou, Y. et al., 2015: A global map of urban extent from nightlights. *Environ. Res. Lett.*, **10**, 2000–2010, doi:10.1088/1748-9326/10/5/054011.

Zhu, J., M. Zhang, Y. Zhang, X. Zeng, and X. Xiao, 2018: Response of tropical terrestrial gross primary production to the super El Niño event in 2015. *J. Geophys. Res. Biogeosciences*, **123**, 3193–3203, doi:10.1029/2018JG004571.

Zhu, P. et al., 2017: Elevated atmospheric CO_2 negatively impacts photosynthesis through radiative forcing and physiology-mediated climate feedback. *Geophys. Res. Lett.*, **44**, 1956–1963, doi:10.1002/2016GL071733.

Zhu, Z. et al., 2016: Greening of the Earth and its drivers. *Nat. Clim. Chang.*, **6**, 791–795, doi:10.1038/nclimate3004.

Zhuang, Q. et al., 2006: CO_2 and CH_4 exchanges between land ecosystems and the atmosphere in northern high latitudes over the 21st century. *Geophys. Res. Lett.*, **33**, 2–6, doi:10.1029/2006GL026972.

Zhuang, Q et al., 2013: Response of global soil consumption of atmospheric methane to changes in atmospheric climate and nitrogen deposition. *Global Biogeochem. Cycles*, **27**, 650–663, doi:10.1002/gbc.20057.

Zilberman, D., 2017: Indirect land use change: much ado about (almost) nothing. *GCB Bioenergy*, **9**, 485–488, doi:10.1111/gcbb.12368.

Zilli, M.T., L.M.V. Carvalho, B. Liebmann, and M.A. Silva Dias, 2017: A comprehensive analysis of trends in extreme precipitation over southeastern coast of Brazil. *Int. J. Climatol.*, **37**, 2269–2279, doi:10.1002/joc.4840.

Zipper, S.C., J. Schatz, C.J. Kucharik, and S.P. Loheide, 2017: Urban heat island-induced increases in evapotranspirative demand. *Geophys. Res. Lett.*, **44**, 873–881, doi:10.1002/2016GL072190.

Zölch, T., J. Maderspacher, C. Wamsler, and S. Pauleit, 2016: Using green infrastructure for urban climate-proofing: An evaluation of heat mitigation measures at the micro-scale. *Urban For. Urban Green.*, **20**, 305–316, doi:10.1016/j.ufug.2016.09.011.

Zomer, R.J. et al., 2016: Global tree cover and biomass carbon on agricultural land: The contribution of agroforestry to global and national carbon budgets. *Sci. Rep.*, **6**, 29987. doi:10.1038/srep29987.

Zomer, R.J., D.A. Bossio, R. Sommer, and L.V. Verchot, 2017: Global sequestration potential of increased organic carbon in cropland soils. *Sci. Rep.*, doi:10.1038/s41598-017-15794-8.

Appendix

This appendix provides all numbers that support Figures 2.14 and 2.17 located in Section 2.5. It lists all model-based studies, with their references, that have been used to create the figures. Studies that examine the effects of historical and future scenarios of changes in anthropogenic land cover are presented in Table A2.1. The responses to idealised latitudinal deforestation and forestation can be found in Table A2.2.

The biophysical effects of changes in anthropogenic land cover reflect the impacts of changes in physical land surface characteristics such as albedo, evapotranspiration, and roughness length. The biogeochemical effects reflect changes in atmospheric CO_2 composition resulting from anthropogenic changes in land cover. The biogeochemical effects are estimated using three different methods:

1. Directly calculated within global climate models (Tables A2.1 and A2.2),
2. Calculated from off-line dynamic global vegetation models (DGVMs) estimates of net changes in the emissions of CO_2 from land (Table A2.1),

3. Calculated from observation-based estimates of net changes in the emissions of CO_2 from land (for historical reconstruction only, Table A2.1).

The mean annual and global temperature change (ΔT) resulting from biogeochemical effects is calculated as follows, for both DGMVs and observation-based estimates:

$$\Delta T = \Delta LCO_2 \times TCRE$$

Where ΔLCO_2 is the cumulative changes in net emissions of CO_2 resulting from anthropogenic land cover changes during the time period considered (in Tera tons of carbon, TtC), and TCRE is the transient climate response to cumulative carbon emissions (Gillett et al. 2013; Matthews et al. 2009). TCRE is a measure of the global temperature response to cumulative emissions of CO_2 and has been identified as a useful and practical tool for evaluating CO_2-induced climate changes (expressed in °C per Tera tons of carbon, °C/TtC). TCRE values have been estimated for a range of Earth system models (Gillett et al. 2013; MacDougall et al. 2016). In the following, we use the 5th percentile, mean and 95th percentile derived from the range of available TCRE values. For each DGVM or observation-based estimate, we then calculate three potential temperature changes to bracket the range of climate sensitivities.

Table A2.1 | Model-based and observation-based estimates of the effects historical and future anthropogenic land cover changes have on mean annual global surface air temperature (°C). BGC and BPH correspond to the change in temperature resulting from respectively biogeochemical processes (e.g., changes in atmospheric CO_2 composition) and biophysical processes (e.g., changes in physical land surface characteristics such as albedo, evapotranspiration, and roughness length).

Reference of the study	Time period	Cumulative CO₂ emissions from anthropogenic land cover change (TtC)	TCRE (°C/TtC)	Change in mean global annual (°C)	
				BGC	BPH
Historical period (global climate models)					
Lawrence et al. (2018)	1850–2005	0.123	1.9	0.23	
Simmons and Matthews (2016)	1750–2000[9]			0.22	−0.24
Devaraju et al. (2016)	1850–2005	0.112	1.9	0.21	
Zhang et al. (2013a)	1850–2005[10]	0.097	1.75	0.17	−0.06
Hua and Chen (2013)	about 1850–2000 (average of two estimates)				−0.015
Jones et al. (2013a)	Preindustrial (no exact dates)				−0.57
Lawrence et al. (2012)	1850–2005	0.120	1.9	0.23	−0.10
De Noblet-Ducoudré et al. (2012)	1972–2002, relative to 1900–1970				−0.042, −0.056, −0.005, −0.041, 0.021, −0.007, −0.005
Pongratz et al. (2010)	20th century			0.16, 0.18	−0.03
Arora and Boer (2010)	1850–2000	0.040, 0.077	2.4	0.1, 0.18	
Strengers et al. (2010)	20th century				−0.06
Kvalevåg et al. (2010)	Preindustrial (no exact dates)				+0.04 (CASE I)
Findell et al. (2009)	1901–2004				+0.02
Findell et al. (2007)	1990 relative to potential vegetation				+0.008
Brovkin et al. (2006)	1700–1992 (5 models)				−0.24, −0.13, −0.14, −0.25, −0.17
Betts et al. (2007), Betts (2001)	1750–1990				−0.02

---

[9] Land-use change + fossil fuel emission simulation values are considered.

[10] Carbon-nitrogen-phosphorous simulation values are considered.

Reference of the study	Time period	Cumulative CO$_2$ emissions from anthropogenic land cover change (TtC)	TCRE (°C/TtC)	Change in mean global annual (°C)	
				BGC	BPH
Hansen et al. (2005)	1880–1990				−0.04
Feddema et al. (2005)	Preindustrial land-cover changes (no exact dates, 'prehuman' simulations)				−0.39
Matthews et al. (2004)	1700–2000 (average of 7 simulations)			0.3	−0.14
Brovkin et al. (2004)	1800–2000			0.18	−0.26
Zhao and Pitman (2002), Chase et al. (2000), (2001)	Preindustrial				+0.06
Hansen et al. (1998)	Preindustrial land-cover changes				−0.14
Mean (± standard deviation) of all studies				**0.2 ± 0.05**	**−0.1 ± 0.14**
Historical period (DGVM/Bookkeeping model results)					
Li et al. (2017a)	1901–2012 (median of models)	0.148	0.88–1.72–2.52	0.13–0.25–0.37	
Peng et al. (2017)	1850–1990 (realistic cases range)	0.087, 0.139	0.88–1.72–2.52	0.1–0.15–0.22, 0.12–0.24–0.35	
Arneth et al. (2017)	1901–2014[11]	0.089	0.88–1.72–2.52	0.1–0.15–0.22	
		0.210	0.88–1.72–2.52	0.18–0.36–0.53	
		0.179	0.88–1.72–2.52	0.16–0.31–0.45	
		0.195	0.88–1.72–2.52	0.17–0.33–0.49	
		0.083	0.88–1.72–2.52	0.1–0.14–0.21	
		0.161	0.88–1.72–2.52	0.14–0.28–0.4	
		0.117	0.88–1.72–2.52	0.1–0.2–0.3	
		0.104	0.88–1.72–2.52	0.1–0.18–0.26	
		0.196	0.88–1.72–2.52	0.17–0.34–0.49	
Pugh et al. (2015)	1850–2012 (gross land clearance flux)	0.157	0.88–1.72–2.52	0.14–0.27–0.39	
Hansis et al. (2015)	1850–2012	0.269	0.88–1.72–2.52	0.19–0.36–0.53	
Houghton et al. (2012), Hansis et al. (2015)	1920–1999 (multi-model range)	0.072, 0.115	0.88–1.72–2.52	0.1–0.12–0.18, 0.1–0.2–0.3	
Mean (± standard deviation) of all studies				**0.24 ± 0.12**	
Historical period (observation-based estimates)					
Li et al. (2017a)	1901–2012	0.155	0.88–1.72–2.52	0.14–0.27–0.39	
Li et al. (2017a), Avitabile et al. (2016), Carvalhais et al. (2014)	1901–2012[12]	0.160, 0.165	0.88–1.72–2.52	0.14–0.27–0.40, 0.14–0.28–0.41	
Liu et al. (2015), Li et al. (2017a)	1901–2012	0.161, 0.163	0.88–1.72–2.52	0.14–0.28–0.41	
Le Quéré et al. (2015)	1870–2014	0.145	0.88–1.72–2.52	0.13–0.25–0.36	
Carvalhais et al. 2014), Li et al. (2017a)	1901–2012	0.152, 0.159	0.88–1.72–2.52	0.13–0.26–0.38, 0.14–0.27–0.4	
Pan et al. (2011), Li et al. (2017a)	1901–2012	0.119, 0.122	0.88–1.72–2.52	0.10–0.20–0.30, 0.11–0.21–0.31	
Mean (± standard deviation) of all studies				**0.25 ± 0.10**	
Future -RCP8.5 (global climate models)					
Tharammal et al. (2018)	2006–2100	0.093	1.9	0.18	
Lawrence et al. (2018)	2006–2100	0.211	1.9	0.40	
Simmons and Matthews (2016)	2000–2100			0.35	−0.34
Hua et al. (2015)	2006–2100	0.032	2.4	0.08	

[11] FLULCC,1 refers to land use change related fluxes accounting for new processes in their study.

[12] Different harmonization methods: method A assumes increase in cropland area in a grid cell taken from forest; method C assumes increase in cropland and pasture taken from forest and then natural grassland if no more forest area available.

Reference of the study	Time period	Cumulative CO$_2$ emissions from anthropogenic land cover change (TtC)	TCRE (°C/TtC)	Change in mean global annual (°C)	
				BGC	BPH
Davies-Barnard et al. (2014)	2005–2100	0.02	2.1	0.04	−0.015
Boysen et al. (2014), Quesada et al. (2017a), Brovkin et al. (2013)	2005–2100	0.034	2.4	0.08	0.04
		0.025	2.1	0.05	0.0
		0.037	1.6	0.06	0.08
		0.062	2.2	0.13	−0.20
		0.205	1.6	0.33	−0.06
Lawrence et al. (2012)	2006–2100	0.256	1.9	0.49	
Mean (± standard deviation) of all studies				**0.20 ± 0.15**	**−0.1 ± 0.14**
Future -RCP8.5 (DGVM results)					
Pugh et al. (2015)	2006–2100	0.169, 0.171	0.88–1.72–2.52	0.15–0.29–0.42, 0.15–0.29–0.43	
IPCC (2013b)	2005–2099	0.151	0.88–1.72–2.52	0.13–0.26–0.38	
Mean (± standard deviation) of all studies				**0.28 ± 0.11**	
Future RCP4.5 (global climate models)					
Tharammal et al. (2018)	2005–2100	−0.029	1.9	−0.05	
Lawrence et al. (2018)	2006–2100	0.053	1.9	0.10	
Simmons and Matthews (2016)	2000–2100			0.37	−0.29
Davies-Barnard et al. (2014)	2005–2100	−0.040	2.1	−0.08	0.14
Lawrence et al. (2012)	2006–2100	0.148	1.9	0.28	
Mean (± standard deviation) of all studies				**0.12 ± 0.17**	**−0.1 ± 0.21**
Future RCP4.5 (DGVM results)					
Pugh et al. (2015)	2006–2100	0.016, −0.018	0.88–1.72–2.52	0.01–0.03–0.04, −0.02–(−0.03)–(−0.045)	
IPCC (2013b)	2005–2099	0.027	0.88–1.72–2.52	0.02–0.05–0.07	
Mean (± standard deviation) of all studies				**0.01 ± 0.04**	
Future RCP2.6 (global climate models)					
Tharammal et al. (2018)	2005–2100	0.039	1.9	0.07	
Simmons and Matthews (2016)	2000–2100			0.42	−0.35
Hua et al. (2015)	2006–2100	0.036	2.4	0.09	
Davies-Barnard et al. (2014)	2005–2100			0.04	−0.01
Brovkin et al. (2013)	2005–2100	0.039	2.4	0.09	
		0.019	2.1	0.04	
		0.065	2.2	0.14	
		0.175	1.6	0.28	
Lawrence et al. (2012)	2006–2100	0.0154	1.9	0.03	
Mean (± standard deviation) of all studies				**0.13 ± 0.12**	**−0.18 ± 0.17**
Future RCP2.6 (DGVM results)					
Pugh et al. (2015)	2006–2100 (no harvest, managed cases)	0.057, 0.084	0.88–1.72–2.52	0.05–0.09–0.14, 0.07–0.14–0.21	
IPCC (2013b)	2005–2099	0.105	0.88–1.72–2.52	0.09–0.18–0.26	
Mean (± standard deviation) of all studies				**0.14 ± 0.06**	

Table A2.2 | Model-based estimates of the effects idealised and latitudinal deforestation or forestation have on mean annual global and latitudinal surface air temperature (°C). BGC and BPH correspond to the change in temperature resulting from respectively biogeochemical processes (e.g., changes in atmospheric CO_2 composition) and biophysical processes (e.g., changes in physical land surface characteristics such as albedo, evapotranspiration and roughness length).

Idealised deforestation/afforestation (global climate models)					
Reference	Change in forest area (Mkm²)	Cumulative LCC flux (TtC)	TCRE (K/TtC)	Mean annual change in surface air temperature, averaged globally (and for the latitudinal band where trees are removed or added) (°C)	
				BGC	BPH
Tropical deforestation					
Devaraju et al. (2018)	36.1				0.02 (1.14)
Longobardi et al. (2016b)	23[13]	0.127	1.72	0.30	0.044 (−0.19)
Devaraju et al. (2015b)	23			1.06	−0.04 (0.20)
Brovkin et al. (2015)					−0.01, −0.13, −0.05
Bathiany et al. (2010)	23.1			0.40	0.18 (0.9)
Snyder (2010)	23				0.2 (1.0)
Bala et al. (2007)	23	0.418	1.72	0.72	0.70
Voldoire (2006)					0.2, 0.4, 0.6
Snyder et al. (2004)	22.7				0.24 (1.2)
Claussen et al. (2001)	7.5			0.19 (0.15)	−0.04 (0.13)
Ganopolski et al. (2001)	7.5				−0.5 (0.5)
Henderson-Sellers and Gornitz (1984)					0.00
Potter et al. (1981), Potter et al. (1975)					−0.2
Sagan et al. (1979)					−0.07
Mean (± standard deviation) of all studies				0.53 ± 0.32	0.1 ± 0.27 (0.61 ± 0.48)
Tropical afforestation					
Wang et al. (2014a) (average of four simulations)					0.925
Bathiany et al. (2010)	23.1				−0.03 (−0.1)
Temperate deforestation					
Devaraju et al. (2018)	18.8				0.18 (0.52)
Longobardi et al. (2016b)	15	0.047	1.72	0.10	-0.077 (-0.22)
Devaraju et al. (2015a)	15.3			0.39	−0.5 (−0.8)
Bala et al. (2007)	15	0.231	1.72	0.40	−0.04
Snyder et al. (2004)	19.1				−0.22 (−1.1)
Mean (± standard deviation) of all studies				0.29 ± 0.13	−0.13 ± 0.22 (−0.4 ± 0.62)
Temperate afforestation					
Laguë and Swann (2016)					0.3 (1.5)
Wang et al. (2014a)					1.14
Swann et al. (2012)	15.3			−0.2, −0.7	0.3
Gibbard et al. (2005)					0.27
Mean (± standard deviation) of all studies				−0.45	0.50 ± 0.36
Boreal afforestation					
Devaraju et al. (2018)	23.5				−0.25 (−1.2)
Longobardi et al. (2016b)	13.7	0.050	1.72	0.11	−0.38 (−0.9)
Devaraju et al. (2015a)	13.7			0.06	−0.9 (−4)
Dass et al. (2013)	18.5			0.12, 0.32	−0.35
Bathiany et al. (2010)	18.5	0.02	2.04	0.04	−0.28 (−1.1)
Bala et al. (2007)	13.7	0.0105	1.72	0.02	−0.8
Snyder et al. (2004)	22.4				−0.77 (−2.8)

---

[13] For some studies that do not provide area deforested, IPSL-CM5 model grids used to calculate the area.

Reference	Change in forest area (Mkm^2)	Cumulative LCC flux (TtC)	TCRE (K/TtC)	Mean annual change in surface air temperature, averaged globally (and for the latitudinal band where trees are removed or added) (°C)	
				BGC	BPH
Idealised deforestation/afforestation (global climate models)					
Caussen et al. (2001)	6			0.09 (0.12)	−0.23 (−0.82)
Ganopolski et al. (2001)	6				−1.0
Mean (± standard deviation) of all studies				0.11 ± 0.09	−0.55 ± 0.29 (−1.8 ± 1.2)
Boreal afforestation					
Bathiany et al. (2010)					0.31 (1.2)

3

Desertification

Coordinating Lead Authors:
Alisher Mirzabaev (Germany/Uzbekistan), Jianguo Wu (China)

Lead Authors:
Jason Evans (Australia), Felipe García-Oliva (Mexico), Ismail Abdel Galil Hussein (Egypt), Muhammad Mohsin Iqbal (Pakistan), Joyce Kimutai (Kenya), Tony Knowles (South Africa), Francisco Meza (Chile), Dalila Nedjraoui (Algeria), Fasil Tena (Ethiopia), Murat Türkeş (Turkey), Ranses José Vázquez (Cuba), Mark Weltz (The United States of America)

Contributing Authors:
Mansour Almazroui (Saudi Arabia), Hamda Aloui (Tunisia), Hesham El-Askary (Egypt), Abdul Rasul Awan (Pakistan), Céline Bellard (France), Arden Burrell (Australia), Stefan van der Esch (The Netherlands), Robyn Hetem (South Africa), Kathleen Hermans (Germany), Margot Hurlbert (Canada), Jagdish Krishnaswamy (India), Zaneta Kubik (Poland), German Kust (The Russian Federation), Eike Lüdeling (Germany), Johan Meijer (The Netherlands), Ali Mohammed (Egypt), Katerina Michaelides (Cyprus/United Kingdom), Lindsay Stringer (United Kingdom), Stefan Martin Strohmeier (Austria), Grace Villamor (The Philippines)

Review Editors:
Mariam Akhtar-Schuster (Germany), Fatima Driouech (Morocco), Mahesh Sankaran (India)

Chapter Scientists:
Chuck Chuan Ng (Malaysia), Helen Berga Paulos (Ethiopia)

This chapter should be cited as:
Mirzabaev, A., J. Wu, J. Evans, F. García-Oliva, I.A.G. Hussein, M.H. Iqbal, J. Kimutai, T. Knowles, F. Meza, D. Nedjraoui, F. Tena, M. Türkeş, R.J. Vázquez, M. Weltz, 2019: Desertification. In: *Climate Change and Land: an IPCC special report on climate change, desertification, land degradation, sustainable land management, food security, and greenhouse gas fluxes in terrestrial ecosystems* [P.R. Shukla, J. Skea, E. Calvo Buendia, V. Masson-Delmotte, H.-O. Pörtner, D.C. Roberts, P. Zhai, R. Slade, S. Connors, R. van Diemen, M. Ferrat, E. Haughey, S. Luz, S. Neogi, M. Pathak, J. Petzold, J. Portugal Pereira, P. Vyas, E. Huntley, K. Kissick, M. Belkacemi, J. Malley, (eds.)]. https://doi.org/10.1017/9781009157988.005

Table of contents

Executive summary 251

3.1 The nature of desertification 254

    3.1.1   Introduction 254

    3.1.2   Desertification in previous IPCC
           and related reports 256

    3.1.3   Dryland populations:
           Vulnerability and resilience 256

    3.1.4   Processes and drivers of desertification
           under climate change 258

3.2 Observations of desertification 260

    3.2.1   Status and trends of desertification ... 260

    3.2.2   Attribution of desertification 265

3.3 Desertification feedbacks to climate 268

    3.3.1   Sand and dust aerosols 268

    3.3.2   Changes in surface albedo 269

    3.3.3   Changes in vegetation and
           greenhouse gas fluxes 270

**3.4 Desertification impacts on natural and
      socio-economic systems under climate change** 270

    3.4.1   Impacts on natural and managed ecosystems ... 270

    3.4.2   Impacts on socio-economic systems ... 272

3.5 Future projections 276

    3.5.1   Future projections of desertification ... 276

    3.5.2   Future projections of impacts 278

**3.6 Responses to desertification
      under climate change** 279

    3.6.1   SLM technologies and practices:
           On-the-ground actions 279

    3.6.2   Socio-economic responses 283

    3.6.3   Policy responses 285

    **Cross-Chapter Box 5 | Policy responses
    to drought** 290

    3.6.4   Limits to adaptation, maladaptation,
           and barriers for mitigation 291

3.7 Hotspots and case studies 292

    3.7.1   Climate change and soil erosion 292

    3.7.2   Green walls and green dams 294

    3.7.3   Invasive plant species 297

    3.7.4   Oases in hyper-arid areas in the
           Arabian Peninsula and northern Africa ... 300

    3.7.5   Integrated watershed management ... 302

3.8 Knowledge gaps and key uncertainties ... 305

Frequently Asked Questions 306

    **FAQ 3.1: How does climate change
    affect desertification?** 306

    **FAQ 3.2: How can climate change
    induced desertification be avoided,
    reduced or reversed?** 306

    **FAQ 3.3: How do sustainable land
    management practices affect ecosystem
    services and biodiversity?** 306

References 307

3

Executive summary

Desertification is land degradation in arid, semi-arid, and dry sub-humid areas, collectively known as drylands, resulting from many factors, including human activities and climatic variations. The range and intensity of desertification have increased in some dryland areas over the past several decades (*high confidence*). Drylands currently cover about 46.2% (±0.8%) of the global land area and are home to 3 billion people. The multiplicity and complexity of the processes of desertification make its quantification difficult. Desertification hotspots, as identified by a decline in vegetation productivity between the 1980s and 2000s, extended to about 9.2% of drylands (±0.5%), affecting about 500 (±120) million people in 2015. The highest numbers of people affected are in South and East Asia, the circum Sahara region including North Africa and the Middle East including the Arabian Peninsula (*low confidence*). Other dryland regions have also experienced desertification. Desertification has already reduced agricultural productivity and incomes (*high confidence*) and contributed to the loss of biodiversity in some dryland regions (*medium confidence*). In many dryland areas, spread of invasive plants has led to losses in ecosystem services (*high confidence*), while over-extraction is leading to groundwater depletion (*high confidence*). Unsustainable land management, particularly when coupled with droughts, has contributed to higher dust-storm activity, reducing human well-being in drylands and beyond (*high confidence*). Dust storms were associated with global cardiopulmonary mortality of about 402,000 people in 2005. Higher intensity of sand storms and sand dune movements are causing disruption and damage to transportation and solar and wind energy harvesting infrastructures (*high confidence*). {3.1.1, 3.1.4, 3.2.1, 3.3.1, 3.4.1, 3.4.2, 3.4.2, 3.7.3, 3.7.4}

Attribution of desertification to climate variability and change, and to human activities, varies in space and time (*high confidence*). Climate variability and anthropogenic climate change, particularly through increases in both land surface air temperature and evapotranspiration, and decreases in precipitation, are *likely* to have played a role, in interaction with human activities, in causing desertification in some dryland areas. The major human drivers of desertification interacting with climate change are expansion of croplands, unsustainable land management practices and increased pressure on land from population and income growth. Poverty is limiting both capacities to adapt to climate change and availability of financial resources to invest in sustainable land management (SLM) (*high confidence*). {3.1.4, 3.2.2, 3.4.2}

Climate change will exacerbate several desertification processes (*medium confidence*). Although CO_2 fertilisation effect is enhancing vegetation productivity in drylands (*high confidence*), decreases in water availability have a larger effect than CO_2 fertilisation in many dryland areas. There is *high confidence* that aridity will increase in some places, but no evidence for a projected global trend in dryland aridity (*medium confidence*). The area at risk of salinisation is projected to increase in the future (*limited evidence, high agreement*). Future climate change is projected to increase the potential for water driven soil erosion in many dryland areas (*medium*

confidence), leading to soil organic carbon decline in some dryland areas. {3.1.1, 3.2.2, 3.5.1, 3.5.2, 3.7.1, 3.7.3}

Risks from desertification are projected to increase due to climate change (*high confidence*). Under shared socio-economic pathway SSP2 ('Middle of the Road') at 1.5°C, 2°C and 3°C of global warming, the number of dryland population exposed (vulnerable) to various impacts related to water, energy and land sectors (e.g., water stress, drought intensity, habitat degradation) is projected to reach 951 (178) million, 1152 (220) million and 1285 (277) million, respectively. While at global warming of 2°C, under SSP1 ('Sustainability'), the exposed (vulnerable) dryland population is 974 (35) million, and under SSP3 ('Fragmented World') it is 1267 (522) million. Around half of the vulnerable population is in South Asia, followed by Central Asia, West Africa and East Asia. {2.2, 3.1.1, 3.2.2, 3.5.1, 3.5.2, 7.2.2}

Desertification and climate change, both individually and in combination, will reduce the provision of dryland ecosystem services and lower ecosystem health, including losses in biodiversity (*high confidence*). Desertification and changing climate are projected to cause reductions in crop and livestock productivity (*high confidence*), modify the composition of plant species and reduce biological diversity across drylands (*medium confidence*). Rising CO_2 levels will favour more rapid expansion of some invasive plant species in some regions. A reduction in the quality and quantity of resources available to herbivores can have knock-on consequences for predators, which can potentially lead to disruptive ecological cascades (*limited evidence, low agreement*). Projected increases in temperature and the severity of drought events across some dryland areas can increase chances of wildfire occurrence (*medium confidence*). {3.1.4, 3.4.1, 3.5.2, 3.7.3}

Increasing human pressures on land, combined with climate change, will reduce the resilience of dryland populations and constrain their adaptive capacities (*medium confidence*). The combination of pressures coming from climate variability, anthropogenic climate change and desertification will contribute to poverty, food insecurity, and increased disease burden (*high confidence*), as well as potentially to conflicts (*low confidence*). Although strong impacts of climate change on migration in dryland areas are disputed (*medium evidence, low agreement*), in some places, desertification under changing climate can provide an added incentive to migrate (*medium confidence*). Women will be impacted more than men by environmental degradation, particularly in those areas with higher dependence on agricultural livelihoods (*medium evidence, high agreement*). {3.4.2, 3.6.2}

Desertification exacerbates climate change through several mechanisms such as changes in vegetation cover, sand and dust aerosols and greenhouse gas fluxes (*high confidence*). The extent of areas in which dryness (rather than temperature) controls CO_2 exchange has increased by 6% between 1948 and 2012, and is projected to increase by at least another 8% by 2050 if the expansion continues at the same rate. In these areas, net carbon uptake is about 27% lower than in other areas (*low confidence*). Desertification also tends to increase albedo, decreasing the energy available at the surface and associated surface temperatures, producing a negative feedback on climate change (*high confidence*). Through its effect on vegetation and soils, desertification changes the absorption and release of associated greenhouse gases (GHGs). Vegetation loss and drying of surface cover due to desertification increases the frequency of dust storms (*high confidence*). Arid ecosystems could be an important global carbon sink, depending on soil water availability (*medium evidence, high agreement*). {3.3.3, 3.4.1, 3.5.2}

Site and regionally-specific technological solutions, based both on new scientific innovations and indigenous and local knowledge (ILK), are available to avoid, reduce and reverse desertification, simultaneously contributing to climate change mitigation and adaptation (*high confidence*). SLM practices in drylands increase agricultural productivity and contribute to climate change adaptation with mitigation co-benefits (*high confidence*). Integrated crop, soil and water management measures can be employed to reduce soil degradation and increase the resilience of agricultural production systems to the impacts of climate change (*high confidence*). These measures include crop diversification and adoption of drought-resilient econogically appropriate plants, reduced tillage, adoption of improved irrigation techniques (e.g., drip irrigation) and moisture conservation methods (e.g., rainwater harvesting using indigenous and local practices), and maintaining vegetation and mulch cover. Conservation agriculture increases the capacity of agricultural households to adapt to climate change (*high confidence*) and can lead to increases in soil organic carbon over time, with quantitative estimates of the rates of carbon sequestration in drylands following changes in agricultural practices ranging between 0.04 and 0.4 t ha^{-1} (*medium confidence*). Rangeland management systems based on sustainable grazing and re-vegetation increase rangeland productivity and the flow of ecosystem services (*high confidence*). The combined use of salt-tolerant crops, improved irrigation practices, chemical remediation measures and appropriate mulch and compost is effective in reducing the impact of secondary salinisation (*medium confidence*). Application of sand dune stabilisation techniques contributes to reducing sand and dust storms (*high confidence*). Agroforestry practices and shelterbelts help reduce soil erosion and sequester carbon. Afforestation programmes aimed at creating windbreaks in the form of 'green walls' and 'green dams' can help stabilise and reduce dust storms, avert wind erosion, and serve as carbon sinks, particularly when done with locally adapted native and other climate resilient tree species (*high confidence*). {3.4.2, 3.6.1, 3.7.2}

Investments into SLM, land restoration and rehabilitation in dryland areas have positive economic returns (*high confidence*). Each USD invested into land restoration can have social returns of about 3–6 USD over a 30-year period. Most SLM practices can become financially profitable within 3 to 10 years (*medium evidence, high agreement*). Despite their benefits in addressing desertification, mitigating and adapting to climate change, and increasing food and economic security, many SLM practices are not widely adopted due to insecure land tenure, lack of access to credit and agricultural advisory services, and insufficient incentives for private land-users (*robust evidence, high agreement*). {3.6.3}

ILK often contributes to enhancing resilience against climate change and combating desertification (*medium confidence*). Dryland populations have developed traditional agroecological practices which are well adapted to resource-sparse dryland environments. However, there is *robust evidence* documenting losses of traditional agroecological knowledge. Traditional agroecological practices are also increasingly unable to cope with growing demand for food. Combined use of ILK and new SLM technologies can contribute to raising the resilience to the challenges of climate change and desertification (*high confidence*). {3.1.3, 3.6.1, 3.6.2}

Policy frameworks promoting the adoption of SLM solutions contribute to addressing desertification as well as mitigating and adapting to climate change, with co-benefits for poverty eradication and food security among dryland populations (*high confidence*). Implementation of Land Degradation Neutrality policies allows populations to avoid, reduce and reverse desertification, thus contributing to climate change adaptation with mitigation co-benefits (*high confidence*). Strengthening land tenure security is a major factor contributing to the adoption of soil conservation measures in croplands (*high confidence*). On-farm and off-farm livelihood diversification strategies increase the resilience of rural households against desertification and extreme weather events, such as droughts (*high confidence*). Strengthening collective action is important for addressing causes and impacts of desertification, and for adapting to climate change (*medium confidence*). A greater emphasis on understanding gender-specific differences over land use and land management practices can help make land restoration projects more successful (*medium confidence*). Improved access to markets raises agricultural profitability and motivates investment into climate change adaptation and SLM (*medium confidence*). Payments for ecosystem services give additional incentives to land users to adopt SLM practices (*medium confidence*). Expanding access to rural advisory services increases the knowledge on SLM and facilitates their wider adoption (*medium confidence*). Developing, enabling and promoting access to cleaner energy sources and technologies can contribute to reducing desertification and mitigating climate change through decreasing the use of fuelwood and crop residues for energy (*medium confidence*). Policy responses to droughts based on proactive drought preparedness and drought risk mitigation are more efficient in limiting drought-caused damages than reactive drought relief efforts (*high confidence*). {3.4.2, 3.6.2, 3.6.3, Cross-Chapter Box 5 in this chapter}

The knowledge on limits of adaptation to the combined effects of climate change and desertification is insufficient. However, the potential for residual risks and maladaptive outcomes is high (*high confidence*). Empirical evidence on the limits to adaptation in dryland areas is limited. Potential limits to adaptation include losses of land productivity due to irreversible forms of desertification. Residual risks can emerge from the inability of SLM measures to fully compensate for yield losses due to climate change impacts. They also arise from foregone reductions in ecosystem services due to soil fertility loss even when applying SLM measures could revert land to initial productivity after some time. Some activities favouring agricultural intensification in dryland areas can become maladaptive due to their negative impacts on the environment (*medium confidence*) Even when solutions are available, social, economic and institutional constraints could pose barriers to their implementation (*medium confidence*). {3.6.4}

Improving capacities, providing higher access to climate services, including local-level early warning systems, and expanding the use of remote sensing technologies are high-return investments for enabling effective adaptation and mitigation responses that help address desertification (*high confidence*). Reliable and timely climate services, relevant to desertification, can aid the development of appropriate adaptation and mitigation options reducing, the impact of desertification on human and natural systems (*high confidence*), with quantitative estimates showing that every USD invested in strengthening hydro-meteorological and early warning services in developing countries can yield between 4 and 35 USD (*low confidence*). Knowledge and flow of knowledge on desertification is currently fragmented. Improved knowledge and data exchange and sharing will increase the effectiveness of efforts to achieve Land Degradation Neutrality (*high confidence*). Expanded use of remotely sensed information for data collection helps in measuring progress towards achieving Land Degradation Neutrality (*low evidence, high agreement*). {3.2.1, 3.6.2, 3.6.3, Cross-Chapter Box 5 in this chapter}

3

1.1   The nature of desertification

1.1.1   Introduction

In this report, desertification is defined as land degradation in arid, semi-arid, and dry sub-humid areas resulting from many factors, including climatic variations and human activities (United Nations Convention to Combat Desertification (UNCCD) 1994). Land degradation is a negative trend in land condition, caused by direct or indirect human-induced processes including anthropogenic climate change, expressed as long-term reduction or loss of at least one of the following: biological productivity, ecological integrity or value to humans (Section 4.1.3). Arid, semi-arid, and dry sub-humid areas, together with hyper-arid areas, constitute drylands (UNEP 1992), home to about 3 billion people (van der Esch et al. 2017). The difference between desertification and land degradation is not process-based but geographic. Although land degradation can occur anywhere across the world, when it occurs in drylands, it is considered desertification (FAQ 1.3). Desertification is not limited to irreversible forms of land degradation, nor is it equated to desert expansion, but represents all forms and levels of land degradation occurring in drylands.

The geographic classification of drylands is often based on the aridity index (AI) – the ratio of average annual precipitation amount (P) to potential evapotranspiration amount (PET, see Glossary) (Figure 3.1). Recent estimates, based on AI, suggest that drylands cover about 46.2% (±0.8%) of the global land area (Koutroulis 2019; Prăvălie 2016) (*low confidence*). Hyper-arid areas, where the aridity index is below 0.05, are included in drylands, but are excluded from the definition of desertification (UNCCD 1994). Deserts are valuable ecosystems (UNEP 2006; Safriel 2009) geographically located in drylands and vulnerable to climate change. However, they are not considered prone to desertification. Aridity is a long-term climatic feature characterised by low average precipitation or available water (Gbeckor-Kove 1989; Türkeş 1999). Thus, aridity is different from drought, which is a temporary climatic event (Maliva and Missimer

2012). Moreover, droughts are not restricted to drylands, but occur both in drylands and humid areas (Wilhite et al. 2014). Following the Synthesis Report (SYR) of the IPCC Fifth Assessment Report (AR5), drought is defined here as "a period of abnormally dry weather long enough to cause a serious hydrological imbalance" (Mach et al. 2014) (Cross-Chapter Box 5 in this chapter).

AI is not an accurate proxy for delineating drylands in an increasing CO_2 environment (Section 3.2.1). The suggestion that most of the world has become more arid, since the AI has decreased, is not supported by changes observed in precipitation, evaporation or drought (Sheffield et al. 2012; Greve et al. 2014). While climate change is expected to decrease the AI due to increases in potential evaporation, the assumptions that underpin the potential evaporation calculation are not consistent with a changing CO_2 environment and the effect this has on transpiration rates (Roderick et al. 2015; Milly and Dunne 2016; Greve et al. 2017) (Section 3.2.1). Given that future climate is characterised by significant increases in CO_2, the usefulness of currently applied AI thresholds to estimate dryland areas is limited under climate change. If instead of the AI, other variables such as precipitation, soil moisture, and primary productivity are used to identify dryland areas, there is no clear indication that the extent of drylands will change overall under climate change (Roderick et al. 2015; Greve et al. 2017; Lemordant et al. 2018). Thus, some dryland borders will expand, while some others will contract (*high confidence*).

Approximately 70% of dryland areas are located in Africa and Asia (Figure 3.2). The biggest land use/cover in terms of area in drylands, if deserts are excluded, are grasslands, followed by forests and croplands (Figure 3.3). The category of 'other lands' in Figure 3.3 includes bare soil, ice, rock, and all other land areas that are not included within the other five categories (FAO 2016). Thus, hyper-arid areas contain mostly deserts, with some small exceptions, for example, where grasslands and croplands are cultivated under oasis conditions with irrigation (Section 3.7.4). Moreover, FAO (2016) defines grasslands as permanent pastures and meadows used continuously for more than five years. In drylands, transhumance, i.e. seasonal migratory grazing,

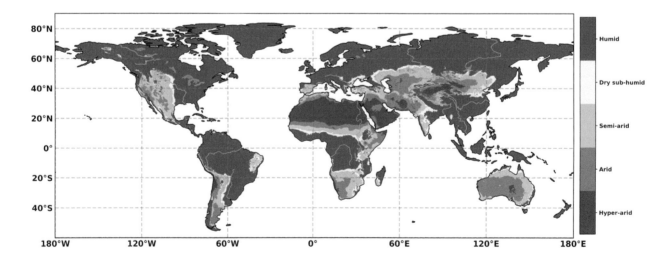

Figure 3.1 | Geographical distribution of drylands, delimited based on the aridity index (AI). The classification of AI is: Humid AI > 0.65, Dry sub-humid 0.50 < AI ≤ 0.65, Semi-arid 0.20 < AI ≤ 0.50, Arid 0.05 < AI ≤ 0.20, Hyper-arid AI < 0.05. Data: TerraClimate precipitation and potential evapotranspiration (1980–2015) (Abatzoglou et al. 2018).

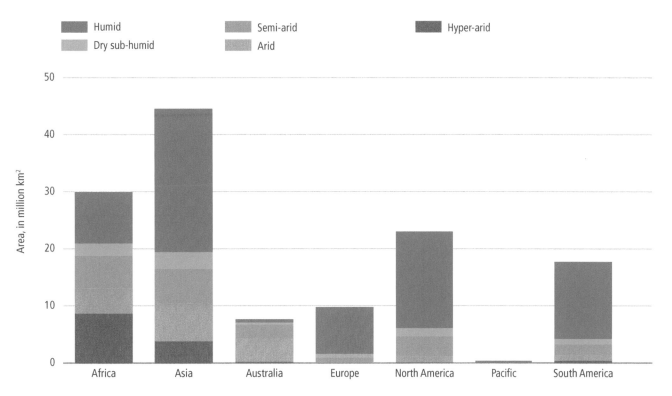

Figure 3.2 | Dryland categories across geographical areas (continents and Pacific region). Data: TerraClimate precipitation and potential evapotranspiration (1980–2015) (Abatzoglou et al. 2018).

often leads to non-permanent pasture systems, thus some of the areas under the 'other land' category are also used as non-permanent pastures (Ramankutty et al. 2008; Fetzel et al. 2017; Erb et al. 2016).

In the earlier global assessments of desertification (since the 1970s), which were based on qualitative expert evaluations, the extent of desertification was found to range between 4% and 70% of the area of drylands (Safriel 2007). More recent estimates, based on remotely sensed data, show that about 24–29% of the global land area experienced reductions in biomass productivity between the 1980s and 2000s (Bai et al. 2008; Le et al. 2016), corresponding to about 9.2% of drylands (±0.5%) experiencing declines in biomass productivity during this period (*low confidence*), mainly due to anthropogenic causes. Both of these studies consider rainfall dynamics, thus, accounting for the effect of droughts. While less than 10% of drylands is undergoing desertification, it is occurring in areas that contain around 20% of dryland population (Klein Goldewijk et al. 2017). In these areas the population has increased from approximately 172 million in 1950 to over 630 million today (Figure 1.1).

Available assessments of the global extent and severity of desertification are relatively crude approximations with considerable uncertainties, for example, due to confounding effects of invasive bush encroachment in some dryland regions. Different indicator sets and approaches have been developed for monitoring and assessment of desertification from national to global scales (Imeson 2012; Sommer et al. 2011; Zucca et al. 2012; Bestelmeyer et al. 2013). Many indicators of desertification only include a single factor or characteristic of desertification, such as the patch size distribution of vegetation (Maestre and Escudero 2009; Kéfi et al. 2010), Normalized Difference Vegetation Index (NDVI) (Piao et al. 2005), drought-tolerant plant

species (An et al. 2007), grass cover (Bestelmeyer et al. 2013), land productivity dynamics (Baskan et al. 2017), ecosystem net primary productivity (Zhou et al. 2015) or Environmentally Sensitive Land Area Index (Symeonakis et al. 2016). In addition, some synthetic indicators of desertification have also been used to assess desertification extent and desertification processes, such as climate, land use, soil, and socio-economic parameters (Dharumarajan et al. 2018), or changes in climate, land use, vegetation cover, soil properties and population as the desertification vulnerability index (Salvati et al. 2009). Current data availability and methodological challenges do not allow for accurately and comprehensively mapping desertification at a global scale (Cherlet et al. 2018). However, the emerging partial evidence points to a lower global extent of desertification than previously estimated (*medium confidence*) (Section 3.2).

This assessment examines the socio-ecological links between drivers (Section 3.1) and feedbacks (Section 3.3) that influence desertification–climate change interactions, and then examines associated observed and projected impacts (Sections 3.4 and 3.5) and responses (Section 3.6). Moreover, this assessment highlights that dryland populations are highly vulnerable to desertification and climate change (Sections 3.2 and 3.4). At the same time, dryland populations also have significant past experience and sources of resilience embodied in indigenous and local knowledge and practices in order to successfully adapt to climatic changes and address desertification (Section 3.6). Numerous site-specific technological response options are also available for SLM in drylands that can help increase the resilience of agricultural livelihood systems to climate change (Section 3.6). However, continuing environmental degradation combined with climate change is straining the resilience of dryland populations. Enabling policy responses for SLM and livelihoods

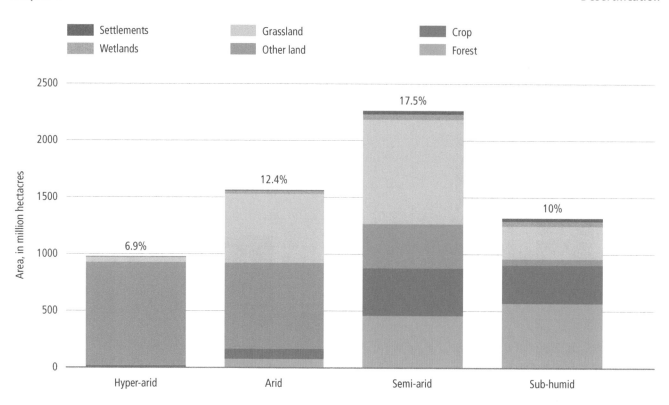

Figure 3.3 | Land use and land cover in drylands and share of each dryland category in global land area. Source: FAO (2016).

diversification can help maintain and strengthen the resilience and adaptive capacities in dryland areas (Section 3.6). The assessment finds that policies promoting SLM in drylands will contribute to climate change adaptation and mitigation, with co-benefits for broader sustainable development (*high confidence*) (Section 3.4).

1.1.2   Desertification in previous IPCC and related reports

The IPCC Fifth Assessment Report (AR5) and Special Report on Global Warming of 1.5°C include a limited discussion of desertification. In AR5 Working Group I desertification is mentioned as a forcing agent for the production of atmospheric dust (Myhre et al. 2013). The same report had *low confidence* in the available projections on the changes in dust loadings due to climate change (Boucher et al. 2013). In AR5 Working Group II desertification is identified as a process that can lead to reductions in crop yields and the resilience of agricultural and pastoral livelihoods (Field et al. 2014; Klein et al. 2015). AR5 Working Group II notes that climate change will amplify water scarcity, with negative impacts on agricultural systems, particularly in semi-arid environments of Africa (*high confidence*), while droughts could exacerbate desertification in southwestern parts of Central Asia (Field et al. 2014). AR5 Working Group III identifies desertification as one of a number of often overlapping issues that must be dealt with when considering governance of mitigation and adaptation (Fleurbaey et al. 2014). The IPCC Special Report on Global Warming of 1.5°C noted that limiting global warming to 1.5°C instead of 2°C is strongly beneficial for land ecosystems and their services (*high confidence*) such as soil conservation, contributing to avoidance of desertification (Hoegh-Guldberg et al. 2018).

The recent Intergovernmental Science-Policy Platform on Biodiversity and Ecosystem Services (IPBES) Land Degradation and Restoration Assessment report (IPBES 2018a) is also of particular relevance. While acknowledging a wide variety of past estimates of the area undergoing degradation, IPBES (2018a) pointed at their lack of agreement about where degradation is taking place. IPBES (2018a) also recognised the challenges associated with differentiating the impacts of climate variability and change on land degradation from the impacts of human activities at a regional or global scale.

The third edition of the World Atlas of Desertification (Cherlet et al. 2018) indicated that it is not possible to deterministically map the global extent of land degradation – and its subset, desertification – pointing out that the complexity of interactions between social, economic, and environmental systems make land degradation not amenable to mapping at a global scale. Instead, Cherlet et al. (2018) presented global maps highlighting the convergence of various pressures on land resources.

1.1.3   Dryland populations: Vulnerability and resilience

Drylands are home to approximately 38.2% (±0.6%) of the global population (Koutroulis 2019; van der Esch et al. 2017), that is about 3 billion people. The highest number of people live in the drylands of South Asia (Figure 3.4), followed by Sub-Saharan Africa and Latin America (van der Esch et al. 2017). In terms of the number of people affected by desertification, Reynolds et al. (2007) indicated that desertification was directly affecting 250 million people. More recent estimates show that 500 (±120) million people lived in 2015 in those dryland areas which experienced significant loss in biomass

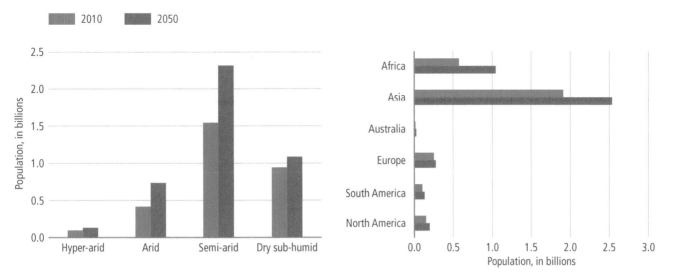

Figure 3.4 | **Current and projected population (under SSP2) in drylands, in billions.** Source: van der Esch et al. (2017).

productivity between the 1980s and 2000s (Bai et al. 2008; Le et al. 2016). The highest numbers of affected people were in South and East Asia, North Africa and the Middle East (*low confidence*). The population in drylands is projected to increase about twice as rapidly as non-drylands, reaching 4 billion people by 2050 (van der Esch et al. 2017). This is due to higher population growth rates in drylands. About 90% of the population in drylands live in developing countries (UN-EMG 2011).

Dryland populations are highly vulnerable to desertification and climate change because their livelihoods are predominantly dependent on agriculture, one of the sectors most susceptible to climate change (Rosenzweig et al. 2014; Schlenker and Lobell 2010). Climate change is projected to have substantial impacts on all types of agricultural livelihood systems in drylands (CGIAR-RPDS 2014) (Sections 3.4.1 and 3.4.2).

One key vulnerable group in drylands are pastoral and agropastoral households.[1] There are no precise figures about the number of people practicing pastoralism globally. Most estimates range between 100 million and 200 million (Rass 2006; Secretariat of the Convention on Biological Diversity 2010), of whom 30–63 million are nomadic pastoralists (Dong 2016; Carr-Hill 2013).[2] Pastoral production systems represent an adaptation to high seasonal climate variability and low biomass productivity in dryland ecosystems (Varghese and Singh 2016; Krätli and Schareika 2010), which require large areas for livestock grazing through migratory pastoralism (Snorek et al. 2014). Grazing lands across dryland environments are being degraded, and/or being converted to crop production, limiting the opportunities for migratory livestock systems, and leading to conflicts with sedentary crop producers (Abbass 2014; Dimelu et al. 2016). These processes, coupled with ethnic differences, perceived security threats, and misunderstanding of pastoral rationality, have led to increasing marginalisation of pastoral

communities and disruption of their economic and cultural structures (Elhadary 2014; Morton 2010). As a result, pastoral communities are not well prepared to deal with increasing weather/climate variability and weather/climate extremes due to changing climate (Dong 2016; López-i-Gelats et al. 2016), and remain amongst the most food insecure groups in the world (FAO 2018).

There is an increasing concentration of poverty in the dryland areas of Sub-Saharan Africa and South Asia (von Braun and Gatzweiler 2014; Barbier and Hochard 2016), where 41% and 12% of the total populations live in extreme poverty, respectively (World Bank 2018). For comparison, the average share of global population living in extreme poverty is about 10% (World Bank 2018). Multidimensional poverty, prevalent in many dryland areas, is a key source of vulnerability (Safriel et al. 2005; Thornton et al. 2014; Fraser et al. 2011; Thomas 2008). Multidimensional poverty incorporates both income-based poverty, and also other dimensions such as poor healthcare services, lack of education, lack of access to water, sanitation and energy, disempowerment, and threat from violence (Bourguignon and Chakravarty 2003; Alkire and Santos 2010, 2014). Contributing elements to this multidimensional poverty in drylands are rapid population growth, fragile institutional environment, lack of infrastructure, geographic isolation and low market access, insecure land tenure systems, and low agricultural productivity (Sietz et al. 2011; Reynolds et al. 2011; Safriel and Adeel 2008; Stafford Smith 2016). Even in high-income countries, those dryland areas that depend on agricultural livelihoods represent relatively poorer locations nationally, with fewer livelihood opportunities, for example in Italy (Salvati 2014). Moreover, in many drylands areas, female-headed households, women and subsistence farmers (both male and female) are more vulnerable to the impacts of desertification and climate change (Nyantakyi-Frimpong and Bezner-Kerr 2015; Sultana 2014; Rahman 2013). Some local cultural traditions and patriarchal

---

[1] Pastoralists derive more than 50% of their income from livestock and livestock products, whereas agropastoralists generate more than 50% of their income from crop production and at least 25% from livestock production (Swift, 1988).

[2] The estimates of the number of pastoralists, and especially of nomadic pastoralists, are very uncertain, because often nomadic pastoralists are not fully captured in national surveys and censuses (Carr-Hill, 2013).

relationships were found to contribute to higher vulnerability of women and female-headed households through restrictions on their access to productive resources (Nyantakyi-Frimpong and Bezner-Kerr 2015; Sultana 2014; Rahman 2013) (Sections 3.4.2 and 3.6.3, and Cross-Chapter Box 11 in Chapter 7).

Despite these environmental, socio-economic and institutional constraints, dryland populations have historically demonstrated remarkable resilience, ingenuity and innovations, distilled into ILK to cope with high climatic variability and sustain livelihoods (Safriel and Adeel 2008; Davis 2016; Davies 2017) (Sections 3.6.1 and 3.6.2, and Cross-Chapter Box 13 in Chapter 7). For example, across the Arabian Peninsula and North Africa, informal community by-laws were successfully used for regulating grazing, collection and cutting of herbs and wood, and which limited rangeland degradation (Gari 2006; Hussein 2011). Pastoralists in Mongolia developed indigenous classifications of pasture resources which facilitated ecologically optimal grazing practices (Fernandez-Gimenez 2000) (Section 3.6.2). Currently, however, indigenous and local knowledge and practices are increasingly lost or can no longer cope with growing demands for land-based resources (Dominguez 2014; Fernández-Giménez and Fillat Estaque 2012; Hussein 2011; Kodirekkala 2017; Moreno-Calles et al. 2012) (Section 3.4.2). Unsustainable land management is increasing the risks from droughts, floods and dust storms (Sections 3.4.2 and 3.5). Policy actions promoting the adoption of SLM practices in dryland areas, based on both indigenous and local knowledge and modern science, and expanding alternative livelihood opportunities outside agriculture can contribute to climate change adaptation and mitigation, addressing desertification, with co-benefits for poverty eradication and food security (*high confidence*) (Cowie et al. 2018; Liniger et al. 2017; Safriel and Adeel 2008; Stafford-Smith et al. 2017).

1.1.4   Processes and drivers of desertification under climate change

1.1.4.1   Processes of desertification and their climatic drivers

Processes of desertification are mechanisms by which drylands are degraded. Desertification consists of both biological and non-biological processes. These processes are classified under broad categories of degradation of physical, chemical and biological properties of terrestrial ecosystems. The number of desertification processes is large and they are extensively covered elsewhere (IPBES 2018a; Lal 2016; Racine 2008; UNCCD 2017). Section 4.2.1 and Tables 4.1 and 4.2 in Chapter 4 highlight those which are particularly relevant for this assessment in terms of their links to climate change and land degradation, including desertification.

Drivers of desertification are factors which trigger desertification processes. Initial studies of desertification during the early-to-mid 20th century attributed it entirely to human activities. In one of the influential publications of that time, Lavauden (1927) stated that: "Desertification is purely artificial. It is only the act of the man…" However, such a uni-causal view of desertification was shown to be invalid (Geist et al. 2004; Reynolds et al. 2007) (Sections 3.1.4.2 and 3.1.4.3). Tables 4.1 and 4.2 in Chapter 4 summarise the drivers,

linking them to the specific processes of desertification and land degradation under changing climate.

Erosion refers to removal of soil by the physical forces of water, wind, or often caused by farming activities such as tillage (Ginoux et al. 2012). The global estimates of soil erosion differ significantly, depending on scale, study period and method used (García-Ruiz et al. 2015), ranging from approximately 20 Gt yr^{-1} to more than 200 Gt yr^{-1} (Boix-Fayos et al. 2006; FAO 2015). There is a significant potential for climate change to increase soil erosion by water, particularly in those regions where precipitation volumes and intensity are projected to increase (Panthou et al. 2014; Nearing et al. 2015). On the other hand, while it is a dominant form of erosion in areas such as West Asia and the Arabian Peninsula (Prakash et al. 2015; Klingmüller et al. 2016), there is *limited evidence* concerning climate change impacts on wind erosion (Tables 4.1 and 4.2 in Chapter 4, and Section 3.5).

Saline and sodic soils (see Glossary) occur naturally in arid, semi-arid and dry sub-humid regions of the world. Climate change or hydrological change can cause soil salinisation by increasing the mineralised groundwater level. However, secondary salinisation occurs when the concentration of dissolved salts in water and soil is increased by anthropogenic processes, mainly through poorly managed irrigation schemes. The threat of soil and groundwater salinisation induced by sea level rise and seawater intrusion are amplified by climate change (Section 4.9.7).

Global warming is expected to accelerate soil organic carbon (SOC) turnover, since the decomposition of the soil organic matter by microbial activity begins with low soil water availability, but this moisture is insufficient for plant productivity (Austin et al. 2004) (Section 3.4.1.1). SOC is also lost due to soil erosion (Lal 2009); therefore, in some dryland areas leading to SOC decline (Sections 3.3.3 and 3.5.2) and the transfer of carbon (C) from soil to the atmosphere (Lal 2009).

Sea surface temperature (SST) anomalies can drive rainfall changes, with implications for desertification processes. North Atlantic SST anomalies are positively correlated with Sahel rainfall anomalies (Knight et al. 2006; Gonzalez-Martin et al. 2014; Sheen et al. 2017). While the eastern tropical Pacific SST anomalies have a negative correlation with Sahel rainfall (Pomposi et al. 2016), a cooler North Atlantic is related to a drier Sahel, with this relationship enhanced if there is a simultaneous relative warming of the South Atlantic (Hoerling et al. 2006). Huber and Fensholt (2011) explored the relationship between SST anomalies and satellite observed Sahel vegetation dynamics, finding similar relationships but with substantial west–east variations in both the significant SST regions and the vegetation response. Concerning the paleoclimatic evidence on aridification after the early Holocene 'Green Sahara' period (11,000 to 5000 years ago), Tierney et al. (2017) indicate that a cooling of the North Atlantic played a role (Collins et al. 2017; Otto-Bliesner et al. 2014; Niedermeyer et al. 2009) similar to that found in modern observations. Besides these SST relationships, aerosols have also been suggested as a potential driver of the Sahel droughts (Rotstayn and Lohmann 2002; Booth et al. 2012; Ackerley et al. 2011). For eastern Africa, both recent droughts and decadal declines have been linked to human-induced warming in the western Pacific (Funk et al. 2018).

Invasive plants contributed to desertification and loss of ecosystem services in many dryland areas in the last century (*high confidence*) (Section 3.7.3). Extensive woody plant encroachment altered runoff and soil erosion across much of the drylands, because the bare soil between shrubs is very susceptible to water erosion, mainly in high-intensity rainfall events (Manjoro et al. 2012; Pierson et al. 2013; Eldridge et al. 2015). Rising CO_2 levels due to global warming favour more rapid expansion of some invasive plant species in some regions. An example is the Great Basin region in western North America where over 20% of ecosystems have been significantly altered by invasive plants, especially exotic annual grasses and invasive conifers, resulting in loss of biodiversity. This land-cover conversion has resulted in reductions in forage availability, wildlife habitat, and biodiversity (Pierson et al. 2011, 2013; Miller et al. 2013).

The wildfire is a driver of desertification, because it reduces vegetation cover, increases runoff and soil erosion, reduces soil fertility and affects the soil microbial community (Vega et al. 2005; Nyman et al. 2010; Holden et al. 2013; Pourreza et al. 2014; Weber et al. 2014; Liu and Wimberly 2016). Predicted increases in temperature and the severity of drought events across some dryland areas (Section 2.2) can increase chances of wildfire occurrence (*medium confidence*) (Jolly et al. 2015; Williams et al. 2010; Clarke and Evans 2018) (Cross-Chapter Box 3 in Chapter 2). In semi-arid and dry sub-humid areas, fire can have a profound influence on observed vegetation and particularly the relative abundance of grasses to woody plants (Bond et al. 2003; Bond and Keeley 2005; Balch et al. 2013).

While large uncertainty exists concerning trends in droughts globally (AR5) (Section 2.2), examining the drought data by Ziese et al. (2014) for drylands only reveals a large inter-annual variability combined with a trend toward increasing dryland area affected by droughts since the 1950s (Figure 1.1). Thus, over the period 1961–2013, the annual area of drylands in drought has increased, on average, by slightly more than 1% per year, with large inter-annual variability.

1.1.4.2   Anthropogenic drivers of desertification under climate change

The literature on the human drivers of desertification is substantial (e.g., D'Odorico et al. 2013; Sietz et al. 2011; Yan and Cai 2015; Sterk et al. 2016; Varghese and Singh 2016) and there have been several comprehensive reviews and assessments of these drivers very recently (Cherlet et al. 2018; IPBES 2018a; UNCCD 2017). IPBES (2018a) identified cropland expansion, unsustainable land management practices including overgrazing by livestock, urban expansion, infrastructure development, and extractive industries as the main drivers of land degradation. IPBES (2018a) also found that the ultimate driver of land degradation is high and growing consumption of land-based resources, e.g., through deforestation and cropland expansion, escalated by population growth. What is particularly relevant in the context of the present assessment is to evaluate if, how and which human drivers of desertification will be modified by climate change effects.

Growing food demand is driving conversion of forests, rangelands, and woodlands into cropland (Bestelmeyer et al. 2015; D'Odorico et al. 2013). Climate change is projected to reduce crop yields across dryland areas (Sections 3.4.1 and 5.2.2), potentially reducing local production of food and feed. Without research breakthroughs mitigating these productivity losses through higher agricultural productivity, and reducing food waste and loss, meeting the increasing food demands of growing populations will require expansion of cropped areas to more marginal areas (with most prime areas in drylands already being under cultivation) (Lambin 2012; Lambin et al. 2013; Eitelberg et al. 2015; Gutiérrez-Elorza 2006; Kapović Solomun et al. 2018). Borrelli et al. (2017) showed that the primary driver of soil erosion in 2012 was cropland expansion. Although local food demands could also be met by importing from other areas, this would mean increasing the pressure on land in those areas (Lambin and Meyfroidt 2011). The net effects of such global agricultural production shifts on land condition in drylands are not known.

Climate change will exacerbate poverty among some categories of dryland populations (Sections 3.4.2 and 3.5.2). Depending on the context, this impact comes through declines in agricultural productivity, changes in agricultural prices and extreme weather events (Hertel and Lobell 2014; Hallegatte and Rozenberg 2017). There is *high confidence* that poverty limits both capacities to adapt to climate change and availability of financial resources to invest into SLM (Gerber et al. 2014; Way 2016; Vu et al. 2014) (Sections 3.5.2, 3.6.2 and 3.6.3).

Labour mobility is another key human driver that will interact with climate change. Although strong impacts of climate change on migration in dryland areas are disputed, in some places, it is *likely* to provide an added incentive to migrate (Section 3.4.2.7). Out-migration will have several contradictory effects on desertification. On one hand, it reduces an immediate pressure on land if it leads to less dependence on land for livelihoods (Chen et al. 2014; Liu et al. 2016a). Moreover, migrant remittances could be used to fund the adoption of SLM practices. Labour mobility from agriculture to non-agricultural sectors could allow land consolidation, gradually leading to mechanisation and agricultural intensification (Wang et al. 2014, 2018). On the other hand, this can increase the costs of labour-intensive SLM practices due to lower availability of rural agricultural labour and/or higher rural wages. Out-migration increases the pressure on land if higher wages that rural migrants earn in urban centres will lead to their higher food consumption. Moreover, migrant remittances could also be used to fund land-use expansion to marginal areas (Taylor et al. 2016; Gray and Bilsborrow 2014). The net effect of these opposite mechanisms varies from place to place (Qin and Liao 2016). There is very little literature evaluating these joint effects of climate change, desertification and labour mobility (Section 7.3.2).

There are also many other institutional, policy and socio-economic drivers of desertification, such as land tenure insecurity, lack of property rights, lack of access to markets, and to rural advisory services, lack of technical knowledge and skills, agricultural price distortions, agricultural support and subsidies contributing to desertification, and lack of economic incentives for SLM (D'Odorico et al. 2013; Geist et al. 2004; Moussa et al. 2016; Mythili and Goedecke 2016; Sow et al.

2016; Tun et al. 2015; García-Ruiz 2010). There is no evidence that these factors will be materially affected by climate change, however, serving as drivers of unsustainable land management practices, they do play a very important role in modulating responses for climate change adaptation and mitigation (Section 3.6.3).

1.1.4.3   Interaction of drivers: Desertification syndrome versus drylands development paradigm

Two broad narratives have historically emerged to describe responses of dryland populations to environmental degradation. The first is 'desertification syndrome' which describes the vicious cycle of resource degradation and poverty, whereby dryland populations apply unsustainable agricultural practices leading to desertification, and exacerbating their poverty, which then subsequently further limits their capacities to invest in SLM (MEA 2005; Safriel and Adeel 2008). The alternative paradigm is one of 'drylands development', which refers to social and technical ingenuity of dryland populations as a driver of dryland sustainability (MEA 2005; Reynolds et al. 2007; Safriel and Adeel 2008). The major difference between these two frameworks is that the 'drylands development' paradigm recognises that human activities are not the sole and/or most important drivers of desertification, but there are interactions of human and climatic drivers within coupled social-ecological systems (Reynolds et al. 2007). This led Behnke and Mortimore (2016), and earlier Swift (1996), to conclude that the concept of desertification as irreversible degradation distorts policy and governance in dryland areas. Mortimore (2016) suggested that instead of externally imposed technical solutions, what is needed is for populations in dryland areas to adapt to this variable environment which they cannot control. All in all, there is *high confidence* that anthropogenic and climatic drivers interact in complex ways in causing desertification. As discussed in Section 3.2.2, the relative influence of human or climatic drivers on desertification varies from place to place (*high confidence*) (Bestelmeyer et al. 2018; D'Odorico et al. 2013; Geist and Lambin 2004; Kok et al. 2016; Polley et al. 2013; Ravi et al. 2010; Scholes 2009; Sietz et al. 2017; Sietz et al. 2011).

1.2   Observations of desertification

1.2.1   Status and trends of desertification

Current estimates of the extent and severity of desertification vary greatly due to missing and/or unreliable information (Gibbs and Salmon 2015). The multiplicity and complexity of the processes of desertification make its quantification difficult (Prince 2016; Cherlet et al. 2018). The most common definition for the drylands is based on defined thresholds of the AI (Figure 3.1; UNEP 1992). While past studies have used the AI to examine changes in desertification or extent of the drylands (Feng and Fu 2013; Zarch et al. 2015; Ji et al. 2015; Spinoni et al. 2015; Huang et al. 2016; Ramarao et al. 2018), this approach has several key limitations: (i) the AI does not measure desertification, (ii) the impact of changes in climate on the land surface and systems is more complex than assumed by AI, and (iii) the relationship between climate change and changes in vegetation is complex due to the influence of CO_2. Expansion of the drylands

does not imply desertification by itself, if there is no long-term loss of at least one of the following: biological productivity, ecological integrity, or value to humans.

The use of the AI to define changing aridity levels and dryland extent in an environment with changing atmospheric CO_2 has been strongly challenged (Roderick et al. 2015; Milly and Dunne 2016; Greve et al. 2017; Liu et al. 2017). The suggestion that most of the world has become more arid, since the AI has decreased, is not supported by changes observed in precipitation, evaporation or drought (*medium confidence*) (Sheffield et al. 2012; Greve et al. 2014). A key issue is the assumption in the calculation of potential evapotranspiration that stomatal conductance remains constant, which is invalid if atmospheric CO_2 changes. Given that atmospheric CO_2 has been increasing over the last century or more, and is projected to continue increasing, this means that AI with constant thresholds (or any other measure that relies on potential evapotranspiration) is not an appropriate way to estimate aridity or dryland extent (Donohue et al. 2013; Roderick et al. 2015; Greve et al. 2017). This issue helps explain the apparent contradiction between the drylands becoming more arid according to the AI and also becoming greener according to satellite observations (Fensholt et al. 2012; Andela et al. 2013) (Figure 3.5). Other climate type classifications based on various combinations of temperature and precipitation (Köppen-Trewartha, Köppen-Geiger) have also been used to examine historical changes in climate zones, finding a tendency toward drier climate types (Feng et al. 2014; Spinoni et al. 2015).

The need to establish a baseline when assessing change in the land area degraded has been extensively discussed in Prince et al. (2018). Desertification is a process, not a state of the system, hence an 'absolute' baseline is not required; however, every study uses a baseline defined by the start of their period of interest.

Depending on the definitions applied and methodologies used in evaluation, the status and extent of desertification globally and regionally still show substantial variations (*high confidence*) (D'Odorico et al. 2013). There is *high confidence* that the range and intensity of desertification has increased in some dryland areas over the past several decades (Sections 3.2.1.1 and 3.2.1.2). The three methodological approaches applied for assessing the extent of desertification: expert judgement, satellite observation of net primary productivity, and use of biophysical models, together provide a relatively holistic assessment but none on its own captures the whole picture (Gibbs and Salmon 2015; Vogt et al. 2011; Prince 2016) (Section 4.2.4).

1.2.1.1   Global scale

Complex human–environment interactions, coupled with biophysical, social, economic and political factors unique to any given location, render desertification difficult to map at a global scale (Cherlet et al. 2018). Early attempts to assess desertification focused on expert knowledge in order to obtain global coverage in a cost-effective manner. **Expert judgement** continues to play an important role because degradation remains a subjective feature whose indicators are different from place to place (Sonneveld and Dent 2007). GLASOD

(Global Assessment of Human-induced Soil Degradation) estimated nearly 2000 million hectares (Mha) (15.3% of the total land area) had been degraded by the early 1990s since the mid-20th century. GLASOD was criticised for perceived subjectiveness and exaggeration (Helldén and Tottrup 2008; Sonneveld and Dent 2007). Dregne and Chou (1992) found 3000 Mha in drylands (i.e. about 50% of drylands) were undergoing degradation. Significant improvements have been made through the efforts of WOCAT (World Overview of Conservation Approaches and Technologies), LADA (Land Degradation Assessment in Drylands) and DESIRE (Desertification Mitigation and Remediation of Land) who jointly developed a mapping tool for participatory expert assessment, with which land experts can estimate current area coverage, type and trends of land degradation (Reed et al. 2011).

A number of studies have used **satellite-based remote sensing** to investigate long-term changes in the vegetation and thus identify parts of the drylands undergoing desertification. Satellite data provides information at the resolution of the sensor, which can be relatively coarse (up to 25 km), and interpretations of the data at sub-pixel levels are challenging. The most widely used remotely sensed vegetation index is the NDVI, providing a measure of canopy greenness that is related to the quantity of standing biomass (Bai et al. 2008; de Jong et al. 2011; Fensholt et al. 2012; Andela et al. 2013; Fensholt et al. 2015; Le et al. 2016) (Figure 3.5). A main challenge associated with NDVI is that although biomass and productivity are closely related in some systems, they can differ widely when looking across land uses and ecosystem types, giving a false positive in some instances (Pattison et al. 2015; Aynekulu et al. 2017). For example, bush encroachment in rangelands and intensive monocropping with high fertiliser application gives an indication of increased productivity in satellite data though these could be considered as land degradation. According to this measure there are regions undergoing desertification, however the drylands are greening on average (Figure 3.6).

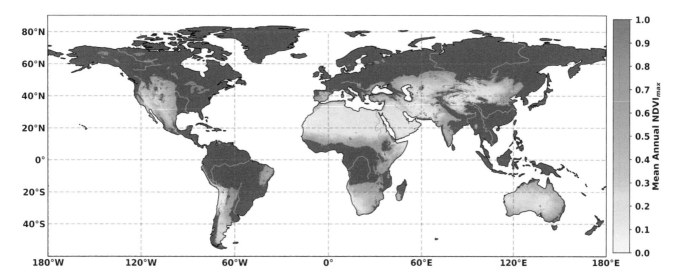

Figure 3.5 | Mean annual maximum NDVI 1982–2015 (Global Inventory Modelling and Mapping Studies NDVI3g v1). Non-dryland regions (aridity index >0.65) are masked in grey.

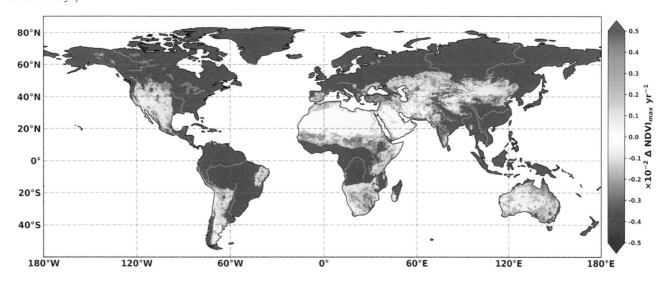

Figure 3.6 | Trend in the annual maximum NDVI 1982–2015 (Global Inventory Modelling and Mapping Studies NDVI3g v1) calculated using the Theil–Sen estimator which is a median based estimator, and is robust to outliers. Non-dryland regions (aridity index >0.65) are masked in grey.

A simple linear trend in NDVI is an unsuitable measure for dryland degradation for several reasons (Wessels et al. 2012; de Jong et al. 2013; Higginbottom and Symeonakis 2014; Le et al. 2016). NDVI is strongly coupled to precipitation in drylands where precipitation has high inter-annual variability. This means that NDVI trend can be dominated by any precipitation trend and is sensitive to wet or dry periods, particularly if they fall near the beginning or end of the time series. Degradation may only occur during part of the time series, while NDVI is stable or even improving during the rest of the time series. This reduces the strength and representativeness of a linear trend. Other factors such as CO_2 fertilisation also influence the NDVI trend. Various techniques have been proposed to address these issues, including the residual trends (RESTREND) method to account for rainfall variability (Evans and Geerken 2004), time-series break point identification methods to find major shifts in the vegetation trends (de Jong et al. 2013; Verbesselt et al. 2010a), and methods to explicitly account for the effect of CO_2 fertilisation (Le et al. 2016).

Using the RESTREND method, Andela et al. (2013) found that human activity contributed to a mixture of improving and degrading regions in drylands. In some locations these regions differed substantially from those identified using the NDVI trend alone, including an increase in the area being desertified in southern Africa and northern Australia, and a decrease in southeast and western Australia and Mongolia. De Jong et al. (2013) examined the NDVI time series for major shifts in vegetation activity and found that 74% of drylands experienced such a shift between 1981 and 2011. This suggests that monotonic linear trends are unsuitable for accurately capturing the changes that have occurred in the majority of the drylands. Le et al. (2016) explicitly accounted for CO_2 fertilisation effect and found that the extent of degraded areas in the world is 3% larger when compared to the linear NDVI trend.

Besides NDVI, there are many vegetation indices derived from satellite data in the optical and infrared wavelengths. Each of these datasets has been derived to overcome some limitation in existing indices. Studies have compared vegetation indices globally (Zhang et al. 2017) and specifically over drylands (Wu 2014). In general, the data from these vegetation indices are available only since around 2000, while NDVI data is available since 1982. With less than 20 years of data, the trend analysis remains problematic with vegetation indices other than NDVI. However, given the various advantages in terms of resolution and other characteristics, these newer vegetation indices will become more useful in the future as more data accumulates.

A major shortcoming of these studies based on vegetation datasets derived from satellite sensors is that they do not account for changes in vegetation composition, thus leading to inaccuracies in the estimation of the extent of degraded areas in drylands. For example, drylands of eastern Africa currently face growing encroachment of invasive plant species, such as *Prosopis juliflora* (Ayanu et al. 2015), which constitutes land degradation since it leads to losses in economic productivity of affected areas but appears as a greening in the satellite data. Another case study in central Senegal found degradation manifested through a reduction in species richness despite satellite observed greening (Herrmann and Tappan 2013). A number of efforts to identify changes in vegetation composition from satellites

have been made (Brandt et al. 2016a, b; Evans and Geerken 2006; Geerken 2009; Geerken et al. 2005; Verbesselt et al. 2010a, b). These depend on well-identified reference NDVI time series for particular vegetation groupings, can only differentiate vegetation types that have distinct spectral phenology signatures, and require extensive ground observations for validation. A recent alternative approach to differentiating woody from herbaceous vegetation involves the combined use of optical/infrared-based vegetation indices, indicating greenness, with microwave based Vegetation Optical Depth (VOD) which is sensitive to both woody and leafy vegetation components (Andela et al. 2013; Tian et al. 2017).

Vegetation Optical Depth (VOD) has been available since the 1980s. VOD is based on microwave measurements and is related to total above-ground biomass water content. Unlike NDVI, which is only sensitive to green canopy cover, VOD is also sensitive to water in woody parts of the vegetation and hence provides a view of vegetation changes that can be complementary to NDVI. Liu et al. (2013) used VOD trends to investigate biomass changes and found that VOD was closely related to precipitation changes in drylands. To complement their work with NDVI, Andela et al. (2013) also applied the RESTREND method to VOD. By interpreting NDVI and VOD trends together they were able to differentiate changes to the herbaceous and woody components of the biomass. They reported that many dryland regions are experiencing an increase in the woody fraction often associated with shrub encroachment and suggest that this was aided by CO_2 fertilisation.

Biophysical models use global datasets that describe climate patterns and soil groups, combined with observations of land use, to define classes of potential productivity and map general land degradation (Gibbs and Salmon 2015). All biophysical models have their own set of assumptions and limitations that contribute to their overall uncertainty, including: model structure; spatial scale; data requirements (with associated errors); spatial heterogeneities of socio-economic conditions; and agricultural technologies used. Models have been used to estimate the vegetation productivity potential of land (Cai et al. 2011) and to understand the causes of observed vegetation changes. Zhu et al. (2016) used an ensemble of ecosystem models to investigate causes of vegetation changes from 1982–2009, using a factorial simulation approach. They found CO_2 fertilisation to be the dominant effect globally, though climate and land-cover change were the dominant effects in various dryland locations. Borrelli et al. (2017) modelled that about 6.1% of the global land area experienced very high soil erosion rates (exceeding 10 $Mg\ ha^{-1}yr^{-1}$) in 2012, particularly in South America, Africa, and Asia.

Overall, improved estimation and mapping of areas undergoing desertification are needed. This requires a combination of rapidly expanding sources of remotely sensed data, ground observations and new modelling approaches. This is a critical gap, especially in the context of measuring progress towards achieving the land degradation-neutrality target by 2030 in the framework of SDGs.

1.2.1.2   Regional scale

While global-scale studies provide information for any region, there are numerous studies that focus on sub-continental scales, providing more in-depth analysis and understanding. Regional and local studies are important to detect location-specific trends in desertification and heterogeneous influences of climate change on desertification. However, these regional and local studies use a wide variety of methodologies, making direct comparisons difficult. For details of the methodologies applied by each study refer to the individual papers.

Africa

It is estimated that 46 of the 54 countries in Africa are vulnerable to desertification, with some already affected (Prǎvǎlie 2016). Moderate or higher severity degradation over recent decades has been identified in many river basins including the Nile (42% of area), Niger (50%), Senegal (51%), Volta (67%), Limpopo (66%) and Lake Chad (26%) (Thiombiano and Tourino-Soto 2007).

The Horn of Africa is getting drier (Damberg and AghaKouchak 2014; Marshall et al. 2012) exacerbating the desertification already occurring (Oroda 2001). The observed decline in vegetation cover is diminishing ecosystem services (Pricope et al. 2013). Based on NDVI residuals, Kenya experienced persistent negative (positive) trends over 21.6% (8.9%) of the country, for the period 1992–2015 (Gichenje and Godinho 2018). Fragmentation of habitats, reduction in the range of livestock grazing, and higher stocking rates are considered to be the main drivers for vegetation structure loss in the rangelands of Kenya (Kihiu 2016; Otuoma et al. 2009).

Despite desertification in the Sahel being a major concern since the 1970s, wetting and greening conditions have been observed in this region over the last three decades (Anyamba and Tucker 2005; Huber et al. 2011; Brandt et al. 2015; Rishmawi et al. 2016; Tian et al. 2016; Leroux et al. 2017; Herrmann et al. 2005; Damberg and AghaKouchak 2014). Cropland areas in the Sahel region of West Africa have doubled since 1975, with settlement area increasing by about 150% (Traore et al. 2014). Thomas and Nigam (2018) found that the Sahara expanded by 10% over the 20th century based on annual rainfall. In Burkina Faso, Dimobe et al. (2015) estimated that from 1984 to 2013, bare soils and agricultural lands increased by 18.8% and 89.7%, respectively, while woodland, gallery forest, tree savannahs, shrub savannahs and water bodies decreased by 18.8%, 19.4%, 4.8%, 45.2% and 31.2%, respectively. In Fakara region in Niger, a 5% annual reduction in herbaceous yield between 1994 and 2006 was largely explained by changes in land use, grazing pressure and soil fertility (Hiernaux et al. 2009). Aladejana et al. (2018) found that between 1986 and 2015, 18.6% of the forest cover around the Owena River basin was lost. For the period 1982–2003, Le et al. (2012) found that 8% of the Volta River basin's landmass had been degraded, with this increasing to 65% after accounting for the effects of CO_2 (and NOx) fertilisation.

Greening has also been observed in parts of southern Africa but it is relatively weak compared to other regions of the continent (Helldén and Tottrup 2008; Fensholt et al. 2012). However, greening can be accompanied by desertification when factors such as decreasing species richness, changes in species composition and shrub encroachment are observed (Smith et al. 2013; Herrmann and Tappan 2013; Kaptué et al. 2015; Herrmann and Sop 2016; Saha et al. 2015) (Sections 3.1.4 and 3.7.3). In the Okavango river Basin in southern Africa, conversion of land towards higher utilisation intensities, unsustainable agricultural practises and overexploitation of the savanna ecosystems have been observed in recent decades (Weinzierl et al. 2016).

In the arid Algerian High Plateaus, desertification due to both climatic and human causes led to the loss of indigenous plant biodiversity between 1975 and 2006 (Hirche et al. 2011). Ayoub (1998) identified 64 Mha in Sudan as degraded, with the Central North Kordofan state being most affected. However, reforestation measures in the last decade sustained by improved rainfall conditions have led to low-medium regrowth conditions in about 20% of the area (Dawelbait and Morari 2012). In Morocco, areas affected by desertification were predominantly on plains with high population and livestock pressure (del Barrio et al. 2016; Kouba et al. 2018; Lahlaoi et al. 2017). The annual costs of soil degradation were estimated at about 1% of Gross Domestic Product (GDP) in Algeria and Egypt, and about 0.5% in Morocco and Tunisia (Réquier-Desjardins and Bied-Charreton 2006).

Asia

Prǎvǎlie (2016) found that desertification is currently affecting 38 of 48 countries in Asia. The changes in drylands in Asia over the period 1982–2011 were mixed, with some areas experiencing vegetation improvement while others showed reduced vegetation (Miao et al. 2015a). Major river basins undergoing salinisation include: Indo-Gangetic Basin in India (Lal and Stewart 2012), Indus Basin in Pakistan (Aslam and Prathapar 2006), Yellow River Basin in China (Chengrui and Dregne 2001), Yinchuan Plain in China (Zhou et al. 2013), Aral Sea Basin of Central Asia (Cai et al. 2003; Pankova 2016; Qadir et al. 2009).

Helldén and Tottrup (2008) highlighted a greening trend in East Asia between 1982 and 2003. Over the past several decades, air temperature and the rainfall increased in the arid and hyper-arid region of Northwest China (Chen et al. 2015; Wang et al. 2017). Within China, rainfall erosivity has shown a positive trend in dryland areas between 1961 and 2012 (Yang and Lu 2015). While water erosion area in Xinjiang, China, has decreased by 23.2%, erosion considered as severe or intense was still increasing (Zhang et al. 2015). Xue et al. (2017) used remote sensing data covering 1975 to 2015 to show that wind-driven desertified land in northern Shanxi in China had expanded until 2000, before contracting again. Li et al. (2012) used satellite data to identify desertification in Inner Mongolia, China and found a link between policy changes and the locations and extent of human-caused desertification. Several oasis regions in China have seen increases in cropland area, while forests, grasslands and available water resources have decreased (Fu et al. 2017; Muyibul et al. 2018; Xie et al. 2014). Between 1990 and 2011 15.3% of Hogno Khaan nature reserve in central Mongolia was subjected to desertification (Lamchin et al. 2016). Using satellite data Liu et al. (2013) found the area of Mongolia undergoing non-climatic desertification was associated with increases in goat density and wildfire occurrence.

In Central Asia, drying up of the Aral Sea is continuing to have negative impacts on regional microclimate and human health (Issanova and Abuduwaili 2017; Lioubimtseva 2015; Micklin 2016; Xi and Sokolik 2015). Half of the region's irrigated lands, especially in the Amudarya and Syrdarya river basins, were affected by secondary salinisation (Qadir et al. 2009). Le et al. (2016) showed that about 57% of croplands in Kazakhstan and about 20% of croplands in Kyrgyzstan had reductions in their vegetation productivity between 1982 and 2006. Chen et al. (2019) indicated that about 58% of the grasslands in the region had reductions in their vegetation productivity between 1999 and 2015. Anthropogenic factors were the main driver of this loss in Turkmenistan and Uzbekistan, while the role of human drivers was smaller than that of climate-related factors in Tajikistan and Kyrgyzstan (Chen et al. 2019). The total costs of land degradation in Central Asia were estimated to equal about 6 billion USD annually (Mirzabaev et al. 2016).

Damberg and AghaKouchak (2014) found that parts of South Asia experienced drying over the last three decades. More than 75% of the area of northern, western and southern Afghanistan is affected by overgrazing and deforestation (UNEP-GEF 2008). Desertification is a serious problem in Pakistan with a wide range of human and natural causes (Irshad et al. 2007; Lal 2018). Similarly, desertification affects parts of India (Kundu et al. 2017; Dharumarajan et al. 2018; Christian et al. 2018). Using satellite data to map various desertification processes, Ajai et al. (2009) found that 81.4 Mha were subject to various processes of desertification in India in 2005, while salinisation affected 6.73 Mha in the country (Singh 2009).

Saudi Arabia is highly vulnerable to desertification (Ministry of Energy Industry and Mineral Resources 2016), with this vulnerability expected to increase in the north-western parts of the country in the coming decades. Yahiya (2012) found that Jazan, south-western Saudi Arabia, lost about 46% of its vegetation cover from 1987 to 2002. Droughts and frequent dust storms were shown to impose adverse impacts over Saudi Arabia, especially under global warming and future climate change (Hasanean et al. 2015). In north-west Jordan, 18% of the area was prone to severe to very severe desertification (Al-Bakri et al. 2016). Large parts of the Syrian drylands have been identified as undergoing desertification (Evans and Geerken 2004; Geerken and Ilaiwi 2004). Moridnejad et al. (2015) identified newly desertified regions in the Middle East based on dust sources, finding that these regions accounted for 39% of all detected dust source points. Desertification has increased substantially in Iran since the 1930s. Despite numerous efforts to rehabilitate degraded areas, it still poses a major threat to agricultural livelihoods in the country (Amiraslani and Dragovich 2011).

Australia

Damberg and AghaKouchak (2014) found that wetter conditions were experienced in northern Australia over the last three decades with widespread greening observed between 1981 and 2006 over much of Australia, except for eastern Australia where large areas

were affected by droughts from 2002 to 2009 based on Advanced High Resolution Radiometer (AVHRR) satellite data (Donohue et al. 2009). For the period 1982–2013, Burrell et al. (2017) also found widespread greening over Australia including eastern Australia over the post-drought period. This dramatic change in the trend found for eastern Australia emphasises the dominant role played by precipitation in the drylands. Degradation due to anthropogenic activities and other causes affects over 5% of Australia, particularly near the central west coast. Jackson and Prince (2016) used a local NPP scaling approach applied with MODIS derived vegetation data to quantify degradation in a dryland watershed in Northern Australia from 2000 to 2013. They estimated that 20% of the watershed was degraded. Salinisation has also been found to be degrading parts of the Murray-Darling Basin in Australia (Rengasamy 2006). Eldridge and Soliveres (2014) examined areas undergoing woody encroachment in eastern Australia and found that rather than degrading the landscape, the shrubs often enhanced ecosystem services.

Europe

Drylands cover 33.8% of northern Mediterranean countries: approximately 69% of Spain, 66% of Cyprus, and between 16% and 62% in Greece, Portugal, Italy and France (Zdruli 2011). The European Environment Agency (EEA) indicated that 14 Mha, that is 8% of the territory of the European Union (mostly in Bulgaria, Cyprus, Greece, Italy, Romania, Spain and Portugal), had a 'very high' and 'high sensitivity' to desertification (European Court of Auditors 2018). This figure increases to 40 Mha (23% of the EU territory) if 'moderately' sensitive areas are included (Prăvălie et al. 2017; European Court of Auditors 2018). Desertification in the region is driven by irrigation developments and encroachment of cultivation on rangelands (Safriel 2009) caused by population growth, agricultural policies, and markets. According to a recent assessment report (ECA 2018), Europe is increasingly affected by desertification leading to significant consequences on land use, particularly in Portugal, Spain, Italy, Greece, Malta, Cyprus, Bulgaria and Romania. Using the Universal Soil Loss Equation, it was estimated that soil erosion can be as high as 300 t ha^{-1} yr^{-1} (equivalent to a net loss of 18 mm yr^{-1}) in Spain (López-Bermúdez 1990). For the badlands region in south-east Spain, however, it was shown that biological soil crusts effectively prevent soil erosion (Lázaro et al. 2008). In Mediterranean Europe, Guerra et al. (2016) found a reduction of erosion due to greater effectiveness of soil erosion prevention between 2001 and 2013. Helldén and Tottrup (2008) observed a greening trend in the Mediterranean between 1982 and 2003, while Fensholt et al. (2012) also show a dominance of greening in Eastern Europe.

In Russia, at the beginning of the 2000s, about 7% of the total area (that is, approximately 130 Mha) was threatened by desertification (Gunin and Pankova 2004; Kust et al. 2011). Turkey is considered highly vulnerable to drought, land degradation and desertification (Türkeş 1999, 2003). About 60% of Turkey's land area is characterised with hydro-climatological conditions favourable for desertification (Türkeş 2013). ÇEMGM (2017) estimated that about half of Turkey's land area (48.6%) is prone to moderate-to-high desertification.

North America

Drylands cover approximately 60% of Mexico. According to Pontifes et al. (2018), 3.5% of the area was converted from natural vegetation to agriculture and human settlements between 2002 and 2011. The region is highly vulnerable to desertification due to frequent droughts and floods (Méndez and Magaña 2010; Stahle et al. 2009; Becerril-Pina Rocio et al. 2015).

For the period 2000–2011 the overall difference between potential and actual NPP in different land capability classes in the south-western United States was 11.8% (Noojipady et al. 2015); reductions in grassland-savannah and livestock grazing area and forests were the highest. Bush encroachment is observed over a fairly wide area of grasslands in the western United States, including Jornada Basin within the Chihuahuan Desert, and is spreading at a fast rate despite grazing restrictions intended to curb the spread (Yanoff and Muldavin 2008; Browning and Archer 2011; Van Auken 2009; Rachal et al. 2012). In comparing sand dune migration patterns and rates between 1995 and 2014, Potter and Weigand (2016) established that the area covered by stable dune surfaces, and sand removal zones, decreased, while sand accumulation zones increased from 15.4 to 25.5 km^2 for Palen Dunes in the Southern California desert, while movement of Kelso Dunes is less clear (Lam et al. 2011). Within the United States, average soil erosion rates on all croplands decreased by about 38% between 1982 and 2003 due to better soil management practices (Kertis 2003).

Central and South America

Morales et al. (2011) indicated that desertification costs between 8% and 14% of gross agricultural product in many Central and South American countries. Parts of the dry Chaco and Caldenal regions in Argentina have undergone widespread degradation over the last century (Verón et al. 2017; Fernández et al. 2009). Bisigato and Laphitz (2009) identified overgrazing as a cause of desertification in the Patagonian Monte region of Argentina. Vieira et al. (2015) found that 94% of northeast Brazilian drylands were susceptible to desertification. It is estimated that up to 50% of the area was being degraded due to frequent prolonged droughts and clearing of forests for agriculture. This land-use change threatens the extinction of around 28 native species (Leal et al. 2005). In Central Chile, dryland forest and shrubland area was reduced by 1.7% and 0.7%, respectively, between 1975 and 2008 (Schulz et al. 2010).

1.2.2   Attribution of desertification

Desertification is a result of complex interactions within coupled social-ecological systems. Thus, the relative contributions of climatic, anthropogenic and other drivers of desertification vary depending on specific socio-economic and ecological contexts. The high natural climate variability in dryland regions is a major cause of vegetation changes but does not necessarily imply degradation. Drought is not degradation as the land productivity may return entirely once the drought ends (Kassas 1995). However, if droughts increase in frequency, intensity and/or duration they may overwhelm

the vegetation's ability to recover (ecosystem resilience, Prince et al. 2018), causing degradation. Assuming a stationary climate and no human influence, rainfall variability results in fluctuations in vegetation dynamics which can be considered temporary, as the ecosystem tends to recover with rainfall, and desertification does not occur (Ellis 1995; Vetter 2005; von Wehrden et al. 2012). Climate change on the other hand, exemplified by a non-stationary climate, can gradually cause a persistent change in the ecosystem through aridification and CO_2 changes. Assuming no human influence, this 'natural' climatic version of desertification may take place rapidly, especially when thresholds are reached (Prince et al. 2018), or over longer periods of time as the ecosystems slowly adjust to a new climatic norm through progressive changes in the plant community composition. Accounting for this climatic variability is required before attributions to other causes of desertification can be made.

For attributing vegetation changes to climate versus other causes, rain use efficiency (RUE – the change in net primary productivity (NPP) per unit of precipitation) and its variations in time have been used (Prince et al. 1998). Global applications of RUE trends to attribute degradation to climate or other (largely human) causes have been performed by Bai et al. (2008) and Le et al. (2016) (Section 3.2.1.1). The RESTREND (residual trend) method analyses the correlation between annual maximum NDVI (or other vegetation index as a proxy for NPP) and precipitation by testing accumulation and lag periods for the precipitation (Evans and Geerken 2004). The identified relationship with the highest correlation represents the maximum amount of vegetation variability that can be explained by the precipitation, and corresponding RUE values can be calculated. Using this relationship, the climate component of the NDVI time series can be reconstructed, and the difference between this and the original time series (the residual) is attributed to anthropogenic and other causes.

The RESTREND method, or minor variations of it, have been applied extensively. Herrmann and Hutchinson (2005) concluded that climate was the dominant causative factor for widespread greening in the Sahel region from 1982–2003, and anthropogenic and other factors were mostly producing land improvements or no change. However, pockets of desertification were identified in Nigeria and Sudan. Similar results were also found from 1982–2007 by Huber et al. (2011). Wessels et al. (2007) applied RESTREND to South Africa and showed that RESTREND produced a more accurate identification of degraded land than RUE alone. RESTREND identified a smaller area undergoing desertification due to non-climate causes compared to the NDVI trends. Liu et al. (2013) extended the climate component of RESTREND to include temperature and applied this to VOD observations of the cold drylands of Mongolia. They found the area undergoing desertification due to non-climatic causes is much smaller than the area with negative VOD trends. RESTREND has also been applied in several other studies to the Sahel (Leroux et al. 2017), Somalia (Omuto et al. 2010), West Africa (Ibrahim et al. 2015), China (Li et al. 2012; Yin et al. 2014), Central Asia (Jiang et al. 2017), Australia (Burrell et al. 2017) and globally (Andela et al. 2013). In each of these studies the extent to which desertification can be attributed to climate versus other causes varies across the landscape.

These studies represent the best regional, remote-sensing based attribution studies to date, noting that RESTREND and RUE have some limitations (Higginbottom and Symeonakis 2014). Vegetation growth (NPP) changes slowly compared to rainfall variations and may be sensitive to rainfall over extended periods (years), depending on vegetation type. Detection of lags and the use of weighted antecedent rainfall can partially address this problem, though most studies do not do this. The method addresses changes since the start of the time series; it cannot identify whether an area is already degraded at the start time. It is assumed that climate, particularly rainfall, is a principal factor in vegetation change which may not be true in more humid regions.

Another assumption in RESTREND is that any trend is linear throughout the period examined. That is, there are no discontinuities (break points) in the trend. Browning et al. (2017) have shown that break points in NDVI time series reflect vegetation changes based on long-term field sites. To overcome this limitation, Burrell et al. (2017) introduced the Time Series Segmentation-RESTREND (TSS-RESTREND) which allows a breakpoint or turning point within the period examined (Figure 3.7). Using TSS-RESTREND over Australia they identified more than double the degrading area than could be identified with a standard RESTREND analysis. The occurrence and drivers of abrupt change (turning points) in ecosystem functioning were also examined by Horion et al. (2016) over the semi-arid Northern Eurasian agricultural frontier. They combined trend shifts in RUE, field data and expert knowledge, to map environmental hotspots of change and attribute them to climate and human activities. One-third of the area showed significant change in RUE, mainly occurring around the fall of the Soviet Union (1991) or as the result of major droughts. Recent human-induced turning points in ecosystem functioning were uncovered around Volgograd (Russia) and around Lake Balkhash (Kazakhstan), attributed to recultivation, increased salinisation, and increased grazing.

Attribution of vegetation changes to human activity has also been done within modelling frameworks. In these methods ecosystem models are used to simulate potential natural vegetation dynamics, and this is compared to the observed state. The difference is attributed to human activities. Applied to the Sahel region during the period of 1982–2002, it showed that people had a minor influence on vegetation changes (Seaquist et al. 2009). Similar model/observation comparisons performed globally found that CO_2 fertilisation was the strongest forcing at global scales, with climate having regionally varying effects (Mao et al. 2013; Zhu et al. 2016). Land-use/ land-cover change was a dominant forcing in localised areas. The use of this method to examine vegetation changes in China (1982–2009) attributed most of the greening trend to CO_2 fertilisation and nitrogen (N) deposition (Piao et al. 2015). However in some parts of northern and western China, which includes large areas of drylands, Piao et al. (2015) found climate changes could be the dominant forcing. In the northern extratropical land surface, the observed greening was consistent with increases in greenhouse gases (notably CO_2) and the related climate change, and not consistent with a natural climate that does not include anthropogenic increase in greenhouse gases (Mao et al. 2016). While many studies found widespread influence of CO_2 fertilisation, it is not ubiquitous; for example, Lévesque et al.

(2014) found little response to CO_2 fertilisation in some tree species in Switzerland/northern Italy.

Using multiple extreme-event attribution methodologies, Uhe et al. (2018) shows that the dominant influence for droughts in eastern Africa during the October–December 'short rains' season is the prevailing tropical SST patterns, although temperature trends mean that the current drought conditions are hotter than they would have been without climate change. Similarly, Funk et al. (2019) found that 2017 March–June East African drought was influenced by Western Pacific SST, with high SST conditions attributed to climate change.

There are numerous local case studies on attribution of desertification, which use different periods, focus on different land uses and covers, and consider different desertification processes. For example, two-thirds of the observed expansion of the Sahara Desert from 1920–2003 has been attributed to natural climate cycles (the cold phase of Atlantic Multi-Decadal Oscillation and Pacific Decadal Oscillation) (Thomas and Nigam 2018). Some studies consider drought to be the main driver of desertification in Africa (e.g., Masih et al. 2014). However, other studies suggest that although droughts may contribute to desertification, the underlying causes are human activities (Kouba et al. 2018). Brandt et al. (2016a) found that woody vegetation trends are negatively correlated with human population density. Changes in land use, water pumping and flow diversion have enhanced drying of wetlands and salinisation of freshwater aquifers in Israel (Inbar 2007). The dryland territory of China has been found to be very sensitive to both climatic variations and land-use/land-cover changes (Fu et al. 2000; Liu and Tian 2010; Zhao et al. 2013, 2006). Feng et al. (2015) shows that socio-economic factors were dominant in causing desertification in north Shanxi, China, between 1983 and 2012, accounting for about 80% of desertification expansion. Successful grass establishment has been impeded by overgrazing and nutrient depletion leading to the encroachment of shrubs into the northern Chihuahuan Desert (USA) since the mid-19th century (Kidron and Gutschick 2017). Human activities led to rangeland degradation in Pakistan and Mongolia during 2000–2011 (Lei et al. 2011). More equal shares of climatic (temperature and precipitation trends) and human factors were attributed for changes in rangeland condition in China (Yang et al. 2016).

This kaleidoscope of local case studies demonstrates how attribution of desertification is still challenging for several reasons. Firstly, desertification is caused by an interaction of different drivers which vary in space and time. Secondly, in drylands, vegetation reacts closely to changes in rainfall so the effect of rainfall changes on biomass needs to be 'removed' before attributing desertification to human activities. Thirdly, human activities and climatic drivers impact vegetation/ ecosystem changes at different rates. Finally, desertification manifests as a gradual change in ecosystem composition and structure (e.g., woody shrub invasion into grasslands). Although initiated at a limited location, ecosystem change may propagate throughout an extensive area via a series of feedback mechanisms. This complicates the attribution of desertification to human and climatic causes, as the process can develop independently once started.

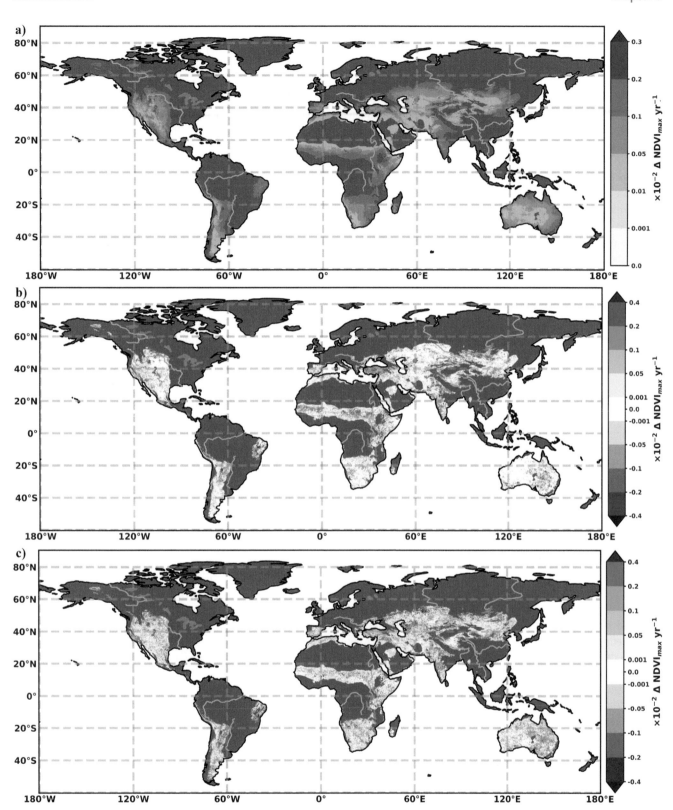

Figure 3.7 | The drivers of dryland vegetation change. The mean annual change in NDVImax between 1982 and 2015 (see Figure 3.6 for total change using Global Inventory Modelling and Mapping Studies NDVI3g v1 dataset) attributable to **(a)** CO_2 fertilisation **(b)** climate and **(c)** land use. The change attributable to CO_2 fertilisation was calculated using the CO_2 fertilisation relationship described in Franks et al. 2013. The Time Series Segmented Residual Trends (TSS-RESTREND) method (Burrell et al. 2017) applied to the CO_2-adjusted NDVI was used to separate Climate and Land Use. A multi-climate dataset ensemble was used to reduce the impact of dataset errors (Burrell et al. 2018). Non-dryland regions (aridity index >0.65) are masked in dark grey. Areas where the change did not meet the multi-run ensemble significance criteria, or are smaller than the error in the sensors (±0.00001) are masked in white.

Rasmussen et al. (2016) studied the reasons behind the overall lack of scientific agreement in trends of environmental changes in the Sahel, including their causes. The study indicated that these are due to differences in conceptualisations and choice of indicators, biases in study site selection, differences in methods, varying measurement accuracy, differences in time and spatial scales. High-resolution, multi-sensor airborne platforms provide a way to address some of these issues (Asner et al. 2012).

The major conclusion of this section is that, with all the shortcomings of individual case studies, relative roles of climatic and human drivers of desertification are location-specific and evolve over time (*high confidence*). Biophysical research on attribution and socio-economic research on drivers of land degradation have long studied the same topic, but in parallel, with little interdisciplinary integration. Interdisciplinary work to identify typical patterns, or typologies, of such interactions of biophysical and human drivers of desertification (not only of dryland vulnerability), and their relative shares, done globally in comparable ways, will help in the formulation of better informed policies to address desertification and achieve land degradation neutrality.

1.3   Desertification feedbacks to climate

While climate change can drive desertification (Section 3.1.4.1), the process of desertification can also alter the local climate, providing a feedback (Sivakumar 2007). These feedbacks can alter the carbon cycle, and hence the level of atmospheric CO_2 and its related global climate change, or they can alter the surface energy and water budgets, directly impacting the local climate. While these feedbacks occur in all climate zones (Chapter 2), here we focus on their effects in dryland regions and assess the literature concerning the major desertification feedbacks to climate. The main feedback pathways discussed throughout Section 3.3 are summarised in Figure 3.8.

Drylands are characterised by limited soil moisture compared to humid regions. Thus, the sensible heat (heat that causes the atmospheric temperature to rise) accounts for more of the surface net radiation than latent heat (evaporation) in these regions (Wang and Dickinson 2013). This tight coupling between the surface energy balance and the soil moisture in semi-arid and dry sub-humid zones makes these regions susceptible to land–atmosphere feedback loops that can amplify changes to the water cycle (Seneviratne et al. 2010). Changes to the land surface caused by desertification can change the surface energy budget, altering the soil moisture and triggering these feedbacks.

1.3.1   Sand and dust aerosols

Sand and mineral dust are frequently mobilised from sparsely vegetated drylands forming 'sand storms' or 'dust storms' (UNEP et al. 2016). The African continent is the most important source of desert dust; perhaps 50% of atmospheric dust comes from the Sahara (Middleton 2017). Ginoux et al. (2012) estimated that 25% of global dust emissions have anthropogenic origins, often in

drylands. These events can play an important role in the local energy balance. Through reducing vegetation cover and drying the surface conditions, desertification can increase the frequency of these events. Biological or structural soil crusts have been shown to effectively stabilise dryland soils. Thus their loss due to intense land use and/or climate change can be expected to cause an increase in sand and dust storms (*high confidence*) (Rajot et al. 2003; Field et al. 2010; Rodriguez-Caballero et al. 2018). These sand and dust aerosols impact the regional climate in several ways (Choobari et al. 2014). The direct effect is the interception, reflection and absorption of solar radiation in the atmosphere, reducing the energy available at the land surface and increasing the temperature of the atmosphere in layers with sand and dust present (Kaufman et al. 2002; Middleton 2017; Kok et al. 2018). The heating of the dust layer can alter the relative humidity and atmospheric stability, which can change cloud lifetimes and water content. This has been referred to as the semi-direct effect (Huang et al. 2017). Aerosols also have an indirect effect on climate through their role as cloud condensation nuclei, changing cloud radiative properties as well as the evolution and development of precipitation (Kaufman et al. 2002). While these indirect effects are more variable than the direct effects, depending on the types and amounts of aerosols present, the general tendency is toward an increase in the number, but a reduction in the size of cloud droplets, increasing the cloud reflectivity and decreasing the chances of precipitation. These effects are referred to as aerosol-radiation and aerosol–cloud interactions (Boucher et al. 2013).

There is *high confidence* that there is a negative relationship between vegetation green-up and the occurrence of dust storms (Engelstaedter et al. 2003; Fan et al. 2015; Yu et al. 2015; Zou and Zhai 2004). Changes in groundwater can affect vegetation and the generation of atmospheric dust (Elmore et al. 2008). This can occur through groundwater processes such as the vertical movement of salt to the surface causing salinisation, supply of near-surface soil moisture, and sustenance of groundwater dependent vegetation. Groundwater dependent ecosystems have been identified in many dryland regions around the world (Decker et al. 2013; Lamontagne et al. 2005; Patten et al. 2008). In these locations declining groundwater levels can decrease vegetation cover. Cook et al. (2009) found that dust aerosols intensified the 'Dust Bowl' drought in North America during the 1930s.

By decreasing the amount of green cover and hence increasing the occurrence of sand and dust storms, desertification will increase the amount of shortwave cooling associated with the direct effect (*high confidence*). There is *medium confidence* that the semi-direct and indirect effects of this dust would tend to decrease precipitation and hence provide a positive feedback to desertification (Huang et al. 2009; Konare et al. 2008; Rosenfeld et al. 2001; Solmon et al. 2012; Zhao et al. 2015). However, the combined effect of dust has also been found to increase precipitation in some areas (Islam and Almazroui 2012; Lau et al. 2009; Sun et al. 2012). The overall combined effect of dust aerosols on desertification remains uncertain with *low agreement* between studies that find positive (Huang et al. 2014), negative (Miller et al. 2004) or no feedback on desertification (Zhao et al. 2015).

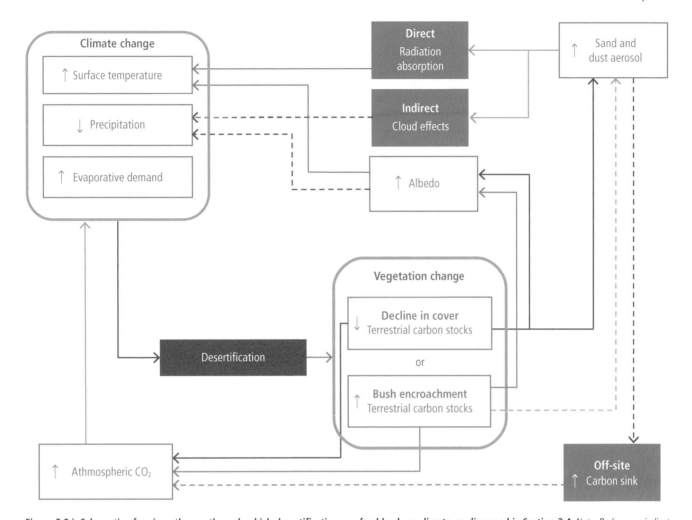

Figure 3.8 | Schematic of main pathways through which desertification can feed back on climate, as discussed in Section 3.4. Note: Red arrows indicate a positive effect. Blue arrows indicate a negative effect. Grey arrows indicate an indeterminate effect (potentially both positive and negative). Solid arrows are direct while dashed arrows are indirect.

1.3.1.1   Off-site feedbacks

Aerosols can act as a vehicle for the long-range transport of nutrients to oceans (Jickells et al. 2005; Okin et al. 2011) and terrestrial land surfaces (Das et al. 2013). In several locations, notably the Atlantic Ocean, the west of northern Africa, and the Pacific Ocean east of northern China, a considerable amount of mineral dust aerosols, sourced from nearby drylands, reaches the oceans. It was estimated that 60% of dust transported off Africa is deposited in the Atlantic Ocean (Kaufman et al. 2005), while 50% of the dust generated in Asia reaches the Pacific Ocean or further (Uno et al. 2009; Zhang et al. 1997). The Sahara is also a major source of dust for the Mediterranean basin (Varga et al. 2014). The direct effect of atmospheric dust over the ocean was found to be a cooling of the ocean surface (*limited evidence, high agreement*) (Evan and Mukhopadhyay 2010; Evan et al. 2009) with the tropical North Atlantic mixed layer cooling by over 1°C (Evan et al. 2009).

It has been suggested that dust may act as a source of nutrients for the upper ocean biota, enhancing the biological activity and related carbon sink (*medium evidence, low agreement*) (Lenes et al. 2001; Shaw et al. 2008; Neuer et al. 2004). The overall response depends on the environmental controls on the ocean biota, the type of aerosols

including their chemical constituents, and the chemical environment in which they dissolve (Boyd et al. 2010).

Dust deposited on snow can increase the amount of absorbed solar radiation leading to more rapid melting (Painter et al. 2018), impacting a region's hydrological cycle (*high confidence*). Dust deposition on snow and ice has been found in many regions of the globe (e.g., Painter et al. 2018; Kaspari et al. 2014; Qian et al. 2015; Painter et al. 2013), however quantification of the effect globally and estimation of future changes in the extent of this effect remain knowledge gaps.

1.3.2   Changes in surface albedo

Increasing surface albedo in dryland regions will impact the local climate, decreasing surface temperature and precipitation, and provide a positive feedback on the albedo (*high confidence*) (Charney et al. 1975). This albedo feedback can occur in desert regions worldwide (Zeng and Yoon 2009). Similar albedo feedbacks have also been found in regional studies over the Middle East (Zaitchik et al. 2007), Australia (Evans et al. 2017; Meng et al. 2014a, b), South America (Lee and Berbery 2012) and the USA (Zaitchik et al. 2013).

Recent work has also found albedo in dryland regions can be associated with soil surface communities of lichens, mosses and cyanobacteria (Rodriguez-Caballero et al. 2018). These communities compose the soil crust in these ecosystems and due to the sparse vegetation cover, directly influence the albedo. These communities are sensitive to climate changes, with field experiments indicating albedo changes greater than 30% are possible. Thus, changes in these communities could trigger surface albedo feedback processes (*limited evidence*, *high agreement*) (Rutherford et al. 2017).

A further pertinent feedback relationship exists between changes in land-cover, albedo, carbon stocks and associated GHG emissions, particularly in drylands with low levels of cloud cover. One of the first studies to focus on the subject was Rotenberg and Yakir (2010), who used the concept of 'radiative forcing' to compare the relative climatic effect of a change in albedo with a change in atmospheric GHGs due to the presence of forest within drylands. Based on this analysis, it was estimated that the change in surface albedo due to the degradation of semi-arid areas has decreased radiative forcing in these areas by an amount equivalent to approximately 20% of global anthropogenic GHG emissions between 1970 and 2005.

1.3.3   Changes in vegetation and greenhouse gas fluxes

Terrestrial ecosystems have the ability to alter atmospheric GHGs through a number of processes (Schlesinger et al. 1990). This may be through a change in plant and soil carbon stocks, either sequestering atmospheric CO_2 during growth or releasing carbon during combustion and respiration, or through processes such as enteric fermentation of domestic and wild ruminants that leads to the release of methane and nitrous oxide (Sivakumar 2007). It is estimated that 241–470 GtC is stored in dryland soils (top 1 m) (Lal 2004; Plaza et al. 2018). When evaluating the effect of desertification, the net balance of all the processes and associated GHG fluxes needs to be considered.

Desertification usually leads to a loss in productivity and a decline in above – and below-ground carbon stocks (Abril et al. 2005; Asner et al. 2003). Drivers such as overgrazing lead to a decrease in both plant and SOC pools (Abdalla et al. 2018). While dryland ecosystems are often characterised by open vegetation, not all drylands have low biomass and carbon stocks in an intact state (Lechmere-Oertel et al. 2005; Maestre et al. 2012). Vegetation types such as the subtropical thicket of South Africa have over 70 tC ha^{-1} in an intact state, greater than 60% of which is released into the atmosphere during degradation through overgrazing (Lechmere-Oertel et al. 2005; Powell 2009). In comparison, semi-arid grasslands and savannahs with similar rainfall, may have only 5–35 tC ha^{-1} (Scholes and Walker 1993; Woomer et al. 2004).

At the same time, it is expected that a decline in plant productivity may lead to a decrease in fuel loads and a reduction in CO_2, nitrous oxide and methane emissions from fire. In a similar manner, decreasing productivity may lead to a reduction in numbers of ruminant animals that in turn would decrease methane emissions. Few studies

have focussed on changes in these sources of emissions due to desertification and it remains a field that requires further research.

In comparison to desertification through the suppression of primary production, the process of woody plant encroachment can result in significantly different climatic feedbacks. Increasing woody plant cover in open rangeland ecosystems leads to an increase in woody carbon stocks both above – and below-ground (Asner et al. 2003; Hughes et al. 2006; Petrie et al. 2015; Li et al. 2016). Within the drylands of Texas, USA, shrub encroachment led to a 32% increase in aboveground carbon stocks over a period of 69 years (3.8 tC ha^{-1} to 5.0 tC ha^{-1}) (Asner et al. 2003). Encroachment by taller woody species can lead to significantly higher observed biomass and carbon stocks. For example, encroachment by *Dichrostachys cinerea* and several Vachellia species in the sub-humid savannahs of north-west South Africa led to an increase of 31–46 tC ha^{-1} over a 50–65 year period (1936–2001) (Hudak et al. 2003). In terms of potential changes in SOC stocks, the effect may be dependent on annual rainfall and soil type. Woody cover generally leads to an increase in SOC stocks in drylands that have less than 800 mm of annual rainfall, while encroachment can lead to a loss of soil carbon in more humid ecosystems (Barger et al. 2011; Jackson et al. 2002).

The suppression of the grass layer through the process of woody encroachment may lead to a decrease in carbon stocks within this relatively small carbon pool (Magandana 2016). Conversely, increasing woody cover may lead to a decrease and even halt in surface fires and associated GHG emissions. In their analysis of drivers of fire in southern Africa, Archibald et al. (2009) note that there is a potential threshold around 40% canopy cover, above which surface grass fires are rare. Whilst there have been a number of studies on changes in carbon stocks due to desertification in North America, southern Africa and Australia, a global assessment of the net change in carbon stocks – as well as fire and ruminant GHG emissions due to woody plant encroachment – has not been done yet.

1.4   Desertification impacts on natural and socio-economic systems under climate change

1.4.1   Impacts on natural and managed ecosystems

1.4.1.1   Impacts on ecosystems and their services in drylands

The Millenium Ecosystem Assessement (2005) proposed four classes of ecosystem services: provisioning, regulating, supporting and cultural services (Cross-Chapter Box 8 in Chapter 6). These ecosystem services in drylands are vulnerable to the impacts of climate change due to high variability in temperature, precipitation and soil fertility (Enfors and Gordon 2008; Mortimore 2005). There is *high confidence* that desertification processes such as soil erosion, secondary salinisation, and overgrazing have negatively impacted provisioning ecosystem services in drylands, particularly food and fodder production (Majeed and Muhammad 2019; Mirzabaev et al. 2016; Qadir et al. 2009; Van Loo et al. 2017; Tokbergenova et al. 2018) (Section 3.4.2.2). Zika and Erb (2009) reported an estimation of NPP losses between 0.8 and 2.0 GtC yr^{-1} due to desertification, comparing the potential NPP and the NPP calculated for the year 2000. In terms of climatic factors, although climatic changes between 1976 and 2016 were found to be favourable for crop yields overall in Russia (Ivanov et al. 2018), yield decreases of up to 40–60% in dryland areas were caused by severe and extensive droughts (Ivanov et al. 2018). Increase in temperature can have a direct impact on animals in the form of increased physiological stress (Rojas-Downing et al. 2017), increased water requirements for drinking and cooling, a decrease in the production of milk, meat and eggs, increased stress during conception and reproduction (Nardone et al. 2010) or an increase in seasonal diseases and epidemics (Thornton et al. 2009; Nardone et al. 2010). Furthermore, changes in temperature can indirectly impact livestock through reducing the productivity and quality of feed crops and forages (Thornton et al. 2009; Polley et al. 2013). On the other hand, fewer days with extreme cold temperatures during winter in the temperate zones are associated with lower livestock mortality. The future projection of impacts on ecosystems is presented in Section 3.5.2.

Over-extraction is leading to groundwater depletion in many dryland areas (*high confidence*) (Mudd 2000; Mays 2013; Mahmod and Watanabe 2014; Jolly et al. 2008). Globally, groundwater reserves have been reduced since 1900, with the highest rate of estimated reductions of 145 km^3 yr^{-1} between 2000 and 2008 (Konikow 2011). Some arid lands are very vulnerable to groundwater reductions, because the current natural recharge rates are lower than during the previous wetter periods (e.g., the Atacama Desert, and Nubian aquifer system in Africa) (Squeo et al. 2006; Mahmod and Watanabe 2014; Herrera et al. 2018).

Among regulating services, desertification can influence levels of atmospheric CO$_2$. In drylands, the majority of carbon is stored below ground in the form of biomass and SOC (FAO 1995) (Section 3.3.3). Land-use changes often lead to reductions in SOC and organic matter inputs into soil (Albaladejo et al. 2013; Almagro et al. 2010; Hoffmann et al. 2012; Lavee et al. 1998; Rey et al. 2011), increasing soil salinity and soil erosion (Lavee et al. 1998; Martinez-Mena et al. 2008). In addition to the loss of soil, erosion reduces soil nutrients and organic matter, thereby impacting land's productive capacity. To illustrate, soil erosion by water is estimated to result in the loss of 23–42 Mt of nitrogen and 14.6–26.4 Mt of phosphorus from soils globally each year (Pierzynski et al. 2017).

Precipitation, by affecting soil moisture content, is considered to be the principal determinant of the capacity of drylands to sequester carbon (Fay et al. 2008; Hao et al. 2008; Mi et al. 2015; Serrano-Ortiz et al. 2015; Vargas et al. 2012; Sharkhuu et al. 2016). Lower annual rainfall resulted in the release of carbon into the atmosphere for a number of sites located in Mongolia, China and North America (Biederman et al. 2017; Chen et al. 2009; Fay et al. 2008; Hao et al. 2008; Mi et al. 2015; Sharkhuu et al. 2016). Low soil water availability promotes soil microbial respiration, yet there is insufficient moisture to stimulate plant productivity (Austin et al. 2004), resulting in net carbon emissions at an ecosystem level. Under even drier conditions, photodegradation of vegetation biomass may often constitute an additional loss of carbon from an ecosystem (Rutledge et al.

2010). In contrast, years of good rainfall in drylands resulted in the sequestration of carbon (Biederman et al. 2017; Chen et al. 2009; Hao et al. 2008). In an exceptionally rainy year (2011) in the southern hemisphere, the semi-arid ecosystems of this region contributed 51% of the global net carbon sink (Poulter et al. 2014). These results suggest that arid ecosystems could be an important global carbon sink, depending on soil water availability (*medium evidence, high agreement*). However, drylands are generally predicted to become warmer with an increasing frequency of extreme drought and high rainfall events (Donat et al. 2016).

When desertification reduces vegetation cover, this alters the soil surface, affecting the albedo and the water balance (Gonzalez-Martin et al. 2014) (Section 3.3). In such situations, erosive winds have no more obstacles, which favours the occurrence of wind erosion and dust storms. Mineral aerosols have an important influence on the dispersal of soil nutrients and lead to changes in soil characteristics (Goudie and Middleton 2001; Middleton 2017). Thereby, the soil formation as a supporting ecosystem service is negatively affected (Section 3.3.1). Soil erosion by wind results in a loss of fine soil particles (silt and clay), reducing the ability of soil to sequester carbon (Wiesmeier et al. 2015). Moreover, dust storms reduce crop yields by loss of plant tissue caused by sandblasting (resulting in loss of plant leaves and hence reduced photosynthetic activity (Field et al. 2010), exposing crop roots, crop seed burial under sand deposits, and leading to losses of nutrients and fertiliser from topsoil (Stefanski and Sivakumar 2009)). Dust storms also impact crop yields by reducing the quantity of water available for irrigation; they can decrease the storage capacity of reservoirs by siltation, and block conveyance canals (Middleton 2017; Middleton and Kang 2017; Stefanski and Sivakumar 2009). Livestock productivity is reduced by injuries caused by dust storms (Stefanski and Sivakumar 2009). Additionally, dust storms favour the dispersion of microbial and plant species, which can make local endemic species vulnerable to extinction and promote the invasion of plant and microbial species (Asem and Roy 2010; Womack et al. 2010). Dust storms increase microbial species in remote sites (*high confidence*) (Kellogg et al. 2004; Prospero et al. 2005; Griffin et al. 2006; Schlesinger et al. 2006; Griffin 2007; De Deckker et al. 2008; Jeon et al. 2011; Abed et al. 2012; Favet et al. 2013; Woo et al. 2013; Pointing and Belnap 2014).

1.4.1.2   Impacts on biodiversity: Plant and wildlife

Plant biodiversity

Over 20% of global plant biodiversity centres are located within drylands (White and Nackoney 2003). Plant species located within these areas are characterised by high genetic diversity within populations (Martínez-Palacios et al. 1999). The plant species within these ecosystems are often highly threatened by climate change and desertification (Millennium Ecosystem Assessment 2005b; Maestre et al. 2012). Increasing aridity exacerbates the risk of extinction of some plant species, especially those that are already threatened due to small populations or restricted habitats (Gitay et al. 2002). Desertification, including through land-use change, already contributed to the loss of biodiversity across drylands (*medium confidence*) (Newbold et al. 2015; Wilting et al. 2017). For example, species richness decreased from 234 species in 1978 to 95 in 2011 following long periods of drought and human driven degradation on the steppe land of south-western Algeria (Observatoire du Sahara et du Sahel 2013). Similarly, drought and overgrazing led to loss of biodiversity in Pakistan to the point that only drought-adapted species can now survive on the arid rangelands (Akhter and Arshad 2006). Similar trends were observed in desert steppes of Mongolia (Khishigbayar et al. 2015). In contrast, the increase in annual moistening of southern European Russia from the late 1980s to the beginning of the 21st century caused the restoration of steppe vegetation, even under conditions of strong anthropogenic pressure (Ivanov et al. 2018). The seed banks of annual species can often survive over the long term, germinating in wet years, suggesting that these species could be resilient to some aspects of climate change (Vetter et al. 2005). Yet, Hiernaux and Houérou (2006) showed that overgrazing in the Sahel tended to decrease the seed bank of annuals, which could make them vulnerable to climate change over time. Perennial species, considered as the structuring element of the ecosystem, are usually less affected as they have deeper roots, xeromorphic properties and physiological mechanisms that increase drought tolerance (Le Houérou 1996). However, in North Africa, long-term monitoring (1978–2014) has shown that important plant perennial species have also disappeared due to drought (*Stipa tenacissima* and *Artemisia herba alba*) (Hirche et al. 2018; Observatoire du Sahara et du Sahel 2013). The aridisation of the climate in the south of Eastern Siberia led to the advance of the steppes to the north and to the corresponding migration of steppe mammal species between 1976 and 2016 (Ivanov et al. 2018). The future projection of impacts on plant biodiversity is presented in Section 3.5.2.

Wildlife biodiversity

Dryland ecosystems have high levels of faunal diversity and endemism (MEA 2005; Whitford 2002). Over 30% of the endemic bird areas are located within these regions, which is also home to 25% of vertebrate species (Maestre et al. 2012; MEA 2005). Yet, many species within drylands are threatened with extinction (Durant et al. 2014; Walther 2016). Habitat degradation and desertification are generally associated with biodiversity loss (Ceballos et al. 2010; Tang et al. 2018; Newbold et al. 2015). The 'grazing value' of land declines with both a reduction in vegetation cover and shrub encroachment, with the former being more detrimental to native vertebrates (Parsons et al. 2017). Conversely, shrub encroachment may buffer desertification by increasing resource and microclimate availability, resulting in an increase in vertebrate species abundance and richness observed in the shrub-encroached arid grasslands of North America (Whitford 1997) and Australia (Parsons et al. 2017). However, compared to historically resilient drylands, these encroached habitats and their new species assemblages may be more sensitive to droughts, which may become more prevalent with climate change (Schooley et al. 2018). Mammals and birds may be particularly sensitive to droughts because they rely on evaporative cooling to maintain their body temperatures within an optimal range (Hetem et al. 2016) and risk lethal dehydration in water limited environments (Albright et al. 2017). The direct effects of reduced rainfall and water availability are *likely* to be exacerbated by the indirect effects

of desertification through a reduction in primary productivity. A reduction in the quality and quantity of resources available to herbivores due to desertification under changing climate can have knock-on consequences for predators and may ultimately disrupt trophic cascades (*limited evidence, low agreement*) (Rey et al. 2017; Walther 2010). Reduced resource availability may also compromise immune response to novel pathogens, with increased pathogen dispersal associated with dust storms (Zinabu et al. 2018). Responses to desertification are species-specific and mechanistic models are not yet able to accurately predict individual species' responses to the many factors associated with desertification (Fuller et al. 2016).

1.4.2   Impacts on socio-economic systems

Combined impacts of desertification and climate change on socio-economic development in drylands are complex. Figure 3.9 schematically represents our qualitative assessment of the magnitudes and the uncertainties associated with these impacts on attainment of the SDGs in dryland areas (UN 2015). The impacts of desertification and climate change are difficult to isolate from the effects of other socio-economic, institutional and political factors (Pradhan et al. 2017). However, there is *high confidence* that climate change will exacerbate the vulnerability of dryland populations to desertification, and that the combination of pressures coming from climate change and desertification will diminish opportunities for reducing poverty, enhancing food and nutritional security, empowering women, reducing disease burden, and improving access to water and sanitation. Desertification is embedded in SDG 15 (Target 15.3) and climate change is under SDG 13. The *high confidence* and high magnitude impacts depicted for these SDGs (Figure 3.9) indicate that the interactions between desertification and climate change strongly affect the achievement of the targets of SDGs 13 and 15.3, pointing at the need for the coordination of policy actions on land degradation neutrality and mitigation and adaptation to climate change. The following subsections present the literature and assessments which serve as the basis for Figure 3.9.

1.4.2.1   Impacts on poverty

Climate change has a high potential to contribute to poverty particularly through the risks coming from extreme weather events (Olsson et al. 2014). However, the evidence rigourously attributing changes in observed poverty to climate change impacts is currently not available. On the other hand, most of the research on links between poverty and desertification (or more broadly, land degradation) focused on whether or not poverty is a cause of land degradation (Gerber et al. 2014; Vu et al. 2014; Way 2016) (Section 4.7.1). The literature measuring the extent to which desertification contributed to poverty globally is lacking: the related literature remains qualitative or correlational (Barbier and Hochard 2016). At the local level, on the other hand, there is *limited evidence* and *high agreement* that desertification increased multidimensional poverty. For example, Diao and Sarpong (2011) estimated that land degradation lowered agricultural incomes in Ghana by 4.2 billion USD between 2006 and 2015, increasing the national poverty rate by 5.4% in 2015. Land degradation increased the probability of households becoming poor

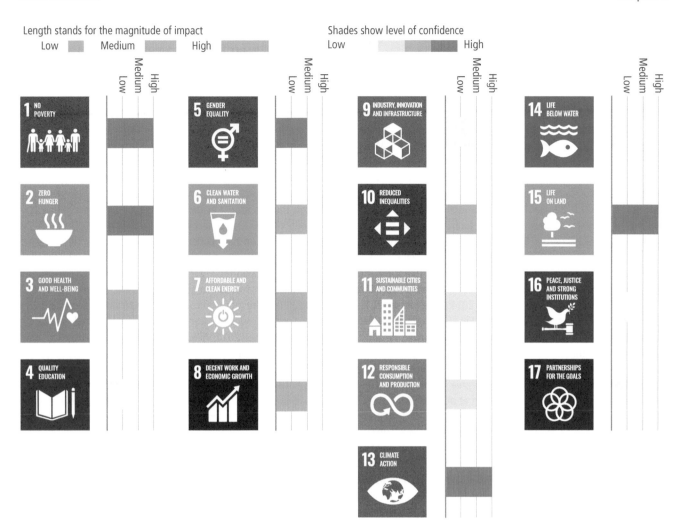

Figure 3.9 | Socio-economic impacts of desertification and climate change with the SDG framework.

by 35% in Malawi and 48% in Tanzania (Kirui 2016). Desertification in China was found to have resulted in substantial losses in income, food production and jobs (Jiang et al. 2014). On the other hand, Ge et al. (2015) indicated that desertification was positively associated with growing incomes in Inner Mongolia in China in the short run since no costs were incurred for SLM, while in the long run higher incomes allowed allocation of more investments to reduce desertification. This relationship corresponds to the Environmental Kuznets Curve, which posits that environmental degradation initially rises and subsequently falls with rising income (e.g., Stern 2017). There is *limited evidence* on the validity of this hypothesis regarding desertification.

1.4.2.2   Impacts on food and nutritional insecurity

About 821 million people globally were food insecure in 2017, of whom 63% in Asia, 31% in Africa and 5% in Latin America and the Caribbean (FAO et al. 2018). The global number of food insecure people rose by 37 million since 2014. Changing climate variability, combined with a lack of climate resilience, was suggested as a key driver of this increase (FAO et al. 2018). Sub-Saharan Africa, East Africa and South Asia had the highest share of undernourished populations in the world in 2017, with 28.8%, 31.4% and 33.7% respectively (FAO et al. 2018). The major mechanism through which

climate change and desertification affect food security is through their impacts on agricultural productivity. There is *robust evidence* pointing to negative impacts of climate change on crop yields in dryland areas (*high agreement*) (Hochman et al. 2017; Nelson et al. 2010; Zhao et al. 2017) (Sections 3.4.1, 5.2.2 and 4.7.2). There is also *robust evidence* and *high agreement* on the losses in agricultural productivity and incomes due to desertification (Kirui 2016; Moussa et al. 2016; Mythili and Goedecke 2016; Tun et al. 2015). Nkonya et al. (2016a) estimated that cultivating wheat, maize, and rice with unsustainable land management practices is currently resulting in global losses of 56.6 billion USD annually, with another 8.7 billion USD of annual losses due to lower livestock productivity caused by rangeland degradation. However, the extent to which these losses affected food insecurity in dryland areas is not known. Lower crop yields and higher agricultural prices worsen existing food insecurity, especially for net food-buying rural households and urban dwellers. Climate change and desertification are not the sole drivers of food insecurity, but especially in the areas with high dependence on agriculture, they are among the main contributors.

1.4.2.3   Impacts on human health through dust storms

The frequency and intensity of dust storms are increasing due to land-use and land-cover changes and climate-related factors (Section 2.4) particularly in some regions of the world such as the Arabian Peninsula (Jish Prakash et al. 2015; Yu et al. 2015; Gherboudj et al. 2017; Notaro et al. 2013; Yu et al. 2013; Alobaidi et al. 2017; Maghrabi et al. 2011; Almazroui et al. 2018) and broader Middle East (Rashki et al. 2012; Türkeş 2017; Namdari et al. 2018) as well as Central Asia (Indoitu et al. 2015; Xi and Sokolik 2015), with growing negative impacts on human health (*high confidence*) (Díaz et al. 2017; Goudarzi et al. 2017; Goudie 2014; Samoli et al. 2011). Dust storms transport particulate matter, pollutants, pathogens and potential allergens that are dangerous for human health over long distances (Goudie and Middleton 2006; Sprigg 2016). Particulate matter (PM; that is, the suspended particles in the air of up to 10 micrometres (PM10) or less in size), have damaging effects on human health (Díaz et al. 2017; Goudarzi et al. 2017; Goudie 2014; Samoli et al. 2011). The health effects of dust storms are largest in areas in the immediate vicinity of their origin, primarily the Sahara Desert, followed by Central and eastern Asia, the Middle East and Australia (Zhang et al. 2016), however, there is *robust evidence* showing that the negative health effects of dust storms reach a much wider area (Bennett et al. 2006; Díaz et al. 2017; Kashima et al. 2016; Lee et al. 2014; Samoli et al. 2011; Zhang et al. 2016). The primary health effects of dust storms include damage to the respiratory and cardiovascular systems (Goudie 2013). Dust particles with a diameter smaller than 2.5μm were associated with global cardiopulmonary mortality of about 402,000 people in 2005, with 3.47 million years of life lost in that single year (Giannadaki et al. 2014). Although globally only 1.8% of cardiopulmonary deaths were caused by dust storms, in the countries of the Sahara region, Middle East, South and East Asia, dust storms were suggested to be the cause of 15–50% of all cardiopulmonary deaths (Giannadaki et al. 2014). A 10 μgm^{-3} increase in PM10 dust particles was associated with mean increases in non-accidental mortality from 0.33% to 0.51% across different calendar seasons in China, Japan and South Korea (Kim et al. 2017). The percentage of all-cause deaths attributed to fine particulate matter in Iranian cities affected by Middle Eastern dust storms (MED) was 0.56–5.02%, while the same percentage for non-affected cities was 0.16–4.13% (Hopke et al. 2018). Epidemics of meningococcal meningitis occur in the Sahelian region during the dry seasons with dusty conditions (Agier et al. 2012; Molesworth et al. 2003). Despite a strong concentration of dust storms in the Sahel, North Africa, the Middle East and Central Asia, there is relatively little research on human health impacts of dust storms in these regions. More research on health impacts and related costs of dust storms, as well as on public health response measures, can help in mitigating these health impacts.

1.4.2.4   Impacts on gender equality

Environmental issues such as desertification and impacts of climate change have been increasingly investigated through a gender lens (Bose (n.d.); Broeckhoven and Cliquet 2015; Kaijser and Kronsell 2014; Kiptot et al. 2014; Villamor and van Noordwijk 2016). There is *medium evidence* and *high agreement* that women will be impacted more than men by environmental degradation (Arora-Jonsson 2011;

Gurung et al. 2006) (Cross-Chapter Box 11 in Chapter 7). Socially structured gender-specific roles and responsibilities, daily activities, access and control over resources, decision-making and opportunities lead men and women to interact differently with natural resources and landscapes. For example, water scarcity affected women more than men in rural Ghana as they had to spend more time in fetching water, which has implications on time allocations for other activities (Ahmed et al. 2016). Despite the evidence pointing to differentiated impact of environmental degradation on women and men, gender issues have been marginally addressed in many land restoration and rehabilitation efforts, which often remain gender-blind. Although there is *robust evidence* on the location-specific impacts of climate change and desertification on gender equality, there is *limited evidence* on the gender-related impacts of land restoration and rehabilitation activities. Women are usually excluded from local decision-making on actions regarding desertification and climate change. Socially constructed gender-specific roles and responsibilities are not static because they are shaped by other factors such as wealth, age, ethnicity and formal education (Kaijser and Kronsell 2014; Villamor et al. 2014). Hence, women's and men's environmental knowledge and priorities for restoration often differ (Sijapati Basnett et al. 2017). In some areas where sustainable land options (e.g., agroforestry) are being promoted, women were not able to participate due to culturally embedded asymmetries in power relations between men and women (Catacutan and Villamor 2016). Nonetheless women, particularly in the rural areas, remain heavily involved in securing food for their households. Food security for them is associated with land productivity and women's contribution to address desertification is crucial.

1.4.2.5   Impacts on water scarcity and use

Reduced water retention capacity of degraded soils amplifies floods (de la Paix et al. 2011), reinforces degradation processes through soil erosion, and reduces annual intake of water to aquifers, exacerbating existing water scarcities (Le Roux et al. 2017; Cano et al. 2018). Reduced vegetation cover and more intense dust storms were found to intensify droughts (Cook et al. 2009). Moreover, secondary salinisation in the irrigated drylands often requires leaching with considerable amounts of water (Greene et al. 2016; Wichelns and Qadir 2015). Thus, different types of soil degradation increase water scarcity both through lower water quantity and quality (Liu et al. 2017; Liu et al. 2016c). All these processes reduce water availability for other needs. In this context, climate change will further intensify water scarcity in some dryland areas and increase the frequency of droughts (*medium confidence*) (IPCC 2013; Zheng et al. 2018) (Section 2.2). Higher water scarcity may imply growing use of wastewater effluents for irrigation (Pedrero et al. 2010). The use of untreated wastewater exacerbates soil degradation processes (Tal 2016; Singh et al. 2004; Qishlaqi et al. 2008; Hanjra et al. 2012), in addition to negative human health impacts (Faour-Klingbeil and Todd 2018; Hanjra et al. 2012). Climate change will thus amplify the need for integrated land and water management for sustainable development.

1.4.2.6   Impacts on energy infrastructure through dust storms

Desertification leads to conditions that favour the production of dust storms (*high confidence*) (Section 3.3.1). There is *robust evidence* and *high agreement* that dust storms negatively affect the operational potential of solar and wind power harvesting equipment through dust deposition, reduced reach of solar radiation and increasing blade-surface roughness, and can also reduce effective electricity distribution in high-voltage transmission lines (Zidane et al. 2016; Costa et al. 2016; Lopez-Garcia et al. 2016; Maliszewski et al. 2012; Mani and Pillai 2010; Mejia and Kleissl 2013; Mejia et al. 2014; Middleton 2017; Sarver et al. 2013; Kaufman et al. 2002; Kok et al. 2018). Direct exposure to desert dust storm can reduce energy generation efficiency of solar panels by 70–80% in one hour (Ghazi et al. 2014). (Saidan et al. 2016) indicated that in the conditions of Baghdad, Iraq, one month's exposure to weather reduced the efficiency of solar modules by 18.74% due to dust deposition. In the Atacama desert, Chile, one month's exposure reduced thin-film solar module performance by 3.7–4.8% (Fuentealba et al. 2015). This has important implications for climate change mitigation efforts using the expansion of solar and wind energy generation in dryland areas for substituting fossil fuels. Abundant access to solar energy in many dryland areas makes them high-potential locations for the installation of solar energy generating infrastructure. Increasing desertification, resulting in higher frequency and intensity of dust storms imposes additional costs for climate change mitigation through deployment of solar and wind energy harvesting facilities in dryland areas. Most frequently used solutions to this problem involve physically wiping or washing the surface of solar devices with water. These result in additional costs and excessive use of already scarce water resources and labour (Middleton 2017). The use of special coatings on the surface of solar panels can help prevent the deposition of dusts (Costa et al. 2016; Costa et al. 2018; Gholami et al. 2017).

1.4.2.7   Impacts on transport infrastructure through dust storms and sand movement

Dust storms and movement of sand dunes often threaten the safety and operation of railway and road infrastructure in arid and hyper-arid areas, and can lead to road and airport closures due to reductions in visibility. For example, the dust storm on 10 March 2009 over Riyadh was assessed to be the strongest in the previous two decades in Saudi Arabia, causing limited visibility, airport shutdown and damages to infrastructure and environment across the city (Maghrabi et al. 2011). There are numerous historical examples of how moving sand dunes led to the forced decommissioning of early railway lines built in Sudan, Algeria, Namibia and Saudi Arabia in the late 19th and early 20th century (Bruno et al. 2018). Currently, the highest concentrations of railways vulnerable to sand movements are located in north-western China, Middle East and North Africa (Bruno et al. 2018; Cheng and Xue 2014). In China, sand dune movements are periodically disrupting the railway transport on the Linhai–Ceke line in north-western China and on the Lanzhou–Xinjiang High-speed Railway in western China, with considerable clean-up and maintenance costs (Bruno et al. 2018; Zhang et al. 2010). There are large-scale plans for expansion of railway networks in arid areas of China, Central Asia, North Africa, the Middle East, and eastern Africa. For example, The Belt and Road Initiative

promoted by China, the Gulf Railway project by the Cooperation Council for the Arab States of the Gulf or Lamu Port–South Sudan–Ethiopia Transport (LAPSSET) Corridor in Eastern Africa. These investments have long-term return and operation periods. Their construction and associated engineering solutions will therefore benefit from careful consideration of potential desertification and climate change effects on sand storms and dune movements.

1.4.2.8   Impacts on conflicts

There is *low confidence* in climate change and desertification leading to violent conflicts. There is *medium evidence* and *low agreement* that climate change and desertification contribute to already existing conflict potentials (Herrero 2006; von Uexkull et al. 2016; Theisen 2017; Olsson 2017; Wischnath and Buhaug 2014) (Section 4.7.3). To illustrate, Hsiang et al. (2013) found that each one standard deviation increase in temperature or rainfall was found to increase interpersonal violence by 4% and intergroup conflict by 14% (Hsiang et al. 2013). However, this conclusion was disputed by Buhaug et al. (2014), who found no evidence linking climate variability to violent conflict after replicating Hsiang et al. (2013) by studying only violent conflicts. Almer et al. (2017) found that a one standard deviation increase in dryness raised the likelihood of riots in Sub-Saharan African countries by 8.3% during the 1990–2011 period. On the other hand, Owain and Maslin (2018) found that droughts and heatwaves were not significantly affecting the level of regional conflict in East Africa. Similarly, it was suggested that droughts and desertification in the Sahel played a relatively minor role in the conflicts in the Sahel in the 1980s, with the major reasons for the conflicts during this period being political, especially the marginalisation of pastoralists (Benjaminsen 2016), corruption and rent-seeking (Benjaminsen et al. 2012). Moreover, the role of environmental factors as the key drivers of conflicts was questioned in the case of Sudan (Verhoeven 2011) and Syria (De Châtel 2014). Selection bias, when the literature focuses on the same few regions where conflicts occurred and relates them to climate change, is a major shortcoming, as it ignores other cases where conflicts did not occur (Adams et al. 2018) despite degradation of the natural resource base and extreme weather events.

1.4.2.9   Impacts on migration

Environmentally induced migration is complex and accounts for multiple drivers of mobility as well as other adaptation measures undertaken by populations exposed to environmental risk (*high confidence*). There is *medium evidence* and *low agreement* that climate change impacts migration. The World Bank (2018) predicted that 143 million people would be forced to move internally by 2050 if no climate action is taken. Focusing on asylum seekers alone, rather than the total number of migrants, Missirian and Schlenker (2017) predict that asylum applications to the European Union will increase from 28% (98,000 additional asylum applications per year) up to 188% (660,000 additional applications per year) depending on the climate scenario by 2100. While the modelling efforts have greatly improved over the years (Hunter et al. 2015; McLeman 2011; Sherbinin and Bai 2018) and in particular, these recent estimates provide an important insight into potential future developments, the quantitative projections are still based on the number of people exposed to risk rather than the number of people who would actually engage in migration as a response to this risk (Gemenne 2011; McLeman 2013) and they do not take into account individual agency in migration decision nor adaptive capacities of individuals (Hartmann 2010; Kniveton et al. 2011; Piguet 2010) (see Section 3.6.2 discussing migration as a response to desertification). Accordingly, the available micro-level evidence suggests that climate-related shocks are one of the many drivers of migration (Adger et al. 2014; London Government Office for Science and Foresight 2011; Melde et al. 2017), but the individual responses to climate risk are more complex than commonly assumed (Gray and Mueller 2012a). For example, despite strong focus on natural disasters, neither flooding (Gray and Mueller 2012b; Mueller et al. 2014) nor earthquakes (Halliday 2006) were found to induce long-term migration; but instead, slow-onset changes, especially those provoking crop failures and heat stress, could affect household or individual migration decisions (Gray and Mueller 2012a; Missirian and Schlenker 2017; Mueller et al. 2014). Out-migration from drought-prone areas has received particular attention (de Sherbinin et al. 2012; Ezra and Kiros 2001). A substantial body of literature suggests that households engage in local or internal migration as a response to drought (Findlay 2011; Gray and Mueller 2012a), while international migration decreases with drought in some contexts (Henry et al. 2004), but might increase in contexts where migration networks are well established (Feng et al. 2010; Nawrotzki and DeWaard 2016; Nawrotzki et al. 2015, 2016). Similarly, the evidence is not conclusive with respect to the effect of environmental drivers, in particular desertification, on mobility. While it has not consistently entailed out-migration in the case of Ecuadorian Andes (Gray 2009, 2010), environmental and land degradation increased mobility in Kenya and Nepal (Gray 2011; Massey et al. 2010), but marginally decreased mobility in Uganda (Gray 2011). These results suggest that in some contexts, environmental shocks actually undermine households' financial capacity to undertake migration (Nawrotzki and Bakhtsiyarava 2017), especially in the case of the poorest households (Barbier and Hochard 2018; Koubi et al. 2016; Kubik and Maurel 2016; McKenzie and Yang 2015). Adding to the complexity, migration, especially to frontier areas, by increasing pressure on land and natural resources, might itself contribute to environmental degradation at the destination (Hugo 2008; IPBES 2018a; McLeman 2017). The consequences of migration can also be salient in the case of migration to urban or peri-urban areas; indeed, environmentally induced migration can add to urbanisation (Section 3.6.2.2), often exacerbating problems related to poor infrastructure and unemployment.

1.4.2.10   Impacts on pastoral communities

Pastoral production systems occupy a significant portion of the world (Rass 2006; Dong 2016). Food insecurity among pastoral households is often high (Gomes 2006) (Section 3.1.3). The Sahelian droughts of the 1970s–1980s provided an example of how droughts could affect livestock resources and crop productivity, contributing to hunger, out-migration and suffering for millions of pastoralists (Hein and De Ridder 2006; Molua and Lambi 2007). During these Sahelian droughts low and erratic rainfall exacerbated desertification processes, leading to ecological changes that forced people to use marginal lands and ecosystems. Similarly, the rate of rangeland

degradation is now increasing because of environmental changes and overexploitation of resources (Kassahun et al. 2008; Vetter 2005). Desertification coupled with climate change is negatively affecting livestock feed and grazing species (Hopkins and Del Prado 2007), changing the composition in favour of species with low forage quality, ultimately reducing livestock productivity (D'Odorico et al. 2013; Dibari et al. 2016) and increasing livestock disease prevalence (Thornton et al. 2009). There is *robust evidence* and *high agreement* that weak adaptive capacity, coupled with negative effects from other climate-related factors, are predisposing pastoralists to increased poverty from desertification and climate change globally (López-i-Gelats et al. 2016; Giannini et al. 2008; IPCC 2007). On the other hand, misguided policies such as enforced sedentarisation, and in certain cases protected area delineation (fencing), which restrict livestock mobility have hampered optimal use of grazing land resources (Du 2012). Such policies have led to degradation of resources and out-migration of people in search of better livelihoods (Gebeye 2016; Liao et al. 2015). Restrictions on the mobile lifestyle are reducing the resilient adaptive capacity of pastoralists to natural hazards including extreme and variable weather conditions, drought and climate change (Schilling et al. 2014). Furthermore, the exacerbation of the desertification phenomenon due to agricultural intensification (D'Odorico et al. 2013) and land fragmentation caused by encroachment of agriculture into rangelands (Otuoma et al. 2009; Behnke and Kerven 2013) is threatening pastoral livelihoods. For example, commercial cotton (*Gossypium hirsutum*) production is crowding out pastoral systems in Benin (Tamou et al. 2018). Food shortages and the urgency to produce enough crop for public consumption are leading to the encroachment of agriculture into productive rangelands and those converted rangelands are frequently prime lands used by pastoralists to produce feed and graze their livestock during dry years (Dodd 1994). The sustainability of pastoral systems is therefore coming into question because of social and political marginalisation of those systems (Davies et al. 2016) and also because of the fierce competition they are facing from other livelihood sources such as crop farming (Haan et al. 2016).

1.5   Future projections

1.5.1   Future projections of desertification

Assessing the impact of climate change on future desertification is difficult as several environmental and anthropogenic variables interact to determine its dynamics. The majority of modelling studies regarding the future evolution of desertification rely on the analysis of specific climate change scenarios and Global Climate Models (GCMs) and their effect on a few processes or drivers that trigger desertification (Cross-Chapter Box 1 in Chapter 1).

With regards to climate impacts, the analysis of global and regional climate models concludes that under all representative concentration pathways (RCPs) potential evapotranspiration (PET) would increase worldwide as a consequence of increasing surface temperatures and surface water vapour deficit (Sherwood and Fu 2014). Consequently, there would be associated changes in aridity indices that depend on this variable (*high agreement*, *robust evidence*) (Cook et al. 2014a;

Dai 2011; Dominguez et al. 2010; Feng and Fu 2013; Ficklin et al. 2016; Fu et al. 2016; Greve and Seneviratne 1999; Koutroulis 2019; Scheff and Frierson 2015). Due to the large increase in PET and decrease in precipitation over some subtropical land areas, aridity index will decrease in some drylands (Zhao and Dai 2015), with one model estimating approximately 10% increase in hyper-arid areas globally (Zeng and Yoon 2009). Increases in PET are projected to continue due to climate change (Cook et al. 2014a; Fu et al. 2016; Lin et al. 2015; Scheff and Frierson 2015). However, as noted in Sections 3.1.1 and 3.2.1, these PET calculations use assumptions that are not valid in an environment with changing CO_2. Evidence from precipitation, runoff or photosynthetic uptake of CO_2 suggest that a future warmer world will be less arid (Roderick et al. 2015). Observations in recent decades indicate that the Hadley cell has expanded poleward in both hemispheres (Fu et al. 2006; Hu and Fu 2007; Johanson et al. 2009; Seidel and Randel 2007), and under all RCPs would continue expanding (Johanson et al. 2009; Lu et al. 2007). This expansion leads to the poleward extension of subtropical dry zones and hence an expansion in drylands on the poleward edge (Scheff and Frierson 2012). Overall, this suggests that while aridity will increase in some places (*high confidence*), there is insufficient evidence to suggest a global change in dryland aridity (*medium confidence*).

Regional modelling studies confirm the outcomes of Global Climate Models (Africa: Terink et al. 2013; China: Yin et al. 2015; Brazil: Marengo and Bernasconi 2015; Cook et al. 2012; Greece: Nastos et al. 2013; Italy: Coppola and Giorgi 2009). According to the IPCC AR5 (IPCC 2013), decreases in soil moisture are detected in the Mediterranean, southwest USA and southern African regions. This is in line with alterations in the Hadley circulation and higher surface temperatures. This surface drying will continue to the end of this century under the RCP8.5 scenario (*high confidence*). Ramarao et al. (2015) showed that a future climate projection based on RCP4.5 scenario indicated the possibility for detecting the summer-time soil drying signal over the Indian region during the 21st century in response to climate change. The IPCC Special Report on Global Warming of 1.5°C (SR15) (Chapter 3; Hoegh-Guldberg et al. 2018) concluded with '*medium confidence*' that global warming by more than 1.5°C increases considerably the risk of aridity for the Mediterranean area and southern Africa. Miao et al. (2015b) showed an acceleration of desertification trends under the RCP8.5 scenario in the middle and northern part of Central Asia and some parts of north-western China. It is also useful to consider the effects of the dynamic–thermodynamical feedback of the climate. Schewe and Levermann (2017) show increases of up to 300% in the Central Sahel rainfall by the end of the century due to an expansion of the West African monsoon. Warming could trigger an intensification of monsoonal precipitation due to increases in ocean moisture availability.

The impacts of climate change on dust storm activity are not yet comprehensively studied and represent an important knowledge gap. Currently, GCMs are unable to capture recent observed dust emission and transport (Evan 2018; Evan et al. 2014), limiting confidence in future projections. Literature suggests that climate change decreases wind erosion/dust emission overall, with regional variation (*low confidence*). Mahowald et al. (2006) and Mahowald (2007) found that climate change led to a decrease in desert dust

source areas globally using CMIP3 GCMs. Wang et al. (2009) found a decrease in sand dune movement by 2039 (increasing thereafter) when assessing future wind-erosion-driven desertification in arid and semi-arid China using a range of SRES scenarios and HadCM3 simulations. Dust activity in the Southern Great Plains in the USA was projected to increase, while in the Northern Great Plains it was projected to decrease under RCP8.5 climate change scenario (Pu and Ginoux 2017). Evan et al. (2016) project a decrease in African dust emission associated with a slowdown of the tropical circulation in the high CO_2 RCP8.5 scenario.

Global estimates of the impact of climate change on soil salinisation show that under the IS92a emissions scenario (a scenario prepared in 1992 that contains 'business as usual' assumptions) (Leggett et al. 1992) the area at risk of salinisation would increase in the future (*limited evidence, high agreement*) (Schofield and Kirkby 2003). Climate change has an influence on soil salinisation that induces further land degradation through several mechanisms that vary in their level of complexity. However, only a few examples can be found to illustrate this range of impacts, including the effect of groundwater table depletion (Rengasamy 2006) and irrigation management (Sivakumar 2007), salt migration in coastal aquifers with decreasing water tables (Sherif and Singh 1999) (Section 4.10.7), and surface hydrology and vegetation that affect wetlands and favour salinisation (Nielsen and Brock 2009).

1.5.1.1   Future vulnerability and risk of desertification

Following the conceptual framework developed in the Special Report on extreme events (SREX) (IPCC 2012), future risks are assessed by examining changes in exposure (that is, presence of people; livelihoods; species or ecosystems; environmental functions, service, and resources; infrastructure; or economic, social or cultural assets; see Glossary), changes in vulnerability (that is, propensity or predisposition to be adversely affected; see Glossary) and changes in the nature and magnitude of hazards (that is, potential occurrence of a natural or human-induced physical event that causes damage; see Glossary). Climate change is expected to further exacerbate the vulnerability of dryland ecosystems to desertification by increasing PET globally (Sherwood and Fu 2014). Temperature increases between 2°C and 4°C are projected in drylands by the end of the 21st century under RCP4.5 and RCP8.5 scenarios, respectively (IPCC 2013). An assessment by Carrão et al. 2017 showed an increase in drought hazards by late-century (2071–2099) compared to a baseline (1971–2000) under high RCPs in drylands around the Mediterranean, south-eastern Africa, and southern Australia. In Latin America, Morales et al. (2011) indicated that areas affected by drought will increase significantly by 2100 under SRES scenarios A2 and B2. The countries expected to be affected include Guatemala, El Salvador, Honduras and Nicaragua. In CMIP5 scenarios, Mediterranean types of climate are projected to become drier (Alessandri et al. 2014; Polade et al. 2017), with the equatorward margins being potentially replaced by arid climate types (Alessandri et al. 2014). Globally, climate change is predicted to intensify the occurrence and severity of droughts (*medium confidence*) (Dai 2013; Sheffield and Wood 2008; Swann et al. 2016; Wang 2005; Zhao and Dai 2015; Carrão et al. 2017; Naumann et al. 2018) (Section 2.2). Ukkola et al. (2018)

showed large discrepancies between CMIP5 models for all types of droughts, limiting the confidence that can be assigned to projections of drought.

Drylands are characterised by high climatic variability. Climate impacts on desertification are not only defined by projected trends in mean temperature and precipitation values but are also strongly dependent on changes in climate variability and extremes (Reyer et al. 2013). The responses of ecosystems depend on diverse vegetation types. Drier ecosystems are more sensitive to changes in precipitation and temperature (Li et al. 2018; Seddon et al. 2016; You et al. 2018), increasing vulnerability to desertification. It has also been reported that areas with high variability in precipitation tend to have lower livestock densities and that those societies that have a strong dependence on livestock that graze natural forage are especially affected (Sloat et al. 2018). Social vulnerability in drylands increases as a consequence of climate change that threatens the viability of pastoral food systems (Dougill et al. 2010; López-i-Gelats et al. 2016). Social drivers can also play an important role with regards to future vulnerability (Máñez Costa et al. 2011). In the arid region of north-western China, Liu et al. (2016b) estimated that under RCP4.5 areas of increased vulnerability to climate change and desertification will surpass those with decreased vulnerability.

Using an ensemble of global climate, integrated assessment and impact models, Byers et al. (2018) investigated 14 impact indicators at different levels of global mean temperature change and socio-economic development. The indicators cover water, energy and land sectors. Of particular relevance to desertification are the water (e.g., water stress, drought intensity) and the land (e.g., habitat degradation) indicators. Under shared socio-economic pathway SSP2 ('Middle of the Road') at 1.5°C, 2°C and 3°C of global warming, the numbers of dryland populations exposed (vulnerable) to various impacts related to water, energy and land sectors (e.g., water stress, drought intensity, habitat degradation) are projected to reach 951 (178) million, 1152 (220) million and 1285 (277) million, respectively. While at global warming of 2°C, under SSP1 ('Sustainability'), the exposed (vulnerable) dryland population is 974 (35) million, and under SSP3 ('Fragmented World') it is 1267 (522) million. Steady increases in the exposed and vulnerable populations are seen for increasing global mean temperatures. However much larger differences are seen in the vulnerable population under different SSPs. Around half the vulnerable population is in South Asia, followed by Central Asia, West Africa and East Asia.

1.5.2   Future projections of impacts

Future climate change is expected to increase the potential for increased soil erosion by water in dryland areas (*medium confidence*). Yang et al. (2003) use a Revised Universal Soil Loss Equation (RUSLE) model to study global soil erosion under historical, present and future conditions of both cropland and climate. Soil erosion potential has increased by about 17%, and climate change will increase this further in the future. In northern Iran, under the SRES A2 emission scenario the mean erosion potential is projected to grow by 45%, comparing the period 1991–2010 with 2031–2050 (Zare et al. 2016).

A strong decrease in precipitation for almost all parts of Turkey was projected for the period 2021–2050 compared to 1971–2000 using Regional Climate Model, RegCM4.4 of the International Centre for Theoretical Physics (ICTP) under RCP4.5 and RCP8.5 scenarios (Türkeş et al. 2019). The projected changes in precipitation distribution can lead to more extreme precipitation events and prolonged droughts, increasing Turkey's vulnerability to soil erosion. In Portugal, a study comparing wet and dry catchments under A1B and B1 emission scenarios showed an increase in erosion in dry catchments (Serpa et al. 2015). In Morocco an increase in sediment load is projected as a consequence of reduced precipitation (Simonneaux et al. 2015). WGII AR5 concluded the impact of increases in heavy rainfall and temperature on soil erosion will be modulated by soil management practices, rainfall seasonality and land cover (Jiménez Cisneros et al. 2014). Ravi et al. (2010) predicted an increase in hydrologic and aeolian soil erosion processes as a consequence of droughts in drylands. However, there are some studies that indicate that soil erosion will be reduced in Spain (Zabaleta et al. 2013), Greece (Nerantzaki et al. 2015) and Australia (Klik and Eitzinger 2010), while others project changes in erosion as a consequence of the expansion of croplands (Borrelli et al. 2017).

Potential dryland expansion implies lower carbon sequestration and higher risk of desertification (Huang et al. 2017), with severe impacts on land usability and threats to food security. At the level of biomes (global-scale zones, generally defined by the type of plant life that they support in response to average rainfall and temperature patterns; see Glossary), soil carbon uptake is determined mostly by weather variability. The area of the land in which dryness controls CO_2 exchange has risen by 6% since 1948 and is projected to expand by at least another 8% by 2050. In these regions net carbon uptake is about 27% lower than elsewhere (Yi et al. 2014). Potential losses of soil carbon are projected to range from 9% to 12% of the total carbon stock in the 0–20 cm layer of soils in southern European Russia by end of this century (Ivanov et al. 2018).

Desertification under climate change will threaten biodiversity in drylands (*medium confidence*). Rodriguez-Caballero et al. (2018) analysed the cover of biological soil crusts under current and future environmental conditions utilising an environmental niche modelling approach. Their results suggest that biological soil crusts currently cover approximately 1600 Mha in drylands. Under RCP scenarios 2.6 to 8.5, 25–40% of this cover will be lost by 2070 with climate and land use contributing equally. The predicted loss is expected to substantially reduce the contribution of biological soil crusts to nitrogen cycling (6.7–9.9 TgN yr^{-1}) and carbon cycling (0.16–0.24 PgC yr^{-1}) (Rodriguez-Caballero et al. 2018). A study in Colorado Plateau, USA showed that changes in climate in drylands may damage the biocrust communities by promoting rapid mortality of foundational species (Rutherford et al. 2017), while in the Southern California deserts climate change-driven extreme heat and drought may surpass the survival thresholds of some desert species (Bachelet et al. 2016). In semi-arid Mediterranean shrublands in eastern Spain, plant species richness and plant cover could be reduced by climate change and soil erosion (García-Fayos and Bochet 2009). The main drivers of species extinctions are land-use change, habitat pollution, over-exploitation, and species invasion, while climate change is

3

indirectly linked to species extinctions (Settele et al. 2014). Malcolm et al. (2006) found that more than 2000 plant species located within dryland biodiversity hotspots could become extinct within 100 years, starting 2004 (within the Cape Floristic Region, Mediterranean Basin and southwest Australia). Furthermore, it is suggested that land use and climate change could cause the loss of 17% of species within shrublands and 8% within hot deserts by 2050 (*low confidence*) (van Vuuren et al. 2006). A study in the semi-arid Chinese Altai Mountains showed that mammal species richness will decline, rates of species turnover will increase, and more than 50% of their current ranges will be lost (Ye et al. 2018).

Changing climate and land use have resulted in higher aridity and more droughts in some drylands, with the rising role of precipitation, wind and evaporation on desertification (Fischlin et al. 2007). In a 2°C world, annual water discharge is projected to decline, and heatwaves are projected to pose risk to food production by 2070 (Waha et al. 2017). However, Betts et al. (2018) found a mixed response of water availability (runoff) in dryland catchments to global temperature increases from 1.5°C to 2°C. The forecasts for Sub-Saharan Africa suggest that higher temperatures, increase in the number of heatwaves, and increasing aridity, will affect the rainfed agricultural systems (Serdeczny et al. 2017). A study by Wang et al. (2009) in arid and semi-arid China showed decreased livestock productivity and grain yields from 2040 to 2099, threatening food security. In Central Asia, projections indicate a decrease in crop yields, and negative impacts of prolonged heat waves on population health (Reyer et al. 2017) (Section 3.7.2). World Bank (2009) projected that, without the carbon fertilisation effect, climate change will reduce the mean yields for 11 major global crops – millet, field pea, sugar beet, sweet potato, wheat, rice, maize, soybean, groundnut, sunflower and rapeseed – by 15% in Sub-Saharan Africa, 11% in Middle East and North Africa, 18% in South Asia, and 6% in Latin America and the Caribbean by 2046–2055, compared to 1996–2005. A separate meta-analysis suggested a similar reduction in yields in Africa and South Asia due to climate change by 2050 (Knox et al. 2012). Schlenker and Lobell (2010) estimated that in sub-Saharan Africa, crop production may be reduced by 17–22% due to climate change by 2050. At the local level, climate change impacts on crop yields vary by location (Section 5.2.2). Negative impacts of climate change on agricultural productivity contribute to higher food prices. The imbalance between supply and demand for agricultural products is projected to increase agricultural prices in the range of 31% for rice, to 100% for maize by 2050 (Nelson et al. 2010), and cereal prices in the range between a 32% increase and a 16% decrease by 2030 (Hertel et al. 2010). In southern European Russia, it is projected that the yields of grain crops will decline by 5–10% by 2050 due to the higher intensity and coverage of droughts (Ivanov et al. 2018).

Climate change can have strong impacts on poverty in drylands (*medium confidence*) (Hallegatte and Rozenberg 2017; Hertel and Lobell 2014). Globally, Hallegatte et al. (2015) project that without rapid and inclusive progress on eradicating multidimensional poverty, climate change could increase the number of the people living in poverty by between 35 million and 122 million people by 2030. Although these numbers are global and not specific to drylands, the highest impacts in terms of the share of the national populations

being affected are projected to be in the drylands areas of the Sahel region, eastern Africa and South Asia (Stephane Hallegatte et al. 2015). The impacts of climate change on poverty vary depending on whether the household is a net agricultural buyer or seller. Modelling results showed that poverty rates would increase by about one-third among the urban households and non-agricultural self-employed in Malawi, Uganda, Zambia and Bangladesh due to high agricultural prices and low agricultural productivity under climate change (Hertel et al. 2010). On the contrary, modelled poverty rates fell substantially among agricultural households in Chile, Indonesia, the Philippines and Thailand, because higher prices compensated for productivity losses (Hertel et al. 2010).

1.6   Responses to desertification under climate change

Achieving sustainable development of dryland livelihoods requires avoiding dryland degradation through SLM and restoring and rehabilitating the degraded drylands due to their potential wealth of ecosystem benefits and importance to human livelihoods and economies (Thomas 2008). A broad suite of on-the-ground response measures exists to address desertification (Scholes 2009), be it in the form of improved fire and grazing management, the control of erosion; integrated crop, soil and water management, among others (Liniger and Critchley 2007; Scholes 2009). These actions are part of the broader context of dryland development and long-term SLM within coupled socio-economic systems (Reynolds et al. 2007; Stringer et al. 2017; Webb et al. 2017). Many of these response options correspond to those grouped under 'land transitions' in the IPCC Special Report on Global Warming of 1.5°C (Coninck et al. 2018) (Table 6.4). It is therefore recognised that such actions require financial, institutional and policy support for their wide-scale adoption and sustainability over time (Sections 3.6.3, 4.8.5 and 6.4.4).

1.6.1   SLM technologies and practices: On-the-ground actions

A broad range of activities and measures can help avoid, reduce and reverse degradation across the dryland areas of the world. Many of these actions also contribute to climate change adaptation and mitigation, with further sustainable development co-benefits for poverty eradication and food security (*high confidence*) (Section 6.3). As preventing desertification is strongly preferable and more cost-effective than allowing land to degrade and then attempting to restore it (IPBES 2018b; Webb et al. 2013), there is a growing emphasis on avoiding and reducing land degradation, following the Land Degradation Neutrality framework (Cowie et al. 2018; Orr et al. 2017) (Section 4.8.5).

An assessment is made of six activities and measures practicable across the biomes and anthromes of the dryland domain (Figure 3.10). This suite of actions is not exhaustive, but rather a set of activities that are particularly pertinent to global dryland ecosystems. They are not necessarily exclusive to drylands and are often implemented across a range of biomes and anthromes (Figure 3.10;

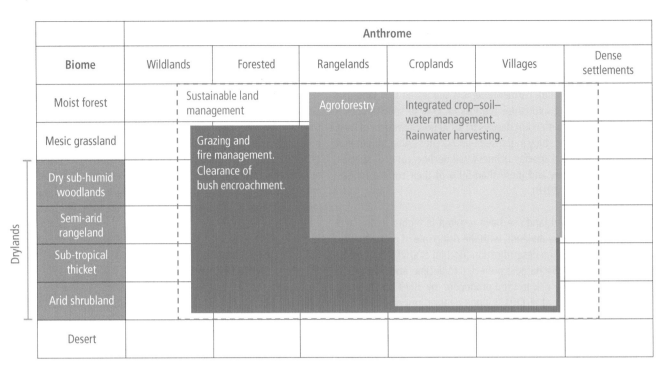

	Anthrome					
Biome	Wildlands	Forested	Rangelands	Croplands	Villages	Dense settlements
Moist forest	Sustainable land management		Agroforestry	Integrated crop–soil–water management. Rainwater harvesting.		
Mesic grassland		Grazing and fire management. Clearance of bush encroachment.				
Dry sub-humid woodlands						
Semi-arid rangeland						
Sub-tropical thicket						
Arid shrubland						
Desert						

Figure 3.10 | The typical distribution of on-the-ground actions across global biomes and anthromes.

for afforestation, see Section 3.7.2, Cross-Chapter Box 2 in Chapter 1, and Chapter 4 (Section 4.8.3)). The use of anthromes as a structuring element for response options is based on the essential role of interactions between social and ecological systems in driving desertification within coupled socio-ecological systems (Cherlet et al. 2018). The concept of the anthromes is defined in the Glossary and explored further in Chapters 1, 4 and 6.

The assessment of each action is twofold: firstly, to assess the ability of each action to address desertification and enhance climate change resilience, and secondly, to assess the potential impact of future climate change on the effectiveness of each action.

1.6.1.1   Integrated crop–soil–water management

Forms of integrated cropland management have been practiced in drylands for thousands of years (Knörzer et al. 2009). Actions include planting a diversity of species including drought-resilient ecologically appropriate plants, reducing tillage, applying organic compost and fertiliser, adopting different forms of irrigation and maintaining vegetation and mulch cover. In the contemporary era, several of these actions have been adopted in response to climate change.

In terms of climate change *adaptation*, the resilience of agriculture to the impacts of climate change is strongly influenced by the underlying health and stability of soils as well as improvements in crop varieties, irrigation efficiency and supplemental irrigation, for example, through rainwater harvesting (*medium evidence, high agreement*) (Altieri et al. 2015; Amundson et al. 2015; Derpsch et al. 2010; Lal 1997; de Vries et al. 2012). Desertification often leads to a reduction in ground cover that in turn results in accelerated water and wind erosion and an associated loss of fertile topsoil that can greatly reduce the resilience of agriculture to climate change (*medium*

evidence, high agreement) (Touré et al. 2019; Amundson et al. 2015; Borrelli et al. 2017; Pierre et al. 2017). Amadou et al. (2011) note that even a minimal cover of crop residues (100 kg ha^{-1}) can substantially decrease wind erosion.

Compared to conventional (flood or furrow) irrigation, drip irrigation methods are more efficient in supplying water to the plant root zone, resulting in lower water requirements and enhanced water use efficiency (*robust evidence, high agreement*) (Ibragimov et al. 2007; Narayanamoorthy 2010; Niaz et al. 2009). For example, in the rainfed area of Fetehjang, Pakistan, the adoption of drip methods reduced water usage by 67–68% during the production of tomato, cucumber and bell peppers, resulting in a 68–79% improvement in water use efficiency compared to previous furrow irrigation (Niaz et al. 2009). In India, drip irrigation reduced the amount of water consumed in the production of sugarcane by 44%, grapes by 37%, bananas by 29% and cotton by 45%, while enhancing yields by up to 29% (Narayanamoorthy 2010). Similarly, in Uzbekistan, drip irrigation increased the yield of cotton by 10–19% while reducing water requirements by 18–42% (Ibragimov et al. 2007).

A prominent response that addresses soil loss, health and cover is altering cropping methods. The adoption of intercropping (inter – and intra-row planting of companion crops) and relay cropping (temporally differentiated planting of companion crops) maintains soil cover over a larger fraction of the year, leading to an increase in production, soil nitrogen, species diversity and a decrease in pest abundance (*robust evidence, medium agreement*) (Altieri and Koohafkan 2008; Tanveer et al. 2017; Wilhelm and Wortmann 2004). For example, intercropping maize and sorghum with *Desmodium* (an insect repellent forage legume) and *Brachiaria* (an insect trapping grass), which is being promoted in drylands of East Africa, led to a two-to-three-fold increase in maize production and an 80% decrease in stem boring insects (Khan et al. 2014). In addition to changes in cropping methods, forms of agroforestry and shelterbelts are often used to reduce erosion and improve soil conditions (Section 3.7.2). For example, the use of tree belts of mixed species in northern China led to a reduction of surface wind speed and an associated reduction in soil temperature of up to 40% and an increase in soil moisture of up to 30% (Wang et al. 2008).

A further measure that can be of increasing importance under climate change is rainwater harvesting (RWH), including traditional *zai* (small basins used to capture surface runoff), earthen bunds and ridges (Nyamadzawo et al. 2013), *fanya juus* infiltration pits (Nyagumbo et al. 2019), contour stone bunds (Garrity et al. 2010) and semi-permeable stone bunds (often referred to by the French term *digue filtrante*) (Taye et al. 2015). RWH increases the amount of water available for agriculture and livelihoods through the capture and storage of runoff, while at the same time reducing the intensity of peak flows following high-intensity rainfall events. It is therefore often highlighted as a practical response to dryness (i.e., long-term aridity and low seasonal precipitation) and rainfall variability, both of which are projected to become more acute over time in some dryland areas (Dile et al. 2013; Vohland and Barry 2009). For example, for drainage in Wadi Al-Lith, Saudi Arabia, the use of rainwater harvesting was suggested as a key climate change adaptation action (Almazroui et al. 2017). There is *robust evidence* and *high agreement* that the implementation of RWH systems leads to an increase in agricultural production in drylands (Biazin et al. 2012; Bouma and Wösten 2016; Dile et al. 2013). A global meta-analysis of changes in crop production due to the adoption of RWH techniques noted an average increase in yields of 78%, ranging from −28% to 468% (Bouma and Wösten 2016). Of particular relevance to climate change in drylands is that the relative impact of RWH on agricultural production generally increases with increasing dryness. Relative yield improvements due to the adoption of RWH were significantly higher in years with less than 330 mm rainfall, compared to years with more than 330 mm (Bouma and Wösten 2016). Despite delivering a clear set of benefits, there are some issues that need to be considered. The impact of RWH may vary at different temporal and spatial scales (Vohland and Barry 2009). At a plot scale, RWH structures may increase available water and enhance agricultural production, SOC and nutrient availability, yet at a catchment scale, they may reduce runoff to downstream uses (Meijer et al. 2013; Singh et al. 2012; Vohland and Barry 2009; Yosef and Asmamaw 2015). Inappropriate storage of water in warm climes can lead to an increase in water related diseases unless managed correctly, for example, schistosomiasis and malaria (Boelee et al. 2013).

Integrated crop–soil–water management may also deliver climate change *mitigation* benefits through avoiding, reducing and reversing the loss of SOC (Table 6.5). Approximately 20–30 Pg of SOC have been released into the atmosphere through desertification processes, for example, deforestation, overgrazing and conventional tillage (Lal 2004). Activities, such as those associated with conservation agriculture (minimising tillage, crop rotation, maintaining organic cover and planting a diversity of species), reduce erosion, improve water use efficiency and primary production, increase inflow of organic material and enhance SOC over time, contributing to climate change mitigation and adaptation (*high confidence*) (Plaza-Bonilla

et al. 2015; Lal 2015; Srinivasa Rao et al. 2015; Sombrero and de Benito 2010). Conservation agriculture practices also lead to increases in SOC (*medium confidence*). However, sustained carbon sequestration is dependent on net primary productivity and on the availability of crop-residues that may be relatively limited and often consumed by livestock or used elsewhere in dryland contexts (Cheesman et al. 2016; Plaza-Bonilla et al. 2015). For this reason, expected rates of carbon sequestration following changes in agricultural practices in drylands are relatively low (0.04–0.4 tC ha^{-1}) and it may take a protracted period of time, even several decades, for carbon stocks to recover if lost (*medium confidence*) (Farage et al. 2007; Hoyle et al. 2013; Lal 2004). This long recovery period enforces the rationale for prioritising the avoidance and reduction of land degradation and loss of C, in addition to restoration activities.

1.6.1.2   Grazing and fire management in drylands

Rangeland management systems such as sustainable grazing approaches and re-vegetation increase rangeland productivity (*high confidence*) (Table 6.5). Open grassland, savannah and woodland are home to the majority of world's livestock production (Safriel et al. 2005). Within these drylands areas, prevailing grazing and fire regimes play an important role in shaping the relative abundance of trees versus grasses (Scholes and Archer 1997; Staver et al. 2011; Stevens et al. 2017), as well as the health of the grass layer in terms of primary production, species richness and basal cover (the propotion of the plant that is in the soil) (Plaza-Bonilla et al. 2015; Short et al. 2003). This in turn influences levels of soil erosion, soil nutrients, secondary production and additional ecosystem services (Divinsky et al. 2017; Pellegrini et al. 2017). A further set of drivers, including soil type, annual rainfall and changes in atmospheric CO_2 may also define observed rangeland structure and composition (Devine et al. 2017; Donohue et al. 2013), but the two principal factors that pastoralists can manage are grazing and fire, by altering their frequency, type and intensity.

The impact of grazing and fire regimes on biodiversity, soil nutrients, primary production and further ecosystem services is not constant and varies between locations (Divinsky et al. 2017; Fleischner 1994; van Oijen et al. 2018). Trade-offs may therefore need to be considered to ensure that rangeland diversity and production are resilient to climate change (Plaza-Bonilla et al. 2015; van Oijen et al. 2018). In certain locations, even light to moderate grazing has led to a significant decrease in the occurrence of particular species, especially forbs (O'Connor et al. 2011; Scott-shaw and Morris 2015). In other locations, species richness is only significantly impacted by heavy grazing and is able to withstand light to moderate grazing (Divinsky et al. 2017). A context specific evaluation of how grazing and fire impact particular species may therefore be required to ensure the persistence of target species over time (Marty 2005). A similar trade-off may need to be considered between soil carbon sequestration and livestock production. As noted by Plaza-Bonilla et al. (2015) increasing grazing pressure has been found to increase SOC stocks in some locations, and decrease them in others. Where it has led to a decrease in soil carbon stocks, for example in Mongolia (Han et al. 2008) and Ethiopia (Bikila et al. 2016), trade-offs between carbon sequestration and the value of livestock to local livelihoods need be considered.

Although certain herbaceous species may be unable to tolerate grazing pressure, a complete lack of grazing or fire may not be desired in terms of ecosystems health. It can lead to a decrease in basal cover and the accumulation of moribund, unpalatable biomass that inhibits primary production (Manson et al. 2007; Scholes 2009). The utilisation of the grass sward through light to moderate grazing stimulates the growth of biomass and basal cover, and allows water services to be sustained over time (Papanastasis et al. 2017; Scholes 2009). Even moderate to heavy grazing in periods of higher rainfall may be sustainable, but constant heavy grazing during dry periods, and especially droughts, can lead to a reduction in basal cover, SOC, biological soil crusts, ecosystem services and an accelerated erosion (*high agreement, robust evidence*) (Archer et al. 2017; Conant and Paustian 2003; D'Odorico et al. 2013; Geist and Lambin 2004; Havstad et al. 2006; Huang et al. 2007; Manzano and Návar 2000; Pointing and Belnap 2012; Weber et al. 2016). For this reason, the inclusion of drought forecasts and contingency planning in grazing and fire management programmes is crucial to avoid desertification (Smith and Foran 1992; Torell et al. 2010). It is an important component of avoiding and reducing early degradation. Although grasslands systems may be relatively resilient and can often recover from a moderately degraded state (Khishigbayar et al. 2015; Porensky et al. 2016), if a tipping point has been exceeded, restoration to a historic state may not be economical or ecologically feasible (D'Odorico et al. 2013).

Together with livestock management (Table 6.5), the use of fire is an integral part of rangeland management, which can be applied to remove moribund and unpalatable forage, exotic weeds and woody species (Archer et al. 2017). Fire has less of an effect on SOC and soil nutrients in comparison to grazing (Abril et al. 2005), yet elevated fire frequency has been observed to lead to a decrease in soil carbon and nitrogen (Abril et al. 2005; Bikila et al. 2016; Bird et al. 2000; Pellegrini et al. 2017). Although the impact of climate change on fire frequency and intensity may not be clear due to its differing impact on fuel accumulation, suitable weather conditions and sources of ignition (Abatzoglou et al. 2018; Littell et al. 2018; Moritz et al. 2012), there is an increasing use of prescribed fire to address several global change phenomena, for example, the spread of invasive species and bush encroachment, as well as the threat of intense runaway fires (Fernandes et al. 2013; McCaw 2013; van Wilgen et al. 2010). Cross-Chapter Box 3 in Chapter 2 provides a further review of the interaction between fire and climate change.

There is often much emphasis on reducing and reversing the degradation of rangelands due to the wealth of benefits they provide, especially in the context of assisting dryland communities to adapt to climate change (Webb et al. 2017; Woollen et al. 2016). The emerging concept of ecosystem-based adaptation has highlighted the broad range of important ecosystem services that healthy rangelands can provide in a resilient manner to local residents and downstream economies (Kloos and Renaud 2016; Reid et al. 2018). In terms of climate change mitigation, the contribution of rangelands, woodland and sub-humid dry forest (e.g., Miombo woodland in south-central Africa) is often undervalued due to relatively low carbon stocks

per hectare. Yet due to their sheer extent, the amount of carbon sequestered in these ecosystems is substantial and can make a valuable contribution to climate change mitigation (Lal 2004; Pelletier et al. 2018).

1.6.1.3   Clearance of bush encroachment

The encroachment of open grassland and savannah ecosystems by woody species has occurred for at least the past 100 years (Archer et al. 2017; O'Connor et al. 2014; Schooley et al. 2018). Dependent on the type and intensity of encroachment, it may lead to a net loss of ecosystem services and be viewed as a form of desertification (Dougill et al. 2016; O'Connor et al. 2014). However, there are circumstances where bush encroachment may lead to a net increase in ecosystem services, especially at intermediate levels of encroachment, where the ability of the landscape to produce fodder for livestock is retained, while the production of wood and associated products increases (Eldridge et al. 2011; Eldridge and Soliveres 2014). This may be particularly important in regions such as southern Africa and India where over 65% of rural households depend on fuelwood from surrounding landscapes as well as livestock production (Komala and Prasad 2016; Makonese et al. 2017; Shackleton and Shackleton 2004).

This variable relationship between the level of encroachment, carbon stocks, biodiversity, provision of water and pastoral value (Eldridge and Soliveres 2014) can present a conundrum to policymakers, especially when considering the goals of three Rio Conventions: UNFCCC, UNCCD and UNCBD. Clearing intense bush encroachment may improve species diversity, rangeland productivity, the provision of water and decrease desertification, thereby contributing to the goals of the UNCBD and UNCCD as well as the adaptation aims of the UNFCCC. However, it would lead to the release of biomass carbon stocks into the atmosphere and potentially conflict with the mitigation aims of the UNFCCC.

For example, Smit et al. (2015) observed an average increase in above-ground woody carbon stocks of 44 tC ha^{-1} in savannahs in northern Namibia. However, since bush encroachment significantly inhibited livestock production, there are often substantial efforts to clear woody species (Stafford-Smith et al. 2017). Namibia has a national programme, currently in its early stages, aimed at clearing woody species through mechanical measures (harvesting of trees) as well as the application of arboricides (Smit et al. 2015). However, the long-term success of clearance and subsequent improved fire and grazing management remains to be evaluated, especially restoration back towards an 'original open grassland state'. For example, in northern Namibia, the rapid reestablishment of woody seedlings has raised questions about whether full clearance and restoration is possible (Smit et al. 2015). In arid landscapes, the potential impact of elevated atmospheric CO_2 (Donohue et al. 2013; Kgope et al. 2010) and opportunity to implement high-intensity fires that remove woody species and maintain rangelands in an open state has been questioned (Bond and Midgley 2000). If these drivers of woody plant encroachment cannot be addressed, a new form of 'emerging ecosystem' (Milton 2003) may need to be explored that includes both improved livestock and fire management as well as the utilisation of biomass as a long-term commodity and source of revenue (Smit et al. 2015). Initial studies in Namibia and South Africa (Stafford-Smith et al. 2017) indicate that there may be good opportunity to produce sawn timber, fencing poles, fuelwood and commercial energy, but factors such as the cost of transport can substantially influence the financial feasibility of implementation.

The benefit of proactive management that prevents land from being degraded (altering grazing systems or treating bush encroachment at early stages before degradation has been initiated) is more cost-effective in the long term and adds more resistance to climate change than treating lands after degradation has occurred (Webb et al. 2013; Weltz and Spaeth 2012). The challenge is getting producers to alter their management paradigm from short-term objectives to long-term objectives.

1.6.1.4 Combating sand and dust storms through sand dune stabilisation

Dust and sand storms have a considerable impact on natural and human systems (Sections 3.4.1 and 3.4.2). Application of sand dune stabilisation techniques contributes to reducing sand and dust storms (*high confidence*). Using a number of methods, sand dune stabilisation aims to avoid and reduce the occurrence of dust and sand storms (Mainguet and Dumay 2011). Mechanical techniques include building palisades to prevent the movement of sand and reduce sand deposits on infrastructure. Chemical methods include the use of calcium bentonite or using silica gel to fix mobile sand (Aboushook et al. 2012; Rammal and Jubair 2015). Biological methods include the use of mulch to stabilise surfaces (Sebaa et al. 2015; Yu et al. 2004) and establishing permanent plant cover using pasture species that improve grazing at the same time (Abdelkebir and Ferchichi 2015; Zhang et al. 2015) (Section 3.7.1.3). When the dune is stabilised, woody perennials are introduced that are selected according to climatic and ecological conditions (FAO 2011). For example, such re-vegetation processes have been implemented on the shifting dunes of the Tengger Desert in northern China leading to the stabilisation of sand and the sequestration of up to 10 tC ha^{-1} over a period of 55 years (Yang et al. 2014).

1.6.1.5 Use of halophytes for the re-vegetation of saline lands

Soil salinity and sodicity can severely limit the growth and productivity of crops (Jan et al. 2017) and lead to a decrease in available arable land. Leaching and drainage provides a possible solution, but can be prohibitively expensive. An alternative, more economical option, is the growth of halophytes (plants that are adapted to grow under highly saline conditions) that allow saline land to be used in a productive manner (Qadir et al. 2000). The biomass produced can be used as forage, food, feed, essential oils, biofuel, timber, or fuelwood (Chughtai et al. 2015; Mahmood et al. 2016; Sharma et al. 2016). A further co-benefit is the opportunity to mitigate climate change through the enhancement of terrestrial carbon stocks as land is re-vegetated (Dagar et al. 2014; Wicke et al. 2013). The combined use of salt-tolerant crops, improved irrigation practices, chemical remediation measures and appropriate mulch and compost is effective in reducing the impact of secondary salinisation (*medium confidence*).

In Pakistan, where about 6.2 Mha of agricultural land is affected by salinity, pioneering work on utilising salt-tolerant plants for the re-vegetation of saline lands (biosaline agriculture) was done in the early 1970s (NIAB 1997). A number of local and exotic varieties were initially screened for salt tolerance in lab – and greenhouse-based studies, and then distributed to similar saline areas (Ashraf et al. 2010). These included tree species (*Acacia ampliceps, Acacia nilotica, Eucalyptus camaldulensis, Prosopis juliflora, Azadirachta indica*) (Awan and Mahmood 2017), forage plants (*Leptochloa fusca, Sporobolus arabicus, Brachiaria mutica, Echinochloa* sp., *Sesbania* and *Atriplex* spp.) and crop species including varieties of barley (*Hordeum vulgare*), cotton, wheat (*Triticum aestivum*) and *Brassica* spp. (Mahmood et al. 2016) as well as fruit crops in the form of date palm (*Phoenix dactylifera*) that has high salt tolerance with no visible adverse effects on seedlings (Yaish and Kumar 2015; Al-Mulla et al. 2013; Alrasbi et al. 2010). Pomegranate (*Punica granatum L.*) is another fruit crop of moderate to high salt tolerance. Through regulating growth form and nutrient balancing, it can maintain water content, chlorophyll fluorescence and enzyme activity at normal levels (Ibrahim 2016; Okhovatian-Ardakani et al. 2010).

In India and elsewhere, tree species including *Prosopis juliflora, Dalbergia sissoo,* and *Eucalyptus tereticornis* have been used to re-vegetate saline land. Certain biofuel crops in the form of *Ricinus communis* (Abideen et al. 2014), *Euphorbia antisyphilitica* (Dagar et al. 2014), *Karelinia caspia* (Akinshina et al. 2016) and *Salicornia* spp. (Sanandiya and Siddhanta 2014) are grown in saline areas, and *Panicum turgidum* (Koyro et al. 2013) and *Leptochloa fusca* (Akhter et al. 2003) have been grown as fodder crop on degraded soils with brackish water. In China, intense efforts are being made on the use of halophytes (Sakai et al. 2012; Wang et al. 2018). These examples reveal that there is great scope for saline areas to be used in a productive manner through the utilisation of halophytes. The most productive species often have yields equivalent to conventional crops, at salinity levels matching even that of seawater.

1.6.2 Socio-economic responses

Socio-economic and policy responses are often crucial in enhancing the adoption of SLM practices (Cordingley et al. 2015; Fleskens and Stringer 2014; Nyanga et al. 2016) and for assisting agricultural households to diversify their sources of income (Barrett et al. 2017; Shiferaw and Djido 2016). Technology and socio-economic responses are not independent, but continuously interact.

1.6.2.1 Socio-economic responses for combating desertification under climate change

Desertification limits the choice of potential climate change mitigation and adaptation response options by reducing climate change adaptive capacities. Furthermore, many additional factors, for example, a lack of access to markets or insecurity of land tenure, hinder the adoption of SLM. These factors are largely beyond the control of individuals or local communities and require broader policy interventions (Section 3.6.3). Nevertheless, local collective action and ILK are still crucial to the ability of households to respond to the combined challenge of climate change and desertification. Raising awareness, capacity building and development to promote collective action and indigenous and local knowledge contribute to avoiding, reducing and reversing desertification under changing climate.

The use of indigenous and local knowledge enhances the success of SLM and its ability to address desertification (Altieri and Nicholls 2017; Engdawork and Bork 2016). Using indigenous and local knowledge for combating desertification could contribute to climate change adaptation strategies (Belfer et al. 2017; Codjoe et al. 2014; Etchart 2017; Speranza et al. 2010; Makondo and Thomas 2018; Maldonado et al. 2016; Nyong et al. 2007). There are abundant examples of how indigenous and local knowledge, which are an important part of broader agroecological knowledge (Altieri 2018), have allowed livelihood systems in drylands to be maintained despite environmental constraints. An example is the numerous traditional water harvesting techniques that are used across the drylands to adapt to dry spells and climate change. These include creating planting pits (*zai, ngoro*) and micro-basins, contouring hill slopes and terracing (Biazin et al. 2012) (Section 3.6.1). Traditional *ndiva* water harvesting systems in Tanzania enable the capture of runoff water from highland areas to downstream community-managed micro-dams for subsequent farm delivery through small-scale canal networks (Enfors and Gordon 2008). A further example are pastoralist communities located in drylands who have developed numerous methods to sustainably manage rangelands. Pastoralist communities in Morocco developed the *agdal* system of seasonally alternating use of rangelands to limit overgrazing (Dominguez 2014) as well as to manage forests in the Moroccan High Atlas Mountains (Auclair et al. 2011). Across the Arabian Peninsula and North Africa, a rotational grazing system, *hema*, was historically practiced by the Bedouin communities (Hussein 2011; Louhaichi and Tastad 2010). The Beni-Amer herders in the Horn of Africa have developed complex livestock breeding and selection systems (Fre 2018). Although well adapted to resource-sparse dryland environments, traditional practices are currently not able to cope with increased demand for food and environmental changes (Enfors and Gordon 2008; Engdawork and Bork 2016). Moreover, there is *robust evidence* documenting the marginalisation or loss of indigenous and local knowledge (Dominguez 2014; Fernández-Giménez and Fillat Estaque 2012; Hussein 2011; Kodirekkala 2017; Moreno-Calles et al. 2012). Combined use of indigenous and local knowledge and new SLM technologies can contribute to raising resilience to the challenges of climate change and desertification (*high confidence*) (Engdawork and Bork 2016; Guzman et al. 2018).

Collective action has the potential to contribute to SLM and climate change adaptation (*medium confidence*) (Adger 2003; Engdawork and Bork 2016; Eriksen and Lind 2009; Ostrom 2009; Rodima-Taylor et al. 2012). Collective action is a result of social capital. Social capital is divided into structural and cognitive forms: structural corresponding to strong networks (including outside one's immediate community); and cognitive encompassing mutual trust and cooperation within communities (van Rijn et al. 2012; Woolcock and Narayan 2000). Social capital is more important for economic growth in settings with weak formal institutions, and less so in those with strong enforcement of formal institutions (Ahlerup et al. 2009). There are cases throughout the drylands showing that community by-laws and collective action successfully limited land degradation and facilitated SLM (Ajayi et al. 2016; Infante 2017; Kassie et al. 2013; Nyangena 2008; Willy and Holm-Müller 2013; Wossen et al. 2015). However, there are also cases when they did not improve SLM where they were not strictly enforced (Teshome et al. 2016). Collective action for implementing responses to dryland degradation is often hindered by local asymmetric power relations and 'elite capture' (Kihiu 2016; Stringer et al. 2007). This illustrates that different levels and types of social capital result in different levels of collective action. In a sample of East, West and southern African countries, structural social capital in the form of access to networks outside one's own community was suggested to stimulate the adoption of agricultural innovations, whereas cognitive social capital, associated with inward-looking community norms of trust and cooperation, was found to have a negative relationship with the adoption of agricultural innovations (van Rijn et al. 2012). The latter is indirectly corroborated by observations of the impact of community-based rangeland management organisations in Mongolia. Although levels of cognitive social capital did not differ between them, communities with strong links to outside networks were able to apply more innovative rangeland management practices in comparison to communities without such links (Ulambayar et al. 2017).

Farmer-led innovations. Agricultural households are not just passive adopters of externally developed technologies, but are active experimenters and innovators (Reij and Waters-Bayer 2001; Tambo and Wünscher 2015; Waters-Bayer et al. 2009). SLM technologies co-generated through direct participation of agricultural households have higher chances of being accepted by them (*medium confidence*) (Bonney et al. 2016; Vente et al. 2016). Usually farmer-driven innovations are more frugal and better adapted to their resource scarcities than externally introduced technologies (Gupta et al. 2016). Farmer-to-farmer sharing of their own innovations and mutual learning positively contribute to higher technology adoption rates (Dey et al. 2017). This innovative ability can be given a new dynamism by combining it with emerging external technologies. For example, emerging low-cost phone applications ('apps') that are linked to soil and water monitoring sensors can provide farmers with previously inaccessible information and guidance (Cornell et al. 2013; Herrick et al. 2017; McKinley et al. 2017; Steger et al. 2017).

Currently, the adoption of SLM practices remains insufficient to address desertification and contribute to climate change adaptation and mitigation more extensively. This is due to the constraints on the use of indigenous and local knowledge and collective action, as well as economic and institutional barriers for SLM adoption (Banadda 2010; Cordingley et al. 2015; Lokonon and Mbaye 2018; Mulinge et al. 2016; Wildemeersch et al. 2015) (Section 3.1.4.2; 3.6.3). Sustainable development of drylands under these socio-economic and environmental (climate change, desertification) conditions will also depend on the ability of dryland agricultural households to diversify their livelihoods sources (Boserup 1965; Safriel and Adeel 2008).

1.6.2.2   Socio-economic responses for economic diversification

Livelihood diversification through non-farm employment increases the resilience of rural households against desertification and extreme weather events by diversifying their income and consumption (*high confidence*). Moreover, it can provide the funds to invest into SLM (Belay et al. 2017; Bryan et al. 2009; Dumenu and Obeng 2016; Salik et al. 2017; Shiferaw et al. 2009). Access to non-agricultural employment is especially important for poorer pastoral households as their small herd sizes make them less resilient to drought (Fratkin 2013; Lybbert et al. 2004). However, access to alternative opportunities is limited in the rural areas of many developing countries, especially for women and marginalised groups who lack education and social networks (Reardon et al. 2008).

Migration is frequently used as an adaptation strategy to environmental change (*medium confidence*). Migration is a form of livelihood diversification and a potential response option to desertification and increasing risk to agricultural livelihoods under climate change (Walther et al. 2002). Migration can be short-term (e.g., seasonal) or long-term, internal within a country or international. There is *medium evidence* showing rural households responding to desertification and droughts through all forms of migration, for example: during the Dust Bowl in the USA in the 1930s (Hornbeck 2012); during droughts in Burkina Faso in the 2000s (Barbier et al. 2009); in Mexico in the 1990s (Nawrotzki et al. 2016); and by the Aymara people of the semi-arid Tarapacá region in Chile between 1820 and 1970, responding to declines in rainfall and growing demands for labour outside the region (Lima et al. 2016). There is *robust evidence* and *high agreement* showing that migration decisions are influenced by a complex set of different factors, with desertification and climate change playing relatively lesser roles (Liehr et al. 2016) (Section 3.4.2). Barrios et al. (2006) found that urbanisation in Sub-Saharan Africa was partially influenced by climatic factors during the 1950–2000 period, in parallel to liberalisation of internal restrictions on labour movements: each 1% reduction in rainfall was associated with a 0.45% increase in urbanisation. This migration favoured more industrially diverse urban areas in Sub-Saharan Africa (Henderson et al. 2017), because they offer more diverse employment opportunities and higher wages. Similar trends were also observed in Iran in response to water scarcity (Madani et al. 2016).

However, migration involves some initial investments. For this reason, reductions in agricultural incomes due to climate change or desertification have the potential to decrease out-migration among the poorest agricultural households, who become less able to afford migration (Cattaneo and Peri 2016), thus increasing social inequalities. There is *medium evidence* and *high agreement* that households with migrant worker members are more resilient against extreme weather events and environmental degradation compared to non-migrant households, who are more dependent on agricultural income (Liehr et al. 2016; Salik et al. 2017; Sikder and Higgins 2017). Remittances from migrant household members potentially contribute to SLM adoptions, however, substantial out-migration was also found to constrain the implementation of labour-intensive land management practices (Chen et al. 2014; Liu et al. 2016a).

1.6.3   Policy responses

The adoption of SLM practices depends on the compatibility of the technology with prevailing socio-economic and biophysical conditions (Sanz et al. 2017). Globally, it was shown that every USD invested into restoring degraded lands yields social returns, including both provisioning and non-provisioning ecosystem services, in the range of 3–6 USD over a 30-year period (Nkonya et al. 2016a). A similar range of returns from land restoration activities was found in Central Asia (Mirzabaev et al. 2016), Ethiopia (Gebreselassie et al. 2016), India (Mythili and Goedecke 2016), Kenya (Mulinge et al. 2016), Niger (Moussa et al. 2016) and Senegal (Sow et al. 2016) (*medium confidence*). Despite these relatively high returns, there is *robust evidence* that the adoption of SLM practices remains low (Cordingley et al. 2015; Giger et al. 2015; Lokonon and Mbaye 2018). Part of the reason for these low adoption rates is that the major share of the returns from SLM are social benefits, namely in the form of non-provisioning ecosystem services (Nkonya et al. 2016a). The adoption of SLM technologies does not always provide implementers with immediate private benefits (Schmidt et al. 2017). High initial investment costs, institutional and governance constraints and a lack of access to technologies and equipment may inhibit their adoption further (Giger et al. 2015; Sanz et al. 2017; Schmidt et al. 2017). However, not all SLM practices have high upfront costs. Analysing the World Overview of Conservation Approaches and Technologies (WOCAT) database, a globally acknowledged reference database for SLM, Giger et al. (2015) found that the upfront costs of SLM technologies ranged from about 20 USD to 5000 USD, with the median cost being around 500 USD. Many SLM technologies are profitable within 3 to 10 years (*medium confidence*) (Djanibekov and Khamzina 2016; Giger et al. 2015; Moussa et al. 2016; Sow et al. 2016). About 73% of 363 SLM technologies evaluated were reported to become profitable within three years, while 97% were profitable within 10 years (Giger et al. 2015). Similarly, it was shown that social returns from investments in restoring degraded lands will exceed their costs within six years in many settings across drylands (Nkonya et al. 2016a). However, even with affordable upfront costs, market failures – in the form of lack of access to credit, input and output markets, and insecure land tenure (Section 3.1.3) – result in the lack of adoption of SLM technologies (Moussa et al. 2016). Payments for ecosystem services, subsidies for SLM, and encouragement of community collective action can lead to a higher level of adoption of SLM and land restoration activities (*medium confidence*) (Bouma and Wösten 2016; Lambin et al. 2014; Reed et al. 2015; Schiappacasse et al. 2012; van Zanten et al. 2014) (Section 3.6.3). Enabling the policy responses discussed in this section will contribute to overcoming these market failures.

Many socio-economic factors shaping individual responses to desertification typically operate at larger scales. Individual households and communities do not exercise control over these factors, such as land tenure insecurity, lack of property rights, lack of access to markets, availability of rural advisory services, and agricultural price distortions. These factors are shaped by national government policies and international markets. As is the case with socio-economic responses, policy responses are classified below in two ways: those which seek to combat desertification under changing climate; and those which seek to provide alternative livelihood sources through economic diversification. These options are mutually complementary and contribute to all the three hierarchical elements of the Land Degradation Neutrality (LDN) framework, namely, avoiding, reducing and reversing land degradation (Cowie et al. 2018; Orr et al. 2017) (Sections 4.8.5 and 7.4.5, and Table 7.2). An enabling policy environment is a critical element for the achievement of LDN (Chasek et al. 2019). Implementation of LDN policies can contribute to climate change adaptation and mitigation (*high confidence*) (Sections 3.6.1 and 3.7.2).

1.6.3.1 Policy responses towards combating desertification under climate change

Policy responses to combat desertification take numerous forms (Marques et al. 2016). Below we discuss major policy responses consistently highlighted in the literature in connection with SLM and climate change, because these response options were found to strengthen adaptation capacities and to contribute to climate change mitigation. They include improving market access, empowering women, expanding access to agricultural advisory services, strengthening land tenure security, payments for ecosystem services, decentralised natural resource management, investing into research and monitoring of desertification and dust storms, and investing into modern renewable energy sources.

Policies aiming at improving market access, that is the ability to access output and input markets at lower costs, help farmers and livestock producers earn more profit from their produce. Increased profits both motivate and enable them to invest more in SLM. Higher access to input, output and credit markets was consistently found as a major factor in the adoption of SLM practices in a wide number of settings across the drylands (*medium confidence*) (Aw-Hassan et al. 2016; Gebreselassie et al. 2016; Mythili and Goedecke 2016; Nkonya and Anderson 2015; Sow et al. 2016). Lack of access to credit limits adjustments and agricultural responses to the impacts of desertification under changing climate, with long-term consequences for the livelihoods and incomes, as was shown during the North American Dust Bowl of the 1930s (Hornbeck 2012). Government policies aimed at improving market access usually involve constructing and upgrading rural–urban transportation infrastructure and agricultural value chains, such as investments into construction of local markets, abattoirs and cold storage warehouses, as well as post-harvest processing facilities (McPeak et al. 2006). However, besides infrastructural constraints, providing improved access often involves relieving institutional constraints to market access (Little 2010), such as improved coordination of cross-border food safety and veterinary regulations (Ait Hou et al. 2015; Keiichiro et al. 2015; McPeak et al. 2006; Unnevehr 2015), and availability and access to market information systems (Bobojonov et al. 2016; Christy et al. 2014; Nakasone et al. 2014).

Women's empowerment. A greater emphasis on understanding gender-specific differences over land use and land management practices as an entry point can make land restoration projects more successful (*medium confidence*) (Broeckhoven and Cliquet 2015; Carr and Thompson 2014; Catacutan and Villamor 2016;

Dah-gbeto and Villamor 2016). In relation to representation and authority to make decisions in land management and governance, women's participation remains lacking particularly in the dryland regions. Thus, ensuring women's rights means accepting women as equal members of the community and citizens of the state (Nelson et al. 2015). This includes equitable access of women to resources (including extension services), networks, and markets. In areas where socio-cultural norms and practices devalue women and undermine their participation, actions for empowering women will require changes in customary norms, recognition of women's (land) rights in government policies, and programmes to assure that their interests are better represented (Section 1.4.2 and Cross-Chapter Box 11 in Chapter 7). In addition, several novel concepts are recently applied for an in-depth understanding of gender in relation to science–policy interface. Among these are the concepts of intersectionality, that is, how social dimensions of identity and gender are bound up in systems of power and social institutions (Thompson-Hall et al. 2016), bounded rationality for gendered decision-making, related to incomplete information interacting with limits to human cognition leading to judgement errors or objectively poor decision making (Villamor and van Noordwijk 2016), anticipatory learning for preparing for possible contingencies and consideration of long-term alternatives (Dah-gbeto and Villamor 2016) and systematic leverage points for interventions that produce, mark, and entrench gender inequality within communities (Manlosa et al. 2018), which all aim to improve gender equality within agroecological landscapes through a systems approach.

Education and expanding access to agricultural services. Providing access to information about SLM practices facilitates their adoption (*medium confidence*) (Kassie et al. 2015; Nkonya et al. 2015; Nyanga et al. 2016). Moreover, improving the knowledge of climate change, capacity building and development in rural areas can help strengthen climate change adaptive capacities (Berman et al. 2012; Chen et al. 2018; Descheemaeker et al. 2018; Popp et al. 2009; Tambo 2016; Yaro et al. 2015). Agricultural initiatives to improve the adaptive capacities of vulnerable populations were more successful when they were conducted through reorganised social institutions and improved communication, for example, in Mozambique (Osbahr et al. 2008). Improved communication and education could be facilitated by wider use of new information and communication technologies (ICTs) (Peters et al. 2015). Investments into education were associated with higher adoption of soil conservation measures, for example, in Tanzania (Tenge et al. 2004). Bryan et al. (2009) found that access to information was the prominent facilitator of climate change adaptation in Ethiopia. However, resource constraints of agricultural services, and disconnects between agricultural policy and climate policy can hinder the dissemination of climate-smart agricultural technologies (Morton 2017). Lack of knowledge was also found to be a significant barrier to implementation of soil rehabilitation programmes in the Mediterranean region (Reichardt 2010). Agricultural services will be able to facilitate SLM best when they also serve as platforms for sharing indigenous and local knowledge and farmer innovations (Mapfumo et al. 2016). Participatory research initiatives conducted jointly with farmers have higher chances of resulting in technology adoption (Bonney et al. 2016; Rusike et al. 2006; Vente et al. 2016). Moreover, rural advisory services are often more successful in disseminating technological innovations when they adopt commodity/value chain approaches, remain open to engagement in input supply, make use of new opportunities presented by ICTs, facilitate mutual learning between multiple stakeholders (Morton 2017), and organise science and SLM information in a location-specific manner for use in education and extension (Bestelmeyer et al. 2017).

Strengthening land tenure security. Strengthening land tenure security is a major factor contributing to the adoption of soil conservation measures in croplands (*high confidence*) (Bambio and Bouayad Agha 2018; Higgins et al. 2018; Holden and Ghebru 2016; Paltasingh 2018; Rao et al. 2016; Robinson et al. 2018), thus contributing to climate change adaptation and mitigation. Moreover, land tenure security can lead to more investment in trees (Deininger and Jin 2006; Etongo et al. 2015). Land tenure recognition policies were found to lead to higher agricultural productivity and incomes, although with inter-regional variations, requiring an improved understanding of overlapping formal and informal land tenure rights (Lawry et al. 2017). For example, secure land tenure increased investments into SLM practices in Ghana, but without affecting farm productivity (Abdulai et al. 2011). Secure land tenure, especially for communally managed lands, helps reduce arbitrary appropriations of land for large-scale commercial farms (Aha and Ayitey 2017; Baumgartner 2017; Dell'Angelo et al. 2017). In contrast, privatisation of rangeland tenures in Botswana and Kenya led to the loss of communal grazing lands and actually increased rangeland degradation (Basupi et al. 2017; Kihiu 2016) as pastoralists needed to graze livestock on now smaller communal pastures. Since food insecurity in drylands is strongly affected by climate risks, there is *robust evidence* and *high agreement* that resilience to climate risks is higher with flexible tenure for allowing mobility for pastoralist communities, and not fragmenting their areas of movement (Behnke 1994; Holden and Ghebru 2016; Liao et al. 2017; Turner et al. 2016; Wario et al. 2016). More research is needed on the optimal tenure mix, including low-cost land certification, redistribution reforms, market-assisted reforms and gender-responsive reforms, as well as collective forms of land tenure such as communal land tenure and cooperative land tenure (see Section 7.6.5 for a broader discussion of land tenure security under climate change).

Payment for ecosystem services (PES) provides incentives for land restoration and SLM (*medium confidence*) (Lambin et al. 2014; Li et al. 2018; Reed et al. 2015; Schiappacasse et al. 2012). Several studies illustrate that the social costs of desertification are larger than its private cost (Costanza et al. 2014; Nkonya et al. 2016a). Therefore, although SLM can generate public goods in the form of provisioning ecosystem services, individual land custodians underinvest in SLM as they are unable to reap these benefits fully. Payment for ecosystem services provides a mechanism through which some of these benefits can be transferred to land users, thereby stimulating further investment in SLM. The effectiveness of PES schemes depends on land tenure security and appropriate design, taking into account specific local conditions (Börner et al. 2017). However, PES has not worked well in countries with fragile institutions (Karsenty and Ongolo 2012). Equity and justice in distributing the payments for ecosystem services were found to be key for the success of the PES programmes in Yunnan,

China (He and Sikor 2015). Yet, when reviewing the performance of PES programmes in the tropics, Calvet-Mir et al. (2015), found that they are generally effective in terms of environmental outcomes, despite being sometimes unfair in terms of payment distribution. It is suggested that the implementation of PES will be improved through decentralised approaches giving local communities a larger role in the decision-making process (He and Lang 2015).

Empowering local communities for decentralised natural resource management. Local institutions often play a vital role in implementing SLM initiatives and climate change adaptation measures (*high confidence*) (Gibson et al. 2005; Smucker et al. 2015). Pastoralists involved in community-based natural resource management in Mongolia had greater capacity to adapt to extreme winter frosts, resulting in less damage to their livestock (Fernandez-Gimenez et al. 2015). Decreasing the power and role of traditional community institutions, due to top-down public policies, resulted in lower success rates in community-based programmes focused on rangeland management in Dirre, Ethiopia (Abdu and Robinson 2017). Decentralised governance was found to lead to improved management in forested landscapes (Dressler et al. 2010; Ostrom and Nagendra 2006). However, there are also cases when local elites were placed in control and this decentralised natural resource management negatively impacted the livelihoods of the poorer and marginalised community members due to reduced access to natural resources (Andersson and Ostrom 2008; Cullman 2015; Dressler et al. 2010).

The success of decentralised natural resource management initiatives depends on increased participation and empowerment of a diverse set of community members, not only local leaders and elites, in the design and management of local resource management institutions (Kadirbeyoglu and Özertan 2015; Umutoni et al. 2016), while considering the interactions between actors and institutions at different levels of governance (Andersson and Ostrom 2008; Carlisle and Gruby 2017; McCord et al. 2017). An example of such programmes where local communities played a major role in land restoration and rehabilitation activities is the cooperative project on The National Afforestation and Erosion Control Mobilization Action Plan in Turkey, initiated by the Turkish Ministry of Agriculture and Forestry (Çalişkan and Boydak 2017), with the investment of 1.8 billion USD between 2008 and 2012. The project mobilised local communities in cooperation with public institutions, municipalities, and non-governmental organisations, to implement afforestation, rehabilitation and erosion control measures, resulting in the afforestation and reforestation of 1.5 Mha (Yurtoglu 2015). Moreover, some 1.75 Mha of degraded forest and 37,880 ha of degraded rangelands were rehabilitated. Finally, the project provided employment opportunities for 300,000 rural residents for six months every year, combining land restoration and rehabilitation activities with measures to promote socio-economic development in rural areas (Çalişkan and Boydak 2017).

Investing in research and development. Desertification has received substantial research attention over recent decades (Turner et al. 2007). There is also a growing research interest on climate change adaptation and mitigation interventions that help address desertification (Grainger 2009). Agricultural research on SLM practices has generated a significant number of new innovations and technologies that increase crop yields without degrading the land, while contributing to climate change adaptation and mitigation (Section 3.6.1). There is *robust evidence* that such technologies help improve the food security of smallholder dryland farming households (Harris and Orr 2014) (Section 6.3.5). Strengthening research on desertification is of high importance not only to meet SDGs but also to manage ecosystems effectively, based on solid scientific knowledge. More investment in research institutes and training the younger generation of researchers is needed for addressing the combined challenges of desertification and climate change (Akhtar-Schuster et al. 2011; Verstraete et al. 2011). This includes improved knowledge management systems that allow stakeholders to work in a coordinated manner by enhancing timely, targeted and contextualised information sharing (Chasek et al. 2011). Knowledge and flow of knowledge on desertification is currently highly fragmented, constraining the effectiveness of those engaged in assessing and monitoring the phenomenon at various levels (Reed et al. 2011). Improved knowledge and data exchange and sharing increase the effectiveness of efforts to address desertification (*high confidence*).

Developing modern renewable energy sources. Transitioning to renewable energy resources contributes to reducing desertification by lowering reliance on traditional biomass in dryland regions (*medium confidence*). This can also have socioeconomic and health benefits, especially for women and children (*high confidence*). Populations in most developing countries continue to rely on traditional biomass, including fuelwood, crop straws and livestock manure, for a major share of their energy needs, with the highest dependence in Sub-Saharan Africa (Amugune et al. 2017; IEA 2013). Use of biomass for energy, mostly fuelwood (especially as charcoal), was associated with deforestation in some dryland areas (Iiyama et al. 2014; Mekuria et al. 2018; Neufeldt et al. 2015; Zulu 2010), while in some other areas there was no link between fuelwood collection and deforestation (Simon and Peterson 2018; Swemmer et al. 2018; Twine and Holdo 2016). Moreover, the use of traditional biomass as a source of energy was found to have negative health effects through indoor air pollution (de la Sota et al. 2018; Lim and Seow 2012), while also being associated with lower female labour force participation (Burke and Dundas 2015). Jiang et al. (2014) indicated that providing improved access to alternative energy sources such as solar energy and biogas could help reduce the use of fuelwood in south-western China, thus alleviating the spread of rocky desertification. The conversion of degraded lands into cultivation of biofuel crops will affect soil carbon dynamics (Albanito et al. 2016; Nair et al. 2011) (Cross-Chapter Box 7 in Chapter 6). The use of biogas slurry as soil amendment or fertiliser can increase soil carbon (Galvez et al. 2012; Negash et al. 2017). Large-scale installation of wind and solar farms in the Sahara Desert was projected to create a positive climate feedback through increased surface friction and reduced albedo, doubling precipitation over the neighbouring Sahel region with resulting increases in vegetation (Li et al. 2018). Transition to renewable energy sources in high-income countries in dryland areas primarily contributes to reducing GHG emissions and mitigating climate change, with some other co-benefits such as diversification of energy sources (Bang 2010), while the impacts on desertification are less evident. The use of renewable energy has been proposed

as an important mitigation option in dryland areas as well (El-Fadel et al. 2003). Transitions to renewable energy are being promoted by governments across drylands (Cancino-Solórzano et al. 2016; Hong et al. 2013; Sen and Ganguly 2017) including in fossil-fuel rich countries (Farnoosh et al. 2014; Dehkordi et al. 2017; Stambouli et al. 2012; Vidadili et al. 2017), despite important social, political and technical barriers to expanding renewable energy production (Afsharzade et al. 2016; Baker et al. 2014; Elum and Momodu 2017; Karatayev et al. 2016). Improving social awareness about the benefits of transitioning to renewable energy resources, and access to hydro-energy, solar and wind energy contributes to their improved adoption (Aliyu et al. 2017; Katikiro 2016).

Developing and strengthening climate services relevant for desertification. Climate services provide climate, drought and desertification-related information in a way that assists decision-making by individuals and organisations. Monitoring desertification, and integrating biogeophysical (climate, soil, ecological factors, biodiversity) and socio-economic (use of natural resources by local population) issues provide a basis for better vulnerability prediction and assessment (OSS, 2012; Vogt et al. 2011). Examples of relevant services include: drought monitoring and early warning systems, often implemented by national climate and meteorological services but also encompassing regional and global systems (Pozzi et al. 2013); and the Sand and Dust Storm Warning Advisory and Assessment System (SDS-WAS), created by WMO in 2007, in partnership with the World Health Organization (WHO) and the United Nations Environment Program (UNEP). Currently, there is also a lack of ecological monitoring in arid and semi-arid regions to study surface winds, dust and sand storms, and their impacts on ecosystems and human health (Bergametti et al. 2018; Marticorena et al. 2010). Reliable and timely climate services, relevant to desertification, can aid the development of appropriate adaptation and mitigation options, reducing the impact of desertification under changing climate on human and natural systems (*high confidence*) (Beegum et al. 2016; Beegum et al. 2018; Cornet 2012; Haase et al. 2018; Sergeant et al. 2012).

1.6.3.2   Policy responses supporting economic diversification

Despite policy responses for combating desertification, other factors will put strong pressures on the land, including climate change and growing food demands, as well as the need to reduce poverty and strengthen food security (Cherlet et al. 2018) (Sections 6.1.4 and 7.2.2). Sustainable development of drylands and their resilience to combined challenges of desertification and climate change will thus also depend on the ability of governments to promote policies for economic diversification within agriculture and in non-agricultural sectors in order make dryland areas less vulnerable to desertification and climate change.

Investing into irrigation. Investments into expanding irrigation in dryland areas can help increase the resilience of agricultural production to climate change, improve labour productivity and boost production and income revenue from agriculture and livestock sectors (Geerts and Raes 2009; Olayide et al. 2016; Oweis and Hachum 2006). This is particularly true for Sub-Saharan Africa, where currently only 6% of the cultivated areas are irrigated (Nkonya et al. 2016b). While renewable groundwater resources could help increase the share of irrigated land to 20.5–48.6% of croplands in the region (Altchenko and Villholth 2015). On the other hand, over-extraction of groundwaters, mainly for irrigating crops, is becoming an important environmental problem in many dryland areas (Cherlet et al. 2018), requiring careful design and planning of irrigation expansion schemes and use of water-efficient irrigation methods (Bjornlund et al. 2017; Woodhouse et al. 2017). For example, in Saudi Arabia, improving the efficiency of water management, for example through the development of aquifers, water recycling and rainwater harvesting, is part of a suite of policy actions to combat desertification (Bazza, et al. 2018; Kingdom of Saudi Arabia 2016). The expansion of irrigation to riverine areas, crucial for dry season grazing of livestock, needs to consider the income from pastoral activities, which is not always lower than income from irrigated crop production (Behnke and Kerven 2013). Irrigation development could be combined with the deployment of clean-energy technologies in economically viable ways (Chandel et al. 2015). For example, solar-powered drip irrigation was found to increase household agricultural incomes in Benin (Burney et al. 2010). The sustainability of irrigation schemes based on solar-powered extraction of groundwaters depends on measures to avoid over-abstraction of groundwater resources and associated negative environmental impacts (Closas and Rap 2017).

Expanding agricultural commercialisation. Faster poverty rate reduction and economic growth enhancement is realised when countries transition into the production of non-staple, high-value commodities and manage to build a robust agro-industry sector (Barrett et al. 2017). Ogutu and Qaim (2019) found that agricultural commercialisation increased incomes and decreased multidimensional poverty in Kenya. Similar findings were earlier reported by Muriithi and Matz (2015) for commercialisation of vegetables in Kenya. Commercialisation of rice production was found to have increased smallholder welfare in Nigeria (Awotide et al. 2016). Agricultural commercialisation contributed to improved household food security in Malawi, Tanzania and Uganda (Carletto et al. 2017). However, such a transition did not improve farmers' livelihoods in all cases (Reardon et al. 2009). High-value cash crop/animal production can be bolstered by wide-scale use of technologies, for example, mechanisation, application of inorganic fertilisers, crop protection and animal health products. Market oriented crop/animal production facilitates social and economic progress, with labour increasingly shifting out of agriculture into non-agricultural

sectors (Cour 2001). Modernised farming, improved access to inputs, credit and technologies enhances competitiveness in local and international markets (Reardon et al. 2009).

Facilitating structural transformations in rural economies implies that the development of non-agricultural sectors encourages the movement of labour from land-based livelihoods, vulnerable to desertification and climate change, to non-agricultural activities (Haggblade et al. 2010). The movement of labour from agriculture to non-agricultural sectors is determined by relative labour productivities in these sectors (Shiferaw and Djido 2016). Given already high underemployment in the farm sector, increasing labour productivity in the non-farm sector was found as the main driver of labour movements from farm sector to non-farm sector (Shiferaw and Djido 2016). More investments into education can facilitate this process (Headey et al. 2014). However, in some contexts, such as pastoralist communities in Xinjiang, China, income diversification was not found to improve the welfare of pastoral households (Liao et al. 2015). Economic transformations also occur through urbanisation, involving the shift of labour from rural areas into gainful employment in urban areas (Jedwab and Vollrath 2015). The majority of world population will be living in urban centres in the 21st century and this will require innovative means of agricultural production with minimum ecological footprint and less dependence on fossil fuels (Revi and Rosenzweig 2013), while addressing the demand of cities (see Section 4.9.1 for discussion on urban green infrastructure). Although there is some evidence of urbanisation leading to the loss of indigenous and local ecological knowledge, however, indigenous and local knowledge systems are constantly evolving, and are also being integrated into urban environments (Júnior et al. 2016; Reyes-García et al. 2013; van Andel and Carvalheiro 2013). Urban areas are attracting an increasing number of rural residents across the developing world (Angel et al. 2011; Cour 2001; Dahiya 2012). Urban development contributes to expedited agricultural commercialisation by providing market outlet for cash crops, high-value crops, and livestock products. At the same time, urbanisation also poses numerous challenges in the form of rapid urban sprawl and pressures on infrastructure and public services, unemployment and associated social risks, which have considerable implications on climate change adaptive capacities (Bulkeley 2013; Garschagen and Romero-Lankao 2015).

3

Cross-Chapter Box 5 | Policy responses to drought

Alisher Mirzabaev (Germany/Uzbekistan), Margot Hurlbert (Canada), Muhammad Mohsin Iqbal (Pakistan), Joyce Kimutai (Kenya), Lennart Olsson (Sweden), Fasil Tena (Ethiopia), Murat Türkeş (Turkey)

Drought is a highly complex natural hazard (for floods, see Box 7.2). It is difficult to precisely identify its start and end. It is usually slow and gradual (Wilhite and Pulwarty 2017), but sometimes can evolve rapidly (Ford and Labosier 2017; Mo and Lettenmaier 2015). It is context-dependent, but its impacts are diffuse, both direct and indirect, short-term and long-term (Few and Tebboth 2018; Wilhite and Pulwarty 2017). Following the Synthesis Report (SYR) of the IPCC Fifth Assessment Report (AR5), drought is defined here as "a period of abnormally dry weather long enough to cause a serious hydrological imbalance" (Mach et al. 2014). Although drought is considered abnormal relative to the water availability under the mean climatic characteristics, it is also a recurrent element of any climate, not only in drylands, but also in humid areas (Cook et al. 2014b; Seneviratne and Ciais 2017; Spinoni et al. 2019; Türkeş 1999; Wilhite et al. 2014). Climate change is projected to increase the intensity or frequency of droughts in some regions across the world (for a detailed assessment see Section 2.2, and IPCC Special Report on Global Warming of 1.5°C (Hoegh-Guldberg et al. 2018)). Droughts often amplify the effects of unsustainable land management practices, especially in drylands, leading to land degradation (Cook et al. 2009; Hornbeck 2012). Especially in the context of climate change, the recurrent nature of droughts requires proactively planned policy instruments both to be well-prepared to respond to droughts when they occur and also undertake *ex ante* actions to mitigate their impacts by strengthening societal resilience against droughts (Gerber and Mirzabaev 2017).

Droughts are among the costliest of natural hazards (*robust evidence*, *high agreement*). According to the International Disaster Database (EM-DAT), droughts affected more than 1.1 billion people between 1994 and 2013, with the recorded global economic damage of 787 billion USD (CRED 2015), corresponding to an average of 41.4 billion USD per year. Drought losses in the agricultural sector alone in developing countries were estimated to equal 29 billion USD between 2005 and 2015 (FAO 2018). Usually, these estimates capture only direct and on-site costs of droughts. However, droughts have also wide-ranging indirect and off-site impacts, which are seldom quantified. These indirect impacts are both biophysical and socio-economic, with poor households and communities being particularly exposed to them (Winsemius et al. 2018). Droughts affect not only water quantity, but also water quality (Mosley 2014). The costs of these water quality impacts are yet to be adequately quantified. Socio-economic indirect impacts of droughts are related to food insecurity, poverty, lowered health and displacement (Gray and Mueller 2012; Johnstone and Mazo 2011; Linke et al. 2015; Lohmann and Lechtenfeld 2015; Maystadt and Ecker 2014; Yusa et al. 2015) (Section 3.4.2.9 and Box 5.5), which are difficult to quantify comprehensively. Research is required for developing methodologies that could allow for more comprehensive assessment of these indirect drought costs. Such methodologies require the collection of highly granular data, which is currently lacking in many countries due to high costs of data collection. However, the opportunities provided by remotely sensed data and novel analytical methods based on big data and artificial intelligence, including use of citizen science for data collection, could help in reducing these gaps.

There are three broad (and sometimes overlapping) policy approaches for responding to droughts (Section 7.4.8). These approaches are often pursued simultaneously by many governments. Firstly, responding to drought when it occurs by providing direct drought relief, known as crisis management. Crisis management is also the costliest among policy approaches to droughts because it often incentivises the continuation of activities vulnerable to droughts (Botterill and Hayes 2012; Gerber and Mirzabaev 2017).

The second approach involves development of drought preparedness plans, which coordinate the policies for providing relief measures when droughts occur. For example, combining resources to respond to droughts at regional level in Sub-Saharan Africa was found to be more cost-effective than separate individual country drought relief funding (Clarke and Hill 2013). Effective drought preparedness plans require well-coordinated and integrated government actions – a key lesson learnt from 2015 to 2017 during drought response in Cape Town, South Africa (Visser 2018). Reliable, relevant and timely climate and weather information helps respond to droughts appropriately (Sivakumar and Ndiang'ui 2007). Improved knowledge and integration of weather and climate information can be achieved by strengthening drought early warning systems at different scales (Verbist et al. 2016). Every USD invested into strengthening hydro-meteorological and early warning services in developing countries was found to yield between 4 and 35 USD (Hallegatte 2012). Improved access and coverage by drought insurance, including index insurance, can help alleviate the impacts of droughts on livelihoods (Guerrero-Baena et al. 2019; Kath et al. 2019; Osgood et al. 2018; Ruiz et al. 2015; Tadesse et al. 2015).

The third category of responses to droughts involves drought risk mitigation. Drought risk mitigation is a set of proactive measures, policies and management activities aimed at reducing the future impacts of droughts (Vicente-Serrano et al. 2012). For example,

Cross-Chapter Box 5 (continued)

policies aimed at improving water use efficiency in different sectors of the economy, especially in agriculture and industry, or public advocacy campaigns raising societal awareness and bringing about behavioural change to reduce wasteful water consumption in the residential sector are among such drought risk mitigation policies (Tsakiris 2017). Public outreach and monitoring of communicable diseases, air and water quality were found to be useful for reducing health impacts of droughts (Yusa et al. 2015). The evidence from household responses to drought in Cape Town, South Africa, between 2015 and 2017, suggests that media coverage and social media could play a decisive role in changing water consumption behaviour, even more so than official water consumption restrictions (Booysen et al. 2019). Drought risk mitigation approaches are less costly than providing drought relief after the occurrence of droughts. To illustrate, Harou et al. (2010) found that establishment of water markets in California considerably reduced drought costs. Application of water saving technologies reduced drought costs in Iran by 282 million USD (Salami et al. 2009). Booker et al. (2005) calculated that inter-regional trade in water could reduce drought costs by 20–30% in the Rio Grande basin, USA. Increasing rainfall variability under climate change can make the forms of index insurance based on rainfall less efficient (Kath et al. 2019). A number of diverse water property instruments, including instruments allowing water transfer, together with the technological and institutional ability to adjust water allocation, can improve timely adjustment to droughts (Hurlbert 2018). Supply-side water management, providing for proportionate reductions in water delivery, prevents the important climate change adaptation option of managing water according to need or demand (Hurlbert and Mussetta 2016). Exclusive use of a water market to govern water allocation similarly prevents the recognition of the human right to water at times of drought (Hurlbert 2018). Policies aiming to secure land tenure, and to expand access to markets, agricultural advisory services and effective climate services, as well as to create off-farm employment opportunities, can facilitate the adoption of drought risk mitigation practices (Alam 2015; Kusunose and Lybbert 2014), increasing resilience to climate change (Section 3.6.3), while also contributing to SLM (Sections 3.6.3 and 4.8.1, and Table 5.7).

The excessive burden of drought relief funding on public budgets is already leading to a paradigm shift towards proactive drought risk mitigation instead of reactive drought relief measures (Verner et al. 2018; Wilhite 2016). Climate change will reinforce the need for such proactive drought risk mitigation approaches. Policies for drought risk mitigation that are already needed now will be even more relevant under higher warming levels (Jerneck and Olsson 2008; McLeman 2013; Wilhite et al. 2014). Overall, there is *high confidence* that responding to droughts through *ex post* drought relief measures is less efficient compared to *ex ante* investments into drought risk mitigation, particularly under climate change.

1.6.4   Limits to adaptation, maladaptation, and barriers for mitigation

Chapter 16 in the IPCC Fifth Assessment Report (AR5) (Klein et al. 2015) discusses the existence of soft and hard limits to adaptation, highlighting that values and perspectives of involved agents are relevant to identify limits (Sections 4.8.5.1 and 7.4.9). In that sense, adaptation limits vary from place to place and are difficult to generalise (Barnett et al. 2015; Dow et al. 2013; Klein et al. 2015). Currently, there is a lack of knowledge on adaptation limits and potential maladaptation to combined effects of climate change and desertification (see Section 4.8.6 for discussion on resilience, thresholds, and irreversible land degradation, also relevant for desertification). However, the potential for residual risks (those risks which remain after adaptation efforts were taken, irrespective of whether they are tolerable or not, tolerability being a subjective concept) and maladaptive outcomes is high (*high confidence*). Some examples of residual risks are illustrated below in this section. Although SLM measures can help lessen the effects of droughts, they cannot fully prevent water stress in crops and resulting lower yields (Eekhout and de Vente 2019). Moreover, although in many cases SLM measures can help reduce and reverse desertification, there would still be short-term losses in land productivity. Irreversible forms of land degradation (for example, loss of topsoil, severe gully erosion) can lead to the complete loss of land productivity. Even when solutions are available, their costs could be prohibitive, presenting the limits to adaptation (Dixon et al. 2013). If warming in dryland areas surpasses human thermal physiological thresholds (Klein et al. 2015; Waha et al. 2013), adaptation could eventually fail (Kamali et al. 2018). Catastrophic shifts in ecosystem functions and services (for example coastal erosion (Chen et al. 2015; Schneider and Kéfi 2016) (Section 4.9.8)) and economic factors can also result in adaptation failure (Evans et al. 2015). Despite the availability of numerous options that contribute to combating desertification, climate change adaptation and mitigation, there are also chances of maladaptive actions (*medium confidence*) (see Glossary). Some activities favouring agricultural intensification in dryland areas can become maladaptive due to their negative impacts on the environment (*medium confidence*). Agricultural expansion to meet food demands can come through deforestation and consequent diminution of carbon sinks (Godfray and Garnett 2014; Stringer et al. 2012). Agricultural insurance programmes encouraging higher agricultural productivity and measures for agricultural intensification can result in detrimental environmental outcomes in some settings (Guodaar et al. 2019; Müller et al. 2017) (Table 6.12). Development of more drought-tolerant crop varieties is considered as a strategy for adaptation to shortening rainy seasons, but this can also lead to a loss of local varieties (Al Hamndou and Requier-Desjardins

2008). Livelihood diversification to collecting and selling firewood and charcoal production can exacerbate deforestation (Antwi-Agyei et al. 2018). Avoiding maladaptive outcomes can often contribute both to reducing the risks from climate change and combating desertification (Antwi-Agyei et al. 2018). Avoiding, reducing and reversing desertification would enhance soil fertility, increase carbon storage in soils and biomass, thus reducing carbon emissions from soils to the atmosphere (Section 3.7.2 and Cross-Chapter Box 2 in Chapter 1). In specific locations, there may be barriers for some of these activities. For example, afforestation and reforestation programmes can contribute to reducing sand storms and increasing carbon sinks in dryland regions (Chu et al. 2019) (Sections 3.6.1 and 3.7.2). However, implementing agroforestry measures in arid locations can be constrained by lack of water (Apuri et al. 2018), leading to a trade-off between soil carbon sequestration and other water uses (Cao et al. 2018). Thus, even when solutions are available, social, economic and institutional constraints could post barriers to their implementation (*medium confidence*).

1.7   Hotspots and case studies

The challenges of desertification and climate change in dryland areas across the world often have very location-specific characteristics. The five case studies in this section present rich experiences and lessons learnt on: (i) soil erosion, (ii) afforestation and reforestation through 'green walls', (iii) invasive plant species, (iv) oases in hyper-arid areas, and (v) integrated watershed management. Although it is impossible to cover all hotspots of desertification and on-the-ground actions from all dryland areas, these case studies present a more focused assessment of these five issues, which emerged as salient in the group discussions and several rounds of review of this chapter. The choice of these case studies was also motivated by the desire to capture a wide diversity of dryland settings.

1.7.1   Climate change and soil erosion

1.7.1.1   Soil erosion under changing climate in drylands

Soil erosion is a major form of desertification occurring in varying degrees in all dryland areas across the world (Section 3.2), with negative effects on dryland ecosystems (Section 3.4). Climate change is projected to increase soil erosion potential in some dryland areas through more frequent heavy rainfall events and rainfall variability (see Section 3.5.2 for a more detailed assessment) (Achite and Ouillon 2007; Megnounif and Ghenim 2016; Vachtman et al. 2013; Zhang and Nearing 2005). There are numerous soil conservation measures that can help reduce soil erosion (Section 3.6.1). Such soil management measures include afforestation and reforestation activities, rehabilitation of degraded forests, erosion control measures, prevention of overgrazing, diversification of crop rotations, and improvement in irrigation techniques, especially in sloping areas (Anache et al. 2018; ÇEMGM 2017; Li and Fang 2016; Poesen 2018; Ziadat and Taimeh 2013). Effective measures for soil conservation can also use spatial patterns of plant cover to reduce sediment connectivity, and the relationships between hillslopes and sediment

transfer in eroded channels (García-Ruiz et al. 2017). The following three examples present lessons learnt from the soil erosion problems and measures to address them in different settings of Chile, Turkey and the Central Asian countries.

1.7.1.2   No-till practices for reducing soil erosion in central Chile

Soil erosion by water is an important problem in Chile. National assessments conducted in 1979, which examined 46% of the continental surface of the country, concluded that very high levels of soil erosion affected 36% of the territory. The degree of soil erosion increases from south to north. The leading locations in Chile are the region of Coquimbo with 84% of eroded soils (Lat. 29°S, semi-arid climate), the region of Valparaíso with 57% of eroded soils (Lat. 33°S, Mediterranean climate) and the region of O'Higgins with 37% of eroded soils (Lat. 34°S, Mediterranean climate). The most important drivers of soil erosion are soil, slope, climate erosivity (i.e., precipitation, intensity, duration and frequency) due to a highly concentrated rainy season, and vegetation structure and cover. In the region of Coquimbo, goat and sheep overgrazing have aggravated the situation (CIREN 2010). Erosion rates reach up to 100 t ha^{-1} annually, having increased substantially over the last 50 years (Ellies 2000). About 10.4% of central Chile exhibits high erosion rates (greater than 1.1 t ha^{-1} annually) (Bonilla et al. 2010).

Over the last few decades there has been an increasing interest in the development of no-till (also called zero tillage) technologies to minimise soil disturbance, reduce the combustion of fossil fuels and increase soil organic matter. No-till, in conjunction with the adoption of strategic cover crops, has positively impacted soil biology with increases in soil organic matter. Early evaluations by Crovetto, (1998) showed that no-till application (after seven years) had doubled the biological activity indicators compared to traditional farming and even surpassed those found in pasture (grown for the previous 15 years). Besides erosion control, additional benefits are an increase of water-holding capacity and reduction in bulk density. Currently, the above no-till farm experiment has lasted for 40 years and continues to report benefits to soil health and sustainable production (Reicosky and Crovetto 2014). The influence of this iconic farm has resulted in the adoption of soil conservation practices – and especially no-till – in dryland areas of the Mediterranean climate region of central Chile (Martínez et al. 2011). Currently, it has been estimated that the area under no-till farming in Chile varies between 0.13 and 0.2 Mha (Acevedo and Silva 2003).

1.7.1.3   Combating wind erosion and deflation in Turkey: The greening desert of Karapınar

In Turkey, the amount of sediment recently released through erosion into seas was estimated to be 168 Mt yr^{-1}, which is considerably lower than the 500 Mt yr^{-1} that was estimated to be lost in the 1970s. The decrease in erosion rates is attributed to an increase in spatial extent of forests, rehabilitation of degraded forests, erosion control, prevention of overgrazing, and improvement in irrigation technologies. Soil conservation measures conducted in the Karapınar district, Turkey, exemplify these activities. The district is characterised

Figure 3.11 | **(1)** A general view of a nearby village of Karapınar town in the early 1960s (Çarkaci 1999). **(2)** A view of the Karapınar wind erosion area in 2013 (Photo: Murat Türkeş, 17 June 2019). **(3)** Construction of cane screens in the early 1960s in order to decrease wind speed and prevent movement of the sand accumulations and dunes; this was one of the physical measures during the prevention and mitigation period (Çarkaci 1999). **(4)** A view of mixed vegetation, which now covers most of the Karapınar wind erosion area in 2013, the main tree species of which were selected for afforestation with respect to their resistance to the arid continental climate conditions along with a warm/hot temperature regime over the district (Photo: Murat Türkeş, 17 June 2013).

by a semi-arid climate and annual average precipitation of 250–300 mm (Türkeş 2003; Türkeş and Tatlı 2011). In areas where vegetation was overgrazed or inappropriately tilled, the surface soil horizon was removed through erosion processes resulting in the creation of large drifting dunes that threatened settlements around Karapınar (Groneman 1968). Such dune movement had begun to affect the Karapınar settlement in 1956 (Kantarcı et al. 2011). Consequently, by the early 1960s, Karapınar town and nearby villages were confronted with the danger of abandonment due to out-migration in the early 1960s (Figure 3.11(1)). The reasons for increasing wind erosion in the Karapınar district can be summarised as follows: sandy material was mobilised following drying of the lake; hot and semi-arid climate conditions; overgrazing and use of pasture plants for fuel; excessive tillage; and strong prevailing winds.

Restoration and mitigation strategies were initiated in 1959, and today 4300 ha of land have been restored (Akay and Yildirim 2010) (Figure 3.11 (2)), using specific measures: (i) physical measures: construction of cane screens to decrease wind speed and prevent sand movement (Figure 3.11(3)); (ii) restoration of cover: increasing grass cover between screens using seeds collected from local pastures or the cultivation of rye (*Secale* sp.) and wheat grass (*Agropyron elongatum*) that are known to grow in arid and hot conditions; and (iii) afforestation: saplings obtained from nursery gardens were planted and grown between these screens. Main tree species selected were oleaster (*Eleagnus* sp.), acacia (*Robinia pseudeaccacia*), ash (*Fraxinus* sp.), elm (*Ulmus* sp.) and maple (*Acer* sp.) (Figure 3.11 (4)). Economic growth occurred after controlling erosion and new tree nurseries have been established with modern irrigation. Potential negative consequences through the excessive use of water can be mitigated through engagement with local stakeholders and transdisciplinary learning processes, as well as by restoring the traditional land uses in the semi-arid Konya closed basin (Akça et al. 2016).

1.7.1.4   Soil erosion in Central Asia under changing climate

Soil erosion is widely acknowledged to be a major form of degradation of Central Asian drylands, affecting a considerable share of croplands and rangelands. However, up-to-date information on the actual extent of eroded soils at the regional or country level is not available. The estimates compiled by Pender et al. (2009), based on the Central Asian Countries Initiative for Land Management (CACILM), indicate that about 0.8 Mha of the irrigated croplands were subject to high degree of soil erosion in Uzbekistan. In Turkmenistan, soil erosion was indicated to be occurring in about 0.7 Mha of irrigated land. In Kyrgyzstan, out of 1 Mha of irrigated land in the foothill zones, 0.76 Mha were subject to soil erosion by water, leading to losses in crop yields of 20–60% in these eroded soils. About 0.65 Mha of arable land were prone to soil erosion by wind (Mavlyanova et al. 2017). Soil erosion is widespread in rainfed and irrigated areas in Kazakhstan (Saparov 2014). About 5 Mha of rainfed croplands were subject to high levels of soil erosion (Pender et al. 2009). Soil erosion by water was indicated to be a major concern in sloping areas in Tajikistan (Pender et al. 2009).

The major causes of soil erosion in Central Asia are related to human factors, primarily excessive water use in irrigated areas (Gupta et al. 2009), deep ploughing and lack of maintenance of vegetative cover in rainfed areas (Suleimenov et al. 2014), and overgrazing in rangelands (Mirzabaev et al. 2016). Lack of good maintenance of watering infrastructure for migratory livestock grazing, and fragmentation of livestock herds led to overgrazing near villages, increasing the soil erosion by wind (Alimaev et al. 2008). Overgrazing in the rangeland areas of the region (e.g., particularly in Kyzylkum) contributes to dust storms, coming primarily from the Ustyurt Plateau, desertified areas of Amudarya and Syrdarya rivers' deltas, the dried seabed of the Aral Sea (now called Aralkum), and the Caspian Sea (Issanova and Abuduwaili 2017; Xi and Sokolik 2015). Xi and Sokolik (2015) estimated that total dust emissions in Central Asia were 255.6 Mt in 2001, representing 10–17% of the global total.

293

Central Asia is one of the regions highly exposed to climate change, with warming levels projected to be higher than the global mean (Hoegh-Guldberg et al. 2018), leading to more heat extremes (Reyer et al. 2017). There is no clear trend in precipitation extremes, with some potential for moderate rise in occurrence of droughts. The diminution of glaciers is projected to continue in the Pamir and Tian Shan mountain ranges, a major source of surface waters along with seasonal snowmelt. Glacier melting will increase the hazards from moraine-dammed glacial lakes and spring floods (Reyer et al. 2017). Increased intensity of spring floods creates favourable conditions for higher soil erosion by water, especially in the sloping areas in Kyrgyzstan and Tajikistan. The continuation of some of the current unsustainable cropland and rangeland management practices may lead to elevated rates of soil erosion, particularly in those parts of the region where climate change projections point to increases in floods (Kyrgyzstan, Tajikistan) or increases in droughts (Turkmenistan, Uzbekistan) (Hijioka et al. 2014). Increasing water use to compensate for higher evapotranspiration due to rising temperatures and heat waves could increase soil erosion by water in the irrigated zones, especially in sloping areas and crop fields with uneven land levelling (Bekchanov et al. 2010). The desiccation of the Aral Sea resulted in a hotter and drier regional microclimate, adding to the growing wind erosion in adjacent deltaic areas and deserts (Kust 1999).

There are numerous sustainable land and water management practices available in the region for reducing soil erosion (Abdullaev et al. 2007; Gupta et al. 2009; Kust et al. 2014; Nurbekov et al. 2016). These include: improved land levelling and more efficient irrigation methods such as drip, sprinkler and alternate furrow irrigation (Gupta et al. 2009); conservation agriculture practices, including no-till methods and maintenance of crop residues as mulch in the rainfed and irrigated areas (Kienzler et al. 2012; Pulatov et al. 2012); rotational grazing; institutional arrangements for pooling livestock for long-distance mobile grazing; reconstruction of watering infrastructure along the livestock migratory routes (Han et al. 2016; Mirzabaev et al. 2016); afforesting degraded marginal lands (Djanibekov and Khamzina 2016; Khamzina et al. 2009; Khamzina et al. 2016); integrated water resource management (Dukhovny et al. 2013; Kazbekov et al. 2009); and planting salt – and drought-tolerant halophytic plants as windbreaks in sandy rangelands (Akinshina et al. 2016; Qadir et al. 2009; Toderich et al. 2009; Toderich et al. 2008), and potentially the dried seabed of the former Aral Sea (Breckle 2013). The adoption of enabling policies, such as those discussed in Section 3.6.3, can facilitate the adoption of these sustainable land and water management practices in Central Asia (*high confidence*) (Aw-Hassan et al. 2016; Bekchanov et al. 2016; Bobojonov et al. 2013; Djanibekov et al. 2016; Hamidov et al. 2016; Mirzabaev et al. 2016).

1.7.2   Green walls and green dams

This case study evaluates the experiences of measures and actions implemented to combat soil erosion, decrease dust storms, and to adapt to and mitigate climate change under the Green Wall and Green Dam programmes in East Asia (e.g., China) and Africa (e.g., Algeria, Sahara and the Sahel region). These measures have also been implemented in other countries, such as Mongolia (Do and

Kang 2014; Lin et al. 2009), Turkey (Yurtoglu 2015; Çalişkan and Boydak 2017) and Iran (Amiraslani and Dragovich 2011), and are increasingly considered as part of many national and international initiatives to combat desertification (Goffner et al. 2019) (Cross-Chapter Box 2 in Chapter 1). Afforestation and reforestation programmes can contribute to reducing sand storms and increasing carbon sinks in dryland regions (*high confidence*). On the other hand, green wall and green dam programmes also decrease the albedo and hence increase the surface absorption of radiation, increasing the surface temperature. The net effect will largely depend on the balance between these and will vary from place to place depending on many factors.

1.7.2.1   The experiences of combating desertification in China

Arid and semi-arid areas of China, including north-eastern, northern and north-western regions, cover an area of more than 509 Mha, with annual rainfall of below 450 mm. Over the past several centuries, more than 60% of the areas in arid and semi-arid regions were used as pastoral and agricultural lands. The coupled impacts of past climate change and human activity have caused desertification and dust storms to become a serious problem in the region (Xu et al. 2010). In 1958, the Chinese government recognised that desertification and dust storms jeopardised the livelihoods of nearly 200 million people, and afforestation programmes for combating desertification have been initiated since 1978. China is committed to go beyond the Land Degradation Neutrality objective, as indicated by the following programmes that have been implemented. The Chinese Government began the Three North's Forest Shelterbelt programme in Northeast China, North China, and Northwest China, with the goal to combat desertification and to control dust storms by improving forest cover in arid and semi-arid regions. The project is implemented in three stages (1978–2000, 2001–2020 and 2021–2050). In addition, the Chinese government launched the Beijing and Tianjin Sandstorm Source Treatment Project (2001–2010), Returning Farmlands to Forest Project (2003–present), and the Returning Grazing Land to Grassland Project (2003–present) to combat desertification, and for adaptation and mitigation of climate change (State Forestry Administration of China 2015; Wang 2014; Wang et al. 2013).

The results of the fifth monitoring period (2010–2014) showed: (i) compared with 2009, the area of degraded land decreased by 12,120 km^2 over a five-year period; (ii) in 2014, the average coverage of vegetation in the sand area was 18.33%, an increase of 0.7% compared with 17.63% in 2009, and the carbon sequestration increased by 8.5%; (iii) compared with 2009, the amount of wind erosion decreased by 33%, the average annual occurrence of sandstorms decreased by 20.3% in 2014; (iv) as of 2014, 203,700 km^2 of degraded land were effectively managed, accounting for 38.4% of the 530,000 km^2 of manageable desertified land; (v) the restoration of degraded land has created an annual output of 53.63 Mt of fresh and dried fruits, accounting for 33.9% of the total national annual output of fresh and dried fruits (State Forestry Administration of China 2015). This has become an important pillar for economic development and a high priority for peasants as a method to eradicate poverty (State Forestry Administration of China 2015).

Stable investment mechanisms for combating desertification have been established along with tax relief policies and financial support policies for guiding the country in its fight against desertification. The investments in scientific and technological innovation for combating desertification have been improved, the technologies for vegetation restoration under drought conditions have been developed, the popularisation and application of new technologies has been accelerated, and the training of technicians to assist farmers and herdsmen has been strengthened. To improve the monitoring capability and technical level of desertification studies, the monitoring network system has been strengthened, and the popularisation and application of modern technologies have been intensified (e.g., information technology and remote sensing) (Wu et al. 2015). Special laws on combating desertification have been decreed by the government. The provincial government's responsibilities for desertification prevention and controlling objectives and laws have been strictly implemented.

Many studies showed that these projects generally played an active role in combating desertification and fighting against dust storms in China over the past several decades (*high confidence*) (Cao et al. 2018; State Forestry Administration of China 2015; Wang et al. 2013; Wang et al. 2014; Yang et al. 2013). At the beginning of the projects, some problems appeared in some places due to lack of enough knowledge and experience (*low confidence*) (Jiang 2016; Wang et al. 2010). For example, some tree species selected were not well suited to local soil and climatic conditions (Zhu et al. 2007), and there was inadequate consideration of the limitation of the amount of available water on the carrying capacity of trees in some arid regions (Dai 2011; Feng et al. 2016) (Section 3.6.4). In addition, at the beginning of the projects, there was an inadequate consideration of the effects of climate change on combating desertification (Feng et al. 2015; Tan and Li 2015). Indeed, climate change and human activities over past years have influenced the desertification and dust storm control effects in China (Feng et al. 2015; Wang et al. 2009; Tan and Li 2015), and future climate change will bring new challenges for combating desertification in China (Wang et al. 2017; Yin et al. 2015; Xu et al. 2019). In particular, the desertification risk in China will be enhanced at 2°C compared to 1.5°C global temperature rise (Ma et al. 2018). Adapting desertification control to climate change involves: improving the adaptation capacity to climate change for afforestation and grassland management by executing SLM practices; optimising the agricultural and animal husbandry structure; and using big data to meet the water resources regulation (Zhang and Huisingh 2018). In particular, improving scientific and technological supports in desertification control is crucial for adaptation to climate change and combating desertification, including protecting vegetation in desertification-prone lands by planting indigenous plant species, facilitating natural restoration of vegetation to conserve biodiversity, employing artificial rain or snow, water-saving irrigation and water storage technologies (Jin et al. 2014; Yang et al. 2013).

1.7.2.2   The Green Dam in Algeria

After independence in 1962, the Algerian government initiated measures to replant forests destroyed by the war, and the steppes affected by desertification, among its top priorities (Belaaz 2003).

In 1972, the government invested in the Green Dam (*Barrage vert*) project. This was the first significant experiment to combat desertification, influence the local climate and decrease the aridity by restoring a barrier of trees. The Green Dam extends across arid and semi-arid zones between the isohyets 300 mm and 200 mm. It is a 3 Mha band of plantation running from east to west (Figure 3.12). It is over 1200 km long (from the Algerian–Moroccan border to the Algerian–Tunisian border) and has an average width of about 20 km. The soils in the area are shallow, low in organic matter and susceptible to erosion. The main objectives of the project were to conserve natural resources, improve the living conditions of local residents and avoid their exodus to urban areas. During the first four decades (1970–2000) the success rate was low (42%) due to lack of participation by the local population and the choice of species (Bensaid 1995).

The Green Dam did not have the desired effects. Despite tree-planting efforts, desertification intensified on the steppes, especially in south-western Algeria, due to the prolonged drought during the 1980s. Rainfall declined in the range from 18% to 27%, and the dry season has increased by two months in the last century (Belala et al. 2018). Livestock numbers in the Green Dam regions, mainly sheep, grew exponentially, leading to severe overgrazing, causing trampling and soil compaction, which greatly increased the risk of erosion. Wind erosion, very prevalent in the region, is due to climatic conditions and the strong anthropogenic action that reduced the vegetation cover. The action of the wind carries fine particles such as sands and clays and leaves on the soil surface a lag-gravel pavement, which is unproductive. Water erosion is largely due to torrential rains in the form of severe thunderstorms that disintegrate the bare soil surface from raindrop impact (Achite et al. 2016). The detached soil and nutrients are transported offsite via runoff, resulting in loss of fertility and water holding capacity. The risk of and severity of water erosion is a function of human land-use activities that increase soil loss through removal of vegetative cover. The National Soil Sensitivity to Erosion Map (Salamani et al. 2012) shows that more than 3 Mha of land in the steppe provinces are currently experiencing intense wind activity (Houyou et al. 2016) and that these areas are at particular risk of soil erosion. Mostephaoui et al. (2013), estimates that each year there is a loss of 7 t ha^{-1} of soils due to erosion. Nearly 0.6 Mha of land in the steppe zone are fully degraded without the possibility of biological recovery.

To combat the effects of erosion and desertification, the government has planned to relaunch the rehabilitation of the Green Dam by incorporating new concepts related to sustainable development, and adaptation to climate change. The experience of previous years has led to integrated rangeland management, improved tree and fodder shrub plantations and the development of water conservation techniques. Reforestation is carried out using several species, including fruit trees, to increase and diversify the sources of income for the population.

The evaluation of the Green Dam from 1972 to 2015 (Merdas et al. 2015) shows that 0.3 Mha of forest plantation have been planted, which represents 10% of the project area. Estimates of the success rate of reforestation vary considerably between 30% and 75%,

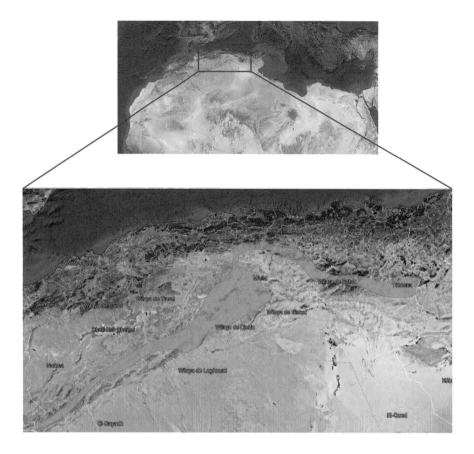

Figure 3.12 | Location of the Green Dam in Algeria (Saifi et al. 2015). Note: The green coloured band represents the location of the Green Dam.

depending on the region. Through demonstration, the Green Dam has inspired several African nations to work together to build a Great Green Wall to combat land degradation, mitigate climate change effects, loss of biodiversity and poverty in a region that stretches from Senegal to Djibouti (Sahara and Sahel Observatory (OSS) 2016) (Section 3.7.2.3).

1.7.2.3   The Great Green Wall of the Sahara and the Sahel Initiative

The Great Green Wall is an initiative of the Heads of State and Government of the Sahelo-Saharan countries to mitigate and adapt to climate change, and to improve the food security of the Sahel and Saharan peoples (Sacande 2018; Mbow 2017). Launched in 2007, this regional project aims to restore Africa's degraded arid landscapes, reduce the loss of biodiversity and support local communities to sustainable use of forests and rangelands. The Great Green Wall focuses on establishing plantations and neighbouring projects, covering a distance of 7775 km from Senegal on the Atlantic coast to Eritrea on the Red Sea coast, with a width of 15 km (Figure 3.13). The wall passes through Djibouti, Eritrea, Ethiopia, Sudan, Chad, Niger, Nigeria, Mali, Burkina Faso, Mauritania and Senegal.

The choice of woody and herbaceous species that will be used to restore degraded ecosystems is based on biophysical and socio-economic criteria, including socio-economic value (food, pastoral, commercial, energetic, medicinal, cultural); ecological importance (carbon sequestration, soil cover, water infiltration);

and resilience to climate change and variability. The Pan-African Agency of the Great Green Wall (PAGGW) was created in 2010 under the auspices of the African Union and CEN-SAD to manage the project. The initiative is implemented at the level of each country by a national structure. A monitoring and evaluation system has been defined, allowing nations to measure outcomes and to propose the necessary adjustments.

In the past, reforestation programmes in the arid regions of the Sahel and North Africa that have been undertaken to stop desertification were poorly studied and cost a lot of money without significant success (Benjaminsen and Hiernaux 2019). Today, countries have changed their strategies and opted for rural development projects that can be more easily funded. Examples of scalable practices for land restoration include managing water bodies for livestock and crop production, and promoting fodder trees to reduce runoff (Mbow 2017).

The implementation of the initiative has already started in several countries. For example, the FAO's Action Against Desertification project was restoring 18,000 hectares of land in 2018 through planting native tree species in Burkina Faso, Ethiopia, The Gambia, Niger, Nigeria and Senegal (Sacande 2018). Berrahmouni et al. (2016) estimated that 166 Mha can be restored in the Sahel, requiring the restoration of 10 Mha per year to achieve Land Degradation Neutrality targets by 2030. Despite these early implementation actions on the ground, the achievement of the planned targets is questionable, and will be challenging without significant additional funding.

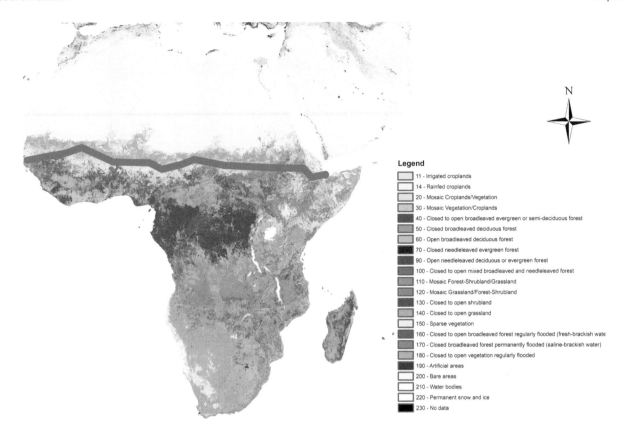

Figure 3.13 | The Great Green Wall of the Sahara and the Sahel. Source for the data layer: This dataset is an extract from the GlobCover 2009 land cover map, covering Africa and the Arabian Peninsula. The GlobCover 2009 land cover map is derived by an automatic and regionally tuned classification of a time series of global MERIS (MEdium Resolution Imaging Spectrometer) FR mosaics for the year 2009. The global land cover map counts 22 land cover classes defined with the United Nations (UN) Land Cover Classification System (LCCS).

1.7.3   Invasive plant species

1.7.3.1   Introduction

The spread of invasive plants can be exacerbated by climate change (Bradley et al. 2010; Davis et al. 2000). In general, it is expected that the distribution of invasive plant species with high tolerance to drought or high temperatures may increase under most climate change scenarios (*medium* to *high confidence*) (Bradley et al. 2010; Settele et al. 2014; Scasta et al. 2015). Invasive plants are considered a major risk to native biodiversity and can disturb the nutrient dynamics and water balance in affected ecosystems (Ehrenfeld 2003). Compared to more humid regions, the number of species that succeed in invading dryland areas is low (Bradley et al. 2012), yet they have a considerable impact on biodiversity and ecosystem services (Le Maitre et al. 2015, 2011; Newton et al. 2011). Moreover, human activities in dryland areas are responsible for creating new invasion opportunities (Safriel et al. 2005).

Current drivers of species introductions include expanding global trade and travel, land degradation and changes in climate (Chytrý et al. 2012; Richardson et al. 2011; Seebens et al. 2018). For example, Davis et al. (2000) suggests that high rainfall variability promotes the success of alien plant species – as reported for semi-arid grasslands and Mediterranean-type ecosystems (Cassidy et al. 2004; Reynolds et al. 2004; Sala et al. 2006). Furthermore, Panda et al. (2018) demonstrated that many invasive species could withstand elevated

temperature and moisture scarcity caused by climate change. Dukes et al. (2011) observed that the invasive plant yellow-star thistle (*Centaurea solstitialis)* grew six time larger under the elevated atmospheric CO_2 expected in future climate change scenarios.

Climate change is *likely* to aggravate the problem as existing species continue to spread unabated and other species develop invasive characteristics (Hellmann et al. 2008). Although the effects of climate change on invasive species distributions have been relatively well explored, the greater impact on ecosystems is less well understood (Bradley et al. 2010; Eldridge et al. 2011).

Due to the time lag between the initial release of invasive species and their impact, the consequence of invasions is not immediately detected and may only be noticed centuries after introduction (Rouget et al. 2016). Climate change and invading species may act in concert (Bellard et al. 2013; Hellmann et al. 2008; Seebens et al. 2015). For example, invasion often changes the size and structure of fuel loads, which can lead to an increase in the frequency and intensity of fire (Evans et al. 2015). In areas where the climate is becoming warmer, an increase in the likelihood of suitable weather conditions for fire may promote invasive species, which in turn may lead to further desertification. Conversely, fire may promote plant invasions via several mechanisms (by reducing cover of competing vegetation, destroying native vegetation and clearing a path for invasive plants or creating favourable soil conditions) (Brooks et al. 2004; Grace et al. 2001; Keeley and Brennan 2012).

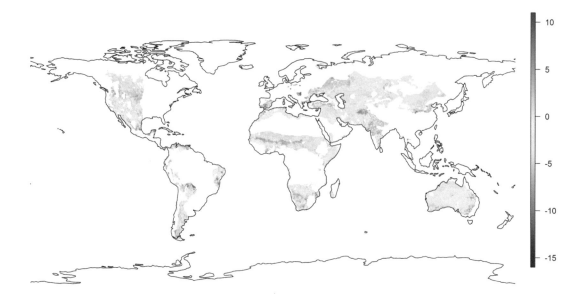

Figure 3.14 | Difference between the number of invasive alien species (n=99, from Bellard et al. (2013)) predicted to occur by 2050 (under A1B scenario) and current period '2000' within the dryland areas.

At a regional scale, Bellard et al. (2013) predicted increasing risk in Africa and Asia, with declining risk in Australia (Figure 3.14). This projection does not represent an exhaustive list of invasive alien species occurring in drylands.

A set of four case studies in Ethiopia, Mexico, the USA and Pakistan is presented below to describe the nuanced nature of invading plant species, their impact on drylands and their relationship with climate change.

1.7.3.2   Ethiopia

The two invasive plants that inflict the heaviest damage to ecosystems, especially biodiversity, are the annual herbaceous weed, *Parthenium hysterophorus* (*Asteraceae*) also known as Congress weed; and the tree species, *Prosopis juliflora* (*Fabaceae*) also called Mesquite, both originating from the southwestern United States to Central/South America (Adkins and Shabbir 2014). *Prosopis* was introduced in the 1970s and has since spread rapidly. *Prosopis*, classified as the highest priority invader in Ethiopia, is threatening livestock production and challenging the sustainability of the pastoral systems. *Parthenium* is believed to have been introduced along with relief aid during the debilitating droughts of the early 1980s, and a recent study reported that it has spread into 32 out of 34 districts in Tigray, the northernmost region of Ethiopia (Teka 2016). A study by Etana et al. (2011) indicated that Parthenium caused a 69% decline in the density of herbaceous species in Awash National Park within a few years of introduction. In the presence of Parthenium, the growth and development of crops is suppressed due to its allelopathic properties. McConnachie et al. (2011) estimated a 28% crop loss across the country, including a 40–90% reduction in sorghum yield in eastern Ethiopia alone (Tamado et al. 2002). The weed is a substantial agricultural and natural resource problem and constitutes a significant health hazard (Fasil 2011). Parthenium causes acute allergic respiratory problems, skin dermatitis, and

reportedly mutagenicity both in humans and livestock (Mekonnen 2017; Patel 2011). The eastern belt of Africa – including Ethiopia – presents a very suitable habitat, and the weed is expected to spread further in the region in the future (Mainali et al. 2015).

There is neither a comprehensive intervention plan nor a clear institutional mandate to deal with invasive weeds, however, there are fragmented efforts involving local communities even though they are clearly inadequate. The lessons learned, related to actions that have contributed to the current scenario, are several. First, lack of coordination and awareness – mesquite was introduced by development agencies as a drought-tolerant shade tree with little consideration of its invasive nature. If research and development institutions had been aware, a containment strategy could have been implemented early on. The second major lesson is the cost of inaction. When research and development organisations did sound the alarm, the warnings went largely unheeded, resulting in the spread and buildup of two of the worst invasive plant species in the world (Fasil 2011).

1.7.3.3   Mexico

Buffelgrass (*Cenchrus ciliaris* L.), a native species from southern Asia and East Africa, was introduced into Texas and northern Mexico in the 1930s and 1940s, as it is highly productive in drought conditions (Cox et al. 1988; Rao et al. 1996). In the Sonoran desert of Mexico, the distribution of buffelgrass has increased exponentially, covering 1 Mha in Sonora State (Castellanos-Villegas et al. 2002). Furthermore, its potential distribution extended to 53% of Sonora State and 12% of semi-arid and arid ecosystems in Mexico (Arriaga et al. 2004). Buffelgrass has also been reported as an aggressive invader in Australia and the USA, resulting in altered fire cycles that enhance further spread of this plant and disrupt ecosystem processes (Marshall et al. 2012; Miller et al. 2010; Schlesinger et al. 2013).

Castellanos et al. (2016) reported that soil moisture was lower in the buffelgrass savannah cleared 35 years ago than in the native semi-arid shrubland, mainly during the summer. The ecohydrological changes induced by buffelgrass can therefore displace native plant species over the long term. Invasion by buffelgrass can also affect landscape productivity, as it is not as productive as native vegetation (Franklin and Molina-Freaner 2010). Incorporation of buffelgrass is considered a good management practice by producers and the government. For this reason, no remedial actions are undertaken.

1.7.3.4   United States of America

Sagebrush ecosystems have declined from 25 Mha to 13 Mha since the late 1800s (Miller et al. 2011). A major cause is the introduction of non-native cheatgrass (*Bromus tectorum*), which is the most prolific invasive plant in the USA. Cheatgrass infests more than 10 Mha in the Great Basin and is expanding every year (Balch et al. 2013). It provides a fine-textured fuel that increases the intensity, frequency and spatial extent of fire (Balch et al. 2013). Historically, wildfire frequency was 60 to 110 years in Wyoming big sagebrush communities and has increased to five years following the introduction of cheatgrass (Balch et al. 2013; Pilliod et al. 2017).

The conversion of the sagebrush steppe biome to annual grassland with higher fire frequencies has severely impacted livestock producers, as grazing is not possible for a minimum of two years after fire. Furthermore, cheatgrass and wildfires reduce critical habitat for wildlife and negatively impact species richness and abundance – for example, the greater sage-grouse (*Centocercus urophasianus*) and pygmy rabbit (*Brachylagus idahoensis*) which are on the verge of being listed for federal protection (Crawford et al. 2004; Larrucea and Brussard 2008; Lockyer et al. 2015).

Attempts to reduce cheatgrass impacts through reseeding of both native and adapted introduced species have occurred for more than 60 years (Hull and Stewart 1949) with little success. Following fire, cheatgrass becomes dominant and recovery of native shrubs and grasses is improbable, particularly in relatively low-elevation sites with minimal annual precipitation (less than 200 mm yr^{-1}) (Davies et al. 2012; Taylor et al. 2014). Current rehabilitation efforts emphasise the use of native and non-native perennial grasses, forbs and shrubs (Bureau of Land Management 2005). Recent literature suggests

that these treatments are not consistently effective at displacing cheatgrass populations or re-establishing sage-grouse habitat, with success varying with elevation and precipitation (Arkle et al. 2014; Knutson et al. 2014). Proper post-fire grazing rest, season-of-use, stocking rates, and subsequent management are essential to restore resilient sagebrush ecosystems before they cross a threshold and become an annual grassland (Chambers et al. 2014; Miller et al. 2011; Pellant et al. 2004). Biological soil crust protection may be an effective measure to reduce cheatgrass germination, as biocrust disturbance has been shown to be a key factor promoting germination of non-native grasses (Hernandez and Sandquist 2011). Projections of increasing temperature (Abatzoglou and Kolden 2011), and observed reductions in and earlier melting of snowpack in the Great Basin region (Harpold and Brooks 2018; Mote et al. 2005) suggest that there is a need to understand current and past climatic variability as this will drive wildfire variability and invasions of annual grasses.

1.7.3.5   Pakistan

The alien plants invading local vegetation in Pakistan include *Brossentia papyrifera* (found in Islamabad Capital territory), *Parthenium hysterophorus* (found in Punjab and Khyber Pakhtunkhwa provinces), *Prosopis juliflora* (found all over Pakistan), *Eucalyptus camaldulensis* (found in Punjab and Sindh provinces), *Salvinia* (aquatic plant widely distributed in water bodies in Sindh), *Cannabis sativa* (found in Islamabad Capital Territory), *Lantana camara* and *Xanthium strumarium* (found in upper Punjab and Khyber Pakhtunkhwa provinces) (Khan et al. 2010; Qureshi et al. 2014). Most of these plants were introduced by the Forest Department decades ago for filling the gap between demand and supply of timber, fuelwood and fodder. These non-native plants have some uses but their disadvantages outweigh their benefits (Marwat et al. 2010; Rashid et al. 2014).

Besides being a source of biological pollution and a threat to biodiversity and habitat loss, the alien plants reduce the land value and cause huge losses to agricultural communities (Rashid et al. 2014). *Brossentia papyrifera*, commonly known as Paper Mulberry, is the root cause of inhalant pollen allergy for the residents of lush green Islamabad during spring. From February to April, the pollen allergy is at its peak, with symptoms of severe persistent coughing, difficulty in breathing, and wheezing. The pollen count, although variable at different times and days, can be as high as 55,000 m^{-3}. Early symptoms of the allergy include sneezing, itching in the eyes and skin, and blocked nose. With changing climate, the onset of disease is getting earlier, and pollen count is estimated to cross 55,000 m^{-3} (Rashid et al. 2014). About 45% of allergic patients in the twin cities of Islamabad and Rawalpindi showed positive sensitivity to the pollens (Marwat et al. 2010). Millions of rupees have been spent by the Capital Development Authority on pruning and cutting of Paper Mulberry trees but because of its regeneration capacity growth is regained rapidly (Rashid et al. 2014). Among other invading plants, *Prosopis juliflora* has allelopathic properties, and *Eucalyptus* is known to transpire huge amounts of water and deplete the soil of its nutrient elements (Qureshi et al. 2014).

Although a Biodiversity Action Plan exists in Pakistan, it is not implemented in letter or spirit. The Quarantine Department focuses only on pests and pathogens but takes no notice of plant and animal species being imported. Also, there is no provision for checking the possible impacts of imported species on the environment (Rashid et al. 2014) or for carrying out bioassays of active allelopathic compounds of alien plants.

1.7.4 Oases in hyper-arid areas in the Arabian Peninsula and northern Africa

Oases are isolated areas with reliable water supply from lakes and springs, located in hyper-arid and arid zones (Figure 3.15). Oasis agriculture has long been the only viable crop production system throughout the hot and arid regions of the Arabian Peninsula and North Africa. Oases in hyper-arid climates are usually subject to water shortage as evapotranspiration exceeds rainfall. This often causes salinisation of soils. While many oases have persisted for several thousand years, many others have been abandoned, often in response to changes in climate or hydrologic conditions (Jones et al. 2019), providing testimony to societies' vulnerability to climatic shifts and raising concerns about similarly severe effects of anthropogenic climate change (Jones et al. 2019).

On the Arabian Peninsula and in North Africa, climate change is projected to have substantial and complex effects on oasis areas (Abatzoglou and Kolden 2011; Ashkenazy et al. 2012; Bachelet et al.

2016; Guan et al. 2018; Iknayan and Beissinger 2018; Ling et al. 2013). To illustrate, by the 2050s, the oases in southern Tunisia are expected to be affected by hydrological and thermal changes, with an average temperature increase of 2.7°C, a 29% decrease in precipitation and a 14% increase in evapotranspiration rate (Ministry of Agriculture and Water Resources of Tunisia and GIZ 2007). In Morocco, declining aquifer recharge is expected to impact the water supply of the Figuig oasis (Jilali 2014), as well as for the Draa Valley (Karmaoui et al. 2016). Saudi Arabia is expected to experience a 1.8°C–4.1°C increase in temperatures by 2050, which is forecast to raise agricultural water demand by 5–15% in order to maintain production levels equal to those of 2011 (Chowdhury and Al-Zahrani 2013). The increase of temperatures and variable pattern of rainfall over the central, north and south-western regions of Saudi Arabia may pose challenges for sustainable water resource management (Tarawneh and Chowdhury 2018). Moreover, future climate scenarios are expected to increase the frequency of floods and flash floods, such as in the coastal areas along the central parts of the Red Sea and the south-southwestern areas of Saudi Arabia (Almazroui et al. 2017).

While many oases are cultivated with very heat-tolerant crops such as date palms, even such crops eventually have declines in their productivity when temperatures exceed certain thresholds or hot conditions prevail for extended periods. Projections so far do not indicate severe losses in land suitability for date palm for the Arabian Peninsula (Aldababseh et al. 2018; Shabani et al. 2015). It is unclear, however, how reliable the climate response parameters in the underlying models are, and actual responses may differ substantially.

Figure 3.15 | Oases across the Arabian Peninsula and North Africa (alphabetically by country). **(a)** Masayrat ar Ruwajah oasis, Ad Dakhiliyah ⬚Governorate, Oman (Photo: Eike Lüdeling). **(b)** Tasselmanet oasis, Ouarzazate Province, Morocco (Photo: Abdellatif Khattabi). **(c)** Al-Ahsa oasis, Al-Ahsa Governarate, Saudi Arabia (Photo: Shijan Kaakkara). **(d)** Zarat oasis, Governorate of Gabes, Tunisia (Photo: Hamda Aloui). The use rights for (a), (b) and (d) were granted by copyright holders; (c) is licensed under the Creative Commons Attribution 2.0 Generic license.

Date palms are routinely assumed to be able to endure very high temperatures, but recent transcriptomic and metabolomic evidence suggests that heat stress reactions already occur at 35°C (Safronov et al. 2017), which is not exceptionally warm for many oases in the region. Given current assumptions about the heat-tolerance of date palm, however, adverse effects are expected to be small (Aldababseh et al. 2018; Shabani et al. 2015). For some other perennial oasis crops, impacts of temperature increases are already apparent. Between 2004/2005 and 2012/2013, high-mountain oases of Al Jabal Al Akhdar in Oman lost almost all fruit and nut trees of temperate-zone origin, with the abundance of peaches, apricots, grapes, figs, pears, apples, and plums dropping by between 86% and 100% (Al-Kalbani et al. 2016). This implies that that the local climate may not remain suitable for species that depend on cool winters to break their dormancy period (Luedeling et al. 2009). A similar impact is very probable in Tunisia and Morocco, as well as in other oasis locations in the Arabian Peninsula and North Africa (Benmoussa et al. 2007). All these studies expect strong decreases in winter chill, raising concerns that many currently well-established species will no longer be viable in locations where they are grown today. The risk of detrimental chill shortfalls is expected to increase gradually, slowly diminishing the economic prospects to produce such species. Without adequate adaptation actions, the consequences of this development for many traditional oasis settlements and other plantations of similar species could be highly negative.

At the same time, population growth and agricultural expansion in many oasis settlements are leading to substantial increases in water demand for human consumption (Al-Kalbani et al. 2014). For example, a large unmet water demand has been projected for future scenarios in the valley of Seybouse in East Algeria (Aoun-Sebaiti et al. 2014), and similar conclusions were drawn for Wadi El Natrun in Egypt (Switzman et al. 2018). Modelling studies have indicated long-term decline in available water and increasing risk of water shortages – for example, for oases in Morocco (Johannsen et al. 2016; Karmaoui et al. 2016), the Dakhla oasis in Egypt's Western Desert (Sefelnasr et al. 2014) and for the large Upper Mega Aquifer of the Arabian Peninsula (Siebert et al. 2016). Mainly due to the risk of water shortages, Souissi et al. (2018) classified almost half of all farmers in Tunisia as non-resilient to climate change, especially those relying on tree crops, which limit opportunities for short-term adaptation actions.

The maintenance of the oasis systems and the safeguarding of their population's livelihoods are currently threatened by continuous water degradation, increasing soil salinisation, and soil contamination (Besser et al. 2017). Waterlogging and salinisation of soils due to rising saline groundwater tables coupled with inefficient drainage systems have become common to all continental oases in Tunisia, most of which are concentrated around saline depressions, known locally as *chotts* (Ben Hassine et al. 2013). Similar processes of salinisation are also occurring in the oasis areas of Egypt due to agricultural expansion, excessive use of water for irrigation and deficiency of the drainage systems (Abo-Ragab 2010; Masoud and Koike 2006). A prime example for this is Siwa oasis (Figure 3.16), a depression extending over 1050 km² in the north-western desert of Egypt in the north of the sand dune belt of the Great Sand Sea (Abo-Ragab and Zaghloul 2017). Siwa oasis has been recognised as

a Globally Important Agricultural Heritage Site (GIAHS) by the FAO for being an *in situ* repository of plant genetic resources, especially of uniquely adapted varieties of date palm, olive and secondary crops that are highly esteemed for their quality and continue to play a significant role in rural livelihoods and diets (FAO 2016).

The population growth in Siwa is leading rapid agricultural expansion and land reclamation. The Siwan farmers are converting the surrounding desert into reclaimed land by applying their old inherited traditional practices. Yet, agricultural expansion in the oasis mainly depends on non-renewable groundwaters. Soil salinisation and vegetation loss have been accelerating since 2000 due to water mismanagement and improper drainage systems (Masoud and Koike 2006). Between 1990 and 2008, the cultivated area increased from 53 to 88 km², lakes from 60 to 76 km², *sabkhas* (salt flats) from 335 to 470 km², and the urban area from 6 to 10 km² (Abo-Ragab 2010). The problem of rising groundwater tables was exacerbated by climatic changes (Askri et al. 2010; Gad and Abdel-Baki 2002; Marlet et al. 2009).

Water supply is *likely* to become even scarcer for oasis agriculture under changing climate in the future than it is today, and viable solutions are difficult to find. While some authors stress the possibility to use desalinated water for irrigation (Aldababseh et al. 2018), the economics of such options, especially given the high evapotranspiration rates in the Arabian Peninsula and North Africa, are debatable. Many oases are located far from water sources that are suitable for desalination, adding further to feasibility constraints. Most authors therefore stress the need to limit water use (Sefelnasr et al. 2014), for example, by raising irrigation efficiency (Switzman et al. 2018), reducing agricultural areas (Johannsen et al. 2016) or imposing water use restrictions (Odhiambo 2017), and to carefully monitor desertification (King and Thomas 2014). Whether adoption of crops with low water demand, such as sorghum (*Sorghum bicolor* (L.) Moench) or jojoba (*Simmondsia chinensis* (Link) C. K. Schneid.) (Aldababseh et al. 2018), can be a viable option for some oases remains to be seen, but given their relatively low profit margins compared to currently grown oasis crops, there are reasons to doubt the economic feasibility of such proposals. While it is currently unclear to what extent oasis agriculture can be maintained in hot locations of the region, cooler sites offer potential for shifting towards new species and cultivars, especially for tree crops, which have particular climatic needs across seasons. Resilient options can be identified, but procedures to match tree species and cultivars with site climate need to be improved to facilitate effective adaptation.

There is *high confidence* that many oases of North Africa and the Arabian Peninsula are vulnerable to climate change. While the impacts of recent climate change are difficult to separate from the consequences of other change processes, it is *likely* that water resources have already declined in many places and the suitability of the local climate for many crops, especially perennial crops, has already decreased. This decline of water resources and thermal suitability of oasis locations for traditional crops is *very likely* to continue throughout the 21st century. In the coming years, the people living in oasis regions across the world will face challenges due to increasing impacts of global environmental change (Chen et al. 2018). Hence, efforts to increase their adaptive capacity to

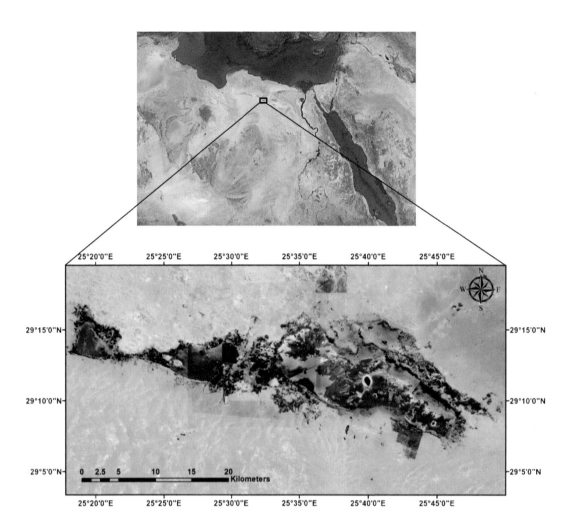

Figure 3.16 | Satellite image of the Siwa Oasis, Egypt. Source: Google Maps.

climate change can facilitate the sustainable development of oasis regions globally. In particular this wil mean addressing the trade-offs between environmental restoration and agricultural livelihoods (Chen et al. 2018). Ultimately, sustainability in oasis regions will depend on policies integrating the provision of ecosystem services and social and human welfare needs (Wang et al. 2017).

1.7.5   Integrated watershed management

Desertification has resulted in significant loss of ecosystem processes and services, as described in detail in this chapter. The techniques and processes to restore degraded watersheds are not linear and integrated watershed management (IWM) must address physical, biological and social approaches to achieve SLM objectives (German et al. 2007).

1.7.5.1   Jordan

Population growth, migration into Jordan and changes in climate have resulted in desertification of the Jordan Badia region. The

Badia region covers more than 80% of the country's area and receives less than 200 mm of rainfall per year, with some areas receiving less than 100 mm (Al-Tabini et al. 2012). Climate analysis has indicated a generally increasing dryness over the West Asia and Middle East region (AlSarmi and Washington 2011; Tanarhte et al. 2015), with reduction in average annual rainfall in Jordan's Badia area (De Pauw et al. 2015). The incidence of extreme rainfall events has not declined over the region. Locally increased incidence of extreme events over the Mediterranean region has been proposed (Giannakopoulos et al. 2009).

The practice of intensive and localised livestock herding, in combination with deep ploughing and unproductive barley agriculture, are the main drivers of severe land degradation and depletion of the rangeland natural resources. This affected both the quantity and the diversity of vegetation as native plants with a high nutrition value were replaced with invasive species with low palatability and nutritional content (Abu-Zanat et al. 2004). The sparsely covered and crusted soils in Jordan's Badia area have a low rainfall interception and infiltration rate, which leads to increased surface runoff and subsequent erosion and gullying, speeding up

Figure 3.17 | (a) Newly prepared micro water harvesting catchment, using the Vallerani system. **(b)** Aerial imaging showing micro water harvesting catchment treatment after planting **(c)** one year after treatment. Source: Stefan Strohmeier.

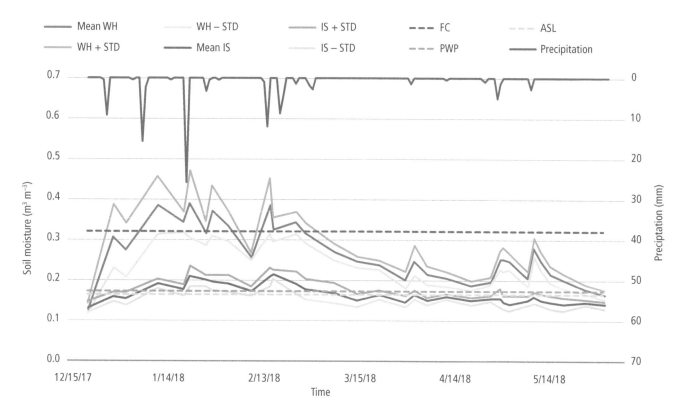

Figure 3.18 | Illustration of enhanced soil water retention in the Mechanized Micro Rainwater Harvesting compared to untreated Badia rangelands in Jordan, showing precipitation (PCP), sustained stress level resulting in decreased production, field capacity and wilting point for available soil moisture, and then measured soil moisture content between the two treatments (degraded rangeland and the restored rangeland with the Vallerani plough).

the drainage of rainwater from the watersheds, which can result in downstream flooding in Amman, Jordan (Oweis 2017).

To restore the desertified Badia an IWM plan was developed using hillslope-implemented water harvesting micro catchments as a targeted restoration approach (Tabieh et al. 2015). Mechanized Micro Rainwater Harvesting (MIRWH) technology using the 'Vallerani plough' (Antinori and Vallerani 1994; Gammoh and Oweis 2011; Ngigi 2003) is being widely applied for rehabilitation of highly degraded rangeland areas in Jordan. A tractor digs out small water harvesting pits on the contour of the slope (Figure 3.17) allowing the retention, infiltration and local storage of surface runoff in the soil (Oweis 2017). The micro catchments are planted with native shrub seedlings, such as saltbush (*Atriplex halimus*), with enhanced survival as a function of increased soil moisture (Figure 3.18) and increased

dry matter yields (>300 kg ha^{-1}) that can serve as forage for livestock (Oweis 2017; Tabieh et al. 2015).

Simultaneously to MIRWH upland measures, the gully erosion is being treated through intermittent stone plug intervention (Figure 3.19), stabilising the gully beds, increasing soil moisture in proximity of the plugs, dissipating the surface runoff's energy, and mitigating further back-cutting erosion and quick drainage of water. Eventually, the treated gully areas silt up and dense vegetation cover can re-establish. In addition, grazing management practices are implemented to increase the longevity of the treatment. Ultimately, the recruitment processes and re-vegetation shall control the watershed's hydrological regime through rainfall interception, surface runoff deceleration and filtration, combined with the less erodible and enhanced infiltration characteristics of the rehabilitated soils.

Figure 3.19 | (a) Gully plug development in September 2017. (b) Post-rainfall event (March 2018). Near Amman, Jordan. Source: Stefan Strohmeier.

In-depth understanding of the Badia's rangeland status transition, coupled with sustainable rangeland management, are still subject to further investigation, development and adoption; a combination of all three is required to mitigate the ongoing degradation of the Middle Eastern rangeland ecosystems.

Oweis (2017) indicated that the cost of the fully automated Vallerani technique was approximately 32 USD ha^{-1}. The total cost of the restoration package included the production, planting and maintenance of the shrub seedlings (11 USD ha^{-1}). Tabieh et al. (2015) calculated a benefit-cost ratio (BCR) of above 1.5 for re-vegetation of degraded Badia areas through MIRWH and saltbush. However, costs vary based on the seedling's costs and availability of trained labour.

Water harvesting is not a recent scientific advancement. Water harvesting is known to have been developed during the Bronze Age and was widely practiced in the Negev Desert during the Byzantine time period (1300–1600 years ago) (Fried et al. 2018; Stavi et al. 2017). Through construction of various structures made of packed clay and stone, water was either held on site in half-circular dam structures (*hafir*) that faced up-slope to capture runoff, or on terraces that slowed water allowing it to infiltrate and to be stored in the soil profile. Numerous other systems were designed to capture water in below-ground cisterns to be used later to provide water to livestock or for domestic use. Other water harvesting techniques divert runoff from hillslopes or *wadis* and spread the water in a systematic manner across *playas* and the toe-slope of a hillslope. These systems allow production of crops in areas with 100 mm of average annual precipitation by harvesting an additional 300+ mm of water (Beckers et al. 2013). Water harvesting is a proven technology to mitigate or adapt to climate change where precipitation may be reduced, and allow for small-scale crop and livestock production to continue supporting local needs.

1.7.5.2 India

The second great challenge after the Green Revolution in India was the low productivity in the rain-fed and semi-arid regions where land degradation and drought were serious concerns. In response to this challenge IWM projects were implemented over large areas in semi-arid biomes over the past few decades. IWM was meant to become a key factor in meeting a range of social development goals in many semi-arid rainfed agrarian landscapes in India (Bouma et al.

2007; Kerr et al. 2002). Over the years, watershed development has become the fulcrum of rural development, and has the potential to achieve the twin objectives of ecosystem restoration and livelihood assurance in the drylands of India (Joy et al. 2004).

Many reports indicate significant improvements in mitigation of drought impacts, raising crops and fodder, livestock productivity, expanding the availability of drinking water and increasing incomes as a result of IWM (Rao 2000), but in some cases overall the positive impact of the programme has been questioned and, except in a few cases, the performance has not lived up to expectations (Joy et al. 2004; JM Kerr et al. 2002). Comparisons of catchments with and without IWM projects using remotely sensed data have sometimes shown no significant enhancement of biomass, in part due to methodological challenges of space for time comparisons (Bhalla et al. 2013). The factors contributing to the successful cases were found to include effective participation of stakeholders in management (Rao 2000; Ratna Reddy et al. 2004).

Attribution of success in soil and water conservation measures was confounded by inadequate monitoring of rainfall variability and lack of catchment hydrologic indicators (Bhalla et al. 2013). Social and economic trade-offs included bias of benefits to downstream crop producers at the expense of pastoralists, women and upstream communities. This biased distribution of IWM benefits could potentially be addressed by compensation for environmental services between communities (Kerr et al. 2002). The successes in some areas also led to increased demand for water, especially groundwater, since there has been no corresponding social regulation of water use after improvement in water regime (Samuel et al. 2007). Policies and management did not ensure water allocation to sectors with the highest social and economic benefits (Batchelor et al. 2003). Limited field evidence of the positive impacts of rainwater harvesting at the local scale is available, but there are several potential negative impacts at the watershed scale (Glendenning et al. 2012). Furthermore, watershed projects are known to have led to more water scarcity, and higher expectations for irrigation water supply, further exacerbating water scarcity (Bharucha et al. 2014).

In summary, the mixed performance of IWM projects has been linked to several factors. These include: inequity in the distribution of benefits (Kerr et al. 2002); focus on institutional aspects rather than application of appropriate watershed techniques and functional aspects of watershed restoration (Joy et al. 2006; Vaidyanathan 2006); mismatch between scales of focus and those that are optimal for catchment processes (Kerr 2007); inconsistencies in criteria used to select watersheds for IWM projects (Bhalla et al. 2011); and in a few cases additional costs and inefficiencies of local non-governmental organisations (Chandrasekhar et al. 2006; Deshpande 2008). Enabling policy responses for improvement of IWM performance include: a greater emphasis on ecological restoration rather than civil engineering; sharper focus on sustainability of livelihoods than just conservation; adoption of 'water justice' as a normative goal and minimising externalities on non-stakeholder communities; rigorous independent biophysical monitoring, with feedback mechanisms and integration with larger schemes for food and ecological security, and maintenance of environmental flows for downstream areas (Bharucha et al. 2014; Calder et al. 2008; Joy et al. 2006). Successful adaptation of IWM to achieve land degradation neutrality would largely depend on how IWM creatively engages with dynamics of large-scale land use and hydrology under a changing climate, involvement of livelihoods and rural incomes in ecological restoration, regulation of groundwater use, and changing aspirations of rural population (*robust evidence*, *high agreement*) (O'Brien et al. 2004; Samuel et al. 2007; Samuel and Joy 2018).

1.7.5.3 Limpopo River Basin

Covering an area of 412,938 km^2, the Limpopo River basin spans parts of Botswana, South Africa, Zimbabwe and Mozambique, eventually entering into the Mozambique Channel. It has been selected as a case study as it provides a clear illustration of the combined effect of desertification and climate change, and why IWM may be a crucial component of reducing exposure to climate change. It is predominantly a semi-arid area with an average annual rainfall of 400 mm (Mosase and Ahiablame 2018). Rainfall is both highly seasonal and variable, with the prominent impact of the El Niño/La Niña phenomena and the Southern Oscillation leading to severe droughts (Jury 2016). It is also exposed to tropical cyclones that sweep in from the Mozambique Channel often leading to extensive casualties and the destruction of infrastructure (Christie and Hanlon 2001). Furthermore, there is good agreement across climate models that the region is going to become warmer and drier, with a change in the frequency of floods and droughts (Engelbrecht et al. 2011; Zhu and Ringler 2012). Seasonality is predicted to increase, which in turn may increase the frequency of flood events in an area that is already susceptible to flooding (Spaliviero et al. 2014).

A clear need exists to both address exposure to flood events as well as predicted decreases in water availability, which are already acute. Without the additional impact of climate change, the basin is rapidly reaching a point where all available water has been allocated to users (Kahinda et al. 2016; Zhu and Ringler 2012). The urgency of the situation was identified several decades ago (FAO 2004), with the countries of the basin recognising that responses are required at several levels, both in terms of system governance and the need to address land degradation.

Recent reviews of the governance and implementation of IWM within the basin recognise that an integrated approach is needed

and that a robust institutional, legal, political, operational, technical and support environment is crucial (Alba et al. 2016; Gbetibouo et al. 2010; Machethe et al. 2004; Spaliviero et al. 2011; van der Zaag and Savenije 1999). Within the scope of emerging lessons, two principal ones emerge. The first is capacity and resource constraints at most levels. Limited capacity within Limpopo Watercourse Commission (LIMCOM) and national water management authorities constrains the implementation of IWM planning processes (Kahinda et al. 2016; Spaliviero et al. 2011). Whereas strategy development is often relatively well-funded and resourced through donor funding, long-term implementation is often limited due to competing priorities. The second is adequate representation of all parties in the process in order to address existing inequalities and ensure full integration of water management. For example, within Mozambique, significant strides have been made towards the decentralisation of river basin governance and IWM. Despite good progress, Alba et al. (2016) found that the newly implemented system may enforce existing inequalities as not all stakeholders, particularly smallholder farmers, are adequately represented in emerging water management structures and are often inhibited by financial and institutional constraints. Recognising economic and socio-political inequalities, and explicitly considering them to ensure the representation of all participants, can increase the chances of successful IWM implementation.

1.8   Knowledge gaps and key uncertainties

- Desertification has been studied for decades and different drivers of desertification have been described, classified, and are generally understood (e.g., overgrazing by livestock or salinisation from inappropriate irrigation) (D'Odorico et al. 2013). However, there are knowledge gaps on the extent and severity of desertification at global, regional, and local scales (Zhang and Huisingh 2018; Zucca et al. 2012). Overall, improved estimation and mapping of areas undergoing desertification is needed. This requires a combination of rapidly expanding sources of remotely sensed data, ground observations and new modelling approaches. This is a critical gap, especially in the context of measuring progress towards achieving the Land Degradation Neutrality target by 2030 in the framework of SDGs.

- Despite numerous relevant studies, consistent indicators for attributing desertification to climatic and/or human causes are still lacking due to methodological shortcomings.

- Climate change impacts on dust and sand storm activity remain a critical gap. In addition, the impacts of dust and sand storms on human welfare, ecosystems, crop productivity and animal health are not measured, particularly in the highly affected regions such as the Sahel, North Africa, the Middle East and Central Asia. Dust deposition on snow and ice has been found in many regions of the globe (e.g., Painter et al. 2018; Kaspari et al. 2014; Qian et al. 2015; Painter et al. 2013), however, the quantification of the effect globally, and estimation of future changes in the extent of this effect, remain knowledge gaps.

- Future projections of combined impacts of desertification and climate change on ecosystem services, fauna and flora, are lacking, even though this topic is of considerable social importance. Available information is mostly on separate, individual impacts of either (mostly) climate change or desertification. Responses to desertification are species-specific and mechanistic models are not yet able to accurately predict individual species responses to the many factors associated with desertification under changing climate.

- Previous studies have focused on the general characteristics of past and current desertification feedbacks to the climate system. However, the information on the future interactions between climate and desertification (beyond changes in the aridity index) are lacking. The knowledge of future climate change impacts on such desertification processes as soil erosion, salinisation, and nutrient depletion remains limited both at the global and at the local levels.

- Further research to develop the technologies and innovations needed to combat desertification is required, but it is also important to gain a better understanding of the reasons for the observed poor adoption of available innovations, to improve adoption rates.

- Desertification under changing climate has a high potential to increase poverty, particularly through the risks coming from extreme weather events (Olsson et al. 2014). However, the evidence rigorously attributing changes in observed poverty to climate change impacts is currently not available.

- The knowledge on the limits to adaptation to the combined effects of climate change and desertification is insufficient. This is an important gap since the potential for residual risks and maladaptive outcomes is high.

- Filling these gaps involves considerable investments in research and data collection. Using Earth observation systems in a standardised approach could help fill some of these gaps. This would increase data comparability and reduce uncertainty in approaches and costs. Systematically collected data would provide far greater insights than incomparable fragmented data.

Frequently Asked Questions

FAQ 3.1 | How does climate change affect desertification?

Desertification is land degradation in drylands. Climate change and desertification have strong interactions. Desertification affects climate change through loss of fertile soil and vegetation. Soils contain large amounts of carbon, some of which could be released to the atmosphere due to desertification, with important repercussions for the global climate system. The impacts of climate change on desertification are complex and knowledge on the subject is still insufficient. On the one hand, some dryland regions will receive less rainfall and increases in temperatures can reduce soil moisture, harming plant growth. On the other hand, the increase of CO_2 in the atmosphere can enhance plant growth if there are enough water and soil nutrients available.

FAQ 3.2 | How can climate change induced desertification be avoided, reduced or reversed?

Managing land sustainably can help avoid, reduce or reverse desertification, and contribute to climate change mitigation and adaptation. Such sustainable land management practices include reducing soil tillage and maintaining plant residues to keep soils covered, planting trees on degraded lands, growing a wider variety of crops, applying efficient irrigation methods, improving rangeland grazing by livestock and many others.

FAQ 3.3 | How do sustainable land management practices affect ecosystem services and biodiversity?

Sustainable land management practices help improve ecosystems services and protect biodiversity. For example, conservation agriculture and better rangeland management can increase the production of food and fibres. Planting trees on degraded lands can improve soil fertility and fix carbon in soils. Sustainable land management practices also support biodiversity through habitat protection. Biodiversity protection allows for the safeguarding of precious genetic resources, thus contributing to human well-being.

References

Abatzoglou, J.T. and C.A. Kolden, 2011: Climate change in western US deserts: Potential for increased wildfire and invasive annual grasses. *Rangel. Ecol. Manag.*, **64**, 471–478, doi:10.2111/REM-D-09-00151.1.

Abatzoglou, J.T., S.Z. Dobrowski, S.A. Parks, and K.C. Hegewisch, 2018: TerraClimate, a high-resolution global dataset of monthly climate and climatic water balance from 1958–2015. *Sci. Data*, **5**, 170191, doi:10.1038/sdata.2017.191.

Abbass, I., 2014: No retreat no surrender conflict for survival between Fulani pastoralists and farmers in Northern Nigeria. *Eur. Sci. J.*, **8**: 331–346, doi:10.19044/esj.2012.v8n1p%25p. DOI: http://dx.doi.org/10.19044/esj.2012.v8n1p%25p,331-346.

Abdalla, M. et al., 2018: Critical review of the impacts of grazing intensity on soil organic carbon storage and other soil quality indicators in extensively managed grasslands. *Agric. Ecosyst. Environ.*, **253**, 62–81, doi:10.1016/j.agee.2017.10.023.

Abdelkebir, T. and A. Ferchichi, 2015: Lutte biologique contre les accumulations sableuses dans la Tunisie aride, *Revue des Régions Arides*, Numéro Spécial 36: 210–220 (ISBN/ISNN 0330-7956).

Abdu, N. and L. Robinson, 2017: *Community-Based Rangeland Management in Dirre Rangeland Unit: Taking Successes in Land Restoration to Scale Project*. ILRI Project Report. Nairobi, Kenya, 43 pp.

Abdulai, A., V. Owusu, and R. Goetz, 2011: Land tenure differences and investment in land improvement measures: Theoretical and empirical analyses. *J. Dev. Econ.*, **96**, 66–78, doi:10.1016/J.JDEVECO.2010.08.002.

Abdullaev, I., M. Ul Hassan, and K. Jumaboev, 2007: Water saving and economic impacts of land leveling: The case study of cotton production in Tajikistan. *Irrig. Drain. Syst.*, **21**, 251–263, doi:10.1007/s10795-007-9034-2.

Abed, R.M.M., A. Ramette, V. Hübner, P. de Deckker, and D. de Beer, 2012: Microbial diversity of eolian dust sources from saline lake sediments and biological soil crusts in arid southern Australia. *FEMS Microbiol. Ecol.*, **80**, 294–304, doi:10.1111/j.1574-6941.2011.01289.x.

Abideen, Z. et al., 2014: Sustainable biofuel production from non-food sources? An overview. *Emirates J. Food Agric.*, **26**, 1057, www.ejfa.me/index.php/journal/article/view/617.

Abo-Ragab, S., Zaghloul, A.Q., 2017: Sand Dunes Movement and its Impact on Development Plans, Siwa Oasis. *International Journal of Research in Economics and Social Sciences*, 5:11-124-133, ISSN 2249-7382.

Abo-Ragab, S., 2010: A desertification impact on Siwa Oasis: Present and future challenges. *Res. J. Agric. Biol. Sci.*, **6**, 791–805. INSInet Publication.

Aboushook, M., M.N. Al Maghrabi, A. Fadol, and G.S. Abdelhaffez, 2012: Different methods for stabilisation of sand dunes using calcium bentonite. *Int. J. Environ. Eng.*, **4**, 79, doi:10.1504/IJEE.2012.048103.

Abril, A., P. Barttfeld, and E.H. Bucher, 2005: The effect of fire and overgrazing disturbes on soil carbon balance in the Dry Chaco forest. *For. Ecol. Manage.*, **206**, 399–405, doi:10.1016/j.foreco.2004.11.014.doi.org/10.1016/j.foreco.2004.11.014.

Abu-Zanat, M.W., G.B. Ruyle, and N.F. Abdel-Hamid, 2004: Increasing range production from fodder shrubs in low rainfall areas. *J. Arid Environ.*, **59**, 205–216, doi:10.1016/J.JARIDENV.2003.12.011.

Acevedo, E., and P. Silva, 2003: Agronomía de la cero labranza. *Ser. Ciencias Agronómicas*, **3**, 118. Universidad de Chile, ISBN: 956-19-0363-6.

Achite, M., and S. Ouillon, 2007: Suspended sediment transport in a semi-arid watershed, Wadi Abd, Algeria (1973–1995). *J. Hydrol.*, **343**, 187–202, doi:10.1016/J.JHYDROL.2007.06.026.

Achite, M., S. Ouillon, H. Quoc Viet, and C. Giay, 2016: Recent changes in climate, hydrology and sediment load in the Wadi Abd, Algeria (1970–2010). *Hydrol. Earth Syst. Sci*, **20**, 1355–1372, doi:10.5194/hess-20-1355-2016.

Ackerley, D. et al., 2011: Sensitivity of 20th century Sahel rainfall to sulfate aerosol and CO_2 forcing. *J. Clim.*, **24**, 4999–5014, doi:10.1175/JCLI-D-11-00019.1.

Adams, C., T. Ide, J. Barnett, and A. Detges, 2018: Sampling bias in climate–conflict research. *Nat. Clim. Chang.*, **8**, 200–203, doi:10.1038/s41558-018-0068-2.

Adger, W.N., 2003: Social capital, collective action, and adaptation to climate change. *Econ. Geogr.*, **79**, 387–404, doi:10.1111/j.1944-8287.2003.tb00220.x.

Adger, W.N., J.M. Pulhin, J. Barnett, G.D. Dabelko, G.K. Hovelsrud, M. Levy, Ú.O. Spring, and C.H. Vogel, 2014: Human security. In: Climate Change 2014: Impacts, Adaptation, and Vulnerability. Part A: Global and Sectoral Aspects. Contribution of Working Group II to the Fifth Assessment Report of the Intergovernmental Panel on Climate Change [Field, C.B., V.R. Barros, D.J. Dokken, K.J. Mach, M.D. Mastrandrea, T.E. Bilir, M. Chatterjee, K.L. Ebi, Y.O. Estrada, R.C. Genova, B. Girma, E.S. Kissel, A.N. Levy, S. MacCracken, P.R. Mastrandrea, and L.L.White (eds.)]. Cambridge University Press, New York, USA, pp. 755–791.

Adkins, S. and A. Shabbir, 2014: Biology, ecology and management of the invasive parthenium weed (Parthenium hysterophorus L.). *Pest Manag. Sci.*, **70**, 1023–1029, doi:10.1002/ps.3708.

Afsharzade, N., A. Papzan, M. Ashjaee, S. Delangizan, S. van Passel, and H. Azadi, 2016: Renewable energy development in rural areas of Iran. *Renew. Sustain. Energy Rev.*, **65**, 743–755, doi:10.1016/j.rser.2016.07.042.

Agier, L., A. Deroubaix, N. Martiny, P. Yaka, A. Djibo, and H. Broutin, 2012: Seasonality of meningitis in Africa and climate forcing: Aerosols stand out. *J.R. Soc. Interface*, **10**, 20120814, doi:10.1098/rsif.2012.0814.

Aha, B., and J.Z. Ayitey, 2017: Biofuels and the hazards of land grabbing: Tenure (in)security and indigenous farmers' investment decisions in Ghana. *Land Use Policy*, **60**, 48–59, doi:10.1016/J.LANDUSEPOL.2016.10.012.

Ahlerup, P., O. Olsson, and D. Yanagizawa, 2009: Social capital vs institutions in the growth process. *Eur. J. Polit. Econ.*, **25**, 1–14, doi:10.1016/J.EJPOLECO.2008.09.008.

Ahmed, A., E.T. Lawson, A. Mensah, C. Gordon, and J. Padgham, 2016: Adaptation to climate change or non-climatic stressors in semi-arid regions? Evidence of gender differentiation in three agrarian districts of Ghana. *Environ. Dev.*, **20**, 45–58, doi:10.1016/J.ENVDEV.2016.08.002.

Ait Hou, M., C. Grazia, and G. Malorgio, 2015: Food safety standards and international supply chain organization: A case study of the Moroccan fruit and vegetable exports. *Food Control*, **55**, 190–199, doi:10.1016/J.FOODCONT.2015.02.023.

Ajai, R.R., A.S. Arya, A.S. Dhinwa, S.K. Pathan, and K.G. Raj, 2009: Desertification/land degradation status mapping of India. *Curr. Sci.*, **97**, 1478–1483.

Ajayi, O.C., F.K. Akinnifesi, and A.O. Ajayi, 2016: How by-laws and collective action influence farmers' adoption of agroforestry and natural resource management technologies: Lessons from Zambia. *For. Trees Livelihoods*, **25**, 102–113, doi:10.1080/14728028.2016.1153435.

Akay, A., and A.I. Yildirim, 2010: *The Greening Desert Of Karapinar: An Example from Turkey*. Soil and Water Resources Research Institute Laboratory Agricultural Engineer, Konya, Turkey, pp. 546–551.

Akça, E., K. Takashi, and T. Sato, 2016: Development and success, for whom and where: The central Anatolian case. IN: Chabay I, Frick M, Helgeson J. *Land Restoration.* Pages 33–541, Boston, Academic Press, doi:10.1016/B978-0-12-801231-4.00034-3.

Akhtar-Schuster, M., R.J. Thomas, L.C. Stringer, P. Chasek, and M. Seely, 2011: Improving the enabling environment to combat land degradation: Institutional, financial, legal and science-policy challenges and solutions. *L. Degrad. Dev.*, **22**, 299–312, doi:10.1002/ldr.1058.

Akhter, J., K. Mahmood, K.A. Malik, S. Ahmed, and R. Murray, 2003: Amelioration of a saline sodic soil through cultivation of a salt-tolerant grass Leptochloa fusca. *Environ. Conserv.*, **30**, 168–174, doi:10.1017/S0376892903000158.

Akhter, R., and M. Arshad, 2006: Arid rangelands in the Cholistan Desert (Pakistan). *Artic. Sci.*, **17**, 210–217.

Akinshina, N., A. Azizov, T. Karasyova, and E. Klose, 2016: On the issue of halophytes as energy plants in saline environment. *Biomass and Bioenergy*, **91**, 306–311, doi:10.1016/J.BIOMBIOE.2016.05.034.

Al-Bakri, J.T. et al., 2016: Modelling desertification risk in the northwest of Jordan using geospatial and remote sensing techniques. *Geomatics, Nat. Hazards Risk*, **7**, 531–549, doi:10.1080/19475705.2014.945102.

Al-Kalbani, M. et al., 2014: Vulnerability assessment of environmental and climate change impacts on water resources in Al Jabal Al Akhdar, Sultanate of Oman. *Water*, **6**, 3118–3135, doi:10.3390/w6103118.

Al-Kalbani, M.S., M.F. Price, T. O'Higgins, M. Ahmed, and A. Abahussain, 2016: Integrated environmental assessment to explore water resources management in Al Jabal Al Akhdar, Sultanate of Oman. *Reg. Environ. Chang.*, **16**, 1345–1361, doi:10.1007/s10113-015-0864-4.

Al-Mulla, L., N.R. Bhat, and M. Khalil, 2013: Salt-tolerance of tissue-cultured date palm cultivars under controlled environment. *World Acad. Sci. Eng. Technol. Int. J. Anim. Vet. Sci.*, **7**, 811–814.

Al-Tabini, R., K. Al-Khalidi, and M. Al-Shudiefat, 2012: Livestock, medicinal plants and rangeland viability in Jordan's Badia: Through the lens of traditional and local knowledge. *Pastor. Res. Policy Pract.*, **2**, 4, doi:10.1186/2041-7136-2-4.

Al Hamndou, D., and M. Requier-Desjardins, 2008: Variabilité climatique, désertification et biodiversité en afrique: S'adapter, une approche intégrée. *VertigO*, 8(2): 1-24, doi:10.4000/vertigo.5356.

Aladejana, O.O., A.T. Salami, and O.-I.O. Adetoro, 2018: Hydrological responses to land degradation in the Northwest Benin Owena River Basin, Nigeria. *J. Environ. Manage.*, **225**, 300–312, doi:10.1016/J.JENVMAN.2018.07.095.

Alam, K., 2015: Farmers' adaptation to water scarcity in drought-prone environments: A case study of Rajshahi District, Bangladesh. *Agric. Water Manag.*, **148**, 196–206, doi:10.1016/J.AGWAT.2014.10.011.

Alba, R., A. Bolding, and R. Ducrot, 2016: The politics of water payments and stakeholder participation in the Limpopo River Basin, Mozambique Raphaëlle Ducrot. *Water Altern.*, **9**, 569–587.

Albaladejo, J., R. Ortiz, N. Garcia-Franco, A.R. Navarro, M. Almagro, J.G. Pintado, and M. Martínez-Mena, 2013: Land use and climate change impacts on soil organic carbon stocks in semi-arid Spain. *J. Soils Sediments*, **13**, 265–277, doi:10.1007/s11368-012-0617-7.

Albanito, F. et al., 2016: Carbon implications of converting cropland to bioenergy crops or forest for climate mitigation: A global assessment. *GCB Bioenergy*, **8**, 81–95, doi:10.1111/gcbb.12242.

Albright, T.P. et al., 2017: Mapping evaporative water loss in desert passerines reveals an expanding threat of lethal dehydration. *Proc. Natl. Acad. Sci.*, **114**, 2283–2288, doi:10.1073/pnas.1613625114.

Aldababseh, A. et al., 2018: Multi-criteria evaluation of irrigated agriculture suitability to achieve food security in an arid environment. *Sustainability*, **10**, 803, doi:10.3390/su10030803.

Alessandri, A. et al., 2014: Robust assessment of the expansion and retreat of Mediterranean climate in the 21st century. *Sci. Rep.*, **4**, 4–11, doi:10.1038/srep07211.

Alimaev, I.I. et al., 2008: The impact of livestock grazing on soils and vegetation around settlements in Southeast Kazakhstan. In: *The Socio-Economic Causes and Consequences of Desertification in Central Asia*, [Behnke R. (ed.)]. Springer Netherlands, Dordrecht, Netherlands, pp. 81–112.

Aliyu, A., B. Modu, and C.W. Tan, 2017: A review of renewable energy development in Africa: A focus in South Africa, Egypt and Nigeria. *Renew. Sustain. Energy Rev.*, **81**, 2502–2518, doi:10.1016/j.rser.2017.06.055.

Alkire, S., and M.E. Santos, 2010: *Acute Multidimensional Poverty: A New Index for Developing Countries*. Working Paper: 38, University of Oxford, Poverty and Human Development Initiative, Oxford, UK, 133 pp.

Alkire, S., and M.E. Santos, 2014: Measuring acute poverty in the developing world: Robustness and scope of the multidimensional poverty index. *World Dev.*, **59**, 251–274, doi:10.1016/J.WORLDDEV.2014.01.026.

Almagro, M., J. López, C. Boix-Fayos, J. Albaladejo, and M. Martínez-Mena, 2010: Below-ground carbon allocation patterns in a dry Mediterranean ecosystem: A comparison of two models. *Soil Biol. Biochem.*, **42**, 1549–1557, doi:10.1016/j.soilbio.2010.05.031.

Almazroui, M., M.N. Islam, K.S. Balkhair, Z. Şen, and A. Masood, 2017: Rainwater harvesting possibility under climate change: A basin-scale case study over western province of Saudi Arabia. *Atmos. Res.*, **189**, 11–23, doi:10.1016/J.ATMOSRES.2017.01.004.

Almazroui, M., M. Alobaidi, S. Saeed, A. Mashat, and M. Assiri, 2018: The possible impact of the circumglobal wave train on the wet season dust storm activity over the northern Arabian Peninsula. *Clim. Dyn.*, **50**, 2257–2268, doi:10.1007/s00382-017-3747-1.

Almer, C., J. Laurent-Lucchetti, and M. Oechslin, 2017: Water scarcity and rioting: Disaggregated evidence from Sub-Saharan Africa. *J. Environ. Econ. Manage.*, **86**, 193–209, doi:10.1016/J.JEEM.2017.06.002.

Alobaidi, M., M. Almazroui, A. Mashat, and P.D. Jones, 2017: Arabian Peninsula wet season dust storm distribution: Regionalization and trends analysis (1983-2013). *Int. J. Climatol.*, **37**, 1356–1373, doi:10.1002/joc.4782.

Alrasbi, S.A.R., N. Hussain, and H. Schmeisky, 2010: Evaluation of the growth of date palm seedlings irrigated with saline water in the Sultanate of Oman. *IV International Date Palm Conference*, 233–246, doi:10.17660/ActaHortic.2010.882.26.

AlSarmi, S., and R. Washington, 2011: Recent observed climate change over the Arabian Peninsula. *J. Geophys. Res.*, **116**, D11109, doi:10.1029/2010JD015459.

Altchenko, Y., and K.G. Villholth, 2015: Mapping irrigation potential from renewable groundwater in Africa – A quantitative hydrological approach. *Hydrol. Earth Syst. Sci.*, **19**, 1055–1067, doi:10.5194/hess-19-1055-2015.

Altieri, M., and P. Koohafkan, 2008: *Enduring Farms: Climate Change, Smallholders and Traditional Farming Communities*. Third World Network, TWN Environment & Development Series no. 6, Penang, Malaysia. ISBN: 978-983-2729-55-6, 63 pp.

Altieri, M.A., 1995: Agroecology: The science of sustainable agriculture. 2nd edition, CRC Press, ISBN 9780813317182, 448 pp.

Altieri, M.A. and C.I. Nicholls, 2017: The adaptation and mitigation potential of traditional agriculture in a changing climate. *Clim. Change*, **140**, 33–45, doi:10.1007/s10584-013-0909-y.

Altieri, M.A., C.I. Nicholls, A. Henao, and M.A. Lana, 2015: Agroecology and the design of climate change-resilient farming systems. *Agron. Sustain. Dev.*, **35**, 869–890, doi:10.1007/s13593-015-0285-2.

Amadou, A. et al., 2011: Impact of very low crop residues cover on wind erosion in the Sahel. *Catena*, **85**, 205–214, doi:10.1016/j.catena.2011.01.002.

Amiraslani, F., and D. Dragovich, 2011: Combating desertification in Iran over the last 50 years: An overview of changing approaches. *J. Environ. Manage.*, **92**, 1–13, doi:10.1016/j.jenvman.2010.08.012.

Amugune, I., P.O. Cerutti, H. Baral, S. Leonard, and C. Martius, 2017: *Small Flame but No Fire: Wood Fuel in the (Intended) Nationally Determined Contributions of Countries in Sub-Saharan Africa*. Center for International Forestry Research (CIFOR), Bogor, Indonesia, 35 pp.

Amundson, R. et al., 2015: Soil science. Soil and human security in the 21st century. *Science*, **348**, doi: 10.1126/science.1261071.

An, P., S. Inanaga, N. Zhu, X. Li, H.M. Fadul, and M. Mars, 2007: Plant species as indicators of the extent of desertification in four sandy rangelands. *Afr. J. Ecol.*, **45**, 94–102, doi:10.1111/j.1365-2028.2006.00681.x.

Anache, J.A.A., D.C. Flanagan, A. Srivastava, and E.C. Wendland, 2018: Land use and climate change impacts on runoff and soil erosion at the hillslope scale in the Brazilian Cerrado. *Sci. Total Environ.*, **622–623**, 140–151, doi:10.1016/J.SCITOTENV.2017.11.257.

van Andel, T., and L.G. Carvalheiro, 2013: Why urban citizens in developing countries use traditional medicines: The case of Suriname. *Evid. Based. Complement. Alternat. Med.*, **2013**, 687197, doi:10.1155/2013/687197.

Andela, N., Y.Y. Liu, A.I.J.M. van Dijk, R.A.M. de Jeu, and T.R. McVicar, 2013: Global changes in Dryland vegetation dynamics (1988-2008) assessed by satellite remote sensing: Comparing a new passive microwave vegetation density record with reflective greenness data. *Biogeosciences*, **10**, 6657–6676, doi:10.5194/bg-10-6657-2013.

Andersson, K.P., and E. Ostrom, 2008: Analyzing decentralized resource regimes from a polycentric perspective. *Policy Sci.*, **41**, 71–93, doi:10.1007/s11077-007-9055-6.

Angel, S., J. Parent, D.L. Civco, and A.M. Blei, 2011: *Making Room for a Planet of Cities*. Lincoln Institute of Land Policy, Cambridge, USA, 76 pp.

Antinori, P., and V. Vallerani, 1994: Experiments in water harvesting technology with the dolphin and train ploughs. In: *Water harvesting for improved agricultural production*, [FAO, (ed.)]. Food and Agriculture Organization of the United Nations, Rome, Italy, pp. 113–132.

Antwi-Agyei, P., A.J. Dougill, L.C. Stringer, and S.N.A. Codjoe, 2018: Adaptation opportunities and maladaptive outcomes in climate vulnerability hotspots of Northern Ghana. *Clim. Risk Manag.*, **19**, 83–93, doi:10.1016/J.CRM.2017.11.003.

Anyamba, A., and C.J. Tucker, 2005: Analysis of Sahelian vegetation dynamics using NOAA-AVHRR NDVI data from 1981–2003. *J. Arid Environ.*, **63**, 596–614, doi:10.1016/J.JARIDENV.2005.03.007.

Aoun-Sebaiti, B., A. Hani, L. Djabri, H. Chaffai, I. Aichouri, and N. Boughrira, 2014: Simulation of water supply and water demand in the valley of Seybouse (East Algeria). *Desalin. Water Treat.*, **52**, 2114–2119, doi:10.1080/19443994.2013.855662.

Apuri, I., K. Peprah, and G.T.W. Achana, 2018: Climate change adaptation through agroforestry: The case of Kassena Nankana West District, Ghana. *Environ. Dev.*, **28**, 32–41, doi:10.1016/J.ENVDEV.2018.09.002.

Archer, S.R. et al., 2017: Woody plant encroachment: Causes and consequences. In: *Rangeland Systems* [Briske, D.D. (ed.)]. Springer Series on Environmental Management, Springer International Publishing, Cham, Switzerland, pp 25–84.

Archibald, S., D.P. Roy, B.W. Van Wilgen, and R.J. Scholes, 2009: What limits fire? An examination of drivers of burnt area in southern Africa. *Glob. Chang. Biol.*, **15**, 613–630, doi:10.1111/j.1365-2486.2008.01754.x.

Arkle, R.S. et al., 2014: Quantifying restoration effectiveness using multi-scale habitat models: Implications for sage-grouse in the Great Basin. *Ecosphere*, **5**, art31, doi:10.1890/ES13-00278.1.

Arora-Jonsson, S., 2011: Virtue and vulnerability: Discourses on women, gender and climate change. *Glob. Environ. Chang.*, **21**, 744–751, doi:10.1016/J.GLOENVCHA.2011.01.005.

Arriaga, L., A.E. Castellanos, E. Moreno, and J. Alarcón, 2004: Potential ecological distribution of alien invasive species and risk assessment: A case study of buffel grass in arid regions of Mexico. *Conserv. Biol.*, **18**, 1504–1514, doi:10.1111/j.1523-1739.2004.00166.x.

Asadi Zarch, M.A., B. Sivakumar, and A. Sharma, 2015: Assessment of global aridity change. *J. Hydrol.*, **520**, 300–313, doi:10.1016/J.JHYDROL.2014.11.033.

Asem, S.O., and W.Y. Roy, 2010: Biodiversity and climate change in Kuwait. *Int. J. Clim. Chang. Strateg. Manag.*, **2**, 68–83, doi:10.1108/17568691011020265.

Ashkenazy, Y., H. Yizhaq, and H. Tsoar, 2012: Sand dune mobility under climate change in the Kalahari and Australian deserts. *Clim. Change*, **112**, 901–923, doi:10.1007/s10584-011-0264-9.

Ashraf, M.Y., M. Ashraf, K. Mahmood, J. Akhter, F. Hussain, and M. Arshad, 2010: Phytoremediation of saline soils for sustainable agricultural productivity. In: *Plant Adaptation and Phytoremediation* [Ashraf M., M. Ozturk and M. Ahmad (eds.)]. Springer Netherlands, Dordrecht, Netherlands, pp. 335–355.

Askri, B., R. Bouhlila, and J.O. Job, 2010: Development and application of a conceptual hydrologic model to predict soil salinity within modern Tunisian oases. *J. Hydrol.*, **380**, 45–61, doi:10.1016/J.JHYDROL.2009.10.022.

Aslam, M., and S.A. Prathapar, 2006: *Strategies to Mitigate Secondary Salinization in the Indus Basin of Pakistan: A Selective Review*. Research Report 97, International Water Management Institute, Colombo, Sri Lanka, 31 pp.

Asner, G.P., S. Archer, R.F. Hughes, R.J. Ansley, and C.A. Wessman, 2003: Net changes in regional woody vegetation cover and carbon storage in Texas drylands, 1937–1999. *Glob. Chang. Biol.*, **9**, 316–335, doi:10.1046/j.1365-2486.2003.00594.x.

Asner, G.P. et al., 2012: Carnegie Airborne Observatory-2: Increasing science data dimensionality via high-fidelity multi-sensor fusion. *Remote Sens. Environ.*, **124**, 454–465, doi:10.1016/J.RSE.2012.06.012.

Auclair, L., P. Baudot, D. Genin, B. Romagny, and R. Simenel, 2011: Patrimony for resilience evidence from the Forest Agdal in the Moroccan High Atlas. *Ecol. Soc.*, **16**, doi:10.5751/ES-04429-160424.

Van Auken, O.W., 2009: Causes and consequences of woody plant encroachment into western North American grasslands. *J. Environ. Manage.*, **90**, 2931–2942, doi:10.1016/J.JENVMAN.2009.04.023.

Austin, A.T. et al., 2004: Water pulses and biogeochemical cycles in arid and semi-arid ecosystems. *Oecologia*, **141**, 221–235, doi:10.1007/s00442-004-1519-1.

Aw-Hassan, A. et al., 2016: Economics of land degradation and improvement in Uzbekistan. In: *Economics of Land Degradation and Improvement – A Global Assessment for Sustainable Development* [Nkonya, E., A. Mirzabaev and J. von Braun (eds.)]. Springer International Publishing, Cham, Switzerland, pp. 651–682.

Awan, A.R., and K. Mahmood, 2017: Tree plantation in problem soils. In: *Text Book of Applied Forestry* [Tahir Siddiqui, M. and F. Nawaz (ed.)]. University of Agriculture, Faisalabad, Pakistan, pp. 140–159.

Awotide, B.A., A.A. Karimov, and A. Diagne, 2016: Agricultural technology adoption, commercialization and smallholder rice farmers' welfare in rural Nigeria. *Agric. Food Econ.*, **4**, 3, doi:10.1186/s40100-016-0047-8.

Ayanu, Y., A. Jentsch, D. Müller-Mahn, S. Rettberg, C. Romankiewicz, and T. Koellner, 2015: Ecosystem engineer unleashed: Prosopis juliflora threatening ecosystem services? *Reg. Environ. Chang.*, **15**, 155–167, doi:10.1007/s10113-014-0616-x.

Aynekulu, E. et al., 2017: *Review of Methodologies for Land Degradation Neutrality Baselines Sub-National Case Studies from Costa Rica and Namibia*. CIAT Publication No. 441, International Center for Tropical Agriculture and World Agroforestry Center, Nairobi, Kenya, 58 pp.

Ayoub, A.T., 1998: Extent, severity and causative factors of land degradation in the Sudan. *J. Arid Environ.*, **38**, 397–409, doi:10.1006/JARE.1997.0346.

Bachelet, D., K. Ferschweiler, T. Sheehan, and J. Strittholt, 2016: Climate change effects on southern California deserts. *J. Arid Environ.*, **127**, 17–29, doi:10.1016/J.JARIDENV.2015.10.003.

Bai, Z.G., D.L. Dent, L. Olsson, and M.E. Schaepman, 2008: Proxy global assessment of land degradation. *Soil Use Manag.*, **24**, 223–234, doi:10.1111/j.1475-2743.2008.00169.x.

Baker, L., P. Newell, and J. Phillips, 2014: The political economy of energy transitions: The case of South Africa. *New Polit. Econ.*, **19**, 791–818, doi:10.1080/13563467.2013.849674.

Balch, J.K., B.A. Bradley, C.M. D'Antonio, and J. Gómez-Dans, 2013: Introduced annual grass increases regional fire activity across the arid western USA (1980–2009). *Glob. Chang. Biol.*, **19**, 173–183, doi:10.1111/gcb.12046.

Bambio, Y., and S. Bouayad Agha, 2018: Land tenure security and investment: Does strength of land right really matter in rural Burkina Faso? *World Dev.*, **111**, 130–147, doi:10.1016/J.WORLDDEV.2018.06.026.

Banadda, N., 2010: Gaps, barriers and bottlenecks to sustainable land management (SLM) adoption in Uganda. *African J. Agric. Res.*, **5**, 3571–3580, doi:10.5897/ajar10.029.

Bang, G., 2010: Energy security and climate change concerns: Triggers for energy policy change in the United States? *Energy Policy*, **38**, 1645–1653, doi:10.1016/J.ENPOL.2009.01.045.

Barbier, B., H. Yacouba, H. Karambiri, M. Zoromé, and B. Somé, 2009: Human vulnerability to climate variability in the Sahel: Farmers' adaptation strategies in northern Burkina Faso. *Environ. Manage.*, **43**, 790–803, doi:10.1007/s00267-008-9237-9.

Barbier, E.B., and J.P. Hochard, 2016: Does land degradation increase poverty in developing countries? *PLoS One*, **11**, e0152973, doi:10.1371/journal.pone.0152973.

Barbier, E.B., and J.P. Hochard, 2018: Land degradation and poverty. *Nat. Sustain.*, **1**, 623–631, doi:10.1038/s41893-018-0155-4.

Barger, N.N. et al., 2011: Woody plant proliferation in North American drylands: A synthesis of impacts on ecosystem carbon balance. *J. Geophys. Res. Biogeosciences*, **116**:1-17, G00K07, doi:10.1029/2010JG001506.

Barnett, J., L.S. Evans, C. Gross, A.S. Kiem, R.T. Kingsford, J.P. Palutikof, C.M. Pickering, and S.G. Smithers, 2015: From barriers to limits to climate change adaptation: path dependency and the speed of change. *Ecol. Soc.*, **20**, 5, doi:10.5751/ES-07698-200305.

Barrett, C.B., L. Christiaensen, M. Sheahan, and A. Shimeles, 2017: On the structural transformation of rural Africa. *J. Afr. Econ.*, **26**, i11–i35, doi:10.1093/jae/ejx009.

del Barrio, G. et al., 2016: Land degradation states and trends in the north-western Maghreb drylands, 1998–2008. *Remote Sens.*, **8**, 603, doi:10.3390/rs8070603.

Barrios, S., L. Bertinelli, and E. Strobl, 2006: Climatic change and rural–urban migration: The case of Sub-Saharan Africa. *J. Urban Econ.*, **60**, 357–371, doi:10.1016/J.JUE.2006.04.005.

Baskan, O., O. Dengiz, and İ.T. Demirag, 2017: The land productivity dynamics trend as a tool for land degradation assessment in a Dryland ecosystem. *Environ. Monit. Assess.*, **189**, 212, doi:10.1007/s10661-017-5909-3.

Basupi, L., C. Quinn, and A.J. Dougill, 2017: Pastoralism and land tenure transformation in Sub-Saharan Africa: Conflicting policies and priorities in Ngamiland, Botswana. *Land*, **6**, 89, doi:10.3390/land6040089.

Batchelor, C.H., M.S. Rama Mohan Rao, and S. Manohar Rao, 2003: Watershed development: A solution to water shortages in semi-arid India or part of the problem? *L. Use Water Resour. Res.*, **3**, 10, doi:10.22004/ag.econ.47866.

Baumgartner, P., 2017: The Impacts of Large-Scale Land-Acquisition in East Africa on Poverty Reduction and the Rural Economy: Studies in Ethiopia and Uganda. PhD Thesis, University of Bonn, Bonn, Germany, 236 pp.

Bazza, M., M. Kay, and C. Knutson, 2018: *Drought Characteristics and Management in North Africa and the Near East*. FAO Water Reports 45, Food and Agriculture Organization of the United Nations, Rome, Italy, 1020–1203 pp.

Becerril-Pina Rocio et al., 2015: Assessing desertification risk in the semi-arid highlands of central Mexico. *J. Arid Environ.*, **120**, 4–13, doi:10.1016/j.jaridenv.2015.04.006.

Beckers, B., J. Berking, and B. Schütt, 2013: Ancient water harvesting methods in the drylands of the Mediterranean and Western Asia. *eTopoi. J. Anc. Stud.*, **2**, 145–164.

Beegum, S.N. et al., 2016: Simulating aerosols over Arabian Peninsula with CHIMERE: Sensitivity to soil, surface parameters and anthropogenic emission inventories. *Atmos. Environ.*, **128**, 185–197, doi:10.1016/J.ATMOSENV.2016.01.010.

Naseema Beegum, S., I. Gherboudj, N. Chaouch, M. Temimi, and H. Ghedira, 2018: Simulation and analysis of synoptic scale dust storms over the Arabian Peninsula. *Atmos. Res.*, **199**, 62–81, doi:10.1016/J.ATMOSRES.2017.09.003.

Behnke, R., 1994: Natural resource management in pastoral Africa. *Dev. Policy Rev.*, **12**, 5–28, doi:10.1111/j.1467-7679.1994.tb00053.x.

Behnke, R., and C. Kerven, 2013: Counting the costs: Replacing pastoralism with irrigated agriculture in the Awash Valley, north-eastern Ethiopia. In: *Pastoralism and Development in Africa: Dynamic Change at the Margins* [Catley, A., J. Lind, and I. Scoones (eds.)]. Routledge, London, UK, pp. 328.

Behnke, R. and M. Mortimore, 2016: Introduction: The end of desertification? In: *The End of Desertification? Disputing Environmental Change in the Drylands* [Behnke R. and M. Mortimore (eds.)]. Springer, Berlin, Heidelberg, pp. 1–34.

Bekchanov, M. et al., 2010: Pros and cons of adopting water-wise approaches in the lower reaches of the Amu Darya: A socio-economic view. *Water*, **2**, 200–216, doi:10.3390/w2020200.

Bekchanov, M., C. Ringler, A. Bhaduri, and M. Jeuland, 2016: Optimizing irrigation efficiency improvements in the Aral Sea Basin. *Water Resour. Econ.*, **13**, 30–45, doi:10.1016/J.WRE.2015.08.003.

Belaaz, M., 2003: *Le barrage vert en tant que patrimoine naturel national et moyen de lutte contre la désertification. Dans actes du XII°*. Proceedings of the XII World Forestry Congress 2003, Quebec, Canada. 0301-B3.

Belala, F. et al., 2018: Rainfall patterns of Algerian steppes and the impacts on natural vegetation in the 20th century. *J. Arid Land*, **10**, 561–573, doi:10.1007/s40333-018-0095-x.

Belay, A., J.W. Recha, T. Woldeamanuel, and J.F. Morton, 2017: Smallholder farmers' adaptation to climate change and determinants of their adaptation decisions in the Central Rift Valley of Ethiopia. *Agric. Food Secur.*, **6**, 24, doi:10.1186/s40066-017-0100-1.

Belfer, E., J.D. Ford, and M. Maillet, 2017: Representation of indigenous peoples in climate change reporting. *Clim. Change*, **145**, 57–70, doi:10.1007/s10584-017-2076-z.

Bellard, C., W. Thuiller, B. Leroy, P. Genovesi, M. Bakkenes, and F. Courchamp, 2013: Will climate change promote future invasions? *Glob. Chang. Biol.*, **19**, 3740–3748, doi:10.1111/gcb.12344.

Ben Hassine, H., A. Ben Slimane, M. Mlawah, L. Albouchi, and A. Gandouzi, 2013: Effects of underground water on soil salinity and dates production in Kebili Oases Area (Tunisia): The case of El Bahaier Oasis. *IOSR J. Environ. Sci. Toxicol. Food Technol.*, **4**, 51–58, doi:10.9790/2402-0445158.

Benjaminsen, T.A., 2016: Does climate change lead to conflicts in the Sahel? In: *The End of Desertification? Disputing Environmental Change in the Drylands* [Behnke R. and M. Mortimore, (eds.)]. Springer, Berlin, Heidelberg, Germany, pp. 99–116.

Benjaminsen, T.A. and P. Hiernaux, 2019: From desiccation to global climate change: A history of the desertification narrative in the West African Sahel, 1900-2018. *Glob. Environ.*, **12**, 206–236, doi:10.3197/ge.2019.120109.

Benjaminsen, T.A., K. Alinon, H. Buhaug, and J.T. Buseth, 2012: Does climate change drive land-use conflicts in the Sahel? *J. Peace Res.*, **49**, 97–111, doi:10.1177/0022343311427343.

Benmoussa, H., E. Luedeling, M. Ghrab, J. Ben Yahmed, and B. Mimoun, 2017: Performance of pistachio (Pistacia vera L.) in warming Mediterranean orchards. *Environ. Exp. Bot.*, **140**, 76–85, doi:10.1016/j.envexpbot.2017.05.007.

Bennett, C.M., I.G. McKendry, S. Kelly, K. Denike, and T. Koch, 2006: Impact of the 1998 Gobi dust event on hospital admissions in the Lower Fraser Valley, British Columbia. *Sci. Total Environ.*, **366**, 918–925, doi:10.1016/j.scitotenv.2005.12.025.

Bensaid, S., 1995: Bilan critique du barrage vert en Algérie. *Sci. Chang. planétaires/Sécheresse*, **6**, 247–255.

Bergametti, G. et al., 2018: Size-resolved dry deposition velocities of dust particles: In situ measurements and parameterizations testing. *J. Geophys. Res. Atmos.*, **123**, 11, 11–80, 99, doi:10.1029/2018JD028964.

Berman, R., C. Quinn, and J. Paavola, 2012: The role of institutions in the transformation of coping capacity to sustainable adaptive capacity. *Environ. Dev.*, **2**, 86–100, doi:10.1016/J.ENVDEV.2012.03.017.

Berrahmouni, N., L. Laestadius, A. Martucci, D. Mollicone, C. Patriarca, and M. Sacande, 2016: *Building Africa's Great Green Wall: Restoring Degraded Drylands for Stronger and More Resilient Communities*. Food and Agriculture Organization of the United Nations, Rome, Italy, 7 pp.

Besser, H. et al., 2017: GIS-based evaluation of groundwater quality and estimation of soil salinization and land degradation risks in an arid Mediterranean site (SW Tunisia). *Arab. J. Geosci.*, **10**, 350, doi:10.1007/s12517-017-3148-0.

Bestelmeyer, B.T., M.C. Duniway, D.K. James, L.M. Burkett, and K.M. Havstad, 2013: A test of critical thresholds and their indicators in a desertification-prone ecosystem: More resilience than we thought. *Ecol. Lett.*, **16**, 339–345, doi:10.1111/ele.12045.

Bestelmeyer, B.T., G.S. Okin, M.C. Duniway, S.R. Archer, N.F. Sayre, J.C. Williamson, and J.E. Herrick, 2015: Desertification, land use, and the transformation of global drylands. *Front. Ecol. Environ.*, **13**, 28–36, doi:10.1890/140162.

Bestelmeyer, B.T. et al., 2018: The grassland–shrubland regime shift in the south-western United States: Misconceptions and their implications for management. *Bioscience*, **68**, 678–690, doi:10.1093/biosci/biy065.

Bestelmeyer, B.T. et al., 2017: *State and Transition Models: Theory, Applications, and Challenges*. Springer International Publishing, Cham, Switzerland, pp. 303–345.

Betts, R.A. et al., 2018: Changes in climate extremes, fresh water availability and vulnerability to food insecurity projected at 1.5°C and 2°C global warming with a higher-resolution global climate model. *Philos. Trans. R. Soc. A Math. Phys. Eng. Sci.*, **376**, 20160452, doi:10.1098/rsta.2016.0452.

Bhalla, R.S., N.W. Pelkey, and K.V. Devi Prasad, 2011: Application of GIS for evaluation and design of watershed guidelines. *Water Resour. Manag.*, **25**, 113–140, doi:10.1007/s11269-010-9690-0.

Bhalla, R.S., K.V. Devi Prasad, and N.W. Pelkey, 2013: Impact of India's watershed development programs on biomass productivity. *Water Resour. Res.*, **49**, 1568–1580, doi:10.1002/wrcr.20133.

Bharucha, Z.P., D. Smith, and J. Pretty, 2014: All paths lead to rain: Explaining why watershed development in India does not alleviate the experience of water scarcity. *J. Dev. Stud.*, **50**, 1209–1225, doi:10.1080/00220388.2014.928699.

Biazin, B., G. Sterk, M. Temesgen, A. Abdulkedir, and L. Stroosnijder, 2012: Rainwater harvesting and management in rainfed agricultural systems in Sub-Saharan Africa – A review. *Phys. Chem. Earth, Parts A/B/C*, **47–48**, 139–151, doi:10.1016/J.PCE.2011.08.015.

Biederman, J.A. et al., 2017: CO_2 exchange and evapotranspiration across Dryland ecosystems of south-western North America. *Glob. Chang. Biol.*, **23**, 4204–4221, doi:10.1111/gcb.13686.

Bikila, N.G., Z.K. Tessema, and E.G. Abule, 2016: Carbon sequestration potentials of semi-arid rangelands under traditional management practices in Borana, southern Ethiopia. *Agric. Ecosyst. Environ.*, **223**, 108–114, doi:10.1016/J.AGEE.2016.02.028.

Bird, M.I., E.M. Veenendaal, C. Moyo, J. Lloyd, and P. Frost, 2000: Effect of fire and soil texture on soil carbon in a sub-humid savanna (Matopos, Zimbabwe). *Geoderma*, **94**, 71–90, doi:10.1016/S0016-7061(99)00084-1.

Bisigato, A.J., and R.M.L. Laphitz, 2009: Ecohydrological effects of grazing-induced degradation in the Patagonian Monte, Argentina. *Austral Ecol.*, **34**, 545–557, doi:10.1111/j.1442-9993.2009.01958.x.

Bjornlund, H., A. van Rooyen, and R. Stirzaker, 2017: Profitability and productivity barriers and opportunities in small-scale irrigation schemes. *Int. J. Water Resour. Dev.*, **33**, 690–704, doi:10.1080/07900627.2016.1263552.

Bobojonov, I., J.P.A. Lamers, M. Bekchanov, N. Djanibekov, J. Franz-Vasdeki, J. Ruzimov, and C. Martius, 2013: Options and constraints for crop diversification: A case study in sustainable agriculture in Uzbekistan. *Agroecol. Sustain. Food Syst.*, **37**, 788–811, doi:10.1080/21683565.2013.775539.

Bobojonov, I., R. Teuber, S. Hasanov, V. Urutyan, and T. Glauben, 2016: Farmers' export market participation decisions in transition economies: A comparative study between Armenia and Uzbekistan. *Dev. Stud. Res.*, **3**, 25–35, doi:10.1080/21665095.2016.1262272.

Boelee, E. et al., 2013: Options for water storage and rainwater harvesting to improve health and resilience against climate change in Africa. *Reg. Environ. Chang.*, **13**, 509–519, doi:10.1007/s10113-012-0287-4.

Boix-Fayos, C., M. Martínez-Mena, E. Arnau-Rosalén, A. Calvo-Cases, V. Castillo, and J. Albaladejo, 2006: Measuring soil erosion by field plots: Understanding the sources of variation. *Earth-Science Rev.*, **78**, 267–285, doi:10.1016/J.EARSCIREV.2006.05.005.

Bond, W.J., and G.F. Midgley, 2000: A proposed CO_2 controlled mechanism of woody plant invasion in grasslands and savannas. *Glob. Chang. Biol.*, **6**, 865–869.

Bond, W.J., and J.E. Keeley, 2005: Fire as a global 'herbivore': The ecology and evolution of flammable ecosystems. *Trends Ecol. Evol.*, **20**, 387–394, doi:10.1016/j.tree.2005.04.025.

Bond, W.J., G.F. Midgley, and F.I. Woodward, 2003: What controls South African vegetation – Climate or fire? *South African J. Bot.*, **69**, 79–91.

Bonilla, C.A., J.L. Reyes, and A. Magri, 2010: Water erosion prediction using the revised universal soil loss equation (RUSLE) in a GIS framework, Central Chile. *Chil. J. Agric. Res.*, **70**, 159–169, doi:10.4067/S0718-58392010000100017.

Bonney, R., T.B. Phillips, H.L. Ballard, and J.W. Enck, 2016: Can citizen science enhance public understanding of science? *Public Underst. Sci.*, **25**, 2–16, doi:10.1177/0963662515607406.

Booker, J.F., A.M. Michelsen, and F.A. Ward, 2005: Economic impact of alternative policy responses to prolonged and severe drought in the Rio Grande Basin. *Water Resour. Res.*, **41**, doi:10.1029/2004WR003486.

Booth, B.B.B., N.J. Dunstone, P.R. Halloran, T. Andrews, and N. Bellouin, 2012: Aerosols implicated as a prime driver of 20th century North Atlantic climate variability. *Nature*, **484**, 228–232, doi:10.1038/nature10946.

Booysen, M.J., M. Visser, and R. Burger, 2019: Temporal case study of household behavioural response to Cape Town's 'Day Zero' using smart meter data. *Water Res.*, **149**, 414–420, doi:10.1016/J.WATRES.2018.11.035.

Börner, J., K. Baylis, E. Corbera, D. Ezzine-de-Blas, J. Honey-Rosés, U.M. Persson, and S. Wunder, 2017: The effectiveness of payments for environmental services. *World Dev.*, **96**, 359–374, doi:10.1016/J.WORLDDEV.2017.03.020.

Borrelli, P. et al., 2017: An assessment of the global impact of 21st century land use change on soil erosion. *Nat. Commun.*, **8**, 2013, doi:10.1038/s41467-017-02142-7.

Bose, P., India's drylands agroforestry: A ten-year analysis of gender and social diversity, tenure and climate variability. *Int. For. Rev.*, **17**, doi:10.1505/146554815816086435.

Boserup, E., 1965: *The Conditions of Agricultural Growth: The Economics of Agrarian Change Under Population Pressure*. George All & Unwin, Ltd., London, UK, 124 pp.

Botterill, L.C., and M.J. Hayes, 2012: Drought triggers and declarations: Science and policy considerations for drought risk management. *Nat. Hazards*, **64**, 139–151, doi:10.1007/s11069-012-0231-4.

Boucher, O., D. Randall, P. Artaxo, C. Bretherton, G. Feingold, P. Forster, V.-M. Kerminen, Y. Kondo, H. Liao, U. Lohmann, P. Rasch, S.K. Satheesh, S. Sherwood, B. Stevens and X.Y. Zhang, 2013: Clouds and aerosols. In: Climate Change 2013: The Physical Science Basis. Contribution of Working Group I to the Fifth Assessment Report of the Intergovernmental Panel on Climate Change [Stocker, T.F., D. Qin, G.-K. Plattner, M. Tignor, S.K. Allen, J. Boschung, A. Nauels, Y. Xia, V. Bex and P.M. Midgley (eds.)]. Cambridge University Press, Cambridge, United Kingdon and New York, NY, USA, pp. 571–657.

Bouma, J., and J.H.M. Wösten, 2016: How to characterize 'good' and 'greening' in the EU Common Agricultural Policy (CAP): The case of clay soils in the Netherlands. *Soil Use Manag.*, **32**, 546–552, doi:10.1111/sum.12289.

Bouma, J., D. Van Soest, and E. Bulte, 2007: How sustainable is participatory watershed development in India? *Agric. Econ.*, **36**, 13–22, doi:10.1111/j.1574-0862.2007.00173.x.

Bourguignon, F., and S.R. Chakravarty, 2003: The measurement of multidimensional poverty. *J. Econ. Inequal.*, **1**, 25–49, doi:10.1023/A:1023913831342.

Boyd, P.W., D.S. Mackie, and K.A. Hunter, 2010: Aerosol iron deposition to the surface ocean – Modes of iron supply and biological responses. *Mar. Chem.*, **120**, 128–143, doi:10.1016/J.MARCHEM.2009.01.008.

Bradley, B.A., D.M. Blumenthal, D.S. Wilcove, and L.H. Ziska, 2010: Predicting plant invasions in an era of global change. *Trends Ecol. Evol.*, **25**, 310–318, doi:10.1016/J.TREE.2009.12.003.

Bradley, B.A. et al., 2012: Global change, global trade, and the next wave of plant invasions. *Front. Ecol. Environ.*, **10**, 20–28, doi:10.1890/110145.

3

Brandt, M. et al., 2015: Ground- and satellite-based evidence of the biophysical mechanisms behind the greening Sahel. *Glob. Chang. Biol.*, **21**, 1610–1620, doi:10.1111/gcb.12807.

Brandt, M. et al., 2016a: Assessing woody vegetation trends in Sahelian drylands using MODIS based seasonal metrics. *Remote Sens. Environ.*, **183**, 215–225, doi:10.1016/J.RSE.2016.05.027.

Brandt, M. et al., 2016b: Woody plant cover estimation in drylands from Earth Observation based seasonal metrics. *Remote Sens. Environ.*, **172**, 28–38, doi:10.1016/J.RSE.2015.10.036.

Von Braun, J., and F.W. Gatzweiler (eds.), 2014: *Marginality: Addressing the nexus of poverty, exclusion and ecology*. Springer Netherlands, Dordrecht, Netherlands, 389 pp.

Breckle, S.-W., 2013: *From Aral Sea to Aralkum: An Ecological Disaster or Halophytes' Paradise*. Springer, Berlin, Germany, pp. 351–398.

Broeckhoven, N., and A. Cliquet, 2015: Gender and ecological restoration: Time to connect the dots. *Restor. Ecol.*, **23**, 729–736, doi:10.1111/rec.12270.

Brooks, M.L. et al., 2004: Effects of invasive alien plants on fire regimes. *Bioscience*, **54**, 677–688, doi:10.1641/0006-3568(2004)054[0677:eoiapo]2.0.co;2.

Browning, D.M., and S.R. Archer, 2011: Protection from livestock fails to deter shrub proliferation in a desert landscape with a history of heavy grazing. *Ecol. Appl.*, **21**, 1629–1642, doi:10.1890/10-0542.1.

Browning, D.M., J.J. Maynard, J.W. Karl, and D.C. Peters, 2017: Breaks in MODIS time series portend vegetation change: Verification using long-term data in an arid grassland ecosystem. *Ecol. Appl.*, **27**, 1677–1693, doi:10.1002/eap.1561.

Bruno, L., M. Horvat, and L. Raffaele, 2018: Windblown sand along railway infrastructures: A review of challenges and mitigation measures. *J. Wind Eng. Ind. Aerodyn.*, **177**, 340–365, doi:10.1016/J.JWEIA.2018.04.021.

Bryan, E., T.T. Deressa, G.A. Gbetibouo and C. Ringler, 2009: Adaptation to climate change in Ethiopia and South Africa: Options and constraints. *Environ. Sci. Policy*, **12**, 413–426, doi:10.1016/j.envsci.2008.11.002.

Buhaug, H. et al., 2014: One effect to rule them all? A comment on climate and conflict. *Clim. Change*, **127**, 391–397, doi:10.1007/s10584-014-1266-1.

Bulkeley, H., 2013: *Cities and Climate Change*. Routledge, London, UK, 268 pp.

Bureau of Land Management, 2005: *Vegetation Treatments on Bureau of Land Management Lands in 17 Western State*. Nevada State Office, Reno, Nevada, USA. 497 pp.

Burke, P.J., and G. Dundas, 2015: Female labour force participation and household dependence on biomass energy: Evidence from National Longitudinal Data. *World Dev.*, **67**, 424–437, doi:10.1016/J.WORLDDEV.2014.10.034.

Burney, J., L. Woltering, M. Burke, R. Naylor, and D. Pasternak, 2010: Solar-powered drip irrigation enhances food security in the Sudano-Sahel. *Proc. Natl. Acad. Sci. U.S.A.*, **107**, 1848–1853, doi:10.1073/pnas.0909678107.

Burrell, A.L., J.P. Evans, and Y. Liu, 2017: Detecting dryland degradation using Time Series Segmentation and Residual Trend analysis (TSS-RESTREND). *Remote Sens. Environ.*, **197**, doi:10.1016/j.rse.2017.05.018.

Burrell, A.L., J.P. Evans, and Y. Liu, 2018: The impact of dataset selection on land degradation assessment. *ISPRS J. Photogramm. Remote Sens.*, **146**, 22–37, doi:10.1016/J.ISPRSJPRS.2018.08.017.

Byers, E. et al., 2018: Global exposure and vulnerability to multi-sector development and climate change hotspots. *Environ. Res. Lett.*, **13**, 55012, doi:10.1088/1748-9326/aabf45.

Cai, X., D.C. McKinney, and M.W. Rosegrant, 2003: Sustainability analysis for irrigation water management in the Aral Sea region. *Agric. Syst.*, **76**, 1043–1066, doi:10.1016/S0308-521X(02)00028-8.

Cai, X., X. Zhang, and D. Wang, 2011: Land availability for biofuel production. *Environ. Sci. Technol.*, **45**, 334–339, doi:10.1021/es103338e.

Calder, I., A. Gosain, M.S.R.M. Rao, C. Batchelor, M. Snehalatha, and E. Bishop, 2008: Watershed development in India. 1. Biophysical and societal impacts. *Environ. Dev. Sustain.*, **10**, 537–557, doi:10.1007/s10668-006-9079-7.

Çalışkan, S., and M. Boydak, 2017: Afforestation of arid and semi-arid ecosystems in Turkey. *Turkish J. Agric. For.*, **41**, 317–330, doi:10.3906/tar-1702-39.

Calvet-Mir, L., E. Corbera, A. Martin, J. Fisher, and N. Gross-Camp, 2015: Payments for ecosystem services in the tropics: A closer look at effectiveness and equity. *Curr. Opin. Environ. Sustain.*, **14**, 150–162, doi:10.1016/J.COSUST.2015.06.001.

Cancino-Solórzano, Y., J.P. Paredes-Sánchez, A.J. Gutiérrez-Trashorras, and J. Xiberta-Bernat, 2016: The development of renewable energy resources in the State of Veracruz, Mexico. *Util. Policy*, **39**, 1–4, doi:10.1016/J.JUP.2016.01.001.

Cano, A. et al., 2018: Current knowledge and future research directions to link soil health and water conservation in the Ogallala Aquifer region. *Geoderma*, **328**, 109–118, doi:10.1016/J.GEODERMA.2018.04.027.

Cao, J., X. Zhang, R. Deo, Y. Gong, and Q. Feng, 2018: Influence of stand type and stand age on soil carbon storage in China's arid and semi-arid regions. *Land Use Policy*, **78**, 258–265, doi:10.1016/J.LANDUSEPOL.2018.07.002.

Çarkaci, D.A., 1999: *Examples of good practices of Turkey for combating wind erosion (examples of karapinar) the blooming desert Karapinar situation of desertification in Turkey, distribution of problem areas in Turkey*, www.mgm.gov.tr/FTPDATA/arastirma/toz/Day3_99_acarkaci.pdf.

Carletto, C., P. Corral, and A. Guelfi, 2017: Agricultural commercialization and nutrition revisited: Empirical evidence from three African countries. *Food Policy*, **67**, 106–118, doi:10.1016/J.FOODPOL.2016.09.020.

Carlisle, K., and R.L. Gruby, 2017: Polycentric systems of governance: A theoretical model for the commons. *Policy Stud. J.*, doi:10.1111/psj.12212.

Carr-Hill, R., 2013: Missing millions and measuring development progress. *World Dev.*, **46**, 30–44, doi:10.1016/J.WORLDDEV.2012.12.017.

Carr, E.R., and M.C. Thompson, 2014: Gender and climate change adaptation in agrarian settings: Current thinking, new directions, and research frontiers. *Geogr. Compass*, **8**, 182–197, doi:10.1111/gec3.12121.

Carrão, H., G. Naumann, and P. Barbosa, 2017: Global projections of drought hazard in a warming climate: A prime for disaster risk management. *Clim. Dyn.*, **50**, 2137–2155, doi:10.1007/s00382-017-3740-8.

Cassidy, T.M., J.H. Fownes, and R.A. Harrington, 2004: Nitrogen limits an invasive perennial shrub in forest understory. *Biol. Invasions*, **6**, 113–121, doi:10.1023/B:BINV.0000010128.44332.0f.

Castellanos-Villegas, A.E., G. Yanes, and D. Valdes, 2002: Drought-tolerant exotic buffel-grass and desertification. In: *Proceedings of a North American Conference Weeds Across Borders Arizona-Sonora Desert Museum* [Tellman, B. (ed.)]. Arizona-Sonora Desert Museum, Tucson, USA. pp. 99–112.

Castellanos, A.E., H. Celaya-Michel, J.C. Rodríguez, and B.P. Wilcox, 2016: Ecohydrological changes in semi-arid ecosystems transformed from shrubland to buffelgrass savanna. *Ecohydrology*, **9**, 1663–1674, doi:10.1002/eco.1756.

Catacutan, D.C., and G.B. Villamor, 2016: Gender roles and land use preferences – Implications to landscape restoration in Southeast Asia. In: *Land Restoration: Reclaiming Landscapes for a Sustainable Future*. [Chabay, I., M. Frick and J. Helgeson (eds.)]. Elsevier, Oxford, UK, pp. 431–440.

Cattaneo, C., and G. Peri, 2016: The migration response to increasing temperatures. *J. Dev. Econ.*, **122**, 127–146, doi:10.1016/J.JDEVECO.2016.05.004.

Ceballos, G. et al., 2010: Rapid decline of a grassland system and its ecological and conservation implications. *PLoS One*, **5**, e8562, doi:10.1371/journal.pone.0008562.

CGIAR-RPDS, 2014: *Annual Report 2014. Pathways to Lasting Impact for Rural Dryland Communities in the Developing World*. Amman, Jordan. Consultative Group for International Agricultural Research (CGIAR). 68 pages.

3

Chambers, J.C. et al., 2014: Resilience to stress and disturbance, and resistance to Bromus tectorum L. Invasion in cold desert shrublands of western North America. *Ecosystems*, **17**, 360–375, doi:10.1007/s10021-013-9725-5.

Chandel, S.S., M. Nagaraju Naik, and R. Chandel, 2015: Review of solar photovoltaic water pumping system technology for irrigation and community drinking water supplies. *Renew. Sustain. Energy Rev.*, **49**, 1084–1099, doi:10.1016/J.RSER.2015.04.083.

Chandrasekhar, C.P., J. Ghosh, and A. Roychowdhury, 2006: The 'demographic dividend' and young India's economic future. *Econ. Polit. Wkly.*, 5055–5064.

Charney, J., P.H. Stone, and W.J. Quirk, 1975: Drought in the Sahara: A biogeophysical feedback mechanism. *Science*, **187**, 434–435, doi:10.1126/science.187.4175.434.

Chasek, P., W. Essahli, M. Akhtar-Schuster, L.C. Stringer, and R. Thomas, 2011: Integrated land degradation monitoring and assessment: Horizontal knowledge management at the national and international levels. *L. Degrad. Dev.*, **22**, 272–284, doi:10.1002/ldr.1096.

Chasek, P., M. Akhtar-Schuster, B.J. Orr, A. Luise, H. Rakoto Ratsimba, and U. Safriel, 2019: Land degradation neutrality: The science-policy interface from the UNCCD to national implementation. *Environ. Sci. Policy*, **92**, 182–190, doi:10.1016/j.envsci.2018.11.017.

de Châtel, F., 2014: The role of drought and climate change in the Syrian uprising: Untangling the triggers of the revolution. *Middle East. Stud.*, **50**, 521–535, doi:10.1080/00263206.2013.850076.

Cheesman, S., C. Thierfelder, N.S. Eash, G.T. Kassie, and E. Frossard, 2016: Soil carbon stocks in conservation agriculture systems of southern Africa. *Soil Tillage Res.*, **156**, 99–109, doi:10.1016/j.still.2015.09.018.

Chen, D. et al., 2015a: Patterns and drivers of soil microbial communities along a precipitation gradient on the Mongolian Plateau. *Landsc. Ecol.*, **30**, 1669–1682, doi:10.1007/s10980-014-9996-z.

Chen, J., S. Yin, H. Gebhardt, and X. Yang, 2018: Farmers' livelihood adaptation to environmental change in an arid region: A case study of the Minqin Oasis, north-western China. *Ecol. Indic.*, **93**, 411–423, doi:10.1016/J.ECOLIND.2018.05.017.

Chen, R., C. Ye, Y. Cai, X. Xing, and Q. Chen, 2014: The impact of rural out-migration on land use transition in China: Past, present and trend. *Land Use Policy*, **40**, 101–110, doi:10.1016/J.LANDUSEPOL.2013.10.003.

Chen, S., G. Lin, J. H, and G.D. Jenerette, 2009: Dependence of carbon sequestration on the differential responses of ecosystem photosynthesis and respiration to rain pulses in a semi-arid steppe. *Glob. Chang. Biol.*, **15**, 2450–2461, doi:10.1111/j.1365-2486.2009.01879.x.

Chen, T. et al., 2019: Disentangling the relative impacts of climate change and human activities on arid and semi-arid grasslands in Central Asia during 1982–2015. *Sci. Total Environ.*, **653**, 1311–1325, doi:10.1016/j.scitotenv.2018.11.058.

Chen, Y., Z. Li, Y. Fan, H. Wang, and H. Deng, 2015b: Progress and prospects of climate change impacts on hydrology in the arid region of Northwest China. *Environ. Res.*, **139**, 11–19, doi:10.1016/J.ENVRES.2014.12.029.

Cheng, J., and C. Xue, 2014: The sand-damage-prevention engineering system for the railway in the desert region of the Qinghai-Tibet plateau. *J. Wind Eng. Ind. Aerodyn.*, **125**, 30–37, doi:10.1016/J.JWEIA.2013.11.016.

Chengrui, M., and H.E. Dregne, 2001: Review article: Silt and the future development of China's Yellow River. *Geogr. J.*, **167**, 7–22, doi:10.1111/1475-4959.00002.

Cherlet, M. et al. (eds.), 2018: *World Atlas of Desertification*. Publication Office of the European Union, Luxembourg, 248 pp.

Choobari, O.A., P. Zawar-Reza, and A. Sturman, 2014: The global distribution of mineral dust and its impacts on the climate system: A review. *Atmos. Res.*, **138**, 152–165, doi:10.1016/J.ATMOSRES.2013.11.007.

Chowdhury, S., and M. Al-Zahrani, 2013: Implications of climate change on water resources in Saudi Arabia. *Arab. J. Sci. Eng.*, **38**, 1959–1971, doi:10.1007/s13369-013-0565-6.

Christian, B.A., P.S. Dhinwa, and Ajai, 2018: Long term monitoring and assessment of desertification processes using medium and high resolution satellite data. *Appl. Geogr.*, **97**, 10–24, doi:10.1016/J.APGEOG.2018.04.010.

Christie, F., and J. Hanlon, 2001: *Mozambique and the Great Flood Of 2000*. Indiana University Press, Indiana, USA. ISBN 0-85255-858-9 (hardback); 0-85255-857-0 (paperback). 176 pp.

Christy, R.D., C.A. da Silva, N. Mhlanga, E. Mabaya, and K. Tihanyi (eds.), 2014: *Innovative Institutions, Public Policies and Private Strategies for Agro-Enterprise Development*. Co-published with Food and Agriculture Organization of the United Nations, Rome, Italy, 368 pp. doi:10.1142/9131.

Chu, X., J. Zhan, Z. Li, F. Zhang, and W. Qi, 2019: Assessment on forest carbon sequestration in the Three-North Shelterbelt Program region, China. *J. Clean. Prod.*, **215**, 382–389, doi:10.1016/J.JCLEPRO.2018.12.296.

Chughtai, M.I., K. Mahmood, and A.R. Awan, 2015: Growth performance of carp species fed on salt-tolerant roughages and formulated feed in brackish water under polyculture system. *Pak. J. Zool.*, **47**, 775–781.

Chytrý, M. et al., 2012: Projecting trends in plant invasions in Europe under different scenarios of future land-use change. *Glob. Ecol. Biogeogr.*, **21**, 75–87, doi:10.1111/j.1466-8238.2010.00573.x.

CIREN, 2010: *Determinación de la erosión potencial y actual de los suelos de Chile*. Centro de Información de Recursos Naturales, Santiago, Chile, 285 pp.

Clarke, D.J., and R.V. Hill, 2013: *Cost-Benefit Analysis of the African Risk Capacity Facility*. No. 01292, Washington, DC, USA, 64 pp.

Clarke, H., and J.P. Evans, 2018: Exploring the future change space for fire weather in Southeast Australia. *Theor. Appl. Climatol.*, **136**, 513–527, doi:10.1007/s00704-018-2507-4.

Closas, A., and E. Rap, 2017: Solar-based groundwater pumping for irrigation: Sustainability, policies, and limitations. *Energy Policy*, **104**, 33–37, doi:10.1016/J.ENPOL.2017.01.035.

Codjoe, S.N.A., G. Owusu, and V. Burkett, 2014: Perception, experience, and indigenous knowledge of climate change and variability: The case of Accra, a Sub-Saharan African city. *Reg. Environ. Chang.*, **14**, 369–383, doi:10.1007/s10113-013-0500-0.

Collins, J.A. et al., 2017: Rapid termination of the African Humid Period triggered by northern high-latitude cooling. *Nat. Commun.*, **8**, 1372, doi:10.1038/s41467-017-01454-y.

Conant, R.T., and K. Paustian, 2003: Potential soil carbon sequestration in overgrazed grassland ecosystems. *Global Biogeochem. Cycles*, **16**, 90–99, doi:10.1029/2001gb001661.

de Coninck, H. et al., 2018: Chapter 4: Strengthening and Implementing the Global Response. In: Global Warming of 1.5°C. An IPCC Special Report on the Impacts of Global Warming of 1.5°C Above Pre-Industrial Levels and Related Global Greenhouse Gas Emission Pathways, in the Context of Strengthening the Global Response to the Threat of Climate Change, Sustainable Development, and Efforts to Eradicate Poverty [Masson-Delmotte, V., P. Zhai, H.O. Pörtner, D. Roberts, J. Skea, P.R. Shukla, A. Pirani, W. Moufouma-Okia, C. Péan, R. Pidcock, S. Connors, J.B.R. Matthews, Y. Chen, X. Zhou, M.I. Gomis, E. Lonnoy, T. Maycock, M. Tignor, T. Waterfield (eds.)]. In Press.

Cook, B., N. Zeng, and J.H. Yoon, 2012: Will Amazonia dry out? Magnitude and causes of change from IPCC climate model projections. *Earth Interact.*, **16**, doi:10.1175/2011EI398.1.

Cook, B.I., R.L. Miller, and R. Seager, 2009: Amplification of the North American 'Dust Bowl' drought through human-induced land degradation. *Proc. Natl. Acad. Sci. U.S.A.*, **106**, 4997–5001, doi:10.1073/pnas.0810200106.

Cook, B.I., J.E. Smerdon, R. Seager, and S. Coats, 2014a: Global warming and 21st century drying. *Clim. Dyn.*, **43**, 2607–2627, doi:10.1007/s00382-014-2075-y.

Cook, B.I., Smerdon, J.E.,, R. Seager, and E.R. Cook, 2014b: Pan-continental droughts in North America over the last millennium. *J. Clim.*, **27**, 383–397, doi:10.1175/JCLI-D-13-00100.1.

3

Coppola, E., and F. Giorgi, 2009: An assessment of temperature and precipitation change projections over Italy from recent global and regional climate model simulations. *Int. J. Climatol.*, **30**: 11-32, doi:10.1002/joc.1867.

Cordingley, J.E., K.A. Snyder, J. Rosendahl, F. Kizito, and D. Bossio, 2015: Thinking outside the plot: Addressing low adoption of sustainable land management in Sub-Saharan Africa. *Curr. Opin. Environ. Sustain.*, **15**, 35–40, doi:10.1016/J.COSUST.2015.07.010.

Cornell, S. et al., 2013: Opening up knowledge systems for better responses to global environmental change. *Environ. Sci. Policy*, **28**, 60–70, doi:10.1016/J.ENVSCI.2012.11.008.

Cornet, A., 2012: Surveillance environnementale et développement Des observations écologiques à la surveillance environnementale: un besoin pour comprendre et agir. *Options Méditerranéennes*, **B**, 1–14.

Costa, S.C.S., A.S.A.C. Diniz, and L.L. Kazmerski, 2016: Dust and soiling issues and impacts relating to solar energy systems: Literature review update for 2012–2015. *Renew. Sustain. Energy Rev.*, **63**, 33–61, doi:10.1016/J.RSER.2016.04.059.

Costa, C.S.C., Suellen, A.S.A.C. Diniz, and L.L. Kazmerski, 2018: Solar energy dust and soiling R&D progress: Literature review update for 2016. *Renew. Sustain. Energy Rev.*, **82**, 2504–2536, doi:10.1016/J.RSER.2017.09.015.

Costanza, R., R. de Groot, and P. Sutton, 2014: Changes in the global value of ecosystem services. *Glob. Environ. Chang.*, **26**, 152–158, doi:10.1016/j.gloenvcha.2014.04.002.

Cour, J.M., 2001: The Sahel in West Africa: Countries in transition to a full market economy. *Glob. Environ. Chang.*, **11**, 31–47, doi:10.1016/S0959-3780(00)00043-1.

Cowie, A.L. et al., 2018: Land in balance: The scientific conceptual framework for. *Environ. Sci. Policy*, **79**, 25–35, doi:10.1016/J.ENVSCI.2017.10.011.

Cox, J.R., M.H. Martin-R, F.A. Ibarra-F, J.H. Fourie, J.F.G. Rethman, and D.G. Wilcox, 1988: The influence of climate and soils on the distribution of four African grasses. *J. Range Manag.*, **41**, 127, doi:10.2307/3898948.

Crawford, J.A., Olson, R.A., West, N.E., Mosley, J.C., Schroeder, M.A., Whitson, T.D., Miller, R.F., Gregg, M.A., Boyd, C.S., 2004: Ecology and management of sage-grouse and sage-grouse habitat. *J. Range Manag.*, **57**, 2–19, doi:10.2111/1551-5028(2004)057[0002:EAMOSA]2.0.CO;2.

CRED, 2015: *Human Cost of Natural Disasters 2015: A global perspective*. CRED and UNISDR. Brussels, Belgium. 59 pp.

Crovetto, C.C., 1998: No-till development in Chequén Farm and its influence on some physical, chemical and biological parameters. *J. soil water Conserv.*, **58**, 194–199. ISSN: 1941–3300.

Cullman, G., 2015: Community forest management as virtualism in north-eastern Madagascar. *Hum. Ecol.*, **43**, 29–41, doi:10.1007/s10745-015-9725-5.

D'Odorico, P. et al., 2013: Global desertification: Drivers and feedbacks. *Adv. Water Resour.*, **51**, 326–344, doi:10.1016/j.advwatres.2012.01.013.

Dagar, J.C., C.B. Pandey, and C.S. Chaturvedi, 2014: Agroforestry: A way forward for sustaining fragile coastal and island agroecosystems. In: *Agroforestry Systems in India: Livelihood Security & Environmental Services Advances in Agroforestry* [Dagar, J.C., A.K. Singh, and A. Arunachalam (eds.)]. Springer, New Delhi, India, pp. 185–232.

Dah-gbeto, A.P., and G.B. Villamor, 2016: Gender-specific responses to climate variability in a semi-arid ecosystem in northern Benin. *Ambio*, **45**, 297–308, doi:10.1007/s13280-016-0830-5.

Dahiya, B., 2012: Cities in Asia, 2012: Demographics, economics, poverty, environment and governance. *Cities*, **29**, S44–S61, doi:10.1016/J.CITIES.2012.06.013.

Dai, A., 2011: Drought under global warming: A review. *Wiley Interdiscip. Rev. Clim. Chang.*, **2**, 45–65, doi:10.1002/wcc.81.

Dai, A., 2013: Increasing drought under global warming in observations and models. *Nat. Clim. Chang.*, **3**, 52–58, doi:10.1038/nclimate1633.

Damberg, L., and A. AghaKouchak, 2014: Global trends and patterns of drought from space. *Theor. Appl. Climatol.*, **117**, 441–448, doi:10.1007/s00704-013-1019-5.

Das, R., A. Evan, and D. Lawrence, 2013: Contributions of long-distance dust transport to atmospheric P inputs in the Yucatan Peninsula. *Global Biogeochem. Cycles*, **27**, 167–175, doi:10.1029/2012GB004420.

Davies, G.M. et al., 2012: Trajectories of change in sagebrush steppe vegetation communities in relation to multiple wildfires. *Ecol. Appl.*, **22**, 1562–1577, doi:10.1890/10-2089.1.

Davies, J., 2017: *The Land in Drylands: Thriving In Uncertainty through Diversity*. Working Paper, Global Land Outlook, Bonn, Germany, 18 pp.

Davies, J. et al., 2016: *Improving Governance of Pastoral Lands: Implementing the Voluntary Guidelines on the Responsible Governance of Tenure Of Land, Fisheries and Forests in the Context of National Food Security*. Governance of Tenure Technical Guide No. 6, Food and Agriculture Organization of the United Nations, Rome, 152 pp.

Davis, D.K., 2016: *The Arid Lands: History, Power, Knowledge*. MIT Press, Cambridge, USA, 271 pp.

Davis, M.A., J.P. Grime, and K. Thompson, 2000: Fluctuating resources in plant communities: A general theory of invasibility. *J. Ecol.*, **88**, 528–534, doi:10.1046/j.1365-2745.2000.00473.x.

Dawelbait, M., and F. Morari, 2012: Monitoring desertification in a savanna region in Sudan using Landsat images and spectral mixture analysis. *J. Arid Environ.*, **80**, 45–55, doi:10.1016/j.jaridenv.2011.12.011.

Decker, M., A.J. Pitman, J.P. Evans, M. Decker, A.J. Pitman, and J.P. Evans, 2013: Groundwater constraints on simulated transpiration variability over south-eastern Australian forests. *J. Hydrometeorol.*, **14**, 543–559, doi:10.1175/JHM-D-12-058.1.

de Deckker, P. et al., 2008: Geochemical and microbiological fingerprinting of airborne dust that fell in Canberra, Australia, in October 2002. *Geochemistry, Geophys. Geosystems*, **9**, doi:10.1029/2008GC002091.

Deininger, K., and S. Jin, 2006: Tenure security and land-related investment: Evidence from Ethiopia. *Eur. Econ. Rev.*, **50**, 1245–1277, doi:10.1016/J.EUROECOREV.2005.02.001.

Dell'Angelo, J., P. D'Odorico, M.C. Rulli, and P. Marchand, 2017: The tragedy of the grabbed commons: Coercion and dispossession in the global land rush. *World Dev.*, **92**, 1–12, doi:10.1016/J.WORLDDEV.2016.11.005.

Derpsch, R., T. Friedrich, A. Kassam, and L. Hongwen, 2010: Current status of adoption of no-till farming in the world and some of its main benefits. *Int. J. Agric. Biol. Eng.*, **3**, 1–25, doi:10.3965/j.issn.1934-6344.2010.01.0-0.

Descheemaeker, K., M. Zijlstra, P. Masikati, O. Crespo, and S. Homann-Kee Tui, 2018: Effects of climate change and adaptation on the livestock component of mixed farming systems: A modelling study from semi-arid Zimbabwe. *Agric. Syst.*, **159**, 282–295, doi:10.1016/J.AGSY.2017.05.004.

Deshpande, R.S., 2008: Watersheds: Putting the cart before the horse. *Econ. Polit. Wkly.*, **43**, 74–76.

Devine, A.P., R.A. McDonald, T. Quaife, and I.M.D. Maclean, 2017: Determinants of woody encroachment and cover in African savannas. *Oecologia*, **183**, 939–951, doi:10.1007/s00442-017-3807-6.

Dey, A., A. Gupta, and G. Singh, 2017: Open innovation at different levels for higher climate risk resilience. *Sci. Technol. Soc.*, **22**, 388–406, doi:10.1177/0971721817723242.

Dharumarajan, S., T.F.A. Bishop, R. Hegde, and S.K. Singh, 2018: Desertification vulnerability index – An effective approach to assess desertification processes: A case study in Anantapur District, Andhra Pradesh, India. *L. Degrad. Dev.*, **29**, 150–161, doi:10.1002/ldr.2850.

Diao, X., and D.B. Sarpong, 2011: Poverty implications of agricultural land degradation in Ghana: An economy-wide, multimarket model assessment. *African Dev. Rev.*, **23**, 263–275, doi:10.1111/j.1467-8268.2011.00285.x.

Díaz, J. et al., 2017: Saharan dust intrusions in Spain: Health impacts and associated synoptic conditions. *Environ. Res.*, **156**, 455–467, doi:10.1016/J.ENVRES.2017.03.047.

Dibari, C., M. Bindi, M. Moriondo, N. Staglianò, S. Targetti, and G. Argenti, 2016: Spatial data integration for the environmental characterisation of pasture macrotypes in the Italian Alps. *Grass Forage Sci.*, **71**, 219–234, doi:10.1111/gfs.12168.

Dile, Y.T., L. Karlberg, M. Temesgen, and J. Rockström, 2013: The role of water harvesting to achieve sustainable agricultural intensification and resilience against water-related shocks in Sub-Saharan Africa. *Agric. Ecosyst. Environ.*, **181**, 69–79, doi:10.1016/j.agee.2013.09.014.

Dimelu, M., E. Salifu, and E. Igbokwe, 2016: Resource use conflict in agrarian communities, management and challenges: A case of farmer-herdsmen conflict in Kogi State, Nigeria. *J. Rural Stud.*, **46**, 147–154.

Dimobe, K., A. Ouédraogo, S. Soma, D. Goetze, S. Porembski, and A. Thiombiano, 2015: Identification of driving factors of land degradation and deforestation in the wildlife reserve of Bontioli (Burkina Faso, West Africa). *Glob. Ecol. Conserv.*, **4**, 559–571, doi:10.1016/j.gecco.2015.10.006.

Divinsky, I., N. Becker, and P. Bar (Kutiel), 2017: Ecosystem service tradeoff between grazing intensity and other services – A case study in Karei-Deshe experimental cattle range in northern Israel. *Ecosyst. Serv.*, **24**, 16–27, doi:10.1016/j.ecoser.2017.01.002.

Dixon, J.A., D.E. James, P.B. Sherman, D.E. James, and P.B. Sherman, 2013: *Economics of Dryland Management*. Routledge, London, UK. 324 pp.

Djanibekov, U., and A. Khamzina, 2016: Stochastic economic assessment of afforestation on marginal land in irrigated farming system. *Environ. Resour. Econ.*, **63**, 95–117, doi:10.1007/s10640-014-9843-3.

Djanibekov, U. et al., 2016: Adoption of sustainable land uses in post-Soviet Central Asia: The case for agroforestry. *Sustainability*, **8**, 1030, doi:10.3390/su8101030.

Do, N., and S. Kang, 2014: Assessing drought vulnerability using soil moisture-based water use efficiency measurements obtained from multi-sensor satellite data in Northeast Asia Dryland regions. *J. Arid Environ.*, **105**, 22–32, doi:10.1016/J.JARIDENV.2014.02.018.

Dodd, J., 1994: Desertification and degradation of Africa's rangelands. *Rangelands*, **16(5)**, 180–183.

Dominguez, F., J. Cañon, and J. Valdes, 2010: IPCC-AR4 climate simulations for the south-western US: The importance of future ENSO projections. *Clim. Change*, **99**, 499–514, doi:10.1007/s10584-009-9672-5.

Dominguez, P., 2014: Current situation and future patrimonializing perspectives for the governance of agropastoral resources in the Ait Ikis transhumants of the High Atlas (Morocco). In: Herrera P, Davies J, Baena P (eds.) The Governance of Rangelands. Collective Action for Sustainable Pastoralism. Routledge, Oxon and New York, pp. 126–144.

Donat, M.G., A.L. Lowry, L.V. Alexander, P.A. O'Gorman, and N. Maher, 2016: More extreme precipitation in the world's dry and wet regions. *Nat. Clim. Chang.*, **6**, 508–513, doi:10.1038/nclimate2941.

Dong, S., 2016: Overview: Pastoralism in the World. In: *Building Resilience of Human-Natural Systems of Pastoralism in the Developing World* [Dong, S., K.A.S. Kassam, J. Tourrand and Boone, R.B. (eds.)]. Springer International Publishing, Cham, Switzerland, pp. 1–37.

Donohue, R.J., T.R. McVicar, and M.L. Roderick, 2009: Climate-related trends in Australian vegetation cover as inferred from satellite observations, 1981–2006. *Glob. Chang. Biol.*, **15**, 1025–1039, doi:10.1111/j.1365-2486.2008.01746.x.

Donohue, R.J., M.L. Roderick, T.R. McVicar, and G.D. Farquhar, 2013: Impact of CO_2 fertilization on maximum foliage cover across the globe's warm, arid environments. *Geophys. Res. Lett.*, **40**, 3031–3035, doi:10.1002/grl.50563.

Dougill, A.J., E.D.G. Fraser, and M.S. Reed, 2010: Anticipating vulnerability to climate change in Dryland pastoral systems: Using dynamic systems models for the Kalahari. *Ecol. Soc.*, **15 (2): 17**. www.ecologyandsociety.org/vol15/iss2/art17/.

Dougill, A.J. et al., 2016: Land use, rangeland degradation and ecological changes in the southern Kalahari, Botswana. *Afr. J. Ecol.*, **54**, 59–67, doi:10.1111/aje.12265.

Dow, K. et al., 2013: Limits to adaptation. *Nat. Clim. Chang.*, **3**, 305–307, doi:10.1038/nclimate1847.

Dregne, H.E., and N.-T. Chou, 1992: Global desertification dimensions and costs. In: *Degradation and restoration of arid lands*. Texas Tech University, Texas, USA, pp. 249–282.

Dressler, W. et al., 2010: From hope to crisis and back again? A critical history of the global CBNRM narrative. *Environ. Conserv.*, **37**, 5–15, doi:10.1017/S0376892910000044.

Du, F., 2012: Ecological resettlement of Tibetan herders in the Sanjiangyuan: A case study in Madoi County of Qinghai. *Nomad. People.*, **16**, 116–133, doi:10.3167/np.2012.160109.

Dukes, J.S., N.R. Chiariello, S.R. Loarie, and C.B. Field, 2011: Strong response of an invasive plant species (*Centaurea solstitialis* L.) to global environmental changes. *Ecol. Appl.*, **21**, 1887–1894, doi:10.1890/11-0111.1.

Dukhovny, V.A., V.I. Sokolov, and D.R. Ziganshina, 2013: Integrated water resources management in Central Asia, as a way of survival in conditions of water scarcity. *Quat. Int.*, **311**, 181–188, doi:10.1016/J.QUAINT.2013.07.003.

Dumenu, W.K. and E.A. Obeng, 2016: Climate change and rural communities in Ghana: Social vulnerability, impacts, adaptations and policy implications. *Environ. Sci. Policy*, **55**, 208–217, doi:10.1016/J.ENVSCI.2015.10.010.

Durant, S.M. et al., 2014: Fiddling in biodiversity hotspots while deserts burn? Collapse of the Sahara's megafauna. *Divers. Distrib.*, **20**, 114–122, doi:10.1111/ddi.12157.

Eekhout, J. and J. de Vente, 2019: Assessing the effectiveness of Sustainable Land Management for large-scale climate change adaptation. *Sci. Total Environ.*, **654**, 85–93, doi:10.1016/J.SCITOTENV.2018.10.350.

Ehrenfeld, J.G., 2003: Effects of Exotic plant invasions on soil nutrient cycling processes. *Ecosystems*, **6**, 503–523, doi:10.1007/s10021-002-0151-3.

Eitelberg, D.A., J. van Vliet, and P.H. Verburg, 2015: A review of global potentially available cropland estimates and their consequences for model-based assessments. *Glob. Chang. Biol.*, **21**, 1236–1248, doi:10.1111/gcb.12733.

El-Fadel, M., R. Chedid, M. Zeinati, and W. Hmaidan, 2003: Mitigating energy-related GHG emissions through renewable energy. *Renew. Energy*, **28**, 1257–1276, doi:10.1016/S0960-1481(02)00229-X.

Eldridge, D.J., and S. Soliveres, 2014: Are shrubs really a sign of declining ecosystem function? Disentangling the myths and truths of woody encroachment in Australia. *Aust. J. Bot.*, **62**, 594–608, doi:10.1071/BT14137.

Eldridge, D.J. et al., 2011: Impacts of shrub encroachment on ecosystem structure and functioning: Towards a global synthesis. *Ecol. Lett.*, **14**, 709–722, doi:10.1111/j.1461-0248.2011.01630.x.

Eldridge, D.J., L. Wang, and M. Ruiz-Colmenero, 2015: Shrub encroachment alters the spatial patterns of infiltration. *Ecohydrology*, **8**, 83–93, doi:10.1002/eco.1490.

Elhadary, Y., 2014: Examining drivers and indicators of the recent changes among pastoral communities of Butana locality, Gedarif State, Sudan. *Am. J. Sociol. Res.*, **4**, 88–101.

Ellies, A., 2000: Soil erosion and its control in Chile. *Acta Geológica Hisp.*, **35**, 279–284.

Ellis, J., 1995: Climate variability and complex ecosystem dynamics: Implications for pastoral development. In: *Living with Uncertainty* [Scoones I. (ed.)]. Practical Action Publishing, London, UK, pp. 37–46.

Elmore, A.J., J.M. Kaste, G.S. Okin, and M.S. Fantle, 2008: Groundwater influences on atmospheric dust generation in deserts. *J. Arid Environ.*, **72**, 1753–1765, doi:10.1016/J.JARIDENV.2008.05.008.

Elum, Z. and A.S. Momodu, 2017: Climate change mitigation and renewable energy for sustainable development in Nigeria: A discourse approach. *Renew. Sustain. Energy Rev.*, **76**, 72–80.

Enfors, E.I., and L.J. Gordon, 2008: Dealing with drought: The challenge of using water system technologies to break Dryland poverty traps. *Glob. Environ. Chang.*, **18**, 607–616, doi:10.1016/J.GLOENVCHA.2008.07.006.

3

Engdawork, A., and H.-R. Bork, 2016: Farmers' perception of land degradation and traditional knowledge in southern Ethiopia – Resilience and stability. *L. Degrad. Dev.*, **27**, 1552–1561, doi:10.1002/ldr.2364.

Engelbrecht, F.A. et al., 2011: Multi-scale climate modelling over southern Africa using a variable-resolution global model. *Water Research Commission 40-year Celebration Conference*, Kempton Park, **37**, 647–658. http://dx.doi.org/10.4314/wsa.v37i5.2.

Engelstaedter, S., K.E. Kohfeld, I. Tegen, and S.P. Harrison, 2003: Controls of dust emissions by vegetation and topographic depressions: An evaluation using dust storm frequency data. *Geophys. Res. Lett.*, **30** 1294, doi:10.1029/2002GL016471.

Erb, K.-H. et al., 2016: Livestock grazing, the neglected land use. In: *Social Ecology* [Haberl, H., M. Fischer-Kowalski and F. Winiwarter (eds.)]. Springer International Publishing, Cham, Switzerland, pp. 295–313.

Eriksen, S., and J. Lind, 2009: Adaptation as a political process: Adjusting to drought and conflict in Kenya's drylands. *Environ. Manage.*, **43**, 817–835, doi:10.1007/s00267-008-9189-0.

van der Esch, S. et al., 2017: *Exploring Future Changes in Land Use and Land Condition and the Impacts on Food, Water, Climate Change and Biodiversity: Scenarios for the UNCCD Global Land Outlook*. Policy Report, PBL Netherlands Environmental Assessment Agency, the Hague, Netherlands, 115 pp.

Etana, A., Kelbessa, E., and Soromessa, T., 2011: Impact of Parthenium hysterophorus L.(Asteraceae) on herbaceous plant biodiversity of Awash National Park (ANP), Ethiopia. *Manag. Biol. Invasions*, **2**, doi:10.3391/mbi.2011.2.1.07.

Etchart, L., 2017: The role of indigenous peoples in combating climate change. *Palgrave Commun.*, **3**, 17085, doi:10.1057/palcomms.2017.85.

Etongo, D., I. Djenontin, M. Kanninen, and K. Fobissie, 2015: Smallholders' tree planting activity in the Ziro Province, southern Burkina Faso: Impacts on livelihood and policy implications. *Forests*, **6**, 2655–2677, doi:10.3390/f6082655.

European Court of Auditors (ECA), 2018: *Desertfication in Europe*. Background paper, European Court of Auditors, Luxembourg, 15 pp.

European Court of Auditors (ECA), 2018: *Combating desertification in the EU: A growing threat in need of more action*. Special Report No. 33, European Court of Auditors, Luxembourg, 66 pp.

Evan, A.T., 2018: Surface winds and dust biases in climate models. *Geophys. Res. Lett.*, **45**, 1079–1085, doi:10.1002/2017GL076353.

Evan, A.T., and S. Mukhopadhyay, 2010: African Dust over the Northern Tropical Atlantic: 1955–2008. *J. Appl. Meteorol. Climatol.*, **49**, 2213–2229, doi:10.1175/2010JAMC2485.1.

Evan, A.T., D.J. Vimont, A.K. Heidinger, J.P. Kossin, and R. Bennartz, 2009: The role of aerosols in the evolution of tropical North Atlantic Ocean temperature anomalies. *Science*, **324**, 778–781, doi:10.1126/science.1167404.

Evan, A.T., C. Flamant, S. Fiedler, and O. Doherty, 2014: An analysis of aeolian dust in climate models. *Geophys. Res. Lett.*, **41**, 5996–6001, doi:10.1002/2014GL060545.

Evan, A.T., C. Flamant, M. Gaetani, and F. Guichard, 2016: The past, present and future of African dust. *Nature*, **531**, 493–495, doi:10.1038/nature17149.

Evans, J., and R. Geerken, 2004: Discrimination between climate and human-induced Dryland degradation. *J. Arid Environ.*, **57**, 535–554, doi:10.1016/S0140-1963(03)00121-6.

Evans, J.P., and R. Geerken, 2006: Classifying rangeland vegetation type and coverage using a Fourier component based similarity measure. *Remote Sens. Environ.*, **105**, 1–8, doi:10.1016/j.rse.2006.05.017.

Evans, J.P., X. Meng, and M.F. McCabe, 2017: Land surface albedo and vegetation feedbacks enhanced the millennium drought in Southeast Australia. *Hydrol. Earth Syst. Sci.*, **21**, 409–422, doi:10.5194/hess-21-409-2017.

Ezra, M., and G.-E. Kiros, 2001: Rural out-migration in the drought prone areas of Ethiopia: A multilevel analysis. *Int. Migr. Rev.*, **35**, 749–771, doi:10.1111/j.1747-7379.2001.tb00039.x.

Fan, B. et al., 2015: Earlier vegetation green-up has reduced spring dust storms. *Sci. Rep.*, **4**, 6749, doi:10.1038/srep06749.

FAO, 1995: *Desertification and Drought – Extent and Consequences Proposal for a Participatory Approach to Combat Desertification*. Proceedings of a conference, Food and Agriculture Organization of the United Nations, Rome, Italy, 15 pp.

FAO, 2004: *Drought Impact Mitigation and Prevention in the Limpopo River Basin, A Situation Analysis*. Land and Water Discussion Paper No. 4, Food and Agriculture Organization of the United Nations, Rome, Italy, 178 pp.

FAO, 2011: *Gestion des Plantations sur Dunes. Document de Travail sur les Forêts et la Foresterie en Zones Arides*. Document de travail sur les les Forêts et la Foresterie en zones arides, n° 3, Food and Agriculture Organization of the United Nations, Rome, Italy, 43 pp.

FAO, 2015: *Global Soil Status, Processes and Trends*. Food and Agriculture Organization of the United Nations, Rome, Italy, 605 pp.

FAO, 2016: *Trees, Forests and Land Use in Drylands: The First Global Assessment Preliminary Findings*. Food and Agriculture Organization of the United Nations, Rome, Italy, 31 pp.

FAO, 2018: *The Impact of Disasters and Crises on Agriculture and Food Security 2017*. Food and Agriculture Organization of the United Nations, Rome, Italy, 143 pp.

FAO, IFAD, UNICEF, WFP, and WHO, 2018: *The State of Food Security and Nutrition in the World 2018: Building Climate Resilience for Food Security and Nutrition*. Food and Agriculture Organization of the United Nations, Rome, Italy, 181 pp.

Faour-Klingbeil, D., and E.C.D. Todd, 2018: The impact of climate change on raw and untreated wastewater use for agriculture, especially in arid regions: A review. *Foodborne Pathog. Dis.*, **15**, 61–72, doi:10.1089/fpd.2017.2389.

Farage, P.K., J. Ardo, L. Olsson, E.A. Rienzi, A.S. Ball, and J.N. Pretty, 2007: The potential for soil carbon sequestration in three tropical Dryland farming systems of Africa and Latin America: A modelling approach. *Soil Tillage Res.*, **94**, 457–472, doi:10.1016/j.still.2006.09.006.

Farnoosh, A., F. Lantz, and J. Percebois, 2014: Electricity generation analyses in an oil-exporting country: Transition to non-fossil fuel based power units in Saudi Arabia. *Energy*, **69**, 299–308, doi:10.1016/J.ENERGY.2014.03.017.

Fasil, R., 2011: *Parthenium Hysterophorus in Ethiopia: Distribution and Importance, and Current Efforts to Manage the Scourge*. Proceedings of Workshop on Noxious Weeds in Production of Certified Seeds, 11–12 July. Food and Agriculture Organization of the United Nations, Accra, Ghana.

Favet, J. et al., 2013: Microbial hitchhikers on intercontinental dust: Catching a lift in Chad. *ISME J.*, **7**, 850–867, doi:10.1038/ismej.2012.152.

Fay, P.A., D.M. Kaufman, J.B. Nippert, J.D. Carlisle, and C.W. Harper, 2008: Changes in grassland ecosystem function due to extreme rainfall events: Implications for responses to climate change. *Glob. Chang. Biol.*, **14**, 1600–1608, doi:10.1111/j.1365-2486.2008.01605.x.

Feng, S., Krueger, A.B., and Oppenheimer, M., 2010: Linkages among climate change, crop yields and Mexico–US cross-border migration. *Natl. Acad Sci.*, **107**, 14257–14262, doi:10.1073/pnas.1002632107.

Feng, Q., H. Ma, X. Jiang, X. Wang, and S. Cao, 2015: What has caused desertification in China? *Sci. Rep.*, **5**, 15998, doi:10.1038/srep15998.

Feng, S., and Q. Fu, 2013: Expansion of global drylands under a warming climate. *Atmos. Chem. Phys.*, **13**, 10081–10094, doi:10.5194/acp-13-10081-2013.

Feng, S., Q. Hu, W. Huang, C.-H. Ho, R. Li, and Z. Tang, 2014: Projected climate regime shift under future global warming from multi-model, multi-scenario CMIP5 simulations. *Glob. Planet. Change*, **112**, 41–52, doi:10.1016/J.GLOPLACHA.2013.11.002.

Feng, X. et al., 2016: Revegetation in China's Loess Plateau is approaching sustainable water resource limits. *Nat. Clim. Chang.*, **6**, 1019–1022, doi:10.1038/nclimate3092.

3

Fensholt, R. et al., 2012: Greenness in semi-arid areas across the globe 1981–2007 – An earth observing satellite based analysis of trends and drivers. *Remote Sens. Environ.*, **121**, 144–158, doi:10.1016/J.RSE.2012.01.017.

Fensholt, R. et al., 2015: Assessment of vegetation trends in drylands from time series of earth observation data. In: *Remote Sensing and Digital Image Processing, Remote Sensing Time Series: Revealing Land Surface Dynamics* [Kuenzer, C., S. Dech and W. Wagner (eds.)]. Springer International Publishing, Cham, Switzerland, pp. 159–182.

Fernandes, P.M. et al., 2013: Prescribed burning in southern Europe: Developing fire management in a dynamic landscape. *Front. Ecol. Environ.*, **11**, e4–e14, doi:10.1890/120298.

Fernandez-Gimenez, M.E., 2000: The role of Mongolian nomadic pastoralists' ecological knowledge in rangeland management. *Ecol. Appl.*, **10**, 1318–1326, doi:10.1890/1051-0761(2000)010[1318:TROMNP]2.0.CO;2.

Fernandez-Gimenez, M.E., B. Batkhishig, and B. Batbuyan, 2015: Lessons from the dzud: Community-based rangeland management increases the adaptive capacity of Mongolian herders to winter disasters. *World Dev.*, **68**, 48–65. doi:10.1016/j.worlddev.2014.11.015.

Fernández-Giménez, M.E., and F. Fillat Estaque, 2012: Pyrenean Pastoralists' Ecological knowledge: Documentation and application to natural resource management and adaptation. *Hum. Ecol.*, **40**, 287–300, doi:10.1007/s10745-012-9463-x.

Fernández, O.A., M.E. Gil, and R.A. Distel, 2009: The challenge of rangeland degradation in a temperate semi-arid region of Argentina: The Caldenal. *L. Degrad. Dev.*, **20**, 431–440, doi:10.1002/ldr.851.

Fetzel, T. et al., 2017: Quantification of uncertainties in global grazing systems assessment. *Global Biogeochem. Cycles*, **31**, 1089–1102, doi:10.1002/2016GB005601.

Few, R., and M.G.L. Tebboth, 2018: Recognising the dynamics that surround drought impacts. *J. Arid Environ.*, **157**, 113–115, doi:10.1016/J.JARIDENV.2018.06.001.

Ficklin, D.L., J.T. Abatzoglou, S.M. Robeson, and A. Dufficy, 2016: The influence of climate model biases on projections of aridity and drought. *J. Clim.*, **29**, 1369–1389, doi:10.1175/JCLI-D-15-0439.1.

Field, J.P. et al., 2010: The ecology of dust. *Front. Ecol. Environ.*, **8**, 423–430, doi:10.1890/090050.

Field, C.B., V.R. Barros, K.J. Mach, M.D. Mastrandrea, M. van Aalst, W.N. Adger, D.J. Arent, J. Barnett, R. Betts, T.E. Bilir, J. Birkmann, J. Carmin, D.D. Chadee, A.J. Challinor, M. Chatterjee, W. Cramer, D.J. Davidson, Y.O. Estrada, J.-P. Gattuso, Y. Hijioka, O. Hoegh-Guldberg, H.Q. Huang, G.E. Insarov, R.N. Jones, R.S. Kovats, P. Romero-Lankao, J.N. Larsen, I.J. Losada, J.A. Marengo, R.F. McLean, L.O. Mearns, R. Mechler, J.F. Morton, I. Niang, T. Oki, J.M. Olwoch, M. Opondo, E.S. Poloczanska, H.-O. Pörtner, M.H. Redsteer, A. Reisinger, A. Revi, D.N. Schmidt, M.R. Shaw, W. Solecki, D.A. Stone, J.M.R. Stone, K.M. Strzepek, A.G. Suarez, P. Tschakert, R. Valentini, S. Vicuña, A. Villamizar, K.E. Vincent, R. Warren, L.L. White, T.J. Wilbanks, P.P. Wong, and G.W. Yohe, 2014: Technical Summary. Climate Change 2014: Impacts, Adaptation, and Vulnerability. Part A: Global and Sectoral Aspects. Contribution of Working Group II to the Fifth Assessment Report of the Intergovernmental Panel on Climate Change [Field, C.B., V.R. Barros, D.J. Dokken, K.J. Mach, M.D. Mastrandrea, T.E. Bilir, M. Chatterjee, K.L. Ebi, Y.O. Estrada, R.C. Genova, E.S. Kissel, A.N. Levy, S. MacCracken, P.R. Mastrandrea and L.L. White (eds.)]. Cambridge University Press, Cambridge, United Kingdom and New York, USA.

Findlay, A.M., 2011: Migrant destinations in an era of environmental change. *Glob. Environ. Chang.*, **21**, S50–S58, doi:10.1016/J.GLOENVCHA.2011.09.004.

Fischlin, A., G.F. Midgley, J. Price, R. Leemans, B. Gopal, C. Turley, M. Rounsevell, P. Dube, J. Tarazona, A. Velichko, 2007: Ecosystems, Their Properties, Goods, and Services. Climate Change 2007: Impacts, Adaptation and Vulnerability. Contribution of Working Group II to the Fourth Assessment Report of the Intergovernmental Panel on Climate Change [Parry, M.L., O.F. Canziani, J.P. Palutikof, P.J. Van der Linden, and C.E. Hanson (eds.)].

Cambridge University Press, Cambridge, United Kingdom and New York, USA, 211–272 pp.

Fleischner, T.L., 1994: Costos ecológicos del pastoreo de ganado en el oeste de Estados Unidos. *Conserv. Biol.*, **8**, 629–644, doi:10.1046/j.1523-1739.1994.08030629.x.

Fleskens, L., and L.C. Stringer, 2014: Land management and policy responses to mitigate desertification and land degradation. *L. Degrad. Dev.*, **25**, 1–4, doi:10.1002/ldr.2272.

Fleurbaey, M., S. Kartha, S. Bolwig, Y.L. Chee, Y. Chen, E. Corbera, F. Lecocq, W. Lutz, M.S. Muylaert, R.B. Norgaard, C. Oker-eke, and A.D. Sagar, 2014: *Sustainable Development and Equity. Climate Change 2014: Mitigation of Climate Change. Contribution of Working Group III to the Fifth Assessment Report of the Intergovernmental Panel on Climate Change* [Edenhofer O., R. Pichs-Madruga, Y. Sokona, E. Farahani, S. Kadner, K. Seyboth, A. Adler, I. Baum, S. Brunner, P. Eickemeier, B. Kriemann, J. Savolainen, S. Schlömer, C. von Stechow, T. Zwickel and J.C. Minx (eds.)]. Cambridge University Press, Cambridge, United Kingdom and New York, NY, USA, 238–350.

Ford, T.W., and C.F. Labosier, 2017: Meteorological conditions associated with the onset of flash drought in the Eastern United States. *Agric. For. Meteorol.*, **247**, 414–423, doi:10.1016/J.AGRFORMET.2017.08.031.

Franklin, K., and F. Molina-Freaner, 2010: Consequences of buffelgrass pasture development for primary productivity, perennial plant richness, and vegetation structure in the drylands of Sonora, Mexico. *Conserv. Biol.*, **24**, 1664–1673, doi:10.1111/j.1523-1739.2010.01540.x.

Franks, P.J. et al., 2013: Sensitivity of plants to changing atmospheric CO_2 concentration: From the geological past to the next century. *New Phytol.*, **197**, 1077–1094, doi:10.1111/nph.12104.

Fraser, E.D.G., A. Dougill, K. Hubacek, C. Quinn, J. Sendzimir, and M. Termansen, 2011: Assessing vulnerability to climate change in dryland livelihood systems: Conceptual challenges and interdisciplinary solutions. *Ecol. Soc.*, **16 (3): 3.** DOI: 10.5751/ES-03402-160303.

Fratkin, E., 2013: Seeking alternative livelihoods in pastoral areas. In: *Pastoralism and Development in Africa Dynamic Change at the Margins* [Catley, A., J. Lind, and I. Scoones (eds.)]. Routledge, London, UK, 328 pp.

Fre, Z., 2018: *Knowledge Sovereignty among African Cattle Herders*. UCL Press, London, UK, 200 pp.

Fried, T., L. Weissbrod, Y. Tepper, and G. Bar-Oz, 2018: A glimpse of an ancient agricultural ecosystem based on remains of micromammals in the Byzantine Negev Desert. *R. Soc. Open Sci.*, **5**, 171528, doi:10.1098/rsos.171528.

Fu, B., L. Chen, K. Ma, H. Zhou, and J. Wang, 2000: The relationships between land use and soil conditions in the hilly area of the Loess Plateau in northern Shaanxi, China. *CATENA*, **39**, 69–78, doi:10.1016/S0341-8162(99)00084-3.

Fu, Q., C.M. Johanson, J.M. Wallace, and T. Reichler, 2006: Enhanced mid-latitude tropospheric warming in satellite measurements. *Science*, **312**, 1179, doi:10.1126/science.1125566.

Fu, Q., L. Lin, J. Huang, S. Feng, and A. Gettelman, 2016: Changes in terrestrial aridity for the period 850–2080 from the community earth system model. *J. Geophys. Res.*, **121**, 2857–2873, doi:10.1002/2015JD024075.

Fu, Q., B. Li, Y. Hou, X. Bi, and X. Zhang, 2017: Effects of land use and climate change on ecosystem services in Central Asia's arid regions: A case study in Altay Prefecture, China. *Sci. Total Environ.*, **607–608**, 633–646, doi:10.1016/J.SCITOTENV.2017.06.241.

Fuentealba, E. et al., 2015: Photovoltaic performance and LCoE comparison at the coastal zone of the Atacama Desert, Chile. *Energy Convers. Manag.*, **95**, 181–186, doi:10.1016/J.ENCONMAN.2015.02.036.

Fuller, A., D. Mitchell, S.K. Maloney, and R.S. Hetem, 2016: Towards a mechanistic understanding of the responses of large terrestrial mammals to heat and aridity associated with climate change. *Clim. Chang. Responses*, **3**, 10, doi:10.1186/s40665-016-0024-1.

Funk, C. et al., 2018: Examining the role of unusually warm Indo-Pacific sea-surface temperatures in recent African droughts. *Q. J. R. Meteorol. Soc.*, **144**, 360–383, doi:10.1002/qj.3266.

3

Funk, C. et al., 2019: Examining the potential contributions of extreme 'Western V' sea surface temperatures to the 2017 March–June East African drought. *Bull. Am. Meteorol. Soc.*, **100**, S55–S60, doi:10.1175/BAMS-D-18-0108.1.

Gad, M.I. M., and A.A. Abdel-Baki, 2002: Estimation of salt balance in the soil-water of the old cultivated lands in Siwa Oasis, Western Desert, Egypt. Int. Journal of Engineering Research and Application Vol. 8, 37–44.Galvez, A., T. Sinicco, M.L. Cayuela, M.D. Mingorance, F. Fornasier, and C. Mondini, 2012: Short-term effects of bioenergy by-products on soil C and N dynamics, nutrient availability and biochemical properties. *Agric. Ecosyst. Environ.*, **160**, 3–14, doi:10.1016/J.AGEE.2011.06.015.

Gammoh, I.A., and T.Y. Oweis, 2011: Performance and adaptation of the Vallerani mechanized water harvesting system in degraded Badia Rangelands. *J. Environ. Sci. Eng.*, **5**, 1370–1380.

García-Fayos, P., and E. Bochet, 2009: Indication of antagonistic interaction between climate change and erosion on plant species richness and soil properties in semi-arid Mediterranean ecosystems. *Glob. Chang. Biol.*, **15**, 306–318, doi:10.1111/j.1365-2486.2008.01738.x.

García-Ruiz, J.M., 2010: The effects of land uses on soil erosion in Spain: A review. *Catena*, **81**, 1–11, doi:10.1016/j.catena.2010.01.001.

García-Ruiza, J.M., S. Beguería, E. Nadal-Romero, J.C. González-Hidalgo, N. Lana-Renault, and Y. Sanjuán, 2015: A meta-analysis of soil erosion rates across the world. *Geomorphology*, **239**, 160–173, doi:10.1016/j.geomorph.2015.03.008.

García-Ruiza, J.M., S. Beguería, N. Lana-Renault, E. Nadal-Romero, and A. Cerdà, 2017: Ongoing and emerging questions in water erosion studies. *L. Degrad. Dev.*, **28**, 5–21, doi:10.1002/ldr.2641.

Gari, L., 2006: A history of the Hima conservation system. *Environ. Hist. Camb.*, **12**, 213–228, doi:10.3197/096734006776680236.

Garrity, D.P. et al., 2010: Evergreen agriculture: A robust approach to sustainable food security in Africa. *Food Secur.*, **2**, 197–214, doi:10.1007/s12571-010-0070-7.

Garschagen, M., and P. Romero-Lankao, 2015: Exploring the relationships between urbanization trends and climate change vulnerability. *Clim. Change*, **133**, 37–52, doi:10.1007/s10584-013-0812-6.

Gbeckor-Kove, N., 1989: *Drought and Desertification*. World Meteorological Organization, Geneva, Switzerland, 41–73, 286 pp.

Gbetibouo, G.A., R.M. Hassan, and C. Ringler, 2010: Modelling farmers' adaptation strategies for climate change and variability: The case of the Limpopo Basin, South Africa. *Agrekon*, **49**, 217–234, doi:10.1080/03031853.2010.491294.

Ge, X., Y. Li, A.E. Luloff, K. Dong, and J. Xiao, 2015: Effect of agricultural economic growth on sandy desertification in Horqin Sandy Land. *Ecol. Econ.*, **119**, 53–63, doi:10.1016/J.ECOLECON.2015.08.006.

Gebeye, B.A., 2016: Unsustain the sustainable: An evaluation of the legal and policy interventions for pastoral development in Ethiopia. *Pastoralism*, **6:2**, doi:10.1186/s13570-016-0049-x.

Gebreselassie, S., O.K. Kirui, and A. Mirzabaev, 2016: Economics of land degradation and improvement in Ethiopia. In: *Economics of Land Degradation and Improvement – A Global Assessment for Sustainable Development* [Nkonya, E., A. Mirzabaev and J. von Braun (eds.)]. Springer International Publishing, Cham, Switzerland, pp. 401–430.

Geerken, R., and M. Ilaiwi, 2004: Assessment of rangeland degradation and development of a strategy for rehabilitation. *Remote Sens. Environ.*, **90**, 490–504, doi:10.1016/J.RSE.2004.01.015.

Geerken, R., B. Zaitchik, and J.P. Evans, 2005: Classifying rangeland vegetation type and coverage from NDVI time series using Fourier filtered cycle similarity. *Int. J. Remote Sens.*, **26**, 5535–5554, doi:10.1080/01431160500300297.

Geerken, R.A., 2009: An algorithm to classify and monitor seasonal variations in vegetation phenologies and their inter-annual change. *ISPRS J. Photogramm. Remote Sens.*, **64**, 422–431, doi:10.1016/J.ISPRSJPRS.2009.03.001.

Geerts, S., and D. Raes, 2009: Deficit irrigation as an on-farm strategy to maximize crop water productivity in dry areas. *Agric. Water Manag.*, **96**, 1275–1284, doi:10.1016/J.AGWAT.2009.04.009.

Geist, H.H.J., and E.F. Lambin, 2004: Dynamic Causal Patterns of Desertification. *Bioscience*, **54**, 817–829, doi:10.1641/0006-3568(2004)054[0817:DCPOD]2.0.CO;2.

Gemenne, F., 2011: Why the numbers don't add up: A review of estimates and predictions of people displaced by environmental changes. *Glob. Environ. Chang.*, **21**, S41–S49, doi:10.1016/J.GLOENVCHA.2011.09.005.

General Directorate of Combating Desertification and Erosion (ÇEMGM), 2017: *Combating Desertification and Erosion Activities in Turkey*. General Directorate of Combating Desertification and Erosion, Ministry of Forestry and Water Affairs, Republic of Turkey. Ankara, Turkey, 44 pp.

Gerber, N., and A. Mirzabaev, 2017: *Benefits of Action and Costs of Inaction: Drought Mitigation and Preparedness – A Literature Review*. Integrated Drought Management Programme Working Paper No. 1., World Meteorological Organization, Geneva, Switzerland and Global Water Partnership, Stockholm, Sweden, 23 pp.

Gerber, N., E. Nkonya, and J. von Braun, 2014: Land degradation, poverty and marginality. In: *Marginality: Addressing the nexus of poverty, exclusion and ecology* [Von Braun, J., and F.W. Gatzweiler (eds.)]. Springer Netherlands, Dordrecht, Netherlands, pp. 181–202.

German, L., H. Mansoor, G. Alemu, W. Mazengia, T. Amede, and A. Stroud, 2007: Participatory integrated watershed management: Evolution of concepts and methods in an ecoregional program of the eastern African highlands. *Agric. Syst.*, **94**, 189–204, doi:10.1016/J.AGSY.2006.08.008.

Ghazi, S., A. Sayigh, and K. Ip, 2014: Dust effect on flat surfaces – A review paper. *Renew. Sustain. Energy Rev.*, **33**, 742–751, doi:10.1016/J.RSER.2014.02.016.

Gherboudj, I., S. Naseema Beegum, and H. Ghedira, 2017: Identifying natural dust source regions over the Middle-East and North-Africa: Estimation of dust emission potential. *Earth-Science Rev.*, **165**, 342–355, doi:10.1016/J.EARSCIREV.2016.12.010.

Gholami, A., A.A. Alemrajabi, and A. Saboonchi, 2017: Experimental study of self-cleaning property of titanium dioxide and nanospray coatings in solar applications. *Sol. Energy*, **157**, 559–565, doi:10.1016/J.SOLENER.2017.08.075.

Giannadaki, D., A. Pozzer, and J. Lelieveld, 2014: Modeled global effects of airborne desert dust on air quality and premature mortality. *Atmos. Chem. Phys.*, **14**, 957–968, doi:10.5194/acp-14-957-2014.

Giannakopoulos, C. et al., 2009: Climatic changes and associated impacts in the Mediterranean resulting from a 2°C global warming. *Glob. Planet. Change*, **68**, 209–224.

Giannini, A., M. Biasutti, I.M. Held, and A. Sobel, 2008: A global perspective on African climate. *Clim. Change*, **90**, 359–383, doi:10.1007/s10584-008-9396-y.

Gibbs, H.K., and J.M. Salmon, 2015: Mapping the world's degraded lands. *Appl. Geogr.*, **57**, 12–21, doi:10.1016/j.apgeog.2014.11.024.

Gibson, C.C., J.T. Williams, and E. Ostrom, 2005: Local enforcement and better forests. *World Dev.*, **33**, 273–284, doi:10.1016/J.WORLDDEV.2004.07.013.

Gichenje, H., and S. Godinho, 2018: Establishing a land degradation neutrality national baseline through trend analysis of GIMMS NDVI Time-series. *L. Degrad. Dev.*, **29**, 2985–2997, doi:10.1002/ldr.3067.

Giger, M., H. Liniger, C. Sauter, and G. Schwilch, 2015: Economic benefits and costs of sustainable land management technologies: An analysis of WOCAT's global data. *L. Degrad. Dev.*, **29**, 962–974, doi:10.1002/ldr.2429.

Ginoux, P., J.M. Prospero, T.E. Gill, N.C. Hsu, and M. Zhao, 2012: Global-scale attribution of anthropogenic and natural dust sources and their emission rates based on MODIS Deep Blue aerosol products. *Rev. Geophys.*, **50**, doi:10.1029/2012RG000388.

Gitay, H., A. Suárez, R.T. Watson, and D.J. Dokken, 2002: *Climate change and biodiversity*. IPCC Technical Paper V, Intergovernmental Panel on Climate Change, World Meteorological Organization, Geneva, Switzerland. 77 pp.

3

Glendenning, C.J., F.F. Van Ogtrop, A.K. Mishra, and R.W. Vervoort, 2012: Balancing watershed and local scale impacts of rainwater harvesting in India – A review. *Agric. Water Manag.*, **107**, 1–13.

Godfray, H.C.J., and T. Garnett, 2014: Food security and sustainable intensification. *Philos. Trans. R. Soc. B Biol. Sci.*, **369**, 20120273, doi:10.1098/rstb.2012.0273.

Goffner, D., H. Sinare, and L.J. Gordon, 2019: The Great Green Wall for the Sahara and the Sahel Initiative as an opportunity to enhance resilience in Sahelian landscapes and livelihoods. *Reg. Environ. Chang.*, **19**, 1417–1428, doi:10.1007/s10113-019-01481-z.

Gomes, N., 2006: *Access to Water, Pastoral Resource Management and Pastoralists' Livelihoods: Lessons Learned from Water Development in Selected Areas of Eastern Africa (Kenya, Ethiopia, Somalia)*. LSP Working Paper 26, Food and Agriculture Organization of the United Nations, Rome, Italy, 47 pp.

Gonzalez-Martin, C., N. Teigell-Perez, B. Valladares, and D.W. Griffin, 2014: The global dispersion of pathogenic microorganisms by dust storms and its relevance to agriculture. *Adv. Agron.*, **127**, 1–41, doi:10.1016/B978-0-12-800131-8.00001-7.

Goudarzi, G. et al., 2017: Health risk assessment of exposure to the Middle-Eastern dust storms in the Iranian megacity of Kermanshah. *Public Health*, **148**, 109–116, doi:10.1016/J.PUHE.2017.03.009.

Goudie, A., 2013: *The Human Impact on the Natural Environment: Past, Present and Future*. Seventh edition, Wiley-Blackwell, West Sussex, UK, 422 pp.

Goudie, A., and N. Middleton, 2006: *Desert Dust in the Global System*. Springer-Verlag, Berlin, Germany, 287 pp.

Goudie, A.S., 2014: Desert dust and human health disorders. *Environ. Int.*, **63**, 101–113, doi:10.1016/J.ENVINT.2013.10.011.

Goudie, A.S., and N.J. Middleton, 2001: Saharan dust storms: Nature and consequences. *Earth-Science Rev.*, **56**, 179–204.

Grace, J.B., M.D. Smith, S.L. Grace, S.L.C., and T.J. Stohlgren, 2001: Interactions between fire and invasive plants in temperate grasslands of North America. In: *Proceedings of the Invasive Species Workshop: The Role of Fire in the Control and Spread of Invasive Species* [Galley, K.E.M. and T.P. Wilson (eds.)]. Tall Timbers Research Station, Florida, USA, pp. 40–65.

Grainger, A., 2009: The role of science in implementing international environmental agreements: The case of desertification. *L. Degrad. Dev.*, **20**, 410–430, doi:10.1002/ldr.898.

Gray, C., and V. Mueller, 2012a: Drought and population mobility in rural Ethiopia. *World Dev.*, **40**, 134–145, doi:10.1016/J.WORLDDEV.2011.05.023.

Gray, C.L., 2009: Environment, land, and rural out-migration in the southern Ecuadorian Andes. *World Dev.*, **37**, 457–468, doi:10.1016/J.WORLDDEV.2008.05.004.

Gray, C.L., 2010: Gender, natural capital, and migration in the southern Ecuadorian Andes. *Environ. Plan. A*, **42**, 678–696, doi:10.1068/a42170.

Gray, C.L., 2011: Soil quality and human migration in Kenya and Uganda. *Glob. Environ. Chang.*, **21**, 421–430, doi:10.1016/J.GLOENVCHA.2011.02.004.

Gray, C.L., and V. Mueller, 2012b: Natural disasters and population mobility in Bangladesh. *Proc. Natl. Acad. Sci. U.S.A.*, **109**, 6000–6005, doi:10.1073/pnas.1115944109.

Gray, C.L., and R.E. Bilsborrow, 2014: Consequences of out-migration for land use in rural Ecuador. *Land Use Policy*, **36**, 182–191, doi:10.1016/J.LANDUSEPOL.2013.07.006.

Greene, R., W. Timms, P. Rengasamy, M. Arshad, and R. Cresswell, 2016: Soil and aquifer salinization: Toward an integrated approach for salinity management of groundwater. In: *Integrated Groundwater Management* [Jakeman, A.J., O. Barreteau, R.J. Hunt, J. Rinaudo and A. Ross (eds.)]. Springer International Publishing, Cham, Switzerland, pp. 377–412.

Greve, P., and S.I. Seneviratne, 1999: Assessment of future changes in water availability and aridity. *Geophys. Res. Lett.*, **42**, 5493–5499, doi:10.1002/2015GL064127.

Greve, P., B. Orlowsky, B. Mueller, J. Sheffield, M. Reichstein, and S.I. Seneviratne, 2014: Global assessment of trends in wetting and drying over land. *Nat. Geosci.*, **7**, 716–721, doi:10.1038/ngeo2247.

Greve, P., M.L. Roderick, and S.I. Seneviratne, 2017: Simulated changes in aridity from the last glacial maximum to 4xCO$_2$. *Environ. Res. Lett.*, **12**, 114021, doi:10.1088/1748-9326/aa89a3.

Griffin, D.W., 2007: Atmospheric movement of microorganisms in clouds of desert dust and implications for human health. *Clin. Microbiol. Rev.*, **20**, 459–477, doi:10.1128/CMR.00039-06.

Griffin, D.W., D.L. Westphal, and M.A. Gray, 2006: Airborne microorganisms in the African desert dust corridor over the mid-Atlantic ridge, Ocean Drilling Program, Leg 209. *Aerobiologia (Bologna).*, **22**, 211–226, doi:10.1007/s10453-006-9033-z.

Groneman, A., 1968: *The Soils of the Wind Erosion Control Camp Area Karapinar, Turkey*. Publication Series No. 472, Wageningen Agricultural University, Wageningen, Netherlands, 161 pp.

Guan, C., X. Li, P. Zhang, and C. Li, 2018: Effect of global warming on soil respiration and cumulative carbon release in biocrust-dominated areas in the Tengger Desert, northern China. *J. Soils Sediments*, **19**, 1161–1170, doi:10.1007/s11368-018-2171-4.

Guerra, C.A., J. Maes, I. Geijzendorffer, and M.J. Metzger, 2016: An assessment of soil erosion prevention by vegetation in Mediterranean Europe: Current trends of ecosystem service provision. *Ecol. Indic.* **60**, 213–222, doi:10.1016/j.ecolind.2015.06.043.

Guerrero-Baena, M., J. Gómez-Limón, M.D. Guerrero-Baena, and J.A. Gómez-Limón, 2019: Insuring water supply in irrigated agriculture: A proposal for hydrological drought index-based insurance in Spain. *Water*, **11**, 686, doi:10.3390/w11040686.

Gunin, P., and E. Pankova, 2004: Contemporary processes of degradation and desertification of ecosystems of the East Asian sector of steppes and forest-steppes. In: *Modern Global Changes of the Natural Environment*. Scientific World, Moscow, Russia, pp. 389–412.

Guodaar, L. et al., 2019: How do climate change adaptation strategies result in unintended maladaptive outcomes? Perspectives of tomato farmers. *Int. J. Veg. Sci.*, 1–17, doi:10.1080/19315260.2019.1573393.

Gupta, A.K. et al., 2016: Theory of open inclusive innovation for reciprocal, responsive and respectful outcomes: Coping creatively with climatic and institutional risks. *J. Open Innov. Technol. Mark. Complex.*, **2**, 16, doi:10.1186/s40852-016-0038-8.

Gupta, R. et al., 2009: *Research prospectus: A vision for sustainable land management research in Central Asia*. ICARDA Central Asia Caucasus program. Sustainable Agriculture in Central Asia and the Caucasus Series No.1, CGIAR-PFU, Tashkent, Uzbekistan, 84 pp.

Gurung, J.D., S. Mwanundu, A. Lubbock, M. Hartl, I. and Firmian, and IFAD, 2006: *Gender and Desertification: Expanding Roles for Women to Restore Dryland Areas*. International Fund for Agricultural Development (IFAD), Rome, Italy, 27 pp.

Gutiérrez-Elorza, M., 2006: Erosión e influencia del cambio climático. *Rev. C G*, **20**, 45–59.

Guzman, C.D. et al., 2018: Developing soil conservation strategies with technical and community knowledge in a degrading sub-humid mountainous landscape. *L. Degrad. Dev.*, **29**, 749–764, doi:10.1002/ldr.2733.

Haan, C. De, E. Dubern, B. Garancher, and C. Quintero, 2016: *Pastoralism Development in the Sahel: A Road to Stability?* World Bank Group, Washington DC, USA, 61 pp.

Haase, P. et al., 2018: The next generation of site-based long-term ecological monitoring: Linking essential biodiversity variables and ecosystem integrity. *Sci. Total Environ.*, **613–614**, 1376–1384, doi:10.1016/J.SCITOTENV.2017.08.111.

Haggblade, S., P. Hazell, and T. Reardon, 2010: The rural non-farm economy: Prospects for growth and poverty reduction. *World Dev.*, **38**, 1429–1441, doi:10.1016/J.WORLDDEV.2009.06.008.

3

Hallegatte, S., 2012: *A Cost Effective Solution to Reduce Disaster Losses in Developing Countries: Hydro-Meteorological Services, Early Warning, and Evacuation*. Policy Research Working Paper no. WPS 6058, The World Bank, Washington DC, USA, 22 pp.

Hallegatte, S., and J. Rozenberg, 2017: Climate change through a poverty lens. *Nat. Clim. Chang.*, **7**, 250–256, doi:10.1038/nclimate3253.

Hallegatte, S. et al., 2016: *Shock Waves: Managing the Impacts of Climate Change on Poverty*. The World Bank, Washington DC, USA, 207 pp.

Halliday, T., 2006: Migration, risk, and liquidity constraints in El Salvador. *Econ. Dev. Cult. Change*, **54**, 893–925, doi:10.1086/503584.

Hamidov, A., K. Helming, and D. Balla, 2016: Impact of agricultural land use in Central Asia: A review. *Agron. Sustain. Dev.*, **36**, 6, doi:10.1007/s13593-015-0337-7.

Han, G., X. Hao, M. Zhao, M. Wang, B.H. Ellert, W. Willms, and M. Wang, 2008: Effect of grazing intensity on carbon and nitrogen in soil and vegetation in a meadow steppe in Inner Mongolia. *Agric. Ecosyst. Environ.*, **125**, 21–32, doi:10.1016/j.agee.2007.11.009.

Han, Q., G. Luo, C. Li, A. Shakir, M. Wu, and A. Saidov, 2016: Simulated grazing effects on carbon emission in Central Asia. *Agric. For. Meteorol.*, **216**, 203–214, doi:10.1016/J.AGRFORMET.2015.10.007.

Hanjra, M.A., J. Blackwell, G. Carr, F. Zhang, and T.M. Jackson, 2012: Wastewater irrigation and environmental health: Implications for water governance and public policy. *Int. J. Hyg. Environ. Health*, **215**, 255–269, doi:10.1016/J.IJHEH.2011.10.003.

Hao, Y. et al., 2008: CO_2, H_2O and energy exchange of an Inner Mongolia steppe ecosystem during a dry and wet year. *Acta Oecologica*, **33**, 133–143, doi:10.1016/j.actao.2007.07.002.

Harou, J.J. et al., 2010: Economic consequences of optimized water management for a prolonged, severe drought in California. *Water Resour. Res.*, **46**, doi:10.1029/2008WR007681.

Harpold, A.A., and P.D. Brooks, 2018: Humidity determines snowpack ablation under a warming climate. *Proc. Natl. Acad. Sci. U.S.A.*, **115**, 1215–1220, doi:10.1073/pnas.1716789115.

Harris, D., and A. Orr, 2014: Is rainfed agriculture really a pathway from poverty? *Agric. Syst.*, **123**, 84–96, doi:10.1016/J.AGSY.2013.09.005.

Hartmann, B., 2010: Rethinking climate refugees and climate conflict: Rhetoric, reality and the politics of policy discourse. *J. Int. Dev.*, **22**, 233–246, doi:10.1002/jid.1676.

Hasanean, H., M. Almazroui, H. Hasanean, and M. Almazroui, 2015: Rainfall: Features and variations over Saudi Arabia, a review. *Climate*, **3**, 578–626, doi:10.3390/cli3030578.

Havstad, K.M., L.F. Huenneke, and W.H. Schlesinger (eds.), 2006: *Structure and Function of a Chihuahuan Desert Ecosystem: The Jornada Basin Long-Term Ecological Research Site*. Oxford University Press, New York, USA, 247 pp.

He, J., and R. Lang, 2015: Limits of state-led programs of payment for ecosystem services: Field evidence from the sloping land conversion program in south-west China. *Hum. Ecol.*, **43**, 749–758, doi:10.1007/s10745-015-9782-9.

He, J., and T. Sikor, 2015: Notions of justice in payments for ecosystem services: Insights from China's sloping land conversion program in Yunnan Province. *Land Use Policy*, **43**, 207–216, doi:10.1016/J.LANDUSEPOL.2014.11.011.

Headey, D., A.S. Taffesse, and L. You, 2014: Diversification and development in pastoralist Ethiopia. *World Dev.*, **56**, 200–213, doi:10.1016/J.WORLDDEV.2013.10.015.

Hein, L., and N. De Ridder, 2006: Desertification in the Sahel: A reinterpretation. *Glob. Chang. Biol.*, **12**, 751–758, doi:10.1111/j.1365-2486.2006.01135.x.

Helldén, U., and C. Tottrup, 2008: Regional desertification: A global synthesis. *Glob. Planet. Change*, **64**, 169–176, doi:10.1016/j.gloplacha.2008.10.006.

Hellmann, J.J., J.E. Byers, B.G. Bierwagen, and J.S. Dukes, 2008: Five potential consequences of climate change for invasive species. *Conserv. Biol.*, **22**, 534–543, doi:10.1111/j.1523-1739.2008.00951.x.

Henderson, J.V., A. Storeygard, and U. Deichmann, 2017: Has climate change driven urbanisation in Africa? *J. Dev. Econ.*, **124**, 60–82, doi:10.1016/J.JDEVECO.2016.09.001.

Henry, S., B. Schoumaker, and C. Beauchemin, 2004: The impact of rainfall on the first out-migration: A multi-level event-history analysis in Burkina Faso. *Popul. Environ.*, **25**, 423–460, doi:10.1023/B:POEN.0000036928.17696.e8.

Hernandez, R.R., and D.R. Sandquist, 2011: Disturbance of biological soil crust increases emergence of exotic vascular plants in California sage scrub. *Plant Ecol.*, **212**, 1709–1721, doi:10.1007/s11258-011-9943-x.

Herrera, C. et al., 2018: Science of the total environment groundwater origin and recharge in the hyper-arid Cordillera de la Costa, Atacama Desert, northern Chile. *Sci. Total Environ.*, **624**, 114–132, doi:10.1016/j.scitotenv.2017.12.134.

Herrero, S.T., 2006: Desertification and environmental security: The case of conflicts between farmers and herders in the arid environments of the Sahel. In: *Desertification in the Mediterranean Region. A Security Issue* [Kepner W.G., J.L. Rubio, D.A. Mouat and F. Pedrazzini (eds.)]. NATO Security Series, Vol. 3, Springer Netherlands, Dordrecht, Netherlands, pp. 109–132.

Herrick, J.E. et al., 2017: Two new mobile apps for rangeland inventory and monitoring by landowners and land managers. *Rangelands*, **39**, 46–55, doi:10.1016/j.rala.2016.12.003.

Herrmann, S.M., and C.F. Hutchinson, 2005: The changing contexts of the desertification debate. *J. Arid Environ.*, **63**, 538–555, doi:10.1016/J.JARIDENV.2005.03.003.

Herrmann, S.M., and G.G. Tappan, 2013: Vegetation impoverishment despite greening: A case study from central Senegal. *J. Arid Environ.*, **90**, 55–66, doi:10.1016/J.JARIDENV.2012.10.020.

Herrmann, S.M., and T.K. Sop, 2016: The map is not the territory: How satellite remote sensing and ground evidence have re-shaped the image of Sahelian desertification. In: *The End of Desertification?* [Behnke, R. and M. Mortimore (eds.)]. Springer Earth System Sciences, Springer, Berlin, Germany, pp. 117–145.

Hertel, T.W., and D.B. Lobell, 2014: Agricultural adaptation to climate change in rich and poor countries: Current modeling practice and potential for empirical contributions. *Energy Econ.*, **46**, 562–575, doi:10.1016/J.ENECO.2014.04.014.

Hertel, T.W., M.B. Burke, and D.B. Lobell, 2010: The poverty implications of climate-induced crop yield changes by 2030. *Glob. Environ. Chang.*, **20**, 577–585, doi:10.1016/J.GLOENVCHA.2010.07.001.

Hetem, R.S., S.K. Maloney, A. Fuller, and D. Mitchell, 2016: Heterothermy in large mammals: Inevitable or implemented? *Biol. Rev.*, **91**, 187–205, doi:10.1111/brv.12166.

Hiernaux, P., and H.N. Le Houérou, 2006: The rangelands of the Sahel. *Sécheresse 17(1)* 17, pp. 51–71.

Hiernaux, P. et al., 2009: Trends in productivity of crops, fallow and rangelands in Southwest Niger: Impact of land use, management and variable rainfall. *J. Hydrol.*, **375**, 65–77, doi:10.1016/J.JHYDROL.2009.01.032.

Higginbottom, T., and E. Symeonakis, 2014: Assessing land degradation and desertification using vegetation index data: Current frameworks and future directions. *Remote Sens.*, **6**, 9552–9575, doi:10.3390/rs6109552.

Higgins, D., T. Balint, H. Liversage, and P. Winters, 2018: Investigating the impacts of increased rural land tenure security: A systematic review of the evidence. *J. Rural Stud.*, **61**, 34–62, doi:10.1016/J.JRURSTUD.2018.05.001.

Hijioka, Y., E. Lin, J.J. Pereira, R.T. Corlett, X. Cui, G.E. Insarov, R.D. Lasco, E. Lindgren, and A. Surjan, 2014: Asia. In: Climate Change 2014: Impacts, Adaptation, and Vulnerability. Part B: Regional Aspects. Contribution of Working Group II to the Fifth Assessment Report of the Intergovernmental Panel on Climate Change [Barros, V.R., C.B. Field, D.J. Dokken, M.D. Mastrandrea, K.J. Mach, T.E. Bilir, M. Chatterjee, K.L. Ebi, Y.O. Estrada, R.C. Genova, B. Girma, E.S. Kissel, A.N. Levy, S. MacCracken, P.R. Mastrandrea, and L.L.White (eds.)]. Cambridge University Press, New York, USA, pp. 1327–1370.

Hirche, A., M. Salamani, A. Abdellaoui, S. Benhouhou, and J.M. Valderrama, 2011: Landscape changes of desertification in arid areas: The case of south-west Algeria. *Environ. Monit. Assess.*, **179**, 403–420, doi:10.1007/s10661-010-1744-5.

Hirche, A. et al., 2018: The Maghreb (North Africa) rangelands evolution for forty years: Regreening or degradation? In: *Desertification: Past, Current and Future Trends* [Squires, V.R. and A. Ariapour (eds.)]. Nova Science Publishers, New York, USA, pp. 73–106.

Hochman, Z., D.L. Gobbett, and H. Horan, 2017: Climate trends account for stalled wheat yields in Australia since 1990. *Glob. Chang. Biol.*, **23**, 2071–2081, doi:10.1111/gcb.13604.

Hoegh-Guldberg, O., D. Jacob, M. Taylor, M. Bindi, S. Brown, I. Camilloni, A. Diedhiou, R. Djalante, K.L. Ebi, F. Engelbrecht, J.Guiot, Y. Hijioka, S. Mehrotra, A. Payne, S.I. Seneviratne, A. Thomas, R. Warren, and G. Zhou, 2018: Impacts of 1.5°C of Global Warming on Natural and Human Systems. In: Global Warming of 1.5°C. An IPCC Special Report on the impacts of global warming of 1.5°C above pre-industrial levels and related global greenhouse gas emission pathways, in the context of strengthening the global response to the threat of climate change, sustainable development, and efforts to eradicate poverty [Masson-Delmotte, V., P. Zhai, H.-O. Pörtner, D. Roberts, J. Skea, P.R. Shukla, A. Pirani, W. Moufouma-Okia, C. Péan, R. Pidcock, S. Connors, J.B.R. Matthews, Y. Chen, X. Zhou, M.I.Gomis, E. Lonnoy, T.Maycock, M.Tignor, and T. Waterfield (eds.)]. In press. pp. 175–311.

Hoerling, M. et al., 2006: Detection and attribution of 20th century northern and southern African. *Clim. Res.*, **19**, 3989–4008, doi:10.1175/JCLI3842.1.

Hoffmann, U., A. Yair, H. Hikel, and N.J. Kuhn, 2012: Soil organic carbon in the rocky desert of northern Negev (Israel). *J. Soils Sediments*, **12**, 811–825, doi:10.1007/s11368-012-0499-8.

Holden, S.R., A. Gutierrez, and K.K. Treseder, 2013: Changes in soil fungal communities, extracellular enzyme activities, and litter decomposition across a fire chronosequence in Alaskan boreal forests. *Ecosystems*, **16**, 34–46, doi:10.1007/s10021-012-9594-3.

Holden, S.T., and H. Ghebru, 2016: Land tenure reforms, tenure security and food security in poor agrarian economies: Causal linkages and research gaps. *Glob. Food Sec.*, **10**, 21–28, doi:10.1016/J.GFS.2016.07.002.

Hong, L., N. Zhou, D. Fridley, and C. Raczkowski, 2013: Assessment of China's renewable energy contribution during the 12th Five Year Plan. *Energy Policy*, **62**, 1533–1543, doi:10.1016/J.ENPOL.2013.07.110.

Hopke, P.K. et al., 2018: Spatial and temporal trends of short-term health impacts of PM 2.5 in Iranian cities; A modelling approach (2013–2016). *Aerosol Air Qual. Res.*, **18**, 497–504, doi:10.4209/aaqr.2017.09.0325.

Hopkins, A., and A. Del Prado, 2007: Implications of climate change for grassland in Europe: Impacts, adaptations and mitigation options: A review. *Grass Forage Sci.*, **62**, 118–126, doi:10.1111/j.1365-2494.2007.00575.x.

Horion, S., A.V. Prishchepov, J. Verbesselt, K. de Beurs, T. Tagesson, and R. Fensholt, 2016: Revealing turning points in ecosystem functioning over the northern Eurasian agricultural frontier. *Glob. Chang. Biol.*, **22**, 2801–2817, doi:10.1111/gcb.13267.

Hornbeck, R., 2012: The enduring impact of the American Dust Bowl: Short- and long-run adjustments to environmental catastrophe. *Am. Econ. Rev.*, **102**, 1477–1507, doi:10.1257/aer.102.4.1477.

Le Houérou, H.N., 1996: Climate change, drought and desertification. *J. Arid Environ.*, **34**, 133–185, doi:10.1006/jare.1996.0099.

Houyou, Z., C.L. Bielders, H.A. Benhorma, A. Dellal, and A. Boutemdjet, 2016: Evidence of strong land degradation by wind erosion as a result of rainfed cropping in the Algerian steppe: A case study at Laghouat. *L. Degrad. Dev.*, **27**, 1788–1796, doi:10.1002/ldr.2295.

Howe, C., H. Suich, P. van Gardingen, A. Rahman, and G.M. Mace, 2013: Elucidating the pathways between climate change, ecosystem services and poverty alleviation. *Curr. Opin. Environ. Sustain.*, **5**, 102–107, doi:10.1016/J.COSUST.2013.02.004.

Hoyle, F.C., M.D. D'Antuono, T. Overheu, and D.V. Murphy, 2013: Capacity for increasing soil organic carbon stocks in Dryland agricultural systems. *Soil Res.*, **51**, 657–667.

Hsiang, S.M., M. Burke, and E. Miguel, 2013: Quantifying the influence of climate on human conflict. *Science*, **341**, doi:10.1126/science.1235367.

Hu, Y., and Q. Fu, 2007: Observed poleward expansion of the Hadley circulation since 1979. *Atmos. Chem. Phys.*, **7**, 5229–5236, doi:10.5194/acp-7-5229-2007.

Huang, D., K. Wang, and W.L. Wu, 2007: Dynamics of soil physical and chemical properties and vegetation succession characteristics during grassland desertification under sheep grazing in an agropastoral transition zone in northern China. *J. Arid Environ.*, **70**, 120–136, doi:10.1016/j.jaridenv.2006.12.009.

Huang, J., Q. Fu, J. Su, Q. Tang, P. Minnis, Y. Hu, Y. Yi, and Q. Zhao, 2009: Taklimakan dust aerosol radiative heating derived from CALIPSO observations using the Fu-Liou radiation model with CERES constraints. *Atmos. Chem. Phys.*, **9**, 4011–4021, doi:10.5194/acp-9-4011-2009.

Huang, J., T. Wang, W. Wang, Z. Li, and H. Yan, 2014: Climate effects of dust aerosols over East Asian arid and semi-arid regions. *J. Geophys. Res. Atmos.*, **119**, 11, 311–398, 416, doi:10.1002/2014JD021796.

Huang, J., H. Yu, X. Guan, G. Wang, and R. Guo, 2016a: Accelerated Dryland expansion under climate change. *Nat. Clim. Chang.*, **6**, 166–171, doi:10.1038/nclimate2837.

Huang, J., H. Yu, A. Dai, Y. Wei, and L. Kang, 2017: Drylands face potential threat under 2°C global warming target. *Nat. Clim. Chang.*, **7**, 417–422, doi:10.1038/nclimate3275.

Huang, J. et al., 2017: Dryland climate change: Recent progress and challenges. *Rev. Geophys.*, **55**, 719–778, doi:10.1002/2016RG000550.

Huber, S., and R. Fensholt, 2011: Analysis of teleconnections between AVHRR-based sea surface temperature and vegetation productivity in the semi-arid Sahel. *Remote Sens. Environ.*, **115**, 3276–3285, doi:10.1016/J.RSE.2011.07.011.

Huber, S., R. Fensholt, and K. Rasmussen, 2011: Water availability as the driver of vegetation dynamics in the African Sahel from 1982 to 2007. *Glob. Planet. Change*, **76**, 186–195, doi:10.1016/J.GLOPLACHA.2011.01.006.

Hudak, A.T., C.A. Wessman, and T.R. Seastedt, 2003: Woody overstorey effects on soil carbon and nitrogen pools in a South African savanna. *Austral Ecol.*, **28**, 173–181. doi:10.1046/j.1442-9993.2003.01265.x.

Hughes, R.F. et al., 2006: Changes in aboveground primary production and carbon and nitrogen pools accompanying woody plant encroachment in a temperate savanna. *Glob. Chang. Biol.*, **12**, 1733–1747, doi:10.1111/j.1365-2486.2006.01210.x.

Hugo, G., 2008: *Migration, Development and Environment*. IOM Migration Research Series, 35, International Organization for Migration, ISSN 1607-338X, Geneva, Switzerland. 63 pp.

Hull, A.C. and G. Stewart, 1949: Replacing Cheatgrass by Reseeding with Perennial Grass on Southern Idaho Ranges. *American Society of Agronomy Journal*, **40**, 694–703.

Hunter, L.M., J.K. Luna, and R.M. Norton, 2015: Environmental dimensions of migration. *Annu. Rev. Sociol.*, **41**, 377–397, doi:10.1146/annurev-soc-073014-112223.

Hurlbert, M. and P. Mussetta, 2016: Creating resilient water governance for irrigated producers in Mendoza, Argentina. *Environ. Sci. Policy*, **58**, 83–94, doi:10.1016/j.envsci.2016.01.004.

Hurlbert, M.A., 2018: *Adaptive Governance of Disaster: Drought and Flood in Rural Areas*. Springer International Publishing, Springer. Cham, Switzerland. ISBN 978-3-319-57801-9, 239 pp.

Hussein, I.A.E., 2011: Desertification process in Egypt. In: *Coping with Global Environmental Change, Disasters and Security: Threats, Challenges, Vulnerabilities and Risks* [Brauch, H.G., U. Oswald Spring, C. Mesjasz, J. Grin, P. Kameri-Mbote, B. Chourou, P. Dunay and J. Brikmann (eds.)]. Springer, Berlin, Germany, pp. 863–874.

3

Ibragimov, N. et al., 2007: Water use efficiency of irrigated cotton in Uzbekistan under drip and furrow irrigation. *Agric. Water Manag.*, **90**, 112–120, doi:10.1016/j.agwat.2007.01.016.

Ibrahim, H.I.M., 2016: Tolerance of two pomegranates cultivars (Punica granatum L.) to salinity stress under hydroponic conditions. *J. Basic Appl. Sci. Res*, **6**, 38–46. ISSN 2090-4304.

Ibrahim, Y., H. Balzter, J. Kaduk, and C. Tucker, 2015: Land degradation assessment using residual trend analysis of GIMMS NDVI3g, soil moisture and rainfall in Sub-Saharan West Africa from 1982 to 2012. *Remote Sens.*, **7**, 5471–5494, doi:10.3390/rs70505471.

IEA, 2013: *World Energy Outlook 2013*. International Energy Agency, Paris, France. ISBN: 978-92-64-20130-9, 687 pp.

Ifejika Speranza, C., B. Kiteme, P. Ambenje, U. Wiesmann, and S. Makali, 2010: Indigenous knowledge related to climate variability and change: Insights from droughts in semi-arid areas of former Makueni District, Kenya. *Clim. Change*, **100**, 295–315, doi:10.1007/s10584-009-9713-0.

Iiyama, M., P. Dobie, M. Njenga, G. Ndegwa, and R. Jamnadass, 2014: The potential of agroforestry in the provision of sustainable woodfuel in Sub-Saharan Africa. *Curr. Opin. Environ. Sustain.*, **6**, 138–147, doi:10.1016/J.COSUST.2013.12.003.

Iknayan, K.J., and S.R. Beissinger, 2018: Collapse of a desert bird community over the past century driven by climate change. *Proc. Natl. Acad. Sci. U.S.A.*, **115**, 8597–8602, doi:10.1073/pnas.1805123115.

Imeson, A., 2012: *Desertification, Land Degradation and Sustainability*. Wiley-Blackwell, Oxford, UK. ISBN:9780470714485, DOI:10.1002/9781119977759, 326 pp.

Inbar, M., 2007: Importance of drought information in monitoring and assessing land degradation. In: *Climate and Land Degradation* [Sivakumar, M.V.K. and N. Ndiang'ui (eds.)]. Springer Berlin Germany, pp. 253–266.

Indoitu, R. et al., 2015: Dust emission and environmental changes in the dried bottom of the Aral Sea. *Aeolian Res.*, **17**, 101–115, doi:10.1016/J.AEOLIA.2015.02.004.

Infante, F., 2017: The role of social capital and labour exchange in the soils of Mediterranean Chile. *Rural Soc.*, **26**, 107–124, doi:10.1080/10371656.2017.1330837.

IPBES, 2018: *Assessment Report on Land Degradation and Restoration* [Montanarella, L., R. Scholes., and A. Brainich. (eds.)]. Secretariate of the Intergovernmental Science-Policy Platform on Biodiversity and Ecosystem Services, Bonn, Germany, 686 pp.

IPCC, 2007: Climate Change 2007: Impacts, Adaptation and Vulnerability. Contribution of Working Group II to the Fourth Assessment Report of the Intergovernmental Panel on Climate Change [Parry, M.L., O.F. Canziani, J.P. Palutikof, P.J. van der Linden and C.E. Hanson (eds.)]. Cambridge University Press, Cambridge, United Kingdom, 976 pp.

IPCC, 2012: Managing the Risks of Extreme Events and Disasters to Advance Climate Change Adaptation. A Special Report of Working Groups I and II of the Intergovernmental Panel on Climate Change [Field, C.B., V. Barros, T.F. Stocker, D. Qin, D.J. Dokken, K.L. Ebi, M.D. Mastrandrea, K.J. Mach, G.-K. Plattner, S.K. Allen, M. Tignor, and P.M. Midgley (eds.)]. Cambridge University Press, Cambridge, UK, and New York, NY, USA. ISBN 978-1-107-02506-6, 582 pp.

IPCC, 2013: Climate Change 2013: The Physical Science Basis. Contribution of Working Group I to the Fifth Assessment Report of the Intergovernmental Panel on Climate Change [Stocker, T.F., D. Qin, G.-K. Plattner, M. Tignor, S.K. Allen, J. Boschung, A. Nauels, Y. Xia, V. Bex and P.M. Midgley (eds.)]. Cambridge University Press, Cambridge, United Kingdom and New York, USA.

IPCC, 2014: Annex II: Glossary. In: Climate Change 2014: Synthesis Report. Contribution of Working Groups I, II and III to the Fifth Assessment Report of the Intergovernmental Panel on Climate Change [R.K. Pachauri and L.A. Meyer (eds.)]. IPCC, Geneva, Switzerland, 117–130.

Irshad, M. et al., 2007: Land desertification – An emerging threat to environment and food security of Pakistan. *J. Appl. Sci.*, **7**, 1199–1205, doi:10.3923/jas.2007.1199.1205.

Islam, M.N., and M. Almazroui, 2012: Direct effects and feedback of desert dust on the climate of the Arabian Peninsula during the wet season: A regional climate model study. *Clim. Dyn.*, **39**, 2239–2250, doi:10.1007/s00382-012-1293-4.

Issanova, G., and J. Abuduwaili, 2017: Relationship between storms and land degradation. In: *Aeolian Process as Dust Storms in the Deserts of Central Asia and Kazakhstan* [Issanova, G. and J. Abuduwaili (eds.)]. Springer, Singapore, pp. 71–86.

Ivanov, A.L. et al., 2018: National report global climate and soil cover of Russia: Assessment of risks and environmental and economic consequences of land degradation. Adaptive systems and technologies of environmental management (agriculture and forestry). In: *Russian: NATSIONAL'N*. Moscow, Russia, 357 pp.

Jackson, H., and S.D. Prince, 2016: Degradation of net primary production in a semi-arid rangeland. *Biogeosciences*, **13**, 4721–4734, doi:10.5194/bg-13-4721-2016.

Jackson, R.B., J.L. Banner, E.G. Jobbagy, W.T. Pockman, and D.H. Wall, 2002: Ecosystem carbon loss with woody plant invasion of grasslands. *Nature*, **418**, 623–626, doi:10.1038/nature00910.

Jan, S.U. et al., 2017: Analysis of salinity tolerance potential in synthetic hexaploid wheat. *Pak. J. Bot*, **49**, 1269–1278.

Jedwab, R., and D. Vollrath, 2015: Urbanization without growth in historical perspective. *Explor. Econ. Hist.*, **58**, 1–21, doi:10.1016/J.EEH.2015.09.002.

Jeon, E.M. et al., 2011: Impact of Asian dust events on airborne bacterial community assessed by molecular analyses. *Atmos. Environ.*, **45**, 4313–4321, doi:10.1016/j.atmosenv.2010.11.054.

Jerneck, A., and L. Olsson, 2008: Adaptation and the poor: Development, resilience and transition. *Clim. Policy*, **8**, 170–182, doi:10.3763/cpol.2007.0434.

Ji, M., J. Huang, Y. Xie, and J. Liu, 2015: Comparison of Dryland climate change in observations and CMIP5 simulations. *Adv. Atmos. Sci.*, **32**, 1565–1574, doi:10.1007/s00376-015-4267-8.

Jiang, H., 2016: Taking down the 'Great Green Wall': The science and policy discourse of desertification and its control in China. In: *The End of Desertification? Disputing Environmental Change in the Drylands* [Behnke R. and M. Mortimore (eds.)]. Springer, Berlin, Germany, pp. 513–536.

Jiang, L., G. Jiapaer, A. Bao, H. Guo, and F. Ndayisaba, 2017: Vegetation dynamics and responses to climate change and human activities in Central Asia. *Sci. Total Environ.*, **599–600**, 967–980, doi:10.1016/J.SCITOTENV.2017.05.012.

Jiang, Z., Y. Lian, and X. Qin, 2014: Rocky desertification in Southwest China: Impacts, causes, and restoration. *Earth-Science Rev.*, **132**, 1–12, doi:10.1016/J.EARSCIREV.2014.01.005.

Jickells, T.D. et al., 2005: Global iron connections between desert dust, ocean biogeochemistry, and climate. *Science*, **308**, 67–71, doi:10.1126/science.1105959.

Jilali, A., 2014: Impact of climate change on the Figuig aquifer using a numerical model: Oasis of Eastern Morocco. *J. Biol. Earth Sci.*, **4**, 16–24.

Jiménez Cisneros, B.E., T. Oki, N.W. Arnell, G. Benito, J.G. Cogley, P. Döll, T. Jiang, and S.S. Mwakalila, 2014: Freshwater resources. In: Climate Change 2014: Impacts, Adaptation, and Vulnerability. Part A: Global and Sectoral Aspects. Contribution of Working Group II to the Fifth Assessment Report of the Intergovernmental Panel on Climate Change [Field, C.B., V.R. Barros, D.J. Dokken, K.J. Mach, M.D. Mastrandrea, T.E. Bilir, M. Chatterjee, K.L. Ebi, Y.O. Estrada, R.C. Genova, B. Girma, E.S. Kissel, A.N. Levy, S. MacCracken, P.R. Mastrandrea, and L.L.White (eds.)]. Cambridge University Press, New York, USA, pp. 229–269.

Jin, Z. et al., 2014: Natural vegetation restoration is more beneficial to soil surface organic and inorganic carbon sequestration than tree plantation on the Loess Plateau of China. *Sci. Total Environ.*, **485–486**, 615–623, doi:10.1016/J.SCITOTENV.2014.03.105.

Jish Prakash, P., G. Stenchikov, S. Kalenderski, S. Osipov, and H. Bangalath, 2015: The impact of dust storms on the Arabian Peninsula and the Red Sea. *Atmos. Chem. Phys.*, **15**, 199–222, doi:10.5194/acp-15-199-2015.

Johannsen, I. et al., 2016: Future of water supply and demand in the middle Drâa Valley, Morocco, under climate and land use change. *Water*, **8**, 313, doi:10.3390/w8080313.

Johanson, C.M., Q. Fu, C.M. Johanson, and Q. Fu, 2009: Hadley cell widening: Model simulations versus observations. *J. Clim.*, **22**, 2713–2725, doi:10.1175/2008JCLI2620.1.

Johnstone, S., and J. Mazo, 2011: Global warming and the Arab Spring. *Survival (Lond).*, **53**, 11–17, doi:10.1080/00396338.2011.571006.

Jolly, I.D., K.L. Mcewan, and K.L. Holland, 2008: A review of groundwater – Surface water interactions in arid/semi-arid wetlands and the consequences of salinity for wetland ecology. *Ecohydrology*, **1**, 43–58, doi:10.1002/eco.6.

Jolly, W.M. et al., 2015: Climate-induced variations in global wildfire danger from 1979 to 2013. *Nat. Commun.*, **6**, 1–11, doi:10.1038/ncomms8537.

Jones, M.D. et al., 2019: 20,000 years of societal vulnerability and adaptation to climate change in Southwest Asia. *Wiley Interdiscip. Rev. Water*, **6**, e1330, doi:10.1002/wat2.1330.

de Jong, R., S. de Bruin, A. de Wit, M.E. Schaepman, and D.L. Dent, 2011: Analysis of monotonic greening and browning trends from global NDVI time-series. *Remote Sens. Environ.*, **115**, 692–702, doi:10.1016/J.RSE.2010.10.011.

de Jong, R., J. Verbesselt, A. Zeileis, and M. Schaepman, 2013: Shifts in global vegetation activity trends. *Remote Sens.*, **5**, 1117–1133, doi:10.3390/rs5031117.

Joy, K.J., S. Paranjape, A.K. Kirankumar, R. Lele, and R. Adagale, 2004: *Watershed Development Review: Issues and Prospects*. Technical Report, Centre for Interdisciplinary Studies in Environment and Development, Bangalore, India, 145 pp.

Joy, K.J., A. Shah, S. Paranjape, S. Badiger, and S. Lélé, 2006: Issues in restructuring. *Econ. Polit. Wkly.*, July 8-15: 2994–2996, doi:10.2307/4418434.

Júnior, W.S.F., F.R. Santoro, I. Vandebroek, and U.P. Albuquerque, 2016: Urbanization, modernization, and nature knowledge. In: *Introduction to Ethnobiology* [Albuquerque, U. and R. Nobrega Alves (eds.)]. Springer International Publishing, Cham, Switzerland, pp. 251–256.

Jury, M.R., 2016: Climate influences on upper Limpopo River flow. *Water SA*, **42**, 63, doi:10.4314/wsa.v42i1.08.

Kadirbeyoglu, Z., and G. Özertan, 2015: Power in the governance of common-pool resources: A comparative analysis of irrigation management decentralization in Turkey. *Environ. Policy Gov.*, **25**, 157–171, doi:10.1002/eet.1673.

Kahinda, J.M., Meissner, R., and Engelbrecht, F.A., 2016: Implementing integrated catchment management in the upper Limpopo River Basin: A situational assessment. *Phys. Chem. Earth*, **93**, 104–118, doi:10.1016/j.pce.2015.10.003.

Kaijser, A., and A. Kronsell, 2014: Climate change through the lens of intersectionality. *Env. Polit.*, **23**, 417–433, doi:10.1080/09644016.2013.835203.

Kamali, B., K.C. Abbaspour, B. Wehrli, and H. Yang, 2018: Drought vulnerability assessment of maize in Sub-Saharan Africa: Insights from physical and social perspectives. *Glob. Planet. Change*, **162**, 266–274, doi:10.1016/J.GLOPLACHA.2018.01.011.

Kantarcı, M.D., Özel, H.B., Ertek, M. and Kırdar, E., 2011: An assessment on the adaptation of six tree species to steppe habitat during Konya-Karapinar sand-dune afforestations. *J. Bartın For. Fac.*, **13**, 107–127.

Kapović Solomun, M., N. Barger, A. Cerda, S. Keesstra, and M. Marković, 2018: Assessing land condition as a first step to achieving land degradation neutrality: A case study of the Republic of Srpska. *Environ. Sci. Policy*, **90**, 19–27, doi:10.1016/j.envsci.2018.09.014.

Kaptué, A.T., L. Prihodko, and N.P. Hanan, 2015: On regreening and degradation in Sahelian watersheds. *Proc. Natl. Acad. Sci. U.S.A.*, **112**, 12133–12138, doi:10.1073/pnas.1509645112.

Karami Dehkordi, M., H. Kohestani, H. Yadavar, R. Roshandel, and M. Karbasioun, 2017: Implementing conceptual model using renewable energies in rural area of Iran. *Inf. Process. Agric.*, **4**, 228–240, doi:10.1016/J.INPA.2017.02.003.

Karatayev, M., S. Hall, Y. Kalyuzhnova, and M.L. Clarke, 2016: Renewable energy technology uptake in Kazakhstan: Policy drivers and barriers in a transitional economy. *Renew. Sustain. Energy Rev.*, **66**, 120–136, doi:10.1016/J.RSER.2016.07.057.

Karmaoui, A., I. Ifaadassan, A. Babqiqi, M. Messouli, and Y.M. Khebiza, 2016: Analysis of the water supply-demand relationship in the Middle Draa Valley, Morocco, under climate change and socio-economic scenarios. *J. Sci. Res. Reports*, **9**, 1–10.

Karsenty, A., and S. Ongolo, 2012: Can 'fragile states' decide to reduce their deforestation? The inappropriate use of the theory of incentives with respect to the REDD mechanism. *For. Policy Econ.*, **18**, 38–45, doi:10.1016/J.FORPOL.2011.05.006.

Kashima, S., T. Yorifuji, S. Bae, Y. Honda, Y.-H. Lim, and Y.-C. Hong, 2016: Asian dust effect on cause-specific mortality in five cities across South Korea and Japan. *Atmos. Environ.*, **128**, 20–27, doi:10.1016/J.ATMOSENV.2015.12.063.

Kaspari, S., T.H. Painter, M. Gysel, S.M. Skiles, and M. Schwikowski, 2014: Seasonal and elevational variations of black carbon and dust in snow and ice in the Solu-Khumbu, Nepal and estimated radiative forcings. *Atmos. Chem. Phys.*, **14**, 8089–8103, doi:10.5194/acp-14-8089-2014.

Kassahun, A., H.A. Snyman, and G.N. Smit, 2008: Impact of rangeland degradation on the pastoral production systems, livelihoods and perceptions of the Somali pastoralists in Eastern Ethiopia. *J. Arid Environ.*, **72**, 1265–1281, doi:10.1016/J.JARIDENV.2008.01.002.

Kassas, M., 1995: Desertification: A general review. *J. Arid Environ.*, **30**, 115–128, doi:10.1016/S0140-1963(05)80063-1.

Kassie, M., M. Jaleta, B. Shiferaw, F. Mmbando, and M. Mekuria, 2013: Adoption of interrelated sustainable agricultural practices in smallholder systems: Evidence from rural Tanzania. *Technol. Forecast. Soc. Change*, **80**, 525–540, doi:10.1016/J.TECHFORE.2012.08.007.

Kassie, M., H. Teklewold, M. Jaleta, P. Marenya, and O. Erenstein, 2015: Understanding the adoption of a portfolio of sustainable intensification practices in eastern and southern Africa. *Land Use Policy*, **42**, 400–411, doi:10.1016/J.LANDUSEPOL.2014.08.016.

Kath, J., S. Mushtaq, R. Henry, A.A. Adeyinka, R. Stone, T. Marcussen, and L. Kouadio, 2019: Spatial variability in regional scale drought index insurance viability across Australia's wheat growing regions. *Clim. Risk Manag.*, **24**, 13–29, doi:10.1016/J.CRM.2019.04.002.

Katikiro, R.E., 2016: Prospects for the uptake of renewable energy technologies in rural Tanzania. *Energy Procedia*, **93**, 229–233, doi:10.1016/J.EGYPRO.2016.07.175.

Kaufman, Y.J., D. Tanré, and O. Boucher, 2002: A satellite view of aerosols in the climate system. *Nature*, **419**, 215–223, doi:10.1038/nature01091.

Kaufman, Y.J. et al., 2005: Dust transport and deposition observed from the Terra-Moderate Resolution Imaging Spectroradiometer (MODIS) spacecraft over the Atlantic Ocean. *J. Geophys. Res.*, **110**, D10S12, doi:10.1029/2003JD004436.

Kazbekov, J., I. Abdullaev, H. Manthrithilake, A. Qureshi, and K. Jumaboev, 2009: Evaluating planning and delivery performance of water user associations (WUAs) in Osh Province, Kyrgyzstan. *Agric. Water Manag.*, **96**, 1259–1267, doi:10.1016/J.AGWAT.2009.04.002.

Keeley, J.E., and T.J. Brennan, 2012: Fire-driven alien invasion in a fire-adapted ecosystem. *Oecologia*, **169**, 1043–1052, doi:10.1007/s00442-012-2253-8.

Kéfi, S., C.L. Alados, R.C.G. Chaves, Y. Pueyo, and M. Rietkerk, 2010: Is the patch size distribution of vegetation a suitable indicator of desertification processes? Comment. *Ecology*, **91**, 3739–3742, doi:10.1890/09-1915.1.

Keiichiro, H., T. Otsuki, and J.S. Wilson, 2015: Food safety standards and international trade: The impact on developing countries' export performance. In: *Food Safety, Market Organization, Trade and Development*

3

[Hammoudi, A., C, Grazia, Y. Surry, J.B. Traversac (eds.)]. Springer International Publishing, Cham, Switzerland, pp. 151–166.

Kellogg, C.A. et al., 2004: Characterization of aerosolized bacteria and fungi from desert dust events in Mali, West Africa. *Aerobiologia*, **20**, 99–110, doi:10.1023/B:AERO.0000032947.88335.bb 99–110.

Kerr, J., 2007: Watershed management: Lessons from common property theory. *Int. J. Commons*, **1**, 89–109, doi:10.18352/ijc.8.

Kerr, J.M., G. Pangare, and V. Pangare, 2002: *Watershed development projects in India: An evaluation*. Research Report 127, International Food Policy Research Institute, Washington, DC, USA.

Kertis, C.A., 2003: *Soil Erosion on Cropland in The United States: Status and Trends for 1982–2003*. Proceedings of the Eight Federal Interagency Sedimentation Conference, April 2006, Nevada, USA.

Kgope, B.S., W.J. Bond, and G.F. Midgley, 2010: Growth responses of African savanna trees implicate atmospheric [CO_2] as a driver of past and current changes in savanna tree cover. *Austral Ecol.*, **35**, 451–463, doi:10.1111/j.1442-9993.2009.02046.x.

Khamzina, A., J.P.A. Lamers, and P.L.G. Vlek, 2009: Nitrogen fixation by Elaeagnus angustifolia in the reclamation of degraded croplands of Central Asia. *Tree Physiol.*, **29**, 799–808, doi:10.1093/treephys/tpp017.

Khamzina, A., J.P.A. Lamers, and C. Martius, 2016: Above- and below-ground litter stocks and decay at a multi-species afforestation site on arid, saline soil. *Nutr. Cycl. Agroecosystems*, **104**, 187–199, doi:10.1007/s10705-016-9766-1.

Khan, M.A. et al., 2010: Invasive species of federal capital area Islamabad, Pakistan. *Pakistan J. Bot.*, **42**, 1529–1534.

Khan, Z.R. et al., 2014: Achieving food security for one million Sub-Saharan African poor through push-pull innovation by 2020. *Philos. Trans. R. Soc. B Biol. Sci.*, **369**, 20120284, doi:10.1098/rstb.2012.0284.

Khishigbayar, J. et al., 2015: Mongolian rangelands at a tipping point? Biomass and cover are stable but composition shifts and richness declines after 20 years of grazing and increasing temperatures. *J. Arid Environ.*, **115**, 100–112, doi:10.1016/j.jaridenv.2015.01.007.

Kidron, G.J., and V.P. Gutschick, 2017: Temperature rise may explain grass depletion in the Chihuahuan Desert. *Ecohydrology*, **10**, e1849, doi:10.1002/eco.1849.

Kienzler, K.M. et al., 2012: Conservation agriculture in Central Asia – What do we know and where do we go from here? *F. Crop. Res.*, **132**, doi:10.1016/j.fcr.2011.12.008.

Kihiu, E., 2016a: *Pastoral Practices, Economics, and Institutions of Sustainable Rangeland Management in Kenya*. PhD Thesis, University of Bonn, Bonn, Germany, 167 pp.

Kihiu, E.N., 2016b: Basic capability effect: Collective management of pastoral resources in south-western Kenya. *Ecol. Econ.*, **123**, 23–34, doi:10.1016/J.ECOLECON.2016.01.003.

Kim, S.E. et al., 2017: Seasonal analysis of the short-term effects of air pollution on daily mortality in Northeast Asia. *Sci. Total Environ.*, **576**, 850–857, doi:10.1016/J.SCITOTENV.2016.10.036.

King, C., and D.S.G. Thomas, 2014: Monitoring environmental change and degradation in the irrigated oases of the Northern Sahara. *J. Arid Environ.*, **103**, 36–45, doi:10.1016/J.JARIDENV.2013.12.009.

Kingdom of Saudi Arabia, 2016: *National Transformation Program 2020, Vision 2030: Delivery Plan 2018–2020*. Government of Saudi Arabia, Riyad, Saudi Arabia, pp. 1–57.

Kiptot, E., S. Franzel, and A. Degrande, 2014: Gender, agroforestry and food security in Africa. *Curr. Opin. Environ. Sustain.*, **6**, 104–109, doi:10.1016/J.COSUST.2013.10.019.

Kirui, O.K., 2016: *Economics of Land Degradation, Sustainable Land Management and Poverty in Eastern Africa – The Extent, Drivers, Costs and Impacts*. PhD Thesis, University of Bonn, Bonn, Germany, 165 pp.

Klein Goldewijk, K., A. Beusen, J. Doelman, and E. Stehfest, 2017: Anthropogenic land use estimates for the Holocene – HYDE 3.2. *Earth Syst. Sci. Data*, **9**, 927–953, doi:10.5194/essd-9-927-2017.

Klein, R.J.T., G.F. Midgley, B.L. Preston, M. Alam, F.G.H. Berkhout, K. Dow, and M.R. Shaw, 2014: Adaptation Opportunities, Constraints, and Limits. In: Climate Change 2014: Impacts, Adaptation, and Vulnerability. Part A: Global and Sectoral Aspects. Contribution of Working Group II to the Fifth Assessment Report of the Intergovernmental Panel on Climate Change [Field, C.B., V.R. Barros, D.J. Dokken, K.J. Mach, M.D. Mastrandrea, T.E. Bilir, M. Chatterjee, K.L. Ebi, Y.O. Estrada, R.C. Genova, B. Girma, E.S. Kissel, A.N. Levy, S. MacCracken, P.R. Mastrandrea, and L.L.White (eds.)]. Cambridge University Press, Cambridge, United Kingdom and New York, NY, USA, pp. 899–943, doi:10.1017/CBO9781107415379.021.

Klik, A., and J. Eitzinger, 2010: Impact of climate change on soil erosion and the efficiency of soil conservation practices in Austria. *J. Agric. Sci.*, **148**, 529–541, doi:10.1017/S0021859610000158.

Klingmüller, K., A. Pozzer, S. Metzger, G.L. Stenchikov, and J. Lelieveld, 2016: Aerosol Optical Depth trend over the Middle East. *Atmos. Chem. Phys.*, **16**, 5063–5073, doi:10.5194/acp-16-5063-2016.

Kloos, J., and F.G. Renaud, 2016: Overview of ecosystem-based approaches to drought risk reduction targeting small-scale farmers in Sub-Saharan Africa. In: *Ecosystem Based Disaster Risk Reduction and Adaptation in Practice* [Renaud, F.G., K. Sudmeier-Rieux, M. Estrella and U. Nehren (eds.)]. Springer, Cham, Switzerland, pp. 199–226.

Knight, J.R., C.K. Folland, and A.A. Scaife, 2006: Climate impacts of the Atlantic Multidecadal Oscillation. *Geophys. Res. Lett.*, **33**, L17706, doi:10.1029/2006GL026242.

Kniveton, D., C. Smith, and S. Wood, 2011: Agent-based model simulations of future changes in migration flows for Burkina Faso. *Glob. Environ. Chang.*, **21**, S34–S40, doi:10.1016/J.GLOENVCHA.2011.09.006.

Knörzer, H., S. Graeff-Hönninger, B. Guo, P. Wang, and W. Claupein, 2009: The rediscovery of intercropping in China: A traditional cropping system for future Chinese agriculture – A review. In: *Climate Change, Intercropping, Pest Control and Beneficial Microorganisms* [Lichtfouse, E (ed.)]. Springer Netherlands, Dordrecht, Netherlands, pp. 13–44.

Knox, J., T. Hess, A. Daccache, and T. Wheeler, 2012: Climate change impacts on crop productivity in Africa and South Asia. *Environ. Res. Lett.*, **7**, 34032, doi:10.1088/1748-9326/7/3/034032.

Knutson, K.C. et al., 2014: Long-term effects of seeding after wildfire on vegetation in Great Basin shrubland ecosystems. *J. Appl. Ecol.*, **51**, 1414–1424, doi:10.1111/1365-2664.12309.

Kodirekkala, K.R., 2017: Internal and external factors affecting loss of traditional knowledge: Evidence from a Horticultural Society in South India. *J. Anthropol. Res.*, **73**, 22–42, doi:10.1086/690524.

Kok, J.F., D.S. Ward, N.M. Mahowald, and A.T. Evan, 2018: Global and regional importance of the direct dust-climate feedback. *Nat. Commun.*, **9**, 241, doi:10.1038/s41467-017-02620-y.

Kok, M. et al., 2016: A new method for analysing socio-ecological patterns of vulnerability. *Reg. Environ. Chang.*, **16**, 229–243, doi:10.1007/s10113-014-0746-1.

Komala, P., and G. Prasad, 2016: Biomass: A key source of energy in rural households of Chamarajanagar district. *Pelagia Res. Libr. Adv. Appl. Sci. Res.*, **7**, 85–89.

Konare, A. et al., 2008: A regional climate modeling study of the effect of desert dust on the West African monsoon. *J. Geophys. Res.*, **113**, D12206, doi:10.1029/2007JD009322.

Konikow, L.F., 2011: Contribution of global groundwater depletion since 1900 to sea-level rise. *Geophys. Res. Lett.*, **38**, 1–5, doi:10.1029/2011GL048604.

Kouba, Y., M. Gartzia, A. El Aich, and C.L. Alados, 2018: Deserts do not advance, they are created: Land degradation and desertification in semi-arid environments in the Middle Atlas, Morocco. *J. Arid Environ.*, **158**, 1–8, doi:10.1016/J.JARIDENV.2018.07.002.

Koubi, V., G. Spilker, L. Schaffer, and T. Bernauer, 2016: Environmental stressors and migration: Evidence from Vietnam. *World Dev.*, **79**, 197–210, doi:10.1016/J.WORLDDEV.2015.11.016.

3

Koutroulis, A.G., 2019: Dryland changes under different levels of global warming. *Sci. Total Environ.*, **655**, 482–511, doi:10.1016/J.SCITOTENV.2018.11.215.

Koyro, H.-W., T. Hussain, B. Huchzermeyer, and M.A. Khan, 2013: Photosynthetic and growth responses of a perennial halophytic grass *Panicum turgidum* to increasing NaCl concentrations. *Environ. Exp. Bot.*, **91**, 22–29, doi:10.1016/J.ENVEXPBOT.2013.02.007.

Krätli, S., and N. Schareika, 2010: Living off uncertainty: The intelligent animal production of Dryland pastoralists. *Eur. J. Dev. Res.*, **22**, 605–622, doi:10.1057/ejdr.2010.41.

Kubik, Z., and M. Maurel, 2016: Weather shocks, agricultural production and migration: Evidence from Tanzania. *J. Dev. Stud.*, **52**, 665–680, doi:10.1080/00220388.2015.1107049.

Kundu, A., N.R. Patel, S.K. Saha, and D. Dutta, 2017: Desertification in western Rajasthan (India): An assessment using remote sensing derived rain-use efficiency and residual trend methods. *Nat. Hazards*, **86**, 297–313, doi:10.1007/s11069-016-2689-y.

Kust, G., 1999: *Desertification: principles of ecological assessment and mapping*. MSU-RAS Institute of Soil Science, Moscow, Russia, 362 pp.

Kust, G., J. Mott, N. Jain, T. Sampath, and A. Armstrong, 2014: SLM oriented projects in Tajikistan: Experience and lessons learned. *Planet@Risk*, **2 (1)**.

Kust, G.S., O.V. Andreeva, and D.V. Dobrynin, 2011: Desertification assessment and mapping in the Russian Federation. *Arid Ecosyst.*, **1**, 14–28, doi:10.1134/S2079096111010057.

Kusunose, Y., and T.J. Lybbert, 2014: Coping with drought by adjusting land tenancy contracts: A model and evidence from rural Morocco. *World Dev.*, **61**, 114–126, doi:10.1016/J.WORLDDEV.2014.04.006.

Lahlaoi, H. et al., 2017: Desertification assessment using MEDALUS model in watershed Oued El Maleh, Morocco. *Geosciences*, **7**, 50, doi:10.3390/geosciences7030050.

Lal, R., Stewart, B.A. (eds.), 2012: *Advances in Soil Science: Soil Water and Agronomic Productivity*. CRC Press, Florida, USA, 320 pp.

Lal, R., 1997: Degradation and resilience of soils. *Philos. Trans. R. Soc. London. Ser. B Biol. Sci.*, **352**, 997–1010, doi:10.1098/rstb.1997.0078.

Lal, R., 2004: Carbon Sequestration in Dryland Ecosystems. *Environ. Manage.*, **33 (4)**, 528–544, doi:10.1007/s00267-003-9110-9.

Lal, R., 2009: Sequestering carbon in soils of arid ecosystems. *L. Degrad. Dev.*, **20**, 441–454, doi:10.1002/ldr.934.

Lal, R., 2015: The soil–peace nexus: Our common future. *Soil Sci. Plant Nutr.*, **61**, 566–578, doi:10.1080/00380768.2015.1065166.

Lal, R., 2018: Managing agricultural soils of Pakistan for food and climate. *Soil Environ.*, **37**, 1–10.

Lal, R., 2016: *Encyclopedia of Soil Science*. CRC Press, Florida, USA, 2653 pp.

Lam, D.K. et al., 2011: Tracking desertification in California using remote sensing: A sand dune encroachment approach. *Remote Sens.*, **3**, 1–13, doi:10.3390/rs3010001.

Lambin, E.F., 2012: Global land availability: Malthus versus Ricardo. *Glob. Food Sec.*, **1**, 83–87, doi:10.1016/J.GFS.2012.11.002.

Lambin, E.F., and P. Meyfroidt, 2011: Global land use change, economic globalization, and the looming land scarcity. *Proc. Natl. Acad. Sci. U.S.A.*, **108**, 3465–3472, doi:10.1073/pnas.1100480108.

Lambin, E.F. et al., 2013: Estimating the world's potentially available cropland using a bottom-up approach. *Glob. Environ. Chang.*, **23**, 892–901, doi:10.1016/J.GLOENVCHA.2013.05.005.

Lambin, E.F. et al., 2014: Effectiveness and synergies of policy instruments for land use governance in tropical regions. *Glob. Environ. Chang.*, **28**, 129–140, doi:10.1016/J.GLOENVCHA.2014.06.007.

Lamchin, M. et al., 2016: Assessment of land cover change and desertification using remote sensing technology in a local region of Mongolia. *Adv. Sp. Res.*, **57**, 64–77, doi:10.1016/J.ASR.2015.10.006.

Lamontagne, S., P.G. Cook, A. O'Grady, and D. Eamus, 2005: Groundwater use by vegetation in a tropical savanna riparian zone (Daly River, Australia). *J. Hydrol.*, **310**, 280–293, doi:10.1016/J.JHYDROL.2005.01.009.

Lands, D., and U. Nations, 2011: Publisher: Arab Center for the Studies of Arid Zones and Dry Lands, Beirut, Lebanon.

Larrucea, E.S., and P.F. Brussard, 2008: Habitat selection and current distribution of the pygmy rabbit in Nevada and California, USA, *J. Mammal.*, **89**, 691–699, doi:10.1644/07-MAMM-A-199R.1.

Lau, K.M., K.M. Kim, Y.C. Sud, and G.K. Walker, 2009: A GCM study of the response of the atmospheric water cycle of West Africa and the Atlantic to Saharan dust radiative forcing. *Ann. Geophys.*, **27**, 4023–4037, doi:10.5194/angeo-27-4023-2009.

Lavauden, L., 1927: *Les forêts du Sahara*. Berger-Levrault, Paris, France, 26 pp.

Lavee, H., A.C. Imeson, and P. Sarah, 1998: The impact of climate change on geomorphology and desertification along a Mediterranean arid transect. *L. Degrad. Dev.*, **9**, 407–422, doi:10.1002/(SICI)1099-145X(199809/10)9:5<407::AID-LDR302>3.0.CO;2-6.

Lawrence, P.G., B.D. Maxwell, L.J. Rew, C. Ellis, and A. Bekkerman, 2018: Vulnerability of dryland agricultural regimes to economic and climatic change. *Ecol. Soc.*, **23**, art34, doi:10.5751/ES-09983-230134.

Lawry, S., C. Samii, R. Hall, A. Leopold, D. Hornby, and F. Mtero, 2017: The impact of land property rights interventions on investment and agricultural productivity in developing countries: A systematic review. *J. Dev. Eff.*, **9**, 61–81, doi:10.1080/19439342.2016.1160947.

Lázaro, R., Y. Cantón, A. Solé-Benet, J. Bevan, R. Alexander, L.G. Sancho, and J. Puigdefábregas, 2008: The influence of competition between lichen colonization and erosion on the evolution of soil surfaces in the Tabernas badlands (SE Spain) and its landscape effects. *Geomorphology*, **102**, 252–266, doi:10.1016/J.GEOMORPH.2008.05.005.

Le, Q., C. Biradar, R. Thomas, C. Zucca, and E. Bonaiuti, 2016a: Socio-ecological context typology to support targeting and upscaling of sustainable land management practices in diverse global Dryland. In: *Proceedings of the International Congress on Environmental Modeling and Software, Paper 45, July 2016, Toulouse, France* [Sauvage, S., J. Sanchez-Perez and A. Rizzoli (eds.)]. Brigham Young University, Utah, USA.

Le, Q.B., L. Tamene and P.L.G. Vlek, 2012: Multi-pronged assessment of land degradation in West Africa to assess the importance of atmospheric fertilization in masking the processes involved. *Glob. Planet. Change*, **92–93**, 71–81, doi:10.1016/J.GLOPLACHA.2012.05.003.

Le, Q.B., E. Nkonya, and A. Mirzabaev, 2016b: Biomass productivity-based mapping of global land degradation hotspots. In: *Economics of Land Degradation and Improvement – A Global Assessment for Sustainable Development* [Nkonya, E., A. Mirzabaev, and J. von Braun (eds.)]. Springer International Publishing, Cham, Switzerland, pp. 55–84.

Leal, I.R., J.M.C. Da Silva, M. Tabarelli, and T.E. Lacher, 2005: Changing the course of biodiversity conservation in the Caatinga of north-eastern Brazil. *Conserv. Biol.*, **19**, 701–706, doi:10.1111/j.1523-1739.2005.00703.x.

Lechmere-Oertel, R.G., G.I.H. Kerley, and R.M. Cowling, 2005: Patterns and implications of transformation in semi-arid succulent thicket, South Africa. *J. Arid Environ.*, **62**, 459–474, doi:10.1016/j.jaridenv.2004.11.016.

Lee, H., Y. Honda, Y.-H. Lim, Y.L. Guo, M. Hashizume, and H. Kim, 2014: Effect of Asian dust storms on mortality in three Asian cities. *Atmos. Environ.*, **89**, 309–317, doi:10.1016/J.ATMOSENV.2014.02.048.

Lee, S.-J., and E.H. Berbery, 2012: Land cover change effects on the climate of the La Plata Basin. *J. Hydrometeorol.*, **13**, 84–102, doi:10.1175/JHM-D-11-021.1.

Leggett, J., W.J. Pepper, R.J. Swart, J. Edmonds, L.G. Meira Filho, I. Mintzer, M.X. Wang, and J. Wasson, 1992: Emissions scenarios for the IPCC: An update. In: Climate Change 1992: The Supplementary Report to IPCC Scientific Assessment [Houghton, J.T., B.A. Callander and S.K. Varney (eds.)]. Cambridge University Press, Cambridge, United Kingdom, pp. 69–95, doi:10.1007/s00213-003-1546-3.

Lei, Y., B. Hoskins, and J. Slingo, 2011: Exploring the interplay between natural decadal variability and anthropogenic climate change in summer rainfall over China. Part I: Observational evidence. *J. Clim.*, **24**, 4584–4599, doi:10.1175/2010JCLI3794.1.

3

Lemordant, L., P. Gentine, A.S. Swann, B.I. Cook, and J. Scheff, 2018: Critical impact of vegetation physiology on the continental hydrologic cycle in response to increasing CO_2. *Proc. Natl. Acad. Sci. U.S.A.*, **115**, 4093–4098, doi:10.1073/pnas.1720712115.

Lenes, J.M. et al., 2001: Iron fertilization and the Trichodesmium response on the West Florida shelf. *Limnol. Oceanogr.*, **46**, 1261–1277, doi:10.4319/lo.2001.46.6.1261.

Leroux, L., A. Bégué, D. Lo Seen, A. Jolivot, and F. Kayitakire, 2017: Driving forces of recent vegetation changes in the Sahel: Lessons learned from regional and local level analyses. *Remote Sens. Environ.*, **191**, 38–54, doi:10.1016/J.RSE.2017.01.014.

Lévesque, M., R. Siegwolf, M. Saurer, B. Eilmann, and A. Rigling, 2014: Increased water-use efficiency does not lead to enhanced tree growth under xeric and mesic conditions. *New Phytol.*, **203**, 94–109, doi:10.1111/nph.12772.

Li, A., J. Wu, and J. Huang, 2012: Distinguishing between human-induced and climate-driven vegetation changes: A critical application of RESTREND in inner Mongolia. *Landsc. Ecol.*, **27**, 969–982, doi:10.1007/s10980-012-9751-2.

Li, D., S. Wu, L. Liu, Y. Zhang, and S. Li, 2018a: Vulnerability of the global terrestrial ecosystems to climate change. *Glob. Chang. Biol.*, **24**, 4095–4106, doi:10.1111/gcb.14327.

Li, H. et al., 2016: Effects of shrub encroachment on soil organic carbon in global grasslands. *Sci. Rep.*, **6**, 28974, doi:10.1038/srep28974.

Li, Y. et al., 2018b: Climate model shows large-scale wind and solar farms in the Sahara increase rain and vegetation. *Science*, **361**, 1019–1022, doi:10.1126/science.aar5629.

Li, Y. et al., 2012: Mongolian pine plantations enhance soil physico-chemical properties and carbon and nitrogen capacities in semi-arid degraded sandy land in China. *Appl. Soil Ecol.*, **56**, 1–9, doi:10.1016/J.APSOIL.2012.01.007.

Li, Z., and H. Fang, 2016: Impacts of climate change on water erosion: A review. *Earth-Science Rev.*, **163**, 94–117, doi:10.1016/J.EARSCIREV.2016.10.004.

Liao, C., C. Barrett, and K.-A. Kassam, 2015: Does diversification improve livelihoods? Pastoral households in Xinjiang, China. *Dev. Change*, **46**, 1302–1330, doi:10.1111/dech.12201.

Liao, C., P.E. Clark, S.D. DeGloria, and C.B. Barrett, 2017: Complexity in the spatial utilization of rangelands: Pastoral mobility in the Horn of Africa. *Appl. Geogr.*, **86**, 208–219, doi:10.1016/J.APGEOG.2017.07.003.

Liehr, S., L. Drees, and D. Hummel, 2016: Migration as societal response to climate change and land degradation in Mali and Senegal. In: *Adaptation to Climate Change and Variability in Rural West Africa* [Yaro, J.A. and J. Hesselberg (eds.)]. Springer International Publishing, Cham, Switzerland, pp. 147–169.

Lim, W.-Y., and A. Seow, 2012: Biomass fuels and lung cancer. *Respirology*, **17**, 20–31, doi:10.1111/j.1440-1843.2011.02088.x.

Lima, M., D.A. Christie, M.C. Santoro, and C. Latorre, 2016: Coupled socio-environmental changes triggered indigenous Aymara depopulation of the semi-arid Andes of Tarapacá-Chile during the late 19th–20th centuries. *PLoS One*, **11**, e0160580, doi:10.1371/journal.pone.0160580.

Lin, L., A. Gettelman, S. Feng, and Q. Fu, 2015: Simulated climatology and evolution of aridity in the 21st century. *J. Geophys. Res. Atmos.*, **120**, 5795–5815, doi:10.1002/2014JD022912.

Lin, M., C. Chen, Q. Wang, Y. Cao, J. Shih, Y. Lee, C. Chen, and S. Wang, 2009: Fuzzy model-based assessment and monitoring of desertification using MODIS satellite imagery. *Eng. Comput.*, **26**, 745–760, doi:10.1108/02644400910985152.

Ling, H., H. Xu, J. Fu, Z. Fan, and X. Xu, 2013: Suitable oasis scale in a typical continental river basin in an arid region of China: A case study of the Manas River Basin. *Quat. Int.*, **286**, 116–125, doi:10.1016/j.quaint.2012.07.027.

Liniger, H., and W. Critchley, 2007: *Case Studies and Analysis of Soil and Water Conservation Initiatives Worldwide*. University of Minnesota, Minnesota, USA, 364 pp.

Liniger, H., R. Mektaschi Studer, P. Moll, and U. Zander, 2017: *Making Sense of Research for Sustainable Land Management*. Centre for Development and Environment, University of Bern, Switzerland and Helmholtz-Centre for Environmental Research GmbH-UFZ, Leipzig, Germany. ISBN 978-3-944280-99-8, 304 pp.

Linke, A.M., J. O'Loughlin, J.T. McCabe, J. Tir, and F.D.W. Witmer, 2015: Rainfall variability and violence in rural Kenya: Investigating the effects of drought and the role of local institutions with survey data. *Glob. Environ. Chang.*, **34**, 35–47, doi:10.1016/J.GLOENVCHA.2015.04.007.

Lioubimtseva, E., 2015: A multi-scale assessment of human vulnerability to climate change in the Aral Sea Basin. *Environ. Earth Sci.*, **73**, 719–729, doi:10.1007/s12665-014-3104-1.

Littell, J., D. McKenzie, H.Y. Wan, and S. Cushman, 2018: Climate change and future wildfire in the western USA: An ecological approach to non-stationarity. *Earth's Futur.*, **6**, 1097–1111, doi:10.1029/2018EF000878.

Little, P., 2010: Unofficial cross-border trade in eastern Africa. In: *Food Security in Africa: Market and Trade Policy for Staple Foods in Eastern and Southern Africa* [Sarris, A., T. Jayne, and J. Morrison, (eds.)]. Edwin Elgar Publishing, Cheltenham, UK, pp. 158–181.

Liu, G., H. Wang, Y. Cheng, B. Zheng, and Z. Lu, 2016a: The impact of rural out-migration on arable land use intensity: Evidence from mountain areas in Guangdong, China. *Land Use Policy*, **59**, 569–579, doi:10.1016/J.LANDUSEPOL.2016.10.005.

Liu, H.-L., P. Willems, A.-M. Bao, L. Wang, and X. Chen, 2016b: Effect of climate change on the vulnerability of a socio-ecological system in an arid area. *Glob. Planet. Change*, **137**, 1–9, doi:10.1016/J.GLOPLACHA.2015.12.014.

Liu, J., Q. Liu, and H. Yang, 2016c: Assessing water scarcity by simultaneously considering environmental flow requirements, water quantity, and water quality. *Ecol. Indic.*, **60**, 434–441, doi:10.1016/J.ECOLIND.2015.07.019.

Liu, J. et al., 2017: Water scarcity assessments in the past, present, and future. *Earth's Futur.*, **5**, 545–559, doi:10.1002/2016EF000518.

Liu, M., and H. Tian, 2010: China's land cover and land use change from 1700 to 2005: Estimations from high-resolution satellite data and historical archives. *Global Biogeochem. Cycles*, **24**, GB3003, 1-18, doi:10.1029/2009GB003687.

Liu, Y., J. Liu, and Y. Zhou, 2017: Spatio-temporal patterns of rural poverty in China and targeted poverty alleviation strategies. *J. Rural Stud.*, **52**, 66–75, doi:10.1016/J.JRURSTUD.2017.04.002.

Liu, Y.Y. et al., 2013: Changing climate and overgrazing are decimating Mongolian steppes. *PLoS One*, **8**, e57599. 1-6, doi:10.1371/journal.pone.0057599.

Liu, Z., and M.C. Wimberly, 2016: Direct and indirect effects of climate change on projected future fire regimes in the western United States. *Sci. Total Environ.*, **542**, 65–75, doi:10.1016/j.scitotenv.2015.10.093.

Lockyer, Z.B., P.S. Coates, M.L. Casazza, S. Espinosa, and D.J. Delehanty, 2015: Nest-site selection and reproductive success of greater sage-grouse in a fire-affected habitat of north-western Nevada. *J. Wildl. Manage.*, **79**, 785–797, doi:10.1002/jwmg.899.

Lohmann, S., and T. Lechtenfeld, 2015: The effect of drought on health outcomes and health expenditures in rural Vietnam. *World Dev.*, **72**, 432–448, doi:10.1016/J.WORLDDEV.2015.03.003.

Lokonon, B.O.K., and A.A. Mbaye, 2018: Climate change and adoption of sustainable land management practices in the Niger Basin of Benin. *Nat. Resour. Forum*, **42**, 42–53, doi:10.1111/1477-8947.12142.

London Government Office for Science, 2011: *Migration and Global Environmental Change: Future Challenges and Opportunities*. Final Project Report, The Government Office for Science, London, UK, 235 pp.

Van Loo, M. et al., 2017: Human induced soil erosion and the implications on crop yield in a small mountainous Mediterranean catchment (SW-Turkey). *CATENA*, **149**, 491–504, doi:10.1016/J.CATENA.2016.08.023.

López-Bermúdez, F., 1990: Soil erosion by water on the desertification of a semi-arid Mediterranean fluvial basin: The Segura Basin, Spain. *Agric. Ecosyst. Environ.*, **33**, 129–145, doi:10.1016/0167-8809(90)90238-9.

3

Lopez-Garcia, J., A. Pozza, and T. Sample, 2016: Long-term soiling of silicon PV modules in a moderate subtropical climate. *Sol. Energy*, **130**, 174–183, doi:10.1016/J.SOLENER.2016.02.025.

López-i-Gelats, F., E.D.G. Fraser, J.F. Morton, and M.G. Rivera-Ferre, 2016: What drives the vulnerability of pastoralists to global environmental change? A qualitative meta-analysis. *Glob. Environ. Chang.*, **39**, 258–274, doi:10.1016/J.GLOENVCHA.2016.05.011.

Louhaichi, M., and A. Tastad, 2010: The Syrian steppe: Past trends, current status, and future priorities. *Rangelands*, **32**, 2–7, doi:10.2111/1551-501X-32.2.2.

Lu, J., G.A. Vecchi, and T. Reichler, 2007: Expansion of the Hadley cell under global warming. *Geophys. Res. Lett.*, **34**, L06805, doi:10.1029/2006GL028443.

Luedeling, E., J. Gebauer, and A. Buerkert, 2009: Climate change effects on winter chill for tree crops with chilling requirements on the Arabian Peninsula. *Clim. Change*, **96**, 219–237, doi:10.1007/s10584-009-9581-7.

Lybbert, T.J., C.B. Barrett, S. Desta, and D. Layne Coppock, 2004: Stochastic wealth dynamics and risk management among a poor population. *Econ. J.*, **114**, 750–777, doi:10.1111/j.1468-0297.2004.00242.x.

Ma, X., C. Zhao, H. Tao, J. Zhu, and Z.W. Kundzewicz, 2018: Projections of actual evapotranspiration under the 1.5°C and 2.0°C global warming scenarios in sandy areas in northern China. *Sci. Total Environ.*, **645**, 1496–1508, doi:10.1016/j.scitotenv.2018.07.253.

Machethe, C.L. et al., 2004: *Smallholder Irrigation and Agricultural Development in the Olifants River Basin of Limpopo Province: Management Transfer, Productivity, Profitability and Food Security Issues*. Report to the Water Research Commission on the Project 'Sustainable Local Management of Smallholder Irrigation', University of the North, Manitoba, Canada, 112 pp.

Madani, K., A. AghaKouchak, and A. Mirchi, 2016: Iran's socio-economic drought: Challenges of a water-bankrupt nation. *Iran. Stud.*, **49**, 997–1016, doi:10.1080/00210862.2016.1259286.

Maestre, F.T., and A. Escudero, 2009: Is the patch size distribution of vegetation a suitable indicator of desertification processes? *Ecology*, **90**, 1729–1735, doi:10.1890/08-2096.1.

Maestre, F.T. et al., 2012: Plant species richness and ecosystems multifunctionality in global drylands. *Science*, **335**, 2014–2017, doi:10.1126/science.1215442.

Magandana, T.P., 2016: Effect of Acacia karoo encroachment on grass production in the semi-arid savannas of the Eastern Cape, South Africa. Thesis, University of Fort Hare, Alice, South Africa, 104 pp.

Maghrabi, A., B. Alharbi, and N. Tapper, 2011: Impact of the March 2009 dust event in Saudi Arabia on aerosol optical properties, meteorological parameters, sky temperature and emissivity. *Atmos. Environ.*, **45**, 2164–2173, doi:10.1016/J.ATMOSENV.2011.01.071.

Mahmod, W.E., and K. Watanabe, 2014: Modified Grey Model and its application to groundwater flow analysis with limited hydrogeological data: A case study of the Nubian Sandstone, Kharga Oasis, Egypt. *Environ. Monit. Assess.*, **186**, 1063–1081, doi:10.1007/s10661-013-3439-1.

Mahmood, R., R.A. Pielke, and C.A. McAlpine, 2016: Climate-relevant land use and land cover change policies. *Bull. Am. Meteorol. Soc.*, **97**, 195–202, doi:10.1175/BAMS-D-14-00221.1.

Mahowald, N.M., 2007: Anthropocene changes in desert area: Sensitivity to climate model predictions. *Geophys. Res. Lett.*, **34**, L18817, doi:10.1029/2007GL030472.

Mahowald, N.M. et al., 2006: Change in atmospheric mineral aerosols in response to climate: Last glacial period, preindustrial, modern, and doubled carbon dioxide climates. *J. Geophys. Res. Atmos.*, **111:1-22**, D10202, n/a–n/a, doi:10.1029/2005JD006653.

Mainali, K.P. et al., 2015: Projecting future expansion of invasive species: Comparing and improving methodologies for species distribution modeling. *Glob. Chang. Biol.*, **21**, 4464–4480, doi:10.1111/gcb.13038.

Mainguet, M. and F. Dumay, 2011: *Fighting Wind Erosion. One Aspect of the Combat Against Desertification*. CSFD/Agropolis International, Les dossiers thématiques du CSFD/Agropolis International, Montpellier, France, 44 pp.

Le Maitre, D.C. et al., 2011: Impacts of invasive Australian acacias: Implications for management and restoration. *Divers. Distrib.*, **17**, 1015–1029, doi:10.1111/j.1472-4642.2011.00816.x.

Le Maitre, D.C. et al., 2015: Impacts of invading alien plant species on water flows at stand and catchment scales. *AoB Plants*, **7**, plv043, doi:10.1093/aobpla/plv043.

Majeed, A. and Z. Muhammad, 2019: Salinity: A major agricultural problem – Causes, impacts on crop productivity and management strategies. In: *Plant Abiotic Stress Tolerance* [Hasanuzzaman, M., K.R. Hakeem, K. Nahar and H. Alharby (eds.)]. Springer International Publishing, Cham, Switzerland, pp. 83–99.

Makondo, C.C., and D.S.G. Thomas, 2018: Climate change adaptation: Linking indigenous knowledge with Western science for effective adaptation. *Environ. Sci. Policy*, **88**, 83–91, doi:10.1016/J.ENVSCI.2018.06.014.

Makonese, T., A. Ifegbesan, and I. Rampedi, 2017: Household cooking fuel use patterns and determinants across southern Africa: Evidence from the demographic and health survey data. *Energy Environ.*, **29**, 29–48, doi:10.1177/0958305X17739475.

Malcolm, J.R., C. Liu, R.P. Neilson, L. Hansen, and L. Hannah, 2006: Global warming and extinctions of endemic species from biodiversity hotspots. *Conserv. Biol.*, **20**, 538–548, doi:10.1111/j.1523-1739.2006.00364.x.

Maldonado, J. et al., 2016: Engagement with indigenous peoples and honoring traditional knowledge systems. *Clim. Change*, **135**, 111–126, doi:10.1007/s10584-015-1535-7.

Maliszewski, P.J., E.K. Larson, and C. Perrings, 2012: Environmental determinants of unscheduled residential outages in the electrical power distribution of Phoenix, Arizona. *Reliab. Eng. Syst. Saf.*, **99**, 161–171, doi:10.1016/J.RESS.2011.10.011.

Maliva, R., and T. Missimer, 2012: Aridity and drought. In: *Arid Lands Water Evaluation and Management* [Maliva, R. and T. Missimer (eds.)]. Springer, Berlin, Germany, pp. 21–39.

Máñez Costa, M.A., E.J. Moors, and E.D.G. Fraser, 2011: Socio-economics, policy, or climate change: What is driving vulnerability in southern Portugal? *Ecol. Soc.*, **16 (1): 28**, doi:10.5751/ES-03703-160128.

Mani, M., and R. Pillai, 2010: Impact of dust on solar photovoltaic (PV) performance: Research status, challenges and recommendations. *Renew. Sustain. Energy Rev.*, **14**, 3124–3131, doi:10.1016/J.RSER.2010.07.065.

Manjoro, M., V. Kakembo, and K.M. Rowntree, 2012: Trends in soil erosion and woody shrub encroachment in Ngqushwa district, Eastern Cape province, South Africa. *Environ. Manage.*, **49**, 570–579, doi:10.1007/s00267-012-9810-0.

Manlosa, A.O., J. Schultner, I. Dorresteijn, and J. Fischer, 2018: Leverage points for improving gender equality and human well-being in a smallholder farming context. *Sustain. Sci.*, 1–13, doi:10.1007/s11625-018-0636-4.

Manson, A.D., D. Jewitt, and A.D. Short, 2007: Effects of season and frequency of burning on soils and landscape functioning in a moist montane grassland. *African J. Range Forage Sci.*, **24**, 9–18.

Manzano, M.G. and J. Návar, 2000: Processes of desertification by goats overgrazing in the Tamaulipan thornscrub (matorral) in north-eastern Mexico. *J. Arid Environ.*, **44**, 1–17, doi:10.1006/JARE.1999.0577.

Mao, J., X. Shi, P. Thornton, F. Hoffman, Z. Zhu, and R. Myneni, 2013: Global latitudinal-asymmetric vegetation growth trends and their driving mechanisms: 1982–2009. *Remote Sens.*, **5**, 1484–1497, doi:10.3390/rs5031484.

Mao, J. et al., 2016: Human-induced greening of the northern extratropical land surface. *Nat. Clim. Chang.*, **6**, 959–963, doi:10.1038/nclimate3056.

Mapfumo, P., F. Mtambanengwe, and R. Chikowo, 2016: Building on indigenous knowledge to strengthen the capacity of smallholder farming communities to adapt to climate change and variability in southern Africa. *Clim. Dev.*, **8**, 72–82, doi:10.1080/17565529.2014.998604.

3

Marengo, J.A., and M. Bernasconi, 2015: Regional differences in aridity/drought conditions over Northeast Brazil: Present state and future projections. *Clim. Change*, **129**, 103–115, doi:10.1007/s10584-014-1310-1.

Marlet, S., F. Bouksila, and A. Bahri, 2009: Water and salt balance at irrigation scheme scale: A comprehensive approach for salinity assessment in a Saharan oasis. *Agric. Water Manag.*, **96**, 1311–1322, doi:10.1016/J. AGWAT.2009.04.016.

Marques, M. et al., 2016: Multifaceted impacts of sustainable land management in drylands: A review. *Sustainability*, **8**, 177, doi:10.3390/su8020177.

Marshall, M., C. Funk, and J. Michaelsen, 2012a: Examining evapotranspiration trends in Africa. *Clim. Dyn.*, **38**, 1849–1865, doi:10.1007/s00382-012-1299-y.

Marshall, V.M., M.M. Lewis, and B. Ostendorf, 2012b: Buffel grass (Cenchrus ciliaris) as an invader and threat to biodiversity in arid environments: A review. *J. Arid Environ.*, **78**, 1–12, doi:10.1016/J.JARIDENV.2011.11.005.

Marticorena, B. et al., 2010: Temporal variability of mineral dust concentrations over West Africa: Analyses of a pluriannual monitoring from the AMMA Sahelian Dust Transect. *Atmos. Chem. Phys.*, **10**, 8899–8915, doi:10.5194/acp-10-8899-2010.

Martinez-Mena, M., J. Lopez, M. Almagro, C. Boix-Fayos, and J. Albaladejo, 2008: Effect of water erosion and cultivation on the soil carbon stock in a semi-arid area of Southeast Spain. *Soil Tillage Res.*, **99**, 119–129, doi:10.1016/j.still.2008.01.009.

Martínez-Palacios, A., L.E. Eguiarte, and G.R. Furnier, 1999: Genetic diversity of the endangered endemic *Agave victoriae-reginae* (Agavaceae) in the Chihuahuan Desert. *Am. J. Bot.*, **86**, 1093–1098, doi:10.2307/2656971.

Martínez, G.I. et al., 2011: Influence of conservation tillage and soil water content on crop yield in Dryland compacted Alfisol of Central Chile. *Chil. J. Agric. Res.*, **71**, 615–622, doi:10.4067/S0718-58392011000400018.

Marty, J., 2005: Effects of cattle grazing on diversity in ephemeral wetlands. *Conserv. Biol.*, **19**, 1626–1632, doi:10.1111/j.1523-1739.2005.00198.x.

Marwat, K.B., S. Hashim, and H. Ali, 2010: Weed management: A case study from north-west Pakistan. *Pakistan J. Bot.*, **42**, 341–353.

Masih, I., S. Maskey, F.E.F. Mussá, and P. Trambauer, 2014: A review of droughts on the African continent: A geospatial and long-term perspective. *Hydrol. Earth Syst. Sci.*, **18**, 3635–3649, doi:10.5194/hess-18-3635-2014.

Masoud, A.A. and K. Koike, 2006: Arid land salinization detected by remotely-sensed landcover changes: A case study in the Siwa region, NW Egypt. *J. Arid Environ.*, **66**, 151–167, doi:10.1016/J.JARIDENV.2005.10.011.

Massey, D.S., W.G. Axinn, and D.J. Ghimire, 2010: Environmental change and out-migration: Evidence from Nepal. *Popul. Environ.*, **32**, 109–136, doi:10.1007/s11111-010-0119-8.

Mavlyanova, N., K. Kulov, and P. Jooshov, 2016: *Euriasian Food Security Center, Moscow, Russia, 18 pages.on, 2016. (in Russian)*.

Mays, L.W., 2013: Groundwater resources sustainability: Past, present, and future. *Water Resour. Manag.*, **27**, 4409–4424, doi:10.1007/s11269-013-0436-7.

Maystadt, J.-F., and O. Ecker, 2014: Extreme weather and civil war: Does drought fuel conflict in Somalia through livestock price shocks? *Am. J. Agric. Econ.*, **96**, 1157–1182, doi:10.1093/ajae/aau010.

Mbow, C., 2017: *The Great Green Wall in the Sahel*. Oxford Research Encyclopedia, Oxford University Press, UK, doi: 10.1093/acrefore/9780190228620.013.559.

McCaw, W.L., 2013: Managing forest fuels using prescribed fire – A perspective from southern Australia. *For. Ecol. Manage.*, **294**, 217–224, doi:10.1016/j.foreco.2012.09.012.

McConnachie, A.J. et al., 2011: Current and potential geographical distribution of the invasive plant *Parthenium hysterophorus* (Asteraceae) in eastern and southern Africa. *Weed Res.*, **51**, 71–84, doi:10.1111/j.1365-3180.2010.00820.x.

McCord, P., J. Dell'Angelo, E. Baldwin, and T. Evans, 2017: Polycentric transformation in Kenyan water governance: A dynamic analysis of institutional and social-ecological change. *Policy Stud. J.*, **45**, 633–658, doi:10.1111/psj.12168.

McKenzie, D., and D. Yang, 2015: Evidence on policies to increase the development impacts of international migration. *World Bank Res. Obs.*, **30**, 155–192, doi:10.1093/wbro/lkv001.

McKinley, D.C. et al., 2017: Citizen science can improve conservation science, natural resource management, and environmental protection. *Biol. Conserv.*, **208**, 15–28, doi:10.1016/J.BIOCON.2016.05.015.

McLeman, R., 2013: Developments in modelling of climate change-related migration. *Clim. Change*, **117**, 599–611, doi:10.1007/s10584-012-0578-2.

McLeman, R., 2017: *Migration and Land Degradation: Recent Experience and Future Trends*. Global Land Outlook Working Paper, UNCCD: GLO, Bonn, Germany, 44 pp.

McLeman, R.A., 2011: Settlement abandonment in the context of global environmental change. *Glob. Environ. Chang.*, **21**, S108–S120, doi:10.1016/J.GLOENVCHA.2011.08.004.

Mcpeak, J., P. Little, and M. Demment, 2006: Policy implications and future research directions. In: *Pastoral Livestock Marketing in Eastern Africa* [McPeak, J. and P. Little (eds.)]. Intermediate Technology Publications, Warwickshire, UK, pp. 247–256.

Megnounif, A. and A.N. Ghenim, 2016: Rainfall irregularity and its impact on the sediment yield in Wadi Sebdou watershed, Algeria. *Arab. J. Geosci.*, **9**, 267, doi:10.1007/s12517-015-2280-y.

Meijer, E., E. Querner, and H. Boesveld, 2013: *Impact of Farm Dams on River Flows; A Case Study in the Limpopo River Basin, Southern Africa*. Alterra report 2394, Alterra Wageningen UR, Wageningen, Netherlands. 61 pages.

Mejia, F., J. Kleissl, and J.L. Bosch, 2014: The effect of dust on solar photovoltaic systems. *Energy Procedia*, **49**, 2370–2376, doi:10.1016/J.EGYPRO.2014.03.251.

Mejia, F.A., and J. Kleissl, 2013: Soiling losses for solar photovoltaic systems in California. *Sol. Energy*, **95**, 357–363, doi:10.1016/J.SOLENER.2013.06.028.

Mekonnen, G., 2017: Threats and management options of Parthenium (Parthenium hysterophorus l.) in Ethiopia. *Agric. Res. Technol.*, **10**, 1–7, doi:10.19080/ARTOAJ.2017.10.555798.

Mekuria, W., M. Yami, M. Haile, K. Gebrehiwot, and E. Birhane, 2018: Impact of exclosures on wood biomass production and fuelwood supply in northern Ethiopia. *J. For. Res.*, **30**, 629–637, doi:10.1007/s11676-018-0643-4.

Melde, S., Laczko, F., and Gemenne, F., 2017: *Making Mobility Work for Adaptation to Environmental Changes: Results from the MECLEP Global Research*. International Organization for Migration, Geneva, Switzerland, 122 pp.

Méndez, M., and V. Magaña, 2010: Regional aspects of prolonged meteorological droughts over Mexico and Central America. *J. Clim.*, **23**, 1175–1188, doi:10.1175/2009JCLI3080.1.

Meng, X.H., J.P. Evans, and M.F. McCabe, 2014a: The influence of inter-annually varying albedo on regional climate and drought. *Clim. Dyn.*, **42**, 787–803, doi:10.1007/s00382-013-1790-0.

Meng X.H., J.P. Evans, and M.F. McCabe, 2014b: The impact of observed vegetation changes on land-atmosphere feedbacks during drought. *J. Hydrometeorol.*, **15**, 759–776, doi:10.1175/JHM-D-13-0130.1.

Meng X.H., J. P. Evans, and M. F. McCabe (2014). The impact of observed vegetation changes on land-atmosphere feedbacks during drought. Journal of Hydrometeorology, 15, 759-776, DOI:10.1175/JHM-D-13-0130.1.

Merdas, S., B. Nouar, and F. Lakhdari, 2015: The Green Dam in Algeria as a tool to combat desertification. *Planet@Risk*, **3**, 3–6.

Mi, J., J. Li, D. Chen, Y. Xie, and Y. Bai, 2015: Predominant control of moisture on soil organic carbon mineralization across a broad range of arid and semi-arid ecosystems on the Mongolia Plateau. *Landsc. Ecol.*, **30**, 1683–1699, doi:10.1007/s10980-014-0040-0.

Miao, L. et al., 2015a: Footprint of research in desertification management in China. *L. Degrad. Dev.*, **26**, 450–457, doi:10.1002/ldr.2399.

3

Miao, L., P. Ye, B. He, L. Chen, and X. Cui, 2015b: Future climate impact on the desertification in the Dryland Asia using AVHRR GIMMS NDVI3g data. *Remote Sens.*, **7**, 3863–3877, doi:10.3390/rs70403863.

Micklin, P., 2016: The future Aral Sea: Hope and despair. *Environ. Earth Sci.*, **75**, 844, doi:10.1007/s12665-016-5614-5.

Middleton, N., and U. Kang, 2017: Sand and dust storms: Impact mitigation. *Sustainability*, **9**, 1053, doi:10.3390/su9061053.

Middleton, N.J., 2017: Desert dust hazards: A global review. *Aeolian Res.*, **24**, 53–63, doi:10.1016/J.AEOLIA.2016.12.001.

Millennium Ecosystem Assessment, 2005: *Ecosystems and Human Well-Being: Desertification Synthesis*. World Resource Institute, Washington, DC, USA, 26 pp.

Miller, R.F. et al., 2011: Characteristics of sagebrush habitats and limitations to long-term conservation. Greater sage-grouse: Ecology and conservation of a landscape species and its habitats. *Stud. Avian Biol.*, **38**, 145–184.

Miller, G., M. Friedel, P. Adam, and V. Chewings, 2010: Ecological impacts of buffel grass (*Cenchrus ciliaris* L.) invasion in Central Australia – Does field evidence support a fire-invasion feedback? *Rangel. J.*, **32**, 353, doi:10.1071/RJ09076.

Miller, R.F., J.C. Chambers, D.A. Pyke, and F.B. Pierson, 2013: *A Review of Fire Effects on Vegetation and Soils in the Great Basin Region: Response and Ecological Site Characteristics*. US Department of Agriculture, Forest Service, Rocky Mountain Research Station, Colorado, USA, 126 pp.

Miller, R.L., I. Tegen, and J. Perlwitz, 2004: Surface radiative forcing by soil dust aerosols and the hydrologic cycle. *J. Geophys. Res. Atmos.*, **109**, D04203: 1–24 D04203,n/a, doi:10.1029/2003JD004085.

Milly, P.C.D. and K.A. Dunne, 2016: Potential evapotranspiration and continental drying. *Nat. Clim. Chang.*, **6**, 946–949, doi:10.1038/nclimate3046.

Milton, S.J., 2003: 'Emerging ecosystems' – a washing-stone for ecologists, economists and sociologists? *S. Afr. J. Sci.*, **99**, 404–406.

Ministry of Agriculture and Water Resources of Tunisia, and GIZ, 2007: *Stratégie nationale d'adaptation de l'agriculture tunisienne et des écosystèmes aux changements climatiques*.Eschborn, Germany, GIZ, 51 pages.

Ministry of Energy Industry and Mineral Resources, 2016: *Third National Communication of the Kingdom of Saudi Arabia*. Submitted to United Nations Framework Convention on Climate Change, Riyadh, Saudi Arabia, pp. 173–174.

Mirzabaev, A., M. Ahmed, J. Werner, J. Pender, and M. Louhaichi, 2016a: Rangelands of Central Asia: Challenges and opportunities. *J. Arid Land*, **8**, 93–108, doi:10.1007/s40333-015-0057-5.

Mirzabaev, A., J. Goedecke, O. Dubovyk, U. Djanibekov, Q.B. Le, and A. Aw-Hassan, 2016b: Economics of land degradation in Central Asia. In: *Economics of Land Degradation and Improvement – A Global Assessment for Sustainable Development* [Nkonya, E., A. Mirzabaev and J. von Braun (eds.)]. Springer International Publishing, Cham, Switzerland, pp. 261–290.

Missirian, A., and W. Schlenker, 2017: Asylum applications respond to temperature fluctuations. *Science*, **358**, 1610–1614, doi:10.1126/science.aao0432.

Mo, K.C., and D.P. Lettenmaier, 2015: Heatwave flash droughts in decline. *Geophys. Res. Lett.*, **42**, 2823–2829, doi:10.1002/2015GL064018.

Molesworth, A.M., L.E. Cuevas, S.J. Connor, A.P. Morse, and M.C. Thomson, 2003: Environmental risk and meningitis epidemics in Africa. *Emerg. Infect. Dis.*, **9**, 1287–1293, doi:10.3201/eid0910.030182.

Molua, E.L., and C.M. Lambi, 2007: *The Economic Impact of Climate Change on Agriculture in Cameroon, Volume 1 of 1*. Policy Research Working Paper, The World Bank, Washington, DC, doi:10.1596/1813-9450-4364.

Morales, C. et al., 2011: *Measuring the Economic Value of Land Degradation/Desertification Considering the Effects of Climate Change. A Study for Latin America and the Caribbean*. Communication au Séminaire, Politiques, programmes et projets de lutte contre la désertification, quelles évaluations, Montpellier, France. 20 pages.

Moreno-Calles, A.I., A. Casas, E. García-Frapolli, and I. Torres-García, 2012: Traditional agroforestry systems of multi-crop 'milpa' and 'chichipera' cactus forest in the arid Tehuacán Valley, Mexico: Their management and role in people's subsistence. *Agrofor. Syst.*, **84**, 207–226, doi:10.1007/s10457-011-9460-x.

Moridnejad, A., N. Karimi, and P.A. Ariya, 2015: Newly desertified regions in Iraq and its surrounding areas: Significant novel sources of global dust particles. *J. Arid Environ.*, **116**, 1–10, doi:10.1016/J.JARIDENV.2015.01.008.

Moritz, M.A., M.-A. Parisien, E. Batllori, M.A. Krawchuk, J. Van Dorn, D.J. Ganz, and K. Hayhoe, 2012: Climate change and disruptions to global fire activity. *Ecosphere*, **3**, art49, doi:10.1890/ES11-00345.1.

Mortimore, M., 2005: Dryland development: Success stories from West Africa. *Environment*, **47**, 8–21, doi:10.3200/ENVT.47.1.8-21.

Mortimore, M., 2016: Changing paradigms for people-centred development in the Sahel. In: *The End of Desertification? Disputing Environmental Change in the Drylands* [Behnke, R., and M. Mortimore (eds.)]. Springer, Berlin, Germany, pp. 65–98.

Morton, J., 2010: Why should governmentality matter for the study of pastoral development? *Nomad. People.*, **14**, 6–30, doi:10.3167/np.2010.140102.

Morton, J., 2017: Climate change and African agriculture. In: *Making Climate Compatible Development Happen* [Nunan, F. (ed.)]. Routledge, London, UK, pp. 87–113.

Mosase, E., and L. Ahiablame, 2018: Rainfall and temperature in the Limpopo River Basin, southern Africa: Means, variations, and trends from 1979 to 2013. *Water*, **10**, 364, doi:10.3390/w10040364.

Mosley, L., 2014: Drought impacts on the water quality of freshwater systems; Review and integration. *Earth-Science Rev.*, **140**, 203–214, doi:10.1016/j.earscirev.2014.11.010.

Mostephaoui, M., Merdas, S., Sakaa, B., Hanafi, M., and Benazzouz, M., 2013: Cartographie des risques d'érosion hydrique par l'application de l'équation universelle de pertes en sol à l'aide d'un système d'information géographique dans le bassin versant d'El hamel (Boussaâda) Algérie *Journal algérien des régions Arid.*, Nspécial, 131–147.

Mote, P.W., A.F. Hamlet, M.P. Clark, and D.P. Lettenmaier, 2005: Declining mountain snowpack in western North America. *Bull. Am. Meteorol. Soc.*, **86**, 39–49, doi:10.1175/BAMS-86-1-39.

Moussa, B., E. Nkonya, S. Meyer, E. Kato, T. Johnson, and J. Hawkins, 2016: Economics of land degradation and improvement in Niger. In: *Economics of Land Degradation and Improvement – A Global Assessment for Sustainable Development* [Nkonya, E., A. Mirzabaev, and J. von Braun (eds.)]. Springer International Publishing, Cham, Switzerland, pp. 499–539.

Mudd, G.M., 2000: Mound springs of the Great Artesian Basin in South Australia: A case study from Olympic Dam. *Environ. Geol.*, **39**, 463–476, doi:10.1007/s002540050452.

Mueller, V., C. Gray, and K. Kosec, 2014: Heat stress increases long-term human migration in rural Pakistan. *Nat. Clim. Chang.*, **4**, 182–185, doi:10.1038/nclimate2103.

Mulinge, W., P. Gicheru, F. Murithi, P. Maingi, E. Kihiu, O.K. Kirui, and A. Mirzabaev, 2016: Economics of land degradation and improvement in Kenya. In: *Economics of Land Degradation and Improvement – A Global Assessment for Sustainable Development* [Nkonya, E., A. Mirzabaev, and J. von Braun (eds.)]. Springer International Publishing, Cham, Switzerland, pp. 471–498.

Müller, B., L. Johnson, and D. Kreuer, 2017: Maladaptive outcomes of climate insurance in agriculture. *Glob. Environ. Chang.*, **46**, 23–33, doi:10.1016/J.GLOENVCHA.2017.06.010.

Muriithi, B.W., and J.A. Matz, 2015: Welfare effects of vegetable commercialization: Evidence from smallholder producers in Kenya. *Food Policy*, **50**, 80–91, doi:10.1016/J.FOODPOL.2014.11.001.

Muyibul, Z., X. Jianxin, P. Muhtar, S. Qingdong, and Z. Run, 2018: Spatiotemporal changes of land use/cover from 1995 to 2015 in an oasis in the middle reaches of the Keriya River, southern Tarim Basin, Northwest China. *CATENA*, **171**, 416–425, doi:10.1016/J.CATENA.2018.07.038.

3

Myhre, G., D. Shindell, F.-M. Breon, W. Collins, J. Fuglestvedt, J. Huang, D. Koch, J.-F. Lamarque, D. Lee, B. Mendoza, T. Nakajima, A. Robock, G. Stephens, T. Takemura, and H. Zhang, 2013: Anthropogenic and Natural Radiative Forcing. In: Climate Change 2013: The Physical Science Basis. Contribution of Working Group I to the Fifth Assessment Report of the Intergovernmental Panel on Climate Change [Stocker, T.F., D. Qin, G.-K. Plattner, M. Tignor, S.K. Allen, J. Boschung, A. Nauels, Y. Xia, V. Bex and P.M. Midgley (eds.)]. Cambridge University Press, Cambridge, United Kingdon and New York, NY, USA.

Mythili, G., and J. Goedecke, 2016: Economics of Land Degradation in India. In: *Economics of Land Degradation and Improvement – A Global Assessment for Sustainable Development* [Nkonya, E., A. Mirzabaev, and J. von Braun (eds.)]. Springer International Publishing, Cham, Switzerland, pp. 431–469.

Nair, P.K.R., S.K. Saha, V.D. Nair, and S.G. Haile, 2011: Potential for greenhouse gas emissions from soil carbon stock following biofuel cultivation on degraded lands. *L. Degrad. Dev.*, **22**, 395–409, doi:10.1002/ldr.1016.

Nakasone, E., M. Torero, and B. Minten, 2014: The power of information: The ICT revolution in agricultural development. *Annu. Rev. Resour. Econ.*, **6**, 533–550, doi:10.1146/annurev-resource-100913-012714.

Namdari, S., N. Karimi, A. Sorooshian, G. Mohammadi, and S. Sehatkashani, 2018: Impacts of climate and synoptic fluctuations on dust storm activity over the Middle East. *Atmos. Environ.*, **173**, 265–276, doi:10.1016/J. ATMOSENV.2017.11.016.

Narayanamoorthy, A., 2010: Can drip method of irrigation be used to achieve the macro objectives of conservation agriculture? *Indian J. Agric. Econ.*, **65**, 428–438.

Nardone, A., B. Ronchi, N. Lacetera, M.S. Ranieri, and U. Bernabucci, 2010: Effects of climate changes on animal production and sustainability of livestock systems. *Livest. Sci.*, **130**, 57–69, doi:10.1016/j.livsci.2010.02.011.

Nastos, P.T., N. Politi, and J. Kapsomenakis, 2013: Spatial and temporal variability of the Aridity Index in Greece. *Atmos. Res.*, **119**, 140–152, doi:10.1016/j.atmosres.2011.06.017.

Naumann, G. et al., 2018: Global changes in drought conditions under different levels of warming. *Geophys. Res. Lett.*, **45**, 3285–3296, doi:10.1002/2017GL076521.

Nawrotzki, R.J., and J. DeWaard, 2016: Climate shocks and the timing of migration from Mexico. *Popul. Environ.*, **38**, 72–100, doi:10.1007/s11111-016-0255-x.

Nawrotzki, R.J., and M. Bakhtsiyarava, 2017: International climate migration: Evidence for the climate inhibitor mechanism and the agricultural pathway. *Popul. Space Place*, **23**, e2033, doi:10.1002/psp.2033.

Nawrotzki, R.J., F. Riosmena, L.M. Hunter, and D.M. Runfola, 2015: Undocumented migration in response to climate change. *Int. J. Popul. Stud.*, **1**, 60–74, doi:10.18063/IJPS.2015.01.004.

Nawrotzki, R.J., D.M. Runfola, L.M. Hunter, and F. Riosmena, 2016: Domestic and international climate migration from rural Mexico. *Hum. Ecol.*, **44**, 687–699, doi:10.1007/s10745-016-9859-0.

Nearing, M.A., C.L. Unkrich, D.C. Goodrich, M.H. Nichols, and T.O. Keefer, 2015: Temporal and elevation trends in rainfall erosivity on a 149 km^2 watershed in a semi-arid region of the American Southwest. *Int. Soil Water Conserv. Res.*, **3**, 77–85, doi:10.1016/j.iswcr.2015.06.008.

Negash, D., A. Abegaz, J.U. Smith, H. Araya, and B. Gelana, 2017: Household energy and recycling of nutrients and carbon to the soil in integrated crop-livestock farming systems: A case study in Kumbursa village, Central Highlands of Ethiopia. *GCB Bioenergy*, **9**, 1588–1601, doi:10.1111/gcbb.12459.

Nelson, G.C. et al., 2010: *Food Security, Farming, and Climate Change to 2050: Scenarios, Results, Policy Options*. International Food Policy Research Institute, Washington, DC, USA. DOI: 10.2499/9780896291867, 131 pp.

Nelson, V., L. Forsythe, and J. Morton, 2015: *Empowering Dryland Women: Capturing Opportunities in Land Rights, Governance and Resilience. A Synthesis of Thematic Papers from the Series 'Women's Empowerment*

in the Drylands. Natural Resource Institute, University of Greenwich, Chatham, UK, 11 pp.

Nerantzaki, S.D. et al., 2015: Modeling suspended sediment transport and assessing the impacts of climate change in a karstic Mediterranean watershed. *Sci. Total Environ.*, **538**, 288–297, doi:10.1016/j. scitotenv.2015.07.092.

Neuer, S., M.E. Torres-Padrón, M.D. Gelado-Caballero, M.J. Rueda, J. Hernández-Brito, R. Davenport, and G. Wefer, 2004: Dust deposition pulses to the eastern subtropical North Atlantic gyre: Does ocean's biogeochemistry respond? *Global Biogeochem. Cycles*, **18**, GB4020, doi:10.1029/2004GB002228.

Neufeldt, H., K. Langford, J. Fuller, M. Iiyama, and P. Dobie, 2015: *From Transition Fuel to Viable Energy Source: Improving Sustainability in the Sub-Saharan Charcoal Sector*. ICRAF Working Paper No. 196, World Agroforestry Centre, Nairobi, Kenya, 20 pp.

Newbold, T. et al., 2015: Global effects of land use on local terrestrial biodiversity. *Nature*, **520**, 45–50, doi:10.1038/nature14324.

Newton, A.C., C. Echeverría, E. Cantarello, and G. Bolados, 2011: Projecting impacts of human disturbances to inform conservation planning and management in a Dryland forest landscape. *Biol. Conserv.*, **144**, 1949–1960, doi:10.1016/J.BIOCON.2011.03.026.

Ngigi, S.N., 2003: What is the limit of up-scaling rainwater harvesting in a river basin? *Phys. Chem. Earth, Parts A/B/C*, **28**, 943–956, doi:10.1016/J. PCE.2003.08.015.

NIAB, 1997: *Economic utilization of salt-affected soils In Twenty Five Years (1992–1997) of NIAB (Silver Jubilee Publication)*. Nuclear Institute for Agriculture and Biology, Faisalabad, Pakistan, pp. 123–151.

Niaz, S., M.A. Ali, S. Ali, and S. Awan, 2009: Comparative water use efficiency of drip and furrow irrigation systems for off-season vegetables under plastic tunnel in rainfed areas. *Life Sci. Int. J.*, **2**, 952–955.

Niedermeyer, E.M., M. Prange, S. Mulitza, G. Mollenhauer, E. Schefuß, and M. Schulz, 2009: Extratropical forcing of Sahel aridity during Heinrich stadials. *Geophys. Res. Lett.*, **36**, L20707, doi:10.1029/2009GL039687.

Nielsen, D.L., and M.A. Brock, 2009: Modified water regime and salinity as a consequence of climate change: Prospects for wetlands of southern Australia. *Clim. Change*, **95**, 523–533, doi:10.1007/s10584-009-9564-8.

Nkonya, E., and W. Anderson, 2015: Exploiting provisions of land economic productivity without degrading its natural capital. *J. Arid Environ.*, **112**, 33–43, doi:10.1016/J.JARIDENV.2014.05.012.

Nkonya, E., F. Place, E. Kato, and M. Mwanjololo, 2015: Climate risk management through sustainable land management in Sub-Saharan Africa. In: *Sustainable Intensification to Advance Food Security and Enhance Climate Resilience in Africa* [Lal, R., B.R. Singh, D.L. Mwaseba, D. Kraybill, D.O. Hansen and L.O. Eik (eds.)]. Springer International Publishing, Cham, Switzerland, pp. 75–111.

Nkonya, E. et al., 2016a: Global cost of land degradation. In: *Economics of Land Degradation and Improvement – A Global Assessment for Sustainable Development* [Nkonya, E., A. Mirzabaev, and J. von Braun (eds.)]. Springer International Publishing, Cham, Switzerland, pp. 117–165.

Nkonya, E, T. Johnson, H.Y. Kwon, and E. Kato, 2016b: Economics of land degradation in Sub-Saharan Africa. In: *Economics of Land Degradation and Improvement – A Global Assessment for Sustainable Development* [Nkonya, E., A. Mirzabaev, and J. von Braun (eds.)]. Springer International Publishing, Cham, Switzerland, pp. 215–259.

Noojipady, P., S.D. Prince, and K. Rishmawi, 2015: Reductions in productivity due to land degradation in the drylands of the south-western united states. *Ecosyst. Heal. Sustain.*, **1**, 1–15, doi:10.1890/EHS15-0020.1.

Notaro, M., F. Alkolibi, E. Fadda, and F. Bakhrjy, 2013: Trajectory analysis of Saudi Arabian dust storms. *J. Geophys. Res. Atmos.*, **118**, 6028–6043, doi:10.1002/jgrd.50346.

Nurbekov, A. et al., 2016: Conservation agriculture for combating land degradation in Central Asia: A synthesis. *AIMS Agric. Food*, **1**, 144–156, doi:10.3934/agrfood.2016.2.144.

3

Nyagumbo, I., G. Nyamadzawo, and C. Madembo, 2019: Effects of three in-field water harvesting technologies on soil water content and maize yields in a semi-arid region of Zimbabwe. *Agric. Water Manag.*, **216**, 206–213, doi:10.1016/J.AGWAT.2019.02.023.

Nyamadzawo, G., M. Wuta, J. Nyamangara, and D. Gumbo, 2013: Opportunities for optimization of in-field water harvesting to cope with changing climate in semi-arid smallholder farming areas of Zimbabwe. *Springerplus*, **2**, 1–9, doi:10.1186/2193-1801-2-100.

Nyanga, A., A. Kessler, and A. Tenge, 2016: Key socio-economic factors influencing sustainable land management investments in the West Usambara Highlands, Tanzania. *Land Use Policy*, **51**, 260–266, doi:10.1016/J.LANDUSEPOL.2015.11.020.

Nyangena, W., 2008: Social determinants of soil and water conservation in rural Kenya. *Environ. Dev. Sustain.*, **10**, 745–767, doi:10.1007/s10668-007-9083-6.

Nyantakyi-Frimpong, H., and R. Bezner-Kerr, 2015: The relative importance of climate change in the context of multiple stressors in semi-arid Ghana. *Glob. Environ. Chang.*, **32**, 40–56, doi:10.1016/J.GLOENVCHA.2015.03.003.

Nyman, P., G. Sheridan, and P.N.J. Lane, 2010: Synergistic effects of water repellency and macropore flow on the hydraulic conductivity of a burned forest soil, south-east Australia. *Hydrol. Process.*, **24**, 2871–2887, doi:10.1002/hyp.7701.

Nyong, A., F. Adesina, and B. Osman Elasha, 2007: The value of indigenous knowledge in climate change mitigation and adaptation strategies in the African Sahel. *Mitig. Adapt. Strateg. Glob. Chang.*, **12**, 787–797, doi:10.1007/s11027-007-9099-0.

O'Brien, K. et al., 2004: Mapping vulnerability to multiple stressors: Climate change and globalization in India. *Glob. Environ. Chang.*, **14**, 303–313, doi:10.1016/J.GLOENVCHA.2004.01.001.

O'Connor, T.G. et al., 2011: Influence of grazing management on plant diversity of Highland Sourveld influence of grazing management on plant diversity of Highland Sourveld grassland. *Rangel. Ecol. Manag.*, **64**, 196–207, doi:10.2111/REM-D-10-00062.1.

O'Connor, T. G, J.R. Puttick, and M.T. Hoffman, 2014: Bush encroachment in southern Africa: Changes and causes. *African J. Range Forage Sci.*, **31**, 67–88, doi:10.2989/10220119.2014.939996.

Observatoire du Sahara et du Sahel, 2013: La Surveillance environnementale dans le circum-Sahara: Synthèse régionale Ecologie (Algérie, Burkina Faso – Kenya – Mali Niger – Sénégal – Tunisie) 2012, Observatoire du Sahara et du Sahel.

Odhiambo, G.O., 2017: Water scarcity in the Arabian Peninsula and socio-economic implications. *Appl. Water Sci.*, **7**, 2479–2492, doi:10.1007/s13201-016-0440-1.

Ogutu, S.O., and M. Qaim, 2019: Commercialization of the small farm sector and multidimensional poverty. *World Dev.*, **114**, 281–293, doi:10.1016/J.WORLDDEV.2018.10.012.

van Oijen, M., G. Bellocchi, and M. Höglind, 2018: Effects of climate change on grassland biodiversity and productivity: The need for a diversity of models. *Agronomy*, **8**, 14, doi:10.3390/agronomy8020014.

Okhovatian-Ardakani, A.R., M. Mehrabanian, F. Dehghani, and A. Akbarzadeh, 2010: Salt tolerance evaluation and relative comparison in cuttings of different pomegranate cultivars. *Plant Soil Environ.*, **2010**, 176–185. DOI: 10.17221/158/2009-PSE.

Okin, G.S. et al., 2011: Impacts of atmospheric nutrient deposition on marine productivity: Roles of nitrogen, phosphorus, and iron. *Global Biogeochem. Cycles*, **25**, n/a–n/a, doi:10.1029/2010GB003858.

Olayide, O.E., I.K. Tetteh, and L. Popoola, 2016: Differential impacts of rainfall and irrigation on agricultural production in Nigeria: Any lessons for climate-smart agriculture? *Agric. Water Manag.*, **178**, 30–36, doi:10.1016/J.AGWAT.2016.08.034.

Olsson, L., 2017: Climate migration and conflicts. In: Climate change, *Migration and Human Rights* [Manou, D., A. Baldwin, D. Cubie, A. Mihr and T. Thorp (eds.)]. Routledge, London, UK, pp. 116–128.

Olsson, L., M. Opondo, P. Tschakert, A. Agrawal, S.H. Eriksen, S. Ma, L.N. Perch, and S.A. Zakieldeen, 2014: Livelihoods and Poverty. In: Climate Change 2014: Impacts, Adaptation, and Vulnerability. Part A: Global and Sectoral Aspects. Contribution of Working Group II to the Fifth Assessment Report of the Intergovernmental Panel on Climate Change [Field, C.B., V.R. Barros, D.J. Dokken, K.J. Mach, M.D. Mastrandrea, T.E. Bilir, M. Chatterjee, K.L. Ebi, Y.O. Estrada, R.C. Genova, B. Girma, E.S. Kissel, A.N. Levy, S. MacCracken, P.R. Mastrandrea, and L.L.White, (eds.)]. Cambridge University Press, New York, USA. pp. 793–832.

Omuto, C.T., R.R. Vargas, M.S. Alim, and P. Paron, 2010: Mixed-effects modelling of time series NDVI-rainfall relationship for detecting human-induced loss of vegetation cover in drylands. *J. Arid Environ.*, **74**, 1552–1563, doi:10.1016/J.JARIDENV.2010.04.001.

Oroda, A.S., 2001: Application of remote sensing to early warning for food security and environmental monitoring in the Horn of Africa. *Int. Arch. Photogramm. Remote Sens. Spat. Inf. Sci.*, **XXXIV**, 66–72.

Orr, B.J. et al., 2017: *Scientific Conceptual Framework for Land Degradation Neutrality. A Report of the Science-Policy Interface.* United Nations Convention to Combat Desertification, Bonn, Germany, 128 pp.

Osbahr, H., C. Twyman, W. Neil Adger, and D.S.G. Thomas, 2008: Effective livelihood adaptation to climate change disturbance: Scale dimensions of practice in Mozambique. *Geoforum*, **39**, 1951–1964, doi:10.1016/J.GEOFORUM.2008.07.010.

Osgood, D. et al., 2018: Farmer perception, recollection, and remote sensing in weather index insurance: An Ethiopia case study. *Remote Sens.*, **10**, 1887, doi:10.3390/rs10121887.

OSS, 2012: *Synthèse Régionale Ecologie(Algérie – Burkina Faso – Kenya – Mali Niger – Sénégal – Tunisie) OSS*.Tunis, Tunisia, 128 pp.

Ostrom, E., 2009: A general framework for analyzing sustainability of social-ecological systems. *Science*, **325**, 419–422, doi:10.1126/science.1172133.

Ostrom, E., and H. Nagendra, 2006: Insights on linking forests, trees, and people from the air, on the ground, and in the laboratory. *Proc. Natl. Acad. Sci. U.S.A.*, **103**, 19224–19231, doi:10.1073/pnas.0607962103.

Otto-Bliesner, B.L. et al., 2014: Coherent changes of south-eastern equatorial and northern African rainfall during the last deglaciation. *Science*, **346**, 1223–1227, doi:10.1126/science.1259531.

Otuoma, J., J. Kinyamario, W. Ekaya, M. Kshatriya, and M. Nyabenge, 2009: Effects of human-livestock-wildlife interactions on habitat in an Eastern Kenya rangeland. *Afr. J. Ecol.*, **47**, 567–573, doi:10.1111/j.1365-2028.2008.01009.x.

Owain, E.L., and M.A. Maslin, 2018: Assessing the relative contribution of economic, political and environmental factors on past conflict and the displacement of people in East Africa. *Palgrave Commun.*, **4**, 47, doi:10.1057/s41599-018-0096-6.

Oweis, T., and A. Hachum, 2006: Water harvesting and supplemental irrigation for improved water productivity of dry farming systems in West Asia and North Africa. *Agric. Water Manag.*, **80**, 57–73, doi:10.1016/J.AGWAT.2005.07.004.

Oweis, T.Y., 2017: Rainwater harvesting for restoring degraded dry agropastoral ecosystems: A conceptual review of opportunities and constraints in a changing climate. *Environ. Rev.*, **25**, 135–149, doi:10.1139/er-2016-0069.

Painter, T.H., M.G. Flanner, G. Kaser, B. Marzeion, R.A. VanCuren, and W. Abdalati, 2013: End of the Little Ice Age in the Alps forced by industrial black carbon. *Proc. Natl. Acad. Sci.*, **110**, 15216–15221, doi:10.1073/PNAS.1302570110.

Painter, T.H., S.M. Skiles, J.S. Deems, W.T. Brandt, and J. Dozier, 2018: Variation in rising limb of Colorado River snowmelt runoff hydrograph controlled by dust radiative forcing in snow. *Geophys. Res. Lett.*, **45**, 797–808, doi:10.1002/2017GL075826.

De la Paix, M.J., L. Lanhai, C. Xi, S. Ahmed, and A. Varenyam, 2011: Soil degradation and altered flood risk as a consequence of deforestation. *L. Degrad. Dev.*, **24**, 478-485, doi:10.1002/ldr.1147, pp. 478–485.

Paltasingh, K.R., 2018: Land tenure security and adoption of modern rice technology in Odisha, Eastern India: Revisiting Besley's hypothesis. *Land Use Policy*, **78**, 236–244, doi:10.1016/J.LANDUSEPOL.2018.06.031.

Panda, R.M., M.D. Behera, and P.S. Roy, 2018: Assessing distributions of two invasive species of contrasting habits in future climate. *J. Environ. Manage.*, **213**, 478–488, doi:10.1016/J.JENVMAN.2017.12.053.

Pankova, E.I., 2016: Salinization of irrigated soils in the Middle-Asian Region: Old and new issues. *Arid Ecosyst.*, **6**, 241–248, doi:10.1134/S2079096116040077.

Panthou, G., T. Vischel, and T. Lebel, 2014: Recent trends in the regime of extreme rainfall in the Central Sahel. *Int. J. Climatol.*, **34**, 3998–4006, doi:10.1002/joc.3984.

Papanastasis, V.P. et al., 2017: Comparative assessment of goods and services provided by grazing regulation and reforestation in degraded Mediterranean rangelands. *L. Degrad. Dev.*, **28**, 1178–1187, doi:10.1002/ldr.2368.

Parsons, S.A., A. Kutt, E.P. Vanderduys, J.J. Perry, and L. Schwarzkopf, 2017: Exploring relationships between native vertebrate biodiversity and grazing land condition. *Rangel. J.*, **39**, 25–37, doi:10.1071/RJ16049.

Patel, S., 2011: Harmful and beneficial aspects of *Parthenium hysterophorus*: An update. *3 Biotech*, **1**, 1–9, doi:10.1007/s13205-011-0007-7.

Patten, D.T., L. Rouse, and J.C. Stromberg, 2008: Isolated spring wetlands in the Great Basin and Mojave Deserts, USA: Potential response of vegetation to groundwater withdrawal. *Environ. Manage.*, **41**, 398–413, doi:10.1007/s00267-007-9035-9.

Pattison, R.R., J.C. Jorgenson, M.K. Raynolds, and J.M. Welker, 2015: Trends in NDVI and Tundra community composition in the Arctic of NE Alaska between 1984 and 2009. *Ecosystems*, **18**, 707–719, doi:10.1007/s10021-015-9858-9.

De Pauw, E., M. Saba, and S. Ali, 2015: *Mapping Climate Change in Iraq and Jordan*. ICARDA Working Paper No. 27, Beirut, Lebanon, 141 pp., doi:10.13140/RG.2.1.1713.1365.

Pedrero, F., I. Kalavrouziotis, J.J. Alarcón, P. Koukoulakis, and T. Asano, 2010: Use of treated municipal wastewater in irrigated agriculture – Review of some practices in Spain and Greece. *Agric. Water Manag.*, **97**, 1233–1241, doi:10.1016/J.AGWAT.2010.03.003.

Pellant, M., B. Abbey, and S. Karl, 2004: Restoring the Great Basin Desert, USA: Integrating science, management, and people. *Environ. Monit. Assess.*, **99**, 169–179, doi:10.1007/s10661-004-4017-3.

Pellegrini, A.F.A. et al., 2017: Fire frequency drives decadal changes in soil carbon and nitrogen and ecosystem productivity. *Nature*, **553**, 194–198, doi:10.1038/nature24668.

Pelletier, J., A. Paquette, K. Mbindo, N. Zimba, A. Siampale, and B. Chendauka, 2018: Carbon sink despite large deforestation in African tropical dry forests (Miombo woodlands). *Environ. Res. Lett.*, **13**, 1–14, doi:10.1088/1748-9326/aadc9a.

Pender, J., A. Mirzabaev, and E. Kato, 2009: *Economic Analysis of Sustainable Land Management Options in Central Asia*. Final Report for ADB. IFPRI/ICARDA.Washington, D.C., USA, 44 pages.

Peters, D.P.C., K.M. Havstad, S.R. Archer, and O.E. Sala, 2015: Beyond desertification: New paradigms for Dryland landscapes. *Front. Ecol. Environ.*, **13**, 4–12, doi:10.1890/140276.

Petrie, M.D., S.L. Collins, A.M. Swann, P.L. Ford, and M.E. Litvak, 2015: Grassland to shrubland state transitions enhance carbon sequestration in the northern Chihuahuan Desert. *Glob. Chang. Biol.*, **21**, 1226–1235, doi:10.1111/gcb.12743.

Piao, S., J. Fang, H. Liu, and B. Zhu, 2005: NDVI-indicated decline in desertification in China in the past two decades. *Geophys. Res. Lett.*, **32**, L06402, doi:10.1029/2004GL021764.

Piao, S. et al., 2015: Detection and attribution of vegetation greening trend in China over the last 30 years. *Glob. Chang. Biol.*, **21**, 1601–1609, doi:10.1111/gcb.12795.

Pierre, C. et al., 2017: Impact of agropastoral management on wind erosion in Sahelian croplands. *L. Degrad. Dev.*, **29**, 800–811, doi:10.1002/ldr.2783.

Pierson, F.B. et al., 2011: Fire, plant invasions, and erosion events on western rangelands. *Rangel. Ecol. Manag.*, **64**, 439–449, doi:10.2111/REM-D-09-00147.1.

Pierson, F.B. et al., 2013: Hydrologic and erosion responses of sagebrush steppe following juniper encroachment, wildfire, and tree cutting. *Rangel. Ecol. Manag.*, **66**, 274–289, doi:10.2111/REM-D-12-00104.1.

Pierzynski, G., Brajendra, L. Caon, and R. Vargas, 2017: *Threats to Soils: Global Trends and Perspectives*. Global Land Outlook Working Paper 28. UNCCD, Bonn, Germany, 27 pp.

Piguet, E., 2010: Linking climate change, environmental degradation, and migration: A methodological overview. *Wiley Interdiscip. Rev. Clim. Chang.*, **1**, 517–524, doi:10.1002/wcc.54.

Pilliod, D.S., J.L. Welty, and R.S. Arkle, 2017: Refining the cheatgrass-fire cycle in the Great Basin: Precipitation timing and fine fuel composition predict wildfire trends. *Ecol. Evol.*, **7**, 8126–8151, doi:10.1002/ece3.3414.

Plaza-Bonilla, D. et al., 2015: Carbon management in dryland agricultural systems. A review. *Agron. Sustain. Dev.*, **35**, 1319–1334, doi:10.1007/s13593-015-0326-x.

Plaza, C. et al., 2018: Soil resources and element stocks in drylands to face global issues. *Sci. Rep.*, **8**, 13788, doi:10.1038/s41598-018-32229-0.

Poesen, J., 2018: Soil erosion in the Anthropocene: Research needs. *Earth Surf. Process. Landforms*, **43**, 64–84, doi:10.1002/esp.4250.

Pointing, S.B., and J. Belnap, 2012: Microbial colonization and controls in Dryland systems. *Nat. Rev. Microbiol.*, **10**, 551–562, doi:10.1038/nrmicro2831.

Pointing, S.B., and J. Belnap, 2014: Disturbance to desert soil ecosystems contributes to dust-mediated impacts at regional scales. *Biodivers. Conserv.*, **23**, 1659–1667, doi:10.1007/s10531-014-0690-x.

Polade, S.D., A. Gershunov, D.R. Cayan, M.D. Dettinger, and D.W. Pierce, 2017: Precipitation in a warming world: Assessing projected hydro-climate changes in California and other Mediterranean climate regions. *Sci. Rep.*, **7**, 1–10, doi:10.1038/s41598-017-11285-y.

Polley, H.W. et al., 2013: Climate change and North American rangelands: Trends, projections, and implications. *Rangel. Ecol. Manag.*, **66**, 493–511, doi:10.2111/REM-D-12-00068.1.

Pomposi, C., A. Giannini, Y. Kushnir, and D.E. Lee, 2016: Understanding Pacific ocean influence on interannual precipitation variability in the Sahel. *Geophys. Res. Lett.*, **43**, 9234–9242, doi:10.1002/2016GL069980.

Pontifes, P.A., P.M. García-Meneses, L. Gómez-Aíza, A.I. Monterroso-Rivas, and M. Caso Chávez, 2018: Land use/land cover change and extreme climatic events in the arid and semi-arid ecoregions of Mexico. *ATMÓSFERA*, **31**, 355–372, doi:10.20937/ATM.2018.31.04.04.

Popp, A., S. Domptail, N. Blaum, and F. Jeltsch, 2009: Landuse experience does qualify for adaptation to climate change. *Ecol. Modell.*, **220**, 694–702, doi:10.1016/J.ECOLMODEL.2008.11.015.

Porensky, L.M., K.E. Mueller, D.J. Augustine, and J.D. Derner, 2016: Thresholds and gradients in a semi-arid grassland: Long-term grazing treatments induce slow, continuous and reversible vegetation change. *J. Appl. Ecol.*, **53**, 1013–1022, doi:10.1111/1365-2664.12630.

Potter, C., and J. Weigand, 2016: Analysis of desert sand dune migration patterns from Landsat image time deries for the southern California desert. *J. Remote Sens. GIS*, **5**, 1–8, doi:10.4172/2469-4134.1000164.

Poulter, B. et al., 2014: Contribution of semi-arid ecosystems to interannual variability of the global carbon cycle. *Nature*, **509**, 600–603, doi:10.1038/nature13376.

Pourreza, M., S.M. Hosseini, A.A. Safari Sinegani, M. Matinizadeh, and W.A. Dick, 2014: Soil microbial activity in response to fire severity in Zagros

3

oak (Quercus brantii Lindl.) forests, Iran, after one year. *Geoderma*, **213**, 95–102, doi:10.1016/j.geoderma.2013.07.024.

Powell, M.J., 2009: Restoration of Degraded Subtropical Thickets in the Baviaanskloof Megareserve, South Africa. Master Thesis, Rhodes University, Grahamstown, South Africa.

Pozzi, W. et al., 2013: Toward global drought early warning capability: Expanding international cooperation for the development of a framework for monitoring and forecasting. *Bull. Am. Meteorol. Soc.*, **94**, 776–785, doi:10.1175/BAMS-D-11-00176.1.

Pradhan, P., L. Costa, D. Rybski, W. Lucht, and J.P. Kropp, 2017: A systematic study of sustainable development goal (SDG) interactions. *Earth's Futur.*, **5**, 1169–1179, doi:10.1002/2017EF000632.

Prăvălie, R., 2016: Drylands extent and environmental issues. A global approach. *Earth-Science Rev.*, **161**, 259–278, doi:10.1016/J.EARSCIREV.2016.08.003.

Prăvălie, R., C. Patriche, and G. Bandoc, 2017: Quantification of land degradation sensitivity areas in southern and central south-eastern Europe. New results based on improving DISMED methodology with new climate data. *Catena*, **158**, 309–320, doi:10.1016/j.catena.2017.07.006.

Pricope, N.G., G. Husak, D. Lopez-Carr, C. Funk, and J. Michaelsen, 2013: The climate-population nexus in the East African Horn: Emerging degradation trends in rangeland and pastoral livelihood zones. *Glob. Environ. Chang.*, **23**, 1525–1541, doi:10.1016/J.GLOENVCHA.2013.10.002.

Prince, S. et al., 2018: Chapter 4: Status and trends of land degradation and restoration and associated changes in biodiversity and ecosystem functions. In: *IPBES (2018): Assessment Report on Land Degradation and Restoration* [Montanarella, L., R. Scholes., and A. Brainich. (eds.)]. Secretariat of the Intergovernmental Science-Policy Platform on Biodiversity and Ecosystem Services, Bonn, Germany, pp. 221–338.

Prince, S.D., 2016: *Where Does Desertification Occur? Mapping Dryland Degradation at Regional to Global Scales*. Springer, Berlin, Germany, pp. 225–263.

Prince, S.D., E.B. De Colstoun, and L.L. Kravitz, 1998: Evidence from rain-use efficiencies does not indicate extensive Sahelian desertification. *Glob. Chang. Biol.*, **4**, 359–374, doi:10.1046/j.1365-2486.1998.00158.x.

Prospero, J.M., E. Blades, G. Mathison, and R. Naidu, 2005: Interhemispheric transport of viable fungi and bacteria from Africa to the Caribbean with soil dust. *Aerobiologia (Bologna).*, **21**, 1–19, doi:10.1007/s10453-004-5872-7.

Pu, B., and P. Ginoux, 2017: Projection of American dustiness in the late 21st century due to climate change. *Sci. Rep.*, **7**, 1–10, doi:10.1038/s41598-017-05431-9.

Pulatov, A. et al., 2012: Introducing conservation agriculture on irrigated meadow alluvial soils (arenosols) in Khorezm, Uzbekistan. In: *Cotton, Water, Salts and Soums* [Martius, C., I. Rudenko, J.P.A. Lamers and P.L.G. Vlek (eds.)]. Springer Netherlands, Dordrecht, Netherlands, pp. 195–217.

Qadir, M., A. Ghafoor, and G. Murtaza, 2000: Amelioration strategies for saline soils: A review. *L. Degrad. Dev.*, **11**, 501–521, doi:10.1002/1099-145X(200011/12)11:6<501::AID-LDR405>3.0.CO;2-S.

Qadir, M., A.D. Noble, A.S. Qureshi, R.K. Gupta, T. Yuldashev, and A. Karimov, 2009: Salt-induced land and water degradation in the Aral Sea Basin: A challenge to sustainable agriculture in Central Asia. *Nat. Resour. Forum*, **33**, 134–149, doi:10.1111/j.1477-8947.2009.01217.x.

Qian, Y. et al., 2015: Light-absorbing particles in snow and ice: Measurement and modeling of climatic and hydrological impact. *Adv. Atmos. Sci.*, **32**, 64–91, doi:10.1007/s00376-014-0010-0.

Qin, H., and T.F. Liao, 2016: Labor out-migration and agricultural change in rural China: A systematic review and meta-analysis. *J. Rural Stud.*, **47**, 533–541, doi:10.1016/J.JRURSTUD.2016.06.020.

Qishlaqi, A., F. Moore, and G. Forghani, 2008: Impact of untreated wastewater irrigation on soils and crops in Shiraz suburban area, SW Iran. *Environ. Monit. Assess.*, **141**, 257–273, doi:10.1007/s10661-007-9893-x.

Qureshi, H., M. Arshad, and Y. Bibi, 2014: Invasive flora of Pakistan: A critical analysis. *Int. J. Biosci.*, **6655**, 407–424, doi:10.12692/ijb/4.1.407-424.

Rachal, D.M., H.C. Monger, G.S. Okin, and D.C. Peters, 2012: Landform influences on the resistance of grasslands to shrub encroachment, Northern Chihuahuan Desert, USA. *J. Maps*, **8**, 507–513, doi:10.1080/17445647.2012.727593.

Racine, C.K., 2008: Soil in the environment: Crucible of terrestrial life: By Daniel Hillel. *Integr. Environ. Assess. Manag.*, **4**, 526. https://doi.org/10.1002/ieam.5630040427.

Rahman, M.S., 2013: Climate change, disaster and gender vulnerability: A study on two divisions of Bangladesh. *Am. J. Hum. Ecol.*, **2**, 72–82, doi:10.11634/216796221504315.

Rajot, J.L., S.C. Alfaro, L. Gomes, and A. Gaudichet, 2003: Soil crusting on sandy soils and its influence on wind erosion. *CATENA*, **53**, 1–16, doi:10.1016/S0341-8162(02)00201-1.

Ramankutty, N., A.T. Evan, C. Monfreda, and J.A. Foley, 2008: Farming the planet: 1. Geographic distribution of global agricultural lands in the year 2000. *Global Biogeochem. Cycles*, **22**, GB1003: 1-19, doi:10.1029/2007GB002952.

Ramarao, M.V.S., R. Krishnan, J. Sanjay, and T.P. Sabin, 2015: Understanding land surface response to changing South Asian monsoon in a warming climate. *Earth Syst. Dyn.*, **6**, 569–582, doi:10.5194/esd-6-569-2015.

Ramarao, M.V.S., J. Sanjay, R. Krishnan, M. Mujumdar, A. Bazaz, and A. Revi, 2018: On observed aridity changes over the semi-arid regions of India in a warming climate. *Theor. Appl. Climatol.*, 1–10, doi:10.1007/s00704-018-2513-6.

Rammal, M.M., and A.A. Jubair, 2015: Sand dunes stabilization using silica gel and cement kiln dust. *Al-Nahrain J. Eng. Sci.*, **18**, 179–191.

Rao, A.S., K.C. Singh, and J.R. Wight, 1996: Productivity of Cenchrus Ciliaris in relation to rainfall and fertilization. *J. Range Manag.*, **49**, 143, doi:10.2307/4002684.

Rao, C.H.H., 2000: Watershed development in India: Recent experience and emerging issues. *Econ. Polit. Wkly.*, **35**, 3943–3947, doi:10.2307/4409924.

Rao, F., M. Spoor, X. Ma, and X. Shi, 2016: Land tenure (in)security and crop-tree intercropping in rural Xinjiang, China. *Land Use Policy*, **50**, 102–114, doi:10.1016/J.LANDUSEPOL.2015.09.001.

Rashid, M., S.H. Abbas, and A. Rehman, 2014: The Status of highly alien invasive plants in Pakistan and their impact on the ecosystem: A Review. *Innovare J. Agric. Sci,.* **2**, 2–5.

Rashki, A., D.G. Kaskaoutis, C.J. deW. Rautenbach, P.G. Eriksson, M. Qiang, and P. Gupta, 2012: Dust storms and their horizontal dust loading in the Sistan region, Iran. *Aeolian Res.*, **5**, 51–62, doi:10.1016/J.AEOLIA.2011.12.001.

Rasmussen, K., S. D'haen, R. Fensholt, B. Fog, S. Horion, J.O. Nielsen, L.V. Rasmussen, and A. Reenberg, 2016: Environmental change in the Sahel: Reconciling contrasting evidence and interpretations. *Reg. Environ. Chang.*, **16**, 673–680, doi:10.1007/s10113-015-0778-1.

Rass, N., 2006: *Policies and Strategies to Address the Vulnerability of Pastoralists in Sub-Saharan Africa*. PPLPI Working Paper No. 37, Food and Agriculture Organization of the United Nations, Rome, Italy, 22 pp.

Ratna Reddy, V., M. Gopinath Reddy, S. Galab, J. Soussan, and O. Springate-Baginski, 2004: Participatory watershed development in India: Can it sustain rural livelihoods? *Dev. Change*, **35**, 297–326, doi:10.1111/j.1467-7660.2004.00353.x.

Ravi, S., D.D. D Breshears, T.E. Huxman, P. D'Odorico, 2010: Land degradation in drylands: Interactions among hydrologic-aeolian erosion and vegetation dynamics. *Geomorphology*, **116**, 236–245, doi:10.1016/J.GEOMORPH.2009.11.023.

Reardon, T., J.E. Taylor, K. Stamoulis, P. Lanjouw, and A. Balisacan, 2008: Effects of non-farm employment on rural income inequality in developing countries: An investment perspective. *J. Agric. Econ.*, **51**, 266–288, doi:10.1111/j.1477-9552.2000.tb01228.x.

Reardon, T. et al., 2009: Agrifood Industry Transformation and Small Farmers in Developing Countries. *World Dev.*, **37**, 1717–1727, doi:10.1016/J.WORLDDEV.2008.08.023.

Reed, M.S., E. Nkonya, M. Winslow, M. Mortimore, and A. Mirzabaev, 2011: Monitoring and assessing the influence of social, economic and policy factors on sustainable land management in drylands. *L. Degrad. Dev.*, **22**, 240–247, doi:10.1002/ldr.1048.

Reed, M.S., 2015: Reorienting land degradation towards sustainable land management: Linking sustainable livelihoods with ecosystem services in rangeland systems. *J. Environ. Manage.*, **151**, 472–485, doi:10.1016/J.JENVMAN.2014.11.010.

Reed, M.S. et al., 2011: Cross-scale monitoring and assessment of land degradation and sustainable land management: A methodological framework for knowledge management. *L. Degrad. Dev.*, **22**, 261–271, doi:10.1002/ldr.1087.

Reichardt, K., 2010: College on soil physics: Soil physical properties and processes under climate change. *Soil and Tillage Research*, **79**, 131–143, doi:10.1016/j.still.2004.07.002.

Reicosky, D., and C. Crovetto, 2014: No-till systems on the Chequen Farm in Chile: A success story in bringing practice and science together. *Int. Soil Water Conserv. Res.*, **2**, 66–77, doi:10.1016/S2095-6339(15)30014-9.

Reid, H., A. Bourne, H. Muller, K. Podvin, S. Scorgie, and V. Orindi, 2018: Chapter 16 – A framework for assessing the effectiveness of ecosystem-based approaches to adaptation. In: *Resilience – The Science of Adaptation to Climate Change* [Zommers Z. and K. Alverson (eds.)]. Elsevier, Amsterdam, Netherlands, pp. 207–216.

Reij, C., and A. Waters-Bayer (eds.), 2001: *Farmer Innovation in Africa: A Source of Inspiration for Agricultural Development*. Earthscan Publications Ltd, London, UK. ISBN 9781853838163, 384 pp.

Rengasamy, P., 2006: World salinization with emphasis on Australia. *J. Exp. Bot.*, **57**, 1017–1023, doi:10.1093/jxb/erj108.

Réquier-Desjardins, M., and M. Bied-Charreton, 2006: *Évaluation économique des coûts économiques et sociaux de la désertification en Afrique*. Centre d'Economie et d'Ethique pour l'Environnement et le Développement, Université de Versailles St Quentin- en- Yvelines, Paris, France, 163 pp.

Revi, A., and C. Rosenzweig, 2013: *The Urban Opportunity to Enable Transformative and Sustainable Development*. Background paper for the High-Level Panel of Eminent Persons on the Post-2015 Development Agenda. Prepared by the Co-Chairs of the Sustainable Development Solutions Network Thematic Group on Sustainable Cities. Paris, France and New York, USA, 47 pp.

Rey, A., E. Pegoraro, C. Oyonarte, A. Were, P. Escribano, and J. Raimundo, 2011: Impact of land degradation on soil respiration in a steppe (*Stipa tenacissima* L.) semi-arid ecosystem in the SE of Spain. *Soil Biol. Biochem.*, **43**, 393–403, doi:10.1016/j.soilbio.2010.11.007.

Rey, B., A. Fuller, D. Mitchell, L.C.R. Meyer, and R.S. Hetem, 2017: Drought-induced starvation of aardvarks in the Kalahari: An indirect effect of climate change. *Biol. Lett.*, **13**, 20170301, doi:10.1098/rsbl.2017.0301.

Reyer, C.P.O. et al., 2013: A plant's perspective of extremes: Terrestrial plant responses to changing climatic variability. *Glob. Chang. Biol.*, **19**, 75–89, doi:10.1111/gcb.12023.

Reyer, C.P.O., 2017: Climate change impacts in Central Asia and their implications for development. *Reg. Environ. Chang.*, **17**, 1639–1650, doi:10.1007/s10113-015-0893-z.

Reyes-García, V., M. Guèze, A.C. Luz, J. Paneque-Gálvez, M.J. Macía, M. Orta-Martínez, J. Pino, and X. Rubio-Campillo, 2013: Evidence of traditional knowledge loss among a contemporary indigenous society. *Evol. Hum. Behav.*, **34**, 249–257, doi:10.1016/J.EVOLHUMBEHAV.2013.03.002.

Reynolds, J.F., P.R. Kemp, K. Ogle, and R.J. Fernández, 2004: Modifying the 'pulse–reserve' paradigm for deserts of North America: Precipitation pulses, soil water, and plant responses. *Oecologia*, **141**, 194–210, doi:10.1007/s00442-004-1524-4.

Reynolds, J.F. et al., 2007: Global desertification: Building a science for Dryland development. *Science*, **316**, 847–851, doi:10.1126/science.1131634.

Reynolds, J.F., 2011: Scientific concepts for an integrated analysis of desertification. *L. Degrad. Dev.*, **22**, 166–183, doi:10.1002/ldr.1104.

Richardson, D.M. et al., 2011: Human-mediated introductions of Australian acacias – A global experiment in biogeography. *Divers. Distrib.*, **17**, 771–787, doi:10.1111/j.1472-4642.2011.00824.x.

van Rijn, F., E. Bulte, and A. Adekunle, 2012: Social capital and agricultural innovation in Sub-Saharan Africa. *Agric. Syst.*, **108**, 112–122, doi:10.1016/J.AGSY.2011.12.003.

Rishmawi, K., S. Prince, K. Rishmawi, and S.D. Prince, 2016: Environmental and anthropogenic degradation of v egetation in the Sahel from 1982 to 2006. *Remote Sens.*, **8**, 948, doi:10.3390/rs8110948.

Robinson, B.E. et al., 2018: Incorporating land tenure security into conservation. *Conserv. Lett.*, **11**, e12383, doi:10.1111/conl.12383.

Roderick, M.L., P. Greve, and G.D. Farquhar, 2015: On the assessment of aridity with changes in atmospheric CO_2. *Water Resour. Res.*, **51**, 5450–5463, doi:10.1002/2015WR017031.

Rodima-Taylor, D., M.F. Olwig, and N. Chhetri, 2012: Adaptation as innovation, innovation as adaptation: An institutional approach to climate change. *Appl. Geogr.*, **33**, 107–111, doi:10.1016/J.APGEOG.2011.10.011.

Rodriguez-Caballero, E., J. Belnap, B. Büdel, P.J. Crutzen, M.O. Andreae, U. Pöschl, and B. Weber, 2018: Dryland photoautotrophic soil surface communities endangered by global change. *Nat. Geosci.*, **11**, 185–189, doi:10.1038/s41561-018-0072-1.

Rojas-Downing, M.M., A.P. Nejadhashemi, T. Harrigan, and S.A. Woznicki, 2017: Climate change and livestock: Impacts, adaptation, and mitigation. *Clim. Risk Manag.*, **16**, 145–163, doi:10.1016/j.crm.2017.02.001.

Rosenfeld, D., Y. Rudich, and R. Lahav, 2001: Desert dust suppressing precipitation: A possible desertification feedback loop. *Proc. Natl. Acad. Sci. U.S.A.*, **98**, 5975–5980, doi:10.1073/pnas.101122798.

Rosenzweig, C. et al., 2014: Assessing agricultural risks of climate change in the 21st century in a global gridded crop model intercomparison. *Proc. Natl. Acad. Sci. U.S.A.*, **111**, 3268–3273, doi:10.1073/pnas.1222463110.

Rotenberg, E. and D. Yakir, 2010: Contribution of semi-arid forests to the climate system. *Science*, **327**, 451–454, doi:10.1126/science.1179998.

Rotstayn, L.D., and U. Lohmann, 2002: Tropical rainfall trends and the indirect aerosol effect. *J. Clim.*, **15**, 2103–2116, doi:10.1175/1520-0442(2002)015<2103:TRTATI>2.0.CO;2.

Rouget, M., M.P. Robertson, J.R.U. Wilson, C. Hui, F. Essl, J.L. Renteria, and D.M. Richardson, 2016: Invasion debt – Quantifying future biological invasions. *Divers. Distrib.*, **22**, 445–456, doi:10.1111/ddi.12408.

le Roux, B., M. van der Laan, T. Vahrmeijer, K.L. Bristow, and J.G. Annandale, 2017: Establishing and testing a catchment water footprint framework to inform sustainable irrigation water use for an aquifer under stress. *Sci. Total Environ.*, **599–600**, 1119–1129, doi:10.1016/J.SCITOTENV.2017.04.170.

Ruiz, J., M. Bielza, A. Garrido, and A. Iglesias, 2015: Dealing with drought in irrigated agriculture through insurance schemes: An application to an irrigation district in southern Spain. *Spanish J. Agric. Res.*, **13**, e0106, doi:10.5424/sjar/2015134-6941.

Rusike, J., S. Twomlow, H.A. Freeman, and G.M. Heinrich, 2006: Does farmer participatory research matter for improved soil fertility technology development and dissemination in southern Africa? *Int. J. Agric. Sustain.*, **4**, 176–192, doi:10.1080/14735903.2006.9684801.

Rutherford, W.A., T.H. Painter, S. Ferrenberg, J. Belnap, G.S. Okin, C. Flagg, and S.C. Reed, 2017: Albedo feedbacks to future climate via climate change impacts on dryland biocrusts. *Sci. Rep.*, **7**, 44188, doi:10.1038/srep44188.

Rutledge, S., D.I. Campbell, D. Baldocchi, and L.A. Schipper, 2010: Photodegradation leads to increased carbon dioxide losses from terrestrial organic matter. *Glob. Chang. Biol.*, **16**, 3065–3074, doi:10.1111/j.1365-2486.2009.02149.x.

Sacande, M., 2018: *Action Against Desertification, Land Restoration*. Food and Agriculture Organization of the United Nations, Rome, Italy.

3

Safriel, U., Adeel, Z. et al., 2005: Dryland systems. In: *Ecosystems and Human Well-Being: Current State and Trends* [Hassan, R., R.J. Scholes, and N. Ash, (eds.)]. Island Press, Washington, DC, USA, pp. 623–662.

Safriel, U., 2009: Deserts and desertification: Challenges but also opportunities. *L. Degrad. Dev.*, **20**, 353–366, doi:10.1002/ldr.935.

Safriel, U., and Z. Adeel, 2008: Development paths of drylands: Thresholds and sustainability. *Sustain. Sci.*, **3**, 117–123, doi:10.1007/s11625-007-0038-5.

Safriel, U.N., 2007: The assessment of global trends in land degradation. In: *Climate and Land Degradation* [Sivakumar, M.V.K. and N. Ndiang'ui (eds.)]. Springer, Berlin, Germany, pp. 1–38.

Safronov, O. et al., 2017: Detecting early signs of heat and drought stress in Phoenix dactylifera (date palm). *PloS One*, **12**, e0177883, doi:10.1371/journal.pone.0177883.

Saha, M.V., T.M. Scanlon, and P. D'Odorico, 2015: Examining the linkage between shrub encroachment and recent greening in water-limited southern Africa. *Ecosphere*, **6**, art156, doi:10.1890/ES15-00098.1.

Sahara and Sahel Observatory (OSS), 2016: *The Great Green Wall, a Development Programme for the Sahara and the Sahel: Projects Monitoring and Evaluation Approach Based on Geospatial Applications.* Sahara and Sahel Observatory, Tunis, Tunisia, 92 pp.

Saidan, M., A.G. Albaali, E. Alasis, and J.K. Kaldellis, 2016: Experimental study on the effect of dust deposition on solar photovoltaic panels in desert environment. *Renew. Energy*, **92**, 499–505, doi:10.1016/J.RENENE.2016.02.031.

Saifi, M., Boulghobra, N. and Fattoum, L., and M. Oesterheld, 2015: The Green Dam in Algeria as a tool to combat desertification. *Planet@Risk*, **3**, 68–71, doi:10.1016/j.jaridenv.2006.01.021.

Sakai, Y. et al., 2012: Phytodesalination of a salt-affected soil with four halophytes in China. *J. Arid L. Stud.*, **22**, 239–302.

Sala, A., D. Verdaguer, and M. Vila, 2006: Sensitivity of the invasive geophyte Oxalis pes-caprae to nutrient availability and competition. *Ann. Bot.*, **99**, 637–645, doi:10.1093/aob/mcl289.

Salamani, M., H.K. Hanifi, A. Hirche, and D. Nedjraoui, 2012: Évaluation de la sensibilitÉ À la dÉsertification en Algerie. *Rev. Écol. (Terre Vie)*, **67**, 71–84.

Salami, H., N. Shahnooshi, and K.J. Thomson, 2009: The economic impacts of drought on the economy of Iran: An integration of linear programming and macroeconometric modelling approaches. *Ecol. Econ.*, **68**, 1032–1039, doi:10.1016/J.ECOLECON.2008.12.003.

Salik, K.M., A. Qaisrani, M.A. Umar, and S.M. Ali, 2017: *Migration Futures in Asia and Africa: Economic Opportunities and Distributional Effects – the Case of Pakistan.* Working Paper, Sustainable Development Policy Institute, Islamabad, Pakistan, 59 pp.

Salvati, L., 2014: A socioeconomic profile of vulnerable land to desertification in Italy. *Sci. Total Environ.*, **466–467**, 287–299. doi.org/10.1016/j.scitotenv.2013.06.091.

Salvati, L., M. Zitti, T. Ceccarelli, and L. Perini, 2009: Developing a synthetic index of land vulnerability to drought and desertification. *Geogr. Res.*, **47**, 280–291, doi:10.1111/j.1745-5871.2009.00590.x.

Samoli, E., E. Kougea, P. Kassomenos, A. Analitis, and K. Katsouyanni, 2011: Does the presence of desert dust modify the effect of PM$_{10}$ on mortality in Athens, Greece? *Sci. Total Environ.*, **409**, 2049–2054, doi:10.1016/J.SCITOTENV.2011.02.031.

Samuel, A. et al., 2007: *Watershed Development in Maharashtra: Present Scenario and Issues for Restructuring the Programme.* www.indiawaterportal.org/articles/watershed-development-maharashtra-present-scenario-and-issues-restructuring-programme. Society for Promoting Participative Ecosystem Management (SOPPECOM), Pune, India, 158 pp.

Samuel, A., and K.J. Joy, 2018: Changing land use, agrarian context and rural transformation. In: *India's Water Futures* [Joy, K. J and N. Janakarajan (eds.)]. Routledge, India, pp. 57–78.

Sanandiya, N.D., and A.K. Siddhanta, 2014: Chemical studies on the polysaccharides of Salicornia brachiata. *Carbohydr. Polym.*, **112**, 300–307, doi:10.1016/J.CARBPOL.2014.05.072.

Sanz, M.J. et al., 2017: *Sustainable Land Management Contribution to Successful Land-Based Climate Change Adaptation and Mitigation. A Report of the Science-Policy Interface.* United Nations Convention to Combat Desertification, Bonn, Germany, 170 pp.

Saparov, A., 2014: Soil resources of the Republic of Kazakhstan: Current status, problems and solutions. In: *Novel Measurement and Assessment Tools for Monitoring and Management of Land and Water Resources in Agricultural Landscapes of Central Asia* [Mueller, L., A. Saparov and G. Lischeid (eds.)]. Springer International Publishing, Cham, Switzerland, pp. 61–73.

Sarver, T., A. Al-Qaraghuli, and L.L. Kazmerski, 2013: A comprehensive review of the impact of dust on the use of solar energy: History, investigations, results, literature, and mitigation approaches. *Renew. Sustain. Energy Rev.*, **22**, 698–733, doi:10.1016/J.RSER.2012.12.065.

Scasta, J.D., D.M. Engle, S.D. Fuhlendorf, D.D. Redfearn, and T.G. Bidwell, 2015: Meta-analysis of exotic forages as invasive plants in complex multi-functioning landscapes. *Invasive Plant Sci. Manag.*, **8**, 292–306, doi:10.1614/ipsm-d-14-00076.1.

Scheff, J., and D.M.W. Frierson, 2012: Robust future precipitation declines in CMIP5 largely reflect the poleward expansion of model subtropical dry zones. *Geophys. Res. Lett.*, **39**, doi:10.1029/2012GL052910.

Scheff, J., and D.M W Frierson, 2015: Terrestrial aridity and its response to greenhouse warming across CMIP5 climate models. *J. Clim.*, **28**, 5583–5600, doi:10.1175/JCLI-D-14-00480.1.

Schewe, J., and A. Levermann, 2017: Non-linear intensification of Sahel rainfall as a possible dynamic response to future warming. *Earth Syst. Dynam*, **8**, 495–505, doi:10.5194/esd-8-495-2017.

Schiappacasse, I., L. Nahuelhual, F. Vásquez, and C. Echeverría, 2012: Assessing the benefits and costs of Dryland forest restoration in central Chile. *J. Environ. Manage.*, **97**, 38–45. https://doi.org/10.1016/j.jenvman.2011.11.007.

Schilling, J., Akuno, M., Scheffran, J., and Weinzierl, T., 2014: On raids and relations: Climate change, pastoral conflict and adaptation in north-western Kenya. In: *Climate Change and Conflict: Where to for Conflict-Sensitive Adaptation to Climate Change in Africa?* [Bronkhorst, S. and U. Bob (eds.)]. Human Research Council, Durban, South Africa. Pages 241–265.

Schlenker, W., and D.B. Lobell, 2010: Robust negative impacts of climate change on African agriculture. *Environ. Res. Lett.*, **5**, 14010, doi:10.1088/1748-9326/5/1/014010.

Schlesinger, C., S. White, and S. Muldoon, 2013: Spatial pattern and severity of fire in areas with and without buffel grass (Cenchrus ciliaris) and effects on native vegetation in central Australia. *Austral Ecol.*, **38**, 831–840, doi:10.1111/aec.12039.

Schlesinger, P., Y. Mamane, and I. Grishkan, 2006: Transport of microorganisms to Israel during Saharan dust events. *Aerobiologia (Bologna).*, **22**, 259–273, doi:10.1007/s10453-006-9038-7.

Schlesinger, W.H. et al., 1990: Biological feedbacks in global desertification. *Science,* **4946**, 1043–1048, doi:10.1126/science.247.4946.1043.

Schmidt, E., P. Chinowsky, S. Robinson, and K. Strzepek, 2017: Determinants and impact of sustainable land management (SLM) investments: A systems evaluation in the Blue Nile Basin, Ethiopia. *Agric. Econ.*, **48**, 613–627, doi:10.1111/agec.12361.

Schneider, F.D., and S. Kéfi, 2016: Spatially heterogeneous pressure raises risk of catastrophic shifts. *Theor. Ecol.*, **9**, 207–217, doi:10.1007/s12080-015-0289-1.

Schofield, R. V, and M.J. Kirkby, 2003: Application of salinization indicators and initial development of potential global soil salinization scenario under climatic change. *Global Biogeochem. Cycles*, **17**, doi:10.1029/2002GB001935.

Scholes, R.J., 2009: Syndromes of dryland degradation in southern Africa. *African J. Range Forage Sci.*, **26**, 113–125, doi:10.2989/AJRF.2009.26.3.2.947.

3

Scholes, R.J., and B.H. Walker and B.H. Walker, 1993: *An African Savanna: Synthesis of the Nylsvley Study*. Cambridge University Press, Cambridge, UK. https://doi.org/10.1017/CBO9780511565472, 306 pp.

Scholes, R.J., and S.R. Archer, 1997: Tree-grass interactions in savannas. *Annu. Rev. Ecol. Syst.*, **28**, 517–544, doi:10.1146/annurev.ecolsys.28.1.517.

Schooley, R.L., B.T. Bestelmeyer, and A. Campanella, 2018: Shrub encroachment, productivity pulses, and core-transient dynamics of Chihuahuan Desert rodents. *Ecosphere*, **9**, doi:10.1002/ecs2.2330.

Schulz, J.J., L. Cayuela, C. Echeverria, J. Salas, and J.M. Rey Benayas, 2010: Monitoring land cover change of the Dryland forest landscape of Central Chile (1975–2008). *Appl. Geogr.*, **30**, 436–447, doi:10.1016/J.APGEOG.2009.12.003.

Scott-shaw, R., and C.D. Morris, 2015: Grazing depletes forb species diversity in the mesic grasslands of KwaZulu-Natal, South Africa. *African J. Range Forage Sci.*, **32**, 37–41, doi:10.2989/10220119.2014.901418.

Seaquist, J.W., T. Hickler, L. Eklundh, J. Ardö, and B.W. Heumann, 2009: Disentangling the effects of climate and people on Sahel vegetation dynamics. *Biogeosciences*, **6**, 469–477, doi:10.5194/bg-6-469-2009.

Sebaa A. et al., 2015: *Guide des Techniques de Lutte Contre l'Ensablement au Sahara Algérien*. CRSTRA, ISBN: 978-9931-438-05-2, Biskra, Algeria, 82 pp.

Secretariat of the Convention on Biological Diversity, 2010: *Pastoralism, Nature Conservation and Development: A Good Practice Guide*. Convention on Biological Diversity, Montreal, Canada, 40 pp.

Seddon, A.W.R., M. Macias-Fauria, P.R. Long, D. Benz, and K.J. Willis, 2016: Sensitivity of global terrestrial ecosystems to climate variability. *Nature*, **531**, 229–232, doi:10.1038/nature16986.

Seebens, H. et al., 2015: Global trade will accelerate plant invasions in emerging economies under climate change. *Glob. Chang. Biol.*, **21**, 4128–4140, doi:10.1111/gcb.13021.

Seebens, H. et al., 2018: Global rise in emerging alien species results from increased accessibility of new source pools. *Proc. Natl. Acad. Sci. U.S.A.*, **115**, E2264–E2273, doi:10.1073/pnas.1719429115.

Sefelnasr, A., W. Gossel, and P. Wycisk, 2014: Three-dimensional groundwater flow modeling approach for the groundwater management options for the Dakhla Oasis, Western Desert, Egypt. *Environ. Earth Sci.*, **72**, 1227–1241, doi:10.1007/s12665-013-3041-4.

Seidel, D.J., and W.J. Randel, 2007: Recent widening of the tropical belt: Evidence from tropopause observations. *J. Geophys. Res.*, **112**, D20113, doi:10.1029/2007JD008861.

Sen, S., and S. Ganguly, 2017: Opportunities, barriers and issues with renewable energy development – A discussion. *Renew. Sustain. Energy Rev.*, **69**, 1170–1181, doi:10.1016/j.rser.2016.09.137.

Seneviratne, S.I., and P. Ciais, 2017: Environmental science: Trends in ecosystem recovery from drought. *Nature*, **548**, 164–165, doi:10.1038/548164a.

Seneviratne, S.I., T. Corti, E.L. Davin, M. Hirschi, E.B. Jaeger, I. Lehner, B. Orlowsky, and A.J. Teuling, 2010: Investigating soil moisture-climate interactions in a changing climate: A review. *Earth-Science Rev.*, **99**, 125–161, doi:10.1016/J.EARSCIREV.2010.02.004.

Serdeczny, O. et al., 2017: Climate change impacts in Sub-Saharan Africa: From physical changes to their social repercussions. *Reg. Environ. Chang.*, **17**, 1585–1600, doi:10.1007/s10113-015-0910-2.

Sergeant, C.J., B.J. Moynahan, and W.F. Johnson, 2012: Practical advice for implementing long-term ecosystem monitoring. *J. Appl. Ecol.*, **49**, 969–973, doi:10.1111/j.1365-2664.2012.02149.x.

Serpa, D. et al., 2015: Impacts of climate and land use changes on the hydrological and erosion processes of two contrasting Mediterranean catchments. *Sci. Total Environ.*, **538**, 64–77, doi:10.1016/j.scitotenv.2015.08.033.

Serrano-Ortiz, P. et al., 2015: Seasonality of net carbon exchanges of Mediterranean ecosystems across an altitudinal gradient. *J. Arid Environ.*, **115**, 1–9, doi:10.1016/j.jaridenv.2014.12.003.

Settele, J., R. Scholes, R. Betts, S. Bunn, P. Leadley, D. Nepstad, J.T. Overpeck, and M.A. Taboada, 2014: Terrestrial and inland water systems. In: *Climate Change 2014: Impacts, Adaptation, and Vulnerability. Part A: Global and Sectoral Aspects. Contribution of Working Group II to the Fifth Assessment Report of the Intergovernmental Panel on Climate Change* [Field, C.B., V.R. Barros, D.J. Dokken, K.J. Mach, M.D. Mastrandrea, T.E. Bilir, M. Chatterjee, K.L. Ebi, Y.O. Estrada, R.C. Genova, B. Girma, E.S. Kissel, A.N. Levy, S. MacCracken, P.R. Mastrandrea, and L.L.White, (eds.)]. Cambridge University Press, New York, USA, pp. 271–359.

Shabani, F., L. Kumar, and S. Taylor, 2015: Distribution of date palms in the Middle East based on future climate scenarios. *Exp. Agric.*, **51**, 244–263, doi:10.1017/S001447971400026X.

Shackleton, C.M., and S.E. Shackleton, 2004: Use of woodland resources for direct household provisioning. In: *Indigenous Forests and Woodlands in South Africa* [Lawes, M.J., H.A.C. Eeley, C.M. Shackleton, and B.G.S. Geach (eds.)]. University of KwaZulu-Natal Press, Pietermaritzburg, South Africa. pp. 195–196.

Sharkhuu, A., A.F. Plante, O. Enkhmandal, C. Gonneau, B.B. Casper, B. Boldgiv, and P.S. Petraitis, 2016: Soil and ecosystem respiration responses to grazing, watering and experimental warming chamber treatments across topographical gradients in northern Mongolia. *Geoderma*, **269**, 91–98, doi:10.1016/j.geoderma.2016.01.041.

Sharma, R., S. Wungrampha, V. Singh, A. Pareek, and M.K. Sharma, 2016: Halophytes as bioenergy crops. *Front. Plant Sci.*, **7**, 1372, doi:10.3389/fpls.2016.01372.

Shaw, E.C., A.J. Gabric, and G.H. McTainsh, 2008: Impacts of aeolian dust deposition on phytoplankton dynamics in Queensland coastal waters. *Mar. Freshw. Res.*, **59**, 951, doi:10.1071/MF08087.

Sheen, K.L., D.M. Smith, N.J. Dunstone, R. Eade, D.P. Rowell, and M. Vellinga, 2017: Skilful prediction of Sahel summer rainfall on inter-annual and multi-year time scales. *Nat. Commun.*, **8**, 14966, doi:10.1038/ncomms14966.

Sheffield, J., and E.F. Wood, 2008: Projected changes in drought occurrence under future global warming from multi-model, multi-scenario, IPCC AR4 simulations. *Clim. Dyn.*, **31**, 79–105, doi:10.1007/s00382-007-0340-z.

Sheffield, J., E.F. Wood, and M.L. Roderick, 2012: Little change in global drought over the past 60 years. *Nature*, **491**, 435–438, doi:10.1038/nature11575.

De Sherbinin, A., and L. Bai, 2018: Geospatial modeling and mapping. In: *Routledge Handbook of Environmental Displacement and Migration*. Routledge, 85–91, doi:10.4324/9781315638843-6.

De Sherbinin, A. et al., 2012: Migration and risk: Net migration in marginal ecosystems and hazardous areas. *Environ. Res. Lett.*, **7**, 45602, doi:10.1088/1748-9326/7/4/045602.

Sherif, M.M., and V.P. Singh, 1999: Effect of climate change on seawater intrusion in coastal aquifers. *Hydrol. Process.*, **13**, 1277–1287, doi:10.1002/(SICI)1099-1085(19990615)13:8<1277::AID-HYP765>3.0.CO;2-W.

Sherwood, S., and Q. Fu, 2014: A drier future? *Science*, **343**, 737–739, doi:10.1126/science.1247620.

Shiferaw, B., and A. Djido, 2016: *Patterns of Labor Productivity and Income Diversification in the Rural Farm and Non-farm Sectors in Sub-Saharan Africa: Partnership for Economic Policy*. Policy Brief No. 143. Nairobi, Kenya, 3 pp.

Shiferaw, B.A., J. Okello, and R.V. Reddy, 2009: Adoption and adaptation of natural resource management innovations in smallholder agriculture: Reflections on key lessons and best practices. *Environ. Dev. Sustain.*, **11**, 601–619, doi:10.1007/s10668-007-9132-1.

Short, A.D., T.G. O'Connor, and C.R. Hurt, 2003: Medium-term changes in grass composition and diversity of Highland Sourveld grassland in the southern Drakensberg in response to fire and grazing management. *African J. Range Forage Sci.*, **20**, 1–10, doi:10.2989/10220110309485792.

Siebert, C. et al., 2016: New tools for coherent information base for IWRM in arid regions: The upper mega aquifer system on the Arabian peninsula. In: *Integrated Water Resources Management: Concept, Research and*

Implementation [Borchardt, D., J.J. Bogardi and R.B. Ibisch (eds.)]. Springer International Publishing, Cham, Switzerland, pp. 85–106.

Sietz, D., M.K.B. Lüdeke and C. Walther, 2011: Categorisation of typical vulnerability patterns in global drylands. *Glob. Environ. Chang.*, **21**, 431–440, doi:10.1016/J.GLOENVCHA.2010.11.005.

Sietz, D. et al., 2017: Nested archetypes of vulnerability in African drylands: Where lies potential for sustainable agricultural intensification? *Environ. Res. Lett.*, **12**, 95006, doi:10.1088/1748-9326/aa768b.

Sijapati Basnett, B., M. Elias, M. Ihalainen, and A.M. Paez Valencia, 2017: *Gender Matters in Forest Landscape Restoration: A framework for Design and Evaluation*. Center for International Forestry Research, Bogor, Indonesia, 12 pp.

Sikder, M.J.U., and V. Higgins, 2017: Remittances and social resilience of migrant households in rural Bangladesh. *Migr. Dev.*, **6**, 253–275, doi:10.1080/21632324.2016.1142752.

Simon, G.L., and C. Peterson, 2018: Disingenuous forests: A historical political ecology of fuelwood collection in South India. *J. Hist. Geogr.*, doi:10.1016/J.JHG.2018.09.003.

Simonneaux, V., A. Cheggour, C. Deschamps, F. Mouillot, O. Cerdan, and Y. Le Bissonnais, 2015: Land use and climate change effects on soil erosion in a semi-arid mountainous watershed (High Atlas, Morocco). *J. Arid Environ.*, **122**, 64–75, doi:10.1016/j.jaridenv.2015.06.002.

Singh, G., 2009: Salinity-related desertification and management strategies: Indian experience. *L. Degrad. Dev.*, **20**, 367–385, doi:10.1002/ldr.933.

Singh, G. et al., 2012: Effects of rainwater harvesting and afforestation on soil properties and growth of Emblica officinalis while restoring degraded hills in western India. *African J. Environ. Sci. Technol.*, **6**, 300–311, doi:10.5897/AJEST11.040.

Singh, K.P., D. Mohan, S. Sinha, and R. Dalwani, 2004: Impact assessment of treated/untreated wastewater toxicants discharged by sewage treatment plants on health, agricultural, and environmental quality in the wastewater disposal area. *Chemosphere*, **55**, 227–255, doi:10.1016/J.CHEMOSPHERE.2003.10.050.

Sivakumar, M.V.K., Ndiang'ui, N. (ed.), 2007: *Climate and Land Degradation (Environmental Science and Engineering)*. Springer-Verlag, Berlin, Germany, 623 pp.

Sivakumar, M.V.K., 2007: Interactions between climate and desertification. *Agric. For. Meteorol.*, **142**, 143–155, doi:10.1016/j.agrformet.2006.03.025.

Sloat, L.L. et al., 2018: Increasing importance of precipitation variability on global livestock grazing lands. *Nat. Clim. Chang.*, **8**, 214–218, doi:10.1038/s41558-018-0081-5.

Smit, G.N., J.N. de Klerk, M.B. Schneider, and J. van Eck, 2015: *Detailed Assessment of the Biomass Resource and Potential Yield in a Selected Bush Encroached Area Of Namibia*. Ministry of Agriculture, Water and Forestry, Windhoek, Namibia, 126 pp.

Smith, M.S., and B. Foran, 1992: An approach to assessing the economic risk of different drought management tactics on a South Australian pastoral sheep station. *Agric. Syst.*, **39**, 83–105.

Smith, P. et al., 2013: How much land-based greenhouse gas mitigation can be achieved without compromising food security and environmental goals? *Glob. Chang. Biol.*, **19**, 2285–2302, doi:10.1111/gcb.12160.

Smucker, T.A. et al., 2015: Differentiated livelihoods, local institutions, and the adaptation imperative: Assessing climate change adaptation policy in Tanzania. *Geoforum*, **59**, 39–50, doi:10.1016/j.geoforum.2014.11.018.

Snorek, J., F.G. Renaud, and J. Kloos, 2014: Divergent adaptation to climate variability: A case study of pastoral and agricultural societies in Niger. *Glob. Environ. Chang.*, **29**, 371–386, doi:10.1016/J.GLOENVCHA.2014.06.014.

Solmon, F., N. Elguindi, and M. Mallet, 2012: Radiative and climatic effects of dust over West Africa, as simulated by a regional climate model. *Clim. Res.*, **52**, 97–113, doi:10.3354/cr01039.

Sombrero, A., and A. De Benito, 2010: Carbon accumulation in soil. Ten-year study of conservation tillage and crop rotation in a semi-arid area of Castile-Leon, Spain. *Soil Tillage Res.*, **107**, 64–70, doi:10.1016/j.still.2010.02.009.

Sommer, S. et al., 2011: Application of indicator systems for monitoring and assessment of desertification from national to global scales. *L. Degrad. Dev*, **22**, 184–197, doi:10.1002/ldr.1084.

Sonneveld, B.G.J.S., and D.L. Dent, 2007: How good is GLASOD? *J. Environ. Manage.*, **90**, 274–283, doi:10.1016/J.JENVMAN.2007.09.008.

de la Sota, C., J. Lumbreras, N. Pérez, M. Ealo, M. Kane, I. Youm, and M. Viana, 2018: Indoor air pollution from biomass cookstoves in rural Senegal. *Energy Sustain. Dev.*, **43**, 224–234, doi:10.1016/J.ESD.2018.02.002.

Souissi, I., J.M. Boisson, I. Mekki, O. Therond, G. Flichman, J. Wery, and H. Belhouchette, 2018: Impact assessment of climate change on farming systems in the South Mediterranean area: A Tunisian case study. *Reg. Environ. Chang.*, **18**, 637–650, doi:10.1007/s10113-017-1130-8.

Sow, S., E. Nkonya, S. Meyer, and E. Kato, 2016: Cost, drivers and action against land degradation in Senegal. In: *Economics of Land Degradation and Improvement – A Global Assessment for Sustainable Development* [Nkonya E., A. Mirzabaev and J. von Braun (eds.)]. Springer International Publishing, Cham, Switzerland, pp. 577–608.

Spaliviero, M., M. De Dapper, C.M. Mannaerts, and A. Yachan, 2011: Participatory approach for integrated basin planning with focus on disaster risk reduction: The case of the Limpopo river. *Water (Switzerland)*, **3**, 737–763, doi:10.3390/w3030737.

Spaliviero, M., M. De Dapper, and S. Maló, 2014: Flood analysis of the Limpopo River Basin through past evolution reconstruction and a geomorphological approach. *Nat. Hazards Earth Syst. Sci.*, **14**, 2027–2039, doi:10.5194/nhess-14-2027-2014.

Spinoni, J., J. Vogt, G. Naumann, H. Carrao, and P. Barbosa, 2015: Towards identifying areas at climatological risk of desertification using the Köppen-Geiger classification and FAO Aridity Index. *Int. J. Climatol.*, **35**, 2210–2222, doi:10.1002/joc.4124.

Spinoni, J. et al., 2019: A new global database of meteorological drought events from 1951 to 2016. *J. Hydrol. Reg. Stud.*, **22**, 100593, doi:10.1016/J.EJRH.2019.100593.

Sprigg, W.A., 2016: *Dust Storms, Human Health and a Global Early Warning System*. Springer, Cham, Switzerland, 59–87 pp.

Squeo, F.A., R. Aravena, E. Aguirre, A. Pollastri, C.B. Jorquera, and J.R. Ehleringer, 2006: Groundwater dynamics in a coastal aquifer in north-central Chile: Implications for groundwater recharge in an arid ecosystem. *J. Arid Environ.*, **67**, 240–254, doi:10.1016/j.jaridenv.2006.02.012.

Srinivasa Rao, C.H., K. Sumantra, and P. Thakur, 2015: Carbon sequestration through conservation agriculture in rainfed systems. In: *Integrated Soil and Water Resource Management for Livelihood and Environmental Security* [Rajkhowa, D.J., A. Das, S.V. Ngachan, A. Sikka, and M. Lyngdoh (eds.)]. ICAR Research Complex for NEH Region, Umiam, Meghalaya, India, pp. 56–67.

Stafford-Smith, M. et al., 2017: Integration: The Key to Implementing the Sustainable Development Goals. *Sustain. Sci.*, **12**, 911–919, doi:10.1007/s11625-016-0383-3.

Stafford Smith, M., 2016: Desertification: Reflections on the Mirage. In: *The End of Desertification? Disputing Environmental Change in the Drylands* [Behnke, R. and M. Mortimore (eds.)]. Springer, Berlin, Germany, pp. 539–560.

Stahle, D.W. et al., 2009: Cool- and warm-season precipitation reconstructions over western New Mexico. *J. Clim.*, **22**, 3729–3750, doi:10.1175/2008JCLI2752.1.

Stambouli, A.B., Z. Khiat, S. Flazi, and Y. Kitamura, 2012: A review on the renewable energy development in Algeria: Current perspective, energy scenario and sustainability issues. *Renew. Sustain. Energy Rev.*, **16**, 4445–4460, doi:10.1016/j.rser.2012.04.031.

State Forestry Administration of China, 2015: *A Bulletin of Desertification and Sandification State of China*. www.documentcloud.org/documents/1237947-state-forestry-administration-desertification.html, State Forestry Administration of China, Beijing, China, 32 pp.

3

Staver, A.C., S. Archibald, and S.A. Levin, 2011: The global extent and determinant of savanna and forest as alternative biome states. *Science*, **334**, 230–233. DOI: 10.1126/science.1210465.

Stavi, I. et al., 2017: Ancient to recent-past runoff harvesting agriculture in recharge playas of the hyper-arid southern Israel. *Water*, **9**, 991, doi:10.3390/w9120991.

Stefanski, R., and M.V.K. Sivakumar, 2009: Impacts of sand and dust storms on agriculture and potential agricultural applications of a SDSWS. *IOP Conf. Ser. Earth Environ. Sci.*, **7**, 12016, doi:10.1088/1755-1307/7/1/012016.

Steger, C., B. Butt, and M.B. Hooten, 2017: Safari science: Assessing the reliability of citizen science data for wildlife surveys. *J. Appl. Ecol.*, **54**, 2053–2062, doi:10.1111/1365-2664.12921.

Sterk, G., J. Boardman, and A. Verdoodt, 2016: Desertification: History, causes and options for its control. *L. Degrad. Dev.*, **27**, 1783–1787, doi:10.1002/ldr.2525.

Stern, D.I., 2017: The environmental Kuznets curve after 25 years. *J. Bioeconomics*, **19**, 7–28, doi:10.1007/s10818-017-9243-1.

Stevens, N., C.E.R. Lehmann, B.P. Murphy, and G. Durigan, 2017: Savanna woody encroachment is widespread across three continents. *Glob. Chang. Biol.*, **23**, 235–244, doi:10.1111/gcb.13409.

Stringer, L.C., C. Twyman, and D.S.G. Thomas, 2007: Combating land degradation through participatory means: The case of Swaziland. *Ambio*, **36**, 387–393. doi.org/10.1579/0044-7447(2007)36[387:CLDTPM]2.0.CO;2.

Stringer, L.C., et al., 2012: Challenges and opportunities for carbon management in Malawi and Zambia. *Carbon Manag.*, **3**, 159–173, doi:10.4155/cmt.12.14.

Stringer, L.C. et al., 2017: A new Dryland development paradigm grounded in empirical analysis of Dryland systems science. *L. Degrad. Dev.*, **28**, 1952–1961, doi:10.1002/ldr.2716.

Suleimenov, M., Z. Kaskarbayev, K. Akshalov, and N. Yushchenko, 2014: Conservation agriculture for long-term soil productivity. In: *Novel Measurement and Assessment Tools for Monitoring and Management of Land and Water Resources in Agricultural Landscapes of Central Asia* [Mueller, L., A. Saparov and G. Lischeid (eds.)]. Springer International Publishing, Cham, Switzerland, pp. 441–454.

Sultana, F., 2014: Gendering climate change: Geographical insights. *Prof. Geogr.*, **66**, 372–381, doi:10.1080/00330124.2013.821730.

Sun, H., Z. Pan, and X. Liu, 2012: Numerical simulation of spatial-temporal distribution of dust aerosol and its direct radiative effects on East Asian climate. *J. Geophys. Res. Atmos.*, **117**, D13206: 1–14, doi:10.1029/2011JD017219.

Swann, A.L.S., F.M. Hoffman, C.D. Koven, and J.T. Randerson, 2016: Plant responses to increasing CO_2 reduce estimates of climate impacts on drought severity. *Proc. Natl. Acad. Sci. U.S.A.*, **113**, 10019–10024, doi:10.1073/pnas.1604581113.

Swemmer, A.M., M. Mashele, and P.D. Ndhlovu, 2018: Evidence for ecological sustainability of fuelwood harvesting at a rural village in South Africa. *Reg. Environ. Chang.*, 1–11, doi:10.1007/s10113-018-1402-y.

Swift, J., 1988: *Major Issues in Pastoral Development with Special Emphasis on Selected African Countries*. Food and Agriculture Organization of the United Nations, Rome, Italy, 91 pp.

Swift J., 1996: Desertification: Narratives, winners and losers. In: *The Lie of the Land – Challenging Received Wisdom on the African Environment* [Leach, M., and R. Mearns (ed.)]. International African Institute, School of Oriental and African Studies, London, UK, pp. 73–90.

Switzman, H., B. Salem, M. Gad, Z. Adeel, and P. Coulibaly, 2018: Conservation planning as an adaptive strategy for climate change and groundwater depletion in Wadi El Natrun, Egypt. *Hydrogeol. J.*, **26**, 689–703, doi:10.1007/s10040-017-1669-y.

Symeonakis, E., N. Karathanasis, S. Koukoulas, and G. Panagopoulos, 2016: Monitoring sensitivity to land degradation and desertification with the environmentally sensitive area index: The case of Lesvos Island. *L. Degrad. Dev.*, **27**, 1562–1573, doi:10.1002/ldr.2285.

Tabieh, M. et al., 2015: Economic analysis of micro-catchment rainwater harvesting techniques in Jordan's arid zones. *Int. J. Appl. Environ. Sci.*, **10**, 1205–1225.

Tadesse, M.A., B.A. Shiferaw, and O. Erenstein, 2015: Weather index insurance for managing drought risk in smallholder agriculture: Lessons and policy implications for Sub-Saharan Africa. *Agric. Food Econ.*, **3**, 26, doi:10.1186/s40100-015-0044-3.

Tal, A., 2016: Rethinking the sustainability of Israel's irrigation practices in the drylands. *Water Res.*, **90**, 387–394, doi:10.1016/J.WATRES.2015.12.016.

Tamado, T., W. Schutz, and P. Milberg, 2002: Germination ecology of the weed *Parthenium hysterophorus* in eastern Ethiopia. *Ann. Appl. Biol.*, **140**, 263–270, doi:10.1111/j.1744-7348.2002.tb00180.x.

Tambo, J.A., 2016: Adaptation and resilience to climate change and variability in north-east Ghana. *Int. J. Disaster Risk Reduct.*, **17**, 85–94, doi:10.1016/J.IJDRR.2016.04.005.

Tambo, J.A., and T. Wünscher, 2015: Identification and prioritization of farmers' innovations in northern Ghana. *Renew. Agric. Food Syst.*, **30**, 537–549, doi:10.1017/S1742170514000374.

Tamou, C., R. Ripoll-Bosch, I.J.M.M. de Boer, and S.J. Oosting, 2018: Pastoralists in a changing environment: The competition for grazing land in and around the W Biosphere Reserve, Benin Republic. *Ambio*, **47**, 340–354, doi:10.1007/s13280-017-0942-6.

Tan, M., and X. Li, 2015: Does the Green Great Wall effectively decrease dust storm intensity in China? A study based on NOAA NDVI and weather station data. *Land Use Policy*, **43**, 42–47, doi:10.1016/J.LANDUSEPOL.2014.10.017.

Tanarhte, M., P. Hadjinicolaou, and J. Lelieveld, 2015: Heatwave characteristics in the eastern Mediterranean and Middle East using extreme value theory. *Clim. Res.*, **63**, 99–113, doi:10.3354/cr01285.

Tang, Z., H. An, G. Zhu, and Z. Shangguan, 2018: Beta diversity diminishes in a chronosequence of desertification in a desert steppe. *L. Degrad. Dev.*, **29**, 543–550, doi:10.1002/ldr.2885.

Tanveer, M., S.A. Anjum, S. Hussain, A. Cerdà, and U. Ashraf, 2017: Relay cropping as a sustainable approach: Problems and opportunities for sustainable crop production. *Environ. Sci. Pollut. Res.*, **24**, 6973–6988, doi:10.1007/s11356-017-8371-4.

Tarawneh, Q., and S. Chowdhury, 2018: Trends of climate change in Saudi Arabia: Implications on water resources. *Climate*, **6**, 8, doi:10.3390/cli6010008.

Taye, G. et al., 2015: Evolution of the effectiveness of stone bunds and trenches in reducing runoff and soil loss in the semi-arid Ethiopian highlands. *Zeitschrift für Geomorphol.*, **59**, 477–493, doi:10.1127/zfg/2015/0166.

Taylor, K., T. Brummer, L.J. Rew, M. Lavin, and B.D. Maxwell, 2014: Bromus tectorum response to fire varies with climate conditions. *Ecosystems*, **17**, 960–973, doi:10.1007/s10021-014-9771-7.

Taylor, M.J., M. Aguilar-Støen, E. Castellanos, M.J. Moran-Taylor, and K. Gerkin, 2016: International migration, land use change and the environment in Ixcán, Guatemala. *Land Use Policy*, **54**, 290–301, doi:10.1016/J.LANDUSEPOL.2016.02.024.

Teka, K., 2016: Parthenium hysterophorus (asteraceae) expansion, environmental impact and controlling strategies in Tigray, Northern Ethiopia: A review. *J. Drylands*, **6**, 434–448.

Tenge, A.J., J. De Graaff, and J.P. Hella, 2004: Social and economic factors affecting the adoption of soil and water conservation in West Usambara highlands, Tanzania. *L. Degrad. Dev.*, **15**, 99–114, doi:10.1002/ldr.606.

Terink, W., W.W. Immerzeel, and P. Droogers, 2013: Climate change projections of precipitation and reference evapotranspiration for the Middle East and northern Africa until 2050. *Int. J. Climatol.*, **33**, 3055–3072, doi:10.1002/joc.3650.

Teshome, A., J. de Graaff, and A. Kessler, 2016: Investments in land management in the north-western highlands of Ethiopia: The role of social capital. *Land Use Policy*, **57**, 215–228, doi:10.1016/J.LANDUSEPOL.2016.05.019.

Theisen, O.M., 2017: Climate change and violence: Insights from political science. *Curr. Clim. Chang. Reports*, **3**, 210–221, doi:10.1007/s40641-017-0079-5.

Thiombiano, L., and I. Tourino-Soto, 2007: Status and trends in land degradation in Africa. In: *Climate and Land Degradation* [Sivakumar, M.V. K and N. Ndiang'ui (eds.)]. Springer, Berlin, Germany, pp. 39–53.

Thomas, N., and S. Nigam, 2018: 20th century climate change over Africa: Seasonal hydroclimate trends and Sahara desert expansion. *J. Clim.*, **31**, 3349–3370, doi:10.1175/JCLI-D-17-0187.1.

Thomas, R., 2008: Opportunities to reduce the vulnerability of Dryland farmers in Central and West Asia and North Africa to climate change. *Agric. Ecosyst. Environ.*, **126**, 36–45, doi:10.1016/j.agee.2008.01.011.

Thompson-Hall, M., E.R. Carr, and U. Pascual, 2016: Enhancing and expanding intersectional research for climate change adaptation in agrarian settings. *Ambio*, **45**, 373–382, doi:10.1007/s13280-016-0827-0.

Thornton, P.K., J. van de Steeg, A. Notenbaert, and M. Herrero, 2009: The impacts of climate change on livestock and livestock systems in developing countries: A review of what we know and what we need to know. *Agric. Syst.*, **101**, 113–127, doi:10.1016/J.AGSY.2009.05.002.

Thornton, P.K., P.J. Ericksen, M. Herrero, and A.J. Challinor, 2014: Climate variability and vulnerability to climate change: A review. *Glob. Chang. Biol.*, **20**, 3313–3328, doi:10.1111/gcb.12581.

Tian, F., M. Brandt, Y.Y. Liu, K. Rasmussen, and R. Fensholt, 2017: Mapping gains and losses in woody vegetation across global tropical drylands. *Glob. Chang. Biol.*, **23**, 1748–1760, doi:10.1111/gcb.13464.

Tian, F. et al., 2016: Remote sensing of vegetation dynamics in drylands: Evaluating vegetation optical depth (VOD) using AVHRR NDVI and in situ green biomass data over West African Sahel. *Remote Sens. Environ.*, **177**, 265–276, doi:10.1016/J.RSE.2016.02.056.

Tierney, J.E., F.S.R. Pausata, and P.B. DeMenocal, 2017: Rainfall regimes of the Green Sahara. *Sci. Adv.*, **3**, e1601503, doi:10.1126/sciadv.1601503.

Toderich, K.N. et al., 2009: Phytogenic resources of halophytes of Central Asia and their role for rehabilitation of sandy desert degraded rangelands. *L. Degrad. Dev.*, **20**, 386–396, doi:10.1002/ldr.936.

Toderich, K.N. et al., 2008: New approaches for biosaline agriculture development, management and conservation of sandy desert ecosystems. In: *Biosaline Agriculture and High Salinity Tolerance* [Abdelly, C., M. Öztürk, M. Ashraf, C. Grignon (eds.)]. Springer, Birkhäuser Basel, Switzerland, pp. 247–264.

Tokbergenova, A., G. Nyussupova, M. Arslan, and S.K.L. Kiyassova, 2018: Causes and impacts of land degradation and desertification: Case study from Kazakhstan. In: *Vegetation of Central Asia and Environs* [Egamberdieva, D. and M. Ozturk (eds.)]. Springer International Publishing, Cham, Switzerland, pp. 291–302.

Torell, L.A.A., S. Murugan, and O.A.A. Ramirez, 2010: Economics of flexible versus conservative stocking strategies to manage climate variability risk. *Rangel. Ecol. Manag.*, **63**, 415–425. doi.org/10.2111/REM-D-09-00131.1.

Touré, A., et al., 2019: Dynamics of wind erosion and impact of vegetation cover and land use in the Sahel: A case study on sandy dunes in south-eastern Niger. *Catene*, **177**, 272–285, doi: 10.1016/j.catena.2019.02.011.

Traore, S.B. et al., 2014: AGRHYMET: A drought monitoring and capacity building center in the West Africa region. *Weather Clim. Extrem.*, **3**, 22–30, doi:10.1016/J.WACE.2014.03.008.

Tsakiris, G., 2017: Facets of modern water resources management: Prolegomena. *Water Resour. Manag.*, **31**, 2899–2904, doi:10.1007/s11269-017-1742-2.

Tun, K.K.K., R.P. Shrestha, and A. Datta, 2015: Assessment of land degradation and its impact on crop production in the dry zone of Myanmar. *Int. J. Sustain. Dev. World Ecol.*, **22**, 533–544, doi:10.1080/13504509.2015.1091046.

Türkeş. M., 1999: Vulnerability of Turkey to desertification with respect to precipitation and aridity conditions. *Turkish J. Eng. Environ. Sci.*, **23**, 363–380.

Türkeş, M., 2003: Spatial and temporal variations in precipitation and aridity index series of Turkey. In: *Mediterranean Climate* [Bolle, H.J. (ed.)]. Springer, Berlin, Germany, pp. 181–213.

Türkeş, M., 2013: *İklim Verileri Kullanılarak Türkiye'nin Çölleşme Haritası Dokümanı Hazırlanması Raporu.* Orman ve Su İşleri Bakanlığı, Çölleşme ve Erozyonla Mücadele Genel Müdürlüğü Yayını, Ankara, Turkey, 57 pp.

Türkeş, M.2017: Recent spatiotemporal variations of synoptic meteorological sand and dust storm events observed over the Middle East and surrounding regions. *Proceedings of the 5th International Workshop on Sand and Dust Storms (SDS): Dust Sources and their Impacts in the Middle East, 23–25 October 2017.* Istanbul, Turkey, pp. 45–59.

Türkeş, M., and H. Tatlı, 2011: Use of the spectral clustering to determine coherent precipitation regions in Turkey for the period 1929–2007. *Int. J. Climatol.*, **31**, 2055–2067, doi:10.1002/joc.2212.

Türkeş, M., M. Turp, T. An, N. Ozturk, and M.L. Kurnaz, 2019: Impacts of climate change on precipitation climatology and variability in Turkey. In: *Water Resources of Turkey* [Harmancioglu, N.B. and D. Altinbilek, (eds.)]. Springer International Publishing, New York, USA. Pages 467–491.

Turner, B.L., E.F. Lambin, and A. Reenberg, 2007: The emergence of land change science for global environmental change and sustainability. *Proc. Natl. Acad. Sci.*, **104**, 20666–20671, doi:10.1073/pnas.0704119104.

Turner, M.D., J.G. McPeak, K. Gillin, E. Kitchell, and N. Kimambo, 2016: Reconciling flexibility and tenure security for pastoral resources: The geography of transhumance networks in Eastern Senegal. *Hum. Ecol.*, **44**, 199–215, doi:10.1007/s10745-016-9812-2.

Twine, W.C., and R.M. Holdo, 2016: Fuelwood sustainability revisited: Integrating size structure and resprouting into a spatially realistic fuelshed model. *J. Appl. Ecol.*, **53**, 1766–1776, doi:10.1111/1365-2664.12713.

von Uexkull, N., M. Croicu, H. Fjelde, and H. Buhaug, 2016: Civil conflict sensitivity to growing-season drought. *Proc. Natl. Acad. Sci. U.S.A.*, **113**, 12391–12396, doi:10.1073/pnas.1607542113.

Uhe, P. et al., 2018: Attributing drivers of the 2016 Kenyan drought. *Int. J. Climatol.*, **38**, e554–e568, doi:10.1002/joc.5389.

Ukkola, A.M. et al., 2018: Evaluating CMIP5 model agreement for multiple drought metrics. *J. Hydrometeorol.*, **19**, 969–988, doi:10.1175/JHM-D-17-0099.1.

Ulambayar, T., M.E. Fernández-Giménez, B. Baival, and B. Batjav, 2017: Social outcomes of community-based rangeland management in Mongolian steppe ecosystems. *Conserv. Lett.*, **10**, 317–327, doi:10.1111/conl.12267.

Umutoni, C., A. Ayantunde, M. Turner, and G.J. Sawadogo, 2016: Community participation in decentralized management of natural resources in the southern region of Mali. *Environ. Nat. Resour. Res.*, **6**, 1, doi:10.5539/enrr.v6n2p1.

UN-EMG, 2011: *Global Drylands: A UN System-wide Response*. United Nations Environment World Conservation Monitoring Centre, Cambridge, UK, 132 pp.

UN, 2015: *Outcomes and Policy-Oriented Recommendations from the UNCCD 3rd Scientific Conference*. United Nations Convention to Combat Desertification, Twelfth Session, Committee on Science and Technology, Ankara, Turkey, 15 pp.

UNCCD, 1994: *Elaboration of an international convention to combat desertification in countries experiencing serious drought and/or desertification, particularly in Africa*. General Assembly, United Nations, 1–58 pp.

UNCCD, 2017: *Global Land Outlook*. First. UNCCD, Bonn, Germany, 336 pp.

UNEP-GEF, 2008: *Desertification, Rangelands and Water Resources Working Group Final Thematic Report*, 47 pp.

UNEP, 1992: *World Atlas of Desertification* [Middleton, N. and D.S.G. Thomas (eds.)]. UNEP.Edward Arnold, London, UK, 69 pp.

UNEP, 2006: *Global Deserts Outlook*. United Nations Environment Programme, Nairobi, Kenya, 148 pp.

UNEP WMO UNCCD, 2016: *Global Assessment of Sand and Dust Storms*. United Nations Environment Programme, Nairobi, Kenya, 139 pp.

3

Unnevehr, L., 2015: Food safety in developing countries: Moving beyond exports. *Glob. Food Sec.*, **4**, 24–29, doi:10.1016/J.GFS.2014.12.001.

Uno, I. et al., 2009: Asian dust transported one full circuit around the globe. *Nat. Geosci.*, **2**, 557–560, doi:10.1038/ngeo583.

Vachtman, D., A. Sandler, N. Greenbaum, and B. Herut, 2013: Dynamics of suspended sediment delivery to the Eastern Mediterranean continental shelf. *Hydrol. Process.*, **27**, 1105–1116, doi:10.1002/hyp.9265.

Vaidyanathan, A., 2006: Restructuring watershed development programmes. *Economic & Political Weekly*, **41**, 2984–2987. www.epw.in/journal/2006/27-28/commentary/restructuring-watershed-development-programmes.html.

Varga, G., G. Újvári, and J. Kovács, 2014: Spatiotemporal patterns of Saharan dust outbreaks in the Mediterranean Basin. *Aeolian Res.*, **15**, 151–160, doi:10.1016/j.aeolia.2014.06.005.

Vargas, R. et al., 2012: Precipitation variability and fire influence the temporal dynamics of soil CO_2 efflux in an arid grassland. *Glob. Chang. Biol.*, **18**, 1401–1411, doi:10.1111/j.1365-2486.2011.02628.x.

Varghese, N., and N.P. Singh, 2016: Linkages between land use changes, desertification and human development in the Thar Desert Region of India. *Land Use Policy*, **51**, 18–25, doi:10.1016/J.LANDUSEPOL.2015.11.001.

Vega, J.A., C. Fernández, and T. Fonturbel, 2005: Throughfall, runoff and soil erosion after prescribed burning in gorse shrubland in Galicia (NW Spain). *L. Degrad. Dev.*, **16**, 37–51, doi:10.1002/ldr.643.

de Vente, J., M.S. Reed, L.C. Stringer, S. Valente, and J. Newig, 2016: How does the context and design of participatory decision-making processes affect their outcomes? Evidence from sustainable land management in global drylands. *Ecol. Soc.*, **21**, 1–24, doi:10.5751/ES-08053-210224.

Verbesselt, J., H.R., A. Zeileis, and D. Culvenor, 2010: Phenological change detection while accounting for abrupt and gradual trends in satellite image time series. *Remote Sens. Environ.*, **114**, 2970–2980, doi:10.1016/J.RSE.2010.08.003.

Verbesselt, J., R. Hyndman, G. Newnham, and D. Culvenor, 2010: Detecting trend and seasonal changes in satellite image time series. *Remote Sens. Environ.*, **114**, 106–115, doi:10.1016/J.RSE.2009.08.014.

Verbist, K., A. Amani, A. Mishra, and B. Jiménez, 2016: Strengthening drought risk management and policy: UNESCO International Hydrological Programme's case studies from Africa and Latin America and the Caribbean. *Water Policy*, **18**, 245–261, doi:10.2166/wp.2016.223.

Verhoeven, H., 2011: Climate change, conflict and development in Sudan: Global neo-malthusian narratives and local power struggles. *Dev. Change*, **42**, 679–707, doi:10.1111/j.1467-7660.2011.01707.x.

Verner, D. et al., 2018: *Climate Variability, Drought, and Drought Management in Morocco's Agricultural Sector*. World Bank Group, Washington, DC, USA. https://doi.org/10.1596/30603, 146 pp.

Verón, S.R., L.J. Blanco, M.A. Texeira, J.G.N. Irisarri, and J.M. Paruelo, 2017: Desertification and ecosystem services supply: The case of the Arid Chaco of South America. *J. Arid Environ.*, **159**, 66–74, doi:10.1016/J.JARIDENV.2017.11.001.

Verstraete, M.M. et al., 2011: Towards a global drylands observing system: Observational requirements and institutional solutions. *L. Degrad. Dev.*, **22**, 198–213, doi:10.1002/ldr.1046.

Vetter, M., C. Wirth, H. Bottcher, G. Churkina, E.-D. Schulze, T. Wutzler, and G. Weber, 2005: Partitioning direct and indirect human-induced effects on carbon sequestration of managed coniferous forests using model simulations and forest inventories. *Glob. Chang. Biol.*, **11**, 810–827, doi:10.1111/j.1365-2486.2005.00932.x.

Vetter, S., 2005: Rangelands at equilibrium and non-equilibrium: Recent developments in the debate. *J. Arid Environ.*, **62**, 321–341, doi:10.1016/J.JARIDENV.2004.11.015.

Vicente-Serrano, S.M. et al., 2012: Challenges for drought mitigation in Africa: The potential use of geospatial data and drought information systems. *Appl. Geogr.*, **34**, 471–486, doi:10.1016/J.APGEOG.2012.02.001.

Vidadili, N., E. Suleymanov, C. Bulut, and C. Mahmudlu, 2017: Transition to renewable energy and sustainable energy development in Azerbaijan.

Renew. Sustain. Energy Rev., **80**, 1153–1161, doi:10.1016/J.RSER.2017.05.168.

Vieira, R.M.S.P. et al., 2015: Identifying areas susceptible to desertification in the Brazilian Northeast. *Solid Earth*, **6**, 347–360, doi:10.5194/se-6-347-2015.

Villamor, G.B., and M. van Noordwijk, 2016: Gender specific land-use decisions and implications for ecosystem services in semi-matrilineal Sumatra. *Glob. Environ. Chang.*, **39**, 69–80, doi:10.1016/J.GLOENVCHA.2016.04.007.

Villamor, G.B., M. van Noordwijk, U. Djanibekov, M.E. Chiong-Javier, and D. Catacutan, 2014: Gender differences in land-use decisions: Shaping multifunctional landscapes? *Curr. Opin. Environ. Sustain.*, **6**, 128–133, doi:10.1016/J.COSUST.2013.11.015.

Visser, W.P., 2018: A perfect storm: The ramifications of Cape Town's drought crisis. *J. Transdiscipl. Res. South. Africa*, **14**, 1–10, doi:10.4102/td.v14i1.567.

Vogt, J.V., U. Safriel, G. Von Maltitz, Y. Sokona, R. Zougmore, G. Bastin, and J. Hill, 2011: Monitoring and assessment of land degradation and desertification: Towards new conceptual and integrated approaches. *L. Degrad. Dev.*, **22**, 150–165, doi:10.1002/ldr.1075.

Vohland, K., and B. Barry, 2009: A review of in situ rainwater harvesting (RWH) practices modifying landscape functions in African drylands. *Agric. Ecosyst. Environ.*, **131**, 119–127, doi:10.1016/j.agee.2009.01.010.

de Vries, F.T., M.E. Liiri, L. Bjørnlund, M. Bowker, S. Christensen, H. Setälä, and R.D. Bardgett, 2012: Land use alters the resistance and resilience of soil food webs to drought. *Nat. Clim. Chang.*, **2**, 276–280, doi:10.1038/nclimate1368.

Vu, Q.M., Q.B. Le, E. Frossard, and P.L.G. Vlek, 2014: Socio-economic and biophysical determinants of land degradation in Vietnam: An integrated causal analysis at the national level. *Land Use Policy*, **36**, 605–617, doi:10.1016/J.LANDUSEPOL.2013.10.012.

van Vuuren, D.P., O.E. Sala, and H.M. Pereira, 2006: The future of vascular plant diversity under four global scenarios. *Ecol. Soc.*, **11** (2): 5, 25. Van Vuuren, D. P., O. E. Sala, and H. M. Pereira. 2006. The future of vascular plant diversity under four global scenarios. Ecology and Society 11(2): 25. [online] URL: www.ecologyandsociety.org/vol11/iss2/art25/.

Waha, K., C. Müller, A. Bondeau, J.P. Dietrich, P. Kurukulasuriya, J. Heinke, and H. Lotze-Campen, 2013: Adaptation to climate change through the choice of cropping system and sowing date in Sub-Saharan Africa. *Glob. Environ. Chang.*, **23**, 130–143, doi:10.1016/J.GLOENVCHA.2012.11.001.

Waha, K. et al., 2017: Climate change impacts in the Middle East and northern Africa (MENA) region and their implications for vulnerable population groups. *Reg. Environ. Chang.*, **17**, 1623–1638, doi:10.1007/s10113-017-1144-2.

Walther, B.A., 2016: A review of recent ecological changes in the Sahel, with particular reference to land use change, plants, birds and mammals. *Afr. J. Ecol.*, **54**, 268–280, doi:10.1111/aje.12350.

Walther, G.-R. et al., 2002: Ecological responses to recent climate change. *Nature*, **416**, 389–395, doi:10.1038/416389a.

Walther, G.R., 2010: Community and ecosystem responses to recent climate change. *Philos. Trans. R. Soc. B Biol. Sci.*, **365**, 2019–2024, doi:10.1098/rstb.2010.0021.

Wang, F., X. Pan, D. Wang, C. Shen, and Q. Lu, 2013: Combating desertification in China: Past, present and future. *Land Use Policy*, **31**, 311–313, doi:10.1016/j.landusepol.2012.07.010.

Wang, G., 2005: Agricultural drought in a future climate: Results from 15 global climate models participating in the IPCC 4th assessment. *Clim. Dyn.*, **25**, 739–753, doi:10.1007/s00382-005-0057-9.

Wang, H., Y. Pan, Y. Chen, and Z. Ye, 2017a: Linear trend and abrupt changes of climate indices in the arid region of north-western China. *Atmos. Res.*, **196**, 108–118. doi.org/10.1016/j.atmosres.2017.06.008.

Wang, K., and R.E. Dickinson, 2013: Contribution of solar radiation to decadal temperature variability over land. *Proc. Natl. Acad. Sci. U.S.A.*, **110**, 14877–14882, doi:10.1073/pnas.1311433110.

Tao, W., 2014: Aeolian desertification and its control in Northern China. *Int. Soil Water Conserv. Res.*, **2**, 34–41, doi:10.1016/S2095-6339(15)30056-3.

Wang, X., Y. Yang, Z. Dong, and C. Zhang, 2009: Responses of dune activity and desertification in China to global warming in the twenty-first century. *Glob. Planet. Change*, **67**, 167–185, doi:10.1016/j.gloplacha.2009.02.004.

Wang, X., F. Yamauchi, K. Otsuka, and J. Huang, 2014: *Wage Growth, Landholding, and Mechanization in Chinese Agriculture*. Policy Research Working Paper No. WPS 7138, World Bank Group, Washington, DC, USA, 43 pp.

Wang, X., T. Hua, L. Lang, and W. Ma, 2017b: Spatial differences of aeolian desertification responses to climate in arid Asia. *Glob. Planet. Change*, **148**, 22–28, doi:10.1016/j.gloplacha.2016.11.008.

Wang, X., F. Yamauchi, J. Huang, and S. Rozelle, 2018a: What constrains mechanization in Chinese agriculture? Role of farm size and fragmentation. *China Econ. Rev.*, in press, doi:10.1016/J.CHIECO.2018.09.002.

Wang, X.M., C.X. Zhang, E. Hasi, and Z.B. Dong, 2010: Has the Three-Norths Forest Shelterbelt Program solved the desertification and dust storm problems in arid and semi-arid China? *J. Arid Environ.*, **74**, 13–22, doi:10.1016/J.JARIDENV.2009.08.001.

Wang, Y., J. Zhang, S. Tong, and E. Guo, 2017c: Monitoring the trends of aeolian desertified lands based on time-series remote sensing data in the Horqin Sandy Land, China. *Catena*, **157**, 286–298, doi:10.1016/j.catena.2017.05.030.

Wang, Y. et al., 2008: The dynamics variation of soil moisture of shelterbelts along the Tarim Desert Highway. *Sci. Bull.*, **53**, 102–108, doi:10.1007/s11434-008-6011-6.

Wang, Z., B. Zhou, L. Pei, J. Zhang, X. He, and H. Lin, 2018b: Controlling threshold in soil salinity when planting spring wheat and sequential cropping silage corn in Northern Xinjiang using drip irrigation. *Int. J. Agric. Biol. Eng.*, **11**, 108–114, doi:10.25165/IJABE.V11I2.3621.

Wario, H.T., H.G. Roba, and B. Kaufmann, 2016: Responding to mobility constraints: Recent shifts in resource use practices and herding strategies in the Borana pastoral system, southern Ethiopia. *J. Arid Environ.*, **127**, 222–234, doi:10.1016/J.JARIDENV.2015.12.005.

Waters-Bayer, A., L. van Veldhuizen, M. Wongtschowski, and C. Wettasinha, 2009: Recognizing and enhancing processes of local innovation. In: *Innovation Africa: Enriching Farmers' Livelihoods* [Sanginga, P., A. Waters-Bayer, S. Kaaria, J. Njuki, and C. Wettasinha (eds.)]. Routledge, London, UK, pp. 239–254.

Way, S.-A., 2016: Examining the links between poverty and land degradation: From blaming the poor toward recognising the rights of the poor. In: *Governing Global Desertification: Linking Environmental Degradation, Poverty and Participation* [Johnson, P.M. (ed.)]. Routledge, London, UK, pp. 47–62, doi:10.4324/9781315253916-13.

Webb, N.P., C.J. Stokes, and N.A. Marshall, 2013: Integrating biophysical and socio-economic evaluations to improve the efficacy of adaptation assessments for agriculture. *Glob. Environ. Chang.*, **23**, 1164–1177, doi:10.1016/j.gloenvcha.2013.04.007.

Webb, N.P., N.A. Marshall, L.C. Stringer, M.S. Reed, A. Chappell, and J.E. Herrick, 2017: Land degradation and climate change: Building climate resilience in agriculture. *Front. Ecol. Environ.*, **15**, 450–459, doi:10.1002/fee.1530.

Weber, B., M. Bowker, Y. Zhang, and J. Belnap, 2016: Natural recovery of biological soil crusts after disturbance. In: *Biological Soil Crusts: An Organizing Principle in Drylands* [Weber, B., B. Büdel, and J. Belnap (eds.)]. Springer International Publishing, New York, USA, pp. 479–498.

Weber, C.F., J.S. Lockhart, E. Charaska, K. Aho, and K.A. Lohse, 2014: Bacterial composition of soils in ponderosa pine and mixed conifer forests exposed to different wildfire burn severity. *Soil Biol. Biochem.*, **69**, 242–250, doi:10.1016/j.soilbio.2013.11.010.

von Wehrden, H., J. Hanspach, P. Kaczensky, J. Fischer, and K. Wesche, 2012: Global assessment of the non-equilibrium concept in rangelands. *Ecol. Appl.*, **22**, 393–399, doi:10.1890/11-0802.1.

Weinzierl, T., J. Wehberg, J. Böhner, and O. Conrad, 2016: Spatial Assessment of Land Degradation Risk for the Okavango River Catchment, southern Africa. *L. Degrad. Dev.*, **27**, 281–294, doi:10.1002/ldr.2426.

Weltz, M., and K. Spaeth, 2012: Estimating effects of targeted conservation on nonfederal rangelands. *Rangelands*, **34**, 35–40, doi:10.2111/RANGELANDS-D-12-00028.1.

Wessels, K.J., S.D. Prince, J. Malherbe, J. Small, P.E. Frost, and D. VanZyl, 2007: Can human-induced land degradation be distinguished from the effects of rainfall variability? A case study in South Africa. *J. Arid Environ.*, **68**, 271–297, doi:10.1016/J.JARIDENV.2006.05.015.

Wessels, K.J., F. van den Bergh, and R.J. Scholes, 2012: Limits to detectability of land degradation by trend analysis of vegetation index data. *Remote Sens. Environ.*, **125**, 10–22, doi:10.1016/J.RSE.2012.06.022.

White, R., and J. Nackoney, 2003: *Drylands, People, and Ecosystem Goods and Services: A Web-Based Geospatial Analysis*. World Resource Institute, Washington, DC, USA, 40 pp.

Whitford, W.G., 1997: Desertification and animal biodiversity in the desert grasslands of North America. *J. Arid Environ.*, **37**, 709–720, doi:10.1006/jare.1997.0313.

Whitford, W., 2002: *Ecology of Desert Systems*. Academic Press, San Diego, USA, 343 pp.

Wichelns, D., and M. Qadir, 2015: Achieving sustainable irrigation requires effective management of salts, soil salinity, and shallow groundwater. *Agric. Water Manag.*, **157**, 31–38, doi:10.1016/J.AGWAT.2014.08.016.

Wicke, B., E.M.W. Smeets, R. Akanda, L. Stille, R.K. Singh, A.R. Awan, K. Mahmood, and A.P.C. Faaij, 2013: Biomass production in agroforestry and forestry systems on salt-affected soils in South Asia: Exploration of the GHG balance and economic performance of three case studies. *J. Environ. Manage.*, **127**, 324–334, doi:10.1016/J.JENVMAN.2013.05.060.

Wiesmeier, M., S. Munro, F. Barthold, M. Steffens, P. Schad, and I. Kögel-Knabner, 2015: Carbon storage capacity of semi-arid grassland soils and sequestration potentials in northern China. *Glob. Chang. Biol.*, **21**, 3836–3845, doi:10.1111/gcb.12957.

Wildemeersch, J.C.J. et al., 2015: Assessing the constraints to adopt water and soil conservation techniques in Tillaberi, Niger. *L. Degrad. Dev.*, **26**, 491–501, doi:10.1002/ldr.2252.

van Wilgen, B.W. et al., 2010: Fire management in Mediterranean-climate shrublands: A case study from the Cape fynbos, South Africa. *J. Appl. Ecol.*, **47**, 631–638, doi:10.1111/j.1365-2664.2010.01800.x.

Wilhelm, W.W., and C.S. Wortmann, 2004: Tillage and rotation interactions for corn and soybean grain yield as affected by precipitation and air temperature. *Agron. J.*, **96**, 425–432, doi:10.2134/agronj2004.4250.

Wilhelm, W.W., and C.S. Wortmann, 2004: Tillage and Rotation Interaction for Corn and Soybean Grain Yield as Affected by Precipitation and Air Temperature. Agronomy Journal, **96**, 425-432, dx.doi.org/10.2134/agronj2004.0425.

Wilhite, D., and R.S. Pulwarty, 2017: *Drought and Water Crises, Integrating Science, Management, and Policy, Second Edition*. CRC Press, Boca Raton, Florida, USA, 542 pp.

Wilhite, D.A., 2016: Drought-management policies and preparedness plans: Changing the paradigm from crisis to risk management. *L. Restor.*, 443–462, doi:10.1016/B978-0-12-801231-4.00007-0.

Wilhite, D.A., M.V.K. Sivakumar, and R. Pulwarty, 2014: Managing drought risk in a changing climate: The role of national drought policy. *Weather Clim. Extrem.*, **3**, 4–13, doi:10.1016/J.WACE.2014.01.002.

Williams, A.P. et al., 2010: Forest responses to increasing aridity and warmth in the south-western United States. *Proc. Natl. Acad. Sci.*, **107**, 21289–21294, doi:10.1073/pnas.0914211107.

Willy, D.K., and K. Holm-Müller, 2013: Social influence and collective action effects on farm level soil conservation effort in rural Kenya. *Ecol. Econ.*, **90**, 94–103, doi:10.1016/J.ECOLECON.2013.03.008.

Wilting, H.C., A.M. Schipper, M. Bakkenes, J.R. Meijer, and M.A.J. Huijbregts, 2017: Quantifying biodiversity losses due to human consumption: A global-scale footprint analysis. *Environ. Sci. Technol.*, **51**, 3298–3306, doi:10.1021/acs.est.6b05296.

3

Winsemius, H.C., B. Jongman, T.I.E. Veldkamp, S. Hallegatte, M. Bangalore, and P.J. Ward, 2018: Disaster risk, climate change, and poverty: Assessing the global exposure of poor people to floods and droughts. *Environ. Dev. Econ.*, **23**, 328–348, doi:10.1017/S1355770X17000444.

Wischnath, G., and H. Buhaug, 2014: Rice or riots: On food production and conflict severity across India. *Polit. Geogr.*, **43**, 6–15, doi:10.1016/J.POLGEO.2014.07.004.

Womack, A.M., B.J.M. Bohannan, and J.L. Green, 2010: Biodiversity and biogeography of the atmosphere. *Philos. Trans. R. Soc. B Biol. Sci.*, **365**, 3645–3653, doi:10.1098/rstb.2010.0283.

Woo, A.C. et al., 2013: Temporal variation in airborne microbial populations and microbially-derived allergens in a tropical urban landscape. *Atmos. Environ.*, **74**, 291–300, doi:10.1016/j.atmosenv.2013.03.047.

Woodhouse, P., G.J. Veldwisch, J.-P. Venot, D. Brockington, H. Komakech, and Â. Manjichi, 2017: African farmer-led irrigation development: Re-framing agricultural policy and investment? *J. Peasant Stud.*, **44**, 213–233, doi:10.1080/03066150.2016.1219719.

Woolcock, M., and D. Narayan, 2000: Social capital: Implications for development theory, research, and policy. *World Bank Res. Obs.*, **15**, 25–249.

Woollen, E. et al., 2016: Supplementary materials for charcoal production in the Mopane woodlands of Mozambique: What are the tradeoffs with other ecosystem services? *Philos. Trans. R. Soc. B-Biological Sci.*, **371**, 1–14, doi:10.1098/rstb.2015.0315.

Woomer, P.L., A. Touré, and M. Sall, 2004: Carbon stocks in Senegal's Sahel transition zone. *J. Arid Environ.*, **59**, 499–510. doi.org/10.1016/j.jaridenv.2004.03.027.

World Bank, 2009: *World Development Report 2010: Development and Climate Change*. World Bank Group, Washington, DC, doi:10.1596/978-0-8213-7987-5, 417 pp.

World Bank, 2018: *Groundswell: Preparing for Internal Climate Migration*. World Bank Group, Washington DC, USA, 256 pp.

Wossen, T., T. Berger, and S. Di Falco, 2015: Social capital, risk preference and adoption of improved farm land management practices in Ethiopia. *Agric. Econ.*, **46**, 81–97, doi:10.1111/agec.12142.

Wu, J., L. Zhou, X. Mo, H. Zhou, J. Zhang, and R. Jia, 2015: Drought monitoring and analysis in China based on the Integrated Surface Drought Index (ISDI). *Int. J. Appl. Earth Obs. Geoinf.*, **41**, 23–33, doi:10.1016/J.JAG.2015.04.006.

Wu, W., 2014: The generalized difference vegetation Index (GDVI) for Dryland characterization. *Remote Sens.*, **6**, 1211–1233, doi:10.3390/rs6021211.

Xi, X., and I.N. Sokolik, 2015: Seasonal dynamics of threshold friction velocity and dust emission in Central Asia. *J. Geophys. Res. Atmos.*, **120**, 1536–1564, doi:10.1002/2014JD022471.

Xie, Y., J. Gong, P. Sun, and X. Gou, 2014: Oasis dynamics change and its influence on landscape pattern on Jinta oasis in arid China from 1963a to 2010a: Integration of multi-source satellite images. *Int. J. Appl. Earth Obs. Geoinf.*, **33**, 181–191, doi:10.1016/j.jag.2014.05.008.

Xu, D., X. You, and C. Xia, 2019: Assessing the spatial-temporal pattern and evolution of areas sensitive to land desertification in North China. *Ecol. Indic.*, **97**, 150–158, doi:10.1016/J.ECOLIND.2018.10.005.

Xu, D.Y., X.W. Kang, D.F. Zhuang, and J.J. Pan, 2010: Multi-scale quantitative assessment of the relative roles of climate change and human activities in desertification – A case study of the Ordos Plateau, China. *J. Arid Environ.*, **74**, 498–507, doi:10.1016/j.jaridenv.2009.09.030.

Xue, Z., Z. Qin, F. Cheng, G. Ding, and H. Li, 2017: Quantitative assessment of aeolian desertification dynamics – A case study in North Shanxi of China (1975 to 2015). *Sci. Rep.*, **7**, 10460, doi:10.1038/s41598-017-11073-8.

Yahiya, A.B., 2012: Environmental degradation and its impact on tourism in Jazan, KSA using remote sensing and GIS. *International Journal Environmental Sciences*, **3**, 421–432, doi:10.6088/ijes.2012030131041.

Yaish, M.W., and P.P. Kumar, 2015: Salt tolerance research in date palm tree (Phoenix dactylifera L.), past, present, and future perspectives. *Front. Plant Sci.*, **6**, 348, doi:10.3389/fpls.2015.00348.

Yan, X., and Y.L. Cai, 2015: Multi-scale anthropogenic driving forces of Karst rocky desertification in south-west China. *L. Degrad. Dev.*, **26**, 193–200, doi:10.1002/ldr.2209.

Yang, D., S. Kanae, T. Oki, T. Koike, and K. Musiake, 2003: Global potential soil erosion with reference to land use and climate changes. *Hydrol. Process.*, **17**, 2913–2928, doi:10.1002/hyp.1441.

Yang, F., and C. Lu, 2015: Spatiotemporal variation and trends in rainfall erosivity in China's Dryland region during 1961–2012. *Catena*, **133**, 362–372, doi:10.1016/j.catena.2015.06.005.

Yang, H. et al., 2014: Carbon sequestration capacity of shifting sand dune after establishing new vegetation in the Tengger Desert, northern China. *Sci. Total Environ.*, **478**, 1–11, doi:10.1016/J.SCITOTENV.2014.01.063.

Yang, L., J. Wu, and P. Shen, 2013: Roles of science in institutional changes: The case of desertification control in China. *Environ. Sci. Policy*, **27**, 32–54, doi:10.1016/j.envsci.2012.10.017.

Yang, Y., Z. Wang, J. Li, C. Gang, Y. Zhang, Y. Zhang, I. Odeh, and J. Qi, 2016: Comparative assessment of grassland degradation dynamics in response to climate variation and human activities in China, Mongolia, Pakistan and Uzbekistan from 2000 to 2013. *J. Arid Environ.*, **135**, 164–172, doi:10.1016/J.JARIDENV.2016.09.004.

Yanoff, S., and E. Muldavin, 2008: Grassland-shrubland transformation and grazing: A century-scale view of a northern Chihuahuan Desert grassland. *J. Arid Environ.*, **72**, 1594–1605, doi:10.1016/J.JARIDENV.2008.03.012.

Yaro, J.A., J. Teye, and S. Bawakyillenuo, 2015: Local institutions and adaptive capacity to climate change/variability in the northern savanna of Ghana. *Clim. Dev.*, **7**, 235–245, doi:10.1080/17565529.2014.951018.

Ye, X., X. Yu, C. Yu, A. Tayibazhaer, F. Xu, A.K. Skidmore, and T. Wang, 2018: Impacts of future climate and land cover changes on threatened mammals in the semi-arid Chinese Altai Mountains. *Sci. Total Environ.*, **612**, 775–787, doi:10.1016/J.SCITOTENV.2017.08.191.

Yi, C., S. Wei, and G. Hendrey, 2014: Warming climate extends dryness-controlled areas of terrestrial carbon sequestration. *Sci. Rep.*, **4**, 1–6, doi:10.1038/srep05472.

Yin, F., X. Deng, Q. Jin, Y. Yuan, and C. Zhao, 2014: The impacts of climate change and human activities on grassland productivity in Qinghai Province, China. *Front. Earth Sci.*, **8**, 93–103, doi:10.1007/s11707-013-0390-y.

Yin, Y., D. Ma, S. Wu, and T. Pan, 2015: Projections of aridity and its regional variability over China in the mid-21st century. *Int. J. Climatol.*, **35**, 4387–4398, doi:10.1002/joc.4295.

Yosef, B.A.A., and D.K.K. Asmamaw, 2015: Rainwater harvesting: An option for dry land agriculture in arid and semi-arid Ethiopia. *Int. J. Water Resour. Environ. Eng.*, **7**, 17–28, doi:10.5897/IJWREE2014.0539.

You, N., J. Meng, and L. Zhu, 2018: Sensitivity and resilience of ecosystems to climate variability in the semi-arid to hyper-arid areas of Northern China: A case study in the Heihe River Basin. *Ecol. Res.*, **33**, 161–174, doi:10.1007/s11284-017-1543-3.

Yu, K., P. D'Odorico, A. Bhattachan, G.S. Okin, and A.T. Evan, 2015: Dust-rainfall feedback in West African Sahel. *Geophys. Res. Lett.*, **42**, 7563–7571, doi:10.1002/2015GL065533.

Yu Qiu, G., I.-B. Lee, H. Shimizu, Y. Gao, and G. Ding, 2004: Principles of sand dune fixation with straw checkerboard technology and its effects on the environment. *J. Arid Environ.*, **56**, 449–464, doi:10.1016/S0140-1963(03)00066-1.

Yu, Y. et al., 2013: Assessing temporal and spatial variations in atmospheric dust over Saudi Arabia through satellite, radiometric, and station data. *J. Geophys. Res. Atmos.*, **118**, 253–266, doi:10.1002/2013JD020677.

Yurtoglu, M.A., 2015: The afforestation and erosion control mobilization action plan in Turkey. In: *Living Land* [Griffiths, J. (ed.)]. UNCCD and Tudor Rose, Leicester, United Kingdom, pp. 37–39.

Yusa, A. et al., 2015: Climate change, drought and human health in Canada. *Int. J. Environ. Res. Public Health*, **12**, 8359–8412, doi:10.3390/ijerph120708359.

van der Zaag, P., and H. Savenije, 1999: The management of international waters in EU and SADC compared. *Phys. Chem. Earth, Part B Hydrol. Ocean. Atmos.*, **24**, 579–589, doi:10.1016/S1464-1909(99)00048-9.

Zabaleta, A., M. Meaurio, E. Ruiz, and I. Antigüedad, 2013: Simulation climate change impact on runoff and sediment yield in a small watershed in the Basque country, northern Spain. *J. Environ. Qual.*, **43**, 235, doi:10.2134/jeq2012.0209.

Zaitchik, B.F. et al., 2007: Climate and vegetation in the Middle East: Interannual variability and drought feedbacks. *J. Clim.*, **20**, 3924–3941, doi:10.1175/JCLI4223.1.

Zaitchik, B.F. et al., 2013: Representation of soil moisture feedbacks during drought in NASA Unified WRF (NU-WRF). *J. Hydrometeorol.*, **14**, 360–367, doi:10.1175/JHM-D-12-069.1.

van Zanten, B.T. et al., 2014: European agricultural landscapes, common agricultural policy and ecosystem services: A review. *Agron. Sustain. Dev.*, **34**, 309–325, doi:10.1007/s13593-013-0183-4.

Zare, M., A.A. Nazari Samani, M. Mohammady, T. Teimurian, and J. Bazrafshan, 2016: Simulation of soil erosion under the influence of climate change scenarios. *Environ. Earth Sci.*, **75**, 1–15, doi:10.1007/s12665-016-6180-6.

Zdruli, P., 2011: *Desertification in the Mediterranean Region*. European Institute of the Mediterranean, Girona, Barcelona, 250–255 pp.

Zeng, N., and J. Yoon, 2009: Expansion of the world's deserts due to vegetation-albedo feedback under global warming. *Geophys. Res. Lett.*, **36**, L17401, doi:10.1029/2009GL039699.

Zhang, J. et al., 2015: Effects of sand dune stabilization on the spatial pattern of Artemisia ordosica population in Mu Us desert, Northwest China. *PLoS One*, **10**, e0129728, doi:10.1371/journal.pone.0129728.

Zhang, K., J. Qu, K. Liao, Q. Niu, and Q. Han, 2010: Damage by wind-blown sand and its control along Qinghai-Tibet railway in China. *Aeolian Res.*, **1**, 143–146, doi:10.1016/J.AEOLIA.2009.10.001.

Zhang, W., J. Zhou, G. Feng, D.C. Weindorf, G. Hu, and J. Sheng, 2015: Characteristics of water erosion and conservation practice in arid regions of Central Asia: Xinjiang Province, China as an example. *Int. Soil Water Conserv. Res.*, **3**, 97–111, doi:10.1016/j.iswcr.2015.06.002.

Zhang, X. et al., 2016: A systematic review of global desert dust and associated human health effects. *Atmosphere (Basel).*, **7**, 158, doi:10.3390/atmos7120158.

Zhang, X.C., and M.A. Nearing, 2005: Impact of climate change on soil erosion, runoff, and wheat productivity in central Oklahoma. *CATENA*, **61**, 185–195, doi:10.1016/J.CATENA.2005.03.009.

Zhang, X.Y., R. Arimoto, and Z.S. An, 1997: Dust emission from Chinese desert sources linked to variations in atmospheric circulation. *J. Geophys. Res. Atmos.*, **102**, 28041–28047, doi:10.1029/97JD02300.

Zhang, Y., C. Song, L.E. Band, G. Sun, and J. Li, 2017: Reanalysis of global terrestrial vegetation trends from MODIS products: Browning or greening? *Remote Sens. Environ.*, **191**, 145–155, doi:10.1016/J.RSE.2016.12.018.

Zhang, Z., and D. Huisingh, 2018: Combating desertification in China: Monitoring, control, management and revegetation. *J. Clean. Prod.*, **182**, 765–775, doi:10.1016/J.JCLEPRO.2018.01.233.

Zhao, C. et al., 2017: Temperature increase reduces global yields of major crops in four independent estimates. *Proc. Natl. Acad. Sci. U.S.A.*, **114**, 9326–9331, doi:10.1073/pnas.1701762114.

Zhao, R., Y. Chen, P. Shi, L. Zhang, J. Pan, and H. Zhao, 2013: Land use and land cover change and driving mechanism in the arid inland river basin: A case study of Tarim River, Xinjiang, China. *Environ. Earth Sci.*, **68**, 591–604, doi:10.1007/s12665-012-1763-3.

Zhao, S., C. Peng, H. Jiang, D. Tian, X. Lei, and X. Zhou, 2006: Land use change in Asia and the ecological consequences. *Ecol. Res.*, **21**, 890–896, doi:10.1007/s11284-006-0048-2.

Zhao, S., H. Zhang, S. Feng, and Q. Fu, 2015: Simulating direct effects of dust aerosol on arid and semi-arid regions using an aerosol-climate coupled system. *Int. J. Climatol.*, **35**, 1858–1866, doi:10.1002/joc.4093.

Zhao, T., and A. Dai, 2015: The magnitude and causes of global drought changes in the twenty-first century under a low-moderate emissions scenario. *J. Clim.*, **28**, 4490–4512, doi:10.1175/JCLI-D-14-00363.1.

Zheng, J., Y. Yu, X. Zhang, and Z. Hao, 2018: Variation of extreme drought and flood in North China revealed by document-based seasonal precipitation reconstruction for the past 300 years. *Clim. Past*, **14**, 1135–1145, doi:10.5194/cp-14-1135-2018.

Zhou, D., Z. Lin, L. Liu, and D. Zimmermann, 2013: Assessing secondary soil salinization risk based on the PSR sustainability framework. *J. Environ. Manage.*, **128**, 642–654, doi:10.1016/J.JENVMAN.2013.06.025.

Zhou, W., C. Gang, F. Zhou, J. Li, X. Dong, and C. Zhao, 2015: Quantitative assessment of the individual contribution of climate and human factors to desertification in Northwest China using net primary productivity as an indicator. *Ecol. Indic.*, **48**, 560–569, doi:10.1016/J.ECOLIND.2014.08.043.

Zhu, J., H. Kang, and M.-L. Xu, 2007: Natural regeneration barriers of Pinus sylvestris var. mongolica plantations in southern Keerqin sandy land, China. *Shengtai Xuebao/ Acta Ecol. Sin.*, **27**, 4086–4095.

Zhu, T., and C. Ringler, 2012: Climate change impacts on water availability and use in the Limpopo River Basin. *Water*, **4**, 63–84, doi:10.3390/w4010063.

Zhu, Z. et al., 2016: Greening of the Earth and its drivers. *Nat. Clim. Chang.*, **6**, 791–795, doi:10.1038/nclimate3004.

Ziadat, F.M., and A.Y. Taimeh, 2013: Effect of rainfall intensity, slope, land use and antecedent soil moisture on soil erosion in an arid environment. *L. Degrad. Dev.*, **24**, 582–590, doi:10.1002/ldr.2239.

Zidane, I.F., K.M. Saqr, G. Swadener, X. Ma, and M.F. Shehadeh, 2016: On the role of surface roughness in the aerodynamic performance and energy conversion of horizontal wind turbine blades: A review. *Int. J. Energy Res.*, **40**, 2054–2077, doi:10.1002/er.3580.

Ziese, M. et al., 2014: The GPCC Drought Index – a new, combined and gridded global drought index. *Earth Syst. Sci. Data*, **6**, 285–295, doi:10.5194/essd-6-285-2014.

Zika, M., and K.H. Erb, 2009: The global loss of net primary production resulting from human-induced soil degradation in drylands. *Ecol. Econ.*, **69**, 310–318, doi:10.1016/j.ecolecon.2009.06.014.

Zinabu, S., A. Kebede, B. Ferede, and J. Dugassa, 2018: Review on the relationship of climate change and prevalence of animal diseases. *World Journal of Veterinary Science*, **6**, 6–18, doi:10.12970/2310-0796.2018.06.02.

Zou, X.K., and P.M. Zhai, 2004: Relationship between vegetation coverage and spring dust storms over northern China. *J. Geophys. Res. Atmos.*, **109**, D03104: 1-9, doi:10.1029/2003JD003913.

Zucca, C., R. Della Peruta, R. Salvia, S. Sommer, and M. Cherlet, 2012: Towards a World Desertification Atlas. Relating and selecting indicators and data sets to represent complex issues. *Ecol. Indic.*, **15**, 157–170, doi:10.1016/J.ECOLIND.2011.09.012.

Zulu, L.C., 2010: The forbidden fuel: Charcoal, urban woodfuel demand and supply dynamics, community forest management and woodfuel policy in Malawi. *Energy Policy*, **38**, 3717–3730, doi:10.1016/J.ENPOL.2010.02.050.

3

4

Land degradation

Coordinating Lead Authors:
Lennart Olsson (Sweden), Humberto Barbosa (Brazil)

Lead Authors:
Suruchi Bhadwal (India), Annette Cowie (Australia), Kenel Delusca (Haiti), Dulce Flores-Renteria (Mexico), Kathleen Hermans (Germany), Esteban Jobbagy (Argentina), Werner Kurz (Canada), Diqiang Li (China), Denis Jean Sonwa (Cameroon), Lindsay Stringer (United Kingdom)

Contributing Authors:
Timothy Crews (The United States of America), Martin Dallimer (United Kingdom), Joris Eekhout (The Netherlands), Karlheinz Erb (Italy), Eamon Haughey (Ireland), Richard Houghton (The United States of America), Muhammad Mohsin Iqbal (Pakistan), Francis X. Johnson (The United States of America), Woo-Kyun Lee (The Republic of Korea), John Morton (United Kingdom), Felipe Garcia Oliva (Mexico), Jan Petzold (Germany), Mohammad Rahimi (Iran), Florence Renou-Wilson (Ireland), Anna Tengberg (Sweden), Louis Verchot (Colombia/ The United States of America), Katharine Vincent (South Africa)

Review Editors:
José Manuel Moreno (Spain), Carolina Vera (Argentina)

Chapter Scientist:
Aliyu Salisu Barau (Nigeria)

This chapter should be cited as:
Olsson, L., H. Barbosa, S. Bhadwal, A. Cowie, K. Delusca, D. Flores-Renteria, K. Hermans, E. Jobbagy, W. Kurz, D. Li, D.J. Sonwa, L. Stringer, 2019: Land Degradation. In: *Climate Change and Land: an IPCC special report on climate change, desertification, land degradation, sustainable land management, food security, and greenhouse gas fluxes in terrestrial ecosystems* [P.R. Shukla, J. Skea, E. Calvo Buendia, V. Masson-Delmotte, H.-O. Pörtner, D. C. Roberts, P. Zhai, R. Slade, S. Connors, R. van Diemen, M. Ferrat, E. Haughey, S. Luz, S. Neogi, M. Pathak, J. Petzold, J. Portugal Pereira, P. Vyas, E. Huntley, K. Kissick, M. Belkacemi, J. Malley, (eds.)]. https://doi.org/10.1017/9781009157988.006

Table of contents

Executive summary 347

4.1 Introduction 349

    4.1.1 Scope of the chapter 349

    4.1.2 Perspectives of land degradation 349

    4.1.3 Definition of land degradation 349

    4.1.4 Land degradation in previous IPCC reports 350

    4.1.5 Sustainable land management (SLM) and sustainable forest management (SFM) 351

    4.1.6 The human dimension of land degradation and forest degradation 353

4.2 Land degradation in the context of climate change 353

    4.2.1 Processes of land degradation 354

    4.2.2 Drivers of land degradation 359

    4.2.3 Attribution in the case of land degradation 360

    4.2.4 Approaches to assessing land degradation 363

4.3 Status and current trends of land degradation 365

    4.3.1 Land degradation 365

    4.3.2 Forest degradation 367

4.4 Projections of land degradation in a changing climate 369

    4.4.1 Direct impacts on land degradation 369

    4.4.2 Indirect impacts on land degradation 373

4.5 Impacts of bioenergy and technologies for CO_2 removal (CDR) on land degradation 373

    4.5.1 Potential scale of bioenergy and land-based CDR 373

    4.5.2 Risks of land degradation from expansion of bioenergy and land-based CDR 373

    4.5.3 Potential contributions of land-based CDR to reducing and reversing land degradation 374

    4.5.4 Traditional biomass provision and land degradation 375

4.6 Impacts of land degradation on climate 375

    4.6.1 Impact on greenhouse gases (GHGs) 376

    4.6.2 Physical impacts 377

4.7 Impacts of climate-related land degradation on poverty and livelihoods 377

4.7.1 Relationships between land degradation, climate change and poverty 378

4.7.2 Impacts of climate-related land degradation on food security 379

4.7.3 Impacts of climate-related land degradation on migration and conflict 380

4.8 Addressing land degradation in the context of climate change 381

    4.8.1 Actions on the ground to address land degradation 381

    4.8.2 Local and indigenous knowledge for addressing land degradation 384

    4.8.3 Reducing deforestation and forest degradation and increasing afforestation 385

    4.8.4 Sustainable forest management (SFM) and CO_2 removal (CDR) technologies 386

    4.8.5 Policy responses to land degradation 387

    4.8.6 Resilience and thresholds 388

    4.8.7 Barriers to implementation of sustainable land management (SLM) 389

4.9 Case studies 391

    4.9.1 Urban green infrastructure 391

    4.9.2 Perennial grains and soil organic carbon (SOC) 393

    4.9.3 Reversing land degradation through reforestation 395

    4.9.4 Degradation and management of peat soils 397

    4.9.5 Biochar 398

    4.9.6 Management of land degradation induced by tropical cyclones 400

    4.9.7 Saltwater intrusion 401

    4.9.8 Avoiding coastal maladaptation 402

4.10 Knowledge gaps and key uncertainties 403

Frequently Asked Questions 404

    FAQ 4.1: How do climate change and land degradation interact with land use? 404

    FAQ 4.2: How does climate change affect land-related ecosystem services and biodiversity? 404

References 405

4

Executive summary

Land degradation affects people and ecosystems throughout the planet and is both affected by climate change and contributes to it. In this report, land degradation is defined as a *negative trend in land condition, caused by direct or indirect human-induced processes including anthropogenic climate change, expressed as long-term reduction or loss of at least one of the following: biological productivity, ecological integrity, or value to humans.* Forest degradation is land degradation that occurs in forest land. Deforestation is the conversion of forest to non-forest land and can result in land degradation. {4.1.3}

Land degradation adversely affects people's livelihoods (*very high confidence*) and occurs over a quarter of the Earth's ice-free land area (*medium confidence*). The majority of the 1.3 to 3.2 billion affected people (*low confidence*) are living in poverty in developing countries (*medium confidence*). Land-use changes and unsustainable land management are direct human causes of land degradation (*very high confidence*), with agriculture being a dominant sector driving degradation (*very high confidence*). Soil loss from conventionally tilled land exceeds the rate of soil formation by >2 orders of magnitude (*medium confidence*). Land degradation affects humans in multiple ways, interacting with social, political, cultural and economic aspects, including markets, technology, inequality and demographic change (*very high confidence*). Land degradation impacts extend beyond the land surface itself, affecting marine and freshwater systems, as well as people and ecosystems far away from the local sites of degradation (*very high confidence*). {4.1.6, 4.2.1, 4.2.3, 4.3, 4.6.1, 4.7, Table 4.1}

Climate change exacerbates the rate and magnitude of several ongoing land degradation processes and introduces new degradation patterns (*high confidence*). Human-induced global warming has already caused observed changes in two drivers of land degradation: increased frequency, intensity and/or amount of heavy precipitation (*medium confidence*); and increased heat stress (*high confidence*). In some areas sea level rise has exacerbated coastal erosion (*medium confidence*). Global warming beyond present day will further exacerbate ongoing land degradation processes through increasing floods (*medium confidence*), drought frequency and severity (*medium confidence*), intensified cyclones (*medium confidence*), and sea level rise (*very high confidence*), with outcomes being modulated by land management (*very high confidence*). Permafrost thawing due to warming (*high confidence*), and coastal erosion due to sea level rise and impacts of changing storm paths (*low confidence*), are examples of land degradation affecting places where it has not typically been a problem. Erosion of coastal areas because of sea level rise will increase worldwide (*high confidence*). In cyclone prone areas, the combination of sea level rise and more intense cyclones will cause land degradation with serious consequences for people and livelihoods (*very high confidence*). {4.2.1, 4.2.2, 4.2.3, 4.4.1, 4.4.2, 4.9.6, Table 4.1}

Land degradation and climate change, both individually and in combination, have profound implications for natural resource-based livelihood systems and societal groups (*high *confidence*).** The number of people whose livelihood depends on degraded lands has been estimated to be about 1.5 billion worldwide (*very low confidence*). People in degraded areas who directly depend on natural resources for subsistence, food security and income, including women and youth with limited adaptation options, are especially vulnerable to land degradation and climate change (*high confidence*). Land degradation reduces land productivity and increases the workload of managing the land, affecting women disproportionally in some regions. Land degradation and climate change act as threat multipliers for already precarious livelihoods (*very high confidence*), leaving them highly sensitive to extreme climatic events, with consequences such as poverty and food insecurity (*high confidence*) and, in some cases, migration, conflict and loss of cultural heritage (*low confidence*). Changes in vegetation cover and distribution due to climate change increase the risk of land degradation in some areas (*medium confidence*). Climate change will have detrimental effects on livelihoods, habitats and infrastructure through increased rates of land degradation (*high confidence*) and from new degradation patterns (*low evidence, high agreement*). {4.1.6, 4.2.1, 4.7}

Land degradation is a driver of climate change through emission of greenhouse gases (GHGs) and reduced rates of carbon uptake (*very high confidence*). Since 1990, globally the forest area has decreased by 3% (*low confidence*) with net decreases in the tropics and net increases outside the tropics (*high confidence*). Lower carbon density in re-growing forests, compared to carbon stocks before deforestation, results in net emissions from land-use change (*very high confidence*). Forest management that reduces carbon stocks of forest land also leads to emissions, but global estimates of these emissions are uncertain. Cropland soils have lost 20–60% of their organic carbon content prior to cultivation, and soils under conventional agriculture continue to be a source of GHGs (*medium confidence*). Of the land degradation processes, deforestation, increasing wildfires, degradation of peat soils, and permafrost thawing contribute most to climate change through the release of GHGs and the reduction in land carbon sinks following deforestation (*high confidence*). Agricultural practices also emit non-CO_2 GHGs from soils and these emissions are exacerbated by climate change (*medium confidence*). Conversion of primary to managed forests, illegal logging and unsustainable forest management result in GHG emissions (*very high confidence*) and can have additional physical effects on the regional climate including those arising from albedo shifts (*medium confidence*). These interactions call for more integrative climate impact assessments. {4.2.2, 4.3, 4.5.4, 4.6}

Large-scale implementation of dedicated biomass production for bioenergy increases competition for land with potentially serious consequences for food security and land degradation (*high confidence*). Increasing the extent and intensity of biomass production, for example, through fertiliser additions, irrigation or monoculture energy plantations, can result in local land degradation. Poorly implemented intensification of land management contributes to land degradation (e.g., salinisation from irrigation) and disrupted livelihoods (*high confidence*). In areas where afforestation and reforestation occur on previously degraded lands, opportunities exist to restore and rehabilitate lands with potentially significant co-benefits (*high confidence*) that depend on whether restoration involves natural or plantation forests. The total area of degraded lands has been estimated at 10–60 Mkm^2 (*very low confidence*). The extent of degraded and marginal lands suitable for dedicated biomass production is highly uncertain and cannot be established without due consideration of current land use and land tenure. Increasing the area of dedicated energy crops can lead to land degradation elsewhere through indirect land-use change (*medium confidence*). Impacts of energy crops can be reduced through strategic integration with agricultural and forestry systems (*high confidence*) but the total quantity of biomass that can be produced through synergistic production systems is unknown. {4.1.6, 4.4.2, 4.5, 4.7.1, 4.8.1, 4.8.3, 4.8.4, 4.9.3}

Reducing unsustainable use of traditional biomass reduces land degradation and emissions of CO$_2$ while providing social and economic co-benefits (*very high confidence*). Traditional biomass in the form of fuelwood, charcoal and agricultural residues remains a primary source of energy for more than one-third of the global population, leading to unsustainable use of biomass resources and forest degradation and contributing around 2% of global GHG emissions (*low confidence*). Enhanced forest protection, improved forest and agricultural management, fuel-switching and adoption of efficient cooking and heating appliances can promote more sustainable biomass use and reduce land degradation, with co-benefits of reduced GHG emissions, improved human health, and reduced workload especially for women and youth (*very high confidence*). {4.1.6, 4.5.4}

Land degradation can be avoided, reduced or reversed by implementing sustainable land management, restoration and rehabilitation practices that simultaneously provide many co-benefits, including adaptation to and mitigation of climate change (*high confidence*). Sustainable land management involves a comprehensive array of technologies and enabling conditions, which have proven to address land degradation at multiple landscape scales, from local farms (*very high confidence*) to entire watersheds (*medium confidence*). Sustainable forest management can prevent deforestation, maintain and enhance carbon sinks and can contribute towards GHG emissions-reduction goals. Sustainable forest management generates socio-economic benefits, and provides fibre, timber and biomass to meet society's growing needs. While sustainable forest management sustains high carbon sinks, the conversion from primary forests to sustainably managed forests can result in carbon emission during the transition and loss of biodiversity (*high confidence*). Conversely, in areas of degraded forests, sustainable forest management can increase carbon stocks and biodiversity (*medium confidence*). Carbon storage in long-lived wood products and reductions of emissions from use of wood products to substitute for emissions-intensive materials also contribute to mitigation objectives. {4.8, 4.9, Table 4.2}

Lack of action to address land degradation will increase emissions and reduce carbon sinks and is inconsistent with the emissions reductions required to limit global warming to 1.5°C or 2°C. (*high confidence*). Better management of soils can offset 5–20% of current global anthropogenic GHG emissions (*medium confidence*). Measures to avoid, reduce and reverse land degradation are available but economic, political, institutional, legal and socio-cultural barriers, including lack of access to resources and knowledge, restrict their uptake (*very high confidence*). Proven measures that facilitate implementation of practices that avoid, reduce, or reverse land degradation include tenure reform, tax incentives, payments for ecosystem services, participatory integrated land-use planning, farmer networks and rural advisory services. Delayed action increases the costs of addressing land degradation, and can lead to irreversible biophysical and human outcomes (*high confidence*). Early actions can generate both site-specific and immediate benefits to communities affected by land degradation, and contribute to long-term global benefits through climate change mitigation (*high confidence*). {4.1.5, 4.1.6, 4.7.1, 4.8, Table 4.2}

Even with adequate implementation of measures to avoid, reduce and reverse land degradation, there will be residual degradation in some situations (*high confidence*). Limits to adaptation are dynamic, site specific and determined through the interaction of biophysical changes with social and institutional conditions. Exceeding the limits of adaptation will trigger escalating losses or result in undesirable changes, such as forced migration, conflicts, or poverty. Examples of potential limits to adaptation due to climate-change-induced land degradation are coastal erosion (where land disappears, collapsing infrastructure and livelihoods due to thawing of permafrost), and extreme forms of soil erosion. {4.7, 4.8.5, 4.8.6, 4.9.6, 4.9.7, 4.9.8}

Land degradation is a serious and widespread problem, yet key uncertainties remain concerning its extent, severity, and linkages to climate change (*very high confidence*). Despite the difficulties of objectively measuring the extent and severity of land degradation, given its complex and value-based characteristics, land degradation represents – along with climate change – one of the biggest and most urgent challenges for humanity (*very high confidence*). The current global extent, severity and rates of land degradation are not well quantified. There is no single method by which land degradation can be measured objectively and consistently over large areas because it is such a complex and value-laden concept (*very high confidence*). However, many existing scientific and locally-based approaches, including the use of indigenous and local knowledge, can assess different aspects of land degradation or provide proxies. Remote sensing, corroborated by other data, can generate geographically explicit and globally consistent data that can be used as proxies over relevant time scales (several decades). Few studies have specifically addressed the impacts of proposed land-based negative emission technologies on land degradation. Much research has tried to understand how livelihoods and ecosystems are affected by a particular stressor – for example, drought, heat stress, or waterlogging. Important knowledge gaps remain in understanding how plants, habitats and ecosystems are affected by the cumulative and interacting impacts of several stressors, including potential new stressors resulting from large-scale implementation of negative emission technologies. {4.10}

1.1 Introduction

1.1.1 Scope of the chapter

This chapter examines the scientific understanding of how climate change impacts land degradation, and vice versa, with a focus on non-drylands. Land degradation of drylands is covered in Chapter 3. After providing definitions and the context (Section 4.1) we proceed with a theoretical explanation of the different processes of land degradation and how they are related to climate and to climate change, where possible (Section 4.2). Two sections are devoted to a systematic assessment of the scientific literature on status and trend of land degradation (Section 4.3) and projections of land degradation (Section 4.4). Then follows a section where we assess the impacts of climate change mitigation options, bioenergy and land-based technologies for carbon dioxide removal (CDR), on land degradation (Section 4.5). The ways in which land degradation can impact on climate and climate change are assessed in Section 4.6. The impacts of climate-related land degradation on human and natural systems are assessed in Section 4.7. The remainder of the chapter assesses land degradation mitigation options based on the concept of sustainable land management: avoid, reduce and reverse land degradation (Section 4.8), followed by a presentation of eight illustrative case studies of land degradation and remedies (Section 4.9). The chapter ends with a discussion of the most critical knowledge gaps and areas for further research (Section 4.10).

1.1.2 Perspectives of land degradation

Land degradation has accompanied humanity at least since the widespread adoption of agriculture during Neolithic time, some 10,000 to 7,500 years ago (Dotterweich 2013; Butzer 2005; Dotterweich 2008) and the associated population increase (Bocquet-Appel 2011). There are indications that the levels of greenhouse gases (GHGs) – particularly carbon dioxide (CO_2) and methane (CH_4) – in the atmosphere already started to increase more than 3,000 years ago as a result of expanding agriculture, clearing of forests, and domestication of wild animals (Fuller et al. 2011; Kaplan et al. 2011; Vavrus et al. 2018; Ellis et al. 2013). While the development of agriculture (cropping and animal husbandry) underpinned the development of civilisations, political institutions and prosperity, farming practices led to conversion of forests and grasslands to farmland, and the heavy reliance on domesticated annual grasses for our food production meant that soils started to deteriorate through seasonal mechanical disturbances (Turner et al. 1990; Steffen et al. 2005; Ojima et al. 1994; Ellis et al. 2013). More recently, urbanisation has significantly altered ecosystems (Cross-Chapter Box 4 in Chapter 2). Since around 1850, about 35% of human-caused CO_2 emissions to the atmosphere has come from land as a combined effect of land degradation and land-use change (Foley et al. 2005) and about 38% of the Earth's land area has been converted to agriculture (Foley et al. 2011). See Chapter 2 for more details.

Not all human impacts on land result in degradation according to the definition of land degradation used in this report (Section 4.1.3). There are many examples of long-term sustainably managed land around the world (such as terraced agricultural systems and sustainably managed forests) although degradation and its management are the focus of this chapter. We also acknowledge that human use of land and ecosystems provides essential goods and services for society (Foley et al. 2005; Millennium Ecosystem Assessment 2005).

Land degradation was long subject to a polarised scientific debate between disciplines and perspectives in which social scientists often proposed that natural scientists exaggerated land degradation as a global problem (Blaikie and Brookfield 1987; Forsyth 1996; Lukas 2014; Zimmerer 1993). The elusiveness of the concept in combination with the difficulties of measuring and monitoring land degradation at global and regional scales by extrapolation and aggregation of empirical studies at local scales, such as the Global Assessment of Soil Degradation database (GLASOD) (Sonneveld and Dent 2009) contributed to conflicting views. The conflicting views were not confined to science only, but also caused tension between the scientific understanding of land degradation and policy (Andersson et al. 2011; Behnke and Mortimore 2016; Grainger 2009; Toulmin and Brock 2016). Another weakness of many land degradation studies is the exclusion of the views and experiences of the land users, whether farmers or forest-dependent communities (Blaikie and Brookfield 1987; Fairhead and Scoones 2005; Warren 2002; Andersson et al. 2011). More recently, the polarised views described above have been reconciled under the umbrella of Land Change Science, which has emerged as an interdisciplinary field aimed at examining the dynamics of land cover and land-use as a coupled human-environment system (Turner et al. 2007). A comprehensive discussion about concepts and different perspectives of land degradation was presented in Chapter 2 of the recent report from the Intergovernmental Science-Policy Platform on Biodiversity and Ecosystem Services (IPBES) on land degradation (Montanarella et al. 2018).

In summary, agriculture and clearing of land for food and wood products have been the main drivers of land degradation for millennia (*high confidence*). This does not mean, however, that agriculture and forestry always cause land degradation (*high confidence*); sustainable management is possible but not always practised (*high confidence*). Reasons for this are primarily economic, political and social.

1.1.3 Definition of land degradation

To clarify the scope of this chapter, it is important to start by defining land itself. The Special Report on Climate Change and Land (SRCCL) defines land as 'the terrestrial portion of the biosphere that comprises the natural resources (soil, near surface air, vegetation and other biota, and water), the ecological processes, topography, and human settlements and infrastructure that operate within that system' (Henry et al. 2018, adapted from FAO 2007; UNCCD 1994).

Land degradation is defined in many different ways within the literature, with differing emphases on biodiversity, ecosystem functions and ecosystem services (e.g., Montanarella et al. 2018). In this report, land degradation is defined as a *negative trend in land condition, caused by direct or indirect human-induced processes including anthropogenic climate change, expressed as long-term reduction or loss of at least one of the following: biological productivity, ecological integrity or value to humans*. This definition applies to forest and non-forest land: forest degradation is land degradation that occurs in forest land. Soil degradation refers to a subset of land degradation processes that directly affect soil.

The SRCCL definition is derived from the IPCC AR5 definition of desertification, which is in turn taken from the United Nations Convention to Combat Desertification (UNCCD): 'Land degradation in arid, semi-arid, and dry sub-humid areas resulting from various factors, including climatic variations and human activities. Land degradation in arid, semi-arid, and dry sub-humid areas is a reduction or loss of the biological or economic productivity and integrity of rainfed cropland, irrigated cropland, or range, pasture, forest, and woodlands resulting from land uses or from a process or combination of processes, including processes arising from human activities and habitation patterns, such as (i) soil erosion caused by wind and/ or water; (ii) deterioration of the physical, chemical, biological, or economic properties of soil; and (iii) long-term loss of natural vegetation' (UNCCD 1994, Article 1).

For this report, the SRCCL definition is intended to complement the more detailed UNCCD definition above, expanding the scope to all regions, not just drylands, providing an operational definition that emphasises the relationship between land degradation and climate. Through its attention to the three aspects – biological productivity, ecological integrity and value to humans – the SRCCL definition is consistent with the Land Degradation Neutrality (LDN) concept, which aims to maintain or enhance the land-based natural capital, and the ecosystem services that flow from it (Cowie et al. 2018).

In the SRCCL definition of land degradation, changes in land condition resulting solely from natural processes (such as volcanic eruptions and tsunamis) are not considered land degradation, as these are not direct or indirect human-induced processes. Climate variability exacerbated by human-induced climate change can contribute to land degradation. Value to humans can be expressed in terms of ecosystem services or Nature's Contributions to People.

The definition recognises the reality presented in the literature that land-use and land management decisions often result in trade-offs between time, space, ecosystem services, and stakeholder groups (e.g., Dallimer and Stringer 2018). The interpretation of a negative trend in land condition is somewhat subjective, especially where there is a trade-off between ecological integrity and value to humans. The definition also does not consider the magnitude of the negative trend or the possibility that a negative trend in one criterion may be an acceptable trade-off for a positive trend in another criterion. For example, reducing timber yields to safeguard biodiversity by leaving on site more wood that can provide habitat, or vice versa, is a trade-off that needs to be evaluated based on context (i.e. the broader landscape) and society's priorities. Reduction of biological productivity *or* ecological integrity *or* value to humans *can* constitute degradation, but any one of these changes need not necessarily be considered degradation. Thus, a land-use change that reduces ecological integrity and enhances sustainable food production at a specific location is not necessarily degradation. Different

stakeholder groups with different world views value ecosystem services differently. As Warren (2002) explained: land degradation is contextual. Further, a decline in biomass carbon stock does not always signify degradation, such as when caused by periodic forest harvest. Even a decline in productivity may not equate to land degradation, such as when a high-intensity agricultural system is converted to a lower-input, more sustainable production system.

In the SRCCL definition, degradation is indicated by a negative trend in land condition during the period of interest, thus the baseline is the land condition at the start of this period. The concept of baseline is theoretically important but often practically difficult to implement for conceptual and methodological reasons (Herrick et al. 2019; Prince et al. 2018; also Sections 4.3.1 and 4.4.1). Especially in biomes characterised by seasonal and interannual variability, the baseline values of the indicators to be assessed should be determined by averaging data over a number of years prior to the commencement of the assessment period (Orr et al. 2017) (Section 4.2.4).

Forest degradation is land degradation in forest remaining forest. In contrast, deforestation refers to the conversion of forest to non-forest that involves a loss of tree cover and a change in land use. Internationally accepted definitions of forest (FAO 2015; UNFCCC 2006) include lands where tree cover has been lost temporarily, due to disturbance or harvest, with an expectation of forest regrowth. Such temporary loss of forest cover, therefore, is not deforestation.

1.1.4   Land degradation in previous IPCC reports

Several previous IPCC assessment reports include brief discussions of land degradation. In AR5 WGIII land degradation is one factor contributing to uncertainties of the mitigation potential of land-based ecosystems, particularly in terms of fluxes of soil carbon (Smith et al. 2014, p. 817). In AR5 WGI, soil carbon was discussed comprehensively but not in the context of land degradation, except forest degradation (Ciais et al. 2013) and permafrost degradation (Vaughan et al. 2013). Climate change impacts were discussed comprehensively in AR5 WGII, but land degradation was not prominent. Land-use and land-cover changes were treated comprehensively in terms of effects on the terrestrial carbon stocks and flows (Settele et al. 2015) but links to land degradation were, to a large extent, missing. Land degradation was discussed in relation to human security as one factor which, in combination with extreme weather events, has been proposed to contribute to human migration (Adger et al. 2014), an issue discussed more comprehensively in this chapter (Section 4.7.3). Drivers and processes of degradation by which land-based carbon is released to the atmosphere and/or the long-term reduction in the capacity of the land to remove atmospheric carbon and to store this in biomass and soil carbon, have been discussed in the methodological reports of IPCC (IPCC 2006, 2014a) but less so in the assessment reports.

The Special Report on Land Use, Land-Use Change and Forestry (SR-LULUCF) (Watson et al. 2000) focused on the role of the biosphere in the global cycles of GHG. Land degradation was not addressed in a comprehensive way. Soil erosion was discussed as a process by which soil carbon is lost and the productivity of the land is reduced. Deposition

4

of eroded soil carbon in marine sediments was also mentioned as a possible mechanism for permanent sequestration of terrestrial carbon (Watson et al. 2000, p. 194). The possible impacts of climate change on land productivity and degradation were not discussed comprehensively. Much of the report was about how to account for sources and sinks of terrestrial carbon under the Kyoto Protocol.

The IPCC Special Report on Managing the Risks of Extreme Events and Disasters to Advance Climate Change Adaptation (SREX) (IPCC 2012) did not provide a definition of land degradation. Nevertheless, it addressed different aspects related to some types of land degradation in the context of weather and climate extreme events. From this perspective, it provided key information on both observed and projected changes in weather and climate (extremes) events that are relevant to extreme impacts on socio-economic systems and on the physical components of the environment, notably on permafrost in mountainous areas and coastal zones for different geographic regions, but few explicit links to land degradation. The report also presented the concept of sustainable land management as an effective risk-reduction tool.

Land degradation has been treated in several previous IPCC reports, but mainly as an aggregated concept associated with GHG emissions, or as an issue that can be addressed through adaptation and mitigation.

1.1.5 Sustainable land management (SLM) and sustainable forest management (SFM)

Sustainable land management (SLM) is defined as 'the stewardship and use of land resources, including soils, water, animals and plants, to meet changing human needs, while simultaneously ensuring the long-term productive potential of these resources and the maintenance of their environmental functions' – adapted from World Overview of Conservation Approaches and Technologies (WOCAT n.d.). Achieving the objective of ensuring that productive potential is maintained in the long term will require implementation of adaptive management and 'triple loop learning', that seeks to monitor outcomes, learn from experience and emerging new knowledge, modifying management accordingly (Rist et al. 2013).

Sustainable Forest Management (SFM) is defined as 'the stewardship and use of forests and forest lands in a way, and at a rate, that maintains their biodiversity, productivity, regeneration capacity, vitality and their potential to fulfill, now and in the future, relevant ecological, economic and social functions, at local, national, and global levels, and that does not cause damage to other ecosystems' (Forest Europe 1993; Mackey et al. 2015). This SFM definition was developed by the Ministerial Conference on the Protection of Forests in Europe and has since been adopted by the Food and Agriculture Organization. Forest management that fails to meet these sustainability criteria can contribute to land degradation.

Land degradation can be reversed through restoration and rehabilitation. These terms are defined in the Glossary, along with other terms that are used but not explicitly defined in this section of the report. While the definitions of SLM and SFM are very similar and could be merged, both are included to maintain the subtle differences in the existing definitions. SFM can be considered a subset of SLM – that is, SLM applied to forest land.

Climate change impacts interact with land management to determine sustainable or degraded outcome (Figure 4.1). Climate change can exacerbate many degradation processes (Table 4.1) and introduce novel ones (e.g., permafrost thawing or biome shifts). To avoid, reduce or reverse degradation, land management activities can be selected to mitigate the impact of, and adapt to, climate change. In some cases, climate change impacts may result in increased productivity and carbon stocks, at least in the short term. For example, longer growing seasons due to climate warming can lead to higher forest productivity (Henttonen et al. 2017; Kauppi et al. 2014; Dragoni et al. 2011), but warming alone may not increase productivity where other factors such a water supply are limiting (Hember et al. 2017).

The types and intensity of human land-use and climate change impacts on lands affect their carbon stocks and their ability to operate as carbon sinks. In managed agricultural lands, degradation can result in reductions of soil organic carbon stocks, which also adversely affects land productivity and carbon sinks (Figure 4.1).

The transition from natural to managed forest landscapes usually results in an initial reduction of landscape-level carbon stocks. The magnitude of this reduction is a function of the differential in frequency of stand-replacing natural disturbances (e.g., wildfires) and harvest disturbances, as well as the age-dependence of these disturbances (Harmon et al. 1990; Kurz et al. 1998; Trofymow et al. 2008).

SFM applied at the landscape scale to existing unmanaged forests can first reduce average forest carbon stocks and subsequently increase the rate at which CO_2 is removed from the atmosphere, because net ecosystem production of forest stands is highest in intermediate stand ages (Kurz et al. 2013; Volkova et al. 2018; Tang et al. 2014). The net impact on the atmosphere depends on the magnitude of the reduction in carbon stocks, the fate of the harvested biomass (i.e. use in short – or long-lived products and for bioenergy, and therefore displacement of emissions associated with GHG-intensive building materials and fossil fuels), and the rate of regrowth. Thus, the impacts of SFM on one indicator (e.g., past reduction in carbon stocks in the forested landscape) can be negative, while those on another indicator (e.g., current forest productivity and rate of CO_2 removal from the atmosphere, avoided fossil fuel emissions) can be positive. Sustainably managed forest landscapes can have a lower biomass carbon density than unmanaged forest, but the younger forests can have a higher growth rate, and therefore contribute stronger carbon sinks than older forests (Trofymow et al. 2008; Volkova et al. 2018; Poorter et al. 2016).

Selective logging and thinning can maintain and enhance forest productivity and achieve co-benefits when conducted with due care for the residual stand and at intensity and frequency that does not exceed the rate of regrowth (Romero and Putz 2018). In contrast, unsustainable logging practices can lead to stand-level degradation. For example, degradation occurs when selective logging

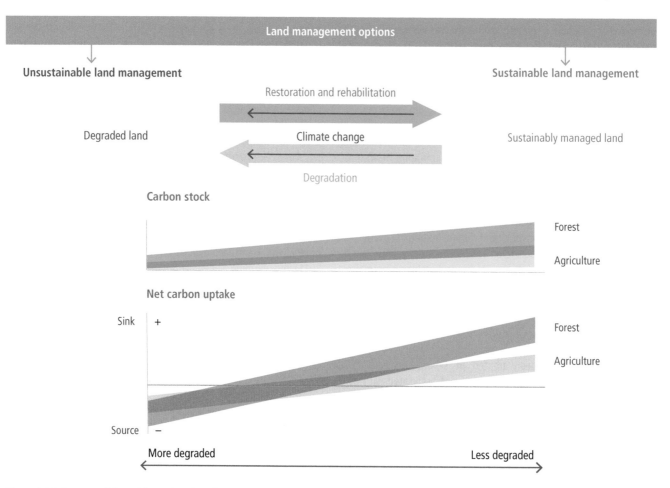

Figure 4.1 | Conceptual figure illustrating that climate change impacts interact with land management to determine sustainable or degraded outcome. Climate change can exacerbate many degradation processes (Table 4.1) and introduce novel ones (e.g., permafrost thawing or biome shifts), hence management needs to respond to climate impacts in order to avoid, reduce or reverse degradation. The types and intensity of human land-use and climate change impacts on lands affect their carbon stocks and their ability to operate as carbon sinks. In managed agricultural lands, degradation typically results in reductions of soil organic carbon stocks, which also adversely affects land productivity and carbon sinks. In forest land, reduction in biomass carbon stocks alone is not necessarily an indication of a reduction in carbon sinks. Sustainably managed forest landscapes can have a lower biomass carbon density but the younger forests can have a higher growth rate, and therefore contribute stronger carbon sinks, than older forests. Ranges of carbon sinks in forest and agricultural lands are overlapping. In some cases, climate change impacts may result in increased productivity and carbon stocks, at least in the short term.

(high-grading) removes valuable large-diameter trees, leaving behind damaged, diseased, non-commercial or otherwise less productive trees, reducing carbon stocks and also adversely affecting subsequent forest recovery (Belair and Ducey 2018; Nyland 1992).

SFM is defined using several criteria (see above) and its implementation will typically involve trade-offs among these criteria. The conversion of primary forests to sustainably managed forest ecosystems increases relevant economic, social and other functions but often with adverse impacts on biodiversity (Barlow et al. 2007). In regions with infrequent or no stand-replacing natural disturbances, the timber yield per hectare harvested in managed secondary forests is typically lower than the yield per hectare from the first harvest in the primary forest (Romero and Putz 2018).

The sustainability of timber yield has been achieved in temperate and boreal forests where intensification of management has resulted in increased growing stocks and increased harvest rates in countries where forests had previously been overexploited (Henttonen et al. 2017; Kauppi et al. 2018). However, intensification of management

to increase forest productivity can be associated with reductions in biodiversity. For example, when increased productivity is achieved by periodic thinning and removal of trees that would otherwise die due to competition, thinning reduces the amount of dead organic matter of snags and coarse woody debris that can provide habitat, and this loss reduces biodiversity (Spence 2001; Ehnström 2001) and forest carbon stocks (Russell et al. 2015; Kurz et al. 2013). Recognition of adverse biodiversity impacts of high-yield forestry is leading to modified management aimed at increasing habitat availability through, for example, variable retention logging and continuous cover management (Roberts et al. 2016) and through the re-introduction of fire disturbances in landscapes where fires have been suppressed (Allen et al. 2002). Biodiversity losses are also observed during the transition from primary to managed forests in tropical regions (Barlow et al. 2007) where tree species diversity can be very high – for example, in the Amazon region, about 16,000 tree species are estimated to exist (ter Steege et al. 2013).

Forest certification schemes have been used to document SFM outcomes (Rametsteiner and Simula 2003) by assessing a set of criteria and indicators (e.g., Lindenmayer et al. 2000). While many of the certified forests are found in temperate and boreal countries (Rametsteiner and Simula 2003; MacDicken et al. 2015), examples from the tropics also show that SFM can improve outcomes. For example, selective logging emits 6% of the tropical GHG annually and improved logging practices can reduce emissions by 44% while maintaining timber production (Ellis et al. 2019). In the Congo Basin, implementing reduced impact logging (RIL-C) practices can cut emissions in half without reducing the timber yield (Umunay et al. 2019). SFM adoption depends on the socio-economic and political context, and its improvement depends mainly on better reporting and verification (Siry et al. 2005).

The successful implementation of SFM requires well-established and functional governance, monitoring, and enforcement mechanisms to eliminate deforestation, illegal logging, arson, and other activities that are inconsistent with SFM principles (Nasi et al. 2011). Moreover, following human and natural disturbances, forest regrowth must be ensured through reforestation, site rehabilitation activities or natural regeneration. Failure of forests to regrow following disturbances will lead to unsustainable outcomes and long-term reductions in forest area, forest cover, carbon density, forest productivity and land-based carbon sinks (Nasi et al. 2011).

Achieving all of the criteria of the definitions of SLM and SFM is an aspirational goal that will be made more challenging where climate change impacts, such as biome shifts and increased disturbances, are predicted to adversely affect future biodiversity and contribute to forest degradation (Warren et al. 2018). Land management to enhance land sinks will involve trade-offs that need to be assessed within their spatial, temporal and societal context.

1.1.6 The human dimension of land degradation and forest degradation

Studies of land and forest degradation are often biased towards biophysical aspects, both in terms of its processes, such as erosion or nutrient depletion, and its observed physical manifestations, such as gullying or low primary productivity. Land users' own perceptions and knowledge about land conditions and degradation have often been neglected or ignored by both policymakers and scientists (Reed et al. 2007; Forsyth 1996; Andersson et al. 2011). A growing body of work is nevertheless beginning to focus on land degradation through the lens of local land users (Kessler and Stroosnijder 2006; Fairhead and Scoones 2005; Zimmerer 1993; Stocking et al. 2001) and the importance of local and indigenous knowledge within land management is starting to be appreciated (Montanarella et al. 2018). Climate change impacts directly and indirectly on the social reality, the land users, and the ecosystem, and vice versa. Land degradation can also have an impact on climate change (Section 4.6).

The use and management of land is highly gendered and is expected to remain so for the foreseeable future (Kristjanson et al. 2017). Women often have less formal access to land than men and less influence over decisions about land, even if they carry out many of the land management tasks (Jerneck 2018a; Elmhirst 2011; Toulmin

2009; Peters 2004; Agarwal 1997; Jerneck 2018b). Many oft-cited general statements about women's subordination in agriculture are difficult to substantiate, yet it is clear that gender inequality persists (Doss et al. 2015). Even if women's access to land is changing formally (Kumar and Quisumbing 2015), the practical outcome is often limited due to several other factors related to both formal and informal institutional arrangements and values (Lavers 2017; Kristjanson et al. 2017; Djurfeldt et al. 2018). Women are also affected differently than men when it comes to climate change, having lower adaptive capacities due to factors such as prevailing land tenure frameworks, less access to other capital assets and dominant cultural practices (Vincent et al. 2014; Antwi-Agyei et al. 2015; Gabrielsson et al. 2013). This affects the options available to women to respond to both land degradation and climate change. Indeed, access to land and other assets (e.g., education and training) is key in shaping land-use and land management strategies (Liu et al. 2018b; Lambin et al. 2001). Young people are also often disadvantaged in terms of access to resources and decision-making power, even though they carry out much of the day-to-day work (Wilson et al. 2017; Kosec et al. 2018; Naamwintome and Bagson 2013).

Land rights differ between places and are dependent on the political-economic and legal context (Montanarella et al. 2018). This means that there is no universally applicable best arrangement. Agriculture in highly erosion-prone regions requires site-specific and long-lasting soil and water conservation measures, such as terraces (Section 4.8.1), which may benefit from secure private land rights (Tarfasa et al. 2018; Soule et al. 2000). Pastoral modes of production and community-based forest management systems are often dominated by, and benefit from, communal land tenure arrangements, which may conflict with agricultural/forestry modernisation policies implying private property rights (Antwi-Agyei et al. 2015; Benjaminsen and Lund 2003; Itkonen 2016; Owour et al. 2011; Gebara 2018).

Cultural ecosystem services, defined as the non-material benefits people obtain from ecosystems through spiritual enrichment, cognitive development, reflection, recreation and aesthetic experiences (Millennium Ecosystem Assessment 2005) are closely linked to land and ecosystems, although often under-represented in the literature on ecosystem services (Tengberg et al. 2012; Hernández-Morcillo et al. 2013). Climate change interacting with land conditions can impact on cultural aspects, such as sense of place and sense of belonging (Olsson et al. 2014).

1.2 Land degradation in the context of climate change

Land degradation results from a complex chain of causes making the clear distinction between direct and indirect drivers difficult. In the context of climate change, an additional complex aspect is brought by the reciprocal effects that both processes have on each other (i.e. climate change influencing land degradation and vice versa). In this chapter, we use the terms 'processes' and 'drivers' with the following meanings:

4

Processes of land degradation are those direct mechanisms by which land is degraded and are similar to the notion of 'direct drivers' in the Millennium Ecosystem Assessment framework (Millennium Ecosystem Assessment, 2005). A comprehensive list of land degradation processes is presented in Table 4.1.

Drivers of land degradation are those indirect conditions which may drive processes of land degradation and are similar to the notion of 'indirect drivers' in the Millennium Ecosystem Assessment framework. Examples of indirect drivers of land degradation are changes in land tenure or cash crop prices, which can trigger land-use or management shifts that affect land degradation.

An exact demarcation between processes and drivers is not possible. Drought and fires are described as drivers of land degradation in the next section but they can also be a process: for example, if repeated fires deplete seed sources, they can affect regeneration and succession of forest ecosystems. The responses to land degradation follow the logic of the LDN concept: avoiding, reducing and reversing land degradation (Orr et al. 2017; Cowie et al. 2018).

In research on land degradation, climate and climate variability are often intrinsic factors. The role of climate change, however, is less articulated. Depending on what conceptual framework is used, climate change is understood either as a process or a driver of land degradation, and sometimes both.

1.2.1   Processes of land degradation

A large array of interactive physical, chemical, biological and human processes lead to what we define in this report as land degradation (Johnson and Lewis 2007). The biological productivity, ecological integrity (which encompasses both functional and structural attributes of ecosystems) or the human value (which includes any benefit that people get from the land) of a given territory can deteriorate as the result of processes triggered at scales that range from a single furrow (e.g., water erosion under cultivation) to the landscape level (e.g., salinisation through raising groundwater levels under irrigation). While pressures leading to land degradation are often exerted on specific components of the land systems (i.e., soils, water, biota), once degradation processes start, other components become affected through cascading and interactive effects. For example, different pressures and degradation processes can have convergent effects, as can be the case of overgrazing leading to wind erosion, landscape drainage resulting in wetland drying, and warming causing more frequent burning; all of which can independently lead to reductions of the soil organic matter (SOM) pools as a second-order process. Still, the reduction of organic matter pools is also a first-order process triggered directly by the effects of rising temperatures (Crowther et al. 2016) as well as other climate changes such as precipitation shifts (Viscarra Rossel et al. 2014). Beyond this complexity, a practical assessment of the major land degradation processes helps to reveal and categorise the multiple pathways in which climate change exerts a degradation pressure (Table 4.1).

Conversion of freshwater wetlands to agricultural land has historically been a common way of increasing the area of arable land. Despite the small areal extent – about 1% of the earth's surface (Hu et al. 2017; Dixon et al. 2016) – freshwater wetlands provide a very large number of ecosystem services, such as groundwater replenishment, flood protection and nutrient retention, and are biodiversity hotspots (Reis et al. 2017; Darrah et al. 2019; Montanarella et al. 2018). The loss of wetlands since 1900 has been estimated at about 55% globally (Davidson 2014) (*low confidence*) and 35% since 1970 (Darrah et al. 2019) (*medium confidence*) which in many situations pose a problem for adaptation to climate change. Drainage causes loss of wetlands, which can be exacerbated by climate change, further reducing the capacity to adapt to climate change (Barnett et al. 2015; Colloff et al. 2016; Finlayson et al. 2017) (*high confidence*).

1.2.1.1   Types of land degradation processes

Land degradation processes can affect the soil, water or biotic components of the land as well as the reactions between them (Table 4.1). Across land degradation processes, those affecting the soil have received more attention. The most widespread and studied land degradation processes affecting soils are water and wind erosion, which have accompanied agriculture since its onset and are still dominant (Table 4.1). Degradation through erosion processes is not restricted to soil loss in detachment areas but includes impacts on transport and deposition areas as well (less commonly, deposition areas can have their soils improved by these inputs). Larger-scale degradation processes related to the whole continuum of soil erosion, transport and deposition include dune field expansion/displacement, development of gully networks and the accumulation of sediments in natural and artificial water-bodies (siltation) (Poesen and Hooke 1997; Ravi et al. 2010). Long-distance sediment transport during erosion events can have remote effects on land systems, as documented for the fertilisation effect of African dust on the Amazon (Yu et al. 2015).

Coastal erosion represents a special case among erosional processes, with reports linking it to climate change. While human interventions in coastal areas (e.g., expansion of shrimp farms) and rivers (e.g., upstream dams cutting coastal sediment supply), and economic activities causing land subsidence (Keogh and Törnqvist 2019; Allison et al. 2016) are dominant human drivers, storms and sea-level rise have already left a significant global imprint on coastal erosion (Mentaschi et al. 2018). Recent projections that take into account geomorphological and socioecological feedbacks suggest that coastal wetlands may not be reduced by sea level rise if their inland growth is accommodated with proper management actions (Schuerch et al. 2018).

Other physical degradation processes in which no material detachment and transport are involved include soil compaction, hardening, sealing and any other mechanism leading to the loss of porous space crucial for holding and exchanging air and water (Hamza and Anderson 2005). A very extreme case of degradation through pore volume loss, manifested at landscape or larger scales, is ground subsidence. Typically caused by the lowering of groundwater or oil levels, subsidence involves a sustained collapse of the ground surface, which can lead to other degradation processes such as salinisation and permanent flooding. Chemical soil degradation processes include relatively simple changes, like nutrient depletion resulting from the imbalance of nutrient extraction on harvested products and fertilisation, and more complex ones, such as acidification and increasing metal toxicity. Acidification in croplands is increasingly driven by excessive nitrogen fertilisation and, to a lower extent, by the depletion of cation like calcium, potassium or magnesium through exports in harvested biomass (Guo et al. 2010). One of the most relevant chemical degradation processes of soils in the context of climate change is the depletion of its organic matter pool. Reduced in agricultural soils through the increase of respiration rates by tillage and the decline of below-ground plant biomass inputs, SOM pools have been diminished also by the direct effects of warming, not only in cultivated land, but also under natural vegetation (Bond-Lamberty et al. 2018). Debate persists, however, on whether in more humid and carbon-rich ecosystems the simultaneous stimulation of decomposition and productivity may result in the lack of effects on soil carbon (Crowther et al. 2016; van Gestel et al. 2018). In the case of forests, harvesting – particularly if it is exhaustive, as in the case of the use of residues for energy generation – can also lead to organic matter declines (Achat et al. 2015). Many other degradation processes (e.g., wildfire increase, salinisation) have negative effects on other pathways of soil degradation (e.g., reduced nutrient availability, metal toxicity). SOM can be considered a 'hub' of degradation processes and a critical link with the climate system (Minasny et al. 2017).

Land degradation processes can also start from alterations in the hydrological system that are particularly important in the context of climate change. Salinisation, although perceived and reported in soils, is typically triggered by water table-level rises, driving salts to the surface under dry to sub-humid climates (Schofield and Kirkby 2003). While salty soils occur naturally under these climates (primary salinity), human interventions have expanded their distribution, secondary salinity with irrigation without proper drainage being the predominant cause of salinisation (Rengasamy 2006). Yet, it has also taken place under non-irrigated conditions where vegetation changes (particularly dry forest clearing and cultivation) have reduced the magnitude and depth of soil water uptake, triggering water table rises towards the surface. Changes in evapotranspiration and rainfall regimes can exacerbate this process (Schofield and Kirkby 2003). Salinisation can also result from the intrusion of sea water into coastal areas, both as a result of sea level rise and ground subsidence (Colombani et al. 2016).

Recurring flood and waterlogging episodes (Bradshaw et al. 2007; Poff 2002), and the more chronic expansion of wetlands over dryland ecosystems, are mediated by the hydrological system, on occasions aided by geomorphological shifts as well (Kirwan et al. 2011). This is also the case for the drying of continental water bodies and wetlands, including the salinisation and drying of lakes and inland seas (Anderson et al. 2003; Micklin 2010; Herbert et al. 2015). In the context of climate change, the degradation of peatland ecosystems is particularly relevant given their very high carbon storage and their sensitivity to changes in soils, hydrology and/or vegetation (Leifeld and Menichetti 2018). Drainage for land-use conversion together

with peat mining are major drivers of peatland degradation, yet other factors such as the extractive use of their natural vegetation and the interactive effects of water table levels and fires (both sensitive to climate change) are important (Hergoualc'h et al. 2017a; Lilleskov et al. 2019).

The biotic components of the land can also be the focus of degradation processes. Vegetation clearing processes associated with land-use changes are not limited to deforestation but include other natural and seminatural ecosystems such as grasslands (the most cultivated biome on Earth), as well as dry steppes and shrublands, which give place to croplands, pastures, urbanisation or just barren land. This clearing process is associated with net carbon losses from the vegetation and soil pool. Not all biotic degradation processes involve biomass losses. Woody encroachment of open savannahs involves the expansion of woody plant cover and/or density over herbaceous areas and often limits the secondary productivity of rangelands (Asner et al. 2004; Anadon et al. 2014). These processes have accelerated since the mid-1800s over most continents (Van Auken 2009). Change in plant composition of natural or semi-natural ecosystems without any significant vegetation structural changes is another pathway of degradation affecting rangelands and forests. In rangelands, selective grazing and its interaction with climate variability and/or fire can push ecosystems to new compositions with lower forage value and a higher proportion of invasive species (Illius and O'Connor 1999; Sasaki et al. 2007), in some cases with higher carbon sequestration potential, yet with very complex interactions between vegetation and soil carbon shifts (Piñeiro et al. 2010). In forests, extractive logging can be a pervasive cause of degradation, leading to long-term impoverishment and, in extreme cases, a full loss of the forest cover through its interaction with other agents such as fires (Foley et al. 2007) or progressive intensification of land use. Invasive alien species are another source of biological degradation. Their arrival into cultivated systems is constantly reshaping crop production strategies, making agriculture unviable on occasions. In natural and seminatural systems such as rangelands, invasive plant species not only threaten livestock production through diminished forage quality, poisoning and other deleterious effects, but have cascading effects on other processes such as altered fire regimes and water cycling (Brooks et al. 2004). In forests, invasions affect primary productivity and nutrient availability, change fire regimes, and alter species composition, resulting in long-term impacts on carbon pools and fluxes (Peltzer et al. 2010).

Other biotic components of ecosystems have been shown as a focus of degradation processes. Invertebrate invasions in continental waters can exacerbate other degradation processes such as eutrophication, which is the over-enrichment of nutrients, leading to excessive algal growth (Walsh et al. 2016a). Shifts in soil microbial and mesofaunal composition – which can be caused by pollution with pesticides or nitrogen deposition and by vegetation or disturbance regime shifts – alter many soil functions, including respiration rates and carbon release to the atmosphere (Hussain et al. 2009; Crowther et al. 2015). The role of the soil biota in modulating the effects of climate change on soil carbon has been recently demonstrated (Ratcliffe et al. 2017), highlighting the importance of this lesser-known component of the biota as a focal point of land degradation. Of special relevance as both

4

indicators and agents of land degradation recovery are mycorrhiza, which are root-associated fungal organisms (Asmelash et al. 2016; Vasconcellos et al. 2016). In natural dry ecosystems, biological soil crusts composed of a broad range of organisms, including mosses, are a particularly sensitive focus for degradation (Field et al. 2010) with evidenced sensitivity to climate change (Reed et al. 2012).

1.2.1.2   Land degradation processes and climate change

While the subdivision of individual processes is challenged by their strong interconnectedness, it provides a useful setting to identify the most important 'focal points' of climate change pressures on land degradation. Among land degradation processes, those responding more directly to climate change pressures include all types of erosion and SOM declines (soil focus), salinisation, sodification and permafrost thawing (soil/water focus), waterlogging of dry ecosystems and drying of wet ecosystems (water focus), and a broad group of biologically-mediated processes like woody encroachment, biological invasions, pest outbreaks (biotic focus), together with biological soil crust destruction and increased burning (soil/biota focus) (Table 4.1). Processes like ground subsidence can be affected by climate change indirectly through sea level rise (Keogh and Törnqvist 2019).

Even when climate change exerts a direct pressure on degradation processes, it can be a secondary driver subordinated to other overwhelming human pressures. Important exceptions are three processes in which climate change is a dominant global or regional pressure and the main driver of their current acceleration. These are: coastal erosion as affected by sea level rise and increased storm frequency/intensity (*high agreement, medium evidence*) (Johnson et al. 2015; Alongi 2015; Harley et al. 2017; Nicholls et al. 2016); permafrost thawing responding to warming (*high agreement, robust evidence*) (Liljedahl et al. 2016; Peng et al. 2016; Batir et al. 2017); and increased burning responding to warming and altered precipitation regimes (*high agreement, robust evidence*) (Jolly et al. 2015; Abatzoglou and Williams 2016; Taufik et al. 2017; Knorr et al. 2016). The previous assessment highlights the fact that climate change not only exacerbates many of the well-acknowledged ongoing land degradation processes of managed ecosystems (i.e., croplands and pastures), but becomes a dominant pressure that introduces novel degradation pathways in natural and seminatural ecosystems. Climate change has influenced species invasions and the degradation that they cause by enhancing the transport, colonisation, establishment, and ecological impact of the invasive species, and also by impairing their control practices (*medium agreement, medium evidence*) (Hellmann et al. 2008).

Table 4.1 | **Major land degradation processes and their connections with climate change.** For each process a 'focal point' (soil, water, biota) on which degradation occurs in the first place is indicated, acknowledging that most processes propagate to other land components and cascade into or interact with some of the other processes listed below. The impact of climate change on each process is categorised based on the proximity (very direct = high, very indirect = low) and dominance (dominant = high, subordinate to other pressures = low) of effects. The major effects of climate change on each process are highlighted together with the predominant pressures from other drivers. Feedbacks of land degradation processes on climate change are categorised according to the intensity (very intense = high, subtle = low) of the chemical (GHG emissions or capture) or physical (energy and momentum exchange, aerosol emissions) effects. Warming effects are indicated in red and cooling effects in blue. Specific feedbacks on climate change are highlighted.

Processes	Focal point	Impacts of climate change				Feedbacks on climate change			
		Proximity	Dominance	Climate change pressures	Other pressures	Intensity of chemical effects	Intensity of physical effects	Global extent	Specific impacts
Wind erosion	Soil	high	medium	Altered wind/drought patterns (*high confidence* on effect, *medium-low confidence* on trend) (1). Indirect effect through vegetation type and biomass production shifts	Tillage, leaving low cover, overgrazing, deforestation/vegetation clearing, large plot sizes, vegetation and fire regime shifts	low	medium	high	Radiative cooling by dust release (*medium confidence*). Ocean and land fertilisation and carbon burial (*medium confidence*). Albedo increase. Dust effect as condensation nuclei (19)
Water erosion	Soil	high	medium	Increasing rainfall intensity (*high confidence* on effect and trend) (2). Indirect effects on fire frequency/intensity, permafrost thawing, biomass production	Tillage, cultivation leaving low cover, overgrazing, deforestation/vegetation clearing, vegetation burning, poorly designed roads and paths	medium	medium	high	Net carbon release. Net release is probably less than site-specific loss due to deposition and burial (*high confidence*). Albedo increase (20)
Coastal erosion	Soil/water	high	high	Sea level rise, increasing intensity/frequency of storm surges (*high confidence* on effects and trends) (3)	Retention of sediments by upstream dams, coastal aquiculture, elimination of mangrove forests, subsidence	high	low	low	Release of old buried carbon pools (*medium confidence*) (21)

Processes	Focal point	Proximity	Dominance	Impacts of climate change		Feedbacks on climate change			
				Climate change pressures	Other pressures	Intensity of chemical effects	Intensity of physical effects	Global extent	Specific impacts
Subsidence	Soil/water	low	low	Indirect through increasing drought leading to higher ground water use. Indirect through enhanced decomposition (e.g., through drainage) in organic soils	Groundwater depletion/overpumping, peatland drainage	low/high	low	low	Unimportant in the case of groundwater depletion. Very high net carbon release in the case of drained peatlands
Compaction/hardening	Soil	low	low	Indirect through reduced organic matter content	Land-use conversion, machinery overuse, intensive grazing, poor tillage/grazing management (e.g., under wet or waterlogged conditions)	low	low	medium	Contradictory effects of reduced aeration on N_2O emissions
Nutrient depletion	Soil	low	low	Indirect (e.g., shifts in cropland distribution, BECCS)	Insufficient replenishment of harvested nutrients	low	low	medium	Net carbon release due shrinking SOC pools. Larger reliance on soil liming with associated CO_2 releases
Acidification/overfertilisation	Soil	low	low	Indirect (e.g., shifts in cropland distribution, BECCS). Sulfidic wetland drying due to increased drought as special direct effect	High nitrogen fertilisation, high cation depletion, acid rain/deposition	medium	low	medium	N_2O release from overfertilised soils, increased by acidification. Inorganic carbon release from acidifying soils (*medium to high confidence*) (22)
Pollution	Soil/biota	low	low	Indirect (e.g., increased pest and weed incidence)	Intensifying chemical control of weed and pests	low	low	medium	Unknown, probably unimportant
Organic matter decline	Soil	high	medium	Warming accelerates soil respiration rates (*medium confidence* on effects and trends) (4). Indirect effects through changing quality of plant litter or fire/waterlogging regimes	Tillage. reduced plant input to soil. Drainage of waterlogged soils. Influenced by most of the other soil degradation processes.	high	low	high	Net carbon release (*high confidence*) (23)
Metal toxicity	Soil	low	low	Indirect	High cation depletion, fertilisation, mining activities	low	low	low	Unknown, probably unimportant
Salinisation	Soil/water	high	low	Sea level rise (*high confidence* on effects and trends) (5). Water balance shifts (*medium confidence* on effects and trends) (6). Indirect effects through irrigation expansion	Irrigation without good drainage infrastructure. Deforestation and water table-level rises under dryland agriculture	low	medium	medium	Reduced methane emissions with high sulfate load. Albedo increase
Sodification (increased sodium and associated physical degradation in soils)	Soil/water	high	low	Water balance shifts (*medium confidence* on effects and trends) (7). Indirect effects through irrigation expansion	Poor water management	low	medium	low	Net carbon release due to soil structure and organic matter dispersion. Albedo increase
Permafrost thawing	Soil/water	high	high	Warming (*very high confidence* on effects and trends) (8), seasonality shifts and accelerated snow melt leading to higher erosivity.		high	low	high	Net carbon release. CH_4 release (*high confidence*) (24)

4

Processes	Focal point	Impacts of climate change				Feedbacks on climate change			
		Proximity	Dominance	Climate change pressures	Other pressures	Intensity of chemical effects	Intensity of physical effects	Global extent	Specific impacts
Waterlogging of dry systems	Water	high	medium	Water balance shifts (*medium confidence* on effects and trends) (9). Indirect effects through vegetation shifts	Deforestation. Irrigation without good drainage infrastructure	medium	medium	low	CH_4 release. Albedo decrease
Drying of continental waters/wetland/lowlands	Water	high	medium	Increasing extent and duration of drought (*high confidence* on effects, *medium confidence* on trends) (10). Indirect effects through vegetation shifts	Upstream surface and groundwater consumption. Intentional drainage. Trampling/overgrazing	medium	medium	medium	Net carbon release. N_2O release. Albedo increase
Flooding	Water	high	medium	Sea level rise, increasing intensity/frequency of storm surges, increasing rainfall intensity causing flash floods (*high confidence* on effects and trends) (11)	Land clearing. Increasing impervious surface. Transport infrastructure	medium	medium	low	CH_4 and N_2O release. Albedo decrease
Eutrophication of continental waters	Water/biota	low	low	Indirect through warming effects on nitrogen losses from the land or climate change effects on erosion rates. Interactive effects of warming and nutrient loads on algal blooms	Excess fertilisation. Erosion. Poor management of livestock/human sewage	medium	low	low	CH_4 and N_2O release
Woody encroachment	Biota	high	medium	Rainfall shifts (*medium confidence* on effects and trends), CO_2 rise (*medium confidence* on effects, *very high confidence* on trends) (12)	Overgrazing. Altered fire regimes, fire suppression. Invasive alien species	high	high	high	Net carbon storage. Albedo decrease
Species loss, compositional shifts	Biota	high	medium	Habitat loss as a result of climate shifts (*medium confidence* on effects and trends) (13)	Selective grazing and logging causing plant species loss, Pesticides causing soil microbial and soil faunal losses, large animal extinctions, interruption of disturbance regimes	low	low	medium	Unknown
Soil microbial and mesofaunal shifts	Biota	high	low	Habitat loss as a result of climate shifts (*medium confidence* on effects and trends) (14)	Altered fire regimes, nitrogen deposition, pesticide pollution, vegetation shifts, disturbance regime shifts	low	low	medium	Unknown
Biological soil crust destruction	Biota/soil	high	medium	Warming. Changing rainfall regimes. (*medium confidence* on effects, high confidence and trends). Indirect through fire regime shifts and/or invasions (15)	Overgrazing and trampling. Land-use conversion	low	high	high	Radiative cooling through albedo rise and dust release (*high confidence*) (25)
Invasions	Biota	high	medium	Habitat gain as a result of climate shifts (*medium confidence* on effects and trends) (16)	Intentional and unintentional species introductions	low	low	medium	Unknown
Pest outbreaks	Biota	high	medium	Habitat gain and accelerated reproduction as a result of climate shifts (*medium confidence* on effects and trends) (17)	Large-scale monocultures. Poor pest management practices	medium	low	medium	Net carbon release

| Processes | Focal point | Impacts of climate change | | | | Feedbacks on climate change | | | |
		Proximity	Dominance	Climate change pressures	Other pressures	Intensity of chemical effects	Intensity of physical effects	Global extent	Specific impacts
Increased burning	Soil/biota	high	high	Warming, drought, shifting precipitation regimes, also wet spells rising fuel load (*high confidence* on effects and trends) (18)	Fire suppression policies increasing wildfire intensity. Increasing use of fire for rangeland management. Agriculture introducing fires in humid climates without previous fire history. Invasions	high	medium	medium	Net carbon release. CO, CH$_4$, N$_2$O release. Albedo increase (*high confidence*). Long-term decline of NPP in non-adapted ecosystems (26)

References in Table 4.1:

(1) Bärring et al. 2003; Munson et al. 2011; Sheffield et al. 2012, (2) Nearing et al. 2004; Shakesby 2011; Panthou et al. 2014, (3) Johnson et al. 2015; Alongi 2015; Harley et al. 2017, (4) Bond-Lamberty et al. 2018; Crowther et al. 2016; van Gestel et al. 2018, (5) Colombani et al. 2016, (6) Schofield and Kirkby 2003; Aragüés et al. 2015; Benini et al. 2016, (7) Jobbágy et al. 2017, (8) Liljedahl et al. 2016; Peng et al. 2016; Batir et al. 2017, (9) Piovano et al. 2004; Osland et al. 2016, (10) Burkett and Kusler 2000; Nielsen and Brock 2009; Johnson et al. 2015; Green et al. 2017, (11) Panthou et al. 2014; Arnell and Gosling 2016; Vitousek et al. 2017, (12) Van Auken 2009; Wigley et al. 2010, (13) Vincent et al. 2014; Gonzalez et al. 2010; Scheffers et al. 2016, (14) Pritchard 2011; Ratcliffe et al. 2017, (15) Reed et al. 2012; Maestre et al. 2013, (16) Hellmann et al. 2008; Hulme 2017, (17) Pureswaran et al. 2015; Cilas et al. 2016; Macfadyen et al. 2018, (18) Jolly et al. 2015; Abatzoglou and Williams 2016; Taufik et al. 2017; Knorr et al. 2016, (19) Davin et al. 2010; Pinty et al. 2011, (20) Wang et al. 2017b; Chappell et al. 2016, (21) Pendleton et al. 2012, (22) Oertel et al. 2016, (23) Houghton et al. 2012; Eglin et al. 2010, (24) Schuur et al. 2015; Christensen et al. 2004; Walter Anthony et al. 2016; Abbott et al. 2016, (25) Belnap, Walker, Munson & Gill, 2014; Rutherford et al. 2017, (26) Page et al. 2002; Pellegrini et al. 2018.

1.2.2   Drivers of land degradation

Drivers of land degradation and land improvement are many and they interact in multiple ways. Figure 4.2 illustrates how some of the most important drivers interact with the land users. It is important to keep in mind that natural and human factors can drive both degradation and improvement (Kiage 2013; Bisaro et al. 2014).

Land degradation is driven by the entire spectrum of factors, from very short and intensive events, such as individual rain storms of 10 minutes removing topsoil or initiating a gully or a landslide (Coppus and Imeson 2002; Morgan 2005b) to century-scale slow depletion of nutrients or loss of soil particles (Johnson and Lewis 2007, pp. 5–6). But, instead of focusing on absolute temporal variations, the drivers of land degradation can be assessed in relation to the rates of possible recovery. Unfortunately, this is impractical to do in a spatially explicit way because rates of soil formation are difficult to measure due to the slow rate, usually <5mm/century (Delgado and Gómez 2016). Studies suggest that erosion rates of conventionally tilled agricultural fields exceed the rate at which soil is generated by one to two orders of magnitude (Montgomery 2007a).

The landscape effects of gully erosion from one short intensive rainstorm can persist for decades and centuries (Showers 2005). Intensive agriculture under the Roman Empire in occupied territories in France is still leaving its marks and can be considered an example of irreversible land degradation (Dupouey et al. 2002).

The climate-change-related drivers of land degradation are gradual changes of temperature, precipitation and wind, as well as changes of the distribution and intensity of extreme events (Lin et al. 2017). Importantly, these drivers can act in two directions: land improvement

and land degradation. Increasing CO$_2$ level in the atmosphere is a driver of land improvement, even if the net effect is modulated by other factors, such as the availability of nitrogen (Terrer et al. 2016) and water (Gerten et al. 2014; Settele et al. 2015; Girardin et al. 2016).

The gradual and planetary changes that can cause land degradation/improvement have been studied by global integrated models and Earth observation technologies. Studies of global land suitability for agriculture suggest that climate change will increase the area suitable for agriculture by 2100 in the Northern high latitudes by 16% (Ramankutty et al. 2002) or 5.6 million km^2 (Zabel et al. 2014), while tropical regions will experience a loss (Ramankutty et al. 2002; Zabel et al. 2014).

Temporal and spatial patterns of tree mortality can be used as an indicator of climate change impacts on terrestrial ecosystems. Episodic mortality of trees occurs naturally even without climate change, but more widespread spatio-temporal anomalies can be a sign of climate-induced degradation (Allen et al. 2010). In the absence of systematic data on tree mortality, a comprehensive meta-analysis of 150 published articles suggests that increasing tree mortality around the world can be attributed to increasing drought and heat stress in forests worldwide (Allen et al. 2010).

Other and more indirect drivers can be a wide range of factors such as demographic changes, technological change, changes of consumption patterns and dietary preferences, political and economic changes, and social changes (Mirzabaev et al. 2016). It is important to stress that there are no simple or direct relationships between underlying drivers and land degradation, such as poverty or high population density, that are necessarily causing land degradation (Lambin et al. 2001). However, drivers of land degradation need to be studied in the context

4

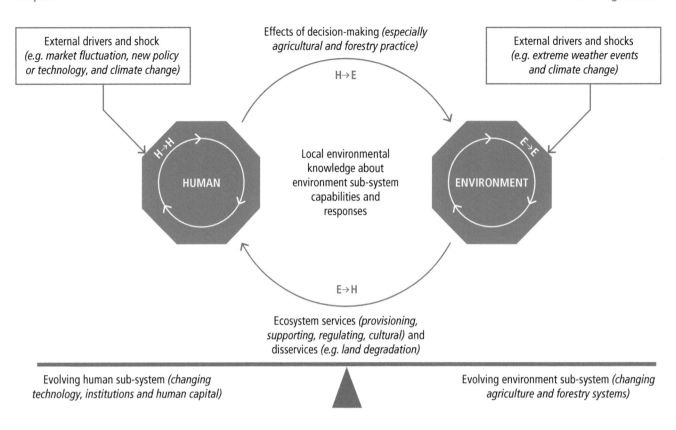

Figure 4.2 | Schematic representation of the interactions between the human (H) and environmental (E) components of the land system showing decision-making and ecosystem services as the key linkages between the components (moderated by an effective system of local and scientific knowledge), and indicating how the rates of change and the way these linkages operate must be kept broadly in balance for functional coevolution of the components. Modified with permission from Stafford Smith et al. (2007).

of spatial, temporal, economic, environmental and cultural aspects (Warren 2002). Some analyses suggest an overall negative correlation between population density and land degradation (Bai et al. 2008) but we find many local examples of both positive and negative relationships (Brandt et al. 2018a, 2017). Even if there are correlations in one or the other direction, causality is not always the same.

Land degradation is inextricably linked to several climate variables, such as temperature, precipitation, wind, and seasonality. This means that there are many ways in which climate change and land degradation are linked. The linkages are better described as a web of causality rather than a set of cause–effect relationships.

1.2.3   Attribution in the case of land degradation

The question here is whether or not climate change can be attributed to land degradation and vice versa. Land degradation is a complex phenomenon often affected by multiple factors such as climatic (rainfall, temperature, and wind), abiotic ecological factors (e.g., soil characteristics and topography), type of land use (e.g., farming of various kinds, forestry, or protected area), and land management practices (e.g., tilling, crop rotation, and logging/thinning). Therefore, attribution of land degradation to climate change is extremely challenging. Because land degradation is highly dependent on land management, it is even possible that climate impacts would trigger land management changes reducing or reversing land degradation,

sometimes called transformational adaptation (Kates et al. 2012). There is not much research on attributing land degradation explicitly to climate change, but there is more on climate change as a threat multiplier for land degradation. However, in some cases, it is possible to infer climate change impacts on land degradation, both theoretically and empirically. Section 4.2.3.1 outlines the potential direct linkages of climate change on land degradation based on current theoretical understanding of land degradation processes and drivers. Section 4.2.3.2 investigates possible indirect impacts on land degradation.

1.2.3.1   Direct linkages with climate change

The most important direct impacts of climate change on land degradation are the results of increasing temperatures, changing rainfall patterns, and intensification of rainfall. These changes will, in various combinations, cause changes in erosion rates and the processes driving both increases and decreases of soil erosion. From an attribution point of view, it is important to note that projections of precipitation are, in general, more uncertain than projections of temperature changes (Murphy et al. 2004; Fischer and Knutti 2015; IPCC 2013a). Precipitation involves local processes of larger complexity than temperature, and projections are usually less robust than those for temperature (Giorgi and Lionello 2008; Pendergrass 2018).

Theoretically the intensification of the hydrological cycle as a result of human-induced climate change is well established (Guerreiro et al. 2018; Trenberth 1999; Pendergrass et al. 2017; Pendergrass and Knutti 2018) and also empirically observed (Blenkinsop et al. 2018; Burt et al. 2016a; Liu et al. 2009; Bindoff et al. 2013). AR5 WGI concluded that heavy precipitation events have increased in frequency, intensity, and/or amount since 1950 (*likely*) and that further changes in this direction are *likely* to *very likely* during the 21st century (IPCC 2013). The IPCC Special Report on 1.5°C concluded that human-induced global warming has already caused an increase in the frequency, intensity and/or amount of heavy precipitation events at the global scale (Hoegh-Guldberg et al. 2018). As an example, in central India, there has been a threefold increase in widespread extreme rain events during 1950–2015 which has influenced several land degradation processes, not least soil erosion (Burt et al. 2016b). In Europe and North America, where observation networks are dense and extend over a long time, it is *likely* that the frequency or intensity of heavy rainfall have increased (IPCC 2013b). It is also expected that seasonal shifts and cycles such as monsoons and El Niño–Southern Oscillation (ENSO) will further increase the intensity of rainfall events (IPCC 2013).

When rainfall regimes change, it is expected to drive changes in vegetation cover and composition, which may be a cause of land degradation in and of itself, as well as impacting on other aspects of land degradation. Vegetation cover, for example, is a key factor in determining soil loss through water (Nearing et al. 2005) and wind erosion (Shao 2008). Changing rainfall regimes also affect below-ground biological processes, such as fungi and bacteria (Meisner et al. 2018; Shuab et al. 2017; Asmelash et al. 2016).

Changing snow accumulation and snow melt alter volume and timing of hydrological flows in and from mountain areas (Brahney et al. 2017; Lutz et al. 2014), with potentially large impacts on downstream areas. Soil processes are also affected by changing snow conditions with partitioning between evaporation and streamflow and between subsurface flow and surface runoff (Barnhart et al. 2016). Rainfall

intensity is a key climatic driver of soil erosion. Early modelling studies and theory suggest that light rainfall events will decrease while heavy rainfall events increase at about 7% per degree of warming (Liu et al. 2009; Trenberth 2011). Such changes result in increased intensity of rainfall, which increases the erosive power of rainfall (erosivity) and hence enhances the likelihood of water erosion. Increases in rainfall intensity can even exceed the rate of increase of atmospheric moisture content (Liu et al. 2009; Trenberth 2011). Erosivity is highly correlated to the product of total rainstorm energy and the maximum 30-minute rainfall intensity of the storm (Nearing et al. 2004) and increased erosivity will exacerbate water erosion substantially (Nearing et al. 2004). However, the effects will not be uniform, but highly variable across regions (Almagro et al. 2017; Mondal et al. 2016). Several empirical studies around the world have shown the increasing intensity of rainfall (IPCC 2013b; Ma et al. 2015, 2017) and also suggest that this will be accentuated with future increased global warming (Cheng and AghaKouchak 2015; Burt et al. 2016b; O'Gorman 2015).

The very comprehensive database of direct measurements of water erosion presented by García-Ruiz et al. (2015) contains 4377 entries (North America: 2776, Europe: 847, Asia: 259, Latin America: 237, Africa: 189, Australia and Pacific: 67), even though not all entries are complete (Figure 4.3).

An important finding from that database is that almost any erosion rate is possible under almost any climatic condition (García-Ruiz et al. 2015). Even if the results show few clear relationships between erosion and land conditions, the authors highlighted four observations (i) the highest erosion rates were found in relation to agricultural activities – even though moderate erosion rates were also found in agricultural settings, (ii) high erosion rates after forest fires were not observed (although the cases were few), (iii) land covered by shrubs showed generally low erosion rates, (iv) pasture land showed generally medium rates of erosion. Some important

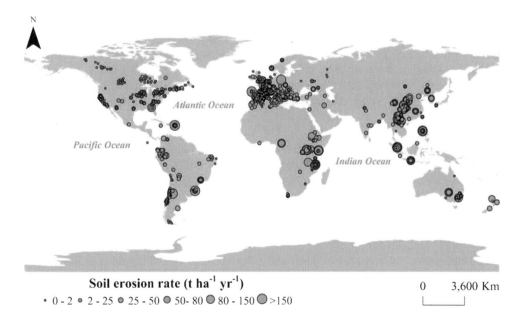

Figure 4.3. | Map of observed soil erosion rates in database of 4,377 entries by García-Ruiz et al. (2015). The map was published by Li and Fang (2016).

findings for the link between soil erosion and climate change can be noted from erosion measurements: erosion rates tend to increase with increasing mean annual rainfall, with a peak in the interval of 1000 to 1400 mm annual rainfall (García-Ruiz et al. 2015) (*low confidence*). However, such relationships are overshadowed by the fact that most rainfall events do not cause any erosion, instead erosion is caused by a few high-intensity rainfall events (Fischer et al. 2016; Zhu et al. 2019). Hence, mean annual rainfall is not a good predictor of erosion (Gonzalez-Hidalgo et al. 2012, 2009). In the context of climate change, it means that the tendency for rainfall patterns to change towards more intensive precipitation events is serious. Such patterns have already been observed widely, even in cases where the total rainfall is decreasing (Trenberth 2011). The findings generally confirm the strong consensus about the importance of vegetation cover as a protection against soil erosion, emphasising how extremely important land management is for controlling erosion.

In the Mediterranean region, the observed and expected decrease in annual rainfall due to climate change is accompanied by an increase of rainfall intensity, and hence erosivity (Capolongo et al. 2008). In tropical and sub-tropical regions, the on-site impacts of soil erosion dominate, and are manifested in very high rates of soil loss, in some cases exceeding 100 t ha^{-1} yr^{-1} (Tadesse 2001; García-Ruiz et al. 2015). In temperate regions, the off-site costs of soil erosion are often a greater concern, for example, siltation of dams and ponds, downslope damage to property, roads and other infrastructure (Boardman 2010). In cases where water erosion occurs, the downstream effects, such as siltation of dams, are often significant and severe in terms of environmental and economic damages (Kidane and Alemu 2015; Reinwarth et al. 2019; Quiñonero-Rubio et al. 2016; Adeogun et al. 2018; Ben Slimane et al. 2016).

The distribution of wet and dry spells also affects land degradation, although uncertainties remain depending on resolution of climate models used for prediction (Kendon et al. 2014). Changes in timing of rainfall events may have significant impacts on processes of soil erosion through changes in wetting and drying of soils (Lado et al. 2004).

Soil moisture content is affected by changes in evapotranspiration and evaporation, which may influence the partitioning of water into surface and subsurface runoff (Li and Fang 2016; Nearing et al. 2004). This portioning of rainfall can have a decisive effect on erosion (Stocking et al. 2001).

Wind erosion is a serious problem in agricultural regions, not only in drylands (Wagner 2013). Near-surface wind speeds over land areas have decreased in recent decades (McVicar and Roderick 2010), partly as a result of changing surface roughness (Vautard et al. 2010). Theoretically (Bakun 1990; Bakun et al. 2015) and empirically (Sydeman et al. 2014; England et al. 2014) average winds along coastal regions worldwide have increased with climate change (*medium evidence, high agreement*). Other studies of wind and wind erosion have not detected any long-term trend, suggesting that climate change has altered wind patterns outside drylands in a way that can significantly affect the risk of wind erosion (Pryor and Barthelmie 2010; Bärring et al. 2003). Therefore, the findings

regarding wind erosion and climate change are inconclusive, partly due to inadequate measurements.

Global mean temperatures are rising worldwide, but particularly in the Arctic region (*high confidence*) (IPCC 2018a). Heat stress from extreme temperatures and heatwaves (multiple days of hot weather in a row) have increased markedly in some locations in the last three decades (*high confidence*), and are *virtually certain* to continue during the 21st century (Olsson et al. 2014a). The IPCC Special Report on Global Warming of 1.5°C concluded that human-induced global warming has already caused more frequent heatwaves in most of land regions, and that climate models project robust differences between present-day and global warming up to 1.5°C and between 1.5°C and 2°C (Hoegh-Guldberg et al. 2018). Direct temperature effects on soils are of two kinds. Firstly, permafrost thawing leads to soil degradation in boreal and high-altitude regions (Yang et al. 2010; Jorgenson and Osterkamp 2005). Secondly, warming alters the cycling of nitrogen and carbon in soils, partly due to impacts on soil microbiota (Solly et al. 2017). There are many studies with particularly strong experimental evidence, but a full understanding of cause and effect is contextual and elusive (Conant et al. 2011a,b; Wu et al. 2011). This is discussed comprehensively in Chapter 2.

Climate change, including increasing atmospheric CO_2 levels, affects vegetation structure and function and hence conditions for land degradation. Exactly how vegetation responds to changes remains a research task. In a comparison of seven global vegetation models under four representative concentration pathways, Friend et al. (2014) found that all models predicted increasing vegetation carbon storage, however, with substantial variation between models. An important insight compared with previous understanding is that structural dynamics of vegetation seems to play a more important role for carbon storage than vegetation production (Friend et al. 2014). The magnitude of CO_2 fertilisation of vegetation growth, and hence conditions for land degradation, is still uncertain (Holtum and Winter 2010), particularly in tropical rainforests (Yang et al. 2016). For more discussion on this topic, see Chapter 2 in this report.

In summary, rainfall changes attributed to human-induced climate change have already intensified drivers of land degradation (*robust evidence, high agreement*) but attributing land degradation to climate change is challenging because of the importance of land management (*medium evidence, high agreement*). Changes in climate variability modes, such as in monsoons and El Niño–Southern Oscillation (ENSO) events, can also affect land degradation (*low evidence, low agreement*).

1.2.3.2   Indirect and complex linkages with climate change

Many important indirect linkages between land degradation and climate change occur via agriculture, particularly through changing outbreaks of pests (Rosenzweig et al. 2001; Porter et al. 1991; Thomson et al. 2010; Dhanush et al. 2015; Lamichhane et al. 2015), which is covered comprehensively in Chapter 5. More negative impacts have been observed than positive ones (IPCC 2014b). After 2050, the risk of yield loss increases as a result of climate change in combination with other drivers (*medium confidence*) and such risks will increase dramatically if global mean temperatures increase by about 4°C (*high confidence*) (Porter et al. 2014). The reduction (or plateauing) in yields in major production areas (Brisson et al. 2010; Lin and Huybers 2012; Grassini et al. 2013) may trigger cropland expansion elsewhere, either into natural ecosystems, marginal arable lands or intensification on already cultivated lands, with possible consequences for increasing land degradation.

Precipitation and temperature changes will trigger changes in land and crop management, such as changes in planting and harvest dates, type of crops, and type of cultivars, which may alter the conditions for soil erosion (Li and Fang 2016).

Much research has tried to understand how plants are affected by a particular stressor, for example, drought, heat, or waterlogging, including effects on below-ground processes. But less research has tried to understand how plants are affected by several simultaneous stressors – which of course is more realistic in the context of climate change (Mittler 2006; Kerns et al. 2016) and from a hazards point of view (Section 7.2.1). From an attribution point of view, such a complex web of causality is problematic if attribution is only done through statistically-significant correlation. It requires a combination of statistical links and theoretically informed causation, preferably integrated into a model. Some modelling studies have combined several stressors with geomorphologically explicit mechanisms – using the Water Erosion Prediction Project (WEPP) model – and realistic land-use scenarios, and found severe risks of increasing erosion from climate change (Mullan et al. 2012; Mullan 2013). Other studies have included various management options, such as changing planting and harvest dates (Zhang and Nearing 2005; Parajuli et al. 2016; Routschek et al. 2014; Nunes and Nearing 2011), type of cultivars (Garbrecht and Zhang 2015), and price of crops (Garbrecht et al. 2007; O'Neal et al. 2005) to investigate the complexity of how new climate regimes may alter soil erosion rates.

In summary, climate change increases the risk of land degradation, both in terms of likelihood and consequence, but the exact attribution to climate change is challenging due to several confounding factors. But since climate change exacerbates most degradation processes, it is clear that, unless land management is improved, climate change will result in increasing land degradation (*very high confidence*).

1.2.4   Approaches to assessing land degradation

In a review of different approaches and attempts to map global land degradation, Gibbs and Salmon (2015) identified four main approaches to map the global extent of degraded lands: expert opinions (Oldeman and van Lynden 1998; Dregne 1998; Reed 2005; Bot et al. 2000); satellite observation of vegetation greenness – for example, remote sensing of Normalized Difference Vegetation Index (NDVI), Enhanced Vegetation Index (EVI), Plant Phenology Index (PPI) – (Yengoh et al. 2015; Bai et al. 2008c; Shi et al. 2017; Abdi et al. 2019; JRC 2018); biophysical models (biogeographical/ topological) (Cai et al. 2011b; Hickler et al. 2005; Steinkamp and Hickler 2015; Stoorvogel et al. 2017); and inventories of land use/ condition. Together they provide a relatively complete evaluation,

but none on its own assesses the complexity of the process (Vogt et al. 2011; Gibbs and Salmon 2015). There is, however, a robust consensus that remote sensing and field-based methods are critical to assess and monitor land degradation, particularly over large areas (such as global, continental and sub-continental) although there are still knowledge gaps to be filled (Wessels et al. 2007, 2004; Prince 2016; Ghazoul and Chazdon 2017) as well as the problem of baseline values (Section 4.1.3).

Remote sensing can provide meaningful proxies of land degradation in terms of severity, temporal development, and areal extent. These proxies of land degradation include several indexes that have been used to assess land conditions, and monitoring changes of land conditions – for example, extent of gullies, severe forms of rill and sheet erosion, and deflation. The presence of open-access, quality controlled and continuously updated global databases of remote sensing data is invaluable, and is the only method for consistent monitoring of large areas over several decades (Sedano et al. 2016; Brandt et al. 2018b; Turner 2014). The NDVI, as a proxy for Net Primary Production (NPP) (see Glossary), is one of the most commonly used methods to assess land degradation, since it indicates land cover, an important factor for soil protection. Although NDVI is not a direct measure of vegetation biomass, there is a close coupling between NDVI integrated over a season and in situ NPP (*high agreement, robust evidence*) (see Higginbottom et al. 2014; Andela et al. 2013; Wessels et al. 2012).

Distinction between land degradation/improvement and the effects of climate variation is an important and contentious issue (Murthy and Bagchi 2018; Ferner et al. 2018). There is no simple and straightforward way to disentangle these two effects. The interaction of different determinants of primary production is not well understood. A key barrier to this is a lack of understanding of the inherent interannual variability of vegetation (Huxman et al. 2004; Knapp and Smith 2001; Ruppert et al. 2012; Bai et al. 2008a; Jobbágy and Sala 2000). One possibility is to compare potential land productivity modelled by vegetation models and actual productivity measured by remote sensing (Seaquist et al. 2009; Hickler et al. 2005; van der Esch et al. 2017), but the difference in spatial resolution, typically 0.5 degrees for vegetation models compared to 0.25–0.5 km for remote sensing data, is hampering the approach. The Moderate Resolution Imaging Spectroradiometer (MODIS) provides higher spatial resolution (up to 0.25 km), delivers data for the EVI, which is calculated in the same way as NDVI, and has showed a robust approach to estimate spatial patterns of global annual primary productivity (Shi et al. 2017; Testa et al. 2018).

Another approach to disentangle the effects of climate and land use/ management is to use the Rain Use Efficiency (RUE), defined as the biomass production per unit of rainfall, as an indicator (Le Houerou 1984; Prince et al. 1998; Fensholt et al. 2015). A variant of the RUE approach is the residual trend (RESTREND) of a NDVI time series, defined as the fraction of the difference between the observed NDVI and the NDVI predicted from climate data (Yengoh et al. 2015; John et al. 2016). These two metrics aim to estimate the NPP, rainfall and the time dimensions. They are simple transformations of the same three variables: RUE shows the NPP relationship with rainfall

4

for individual years, while RESTREND is the interannual change of RUE; also, both consider that rainfall is the only variable that affects biomass production. They are legitimate metrics when used appropriately, but in many cases they involve oversimplifications and yield misleading results (Fensholt et al. 2015; Prince et al. 1998).

Furthermore, increases in NPP do not always indicate improvement in land condition/reversal of land degradation, since this does not account for changes in vegetation composition. It could, for example, result from conversion of native forest to plantation, or due to bush encroachment, which many consider to be a form of land degradation (Ward 2005). Also, NPP may be increased by irrigation, which can enhance productivity in the short to medium term while increasing risk of soil salinisation in the long term (Niedertscheider et al. 2016).

Recent progress and expanding time series of canopy characterisations based on passive microwave satellite sensors have offered rapid progress in regional and global descriptions of forest degradation and recovery trends (Tian et al. 2017). The most common proxy is vertical optical depth (VOD) and has already been used to describe global forest/savannah carbon stock shifts over two decades, highlighting strong continental contrasts (Liu et al. 2015a) and demonstrating the value of this approach to monitor forest degradation at large scales. Contrasting with NDVI, which is only sensitive to vegetation 'greenness', from which primary production can be modelled, VOD is also sensitive to water in woody parts of the vegetation and hence provides a view of vegetation dynamics that can be complementary to NDVI. As well as the NDVI, VOD also needs to be corrected to take into account the rainfall variation (Andela et al. 2013).

Even though remote sensing offers much potential, its application to land degradation and recovery remains challenging as structural changes often occur at scales below the detection capabilities of most remote-sensing technologies. Additionally, if the remote sensing is based on vegetation index data, other forms of land degradation, such as nutrient depletion, changes of soil physical or biological properties, loss of values for humans, among others, cannot be inferred directly by remote sensing. The combination of remotely sensed images and field-based approach can give improved estimates of carbon stocks and tree biodiversity (Imai et al. 2012; Fujiki et al. 2016).

Additionally, the majority of trend techniques employed would be capable of detecting only the most severe of degradation processes, and would therefore not be useful as a degradation early-warning system (Higginbottom et al. 2014; Wessels et al. 2012). However, additional analyses using higher-resolution imagery, such as the Landsat and SPOT satellites, would be well suited to providing further localised information on trends observed (Higginbottom et al. 2014). New approaches to assess land degradation using high spatial resolution are developing, but the need for time series makes progress slow. The use of synthetic aperture radar (SAR) data has been shown to be advantageous for the estimation of soil surface characteristics, in particular, surface roughness and soil moisture (Gao et al. 2017; Bousbih et al. 2017), and detecting and quantifying selective logging (Lei et al. 2018). Continued research effort is required to enable full assessment of land degradation using remote sensing.

Computer simulation models can be used alone or combined with the remote sensing observations to assess land degradation. The Revised Universal Soil Loss Equation (RUSLE) can be used, to some extent, to predict the long-term average annual soil loss by water erosion. RUSLE has been constantly revisited to estimate soil loss based on the product of rainfall–runoff erosivity, soil erodibility, slope length and steepness factor, conservation factor, and support practice parameter (Nampak et al. 2018). Inherent limitations of RUSLE include data-sparse regions, inability to account for soil loss from gully erosion or mass wasting events, and that it does not predict sediment pathways from hillslopes to water bodies (Benavidez et al. 2018). Since RUSLE models only provide gross erosion, the integration of a further module in the RUSLE scheme to estimate the sediment yield from the modelled hillslopes is needed. The spatially distributed sediment delivery model, WaTEM/SEDEM, has been widely tested in Europe (Borrelli et al. 2018). Wind erosion is another factor that needs to be taken into account in the modelling of soil erosion (Webb et al. 2017a, 2016). Additional models need to be developed to include the limitations of the RUSLE models.

Regarding the field-based approach to assess land degradation, there are multiple indicators that reflect functional ecosystem processes linked to ecosystem services and thus to the value for humans. These indicators are a composite set of measurable attributes from different factors, such as climate, soil, vegetation, biomass, management, among others, that can be used together or separately to develop indexes to better assess land degradation (Allen et al. 2011; Kosmas et al. 2014).

Declines in vegetation cover, changes in vegetation structure, decline in mean species abundances, decline in habitat diversity, changes in abundance of specific indicator species, reduced vegetation health and productivity, and vegetation management intensity and use, are the most common indicators in the vegetation condition of forest and woodlands (Stocking et al. 2001; Wiesmair et al. 2017; Ghazoul and Chazdon 2017; Alkemade et al. 2009).

Several indicators of the soil quality (SOM, depth, structure, compaction, texture, pH, C:N ratio, aggregate size distribution and stability, microbial respiration, soil organic carbon, salinisation, among others) have been proposed (Schoenholtz et al. 2000) (Section 2.2). Among these, SOM directly and indirectly drives the majority of soil functions. Decreases in SOM can lead to a decrease in fertility and biodiversity, as well as a loss of soil structure, causing reductions in water-holding capacity, increased risk of erosion (both wind and water) and increased bulk density and hence soil compaction (Allen et al. 2011; Certini 2005; Conant et al. 2011a). Thus, indicators related with the quantity and quality of the SOM are necessary to identify land degradation (Pulido et al. 2017; Dumanski and Pieri 2000). The composition of the microbial community is *very likely* to be positive impacted by both climate change and land degradation processes (Evans and Wallenstein 2014; Wu et al. 2015; Classen et al. 2015), thus changes in microbial community composition can be very useful to rapidly reflect land degradation (e.g., forest degradation increased the bacterial alpha-diversity indexes) (Flores-Rentería et al. 2016; Zhou et al. 2018). These indicators might be used as a set of site-dependent indicators, and in a plant-soil system (Ehrenfeld et al. 2005).

4

Useful indicators of degradation and improvement include changes in ecological processes and disturbance regimes that regulate the flow of energy and materials and that control ecosystem dynamics under a climate change scenario. Proxies of dynamics include spatial and temporal turnover of species and habitats within ecosystems (Ghazoul et al. 2015; Bahamondez and Thompson 2016). Indicators in agricultural lands include crop yield decreases and difficulty in maintaining yields (Stocking et al. 2001). Indicators of landscape degradation/improvement in fragmented forest landscapes include the extent, size and distribution of remaining forest fragments, an increase in edge habitat, and loss of connectivity and ecological memory (Zahawi et al. 2015; Pardini et al. 2010).

In summary, as land degradation is such a complex and global process, there is no single method by which land degradation can be estimated objectively and consistently over large areas (*very high confidence*). However, many approaches exist that can be used to assess different aspects of land degradation or provide proxies of land degradation. Remote sensing, complemented by other kinds of data (i.e., field observations, inventories, expert opinions), is the only method that can generate geographically explicit and globally consistent data over time scales relevant for land degradation (several decades).

1.3   Status and current trends of land degradation

The scientific literature on land degradation often excludes forest degradation, yet here we attempt to assess both issues. Because of the different bodies of scientific literature, we assess land degradation and forest degradation under different sub-headings and, where possible, draw integrated conclusions.

1.3.1   Land degradation

There are no reliable global maps of the extent and severity of land degradation (Gibbs and Salmon 2015; Prince et al. 2018; van der Esch et al. 2017), despite the fact that land degradation is a severe problem (Turner et al. 2016). The reasons are both conceptual – that is, how land degradation is defined, using what baseline (Herrick et al. 2019) or over what time period – and methodological – that is, how it can be measured (Prince et al. 2018). Although there is a strong consensus that land degradation is a reduction in productivity of the land or soil, there are diverging views regarding the spatial and temporal scales at which land degradation occurs (Warren 2002), and how this can be quantified and mapped. Proceeding from the definition in this report, there are also diverging views concerning ecological integrity and the value to humans. A comprehensive treatment of the conceptual discussion about land degradation is provided by the recent report on land degradation from the Intergovernmental Science-Policy Platform on Biodiversity and Ecosystem Services (IPBES) (Montanarella et al. 2018).

A review of different attempts to map global land degradation, based on expert opinion, satellite observations, biophysical models and a database of abandoned agricultural lands, suggested that between <10 Mkm^2 to 60 Mkm^2 (corresponding to 8–45% of the ice-free land area) have been degraded globally (Gibbs and Salmon, 2015) (*very low confidence*).

One often-used global assessment of land degradation uses trends in NDVI as a proxy for land degradation and improvement during the period 1983 to 2006 (Bai et al. 2008b,c) with an update to 2011 (Bai et al. 2015). These studies, based on very coarse resolution satellite data (NOAA AVHRR data with a resolution of 8 km), indicated that, between 22% and 24% of the global ice-free land area was subject to a downward trend, while about 16% showed an increasing trend. The study also suggested, contrary to earlier assessments (Middleton and Thomas 1997), that drylands were not among the most affected regions. Another study using a similar approach for the period 1981–2006 suggested that about 29% of the global land area is subject to 'land degradation hotspots', that is, areas with acute land degradation in need of particular attention. These hotspot areas were distributed over all agro-ecological regions and land cover types. Two different studies have tried to link land degradation, identified by NDVI as a proxy, and number of people affected: Le et al. (2016) estimated that at least 3.2 billion people were affected, while Barbier and Hochard (2016, 2018) estimated that 1.33 billion people were affected, of which 95% were living in developing countries.

Yet another study, using a similar approach and type of remote-sensing data, compared NDVI trends with biomass trends calculated by a global vegetation model over the period 1982–2010 and found that 17–36% of the land areas showed a negative NDVI trend, while a positive or neutral trend was predicted in modelled vegetation (Schut et al. 2015). The World Atlas of Desertification (3rd edition) includes a global map of land productivity change over the period 1999 to 2013, which is one useful proxy for land degradation (Cherlet et al. 2018). Over that period, about 20% of the global ice-free land area shows signs of declining or unstable productivity, whereas about 20% shows increasing productivity. The same report also summarised the productivity trends by land categories and found that most forest land showed increasing trends in productivity, while rangelands had more declining trends than increasing trends (Figure 4.4). These productivity assessments, however, do not distinguish between trends due to climate change and trends due to other factors. A recent analysis of 'greening' of the world using MODIS time series of NDVI 2000–2017, shows a striking increase in the greening over China and India. In China the greening is seen over forested areas, 42%, and cropland areas, in which 32% is increasing (Section 4.9.3). In India, the greening is almost entirely associated with cropland (82%) (Chen et al. 2019).

All these studies of vegetation trends show that there are regionally differentiated trends of either decreasing or increasing vegetation. When comparing vegetation trends with trends in climatic variables, Schut et al. (2015) found very few areas (1–2%) where an increase in vegetation trend was independent of the climate drivers, and that study suggested that positive vegetation trends are primarily caused by climatic factors.

4

In an attempt to go beyond the mapping of global vegetation trends for assessing land degradation, Borelli et al. (2017) used a soil erosion model (RUSLE) and suggested that soil erosion is mainly caused in areas of cropland expansion, particularly in Sub-Saharan Africa, South America and Southeast Asia. The method is controversial for conceptual reasons (i.e., the ability of the model to capture the most important erosion processes) and data limitations (i.e., the availability of relevant data at regional to global scales), and its validity for assessing erosion over large areas has been questioned by several studies (Baveye 2017; Evans and Boardman 2016a,b; Labrière et al. 2015).

An alternative to using remote sensing for assessing the state of land degradation is to compile field-based data from around the globe (Turner et al. 2016). In addition to the problems of definitions and baselines, this approach is also hampered by the lack of standardised methods used in the field. An assessment of the global severity of soil erosion in agriculture, based on 1673 measurements around the world (compiled from 201 peer-reviewed articles), indicated that the global net median rate of soil formation (i.e., formation minus erosion) is about 0.004 mm yr^{-1} (about 0.05 t ha^{-1} yr^{-1}) compared with the median net rate of soil loss in agricultural fields, 1.52 mm yr^{-1} (about 18 t ha^{-1} yr^{-1}) in tilled fields and 0.065 mm yr^{-1} (about 0.8 t ha^{-1} yr^{-1}) in no-till fields (Montgomery 2007a). This means that the rate of soil erosion from agricultural fields is between 380 (conventional tilling) and 16 times (no-till) the natural rate of soil formation (*medium agreement, limited evidence*). These approximate figures are supported by another large meta-study including over 4000 sites

around the world (see Figure 4.4) where the average soil loss from agricultural plots was about 21 t ha^{-1} yr^{-1} (García-Ruiz et al. 2015). Climate change, mainly through the intensification of rainfall, will further increase these rates unless land management is improved (*high agreement, medium evidence*).

Soils contain about 1500 Gt of organic carbon (median across 28 different estimates presented by Scharlemann et al. (2014)), which is about 1.8 times more carbon than in the atmosphere (Ciais et al. 2013) and 2.3–3.3 times more than what is held in the terrestrial vegetation of the world (Ciais et al. 2013). Hence, land degradation, including land conversion leading to soil carbon losses, has the potential to impact on the atmospheric concentration of CO_2 substantially. When natural ecosystems are cultivated they lose soil carbon that accumulated over long time periods. The loss rate depends on the type of natural vegetation and how the soil is managed. Estimates of the magnitude of loss vary but figures between 20% and 59% have been reported in several meta studies (Poeplau and Don 2015; Wei et al. 2015; Li et al. 2012; Murty et al. 2002; Guo and Gifford 2002). The amount of soil carbon lost explicitly due to land degradation after conversion is hard to assess due to large variation in local conditions and management, see also Chapter 2.

From a climate change perspective, land degradation plays an important role in the dynamics of nitrous oxide (N_2O) and methane (CH_4). N_2O is produced by microbial activity in the soil and the dynamics are related to both management practices and weather

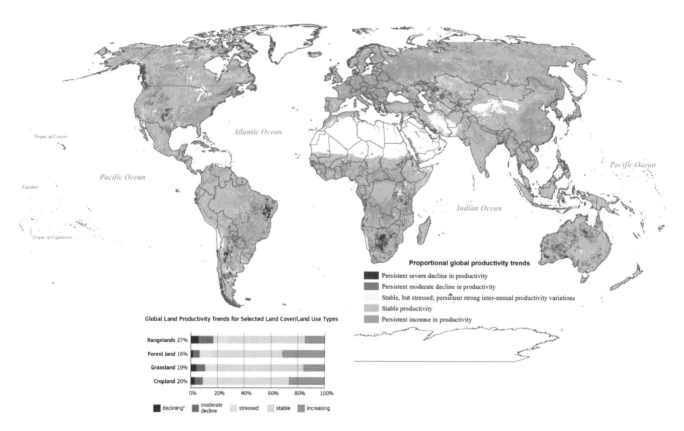

Figure 4.4 | Proportional global land productivity trends by land-cover/land-use class. (Cropland includes arable land, permanent crops and mixed classes with over 50% crops; grassland includes natural grassland and managed pasture land; rangelands include shrubland, herbaceous and sparsely vegetated areas; forest land includes all forest categories and mixed classes with tree cover greater than 40%.) Data source: Copernicus Global Land SPOT VGT, 1999–2013, adapted from (Cherlet et al. 2018).

conditions, while CH_4 dynamics are primarily determined by the amount of soil carbon and to what extent the soil is subject to waterlogging (Palm et al. 2014), see also Chapter 2.

Several attempts have been made to map the human footprint on the planet (Čuček et al. 2012; Venter et al. 2016) but, in some cases, they confuse human impact on the planet with degradation. From our definition it is clear that human impact (or pressure) is not synonymous with degradation, but information on the human footprint provides a useful mapping of potential non-climatic drivers of degradation.

In summary, there are no uncontested maps of the location, extent and severity of land degradation. Proxy estimates based on remote sensing of vegetation dynamics provide one important information source, but attribution of the observed changes in productivity to climate change, human activities, or other drivers is hard. Nevertheless, the different attempts to map the extent of global land degradation using remotely sensed proxies show some convergence and suggest that about a quarter of the ice-free land area is subject to some form of land degradation (*limited evidence, medium agreement*) affecting about 3.2 billion people (*low confidence*). Attempts to estimate the severity of land degradation through soil erosion estimates suggest that soil erosion is a serious form of land degradation in croplands closely associated with unsustainable land management in combination with climatic parameters, some of which are subject to climate change (*limited evidence, high agreement*). Climate change is one among several causal factors in the status and current trends of land degradation (*limited evidence, high agreement*).

1.3.2 Forest degradation

Quantifying degradation in forests has also proven difficult. Remote sensing based inventory methods can measure reductions in canopy cover or carbon stocks more easiliy than reductions in biological productivity, losses of ecological integrity or value to humans. However, the causes of reductions in canopy cover or carbon stocks can be many (Curtis et al. 2018), including natural disturbances (e.g., fires, insects and other forest pests), direct human activities (e.g., harvest, forest management) and indirect human impacts (such as climate change) and these may not reduce long-term biological productivity. In many boreal, some temperate and other forest types natural disturbances are common, and consequently these disturbance-adapted forest types are comprised of a mosaic of stands of different ages and stages of stand recovery following natural disturbances. In those managed forests where natural disturbances are uncommon or suppressed, harvesting is the primary determinant of forest age-class distributions.

Quantifying forest degradation as a reduction in productivity, carbon stocks or canopy cover also requires that an initial condition (or baseline) is established, against which this reduction is assessed (Section 4.1.4). In forest types with rare stand-replacing disturbances, the concept of 'intact' or 'primary' forest has been used to define the initial condition (Potapov et al. 2008) but applying a single metric can be problematic (Bernier et al. 2017). Moreover, forest types with

frequent stand-replacing disturbances, such as wildfires, or with natural disturbances that reduce carbon stocks, such as some insect outbreaks, experience over time a natural variability of carbon stocks or canopy density, making it more difficult to define the appropriate baseline carbon density or canopy cover against which to assess degradation. In these systems, forest degradation cannot be assessed at the stand level, but requires a landscape-level assessment that takes into consideration the stand age-class distribution of the landscape, which reflects natural and human disturbance regimes over past decades to centuries and also considers post-disturbance regrowth (van Wagner 1978; Volkova et al. 2018; Lorimer and White 2003).

The lack of a consistent definition of forest degradation also affects the ability to establish estimates of the rates or impacts of forest degradation because the drivers of degradation are not clearly defined (Sasaki and Putz 2009). Moreover, the literature at times confounds estimates of forest degradation and deforestation (i.e., the conversion of forest to non-forest land uses). Deforestation is a change in land use, while forest degradation is not, although severe forest degradation can ultimately lead to deforestation.

Based on empirical data provided by 46 countries, the drivers for deforestation (due to commercial agriculture) and forest degradation (due to timber extraction and logging) are similar in Africa, Asia and Latin America (Hosonuma et al. 2012). More recently, global forest disturbance over the period 2001–2015 was attributed to commodity-driven deforestation (27 ± 5%), forestry (26 ± 4%), shifting agriculture (24 ± 3%) and wildfire (23 ± 4%). The remaining 0.6 ± 0.3% was attributed to the expansion of urban centres (Curtis et al. 2018).

The trends of productivity shown by several remote-sensing studies (see previous section) are largely consistent with mapping of forest cover and change using a 34-year time series of coarse resolution satellite data (NOAA AVHRR) (Song et al. 2018). This study, based on a thematic classification of satellite data, suggests that (i) global tree canopy cover increased by 2.24 million km² between 1982 and 2016 (corresponding to +7.1%) but with regional differences that contribute a net loss in the tropics and a net gain at higher latitudes, and (ii) the fraction of bare ground decreased by 1.16 million km² (corresponding to –3.1%), mainly in agricultural regions of Asia (Song et al. 2018), see Figure 4.5. Other tree or land cover datasets show opposite global net trends (Li et al. 2018b), but high agreement in terms of net losses in the tropics and large net gains in the temperate and boreal zones (Li et al. 2018b; Song et al. 2018; Hansen et al. 2013). Differences across global estimates are further discussed in Chapter 1 (Section 1.1.2.3) and Chapter 2.

The changes detected from 1982 to 2016 were primarily linked to direct human action, such as land-use changes (about 60% of the observed changes), but also to indirect effects, such as human-induced climate change (about 40% of the observed changes) (Song et al. 2018), a finding also supported by a more recent study (Chen et al. 2019). The climate-induced effects were clearly discernible in some regions, such as forest decline in the US Northwest due to increasing pest infestation and increasing fire frequency (Lesk et al. 2017; Abatzoglou and Williams 2016; Seidl et al. 2017), warming-induced vegetation increase in the Arctic region, general greening in the Sahel probably as a result of increasing rainfall and atmospheric CO_2, and advancing treelines in mountain regions (Song et al. 2018).

Keenan et al. (2015) and Sloan and Sayer (2015) studied the 2015 Forest Resources Assessment (FRA) of the Food and Agriculture Organization of the United Nations (FAO) (FAO 2016) and found that the total forest area from 1990 to 2015 declined by 3%, an estimate that is supported by a global remote-sensing assessment of forest area change that found a 2.8% decline between 1990–2010 (D'Annunzio et al. 2017; Lindquist and D'Annunzio 2016). The trend in deforestation is, however, contradicted between these two global assessments, with FAO (2016) suggesting that deforestation is slowing down, while the remote sensing assessments finds it to be accelerating (D'Annunzio et al. 2017). Recent estimates (Song et al. 2018) owing to semantic and methodological differences (see Chapter 1, Section 1.1.2.3) suggest that global tree cover has increased over the period 1982–2016, which contradicts the forest area dynamics assessed by FAO (2016) and Lindquist and D'Annunzio (2016). The loss rate in tropical forest areas from 2010 to 2015 is 55,000 km^2 yr^{-1}. According to the FRA, the global natural forest area also declined from 39.61 Mkm^2 to 37.21 Mkm^2 during the period 1990 to 2015 (Keenan et al. 2015).

Since 1850, deforestation globally contributed 77% of the emissions from land-use and land-cover change while degradation contributed 10% (with the remainder originating from non-forest land uses) (Houghton and Nassikas 2018). That study also showed large temporal and regional differences with northern mid-latitude forests currently contributing to carbon sinks due to increasing forest area and forest management. However, the contribution to carbon emissions of degradation as percentage of total forest emissions (degradation and deforestation) are uncertain, with estimates varying from 25% (Pearson et al. 2017) to nearly 70% of carbon losses (Baccini et al. 2017). The 25% estimate refers to an analysis of 74 developing countries within tropical and subtropical regions covering 22 million km^2 for the period 2005–2010, while the 70% estimate refers to an analysis of the tropics for the period 2003–2014, but, by and large, the scope of these studies is the same. Pearson et al. (2017) estimated annual gross emissions of 2.1 $GtCO_2$, of which 53% were derived from timber harvest, 30% from woodfuel harvest and 17% from forest fire. Estimating gross emissions only, creates a distorted representation of human impacts on the land sector carbon cycle. While forest harvest for timber and fuelwood and land-use change (deforestation) contribute to gross emissions, to quantify impacts on the atmosphere, it is necessary to estimate net emissions, that is, the balance of gross emissions and gross removals of carbon from the atmosphere through forest regrowth (Chazdon et al. 2016a; Poorter et al. 2016; Sanquetta et al. 2018).

Current efforts to reduce atmospheric CO_2 concentrations can be supported by reductions in forest-related carbon emissions and increases in sinks, which requires that the net impact of forest management on the atmosphere be evaluated (Griscom et al. 2017). Forest management and the use of wood products in GHG mitigation strategies result in changes in forest ecosystem carbon stocks, changes in harvested wood product carbon stocks, and potential changes in emissions resulting from the use of wood products and forest biomass that substitute for other emissions-intensive materials such as concrete, steel and fossil fuels (Kurz et al. 2016; Lemprière et al. 2013; Nabuurs et al. 2007). The net impact of these changes on GHG emissions and removals, relative to a scenario without forest mitigation actions, needs to be quantified, (e.g., Werner et al. 2010; Smyth et al. 2014; Xu et al. 2018). Therefore, reductions in forest ecosystem carbon stocks alone are an incomplete estimator of the impacts of forest management on the atmosphere (Nabuurs et al. 2007; Lemprière et al. 2013; Kurz et al. 2016; Chen et al. 2018b). The impacts of forest management and the carbon storage in long-lived products and landfills vary greatly by region, however, because of the typically much shorter lifespan of wood products produced from tropical regions compared to temperate and boreal regions (Earles et al. 2012; Lewis et al. 2019; Iordan et al. 2018) (Section 4.8.4).

Assessments of forest degradation based on remote sensing of changes in canopy density or land cover, (e.g., Hansen et al. 2013; Pearson et al. 2017) quantify changes in above-ground biomass carbon stocks and require additional assumptions or model-based analyses to also quantify the impacts on other ecosystem carbon pools including below-ground biomass, litter, woody debris and soil carbon. Depending on the type of disturbance, changes in above-ground biomass may lead to decreases or increases in other carbon pools, for example, windthrow and insect-induced tree mortality may result in losses in above-ground biomass that are (initially) offset by corresponding increases in dead organic matter carbon pools (Yamanoi et al. 2015; Kurz et al. 2008), while deforestation will reduce the total ecosystem carbon pool (Houghton et al. 2012).

A global study of current vegetation carbon stocks (450 Gt C), relative to a hypothetical condition without land use (916 Gt C), attributed 42–47% of carbon stock reductions to land management effects without land-use change, while the remaining 53–58% of carbon stock reductions were attributed to deforestation and other land-use changes (Erb et al. 2018). While carbon stocks in European forests are lower than hypothetical values in the complete absence of human land use, forest area and carbon stocks have been increasing over recent decades (McGrath et al. 2015; Kauppi et al. 2018). Studies by Gingrich et al. (2015) on the long-term trends in land use over nine European countries (Albania, Austria, Denmark, Germany, Italy, the Netherlands, Romania, Sweden and the United Kingdom) also show an increase in forest land and reduction in cropland and grazing land from the 19th century to the early 20th century. However, the extent to which human activities have affected the productive capacity of forest lands is poorly understood. Biomass Production Efficiency (BPE), i.e. the fraction of photosynthetic production used for biomass production, was significantly higher in managed forests (0.53) compared to natural forests (0.41) (and it was also higher in managed (0.63) compared to natural (0.44) grasslands) (Campioli et al. 2015). Managing lands for production may involve trade-offs. For example, a larger proportion of NPP in managed forests is allocated to biomass carbon storage, but lower allocation to fine roots is hypothesised to reduce soil carbon stocks in the long term (Noormets et al. 2015). Annual volume increment in Finnish forests has more than doubled over the last century, due to increased growing stock, improved forest management and environmental changes (Henttonen et al. 2017).

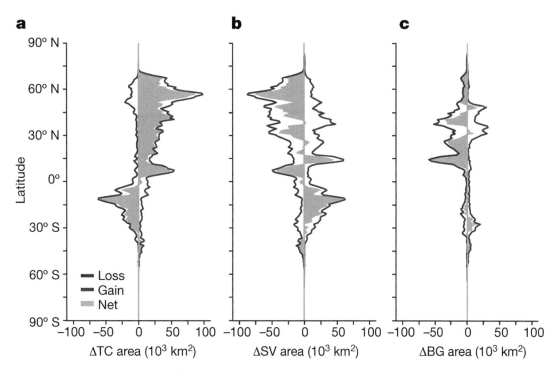

Figure 4.5 | Diagrams showing latitudinal profiles of land cover change over the period 1982 to 2016 based on analysis of time-series of NOAA AVHRR imagery: a) tree canopy cover change (ΔTC); **b)** short vegetation cover change (ΔSV); **c)** bare ground cover change (ΔBG). Area statistics were calculated for every 1° of latitude (Song et al. 2018). Source of data: NOAA AVHRR.

As economies evolve, the patterns of land-use and carbon stock changes associated with human expansion into forested areas often include a period of rapid decline of forest area and carbon stocks, recognition of the need for forest conservation and rehabilitation, and a transition to more sustainable land management that is often associated with increasing carbon stocks, (e.g., Birdsey et al. 2006). Developed and developing countries around the world are in various stages of forest transition (Kauppi et al. 2018; Meyfroidt and Lambin 2011). Thus, opportunities exist for SFM to contribute to atmospheric carbon targets through reduction of deforestation and degradation, forest conservation, forest restoration, intensification of management, and enhancements of carbon stocks in forests and harvested wood products (Griscom et al. 2017) (*medium evidence, medium agreement*).

1.4 Projections of land degradation in a changing climate

Land degradation will be affected by climate change in both direct and indirect ways, and land degradation will, to some extent, also feed back into the climate system. The direct impacts are those in which climate and land interact directly in time and space. Examples of direct impacts are when increasing rainfall intensity exacerbates soil erosion, or when prolonged droughts reduce the vegetation cover of the soil, making it more prone to erosion and nutrient depletion. The indirect impacts are those where climate change impacts and land degradation are separated in time and/or space. Examples of such impacts are when declining agricultural productivity due to climate change drives an intensification of agriculture elsewhere, which may cause land degradation. Land degradation, if sufficiently widespread,

may also feed back into the climate system by reinforcing ongoing climate change.

Although climate change is exacerbating many land degradation processes (*high to very high confidence*), prediction of future land degradation is challenging because land management practices determine, to a very large extent, the state of the land. Scenarios of climate change in combination with land degradation models can provide useful knowledge on what kind and extent of land management will be necessary to avoid, reduce and reverse land degradation.

1.4.1 Direct impacts on land degradation

There are two main levels of uncertainty in assessing the risks of future climate-change-induced land degradation. The first level, where uncertainties are comparatively low, involves changes of the degrading agent, such as erosive power of precipitation, heat stress from increasing temperature extremes (Hüve et al. 2011), water stress from droughts, and high surface wind speed. The second level of uncertainties, and where the uncertainties are much larger, relates to the above – and below-ground ecological changes as a result of changes in climate, such as rainfall, temperature, and increasing level of CO_2. Vegetation cover is crucial to protect against erosion (Mullan et al. 2012; García-Ruiz et al. 2015).

Changes in rainfall patterns, such as distribution in time and space, and intensification of rainfall events will increase the risk of land degradation, both in terms of likelihood and consequences (*high agreement, medium evidence*). Climate-induced vegetation changes will increase the risk of land degradation in some areas (where vegetation cover will decline) (*medium confidence*). Landslides are a form of land degradation, induced by extreme rainfall events. There is a strong theoretical reason for increasing landslide activity due to intensification of rainfall, but so far, the empirical evidence that climate change has contributed to landslides is lacking (Crozier 2010; Huggel et al. 2012; Gariano and Guzzetti 2016). Human disturbance may be a more important future trigger than climate change (Froude and Petley 2018).

Erosion of coastal areas as a result of sea level rise will increase worldwide (*very high confidence*). In cyclone-prone areas (such as the Caribbean, Southeast Asia, and the Bay of Bengal) the combination of sea level rise and more intense cyclones (Walsh et al. 2016b) and, in some areas, land subsidence (Yang et al. 2019; Shirzaei and Bürgmann 2018; Wang et al. 2018; Fuangswasdi et al. 2019; Keogh and Törnqvist 2019), will pose a serious risk to people and livelihoods (*very high confidence*), in some cases even exceeding limits to adaption (Sections 4.8.4.1, 4.9.6 and 4.9.8).

1.4.1.1   Changes in water erosion risk due to precipitation changes

The hydrological cycle is intensifying with increasing warming of the atmosphere. The intensification means that the number of heavy rainfall events is increasing, while the total number of rainfall events tends to decrease (Trenberth 2011; Li and Fang 2016; Kendon et al. 2014; Guerreiro et al. 2018; Burt et al. 2016a; Westra et al. 2014; Pendergrass and Knutti 2018) (*robust evidence, high agreement*). Modelling of the changes in land degradation that are a result of climate change alone is hard because of the importance of local contextual factors. As shown above, actual erosion rate is extremely dependent on local conditions, primarily vegetation cover and topography (García-Ruiz et al. 2015). Nevertheless, modelling of soil erosion risks has advanced substantially in recent decades, and such studies are indicative of future changes in the risk of soil erosion, while actual erosion rates will still primarily be determined by land management. In a review article, Li and Fang (2016) summarised 205 representative modelling studies around the world where erosion models were used in combination with downscaled climate models to assess future (between 2030 to 2100) erosion rates. The meta-study by Li and Fang, where possible, considered climate change in terms of temperature increase and changing rainfall regimes and their impacts on vegetation and soils. Almost all of the sites had current soil loss rates above 1 t ha⁻¹ (assumed to be the upper limit for acceptable soil erosion in Europe) and 136 out of 205 studies predicted increased soil erosion rates. The percentage increase in erosion rates varied between 1.2% to as much as over 1600%, whereas 49 out of 205 studies projected more than 50% increase. Projected soil erosion rates varied substantially between studies because the important of local factors, hence climate change impacts on soil erosion, should preferably be assessed at the local to regional scale, rather than the global (Li and Fang 2016).

Mesoscale convective systems (MCS), typically thunder storms, have increased markedly in the last three to four decades in the USA and Australia and they are projected to increase substantially (Prein et al. 2017). Using a climate model with the ability to represent MCS, Prein

and colleagues were able to predict future increases in frequency, intensity and size of such weather systems. Findings include the 30% decrease in number of MCS of <40 mm h⁻¹, but a sharp increase of 380% in the number of extreme precipitation events of >90 mm h⁻¹ over the North American continent. The combined effect of increasing precipitation intensity and increasing size of the weather systems implies that the total amount of precipitation from these weather systems is expected to increase by up to 80% (Prein et al. 2017), which will substantially increase the risk of land degradation in terms of landslides, extreme erosion events, flashfloods, and so on.

The potential impacts of climate change on soil erosion can be assessed by modelling the projected changes in particular variables of climate change known to cause erosion, such as erosivity of rainfall. A study of the conterminous United States based on three climate models and three scenarios (A2, A1B, and B1) found that rainfall erosivity will increase in all scenarios, even if there are large spatial differences – a strong increase in the north-east and north-west, and either weak or inconsistent trends in the south-west and mid-west (Segura et al. 2014).

In a study of how climate change will impact on future soil erosion processes in the Himalayas, Gupta and Kumar (2017) estimated that soil erosion will increase by about 27% in the near term (2020s) and 22% in the medium term (2080s), with little difference between scenarios. A study from Northern Thailand estimated that erosivity will increase by 5% in the near term (2020s) and 14% in the medium term (2080s), which would result in a similar increase of soil erosion, all other factors being constant (Plangoen and Babel 2014). Observed rainfall erosivity has increased significantly in the lower Niger Basin (Nigeria) and is predicted to increase further based on statistical downscaling of four General Circulation Models (GCM) scenarios, with an estimated increase of 14%, 19% and 24% for the 2030s, 2050s, and 2070s respectively (Amanambu et al. 2019).

Many studies from around the world where statistical downscaling of GCM results have been used in combination with process-based erosion models show a consistent trend of increasing soil erosion.

Using a comparative approach, Serpa et al. (2015) studied two Mediterranean catchments (one dry and one humid) using a spatially explicit hydrological model – soil and water assessment tool (SWAT) – in combination with land-use and climate scenarios for 2071–2100. Climate change projections showed, on the one hand, decreased rainfall and streamflow for both catchments, whereas sediment export decreased only for the humid catchment; projected land-use change, from traditional to more profitable, on the other hand, resulted in increase in streamflow. The combined effect of climate and land-use change resulted in reduced sediment export for the humid catchment (−29% for A1B; −22% for B1) and increased sediment export for the dry catchment (+222% for A1B; +5% for B1). Similar methods have been used elsewhere, also showing the dominant effect of land-use/land cover for runoff and soil erosion (Neupane and Kumar 2015).

A study of future erosion rates in Northern Ireland, using a spatially explicit erosion model in combination with downscaled climate projections (with and without sub-daily rainfall intensity changes), showed that erosion rates without land management changes would decrease by the 2020s, 2050s and 2100s, irrespective of changes in intensity, mainly as a result of a general decline in rainfall (Mullan et al. 2012). When land management scenarios were added to the modelling, the erosion rates started to vary dramatically for all three time periods, ranging from a decrease of 100% for no-till land use, to an increase of 3621% for row crops under annual tillage and sub-days intensity changes (Mullan et al. 2012). Again, it shows how crucial land management is for addressing soil erosion, and the important role of rainfall intensity changes.

There is a large body of literature based on modelling future land degradation due to soil erosion concluding that, in spite of the increasing trend of erosive power of rainfall, (*medium evidence, high agreement*) land degradation is primarily determined by land management (*very high confidence*).

1.4.1.2   Climate-induced vegetation changes, implications for land degradation

The spatial mosaic of vegetation is determined by three factors: the ability of species to reach a particular location, how species tolerate the environmental conditions at that location (e.g., temperature, precipitation, wind, the topographic and soil conditions), and the interaction between species (including above/below ground species (Settele et al. 2015). Climate change is projected to alter the conditions and hence impact on the spatial mosaic of vegetation, which can be considered a form of land degradation. Warren et al. (2018) estimated that only about 33% of globally important biodiversity conservation areas will remain intact if global mean temperature increases to 4.5°C, while twice that area (67%) will remain intact if warming is restricted to 2°C. According to AR5, the clearest link between climate change and ecosystem change is when temperature is the primary driver, with changes of Arctic tundra as a response to significant warming as the best example (Settele et al. 2015). Even though distinguishing climate-induced changes from land-use changes is challenging, Boit et al. (2016) suggest that 5–6% of biomes in South America will undergo biome shifts until 2100, regardless of scenario, attributed to climate change. The projected biome shifts are primarily forests shifting to shrubland and dry forests becoming fragmented and isolated from more humid forests (Boit et al. 2016). Boreal forests are subject to unprecedented warming in terms of speed and amplitude (IPCC 2013b), with significant impacts on their regional distribution (Juday et al. 2015). Globally, tree lines are generally expanding northward and to higher elevations, or remaining stable, while a reduction in tree lines was rarely observed, and only where disturbances occurred (Harsch et al. 2009). There is *limited evidence* of a slow northward migration of the boreal forest in eastern North America (Gamache and Payette 2005). The thawing of permafrost may increase drought-induced tree mortality throughout the circumboreal zone (Gauthier et al. 2015).

Forests are a prime regulator of hydrological cycling, both fluxes of atmospheric moisture and precipitation, hence climate and forests are inextricably linked (Ellison et al. 2017; Keys et al. 2017). Forest management influences the storage and flow of water in forested

watersheds. In particular, harvesting, forest thinning and the construction of roads increase the likelihood of floods as an outcome of extreme climate events (Eisenbies et al. 2007). Water balance of at least partly forested landscapes is, to a large extent, controlled by forest ecosystems (Sheil and Murdiyarso 2009; Pokam et al. 2014). This includes surface runoff, as determined by evaporation and transpiration and soil conditions, and water flow routing (Eisenbies et al. 2007). Water-use efficiency (i.e., the ratio of water loss to biomass gain) is increasing with increased CO_2 levels (Keenan et al. 2013), hence transpiration is predicted to decrease which, in turn, will increase surface runoff (Schlesinger and Jasechko 2014). However, the interaction of several processes makes predictions challenging (Frank et al. 2015; Trahan and Schubert 2016). Surface runoff is an important agent in soil erosion.

Generally, removal of trees through harvesting or forest death (Anderegg et al. 2012) will reduce transpiration and hence increase the runoff during the growing season. Management-induced soil disturbance (such as skid trails and roads) will affect water flow routing to rivers and streams (Zhang et al. 2017; Luo et al. 2018; Eisenbies et al. 2007).

Climate change affects forests in both positive and negative ways (Trumbore et al. 2015; Price et al. 2013) and there will be regional and temporal differences in vegetation responses (Hember et al. 2017; Midgley and Bond 2015). Several climate-change-related drivers interact in complex ways, such as warming, changes in precipitation and water balance, CO_2 fertilisation, and nutrient cycling, which makes projections of future net impacts challenging (Kurz et al. 2013; Price et al. 2013) (Section 2.3.1.2). In high latitudes, a warmer climate will extend the growing seasons. However, this could be constrained by summer drought (Holmberg et al. 2019), while increasing levels of atmospheric CO_2 will increase water-use efficiency but not necessarily tree growth (Giguère-Croteau et al. 2019). Improving one growth-limiting factor will only enhance tree growth if other factors are not limiting (Norby et al. 2010; Trahan and Schubert 2016; Xie et al. 2016; Frank et al. 2015). Increasing forest productivity has been observed in most of Fennoscandia (Kauppi et al. 2014; Henttonen et al. 2017), Siberia and the northern reaches of North America as a response to a warming trend (Gauthier et al. 2015) but increased warming may also decrease forest productivity and increase risk of tree mortality and natural disturbances (Price et al. 2013; Girardin et al. 2016; Beck et al. 2011; Hember et al. 2016; Allen et al. 2011). The climatic conditions in high latitudes are changing at a magnitude faster than the ability of forests to adapt with detrimental, yet unpredictable, consequences (Gauthier et al. 2015).

Negative impacts dominate, however, and have already been documented (Lewis et al. 2004; Bonan et al. 2008; Beck et al. 2011) and are predicted to increase (Miles et al. 2004; Allen et al. 2010; Gauthier et al. 2015; Girardin et al. 2016; Trumbore et al. 2015). Several authors have emphasised a concern that tree mortality (forest dieback) will increase due to climate-induced physiological stress as well as interactions between physiological stress and other stressors, such as insect pests, diseases, and wildfires (Anderegg et al. 2012; Sturrock et al. 2011; Bentz et al. 2010; McDowell et al. 2011). Extreme events such as extreme heat and drought, storms, and floods also pose increased threats to forests in both high – and low-latitude forests (Lindner et al. 2010; Mokria et al. 2015). However, comparing observed forest dieback with modelled climate-induced damages did not show a general link between climate change and forest dieback (Steinkamp and Hickler 2015). Forests are subject to increasing frequency and intensity of wildfires which is projected to increase substantially with continued climate change (Price et al. 2013) (Cross-Chapter Box 3 in Chapter 2, and Chapter 2). In the tropics, interaction between climate change, CO_2 and fire could lead to abrupt shifts between woodland – and grassland-dominated states in the future (Shanahan et al. 2016).

Within the tropics, much research has been devoted to understanding how climate change may alter regional suitability of various crops. For example, coffee is expected to be highly sensitive to both temperature and precipitation changes, both in terms of growth and yield, and in terms of increasing problems of pests (Ovalle-Rivera et al. 2015). Some studies conclude that the global area of coffee production will decrease by 50% (Bunn et al. 2015). Due to increased heat stress, the suitability of Arabica coffee is expected to deteriorate in Mesoamerica, while it can improve in high-altitude areas in South America. The general pattern is that the climatic suitability for Arabica coffee will deteriorate at low altitudes of the tropics as well as at the higher latitudes (Ovalle-Rivera et al. 2015). This means that climate change in and of itself can render unsustainable previously sustainable land-use and land management practices, and vice versa (Laderach et al. 2011).

Rangelands are projected to change in complex ways due to climate change. Increasing levels of atmospheric CO_2 directly stimulate plant growth and can potentially compensate for negative effects from drying by increasing rain-use efficiency. But the positive effect of increasing CO_2 will be mediated by other environmental conditions, primarily water availability, but also nutrient cycling, fire regimes and invasive species. Studies over the North American rangelands suggest, for example, that warmer and dryer climatic conditions will reduce NPP in the southern Great Plains, the Southwest, and northern Mexico, but warmer and wetter conditions will increase NPP in the northern Plains and southern Canada (Polley et al. 2013).

1.4.1.3 Coastal erosion

Coastal erosion is expected to increase dramatically by sea level rise and, in some areas, in combination with increasing intensity of cyclones (highlighted in Section 4.9.6) and cyclone-induced coastal erosion. Coastal regions are also characterised by high population density, particularly in Asia (Bangladesh, China, India, Indonesia, Vietnam), whereas the highest population increase in coastal regions is projected in Africa (East Africa, Egypt, and West Africa) (Neumann et al. 2015). For coastal regions worldwide, and particularly in developing countries with high population density in low-lying coastal areas, limiting the warming to 1.5°C to 2.0°C will have major socio-economic benefits compared with higher temperature scenarios (IPCC 2018a; Nicholls et al. 2018). For more in-depth discussions on coastal process, please refer to Chapter 4 of the IPCC Special Report on the Ocean and Cryosphere in a Changing Climate (IPCC SROCC).

Despite the uncertainty related to the responses of the large ice sheets of Greenland and west Antarctica, climate-change-induced sea level rise is largely accepted and represents one of the biggest threats faced by coastal communities and ecosystems (Nicholls et al. 2011; Cazenave and Cozannet 2014; DeConto and Pollard 2016; Mengel et al. 2016). With significant socio-economic effects, the physical impacts of projected sea level rise, notably coastal erosion, have received considerable scientific attention (Nicholls et al. 2011; Rahmstorf 2010; Hauer et al. 2016).

Rates of coastal erosion or recession will increase due to rising sea levels and, in some regions, also in combination with increasing oceans waves (Day and Hodges 2018; Thomson and Rogers 2014; McInnes et al. 2011; Mori et al. 2010), lack or absence of sea-ice (Savard et al. 2009; Thomson and Rogers 2014) thawing of permafrost (Hoegh-Guldberg et al. 2018), and changing cyclone paths (Tamarin-Brodsky and Kaspi 2017; Lin and Emanuel 2016a). The respective role of the different climate factors in the coastal erosion process will vary spatially. Some studies have shown that the role of sea level rise on the coastal erosion process can be less important than other climate factors, like wave heights, changes in the frequency of the storms, and the cryogenic processes (Ruggiero 2013; Savard et al. 2009). Therefore, in order to have a complete picture of the potential effects of sea level rise on rates of coastal erosion, it is crucial to consider the combined effects of the aforementioned climate controls and the geomorphology of the coast under study.

Coastal wetlands around the world are sensitive to sea level rise. Projections of the impacts on global coastlines are inconclusive, with some projections suggesting that 20% to 90% (depending on sea level rise scenario) of present day wetlands will disappear during the 21st century (Spencer et al. 2016). Another study, which included natural feedback processes and management responses, suggested that coastal wetlands may actually increase (Schuerch et al. 2018).

Low-lying coastal areas in the tropics are particularly subject to the combined effect of sea level rise and increasing intensity of tropical cyclones, conditions that, in many cases, pose limits to adaptation (Section 4.8.5.1).

Many large coastal deltas are subject to the additional stress of shrinking deltas as a consequence of the combined effect of reduced sediment loads from rivers due to damming and water use, and land subsidence resulting from extraction of ground water or natural gas, and aquaculture (Higgins et al. 2013; Tessler et al. 2016; Minderhoud et al. 2017; Tessler et al. 2015; Brown and Nicholls 2015; Szabo et al. 2016; Yang et al. 2019; Shirzaei and Bürgmann 2018; Wang et al. 2018; Fuangswasdi et al. 2019). In some cases the rate of subsidence can outpace the rate of sea level rise by one order of magnitude (Minderhoud et al. 2017) or even two (Higgins et al. 2013). Recent findings from the Mississippi Delta raise the risk of a systematic underestimation of the rate of land subsidence in coastal deltas (Keogh and Törnqvist 2019).

In sum, from a land degradation point of view, low-lying coastal areas are particularly exposed to the nexus of climate change and increasing concentration of people (Elliott et al. 2014) (*robust evidence, high agreement*) and the situation will become particularly acute in delta areas shrinking from both reduced sediment loads and land subsidence (*robust evidence, high agreement*).

1.4.2   Indirect impacts on land degradation

Indirect impacts of climate change on land degradation are difficult to quantify because of the many conflating factors. The causes of land-use change are complex, combining physical, biological and socio-economic drivers (Lambin et al. 2001; Lambin and Meyfroidt 2011). One such driver of land-use change is the degradation of agricultural land, which can result in a negative cycle of natural land being converted to agricultural land to sustain production levels. The intensive management of agricultural land can lead to a loss of soil function, negatively impacting on the many ecosystem services provided by soils, including maintenance of water quality and soil carbon sequestration (Smith et al. 2016a). The degradation of soil quality due to cropping is of particular concern in tropical regions, where it results in a loss of productive potential of the land, affecting regional food security and driving conversion of non-agricultural land, such as forestry, to agriculture (Lambin et al. 2003; Drescher et al. 2016; Van der Laan et al. 2017). Climate change will exacerbate these negative cycles unless sustainable land management practices are implemented.

Climate change impacts on agricultural productivity (see Chapter 5) will have implications for the intensity of land use and hence exacerbate the risk of increasing land degradation. There will be both localised effects (i.e., climate change impacts on productivity affecting land use in the same region) and teleconnections (i.e., climate change impacts and land-use changes that are spatially and temporally separate) (Wicke et al. 2012; Pielke et al. 2007). If global temperature increases beyond 3°C it will have negative yield impacts on all crops (Porter et al. 2014) which, in combination with a doubling of demands by 2050 (Tilman et al. 2011), and increasing competition for land from the expansion of negative emissions technologies (IPCC 2018a; Schleussner et al. 2016), will exert strong pressure on agricultural lands and food security.

In sum, reduced productivity of most agricultural crops will drive land-use changes worldwide (*robust evidence, medium agreement*), but predicting how this will impact on land degradation is challenging because of several conflating factors. Social change, such as widespread changes in dietary preferences, will have a huge impact on agriculture and hence land degradation (*medium evidence, high agreement*).

1.5   Impacts of bioenergy and technologies for CO₂ removal (CDR) on land degradation

1.5.1   Potential scale of bioenergy and land-based CDR

In addition to the traditional land-use drivers (e.g., population growth, agricultural expansion, forest management), a new driver will interact to increase competition for land throughout this century:

the potential large-scale implementation of land-based technologies for CO_2 removal (CDR). Land-based CDR includes afforestation and reforestation, bioenergy with carbon capture and storage (BECCS), soil carbon management, biochar and enhanced weathering (Smith et al. 2015; Smith 2016).

Most scenarios, including two of the four pathways in the IPCC Special Report on 1.5°C (IPCC 2018a), compatible with stabilisation at 2°C involve substantial areas devoted to land-based CDR, specifically afforestation/reforestation and BECCS (Schleussner et al. 2016; Smith et al. 2016b; Mander et al. 2017). Even larger land areas are required in most scenarios aimed at keeping average global temperature increases to below 1.5°C, and scenarios that avoid BECCS also require large areas of energy crops in many cases (IPCC 2018b), although some options with strict demand-side management avoid this need (Grubler et al. 2018). Consequently, the addition of carbon capture and storage (CCS) systems to bioenergy facilities enhances mitigation benefits because it increases the carbon retention time and reduces emissions relative to bioenergy facilities without CCS. The IPCC SR15 states that, 'When considering pathways limiting warming to 1.5°C with no or limited overshoot, the full set of scenarios shows a conversion of 0.5–11 Mkm² of pasture into 0–6 Mkm² for energy crops, a 2 Mkm² reduction to 9.5 Mkm² increase [in] forest, and a 4 Mkm² decrease to a 2.5 Mkm² increase in non-pasture agricultural land for food and feed crops by 2050 relative to 2010.' (Rogelj et al. 2018, p. 145). For comparison, the global cropland area in 2010 was 15.9 Mkm² (Table 1.1), and Woods et al. (2015) estimate that the area of abandoned and degraded land potentially available for energy crops (or afforestation/reforestation) exceeds 5 Mkm². However, the area of available land has long been debated, as much marginal land is subject to customary land tenure and used informally, often by impoverished communities (Baka 2013, 2014; Haberl et al. 2013; Young 1999). Thus, as noted in SR15, 'The implementation of land-based mitigation options would require overcoming socio-economic, institutional, technological, financing and environmental barriers that differ across regions.' (IPCC, 2018a, p. 18).

The wide range of estimates reflects the large differences among the pathways, availability of land in various productivity classes, types of negative emission technology implemented, uncertainties in computer models, and social and economic barriers to implementation (Fuss et al. 2018; Nemet et al. 2018; Minx et al. 2018).

1.5.2   Risks of land degradation from expansion of bioenergy and land-based CDR

The large-scale implementation of high-intensity dedicated energy crops, and harvest of crop and forest residues for bioenergy, could contribute to increases in the area of degraded lands: intensive land management can result in nutrient depletion, over-fertilisation and soil acidification, salinisation (from irrigation without adequate drainage), wet ecosystems drying (from increased evapotranspiration), as well as novel erosion and compaction processes (from high-impact biomass harvesting disturbances) and other land degradation processes described in Section 4.2.1.

Global integrated assessment models used in the analysis of mitigation pathways vary in their approaches to modelling CDR (Bauer et al. 2018) and the outputs have large uncertainties due to their limited capability to consider site-specific details (Krause et al. 2018). Spatial resolutions vary from 11 world regions to 0.25 degrees gridcells (Bauer et al. 2018). While model projections identify potential areas for CDR implementation (Heck et al. 2018), the interaction with climate-change-induced biome shifts, available land and its vulnerability to degradation are unknown. The crop/forest types and management practices that will be implemented are also unknown, and will be influenced by local incentives and regulations. While it is therefore currently not possible to project the area at risk of degradation from the implementation of land-based CDR, there is a clear risk that expansion of energy crops at the scale anticipated could put significant strain on land systems, biosphere integrity, freshwater supply and biogeochemical flows (Heck et al. 2018). Similarly, extraction of biomass for energy from existing forests, particularly where stumps are utilised, can impact on soil health (de Jong et al. 2017). Reforestation and afforestation present a lower risk of land degradation and may in fact reverse degradation (Section 4.5.3) although potential adverse hydrological and biodiversity impacts will need to be managed (Caldwell et al. 2018; Brinkman et al. 2017). Soil carbon management can deliver negative emissions while reducing or reversing land degradation. Chapter 6 discusses the significance of context and management in determining environmental impacts of implementation of land-based options.

1.5.3   Potential contributions of land-based CDR to reducing and reversing land degradation

Although large-scale implementation of land-based CDR has significant potential risks, the need for negative emissions and the anticipated investments to implement such technologies can also create significant opportunities. Investments into land-based CDR can contribute to halting and reversing land degradation, to the restoration or rehabilitation of degraded and marginal lands (Chazdon and Uriarte 2016; Fritsche et al. 2017) and can contribute to the goals of LDN (Orr et al. 2017).

Estimates of the global area of degraded land range from less than 10 to 60 Mkm^2 (Gibbs and Salmon 2015) (Section 4.3.1). Additionally, large areas are classified as marginal lands and may be suitable for the implementation of bioenergy and land-based CDR (Woods et al. 2015). The yield per hectare of marginal and degraded lands is lower than on fertile lands, and if CDR will be implemented on marginal and degraded lands, this will increase the area demand and costs per unit area of achieving negative emissions (Fritsche et al. 2017). The selection of lands suitable for CDR must be considered carefully to reduce conflicts with existing users, to assess the possible trade-offs in biodiversity contributions of the original and the CDR land uses, to quantify the impacts on water budgets, and to ensure sustainability of the CDR land use.

Land use and land condition prior to the implementation of CDR affect climate change benefits (Harper et al. 2018). Afforestation/reforestation on degraded lands can increase carbon stocks in vegetation and soil, increase carbon sinks (Amichev et al. 2012), and deliver co-benefits for biodiversity and ecosystem services, particularly if a diversity of local species are used. Afforestation and reforestation on native grasslands can reduce soil carbon stocks, although the loss is typically more than compensated by increases in biomass and dead organic matter carbon stocks (Bárcena et al. 2014; Li et al. 2012; Ovalle-Rivera et al. 2015; Shi et al. 2013), and may impact on biodiversity (Li et al. 2012).

Strategic incorporation of energy crops into agricultural production systems, applying an integrated landscape management approach, can provide co-benefits for management of land degradation and other environmental objectives. For example, buffers of Miscanthus and other grasses can enhance soil carbon and reduce water pollution (Cacho et al. 2018; Odgaard et al. 2019), and strip-planting of short-rotation tree crops can reduce the water table where crops are affected by dryland salinity (Robinson et al. 2006). Shifting to perennial grain crops has the potential to combine food production with carbon sequestration at a higher rate than annual grain crops and avoid the trade-off between food production and climate change mitigation (Crews et al. 2018; de Olivera et al. 2018; Ryan et al. 2018) (Section 4.9.2).

Changes in land cover can affect surface reflectance, water balances and emissions of volatile organic compounds and thus the non-GHG impacts on the climate system from afforestation/reforestation or planting energy crops (Anderson et al. 2011; Bala et al. 2007; Betts 2000; Betts et al. 2007) (see Section 4.6 for further details). Some of these impacts reinforce the GHG mitigation benefits, while others offset the benefits, with strong local (slope, aspect) and regional (boreal vs. tropical biomes) differences in the outcomes (Li et al. 2015). Adverse effects on albedo from afforestation with evergreen conifers in boreal zones can be reduced through planting of broadleaf deciduous species (Astrup et al. 2018; Cai et al. 2011a; Anderson et al. 2011).

Combining CDR technologies may prove synergistic. Two soil management techniques with an explicit focus on increasing the soil carbon content rather than promoting soil conservation more broadly have been suggested: addition of biochar to agricultural soils (Section 4.9.5) and addition of ground silicate minerals to soils in order to take up atmospheric CO_2 through chemical weathering (Taylor et al. 2017; Haque et al. 2019; Beerling 2017; Strefler et al. 2018). The addition of biochar is comparatively well understood and also field tested at large scale, see Section 4.9.5 for a comprehensive discussion. The addition of silicate minerals to soils is still highly uncertain in terms of its potential (from 95 GtCO$_2$ yr^{-1} (Strefler et al. 2018) to only 2–4 GtCO$_2$ yr^{-1} (Fuss et al. 2018)) and costs (Schlesinger and Amundson 2018).

Effectively addressing land degradation through implementation of bioenergy and land-based CDR will require site-specific local knowledge, matching of species with the local land, water balance, nutrient and climatic conditions, ongoing monitoring and, where necessary, adaptation of land management to ensure sustainability under global change (Fritsche et al. 2017). Effective land governance mechanisms including integrated land-use planning, along with strong sustainability standards could support deployment of

4

energy crops and afforestation/reforestation at appropriate scales and geographical contexts (Fritsche et al. 2017). Capacity-building and technology transfer through the international cooperation mechanisms of the Paris Agreement could support such efforts. Modelling to inform policy development is most useful when undertaken with close interaction between model developers and other stakeholders including policymakers to ensure that models account for real world constraints (Dooley and Kartha 2018).

International initiatives to restore lands, such as the Bonn Challenge (Verdone and Seidl 2017) and the New York Declaration on Forests (Chazdon et al. 2017), and interventions undertaken for LDN and implementation of NDCs (see Glossary) can contribute to NET objectives. Such synergies may increase the financial resources available to meet multiple objectives (Section 4.8.4).

1.5.4   Traditional biomass provision and land degradation

Traditional biomass (fuelwood, charcoal, agricultural residues, animal dung) used for cooking and heating by some 2.8 billion people (38% of global population) in non-OECD countries accounts for more than half of all bioenergy used worldwide (IEA 2017; REN21 2018) (Cross-Chapter Box 7 in Chapter 6). Cooking with traditional biomass has multiple negative impacts on human health, particularly for women, children and youth (Machisa et al. 2013; Sinha and Ray 2015; Price 2017; Mendum and Njenga 2018; Adefuye et al. 2007) and on household productivity, including high workloads for women and youth (Mendum and Njenga 2018; Brunner et al. 2018; Hou et al. 2018; Njenga et al. 2019). Traditional biomass is land-intensive due to reliance on open fires, inefficient stoves and overharvesting of woodfuel, contributing to land degradation, losses in biodiversity and reduced ecosystem services (IEA 2017; Bailis et al. 2015; Masera et al. 2015; Specht et al. 2015; Fritsche et al. 2017; Fuso Nerini et al. 2017). Traditional woodfuels account for 1.9–2.3% of global GHG emissions, particularly in 'hotspots' of land degradation and fuelwood depletion in eastern Africa and South Asia, such that one-third of traditional woodfuels globally are harvested unsustainably (Bailis et al. 2015). Scenarios to significantly reduce reliance on traditional biomass in developing countries present multiple co-benefits (*high evidence, high agreement*), including reduced emissions of black carbon, a short-lived climate forcer that also causes respiratory disease (Shindell et al. 2012).

A shift from traditional to modern bioenergy, especially in the African context, contributes to improved livelihoods and can reduce land degradation and impacts on ecosystem services (Smeets et al. 2012; Gasparatos et al. 2018; Mudombi et al. 2018). In Sub-Saharan Africa, most countries mention woodfuel in their Nationally Determined Contribution (NDC) but fail to identify transformational processes to make fuelwood a sustainable energy source compatible with improved forest management (Amugune et al. 2017). In some regions, especially in South and Southeast Asia, a scarcity of woody biomass may lead to excessive removal and use of agricultural wastes and residues, which contributes to poor soil quality and land degradation (Blanco-Canqui and Lal 2009; Mateos et al. 2017).

In Sub-Saharan Africa, forest degradation is widely associated with charcoal production, although in some tropical areas rapid re-growth can offset forest losses (Hoffmann et al. 2017; McNicol et al. 2018). Overharvesting of wood for charcoal contributes to the high rate of deforestation in Sub-Saharan Africa, which is five times the world average, due in part to corruption and weak governance systems (Sulaiman et al. 2017). Charcoal may also be a by-product of forest clearing for agriculture, with charcoal sale providing immediate income when the land is cleared for food crops (Kiruki et al. 2017; Ndegwa et al. 2016). Besides loss of forest carbon stock, a further concern for climate change is methane and black carbon emissions from fuelwood burning and traditional charcoal-making processes (Bond et al. 2013; Patange et al. 2015; Sparrevik et al. 2015).

A fundamental difficulty in reducing environmental impacts associated with charcoal lies in the small-scale nature of much charcoal production in Sub-Saharan Africa, leading to challenges in regulating its production and trade, which is often informal, and in some cases illegal, but nevertheless widespread since charcoal is the most important urban cooking fuel (Zulu 2010; Zulu and Richardson 2013; Smith et al. 2015; World Bank 2009). Urbanisation combined with population growth has led to continuously increasing charcoal production. Low efficiency of traditional charcoal production results in a four-fold increase in raw woody biomass required and thus much greater biomass harvest (Hojas-Gascon et al. 2016; Smeets et al. 2012). With continuing urbanisation anticipated, increased charcoal production and use will probably contribute to increasing land pressures and increased land degradation, especially in Sub-Saharan Africa (*medium evidence, high agreement*).

Although it could be possible to source this biomass more sustainably, the ecosystem and health impacts of this increased demand for cooking fuel would be reduced through use of other renewable fuels or, in some cases, non-renewable fuels (LPG), as well as through improved efficiency in end-use and through better resource and supply chain management (Santos et al. 2017; Smeets et al. 2012; Hoffmann et al. 2017). Integrated response options such as agro-forestry (Chapter 6) and good governance mechanisms for forest and agricultural management (Chapter 7) can support the transition to sustainable energy for households and reduce the environmental impacts of traditional biomass.

1.6   Impacts of land degradation on climate

While Chapter 2 has its focus on land cover changes and their impacts on the climate system, this chapter focuses on the influences of individual land degradation processes on climate (see Table 4.1) which may or may not take place in association with land cover changes. The effects of land degradation on CO_2 and other GHGs as well as those on surface albedo and other physical controls of the global radiative balance are discussed.

1.6.1   Impact on greenhouse gases (GHGs)

Land degradation processes with direct impact on soil and terrestrial biota have great relevance in terms of CO_2 exchange with the atmosphere, given the magnitude and activity of these reservoirs in the global carbon cycle. As the most widespread form of soil degradation, erosion detaches the surface soil material, which typically hosts the highest organic carbon stocks, favouring the mineralisation and release as CO_2. Yet complementary processes such as carbon burial may compensate for this effect, making soil erosion a long-term carbon sink (*low agreement, limited evidence*), (Wang et al. (2017b), but see also Chappell et al. (2016)). Precise estimation of the CO_2 released from eroded lands is challenged by the fact that only a fraction of the detached carbon is eventually lost to the atmosphere. It is important to acknowledge that a substantial fraction of the eroded material may preserve its organic carbon load in field conditions. Moreover, carbon sequestration may be favoured through the burial of both the deposited material and the surface of its hosting soil at the deposition location (Quinton et al. 2010). The cascading effects of erosion on other environmental processes at the affected sites can often cause net CO_2 emissions through their indirect influence on soil fertility, and the balance of organic carbon inputs and outputs, interacting with other non-erosive soil degradation processes (such as nutrient depletion, compaction and salinisation), which can lead to the same net carbon effects (see Table 4.1) (van de Koppel et al. 1997).

As natural and human-induced erosion can result in net carbon storage in very stable buried pools at the deposition locations, degradation in those locations has a high C-release potential. Coastal ecosystems such as mangrove forests, marshes and seagrasses are at typical deposition locations, and their degradation or replacement with other vegetation is resulting in a substantial carbon release (0.15 to 1.02 GtC yr^{-1}) (Pendleton et al. 2012), which highlights the need for a spatially integrated assessment of land degradation impacts on climate that considers in-situ but also ex-situ emissions.

Cultivation and agricultural management of cultivated land are relevant in terms of global CO_2 land–atmosphere exchange (Section 4.8.1). Besides the initial pulse of CO_2 emissions associated with the onset of cultivation and associated vegetation clearing (Chapter 2), agricultural management practices can increase or reduce carbon losses to the atmosphere. Although global croplands are considered to be at a relatively neutral stage in the current decade (Houghton et al. 2012), this results from a highly uncertain balance between coexisting net losses and gains. Degradation losses of soil and biomass carbon appear to be compensated by gains from soil protection and restoration practices such as cover crops, conservation tillage and nutrient replenishment favouring organic matter build-up. Cover crops, increasingly used to improve soils, have the potential to sequester 0.12 GtC yr^{-1} on global croplands with a saturation time of more than 150 years (Poeplau and Don 2015). No-till practices (i.e., tillage elimination favouring crop residue retention in the soil surface) which were implemented to protect soils from erosion and reduce land preparation times, were also seen with optimism as a carbon sequestration option, which today is considered more modest globally and, in some systems, even less

certain (VandenBygaart 2016; Cheesman et al. 2016; Powlson et al. 2014). Among soil fertility restoration practices, lime application for acidity correction, increasingly important in tropical regions, can generate a significant net CO_2 source in some soils (Bernoux et al. 2003; Desalegn et al. 2017).

Land degradation processes in seminatural ecosystems driven by unsustainable uses of their vegetation through logging or grazing lead to reduced plant cover and biomass stocks, causing net carbon releases from soils and plant stocks. Degradation by logging activities is particularly prevalent in developing tropical and subtropical regions, involving carbon releases that exceed by far the biomass of harvested products, including additional vegetation and soil sources that are estimated to reach 0.6 GtC yr^{-1} (Pearson et al. 2014, 2017). Excessive grazing pressures pose a more complex picture with variable magnitudes and even signs of carbon exchanges. A general trend of higher carbon losses in humid overgrazed rangelands suggests a high potential for carbon sequestration following the rehabilitation of those systems (Conant and Paustian 2002) with a global potential sequestration of 0.045 GtC yr^{-1}. A special case of degradation in rangelands is the process leading to the woody encroachment of grass-dominated systems, which can be responsible for declining animal production but high carbon sequestration rates (Asner et al. 2003; Maestre et al. 2009).

Fire regime shifts in wild and seminatural ecosystems can become a degradation process in itself, with high impact on net carbon emission and with underlying interactive human and natural drivers such as burning policies (Van Wilgen et al. 2004), biological invasions (Brooks et al. 2009), and plant pest/disease spread (Kulakowski et al. 2003). Some of these interactive processes affecting unmanaged forests have resulted in massive carbon release, highlighting how degradation feedbacks on climate are not restricted to intensively used land but can affect wild ecosystems as well (Kurz et al. 2008).

Agricultural land and wetlands represent the dominant source of non-CO_2 greenhouse gases (GHGs) (Chen et al. 2018d). In agricultural land, the expansion of rice cultivation (increasing CH_4 sources), ruminant stocks and manure disposal (increasing CH_4, N_2O and NH_3 fluxes) and nitrogen over-fertilisation combined with soil acidification (increasing N_2O fluxes) are introducing the major impacts (*medium agreement, medium evidence*) and their associated emissions appear to be exacerbated by global warming (*medium agreement, medium evidence*) (Oertel et al. 2016).

As the major sources of global N_2O emissions, over-fertilisation and manure disposal are not only increasing in-situ sources but also stimulating those along the pathway of dissolved inorganic nitrogen transport all the way from draining waters to the ocean (*high agreement, medium evidence*). Current budgets of anthropogenically fixed nitrogen on the Earth System (Tian et al. 2015; Schaefer et al. 2016; Wang et al. 2017a) suggest that N_2O release from terrestrial soils and wetlands accounts for 10–15% of the emissions, yet many further release fluxes along the hydrological pathway remain uncertain, with emissions from oceanic 'dead-zones' being a major aspect of concern (Schlesinger 2009; Rabalais et al. 2014).

Environmental degradation processes focused on the hydrological system, which are typically manifested at the landscape scale, include both drying (as in drained wetlands or lowlands) and wetting trends (as in waterlogged and flooded plains). Drying of wetlands reduces CH_4 emissions (Turetsky et al. 2014) but favours pulses of organic matter mineralisation linked to high N_2O release (Morse and Bernhardt 2013; Norton et al. 2011). The net warming balance of these two effects is not resolved and may be strongly variable across different types of wetlands. In the case of flooding of non-wetland soils, a suppression of CO_2 release is typically overcompensated in terms of net greenhouse impact by enhanced CH_4 fluxes that stem from the lack of aeration but are aided by the direct effect of extreme wetting on the solubilisation and transport of organic substrates (McNicol and Silver 2014). Both wetlands rewetting/restoration and artificial wetland creation can increase CH_4 release (Altor and Mitsch 2006; Fenner et al. 2011). Permafrost thawing is another major source of CH_4 release, with substantial long-term contributions to the atmosphere that are starting to be globally quantified (Christensen et al. 2004; Schuur et al. 2015; Walter Anthony et al. 2016).

1.6.2   Physical impacts

Among the physical effects of land degradation, surface albedo changes are those with the most evident impact on the net global radiative balance and net climate warming/cooling. Degradation processes affecting wild and semi-natural ecosystems, such as fire regime changes, woody encroachment, logging and overgrazing, can trigger strong albedo changes before significant biogeochemical shifts take place. In most cases these two types of effects have opposite signs in terms of net radiative forcing, making their joint assessment critical for understanding climate feedbacks (Bright et al. 2015).

In the case of forest degradation or deforestation, the albedo impacts are highly dependent on the latitudinal/climatic belt to which they belong. In boreal forests, the removal or degradation of the tree cover increases albedo (net cooling effect) (*medium evidence, high agreement*) as the reflective snow cover becomes exposed, which can exceed the net radiative effect of the associated carbon release to the atmosphere (Davin et al. 2010; Pinty et al. 2011). On the other hand, progressive greening of boreal and temperate forests has contributed to net albedo declines (*medium agreement, medium evidence*) (Planque et al. 2017; Li et al. 2018a). In the northern treeless vegetation belt (tundra), shrub encroachment leads to the opposite effect as the emergence of plant structures above the snow cover level reduce winter-time albedo (Sturm 2005).

The extent to which albedo shifts can compensate for carbon storage shifts at the global level has not been estimated. A significant but partial compensation takes place in temperate and subtropical dry ecosystems in which radiation levels are higher and carbon stocks smaller compared to their more humid counterparts (*medium agreement, medium evidence*). In cleared dry woodlands, half of the net global warming effect of net carbon release has been compensated by albedo increase (Houspanossian et al. 2013), whereas in afforested dry rangelands, albedo declines cancelled one-fifth of the net carbon sequestration (Rotenberg and Yakir 2010). Other important cases

in which albedo effects impose a partial compensation of carbon exchanges are the vegetation shifts associated with wildfires, as shown for the savannahs, shrublands and grasslands of Sub-Saharan Africa (Dintwe et al. 2017). Besides the net global effects discussed above, albedo shifts can play a significant role in local climate (*high agreement, medium evidence*), as exemplified by the effect of no-till agriculture reducing local heat extremes in European landscapes (Davin et al. 2014) and the effects of woody encroachment causing precipitation rises in the North American Great Plains (Ge and Zou 2013). Modelling efforts that integrate ground data from deforested areas worldwide accounting for both physical and biogeochemical effects, indicate that massive global deforestation would have a net warming impact (Lawrence and Vandecar 2015) at both local and global levels with highlight non-linear effects of forest loss on climate variables.

Beyond the albedo effects presented above, other physical impacts of land degradation on the atmosphere can contribute to global and regional climate change. Of particular relevance, globally and continentally, are the net cooling effects of dust emissions (*low agreement, medium evidence*) (Lau and Kim (2007), but see also Huang et al. (2014)). Anthropogenic emission of mineral particles from degrading land appear to have a similar radiative impact than all other anthropogenic aerosols (Sokolik and Toon 1996). Dust emissions may explain regional climate anomalies through reinforcing feedbacks, as suggested for the amplification of the intensity, extent and duration of the low precipitation anomaly of the North American Dust Bowl in the 1930s (Cook et al. 2009). Another source of physical effects on climate are surface roughness changes which, by affecting atmospheric drag, can alter cloud formation and precipitation (*low agreement, low evidence*), as suggested by modelling studies showing how the massive deployment of solar panels in the Sahara could increase rainfall in the Sahel (Li et al. 2018c), or how woody encroachment in the Arctic tundra could reduce cloudiness and raise temperature (Cho et al. 2018). The complex physical effects of deforestation, as explored through modelling, converge into general net regional precipitation declines, tropical temperature increases and boreal temperature declines, while net global effects are less certain (Perugini et al. 2017). Integrating all the physical effects of land degradation and its recovery or reversal is still a challenge, yet modelling attempts suggest that, over the last three decades, the slow but persistent net global greening caused by the average increase of leaf area in the land has caused a net cooling of the Earth, mainly through the rise in evapotranspiration (Zeng et al. 2017) (*low confidence*).

1.7   Impacts of climate-related land degradation on poverty and livelihoods

Unravelling the impacts of climate-related land degradation on poverty and livelihoods is highly challenging. This complexity is due to the interplay of multiple social, political, cultural and economic factors, such as markets, technology, inequality, population growth, (Barbier and Hochard 2018) each of which interact and shape the ways in which social-ecological systems respond (Morton 2007). We find *limited evidence* attributing the impacts of climate-related land degradation to poverty and livelihoods, with climate often not distinguished from any other driver of land degradation. Climate is nevertheless frequently noted as a risk multiplier for both land degradation and poverty (*high agreement, robust evidence*) and is one of many stressors people live with, respond to and adapt to in their daily lives (Reid and Vogel 2006). Climate change is considered to exacerbate land degradation and potentially accelerate it due to heat stress, drought, changes to evapotranspiration rates and biodiversity, as well as a result of changes to environmental conditions that allow new pests and diseases to thrive (Reed and Stringer 2016). In general terms, the climate (and climate change) can increase human and ecological communities' sensitivity to land degradation. Land degradation then leaves livelihoods more sensitive to the impacts of climate change and extreme climatic events (*high agreement, robust evidence*). If human and ecological communities exposed to climate change and land degradation are sensitive and cannot adapt, they can be considered vulnerable to it; if they are sensitive and can adapt, they can be considered resilient (Reed and Stringer 2016). The impacts of land degradation will vary under a changing climate, both spatially and temporally, leading some communities and ecosystems to be more vulnerable or more resilient than others under different scenarios. Even within communities, groups such as women and youth are often more vulnerable than others.

1.7.1 Relationships between land degradation, climate change and poverty

This section sets out the relationships between land degradation and poverty, and climate change and poverty, leading to inferences about the three-way links between them. Poverty is multidimensional and includes a lack of access to the whole range of capital assets that can be used to pursue a livelihood. Livelihoods constitute the capabilities, assets and activities that are necessary to make a living (Chambers and Conway 1992; Olsson et al. 2014b).

The literature shows *high agreement* in terms of speculation that there are *potential* links between land degradation and poverty. However, studies have not provided robust quantitative assessments of the extent and incidence of poverty within populations affected by land degradation (Barbier and Hochard 2016). Some researchers, for example, Nachtergaele et al. (2011) estimate that 1.5 billion people were dependent upon degraded land to support their livelihoods in 2007, while >42% of the world's poor population inhabit degraded areas. However, there is overall *low confidence* in the evidence base, a lack of studies that look beyond the past and present, and the literature calls for more in-depth research to be undertaken on these issues (Gerber et al. 2014). Recent work by Barbier and Hochard (2018) points to biophysical constraints such as poor soils and limited rainfall, which interact to limit land productivity, suggesting that those farming in climatically less-favourable agricultural areas are challenged by poverty. Studies such as those by Coomes et al. (2011), focusing on an area in the Amazon, highlight the importance of the initial conditions of land holding in the dominant (shifting) cultivation system in terms of long-term effects on household poverty and future forest cover, showing that initial land tenure and socio-economic aspects can make some areas less favourable too.

Much of the qualitative literature is focused on understanding the livelihood and poverty impacts of degradation through a focus on subsistence agriculture, where farms are small, under traditional or informal tenure and where exposure to environmental (including climate) risks is high (Morton 2007). In these situations, poorer people lack access to assets (financial, social, human, natural and physical) and in the absence of appropriate institutional supports and social protection, this leaves them sensitive and unable to adapt, so a vicious cycle of poverty and degradation can ensue. To further illustrate the complexity, livelihood assessments often focus on a single snapshot in time. Livelihoods are dynamic and people alter their livelihood activities and strategies depending on the internal and external stressors to which they are responding (O'Brien et al. 2004). When certain livelihood activities and strategies are no longer tenable as a result of land degradation (and may push people into poverty), land degradation can have further effects on issues such as migration (Lee 2009), as people adapt by moving (Section 4.7.3); and may result in conflict (Section 4.7.3), as different groups within society compete for scarce resources, sometimes through non-peaceful actions. Both migration and conflict can lead to land-use changes elsewhere that further fuel climate change through increased emissions.

Similar challenges as for understanding land degradation–poverty linkages are experienced in unravelling the relationship between climate change and poverty. A particular issue in examining climate change–poverty links relates to the common use of aggregate economic statistics like GDP, as the assets and income of the poor constitute a minor proportion of national wealth (Hallegatte et al. 2018). Aggregate quantitative measures also fail to capture the distributions of costs and benefits from climate change. Furthermore, people fall into and out of poverty, with climate change being one of many factors affecting these dynamics, through its impacts on livelihoods. Much of the literature on climate change and poverty tends to look backward rather than forward (Skoufias et al. 2011), providing a snapshot of current or past relationships (for example, Dell et al. (2009) who examine the relationship between temperature and income (GDP) using cross-sectional data from countries in the Americas). Yet, simulations of future climate change impacts on income or poverty are largely lacking.

Noting the *limited evidence* that exists that explicitly focuses on the relationship between land degradation, climate change and poverty, Barbier and Hochard (2018b) suggest that those people living in less-favoured agricultural areas face a poverty–environment trap that can result in increased land degradation under climate change conditions. The emergent relationships between land degradation, climate change and poverty are shown in Figure 4.6 (see also Figure 6.1).

The poor have access to few productive assets – so land, and the natural resource base more widely, plays a key role in supporting the livelihoods of the poor. It is, however, hard to make generalisations about how important income derived from the natural resource base is for rural livelihoods in the developing world (Angelsen et al. 2014). Studies focusing on forest resources have shown that approximately one quarter of the total rural household income in developing countries stems from forests, with forest-based income shares being tentatively higher for low-income households (Vedeld et al. 2007; Angelsen et al. 2014). Different groups use land in different ways within their overall livelihood portfolios and are, therefore, at different levels of exposure and sensitivity to climate shocks and stresses. The literature nevertheless displays *high evidence* and *high agreement* that those populations whose livelihoods are more sensitive to climate change and land degradation are often more dependent on environmental assets, and these people are often the poorest members of society. There is further *high evidence* and *high agreement* that both climate change and land degradation can affect livelihoods and poverty through their threat multiplier effect. Research in Bellona, in the Solomon Islands in the South Pacific (Reenberg et al. 2008) examined event-driven impacts on livelihoods, taking into account weather events as one of many drivers of land degradation and links to broader land use and land cover changes that have taken place. Geographical locations experiencing land degradation are often the same locations that are directly affected by poverty, and also by extreme events linked to climate change and variability.

Much of the assessment presented above has considered place-based analyses examining the relationships between poverty, land degradation and climate change in the locations in which these outcomes have occurred. Altieri and Nicholls (2017) note that, due to the globalised nature of markets and consumption systems, the impacts of changes in crop yields linked to climate-related land degradation (manifest as lower yields) will be far reaching, beyond the sites and livelihoods experiencing degradation. Despite these teleconnections, farmers living in poverty in developing countries will be especially vulnerable due to their exposure, dependence on the environment for income and limited options to engage in other ways to make a living (Rosenzweig and Hillel 1998). In identifying ways in which these interlinkages can be addressed, Scherr (2000) observes that key actions that can jointly address poverty and environmental improvement often seek to increase access to natural resources, enhance the productivity of the natural resource assets of the poor, and engage stakeholders in addressing public natural resource management issues. In this regard, it is increasingly recognised that those suffering from, and being vulnerable to, land degradation and poverty need to have a voice and play a role in the development of solutions, especially where the natural resources and livelihood activities they depend on are further threatened by climate change.

1.7.2 Impacts of climate-related land degradation on food security

How and where we grow food, compared to where and when we need to consume it, is at the crux of issues surrounding land degradation, climate change and food security, especially because more than 75% of the global land surface (excluding Antarctica) faces rain-fed crop production constraints (Fischer et al. 2009), see also Chapter 5. Taken separately, knowledge on land degradation processes and human-induced climate change has attained a great level of maturity. However, their combined effects on food security, notably food supply, remain underappreciated (Webb et al. 2017b), and quantitative information is lacking. Just a few studies have shown how the interactive effects of the aforementioned challenging, interrelated phenomena can impact on crop productivity and hence food security and quality (Karami et al. 2009; Allen et al. 2001; Högy and Fangmeier 2008) (*low evidence*). Along with socio-economic drivers, climate change accelerates land degradation due to its influence on land-use systems (Millennium Ecosystem Assessment 2005; UNCCD 2017), potentially leading to a decline in agri-food system productivity, particularly on the supply side. Increases in temperature and changes in precipitation patterns are expected to have impacts on soil quality, including nutrient availability and assimilation (St.Clair and Lynch 2010). Those climate-related changes are expected to have net negative impacts on agricultural productivity, particularly in tropical regions, though the magnitude of impacts depends on the models used. Anticipated supply-side issues linked to land and climate relate to biocapacity factors (including e.g., whether there is enough water to support agriculture); production factors (e.g., chemical pollution of soil and water resources or lack of soil nutrients) and distribution issues (e.g., decreased availability of and/or accessibility to the necessary diversity of quality food where and when it is needed) (Stringer et al. 2011). Climate-sensitive transport infrastructure is

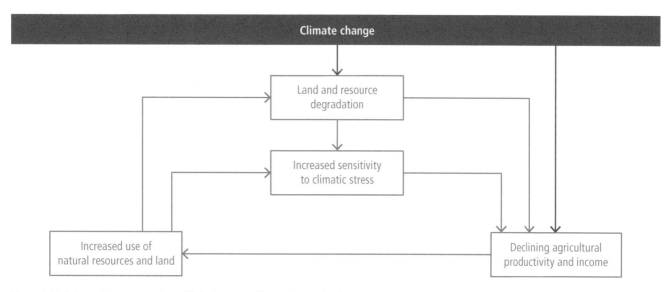

Figure 4.6 | Schematic representation of links between climate change, land management and socio-economic conditions.

also problematic for food security (Islam et al. 2017), and can lead to increased food waste, while poor siting of roads and transport links can lead to soil erosion and forest loss (Xiao et al. 2017), further feeding back into climate change.

Over the past decades, crop models have been useful tools for assessing and understanding climate change impacts on crop productivity and food security (White et al. 2011; Rosenzweig et al. 2014). Yet, the interactive effects of soil parameters and climate change on crop yields and food security remain limited, with *low evidence* of how they play out in different economic and climate settings (e.g., Sundström et al. 2014). Similarly, there have been few meta-analyses focusing on the adaptive capacity of land-use practices such as conservation agriculture in light of climate stress (see e.g., Steward et al. 2018), as well as *low evidence* quantifying the role of wild foods and forests (and, by extension, forest degradation) in both the global food basket and in supporting household-scale food security (Bharucha and Pretty 2010; Hickey et al. 2016).

To be sustainable, any initiative aimed at addressing food security – encompassing supply, diversity and quality – must take into consideration the interactive effects between climate and land degradation in a context of other socio-economic stressors. Such socio-economic factors are especially important if we look at demand-side issues too, which include lack of purchasing power, large-scale speculation on global food markets, leading to exponential price rises (Tadesse et al. 2014), competition in appropriation of supplies and changes to per capita food consumption (Stringer et al. 2011) (Chapter 5). Lack of food security, combined with lack of livelihood options, is often an important manifestation of vulnerability, and can act as a key trigger for people to migrate. In this way, migration becomes an adaptation strategy.

1.7.3   Impacts of climate-related land degradation on migration and conflict

Land degradation may trigger competition for scarce natural resources, potentially leading to migration and/or conflict, though, even with *medium evidence,* there is *low agreement* in the literature. Linkages between land degradation and migration occur within a larger context of multi-scale interaction of environmental and non-environmental drivers and processes, including resettlement projects, searches for education and/or income, land shortages, political turmoil, and family-related reasons (McLeman 2017; Hermans and Ide 2019). The complex contribution of climate to migration and conflict hampers retrieving any level of confidence on climate-migration and climate-conflict linkages, therefore constituting a major knowledge gap (Cramer et al. 2014; Hoegh-Guldberg et al. 2018).

There is *low evidence* on the causal linkages between climate change, land degradation processes (other than desertification) and migration. Existing studies on land degradation and migration – particularly in drylands – largely focus on the effect of rainfall variability and drought, and show how migration serves as adaptation strategy (Piguet et al. 2018; McLeman 2017; Chapter 3). For example, in the Ethiopian highlands, severe topsoil erosion and forest degradation

is a major environmental stressor which is amplified by recurring droughts, with migration being an important household adaptation strategy (Morrissey 2013). In the humid tropics, land degradation, mainly as a consequence of deforestation, has been a reported reason for people leaving their homes during the Amazonian colonisation (Hecht 1983) but was also observed more recently, for example in Guatemala, where soil degradation was one of the most frequently cited migration pushes (López-Carr 2012) and Kenya, where households respond to low soil quality by sending temporary migrants for additional income generation (Gray 2011). In contrast, in the Andean highlands and the Pacific coastal plain, migration increased with land quality, probably because revenues from additional agricultural production was invested in costly forms of migration (Gray and Bilsborrow 2013). These mixed results illustrate the complex, non-linear relationship of land degradation–migration linkages and suggest that explaining land degradationand migration linkages requires considering a broad range of socio-ecological conditions (McLeman 2017).

In addition to people moving away from an area due to 'lost' livelihood activities, climate-related land degradation can also reduce the availability of livelihood safety nets – environmental assets that people use during times of shocks or stress. For example, Barbier (2000) notes that wetlands in north-east Nigeria around Hadejia–Jama'are floodplain provide dry season pastures for seminomadic herders, agricultural surpluses for Kano and Borno states, groundwater recharge of the Chad formation aquifer and 'insurance' resources in times of drought. The floodplain also supports many migratory bird species. As climate change and land degradation combine, delivery of these multiple services can be undermined, particularly as droughts become more widespread, reducing the utility of this wetland environment as a safety net for people and wildlife alike.

Early studies conducted in Africa hint at a significant causal link between land degradation and violent conflict (Homer-Dixon et al. 1993). For example, Percival and Homer-Dixon (1995) identified land degradation as one of the drivers of the crisis in Rwanda in the early 1990s, which allowed radical forces to stoke ethnic rivalries. With respect to the Darfur conflict, some scholars and United Nations Environment Programme (UNEP) concluded that land degradation, together with other environmental stressors, constitute a major security threat for the Sudanese people (Byers and Dragojlovic 2004; Sachs 2007; UNEP 2007). Recent studies show *low agreement*, suggesting that climate change can increase the likelihood of civil violence if certain economic, political and social factors, including low development and weak governance mechanisms, are present (Scheffran et al. 2012; Benjaminsen et al. 2012). In contrast, Raleigh and Urdal (2007) found in a global study that land degradation is a weak predictor for armed conflict. As such, studies addressing possible linkages between climate change – a key driver of land degradation – and the risks of conflict have yielded contradictory results, and it remains largely unclear whether land degradation resulting from climate change leads to conflict or cooperation (Salehyan 2008; Solomon et al. 2018).

Land degradation–conflict linkages can be bi-directional. Research suggests that households experiencing natural resource degradation often engage in migration for securing livelihoods (Kreamer 2012), which potentially triggers land degradation at the destination, leading to conflict there (Kassa et al. 2017). While this indeed holds true for some cases, it may not for others, given the complexity of processes, contexts and drivers. Where conflict and violence do ensue, it is often as a result of a lack of appreciation for the cultural practices of others.

1.8 Addressing land degradation in the context of climate change

Land degradation in the form of soil carbon loss is estimated to have been ongoing for at least 12,000 years, but increased exponentially in the last 200 years (Sanderman et al. 2017). Before the advent of modern sources of nutrients, it was imperative for farmers to maintain and improve soil fertility through the prevention of runoff and erosion, and management of nutrients through vegetation residues and manure. Many ancient farming systems were sustainable for hundreds and even thousands of years, such as raised-field agriculture in Mexico (Crews and Gliessman 1991), tropical forest gardens in Southeast Asia and Central America (Ross 2011; Torquebiau 1992; Turner and Sabloff 2012), terraced agriculture in East Africa, Central America, Southeast Asia and the Mediterranean basin (Turner and Sabloff 2012; Preti and Romano 2014; Widgren and Sutton 2004; Håkansson and Widgren 2007; Davies and Moore 2016; Davies 2015), and integrated rice–fish cultivation in East Asia (Frei and Becker 2005).

Such long-term sustainable farming systems evolved in very different times and geographical contexts, but they share many common features, such as: the combination of species and structural diversity in time and space (horizontally and vertically) in order to optimise the use of available land; recycling of nutrients through biodiversity of plants, animals and microbes; harnessing the full range of site-specific micro-environments (e.g., wet and dry soils); biological interdependencies which help suppression of pests; reliance on mainly local resources; reliance on local varieties of crops, and sometimes incorporation of wild plants and animals; the systems are often labour and knowledge intensive (Rudel et al. 2016; Beets 1990; Netting 1993; Altieri and Koohafkan 2008). Such farming systems have stood the test of time and can provide important knowledge for adapting farming systems to climate change (Koohafkann and Altieri 2011).

In modern agriculture, the importance of maintaining the biological productivity and ecological integrity of farmland has not been a necessity in the same way as in pre-modern agriculture because nutrients and water have been supplied externally. The extreme land degradation in the US Midwest during the Dust Bowl period in the 1930s became an important wake-up call for agriculture and agricultural research and development, from which we can still learn much in order to adapt to ongoing and future climate change (McLeman et al. 2014; Baveye et al. 2011; McLeman and Smit 2006).

SLM is a unifying framework for addressing land degradation and can be defined as the stewardship and use of land resources, including soils, water, animals and plants, to meet changing human needs, while simultaneously ensuring the long-term productive potential of these resources and the maintenance of their environmental functions. It is a comprehensive approach comprising technologies combined with social, economic and political enabling conditions (Nkonya et al. 2011). It is important to stress that farming systems are informed by both scientific and local/traditional knowledge. The power of SLM in small-scale diverse farming was demonstrated effectively in Nicaragua after the severe cyclone Mitch in 1998 (Holt-Giménez 2002). Pairwise analysis of 880 fields with and without implementation of SLM practices showed that the SLM fields systematically fared better than the fields without SLM in terms of more topsoil remaining, higher field moisture, more vegetation, less erosion and lower economic losses after the cyclone. Furthermore, the difference between fields with and without SLM increased with increasing levels of storm intensity, slope gradient, and age of SLM (Holt-Giménez 2002).

When addressing land degradation through SLM and other approaches, it is important to consider feedbacks that impact on climate change. Table 4.2 shows some of the most important land degradation issues, their potential solutions, and their impacts on climate change. This table provides a link between the comprehensive lists of land degradation processes (Table 4.1) and land management solutions.

1.8.1 Actions on the ground to address land degradation

Concrete actions on the ground to address land degradation are primarily focused on soil and water conservation. In the context of adaptation to climate change, actions relevant for addressing land degradation are sometimes framed as ecosystem-based adaptation (Scarano 2017) or Nature-Based Solutions (Nesshöver et al. 2017), and in an agricultural context, agroecology (see Glossary) provides an important frame. The site-specific biophysical and social conditions, including local and indigenous knowledge, are important for successful implementation of concrete actions.

Responses to land degradation generally take the form of agronomic measures (methods related to managing the vegetation cover), soil management (methods related to tillage, nutrient supply), and mechanical methods (methods resulting in durable changes to the landscape) (Morgan 2005a). Measures may be combined to reinforce benefits to land quality, as well as improving carbon sequestration that supports climate change mitigation. Some measures offer adaptation options and other co-benefits, such as agroforestry, involving planting fruit trees that can support food security in the face of climate change impacts (Reed and Stringer 2016), or application of compost or biochar that enhances soil water-holding capacity, so increases resilience to drought.

There are important differences in terms of labour and capital requirements for different technologies, and also implications for land tenure arrangements. Agronomic measures and soil management require generally little extra capital input and comprise activities repeated annually, so have no particular implication for land tenure

4

Table 4.2 | **Interaction of human and climate drivers can exacerbate desertification and land degradation.** Climate change exacerbates the rate and magnitude of several ongoing land degradation and desertification processes. Human drivers of land degradation and desertification include expanding agriculture, agricultural practices and forest management. In turn, land degradation and desertification are also drivers of climate change through GHG emissions, reduced rates of carbon uptake, and reduced capacity of ecosystems to act as carbon sinks into the future. Impacts on climate change are either warming (in red) or cooling (in blue).

Issue/ syndrome	Impact on climate change	Human driver	Climate driver	Land management options	References
Erosion of agricultural soils	Emission: CO_2, N_2O	(agriculture practice, expansion of agriculture)	(warming trend, drying trend, extreme rainfall, shifting rains)	Increase soil organic matter, no-till, perennial crops, erosion control, agroforestry, dietary change	3.1.4, 3.4.1, 3.5.2, 3.7.1, 4.8.1, 4.8.5, 4.9.2, 4.9.5
Deforestation	Emission of CO_2	(forest clearing, expansion of agriculture)		Forest protection, sustainable forest management and dietary change	4.1.5, 4.5, 4.8.3, 4.8.4, 4.9.3
Forest degradation	Emission of CO_2 Reduced carbon sink	(forest clearing)		Forest protection, sustainable forest management	4.1.5, 4.5, 4.8.3, 4.8.4, 4.9.3
Overgrazing	Emission: CO_2, CH_4 Increasing albedo	(grazing pressure)	(warming trend, drying trend)	Controlled grazing, rangeland management	3.1.4.2, 3.4.1, 3.6.1, 3.7.1, 4.8.1.4
Firewood and charcoal production	Emission: CO_2, CH_4 Increasing albedo	(wood fuel)		Clean cooking (health co-benefits, particularly for women and children)	3.6.3, 4.5.4, 4.8.3, 4.8.4
Increasing fire frequency and intensity	Emission: CO_2, CH_4, N_2O Emission: aerosols, increasing albedo		(warming trend, extreme temperature, drying trend)	Fuel management, fire management	3.1.4, 3.6.1, 4.1.5, 4.8.3, Cross-Chapter Box 3 in Chp 2
Degradation of tropical peat soils	Emission: CO_2, CH_4	(agriculture practice, expansion of agriculture)	(drying trend)	Peatland restoration, erosion control, regulating the use of peat soils	4.9.4
Thawing of permafrost	Emission: CO_2, CH_4		(warming trend, extreme temperature)	Relocation of settlement and infrastructure	4.8.5.1
Coastal erosion	Emission: CO_2, CH_4		(intensifying cyclones, sea level rise, extreme rainfall)	Wetland and coastal restoration, mangrove conservation, long-term land-use planning	4.9.6, 4.9.7, 4.9.8
Sand and dust storms, wind erosion	Emission: aerosols	(grazing pressure, agriculture practice, expansion of agriculture, wood fuel)	(drying trend, shifting rains)	Vegetation management, afforestation, windbreaks	3.3.1, 3.4.1, 3.6.1, 3.7.1, 3.7.2
Bush encroachment	Capturing: CO_2, Decreasing albedo	(grazing pressure, wood fuel)	(shifting rains)	Grazing land management, fire management	3.6.1.3, 3.7.3.2

Human driver	Climate driver
Grazing pressure	Warming trend
Agriculture practice	Extreme temperature
Expansion of agriculture	Drying trend
Forest clearing	Extreme rainfall
Wood fuel	Shifting rains
	Intensifying cyclones
	Sea level rise

arrangements. Mechanical methods require substantial upfront investments in terms of capital and labour, resulting in long-lasting structural change, requiring more secure land tenure arrangements (Mekuriaw et al. 2018). Agroforestry is a particularly important strategy for SLM in the context of climate change because of the large potential to sequester carbon in plants and soil and enhance resilience of agricultural systems (Zomer et al. 2016).

Implementation of SLM practices has been shown to increase the productivity of land (Branca et al. 2013) and to provide good economic returns on investment in many different settings around the world (Mirzabaev et al. 2015). Giger et al. (2018) showed, in a meta study of 363 SLM projects over the period 1990 to 2012, that 73% of the projects were perceived to have a positive or at least neutral cost-benefit ratio in the short term, and 97% were perceived to have a positive or very positive cost-benefit ratio in the long term (*robust evidence, high agreement*). Despite the positive effects, uptake is far from universal. Local factors, both biophysical conditions (e.g., soils,

drainage, and topography) and socio-economic conditions (e.g., land tenure, economic status, and land fragmentation) play decisive roles in the interest in, capacity to undertake, and successful implementation of SLM practices (Teshome et al. 2016; Vogl et al. 2017; Tesfaye et al. 2016; Cerdà et al. 2018; Adimassu et al. 2016). From a landscape perspective, SLM can generate benefits, including adaptation to and mitigation of climate change, for entire watersheds, but challenges remain regarding coordinated and consistent implementation (*medium evidence, medium agreement*) (Kerr et al. 2016; Wang et al. 2016a).

1.8.1.1   Agronomic and soil management measures

Rebuilding soil carbon is an important goal of SLM, particularly in the context of climate change (Rumpel et al. 2018). The two most important reasons why agricultural soils have lost 20–60% of the soil carbon they contained under natural ecosystem conditions are the frequent disturbance through tillage and harvesting, and the change from deep-rooted perennial plants to shallow-rooted annual plants (Crews and

Rumsey 2017). Practices that build soil carbon are those that increase organic matter input to soil, or reduce decomposition of SOM.

Agronomic practices can alter the carbon balance significantly, by increasing organic inputs from litter and roots into the soil. Practices include retention of residues, use of locally adapted varieties, inter-cropping, crop rotations, and green manure crops that replace the bare field fallow during winter and are eventually ploughed before sowing the next main crop (Henry et al. 2018). Cover crops (green manure crops and catch crops that are grown between the main cropping seasons) can increase soil carbon stock by between 0.22 and 0.4 t C ha^{-1}yr^{-1} (Poeplau and Don 2015; Kaye and Quemada 2017).

Reduced tillage (or no-tillage) is an important strategy for reducing soil erosion and nutrient loss by wind and water (Van Pelt et al. 2017; Panagos et al. 2015; Borrelli et al. 2016). But the evidence that no-till agriculture also sequesters carbon is not compelling (VandenBygaart 2016). Soil sampling of only the upper 30 cm can give biased results, suggesting that soils under no-till practices have higher carbon content than soils under conventional tillage (Baker et al. 2007; Ogle et al. 2012; Fargione et al. 2018; VandenBygaart 2016).

Changing from annual to perennial crops can increase soil carbon content (Culman et al. 2013; Sainju et al. 2017). A perennial grain crop (intermediate wheatgrass) was, on average, over four years a net carbon sink of about 13.5 tCO$_2$ ha^{-1} yr^{-1} (de Oliveira et al. 2018). Sprunger et al. (2018) compared an annual winter wheat crop with a perennial grain crop (intermediate wheatgrass) and found that the perennial grain root biomass was 15 times larger than winter wheat, however, there was no significant difference in soil carbon pools after the four-year experiment. Exactly how much, and over what time period, carbon can be sequestered through changing from annual to perennial crops depends on the degree of soil carbon depletion and other local biophysical factors (Section 4.9.2).

Integrated soil fertility management is a sustainable approach to nutrient management that uses a combination of chemical and organic amendments (manure, compost, biosolids, biochar), rhizobial nitrogen fixation, and liming materials to address soil chemical constraints (Henry et al. 2018). In pasture systems, management of grazing pressure, fertilisation, diverse species including legumes and perennial grasses can reduce erosion and enhance soil carbon (Conant et al. 2017).

1.8.1.2   Mechanical soil and water conservation

In hilly and mountainous terrain, terracing is an ancient but still practised soil conservation method worldwide (Preti and Romano 2014) in climatic zones from arid to humid tropics (Balbo 2017). By reducing the slope gradient of hillsides, terraces provide flat surfaces. Deep, loose soils that increase infiltration, reduce erosion and thus sediment transport. They also decrease the hydrological connectivity and thus reduce hillside runoff (Preti et al. 2018; Wei et al. 2016; Arnáez et al. 2015; Chen et al. 2017). In terms of climate change, terraces are a form of adaptation that helps in cases where rainfall is increasing or intensifying (by reducing slope gradient and the hydrological connectivity), and where rainfall is decreasing (by

increasing infiltration and reducing runoff) (*robust evidence, high agreement*). There are several challenges, however, to continued maintenance and construction of new terraces, such as the high costs in terms of labour and/or capital (Arnáez et al. 2015) and disappearing local knowledge for maintaining and constructing new terraces (Chen et al. 2017). The propensity of farmers to invest in mechanical soil conservation methods varies with land tenure; farmers with secure tenure arrangements are more willing to invest in durable practices such as terraces (Lovo 2016; Sklenicka et al. 2015; Haregeweyn et al. 2015). Where the slope is less severe, erosion can be controlled by contour banks, and the keyline approach (Duncan 2016; Stevens et al. 2015) to soil and water conservation.

1.8.1.3   Agroforestry

Agroforestry is defined as a collective name for land-use systems in which woody perennials (trees, shrubs, etc.) are grown in association with herbaceous plants (crops, pastures) and/or livestock in a spatial arrangement, a rotation, or both, and in which there are both ecological and economic interactions between the tree and non-tree components of the system (Young, 1995, p. 11). At least since the 1980s, agroforestry has been widely touted as an ideal land management practice in areas vulnerable to climate variations and subject to soil erosion. Agroforestry holds the promise of improving soil and climatic conditions, while generating income from wood energy, timber and non-timber products – sometimes presented as a synergy of adaptation and mitigation of climate change (Mbow et al. 2014).

There is strong scientific consensus that a combination of forestry with agricultural crops and/or livestock, agroforestry systems can provide additional ecosystem services when compared with monoculture crop systems (Waldron et al. 2017; Sonwa et al. 2011, 2014, 2017; Charles et al. 2013). Agroforestry can enable sustainable intensification by allowing continuous production on the same unit of land with higher productivity without the need to use shifting agriculture systems to maintain crop yields (Nath et al. 2016). This is especially relevant where there is a regional requirement to find a balance between the demand for increased agricultural production and the protection of adjacent natural ecosystems such as primary and secondary forests (Mbow et al. 2014). For example, the use of agroforestry for perennial crops such as coffee and cocoa is increasingly promoted as offering a route to sustainable farming, with important climate change adaptation and mitigation co-benefits (Sonwa et al. 2001; Kroeger et al. 2017). Reported co-benefits of agroforestry in cocoa production include increased carbon sequestration in soils and biomass, improved water and nutrient use efficiency and the creation of a favourable micro-climate for crop production (Sonwa et al. 2017; Chia et al. 2016). Importantly, the maintenance of soil fertility using agroforestry has the potential to reduce the practice of shifting agriculture (of cocoa) which results in deforestation (Gockowski and Sonwa 2011). However, positive interactions within these systems can be ecosystem and/or species specific, but co-benefits such as increased resilience to extreme climate events, or improved soil fertility are not always observed (Blaser et al. 2017; Abdulai et al. 2018). These contrasting outcomes indicate the importance of field-scale research programmes to inform agroforestry system design, species selection and management practices (Sonwa et al. 2014).

Despite the many proven benefits, adoption of agroforestry has been low and slow (Toth et al. 2017; Pattanayak et al. 2003; Jerneck and Olsson 2014). There are several reasons for the slow uptake, but the perception of risks and the time lag between adoption and realisation of benefits are often important (Pattanayak et al. 2003; Mercer 2004; Jerneck and Olsson 2013).

An important question for agroforestry is whether it supports poverty alleviation, or if it favours comparatively affluent households. Experiences from India suggest that the overall adoption is low, with a differential between rich and poor households. Brockington el al. (2016), studied agroforestry adoption over many years in South India and found that, overall, only 18% of the households adopted agroforestry. However, among the relatively rich households who adopted agroforestry, 97% were still practising it after six to eight years, and some had expanded their operations. Similar results were obtained in Western Kenya, where food-secure households were much more willing to adopt agroforestry than food-insecure households (Jerneck and Olsson 2013, 2014). Other experiences from Sub-Saharan Africa illustrate the difficulties (such as local institutional support) of having a continued engagement of communities in agroforestry (Noordin et al. 2001; Matata et al. 2013; Meijer et al. 2015).

1.8.1.4 Crop–livestock interaction as an approach to managing land degradation

The integration of crop and livestock production into 'mixed farming' for smallholders in developing countries became an influential model, particularly for Africa, in the early 1990s (Pritchard et al. 1992; McIntire et al. 1992). Crop–livestock integration under this model was seen as founded on three pillars: improved use of manure for crop fertility management; expanded use of animal traction (draught animals); and promotion of cultivated fodder crops. For Asia, emphasis was placed on draught power for land preparation, manure for soil fertility enhancement, and fodder production as an entry point for cultivation of legumes (Devendra and Thomas 2002). Mixed farming was seen as an evolutionary process to expand food production in the face of population increase, promote improvements in income and welfare, and protect the environment. The process could be further facilitated and steered by research, agricultural advisory services and policy (Pritchard et al. 1992; McIntire et al. 1992; Devendra 2002).

Scoones and Wolmer (2002) place this model in historical context, including concern about population pressure on resources and the view that mobile pastoralism was environmentally damaging. The latter view had already been critiqued by developing understandings of pastoralism, mobility and communal tenure of grazing lands (e.g., Behnke 1994; Ellis 1994). They set out a much more differentiated picture of crop–livestock interactions, which can take place either within a single-farm household, or between crop and livestock producers, in which case they will be mediated by formal and informal institutions governing the allocation of land, labour and capital, with the interactions evolving through multiple place-specific pathways (Ramisch et al. 2002; Scoones and Wolmer 2002). Promoting a diversity of approaches to crop–livestock interactions does not imply that the integrated model necessarily leads to land

degradation, but increases the space for institutional support to local innovation (Scoones and Wolmer 2002).

However, specific managerial and technological practices that link crop and livestock production will remain an important part of the repertoire of on-farm adaptation and mitigation. Howden and coauthors (Howden et al. 2007) note the importance of innovation within existing integrated systems, including use of adapted forage crops. Rivera-Ferre et al. (2016) list as adaptation strategies with high potential for grazing systems, mixed crop–livestock systems or both: crop–livestock integration in general; soil management, including composting; enclosure and corralling of animals; improved storage of feed. Most of these are seen as having significant co-benefits for mitigation, and improved management of manure is seen as a mitigation measure with adaptation co-benefits.

1.8.2 Local and indigenous knowledge for addressing land degradation

In practice, responses are anchored in scientific research, as well as local, indigenous and traditional knowledge and know-how. For example, studies in the Philippines by Camacho et al. (2016) examine how traditional integrated watershed management by indigenous people sustain regulating services vital to agricultural productivity, while delivering co-benefits in the form of biodiversity and ecosystem resilience at a landscape scale. Although responses can be site specific and sustainable at a local scale, the multi-scale interplay of drivers and pressures can nevertheless cause practices that have been sustainable for centuries to become less so. Siahaya et al. (2016) explore the traditional knowledge that has informed rice cultivation in the uplands of East Borneo, grounded in sophisticated shifting cultivation methods (*gilir balik*) which have been passed on for generations (more than 200 years) in order to maintain local food production. Gilir balik involves temporary cultivation of plots, after which, abandonment takes place as the land user moves to another plot, leaving the natural (forest) vegetation to return. This approach is considered sustainable if it has the support of other subsistence strategies, adapts to and integrates with the local context, and if the carrying capacity of the system is not surpassed (Siahaya et al. 2016). Often gilir balik cultivation involves intercropping of rice with bananas, cassava and other food crops. Once the abandoned plot has been left to recover such that soil fertility is restored, clearance takes place again and the plot is reused for cultivation. Rice cultivation in this way plays an important role in forest management, with several different types of succession forest being found in the study by Siahaya et al. (2016). Nevertheless, interplay of these practices with other pressures (large-scale land acquisitions for oil palm plantation, logging and mining), risk their future sustainability. Use of fire is critical in processes of land clearance, so there are also trade-offs for climate change mitigation, which have been sparsely assessed.

Interest appears to be growing in understanding how indigenous and local knowledge inform land users' responses to degradation, as scientists engage farmers as experts in processes of knowledge co-production and co-innovation (Oliver et al. 2012; Bitzer and Bijman 2015). This can help to introduce, implement, adapt and

4

promote the use of locally appropriate responses (Schwilch et al. 2011). Indeed, studies strongly agree on the importance of engaging local populations in both sustainable land and forest management. Meta-analyses in tropical regions that examined both forests in protected areas and community-managed forests suggest that deforestation rates are lower, with less variation in deforestation rates presenting in community-managed forests compared to protected forests (Porter-Bolland et al. 2012). This suggests that consideration of the social and economic needs of local human populations is vital in preventing forest degradation (Ward et al. 2018). However, while disciplines such as ethnopedology seek to record and understand how local people perceive, classify and use soil, and draw on that information to inform its management (Barrera-Bassols and Zinck 2003), links with climate change and its impacts (perceived and actual) are not generally considered.

1.8.3 Reducing deforestation and forest degradation and increasing afforestation

Improved stewardship of forests through reduction or avoidance of deforestation and forest degradation, and enhancement of forest carbon stocks can all contribute to land-based natural climate solutions (Angelsen et al. 2018; Sonwa et al. 2011; Griscom et al. 2017). While estimates of annual emissions from tropical deforestation and forest degradation range widely from 0.5 to 3.5 GtC yr^{-1} (Baccini et al. 2017; Houghton et al. 2012; Mitchard 2018; see also Chapter 2), they all indicate the large potential to reduce annual emissions from deforestation and forest degradation. Recent estimates of forest extent for Africa in 1900 may result in downward adjustments of historic deforestation and degradation emission estimates (Aleman et al. 2018). Emissions from forest degradation in non-Annex I countries have declined marginally from 1.1 $GtCO_2$ yr^{-1} in 2001–2010 to 1 $GtCO_2$ yr^{-1} in 2011–2015, but the relative emissions from degradation compared to deforestation have increased from a quarter to a third (Federici et al. 2015). Forest sector activities in developing countries were estimated to represent a technical mitigation potential in 2030 of 9 $GtCO_2$ (Miles et al. 2015). This was partitioned into reduction of deforestation (3.5 $GtCO_2$), reduction in degradation and forest management (1.7 $GtCO_2$) and afforestation and reforestation (3.8 $GtCO_2$). The economic mitigation potential will be lower than the technical potential (Miles et al. 2015).

Natural regeneration of second-growth forests enhances carbon sinks in the global carbon budget (Chazdon and Uriarte 2016). In Latin America, Chazdon et al. (2016) estimated that, in 2008, second-growth forests (up to 60 years old) covered 2.4 Mkm^2 of land (28.1% of the total study area). Over 40 years, these lands can potentially accumulate 8.5 GtC in above-ground biomass via low-cost natural regeneration or assisted regeneration, corresponding to a total CO_2 sequestration of 31.1 $GtCO_2$ (Chazdon et al. 2016b). While above-ground biomass carbon stocks are estimated to be declining in the tropics, they are increasing globally due to increasing stocks in temperate and boreal forests (Liu et al. 2015b), consistent with the observations of a global land sector carbon sink (Le Quéré et al. 2013; Keenan et al. 2017; Pan et al. 2011).

Moving from technical mitigation potentials (Miles et al. 2015) to real reduction of emissions from deforestation and forest degradation required transformational changes (Korhonen-Kurki et al. 2018). This transformation can be facilitated by two enabling conditions: the presence of already initiated policy change; or the scarcity of forest resources combined with an absence of any effective forestry framework and policies. These authors and others (Angelsen et al. 2018) found that the presence of powerful transformational coalitions of domestic pro-REDD+ (the United Nations Collaborative Programme on Reducing Emissions from Deforestation and Forest Degradation in Developing Countries) political actors combined with strong ownership and leadership, regulations and law enforcement, and performance-based funding, can provide a strong incentive for achieving REDD+ goals.

Implementing schemes such as REDD+ and various projects related to the voluntary carbon market is often regarded as a no-regrets investment (Seymour and Angelsen 2012) but the social and ecological implications (including those identified in the Cancun Safeguards) must be carefully considered for REDD+ projects to be socially and ecologically sustainable (Jagger et al. 2015). In 2018, 34 countries have submitted a REDD+ forest reference level and/ or forest reference emission level to the United Nations Framework Convention on Climate Change (UNFCCC). Of these REDD+ reference levels, 95% included the activity 'reducing deforestation' while 34% included the activity 'reducing forest degradation' (FAO 2018). Five countries submitted REDD+ results in the technical annex to their Biennial Update Report totalling an emission reduction of 6.3 $GtCO_2$ between 2006 and 2015 (FAO 2018).

Afforestation is another mitigation activity that increases carbon sequestration (Cross-Chapter Box 2 in Chapter 1). Yet, it requires careful consideration about where to plant trees to achieve potential climatic benefits, given an altering of local albedo and turbulent energy fluxes and increasing night-time land surface temperatures (Peng et al. 2014). A recent hydro-climatic modelling effort has shown that forest cover can account for about 40% of the observed decrease in annual runoff (Buendia et al. 2016). A meta-analysis of afforestation in Northern Europe (Bárcena et al. 2014) concluded that significant soil organic carbon sequestration in Northern Europe occurs after afforestation of croplands but not grasslands. Additional sequestration occurs in forest floors and biomass carbon stocks. Successful programmes of large-scale afforestation activities in South Korea and China are discussed in-depth in a special case study (Section 4.9.3).

The potential outcome of efforts to reduce emissions from deforestation and degradation in Indonesia through a 2011 moratorium on concessions to convert primary forests to either timber or palm oil uses was evaluated against rates of emissions over the period 2000 to 2010. The study concluded that less than 7% of emissions would have been avoided had the moratorium been implemented in 2000 because it only curtailed emissions due to a subset of drivers of deforestation and degradation (Busch et al. 2015).

4

In terms of ecological integrity of tropical forests, the policy focus on carbon storage and tree cover can be problematic if it leaves out other aspects of forests ecosystems, such as biodiversity – and particularly fauna (Panfil and Harvey 2016; Peres et al. 2016; Hinsley et al. 2015). Other concerns of forest-based projects under the voluntary carbon market are potential negative socio-economic side effects (Edstedt and Carton 2018; Carton and Andersson 2017; Osborne 2011; Scheidel and Work 2018; Richards and Lyons 2016; Borras and Franco 2018; Paladino and Fiske 2017) and leakage (particularly at the subnational scale), that is, when interventions to reduce deforestation or degradation at one site displace pressures and increase emissions elsewhere (Atmadja and Verchot 2012; Phelps et al. 2010; Lund et al. 2017; Balooni and Lund 2014).

Maintaining and increasing forest area, in particular native forests rather than monoculture and short-rotation plantations, contributes to the maintenance of global forest carbon stocks (Lewis et al. 2019) (*robust evidence*, *high agreement*).

1.8.4   Sustainable forest management (SFM) and CO_2 removal (CDR) technologies

While reducing deforestation and forest degradation may directly help to meet mitigation goals, SFM aimed at providing timber, fibre, biomass and non-timber resources can provide long-term livelihood for communities, reduce the risk of forest conversion to non-forest uses (settlement, crops, etc.), and maintain land productivity, thus reducing the risks of land degradation (Putz et al. 2012; Gideon Neba et al. 2014; Sufo Kankeu et al. 2016; Nitcheu Tchiadje et al. 2016; Rossi et al. 2017).

Developing SFM strategies aimed at contributing towards negative emissions throughout this century requires an understanding of forest management impacts on ecosystem carbon stocks (including soils), carbon sinks, carbon fluxes in harvested wood, carbon storage in harvested wood products, including landfills and the emission reductions achieved through the use of wood products and bioenergy (Nabuurs et al. 2007; Lemprière et al. 2013; Kurz et al. 2016; Law et al. 2018; Nabuurs et al. 2017). Transitions from natural to managed forest landscapes can involve a reduction in forest carbon stocks, the magnitude of which depends on the initial landscape conditions, the harvest rotation length relative to the frequency and intensity of natural disturbances, and on the age-dependence of managed and natural disturbances (Harmon et al. 1990; Kurz et al. 1998). Initial landscape conditions, in particular the age-class distribution and therefore carbon stocks of the landscape, strongly affect the mitigation potential of forest management options (Ter-Mikaelian et al. 2013; Kilpeläinen et al. 2017). Landscapes with predominantly mature forests may experience larger reductions in carbon stocks during the transition to managed landscapes (Harmon et al. 1990; Kurz et al. 1998; Lewis et al. 2019). In landscapes with predominantly young or recently disturbed forests, SFM can enhance carbon stocks (Henttonen et al. 2017).

Forest growth rates, net primary productivity, and net ecosystem productivity are age-dependent, with maximum rates of CO_2 removal

(CDR) from the atmosphere occurring in young to medium-aged forests and declining thereafter (Tang et al. 2014). In boreal forest ecosystem, estimation of carbon stocks and carbon fluxes indicate that old growth stands are typically small carbon sinks or carbon sources (Gao et al. 2018; Taylor et al. 2014; Hadden and Grelle 2016). In tropical forests, carbon uptake rates in the first 20 years of forest recovery were 11 times higher than uptake rates in old-growth forests (Poorter et al. 2016). Age-dependent increases in forest carbon stocks and declines in forest carbon sinks mean that landscapes with older forests have accumulated more carbon but their sink strength is diminishing, while landscapes with younger forests contain less carbon but they are removing CO_2 from the atmosphere at a much higher rate (Volkova et al. 2017; Poorter et al. 2016). The rates of CDR are not just age-related but also controlled by many biophysical factors and human activities (Bernal et al. 2018). In ecosystems with uneven-aged, multispecies forests, the relationships between carbon stocks and sinks are more difficult and expensive to quantify.

Whether or not forest harvest and use of biomass is contributing to net reductions of atmospheric carbon depends on carbon losses during and following harvest, rates of forest regrowth, and the use of harvested wood and carbon retention in long-lived or short-lived products, as well as the emission reductions achieved through the substitution of emissions-intensive products with wood products (Lemprière et al. 2013; Lundmark et al. 2014; Xu et al. 2018b; Olguin et al. 2018; Dugan et al. 2018; Chen et al. 2018b; Pingoud et al. 2018; Seidl et al. 2007). Studies that ignore changes in forest carbon stocks (such as some lifecycle analyses that assume no impacts of harvest on forest carbon stocks), ignore changes in wood product pools (Mackey et al. 2013) or assume long-term steady state (Pingoud et al. 2018), or ignore changes in emissions from substitution benefits (Mackey et al. 2013; Lewis et al. 2019) will arrive at diverging conclusions about the benefits of SFM. Moreover, assessments of climate benefits of any mitigation action must also consider the time dynamics of atmospheric impacts, as some actions will have immediate benefits (e.g., avoided deforestation), while others may not achieve net atmospheric benefits for decades or centuries. For example, the climate benefits of woody biomass use for bioenergy depend on several factors, such as the source and alternate fate of the biomass, the energy type it substitutes, and the rates of regrowth of the harvested forest (Laganière et al. 2017; Ter-Mikaelian et al. 2014; Smyth et al. 2017). Conversion of primary forests in regions of very low stand-replacing disturbances to short-rotation plantations where the harvested wood is used for short-lived products with low displacement factors will increase emissions. In general, greater mitigation benefits are achieved if harvested wood products are used for products with long carbon retention time and high displacement factors.

With increasing forest age, carbon sinks in forests will diminish until harvest or natural disturbances, such as wildfire, remove biomass carbon or release it to the atmosphere (Seidl et al. 2017). While individual trees can accumulate carbon for centuries (Köhl et al. 2017), stand-level carbon accumulation rates depend on both tree growth and tree mortality rates (Hember et al. 2016; Lewis et al. 2004). SFM, including harvest and forest regeneration, can help maintain active carbon sinks by maintaining a forest age-class distribution that includes a share of young, actively growing stands (Volkova et al.

2018; Nabuurs et al. 2017). The use of the harvested carbon in either long-lived wood products (e.g., for construction), short-lived wood products (e.g., pulp and paper), or biofuels affects the net carbon balance of the forest sector (Lemprière et al. 2013; Matthews et al. 2018). The use of these wood products can further contribute to GHG emission-reduction goals by avoiding the emissions from the products with higher embodied emissions that have been displaced (Nabuurs et al. 2007; Lemprière et al. 2013). In 2007 the IPCC concluded that '[i]n the long term, a sustainable forest management strategy aimed at maintaining or increasing forest carbon stocks, while producing an annual sustained yield of timber, fibre or energy from the forest, will generate the largest sustained mitigation benefit' (Nabuurs et al. 2007). The apparent trade-offs between maximising forest carbon stocks and maximising ecosystem carbon sinks are at the origin of ongoing debates about optimum management strategies to achieve negative emissions (Keith et al. 2014; Kurz et al. 2016; Lundmark et al. 2014). SFM, including the intensification of carbon-focused management strategies, can make long-term contributions towards negative emissions if the sustainability of management is assured through appropriate governance, monitoring and enforcement. As specified in the definition of SFM, other criteria such as biodiversity must also be considered when assessing mitigation outcomes (Lecina-Diaz et al. 2018). Moreover, the impacts of changes in management on albedo and other non-GHG factors also need to be considered (Luyssaert et al. 2018) (Chapter 2). The contribution of SFM for negative emissions is strongly affected by the use of the wood products derived from forest harvest and the time horizon over which the carbon balance is assessed. SFM needs to anticipate the impacts of climate change on future tree growth, mortality and disturbances when designing climate change mitigation and adaptation strategies (Valade et al. 2017; Seidl et al. 2017).

1.8.5   Policy responses to land degradation

The 1992 United Nations Conference on Environment and Development (UNCED), also known as the Rio de Janeiro Earth Summit, recognised land degradation as a major challenge to sustainable development, and led to the establishment of the UNCCD, which specifically addressed land degradation in the drylands. The UNCCD emphasises sustainable land use to link poverty reduction on one hand and environmental protection on the other. The two other 'Rio Conventions' emerging from the UNCED – the UNFCCC and the Convention on Biological Diversity (CBD) – focus on climate change and biodiversity, respectively. The land has been recognised as an aspect of common interest to the three conventions, and SLM is proposed as a unifying theme for current global efforts on combating land degradation, climate change and loss of biodiversity, as well as facilitating land-based adaptation to climate change and sustainable development.

The Global Environmental Facility (GEF) funds developing countries to undertake activities that meet the goals of the conventions and deliver global environmental benefits. Since 2002, the GEF has invested in projects that support SLM through its Land Degradation Focal Area Strategy, to address land degradation within and beyond the drylands.

Under the UNFCCC, parties have devised National Adaptation Plans (NAPs) that identify medium- and long-term adaptation needs. Parties have also developed their climate change mitigation plans, presented as NDCs. These programmes have the potential of assisting the promotion of SLM. It is understood that the root causes of land degradation and successful adaptation will not be realised until holistic solutions to land management are explored. SLM can help address root causes of low productivity, land degradation, loss of income-generating capacity, as well as contribute to the amelioration of the adverse effects of climate change.

The '4 per 1000' (4p1000) initiative (Soussana et al. 2019) launched by France during the UNFCCC COP21 in 2015 aims at capturing CO_2 from the atmosphere through changes to agricultural and forestry practices at a rate that would increase the carbon content of soils by 0.4% per year (Rumpel et al. 2018). If global soil carbon content increases at this rate in the top 30–40 cm, the annual increase in atmospheric CO_2 would be stopped (Dignac et al. 2017). This is an illustration of how extremely important soils are for addressing climate change. The initiative is based on eight steps: stop carbon loss (priority #1 is peat soils); promote carbon uptake; monitor, report and verify impacts; deploy technology for tracking soil carbon; test strategies for implementation and upscaling; involve communities; coordinate policies; and provide support (Rumpel et al. 2018). Questions remain, however, about the extent that the 4p1000 is achievable as a universal goal (van Groenigen et al. 2017; Poulton et al. 2018; Schlesinger and Amundson 2018).

LDN was introduced by the UNCCD at Rio +20, and adopted at UNCCD COP12 (UNCCD 2016a). LDN is defined as 'a state whereby the amount and quality of land resources necessary to support ecosystem functions and services and enhance food security remain stable or increase within specified temporal and spatial scales and ecosystems' (Cowie et al. 2018). Pursuit of LDN requires effort to avoid further net loss of the land-based natural capital relative to a reference state, or baseline. LDN encourages a dual-pronged effort involving SLM to reduce the risk of land degradation, combined with efforts in land restoration and rehabilitation, to maintain or enhance land-based natural capital, and its associated ecosystem services (Orr et al. 2017; Cowie et al. 2018). Planning for LDN involves projecting the expected cumulative impacts of land-use and land management decisions, then counterbalancing anticipated losses with measures to achieve equivalent gains, within individual land types (where land type is defined by land potential). Under the LDN framework developed by UNCCD, three primary indicators are used to assess whether LDN is achieved by 2030: land cover change; net primary productivity; and soil organic carbon (Cowie et al. 2018; Sims et al. 2019). Achieving LDN therefore requires integrated landscape management that seeks to optimise land use to meet multiple objectives (ecosystem health, food security, human well-being) (Cohen-Shacham et al. 2016). The response hierarchy of Avoid > Reduce > Reverse land degradation articulates the priorities in planning LDN interventions. LDN provides the impetus for widespread adoption of SLM and efforts to restore or rehabilitate land. Through its focus, LDN ultimately provides tremendous potential for mitigation of, and adaptation to, climate change by halting and reversing land degradation and transforming land from a carbon source to a sink. There are strong synergies

4

between the concept of LDN and the NDCs of many countries, with linkages to national climate plans. LDN is also closely related to many Sustainable Development Goals (SDG) in the areas of poverty, food security, environmental protection and sustainable use of natural resources (UNCCD 2016b). The GEF is supporting countries to set LDN targets and implement their LDN plans through its land degradation focal area, which encourages application of integrated landscape approaches to managing land degradation (GEF 2018).

The 2030 Agenda for Sustainable Development, adopted by the United Nations in 2015, comprises 17 SDGs. Goal 15 is of direct relevance to land degradation, with the objective to protect, restore and promote sustainable use of terrestrial ecosystems, sustainably manage forests, combat desertification and halt and reverse land degradation and halt biodiversity loss. Target 15.3 specifically addresses LDN. Other goals that are relevant for land degradation include Goal 2 (Zero hunger), Goal 3 (Good health and well-being), Goal 7 (Affordable and clean energy), Goal 11 (Sustainable cities and communities), and Goal 12 (Responsible production and consumption). Sustainable management of land resources underpins the SDGs related to hunger, climate change and environment. Further goals of a cross-cutting nature include 1 (No poverty), 6 (Clean water and sanitation) and 13 (Climate action). It remains to be seen how these interconnections are dealt with in practice.

With a focus on biodiversity, IPBES published a comprehensive assessment of land degradation in 2018 (Montanarella et al. 2018). The IPBES report, together with this report focusing on climate change, may contribute to creating a synergy between the two main global challenges for addressing land degradation in order to help achieve the targets of SDG 15 (protect, restore and promote sustainable use of terrestrial ecosystems, sustainably manage forests, combat desertification, and halt and reverse land degradation and halt biodiversity loss).

Market-based mechanisms like the Clean Development Mechanism (CDM) under the UNFCCC and the voluntary carbon market provide incentives to enhance carbon sinks on the land through afforestation and reforestation. Implications for local land use and food security have been raised as a concern and need to be assessed (Edstedt and Carton 2018; Olsson et al. 2014b). Many projects aimed at reducing emissions from deforestation and forest degradations (not to be confused with the national REDD+ programmes in accordance with the UNFCCC Warsaw Framework) are being planned and implemented to primarily target countries with high forest cover and high deforestation rates. Some parameters of incentivising emissions reduction, quality of forest governance, conservation priorities, local rights and tenure frameworks, and sub-national project potential are being looked into, with often very mixed results (Newton et al. 2016; Gebara and Agrawal 2017).

Besides international public initiatives, some actors in the private sector are increasingly aware of the negative environmental impacts of some global value chains producing food, fibre, and energy products (Lambin et al. 2018; van der Ven and Cashore 2018; van der Ven et al. 2018; Lyons-White and Knight 2018). While improvements are underway in many supply chains, measures implemented so far are often insufficient to be effective in reducing or stopping deforestation and forest degradation (Lambin et al. 2018). The GEF is investing in actions to reduce deforestation in commodity supply chains through its Food Systems, Land Use, and Restoration Impact Program (GEF 2018).

1.8.5.1   Limits to adaptation

SLM can be deployed as a powerful adaptation strategy in most instances of climate change impacts on natural and social systems, yet there are limits to adaptation (Klein et al. 2014; Dow, Berhout and Preston 2013). Such limits are dynamic and interact with social and institutional conditions (Barnett et al. 2015; Filho and Nalau 2018). Exceeding adaptation limits will trigger escalating losses or require undesirable transformational change, such as forced migration. The rate of change in relation to the rate of possible adaptation is crucial (Dow et al. 2013). How limits to adaptation are defined, and how they can be measured, is contextual and contested. Limits must be assessed in relation to the ultimate goals of adaptation, which is subject to diverse and differential values (Dow et al. 2013; Adger et al. 2009). A particularly sensitive issue is whether migration is accepted as adaptation or not (Black et al. 2011; Tacoli 2009; Bardsley and Hugo 2010). If migration were understood and accepted as a form of successful adaptation, it would change the limits to adaptation by reducing, or even avoiding, future humanitarian crises caused by climate extremes (Adger et al. 2009; Upadhyay et al. 2017; Nalau et al. 2018).

In the context of land degradation, potential limits to adaptation exist if land degradation becomes so severe and irreversible that livelihoods cannot be maintained, and if migration is either not acceptable or not possible. Examples are coastal erosion where land disappears (Gharbaoui and Blocher 2016; Luetz 2018), collapsing livelihoods due to thawing of permafrost (Landauer and Juhola 2019), and extreme forms of soil erosion, (e.g., landslides (Van der Geest and Schindler 2016) and gully erosion leading to badlands (Poesen et al. 2003)).

1.8.6   Resilience and thresholds

Resilience refers to the capacity of interconnected social, economic and ecological systems, such as farming systems, to absorb disturbance (e.g., drought, conflict, market collapse), and respond or reorganise, to maintain their essential function, identity and structure. Resilience can be described as 'coping capacity'. The disturbance may be a shock – sudden events such as a flood or disease epidemic – or it may be a trend that develops slowly, like a drought or market shift. The shocks and trends anticipated to occur due to climate change are expected to exacerbate risk of land degradation. Therefore, assessing and enhancing resilience to climate change is a critical component of designing SLM strategies.

Resilience as an analytical lens is particularly strong in ecology and related research on natural resource management (Folke et al. 2010; Quinlan et al. 2016) while, in the social sciences, the relevance of resilience for studying social and ecological interactions is contested

(Cote and Nightingale 2012; Olsson et al. 2015; Cretney 2014; Béné et al. 2012; Joseph 2013). In the case of adaptation to climate change (and particularly regarding limits to adaptation), a crucial ambiguity of resilience is the question of whether resilience is a normative concept (i.e., resilience is good or bad) or a descriptive characteristic of a system (i.e., neither good nor bad). Previous IPCC reports have defined resilience as a normative (positive) attribute (see AR5 Glossary), while the wider scientific literature is divided on this (Weichselgartner and Kelman 2015; Strunz 2012; Brown 2014; Grimm and Calabrese 2011; Thorén and Olsson 2018). For example, is outmigration from a disaster-prone area considered a successful adaptation (high resilience) or a collapse of the livelihood system (lack of resilience) (Thorén and Olsson 2018)? In this report, resilience is considered a positive attribute when it maintains capacity for adaptation, learning and/or transformation.

Furthermore, 'resilience' and the related terms 'adaptation' and 'transformation' are defined and used differently by different communities (Quinlan et al. 2016). The relationship and hierarchy of resilience with respect to vulnerability and adaptive capacity are also debated, with different perspectives between disaster management and global change communities, (e.g., Cutter et al. 2008). Nevertheless, these differences in usage need not inhibit the application of 'resilience thinking' in managing land degradation; researchers using these terms, despite variation in definitions, apply the same fundamental concepts to inform management of human-environment systems, to maintain or improve the resource base, and sustain livelihoods.

Applying resilience concepts involves viewing the land as a component of an interlinked social-ecological system; identifying key relationships that determine system function and vulnerabilities of the system; identifying thresholds or tipping points beyond which the system transitions to an undesirable state; and devising management strategies to steer away from thresholds of potential concern, thus facilitating healthy systems and sustainable production (Walker et al. 2009).

A threshold is a non-linearity between a controlling variable and system function, such that a small change in the variable causes the system to shift to an alternative state. Bestelmeyer et al. (2015) and Prince et al. (2018) illustrate this concept in the context of land degradation. Studies have identified various biophysical and socio-economic thresholds in different land-use systems. For example, 50% ground cover (living and dead plant material and biological crusts) is a recognised threshold for dryland grazing systems (e.g., Tighe et al. 2012); below this threshold, the infiltration rate declines, risk of erosion causing loss of topsoil increases, a switch from perennial to annual grass species occurs and there is a consequential sharp decline in productivity. This shift to a lower-productivity state cannot be reversed without significant human intervention. Similarly, the combined pressure of water limitations and frequent fire can lead to transition from closed forest to savannah or grassland: if fire is too frequent, trees do not reach reproductive maturity and post-fire regeneration will fail; likewise, reduced rainfall/increased drought prevents successful forest regeneration (Reyer et al. 2015; Thompson et al. 2009) (Cross-Chapter Box 3 in Chapter 2).

In managing land degradation, it is important to assess the resilience of the existing system, and the proposed management interventions. If the existing system is in an undesirable state or considered unviable under expected climate trends, it may be desirable to promote adaptation or even transformation to a different system that is more resilient to future changes. For example, in an irrigation district where water shortages are predicted, measures could be implemented to improve water use efficiency, for example, by establishing drip irrigation systems for water delivery, although transformation to pastoralism or mixed dryland cropping/livestock production may be more sustainable in the longer term, at least for part of the area. Application of SLM practices, especially those focused on ecological functions (e.g., agroecology, ecosystem-based approaches, regenerative agriculture, organic farming), can be effective in building resilience of agro-ecosystems (Henry et al. 2018). Similarly, the resilience of managed forests can be enhanced by SFM that protects or enhances biodiversity, including assisted migration of tree species within their current range limit (Winder et al. 2011; Pedlar et al. 2012) or increasing species diversity in plantation forests (Felton et al. 2010; Liu et al. 2018a). The essential features of a resilience approach to management of land degradation under climate change are described by O'Connell et al. (2016) and Simonsen et al. (2014).

Consideration of resilience can enhance effectiveness of interventions to reduce or reverse land degradation (*medium agreement, limited evidence*). This approach will increase the likelihood that SLM/SFM and land restoration/rehabilitation interventions achieve long-term environmental and social benefits. Thus, consideration of resilience concepts can enhance the capacity of land systems to cope with climate change and resist land degradation, and assist land-use systems to adapt to climate change.

1.8.7   Barriers to implementation of sustainable land management (SLM)

There is a growing recognition that addressing barriers and designing solutions to complex environmental problems, such as land degradation, requires awareness of the larger system into which the problems and solutions are embedded (Laniak et al. 2013). An ecosystem approach to sustainable land management (SLM) based on an understanding of land degradation processes has been recommended to separate multiple drivers, pressures and impacts (Kassam et al. 2013), but large uncertainty in model projections of future climate, and associated ecosystem processes (IPCC 2013a) pose additional challenges to the implementation of SLM. As discussed earlier in this chapter, many SLM practices, including technologies and approaches, are available that can increase yields and contribute to closing the yield gap between actual and potential crop or pasture yield, while also enhancing resilience to climate change (Yengoh and Ardö 2014; WOCAT n.d.). However, there are often systemic barriers to adoption and scaling up of SLM practices, especially in developing countries.

Uitto (2016) identified areas that the GEF, the financial mechanism of the UNCCD, UNFCCC and other multilateral environmental agreements, can address to solve global environmental problems. These include:

removal of barriers related to knowledge and information; strategies for implementation of technologies and approaches; and institutional capacity. Strengthening these areas would drive transformational change, leading to behavioural change and broader adoption of sustainable environmental practices. Detailed analysis of barriers as well as strategies, methods and approaches to scale up SLM have been undertaken for GEF programmes in Africa, China and globally (Tengberg and Valencia 2018; Liniger et al. 2011; Tengberg et al. 2016). A number of interconnected barriers and bottlenecks to the scaling up of SLM have been identified in this context and are related to:

- limited access to knowledge and information, including new SLM technologies and problem-solving capacities
- weak enabling environment, including the policy, institutional and legal framework for SLM, and land tenure and property rights
- inadequate learning and adaptive knowledge management in the project cycle, including monitoring and evaluation of impacts
- limited access to finance for scaling up, including public and private funding, innovative business models for SLM technologies and financial mechanisms and incentives, such as payment for ecosystem services (PES), insurance and micro-credit schemes (see also Shames et al. 2014).

Adoption of innovations and new technologies are increasingly analysed using the transition theory framework (Geels 2002), the starting point being the recognition that many global environmental problems cannot be solved by technological change alone, but require more far-reaching change of social-ecological systems. Using transition theory makes it possible to analyse how adoption and implementation follow the four stages of sociotechnical transitions,

from predevelopment of technologies and approaches at the niche level, take-off and acceleration, to regime shift and stabilisation at the landscape level. According to a recent review of sustainability transitions in developing countries (Wieczorek 2018), three internal niche processes are important, including the formation of networks that support and nurture innovation, the learning process, and the articulation of expectations to guide the learning process. While technologies are important, institutional and political aspects form the major barriers to transition and upscaling. In developing and transition economies, informal institutions play a pivotal role, and transnational linkages are also important, such as global value chains. In these countries, it is therefore more difficult to establish fully coherent regimes or groups of individuals who share expectations, beliefs or behaviour, as there is a high level of uncertainty about rules and social networks or dominance of informal institutions, which creates barriers to change. This uncertainty is further exacerbated by climate change. Landscape forces comprise a set of slow-changing factors, such as broad cultural and normative values, long-term economic effects such as urbanisation, and shocks such as war and crises that can lead to change.

A study on SLM in the Kenyan highlands using transition theory concluded that barriers to adoption of SLM included high poverty levels, a low-input/low-output farming system with limited potential to generate income, diminishing land sizes, and low involvement of the youth in farming activities. Coupled with a poor coordination of government policies for agriculture and forestry, these barriers created negative feedbacks in the SLM transition process. Other factors to consider include gender issues and lack of secure land tenure. Scaling up of SLM technologies would require collaboration of

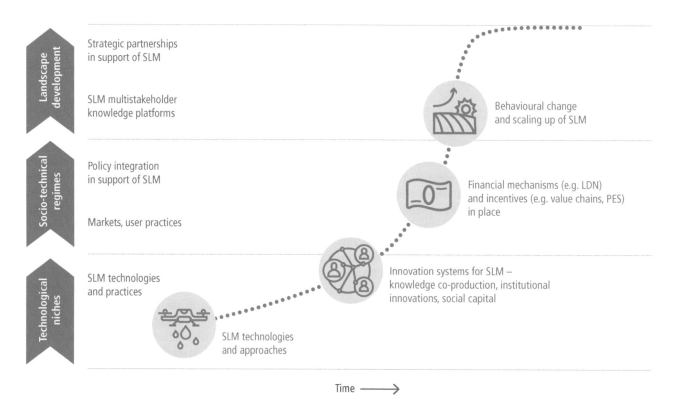

Figure 4.7 | The transition from SLM niche adoption to regime shift and landscape development. Figure draws inspiration from Geels (2002), adapted from Tengberg and Valencia (2018).

diverse stakeholders across multiple scales, a more supportive policy environment and substantial resource mobilisation (Mutoko et al. 2014). Tengberg and Valencia (2018) analysed the findings from a review of the GEF's integrated natural resources management portfolio of projects using the transition theory framework (Figure 4.7). They concluded that to remove barriers to SLM, an agricultural innovations systems approach that supports co-production of knowledge with multiple stakeholders, institutional innovations, a focus on value chains and strengthening of social capital to facilitate shared learning and collaboration could accelerate the scaling up of sustainable technologies and practices from the niche to the landscape level. Policy integration and establishment of financial mechanisms and incentives could contribute to overcoming barriers to a regime shift. The new SLM regime could, in turn, be stabilised and sustained at the landscape level by multi-stakeholder knowledge platforms and strategic partnerships. However, transitions to more sustainable regimes and practices are often challenged by lock-in mechanisms in the current system (Lawhon and Murphy 2012) such as economies of scale, investments already made in equipment, infrastructure and competencies, lobbying, shared beliefs, and practices, that could hamper wider adoption of SLM.

Adaptive, multi-level and participatory governance of social-ecological systems is considered important for regime shifts and transitions to take place (Wieczorek 2018) and essential to secure the capacity of environmental assets to support societal development over longer time periods (Folke et al. 2005). There is also recognition that effective environmental policies and programmes need to be informed by a comprehensive understanding of the biophysical, social, and economic components and processes of a system, their complex interactions, and how they respond to different changes (Kelly (Letcher) et al. 2013). But blueprint policies will not work, due to the wide diversity of rules and informal institutions used across sectors and regions of the world, especially in traditional societies (Ostrom 2009).

The most effective way of removing barriers to funding of SLM has been mainstreaming of SLM objectives and priorities into relevant policy and development frameworks, and combining SLM best practices with economic incentives for land users. As the short-term costs for establishing and maintaining SLM measures are generally high and constitute a barrier to adoption, land users may need to be compensated for generation of longer-term public goods, such as ecosystem services. Cost-benefit analyses can be conducted on SLM interventions to facilitate such compensations (Liniger et al. 2011; Nkonya et al. 2016; Tengberg et al. 2016). The landscape approach is a means to reconcile competing demands on the land and remove barriers to implementation of SLM (e.g., Sayer et al. 2013; Bürgi et al. 2017). It involves an increased focus on participatory governance, development of new SLM business models, and innovative funding schemes, including insurance (Shames et al. 2014). The LDN Fund takes a landscape approach and raises private finance for SLM and promotes market-based instruments, such as PES, certification and carbon trading, that can support scaling up of SLM to improve local livelihoods, sequester carbon and enhance the resilience to climate change (Baumber et al. 2019).

1.9   Case studies

Climate change impacts on land degradation can be avoided, reduced or even reversed, but need to be addressed in a context-sensitive manner. Many of the responses described in this section can also provide synergies of adaptation and mitigation. In this section we provide more in-depth analysis of a number of salient aspects of how land degradation and climate change interact. Table 4.3 is a synthesis of how of these case studies relate to climate change and other broader issues in terms of co-benefits.

1.9.1   Urban green infrastructure

Over half of the world's population now lives in towns and cities, a proportion that is predicted to increase to about 70% by the middle of the century (United Nations 2015). Rapid urbanisation is a severe threat to land and the provision of ecosystem services (Seto et al. 2012). However, as cities expand, the avoidance of land degradation, or the maintenance/enhancement of ecosystem services is rarely considered in planning processes. Instead, economic development and the need for space for construction is prioritised, which can result in substantial pollution of air and water sources, the degradation of existing agricultural areas and indigenous, natural or semi-natural ecosystems both within and outside of urban areas. For instance, urban areas are characterised by extensive impervious surfaces. Degraded, sealed soils beneath these surfaces do not provide the same quality of water retention as intact soils. Urban landscapes comprising 50–90% impervious surfaces can therefore result in 40–83% of rainfall becoming surface water runoff (Pataki et al. 2011). With rainfall intensity predicted to increase in many parts of the world under climate change (Royal Society 2016), increased water runoff is going to get worse. Urbanisation, land degradation and climate change are therefore strongly interlinked, suggesting the need for common solutions (Reed and Stringer 2016).

There is now a large body of research and application demonstrating the importance of retaining urban green infrastructure (UGI) for the delivery of multiple ecosystem services (DG Environment News Alert Service, 2012; Wentworth, 2017) as an important tool to mitigate and adapt to climate change. UGI can be defined as all green elements within a city, including, but not limited to, retained indigenous ecosystems, parks, public greenspaces, green corridors, street trees, urban forests, urban agriculture, green roofs/walls and private domestic gardens (Tzoulas et al. 2007). The definition is usually extended to include 'blue' infrastructure, such as rivers, lakes, bioswales and other water drainage features. The related concept of Nature-Based Solutions (defined as: *living solutions inspired by, continuously supported by and using nature, which are designed to address various societal challenges in a resource-efficient and adaptable manner and to provide simultaneously economic, social, and environmental benefits*) has gained considerable traction within the European Commission as one approach to mainstreaming the importance of UGI (Maes and Jacobs 2017; European Union 2015).

4

Table 4.3 | Synthesis of how the case studies interact with climate change and a broader set of co-benefits.

Case studies (4.9)	Mitigation benefits and potential	Adaptation benefits	Co-benefits	Legend	
Urban green infrastructure (4.9.1) An increasing majority of the world population live in cities and land degradation is an urgent matter for urban areas	↓	≋↓ 🌡	human health, recreation	↓	carbon sink
Perennial grains (4.9.2) After 40 years of breeding, perennial grains now seem to have the potential of reducing climate impacts of agriculture while increasing its overall sustainability	↓ ↑	☁ ≋↓	reduced use of herbicides, reduced soil erosion and nutrient leakage	↑	reduced emission
Reforestation (4.9.3) Two cases of successful reforestation serve as illustrations of the potential of sustained efforts into reforestation	↓	≋↓	economic return from sustainable forestry, reduced flood risk downstream	≋↓	reduced flood risk
Management of peat soils (4.9.4) Degradation of peat soils in tropical and Arctic regions is a major source of greenhouse gases, hence an urgent mitigation option	↑		improved air quality in tropical regions	🌡	reduced heat stress
Biochar (4.9.5) Biochar is a land-management technique of high potential, but controversial	↓ ↑	☁	improved soil fertility	☁	drought resistance
Protection against hurricane damages (4.9.6) More severe tropical cyclones increase the risk of land degradation in some areas, hence the need for increased adaptation		≋↓ 🌀	reduced losses (human lives, livelihoods, and assets)	🌀	storm protection
Responses to saltwater intrusion (4.9.7) The combined effect of climate-induced sea level rise and land-use change in coastal regions increases the risk of saltwater intrusion in many coastal regions		≋ 🌀	improved food and water security	≋	protection against sea level rise
Avoiding coastal maladaptation (4.9.8) Low-lying coastal areas are in urgent need of adaptation, but examples have resulted in maladaptation		≋ 🌀	reduced losses (human lives, livelihoods, and assets)		

Through retaining existing vegetation and ecosystems, revegetating previous developed land or integrating vegetation into buildings in the form of green walls and roofs, UGI can play a direct role in mitigating climate change through carbon sequestration. However, compared to overall carbon emissions from cities, effects will be small. Given that UGI necessarily involves the retention and management of non-sealed surfaces, co-benefits for land degradation (e.g., soil compaction avoidance, reduced water runoff, carbon storage and vegetation productivity (Davies et al. 2011; Edmondson et al. 2011, 2014; Yao et al. 2015) will also be apparent. Although not currently a priority, its role in mitigating land degradation could be substantial. For instance, appropriately managed innovative urban agricultural production systems, such as vertical farms, could have the potential to meet some of the food needs of cities and reduce the production (and therefore degradation) pressure on agricultural land in rural areas, although thus far this is unproven (for a recent review, see Wilhelm and Smith 2018).

The importance of UGI as part of a climate change adaptation approach has received greater attention and application (Gill et al. 2007; Fryd et al. 2011; Demuzere et al. 2014; Sussams et al. 2015). The EU's Adapting to Climate Change white paper emphasises the 'crucial role in adaptation in providing essential resources for social and economic purposes under extreme climate conditions' (CEC, 2009, p. 9). Increasing vegetation cover, planting street trees and maintaining/expanding public parks reduces temperatures (Cavan et al. 2014; Di Leo et al. 2016; Feyisa et al. 2014; Tonosaki and Kawai 2014; Zölch et al. 2016). Further, the appropriate design and spatial distribution of greenspaces within cities can help to alter urban climates to improve human health and comfort (e.g., Brown and Nicholls 2015; Klemm et al. 2015). The use of green walls and roofs can also reduce energy use in buildings (e.g., Coma et al. 2017). Similarly, natural flood management and ecosystem-based approaches of providing space for water, renaturalising rivers and reducing surface runoff through the presence of permeable surfaces and vegetated features (including walls and roofs) can manage flood risks, impacts and vulnerability (e.g., Gill et al. 2007; Munang et al. 2013). Access to UGI in times of environmental stresses and shock can provide safety nets for people, and so can be an important adaptation mechanism, both to climate change (Potschin et al. 2016) and land degradation.

Most examples of UGI implementation as a climate change adaptation strategy have centred on its role in water management for flood risk reduction. The importance for land degradation is either not stated, or not prioritised. In Beira, Mozambique, the government is using UGI to mitigate increased flood risks predicted to occur under climate change and urbanisation, which will be done by improving the natural water capacity of the Chiveve River. As part of the UGI approach, mangrove habitats have been restored, and future phases include developing new multi-functional urban green spaces along the river (World Bank 2016). The retention of green spaces within the city will have the added benefit of halting further degradation in those areas. Elsewhere, planning mechanisms promote the retention and expansion of green areas within cities to ensure ecosystem service delivery, which directly halts land degradation, but are largely viewed and justified in the context of climate change adaptation and mitigation. For instance, the Berlin Landscape Programme includes five plans, one of which covers adapting to climate change through the recognition of the role of UGI (Green Surge 2016). Major climate-related challenges facing Durban, South Africa, include sea level rise, urban heat island, water runoff and conservation (Roberts and O'Donoghue 2013). Now considered a global leader in climate adaptation planning (Roberts 2010), Durban's Climate Change Adaptation plan includes the retention and maintenance of natural ecosystems, in particular those that are important for mitigating flooding, coastal erosion, water pollution, wetland siltation and climate change (eThekwini Municipal Council 2014).

1.9.2 Perennial grains and soil organic carbon (SOC)

The severe ecological perturbation that is inherent in the conversion of native perennial vegetation to annual crops, and the subsequent high frequency of perturbation required to maintain annual crops, results in at least four forms of soil degradation that will be exacerbated by the effects of climate change (Crews et al. 2016). First, soil erosion is a very serious consequence of annual cropping, with median losses exceeding rates of formation by one to two orders of magnitude in conventionally plowed agroecosystems, and while erosion is reduced with conservation tillage, median losses still exceed formation by several fold (Montgomery 2007). More severe storm intensity associated with climate change is expected to cause even greater losses to wind and water erosion (Nearing et al. 2004). Second, the periods of time in which live roots are reduced or altogether absent from soils in annual cropping systems allow for substantial losses of nitrogen from fertilised croplands, averaging 50% globally (Ladha et al. 2005). This low retention of nitrogen is also expected to worsen with more intense weather events (Bowles et al. 2018). A third impact of annual cropping is the degradation of soil structure caused by tillage, which can reduce infiltration of precipitation, and increase surface runoff. It is predicted that the percentage of precipitation that infiltrates into agricultural soils will decrease further under climate-change scenarios (Basche and DeLonge 2017; Wuest et al. 2006). The fourth form of soil degradation that results from annual cropping is the reduction of soil organic matter (SOM), a topic of particular relevance to climate change mitigation and adaptation.

Undegraded cropland soils can theoretically hold far more SOM (which is about 58% carbon) than they currently do (Soussana et al. 2006). We know this deficiency because, with few exceptions, comparisons between cropland soils and those of proximate mature native ecosystems commonly show a 40–75% decline in soil carbon attributable to agricultural practices. What happens when native ecosystems are converted to agriculture that induces such significant losses of SOM? Wind and water erosion commonly results in preferential removal of light organic matter fractions that can accumulate on or near the soil surface (Lal 2003). In addition to the effects of erosion, the fundamental practices of growing annual food and fibre crops alters both inputs and outputs of organic matter from most agroecosystems, resulting in net reductions in soil carbon equilibria (Soussana et al. 2006; McLauchlan 2006; Crews et al. 2016). Native vegetation of almost all terrestrial ecosystems is dominated

by perennial plants, and the below-ground carbon allocation of these perennials is a key variable in determining formation rates of stable soil organic carbon (SOC) (Jastrow et al. 2007; Schmidt et al. 2011). When perennial vegetation is replaced by annual crops, inputs of root-associated carbon (roots, exudates, mycorrhizae) decline substantially. For example, perennial grassland species allocate around 67% of productivity to roots, whereas annual crops allocate between 13–30% (Saugier 2001; Johnson et al. 2006).

At the same time, inputs of SOC are reduced in annual cropping systems, and losses are increased because of tillage, compared to native perennial vegetation. Tillage breaks apart soil aggregates which, among other functions, are thought to inhibit soil bacteria, fungi and other microbes from consuming and decomposing SOM (Grandy and Neff 2008). Aggregates reduce microbial access to organic matter by restricting physical access to mineral-stabilised organic compounds as well as reducing oxygen availability (Cotrufo et al. 2015; Lehmann and Kleber 2015). When soil aggregates are broken open with tillage in the conversion of native ecosystems to agriculture, microbial consumption of SOC and subsequent respiration of CO_2 increase dramatically, reducing soil carbon stocks (Grandy and Robertson 2006; Grandy and Neff 2008).

Many management approaches are being evaluated to reduce soil degradation in general, especially by increasing mineral-protected forms of SOC in the world's croplands (Paustian et al. 2016). The menu of approaches being investigated focuses either on increasing below-ground carbon inputs, usually through increases in total crop productivity, or by decreasing microbial activity, usually through reduced soil disturbance (Crews and Rumsey 2017). However, the basic biogeochemistry of terrestrial ecosystems managed for production of annual crops presents serious challenges to achieving the standing stocks of SOC accumulated by native ecosystems that preceded agriculture. A novel new approach that is just starting to receive significant attention is the development of perennial cereal, legume and oilseed crops (Glover et al. 2010; Baker 2017).

There are two basic strategies that plant breeders and geneticists are using to develop new perennial grain crop species. The first involves making wide hybrid crosses between existing elite lines of annual crops, such as wheat, sorghum and rice, with related wild perennial species in order to introgress perennialism into the genome of the annual (Cox et al. 2018; Huang et al. 2018; Hayes et al. 2018). The other approach is *de novo* domestication of wild perennial species that have crop-like traits of interest (DeHaan et al. 2016; DeHaan and Van Tassel 2014). New perennial crop species undergoing *de novo* domestication include intermediate wheatgrass, a relative of wheat that produces grain known as Kernza (DeHaan et al. 2018; Cattani and Asselin 2018) and *Silphium integrifolium*, an oilseed crop in the sunflower family (Van Tassel et al. 2017). Other grain crops receiving attention for perennialisation include pigeon pea, barley, buckwheat and maize (Batello et al. 2014; Chen et al. 2018c) and a number of legume species (Schlautman et al. 2018). In most cases, the seed yields of perennial grain crops under development are well below those of elite modern grain varieties. During the period that it will take for intensive breeding efforts to close the yield and other trait gaps between annual and perennial grains, perennial

4

proto-crops may be used for purposes other than grain, including forage production (Ryan et al. 2018). Perennial rice stands out as a high-yielding exception, as its yields matched those of elite local varieties in the Yunnan Province for six growing seasons over three years (Huang et al. 2018).

In a perennial agroecosystem, the biogeochemical controls on SOC accumulation shift dramatically, and begin to resemble the controls that govern native ecosystems (Crews et al. 2016). When erosion is reduced or halted, and crop allocation to roots increases by 100–200%, and when soil aggregates are not disturbed thus reducing microbial respiration, SOC levels are expected to increase (Crews and Rumsey 2017). Deep roots growing year round are also effective at increasing nitrogen retention (Culman et al. 2013; Jungers et al. 2019). Substantial increases in SOC have been measured where croplands that had historically been planted to annual grains were converted to perennial grasses, such as in the US Conservation Reserve Program or in plantings of second-generation perennial biofuel crops. Two studies have assessed carbon accumulation in soils when croplands were converted to the perennial grain Kernza. In one, researchers found no differences in soil labile (permanganate-oxidisable) carbon after four years of cropping to perennial Kernza versus annual wheat in a sandy textured soil. Given that coarse textured soils do not offer the same physicochemical protection against microbial attack as many finer textured soils, these results are not surprising, but these results do underscore how variable the rates of carbon accumulation can be (Jastrow et al. 2007). In the second study, researchers assessed the carbon balance of a Kernza field in Kansas, USA over 4.5 years using eddy covariance observations (de Oliveira et al. 2018). They found that the net carbon accumulation rate of about 1500 gC m^{-2} yr^{-1} in the first year of the study corresponding to the biomass of Kernza, increasing to about 300 gC m^{-2} yr^{-1} in the final year, where CO_2 respiration losses from the decomposition of roots and SOM approached new carbon inputs from photosynthesis. Based on measurements of soil carbon accumulation in restored grasslands in this part of the USA, the net carbon accumulation in stable organic matter under a perennial grain crop might be expected to

sequester 30–50 gC m^{-2} yr^{-1} (Post and Kwon 2000) until a new equilibrium is reached. Sugar cane, a highly productive perennial, has been shown to accumulate a mean of 187 gC m^{-2} yr^{-1} in Brazil (La Scala Júnior et al. 2012).

Reduced soil erosion, increased nitrogen retention, greater water uptake efficiency and enhanced carbon sequestration represent improved ecosystem functions, made possible in part by deep and extensive root systems of perennial crops (Figure 4.8).

When compared to annual grains like wheat, single species stands of deep-rooted perennial grains such as Kernza are expected to reduce soil erosion, increase nitrogen retention, achieve greater water uptake efficiency and enhance carbon sequestration (Crews et al. 2018) (Figure 4.8). An even higher degree of ecosystem services can, at least theoretically, be achieved by strategically combining different functional groups of crops such as a cereal and a nitrogen-fixing legume (Soussana and Lemaire 2014). Not only is there evidence from plant-diversity experiments that communities with higher species richness sustain higher concentrations of SOC (Hungate et al. 2017; Sprunger and Robertson 2018; Chen, S. 2018; Yang et al. 2019), but other valuable ecosystem services such as pest suppression, lower GHG emissions, and greater nutrient retention may be enhanced (Schnitzer et al. 2011; Culman et al. 2013).

Similar to perennial forage crops such as alfalfa, perennial grain crops are expected to have a definite productive lifespan, probably in the range of three to 10 years. A key area of research on perennial grains cropping systems is to minimise losses of SOC during conversion of one stand of perennial grains to another. Recent work demonstrates that no-till conversion of a mature perennial grassland to another perennial crop will experience several years of high net CO_2 emissions as decomposition of copious crop residues exceed ecosystem uptake of carbon by the new crop (Abraha et al. 2018). Most, if not all, of this lost carbon will be recaptured in the replacement crop. It is not known whether mineral-stabilised carbon that is protected in soil aggregates is vulnerable to loss in perennial crop succession.

Figure 4.8 | Comparison of root systems between the newly domesticated intermediate wheatgrass (left) and annual wheat (right). Photo: Copyright © Jim Richardson.

Perennial grains hold promises of agricultural practices, which can significantly reduce soil erosion and nutrient leakage while sequestering carbon. When cultivated in mixes with N-fixing species (legumes) such polycultures also reduce the need for external inputs of nitrogen – a large source of GHG from conventional agriculture.

1.9.3   Reversing land degradation through reforestation

1.9.3.1   South Korea case study on reforestation success

In the first half of the 20th century, forests in the Republic of Korea (South Korea) were severely degraded and deforested during foreign occupations and the Korean War. Unsustainable harvest for timber and fuelwood resulted in severely degraded landscapes, heavy soil erosion and large areas denuded of vegetation cover. Recognising that South Korea's economic health would depend on a healthy environment, South Korea established a national forest service (1967) and embarked on the first phase of a 10-year reforestation programme in 1973 (Forest Development Program), which was followed by subsequent reforestation programmes that ended in 1987, after 2.4 Mha of forests were restored (Figure 4.9).

As a consequence of reforestation, forest volume increased from 11.3 m^3 ha^{-1} in 1973 to 125.6 m^3 ha^{-1} in 2010 and 150.2 m^3 ha^{-1} in 2016 (Korea Forest Service 2017). Increases in forest volume had significant co-benefits such as increasing water yield by 43% and reducing soil losses by 87% from 1971 to 2010 (Kim et al. 2017).

The forest carbon density in South Korea has increased from 5–7 MgC ha^{-1} in the period 1955–1973 to more than 30 MgC ha^{-1} in the late 1990s (Choi et al. 2002). Estimates of carbon uptake rates in the late 1990s were 12 TgC yr^{-1} (Choi et al. 2002). For the period 1954 to 2012, carbon uptake was 8.3 TgC yr^{-1} (Lee et al. 2014), lower than other estimates because reforestation programmes did not start until 1973. Net ecosystem production in South Korea was 10.55 $\pm$ 1.09 TgC yr^{-1} in the 1980s, 10.47 $\pm$ 7.28 Tg C yr^{-1} in the 1990s, and 6.32 $\pm$ 5.02 Tg C yr^{-1} in the 2000s, showing a gradual decline as average forest age increased (Cui et al. 2014). The estimated past and projected future increase in the carbon content of South Korea's forest area during 1992–2034 was 11.8 TgC yr^{-1} (Kim et al. 2016).

During the period of forest restoration, South Korea also promoted inter-agency cooperation and coordination, especially between the energy and forest sectors, to replace firewood with fossil fuels, and to reduce demand for firewood to help forest recovery (Bae et al. 2012). As experience with forest restoration programmes has increased, emphasis has shifted from fuelwood plantations, often

4

Figure 4.9 | Example of severely degraded hills in South Korea and stages of forest restoration. The top two photos are taken in the early 1970s, before and after restoration, the third photo about five years after restoration, and the bottom photo was taken about 20 years after restoration. Many examples of such restoration success exist throughout South Korea. (Photos: Copyright © Korea Forest Service)

with exotic species and hybrid varieties to planting more native species and encouraging natural regeneration (Kim and Zsuffa 1994; Lee et al. 2015). Avoiding monocultures in reforestation programmes can reduce susceptibility to pests (Kim and Zsuffa 1994). Other important factors in the success of the reforestation programme were that private landowners were heavily involved in initial efforts (both corporate entities and smallholders) and that the reforestation programme was made part of the national economic development programme (Lamb 2014).

The net present value and the cost-benefit ratio of the reforestation programme were 54.3 billion and 5.84 billion USD in 2010, respectively. The breakeven point of the reforestation investment appeared within two decades. Substantial benefits of the reforestation programme included disaster risk reduction and carbon sequestration (Lee et al. 2018a).

In summary, the reforestation programme was a comprehensive technical and social initiative that restored forest ecosystems, enhanced the economic performance of rural regions, contributed to disaster risk reduction, and enhanced carbon sequestration (Kim et al. 2017; Lee et al. 2018a; UNDP 2017).

The success of the reforestation programme in South Korea and the associated significant carbon sink indicate a high mitigation potential that might be contributed by a potential future reforestation programme in the Democratic People's Republic of Korea (North Korea) (Lee et al. 2018b).

1.9.3.2   China case study on reforestation success

The dramatic decline in the quantity and quality of natural forests in China resulted in land degradation, such as soil erosion, floods, droughts, carbon emission, and damage to wildlife habitat (Liu and Diamond 2008). In response to failures of previous forestry and land policies, the severe droughts in 1997, and the massive floods in 1998, the central government decided to implement a series of land degradation control policies, including the National Forest Protection Program (NFPP), Grain for Green or the Conversion of Cropland to Forests and Grassland Program (GFGP) (Liu et al. 2008; Yin 2009; Tengberg et al. 2016; Zhang et al. 2000). The NFPP aimed to completely ban logging of natural forests in the upper reaches of the Yangtze and Yellow rivers as well as in Hainan Province by 2000 and to substantially reduce logging in other places (Xu et al. 2006). In 2011, NFPP was renewed for the 10-year second phase, which also added another 11 counties around Danjiangkou Reservoir in Hubei and Henan Provinces, the water source for the middle route of the South-to-North Water Diversion Project (Liu et al. 2013). Furthermore, the NFPP afforested 31 Mha by 2010 through aerial seeding, artificial planting, and mountain closure (i.e., prohibition of human activities such as fuelwood collection and lifestock grazing) (Xu et al. 2006). China banned commercial logging in all natural forests by the end of 2016, which imposed logging bans and harvesting reductions in 68.2 Mha of forest land – including 56.4 Mha of natural forest (approximately 53% of China's total natural forests).

GFGP became the most ambitious of China's ecological restoration efforts, with more than 45 billion USD devoted to its implementation since 1990 (Kolinjivadi and Sunderland 2012) The programme involves the conversion of farmland on slopes of 15–25° or greater to forest or grassland (Bennett 2008). The pilot programme started in three provinces – Sichuan, Shaanxi and Gansu – in 1999 (Liu and Diamond 2008). After its initial success, it was extended to 17 provinces by 2000 and finally to all provinces by 2002, including the headwaters of the Yangtze and Yellow rivers (Liu et al. 2008).

NFPP and GFGP have dramatically improved China's land conditions and ecosystem services, and thus have mitigated the unprecedented land degradation in China (Liu et al. 2013; Liu et al. 2002; Long et al. 2006; Xu et al. 2006). NFPP protected 107 Mha forest area and increased forest area by 10 Mha between 2000 and 2010. For the second phase (2011–2020), the NFPP plans to increase forest cover by a further 5.2 Mha, capture 416 million tons of carbon, provide 648,500 forestry jobs, further reduce land degradation, and enhance biodiversity (Liu et al. 2013). During 2000–2007, sediment concentration in the Yellow River had declined by 38%. In the Yellow River basin, it was estimated that surface runoff would be reduced by 450 million m^3 from 2000 to 2020, which is equivalent to 0.76% of the total surface water resources (Jia et al. 2006). GFGP had cumulatively increased vegetative cover by 25 Mha, with 8.8 Mha of cropland being converted to forest and grassland, 14.3 Mha barren land being afforested, and 2.0 Mha of forest regeneration from mountain closure. Forest cover within the GFGP region has increased 2% during the first eight years (Liu et al. 2008). In Guizhou Province, GFGP plots had 35–53% less loss of phosphorus than non-GFGP plots (Liu et al. 2002). In Wuqi County of Shaanxi Province, the Chaigou Watershed had 48% and 55% higher soil moisture and moisture-holding capacity in GFGP plots than in non-GFGP plots, respectively (Liu et al. 2002). According to reports on China's first national ecosystem assessment (2000–2010), for carbon sequestration and soil retention, coefficients for the GFGP targeting forest restoration and NFPP are positive and statistically significant. For sand fixation, GFGP targeting grassland restoration is positive and statistically significant. Remote sensing observations confirm that vegetation cover increased and bare soil declined in China over the period 2001 to 2015 (Qiu et al. 2017). But, where afforestation is sustained by drip irrigation from groundwater, questions about plantation sustainability arise (Chen et al. 2018a). Moreover, greater gains in biodiversity could be achieved by promoting mixed forests over monocultures (Hua et al. 2016).

NFPP-related activities received a total commitment of 93.7 billion yuan (about 14 billion USD at 2018 exchange rate) between 1998 and 2009. Most of the money was used to offset economic losses of forest enterprises caused by the transformation from logging to tree plantations and forest management (Liu et al. 2008). By 2009, the cumulative total investment through the NFPP and GFGP exceeded 50 billion USD2009 and directly involved more than 120 million farmers in 32 million households in the GFGP alone (Liu et al. 2013). All programmes reduce or reverse land degradation and improve human well-being. Thus, a coupled human and natural systems perspective (Liu et al. 2008) would be helpful to understand the complexity of policies and their impacts, and to establish long-term

4

management mechanisms to improve the livelihood of participants in these programmes and other land management policies in China and many other parts of the world.

1.9.4   Degradation and management of peat soils

Globally, peatlands cover 3–4% of the Earth's land area (about 430 Mha) (Xu et al. 2018a) and store 26–44% of estimated global SOC (Moore 2002). They are most abundant in high northern latitudes, covering large areas in North America, Russia and Europe. At lower latitudes, the largest areas of tropical peatlands are located in Indonesia, the Congo Basin and the Amazon Basin in the form of peat swamp forests (Gumbricht et al. 2017; Xu et al. 2018a). It is estimated that, while 80–85% of the global peatland areas is still largely in a natural state, they are such carbon-dense ecosystems that degraded peatlands (0.3% of the terrestrial land) are responsible for a disproportional 5% of global anthropogenic CO_2 emissions – that is, an annual addition of 0.9–3 $GtCO_2$ to the atmosphere (Dommain et al. 2012; IPCC 2014c).

Peatland degradation is not well quantified globally, but regionally peatland degradation can involve a large percentage of the areas. Land-use change and degradation in tropical peatlands have primarily been quantified in Southeast Asia, where drainage and conversion to plantation crops is the dominant transition (Miettinen et al. 2016). Degradation of peat swamps in Peru is also a growing concern and one pilot survey showed that more than 70% of the peat swamps were degraded in one region surveyed (Hergoualc'h et al. 2017a). Around 65,000 km^2 or 10% of the European peatland area has been lost and 44% of the remaining European peatlands are degraded (Joosten, H., Tanneberger 2017). Large areas of fens have been entirely 'lost' or greatly reduced in thickness due to peat wastage (Lamers et al. 2015).

The main drivers of the acceleration of peatland degradation in the 20th century were associated with drainage for agriculture, peat extraction and afforestation related activities (burning, over-grazing, fertilisation) with a variable scale and severity of impact depending on existing resources in the various countries (O'Driscoll et al. 2018; Cobb, A.R. et al. Dommain et al. 2018; Lamers et al. 2015). New drivers include urban development, wind farm construction (Smith et al. 2012), hydroelectric development, tar sands mining and recreational uses (Joosten and Tanneberger 2017). Anthropogenic pressures are now affecting peatlands in previously geographically isolated areas with consequences for global environmental concerns and impacts on local livelihoods (Dargie et al. 2017; Lawson et al. 2015; Butler et al. 2009).

Drained and managed peatlands are GHG-emission hotspots (Swails et al. 2018; Hergoualc'h et al. 2017a, 2017b; Roman-Cuesta et al. 2016). In most cases, lowering of the water table leads to direct and indirect CO_2 and N_2O emissions to the atmosphere, with rates dependent on a range of factors, including the groundwater level and the water content of surface peat layers, nutrient content, temperature, and vegetation communities. The exception is nutrient-limited boreal peatlands (Minkkinen et al. 2018; Ojanen et al. 2014). Drainage also increases erosion and dissolved organic carbon loss, removing stored carbon into streams as dissolved and particulate organic carbon, which ultimately returns to the atmosphere (Moore et al. 2013; Evans et al. 2016).

In tropical peatlands, oil palm is the most widespread plantation crop and, on average, it emits around 40 tCO_2 ha^{-1} yr^{-1}; Acacia plantations for pulpwood are the second most widespread plantation crop and emit around 73 tCO_2 ha^{-1} yr^{-1} (Drösler et al. 2013). Other land uses typically emit less than 37 tCO_2 ha^{-1} yr^{-1}. Total emissions from peatland drainage in the region are estimated to be between 0.07 and 1.1 $GtCO_2$ yr^{-1} (Houghton and Nassikas 2017; Frolking et al. 2011). Land-use change also affects the fluxes of N_2O and CH_4. Undisturbed tropical peatlands emit about 0.8 $MtCH_4$ yr^{-1} and 0.002 MtN_2O yr^{-1}, while disturbed peatlands emit 0.1 $MtCH_4$ yr^{-1} and 0.2 MtN_2O–N yr^{-1} (Frolking et al. 2011). These N_2O emissions are probably low, as new findings show that emissions from fertilised oil palm can exceed 20 kgN_2O–N ha^{-1} yr^{-1} (Oktarita et al. 2017).

In the temperate and boreal zones, peatland drainage often leads to emissions in the order of 0.9 to 9.5 tCO_2 ha^{-1} y^{-1} in forestry plantations and 21 to 29 tCO_2 ha^{-1} y^{-1} in grasslands and croplands. Nutrient-poor sites often continue to be CO_2 sinks for long periods (e.g., 50 years) following drainage and, in some cases, sinks for atmospheric CH_4, even when drainage ditch emissions are considered (Minkkinen et al. 2018; Ojanen et al. 2014). Undisturbed boreal and temperate peatlands emit about 30 $MtCH_4$ yr^{-1} and 0.02 MtN_2O–N yr^{-1}, while disturbed peatlands emit 0.1 $MtCH_4$ yr^{-1} and 0.2 MtN_2O–N yr^{-1} (Frolking et al. 2011).

Fire emissions from tropical peatlands are only a serious issue in Southeast Asia, where they are responsible for 634 (66–4070) $MtCO_2$ yr^{-1} (van der Werf et al. 2017). Much of the variability is linked with the El Niño–Southern Oscillation (ENSO), which produces drought conditions in this region. Anomalously active fire seasons have also been observed in non-drought years and this has been attributed to the increasing effect of high temperatures that dry vegetation out during short dry spells in otherwise normal rainfall years (Fernandes et al. 2017; Gaveau et al. 2014). Fires have significant societal impacts; for example, the 2015 fires caused more than 100,000 additional deaths across Indonesia, Malaysia and Singapore, and this event was more than twice as deadly as the 2006 El Niño event (Koplitz et al. 2016).

Peatland degradation in other parts of the world differs from Asia. In Africa, for large peat deposits like those found in the Cuvette Centrale in the Congo Basin or in the Okavango inland delta, the principle threat is changing rainfall regimes due to climate variability and change (Weinzierl et al. 2016; Dargie et al. 2017). Expansion of agriculture is not yet a major factor in these regions. In the Western Amazon, extraction of non-timber forest products like the fruits of *Mauritia flexuosa* (moriche palm) and Suri worms are major sources of degradation that lead to losses of carbon stocks (Hergoualc'h et al. 2017a).

The effects of peatland degradation on livelihoods have not been systematically characterised. In places where plantation crops are driving the conversion of peat swamps, the financial benefits can be considerable. One study in Indonesia found that the net present value of an oil palm plantation is between 3,835 and 9,630 USD per ha to land owners (Butler et al. 2009). High financial returns are creating incentives for the expansion of smallholder production in peatlands. Smallholder plantations extend over 22% of the peatlands in insular Southeast Asia compared to 27% for industrial plantations (Miettinen et al. 2016). In places where income is generated from extraction of marketable products, ecosystem degradation probably has a negative effect on livelihoods. For example, the sale of fruits of *M. flexuosa* in some parts of the western Amazon constitutes as much as 80% of the winter income of many rural households, but information on trade values and value chains of *M. flexuosa* is still sparse (Sousa et al. 2018; Virapongse et al. 2017).

There is little experience with peatland restoration in the tropics. Experience from northern latitudes suggests that extensive damage and changes in hydrological conditions mean that restoration in many cases is unachievable (Andersen et al. 2017). In the case of Southeast Asia, where peatlands form as raised bogs, drainage leads to collapse of the dome, and this collapse cannot be reversed by rewetting. Nevertheless, efforts are underway to develop solutions, or at least partial solutions in Southeast Asia, for example, by the Indonesian Peatland Restoration Agency. The first step is to restore the hydrological regime in drained peatlands, but so far experiences with canal blocking and reflooding of the peat have been only partially successful (Ritzema et al. 2014). Market incentives with certification through the Roundtable on Sustainable Palm Oil have also not been particularly successful as many concessions seek certification only after significant environmental degradation has occurred (Carlson et al. 2017). Certification had no discernible effect on forest loss or fire detection in peatlands in Indonesia. To date there is no documentation of restoration methods or successes in many other parts of the tropics. However, in situations where degradation does not involve drainage, ecological restoration may be possible. In South America, for example, there is growing interest in restoration of palm swamps, and as experiences are gained it will be important to document success factors to inform successive efforts (Virapongse et al. 2017).

In higher latitudes where degraded peatlands have been drained, the most effective option to reduce losses from these large organic carbon stocks is to change hydrological conditions and increase soil moisture and surface wetness (Regina et al. 2015). Long-term GHG monitoring in boreal sites has demonstrated that rewetting and restoration noticeably reduce emissions compared to degraded drained sites and can restore the carbon sink function when vegetation is re-established (Wilson et al. 2016; IPCC 2014a; Nugent et al. 2018) although, restored ecosystems may not yet be as resilient as their undisturbed counterparts (Wilson et al. 2016). Several studies have demonstrated the co-benefits of rewetting specific degraded peatlands for biodiversity, carbon sequestration, (Parry et al. 2014; Ramchunder et al. 2012; Renou-Wilson et al. 2018) and other ecosystem services, such as improvement of water storage and quality (Martin-Ortega et al. 2014) with beneficial consequences for human well-being (Bonn et al. 2016; Parry et al. 2014).

1.9.5   Biochar

Biochar is organic matter that is carbonised by heating in an oxygen-limited environment, and used as a soil amendment. The properties of biochar vary widely, dependent on the feedstock and the conditions of production. Biochar could make a significant contribution to mitigating both land degradation and climate change, simultaneously.

1.9.5.1   Role of biochar in climate change mitigation

Biochar is relatively resistant to decomposition compared with fresh organic matter or compost, so represents a long-term carbon store (*very high confidence*). Biochars produced at higher temperature (>450°C) and from woody material have greater stability than those produced at lower temperature (300–450°C), and from manures (*very high confidence*) (Singh et al. 2012; Wang et al. 2016b). Biochar stability is influenced by soil properties: biochar carbon can be further stabilised by interaction with clay minerals and native SOM (*medium evidence*) (Fang et al. 2015). Biochar stability is estimated to range from decades to thousands of years, for different biochars in different applications (Singh et al. 2015; Wang et al. 2016). Biochar stability decreases as ambient temperature increases (*limited evidence*) (Fang et al. 2017).

Biochar can enhance soil carbon stocks through 'negative priming', in which rhizodeposits are stabilised through sorption of labile carbon on biochar, and formation of biochar-organo-mineral complexes (Weng et al. 2015, 2017, 2018; Wang et al. 2016b). Conversely, some studies show increased turnover of native soil carbon ('positive priming') due to enhanced soil microbial activity induced by biochar. In clayey soils, positive priming is minor and short-lived compared to negative priming effects, which dominate in the medium to long term (Singh and Cowie 2014; Wang et al. 2016b). Negative priming has been observed particularly in loamy grassland soil (Ventura et al. 2015) and clay-dominated soils, whereas positive priming is reported in sandy soils (Wang et al. 2016b) and those with low carbon content (Ding et al. 2018).

Biochar can provide additional climate-change mitigation by decreasing nitrous oxide (N_2O) emissions from soil, due in part to decreased substrate availability for denitrifying organisms, related to the molar H/C ratio of the biochar (Cayuela et al. 2015). However, this impact varies widely: meta-analyses found an average decrease in N_2O emissions from soil of 30–54%, (Cayuela et al. 2015; Borchard et al. 2019; Moore 2002), although another study found no significant reduction in field conditions when weighted by the inverse of the number of observations per site (Verhoeven et al. 2017). Biochar has been observed to reduce methane emissions from flooded soils, such as rice paddies, though, as for N_2O, results vary between studies and increases have also been observed (He et al. 2017; Kammann et al. 2017). Biochar has also been found to reduce methane uptake by dryland soils, though the effect is small in absolute terms (Jeffery et al. 2016).

Additional climate benefits of biochar can arise through: reduced nitrogen fertiliser requirements, due to reduced losses of nitrogen through leaching and/or volatilisation (Singh et al. 2010) and enhanced biological nitrogen fixation (Van Zwieten et al. 2015); increased yields of crop, forage, vegetable and tree species (Biederman and Harpole 2013), particularly in sandy soils and acidic tropical soils (Simon et al. 2017); avoided GHG emissions from manure that would otherwise be stockpiled, crop residues that would be burned or processing residues that would be landfilled; and reduced GHG emissions from compost when biochar is added (Agyarko-Mintah et al. 2017; Wu et al. 2017a).

Climate benefits of biochar could be substantially reduced through reduction in albedo if biochar is surface-applied at high rates to light-coloured soils (Genesio et al. 2012; Bozzi et al. 2015; Woolf et al. 2010), or if black carbon dust is released (Genesio et al. 2016). Pelletising or granulating biochar, and applying below the soil surface or incorporating into the soil, minimises the release of black carbon dust and reduces the effect on albedo (Woolf et al. 2010).

Biochar is a potential 'negative emissions' technology: the thermochemical conversion of biomass to biochar slows mineralisation of the biomass, delivering long-term carbon storage; gases released during pyrolysis can be combusted for heat or power, displacing fossil energy sources, and could be captured and sequestered if linked with infrastructure for CCS (Smith 2016). Studies of the lifecycle climate change impacts of biochar systems generally show emissions reduction in the range $0.4-1.2$ tCO$_2$e t^{-1} (dry) feedstock (Cowie et al. 2015). Use of biomass for biochar can deliver greater benefits than use for bioenergy, if applied in a context where it delivers agronomic benefits and/or reduces non-CO$_2$ GHG emissions (Ji et al. 2018; Woolf et al. 2010, 2018; Xu et al. 2019). A global analysis of technical potential, in which biomass supply constraints were applied to protect against food insecurity, loss of habitat and land degradation, estimated technical potential abatement of $3.7-6.6$ GtCO$_2$e yr^{-1} (including $2.6-4.6$ GtCO$_2$e yr^{-1} carbon stabilisation), with theoretical potential to reduce total emissions over the course of a century by $240-475$ GtCO$_2$e (Woolf et al. 2010). Fuss et al. (2018) propose a range of $0.5-2$ GtCO$_2$e per year as the sustainable potential for negative emissions through biochar. Mitigation potential of biochar is reviewed in Chapter 2.

1.9.5.2   Role of biochar in management of land degradation

Biochars generally have high porosity, high surface area and surface-active properties that lead to high absorptive and adsorptive capacity, especially after interaction in soil (Joseph et al. 2010). As a result of these properties, biochar could contribute to avoiding, reducing and reversing land degradation through the following documented benefits:

- Improved nutrient use efficiency due to reduced leaching of nitrate and ammonium (e.g., Haider et al. 2017) and increased availability of phosphorus in soils with high phosphorus fixation capacity (Liu et al. 2018c), potentially reducing nitrogen and phosphorus fertiliser requirements.
- Management of heavy metals and organic pollutants: through reduced bioavailability of toxic elements (O'Connor et al. 2018; Peng et al. 2018), by reducing availability, through immobilisation due to increased pH and redox effects (Rizwan et al. 2016) and adsorption on biochar surfaces (Zhang et al. 2013) thus providing

a means of remediating contaminated soils, and enabling their utilisation for food production.

- Stimulation of beneficial soil organisms, including earthworms and mycorrhizal fungi (Thies et al. 2015).
- Improved porosity and water-holding capacity (Quin et al. 2014), particularly in sandy soils (Omondi et al. 2016), enhancing microbial function during drought (Paetsch et al. 2018).
- Amelioration of soil acidification, through application of biochars with high pH and acid-neutralising capacity (Chan et al. 2008; Van Zwieten et al. 2010).

Biochar systems can deliver a range of other co-benefits, including destruction of pathogens and weed propagules, avoidance of landfill, improved handling and transport of wastes such as sewage sludge, management of biomass residues such as environmental weeds and urban greenwaste, reduction of odours and management of nutrients from intensive livestock facilities, reduction in environmental nitrogen pollution and protection of waterways. As a compost additive, biochar has been found to reduce leaching and volatilisation of nutrients, increasing nutrient retention through absorption and adsorption processes (Joseph et al. 2018).

While many studies report positive responses, some studies have found negative or zero impacts on soil properties or plant response (e.g., Kuppusamy et al. 2016). The risk that biochar may enhance polycyclic aromatic hydrocarbon (PAH) in soil or sediments has been raised (Quilliam et al. 2013; Ojeda et al. 2016), but bioavailability of PAH in biochar has been shown to be very low (Hilber et al. 2017) Pyrolysis of biomass leads to losses of volatile nutrients, especially nitrogen. While availability of nitrogen and phosphorus in biochar is lower than in fresh biomass, (Xu et al. 2016) the impact of biochar on plant uptake is determined by the interactions between biochar, soil minerals and activity of microorganisms (e.g., Vanek and Lehmann 2015; Nguyen et al. 2017). To avoid negative responses, it is important to select biochar formulations to address known soil constraints, and to apply biochar prior to planting (Nguyen et al. 2017). Nutrient enrichment improves the performance of biochar from low nutrient feedstocks (Joseph et al. 2013). While there are many reports of biochar reducing disease or pest incidence, there are also reports of nil or negative effects (Bonanomi et al. 2015). Biochar may induce systemic disease resistance (e.g., Elad et al. 2011), though Viger et al. (2015) reported down-regulation of plant defence genes, suggesting increased susceptibility to insect and pathogen attack. Disease suppression where biochar is applied is associated with increased microbial diversity and metabolic potential of the rhizosphere microbiome (Kolton et al. 2017). Differences in properties related to feedstock (Bonanomi et al. 2018) and differential response to biochar dose, with lower rates more effective (Frenkel et al. 2017), contribute to variable disease responses.

The constraints on biochar adoption include: the high cost and limited availability due to limited large-scale production; limited amount of unutilised biomass; and competition for land for growing biomass. While early biochar research tended to use high rates of application (10 t ha^{-1} or more) subsequent studies have shown that biochar can be effective at lower rates, especially when combined

4

with chemical or organic fertilisers (Joseph et al. 2013). Biochar can be produced at many scales and levels of engineering sophistication, from simple cone kilns and cookstoves to large industrial-scale units processing several tonnes of biomass per hour (Lehmann and Joseph 2015). Substantial technological development has occurred recently, though large-scale deployment is limited to date.

Governance of biochar is required to manage climate, human health and contamination risks associated with biochar production in poorly designed or operated facilities that release methane or particulates (Downie et al. 2012; Buss et al. 2015), to ensure quality control of biochar products, and to ensure that biomass is sourced sustainably and is uncontaminated. Measures could include labelling standards, sustainability certification schemes and regulation of biochar production and use. Governance mechanisms should be tailored to context, commensurate with risks of adverse outcomes.

In summary, application of biochar to soil can improve soil chemical, physical and biological attributes, enhancing productivity and resilience to climate change, while also delivering climate-change mitigation through carbon sequestration and reduction in GHG emissions (*medium agreement, robust evidence*). However, responses to biochar depend on the biochar's properties, which are in turn dependent on feedstock and biochar production conditions, and the soil and crop to which it is applied. Negative or nil results have been recorded. Agronomic and methane-reduction benefits appear greatest in tropical regions, where acidic soils predominate and suboptimal rates of lime and fertiliser are common, while carbon stabilisation is greater in temperate regions. Biochar is most effective when applied in low volumes to the most responsive soils and when properties are matched to the specific soil constraints and plant needs. Biochar is thus a practice that has potential to address land degradation and climate change simultaneously, while also supporting sustainable development. The potential of biochar is limited by the availability of biomass for its production. Biochar production and use requires regulation and standardisation to manage risks (*strong agreement*).

1.9.6 Management of land degradation induced by tropical cyclones

Tropical cyclones are normal disturbances that natural ecosystems have been affected by and recovered from for millennia. Climate models mostly predict decreasing frequency of tropical cyclones, but dramatically increasing intensity of the strongest storms, as well as increasing rainfall rates (Bacmeister et al. 2018; Walsh et al. 2016b). Large amplitude fluctuations in the frequency and intensity complicate both the detection and attribution of tropical cyclones to climate change (Lin and Emanuel 2016b). Yet, the force of high-intensity cyclones has increased and is expected to escalate further due to global climate change (*medium agreement, robust evidence*) (Knutson et al. 2010; Bender et al. 2010; Vecchi et al. 2008; Bhatia et al. 2018; Tu et al. 2018; Sobel et al. 2016). Tropical cyclone paths are also shifting towards the poles, increasing the area subject to tropical cyclones (Sharmila and Walsh 2018; Lin and Emanuel 2016b). Climate change alone will affect the hydrology of individual wetland ecosystems, mostly through changes in precipitation and temperature

regimes with great global variability (Erwin 2009). Over the last seven decades, the speed at which tropical cyclones move has decreased significantly, as expected from theory, exacerbating the damage on local communities from increasing rainfall amounts and high wind speed (Kossin 2018). Tropical cyclones will accelerate changes in coastal forest structure and composition. The heterogeneity of land degradation at coasts that are affected by tropical cyclones can be further enhanced by the interaction of its components (for example, rainfall, wind speed, and direction) with topographic and biological factors (for example, species susceptibility) (Luke et al. 2016).

Small Island Developing States (SIDS) are particularly affected by land degradation induced by tropical cyclones; recent examples are Matthew (2016) in the Caribbean, and Pam (2015) and Winston (2016) in the Pacific (Klöck and Nunn 2019; Handmer and Nalau 2019). Even if the Pacific Ocean has experienced cyclones of unprecedented intensity in recent years, their geomorphological effects may not be unprecedented (Terry and Lau 2018).

Cyclone impacts on coastal areas is not restricted to SIDS, but a problem for all low-lying coastal areas (Petzold and Magnan 2019). The Sundarbans, one of the world's largest coastal wetlands, covers about one million hectares between Bangladesh and India. Large areas of the Sundarbans mangroves have been converted into paddy fields over the past two centuries and, more recently, into shrimp farms (Ghosh et al. 2015). In 2009, cyclone Aila caused incremental stresses on the socio-economic conditions of the Sundarbans coastal communities through rendering huge areas of land unproductive for a long time (Abdullah et al. 2016). The impact of Aila was widespread throughout the Sundarbans mangroves, showing changes between the pre- and post-cyclonic period of 20–50% in the enhanced vegetation index (Dutta et al. 2015), although the magnitude of the effects of the Sundarbans mangroves derived from climate change is not yet defined (Payo et al. 2016; Loucks et al. 2010; Gopal and Chauhan 2006; Ghosh et al. 2015; Chaudhuri et al. 2015). There is *high agreement* that the joint effect of climate change and land degradation will be very negative for the area, strongly affecting the environmental services provided by these forests, including the extinction of large mammal species (Loucks et al. 2010). The changes in vegetation are mainly due to inundation and erosion (Payo et al. 2016).

Tropical cyclone Nargis unexpectedly hit the Ayeyarwady River delta (Myanmar) in 2008 with unprecedented and catastrophic damages to livelihoods, destruction of forests and erosion of fields (Fritz et al. 2009) as well as eroding the shoreline 148 m compared with the long-term average (1974–2015) of 0.62 m yr^{-1}. This is an example of the disastrous effects that changing cyclone paths can have on areas previously not affected by cyclones (Fritz et al. 2010).

1.9.6.1 Management of coastal wetlands

Tropical cyclones mainly, but not exclusively, affect coastal regions, threatening maintenance of the associated ecosystems, mangroves, wetlands, seagrasses, and so on. These areas not only provide food, water and shelter for fish, birds and other wildlife, but also provide important ecosystem services such as water-quality improvement, flood abatement and carbon sequestration (Meng et al. 2017).

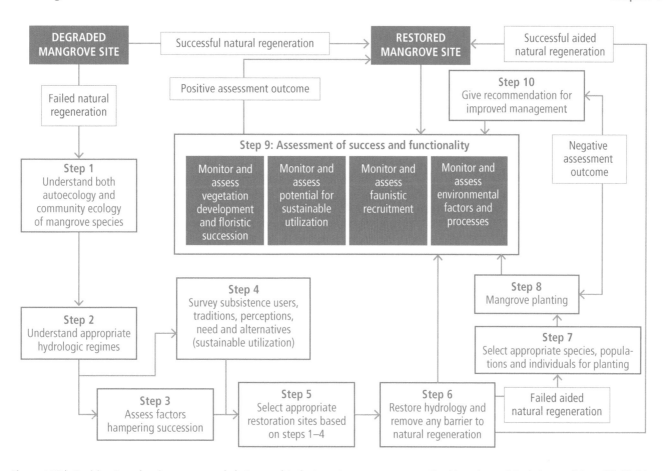

Figure 4.10 | Decision tree showing recommended steps and tasks to restore a mangrove wetland based on original site conditions. (Modified from Bosire et al. 2008.)

Despite their importance, coastal wetlands are listed amongst the most heavily damaged of natural ecosystems worldwide. Starting in the 1990s, wetland restoration and re-creation became a 'hotspot' in the ecological research fields (Zedler 2000). Coastal wetland restoration and preservation is an extremely cost-effective strategy for society, for example, the preservation of coastal wetlands in the USA provides storm protection services, with a cost of 23.2 billion USD yr^{-1} (Costanza et al. 2008).

There is a *high agreement* with *medium evidence* that the success of wetland restoration depends mainly on the flow of the water through the system, the degree to which re-flooding occurs, disturbance regimes, and the control of invasive species (Burlakova et al. 2009; López-Rosas et al. 2013). The implementation of the Ecological Mangrove Rehabilitation protocol (López-Portillo et al. 2017) that includes monitoring and reporting tasks, has been proven to deliver successful rehabilitation of wetland ecosystem services.

1.9.7   Saltwater intrusion

Current environmental changes, including climate change, have caused sea levels to rise worldwide, particularly in tropical and subtropical regions (Fasullo and Nerem 2018). Combined with scarcity of water in river channels, such rises have been instrumental in the intrusion of highly saline seawater inland, posing a threat

to coastal areas and an emerging challenge to land managers and policymakers. Assessing the extent of salinisation due to sea water intrusion at a global scale nevertheless remains challenging. Wicke et al. (2011) suggest that across the world, approximately 1.1 Gha of land is affected by salt, with 14% of this categorised as forest, wetland or some other form of protected area. Seawater intrusion is generally caused by (i) increased tidal activity, storm surges, cyclones and sea storms due to changing climate, (ii) heavy groundwater extraction or land-use changes as a result of changes in precipitation, and droughts/floods, (iii) coastal erosion as a result of destruction of mangrove forests and wetlands, (iv) construction of vast irrigation canals and drainage networks leading to low river discharge in the deltaic region; and (v) sea level rise contaminating nearby freshwater aquifers as a result of subsurface intrusion (Uddameri et al. 2014).

The Indus Delta, located in the south-eastern coast of Pakistan near Karachi in the North Arabian Sea, is one of the six largest estuaries in the world, spanning an area of 600,000 ha. The Indus delta is a clear example of seawater intrusion and land degradation due to local as well as up-country climatic and environmental conditions (Rasul et al. 2012). Salinisation and waterlogging in the up-country areas including provinces of Punjab and Sindh is, however, caused by the irrigation network and over-irrigation (Qureshi 2011).

Such degradation takes the form of high soil salinity, inundation and waterlogging, erosion and freshwater contamination. The interannual variability of precipitation with flooding conditions in some years and drought conditions in others has caused variable river flows and sediment runoff below Kotri Barrage (about 200 km upstream of the Indus delta). This has affected hydrological processes in the lower reaches of the river and the delta, contributing to the degradation (Rasul et al. 2012).

Over 480,000 ha of fertile land is now affected by sea water intrusion, wherein eight coastal subdivisions of the districts of Badin and Thatta are mostly affected (Chandio et al. 2011). A very high intrusion rate of 0.179 ± 0.0315 km yr^{-1}, based on the analysis of satellite data, was observed in the Indus delta during the 10 years between 2004 and 2015 (Kalhoro et al. 2016). The area of agricultural crops under cultivation has been declining, with economic losses of millions of USD (IUCN 2003). Crop yields have reduced due to soil salinity, in some places failing entirely. Soil salinity varies seasonally, depending largely on the river discharge: during the wet season (August 2014), salinity (0.18 mg L^{-1}) reached 24 km upstream, while during the dry season (May 2013), it reached 84 km upstream (Kalhoro et al. 2016). The freshwater aquifers have also been contaminated with sea water, rendering them unfit for drinking or irrigation purposes. Lack of clean drinking water and sanitation causes widespread diseases, of which diarrhoea is most common (IUCN 2003).

Lake Urmia in northwest Iran, the second-largest saltwater lake in the world and the habitat for endemic Iranian brine shrimp, *Artemia urmiana*, has also been affected by salty water intrusion. During a 17-year period between 1998 and 2014, human disruption, including agriculture and years of dam building affected the natural flow of freshwater as well as salty sea water in the surrounding area of Lake Urmia. Water quality has also been adversely affected, with salinity fluctuating over time, but in recent years reaching a maximum of 340 g L^{-1} (similar to levels in the Dead Sea). This has rendered the underground water unfit for drinking and agricultural purposes and risky to human health and livelihoods. Adverse impacts of global climate change as well as direct human impacts have caused changes in land use, overuse of underground water resources and construction of dams over rivers, which resulted in the drying-up of the lake in large part. This condition created sand, dust and salt storms in the region which affected many sectors including agriculture, water resources, rangelands, forests and health, and generally presented desertification conditions around the lake (Karbassi et al. 2010; Marjani and Jamali 2014; Shadkam et al. 2016).

Rapid irrigation expansion in the basin has, however, indirectly contributed to inflow reduction. Annual inflow to Lake Urmia has dropped by 48% in recent years. About three-fifths of this change was caused by climate change and two-fifths by water resource development and agriculture (Karbassi et al. 2010; Marjani and Jamali 2014; Shadkam et al. 2016).

In the drylands of Mexico, intensive production of irrigated wheat and cotton using groundwater (Halvorson et al. 2003) resulted in sea water intrusion into the aquifers of La Costa de Hermosillo, a coastal agricultural valley at the centre of Sonora Desert in Northwestern Mexico. Production of these crops in 1954 was on 64,000 ha of cultivated area, increasing to 132,516 ha in 1970,

but decreasing to 66,044 ha in 2009 as a result of saline intrusion from the Gulf of California (Romo-Leon et al. 2014). In 2003, only 15% of the cultivated area was under production, with around 80,000 ha abandoned due to soil salinisation whereas in 2009, around 40,000 ha was abandoned (Halvorson et al. 2003; Romo-Leon et al. 2014). Salinisation of agricultural soils could be exacerbated by climate change, as Northwestern Mexico is projected to be warmer and drier under climate change scenarios (IPCC 2013a).

In other countries, intrusion of seawater is exacerbated by destruction of mangrove forests. Mangroves are important coastal ecosystems that provide spawning bed for fish, timber for building, and livelihoods to dependent communities. They also act as barriers against coastal erosion, storm surges, tropical cyclones and tsunamis (Kalhoro et al. 2017) and are among the most carbon-rich stocks on Earth (Atwood et al. 2017). They nevertheless face a variety of threats: climatic (storm surges, tidal activities, high temperatures) and human (coastal developments, pollution, deforestation, conversion to aquaculture, rice culture, oil palm plantation), leading to declines in their areas. In Pakistan, using remote sensing, the mangrove forest cover in the Indus delta decreased from 260,000 ha in 1980s to 160,000 ha in 1990 (Chandio et al. 2011). Based on remotely sensed data, a sharp decline in the mangrove area was also found in the arid coastal region of Hormozgan province in southern Iran during 1972, 1987 and 1997 (Etemadi et al. 2016). Myanmar has the highest rate (about 1% yr^{-1}) of mangrove deforestation in the world (Atwood et al. 2017). Regarding global loss of carbon stored in the mangrove due to deforestation, four countries exhibited high levels of loss: Indonesia (3410 GgCO$_2$ yr^{-1}), Malaysia (1288 GgCO$_2$ yr^{-1}), US (206 GgCO$_2$ yr^{-1}) and Brazil (186 GgCO$_2$ yr^{-1}). Only in Bangladesh and Guinea Bissau was there no decline in the mangrove area from 2000 to 2012 (Atwood et al. 2017).

Frequency and intensity of average tropical cyclones will continue to increase (Knutson et al. 2015) and global sea level will continue to rise. The IPCC (2013) projected with *medium confidence* that the sea level in the Asia Pacific region will rise from 0.4 to 0.6 m, depending on the emission pathway, by the end of this century. Adaptation measures are urgently required to protect the world's coastal areas from further degradation due to saline intrusion. A viable policy framework is needed to ensure that the environmental flows to deltas in order to repulse the intruding seawater.

1.9.8   Avoiding coastal maladaptation

Coastal degradation – for example, beach erosion, coastal squeeze, and coastal biodiversity loss – as a result of rising sea levels is a major concern for low lying coasts and small islands (*high confidence*). The contribution of climate change to increased coastal degradation has been well documented in AR5 (Nurse et al. 2014; Wong et al. 2014) and is further discussed in Section 4.4.1.3 as well as in the IPCC Special Report on the Ocean and Cryosphere in a Changing Climate (SROCC). However, coastal degradation can also be indirectly induced by climate change as the result of adaptation measures that involve changes to the coastal environment, for example, coastal protection measures against increased flooding and erosion due to sea level rise, and storm surges transforming the natural coast to a 'stabilised' coastline (Cooper and Pile 2014; French 2001). Every kind of adaptation response option is context-dependent, and, in fact, sea walls play an important role for adaptation in many places. Nonetheless, there are observed cases where the construction of sea walls can be considered 'maladaptation' (Barnett and O'Neill 2010; Magnan et al. 2016) by leading to increased coastal degradation, such as in the case of small islands where, due to limitations of space, coastal retreat is less of an option than in continental coastal zones. There is emerging literature on the implementation of alternative coastal protection measures and mechanisms on small islands to avoid coastal degradation induced by sea walls (e.g., Mycoo and Chadwick 2012; Sovacool 2012).

In many cases, increased rates of coastal erosion due to the construction of sea walls are the result of the negligence of local coastal morphological dynamics and natural variability as well as the interplay of environmental and anthropogenic drivers of coastal change (*medium evidence, high agreement*). Sea walls in response to coastal erosion may be ill-suited for extreme wave heights under cyclone impacts and can lead to coastal degradation by keeping overflowing sea water from flowing back into the sea, and therefore affect the coastal vegetation through saltwater intrusion, as observed in Tuvalu (Government of Tuvalu 2006; Wairiu 2017). Similarly, in Kiribati, poor construction of sea walls has resulted in increased erosion and inundation of reclaimed land (Donner 2012; Donner and Webber 2014). In the Comoros and Tuvalu, sea walls have been constructed from climate change adaptation funds and 'often by international development organisations seeking to leave tangible evidence of their investments' (Marino and Lazrus 2015, p. 344). In these cases, they have even increased coastal erosion, due to poor planning and the negligence of other causes of coastal degradation, such as sand mining (Marino and Lazrus 2015; Betzold and Mohamed 2017; Ratter et al. 2016). On the Bahamas, the installation of sea walls as a response to coastal erosion in areas with high wave action has led to the contrary effect and has even increased sand loss in those areas (Sealey 2006). The reduction of natural buffer zones – such as beaches and dunes – due to vertical structures, such as sea walls, increased the impacts of tropical cyclones on Reunion Island (Duvat et al. 2016). Such a process of 'coastal squeeze' (Pontee 2013) also results in the reduction of intertidal habitat zones, such as wetlands and marshes (Zhu et al. 2010). Coastal degradation resulting from the construction of sea walls, however, is not only observed in SIDS, as described above, but also on islands in the Global North, for example, the North Atlantic (Muir et al. 2014; Young et al. 2014; Cooper and Pile 2014; Bush 2004).

The adverse effects of coastal protection measures may be avoided by the consideration of local social-ecological dynamics, including critical study of the diverse drivers of ongoing shoreline changes, and the appropriate implementation of locally adequate coastal protection options (French 2001; Duvat 2013). Critical elements for avoiding maladaptation include profound knowledge of local tidal regimes, availability of relative sea level rise scenarios and projections for extreme water levels. Moreover, the downdrift effects of sea walls need to be considered, since undefended coasts may be exposed to increased erosion (Zhu et al. 2010). In some cases, it may be possible to keep intact and restore natural buffer zones as an alternative to the construction of hard engineering solutions. Otherwise, changes in land use, building codes, or even coastal realignment can be an option in order to protect and avoid the loss of the buffer function of beaches (Duvat et al. 2016; Cooper and Pile 2014). Examples in Barbados show that combinations of hard and soft coastal protection approaches can be sustainable and reduce the risk of coastal ecosystem degradation while keeping the desired level of protection for coastal users (Mycoo and Chadwick 2012). Nature-based solutions and approaches such as 'building with nature' (Slobbe et al. 2013) may allow for more sustainable coastal protection mechanisms and avoid coastal degradation. Examples from the Maldives, several Pacific islands and the North Atlantic show the importance of the involvement of local communities in coastal adaptation projects, considering local skills, capacities, as well as demographic and socio-political dynamics, in order to ensure the proper monitoring and maintenance of coastal adaptation measures (Sovacool 2012; Muir et al. 2014; Young et al. 2014; Buggy and McNamara 2016; Petzold 2016).

1.10   Knowledge gaps and key uncertainties

The co-benefits of improved land management, such as mitigation of climate change, increased climate resilience of agriculture, and impacts on rural areas/societies are well known in theory, but there is a lack of a coherent and systematic global inventory of such integrated efforts. Both successes and failures are important to document systematically.

Efforts to reduce climate change through land-demanding mitigation actions aimed at removing atmospheric carbon, such as afforestation, reforestation, bioenergy crops, intensification of land management and plantation forestry can adversely affect land conditions and lead to degradation. However, they may also lead to avoidance, reduction and reversal of degradation. Regionally differentiated, socially and ecologically appropriate SLM strategies need to be identified, implemented, monitored and the results communicated widely to ensure climate effective outcomes.

Impacts of new technologies on land degradation and their social and economic ramifications need more research.

Improved quantification of the global extent, severity and rates of land degradation by combining remote sensing with a systematic use of ancillary data is a priority. The current attempts need better scientific underpinning and appropriate funding.

Land degradation is defined using multiple criteria but the definition does not provide thresholds or the magnitude of acceptable change. In practice, human interactions with land will result in a variety of changes; some may contribute positively to one criterion while adversely affecting another. Research is required on the magnitude of impacts and the resulting trade-offs. Given the urgent need to remove carbon from the atmosphere and to reduce climate change impacts, it is important to reach agreement on what level of reduction in one criterion (biological productivity, ecological integrity) may be acceptable for a given increase in another criterion (ecological integrity, biological productivity).

4

Attribution of land degradation to the underlying drivers is a challenge because it is a complex web of causality rather than a simple cause–effect relationship. Also, diverging views on land degradation in relation to other challenges is hampering such efforts.

A more systematic treatment of the views and experiences of land users would be useful in land degradation studies.

Much research has tried to understand how social and ecological systems are affected by a particular stressor, for example, drought, heat, or waterlogging. But less research has tried to understand how such systems are affected by several simultaneous stressors – which is more realistic in the context of climate change (Mittler 2006).

More realistic modelling of carbon dynamics, including better appreciation of below-ground biota, would help us to better quantify the role of soils and soil management for soil carbon sequestration.

Frequently Asked Questions

FAQ 4.1 | How do climate change and land degradation interact with land use?

Climate change, land degradation and land use are linked in a complex web of causality. One important impact of climate change on land degradation is that increasing global temperatures intensify the hydrological cycle, resulting in more intense rainfall, which is an important driver of soil erosion. This means that sustainable land management (SLM) becomes even more important with climate change. Land-use change in the form of clearing of forest for rangeland and cropland (e.g., for provision of bio-fuels), and cultivation of peat soils, is a major source of greenhouse gas (GHG) emission from both biomass and soils. Many SLM practices (e.g., agroforestry, perennial crops, organic amendments, etc.) increase carbon content of soil and vegetation cover and hence provide both local and immediate adaptation benefits, combined with global mitigation benefits in the long term, while providing many social and economic co-benefits. Avoiding, reducing and reversing land degradation has a large potential to mitigate climate change and help communities to adapt to climate change.

FAQ 4.2 | How does climate change affect land-related ecosystem services and biodiversity?

Climate change will affect land-related ecosystem services (e.g., pollination, resilience to extreme climate events, water yield, soil conservation, carbon storage, etc.) and biodiversity, both directly and indirectly. The direct impacts range from subtle reductions or enhancements of specific services, such as biological productivity, resulting from changes in temperature, temperature variability or rainfall, to complete disruption and elimination of services. Disruptions of ecosystem services can occur where climate change causes transitions from one biome to another, for example, forest to grassland as a result of changes in water balance or natural disturbance regimes. Climate change will result in range shifts and, in some cases, extinction of species. Climate change can also alter the mix of land-related ecosystem services, such as groundwater recharge, purification of water, and flood protection. While the net impacts are specific to time as well as ecosystem types and services, there is an asymmetry of risk such that overall impacts of climate change are expected to reduce ecosystem services. Indirect impacts of climate change on land-related ecosystem services include those that result from changes in human behaviour, including potential large-scale human migrations or the implementation of afforestation, reforestation or other changes in land management, which can have positive or negative outcomes on ecosystem services.

4

References

Abatzoglou, J.T. and A.P. Williams, 2016: Impact of anthropogenic climate change on wildfire across western US forests. *Proc. Natl. Acad. Sci. U.S.A.*, **113**, 11770–11775, doi:10.1073/pnas.1607171113.

Abbott, B.W. et al. 2016: Biomass offsets little or none of permafrost carbon release from soils, streams, and wildfire: an expert assessment. *Environ. Res. Lett.*, **11**, 034014, doi:10.1088/1748-9326/11/3/034014.

Abdi, A.M. et al. 2019: First assessment of the plant phenology index (PPI) for estimating gross primary productivity in African semi-arid ecosystems. *Int. J. Appl. Earth Obs.* Geoinf., **78**, 249–260, doi:10.1016/J.JAG.2019.01.018.

Abdulai, I. et al. 2018: Cocoa agroforestry is less resilient to sub-optimal and extreme climate than cocoa in full sun. *Glob. Chang. Biol.*, **24**, 273–286, doi:10.1111/gcb.13885.

Abdullah, A.N.M., K.K. Zander, B. Myers, N. Stacey, and S.T. Garnett, 2016: A short-term decrease in household income inequality in the Sundarbans, Bangladesh, following Cyclone Aila. *Nat. Hazards*, **83**, 1103–1123, doi:10.1007/s11069-016-2358-1.

Abraha, M., S.K. Hamilton, J. Chen, and G.P. Robertson, 2018: Ecosystem carbon exchange on conversion of Conservation Reserve Program grasslands to annual and perennial cropping systems. *Agric. For. Meteorol.*, **253–254**, 151–160, doi:10.1016/J.AGRFORMET.2018.02.016.

Achat, D.L., M. Fortin, G. Landmann, B. Ringeval, and L. Augusto, 2015: Forest soil carbon is threatened by intensive biomass harvesting. *Sci. Rep.*, **5**, 15991, doi:10.1038/srep15991.

Adefuye, B.O. et al. 2007: Practice and perception of biomass fuel use and its health effects among residents in a sub urban area of southern Nigeria: A qualitative study. *Niger. Hosp. Pract.*, **22**, 48–54.

Adeogun, A.G., B.A. Ibitoye, A.W. Salami, and G.T. Ihagh, 2018: Sustainable management of erosion prone areas of upper watershed of Kainji hydropower dam, Nigeria. *J. King Saud Univ. – Eng. Sci.*, doi:10.1016/J.JKSUES.2018.05.001.

Adger, N.W. et al. 2014: Human Security. In: *Climate Change 2014 Impacts, Adaptation, and Vulnerability*, [C.B. Field and V.R. Barros, (eds.)]. Cambridge University Press, Cambridge, UK and New York, USA, pp. 755–791.

Adger, W.N. et al. 2009: Are there social limits to adaptation to climate change? *Clim. Change*, **93**, 335–354, doi:10.1007/s10584-008-9520-z.

Adimassu, Z., S. Langan, and R. Johnston, 2016: Understanding determinants of farmers' investments in sustainable land management practices in Ethiopia: review and synthesis. *Environ. Dev. Sustain.*, **18**, 1005–1023, doi:10.1007/s10668-015-9683-5.

Agarwal, B., 1997: Environmental action, gender equity and women's participation. *Dev. Change*, **28**, 1–44, doi:10.1111/1467-7660.00033.

Agyarko-Mintah, E. et al. 2017: Biochar increases nitrogen retention and lowers greenhouse gas emissions when added to composting poultry litter. *Waste Manag.*, **61**, 138–149, doi:10.1016/j.wasman.2016.11.027.

Aleman, J.C., M.A. Jarzyna, and A.C. Staver, 2018: Forest extent and deforestation in tropical Africa since. *Nat. Ecol. Evol.*, **2**, 26–33, doi:10.1038/s41559-017-0406-1.

Alkemade, R. et al. 2009: GLOBIO3: A framework to investigate options for reducing global terrestrial biodiversity loss. *Ecosystems*, **12**, 374–390, doi:10.1007/s10021-009-9229-5.

Allen, C.D. et al. 2002: Ecological restoration of southwestern ponderosa pine ecosystems: A broad perspective. *Ecol. Appl.*, **12**, 1418–1433, doi:10.2307/3099981.

Allen, C.D. et al. 2010: A global overview of drought and heat-induced tree mortality reveals emerging climate change risks for forests. *For. Ecol. Manage.*, **259**, 660–684, doi:10.1016/J.FORECO.2009.09.001.

Allen, D.E., B.P. Singh, and R.C. Dalal, 2011: *Soil Health Indicators Under Climate Change: A Review of Current Knowledge*. Springer, Berlin, Heidelberg, Germany, pp. 25–45.

Allen, H.M., J.K. Pumpa, and G.D. Batten, 2001: Effect of frost on the quality of samples of Janz wheat. *Aust. J. Exp. Agric.*, **41**, 641, doi:10.1071/EA00187.

Allison, M. et al. 2016: Global risks and research priorities for coastal subsidence. *Eos Earth and Space Science News*, (Washington. DC)., **97**, doi:org/10.1029/2016EO055013.

Almagro, A., P.T.S. Oliveira, M.A. Nearing, and S. Hagemann, 2017: Projected climate change impacts in rainfall erosivity over Brazil. *Sci. Rep.*, **7**, 8130, doi:10.1038/s41598-017-08298-y.

Alongi, D.M., 2015: The impact of climate change on mangrove forests. *Curr. Clim. Chang. Reports*, **1**, 30–39, doi:10.1007/s40641-015-0002-x.

Altieri, M.A. and P. Koohafkan, 2008: *Enduring Farms: Climate Change, Smallholders and Traditional Farming Communities*. Penang, Malaysia, 72 pp.

Altieri, M.A. and C.I. Nicholls, 2017: The adaptation and mitigation potential of traditional agriculture in a changing climate. *Clim. Change*, **140**, 33–45, doi:10.1007/s10584-013-0909-y.

Altor, A.E. and W.J. Mitsch, 2006: Methane flux from created riparian marshes: Relationship to intermittent versus continuous inundation and emergent macrophytes. *Ecol. Eng.*, **28**, 224–234, doi:10.1016/j.ecoleng.2006.06.006.

Amanambu, A.C. et al. 2019: Spatio-temporal variation in rainfall-runoff erosivity due to climate change in the Lower Niger Basin, West Africa. *CATENA*, **172**, 324–334, doi:10.1016/J.CATENA.2018.09.003.

Amichev, B.Y., W.A. Kurz, C. Smyth, and K.C.J. Van Rees, 2012: The carbon implications of large-scale afforestation of agriculturally marginal land with short-rotation willow in Saskatchewan. *GCB Bioenergy*, **4**, doi:10.1111/j.1757-1707.2011.01110.x.

Amugune, I., P. Cerutti, H. Baral, S. Leonard, and C. Martius, 2017: Small flame but no fire: Wood fuel in the (Intended) Nationally Determined Contributions of countries in Sub-Saharan Africa. Amugune, I., Cerutti, P.O., Baral, H., Leonard, S., Martius, C., Working Paper 232, [Center for International Forestry Research] Bogor, Indonesia, 35 pp.

Andela, N. et al. 2013:. Global changes in dryland vegetation dynamics (1988–2008) assessed by satellite remote sensing: Comparing a new passive microwave vegetation density record with reflective greenness data. *Biogeosciences*, **10**, 6657–6676, doi.org/10.5194/bg-10-6657-2013, 2013.

Anderegg, W.R.L., J.A. Berry, and C.B. Field, 2012: Linking definitions, mechanisms, and modeling of drought-induced tree death. *Trends Plant Sci.*, **17**, 693–700, doi:10.1016/J.TPLANTS.2012.09.006.

Andersen, R. et al. 2017: An overview of the progress and challenges of peatland restoration in Western Europe. *Restor. Ecol.*, **25**, 271–282, doi:10.1111/rec.12415.

Anderson, R.G. et al. 2011: Biophysical considerations in forestry for climate protection. *Front. Ecol. Environ.*, **9**, 174–182, doi:10.1890/090179.

Anderson, R.L., D.R. Foster, and G. Motzkin, 2003: Integrating lateral expansion into models of peatland development in temperate New England. *J. Ecol.*, **91**, 68–76, doi:10.1046/j.1365-2745.2003.00740.x.

Andersson, E., S. Brogaard, and L. Olsson, 2011: The political ecology of land degradation. *Annu. Rev. Environ. Resour.*, **36**, 295–319, doi:10.1146/annurev-environ-033110-092827.

Angelsen, A. et al. 2014: Environmental income and rural livelihoods: A global-comparative analysis. *World Dev.*, **64**, S12–S28, doi:10.1016/J.WORLDDEV.2014.03.006.

Angelsen, A. et al. 2018: *Transforming REDD+: Lessons and new directions*. Center for International Forestry Research (CIFOR),, Bonn, Germany, 229 p.

Antwi-Agyei, P., A.J. Dougill, and L.C. Stringer, 2015: Impacts of land tenure arrangements on the adaptive capacity of marginalized groups: The case of Ghana's Ejura Sekyedumase and Bongo districts. *Land use policy*, **49**, 203–212, doi:10.1016/J.LANDUSEPOL.2015.08.007.

4

Aragüés, R. et al. 2015: Soil salinization as a threat to the sustainability of deficit irrigation under present and expected climate change scenarios. *Irrig. Sci.*, **33**, 67–79, doi:10.1007/s00271-014-0449-x.

Arnáez, J., N. Lana-Renault, T. Lasanta, P. Ruiz-Flaño, and J. Castroviejo, 2015: Effects of farming terraces on hydrological and geomorphological processes. A review. *CATENA*, **128**, 122–134, doi:10.1016/J.CATENA.2015.01.021.

Arnell, N.W., and S.N. Gosling, 2016: The impacts of climate change on river flood risk at the global scale. *Clim. Change*, **134**, 387–401, doi:10.1007/s10584-014-1084-5.

Asmelash, F., T. Bekele, and E. Birhane, 2016: The potential role of arbuscular mycorrhizal fungi in the restoration of degraded lands. *Front. Microbiol.*, **7**, 1095, doi:10.3389/fmicb.2016.01095.

Asner, G.P., A.J. Elmore, L.P. Olander, R.E. Martin, and A.T. Harris, 2004: Grazing systems, ecosystem responses, and global change. *Annu. Rev. Environ. Resour.*, **29**, 261–299, doi:10.1146/annurev.energy.29.062403.102142.

Astrup, R., P.Y. Bernier, H. Genet, D.A. Lutz, and R.M. Bright, 2018: A sensible climate solution for the boreal forest. *Nat. Clim. Chang.*, doi:10.1038/s41558-017-0043-3.

Atmadja, S., and L. Verchot, 2012: A review of the state of research, policies and strategies in addressing leakage from reducing emissions from deforestation and forest degradation (REDD+). *Mitig. Adapt. Strateg. Glob. Chang.*, **17**, 311–336, doi:10.1007/s11027-011-9328-4.

Atwood, T.B. et al. 2017: Global patterns in mangrove soil carbon stocks and losses. *Nat. Clim. Chang.*, **7**, 523–528, doi:10.1038/nclimate3326.

Van Auken, O.W., 2009: Causes and consequences of woody plant encroachment into western North American grasslands. *J. Environ. Manage.*, **90**, 2931–2942, doi:10.1016/j.jenvman.2009.04.023.

Baccini, A. et al. 2017: Tropical forests are a net carbon source based on aboveground measurements of gain and loss. *Science*, **358**, 230–234, doi:10.1126/science.aam5962.

Bacmeister, J.T. et al. 2018: Projected changes in tropical cyclone activity under future warming scenarios using a high-resolution climate model. *Clim. Change*, **146**, 547–560, doi:10.1007/s10584-016-1750-x.

Bae, J.S., R.W. Joo, Y.-S. Kim, and Y.S. Kim Yeon-Su., 2012: Forest transition in South Korea: Reality, path and drivers. *Land use policy*, **29**, 198–207, doi:10.1016/j.landusepol.2011.06.007.

Bahamondez, C., and I.D. Thompson, 2016: Determining forest degradation, ecosystem state and resilience using a standard stand stocking measurement diagram: Theory into practice. *Forestry*, **89**, 290–300, doi:10.1093/forestry/cpv052.

Bai, Y. et al. 2008a: Primary production and rain use efficiency across a precipitation gradient on the mongolia plateau. *Ecology*, **89**, 2140–2153, doi:10.1890/07-0992.1.

Bai, Z., D. Dent, L. Olsson, and M.E. Schaepman, 2008b: *Global Assessment of Land Degradation and Improvement. 1. Identification by Remote Sensing.* Wageningen, The Netherlands, 78 pp.

Bai, Z.G., D.L. Dent, L. Olsson, and M.E. Schaepman, 2008c: Proxy global assessment of land degradation. *Soil Use Manag.*, **24**, 223–234, doi:10.1111/j.1475-2743.2008.00169.x.

Bai, Z.G. et al. 2015: A longer, closer, look at land degradation. *Agric. Dev.*, **24**, 3–9.

Bailis, R., R. Drigo, A. Ghilardi, and O. Masera, 2015: The carbon footprint of traditional woodfuels. *Nat. Clim. Chang.*, **5**, 266–272, doi:10.1038/nclimate2491.

Baka, J., 2013: The political construction of wasteland: Governmentality, land acquisition and social inequality in South India. *Dev. Change*, **44**, 409–428, doi:10.1111/dech.12018.

Baka, J., 2014: What wastelands? A critique of biofuel policy discourse in South India. *Geoforum*, **54**, 315–323, doi:10.1016/J.GEOFORUM.2013.08.007.

Baker, B., 2017: Can modern agriculture be sustainable? *Bioscience*, **67**, 325–331, doi:10.1093/biosci/bix018.

Baker, J.M., T.E. Ochsner, R.T. Venterea, and T.J. Griffis, 2007: Tillage and soil carbon sequestration: What do we really know? *Agric. Ecosyst. Environ.*, **118**, 1–5, doi:10.1016/J.AGEE.2006.05.014.

Bakun, A., 1990: Global climate change and intensification of coastal ocean upwelling. *Science*, **247**, 198–201, doi:10.1126/science.247.4939.198.

Bakun, A. et al. 2015: Anticipated effects of climate change on coastal upwelling ecosystems. *Curr. Clim. Chang. Reports*, **1**, 85–93, doi:10.1007/s40641-015-0008-4.

Bala, G. et al. 2007: Combined climate and carbon-cycle effects of large-scale deforestation. *Proc. Natl. Acad. Sci.*, **104**, 6550–6555, doi:10.1073/pnas.0608998104.

Balbo, A.L., 2017: Terrace landscapes. Editorial to the special issue. *J. Environ. Manage.*, **202**, 495–499, doi:10.1016/J.JENVMAN.2017.02.001.

Balooni, K., and J.F. Lund, 2014: Forest rights: The hard currency of REDD+. *Conserv. Lett.*, **7**, 278–284, doi:10.1111/conl.12067.

Barbier, E.B., 2000: Valuing the environment as input: Review of applications to mangrove-fishery linkages. *Ecol. Econ.*, **35**, 47–61, doi:10.1016/S0921-8009(00)00167-1.

Barbier, E.B. and J.P. Hochard, 2016: Does land degradation increase poverty in developing countries? *PLoS One*, **11**, e0152973, doi:10.1371/journal.pone.0152973.

Barbier, E.B. and J.P. Hochard, 2018: Land degradation and poverty. *Nat. Sustain.*, **1**, 623–631, doi:10.1038/s41893-018-0155-4.

Bárcena, T.G. et al. 2014: Soil carbon stock change following afforestation in Northern Europe: A meta-analysis. *Glob. Chang. Biol.*, **20**, 2393–2405, doi:10.1111/gcb.12576.

Bardsley, D.K., and G.J. Hugo, 2010: Migration and climate change: examining thresholds of change to guide effective adaptation decision-making. *Popul. Environ.*, **32**, 238–262, doi:10.1007/s11111-010-0126-9.

Barlow, J. et al. 2007: Quantifying the biodiversity value of tropical primary, secondary, and plantation forests. *Proc. Natl. Acad. Sci. U.S.A.*, **104**, 18555–18560, doi:10.1073/pnas.0703333104.

Barnett, J., and S. O'Neill, 2010: Maladaptation. *Glob. Environ. Chang.*, **20**, 211–213, doi:10.1016/j.gloenvcha.2009.11.004.

Barnett, J. et al. 2015: From barriers to limits to climate change adaptation: Path dependency and the speed of change. *Ecol. Soc.*, **20**, art5, doi:10.5751/ES-07698-200305.

Barnhart, T.B. et al. 2016: Snowmelt rate dictates streamflow. *Geophys. Res. Lett.*, **43**, 8006–8016, doi:10.1002/2016GL069690.

Barrera-Bassols, N., and J.A. Zinck, 2003: Ethnopedology: a worldwide view on the soil knowledge of local people. *Geoderma*, **111**, 171–195, doi:10.1016/S0016-7061(02)00263-X.

Bärring, L., P. Jönsson, J.O. Mattsson, and R. Åhman, 2003: Wind erosion on arable land in Scania, Sweden and the relation to the wind climate: A review. *CATENA*, **52**, 173–190, doi:10.1016/S0341-8162(03)00013-4.

Basche, A., and M. DeLonge, 2017: The impact of continuous living cover on soil hydrologic properties: A meta-analysis. *Soil Sci. Soc. Am. J.*, **81**, 1179, doi:10.2136/sssaj2017.03.0077.

Batello, C. et al. 2014: *Perennial Crops for Food Security.* Food and Agriculture Organization of the United Nations (FAO), Rome, Italy, 390 p.

Batir, J.F., M.J. Hornbach, and D.D. Blackwell, 2017: Ten years of measurements and modeling of soil temperature changes and their effects on permafrost in Northwestern Alaska. *Glob. Planet. Change*, **148**, 55–71, doi:10.1016/J.GLOPLACHA.2016.11.009.

Bauer, N. et al. 2018: Global energy sector emission reductions and bioenergy use: Overview of the bioenergy demand phase of the EMF-33 model comparison. *Clim. Change*, 1–16, doi:10.1007/s10584-018-2226-y.

Baumber, A., E. Berry, and G. Metternicht, 2019: Synergies between land degradation neutrality goals and existing market-based instruments. *Environ. Sci. Policy*, **94**, 174–181, doi:10.1016/J.ENVSCI.2019.01.012.

Baveye, P.C., 2017: Quantification of ecosystem services: Beyond all the "guesstimates", how do we get real data? *Ecosyst. Serv.*, **24**, 47–49, doi:10.1016/J.ECOSER.2017.02.006.

Baveye, P.C. et al. 2011: From dust bowl to dust bowl: Soils are still very much a frontier of science. *Soil Sci. Soc. Am. J.*, **75**, 2037, doi:10.2136/sssaj2011.0145.

Beck, P.S.A. et al. 2011: Changes in forest productivity across Alaska consistent with biome shift. *Ecol. Lett.*, 14, 373–379, doi:10.1111/j.1461-0248.2011.01598.x.

Beerling, D.J., 2017: Enhanced rock weathering: Biological climate change mitigation with co-benefits for food security? *Biol. Lett.*, **13**, 20170149, doi:10.1098/rsbl.2017.0149.

Beets, W.C., 1990: *Raising and Sustaining Productivity of Smallholder Farming Systems in the Tropics: A handbook of sustainable agricultural development*. Agbe Publishing, Alkmaar, The Netherlands, 754 pp.

Behnke, R., 1994: Natural resource management in pastoral Africa. *Dev. Policy Rev.*, **12**, 5–28, doi:10.1111/j.1467-7679.1994.tb00053.x.

Behnke, R., and M. Mortimore, 2016: *Introduction: The End of Desertification?* Springer, Berlin, Heidelberg, pp. 1–34.

Belair, E.P., and M.J. Ducey, 2018: Patterns in forest harvesting in New England and New York: Using FIA data to evaluate silvicultural outcomes. *J. For.*, **116**, 273–282, doi:10.1093/jofore/fvx019.

Bellamy, P.H., P.J. Loveland, R.I. Bradley, R.M. Lark, and G.J.D. Kirk, 2005: Carbon losses from all soils across England and Wales 1978–2003. *Nature*, **437**, 245–248, doi:10.1038/nature04038.

Belnap, J., B.J. Walker, S.M. Munson, and R.A. Gill, 2014: Controls on sediment production in two U.S. deserts. *Aeolian Res.*, **14**, 15–24, doi:10.1016/J.AEOLIA.2014.03.007.

Benavidez, R., B. Jackson, D. Maxwell, and K. Norton, 2018: A review of the (Revised) Universal Soil Loss Equation (R/USLE): with a view to increasing its global applicability and improving soil loss estimates. *Hydrol. Earth Syst. Sci.* **22**, 6059–6086, doi:10.5194/hess-22-6059-2018.

Bender, M.A. et al. 2010: Modeled impact of anthropogenic warming on the frequency of intense Atlantic hurricanes. *Science*, **327**, 454 LP-458.

Béné, C., R.G. Wood, A. Newsham, and M. Davies, 2012: Resilience: New utopia or new tyranny? Reflection about the potentials and limits of the concept of resilience in relation to vulnerability reduction programmes. *IDS Work. Pap.*, **2012**, 1–61, doi:10.1111/j.2040-0209.2012.00405.x.

Benini, L., M. Antonellini, M. Laghi, and P.N. Mollema, 2016: Assessment of water resources availability and groundwater salinization in future climate and land use change scenarios: A case study from a coastal drainage basin in Italy. *Water Resour. Manag.*, **30**, 731–745, doi:10.1007/s11269-015-1187-4.

Benjaminsen, T.A., and C. Lund, 2003: *Securing Land Rights in Africa*. Frank Cass Publishers, London, UK, 175 pp.

Benjaminsen, T.A., K. Alinon, H. Buhaug, and J.T. Buseth, 2012: Does climate change drive land-use conflicts in the Sahel? *J. Peace Res.*, **49**, 97–111, doi:10.1177/0022343311427343.

Bennett, M.T., 2008: China's sloping land conversion program: Institutional innovation or business as usual? *Ecol. Econ.*, **65**(4), 699–711 doi:10.1016/j.ecolecon.2007.09.017.

Bentz, B.J. et al. 2010: Climate change and bark beetles of the western United States and Canada: Direct and indirect effects. *Bioscience*, **60**, 602–613, doi:10.1525/bio.2010.60.8.6.

Bernal, B., L.T. Murray, and T.R.H. Pearson, 2018: Global carbon dioxide removal rates from forest landscape restoration activities. *Carbon Balance Manag.*, **13**, doi:10.1186/s13021-018-0110-8.

Bernier, P.Y. et al. 2017: Moving beyond the concept of "primary forest" as a metric of forest environment quality. *Ecol. Appl.*, **27**, 349–354, doi:10.1002/eap.1477.

Bernoux, M., B. Volkoff, M. da C.S. Carvalho, and C.C. Cerri, 2003: CO_2 emissions from liming of agricultural soils in Brazil. *Global Biogeochem. Cycles*, **17**, n/a-n/a, doi:10.1029/2001GB001848.

Betts, R.A., 2000: Offset of the potential carbon sink from boreal forestation by decreases in surface albedo. *Nature*, **408**, 187–190, doi:10.1038/35041545.

Betts, R.A. et al. 2007: Projected increase in continental runoff due to plant responses to increasing carbon dioxide. *Nature*, **448**, 1037–1041, doi:10.1038/nature06045.

Betzold, C., and I. Mohamed, 2017: Seawalls as a response to coastal erosion and flooding: A case study from Grande Comore, Comoros (West Indian Ocean). *Reg. Environ. Chang.*, **17**, 1077–1087, doi:10.1007/s10113-016-1044-x.

Bharucha, Z., and J. Pretty, 2010: The roles and values of wild foods in agricultural systems. *Philos. Trans. R. Soc. B Biol. Sci.*, **365**, 2913–2926, doi:10.1098/rstb.2010.0123.

Bhatia, K. et al. 2018: Projected response of tropical cyclone intensity and intensification in a global climate model. *J. Clim.*, **31**, 8281–8303, doi:10.1175/JCLI-D-17-0898.1.

Biederman, L.A., and W. Stanley Harpole, 2013: Biochar and its effects on plant productivity and nutrient cycling: A meta-analysis. *GCB Bioenergy*, **5**(2), 202–214 doi:10.1111/gcbb.12037.

Bindoff, N.L., P.A. Stott, K.M. AchutaRao, M.R. Allen, N. Gillett, D. Gutzler, K. Hansingo, G. Hegerl, Y. Hu, S. Jain, I.I. Mokhov, J. Overland, J. Perlwitz, R. Sebbari and X. Zhang, 2013: Detection and Attribution of Climate Change: from Global to Regional. In: Climate Change 2013: The Physical Science Basis. Contribution of Working Group I to the Fifth Assessment Report of the Intergovernmental Panel on Climate Change, [T.F. Stocker et al. (eds.)]. Cambridge University Press, Cambrdige, UK and New York, USA, pp. 867–940.

Birdsey, R., K. Pregitzer, and A. Lucier, 2006: Forest carbon management in the United States. *J. Environ. Qual.*, **35**, 1461, doi:10.2134/jeq2005.0162.

Bisaro, A., M. Kirk, P. Zdruli, and W. Zimmermann, 2014: Global drivers setting desertification research priorities: Insights from a stakeholder consultation forum. *L. Degrad. Dev.*, **25**, 5–16, doi:10.1002/ldr.2220.

Bitzer, V., and J. Bijman, 2015: From innovation to co-innovation? An exploration of African agrifood chains. *Br. Food J.*, **117**, 2182–2199, doi:10.1108/BFJ-12-2014-0403.

Black, R., S.R.G. Bennett, S.M. Thomas, and J.R. Beddington, 2011: Climate change: Migration as adaptation. *Nat.* **2011** 4787370.

Blaikie, P.M., and H.C. Brookfield, 1987: Land degradation and society. [P.M. Blaikie and H.C. Brookfield, (eds.)]. Methuen, Milton Park, Abingdon, UK, 222 pp.

Blanco-Canqui, H., and R. Lal, 2009: Crop residue removal impacts on soil productivity and environmental quality. *CRC. Crit. Rev. Plant Sci.*, **28**, 139–163, doi:10.1080/07352680902776507.

Blaser, W.J., J. Oppong, E. Yeboah, and J. Six, 2017: Shade trees have limited benefits for soil fertility in cocoa agroforests. *Agric. Ecosyst. Environ.*, **243**, 83–91, doi:10.1016/J.AGEE.2017.04.007.

Blenkinsop, S. et al. 2018: The INTENSE project: using observations and models to understand the past, present and future of sub-daily rainfall extremes. *Adv. Sci. Res.*, **15**, 117–126, doi:10.5194/asr-15-117-2018.

Boardman, J., 2010: A short history of muddy floods. *L. Degrad. Dev.*, **21**, 303–309, doi:10.1002/ldr.1007.

Bocquet-Appel, J.-P., 2011: When the world's population took off: The springboard of the Neolithic Demographic Transition. *Science*, **333**, 560–561, doi:10.1126/science.1208880.

Boit, A. et al. 2016: Large-scale impact of climate change vs. land-use change on future biome shifts in Latin America. *Glob. Chang. Biol.*, **22**, 3689–3701, doi:10.1111/gcb.13355.

Bonan, G.B. et al. 2008: Forests and climate change: Forcings, feedbacks, and the climate benefits of forests. *Science*, **320**, 1444–1449, doi:10.1126/science.1155121.

Bonanomi, G. et al. 2018: Biochar chemistry defined by13C-CPMAS NMR explains opposite effects on soilborne microbes and crop plants. *Appl. Soil Ecol.*, **124**, 351–361, doi:10.1016/j.apsoil.2017.11.027.

Bond-Lamberty, B., V.L. Bailey, M. Chen, C.M. Gough, and R. Vargas, 2018: Globally rising soil heterotrophic respiration over recent decades. *Nature*, **560**, 80–83, doi:10.1038/s41586-018-0358-x.

4

Bond, T.C. et al. 2013: Bounding the role of black carbon in the climate system: A scientific assessment. *J. Geophys. Res. Atmos.*, **118**, 5380–5552, doi:10.1002/jgrd.50171.

Bonn, A., Allott, T., Evans, M., Joosten, H., Stoneman, R., 2016: Peatland restoration and ecosystem services: nature-based solutions for societal goals. In: *Peatland restoration and ecosystem services: Science, policy and practice*. [Bonn A., Allott T., Evans M., Joosten H., Stoneman R., (eds.)]. Cambridge University Press. pp. 402–417.

Borchard, N. et al. 2019: Biochar, soil and land-use interactions that reduce nitrate leaching and N2O emissions: A meta-analysis. *Sci. Total Environ.*, **651**, 2354–2364, doi:10.1016/J.SCITOTENV.2018.10.060.

Borras, S.M., and J.C. Franco, 2018: The challenge of locating land-based climate change mitigation and adaptation politics within a social justice perspective: Towards an idea of agrarian climate justice. *Third World Q.*, **39**, 1308–1325, doi:10.1080/01436597.2018.1460592.

Borrelli, P. et al. 2016: Effect of good agricultural and environmental conditions on erosion and soil organic carbon balance: A national case study. *Land use policy*, **50**, 408–421, doi:10.1016/J.LANDUSEPOL.2015.09.033.

Borrelli, P. et al. 2017: An assessment of the global impact of 21st century land use change on soil erosion. *Nat. Commun.*, **8**, 2013, doi:10.1038/s41467-017-02142-7.

Borrelli, P. et al. 2018: A step towards a holistic assessment of soil degradation in Europe: Coupling on-site erosion with sediment transfer and carbon fluxes. *Environ. Res.*, **161**, 291–298, doi:10.1016/J.ENVRES.2017.11.009.

Bosire, J.O. et al. 2008: Functionality of restored mangroves: A review. *Aquat. Bot.*, **89**, 251–259, doi:10.1016/J.AQUABOT.2008.03.010.

Bot, A., F. Nachtergaele, and A. Young, 2000: *Land Resource Potential and Constraints at Regional and Country Levels*. Land and Water Development Division, Food and Agriculture Organization of the United Nations, Rome, Italy, 114 pp.

Bousbih, S. et al. 2017: Potential of Sentinel-1 radar data for the assessment of soil and cereal cover parameters. *Sensors*, **17**, 2617, doi:10.3390/s17112617.

Bowles, T.M. et al. 2018: Addressing agricultural nitrogen losses in a changing climate. *Nat. Sustain.*, **1**, 399–408, doi:10.1038/s41893-018-0106-0.

Bozzi, E., L. Genesio, P. Toscano, M. Pieri, and F. Miglietta, 2015: Mimicking biochar-albedo feedback in complex Mediterranean agricultural landscapes. *Environ. Res. Lett.*, **10**, 084014, doi:10.1088/1748-9326/10/8/084014.

Bradshaw, C.J.A., N.S. Sodhi, K.S.-H. Peh, and B.W. Brook, 2007: Global evidence that deforestation amplifies flood risk and severity in the developing world. *Glob. Chang. Biol.*, **13**, 2379–2395, doi:10.1111/j.1365-2486.2007.01446.x.

Brahney, J., F. Weber, V. Foord, J. Janmaat, and P.J. Curtis, 2017: Evidence for a climate-driven hydrologic regime shift in the Canadian Columbia Basin. *Can. Water Resour. J. / Rev. Can. des ressources hydriques*, **42**, 179–192, doi:10.1080/07011784.2016.1268933.

Branca, G., L. Lipper, N. McCarthy, and M.C. Jolejole, 2013: Food security, climate change, and sustainable land management. A review. *Agron. Sustain. Dev.*, **33**, 635–650, doi:10.1007/s13593-013-0133-1.

Brandt, M. et al. 2017: Human population growth offsets climate-driven increase in woody vegetation in sub-Saharan Africa. *Nat. Ecol. Evol.*, **1**, 0081, doi:10.1038/s41559-017-0081.

Brandt, M. et al. 2018a: Reduction of tree cover in West African woodlands and promotion in semi-arid farmlands. *Nat. Geosci.*, **11**, 328–333, doi:10.1038/s41561-018-0092-x.

Brandt, M. et al. 2018b: Satellite passive microwaves reveal recent climate-induced carbon losses in African drylands. *Nat. Ecol. Evol.*, **2**, 827–835, doi:10.1038/s41559-018-0530-6.

Bright, R.M., K. Zhao, R.B. Jackson, and F. Cherubini, 2015: Quantifying surface albedo and other direct biogeophysical climate forcings of forestry activities. *Glob. Chang. Biol.*, **21**, 3246–3266, doi:10.1111/gcb.12951.

Brinkman, E.P., R. Postma, W.H. van der Putten, and A.J. Termorshuizen, 2017: *Influence of Growing Eucalyptus Trees for Biomass on Soil Quality*. 1–39.

Brisson, N. et al. 2010: Why are wheat yields stagnating in Europe? A comprehensive data analysis for France. *F. Crop. Res.*, **119**, 201–212, doi:10.1016/J.FCR.2010.07.012.

Brockington, J.D., I.M. Harris, and R.M. Brook, 2016: Beyond the project cycle: A medium-term evaluation of agroforestry adoption and diffusion in a south Indian village. *Agrofor. Syst.*, **90**, 489–508, doi:10.1007/s10457-015-9872-0.

Brooks, M.L. et al. 2004: Effects of invasive alien plants on fire regimes. *Bioscience*, **54**, 677–688, doi:10.1641/0006-3568(2004)054[0677:eoiapo]2.0.co;2.

Brown, K., 2014: Global environmental change I: A social turn for resilience? *Prog. Hum. Geogr.*, **38**, 107–117, doi:10.1177/0309132513498837.

Brown, S., and R.J. Nicholls, 2015: Subsidence and human influences in mega deltas: The case of the Ganges–Brahmaputra–Meghna. *Sci. Total Environ.*, **527–528**, 362–374, doi:10.1016/J.SCITOTENV.2015.04.124.

Brunner, K.-M., S. Mandl, and H. Thomson, 2018: Energy Poverty: Energy equity in a world of high demand and low supply. In: *The Oxford Handbook of Energy and Society*, [D.J. Davidson and M. Gross, (eds.)]. Oxford University Press, New York, United States, pp. 297–316.

Buendia, C., R.J. Batalla, S. Sabater, A. Palau, and R. Marcé, 2016: Runoff trends driven by climate and afforestation in a Pyrenean Basin. *L. Degrad. Dev.*, **27**, 823–838, doi:10.1002/ldr.2384.

Buggy, L. and K.E. McNamara, 2016: The need to reinterpret "community" for climate change adaptation: A case study of Pele Island, Vanuatu. *Clim. Dev.*, **8**, 270–280, doi:10.1080/17565529.2015.1041445.

Bunn, C., P. Läderach, O. Ovalle Rivera, and D. Kirschke, 2015: A bitter cup: Climate change profile of global production of Arabica and Robusta coffee. *Clim. Change*, **129**, 89–101, doi:10.1007/s10584-014-1306-x.

Bürgi, M. et al. 2017: Integrated Landscape Approach: Closing the gap between theory and application. *Sustainability*, **9**, 1371, doi:10.3390/su9081371.

Burkett, V., and J. Kusler, 2000: Climate change: Potential impacts and interactions in wetlands in the United States. JAWRA *J. Am. Water Resour. Assoc.*, **36**, 313–320, doi:10.1111/j.1752-1688.2000.tb04270.x.

Burlakova, L.E., A.Y. Karatayev, D.K. Padilla, L.D. Cartwright, and D.N. Hollas, 2009: Wetland restoration and invasive species: Apple snail (Pomacea insularum) feeding on native and invasive aquatic plants. *Restor. Ecol.*, **17**, 433–440, doi:10.1111/j.1526-100X.2008.00429.x.

Burt, T., J. Boardman, I. Foster, and N. Howden, 2016a: More rain, less soil: long-term changes in rainfall intensity with climate change. *Earth Surf. Process. Landforms*, **41**, 563–566, doi:10.1002/esp.3868.

Busch, J. et al. 2015: Reductions in emissions from deforestation from Indonesia's moratorium on new oil palm, timber, and logging concessions. *Proc. Natl. Acad. Sci. USA*, **112**, 1328–1333, doi:10.7910/DVN/28615.

Bush, D.M., 2004: *Living with Florida's Atlantic beaches: coastal hazards from Amelia Island to Key West*. Duke University Press, 338 pp.

Buss, W., O. Mašek, M. Graham, and D. Wüst, 2015: Inherent organic compounds in biochar: Their content, composition and potential toxic effects. *J. Environ. Manage.*, **156**, 150–157, doi:10.1016/j.jenvman.2015.03.035.

Butler, R.A., L.P. Koh, and J. Ghazoul, 2009: REDD in the red: Palm oil could undermine carbon payment schemes. *Conserv. Lett.*, **2**, 67–73, doi:10.1111/j.1755-263X.2009.00047.x.

Butzer, K.W., 2005: Environmental history in the Mediterranean world: Cross-disciplinary investigation of cause-and-effect for degradation and soil erosion. *J. Archaeol. Sci.*, **32**, 1773–1800, doi:10.1016/J.JAS.2005.06.001.

Byers, M., and N. Dragojlovic, 2004: Darfur: A climate change-induced humanitarian crisis? *Hum. Secur. Bull.*, October 2004, 16–18.

Cacho, J.F., M.C. Negri, C.R. Zumpf, and P. Campbell, 2018: Introducing perennial biomass crops into agricultural landscapes to address water quality challenges and provide other environmental services. *Wiley Interdiscip. Rev. Energy Environ.*, **7**, e275, doi:10.1002/wene.275.

Cai, T., D.T. Price, A.L. Orchansky, and B.R. Thomas, 2011a: Carbon, water, and energy exchanges of a hybrid poplar plantation during the first five years following planting. *Ecosystems*, **14**, 658–671, doi:10.1007/s10021-011-9436-8.

Cai, X., X. Zhang, and D. Wang, 2011b: Land availability for biofuel production. *Environ. Sci. Technol.*, **45**, 334–339, doi:10.1021/es103338e.

Caldwell, P.V. et al. 2018: Woody bioenergy crop selection can have large effects on water yield: A southeastern United States case study. *Biomass and Bioenergy*, **117**, 180–189, doi:10.1016/J.BIOMBIOE.2018.07.021.

Camacho, L.D., D.T. Gevaña, Antonio P. Carandang, and S.C. Camacho, 2016: Indigenous knowledge and practices for the sustainable management of Ifugao forests in Cordillera, Philippines. *Int. J. Biodivers. Sci. Ecosyst. Serv. Manag.*, **12**, 5–13, doi:10.1080/21513732.2015.1124453.

Campioli, M. et al. 2015: Biomass production efficiency controlled by management in temperate and boreal ecosystems. *Nat. Geosci.*, **8**, 843–846, doi:10.1038/ngeo2553.

Capolongo, D., N. Diodato, C.M. Mannaerts, M. Piccarreta, and R.O. Strobl, 2008: Analyzing temporal changes in climate erosivity using a simplified rainfall erosivity model in Basilicata (southern Italy). *J. Hydrol.*, **356**, 119–130, doi:10.1016/J.JHYDROL.2008.04.002.

Carlson, K.M. et al. 2017: Effect of oil palm sustainability certification on deforestation and fire in Indonesia. *Proc. Natl. Acad. Sci.*, **115**, 201704728, doi:10.1073/pnas.1704728114.

Carton, W., and E. Andersson, 2017: Where forest carbon meets its maker: Forestry-based offsetting as the subsumption of nature. *Soc. Nat. Resour.*, **30**, 829–843, doi:10.1080/08941920.2017.1284291.

Cattani, D., and S. Asselin, 2018: Has selection for grain yield altered intermediate wheatgrass? *Sustainability*, **10**, 688, doi:10.3390/su10030688.

Cavan, G. et al. 2014: Urban morphological determinants of temperature regulating ecosystem services in two African cities. *Ecol. Indic.*, **42**, 43–57, doi:10.1016/J.ECOLIND.2014.01.025.

Cayuela, M.L., S. Jeffery, and L. Van Zwieten, 2015: The molar H: Corg ratio of biochar is a key factor in mitigating N_2O emissions from soil. *Agric. Ecosyst. Environ.*, **202**, 135–138.

Cazenave, A., and G. Le Cozannet, 2014: Sea level rise and its coastal impacts. *Earth's Futur.*, **2**, 15–34, doi:10.1002/2013EF000188.

CEC, 2009: *Adapting to Climate Change: Towards a European Framework for Action*. White Paper, European Commission, Brussels., 17 pp.

Cerdà, A., J. Rodrigo-Comino, A. Giménez-Morera, and S.D. Keesstra, 2018: Hydrological and erosional impact and farmer's perception on catch crops and weeds in citrus organic farming in Canyoles river watershed, Eastern Spain. *Agric. Ecosyst. Environ.*, **258**, 49–58, doi:10.1016/J.AGEE.2018.02.015.

Certini, G., 2005: Effects of fire on properties of forest soils: A review. *Oecologia*, **143**, 1–10, doi:10.1007/s00442-004-1788-8.

Chambers, R., and G. Conway, 1992: *Sustainable Rural Livelihoods: Practical concepts for the 21st Century*. Institute of Development Studies, Brighton, UK,, 42 pp.

Chan, K.Y., L. Van Zwieten, I. Meszaros, A. Downie, and S. Joseph, 2008: Using poultry litter biochars as soil amendments. *Soil Res.*, **46**, 437, doi:10.1071/SR08036.

Chandio, N.H., M.M. Anwar, and A.A. Chandio, 2011: Degradation of Indus Delta, removal of mangroves forestland; its causes. A case study of Indus River Delta. *Sindh Univ. Res. Jour.* (Sci. Ser.), **43**. 1.

Changjin, S., and C.L., 2007: A review of watershed environmental services from forest in China. Chapter 3. In: *Payment for Watershed Services In China: The Role of Government and Market*. L.Xiaoyun, J. Leshan, Z. Ting, and I. Bond, (eds.)]. Social Sciences Academic Press (China), Beijing, China.

Chappell, A., J. Baldock, and J. Sanderman, 2016: The global significance of omitting soil erosion from soil organic carbon cycling schemes. *Nat. Clim. Chang.*, **6**, 187–191, doi:10.1038/nclimate2829.

Charles, R.L., P.K.T. Munushi, and E.F. Nzunda, 2013: Agroforestry as adaptation strategy under climate change in Mwanga District, Kilimanjaro, Tanzania. *Int. J. Environ. Prot.*, **3**, 29–38.

Chaudhuri, P., S. Ghosh, M. Bakshi, S. Bhattacharyya, and B. Nath, 2015: A review of threats and vulnerabilities to mangrove habitats: With special emphasis on East Coast of India. *J. Earth Sci. Clim. Change*, **06**, 270, doi:10.4172/2157-7617.1000270.

Chazdon, R.L., and M. Uriarte, 2016: Natural regeneration in the context of large-scale forest and landscape restoration in the tropics. *Biotropica*, **48**, 709–715, doi:10.1111/btp.12409.

Chazdon, R.L. et al. 2016a: Carbon sequestration potential of second-growth forest regeneration in the Latin American tropics. *Sci. Adv.*, **2**, e1501639–e1501639, doi:10.1126/sciadv.1501639.

Chazdon, R.L. et al. 2017: A policy-driven knowledge agenda for global forest and landscape restoration. *Conserv. Lett.*, **10**, 125–132, doi:10.1111/conl.12220.

Cheesman, S., C. Thierfelder, N.S. Eash, G.T. Kassie, and E. Frossard, 2016: Soil carbon stocks in conservation agriculture systems of Southern Africa. *Soil Tillage Res.*, **156**, 99–109, doi:10.1016/J.STILL.2015.09.018.

Chen, C. et al. 2019: China and India lead in greening of the world through land-use management. *Nat. Sustain.*, **2**, 122–129, doi:10.1038/s41893-019-0220-7.

Chen, D., W. Wei, and L. Chen, 2017: Effects of terracing practices on water erosion control in China: A meta-analysis. *Earth-Science Rev.*, **173**, 109–121, doi:10.1016/J.EARSCIREV.2017.08.007.

Chen, J. et al. 2018a: Prospects for the sustainability of social-ecological systems (SES) on the Mongolian plateau: Five critical issues. *Environ. Res. Lett.*, **13**, 123004, doi:10.1088/1748-9326/aaf27b.

Chen, J., M.T. Ter-Mikaelian, P.Q. Ng, and S.J. Colombo, 2018b: Ontario's managed forests and harvested wood products contribute to greenhouse gas mitigation from 2020 to 2100. *For. Chron.*, **43**, 269–282, doi:10.5558/tfc2018-040.

Chen, S. et al. 2018: Plant diversity enhances productivity and soil carbon storage. *Proc. Natl. Acad. Sci. U.S.A.*, **115**, 4027–4032, doi:10.1073/pnas.1700298114.

Chen, Z. et al. 2018d: Source partitioning of methane emissions and its seasonality in the U.S. Midwest. *J. Geophys. Res. Biogeosciences*, **123**, 646–659, doi:10.1002/2017JG004356.

Cheng, L., and A. AghaKouchak, 2015: Nonstationary precipitation intensity-duration-frequency curves for infrastructure design in a changing climate. *Sci. Rep.*, **4**, 7093, doi:10.1038/srep07093.

Cherlet, M. et al. 2018: *World Atlas of Desertification*. 3rd edition. Publication Office of the European Union, Luxemburg, 248 pp.

Chia, E. et al. 2016: Exploring opportunities for promoting synergies between climate change adaptation and mitigation in forest carbon initiatives. *Forests*, **7**, 24, doi:10.3390/f7010024.

Cho, M.-H. et al. 2018: Vegetation-cloud feedbacks to future vegetation changes in the Arctic regions. *Clim. Dyn.*, **50**, 3745–3755, doi:10.1007/s00382-017-3840-5.

Choi, S.-D., K. Lee, and Y.-S. Chang, 2002: Large rate of uptake of atmospheric carbon dioxide by planted forest biomass in Korea. *Global Biogeochem. Cycles*, **16**, 1089, doi:10.1029/2002GB001914.

Christensen, T.R. et al. 2004: Thawing sub-arctic permafrost: Effects on vegetation and methane emissions. *Geophys. Res. Lett.*, **31**, L04501, doi:10.1029/2003GL018680.

Ciais, P. et al. 2013: Carbon and Other Biogeochemical Cycles. Climate Change 2013: The Physical Science Basis. In: *Contribution of Working Group I to the Fifth Assessment Report of the Intergovernmental Panel on Climate Change*, [J. Stocker, T.F. et al. (eds.)]. Cambridge University Press, Cambridge, UK and New York, USA, pp. 467–570.

Cilas, C., F.-R. Goebel, R. Babin, and J. Avelino, 2016: *Tropical Crop Pests and Diseases in a Climate Change Setting—A Few Examples*. Climate Change and Agriculture Worldwide, Springer Netherlands, Dordrecht, pp. 73–82.

Classen, A.T. et al. 2015: Direct and indirect effects of climate change on soil microbial and soil microbial-plant interactions: What lies ahead? *Ecosphere*, **6**, art130, doi:10.1890/ES15-00217.1.

Cobb, A.R. et al. 2017: How temporal patterns in rainfall determine the geomorphology and carbon fluxes of tropical peatlands. *Proc. Natl. Acad. Sci.,* **114**(26), E5187-E5196, , doi:10.1073/pnas.1701090114.

Cohen-Shacham, E., Walters, G., Janzen, C., Maginnis, S., 2016: *Nature-based Solutions to Address Global Societal Challenges*. Gland, Switzerland, xiii + 97 pp.

Colloff, M.J. et al. 2016: Adaptation services of floodplains and wetlands under transformational climate change. *Ecol. Appl.,* **26**, 1003–1017, doi:10.1890/15-0848.

Colombani, N., A. Osti, G. Volta, and M. Mastrocicco, 2016: Impact of climate change on salinization of coastal water resources. *Water Resour. Manag.,* **30**, 2483–2496, doi:10.1007/s11269-016-1292-z.

Coma, J. et al. 2017: Vertical greenery systems for energy savings in buildings: A comparative study between green walls and green facades. *Build. Environ.,* **111**, 228–237, doi:10.1016/j.buildenv.2016.11.014.

Conant, R.T., and K. Paustian, 2002: Potential soil carbon sequestration in overgrazed grassland ecosystems. *Global Biogeochem. Cycles*, **16**, 90–1-90–99, doi:10.1029/2001GB001661.

Conant, R.T., S.M. Ogle, E.A. Paul, and K. Paustian, 2011a: Temperature and soil organic matter decomposition rates – synthesis of current knowledge and a way forward. *Glob. Chang. Biol.,* **17**, 3392–3404, doi:10.1111/j.1365-2486.2011.02496.x.

Conant, R.T., S.M. Ogle, E.A. Paul, and K. Paustian, 2011b: Measuring and monitoring soil organic carbon stocks in agricultural lands for climate mitigation. *Front. Ecol. Environ.,* **9**, 169–173, doi:10.1890/090153.

Conant, R.T., C.E.P. Cerri, B.B. Osborne, and K. Paustian, 2017: Grassland management impacts on soil carbon stocks: A new synthesis. *Ecol. Appl.,* **27**, 662–668, doi:10.1002/eap.1473.

Coomes, O.T., Y. Takasaki, and J.M. Rhemtulla, 2011: Land-use poverty traps identified in shifting cultivation systems shape long-term tropical forest cover. *Proc. Natl. Acad. Sci.,* **108**, 13925–13930, doi:10.1073/PNAS.1012973108.

Cooper, J.A.G., and J. Pile, 2014: The adaptation-resistance spectrum: A classification of contemporary adaptation approaches to climate-related coastal change. *Ocean Coast. Manag.,* **94**, 90–98, doi:10.1016/j.ocecoaman.2013.09.006.

Coppus, R., and A.C. Imeson, 2002: Extreme events controlling erosion and sediment transport in a semi-arid sub-andean valley. *Earth Surf. Process. Landforms,* **27**, 1365–1375, doi:10.1002/esp.435.

Costanza, R. et al. 2008: The value of coastal wetlands for hurricane protection. AMBIO A *J. Hum. Environ.,* **37**, 241–248, doi:10.1579/0044-7447(2008)37[241:TVOCWF]2.0.CO;2.

Cote, M., and A.J. Nightingale, 2012: Resilience thinking meets social theory. *Prog. Hum. Geogr.,* **36**, 475–489, doi:10.1177/0309132511425708.

Cotrufo, M.F. et al. 2015: Formation of soil organic matter via biochemical and physical pathways of litter mass loss. *Nat. Geosci.,* **8**, 776–779, doi:10.1038/ngeo2520.

Cowie, A. et al. 2015: Biochar, carbon accounting and climate change. In: *Biochar for Environmental Management Science, Technology and Implementation*. [Joseph, S., Lehmann, J., (eds.)]. Taylor and Francis, London, UK, pp. 763–794.

Cowie, A.L. et al. 2018: Land in balance: The scientific conceptual framework for land degradation neutrality. *Environ. Sci. Policy,* **79**, 25–35.

Cox, S., P. Nabukalu, A. Paterson, W. Kong, and S. Nakasagga, 2018: Development of Perennial Grain Sorghum. *Sustainability,* **10**, 172, doi:10.3390/su10010172.

Cramer, W. et al. 2014: Detection and attribution of observed impacts. In: Climate Change 2014: Impacts, Adaptation, and Vulnerability. Part A: Global and Sectoral Aspects. Contribution of Working Group II to the Fifth Assessment Report of the Intergovernmental Panel on Climate Change, in [Field, C.B. et al. (eds.)]. Cambridge University Press, Cambrdige, UK and New York, USA, 979–1037.

Cretney, R., 2014: Resilience for whom? Emerging critical geographies of socio-ecological resilience. *Geogr. Compass,* **8**, 627–640, doi:10.1111/gec3.12154.

Crews, T., W. Carton, and L. Olsson, 2018: Is the future of agriculture perennial? Imperatives and opportunities to reinvent agriculture by shifting from annual monocultures to perennial polycultures. *Glob. Sustain.,* **1**, e11, doi:10.1017/sus.2018.11.

Crews, T.E., and S.R. Gliessman, 1991: Raised field agriculture in Tlaxcala, Mexico: An ecosystem perspective on maintenance of soil fertility. *Am. J. Altern. Agric.,* **6**, 9, doi:10.1017/S088918930000374X.

Crews, T.E., and B.E. Rumsey, 2017: What agriculture can learn from native ecosystems in building soil organic matter: A review. *Sustain.,* **9**, 1–18, doi:10.3390/su9040578.

Crews, T.E. et al. 2016: Going where no grains have gone before: From early to mid-succession. *Agric. Ecosyst. Environ.,* **223**, 223–238, doi:10.1016/j.agee.2016.03.012.

Crowther, T.W. et al. 2015: Biotic interactions mediate soil microbial feedbacks to climate change. *Proc. Natl. Acad. Sci.,* **112**, 7033–7038, doi:10.1073/pnas.1502956112.

Crowther, T.W. et al. 2016: Quantifying global soil carbon losses in response to warming. *Nature,* **540**, 104–108, doi:10.1038/nature20150.

Crozier, M.J., 2010: Deciphering the effect of climate change on landslide activity: A review. *Geomorphology,* **124**, 260–267, doi:10.1016/J.GEOMORPH.2010.04.009.

Čuček, L., J.J. Klemeš, and Z. Kravanja, 2012: A review of footprint analysis tools for monitoring impacts on sustainability. *J. Clean. Prod.,* **34**, 9–20, doi:10.1016/J.JCLEPRO.2012.02.036.

Cui, G. et al. 2014: Estimation of forest carbon budget from land cover change in South and North Korea between 1981 and 2010. *J. Plant Biol.,* **57**, 225–238, doi:10.1007/s12374-014-0165-3.

Culman, S.W., S.S. Snapp, M. Ollenburger, B. Basso, and L.R. DeHaan, 2013: Soil and water quality rapidly responds to the perennial grain Kernza wheatgrass. *Agron. J.,* **105**, 735–744, doi:10.2134/agronj2012.0273.

Curtis, P.G., C.M. Slay, N.L. Harris, A. Tyukavina, and M.C. Hansen, 2018: Classifying drivers of global forest loss. *Science,* **361**, 1108–1111, doi:10.1126/science.aau3445.

Cutter, S.L. et al. 2008: A place-based model for understanding community resilience to natural disasters. *Glob. Environ. Chang.,* **18**, 598–606, doi:10.1016/J.GLOENVCHA.2008.07.013.

D'Annunzio, R., E.J. Lindquist, and K.G. MacDicken, 2017: *Global Forest Land-Use Change from 1990 to 2010: An Update to a Global Remote Sensing Survey Of Forests*. FAO, Rome, Italy., 6 pp.

Dallimer, M., and L.C. Stringer, 2018: Informing investments in land degradation neutrality efforts: A triage approach to decision making. *Environ. Sci. Policy,* **89**, 198–205, doi:10.1016/j.envsci.2018.08.004.

Dargie, G.C. et al. 2017: Age, extent and carbon storage of the central Congo Basin peatland complex. *Nature,* **542**, 86–90, doi:10.1038/nature21048.

Darrah, S.E. et al. 2019: Improvements to the Wetland Extent Trends (WET) index as a tool for monitoring natural and human-made wetlands. *Ecol. Indic.,* **99**, 294–298, doi:10.1016/J.ECOLIND.2018.12.032.

Davidson, N.C., 2014: How much wetland has the world lost? Long-term and recent trends in global wetland area. *Mar. Freshw. Res.,* **65**, 934, doi:10.1071/MF14173.

Davies, M.I.J., 2015: Economic Specialisation, resource variability, and the origins of intensive agriculture in Eastern Africa. *Rural Landscapes Soc. Environ. Hist.,* **2**, 1 doi:10.16993/rl.af.

Davies, M.I.J. and H.L. Moore, 2016: Landscape, time and cultural resilience: A brief history of agriculture in Pokot and Marakwet, Kenya. *J. East. African Stud.,* **10**, 67–87, doi:10.1080/17531055.2015.1134417.

Davies, Z.G., J.L. Edmondson, A. Heinemeyer, J.R. Leake, and K.J. Gaston, 2011: Mapping an urban ecosystem service: Quantifying above-ground carbon storage at a city-wide scale. *J. Appl. Ecol.,* **48**, 1125–1134, doi:10.1111/j.1365-2664.2011.02021.x.

Davin, E.L., N. de Noblet-Ducoudré, E.L. Davin, and N. de Noblet-Ducoudré, 2010: Climatic impact of global-scale deforestation: Radiative versus nonradiative processes. *J. Clim.*, **23**, 97–112, doi:10.1175/2009JCLI3102.1.

Davin, E.L., S.I. Seneviratne, P. Ciais, A. Olioso, and T. Wang, 2014: Preferential cooling of hot extremes from cropland albedo management. *Proc. Natl. Acad. Sci. U.S.A.*, **111**, 9757–9761, doi:10.1073/pnas.1317323111.

Day, J.J., and K.I. Hodges, 2018: Growing land-sea temperature contrast and the intensification of arctic cyclones. *Geophys. Res. Lett.*, **45**, 3673–3681, doi:10.1029/2018GL077587.

DeConto, R.M., and D. Pollard, 2016: Contribution of Antarctica to past and future sea-level rise. *Nature*, **531**, 591–597, doi:10.1038/nature17145.

DeHaan, L., M. Christians, J. Crain, and J. Poland, 2018: Development and evolution of an intermediate wheatgrass domestication program. *Sustainability*, **10**, 1499, doi:10.3390/su10051499.

DeHaan, L.R., and D.L. Van Tassel, 2014: Useful insights from evolutionary biology for developing perennial grain crops. *Am. J. Bot.*, **101**, 1801–1819, doi:10.3732/ajb.1400084.

DeHaan, L.R. et al. 2016: A pipeline strategy for grain crop domestication. *Crop Sci.*, **56**, 917–930, doi:10.2135/cropsci2015.06.0356.

Delgado, A., and J.A. Gómez, 2016: *The Soil. Physical, Chemical and Biological Properties. Principles of Agronomy for Sustainable Agriculture*, Springer International Publishing, Cham, Switzerland, pp. 15–26.

Dell, M., B.F. Jones, and B.A. Olken, 2009: Temperature and income: Reconciling new cross-sectional and panel estimates. *Am. Econ. Rev.*, **99**, 198–204, doi:10.1257/aer.99.2.198.

Demuzere, M. et al. 2014: Mitigating and adapting to climate change: Multi-functional and multi-scale assessment of green urban infrastructure. *J. Environ. Manage.*, **146**, 107–115, doi:10.1016/J.JENVMAN.2014.07.025.

Desalegn, T., Alemu, G., Adella, A., & Debele, T. (2017). Effect of lime and phosphorus fertilizer on acid soils and barley (Hordeum vulgare L.) performance in the central highlands of Ethiopia. *Experimental Agriculture*, **53(3)**, 432–444., doi:10.1017/S0014479716000491.

Devendra, C., 2002: Crop–animal systems in Asia: Implications for research. *Agric. Syst.*, **71**, 169–177, doi:10.1016/S0308-521X(01)00042-7.

Devendra, C. and D. Thomas, 2002: Crop–animal systems in Asia: importance of livestock and characterisation of agro-ecological zones. *Agric. Syst.*, **71**, 5–15, doi:10.1016/S0308-521X(01)00032-4.

DG Environment News Alert Service, 2012: *The Multifunctionality of Green Infrastructure*. European Commission, Brussels, Belgium, 40 pp.

Dhanush, D. et al. 2015: *Impact of climate change on African agriculture: focus on pests and diseases*. CCAFS Info Note, Copenhagen, Denmark, 4 pp.

Dignac, M.-F. et al. 2017: Increasing soil carbon storage: Mechanisms, effects of agricultural practices and proxies. A review. *Agron. Sustain. Dev.*, **37**, 14, doi:10.1007/s13593-017-0421-2.

Di Leo, N., F.J. Escobedo, and M. Dubbeling, 2016: The role of urban green infrastructure in mitigating land surface temperature in Bobo-Dioulasso, Burkina Faso. *Environ. Dev. Sustain.*, **18**, 373–392, doi:10.1007/s10668-015-9653-y.

Ding, F. et al. 2018: A meta-analysis and critical evaluation of influencing factors on soil carbon priming following biochar amendment. *J. Soils Sediments*, **18(4)**, 1507–1517, doi:10.1007/s11368-017-1899-6.

Dintwe, K., G.S. Okin, and Y. Xue, 2017: Fire-induced albedo change and surface radiative forcing in sub-Saharan Africa savanna ecosystems: Implications for the energy balance. *J. Geophys. Res. Atmos.*, **122**, 6186–6201, doi:10.1002/2016JD026318.

Dixon, M.J.R. et al. 2016: Tracking global change in ecosystem area: The Wetland Extent Trends index. *Biol. Conserv.*, **193**, 27–35, doi:10.1016/j.biocon.2015.10.023.

Djurfeldt, A.A., E. Hillbom, W.O. Mulwafu, P. Mvula, and G. Djurfeldt, 2018: "The family farms together, the decisions, however are made by the man" – Matrilineal land tenure systems, welfare and decision making in rural Malawi. *Land use policy*, **70**, 601–610, doi:10.1016/j.landusepol.2017.10.048.

Dommain, R. et al. 2012: *Peatlands – Guidance For Climate Change Mitigation Through Conservation, Rehabilitation And Sustainable Use*. Mitigation of Climate Change in Agriculture (MICCA) Programme series 5, Food and Agriculture Organization of the United Nations and Wetlands International, Rome, Italy, 114 pp.

Dommain, R. et al. 2018: A radiative forcing analysis of tropical peatlands before and after their conversion to agricultural plantations. *Glob. Chang. Biol.*, **24**, 5518–5533, doi:10.1111/gcb.14400.

Donner, S., 2012: Sea level rise and the ongoing battle of Tarawa. *Eos, Trans. Am. Geophys. Union*, **93**, 169–170, doi:10.1029/2012EO170001.

Donner, S.D., and S. Webber, 2014: Obstacles to climate change adaptation decisions: A case study of sea-level rise and coastal protection measures in Kiribati. *Sustain. Sci.*, **9**, 331–345, doi:10.1007/s11625-014-0242-z.

Dooley, K., and S. Kartha, 2018: Land-based negative emissions: Risks for climate mitigation and impacts on sustainable development. *Int. Environ. Agreements Polit. Law Econ.*, **18**, 79–98, doi:10.1007/s10784-017-9382-9.

Doss, C., C. Kovarik, A. Peterman, A. Quisumbing, and M. van den Bold, 2015: Gender inequalities in ownership and control of land in Africa: Myth and reality. *Agric. Econ.*, **46**, 403–434, doi:10.1111/agec.12171.

Dotterweich, M., 2008: The history of soil erosion and fluvial deposits in small catchments of central Europe: Deciphering the long-term interaction between humans and the environment – A review. *Geomorphology*, **101**, 192–208, doi:10.1016/J.GEOMORPH.2008.05.023.

Dotterweich, M., 2013: The history of human-induced soil erosion: Geomorphic legacies, early descriptions and research, and the development of soil conservation – A global synopsis. *Geomorphology*, **201**, 1–34, doi:10.1016/J.GEOMORPH.2013.07.021.

Dow, K., F. Berkhout, and B.L. Preston, 2013a: Limits to adaptation to climate change: A risk approach. *Curr. Opin. Environ. Sustain.*, **5**, 384–391, doi:10.1016/J.COSUST.2013.07.005.

Dow, K. et al. 2013b: Limits to adaptation. *Nat. Clim. Chang.*, **3**, 305–307, doi:10.1038/nclimate1847.

Downie, A., P. Munroe, A. Cowie, L. Van Zwieten, and D.M.S. Lau, 2012: Biochar as a geoengineering climate solution: Hazard identification and risk management. *Crit. Rev. Environ. Sci. Technol.*, **42**, 225–250, doi:10.1080/10643389.2010.507980.

Dragoni, D. et al. 2011: Evidence of increased net ecosystem productivity associated with a longer vegetated season in a deciduous forest in south-central Indiana, USA. *Glob. Chang. Biol.*, 17, 886–897, doi:10.1111/j.1365-2486.2010.02281.x.

Dregne, H.E., 1998: Desertification Assessment. In: *Methods for Assessment Of Soil Degradation*, [Lal, R., (ed.)]. CRC Press, Boca Raton, London, New York, Washington DC, pp. 441–458.

Drescher, J. et al. 2016: Ecological and socio-economic functions across tropical land use systems after rainforest conversion. *Philos. Trans. R. Soc. B Biol. Sci.*, **371**, 20150275, doi:10.1098/rstb.2015.0275.

Drösler, M., L.V. Verchot, A. Freibauer, G. Pan, 2014. Drained Inland Organic Soils, 2013. In: Task Force on National Greenhouse Gas Inventories of the IPCC, [Hiraishi T., T. Krug, K. Tanabe, N. Srivastava, B. Jamsranjav, M. Fukuda, and T. Troxler (eds.)]. Supplement to the 2006 IPCC Guidelines: Wetlands. Hayama, Japan: Institute for Global Environmental Strategies (IGES) on behalf of the Intergovernmental Panel on Climate Change (IPCC).

Dugan, A.J. et al. 2018: A systems approach to assess climate change mitigation options in landscapes of the United States forest sector. *Carbon Balance Manag.*, **13**, doi:10.1186/s13021-018-0100-x.

Dumanski, J., and C. Pieri, 2000: Land quality indicators: Research plan. *Agric. Ecosyst. Environ.*, **81**, 93–102, doi:10.1016/S0167-8809(00)00183-3.

Duncan, T., 2016: Case Study: Taranaki farm regenerative agriculture. Pathways to integrated ecological farming. *L. Restor.*, **2016**, 271–287, doi:10.1016/B978-0-12-801231-4.00022-7.

Dupouey, J.L., E. Dambrine, J.D. Laffite, and C. Moares, 2002: Irreversible impact of past land use on forest soils and biodiversity. *Ecology*, **83**, 2978–2984, doi:10.1890/0012-9658(2002)083[2978:IIOPLU]2.0.CO;2.

Dutta, D., P.K. Das, S. Paul, J.R. Sharma, and V.K. Dadhwal, 2015: Assessment of ecological disturbance in the mangrove forest of Sundarbans caused by cyclones using MODIS time-series data (2001–2011). *Nat. Hazards*, **79**, 775–790, doi:10.1007/s11069-015-1872-x.

Duvat, V., 2013: Coastal protection structures in Tarawa Atoll, Republic of Kiribati. *Sustain. Sci.*, **8**, 363–379, doi:10.1007/s11625-013-0205-9.

Duvat, V.K.E., A.K. Magnan, S. Etienne, C. Salmon, and C. Pignon-Mussaud, 2016: Assessing the impacts of and resilience to Tropical Cyclone Bejisa, Reunion Island (Indian Ocean). *Nat. Hazards*, **83**, 601–640, doi:10.1007/s11069-016-2338-5.

Earles, J.M., S. Yeh, and K.E. Skog, 2012: Timing of carbon emissions from global forest clearance. *Nat. Clim. Chang.*, **2**, 682–685, doi:10.1038/nclimate1535.

Edmondson, J.L., Z.G. Davies, S.A. McCormack, K.J. Gaston, and J.R. Leake, 2011: Are soils in urban ecosystems compacted? A citywide analysis. *Biol. Lett.*, **7**(5), 771–774, doi:10.1098/rsbl.2011.0260.

Edmondson, J.L., Z.G. Davies, S.A. McCormack, K.J. Gaston, and J.R. Leake, 2014: Land-cover effects on soil organic carbon stocks in a European city. *Sci. Total Environ.*, **472**, 444–453, doi:10.1016/j.scitotenv.2013.11.025.

Edstedt, K., and W. Carton, 2018: The benefits that (only) capital can see? Resource access and degradation in industrial carbon forestry, lessons from the CDM in Uganda. *Geoforum*, **97**, 315–323, doi:10.1016/J.GEOFORUM.2018.09.030.

Eglin, T. et al. 2010: Historical and future perspectives of global soil carbon response to climate and land-use changes. *Tellus B Chem. Phys. Meteorol.*, **62**, 700–718, doi:10.1111/j.1600-0889.2010.00499.x.

Ehnström, B., 2001: Leaving dead wood for insects in boreal forests – suggestions for the future. *Scand. J. For. Res.*, **16**, 91–98, doi:10.1080/028275801300090681.

Ehrenfeld, J.G., B. Ravit, and K. Elgersma, 2005: Feedback in the plant-soil system. *Annu. Rev. Environ. Resour.*, **30**, 75–115, doi:10.1146/annurev.energy.30.050504.144212.

Eisenbies, M.H., W.M. Aust, J.A. Burger, and M.B. Adams, 2007: Forest operations, extreme flooding events, and considerations for hydrologic modeling in the Appalachians – A review. *For. Ecol. Manage.*, **242**, 77–98, doi:10.1016/j.foreco.2007.01.051.

Elad, Y. et al. 2011: The biochar effect: Plant resistance to biotic stresses. *Phytopathol. Mediterr.*, **50**(3), 335–349, doi:10.14601/Phytopathol_Mediterr-9807.

Elbehri, A., A. Challinor, L. Verchot, A. Angelsen, T. Hess, A. Ouled Belgacem, H. Clark, M. Badraoui, A. Cowie, S. De Silva, J. Joar Hegland Erickson, A. Iglesias, D. Inouye, A. Jarvis, E. Mansur, A. Mirzabaev, L. Montanarella, D. Murdiyarso, A. Notenbaert, M. Obersteiner, K. Paustian, D. Pennock, A. Reisinger, D. Soto, J.-F. Soussana, R. Thomas, R. Vargas, M. Van Wijk, and R. Walker, 2017: In: FAO-IPCC Expert Meeting on Climate Change, Land Use and Food Security: Final Meeting Report. January 23–25, 2017. Food and Agriculture Organization of the UN (FAO)/Intergovernmental Panel on Climate Change (IPCC) Rome, Italy, 156 pp.

Elliott, M., N.D. Cutts, and A. Trono, 2014: A typology of marine and estuarine hazards and risks as vectors of change: A review for vulnerable coasts and their management. *Ocean Coast. Manag.*, **93**, 88–99, doi:10.1016/J.OCECOAMAN.2014.03.014.

Ellis, E.C. et al. 2013: Used planet: a global history. *Proc. Natl. Acad. Sci. U.S.A.*, **110**, 7978–7985, doi:10.1073/pnas.1217241110.

Ellis, J.E., 1994: Climate variability and complex ecosystem dynamics: Implications for pastoral development. In: *Living with uncertainty: New directions in pastoral development in Africa*, [Scoones, I. (ed.)]. Internmediate Technology Publications, London, UK, pp. 37–46.

Ellis, P.W. et al. 2019: Reduced-impact logging for climate change mitigation (RIL-C) can halve selective logging emissions from tropical forests. *For. Ecol. Manage.*, **438**, 255–266, doi:10.1016/J.FORECO.2019.02.004.

Ellison, D. et al. 2017: Trees, forests and water: Cool insights for a hot world. *Glob. Environ. Chang.*, **43**, 51–61, doi:10.1016/J.GLOENVCHA.2017.01.002.

Elmhirst, R., 2011: Introducing new feminist political ecologies. *Geoforum*, **42**, 129–132, doi:10.1016/j.geoforum.2011.01.006.

England, M.H. et al. 2014: Recent intensification of wind-driven circulation in the Pacific and the ongoing warming hiatus. *Nat. Clim. Chang.*, **4**, 222–227, doi:10.1038/nclimate2106.

Erb, K.-H.H. et al. 2018: Unexpectedly large impact of forest management and grazing on global vegetation biomass. *Nature*, **553**, 73–76, doi:10.1038/nature25138.

Erwin, K.L., 2009: Wetlands and global climate change: The role of wetland restoration in a changing world. *Wetl. Ecol. Manag.*, **17**, 71–84, doi:10.1007/s11273-008-9119-1.

van der Esch, S. et al. 2017: *Exploring future changes in land use and land condition and the impacts on food, water, climate change and biodiversity: Scenarios for the Global Land Outlook.* Netherlands Environmental Assessment Agency, The Hague, The Netherlands, 116 pp.

Etemadi, H., S.Z. Samadi, M. Sharifikia, and J.M. Smoak, 2016: Assessment of climate change downscaling and non-stationarity on the spatial pattern of a mangrove ecosystem in an arid coastal region of southern Iran. *Theor. Appl. Climatol.*, **126**, 35–49, doi:10.1007/s00704-015-1552-5.

eThekwini Municipal Council, 2014: *The Durban Climate Change Strategy.* Environmental Planning and Climate Protection Department (EPCPD) and the Energy Office (EO) of eThekwini Municipality, Durban, South Africa. 54 pp.

European Union, 2015: Towards an EU research and innovation policy agenda for nature-based solutions & re-naturing cities. *Nature-Based Solut. Re-Naturing Cities*, Directorate-General for Research and Innovation, Brussels, Belgium, 74 pp. doi:10.2777/765301.

Evans, C.D., F. Renou-Wilson, and M. Strack, 2016: The role of waterborne carbon in the greenhouse gas balance of drained and re-wetted peatlands. *Aquat. Sci.*, **78**(3), 573–590, doi:10.1007/s00027-015-0447-y.

Evans, R., and J. Boardman, 2016a: A reply to panagos et al. 2016. (*Environ. Sci. Policy* 59, 2016, 53–57). *Environ. Sci. Policy*, **60**, 63–68, doi:10.1016/J.ENVSCI.2016.03.004.

Evans, R., and J. Boardman, 2016b: The new assessment of soil loss by water erosion in Europe. Panagos P. et al. 2015 *Environmental Science & Policy* 54, 438–447—A response. *Environ. Sci. Policy*, **58**, 11–15, doi:10.1016/j.envsci.2015.12.013.

Evans, S.E., and M.D. Wallenstein, 2014: Climate change alters ecological strategies of soil bacteria. *Ecol. Lett.*, **17**, 155–164, doi:10.1111/ele.12206.

Fairhead, J., and I. Scoones, 2005: Local knowledge and the social shaping of soil investments: Critical perspectives on the assessment of soil degradation in Africa. *Land use policy*, **22**, 33–41, doi:10.1016/J.LANDUSEPOL.2003.08.004.

Fang, Y., B.P.B. Singh, and B.P.B. Singh, 2015: Effect of temperature on biochar priming effects and its stability in soils. *Soil Biol. Biochem.*, **80**, 136–145, doi:10.1016/j.soilbio.2014.10.006.

Fang, Y., B.P. Singh, P. Matta, A.L. Cowie, and L. Van Zwieten, 2017: Temperature sensitivity and priming of organic matter with different stabilities in a Vertisol with aged biochar. *Soil Biol. Biochem.*, **115**, 346–356, doi:10.1016/j.soilbio.2017.09.004.

FAO, 2007: *Land Evaluation: Towards a Revised Framework.* Land and Water Discussion Paper No. 6. Food and Agricultural Organization of the UN, Rome, Italy, 124 pp.

FAO, 2015: *FRA 2015 Terms and Definitions.* Food and Agricultural Organization of the UN, Rome, Italy, 1–81 pp.

FAO, 2016: *Global Forest Resources Assessment 2015 : How Are The World's Forests Changing?* K. MacDicken, Ö. Jonsson, L. Pina, and S. Maulo, Eds. Food and Agricultural Organization of the UN, Rome, Italy, 44 pp.

FAO, 2018: *From Reference Levels to Results Reporting: REDD+ Under the UNFCCC*, Update. Food and Agricultural Organization of the UN, Rome, Italy. 47 pp.

Fargione, J.E. et al. 2018: Natural climate solutions for the United States. *Sci. Adv.*, **4**, eaat1869, doi:10.1126/sciadv.aat1869.

4

Fasullo, J.T., and R.S. Nerem, 2018: Altimeter-era emergence of the patterns of forced sea-level rise in climate models and implications for the future. *Proc. Natl. Acad. Sci. U.S.A.*, **115**, 12944–12949, doi:10.1073/pnas.1813233115.

Federici, S., F.N. Tubiello, M. Salvatore, H. Jacobs, and J. Schmidhuber, 2015: Forest ecology and management new estimates of CO_2 forest emissions and removals: 1990 – 2015. *For. Ecol. Manage.*, **352**, 89–98, doi:10.1016/j.foreco.2015.04.022.

Felton, A., M. Lindbladh, J. Brunet, and Ö. Fritz, 2010: Replacing coniferous monocultures with mixed-species production stands: An assessment of the potential benefits for forest biodiversity in northern Europe. *For. Ecol. Manage.*, **260**, 939–947, doi:10.1016/j.foreco.2010.06.011.

Fenner, N. et al. 2011: Decomposition 'hotspots' in a rewetted peatland: Implications for water quality and carbon cycling. *Hydrobiologia*, **674**, 51–66, doi:10.1007/s10750-011-0733-1.

Fensholt, R. et al. 2015: *Assessing Drivers of Vegetation Changes in Drylands from Time Series of Earth Observation Data*. Springer, Cham, Switzerland, pp. 183–202.

Fernandes, K. et al. 2017: Heightened fire probability in Indonesia in non-drought conditions: The effect of increasing temperatures. *Environ. Res. Lett.*, **12**, doi:10.1088/1748-9326/aa6884.

Ferner, J., S. Schmidtlein, R.T. Guuroh, J. Lopatin, and A. Linstädter, 2018: Disentangling effects of climate and land-use change on West African drylands' forage supply. *Glob. Environ. Chang.*, **53**, 24–38, doi:10.1016/J.GLOENVCHA.2018.08.007.

Feyisa, G.L., K. Dons, and H. Meilby, 2014: Efficiency of parks in mitigating urban heat island effect: An example from Addis Ababa. *Landsc. Urban Plan.*, **123**, 87–95, doi:10.1016/j.landurbplan.2013.12.008.

Field, J.P. et al. 2010: The ecology of dust. *Front. Ecol. Environ.*, **8**, 423–430, doi:10.1890/090050.

Filho, W.L., and J. Nalau, (eds.) 2018: *Limits to Climate Change Adaptation*. Springer International Publishing, Berlin, Heidelberg, 1–410 pp.

Finlayson, C.M. et al. 2017: Policy considerations for managing wetlands under a changing climate. *Mar. Freshw. Res.*, **68**, 1803, doi:10.1071/MF16244.

Fischer, E.M., and R. Knutti, 2015: Anthropogenic contribution to global occurrence of heavy-precipitation and high-temperature extremes. *Nat. Clim. Chang.*, **5**, 560–564, doi:10.1038/nclimate2617.

Fischer, F. et al. 2016: Spatio-temporal variability of erosivity estimated from highly resolved and adjusted radar rain data (RADOLAN). *Agric. For. Meteorol.*, **223**, 72–80, doi:10.1016/J.AGRFORMET.2016.03.024.

Fischer, G., M. Shah, H. van Velthuizen, and F. Nachtergaele, 2009: *Agro-ecological Zones Assessment. Land Use, Land Cover and Soil Sciences – Volume III*: Land Use Planning, Eolss Publishers Co., Oxford, UK, pp. 61–81.

Flores-Rentería, D., A. Rincón, F. Valladares, and J. Curiel Yuste, 2016: Agricultural matrix affects differently the alpha and beta structural and functional diversity of soil microbial communities in a fragmented Mediterranean holm oak forest. *Soil Biol. Biochem.*, **92**, doi:10.1016/j.soilbio.2015.09.015.

Foley, J. et al. 2007, *Amazonia Revealed: Forest Degradation And Loss Of Ecosystem Goods And Services in the Amazon Basin*. Frontiers in Ecology and the Environment, 5(1), 25–32, doi:10.1890/1540-9295(2007)5[25:ARFDAL]2.0.CO;2

Foley, J.A. et al. 2005: Global consequences of land use. *Science*, **309**, 570–574, doi:10.1126/science.1111772.

Foley, J.A. et al. 2011: Solutions for a cultivated planet. *Nature*, **478**, 337–342, doi:10.1038/nature10452.

Folke, C., T. Hahn, P. Olsson, and J. Norberg, 2005: Adaptive governance of social-ecological systems. *Annu. Rev. Environ. Resour.*, **30**, 441–473, doi:10.1146/annurev.energy.30.050504.144511.

Folke, C. et al. 2010: Resilience thinking: Integrating resilience, adaptability and transformability. *Ecol. Soc.*, **15**.

Forest Europe, 1993: Resolution H1: General Guidelines for the Sustainable Management of Forests in Europe. Second Ministerial Conference on the Protection of Forests in Europe 16–17 June 1993, Helsinki. 5 pp.

Forsyth, T., 1996: Science, myth and knowledge: Testing himalayan environmental degradation in Thailand. *Geoforum*, **27**, 375–392, doi:10.1016/S0016-7185(96)00020-6.

Frank, D.C. et al. 2015: Water-use efficiency and transpiration across European forests during the Anthropocene. *Nat. Clim. Chang.*, **5**, 579–583, doi:10.1038/nclimate2614.

Frei, M., and K. Becker, 2005: Integrated rice-fish culture: Coupled production saves resources. *Nat. Resour. Forum*, **29**, 135–143, doi:10.1111/j.1477-8947.2005.00122.x.

French, P.W., 2001: *Coastal Defences: Processes, Problems and Solutions*. Routledge, London, 350 pp.

Frenkel, O. et al. 2017: The effect of biochar on plant diseases: What should we learn while designing biochar substrates? *J. Environ. Eng. Landsc. Manag.*, **25**(2), 105–113, doi:10.3846/16486897.2017.1307202.

Friend, A.D. et al. 2014: Carbon residence time dominates uncertainty in terrestrial vegetation responses to future climate and atmospheric CO_2. *Proc. Natl. Acad. Sci. U.S.A.*, **111**, 3280–3285, doi:10.1073/pnas.1222477110.

Fritsche, U.R. et al. 2017: *Energy and Land Use*. Working Paper for the UNCCD Global Land Outlook, Darmstadt, Germany, 61 pp.

Fritz, H.M., C.D. Blount, S. Thwin, M.K. Thu, and N. Chan, 2009: Cyclone Nargis storm surge in Myanmar. *Nat. Geosci.*, **2**, 448–449, doi:10.1038/ngeo558.

Fritz, H.M., C. Blount, S. Thwin, M.K. Thu, and N. Chan, 2010: *Cyclone Nargis Storm Surge Flooding in Myanmar's Ayeyarwady River Delta. Indian Ocean Tropical Cyclones and Climate Change*, Springer Netherlands, Dordrecht, pp. 295–303.

Frolking, S. et al. 2011: Peatlands in the Earth's 21st century climate system. *Environ. Rev.*, **19**, 371–396, doi:10.1139/a11-014.

Froude, M.J., and D.N. Petley, 2018: Global fatal landslide occurrence from 2004 to 2016. *Nat. Hazards Earth Syst. Sci.*, **18**, 2161–2181, doi:10.5194/nhess-18-2161-2018.

Fryd, O., S. Pauleit, and O. Bühler, 2011: The role of urban green space and trees in relation to climate change. CAB Rev. *Perspect. Agric. Vet. Sci. Nutr. Nat. Resour.*, **6**, 1–18, doi:10.1079/PAVSNNR20116053.

Fuangswasdi, A., S. Worakijthamrong, and S.D. Shah, 2019: *Addressing Subsidence in Bangkok, Thailand and Houston, Texas: Scientific Comparisons and Data-Driven Groundwater Policies for Coastal Land-Surface Subsidence. IAEG/AEG Annual Meeting Proceedings, San Francisco, California, 2018 – Volume 5*, Springer International Publishing, Cham, Switzerland, 51–60.

Fujiki, S. et al. 2016: Large-scale mapping of tree-community composition as a surrogate of forest degradation in Bornean tropical rain forests. *Land*, **5**, 45, doi:10.3390/land5040045.

Fuller, D.Q. et al. 2011: The contribution of rice agriculture and livestock pastoralism to prehistoric methane levels. *The Holocene*, **21**, 743–759, doi:10.1177/0959683611398052.

Fuso Nerini, F., C. Ray, and Y. Boulkaid, 2017: The cost of cooking a meal. The case of Nyeri County, Kenya. *Environ. Res. Lett.*, **12**, 65007, doi:10.1088/1748-9326/aa6fd0.

Fuss, S. et al. 2018: Negative emissions – Part 2 : Costs, potentials and side effects, *Environmental Research Letters* 13, no. 6 (2018): 063002, doi:10.1088/1748-9326/aabf9f.

Gabrielsson, S., S. Brogaard, and A. Jerneck, 2013: Living without buffers – illustrating climate vulnerability in the Lake Victoria basin. *Sustain. Sci.*, **8**, 143–157, doi:10.1007/s11625-012-0191-3.

Gamache, I., and S. Payette, 2005: Latitudinal response of subarctic tree lines to recent climate change in eastern Canada. *J. Biogeogr.*, **32**, 849–862.

Gao, B. et al. 2018: Carbon storage declines in old boreal forests irrespective of succession pathway. *Ecosystems*, **21**, 1–15, doi:10.1007/s10021-017-0210-4.

Gao, Q. et al. 2017: Synergetic Use of Sentinel-1 and Sentinel-2 data for soil moisture mapping at 100 m resolution. *Sensors*, **17**, 1966, doi:10.3390/s17091966.

Garbrecht, J.D., and X.C. Zhang, 2015: Soil erosion from winter wheat cropland under climate change in central Oklahoma. *Appl. Eng. Agric.*, **31**, 439–454, doi:10.13031/aea.31.10998.

Garbrecht, J.D., J.L. Steiner, and A. Cox, Craig, 2007: The times they are changing: Soil and water conservation in the 21st century. *Hydrol. Process.*, **21**, 2677–2679.

García-Ruiz, J.M. et al. 2015: A meta-analysis of soil erosion rates across the world. *Geomorphology*, 239, 160–173, doi:10.1016/j.geomorph.2015.03.008.

Gariano, S.L., and F. Guzzetti, 2016: Landslides in a changing climate. *Earth-Science Rev.*, **162**, 227–252, doi:10.1016/J.EARSCIREV.2016.08.011.

Gasparatos, A. et al. 2018: Survey of local impacts of biofuel crop production and adoption of ethanol stoves in southern Africa. *Sci. data*, **5**, 180186, doi:10.1038/sdata.2018.186.

Gauthier, S., P. Bernier, T. Kuuluvainen, A.Z. Shvidenko, and D.G. Schepaschenko, 2015: Boreal forest health and global change. *Science*, **349**, 819–822, doi:10.1126/science.aaa9092.

Gaveau, D.L. a et al. 2014: Major atmospheric emissions from peat fires in Southeast Asia during non-drought years: Evidence from the 2013 Sumatran fires. *Sci. Rep.*, **4**, 1–7, doi:10.1038/srep06112.

Ge, J., and C. Zou, 2013: Impacts of woody plant encroachment on regional climate in the southern Great Plains of the United States. *J. Geophys. Res. Atmos.*, **118**, 9093–9104, doi:10.1002/jgrd.50634.

Gebara, M., and A. Agrawal, 2017: Beyond rewards and punishments in the Brazilian Amazon: practical implications of the REDD+ discourse. *Forests*, **8**, 66, doi:10.3390/f8030066.

Gebara, M.F., 2018: Tenure reforms in indigenous lands: Decentralized forest management or illegalism? *Curr. Opin. Environ. Sustain.*, **32**, 60–67, doi:10.1016/J.COSUST.2018.04.008.

Geels, F.W., 2002: Technological transitions as evolutionary reconfiguration processes: A multi-level perspective and a case-study. *Res. Policy*, **31**, 1257–1274, doi:10.1016/S0048-7333(02)00062-8.

Van der Geest, K., and M. Schindler, 2016: *Case Study Report: Loss And Damage From a Catastrophic Landslide in Sindhupalchok District, Nepal*. Report No.1. United Nations University Institute for Environment and Human Security (UNU-EHS), Bonn, Germany, 1–96 pp.

GEF, 2018: *GEF-7 Replenishment, Programming Directions*. Global Environment Facility (GEF), Washington DC,, 155 pp.

Genesio, L. et al. 2012: Surface albedo following biochar application in durum wheat. *Environ. Res. Lett.*, **7**(1) 014025, doi:10.1088/1748-9326/7/1/014025.

Genesio, L., F.P. Vaccari, and F. Miglietta, 2016: Black carbon aerosol from biochar threats its negative emission potential. *Glob. Chang. Biol.*, **22**(7), 2313–2314, doi:10.1111/gcb.13254.

Gerber, N., E. Nkonya, and J. von Braun, 2014: *Land Degradation, Poverty and Marginality*. Springer Netherlands, Dordrecht, pp. 181–202.

Gerten, D., R. Betts, and P. Döll, 2014: Cross-Chapter Box on the Active Role of Vegetation in Altering Water Flows Under Climate Change. In: Climate Change 2014: Impacts, Adaptation, and Vulnerability. Part A: Global and Sectoral Aspects. Contribution of Working Group II to the Fifth Assessment Report of the Intergovernmental Panel on Climate Change, [Field, C.B., V.R. Barros, D.J. Dokken, K.J. Mach, M.D. Mastrandrea, T.E. Bilir, M. Chatterjee, K.L. Ebi, Y.O. Estrada, R.C. Genova, B. Girma, E.S. Kissel, A.N. Levy, S. MacCracken, P.R. Mastrandrea, and L.L.White (eds.)]. Cambridge University Press, Cambridge, UK and New York, USA, pp. 157–161.

van Gestel, N. et al. 2018: Predicting soil carbon loss with warming. *Nature*, **554**, E4–E5, doi:10.1038/nature25745.

Gharbaoui, D., and J. Blocher, 2016: *The Reason Land Matters: Relocation as Adaptation to Climate Change in Fiji Islands*. Springer, Cham, Switzerland, pp. 149–173.

Ghazoul, J., and R. Chazdon, 2017: Degradation and recovery in changing forest landscapes: A multiscale conceptual framework. *Annu. Rev. Environ. Resour.*, **42**, 161–188, doi:10.1146/annurev-environ.

Ghazoul, J., Z. Burivalova, J. Garcia-Ulloa, and L.A. King, 2015: Conceptualizing forest degradation. *Trends Ecol. Evol.*, **30**, 622–632, doi:10.1016/j.tree.2015.08.001.

Ghosh, A., S. Schmidt, T. Fickert, and M. Nüsser, 2015: The Indian sundarbans mangrove forests: History, utilization, conservation strategies and local perception. *Diversity*, **7**, 149–169, doi:10.3390/d7020149.

Gibbs, H.K., and J.M. Salmon, 2015: Mapping the world's degraded lands. *Appl. Geogr.*, **57**, 12–21, doi:10.1016/J.APGEOG.2014.11.024.

Gideon Neba, S., M. Kanninen, R. Eba'a Atyi, and D.J. Sonwa, 2014: Assessment and prediction of above-ground biomass in selectively logged forest concessions using field measurements and remote sensing data: Case study in South East Cameroon. *For. Ecol. Manage.*, **329**, 177–185, doi:10.1016/J.FORECO.2014.06.018.

Giger, M., H. Liniger, C. Sauter, and G. Schwilch, 2018: Economic benefits and costs of sustainable land management technologies: An analysis of WOCAT's global data. *L. Degrad. Dev.*, **29**, 962–974, doi:10.1002/ldr.2429.

Giguère-Croteau, C. et al. 2019: North America's oldest boreal trees are more efficient water users due to increased CO_2, but do not grow faster. *Proc. Natl. Acad. Sci.*, **116**, 2749–2754, doi:10.1073/pnas.1816686116.

Gill, S., J. Handley, A. Ennos, and S. Pauleit, 2007: Adapting cities for climate change: The role of the green infrastructure. *Built Environ.*, **33**, 115–133, doi:10.2148/benv.33.1.115.

Gingrich, S. et al. 2015: Exploring long-term trends in land use change and aboveground human appropriation of net primary production in nine European countries. *Land use policy*, **47**, 426–438, doi:10.1016/J.LANDUSEPOL.2015.04.027.

Giorgi, F., and P. Lionello, 2008: Climate change projections for the Mediterranean region. *Glob. Planet. Change*, **63**, 90–104, doi:10.1016/J.GLOPLACHA.2007.09.005.

Girardin, M.P. et al. 2016: No growth stimulation of Canada's boreal forest under half-century of combined warming and CO_2 fertilization. *Proc. Natl. Acad. Sci. U.S.A.*, **113**, E8406–E8414, doi:10.1073/pnas.1610156113.

Glover, J.D. et al. 2010: Harvested perennial grasslands provide ecological benchmarks for agricultural sustainability. *Agric. Ecosyst. Environ.*, **137**, 3–12, doi:10.1016/j.agee.2009.11.001.

Gockowski, J., and D. Sonwa, 2011: Cocoa intensification scenarios and their predicted impact on CO_2 emissions, biodiversity conservation, and rural livelihoods in the Guinea rain forest of West Africa. *Environ. Manage.*, **48**, 307–321, doi:10.1007/s00267-010-9602-3.

Gonzalez-Hidalgo, J.C., M. de Luis, and R.J. Batalla, 2009: Effects of the largest daily events on total soil erosion by rainwater. An analysis of the USLE database. *Earth Surf. Process. Landforms*, **34**, 2070–2077, doi:10.1002/esp.1892.

Gonzalez-Hidalgo, J.C., R.J. Batalla, A. Cerda, and M. de Luis, 2012: A regional analysis of the effects of largest events on soil erosion. *CATENA*, **95**, 85–90, doi:10.1016/J.CATENA.2012.03.006.

Gonzalez, P., R.P. Neilson, J.M. Lenihan, and R.J. Drapek, 2010: Global patterns in the vulnerability of ecosystems to vegetation shifts due to climate change. *Glob. Ecol. Biogeogr.*, **19**, 755–768, doi:10.1111/j.1466-8238.2010.00558.x.

Gopal, B., and M. Chauhan, 2006: Biodiversity and its conservation in the Sundarbans Mangrove Ecosystem. *Aquat. Sci.*, **68**, 338–354, doi:10.1007/s00027-006-0868-8.

Government of Tuvalu, 2006: *National Action Plan to Combat Land Degradation and Drought*. Funafuti, Tuvalu,. 38 pp.

Grainger, A., 2009: The role of science in implementing international environmental agreements: The case of desertification. *L. Degrad. Dev.*, **20**, 410–430, doi:10.1002/ldr.898.

Grandy, A.S., and G.P. Robertson, 2006: Aggregation and organic matter protection following tillage of a previously uncultivated soil. *Soil Sci. Soc. Am. J.*, **70**, 1398, doi:10.2136/sssaj2005.0313.

Grandy, A.S., and J.C. Neff, 2008: Molecular C dynamics downstream: The biochemical decomposition sequence and its impact on soil organic matter structure and function. *Sci. Total Environ.*, **404**, 297–307, doi:10.1016/j.scitotenv.2007.11.013.

Grassini, P., K.M. Eskridge, and K.G. Cassman, 2013: Distinguishing between yield advances and yield plateaus in historical crop production trends. *Nat. Commun.*, **4**, 2918, doi:10.1038/ncomms3918.

Gray, C., and R. Bilsborrow, 2013: Environmental influences on human migration in rural Ecuador. *Demography*, **50**, 1217–1241, doi:10.1007/s13524-012-0192-y.

Gray, C.L., 2011: Soil quality and human migration in Kenya and Uganda. *Glob. Environ. Chang.*, **21**, 421–430, doi:10.1016/J.GLOENVCHA.2011.02.004.

Green, A.J. et al. 2017: Creating a safe operating space for wetlands in a changing climate. *Front. Ecol. Environ.*, **15**, 99–107, doi:10.1002/fee.1459.

Green Surge, 2016: *Advancing Approaches And Strategies For UGI Planning And Implementation*. Green Surge Report D5.2, Brussels, 204 pp.

Grimm, V., and J.M. Calabrese, 2011: *What Is Resilience? A Short Introduction*. Springer, Berlin, Heidelberg, Germany, pp. 3–13.

Griscom, B.W. et al. 2017: Natural climate solutions. *Proc. Natl. Acad. Sci.*, **114**, 11645–11650, doi:10.1073/pnas.1710465114.

van Groenigen, J.W. et al. 2017: Sequestering soil organic carbon: A nitrogen dilemma. *Environ. Sci. Technol.*, **51**, 4738–4739, doi:10.1021/acs.est.7b01427.

Groisman, P.Y. et al. 2005: Trends in intense precipitation in the climate record. *J. Clim.*, **18**, 1326–1350, doi:10.1175/JCLI3339.1.

Grubler, A. et al. 2018: A low energy demand scenario for meeting the 1.5 °C target and sustainable development goals without negative emission technologies. *Nat. Energy*, **3**, 515–527, doi:10.1038/s41560-018-0172-6.

Guerreiro, S.B. et al. 2018: Detection of continental-scale intensification of hourly rainfall extremes. *Nat. Clim. Chang.*, **8**, 803–807, doi:10.1038/s41558-018-0245-3.

Gumbricht, T. et al. 2017: An expert system model for mapping tropical wetlands and peatlands reveals South America as the largest contributor. *Glob. Chang. Biol.*, **23**, 3581–3599, doi:10.1111/gcb.13689.

Guo, J.H. et al. 2010: Significant acidification in major Chinese croplands. *Science*, **327**, 1008–1010, doi:10.1126/science.1182570.

Guo, L.B., and R.M. Gifford, 2002: Soil carbon stocks and land use change: A meta analysis. *Glob. Chang. Biol.*, **8**, 345–360, doi:10.1046/j.1354-1013.2002.00486.x.

Gupta, S., and S. Kumar, 2017: Simulating climate change impact on soil erosion using RUSLE model – A case study in a watershed of mid-Himalayan landscape. *J. Earth Syst. Sci.*, **126**, 43, doi:10.1007/s12040-017-0823-1.

Haberl, H. et al. 2013: Bioenergy: How much can we expect for 2050? *Environ. Res. Lett.*, **8**, 031004, doi:10.1088/1748-9326/8/3/031004.

Hadden, D., and A. Grelle, 2016: Changing temperature response of respiration turns boreal forest from carbon sink into carbon source. *Agric. For. Meteorol.*, **223**, 30–38, doi:10.1016/j.agrformet.2016.03.020.

Haider, G., D. Steffens, G. Moser, C. Müller, and C.I. Kammann, 2017: Biochar reduced nitrate leaching and improved soil moisture content without yield improvements in a four-year field study. *Agric. Ecosyst. Environ.*, **237**, 80–94, doi:10.1016/j.agee.2016.12.019.

Håkansson, N.T., and M. Widgren, 2007: Labour and landscapes: The political economy of landesque capital in nineteenth century Tanganyika. *Geogr. Ann. Ser. B, Hum. Geogr.*, **89**, 233–248, doi:10.1111/j.1468-0467.2007.00251.x.

Hallegatte, S., M. Fay, and E.B. Barbier, 2018: Poverty and climate change: Introduction. *Environ. Dev. Econ.*, **23**, 217–233, doi:10.1017/S1355770X18000141.

Halvorson, W.L., A.E. Castellanos, and J. Murrieta-Saldivar, 2003: Sustainable land use requires attention to ecological signals. *Environ. Manage.*, **32**, 551–558, doi:10.1007/s00267-003-2889-6.

Hamza, M.A., and W.K. Anderson, 2005: Soil compaction in cropping systems. *Soil Tillage Res.*, **82**, 121–145, doi:10.1016/j.still.2004.08.009.

Handmer, J., and J. Nalau, 2019: *Understanding Loss and Damage in Pacific Small Island Developing States*. Springer, Cham, Switzerland, pp. 365–381

Hansen, M.C. et al. 2013: High-resolution global maps of 21st-century forest cover change. *Science*, **342**, 850–853.

Haque, F., R.M. Santos, A. Dutta, M. Thimmanagari, and Y.W. Chiang, 2019: Co-benefits of wollastonite weathering in agriculture: CO_2 sequestration and promoted plant growth. *ACS Omega*, **4**, 1425–1433, doi:10.1021/acsomega.8b02477.

Haregeweyn, N. et al. 2015: Soil erosion and conservation in Ethiopia. *Prog. Phys. Geogr. Earth Environ.*, **39**, 750–774, doi:10.1177/0309133315598725.

Harley, M.D. et al. 2017: Extreme coastal erosion enhanced by anomalous extratropical storm wave direction. *Sci. Rep.*, **7**, 6033, doi:10.1038/s41598-017-05792-1.

Harmon, M.E., W.K. Ferrell, and J.F. Franklin, 1990: Effects on carbon storage of conversion of old-growth forests to young forests. *Science*, **247**, 699–702.

Harper, A.B. et al. 2018: Land-use emissions play a critical role in land-based mitigation for Paris climate targets. *Nat. Commun.*, **9**, doi:10.1038/s41467-018-05340-z.

Harsch, M.A., P.E. Hulme, M.S. McGlone, and R.P. Duncan, 2009: Are treelines advancing? A global meta-analysis of treeline response to climate warming. *Ecol. Lett.*, **12**, 1040–1049, doi:10.1111/j.1461-0248.2009.01355.x.

Hauer, M.E., J.M. Evans, and D.R. Mishra, 2016: Millions projected to be at risk from sea-level rise in the continental United States. *Nat. Clim. Chang.*, **6**, 691–695, doi:10.1038/nclimate2961.

Hayes, R. et al. 2018: The performance of early-generation perennial winter cereals at 21 sites across four continents. *Sustainability*, **10**, 1124, doi:10.3390/su10041124.

He, Y. et al. 2017: Effects of biochar application on soil greenhouse gas fluxes: A meta-analysis. *GCB Bioenergy*, **9**, 743–755, doi:10.1111/gcbb.12376.

Hecht, and S.B., 1983: Cattle Ranching in the eastern Amazon: Environmental and Social Implications. In: *The Dilemma of Amazonian Development*, [Moran, E.F. (ed.)]. Westview Press, Boulder, CO, USA, pp. 155–188.

Heck, V., D. Gerten, W. Lucht, and A. Popp, 2018: Biomass-based negative emissions difficult to reconcile with planetary boundaries. *Nat. Clim. Chang.*, **8**, 151–155, doi:10.1038/s41558-017-0064-y.

Hellmann, J.J., J.E. Byers, B.G. Bierwagen, and J.S. Dukes, 2008: Five potential consequences of climate change for invasive species. *Conserv. Biol.*, **22**, 534–543, doi:10.1111/j.1523-1739.2008.00951.x.

Hember, R.A., W.A. Kurz, and N.C. Coops, 2016: Relationships between individual-tree mortality and water-balance variables indicate positive trends in water stress-induced tree mortality across North America. *Glob. Chang. Biol.*, **23**, 1691–1710, doi:10.1111/gcb.13428.

Hember, R.A., W.A. Kurz, and N.C. Coops, 2017: Increasing net ecosystem biomass production of Canada's boreal and temperate forests despite decline in dry climates. *Global Biogeochem. Cycles*, **31**, 134–158, doi:10.1002/2016GB005459.

Henry, B., B. Murphy, and A. Cowie, 2018: *Sustainable Land Management for Environmental Benefits and Food Security*. A synthesis report for the GEF. Washington DC, USA, 127 pp.

Henttonen, H.M., P. Nöjd, and H. Mäkinen, 2017: Environment-induced growth changes in the Finnish forests during 1971–2010 – An analysis based on national forest inventory. *For. Ecol. Manage.*, **386**, 22–36, doi:10.1016/j.foreco.2016.11.044.

Herbert, E.R. et al. 2015: A global perspective on wetland salinization: Ecological consequences of a growing threat to freshwater wetlands. *Ecosphere*, **6**, art206, doi:10.1890/ES14-00534.1.

Hergoualc'h, K., V.H. Gutiérrez-vélez, M. Menton, and L. V Verchot, 2017a: Forest ecology and management characterizing degradation of palm swamp peatlands from space and on the ground: An exploratory study in the Peruvian Amazon. *For. Ecol. Manage.*, **393**, 63–73, doi:10.1016/j.foreco.2017.03.016.

Hergoualc'h, K., D.T. Hendry, D. Murdiyarso, and L.V. Verchot, 2017b: Total and heterotrophic soil respiration in a swamp forest and oil palm plantations on peat in Central Kalimantan, Indonesia. *Biogeochemistry*, **135**, 203–220, doi:10.1007/s10533-017-0363-4.

Hermans, K., and T. Ide, 2019: *Advancing Research on Climate Change, Conflict and Migration*. Die Erde, **150**(1), 40–44, doi:10.12854/erde-2019-411.

Hernández-Morcillo, M., T. Plieninger, and C. Bieling, 2013: An empirical review of cultural ecosystem service indicators. *Ecol. Indic.*, **29**, 434–444, doi:10.1016/J.ECOLIND.2013.01.013.

Herrick, J.E. et al. 2019: A strategy for defining the reference for land health and degradation assessments. *Ecol. Indic.*, **97**, 225–230, doi:10.1016/J.ECOLIND.2018.06.065.

Hickey, G.M., M. Pouliot, C. Smith-Hall, S. Wunder, and M.R. Nielsen, 2016: Quantifying the economic contribution of wild food harvests to rural livelihoods: A global-comparative analysis. *Food Policy*, **62**, 122–132, doi:10.1016/J.FOODPOL.2016.06.001.

Hickler, T. et al. 2005: Precipitation controls Sahel greening trend. *Geophys. Res. Lett.*, **32**, L21415, doi:10.1029/2005GL024370.

Higginbottom, T., E. Symeonakis, T.P. Higginbottom, and E. Symeonakis, 2014: Assessing land degradation and desertification using vegetation index data: current frameworks and future directions. *Remote Sens.*, **6**, 9552–9575, doi:10.3390/rs6109552.

Higgins, S., I. Overeem, A. Tanaka, and J.P.M. Syvitski, 2013: Land subsidence at aquaculture facilities in the Yellow River delta, China. *Geophys. Res. Lett.*, **40**, 3898–3902, doi:10.1002/grl.50758.

Hilber, I. et al. 2017: Bioavailability and bioaccessibility of polycyclic aromatic hydrocarbons from (post-pyrolytically treated) biochars. *Chemosphere*, **174**, 700–707, doi:10.1016/J.CHEMOSPHERE.2017.02.014.

Hinsley, A., A. Entwistle, and D.V. Pio, 2015: Does the long-term success of REDD+ also depend on biodiversity? *Oryx*, **49**, 216–221, doi:10.1017/S0030605314000507.

Hoegh-Guldberg, O. et al. 2018: Impacts of 1.5°C global warming on natural and human systems. In: Global Warming of 1.5°C: An IPCC special report on the impacts of global warming of 1.5°C above pre-industrial levels and related global greenhouse gas emission pathways, in the context of strengthening the global response to the threat of climate change [V. Masson-Delmotte, P. Zhai, H.-O. Pörtner, D. Roberts, J. Skea, P.R. Shukla, A. Pirani, W. Moufouma-Okia, C. Péan, R. Pidcock, S. Connors, J.B.R. Matthews, Y. Chen, X. Zhou, M.I. Gomis, E. Lonnoy, T. Maycock, M. Tignor, and T. Waterfield (eds.)]. In press.

Hoffmann, H.K., K. Sander, M. Brüntrup, and S. Sieber, 2017: Applying the water-energy-food nexus to the charcoal value chain. *Front. Environ. Sci.*, 5, 84.

Högy, P. and A. Fangmeier, 2008: Effects of elevated atmospheric CO_2 on grain quality of wheat. *J. Cereal Sci.*, **48**, 580–591, doi:10.1016/J.JCS.2008.01.006.

Hojas-Gascon, L. et al. 2016: *Urbanization and Forest Degradation In East Africa: A Case Study Around Dar es Salaam, Tanzania*. IEEE International Geoscience and Remote Sensing Symposium (IGARSS), Beijing, pp. 7293–7295.

Holmberg, M. et al. 2019: Ecosystem services related to carbon cycling – modeling present and future impacts in boreal forests. *Front. Plant Sci.*, **10**.

Holt-Giménez, E., 2002: Measuring farmers' agroecological resistance after Hurricane Mitch in Nicaragua: A case study in participatory, sustainable land management impact monitoring. *Agric. Ecosyst. Environ.*, **93**, 87–105, doi:10.1016/S0167-8809(02)00006-3.

Holtum, J.A.M., and K. Winter, 2010: Elevated CO_2 and forest vegetation: More a water issue than a carbon issue? *Funct. Plant Biol.*, **37**, 694, doi:10.1071/FP10001.

Homer-Dixon, T.F., J.H. Boutwell, and G.W. Rathjens, 1993: Environmental change and violent conflict. *Sci. Am.*, **268**, 38–45, doi:10.2307/24941373.

Hosonuma, N. et al. 2012: An assessment of deforestation and forest degradation drivers in developing countries. *Environ. Res. Lett.*, **7**, 044009, doi:10.1088/1748-9326/7/4/044009.

Hou, B., H. Liao, and J. Huang, 2018: Household cooking fuel choice and economic poverty: Evidence from a nationwide survey in China. *Energy Build.*, **166**, 319–329, doi:10.1016/J.ENBUILD.2018.02.012.

Le Houerou, H.N., 1984: Rain use efficiency: a unifying concept in arid-land ecology. *J. Arid Environ.*, **7**, 213–247.

Houghton, R.A. and A.A. Nassikas, 2017: Global and regional fluxes of carbon from land use and land cover change 1850–2015. *Global Biogeochem. Cycles*, **31**, 456–472, doi:10.1002/2016GB005546.

Houghton, R.A. and A.A. Nassikas, 2018: Negative emissions from stopping deforestation and forest degradation, globally. *Glob. Chang. Biol.*, **24**, 350–359, doi:10.1111/gcb.13876.

Houghton, R.A. et al. 2012: Carbon emissions from land use and land-cover change. *Biogeosciences*, **9**, 5125–5142, doi:10.5194/bg-9-5125-2012.

Houspanossian, J., M. Nosetto, and E.G. Jobbágy, 2013: Radiation budget changes with dry forest clearing in temperate Argentina. *Glob. Chang. Biol.*, **19**, 1211–1222, doi:10.1111/gcb.12121.

Howden, S.M. et al. 2007: Adapting agriculture to climate change. *Proc. Natl. Acad. Sci.*, **104**, 19691–19696, doi:10.1073/pnas.0701890104.

Hu, S., Z. Niu, Y. Chen, L. Li, and H. Zhang, 2017: Global wetlands: Potential distribution, wetland loss, and status. *Sci. Total Environ.*, **586**, 319–327, doi:10.1016/j.scitotenv.2017.02.001.

Hua, F. et al. 2016: Opportunities for biodiversity gains under the world's largest reforestation programme. *Nat. Commun.*, **7**, doi:10.1038/ncomms12717.

Huang, G. et al. 2018: Performance, economics and potential impact of perennial rice PR23 relative to annual rice cultivars at multiple locations in Yunnan Province of China. *Sustainability*, **10**, 1086, doi:10.3390/su10041086.

Huang, J., T. Wang, W. Wang, Z. Li, and H. Yan, 2014: Climate effects of dust aerosols over East Asian arid and semiarid regions. *J. Geophys. Res. Atmos.*, **119**, 11,398–11,416, doi:10.1002/2014JD021796.

Huggel, C., J.J. Clague, and O. Korup, 2012: Is climate change responsible for changing landslide activity in high mountains? *Earth Surf. Process. Landforms*, **37**, 77–91, doi:10.1002/esp.2223.

Hulme, P.E., 2017: Climate change and biological invasions: evidence, expectations, and response options. *Biol. Rev.*, **92**, 1297–1313, doi:10.1111/brv.12282.

Hungate, B.A. et al. 2017: The economic value of grassland species for carbon storage. *Sci. Adv.*, **3**, e1601880, doi:10.1126/sciadv.1601880.

Hussain, S., T. Siddique, M. Saleem, M. Arshad, and A. Khalid, 2009: Impact of pesticides on soil microbial diversity, enzymes, and biochemical reactions. *Adv. Agron.*, **102**, 159–200, doi:10.1016/S0065-2113(09)01005-0.

Hüve, K., I. Bichele, B. Rasulov, and Ü. Niinemets, 2011: When it is too hot for photosynthesis: Heat-induced instability of photosynthesis in relation to respiratory burst, cell permeability changes and H_2O_2 formation. *Plant. Cell Environ.*, **34**, 113–126, doi:10.1111/j.1365-3040.2010.02229.x.

Huxman, T.E. et al. 2004: Convergence across biomes to a common rain-use efficiency. *Nature*, **429**, 651–654, doi:10.1038/nature02561.

IEA, 2017: *World Energy Outlook 2017*. International Energy Agency, Paris. 763 pp.

Illius, A.W., O'Connor, T.G., 1999. On the relevance of nonequilibrium concepts to arid and semiarid grazing systems. *Ecological Applications* 9(3), 798–813, doi: 10.1890/1051-0761(1999)009[0798:OTRONC]2.0.CO;2.

Imai, N. et al. 2012: Effects of selective logging on tree species diversity and composition of Bornean tropical rain forests at different spatial scales. *Plant Ecol.*, **213**, 1413–1424, doi:10.1007/s11258-012-0100-y.

Iordan, C.M., X. Hu, A. Arvesen, P. Kauppi, and F. Cherubini, 2018: Contribution of forest wood products to negative emissions: Historical comparative analysis from 1960 to 2015 in Norway, Sweden and Finland. *Carbon Balance Manag.*, **13**, doi:10.1186/s13021-018-0101-9.

IPCC, 2006: 2006 IPCC Guidelines for National Greenhouse Gas Inventories – A Primer. [Eggleston H.S., K. Miwa, N. Srivastava, and K. Tanabe (eds.)]. Institute for Global Environmental Strategies (IGES) for the Intergovernmental Panel on Climate Change. IGES, Japan, 20 pp.

IPCC, 2012: Managing the Risks of Extreme Events and Disasters to Advance Climate Change Adaptation: Special report of the Intergovernmental Panel on Climate Change. Cambridge University Press, Cambridge, UK and New York, USA, 582 pp.

IPCC, 2013a: Annex I: Atlas of Global and Regional Climate Projections. In: Climate Change 2013: The Physical Science Basis. Contribution of Working Group I to the Fifth Assessment Report of the Intergovernmental Panel on Climate Change, [Stocker, T.F., D. Qin, G.-K. Plattner, M. Tignor, S.K. Allen, J. Boschung, A. Nauels, Y. Xia, V. Bex and P.M. Midgley (eds.)]. Cambridge University Press, Cambridge, UK and New York, NY, USA, 1313–1390 pp.

IPCC, 2013b: Summary for Policy Makers. In: Climate Change 2013: The Physical Science Basis. Contribution of Working Group I to the Fifth Assessment Report of the Intergovernmental Panel on Climate Change, [Stocker, T.F., D. Qin, G.-K. Plattner, M. Tignor, S.K. Allen, J. Boschung, A. Nauels, Y. Xia, V. Bex and P.M. Midgley (eds.)]. Cambridge University Press, Cambridge, UK and New York, USA, p. 1535.

IPCC, 2014a: 2013 Supplement to the 2006 IPCC Guidelines for National Greenhouse Gas Inventories: Wetlands. [Blain, D. Boer, R., Eggleston S., Gonzalez, S., Hiraishi, T., Irving, W., Krug, T., Krusche, A., Mpeta, E.J., Penman, J., Pipatti, R., Sturgiss, R., Tanabe, K., Towprayoon, S.], IPCC Geneva, 354 pp.

IPCC, 2014b: Summary for Policymakers. In: Climate Change 2014: Mitigation of Climate Change. Contribution of Working Group III to the Fifth Assessment Report of the Intergovernmental Panel on Climate Change [Edenhofer, O., R. Pichs-Madruga, Y. Sokona, E. Farahani, S. Kadner, K. Seyboth, A. Adler, I. Baum, S. Brunner, P. Eickemeier, B. Kriemann, J. Savolainen, S. Schlömer, C. von Stechow, T. Zwickel and J.C. Minx (eds.)]. Cambridge University Press, Cambridge, United Kingdom and New York, NY, USA., pp. 1–34.

IPCC, 2018a: Summary for Policymakers. In: Global Warming of 1.5°C: An IPCC special report on the impacts of global warming of 1.5°C above pre-industrial levels and related global greenhouse gas emission pathways, in the context of strengthening the global response to the threat of climate change. [V. Masson-Delmotte, P. Zhai, H.-O. Pörtner, D. Roberts, J. Skea, P.R. Shukla, A. Pirani, W. Moufouma-Okia, C. Péan, R. Pidcock, S. Connors, J.B.R. Matthews, Y. Chen, X. Zhou, M.I. Gomis, E. Lonnoy, T. Maycock, M. Tignor, and T. Waterfield (eds.)]. In press.

IPCC, 2018b: Global Warming of 1.5°C an IPCC special report on the impacts of global warming of 1.5°C above pre-industrial levels and related global greenhouse gas emission pathways, in the context of strengthening the global response to the threat of climate change. [V. Masson-Delmotte, P. Zhai, H.-O. Pörtner, D. Roberts, J. Skea, P.R. Shukla, A. Pirani, W. Moufouma-Okia, C. Péan, R. Pidcock, S. Connors, J.B.R. Matthews, Y. Chen, X. Zhou, M.I. Gomis, E. Lonnoy, T. Maycock, M. Tignor, and T. Waterfield (eds.)]. In press.

IPCC, 2014: Annex II, 2014c: Glossary. In: Climate Change 2014: Synthesis Report. Contribution of Working Groups I, II and III to the Fifth Assessment Report of the Intergovernmental Panel on Climate Change. Mach, K.J., S. Planton and C. von Stechow (eds.). Cambridge University Press, Cambridge, United Kingdom and New York, NY, USA, 117–130 pp.

Islam, M.S., A.T. Wong, M.S. Islam, and A.T. Wong, 2017: Climate change and food in/security: A critical nexus. *Environments*, **4**, 38, doi:10.3390/environments4020038.

Itkonen, P., 2016: Land rights as the prerequisite for Sámi culture: Skolt Sámi's changing relation to nature in Finland. In: *Indigenous Rights in Modern Landscapes*, Elenius, L., Allard, C. and Sandström, C. (eds.)]. Routledge, Abingdon, Oxfordshire, UK, 94–105.

IUCN, 2003: *Environmental Degradation And Impacts On Livelihood: Sea Intrusion – A Case Study*. International Union for Conservation of Nature, Sindh Programme Office, Pakistan. IUCN, Gland, Switzerland, 77 pp.

Jagger, P. et al. 2015: REDD+ safeguards in national policy discourse and pilot projects. In: *Analysing REDD+: Challenges and Choices*, [Angelsen, L., Brockhaus, A., Sunderlin, M., Verchot, W.D. (eds.)]. CIFOR, Bogor, Indonesia. 301–316.

Jastrow, J.D., J.E. Amonette, and V.L. Bailey, 2007: Mechanisms controlling soil carbon turnover and their potential application for enhancing carbon sequestration. *Clim. Change*, **80**, 5–23, doi:10.1007/s10584-006-9178-3.

Jeffery, S., F.G.A. Verheijen, C. Kammann, and D. Abalos, 2016: Biochar effects on methane emissions from soils: A meta-analysis. *Soil Biol. Biochem.*, **101**, 251–258, doi:10.1016/J.SOILBIO.2016.07.021.

Jerneck, A., 2018a: What about gender in climate change? Twelve feminist lessons from development. *Sustainability*, **10**, 627, doi:10.3390/su10030627.

Jerneck, A., 2018b: Taking gender seriously in climate change adaptation and sustainability science research: views from feminist debates and sub-Saharan small-scale agriculture. *Sustain. Sci.*, **13**, 403–416, doi:10.1007/s11625-017-0464-y.

Jerneck, A. and L. Olsson, 2013: More than trees! Understanding the agroforestry adoption gap in subsistence agriculture: Insights from narrative walks in Kenya. *J. Rural Stud.*, **32**, 114–125, doi:10.1016/J.JRURSTUD.2013.04.004.

Jerneck, A. and L. Olsson, 2014: Food first! Theorising assets and actors in agroforestry: Risk evaders, opportunity seekers and 'the food imperative' in sub-Saharan Africa. *Int. J. Agric. Sustain.*, **12**, 1–22, doi:10.1080/14735903.2012.751714.

Ji, C., K. Cheng, D. Nayak, and G. Pan, 2018: Environmental and economic assessment of crop residue competitive utilization for biochar, briquette fuel and combined heat and power generation. *J. Clean. Prod.*, **192**, 916–923, doi:10.1016/J.JCLEPRO.2018.05.026.

Jia, Y., Zhou, Z., Qiu, Y., 2006: The potential effect of water yield reduction caused by land conversion in the Yellow River basin. In: *Study of Sustainable Use of Land Resources in Northwestern China*. [Zhang L., J. Bennett J, X.H. Wang, C. Xie, and A. Zhao. (eds.)]. China National Forestry Economics and Development Research Center, Beijing, China.

Jobbágy, E.G. and O.E. Sala, 2000: Controls of grass and shrub aboveground production in the Patagonian steppe. *Ecol. Appl.*, **10**, 541–549, doi:10.1890/1051-0761(2000)010[0541:COGASA]2.0.CO;2.

Jobbágy, E.G., T. Tóth, M.D. Nosetto, and S. Earman, 2017: On the fundamental causes of high environmental alkalinity (pH ≥ 9): An assessment of its drivers and global distribution. *L. Degrad. Dev.*, **28**, 1973–1981, doi:10.1002/ldr.2718.

John, R. et al. 2016: Differentiating anthropogenic modification and precipitation-driven change on vegetation productivity on the Mongolian Plateau. *Landsc. Ecol.*, **31**, 547–566, doi:10.1007/s10980-015-0261-x.

Johnson, D.L. and L.A. Lewis, 2007: Land degradation: Creation and destruction. Rowman & Littlefield, Lanham, MD, 303 pp.

Johnson, J.M.-F., R.R. Allmaras, and D.C. Reicosky, 2006: Estimating source carbon from crop residues, roots and rhizodeposits using the National Grain-Yield database. *Agron. J.*, **98**, 622, doi:10.2134/agronj2005.0179.

Johnson, J.M. et al. 2015: Recent shifts in coastline change and shoreline stabilization linked to storm climate change. *Earth Surf. Process. Landforms*, **40**, 569–585, doi:10.1002/esp.3650.

Jolly, W.M. et al. 2015: Climate-induced variations in global wildfire danger from 1979 to 2013. *Nat. Commun.*, **6**, 7537, doi:10.1038/ncomms8537.

de Jong, J., C. Akselsson, G. Egnell, S. Löfgren, and B.A. Olsson, 2017: Realizing the energy potential of forest biomass in Sweden – How much is environmentally sustainable? *For. Ecol. Manage.*, **383**, 3–16, doi:10.1016/J.FORECO.2016.06.028.

Joosten, H., Tanneberger, F., 2017: Peatland use in Europe. In: *Mires and Peatlands of Europe: Status, Distribution And Conservation*. [Joosten H.,

4

Tanneberger F., Moen A., (eds.)]. Schweizerbart Science Publisher. Stuttgart, Germany, pp. 151–173.

Jorgenson, M.T. and T.E. Osterkamp, 2005: Response of boreal ecosystems to varying modes of permafrost degradation. *Can. J. For. Res.*, **35**, 2100–2111, doi:10.1139/x05-153.

Joseph, J., 2013: Resilience as embedded neoliberalism: A governmentality approach. *Resilience*, **1**, 38–52, doi:10.1080/21693293.2013.765741.

Joseph, S. et al. 2013: Shifting paradigms: Development of high-efficiency biochar fertilizers based on nano-structures and soluble components. *Carbon Manag.*, **4**, 323–343, doi:10.4155/cmt.13.23.

Joseph, S. et al. 2018: Microstructural and associated chemical changes during the composting of a high temperature biochar: Mechanisms for nitrate, phosphate and other nutrient retention and release. *Sci. Total Environ.*, **618**, 1210–1223, doi:10.1016/j.scitotenv.2017.09.200.

Joseph, S.D. et al. 2010: An investigation into the reactions of biochar in soil. *Soil Research*. **48**(7), 501–515. doi: 10.1071/SR10009.

JRC, 2018: World Atlas of Desertification. [Cherlet, M., Hutchinson, C., Reynolds, J., Hill, J., Sommer, S., von Maltitz, G. (Eds.)] Publication Office of the European Union, Luxembourg, 2018. 237 pp. doi: 10.2760/06292.

Juday, G.P., C. Alix, and T.A. Grant, 2015: Spatial coherence and change of opposite white spruce temperature sensitivities on floodplains in Alaska confirms early-stage boreal biome shift. *For. Ecol. Manage.*, **350**, 46–61, doi:10.1016/J.FORECO.2015.04.016.

Kalhoro, N. et al. 2016: Vulnerability of the Indus River Delta of the North Arabian Sea, Pakistan. *Glob. Nest J.* (gnest). **18**, 599–610.

Kalhoro, N.A., Z. He, D. Xu, I. Muhammad, and A.F. Sohoo, 2017: Seasonal variation of oceanographic processes in Indus River estuary. *Seas. Var. Oceanogr. Process. Indus river estuary.* MAUSAM, **68**, 643–654.

Kammann, C. et al. 2017: Biochar as a tool to reduce the agricultural greenhouse-gas burden – knowns, unknowns and future research needs. *J. Environ. Eng. Landsc. Manag.*, **25**, 114–139, doi:10.3846/16486897.2017. 1319375.

Kaplan, J.O. et al. 2011: Holocene carbon emissions as a result of anthropogenic land cover change. *The Holocene*, **21**, 775–791, doi:10.1177/0959683610386983.

Karami, M., M. Afyuni, A.H. Khoshgoftarmanesh, A. Papritz, and R. Schulin, 2009: Grain zinc, iron, and copper concentrations of wheat grown in central Iran and their relationships with soil and climate variables. *J. Agric. Food Chem.*, **57**, 10876–10882, doi:10.1021/jf902074f.

Karbassi, A., G.N. Bidhendi, A. Pejman, and M.E. Bidhendi, 2010: Environmental impacts of desalination on the ecology of Lake Urmia. *J. Great Lakes Res.*, **36**, 419–424, doi:10.1016/j.jglr.2010.06.004.

Kassa, H., S. Dondeyne, J. Poesen, A. Frankl, and J. Nyssen, 2017: Transition from forest-based to cereal-based agricultural systems: A review of the drivers of land use change and degradation in Southwest Ethiopia. *L. Degrad. Dev.*, **28**, 431–449, doi:10.1002/ldr.2575.

Kassam, A. et al. 2013: Sustainable soil management is more than what and how crops are grown. In: *Principles of Sustainable Soil Management in Agroecosystems*, [R. Lal and B.A. Stewart, (eds.)]. CRC Press, Boca Raton, Fl, USA, 337–400.

Kates, R.W., W.R. Travis, and T.J. Wilbanks, 2012: Transformational adaptation when incremental adaptations to climate change are insufficient. *Proc. Natl. Acad. Sci. U.S.A.*, **109**, 7156–7161, doi:10.1073/pnas.1115521109.

Kauppi, P.E., M. Posch, and P. Pirinen, 2014: Large impacts of climatic warming on growth of boreal forests since 1960. *PLoS One*, **9**, 1–6, doi:10.1371/journal.pone.0111340.

Kauppi, P.E., V. Sandström, and A. Lipponen, 2018: Forest resources of nations in relation to human well-being. *PLoS One,* **13**, e0196248, doi:10.1371/journal.pone.0196248.

Kaye, J.P., and M. Quemada, 2017: Using cover crops to mitigate and adapt to climate change. A review. *Agron. Sustain. Dev.*, **37**, 4, doi:10.1007/s13593-016-0410-x.

Keenan, R.J. et al. 2015: Dynamics of global forest area: Results from the FAO Global Forest Resources Assessment 2015. *For. Ecol. Manage.*, **352**, 9–20, doi:10.1016/J.FORECO.2015.06.014.

Keenan, T.F. et al. 2013: Increase in forest water-use efficiency as atmospheric carbon dioxide concentrations rise. *Nature*, **499**, 324–327, doi:10.1038/nature12291.

Keenan, T.F. et al. 2017: Corrigendum: Recent pause in the growth rate of atmospheric CO_2 due to enhanced terrestrial carbon uptake. *Nat. Commun.*, **8**, 16137, doi:10.1038/ncomms16137.

Keith, H. et al. 2014: Managing temperate forests for carbon storage: Impacts of logging versus forest protection on carbon stocks. *Ecosphere*, **5**, art75, doi:10.1890/ES14-00051.1.

Kelly (Letcher), R.A. et al. 2013: Selecting among five common modelling approaches for integrated environmental assessment and management. *Environ. Model. Softw.*, **47**, 159–181, doi:10.1016/J.ENVSOFT.2013.05.005.

Kendon, E.J. et al. 2014: Heavier summer downpours with climate change revealed by weather forecast resolution model. *Nat. Clim. Chang.*, **4**, 570–576, doi:10.1038/nclimate2258.

Keogh, M.E., and T.E. Törnqvist, 2019: Measuring rates of present-day relative sea-level rise in low-elevation coastal zones: A critical evaluation. *Ocean Sci.*, **15**, 61–73, doi:10.5194/os-15-61-2019.

Kerns, B.K., J.B. Kim, J.D. Kline, and M.A. Day, 2016: US exposure to multiple landscape stressors and climate change. *Reg. Environ. Chang.*, **16**, 2129–2140, doi:10.1007/s10113-016-0934-2.

Kerr, J.M., J.V. DePinto, D. McGrath, S.P. Sowa, and S.M. Swinton, 2016: Sustainable management of Great Lakes watersheds dominated by agricultural land use. *J. Great Lakes Res.*, **42**, 1252–1259, doi:10.1016/J.JGLR.2016.10.001.

Kessler, C.A. and L. Stroosnijder, 2006: Land degradation assessment by farmers in Bolivian mountain valleys. *L. Degrad. Dev.*, **17**, 235–248, doi:10.1002/ldr.699.

Keys, P.W., L. Wang-Erlandsson, L.J. Gordon, V. Galaz, and J. Ebbesson, 2017: Approaching moisture recycling governance. *Glob. Environ. Chang.*, **45**, 15–23, doi:10.1016/J.GLOENVCHA.2017.04.007.

Kiage, L.M., 2013: Perspectives on the assumed causes of land degradation in the rangelands of Sub-Saharan Africa. *Prog. Phys. Geogr.*, **37**, 664–684, doi:10.1177/0309133313492543.

Kidane, D. and B. Alemu, 2015: The effect of upstream land use practices on soil erosion and sedimentation in the Upper Blue Nile Basin, Ethiopia. *Res. J. Agric. Environ. Manag.*, **4**, 55–68.

Kilpeläinen, A. et al. 2017: Effects of initial age structure of managed Norway spruce forest area on net climate impact of using forest biomass for energy. *Bioenergy Res.*, **10**, 499–508, doi:10.1007/s12155-017-9821-z.

Kim, G.S. et al. 2017: Effect of national-scale afforestation on forest water supply and soil loss in South Korea, 1971–2010. *Sustainability*, **9**, 1017, doi:10.3390/su9061017.

Kim, K.H. and L. Zsuffa, 1994: Reforestation of South Korea: The history and analysis of a unique case in forest tree improvement and forestry. *For. Chron.*, **70**, 58–64, doi:10.5558/tfc70058-1.

Kim, M. et al. 2016: Estimating carbon dynamics in forest carbon pools under IPCC standards in South Korea using CBM-CFS3. *iForest – Biogeosciences For.*, **10**, 83–92, doi:10.3832/ifor2040-009.

Kiruki, H.M., E.H. van der Zanden, Ž. Malek, and P.H. Verburg, 2017: Land cover change and woodland degradation in a charcoal producing semi-arid area in Kenya. *L. Degrad. Dev.*, **28**, 472–481, doi:10.1002/ldr.2545.

Kirwan, M.L., A.B. Murray, J.P. Donnelly, and D.R. Corbett, 2011: Rapid wetland expansion during European settlement and its implication for marsh survival under modern sediment delivery rates. *Geology*, **39**, 507–510, doi:10.1130/G31789.1.

Klein, R.J.T. et al. 2014: Adaptation opportunities, constraints, and limits. In: Climate Change 2014: Impacts, Adaptation, and Vulnerability. Part A: Global and Sectoral Aspects. Contribution of Working Group II to the Fifth Assessment Report of the Intergovernmental Panel on Climate Change.

4

[Field, C.B., V.R. Barros, D.J. Dokken, K.J. Mach, M.D. Mastrandrea, T.E. Bilir, M. Chatterjee, K.L. Ebi, Y.O. Estrada, R.C. Genova, B. Girma, E.S. Kissel, A.N. Levy, S. MacCracken, P.R. Mastrandrea, and L.L.White (eds.)]. Cambridge University Press, Cambridge, UK and New York, USA, 899–943.

Klemm, W., B.G. Heusinkveld, S. Lenzholzer, and B. van Hove, 2015: Street greenery and its physical and psychological impact on thermal comfort. *Landsc. Urban Plan.*, **138**, 87–98, doi:10.1016/J.LANDURBPLAN.2015.02.009.

Klöck, C. and P.D. Nunn, 2019: Adaptation to climate change in small island developing states: A systematic literature review of academic research. *J. Environ. Dev.*, **28**(2), 196–218, 107049651983589, doi:10.1177/1070496519835895.

Knapp, A.K. and M.D. Smith, 2001: Variation among biomes in temporal dynamics of aboveground primary production. *Science*, 291, 481–484, doi:10.1126/science.291.5503.481.

Knorr, W., L. Jiang, and A. Arneth, 2016: Climate, CO_2 and human population impacts on global wildfire emissions. *Biogeosciences*, **13**, 267–282, doi:10.5194/bg-13-267-2016.

Knutson, T.R. et al. 2010: Tropical cyclones and climate change. *Nat. Geosci.*, **3**, 157–163, doi:10.1038/ngeo779.

Knutson, T.R. et al. 2015: Global projections of intense tropical cyclone activity for the late twenty-first century from dynamical downscaling of CMIP5/RCP4.5 scenarios. *J. Clim.*, **28**, 7203–7224, doi:10.1175/JCLI-D-15-0129.1.

Köhl, M., P.R. Neupane, and N. Lotfiomran, 2017: The impact of tree age on biomass growth and carbon accumulation capacity: A retrospective analysis using tree ring data of three tropical tree species grown in natural forests of Suriname. *PLoS One*, **12**, e0181187, doi:10.1371/journal.pone.0181187.

Kolinjivadi, V.K. and T. Sunderland, 2012: A review of two payment schemes for watershed services from China and Vietnam: The interface of government control and PES theory. *Ecol. Soc.*, **17**(4), doi:10.5751/ES-05057-170410.

Kolton, M., E.R. Graber, L. Tsehansky, Y. Elad, and E. Cytryn, 2017: Biochar-stimulated plant performance is strongly linked to microbial diversity and metabolic potential in the rhizosphere. *New Phytol.*, **213**(3), 1393–1404, doi:10.1111/nph.14253.

Koohafkann, P., and M.A. Altieri, 2011: *Agricultural Heritage Systems: A legacy for the Future*. Food and Agriculture Organization of the United Nations, Rome, Italy, 49 pp.

Koplitz, S.N. et al. 2016: Public health impacts of the severe haze in Equatorial Asia in September –October 2015: Demonstration of a new framework for informing fire management strategies to reduce downwind smoke exposure. *Environ. Res. Lett.*, **11**, doi:10.1088/1748-9326/11/9/094023.

Korea Forest Service, 2017: *Statistical Yearbook of Forestry*. Korea Forest Service, Ed. Deajeon, Korea.

Korhonen-Kurki, K. et al. 2018: What drives policy change for REDD+? A qualitative comparative analysis of the interplay between institutional and policy arena factors. *Clim. Policy*, 1–14, doi:10.1080/14693062.2018.1507897.

Kosec, K., H. Ghebru, B. Holtemeyer, V. Mueller, and E. Schmidt, 2018: The effect of land access on youth employment and migration decisions: Evidence from rural Ethiopia. *Am. J. Agric. Econ.*, **100**, 931–954, doi:10.1093/ajae/aax087.

Kosmas, C. et al. 2014: Evaluation and selection of indicators for land degradation and desertification monitoring: Methodological approach. *Environ. Manage.*, **54**, 951–970, doi:10.1007/s00267-013-0109-6.

Kossin, J.P., 2018: A global slowdown of tropical-cyclone translation speed. *Nature*, **558**, 104–107, doi:10.1038/s41586-018-0158-3.

Krause, A. et al. 2018: Large uncertainty in carbon uptake potential of land-based climate-change mitigation efforts. *Glob. Chang. Biol.*, **24**, 3025–3038, doi:10.1111/gcb.14144.

Kreamer, D.K., 2012: The past, present, and future of water conflict and international security. *J. Contemp. Water Res. Educ.*, **149**, 87–95, doi:10.1111/j.1936-704X.2012.03130.x.

Kristjanson, P. et al. 2017: Addressing gender in agricultural research for development in the face of a changing climate: Where are we and where should we be going? *Int. J. Agric. Sustain.*, **15**, 482–500, doi:10.1080/14735903.2017.1336411.

Kroeger, A. et al. 2017: *Forest – and Climate-Smart Cocoa in Côte d'Ivoire and Ghana, Aligning Stakeholders to Support Smallholders in Deforestation-Free Cocoa*. World Bank, Washington, DC, USA, 57 pp.

Kulakowski, D., T.T. Veblen, and P. Bebi, 2003: Effects of fire and spruce beetle outbreak legacies on the disturbance regime of a subalpine forest in Colorado. *J. Biogeogr.*, **30**, 1445–1456, doi:10.1046/j.1365-2699.2003.00912.x.

Kumar, N. and A.R. Quisumbing, 2015: Policy reform toward gender equality in Ethiopia: Little by little the egg begins to walk. *World Dev.*, **67**, 406–423, doi:10.1016/J.WORLDDEV.2014.10.029.

Kuppusamy, S., P. Thavamani, M. Megharaj, K. Venkateswarlu, and R. Naidu, 2016: Agronomic and remedial benefits and risks of applying biochar to soil: Current knowledge and future research directions. *Environ. Int.*, **87**, 1–12, doi:10.1016/J.ENVINT.2015.10.018.

Kurz, W.A., S.J. Beukema, and M.J. Apps, 1998: Carbon budget implications of the transition from natural to managed disturbance regimes in forest landscapes. *Mitig. Adapt. Strateg. Glob. Chang.*, **2**, 405–421, doi:10.1023/b:miti.0000004486.62808.29.

Kurz, W.A. et al. 2008: Mountain pine beetle and forest carbon feedback to climate change. *Nature*, **452**, 987–990, doi:10.1038/nature06777.

Kurz, W.A. et al. 2013: Carbon in Canada's boreal forest – A synthesis. *Environ. Rev.*, **21**, 260–292, doi:10.1139/er-2013-0041.

Kurz, W.A., C. Smyth, and T. Lemprière, 2016: Climate change mitigation through forest sector activities: Principles, potential and priorities. *Unasylva*, **67**, 61–67.

Labrière, N., B. Locatelli, Y. Laumonier, V. Freycon, and M. Bernoux, 2015: Soil erosion in the humid tropics: A systematic quantitative review. *Agric. Ecosyst. Environ.*, **203**, 127–139, doi:10.1016/J.AGEE.2015.01.027.

Laderach, P. et al. 2011: *Predicted Impact of Climate Change on Coffee Supply Chains*. Springer, Berlin, Heidelberg, Germany, pp. 703–723.

Ladha, J.K., H. Pathak, T.J. Krupnik, J. Six, and C. van Kessel, 2005: Efficiency of fertilizer nitrogen in cereal production: Retrospects and prospects. *Adv. Agron.*, **87**, 85–156, doi:10.1016/S0065-2113(05)87003-8.

Lado, M., M. Ben-Hur, and I. Shainberg, 2004: Soil wetting and texture effects on aggregate stability, seal formation, and erosion. *Soil Sci. Soc. Am. J.*, **68**, 1992, doi:10.2136/sssaj2004.1992.

Laganière, J., D. Paré, E. Thiffault, and P.Y. Bernier, 2017: Range and uncertainties in estimating delays in greenhouse gas mitigation potential of forest bioenergy sourced from Canadian forests. *GCB Bioenergy*, **9**, 358–369, doi:10.1111/gcbb.12327.

Lal, R., 2003: Soil erosion and the global carbon budget. *Environ. Int.*, **29**, 437–450, doi:10.1016/S0160-4120(02)00192-7.

Lamb, D., 2014: *Large-scale Forest Restoration*. Routledge, London, UK, 302 pp.

Lambin, E.F. and P. Meyfroidt, 2011: Global land use change, economic globalization, and the looming land scarcity. *Proc. Natl. Acad. Sci. U.S.A.*, **108**, 3465–3472, doi:10.1073/pnas.1100480108.

Lambin, E.F. et al. 2001: The causes of land-use and land-cover change: Moving beyond the myths. *Glob. Environ. Chang.*, **11**, 261–269, doi:10.1016/S0959-3780(01)00007-3.

Lambin, E.F., H.J. Geist, and E. Lepers, 2003: Dynamics of land – use and land – cover change in tropical regions. *Annu. Rev. Environ. Resour.*, **28**, 205–241, doi:10.1146/annurev.energy.28.050302.105459.

Lambin, E.F. et al. 2018: The role of supply-chain initiatives in reducing deforestation. *Nat. Clim. Chang.*, **8**, 109–116, doi:10.1038/s41558-017-0061-1.

Lamers, L.P.M. et al. 2015: Ecological restoration of rich fens in Europe and North America: From trial and error to an evidence-based approach. *Biol. Rev. Camb. Philos. Soc.*, **90**(1), 182–203, doi:10.1111/brv.12102.

Lamichhane, J.R. et al. 2015: Robust cropping systems to tackle pests under climate change. A review. *Agron. Sustain. Dev.*, **35**, 443–459, doi:10.1007/s13593-014-0275-9.

Landauer, M. and S. Juhola, 2019: *Loss and Damage in the Rapidly Changing Arctic.* Springer, Cham, Switzerland, pp. 425–447.

Laniak, G.F. et al. 2013: Integrated environmental modeling: A vision and roadmap for the future. *Environ. Model. Softw.*, **39**, 3–23, doi:10.1016/j.envsoft.2012.09.006.

Lau, K.M. and K.M. Kim, 2007: Cooling of the Atlantic by Saharan dust. *Geophys. Res. Lett.*, **34**, n/a-n/a, doi:10.1029/2007GL031538.

Lavers, T., 2017: Land registration and gender equality in Ethiopia: How state-society relations influence the enforcement of institutional change. *J. Agrar. Chang.*, **17**, 188–207, doi:10.1111/joac.12138.

Law, B.E. et al. 2018: Land use strategies to mitigate climate change in carbon dense temperate forests. *Proc. Natl. Acad. Sci. U.S.A.*, **115**, 3663–3668, doi:10.1073/pnas.1720064115.

Lawhon, M. and J.T. Murphy, 2012: Socio-technical regimes and sustainability transitions. *Prog. Hum. Geogr.*, **36**, 354–378, doi:10.1177/0309132511427960.

Lawrence, D. and K. Vandecar, 2015: Effects of tropical deforestation on climate and agriculture. *Nat. Clim. Chang.*, **5**, 27–36, doi:10.1038/nclimate2430.

Lawson, I.T. et al. 2015: Improving estimates of tropical peatland area, carbon storage, and greenhouse gas fluxes. *Wetl. Ecol. Manag.*, doi:10.1007/s11273-014-9402-2.

Le, Q.B., E. Nkonya and A. Mirzabaev, 2016: Biomass Productivity-Based Mapping of Global Land Degradation Hotspots. In: *Economics of Land Degradation and Improvement – A Global Assessment for Sustainable Development* [E. Nkonya, A. Mirzabaev, and J. Von Braun, (eds.)]. Springer International Publishing, Cham, Switzerland, pp. 55–84.

Lecina-Diaz, J. et al. 2018: The positive carbon stocks-biodiversity relationship in forests: Co-occurrence and drivers across five subclimates. *Ecol. Appl.*, **28**, 1481–1493, doi:10.1002/eap.1749.

Lee, D.K., P.S. Park, and Y.D. Park, 2015: Forest Restoration and Rehabilitation in the Republic of Korea. In: *Restoration of Boreal and Temperate Forests* [J.A. Stanturf, (ed.)]. CRC Press, Boca Raton, Florida, USA, pp. 230–245.

Lee, H.-L., 2009: The impact of climate change on global food supply and demand, food prices, and land use. *Paddy Water Environ.*, **7**, 321–331, doi:10.1007/s10333-009-0181-y.

Lee, J. et al. 2014: Estimating the carbon dynamics of South Korean forests from 1954 to 2012. *Biogeosciences*, **11**, 4637–4650, doi:10.5194/bg-11-4637-2014.

Lee, J. et al. 2018a: Economic viability of the national-scale forestation program: The case of success in the Republic of Korea. *Ecosyst. Serv.*, **29**, 40–46, doi:10.1016/j.ecoser.2017.11.001.

Lee, S.G. et al. 2018b: Restoration plan for degraded forest in the democratic people's republic of Korea considering suitable tree species and spatial distribution. *Sustain.*, **10**, 856, doi:10.3390/su10030856.

Lehmann, J. and M. Kleber, 2015: The contentious nature of soil organic matter. *Nature*, **528**, 60, doi:10.1038/nature16069.

Lehmann, J., and S. Joseph, 2015: *Biochar for Environmental Management: Science, Technology and Implementation*. Routledge, Abingdon, 907 pp.

Lei, Y. et al. 2018: Quantification of selective logging in tropical forest with spaceborne SAR interferometry. *Remote Sens. Environ.*, **211**, 167–183, doi:10.1016/J.RSE.2018.04.009.

Leifeld, J., and L. Menichetti, 2018: The underappreciated potential of peatlands in global climate change mitigation strategies. *Nat. Commun.*, **9**, 1071, doi:10.1038/s41467-018-03406-6.

Lemprière, T.C. et al. 2013: Canadian boreal forests and climate change mitigation. *Environ. Rev.*, **21**, 293–321, doi:10.1139/er-2013-0039.

Lesk, C., E. Coffel, A.W. D'Amato, K. Dodds, and R. Horton, 2017: Threats to North American forests from southern pine beetle with warming winters. *Nat. Clim. Chang.*, **7**, 713–717, doi:10.1038/nclimate3375.

Lewis, S.L., Y. Malhi, and O.L. Phillips, 2004: Fingerprinting the impacts of global change on tropical forests. *Philos. Trans. R. Soc. Lond. B. Biol. Sci.*, **359**, 437–462, doi:10.1098/rstb.2003.1432.

Lewis, S.L., C.E. Wheeler, E.T.A. Mitchard, and A. Koch, 2019: Restoring natural forests is the best way to remove atmospheric carbon. *Nature*, 568, 25–28.

Li, D., S. Niu, and Y. Luo, 2012: Global patterns of the dynamics of soil carbon and nitrogen stocks following afforestation: A meta-analysis. *New Phytol.*, **195**, 172–181, doi:10.1111/j.1469-8137.2012.04150.x.

Li, Q., M. Ma, X. Wu, and H. Yang, 2018a: Snow cover and vegetation-induced decrease in global albedo from 2002 to 2016. *J. Geophys. Res. Atmos.*, **123**, 124–138, doi:10.1002/2017JD027010.

Li, W. et al. 2018b: Gross and net land cover changes in the main plant functional types derived from the annual ESA CCI land cover maps (1992–2015). *Earth Syst. Sci. Data*, **10**, 219–234, doi:10.5194/essd-10-219-2018.

Li, Y. et al. 2015: Local cooling and warming effects of forests based on satellite observations. *Nat. Commun.*, **6**, doi:10.1038/ncomms7603.

Li, Y. et al. 2018c: Climate model shows large-scale wind and solar farms in the Sahara increase rain and vegetation. *Science*, **361**, 1019–1022, doi:10.1126/SCIENCE.AAR5629.

Li, Z., and H. Fang, 2016: Impacts of climate change on water erosion: A review. *Earth-Science Rev.*, **163**, 94–117, doi:10.1016/J.EARSCIREV.2016.10.004.

Liljedahl, A.K. et al. 2016: Pan-Arctic ice-wedge degradation in warming permafrost and its influence on tundra hydrology. *Nat. Geosci.*, **9**, 312–318, doi:10.1038/ngeo2674.

Lilleskov, E. et al. 2019: Is Indonesian peatland loss a cautionary tale for Peru? A two-country comparison of the magnitude and causes of tropical peatland degradation. *Mitig. Adapt. Strateg. Glob. Chang.*, **24**, 591–623, doi:10.1007/s11027-018-9790-3.

Lin, K.-C. et al. 2017: Impacts of increasing typhoons on the structure and function of a subtropical forest: Reflections of a changing climate. *Sci. Rep.*, **7**, 4911, doi:10.1038/s41598-017-05288-y.

Lin, M., and P. Huybers, 2012: Reckoning wheat yield trends. *Environ. Res. Lett.*, **7**, 024016, doi:10.1088/1748-9326/7/2/024016.

Lin, N., and K. Emanuel, 2016a: Grey swan tropical cyclones. *Nat. Clim. Chang.*, **6**, 106–111, doi:10.1038/nclimate2777.

Lindenmayer, D.B., C.R. Margules, and D.B. Botkin, 2000: Indicators of biodiversity for ecologically sustainable forest management. *Conserv. Biol.*, **14**, 941–950, doi:10.1046/j.1523-1739.2000.98533.x.

Lindner, M. et al. 2010: Climate change impacts, adaptive capacity, and vulnerability of European forest ecosystems. *For. Ecol. Manage.*, **259**, 698–709, doi:10.1016/J.FORECO.2009.09.023.

Lindquist, E.J. and R. D'Annunzio, 2016: Assessing global forest land-use change by object-based image analysis. *Remote Sens.*, **8**, doi:10.3390/rs8080678.

Liniger, H.P., M. Studer, R.C. Hauert, and M. Gurtner, 2011: *Sustainable Land Management in Practice. Guidelines and Best Practices for Sub-Saharan Africa*. Food and Agricultural Organization of the United Nations, Rome, Italy, 13 pp.

Liu, F. et al. 2002: Role of Grain to Green Program in reducing loss of phosphorus from yellow soil in hilly areas. *J. Soil Water Conserv.* **16**, 20–23.

Liu, C.L.C., O. Kuchma, and K.V. Krutovsky, 2018a: Mixed-species versus monocultures in plantation forestry: Development, benefits, ecosystem services and perspectives for the future. *Glob. Ecol. Conserv.*, **15**, e00419, doi:10.1016/j.gecco.2018.e00419.

Liu, J., and J. Diamond, 2008: Science and government: Revolutionizing China's environmental protection. *Science*, doi:10.1126/science.1150416.

Liu, J., S. Li, Z. Ouyang, C. Tam, and X. Chen, 2008: Ecological and socioeconomic effects of China's policies for ecosystem services. *Proc. Natl. Acad. Sci.*, **105**(28), 9477–9482, doi:10.1073/pnas.0706436105.

Liu, J.Z., W. Ouyang, W. Yang, and S.L. Xu, 2013: *Encyclopedia of Biodiversity*, S.A. Levin, Academic Press, Waltham, MA, ed. 2. 372–384.

4

Liu, S.C., C. Fu, C.-J. Shiu, J.-P. Chen, and F. Wu, 2009: Temperature dependence of global precipitation extremes. *Geophys. Res. Lett.*, 36, L17702, doi:10.1029/2009GL040218.

Liu, T., R. Bruins, and M. Heberling, 2018b: Factors influencing farmers' adoption of best management practices: A review and synthesis. *Sustainability*, 10, 432, doi:10.3390/su10020432.

Liu, Y. et al. 2018c: Mechanisms of rice straw biochar effects on phosphorus sorption characteristics of acid upland red soils. *Chemosphere*, **207**, 267–277, doi:10.1016/j.chemosphere.2018.05.086.

Liu, Y.Y. et al. 2015: Recent reversal in loss of global terrestrial biomass. *Nat. Clim. Chang.*, **5**, 470–474, doi:10.1038/nclimate2581.

Long, H.L. et al. 2006: Land use and soil erosion in the upper reaches of the Yangtze River: Some socio-economic considerations on China's Grain-for-Green Programme. *L. Degrad. Dev.*, 17(6), 589–603, doi:10.1002/ldr.736.

López-Carr, D., 2012: Agro-ecological drivers of rural out-migration to the Maya Biosphere Reserve, Guatemala. *Environ. Res. Lett.*, **7**, 045603, doi:10.1088/1748-9326/7/4/045603.

López-Portillo, J. et al. 2017: *Mangrove Forest Restoration and Rehabilitation. Mangrove Ecosystems: A Global Biogeographic Perspective*, Springer International Publishing, Cham, Switzerland, pp. 301–345.

López-Rosas, H. et al. 2013: Interdune wetland restoration in central Veracruz, Mexico: plant diversity recovery mediated by the hydroperiod. *Restoration of Coastal Dunes*, Springer, Berlin, Heidelberg, 255–269.

Lorimer, C.G. and A.S. White, 2003: Scale and frequency of natural disturbances in the northeastern US: Implications for early successional forest habitats and regional age distributions. *For. Ecol. Manage.*, **185**, 41–64, doi:10.1016/S0378-1127(03)00245-7.

Loucks, C., S. Barber-Meyer, M.A.A. Hossain, A. Barlow, and R.M. Chowdhury, 2010: Sea level rise and tigers: Predicted impacts to Bangladesh's Sundarbans mangroves. *Clim. Change*, 98, 291–298, doi:10.1007/s10584-009-9761-5.

Lovo, S., 2016: Tenure insecurity and investment in soil conservation. Evidence from Malawi. *World Dev.*, **78**, 219–229, doi:10.1016/J.WORLDDEV.2015.10.023.

Luetz, J., 2018: Climate Change and Migrationin Bangladesh: Empirically Derived Lessons and Opportunities for Policy Makers and Practitioners. In: *Limits to Climate Change Adaptation* [W.L. Filho and J. Nalau (eds.)]. Springer International Publishing, Berlin, Heidelberg, 59–105.

Lukas, M.C., 2014: Eroding battlefields: Land degradation in Java reconsidered. *Geoforum*, 56, 87–100, doi:10.1016/J.GEOFORUM.2014.06.010.

Luke, D., K. McLaren, and B. Wilson, 2016: Modeling hurricane exposure in a Caribbean lower montane tropical wet forest: The effects of frequent, intermediate disturbances and topography on forest structural dynamics and composition. *Ecosystems*, 19, 1178–1195, doi:10.1007/s10021-016-9993-y.

Lund, J.F., E. Sungusia, M.B. Mabele, and A. Scheba, 2017: Promising change, delivering continuity: REDD+ as conservation fad. *World Dev.*, **89**, 124–139, doi:10.1016/J.WORLDDEV.2016.08.005.

Lundmark, T. et al. 2014: Potential roles of Swedish forestry in the context of climate change mitigation. *Forests*, **5**, 557–578, doi:10.3390/f5040557.

Luo, P. et al. 2018: Impact of forest maintenance on water shortages: Hydrologic modeling and effects of climate change. *Sci. Total Environ.*, 615, 1355–1363, doi:10.1016/J.SCITOTENV.2017.09.044.

Lutz, A.F., W.W. Immerzeel, A.B. Shrestha, and M.F.P. Bierkens, 2014: Consistent increase in high Asia's runoff due to increasing glacier melt and precipitation. *Nat. Clim. Chang.*, **4**, 587–592, doi:10.1038/nclimate2237.

Luyssaert, S. et al. 2018: Trade-offs in using European forests to meet climate objectives. *Nature*, **562**, 259–262, doi:10.1038/s41586-018-0577-1.

Lyons-White, J. and A.T. Knight, 2018: Palm oil supply chain complexity impedes implementation of corporate no-deforestation commitments. *Glob. Environ. Chang.*, **50**, 303–313, doi:10.1016/J.GLOENVCHA.2018.04.012.

MA (Millennium Ecosystem Assessment), 2005: *Millennium Ecosystem Assessment Synthesis Report*. Island Press, Washington DC, 160 pp.

Ma, S. et al. 2015: Observed changes in the distributions of daily precipitation frequency and amount over China from 1960 to 2013. *J. Clim.*, **28**, 6960–6978, doi:10.1175/JCLI-D-15-0011.1.

Ma, S. et al. 2017: Detectable anthropogenic shift toward heavy precipitation over eastern China. *J. Clim.*, **30**, 1381–1396, doi:10.1175/JCLI-D-16-0311.1.

MacDicken, K.G. et al. 2015: Global progress toward sustainable forest management. *For. Ecol. Manage.*, **352**, 47–56, doi:10.1016/j.foreco.2015.02.005.

Macfadyen, S., G. McDonald, and M.P. Hill, 2018: From species distributions to climate change adaptation: Knowledge gaps in managing invertebrate pests in broad-acre grain crops. *Agric. Ecosyst. Environ.*, **253**, 208–219, doi:10.1016/J.AGEE.2016.08.029.

Machisa, M., J. Wichmann, and P.S. Nyasulu, 2013: Biomass fuel use for household cooking in Swaziland: Is there an association with anaemia and stunting in children aged 6–36 months? *Trans. R. Soc. Trop. Med. Hyg.*, **107**, 535–544, doi:10.1093/trstmh/trt055.

Mackey, B. et al. 2013: Untangling the confusion around land carbon science and climate change mitigation policy. *Nat. Clim. Chang.*, **3**, 552–557, doi:10.1038/nclimate1804.

Mackey, B. et al. 2015: Policy options for the world's primary forests in multilateral environmental agreements. *Conserv. Lett.*, **8**, 139–147, doi:10.1111/conl.12120.

Maes, J., and S. Jacobs, 2017: Nature-based solutions for europe's sustainable development. *Conserv. Lett.*, doi:10.1111/conl.12216.

Maestre, F.T. et al. 2009: Shrub encroachment can reverse desertification in semi-arid Mediterranean grasslands. *Ecol. Lett.*, **12**, 930–941, doi:10.1111/j.1461-0248.2009.01352.x.

Maestre, F.T. et al. 2013: Changes in biocrust cover drive carbon cycle responses to climate change in drylands. *Glob. Chang. Biol.*, **19**, 3835–3847, doi:10.1111/gcb.12306.

Magnan, A.K. et al. 2016: Addressing the risk of maladaptation to climate change. *Wiley Interdiscip. Rev. Clim. Chang.*, **7**, 646–665, doi:10.1002/wcc.409.

Mander, S., K. Anderson, A. Larkin, C. Gough, and N. Vaughan, 2017: The role of bio-energy with carbon capture and storage in meeting the climate mitigation challenge: A whole system perspective. *Energy Procedia*, **114**, 6036–6043, doi:10.1016/J.EGYPRO.2017.03.1739.

Marino, E., and H. Lazrus, 2015: Migration or Forced Displacement? The complex choices of climate change and disaster migrants in Shishmaref, Alaska and Nanumea, Tuvalu. *Hum. Organ.*, **74**, 341–350, doi:10.17730/0018-7259-74.4.341.

Marjani, A. and M. Jamali, 2014: Role of exchange flow in salt water balance of Urmia Lake. *Dyn. Atmos. Ocean.*, **65**, 1–16, doi:10.1016/j.dynatmoce.2013.10.001.

Martin-Ortega, J., T.E.H. Allott, K. Glenk, and M. Schaafsma, 2014: Valuing water quality improvements from peatland restoration: Evidence and challenges. *Ecosyst. Serv.*, **9**, 34–43, doi:10.1016/j.ecoser.2014.06.00.

Masera, O.R., R. Bailis, R. Drigo, A. Ghilardi, and I. Ruiz-Mercado, 2015: Environmental burden of traditional bioenergy use. *Annu. Rev. Environ. Resour.*, **40**, 121–150, doi:10.1146/annurev-environ-102014-021318.

Matata, P.Z., L.W. Masolwa, S. Ruvuga, and F.M. Bagarama, 2013: Dissemination pathways for scaling-up agroforestry technologies in western Tanzania. *J. Agric. Ext. Rural Dev.*, **5**, 31–36.

Mateos, E., J.M. Edeso, and L. Ormaetxea, 2017: Soil erosion and forests biomass as energy resource in the basin of the Oka River in Biscay, Northern Spain. *Forests*, **8**, 258, doi:10.3390/f8070258.

Matthews, R., G. Hogan, and E. Mackie, 2018: Carbon impacts of biomass consumed in the EU. *Res. Agency For. Comm.*, **61**.

Mbow, C., P. Smith, D. Skole, L. Duguma, and M. Bustamante, 2014: Achieving mitigation and adaptation to climate change through sustainable agroforestry practices in Africa. *Curr. Opin. Environ. Sustain.*, **6**, 8–14, doi:10.1016/j.cosust.2013.09.002.

4

McDowell, N.G. et al. 2011: The interdependence of mechanisms underlying climate-driven vegetation mortality. *Trends Ecol. Evol.*, **26**, 523–532, doi:10.1016/J.TREE.2011.06.003.

McGrath, M.J. et al. 2015: Reconstructing European forest management from 1600 to 2010. *Biogeosciences*, **12**, 4291–4316, doi:10.5194/bg-12-4291-2015.

McInnes, K.L., T.A. Erwin, and J.M. Bathols, 2011: Global Climate Model projected changes in 10m wind speed and direction due to anthropogenic climate change. *Atmos. Sci. Lett.*, **12**, 325–333, doi:10.1002/asl.341.

McIntire, J., D. (Daniel) Bourzat, and P.L. Pingali, 1992: *Crop-livestock interaction in Sub-Saharan Africa*. World Bank, Washington DC, USA, 246 pp.

McLauchlan, K., 2006: The nature and longevity of agricultural impacts on soil carbon and nutrients: A review. *Ecosystems*, **9**, 1364–1382, doi:10.1007/s10021-005-0135-1.

McLeman, R., 2017: *Migration and Land Degradation: Recent Experience and Future Trends*. Working paper for the Global Land Outlook, 1st edition. UNCCD, Bonn, Germany, 45 pp.

McLeman, R. and B. Smit, 2006: Migration as an adaptation to climate change. *Clim. Change*, **76**, 31–53, doi:10.1007/s10584-005-9000-7.

McLeman, R.A. et al. 2014: What we learned from the Dust Bowl: Lessons in science, policy, and adaptation. *Popul. Environ.*, **35**, 417–440, doi:10.1007/s11111-013-0190-z.

McNicol, G. and W.L. Silver, 2014: Separate effects of flooding and anaerobiosis on soil greenhouse gas emissions and redox sensitive biogeochemistry. *J. Geophys. Res. Biogeosciences*, **119**, 557–566, doi:10.1002/2013JG002433.

McNicol, I.M., C.M. Ryan, and E.T.A. Mitchard, 2018: Carbon losses from deforestation and widespread degradation offset by extensive growth in African woodlands. *Nat. Commun.*, **9**, 3045, doi:10.1038/s41467-018-05386-z.

McVicar, T.R. and M.L. Roderick, 2010: Winds of change. *Nat. Geosci.*, **3**, 747–748, doi:10.1038/ngeo1002.

Meijer, S.S., D. Catacutan, O.C. Ajayi, G.W. Sileshi, and M. Nieuwenhuis, 2015: The role of knowledge, attitudes and perceptions in the uptake of agricultural and agroforestry innovations among smallholder farmers in sub-Saharan Africa. *Int. J. Agric. Sustain.*, **13**, 40–54, doi:10.1080/14735903.2014.912493.

Meisner, A., S. Jacquiod, B.L. Snoek, F.C. ten Hooven, and W.H. van der Putten, 2018: Drought legacy effects on the composition of soil fungal and prokaryote communities. *Front. Microbiol.*, **9**, 294, doi:10.3389/fmicb.2018.00294.

Mekuriaw, A., A. Heinimann, G. Zeleke, and H. Hurni, 2018: Factors influencing the adoption of physical soil and water conservation practices in the Ethiopian highlands. *Int. Soil Water Conserv. Res.*, **6**, 23–30, doi:10.1016/J.ISWCR.2017.12.006.

Mendum, R. and M. Njenga, 2018: *Recovering bioenergy in Sub-Saharan Africa: Gender Dimensions, Lessons and Challenges*. International Water Management Institute (IWMI), Colombo, Sri Lanka, pp. 1–83.

Meng, W. et al. 2017: Status of wetlands in China: A review of extent, degradation, issues and recommendations for improvement. *Ocean Coast. Manag.*, **146**, 50–59, doi:10.1016/J.OCECOAMAN.2017.06.003.

Mengel, M. et al. 2016: Future sea level rise constrained by observations and long-term commitment. *Proc. Natl. Acad. Sci.*, **113**, 2597–2602, doi:10.1073/PNAS.1500515113.

Mentaschi, L., M.I. Vousdoukas, J.-F. Pekel, E. Voukouvalas, and L. Feyen, 2018: Global long-term observations of coastal erosion and accretion. *Sci. Rep.*, **8**, 12876, doi:10.1038/s41598-018-30904-w.

Mercer, D.E., 2004: Adoption of agroforestry innovations in the tropics: A review. *Agrofor. Syst.*, **61–62**, 311–328, doi:10.1023/B:AGFO.0000029007.85754.70.

Meyfroidt, P. and E.F. Lambin, 2011: *Global Forest Transition: Prospects for an End to Deforestation*. Ann. Rev. Env. Res., **36**, 343–371 pp.

Micklin, P., 2010: The past, present, and future Aral Sea. *Lakes Reserv. Res. Manag.*, **15**, 193–213, doi:10.1111/j.1440-1770.2010.00437.x.

Middleton, N.J. and D.S. Thomas, 1997: *World Atlas of Desertification*. Arnold, London, 181 pp.

Midgley, G.F. and W.J. Bond, 2015: Future of African terrestrial biodiversity and ecosystems under anthropogenic climate change. *Nat. Clim. Chang.*, **5**, 823–829, doi:10.1038/nclimate2753.

Miettinen, J., C. Shi, and S.C. Liew, 2016: Land cover distribution in the peatlands of Peninsular Malaysia, Sumatra and Borneo in 2015 with changes since 1990. *Glob. Ecol. Conserv.*, **6**, 67–78, doi:10.1016/j.gecco.2016.02.004.

Miles, L. et al. 2015: *Mitigation Potential from Forest-Related Activities and Incentives for Enhanced Action in Developing Countries*. United Nations Environment Programme, Nairobi, Kenya, 44–50 pp.

Millennium Ecosystem Assessment, 2005: *Ecosystems and Human Well-being, Synthesis*. Island Press, Washington DC, USA, 155 pp.

Minasny, B. et al. 2017: Soil carbon 4 per mille. *Geoderma*, **292**, 59–86, doi:10.1016/J.GEODERMA.2017.01.002.

Minderhoud, P.S.J. et al. 2017: Impacts of 25 years of groundwater extraction on subsidence in the Mekong delta, Vietnam. *Environ. Res. Lett.*, **12**, 064006, doi:10.1088/1748-9326/aa7146.

Minkkinen, K. et al. 2018: Persistent carbon sink at a boreal drained bog forest. *Biogeosciences*, **15**, 3603–3624, doi:10.5194/bg-15-3603-2018.

Minx, J.C. et al. 2018: Negative emissions – Part 1: Research landscape and synthesis. *Environ. Res. Lett.*, **13**, doi:10.1088/1748-9326/aabf9b.

Mirzabaev, A., E. Nkonya, and J. von Braun, 2015: Economics of sustainable land management. *Curr. Opin. Environ. Sustain.*, **15**, 9–19, doi:10.1016/J.COSUST.2015.07.004.

Mirzabaev, A., E. Nkonya, J. Goedecke, T. Johnson, and W. Anderson, 2016: *Global Drivers of Land Degradation and Improvement. Economics of Land Degradation and Improvement – A Global Assessment for Sustainable Development*, Springer International Publishing, Cham, Switzerland, pp. 167–195.

Mitchard, E.T.A., 2018: The tropical forest carbon cycle and climate change. *Nature*, **559**, 527–534, doi:10.1038/s41586-018-0300-2.

Mittler, R., 2006: Abiotic stress, the field environment and stress combination. *Trends Plant Sci.*, **11**, 15–19, doi:10.1016/J.TPLANTS.2005.11.002.

Mokria, M., A. Gebrekirstos, E. Aynekulu, and A. Bräuning, 2015: Tree dieback affects climate change mitigation potential of a dry afromontane forest in northern Ethiopia. *For. Ecol. Manage.*, **344**, 73–83, doi:10.1016/j.foreco.2015.02.008.

Mondal, A., D. Khare, and S. Kundu, 2016: Change in rainfall erosivity in the past and future due to climate change in the central part of India. *Int. Soil Water Conserv. Res.*, **4**, 186–194, doi:10.1016/J.ISWCR.2016.08.004.

Montanarella, L., R. Scholes and A. Brainich, 2018: *The IPBES Assessment Report on Land Degradation and Restoration*. Secretariat of the Intergovernmental Science-Policy Platform on Biodiversity and Ecosystem Services, Bonn, Germany. 744 pp. doi: 10.5281/zenodo.3237392.

Montgomery, D.R., 2007a: Soil erosion and agricultural sustainability. *Proceedings of the National Academy of Sciences*, **104**(33), 13268–13272, doi: 10.1073/pnas.0611508104.

Montgomery, D.R., 2007b: *Dirt: the Erosion of Civilizations*. University of California Press, 285 pp.

Moore, P.D., 2002: The future of cool temperate bogs. *Environmental Conservation*. 29, 3–20.

National Climate Commission. 2009. *Belgium's Fifth National Communication under the United Nations Framework Convention on Climate Change*. Federal Public Service Health, Food Chain Safety and Environment, Brussels, 206 pp.

Moore, S. et al. 2013: Deep instability of deforested tropical peatlands revealed by fluvial organic carbon fluxes. *Nature*, **493**, 660–663, doi:10.1038/nature11818.

Morgan, R.P.C., 2005a: *Soil Erosion and Conservation*. 3rd ed. Blackwell Science Ltd, Malden, USA.

Morgan, R.P.C., and Royston P.C., 2005b: *Soil Erosion And Conservation*. 2nd ed. Blackwell Publishing, Harlow, Essex, UK, 198 pp.

Mori, N., T. Yasuda, H. Mase, T. Tom, and Y. Oku, 2010: Projection of extreme wave climate change under global warming. *Hydrol. Res. Lett.*, **4**, 15–19, doi:10.3178/hrl.4.15.

Morrissey, J.W., 2013: Understanding the relationship between environmental change and migration: The development of an effects framework based on the case of northern Ethiopia. *Glob. Environ. Chang.*, **23**, 1501–1510, doi:10.1016/J.GLOENVCHA.2013.07.021.

Morse, J.L. and E.S. Bernhardt, 2013: Using15N tracers to estimate N_2O and N_2 emissions from nitrification and denitrification in coastal plain wetlands under contrasting land-uses. *Soil Biol. Biochem.*, **55**, 635–643, doi:10.1016/j.soilbio.2012.07.025.

Morton, J.F., 2007: The impact of climate change on smallholder and subsistence agriculture. *Proc. Natl. Acad. Sci. U.S.A.*, **104**, 19680–19685, doi:10.1073/pnas.0701855104.

Mudombi, S. et al. 2018: Multi-dimensional poverty effects around operational biofuel projects in Malawi, Mozambique and Swaziland. *Biomass and Bioenergy*, **114**, 41–54, doi:https://doi.org/10.1016/j.biombioe.2016.09.003.

Muir, D., J.A.G. Cooper, and G. Pétursdóttir, 2014: Challenges and opportunities in climate change adaptation for communities in Europe's northern periphery. *Ocean Coast. Manag.*, **94**, 1–8, doi:10.1016/j.ocecoaman.2014.03.017.

Mullan, D., 2013: Soil erosion under the impacts of future climate change: Assessing the statistical significance of future changes and the potential on-site and off-site problems. *CATENA*, **109**, 234–246, doi:10.1016/J.CATENA.2013.03.007.

Mullan, D., D. Favis-Mortlock, and R. Fealy, 2012: Addressing key limitations associated with modelling soil erosion under the impacts of future climate change. *Agric. For. Meteorol.*, **156**, 18–30, doi:10.1016/j.agrformet.2011.12.004.

Munang, R., I. Thiaw, K. Alverson, and Z. Han, 2013: The role of ecosystem services in climate change adaptation and disaster risk reduction. *Curr. Opin. Environ. Sustain.*, **5**, 47–52, doi:10.1016/J.COSUST.2013.02.002.

Munson, S.M., J. Belnap, and G.S. Okin, 2011: Responses of wind erosion to climate-induced vegetation changes on the Colorado Plateau. *Proc. Natl. Acad. Sci.*, **108**, 3854–3859, doi:10.1073/pnas.1014947108.

Murphy, J.M. et al. 2004: Quantification of modelling uncertainties in a large ensemble of climate change simulations. *Nature*, **430**, 768–772, doi:10.1038/nature02771.

Murthy, K. and S. Bagchi, 2018: Spatial patterns of long-term vegetation greening and browning are consistent across multiple scales: Implications for monitoring land degradation. *L. Degrad. Dev.*, **29**, 2485–2495, doi:10.1002/ldr.3019.

Murty, D., M.U.F. Kirschbaum, R.E. Mcmurtrie, and H. Mcgilvray, 2002: Does conversion of forest to agricultural land change soil carbon and nitrogen? A review of the literature. *Glob. Chang. Biol.*, **8**, 105–123, doi:10.1046/j.1354-1013.2001.00459.x.

Mutoko, M.C., C.A. Shisanya, and L. Hein, 2014: Fostering technological transition to sustainable land management through stakeholder collaboration in the western highlands of Kenya. *Land use policy*, **41**, 110–120, doi:10.1016/J.LANDUSEPOL.2014.05.005.

Mycoo, M. and A. Chadwick, 2012: Adaptation to climate change: The coastal zone of Barbados. *Proc. Inst. Civ. Eng. – Marit. Eng.*, **165**, 159–168, doi:10.1680/maen.2011.19.

Naamwintome, B.A. and E. Bagson, 2013: Youth in agriculture: Prospects and challenges in the Sissala area of Ghana. *Net J. Agric. Sci.*, **1**, 60–68.

Nabuurs, G.J., O. Masera, K. Andrasko, P. Benitez-Ponce, R. Boer, M. Dutschke, E. Elsiddig, J. Ford-Robertson, P. Frumhoff, T. Karjalainen, O. Krankina, W.A. Kurz, M. Matsumoto, W. Oyhantcabal, N.H. Ravindranath, M.J. Sanz Sanchez, X. Zhang, 2007: In Climate Change 2007: Mitigation. Contribution of Working Group III to the Fourth Assessment Report of the Intergovernmental Panel on Climate Change [Metz, B., O.R. Davidson, P.R. Bosch, R. Dave, L.A. Meyer (eds)]. Cambridge University Press, Cambridge, United Kingdom and New York, NY, USA, pp. 541–584.

Nabuurs, G.J. et al. 2017: By 2050 the mitigation effects of EU forests could nearly double through climate smart forestry. *Forests*, **8**, 1–14, doi:10.3390/f8120484.

Nachtergaele, F.O. et al. 2011: *Global Land Degradation Information System (GLADIS) Version 1.0. An Information database for Land Degradation Assessment at Global Level*. LADA Technical Report 17. Rome, Italy, pp. 1–110.

Nalau, J., J. Handmer, J. Nalau, and J. Handmer, 2018: Improving development outcomes and reducing disaster risk through planned community relocation. *Sustainability*, **10**, 3545, doi:10.3390/su10103545.

Nampak, H., B. Pradhan, H. Mojaddadi Rizeei, and H.-J. Park, 2018: Assessment of land cover and land use change impact on soil loss in a tropical catchment by using multitemporal SPOT-5 satellite images and Revised Universal Soil Loss Equation model. *L. Degrad. Dev.*, **29**(10), 3440–3455, doi:10.1002/ldr.3112.

Nasi, R., F.E. Putz, P. Pacheco, S. Wunder, and S. Anta, 2011: Sustainable forest management and carbon in tropical Latin America: The case for REDD+. *Forests*, **2**, 200–217, doi:10.3390/f2010200.

Nath, T.K., M. Jashimuddin, M. Kamrul Hasan, M. Shahjahan, and J. Pretty, 2016: The sustainable intensification of agroforestry in shifting cultivation areas of Bangladesh. *Agrofor. Syst.*, **90**, 405–416, doi:10.1007/s10457-015-9863-1.

Ndegwa, G.M., U. Nehren, F. Grüninger, M. Iiyama, and D. Anhuf, 2016: Charcoal production through selective logging leads to degradation of dry woodlands: A case study from Mutomo District, Kenya. *J. Arid Land*, **8**, 618–631, doi:10.1007/s40333-016-0124-6.

Nearing, M.A., F.F. Pruski, and M.R. O'Neal, 2004: Expected climate change impacts on soil erosion rates: A review. *J. Soil Water Conserv.*, **59**, 43–50.

Nearing, M.A. et al. 2005: Modeling response of soil erosion and runoff to changes in precipitation and cover. *CATENA*, **61**, 131–154, doi:10.1016/j.catena.2005.03.007.

Nemet, G.F. et al. 2018: Negative emissions – Part 3: Innovation and upscaling. *Environ. Res. Lett.*, **13**, doi:10.1088/1748-9326/aabff4.

Nesshöver, C. et al. 2017: The science, policy and practice of nature-based solutions: An interdisciplinary perspective. *Sci. Total Environ.*, **579**, 1215–1227, doi:10.1016/J.SCITOTENV.2016.11.106.

Netting, R.M., 1993: *Smallholders, Householders: Farm Families and the Ecology of Intensive, Sustainable Agriculture*. Stanford University Press, Stanford, CA, 389 pp.

Neumann, B., A.T. Vafeidis, J. Zimmermann, and R.J. Nicholls, 2015: Future coastal population growth and exposure to sea-level rise and coastal flooding – A global assessment. *PLoS One*, **10**, e0118571, doi:10.1371/journal.pone.0118571.

Neupane, R.P., and S. Kumar, 2015: Estimating the effects of potential climate and land use changes on hydrologic processes of a large agriculture dominated watershed. *J. Hydrol.*, **529**, 418–429, doi:10.1016/J.JHYDROL.2015.07.050.

Newton, P., J. A Oldekop, G. Brodnig, B.K. Karna, and A. Agrawal, 2016: Carbon, biodiversity, and livelihoods in forest commons: Synergies, trade-offs, and implications for REDD+. *Environ. Res. Lett.*, **11**, 044017, doi:10.1088/1748-9326/11/4/044017.

Nguyen, T.T.N. et al. 2017: Effects of biochar on soil available inorganic nitrogen: A review and meta-analysis. *Geoderma*, **288**, 79–96 doi:10.1016/j.geoderma.2016.11.004.

Nicholls, R.J. et al. 2011: Sea-level rise and its possible impacts given a "beyond 4°C world" in the twenty-first century. *Philos. Trans. A. Math. Phys. Eng. Sci.*, **369**, 161–181, doi:10.1098/rsta.2010.0291.

4

Nicholls, R.J., C. Woodroffe, and V. Burkett, 2016: Chapter 20 – Coastline degradation as an indicator of global change. *Clim. Chang.*, 309–324, doi:10.1016/B978-0-444-63524-2.00020-8.

Nicholls, R.J. et al. 2018: Stabilization of global temperature at 1.5°C and 2.0°C: Implications for coastal areas. *Philos. Trans. R. Soc. A Math. Eng. Sci.*, 376, 20160448, doi:10.1098/rsta.2016.0448.

Niedertscheider, M. et al. 2016: Mapping and analysing cropland use intensity from a NPP perspective. *Environ. Res. Lett.*, 11, 014008, doi:10.1088/1748-9326/11/1/014008.

Nielsen, D.L. and M.A. Brock, 2009: Modified water regime and salinity as a consequence of climate change: Prospects for wetlands of Southern Australia. *Clim. Change*, 95, 523–533, doi:10.1007/s10584-009-9564-8.

Nitcheu Tchiadje, S., D.J. Sonwa, B.-A. Nkongmeneck, L. Cerbonney, and R. Sufo Kankeu, 2016: Preliminary estimation of carbon stock in a logging concession with a forest management plan in East Cameroon. *J. Sustain. For.*, 35, 355–368, doi:10.1080/10549811.2016.1190757.

Njenga, M. et al. 2019: Innovative biomass cooking approaches for sub-Saharan Africa. *African J. Food, Agric. Nutr. Dev.*, 19, 14066–14087.

Nkonya, E., M. Winslow, M.S. Reed, M. Mortimore, and A. Mirzabaev, 2011: Monitoring and assessing the influence of social, economic and policy factors on sustainable land management in drylands. *L. Degrad. Dev.*, 22, 240–247, doi:10.1002/ldr.1048.

Nkonya, E., A. Mirzabaev, and J. von Braun, 2016: Economics of Land Degradation and Improvement – A Global Assessment for Sustainable Development. E. Nkonya, A. Mirzabaev, and J. Von Braun (Eds.) Springer, Heidelberg, New York, Dordrecht, London, 695 pp.

Noordin, Q., A. Niang, B. Jama, and M. Nyasimi, 2001: Scaling up adoption and impact of agroforestry technologies: Experiences from western Kenya. *Dev. Pract.*, 11, 509–523, doi:10.1080/09614520120066783.

Noormets, A. et al. 2015: Effects of forest management on productivity and carbon sequestration: A review and hypothesis. *For. Ecol. Manage.*, 355, 124–140, doi:10.1016/j.foreco.2015.05.019.

Norby, R.J., J.M. Warren, C.M. Iversen, B.E. Medlyn, and R.E. McMurtrie, 2010: CO_2 enhancement of forest productivity constrained by limited nitrogen availability. *Proc. Natl. Acad. Sci. USA*, 107, 19368–19373, doi: 10.1073/pnas.1006463107.

Norton, J.B. et al. 2011: Soil carbon and nitrogen storage in upper montane riparian meadows. *Ecosystems*, 14, 1217–1231, doi:10.1007/s10021-011-9477-z.

Nugent, K.A., I.B. Strachan, M. Strack, N.T. Roulet, and L. Rochefort, 2018: Multi-year net ecosystem carbon balance of a restored peatland reveals a return to carbon sink. *Glob. Chang. Biol.*, 24(12), 5751–5768, doi:10.1111/gcb.14449.

Nunes, J.P., and M.A. Nearing, 2011: Modelling Impacts of Climate Change: Case studies using the new generation of erosion models. *Handbook of Erosion Modelling*, [R.P.C. Morgan and M.A. Nearing, (eds.)]. Wiley, Chichester, West Sussex, UK, p. 400.

Nurse, L.A. et al. 2014: Small Islands. Climate Change 2014: Impacts, Adaptation, and Vulnerability [V.R. Barros et al. (eds.)]. Cambridge University Press, Cambridge; New York, 1613–1654.

Nyland, R.D., 1992: Exploitation and greed in eastern hardwood forests. *J. For.*, 90, 33–37, doi:10.1093/jof/90.1.33.

O'Brien, K. et al. 2004: Mapping vulnerability to multiple stressors: Climate change and globalization in India. *Glob. Environ. Chang.*, 14, 303–313, doi:10.1016/J.GLOENVCHA.2004.01.001.

O'Connell, D. et al. 2016: Designing projects in a rapidly changing world: Guidelines for embedding resilience, adaptation and transformation into sustainable development projects.C Global Environment Facility, Washington, D.C.

O'Connor, D. et al. 2018: Biochar application for the remediation of heavy metal polluted land: A review of in situ field trials. *Sci. Total Environ.*, 619, 815–826, doi:10.1016/j.scitotenv.2017.11.132.

O'Driscoll, C. et al. 2018: National scale assessment of total trihalomethanes in Irish drinking water. *J. Environ. Manage.*, 212, 131–141, doi:10.1016/j.jenvman.2018.01.070.

O'Gorman, P.A., 2015: Precipitation extremes under climate change. *Curr. Clim. Chang. Reports*, 1, 49–59, doi:10.1007/s40641-015-0009-3.

O'Neal, M.R., M.A. Nearing, R.C. Vining, J. Southworth, and R.A. Pfeifer, 2005: Climate change impacts on soil erosion in Midwest United States with changes in crop management. *CATENA*, 61, 165–184, doi:10.1016/J.CATENA.2005.03.003.

Odgaard, M.V., M.T. Knudsen, J.E. Hermansen, and T. Dalgaard, 2019: Targeted grassland production – A Danish case study on multiple benefits from converting cereal to grasslands for green biorefinery. *J. Clean. Prod.*, 223, 917–927, doi:10.1016/J.JCLEPRO.2019.03.072.

Oertel, C., J. Matschullat, K. Zurba, F. Zimmermann, and S. Erasmi, 2016: Greenhouse gas emissions from soils – A review. *Chemie der Erde – Geochemistry*, 76, 327–352, doi:10.1016/J.CHEMER.2016.04.002.

Ogle, S.M., A. Swan, and K. Paustian, 2012: No-till management impacts on crop productivity, carbon input and soil carbon sequestration. *Agric. Ecosyst. Environ.*, 149, 37–49, doi:10.1016/J.AGEE.2011.12.010.

Ojanen, P., A. Lehtonen, J. Heikkinen, T. Penttilä, and K. Minkkinen, 2014: Soil CO_2 balance and its uncertainty in forestry-drained peatlands in Finland. *For. Ecol. Manage.*, 325, 60–73, doi:10.1016/j.foreco.2014.03.049.

Ojeda, G., J. Patrício, S. Mattana, and A.J.F.N. Sobral, 2016: Effects of biochar addition to estuarine sediments. *J. Soils Sediments*, 16, 2482–2491, doi:10.1007/s11368-016-1493-3.

Ojima, D.S., K.A. Galvin, and B.L. Turner, 1994: The global impact of land-use change. *Bioscience*, 44, 300–304, doi:10.2307/1312379.

Oktarita, S., K. Hergoualc'h, S. Anwar, and L. V Verchot, 2017: Environmental Research Letters Substantial N_2O emissions from peat decomposition and N fertilization in an oil palm plantation exacerbated by hotspots Substantial N_2O emissions from peat decomposition and N fertilization in an oil palm plantation exac. *Environ. Res. Lett*, 12.

Oldeman, L.R. and G.W.J. van Lynden, 1998: Revisiting the Glasod Methodology. Methods for Assessment of Soil Degradation [R. Lal, W.H. Blum, C. Valentine, and B.A. Stewart, (eds.)]. CRC Press, Boca Raton, London, New York, Washington D.C., 423–440.

Olguin, M. et al. 2018: Applying a systems approach to assess carbon emission reductions from climate change mitigation in Mexico's forest sector. *Environ. Res. Lett.*, 13, doi:10.1088/1748-9326/aaaa03.

de Oliveira, G., N.A. Brunsell, C.E. Sutherlin, T.E. Crews, and L.R. DeHaan, 2018: Energy, water and carbon exchange over a perennial Kernza wheatgrass crop. *Agric. For. Meteorol.*, 249, 120–137, doi:10.1016/J.AGRFORMET.2017.11.022.

Oliver, D.M. et al. 2012: Valuing local knowledge as a source of expert data: Farmer engagement and the design of decision support systems. *Environ. Model. Softw.*, 36, 76–85, doi:10.1016/J.ENVSOFT.2011.09.013.

Olsson, L. et al. 2014a: Cross-Chapter Box on Heat Stress and Heat Waves. In: Climate Change 2014: Impacts, Adaptation, and Vulnerability. Part A: Global and Sectoral Aspects. Contribution of Working Group II to the Fifth Assessment Report of the Intergovernmental Panel on Climate Change, [Field, C.B., V.R. Barros, D.J. Dokken, K.J. Mach, M.D. Mastrandrea, T.E. Bilir, M. Chatterjee, K.L. Ebi, Y.O. Estrada, R.C. Genova, B. Girma, E.S. Kissel, A.N. Levy, S. MacCracken, P.R. Mastrandrea, and L.L.White (eds.)]. Cambridge University Press, Cambrdige, UK and New York, USA, pp. 109–111.

Olsson, L. et al. 2014b: Livelihoods and Poverty. In: Climate Change 2014: Impacts, Adaptation, and Vulnerability: Contribution of Working Group II to the Fifth Assessment Report of the Intergovernmental Panel on Climate Change, [Field, C.B., V.R. Barros, D.J. Dokken, K.J. Mach, M.D. Mastrandrea, T.E. Bilir, M. Chatterjee, K.L. Ebi, Y.O. Estrada, R.C. Genova, B. Girma, E.S. Kissel, A.N. Levy, S. MacCracken, P.R. Mastrandrea, and L.L.White (eds.)]. Cambridge University Press, Cambridge, UK, and New York, USA, pp. 793–832.

4

Olsson, L., A. Jerneck, H. Thoren, J. Persson, and D. O'Byrne, 2015: Why resilience is unappealing to social science: Theoretical and empirical investigations of the scientific use of resilience. *Sci. Adv.*, **1**, e1400217–e1400217, doi:10.1126/sciadv.1400217.

Omondi, M.O. et al. 2016: Quantification of biochar effects on soil hydrological properties using meta-analysis of literature data. *Geoderma*, **274**, 28–34, doi:10.1016/J.GEODERMA.2016.03.029.

Orr, B.J. et al. 2017: *Scientific Conceptual Framework For Land Degradation Neutrality*. A Report of the Science-Policy Interface. United Nations Convention to Combat Desertification (UNCCD), Bonn, Germany. 136 pp.

Osborne, T.M., 2011: Carbon forestry and agrarian change: Access and land control in a Mexican rainforest. *J. Peasant Stud.*, **38**, 859–883, doi:10.1080/03066150.2011.611281.

Osland, M.J. et al. 2016: Beyond just sea-level rise: Considering macroclimatic drivers within coastal wetland vulnerability assessments to climate change. *Glob. Chang. Biol.*, **22**, 1–11, doi:10.1111/gcb.13084.

Ostrom, E., 2009: A general framework for analyzing sustainability of social-ecological systems. *Science*, **325**, 419–422, doi:10.1126/science.1172133.

Ovalle-Rivera, O., P. Läderach, C. Bunn, M. Obersteiner, and G. Schroth, 2015: Projected shifts in coffea arabica suitability among major global producing regions due to climate change. *PLoS One*, **10**, e0124155, doi:10.1371/journal.pone.0124155.

Owour, B., W. Mauta, and S. Eriksen, 2011: Sustainable adaptation and human security: Interactions between pastoral and agropastoral groups in dryland Kenya. *Clim. Dev.*, **3**, 42–58, doi:10.3763/cdev.2010.0063.

Paetsch, L. et al. 2018: Effect of in-situ aged and fresh biochar on soil hydraulic conditions and microbial C use under drought conditions. *Sci. Rep.*, **8**(1), 6852, doi:10.1038/s41598-018-25039-x.

Page, S.E. et al. 2002: The amount of carbon released from peat and forest fires in Indonesia during 1997. *Nature*, **420**, 61–65, doi:10.1038/nature01131.

Paladino, S., and S.J. Fiske, 2017: *The Carbon Fix: Forest Carbon, Social Justice, and Environmental Governance*. Routledge, New York, United States, 320 pp.

Palm, C., H. Blanco-Canqui, F. DeClerck, L. Gatere, and P. Grace, 2014: Conservation agriculture and ecosystem services: An overview. *Agric. Ecosyst. Environ.*, **187**, 87–105, doi:10.1016/J.AGEE.2013.10.010.

Pan, Y. et al. 2011: A large and persistent carbon sink in the world's forests. *Science*, **333**, 988–993, doi:10.1126/science.1201609.

Panagos, P. et al. 2015: Estimating the soil erosion cover-management factor at the European scale. *Land use policy*, **48**, 38–50, doi:10.1016/J.LANDUSEPOL.2015.05.021.

Panfil, S.N. and C.A. Harvey, 2016: REDD+ and biodiversity conservation: A review of the biodiversity goals, monitoring methods, and impacts of 80 REDD+ projects. *Conserv. Lett.*, **9**, 143–150, doi:10.1111/conl.12188.

Panthou, G., T. Vischel, and T. Lebel, 2014: Recent trends in the regime of extreme rainfall in the Central Sahel. *Int. J. Climatol.*, **34**, 3998–4006, doi:10.1002/joc.3984.

Parajuli, P.B., P. Jayakody, G.F. Sassenrath, and Y. Ouyang, 2016: Assessing the impacts of climate change and tillage practices on stream flow, crop and sediment yields from the Mississippi River Basin. *Agric. Water Manag.*, **168**, 112–124, doi:10.1016/J.AGWAT.2016.02.005.

Pardini, R., A. de A. Bueno, T.A. Gardner, P.I. Prado, and J.P. Metzger, 2010: Beyond the fragmentation threshold hypothesis: Regime shifts in biodiversity across fragmented landscapes. *PLoS One*, **5**, e13666, doi:10.1371/journal.pone.0013666.

Parry, L.E., J. Holden, and P.J. Chapman, 2014: Restoration of blanket peatlands. *J. Environ. Manage.*, **133**, 193–205, doi:10.1016/j.jenvman.2013.11.033.

Pataki, D.E. et al. 2011: Coupling biogeochemical cycles in urban environments: Ecosystem services, green solutions, and misconceptions. *Front. Ecol. Environ.*, **9**, 27–36, doi:10.1890/090220.

Patange, O.S. et al. 2015: Reductions in indoor black carbon concentrations from improved biomass stoves in rural India. *Environ. Sci. Technol.*, **49**, 4749–4756, doi:10.1021/es506208x.

Pattanayak, S.K., D. Evan Mercer, E. Sills, and J.-C. Yang, 2003: Taking stock of agroforestry adoption studies. *Agrofor. Syst.*, **57**, 173–186, doi:10.1023/A:1024809108210.

Paustian, K. et al. 2016: Climate-smart soils. *Nature*, **532**, 49–57, doi:10.1038/nature17174.

Payo, A. et al. 2016: Projected changes in area of the Sundarbans mangrove forest in Bangladesh due to SLR by 2100. *Clim. Change*, **139**, 279–291, doi:10.1007/s10584-016-1769-z.

Pearson, T.R.H., S. Brown, and F.M. Casarim, 2014: Carbon emissions from tropical forest degradation caused by logging. *Environ. Res. Lett.*, **9**, 034017, doi:10.1088/1748-9326/9/3/034017.

Pearson, T.R.H., S. Brown, L. Murray, and G. Sidman, 2017: Greenhouse gas emissions from tropical forest degradation: An underestimated source. *Carbon Balance Manag.*, **12**, 3, doi:10.1186/s13021-017-0072-2.

Pedlar, J.H. et al. 2012: Placing forestry in the assisted migration debate. *Bioscience*, **62**, 835–842, doi:10.1525/bio.2012.62.9.10.

Pellegrini, A.F.A. et al. 2018: Fire frequency drives decadal changes in soil carbon and nitrogen and ecosystem productivity. *Nature*, **553**, 194–198. doi:10.1038/nature24668.

Van Pelt, R.S. et al. 2017: The reduction of partitioned wind and water erosion by conservation agriculture. *CATENA*, **148**, 160–167, doi:10.1016/J.CATENA.2016.07.004.

Peltzer, D.A., R.B. Allen, G.M. Lovett, D. Whitehead, and D.A. Wardle, 2010: Effects of biological invasions on forest carbon sequestration. *Glob. Chang. Biol.*, **16**, 732–746, doi:10.1111/j.1365-2486.2009.02038.x.

Pendergrass, A.G., 2018: What precipitation is extreme? *Science*, **360**, 1072–1073, doi:10.1126/science.aat1871.

Pendergrass, A.G. and R. Knutti, 2018: The uneven nature of daily precipitation and its change. *Geophys. Res. Lett.*, **45**, 11,980–11,988, doi:10.1029/2018GL080298.

Pendergrass, A.G., R. Knutti, F. Lehner, C. Deser, and B.M. Sanderson, 2017: Precipitation variability increases in a warmer climate. *Sci. Rep.*, **7**, 17966, doi:10.1038/s41598-017-17966-y.

Pendleton, L. et al. 2012: Estimating global "blue carbon" emissions from conversion and degradation of vegetated coastal ecosystems. *PLoS One*, **7**, e43542, doi:10.1371/journal.pone.0043542.

Peng, S.-S. et al. 2014: Afforestation in China cools local land surface temperature. *Proc. Natl. Acad. Sci. U.S.A.*, **111**, 2915–2919, doi:10.1073/pnas.1315126111.

Peng, X. et al. 2016: Response of changes in seasonal soil freeze/thaw state to climate change from 1950 to 2010 across China. *J. Geophys. Res. Earth Surf.*, **121**, 1984–2000, doi:10.1002/2016JF003876.

Peng, X., Y. Deng, Y. Peng, and K. Yue, 2018: Effects of biochar addition on toxic element concentrations in plants: A meta-analysis. *Sci. Total Environ.*, **616–617**, 970–977.

Percival, V. and T. Homer-Dixon, 1995: Environmental Scarcity and Violent Conflict: The Case of Rwanda. *J. Env. Dev.*, **5**(3), 270–291, doi:10.1177/107049659600500302.

Peres, C.A., T. Emilio, J. Schietti, S.J.M. Desmoulière, and T. Levi, 2016: Dispersal limitation induces long-term biomass collapse in overhunted Amazonian forests. *Proc. Natl. Acad. Sci. U.S.A.*, **113**, 892–897, doi:10.1073/pnas.1516525113.

Perugini, L. et al. 2017: Biophysical effects on temperature and precipitation due to land cover change. *Environ. Res. Lett.*, **12**, 053002, doi:10.1088/1748-9326/aa6b3f.

Peters, P.E., 2004: Inequality and social conflict over land in Africa. *J. Agrar. Chang.*, **4**, 269–314, doi:10.1111/j.1471-0366.2004.00080.x.

Petzold, J., 2016: Limitations and opportunities of social capital for adaptation to climate change: A case study on the Isles of Scilly. *Geogr. J.*, **182**, 123–134, doi:10.1111/geoj.12154.

Petzold, J. and A.K. Magnan, 2019: Climate change: Thinking small islands beyond Small Island Developing States (SIDS). *Clim. Change,* **152,** 145–165, doi:10.1007/s10584-018-2363-3.

Phelps, J., E.L. Webb, and A. Agrawal, 2010: Does REDD+ threaten to recentralize forest governance? *Science,* **328,** 312–313, doi:10.1126/ science.1187774.

Pielke, R.A. et al. 2007: An overview of regional land-use and land-cover impacts on rainfall. *Tellus B Chem. Phys. Meteorol.,* **59,** 587–601, doi:10.1111/j.1600-0889.2007.00251.x.

Piguet, E., R. Kaenzig, and J. Guélat, 2018: The uneven geography of research on "environmental migration." *Popul. Environ.,* **1–27,** doi:10.1007/ s11111-018-0296-4.

Piñeiro, G., J.M. Paruelo, M. Oesterheld, and E.G. Jobbágy, 2010: Pathways of grazing effects on soil organic carbon and nitrogen. *Rangel. Ecol. Manag.,* **63,** 109–119, doi:10.2111/08-255.1.

Pingoud, K., T. Ekholm, R. Sievänen, S. Huuskonen, and J. Hynynen, 2018: Trade-offs between forest carbon stocks and harvests in a steady state – A multi-criteria analysis. *J. Environ. Manage.,* **210,** 96–103, doi:10.1016/j. jenvman.2017.12.076.

Pinty, B. et al. 2011: Snowy backgrounds enhance the absorption of visible light in forest canopies. *Geoph. Res. Lett.* **38**(6), 1–5 doi:10.1029/2010GL046417.

Piovano, E.L., D. Ariztegui, S.M. Bernasconi, and J.A. McKenzie, 2004: Stable isotopic record of hydrological changes in subtropical Laguna Mar Chiquita (Argentina) over the last 230 years. *The Holocene,* **14,** 525–535, doi:10.1191/0959683604hl729rp.

Plangoen, P. and M.S. Babel, 2014: Projected rainfall erosivity changes under future climate in the Upper Nan Watershed, Thailand. *J. Earth Sci. Clim. Change,* **5.**

Planque, C., D. Carrer, and J.-L. Roujean, 2017: Analysis of MODIS albedo changes over steady woody covers in France during the period of 2001–2013. *Remote Sens. Environ.,* **191,** 13–29, doi:10.1016/J. RSE.2016.12.019.

Poeplau, C., and A. Don, 2015: Carbon sequestration in agricultural soils via cultivation of cover crops – A meta-analysis. *Agric. Ecosyst. Environ.,* **200,** 33–41, doi:10.1016/J.AGEE.2014.10.024.

Poesen, J., J. Nachtergaele, G. Verstraeten, and C. Valentin, 2003: Gully erosion and environmental change: Importance and research needs. *CATENA,* **50,** 91–133, doi:10.1016/S0341-8162(02)00143-1.

Poesen, J.W.A. and J.M. Hooke, 1997: Erosion, flooding and channel management in Mediterranean environments of southern Europe. *Prog. Phys. Geogr.,* **21,** 157–199, doi:10.1177/030913339702100201.

Poff, N.L., 2002: Ecological response to and management of increased flooding caused by climate change. *Philos. Trans. R. Soc. A Math. Phys. Eng. Sci.,* **360,** 1497–1510, doi:10.1098/rsta.2002.1012.

Pokam, W.M. et al. 2014: Identification of processes driving low-level westerlies in West Equatorial Africa. *J. Clim.,* **27,** 4245–4262, doi:10.1175/ JCLI-D-13-00490.1.

Polley, H.W. et al. 2013: Climate change and North American rangelands: Trends, projections, and implications. *Rangel. Ecol. Manag.,* **66,** 493–511, doi:10.2111/REM-D-12-00068.1.

Pontee, N., 2013: Defining coastal squeeze: A discussion. *Ocean Coast. Manag.,* **84,** 204–207, doi:10.1016/J.OCECOAMAN.2013.07.010.

Poorter, L. et al. 2016: Biomass resilience of neotropical secondary forests. *Nature,* **530,** 211–214.

Porter-Bolland, L. et al. 2012: Community managed forests and forest protected areas: An assessment of their conservation effectiveness across the tropics. *For. Ecol. Manage.,* **268,** 6–17, doi:10.1016/J.FORECO.2011.05.034.

Porter, J. et al. 2014: In: Climate Change 2014: Impacts, Adaptation and Vulnerability. Part A: Global and Sectoral Aspects. Contribution of Working Group II to the Fifth Assessment Report of the Intergovernmental Panel on Climate Change [Field, C.B., V.R. Barros, D.J. Dokken, K.J. Mach, M.D. Mastrandrea, T.E. Bilir, M. Chatterjee, K.L. Ebi, Y.O. Estrada, R.C. Genova, B. Girma, E.S. Kissel, A.N. Levy, S. MacCracken, P.R. Mastrandrea, and L.L.White

(eds.)]. Cambridge University Press, Cambridge, United Kingdom and New York, USA, pp. 485–533.

Porter, J.H., M.L. Parry, and T.R. Carter, 1991: The potential effects of climatic change on agricultural insect pests. *Agric. For. Meteorol.,* **57,** 221–240, doi:10.1016/0168-1923(91)90088-8.

Post, W.M. and K.C. Kwon, 2000: Soil carbon sequestration and land-use change: Processes and potential. *Glob. Chang. Biol.,* **6,** 317–327, doi:10.1046/j.1365-2486.2000.00308.x.

Potapov, P. et al. 2008: Mapping the world's intact forest landscapes by remote sensing. *Ecol. Soc.,* **13,** art51, doi:10.5751/ES-02670-130251.

Potschin, M., R.H. 2016: *Routledge Handbook Of Ecosystem Services.* [Roy H. . Haines-Young, R. Fish, and R.K. Turner, (eds.)]. Routledge, Abingdon, Oxfordshire, UK, 629 pp.

Poulton, P., J. Johnston, A. Macdonald, R. White, and D. Powlson, 2018: Major limitations to achieving "4 per 1000" increases in soil organic carbon stock in temperate regions: Evidence from long-term experiments at Rothamsted Research, United Kingdom. *Glob. Chang. Biol.,* **24,** 2563–2584, doi:10.1111/gcb.14066.

Powell, B., J. Hall, and T. Johns, 2011: Forest cover, use and dietary intake in the East Usambara Mountains, Tanzania. *Int. For. Rev.,* **13,** 305–317, doi:1 0.1505/146554811798293944.

Powlson, D.S. et al. 2014: Limited potential of no-till agriculture for climate change mitigation. *Nat. Clim. Chang.,* **4,** 678–683, doi:10.1038/ nclimate2292.

Prein, A.F. et al. 2017: Increased rainfall volume from future convective storms in the US. *Nat. Clim. Chang.,* **7,** 880–884, doi:10.1038/s41558-017-0007-7.

Preti, F. and N. Romano, 2014: Terraced landscapes: From an old best practice to a potential hazard for soil degradation due to land abandonment. *Anthropocene,* **6,** 10–25, doi:10.1016/J.ANCENE.2014.03.002.

Preti, F. et al. 2018: Conceptualization of water flow pathways in agricultural terraced landscapes. *L. Degrad. Dev.,* **29,** 651–662, doi:10.1002/ldr.2764.

Price, D.T. et al. 2013: Anticipating the consequences of climate change for Canada' s boreal forest ecosystems. *Environ. Rev.,* **21,** 322–365, doi:10.1139/er-2013-0042.

Price, R., 2017: *"Clean" Cooking Energy in Uganda – Technologies, Impacts, and Key Barriers and Enablers to Market Acceleration.* Institute of Development Studies, Brighton, UK.

Prince, S. et al. 2018: Status and trends of land degradation and restoration and associated changes in biodiversity and ecosystem fundtions. *The IPBES Assessment Report On Land Degradation And Restoration,* [L. Montanarella, R. Scholes, and A. Brainich, (eds.)]. Bonn, Germany, pp. 221–338.

Prince, S.D., 2016: Where Does Desertification Occur? Mapping Dryland Degradation at Regional to Global Scales. The End of Desertification, [R. Behnke and M. Mortimore, (eds.)]. Springer, Berlin, Heidelberg, Gernmany, pp. 225–263.

Prince, S.D., E.B. De Colstoun, and L.L. Kravitz, 1998: Evidence from rain-use efficiencies does not indicate extensive Sahelian desertification. *Glob. Chang. Biol.,* **4,** 359–374, doi:10.1046/j.1365-2486.1998.00158.x.

Pritchard, S.G., 2011: Soil organisms and global climate change. *Plant Pathol.,* **60,** 82–99, doi:10.1111/j.1365-3059.2010.02405.x.

Pritchard, W.R. et al. 1992: *Assessment of Animal Agriculture in Sub-Saharan Africa.* Morrilton: Winrock International, Washington D.C., 169 p.

Pryor, S.C., and R.J. Barthelmie, 2010: Climate change impacts on wind energy: A review. *Renew. Sustain. Energy Rev.,* **14,** 430–437, doi:10.1016/J. RSER.2009.07.028.

Pulido, M., S. Schnabel, J.F.L. Contador, J. Lozano-Parra, and Á. Gómez-Gutiérrez, 2017: Selecting indicators for assessing soil quality and degradation in rangelands of Extremadura (SW Spain). *Ecol. Indic.,* **74,** 49–61, doi:10.1016/J.ECOLIND.2016.11.016.

Pureswaran, D.S. et al. 2015: Climate-induced changes in host tree–insect phenology may drive ecological state-shift in boreal forests. *Ecology,* **96,** 1480–1491, doi:10.1890/13-2366.1.

4

Putz, F.E. et al. 2012: Sustaining conservation values in selectively logged tropical forests: The attained and the attainable. *Conserv. Lett.*, **5**, 296–303, doi:10.1111/j.1755-263X.2012.00242.x.

Qiu, B. et al. 2017: Assessing the Three-North Shelter Forest Program in China by a novel framework for characterizing vegetation changes. *ISPRS J. Photogramm. Remote Sens.*, **133**, 75–88, doi:10.1016/j.isprsjprs.2017.10.003.

Le Quéré, C. et al. 2013: The global carbon budget 1959–2011. *Earth Syst. Sci. Data*, **5**, 165–185, doi:10.5194/essd-5-165-2013.

Quilliam, R.S., S. Rangecroft, B.A. Emmett, T.H. Deluca, and D.L. Jones, 2013: Is biochar a source or sink for polycyclic aromatic hydrocarbon (PAH) compounds in agricultural soils? *GCB Bioenergy*, **5**, 96–103, doi:10.1111/gcbb.12007.

Quin, P.R. et al. 2014: Oil mallee biochar improves soil structural properties – A study with x-ray micro-CT. *Agric. Ecosyst. Environ.*, **191**, 142–149, doi:10.1016/j.agee.2014.03.022.

Quinlan, A.E., M. Berbés-Blázquez, L.J. Haider, and G.D. Peterson, 2016: Measuring and assessing resilience: Broadening understanding through multiple disciplinary perspectives. *J. Appl. Ecol.*, **53**, 677–687, doi:10.1111/1365-2664.12550.

Quiñonero-Rubio, J.M., E. Nadeu, C. Boix-Fayos, and J. de Vente, 2016: Evaluation of the effectiveness of forest restoration and check-dams to reduce catchment sediment yield. *L. Degrad. Dev.*, **27**, 1018–1031, doi:10.1002/ldr.2331.

Quinton, J.N. et al. 2010: The impact of agricultural soil erosion on biogeochemical cycling. *Nat. Geosci.*, **3**, 311–314, doi:10.1038/ngeo838.

Qureshi, A.S., 2011: Water management in the Indus Basin in Pakistan: Challenges and opportunities. *Mt. Res. Dev.*, **31**, 252–260, doi:10.1659/MRD-JOURNAL-D-11-00019.1.

Rabalais, N.N. et al. 2014: Eutrophication-driven deoxygenation in the Coastal Ocean. *Oceanography*, **27**, 172–183, doi:10.2307/24862133.

Rahmstorf, S., 2010: A new view on sea level rise. *Nat. Reports Clim. Chang.*, **44–45**, doi:10.1038/climate.2010.29.

Raleigh, C. and H. Urdal, 2007: Climate change, environmental degradation and armed conflict. *Polit. Geogr.*, **26**, 674–694, doi:10.1016/J.POLGEO.2007.06.005.

Ramankutty, N., J.A. Foley, J. Norman, and K. McSweeney, 2002: The global distribution of cultivable lands: Current patterns and sensitivity to possible climate change. *Glob. Ecol. Biogeogr.*, **11**, 377–392, doi:10.1046/j.1466-822x.2002.00294.x.

Ramchunder, S.J., L.E. Brown, and J. Holden, 2012: Catchment-scale peatland restoration benefits stream ecosystem biodiversity. *J. Appl. Ecol.*, **49**(1), 182–191, doi:10.1111/j.1365-2664.2011.02075.x.

Rametsteiner, E. and M. Simula, 2003: Forest certification – An instrument to promote sustainable forest management? *J. Environ. Manage.*, **67**, 87–98, doi:10.1016/S0301-4797(02)00191-3.

Ramisch, J., J. Keeley, I. Scoones, and W. Wolmer, 2002: Crop-Livestock policy in Africa: What is to be done? In: Pathways of change in Africa: crops, livestock & livelihoods in Mali, Ethiopia & Zimbabwe, [I. Scoones and W. Wolmer, (eds.)], James Currey Ltd., Oxford, pp. 183–210.

Rasul, G., A Mahmood, A Sadiq, and S.I. Khan, 2012: Vulnerability of the Indus Delta to Climate Change in Pakistan. *Pakistan J. Meteorol.*, **8**, 89–107.

Ratcliffe, S. et al. 2017: Biodiversity and ecosystem functioning relations in European forests depend on environmental context. *Ecol. Lett.*, **20**, 1414–1426, doi:10.1111/ele.12849.

Ratter, B.M.W., J. Petzold, and K. Sinane, 2016: Considering the locals: Coastal construction and destruction in times of climate change on Anjouan, Comoros. *Nat. Resour. Forum*, **40**, 112–126, doi:10.1111/1477-8947.12102.

Ravi, S., D.D. Breshears, T.E. Huxman, and P. D'Odorico, 2010: Land degradation in drylands: Interactions among hydrologic–aeolian erosion and vegetation dynamics. *Geomorphology*, **116**, 236–245, doi:10.1016/j.geomorph.2009.11.023.

Reed, M.S., 2005: *Participatory Rangeland Monitoring and Management in the Kalahari, Botswana*. University of Leeds, Leeds, UK, 267 pp.

Reed, M.S., and L. Stringer, 2016: *Land Degradation, Desertification and Climate Change: Anticipating, Assessing and Adapting to Future Change*. New York, NY: Routledge,178 pp.

Reed, M.S., A.J. Dougill, and M.J. Taylor, 2007: Integrating local and scientific knowledge for adaptation to land degradation: Kalahari rangeland management options. *L. Degrad. Dev.*, **18**, 249–268, doi:10.1002/ldr.777.

Reed, S.C. et al. 2012: Changes to dryland rainfall result in rapid moss mortality and altered soil fertility. *Nat. Clim. Chang.*, **2**, 752–755, doi:10.1038/nclimate1596.

Reenberg, A., T. Birch-Thomsen, O. Mertz, B. Fog, and S. Christiansen, 2008: Adaptation of human coping strategies in a small island society in the SW Pacific—50 years of change in the coupled human-environment system on Bellona, Solomon Islands. *Hum. Ecol.*, **36**, 807–819, doi:10.1007/s10745-008-9199-9.

Regina, K., J. Sheehy, and M. Myllys, 2015: Mitigating greenhouse gas fluxes from cultivated organic soils with raised water table. *Mitig. Adapt. Strateg. Glob. Chang.*, **20**, 1529–1544, doi:10.1007/s11027-014-9559-2.

Reid, P. and C. Vogel, 2006: Living and responding to multiple stressors in South Africa – Glimpses from KwaZulu-Natal. *Glob. Environ. Chang.*, **16**, 195–206, doi:10.1016/J.GLOENVCHA.2006.01.003.

Reinwarth, B., R. Petersen, and J. Baade, 2019: Inferring mean rates of sediment yield and catchment erosion from reservoir siltation in the Kruger National Park, South Africa: An uncertainty assessment. *Geomorphology*, **324**, 1–13, doi:10.1016/J.GEOMORPH.2018.09.007.

Reis, V. et al. 2017: A global assessment of inland wetland conservation status. *Bioscience*, **67**, 523–533, doi:10.1093/biosci/bix045.

REN21, 2018: *Renewables 2018: Global Status Report*. Renewable Energy for the 21st Century Policy Network, Paris, France, 325 pp.

Rengasamy, P., 2006: World salinization with emphasis on Australia. *J. Exp. Bot.*, **57**, 1017–1023, doi:10.1093/jxb/erj108.

Renou-Wilson, F. et al. 2018: Rewetting degraded peatlands for climate and biodiversity benefits: Results from two raised bogs *Ecological Engineering*, **127**, 547–560, doi: 10.1016/j.ecoleng.2018.02.014.

Reyer, C.P.O. et al. 2015: Forest resilience and tipping points at different spatio-temporal scales: Approaches and challenges. *J. Ecol.*, **103**, 5–15, doi:10.1111/1365-2745.12337.

Richards, C. and K. Lyons, 2016: The new corporate enclosures: Plantation forestry, carbon markets and the limits of financialised solutions to the climate crisis. *Land use policy*, **56**, 209–216, doi:10.1016/J.LANDUSEPOL.2016.05.013.

Rist, L., A. Felton, L. Samuelsson, C. Sandström, and O. Rosvall, 2013: A new paradigm for adaptive management. *Ecol. Soc.*, **18**, doi:10.5751/ES-06183-180463.

Ritzema, H., S. Limin, K. Kusin, J. Jauhiainen, and H. Wösten, 2014: Canal blocking strategies for hydrological restoration of degraded tropical peatlands in Central Kalimantan, Indonesia. *Catena*, **114**, 11–20, doi:10.1016/j.catena.2013.10.009.

Rivera-Ferre, M.G. et al. 2016: Re-framing the climate change debate in the livestock sector: Mitigation and adaptation options. *Wiley Interdiscip. Rev. Clim. Chang.*, **7**, 869–892, doi:10.1002/wcc.421.

Rizwan, M. et al. 2016: Mechanisms of biochar-mediated alleviation of toxicity of trace elements in plants: A critical review. *Environ. Sci. Pollut. Res.*, **23**, 2230–2248, doi:10.1007/s11356-015-5697-7.

Roberts, D., 2010: Prioritizing climate change adaptation and local level resilience in Durban, South Africa. *Environ. Urban.* **22**(2), 397–413, doi:10.1177/0956247810379948.

Roberts, D. and S. O'Donoghue, 2013: Urban environmental challenges and climate change action in Durban, South Africa. *Environ. Urban.*, **25**, 299–319, doi:10.1177/0956247813500904.

Roberts, M.W., A.W. D'Amato, C.C. Kern, and B.J. Palik, 2016: Long-term impacts of variable retention harvesting on ground-layer plant communities in Pinus resinosa forests. *J. Appl. Ecol.*, **53**, 1106–1116, doi:10.1111/1365-2664.12656.

Robinson, N., R.J. Harper, and K.R.J. Smettem, 2006: Soil water depletion by *Eucalyptus* spp. integrated into dryland agricultural systems. *Plant Soil*, **286**, 141–151, doi:10.1007/s11104-006-9032-4.

Rogelj, J. et al. 2018: Pathways Compatible With 1.5°C in the Context of Sustainable Development. In: Global Warming of 1.5°C: An IPCC special report on the impacts of global warming of 1.5°C above pre-industrial levels and related global greenhouse gas emission pathways, in the context of strengthening the global response to the threat of climate change [V. Masson-Delmotte, P. Zhai, H.-O. Pörtner, D. Roberts, J. Skea, P.R. Shukla, A. Pirani, W. Moufouma-Okia, C. Péan, R. Pidcock, S. Connors, J.B.R. Matthews, Y. Chen, X. Zhou, M.I. Gomis, E. Lonnoy, T. Maycock, M. Tignor, and T. Waterfield (eds.)]. In press, pp. 93–174.

Roman-Cuesta, R.M. et al. 2016: Hotspots of gross emissions from the land use sector: Patterns, uncertainties, and leading emission sources for the period 2000–2005 in the tropics. *Biogeosciences*, **13**, 4253–4269, doi:10.5194/bg-13-4253-2016.

Romero, C. and F.E. Putz, 2018: Theory-of-change development for the evaluation of forest stewardship council certification of sustained timber yields from natural forests in Indonesia. *Forests*, **9**, doi:10.3390/f9090547.

Romo-Leon, J.R., W.J.D. van Leeuwen, and A. Castellanos-Villegas, 2014: Using remote sensing tools to assess land use transitions in unsustainable arid agro-ecosystems. *J. Arid Environ.*, **106**, 27–35, doi:10.1016/j.jaridenv.2014.03.002.

Rosenzweig, C. and D. Hillel, 1998: Climate change and the global harvest: Potential impacts of the greenhouse effect on agriculture. Oxford University Press, Oxford, UK, 324 pp.

Rosenzweig, C., A. Iglesias, X.B. Yang, P.R. Epstein, and E. Chivian, 2001: Climate change and extreme weather events; implications for food production, plant diseases, and pests. *Glob. Chang. Hum. Heal.*, **2**, 90–104, doi:10.1023/A:1015086831467.

Rosenzweig, C. et al. 2014: Assessing agricultural risks of climate change in the 21st century in a global gridded crop model intercomparison. *Proc. Natl. Acad. Sci. U.S.A.*, **111**, 3268–3273, doi:10.1073/pnas.1222463110.

Ross, N.J., 2011: Modern tree species composition reflects ancient Maya "forest gardens" in northwest Belize. *Ecol. Appl.*, **21**, 75–84, doi:10.1890/09-0662.1.

Rossi, V. et al. 2017: Could REDD+ mechanisms induce logging companies to reduce forest degradation in Central Africa? *J. For. Econ.*, **29**, 107–117, doi:10.1016/J.JFE.2017.10.001.

Rotenberg, E. and D. Yakir, 2010: Contribution of semi-arid forests to the climate system. *Science*, **327**, 451–454, doi:10.1126/science.1179998.

Routschek, A., J. Schmidt, and F. Kreienkamp, 2014: Impact of climate change on soil erosion – A high-resolution projection on catchment scale until 2100 in Saxony/Germany. *CATENA*, **121**, 99–109, doi:10.1016/J.CATENA.2014.04.019.

Royal Society, 2016: *Resilience to Extreme Weather*. The Royal Society, London, 124 pp.

Rudel, T. et al. 2016: Do smallholder, mixed crop-livestock livelihoods encourage sustainable agricultural practices? A meta-analysis. *Land*, **5, 6**, doi:10.3390/land5010006.

Ruggiero, P., 2013: Is the intensifying wave climate of the u.s. pacific northwest increasing flooding and erosion risk faster than sea-level rise? *J. Waterw. Port, Coastal, Ocean Eng.*, **139**, 88–97, doi:10.1061/(ASCE)WW.1943-5460.0000172.

Rumpel, C. et al. 2018: Put more carbon in soils to meet Paris climate pledges. *Nature*, **564**, 32–34, doi:10.1038/d41586-018-07587-4.

Ruppert, J.C. et al. 2012: Meta-analysis of ANPP and rain-use efficiency confirms indicative value for degradation and supports non-linear response along precipitation gradients in drylands. *J. Veg. Sci.*, **23**, 1035–1050, doi:10.1111/j.1654-1103.2012.01420.x.

Russell, M.B. et al. 2015: Quantifying carbon stores and decomposition in dead wood: A review. *For. Ecol. Manage.*, **350**, 107–128, doi:10.1016/j.foreco.2015.04.033.

Rutherford, W.A. et al. 2017: Albedo feedbacks to future climate via climate change impacts on dryland biocrusts. *Sci. Rep.*, **7**, 44188, doi:10.1038/srep44188.

Ryan, M.R. et al. 2018: Managing for multifunctionality in perennial grain crops. *Bioscience*, **68**, 294–304, doi:10.1093/biosci/biy014.

Sachs, J.D., 2007: Poverty and environmental stress fuel Darfur crisis. *Nature*, **449**, 24–24, doi:10.1038/449024a.

Sainju, U.M., B.L. Allen, A.W. Lenssen, and R.P. Ghimire, 2017: Root biomass, root/shoot ratio, and soil water content under perennial grasses with different nitrogen rates. *F. Crop. Res.*, **210**, 183–191, doi:10.1016/J.FCR.2017.05.029.

Salehyan, I., 2008: From climate change to conflict? No consensus yet. *J. Peace Res.*, **45**, 315–326, doi:10.1177/0022343308088812.

Sanderman, J., T. Hengl, and G.J. Fiske, 2017: Soil carbon debt of 12,000 years of human land use. *Proc. Natl. Acad. Sci. U.S.A.*, **114**, 9575–9580, doi:10.1073/pnas.1706103114.

Sanquetta, C.R. et al. 2018: Dynamics of carbon and CO_2 removals by Brazilian forest plantations during 1990–2016. *Carbon Balance Manag.*, **13**, doi:10.1186/s13021-018-0106-4.

Santos, M.J., S.C. Dekker, V. Daioglou, M.C. Braakhekke, and D.P. van Vuuren, 2017: Modeling the effects of future growing demand for charcoal in the tropics. *Front. Environ. Sci.*, **5**, 28.

Sasaki, N., and F.E. Putz, 2009: Critical need for new definitions of "forest" and "forest degradation" in global climate change agreements. *Conserv. Lett.*, **2**, 226–232, doi:10.1111/j.1755-263X.2009.00067.x.

Sasaki, T., T. Okayasu, U. Jamsran, and K. Takeuchi, 2007: Threshold changes in vegetation along a grazing gradient in Mongolian rangelands. *J. Ecol.*, **96**(1), 145–154, doi:10.1111/j.1365-2745.2007.01315.x.

Saugier, B., 2001: Estimations of Global Terrestrial Productivity: Converging Toward a Single Number? *Terrestrial Global Productivity*, [J. Roy, (ed.)]. Academic Press, San Diego, CA, CA, USA, pp. 543–556.

Savard, J.-P., P. Bernatchez, F. Morneau, and F. Saucier, 2009: Vulnérabilité des communautés côtières de l'est du Québec aux impacts des changements climatiques. *La Houille Blanche*, **59–66**, doi:10.1051/lhb/2009015.

Sayer, J. et al. 2013: Ten principles for a landscape approach to reconciling agriculture, conservation, and other competing land uses. *Proc. Natl. Acad. Sci. U.S.A.*, **110**, 8349–8356, doi:10.1073/pnas.1210595110.

La Scala Júnior, N., E. De Figueiredo, and A. Panosso, 2012: A review on soil carbon accumulation due to the management change of major Brazilian agricultural activities. *Brazilian J. Biol.*, **72**, 775–785, doi:10.1590/S1519-69842012000400012.

Scarano, F.R., 2017: Ecosystem-based adaptation to climate change: concept, scalability and a role for conservation science. *Perspect. Ecol. Conserv.*, **15**, 65–73, doi:10.1016/J.PECON.2017.05.003.

Schaefer, H. et al. 2016: A 21st-century shift from fossil-fuel to biogenic methane emissions indicated by 13CH. *Science*, **352**, 80–84, doi:10.1126/science.aad2705.

Scharlemann, J.P., E.V. Tanner, R. Hiederer, and V. Kapos, 2014: Global soil carbon: Understanding and managing the largest terrestrial carbon pool. *Carbon Manag.*, **5**, 81–91, doi:10.4155/cmt.13.77.

Scheffers, B.R. et al. 2016: The broad footprint of climate change from genes to biomes to people. *Science*, **354**, aaf7671, doi:10.1126/science.aaf7671.

Scheffran, J., M. Brzoska, J. Kominek, P.M. Link, and J. Schilling, 2012: Disentangling the climate-conflict nexus: Empirical and theoretical assessment of vulnerabilities and pathways. *Rev. Eur. Stud.*, **4**, 1.

Scheidel, A. and C. Work, 2018: Forest plantations and climate change discourses: New powers of 'green' grabbing in Cambodia. *Land use policy*, **77**, 9–18, doi:10.1016/J.LANDUSEPOL.2018.04.057.

Scherr, S.J., 2000: A downward spiral? Research evidence on the relationship between poverty and natural resource degradation. *Food Policy*, **25**, 479–498, doi:10.1016/S0306-9192(00)00022-1.

Schlautman, B., S. Barriball, C. Ciotir, S. Herron, and A. Miller, 2018: Perennial grain legume domestication phase I: Criteria for candidate species selection. *Sustainability*, **10**, 730, doi:10.3390/su10030730.

Schlesinger, W.H., 2009: On the fate of anthropogenic nitrogen. *Proc. Natl. Acad. Sci.*, **106**, 203 LP-208.

Schlesinger, W.H. and S. Jasechko, 2014: Transpiration in the global water cycle. *Agric. For. Meteorol.*, **189–190**, 115–117, doi:10.1016/J.AGRFORMET.2014.01.011.

Schlesinger, W.H. and R. Amundson, 2018: Managing for soil carbon sequestration: Let's get realistic. *Glob. Chang. Biol.*, **25**, gcb.14478, doi:10.1111/gcb.14478.

Schleussner, C.-F. et al. 2016: Science and policy characteristics of the Paris Agreement temperature goal. *Nat. Clim. Chang.*, **6**, 827–835, doi:10.1038/nclimate3096.

Schmidt, M.W.I. et al. 2011: Persistence of soil organic matter as an ecosystem property. *Nature*, **478**, 49–56, doi:10.1038/nature10386.

Schnitzer, S.A. et al. 2011: Soil microbes drive the classic plant diversity–productivity pattern. *Ecology*, **92**, 296–303.

Schoenholtz, S., H.V. Miegroet, and J. Burger, 2000: A review of chemical and physical properties as indicators of forest soil quality: Challenges and opportunities. *For. Ecol. Manage.*, **138**, 335–356, doi:10.1016/S0378-1127(00)00423-0.

Schofield, R.V. and M.J. Kirkby, 2003: Application of salinization indicators and initial development of potential global soil salinization scenario under climatic change. *Global Biogeochem. Cycles*, **17**, doi:10.1029/2002GB001935.

Schuerch, M. et al. 2018: Future response of global coastal wetlands to sea-level rise. *Nature*, **561**, 231–234, doi:10.1038/s41586-018-0476-5.

Schut, A.G.T., E. Ivits, J.G. Conijn, B. ten Brink, and R. Fensholt, 2015: Trends in global vegetation activity and climatic drivers indicate a decoupled response to climate change. *PLoS One*, **10**, e0138013, doi:10.1371/journal.pone.0138013.

Schuur, E.A.G. et al. 2015: Climate change and the permafrost carbon feedback. *Nature*, **520**, 171–179, doi:10.1038/nature14338.

Schwilch, G. et al. 2011: Experiences in monitoring and assessment of sustainable land management. *L. Degrad. Dev.*, **22**, 214–225, doi:10.1002/ldr.1040.

Scoones, I. and W. Wolmer, 2002: Pathways of Change: Crop-Livestock Integration in Africa. *Pathways of Change In Africa: Crops, Livestock & Livelihoods in Mali, Ethiopia & Zimbabwe*, [I. Scoones and W. Wolmer, (eds.)]. James Currey Ltd. Oxford, 236 p.

Sealey, N.E., 2006: The cycle of casuarina-induced beach erosion – A case study from Andros, Bahamas. 12th Symp. Geol. Bahamas other Carbonate Reg. San Salvador. Bahamas, 196–204.

Seaquist, J.W., T. Hickler, L. Eklundh, J. Ardö, and B.W. Heumann, 2009: Disentangling the effects of climate and people on Sahel vegetation dynamics. *Biogeosciences*, **6**, 469–477, doi:10.5194/bg-6-469-2009.

Sedano, F. et al. 2016: The impact of charcoal production on forest degradation: A case study in Tete, Mozambique. *Environ. Res. Lett.*, **11**, 094020, doi:10.1088/1748-9326/11/9/094020.

Segura, C., G. Sun, S. McNulty, and Y. Zhang, 2014: Potential impacts of climate change on soil erosion vulnerability across the conterminous United States. *J. Soil Water Conserv.*, **69**, 171–181, doi:10.2489/jswc.69.2.171.

Seidl, R., W. Rammer, D. Jäger, W.S. Currie, and M.J. Lexer, 2007: Assessing trade-offs between carbon sequestration and timber production within a framework of multi-purpose forestry in Austria. *For. Ecol. Manage.*, **248**, 64–79, doi:https://doi.org/10.1016/j.foreco.2007.02.035.

Seidl, R. et al. 2017: Forest disturbances under climate change. *Nat. Clim. Chang.*, **7**, 395–402, doi:10.1038/nclimate3303.

Serpa, D. et al. 2015: Impacts of climate and land use changes on the hydrological and erosion processes of two contrasting Mediterranean catchments. *Sci. Total Environ.*, **538**, 64–77, doi:10.1016/J.SCITOTENV.2015.08.033.

Seto, K.C. et al. 2012: Urban land teleconnections and sustainability. *Proc. Natl. Acad. Sci.*, **109**, 7687–7692, doi:10.1073/pnas.1117622109.

Settele, J. et al. 2015: Terrestrial and Inland Water Systems. In: Climate Change 2014: Impacts, Adaptation and Vulnerability. Part A: Global and Sectoral Aspects. Contribution of Working Group II to the Fifth Assessment Report of the Intergovernmental Panel on Climate Change [Field, C.B., V.R. Barros, D.J. Dokken, K.J. Mach, M.D. Mastrandrea, T.E. Bilir, M. Chatterjee, K.L. Ebi, Y.O. Estrada, R.C. Genova, B. Girma, E.S. Kissel, A.N. Levy, S. MacCracken, P.R. Mastrandrea, and L.L.White (eds.)]. Cambridge University Press, Cambridge, United Kingdom and New York, NY, USA, pp. 271–360.

Seymour, F. and A. Angelsen, 2012: *Summary and Conclusions: REDD+ without regrets. Analysing REDD+: Challenges and Choices*, Center for International Forestry Research (CIFOR), Bogor, Indonesia, 317–334.

Shadkam, S., F. Ludwig, P. van Oel, Ç Kirmit, and P. Kabat, 2016: Impacts of climate change and water resources development on the declining inflow into Iran's Urmia Lake. *J. Great Lakes Res.*, **42**, 942–952, doi:10.1016/j.jglr.2016.07.033.

Shakesby, R.A., 2011: Post-wildfire soil erosion in the Mediterranean: Review and future research directions. *Earth-Science Rev.*, **105**, 71–100, doi:10.1016/J.EARSCIREV.2011.01.001.

Shames, S., M. Hill Clarvis, and G. Kissinger, 2014: *Financing Strategies for Integrated Landscape Investment*. EcoAgriculture Partners, Washington DC, USA, 1–60 pp.

Shanahan, T.M. et al. 2016: CO_2 and fire influence tropical ecosystem stability in response to climate change. *Sci. Rep.*, **6**, 29587, doi:10.1038/srep29587.

Shao, Y., 2008: *Physics and Modelling Of Wind Erosion*. Springer, Berlin, Germany, 452 pp.

Sharmila, S. and K.J.E. Walsh, 2018: Recent poleward shift of tropical cyclone formation linked to Hadley cell expansion. *Nat. Clim. Chang.*, **8**, 730–736, doi:10.1038/s41558-018-0227-5.

Sheffield, J., E.F. Wood, and M.L. Roderick, 2012: Little change in global drought over the past 60 years. *Nature*, **491**, 435–438, doi:10.1038/nature11575.

Sheil, D. and D. Murdiyarso, 2009: How forests attract rain: An examination of a new hypothesis. *Bioscience*, **59**, 341–347, doi:10.1525/bio.2009.59.4.12.

Shi, H. et al. 2017: Assessing the ability of MODIS EVI to estimate terrestrial ecosystem gross primary production of multiple land cover types. *Ecol. Indic.*, **72**, 153–164, doi:10.1016/J.ECOLIND.2016.08.022.

Shi, S., W. Zhang, P. Zhang, Y. Yu, and F. Ding, 2013: A synthesis of change in deep soil organic carbon stores with afforestation of agricultural soils. *For. Ecol. Manage.*, **296**, 53–63, doi:10.1016/j.foreco.2013.01.026.

Shindell, D. et al. 2012: Simultaneously mitigating near-term climate change and improving human health and food security. *Science*, **335**, 183 LP-189, doi:10.1126/science.1210026.

Shirzaei, M. and R. Bürgmann, 2018: Global climate change and local land subsidence exacerbate inundation risk to the San Francisco Bay Area. *Sci. Adv.*, **4**, eaap9234, doi:10.1126/sciadv.aap9234.

Showers, K.B., 2005: *Imperial Gullies: Soil Erosion and Conservation in Lesotho*. Ohio University Press, Athens, Ohio, USA, 346 pp.

Shuab, R., R. Lone, J. Ahmad, and Z.A. Reshi, 2017: *Arbuscular Mycorrhizal Fungi: A Potential Tool for Restoration of Degraded Land*. Mycorrhiza – Nutrient Uptake, Biocontrol, Ecorestoration, Springer International Publishing, Cham, Switzerland, pp. 415–434.

Siahaya, M.E., T.R. Hutauruk, H.S.E.S. Aponno, J.W. Hatulesila, and A.B. Mardhanie, 2016: Traditional ecological knowledge on shifting cultivation and forest management in East Borneo, Indonesia. *Int. J. Biodivers. Sci. Ecosyst. Serv. Manag.*, **12**, 14–23, doi:10.1080/21513732.2016.1169559.

Simon, J. et al. 2017: Biochar boosts tropical but not temperate crop yields. *Environ. Res. Lett.*, **12**, 53001.

Simonsen, S.H. et al. 2014: *Applying resilience thinking: Seven principles for building resilience in social-ecological systems*. Stockholm Resilience Centre, 20 p.

Sims, N.C. et al. 2019: Developing good practice guidance for estimating land degradation in the context of the United Nations Sustainable Development Goals. *Environ. Sci. Policy*, **92**, 349–355, doi:10.1016/J.ENVSCI.2018.10.014.

Singh, B.P. and A.L. Cowie, 2014: Long-term influence of biochar on native organic carbon mineralisation in a low-carbon clayey soil. *Sci. Rep.*, **4**, 3687, doi:10.1038/srep03687.

Singh, B.P., B.J. Hatton, S. Balwant, and A.L. Cowie, 2010: The role of biochar in reducing nitrous oxide emissions and nitrogen leaching from soil. Proceedings from the *19th World Congress of Soil Science Soil Solutions for a Changing World*, Brisbane, Australia. 3 p.

Singh, B.P., A.L. Cowie, and R.J. Smernik, 2012: Biochar carbon stability in a clayey soil as a function of feedstock and pyrolysis temperature. *Environ. Sci. Technol.*, **46**(21), 11770–11778 doi:10.1021/es302545b.

Singh, B.P. et al. 2015: In situ persistence and migration of biochar carbon and its impact on native carbon emission in contrasting soils under managed temperate pastures. *PLoS One*, **10**, e0141560, doi:10.1371/journal.pone.0141560.

Sinha, D. and M.R. Ray, 2015: *Health Effects of Indoor Air Pollution Due to Cooking with Biomass Fuel*. Humana Press, Cham, Switzerland, pp. 267–302.

Siry, J.P., F.W. Cubbage, and M.R. Ahmed, 2005: Sustainable forest management: Global trends and opportunities. *For. Policy Econ.*, **7**, 551–561, doi:10.1016/j.forpol.2003.09.003.

Sklenicka, P. et al. 2015: Owner or tenant: Who adopts better soil conservation practices? *Land use policy*, **47**, 253–261, doi:10.1016/J.LANDUSEPOL.2015.04.017.

Skoufias, E., M. Rabassa, and S. Olivieri, 2011: *The Poverty Impacts Of Climate Change: A Review of the Evidence*. The World Bank, Washington DC, USA, 38 pp.doi: 10.1596/1813-9450-5622.

Ben Slimane, A. et al. 2016: Relative contribution of rill/interrill and gully/channel erosion to small reservoir siltation in mediterranean environments. *L. Degrad. Dev.*, **27**, 785–797, doi:10.1002/ldr.2387.

Sloan, S. and J.A. Sayer, 2015: Forest resources assessment of 2015 shows positive global trends but forest loss and degradation persist in poor tropical countries. *For. Ecol. Manage.*, **352**, 134–145, doi:10.1016/J.FORECO.2015.06.013.

Slobbe, E. et al. 2013: Building with nature: In search of resilient storm surge protection strategies. *Nat. Hazards,* **65**, 947–966, doi:10.1007/s11069-012-0342-y.

Smeets, E., F.X. Johnson, and G. Ballard-Tremeer, 2012: *Traditional and Improved Use of Biomass for Energy in Africa. Bioenergy for Sustainable Development in Africa*, [R. Janssen and D. Rutz, (eds.)]. Springer Netherlands, Dordrecht, pp. 3–12.

Smith, H.E., F. Eigenbrod, D. Kafumbata, M.D. Hudson, and K. Schreckenberg, 2015: Criminals by necessity: The risky life of charcoal transporters in Malawi. *For. Trees Livelihoods*, **24**, 259–274, doi:10.1080/14728028.2015.1062808.

Smith, J., D.R. Nayak, and P. Smith, 2012: Renewable energy: Avoid constructing wind farms on peat. *Nature*, **489**(7414), p.33. doi:10.1038/489033d.

Smith, P., 2016: Soil carbon sequestration and biochar as negative emission technologies. *Glob. Chang. Biol.*, **22**, 1315–1324, doi:10.1111/gcb.13178.

Smith, P. et al. 2016a: Global change pressures on soils from land use and management. *Glob. Chang. Biol.*, **22**, 1008–1028, doi:10.1111/gcb.13068.

Smith, P. et al. 2016b: Biophysical and economic limits to negative CO_2 emissions. *Nat. Clim. Chang.*, **6**, 42–50, doi:10.1038/nclimate2870.

Smith, P. et al. 2014: Agriculture, Forestry and Other Land Use (AFOLU). In: Climate Change 2014: Mitigation of Climate Change. Contribution of Working Group III to the Fifth Assessment Report of the Intergovernmental Panel on Climate Change [Edenhofer, O., R. Pichs-Madruga, Y. Sokona,

E. Farahani, S. Kadner, K. Seyboth, A. Adler, I. Baum, S. Brunner, P. Eickemeier, B. Kriemann, J. Savolainen, S. Schlömer, C. von Stechow, T. Zwickel and J.C. Minx (eds.)]. Cambridge University Press, Cambridge, United Kingdom and New York, NY, USA, p. 83.

Smyth, C., W.A. Kurz, G. Rampley, T.C. Lemprière, and O. Schwab, 2017: Climate change mitigation potential of local use of harvest residues for bioenergy in Canada. *GCB Bioenergy*, **9**, 817–832, doi:10.1111/gcbb.12387.

Smyth, C.E. et al., 2014: Quantifying the biophysical climate change mitigation potential of Canada's forest sector. *Biogeosciences*, **11**, 3515–3529, doi:10.5194/bg-11-3515-2014.

Sobel, A.H. et al. 2016: Human influence on tropical cyclone intensity. *Science*, **353**, 242–246, doi:10.1126/science.aaf6574.

Sokolik, I.N. and O.B. Toon, 1996: Direct radiative forcing by anthropogenic airborne mineral aerosols. *Nature*, **381**, 681–683, doi:10.1038/381681a0.

Solly, E.F. et al. 2017: Experimental soil warming shifts the fungal community composition at the alpine treeline. *New Phytol.*, **215**, 766–778, doi:10.1111/nph.14603.

Solomon, N. et al. 2018: Environmental impacts and causes of conflict in the Horn of Africa: A review. *Earth-Science Rev.*, **177**, 284–290, doi:10.1016/J.EARSCIREV.2017.11.016.

Song, X.-P. et al. 2018: Global land change from 1982 to 2016. *Nature*, **560**, 639–643, doi:10.1038/s41586-018-0411-9.

Sonneveld, B.G.J.S., and D.L. Dent, 2009: How good is GLASOD? *J. Environ. Manage.*, **90**, 274–283, doi:10.1016/J.JENVMAN.2007.09.008.

Sonwa, D.J., S. Walker, R. Nasi, and M. Kanninen, 2011: Potential synergies of the main current forestry efforts and climate change mitigation in Central Africa. *Sustain. Sci.*, **6**, 59–67, doi:10.1007/s11625-010-0119-8.

Sonwa, D.J., S.F. Weise, G. Schroth, M.J.J. Janssens, and Howard-Yana Shapiro, 2014: Plant diversity management in cocoa agroforestry systems in West and Central Africa – effects of markets and household needs. *Agrofor. Syst.*, **88**, 1021–1034, doi:10.1007/s10457-014-9714-5.

Sonwa, D.J., S.F. Weise, B.A. Nkongmeneck, M. Tchatat, and M.J.J. Janssens, 2017: Structure and composition of cocoa agroforests in the humid forest zone of Southern Cameroon. *Agrofor. Syst.*, **91**, 451–470, doi:10.1007/s10457-016-9942-y.

Sonwa, F. et al. 2001: *The Role of Cocoa Agroforests in Rural and Community Forestry in Southern Cameroon*. Overseas Development Institute, London, UK, 1–10 pp.

Soule, M.J., A. Tegene, and K.D. Wiebe, 2000: Land tenure and the adoption of conservation practices. *Am. J. Agric. Econ.*, **82**, 993–1005, doi:10.1111/0002-9092.00097.

Sousa, F.F. de, C. Vieira-da-Silva, and F.B. Barros, 2018: The (in)visible market of miriti (Mauritia flexuosa L.f.) fruits, the "winter acai", in Amazonian riverine communities of Abaetetuba, Northern Brazil. *Glob. Ecol. Conserv.*, **14**, e00393, doi:10.1016/j.gecco.2018.e00393.

Soussana, J.-F. and G. Lemaire, 2014: Coupling carbon and nitrogen cycles for environmentally sustainable intensification of grasslands and crop-livestock systems. *Agric. Ecosyst. Environ.*, **190**, 9–17, doi:10.1016/J.AGEE.2013.10.012.

Soussana, J.-F. et al. 2006: Carbon cycling and sequestration opportunities in temperate grasslands. *Soil Use Manag.*, **20**, 219–230, doi:10.1111/j.1475-2743.2004.tb00362.x.

Soussana, J.-F. et al. 2019: Matching policy and science: Rationale for the '4 per 1000 – soils for food security and climate' initiative. *Soil Tillage Res.*, **188**, 3–15, doi:10.1016/J.STILL.2017.12.002.

Sovacool, B.K., 2012: Perceptions of climate change risks and resilient island planning in the Maldives. *Mitig. Adapt. Strateg. Glob. Chang.*, **17**, 731–752, doi:10.1007/s11027-011-9341-7.

Sparrevik, M., C. Adam, V. Martinsen, Jubaedah, and G. Cornelissen, 2015: Emissions of gases and particles from charcoal/biochar production in rural areas using medium-sized traditional and improved "retort" kilns. *Biomass and Bioenergy*, **72**, 65–73, doi:10.1016/j.biombioe.2014.11.016.

Specht, M.J., S.R.R. Pinto, U.P. Albuquerque, M. Tabarelli, and F.P.L. Melo, 2015: Burning biodiversity: Fuelwood harvesting causes forest degradation in human-dominated tropical landscapes. *Glob. Ecol. Conserv.*, **3**, 200–209, doi:10.1016/j.gecco.2014.12.002.

Spence, J.R., 2001: The new boreal forestry: Adjusting timber management to accommodate biodiversity. *Trends Ecol. Evol.*, **16**, 591–593, doi:10.1016/S0169-5347(01)02335-7.

Spencer, T. et al. 2016: Global coastal wetland change under sea-level rise and related stresses: The DIVA Wetland Change Model. *Glob. Planet. Change,* **139**, 15–30, doi:10.1016/J.GLOPLACHA.2015.12.018.

Sprunger, C.D., S.W. Culman, G.P. Robertson, and S.S. Snapp, 2018: Perennial grain on a Midwest Alfisol shows no sign of early soil carbon gain. *Renew. Agric. Food Syst.*, **33**, 360–372, doi:10.1017/S1742170517000138.

St.Clair, S.B. and J.P. Lynch, 2010: The opening of Pandora's Box: Climate change impacts on soil fertility and crop nutrition in developing countries. *Plant Soil*, **335**, 101–115, doi:10.1007/s11104-010-0328-z.

Stafford Smith, D.M. et al. 2007: Learning from episodes of degradation and recovery in variable Australian rangelands. *Proc. Natl. Acad. Sci. U.S.A.*, **104**, 20690–20695, doi:10.1073/pnas.0704837104.

Steffen, W.L. et al. 2005: Global Change and The Earth System: A Planet Under Pressure. Springer, Berlin, Germany, 336 pp.

Steinkamp, J. and T. Hickler, 2015: Is drought-induced forest dieback globally increasing? *J. Ecol.*, **103**, 31–43, doi:10.1111/1365-2745.12335.

Stevens, P., T. Roberts, and S. Lucas, 2015: *Life on Mars: Using Micro-topographic Relief to Secure Soil, Water and Biocapacity.* Engineers Australia, Barton ACT, Australia, pp. 505–519.

Steward, P.R. et al. 2018: The adaptive capacity of maize-based conservation agriculture systems to climate stress in tropical and subtropical environments: A meta-regression of yields. *Agric. Ecosyst. Environ.*, **251**, 194–202, doi:10.1016/J.AGEE.2017.09.019.

Stocking, M.A., N. Murnaghan, and N. Murnaghan, 2001: *A Handbook for the Field Assessment of Land Degradation*. Routledge, London, UK, 169 p.

Stoorvogel, J.J., M. Bakkenes, A.J.A.M. Temme, N.H. Batjes, and B.J.E. ten Brink, 2017: S-World: A global soil map for environmental modelling. *L. Degrad. Dev.*, **28**, 22–33, doi:10.1002/ldr.2656.

Strefler, J., T. Amann, N. Bauer, E. Kriegler, and J. Hartmann, 2018: Potential and costs of carbon dioxide removal by enhanced weathering of rocks. *Environ. Res. Lett.*, **13**, 034010, doi:10.1088/1748-9326/aaa9c4.

Stringer, L.C. et al. 2011: Combating land degradation and desertification and enhancing food security: Towards integrated solutions. Ann. *Arid Zone*, **50**, 1–23.

Strunz, S., 2012: Is conceptual vagueness an asset? Arguments from philosophy of science applied to the concept of resilience. *Ecol. Econ.*, **76**, 112–118, doi:10.1016/J.ECOLECON.2012.02.012.

Sturm, M., 2005: Changing snow and shrub conditions affect albedo with global implications. *J. Geophys. Res.*, **110**, G01004, doi:10.1029/2005JG000013.

Sturrock, R.N. et al. 2011: Climate change and forest diseases. *Plant Pathol.*, **60**, 133–149, doi:10.1111/j.1365-3059.2010.02406.x.

Sufo Kankeu, R., D.J. Sonwa, R. Eba'a Atyi, and N.M. Moankang Nkal, 2016: Quantifying post logging biomass loss using satellite images and ground measurements in Southeast Cameroon. *J. For. Res.*, **27**, 1415–1426, doi:10.1007/s11676-016-0277-3.

Sulaiman, C., A.S. Abdul-Rahim, H.O. Mohd-Shahwahid, and L. Chin, 2017: Wood fuel consumption, institutional quality, and forest degradation in sub-Saharan Africa: Evidence from a dynamic panel framework. *Ecol. Indic.*, **74**, 414–419, doi:https://doi.org/10.1016/j.ecolind.2016.11.045.

Sundström, J.F. et al. 2014: Future threats to agricultural food production posed by environmental degradation, climate change, and animal and plant diseases – a risk analysis in three economic and climate settings. *Food Secur.*, **6**, 201–215, doi:10.1007/s12571-014-0331-y.

Sussams, L.W., W.R. Sheate, and R.P. Eales, 2015: Green infrastructure as a climate change adaptation policy intervention: Muddying the waters or clearing a path to a more secure future? *J. Environ. Manage.*, **147**, 184–193, doi:10.1016/J.JENVMAN.2014.09.003.

Swails, E. et al. 2018: Will CO_2 emissions from drained tropical peatlands decline over time? Links between soil organic matter quality, nutrients, and C mineralization rates. *Ecosystems*, **21**, 868–885, doi:10.1007/s10021-017-0190-4.

Sydeman, W.J. et al. 2014: Climate change and wind intensification in coastal upwelling ecosystems. *Science*, **345**, 77–80, doi:10.1126/science.1251635.

Szabo, S. et al. 2016: Population dynamics, delta vulnerability and environmental change: Comparison of the Mekong, Ganges–Brahmaputra and Amazon delta regions. *Sustain. Sci.*, **11**, 539–554, doi:10.1007/s11625-016-0372-6.

Tacoli, C., 2009: Crisis or adaptation? Migration and climate change in a context of high mobility. *Environ. Urban.*, **21**, 513–525, doi:10.1177/0956247809342182.

Tadesse, G., 2001: Land degradation: A challenge to Ethiopia. *Environ. Manage.*, **27**, 815–824, doi:10.1007/s002670010190.

Tadesse, G., B. Algieri, M. Kalkuhl, and J. von Braun, 2014: Drivers and triggers of international food price spikes and volatility. *Food Policy,* **47**, 117–128, doi:10.1016/J.FOODPOL.2013.08.014.

Tamarin-Brodsky, T., and Y. Kaspi, 2017: Enhanced poleward propagation of storms under climate change. *Nat. Geosci.,* **10**, 908–913, doi:10.1038/s41561-017-0001-8.

Tang, J., S. Luyssaert, A.D. Richardson, W. Kutsch, and I.A. Janssens, 2014: Steeper declines in forest photosynthesis than respiration explain age-driven decreases in forest growth. *Proc. Natl. Acad. Sci.,* **111**, 8856–8860, doi:10.1073/pnas.1320761111.

Tarfasa, S. et al. 2018: Modeling smallholder farmers' preferences for soil management measures: A case study from South Ethiopia. *Ecol. Econ.,* **145**, 410–419, doi:10.1016/j.ecolecon.2017.11.027.

Taufik, M. et al. 2017: Amplification of wildfire area burnt by hydrological drought in the humid tropics. *Nat. Clim. Chang.,* **7**, 428–431, doi:10.1038/nclimate3280.

Taylor, A.R., M. Seedre, B.W. Brassard, and H.Y.H. Chen, 2014: Decline in net ecosystem productivity following canopy transition to late-succession forests. *Ecosystems*, **17**, 778–791, doi:10.1007/s10021-014-9759-3.

Taylor, L.L., D.J. Beerling, S. Quegan, and S.A. Banwart, 2017: Simulating carbon capture by enhanced weathering with croplands: An overview of key processes highlighting areas of future model development. *Biol. Lett.,* **13**, 20160868, doi:10.1098/rsbl.2016.0868.

Tengberg, A., and S. Valencia, 2018: Integrated approaches to natural resources management – Theory and practice. *L. Degrad. Dev.*, **29**, 1845–1857, doi:10.1002/ldr.2946.

Tengberg, A., S. Fredholm, I. Eliasson, I. Knez, K.Saltzman, and O. Wetterberg, 2012: Cultural ecosystem services provided by landscapes: Assessment of heritage values and identity. *Ecosyst. Serv.*, **2**, 14–26.

Tengberg, A., F. Radstake, K. Zhang, and B. Dunn, 2016: Scaling up of sustainable land management in the Western People's Republic of China: Evaluation of a 10-year partnership. *L. Degrad. Dev.*, **27**(2), 134–144, doi:10.1002/ldr.2270.

Ter-Mikaelian, M.T., S.J. Colombo, and J. Chen, 2013: Effects of harvesting on spatial and temporal diversity of carbon stocks in a boreal forest landscape. *Ecol. Evol.*, **3**, 3738–3750, doi:10.1002/ece3.751.

Ter-Mikaelian, M.T., S.J. Colombo, and J. Chen, 2014: The burning question: Does forest bioenergy reduce carbon emissions? A review of common misconceptions about forest carbon accounting. *J. For.*, **113**, 57–68, doi:10.5849/jof.14-016.

ter Steege, H. et al. 2013: Hyperdominance in the Amazonian tree flora. *Science* **342**, 1243092–1243092, doi:10.1126/science.1243092.

Terrer, C., S. Vicca, B.A. Hungate, R.P. Phillips, and I.C. Prentice, 2016: Mycorrhizal association as a primary control of the CO_2 fertilization effect. *Science*, **353**, 72–74, doi:10.1126/science.aaf4610.

4

Terry, J.P. and A.Y.A. Lau, 2018: Magnitudes of nearshore waves generated by tropical cyclone Winston, the strongest landfalling cyclone in South Pacific records. Unprecedented or unremarkable? *Sediment. Geol.*, **364**, 276–285, doi:10.1016/J.SEDGEO.2017.10.009.

Tesfaye, A., R. Brouwer, P. van der Zaag, and W. Negatu, 2016: Assessing the costs and benefits of improved land management practices in three watershed areas in Ethiopia. *Int. Soil Water Conserv. Res.*, **4**, 20–29, doi:10.1016/J.ISWCR.2016.01.003.

Teshome, A., J. de Graaff, C. Ritsema, and M. Kassie, 2016: Farmers' perceptions about the influence of land quality, land fragmentation and tenure systems on sustainable land management in the north western Ethiopian highlands. *L. Degrad. Dev.*, **27**, 884–898, doi:10.1002/ldr.2298.

Tessler, Z.D. et al. 2015: Environmental science. Profiling risk and sustainability in coastal deltas of the world. *Science*, **349**, 638–643, doi:10.1126/science.aab3574.

Tessler, Z.D., C.J. Vörösmarty, M. Grossberg, I. Gladkova, and H. Aizenman, 2016: A global empirical typology of anthropogenic drivers of environmental change in deltas. *Sustain. Sci.*, **11**, 525–537, doi:10.1007/s11625-016-0357-5.

Testa, S., K. Soudani, L. Boschetti, and E. Borgogno Mondino, 2018: MODIS-derived EVI, NDVI and WDRVI time series to estimate phenological metrics in French deciduous forests. *Int. J. Appl. Earth Obs. Geoinf.*, **64**, 132–144, doi:10.1016/j.jag.2017.08.006.

Thies, J.E., M.C. Rillig, and E.R. Graber, 2015: Biochar effects on the abundance, activity and diversity of the soil biota. In Biochar for Environmental Management: Science, Technology and Implementation [Lehmann, J., Joseph, S. Eds.], Routledge, Abingdon, 907 p.

Thompson, I., B. Mackey, S. McNulty, and A. Mosseler, 2009: Forest resilience, biodiversity, and climate change. A synthesis of the biodiversity/resilience/stability relationship in forest ecosystems. Secretariat of the Convention on Biological Diversity, Montreal, Technical Series, Vol. 43 of, 67.

Thomson, J. and W.E. Rogers, 2014: Swell and sea in the emerging Arctic Ocean. *Geophys. Res. Lett.*, **41**, 3136–3140, doi:10.1002/2014GL059983.

Thomson, L.J., S. Macfadyen, and A.A. Hoffmann, 2010: Predicting the effects of climate change on natural enemies of agricultural pests. *Biol. Control*, **52**, 296–306, doi:10.1016/J.BIOCONTROL.2009.01.022.

Thorén, H., and L. Olsson, 2018: Is resilience a normative concept? *Resilience*, **6**, 112–128, doi:10.1080/21693293.2017.1406842.

Tian, F., M. Brandt, Y.Y. Liu, K. Rasmussen, and R. Fensholt, 2017: Mapping gains and losses in woody vegetation across global tropical drylands. *Glob. Chang. Biol.*, **23**, 1748–1760, doi:10.1111/gcb.13464.

Tian, H. et al. 2015: North American terrestrial CO_2 uptake largely offset by CH_4 and N_2O emissions: toward a full accounting of the greenhouse gas budget. *Clim. Change*, **129**, 413–426, doi:10.1007/s10584-014-1072-9.

Tighe, M., C. Muñoz-Robles, N. Reid, B. Wilson, and S. V Briggs, 2012: Hydrological thresholds of soil surface properties identified using conditional inference tree analysis. *Earth Surf. Process. Landforms*, **37**, 620–632, doi:10.1002/esp.3191.

Tilman, D., C. Balzer, J. Hill, and B.L. Befort, 2011: Global food demand and the sustainable intensification of agriculture. *Proc. Natl. Acad. Sci. U.S.A.*, **108**, 20260–20264, doi:10.1073/pnas.1116437108.

Tonosaki K, Kawai S, T.K., 2014: Cooling Potential of Urban Green Spaces in Summer. Designing Low Carbon Societies in Landscapes. [Nakagoshi N, Mabuhay AJ. (eds.)]. Springer, Tokyo, pp. 15–34.

Torquebiau, E., 1992: Are tropical agroforestry home gardens sustainable? *Agric. Ecosyst. Environ.*, **41**, 189–207, doi:10.1016/0167-8809(92)90109-O.

Toth, G.G., P.K. Ramachandran Nair, M. Jacobson, Y. Widyaningsih, and C.P. Duffy, 2017: Malawi's energy needs and agroforestry: Adoption potential of woodlots. *Hum. Ecol.*, **45**, 735–746, doi:10.1007/s10745-017-9944-z.

Toulmin, C., 2009: Securing land and property rights in sub-Saharan Africa: The role of local institutions. *Land use policy*, **26**, 10–19, doi:10.1016/J.LANDUSEPOL.2008.07.006.

Toulmin, C. and K. Brock, 2016: *Desertification in the Sahel: Local Practice Meets Global Narrative*. Springer, Berlin, Heidelberg, pp. 37–63.

Trahan, M.W. and B.A. Schubert, 2016: Temperature-induced water stress in high-latitude forests in response to natural and anthropogenic warming. *Glob. Chang. Biol.*, **22**, 782–791, doi:10.1111/gcb.13121.

Trenberth, K.E., 1999: *Conceptual Framework for Changes of Extremes of the Hydrological Cycle With Climate Change*. Weather and Climate Extremes, Springer Netherlands, Dordrecht, pp. 327–339.

Trenberth, K.E., 2011: Changes in precipitation with climate change. *Clim. Res.*, **47**, 123–138, doi:10.2307/24872346.

Trofymow, J.A., G. Stinson, and W.A. Kurz, 2008: Derivation of a spatially explicit 86-year retrospective carbon budget for a landscape undergoing conversion from old-growth to managed forests on Vancouver Island, BC. *For. Ecol. Manage.*, **256**, doi:10.1016/j.foreco.2008.02.056.

Trumbore, S., P. Brando, and H. Hartmann, 2015: Forest health and global change. *Science*, **349**, 814–818, doi:10.1126/science.aac6759.

Tu, S., F. Xu, and J. Xu, 2018: Regime shift in the destructiveness of tropical cyclones over the western North Pacific. *Environ. Res. Lett.*, **13**, 094021, doi:10.1088/1748-9326/aade3a.

Turetsky, M.R. et al. 2014: A synthesis of methane emissions from 71 northern, temperate, and subtropical wetlands. *Glob. Chang. Biol.*, **20**, 2183–2197, doi:10.1111/gcb.12580.

Turner, B.L. and J.A. Sabloff, 2012: Classic Period collapse of the Central Maya Lowlands: Insights about human–environment relationships for sustainability. *Proc. Natl. Acad. Sci. U.S.A.*, **109**, 13908–13914, doi:10.1073/pnas.1210106109.

Turner, B.L., E.F. Lambin, and A. Reenberg, 2007: The emergence of land change science for global environmental change and sustainability. *Proc. Natl. Acad. Sci. U.S.A.*, **104**, 20666–20671, doi:10.1073/pnas.0704119104.

Turner, B.L. (Billie L., W.C. Clark, R.W. Kates, J.F. Richards, T. Mathews, Jessica, and W.B. Meyer, eds.,) 1990: *The Earth as transformed by human action: global and regional changes in the biosphere over the past 300 years*. Cambridge University Press with Clark University, Cambrdige, UK and New York, USA, 713 pp.

Turner, K.G. et al. 2016: A review of methods, data, and models to assess changes in the value of ecosystem services from land degradation and restoration. *Ecol. Modell.*, **319**, 190–207, doi:10.1016/J.ECOLMODEL.2015.07.017.

Turner, W., 2014: Conservation. Sensing biodiversity. *Science*, **346**, 301–302, doi:10.1126/science.1256014.

Tzoulas, K. et al. 2007: Promoting ecosystem and human health in urban areas using Green Infrastructure: A literature review. Landsc. *Urban Plan.*, **81**, 167–178, doi:10.1016/J.LANDURBPLAN.2007.02.001.

Uddameri, V., S. Singaraju, and E.A. Hernandez, 2014: Impacts of sea-level rise and urbanization on groundwater availability and sustainability of coastal communities in semi-arid South Texas. *Environ. Earth Sci.*, **71**, 2503–2515, doi:10.1007/s12665-013-2904-z.

Uitto, J.I., 2016: Evaluating the environment as a global public good. *Evaluation*, **22**, 108–115, doi:10.1177/1356389015623135.

Umunay, P.M., T.G. Gregoire, T. Gopalakrishna, P.W. Ellis, and F.E. Putz, 2019: Selective logging emissions and potential emission reductions from reduced-impact logging in the Congo Basin. *For. Ecol. Manage.*, **437**, 360–371, doi:10.1016/j.foreco.2019.01.049.

UNCCD, 1994: United Nations Convention to Combat Desertification. United Nations General Assembly, New York City, 54 p.

UNCCD, 2016a: *Report of the Conference of the Parties on its twelfth Session, held in Ankara from 12 to 23 October 2015*. United Nations Convention to Combat Desertification, Bonn, Germany,

UNCCD, 2016b: *Land Degradation Neutrality Target Setting – A Technical Guide*. Land Degradation Neutrality Target Setting Programme. United Nations Convention to Combat Desertification, Bonn, Germany.

UNCCD, 2017: *The Global Land Outlook*. 1st ed. United Nations Convention to Combat Desertification, Bonn, Germany, 340 pp.

4

UNDP, 2017: *Valuation of Reforestation in Terms of Disaster Risk Reduction: A Technical Study From the Republic of Korea*. Sustainable Development Goals Policy Brief Series No. 1, United Nations Development Programme, New York, USA, 80 pp.

UNEP, 2007: *Sudan Post-Conflict Environmental Assessment*. United Nations Environment Programme, Nairobi, Kenya, 358 pp.

UNFCCC, 2006: *Report of the Conference of the Parties serving as the meeting of the Parties to the Kyoto Protocol on its first session, held at Montreal from 28 November to 10 December 2005*. United Nations Framework Convention on Climate Change, Bonn, Germany, 103 pp.

United Nations, 2015: *World Urbanization Prospects: The 2014 Revision*. Department of Economic and Social Affairs, NewYork, 517 pp.

Upadhyay, H., D. Mohan, and D. Mohan, 2017: *Migrating to adapt? Climate Change, Vulnerability and Migration*, Routledge India, pp. 43–58.

Valade, A., V. Bellassen, C. Magand, and S. Luyssaert, 2017: Sustaining the sequestration efficiency of the European forest sector. *For. Ecol. Manage.*, **405**, 44–55, doi:10.1016/j.foreco.2017.09.009.

van de Koppel, J., M. Rietkerk, and F.J. Weissing, 1997: Catastrophic vegetation shifts and soil degradation in terrestrial grazing systems. *Trends Ecol. Evol.*, **12**, 352–356, doi:10.1016/S0169-5347(97)01133-6.

VandenBygaart, A.J., 2016: The myth that no-till can mitigate global climate change. *Agric. Ecosyst. Environ.*, **216**, 98–99, doi:10.1016/J.AGEE.2015.09.013.

Van der Laan, C., B. Wicke, P.A. Verweij, and A.P.C. Faaij, 2017: Mitigation of unwanted direct and indirect land-use change – an integrated approach illustrated for palm oil, pulpwood, rubber and rice production in North and East Kalimantan, Indonesia. *GCB Bioenergy*, **9**, 429–444, doi:10.1111/gcbb.12353.

Vanek, S.J., and J. Lehmann, 2015: Phosphorus availability to beans via interactions between mycorrhizas and biochar. *Plant Soil*, **395**(1–2), 105–123, doi:10.1007/s11104-014-2246-y.

Van Tassel, D.L. et al. 2017: Accelerating silphium domestication: An opportunity to develop new crop ideotypes and breeding strategies informed by multiple disciplines. *Crop Sci.*, **57**, 1274–1284, doi:10.2135/cropsci2016.10.0834.

Vasconcellos, R.L.F., J.A. Bonfim, D. Baretta, and E.J.B.N. Cardoso, 2016: Arbuscular mycorrhizal fungi and glomalin-related soil protein as potential indicators of soil quality in a recuperation gradient of the Atlantic Forest in Brazil. *L. Degrad. Dev.*, **27**, 325–334, doi:10.1002/ldr.2228.

Vaughan, D.G. et al. 2013: Observations: Cryosphere. In: Climate Change 2013: The Physical Science Basis. Contribution of Working Group I to the Fifth Assessment Report of the Intergovernmental Panel on Climate Change, [Stocker, T.F., D. Qin, G.-K. Plattner, M. Tignor, S.K. Allen, J. Boschung, A. Nauels, Y. Xia, V. Bex and P.M. Midgley (eds.)]. Cambridge University Press, Cambridge, United Kingdom and New York, NY, USA, p. 317.

Vautard, R., J. Cattiaux, P. Yiou, J.-N. Thépaut, and P. Ciais, 2010: Northern Hemisphere atmospheric stilling partly attributed to an increase in surface roughness. *Nat. Geosci.*, **3**, 756–761, doi:10.1038/ngeo979.

Vavrus, S.J., F. He, J.E. Kutzbach, W.F. Ruddiman, and P.C. Tzedakis, 2018: Glacial inception in marine isotope stage 19: An orbital analog for a natural holocene climate. *Sci. Rep.*, **8**, 10213, doi:10.1038/s41598-018-28419-5.

Vecchi, G.A. et al. 2008: Climate change. Whither hurricane activity? *Science*, **322**, 687–689, doi:10.1126/science.1164396.

Vedeld, P., A. Angelsen, J. Bojö, E. Sjaastad, and G. Kobugabe Berg, 2007: Forest environmental incomes and the rural poor. *For. Policy Econ.*, **9**, 869–879, doi:10.1016/J.FORPOL.2006.05.008.

van der Ven, H. and B. Cashore, 2018: Forest certification: The challenge of measuring impacts. *Curr. Opin. Environ. Sustain.*, **32**, 104–111, doi:10.1016/J.COSUST.2018.06.001.

van der Ven, H., C. Rothacker, and B. Cashore, 2018: Do eco-labels prevent deforestation? Lessons from non-state market driven governance in the soy, palm oil, and cocoa sectors. *Glob. Environ. Chang.*, **52**, 141–151, doi:10.1016/J.GLOENVCHA.2018.07.002.

Venter, O. et al. 2016: Sixteen years of change in the global terrestrial human footprint and implications for biodiversity conservation. *Nat. Commun.*, **7**, 12558, doi:10.1038/ncomms12558.

Ventura, M. et al. 2015: Biochar mineralization and priming effect on SOM decomposition in two European short rotation coppices. *GCB Bioenergy,* **7**(5), 1150–1160, doi:10.1111/gcbb.12219.

Verdone, M. and A. Seidl, 2017: Time, space, place, and the Bonn Challenge global forest restoration target. *Restor. Ecol.*, **25**, 903–911, doi:10.1111/rec.12512.

Verhoeven, E. et al. 2017: Toward a better assessment of biochar–nitrous oxide mitigation potential at the field scale. *J. Environ. Qual.*, **46**(2), 237–246, doi:10.2134/jeq2016.10.0396.

Viger, M., R.D. Hancock, F. Miglietta, and G. Taylor, 2015: More plant growth but less plant defence? First global gene expression data for plants grown in soil amended with biochar. *GCB Bioenergy*, **7**(4), 658–672, doi:10.1111/gcbb.12182.

Vincent, K.E., P. Tschakert, J. Barnett, M.G. Rivera-Ferre, and A. Woodward, 2014: Cross-Chapter Box on Gender and Climate Change. In: Climate Change 2014: Impacts, Adaptation, and Vulnerability. Part A: Global and Sectoral Aspects. Contribution of Working Group II to the Fifth Assessment Report of the Intergovernmental Panel on Climate Change, [Field, C.B., V.R. Barros, D.J. Dokken, K.J. Mach, M.D. Mastrandrea, T.E. Bilir, M. Chatterjee, K.L. Ebi, Y.O. Estrada, R.C. Genova, B. Girma, E.S. Kissel, A.N. Levy, S. MacCracken, P.R. Mastrandrea, and L.L.White (eds.)]. Cambridge University Press, Cambridge, UK and New York, NY, USA, 105–107.

Virapongse, A., B.A. Endress, M.P. Gilmore, C. Horn, and C. Romulo, 2017: Ecology, livelihoods, and management of the Mauritia flexuosa palm in South America. *Glob. Ecol. Conserv.*, **10**, 70–92, doi:10.1016/j.gecco.2016.12.005.

Viscarra Rossel, R.A., R. Webster, E.N. Bui, and J.A. Baldock, 2014: Baseline map of organic carbon in Australian soil to support national carbon accounting and monitoring under climate change. *Glob. Chang. Biol.*, **20**, 2953–2970, doi:10.1111/gcb.12569.

Vitousek, S. et al. 2017: Doubling of coastal flooding frequency within decades due to sea-level rise. *Sci. Rep.*, **7**, 1399, doi:10.1038/s41598-017-01362-7.

Vogl, A.L. et al. 2017: Valuing investments in sustainable land management in the Upper Tana River basin, Kenya. *J. Environ. Manage.*, **195**, 78–91, doi:10.1016/J.JENVMAN.2016.10.013.

Vogt, J.V. et al. 2011: Monitoring and assessment of land degradation and desertification: Towards new conceptual and integrated approaches. *L. Degrad. Dev.*, **22**, 150–165, doi:10.1002/ldr.1075.

Volkova, L., H. Bi, J. Hilton, and C.J. Weston, 2017: Impact of mechanical thinning on forest carbon, fuel hazard and simulated fire behaviour in *Eucalyptus delegatensis* forest of south-eastern Australia. *For. Ecol. Manage.*, **405**, 92–100, doi:10.1016/j.foreco.2017.09.032.

Volkova, L. et al. 2018: Importance of disturbance history on net primary productivity in the world's most productive forests and implications for the global carbon cycle. *Glob. Chang. Biol.*, **24**, 4293–4303, doi:10.1111/gcb.14309.

van Wagner, C.E., 1978: Age-class distribution and the forest fire cycle. *Can. J. For. Res.*, **8**, 220–227, doi:10.1139/x78-034.

Wagner, L.E., 2013: A history of Wind Erosion Prediction Models in the United States Department of Agriculture: The Wind Erosion Prediction System (WEPS). *Aeolian Res.*, **10**, 9–24, doi:10.1016/J.AEOLIA.2012.10.001.

Wairiu, M., 2017: Land degradation and sustainable land management practices in Pacific Island Countries. *Reg. Environ. Chang.*, **17**, 1053–1064, doi:10.1007/s10113-016-1041-0.

Waldron, A. et al. 2017: Agroforestry can enhance food security while meeting other sustainable development goals. *Trop. Conserv. Sci.*, **10**, 194008291772066, doi:10.1177/1940082917720667.

Walsh, J.R., S.R. Carpenter, and M.J. Vander Zanden, 2016a: Invasive species triggers a massive loss of ecosystem services through a trophic cascade. *Proc. Natl. Acad. Sci.*, **113**, 4081–4085, doi:10.1073/pnas.1600366113.

Walsh, K.J.E. et al. 2016b: Tropical cyclones and climate change. *Wiley Interdiscip. Rev. Clim. Chang.*, **7**, 65–89, doi:10.1002/wcc.371.

Walter Anthony, K. et al. 2016: Methane emissions proportional to permafrost carbon thawed in Arctic lakes since the 1950s. *Nat. Geosci.*, **9**, 679–682, doi:10.1038/ngeo2795.

Wang, G. et al. 2016a: Integrated watershed management: Evolution, development and emerging trends. *J. For. Res.*, **27**, 967–994, doi:10.1007/s11676-016-0293-3.

Wang, J., Z. Xiong, and Y. Kuzyakov, 2016b: Biochar stability in soil: Meta-analysis of decomposition and priming effects. *GCB Bioenergy*, **8**, 512–523, doi:10.1111/gcbb.12266.

Wang, J., S. Yi, M. Li, L. Wang, and C. Song, 2018: Effects of sea level rise, land subsidence, bathymetric change and typhoon tracks on storm flooding in the coastal areas of Shanghai. *Sci. Total Environ.*, **621**, 228–234, doi:10.1016/J.SCITOTENV.2017.11.224.

Wang, W. et al. 2017a: Relationships between the potential production of the greenhouse gases CO_2, CH_4 and N_2O and soil concentrations of C, N and P across 26 paddy fields in southeastern China. *Atmos. Environ.*, **164**, 458–467, doi:10.1016/J.ATMOSENV.2017.06.023.

Wang, Z. et al. 2017b: Human-induced erosion has offset one-third of carbon emissions from land cover change. *Nat. Clim. Chang.*, **7**, 345–349, doi:10.1038/nclimate3263.

Ward, C., L. Stringer, and G. Holmes, 2018: Changing governance, changing inequalities: Protected area co-management and access to forest ecosystem services – a Madagascar case study. *Ecosyst. Serv.*, **30**, 137–148, doi:10.1016/J.ECOSER.2018.01.014.

Ward, D., 2005: Do we understand the causes of bush encroachment in African savannas? *African J. Range Forage Sci.*, **22**, 101–105, doi:10.2989/10220110509485867.

Warren, A., 2002: Land degradation is contextual. *L. Degrad. Dev.*, **13**, 449–459, doi:10.1002/ldr.532.

Warren, R., J. Price, J. VanDerWal, S. Cornelius, and H. Sohl, 2018: The implications of the United Nations Paris Agreement on climate change for globally significant biodiversity areas. *Clim. Change*, **147**, 395–409, doi:10.1007/s10584-018-2158-6.

Watson, R.T. et al. (eds.) 2000: *Land Use, Land-Use Change, and Forestry*. Cambridge University Press, Cambridge, UK, 370 pp.

Webb, N.P. et al. 2016: The National Wind Erosion Research Network: Building a standardized long-term data resource for aeolian research, modeling and land management. *Aeolian Res.*, 22, 23–**36**, doi:10.1016/J.AEOLIA.2016.05.005.

Webb, N.P. et al. 2017a: Enhancing wind erosion monitoring and assessment for U.S. rangelands. *Rangelands*, **39**, 85–96, doi:10.1016/J.RALA.2017.04.001.

Webb, N.P. et al. 2017b: Land degradation and climate change: Building climate resilience in agriculture. *Front. Ecol. Environ.*, **15**, 450–459, doi:10.1002/fee.1530.

Wei, W. et al. 2016: Global synthesis of the classifications, distributions, benefits and issues of terracing. *Earth-Science Rev.*, **159**, 388–403, doi:10.1016/J.EARSCIREV.2016.06.010.

Wei, X., M. Shao, W. Gale, and L. Li, 2015: Global pattern of soil carbon losses due to the conversion of forests to agricultural land. *Sci. Rep.*, **4**, 4062, doi:10.1038/srep04062.

Weichselgartner, J., and I. Kelman, 2015: Geographies of resilience. *Prog. Hum. Geogr.*, **39**, 249–267, doi:10.1177/0309132513518834.

Weinzierl, T., J. Wehberg, J. Böhner, and O. Conrad, 2016: Spatial assessment of land degradation risk for the Okavango river catchment, Southern Africa. *L. Degrad. Dev.*, **27**, 281–294, doi:10.1002/ldr.2426.

Weng, Z. et al. 2017: Biochar built soil carbon over a decade by stabilizing rhizodeposits. *Nat. Clim. Chang.*, **7**, 371–376, doi:10.1038/nclimate3276

Weng, Z. (Han) et al. 2018: The accumulation of rhizodeposits in organo-mineral fractions promoted biochar-induced negative priming of native soil organic carbon in Ferralsol. *Soil Biol. Biochem.*, **118**, 91–96, doi:10.1016/j.soilbio.2017.12.008.

Weng, Z.H. (Han) et al. 2015: Plant-biochar interactions drive the negative priming of soil organic carbon in an annual ryegrass field system. *Soil Biol. Biochem.*, **90**, 111–121, doi:10.1016/j.soilbio.2015.08.005.

Wentworth, J., 2017: *Urban Green Infrastructure and Ecosystem Services*. POSTbrief from UK Parliamentary Office of Science and Technology London, UK, 26 pp.

van der Werf, G.R. et al. 2017: Global fire emissions estimates during 1997–2016. *Earth Syst. Sci.* Data, **9**, 697–720, doi:10.5194/essd-9-697-2017.

Werner, F., R. Taverna, P. Hofer, E. Thürig, and E. Kaufmann, 2010: National and global greenhouse gas dynamics of different forest management and wood use scenarios: A model-based assessment. *Environ. Sci. Policy*, **13**, 72–85, doi:10.1016/j.envsci.2009.10.004.

Wessels, K., S. Prince, P. Frost, and D. van Zyl, 2004: Assessing the effects of human-induced land degradation in the former homelands of northern South Africa with a 1 km AVHRR NDVI time-series. *Remote Sens. Environ.*, **91**, 47–67, doi:10.1016/j.rse.2004.02.005.

Wessels, K.J. et al. 2007: Can human-induced land degradation be distinguished from the effects of rainfall variability? A case study in South Africa. *J. Arid Environ.*, **68**, 271–297, doi:10.1016/j.jaridenv.2006.05.015.

Wessels, K.J., F. van den Bergh, and R.J. Scholes, 2012: Limits to detectability of land degradation by trend analysis of vegetation index data. *Remote Sens. Environ.*, **125**, 10–22, doi:10.1016/j.rse.2012.06.022.

Westra, S. et al. 2014: Future changes to the intensity and frequency of short-duration extreme rainfall. *Rev. Geophys.*, **52**, 522–555, doi:10.1002/2014RG000464.

White, J.W., G. Hoogenboom, B.A. Kimball, and G.W. Wall, 2011: Methodologies for simulating impacts of climate change on crop production. *F. Crop. Res.*, **124**, 357–368, doi:10.1016/J.FCR.2011.07.001.

Wicke, B. et al. 2011: The global technical and economic potential of bioenergy from salt-affected soils. *Energy Environ. Sci.*, **4**, 2669, doi:10.1039/c1ee01029h.

Wicke, B., P. Verweij, H. van Meijl, D.P. van Vuuren, and A.P. Faaij, 2012: Indirect land use change: Review of existing models and strategies for mitigation. *Biofuels*, **3**, 87–100, doi:10.4155/bfs.11.154.

Widgren, M., and J.E.G. Sutton, 2004: Islands of intensive agriculture in Eastern Africa: Past & present. Ohio University Press, Athens, Ohio, USA, 160 pp.

Wieczorek, A.J., 2018: Sustainability transitions in developing countries: Major insights and their implications for research and policy. *Environ. Sci. Policy*, **84**, 204–216, doi:10.1016/J.ENVSCI.2017.08.008.

Wiesmair, M., A. Otte, and R. Waldhardt, 2017: Relationships between plant diversity, vegetation cover, and site conditions: Implications for grassland conservation in the Greater Caucasus. *Biodivers. Conserv.*, **26**, 273–291, doi:10.1007/s10531-016-1240-5.

Wigley, B.J., W.J. Bond, and M.T. Hoffman, 2010: Thicket expansion in a South African savanna under divergent land use: Local vs. global drivers? *Glob. Chang. Biol.*, **16**, 964–976, doi:10.1111/j.1365-2486.2009.02030.x.

Van Wilgen, B.W., N. Govender, H.C. Biggs, D. Ntsala, and X.N. Funda, 2004: Response of savanna fire regimes to changing fire-management policies in a large African national park. *Conserv. Biol.*, **18**, 1533–1540, doi:10.1111/j.1523-1739.2004.00362.x.

Wilhelm, J.A., and R.G. Smith, 2018: Ecosystem services and land sparing potential of urban and peri-urban agriculture: A review. *Renew. Agric. Food Syst.*, **33**, 481–494, doi:10.1017/S1742170517000205.

Wilson, D. et al. 2016: Greenhouse gas emission factors associated with rewetting of organic soils. *Mires Peat*, **17**(4), 1–28, doi:10.19189/MaP.2016.OMB.222.

Wilson, G.A. et al. 2017: Social memory and the resilience of communities affected by land degradation. *L. Degrad. Dev.*, 28, 383–400, doi:10.1002/ldr.2669.

Winder, R., E. Nelson, and T. Beardmore, 2011: Ecological implications for assisted migration in Canadian forests. *For. Chron.*, **87**, 731–744, doi:10.5558/tfc2011-090.

WOCAT n.d., WOCAT (World Overview of Conservation Approaches and Technologies). Glossary.

Wong, P.P. et al. 2014: Coastal Systems and Low-Lying Areas. In: Climate Change 2014: Impacts, Adaptation, and Vulnerability, Part A: Global and Sectoral Aspects. Contribution of Working Group II to the Fifth Assessment Report of the Intergovernmental Panel on Climate Change [Field, C.B., V.R. Barros, D.J. Dokken, K.J. Mach, M.D. Mastrandrea, T.E. Bilir, M. Chatterjee, K.L. Ebi, Y.O. Estrada, R.C. Genova, B. Girma, E.S. Kissel, A.N. Levy, S. MacCracken, P.R. Mastrandrea, and L.L.White (eds.)]. C.B. Field et al. Eds., Cambridge University Press, Cambridge, United Kingdom and New York, 361–409.

Woods, J. et al. 2015: Land and Bioenergy. Bioenergy & Sustainability: Bridging the gaps – Scope Bioenergy, [G.M. Souza, R.L. Victoria, C.A. Joly, and L.M. Verdade, (eds.)]. SCOPE, Paris, 258–300.

Woolf, D., J.E. Amonette, F.A. Street-Perrott, J. Lehmann, and S. Joseph, 2010: Sustainable biochar to mitigate global climate change. *Nat. Commun.*, **1**, 56.

Woolf, D. et al. 2018: *Biochar for Climate Change Mitigation. Soil and Climate*, Series: Advances in Soil Science, CRC Press, Taylor & Francis Group, Boca Raton, Florida, USA, pp. 219–248.

World Bank, 2009: *Environmental Crisis or Sustainable Development Opportunity? Transforming the Charcoal Sector in Tanzania*. World Bank, Washington, DC, USA, 72 pp.

World Bank, 2016: *The Role of Green Infrastructure Solutions in Urban Flood Risk Management*. World Bank, Washington DC, USA, 18 pp.

Wu, H. et al. 2017a: The interactions of composting and biochar and their implications for soil amendment and pollution remediation: A review. *Crit. Rev. Biotechnol.*, **37**, 754–764, doi:10.1080/07388551.2016.1232696.

Wu, J. et al. 2015: Temperature sensitivity of soil bacterial community along contrasting warming gradient. *Appl. Soil Ecol.*, **94**, 40–48, doi:10.1016/J.APSOIL.2015.04.018.

Wu, Z., P. Dijkstra, G.W. Koch, J. Peñuelas, and B.A. Hungate, 2011: Responses of terrestrial ecosystems to temperature and precipitation change: A meta-analysis of experimental manipulation. *Glob. Chang. Biol.*, **17**, 927–942, doi:10.1111/j.1365-2486.2010.02302.x.

Wuest, S.B., J.D. Williams, and H.T. Gollany, 2006: Tillage and perennial grass effects on ponded infiltration for seven semi-arid loess soils. *J. Soil Water Conserv.*, **61**, 218–223.

Xiao, L. et al. 2017: The indirect roles of roads in soil erosion evolution in Jiangxi Province, China: A large scale perspective. *Sustainability*, **9**, 129, doi:10.3390/su9010129.

Xie, J. et al. 2016: Ten-year variability in ecosystem water use efficiency in an oak-dominated temperate forest under a warming climate. *Agric. For. Meteorol.*, **218–219**, 209–217, doi:10.1016/J.AGRFORMET.2015.12.059.

Xu, G., Y. Zhang, H. Shao, and J. Sun, 2016: Pyrolysis temperature affects phosphorus transformation in biochar: Chemical fractionation and 31P NMR analysis. *Sci. Total Environ.*, **569**, 65–72, doi:10.1016/j.scitotenv.2016.06.081.

Xu, J., R. Yin, Z. Li, and C. Liu, 2006: China's ecological rehabilitation: Unprecedented efforts, dramatic impacts, and requisite policies. *Ecol. Econ.*, doi:10.1016/j.ecolecon.2005.05.008.

Xu, J., P.J. Morris, J. Liu, and J. Holden, 2018a: PEATMAP: Refining estimates of global peatland distribution based on a meta-analysis. *Catena*, **160**, 134–140, doi:10.1016/j.catena.2017.09.010.

Xu, X. et al. 2019: Greenhouse gas mitigation potential in crop production with biochar soil amendment – a carbon footprint assessment for cross-site field experiments from China. *GCB Bioenergy*, **11**, 592–605, doi:10.1111/gcbb.12561.

Xu, Z., C.E. Smyth, T.C. Lemprière, G.J. Rampley, and W.A. Kurz, 2018: Climate change mitigation strategies in the forest sector: Biophysical impacts and economic implications in British Columbia, Canada. *Mitig. Adapt. Strateg. Glob. Chang.*, **23**, 257–290, doi:10.1007/s11027-016-9735-7.

Yamanoi, K., Y. Mizoguchi, and H. Utsugi, 2015: Effects of a windthrow disturbance on the carbon balance of a broadleaf deciduous forest in Hokkaido, Japan. *Biogeosciences*, **12**, 6837–6851, doi:10.5194/bg-12-6837-2015.

Yang, J. et al. 2019: Deformation of the aquifer system under groundwater level fluctuations and its implication for land subsidence control in the Tianjin coastal region. *Environ. Monit. Assess.*, **191**, 162, doi:10.1007/s10661-019-7296-4.

Yang, M., F.E. Nelson, N.I. Shiklomanov, D. Guo, and G. Wan, 2010: Permafrost degradation and its environmental effects on the Tibetan Plateau: A review of recent research. *Earth-Science Rev.*, **103**, 31–44, doi:10.1016/J.EARSCIREV.2010.07.002.

Yang, Y., R.J. Donohue, T.R. McVicar, M.L. Roderick, and H.E. Beck, 2016: Long-term CO_2 fertilization increases vegetation productivity and has little effect on hydrological partitioning in tropical rainforests. *J. Geophys. Res. Biogeosciences*, **121**, 2125–2140, doi:10.1002/2016JG003475.

Yao, L., L. Chen, W. Wei, and R. Sun, 2015: Potential reduction in urban runoff by green spaces in Beijing: A scenario analysis. *Urban For. Urban Green.*, **14**(2), 300–308, doi:10.1016/j.ufug.2015.02.014.

Yengoh, G.T., and J. Ardö, 2014: Crop Yield Gaps in Cameroon. *Ambio*, **43**, 175–190, doi:10.1007/s13280-013-0428-0.

Yengoh, G.T., D. Dent, L. Olsson, A. Tengberg, and C.J. Tucker, 2015: *Use of the Normalized Difference Vegetation Index (NDVI) to Assess Land Degradation at Multiple Scales: Current Status, Future Trends, and Practical Considerations*. Springer, Heidelberg, New York, Dordrecht, London, 110 pp.

Yin, R., 2009: *An Integrated Assessment of China's Ecological Restoration Programs*. Springer, New York, USA, 254 pp.

Young, A., 1995: *Agroforestry for Soil Conservation*. CTA, Wageningen, The Netherlands, 194 pp.

Young, A., 1999: Is there really spare land? A critique of estimates of available cultivable land in developing countries. *Environ. Dev. Sustain.*, **1**, 3–18, doi:10.1023/A:1010055012699.

Young, E., D. Muir, A. Dawson, and S. Dawson, 2014: Community driven coastal management: An example of the implementation of a coastal defence bund on South Uist, Scottish Outer Hebrides. *Ocean Coast. Manag.*, **94**, 30–37, doi:10.1016/j.ocecoaman.2014.01.001.

Yu, H. et al. 2015: The fertilizing role of African dust in the Amazon rainforest: A first multiyear assessment based on data from Cloud-Aerosol Lidar and Infrared Pathfinder Satellite Observations. *Geophys. Res. Lett.*, **42**, 1984–1991, doi:10.1002/2015GL063040.

Zabel, F., B. Putzenlechner, and W. Mauser, 2014: Global agricultural land resources – a high resolution suitability evaluation and its perspectives until 2100 under climate change conditions. *PLoS One*, **9**, e107522, doi:10.1371/journal.pone.0107522.

Zahawi, R.A., G. Duran, and U. Kormann, 2015: Sixty-seven years of land-use change in Southern Costa Rica. *PLoS One*, **10**, e0143554, doi:10.1371/journal.pone.0143554.

Zedler, J.B., 2000: Progress in wetland restoration ecology. *Trends Ecol. Evol.*, **15**, 402–407, doi:10.1016/S0169-5347(00)01959-5.

Zeng, Z. et al. 2017: Climate mitigation from vegetation biophysical feedbacks during the past three decades. *Nat. Clim. Chang.*, **7**, 432–436, doi:10.1038/nclimate3299.

Zhang, M. et al. 2017: A global review on hydrological responses to forest change across multiple spatial scales: Importance of scale, climate, forest type and hydrological regime. *J. Hydrol.*, **546**, 44–59, doi:10.1016/J.JHYDROL.2016.12.040.

Zhang, P. et al. 2000: China's forest policy for the 21st century. *Science*, doi:10.1126/science.288.5474.2135.

Zhang, X. et al. 2013: Using biochar for remediation of soils contaminated with heavy metals and organic pollutants. *Environ. Sci. Pollut. Res.*, **20**(12), 8472–8443, doi:10.1007/s11356-013-1659-0.

Zhang, X.C. and M.A. Nearing, 2005: Impact of climate change on soil erosion, runoff, and wheat productivity in central Oklahoma. *CATENA*, **61**, 185–195, doi:10.1016/J.CATENA.2005.03.009.

Zhou, Z., C. Wang, and Y. Luo, 2018: Effects of forest degradation on microbial communities and soil carbon cycling: A global meta-analysis. *Glob. Ecol. Biogeogr.*, **27**, 110–124, doi:10.1111/geb.12663.

Zhu, Q. et al. 2019: Estimation of event-based rainfall erosivity from radar after wildfire. *L. Degrad. Dev.*, **30**, 33–48, doi:10.1002/ldr.3146.

Zhu, X., M.M. Linham, and R.J. Nicholls, 2010: *Technologies for Climate Change Adaptation – Coastal Erosion and Flooding.* Roskilde: Danish Technical University, Risø National Laboratory for Sustainable Energy. TNA Guidebook Series.

Zimmerer, K.S., 1993: Soil erosion and social (dis)courses in Cochabamba, Bolivia: Perceiving the nature of environmental degradation. *Econ. Geogr.*, **69**, 312, doi:10.2307/143453.

Zölch, T., J. Maderspacher, C. Wamsler, and S. Pauleit, 2016: Using green infrastructure for urban climate-proofing: An evaluation of heat mitigation measures at the micro-scale. *Urban For. Urban Green.*, **20**, 305–316, doi:10.1016/J.UFUG.2016.09.011.

Zomer, R.J. et al. 2016: Global tree cover and biomass carbon on agricultural land: The contribution of agroforestry to global and national carbon budgets. *Sci. Rep.*, **6**, 29987, doi:10.1038/srep29987.

Zulu, L.C., 2010: The forbidden fuel: Charcoal, urban woodfuel demand and supply dynamics, community forest management and woodfuel policy in Malawi. *Energy Policy*, **38**, 3717–3730, doi:https://doi.org/10.1016/j.enpol.2010.02.050.

Zulu, L.C. and R.B. Richardson, 2013: Charcoal, livelihoods, and poverty reduction: Evidence from sub-Saharan Africa. *Energy Sustain. Dev.*, **17**, 127–137, doi:https://doi.org/10.1016/j.esd.2012.07.007.

Van Zwieten, L. et al. 2010: Effects of biochar from slow pyrolysis of papermill waste on agronomic performance and soil fertility. *Plant Soil,* **327**, 235–246, doi:10.1007/s11104-009-0050-x.

Van Zwieten, L. et al. 2015: Enhanced biological N2 fixation and yield of faba bean (*Vicia faba L.*) in an acid soil following biochar addition: Dissection of causal mechanisms. *Plant Soil*, **395**, 7–20, doi:10.1007/s11104-015-2427-3.

4

5

Food security

Coordinating Lead Authors:
Cheikh Mbow (Senegal), Cynthia Rosenzweig (The United States of America)

Lead Authors:
Luis G. Barioni (Brazil), Tim G. Benton (United Kingdom), Mario Herrero (Australia/ Costa Rica), Murukesan Krishnapillai (Micronesia/India), Emma Liwenga (Tanzania), Prajal Pradhan (Germany/Nepal), Marta G. Rivera-Ferre (Spain), Tek Sapkota (Canada/Nepal), Francesco N. Tubiello (The United States of America/Italy), Yinlong Xu (China)

Contributing Authors:
Erik Mencos Contreras (The United States of America/Mexico), Joana Portugal Pereira (United Kingdom), Julia Blanchard (Australia), Jessica Fanzo (The United States of America), Stefan Frank (Austria), Steffen Kriewald (Germany), Gary Lanigan (Ireland), Daniel López (Spain), Daniel Mason-D'Croz (The United States of America), Peter Neofotis (The United States of America), Laxmi Pant (Canada), Renato Rodrigues (Brazil), Alex C. Ruane (The United States of America), Katharina Waha (Australia)

Review Editors:
Noureddine Benkeblia (Jamaica), Andrew Challinor (United Kingdom), Amanullah Khan (Pakistan), John R. Porter (United Kingdom)

Chapter Scientists:
Erik Mencos Contreras (The United States of America/Mexico), Abdoul Aziz Diouf (Senegal)

This chapter should be cited as:
Mbow, C., C. Rosenzweig, L.G. Barioni, T.G. Benton, M. Herrero, M. Krishnapillai, E. Liwenga, P. Pradhan, M.G. Rivera-Ferre, T. Sapkota, F.N. Tubiello, Y. Xu, 2019: Food Security. In: *Climate Change and Land: an IPCC special report on climate change, desertification, land degradation, sustainable land management, food security, and greenhouse gas fluxes in terrestrial ecosystems* [P.R. Shukla, J. Skea, E. Calvo Buendia, V. Masson-Delmotte, H.-O. Pörtner, D.C. Roberts, P. Zhai, R. Slade, S. Connors, R. van Diemen, M. Ferrat, E. Haughey, S. Luz, S. Neogi, M. Pathak, J. Petzold, J. Portugal Pereira, P. Vyas, E. Huntley, K. Kissick, M. Belkacemi, J. Malley, (eds.)]. https://doi.org/10.1017/9781009157988.007

Table of contents

Executive summary 439

5.1   Framing and context 441

    5.1.1   Food security and insecurity, the
            food system and climate change 442

    5.1.2   Status of the food system,
            food insecurity and malnourishment 445

    5.1.3   Climate change, gender and equity 446

    **Box 5.1: Gender, food security
    and climate change** 447

    5.1.4   Food systems in AR5, SR15,
            and the Paris Agreement 448

5.2   Impacts of climate change on food systems 450

    5.2.1   Climate drivers important to food security 450

    5.2.2   Climate change impacts on food availability ... 451

    5.2.3   Climate change impacts on access 460

    5.2.4   Climate change impacts on food utilisation 462

    5.2.5   Climate change impacts on food stability 464

**5.3   Adaptation options, challenges
        and opportunities** 464

    5.3.1   Challenges and opportunities 464

    **Box 5.2: Sustainable solutions for
    food systems and climate change in Africa** 465

    5.3.2   Adaptation framing and key concepts 466

    **Box 5.3: Climate change and indigenous food
    systems in the Hindu-Kush Himalayan Region** 469

    5.3.3   Supply-side adaptation 470

    5.3.4   Demand-side adaptation 472

    5.3.5   Institutional measures 473

    5.3.6   Tools and finance 475

5.4   Impacts of food systems on climate change 475

    5.4.1   Greenhouse gas emissions from food systems 475

    5.4.2   Greenhouse gas emissions
            from croplands and soils 476

    5.4.3   Greenhouse gas emissions from livestock 477

    5.4.4   Greenhouse gas emissions from aquaculture 478

    5.4.5   Greenhouse gas emissions from inputs,
            processing, storage and transport 478

    5.4.6   Greenhouse gas emissions associated
            with different diets 479

**5.5   Mitigation options, challenges
        and opportunities** 480

    5.5.1   Supply-side mitigation options 480

    **Box 5.4: Towards sustainable
    intensification in South America** 481

    5.5.2   Demand-side mitigation options 487

**5.6   Mitigation, adaptation, food security and
        land use: Synergies, trade-offs and co-benefits** ...
        492

    5.6.1   Land-based carbon dioxide removal (CDR)
            and bioenergy 492

    5.6.2   Mitigation, food prices, and food security 494

    5.6.3   Environmental and health effects of
            adopting healthy and sustainable diets 497

    5.6.4   Sustainable integrated agricultural systems 499

    **Cross-Chapter Box 6 | Agricultural intensification:
    Land sparing, land sharing and sustainability** 502

    5.6.5   Role of urban agriculture 505

    5.6.6   Links to the Sustainable Development Goals 507

5.7   Enabling conditions and knowledge gaps 507

    5.7.1   Enabling policy environments 508

    5.7.2   Enablers for changing markets and trade 511

    5.7.3   Just Transitions to sustainability 511

    5.7.4   Mobilising knowledge 512

    5.7.5   Knowledge gaps and key research areas 513

5.8   Future challenges to food security 514

    5.8.1   Food price spikes 514

    **Box 5.5: Market drivers and the consequences
    of extreme weather in 2010–2011** 516

    5.8.2   Migration and conflict 516

    **Box 5.6: Migration in the Pacific region:
    Impacts of climate change on food security** 517

Frequently Asked Questions 519

    **FAQ 5.1: How does climate change
    affect food security?** 519

    **FAQ 5.2: How can changing diets
    help address climate change?** 519

References 520

Executive summary

The current food system (production, transport, processing, packaging, storage, retail, consumption, loss and waste) feeds the great majority of world population and supports the livelihoods of over 1 billion people. Since 1961, food supply per capita has increased more than 30%, accompanied by greater use of nitrogen fertilisers (increase of about 800%) and water resources for irrigation (increase of more than 100%). However, an estimated 821 million people are currently undernourished, 151 million children under five are stunted, 613 million women and girls aged 15 to 49 suffer from iron deficiency, and 2 billion adults are overweight or obese. The food system is under pressure from non-climate stressors (e.g., population and income growth, demand for animal-sourced products), and from climate change. These climate and non-climate stresses are impacting the four pillars of food security (availability, access, utilisation, and stability). {5.1.1, 5.1.2}

Observed climate change is already affecting food security through increasing temperatures, changing precipitation patterns, and greater frequency of some extreme events (*high confidence*). Studies that separate out climate change from other factors affecting crop yields have shown that yields of some crops (e.g., maize and wheat) in many lower-latitude regions have been affected negatively by observed climate changes, while in many higher-latitude regions, yields of some crops (e.g., maize, wheat, and sugar beets) have been affected positively over recent decades. Warming compounded by drying has caused large negative effects on yields in parts of the Mediterranean. Based on indigenous and local knowledge (ILK), climate change is affecting food security in drylands, particularly those in Africa, and high mountain regions of Asia and South America. {5.2.2}

Food security will be increasingly affected by projected future climate change (*high confidence*). Across Shared Socio-economic Pathways (SSPs) 1, 2, and 3, global crop and economic models projected a 1–29% cereal price increase in 2050 due to climate change (RCP 6.0), which would impact consumers globally through higher food prices; regional effects will vary (*high confidence*). Low-income consumers are particularly at risk, with models projecting increases of 1–183 million additional people at risk of hunger across the SSPs compared to a no climate change scenario (*high confidence*). While increased CO_2 is projected to be beneficial for crop productivity at lower temperature increases, it is projected to lower nutritional quality (*high confidence*) (e.g., wheat grown at 546–586 ppm CO_2 has 5.9–12.7% less protein, 3.7–6.5% less zinc, and 5.2–7.5% less iron). Distributions of pests and diseases will change, affecting production negatively in many regions (*high confidence*). Given increasing extreme events and interconnectedness, risks of food system disruptions are growing (*high confidence*). {5.2.3, 5.2.4}

Vulnerability of pastoral systems to climate change is very high (*high confidence*). Pastoralism is practiced in more than 75% of countries by between 200 and 500 million people, including nomadic communities, transhumant herders, and agropastoralists. Impacts in pastoral systems in Africa include lower pasture and animal productivity, damaged reproductive function, and biodiversity loss.

Pastoral system vulnerability is exacerbated by non-climate factors (land tenure, sedentarisation, changes in traditional institutions, invasive species, lack of markets, and conflicts). {5.2.2}

Fruit and vegetable production, a key component of healthy diets, is also vulnerable to climate change (*medium evidence, high agreement*). Declines in yields and crop suitability are projected under higher temperatures, especially in tropical and semi-tropical regions. Heat stress reduces fruit set and speeds up development of annual vegetables, resulting in yield losses, impaired product quality, and increasing food loss and waste. Longer growing seasons enable a greater number of plantings to be cultivated and can contribute to greater annual yields. However, some fruits and vegetables need a period of cold accumulation to produce a viable harvest, and warmer winters may constitute a risk. {5.2.2}

Food security and climate change have strong gender and equity dimensions (*high confidence*). Worldwide, women play a key role in food security, although regional differences exist. Climate change impacts vary among diverse social groups depending on age, ethnicity, gender, wealth, and class. Climate extremes have immediate and long-term impacts on livelihoods of poor and vulnerable communities, contributing to greater risks of food insecurity that can be a stress multiplier for internal and external migration (*medium confidence*). {5.2.6} Empowering women and rights-based approaches to decision-making can create synergies among household food security, adaptation, and mitigation. {5.6.4}

Many practices can be optimised and scaled up to advance adaptation throughout the food system (*high confidence*). Supply-side options include increased soil organic matter and erosion control, improved cropland, livestock, grazing land management, and genetic improvements for tolerance to heat and drought. Diversification in the food system (e.g., implementation of integrated production systems, broad-based genetic resources, and heterogeneous diets) is a key strategy to reduce risks (*medium confidence*). Demand-side adaptation, such as adoption of healthy and sustainable diets, in conjunction with reduction in food loss and waste, can contribute to adaptation through reduction in additional land area needed for food production and associated food system vulnerabilities. ILK can contribute to enhancing food system resilience (*high confidence*). {5.3, 5.6.3 Cross-Chapter Box 6 in Chapter 5}

About 21–37% of total greenhouse gas (GHG) emissions are attributable to the food system. These are from agriculture and land use, storage, transport, packaging, processing, retail, and consumption (*medium confidence*). This estimate includes emissions of 9–14% from crop and livestock activities within the farm gate and 5–14% from land use and land-use change including deforestation and peatland degradation (*high confidence*); 5–10% is from supply chain activities (*medium confidence*). This estimate includes GHG emissions from food loss and waste. Within the food system, during the period 2007–2016, the major sources of emissions from the supply side were agricultural production, with crop and livestock activities within the farm gate generating respectively 142 ± 42 $TgCH_4$ yr^{-1} (*high confidence*) and 8.0 ± 2.5 TgN_2O yr^{-1} (*high confidence*), and CO_2 emissions linked to relevant land-use change dynamics such as deforestation and peatland degradation, generating 4.9 ± 2.5 $GtCO_2$ yr^{-1}. Using 100-year GWP values (no climate feedback) from the IPCC AR5, this implies that total GHG emissions from agriculture were 6.2 ± 1.4 $GtCO_2$-eq yr^{-1}, increasing to 11.1 ± 2.9 $GtCO_2$-eq yr^{-1} including relevant land use. Without intervention, these are likely to increase by about 30–40% by 2050, due to increasing demand based on population and income growth and dietary change (*high confidence*). {5.4}

Supply-side practices can contribute to climate change mitigation by reducing crop and livestock emissions, sequestering carbon in soils and biomass, and by decreasing emissions intensity within sustainable production systems (*high confidence*). Total technical mitigation potential from crop and livestock activities and agroforestry is estimated as 2.3–9.6 GtCO2-eq yr^{-1} by 2050 (*medium confidence*). Options with large potential for GHG mitigation in cropping systems include soil carbon sequestration (at decreasing rates over time), reductions in N_2O emissions from fertilisers, reductions in CH_4 emissions from paddy rice, and bridging of yield gaps. Options with large potential for mitigation in livestock systems include better grazing land management, with increased net primary production and soil carbon stocks, improved manure management, and higher-quality feed. Reductions in GHG emissions intensity (emissions per unit product) from livestock can support reductions in absolute emissions, provided appropriate governance to limit total production is implemented at the same time (*medium confidence*). {5.5.1}

Consumption of healthy and sustainable diets presents major opportunities for reducing GHG emissions from food systems and improving health outcomes (*high confidence*). Examples of healthy and sustainable diets are high in coarse grains, pulses, fruits and vegetables, and nuts and seeds; low in energy-intensive animal-sourced and discretionary foods (such as sugary beverages); and with a carbohydrate threshold. Total technical mitigation potential of dietary changes is estimated as 0.7–8.0 $GtCO_2$-eq yr^{-1} by 2050 (*medium confidence*). This estimate includes reductions in emissions from livestock and soil carbon sequestration on spared land, but co-benefits with health are not taken into account. Mitigation potential of dietary change may be higher, but achievement of this potential at broad scales depends on consumer choices and dietary preferences that are guided by social, cultural, environmental, and traditional factors, as well as income growth. Meat analogues such as imitation meat (from plant products), cultured meat, and insects may help in the transition to more healthy and sustainable diets, although their carbon footprints and acceptability are uncertain. {5.5.2, 5.6.5}

Reduction of food loss and waste could lower GHG emissions and improve food security (*medium confidence*). Combined food loss and waste amount to 25–30% of total food produced (*medium confidence*). During 2010–2016, global food loss and waste equalled 8–10% of total anthropogenic GHG emissions (*medium confidence*); and cost about 1 trillion USD2012 per year (*low confidence*). Technical options for reduction of food loss and waste include improved harvesting techniques, on-farm storage, infrastructure, and packaging. Causes of food loss (e.g., lack of refrigeration) and waste (e.g., behaviour) differ substantially in developed and developing countries, as well as across regions (*robust evidence, medium agreement*). {5.5.2}

Agriculture and the food system are key to global climate change responses. Combining supply-side actions such as efficient production, transport, and processing with demand-side interventions such as modification of food choices, and reduction of food loss and waste, reduces GHG emissions and enhances food system resilience (*high confidence*). Such combined measures can enable the implementation of large-scale land-based adaptation and mitigation strategies without threatening food security from increased competition for land for food production and higher food prices. Without combined food system measures in farm management, supply chains, and demand, adverse effects would include increased numbers of malnourished people and impacts on smallholder farmers (*medium evidence, high agreement*). Just transitions are needed to address these effects. {5.5, 5.6, 5.7}

For adaptation and mitigation throughout the food system, enabling conditions need to be created through policies, markets, institutions, and governance (*high confidence*). For adaptation, resilience to increasing extreme events can be accomplished through risk sharing and transfer mechanisms such as insurance markets and index-based weather insurance (*high confidence*). Public health policies to improve nutrition – such as school procurement, health insurance incentives, and awareness-raising campaigns – can potentially change demand, reduce healthcare costs, and contribute to lower GHG emissions (*limited evidence, high agreement*). Without inclusion of comprehensive food system responses in broader climate change policies, the mitigation and adaptation potentials assessed in this chapter will not be realised and food security will be jeopardised (*high confidence*). {5.7, 5.8}

1.1 Framing and context

The current food system (production, transport, processing, packaging, storage, retail, consumption, loss and waste) feeds the great majority of world population and supports the livelihoods of over 1 billion people. Agriculture as an economic activity generates between 1% and 60% of national GDP in many countries, with a world average of about 4% in 2017 (World Bank 2019). Since 1961, food supply per capita has increased more than 30%, accompanied by greater use of nitrogen fertiliser (increase of about 800%) and water resources for irrigation (increase of more than 100%).

The rapid growth in agricultural productivity since the 1960s has underpinned the development of the current global food system that is both a major driver of climate change, and increasingly vulnerable to it (from production, transport, and market activities). Given the current food system, the UN Food and Agriculture Organization (FAO) estimates that there is a need to produce about 50% more food by 2050 in order to feed the increasing world population (FAO 2018a). This would engender significant increases in GHG emissions and other environmental impacts, including loss of biodiversity. FAO (2018a) projects that by 2050 cropland area will increase

90–325 Mha, between 6% and 21% more than the 1567 Mha cropland area of 2010, depending on climate change scenario and development pathway (the lowest increase arises from reduced food loss and waste and adoption of more sustainable diets).

Climate change has direct impacts on food systems, food security, and, through the need to mitigate, potentially increases the competition for resources needed for agriculture. Responding to climate change through deployment of land-based technologies for negative emissions based on biomass production would increasingly put pressure on food production and food security through potential competition for land.

Using a food system approach, this chapter addresses how climate change affects food security, including nutrition, the options for the food system to adapt and mitigate, synergies and trade-offs among these options, and enabling conditions for their adoption. The chapter assesses the role of incremental and transformational adaptation, and the potential for combinations of supply-side measures such as sustainable intensification (increasing productivity per hectare) and demand-side measures (e.g., dietary change and waste reduction) to contribute to climate change mitigation.

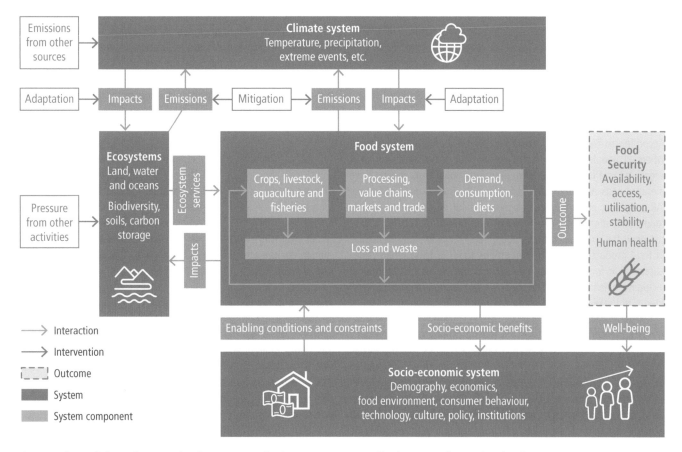

Figure 5.1 | Interlinkages between the climate system, food system, ecosystems (land, water and oceans) and socio-economic system. These systems operate at multiple scales, both global and regional. Food security is an outcome of the food system leading to human well-being, which is also indirectly linked with climate and ecosystems through the socio-economic system. Adaptation measures can help to reduce negative impacts of climate change on the food system and ecosystems. Mitigation measures can reduce GHG emissions coming from the food system and ecosystems.

1.1.1   Food security and insecurity, the food system and climate change

The **food system** encompasses all the activities and actors in the production, transport, manufacturing, retailing, consumption, and waste of food, and their impacts on nutrition, health and well-being, and the environment (Figure 5.1).

1.1.1.1   Food security as an outcome of the food system

The activities and the actors in the food system lead to outcomes such as food security and generate impacts on the environment. As part of the environmental impacts, food systems are a considerable contributor to GHG emissions, and thus climate change (Section 5.4). In turn, climate change has complex interactions with food systems, leading to food insecurity through impacts on food availability, access, utilisation and stability (Table 5.1 and Section 5.2).

We take a **food systems lens** in the Special Report on Climate Change and Land (SRCCL) to recognise that demand for and supply of food are interlinked and need to be jointly assessed in order to identify the challenges of mitigation and adaptation to climate change. Outcomes cannot be disaggregated solely to, for example, agricultural production, because the demand for food shapes what is grown, where it is grown, and how much is grown. Thus, GHG emissions from agriculture result, in large part, from 'pull' from the demand side. Mitigation and adaptation involve modifying production, supply chain, and demand practices (through, for example, dietary choices, market incentives, and trade relationships), so as to evolve to a more sustainable and healthy food system.

According to FAO (2001a), **food security** is a situation that exists when all people, at all times, have physical, social, and economic access to sufficient, safe, and nutritious food that meets their dietary needs and food preferences for an active and healthy life. 'All people at all times' implies the need for equitable and stable food distribution, but it is increasingly recognised that it also covers the need for inter-generational equity, and therefore 'sustainability' in food production. 'Safe and nutritious food … for a healthy life' implies that food insecurity can occur if the diet is not nutritious, including when there is consumption of an excess of calories, or if food is not safe, meaning free from harmful substances.

A prime impact of food insecurity is **malnourishment** (literally 'bad nourishment') leading to **malnutrition**, which refers to deficiencies, excesses, or imbalances in a person's intake of energy and/or nutrients. As defined by FAO et al. (2018), undernourishment occurs when an individual's habitual food consumption is insufficient to provide the amount of dietary energy required to maintain a normal, active, healthy life. In addition to undernourishment in the sense of insufficient calories ('hunger'), undernourishment occurs in terms of nutritional deficiencies in vitamins (e.g., vitamin A) and minerals

(e.g., iron, zinc, iodine), so-called 'hidden hunger'. Hidden hunger tends to be present in countries with high levels of undernourishment (Muthayya et al. 2013), but micronutrient deficiency can occur in societies with low prevalence of undernourishment. For example, in many parts of the world teenage girls suffer from iron deficiency (Whitfield et al. 2015) and calcium deficiency is common in Western-style diets (Aslam and Varani 2016). Food security is related to nutrition, and conversely food insecurity is related to malnutrition. Not all malnourishment arises from food insecurity, as households may have access to healthy diets but choose to eat unhealthily, or it may arise from illness. However, in many parts of the world, poverty is linked to poor diets (FAO et al. 2018). This may be through lack of resources to produce or access food in general, or healthy food, in particular, as healthier diets are more expensive than diets rich in calories but poor in nutrition *(high confidence)* (see meta-analysis by Darmon and Drewnowski 2015). The relationship between poverty and poor diets may also be linked to unhealthy 'food environments,' with retail outlets in a locality only providing access to foods of low nutritional quality (Gamba et al. 2015) – such areas are sometimes termed 'food deserts' (Battersby 2012).

Whilst conceptually the definition of food security is clear, it is not straightforward to measure in a simple way that encompasses all its aspects. Although there are a range of methods to assess food insecurity, they all have some shortcomings. For example, the FAO has developed the Food Insecurity Experience Scale (FIES), a survey-based tool to measure the severity of overall households' inability to access food. While it provides reliable estimates of the prevalence of food insecurity in a population, it does not reveal whether actual diets are adequate or not with respect to all aspects of nutrition (Section 5.1.2.1).

1.1.1.2   Effects of climate change on food security

Climate change is projected to negatively impact the four pillars of food security – availability, access, utilisation and stability – and their interactions (FAO et al. 2018) *(high confidence)*. This chapter assesses recent work since AR5 that has strengthened understanding of how climate change affects each of these pillars across the full range of food system activities (Table 5.1 and Section 5.2).

While most studies continue to focus on availability via impacts on food production, more studies are addressing related issues of access (e.g., impacts on food prices), utilisation (e.g., impacts on nutritional quality), and stability (e.g., impacts of increasing extreme events) as they are affected by a changing climate (Bailey et al. 2015). Low-income producers and consumers are likely to be most affected because of a lack of resources to invest in adaptation and diversification measures (UNCCD 2017; Bailey et al. 2015).

Table 5.1 | Relationships between food security, the food system, and climate change, and guide to chapter.

Food security pillar	Examples of observed and projected climate change impacts	Sections	Examples of adaptation and mitigation	Section
Availability *Production of food and its readiness for use through storage, processing, distribution, sale and/or exchange*	Reduced yields in crop and livestock systems	5.2.2.1, 5.2.2.2	Development of adaptation practices	5.3
	Reduced yields from lack of pollinators; pests and diseases	5.2.2.3, 5.2.2.4	Adoption of new technologies, new and neglected varieties	5.3.2.3, 5.3.3.1,
	Reduced food quality affecting availability (e.g., food spoilage and loss from mycotoxins)	5.2.4.1, 5.5.2.5	Enhanced resilience by integrated practices, better food storage	5.3.2.3, 5.3.3.4, 5.6.4
	Disruptions to food storage and transport networks from change in climate, including extremes	5.2.5.1, 5.3.3.4, 5.8.1, Box 5.5	Reduction of food demand by reducing waste, modifying diets	5.3.4, 5.5.2, 5.7
			Closing of crop yield and livestock productivity gaps	5.6.4.4, 5.7
			Risk management, including marketing mechanisms, financial insurance	5.3.2, 5.7
Access *Ability to obtain food, including effects of price*	Yield reductions, changes in farmer livelihoods, limitations on ability to purchase food	5.2.2.1, 5.2.2.2	Integrated agricultural practices to build resilient livelihoods	5.6.4
	Price rise and spike effects on low-income consumers, in particular women and children, due to lack of resources to purchase food	5.1.3, 5.2.3.1, 5.2.5.1, Box 5.1	Increased supply chain efficiency (e.g., reducing loss and waste)	5.3.3, 5.3.4
	Effects of increased extreme events on food supplies, disruption of agricultural trade and transportation infrastructure	5.8.1	More climate-resilient food systems, shortened supply chains, dietary change, market change	5.7
Utilisation *Achievement of food potential through nutrition, cooking, health*	Impacts on food safety due to increased prevalence of microorganisms and toxins	5.2.4.1	Improved storage and cold chains	5.3.3, 5.3.4
	Decline in nutritional quality resulting from increasing atmospheric CO_2	5.2.4.2	Adaptive crop and livestock varieties, healthy diets, better sanitation	5.3.4, 5.5.2, 5.7
	Increased exposure to diarrheal and other infectious diseases due to increased risk of flooding	5.2.4.1		
Stability *Continuous availability and access to food without disruption*	Greater instability of supply due to increased frequency and severity of extreme events; food price rises and spikes; instability of agricultural incomes	5.2.5, 5.8.1	Resilience via integrated systems and practices, diversified local agriculture, infrastructure investments, modifying markets and trade, reducing food loss and waste	5.6.4, 5.7, 5.8.1
	Widespread crop failure contributing to migration and conflict	5.8.2	Crop insurance for farmers to cope with extreme events	5.3.2.2, 5.7
			Capacity building to develop resilient systems	5.3.6, 5.7.4
Combined *Systemic impacts from interactions of all four pillars*	Increasing undernourishment as food system is impacted by climate change	5.1	Increased food system productivity and efficiency (e.g., supply side mitigation, reducing waste, dietary change)	5.5.1, 5.7
	Increasing obesity and ill health through narrow focus on adapting limited number of commodity crops	5.1	Increased production of healthy food and reduced consumption of energy-intensive products	5.5.2, 5.7
	Increasing environmental degradation and GHG emissions	Cross-Chapter Box 6	Development of climate smart food systems by reducing GHG emissions, building resilience, adapting to climate change	5.3.3, 5.7
	Increasing food insecurity due to competition for land and natural resources (e.g., for land-based mitigation)	5.6.1	Governance and institutional responses (including food aid) that take into consideration gender and equity.	5.2.5, 5.7

5

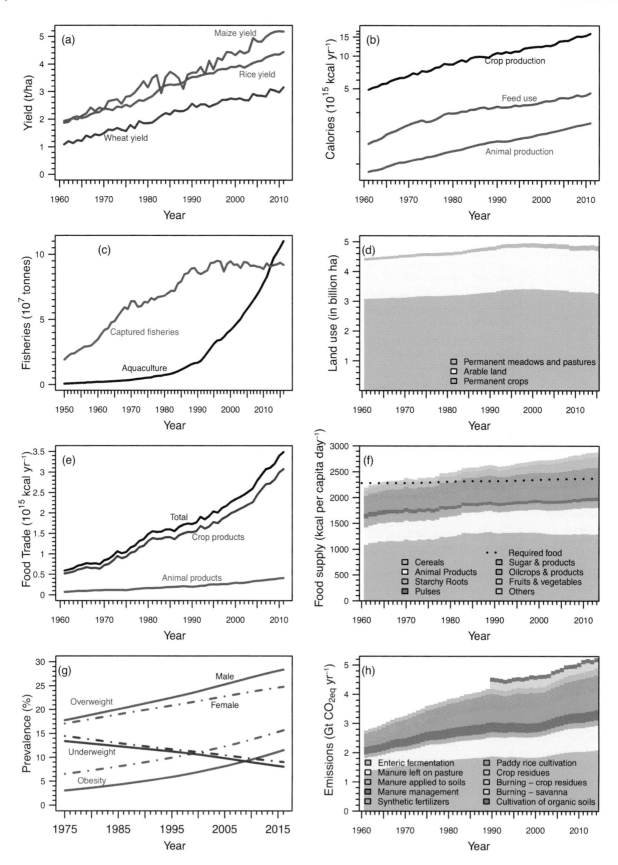

Figure 5.2 | Global trends in (a) yields of maize, rice, and wheat (FAOSTAT 2018) – the top three crops grown in the world; **(b)** production of crop and animal calories and use of crop calories as livestock feed (FAOSTAT 2018); **(c)** production from marine and aquaculture fisheries (FishStat 2019); **(d)** land used for agriculture (FAOSTAT 2018); **(e)** food trade in calories (FAOSTAT 2018); **(f)** food supply and required food (i.e., based on human energy requirements for medium physical activities) from 1961–2012 (FAOSTAT 2018; Hiç et al. 2016); **(g)** prevalence of overweight, obesity and underweight from 1975–2015 (Abarca-Gómez et al. 2017); and **(h)** GHG emissions for the agriculture sector, excluding land-use change (FAOSTAT 2018). For figures (b) and (e), data provided in mass units were converted into calories using nutritive factors (FAO 2001b). Data on emissions due to burning of savanna and cultivation of organic soils is provided only after 1990 (FAOSTAT 2018).

1.1.2   Status of the food system, food insecurity and malnourishment

1.1.2.1   Trends in the global food system

Food is predominantly produced on land, with, on average, 83% of the 697 kg of food consumed per person per year, 93% of the 2884 kcal per day, and 80% of the 81 g of protein eaten per day coming from terrestrial production in 2013 (FAOSTAT 2018).[1] With increases in crop yields and production (Figure 5.2), the absolute supply of food has been increasing over the last five decades. Growth in production of animal-sourced food is driving crop utilisation for livestock feed (FAOSTAT 2018; Pradhan et al. 2013a). Global trade of crop and animal-sourced food has increased by around 5 times between 1961 and 2013 (FAOSTAT 2018). During this period, global food availability has increased from 2200 kcal/cap/day to 2884 kcal/cap/day, making a transition from a food deficit to a food surplus situation (FAOSTAT 2018; Hiç et al. 2016).

The availability of cereals, animal products, oil crops, and fruits and vegetables has mainly grown (FAOSTAT 2018), reflecting shifts towards more affluent diets. This, in general, has resulted in a decrease in prevalence of underweight and an increase in prevalence of overweight and obesity among adults (Abarca-Gómez et al. 2017). During the period 1961–2016, anthropogenic GHG emissions associated with agricultural production has grown from 3.1 GtCO2-eq yr^{-1} to 5.8 GtCO$_2$-eq yr^{-1} (Section 5.4.2 and Chapter 2). The increase in emissions is mainly from the livestock sector (from enteric fermentation and manure left on pasture), use of synthetic fertiliser, and rice cultivation (FAOSTAT 2018).

1.1.2.2   Food insecurity status and trends

In addressing food security the dual aspects of malnutrition – under-nutrition and micro-nutrient deficiency, as well as over-consumption, overweight, and obesity – need to be considered (Figure 5.2 (g) and Table 5.2 | Global prevalence of various forms of malnutrition.5.2). The UN agencies' State of Food Security and Nutrition 2018 report (FAO et al. 2018) and the Global Nutrition Report 2017 (Development Initiatives 2017) summarise the global data. The *State of Food Security* report's estimate for undernourished people on a global basis is 821 million, up from 815 million the previous year and 784 million the year before that. Previous to 2014/2015 the prevalence of hunger had been declining over the last three decades. The proportion of young children (under five) who are stunted (low height-for-age), has been gradually declining, and was 22% in 2017 compared to 31% in 2012 (150.8 million, down from 165.2 million in 2012). In 2017, 50.5 million children (7.5%) under five were wasted (low weight-for-height). Since 2014, undernutrition has worsened, particularly in parts of Sub-Saharan Africa, south-eastern Asia and Western Asia, and recently Latin America. Deteriorations have been observed most notably in situations of conflict and conflict combined with droughts or floods (FAO et al. 2018).

Regarding micronutrient deficiencies known as 'hidden hunger', reporting suggests a prevalence of one in three people globally (FAO 2013a; von Grebmer et al. 2014; Tulchinsky 2010) (Table 5.2). In the last decades, hidden hunger (measured through proxies targeting iron, vitamin A, and zinc deficiencies) worsened in Africa, while it mainly improved in the Asia and Pacific regions (Ruel-Bergeron et al. 2015). In 2016, 613 million women and girls aged 15 to 49 suffered

Table 5.2 | Global prevalence of various forms of malnutrition.

	HLPE 2017 (UN)	SOFI 2017 (FAO)	GNR 2017	SOFI 2018 (FAO)	GNR2018
Overweight but not obese[a]	1.3 billion		1.29 billion		1.34 billion (38,9%)[c]
Overweight under five	41 million	41 million	41 million	38 million	38 million
Obesity[b]	600 million	600 million (13%)	641 million	672 million	678 million (13,1%)[c]
Undernourishment	800 million	815 million	815 million	821 million	
Stunting under five	155 million	155 million	155 million[d]	151 million	151 million[d] (22%)
Wasting under five	52 million	52 million (8%)	52 million[d]	50 million	51 million[d] (7%)
MND (iron)	19.2% of pregnant women[e]	33% women of reproductive age	613 million women and girls aged 15 to 49[f]	613 million (32.8%) women and girls aged 15 to 49[f]	613 million (32.8%) women and girls aged 15 to 49[f]

HLPE: High Level Panel of Experts on Food Security and Nutrition; SOFI: The State of Food Security and Nutrition in the World; GNR: Global Nutrition Report; MND: Micro nutrient deficiency (iron deficiency for year 2016, uses anaemia as a proxy (percentage of pregnant women whose haemoglobin level is less than 110 grams per litre at sea level and percentage of non-pregnant women whose haemoglobin level is less than 120 grams per litre at sea level).

[a] Body mass index between 25 kg m^{-2} and 29.9 kg m^{-2}.

[b] Body mass index greater than 30 kg m^{-2}.

[c] Prevalence of overweight/obesity among adults (age ≥18) in year 2016. Data from NCD Risc data source.

[d] UNICEF WHO Joint Malnutrition.

[e] In 2011.

[f] Anaemia prevalence in girls and women aged 15 to 49.

---

[1]   Does not take into account terrestrial production of feed.

from iron deficiency (Development Initiatives 2018); in 2013, 28.5% of the global population suffered from iodine deficiency; and in 2005, 33.3% of children under five and 15.3% of pregnant women suffered from vitamin A deficiency, and 17.3% of the global population suffered from zinc deficiency (HLPE 2017).

Globally, as the availability of inexpensive calories from commodity crops increases, so does per capita consumption of calorie-dense foods (Ng et al. 2014; NCD-RisC 2016a; Abarca-Gómez et al. 2017 and Doak and Popkin 2017). As a result, in every region of the world, the prevalence of obesity (body mass index >30 kg m^{-2}) and overweight (body mass index range between normality [18.5–24.9] and obesity) is increasing. There are now more obese adults in the world than underweight adults (Ng et al. 2014; NCD-RisC 2016a; Abarca-Gómez et al. 2017 and Doak and Popkin 2017). In 2016, around two billion adults were overweight, including 678 million suffering from obesity (NCD-RisC 2016a; Abarca-Gómez et al. 2017). The prevalence of overweight and obesity has been observed in all age groups.

Around 41 million children under five years and 340 million children and adolescents aged 5–19 years were suffering from overweight or obesity in 2016 (NCD-RisC 2016a; FAO et al. 2017; WHO 2015). In many high-income countries, the rising trends in children and adolescents suffering from overweight or obesity have stagnated at high levels; however, these have accelerated in parts of Asia and have very slightly reduced in European and Central Asian lower and middle-income countries (Abarca-Gómez et al. 2017; Doak and Popkin 2017; Christmann et al. 2009).

There are associations between obesity and non-communicable diseases such as diabetes, dementia, inflammatory diseases (Saltiel and Olefsky 2017), cardiovascular disease (Ortega et al. 2016) and some cancers, for example, of the colon, kidney, and liver (Moley and Colditz 2016). There is a growing recognition of the rapid rise in overweight and obesity on a global basis and its associated health burden created through non-communicable diseases (NCD-RisC 2016a; HLPE 2017).

Analyses reported in FAO et al. (2018) highlight the link between food insecurity, as measured by the FIES scale, and malnourishment (*medium agreement, robust evidence*). This varies by malnourishment measure as well as country (FAO et al. 2018). For example, there is *limited evidence* (*low agreement* but multiple studies) that food insecurity and childhood wasting (i.e., or low weight for height) are closely related, but it is very likely (*high agreement, robust evidence*) that childhood stunting and food insecurity are related (FAO et al. 2018). With respect to adult obesity there is *robust evidence, with medium agreement*, that food insecurity, arising from poverty reducing access to nutritious diets, is related to the prevalence of obesity, especially in high-income countries and adult females. An additional meta-analysis (for studies in Europe and North America) also finds a negative relationship between income and obesity, with some support for an effect of obesity causing low income (as well as vice versa) (Kim and von dem Knesebeck 2018).

As discussed in Section 5.1.1.1, different methods of assessing food insecurity can provide differential pictures. Of particular note is the

spatial distribution of food insecurity, especially in higher-income countries. FAO et al. (2018) reports FIES estimates of severe food insecurity in Africa, Asia and Latin America of 29.8%, 6.9% and 9.8% of the population, respectively, but of 1.4% of the population (i.e., about 20 million in total; pro rata <5 million for US, <1 million for UK) in Europe and North America. However, in the USA, USDA estimates 40 million people were exposed to varying degrees of food insecurity, from mild to severe (overall prevalence about 12%) (Coleman-Jensen et al. 2018). In the UK, estimates from 2017 and 2018 indicate about 4 million adults are moderately to severely food insecure (prevalence 8%) (End Hunger UK 2018; Bates et al. 2017). The UK food bank charity, the Trussell Trust, over a year in 2017/18, distributed 1,332,952 three-day emergency food parcels to people referred to the charity as being in food crisis. Furthermore, a 2003 study in the UK (Schenker 2003) estimated that 40% of adults, and 15% of children admitted to hospitals were malnourished, and that 70% of undernourishment in the UK was unreported.

In total, more than half the world's population are underweight or overweight (NCD-RisC 2017a), so their diets do not provide the conditions for 'an active and healthy life'. This will be more compromised under the impacts of climate change by changing the availability, access, utilisation, and stability of diets of sufficient nutritional quality as shown in Table 5.2 and discussed in detail below (Section 5.2).

1.1.3   Climate change, gender and equity

Throughout, the chapter considers many dimensions of gender and equity in regard to climate change and the food system (Box 5.1). Climate change impacts differ among diverse social groups depending on factors such as age, ethnicity, ability/disability, sexual orientation, gender, wealth, and class (*high confidence*) (Vincent and Cull 2014; Kaijser and Kronsell 2014). Poverty, along with socio-economic and political marginalisation, cumulatively put women, children and the elderly in a disadvantaged position in coping with the adverse impacts of the changing climate (UNDP 2013; Skoufias et al. 2011). The contextual vulnerability of women is higher due to their differentiated relative power, roles, and responsibilities at the household and community levels (Bryan and Behrman 2013; Nelson et al. 2002). They often have a higher reliance on subsistence agriculture, which will be severely impacted by climate change (Aipira et al. 2017).

Through impacts on food prices (Section 5.2.3.1) poor people's food security is particularly threatened. Decreased yields can impact nutrient intake of the poor by decreasing supplies of highly nutritious crops and by promoting adaptive behaviours that may substitute crops that are resilient but less nutritious (Thompson et al. 2012; Lobell and Burke 2010). In Guatemala, food prices and poverty have been correlated with lower micronutrient intakes (Iannotti et al. 2012). In the developed world, poverty is more typically associated with calorically-dense but nutrient-poor diets, obesity, overweight, and other related diseases (Darmon and Drewnowski 2015).

5

Rural areas are especially affected by climate change (Dasgupta et al. 2014), through impacts on agriculture-related livelihoods and rural income (Mendelsohn et al. 2007) and through impacts on employment. Jessoe et al. (2018) using a 28-year panel on individual employment in rural Mexico, found that years with a high occurrence of heat lead to a reduction in local employment by up to 1.4% with a medium emissions scenario, particularly for wage work and non-farm labour, with impacts on food access. Without employment opportunities in areas where extreme poverty is prevalent, people may be forced to migrate, exacerbating potential for ensuing conflicts (FAO 2018a).

Finally, climate change can affect human health in other ways that interact with food utilisation. In many parts of the world where agriculture relies still on manual labour, projections are that heat stress will reduce the hours people can work, and increase their risk (Dunne et al. 2013). For example, Takakura et al. (2017) estimates that under RCP8.5, the global economic loss from people working shorter hours to mitigate heat loss may be 2.4–4% of GDP. Furthermore, as

discussed by Watts et al. (2018); people's nutritional status interacts with other stressors and affects their susceptibility to ill health (the 'utilisation pillar' of food security): so food-insecure people are more likely to be adversely affected by extreme heat, for example.

In the case of food price hikes, those more vulnerable are more affected (Uraguchi 2010), especially in urban areas (Ruel et al. 2010), where livelihood impacts are particularly severe for the individuals and groups that have scarce resources or are socially isolated (Revi et al. 2014; Gasper et al. 2011) (*high confidence*). These people often lack power and access to resources, adequate urban services and functioning infrastructure. As climate events become more frequent and intense, this can increase the scale and depth of urban poverty (Rosenzweig et al. 2018b). Urban floods and droughts may result in water contamination increasing the incidence of diarrhoeal illness in poor children (Bartlett 2008). In the near destruction of New Orleans by Hurricane Katrina, about 40,000 jobs were lost (Rosemberg 2010).

Box 5.1 | Gender, food security and climate change

Differentiated impacts, vulnerability, risk perception, behaviours and coping strategies for climate change related to food security derive from cultural (gendered) norms. That is, the behaviours, tasks, and responsibilities a society defines as 'male' or 'female', and the differential gendered access to resources (Paris and Rola-Rubzen 2018; Aberman and Tirado 2014; Lebel et al. 2014; Bee 2016). In many rural areas women often grow most of the crops for domestic consumption and are primarily responsible for storing, processing, and preparing food; handling livestock; gathering food, fodder and fuelwood; managing domestic water supply; and providing most of the labour for post-harvest activities (FAO 2011a). They are mostly impacted through increased hardship, implications for household roles, and subsequent organisational responsibilities (Boetto and McKinnon 2013; Jost et al. 2016). Water scarcity can particularly affect women because they need to spend more time and energy to collect water, where they may be more exposed to physical and sexual violence (Sommer et al. 2015; Aipira et al. 2017). They may be forced to use unsafe water in the household increasing risk of water-borne diseases (Parikh 2009). Climate change also has differentiated gendered impacts on livestock-holders' food security (McKune et al. 2015; Ongoro and Ogara 2012; Fratkin et al. 2004) (Supplementary Material Table SM5.1).

Gender dimensions of the four pillars
Worldwide, women play a key role in food security (World Bank 2015) and the four pillars of food security have strong gender dimensions (Thompson 2018). In terms of **food availability**, women tend to have less access to productive resources, including land, and thus less capacity to produce food (Cross-Chapter Box 11 in Chapter 7).

In terms of **food access**, gendered norms in how food is

divided at mealtimes may lead to smaller food portions for women and girls. Women's intra-household inequity limits their ability to purchase food; limitations also include lack of women's mobility impacting trips to the market and lack of decision-making within the household (Ongoro and Ogara 2012; Mason et al. 2017; Riley and Dodson 2014).

In terms of **food utilisation**, men, women, children and the elderly have different nutritional needs (e.g., during pregnancy or breast-feeding).

In terms of **food stability**, women are more likely to be disproportionately affected by price spikes (Vellakkal et al. 2015; Arndt et al. 2016; Hossain and Green 2011; Darnton-Hill and Cogill 2010; Cohen and Garrett 2010; Kumar and Quisumbing 2013) because when food is scarce women reduce food consumption relative to other family members, although these norms vary according to age, ethnicity, culture, region, and social position, as well as by location in rural or urban areas (Arora-Jonsson 2011; Goh 2012; Niehof 2016; Ongoro and Ogara 2012).

Integrating gender into adaptation
Women have their own capabilities to adapt to climate change. In the Pacific Islands, women hold critical knowledge on where or how to find clean water; which crops to grow in a wet or dry season; how to preserve and store food and seeds ahead of approaching storms,

5

Box 5.1 (continued)

floods or droughts; and how to carry their families through the recovery months. They also play a pivotal role in managing household finances and investing their savings in education, health, livelihoods, and other activities that assist their families to adapt and respond to climate effects (Aipira et al. 2017). Decreasing women's capacity to adapt to the impacts of climate change also decreases that of the household (Bryan and Behrman 2013).

However, gender norms and power inequalities also shape the ability of men, women, boys, girls and the elderly to adapt to climate risks (Rossi and Lambrou 2008). For example, women pastoralists in the Samburu district of Kenya cannot make decisions affecting their lives, limiting their adaptive capacity (Ongoro and Ogara 2012).

Participation in decision-making and politics, division of labour, resource access and control, and knowledge and skills (Nelson and Stathers 2009) are some of the barriers to adaptation. Women's adaptive capacity is also diminished because their work often goes unrecognised (Rao 2005; Nelson and Stathers 2009). Many of women's activities are not defined as 'economically active employment' in national accounts (FAO 2011a). This non-economic status of women's activities implies that they are not included in wider discussions of priorities or interventions for climate change. Their perspectives and needs are not met; and thus, interventions, information, technologies, and tools promoted are potentially not relevant, and even can increase discrimination (Alston 2009; Edvardsson Björnberg and Hansson 2013; Huynh and Resurreccion 2014).

Where gender-sensitive policies to climate change may exist, effective implementation in practice of gender equality and empowerment may not be achieved on the ground due to lack of technical capacity, financial resources and evaluation criteria, as shown in the Pacific Islands (Aipira et al. 2017). Thus, corresponding institutional frameworks that are well-resourced, coordinated, and informed are required, along with adequate technical capacity within government agencies, NGOs and project teams, to strengthen collaboration and promote knowledge sharing (Aipira et al. 2017).

Women's empowerment: Synergies among adaptation,

mitigation, and food security

Empowering and valuing women in their societies increases their capacity to improve food security under climate change and make substantial contributions to their own well-being, to that of their families and of their communities (Langer et al. 2015; Ajani et al. 2013 and Alston 2014) (*high confidence*). Women's empowerment includes economic, social and institutional arrangements and may include targeting men in

integrated agriculture programmes to change gender norms and improve nutrition (Kerr et al. 2016). Empowerment through collective action and groups-based approaches in the near-term has the potential to equalise relationships on the local, national and global scale (Ringler et al. 2014). Empowered women are crucial to creating effective synergies among adaptation, mitigation, and food security.

In Western Kenya, widows in their new role as main livelihood providers invested in sustainable innovations like rainwater harvesting systems and agroforestry (this can serve as both adaptation and mitigation), and worked together in formalised groups of collective action (Gabrielsson and Ramasar 2013) to ensure food and water security. In Nepal, women's empowerment had beneficial outcomes in maternal and children nutrition, reducing the negative effect of low production diversity (Malapit et al. 2015). Integrated nutrition and agricultural programmes have increased women's decision-making power and control over home gardens in Burkina Faso (van den Bold et al. 2015) with positive impacts on food security.

1.1.4   Food systems in AR5, SR15, and the Paris Agreement

Food, and its relationship to the environment and climate change, has grown in prominence since the Rio Declaration in 1992, where food production is Chapter 14 of Agenda 21, to the Paris Agreement of 2015, which includes the need to ensure food security under the threat of climate change on its first page. This growing prominence of food is reflected in recent IPCC reports, including its Fifth Assessment Report (IPCC 2014a) and the Special Report on global warming of 1.5°C (SR15) (IPCC 2018a).

1.1.4.1   Food systems in AR5 and SR15

The IPCC Working Group (WG) II AR5 chapter on Food Security and Food Production Systems broke new ground by expanding its focus beyond the effects of climate change primarily on agricultural production (crops, livestock and aquaculture) to include a food systems approach as well as directing attention to undernourished people (Porter et al. 2014). However, it focused primarily on food production systems due to the prevalence of studies on that topic (Porter et al. 2017). It highlighted that a range of potential adaptation options exist across all food system activities, not just in food production, and that benefits from potential innovations in food processing, packaging, transport, storage, and trade were insufficiently researched at that time.

The IPCC WG III AR5 chapter on Agriculture, Forestry and Other Land Use (AFOLU) (Smith et al. 2014) assessed mitigation potential considering not only the supply, but also the demand side of land uses, by consideration of changes in diets; it also included food loss and waste. AR5 focused on crop and livestock activities within the farm gate and land use and land-use change dynamics associated with agriculture. It did not take a full food system approach to emissions estimates that include processing, transport, storage, and retail.

The IPCC WG II AR5 Rural Areas chapter (Revi et al. 2014) found that farm households in developing countries are vulnerable to climate change due to socio-economic characteristics and non-climate stressors, as well as climate risks (Dasgupta et al. 2014). They also found that a wide range of on-farm and off-farm climate change adaptation measures are already being implemented and that the local social and cultural context played a prominent role in the success or failure of different adaptation strategies for food security, such as trade, irrigation or diversification. The IPCC WG II AR5 Urban Areas chapter found that food security for people living in cities was severely affected by climate change through reduced supplies, including urban-produced food, and impacts on infrastructure, as well as a lack of access to food. Poor urban dwellers are more vulnerable to rapid changes of food prices due to climate change.

Many climate change response options in IPCC WG II and WG III AR5 (IPCC 2014b) address incremental adaptation or mitigation responses separately rather than being inclusive of more systemic or transformational changes in multiple food systems that are large-scale, in depth, and rapid, requiring social, technological, organisational and system responses (Rosenzweig and Solecki 2018; Mapfumo et al. 2017; Termeer et al. 2017). In many cases, transformational change will require integration of resilience and mitigation across all parts of the food system including production, supply chains, social aspects, and dietary choices. Further, these transformational changes in the food system need to encompass linkages to ameliorative responses to land degradation (Chapter 4), desertification (Chapter 3), and declines in quality and quantity of water resources throughout the food-energy-water nexus (Chapter 2 and Section 5.7).

The IPCC Special Report on global warming of 1.5°C found that climate-related risks to food security are projected to increase with global warming of 1.5°C and increase further with 2°C (IPCC 2018a).

1.1.4.2   Food systems and the Paris Agreement

To reach the temperature goal put forward in the Paris Agreement of limiting warming to well below 2°C, and pursuing efforts to limit warming to 1.5°C, representatives from 196 countries signed the United Nations Framework Convention on Climate Change (UNFCCC) Paris Agreement (UNFCCC 2015) in December 2015. The Agreement put forward a temperature target of limiting warming to well below 2°C, and pursuing efforts to limit warming to 1.5°C. Under the Paris Agreement, Parties are expected to put forward their best efforts through nationally determined contributions (NDCs) and to strengthen these efforts in the years ahead. Article 2 of the Agreement makes clear the agreement is within 'the context of sustainable development' and states actions should be 'in a manner that does not threaten food production' to ensure food security.

Many countries have included food systems in their mitigation and adaptation plans as found in their NDCs for the Paris Agreement (Rosenzweig et al. 2018a). Richards et al. (2015) analysed 160 Party submissions and found that 103 include agricultural mitigation; of the 113 Parties that include adaptation in their NDCs, almost all (102) include agriculture among their adaptation priorities. There is much attention to conventional agricultural practices that can be climate-smart and sustainable (e.g., crop and livestock management), but less to the enabling services that can facilitate uptake (e.g., climate information services, insurance, credit). Considerable finance is needed for agricultural adaptation and mitigation by the least developed countries – in the order of 3 billion USD annually for adaptation and 2 billion USD annually for mitigation, which may be an underestimate due to a small sample size (Richards et al. 2015). On the mitigation side, none of the largest agricultural emitters included sector-specific contributions from the agriculture sector in their NDCs, but most included agriculture in their economy-wide targets (Richards et al. 2018).

Carbon dioxide removal (CDR). A key aspect regarding the implementation of measures to achieve the Paris Agreement goals involves measures related to carbon dioxide removal (CDR) through bioenergy (Sections 5.5 and 5.6). To reach the temperature target of limiting warming to well below 2°C, and pursuing efforts to limit warming to 1.5°C, large investments and abrupt changes in land use will be required to advance bioenergy with carbon capture and sequestration (BECCS), afforestation and reforestation (AR), and biochar technologies. Existing scenarios estimate the global area required for energy crops to help limit warming to 1.5°C in the range of 109–990 Mha, most commonly around 380–700 Mha.

Most scenarios assume very rapid deployment between 2030 and 2050, reaching rates of expansion in land use in 1.5°C scenarios exceeding 20 Mha yr^{-1}, which are unprecedented for crops and forestry reported in the FAO database from 1961. Achieving the 1.5°C target would thus result in major competing demands for land between climate change mitigation and food production, with cascading impacts on food security.

This chapter assesses how the potential conflict for land could be alleviated by sustainable intensification to produce food with a lower land footprint (Cross-Chapter Box 6 in Section 5.6). To accomplish this, farmers would need to produce the same amount of food with lower land requirement, which depends on technology, skills, finance, and markets. Achieving this would also rely on demand-side changes including dietary choices that enable reduction of the land footprint for food production while still meeting dietary needs. Transitions required for such transformative changes in food systems are addressed in Section 5.7.

1.1.4.3   Charting the future of food security

This chapter utilises the common framework of the Representative Concentration Pathways (RCPs) and the Shared Socio-economic Pathways (SSPs) (Popp et al. 2017; Riahi et al. 2017 and Doelman et al. 2018) to assess the impacts of future GHG emissions, mitigation measures, and adaptation on food security (Cross-Chapter Box 1 in Chapter 1, Sections 5.2 and 5.6).

New work utilising these scenario approaches has shown that the food system externalises costs onto human health and the environment (Springmann et al. 2018a; Swinburn et al. 2019; Willett et al. 2019), leading to calls for transforming the food system to deliver better human and sustainability outcomes (Willett et al. 2019; IAP 2018; Development Initiatives 2018; Lozano et al. 2018). Such a transformation could be an important lever to address the complex interactions between climate change and food security. Through acting on mitigation and adaptation in regard to both food demand and food supply we assess the potential for improvements to both human health and the Sustainable Development Goals (Section 5.6).

This chapter builds on the food system and scenario approaches followed by AR5 and its focus on climate change and food security, but new work since AR5 has extended beyond production to how climate change interacts with the whole food system. The analysis of climate change and food insecurity has expanded beyond undernutrition to include the over-consumption of unhealthy mass-produced food high in sugar and fat, which also threatens health in different but highly damaging ways, as well as the role of dietary choices and consumption in GHG emissions. It focuses on land-based food systems, though highlighting in places the contributions of freshwater and marine production.

The chapter assesses new work on the observed and projected effects of CO_2 concentrations on the nutritional quality of crops (Section 5.2.4.2) emphasising the role of extreme climate events (Section 5.2.5.1), social aspects including gender and equity (Box 5.1, and Cross-Chapter Box 11 in Chapter 7), and dietary choices (Section 5.4.6, 5.5.2). Other topics with considerable new literature include impacts on smallholder farming systems (Section 5.2.2.6), food loss and waste (Section 5.5.2.5), and urban and peri-urban agriculture (Section 5.6.5). The chapter explores the potential competing demands for land that mitigation measures to achieve temperature targets may engender, with cascading impacts on food production, food security, and farming systems (Section 5.6), and the enabling conditions for achieving mitigation and adaptation in equitable and sustainable ways (Section 5.7). Section 5.8 presents challenges to future food security, including food price spikes, migration, and conflict.

1.2   Impacts of climate change on food systems

There are many routes by which climate change can impact food security and thus human health (Watts et al. 2018; Fanzo et al. 2017). One major route is via climate change affecting the amount of food,

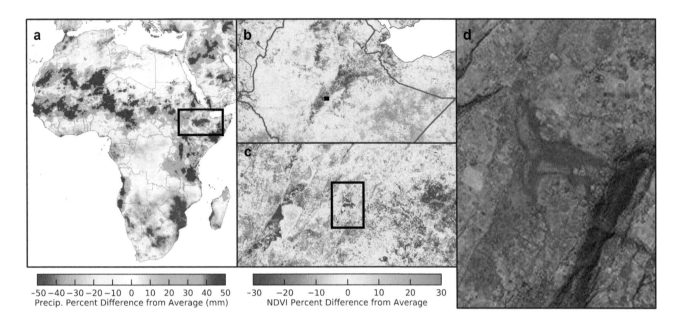

Figure 5.3 | Precipitation anomaly and vegetation response in eastern Africa. (a) Sep 2015–Feb 2016 Climate Hazards Group Infrared Precipitation with Station (CHIRPS) precipitation anomaly over Africa relative to the 1981–2010 average shows that large areas of Ethiopia received less than half of normal precipitation. Consequently, widespread impacts to agricultural productivity, especially within pastoral regions, were present across Ethiopia as evidenced by **(d)** reduced greenness in remote sensing images. **(b)** MODIS NDVI anomalies for Sep 2015–Feb 2016 relative to 2000–2015 average are shown for the inset box in (a). **(c)** Landsat NDVI anomalies for Sep 2015–Feb 2016 relative to 2000–2015 average are shown for the inset box in (b) (Huntington et al. 2017).

both from direct impacts on yields (Section 5.2.2.1) and indirect effects through climate change's impacts on water availability and quality, pests and diseases (Section 5.2.2.3), and pollination services (Section 5.2.2.4). Another route is via changing CO_2 in the atmosphere, affecting biomass and nutritional quality (Section 5.2.4.2). Food safety risks during transport and storage can also be exacerbated by changing climate (Section 5.2.4.1).

Further, the direct impacts of changing weather can affect human health through the agricultural workforce's exposure to extreme temperatures (Section 5.2.5.1). Through changing metabolic demands and physiological stress for people exposed to extreme temperatures, there is also the potential for interactions with food availability; people may require more food to cope, whilst at the same time being impaired from producing it (Watts et al. 2018). All these factors have the potential to alter both physical health as well as cultural health, through changing the amount, safety and quality of food available for individuals within their cultural context.

This section assesses recent literature on climate change impacts on the four pillars of food security: availability (Section 5.2.2), access (Section 5.2.3), utilisation (Section 5.2.4), and stability (Section 5.2.5). It considers impacts on the food system from climate changes that are already taking place and how impacts are projected to occur in the future. See Supplementary Material Section SM5.2 for discussion of detection and attribution and improvement in projection methods.

1.2.1   Climate drivers important to food security

Climate drivers relevant to food security and food systems include temperature-related, precipitation-related, and integrated metrics that combine these and other variables. These are projected to affect many aspects of the food security pillars (FAO 2018b) (see Supplementary Material Table SM5.2, and Chapter 6 for assessment of observed and projected climate impacts). Climate drivers relevant to food production and availability may be categorised as modal climate changes (e.g., shifts in climate envelopes causing shifts in cropping varieties planted), seasonal changes (e.g., warming trends extending growing seasons), extreme events (e.g., high temperatures affecting critical growth periods, flooding/droughts), and atmospheric conditions for example, CO_2 concentrations, short-lived climate pollutants (SLCPs), and dust. Water resources for food production will be affected through changing rates of precipitation and evaporation, ground water levels, and dissolved oxygen content (Cruz-Blanco et al. 2015; Sepulcre-Canto et al. 2014; Huntington et al. 2017; Schmidtko et al. 2017). Potential changes in major modes of climate variability can also have widespread impacts such as those that occurred during late 2015 to early 2016 when a strong El Niño contributed to regional shifts in precipitation in the Sahel region. Significant drought across Ethiopia resulted in widespread crop failure and more than 10 million people in Ethiopia requiring food aid (U.S. Department of State 2016; Huntington et al. 2017) (Figure 5.3).

Other variables that affect agricultural production, processing, and/or transport are solar radiation, wind, humidity, and (in coastal areas) salinisation and storm surge (Mutahara et al. 2016; Myers et al. 2017). Extreme climate events resulting in inland and coastal flooding, can affect the ability of people to obtain and prepare food (Rao et al. 2016; FAO et al. 2018). For direct effects of atmospheric CO_2 concentrations on crop nutrient status see Section 5.2.4.2.

1.2.1.1   Short-lived climate pollutants

The important role of short-lived climate pollutants such as ozone and black carbon is increasingly emphasised since they affect agricultural production through direct effects on crops and indirect effects on climate (Emberson et al. 2018; Lal et al. 2017; Burney and Ramanathan 2014; Ghude et al. 2014) (Chapters 2 and 4). Ozone causes damage to plants through damages to cellular metabolism that influence leaf-level physiology to whole-canopy and root-system processes and feedbacks; these impacts affect leaf-level photosynthesis senescence and carbon assimilation, as well as whole-canopy water and nutrient acquisition and ultimately crop growth and yield (Emberson et al. 2018).

Using atmospheric chemistry and a global integrated assessment model, Chuwah et al. (2015) found that without a large decrease in air pollutant emissions, high ozone concentration could lead to an increase in crop damage of up to 20% in agricultural regions in 2050 compared to projections in which changes in ozone are not accounted for. Higher temperatures are associated with higher ozone concentrations; C3 crops are sensitive to ozone (e.g., soybeans, wheat, rice, oats, green beans, peppers, and some types of cottons) and C4 crops are moderately sensitive (Backlund et al. 2008).

Methane increases surface ozone which augments warming-induced losses and some quantitative analyses now include climate, long-lived (CO_2) and multiple short-lived pollutants (CH_4, O_3) simultaneously (Shindell et al. 2017; Shindell 2016). Reduction of tropospheric ozone and black carbon can avoid premature deaths from outdoor air pollution and increases annual crop yields (Shindell et al. 2012). These actions plus methane reduction can influence climate on shorter time scales than those of carbon dioxide reduction measures. Implementing them substantially reduces the risks of crossing the 2°C threshold and contributes to achievement of the SDGs (Haines et al. 2017; Shindell et al. 2017).

1.2.2 Climate change impacts on food availability

Climate change impacts food availability through its effect on the production of food and its storage, processing, distribution, and exchange.

1.2.2.1 Impacts on crop production

Observed impacts. Since AR5, there have been further studies that document impacts of climate change on crop production and related variables (Supplementary Material Table SM5.3). There have also been a few studies that demonstrate a strengthening relationship between observed climate variables and crop yields that indicate future expected warming will have severe impacts on crop production (Mavromatis 2015; Innes et al. 2015). At the global scale, Iizumi et al. (2018) used a counterfactual analysis and found that climate change between 1981 and 2010 has decreased global mean yields of maize, wheat, and soybeans by 4.1, 1.8 and 4.5%, respectively, relative to preindustrial climate, even when CO_2 fertilisation and agronomic adjustments are considered. Uncertainties (90% probability interval) in the yield impacts are −8.5 to +0.5% for maize, −7.5 to +4.3% for wheat, and −8.4 to −0.5% for soybeans. For rice, no significant impacts were detected. This study suggests that climate change has

modulated recent yields on the global scale and led to production losses, and that adaptations to date have not been sufficient to offset the negative impacts of climate change, particularly at lower latitudes.

Dryland settlements are perceived as vulnerable to climate change with regard to food security, particularly in developing countries; such areas are known to have low capacities to cope effectively with decreasing crop yields (Shah et al. 2008; Nellemann et al. 2009). This is of concern because drylands constitute over 40% of the earth's land area, and are home to 2.5 billion people (FAO et al. 2011).

Australia

In Australia, declines in rainfall and rising daily maximum temperatures based on simulations of 50 sites caused water-limited yield potential to decline by 27% from 1990 to 2015, even though elevated atmospheric CO_2 concentrations had a positive effect (Hochman et al. 2017). In New South Wales, high-temperature episodes during the reproduction stage of crop growth were found to have negative effects on wheat yields, with combinations of low rainfall and high temperatures being the most detrimental (Innes et al. 2015).

Asia

There are numerous studies demonstrating that climate change is affecting agriculture and food security in Asia. Several studies with remote sensing and statistical data have examined rice areas in north-eastern China, the northernmost region of rice cultivation, and found expansion over various time periods beginning in the 1980s, with most of the increase occurring after 2000 (Liu et al. 2014; Wang et al. 2014; Zhang et al. 2017). Rice yield increases have also been found over a similar period (Wang et al. 2014). Multiple factors, such as structural adjustment, scientific and technological progress, and government policies, along with regional warming (1.43°C in the past century) (Fenghua et al. 2006) have been put forward as contributing to the observed expanded rice areas and yield in the region. Shi et al. (2013) indicate that there is a partial match between climate change patterns and shifts in extent and location of the rice-cropping area (2000–2010).

There have also been documented changes in winter wheat phenology in Northwest China (He 2015). Consistent with this finding, dates of sowing and emergence of spring and winter wheat were delayed, dates of anthesis and maturity was advanced, and length of reproductive growth period was prolonged from 1981–2011 in a study looking at these crops across China (Liu et al. 2018b). Another study looking in Northwest China demonstrated that there have been changes in the phenology and productivity of spring cotton (Huang and Ji 2015). A counterfactual study looking at wheat growth and yield in different climate zones of China from 1981–2009 found that impacts were positive in northern China and negative in southern China (Tao et al. 2014). Temperature increased across the zones while precipitation changes were not consistent (Tao et al. 2014).

Similar crop yield studies focusing on India have found that warming has reduced wheat yields by 5.2% from 1981 to 2009, despite adaptation

5

(Gupta et al. 2017), and that maximum daytime temperatures have risen along with some night-time temperatures (Jha and Tripathi 2017).

Agriculture in Pakistan has also been affected by climate change. From 1980 to 2014, spring maize growing periods have shifted an average of 4.6 days per decade earlier, while sowing of autumn maize has been delayed 3.0 days per decade (Abbas et al. 2017). A similar study with sunflower showed that increases in mean temperature from 1980 to 2016 were highly correlated with shifts in sowing, emergence, anthesis, and maturity for fall and spring crops (Tariq et al. 2018).

Mountain people in the Hindu-Kush Himalayan region encompassing parts of Pakistan, India, Nepal, and China, are particularly vulnerable to food insecurity related to climate change because of poor infrastructure, limited access to global markets, physical isolation, low productivity, and hazard exposure, including Glacial Lake Outburst Floods (GLOFs) (Rasul et al. 2019; Rasul 2010; Tiwari and Joshi 2012; Huddleston et al. 2003; Ward et al. 2013; FAO 2008; Nautiyal et al. 2007; Din et al. 2014). Surveys have been conducted to determine how climate-related changes have affected food security (Hussain et al. 2016; Shrestha and Nepal 2016) with results showing that the region is experiencing an increase in extremes, with farmers facing more frequent floods as well as prolonged droughts with ensuing negative impacts on agricultural yields and increases in food insecurity (Hussain et al. 2016; Manzoor et al. 2013).

South America

In another mountainous region, the Andes, inhabitants are also beginning to experience changes in the timing, severity, and patterns of the annual weather cycle. Data collected through participatory workshops, semi-structured interviews with agronomists, and qualitative fieldwork from 2012 to 2014 suggest that in Colomi, Bolivia, climate change is affecting crop yields and causing farmers to alter the timing of planting, their soil management strategies, and the use and spatial distribution of crop varieties (Saxena et al. 2016). In Argentina, there has also been an increase in yield variability of

maize and soybeans (Iizumi and Ramankutty 2016). These changes have had important implications for the agriculture, human health, and biodiversity of the region (Saxena et al. 2016).

Africa

In recent years, yields of staple crops such as maize, wheat, sorghum, and fruit crops, such as mangoes, have decreased across Africa, widening food insecurity gaps (Ketiem et al. 2017). In Nigeria, there have been reports of climate change having impacts on the livelihoods of arable crop farmers (Abiona et al. 2016; Ifeanyi-obi et al. 2016; Onyeneke 2018). The Sahel region of Cameroon has experienced an increasing level of malnutrition. This is partly due to the impact of climate change since harsh climatic conditions leading to extreme drought have a negative influence on agriculture (Chabejong 2016).

Utilising farmer interviews in Abia State, Nigeria, researchers found that virtually all responders agreed that the climate was changing in their area (Ifeanyi-obi et al. 2016). With regard to management responses, a survey of farmers from Anambra State, Nigeria, showed that farmers are adapting to climate change by utilising such techniques as mixed cropping systems, crop rotation, and fertiliser application (Onyeneke et al. 2018). In Ebonyi State, Nigeria, Eze (2017) interviewed 160 women cassava farmers and found the major climate change risks in production to be severity of high temperature stress, variability in relative humidity, and flood frequency.

Europe

The impacts of climate change are varied across the continent. Moore and Lobell (2015) showed via counterfactual analysis that climate trends are affecting European crop yields, with long-term temperature and precipitation trends since 1989 reducing continent-wide wheat and barley yields by 2.5% and 3.8%, respectively, and having slightly increased maize and sugar beet yields. Though these aggregate affects appear small, the impacts are not evenly distributed. In cooler regions such as the United Kingdom and Ireland, the effect of increased warming has been ameliorated by an increase in rainfall. Warmer regions, such

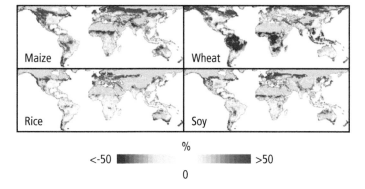

GGCMs with explicit N stress

Maize   Wheat

Rice   Soy

%

<-50 ■■■■■■ >50

0

Figure 5.4 | AgMIP median yield changes (%) for RCP8.5 (2070–2099 in comparison to 1980–2010 baseline) with CO_2 effects and explicit nitrogen stress over five GCMs x four Global Gridded Crop Models (GGCMs) for rainfed maize, wheat, rice, and soy (20 ensemble members from EPIC, GEPIC, pDSSAT, and PEGASUS; except for rice which has 15). Grey areas indicate historical areas with little to no yield capacity. All models use a 0.5°C grid, but there are differences in grid cells simulated to represent agricultural land. While some models simulated all land areas, others simulated only potential suitable cropland area according to evolving climatic conditions. Others utilised historical harvested areas in 2000 according to various data sources (Rosenzweig et al. 2014).

as Southern Europe, have suffered more from the warming; in Italy this effect has been amplified by a drying trend, leading to yield declines of 5% or greater.

Another study examining the impacts of recent climate trends on cereals in Greece showed that crops are clearly responding to changes in climate – and demonstrated (via statistical analysis) that significant impacts on wheat and barley production are expected at the end of the 21st century (Mavromatis 2015). In the Czech Republic, a study documented positive long-term impacts of recent warming on yields of fruiting vegetables (cucumbers and tomatoes) from 4.9 to 12% per 1°C increase in local temperature, but decreases in yield stability of traditionally grown root vegetables in the warmest areas of the country (Potopová et al. 2017). A study in Hungary also indicated the increasingly negative impacts of temperature on crops and indicated that a warming climate is at least partially responsible for the stagnation in crop yields since the mid-1980s in Eastern Europe (Pinke and Lövei 2017).

In summary, climate change is already affecting food security (*high confidence*). Recent studies in both large-scale and smallholder farming systems document declines in crop productivity related to rising temperatures and changes in precipitation. Evidence for climate change impacts (e.g., declines and stagnation in yields, changes in sowing and harvest dates, increased infestation of pests and diseases, and declining viability of some crop varieties) is emerging from detection and attribution studies and ILK in Australia, Europe, Asia, Africa, North America, and South America (*medium evidence, robust agreement*).

Projected impacts

Climate change effects have been studied on a global scale following a variety of methodologies that have recently been compared (Lobell and Asseng 2017; Zhao et al. 2017a and Liu et al. 2016). Approaches to study global and local changes include global gridded crop model simulations (e.g., Deryng et al. 2014), point-based crop model simulations (e.g., Asseng et al. 2015), analysis of point-based observations in the field (e.g., Zhao et al. 2016), and temperature-yield regression models (e.g., Auffhammer and Schlenker 2014). For an evaluation of model skills see example used in AgMIP (Müller et al. 2017b).

Results from Zhao et al. (2017a) across different methods consistently showed negative temperature impacts on crop yield at the global scale, generally underpinned by similar impacts at country and site scales. A limitation of Zhao et al. (2017a) is that it is based on the assumption that yield responses to temperature increase are linear, while yield response differs depending on growing season temperature levels. Iizumi et al. (2017) showed that the projected global mean yields of maize and soybean at the end of this century do decrease monotonically with warming, whereas those of rice and wheat increase with warming but level off at about 3°C (2091–2100 relative to 1850–1900).

Empirical statistical models have been applied widely to different cropping systems, at multiple scales. Analyses using statistical models for maize and wheat tested with global climate model scenarios found that the RCP4.5 scenario reduced the size of average yield impacts,

risk of major slowdowns, and exposure to critical heat extremes compared to RCP8.5 in the latter decades of the 21st century (Tebaldi and Lobell 2018). Impacts on crops grown in the tropics are projected to be more negative than in mid – to high-latitudes as stated in AR5 and confirmed by recent studies (e.g., Levis et al. 2018). These projected negative effects in the tropics are especially pronounced under conditions of explicit nitrogen stress (Rosenzweig et al. 2014) (Figure 5.4).

Reyer et al. (2017b) examined biophysical impacts in five world regions under different warming scenarios: 1°C, 1.5°C, 2°C, and 4°C warming. For the Middle East and northern African region a significant correlation between crop yield decrease and temperature increase was found, regardless of whether the effects of CO_2 fertilisation or adaptation measures are taken into account (Waha et al. 2017). For Latin America and the Caribbean the relationship between temperature and crop yield changes was only significant when the effect of CO_2 fertilisation is considered (Reyer et al. 2017a).

A review of recent scientific literature found that projected yield loss for West Africa depends on the degree of wetter or drier conditions and elevated CO_2 concentrations (Sultan and Gaetani 2016). Faye et al. (2018b) in a crop modelling study with RCPs 4.5 and 8.5 found that climate change could have limited effects on peanut yield in Senegal due to the effect of elevated CO_2 concentrations.

Crop productivity changes for 1.5°C and 2.0°C. The IPCC Special Report on global warming of 1.5°C found that climate-related risks to food security are projected to increase with global warming of 1.5°C and increase further with 2°C (IPCC 2018b). These findings are based among others on Schleussner et al. (2018); Rosenzweig et al. (2018a); Betts et al. (2018), Parkes et al. (2018) and Faye et al. (2018a). The importance of assumptions about CO_2 fertilisation was found to be significant by Ren et al. (2018) and Tebaldi and Lobell (2018).

AgMIP coordinated global and regional assessment (CGRA) results confirm that at the global scale, positive and negative changes are mixed in simulated wheat and maize yields, with declines in some breadbasket regions, at both 1.5°C and 2.0°C (Rosenzweig et al. 2018a). In conjunction with price changes from the global economics models, productivity declines in the Punjab, Pakistan resulted in an increase in vulnerable households and poverty rate (Rosenzweig et al. 2018a).

Crop suitability. Another method of assessing the effects of climate change on crop yields that combined observations of current maximum-attainable yield with climate analogues also found strong reductions in attainable yields across a large fraction of current cropland by 2050 (Pugh et al. 2016). However, the study found the projected total land area in 2050, including regions not currently used for crops, climatically suitable for a high attainable yield similar to today. This indicates that large shifts in land-use patterns and crop choice will likely be necessary to sustain production growth and keep pace with current trajectories of demand.

Fruits and vegetables. Understanding the full range of climate impacts on fruits and vegetables is important for projecting future food security, especially related to dietary diversity and healthy diets.

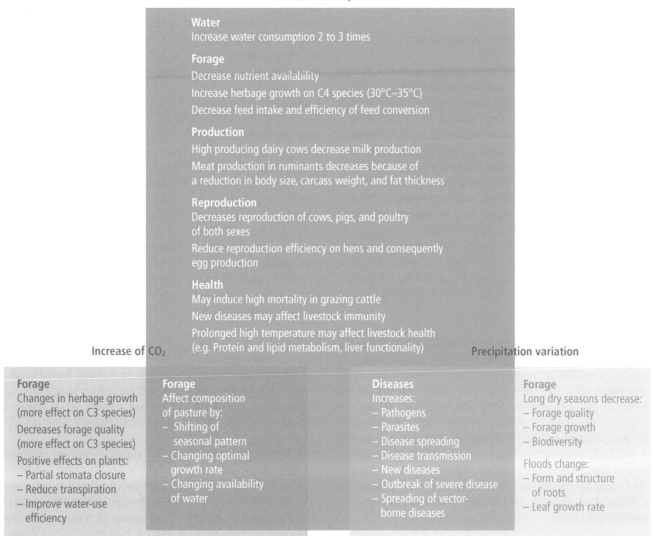

Figure 5.5 | **Impacts of climate change on livestock** (based on Rojas-Downing et al. 2017).

However, studies for vegetables are very limited (Bisbis et al. 2018). Of the 174 studies considered in a recent review, only 14 described results of field or greenhouse experiments studying impacts of increased temperatures on yields of different root and leafy vegetables, tomatoes and legumes (Scheelbeek et al. 2018). Bisbis et al. (2018) found similar effects for vegetables as have been found for grain crops. That is, the effect of increased CO_2 on vegetables is mostly beneficial for production, but may alter internal product quality, or result in photosynthetic down-regulation. Heat stress reduces fruit set of fruiting vegetables, and speeds up development of annual vegetables, shortening their time for photoassimilation. Yield losses and impaired product quality result, thereby increasing food loss and waste. On the other hand, a longer growing season due to warmer temperatures enables a greater number of plantings and can contribute to greater annual yields. However, some vegetables, such as cauliflower and asparagus, need a period of cold accumulation to produce a harvest and warmer winters may not provide those requirements.

For vegetables growing in higher baseline temperatures (>20°C), mean yield declines caused by 4°C warming were 31.5%; for vegetables growing in cooler environments (≤20°C), yield declines caused by 4°C were much less, on the order of about 5% (Scheelbeek et al. 2018). Rippke et al. (2016) found that 30–60% of the common bean growing area and 20–40% of the banana growing areas in Africa will lose viability in 2078–2098 with a global temperature increase of 2.6°C and 4°C respectively. Tripathi et al. (2016) found fruits and vegetable production to be highly vulnerable to climate change at their reproductive stages and also due to potential for greater disease pressure.

In summary, studies assessed find that climate change will increasingly be detrimental to crop productivity as levels of warming progress (*high confidence*). Impacts will vary depending on CO_2 concentrations, fertility levels, and region. Productivity of major commodity crops as well as crops such as millet and sorghum yields will be affected. Studies on fruits and vegetables find similar effects to those projected for grain crops in regard to temperature and CO_2 effects. Total land area climatically suitable for high attainable yield, including regions not currently used for crops, will be similar in 2050 to today.

1.2.2.2   Impacts on livestock production systems

Livestock systems are impacted by climate change mainly through increasing temperatures and precipitation variation, as well as atmospheric carbon dioxide (CO_2) concentration and a combination of these factors. Temperature affects most of the critical factors of livestock production, such as water availability, animal production and reproduction, and animal health (mostly through heat stress) (Figure 5.5). Livestock diseases are mostly affected by increases in temperature and precipitation variation (Rojas-Downing et al. 2017). Impacts of climate change on livestock productivity, particularly of mixed and extensive systems, are strongly linked to impacts on rangelands and pastures, which include the effects of increasing CO_2 on their biomass and nutritional quality. This is critical considering the very large areas concerned and the number of vulnerable people affected (Steinfeld 2010; Morton 2007). Pasture quality and quantity are mainly affected through increases in temperature and CO_2, and precipitation variation.

Among livestock systems, pastoral systems are particularly vulnerable to climate change (Dasgupta et al. 2014) (see Section 5.2.2.6 for impacts on smallholder systems that combine livestock and crops). Industrial systems will suffer most from indirect impacts leading to rises in the costs of water, feeding, housing, transport and the destruction of infrastructure due to extreme events, as well as an increasing volatility of the price of feedstuff which increases the level of uncertainty in production (Rivera-Ferre et al. 2016b; Lopez-i-Gelats 2014). Mixed systems and industrial or landless livestock systems could encounter several risk factors mainly due to the variability of grain availability and cost, and low adaptability of animal genotypes (Nardone et al. 2010).

Considering the diverse typologies of animal production, from grazing to industrial, Rivera-Ferre et al. (2016b) distinguished impacts of climate change on livestock between those related to extreme events and those related to more gradual changes in the average of climate-related variables. Considering vulnerabilities, they grouped the impacts as those impacting the animal directly, such as heat and cold stress, water stress, physical damage during extremes; and others impacting their environment, such as modification in the geographical distribution of vector-borne diseases, location, quality and quantity of feed and water and destruction of livestock farming infrastructures.

With severe negative impacts due to drought and high frequency of extreme events, the average gain of productivity might be cancelled by the volatility induced by increasing variability in the weather. For instance, semi-arid and arid pasture will likely have reduced livestock productivity, while nutritional quality will be affected by CO_2 fertilisation (Schmidhuber and Tubiello 2007).

Observed impacts. Pastoralism is practiced in more than 75% of countries by between 200 and 500 million people, including nomadic communities, transhumant herders, and agropastoralists (McGahey et al. 2014). Observed impacts in pastoral systems reported in

the literature include decreasing rangelands, decreasing mobility, decreasing livestock numbers, poor animal health, overgrazing, land degradation, decreasing productivity, decreasing access to water and feed, and increasing conflicts for the access to pasture land (*high confidence*) (López-i-Gelats et al. 2016; Batima et al. 2008; Njiru 2012; Fjelde and von Uexkull 2012; Raleigh and Kniveton 2012; Egeru 2016).

Pastoral systems in different regions have been affected differently. For instance, in China changes in precipitation were a more important factor in nomadic migration than temperature (Pei and Zhang 2014). There is some evidence that recent years have already seen an increase in grassland fires in parts of China and tropical Asia (IPCC 2012). In Mongolia, grassland productivity has declined by 20–30% over the latter half of the 20th century, and ewe average weight reduced by 4 kg on an annual basis, or about 8% since 1980 (Batima et al. 2008). Substantial decline in cattle herd sizes can be due to increased mortality and forced off-take (Megersa et al. 2014). Important, but less studied, is the impact of the interaction of grazing patterns with climate change on grassland composition. Spence et al. (2014) showed that climate change effects on Mongolia mountain steppe could be contingent on land use.

Conflicts due to resource scarcity, as well as other socio-political factors (Benjaminsen et al. 2012) aggravated by climate change, has differentiated impact on women. In Turkana, female-headed households have lower access to decision-making on resource use and allocation, investment and planning (Omolo 2011), increasing their vulnerability (Cross-Chapter Box 11 in Chapter 7, Section 5.1.3).

Non-climate drivers add vulnerability of pastoral systems to climate change (McKune and Silva 2013). For instance, during environmental disasters, livestock holders have been shown to be more vulnerable to food insecurity than their crop-producing counterparts because of limited economic access to food and unfavourable market exchange rates (Nori et al. 2005). Sami reindeer herders in Finland showed reduced freedom of action in response to climate change due to loss of habitat, increased predation, and presence of economic and legal constraints (Tyler et al. 2007; Pape and Löffler 2012). In Tibet, emergency aid has provided shelters and privatised communally owned rangeland, which have increased the vulnerability of pastoralists to climate change (Yeh et al. 2014; Næss 2013).

Projected impacts. The impacts of climate change on global rangelands and livestock have received comparatively less attention than the impacts on crop production. Projected impacts on grazing systems include changes in herbage growth (due to changes in atmospheric CO_2 concentrations and rainfall and temperature regimes) and changes in the composition of pastures and in herbage quality, as well as direct impacts on livestock (Herrero et al. 2016b). Droughts and high temperatures in grasslands can also be a predisposing factor for fire occurrence (IPCC 2012).

Net primary productivity, soil organic carbon, and length of growing period. There are large uncertainties related to grasslands and grazing lands (Erb et al. 2016), especially in regard to net primary productivity (NPP) (Fetzel et al. 2017; Chen et al. 2018). Boone et al. (2017) estimated that the mean global annual net primary production

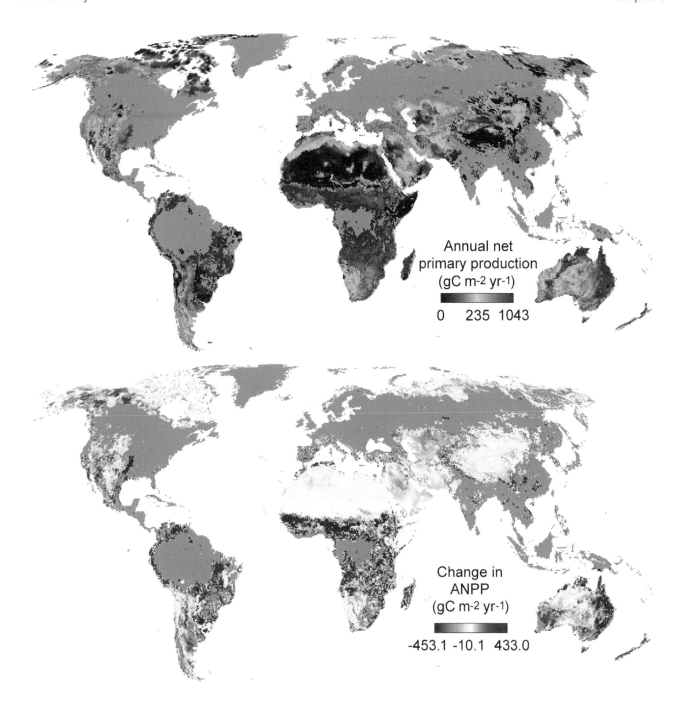

Figure 5.6 | Ensemble simulation results for projected annual net primary productivity of rangelands as simulated in 2000 (top) and their change in 2050 (bottom) under emissions scenario RCP 8.5, with plant responses enhanced by CO$_2$ fertilisation. Results from RCP 4.5 and 8.5, with and without positive effects of atmospheric CO$_2$ on plant production, differed considerably in magnitude but had similar spatial patterns, and so results from RCP 8.5 with increasing production are portrayed spatially here and in other figures. Scale bar labels and the stretch applied to colours are based on the spatial mean value plus or minus two standard deviations (Boone et al. 2017).

(NPP) in rangelands may decline by 10 gC m^{-2} yr^{-1} in 2050 under RCP8.5, but herbaceous NPP is likely to increase slightly (i.e., average of 3 gC m^{-2} yr^{-1}) (Figure 5.6). Results of a similar magnitude were obtained by Havlík et al. (2015), using EPIC and LPJmL on a global basis. According to Rojas-Downing et al. (2017), an increase of 2°C is estimated to negatively impact pasture and livestock production in arid and semi-arid regions and positively impact humid temperate regions.

Boone et al. (2017) identified significant regional heterogeneity in responses, with large increases in annual productivity projected in northern regions (e.g., a 21% increase in productivity in the USA and Canada) and large declines in western Africa (–46% in Sub-Saharan western Africa) and Australia (–17%). Regarding the length of growing period (LGP, average number of growing days per year) Herrero et al. (2016b) projected reductions in lower latitudes due to changes in rainfall patterns and increases in temperatures, which indicate increasing limitations of water. They identified 35°C as a critical threshold for rangeland vegetation and heat tolerance in some livestock species.

Rangeland composition. According to Boone et al. (2017), the composition of rangelands is projected to change as well (Chapter 3). Bare ground cover is projected to increase, averaging 2.4% across rangelands, with increases projected for the eastern Great Plains, eastern Australia, parts of southern Africa, and the southern Tibetan Plateau. Herbaceous cover declines are projected in the Tibetan Plateau, the eastern Great Plains, and scattered parts of the Southern Hemisphere. Shrub cover is projected to decline in eastern Australia, parts of southern Africa, the Middle East, the Tibetan Plateau, and the eastern Great Plains. Shrub cover could also increase in much of the Arctic and some parts of Africa. In mesic and semi-arid savannas south of the Sahara, both shrub and tree cover are projected to increase, albeit at lower productivity and standing biomass. Rangelands in western and south-western parts of the Isfahan province in Iran were found to be more vulnerable to future drying–warming conditions (Saki et al. 2018; Jaberalansar et al. 2017).

Soil degradation and expanding woody cover suggest that climate-vegetation-soil feedbacks catalysing shifts toward less productive, possibly stable states (Ravi et al. 2010) may threaten mesic and semi-arid savannas south of the Sahara (Chapters 3 and 4). This will also change their suitability for grazing different animal species; switches from cattle, which mainly consume herbaceous plants, to goats or camels are likely to occur as increases in shrubland occur.

Direct and indirect effects on livestock. Direct impacts of climate change in mixed and extensive production systems are linked to increased water and temperature stress on the animals potentially leading to animal morbidity, mortality and distress sales. Most livestock species have comfort zones between 10°C–30°C, and at temperatures above this animals reduce their feed intake 3–5% per additional degree of temperature (NRC 1981). In addition to reducing animal production, higher temperatures negatively affect fertility (HLPE 2012).

Indirect impacts to mixed and extensive systems are mostly related to the impacts on the feed base, whether pastures or crops, leading to increased variability and sometimes reductions in availability and quality of the feed for the animals (Rivera-Ferre et al. 2016b). Reduced forage quality can increase CH_4 emissions per unit of gross energy consumed. Increased risk of animal diseases is also an important impact to all production systems (Bett et al. 2017). These depend on the geographical region, land-use type, disease characteristics, and animal susceptibility (Thornton et al. 2009). Also important is the interaction of grazing intensity with climate change. Pfeiffer et al. (2019) estimated that, in a scenario of mean annual precipitation below 500 mm, increasing grazing intensity reduced rangeland productivity and increased annual grass abundance.

Pastoral systems. In Kenya, some 1.8 million extra cattle could be lost by 2030 because of increased drought frequency, the value of the lost animals and production foregone amounting to 630 million USD (Herrero et al. 2010). Martin et al. (2014) assessed impacts of changing precipitation regimes to identify limits of tolerance beyond which pastoral livelihoods could not be secured and found that reduced mean annual precipitation always had negative effects as opposed to increased rainfall variability. Similarly, Martin et al. (2016) found that drought effects on pastoralists in High Atlas in Morocco depended on income needs and mobility options (see Section 5.2.2.6 for additional information about impacts on smallholder farmers).

In summary, observed impacts in pastoral systems include changes in pasture productivity, lower animal growth rates and productivity, damaged reproductive functions, increased pests and diseases, and loss of biodiversity (*high confidence*). Livestock systems are projected to be adversely affected by rising temperatures, depending on the extent of changes in pasture and feed quality, spread of diseases, and water resource availability (*high confidence*). Impacts will differ for different livestock systems and for different regions (*high confidence*). Vulnerability of pastoral systems to climate change is very high (*high confidence*), and mixed systems and industrial or landless livestock systems could encounter several risk factors mainly due to variability of grain availability and cost, and low adaptability of animal genotypes. Pastoral system vulnerability is exacerbated by non-climate factors (land tenure issues, sedentarisation programmes, changes in traditional institutions, invasive species, lack of markets, and conflicts) (*high confidence*).

1.2.2.3   Impacts on pests and diseases

Climate change is changing the dynamics of pests and diseases of both crops and livestock. The nature and magnitude of future changes is likely to depend on local agroecological and management context. This is because of the many biological and ecological mechanisms by which climate change can affect the distribution, population size, and impacts of pests and diseases on food production (Canto et al. 2009; Gale et al. 2009; Thomson et al. 2010; Pangga et al. 2011; Juroszek and von Tiedemann 2013; Bett et al. 2017).

These mechanisms include changes in host susceptibility due to CO_2 concentration effects on crop composition and climate stresses; changes in the biology of pests and diseases or their vectors (e.g., more generational cycles, changes in selection pressure driving evolution); mismatches in timing between pests or vectors and their 'natural enemies'; changes in survival or persistence of pests or disease pathogens (e.g., changes in crop architecture driven by CO_2 fertilisation and increased temperature, providing a more favourable environment for persistence of pathogens like fungi), and changes in pest distributions as their 'climate envelopes' shift. Such processes may affect pathogens, and their vectors, as well as plant, invertebrate and vertebrate pests (Latham et al. 2015).

Furthermore, changes in diseases and their management, as well as changing habitat suitability for pests and diseases in the matrix surrounding agricultural fields, have the ability to reduce or exacerbate impacts (Bebber 2015). For example, changes in water storage and irrigation to adapt to rainfall variation have the potential to enhance disease vector populations and disease occurrence (Bett et al. 2017).

There is *robust evidence* that pests and diseases have already responded to climate change (Bebber et al. 2013), and many studies have now built predictive models based on current incidence of pests, diseases or vectors that indicate how they may respond in future (e.g., Caminade et al. 2015; Kim et al. 2015; Kim and Cho 2016; Samy and Peterson 2016; Yan et al. 2017). Warren et al. (2018) estimate that about 50% of insects, which are often pests or disease vectors, will change ranges by about 50% by 2100 under current GHG emissions trajectories. These changes will lead to crop losses due to changes in insect pests (Deutsch et al. 2018) and weed pressure (Ziska et al. 2018), and thus affect pest and disease management at the farm level (Waryszak et al. 2018). For example, Samy and Peterson (2016) modelled bluetongue virus (BTV), which is spread by biting *Culicodes* midges, finding that the distribution of BTV is likely to be extended, particularly in Central Africa, the USA, and Western Russia.

There is some evidence (*medium confidence*) that exposure will, on average, increase (Bebber and Gurr 2015; Yan et al. 2017), although there are a few examples where changing stresses may limit the range of a vector. There is also a general expectation that perturbations may increase the likelihood of pest and disease outbreaks by disturbing processes that may currently be at some quasi-equilibrium (Canto et al. 2009; Thomson et al. 2010; Pangga et al. 2011). However, in some places, and for some diseases, risks may decrease as well as increase (e.g., drying out may reduce the ability of fungi to survive) (Kim et al. 2015; Skelsey and Newton 2015), or tsetse fly's range may decrease (Terblanche et al. 2008; Thornton et al. 2009).

Pests, diseases, and vectors for both crop and livestock diseases are likely to be altered by climate change (*high confidence*). Such changes are likely to depend on specifics of the local context, including management, but perturbed agroecosystems are more likely, on theoretical grounds, to be subject to pest and disease outbreaks (*low confidence*). Whilst specific changes in pest and disease pressure will vary with geography, farming system, pest/pathogen – increasing in some situations decreasing in others – there is robust evidence, with *high agreement*, that pest and disease pressures are likely to change; such uncertainty requires robust strategies for pest and disease mitigation.

1.2.2.4   Impacts on pollinators

Pollinators play a key role on food security globally (Garibaldi et al. 2016). Pollinator-dependent crops contribute up to 35% of global crop production volume and are important contributors to healthy human diets and nutrition (IPBES 2016). On a global basis, some 1500 crops require pollination (typically by insects, birds and bats) (Klein et al. 2007). Their importance to nutritional security is therefore perhaps under-rated by valuation methodologies, which, nonetheless, include estimates of the global value of pollination services at over 225 billion USD2010 (Hanley et al. 2015). As with other ecosystem processes affected by climate change (e.g., changes in pests and diseases), how complex systems respond is highly context dependent. Thus, predicting the effects of climate on pollination services is difficult (Tylianakis et al. 2008; Schweiger et al. 2010) and uncertain, although there is *limited evidence* that impacts are occurring already (Section 5.2.2.4), and *medium evidence* that there will be an effect.

Pollination services arise from a mutualistic interaction between an animal and a plant – which can be disrupted by climate's impacts on one or the other or both (Memmott et al. 2007). Disruption can occur through changes in species' ranges or by changes in timing of growth stages (Settele et al. 2016). For example, if plant development responds to different cues (e.g., day length) from insects (e.g., temperature), the emergence of insects may not match the flowering times of the plants, causing a reduction in pollination. Climate change will affect pollinator ranges depending on species, life-history, dispersal ability and location. Warren et al. (2018) estimate that under a 3.2°C warming scenario, the existing range of about 49% of insects will be reduced by half by 2100, suggesting either significant range changes (if dispersal occurs) or extinctions (if it does not). However, in principle, ecosystem changes caused by invasions, in some cases, could compensate for the decoupling generated between native pollinators and pollinated species (Schweiger et al. 2010).

Other impacts include changes in distribution and virulence of pathogens affecting pollinators, such as the fungus *Nosema cerana*, which can develop at a higher temperature range than the less-virulent *Nosema apis*; increased mortality of pollinators due to higher frequency of extreme weather events; food shortage for pollinators due to reduction of flowering length and intensity; and aggravation of other threats, such as habitat loss and fragmentation (González-Varo et al. 2013; Goulson et al. 2015; Le Conte and Navajas 2008; Menzel et al. 2006; Walther et al. 2009; IPBES, 2016). The increase in atmospheric CO_2 is also reducing the protein content of pollen, with potential impact on pollination population biology (Ziska et al. 2016).

In summary, as with other complex agroecosystem processes affected by climate change (e.g., changes in pests and diseases), how pollination services respond will be highly context dependent. Thus, predicting the effects of climate on pollination services is difficult and uncertain, although there is *medium evidence* that there will be an effect.

1.2.2.5   Impacts on aquaculture

This report focuses on land-based aquaculture; for assessment of impacts on marine fisheries both natural and farmed see the IPCC Special Report on the ocean and cryosphere in a changing climate (SROCC).

Aquaculture will be affected by both direct and indirect climate change drivers, both in the short and the long-term. Barange et al. (2018) provides some examples of short-term loss of production or infrastructure due to extreme events such as floods, increased risk of diseases, toxic algae and parasites; and decreased productivity due to suboptimal farming conditions. Long-term impacts may include scarcity of wild seed, limited access to freshwater for farming due to reduced precipitation, limited access to feeds from marine and terrestrial sources, decreased productivity due to suboptimal farming conditions, eutrophication and other perturbations.

FAO (2014a) assessed the vulnerability of aquaculture stakeholders to non-climate change drivers, which add to climate change hazards. Vulnerability arises from discrimination in access to inputs and decision-making; conflicts; infrastructure damage; and dependence on global markets and international pressures. Other non-climate drivers identified by McClanahan et al. (2015) include: declining fishery resources; a North–South divide in investment; changing consumption patterns; increasing reliance on fishery resources for coastal communities; and inescapable poverty traps created by low net resource productivity and few alternatives. In areas where vulnerability to climate change is heightened, increased exposure to climate change variables and impacts is likely to exacerbate current inequalities in the societies concerned, penalising further already disadvantaged groups such as migrant fishers (e.g., Lake Chad) or women (e.g., employees in Chile's processing industry) (FAO 2014a).

In many countries the projected declines co-occur across both marine fisheries and agricultural crops (Blanchard et al. 2017), both of which will impact the aquaculture and livestock sectors (Supplementary Material Figure SM5.1). Countries with low Human Development Index, trade opportunities and aquaculture technologies are likely to face greater challenges. These cross-sectoral impacts point to the need for a more holistic account of the inter-connected vulnerabilities of food systems to climate and global change.

1.2.2.6 Impacts on smallholder farming systems

New work has developed farming system approaches that take into account both biophysical and economic processes affected by climate change and multiple activities. Farm households in the developing world often rely on a complex mix of crops, livestock, aquaculture, and non-agricultural activities for their livelihoods (Rosenzweig and Hillel 2015; Antle et al. 2015). Across the world, smallholder farmers are considered to be disproportionately vulnerable to climate change because changes in temperature, rainfall and the frequency or intensity of extreme weather events directly affect their crop and animal productivity as well as their household's food security, income and well-being (Vignola et al. 2015; Harvey et al. 2014b). For example, smallholder farmers in the Philippines, whose survival and livelihood largely depend on the environment, constantly face risks and bear the impacts of the changing climate (Peria et al. 2016).

Smallholder farming systems have been recognised as highly vulnerable to climate change (Morton, 2007) because they are highly dependent on agriculture and livestock for their livelihood (*high confidence*) (Dasgupta et al. 2014). In Zimbabwe, farmers were found vulnerable due to their marginal location, low levels of technology, and lack of other essential farming resources. Farmers observed high frequency and severity of drought; excessive precipitation; drying of rivers, dams and wells; and changes in timing and pattern of seasons as evidence of climate change, and indicated that prolonged wet, hot, and dry weather conditions resulted in crop damage, death of livestock, soil erosion, bush fires, poor plant germination, pests, lower incomes, and deterioration of infrastructure (Mutekwa 2009).

In Madagascar, Harvey et al. (2014b) surveyed 600 small farmers and found that chronic food insecurity, physical isolation and lack of access to formal safety nets increased Malagasy farmers' vulnerability to any shocks to their agricultural system, particularly extreme events. In Chitwan, Nepal, occurrence of extreme events and

increased variability in temperature has increased the vulnerability of crops to biotic and abiotic stresses and altered the timing of agricultural operations; thereby affecting crop production (Paudel et al. 2014). In Lesotho, a study on subsistence farming found that food crops were the most vulnerable to weather, followed by soil and livestock. Climate variables of major concern were hail, drought and dry spells which reduced crop yields. In the Peruvian Altiplano, Sietz et al. (2012) evaluated smallholders' vulnerability to weather extremes with regard to food security and found that resource scarcity (livestock, land area), diversification of activities (lack of alternative income, education deprivation) and income restrictions (harvest failure risk) shaped the vulnerability of smallholders. See Section 5.2.2.2 for observed impacts on smallholder pastoral systems.

Projected impacts. By including regional economic models, integrated methods take into account the potential for yield declines to raise prices and thus livelihoods (up to a certain point) in some climate change scenarios. Regional economic models of farming systems can be used to examine the potential for switching to other crops and livestock, as well as the role that non-farm income can play in adaptation (Valdivia et al. 2015 Antle et al. 2015). On the other hand, lost income for smallholders from climate change-related declines (for example, in coffee production), can decrease their food security (Hannah et al. 2017).

Farming system methods developed by AgMIP (Rosenzweig et al. 2013) have been used in regional integrated assessments in Sub-Saharan Africa (Kihara et al. 2015), West Africa (Adiku et al. 2015); East Africa (Rao et al. 2015), South Africa (Beletse et al. 2015), Zimbabwe (Masikati et al. 2015), South Asia (McDermid et al. 2015), Pakistan (Ahmad et al. 2015), the Indo-Gangetic Basin (Subash et al. 2015), Tamil Nadu (Ponnusamy et al. 2015) and Sri Lanka (Zubair et al. 2015). The assessments found that climate change adds pressure to smallholder farmers across Sub-Saharan Africa and South Asia, with winners and losers within each area studied. Temperatures are expected to increase in all locations, and rainfall decreases are projected for the western portion of West Africa and southern Africa, while increases in rainfall are projected for eastern West Africa and all studied regions of South Asia. The studies project that climate change will lead to yield decreases in most study regions except South India and areas in central Kenya, as detrimental temperature effects overcome the positive effects of CO_2.

These studies use AgMIP representative agricultural pathways (RAPs) as a way to involve stakeholders in regional planning and climate resilience (Valdivia et al. 2015). RAPs are consistent with and complement the RCP/SSP approaches for use in agricultural model intercomparisons, improvement, and impact assessments.

New methods have been developed for improving analysis of climate change impacts and adaptation options for the livestock component of smallholder farming systems in Zimbabwe (Descheemaeker et al. 2018). These methods utilised disaggregated climate scenarios, as well as differentiating farms with larger stocking rates compared to less densely stocked farms. By disaggregating climate scenarios, impacts, and smallholder farmer attributes, such assessments can more effectively inform decision-making towards climate change adaptation.

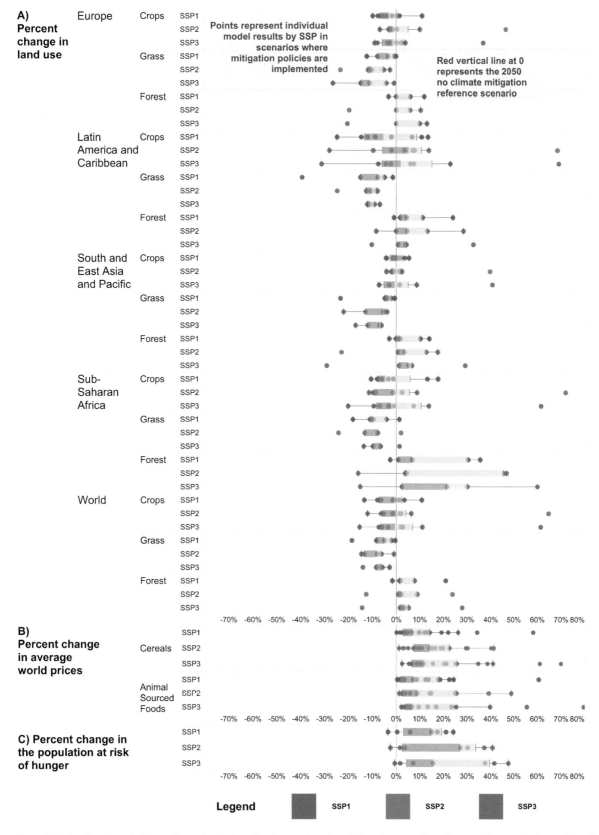

A) Percent change in land use

B) Percent change in average world prices

C) Percent change in the population at risk of hunger

Figure 5.7 | Implications of climate change by 2050 on land-use, selected agricultural commodity prices, and the population at risk of hunger based on AgMIP Global Economic Model analysis. (A) Projected % change in land-use by 2050 by land type (cropland, grassland, and forest) and SSP. (B) Projected % changes in average world prices by 2050 for cereals (rice, wheat, and coarse grains) and animal sourced foods (ruminant meat, monogastric, and dairy) by SSP. (C) Percentage change by 2050 in the global population at risk of hunger by SSP. (Hasegawa et al. 2018).

In Central Asia, a study using the bio-economic farm model (BEFM) found large differences in projected climate change impact ranging from positive income gains in large-scale commercial farms in contrast to negative impacts in small-scale farms (Bobojonov and Aw-Hassan 2014). Negative impacts may be exacerbated if irrigation water availability declines due to climate change and increased water demand in upstream regions. In Iran, changes in rainfall and water endowments are projected to significantly impact crop yield and water requirements, as well as income and welfare of farm families (Karimi et al. 2018).

Climate change impacts on food, feed and cash crops other than cereals, often grown in smallholder systems or family farms are less often studied, although impacts can be substantial. For example, areas suitable for growing coffee are expected to decrease by 21% in Ethiopia with global warming of 2.4°C (Moat et al. 2017) and more than 90% in Nicaragua (Läderach et al. 2017) with 2.2°C local temperature increase.

Climate change can modify the relationship between crops and livestock in the landscape, affecting mixed crop-livestock systems in many places. Where crop production will become marginal, livestock may provide an alternative to cropping. Such transitions could occur in up to 3% of the total area of Africa, largely as a result of increases in the probability of season failure in the drier mixed crop–livestock systems of the continent (Thornton et al. 2014).

In Mexico, subsistence agriculture is expected to be the most vulnerable to climate change, due to its intermittent production and reliance on maize and beans (Monterroso et al. 2014). Overall, a decrease in suitability and yield is expected in Mexico and Central America for beans, coffee, maize, plantain and rice (Donatti et al. 2018). Municipalities with a high proportional area under subsistence crops in Central America tend to have less resources to promote innovation and action for adaptation (Bouroncle et al. 2017).

In summary, smallholder farmers are especially vulnerable to climate change because their livelihoods often depend primarily on agriculture. Further, smallholder farmers often suffer from chronic food insecurity (*high confidence*). Climate change is projected to exacerbate risks of pests and diseases and extreme weather events in smallholder farming systems.

1.2.3   Climate change impacts on access

Access to food involves the ability to obtain food, including the ability to purchase food at affordable prices.

1.2.3.1   Impacts on prices and risk of hunger

A protocol-based analysis based on AgMIP methods tested a combination of RCPs and SSPs to provide a range of projections for prices, risk of hunger, and land-use change (Hasegawa et al. 2018) (Figure 5.7 and Supplementary Material Table SM5.4.). Previous studies have found that decreased agricultural productivity will depress agricultural supply, leading to price increases. Despite different economic models with various representations of the global food system (Valin et al. 2014; Robinson et al. 2014; Nelson et al. 2013; Schmitz et al. 2014), as well as having represented the SSPs in different ways, for example, technological change, land-use policies, and sustainable diets (Stehfest et al. 2019; Hasegawa et al. 2018), the ensemble of participating models projected a 1–29% cereal price increase in 2050 across SSPs 1, 2 and 3 due to climate change (RCP 6.0). This would impact consumers globally through higher food prices, though regional effects will vary. The median cereal price increase was 7%, given current projections of demand. In all cases (across SSPs and global economic models), prices are projected to increase for rice and coarse grains, with only one instance of a price decline (–1%) observed for wheat in SSP1, with price increases projected in all other cases. Animal-sourced foods (ASFs) are also projected to see price increases (1%), but the range of projected price changes are about half those of cereals, highlighting that the climate impacts on ASFs will be felt indirectly, through the cost and availability of feed, and that there is significant scope for feed substitution within the livestock sector.

Declining food availability caused by climate change is likely to lead to increasing food cost impacting consumers globally through higher prices and reduced purchasing power, with low-income consumers particularly at risk from higher food prices (Nelson et al. 2010; Springmann et al. 2016a and Nelson et al. 2018). Higher prices depress consumer demand, which in turn will not only reduce energy intake (calories) globally (Hasegawa et al. 2015; Nelson et al. 2010; Springmann et al. 2016a and Hasegawa et al. 2018), but will also likely lead to less healthy diets with lower availability of key micronutrients (Nelson et al. 2018) and increase diet-related mortality in lower and middle-income countries (Springmann et al. 2016a). These changes will slow progress towards the eradication of malnutrition in all its forms.

The extent that reduced energy intake leads to a heightened risk of hunger varies by global economic model. However, all models project an increase in the risk of hunger, with the median projection of an increase in the population at risk of insufficient energy intake by 6, 14, and 12% in 2050 for SSPs 1, 2 and 3 respectively compared to a no climate change reference scenario. This median percentage increase would be the equivalent of 8, 24 and 80 million (full range 1–183 million) additional people at risk of hunger due to climate change (Hasegawa et al. 2018).

1.2.3.2   Impacts on land use

Climate change is likely to lead to changes in land use globally (Nelson et al. 2014; Schmitz et al. 2014 and Wiebe et al. 2015). Hasegawa et al. (2018) found that declining agricultural productivity broadly leads to the need for additional cropland, with 7 of 8 models projecting increasing cropland and the median increase by 2050 projected across all models of 2% compared to a no climate change reference (Figure 5.7). Not all regions will respond to climate impacts equally, with more uncertainty on regional land-use change across the model ensemble than the global totals might suggest. For example, the median land-use change for Latin America is an increase of cropland by 3%, but the range across the model ensemble is significant, with three models projecting declines in cropland (–25 to –1%) compared to the five models projecting cropland increase (0–5%). For further discussion on land-use change and food security see Section 5.6.

1.2.4   Climate change impacts on food utilisation

Food utilisation involves nutrient composition of food, its preparation, and overall state of health. Food safety and quality affects food utilisation.

1.2.4.1   Impacts on food safety and human health

Climate change can influence food safety through changing the population dynamics of contaminating organisms due to, for example, changes in temperature and precipitation patterns, humidity, increased frequency and intensity of extreme weather events, and changes in contaminant transport pathways. Changes in food and farming systems, for example, intensification to maintain supply under climate change, may also increase vulnerabilities as the climate changes (Tirado et al. 2010).

Climate-related changes in the biology of contaminating organisms include changing the activity of mycotoxin-producing fungi, changing the activity of microorganisms in aquatic food chains that cause disease (e.g., dinoflagellates, bacteria like *Vibrio*), and increasingly heavy rainfall and floods causing contamination of pastures with enteric microbes (like *Salmonella*) that can enter the human food chain. Degradation and spoilage of products in storage and transport can also be affected by changing humidity and temperature outside of cold chains, notably from microbial decay but also from potential changes in the population dynamics of stored product pests (e.g., mites, beetles, moths) (Moses et al. 2015).

Mycotoxin-producing fungi occur in specific conditions of temperature and humidity, so climate change will affect their range, increasing risks in some areas (such as mid-temperate latitudes) and reducing them in others (e.g., the tropics) (Paterson and Lima 2010). There is *robust evidence* from process-based models of particular species (*Aspergillus*/Aflatoxin B1, *Fusarium*/deoxynivalenol), which include projections of future climate that show that aflatoxin contamination of maize in Southern Europe will increase significantly (Battilani et al. 2016), and deoxynivalenol contamination of wheat in Northwestern Europe will increase by up to three times current levels (van der Fels-Klerx et al. 2012b, a).

Whilst downscaled climate models make any specific projection for a given geography uncertain (Van der Fels-Klerx et al. 2013), experimental evidence on the small scale suggests that the combination of rising CO_2 levels, affecting physiological processes in photosynthetic organisms, and temperature changes, can be significantly greater than temperature alone (Medina et al. 2014). Risks related to aflatoxins are likely to change, but detailed projections are difficult because they depend on local conditions (Vaughan et al. 2016).

Foodborne pathogens in the terrestrial environment typically come from enteric contamination (from humans or animals), and can be spread by wind (blowing contaminated soil) or flooding – the incidence of both of which are likely to increase with climate change (Hellberg and Chu 2016). Furthermore, water stored for irrigation, which may be increased in some regions as an adaptation strategy, can become an important route for the spread of pathogens (as well as other pollutants). Contaminated water and diarrheal diseases are acute threats to food security (Bond et al. 2018). Whilst there is little direct evidence (in terms of modelled projections) the results of a range of reviews, as well as expert groups, suggest that risks from foodborne pathogens are likely to increase through multiple mechanisms (Tirado et al. 2010; van der Spiegel et al. 2012; Liu et al. 2013; Kirezieva et al. 2015; Hellberg and Chu 2016).

An additional route to climate change impacts on human health can arise from the changing biology of plants altering human exposure levels. This may include climate changing how crops sequester heavy metals (Rajkumar et al. 2013), or how they respond to changing pest pressure (e.g., cassava produces hydrogen cyanide as a defence against herbivore attack).

All of these factors will lead to regional differences regarding food safety impacts (Paterson and Lima 2011). For instance, in Europe it is expected that most important food safety-related impacts will be mycotoxins formed on plant products in the field or during storage; residues of pesticides in plant products affected by changes in pest pressure; trace elements and/or heavy metals in plant products depending on changes in abundance and availability in soils; polycyclic aromatic hydrocarbons in foods following changes in long-range atmospheric transport and deposition; and presence of pathogenic bacteria in foods following more frequent extreme weather, such as flooding and heat waves (Miraglia et al. 2009).

In summary, there is *medium evidence*, with *high agreement* that food utilisation via changes in food safety (and potentially food access from food loss) will be impacted by climate change, mostly by increasing risks, but there is *low confidence*, exactly how they may change for any given place.

1.2.4.2   Impacts on food quality

There are two main routes by which food quality may change. First, the direct effects of climate change on plant and animal biology, such as through changing temperatures changing the basic metabolism of plants. Secondly, by increasing carbon dioxide's effect on biology through CO_2 fertilisation.

Direct effects on plant and animal biology. Climate affects a range of biological processes, including the metabolic rate in plants and ectothermic animals. Changing these processes can change growth rates, and therefore yields, but can also cause organisms to change relative investments in growth vs reproduction, and therefore change the nutrients assimilated. This may decrease protein and mineral nutrient concentrations, as well as alter lipid composition (DaMatta et al. 2010). For example, apples in Japan have been exposed to higher temperatures over 3–4 decades and have responded by blooming earlier. This has led to changes in acidity, firmness, and water content, reducing quality (Sugiura et al. 2013). In other fruit, such as grapes, warming-induced changes in sugar composition affect both colour and aroma (Mira de Orduña 2010). Changing heat stress in poultry can affect yield as well as meat quality (by altering fat deposition and chemical constituents), shell quality of eggs, and immune systems (Lara and Rostagno 2013).

Effects of rising CO_2 concentrations. Climate change is being driven by rising concentrations of carbon dioxide and other GHG's in the atmosphere. As plants use CO_2 in photosynthesis to form sugar, rising CO_2 levels, all things being equal, enhances the process unless limited by water or nitrogen availability. This is known as 'CO_2 fertilisation'. Furthermore, increasing CO_2 allows stomata to partially close during gas exchange, reducing water loss through transpiration. These two factors affect the metabolism of plants, and, as with changing temperatures, affects plant growth rates, yields and their nutritional quality. Studies of these effects include meta-analyses, modelling, and small-scale experiments (Franzaring et al. 2013; Mishra and Agrawal 2014; Myers et al. 2014; Ishigooka et al. 2017; Zhu et al. 2018; Loladze 2014 and Yu et al. 2014).

With regard to nutrient quality, a meta-analysis from seven Free-Air Carbon dioxide Enrichment (FACE), (with elevated atmospheric CO_2 concentration of 546–586 ppm) experiments (Myers et al. 2014), found that wheat grains had 9.3% lower zinc (CI 5.9–12.7%), 5.1% lower iron (CI 3.7–6.5%) and 6.3% lower protein (CI 5.2–7.5%), and rice grains had 7.8% lower protein content (CI 6.8–8.9%). Changes in nutrient concentration in field pea, soybean and C4 crops such as sorghum and maize were small or insignificant. Zhu et al. (2018) report a meta-analysis of FACE trials on a range of rice cultivars. They show that protein declines by an average of 10% under elevated CO_2, iron and zinc decline by 8% and 5% respectively. Furthermore, a range of vitamins show large declines across all rice cultivars, including B1 (–17%), B2 (–17%), B5 (–13%) and B9 (–30%), whereas vitamin E increased. As rice underpins the diets of many of the world's poorest people in low-income countries, especially in Asia, Zhu et al. (2018) estimate that these changes under high CO_2 may affect the nutrient status of about 600 million people.

Decreases in protein concentration with elevated CO_2 are related to reduced nitrogen concentration possibly caused by nitrogen uptake not keeping up with biomass growth, an effect called 'carbohydrate dilution' or 'growth dilution', and by inhibition of photorespiration which can provide much of the energy used for assimilating nitrate into proteins (Bahrami et al. 2017). Other mechanisms have also been postulated (Feng et al. 2015; Bloom et al. 2014; Taub and Wang 2008). Together, the impacts on protein availability may take as many as 150 million people into protein deficiency by 2050 (Medek et al. 2017). Legume and vegetable yields increased with elevated CO_2 concentration of 250 ppm above ambient by 22% (CI 11.6–32.5%), with a stronger effect on leafy vegetables than on legumes and no impact for changes in iron, vitamin C or flavonoid concentration (Scheelbeek et al. 2018).

Increasing concentrations of atmospheric CO_2 lower the content of zinc and other nutrients in important food crops. Dietary deficiencies of zinc and iron are a substantial global public health problem (Myers et al. 2014). An estimated two billion people suffer these deficiencies (FAO 2013a), causing a loss of 63 million life-years annually (Myers et al. 2014). Most of these people depend on C3 grain legumes as their primary dietary source of zinc and iron. Zinc deficiency is currently responsible for large burdens of disease globally, and the populations who are at highest risk of zinc deficiency receive most of their dietary zinc from crops (Myers et al. 2015). The total number of people estimated to be placed at new risk of zinc deficiency by 2050 is 138 million. The people likely to be most affected live in Africa and South Asia, with nearly 48 million residing in India alone. Differences between cultivars of a single crop suggest that breeding for decreased sensitivity to atmospheric CO_2 concentration could partly address these new challenges to global health (Myers et al. 2014).

In summary, while increased CO_2 is projected to be beneficial for crop productivity at lower temperature increases, it is projected to lower nutritional quality (e.g., less protein, zinc, and iron) (*high confidence*).

1.2.5 Climate change impacts on food stability

Food stability is related to people's ability to access and use food in a steady way, so that there are not intervening periods of hunger. Increasing extreme events associated with climate change can disrupt food stability (see Section 5.8.1 for assessment of food price spikes).

1.2.5.1 Impacts of extreme events

FAO et al. (2018) conducted an analysis of the prevalence of undernourishment (PoU) and found that in 2017, the average of the PoU was 15.4% for all countries exposed to climate extremes (Supplementary Material Figure SM5.2). At the same time, the PoU was 20% for countries that additionally show high vulnerability of agriculture production/yields to climate variability, or 22.4% for countries with high PoU vulnerability to severe drought. When there is both high vulnerability of agriculture production/yields and high PoU sensitivity to severe drought, the PoU is 9.8 points higher (25.2%). These vulnerabilities were found to be higher when countries had a high dependence on agriculture as measured by the number of people employed in the sector. Bangkok experienced severe flooding in 2011–2012 with large-scale disruption of the national food supply chains since they were centrally organised in the capital city (Allen et al. 2017).

The IPCC projects that frequency, duration, and intensity of some extreme events will increase in the coming decades (IPCC 2018a, 2012). To test these effects on food security, Tigchelaar et al. (2018) showed rising instability in global grain trade and international grain prices, affecting especially the about 800 million people living in extreme poverty who are most vulnerable to food price spikes

(Section 5.8.1). They used global datasets of maize production and climate variability combined with future temperature projections to quantify how yield variability will change in the world's major maize-producing and exporting countries under 2°C and 4°C of global warming.

Tesfaye et al. (2017) projected that the extent of heat-stressed areas in South Asia could increase by up to 12% in 2030 and 21% in 2050 relative to the baseline (1950–2000). Another recent study found that drier regions are projected to dry earlier, more severely and to a greater extent than humid regions, with the population of Sub-Saharan Africa most vulnerable (Lickley and Solomon 2018).

1.2.5.2   Food aid

Food aid plays an important role in providing food security and saving lives after climate disasters. In 2015, 14.5 million people were assisted through disaster-risk reduction, climate change and/ or resilience building activities (WFP 2018). However, there is no agreement on how to better use emergency food aid, since it can come with unintended consequences for individuals, groups, regions, and countries (Barrett 2006). These may include negative dependency of food recipients (Lentz et al. 2005) or price increases, among others.

Some authors state that tied food aid provided as 'in kind' by the donor country hampers local food production (Clay 2006), although others found no evidence of this (Ferrière and Suwa-Eisenmann 2015). Untied cash aid can be used to buy food locally or in neighbouring countries, which is cheaper and can contribute to improving the livelihoods of local farmers (Clay 2006).

Ahlgren et al. (2014) found that food aid dependence of Marshall Islands due to climate change impacts can result in poor health outcomes due to the poor nutritional quality of food aid, which may result in future increases of chronic diseases. In this regard, Mary et al. (2018) showed that nutrition-sensitive aid can reduce the prevalence of undernourishment.

In summary, based on AR5 and SR15 assessments that the likelihood of extreme weather events will increase, (e.g., increases in heatwaves, droughts, inland flooding, and coastal flooding due to rising sea levels, depending on region) in both frequency and magnitude, decreases in food stability and thus increases in food insecurity will likely rise as well (*medium evidence, high agreement*).

1.3   Adaptation options, challenges and opportunities

This section assesses the large body of literature on food system adaptation to climate change, including increasing extreme events, within a framework of autonomous, incremental, and transformational adaptation. It focuses primarily on regional and local considerations and adaptation options for both the supply side (production, storage, transport, processing, and trade) and the demand side (consumption and diets) of the food system. Agroecological, social, and cultural contexts are considered throughout. Finally, the section assesses the role of institutional measures at global, regional (multiple countries), national, and local scales and capacity-building.

1.3.1   Challenges and opportunities

By formulating effective adaptation strategies, it is possible to reduce or even avoid some of the negative impacts of climate change on food security (Section 5.2). However, if unabated climate change continues, limits to adaptation will be reached (SR15). In the food system, adaptation actions involve any activities designed to reduce vulnerability and enhance resilience of the system to climate change. In some areas, expanded climate envelopes will alter agroecological zones, with opportunity for expansion towards higher latitudes and altitudes, soil and water resources permitting (Rosenzweig and Hillel 2015).

More extreme climatic events are projected to lead to more agrometeorological disasters with associated economic and social losses. There are many options for adapting the food system to extreme events reported in IPCC (2012), highlighting measures that reduce exposure and vulnerability and increase resilience, even though risks cannot fully be eliminated (IPCC 2012). Adaptation responses to extreme events aim to minimise damages, modify threats, prevent adverse impacts, or share losses, thus making the system more resilient (Harvey et al. 2014a).

With current and projected climate change (higher temperature, changes in precipitation, flooding and extremes events), achieving adaptation will require both technological (e.g., recovering and improving orphan crops, new cultivars from breeding or biotechnology) and non-technological (e.g., markets, land management, dietary change) solutions. Climate interacts with other factors such as food supplied over longer distances and policy drivers (Mbow et al. 2008; Howden et al. 2007), as well as local agricultural productivity.

Given the site-specific nature of climate change impacts on food system components together with wide variation in agroecosystems types and management, and socio-economic conditions, it is widely understood that adaptation strategies are linked to environmental and cultural contexts at the regional and local levels (*high confidence*). Developing systemic resilience that integrates climate drivers with social and economic drivers would reduce the impact on food security, particularly in developing countries. For example, in Africa, improving food security requires evolving food systems to be highly climate resilient, while supporting the need for increasing yield to feed the growing population (Mbow et al. 2014b) (Box 5.2).

Adaptation involves producing more food where needed, moderating demand, reducing waste, and improving governance (Godfray and Garnett 2014) (see Section 5.6 for the significant synergies between adaptation and mitigation through specific practices, actions and strategies).

Box 5.2 | Sustainable solutions for food systems and climate change in Africa

Climate change, land-use change, and food security are   important aspects of sustainability policies in Africa.

Table 5.3 | Synthesis of food security related adaptation options to address climate risks (IPCC 2014b; Vermeulen et al. 2013, 2018; Burnham and Ma 2016; Bhatta and Aggarwal 2016).

Key climate drivers and risks	Incremental adaptation	Transformational adaptation	Enabling conditions
– Extreme events and short-term climate variability – Stress on water resources, drought stress, dry spells, heat extremes, flooding, shorter rainy seasons, pests	– Change in variety, water management, water harvesting, supplemental irrigation during dry spells – Planting dates, pest control, feed banks – Transhumance, other sources of revenue (e.g., charcoal, wild fruits, wood, temporary work) – Soil management, composting	– Early Warning Systems – Planning for and prediction of seasonal to intra-seasonal climate risks to transition to safer food conditions – Abandonment of monoculture, diversification – Crop and livestock insurance – Alternate cropping, intercropping – Erosion control	– Establishment of climate services – Integrated water management policies, integrated land and water governance – Seed banks, seed sovereignty and seed distribution policies – Capacity building and extension programmes
– Warming trend, drying trend – Reduced crop productivity due to persistent heat, long drought cycles, deforestation and land degradation with strong adverse effects on food production and nutrition quality, increased pest and disease damage	– Strategies to reduce effects of recurring food challenges – Sustainable intensification, agroforestry, conservation agriculture, SLM – Adoption of existing drought-tolerant crop and livestock species – Counter season crop production – Livestock fattening – New ecosystem-based adaptation (e.g., bee keeping, woodlots) – Farmers management of natural resources – Labour redistribution (e.g., mining, development projects, urban migration) – Adjustments to markets and trade pathways already in place	– Climate services for new agricultural programmes (e.g., sustainable irrigation districts) – New technology (e.g., new farming systems, new crops and livestock breeds) – Switches between cropping and transhumant livelihoods, replacement of pasture or forest to irrigated/rainfed crops – Shifting to small ruminants or drought resistant livestock or fish farming – Food storage infrastructures, food transformation – Changes in cropping area, land rehabilitation (enclosures, afforestation) perennial farming – New markets and trade pathways	– Climate information in local development policies – Stallholders' access to credit and production resources – National food security programme based on increased productivity, diversification, transformation and trade – Strengthening (budget, capacities, expertise) of local and national institutions to support agriculture and livestock breeding – Devolution to local communities, women's empowerment, market opportunities – Incentives for establishing new markets and trade pathways

According to the McKinsey Global Institute (2010), Africa has around 60% of the global uncultivated arable land; thus the continent has a high potential for transformative change in food production. With short and long-term climate change impacts combined with local poverty conditions, land degradation and poor farming practices, Africa cannot grow enough food to feed its rapidly growing population. Sustainable improvement of productivity is essential, even as the impacts of climate change on food security in Africa are projected to be multiple and severe.

Sustainable Land Management (SLM) of farming systems is important to address climate change while dealing with these daunting food security needs and the necessity to improve access to nutritious food to maintain healthy and active lives in Africa (AGRA 2017). SLM has functions beyond the production of food, such as delivery of water, protection against disease (especially zoonotic diseases), the delivery of energy, fibre and building materials.

Commodity-based systems – driven by external markets – are increasing in Africa (cotton, cocoa, coffee, palm oil, groundnuts) with important impacts on the use of land and climate. Land degradation, decreasing water resources, loss of biodiversity, excessive use of synthetic fertilisers and pesticides are some of the environmental challenges that influence preparedness to adapt to climate change (Pretty and Bharucha 2015). A balanced strategy on African agriculture can be based on SLM and multifunctional land-use approaches combining food production, cash crops, ecosystem services, biodiversity conservation, ecosystem

services delivery, and ILK.

Box 5.2 (continued)

Thus, sustainable food systems in Africa entail multiple dimensions as shown in Figure 5.8.

With rapid urbanisation, it is important to integrate strategies (e.g., zero-carbon energy, smart irrigation systems, and climate-resilient agriculture) to minimise the negative effects of climate change while securing quality food for a growing population.

Building resilience into productivity and production can be based on simultaneous attention to the following five overarching issues:

1. Closing yield gaps through adapted cultivars, sustainable land management combining production and preservation of ecosystems essential functions, such as sustainable intensification approaches based on conservation agriculture and community-based adaptation with functioning support services and market access (Mbow et al. 2014a).
2. Identifying sustainable land management practices (agroecology, agroforestry, etc.) addressing different ecosystem services (food production, biodiversity, reduction of GHG emissions, soil carbon sequestration) for improved land-based climate change adaptation and mitigation (Sanz et al. 2017; Francis 2016).
3. Paying attention to the food-energy-water nexus,

especially water use and reutilisation efficiency but also management of rainwater (Albrecht et al. 2018).

4. Implementing institutional designs focused on youth and women through new economic models that help enable access to credit and loans to support policies that balance cash and food crops.

5. Building on local knowledge, culture and traditions while seeking innovations for food waste reduction and transformation of agricultural products.

These aspects suppose both incremental and transformational adaptation that may stem from better infrastructure (storage and food processing), adoption of harvest and post-harvest technologies that minimise food waste, and development of new opportunities for farmers to respond to environmental, economic and social shocks that affect their livelihoods (Morton 2017).

Agriculture in Africa offers a unique opportunity for merging adaption to and mitigation of climate change with sustainable production to ensure food security (CCAFS 2012; FAO 2012). Initiatives throughout the food system on both the supply and demand sides can lead to positive outcomes.

1.3.2 Adaptation framing and key concepts

1.3.2.1 Autonomous, incremental, and transformational adaptation

Framing of adaptation in this section categorises and assesses adaptation measures as autonomous, incremental, and transformational (Glossary and Table 5.3). Adaptation responses can be reactive or anticipatory.

Autonomous. Autonomous adaptation in food systems does not constitute a conscious response to climatic stimuli but is triggered by changes in agroecosystems, markets, or welfare changes. It is also referred to as spontaneous adaptation (IPCC 2007). Examples of autonomous adaptation of rural populations have been documented in the Sahel (IRD 2017). In India, farmers are changing sowing and harvesting timing, cultivating short duration varieties, inter-cropping, changing cropping patterns, investing in irrigation, and establishing agroforestry. These are considered as passive responses or autonomous adaptation, because they do not acknowledge that these steps are taken in response to perceived climatic changes (Tripathi and Mishra 2017).

Incremental. Incremental adaptation maintains the essence and integrity of a system or process at a given scale (Park et al. 2012). Incremental adaptation focuses on improvements to existing resources and management practices (IPCC 2014a).

Transformational. Transformational adaptation changes the fundamental attributes of a socio-ecological system either in anticipation of, or in response to, climate change and its impacts

(IPCC 2014a). Transformational adaptation seeks alternative livelihoods and land-use strategies needed to develop new farming systems (Termeer et al. 2016). For example, limitations in incremental adaptation among smallholder rice farmers in Northwest Costa Rica led to a shift from rice to sugarcane production due to decreasing market access and water scarcity (Warner et al. 2015). Migration from the Oldman River Basin has been described as a transformational adaption to climate change in the Canadian agriculture sector (Hadarits et al. 2017). If high-end scenarios of climate change eventuate, the food security of farmers and consumers will depend on how transformational change in food systems is managed. An integrated framework of adaptive transition – management of socio-technical transitions and adaptation to socio-ecological changes – may help build transformational adaptive capacity (Mockshell and Kamanda 2018 and Pant et al. 2015). Rippke et al. (2016) has suggested overlapping phases of adaptation needed to support transformational change in Africa.

1.3.2.2 Risk management

Climate risks affect all pillars of food security, particularly stability because extreme events lead to strong variation to food access. The notion of risk is widely treated in IPCC reports (IPCC 2014c) (see also Chapter 7 in this report). With food systems, many risks co-occur or reinforce each other, and this can limit effective adaptation planning as they require a comprehensive and dynamic policy approach covering a range of drivers and scales. For example, from the understanding by farmers of change in risk profiles to the establishment of efficient markets that facilitate response strategies will require more than systemic reviews of risk factors (Howden et al. 2007).

Integration of Climate Change Adaptation (CCA) and Disaster Risk Reduction (DRR) helps to minimise the overlap and duplication of projects and programmes (Nalau et al. 2016). Recently, countries started integrating the concept of DRR and CCA. For instance, the Philippines introduced new legislation calling for CCA and DRR integration, as current policy instruments had been largely unsuccessful in combining agencies and experts across the two areas (Leon and Pittock 2016).

Studies reveal that the amplitude of interannual growing-season temperature variability is in general larger than that of long-term temperature change in many locations. Responding better to seasonal climate-induced food supply shocks therefore increases society's capability to adapt to climate change. Given these backgrounds, seasonal crop forecasting and early response recommendations (based on seasonal climate forecasts), are emerging to strengthen existing operational systems for agricultural monitoring and forecasting (FAO 2016a; Ceglar et al. 2018 and Iizumi et al. 2018).

While adaptation and mitigation measures are intended to reduce the risk from climate change impacts in food systems, they can also be sources of risk themselves (e.g., investment risk, political risk) (IPCC 2014b). Climate-related hazards are a necessary element of risks related to climate impacts but may have little or nothing to do with risks related to some climate policies/responses.

Box 5.3 | Climate change and indigenous food systems in the Hindu-Kush Himalayan Region

Diversification of production systems through promotion of Neglected and Underutilised Species (NUS; also known as understudied, neglected, orphan, lost or disadvantaged crops) offers adaptation opportunities to climate change, particularly in mountains. Neglected and Underutilised Species (NUS) have a potential to improve food security and at the same time help protect and conserve traditional knowledge and biodiversity. Scaling-up NUS requires training farmers and other stakeholders on ways to adopt adequate crop management, quality seed, select varieties, farming systems, soil management, development of new products, and market opportunities (Padulosi et al. 2013). Farmers in the Rasuwa district, in the mid-hills of Nepal, prefer to cultivate local bean, barley, millet and local maize, rather than commodity crops because they are more tolerant to water stress and extremely cold conditions (Adhikari et al. 2017). Farmers in the high-altitude, cold climate of Nepal prefer local barley with its short growing period because of a shorter growing window. Buckwheat is commonly grown in the Hindu-Kush Himalayan (HKH) region mainly because it grows fast and suppresses weeds. In Pakistan, quinoa (*Chenopodium quinoa*) grew and produced well under saline and marginal soil where other crops would not grow (Adhikari et al. 2017).

At the same time, in many parts of the HKH region, a substantial proportion of the population is facing malnutrition. Various factors are responsible for this, and lack of diversity in food and nutrition resulting from production and consumption of few crops is one of them. In the past, food baskets in this region consisted of many different edible plant species, many of which are now neglected and underutilised. This is because almost all the efforts of the Green Revolution after 1960 focused on major crops. Four crops, namely rice, wheat, maize and potato, account for about 60% of global plant-derived energy supply (Padulosi et al. 2013).

While the Green Revolution technologies substantially increased the yield of few crops and allowed countries to reduce hunger, they also resulted in inappropriate and excessive use of agrochemicals, inefficient water use, loss of beneficial biodiversity, water and soil pollution and significantly reduced crop and varietal diversity. With farming systems moving away from subsistence-based to commercial farming, farmers are also reluctant to grow these local crops because of low return, poor market value and lack of knowledge about their nutritional environmental value.

However, transition from traditional diets based on local foods to a commercial crop-based diet with high fats, salt, sugar and processed foods, increased the incidence of non-communicable diseases, such as diabetes, obesity, heart diseases and certain types of cancer (Abarca-Gómez et al. 2017; NCD-RisC 2016b, 2017b). This 'hidden hunger' – enough calories, but insufficient vitamins – is increasingly evident in mountainous communities including the HKH region.

Internationally, there is rising interest in NUS, not only because they present opportunities for fighting poverty, hunger and malnutrition, but also because of their role in mitigating climate risk in agricultural production systems. NUS play an important role in mountain agroecosystems because mountain agriculture is generally low-input agriculture, for which many NUS are well adapted.

In the HKH region, mountains are agroecologically suitable for cultivation of traditional food crops, such as barley, millet, sorghum, buckwheat, bean, grams, taro, yam and a vast range of wild fruits, vegetables and medicinal plants. In one study carried out in two villages of mid-hills in Nepal, Khanal et al. (2015) reported 52 indigenous crop species belonging to 27 families with their various uses. Farming communities continue to grow various indigenous crops, albeit in marginal land, because of their value on traditional food and associated culture. Nepal Agricultural Research Council (NARC) has identified a list of indigenous crops based on their nutritional, medicinal, cultural and other values.

Many indigenous crops supply essential micronutrients to the human body, and need to be conserved in mountain food systems. Farmers in HKH region are cultivating and maintaining various indigenous crops such as Amaranthus, barley, black gram, horse gram, yam, and sesame. because of their nutritional value. Most of these indigenous crops are comparable with commercial cereals in terms of dietary energy and protein content, but are also rich in micronutrients. For example, pearl millet has higher content of calcium, iron, zinc, riboflavin and folic acid than rice or maize (Adhikari et al. 2017).

NUS can provide both climate resilience and more options for dietary diversity to the farming communities of mountain ecosystems. Some of these indigenous crops have high medical importance. For example, mountain people in the HKH region have been using *jammun* (i.e., *Syzygium cumini*) to treat diabetes. In the Gilgit-Baltistan province of Pakistan, realising the importance of sea-buckthorn for nutritional and medicinal purposes, local communities have expanded its cultivation to larger areas. Many of these crops can be cultivated in marginal and/or fallow land which otherwise remains fallow. Most of these species are drought resistant and can be easily grown in rainfed conditions in non-irrigated land.

5

Adoption of agroecological practices could provide resilience for future shocks, spread farmer risk and mitigate the impact of droughts (Niles et al. 2018) (Section 5.3.2.3). Traditionally, risk management is performed through multifunctional landscape approaches in which resource utilisation is planned across wide areas and local agreements on resource access. Multifunctionality permits vulnerable communities to access various resources at various times and under various risk conditions (Minang et al. 2015).

In many countries, governmental compensation for crop-failure and financial losses are used to protect against risk of severe yield reductions. Both public and private sector groups develop insurance markets and improve and disseminate index-based weather insurance programmes. Catastrophe bonds, microfinance, disaster contingency funds, and cash transfers are other available mechanisms for risk management.

In summary, risk management can be accomplished through agroecological landscape approaches and risk sharing and transfer mechanisms, such as development of insurance markets and improved index-based weather insurance programmes (*high confidence*).

1.3.2.3   Role of agroecology and diversification

Agroecological systems are integrated land-use systems that maintain species diversity in a range of productive niches. Diversified cropping systems and practicing traditional agroecosystems of crop production where a wide range of crop varieties are grown in various spatial and temporal arrangements, are less vulnerable to catastrophic loss (Zhu et al. 2011). The use of local genetic diversity, soil organic matter enhancement, multiple-cropping or poly-culture systems, home gardening, and agroecological approaches can build resilience against extreme climate events (Altieri and Koohafkan 2008).

However, Nie et al. (2016) argued that while integrated crop-livestock systems present some opportunities such as control of weeds, pests and diseases, and environmental benefits, there are some challenges, including yield reduction, difficulty in pasture-cropping, grazing, and groundcover maintenance in high rainfall zones, and development of persistent weeds and pests.

Adaptation measures based on agroecology entail enhancement of agrobiodiversity; improvement of ecological processes and delivery of ecosystem services. They also entail strengthening of local communities and recognition of the role and value of ILK. Such practices can enhance the sustainability and resilience of agricultural systems by buffering climate extremes, reducing degradation of soils, and reversing unsustainable use of resources; outbreak of pests and diseases and consequently increase yield without damaging biodiversity. Increasing and conserving biological diversity such as soil microorganisms can promote high crop yields and sustain the environment (Schmitz et al. 2015; Bhattacharyya et al. 2016; Garibaldi et al. 2017).

Diversification of many components of the food system is a key element for increasing performance and efficiency that may translate into increased resilience and reduced risks (integrated land management systems, agrobiodiversity, ILK, local food systems,

dietary diversity, the sustainable use of indigenous fruits, neglected and underutilised crops as a food source) (*medium confidence*) (Makate et al. 2016; Lin 2011; Awodoyin et al. 2015).

The more diverse the food systems are, the more resilient they are in enhancing food security in the face of biotic and abiotic stresses. Diverse production systems are important for providing regulatory ecosystem services such as nutrient cycling, carbon sequestration, soil erosion control, reduction of GHG emissions and control of hydrological processes (Chivenge et al. 2015). Further options for adapting to change in both mean climate and extreme events are livelihood diversification (Michael 2017; Ford et al. 2015), and production diversity (Sibhatu et al. 2015).

Crop diversification, maintaining local genetic diversity, animal integration, soil organic matter management, water conservation, and harvesting the role of microbial assemblages. These types of farm management significantly affect communities in soil, plant structure, and crop growth in terms of number, type, and abundance of species (Morrison-Whittle et al. 2017). Complementary strategies towards sustainable agriculture (ecological intensification, strengthening existing diverse farming systems and investment in ecological infrastructure) also address important drivers of pollinator decline (IPBES 2016).

Evidence also shows that, together with other factors, on-farm agricultural diversity can translate into dietary diversity at the farm level and beyond (Pimbert and Lemke 2018; Kumar et al. 2015; Sibhatu et al. 2015). Dietary diversity is important but not enough as an adaptation option, but results in positive health outcomes by increasing the variety of healthy products in people's diets and reducing exposure to unhealthy environments.

Locally developed seeds and the concept of seed sovereignty can both help protect local agrobiodiversity and can often be more climate resilient than generic commercial varieties (Wattnem 2016; Coomes et al. 2015; van Niekerk and Wynberg 2017; Vasconcelos et al. 2013). Seed exchange networks and banks protect local agrobiodiversity and landraces, and can provide crucial lifelines when crop harvests fail (Coomes et al. 2015; van Niekerk and Wynberg 2017; Vasconcelos et al. 2013).

Related to locally developed seeds, neglected and underutilised species (NUS) can play a key role in increasing dietary diversity (*high confidence*) (Baldermann et al. 2016; van der Merwe et al. 2016; Kahane et al. 2013; Muhanji et al. 2011) (Box 5.3). These species can also improve nutritional and economic security of excluded social groups, such as tribals (Nandal and Bhardwaj 2014; Ghosh-Jerath et al. 2015), indigent (Kucich and Wicht 2016) or rural populations (Ngadze et al. 2017).

Dietary diversity has also been correlated (*medium evidence, medium agreement*) to agricultural diversity in small-holder and subsistence farms (Ayenew et al. 2018; Jones et al. 2014; Jones 2017; Pimbert and Lemke 2018), including both crops and animals, and has been proposed as a strategy to reduce micronutrient malnutrition in developing countries (Tontisirin et al. 2002). In this regard, the capacity of subsistence farming to supply essential nutrients in reasonable balance to the people dependent on them has been considered as a means of overcoming their nutrient limitations in sound agronomic and sustainable ways (Graham et al. 2007).

Ecosystem-based adaptation (EbA). EbA is a set of nature-based methods addressing climate change adaptation and food security by strengthening and conserving natural functions, goods and services that benefit people. EbA approaches to address food security provide co-benefits such as contributions to health and improved diet, sustainable land management, economic revenue and water security. EbA practices can reduce GHG emissions and increase carbon storage (USAID 2017).

For example, agroforestry systems can contribute to improving food productivity while enhancing biodiversity conservation, ecological balance and restoration under changing climate conditions (Mbow et al. 2014a; Paudela et al. 2017; Newaj et al. 2016; Altieri et al. 2015). Agroforestry systems have been shown to reduce erosion through their canopy cover and their contribution to the micro-climate and erosion control (Sida et al. 2018). Adoption of conservation farming practices such as removing weeds from and dredging irrigation canals, draining and levelling land, and using organic fertilisation were among the popular conservation practices in small-scale paddy rice farming community of northern Iran (Ashoori and Sadegh 2016).

Adaptation potential of ecologically-intensive systems includes also forests and river ecosystems, where improved resource management such as soil conservation, water cycling and agrobiodiversity support the function of food production affected by severe climate change (Muthee et al. 2017). The use of non-crop plant resources in agroecosystems (permaculture, perennial polyculture) can improve ecosystem conservation and may lead to increased crop productivity (Balzan et al. 2016; Crews et al. 2018; Toensmeier 2016).

In summary, increasing the resilience of the food system through agroecology and diversification is an effective way to achieve climate change adaptation (*robust evidence, high agreement*). Diversification in the food system is a key adaptation strategy to reduce risks (e.g., implementation of integrated production systems at landscape scales, broad-based genetic resources, and heterogeneous diets) (*medium confidence*).

1.3.2.4   Role of cultural values

Food production and consumption are strongly influenced by cultures and beliefs. Culture, values and norms are primary factors in most climate change and food system policies. The benefits of integrating cultural beliefs and ILK into formal climate change mitigation and adaptation strategies can add value to the development of sustainable climate change, rich in local aspirations, planned with, and for, local people (Nyong et al. 2007).

Cultural dimensions are important in understanding how societies establish food production systems and respond to climate change, since they help to explain differences in responses across populations to the same environmental risks (Adger et al. 2013). There is an inherent adaptability of indigenous people who are particularly connected to land use, developed for many centuries to produce specific solutions to particular climate change challenges. Acknowledging that indigenous cultures across the world are supporting many string strategies and beliefs that offer sustainable systems with pragmatic solutions will help move forward the food and climate sustainability policies. For instance, in the Sahel, the local populations have developed and implemented various adaptation strategies that sustain their resilience despite many threats (Nyong et al. 2007). There is an increased consideration of local knowledge and cultural values and norms in the design and implementation of modern mitigation and adaptation strategies.

5

There are some entrenched cultural beliefs and values that may be barriers to climate change adaptation. For instance, culture has been shown to be a major barrier to adaptation for the Fulbe ethnic group of Burkina Faso (Nielsen and Reenberg 2010). Thus, it is important to understand how beliefs, values, practices and habits interact with the behaviour of individuals and collectivities that have to confront climate change (Heyd and Thomas 2008). Granderson (2014) suggests that making sense of climate change and its responses at the community level demands attention to the cultural and political processes that shape how risk is conceived, prioritised and managed. For a discussion of gender issues related to climate change, see Section 5.2.

Culturally sensitive risk analysis can deliver a better understanding of what climate change means for society (O'Brien and Wolf 2010; Persson et al. 2015) and thus, how to better adapt. Murphy et al. (2016) stated that culture and beliefs play an important role in adaptive capacity but that they are not static. In the work done by Elum et al. (2017) in South Africa (about farmers' perception of climate change), they concluded that perceptions and beliefs often have negative effects on adaptation options.

Culture is a key issue in food systems and the relation of people with nature. Food is an intrinsically cultural process: food production shapes landscapes, which in turn are linked to cultural heritages and identities (Koohafkan and Altieri 2011; Fuller and Qingwen 2013), and food consumption has a strong cultural dimension. The loss of subsistence practices in modern cultures and their related ILK, has resulted in a loss of valuable adaptive capacities (Hernández-Morcillo et al. 2014). This is so because these systems are often characterised by livelihood strategies linked to the management of natural resources that have been evolved to reduce overall vulnerability to climate shocks ('adaptive strategies') and to manage their impacts ex-post ('coping strategies') (Morton 2007; López-i-Gelats et al. 2016).

1.3.3   Supply-side adaptation

Supply-side adaptation takes place in the production (of crops, livestock, and aquaculture), storage, transport, processing, and trade of food.

1.3.3.1   Crop production

There are many current agricultural management practices that can be optimised and scaled up to advance adaptation. Among the often-studied adaptation options are increased soil organic matter, improved cropland management, increased food productivity, prevention and reversal of soil erosion (see Chapter 6 for evaluation of these practices in regard to desertification and land degradation). Many analyses have demonstrated the effectiveness of soil management and changing sowing date, crop type or variety (Waongo et al. 2015; Bodin et al. 2016; Teixeira et al. 2017; Waha et al. 2013; Zimmermann et al. 2017; Chalise and Naranpanawa 2016; Moniruzzaman 2015; Sanz et al. 2017). Biophysical adaptation options also include pest and disease management (Lamichhane

et al. 2015) and water management (Palmer et al. 2015; Korbeľová and Kohnová 2017).

In Africa, Scheba (2017) found that conservation agriculture techniques were embedded in an agriculture setting based on local traditional knowledge, including crop rotation, no or minimum tillage, mulching, and cover crops. Cover cropping and no-tillage also improved soil health in a highly commercialised arid irrigated system in California's San Joaquin Valley, USA (Mitchell et al. 2017). Biofertilisers can enhance rice yields (Kantachote et al. 2016), and Amanullah and Khalid (2016) found that manure and biofertiliser improve maize productivity under semi-arid conditions.

Adaptation also involves use of current genetic resources as well as breeding programmes for both crops and livestock. More drought, flood and heat-resistant crop varieties (Atlin et al. 2017; Mickelbart et al. 2015; Singh et al. 2017) and improved nutrient and water use efficiency, including overabundance as well as water quality (such as salinity) (Bond et al. 2018) are aspects to factor into the design of adaptation measures. Both availability and adoption of these varieties is a possible path for adaptation and can be facilitated by new outreach policy and capacity building.

Water management is another key area for adaptation. Increasing water availability and reliability of water for agricultural production using different techniques of water harvesting, storage, and its judicious utilisation through farm ponds, dams, and community tanks in rainfed agriculture areas have been presented by Rao et al. (2017) and Rivera-Ferre et al. (2016a). In addition, improved drainage systems (Thiel et al. 2015), and Alternate Wetting and Drying (AWD) techniques for rice cultivation (Howell et al. 2015; Rahman and Bulbul 2015) have been proposed. Efficient irrigation systems have been also analysed and proposed by Jägermeyr et al. (2016), Naresh et al. (2017), Gunarathna et al. (2017) and Chartzoulakis and Bertaki (2015). Recent innovation includes using farming systems with low usage of water such as drip-irrigation or hydroponic systems mostly in urban farming.

1.3.3.2   Livestock production systems

Considering the benefits of higher temperature in temperate climates and the increase of pasture with incremental warming in some humid and temperate grasslands, as well as potential negative effects, can be useful in planning adaptation strategies to future climate change. Rivera-Ferre et al. (2016b) characterize adaptation for different livestock systems as managerial, technical, behavioural and policy-related options. Managerial included production adjustments (e.g., intensification, integration with crops, shifting from grazing to browsing species, multispecies herds, mobility, soil and nutrient management, water management, pasture management, corralling, feed and food storage, farm diversification or cooling systems); and changes in labour allocation (diversifying livelihoods, shifting to irrigated farming, and labour flexibility). Technological options included breeding strategies and information technology research. Behavioural options are linked to cultural patterns and included encouraging social collaboration and reciprocity, for example,

5

livestock loans, communal planning, food exchanges, and information sharing. Policy options are discussed in Section 5.7 and Chapter 7.

1.3.3.3 Aquaculture, fisheries, and agriculture interactions

Options may include livelihood diversification within and across sectors of fisheries, aquaculture and agriculture. Thus, adaptation options need to provide management approaches and policies that build the livelihood asset base, reducing vulnerability to multiple stressors with a multi-sector perspective (Badjeck et al. 2010). In Bangladesh, fishing pressure on post-larval prawns has increased as displaced farmers have shifted to fishing following salt-water intrusion of agricultural land (Ahmed et al. 2013). In West Africa, strategies to cope with sudden shifts in fisheries are wider-reaching and have included turning to seafood import (Gephart et al. 2017) or terrestrial food production, including farming and bush-meat hunting on land (Brashares et al. 2004).

Proposed actions for adaptation include effective governance, improved management and conservation, efforts to maximise societal and environmental benefits from trade, increased equitability of distribution and innovation in food production, and the continued development of low-input and low-impact aquaculture (FAO 2018c).

Particular adaptation strategies proposed by FAO (2014a) include diverse and flexible livelihood strategies, such as introduction of fish ponds in areas susceptible to intermittent flood/drought periods; flood-friendly small-scale homestead bamboo pens with trap doors allowing seasonal floods to occur without loss of stocked fish; cage fish aquaculture development using plankton feed in reservoirs created by dam building; supporting the transition to different species, polyculture and integrated systems, allowing for diversified and more resilient systems; promotion of combined rice and fish farming systems that reduce overall water needs and provide integrated pest management; and supporting transitions to alternative livelihoods.

Risk reduction initiatives include innovative weather-based insurance schemes being tested for applicability in aquaculture and fisheries and climate risk assessments introduced for integrated coastal zone management. For aquaculture's contribution to building resilient food systems, Troell et al. (2014) found that aquaculture could potentially enhance resilience through improved resource use efficiencies and increased diversification of farmed species, locales of production, and feeding strategies. Yet, its high reliance on terrestrial crops and wild fish for feeds, its dependence on freshwater and land for culture sites and its environmental impacts reduce this potential. For instance, the increase in aquaculture worldwide may enhance land competition for feed crops, increasing price levels and volatility and worsening food insecurity among the most vulnerable populations.

1.3.3.4 Transport and storage

Fewer studies have been done on adaptation of food system transport and storage compared to the many studies on adaptation to climate in food production.

Transport. One transport example is found in Bangkok. Between mid-November 2011 and early January 2012, Bangkok, the capital city of Thailand, faced its most dramatic flood in approximately 70 years with most transport networks cut-off or destroyed. This caused large-scale disruption of the national food supply chains since they were centrally organised in the capital city (Allen et al. 2017). From this experience, the construction and management of 'climate-proof' rural roads and transport networks is argued as one the most important adaptation strategies for climate change and food security in Thailand (Rattanachot et al. 2015).

Similarly in Africa, it has been shown that enhanced transportation networks combined with other measures could reduce the impact of climate change on food and nutrition security (Brown et al. 2017b). This suggests that strengthening infrastructure and logistics for transport would significantly enhance resilience to climate change, while improving food and nutrition security in developing counties.

Storage. Storage refers to both structures and technologies for storing seed as well as produce. Predominant storage methods used in Uganda are single-layer woven polypropylene bags (popularly called 'kavera' locally), chemical insecticides and granaries. Evidence from Omotilewa et al. (2018) showed that the introduction of new storage technology called Purdue Improved Crop Storage (PICS) could contribute to climate change adaptation. PICS is a chemical-free airtight triple-layered technology consisting of two high-density polyethylene inner liners and one outer layer of woven polypropylene bag. Its adoption has increased the number of households planting hybrid maize varieties that are more susceptible to insect pests in storage than traditional lower-yielding varieties. Such innovations could help to protect crops more safely and for longer periods from postharvest insect pests that are projected to increase as result of climate change, thus contributing to food security.

In the Indo-Gangetic Plain many different storage structures based on ILK provide reliable and low-cost options made of local materials. For example, elevated grain stores protect harvested cereals from floods, but also provide for air circulation to prevent rot and to control insects and other vermin (Rivera-Ferre et al. 2013).

1.3.3.5 Trade and processing

Adaptation measures are also being considered in trade, processing and packaging, other important components of the food system. These will enable availability, stability, and safety of food under changing climate conditions.

Trade. Brooks and Matthews (2015) found that food trade increases the availability of food by enabling products to flow from surplus to deficit areas, raises incomes and favours access to food, improves utilisation by increasing the diversity of national diets while pooling production risks across individual markets to maintain stability.

Processing. Growth of spoilage bacteria of red meat and poultry during storage due to increasing temperature has been demonstrated by European Food Safety Authority (EFSA Panel on Biological Hazards 2016). In a recent experiment conducted on the optimisation of processing conditions of Chinese traditional smoke-cured bacon, Larou, Liu et al. (2018a) showed that the use of a new natural coating solution composed of lysozyme, sodium alginate, and chitosan during the storage period resulted in 99.69% rate of reducing deterioration after 30-day storage. Also, the use of High Hydrostatic Pressure (HHP) technology to inactivate pathogenic, spoilage microorganisms and enzymes (with little or no effects on the nutritional and sensory quality of foods) have been described by Wang et al. (2016) and Ali et al. (2018) as new advances in processing and packaging fruits, vegetables, meats, seafood, dairy, and egg products.

In summary, there are many practices that can be optimised and scaled up to advance supply-side adaptation. On-farm adaptation options include increased soil organic matter and erosion control in cropland, improved livestock and grazing land management, and transition to different species, polyculture and integrated systems in aquaculture. Crop and livestock genetic improvements include tolerance to heat, drought, and pests and diseases. Food transport, storage, trade, and processing will likely play increasingly important roles in adapting to climate change-induced food insecurity.

1.3.4   Demand-side adaptation

Adaptation in the demand side of the food system involves consumption practices, diets, and reducing food loss and waste. Recent studies showed that supply-side adaptation measures alone will not be sufficient to sustainably achieve food security under climate change (Springmann et al. 2018b; Swinburn et al. 2019; Bajželj et al. 2014). As noted by Godfray (2015), people with higher income demand more varied diets, and typically ones that are richer in meat and other food types that require more resources to produce. Therefore, both supply-side (production, processing, transport, and trade) and demand-side solutions (for example, changing diets, food loss and waste reduction) can be effective in adapting to climate change (Creutzig et al. 2016) (see Section 5.5.2.5 for food loss and waste).

The implications of dietary choice can have severe consequences for land. For example, Alexander et al. (2016), found that if every country were to adopt the UK's 2011 average diet and meat consumption, 95% of global habitable land area would be needed for agriculture – up from 50% of land currently used. For the average USA diet, 178% of global land would be needed (relative to 2011) (Alexander et al. 2016); and for 'business as usual' dietary trends and existing rates of improvement in yields, 55% more land would be needed above baseline (2009) (Bajželj et al. 2014). Changing dietary habits have been suggested as an effective food route to affect land use (Beheshti et al. 2017) and promote adaptation to climate change through food demand.

Most literature has focused on demand-side options that analyse the effects on climate change mitigation by dietary changes. Little focus has been brought on demand-side adaptation measures to adjust the demand to the food challenges related to drivers such as market, climate change, inputs limitations (for example, fossil fuels, nitrogen, phosphorus), food access, and quality. Adding to that, the high cost of nutritious foods contributes to a higher risk of overweight and obesity (FAO 2018d). Adaptation measures relate also to the implications of easy access to inexpensive, high-calorie, low-nutrition foods which have been shown to lead to malnutrition (Section 5.1). Therefore, adaptation related to diet may be weighed against the negative side effects on health of current food choices.

Reduction in the demand for animal-based food products and increasing proportions of plant-based foods in diets, particularly pulses and nuts; and replacing red meat with other more efficient protein sources are demand-side adaptation measures (Machovina et al. 2015) (Section 5.5.2). For example, replacing beef in the USA diet with poultry can meet caloric and protein demands of about 120 to 140 million additional people consuming the average American diet (Shepon et al. 2016). Similar suggestions are made for adopting the benefits of moving to plant-based protein, such as beans (Harwatt et al. 2017).

The main reason why reducing meat consumption is an adaptation measure is because it reduces pressure on land and water and thus our vulnerability to climate change and inputs limitations (Vanham et al. 2013). For animal feed, ruminants can have positive ecological effects (species diversity, soil carbon) if they are fed extensively on existing grasslands. Similarly, reducing waste at all points along the entire food chain is a significant opportunity for improving demand-side adaptation measures (Godfray 2015).

It is important to highlight the opportunities for improving the feed-to-meat conversion considered as a form of food loss. However, the unique capacity of ruminants to produce high-quality food from low-quality forage, in particular from landscapes that cannot be cropped and from cellulosic biomass that humans cannot digest could be seen as an effective way to improve the feed:meat ratio (Cawthorn and Hoffman 2015).

In summary, there is potential for demand-side adaptation, such as adoption of diets low in animal-sourced products, in conjunction with reduction in food loss and waste to contribute to reduction in food demand, land sparing, and thus need for adaptation.

1.3.5   Institutional measures

To facilitate the scaling up of adaptation throughout the food system, institutional measures are needed at global, regional, national, and local levels (Section 5.7). Institutional aspects, including policies and laws, depend on scale and context. International institutions (financial and policies) are driving many aspects of global food systems (for example, UN agencies, international private sector agribusinesses and retailers). Many others operate at local level and strongly influence livelihoods and markets of smallholder farmers. Hence, differentiation in the roles of the organisations, their missions and outcomes related to food and climate change action need to be clearly mapped and understood.

Awareness about the institutional context within which adaptation planning decisions are made is essential for the usability of climate change projection (Lorenz 2017) (Chapter 7). In the planning and operational process of food production, handling and consumption, the environment benefits and climate change goals can be mainstreamed under sustainable management approaches that favour alternative solutions for inputs, energy consumption, transformation and diet. For instance, land-use planning would guide current and future decision-making and planners in exploring uncertainty to increase the resilience of communities (Berke and Stevens 2016). One of the important policy implications for enhanced food security are the trade-offs between agricultural production and environmental concerns, including the asserted need for global land-use expansion, biodiversity and ecological restoration (Meyfroidt 2017) (Section 5.6).

There are a number of adaptation options in agriculture in the form of policy, planning, governance and institutions (Lorenz 2017). For example, early spatial planning action is crucial to guide decision-making processes and foster resilience in highly uncertain future climate change (Brunner and Grêt-Regamey, 2016). Institutions may develop new capacities to empower value chain actors, take climate change into account as they develop quality products, promote adoption of improved diet for healthier lifestyles, aid the improvement of livelihoods of communities, and further socioeconomic development (Sehmi et al. 2016). Other adaptation policies include property rights and land tenure security as legal and institutional reforms to ensure transparency and access to land that could stimulate adaptation to climate change (Antwi-Agyei et al. 2015).

1.3.5.1   Global initiatives

Climate change poses serious wide-ranging risks, requiring a broader approach in fighting the phenomenon. The United Nations Framework Convention on Climate Change (UNFCCC) and its annual Conferences of the Parties (COPs) has been instrumental in ensuring international cooperation in the field of tackling the impacts of climate change in a broader framework (Clémençon 2016). The National Adaptation Plan (NAP) programme under the UNFCCC was established to: identify vulnerable regions; assess the impacts of climate change on food security; and prioritise adaptation measures for implementation to increase resilience. The National Adaptation Programs of Action (NAPAs) was also established to support least-developed countries (LDCs) in addressing their particular challenges in adaptation, to enhance food security among other priorities.

The Paris Agreement (UNFCCC 2015) is a major victory for small island states and vulnerable nations that face climate change-related impacts of floods and droughts resulting in food security challenges. Adaptation and mitigation targets set by the parties through their nationally determined commitments (NDCs) are reviewed internationally to ensure consistency and progress towards actions (Falkner 2016).

The Food and Agriculture Organization of the United Nations (FAO) also plays a significant role in designing and coordinating national policies to increase adaptation and food security. The five key strategic objectives of FAO (help eliminate hunger, food insecurity and malnutrition; make agriculture, forestry and fisheries more productive and sustainable; reduce rural poverty; enable inclusive and efficient agricultural and food systems; and increase the resilience of livelihoods to climate threats) (FAO 2018e), all relate to building resilience and increasing global adaptation to climate variability.

In support of the Paris Agreement, FAO launched a global policy, 'Tracking Adaptation' with the aim of monitoring the adaptation processes and outcomes of the parties to increase food security and of making available technical information for evaluation by stakeholders. In response to the estimated world population of 9.7 billion by 2050, FAO adopted the Climate Smart Agriculture (CSA) approach to increase global food security without compromising environmental quality (Section 5.6). FAO supports governments at the national level to plan CSA programmes and to seek climate finance to fund their adaptation programmes.

The Global Commission on Adaptation, co-managed by World Resources Institute (WRI) and the Global Center on Adaptation, seeks to accelerate adaptation action by elevating the political visibility of adaptation and focusing on concrete solutions (Global Commission on Adaptation 2019). The Commission works to demonstrate that adaptation is a cornerstone of better development, and can help improve lives, reduce poverty, protect the environment, and enhance resilience around the world. The Commission is led by Ban Ki-moon, 8th Secretary-General of the United Nations, Bill Gates, co-chair of the Bill & Melinda Gates Foundation, and Kristalina Georgieva, CEO, World Bank. It is convened by 17 countries and guided by 28 commissioners. A global network of research partners and advisors provide scientific, economic, and policy analysis.

1.3.5.2   National policies

The successful development of food systems under climate change conditions requires a national-level management that involves the cooperation of a number of institutions and governance entities to enable more sustainable and beneficial production and consumption practices.

For example, Nepal has developed a novel multi-level institutional partnership, under the Local Adaptation Plan of Action (LAPA), which is an institutional innovation that aims to better integrate local adaptation planning processes and institutions into national adaptation processes. That includes collaboration with farmers and other non-governmental organisations (Chhetri et al. 2012). By combining conventional technological innovation process with the tacit knowledge of farmers, this new alliance has been instrumental in the innovation of location-specific technologies thereby facilitating the adoption of technologies in a more efficient manner.

National Adaptation Planning of Indonesia was officially launched in 2014 and was an important basis for ministries and local governments to mainstream climate change adaptation into their respective sectoral and local development plans (Kawanishi et al. 2016). Crop land-use policy – to switch from crops that are highly impacted by climate change to those that are less vulnerable – were suggested for improving climate change adaptation policy processes and outcomes in Nepal (Chalise and Naranpanawa 2016).

Enhancement of representation, democratic and inclusive governance, as well as equity and fairness for improving climate change adaptation policy processes and outcomes in Nepal were also suggested as institutional measures by Ojha et al. (2015). Further,

5

Table 5.4 | GHG emissions (GtCO$_2$-eq yr^{-1}) from the food system and their contribution (%) to total anthropogenic emissions. Mean of 2007–2016 period.

Food system component	Emissions (Gt CO$_2$eq yr^{-1})	Share in mean total emissions (%)
Agriculture	6.2 ± 1.4 [a,b]	10–14%
Land use	4.9 ± 2.5 [a]	5–14%
Beyond farm gate	2.6 [c] – 5.2 [d]	5–10% [e]
Food system (total)	10.8 – 19.1	21–37%

Notes: Food system emissions are estimated from a) FAOSTAT (2018), b) US EPA (2012), c) Poore and Nemecek (2018) and d) Fischedick et al. (2014) (using square root of sum of squares of standard deviations when adding uncertainty ranges; see also Chapter 2); e) rounded to nearest fifth percentile due to assessed uncertainty in estimates. Percentage shares were computed by using a total emissions value for the period 2007–2016 of nearly 52 GtCO$_2$-eq yr^{-1} (Chapter 2), using GWP values of the IPCC AR5 with no climate feedback (GWP-CH$_4$=28; GWP-N$_2$O=265).

food, nutrition, and health policy adaptation options such as social safety nets and social protection have been implemented in India, Pakistan, Middle East and North Africa (Devereux 2015; Mumtaz and Whiteford 2017; Narayanan and Gerber 2017).

Financial incentives policies at the national scale used as adaptation options include taxes and subsidies; index-based weather insurance schemes; and catastrophe bonds (Zilberman et al. 2018; Linnerooth-Bayer and Hochrainer-Stigler 2015; Ruiter et al. 2017 and Campillo et al. 2017). Microfinance, disaster contingency funds, and cash transfers are other mechanisms (Ozaki 2016 and Kabir et al. 2016).

1.3.5.3 Community-based adaptation

Community-based adaptation (CBA) builds on social organisational capacities and resources to address food security and climate change. CBA represents bottom-up approaches and localised adaptation measures where social dynamics serve as the power to respond to the impacts of climate change (Ayers and Forsyth 2009). It identifies, assists, and implements development activities that strengthen the capacity of local people to adapt to living in a riskier and less predictable climate, while ensuring their food security.

Klenk et al. (2017) found that mobilisation of local knowledge can inform adaptation decision-making and may facilitate greater flexibility in government-funded research. As an example, rural innovation in terrace agriculture developed on the basis of a local coping mechanism and adopted by peasant farmers in Latin America may serve as an adaptation option to climate change (Bocco and Napoletano, 2017). Clemens et al. (2015) indicated that learning alliances provided social learning and knowledge-sharing in Vietnam through an open dialogue platform that provided incentives and horizontal exchange of ideas.

Community-based adaptation generates strategies through participatory processes, involving local stakeholders and development and disaster risk reduction practitioners. Fostering collaboration and community stewardship is central to the success of CBA (Scott et al. 2017). Preparedness behaviours that are encouraged include social connectedness, education, training, and messaging; CBA also can encompass beliefs that might improve household preparedness to climate disaster risk (Thomas et al. 2015). Reliance on social networks, social groups connectivities, or moral economies reflect the importance of collaboration within communities (Reuter 2018; Schramski et al. 2017).

Yet, community-based adaptation also needs to consider methods that engage with the drivers of vulnerability as part of community-based approaches, particularly questions of power, culture, identity and practice (Ensor et al. 2018). The goal is to avoid maladaptation or exacerbation of existing inequalities within the communities (Buggy and McNamara 2016). For example, in the Pacific Islands, elements considered in a CBA plan included people's development aspirations; immediate economic, social and environmental benefits; dynamics of village governance, social rules and protocols; and traditional forms of knowledge that could inform sustainable solutions (Remling and Veitayaki 2016).

With these considerations, community-based adaptation can help to link local adaptation with international development and climate change policies (Forsyth 2013). In developing CBA programmes, barriers exist that may hinder implementation. These include poor coordination within and between organisations implementing adaptation options, poor skills, poor knowledge about climate change, and inadequate communication among stakeholders (Spires et al. 2014). A rights-based approach has been suggested to address issues of equality, transparency, accountability and empowerment in adaptation to climate change (Ensor et al. 2015).

In summary, institutional measures, including risk management, policies, and planning at global, national, and local scales can support adaptation. Advance planning and focus on institutions can aid in guiding decision-making processes and foster resilience. There is evidence that institutional measures can support the scaling up of adaptation and thus there is reason to believe that systemic resilience is achievable.

1.3.6   Tools and finance

1.3.6.1   Early warning systems

Many countries and regions in the world have adopted early warning systems (EWS) to cope with climate variability and change as it helps to reduce interruptions and improve response times before and after extreme weather events (Ibrahim and Kruczkiewicz 2016). The Early Warning and Early Action (EW/EA) framework has been implemented in West Africa (Red Cross 2011) and Mozambique (DKNC 2012). Bangladesh has constructed cyclone shelters where cyclone warnings are disseminated and responses organised (Mallick et al. 2013). In Benin, a Standard Operating Procedure is used to issue early

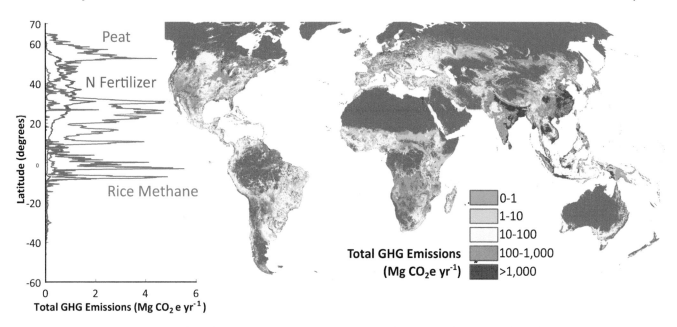

Figure 5.9 | Cropland GHGs consist of CH₄ from rice cultivation, CO₂, N₂O, and CH₄ from peatland draining, and N₂O from N fertiliser application. Total emissions from each grid cell are concentrated in Asia, and are distinct from patterns of production intensity (Carlson et al. 2017).

warnings through the UNDP Climate Information and Early Warning Systems Project (UNDP 2016).

However, there are some barriers to building effective early warning systems in Africa, such as lack of reliable data and distribution systems, lack of credibility, and limited relationships with media and government agencies (UNDP 2016). Mainstreaming early warning systems in adaptation planning could present a significant opportunity for climate disaster risk reduction (Zia and Wagner 2015). Enenkel et al. (2015) suggested that the use of smartphone applications that concentrate on food and nutrition security could help with more frequent and effective monitoring of food prices, availability of fertilisers and drought-resistant seeds, and could help to turn data streams into useful information for decision support and resilience building.

GIS and remote sensing technology are used for monitoring and risk quantification for broad-spectrum stresses such as drought, heat, cold, salinity, flooding, and pests (Skakun et al. 2017; Senay et al. 2015; Hossain et al. 2015 and; Brown 2016), while site-specific applications, such as drones, for nutrient management, precision fertilisers, and residue management can help devise context-specific adaptations (Campbell et al. 2016 and; Baker et al. 2016). Systematic monitoring and remote sensing options, as argued by Aghakouchak et al. (2015), showed that satellite observations provide opportunities to improve early drought warning. Waldner et al. (2015) found that cropland mapping allows strategic food and nutrition security monitoring and climate modelling.

Access to a wide range of adaptation technologies for precipitation change is important, such as rainwater harvesting, wastewater treatment, stormwater management and bioswales, water demand reduction, water-use efficiency, water recycling and reuse, aquifer recharge, inter-basin water transfer, desalination, and surface-water storage (ADB 2014).

1.3.6.2   Financial resources

Financial instruments such as micro-insurance, index-based insurance, provision of post-disaster finances for recovery and pre-disaster payment are fundamental means to reduce lower and medium level risks (Linnerooth-Bayer and Hochrainer-Stigler 2014). Fenton & Paavola, 2015; Dowla, 2018). Hammill et al. (2010) found that microfinance services (MFS) are especially helpful for the poor. MFS can provide poor people with the means to diversify, accumulate and manage the assets needed to become less susceptible to shocks and stresses. As a result, MFS plays an important role in vulnerability reduction and climate change adaptation among some of the poor. The provision of small-scale financial products to low-income and otherwise disadvantaged groups by financial institutions can serve as adaptation to climate change. Access to finance in the context of climate change adaptation that focuses on poor households and women in particular is bringing encouraging results (Agrawala and Carraro 2010).

In summary, effective adaptation strategies can reduce the negative impacts of climate change. Food security under changing climate conditions depends on adaptation throughout the entire food system – production, supply chain, and consumption/demand, as well as reduction of food loss and waste. Adaptation can be autonomous, incremental, or transformative, and can reduce vulnerability and enhance resilience. Local food systems are embedded in culture, beliefs and values, and ILK can contribute to enhancing food system resilience to climate change (*high confidence*). Institutional and capacity-building measures are needed to scale up adaptation measures across local, national, regional, and global scales.

1.4   Impacts of food systems on climate change

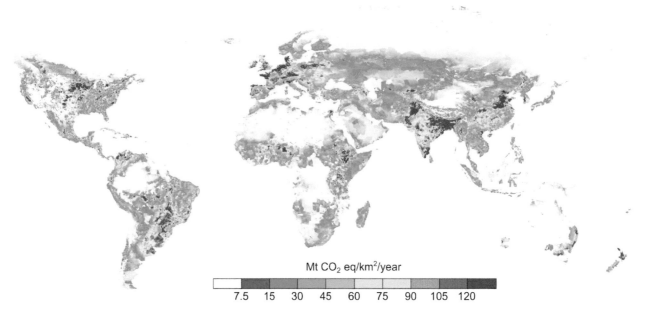

Mt CO$_2$ eq/km^2/year

7.5   15   30   45   60   75   90   105   120

Figure 5.10 | Global GHG emissions from livestock for 1995–2005 (adapted from Herrero et al. 2016a).

1.4.1   Greenhouse gas emissions from food systems

This chapter assesses the contributions of the entire food system to greenhouse gas (GHG) emissions. Food systems emissions include CO$_2$ and non-CO$_2$ gases, specifically those generated from: (i) crop and livestock activities within the farm gate (Table 5.4, category 'Agriculture'); (ii) land use and land-use change dynamics associated with agriculture (Table 5.4, category 'Land Use'); and (iii) food processing, retail and consumption patterns, including upstream and downstream processes such as manufacture of chemical fertilisers and fuel (Table 5.4, category 'Beyond Farm Gate'). The first two categories comprise emissions reported by countries in the AFOLU (agriculture, forestry, and other land use) sectors of national GHG inventories; the latter comprises emissions reported in other sectors of the inventory, as appropriate. For instance, industrial processes, energy use, and food loss and waste.

The first two components (agriculture and land use) identified above are well quantified and supported by an ample body of literature (Smith et al. 2014). During the period 2007–2016, global agricultural non-CO$_2$ emissions from crop and livestock activities within the farm gate were 6.2 ± 1.4 GtCO$_2$-eq yr^{-1} during 2007–2016, with methane (142 ± 42 MtCH$_4$ yr^{-1}, or 4.0 ± 1.2 GtCO$_2$-eq yr^{-1}) contributing in CO$_2$-eq about twice as much as nitrous oxide (8.3 ± 2.5 MtN$_2$O yr^{-1}, or 2.2 ± 0.7 GtCO$_2$-eq yr^{-1}) to this total (Table 2.2 in Chapter 2). Emissions from land use associated with agriculture in some regions, such as from deforestation and peatland degradation (both processes involved in preparing land for agricultural use), added another 4.9 ± 2.5 GtCO$_2$-eq yr^{-1} (Chapter 2) globally during the same period. These estimates are associated with uncertainties of about 30% (agriculture) and 50% (land use), as per IPCC AR5 (Smith et al. 2014).

Agriculture activities within the farm gate and associated land-use dynamics are therefore responsible for about 11.1 ± 2.9 GtCO$_2$-eq yr^{-1}, or some 20% of total anthropogenic emissions (Table 5.4), consistent with post-AR5 findings (for example, Tubiello et al. 2015). In terms of individual gases, the contributions of agriculture to total emissions by gas are significantly larger. For instance, over the period 2010–2016, methane gas emissions within the farm gate represented about half of the total CH$_4$ emitted by all sectors, while nitrous dioxide gas emissions within the farm gate represented about three-quarters of the total N$_2$O emitted by all sectors (Tubiello 2019). In terms of carbon, CO$_2$ emissions from deforestation and peatland degradation linked to agriculture contributed about 10% of the CO$_2$ emitted by all sectors in 2017 (Le Quéré et al. 2018).

Food systems emissions beyond the farm gate, such as those upstream from manufacturing of fertilisers, or downstream such as food processing, transport and retail, and food consumption, generally add to emissions from agriculture and land use, but their estimation is very uncertain due to lack of sufficient studies. The IPCC AR5 (Fischedick et al. 2014) provided some information on these other food system components, noting that emissions beyond the farm gate in developed countries may equal those within the farm gate, and cited one study estimating world total food system emissions to be up to 30% of total anthropogenic emissions (Garnett 2011). More recently, Poore and Nemecek (2018), by looking at a database of farms and using a combination of modelling approaches across relevant processes, estimated a total contribution of food systems around 26% of total anthropogenic emissions. Total emissions from food systems may account for 21–37% of total GHG emissions (*medium confidence*).

Based on the available literature, a break-down of individual contributions of food systems emissions is show in Table 5.4, between those from agriculture within the farm gate (10–14%) (*high confidence*); emissions from land use and land-use change dynamics such as deforestation and peatland degradation, which are associated with agriculture in many regions (5–14%) (*high confidence*); and those from food supply chain activities past the farm gate, such as storage, processing, transport, and retail (5–10%) (*limited evidence, medium agreement)*. Note that the corresponding lower range of emissions past the farm gate, for example, 2.6 GtCO$_2$-eq yr^{-1} (Table 5.4), is consistent with recent estimates made by Poore and Nemecek (2018). Contributions from food loss and waste are implicitly included in these estimates of total emissions from food systems (Section 5.5.2.5). They may account for 8–10% of total anthropogenic GHG emissions (*low confidence*) (FAO 2013b).

1.4.2   Greenhouse gas emissions from croplands and soils

Since AR5, a few studies have quantified separate contributions of crops and soils on the one hand, and livestock on the other, to the total emissions from agriculture and associated land use. For instance, Carlson et al. (2017) estimated emissions from cropland to be in the range of 2–3 GtCO$_2$-eq yr^{-1}, including methane emissions from rice, CO$_2$ emissions from peatland cultivation, and N$_2$O emissions from fertiliser applications. Data from FAOSTAT (2018), recomputed to use AR5 GWP values, indicated that cropland emissions from these categories were 3.6 ± 1.2 GtCO$_2$-eq yr^{-1} over the period 2010–2016. Two-thirds of this were related to peatland degradation, followed by N$_2$O emissions from synthetic fertilisers and methane emissions from paddy rice fields (Tubiello 2019). These figures are a subset of the total emissions from agriculture and land use reported in Table 5.4. Asia, especially India, China and Indonesia accounted for roughly 50% of global emissions from croplands. Figure 5.9 shows the spatial distribution of emissions from cropland according to Carlson et al. (2017), not including emissions related to deforestation or changes in soil carbon.

1.4.3   Greenhouse gas emissions from livestock

Emissions from livestock include non-CO$_2$ gases from enteric fermentation from ruminant animals and from anaerobic fermentation in manure management processes, as well as non-CO$_2$ gases from manure deposited on pastures (Smith et al. 2014). Estimates after the AR5 include those from Herrero et al. (2016), who quantified non-CO$_2$ emissions from livestock to be in the range of 2.0–3.6 GtCO$_2$-eq yr^{-1}, with enteric fermentation from ruminants being the main contributor. FAOSTAT (2018) estimates of these emissions, renormalized to AR5 GWP values, were 4.1 ± 1.2 GtCO$_2$-eq yr^{-1} over the period 2010–2016.

These estimates of livestock emissions are for those generated within the farm gate. Adding emissions from relevant land-use change, energy use, and transportation processes, FAO (2014a) and Gerber et al. (2013) estimated livestock emissions of up to 5.3 ±1.6 GtCO$_2$-eq yr^{-1} circa the year 2010. This data came from original papers, but was scaled to SAR global warming potential (GWP) values for methane, for comparability with previous results.

All estimates agree that cattle are the main source of global livestock emissions (65–77%). Livestock in low and middle-income countries contribute 70% of the emissions from ruminants and 53% from monogastric livestock (animals without ruminant digestion processes such as pigs and poultry), and these are expected to increase as demand for livestock products increases in these countries (Figure 5.10). In contrast to the increasing trend in absolute GHG emissions, GHG emissions intensities, defined as GHG emissions per unit produced, have declined globally and are about 60% lower today than in the 1960s. This is largely due to improved meat and milk productivity of cattle breeds (FAOSTAT 2018; Davis et al. 2015).

Still, products like red meat remain the most inefficient in terms of emissions per kg of protein produced in comparison to milk, pork, eggs and all crop products (IPCC 2014b). Yet, the functional unit used in these measurements is highly relevant and may produce different results (Salou et al. 2017). For instance, metrics based on products tend to rate intensive livestock systems as efficient, while metrics based on area or resources used tend to rate extensive systems as efficient (Garnett 2011). In ruminant dairy systems, less intensified farms show higher emissions if expressed by product, and lower emissions if expressed by Utilizable Agricultural Land (Gutiérrez-Peña et al. 2019; Salvador et al. 2017; Salou et al. 2017).

Furthermore, if other variables are used in the analysis of GHG emissions of different ruminant production systems, such as human-edible grains used to feed animals instead of crop waste and pastures of marginal lands, or carbon sequestration in pasture systems in degraded lands, then the GHG emissions of extensive systems are reduced. Reductions of 26% and 43% have been shown in small ruminants, such as sheep and goats (Gutiérrez-Peña et al. 2019; Salvador et al. 2017; Batalla et al. 2015 and Petersen et al. 2013). In this regard, depending on what the main challenge is in different regions (for example, undernourishment, over-consumption, natural resources degradation), different metrics could be used as reference. Other metrics that consider nutrient density have been proposed because they provide potential for addressing both mitigation and health targets (Doran-Browne et al. 2015).

Uncertainty in worldwide livestock population numbers remains the main source of variation in total emissions of the livestock sector, while at the animal level, feed intake, diet regime, and nutritional composition are the main sources of variation through their impacts on enteric fermentation and manure N excretion.

Increases in economies of scale linked to increased efficiencies and decreased emission intensities may lead to more emissions, rather than less, an observed dynamic referred to by economists as a 'rebound effect'. This is because increased efficiency allows production processes to be performed using fewer resources and often at lower cost. This in turn influences consumer behaviour and product use, increasing demand and leading to increased production. In this way, the expected gains from new technologies that increase the efficiency of resource use may be reduced (for example, increase in the total production of livestock despite increased efficiency of production due to increased demand for meat sold at lower prices). Thus, in order for the livestock sector to provide a contribution to GHG mitigation, reduction in emissions intensities need to be accompanied by appropriate governance and incentive mechanisms to avoid rebound effects, such as limits on total production.

5

Variation in estimates of N$_2$O emissions are due to differing (i) climate regimes, (ii) soil types, and (iii) N transformation pathways (Charles et al. 2017 and Fitton et al. 2017). It was recently suggested that N$_2$O soil emissions linked to livestock through manure applications could be 20–40% lower than previously estimated in some regions. For instance, in Sub-Saharan Africa and Eastern Europe (Gerber et al. 2016) and from smallholder systems in East Africa (Pelster et al. 2017). Herrero et al. (2016a) estimated global livestock enteric methane to range from 1.6–2.7 Gt CO$_2$-eq, depending on assumptions of body weight and animal diet.

1.4.4 Greenhouse gas emissions from aquaculture

Emissions from aquaculture and fisheries may represent some 10% of total agriculture emissions, or about 0.58 GtCO$_2$-eq yr^{-1} (Barange et al. 2018), with two-thirds being non-CO$_2$ emissions from aquaculture (Hu et al. 2013; Yang et al. 2015) and the rest due to fuel use in fishing vessels. They were not included in Table 5.4 under agriculture emissions, as these estimates are not included in national GHG inventories and global numbers are small as well as uncertain.

Methodologies to measure aquaculture emissions are still being developed (Vasanth et al. 2016). N$_2$O emissions from aquaculture are partly linked to fertiliser use for feed as well as aquatic plant growth, and depend on the temperature of water as well as on fish production (Paudel et al. 2015). Hu et al. (2012) estimated the global N$_2$O emissions from aquaculture in 2009 to be 0.028 GtCO$_2$-eq yr^{-1}, but could increase to 0.114 GtCO$_2$-eq yr^{-1} (that is 5.72% of anthropogenic N$_2$O–N emissions) by 2030 for an estimated 7.10% annual growth rate of the aquaculture industry. Numbers estimated by Williams and Crutzen (2010) were around 0.036 GtCO$_2$-eq yr^{-1}, and suggested that this may rise to more than 0.179 GtCO$_2$-eq yr^{-1} within 20 years for an estimated annual growth of 8.7%. Barange et al. (2018) assessed the contribution of aquaculture to climate change as 0.38 GtCO$_2$-eq yr^{-1} in 2010, around 7% of those from agriculture.

CO$_2$ emissions coming from the processing and transport of feed for fish raised in aquaculture, and also the emissions associated with the manufacturing of floating cultivation devices (e.g., rafts or floating fish-farms), connecting or mooring devices, artificial fishing banks or reefs, and feeding devices (as well as their energy consumption) may be considered within the emissions from the food system. Indeed, most of the GHG emissions from aquaculture are associated with the production of raw feed materials and secondarily, with the transport of raw materials to mills and finished feed to farms (Barange et al. 2018).

1.4.5 Greenhouse gas emissions from inputs, processing, storage and transport

Apart from emissions from agricultural activities within the farm gate, food systems also generate emissions from the pre- and post-production stages in the form of input manufacturing (fertilisers, pesticides, feed production) and processing, storage, refrigeration, retail, waste disposal, food service, and transport. The total contribution of these combined activities outside the farm gate is not well documented.

Based on information reported in the AR5 (Fischedick et al. 2014) and Poore and Nemecek (2018), we estimate their total contribution to be roughly 5-10% of total anthropogenic emissions (Table 5.4). There is no post-AR5 assessment at the global level in terms of absolute emissions. Rather, several studies have recently investigated how the combined emissions within and outside the farm gate are embedded in food products and thus associated with specific dietary choices (see next section). Below important components of food systems emissions beyond the farm gate are discussed based on recent literature.

Refrigerated trucks, trailers, shipping containers, warehouses, and retail displays that are vital parts of food supply chains all require energy and are direct sources of GHG emissions. Upstream emissions in terms of feed and fertiliser manufacture and downstream emissions (transport, refrigeration) in intensive livestock production (dairy, beef, pork) can account for up to 24–32% of total livestock emissions, with the higher fractions corresponding to commodities produced by monogastric animals (Weiss and Leip 2012). The proportion of upstream/downstream emissions fall significantly for less-intensive and more-localised production systems (Mottet et al. 2017a).

Transport and processing. Recent globalisation of agriculture has promoted industrial agriculture and encouraged value-added processing and more distant transport of agricultural commodities, all leading to increased GHG emissions. Although often GHG-intensive, food transportation plays an important role in food chains: it delivers food from producers to consumers at various distances, particularly to feed people in food-shortage zones from food-surplus zones. (Section 5.5.2.6 for assessment of local food production.)

To some extent, processing is necessary in order to make food supplies more stable, safe, long-lived, and in some cases, nutritious (FAO 2007). Agricultural production within the farm gate may contribute 80–86% of total food-related emissions in many countries, with emissions from other processes such as processing and transport being small (Vermeulen et al. 2012). However, in net food-importing countries where consumption of processed food is common, emissions from other parts of the food lifecycle generated in other locations are much higher (Green et al. 2015).

A study conducted by Wakeland et al. (2012) in the USA found that the transportation-related carbon footprint varies from a few percent to more than half of the total carbon footprint associated with food production, distribution, and storage. Most of the GHGs emitted from food processing are a result of the use of electricity, natural gas, coal, diesel, gasoline or other energy sources. Cookers, boilers, and furnaces emit carbon dioxide, and wastewater emits methane and nitrous oxide. The most energy-intensive processing is wet milling of maize, which requires 15% of total USA food industry energy (Bernstein et al. 2008); processing of sugar and oils also requires large amounts of energy.

1.4.6   Greenhouse gas emissions associated with different diets

There is now extensive literature on the relationship between food products and emissions, although the focus of the studies has been on high-income countries. Godfray et al. (2018) updated Nelson et al. (2016), a previous systematic review of the literature on environmental impacts associated with food, and concluded that higher consumption of animal-based foods was associated with higher estimated environmental impacts, whereas increased consumption of plant-based foods was associated with estimated lower environmental impact. Assessment of individual foods within these broader categories showed that meat – sometimes specified as ruminant meat (mainly beef) – was consistently identified as the single food with the greatest impact on the environment, most often in terms of GHG emissions and/or land use per unit commodity. Similar hierarchies, linked to well-known energy losses along trophic chains, from roots to beef were found in another recent review focussing exclusively on GHG emissions (Clune et al. 2017), and one on life-cycle assessments (Poore and Nemecek 2018). Poore and Nemecek (2018) amassed an extensive database that specifies both the hierarchy of emissions intensities and the variance with the production context (for example, by country and farming system).

The emissions intensities of red meat mean that its production has a disproportionate impact on total emissions (Godfray et al. 2018). For example, in the USA 4% of food sold (by weight) is beef, which accounts for 36% of food-related emissions (Heller and Keoleian 2015). Food-related emissions are therefore very sensitive to the amount and type of meat consumed. However, 100 g of beef has twice as much protein as the equivalent in cooked weight of beans, for example, and 2.5 times more iron. One can ingest only about 2.5 kg of food per day and not all food items are as dense in nutrition.

There is therefore *robust evidence with high agreement* that the mixture of foods eaten can have a highly significant impact on per capita carbon emissions, driven particularly through the amount of (especially grain-fed) livestock and products.

Given the rising costs of malnutrition in all its forms, a legitimate question is often asked: would a diet that promotes health through good nutrition also be one that mitigates GHG emissions? Whilst sustainable diets need not necessarily provide more nutrition, there is certainly significant overlap between those that are healthier (e.g., via eating more plant-based material and less livestock-based material), and eating the appropriate level of calories. In their systematic review, Nelson et al. (2016) conclude that, in general, a dietary pattern that is higher in plant-based foods, such as vegetables, fruits, whole grains, legumes, nuts, and seeds, and lower in animal-based foods is more health-promoting and is associated with lesser environmental impact (GHG emissions and energy, land, and water use) than is the current average 'meat-based' diet.

Recent FAO projections of food and agriculture to 2050 under alternative scenarios characterised by different degrees of sustainability, provide global-scale evidence that rebalancing diets is key to increasing the overall sustainability of food and agricultural

systems world-wide. A 15% reduction of animal products in the diets of high-income countries by 2050 would contribute to containing the need to expand agricultural output due to upward global demographic trends. Not only would GHG emissions and the pressure on land and water be significantly reduced but the potential for low-income countries to increase the intake of animal-based food, with beneficial nutritional outcomes, could be enhanced (FAO 2018a). Given that higher-income countries typically have higher emissions per capita, results are particularly applicable in such places.

However, Springmann et al. (2018a) found that there are locally applicable upper bounds to the footprint of diets around the world, and for lower-income countries undergoing a nutrition transition, adopting 'Westernised' consumption patterns (over-consumption, large amounts of livestock produce, sugar and fat), even if in culturally applicable local contexts, would increase emissions. The global mitigation potential of healthy but low-emissions diets is discussed in detail in Section 5.5.2.1.

In summary, food system emissions are growing globally due to increasing population, income, and demand for animal-sourced products (*high confidence*). Diets are changing on average toward greater consumption of animal-based foods, vegetable oils and sugar/sweeteners (*high confidence*) (see also Chapter 2), with GHG emissions increasing due to greater amounts of animal-based products in diets (*robust evidence, medium agreement*).

1.5   Mitigation options, challenges and opportunities

The IPCC AR5 WG III concluded that mitigation in agriculture, forestry, and land use (AFOLU) is key to limit climate change in the 21st century, in terms of mitigation of non-CO_2 GHGs, which are predominately emitted in AFOLU, as well as in terms of land-based carbon sequestration. Wollenberg et al. (2016) highlighted the need to include agricultural emissions explicitly in national mitigation targets and plans, as a necessary strategy to meet the 2°C goal of the Paris Agreement. This chapter expands on these key findings to document how mitigation in the entire food system, from farm gate to consumer, can contribute to reaching the stated global mitigation goals, but in a context of improved food security and nutrition. To put the range of mitigation potential of food systems in context, it is worth noting that emissions from crop and livestock are expected to increase by 30–40% from present to 2050, under business-as-usual scenarios that include efficiency improvements as well as dietary changes linked to increased income per capita (FAO 2018a; Tubiello et al. 2014). Using current emissions estimates in this chapter and Chapter 2, these increases translate into projected GHG emissions from agriculture of 8–9 Gt CO_2-eq yr^{-1} by 2050 (*medium confidence*).

The AR5 ranked mitigation measures from simple mechanisms such as improved crop and livestock management (Smith et al. 2014) to more complex carbon dioxide reduction interventions, such as afforestation, soil carbon storage and biomass energy projects with carbon capture and storage (BECCS). The AR5 WGIII AFOLU chapter

(Smith et al. 2014) identified two primary categories of mitigation pathways from the food system:

Supply side: Emissions from agricultural soils, land-use change, land management, and crop and livestock practices can be reduced and terrestrial carbon stocks can be increased by increased production efficiencies and carbon sequestration in soils and biomass, while emissions from energy use at all stages of the food system can be reduced through improvements in energy efficiency and fossil fuel substitution with carbon-free sources, including biomass.

Demand side: GHG emissions could be mitigated by changes in diet, reduction in food loss and waste, and changes in wood consumption for cooking.

In this chapter, supply-side mitigation practices include land-use change and carbon sequestration in soils and biomass in both crop and livestock systems. Cropping systems practices include improved land and fertiliser management, land restoration, biochar applications, breeding for larger root systems, and bridging yield gaps (Dooley and Stabinsky 2018). Options for mitigation in livestock systems include better manure management, improved grazing land management, and better feeding practices for animals. Agroforestry also is a supply-side mitigation practice. Improving efficiency in supply chains is a supply-side mitigation measure.

Demand-side mitigation practices include dietary changes that lead to reduction of GHG emissions from production and changes in land use that sequester carbon. Reduction of food loss and waste can contribute to mitigation of GHGs on both the supply and demand sides. See Section 5.7 and Chapter 7 for the enabling conditions needed to ensure that these food system measures would deliver their potential mitigation outcomes.

1.5.1 Supply-side mitigation options

The IPCC AR5 identified options for GHG mitigation in agriculture, including cropland management, restoration of organic soils, grazing land management and livestock, with a total mitigation potential of 1.6–4.6 $GtCO_2$-eq yr^{-1} by 2030 (compared to baseline emissions in the same year), at carbon prices from 20 to 100 USD per tCO_2-eq (Smith et al. 2014). Reductions in GHG emissions intensity (emissions per unit product) from livestock and animal products can also be a means to achieve reductions in absolute emissions in specific contexts and with appropriate governance (*medium confidence*). Agroforestry mitigation practices include rotational woodlots, long-term fallow, and integrated land use.

Emissions from food systems can be reduced significantly by the implementation of practices that reduce carbon dioxide, methane, and nitrous oxide emissions from agricultural activities related to the production of crops, livestock, and aquaculture. These include implementation of more sustainable and efficient crop and livestock production practices aimed at reducing the amount of land needed per output (reductions in GHG emissions intensity from livestock and animal production can support reductions in absolute emissions if total production is constrained), bridging yield gaps, implementing better feeding practices for animals and fish in aquaculture, and better manure management (FAO 2019a). Practices that promote soil improvements and carbon sequestration can also play an important role. In the South America region, reduction of deforestation, restoration of degraded pasture areas, and adoption of agroforestry and no-till agricultural techniques play a major role in the nation's voluntary commitments to reduce GHG emissions in the country's mitigation activities (Box 5.4).

The importance of supply-side mitigation options is that these can be directly applied by food system actors (farmers, processors, retailers) and can contribute to improved livelihoods and income generation. Recognising and empowering farming system actors with the right incentives and governance systems will be crucial to increasing the adoption rates of effective mitigation practices and to build convincing cases for enabling GHG mitigation (Section 5.7 and Chapter 7).

Box 5.4 | Towards sustainable intensification in South America

Reconciling the increasing global food demand with limited land resources and low environmental impact is a major global challenge (FAO 2018a; Godfray and Garnett 2014; Yao et al. 2017). South America has been a significant contributor of the world's agricultural production growth in the last three decades (OECD and FAO 2015), driven partly by increased export opportunities for specific commodities, mainly soybeans and meat (poultry, beef and pork).

Agricultural expansion, however, has driven profound landscape transformations in the region, particularly between the 1970s and early 2000s, contributing to increased deforestation rates and associated GHG emissions. High rates of native vegetation conversion were found in Argentina, Bolivia, Brazil, Colombia, Ecuador, Paraguay and Peru (FAO

2016b; Graesser et al. 2015), threatening ecologically important biomes, such as the Amazon, the savannas (Cerrado, Chacos and Lannos), the Atlantic Rainforest, the Caatinga, and the Yungas. The Amazon biome is a particularly sensitive biome as it provides crucial ecosystem services including biodiversity, hydrological processes (through evapotranspiration, cloud formation, and precipitation), and biogeochemical cycles (including carbon) (Bogaerts et al. 2017; Fearnside 2015; Beuchle et al. 2015; Grecchi et al. 2014; Celentano et al. 2017; Soares-Filho et al. 2014; Nogueira et al. 2018). Further, deforestation associated with commodity exports has not led to inclusive socioeconomic development, but rather has exacerbated social inequality and created more challenging living conditions for lower-income people (Celentano et al. 2017). Nor has it avoided increased hunger of local populations in the last few years (FAO 2018b).

In the mid-2000s, governments, food industries, NGOs, and international programmes joined forces to put in place important initiatives to respond to the growing concerns about the environmental impacts of agricultural expansion in the region (Negra et al. 2014; Finer et al. 2018). Brazil led regional action by launching the Interministerial Plan of Action for Prevention and Control of Deforestation of the Legal Amazon[2] (PPCDAm), associated with development of a real-time deforestation warning system. Further, Brazil built capacity to respond to alerts by coordinated efforts of ministries, the federal police, the army and public prosecution (Negra et al. 2014; Finer et al. 2018).

Other countries in the region have also launched similar strategies, including a zero-deforestation plan in Paraguay in 2004 (Gasparri and de Waroux 2015), and no-deforestation zones in Argentina in 2007 (Garcia Collazo et al. 2013). Peru also developed the National System of Monitoring and Control, led by the National Forest Service and Wildlife Authority (SERFOR), to provide information and coordinate response to deforestation events, and Colombia started producing quarterly warning reports on active fronts of deforestation in the country (Finer et al. 2018).

Engagement of the food industry and NGOs, particularly through the Soy Moratorium (from 2006) and Beef Moratorium (from 2009) also contributed effectively to keep deforestation at low historical rates in the regions where they were implemented (Nepstad et al. 2014 and Gibbs et al. 2015). In 2012, Brazil also created the national land registry system (SICAR), a georeferenced database, which allows monitoring of farms' environmental liability in order to grant access to rural credit. Besides the governmental schemes, funding agencies and the Amazon Fund provide financial resources to assist smallholder farmers to comply with environmental regulations (Jung et al. 2017).

Box 5.4 (continued)

Nevertheless, Azevedo et al. (2017) argue that the full potential of these financial incentives has not been achieved, due to weak enforcement mechanisms and limited supporting public policies. Agricultural expansion and intensification have complex interactions with deforestation. While mechanisms have been implemented in the region to protect native forests and ecosystems, control of deforestation rates require stronger governance of natural resources (Ceddia et al. 2013 and Oliveira and Hecht 2016), including monitoring programmes to evaluate fully the results of land-use policies in the region.

Public and private sector actions resulted in a reduction of the Brazilian legal Amazon deforestation rate from 2.78 Mha yr^{-1} in 2004, to about 0.75 Mha yr^{-1} (ca. 0.15%) in 2009 (INPE 2015), oscillating from 0.46 Mha and 0.79 Mha (2016) since then (INPE 2018; Boucher and Chi 2018). The governmental forest protection scheme was also expanded to other biomes. As a result, the Brazilian Cerrado deforestation was effectively reduced from 2.9 Mha yr^{-1} in 2004 to an average of 0.71 Mha yr^{-1} in 2016–2017 (INPE 2018).

Overall, deforestation rates in South America have declined significantly, with current deforestation rates being about half of rates in the early 2000s (FAOSTAT 2018). However, inconsistent conservation policies across countries (Gibbs et al. 2015) and recent hiccups (Curtis et al. 2018) indicate that deforestation control still requires stronger reinforcement mechanisms (Tollefson 2018). Further, there are important spill-over effects that need coordinated international governance. Curtis et al. (2018) and Dou et al. (2018) point out that, although the Amazon deforestation rate decreased in Brazil, it has increased in other regions, particularly in South Asia, and in other countries in South America, resulting in nearly constant deforestation rates worldwide.

Despite the reduced expansion rates into forest land, agricultural production continues to rise steadily in South America, relying on increasing productivity and substitution of extensive pastureland by crops. The average soybean and maize productivity in the region increased from 1.8 and 2.0 t ha^{-1} in 1990 to 3.0 and 5.0 t ha^{-1}, respectively, in 2015 (FAOSTAT 2018). Yet, higher crop productivity was not enough to meet growing demand for cereals and oilseeds and cultivation continued to expand, mainly on grasslands (Richards 2015). The reconciliation of this expansion with higher demand for meat and dairy products was carried out through the intensification of livestock systems (Martha et al. 2012). Nevertheless, direct and indirect deforestation still occurs, and recently deforestation rates have increased (INPE

---

2   The Legal Amazon is a Brazilian region of 501.6 Mha (about 59% of the Brazilian territory) that contains all the Amazon but also 40% of the Cerrado and 40% of the Pantanal biomes, with a total population of 25.47 million inhabitants.

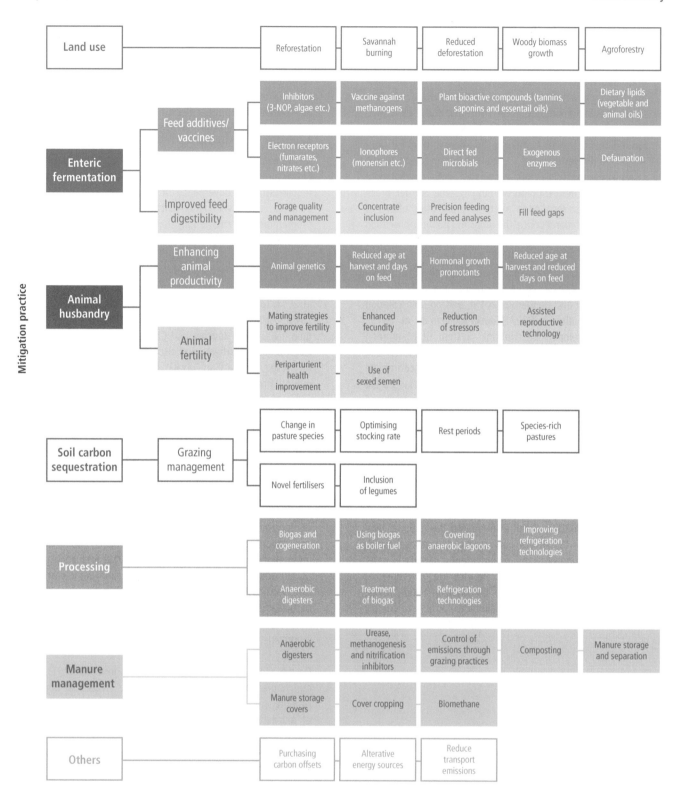

Figure 5.11 | Technical supply-side mitigation practices in the livestock sector (adapted from Hristov et al. 2013b; Herrero et al. 2016b and Smith et al. 2014).

2018), albeit they remain far smaller than observed in the 2000–2010 period.

The effort towards sustainable intensification has also been incorporated in agricultural policies. In Brazil, for instance, the reduction of deforestation, the restoration of degraded pasture areas, the adoption of integrated agroforestry systems[3] and no-till agricultural techniques play a major role in the nation's voluntary commitments to reduce GHG emissions in the country's NAMAs (Mozzer 2011) and NDCs (Silva Oliveira et al. 2017; Rochedo et al. 2018). Such commitment under the UNFCCC is operationalised through the Low Carbon Agriculture Plan (ABC),[4] which is based on low interest credit for investment in sustainable agricultural technologies (Mozzer 2011). Direct pasture restoration and integrated systems reduce area requirements (Strassburg et al. 2014), and increase organic matter (Gil et al. 2015; Bungenstab 2012; Maia et al. 2009), contributing to overall lifecycle emissions reduction (Cardoso et al. 2016; de Oliveira Silva et al. 2016). Also, increased adoption of supplementation and feedlots, often based on agroindustrial co-products and agricultural crop residues are central to improve productivity and increase climate resilience of livestock systems (Mottet et al. 2017a; van Zanten et al. 2018).

Despite providing clear environmental and socio-economic co-benefits, including improved resource productivity, socio-environmental sustainability and higher economic competitiveness, implementation of the Brazilian Low Carbon Agriculture Plan is behind schedule (Köberle et al. 2016). Structural inefficiencies related to the allocation and distribution of resources need to be addressed to put the plan on track to meet its emissions reduction targets. Monitoring and verification are fundamental tools to guarantee the successful implementation of the plan.

Overall, historical data and projections show that South America is one of the regions of the world with the highest potential to increase crop and livestock production in the coming decades in a sustainable manner (Cohn et al. 2014), increasing food supply to more densely populated regions in Asia, Middle East and Europe. However, a great and coordinated effort is required from governments, industry, traders, scientists and the international community to improve planning, monitoring and innovation to guarantee sustainable intensification of its agricultural systems, contribution to GHG mitigation, and conservation of the surrounding environment (Negra et al. 2014; Curtis et al. 2018 and Lambin et al. 2018).

1.5.1.1   Greenhouse gas mitigation in croplands and soils

The mitigation potential of agricultural soils, cropland and grazing land management has been the subject of much research and was thoroughly summarised in the AR5 (Smith et al. 2014) (see also Chapter 2, Section 2.5.1 and Chapter 6, Section 6.3.1). Key mitigation pathways are related to practices reducing nitrous oxide emissions from fertiliser applications, reducing methane emissions from paddy rice, reducing both gases through livestock manure management and applications, and sequestering carbon or reducing its losses, with practices for improving grassland and cropland management identified as the largest mitigation opportunities. Better monitoring reporting and verification (MRV) systems are currently needed for reducing uncertainties and better quantifying the actual mitigation outcomes of these activities.

---

3   Integrated agroforestry systems are agricultural systems that strategically integrate two or more components among crops, livestock and forestry. The activities can be in consortium, succession or rotation in order to achieve overall synergy.

4   ABC – *Agricultura de Baixo Carbono* in Portuguese.

Table 5.5 | Carbon sequestration potential for agroforestry (Mbow et al. 2014b).

Source	Carbon sequestration (tCO$_2$ km^{-2} yr^{-1}) (range)	Carbon stock (tCO$_2$ km^{-2}) (range)	Maximum rotation period (years)
Dominant parklands	183 (73–293)	12,257 (2091–25,983)	50
Rotational woodlots[a]	1,431 (807–2128)	6,789 (4257–9358)	5
Tree planting-windrows-home gardens	220.2 (146–293)	6,973 (–)	25
Long-term fallows, regrowth of wood-lands in abandoned farms[b]	822 (80–2128)	5,761 (–)	25
Integrated land use	1,145 (367–2458)	28,589 (4404–83,676)	50
Soil carbon	330 (91–587)	33,286 (4771–110,100)	–

[a] May be classified as forestry on forest land, depending on the spatial and temporal characteristics of these activities.

[b] This is potentially not agroforestry, but forestry following abandonment of agricultural land.

New work since AR5 has focused on identifying pathways for the reductions of GHG emissions from agriculture to help meet Paris Agreement goals (Paustian et al. 2016 and Wollenberg et al. 2016). Altieri and Nicholls (2017) have characterised mitigation potentials from traditional agriculture. Zomer et al. (2017) have updated previous estimates of global carbon sequestration potential in cropland soils. Mayer et al. (2018) converted soil carbon sequestration potential through agricultural land management into avoided temperature reductions. Fujisaki et al. (2018) identify drivers to increase soil organic carbon in tropical soils. For discussion of integrated practices such as sustainable intensification, conservation agriculture and agroecology, see Section 5.6.4.

Paustian et al. (2016) developed a decision-tree for facilitating implementation of mitigation practices on cropland and described the features of key practices. They observed that most individual mitigation practices will have a small effect per unit of land, and hence they need to be combined and applied at large scales for their impact to be significant. Examples included aggregation of cropland practices (for example, organic amendments, improved crop rotations and nutrient management and reduced tillage) and grazing land practices (e.g., grazing management, nutrient and fire management and species introduction) that could increase net soil carbon stocks while reducing emissions of N_2O and CH_4.

However, it is well-known that the portion of projected mitigation from soil carbon stock increase (about 90% of the total technical potential) is impermanent. It would be effective for only 20–30 years due to saturation of the soil capacity to sequester carbon, whereas non-CO_2 emission reductions could continue indefinitely. '**Technical potential**' is the maximum amount of GHG mitigation achievable through technology diffusion.

Biochar application and management towards enhanced root systems are mitigation options that have been highlighted in recent literature (Dooley and Stabinsky 2018; Hawken 2017; Paustian et al. 2016; Woolf et al. 2010 and Lenton 2010).

1.5.1.2   Greenhouse gas mitigation in livestock systems

The technical options for mitigating GHG emissions in the livestock sector have been the subject of recent reviews (Mottet et al. 2017b; Hristov et al. 2013a,b; Smithers 2015; Herrero et al. 2016a; Rivera-Ferre et al. 2016b) (Figure 5.11). They can be classified as either targeting reductions in enteric methane; reductions in nitrous oxide through manure management; sequestering carbon in pastures; implementation of best animal husbandry and management practices, which would have an effect on most GHG; and land-use practices that also help sequester carbon. Excluding land-use practices, these options have a technical mitigation potential ranging 0.2–2.4 $GtCO_2$-eq yr^{-1} (Herrero et al. 2016a; FAO 2007) (Chapters 2 and 6.)

The opportunities for carbon sequestration in grasslands and rangelands may be significant (Conant 2010), for instance, through changes in grazing intensity or manure recycling aimed at maintaining grassland productivity (Hirata et al. 2013). Recent studies have questioned the economic potential of such practices in regard to whether they could be implement at scale for economic gain (Garnett et al. 2017; Herrero et al. 2016a and Henderson et al. 2015). For instance, Henderson et al. (2015) found economic potentials below 200 $MtCO_2$-eq yr^{-1}. Carbon sequestration can occur in situations where grasslands are highly degraded (Garnett 2016). Carbon sequestration linked to livestock management could thus be considered as a co-benefit of well-managed grasslands, as well as a mitigation practice.

Different production systems will require different strategies, including the assessment of impacts on food security, and this has been the subject of significant research (e.g., Rivera-Ferre et al. 2016b). Livestock systems are heterogeneous in terms of their agroecological orientation (arid, humid or temperate/highland locations), livestock species (cattle, sheep, goats, pigs, poultry and others), structure (grazing only, mixed-crop-livestock systems, industrial systems, feedlots and others), level of intensification, and resource endowment (Robinson 2011).

The implementation of strategies presented in Figure 5.11 builds on this differentiation, providing more depth compared to the previous AR5 analysis. Manure management strategies are more applicable in confined systems, where manure can be easily collected, such as in pigs and poultry systems or in smallholder mixed crop-livestock systems. More intensive systems, with strong market orientation, such as dairy in the US, can implement a range of sophisticated practices like feed additives and vaccines, while many market-oriented dairy systems in tropical regions can improve feed digestibility by improving forage quality and adding larger quantities of concentrate to the rations. Many of these strategies can be implemented as packages in different systems, thus maximising the synergies between different options (Mottet et al. 2017b).

See the Supplementary Material Section SM5.3 for a detailed description of livestock mitigation strategies; synergies and trade-offs with other mitigation and adaptation options are discussed in Section 5.6.

1.5.1.3   Greenhouse gas mitigation in agroforestry

Agroforestry can curb GHG emissions of CO_2, CH_4, and N_2O in agricultural systems in both developed and developing countries (see Glossary for definition) (see Chapter 2, Section 2.5.1 and Figure 2.24). Soil carbon sequestration, together with biological N fixation, improved land health and underlying ecosystem services may be enhanced through agricultural lands management practices used by large-scale and smallholder farmers, such as incorporation of trees within farms or in hedges (manure addition, green manures, cover crops, etc.), whilst promoting greater soil organic matter and nutrients (and thus soil organic carbon) content and improve soil structure (Mbow et al. 2014b) (Table 5.5). The tree cover increases the microbial activity of the soil and increases the productivity of the grass under cover. CO_2 emissions are furthermore lessened indirectly, through lower rates of erosion due to better soil structure and more plant cover in diversified farming systems than in monocultures. There is great potential for increasing above-ground and soil carbon stocks, reducing soil erosion and degradation, and mitigating GHG emissions.

These practices can improve food security through increases in productivity and stability since they contribute to increased soil quality and water-holding capacity. Agroforestry provides economic, ecological, and social stability through diversification of species and products. On the other hand, trade-offs are possible when cropland is taken out of production mainly as a mitigation strategy.

Meta-analyses have been done on carbon budgets in agroforestry systems (Zomer et al. 2016; Chatterjee et al. 2018). In a review of 42 studies, (Ramachandran Nair et al. 2009) estimated carbon sequestration potentials of differing agroforestry systems. These include sequestration rates ranging from 954 (semi-arid); to 1431 (temperate); 2238 (sub-humid) and 3670 tCO_2 km^{-2} yr^{-1} (humid). The global technical potential for agroforestry is 0.1–5.7 Gt CO_2e yr^{-1} (Griscom et al. 2017; Zomer et al. 2016; Dickie et al. 2014) (Chapter 2, Section 2.5.1). Agroforestry-based carbon sequestration can be used to offset N_2O and CO_2 emissions from soils and increase methane sink strength compared to annual cropping systems (Rosenstock et al. 2014).

Agroforestry systems with perennial crops, such as coffee and cacao, may be more important carbon sinks than those that combine trees with annual crops. Brandt et al. (2018) showed that farms in semi-arid regions (300–600 mm precipitation) were increasing in tree cover due to natural regeneration and that the increased application of agroforestry systems were supporting production and reducing GHG emissions.

1.5.1.4   Integrated approaches to crop and livestock mitigation

Livestock mitigation in a circular economy. Novel technologies for increasing the integration of components in the food system are being devised to reduce GHG emissions. These include strategies that help decoupling livestock from land use. Work by van Zanten et al. (2018) shows that 7–23 g of animal protein per capita per day could be produced without livestock competing for vital arable land. This would imply a contraction of the land area utilised by the livestock sector, but also a more efficient use of resources, and would lead to land sparing and overall emissions reductions.

Pikaar et al. (2018) demonstrated the technical feasibility of producing microbial protein as a feedstuff from sewage that could replace use of feed crops such as soybean. The technical potential of this novel practice could replace 10–19% of the feed protein required, and would reduce cropland demand and associated emissions by 6–7%. These practices are, however, not economically feasible nor easily upscalable in most systems. Nonetheless, significant progress in Japan and South Korea in the reduction and use of food waste to increase efficiencies in livestock food chains has been achieved, indicating a possible pathway to progress elsewhere (FAO 2017; zu Ermgassen et al. 2016). Better understanding of biomass and food and feed wastes, value chains, and identification of mechanisms for reducing the transport and processing costs of these materials is required to facilitate larger-scale implementation.

Waste streams into energy. Waste streams from manure and food waste can be used for energy generation and thus reduction in overall GHG emissions in terms of recovered methane (for instance through anaerobic digestion) production (De Clercq et al. 2016) or for the production of microbial protein (Pikaar et al. 2018). Second-generation biorefineries, once the underlying technology is improved, may enable the generation of hydro-carbon from agricultural residues, grass, and woody biomass in ways that do not compete with food and can generate, along with biofuel, high-value products such as plastics (Nguyen et al. 2017). Second-generation energy biomass from residues may constitute a complementary income source for farmers that can increase their incentive to produce. Technologies include CHP (combined heat and power) or gas turbines, and fuel types such as biodiesel, biopyrolysis (i.e., high temperature chemical transformation of organic material in the absence of oxygen), torrefaction of biomass, production of cellulosic bioethanol and of bioalcohols produced by other means than fermentation, and the production of methane by anaerobic fermentation. (Nguyen et al. 2017).

Technology for reducing fossil fuel inputs. Besides biomass and bioenergy, other forms of renewable energy substitution for fossil fuels (e.g., wind, solar, geothermal, hydro) are already being applied on farms throughout the supply chain. Energy efficiency measures are being developed for refrigeration, conservation tillage, precision farming (e.g., fertiliser and chemical application and precision irrigation).

Novel technologies. Measures that can reduce livestock emissions given continued research and development include methane and nitrification inhibitors, methane vaccines, targeted breeding of lower-emitting animals, and genetically modified grasses with higher sugar content. New strategies to reduce methanogenesis include supplementing animal diets with antimethanogenic agents (e.g., 3-NOP, algae, chemical inhibitors such as chloroform) or supplementing with electron acceptors (e.g., nitrate) or dietary lipids. These could potentially contribute, once economically feasible at scale, to significant reductions of methane emissions from ruminant livestock. A well-tested compound is 3-nitrooxypropanol

5

Demand-side mitigation
GHG mitigation potential of different diets

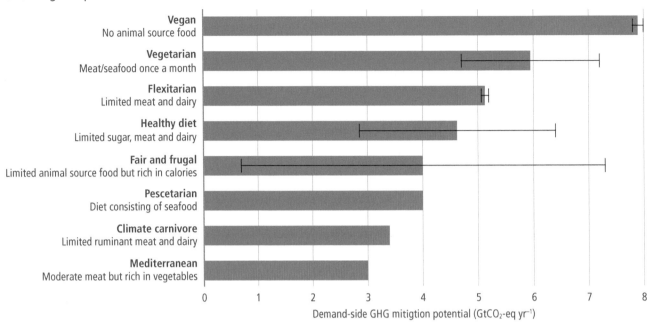

Figure 5.12 | **Technical mitigation potential of changing diets by 2050 according to a range of scenarios examined in the literature.** Estimates indicate technical potential only and include additional effects of carbon sequestration from land-sparing. Data without error bars are from one study only.

All diets need to provide a full complement of nutritional quality, including micronutrients (FAO et al. 2018).

Vegan: Completely plant-based (Springmann et al. 2016b; Stehfest et al. 2009).

Vegetarian: Grains, vegetables, fruits, sugars, oils, eggs and dairy, and generally at most one serving per month of meat or seafood (Springmann et al. 2016b; Tilman and Clark 2014; Stehfest et al. 2009).

Flexitarian: 75% of meat and dairy replaced by cereals and pulses; at least 500 g per day fruits and vegetables; at least 100 g per day of plant-based protein sources; modest amounts of animal-based proteins and limited amounts of red meat (one portion per week), refined sugar (less than 5% of total energy), vegetable oils high in saturated fat, and starchy foods with relatively high glycaemic index (Springmann et al. 2018a; Hedenus et al. 2014).

Healthy diet: Based on global dietary guidelines for consumption of red meat, sugar, fruits and vegetables, and total energy intake (Springmann et al. 2018a; Bajželj et al. 2014).

Fair and frugal: Global daily per-capita calorie intake of 2800 kcal/cap/day (11.7 MJ/cap/day), paired with relatively low level of animal products (Smith et al. 2013).

Pescetarian: Vegetarian diet that includes seafood (Tilman and Clark 2014).

Climate carnivore: 75% of ruminant meat and dairy replaced by other meat (Hedenus et al. 2014).

Mediterranean: Vegetables, fruits, grains, sugars, oils, eggs, dairy, seafood, moderate amounts of poultry, pork, lamb and beef (Tilman and Clark 2014).

(3-NOP), which was shown to decrease methane by up to 40% when incorporated in diets for ruminants (Hristov et al. 2015).

Whilst these strategies may become very effective at reducing methane, they can be expensive and also impact on animal performance and/or welfare (Llonch et al. 2017). The use of novel fertilisers and/or plant species that secrete biological nitrification inhibitors also have the potential to significantly reduce N_2O emissions from agricultural soils (Subbarao et al. 2009; Rose et al. 2018).

Economic mitigation potentials of crop and livestock sectors. Despite the large technical mitigation potential of the agriculture sector in terms of crop and livestock activities, its economic potential is relatively small in the short term (2030) and at modest carbon prices (less than 20 USD tC^{-1}). For crop and soil management practices, it is estimated that 1.0–1.5 GtCO$_2$-eq yr^{-1} could be a feasible mitigation target at a carbon price of 20 USD tC^{-1} (Frank et al. 2018, 2017;

Griscom et al. 2016; Smith et al. 2013; Wollenberg et al. 2016). For the livestock sector, these estimates range from 0.12–0.25 GtCO$_2$-eq yr^{-1} at similar carbon prices (Herrero et al. 2016c; Henderson et al. 2017). But care is needed in comparing crop and livestock economic mitigation potentials due to differing assumptions.

Frank et al. (2018) recently estimated that the economic mitigation potential of non-CO$_2$ emissions from agriculture and livestock to 2030 could be up to four times higher than indicated in the AR5, if structural options such as switching livestock species from ruminants to monogastrics, or allowing for flexibility to relocate production to more efficient regions were implemented, at the same time as the technical options such as those described above. At higher carbon prices (i.e., at about 100 USD tC^{-1}), they found a mitigation potential of supply-side measures of 2.6 GtCO$_2$-eq yr^{-1}.

In this scenario, technical options would account for 38% of the abatement, while another 38% would be obtained through structural changes, and a further 24% would be obtained through shifts in consumption caused by food price increases. Key to the achievement of this mitigation potential lay in the livestock sector, as reductions in livestock consumption, structural changes and implementation of technologies in the sector had some of the highest impacts. Regions with the highest mitigation potentials were Latin America, China and Sub-Saharan Africa. The large-scale implementability of such proposed sweeping changes in livestock types and production systems is likely very limited as well as constrained by long-established socio-economic, traditional and cultural habits, requiring significant incentives to generate change.

In summary, supply-side practices can contribute to climate change mitigation by reducing crop and livestock emissions, sequestering carbon in soils and biomass, and by decreasing emissions intensity within sustainable production systems (*high confidence*). The AR5 estimated the total economic mitigation potential of crop and livestock activities as 1.5–4.0 $GtCO_2$-eq yr^{-1} by 2030 at prices ranging from 20–100 USD tCO_2-eq (*high confidence*). Options with large potential for GHG mitigation in cropping systems include soil carbon sequestration (at decreasing rates over time), reductions in N_2O emissions from fertilisers, reductions in CH_4 emissions from paddy rice, and bridging of yield gaps. Options with large potential for mitigation in livestock systems include better grazing land management, with increased net primary production and soil carbon stocks, improved manure management, and higher-quality feed. Reductions in GHG emissions intensity (emissions per unit product) from livestock can support reductions in absolute emissions, provided appropriate governance structures to limit total production are implemented at the same time (*medium confidence*).

1.5.1.5   Greenhouse gas mitigation in aquaculture

Barange et al. (2018) provide a synthesis of effective options for GHG emissions reduction in aquaculture, including reduction of emissions from production of feed material, replacement of fish-based feed ingredients with crop-based ingredients; reduction of emissions from feed mill energy use, improvement of feed conversion rates, improvement of input use efficiency, shift of energy supply (from high-carbon fossil fuels to low-carbon fossil fuels or renewables), and improvement of fish health. Conversion of 25% of total aquaculture area to integrated aquaculture-agriculture ponds (greening aquaculture) has the potential to sequester 95.4 million tonnes of carbon per year (Ahmed et al. 2017).

Proposed mitigation in aquaculture includes avoided deforestation. By halting annual mangrove deforestation in Indonesia, associated total emissions would be reduced by 10–31% of estimated annual emissions from the land-use sector at present (Murdiyarso et al. 2015). Globally, 25% mangrove regeneration could sequester 0.54–0.65 million tonnes of carbon per year (Ahmed et al. 2017) of which 0.17–0.21 million tonnes could be through integrated or organic shrimp culture (Ahmed et al. 2018).

1.5.1.6   Cellular agriculture

The technology for growing muscle tissue in culture from animal stem cells to produce meat, for example, 'cultured', 'synthetic', 'in vitro' or 'hydroponic' meat could, in theory, be constructed with different characteristics and be produced faster and more efficiently than traditional meat (Kadim et al. 2015). Cultured meat (CM) is part of so-called cellular agriculture, which includes production of milk, egg white and leather from industrial cell cultivation (Stephens et al. 2018). CM is produced from muscle cells extracted from living animals, isolation of adult skeletal muscle stem cells (myosatellite cells), placement in a culture medium which allow their differentiation into myoblasts and then, through another medium, generation of myocytes which coalesce into myotubes and grow into strands in a stirred-tank bioreactor (Mattick et al. 2015).

Current technology enables the creation of beef hamburgers, nuggets, steak chips or similar products from meat of other animals, including wild species, although production currently is far from being economically feasible. Nonetheless, by allowing bioengineering from the manipulation of the stem cells and nutritive culture, CM allows for reduction of harmful fatty acids, with advantages such as reduced GHG emissions, mostly indirectly through reduced land use (Bhat et al. 2015; Kumar et al. 2017b).

Tuomisto and de Mattos (2011) made optimistic technological assumptions, relying on cyanobacteria hydrolysate nutrient source, and produced the lowest estimates on energy and land use. Tuomisto and de Mattos (2011) conducted a lifecycle assessment that indicates that cultured meat could have less than 60% of energy use and 1% of land use of beef production and it would have lower GHG emissions than pork and poultry as well. Newer estimates (Alexander et al. 2017; Mattick et al. 2015) indicate a trade-off between industrial energy consumption and agricultural land requirements of conventional and cultured meat and possibly higher GWP than pork or poultry due to higher energy use. The change in proportion of CO_2 versus CH_4 could have important implications in climate change projections and, depending on decarbonisation of the energy sources and climate change targets, cultured meat may be even more detrimental than exclusive beef production (Lynch and Pierrehumbert 2019).

Overall, as argued by Stephens et al. (2018), cultured meat is an 'as-yet undefined ontological object' and, although marketing targets people who appreciate meat but are concerned with animal welfare and environmental impacts, its market is largely unknown (Bhat et al. 2015 and Slade 2018). In this context it will face the competition of imitation meat (meat analogues from vegetal protein) and insect-derived products, which have been evaluated as more environmentally friendly (Alexander et al. 2017) and it may be considered as being an option for a limited resource world, rather than a mainstream solution. Besides, as the commercial production process is still largely undefined, its actual contribution to climate change mitigation and food security is largely uncertain and challenges are not negligible. Finally, it is important to understand the systemic nature of these challenges and evaluate their social impacts on rural populations due to transforming animal agriculture into an industrialised activity and its possible rebound effects on food security, which are still understudied in the literature.

5

Studies are needed to improve quantification of mitigation options for supply chain activities.

1.5.2   Demand-side mitigation options

Although population growth is one of the drivers of global food demand and the resulting environmental burden, demand-side management of the food system could be one of the solutions to curb climate change. Avoiding food waste during consumption, reducing over-consumption, and changing dietary preferences can contribute significantly to providing healthy diets for all, as well as reducing the environmental footprint of the food system. The number of studies addressing this issue have increased in the last few years (Chapter 2). (See Section 5.6 for synergies and trade-offs with health and Section 5.7 for discussion of Just Transitions.)

1.5.2.1   Mitigation potential of different diets

A systematic review found that higher consumption of animal-based foods was associated with higher estimated environmental impact, whereas increased consumption of plant-based foods was associated with an estimated lower environmental impact (Nelson et al. 2016). Assessment of individual foods within these broader categories showed that meat – especially ruminant meat (beef and lamb) – was consistently identified as the single food with the greatest impact on the environment, on a global basis, most often in terms of GHG emissions and/or land use.

Figure 5.12 shows the technical mitigation potentials of some scenarios of alternative diets examined in the literature. Stehfest et al. (2009) were among the first to examine these questions. They found that under the most extreme scenario, where no animal products are consumed at all, adequate food production in 2050 could be achieved on less land than is currently used, allowing considerable forest regeneration, and reducing land-based GHG emissions to one third of the reference 'business-as-usual' case for 2050, a reduction of 7.8 $GtCO_2$-eq yr^{-1}. Springmann et al. (2016b) recently estimated similar emissions reduction potential of 8 $GtCO_2$-eq yr^{-1} from a vegan diet without animal-sourced foods. This defines the upper bound of the technical mitigation potential of demand side measures.

Herrero et al. (2016a) reviewed available options, with a specific focus on livestock products, assessing technical mitigation potential across a range of scenarios, including 'no animal products', 'no meat', 'no ruminant meat', and 'healthy diet' (reduced meat consumption). With regard to 'credible low-meat diets', where reduction in animal protein intake was compensated by higher intake of pulses, emissions reductions by 2050 could be in the 4.3–6.4 $GtCO_2$-eq yr^{-1}, compared to a business-as-usual scenario. Of this technical potential, 1–2 $GtCO_2$-eq yr^{-1} come from reductions of mostly non-CO_2 GHG within the farm gate, while the remainder was linked to carbon sequestration on agricultural lands no longer needed for livestock production. When the transition to a low-meat diet reduces the agricultural area required, land is abandoned, and the re-growing vegetation can take up carbon until a new equilibrium is reached. This is known as the land-sparing effect.

Other studies have found similar results for potential mitigation linked to diets. For instance, Smith et al. (2013) analysed a dietary change scenario that assumed a convergence towards a global daily per-capita calorie intake of 2800 kcal per person per day (11.7 MJ per person per day), paired with a relatively low level of animal product supply, estimated technical mitigation potential in the range 0.7–7.3 $GtCO_2$-eq yr^{-1} for additional variants including low or high-yielding bioenergy, 4.6 $GtCO_2$-eq yr^{-1} if spare land is afforested.

Bajželj et al. (2014) developed different scenarios of farm systems change, waste management, and dietary change on GHG emissions coupled to land use. Their dietary scenarios were based on target kilocalorie consumption levels and reductions in animal product consumption. Their scenarios were 'healthy diet'; healthy diet with 2500 kcal per person per day in 2050; corresponding to technical mitigation potentials in the range 5.8 and 6.4 $GtCO_2$-eq yr^{-1}.

Hedenus et al. (2014) explored further dietary variants based on the type of livestock product. 'climate carnivore', in which 75% of the baseline-consumption of ruminant meat and dairy was replaced by pork and poultry meat, and 'flexitarian', in which 75% of the baseline-consumption of meat and dairy was replaced by pulses and cereal products. Their estimates of technical mitigation potentials by 2050 ranged 3.4–5.2 $GtCO_2$-eq yr^{-1}, the high end achieved under the flexitarian diet. Finally, Tilman and Clark (2014) used stylised diets as variants that included 'peseatarian', 'Mediterranean', 'vegetarian', compared to a reference diet, and estimated technical mitigation potentials within the farm gate of 1.2–2.3 $GtCO_2$-eq yr^{-1}, with additional mitigation from carbon sequestration on spared land ranging 1.8–2.4 $GtCO_2$-eq yr^{-1}.

Studies have defined dietary mitigation potential as, for example, 20 kg per person per week CO_2-eq for Mediterranean diet, versus 13 kg per person per week CO_2-eq for vegan (Castañé and Antón 2017). Rosi et al. (2017) developed seven-day diets in Italy for about 150 people defined as omnivore 4.0 ± 1.0; ovo-lacto-veggie 2.6 ± 0.6; and vegan 2.3 ± 0.5 kg CO_2-eq per capita per day.

Importantly, many more studies that compute the economic and calorie costs of these scenarios are needed. Herrero et al. (2016a) estimated that once considerations of economic and calorie costs of their diet-based solutions were included, the technical range of 4.3–6.4 $GtCO_2$-eq yr^{-1} in 2050 was reduced to 1.8–3.4 $GtCO_2$-eq yr^{-1} when implementing a GHG tax ranging from 20–100 USD tCO_2. While caloric costs where low below 20 USD tCO_2, they ranged from 27–190 kcal per person per day under the higher economic potential, thus indicating possible negative trade-offs with food security.

In summary, demand-side changes in food choices and consumption can help to achieve global GHG mitigation targets (*high confidence*). Low-carbon diets on average tend to be healthier and have smaller land footprints. By 2050, technical mitigation potential of dietary changes range from 2.7–6.4 $GtCO_2$-eq yr^{-1} for assessed diets (*high confidence*). At the same time, the economic potential of such solutions is lower, ranging from 1.8–3.4 $GtCO_2$-eq yr^{-1} at prices of 20–100 USD tCO_2, with caloric costs up to 190 kcal per person per day. The feasibility of how to create economically viable transitions

5

to more sustainable and healthy diets that also respect food security requirements needs to be addressed in future research.

1.5.2.2   Role of dietary preferences

Food preference is an inherently cultural dimension that can ease or hinder transformations to food systems that contribute to climate change mitigation. Consumer choice and dietary preferences are guided by social, cultural, environmental, and traditional factors as well as economic growth. The food consumed by a given group conveys cultural significance about social hierarchy, social systems and human-environment relationships (Herforth and Ahmed 2015).

As suggested by Springmann et al. (2018a), per capita dietary emissions will translate into different realised diets, according to regional contexts including cultural and gendered norms (e.g., among some groups, eating meat is perceived as more masculine (Ruby and Heine 2011). In some cases, women and men have different preferences in terms of food, with women reporting eating healthier food (Imamura et al. 2015; Kiefer et al. 2005; Fagerli and Wandel 1999): these studies found that men tend to eat more meat, while women eat more vegetables, fruits and dairy products (Kanter and Caballero 2012).

Food preferences can change over time, with the nutrition transition from traditional diets to high-meat, high-sugar, high-saturated fat diets being a clear example of significant changes occurring in a short period of time. Meat consumption per capita consistently responds to income with a saturating trend at high income levels (Sans and Combris 2015; Vranken et al. 2014). Some emerging economies have rapidly increased demand for beef, leading to pressure on natural resources (Bowles et al. 2019). In another example, by reducing beef consumption between 2005 and 2014, Americans avoided approximately 271 million metric tonnes of emissions (CO_2-eq) (NRDC 2017). Attending farmers markets or buying directly from local producers has been shown to change worldviews (Kerton and Sinclair 2010), and food habits towards healthier diets (Pascucci et al. 2011) can be advanced through active learning (Milestad et al. 2010).

Regarding the options to reduce meat intake in developed countries, research shows that there is an apparent sympathy of consumers for meat reduction due to environmental impacts (Dagevos and Voordouw 2013), which has not been exploited. Social factors that influence reducing meat consumption in New Zealand include the need for better education or information dispersal regarding perceived barriers to producing meat-reduced/less meals; ensuring there is sensory or aesthetic appeal; and placing emphasis on human health or nutritional benefits (Tucker 2018).

Different and complementary strategies can be used in parallel for different consumer's profiles to facilitate step-by-step changes in the amounts and the sources of protein consumed. In the Netherlands, a nationwide sample of 1083 consumers were used to study their dietary choices toward smaller portions of meat, smaller portions using meat raised in a more sustainable manner, smaller portions and eating more vegetable protein, and meatless meals with or without meat substitutes. Results showed that strategies to change meat eating frequencies and meat portion sizes appeared to overlap and that these strategies can be applied to address consumers in terms of their own preferences (de Boer et al. 2014).

1.5.2.3   Uncertainties in demand-side mitigation potential

Both reducing ruminant meat consumption and increasing its efficiency are often identified as the main options to reduce GHG emissions (GHGE) and to lessen pressure on land (Westhoek et al. 2014) (see Section 5.6 for synergies and trade-offs with health and Section 5.7 for discussion of Just Transitions). However, analysing ruminant meat production is highly complex because of the extreme heterogeneity of production systems and due to the numerous products and services associated with ruminants (Gerber et al. 2015). See Supplementary Material Section SM5.3 for further discussion of uncertainties in estimates of livestock mitigation technical potential. Further, current market mechanisms are regarded as insufficient to decrease consumption or increase efficiency, and governmental intervention is often suggested to encourage mitigation in both the supply-side and demand-side of the food system (Section 5.7) (Wirsenius et al. 2011; Henderson et al. 2018).

Minimising GHG emissions through mathematical programming with near-minimal acceptability constraints can be understood as a reference or technical potential for mitigation through diet shifts. In this context (Macdiarmid et al. 2012) found up to 36% reduction in emissions in UK with similar diet costs applying fixed lifecycle analyses (LCA) carbon footprints (i.e., no rebound effects considered). Westhoek et al. (2014) found 25–40% in emissions by halving meat, dairy and egg intake in the EU, applying standard IPCC fixed emission intensity factors. Uncertainty about the consequences of on-the-ground implementation of policies towards low ruminant meat consumption in the food system and their externalities remain noteworthy.

Often, all emissions are allocated only to human edible meat and the boundaries are set only within the farm gate (Henderson et al. 2018; Gerber et al. 2013). However, less than 50% of slaughtered cattle weight is human edible meat, and 1–10% of the mass is lost or incinerated, depending on specified risk materials legislation. The remaining mass provide inputs to multiple industries, for example clothing, furniture, vehicle coating materials, biofuel, gelatine, soap, cosmetics, chemical and pharmaceutical industrial supplies, pet feed ingredients and fertilisers (Marti et al. 2011; Mogensen et al. 2016; Sousa et al. 2017). This makes ruminant meat production one of the most complex problems for LCA in the food system (Place and Mitloehner 2012; de Boer et al. 2011). There are only a few examples taking into account slaughter by-products (Mogensen et al. 2016).

1.5.2.4   Insect-based diets

Edible insects are, in general, rich in protein, fat, and energy and can be a significant source of vitamins and minerals (Rumpold and Schlüter 2015). Approximately 1900 insect species are eaten worldwide, mainly in developing countries (van Huis 2013). The development of safe rearing and effective processing methods are mandatory for utilisation of insects in food and feed. Some insect species can be grown on organic side streams, reducing environmental contamination and transforming waste into high-protein feed. Insects are principally considered as meat substitutes, but worldwide meat substitute consumption is still very low, principally due to differences in food culture, and will require transition phases such as powdered forms (Megido et al. 2016 and Smetana et al. 2015). Wider consumer acceptability will relate to pricing, perceived environmental benefits, and the development of tasty insect-derived protein products (van Huis et al. 2015; van Huis 2013). Clearly, increasing the share of insect-derived protein has the potential to reduce GHG emissions otherwise associated with livestock production. However, no study to date has quantified such potential.

1.5.2.5 Food loss and waste, food security, and land use

Food loss and waste impacts food security by reducing global and local food availability, limiting food access due to an increase in food prices and a decrease of producer income, affecting future food production due to the unstainable use of natural resources (HLPE 2014). Food loss is defined as the reduction of edible food during production, postharvest, and processing, whereas food discarded by consumers is considered as food waste (FAO 2011b). Combined food loss and waste amount to 25–30% of total food produced (*medium confidence*). During 2010–2016, global food loss and waste equalled 8–10% of total GHG emissions (*medium confidence*); and cost about 1 trillion USD per year (*low confidence*) (FAO 2014b).

A large share of produced food is lost in developing countries due to poor infrastructure, while a large share of produced food is wasted in developed countries (Godfray et al. 2010). Changing consumer behaviour to reduce per capita over-consumption offers substantial potential to improve food security by avoiding related health burdens (Alexander et al. 2017; Smith 2013) and reduce emissions associated with the extra food (Godfray et al. 2010). In 2007, around 20% of the food produced went to waste in Europe and North America, while around 30% of the food produced was lost in Sub-Saharan Africa (FAO 2011b). During the last 50 years, the global food loss and waste increased from around 540 Mt in 1961 to 1630 Mt in 2011 (Porter et al. 2016).

In 2011, food loss and waste resulted in about 8–10% of total anthropogenic GHG emissions. The mitigation potential of reduced food loss and waste from a full life-cycle perspective, for example, considering both food supply chain activities and land-use change, was estimated as 4.4 GtCO$_2$-eq yr^{-1} (FAO 2015a, 2013b). At a global scale, loss and waste of milk, poultry meat, pig meat, sheep meat, and potatoes are associated with 3% of the global agricultural N$_2$O emissions (more than 200 Gg N$_2$O-N yr^{-1} or 0.06 GtCO$_2$-eq yr^{-1}) in 2009 (Reay et al. 2012). For the USA, 35% of energy use, 34% of blue water use, 34% of GHG emissions, 31% of land use, and 35% of fertiliser use related to an individual's food-related resource consumption were accounted for as food waste and loss in 2010 (Birney et al. 2017).

Similar to food waste, over-consumption (defined as food consumption in excess of nutrient requirements), leads to GHG emissions (Alexander et al. 2017). In Australia for example, over-consumption accounts for about 33% GHGs associated with food (Hadjikakou

2017). In addition to GHG emissions, over-consumption can also lead to severe health conditions such as obesity or diabetes. Over-eating was found to be at least as large a contributor to food system losses (Alexander et al. 2017). Similarly, food system losses associated with consuming resource-intensive animal-based products instead of nutritionally comparable plant-based alternatives are defined as 'opportunity food losses'. These were estimated to be 96, 90, 75, 50, and 40% for beef, pork, dairy, poultry, and eggs, respectively, in the USA (Shepon et al. 2018).

Avoiding food loss and waste will contribute to reducing emissions from the agriculture sector. By 2050, agricultural GHG emissions associated with production of food that might be wasted may increase to 1.9–2.5 GtCO$_2$-eq yr^{-1} (Hiç et al. 2016). When land-use change for agriculture expansion is also considered, halving food loss and waste reduces the global need for cropland area by around 14% and GHG emissions from agriculture and land-use change by 22–28% (4.5 GtCO$_2$-eq yr^{-1}) compared to the baseline scenarios by 2050 (Bajželj et al. 2014). The GHG emissions mitigation potential of food loss and waste reduction would further increase when lifecycle analysis accounts for emissions throughout food loss and waste through all food system activities.

Reducing food loss and waste to zero might not be feasible. Therefore, appropriate options for the prevention and management of food waste can be deployed to reduce food loss and waste and to minimise its environmental consequences. Papargyropoulou et al. (2014) proposed the Three Rs (i.e., reduction, recovery and recycle) options to prevent and manage food loss and waste. A wide range of approaches across the food supply chain is available to reduce food loss and waste, consisting of technical and non-technical solutions (Lipinski et al. 2013). However, technical solutions (e.g., improved harvesting techniques, on-farm storage, infrastructure, packaging to keep food fresher for longer, etc.) include additional costs (Rosegrant et al. 2015) and may have impacts on local environments (FAO 2018b). Additionally, all parts of food supply chains need to become efficient to achieve the full reduction potential of food loss and waste (Lipinski et al. 2013).

Together with technical solutions, approaches (i.e., non-technical solutions) to changes in behaviours and attitudes of a wide range of stakeholders across the food system will play an important role in reducing food loss and waste. Food loss and waste can be recovered by distributing food surplus to groups affected by food poverty or converting food waste to animal feed (Vandermeersch et al. 2014). Unavoidable food waste can also be recycled to produce energy based on biological, thermal and thermochemical technologies (Pham et al. 2015). Additionally, strategies for reducing food loss and waste also need to consider gender dynamics with participation of females throughout the food supply chain (FAO 2018f).

In summary, reduction of food loss and waste can be considered as a climate change mitigation measure that provides synergies with food security and land use (*robust evidence, medium agreement*). Reducing food loss and waste reduces agricultural GHG emissions and the need for agricultural expansion for producing excess food. Technical options for reduction of food loss and waste include

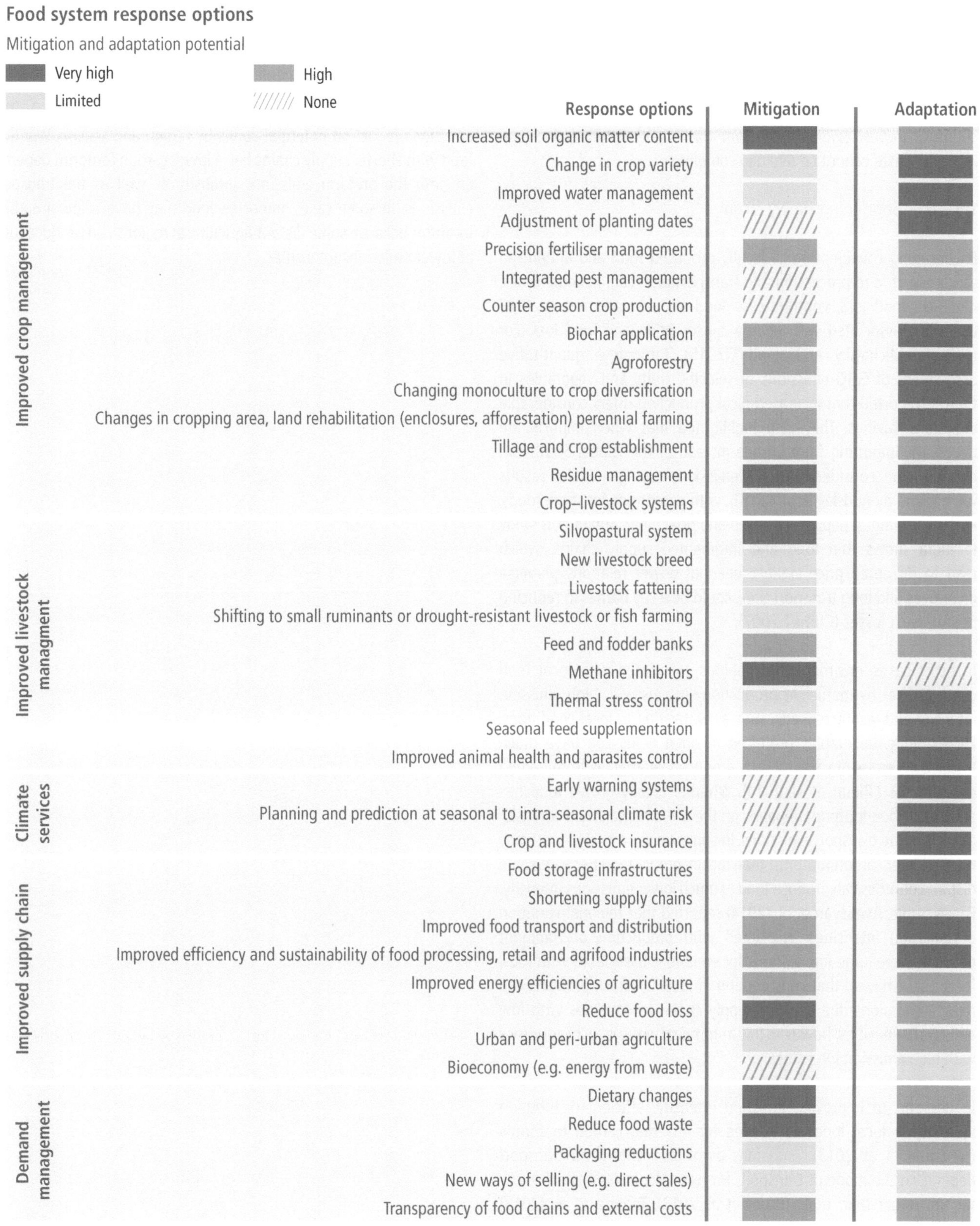

Food system response options

Mitigation and adaptation potential

- Very high
- High
- Limited
- None

Figure 5.13 | Response options related to food system and their potential impacts on mitigation and adaptation. Many response options offer significant potential for both mitigation and adaptation.

improved harvesting techniques, on-farm storage, infrastructure, and packaging. However, the beneficial effects of reducing food loss and waste will vary between producers and consumers, and across regions. Causes of food loss (e.g., lack of refrigeration) and waste (e.g., behaviour) differ substantially in developed and developing countries (*robust evidence, medium agreement*). Additionally, food loss and waste cannot be avoided completely.

1.5.2.6   Shortening supply chains

Encouraging consumption of locally produced food and enhancing efficiency of food processing and transportation can, in some cases, minimise food loss, contribute to food security, and reduce GHG emissions associated with energy consumption and food loss. For example, Michalský and Hooda (2015), through a quantitative assessment of GHG emissions of selected fruits and vegetables in the UK, reported that increased local production offers considerable emissions savings. They also highlighted that when imports are necessary, importing from Europe instead of the Global South can contribute to considerable GHG emissions savings. Similar results were found by Audsley et al. (2010), with exceptions for some foods, such as tomatoes, peppers or sheep and goat meat. Similarly, a study in India shows that long and fragmented supply chains, which lead to disrupted price signals, unequal power relations perverse incentives and long transport time, could be a key barrier to reducing post-harvest losses (CIPHET 2007).

In other cases, environmental benefits associated with local food can be offset by inefficient production systems with high emission intensity and resource needs, such as water, due to local conditions. For example, vegetables produced in open fields can have much lower GHG emissions than locally produced vegetables from heated greenhouses (Theurl et al. 2014). Whether locally grown food has a lower carbon footprint depends on the on-farm emissions intensity as well as the transport emissions. In some cases, imported food may have a lower carbon footprint than locally grown food because some distant countries can produce food at much lower emissions intensity. For example, Avetisyan et al. (2014) reported that regional variation of emission intensities associated with production of ruminant products have large implications for emissions associated with local food. They showed that consumption of local livestock products can reduce emissions due to short supply chains in countries with low emission intensities; however, this might not be the case in countries with high emission intensities.

In addition to improving emission intensity, efficient distribution systems for local food are needed for lowering carbon footprints (Newman et al. 2013). Emissions associated with food transport depend on the mode of transport, for example, emissions are lower for rail rather than truck (Brodt et al. 2013). Tobarra et al. (2018) reported that emissions saving from local food may vary across seasons and regions of import. They highlighted that, in Spain, local production of fruits and vegetables can reduce emissions associated with imports from Africa but imports from France and Portugal can save emissions in comparison to production in Spain. Additionally, local production of seasonal products in Spain reduces emissions, while imports of out-of-season products can save emissions rather than producing them locally.

In summary, consuming locally grown foods can reduce GHG emissions, if they are grown efficiently (*high confidence*). The emissions reduction potential varies by region and season. Whether food with shorter supply chains has a lower carbon footprint depends on both the on-farm emissions intensity as well as the transport emissions. In some cases, imported food may have a lower carbon footprint because some distant agricultural regions can produce food at lower emissions intensities.

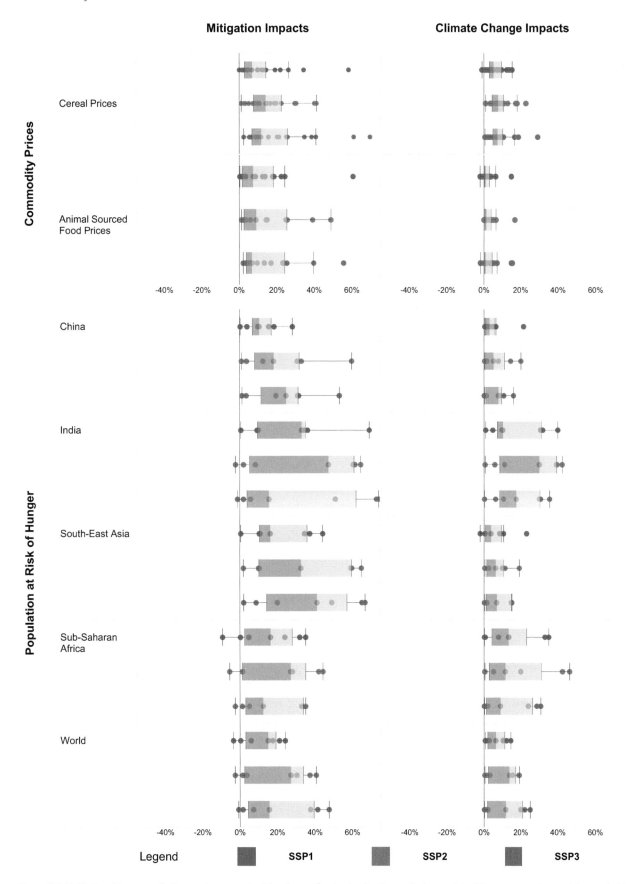

Mitigation Impacts   **Climate Change Impacts**

Figure 5.14 | Regional impacts of climate change and mitigation on food price (top), population at risk of hunger or undernourishment (middle), GHG emissions (bottom) in 2050 under different socio-economic scenarios (SSP1, SSP2 and SSP3) based on AgMIP Global Economic Model analysis. Values indicate changes from no climate change and no climate change mitigation scenario. MAgPIE, a global land-use allocation model, is excluded due to inelastic food demand. The value of India includes that of Other Asia in MAGNET, a global general equilibrium model (Hasegawa et al. 2018).

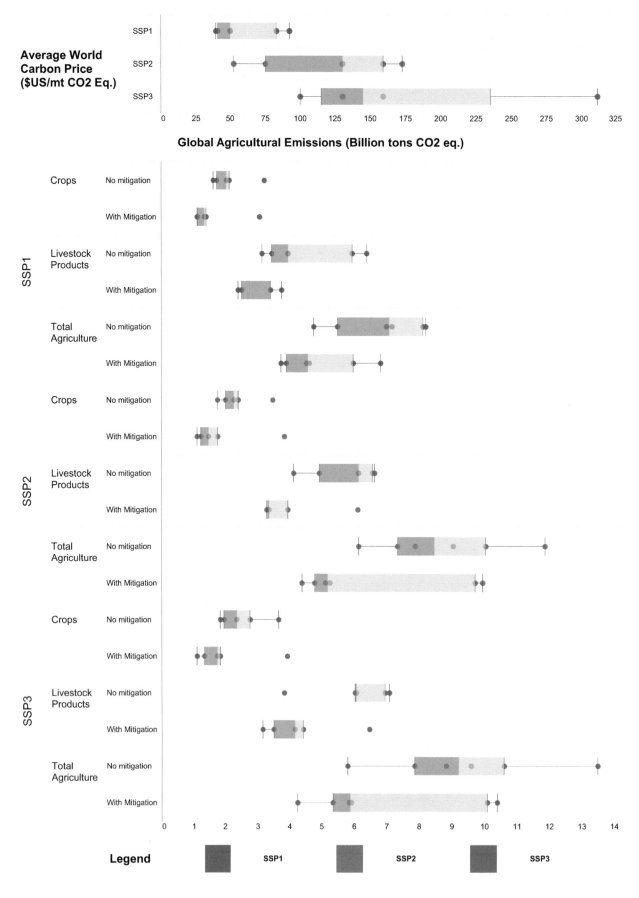

Figure 5.14 (continued).

1.6 Mitigation, adaptation, food security and land use: Synergies, trade-offs and co-benefits

Food systems will need to adapt to changing climates and also reduce their GHG emissions and sequester carbon if Paris Agreement goals are to be met (Springmann et al. 2018a and van Vuuren et al. 2014). The synergies and trade-offs between the food system mitigation and adaptation options described in Sections 5.3 and 5.5 are of increasing importance in both scientific and policy communities because of the necessity to ensure food security, i.e., providing nutritious food for growing populations while responding to climate change (Rosenzweig and Hillel 2015). A special challenge involves interactions between land-based non-food system mitigation, such as negative emissions technologies, and food security. Response options for the food system have synergies and trade-offs between climate change mitigation and adaptation (Figure 5.13; Chapter 6).

Tirado et al. (2013) suggest an integrated approach to address the impacts of climate change to food security that considers a combination of nutrition-sensitive adaptation and mitigation measures, climate-resilient and nutrition-sensitive agricultural development, social protection, improved maternal and child care and health, nutrition-sensitive risk reduction and management, community development measures, nutrition-smart investments, increased policy coherence, and institutional and cross-sectoral collaboration. These measures are a means to achieve both short-term and long-term benefits in poor and marginalised groups.

This section assesses the synergies and trade-offs for land-based atmospheric carbon dioxide removal measures, effects of mitigation measures on food prices, and links between dietary choices and human health. It then evaluates a range of integrated agricultural systems and practices that combine mitigation and adaptation measures, including the role of agricultural intensification. The role urban agriculture is examined, as well as interactions between SDG 2 (zero hunger) and SDG 13 (climate action).

1.6.1 Land-based carbon dioxide removal (CDR) and bioenergy

Large-scale deployment of negative emission technologies (NETs) in emission scenarios has been identified as necessary for avoiding unacceptable climate change (IPCC 2018b). Among the available NETs, carbon dioxide removal (CDR) technologies are receiving increasing attention. Land-based CDRs include afforestation and reforestation (AR), sustainable forest management, biomass energy with carbon capture and storage (BECCS), and biochar (BC) production (Minx et al. 2018). Most of the literature on global land-based mitigation potential relies on CDRs, particularly on BECCS, as a major mitigation action (Kraxner et al. 2014; Larkin et al. 2018 and Rogelj et al. 2018, 2015, 2011). BECCS is not yet deployable at a significant scale, as it faces challenges similar to fossil fuel carbon capture and storage (CCS) (Fuss et al. 2016; Vaughan and Gough 2016; Nemet et al. 2018). Regardless, the effectiveness of large-scale BECCS to meet Paris Agreement goals has been questioned and other pathways to mitigation have been proposed (Anderson and Peters 2016; van Vuuren et al. 2017, 2018; Grubler et al. 2018; Vaughan and Gough 2016).

Atmospheric CO_2 removal by storage in vegetation depends on achieving net organic carbon accumulation in plant biomass over decadal time scales (Kemper 2015) and, after plant tissue decay, in soil organic matter (Del Grosso et al. 2019). AR, BECCS and BC differ in the use and storage of plant biomass. In BECCS, biomass carbon from plants is used in industrial processes (e.g., for electricity, hydrogen, ethanol, and biogas generation), releasing CO_2, which is then captured and geologically stored (Greenberg et al. 2017; Minx et al. 2018).

Afforestation and reforestation result in long-term carbon storage in above and belowground plant biomass on previously unforested areas, and is effective as a carbon sink during the AR establishment period, in contrast to thousands of years for geological carbon storage (Smith et al. 2016).

Biochar is produced from controlled thermal decomposition of biomass in absence of oxygen (pyrolysis), a process that also yields combustible oil and combustible gas in different proportions. Biochar is a very stable carbon form, with storage on centennial time scales (Lehmann et al. 2006) (Chapter 4). Incorporated in soils, some authors suggest it may lead to improved water-holding capacity, nutrient retention, and microbial processes (Lehmann et al. 2015). There is, however, uncertainty about the benefits and risks of this practice (The Royal Society 2018).

Land-based CDRs require high biomass-producing crops. Since not all plant biomass is harvested (e.g., roots and harvesting losses), it can produce co-benefits related to soil carbon sequestration, crop productivity, crop quality, as well improvements in air quality, but the overall benefits strongly depend on the previous land-use and soil management practices (Smith et al. 2016; Wood et al. 2018). In addition, CDR effectiveness varies widely depending on type of biomass, crop productivity, and emissions offset in the energy system. Importantly, its mitigation benefits can be easily lost due to land-use change interactions (Harper et al. 2018; Fuss et al. 2018; Daioglou et al. 2019).

Major common challenges of implementing these large-scale CDR solutions, as needed to stabilise global temperature at 'well-below' 2°C by the end of the century, are the large investments and the associated significant changes in land use required. Most of the existing scenarios estimate the global area required for energy crops in the range of 109–990 Mha (IPCC 2018a), most commonly around 380–700 Mha (Smith et al. 2016), reaching net area expansion rates of up to 23.7 Mha yr^{-1} (IPCC 2018b). The upper limit implies unprecedented rates of area expansion for crops and forestry observed historically, for instance, as reported by FAO since 1961 (FAOSTAT 2018). By comparison, the sum of recent worldwide rates of expansion in the harvested area of soybean and sugarcane has not exceeded 3.5 Mha yr^{-1} on average. Even at this rate, they have been the source of major concerns for their possible negative environmental and food security impacts (Boerema et al. 2016; Popp et al. 2014).

5

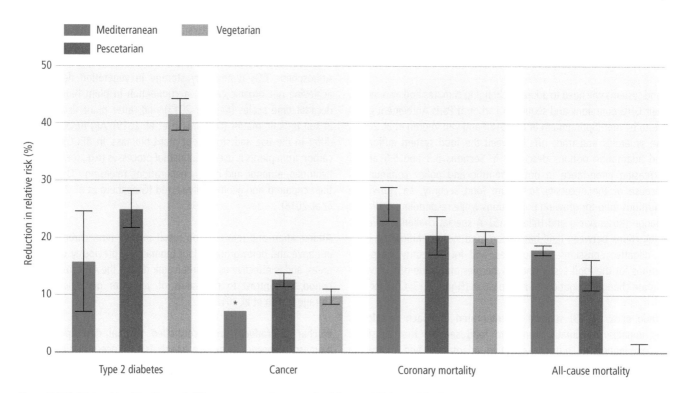

Figure 5.15 | Diet and health effects of different consumption scenarios (Tilman and Clark 2014) (*reflects data from a single study, hence no error bars).

Most land area available for CDR is currently pasture, estimated at 3300 Mha globally (FAOSTAT 2018). However, there is *low confidence* about how much low-productivity land is actually available for CDR (Lambin et al. 2013 and Gibbs and Salmon 2015). There is also *low confidence* as to whether the transition to BECCS will take place directly on low-productivity grasslands (Johansson and Azar 2007), and uncertainty on the governance mechanisms required to avoid unwanted spill-over effects, for instance causing additional deforestation (Keles et al. 2018).

Further, grasslands and rangelands may often occur in marginal areas, in which case, they may be exposed to climate risks, including periodic flooding. Grasslands and especially rangelands and savannas tend to predominate in less-developed regions, often bordering areas of natural vegetation with little infrastructure available for transport and processing of large quantities of CDR-generated biomass (O'Mara 2012; Beringer et al. 2011; Haberl et al. 2010; Magdoff 2007).

CDR-driven reductions in the available pastureland area is a scenario of constant or increasing global animal protein output as proposed by Searchinger et al. (2018). However, despite the recent reduction in meat consumption in western countries, this will require productivity improvements (Cohn et al. 2014; Strassburg et al. 2014). It would also result in lower emission intensities and create conditions for increased soil carbon stocks (de Oliveira Silva et al. 2016; Searchinger et al. 2018; Soussana et al. 2019, 2013). At the same time, food security may be threatened if land-based mitigation displaced crops elsewhere, especially if to regions of lower productivity potential, higher climatic risk, and higher vulnerability.

There is *low agreement* about what are the more competitive regions of the world for CDRs. Smith et al. (2016) and Vaughan et al. (2018)

identify as candidates relatively poor countries in Latin America, Africa and Asia (except China and India). Others indicate those regions may be more competitive for food production, placing Europe as a major BECCS exporter (Muratori et al. 2016). Economically feasible CDR investments are forecast to be directed to regions with high biomass production potential, demand for extra energy production, low leakage potential for deforestation and low competition for food production (Vaughan et al. 2018). Latin America and Africa, for instance, although having high biomass production potential, still have low domestic energy consumption (589 and 673 MTOE – 24.7 and 28.2 EJ, respectively), with about 30% of primary energy from renewable sources (reaching 50% in Brazil), mainly hydropower and traditional biomass.

There is *high confidence* that deployment of BECCS will require ambitious investments and policy interventions (Peters and Geden 2017) with strong regulation and governance of bioenergy production to ensure protection of forests, maintain food security and enhance climate benefits (Burns and Nicholson 2017; Vaughan et al. 2018; Muratori et al. 2016), and that such conditions may be challenging for developing countries. Increased value of bioenergy puts pressure on land, ecosystem services, and the prices of agricultural commodities, including food (*high confidence*).

There is *medium confidence* for the impact of CDR technologies on increased food prices and reduced food security, as these depend on several assumptions. Nevertheless, those impacts could be strong, with food prices doubling under certain scenario combinations (Popp et al. 2017). The impacts of land-mitigation policies on the reduction of dietary energy availability alone (without climate change impacts) is estimated at over 100 kcal per person per day by 2050, with highest regional impacts in South Asia and Sub-Saharan Africa (Hasegawa et al.

2018) (Section 5.2). However, only limited pilot BECCS projects have been implemented to date (Lenzi et al. 2018). Integrated assessment models (IAMs) use theoretical data based on high-level studies and limited regional data from the few on-the-ground BECCS projects.

Furthermore, it has been suggested that several BECCS IAM scenarios rely on unrealistic assumptions regarding regional climate, soils and infrastructure suitability (Anderson and Peters 2016), as well as international bioenergy trade (Lamers et al. 2011). Current global IAMs usually consider major trends in production potential and projected demand, overlooking major challenges for the development of a reliable international market. Such a market will have to be created from scratch and overcome a series of constraints, including trade barriers, logistics, and supply chains, as well as social, ecological and economic impacts (Matzenberger et al. 2015).

In summary, there is *high agreement* that better assessment of BECCS mitigation potential would need to be based on increased regional, bottom-up studies of biomass potentials, socio-economic consequences (including on food security), and environmental impacts in order to develop more realistic estimates (IPCC 2018a).

1.6.2   Mitigation, food prices, and food security

Food prices are the result of supply, demand and trade relations. Earlier studies (e.g., Nelson et al. 2009) showed that recent climate impacts that reduced crop productivity led to higher prices and increased trade of commodities between regions, with asymmetric impacts on producers and consumers. In terms of published scenario analyses, the most affected regions tend to be Sub-Saharan Africa and parts of Asia, but there is significant heterogeneity in results between countries. Relocation of production to less affected areas buffers these impacts to a certain extent, and offers potential for improvements in food production technologies (Hasegawa et al. 2018; van Meijl et al. 2017; Wiebe et al. 2015; Lotze-Campen et al. 2014; Valin et al. 2014; Robinson et al. 2014).

A newer, less studied impact of climate change on prices and their impacts on food security is the level of land-based mitigation necessary to stabilise global temperature. Hasegawa et al. (2018), using an ensemble of seven global economic models across a range of GHG emissions pathways and socioeconomic trajectories, suggested that the level of mitigation effort needed to reduce emissions can have a more significant impact on prices than the climate impacts themselves on reduced crop yields (Figure 5.14). This occurs because in the models, taxing GHG emissions leads to higher crop and livestock prices, while land-based mitigation leads to less land availability for food production, potentially lower food supply, and therefore food price increases.

Price increases in turn lead to reduced consumption, especially by vulnerable groups, or to shifts towards cheaper food, which are often less nutritious. This leads to significant increases in the number of malnourished people. Frank et al. (2017) and Fujimori et al. (2017) arrived at the same conclusions for the 1.5°C mitigation scenario using the IAM Globiom and ensembles of AgMIP global economic

models. While the magnitude of the response differs between models, the results are consistent between them. In contrast, a study based on five global agroeconomic models highlights that the global food prices may not increase much when the required land for bioenergy is accessible on the margin of current cropland, or the feedstock does not have a direct completion with agricultural land (Lotze-Campen et al. 2014).

These studies highlight the need for careful design of emissions mitigation policies in upcoming decades – for example, targeted schemes encouraging more productive and resilient agricultural production systems and the importance of incorporating complementary policies (such as safety-net programmes for poverty alleviation) that compensate or counteract the impacts of climate change mitigation policies on vulnerable regions (Hasegawa et al. 2018). Fujimori et al. (2018) showed how an inclusive policy design can avoid adverse side effects on food security through international aid, bioenergy taxes, or domestic reallocation of income. These strategies can shield impoverished and vulnerable people from the additional risk of hunger that would be caused by the economic effects of policies narrowly focussing on climate objectives only.

In summary, food security will be threatened through increasing numbers of malnourished people if land-based mitigation raises prices, unless other policy mechanisms reduce its impact (*high confidence*). Inclusive policy design can avoid adverse side effects on food security by shielding vulnerable people from the additional risk of hunger that would be caused by the economic effects of policies narrowly focusing on climate objectives (*medium confidence*).

1.6.3   Environmental and health effects of adopting healthy and sustainable diets

Two key questions arise from the potentially significant mitigation potential of dietary change: (i) Are 'low-GHG emission diets' likely to be beneficial for health? and (ii) Would changing diets at scale provide substantial benefits? In short, what are the likely synergies and trade-offs between low-GHG emissions diets and food security, health, and climate change? See Supplementary Material Section SM5.4 for further discussion.

Are 'low GHG emission diets' healthy? Consistent evidence indicates that, in general, a dietary pattern that is higher in plant-based foods, such as vegetables, fruits, whole grains, legumes, nuts, and seeds, and lower in animal-based foods, is more health-promoting and associated with lower environmental impact (GHG emissions and energy, land and water use) than either the current global average diets (Swinburn et al. 2019; Willett et al. 2019; Springmann et al. 2016b), or the current average USA diet (Nelson et al. 2016). Another study (Van Mierlo et al. 2017) showed that nutritionally-equivalent diets can substitute plant-based foods for meat and provide reductions in GHG emissions.

There are several studies that estimate health adequacy and sustainability and conclude that healthy sustainable diets are possible. These include global studies (e.g., Willett et al. 2019; Swinburn et al.

2019), as well as localised studies (e.g., Van Dooren et al. 2014). For example, halving consumption of meat, dairy products and eggs in the European Union would achieve a 40% reduction in ammonia emissions, 25–40% reduction in non-CO_2 GHG emissions (primarily from agriculture) and 23% per capita less use of cropland for food production, with dietary changes lowering health risks (Westhoek et al. 2014). In China, diets were designed that could meet dietary guidelines while creating significant reductions in GHG emissions (between 5% and 28%, depending on scenario) (Song et al. 2017). Changing diets can also reduce non-dietary related health issues caused by emissions of air pollutants. For example, specific changes in diets were assessed for their potential to mitigate PM 2.5 in China (Zhao et al. 2017b).

Some studies are starting to estimate both health and environmental benefits from dietary shifts. For example, Farchi et al. (2017) estimate health (colorectal cancer, cardiovascular disease) and GHG outcomes of 'Mediterranean' diets in Italy, and found the potential to reduce deaths from colorectal cancer of 7–10% and CVD from 9–10%, as well as potential savings of up to 263 CO_2-eq per person per year. In the USA, Hallström et al. (2017) found that adoption of healthier diets (consistent with dietary guidelines, and reducing amounts of red and processed meats) could reduce relative risk of coronary heart disease, colorectal cancer, and type 2 diabetes by 20–45%, USA healthcare costs by 77–93 billion USD per year, and direct GHG emissions by 222–826 kg CO_2-eq per person per year (69–84 kg from the healthcare system, 153–742 kg from the food system). Broadly similar conclusions were found for the Netherlands (Biesbroek et al. 2014); and the UK (Friel et al. 2009 and Milner et al. 2015).

Whilst for any given disease, there are a range of factors, including diet, that can affect it, and evidence is stronger for some diseases than others, a recent review found that an overall trend toward increased cancer risk was associated with unhealthy dietary patterns, suggesting that diet-related choices could significantly affect the risk of cancer (Grosso et al. 2017). Tilman and Clark (2014) found significant benefits in terms of reductions in relative risk of key diseases: type 2 diabetes, cancer, coronary mortality and all causes of mortality (Figure 5.15).

1.6.3.1   Can dietary shifts provide significant benefits?

Many studies now indicate that dietary shifts can significantly reduce GHG emissions. For instance, several studies highlight that if current dietary trends are maintained, this could lead to emissions from agriculture of approximately 20 $GtCO_2$-eq yr^{-1} by 2050, creating significant mitigation potential (Pradhan et al. 2013b; Bajželj et al. 2014; Hedenus et al. 2014; Bryngelsson et al. 2017). Additionally in the USA, a shift in consumption towards a broadly healthier diet, combined with meeting the USDA and Environmental Protection Agency's 2030 food loss and waste reduction goals, could increase *per capita* food-related energy use by 12%, decrease blue water consumption by 4%, decrease green water use by 23%, decrease GHG emissions from food production by 11%, decrease GHG emissions from landfills by 20%, decrease land use by 32%, and increase fertiliser use by 12% (Birney et al. 2017). This study, however, does not account for all potential routes to emissions, ignoring, for

example, fertiliser use in feed production. Similar studies have been conducted, for China (Li et al. 2016), where adoption of healthier diets and technology improvements have the potential to reduce food systems GHG emissions by >40% relative to those in 2010; and India (Green et al. 2017; Vetter et al. 2017), where alternative diet scenarios can affect emissions from the food system by –20 to +15%.

Springmann et al.(2018a) modelled the role of technology, waste reduction and dietary change in living within planetary boundaries (Rockström et al. 2009), with the climate change boundary being a 66% chance of limiting warming to less than 2°C. They found that all are necessary for the achievement of a sustainable food system. Their principal conclusion is that only by adopting a 'flexitarian diet', as a global average, would climate change be limited to under two degrees. Their definition of a flexitarian diet is fruits and vegetables, plant-based proteins, modest amounts of animal-based proteins, and limited amounts of red meat, refined sugar, saturated fats, and starchy foods.

Healthy and sustainable diets address both health and environmental concerns (Springmann et al. 2018b). There is high agreement that there are significant opportunities to achieve both objectives simultaneously. Contrasting results of marginal GHG emissions, that is, variations in emissions as a result of variation in one or more dietary components, are found when comparing low to high emissions in self-selected diets (diets freely chosen by consumers). Vieux et al. (2013) found self-selected healthier diets with higher amounts of plant-based food products did not result in lower emissions, while (Rose et al. 2019) found that the lowest emission diets analysed were lower in meat but higher in oil, refined grains and added sugar. Vieux et al. (2018) concluded that setting nutritional goals with no consideration for the environment may increase GHG emissions.

Tukker et al. (2011) also found a slight increase in emissions by shifting diets towards the European dietary guidelines, even with lower meat consumption. Heller and Keoleian (2015) found a 12% increase in GHG emissons when shifting to iso-caloric diets, defined as diets with the same caloric intake of diets currently consumed, following the USA guidelines and a 1% decrease in GHG emissions when adjusting caloric intake to recommended levels for moderate activity. There is scarce information on the marginal GHG emissions that would be associated with following dietary guidelines in developing countries.

Some studies have found a modest mitigation potential of diet shifts when economic and biophysical systems effects are taken into account in association with current dietary guidelines. Tukker et al. (2011), considering economic rebound effects of diet shifts (i.e., part of the gains would be lost due to increased use at lower prices), found maximum changes in emissions of the EU food system of 8% (less than 2% of total EU emissions) when reducing meat consumption by 40 to 58%. Using an economic optimisation model for studying carbon taxation in food but with adjustments of agricultural production systems and commodity markets in Europe, Zech and Schneider (2019) found a reduction of 0.41% in GHG emissions at a tax level of 50 USD per tCO_2-eq. They estimate a leakage of 43% of the GHG emissions reduced by domestic consumption, (i.e., although reducing emissions due to reducing consumption, around 43% of the emissions would not be reduced because part of the production would be directed to exports).

Studying optimised beef production systems intensification technologies in a scenario of no grasslands area expansion de Oliveira Silva et al. (2016) found marginal GHG emissions to be negligible in response to beef demand in the Brazilian Cerrado. This was because reducing productivity would lead to increased emission intensities, cancelling out the effect of reduced consumption.

In summary, there is significant potential mitigation (*high confidence*) arising from the adoption of diets in line with dietary recommendations made on the basis of health. These are broadly similar across most countries. These are typically capped at the number of calories and higher in plant-based foods, such as vegetables, fruits, whole grains, legumes, nuts and seeds, and lower in animal-sourced foods, fats and sugar. Such diets have the potential to be both more sustainable and healthier than alternative diets (but healthy diets are not necessarily sustainable and vice versa). The extent to which the mitigation potential of dietary choices can be realised requires both climate change and health being considered together. Socio-economic (prices, rebound effects), political, and cultural contexts would require significant consideration to enable this mitigation potential to be realised.

1.6.4   Sustainable integrated agricultural systems

A range of integrated agricultural systems are being tested to evaluate synergies between mitigation and adaptation and lead to low-carbon and climate-resilient pathways for sustainable food security and ecosystem health (*robust evidence, medium agreement*). Integration refers to the use of practices that enhance an agroecosystem's mitigation, resilience, and sustainability functions. These systems follow holistic approaches with the objective of achieving biophysical, socio-cultural, and economic benefits from land management systems (Sanz et al. 2017). These integrated systems may include agroecology (FAO et al. 2018; Altieri et al. 2015), climate smart agriculture (FAO 2011c; Lipper et al. 2014; Aggarwal et al. 2018), conservation agriculture (Aryal et al. 2016; Sapkota et al. 2015), and sustainable intensification (FAO 2011d; Godfray 2015), amongst others.

Many of these systems are complementary in some of their practices, although they tend to be based on different narratives (Wezel et al. 2015; Lampkin et al. 2015; Pimbert 2015). They have been tested in various production systems around the world (Dinesh et al. 2017; Jat et al. 2016; Sapkota et al. 2015 and Neufeldt et al. 2013). Many technical innovations, for example, precision nutrient management (Sapkota et al. 2014) and precision water management (Jat et al. 2015), can lead to both adaptation and mitigation outcomes and even synergies; although negative adaptation and mitigation outcomes (i.e., trade-offs) are often overlooked. Adaptation potential of ecologically intensive systems includes crop diversification, maintaining local genetic diversity, animal integration, soil organic management, water conservation and harvesting the role of microbial assemblages (Section 5.3). Technical innovations may encompass not

only inputs reduction, but complete redesign of agricultural systems (Altieri et al. 2017) and how knowledge is generated (Levidow et al. 2014), including social and political transformations.

1.6.4.1   Agroecology

Agroecology (see Glossary) (Francis et al. 2003; Gliessman and Engles 2014; Gliessman 2018), provides knowledge for their design and management, including social, economic, political, and cultural dimensions (Dumont et al. 2016). It started with a focus at the farm level but has expanded to include the range of food system activities (Benkeblia 2018). Agroecology builds systems resilience through knowledge-intensive practices relying on traditional farming systems and co-generation of new insights and information with stakeholders through participatory action research (Menéndez et al. 2013). It provides a multidimensional view of food systems within ecosystems, building on ILK and co-evolving with the experiences of local people, available natural resources, access to these resources, and ability to share and pass on knowledge among communities and generations, emphasising the inter-relatedness of all agroecosystem components and the complex dynamics of ecological processes (Vandermeer 1995).

At the farm level, agroecological practices recycle biomass and regenerate soil biotic activities. They strive to attain balance in nutrient flows to secure favourable soil and plant growth conditions, minimise loss of water and nutrients, and improve use of solar radiation. Practices include efficient microclimate management, soil cover, appropriate planting time and genetic diversity. They seek to promote ecological processes and services such as nutrient cycling, balanced predator/prey interactions, competition, symbiosis, and successional changes. The overall goal is to benefit human and non-human communities in the ecological sphere, with fewer negative environmental or social impacts and fewer external inputs (Vandermeer et al. 1998; Altieri et al. 1998). From a food system focus, agroecology provides management options in terms of commercialisation and consumption through the promotion of short food chains and healthy diets (Pimbert and Lemke 2018; Loconto et al. 2018).

Agroecology has been proposed as a key set of practices in building climate resilience (FAO et al. 2018; Altieri et al. 2015). These can enhance on-farm diversity (of genes, species, and ecosystems) through a landscape approach (FAO 2018g). Outcomes include soil conservation and restoration and thus soil carbon sequestration, reduction of the use of mineral and chemical fertilisers, watershed protection, promotion of local food systems, waste reduction, and fair access to healthy food through nutritious and diversified diets (Pimbert and Lemke 2018; Kremen et al. 2012; Goh 2011; Gliessman and Engles 2014).

A principle in agroecology is to contribute to food production by smallholder farmers (Altieri 2002). Since climatic events can severely impact smallholder farmers, there is a need to better understand the heterogeneity of small-scale agriculture in order to consider the diversity of strategies that traditional farmers have used and still use to deal with climatic variability. In Africa, many smallholder farmers cope with and even prepare for climate extremes, minimising crop failure through a series of agroecological practices (e.g., biodiversification, soil management, and water harvesting) (Mbow et al. 2014a). Resilience to extreme climate events is also linked to on-farm biodiversity, a typical feature of traditional farming systems (Altieri and Nicholls 2017).

Critiques of agroecology refer to its explicit exclusion of modern biotechnology (Kershen 2013) and the assumption that smallholder farmers are a uniform unit with no heterogeneity in power (and thus gender) relationships (Neira and Montiel 2013; Siliprandi and Zuluaga Sánchez 2014).

1.6.4.2   Climate-smart agriculture

'Climate-smart agriculture' (CSA) is an approach developed to tackle current food security and climate change challenges in a joint and synergistic fashion (Lipper et al. 2014; Aggarwal et al. 2018; FAO 2013c). CSA is designed to be a pathway towards development and food security built on three pillars: increasing productivity and incomes, enhancing resilience of livelihoods and ecosystems and reducing, and removing GHG emissions from the atmosphere (FAO 2013c). Climate-smart agricultural systems are integrated approaches to the closely linked challenges of food security, development, and climate change adaptation/mitigation to enable countries to identify options with maximum benefits and those where trade-offs need management.

Many agricultural practices and technologies already provide proven benefits to farmers' food security, resilience and productivity (Dhanush and Vermeulen 2016). In many cases, these can be implemented by changing the suites of management practices. For example, enhancing soil organic matter to improve the water-holding capacity of agricultural landscapes also sequesters carbon. In annual cropping systems, changes from conventional tillage practices to minimum tillage can convert the system from one that either provides adaptation or mitigation benefits or neither to one that provides both adaptation and mitigation benefits (Sapkota et al. 2017a; Harvey et al. 2014a).

Increasing food production by using more fertilisers in agricultural fields could maintain crop yield in the face of climate change, but may result in greater overall GHG emissions. But increasing or maintaining the same level of yield by increasing nutrient-use-efficiency through adoption of better fertiliser management practices could contribute to both food security and climate change mitigation (Sapkota et al. 2017a).

Mixed farming systems integrating crops, livestock, fisheries and agroforestry could maintain crop yield in the face of climate change, help the system to adapt to climatic risk, and minimise GHG emissions by increasingly improving the nutrient flow in the system (Mbow et al. 2014a; Newaj et al. 2016; Bioversity International 2016). Such systems can help diversify production and/or incomes and support efficient and timely use of inputs, thus contributing to increased resilience, but they require local seed and input systems and extension services. Recent whole farm modelling exercises have shown the economic and environmental (reduced GH emissions, reduced land use) benefits of integrated crop-livestock systems (Gil

et al. 2018) compared different soy-livestock systems across multiple economic and environmental indicators, including climate resilience. However, it is important to note that potential benefits are very context specific.

Although climate-smart agriculture involves a holistic approach, some argue that it narrowly focuses on technical aspects at the production level (Taylor 2018; Newell and Taylor 2018). Studying barriers to the adoption and diffusion of technological innovations for climate-smart agriculture in Europe, Long et al. (2016) found that there was incompatibility between existing policies and climate-smart agriculture objectives, including barriers to the adoption of technological innovations.

Climate-smart agricultural systems recognise that the implementation of the potential options will be shaped by specific country contexts and capacities, as well as enabled by access to better information, aligned policies, coordinated institutional arrangements and flexible incentives and financing mechanisms (Aggarwal et al. 2018). Attention to underlying socio-economic factors that affect adoption of practices and access to technologies is crucial for enhancing biophysical processes, increasing productivity, and reducing GHG emissions at scale. The Government of India, for example, has started a programme of climate resilient villages (CRV) as a learning platform to design, implement, evaluate and promote various climate-smart agricultural interventions, with the goal of ensuring enabling mechanisms at the community level (Srinivasa Rao et al. 2016).

1.6.4.3   Conservation agriculture

Conservation agriculture (CA) is based on the principles of minimum soil disturbance and permanent soil cover, combined with appropriate crop rotation (Jat et al. 2014; FAO 2011e). CA has been shown to respond with positive benefits to smallholder farmers under both economic and environmental pressures (Sapkota et al. 2017a, 2015). This agricultural production system uses a body of soil and residues management practices that control erosion (Blanco Sepúlveda and Aguilar Carrillo 2016) and at the same time improve soil quality, by increasing organic matter content and improving porosity, structural stability, infiltration and water retention (Sapkota et al. 2017a, 2015 and Govaerts et al. 2009).

Intensive agriculture during the second half of the 20th century led to soil degradation and loss of natural resources and contributed to climate change. Sustainable soil management practices can address both food security and climate change challenges faced by these agricultural systems. For example, sequestration of soil organic carbon (SOC) is an important strategy to improve soil quality and to mitigation of climate change (Lal 2004). CA has been reported to increase farm productivity by reducing costs of production (Aryal et al. 2015; Sapkota et al. 2015; Indoria et al. 2017) as well as to reduce GHG emission (Pratibha et al. 2016).

Conservation agriculture brings favourable changes in soil properties that affect the delivery of nature's contribution to people (NCPs) or ecosystem services, including climate regulation through carbon sequestration and GHG emissions (Palm et al. 2013; Sapkota et al.

2017a). However, by analysing datasets for soil carbon in the tropics, Powlson et al. (2014, 2016) argued that the rate of SOC increase

labour (Aryal et al. 2015) although a gender shift of the labour burden to women have also been described (Giller et al. 2009).

Cross-Chapter Box 6, Table 1 | Approaches to sustainable intensification of agriculture (Pretty et al. 2018; Hill 1985).

Approach	Sub-category	Examples/notes
Improving efficiency	Precision agriculture	High- and low-technology options to optimise resource use.
	Genetic improvements	Improved resource use efficiency through crop or livestock breeding.
	Irrigation technology	Increased production in areas currently limited by precipitation (sustainable water supply required).
	Organisational scale-up	Increasing farm organisational scale (e.g., cooperative schemes) can increase efficiency via facilitation of mechanisation and precision techniques.
Substitution	Green fertiliser	Replacing chemical fertiliser with green manures, compost (including vermicompost), biosolids and digestate (by-product of anaerobic digestion) to maintain and improve soil fertility.
	Biological control	Pest control through encouraging natural predators.
	Alternative crops	Replacment of annual with perennial crops reducing the need for soil disturbance and reducing erosion.
	Premium products	Increase farm-level income for less output by producing a premium product.
System redesign	System diversification	Implementation of alternative farming systems: organic, agroforestry and intercropping (including the use of legumes).
	Pest management	Implementing integrated pest and weed management to reduce the quantities of inputs required.
	Nutrient management	Implementing integrated nutrient management by using crop and soil specific nutrient management – guided by soil testing.
	Knowledge transfer	Using knowledge sharing and technology platforms to accelerate the uptake of good agricultural practices.

and resulting GHG mitigation in CA systems, from zero-tillage in particular, has been overstated (Chapter 2).

However, there is unanimous agreement that the gain in SOC and its contribution to GHG mitigation by CA in any given soil is largely determined by the quantity of organic matter returned to the soil (Giller et al. 2009; Virto et al. 2011; Sapkota et al. 2017b). Thus, a careful analysis of the production system is necessary to minimise the trade-offs among the multiple use of residues, especially where residues remain an integral part of livestock feeding (Sapkota et al. 2017b). Similarly, replacing mono-cropping systems with more diversified cropping systems and agroforestry, as well as afforestation and deforestation, can buffer temperatures as well as increase carbon storage (Mbow et al. 2014a; Bioversity International 2016), and provide diversified and healthy diets in the face of climate change.

Adoption of conservation agriculture in Africa has been low despite more than three decades of implementation (Giller et al. 2009), although there is promising uptake recently in east and southern Africa. This calls for a better understanding of the social and institutional aspects around CA adoption. Brown et al. (2017a) found that institutional and community constraints hampered the use of financial, physical, human and informational resources to implement CA programmes.

Gender plays an important role at the intra-household level in regard to decision-making and distributing benefits. Conservation agriculture interventions have implications for labour requirements, labour allocation, and investment decisions, all of which impact the roles of men and women (Farnworth et al. 2016) (Section 5.1.3). For example, in the Global South, CA generally reduces labour and production costs and generally leads to increased returns to family

1.6.4.4   Sustainable intensification

The need to produce about 50% more food by 2050, required to feed the increasing world population (FAO 2018a), may come at the price of significant increases in GHG emissions and environmental impacts, including loss of biodiversity. For instance, land conversion for agriculture is responsible for an estimated 8–10% of all anthropogenic GHG emissions currently (Section 5.4). Recent calls for sustainable intensification (SI) are based on the premise that damage to the environment through extensification outweighs benefits of extra food produced on new lands (Godfray 2015). However, increasing the net production area by restoring already degraded land may contribute to increased production on the one hand and increased carbon sequestration on the other (Jat et al. 2016), thereby contributing to both increased agricultural production and improved natural capital outcomes (Pretty et al. 2018).

Sustainable intensification is a goal but does not specify *a priori* how it could be attained, for example, which agricultural techniques to deploy (Garnett et al. 2013). It can be combined with selected other improved management practices, for example, conservation agriculture (see above), or agroforestry, with additional economic, ecosystem services, and carbon benefits. Sustainable intensification, by improving nutrient, water, and other input-use efficiency, not only helps to close yield gaps and contribute to food security (Garnett et al. 2013), but also reduces the loss of such production inputs and associated emissions (Sapkota et al. 2017c; Wollenberg et al. 2016). Closing yield gaps is a way to become more efficient in use of land per unit production. Currently, most regions in Africa and South Asia have attained less than 40% of their potential crop production (Pradhan et al. 2015). Integrated farming systems (e.g., mixed crop/livestock, crop/aquaculture) are strategies to produce more products per unit land, which in regard to food security, becomes highly relevant.

for high population density and small farm sizes, attaining food security and reducing GHG emissions require the use of more

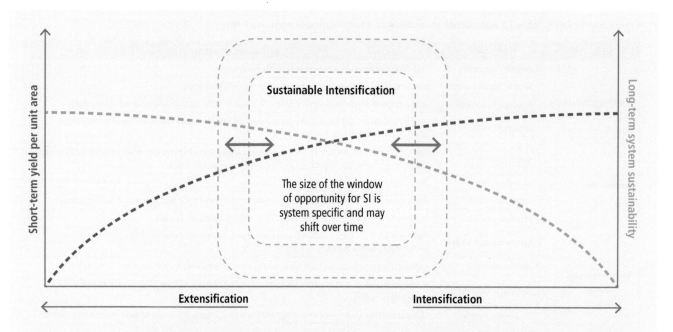

Cross-Chapter Box 6, Figure 1 | There is a need to balance increasing demands for food, fuel and fibre with long-term sustainability of land use. Sustainable intensification can, in theory, offer a window of opportunity for the intensification of land use without causing degradation. This potentially allows the sparing of land to provide other ecosystem services, including carbon sequestration and the protection of biodiversity. However, the potential for SI is system specific and may change through time (indicated by grey arrows). Current practice may already be outside of this window and be unsustainable in terms of negative impacts on the long-term sustainability of the system.

Sustainable intensification acknowledges that enhanced productivity needs to be accompanied by maintenance of other ecosystem services and enhanced resilience to shocks (Vanlauwe et al. 2014). SI in intensively farmed areas may require a reduction in production in favour of increasing sustainability in the broad sense (Buckwell et al. 2014) (Cross-Chapter Box 6 in Chapter 5). Hence, moving towards sustainability may imply lower yield growth rates than those maximally attainable in such situations. For areas that contain valuable natural ecosystems, such as the primary forest in the Congo basin, intensification of agriculture is one of the pillars of the strategy to conserve forest (Vanlauwe et al. 2014). Intensification in agriculture is recognised as one of the pathways to meet food security and climate change adaptation and mitigation goals (Sapkota et al. 2017c).

However, SI does not always confer co-benefits in terms of food security and climate change adaption/mitigation. For example, in the case of Vietnam, intensified production of rice and pigs reduced GHG emissions in the short term through land sparing, but after two decades, the emissions associated with higher inputs were likely to outweigh the savings from land sparing (Thu Thuy et al. 2009). Intensification needs to be sustainable in all components of food system by curbing agricultural sprawl, rebuilding soils, restoring degraded lands, reducing agricultural pollution, increasing water use efficiency, and decreasing the use of external inputs (Cook et al. 2015).

A study conducted by Palm et al. (2010) in Sub-Saharan Africa, reported that, at low population densities and high land availability, food security and climate mitigation goals can be met with intensification scenarios, resulting in surplus crop area for reforestation. In contrast,

mineral fertilisers to make land available for reforestation. However, some forms of intensification in drylands can increase rather than reduce vulnerability due to adverse effects such as environmental degradation and increased social inequity (Robinson et al. 2015).

Sustainable intensification has been critiqued for considering food security only from the supply side, whereas global food security requires attention to all aspects of food system, including access, utilisation, and stability (Godfray 2015). Further, adoption of high-input forms of agriculture under the guise of simultaneously improving yields and environmental performance will attract more investment leading to higher rate of adoption but with the environmental component of SI quickly abandoned (Godfray 2015). Where adopted, SI needs to engage with the sustainable development agenda to (i) identify SI agricultural practices that strengthen rural communities, improve smallholder livelihoods and employment, and avoid negative social and cultural impacts, including loss of land tenure and forced migration; (ii) invest in the social, financial, natural, and physical capital needed to facilitate SI implementation; and (iii) develop mechanisms to pay poor farmers for undertaking sustainability measures (e.g., GHG emissions mitigation or biodiversity protection) that may carry economic costs (Garnett et al. 2013).

In summary, integrated agricultural systems and practices can enhance food system resilience to climate change and reduce GHG emissions, while helping to achieve sustainability (*high confidence*).

Cross-Chapter Box 6 | Agricultural intensification: Land sparing, land sharing and sustainability

Eamon Haughey (Ireland), Tim Benton (United Kingdom), Annette Cowie (Australia), Lennart Olsson (Sweden), Pete Smith (United Kingdom)

Introduction

The projected demand for more food, fuel and fibre for a growing human population necessitates intensification of current land use to avoid conversion of additional land to agriculture and potentially allow the sparing of land to provide other ecosystem services, including carbon sequestration, production of biomass for energy, and the protection of biodiversity (Benton et al. 2018; Garnett et al. 2013). Land-use intensity may be defined in terms of three components; (i) intensity of system inputs (land/soil, capital, labour, knowledge, nutrients and other chemicals), (ii) intensity of system outputs (yield per unit land area or per specific input) and (iii) the impacts of land use on ecosystem services such as changes in soil carbon or biodiversity (Erb et al. 2013). Intensified land use can lead to ecological damage as well as degradation of soil, resulting in a loss of function which underpins many ecosystem services (Wilhelm and Smith 2018; Smith et al. 2016). Therefore, there is a risk that increased agricultural intensification could deliver short-term production goals at the expense of future productive potential, jeopardising long term food security (Tilman et al. 2011).

Agroecosystems which maintain or improve the natural and human capital and services they provide may be defined as sustainable systems, while those which deplete these assets as unsustainable (Pretty and Bharucha 2014). Producing more food, fuel and fibre without the conversion of additional non-agricultural land while simultaneously reducing environmental impacts requires what has been termed sustainable intensification (Godfray et al. 2010; FAO 2011e) (Glossary and Figure 1 in this Cross-Chapter Box). Sustainable intensification (SI) may be achieved through a wide variety of means; from improved nutrient and water use efficiency via plant and animal breeding programmes, to the implementation of integrated soil fertility and pest management practices, as well as by smarter land-use allocation at a larger spatial scale: for example, matching land use to the context and specific capabilities of the land (Benton et al. 2018). However, implementation of SI is broader than simply increasing the technical efficiency of agriculture ('doing more with less'). It sometimes may require a reduction of yields to raise sustainability, and successful implementation can be dependent on place and scale. Pretty et al. (2018), following Hill (1985), highlights three elements to SI: (i) increasing efficiency, (ii) substitution of less beneficial or efficient practices for better ones, and (iii) system redesign to adopt new practices and farming systems (Table 1 in this Cross-Chapter Box).

Under a land sparing strategy, intensification of land use in some areas, generating higher productivity per unit area of land, can allow other land to provide other ecosystem services, such as increased carbon sequestration and the conservation of natural ecosystems and biodiversity (Balmford et al. 2018 and Strassburg et al. 2014). Conversely under a land sharing strategy, less, or no, land is set aside, but lower levels of intensification are applied to agricultural land, providing a combination of provisioning and other functions such as biodiversity conservation from the same land (Green et al. 2005). The two approaches are not mutually exclusive and the suitability of their application is generally system-, scale- and/or location-specific (Fischer et al. 2014). One crucial issue for the success of a land sparing strategy is that spared land is protected from further conversion. As the profits from the intensively managed land increase, there is an incentive for conversion of additional land for production (Byerlee et al. 2014). Furthermore, it is implicit that there are limits to the SI of land at a local and also planetary boundary level (Rockström et al. 2009). These may relate to the 'health' of soil, the presence of supporting services, such as pollination, local limits to water availability, or limits on air quality. This implies that it may not be possible to meet demand 'sustainably' if demand exceeds local and global limits. There are no single global solutions to these challenges and specific in situ responses for different farming systems and locations are required. Bajželj et al. (2014) showed that implementation of SI, primarily through yield gap closure, had better environmental outcomes compared with 'business as usual' trajectories. However, SI alone will not be able to deliver the necessary environmental outcomes from the food system – dietary change and reduced food waste are also required (Springmann et al. 2018a; Bajželj et al. 2014).

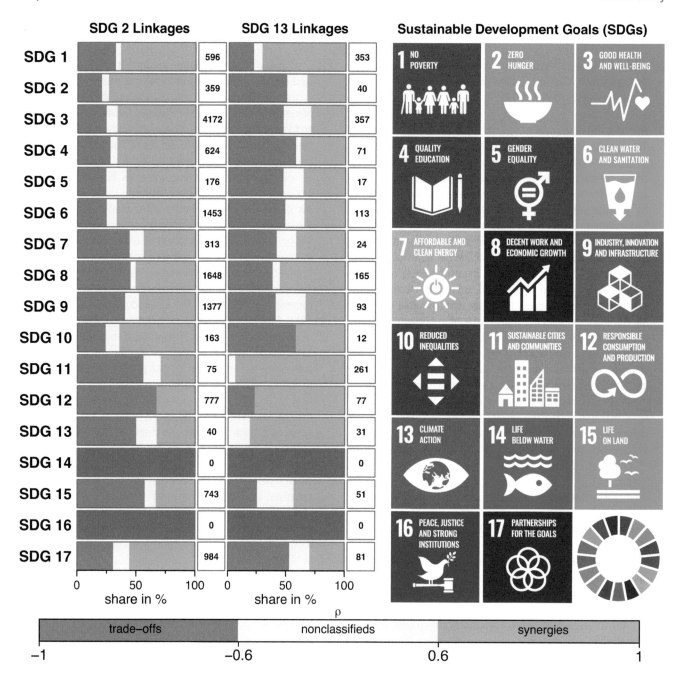

Figure 5.16. | Intra and inter-linkages for SDG 2 (Zero Hunger) and SDG 13 (Climate Action) at the global level using the official indicators of Sustainable Development Goals that consist of data for 122 indicators for a total of 227 countries between the years 1983 and 2016 (United Nations Statistics Division 2016). Synergies and trade-offs defined as significant positive ($\rho > 0.6$, red bar) and negative ($\rho < -0.6$, green bar) Spearman's correlation between SDG indicators, respectively; ρ between 0.6 and −0.6 is considered as nonclassifieds (yellow bar) (Pradhan et al. 2017). Grey bars show insufficient data for analysis; white box shows number of data pairs used in analysis. The correlation between unique pairs of indicator time-series is carried based on country data. For example, between 'prevalence of undernourishment' (an indicator for SDG 2.1) and 'maternal mortality ratio' (an indicator for SDG 3.1). The data pairs can belong to the same goal or to two distinct goals. At the global level, intra-linkages of SDGs are quantified by the percentage of synergies, trade-offs, and nonclassifieds of indicator pairs belonging to the same SDG for all the countries. Similarly, SDG interlinkages are estimated by the percentage of synergies, trade-offs, and nonclassifieds between indicator pairs that fall into two distinct goals for all the countries.

Cross-Chapter Box 6 (continued)

Improved efficiency – example of precision agriculture

Precision farming usually refers to optimising production in fields through site-specific choices of crop varieties, agrochemical application, precise water management (e.g., in given areas or threshold moistures) and management of crops at a small scale (or livestock as individuals) (Hedley 2015). Precision agriculture has the potential to achieve higher yields in a more efficient and sustainable manner compared with traditional low-precision methods.

Precision agriculture

Precision agriculture is a technologically advanced approach that uses continual monitoring of crop and livestock performance to actively inform management practices. Precise monitoring of crop performance over the course of the growing season will enable farmers to economise on their inputs in terms of water, nutrients and pest management. Therefore, it can contribute to both the food security (by maintaining yields), sustainability (by reducing unnecessary inputs) and land sparing goals associated with SI. The site-specific management of weeds allows a more efficient application of herbicide to specific weed patches within crops (Jensen et al. 2012). Such precision weed control has resulted in herbicide savings of 19–22% for winter oilseed rape, 46–57% for sugar beet and 60–77% for winter wheat production (Gutjahr and Gerhards 2010). The use of on-farm sensors for real time management of crop and livestock performance can enhance farm efficiency (Aqeel-Ur-Rehman et al. 2014). Mapping soil nutrition status can allow for more targeted, and therefore more effective, nutrient management practices (Hedley 2015). Using wireless sensors to monitor environmental conditions, such as soil moisture, has the potential to allow more efficient crop irrigation (Srbinovska et al. 2015). Controlled traffic farming, where farm machinery is confined to permanent tracks, using automatic steering and satellite guidance, increases yields by minimising soil compaction. However, barriers to the uptake of many of these high-tech precision agriculture technologies remain. In what is described as the 'implementation problem', despite the potential to collect vast quantities of data on crop or livestock performance, applying these data to inform management decisions remains a challenge (Lindblom et al. 2017).

Low-tech precision agriculture

The principle of precision agriculture can be applied equally to low capital-input farming, in the form of low-tech precision agriculture (Conway 2013). The principle is the same, but instead of adopting capital-heavy equipment (such as sensor technology connected to the 'internet of things', or large machinery and expensive inputs), farmers use knowledge and experience and re-purposed innovative approaches, such as a bottle cap as a fertiliser measure for each plant, applied by hand (Mondal and Basu 2009). This type of precision agriculture

is particularly relevant to small-scale farming in the Global South, where capital investment is major limiting factor. For example, the application of a simple seed priming technique resulted in a 20 to 30% increase in yields of pearl millet and sorghum in semi-arid West Africa (Aune et al. 2017). Low-tech precision agriculture has the potential to increase the economic return per unit land area while also creating new employment opportunities.

Cross-Chapter Box 6 (continued)

Sustainable intensification through farming system redesign

Sustainable intensification requires equal weight to be placed on the sustainability and intensification components (Benton 2016; Garnett et al. 2013). Figure 1 in this Cross-Chapter Box outlines the trade-offs which SI necessitates between the intensity of land use against long-term sustainability. One approach to this challenge is through farming system redesign, including increased diversification.

Diversification of intensively managed systems

Incorporating higher levels of plant diversity in agroecosystems can improve the sustainability of farming systems (Isbell et al. 2017). Where intensive land use has led to land degradation, more diverse land-use systems, such as intercropping, can provide a more sustainable land-use option with co-benefits for food security, adaptation and mitigation objectives. For example, in temperate regions, highly productive agricultural grasslands used to produce meat and dairy products are characterised by monoculture pastures with high agrochemical inputs. Multi-species grasslands may provide a route to SI, as even a modest increase in species richness in intensively managed grasslands can result in higher forage yields without increased inputs, such as chemical fertiliser (Finn et al. 2013; Sanderson et al. 2013; Tilman et al. 2011). Recent evidence also indicates multispecies grasslands have greater resilience to drought, indicating co-benefits for adaptation (Hofer et al. 2016; Haughey et al. 2018).

Diversification of production systems

Agroforestry systems (see Glossary) can promote regional food security and provide many additional ecosystem services when compared with monoculture crop systems. Co-benefits for mitigation and adaptation include increased carbon sequestration in soils and biomass, improved water and nutrient use efficiency and the creation of favourable micro-climates (Waldron et al. 2017). Silvopasture systems, which combine grazing of livestock and forestry, are particularly useful in reducing land degradation where the risk of soil erosion is high (Murgueitio et al. 2011). Crop and livestock systems can also be combined to provide multiple services. Perennial wheat derivatives produced both high quality forage and substantial volumes of cereal grains

5

(Newell and Hayes 2017), and show promise for integrating cereal and livestock production while sequestering soil carbon (Ryan et al. 2018). A key feature of diverse production systems is the provision of multiple income streams for farming households, providing much needed economic resilience in the face of fluctuation of crop yields and prices.

Cross-Chapter Box 6 (continued)

Landscape approaches
The land sparing and land sharing approaches which may be used to implement SI are inherently 'landscape approaches' (e.g., Hodgson et al. 2010). While the term landscape is by no means precise (Englund et al. 2017), landscape approaches, focused, for example, at catchment scale, are generally agreed to be the best way to tackle competing demands for land (e.g., Sayer et al. 2013), and are the appropriate scale at which to focus the implementation of sustainable intensification. The landscape approach allots land to various uses – cropping, intensive and extensive grazing, forestry, mining, conservation, recreation, urban, industry, infrastructure – through a planning process that seeks to balance conservation and production objectives. With respect to SI, a landscape approach is pertinent to achieving potential benefits for biodiversity conservation, ensuring that land 'spared' through SI remains protected, and that adverse impacts of agriculture on conservation land are minimised. Depending on the land governance mechanisms applied in the jurisdiction, different approaches will be appropriate/required. However, benefits are only assured if land-use restrictions are devised and enforced.

Summary
Intensification needs to be achieved sustainably, necessitating a balance between productivity today and future potential (*high agreement, medium evidence*). Improving the efficiency of agriculture systems can increase production per unit of land through more effective resource use. To achieve SI, some intensively managed agricultural systems may have to be diversified as they cannot be further intensified without land degradation. A combination of land sparing and sharing options can be utilised to achieve SI – their application is most likely to succeed if applied using a landscape approach.

1.6.5   Role of urban agriculture

Cities are an important actor in the food system through demand for food by urban dwellers and production of food in urban and peri-urban areas (Cross-Chapter Box 4 in Chapter 2). Both the demand side and supply side roles are important relative to climate change mitigation and adaptation strategies. Urban areas are home to more than half of the world's population, and a minimal proportion of the production. Thus, they are important drivers for the development of the complex food systems in place today, especially with regard to supply chains and dietary preferences.

The increasing separation of urban and rural populations with regard to territory and culture is one of the factors favouring the nutrition transition towards urban diets (Weber and Matthews 2008; Neira et al. 2016). These are primarily based on a high diversity of food products, independent of season and local production, and on the

Table 5.6 | Potential policy 'families' for food-related adaptation and mitigation of climate change. The column 'scale' refers to scale of implementation: International (I), national (N), sub-national-regional (R), and local (L).

Family	Sub-family	Scale	Interventions	Examples
Supply-side efficiency	Increasing agricultural efficiency and yields	I, N	Agricultural R&D	Investment in research, innovation, knowledge exchange, e.g., on genetics, yield gaps, resilience
		I, N	Supporting precision agriculture	Agricultural engineering, robotics, big data, remote sensing, inputs
		I, N	Sustainable intensification projects	Soils, nutrients, capital, labour (Cross-Chapter Box 6)
		N, R	Improving farmer training and knowledge sharing	Extension services, online access, farmer field schools, farmer-to-farmer networks (CABI 2019)
	Land-use planning	N, R, L	Land-use planning for ecosystem services (remote sensing, ILK)	Zoning, protected area networks, multifunctional landscapes, 'land sparing' (Cross-Chapter Box 6; Benton et al. 2018; Jones et al. 2013)
		N, R, L	Conservation agriculture programmes	Soil and water erosion control, soil quality improvement (Conservation Evidence 2019)
		N	Payment for ecosystem services	Incentives for farmers/landowners to choose lower-profit but environmentally benign resource use, e.g., Los Negros Valley in Bolivia (Ezzine-de-Blas et al. 2016)
	Market approaches	I, N	Mandated carbon cost reporting in supply chains; public/private incentivised insurance products	Carbon and natural capital accounts (CDP 2019), crop insurance (Müller et al. 2017a)
	Trade	I	Liberalising trade flows; green trade	Reduction in GHG emissions from supply chains (Neumayer 2001)
Raising profitability and quality	Stimulating markets for premium goods	N, R	Sustainable farming standards, agroecology projects, local food movements	Regional policy development, public procurement of sustainable food (Mairie de Paris 2015)
Modifying demand	Reducing food waste	I, N, L	Regulations, taxes	'Pay-As-You-Throw (PAYT)' schemes; EU Landfill Directives; Japan Food Waste Recycling Law 2008; South Africa Draft Waste Classification and Management Regulations 2010 (Chalak et al. 2016)
		I, N, L	Awareness campaigns, education	FAO Global Initiative on Food Loss and Waste Reduction (FAO 2019b)
		I, N	Funding for reducing food waste	Research and investment for shelf life, processing, packaging, cold storage (MOFPI 2019)
		I, N, L	Circular economy using waste as inputs	Biofuels, distribution of excess food to charities (Baglioni et al. 2017)
	Reducing consumption of carbon-intensive food	I, N, L	Carbon pricing for selected food commodities	Food prices reflective of GHG gas emissions throughout production and supply chain (Springmann et al. 2017; Hasegawa et al. 2018)
		I, N, L	Changing food choice through education	Nutritional and portion-size labelling, 'nudge' strategies (positive reinforcement, indirect suggestion) (Arno and Thomas 2016)
		I, N, L	Changing food choices through money transfers	Unconditional cash transfers; e-vouchers exchanged for set quantity or value of specific, pre-selected goods (Fenn 2018)
		N, L	Changing food environments through planning	Farmers markets, community food production, addressing 'food deserts' (Ross et al. 2014)
	Combining carbon and health objectives	I, N, L	Changing subsidies, standards, regulations to healthier and more sustainably produced foods	USDA's 'Smart Snacks for School' regulation mandating nutritional guidelines (USDA 2016) Incentivising production via subsidies (direct to producer based on output or indirect via subsidising inputs)
		N	Preventative versus curative public healthcare incentives	Health insurance cost reductions for healthy and sustainable diets
		I, N, L	Food system labelling	Organic certification, nutrition labels, blockchain ledgers (Chadwick 2017)
		N, L	Education and awareness campaigns	School curricula; public awareness campaigns
		N, L	Investment in disruptive technologies (e.g., cultured meat)	Tax breaks for R&D, industrial strategies (European Union 2018)
		N, L	Public procurement	For health: Public Procurement of Food for Health (Caldeira et al. 2017) For environment: Paris Sustainable Food Plan 2015–2020 Public Procurement Code (Mairie de Paris 2015)

extension of the distances that food travels between production and consumption. The transition of traditional diets to more homogeneous diets has also become tied to consumption of animal protein, which has increased GHG emissions globally (Section 5.4.6).

Cities are becoming key actors in developing strategies of mitigation to climate change, in their food procurement and in sustainable urban food policies alike (McPhearson et al. 2018). These are being developed by big and medium-sized cities in the world, often integrated within climate change policies (Moragues et al. 2013 and Calori and Magarini 2015). A review of 100 cities shows that urban food consumption is one of the largest sources of urban material flows, urban carbon footprint, and land footprint (Goldstein et al. 2017). Additionally, the urban poor have limited capacity to adapt to climate-related impacts, which place their food security at risk under climate change (Dubbeling and de Zeeuw 2011).

Urban and peri-urban areas. In 2010, around 14% of the global population was nourished by food grown in urban and peri-urban areas (Kriewald et al. 2019). A review study on Sub-Saharan Africa shows that urban and peri-urban agriculture contributes to climate change adaptation and mitigation (Lwasa et al. 2014, 2015). Urban and peri-urban agriculture reduces the food carbon footprint by avoiding long distance food transport. These types of agriculture also limit GHG emissions by recycling organic waste and wastewater that would otherwise release methane from landfills and dumping sites (Lwasa et al. 2014). Urban and peri-urban agriculture also contribute in adapting to climate change, including extreme events, by reducing the urban heat island effect, increasing water infiltration and slowing down run-offs to prevent flooding, etc. (Lwasa et al. 2014, 2015; Kumar et al. 2017a). For example, a scenario analysis shows that urban gardens reduce the surface temperature up to 10°C in comparison to the temperature without vegetation (Tsilini et al. 2015). Urban agriculture can also improve biodiversity and strengthen associated ecosystem services (Lin et al. 2015).

Urban and peri-urban agriculture is exposed to climate risks and urban growth that may undermine its long-term potential to address urban food security (Padgham et al. 2015). Therefore, there is a need to better understand the impact of urban sprawl on peri-urban agriculture; the contribution of urban and peri-urban agriculture to food self-sufficiency of cities; the risks posed by pollutants from urban areas to agriculture and vice-versa; the global and regional extent of urban agriculture; and the role that urban agriculture could play in climate resilience and abating malnutrition (Mok et al. 2014; Hamilton et al. 2014). Globally, urban sprawl is projected to consume 1.8–2.4% and 5% of the current cultivated land by 2030 and 2050 respectively, leading to crop calorie loss of 3–4% and 6–7%, respectively (Pradhan et al. 2014 and Bren d'Amour et al. 2017). Kriewald et al. 2019 shows that the urban growth has the largest impact in many sub-continental regions (e.g., Western, Central, and Eastern Africa), while climate change will mostly reduce potential of urban and peri-urban agriculture in Southern Europe and North Africa.

In summary, urban and peri-urban agriculture can contribute to improving urban food security, reducing GHG emissions, and adapting to climate change impacts (*robust evidence, medium agreement*).

1.6.6 Links to the Sustainable Development Goals

In 2015, the Sustainable Development Goals (SDGs) and the Paris Agreement were two global major international policies adopted by all countries to guide the world to overall sustainability, within the 2030 Sustainable Development Agenda and UNFCCC processes respectively. The 2030 Sustainable Development agenda includes 17 goals and 169 targets, including zero hunger, sustainable agriculture and climate action (United Nations 2015).

This section focuses on intra – and inter-linkages of SDG 2 and SDG 13 based on the official SDG indicators (Figure 5.16), showing the current conditions (Roy et al. (2018) and Chapter 7 for further discussion). The second goal (Zero Hunger – SDG 2) aims to end hunger and all forms of malnutrition by 2030 and commits to universal access to safe, nutritious and sufficient food at all times of the year. SDG 13 (Climate Action) calls for urgent action to combat climate change and its impacts. Integrating the SDGs into the global food system can provide opportunities for mitigation and adaptation and enhancement of food security.

Ensuring food security (SDG 2) shows positive relations (synergies) with most goals, according to Pradhan et al. (2017) and the International Council for Science (ICSU) (2017), but has trade-offs with SDG 12 (Responsible Consumption and Production) and SDG 15 (Life on Land) under current development paradigms (Pradhan et al. 2017). Sustainable transformation of traditional consumption and production approaches can overcome these trade-offs based on several innovative methods (Shove et al. 2012). For example, sustainable intensification and reduction of food waste can minimise the observed negative relations between SDG 2 and other goals (Obersteiner et al. 2016) (Cross-Chapter Box 6 in Chapter 5 and Section 5.5.2). Achieving target 12.3 of SDG 12 'by 2030, to halve per capita global food waste at the retail and consumer levels and reduce food losses along production and supply chains, including post-harvest losses' will contribute to climate change mitigation.

Doubling productivity of smallholder farmers and halving food loss and waste by 2030 are targets of SDG 2 and SDG 12, respectively (United Nations Statistics Division 2016). Agroforestry that promotes biodiversity and sustainable land management also contributes to food security (Montagnini and Metzel 2017). Land restoration and protection (SDG 15) can increase crop productivity (SDG 2) (Wolff et al. 2018). Similarly, efficient irrigation practices can reduce water demand for agriculture that could improve the health of the freshwater ecosystem (SDG 6 and SDG 15) without reducing food production (Jägermeyr et al. 2017).

Climate action (SDG 13) shows negative relations (trade-offs) with most goals and is antagonistic to the 2030 development agenda under the current development paradigm (Figure 5.16) (Lusseau and Mancini 2019 and Pradhan 2019). The targets for SDG 13 have a strong focus on climate change adaptation, and the data for the SDG 13 indicators are limited. SDG 13 shares two indicators with SDG 1 and SDG 11 (United Nations 2017) and therefore, has mainly positive linkages with these two goals. Trade-offs were observed between SDG 2 and SDG 13 for around 50% of the linkages analysed (Pradhan et al. 2017).

Transformation from current development paradigms and the breaking of these lock-in effects can protect climate and achieve food security in future. Sustainable agriculture practices can provide climate change adaptation and mitigation synergies, linking SDG 2 and SDG 13 more positively, according to the International Council for Science (ICSU) (2017). IPCC found that most of the current observed trade-offs between SDG 13 and other SDGs can be converted into synergies based on various mitigation options that can be deployed to limit the global warming well below 1.5°C (IPCC 2018b).

In summary, there are fundamental synergies that can facilitate the joint implementation of strategies to achieve SDGs and climate action, with particular reference to those climate response strategies related to both supply side (production and supply chains) and demand side (consumption and dietary choices) described in this chapter (*high agreement and medium evidence*).

1.7   Enabling conditions and knowledge gaps

To achieve mitigation and adaptation to climate change in food systems, enabling conditions are needed to scale up the adoption of effective strategies (such as those described in Sections 5.3 to 5.6 and Chapter 6). These enabling conditions include multi-level governance and multi-sector institutions (Supplementary Material Section SM5.5) and multiple policy pathways (Sections 5.7.1 and 5.7.2). In this regard, the subnational level is gaining relevance both in food systems and climate change. Just Transitions are needed to address both climate change and food security (Section 5.7.3). Mobilisation of knowledge, education, and capacity will be required (Section 5.7.4) to fill knowledge gaps (Section 5.7.5).

Effective governance of food systems and climate change requires the establishment of institutions responsible for coordinating among multiple sectors (education, agriculture, environment, welfare, consumption, economic, health), levels (local, regional, national, global) and actors (governments, CSO, public sector, private sector, international bodies). Positive outcomes will be engendered by participation, learning, flexibility, and cooperation. See Supplementary Material SM5.5 for further discussion.

1.7.1   Enabling policy environments

The scope for responses to make sustainable land use inclusive of climate change mitigation and adaptation, and the policies to implement them, are covered in detail in Chapters 6 and 7. Here we highlight some of the major policy areas that have shaped the food system, and might be able to shape responses in future. Although two families of policy – agriculture and trade – have been instrumental in shaping the food system in the past (and potentially have led to conditions that increase climate vulnerability) (Benton and Bailey 2019), a much wider family of policy instruments can be deployed to reconfigure the food system to deliver healthy diets in a sustainable way.

1.7.1.1   Agriculture and trade policy

Agriculture. The thrust of agricultural policies over the last 50 years has been to increase productivity, even if at the expense of environmental sustainability (Benton and Bailey 2019). For example, in 2007–2009, 46% of OECD support for agriculture was based on measures of output (price support or payments based on yields), 37% of support was based on the current or historical area planted, herd size (or correlated measures of the notional costs of farming), and 13% was payments linked to input prices. In a similar vein, non-OECD countries have promoted productivity growth for their agricultural sectors.

Trade. Along with agricultural policy to grow productivity, the development of frameworks to liberalise trade (such as the General Agreement on Tariffs and Trade – GATT – Uruguay Round, now incorporated into the World Trade Organization) have been essential in stimulating the growth of a globalised food system. Almost every country has a reliance on trade to fulfil some or all of its local food needs, and trade networks have grown to be highly complex (Puma et al. 2015; MacDonald et al. 2015; Fader et al. 2013 and Ercsey-Ravasz et al. 2012). This is because many countries lack the capacity to produce sufficient food due to climatic conditions, soil quality, water constraints, and availability of farmland (FAO 2015b). In a world of liberalised trade, using comparative advantage to maximise production in high-yielding commodities, exporting excess production, and importing supplies of other goods supports economic growth.

City states as well as many small island states, do not have adequate farmland to feed their populations, while Sub-Saharan African countries are projected to experience high population growth as well as to be negatively impacted by climate change, and thus will likely find it difficult to produce all of their own food supplies (Agarwal et al. 2002). One study estimates that some 66 countries are currently incapable of being self-sufficient in food (Pradhan et al. 2014). Estimates of the proportion of people relying on trade for basic food security vary from about 16% to about 22% (Fader et al. 2013; Pradhan et al. 2014), with this figure rising to between 1.5 and 6 billion people by 2050, depending on dietary shifts, agricultural gains, and climate impacts (Pradhan et al. 2014).

Global trade is therefore essential for achieving food and nutrition security under climate change because it provides a mechanism for enhancing the efficiency of supply chains, reducing the vulnerability of food availability to changes in local weather, and moving production from areas of surplus to areas of deficit (FAO 2018d). However, the benefits of trade will only be realised if trade is managed in ways that maximise broadened access to new markets while minimising the risks of increased exposure to international competition and market volatility (Challinor et al. 2018; Brown et al. 2017b).

As described in Section 5.8.1, trade acts to buffer exposure to climate risks when the market works well. Under certain conditions – such as shocks, or the perception of a shock, coupled with a lack of food stocks or lack of transparency about stocks (Challinor et al. 2018; Marchand et al. 2016) – the market can fail and trade can expose countries to food price shocks.

Furthermore, Clapp (2016) showed that trade, often supported by high levels of subsidy support to agriculture in some countries, can depress world prices and reduce incomes for other agricultural exporters. Lower food prices that result from subsidy support may benefit urban consumers in importing countries, but at the same time they may hurt farmers' incomes in those same countries. The outmigration of smallholder farmers from the agriculture sector across the Global South is significantly attributed to these trade patterns of cheap food imports (Wittman 2011; McMichael 2014; Akram-Lodhi et al. 2013). Food production and trade cartels, as well as financial speculation on food futures markets, affect low-income market-dependent populations.

Food sovereignty is a framing developed to conceptualise these issues (Reuter 2015). They directly relate to the ability of local communities and nations to build their food systems, based, among other aspects, on diversified crops and ILK. If a country enters international markets by growing more commodity crops and reducing local crop varieties, it may get economic benefits, but may also expose itself to climate risks and food insecurity by increasing reliance on trade, which may be increasingly disrupted by climate risks. These include a local lack of resilience from reduced diversity of products, but also exposure to food price spikes, which can become amplified by market mechanisms such as speculation.

In summary, countries must determine the balance between locally produced versus imported food (and feed) such that it both minimises climate risks and ensures sustainable food security. There is *medium evidence* that trade has positive benefits but also creates exposure to risks (Section 5.3).

1.7.1.2   Scope for expanded policies

There are a range of ways that policy can intervene to stimulate change in the food system – through agriculture, research and development, food standards, manufacture and storage, changing the food environment and access to food, changing practices to encourage or discourage trade (Table 5.6). Novel incentives can stimulate the market, for example, through reduction in waste or changes in diets to gain benefits from a health or sustainability direction. Different contexts with different needs will require different set of policies at local, regional and national levels. See Supplementary Material Section SM5.5 for further discussion on expanded policies.

In summary, although agriculture is often thought to be shaped predominantly by agriculture and trade policies, there are over twenty families of policy areas that can shape agricultural production directly or indirectly (through environmental regulations or through markets, including by shaping consumer behaviour). Thus, delivering outcomes promoting climate change adaptation and mitigation can arise from policies across many departments, if suitably designed and aligned.

1.7.1.3   Health-related policies and cost savings

The co-benefits arising from mitigating climate change through changing dietary patterns, and thus demand, have potentially important economic impacts (*high confidence*). The gross value added from agriculture to the global economy (GVA) was 1.9 trillion USD2013 (FAO 2015c), from a global agriculture economy (GDP) of 2.7 trillion USD2016. In 2013, the FAO estimated an annual cost of 3.5 trillion USD for malnutrition (FAO 2013a).

However, this is likely to be an underestimate of the economic health costs of current food systems for several reasons: (i) lack of data – for example there is little robust data in the UK on the prevalence of malnutrition in the general population (beyond estimates of obesity and surveys of malnourishment of patients in hospital and care homes, from which estimates over 3 million people in the UK are undernourished (BAPEN 2012); (ii) lack of robust methodology to determine, for example, the exact relationship between over-consumption of poor diets, obesity and non-communicable diseases like diabetes, cardiovascular disease, a range of cancers or Alzheimer's disease (Pedditizi et al. 2016), and (iii) unequal healthcare spending around the world.

In the USA, the economic cost of diabetes, a disease strongly associated with obesity and affecting about 23 million Americans, is estimated at 327 billion USD2017 (American Diabetes Association 2018), with direct healthcare costs of 9600 USD per person. By 2025, it is estimated that, globally, there will be over 700 million people with diabetes (NCD-RisC 2016b), over 30 times the number in the USA. Even if a global average cost of diabetes per capita were a quarter of that in the USA, the total economic cost of diabetes would be approximately the same as global agricultural GDP. Finally, (iv) the role of agriculture in causing ill-health beyond dietary health, such as through degrading air quality (e.g., Paulot and Jacob 2014).

Whilst data of the healthcare costs associated with the food system and diets are scattered and the proportion of costs directly attributable to diets and food consumption is uncertain, there is potential for more preventative healthcare systems to save significant costs that could incentivise agricultural business models to change what is grown, and how. The potential of moving towards more preventative healthcare is widely discussed in health economics literature, particularly in order to reduce the life-style-related (including dietary-related) disease component in aging populations (e.g., Bloom et al. 2015).

1.7.1.4   Multiple policy pathways

As discussed in more detail in Chapters 6 and 7, there is a wide potential suite of interventions and policies that can potentially enhance the adaptation of food systems to climate change, as well as enhance the mitigation potential of food systems on climate change. There is an increasing number of studies that argue that the key to sustainable land management is not in land management practices but in the factors that determine the demand for products from land (such as food). Public health policy, therefore, has the potential to affect dietary choice and thus the demand for different amounts of, and types of, food.

Obersteiner et al. (2016) show that increasing the average price of food is an important policy lever that, by reducing demand, reduces food waste, pressure on land and water, impacts on biodiversity and through reducing emissions, mitigates climate change and potentially helps to achieve multiple SDGs. Whilst such policy responses – such as a carbon tax applied to goods including food – has the potential to be regressive, affecting the poor differentially (Frank et al. 2017; Hasegawa et al. 2018 and Kehlbacher et al. 2016), and increasing food insecurity – further development of social safety nets can help to avoid the regressive nature (Hasegawa et al. 2018). Hasegawa et al. (2018) point out that such safety nets for vulnerable populations could be funded from the revenues arising from a carbon tax.

The evidence suggests, as with SR15 (IPCC 2018a) and its multiple pathways to climate change solutions, that there is no single solution that will address the problems of food and climate change, but instead there is a need to deploy many solutions, simultaneously adapted to the needs and options available in a given context. For example, Springmann et al. (2018a) indicate that maintaining the food system within planetary boundaries at mid-century, including equitable climate, requires increasing the production (and resilience) of agricultural outputs (i.e., closing yield gaps), reducing waste, and changes in diets towards ones often described as flexitarian (low-meat dietary patterns that are in line with available evidence on healthy eating). Such changes can have significant co-benefits for public health, as well as facing significant challenges to ensure equity (in terms of affordability for those in poverty).

Significant changes in the food system require them to be acceptable to the public ('public license'), or they will be rejected. Focus groups with members of the public around the world, on the issue of changing diets, have shown that there is a general belief that the government plays a key role in leading efforts for change in consumption patterns (Wellesley et al. 2015). If governments are not leading on an issue, or indicating the need for it through leading public dialogue, it signals to their citizens that the issue is unimportant or undeserving of concern.

In summary, there is significant potential (*high confidence*) that, through aligning multiple policy goals, multiple benefits can be realised that positively impact public health, mitigation and adaptation (e.g., adoption of healthier diets, reduction in waste, reduction in environmental impact). These benefits may not occur without the alignment across multiple policy areas (*high confidence*).

1.7.2   Enablers for changing markets and trade

'Demand' for food is not an exogenous variable to the food system but is shaped crucially by its ability to produce, market, and supply food of different types and prices. These market dynamics can be influenced by a variety of factors beyond consumer preferences (e.g., corporate power and marketing, transparency, the food environment more generally), and the ability to reshape the market can also depend on its internal resilience and/or external shocks (Challinor et al. 2018; Oliver et al. 2018).

1.7.2.1   Capital markets

Two areas are often discussed regarding the role of capital markets in shaping the food system. First, investment in disruptive technologies might stimulate climate-smart food systems (WEF/McKinsey & Company 2018 and Bailey and Wellesley 2017), including alternative proteins, such as laboratory or 'clean meat' (which has significant ability to impact on land-use requirements) (Alexander et al. 2017) (Section 5.5.1.6). An innovation environment through which disruptive technology can emerge typically requires the support of public policy, whether in directly financing small and emerging enterprises, or funding research and development via reducing tax burdens.

Second, widespread adoption of (and perhaps underpinned by regulation for) natural capital accounting as well as financial accounting are needed. Investors can then be aware of the risk exposure of institutions, which can undermine sustainability through externalising costs onto the environment. The prime example of this in the realm of climate change is the Carbon Disclosure Project, with around 2500 companies voluntarily disclosing their carbon footprint, representing nearly 60% of the world's market capital (CDP 2018).

1.7.2.2   Insurance and re-insurance

The insurance industry can incentivise actors' behaviour towards greater climate mitigation or adaptation, including building resilience. For example, Lloyd's of London analysed the implications of extreme weather for the insurance market, and conclude that the insurance industry needs to examine their exposure to risks through the food supply chain and develop innovative risk-sharing products that can make an important contribution to resilience of the global food system (Lloyd's 2015).

Many of these potential areas for enabling healthy and sustainable food systems are also knowledge gaps, in that, whilst the levers are widely known, their efficacy and the ability to scale-up, in any given context, are poorly understood.

5

1.7.3   Just Transitions to sustainability

Research is limited on how land-use transitions would proceed from ruminant production to other socio-ecological farming systems. Ruminants have been associated with humans since the early development of agriculture, and the role of ruminants in many agricultural systems and smallholder communities is substantial. Ruminant production systems have been adapted to a wide range of socioeconomic and environmental conditions in crop, forestry, and food processing settings (Čolović et al. 2019), bioenergy production (de Souza et al. 2019), and food waste recycling (Westendorf 2000). Pasture cultivation in succession to crops is recognised as important to management of pest and diseases cycles and to improve soil carbon stocks and soil quality (Carvalho and Dedieu 2014). Grazing livestock is important as a reserve of food and economic stocks for some smallholders (Ouma et al. 2003).

Possible land-use options for transitions away from livestock production in a range of systems include (a) retain land but reduce investments to run a more extensive production system; (b) change land use by adopting a different production activity; (c) abandon land (or part of the farm) to allow secondary vegetation regrowth (Carvalho et al. 2019 and Laue and Arima 2016); and (d) invest in afforestation or reforestation (Baynes et al. 2017). The extensification option could lead to increases rather than decreases in GHG emissions related to reduction in beef consumption. Large-scale abandonment, afforestation, or reforestation would probably have more positive environmental outcomes, but could result in economic and social issues that would require governmental subsidies to avoid decline and migration in some regions (Henderson et al. 2018).

Alternative economic use of land, such as bioenergy production, could balance the negative socioeconomic impact of reducing beef output, reduce the tax values needed to reduce consumption, and avoid extensification of ruminant production systems (Wirsenius et al. 2011). However, the analysis of the transition of land use for ruminants to other agricultural production systems is still a literature gap (Cross-Chapter Box 7 in Chapter 6).

Finally, it is important to recognise that, while energy alternatives produce the same function for the consumer, it is questionable that providing the same nutritional value through an optimised mix of dietary ingredients provides the same utility for humans. Food has a central role in human pleasure, socialisation, cultural identity, and health (Röös et al. 2017), including some of the most vulnerable groups, so Just Transitions and their costs need to be taken into account. Pilot projects are important to provide greater insights for large-scale policy design, implementation, and enforcement.

In summary, more research is needed on how land-use transitions would proceed from ruminant production to other farming systems and affect the farmers and other food system actors involved. There is *limited evidence* on what the decisions of farmers under lower beef demand would be.

1.7.4   Mobilising knowledge

Addressing climate change-related challenges and ensuring food security requires all types of knowledge (formal/non-formal, scientific/indigenous, women, youth, technological). Miles et al. (2017) stated that a research and policy feedback that allows transitions to sustainable food systems must take a whole system approach. Currently, in transmitting knowledge for food security and land sustainability under climate change there are three major approaches: (i) public technology transfer with demonstration (extension agents); (ii) public and private advisory services (for intensification techniques) and; (iii) non-formal education with many different variants such as farmer field schools, rural resource centres; facilitation extension where front-line agents primarily work as 'knowledge brokers' in facilitating the teaching-learning process among all types of farmers (including women and rural young people), or farmer-to-farmer, where farmers act themselves as knowledge transfer and sharing actors through peer processes.

1.7.4.1   Indigenous and local knowledge

Recent discourse has a strong orientation towards scaling-up innovation and adoption by local farmers. However, autonomous adaptation, indigenous knowledge and local knowledge are both important for agricultural adaptation (Biggs et al. 2013) (Section 5.3). These involve the promotion of farmer participation in governance structures, research, and the design of systems for the generation and dissemination of knowledge and technology, so that farmers' needs and knowledge can be taken into consideration. Klenk et al. (2017) found that mobilisation of local knowledge can inform adaptation decision-making and may facilitate greater flexibility in government-funded research. As an example, rural innovation in terrace agriculture developed on the basis of a local coping mechanism and adopted by peasant farmers in Latin America may serve as an adaptation option or starting place for learning about climate change responses (Bocco and Napoletano 2017). Clemens et al. (2015) found that an open dialogue platform enabled horizontal exchange of ideas and alliances for social learning and knowledge-sharing in Vietnam. Improving local technologies in a participatory manner, through on-farm experimentation, farmer-to-farmer exchange, consideration of women and youths, is also relevant in mobilising knowledge and technologies.

1.7.4.2   Citizen science

Citizen science has been tested as a useful tool with potential for biodiversity conservation (Schmitz et al. 2015) and mobilising knowledge from society. In food systems, knowledge-holders (e.g., farmers and pastoralists) are trained to gather scientific data in order to promote conservation and resource management (Fulton et al. 2019) or to conserve and use traditional knowledge in developed countries relevant to climate change adaptation and mitigation through the use of ICT (Calvet-Mir et al. 2018).

Relation of climate shocks to food price spikes

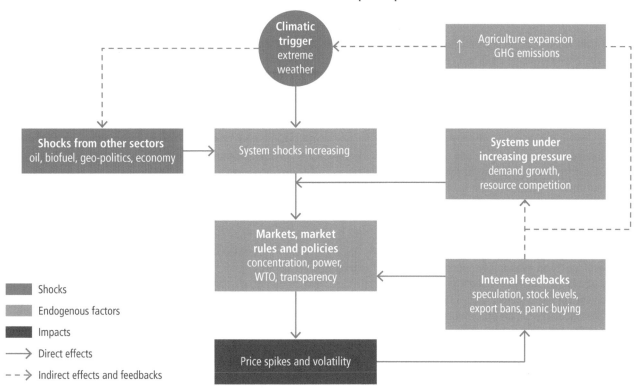

Figure 5.17 | Underlying processes that affect the development of a food price spike in agricultural commodity markets (Challinor et al. 2018).

1.7.4.3   Capacity building and education

Mobilising knowledge may also require significant efforts on capacity building and education to scale up food system responses to climate change. This may involve increasing the capacity of farmers to manage current climate risks and to mitigate and adapt in their local contexts, and of citizens and consumers to understand the links between food demand and climate change emissions and impacts, as well as policy makers to take a systemic view of the issues. Capacity building may also require institutional change. For example, alignment of policies towards sustainable and healthy food systems may require building institutional capacity across policy silos.

As a tool for societal transformation, education is a powerful strategy to accelerate changes in the way we produce and consume food. Education refers to early learning and lifelong acquisition of skills for higher awareness and actions for solving food system challenges (FAO 2005). Education also entails vocational training, research and institutional strengthening (Hollinger 2015). Educational focus changes according to the supply side (e.g., crop selection, input resource management, yield improvement, and diversification) and the demand since (nutrition and dietary health implications). Education on food loss and waste spans both the supply and demand sides.

In developing countries, extension learning such as farmer field schools – also known asrural resources centers – are established to promote experiential learning on improved production and food transformation (FAO 2016c). In developed countries, education

campaigns are being undertaken to reduce food waste, improve diets and redefine acceptable food (e.g., "less than perfect" fruits and vegetables), and ultimately can contribute to changes in the structure of food industries (Heller 2019; UNCCD 2017).

The design of new education modules from primary to secondary to tertiary education could help create new jobs in the realm of sustainability (e.g., certification programmes). For example, one area could be educating managers of recycling programmes for food-efficient cities where food and organic waste are recycled to become fertilisers (Jara-Samaniego et al. 2017). Research and education need to be coordinated so that knowledge gaps can be filled and greater trust established in shifting behaviour of individuals to be more sustainable. Education campaigns can also influence policy and legislation, and help to advance successful outcomes for climate change mitigation and adaptation regarding supply-side innovations, technologies, trade, and investment, and demand-side evolution of food choices for health and sustainability, and greater gender equality throughout the entire food system (Heller 2019).

1.7.5   Knowledge gaps and key research areas

Knowledge gaps around options and solutions and their (co-)benefits and trade-offs are increasingly important now that implementation of mitigation and adaptation measures is scaling up.

Research is needed on how a changing climate and interventions to respond to it will affect all aspects of food security, including access, utilisation and stability, not just availability. Knowledge gaps across all the food security pillars are one of the barriers hindering mitigation and adaptation to climate change in the food system and its capacity to deliver food security. The key areas for climate change, food systems, and food security research are enlisted below.

1.7.5.1   Impacts and adaptation

Climate Services (food availability). Agriculture and food security is a priority area for the Global Framework for Climate Services (GFCS) a programme of the World Meteorological Organization (WMO). The GFCS enables vulnerable sectors and populations to better manage climate variability and adapt to climate change (Hansen et al. 2018). Global precipitation datasets and remote sensing technologies can be used to detect local to regional anomalies in precipitation as a tool for devising early-warning systems for drought-related impacts, such as famine (Huntington et al. 2017).

Crop and livestock genetics (food availability, utilisation). Advances in plant breeding are crucial for enhancing food security under changing climate for a wide variety of crops including fruits and vegetables as well as staples. Genetics improvement is needed in order to breed crops and livestock that can both reduce GHG emissions, increase drought and heat tolerance (e.g., rice), and enhance nutrition and food security (Nankishore and Farrell 2016; Kole et al. 2015). Many of these characteristics already exist in traditional varieties, including orphan crops and indigenous and local breeds, so research is needed to recuperate such varieties and evaluate their potential for adaptation and mitigation.

Phenomics-assisted breeding appears to be a promising tool for deciphering the stress responsiveness of crop and animal species (Papageorgiou 2017; Kole et al. 2015; Lopes et al. 2015; Boettcher et al. 2015). Initially discovered in bacteria and archaea, CRISPR–Cas9 is an adaptive immune system found in prokaryotes and since 2013 has been used as a genome editing tool in plants. The main use of CRISPR systems is to achieve improved yield performance, biofortification, biotic and abiotic stress tolerance, with rice (Oryza sativa) being the most studied crop (Gao 2018 and Ricroch et al. 2017).

Climate impact models (food availability). Understanding the full range of climate impacts on staple crops (especially those important in developing countries, such as fruits and vegetables) is missing in the current climate impact models. Further, the CO_2 effects on nutrition quality of different crops are just beginning to be parameterised in the models (Müller et al. 2014). Bridging these gaps is essential for projecting future dietary diversity, healthy diets, and food security (Bisbis et al. 2018). Crop model improvements are

needed for simulation of evapotranspiration to guide crop water management in future climate conditions (Cammarano et al. 2016). Similarly, mores studies are needed to understand the impacts of climate change on global rangelands, livestock and aquaculture, which have received comparatively less attention than the impacts on crop production.

Resilience to extreme events (food availability, access, utilisation, and stability). On the adaptation side, knowledge gaps include impacts of climate shocks (Rodríguez Osuna et al. 2014) as opposed to impacts of slow-onset climate change, how climate-related harvest failures in one continent may influence food security outcomes in others, impacts of climate change on fruits and vegetables and their nutrient contents.

1.7.5.2   Emissions and mitigation

GHG emissions inventory techniques (food utilisation). Knowledge gaps include food consumption-based emissions at national scales, embedded emissions (overseas footprints) of food systems, comparison of GHG emissions per type of food systems (e.g., smallholder and large-scale commercial food systems), and GHG emissions from land-based aquaculture. An additional knowledge gap is the need for more socio-economic assessments of the potential of various integrated practices to deliver the mitigation potential estimated from a biophysical perspective. This needs to be effectively monitored, verified, and implemented, once barriers and incentives to adoption of the techniques, practices, and technologies are considered. Thus, future research needs fill the gaps on evaluation of climate actions in the food system.

Food supply chains (food availability). The expansion of the cold chain into developing economies means increased energy consumption and GHG emissions at the consumer stages of the food system, but its net impact on GHG emissions for food systems as a whole, is complex and uncertain (Heard and Miller 2016). Further understanding of negative side effects in intensive food processing systems is still needed.

Blockchains, as a distributed digital ledger technology which ensures transparency, traceability, and security, is showing promise for easing some global food supply chain management challenges, including the need for documentation of sustainability and the circular economy for stakeholders including governments, communities, and consumers to meet sustainability goals. Blockchain-led transformation of food supply chains is still in its early stages; research is needed on overcoming barriers to adoption (Tripoli and Schmidhuber 2018; Casado-Vara et al. 2018; Mao et al. 2018; Saberi et al. 2019).

1.7.5.3   Synergies and trade-offs

Supply-side and demand-side mitigation and adaptation (food availability, utilisation). Knowledge gaps exist in characterising the potential and risks associated with novel mitigation technologies on the supply side (e.g., inhibitors, targeted breeding, cellular agriculture, etc.). Additionally, most integrated assessment models

Box 5.6 | Migration in the Pacific region: Impacts of climate change on food security

Climate change-induced displacement and migration in the Pacific has received wide attention in the scientific discourse (Fröhlich and Klepp 2019). The processes of climate change and their effects in the region have serious implications for Pacific Island nations as they influence the environments that are their 'life-support systems' (Campbell 2014). Climate variability poses significant threats to both agricultural production and food security. Rising temperatures and reductions in groundwater availability, as well as increasing frequency and severity of disaster events translate into substantial impacts on food security, causing human displacement, a trend that will be aggravated by future climate impacts (ADB 2017). Declining soil productivity, groundwater depletion, and non-availability of freshwater threatens agricultural production in many remote atolls.

Many countries in the Pacific devote a large share of available land area to agricultural production. For example, more than 60% of land area is cultivated in the Marshall Islands and Tuvalu and more than 40% in Kiribati and Tonga. With few options to expand agricultural area, the projected impacts of climate change on food production are of particular concern (ADB 2013, 2017). The degradation of available land area for traditional agriculture, adverse disruptions of agricultural productivity and diminishing livelihood opportunities through climate change impacts leads to increasing poverty and food insecurity, incentivising migration to urban agglomerations (ADB 2017; FAO et al. 2018).

Campbell (2014) describe the trends that lead to migration. First, climate change, including rising sea levels, affects communities' land security, which is the physical presence on which to live and sustain livelihoods. Second, they impinge on livelihood security (especially food security) of island communities where the productivity of both subsistence and commercial food production systems is reduced. Third, the effects of climate change are especially severe on small-island environments since they result in declining ecological habitat. The effects on island systems are mostly manifested in atolls through erosion and inundation, and on human populations through migration. Population growth and scenarios of climate change are *likely* to further induce food stress as impacts unfold in the coming decades (Campbell 2015).

While the populations of several islands and island groups in the Pacific (e.g., Tuvalu, Carteret Islands, and Kiribati) have been perceived as the first probable victims of rising seas so that their inhabitants would become, and in some quarters already are seen to be, the first 'environmental' or 'climate change refugees', migration patterns vary. Especially in small islands, the range and nature of the interactions among economic, social, and/or political drivers are complex. For example, in the Maldives, Stojanov et al. (2017) show that while collective perceptions support climate change impacts as being one of the key factors prompting migration, individual perceptions give more credence to other cultural, religious, economic or social factors.

In the Pacific, Tuvalu has long been a prime candidate to disappear due to rising sea levels, forcing human migration. However, results of a recent study (Kench et al. 2018) challenge perceptions of island loss in Tuvalu, reporting that there is a net increase in land area of 73.5 ha. The findings suggest that islands are dynamic features likely to persist as habitation sites over the next century, presenting opportunities for adaptation that embrace the heterogeneity of island types and processes. Farbotko (2010) and Farbotko and Lazrus (2012) present Tuvalu as a site of 'wishful sinking', in the climate change discourse. These authors argue that representations of Tuvalu as a laboratory for global climate change migration are visualisations by non-locals.

In Nanumea (Tuvalu), forced displacements and voluntary migrations are complex decisions made by individuals, families and communities in response to discourses on risk, deteriorating infrastructure and other economic and social pressures (Marino and Lazrus 2015). In many atoll nations in the Western Pacific, migration has increasingly become a sustainable livelihood strategy, irrespective of climate change (Connell 2015).

In Lamen Bay, Vanuatu, migration is both a cause and consequence of local vulnerabilities. While migration provides an opportunity for households to meet their immediate economic needs, it limits the ability of the community to foster longer-term economic development. At the same time, migration adversely affects the ability of the community to maintain food security due to lost labour and changing attitudes towards traditional ways of life among community members (Craven 2015).

5

(IAMs) currently have limited regional data on BECCS projects because of little BECCS implementation (Lenzi et al. 2018). Hence, several BECCS scenarios rely on assumptions regarding regional climate, soils and infrastructure suitability (Köberle et al. 2019) as well as international trade (Lamers et al. 2011).

Areas for study include how to incentivise, regulate, and raise awareness of the co-benefits of healthy consumption patterns and climate change mitigation and adaptation; to improve access to healthy diets for vulnerable groups through food assistance programmes; and to implement policies and campaigns to reduce food loss and food waste. Knowledge gaps also exist on the role of different policies, and underlying uncertainties, to promote changes in food habits towards climate resilience and healthy diets.

Food systems, land-use change, and telecoupling (food availability, access, utilisation). The analytical framework of telecoupling has recently been proposed to address this complexity, particularly the connections, flows, and feedbacks characterising food systems (Friis et al. 2016; Easter et al. 2018). For example, how will climate-induced shifts in livestock and crop diseases affect food production and consumption in the future. Investigating the social and ecological consequences of these changes will contribute to decision-making under uncertainty in the future. Research areas include food systems and their boundaries, hierarchies, and scales through metabolism studies, political ecology and cultural anthropology.

Food-Energy-Water Nexus (food availability, utilisation, stability). Emerging interdisciplinary science efforts are providing new understanding of the interdependence of food, energy, and water systems. These interdependencies are beginning to take into account climate change, food security, and AFOLU assessments (Scanlon et al. 2017; Liu et al. 2017). These science advances, in turn, provide critical information for coordinated management to improve the affordability, reliability, and environmental sustainability of food, energy, and water systems. Despite significant advances within the past decade, there are still many challenges for the scientific community. These include the need for interdisciplinary science related to the food-energy-water nexus; ground-based monitoring and modelling at local-to-regional scales (Van Gaelen et al. 2017); incorporating human and institutional behaviour in models; partnerships among universities, industry, and government to develop policy-relevant data; and systems modelling to evaluate trade-offs associated with food-energy-water decisions (Scanlon et al. 2017).

However, the nexus approach, as a conceptual framework, requires the recognition that, although land and the goods and services it provides is finite, potential demand for the goods and services may be greater than the ability to supply them sustainably (Benton et al. 2018). By addressing demand-side issues, as well as supply-side efficiencies, it provides a potential route for minimising trade-offs for different goods and services (Benton et al. 2018) (Section 5.6).

1.8   Future challenges to food security

A particular concern in regard to the future of food security is the potential for the impacts of increasing climate extremes on food production to contribute to multi-factored complex events such as food price spikes. In this section, we assess literature on food price spikes and potential strategies for increasing resilience to such occurrences. We then assess the potential for such food system events to affect migration and conflict.

1.8.1   Food price spikes

Under average conditions, global food system markets may function well, and equilibrium approaches can estimate demand and supply with some confidence; however, if there is a significant shock, the market can fail to smoothly link demand and supply through price, and a range of factors can act to amplify the effects of the shock, and transmit it across the world (Box 5.5). Given the potential for shocks driven by changing patterns of extreme weather to increase with climate change, there is the potential for market volatility to disrupt food supply through creating food price spikes. This potential is exacerbated by the interconnectedness of the food system (Puma et al. 2015) with other sectors (i.e., the food system depends on water, energy, and transport) (Homer-Dixon et al. 2015), so the impact of shocks can propagate across sectors and geographies (Homer-Dixon et al. 2015). There is also less spare land globally than there has been in the past, such that if prices spike, there are fewer options to bring new production on stream (Marianela et al. 2016).

Increasing extreme weather events can disrupt production and transport logistics. For example, in 2012 the USA Corn Belt suffered a widespread drought; USA corn yield declined 16% compared to 2011 and 25% compared to 2009. In 2016, a record yield loss in France that is attributed to a conjunction of abnormal warmness in late autumn and abnormal wet in the following spring (Ben-Ari et al. 2018) is another well-documented example. To the extent that such supply shocks are associated with climate change, they may become more frequent and contribute to greater instability in agricultural markets in the future.

Furthermore, analogue conditions of past extremes might create significantly greater impacts in a warmer world. A study simulating analogous conditions to the Dust Bowl drought in today's agriculture suggests that Dust Bowl-type droughts today would have unprecedented consequences, with yield losses about 50% larger than the severe drought of 2012 (Glotter and Elliott 2016). Damages at these extremes are highly sensitive to temperature, worsening by about 25% with each degree centigrade of warming. By mid-century, over 80% of summers are projected to have average temperatures that are likely to exceed the hottest summer in the Dust Bowl years (1936) (Glotter and Elliott 2016).

How a shortfall in production – or an interruption in trade due to an event affecting a logistics choke-point (Wellesley et al. 2017) – of any given magnitude may create impacts depends on many interacting factors (Homer-Dixon et al. 2015; Tadasse et al. 2016; Challinor et al. 2018). The principal route is by affecting agricultural commodity markets, which respond to a perturbation through multiple routes as in Figure 5.17. This includes pressures from other sectors (such as, if biofuels policy is incentivising crops for the production of ethanol, as happened in 2007–2008). The market response can be amplified by poor policies, setting up trade and non-trade barriers to exports, from countries seeking to ensure their local food security (Bailey et al. 2015). Furthermore, the perception of problems can fuel panic buying on the markets that in turn drives up prices.

Thus, the impact of an extreme weather event on markets has both a *trigger* component (the event) and a risk *perception* component (Challinor et al. 2016, 2018). Through commodity markets, prices change across the world because almost every country depends, to a greater or lesser extent, on trade to fulfil local needs. Commodity prices can also affect local market prices by altering input prices, changing the cost of food aid, and through spill-over effects. For example, in 2007–2008 the grain affected by extreme weather was wheat, but there was a significant price spike in rice markets (Dawe 2010).

5

As discussed by Bailey et al. (2015), there are a range of adaptation measures that can be put in place to reduce the impact of climate-related production shortfalls. These include (i) ensuring transparency of public and private stocks, as well as improved seasonal forecasting to signal forthcoming yield shortfalls (FAO 2016a; Ceglar et al. 2018; Iizumi et al. 2018), (ii) building real or virtual stockholdings, (iii) increasing local productivity and diversity (as a hedge against a reliance on trade) and (iv) ensuring smoother market responses, through, for example, avoiding the imposition of export bans.

In summary, given the likelihood that extreme weather will increase, in both frequency and magnitude (Hansen et al. 2012; Coumou et al. 2014; Mann et al. 2017; Bailey et al. 2015), and the current state of global and cross-sectoral interconnectedness, the food system is at increasing risk of disruption (*medium evidence, medium agreement*), with large uncertainty about how this could manifest. There is, therefore, a need to build resilience into international trade as well as local supplies.

5

Box 5.5 | Market drivers and the consequences of extreme weather in 2010–2011

The 2010–2011 food price spike was initially triggered by the exceptional heat in summer 2010, with an extent from Europe to the Ukraine and Western Russia (Barriopedro et al. 2011; Watanabe et al. 2013; Hoag 2014). The heatwave in Russia was extreme in both temperature (over 40°C) and duration (from July to mid-August in 2010). This reduced wheat yields by approximately one third (Wegren 2011; Marchand et al. 2016). Simultaneously, in the Indus Valley in Pakistan, unprecedented rainfall led to flooding, affecting the lives and livelihoods of 20 million people. There is evidence that these effects were both linked and made more likely through climate change (Mann et al. 2017).

In response to its shortfall in yields, Russia imposed an export ban in order to maintain local food supplies. Other countries responded in a largely uncoordinated ways, each of them driven by internal politics as well as national self-interests (Jones and Hiller 2017). Overall, these measures led to rapid price rises on the global markets (Welton 2011), partly through panic buying, but also through financial speculation (Spratt 2013).

Analysis of responses to higher food prices in the developing world showed that lower-income groups responded by taking on more employment, reducing food intake, limiting expenditures, spending savings (if available), and participating in demonstrations. People often identified their problems as stemming from collusion between powerful incumbent interests (e.g., of politicians and big business) and disregard for the poor (Hossain and Green 2011). This politicised social response helped spark food-related civil protest, including riots, across a range of countries in 2010–2011 (Natalini et al. 2017). In Pakistan, food price rises were exacerbated by the economic impacts of the floods, which further contributed to food-related riots in 2010.

Price spikes also impact on food security in the developed world. In the UK, global commodity price inflation influenced local food prices, increasing food-price inflation by about five times at the end of 2010. Comparing household purchases over the five-year period from 2007 to 2011 showed that the amount of food bought declined, on average, by 4.2%, whilst paying 12% more for it. The lowest income decile spent 17% more by 2011 than they did in 2007 (Holding et al. 2013; Tadasse et al. 2016). Consumers also saved money by trading down for cheaper alternatives. For the poorest, in the extreme situation, food became unaffordable: the Trussell Trust, a charity supplying emergency food handouts for people in crisis, noted a 50% increase in handouts in 2010.

1.8.2 Migration and conflict

Since the IPCC AR5 (Porter et al. 2014; Cramer et al. 2014), new work has advanced multi-factor methodological issues related to migration and conflict (e.g., Kelley et al. 2015, 2017; Werrell et al. 2015; Challinor et al. 2018; Pasini et al. 2018). These in particular have addressed systemic risks to food security that result from cascading impacts triggered by droughts and floods and how these are related to a broad range of societal influences.

Climate variability and extremes have short-, medium – and long-term impacts on livelihoods and livelihood assets – especially of the poor – contributing to greater risk of food insecurity and malnutrition (FAO et al. 2018). Drought threatens local food security and nutrition and aggravates humanitarian conditions, which can trigger large-scale human displacement and create a breeding ground for conflict (Maystadt and Ecker 2014). There is *medium agreement* that existing patterns of conflict could be reinforced under climate change, affecting food security and livelihood opportunities, for example, in already fragile regions with ethnic divides such as North and Central Africa as well as Central Asia (Buhaug 2016; Schleussner et al. 2016) (Box 5.6).

Challinor et al. (2018) have developed a typology for transboundary and transboundary risk transmission that distinguishes the roles of climate and social and economic systems. To understand these complex interactions, they recommend a combination of methods that include expert judgement; interactive scenario building; global systems science and big data; and innovative use of climate and integrated assessment models; and social science techniques (e.g., surveys, interviews, and focus groups).

1.8.2.1 Migration

There has been a surge in international migration in recent years, with around five million people migrating permanently in 2016 (OECD 2017). Though the initial driver of migration may differ across populations, countries and contexts, migrants tend to seek the same fundamental objective: to provide security and adequate living conditions for their families and themselves. Food insecurity is a critical 'push' factor driving international migration, along with conflict, income inequality, and population growth. The act of migration itself causes food insecurity, given the lack of income opportunities and adverse conditions compounded by conflict situations.

Warner et al. (2012) found the interrelationships between changing rainfall patterns, food and livelihood security in eight countries in Asia, Africa and Latin America. Several studies in Africa have found that persistent droughts and land degradation contributed to both seasonal and permanent migration (Gray 2011; Gray and Mueller 2012; Hummel 2015; Henry et al. 2004; Folami and Folami 2013), worsening the vulnerability of different households (Dasgupta et al. 2014).

Dependency on rainfed agriculture ranges from 13% in Mexico to more than 30% in Guatemala, Honduras, and Nicaragua, suggesting a high degree of sensitivity to climate variability and change, and undermined food security (Warner et al. 2009). Studies have demonstrated that Mexican migration (Feng et al. 2010; Nawrotzki et al. 2013) and Central American migration (WFP 2017) fluctuate in response to climate variability. The food system is heavily dependent on maize and bean production and long-term climate change and variability significantly affect the productivity of these crops and the livelihoods of smallholder farmers (WFP 2017). In rural Ecuador, adverse environmental conditions prompt out-migration, although households respond to these challenges in diverse ways resulting in complex migratory responses (Gray and Bilsborrow 2013).

Migration patterns have been linked to heat stress in Pakistan (Mueller et al. 2014) and climate variability in the Sundarbans due to decline in food security (Guha and Roy 2016). In Bangladesh, the impacts of climate change have been on the rise throughout the last three decades with increasing migration, mostly of men leaving women and children to cope with increasing effects of natural disasters (Rabbani et al. 2015).

Small islands are very sensitive to climate change impacts (*high confidence*) (Nurse et al. 2014) and impacted by multiple climatic stressors (IPCC 2018a and SROCC). Food security in the Pacific, especially in Micronesia, has worsened in the past half century and climate change is *likely* to further hamper local food production, especially in low-lying atolls (Connell 2016). Migration in small islands (internally and internationally) occurs for multiple reasons and purposes, mostly for better livelihood opportunities (*high confidence*).

Beyond rising sea levels, the effects of increasing frequency and intensity of extreme events such as severe tropical cyclones are *likely* to affect human migration in the Pacific (Connell 2015; Krishnapillai and Gavenda 2014; Charan et al. 2017; Krishnapillai 2017). On Yap Island, extreme weather events are affecting every aspect of atoll communities' existence, mainly due to the islands' small size, their low elevation, and extensive coastal areas (Krishnapillai 2018). Displaced atoll communities on Yap Island grow a variety of nutritious vegetables and use alternative crop production methods such as small-plot intensive farming, raised bed gardening, as part of a community-based adaptation programme (Krishnapillai and Gavenda 2014; Krishnapillai 2018).

Recurrences of natural disasters and crises threaten food security through impacts on traditional agriculture, causing the forced migration and displacement of coastal communities to highlands in search of better living conditions. Although considerable differences occur in the physical manifestations of severe storms, such climate stressors threaten the life-support systems of many atoll communities (Campbell et al. 2014). The failure of these systems resulting from climate disasters propel vulnerable atoll communities into poverty traps, and low adaptive capacity could eventually force these communities to migrate.

1.8.2.2   Conflict

While climate change will not alone cause conflict, it is often acknowledged as having the potential to exacerbate or catalyse conflict in conjunction with other factors. Increased resource competition can aggravate the potential for migration to lead to conflict. When populations continue to increase, competition for resources will also increase, and resources will become even scarcer due to climate change (Hendrix and Glaser 2007). In agriculture-dependent communities in low-income contexts, droughts have been found to increase the likelihood of violence and prolonged conflict at the local level, which eventually pose a threat to societal stability and peace (FAO et al. 2017). In contrast, conflicts can also have diverging effects on agriculture due to land abandonment, resulting in forest growth, or agriculture expansion causing deforestation, for example, in Colombia (Landholm et al. 2019).

Several studies have explored the causal links among climate change, drought, impacts on agricultural production, livelihoods, and civil unrest in Syria from 2007–2010, but without agreement as to the role played by climate in subsequent migration (Kelley et al. 2015, 2017; Challinor et al. 2018; Selby et al. 2017; Hendrix 2018). Contributing factors that have been examined include rainfall deficits, population growth, agricultural policies, and the influx of refugees that had placed burdens on the region's water resources (Kelley et al. 2015). Drought may have played a role as a trigger, as this drought was the longest and the most intense in the last 900 years (Cook et al. 2016; Mathbout et al. 2018). Some studies linked the drought to widespread crop failure, but the climate hypothesis has been contested (Selby et al. 2017; Hendrix 2018). Recent evidence shows that the severe drought triggered agricultural collapse and displacement of rural farm families, with approximately 300,000 families going to Damascus, Aleppo and other cities (Kelley et al. 2017).

Persistent drought in Morocco during the early 1980s resulted in food riots and contributed to an economic collapse (El-Said and Harrigan 2014). A drought in Somalia that fuelled conflict through livestock price changes, establishing livestock markets as the primary channel of impact (Maystadt and Ecker 2014). Cattle raiding as a normal means of restocking during drought in the Great Horn of Africa led to conflict (ICPAC and WFP 2017) whereas a region-wide drought in northern Mali in 2012 wiped out thousands of livestock and devastated the livelihoods of pastoralists, in turn swelling the ranks of armed rebel factions and forcing others to steal and loot for survival (Breisinger et al. 2015).

On the other hand, inter-annual adjustments in international trade can play an important role in shifting supplies from food surplus regions to regions facing food deficits which emerge as a consequence of extreme weather events, civil strife, and/or other disruptions (Baldos and Hertel 2015). A more freely functioning global trading system is tested for its ability to deliver improved long run food security in 2050.

In summary, given increasing extreme events and global and cross-sectoral interconnectedness, the food system is at increasing risk of disruption, for example, via migration and conflict (*high confidence*). {5.2.3, 5.2.4}

FAQ 5.1 | How does climate change affect food security?

Climate change negatively affects all four pillars of food security: availability, access, utilisation and stability. Food availability may be reduced by negative climate change impacts on productivity of crops, livestock and fish, due, for instance, to increases in temperature and changes in rainfall patterns. Productivity is also negatively affected by increased pests and diseases, as well as changing distributions of pollinators under climate change. Food access and its stability may be affected through disruption of markets, prices, infrastructure, transport, manufacture, and retail, as well as direct and indirect changes in income and food purchasing power of low-income consumers. Food utilisation may be directly affected by climate change due to increases in mycotoxins in food and feed with rising temperatures and increased frequencies of extreme events, and indirectly through effects on health. Elevated atmospheric CO_2 concentrations can increase yields at lower temperature increases, but tend to decrease protein content in many crops, reducing their nutritional values. Extreme events, for example, flooding, will affect the stability of food supply directly through disruption of transport and markets.

FAQ 5.2 | How can changing diets help address climate change?

Agricultural activities emit substantial amounts of greenhouse gases (GHGs). Food supply chain activities past the farm gate (e.g., transportation, storage, packaging) also emit GHGs, for instance due to energy use. GHG emissions from food production vary across food types. Producing animal-sourced food (e.g., meat and dairy) emits larger amount of GHGs than growing crops, especially in intensive, industrial livestock systems. This is mainly true for commodities produced by ruminant livestock such as cattle, due to enteric fermentation processes that are large emitters of methane. Changing diets towards a lower share of animal-sourced food, once implemented at scale, reduces the need to raise livestock and changes crop production from animal feed to human food. This reduces the need for agricultural land compared to present and thus generates changes in the current food system. From field to consumer this would reduce overall GHG emissions. Changes in consumer behaviour beyond dietary changes, such as reduction of food waste, can also have, at scale, effects on overall GHG emissions from food systems. Consuming regional and seasonal food can reduce GHG emissions, if they are grown efficiently.

5

References

Abarca-Gómez, L. et al., 2017: Worldwide trends in body-mass index, underweight, overweight, and obesity from 1975 to 2016: A pooled analysis of 2416 population-based measurement studies in 128·9 million children, adolescents, and adults. *Lancet*, **390**, 2627–2642, doi:10.1016/S0140-6736(17)32129-3.

Abbas, G. et al., 2017: Quantification the impacts of climate change and crop management on phenology of maize-based cropping system in Punjab, Pakistan. *Agric. For. Meteorol.*, **247**, 42–55, doi:10.1016/j.agrformet.2017.07.012.

Aberman, N.-L., and C. Tirado, 2014: Impacts of climate change on food utilization. In: *Glob. Environ. Chang. [Freedman, B. (ed.)]*. Springer, Dordrecht, Holland, pp. 717–724, doi:10.1007/978-94-007-5784-4.

Abiona, B.G., E.O. Fakoya, and J. Esun, 2016: The impacts of climate change on the livelihood of arable crop farmers in Southwest, Nigeria. In: *Innovation in Climate Change Adaptation* [Leal Filho, W. (ed.)]. Springer, Cham, Switzerland, 289–296 pp.

ADB, 2013: Food Security Challenges in Asia. Working Paper from the Independent Evaluation Department (IED) of the Asian Development Bank (ADB). Manila, Phillipines. 59 pp.

ADB, 2014: Technologies to support climate change adaptation in developing Asia. Asian Development Bank. Mandaluyong City, Philippines. 206 pp.

ADB, 2017: Food Insecurity in Asia: Why Institutions Matter [Zhang-Yue, Z. and G. Wan, (eds.)]. Asian Development Bank Institute. Tokyo, Japan, 415 pp.

Adger, W.N., J. Barnett, K. Brown, N. Marshall, and K. O'Brien, 2013: Cultural dimensions of climate change impacts and adaptation. *Nat. Clim. Chang.*, **3**, 112–117, doi:10.1038/nclimate1666.

Adhikari, L., A. Hussain, and G. Rasul, 2017: Tapping the potential of neglected and underutilized food crops for sustainable nutrition security in the mountains of Pakistan and Nepal. *Sustain.*, **9**, 291, doi:10.3390/su9020291.

Adiku, S.G.K., and coauthors, 2015: Climate change impacts on West African agriculture: An integrated regional assessment (CIWARA). In: *Handbook of Climate Change and Agroecosystems: The Agricultural Model Intercomparison and Improvement Project Integrated Crop and Economic Assessments – Joint Publication with American Society of Agronomy, Crop Science Society of America, and Soil Science Society of America (In 2 Parts) (Vol. 3)*. Imperial College Press, London, UK. [Rosenzweig, C. and D. Hillel (eds.)]. Part 2, pp. 25–73, doi:10.1142/p970.

Agarwal, C., G.M. Green, J.M. Grove, T.P. Evans, and C.M. Schweik, 2002: *A Review and Assessment of Land-use Change Models: Dynamics of Space, Time, And Human Choice*. General Technical Report NE-297, Department of Agriculture, Forest Service, Northeastern Research Station Newton Square, Pennsylvania, USA, 61 pp., doi:10.2737/NE-GTR-297.

Aggarwal, P.K. et al., 2018: The climate-smart village approach: Framework of an integrative strategy for scaling up adaptation options in agriculture. *Ecol. Soc.*, **23**, 14, doi: 10.5751/ES-09844-230114.

Aghakouchak, A., A. Farahmand, F.S. Melton, J. Teixeira, M.C. Anderson, B.D. Wardlow, and C.R. Hain, 2015: Reviews of Geophysics Remote sensing of drought: Progress, challenges. *Rev. Geophys*, **53**, 452–480, doi:10.1002/2014RG000456.

AGRA, 2017: *Africa Agriculture Status Report: The Business of Smallholder Agriculture in Sub-Saharan Africa (Issue 5)*. Nairobi, Kenya: Alliance for a Green Revolution in Africa (AGRA). Issue No. 5. [Sumba, D. (ed.)]. 180 pp.

Agrawala, S., and M. Carraro, 2010: *Assessing the Role of Microfinance in Fostering Adaptation to Climate Change*. FEEM Working Paper No. 82.2010; CMCC Research Paper No. 91, *Ssrn*, Paris, France. doi:10.2139/ssrn.1646883.

Ahlgren, I., S. Yamada, and A. Wong, 2014: Rising oceans, climate change, food aid and human rights in the Marshall Islands. *Heal. Hum. Rights J.*, **16**, 69–81, doi:10.2307/healhumarigh.16.1.69.

Ahmad, A. et al., 2015: Impact of climate change on the rice-wheat cropping system of Pakistan. In: *Handbook of Climate Change and Agroecosystems: The Agricultural Model Intercomparison and Improvement Project Integrated Crop and Economic Assessments – Joint Publication with American Society of Agronomy, Crop Science Society of America, and Soil Science Society of America (In 2 Parts) (Vol. 3)*. Imperial College Press, London, UK. [Rosenzweig, C. and D. Hillel (eds.)]. Part 2, pp. 219–258, doi:10.1142/9781783265640_0019.

Ahmed, N., A. Occhipinti-Ambrogi, and J.F. Muir, 2013: The impact of climate change on prawn postlarvae fishing in coastal Bangladesh: Socioeconomic and ecological perspectives. *Mar. Policy*, **39**, 224–233, doi:10.1016/j.marpol.2012.10.008.

Ahmed, N., S.W. Bunting, M. Glaser, M.S. Flaherty, and J.S. Diana, 2017: Can greening of aquaculture sequester blue carbon? *Ambio*, **46**, 468–477, doi:10.1007/s13280-016-0849-7.

Ahmed, N., S. Thompson, and M. Glaser, 2018: Integrated mangrove-shrimp cultivation: Potential for blue carbon sequestration. *Ambio*, **47**, 441–452, doi:10.1007/s13280-017-0946-2.

Aipira, C., A. Kidd, and K. Morioka, 2017: Climate change adaptation in pacific countries: Fostering resilience through gender equality. In: *Climate Change Adaptation in Pacific Countries. Climate Change Management* [Leal Filho W. (ed.)]. Springer International Publishing, Cham, Switzerland, pp. 225–239.

Ajani, E., E. Onwubuya, and R. Mgbenka, 2013: Approaches to economic empowerment of rural women for climate change mitigation and adaptation: Implications for policy. *J. Agric. Ext.*, **17**, 23, doi:10.4314/jae.v17i1.3.

Akram-Lodhi, A.H. (ed.), 2013: *Hungry for Change: Farmers, Food Justice and the Agrarian Question*. Kumarian Press Inc., Colorado, USA, 194 pp.

Albrecht, T.R., A. Crootof, and C.A. Scott, 2018: The water-energy-food nexus: A systematic review of methods for nexus assessment. *Environ. Res. Lett.*, **13**, 43002, doi:10.1088/1748-9326/aaa9c6.

Alexander, P., C. Brown, A. Arneth, J. Finnigan, and M.D.A. Rounsevell, 2016: Human appropriation of land for food: The role of diet. *Glob. Environ. Chang.*, **41**, 88–98, doi:10.1016/J.GLOENVCHA.2016.09.005.

Alexander, P., et al., 2017: Losses, inefficiencies and waste in the global food system. *Agric. Syst.*, **153**, 190–200, doi:10.1016/j.agsy.2017.01.014.

Ali, A., W.K. Yeoh, C. Forney, and M.W. Siddiqui, 2018: Advances in postharvest technologies to extend the storage life of minimally processed fruits and vegetables. *Crit. Rev. Food Sci. Nutr.*, **58**, 2632–2649, doi:10.1080/10408398.2017.1339180.

Allen, A., L. Griffin, and C. Johnson (eds.), 2017: *Environmental Justice and Urban Resilience in the Global South*. Palgrave Macmillan US, New York, 1–307 pp, doi:10.1057/978-1-137-47354-7.

Alston, M., 2009: Drought policy in Australia: Gender mainstreaming or gender blindness? *Gender, Place Cult.*, **16**, 139–154, doi:10.1080/09663690902795738.

Alston, M., 2014: Gender mainstreaming and climate change. *Womens. Stud. Int. Forum*, **47**, 287–294, doi:10.1016/J.WSIF.2013.01.016.

Altieri, M. et al., 2017: Technological approaches to sustainable agriculture at a crossroads: An agroecological perspective. *Sustainability*, **9**, 349, doi:10.3390/su9030349.

Altieri, M.A., 2002: Agroecology: The science of natural resource management for poor farmers in marginal environments. *Agric. Ecosyst. Environ.*, **93**, 1–24, doi:10.1016/S0167-8809(02)00085-3.

Altieri, M.A., and P. Koohafkan, 2008: *Enduring Farms: Climate Change, Smallholders and Traditional Farming Communities*. Third World Network, Penang, Malaysia, 72 pp.

Altieri, M.A. and C.I. Nicholls, 2017: The adaptation and mitigation potential of traditional agriculture in a changing climate. *Clim. Change*, **140**, 33–45, doi:10.1007/s10584-013-0909-y.

Altieri, M.A., P. Rosset, and L.A. Thrupp, 1998: *The Potential of Agroecology to Combat Hunger in the Developing World*. International Food Policy Research Institute (IFPRI), Washington, DC, USA, 1–2 pp.

Altieri, M.A., C.I. Nicholls, A. Henao, and M.A. Lana, 2015: Agroecology and the design of climate change-resilient farming systems. *Agron. Sustain. Dev.*, **35**, 869–890, doi:10.1007/s13593-015-0285-2.

Amanullah, A., and S. Khalid, 2016: Integrated use of phosphorus, animal manures and biofertilizers improve maize productivity under semi-arid condition. In: *Organic Fertilizers – From Basic Concepts to Applied Outcomes* [Larramendy, M.L. and S. Soloneski (eds.)]. IntechOpen Limited, London, UK, pp. 20, doi:10.5772/62388.

American Diabetes Association, 2018: Economic costs of diabetes in the US in 2017. *Diabetes Care*, **41**, 917, doi:10.2337/dci18-0007.

Anderson, K., and G. Peters, 2016: The trouble with negative emissions. *Science.*, **354**, 182–183, doi:10.1126/science.aah4567.

Antle, J.M. et al. 2015: AgMIP's transdisciplinary agricultural systems approach to regional integrated assessment of climate impacts, vulnerability, and adaptation. In: *Handbook of Climate Change and Agroecosystems: The Agricultural Model Intercomparison and Improvement Project Integrated Crop and Economic Assessments – Joint Publication with American Society of Agronomy, Crop Science Society of America, and Soil Science Society of America (In 2 Parts) (Vol. 3).* Imperial College Press, London, UK. [Rosenzweig, C. and D. Hillel (eds.)]. Part 1, pp. 27–44, doi:10.1142/9781783265640_0002.

Antwi-Agyei, P., A.J. Dougill, and L.C. Stringer, 2015: Impacts of land tenure arrangements on the adaptive capacity of marginalized groups: The case of Ghana's Ejura Sekyedumase and Bongo districts. Elsevier, **49**, 203–212, doi:10.1016/j.landusepol.2015.08.007.

Aqeel-Ur-Rehman, A.Z. Abbasi, N. Islam, and Z.A. Shaikh, 2014: A review of wireless sensors and networks' applications in agriculture. *Comput. Stand. Interfaces*, **36**, 263–270, doi:10.1016/j.csi.2011.03.004.

Arndt, C., M.A. Hussain, V. Salvucci, and L.P. Østerdal, 2016: Effects of food price shocks on child malnutrition: The Mozambican experience 2008/2009. *Econ. Hum. Biol.*, **22**, 1–13, doi:10.1016/j.ehb.2016.03.003.

Arno, A., and S. Thomas, 2016: The efficacy of nudge theory strategies in influencing adult dietary behaviour: A systematic review and meta-analysis. *BMC Public Health*, **16**, 676, doi:10.1186/s12889-016-3272-x.

Arora-Jonsson, S., 2011: Virtue and vulnerability: Discourses on women, gender and climate change. *Glob. Environ. Chang.*, **21**, 744–751, doi:10.1016/J.GLOENVCHA.2011.01.005.

Aryal, J.P., T.B. Sapkota, M.L. Jat, and D.K. Bishnoi, 2015: On-farm economic and environmental impact of zero-tillage wheat: A case of North-West India. *Exp. Agric.*, **51**, 1–16, doi:10.1017/S001447971400012X.

Aryal, J.P. et al., 2016: Conservation agriculture-based wheat production better copes with extreme climate events than conventional tillage-based systems: A case of untimely excess rainfall in Haryana, India. *Agric. Ecosyst. Environ.*, **233**, 325–335, doi:10.1016/j.agee.2016.09.013.

Ashoori, D., and M. Sadegh, 2016: Adoption of conservation farming practices for sustainable rice production among small-scale paddy farmers in northern Iran. *Paddy Water Environ.*, **15**, 237–248, doi:10.1007/s10333-016-0543-1.

Aslam, N.M., and J. Varani, 2016: The Western-style diet, calcium deficiency and chronic disease. *J. Nutr. Food Sci.*, **6**, 1–6, doi:10.4172/2155-9600.1000496.

Asseng, S. et al., 2015: Rising temperatures reduce global wheat production. *Nat. Clim. Chang.*, **5**, 143–147, doi:10.1038/nclimate2470.

Atlin, G.N., J.E. Cairns, and B. Das, 2017: Rapid breeding and varietal replacement are critical to adaptation of cropping systems in the developing world to climate change. *Glob. Food Sec.*, **12**, 31–37, doi:10.1016/j.gfs.2017.01.008.

Audsley, E., M. Brander, J.C. Chatterton, D. Murphy-Bokern, C. Webster, and A.G. Williams, 2010: *How low can we go? An assessment of greenhouse gas emissions from the UK foodsystem and the scope reduction by 2050*. Report for the WWF and Food ClimateResearch Network. WWF-UK.

Auffhammer, M., and W. Schlenker, 2014: Empirical studies on agricultural impacts and adaptation. *Energy Econ.*, **46**, 555–561, doi:10.1016/J.ENECO.2014.09.010.

Aune, J.B., A. Coulibaly, and K.E. Giller, 2017: Precision farming for increased land and labour productivity in semi-arid West Africa. A review. *Agron. Sustain. Dev.*, **37**, 16. doi:10.1007/s13593-017-0424-z.

Avetisyan, M., T. Hertel, and G. Sampson, 2014: Is local food more environmentally friendly? The GHG emissions impacts of consuming imported versus domestically produced food. *Environ. Resour. Econ.*, **58**, 415–462, doi:10.1007/s10640-013-9706-3.

Awodoyin, R. et al., 2015: Indigenous fruit trees of tropical Africa: Status, opportunity for development and biodiversity management. *Agric. Sci.*, **6**, 31–41, doi:10.4236/as.2015.61004.

Ayenew, H.Y., S. Biadgilign, L. Schickramm, G. Abate-Kassa, and J. Sauer, 2018: Production diversification, dietary diversity and consumption seasonality: Panel data evidence from Nigeria. *BMC Public Health*, **18**, 988, doi:10.1186/s12889-018-5887-6.

Ayers, J. and T. Forsyth, 2009: Community-based adaptation to climate change. *Environ. Sci. Policy Sustain. Dev.*, **51**, 22–31, doi:10.3200/ENV.51.4.22-31.

Azevedo, A.A. et al., 2017: Limits of Brazil's forest code as a means to end illegal deforestation. *Proc. Natl. Acad. Sci.*, **114**, 7653–7658, doi:10.1073/pnas.1604768114.

Backlund, P., A. Janetos, and D. Schimel, 2008: *The Effects of Climate Change on Agriculture, Land Resources, Water Resources, and Biodiversity in the United States*. U.S. Department of Agriculture, Washington, DC, USA, 362 pp.

Badjeck, M.-C., E.H. Allison, A.S. Halls, and N.K. Dulvy, 2010: Impacts of climate variability and change on fishery-based livelihoods. *Mar. Policy*, **34**, 375–383, doi:10.1016/J.MARPOL.2009.08.007.

Baglioni, S., B. De Pieri, U. Soler, J. Rosell, and T. Tallarico, 2017: Public policy interventions in surplus food recovery in France, Germany, Italy and Spain. In: *Foodsaving in Europe* [Baglioni, S., F. CalòPaola Garrone, and M. Molteni (eds.)], Springer International Publishing, Cham, Switzerland, pp. 37–48.

Bahrami, H. et al., 2017: The proportion of nitrate in leaf nitrogen, but not changes in root growth, are associated with decreased grain protein in wheat under elevated CO_2. *J. Plant Physiol.*, **216**, 44–51, doi:10.1016/j.jplph.2017.05.011.

Bailey, R., and L. Wellesley, 2017: *Chokepoints and Vulnerabilities in Global Food Trade*. The Royal Institute of International Affairs, Chatham House, London, UK 124 pp.

Bailey, R. et al., 2015: *Extreme Weather and Resilience of the Global Food System. Final Project Report from the UK-US Taskforce on Extreme Weather and Global Food System Resilience*. Final Project Report from the UK-US Taskforce on Extreme Weather and Global Food System Resilience, The Global Food Security programme, UK, 17 pp.

Bajželj, B. et al., 2014: Importance of food-demand management for climate mitigation. *Nat. Clim. Chang.*, **4**, 924–929, doi:10.1038/nclimate2353.

Baker, B.H. et al., 2016: A field-scale investigation of nutrient and sediment reduction efficiencies of a low-technology best management practice: Low-grade weirs. *Ecol. Eng.*, **91**, 240–248, doi:10.1016/j.ecoleng.2016.02.038.

Baldermann, S. et al., 2016: Are neglected plants the food for the future? *CRC. Crit. Rev. Plant Sci.*, **35**, 106–119, doi:10.1080/07352689.2016.1201399.

Baldos, U.L.C., and T.W. Hertel, 2015: The role of international trade in managing food security risks from climate change. *Food Secur.*, **7**, 275–290, doi:10.1007/s12571-015-0435-z.

Balmford, A. et al., 2018: The environmental costs and benefits of high-yield farming. *Nat. Sustain.*, **1**, 477–485, doi:10.1038/s41893-018-0138-5.

Balzan, M.V., G. Bocci, and A.C. Moonen, 2016: Utilisation of plant functional diversity in wildflower strips for the delivery of multiple agroecosystem services. *Entomol. Exp. Appl.*, **158**, 304–319, doi:10.1111/eea.12403.

Russell, C.A. and M. Elia, 2012: *Nutrition screening survey in the uk and republic of ireland in 2011*. A Report by the British Association for Parenteral and Enteral Nutrition (BAPEN), Redditch, UK, 1–76 pp.

Barange, M. et al., 2018: *Impacts of Climate Change on Fisheries and Aquaculture: Synthesis of Current Knowledge, Adaptation and Mitigation Options*. Fisheries and Aquaculture Technical Paper No. 627, Rome, Italy, 628 pp.

Barrett, C.B., 2006: *Food Aid's Intended and Unintended Consequences*. ESA Working Paper No. 06–05, Agricultural and Development Economics Division, The Food and Agriculture Organization of the United Nations, Rome, Italy, 27 pp.

Barriopedro, D., E.M. Fischer, J. Luterbacher, R.M. Trigo, and R. García-Herrera, 2011: The hot summer of 2010: Redrawing the temperature record map of Europe. *Science.*, **332**, 220–224, doi:10.1126/science.1201224.

Bartlett, S., 2008: Climate change and urban children: Impacts and implications for adaptation in low – and middle-income countries. *Environ. Urban.*, **20**, 501–519, doi:10.1177/0956247808096125.

Batalla, I., M.T. Knudsen, L. Mogensen, Ó. del Hierro, M. Pinto, and J.E. Hermansen, 2015: Carbon footprint of milk from sheep farming systems in Northern Spain including soil carbon sequestration in grasslands. *J. Clean. Prod.*, **104**, 121–129, doi:10.1016/J.JCLEPRO.2015.05.043.

Bates, B., C. Roberts, H. Lepps, and L. Porter, 2017: *The Food and You Survey Wave 4 Combined Report for England, Wales and Northern Ireland*. NatCen Social Research, London, UK.

Batima, P., L. Natsagdorj, and N. Batnasan, 2008: Vulnerability of Mongolia's pastoralists to climate extremes and changes. In: *Climate Change and Vulnerability and Adaptation* [N. Leary, C. Conde, J. Kulkarni, A. Nyong, and J. Pulin (eds.)]. Earthscan/Taylor and Francis, London, UK, pp. 67–87.

Battersby, J., 2012: Beyond the food desert: Finding ways to speak about urban food security in South Africa. *Geogr. Ann. Ser. B, Hum. Geogr.*, **94**, 141–159, doi:10.1111/j.1468-0467.2012.00401.x.

Battilani, P. et al., 2016: Aflatoxin B1 contamination in maize in Europe increases due to climate change. *Sci. Rep.*, **6**, 24328, doi:10.1038/srep24328.

Baynes, J., J. Herbohn, and W. Unsworth, 2017: Reforesting the grasslands of Papua New Guinea: The importance of a family-based approach. *J. Rural Stud.*, **56**, 124–131, doi:10.1016/J.JRURSTUD.2017.09.012.

Bebber, D.P., 2015: Range-expanding pests and pathogens in a warming world. *Annual Review of Phytopathology*, **53**, 335–356.

Bebber, D.P., and S.J. Gurr, 2015: Crop-destroying fungal and oomycete pathogens challenge food security. *Fungal Genet. Biol.*, **74**, 62–64, doi:10.1016/j.fgb.2014.10.012.

Bebber, D.P., T. Holmes, and S.J. Gurr, 2014: The global spread of crop pests and pathogens. *Glob. Ecol. Biogeogr.*, **23**, 1398–1407, doi:10.1111/geb.12214.

Bee, B.A., 2016: Power, perception, and adaptation: Exploring gender and social-environmental risk perception in northern Guanajuato, Mexico. *Geoforum*, **69**, 71–80, doi:10.1016/J.GEOFORUM.2015.12.006.

Beheshti, R., J.C. Jones-Smith, and T. Igusa, 2017: Taking dietary habits into account: A computational method for modeling food choices that goes beyond price. *PLoS One*, **12**, 1–13, doi:10.1371/journal.pone.0178348.

Beletse, Y.G. et al., 2015: Projected impacts of climate change scenarios on the production of maize in southern Africa: An integrated assessment case study of the Bethlehem district, Central Free State, South Africa. In: *Handbook of Climate Change and Agroecosystems: The Agricultural Model Intercomparison and Improvement Project Integrated Crop and Economic Assessments – Joint Publication with American Society of Agronomy, Crop Science Society of America, and Soil Science Society of America (In 2 Parts) (Vol. 3)*. Imperial College Press, London, UK. [Rosenzweig, C. and D. Hillel (eds.)]. Part 2, pp. 125–157.

Ben-Ari, T., J. Boé, P. Ciais, R. Lecerf, M. Van der Velde, and D. Makowski, 2018: Causes and implications of the unforeseen 2016 extreme yield loss in the breadbasket of France. *Nat. Commun.*, **9**, 1627, doi:10.1038/s41467-018-04087-x.

Benjaminsen, T.A., K. Alinon, H. Buhaug, and J.T. Buseth, 2012: Does climate change drive land-use conflicts in the Sahel? *J. Peace Res.*, **49**, 97–111, doi:10.1177/0022343311427343.

Benkeblia, N., 2018: *Climate Change and Crop Production. Foundations for Agroecosystem Resilience*. CRC Press, Florida, USA, 207 pp.

Benton, T.G., and T. Bailey, 2019: The paradox of productivity: Agricultural productivity promotes food system inefficiency. *Glob. Sustain.*, **2**, E6.

Benton, T.G. et al., 2018: Designing sustainable landuse in a 1.5° world: The complexities of projecting multiple ecosystem services from land. *Curr. Opin. Environ. Sustain.*, **31**, 88–95, doi:10.1016/j.cosust.2018.01.011.

Beringer, T., W. Lucht, and S. Schaphoff, 2011: Bioenergy production potential of global biomass plantations under environmental and agricultural constraints. *GCB Bioenergy*, **3**, 299–312, doi:10.1111/j.1757-1707.2010.01088.x.

Berke, P.R., and M.R. Stevens, 2016: Land use planning for climate adaptation: Theory and practice. *J. Plan. Educ. Res.*, **36**(3): 283–289, doi:10.1177/0739456X16660714.

Bernstein, L.E., E.T. Auer, M. Wagner, and C.W. Ponton, 2008: Spatiotemporal dynamics of audiovisual speech processing. *Neuroimage*, **39**, 423–435, doi:10.1016/j.neuroimage.2007.08.035.

Bett, B. et al., 2017: Effects of climate change on the occurrence and distribution of livestock diseases. *Prev. Vet. Med.*, **137**, 119–129, doi:10.1016/j.prevetmed.2016.11.019.

Betts, R.A. et al., 2018: Changes in climate extremes, fresh water availability and vulnerability to food insecurity projected at 1.5°C and 2°C global warming with a higher-resolution global climate model. *Philos. Trans. R. Soc. A Math. Eng. Sci.*, **376**, 20160452, doi:10.1098/rsta.2016.0452.

Beuchle, R. et al., 2015: Land cover changes in the Brazilian Cerrado and Caatinga biomes from 1990 to 2010 based on a systematic remote sensing sampling approach. *Appl. Geogr.*, **58**, 116–127, doi:10.1016/J.APGEOG.2015.01.017.

Bhat, Z.F., S. Kumar, and H. Fayaz, 2015: In vitro meat production: Challenges and benefits over conventional meat production. *J. Integr. Agric.*, **14**, 241–248, doi:10.1016/S2095-3119(14)60887-X.

Bhatta, G.D. and P.K. Aggarwal, 2016: Coping with weather adversity and adaptation to climatic variability: A cross-country study of smallholder farmers in South Asia. *Clim. Dev.*, **8**, 145–157, doi:10.1080/17565529.2015.1016883.

Bhattacharyya, P.N., M.P. Goswami, and L.H. Bhattacharyya, 2016: Perspective of beneficial microbes in agriculture under changing climatic scenario: A review. *J. Phytol.*, **8**, 26–41, doi:10.19071/jp.2016.v8.3022.

Biesbroek, S., H.B. et al., 2014: Reducing our environmental footprint and improving our health: Greenhouse gas emission and land use of usual diet and mortality in EPIC-NL: A prospective cohort study. *Environ. Health*, **13**, 27, doi:10.1186/1476-069X-13-27.

Biggs, E.M., E.L. Tompkins, J. Allen, C. Moon, and R. Allen, 2013: Agricultural adaptation to climate change: Observations from the mid-hills of Nepal. *Clim. Dev.*, **5**, 165–173, doi:10.1080/17565529.2013.789791.

Bioversity International, 2016: *Mainstreaming Agrobiodiversity in Sustainable Food Systems: Scientific Foundations for an Agrobiodiversity Index-Summary*. [Bailey, A. (Ed.)]. Bioversity International, Rome. Italy, pp. 30.

Birney, C.I., K.F. Franklin, F.T. Davidson, and M.E. Webber, 2017: An assessment of individual foodprints attributed to diets and food waste in the United States. *Environ. Res. Lett.*, **12**, 105008, doi:10.1088/1748-9326/aa8494.

Bisbis, M.B., N. Gruda, and M. Blanke, 2018: Potential impacts of climate change on vegetable production and product quality – A review. *J. Clean. Prod.*, **170**, 1602–1620, doi:10.1016/j.jclepro.2017.09.224.

Blanchard, J.L. et al., 2017: Linked sustainability challenges and trade-offs among fisheries, aquaculture and agriculture. *Nat. Ecol. Evol.*, **1**, 1240–1249, doi:10.1038/s41559-017-0258-8.

Blanco Sepúlveda, R. and A. Aguilar Carrillo, 2016: The erosion threshold for a sustainable agriculture in cultures of bean (*Phaseolus vulgaris* L.) under conventional tillage and no-tillage in Northern Nicaragua. *Soil Use Manag.*, **32**, 368–380, doi:10.1111/sum.12271.

Bloom, A.J., M. Burger, B.A. Kimball, and P.J. Pinter, 2014: Nitrate assimilation is inhibited by elevated CO_2 in field-grown wheat. *Nat. Clim. Chang.*, **4**, 477–480, doi:10.1038/nclimate2183.

Bloom, D.E. et al., 2015: Macroeconomic implications of population ageing and selected policy responses. *Lancet*, **385**, 649–657, doi:10.1016/S0140-6736(14)61464-1.

Bobojonov, I., and A. Aw-Hassan, 2014: Impacts of climate change on farm income security in Central Asia: An integrated modeling approach. *Agric. Ecosyst. Environ.*, **188**, 245–255, doi:10.1016/j.agee.2014.02.033.

Bocco, G., and B.M. Napoletano, 2017: The prospects of terrace agriculture as an adaptation to climate change in Latin America. *Geogr. Compass*, **11**, 1–13, doi:10.1111/gec3.12330.

Bodin, P., S. Olin, T.A.M. Pugh, and A. Arneth, 2016: Accounting for interannual variability in agricultural intensification: The potential of crop selection in Sub-Saharan Africa. *Agric. Syst.*, **148**, 159–168, doi:10.1016/j.agsy.2016.07.012.

de Boer, I. et al., 2011: Greenhouse gas mitigation in animal production: Towards an integrated lifecycle sustainability assessment. *Curr. Opin. Environ. Sustain.*, **3**, 423–431, doi:10.1016/J.COSUST.2011.08.007.

de Boer, J., H. Schösler, and H. Aiking, 2014: 'Meatless days' or 'less but better'? Exploring strategies to adapt Western meat consumption to health and sustainability challenges. *Appetite*, **76**, 120–128, doi:10.1016/J.APPET.2014.02.002.

Boerema, A. et al., 2016: Soybean trade: Balancing environmental and socio-economic impacts of an intercontinental market. *PLoS One*, **11**, e0155222, doi:10.1371/journal.pone.0155222.

Boettcher, P.J. et al., 2015: Genetic resources and genomics for adaptation of livestock to climate change. *Front. Genet.*, **5**, 2014–2016, doi:10.3389/fgene.2014.00461.

Boetto, H., and J. McKinnon, 2013: Rural women and climate change: A gender-inclusive perspective. *Aust. Soc. Work*, **66**, 234–247, doi:10.1080/0312407X.2013.780630.

Bogaerts, M. et al., 2017: Climate change mitigation through intensified pasture management: Estimating greenhouse gas emissions on cattle farms in the Brazilian Amazon. *J. Clean. Prod.*, **162**, 1539–1550, doi:10.1016/J.JCLEPRO.2017.06.130.

van den Bold, M. et al., 2015: Can integrated agriculture-nutrition programmes change gender norms on land and asset ownership? Evidence from Burkina Faso. *J. Dev. Stud.*, **51**, 1155–1174, doi:10.1080/00220388.2015.1036036.

Bond, H. et al., 2018: *Developing a Global Compendium on Water Quality Guidelines*. Report of the International Water Resources Association (IWRA), 1–79 pp.

Boone, R.B., R.T. Conant, J. Sircely, P.K. Thornton, and M. Herrero, 2017: Climate change impacts on selected global rangeland ecosystem services. *Glob. Chang. Biol.*, **24**, 1382–1393, doi:10.1111/gcb.13995.

Boucher, D., and D. Chi, 2018: Amazon deforestation in Brazil: What has not happened and how the global media covered it. *Trop. Conserv. Sci.*, **11**, 1940082918794325, doi:10.1177/1940082918794325.

Bouroncle, C. et al., 2017: Mapping climate change adaptive capacity and vulnerability of smallholder agricultural livelihoods in Central America: Ranking and descriptive approaches to support adaptation strategies. *Clim. Change*, **141**, 123–137, doi:10.1007/s10584-016-1792-0.

Bowles, N., S. Alexander, and M. Hadjikakou, 2019: The livestock sector and planetary boundaries: A 'limits to growth' perspective with dietary implications. *Ecol. Econ.*, **160**, 128–136, doi:10.1016/J.ECOLECON.2019.01.033.

Brandt, M. et al., 2018: Reduction of tree cover in West African woodlands and promotion in semi-arid farmlands. *Nat. Geosci.*, **11**, 328–333, doi:10.1038/s41561-018-0092-x.

Brashares, J.S. et al., 2004: Bushmeat hunting, wildlife declines, and fish supply in West Africa. *Science*, **306**, 1180–1183, doi:10.1126/science.1102425.

Breisinger, C. et al., 2015: Conflict and food insecurity: How do we break the links? In *2014–2015 Global food policy report*. Washington, D.C.: International Food Policy Research Institute (IFPRI). Chapter 7, 51–59 pp.

Bren d'Amour, C. et al., 2017: Future urban land expansion and implications for global croplands. *Proc. Natl. Acad. Sci.*, **114**, 8939 LP-8944.

Brodt, S., K.J. Kramer, A. Kendall, and G. Feenstra, 2013: Comparing environmental impacts of regional and national-scale food supply chains: A case study of processed tomatoes. *Food Policy*, **42**, 106–114, doi:10.1016/j.foodpol.2013.07.004.

Brooks, J., and A. Matthews, 2015: Trade Dimensions of Food Security. *OECD Food, Agric. Fish. Pap.*, 1–45, doi:10.1787/5js65xn790nv-en.

Brown, B., I. Nuberg, and R. Llewellyn, 2017a: Negative evaluation of conservation agriculture: perspectives from African smallholder farmers. *Int. J. Agric. Sustain.*, **15**, 467–481, doi:10.1080/14735903.2017.1336051.

Brown, M.E., 2016: Remote sensing technology and land use analysis in food security assessment security assessment. *Journal of Land Use Science*, **11**, 623-641, doi:10.1080/1747423X.2016.1195455.

Brown, M.E. et al., 2017b: Do markets and trade help or hurt the global food system adapt to climate change? *Food Policy*, **68**, 154–159, doi:10.1016/j.foodpol.2017.02.004.

Brunner, S.H., and A. Grêt-Regamey, 2016: Policy strategies to foster the resilience of mountain social-ecological systems under uncertain global change. *Environ. Sci. Policy*, **66**, 129–139, doi:10.1016/j.envsci.2016.09.003.

Bryan, E., and J.A. Behrman, 2013: *Community-Based Adaptation to Climate Change: A Theoretical Framework, Overview of Key Issues and Discussion of Gender Differentiated Priorities and Participation*. CAPRi Working Paper No. 109, International Food Policy Research Institute, Washington, DC, USA, pp 36.

Bryngelsson, D., F. Hedenus, D. Johansson, C. Azar, and S. Wirsenius, 2017: How do dietary choices influence the energy-system cost of stabilizing the climate? *Energies*, **10**, 182, doi:10.3390/en10020182.

Buckwell, A. et al., 2014: *The Sustainable Intensification of European Agriculture: A review sponsored by The Rise Foundation*. RISE Foundation, Brussels, Belgium, 98 pp.

Buggy, L., and K.E. McNamara, 2016: The need to reinterpret 'community' for climate change adaptation: A case study of Pele Island, Vanuatu. *Clim. Dev.*, **8**, 270–280, doi:10.1080/17565529.2015.1041445.

Buhaug, H., 2016: Climate change and conflict: Taking stock. *Peace Econ. Peace Sci. Public Policy*, **22**, 331–338, doi:10.1515/peps-2016-0034.

Bungenstab, D.J., 2012: *Pecuária de corte brasileira: Redução do aquecimento global pela eficiência dos sistemas de produção*. Brasilia, Brasil: Embrapa Gado de Corte. Documentos 192, Embrapa Gado de Cort, Ministério da Agricultura, Pecuária e Abasteciment, 46 pp.

Burney, J., and V. Ramanathan, 2014: Recent climate and air pollution impacts on Indian agriculture. *PNAS*, **111** (46) 16319-16324 doi:10.1073/pnas.1317275111.

Burnham, M., and Z. Ma, 2016: Linking smallholder farmer climate change adaptation decisions to development. *Clim. Dev.*, **8**, 289–311, doi:10.1080/17565529.2015.1067180.

Burns, W., and S. Nicholson, 2017: Bioenergy and carbon capture with storage (BECCS): The prospects and challenges of an emerging climate policy response. *J. Environ. Stud. Sci.*, **7**, 527–534, doi:10.1007/s13412-017-0445-6.

Byerlee, D., J. Stevenson, and N. Villoria, 2014: Does intensification slow crop land expansion or encourage deforestation? *Glob. Food Sec.*, **3**, 92–98, doi:10.1016/J.GFS.2014.04.001.

CABI, 2019: About Plantwise: Increasing food security and improving rural livelihoods by reducing crop losses CAB International. www.plantwise.org/about/.

Caldeira, S. et al., 2017: *Public Procurement of Food for Health*, Technical Report On The School Setting, Publications Office of the European Commission, Brussels, Belgium, 86 pp.

Calori, A., and A. Magarini, 2015: *Food and the cities: Politiche del cibo per città sostenibili*. Edizioni Ambiente, Milan, Italy. 200 pp.

Calvet-Mir, L. et al., 2018: The contribution of traditional agroecological knowledge as a digital commons to agroecological transitions: The case of the CONECT-e platform. *Sustainability*, **10**, 3214, doi:10.3390/su10093214.

Caminade, C., J. van Dijk, M. Baylis, and D. Williams, 2015: Modelling recent and future climatic suitability for fasciolosis in Europe. *Geospat. Health*, **2**, 301–308.

Cammarano, D. et al., 2016: Uncertainty of wheat water use: Simulated patterns and sensitivity to temperature and CO_2. *F. Crop. Res.*, **198**, 80–92, doi:10.1016/j.fcr.2016.08.015.

Campbell, B.M. et al., 2016: Reducing risks to food security from climate change. *Glob. Food Sec.*, **11**, 34–43, doi:10.1016/j.gfs.2016.06.002.

Campbell, J., R. Bedford, and R. Bedford, 2014: Migration and climate change in Oceania. In: *People on the Move in a Changing Climate: The Regional Impact of Environmental Change on Migration* [Piguet E. and F. Laczko, (eds.)]. Springer Netherlands, Dordrecht, Netherlands, pp. 177–204, doi:10.1007/978-94-007-6985-4_8.

Campbell, J.R., 2014: Climate-change migration in the Pacific. *Contemp. Pac.*, **26**, 1–28, doi:10.1353/cp.2014.0023.

Campbell, J.R., 2015: Development, global change and traditional food security in Pacific Island countries. *Reg. Environ. Chang.*, **15**, 1313–1324, doi:10.1007/s10113-014-0697-6.

Campillo, G., M. Mullan, and L. Vallejo, 2017: *Climate Change Adaptation and Financial Protection*. OECD Environment Working Papers, No. 120, OECD Publishing, Paris, doi:10.1787/0b3dc22a-en.

Canto, T., M.A. Aranda, and A. Fereres, 2009: Climate change effects on physiology and population processes of hosts and vectors that influence the spread of hemipteran-borne plant viruses. *Glob. Chang. Biol.*, **15**, 1884–1894, doi:10.1111/j.1365-2486.2008.01820.x.

Cardoso, A.S. et al., 2016: Impact of the intensification of beef production in Brazil on greenhouse gas emissions and land use. *Agric. Syst.*, **143**, 86–96, doi:10.1016/j.agsy.2015.12.007.

Carlson, K.M. et al., 2017: Greenhouse gas emissions intensity of global croplands. *Nat. Clim. Chang.*, **7**, 63–72, doi:10.1038/nclimate3158.

Carvalho, P.C. de F., and B. Dedieu, 2014: Integrated crop–livestock systems: Strategies to achieve synergy between agricultural production and environmental quality. *Agric. Ecosyst. Environ.*, **190**, 4–8, doi:10.1016/J.AGEE.2013.08.009.

Carvalho, R., M. Adami, S. Amaral, F.G. Bezerra, and A.P.D. de Aguiar, 2019: Changes in secondary vegetation dynamics in a context of decreasing deforestation rates in Pará, Brazilian Amazon. *Appl. Geogr.*, **106**, 40–49, doi:10.1016/J.APGEOG.2019.03.001.

Casado-Vara, R., J. Prieto, F. De la Prieta, and J.M. Corchado, 2018: How blockchain improves the supply chain: Case study alimentary supply chain. *Procedia Comput. Sci.*, **134**, 393–398, doi:10.1016/J.PROCS.2018.07.193.

Castañé, S., and A. Antón, 2017: Assessment of the nutritional quality and environmental impact of two food diets: A Mediterranean and a vegan diet. *J. Clean. Prod.*, **167**, 929–937, doi:10.1016/j.jclepro.2017.04.121.

Cawthorn, D.-M., and L.C. Hoffman, 2015: The bushmeat and food security nexus: A global account of the contributions, conundrums and ethical collisions. *Food Res. Int.*, **76**, 906–925–2015, doi:10.1016/j.foodres.2015.03.025.

CCAFS, 2012: Agriculture's role in both adaptation and mitigation discussed during climate talks. Blog post by Schubert, C. https://ccafs.cgiar.org/blog/agricultures-role-both-adaptation-and-mitigation-discussed-during-climate-talks.

CDP, 2018: Tracking climate progress 2017 – CDP. Second annual analysis. www.cdp.net/en/research/global-reports/tracking-climate-progress-2017. CDP Worldwide, UK.

PWC, 2019: The Carbon Disclosure Project. www.pwc.com/gx/en/services/sustainability/publications/carbon-disclosure-project.html.

Ceddia, M.G., S. Sedlacek, N.O. Bardsley, and S. Gomez-y-Paloma, 2013: Sustainable agricultural intensification or Jevons paradox? The role of public governance in tropical South America. *Glob. Environ. Chang.*, **23**, 1052–1063, doi:10.1016/j.gloenvcha.2013.07.005.

Ceglar, A., A. Toreti, C. Prodhomme, M. Zampieri, M. Turco, and F.J. Doblas-Reyes, 2018: Land-surface initialisation improves seasonal climate prediction skill for maize yield forecast. *Sci. Rep.*, **8**, 1322, doi:10.1038/s41598-018-19586-6. doi:10.1038/s41598-018-19586-6.

Celentano, D. et al., 2017: Towards zero deforestation and forest restoration in the Amazon region of Maranhão state, Brazil. *Land use policy*, **68**, 692–698, doi:10.1016/J.LANDUSEPOL.2017.07.041.

Chabejong, N.E., 2016: A review on the impact of climate change on food security and malnutrition in the Sahel region of Cameroon. In: *Climate Change and Health* [Leal Filho W., U.Azeiteiro, F. Alves (eds.)]., Springer International Publishing, Cham, Switzerland, pp. 133–148.

Chadwick, V., 2017: Banks to incentivize sustainable supply chains under DFID-backed pilot, Devex, Paris www.devex.com/news/banks-to-incentivize-sustainable-supply-chains-under-dfid-backed-pilot-91730.

Chalak, A., C. Abou-Daher, J. Chaaban, and M.G. Abiad, 2016: The global economic and regulatory determinants of household food waste generation: A cross-country analysis. *Waste Manag.*, **48**, 418–422, doi:10.1016/J.WASMAN.2015.11.040.

Chalise, S., and A. Naranpanawa, 2016: Climate change adaptation in agriculture: A computable general equilibrium analysis of land-use change in Nepal. *Land use policy*, **59**, 241–250, doi:10.1016/j.landusepol.2016.09.007.

Challinor, A., W.N. Adger, T.G. Benton, D. Conway, M. Joshi, and D. Frame, 2018: Transmission of climate risks across sectors and borders. *Philos. Trans. R. Soc. A Math. Phys. Eng. Sci.*, **376**(2121), 20170301. doi:10.1098/rsta.2017.0301.

Challinor, A.J. et al., 2016: *UK Climate Change Risk Assessment Evidence Report: Chapter 7, International Dimensions.* Report prepared for the Adaptation Sub-Committee of the Committee on Climate Change, London, UK, 85 pp.

Charan, D., M. Kaur, and P. Singh, 2017: Customary land and climate change induced relocation – A case study of Vunidogoloa village, Vanua Levu, Fiji. In: *Climate Change Adaptation in Pacific Countries* [Leal Filho W. (ed.)]. Springer International Publishing, Cham, Switzerland, pp. 19–34.

Charles, A. et al., 2017: Global nitrous oxide emission factors from agricultural soils after addition of organic amendments: A meta-analysis. *Agric. Ecosyst. Environ.*, **236**, 88–98, doi:10.1016/J.AGEE.2016.11.021.

Chartzoulakis, K., and M. Bertaki, 2015: Sustainable water management in agriculture under climate change. *Agric. Agric. Sci. Procedia*, **4**, 88–98, doi:10.1016/j.aaspro.2015.03.011.

Chatterjee, N., P.K.R. Nair, S. Chakraborty, and V.D. Nair, 2018: Changes in soil carbon stocks across the Forest-Agroforest-Agriculture/Pasture continuum in various agroecological regions: A meta-analysis. *Agric. Ecosyst. Environ.*, **266**, 55–67, doi:10.1016/j.agee.2018.07.014.

Chen, Y. et al., 2018: Great uncertainties in modeling grazing impact on carbon sequestration: A multi-model inter-comparison in temperate Eurasian Steppe. *Environ. Res. Lett.*, **13**, 75005, doi:10.1088/1748-9326/aacc75.

Chhetri, N., P. Chaudhary, P.R. Tiwari, and R.B. Yadaw, 2012: Institutional and technological innovation: Understanding agricultural adaptation to climate change in Nepal. *Appl. Geogr.*, **33**, 142–150, doi:10.1016/J.APGEOG.2011.10.006.

Chivenge, P., T. Mabhaudhi, A.T. Modi, and P. Mafongoya, 2015: The potential role of neglected and underutilised crop species as future crops under water scarce conditions in Sub-Saharan Africa. *Int. J. Environ. Res. Public Health*, **12**, 5685–5711, doi:10.3390/ijerph120605685.

5

Christmann, S. et al., 2009: *Food Security and Climate Change in Central Asia and the Caucasus*. Food Security and Climate Change in Central Asia & Caucasus, International Center for Argricultural Research in the Dry Areas (ICARDA), Tashkent, Uzbekistan, 75 pp.

Chuwah, C., T. van Noije, D.P. van Vuuren, E. Stehfest, and W. Hazeleger, 2015: Global impacts of surface ozone changes on crop yields and land use. *Atmos. Environ.*, **106**, 11–23, doi:10.1016/J.atmosenv.2015.01.062.

CIPHET, 2007: *Vision 2025 – CIPHET Perspective Plan, Central Institute of Post Harvest Engineering and Technology*. Central Institute of Post Harvest Engineering and Technology, Ludhiana, India, 74 pp.

Clapp, J., 2016: *Food Security and International Trade – Unpacking Disputed Narratives*. Background paper prepared for The State of Agricultural Commodity Markets, Rome, Italy 51 pp.

Clay, E., 2006: Is food aid effective? *Id21 insights*, **61**, 6.

Clémençon, R., 2016: The two sides of the Paris climate agreement: Dismal failure or historic breakthrough? *Journal of Environment & Development*, **25**, 3–24, doi:10.1177/1070496516631362.

Clemens, M., J. Rijke, A. Pathirana, and J. Evers, 2015: Social learning for adaptation to climate change in developing countries: insights from Vietnam. *J. Water Clim. Chang.*, 1–13, doi:10.2166/wcc.2015.004.

De Clercq, D., Z. Wen, F. Fan, and L. Caicedo, 2016: Biomethane production potential from restaurant food waste in megacities and project level-bottlenecks: A case study in Beijing. *Renew. Sustain. Energy Rev.*, **59**, 1676–1685, doi:10.1016/J.RSER.2015.12.323.

Clune, S., E. Crossin, and K. Verghese, 2017: Systematic review of greenhouse gas emissions for different fresh food categories. *J. Clean. Prod.*, **140**, 766–783, doi:10.1016/j.jclepro.2016.04.082.

Cohen, M.J., and J.L. Garrett, 2010: The food price crisis and urban food (in) security. *Environ. Urban.*, **22**, 467–482, doi:10.1177/0956247810380375.

Cohn, A.S. et al., 2014: Cattle ranching intensification in Brazil can reduce global greenhouse gas emissions by sparing land from deforestation. *Proc. Natl. Acad. Sci.*, **111**, 7236–7241, doi:10.1073/pnas.1307163111.

Coleman-Jensen, A., M.P. Rabbitt, C.A. Gregory, and A. Singh, 2018: *Household Food Security in the United States in 2017*. Economic Research Report Number 256, US Department of Agriculture, Economic Research Service, Washington, DC, USA, pp. 44, www.ers.usda.gov.

Čolović, D., S. Rakita, V. Banjac, O. Đuragić, and I. Čabarkapa, 2019: Plant food by-products as feed: Characteristics, possibilities, environmental benefits, and negative sides. *Food Rev. Int.*, **35**, 1–27, doi:10.1080/87559129.2019.1573431.

Conant, R.T., 2010: *Challenges and Opportunities for Carbon Sequestration in Grassland Systems: A Technical Report on Grassland Management and Climate Change Mitigation*. A Technical Report On Grassland Management And Climate Change Mitigation, Food And Agriculture Organization Of The United Nations, Rome, Italy, 1–65 pp.

Connell, J., 2015: Vulnerable islands: Climate change, tectonic change, and changing livelihoods in the Western Pacific. *Contemp. Pac.*, **27**, 1–36, doi:10.1353/cp.2015.0014.

Connell, J., 2016: Last days in the Carteret Islands? Climate change, livelihoods and migration on coral atolls. *Asia Pac. Viewp.*, **57**, 3–15, doi:10.1111/apv.12118.

Conservation Evidence, 2019: Conservation Evidence Synopses. www.conservationevidence.com/synopsis/index. Department of Zoology, University of Cambridge, UK.

Le Conte, Y., and M. Navajas, 2008: Climate change: Impact on honey bee populations and diseases. *Rev. Sci. Tech.*, **27**, 485–497, 499–510.

Conway, G., 2013: *Sustainable Intensification: A New Paradigm for African Agriculture*. Montpelllier Panel, London, UK 36 pp.

Cook, B.I., K.J. Anchukaitis, R. Touchan, D.M. Meko, and E.R. Cook, 2016: Spatiotemporal drought variability in the Mediterranean over the last 900 years. *J. Geophys. Res. Atmos.*, **121**, 2060–2074, doi:10.1002/2015JD023929.

Cook, S., L. Silici, B. Adolph, and S. Walker, 2015: *Sustainable Intensification Revisited*. Issue Paper, IIED, London, UK, 32 pp.

Coomes, O.T. et al., 2015: Farmer seed networks make a limited contribution to agriculture? Four common misconceptions. *Food Policy*, **56**, 41–50, doi:10.1016/j.foodpol.2015.07.008.

Coumou, D., V. Petoukhov, S. Rahmstorf, S. Petri, and H.J. Schellnhuber, 2014: Quasi-resonant circulation regimes and hemispheric synchronization of extreme weather in boreal summer. *Proc. Natl. Acad. Sci.*, **111**, 12331–12336, doi:10.1073/pnas.1412797111.

Cramer, W., G.W. Yohe, M. Auffhammer, C. Huggel, U. Molau, M.A.F. da S. Dias, and R. Leemans, 2014: Detection and attribution of observed impacts. In: Climate Climate Change 2014: Impacts, Adaptation, and Vulnerability [Field, C.B., V.R. Barros, D.J. Dokken, K.J. Mach, M.D. Mastrandrea, T.E. Bilir, M. Chatterjee, K.L. Ebi, Y.O. Estrada, S. MacCracken, P.R. Mastrandrea, L.L. White, R.C. Genova, B. Girma, E.S. Kissel, A.N. Levy (eds.)]. Cambridge University Press, United Kingdom and New York, NY, USA, 979–1038.

Craven, L.K., 2015: Migration-affected change and vulnerability in rural Vanuatu. *Asia Pac. Viewp.*, **56**, 223–236, doi:10.1111/apv.12066.

Creutzig, F., B. Fernandez, H. Haberl, R. Khosla, Y. Mulugetta, and K.C. Seto, 2016: Beyond technology: Demand-side solutions for climate change mitigation. *Annu. Rev. Environ. Resour.*, **41**, 173–198, doi:10.1146/annurev-environ-110615-085428.

Crews, T.E., W. Carton, and L. Olsson, 2018: Is the future of agriculture perennial? Imperatives and opportunities to reinvent agriculture by shifting from annual monocultures to perennial polycultures. *Glob. Sustain.*, **1**, e11, doi:10.1017/sus.2018.11.

Cruz-Blanco, M., C. Santos, P. Gavilán, and I.J. Lorite, 2015: Uncertainty in estimating reference evapotranspiration using remotely sensed and forecasted weather data under the climatic conditions of southern Spain. *Int. J. Climatol.*, **35**, 3371–3384, doi:10.1002/joc.4215.

Curtis, P.G., C.M. Slay, N.L. Harris, A. Tyukavina, and M.C. Hansen, 2018: Classifying drivers of global forest loss. *Science.*, **361**, 1108–1111, doi:10.1126/science.aau3445.

Dagevos, H., and J. Voordouw, 2013: Sustainability and meat consumption: Is reduction realistic? *Sustain. Sci. Pract. Policy*, **9**, 60–69, doi:10.1080/15487733.2013.11908115.

Daioglou, V., J.C. Doelman, B. Wicke, A. Faaij, and D.P. van Vuuren, 2019: Integrated assessment of biomass supply and demand in climate change mitigation scenarios. *Glob. Environ. Chang.*, **54**, 88–101, doi:10.1016/J.GLOENVCHA.2018.11.012.

DaMatta, F.M., A. Grandis, B.C. Arenque, and M.S. Buckeridge, 2010: Impacts of climate changes on crop physiology and food quality. *Food Res. Int.*, **43**, 1814–1823, doi:10.1016/J.FOODRES.2009.11.001.

Darmon, N., and A. Drewnowski, 2015: Contribution of food prices and diet cost to socioeconomic disparities in diet quality and health: A systematic review and analysis. *Nutr. Rev.*, **73**, 643–660, doi:10.1093/nutrit/nuv027.

Darnton-Hill, I., and B. Cogill, 2010: Maternal and young child nutrition adversely affected by external shocks such as increasing global food prices. *J. Nutr.*, **140**, 162S–169S, doi:10.3945/jn.109.111682.

Dasgupta, P. et al., 2014: Rural Areas. In: *Climate Change 2014: Impacts, Adaptation, and Vulnerability. Part A: Global and Sectoral Aspects. Contribution of Working Group II to the Fifth Assessment Report of the Intergovernmental Panel on Climate Change* [Field, C.B. et al. (eds.)]. Cambridge University Press, Cambridge, UK and New York, 613–657.

Davis, K.F., K. Yu, M. Herrero, P. Havlik, J.A. Carr, and P. D'Odorico, 2015: Historical trade-offs of livestock's environmental impacts. *Environ. Res. Lett.*, **10**, 125013.

Dawe, D., 2010: *The Rice Crisis: Markets, Policies and Food Security*. FAO and Earthscan, London, UK and Rome, Italy, 393 pp.

Deryng, D., D. Conway, N. Ramankutty, J. Price, and R. Warren, 2014: Global crop yield response to extreme heat stress under multiple climate change futures. *Environ. Res. Lett.*, **9**, 34011, doi:10.1088/1748-9326/9/3/034011.

5

Descheemaeker, K., M. Zijlstra, P. Masikati, O. Crespo, and S. Homann-Kee Tui, 2018: Effects of climate change and adaptation on the livestock component of mixed farming systems: A modelling study from semi-arid Zimbabwe. *Agric. Syst.*, **159**, 282–295, doi:10.1016/j.agsy.2017.05.004.

Deutsch, C.A., J.J. Tewksbury, M. Tigchelaar, D.S. Battisti, S.C. Merrill, R.B. Huey, and R.L. Naylor, 2018: Increase in crop losses to insect pests in a warming climate. *Science*, **361**, 916–919, doi:10.1126/science.aat3466.

Development Initiatives, 2017: *Global Nutrition Report 2017: Nourishing the SDGs.*Produced by Hawkes, C., J. Fanzo, E. Udomkesmalee and the GNR Independent Expert Group. Bristol, UK, 115 pp.

Development Initiatives, 2018: *Global Nutrition Report 2018: Shining a light to spur action on nutrition*. Produced by Fanzo, J., C. Hawkes, E. Udomkesmalee and the GNR Independent Expert Group. Bristol, UK, 161 pp.

Devereux, S., 2015: *Social Protection and Safety Nets in the Middle East and North Africa*. IDS Research Report, Brighton, UK, issue 80, pp 128. www.ids.ac.uk/publications.

Dhanush, D., and S. Vermeulen, 2016: Climate change adaptation in agriculture: Practices and technologies. Opportunities for climate action in agricultural systems. CCAFS, Copenhagen, Denmark, 1–7 pp., doi:10.13140/RG.2.1.4269.0321.

Dickie, I.A. et al., 2014: Conflicting values: Ecosystem services and invasive tree management. *Biol. Invasions*, **16**, 705–719, doi:10.1007/s10530-013-0609-6.

Din, K., S. Tariq, A. Mahmood, and G. Rasul, 2014: Temperature and Precipitation: GLOF Triggering Indicators in Gilgit-Baltistan, Pakistan. *Pakistan J. Meteorol.*, **10**(20), 39–56 pp.

Dinesh, D. et al., 2017: The rise in Climate-Smart Agriculture strategies, policies, partnerships and investments across the globe. *Agric. Dev.*, **30**, 4–9.

DKNC, 2012: Managing Climate Extremes and Disasters in the Agriculture Sector: Lessons from the IPCC SREX Report Contents. Climate and Development Knowledge Network, 36 pp. www.africaportal.org/dspace/articles/managing-climate-extremes-and-disasters-agriculture-sector-lessons-ipcc-srex-report.

Doak, C.M., and B.M. Popkin, 2017: Overweight and obesity. In: *Nutrition and Health in a Developing World* [De Pee M.W., T. Saskia, B. Douglas, (eds.)]. Springer, Humana Press, Cham, Switzerland, pp. 143–158.

Doelman, J.C. et al., 2018: Exploring SSP land-use dynamics using the IMAGE model: Regional and gridded scenarios of land-use change and land-based climate change mitigation. *Glob. Environ. Chang.*, **48**, 119–135, doi:10.1016/j.gloenvcha.2017.11.014.

Donatti, C.I., C.A. Harvey, M.R. Martinez-Rodriguez, R. Vignola, and C.M. Rodriguez, 2018: Vulnerability of smallholder farmers to climate change in Central America and Mexico: Current knowledge and research gaps. *Clim. Dev.*, 1–23, doi:10.1080/17565529.2018.1442796.

Dooley, K., and D. Stabinsky, 2018: *Missing Pathways to 1.5°C: The role of the land sector in ambitious climate action*. [Olden, M. (ed.)]. Climate Land Ambition and Rights Alliance (CLARA). www.climatelandambitionrightsalliance.org/report.

Van Dooren, C., M. Marinussen, H. Blonk, H. Aiking, and P. Vellinga, 2014: Exploring dietary guidelines based on ecological and nutritional values: A comparison of six dietary patterns. *Food Policy*, **44**, 36–46, doi:10.1016/j.foodpol.2013.11.002.

Doran-Browne, N.A., R.J. Eckard, R. Behrendt, and R.S. Kingwell, 2015: Nutrient density as a metric for comparing greenhouse gas emissions from food production. *Clim. Change*, **129**, 73–87, doi:10.1007/s10584-014-1316-8.

Dou, Y., R. Silva, H. Yang, and J. Lliu, 2018: Spillover effect offsets the conservation effort in the Amazon. *Geogr. Sci.*, **28**, 1715–1732, doi:10.1007/s11442-018-1539-0.

Dowla, A., 2018: Climate change and microfinance. *Bus. Strateg. Dev.*, **1**, 78–87, doi:10.1002/bsd2.13.

Dubbeling, M., and H. de Zeeuw, 2011: Urban Agriculture and Climate Change Adaptation: Ensuring Food Security Through Adaptation BT – Resilient Cities. Springer, Dordrecht, Netherlands, 441–449 pp.

Dumont, A.M., G. Vanloqueren, P.M. Stassart, and P.V. Baret, 2016: Clarifying the socioeconomic dimensions of agroecology: Between principles and practices. *Agroecol. Sustain. Food Syst.*, **40**, 24–47, doi:10.1080/21683565.2015.1089967.

Dunne, J.P., R.J. Stouffer, and J.G. John, 2013: Reductions in labour capacity from heat stress under climate warming. *Nat. Clim. Chang.*, **3**, 563, doi:10.1038/nclimate1827.

Easter, T.S., A.K. Killion, and N.H. Carter, 2018: Climate change, cattle, and the challenge of sustainability in a telecoupled system in Africa. *Ecol. Soc.*, **23**, art10, doi:10.5751/ES-09872-230110.

Edvardsson Björnberg, K., and S.O. Hansson, 2013: Gendering local climate adaptation. *Local Environ.*, **18**, 217–232, doi:10.1080/13549839.2012.729571.

EFSA Panel on Biological Hazards, 2016: Growth of spoilage bacteria during storage and transport of meat. *EFSA J.*, **14**, doi:10.2903/j.efsa.2016.4523.

Egeru, A., 2016: Climate risk management information, sources and responses in a pastoral region in East Africa. *Clim. Risk Manag.*, **11**, 1–14, doi:10.1016/J.CRM.2015.12.001.

El-Said, H., and J. Harrigan, 2014: Economic reform, social welfare, and instability: Jordan, Egypt, Morocco, and Tunisia, 1983–2004. *Middle East J.*, **68**, 99–121.

Elum, Z.A., D.M. Modise, and A. Marr, 2017: Farmer's perception of climate change and responsive strategies in three selected provinces of South Africa. *Clim. Risk Manag.*, **16**, 246–257, doi:10.1016/j.crm.2016.11.001.

Emberson, L.D. et al., 2018: Ozone effects on crops and consideration in crop models. *Eur. J. Agron.*, 0–1, doi:10.1016/j.eja.2018.06.002.

End Hunger UK, 2018: Shocking figures showing hidden hunger show why we need to find out more – End Hunger UK. http://endhungeruk.org/shocking-figures-showing-hidden-hunger-show-need-find/#more-274.

Enenkel, M. et al., 2015: Drought and food security – Improving decision-support via new technologies and innovative collaboration. *Glob. Food Sec.*, **4**, 51–55, doi:10.1016/j.gfs.2014.08.005.

Englund, O., G. Berndes, and C. Cederberg, 2017: How to analyse ecosystem services in landscapes – A systematic review. *Ecol. Indic.*, doi:10.1016/j.ecolind.2016.10.009.

Ensor, J.E., S.E. Park, E.T. Hoddy, and B.D. Ratner, 2015: A rights-based perspective on adaptive capacity. *Glob. Environ. Chang.*, **31**, 38–49, doi:10.1016/J.GLOENVCHA.2014.12.005.

Ensor, J.E., S.E. Park, S.J. Attwood, A.M. Kaminski, and J.E. Johnson, 2018: Can community-based adaptation increase resilience? *Clim. Dev.*, **10**, 134–151, doi:10.1080/17565529.2016.1223595.

Erb, K.-H. et al., 2016: Livestock Grazing, the Neglected Land Use. *Social Ecology*, Springer International Publishing, Cham, Switzerland, 295–313 pp.

Erb, K.H., H. Haberl, M.R. Jepsen, T. Kuemmerle, M. Lindner, D. Müller, P.H. Verburg, and A. Reenberg, 2013: A conceptual framework for analysing and measuring land-use intensity. *Curr. Opin. Environ. Sustain.*, doi:10.1016/j.cosust.2013.07.010.

Ercsey-Ravasz, M., Z. Toroczkai, Z. Lakner, and J. Baranyi, 2012: Complexity of the international agro-food trade network and its impact on food safety. *PLoS One*, **7**, e37810, doi:10.1371/annotation/5fe23e20-573f-48d7-b284-4fa0106b8c42.

European Union, 2018: Public-private startup accelerators in regional business support ecosystems: A Policy Brief from the Policy Learning Platform on SME competitiveness. European Regional Development Fund. 13 pp.

Eze, S.O., 2017: Constraints to climate change adaptation among cassava women farmers: Implications for agricultural transformation and food security in ebonyi state, nigeria. *Int. J. Ecosyst. Ecol. Sci.*, **7**, 219–228.

Ezzine-de-Blas, D., S. Wunder, M. Ruiz-Pérez, and R. del P. Moreno-Sanchez, 2016: Global patterns in the implementation of payments for environmental services. *PLoS One*, **11**, e0149847, doi:10.1371/journal.pone.0149847.

Fader, M., D. Gerten, M. Krause, W. Lucht, and W. Cramer, 2013: Spatial decoupling of agricultural production and consumption: Quantifying dependences of countries on food imports due to domestic land and water constraints. *Environ. Res. Lett.*, **8**, 14046, doi:10.1088/1748-9326/8/1/014046.

Fagerli, R.A., and M. Wandel, 1999: Gender Differences in opinions and practices with regard to a 'healthy diet'. *Appetite*, **32**, 171–190, doi:10.1006/APPE.1998.0188.

Falkner, R., 2016: The Paris agreement and the new logic of international climate politics. **5**, 1107–1125, doi:10.1111/1468-2346.12708.

Fanzo, J., R. McLaren, C. Davis, and J. Choufani, 2017: Climate change and variability: What are the risks for nutrition, diets, and food systems? *IFPRI*, 1–128.

FAO, 2001a: *Food Insecurity in the World 2001*. www.fao.org/3/a-y1500e.pdf. Food and Agriculture Organization of the United Nations, Rome, Italy, 8 pp.

FAO, 2001b: *Food Balance Sheets: A Handbook (Rome: FAO)*. Food and Agriculture Organization of the United Nations, Rome, Italy, 99 pp. www.fao.org/3/a-x9892e.pdf.

FAO, 2005: *Voluntary Guidelines to Support the Progressive Realization of the Right to Adequate Food in the Context of the National Food Security*. The Right to Food. Food and Agriculture Organization of the United Nations, Rome, Italy, 48 pp. www.fao.org/right-to-food/resources/resources-detail/en/c/44965/.

FAO, 2007: *The State of Food and Agriculture 2007*. Food and Agriculture Organization of the United Nations, Rome, Italy, 222 pp.

FAO, 2008: *Food Security in Mountains – High Time for Action*. Food and Agriculture Organization of the United Nations, Rome, Italy.

FAO, 2011a: *The State of Food and Agriculture. Women in Agriculture: Closing the Gender Gap for Development*. Food and Agriculture Organization of the United Nations, Rome, Italy, 160 pp.

FAO, 2011b: *Global Food Losses and Food Waste: Extent, Causes and Prevention*. Food and Agriculture Organization of the United Nations, Rome, Italy, 37 pp.

FAO, 2011c: *Dietary Protein Quality Evaluation in Human Nutrition*. Food and Agriculture Organization of the United Nations, Rome, Italy, 1–79 pp.

FAO, 2011d: *Save and Grow. A Policymaker's Guide to the Sustainable Intensification of Smallholder Crop Production*. Food and Agriculture Organization of the United Nations, Rome, Italy, 1–116 pp.

FAO, 2011e: *Save and Grow. A Policymaker's Guide to the Sustainable Intensification of Smallholder Crop Production*. Food and Agriculture Organization of the United Nations, Rome, Italy, 116 pp.

FAO, 2012: *Incorporating Climate Change Considerations into Agricultural Investment*. Food and Agriculture Organization of the United Nations, Rome, Italy, 148 pp.

FAO, 2013a: *The State of Food and Agriculture*. Food and Agriculture Organization of the United Nations, Rome, Italy, 114 pp.

FAO, 2013b: *Food Wastage Footprint. Impacts on Natural Resources*. Food and Agriculture Organization of the United Nations, Rome, Italy, 63 pp.

FAO, 2013c: *Climate-Smart Agriculture: Managing Ecosystems for Sustainable Livelihoods*. Food and Agriculture Organization of the United Nations, Rome, Italy, 14 pp.

FAO, 2014a: *The State of Food Insecurity in the World (SOFI) 2014*. Food and Agriculture Organization of the United Nations, Rome, Italy, 57 pp.

FAO, 2014b: *Food Wastage Footprint: Full-cost Accounting*. Food and Agriculture Organization of the United Nations, Rome, Italy, 98 pp.

FAO, 2015a: *Food Wastage Footprint and Climate Change*. Food and Agriculture Organization of the United Nations, Rome, Italy, 4 pp.

FAO, 2015b: *The State of Food Insecurity in the World 2015. Meeting the 2015 international hunger targets: taking stock of uneven progress* Food and Agriculture Organization of the United Nations, Rome, Italy, 62 pp.

FAO, 2015c: *Global Trends in GDP and Agriculture Value Added (1970–2013)*. Food and Agriculture Organization of the United Nations, Rome, Italy, 1–6 pp.

FAO, 2016a: *2015–2016 El Niño – Early Action and Response for Agriculture, Food Security and Nutrition – UPDATE #10*. Food and Agriculture Organization of the United Nations, Rome, Italy, 43 pp.

FAO, 2016b: *State of the World's forests*. Food and Agriculture Organization of the United Nations, Rome, Italy, 126 pp.

FAO, 2016c: *Farmer Field School: Guidance Document: Planning for Quality Programmes*. Food and Agriculture Organization of the United Nations, Rome, Italy, 112 pp.

FAO, 2017: *Livestock Solutions for Climate Change*. Food and Agriculture Organization of the United Nations, Rome, Italy, 1–8 pp.

FAO, 2018a: *The Future of Food and Agriculture: Alternative Pathways to 2050*. Food and Agriculture Organization of the United Nations, Rome, Italy, 228 pp.

FAO, 2018b: *The Future of Food and Agriculture: Trends and Challenges*. Food and Agriculture Organization of the United Nations, Rome, Italy, 180 pp.

FAO, 2018c: *The State of World Fisheries and Aquaculture 2018: Meeting the Sustainable Development Goals*. Food and Agriculture Organization of the United Nations, Rome, Italy, 210 pp.

FAO, 2018d: *The State of Agricultural Commodity Markets 2018*. Food and Agriculture Organization of the United Nations, Rome, Italy, 112 pp.

FAO, 2018e: *The FAO Strategic Objectives*. Food and Agriculture Organization of the United Nations, Rome, Italy,www.fao.org/3/a-mg994e.pdf.

FAO, 2018f: *Gender and Food Loss in Sustainable Food Value Chains: A Guiding Note*. Food and Agriculture Organization of the United Nations, Rome, Italy, 1–56 pp.

FAO, 2018g: *Globally Important Agricultural Heritage Systems (GIAHS). Combining Agricultural Biodiversity, Resilient Ecosystems, Traditional Farming Practices and Cultural Identity*. Food and Agriculture Organization of the United Nations, Rome, Italy, 1–48 pp.

FAO, 2019a: Chapter 10: Context for sustainable intensification of agriculture. In: *Sustainable Food and Agriculture: An Integrated Approach*. Academic Press, Food and Agriculture Organization of the United Nations, Rome, Italy, pp. 171–172, doi:10.1016/B978-0-12-812134-4.00010-8.

FAO, 2019b: *SAVE FOOD: Global Initiative on Food Loss and Waste Reduction. Food and Agriculture Organization of the United Nations*, Rome, Italy, www.fao.org/save-food/en/.

FAO, Mountain Partnership, UNCCD, SDC, and CDE, 2011: *Highlands and Drylands Mountains, A Source of Resilience in Arid Regions*. Food and Agriculture Organization of FAO, UNCCD, Mountain Partnership, Swiss Agency for Development and Cooperation, and CDE, Rome, 115 pp.

FAO, IFAD, UNICEF, WFP, and WHO, 2017: *The State of Food Security and Nutrition in the World 2017. Building resilience for peace and food security*. Food and Agriculture Organization of the United Nations, Rome, Italy, 132 pp.

FAO, IFAD, UNICEF, WFP and WHO, 2018: *The State of Food Security and Nutrition in the World 2018. Building climate resilience for food security and nutrition*. Food and Agriculture Organization of the United Nations, Rome, Italy, 202 pp.

FAOSTAT, 2018: FAOSTAT. Food and Agriculture Organization Corporate Statistical Database. www.fao.org/faostat/en/#home.

Farbotko, C., 2010: Wishful sinking: Disappearing islands, climate refugees and cosmopolitan experimentation. *Asia Pac. Viewp.*, **51**, 47–60, doi:10.1111/j.1467-8373.2010.001413.x.

Farbotko, C., and H. Lazrus, 2012: The first climate refugees? Contesting global narratives of climate change in Tuvalu. *Glob. Environ. Chang.*, **22**, 382–390, doi:10.1016/J.GLOENVCHA.2011.11.014.

Farchi, S., M. De Sario, E. Lapucci, M. Davoli, and P. Michelozzi, 2017: Meat consumption reduction in Italian regions: Health co-benefits and decreases in GHG emissions. *PLoS One*, **12**, 1–19, doi:10.1371/journal.pone.0182960.

5

Farnworth, C.R., F. Baudron, J.A. Andersson, M. Misiko, L. Badstue, and C.M. Stirling, 2016: Gender and conservation agriculture in east and southern Africa: Towards a research agenda. *Int. J. Agric. Sustain.*, **14**, 142–165, doi: 10.1080/14735903.2015.1065602.

Faye, B. et al., 2018a: Impacts of 1.5 versus 2.0°C on cereal yields in the West African Sudan Savanna. *Environ. Res. Lett.*, **13**, 3, doi:10.1088/1748-9326/aaab40.

Babacar, F., H. Webber, M. Diop, M.L. Mbaye, J.D. Owusu-Sekyere, J.B. Naab, and T. Gaiser, 2018b: Potential impact of climate change on peanut yield in Senegal, West Africa. *F. Crop. Res.*, **219**, 148–159, doi:10.1016/j.fcr.2018.01.034.

Fearnside, P.M., 2015: Deforestation soars in the Amazon. *Nature.* **521**, 423, doi: 10.1038/521423b.

van der Fels-Klerx, H.J., J.E. Olesen, M.S. Madsen, and P.W. Goedhart, 2012a: Climate change increases deoxynivalenol contamination of wheat in north-western Europe. *Food Addit. Contam. Part a-Chemistry Anal. Control Expo. Risk Assess.*, **29**, 1593–1604, doi:10.1080/19440049.2012.691555.

van der Fels-Klerx, H., J.E. Olesen, L.J. Naustvoll, Y. Friocourt, M.J.B. Mengelers, and J.H. Christensen, 2012b: Climate change impacts on natural toxins in food production systems, exemplified by deoxynivalenol in wheat and diarrhetic shellfish toxins. *Food Addit. Contam. Part a-Chemistry Anal. Control Expo. Risk Assess.*, **29**, 1647–1659, doi:10.1080/19440049.2012.714080.

Van der Fels-Klerx, H.J., E.D. van Asselt, M.S. Madsen, and J.E. Olesen, 2013: Impact of climate change effects on contamination of cereal grains with deoxynivalenol. *PLoS One*, **8**, doi:10.1371/journal.pone.0073602.

Feng, S., A.B. Krueger, and M. Oppenheimer, 2010: Linkages among climate change, crop yields and Mexico–US cross-border migration. *Proc. Natl. Acad. Sci.*, **107**, 14257–14262.

Feng, Z., et al., 2015: Constraints to nitrogen acquisition of terrestrial plants under elevated CO_2. *Glob. Chang. Biol.*, **21**, 3152–3168, doi:10.1111/gcb.12938.

Fenghua, S., Y. Xiuqun, L. Shuang, and others, 2006: The contrast analysis on the average and extremum temperature trend in northeast China. *Sci. Meteorol. Sin.*, **26**, 157–163.

Fenn, B., 2018: *Impacts of CASH on Nutrition Outcomes: From Available Scientific Evidence to Informed Actions*. Research 4 Action. 24 pp. www.cashlearning.org/downloads/user-submitted-resources/2018/06/1529400438.WFP-0000071735.pdf.

Fenton, A., and J. Paavola, 2015: Microfinance and climate change adaptation: An overview of the current literature. *Enterprise Development and Microfinance*, **26**, 262–273.

Ferrière, N., and A. Suwa-Eisenmann, 2015: Does Food Aid Disrupt Local Food Market? Evidence from Rural Ethiopia. *World Dev.*, **76**, 114–131, doi:10.1016/J.WORLDDEV.2015.07.002.

Fetzel, T. et al., 2017: Quantification of uncertainties in global grazing systems assessment. *Global Biogeochem. Cycles*, **31**, 1089–1102, doi:10.1002/2016GB005601.

Finer, M., S. Novoa, M.J. Weisse, R. Petersen, J. Mascaro, T. Souto, F. Stearns, and R.G. Martinez, 2018: Combating deforestation: From satellite to intervention. *Science.*, **360**, 1303–1305, doi:10.1126/science.aat1203.

Finn, J.A. et al., 2013: Ecosystem function enhanced by combining four functional types of plant species in intensively managed grassland mixtures: A 3-year continental-scale field experiment. *J. Appl. Ecol.*, **50**, 365–375, doi:10.1111/1365-2664.12041.

Fischedick, M., and Coauthors, 2014: Industry. *Climate Change 2014: Mitigation of Climate Change. Contribution of Working Group III to the Fifth Assessment Report of the Intergovernmental Panel on Climate Change*, [Edenhofer, O., J.C. Minx, A. Adler, P. Eickemeier, S. Schlömer, R. Pichs-Madruga, Y. Sokona, K. Seyboth, S. Brunner, J. Savolainen, T. Zwickel, E. Farahani, S. Kadner, I.B.B. Kriemann, and C. von Stechow (eds.)]. Cambridge University Press, Cambridge, United Kingdom and New York, NY, USA.

Fischer, J. et al., 2014: Land sparing versus land sharing: Moving forward. *Conserv. Lett.*, **7**, 149–157, doi:10.1111/conl.12084.

FishStat, 2019: FishStatJ – Software for Fishery and Aquaculture Statistical Time Series. Food and Agriculture Organization of the United Nations. www.fao.org/fishery/statistics/software/fishstatj/en.

Fitton, N. et al., 2017: Modelling spatial and inter-annual variations of nitrous oxide emissions from UK cropland and grasslands using DailyDayCent. *Agric. Ecosyst. Environ.*, **250**, 1–11, doi:10.1016/J.AGEE.2017.08.032.

Fjelde, H., and N. von Uexkull, 2012: Climate triggers: Rainfall anomalies, vulnerability and communal conflict in Sub-Saharan Africa. *Polit. Geogr.*, **31**, 444–453, doi:10.1016/J.POLGEO.2012.08.004.

Folami, O.M., and A.O. Folami, 2013: Climate change and inter-ethnic conflict in Nigeria. *Peace Rev.*, **25**, 104–110, doi:10.1080/10402659.2013.759783.

Ford, J.D., L. Berrang-Ford, A. Bunce, C. McKay, M. Irwin, and T. Pearce, 2015: The status of climate change adaptation in Africa and Asia. *Reg. Environ. Chang.*, **15**, 801–814, doi:10.1007/s10113-014-0648-2.

Forsyth, T., 2013: Community-based adaptation: A review of past and future challenges. *Wiley Interdiscip. Rev. Clim. Chang.*, **4**, 439–446, doi:10.1002/wcc.231.

Francis, C., 2016: The carbon farming solution: A global toolkit of perennial crops and regenerative agriculture practices for climate change mitigation and food security, by Eric Toensmeier. *Agroecol. Sustain. Food Syst.*, **40**, 1039–1040, doi:10.1080/21683565.2016.1214861.

Francis, C. et al., 2003: Agroecology: The ecology of food systems. *J. Sustain. Agric.*, **22**, 99–118, doi:10.1300/J064v22n03_10.

Frank, S. et al., 2017: Reducing greenhouse gas emissions in agriculture without compromising food security? *Environ. Res. Lett.*, **12**, 105004, doi:10.1088/1748-9326/aa8c83.

Frank, S. et al., 2018: Structural change as a key component for agricultural non-CO_2 mitigation efforts. *Nat. Commun.*, **9**, 1060, doi:10.1038/s41467-018-03489-1.

Franzaring, J., I. Holz, and A. Fangmeier, 2013: Responses of old and modern cereals to CO_2-fertilisation. *Crop Pasture Sci.*, **64**, 943–956, doi:10.1071/cp13311.

Fratkin, E., E.A. Roth, and M.A. Nathan, 2004: Pastoral sedentarization and its effects on children?s diet, health, and growth among rendille of northern Kenya. *Hum. Ecol.*, **32**, 531–559, doi:10.1007/s10745-004-6096-8.

Friel, S. et al., 2009: Public health benefits of strategies to reduce greenhouse-gas emissions: Food and agriculture. *Lancet*, **374**, 2016–2025, doi:10.1016/S0140-6736(09)61753-0.

Friis, C., J.Ø. Nielsen, I. Otero, H. Haberl, J. Niewöhner, and P. Hostert, 2016: From teleconnection to telecoupling: Taking stock of an emerging framework in land system science. *J. Land Use Sci.*, **11**, 131–153, doi:10.1080/1747423X.2015.1096423.

Fröhlich, C., and S. Klepp, 2019: Effects of climate change on migration crises in Oceania. In: *The Oxford Handbook of Migration Crises* [C. Menjívar, M. Ruiz, and I. Ness, (eds.)]. Oxford University Press, Oxford, UK, pp. 330–346.

Fujimori, S. et al., 2017: SSP3: AIM implementation of shared socioeconomic pathways. *Glob. Environ. Chang.*, **42**, 268–283, doi:10.1016/j.gloenvcha.2016.06.009.

Fujimori, S., T. Hasegawa, J. Rogelj, X. Su, P. Havlik, V. Krey, K. Takahashi, and K. Riahi, 2018: Inclusive climate change mitigation and food security policy under 1.5°C climate goal. *Environ. Res. Lett.*, **13**, 74033, doi:10.1088/1748-9326/aad0f7.

Fujisaki, K. et al., 2018: Soil carbon stock changes in tropical croplands are mainly driven by carbon inputs: A synthesis. *Agric. Ecosyst. Environ.*, **259**, 147–158, doi:10.1016/j.agee.2017.12.008.

Fuller, T., and M. Qingwen, 2013: Understanding agricultural heritage sites as complex adaptive systems: The challenge of complexity. *J. Resour. Ecol.*, **4**, 195–201, doi:10.5814/j.issn.1674-764x.2013.03.002.

Fulton, S. et al., 2019: From Fishing Fish to Fishing Data: The Role of Artisanal Fishers in Conservation and Resource Management in Mexico. In: *Viability and Sustainability of Small-Scale Fisheries in Latin America and The*

Caribbean [Salas, S., M.J. Barragán-Paladines, and R. Chuenpagdee (eds.)]. Springer International Publishing, Cham, Switzerland, pp. 151–175.

Fuss, S., et al., 2016: Research priorities for negative emissions. *Environ. Res. Lett.*, **11**, 115007, doi:10.1088/1748-9326/11/11/115007.

Fuss, S. et al., 2018: Negative emissions – Part 2: Costs, potentials and side effects. *Environ. Res. Lett.*, **13**, 63002, doi:10.1088/1748-9326/aabf9f.

Gabrielsson, S., and V. Ramasar, 2013: Widows: Agents of change in a climate of water uncertainty. *J. Clean. Prod.*, **60**, 34–42, doi:10.1016/J.JCLEPRO.2012.01.034.

Van Gaelen, H., E. Vanuytrecht, P. Willems, J. Diels, and D. Raes, 2017: Bridging rigorous assessment of water availability from field to catchment scale with a parsimonious agro-hydrological model. *Environ. Model. Softw.*, **94**, 140–156, doi:10.1016/j.envsoft.2017.02.014.

Gale, P., T. Drew, L.P. Phipps, G. David, and M. Wooldridge, 2009: The effect of climate change on the occurrence and prevalence of livestock diseases in Great Britain: A review. *J. Appl. Microbiol.*, **106**, 1409–1423, doi:10.1111/j.1365-2672.2008.04036.x.

Gamba, R.J., J. Schuchter, C. Rutt, and E.Y.W. Seto, 2015: Measuring the food environment and its effects on obesity in the United States: A systematic review of methods and results. *J. Community Health*, **40**, 464–475, doi:10.1007/s10900-014-9958-z.

Gao, C., 2018: The future of CRISPR technologies in agriculture. *Nat. Rev. Mol. Cell Biol.*, **19**, 275–276, doi:10.1038/nrm.2018.2.

Garcia Collazo, M.A., A. Panizza, and J.M. Paruelo, 2013: Ordenamiento territorial de bosques nativos: Resultados de la zonificación realizada por provincias del norte. *Ecol. Aust.*, **23**, 97–107.

Garibaldi, L.A. et al., 2016: Mutually beneficial pollinator diversity and crop yield outcomes in small and large farms. *Science.*, **351**, 388–391, doi:10.1126/SCIENCE.AAC7287.

Garibaldi, L.A., B. Gemmill-Herren, R. D'Annolfo, B.E. Graeub, S.A. Cunningham, and T.D. Breeze, 2017: Farming approaches for greater biodiversity, livelihoods, and food security. *Trends Ecol. Evol.*, **32**, 68–80, doi:10.1016/j.tree.2016.10.001.

Garnett, T., 2011: Where are the best opportunities for reducing greenhouse gas emissions in the food system (including the food chain)? *Food Policy*, **36**, S23–S32, doi:10.1016/J.FOODPOL.2010.10.010.

Garnett, T., 2016: Plating up solutions. *Science.*, **353**, 1202–1204, doi:10.1126/science.aah4765.

Garnett, T. et al., 2013: Sustainable intensification in agriculture: Premises and policies. *Science.*, **341**, 33–34, doi:10.1126/science.1234485.

Garnett, T. et al., 2017: *Grazed and confused?* Ruminating on cattle, grazing systems, methane, nitrous oxide, the soil carbon sequestration question – and what it all means for greenhouse gas emissions. Food Climate Research Network, Oxford, UK. 127 pp.

Gasparri, N.I., and Y. le P. de Waroux, 2015: The coupling of South American soybean and cattle production frontiers: New challenges for conservation policy and land change science. *Conserv. Lett.*, **8**, 290–298, doi:10.1111/conl.12121.

Gasper, R., A. Blohm, and M. Ruth, 2011: Social and economic impacts of climate change on the urban environment. *Curr. Opin. Environ. Sustain.*, **3**, 150–157, doi:10.1016/J.COSUST.2010.12.009.

Gephart, J.A., L. Deutsch, M.L. Pace, M. Troell, and D.A. Seekell, 2017: Shocks to fish production: Identification, trends, and consequences. *Glob. Environ. Chang.*, **42**, 24–32, doi:10.1016/j.gloenvcha.2016.11.003.

Gerber, J.S. et al., 2016: Spatially explicit estimates of N_2O emissions from croplands suggest climate mitigation opportunities from improved fertilizer management. *Glob. Chang. Biol.*, **22**, 3383–3394, doi:10.1111/gcb.13341.

Gerber, P.J., et al., 2013: *Tackling Climate Change Through Livestock: A global assessment of emissions and mitigation opportunities*. Food and Agriculture Organization of the United Nations, Rome, Italy, 139 pp.

Gerber, P.J., A. Mottet, C.I. Opio, A. Falcucci, and F. Teillard, 2015: Environmental impacts of beef production: Review of challenges and perspectives for durability. *Meat Sci.*, **109**, 2–12, doi:10.1016/J.MEATSCI.2015.05.013.

Ghosh-Jerath, S., A. Singh, P. Kamboj, G. Goldberg, and M.S. Magsumbol, 2015: Traditional knowledge and nutritive value of indigenous foods in the Oraon tribal community of Jharkhand: An exploratory cross-sectional study. *Ecol. Food Nutr.*, **54**, 493–519, doi:10.1080/03670244.2015.1017758.

Ghude, S.D., C. Jena, D.M. Chate, G. Beig, G.G. Pfister, R. Kumar, and V. Ramanathan, 2014: Reductions in India's crop yield due to ozone. *Geophys. Res. Lett.*, **41**, 5685–5691, doi:10.1002/2014GL060930.

Gibbs, H.K., and J.M. Salmon, 2015: Mapping the world's degraded lands. *Appl. Geogr.*, **57**, 12–21, doi:10.1016/J.APGEOG.2014.11.024.

Gibbs, H.K. et al., 2015: Brazil's Soy Moratorium. *Science.*, **347**, 377–378, doi:10.1126/science.aaa0181.

Gil, J., M. Siebold, and T. Berger, 2015: Adoption and development of integrated crop-livestock-forestry systems in Mato Grosso, Brazil. *Agric. Ecosyst. Environ.*, **199**, 394–406, doi:10.1016/j.agee.2014.10.008.

Gil, J.D.B. et al., 2018: Trade-offs in the quest for climate smart agricultural intensification in Mato Grosso, Brazil. *Environ. Res. Lett.*, **13**, 64025, doi:10.1088/1748-9326/aac4d1.

Giller, K.E., E. Witter, M. Corbeels, and P. Tittonell, 2009: Conservation agriculture and smallholder farming in Africa: The heretics' view. *F. Crop. Res.*, **114**, 23–34, doi:10.1016/j.fcr.2009.06.017.

Gliessman, S., 2018: Defining agroecology. *Agroecol. Sustain. Food Syst.*, **42**, 599–600, doi:10.1080/21683565.2018.1432329.

Gliessman, S.R., and E. Engles (eds.), 2014: *Agroecology: The Ecology of Sustainable Food Systems*. CRC Press, Taylor & Francis Group, Florida, US, 405 pp.

Global Commission on Adaptation, 2019: Adapt now: A global call for leadership on climate resilience. Global Commission on Adaptation Rotterdam, The Netherlands, 90pp.

Glotter, M., and J. Elliott, 2016: Simulating US agriculture in a modern Dust Bowl drought. *Nat. Plants*, **3**, 16193, doi:10.1038/nplants.2016.193.

Godfray, H.C.J., 2015: The debate over sustainable intensification. *Food Secur.*, **7**, 199–208, doi:10.1007/s12571-015-0424-2.

Godfray, H.C.J. and T. Garnett, 2014: Food security and sustainable intensification. *Philos. Trans. R. Soc. B Biol. Sci.*, **369**, 20120273–20120273, doi:10.1098/rstb.2012.0273.

Godfray, H.C.J. et al., 2010: Food security: The challenge of feeding 9 billion people. *Science*, **327**, 812–818, doi:10.1126/science.1185383.

Godfray, H.C.J. et al., 2018: Meat consumption, health, and the environment. *Science.*, **361**, eaam5324, doi:10.1126/science.aam5324.

Goh, A.H.X., 2012: *A Literature Review of the Gender-Differentiated Impacts of Climate Change on Women's and Men's Assets and Well-Being in Developing Countries*. CAPRi Working Paper No. 106, International Food Policy Research Institute, Washington, DC, USA, 43 pp.

Goh, K.M., 2011: Greater mitigation of climate change by organic than conventional agriculture: A review. *Biol. Agric. Hortic.*, **27**, 205–229, doi:10.1080/01448765.2011.9756648.

Goldstein, B., M. Birkved, J. Fernández, and M. Hauschild, 2017: Surveying the environmental footprint of urban food consumption. *J. Ind. Ecol.*, **21**, 151–165, doi:10.1111/jiec.12384.

González-Varo, J.P. et al., 2013: Combined effects of global change pressures on animal-mediated pollination. *Trends Ecol. Evol.*, **28**, 524–530, doi:10.1016/j.tree.2013.05.008.

Goulson, D., E. Nicholls, C. Botias, and E.L. Rotheray, 2015: Bee declines driven by combined stress from parasites, pesticides, and lack of flowers. *Science.*, **347**, 1255957–1255957, doi:10.1126/science.1255957.

Govaerts, B., N. Verhulst, A. Castellanos-Navarrete, K.D. Sayre, J. Dixon, and L. Dendooven, 2009: Conservation agriculture and soil carbon sequestration: Between myth and farmer reality. *CRC. Crit. Rev. Plant Sci.*, **28**, 97–122, doi:10.1080/07352680902776358.

Graesser, J., T.M. Aide, H.R. Grau, and N. Ramankutty, 2015: Cropland/pastureland dynamics and the slowdown of deforestation in Latin America. *Environ. Res. Lett.*, **10**, 34017, doi:10.1088/1748-9326/10/3/034017.

Graham, R.D. et al., 2007: Nutritious subsistence food systems. *Adv. Agron.*, **92**, 1–74, doi:10.1016/S0065-2113(04)92001-9.

Granderson, A.A., 2014: Making sense of climate change risks and responses at the community level: A cultural-political lens. *Clim. Risk Manag.*, **3**, 55–64, doi:10.1016/J.CRM.2014.05.003.

Gray, C., and V. Mueller, 2012: Drought and population mobility in rural Ethiopia. *World Dev.*, **40**, 134–145, doi:10.1016/j.worlddev.2011.05.023.

Gray, C.L., 2011: Soil quality and human migration in Kenya and Uganda. *Glob. Environ. Chang.*, **21**, 421–430, doi:10.1016/j.gloenvcha.2011.02.004.

Gray, C.L., and R. Bilsborrow, 2013: Environmental influences on human migration in rural Ecuador. *Demography*, **50**, 1217–1241, doi:10.1007/s13524-012-0192-y.

von Grebmer, K. et al., 2014: *2014 Global Hunger Index The Challenge of Hidden Hunger*. Welthungerhilfe, International Food Policy Research Institute, and Concern Worldwide, Washington, DC, USA, and Dublin, Ireland, 56 pp.

Grecchi, R.C., Q.H.J. Gwyn, G.B. Bénié, A.R. Formaggio, and F.C. Fahl, 2014: Land use and land cover changes in the Brazilian Cerrado: A multidisciplinary approach to assess the impacts of agricultural expansion. *Appl. Geogr.*, **55**, 300–312, doi:10.1016/J.APGEOG.2014.09.014.

Green, R. et al., 2015: The potential to reduce greenhouse gas emissions in the UK through healthy and realistic dietary change. *Clim. Change*, **129**, 253–265.

Green, R. et al., 2017: Environmental Impacts of Typical Dietary Patterns in India. *FASEB J.*, **31**, 651.2–651.2.

Green, R.E., S.J. Cornell, J.P.W. Scharlemann, and A. Balmford, 2005: Farming and the fate of wild nature. *Science*, **307**, 550–555, doi:10.1126/science.1106049.

Greenberg, S.E. et al., 2017: Geologic carbon storage at a one million tonne demonstration project: Lessons learned from the Illinois Basin-Decatur Project. *Energy Procedia*, **114**, 5529–5539.

Griscom, B.W. et al., 2016: Natural climate solutions. *Proc. Natl. Acad. Sci.*, 11–12, doi:10.1073/pnas.1710465114.

Griscom, B.W. et al., 2017: Natural climate solutions. *Proc. Natl. Acad. Sci. U.S.A.*, **114**, 11645–11650, doi:10.1073/pnas.1710465114.

Grosso, G., F. Bella, J. Godos, S. Sciacca, D. Del Rio, S. Ray, F. Galvano, and E.L. Giovannucci, 2017: Possible role of diet in cancer: Systematic review and multiple meta-analyses of dietary patterns, lifestyle factors, and cancer risk. *Nutr. Rev.*, **75**, 405–419, doi:10.1093/nutrit/nux012.

Del Grosso, S.J., W.J. Parton, O. Wendroth, R.J. Lascano, and L. Ma, 2019: History of Ecosystem Model Development at Colorado State University and Current Efforts to Address Contemporary Ecological Issues. In: *Bridging Among Disciplines by Synthesizing Soil and Plant Processes* [Ahuja, L.R., L. Mab and S.S. Anapallic (eds.)]. American Society of Agronomy, Crop Science Society of America, and Soil Science Society of America, Inc., Madison, USA.

Grubler, A. et al., 2018: A low energy demand scenario for meeting the 1.5°C target and sustainable development goals without negative emission technologies. *Nat. Energy*, **3**, 515–527, doi:10.1038/s41560-018-0172-6.

Guha, I., and C. Roy, 2016: Climate change, migration and food security: Evidence from Indian Sundarbans. *Int. J. Theor. Appl. Sci.*, **8**, 45–49.

Gunarathna, M.H.J.P. et al., 2017: Optimized subsurface irrigation system (OPSIS): Beyond traditional subsurface irrigation. *Water*, **9**, 8, 599, doi:10.3390/w9080599.

Gupta, R., E. Somanathan, and S. Dey, 2017: Global warming and local air pollution have reduced wheat yields in India. *Clim. Change*, **140**, 593–604, doi:10.1007/s10584-016-1878-8.

Gutiérrez-Peña, R., Y. Mena, I. Batalla, and J.M. Mancilla-Leytón, 2019: Carbon footprint of dairy goat production systems: A comparison of three contrasting grazing levels in the Sierra de Grazalema Natural Park (Southern Spain). *J. Environ. Manage.*, **232**, 993–998, doi:10.1016/J.JENVMAN.2018.12.005.

Gutjahr, C., and R. Gerhards, 2010: Decision rules for site-specific weed management. In: *Precision Crop Protection – The Challenge and Use of Heterogeneity* [Oerke, E.-C., R. Gerhards, G. Menz, and R.A. Sikora, (eds.)]. Springer, Dordrecht, pp. 223–240.

Haberl, H., T. Beringer, S.C. Bhattacharya, K.-H. Erb, and M. Hoogwijk, 2010: The global technical potential of bio-energy in 2050 considering sustainability constraints. *Curr. Opin. Environ. Sustain.*, **2**, 394–403, doi:10.1016/J.COSUST.2010.10.007.

Hadarits, M., J. Pittman, D. Corkal, H. Hill, K. Bruce, and A. Howard, 2017: The interplay between incremental, transitional, and transformational adaptation: A case study of Canadian agriculture. *Reg. Environ. Chang.*, **17**, 1515–1525, doi:10.1007/s10113-017-1111-y.

Hadjikakou, M., 2017: Trimming the excess: Environmental impacts of discretionary food consumption in Australia. *Ecol. Econ.*, **131**, 119–128, doi:10.1016/j.ecolecon.2016.08.006.

Haines, A. et al., 2017: Short-lived climate pollutant mitigation and the Sustainable Development Goals. *Nat. Clim. Chang.*, **7**, 863–869, doi:10.1038/s41558-017-0012-x.

Hallström, E., Q. Gee, P. Scarborough, and D.A. Cleveland, 2017: A healthier US diet could reduce greenhouse gas emissions from both the food and healthcare systems. *Clim. Change*, **142**, 199–212, doi:10.1007/s10584-017-1912-5.

Hamilton, A.J., K. Burry, H.F. Mok, S.F. Barker, J.R. Grove, and V.G. Williamson, 2014: Give peas a chance? Urban agriculture in developing countries. A review. *Agron. Sustain. Dev.*, **34**, 45–73, doi:10.1007/s13593-013-0155-8.

Hammill, A., R. Matthew, and E. McCarter, 2010: Microfinance and climate change adaptation. *IDS Bull.*, **39**, 113–122, doi:10.1111/j.1759-5436.2008.tb00484.x.

Hanley, N., T.D. Breeze, C. Ellis, and D. Goulson, 2015: Measuring the economic value of pollination services: Principles, evidence and knowledge gaps. *Ecosyst. Serv.*, **14**, 124–132, doi:10.1016/j.ecoser.2014.09.013.

Hannah, L. et al., 2017: Regional modeling of climate change impacts on smallholder agriculture and ecosystems in Central America. *Clim. Change*, **141**, 29–45, doi:10.1007/s10584-016-1867-y.

Hansen, J., M. Sato, and R. Ruedy, 2012: Perception of climate change. *Proc. Natl. Acad. Sci. U.S.A.*, **109**, E2415–E2423, doi:10.1073/pnas.1205276109.

Hansen, J., K. Fara, K. Milliken, C. Boyce, L. Chang'a, and E. Allis, 2018: *Strengthening climate services for the food security sector*. World Meteorological Organization Bulletin 67(2):20–26, Geneva, Switzerland.

Harper, A.B. et al., 2018: Land-use emissions play a critical role in land-based mitigation for Paris climate targets. *Nat. Commun.*, **9**, 2938.

Harvey, C.A. et al., 2014a: Climate-smart landscapes: Opportunities and challenges for integrating adaptation and mitigation in tropical agriculture. *Conserv. Lett.*, **7**, 77–90, doi:10.1111/conl.12066.

Harvey, C.A. et al., 2014b: Extreme vulnerability of smallholder farmers to agricultural risks and climate change in Madagascar. *Philos. Trans. R. Soc. Lond. B. Biol. Sci.*, **369**, 20130089, doi:10.1098/rstb.2013.0089.

Harwatt, H., J. Sabaté, G. Eshel, S. Soret, and W. Ripple, 2017: Substituting beans for beef as a contribution toward US climate change targets. *Clim. Change*, **143**, 1–10, doi:10.1007/s10584-017-1969-1. doi:10.1007/s10584-017-1969-1.

Hasegawa, T., S. Fujimori, K. Takahashi, and T. Masui, 2015: Scenarios for the risk of hunger in the 21st century using Shared Socioeconomic Pathways. *Environ. Res. Lett.*, **10**, 14010, doi:10.1088/1748-9326/10/1/014010.

Hasegawa, T et al., 2018: Risk of increased food insecurity under stringent global climate change mitigation policy. *Nat. Clim. Chang.*, **8**, 699–703, doi:10.1038/s41558-018-0230-x.

Haughey, E. et al., 2018: Higher species richness enhances yield stability in intensively managed grasslands with experimental disturbance. *Sci. Rep.*, **8**, 15047.

Havlík, P. et al., 2015: Global climate change, food supply and livestock production systems: A bioeconomic analysis. In: *Climate Change and Food Systems: Global Assessments and Implications for Food Security and Trade*, Food and Agriculture Organization of the United Nations, Rome, Italy, 176–196 pp.

Hawken, P., 2017: *Drawdown: The most comprehensive plan ever proposed to reverse global warming*. Penguin Books, New York, USA, 256 pp.

He, J., 2015: Chinese public policy on fisheries subsidies: Reconciling trade, environmental and food security stakes. *Mar. Policy*, **56**, 106–116, doi:10.1016/j.marpol.2014.12.021.

Heard, B.R., and S.A. Miller, 2016: Critical research needed to examine the environmental Impacts of expanded refrigeration on the food system. *Environ. Sci. Technol.*, **50**, 12060–12071.

Hedenus, F., S. Wirsenius, and D.J.A. Johansson, 2014: The importance of reduced meat and dairy consumption for meeting stringent climate change targets. *Clim. Change*, **124**, 79–91, doi:10.1007/s10584-014-1104-5.

Hedley, C., 2015: The role of precision agriculture for improved nutrient management on farms. *J. Sci. Food Agric.*, doi:10.1002/jsfa.6734.

Hellberg, R.S., and E. Chu, 2016: Effects of climate change on the persistence and dispersal of foodborne bacterial pathogens in the outdoor environment: A review. *Crit. Rev. Microbiol.*, **42**, 548–572, doi:10.3109/1040841x.2014.972335.

Heller, M., 2019: *Waste Not, Want Not: Reducing Food Loss and Waste in North America Through Life Cycle-Based Approaches*. One planet. United Nations Environment Programme (United Nations/ intergovernmental organizations), Washington, DC, USA, 83 pp.

Heller, M.C., and G.A. Keoleian, 2015: Greenhouse gas emission estimates of US dietary choices and food loss. *J. Ind. Ecol.*, **19**, 391–401, doi:10.1111/jiec.12174.

Henderson, B., A. Falcucci, A. Mottet, L. Early, B. Werner, H. Steinfeld, and P. Gerber, 2017: Marginal costs of abating greenhouse gases in the global ruminant livestock sector. *Mitig. Adapt. Strateg. Glob. Chang.*, **22**, 199–224, doi:10.1007/s11027-015-9673-9. http://link.springer.com/10.1007/s11027-015-9673-9.

Henderson, B et al., 2018: The power and pain of market-based carbon policies: A global application to greenhouse gases from ruminant livestock production. *Mitig. Adapt. Strateg. Glob. Chang.*, **23**, 349–369, doi:10.1007/s11027-017-9737-0.

Henderson, B.B., P.J. Gerber, T.E. Hilinski, A. Falcucci, D.S. Ojima, M. Salvatore, and R.T. Conant, 2015: Greenhouse gas mitigation potential of the world's grazing lands: Modeling soil carbon and nitrogen fluxes of mitigation practices. *Agric. Ecosyst. Environ.*, **207**, 91–100.

Hendrix, C., and S. Glaser, 2007: Trends and triggers: Climate, climate change and civil conflict in Sub-Saharan Africa. *Polit. Geogr.*, **26**, 695–715, doi:10.1016/j.polgeo.2007.06.006.

Hendrix, C.S., 2018: Searching for climate–conflict links. *Nat. Clim. Chang.*, **8**, 190–191, doi:10.1038/s41558-018-0083-3.

Henry, S., B. Schoumaker, C. Beauchemin, S. Population, and N. May, 2004: The impact of rainfall on the first out-migration: A multi-level event-history analysis in Burkina Faso. All use subject to JSTOR Terms and Conditions The Impact of Rainfall on the First Out-Migration: A Multi-level Event-History Analysis in Burkina Faso. *Popul. Environ.*, **25**, 423–460.

Herforth, A., and S. Ahmed, 2015: The food environment, its effects on dietary consumption, and potential for measurement within agriculture-nutrition interventions. *Food Secur.*, **7**, 505–520, doi:10.1007/s12571-015-0455-8.

Hernández-Morcillo, M. et al., 2014: Traditional ecological knowledge in Europe: Status quo and insights for the environmental policy agenda. *Environ. Sci. Policy Sustain. Dev.*, **56**, 3–17, doi:10.1080/00139157.2014.861673.

Herrero, M. et al., 2016a: Greenhouse gas mitigation potentials in the livestock sector. *Nat. Clim. Chang.*, **6**, 452–461, doi:10.1038/nclimate2925.

Herrero, M., J. Addison, Bedelian, E. Carabine, P. Havlik, B. Henderson, J. Van de Steeg, and P.K. Thornton, 2016b: Climate change and pastoralism: impacts, consequences and adaptation. *Rev. Sci. Tech. l'OIE*, **35**, 417–433, doi:10.20506/rst.35.2.2533.

Herrero, M., D. Mayberry, J. Van De Steeg, D. Phelan, A. Ash, and K. Diyezee, 2016c: *Understanding Livestock Yield Gaps for Poverty Alleviation, Food Security and the Environment*. The LiveGAPS Project. CSIRO, Brisbane, Australia, 133 pp.

Herrero, M.T. et al., 2010: Climate variability and climate change and their impacts on Kenya's agricultural sector. Report of the project 'Adaptation of Smallholder Agriculture to Climate Change in Kenya', Nairobi, Kenya, 65 pp.

Heyd, T., and Thomas, 2008: Cultural responses to natural changes such as climate change. *Espac. Popul. sociétés*, 83–88, doi:10.4000/eps.2397.

Hiç, C., P. Pradhan, D. Rybski, and J.P. Kropp, 2016: Food surplus and its climate burdens. *Environ. Sci. Technol.*, **50**, 4269–4277, doi:10.1021/acs.est.5b05088.

Hill, S.B., 1985: Redesigning the food system for sustainability. *Alternatives*, **12**, 32–36.

Hirata, R. et al., 2013: Carbon dioxide exchange at four intensively managed grassland sites across different climate zones of Japan and the influence of manure application on ecosystem carbon and greenhouse gas budgets. *Agric. For. Meteorol.*, **177**, 57–68, doi:10.1016/J.AGRFORMET.2013.04.007.

HLPE, 2012: *Food security and climate change. A report by the High Level Panel of Experts on Food Security and Nutrition of the Committee on World Food Security*. Food and Agriculture Organization of the United Nations, Rome, Italy, 102 pp.

HLPE, 2014: *Food losses and waste in the context of sustainable food systems. A report by The High Level Panel of Experts on Food Security and Nutrition of the Committee on World Fod Security*. Food and Agriculture Organization of the United Nations, Rome, Italy, 1–6 pp. www.fao.org/cfs/cfs-hlpe.

HLPE, 2017: *Nutrition and food systems. A report by the High Level Panel of Experts on Food Security and Nutrition of the Committee on World Food Security*. Food and Agriculture Organization of the United Nations, Rome, Italy, 152 pp. www.fao.org/3/a-i7846e.pdf.

Hoag, H., 2014: Russian summer tops 'universal' heatwave index. *Nature*, **16**, 4 pp., doi:10.1038/nature.2014.16250.

Hochman, Z., D.L. Gobbett, and H. Horan, 2017: Climate trends account for stalled wheat yields in Australia since 1990. *Glob. Chang. Biol.*, **23**, 2071–2081, doi:10.1111/gcb.13604.

Hodgson, J.A., W.E. Kunin, C.D. Thomas, T.G. Benton, and D. Gabriel, 2010: Comparing organic farming and land sparing: Optimizing yield and butterfly populations at a landscape scale. *Ecol. Lett.*, **13**, 1358–1367, doi:10.1111/j.1461-0248.2010.01528.x.

Hofer, D., M. Suter, E. Haughey, J.A. Finn, N.J. Hoekstra, N. Buchmann, and A. Lüscher, 2016: Yield of temperate forage grassland species is either largely resistant or resilient to experimental summer drought. *J. Appl. Ecol.*, **53**, 1023–1034, doi:10.1111/1365-2664.12694.

Holding, J., J. Carr, and K. Stark, 2013: Food Statistics Pocketbook 2012. Department for Environment, Food and Rural Affairs, York, UK, 86 pp.

Hollinger, F., 2015: *Agricultural growth in West Africa market and policy drivers*. Food and Agriculture Organization of the United Nations, Rome, Italy, 406 pp.

Homer-Dixon, T. et al., 2015: Synchronous failure: The emerging causal architecture of global crisis. *Ecol. Soc.*, **20**, doi:10.5751/ES-07681-200306.

Hossain, M.A., J. Canning, S. Ast, P.J. Rutledge, and A. Jamalipour, 2015: Early warning smartphone diagnostics for water security and analysis using real-time pH mapping. *Photonic Sensors*, **5**, 289–297, doi:10.1007/s13320-015-0256-x.

Hossain, N., and D. Green, 2011: Living on a Spike: How is the 2011 food price crisis affecting poor people? *Oxfam Policy Pract. Agric. Food L.*, **11**, 9–56.

Howden, S.M., J.-F. Soussana, F.N. Tubiello, N. Chhetri, M. Dunlop, and H. Meinke, 2007: Adapting agriculture to climate change. *Proc. Natl. Acad. Sci.*, **104**, 19691–19696, doi:10.1073/pnas.0701890104.

Howell, K.R., P. Shrestha, and I.C. Dodd, 2015: Alternate wetting and drying irrigation maintained rice yields despite half the irrigation volume, but is currently unlikely to be adopted by smallholder lowland rice farmers in Nepal. *Food Energy Secur.*, **4**, 144–157, doi:10.1002/fes3.58.

5

Hristov, A.N. et al., 2013a: SPECIAL TOPICS – Mitigation of methane and nitrous oxide emissions from animal operations: III. A review of animal management mitigation options. *J. Anim. Sci.*, **91**, 5095–5113, doi:10.2527/jas2013-6585.

Hristov, A.N. et al., 2013b: SPECIAL TOPICS – Mitigation of methane and nitrous oxide emissions from animal operations: I. A review of enteric methane mitigation options. *J. Anim. Sci.*, **91**, 5045–5069, doi:10.2527/jas2013-6583.

Hristov, A.N. et al., 2015: An inhibitor persistently decreased enteric methane emission from dairy cows with no negative effect on milk production. *Proc. Natl. Acad. Sci.*, **112**, 10663–10668.

Hu, Z., J.W. Lee, K. Chandran, S. Kim, and S.K. Khanal, 2012: Nitrous Oxide (N_2O) Emission from Aquaculture: A Review. *Environ. Sci. Technol.*, **46**, 6470–6480, doi:10.1021/es300110x.

Hu, Z. et al., 2013: Nitrogen transformations in intensive aquaculture system and its implication to climate change through nitrous oxide emission. *Bioresour. Technol.*, **130**, 314–320, doi:10.1016/J.BIORTECH.2012.12.033.

Huang, J., and F. Ji, 2015: Effects of climate change on phenological trends and seed cotton yields in oasis of arid regions. *Int. J. Biometeorol.*, **59**, 877–888, doi:10.1007/s00484-014-0904-7.

Huddleston, B., E. Ataman, P. De Salvo, M. Zanetti, M. Bloise, J. Bel, G. Franicheschini, and L. Fed'Ostiani, 2003: *Towards a GIS-Based Analysis of Mountain Environments and Populations*. Food and Agriculture Organization of the United Nations, Rome, Italy, 34 pp.

van Huis, A., 2013: Potential of insects as food and feed in assuring food security. *Annu. Rev. Entomol.*, **58**, 563–583, doi:10.1146/annurev-ento-120811-153704.

van Huis, A., M. Dicke, and J.J.A. van Loon, 2015: Insects to feed the world. *J. Insects as Food Feed*, **1**, 3–5, doi:10.3920/JIFF2015.x002.

Hummel, D., 2015: Climate change, environment and migration in the Sahel. *Rural 21*, 40–43.

Huntington, J.L., K.C. Hegewisch, B. Daudert, C.G. Morton, J.T. Abatzoglou, D.J. McEvoy, and T. Erickson, 2017: Climate engine: Cloud computing and visualization of climate and remote sensing data for advanced natural resource monitoring and process understanding. *Bull. Am. Meteorol. Soc.*, **98**, 2397–2409, doi:10.1175/BAMS-D-15-00324.1.

Hussain, A., G. Rasul, B. Mahapatra, and S. Tuladhar, 2016: Household food security in the face of climate change in the Hindu-Kush Himalayan region. *Food Secur.*, **8**, 921–937, doi:10.1007/s12571-016-0607-5.

Huynh, P.T.A., and B.P. Resurreccion, 2014: Women's differentiated vulnerability and adaptations to climate-related agricultural water scarcity in rural Central Vietnam. *Clim. Dev.*, **6**, 226–237, doi:10.1080/17565529.2014.886989.

Iannotti, L.L., M. Robles, H. Pachón, and C. Chiarella, 2012: Food prices and poverty negatively affect micronutrient intakes in Guatemala. *J. Nutr.*, **142**, 1568–1576, doi:10.3945/jn.111.157321.

IAP, 2018: *Opportunities for future research and innovation on food and nutrition security and agriculture: The InterAcademy Partnership's global perspective*. The IAP-Policy Secretariat, The U.S. National Academies of Science, Engineering, and Medicine, Washington, DC, USA, 1–94 pp.

Ibrahim, M., and A. Kruczkiewicz, 2016: *Learning from Experience: A Review of Early Warning Systems – Moving Towards Early Action 2016*. World Vision House, Milton Keynes, UK, 30 pp.

ICPAC, and WFP, 2017: *Greater Horn of Africa Climate Risk and Food Security Atlas – Part 1: Regional Analysis*. Climate Adaptation Management and Innovation Initiative, 93 pp.

Ifeanyi-obi, C.C., A.O. Togun, and R. Lamboll, 2016: Influence of climate change on cocoyam production in Aba agricultural zone of Abia State, Nigeria. In: *Innovation in Climate Change Adaptation*. Springer International Publishing, Cham, Switzerland, pp. 261–273.

Iizumi, T., and N. Ramankutty, 2016: Changes in yield variability of major crops for 1981–2010 explained by climate change. *Environ. Res. Lett.*, **11**, 34003, doi:10.1088/1748-9326/11/3/034003.

Iizumi, T. et al., 2017: Responses of crop yield growth to global temperature and socioeconomic changes. *Sci. Rep.*, **7**, 7800, doi:10.1038/s41598-017-08214-4.

Iizumi, T. et al., 2018: Crop production losses associated with anthropogenic climate change for 1981–2010 compared with preindustrial levels. *Int. J. Climatol.*, **38**, 5405–5417, doi:10.1002/joc.5818.

Imamura, F. et al., 2015: Dietary quality among men and women in 187 countries in 1990 and 2010: A systematic assessment. *Lancet Glob. Heal.*, **3**, e132–e142, doi:10.1016/S2214-109X(14)70381-X.

Indoria, A.K., C. Srinivasa Rao, K.L. Sharma, and K. Sammi Reddy, 2017: Conservation agriculture – A panacea to improve soil physical health. *Curr. Sci.*, **112**, 52–61, doi:10.18520/cs/v112/i01/52-61.

Innes, P.J., D.K.Y. Tan, F. Van Ogtrop, and J.S. Amthor, 2015: Effects of high-temperature episodes on wheat yields in New South Wales, Australia. *Agric. For. Meteorol.*, **208**, 95–107, doi:10.1016/J.AGRFORMET.2015.03.018.

INPE, 2015: *Nota tecnica: INPE apresenta taxa de desmatamento consolidada do PRODES 2015*. Brazil, 3 pp.

INPE, 2018: Projeto de Monitoramento do Desmatamento na Amazônia Legal – PRODES (INPE). Brazil, www.obt.inpe.br/prodes/index.php.

International Council for Science (ICSU), 2017: *A Guide to SDG Interactions: From Science to Implementation*. International Council for Science, Paris, France, 239 pp.

IPBES, 2016: Summary for Policymakers of the Assessment Report of the Intergovernmental Science–Policy Platform on Biodiversity and Ecosystem Services on Pollinators, Pollination and Food Production. Secretariat of the Intergovernmental Science–Policy Platform on Biodiversity and Ecosystem Services (IPBES), Bonn, Germany, 1–30 pp.

IPCC, 2007: Climate Change 2007: Impacts, Adaptation and Vulnerability. Contribution of Working Group II to the Fourth Assessment Report of the Intergovernmental Panel on Climate Change. M.L. Parry, O.F. Canziani, J.P. Palutikof, P.J. Van der Linden, and C.E. Hanson, Eds. Cambridge University Press, Cambridge, UK, 976 pp.

IPCC, 2012: Managing the Risks of Extreme Events and Disasters to Advance Climate Change Adaptation. A Special Report of Working Groups I and II of the Intergovernmental Panel on Climate Change. [Field, C.B., V. Barros, T.F. Stocker, D. Qin, D.J. Dokken, K.L. Ebi, M.D. Mastrandrea, K.J. Mach, G.-K. Plattner, S.K. Allen, M. Tignor, and P.M. Midgley (eds.)]. Cambridge University Press, Cambridge, UK, 582 pp.

IPCC, 2014a: Climate Change 2014: Mitigation of Climate Change: Working Group III Contribution to the IPCC Fifth Assessment Report. Cambridge University Press, Cambridge, UK, 1435 pp.

IPCC, 2014b: Climate Change 2014: Synthesis Report. Contribution of Working Groups I, II and III to the Fifth Assessment Report of the Intergovernmental Panel on Climate Change [R.K. Pachauri and L.A. Meyer (eds.)]. IPCC, Geneva, Switzerland, 151 pp.

IPCC, 2014c: Climate Change 2014: Impacts, Adaptation, and Vulnerability: Working Group II Contribution to the Fifth Assessment Report of the Intergovernmental Panel on Climate Change [Field, C.B., D.J. Dokken, T.E. Bilir, M. Chatterjee, E.S. Kissel, R.C. Genova, B. Girma, Andrew N. Levy, K.J. Mach, M.D. Mastrandrea, K.L. Ebi, Y.O. Estrada, S. MacCracken, P.R. Mastrandrea, Leslie L. White (eds.)]. Cambridge University Press, Cambridge, UK.

IPCC, 2018a: Global Warming of 1.5°C an IPCC Special Report on the Impacts of Global Warming of 1.5°C Above Pre-Industrial Levels and Related Global Greenhouse Gas Emission Pathways, in the Context of Strengthening the Global Response to the Threat of Climate Change [Masson-Delmotte, V., P. Zhai, H.-O. Pörtner, D. Roberts, J. Skea, P.R. Shukla, A. Pirani, W. Moufouma-Okia, C. Péan, R. Pidcock, S. Connors, J.B.R. Matthews, Y. Chen, X. Zhou, M.I. Gomis, E. Lonnoy, T. Maycock, M. Tignor, and T. Waterfield (eds.)]. In press.

IPCC 2018b: Summary for Policymakers. In: Global Warming of 1.5°C an IPCC Special Report on the Impacts of Global Warming of 1.5°C Above Pre-Industrial Levels and Related Global Greenhouse Gas Emission Pathways, in the Context of Strengthening the Global Response to the Threat of Climate Change [Masson-Delmotte, V., P. Zhai, H.-O. Pörtner, D. Roberts, J. Skea, P.R. Shukla, A. Pirani, W. Moufouma-Okia, C. Péan, R. Pidcock, S. Connors, J.B.R. Matthews, Y. Chen, X. Zhou, M.I. Gomis, E. Lonnoy, T. Maycock, M. Tignor, and T. Waterfield (eds.)]. In press.

IRD, 2017: *Rural Societies in the Face of Climatic and Environmental Changes in West Africa*. [Sultan, B., R. Lalou, M.A. Sanni, A. Oumarou, M.A. Soumare (eds.)]. IRD Éditions, Marseille, France, 432 pp.

Isbell, F. et al., 2017: Benefits of increasing plant diversity in sustainable agroecosystems. *J. Ecol.*, doi:10.1111/1365-2745.12789.

Ishigooka, Y., S. Fukui, T. Hasegawa, T. Kuwagata, M. Nishimori, and M. Kondo, 2017: Large-scale evaluation of the effects of adaptation to climate change by shifting transplanting date on rice production and quality in Japan. *J. Agric. Meteorol.*, **73**, 156–173, doi:10.2480/agrmet.D-16-00024.

Jaberalansar, Z., M. Tarkesh, M. Bassiri, and S. Pourmanafi, 2017: Modelling the impact of climate change on rangeland forage production using a generalized regression neural network: A case study in Isfahan Province, Central Iran. *J. Arid Land*, **9**, 489–503, doi:10.1007/s40333-017-0058-7.

Jägermeyr, J., D. Gerten, S. Schaphoff, J. Heinke, W. Lucht, and J. Rockström, 2016: Integrated crop water management might sustainably halve the global food gap. *Environ. Res. Lett.*, **11**, 25002, doi:10.1088/1748-9326/11/2/025002.

Jägermeyr, J., A. Pastor, H. Biemans, and D. Gerten, 2017: Reconciling irrigated food production with environmental flows for Sustainable Development Goals implementation. *Nat. Commun.*, **8**, 15900, doi:10.1038/ncomms15900.

Jara-Samaniego, J. et al., 2017: Development of organic fertilizers from food market waste and urban gardening by composting in Ecuador. *PLoS One*, **12**, e0181621, doi:10.1371/journal.pone.0181621.

Jat, M.L., Yadvinder-Singh, G. Gill, H.S. Sidhu, J.P. Aryal, C. Stirling, and B. Gerard, 2015: Laser-assisted precision land leveling impacts in irrigated intensive production systems of South Asia. *Advances in Soil Science*, 323–352.

Jat, M.L. et al., 2016: Climate change and agriculture: Adaptation strategies and mitigation opportunities for food security in South Asia and Latin America. *Advances in Agronomy*, Vol. 137 of, 127–235, doi:10.1016/bs.agron.2015.12.005.

Jat, R.K., T.B. Sapkota, R.G. Singh, M.L. Jat, M. Kumar, and R.K. Gupta, 2014: Seven years of conservation agriculture in a rice–wheat rotation of Eastern Gangetic Plains of South Asia: Yield trends and economic profitability. *F. Crop. Res.*, **164**, 199–210, doi:10.1016/j.fcr.2014.04.015.

Jensen, H.G., L.B. Jacobsen, S.M. Pedersen, and E. Tavella, 2012: Socioeconomic impact of widespread adoption of precision farming and controlled traffic systems in Denmark. *Precis. Agric.*, **13**, 661–677, doi:10.1007/s11119-012-9276-3.

Jessoe, K., D.T. Manning, and J.E. Taylor, 2018: Climate change and labour allocation in rural Mexico: Evidence from annual fluctuations in weather. *Econ. J.*, **128**, 230–261, doi:10.1111/ecoj.12448.

Jha, B., and A. Tripathi, 2017: How susceptible is India's food basket to climate change? *Soc. Change*, **47**, 11–27, doi:10.1177/0049085716681902.

Johansson, D.J.A., and C. Azar, 2007: A scenario based analysis of land competition between food and bioenergy production in the US. *Clim. Change*, **82**, 267–291, doi:10.1007/s10584-006-9208-1.

Jones, A., and B. Hiller, 2017: Exploring the dynamics of responses to food production shocks. *Sustainability*, **9**, 960, doi:10.3390/su9060960.

Jones, A.D., 2017: On-farm crop species richness is associated with household diet diversity and quality in subsistence – and market-oriented farming households in Malawi. *J. Nutr.*, **147**, 86–96, doi:10.3945/jn.116.235879.

Jones, A.D., F.M. Ngure, G. Pelto, and S.L. Young, 2013: What are we assessing when we measure food security? A compendium and review of current metrics. *Am. Soc. Nutr. Adv. Nutr*, 481–505, doi:10.3945/an.113.004119. disciplines.

Jones, A.D., A. Shrinivas, and R. Bezner-Kerr, 2014: Farm production diversity is associated with greater household dietary diversity in Malawi: Findings from nationally representative data. *Food Policy*, **46**, 1–12, doi:10.1016/J.FOODPOL.2014.02.001.

Jost, C. et al., 2016: Understanding gender dimensions of agriculture and climate change in smallholder farming communities. *Clim. Dev.*, **8**, 133–144, doi:10.1080/17565529.2015.1050978.

Jung, S., L.V. Rasmussen, C. Watkins, P. Newton, and A. Agrawal, 2017: Brazil's national environmental registry of rural properties: Implications for livelihoods. *Ecol. Econ.*, **136**, 53–61, doi:10.1016/j.ecolecon.2017.02.004.

Juroszek, P., and A. von Tiedemann, 2013: Climate change and potential future risks through wheat diseases: A review. *Eur. J. Plant Pathol.*, **136**, 21–33, doi:10.1007/s10658-012-0144-9.

Kabir, M., S. Rafiq, S.M.K. Salema, and H. Scheyvens, 2016: *Approaches of MFIs to Disasters and Climate Change Adaptation in Bangladesh*. Institute for Inclusive Finance and Development, Working Paper No. 49, Bangladesh, India, 38 pp.

Kadim, I.T., O. Mahgoub, S. Baqir, B. Faye, and R. Purchas, 2015: Cultured meat from muscle stem cells: A review of challenges and prospects. *J. Integr. Agric.*, **14**, 222–233, doi:10.1016/S2095-3119(14)60881-9.

Kahane, R. et al., 2013: Agrobiodiversity for food security, health and income. *Agron. Sustain. Dev.*, **33**, 671–693, doi:10.1007/s13593-013-0147-8.

Kaijser, A., and A. Kronsell, 2014: Climate change through the lens of intersectionality. *Env. Polit.*, **23**, 417–433, doi:10.1080/09644016.2013.835203.

Kantachote, D., T. Nunkaew, T. Kantha, and S. Chaiprapat, 2016: Biofertilizers from *Rhodopseudomonas palustris* strains to enhance rice yields and reduce methane emissions. *Appl. Soil Ecol.*, **100**, 154–161, doi:10.1016/j.apsoil.2015.12.015.

Kanter, R., and B. Caballero, 2012: Global gender disparities in obesity: A review. *Adv. Nutr.*, **3**, 491–498, doi:10.3945/an.112.002063.

Karimi, V., E. Karami, and M. Keshavarz, 2018: Climate change and agriculture: Impacts and adaptive responses in Iran. *J. Integr. Agric.*, **17**, 1–15, doi:10.1016/S2095-3119(17)61794-5.

Kawanishi, M., B.L. Preston, and N.A. Ridwan, 2016: Evaluation of national adaptation planning: A case study in Indonesia. In: *Climate Change Policies and Challenges in Indonesia* [Kaneko, S., M. Kawanishi (eds.)]. Springer, Tokyo, Japan, pp. 85–107, doi:10.1007/978-4-431-55994-8.

Kehlbacher, A., R. Tiffin, A. Briggs, M. Berners-Lee, and P. Scarborough, 2016: The distributional and nutritional impacts and mitigation potential of emission-based food taxes in the UK. *Clim. Change*, **137**, 121–141, doi:10.1007/s10584-016-1673-6.

Keles, D., J. Choumert-Nkolo, P. Combes Motel, and E. Nazindigouba Kéré, 2018: Does the expansion of biofuels encroach on the forest? *J. For. Econ.*, **33**, 75–82, doi:10.1016/J.JFE.2018.11.001.

Kelley, C., S. Mohtadi, M. Cane, R. Seager, and Y. Kushnir, 2017: Commentary on the Syria case: Climate as a contributing factor. *Polit. Geogr.*, **60**, 245–247, doi:10.1016/j.polgeo.2017.06.013.

Kelley, C.P., S. Mohtadi, M.A. Cane, R. Seager, and Y. Kushnir, 2015: Climate change in the Fertile Crescent and implications of the recent Syrian drought. *Proc. Natl. Acad. Sci.*, **112**, 3241–3246, doi:10.1073/pnas.1421533112.

Kemper, J., 2015: Biomass and carbon dioxide capture and storage: A review. *Int. J. Greenh. Gas Control*, **40**, 401–430, doi:10.1016/j.ijggc.2015.06.012.

Kench, P.S., M.R. Ford, and S.D. Owen, 2018: Patterns of island change and persistence offer alternate adaptation pathways for atoll nations. *Nat. Commun.*, **9**, 605, doi:10.1038/s41467-018-02954-1.

Kerr, R.B., E. Chilanga, H. Nyantakyi-Frimpong, I. Luginaah, and E. Lupafya, 2016: Integrated agriculture programs to address malnutrition in northern Malawi. *BMC Public Health*, **16**, 1197, doi:10.1186/s12889-016-3840-0.

5

Kershen, D.L., 2013: The contested vision for agriculture's future: Sustainable intensive agriculture and agroecology agriculture. *Creighton Law Rev.*, **46**, 591–618.

Kerton, S., and A.J. Sinclair, 2010: Buying local organic food: A pathway to transformative learning. *Agric. Human Values*, **27**, 401–413, doi:10.1007/s10460-009-9233-6.

Ketiem, P., P.M. Makeni, E.K. Maranga, and P.A. Omondi, 2017: Integration of climate change information into drylands crop production practices for enhanced food security: A case study of Lower Tana Basin in Kenya. *African J. Agric. Res.*, **12**, 1763–1771.

Khanal, R., A. Timilsina, P. Pokhrel, and R.K.P. Yadav, 2015: Documenting abundance and use of underutilized plant species in the Mid Hill region of Nepal. *Ecoprint An Int. J. Ecol.*, **21**, 63–71, doi:10.3126/eco.v21i0.11906.

Kiefer, I., T. Rathmanner, and M. Kunze, 2005: Eating and dieting differences in men and women. *J. Men's Heal. Gend.*, **2**, 194–201, doi:10.1016/j.jmhg.2005.04.010.

Kihara, J. et al., 2015: Perspectives on climate effects on agriculture: The international efforts of AgMIP in Sub-Saharan Africa. *In: Handbook of Climate Change and Agroecosystems: The Agricultural Model Intercomparison and Improvement Project Integrated Crop and Economic Assessments – Joint Publication with American Society of Agronomy, Crop Science Society of America, and Soil Science Society of America (In 2 Parts) (Vol. 3).* Imperial College Press, London, UK. [Rosenzweig, C. and D. Hillel (eds.)]. Part 2, pp. 3–23, doi:10.1142/p876.

Kim, K.H., and J. Cho, 2016: Predicting potential epidemics of rice diseases in Korea using multi-model ensembles for assessment of climate change impacts with uncertainty information. *Clim. Change*, **134**, 327–339, doi:10.1007/s10584-015-1503-2.

Kim, K.-H., J. Cho, Y.H. Lee, and W.S. Lee, 2015: Predicting potential epidemics of rice leaf blast and sheath blight in South Korea under the RCP 4.5 and RCP 8.5 climate change scenarios using a rice disease epidemiology model, EPIRICE. *Agric. For. Meteorol.*, **203**, 191–207, doi:10.1016/j.agrformet.2015.01.011.

Kim, T.J., and O. von dem Knesebeck, 2018: Income and obesity: What is the direction of the relationship? A systematic review and meta-analysis. *BMJ Open*, **8**.

Kirezieva, K., L. Jacxsens, M. van Boekel, and P.A. Luning, 2015: Towards strategies to adapt to pressures on safety of fresh produce due to climate change. *Food Res. Int.*, **68**, 94–107, doi:10.1016/j.foodres.2014.05.077.

Klein, A.M., B.E. Vaissiere, J.H. Cane, I. Steffan-Dewenter, S.A. Cunningham, C. Kremen, and T. Tscharntke, 2007: Importance of pollinators in changing landscapes for world crops. *Proc. R. Soc. B-Biological Sci.*, **274**, 303–313, doi:10.1098/rspb.2006.3721.

Klenk, N., A. Fiume, K. Meehan, and C. Gibbes, 2017: Local knowledge in climate adaptation research: Moving knowledge frameworks from extraction to co-production. *Wiley Interdiscip. Rev. Clim. Chang.*, **8**, doi:10.1002/wcc.475.

Köberle, A.C., I. Schmidt Tagomori, J. Portugal-Pereira, and R. Schaeffer, 2016: *Policy Case Study: Evaluation of the Plano ABC in Brazil.* Brazil, 105–133 pp.

Köberle, A.C., J. Portugal-Pereira, B. Cunha, R. Garaffa, A.F. Lucena, A. Szklo, and R. Schaeffer, 2019: Brazilian ethanol expansion subject to limitations. *Nat. Clim. Chang.*, 9, 209–210.

Kole, C. et al., 2015a: Application of genomics-assisted breeding for generation of climate resilient crops: Progress and prospects. *Front. Plant Sci.*, **6**, 1–16, doi:10.3389/fpls.2015.00563.

Kole, C. et al., 2015b: Application of genomics-assisted breeding for generation of climate resilient crops: Progress and prospects. *Front. Plant Sci.*, **6**, 1–16, doi:10.3389/fpls.2015.00563.

Koohafkan, P., and M.A. Altieri, 2011: *Globally important agricultural heritage systems: A legacy for the future.* Food and Agriculture Organization of the United Nations, Rome, Italy, 1–49 pp.

Korbeľová, L., and S. Kohnová, 2017: Methods for improvement of the ecosystem services of soil by sustainable land management in the Myjava River Basin. *Slovak J. Civ. Eng.*, **25**, 29–36, doi:10.1515/sjce-2017-0005.

Kraxner, F. et al., 2014: BECCS in South Korea – Analyzing the negative emissions potential of bioenergy as a mitigation tool. *Renew. Energy*, **61**, 102–108, doi:10.1016/j.renene.2012.09.064.

Kremen, C., A. Iles, and C. Bacon, 2012: Diversified farming systems: An agroecological, systems-based alternative to modern industrial agriculture. *Ecol. Soc.*, **17**, art44, doi:10.5751/ES-05103-170444.

Kriewald, S., P. Pradhan, L. Costa, R.A. Cantu, and J. Kropp, 2019: Hungry cities: How local food self-sufficiency relates to climate change, life styles and demographic development. *Environ. Res. Lett.*, **14**, 094007.

Krishnapillai, M., 2018: Enhancing adaptive capacity and climate change resilience of coastal communities in Yap. In: *Climate Change Impacts and Adaptation Strategies for Coastal Communities* [Filho, W.L. (ed.)]. Springer International Publishing AG, Cham, Switzerland, pp. 87–118.

Krishnapillai, M., and R. Gavenda, 2014: *From Barren Land to Biodiverse Home Gardens.* ILEIA, Wageningen, The Netherlands, pp 26–28.

Krishnapillai, M. V, 2017: Climate-friendly adaptation strategies for the displaced Atoll population in Yap. In: *Climate Change Adaptation in Pacific Countries: Fostering Resilience and Improving the Quality of Life* [Filho, W.L., (ed.)]. Springer International Publishing, Cham, Switzerland, 101–117, doi:10.1007/978-3-319-50094-2_6.

Kucich, D.A., and M.M. Wicht, 2016: South African indigenous fruits – Underutilized resource for boosting daily antioxidant intake among local indigent populations? *South African J. Clin. Nutr.*, **29**, 150–156, doi:10.1080/16070658.2016.1219470.

Kumar, N., and A.R. Quisumbing, 2013: Gendered impacts of the 2007–2008 food price crisis: Evidence using panel data from rural Ethiopia. *Food Policy*, **38**, 11–22, doi:10.1016/J.FOODPOL.2012.10.002.

Kumar, N., J. Harris, and R. Rawat, 2015: If they grow it, will they eat and grow? Evidence from Zambia on agricultural diversity and child undernutrition. *J. Dev. Stud.*, **51**, 1060–1077, doi:10.1080/00220388.2015.1018901.

Kumar, R., V. Mishra, J. Buzan, R. Kumar, D. Shindell, and M. Huber, 2017a: Dominant control of agriculture and irrigation on urban heat island in India. *Sci. Rep.*, **7**, 14054, doi:10.1038/s41598-017-14213-2.

Kumar, Y., R. Berwal, A. Pandey, A. Sharma, and V. Sharma, 2017b: Hydroponics meat: An envisaging boon for sustainable meat production through biotechnological approach – A review. *Int. J. Vet. Sci. Anim. Husb.*, **2**, 34–39.

Läderach, P., J.R. Villegas, C. Navarro-racines, C. Zelaya, A.M. Valle, and A. Jarvis, 2017: Climate change adaptation of coffee production in space and time. *Clim. Change*, **141**, 47–62, doi:10.1007/s10584-016-1788-9.

Lal, R., 2004: Soil carbon sequestration impacts on global climate change and food security. *Science*, **304**, 1623–1627.

Lal, S. et al., 2017: Loss of crop yields in India due to surface ozone: An estimation based on a network of observations. *Environ. Sci. Pollut. Res.*, **24**, 20972–20981, doi:10.1007/s11356-017-9729-3.

Lambin, E.F. et al., 2013: Estimating the world's potentially available cropland using a bottom-up approach. *Glob. Environ. Chang.*, **23**, 892–901, doi:10.1016/j.gloenvcha.2013.05.005.

Lambin, E.F. et al., 2018: The role of supply-chain initiatives in reducing deforestation. *Nat. Clim. Chang.*, **8**, 109–116, doi:10.1038/s41558-017-0061-1.

Lamers, P., C. Hamelinck, M. Junginger, and A. Faaij, 2011: International bioenergy trade – A review of past developments in the liquid biofuel market. *Renew. Sustain. Energy Rev.*, **15**, 2655–2676, doi:10.1016/J.RSER.2011.01.022.

Lamichhane, J.R. et al., 2015: Robust cropping systems to tackle pests under climate change. A review. *Agron. Sustain. Dev.*, **35**, 443–459, doi:10.1007/s13593-014-0275-9.

Lampkin, N. et al., 2015: The Role of Agroecology in Sustainable Intensification. *Report for the Land Use Policy Group.* Organic Research Centre, Elm Farm and Game & Wildlife Conservation Trust, http://orgprints.org/33067/.

Landholm, D.M., P. Pradhan, and J.P. Kropp, 2019: Diverging forest land use dynamics induced by armed conflict across the tropics. *Glob. Environ. Chang.*, **56**, 86–94, doi:10.1016/j.gloenvcha.2019.03.006.

Langer, A. et al., 2015: Women and Health: The key for sustainable development. *Lancet*, London, UK, **386**, 1165–1210, doi:10.1016/S0140-6736(15)60497-4.

Lara, L., and M. Rostagno, 2013: Impact of Heat Stress on Poultry Production. *Animals*, **3**, 356, doi:10.3390/ani3020356.

Larkin, A., J. Kuriakose, M. Sharmina, and K. Anderson, 2018: What if negative emission technologies fail at scale? Implications of the Paris Agreement for big emitting nations. *Clim. Policy*, **18**, 690–714, doi:10.1080/146930 62.2017.1346498.

Latham, A.D.M., M.C. Latham, E. Cieraad, D.M. Tompkins, and B. Warburton, 2015: Climate change turns up the heat on vertebrate pest control. *Biol. Invasions*, **17**, 2821–2829, doi:10.1007/s10530-015-0931-2.

Laue, J.E., and E.Y. Arima, 2016: Spatially explicit models of land abandonment in the Amazon. *J. Land Use Sci.*, **11**, 48–75, doi:10.1080/1 747423X.2014.993341.

Lebel, L., P. Lebel, and B. Lebel, 2014: Gender and the management of climate-related risks in Northern Thailand. *Int. Soc. Sci. J.*, **65**, 147–158, doi:10.1111/issj.12090.

Lehmann, J., J. Gaunt, and M. Rondon, 2006: Bio-char Sequestration in Terrestrial Ecosystems – A Review. *Mitig. Adapt. Strateg. Glob. Chang.*, **11**, 403–427, doi:10.1007/s11027-005-9006-5.

Lehmann, J., Y. Kuzyakov, G. Pan, and Y.S. Ok, 2015: Biochars and the plant-soil interface. *Plant Soil*, **395**, 1–5, doi:10.1007/s11104-015-2658-3.

Lenton, T.M., 2010: The potential for land-based biological CO_2 removal to lower future atmospheric CO_2 concentration. *Carbon Manag.*, **1**, 145–160, doi:10.4155/cmt.10.12.

Lentz, E., C.B. Barrett, and J. Hoddinott, 2005: *Food Aid and Dependency: Implications for Emergency Food Security Assessments*. IFPRI Discussion Paper No. 12–2, SSRN, New York, US, 1–50 pp.

Lenzi, D., W.F. Lamb, J. Hilaire, M. Kowarsch, and J.C. Minx, 2018: Don't deploy negative emissions technologies without ethical analysis. *Nature*, **561**, 303–305, doi:10.1038/d41586-018-06695-5.

Leon, E.G. de, and J. Pittock, 2016: Integrating climate change adaptation and climate-related disaster risk-reduction policy in developing countries: A case study in the Philippines. *Clim. Dev.*, **5529**, doi:10.1080/17565529.2016.1174659.

Levidow, L., M. Pimbert, and G. Vanloqueren, 2014: Agroecological Research: Conforming – Or transforming the dominant agro-food regime? *Agroecol. Sustain. Food Syst.*, **38**, 1127–1155, doi:10.1080/21683565.2014.951459.

Levis, S., A. Badger, B. Drewniak, C. Nevison, and X. Ren, 2018: CLMcrop yields and water requirements: Avoided impacts by choosing RCP 4.5 over 8.5. *Clim. Change*, **146**, 501–515, doi:10.1007/s10584-016-1654-9.

Li, H., T. Wu, X. Wang, and Y. Qi, 2016: The greenhouse gas footprint of China's food system: An analysis of recent trends and future scenarios. *J. Ind. Ecol.*, **20**, 803–817, doi:10.1111/jiec.12323.

Lickley, M., and S. Solomon, 2018: Drivers, timing and some impacts of global aridity change. *Environ. Res. Lett.*, **13**, 104010, doi:10.1088/1748-9326/aae013.

Lin, B.B., 2011: Resilience in Agriculture through crop diversification: Adaptive management for environmental change. *Bioscience*, **61**, 183–193, doi:10.1525/bio.2011.61.3.4.

Lin, B.B., S.M. Philpott, and S. Jha, 2015: The future of urban agriculture and biodiversity-ecosystem services: Challenges and next steps. *Basic Appl. Ecol.*, **16**, 189–201, doi:10.1016/j.baae.2015.01.005.

Lindblom, J., C. Lundström, M. Ljung, and A. Jonsson, 2017: Promoting sustainable intensification in precision agriculture: Review of decision-support systems development and strategies. *Precis. Agric.*, **18**, 309–331, doi:10.1007/s11119-016-9491-4.

Linnerooth-bayer, J., and S. Hochrainer-stigler, 2014: Financial instruments for disaster risk management and climate change adaptation. *Clim. Change*, doi:10.1007/s10584-013-1035-6.

Linnerooth-Bayer, J., and S. Hochrainer-Stigler, 2015: Financial instruments for disaster risk management and climate change adaptation. *Clim. Change*, **133**, 85–100, doi:10.1007/s10584-013-1035-6.

Lipinski, B., C. Hanson, R. Waite, T. Searchinger, J. Lomax, and L. Kitinoja, 2013: *Reducing Food Loss and Waste: Creating a Sustainable Food Future, Installment Two*. World Resources Institute, Washington, DC, USA, 1–40 pp.

Lipper, L. et al., 2014: Climate-smart agriculture for food security. *Nat. Clim. Chang.*, **4**, doi:10.1038/nclimate2437.

Liu, B. et al., 2016: Similar estimates of temperature impacts on global wheat yield by three independent methods. *Nat. Clim. Chang.*, **6**, 1130–1136, doi:10.1038/nclimate3115.

Liu, C., N. Hofstra, and E. Franz, 2013: Impacts of climate change on the microbial safety of pre-harvest leafy green vegetables as indicated by Escherichia coli O157 and Salmonella spp. *Int. J. Food Microbiol.*, **163**, 119–128, doi:10.1016/j.ijfoodmicro.2013.02.026.

Liu, J. et al., 2017: Challenges in operationalizing the water–energy–food nexus. *Hydrol. Sci. J.*, **62**, 1714–1720, doi:10.1080/02626667.2017.1353695.

Liu, N., Q. Zhu, X. Zeng, B. Yang, and M. Liang, 2018a: Optimization of processing conditions of Chinese smoke-cured bacon (Larou) with a new natural coating solution during storage period. **38**, 636–652, doi:10.5851/kosfa.2018.38.3.636.

Liu, Y., Q. Chen, Q. Ge, J. Dai, Y. Qin, L. Dai, X. Zou, and J. Chen, 2018b: Modelling the impacts of climate change and crop management on phenological trends of spring and winter wheat in China. *Agric. For. Meteorol.*, **248**, 518–526, doi:10.1016/J.AGRFORMET.2017.09.008.

Liu, Z., P. Yang, H. Tang, W. Wu, L. Zhang, Q. Yu, and Z. Li, 2014: Shifts in the extent and location of rice cropping areas match the climate change pattern in China during 1980–2010. *Reg. Environ. Chang.*, **15**, 919–929, doi:10.1007/s10113-014-0677-x.

Llonch, P., M.J. Haskell, R.J. Dewhurst, and S.P. Turner, 2017: Current available strategies to mitigate greenhouse gas emissions in livestock systems: An animal welfare perspective. *animal*, **11**, 274–284, doi:10.1017/S1751731116001440.

Lloyd's, 2015: *Food System Shock: The Insurance Impacts of Acute Disruption to Global Food Supply*. Lloyd's Emerging Risk Report – 2015, London, UK, www.lloyds.com/news-and-risk-insight/risk-reports/library/society-and-security/food-system-shock.

Lobell, D., and M. Burke, 2010: *Climate Change and Food Security: Adapting Agriculture to a Warmer World*. Springer, Netherlands, Dordrecht, 197 pp.

Lobell, D.B., and S. Asseng, 2017: Comparing estimates of climate change impacts from process-based and statistical crop models. *Environ. Res. Lett.*, **12**, 15001, doi:10.1088/1748-9326/aa518a.

Loconto, A.M., A. Jimenez, E. Vandecandelaere, and F. Tartanac, 2018: Agroecology, local food systems and their markets. *AGER J. Depopulation Rural Dev. Stud.*, **25**, 13–42, doi:10.4422/AGER.2018.15.

Loladze, I., 2014: Hidden shift of the ionome of plants exposed to elevated CO_2 depletes minerals at the base of human nutrition. *Elife*, **2014**, doi:10.7554/eLife.02245.

Long, T.B., V. Blok, and I. Coninx, 2016: Barriers to the adoption and diffusion of technological innovations for climate-smart agriculture in Europe: Evidence from the Netherlands, France, Switzerland and Italy. *J. Clean. Prod.*, **112**, 9–21, doi:10.1016/j.jclepro.2015.06.044.

Lopes, M.S. et al., 2015: Exploiting genetic diversity from landraces in wheat breeding for adaptation to climate change. *J. Exp. Bot.*, **66**, 3477–3486, doi:10.1093/jxb/erv122.

Lopez-i-Gelats, F., 2014: Impacts of Climate Change on Food Availability: Livestock. In: *Global Environmental Change* [Brimblecombe, P., R. Lal, R. Stanley, J. Trevors (eds.)]. Springer Netherlands, Dordrecht, Netherlands, pp. 689–694.

López-i-Gelats, F., E.D.G. Fraser, J.F. Morton, and M.G. Rivera-Ferre, 2016: What drives the vulnerability of pastoralists to global environmental change? A qualitative meta-analysis. *Glob. Environ. Chang.*, **39**, 258–274, doi:10.1016/J.GLOENVCHA.2016.05.011.

Lorenz, S., 2017: Adaptation planning and the use of climate change projections in local government in England and Germany. *Reg. Environ. Chang.*, **17**, 425–435, doi:10.1007/s10113-016-1030-3.

Lotze-Campen, H. et al., 2014: Impacts of increased bioenergy demand on global food markets: An AgMIP economic model intercomparison. *Agric. Econ.*, **45**, 103–116, doi:10.1111/agec.12092.

Lozano, R. et al., 2018: Measuring progress from 1990 to 2017 and projecting attainment to 2030 of the health-related Sustainable Development Goals for 195 countries and territories: A systematic analysis for the Global Burden of Disease Study 2017. The Lancet, **392**, 2091–2138, doi:10.1016/S0140-6736(18)32281-5.

Lusseau, D., and F. Mancini, 2019: Income-based variation in Sustainable Development Goal interaction networks. *Nat. Sustain.*, **2**, 242–247, doi:10.1038/s41893-019-0231-4.

Lwasa, S., F. Mugagga, B. Wahab, D. Simon, J. Connors, and C. Griffith, 2014: Urban and peri-urban agriculture and forestry: Transcending poverty alleviation to climate change mitigation and adaptation. *Urban Clim.*, **7**, 92–106, doi:10.1016/j.uclim.2013.10.007.

Lwasa, S., F. Mugagga, B. Wahab, D. Simon, J.P. Connors, and C. Griffith, 2015: A meta-analysis of urban and peri-urban agriculture and forestry in mediating climate change. *Curr. Opin. Environ. Sustain.*, **13**, 68–73, doi:10.1016/j.cosust.2015.02.003.

Lynch, J. and R. Pierrehumbert, 2019: Climate impacts of cultured meat and beef cattle. *Frontiers in Sustainable Food Systems*, **3**, p. 5, doi 10.3389/fsufs.2019.00005

Macdiarmid, J.I., J. Kyle, G.W. Horgan, J. Loe, C. Fyfe, A. Johnstone, and G. McNeill, 2012: Sustainable diets for the future: Can we contribute to reducing greenhouse gas emissions by eating a healthy diet? *Am. J. Clin. Nutr.*, **96**, 632–639, doi:10.3945/ajcn.112.038729.

MacDonald, G.K., K.A. Brauman, S. Sun, K.M. Carlson, E.S. Cassidy, J.S. Gerber, and P.C. West, 2015: Rethinking agricultural trade relationships in an era of globalization. *Bioscience*, **65**, 275–289, doi:10.1093/biosci/biu225.

Machovina, B., K.J. Feeley, and W.J. Ripple, 2015: Biodiversity conservation: The key is reducing meat consumption. *Sci. Total Environ.*, **536**, 419–431, doi:10.1016/j.scitotenv.2015.07.022.

Magdoff, F., 2007: Ecological agriculture: Principles, practices, and constraints. *Renew. Agric. Food Syst.*, **22**, 109–117, doi:10.1017/S1742170507001846.

Maia, S.M.F., S.M. Ogle, C.E.P. Cerri, and C.C. Cerri, 2009: Effect of grassland management on soil carbon sequestration in Rondônia and Mato Grosso states, Brazil. *Geoderma*, **149**, 84–91, doi:10.1016/J.GEODERMA.2008.11.023.

Mairie de Paris, 2015: *Sustainable Food Plan 2015–2020*. Direction des espaces verts et de l'environnement, Paris, France, 29 pp, https://api-site-cdn.paris.fr/images/76336.

Makate, C., R. Wang, M. Makate, and N. Mango, 2016: Crop diversification and livelihoods of smallholder farmers in Zimbabwe: Adaptive management for environmental change. *Springerplus*, **5**, 1135, doi:10.1186/s40064-016-2802-4.

Malapit, H.J.L., S. Kadiyala, A.R. Quisumbing, K. Cunningham, and P. Tyagi, 2015: Women's empowerment mitigates the negative effects of low production diversity on maternal and child nutrition in Nepal. *J. Dev. Stud.*, **51**, 1097–1123, doi:10.1080/00220388.2015.1018904.

Mallick, J., A. Rahman, and C.K. Singh, 2013: Modeling urban heat islands in heterogeneous land surface and its correlation with impervious surface area by using night-time ASTER satellite data in highly urbanizing city, Delhi-India. *Adv. Sp. Res.*, **52**, 639–655, doi:10.1016/J.ASR.2013.04.025.

Mann, M.E., S. Rahmstorf, K. Kornhuber, B.A. Steinman, S.K. Miller, and D. Coumou, 2017: Influence of anthropogenic climate change on planetary wave resonance and extreme weather events. *Sci. Rep.*, **7**, 45242, doi:10.1038/srep45242.

Manzoor, M., S. Bibi, M. Manzoor, and R. Jabeen, 2013: Historical analysis of flood information and impacts assessment and associated response in Pakistan (1947–2011). *Res. J. Environ. Earth Sci.*, **5**, 139–146.

Mao, D., F. Wang, Z. Hao, and H. Li, 2018: Credit evaluation system based on blockchain for multiple stakeholders in the food supply chain. *Int. J. Environ. Res. Public Health*, **15**, 1627, doi:10.3390/ijerph15081627.

Mapfumo, P. et al., 2017: Pathways to transformational change in the face of climate impacts: An analytical framework. *Clim. Dev.*, **9**, 439–451, doi:10.1080/17565529.2015.1040365.

Marchand, P. et al., 2016: Reserves and trade jointly determine exposure to food supply shocks. *Environ. Res. Lett.*, **11**, 95009, doi:10.1088/1748-9326/11/9/095009.

Marianela, F. et al., 2016: Past and present biophysical redundancy of countries as a buffer to changes in food supply. *Environ. Res. Lett.*, **11**, 55008, doi:10.1088/1748-9326/11/5/055008.

Marino, E., and H. Lazrus, 2015: Migration or forced displacement?: The complex choices of climate change and disaster migrants in Shishmaref, Alaska and Nanumea, Tuvalu. *Hum. Organ.*, **74**, 341–350, doi:10.17730/0018-7259-74.4.341.

Martha, G.B., E. Alves, and E. Contini, 2012: Land-saving approaches and beef production growth in Brazil. *Agric. Syst.*, **110**, 173–177, doi:10.1016/J.AGSY.2012.03.001.

Marti, D.L., R. Johnson, and K.H. Mathews, 2011: *Where's the (Not) Meat? By-products From Beef and Pork Production*. Livestock, Dairy, and Poultry Outlook No. (LDPM-209-01), United States Department of Agriculture, Economic Research Service, USA, 30 pp.

Martin, R., B. Müller, A. Linstädter, and K. Frank, 2014: How much climate change can pastoral livelihoods tolerate? Modelling rangeland use and evaluating risk. *Glob. Environ. Chang.*, **24**, 183–192, doi:10.1016/J.GLOENVCHA.2013.09.009.

Martin, R., A. Linstädter, K. Frank, and B. Müller, 2016: Livelihood security in face of drought – Assessing the vulnerability of pastoral households. *Environ. Model. Softw.*, **75**, 414–423, doi:10.1016/j.envsoft.2014.10.012.

Mary, S., S. Saravia-Matus, and S. Gomez y Paloma, 2018: Does nutrition-sensitive aid reduce the prevalence of undernourishment? *Food Policy*, **74**, 100–116, doi:10.1016/j.foodpol.2017.11.008.

Masikati, P. et al., 2015: Crop-livestock intensification in the face of climate change: Exploring opportunities to reduce risk and increase resilience in southern Africa by using an integrated multi-modeling approach. In: *Handbook of Climate Change and Agroecosystems: The Agricultural Model Intercomparison and Improvement Project Integrated Crop and Economic Assessments – Joint Publication with American Society of Agronomy, Crop Science Society of America, and Soil Science Society of America (In 2 Parts) (Vol. 3)*. Imperial College Press, London, UK. [Rosenzweig, C. and D. Hillel (eds.)]. Part 2, pp. 159–198.

Mason, R., J.R. Parkins, and A. Kaler, 2017: Gendered mobilities and food security: Exploring possibilities for human movement within hunger prone rural Tanzania. *Agric. Human Values*, **34**, 423–434, doi:10.1007/s10460-016-9723-2.

Mathbout, S., J.A. Lopez-Bustins, J. Martin-Vide, J. Bech, and F.S. Rodrigo, 2018: Spatial and temporal analysis of drought variability at several time scales in Syria during 1961–2012. *Atmos. Res.*, **200**, 153–168, doi:10.1016/J.ATMOSRES.2017.09.016.

Mattick, C.S., A.E. Landis, B.R. Allenby, and N.J. Genovese, 2015: Anticipatory lifecycle analysis of in vitro biomass cultivation for cultured meat production in the United States. *Environ. Sci. Technol.*, **49**, 11941–11949, doi:10.1021/acs.est.5b01614.

Matzenberger, J., L. Kranzl, E. Tromborg, M. Junginger, V. Daioglou, C. Sheng Goh, and K. Keramidas, 2015: Future perspectives of international bioenergy trade. *Renew. Sustain. Energy Rev.*, **43**, 926–941, doi:10.1016/J.RSER.2014.10.106.

Mavromatis, T., 2015: Crop–climate relationships of cereals in Greece and the impacts of recent climate trends. *Theor. Appl. Climatol.*, **120**, 417–432, doi:10.1007/s00704-014-1179-y.

Mayer, A., Z. Hausfather, A.D. Jones, and W.L. Silver, 2018: The potential of agricultural land management to contribute to lower global surface temperatures. *Sci. Adv.*, **4**, 1–9, doi:10.1126/sciadv.aaq0932.

Maystadt, J.-F., and O. Ecker, 2014: Extreme weather and civil war: Does drought fuel conflict in Somalia through livestock price shocks? *Am. J. Agric. Econ.*, **96**, 1157–1182.

Mbow, C., O. Mertz, A. Diouf, K. Rasmussen, and A. Reenberg, 2008: The history of environmental change and adaptation in eastern Saloum–Senegal – Driving forces and perceptions. *Glob. Planet. Change*, **64**, 210–221, doi:10.1016/J.GLOPLACHA.2008.09.008.

Mbow, C., M. Van Noordwijk, E. Luedeling, H. Neufeldt, P.A. Minang, and G. Kowero, 2014a: Agroforestry solutions to address food security and climate change challenges in Africa. *Curr. Opin. Environ. Sustain.*, **6**, 61–67, doi:10.1016/j.cosust.2013.10.014.

Mbow, C., P. Smith, D. Skole, L. Duguma, and M. Bustamante, 2014b: Achieving mitigation and adaptation to climate change through sustainable agroforestry practices in Africa. *Curr. Opin. Environ. Sustain.*, **6**, 8–14, doi:10.1016/J.COSUST.2013.09.002.

McClanahan, T., E.H. Allison, and J.E. Cinner, 2015: Managing fisheries for human and food security. *Fish Fish.*, **16**, 78–103, doi:10.1111/faf.12045.

McDermid, S.P. et al., 2015: Integrated assessments of the impact of climate change on agriculture: An overview of AgMIP regional research in South Asia. In: *Handbook of Climate Change and Agroecosystems: The Agricultural Model Intercomparison and Improvement Project Integrated Crop and Economic Assessments – Joint Publication with American Society of Agronomy, Crop Science Society of America, and Soil Science Society of America (In 2 Parts) (Vol. 3)*. Imperial College Press, London, UK. [Rosenzweig, C. and D. Hillel (eds.)]. Part 2, pp. 201–217.

McGahey, D., J. Davies, N. Hagelberg, and R. Ouedraogo, 2014: *Pastoralism and the green economy: A natural nexus? – Status, challenges and policy implications*. IUCN and UNEP, Nairobi, Kenya, 1–72 pp.

McKinsey Global Institute, 2010: *Lions on the move: The progress and potential of African economies The McKinsey Global Institute*. McKinsey Global Institute, New York, USA, 82 pp.

McKune, S.L., and J.A. Silva, 2013: Pastoralists under pressure: Double exposure to economic and environmental change in Niger. *J. Dev. Stud.*, **49**, 1711–1727, doi:10.1080/00220388.2013.822067.

McKune, S.L.E.C. Borresen, A.G. Young, T.D. Auria Ryley, S.L. Russo, A. Diao Camara, M. Coleman, and E.P. Ryan, 2015: Climate change through a gendered lens: Examining livestock holder food security. *Glob. Food Sec.*, **6**, 1–8, doi:10.1016/J.GFS.2015.05.001.

McMichael, P., 2014: Historicizing food sovereignty. *J. Peasant Stud.*, **41**, 933–957, doi:10.1080/03066150.2013.876999.

McPhearson, T. et al., 2018: Urban ecosystems and biodiversity. In: *Climate Change and Cities: Second Assessment Report of the Urban Climate Change Research Network* [Rosenzweig, C., W. Solecki, P. Romero-Lankao, S. Mehrotra, S. Dhakal, and S. Ali Ibrahim (eds.)]. Cambridge University Press, New York, NY, pp. 257–318.

Medek, D.E., J. Schwartz, and S.S. Myers, 2017: Estimated effects of future atmospheric CO_2 concentrations on protein intake and the risk of protein deficiency by country and region. *Environ. Health Perspect.*, **125**, doi:10.1289/ehp41.

Medina, A., A. Rodriguez, and N. Magan, 2014: Effect of climate change on Aspergillus flavus and aflatoxin B1 production. *Front. Microbiol.*, **5**, 348, doi:10.3389/fmicb.2014.00348.

Megersa, B., A. Markemann, A. Angassa, J.O. Ogutu, H.-P. Piepho, and A. Valle Zaráte, 2014: Impacts of climate change and variability on cattle production in southern Ethiopia: Perceptions and empirical evidence. *Agric. Syst.*, **130**, 23–34, doi:10.1016/J.AGSY.2014.06.002.

Megido, R.C., C. Gierts, C. Blecker, Y. Brostaux, É. Haubruge, T. Alabi, and F. Francis, 2016: Consumer acceptance of insect-based alternative meat products in Western countries. *Food Qual. Prefer.*, **52**, 237–243, doi:10.1016/J.FOODQUAL.2016.05.004.

van Meijl, H. et al., 2017: *Challenges of Global Agriculture in a Climate Change Context by 2050*. JRC Science for Policy Report, Publications Office of the European Union, Luxembourg, 64 pp, doi:10.2760/772445.

Memmott, J., P.G. Craze, N.M. Waser, and M. V Price, 2007: Global warming and the disruption of plant-pollinator interactions. *Ecol. Lett.*, **10**, 710–717, doi:10.1111/j.1461-0248.2007.01061.x.

Mendelsohn, R., A. Basist, P. Kurukulasuriya, and A. Dinar, 2007: Climate and rural income. *Clim. Change*, **81**, 101–118, doi:10.1007/s10584-005-9010-5.

Menéndez, E., C.M. Bacon, and R. Cohen, 2013: Agroecology as a transdisciplinary, participatory, and action-oriented approach. *Agroecol. Sustain. Food Syst.*, **37**, 3–18.

Menzel, A. et al., 2006: European phenological response to climate change matches the warming pattern. *Glob. Chang. Biol.*, **12**, 1969–1976, doi:10.1111/j.1365-2486.2006.01193.x.

van der Merwe, J.D., P.C. Cloete, and M. van der Hoeven, 2016: Promoting food security through indigenous and traditional food crops. *Agroecol. Sustain. Food Syst.*, **40**, 830–847, doi:10.1080/21683565.2016.1159642.

Meyfroidt, P., 2017: Trade-offs between environment and livelihoods: Bridging the global land use and food security discussions. *Elsevier*, **16**, 9–16, doi:10.1016/j.gfs.2017.08.001.

Michael, Y.G., 2017: Vulnerability and local innovation in adaptation to climate change among the pastoralists: Harshin District, Somali Region, Ethiopia. *Environ. Manag. Sustain. Dev.*, **6**, doi:10.5296/emsd.v6i2.11211.

Michalský, M., and P.S. Hooda, 2015: Greenhouse gas emissions of imported and locally produced fruit and vegetable commodities: A quantitative assessment. *Environ. Sci. Policy*, **48**, 32–43, doi:10.1016/j.envsci.2014.12.018.

Mickelbart, M.V., P.M. Hasegawa, and J. Bailey-Serres, 2015: Genetic mechanisms of abiotic stress tolerance that translate to crop yield stability. *Nat. Publ. Gr.*, **16**, 237–251, doi:10.1038/nrg3901.

Van Mierlo, K., S. Rohmer, and J.C. Gerdessen, 2017: A model for composing meat replacers: Reducing the environmental impact of our food consumption pattern while retaining its nutritional value. *J. Clean. Prod.*, **165**, 930–950, doi:10.1016/j.jclepro.2017.07.098.

Miles, A., M.S. DeLonge, and L. Carlisle, 2017: Triggering a positive research and policy feedback cycle to support a transition to agroecology and sustainable food systems. *Agroecol. Sustain. Food Syst.*, **41**, 855–879, doi: 10.1080/21683565.2017.1331179.

Milestad, R., L. Westberg, U. Geber, and J. Björklund, 2010: Enhancing adaptive capacity in food systems: Learning at farmers' markets in Sweden. *Ecol. Soc.*, **15**, 18, doi:10.5751/ES-03543-150329.

Milner, J., R. Green, A.D. Dangour, A. Haines, Z. Chalabi, J. Spadaro, A. Markandya, and P. Wilkinson, 2015: Health effects of adopting low greenhouse gas emission diets in the UK. *BMJ Open*, **5**, e007364–e007364, doi:10.1136/bmjopen-2014-007364.

Minang, P.A., M. Van Noordwijk, O.E. Freeman, C. Mbow, J. de Leeuw, and D. Catacutan, 2015: *Climate-Smart Landscapes: Multifunctionality in Practice*. World Agroforestry Centre, Nairobi, Kenya, Africa, 404 pp.

Minx, J.C. et al., 2018a: Negative emissions – Part 1: Research landscape and synthesis. *Environ. Res. Lett.*, **13**, 63001, doi:10.1088/1748-9326/aabf9b.

Minx, J.C. et al., 2018b: Negative emissions – Part 1: Research landscape and synthesis. *Environ. Res. Lett.*, **13**, 63001, doi:10.1088/1748-9326/aabf9b.

Mira de Orduña, R., 2010: Climate change associated effects on grape and wine quality and production. *Food Res. Int.*, **43**, 1844–1855, doi:10.1016/J.FOODRES.2010.05.001.

Miraglia, M. et al., 2009: Climate change and food safety: An emerging issue with special focus on Europe. *Food Chem. Toxicol.*, **47**, 1009–1021, doi:10.1016/J.FCT.2009.02.005.

5

Mishra, A.K., and S.B. Agrawal, 2014: Cultivar specific response of CO_2 fertilization on two tropical mung bean (Vigna radiata L.) cultivars: ROS Generation, antioxidant status, physiology, growth, yield and seed quality. *J. Agron. Crop Sci.*, **200**, 273–289, doi:10.1111/jac.12057.

Mitchell, J.P. et al., 2017: Cover cropping and no-tillage improve soil health in an arid irrigated cropping system in California's San Joaquin Valley, USA. *Soil Tillage Res.*, **165**, 325–335, doi:10.1016/j.still.2016.09.001.

Moat, J., J. Williams, S. Baena, T. Wilkinson, T.W. Gole, Z.K. Challa, S. Demissew, and A.P. Davis, 2017: Resilience potential of the Ethiopian coffee sector under climate change. *Nat. Plants*, **17081**, doi:10.1038/nplants.2017.81.

Mockshell, J., and J. Kamanda, 2018: Beyond the agroecological and sustainable agricultural intensification debate: Is blended sustainability the way forward? *Int. J. Agric. Sustain.*, **16**, 127–149, doi:10.1080/1473 5903.2018.1448047.

MOFPI, 2019: *Cold Chain*. Ministry of Food Processing Industries, GOI, New Delhi, India. http://mofpi.nic.in/Schemes/cold-chain.

Mogensen, L., T.L.T. Nguyen, N.T. Madsen, O. Pontoppidan, T. Preda, and J.E. Hermansen, 2016: Environmental impact of beef sourced from different production systems – Focus on the slaughtering stage: Input and output. *J. Clean. Prod.*, **133**, 284–293, doi:10.1016/J.JCLEPRO.2016.05.105.

Mok, H.F., V.G. Williamson, J.R. Grove, K. Burry, S.F. Barker, and A.J. Hamilton, 2014: Strawberry fields forever? Urban agriculture in developed countries: A review. *Agron. Sustain. Dev.*, **34**, 21–43, doi:10.1007/s13593-013-0156-7.

Moley, K.H., and G.A. Colditz, 2016: Effects of obesity on hormonally driven cancer in women. *Sci. Transl. Med.*, **8**, 323ps3, doi:10.1126/scitranslmed. aad8842.

Mondal, P., and M. Basu, 2009: Adoption of precision agriculture technologies in India and in some developing countries: Scope, present status and strategies. *Prog. Nat. Sci.*, **19**, 659–666, doi:10.1016/J.PNSC.2008.07.020.

Moniruzzaman, S., 2015: Crop choice as climate change adaptation: Evidence from Bangladesh. *Ecol. Econ.*, **118**, 90–98, doi:10.1016/j. ecolecon.2015.07.012.

Montagnini, F., and R. Metzel, 2017: *The Contribution of Agroforestry to Sustainable Development Goal 2: End Hunger, Achieve Food Security and Improved Nutrition, and Promote Sustainable Agriculture*. Springer International Publishing, Cham, Switzerland, 11–45 pp.

Monterroso, A., C. Conde, C. Gay, D. Gómez, and J. López, 2014: Two methods to assess vulnerability to climate change in the Mexican agricultural sector. *Mitig. Adapt. Strateg. Glob. Chang.*, **19**, 445–461, doi:10.1007/s11027-012-9442-y.

Moore, F.C., and D.B. Lobell, 2015: The fingerprint of climate trends on European crop yields. *Proc. Natl. Acad. Sci.*, **9**, 2670–2675, 201409606, doi:10.1073/pnas.1409606112.

Moragues, A. et al., 2013: *Urban Food Strategies. The rough guide to sustainable food systems*. Document developed in the framework of the FP7 project FOODLINKS (GA No. 265287), Research Institute of Organic Agriculture (FiBL), Frick, Switzerland, 26 pp.

Morrison-Whittle, P., S.A. Lee, and M.R. Goddard, 2017: Fungal communities are differentially affected by conventional and biodynamic agricultural management approaches in vineyard ecosystems. *Agric. Ecosyst. Environ.*, **246**, 306–313, doi:10.1016/j.agee.2017.05.022.

Morton, J., 2017: Climate change and African agriculture: Unlocking the potential of research and advisory services. In: *Making Climate Compatible Development Happen* [Nunan, F., (ed.)]. Routledge, London, UK, pp. 87–113.

Morton, J.F., 2007: The impact of climate change on smallholder and subsistence agriculture. *Proc. Natl. Acad. Sci.*, **104**, 19680–19685, doi:10.1073/pnas.0701855104.

Moses, J.A., D.S. Jayas, and K. Alagusundaram, 2015: Climate change and its implications on stored food grains. *Agric. Res.*, **4**, 21–30, doi:10.1007/ s40003-015-0152-z.

Mottet, A. et al., 2017a: Livestock: On our plates or eating at our table? A new analysis of the feed/food debate. *Glob. Food Sec.*, **14**, 1–8, doi:10.1016/J. GFS.2017.01.001.

Mottet, A. et al., 2017b: Climate change mitigation and productivity gains in livestock supply chains: Insights from regional case studies. *Reg. Environ. Chang.*, **17**, 129–141, doi:10.1007/s10113-016-0986-3.

Mozzer, G., 2011: Agriculture and cattle raising in the context of a low carbon economy. In: *Climate Change in Brazil: Economic, social and regulatory aspects* [Motta, R., J. Hargrave, G. Luedemann, and M. Gutierrez (eds.)]. IPEA, Brasilia, Brazil, pp. 107–122.

Mueller, V., C. Gray, and K. Kosec, 2014: Heat stress increases long-term human migration in rural Pakistan. *Nat. Clim. Chang.*, **4**, 182–185, doi:10.1038/nclimate2103.

Muhanji, G., R.L. Roothaert, C. Webo, and M. Stanley, 2011: African indigenous vegetable enterprises and market access for small-scale farmers in East Africa. *Int. J. Agric. Sustain.*, **9**, 194–202, doi:10.3763/ijas.2010.0561.

Müller, B., L. Johnson, and D. Kreuer, 2017a: Maladaptive outcomes of climate insurance in agriculture. *Glob. Environ. Chang.*, **46**, 23–33, doi:10.1016/J. GLOENVCHA.2017.06.010.

Müller, C., J. Elliott, and A. Levermann, 2014: Food security: Fertilizing hidden hunger. *Nat. Clim. Chang.*, **4**, 540–541, doi:10.1038/nclimate2290.

Müller, C. et al., 2017b: Global gridded crop model evaluation: Benchmarking, skills, deficiencies and implications. *Geosci. Model Dev.*, **10**, 1403–1422, doi:10.5194/gmd-10-1403-2017.

Mumtaz, Z., and P. Whiteford, 2017: Social safety nets in the development of a welfare system in Pakistan: An analysis of the Benazir Income Support Programme. *Asia Pacific J. Public Adm.*, **39**, 16–38, doi:10.1080/2327666 5.2017.1290902.

Muratori, M., K. Calvin, M. Wise, P. Kyle, and J. Edmonds, 2016: Global economic consequences of deploying bioenergy with carbon capture and storage (BECCS). *Environ. Res. Lett.*, **11**, 95004.

Murdiyarso, D. et al., 2015: The potential of Indonesian mangrove forests for global climate change mitigation. *Nat. Clim. Chang.*, **5**, 1089–1092, doi:10.1038/nclimate2734.

Murgueitio, E., Z. Calle, F. Uribe, A. Calle, and B. Solorio, 2011: Native trees and shrubs for the productive rehabilitation of tropical cattle ranching lands. *For. Ecol. Manage.*, doi:10.1016/j.foreco.2010.09.027.

Murphy, C., M. Tembo, A. Phiri, O. Yerokun, and B. Grummell, 2016: Adapting to climate change in shifting landscapes of belief. *Clim. Change*, **134**, 101–114, doi:10.1007/s10584-015-1498-8.

Mutahara, M., A. Haque, M.S.A. Khan, J.F. Warner, and P. Wester, 2016: Development of a sustainable livelihood security model for storm-surge hazard in the coastal areas of Bangladesh. *Stoch. Environ. Res. Risk Assess.*, **30**, 1301–1315, doi:10.1007/s00477-016-1232-8.

Mutekwa, V.T., 2009: *Climate Change Impacts and Adaptation in the Agricultural Sector: The Case of Smallholder Farmers in Zimbabwe*. Southern University and A & M College, Louisiana, USA, 237–256 pp.

Muthayya, S. et al., 2013: The global hidden hunger indices and maps: An advocacy tool for action. *PLoS One*, **8**, e67860, doi:10.1371/journal. pone.0067860.

Muthee, K., C. Mbow, G. Macharia, and W. Leal Filho, 2017: Ecosystem-based adaptation (EbA) as an adaptation strategy in Burkina Faso and Mali. In: *Climate Change Adaptation in Africa* [Leal Filho, W., S. Belay, J. Kalangu Wuta Menas, P. Munishi, K. Musiyiwa (eds.)]. Springer International Publishing, Cham, Switzerland, pp. 205–215, doi:10.1007/978-3-319-49520-0.

Mutuo, P.K., G. Cadisch, A. Albrecht, C.A. Palm, and L. Verchot, 2005: Potential of agroforestry for carbon sequestration and mitigation of greenhouse gas emissions from soils in the tropics. *Nutr. Cycl. Agroecosystems*, **71**, 43–54, doi:10.1007/s10705-004-5285-6.

Myers, S.S. et al., 2014: Increasing CO_2 threatens human nutrition. *Nature*, **510**, 139–142, doi:10.1038/nature13179.

Myers, S.S., K.R. Wessells, I. Kloog, A. Zanobetti, and J. Schwartz, 2015: Effect of increased concentrations of atmospheric carbon dioxide on the global threat of zinc deficiency: A modelling study. *Lancet Glob. Heal.*, **3**, e639–e645, doi:10.1016/S2214-109X(15)00093-5.

5

Myers, S.S. et al., 2017: Climate change and global food systems: Potential impacts on food security and undernutrition. *Annu. Rev. Public Health*, **38**, 259–277, doi:10.1146/annurev-publhealth-031816-044356.

Næss, M.W., 2013: Climate change, risk management and the end of *Nomadic* pastoralism. *Int. J. Sustain. Dev. World Ecol.*, **20**, 123–133, doi:10.1080/13 504509.2013.779615.

Nalau, J. et al., 2016: The practice of integrating adaptation and disaster risk reduction in the south-west Pacific. *Clim. Dev.*, **8**, 365–375, doi:10.1080/1 7565529.2015.1064809.

Nandal, U., and R.L. Bhardwaj, 2014: The role of underutilized fruits in nutritional and economic security of tribals: A review. *Crit. Rev. Food Sci. Nutr.*, **54**, 880–890, doi:10.1080/10408398.2011.616638.

Nankishore, A., and A.D. Farrell, 2016: The response of contrasting tomato genotypes to combined heat and drought stress. *J. Plant Physiol.*, **202**, 75–82, doi:10.1016/j.jplph.2016.07.006.

Narayanan, S., and N. Gerber, 2017: Social safety nets for food and nutrition security in India. *Glob. Food Sec.*, doi:10.1016/j.gfs.2017.05.001.

Nardone, A., B. Ronchi, N. Lacetera, M.S. Ranieri, and U. Bernabucci, 2010: Effects of climate changes on animal production and sustainability of livestock systems. *Livest. Sci.*, **130**, 57–69, doi:10.1016/J.LIVSCI.2010.02.011.

Naresh, R.K. et al., 2017: Water-Smart-Agriculture to Cope With Changing Climate in Smallholders Farming Areas of Subtropical India: A Review. *Int. J. Pure App. Biosci.*, **5**, 400–416.

Natalini, D., G. Bravo, and A.W. Jones, 2017: Global food security and food riots – An agent-based modelling approach. *Food Secur.*, 1–21, doi:10.1007/s12571-017-0693-z.

Nautiyal, S., H. Kaechele, K.S. Rao, R.K. Maikhuri, and K.G. Saxena, 2007: Energy and economic analysis of traditional versus introduced crops cultivation in the mountains of the Indian Himalayas: A case study. *Energy*, **32**, 2321–2335, doi:10.1016/j.energy.2007.07.011.

Nawrotzki, R.J., F. Riosmena, and L.M. Hunter, 2013: Do rainfall deficits predict US-bound migration from rural Mexico? Evidence from the Mexican census. *Popul. Res. Policy Rev.*, **32**, 129–158, doi:10.1007/s11113-012-9251-8.

NCD-RisC, 2016a: Trends in adult body-mass index in 200 countries from 1975 to 2014: A pooled analysis of 1698 population-based measurement studies with 19.2 million participants. *Lancet*, **387**, 1377–1396, doi:10.1016/ S0140-6736(16)30054-X.

NCD-RisC, 2016b: Worldwide trends in diabetes since 1980: A pooled analysis of 751 population-based studies with 4·4 million participants. *Lancet*, **387**, 1513–1530, doi:10.1016/S0140-6736(16)00618-8.

NCD-RisC, 2017a: Worldwide trends in body-mass index, underweight, overweight, and obesity from 1975 to 2016: A pooled analysis of 2416 population-based measurement studies in 128·9 million children, adolescents, and adults. *Lancet*, **390**, 2627–2642, doi:10.1016/S0140-6736(17)32129-3.

NCD-RisC, 2017b: Worldwide trends in blood pressure from 1975 to 2015: A pooled analysis of 1479 population-based measurement studies with 19·1 million participants. *Lancet (London, England)*, **389**, 37–55, doi:10.1016/S0140-6736(16)31919-5.

Negra, C. et al., 2014: Brazil, Ethiopia, and New Zealand lead the way on climate-smart agriculture. *BioMed Central*, 10–15, doi:10.1186/s40066-014-0019-8.

Neira, D.P., and M.S. Montiel, 2013: *Agroecología y Ecofeminismo para descolonizar y despatriarcalizar la alimentación globalizada*. Revista Internacional de Pensamiento Politico, **8**, 95–113 pp.

Neira, D.P., X.S. Fernández, D.C. Rodríguez, M.S. Montiel, and M.D. Cabeza, 2016: Analysis of the transport of imported food in Spain and its contribution to global warming. *Renew. Agric. Food Syst.*, **31**, 37–48, doi:10.1017/S1742170514000428.

Nellemann, C., M. MacDevette, T. Manders, B. Eickhout, B. Svihus, A.G. Prins, and B.P. Kaltenborn, 2009: *The environmental food crisis: the environment's role in averting future food crises: A UNEP rapid response assessment*. UNEP/Earthprint, Stevenage, UK, 104 pp.

Nelson, G. et al., 2010: *Food Security, Farming, and Climate Change to 2050: Scenarios, Results, Policy Options*. IFPRI Research Monograph, Washington, DC, USA, 155 pp.

Nelson, G.C. et al., 2009: *Climate Change Impact on Agriculture and Costs of Adaptation*. International Food Policy Research Institute Washington, DC, USA, 30 pp.

Nelson, G.C. et al., 2013: Agriculture and climate change in global scenarios: Why don't the models agree. *Agric. Econ.*, **45**, 85–101, doi:10.1111/agec.12091.

Nelson, G. et al., 2018: Income growth and climate change effects on global nutrition security to mid-century. *Nat. Sustain.*, **1**, doi:10.1038/s41893-018-0192-z.

Nelson, G.C.G. et al., 2014: Climate change effects on agriculture: Economic responses to biophysical shocks. *Proc. Natl. Acad. Sci.*, **111**, 3274–3279, doi:10.1073/pnas.1222465110.

Nelson, M.E., M.W. Hamm, F.B. Hu, S.A. Abrams, and T.S. Griffin, 2016: Alignment of healthy dietary patterns and environmental sustainability: A systematic review. *Adv. Nutr. An Int. Rev. J.*, **7**, 1005–1025, doi:10.3945/an.116.012567.

Nelson, V., and T. Stathers, 2009: Resilience, power, culture, and climate: A case study from semi-arid Tanzania, and new research directions. *Gend. Dev.*, **17**, 81–94, doi:10.1080/13552070802696946.

Nelson, V, K. Meadows, T. Cannon, J. Morton, and A. Martin, 2002: Uncertain predictions, invisible impacts, and the need to mainstream gender in climate change adaptations. *Gend. Dev.*, **10**, 51–59, doi:10.1080/13552070215911.

Nemet, G.F. et al., 2018: Negative emissions – Part 3: Innovation and upscaling. *Environ. Res. Lett.*, **13**, 63003, doi:10.1088/1748-9326/aabff4.

Nepstad, D. et al., 2014: Slowing Amazon deforestation through public policy and interventions in beef and soy supply chains. *Science.*, **344**, 1118–1123, doi:10.1126/science.1248525.

Neufeldt, H. et al., 2013: Beyond climate-smart agriculture: Toward safe operating spaces for global food systems. *Agric. Food Secur.*, **2**, 12, doi:10.1186/2048-7010-2-12.

Neumayer, E., 2001: *Greening Trade and Investment: Environmental Protection Without Protectionism*. Routledge, London, UK, 240 pp.

Newaj, R., O.P. Chaturvedi, and A.K. Handa, 2016: Recent development in agroforestry research and its role in climate change adaptation and mitigation change adaptation and mitigation. *Indian J. Agrofor.*, **18**, 1–9.

Newell, M.T., and R.C. Hayes, 2017: An initial investigation of forage production and feed quality of perennial wheat derivatives. *Crop Pasture Sci.*, **68**, 1141-1148, doi:10.1071/CP16405.

Newell, P., and O. Taylor, 2018: Contested landscapes: The global political economy of climate-smart agriculture. *J. Peasant Stud.*, **45**, 108–129, doi: 10.1080/03066150.2017.1324426.

Newman, L., C. Ling, and K. Peters, 2013: Between field and table: Environmental implications of local food distribution. *Int. J. Sustain. Soc.*, **5**, 11–23, doi:10.1504/IJSSOC.2013.050532.

Ng, M. et al., 2014: Global, regional, and national prevalence of overweight and obesity in children and adults during 1980–2013: A systematic analysis for the Global Burden of Disease Study 2013. *Lancet*, **384**, 766–781, doi:10.1016/S0140-6736(14)60460-8.

Ngadze, R.T., A.R. Linnemann, L.K. Nyanga, V. Fogliano, and R. Verkerk, 2017: Local processing and nutritional composition of indigenous fruits: The case of monkey orange (*Strychnos* spp.) from southern Africa. *Food Rev. Int.*, **33**, 123–142, doi:10.1080/87559129.2016.1149862.

Nguyen, Q., J. Bow, J. Howe, S. Bratkovich, H. Groot, E. Pepke, and K. Fernholz, 2017: *Global Production of Second Generation Biofuels: Trends and Influences*. Dovetail Partners, Inc., Minneapolis, US, 16 pp.

Nie, Z. et al., 2016: Benefits, challenges and opportunities of integrated crop-livestock systems and their potential application in the high rainfall zone of southern Australia: A review. *Agric. Ecosyst. Environ.*, **235**, 17–31, doi:10.1016/j.agee.2016.10.002.

Niehof, A., 2016: Food and nutrition security as gendered social practice. *Appl. Stud. Agribus. Commer.*, **10**, 59–66, doi:10.19041/APSTRACT/2016/2-3/7.

van Niekerk, J., and R. Wynberg, 2017: Traditional seed and exchange systems cement social relations and provide a safety net: A case study from KwaZulu-Natal, South Africa. *Agroecol. Sustain. Food Syst.*, 1–25, doi:10.1080/21683565.2017.1359738.

Nielsen, J.Ø., and A. Reenberg, 2010: Cultural barriers to climate change adaptation: A case study from Northern Burkina Faso. *Glob. Environ. Chang.*, **20**, 142–152, doi:10.1016/J.GLOENVCHA.2009.10.002.

Niles, M.T. et al., 2018: Climate change mitigation beyond agriculture: A review of food system opportunities and implications. *Renew. Agric. Food Syst.*, **33**, 297–308, doi:10.1017/S1742170518000029.

Njiru, B.N., 2012: *Climate Change, Resource Competition, and Conflict amongst Pastoral Communities in Kenya*. Springer, Berlin, Heidelberg, 513–527 pp.

Nogueira, E.M., A.M. Yanai, S.S. de Vasconcelos, P.M.L. de Alencastro Graça, and P.M. Fearnside, 2018: Brazil's Amazonian protected areas as a bulwark against regional climate change. *Reg. Environ. Chang.*, **18**, 573–579, doi:10.1007/s10113-017-1209-2.

Nori, M., J. Switzer, and A. Crawford, 2005: *Herding on the Brink Towards a Global Survey of Pastoral Communities and Conflict*. Commission on Environmental, Economic and Social Policy. International Institute for Sustainable Development (IISD), 1–33 pp.

NRC, 1981: *Effect of environment on nutrient requirements of domestic animals. National Research Council (U.S.). Committee on Animal Nutrition. Subcommittee on Environmental Stress*. National Academy Press, Washington, DC, USA, 152 pp.

NRDC, 2017: *Less beef, less carbon: Americans shrink their diet-related carbon footprint by 10 percent between 2005 and 2014*. Issue Paper 16-11-B, New York, USA, 8 pp.

Nurse, L.A., R.F. McLean, J. Agard, L.P. Briguglio, V. Duvat-Magnan, N. Pelesikoti, E. Tompkins, and A. Webb, 2014: Small islands. In: *Climate Change 2014: Impacts, Adaptation, and Vulnerability. Part B: Regional Aspects. Contribution of Working Group II to the Fifth Assessment Report of the Intergovernmental Panel on Climate Change* [Field, C.B., D.J. Dokken, T.E. Bilir, M. Chatterjee, E.S. Kissel, R.C. Genova, B. Girma, Andrew N. Levy, K.J. Mach, M.D. Mastrandrea, K.L. Ebi, Y.O. Estrada, S. MacCracken, P.R. Mastrandrea, Leslie L. White (eds.)]. Cambridge University Press, Cambridge, UK and New York, USA, 1613–1654 pp.

Nyong, A., F. Adesina, and B. Osman Elasha, 2007: The value of indigenous knowledge in climate change mitigation and adaptation strategies in the African Sahel. *Mitig. Adapt. Strateg. Glob. Chang.*, **12**, 787–797, doi:10.1007/s11027-007-9099-0.

O'Brien, K.L., and J. Wolf, 2010: A values-based approach to vulnerability and adaptation to climate change. *Wiley Interdiscip. Rev. Clim. Chang.*, **1**, 232–242, doi:10.1002/wcc.30.

O'Mara, F.P., 2012: The role of grasslands in food security and climate change. *Ann. Bot.*, **110**, 1263–1270, doi:10.1093/aob/mcs209.

Obersteiner, M. et al., 2016: Assessing the land resource–food price nexus of the Sustainable Development Goals. *Sci. Adv.*, **2**, e1501499, doi:10.1126/sciadv.1501499.

OECD, 2017: *International Migration Outlook 2017*. OECD Publishing, Paris, France, 366 pp.

OECD and FAO, 2015: *OECD-FAO Agricultural Outlook 2015*. Paris, France, 148 pp. www.fao.org/3/a-i4738e.pdf.

Ojha, H.R., S. Ghimire, A. Pain, A. Nightingale, B. Khatri, and H. Dhungana, 2015: Policy without politics: Technocratic control of climate change adaptation policy making in Nepal. *Climate Policy*, 16, 415–433, doi:10.1080/14693062.2014.1003775.

Oliveira, G., and S. Hecht, 2016: Sacred groves, sacrifice zones and soy production: Globalization, intensification and neo-nature in South America. *J. Peasant Stud.*, **43**, 251–285, doi:10.1080/03066150.2016.1146705.

de Oliveira Silva, R., L.G. Barioni, J.A.J. Hall, M. Folegatti Matsuura, T. Zanett Albertini, F.A. Fernandes, and D. Moran, 2016a: Increasing beef production could lower greenhouse gas emissions in Brazil if decoupled from deforestation. *Nat. Clim. Chang.*, **6**, 493–497, doi:10.1038/nclimate2916.

de Oliveira Silva, R., 2016b: Increasing beef production could lower greenhouse gas emissions in Brazil if decoupled from deforestation. *Nat. Clim. Chang.*, **6**, 493–497, doi:10.1038/nclimate2916.

Oliver, T.H. et al., 2018: Overcoming undesirable resilience in the global food system. *Glob. Sustain.*, **1**, e9, 1–9, doi.org:10.1017/sus.2018.9.

Omolo, N., 2011: Gender and climate change-induced conflict in pastoral communities: Case study of Turkana in north-western Kenya. *African J. Confl. Resolut.*, **10**, 2, pp 81–102, doi:10.4314/ajcr.v10i2.63312.

Omotilewa, O.J., J. Ricker-Gilbert, J.H. Ainembabazi, and G.E. Shively, 2018: Does improved storage technology promote modern input use and food security? Evidence from a randomized trial in Uganda. *J. Dev. Econ.*, **135**, 176–198, doi:10.1016/j.jdeveco.2018.07.006.

Ongoro, E., and W. Ogara, 2012: Impact of climate change and gender roles in community adaptation: A case study of pastoralists in Samburu East District, Kenya. *Int. J. Biodivers. Conserv.*, **42**, 78–89.

Onyeneke, R.U., C.O. Igberi, C.O. Uwadoka, and J.O. Aligbe, 2018: Status of climate-smart agriculture in Southeast Nigeria. *GeoJournal*, **83**, 333–346, doi:10.1007/s10708-017-9773-z.

Ortega, F.B., C.J. Lavie, and S.N. Blair, 2016: Obesity and cardiovascular disease. *Circ. Res.*, **118**, 1752–1770, doi:10.1161/CIRCRESAHA.115.306883.

Ouma, E.A., G.A. Obare, and S.J. Staal, 2003: *Cattle As Assets: Assessment Of Non-Market Benefits From Cattle In Smallholder Kenyan Crop-Livestock Systems*. Proceedings of the 25th International Conference of Agricultural Economists (IAAE), Durban, South Africa, 7 pp.

Ozaki, M., 2016: *Disaster Risk Financing in Bangladesh Disaster Risk Financing in Bangladesh*. ADB South Asia Working Paper Series, Working Paper No. 46., Manila, Philippines, 35 pp.

Padgham, J., J. Jabbour, and K. Dietrich, 2015: Managing change and building resilience: A multi-stressor analysis of urban and peri-urban agriculture in Africa and Asia. *Urban Clim.*, **12**, 183–204, doi:10.1016/j.uclim.2015.04.003.

Padulosi, S., J. Thompson, and P. Ruderbjer, 2013: *Fighting Poverty, Hunger and Malnutrition with Neglected and Underutilized Species (NUS): Needs, Challenges and the Way Forward*. Bioversity International, Rome, Italy, 60 pp.

Palm, C., H. Blanco-canqui, F. Declerck, L. Gatere, and P. Grace, 2013: Conservation agriculture and ecosystem services: An overview. *Agriculture, Ecosyst. Environ.*, **187**, 87–105, doi:10.1016/j.agee.2013.10.010.

Palm, C.A. et al., 2010: Identifying potential synergies and trade-offs for meeting food security and climate change objectives in Sub-Saharan Africa. *Proc. Natl Acad. Sci.*, **107**, 19661, doi:10.1073/pnas.0912248107.

Palmer, M.A., J. Liu, J.H. Matthews, M. Mumba, and P. D'Odorico, 2015: Manage water in a green way. *Science.*, **349**, 584 LP-585, doi:10.1126/science.aac7778.

Pangga, I.B., J. Hanan, and S. Chakraborty, 2011: Pathogen dynamics in a crop canopy and their evolution under changing climate. *Plant Pathol.*, **60**, 70–81, doi:10.1111/j.1365-3059.2010.02408.x.

Pant, L.P., B. Adhikari, and K.K. Bhattarai, 2015: Adaptive transition for transformations to sustainability in developing countries. *Curr. Opin. Environ. Sustain.*, **14**, 206–212, doi:10.1016/j.cosust.2015.07.006.

Papageorgiou, M., 2017: iMedPub journals genomics-centric approach: An insight into the role of genomics in assisting GM-rice varieties within a paradigm of future climate change mitigation, food security, and GM regulation keywords: Genomic tools used in rice improvement. *J. Plant Sci. Agric. Res.*, **1**, 1–10.

Papargyropoulou, E., R. Lozano, J.K. Steinberger, N. Wright, and Z. bin Ujang, 2014: The food waste hierarchy as a framework for the management of food surplus and food waste. *J. Clean. Prod.*, **76**, 106–115, doi:10.1016/j.jclepro.2014.04.020.

5

Pape, R., and J. Löffler, 2012: Climate change, land use conflicts, predation and ecological degradation as challenges for reindeer husbandry in northern Europe: What do we really know after half a century of research? *Ambio*, **41**, 421–434, doi:10.1007/s13280-012-0257-6.

Parikh, J., 2009: *Towards a gender-sensitive agenda for energy, environment and climate change*. Division for the Advancement of Women, United Nations, Geneva, 7 pp.

Paris, T., and M.F. Rola-Rubzen, 2018: *Gender dimension of climate change research in agriculture: Case studies in Southeast Asia*. CGIAR research programme on Climate Change, Agriculture and Food Security, Wageningen, 216 pp.

Park, S.E., N.A. Marshall, E. Jakku, A.M. Dowd, S.M. Howden, E. Mendham, and A. Fleming, 2012: Informing adaptation responses to climate change through theories of transformation. *Glob. Environ. Chang.*, **22**, 115–126, doi:10.1016/J.GLOENVCHA.2011.10.003.

Parkes, B., D. Defrance, B. Sultan, P. Ciais, and X. Wang, 2018: Projected changes in crop yield mean and variability over West Africa in a world 1.5 K warmer than the pre-industrial. *Earth Syst. Dynam.*, **9**, 1–24, doi:10.5194/esd-2017-66.

Pascucci, S., C. Cicatiello, S. Franco, B. Pancino, D. Marinov, and M. Davide, 2011: Back to the future? Understanding change in food habits of farmers' market customers. *The International Food and Agribusiness Management Review*, **14**, 105 – 126, doi:10.1016/j.foodres.2009.07.010.

Pasini, A., G. Mastrojeni, and F.N. Tubiello, 2018: Climate actions in a changing world. *Anthr. Rev.*, **5**, 237–241, doi:10.1177/2053019618794213.

Paterson, R.R.M., and N. Lima, 2010: How will climate change affect mycotoxins in food? *Food Res. Int.*, **43**, 1902–1914.

Paterson, R.R.M., and N. Lima, 2011: Further mycotoxin effects from climate change. *Food Res. Int.*, **44**, 2555–2566, doi:10.1016/J.FOODRES.2011.05.038.

Paudel, B., B.S. Acharya, R. Ghimire, K.R. Dahal, and P. Bista, 2014: Adapting agriculture to climate change and variability in Chitwan: Long-term trends and farmers' perceptions. *Agric. Res.*, **3**, 165–174, doi:10.1007/s40003-014-0103-0.

Paudel, S.R. et al., 2015: Effects of temperature on nitrous oxide (N_2O) emission from intensive aquaculture system. *Sci. Total Environ.*, **518–519**, 16–23, doi:10.1016/J.SCITOTENV.2015.02.076.

Paudela, D., K.R. Tiwaria, R.M. Bajracharyab, N. Rautb, and B.K. Sitaulac, 2017: Agroforestry system: An opportunity for carbon sequestration and climate change adaptation in the Mid-Hills of Nepal. *Octa J. Environ. Res.*, **5**, 10 pp.

Paulot, F., and D.J. Jacob, 2014: Hidden cost of U.S. agricultural exports: Particulate matter from ammonia emissions. *Env. Sci Technol*, **48**, 903–908, doi:10.1021/es4034793.

Paustian, K., J. Lehmann, S. Ogle, D. Reay, G.P. Robertson, and P. Smith, 2016: Climate-smart soils. *Nature*, **532**, 49–57, doi:10.1038/nature17174.

Pedditizi, E., R. Peters, and N. Beckett, 2016: The risk of overweight/obesity in mid-life and late life for the development of dementia: A systematic review and meta-analysis of longitudinal studies. *Age Ageing*, **45**, 14–21, doi:10.1093/ageing/afv151.

Pei, Q., and D.D. Zhang, 2014: Long-term relationship between climate change and nomadic migration in historical China. *Ecol. Soc.*, **19**, art68, doi:10.5751/ES-06528-190268.

Pelster, D., M. Rufino, T. Rosenstock, J. Mango, G. Saiz, E. Diaz-Pines, G. Baldi, and K. Butterbach-Bahl, 2017: Smallholder farms in eastern African tropical highlands have low soil greenhouse gas fluxes. *Biogeosciences*, **14**, 187–202, doi:10.5194/bg-14-187-2017.

Peria, A.S., J.M. Pulhin, M.A. Tapia, C.D. Predo, R.J.J. Peras, R.J.P. Evangelista, R.D. Lasco, and F.B. Pulhin, 2016: Knowledge, Risk Attitudes and Perceptions on Extreme Weather Events of Smallholder Farmers in Ligao City, Albay, Bicol, Philippines. *J. Environ. Sci. Manag.*, **1**, 31–41.

Persson, J., N.-E. Sahlin, and A. Wallin, 2015: Climate change, values, and the cultural cognition thesis. *Environ. Sci. Policy*, **52**, 1–5, doi:10.1016/J.ENVSCI.2015.05.001.

Peters, G.P., and O. Geden, 2017: Catalysing a political shift from low to negative carbon. *Nat. Clim. Chang.*, **7**, 619.

Petersen, B.M., M.T. Knudsen, J.E. Hermansen, and N. Halberg, 2013: An approach to include soil carbon changes in lifecycle assessments. *J. Clean. Prod.*, **52**, 217–224, doi:10.1016/J.JCLEPRO.2013.03.007.

Pfeiffer, M. et al., 2019: Grazing and aridity reduce perennial grass abundance in semi-arid rangelands – Insights from a trait-based dynamic vegetation model. *Ecol. Modell.*, **395**, 11–22, doi:10.1016/j.ecolmodel.2018.12.013.

Pham, T.P.T., R. Kaushik, G.K. Parshetti, R. Mahmood, and R. Balasubramanian, 2015: Food waste-to-energy conversion technologies: Current status and future directions. *Waste Manag.*, **38**, 399–408, doi:10.1016/J.WASMAN.2014.12.004.

Pikaar, I., S. Matassa, K. Rabaey, B. Laycock, N. Boon, and W. Verstraete, 2018: The urgent need to re-engineer nitrogen-efficient food production for the planet. In: *Managing Water, Soil and Waste Resources to Achieve Sustainable Development Goals*, Springer International Publishing, Cham, Switzerland, 35–69 pp.

Pimbert, M., 2015: Agroecology as an alternative vision to conventional development and climate-smart agriculture. *Development*, **58**, 286–298, doi:10.1057/s41301-016-0013-5.

Pimbert, M., and S. Lemke, 2018: Using agroecology to enhance dietary diversity. *UNSCN News*, **43**, 33–42.

Pinke, Z., and G.L. Lövei, 2017: Increasing temperature cuts back crop yields in Hungary over the last 90 years. *Glob. Chang. Biol.*, **23**, 5426–5435, doi:10.1111/gcb.13808.

Place, S.E., and F.M. Mitloehner, 2012: Beef production in balance: Considerations for lifecycle analyses. *Meat Sci.*, **92**, 179–181, doi:10.1016/J.MEATSCI.2012.04.013.

Ponnusamy, P. et al., 2015: Integrated assessment of climate change impacts on maize farms and farm household incomes in South India: A case study from Tamil Nadu. In: *Handbook of Climate Change and Agroecosystems: The Agricultural Model Intercomparison and Improvement Project Integrated Crop and Economic Assessments – Joint Publication with American Society of Agronomy, Crop Science Society of America, and Soil Science Society of America (In 2 Parts) (Vol. 3)*. Imperial College Press, London, UK. [Rosenzweig, C. and D. Hillel (eds.)]. Part 2, pp 281–314, doi:10.1142/9781783265640_0021.

Poore, J., and T. Nemecek, 2018: Reducing food's environmental impacts through producers and consumers. *Science*, **360**, 987–992, doi:10.1126/science.aaq0216.

Popp, A. et al., 2017: Land-use futures in the shared socio-economic pathways. *Glob. Environ. Chang.*, **42**, 331–345, doi:10.1016/j.gloenvcha.2016.10.002.

Popp, J., Z. Lakner, M. Harangi-Rákos, and M. Fári, 2014: The effect of bioenergy expansion: Food, energy, and environment. *Renew. Sustain. Energy Rev.*, **32**, 559–578, doi:10.1016/J.RSER.2014.01.056.

Porter, J.R., L. Xie, A.J. Challinor, K. Cochrane, S.M. Howden, M.M. Iqbal, D.B. Lobell, and M.I. Travasso, 2014: Food security and food production systems. *Clim. Chang. 2014 Impacts, Adapt. Vulnerability. Part A Glob. Sect. Asp. Contrib. Work. Gr. II to Fifth Assess. Rep. Intergov. Panel Clim. Chang.*, **2**, 485–533, doi:10.1111/j.1728-4457.2009.00312.x.

Porter, J.R., M. Howden, and P. Smith, 2017: Considering agriculture in IPCC assessments. *Nat. Clim. Chang.*, **7**, 680–683, doi:10.1038/nclimate3404.

Porter, S.D., D.S. Reay, P. Higgins, and E. Bomberg, 2016: A half-century of production-phase greenhouse gas emissions from food loss & waste in the global food supply chain. *Sci. Total Environ.*, **571**, 721–729, doi:10.1016/j.scitotenv.2016.07.041.

Potopová, V., P. Zahradníček, P. Štěpánek, L. Türkott, A. Farda, and J. Soukup, 2017: The impacts of key adverse weather events on the field-grown vegetable yield variability in the Czech Republic from 1961 to 2014. *Int. J. Climatol.*, **37**, 1648–1664, doi:10.1002/joc.4807.

Powlson, D.S., C.M. Stirling, M.L. Jat, B.G. Gerard, C.A. Palm, P.A. Sanchez, and K.G. Cassman, 2014: Limited potential of no-till agriculture for climate

5

change mitigationchange mitigation. *Nat. Clim. Chang.*, **4**, 678–683, doi:10.1038/NCLIMATE2292.

Powlson, D.S., C.M. Stirling, C. Thierfelder, R.P. White, and M.L. Jat, 2016: Does conservation agriculture deliver climate change mitigation through soil carbon sequestration in tropical agro-ecosystems? *Agric. Ecosyst. Environ.*, **220**, 164–174, doi:10.1016/j.agee.2016.01.005.

Pradhan, P., 2019: Antagonists to meeting the 2030 Agenda. *Nat. Sustain.*, **2**, 171–172, doi:10.1038/s41893-019-0248-8.

Pradhan, P., M.K.B. Lüdeke, D.E. Reusser, and J.P. Kropp, 2013a: Embodied crop calories in animal products. *Environ. Res. Lett.*, **8**, 44044, doi:10.1088/1748-9326/8/4/044044.

Pradhan, P., D.E. Reusser, and J.P. Kropp, 2013b: Embodied greenhouse gas emissions in diets. *PLoS One*, **8**, 1–8, doi:10.1371/journal.pone.0062228.

Pradhan, P., M.K.B. Lüdeke, D.E. Reusser, and J.P. Kropp, 2014: Food Self-Sufficiency across Scales: How Local Can We Go? *Environ. Sci. Technol.*, **15**, 9779, doi:10.1021/es5005939.

Pradhan, P., G. Fischer, H. Van Velthuizen, D.E. Reusser, and J.P. Kropp, 2015: Closing yield gaps: How sustainable can we be? *PLoS One*, **10**, 1–18, doi:10.1371/journal.pone.0129487.

Pradhan, P., L. Costa, D. Rybski, W. Lucht, and J.P. Kropp, 2017: A systematic study of Sustainable Development Goal (SDG) interactions. *Earth's Futur.*, **5**, 1169–1179, doi:10.1002/2017EF000632.

Pratibha, G. et al., 2016: Net global warming potential and greenhouse gas intensity of conventional and conservation agriculture system in rainfed semi arid tropics of India. *Atmos. Environ.*, **145**, 239–250, doi:10.1016/J.ATMOSENV.2016.09.039.

Pretty, J., and Z.P. Bharucha, 2014: Sustainable intensification in agricultural systems. *Ann. Bot.*, **114**, 1571–1596, doi:10.1093/aob/mcu205.

Pretty, J., and Z.P. Bharucha, 2015: Integrated pest management for sustainable intensification of agriculture in Asia and Africa. *Insects*, **6**, 152–182, doi:10.3390/insects6010152.

Pretty, J. et al., 2018: Global assessment of agricultural system redesign for sustainable intensification. *Nat. Sustain.*, **1**, 441–446, doi:10.1038/s41893-018-0114-0.

Pugh, T.A.M. et al., 2016: Climate analogues suggest limited potential for intensification of production on current croplands under climate change. *Nat. Commun.*, **7**, 1–8, doi:10.1038/ncomms12608.

Puma, M.J., S. Bose, S.Y. Chon, and B.I. Cook, 2015: Assessing the evolving fragility of the global food system. *Environ. Res. Lett.*, **10**, 24007, doi:10.1088/1748-9326/10/2/024007.

Le Quéré, C. et al., 2018: Global carbon budget 2018. *Earth Syst. Sci. Data*, **10**, 2141–2194, doi:10.5194/essd-10-2141-2018.

Rabbani, M.G., Z.M. Khan, and M.H. Tuhin, 2015: *Climate change and migration in Bangladesh: A gender perspective*. Bangladesh Centre for Advanced Studies for UN Women, Dhaka, Bangladesh, 49 pp.

Rahman, M.R., and S.H. Bulbul, 2015: Adoption of water saving irrigation techniques for sustainable rice production in Bangladesh. *Environ. Ecol. Res.*, **3**, 1–8, doi:10.13189/eer.2015.030101.

Rajkumar, M., M.N. V Prasad, S. Swaminathan, and H. Freitas, 2013: Climate change driven plant-metal-microbe interactions. *Environ. Int.*, **53**, 74–86, doi:10.1016/j.envint.2012.12.009.

Raleigh, C., and D. Kniveton, 2012: Come rain or shine: An analysis of conflict and climate variability in East Africa. *J. Peace Res.*, **49**, 51–64, doi:10.1177/0022343311427754.

Ramachandran Nair, P.K., B. Mohan Kumar, and V.D. Nair, 2009: Agroforestry as a strategy for carbon sequestration. *J. Plant Nutr. Soil Sci.*, **172**, 10–23, doi:10.1002/jpln.200800030.

Rao, C.A.R. et al., 2016: A district level assessment of vulnerability of Indian agriculture to climate change. *Current Science*, **110**, 1939–1946, doi:10.18520/cs/v110/i10/1939-1946.

Rao, K.P.C., G. Sridhar, R.M. Mulwa, M.N. Kilavi, A. Esilaba, I.N. Athanasiadis, and R.O. Valdivia, 2015: Impacts of climate variability and change on agricultural systems in East Africa. In: *Handbook of Climate Change and Agroecosystems: The Agricultural Model Intercomparison and Improvement Project Integrated Crop and Economic Assessments – Joint Publication with American Society of Agronomy, Crop Science Society of America, and Soil Science Society of America (In 2 Parts) (Vol. 3)*. Imperial College Press, London, UK. [Rosenzweig, C. and D. Hillel (eds.)]. Part 2, pp. 75–124.

A.K. Sikka, A. Islam, and K.V. Rao, 2017: Climate-Smart Land and Water Management for Sustainable Agriculture. *Irrig. Drain.*, **67**, 72–81, doi:10.1002/ird.2162.

Rao, N., 2005: Gender equality, land rights and household food security: Discussion of rice farming systems. *Econ. Polit. Wkly.*, **40**, 2513–2521, doi:10.2307/4416780.

Rasul, G., 2010: The role of the Himalayan mountain systems in food security and agricultural sustainability in South Asia. *Int. J. Rural Manag.*, **6**, 95–116, doi:10.1177/097300521100600105.

Rasul, G., A. Saboor, P.C. Tiwari, A. Hussain, N. Ghosh, and G.B. Chettri, 2019: Food and nutritional security in the Hindu Kush Himalaya: Unique challenges and niche opportunities. In: *The Hindu Kush Himalaya Assessment: Mountains, Climate Change, Sustainability and People* [P. Wester, A. Mishra, A. Mukherji, and A.B. Shrestha, (eds.)]. Springer Netherlands, Dordrecht, Netherlands, pp. 627.

Rattanachot, W., Y. Wang, D. Chong, and S. Suwansawas, 2015: Adaptation strategies of transport infrastructures to global climate change. *Transp. Policy*, **41**, 159–166, doi:10.1016/j.tranpol.2015.03.001.

Ravi, S., D.D. Breshears, T.E. Huxman, and P. D'Odorico, 2010: Land degradation in drylands: Interactions among hydrologic–aeolian erosion and vegetation dynamics. *Geomorphology*, **116**, 236–245, doi:10.1016/J.GEOMORPH.2009.11.023.

Reay, D.S., E.A. Davidson, K.A. Smith, P. Smith, J.M. Melillo, F. Dentener, and P.J. Crutzen, 2012: Global agriculture and nitrous oxide emissions. *Nat. Clim. Chang.*, **2**, 410–416, doi:10.1038/nclimate1458.

Red Cross, 2011: *West and Central Africa: Early Warning/ Early Action*. International Federation of Red Cross and Red Crescent Societies, Dakar, Senegal, 8 pp, https://reliefweb.int/sites/reliefweb.int/files/resources/4BA0944266708735C125771F0036408C-Full_Report.pdf.

Remling, E., and J. Veitayaki, 2016: Community-based action in Fiji's Gau Island: A model for the Pacific? *Int. J. Clim. Chang. Strateg. Manag.*, **8**, 375–398, doi:10.1108/IJCCSM-07-2015-0101.

Ren, X., M. Weitzel, B.C. O'Neill, P. Lawrence, P. Meiyappan, S. Levis, E.J. Balistreri, and M. Dalton, 2018: Avoided economic impacts of climate change on agriculture: Integrating a land surface model (CLM) with a global economic model (iPETS). *Clim. Change*, **146**, 517–531, doi:10.1007/s10584-016-1791-1.

Reuter, T., 2018: Understanding food system resilience in Bali, Indonesia: A moral economy approach. *Cult. Agric. Food Environ.*, **0**, doi:10.1111/cuag.12135.

Reuter, T.A., 2015: The Struggle for Food Sovereignty: A Global Perspective. In: *Averting a Global Environmental Collapse* [Reuter, T.A. (ed.)]. Cambridge Scholars Publishing, Newcastle upon Tyne, UK, pp. 127–147.

Revi, A., D.E. Satterthwaite, F. Aragón-Durand, J. Corfee-Morlot, R.B.R. Kiunsi, M. Pelling, D.C. Roberts, and W. Solecki, 2014: Urban areas Climate Change 2014: Impacts, Adaptation, and Vulnerability. Part A: Global and Sectoral Aspects. In: Contribution of Working Group II to the Fifth Assessment Report of the Intergovernmental Panel on Climate Change [Field, C.B., Barros, V.R., Dokken, D.J. (eds.)]. Cambridge University Press, New York, NY, USA, 535–612.

Reyer, C.P. et al., 2017a: Climate change impacts in Latin America and the Caribbean and their implications for development. *Reg. Environ. Chang.*, **17**, 1601–1621, doi:10.1007/s10113-015-0854-6.

Reyer, C.P.O., Kanta, K. Rigaud, E. Fernandes, W. Hare, O. Serdeczny, and H.J. Schellnhuber, 2017b: Turn down the heat: Regional climate change impacts on development. *Reg. Environ. Chang.*, **17**, 1563–1568, doi:10.1007/s10113-017-1187-4.

5

Riahi, K. et al., 2017: The Shared Socioeconomic Pathways and their energy, land use, and greenhouse gas emissions implications: An overview. *Glob. Environ. Chang.*, **42**, 153–168, doi:10.1016/j.gloenvcha.2016.05.009.

Richards, M. et al., 2015: *How Countries Plan to Address Agricultural Adaptation and Mitigation: An Analysis of Intended Nationally Determined Contributions*. CCAFS dataset version 1.2. Copenhagen, Denmark: CGIAR Research Program on Climate Change, Agriculture and Food Security (CCAFS), 8 pp.

Richards, M.B., E. Wollenberg, and D. van Vuuren, 2018: National contributions to climate change mitigation from agriculture: Allocating a global target. *Clim. Policy*, **18**, 1271–1285, doi:10.1080/14693062.2018.1430018.

Richards, P., 2015: What drives indirect land use change? How Brazil's agriculture sector influences frontier deforestation. *Ann. Assoc. Am. Geogr.*, **105**, 1026–1040, doi:10.1080/00045608.2015.1060924.

Ricroch, A., P. Clairand, and W. Harwood, 2017: Use of CRISPR systems in plant genome editing: Toward new opportunities in agriculture. *Emerg. Top. Life Sci.*, **1**, 169–182, doi:10.1042/ETLS20170085.

Riley, L., and B. Dodson, 2014: Gendered mobilities and food access in Blantyre, Malawi. *Urban Forum*, **25**, 227–239, doi:10.1007/s12132-014-9223-7.

Ringler, C., A.R. Quisumbing, E. Bryan, and R. Meinzen-Dick, 2014: *Enhancing Women's Assets to Manage Risk Under Climate Change: Potential for Group-based Approaches*. International Food Policy Research Institute, Washington, DC, USA, 65 pp.

Rippke, U. et al., 2016: Time scales of transformational climate change adaptation in Sub-Saharan African agriculture. *Nat. Clim. Chang.*, **6**, 605–609, doi:10.1038/nclimate2947.

Rivera-Ferre et al., 2013: *Understanding the Role of Local and Traditional Agricultural Knowledge in a Changing World Climate: The Case of the Indo-Gangetic Plains*. Technical Report, CGIAR-CCAFS Program, Nepal, 98 pp.

Rivera-Ferre, M.G. et al., 2016a: Local agriculture traditional knowledge to ensure food availability in a changing climate: Revisiting water management practices in the Indo-Gangetic Plains. *Agroecol. Sustain. Food Syst.*, **40**, 965–987, doi:10.1080/21683565.2016.1215368.

Rivera-Ferre, M.G. et al., 2016b: Re-framing the climate change debate in the livestock sector: Mitigation and adaptation options. *Wiley Interdiscip. Rev. Clim. Chang.*, **7**, 869–892, doi:10.1002/wcc.421.

Robinson, L.W., P.J. Ericksen, S. Chesterman, and J.S. Worden, 2015: Sustainable intensification in drylands: What resilience and vulnerability can tell us. *Agric. Syst.*, **135**, 133–140, doi:10.1016/J.AGSY.2015.01.005.

Robinson, S. et al., 2014: Comparing supply-side specifications in models of global agriculture and the food system. *Agric. Econ.*, **45**, 21–35, doi:10.1111/agec.12087.

Robinson, W.I., 2011: Globalization and the sociology of Immanuel Wallerstein: A critical appraisal. *Int. Sociol.*, **26**, 723–745, doi:10.1177/0268580910393372.

Rochedo, P.R.R. et al., 2018: The threat of political bargaining to climate mitigation in Brazil. *Nat. Clim. Chang.*, **8**, 695–698, doi:10.1038/s41558-018-0213-y.

Rockström, J. et al., 2009: Planetary boundaries: Exploring the safe operating space for humanity. *Ecol. Soc.*, **14**, 2, 32, doi:10.5751/ES-03180-140232.

Rodríguez Osuna, V., J. Börner, and M. Cunha, 2014: Scoping adaptation needs for smallholders in the Brazilian Amazon: A municipal level case study. *Chang. Adapt. Socio-Ecological Syst.*, **1**, doi:10.2478/cass-2014-0002.

Rogelj, J. et al., 2011: Emission pathways consistent with a 2°C global temperature limit. *Nat. Clim. Chang.*, **1**, 413–418, doi:10.1038/nclimate1258.

Rogelj, J. et al., 2015: Energy system transformations for limiting end-of-century warming to below 1.5°C. *Nat. Clim. Chang.*, **5**, 519–527, doi:10.1038/nclimate2572.

Rogelj, J. et al., 2018: Scenarios towards limiting global mean temperature increase below 1.5°C. *Nat. Clim. Chang.*, **8**, 325–332, doi:10.1038/s41558-018-0091-3.

Rojas-Downing, M.M., A.P. Nejadhashemi, T. Harrigan, and S.A. Woznicki, 2017: Climate change and livestock: Impacts, adaptation, and mitigation. *Clim. Risk Manag.*, **16**, 145–163, doi:10.1016/J.CRM.2017.02.001.

Röös, E., B. Bajželj, P. Smith, M. Patel, D. Little, and T. Garnett, 2017: Greedy or needy? Land use and climate impacts of food in 2050 under different livestock futures. *Glob. Environ. Chang.*, **47**, 1–12, doi:10.1016/j.gloenvcha.2017.09.001.

Rose, D., M.C. Heller, A.M. Willits-Smith, and R.J. Meyer, 2019: Carbon footprint of self-selected US diets: Nutritional, demographic, and behavioral correlates. *Am. J. Clin. Nutr.*, **109**, 526–534, doi:10.1093/ajcn/nqy327.

Rose, T.J., R.H. Wood, M.T. Rose, and L. Van Zwieten, 2018: A re-evaluation of the agronomic effectiveness of the nitrification inhibitors DCD and DMPP and the urease inhibitor NBPT. *Agric. Ecosyst. Environ.*, **252**, 69–73, doi:10.1016/J.AGEE.2017.10.008.

Rosegrant, M.W., E. Magalhaes, R.A. Valmonte-Santos, and D. Mason-D'Croz, 2015: *Returns to investment in reducing postharvest food losses and increasing agricultural productivity growth: Post-2015 consensus*. Copenhagen Consensus Center, Lowell, USA, 46 pp.

Rosemberg, A., 2010: Building a just transition: The linkages between climate change and employment. *Int. J. Labour Res.*, **2**, 125–161.

Rosenstock, T. et al., 2014: Agroforestry with N_2-fixing trees: Sustainable development's friend or foe? *Curr. Opin. Environ. Sustain.*, **6**, 15–21, doi:10.1016/j.cosust.2013.09.001.

Rosenzweig, C., et al., 2013: The agricultural model intercomparison and improvement project (AgMIP): protocols and pilot studies. *Agricultural and Forest Meteorology.* **170**, 166–182, doi: 10.1016/j.agrformet.2012.09.011.

Rosenzweig, C., and D. Hillel, 2015: The Role of the Agricultural Model Intercomparison and Improvement Project. *In: Handbook of Climate Change and Agroecosystems: The Agricultural Model Intercomparison and Improvement Project Integrated Crop and Economic Assessments – Joint Publication with American Society of Agronomy, Crop Science Society of America, and Soil Science Society of America (In 2 Parts) (Vol. 3)*. Imperial College Press, London, UK. [Rosenzweig, C. and D. Hillel (eds.)]. Part 1, pp. xxiii–xxvii.

Rosenzweig, C., and W. Solecki, 2018: Action pathways for transforming cities. *Nat. Clim. Chang.*, **8**, 756–759, doi:10.1038/s41558-018-0267-x.

Rosenzweig, C. et al., 2014: Assessing agricultural risks of climate change in the 21st century in a global gridded crop model intercomparison. *Proc. Natl. Acad. Sci. U.S.A.*, **111**, 3268–3273, doi:10.1073/pnas.1222463110.

Rosenzweig, C. et al., 2018a: Coordinating AgMIP data and models across global and regional scales for 1.5°C and 2.0°C assessments. *Phil. Trans. R. Soc.*, doi:10.1098/rsta.2016.0455.

Rosenzweig, C. et al., 2018b: *Climate Change and Cities: Second Assessment Report of the Urban Climate Change Research Network*. Cambridge University Press, Cambridge, United Kingdom and New York, NY, USA.

Rosi, A. et al., 2017: Environmental impact of omnivorous, ovo-lacto-vegetarian, and vegan diet. *Sci. Rep.*, **7**, 1–9, doi:10.1038/s41598-017-06466-8.

Ross, C.L., M. Orenstein, and N. Botchwey, 2014: Public health and community planning 101. In: *Health Impact Assessment in the United States* [Ross, C.L., M. Orenstein and N. Botchwey (eds.)]. Springer New York, New York, NY, 15–32.

Rossi, A., and Y. Lambrou, 2008: *Gender and equity issues in liquid biofuels production: Minimizing the risks to maximize the opportunities*. Food and Agriculture Organization of the United Nations, Rome, Italy, 33 pp.

Roy, J., P. Tschakert, and H. Waisman, 2018: Sustainable Development, Poverty eradication and Reducing Inequalities. In: Global Warming of 1.5°C an IPCC Special Report on the Impacts of Global Warming of 1.5°C Above Pre-Industrial Levels and Related Global Greenhouse Gas Emission Pathways, in the Context of Strengthening the Global Response to the Threat of Climate Change [Masson-Delmotte, V., P. Zhai, H.-O. Pörtner, D. Roberts, J. Skea, P.R. Shukla, A. Pirani, W. Moufouma-Okia, C. Péan,

R. Pidcock, S. Connors, J.B.R. Matthews, Y. Chen, X. Zhou, M.I. Gomis, E. Lonnoy, T. Maycock, M. Tignor, and T. Waterfield (eds.)]. In press.

Ruby, M.B., and S.J. Heine, 2011: Meat, morals, and masculinity. *Appetite*, **56**, 447–450, doi:10.1016/J.APPET.2011.01.018.

Ruel-Bergeron, J.C., G.A. Stevens, J.D. Sugimoto, F.F. Roos, M. Ezzati, R.E. Black, and K. Kraemer, 2015: Global update and trends of hidden hunger, 1995–2011: The hidden hunger Index. *PLoS One*, **10**, 1–13, doi:10.1371/journal.pone.0143497.

Ruel, M.T., J.L. Garrett, C. Hawkes, and M.J. Cohen, 2010: The food, fuel, and financial crises affect the urban and rural poor disproportionately: A review of the evidence. *J. Nutr.*, **140**, 170S–176S, doi:10.3945/jn.109.110791.

de Ruiter, M., P. Hudson, L. De Ruig, O. Kuik, and W. Botzen, 2017: A comparative study of European insurance schemes for extreme weather risks and incentives for risk reduction. *EGU General Assembly*, **19**, EGU2017-14716.

Rumpold, B.A., and O. Schlüter, 2015: Insect-based protein sources and their potential for human consumption: Nutritional composition and processing. *Anim. Front.*, **5**, 20–24, doi:10.2527/af.2015-0015.

Ryan, M.R., T.E. Crews, S.W. Culman, L.R. Dehaan, R.C. Hayes, J.M. Jungers, and M.G. Bakker, 2018: Managing for multifunctionality in perennial grain crops. *Bioscience*, **68**, 294–304 doi:10.1093/biosci/biy014.

Saberi, S., M. Kouhizadeh, J. Sarkis, and L. Shen, 2019: Blockchain technology and its relationships to sustainable supply chain management. *Int. J. Prod. Res.*, **57**, 2117–2135, doi:10.1080/00207543.2018.1533261.

Saki, M., M. Tarkesh Esfahani, and S. Soltani, 2018: A scenario-based modeling of climate change impacts on the above-ground net primary production in rangelands of central Iran. *Environ. Earth Sci.*, **77**, 670, doi:10.1007/s12665-018-7864-x.

Salou, T., C. Le Mouël, and H.M.G. van der Werf, 2017: Environmental impacts of dairy system intensification: The functional unit matters! *J. Clean. Prod.*, **140**, 445–454, doi:10.1016/J.JCLEPRO.2016.05.019.

Saltiel, A.R., and J.M. Olefsky, 2017: Inflammatory mechanisms linking obesity and metabolic disease. *J. Clin. Invest.*, **127**, 1–4, doi:10.1172/JCI92035.

Salvador, S., M. Corazzin, A. Romanzin, and S. Bovolenta, 2017: Greenhouse gas balance of mountain dairy farms as affected by grassland carbon sequestration. *J. Environ. Manage.*, **196**, 644–650, doi:10.1016/J.JENVMAN.2017.03.052.

Samy, A.M., and A.T. Peterson, 2016: Climate change influences on the global potential distribution of bluetongue virus. *PLoS One*, **11**, e0150489, doi:10.1371/journal.pone.0150489.

Sanderson, M.A., G. Brink, R. Stout, and L. Ruth, 2013: Grass–legume proportions in forage seed mixtures and effects on herbage yield and weed abundance. *Agron. J.*, **105**, 1289, doi:10.2134/agronj2013.0131.

Sans, P., and P. Combris, 2015: World meat consumption patterns: An overview of the last fifty years (1961–2011). *Meat Sci.*, **109**, 106–111, doi:10.1016/j.meatsci.2015.05.012.

Sanz, M.J. et al., 2017: *Sustainable Land Management Contribution to Successful Land-Based Climate Change Adaptation and Mitigation. A Report of the Science–Policy Interface*. A Report of the Science–Policy Interface, United Nations Convention to Combat Desertification (UNCCD), Bonn, Germany, 178 pp.

Sapkota, T. et al., 2017a: Soil organic carbon changes after seven years of conservation agriculture based rice-wheat cropping system in the eastern Indo-Gangetic Plain of India. *Soil Use Manag.*, 1–9, doi:10.1111/sum.12331.

Sapkota, T.B., M.L. Jat, J.P. Aryal, R.K. Jat, and A. Khatri-Chhetri, 2015: Climate change adaptation, greenhouse gas mitigation and economic profitability of conservation agriculture: Some examples from cereal systems of Indo-Gangetic Plains. *J. Integr. Agric.*, **14**, 1524–1533, doi:10.1016/S2095-3119(15)61093-0.

Sapkota, T.B., J.P. Aryal, A. Khatri-chhetri, P.B. Shirsath, P. Arumugam, and C.M. Stirling, 2017b: Identifying high-yield low-emission pathways for the cereal production in South Asia. *Mitig. Adapt. Strateg. Glob. Chang.*, doi:10.1007/s11027-017-9752-1.

Sapkota, T.B., V. Shankar, M. Rai, M.L. Jat, C.M. Stirling, L.K. Singh, H.S. Jat, and M.S. Grewal, 2017c: Reducing global warming potential through sustainable intensification of Basmati rice-wheat systems in India. *Sustain.*, **9**, 1–17, doi:10.3390/su9061044.

Sapkota, T.B.T.B., K. Majumdar, M.L. Jat, A. Kumar, D.K. Bishnoi, A.J. Mcdonald, and M. Pampolino, 2014: Precision nutrient management in conservation agriculture based wheat production of Northwest India: Profitability, nutrient use efficiency and environmental footprint. *F. Crop. Res.*, **155**, 233–244, doi:10.1016/j.fcr.2013.09.001.

Saxena, A.K., X.C. Fuentes, R.G. Herbas, and D.L. Humphries, 2016: Indigenous food systems and climate change: Impacts of climatic shifts on the production and processing of native and traditional crops in the Bolivian Andes. *Front. public Heal.*, **4**, 20, doi: 10.3389/fpubh.2016.00020.

Sayer, J. et al., 2013: Ten principles for a landscape approach to reconciling agriculture, conservation, and other competing land uses. *Proc. Natl. Acad. Sci.*, **110**, 21,, 8349-8356, doi:10.1073/pnas.1210595110.

Scanlon, B.R., B.L. Ruddell, P.M. Reed, R.I. Hook, C. Zheng, V.C. Tidwell, and S. Siebert, 2017: The food-energy-water nexus: Transforming science for society. *Water Resour. Res.*, **53**, 3550–3556, doi:10.1002/2017WR020889.

Scheba, A., 2017: Conservation agriculture and sustainable development in Africa: Insights from Tanzania. *Nat. Resour. Forum*, doi:10.1111/1477-8947.12123.

Scheelbeek, P.F.D. et al., 2018: Effect of environmental changes on vegetable and legume yields and nutritional quality. *Proc. Natl. Acad. Sci.*, **115**, 6804 LP-6809. www.pnas.org/content/115/26/6804.abstract.

Schenker, S., 2003: Undernutrition in the UK. *Bull. Nutr./Br. Nutr. Found.*, **28**, 87–120, doi:10.1046/j.1467-3010.2003.00303.x.

Schleussner, C.-F., J.F. Donges, R.V. Donner, and H.J. Schellnhuber, 2016: Armed-conflict risks enhanced by climate-related disasters in ethnically fractionalized countries. *Proc. Natl. Acad. Sci.*, **113**, 9216–9221, doi:10.1073/PNAS.1601611113.

Schleussner, C.-F., et al., 2018: Crop productivity changes in 1.5°C and 2°C worlds under climate sensitivity uncertainty. *Environ. Res. Lett.*, **13**, 64007, doi:10.1088/1748-9326/aab63b.

Schmidhuber, J., and F.N. Tubiello, 2007: Global food security under climate change. *Proc. Natl. Acad. Sci. USA.*, **104**, 19703–19708, doi:10.1073/pnas.0701976104.

Schmidtko, S., L. Stramma, and M. Visbeck, 2017: Decline in global oceanic oxygen content during the past five decades. *Nature*, **542**, 335, doi:10.1038/nature21399.

Schmitz, C. et al., 2014: Land-use change trajectories up to 2050: Insights from a global agro-economic model comparison. *Agric. Econ.*, **45**, 69–84, doi:10.1111/agec.12090.

Schmitz, O.J. et al., 2015: Conserving biodiversity: Practical guidance about climate change adaptation approaches in support of land-use planning. *Nat. Areas J.*, **1**, 190–203, doi:10.3375/043.035.0120.

Schramski, S., C. Mccarty, and G. Barnes, 2017: Household adaptive capacity: A social networks approach in rural South Africa. *Clim. Dev.*, **0**, 1–13, doi:10.1080/17565529.2017.1301861.

Schweiger, O. et al., 2010: Multiple stressors on biotic interactions: How climate change and alien species interact to affect pollination. *Biol. Rev.*, **85**, 777–795, doi:10.1111/j.1469-185X.2010.00125.x.

Scott, P., B. Tayler, and D. Walters, 2017: Lessons from implementing integrated water resource management: A case study of the North Bay-Mattawa Conservation Authority, Ontario. *Int. J. Water Resour. Dev.*, **33**, 393–407, doi:10.1080/07900627.2016.1216830.

Searchinger, T.D., S. Wirsenius, T. Beringer, and P. Dumas, 2018: Assessing the efficiency of changes in land use for mitigating climate change. *Nature*, **564**, 249–253, doi:10.1038/s41586-018-0757-z.

Sehmi, R., C. Mbow, S. Pitkanen, H. Cross, N. Berry, M. Riddell, J. Heiskanen, and E. Aynekulu, 2016: *Replicable Tools and Frameworks for Biocarbon Development in West Africa*. ICRAF Working Paper 237, World Agroforestry Centre, Nairobi, Kenya, 74 pp.

Selby, J., O.S. Dahi, C. Fröhlich, and M. Hulme, 2017: Climate change and the Syrian civil war revisited. *Polit. Geogr.*, **60**, 232–244, doi:10.1016/J.POLGEO.2017.05.007.

Senay, G.B. et al., 2015: Drought monitoring and assessment: Remote sensing and modeling approaches for the famine early warning systems network. In: *Hydro-Meteorological Hazards, Risks and Disasters* [Shroder, J.F., P. Paron, G. Di Baldassarre (eds.)]. Elsevier, Oxford, UK, pp. 233–262, doi:10.1016/B978-0-12-394846-5.00009-6.

Sepulcre-Canto, G., J. Vogt, A. Arboleda, and T. Antofie, 2014: Assessment of the EUMETSAT LSA-SAF evapotranspiration product for drought monitoring in Europe. *Int. J. Appl. Earth Obs. Geoinf.*, **30**, 190–202, doi:10.1016/j.jag.2014.01.021.

Settele, J., J. Bishop, and S.G. Potts, 2016: Climate change impacts on pollination. *Nat. Plants*, **2**, 16092, doi:10.1038/nplants.2016.92.

Shah, M., G. Fischer, and H. Van Velthuizen, 2008: *Food Security and Sustainable Agriculture: The Challenges of Climate Change in Sub-Saharan Africa*. International Institute for Applied Systems Analysis, Laxenburg, Austria, 40 pp.

Shepon, A., G. Eshel, E. Noor, and R. Milo, 2016: Energy and protein feed-to-food conversion efficiencies in the US and potential food security gains from dietary changes. *Environ. Res. Lett.*, **11**, 10, 105002, doi:10.1088/1748-9326/11/10/105002.

Shepon, A., G. Eshel, E. Noor, and R. Milo, 2018: The opportunity cost of animal-based diets exceeds all food losses. *Proc. Natl. Acad. Sci.*, **115**, 3804 LP-3809. www.pnas.org/content/115/15/3804.abstract.

Shi, Q., Y. Lin, E. Zhang, H. Yan, and J. Zhan, 2013: Impacts of Cultivated Land Reclamation on the Climate and Grain Production in Northeast China in the Future 30 Years. *Adv. inMeteorology*, **2013**, 1–8, doi:10.1155/2013/853098.

Shindell, D. et al., 2012: Simultaneously mitigating near-term climate change and improving human health and food security. *Science*, **335**, 183–189, doi:10.1126/science.1210026.

Shindell, D.T., 2016: Crop yield changes induced by emissions of individual climate-altering pollutants. *Earth's Futur.*, **4**, 373–380, doi:10.1002/2016EF000377.

Shindell, D.T., J.S. Fuglestvedt, and W.J. Collins, 2017: The social cost of methane: Theory and applications. *Faraday Discuss.*, **200**, 429–451, doi:10.1039/C7FD00009J.

Shove, E., M. Pantzar, and M. Watson, 2012: *The Dynamics of Social Practice: Everyday Life and How it Changes*. SAGE Publications Ltd, London, UK, 183 pp.

Shrestha, R.P., and N. Nepal, 2016: An assessment by subsistence farmers of the risks to food security attributable to climate change in Makwanpur, Nepal. *Food Secur.*, **8**, 415–425, doi:10.1007/s12571-016-0554-1.

Sibhatu, K.T., V.V. Krishna, and M. Qaim, 2015a: Production diversity and dietary diversity in smallholder farm households. *Proc. Natl. Acad. Sci.*, **112**, 10657–10662, doi:10.1073/pnas.1510982112.

Sibhatu, K.T., V. V Krishna, and M. Qaim, 2015b: Production diversity and dietary diversity in smallholder farm households. *PNAS*, **2015**, doi:10.1073/pnas.1510982112.

Sida, T.S., F. Baudron, H. Kim, and K.E. Giller, 2018: Climate-smart agroforestry: Faidherbia albida trees buffer wheat against climatic extremes in the Central Rift Valley of Ethiopia. *Agric. For. Meteorol.*, **248**, 339–347, doi:10.1016/J.AGRFORMET.2017.10.013.

Sietz, D., S.E. Mamani Choque, and M.K.B. Lüdeke, 2012: Typical patterns of smallholder vulnerability to weather extremes with regard to food security in the Peruvian Altiplano. *Reg. Environ. Chang.*, **12**, 489–505, doi:10.1007/s10113-011-0246-5.

Siliprandi, E., and G.P. Zuluaga Sánchez, 2014: *Género, agroecología y soberanía alimentaria: perspectivas ecofeministas*. Icaria, Barcelona, Spain, 240 pp.

Silva Oliveira, R. et al., 2017: Sustainable intensification of Brazilian livestock production through optimized pasture restoration. *Agric. Syst.*, **153**, 201–211, doi:10.1016/J.AGSY.2017.02.001.

Singh, P., K.J. Boote, M.D.M. Kadiyala, S. Nedumaran, S.K. Gupta, K. Srinivas, and M.C.S. Bantilan, 2017: An assessment of yield gains under climate change due to genetic modification of pearl millet. *Sci. Total Environ.*, **601–602**, 1226–1237, doi:10.1016/j.scitotenv.2017.06.002.

Skakun, S., N. Kussul, A. Shelestov, and O. Kussul, 2017: The use of satellite data for agriculture drought risk quantification in Ukraine. *Geomatics, Natural Hazards and Risk*, **7**, 3, 901–917 [, doi:10.1080/19475705.2015.1016555.

Skelsey, P., and A.C. Newton, 2015: Future environmental and geographic risks of Fusarium head blight of wheat in Scotland. *Eur. J. Plant Pathol.*, **142**, 133–147, doi:10.1007/s10658-015-0598-7.

Skoufias, E., M. Rabassa, and S. Olivieri, 2011: *The Poverty Impacts of Climate Change: A Review of the Evidence*. Policy Research Working Paper 5622, The World Bank, Washington, DC, USA.

Slade, P., 2018: If you build it, will they eat it? Consumer preferences for plant-based and cultured meat burgers. *Appetite*, **125**, 428–437, doi:10.1016/J.APPET.2018.02.030.

Smetana, S., A. Mathys, A. Knoch, and V. Heinz, 2015: Meat alternatives: Lifecycle assessment of most known meat substitutes. *Int. J. Life Cycle Assess.*, **20**, 1254–1267, doi:10.1007/s11367-015-0931-6.

Smith, P., 2013: Delivering food security without increasing pressure on land. *Glob. Food Sec.*, **2**, 18–23, doi:10.1016/j.gfs.2012.11.008.

Smith, P. et al., 2013: How much land-based greenhouse gas mitigation can be achieved without compromising food security and environmental goals? *Glob. Chang. Biol.*, **19**, 2285–2302, doi:10.1111/gcb.12160.

Smith P., M. Bustamante, H. Ahammad, H. Clark, H. Dong, E.A. Elsiddig, H. Haberl, R. Harper, J. House, M. Jafari, O. Masera, C. Mbow, N.H. Ravindranath, C.W. Rice, C. Robledo Abad, A. Romanovskaya, F. Sperling, and F. Tubiello, 2014: Agriculture, forestry and other land use (AFOLU). In: *Climate Change 2014: Mitigation of Climate Change. Contribution of Working Group III to the Fifth Assessment Report of the Intergovernmental Panel on Climate Change* [Edenhofer, O., R. Pichs-Madruga, Y. Sokona, E. Farahani, S. Kadner, K. Seyboth, A. Adler, I. Baum, S. Brunner, P. Eickemeier, B. Kriemann, J. Savolainen, S. Schlömer, C. von Stechow, T. Zwickel and J.C. Minx (eds.)]. Cambridge University Press, Cambridge, United Kingdom and New York, NY, USA 811–922.

Smith, P. et al., 2016: Global change pressures on soils from land use and management. *Glob. Chang. Biol.*, **22**, 1008–1028, doi:10.1111/gcb.13068.

Smithers, G.W., 2015: Food science – Yesterday, today and tomorrow. In: *Reference Module in Food Science*. Elsevier, Oxford, UK 31 pp, doi:10.1016/B978-0-08-100596-5.03337-0.

Soares-Filho, B., R. Rajão, M. Macedo, A. Carneiro, W. Costa, M. Coe, H. Rodrigues, and A. Alencar, 2014: Land use. Cracking Brazil's forest code. *Science*, **344**, 363–364, doi:10.1126/science.1246663.

Sommer, M., S. Ferron, S. Cavill, and S. House, 2015: Violence, gender and WASH: Spurring action on a complex, under-documented and sensitive topic. *Environ. Urban.*, **27**, 105–116, doi:10.1177/0956247814564528.

Song, G., M. Li, P. Fullana-i-Palmer, D. Williamson, and Y. Wang, 2017: Dietary changes to mitigate climate change and benefit public health in China. *Sci. Total Environ.*, **577**, 289–298, doi:10.1016/j.scitotenv.2016.10.184.

Sousa, V.M.Z., S.M. Luz, A. Caldeira-Pires, F.S. Machado, and C.M. Silveira, 2017: Lifecycle assessment of biodiesel production from beef tallow in Brazil. *Int. J. Life Cycle Assess.*, **22**, 1837–1850, doi:10.1007/s11367-017-1396-6.

Soussana, J. et al., 2013: Managing grassland systems in a changing climate: The search for practical solutions. In: *Revitalising grasslands to sustain our communities: 22nd International Grassland Congress, 15 – 19 September 2013, Sydney, Australia* [Michalk, D.L., G.D. Millar, W.B. Badgery, K.M. Broadfoot, (eds.)]. New South Wales Department of Primary Industry, Orange, Australia, 18 pp.

Soussana, J.-F. et al., 2019: Matching policy and science: Rationale for the '4 per 1000 – soils for food security and climate' initiative. *Soil Tillage Res.*, **188**, 3–15, doi:10.1016/J.STILL.2017.12.002.

5

de Souza, N.R.D. et al., 2019: Sugarcane ethanol and beef cattle integration in Brazil. *Biomass and Bioenergy*, **120**, 448–457, doi:10.1016/J.BIOMBIOE.2018.12.012.

Spence, L.A., P. Liancourt, B. Boldgiv, P.S. Petraitis, and B.B. Casper, 2014: Climate change and grazing interact to alter flowering patterns in the Mongolian steppe. *Oecologia*, **175**, 251–260, doi:10.1007/s00442-014-2884-z.

van der Spiegel, M., H.J. van der Fels-Klerx, and H.J.P. Marvin, 2012: Effects of climate change on food safety hazards in the dairy production chain. *Food Res. Int.*, **46**, 201–208, doi:10.1016/j.foodres.2011.12.011.

Spires, M., S. Shackleton, and G. Cundill, 2014: Barriers to implementing planned community-based adaptation in developing countries: A systematic literature review. *Clim. Dev.*, **6**, 277–287, doi:10.1080/17565529.2014.886995.

Spratt, S., 2013: *Food Price Volatility and Financial Speculation*. Working Paper 047, Institute of Development Studies, Future Agricultures Consortium Secretariat at the University of Sussex, Brighton, UK, 22 pp.

Springmann, M. et al., 2016a: Global and regional health effects of future food production under climate change: A modelling study. *Lancet*, **387**, 1937–1946, doi:10.1016/S0140-6736(15)01156-3.

Springmann, M., H.C.J. Godfray, M. Rayner, and P. Scarborough, 2016b: Analysis and valuation of the health and climate change co-benefits of dietary change. *Proc. Natl. Acad. Sci.*, **113**, 4146–4151, doi:10.1073/pnas.1523119113.

Springmann, M., D. Mason-D'Croz, S. Robinson, K. Wiebe, H.C.J. Godfray, M. Rayner, and P. Scarborough, 2017: Mitigation potential and global health impacts from emissions pricing of food commodities. *Nat. Clim. Chang.*, **7**, 69–74, doi:10.1038/nclimate3155.

Springmann, M. et al., 2018a: Options for keeping the food system within environmental limits. *Nature*, **562**, 519–525, doi:10.1038/s41586-018-0594-0.

Springmann, M. et al., 2018b: Health and nutritional aspects of sustainable diet strategies and their association with environmental impacts: A global modelling analysis with country-level detail. *Lancet. Planet. Heal.*, **2**, e451–e461, doi:10.1016/S2542-5196(18)30206-7.

Srbinovska, M., C. Gavrovski, V. Dimcev, A. Krkoleva, and V. Borozan, 2015: Environmental parameters monitoring in precision agriculture using wireless sensor networks. *J. Clean. Prod.*, **88**, 297–307, doi:10.1016/j.jclepro.2014.04.036.

Srinivasa Rao, C., K.A. Gopinath, J.V.N.S. Prasad, Prasannakumar, and A.K. Singh, 2016: Climate resilient villages for sustainable food security in tropical India: Concept, process, technologies, institutions, and impacts. *Adv. Agron.*, **140**, 101–214, doi:10.1016/BS.AGRON.2016.06.003.

Stehfest, E., et al., 2009: Climate benefits of changing diet. *Clim. Change*, **95**, 83–102, doi:10.1007/s10584-008-9534-6.

Stehfest, E. et al., 2019: Key determinants of global land-use futures. *Nat. Commun.*, **10**, 2166, doi:10.1038/s41467-019-09945-w.

Steinfeld, H., H.A. Mooney, F. Schneider, and L.E. Neville, 2010: *Livestock in a Changing Landscape. Volume 1, Drivers, Consequences, and Responses*. Island Press, Washington, DC, USA, 396 pp.

Stephens, N., L. Di Silvio, I. Dunsford, M. Ellis, A. Glencross, and A. Sexton, 2018: Bringing cultured meat to market: Technical, socio-political, and regulatory challenges in cellular agriculture. *Trends Food Sci. Technol.*, **78**, 155–166, doi:10.1016/J.TIFS.2018.04.010.

Stojanov, R., B. Duží, I. Kelman, D. Němec, and D. Procházka, 2017: Local perceptions of climate change impacts and migration patterns in Malé, Maldives. *Geogr. J.*, **183**, 370–385, doi:10.1111/geoj.12177.

Strassburg, B.B.N. et al., 2014: When enough should be enough: Improving the use of current agricultural lands could meet production demands and spare natural habitats in Brazil. *Glob. Environ. Chang.*, **28**, 84–97, doi:10.1016/j.gloenvcha.2014.06.001.

Subash, N., H. Singh, B. Gangwar, G. Baigorria, A.K. Sikka, and R.O. Valdivia, 2015: Integrated climate change assessment through linking crop simulation with economic modeling – Results from the Indo-Gangetic Basin. In: *Handbook of Climate Change and Agroecosystems: The Agricultural Model Intercomparison and Improvement Project Integrated Crop and Economic Assessments – Joint Publication with American Society of Agronomy, Crop Science Society of America, and Soil Science Society of America (In 2 Parts) (Vol. 3)*. Imperial College Press, London, UK. [Rosenzweig, C. and D. Hillel (eds.)]. Part 2, pp. 259–280.

Subbarao, G. V et al., 2009: Evidence for biological nitrification inhibition in Brachiaria pastures. *Proc. Natl. Acad. Sci.*, **106**, 17302–17307, doi:10.1073/pnas.0903694106.

Sugiura, T., H. Ogawa, N. Fukuda, and T. Moriguchi, 2013: Changes in the taste and textural attributes of apples in response to climate change. *Sci. Rep.*, **3**, 2418, doi:10.1038/srep02418.

Sultan, B., and M. Gaetani, 2016: Agriculture in West Africa in the 21st century: Climate change and impacts scenarios, and potential for adaptation. *Front. Plant Sci.*, **7**, 1–20, doi:10.3389/fpls.2016.01262.

Swinburn, B.A. et al., 2019: The global syndemic of obesity, undernutrition, and climate change: The Lancet commission report. *Lancet (London, England)*, **393**, 791–846, doi:10.1016/S0140-6736(18)32822-8.

Tadasse, G., B. Algieri, M. Kalkuhl, and J. von Braun, 2016: Drivers and Triggers of International Food Price Spikes and Volatility. In: *Food Price Volatility and Its Implications for Food Security and Policy* [Kalkuhl, M., J. von Braun, M. Torero (eds.)]. Springer International Publishing, Cham, Switzerland, pp. 59–82.

Takakura, J., S. Fujimori, K. Takahashi, and M. Hijioka, Yasuaki and Tomoko Hasegawa and Yasushi HondaToshihiko, 2017: Cost of preventing workplace heat-related illness through worker breaks and the benefit of climate-change mitigation. *Environ. Res. Lett.*, **12**, 64010.

Tao, F., et al., 2014: Responses of wheat growth and yield to climate change in different climate zones of China, 1981–2009. *Agric. For. Meteorol.*, **189–190**, 91–104, doi:10.1016/J.AGRFORMET.2014.01.013.

Tariq, M. et al., 2018: The impact of climate warming and crop management on phenology of sunflower-based cropping systems in Punjab, Pakistan. *Agric. For. Meteorol.*, **256–257**, 270–282, doi:10.1016/J.AGRFORMET.2018.03.015.

Taub, D.R., and X. Wang, 2008: Why are nitrogen concentrations in plant tissues lower under elevated CO_2? A critical examination of the hypotheses. *J. Integr. Plant Biol.*, **50**, 1365–1374, doi:10.1111/j.1744-7909.2008.00754.x.

Taylor, M., 2018: Climate-smart agriculture: What is it good for? *J. Peasant Stud.*, **45**, 89–107, doi:10.1080/03066150.2017.1312355.

Tebaldi, C., and D. Lobell, 2018: Estimated impacts of emission reductions on wheat and maize crops. *Clim. Change*, **146**, 533–545, doi:10.1007/s10584-015-1537-5.

Teixeira, E.I., J. De Ruiter, A. Ausseil, A. Daigneault, P. Johnstone, A. Holmes, A. Tait, and F. Ewert, 2017: Adapting crop rotations to climate change in regional impact modelling assessments. *Sci. Total Environ.*, **616–617**, 785–795, doi:10.1016/j.scitotenv.2017.10.247.

Terblanche, J.S., S. Clusella-Trullas, J.A. Deere, and S.L. Chown, 2008: Thermal tolerance in a south-east African population of the tsetse fly Glossina pallidipes (Diptera, Glossinidae): Implications for forecasting climate change impacts. *J. Insect Physiol.*, **54**, 114–127, doi:10.1016/J.JINSPHYS.2007.08.007.

Termeer, C.J.A.M., A. Dewulf, G.R. Biesbroek, C.J.A.M. Termeer, A. Dewulf, and G.R. Biesbroek, 2016: Transformational change: Governance interventions for climate change adaptation from a continuous change perspective. *J. Environ. Plan. Manag.*, **568**, doi:10.1080/09640568.2016.1168288.

Termeer, C.J.A.M., A. Dewulf, and G.R. Biesbroek, 2017: Transformational change: Governance interventions for climate change adaptation from a continuous change perspective. *J. Environ. Plan. Manag.*, **60**, 558–576, doi:10.1080/09640568.2016.1168288.

5

6

Interlinkages between desertification, land degradation, food security and greenhouse gas fluxes: Synergies, trade-offs and integrated response options

Coordinating Lead Authors:
Pete Smith (United Kingdom), Johnson Nkem (Cameroon), Katherine Calvin (The United States of America)

Lead Authors:
Donovan Campbell (Jamaica), Francesco Cherubini (Norway/Italy), Giacomo Grassi (Italy/European Union), Vladimir Korotkov (The Russian Federation), Anh Le Hoang (Viet Nam), Shuaib Lwasa (Uganda), Pamela McElwee (The United States of America), Ephraim Nkonya (Tanzania), Nobuko Saigusa (Japan), Jean-Francois Soussana (France), Miguel Angel Taboada (Argentina)

Contributing Authors:
Cristina Arias-Navarro (Spain), Otavio Cavalett (Brazil), Annette Cowie (Australia), Joanna House (United Kingdom), Daniel Huppmann (Austria), Jagdish Krishnaswamy (India), Alexander Popp (Germany), Stephanie Roe (The Philippines/The United States of America), Raphael Slade (United Kingdom), Lindsay Stringer (United Kingdom), Matteo Vizzarri (Italy)

Review Editors:
Amjad Abdulla (Maldives), Ian Noble (Australia), Yoshiki Yamagata (Japan), Taha Zatari (Saudi Arabia)

Chapter Scientists:
Frances Manning (United Kingdom), Dorothy Nampanzira (Uganda)

This chapter should be cited as:
Smith, P., J. Nkem, K. Calvin, D. Campbell, F. Cherubini, G. Grassi, V. Korotkov, A.L. Hoang, S. Lwasa, P. McElwee, E. Nkonya, N. Saigusa, J.-F. Soussana, M.A. Taboada, 2019: Interlinkages Between Desertification, Land Degradation, Food Security and Greenhouse Gas Fluxes: Synergies, Trade-offs and Integrated Response Options. In: *Climate Change and Land: an IPCC special report on climate change, desertification, land degradation, sustainable land management, food security, and greenhouse gas fluxes in terrestrial ecosystems* [P.R. Shukla, J. Skea, E. Calvo Buendia, V. Masson-Delmotte, H.- O. Portner, D. C. Roberts, P. Zhai, R. Slade, S. Connors, R. van Diemen, M. Ferrat, E. Haughey, S. Luz, S. Neogi, M. Pathak, J. Petzold, J. Portugal Pereira, P. Vyas, E. Huntley, K. Kissick, M. Belkacemi, J. Malley, (eds.)]. https://doi.org/10.1017/9781009157988.008

Table of contents

Executive summary 553

6.1   Introduction 556

   6.1.1   Context of this chapter 556

   6.1.2   Framing social challenges and
           acknowledging enabling factors 556

   6.1.3   Challenges and response options in current
           and historical interventions 558

   **Box 6.1 | Case studies by anthrome type
   showing historical interlinkages between
   land-based challenges and the development
   of local responses** 561

   6.1.4   Challenges represented in future scenarios 564

**6.2   Response options, co-benefits and adverse
       side effects across the land challenges** 565

   6.2.1   Integrated response options based
           on land management 569

   6.2.2   Integrated response options based
           on value chain management 576

   6.2.3   Integrated response options based
           on risk management 576

   **Cross-Chapter Box 7 | Bioenergy
   and bioenergy with carbon capture and
   storage (BECCS) in mitigation scenarios** 580

6.3   Potentials for addressing the land challenges 583

   6.3.1   Potential of the integrated response
           options for delivering mitigation 583

   6.3.2   Potential of the integrated response
           options for delivering adaptation 589

   6.3.3   Potential of the integrated response
           options for addressing desertification 595

   6.3.4   Potential of the integrated response
           options for addressing land degradation 599

   6.3.5   Potential of the integrated response
           options for addressing food security 603

   6.3.6   Summarising the potential of the
           integrated response options across
           mitigation, adaptation, desertification
           land degradation and food security 609

6.4   Managing interactions and interlinkages 618

   6.4.1   Feasibility of the integrated response
           options with respect to costs, barriers,
           saturation and reversibility 618

   6.4.2   Sensitivity of the integrated response
           options to climate change impacts 623

   **Cross-Chapter Box 8 | Ecosystem services
   and Nature's Contributions to People, and
   their relation to the land–climate system** 625

   6.4.3   Impacts of integrated response
           options on Nature's Contributions
           to People (NCP) and the UN Sustainable
           Development Goals (SDGs) 627

   6.4.4   Opportunities for implementing
           integrated response options 633

   **Cross-Chapter Box 9 | Climate and
   land pathways** 641

   6.4.5   Potential consequences of delayed action 644

Frequently Asked Questions 646

   **FAQ 6.1: What types of land-based options can
   help mitigate and adapt to climate change?** 646

   **FAQ 6.2 Which land-based mitigation
   measures could affect desertification,
   land degradation or food security?** 646

   **FAQ 6.3: What is the role of bioenergy
   in climate change mitigation, and what
   are its challenges?** 646

References 647

Executive summary

The land challenges, in the context of this report, are climate change mitigation, adaptation, desertification, land degradation, and food security. The chapter also discusses implications for Nature's Contributions to People (NCP), including biodiversity and water, and sustainable development, by assessing intersections with the Sustainable Development Goals (SDGs). The chapter assesses response options that could be used to address these challenges. These response options were derived from the previous chapters and fall into three broad categories: land management, value chain, and risk management.

The land challenges faced today vary across regions; climate change will increase challenges in the future, while socio-economic development could either increase or decrease challenges (*high confidence*). Increases in biophysical impacts from climate change can worsen desertification, land degradation, and food insecurity (*high confidence*). Additional pressures from socio-economic development could further exacerbate these challenges; however, the effects are scenario dependent. Scenarios with increases in income and reduced pressures on land can lead to reductions in food insecurity; however, all assessed scenarios result in increases in water demand and water scarcity (*medium confidence*). {6.1}

The applicability and efficacy of response options are region and context specific; while many value chain and risk management options are potentially broadly applicable, many land management options are applicable on less than 50% of the ice-free land surface (*high confidence*). Response options are limited by land type, bioclimatic region, or local food system context (*high confidence*). Some response options produce adverse side effects only in certain regions or contexts; for example, response options that use freshwater may have no adverse side effects in regions where water is plentiful, but large adverse side effects in regions where water is scarce (*high confidence*). Response options with biophysical climate effects (e.g., afforestation, reforestation) may have different effects on local climate, depending on where they are implemented (*medium confidence*). Regions with more challenges have fewer response options available for implementation (*medium confidence*). {6.1, 6.2, 6.3, 6.4}

Nine options deliver medium-to-large benefits for all five land challenges (*high confidence*). The options with medium-to-large benefits for all challenges are increased food productivity, improved cropland management, improved grazing land management, improved livestock management, agroforestry, forest management, increased soil organic carbon content, fire management and reduced post-harvest losses. A further two options, dietary change and reduced food waste, have no global estimates for adaptation but have medium-to-large benefits for all other challenges (*high confidence*). {6.3, 6.4}

Five options have large mitigation potential (>3 GtCO$_2$e yr^{-1}) without adverse impacts on the other challenges (*high confidence*). These are: increased food productivity; reduced deforestation and forest degradation; increased soil organic carbon content; fire management; and reduced post-harvest losses.

Two further options with large mitigation potential, dietary change and reduced food waste, have no global estimates for adaptation but show no negative impacts across the other challenges. Five options: improved cropland management; improved grazing land managements; agroforestry; integrated water management; and forest management, have moderate mitigation potential, with no adverse impacts on the other challenges (*high confidence*). {6.3.6}

Sixteen response options have large adaptation potential (more than 25 million people benefit), without adverse side effects on other land challenges (*high confidence*). These are increased food productivity, improved cropland management, agroforestry, agricultural diversification, forest management, increased soil organic carbon content, reduced landslides and natural hazards, restoration and reduced conversion of coastal wetlands, reduced post-harvest losses, sustainable sourcing, management of supply chains, improved food processing and retailing, improved energy use in food systems, livelihood diversification, use of local seeds, and disaster risk management (*high confidence*). Some options (such as enhanced urban food systems or management of urban sprawl) may not provide large global benefits but may have significant positive local effects without adverse effects (*high confidence*). {6.3, 6.4}

Seventeen of 40 options deliver co-benefits or no adverse side effects for the full range of NCPs and SDGs; only three options (afforestation, bioenergy and bioenergy with carbon capture and storage (BECCS), and some types of risk sharing instruments, such as insurance) have potentially adverse side effects for five or more NCPs or SDGs (*medium confidence*). The 17 options with co-benefits and no adverse side effects include most agriculture- and soil-based land management options, many ecosystem-based land management options, forest management, reduced post-harvest losses, sustainable sourcing, improved energy use in food systems, and livelihood diversification (*medium confidence*). Some of the synergies between response options and SDGs include positive poverty eradication impacts from activities like improved water management or improved management of supply chains. Examples of synergies between response options and NCPs include positive impacts on habitat maintenance from activities like invasive species management and agricultural diversification. However, many of these synergies are not automatic, and are dependent on well-implemented activities requiring institutional and enabling conditions for success. {6.4}

Most response options can be applied without competing for available land; however, seven options result in competition for land (*medium confidence*). A large number of response options do not require dedicated land, including several land management options, all value chain options, and all risk management options. Four options could greatly increase competition for land if applied at scale: afforestation, reforestation, and land used to provide feedstock for BECCS or biochar, with three further options: reduced grassland conversion to croplands, restoration and reduced conversion of peatlands and restoration, and reduced conversion of coastal wetlands having smaller or variable impacts on competition for land. Other options such as reduced deforestation and forest degradation, restrict land conversion for other options and uses.

6

Expansion of the current area of managed land into natural ecosystems could have negative consequences for other land challenges, lead to the loss of biodiversity, and adversely affect a range of NCPs (*high confidence*). {6.3.6, 6.4}

Some options, such as bioenergy and BECCS, are scale dependent. The climate change mitigation potential for bioenergy and BECCS is large (up to 11 GtCO$_2$ yr^{-1}); however, the effects of bioenergy production on land degradation, food insecurity, water scarcity, greenhouse gas (GHG) emissions, and other environmental goals are scale- and context-specific (*high confidence*). These effects depend on the scale of deployment, initial land use, land type, bioenergy feedstock, initial carbon stocks, climatic region and management regime (*high confidence*). Large areas of monoculture bioenergy crops that displace other land uses can result in land competition, with adverse effects for food production, food consumption, and thus food security, as well as adverse effects for land degradation, biodiversity, and water scarcity (*medium confidence*). However, integration of bioenergy into sustainably managed agricultural landscapes can ameliorate these challenges (*medium confidence*). {6.2, 6.3, 6.4, Cross-Chapter Box 7 in this chapter}

Response options are interlinked; some options (e.g., land sparing and sustainable land management options) can enhance the co-benefits or increase the potential for other options (*medium confidence*). Some response options can be more effective when applied together (*medium confidence*); for example, dietary change and waste reduction expand the potential to apply other options by freeing as much as 5.8 Mkm^2 (0.8–2.4 Mkm^2 for dietary change; about 2 Mkm^2 for reduced post-harvest losses, and 1.4 Mkm^2 for reduced food waste) of land (*low confidence*). Integrated water management and increased soil organic carbon can increase food productivity in some circumstances. {6.4}

Other response options (e.g., options that require land) may conflict; as a result, the potentials for response options are not all additive, and a total potential from the land is currently unknown (*high confidence*). Combining some sets of options (e.g., those that compete for land) may mean that maximum potentials cannot be realised, for example, reforestation, afforestation, and bioenergy and BECCS, all compete for the same finite land resource so the combined potential is much lower than the sum of potentials of each individual option, calculated in the absence of alternative uses of the land (*high confidence*). Given the interlinkages among response options and that mitigation potentials for individual options assume that they are applied to all suitable land, the total mitigation potential is much lower than the sum of the mitigation potential of the individual response options (*high confidence*). {6.4}

The feasibility of response options, including those with multiple co-benefits, is limited due to economic, technological, institutional, socio-cultural, environmental and geophysical barriers (*high confidence*). A number of response options (e.g., most agriculture-based land management options, forest management, reforestation and restoration) have already been implemented widely to date (*high confidence*). There is robust evidence that many other response options can deliver co-benefits across the range of

land challenges, yet these are not being implemented. This limited application is evidence that multiple barriers to implementation of response options exist (*high confidence*). {6.3, 6.4}

Coordinated action is required across a range of actors, including business, producers, consumers, land managers, indigenous peoples and local communities and policymakers to create enabling conditions for adoption of response options (*high confidence*). The response options assessed face a variety of barriers to implementation (economic, technological, institutional, socio-cultural, environmental and geophysical) that require action across multiple actors to overcome (*high confidence*). There are a variety of response options available at different scales that could form portfolios of measures applied by different stakeholders – from farm to international scales. For example, agricultural diversification and use of local seeds by smallholders can be particularly useful poverty eradication and biodiversity conservation measures, but are only successful when higher scales, such as national and international markets and supply chains, also value these goods in trade regimes, and consumers see the benefits of purchasing these goods. However, the land and food sectors face particular challenges of institutional fragmentation, and often suffer from a lack of engagement between stakeholders at different scales (*medium confidence*). {6.3, 6.4}

Delayed action will result in an increased need for response to land challenges and a decreased potential for land-based response options due to climate change and other pressures (*high confidence*). For example, failure to mitigate climate change will increase requirements for adaptation and may reduce the efficacy of future land-based mitigation options (*high confidence*). The potential for some land management options decreases as climate change increases; for example, climate alters the sink capacity for soil and vegetation carbon sequestration, reducing the potential for increased soil organic carbon (*high confidence*). Other options (e.g., reduced deforestation and forest degradation) prevent further detrimental effects to the land surface; delaying these options could lead to increased deforestation, conversion, or degradation, serving as increased sources of GHGs and having concomitant negative impacts on NCPs (*medium confidence*). Carbon dioxide removal (CDR) options – such as reforestation, afforestation, bioenergy and BECCS – are used to compensate for unavoidable emissions in other sectors; delayed action will result in larger and more rapid deployment later (*high confidence*). Some response options will not be possible if action is delayed too long; for example, peatland restoration might not be possible after certain thresholds of degradation have been exceeded, meaning that peatlands could not be restored in certain locations (*medium confidence*). {6.2, 6.3, 6.4}

Early action, however, has challenges including technological readiness, upscaling, and institutional barriers (*high confidence*). Some of the response options have technological barriers that may limit their wide-scale application in the near term (*high confidence*). Some response options, for example, BECCS, have only been implemented at small-scale demonstration facilities; challenges exist with upscaling these options to the levels discussed in this Chapter (*medium confidence*). Economic and institutional barriers, including governance, financial incentives and financial resources,

6

limit the near-term adoption of many response options, and 'policy lags', by which implementation is delayed by the slowness of the policy implementation cycle, are significant across many options (*medium confidence*). Even some actions that initially seemed like 'easy wins' have been challenging to implement, with stalled policies for reducing emissions from deforestation and forest degradation and fostering conservation (REDD+) providing clear examples of how response options need sufficient funding, institutional support, local buy-in, and clear metrics for success, among other necessary enabling conditions. {6.2, 6.4}

Some response options reduce the consequences of land challenges, but do not address underlying drivers (*high confidence*). For example, management of urban sprawl can help reduce the environmental impact of urban systems; however, such management does not address the socio-economic and demographic changes driving the expansion of urban areas. By failing to address the underlying drivers, there is a potential for the challenge to re-emerge in the future (*high confidence*). {6.4}

Many response options have been practised in many regions for many years; however, there is limited knowledge of the efficacy and broader implications of other response options (*high confidence*). For the response options with a large evidence base and ample experience, further implementation and upscaling would carry little risk of adverse side effects (*high confidence*). However, for other options, the risks are larger as the knowledge gaps are greater; for example, uncertainty in the economic and social aspects of many land response options hampers the ability to predict their effects (*medium confidence*). Furthermore, Integrated Assessment Models, like those used to develop the pathways in the IPCC Special Report on Global Warming of 1.5°C (SR15), omit many of these response options and do not assess implications for all land challenges (*high confidence*). {6.4}

6.1   Introduction

6.1.1   Context of this chapter

This chapter focuses on the interlinkages between response options[1] to deliver climate mitigation and adaptation, to address desertification and land degradation, and to enhance food security. It also assesses reported impacts on Nature's Contributions to People (NCP) and contributions to the UN Sustainable Development Goals (SDGs). By identifying which options provide the most co-benefits with the fewest adverse side effects, this chapter aims to provide *integrated response options* that could co-deliver across the range of challenges. This chapter *does not consider* response options that affect only one aspect of climate mitigation, adaptation, desertification, land degradation, or food security in isolation, since these are the subjects of Chapters 2–5; this chapter *considers only* interlinkages between response options, and two or more of these challenges in the land sector.

Since the aim is to assess and provide guidance on integrated response options, each response option is first described and categorised, drawing on previous chapters 2–5 (Section 6.2), and their impact on climate mitigation/adaptation, desertification, land degradation, and food security is quantified (Section 6.3). The feasibility of each response option, respect to costs, barriers, saturation and reversibility is then assessed (Section 6.4.1), before considering their sensitivity to future climate change (Section 6.4.2).

The *co-benefits* and *adverse side effects*[2] of each integrated response option across the five land challenges, and their impacts on the NCP and the SDGs, are then assessed in Section 6.4.3. In section 6.4.4, the spatial applicability of these integrated response options is assessed in relation to the location of the challenges, with the aim of identifying which options have the greatest potential to co-deliver across the challenges, and the contexts and circumstances in which they do so. Interlinkages among response options and challenges in future scenarios are also assessed in Section 6.4.4. Finally, Section 6.4.5 discusses the potential consequences of delayed action.

In providing this evidence-based assessment, drawing on the relevant literature, this chapter does not assess the merits of policies to deliver these integrated response options – Chapter 7 assesses the various policy options currently available to deliver these interventions. Rather, this chapter provides an assessment of the integrated response options and their ability to co-deliver across the multiple challenges addressed in this Special Report.

6.1.2   Framing social challenges and acknowledging enabling factors

In this section we outline the approach used in assessing the evidence for interactions between response options to deliver climate mitigation and adaptation, to prevent desertification and land degradation, and to enhance food security. Overall, while defining and presenting the response options to meet these goals is the primary goal of this chapter, we note that these options must not be considered only as technological interventions, or one-off actions. Rather, they need to be understood as responses to socio-ecological challenges whose success will largely depend on external enabling factors. There have been many previous efforts at compiling positive response options that meet numerous SDGs, but which have not resulted in major shifts in implementation; for example, online databases of multiple response options for sustainable land management (SLM), adaptation, and other objectives have been compiled by many donor agencies, including World Overview of Conservation Approaches and Technologies (WOCAT), Climate Adapt, and the Adaptation Knowledge Portal (Schwilch et al. 2012b).[3] Yet, clearly barriers to adoption remain, or these actions would have been more widely used by now. Much of the scientific literature on barriers to implementing response options focuses on the individual and household level, and discusses limits to adoption, often primarily identified as economic factors (Nigussie et al. 2017; Dallimer et al. 2018). While a useful approach, such studies are often unable to account for the larger enabling factors that might assist in more wide-scale implementation (Chapter 7 discusses these governance factors and associated barriers in more detail).

Instead, this chapter proposes that each response option identified and assessed needs to be understood as an intervention within complex socio-ecological systems (SES) (introduced in Chapter 1). In this understanding, physical changes affect human decision-making over land and risk management options, as do economics, policies, and cultural factors, which in turn may drive additional ecological change (Rawlins and Morris 2010). This co-evolution of responses within an SES provides a more nuanced understanding of the dynamics between drivers of change and impacts of interventions. Thus, in discussions of the 40 specific response options in this chapter, it must be kept in mind that all need to be contextualised within the specific SES in which they are deployed (Figure 6.1). Framing response options within SESs also recognises the interactions *between* different response options. However, a major problem within SESs is that the choice and use of different response options requires knowledge of the problems they are aimed at solving, which may be unclear, contested, or not shared equally among stakeholders (Carmenta et al. 2017). Drivers of environmental change often have primarily social or economic, rather than technological roots, which requires acknowledgement that the response options not aimed at reducing the drivers of change may thus be less successful (Schwilch et al. 2014).

---

[1]   Many of the response options considered are *sustainable land management* options, but several response options are not based on land management – for example, those based on value chain management and governance and risk management options.

[2]   We use the IPCC Fifth Assessment Report Working Group III definitions of co-benefits and adverse side effect – see Glossary. Co-benefits and adverse side effects can be biophysical and/or socio-economic in nature, and all are assessed as far as the literature allows.

[3]   For example, see https://qcat.wocat.net/en/wocat/; https://climate-adapt.eea.europa.eu; https://www4.unfccc.int/sites/NWPStaging/Pages/Home.aspx.

6

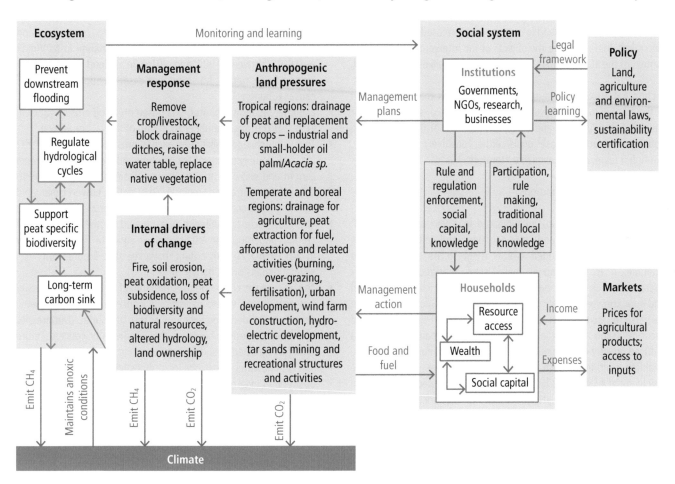

Figure 6.1 | Model to represent a social-ecological system of one of the integrated response options in this chapter, using restoration and reduced impact of peatlands as an example. The boxes show systems (ecosystem, social system), external and internal drivers of change and the management response – here enacting the response option. Unless included in the 'internal drivers of change' box, all other drivers of change are external (e.g., climate, policy, markets, anthropogenic land pressures). The arrows represent how the systems can influence each other, with key drivers of impact written in the arrow in the direction of effect.

Response options must also account for the uneven distribution of impacts among populations of both environmental change and intervention responses to this change. Understanding the integrated response options available in a given context requires an understanding of the specificities of social vulnerability, adaptive capacity, and institutional support to assist communities, households and regions to reach their capabilities and achievement of the SDG and other social and land management goals. Vulnerability reflects how assets are distributed within and among communities, shaped by factors that are not easily overcome with technical solutions, including inequality and marginalisation, poverty, and access to resources (Adger et al. 2004; Hallegate et al. 2016). Understanding why some people are vulnerable, and what structural factors perpetuate this vulnerability requires attention to both micro and meso scales (Tschakert et al. 2013). These vulnerabilities create barriers to adoption of even low-cost high-return response options, such as soil carbon management, that may seem obviously beneficial to implement (Mutoko et al. 2014; Cavanagh et al. 2017). Thus, assessment of the differentiated vulnerabilities that may prevent the adoption of a response option need to be considered as part of any package of interventions.

Adaptive capacity relates to the ability of institutions or people to modify or change characteristics or behaviour so as to cope better with existing or anticipated external stresses (Moss et al. 2001;

Brenkert and Malone 2005; Brooks et al. 2005). Adaptive capacity reflects institutional and policy support networks, and has often been associated at the national level with strong developments in the fields of economics, education, health, and governance and political rights (Smit et al. 2001). Areas with low adaptive capacity, as reflected in low Human Development Index scores, might constrain the ability of communities to implement response options (Section 6.4.4.1 and Figure 6.7).

Further, while environmental changes like land degradation have obvious social and cultural impacts, (as discussed in the preceding chapters), so do response options. Therefore, careful thought is needed about what impacts are expected and what trade-offs are acceptable. One potential way to assess the impact of response interventions relates to the idea of capabilities, a concept first proposed by economist Amartya Sen (Sen 1992). Understanding capability as the 'freedom to achieve well-being' frames a problem as being a matter of facilitating what people aspire to do and be, rather than telling them to achieve a standardised or predetermined outcome (Nussbaum and Sen 1993). Thus a capability approach is generally a more flexible and multi-purpose framework, appropriate to an SES understanding because of its open-ended approach (Bockstael and Berkes 2017). Thus, one question for any decision-maker approaching schematics of response options is to determine

6

which response options lead to increased or decreased capabilities for the stakeholders who are the objects of the interventions, given the context of the SES in which the response option will be implemented.

Section 6.4.3 examines some of the capabilities that are reflected in the UN Sustainable Development Goals (SDGs), such as gender equality and education, and assesses how each of the 40 response options may affect those goals, either positively or negatively, through a review of the available literature.

6.1.2.1   Enabling conditions

Response options are not implemented in a vacuum and rely on knowledge production and socio-economic and cultural strategies and approaches embedded within them to be successful. For example, it is well known that "Weak grassroots institutions characterised by low capacity, failure to exploit collective capital and poor knowledge sharing and access to information, are common barriers to sustainable land management and improved food security" (Oloo and Omondi 2017). Achieving broad goals such as reduced poverty or sustainable land management requires conducive enabling conditions, such as attention to gender issues and the involvement of stakeholders, such as indigenous peoples and local communities, as well as attention to governance, including adaptive governance, stakeholder engagement, and institutional facilitation (Section 6.4.4.3). These enabling conditions – such as gender-sensitive programming or community-based solutions – are not categorised as individual response options in subsequent sections of this chapter because they are conditions that can potentially help improve *all* response options when used in tandem to produce more sustainable outcomes. Chapter 7 picks up on these themes and discusses the ways various policies to implement response options have tried to minimise unwanted social and economic impacts on participants in more depth, through deeper analysis of concepts such as citizen science and adaptive governance. Here we simply note the importance of assessing the contexts in which response options will be delivered, as no two situations are the same, and no single response option is likely to be a 'silver bullet' to solve all land–climate problems; each option comes with potential challenges and trade-offs (Section 6.2), barriers to implementation (Section 6.4.1), interactions with other sectors of society (Section 6.4.3), and potential environmental limitations (Section 6.4.4).

6.1.3   Challenges and response options in current and historical interventions

Land-based systems are exposed to multiple overlapping challenges, including climate change (adaptation and mitigation), desertification (Chapter 3), land degradation (Chapter 4) and food insecurity (Chapter 5), as well as loss of biodiversity, groundwater stress (from over-abstraction) and water quality. The spatial distribution of these individual land-based challenges is shown in Figure 6.2, based on recent studies and using the following indicators:

- Desertification attributed to land use is estimated from vegetation remote sensing (Figure 3.7c), mean annual change in NDVImax <−0.001 (between 1982 and 2015) in dryland areas (Aridity Index >0.65), noting, however, that desertification has multiple causes (Chapter 3).
- Land degradation (Chapter 4) is based on a soil erosion (Borrelli et al. 2017) proxy (annual erosion rate of 3 t ha^{-1} or above).
- The climate change challenge for adaptation is based on a dissimilarity index of monthly means of temperature and precipitation between current and end-of-century scenarios (dissimilarity index equal to 0.7 or above; Netzel and Stepinski 2018), noting, however, that rapid warming could occur in all land regions (Chapter 2).
- The food security challenge is estimated as the prevalence of chronic undernourishment (higher or equal to 5%) by country in 2015 (FAO 2017a), noting, however, that food security has several dimensions (Chapter 5).
- The biodiversity challenge uses threatened terrestrial biodiversity hotspots (areas where exceptional concentrations of endemic species are undergoing exceptional loss of habitat, (Mittermeier et al. 2011), noting, however, that biodiversity concerns more than just threatened endemic species.
- The groundwater stress challenge is estimated as groundwater abstraction over recharge ratios above one (Gassert et al. 2014) in agricultural areas (croplands and villages).
- The water quality challenge is estimated as critical loads (higher or equal to 1000 kg N km^{-2} or 50 kg P km^{-2}) of nitrogen (N) and phosphorus (P) (Xie and Ringler 2017).

Overlapping land-based challenges affect all land-use categories: croplands, rangelands, semi-natural forests, villages, dense settlements, wild forests and sparse trees and barren lands. These land-use categories can be defined as anthropogenic biomes, or anthromes, and their global distribution was mapped by Ellis and Ramankutty (2008) (Figure 6.2).

The majority of the global population is concentrated in dense settlements and villages, accounting for less than 7% of the global ice-free land area, while croplands and rangelands use 39% of land. The remainder of the ice-free land area (more than half) is used by semi-natural forests, by wild forests, sparse trees and barren lands (Table 6.1).

Land-use types (or anthromes) are exposed to multiple overlapping challenges. Climate change could induce rapid warming in all land areas (Chapter 2). In close to 70% of the ice-free land area, the climate change adaptation challenge could be reinforced by a strong dissimilarity between end-of-century and current temperature and precipitation seasonal cycles (Netzel and Stepinski 2018). Chronic undernourishment (a component of food insecurity) is concentrated in 20% of global ice-free land area. Severe soil erosion (a proxy of land degradation) and desertification from land use affect 13% and 3% of ice-free land area, respectively. Both groundwater stress and severe water-quality decline (12% and 10% of ice-free land area, respectively) contribute to the water challenge. Threatened biodiversity hot-spots (15% of ice-free land area) are significant for the biodiversity challenge (Table 6.1).

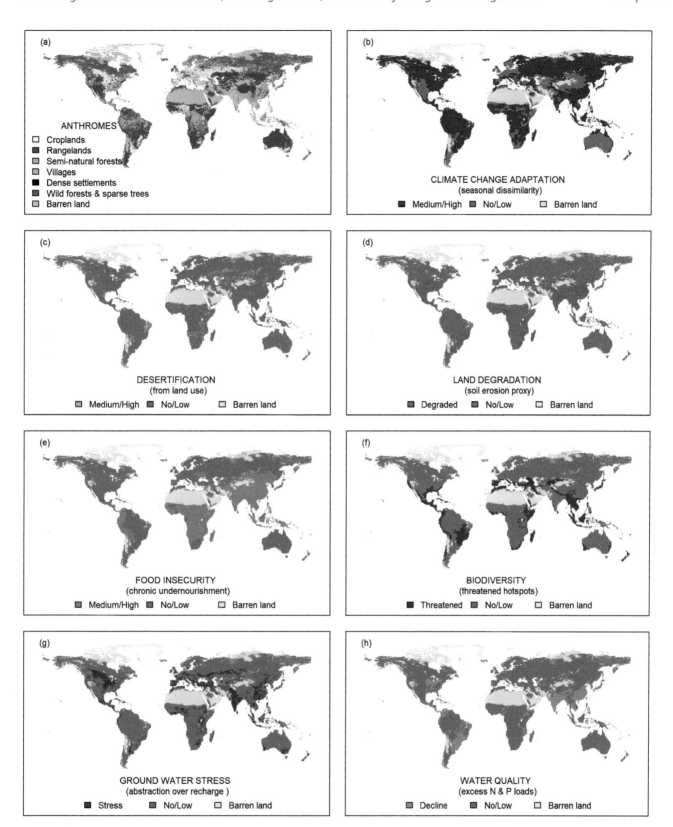

Figure 6.2 | Global distributions of land-use types and individual land-based challenges. (a) Land-use types (or anthromes, after Ellis and Ramankutty 2008); **(b)** Climate change adaptation challenge (estimated from the dissimilarity between current and end-of-century climate scenarios, Netzel and Stepinski 2018); **(c)** Desertification challenge (after Chapter 3, Figure 3.7c); **(d)** Land degradation challenge (estimated from a soil erosion proxy, one indicator of land degradation; Borrelli et al. 2017); **(e)** Food security challenge (estimated from chronic undernourishment, a component of food security, FAO 2017a); **(f)** biodiversity challenge (estimated from threatened biodiversity hotspots, a component of biodiversity, Mittermeier et al. 2011); **(g)** Groundwater stress challenge (estimated from water over-abstraction, Gassert et al. 2014); **(h)** Water quality challenge (estimated from critical nitrogen and phosphorus loads of water systems, Xie and Ringler 2017).

Table 6.1 | Global area of land-use types (or anthromes) and current percentage area exposure to individual (overlapping) land-based challenges. See Figure 6.2 and text for further details on criteria for individual challenges.

Land-use type (anthrome[a])	Anthrome area	Climate change adaptation (dissimilarity index proxy)[b]	Land degradation (soil erosion proxy)[c]	Desertifica-tion (ascribed to land use)[d]	Food security (chronic undernourish-ment)[e]	Biodiversity (threatened hotspot)[f]	Groundwater stress (over abstraction)[g]	Water quality (critical N-P loads)[h]
	% of ice-free land area[i]		% anthrome area exposed to an individual challenge					
Dense settlement	1	76	20	3	30	32	–	30
Village	5	70	49	3	78	28	77	59
Cropland	13	68	21	7	28	27	65	20
Rangeland	26	46	14	7	43	21	–	10
Semi-natural forests	14	91	17	0.7	–	21	–	7
Wild forests and sparse trees	17	98	4	0.5	–	2	–	0.3
Barren	19	53	6	0.9	2	4	–	0.4
*Organic soils	4	95	10	2	9	13	–	6
*Coastal wetlands	0.6	74	11	2	24	33	–	26
All anthromes	100	69	13	3.2	20	15	12	10

[a] Ellis and Ramankutty (2008); [b] Borrelli et al. 2017; [c] Netzel and Stepinski 2018; [d] from Figure 3.7c in Chapter 3; [e] FAO 2017a; [f] Mittermeier et al. 2011; [g] Gassert et al. 2014; [h] Xie and Ringler 2017; [i] the global ice-free land area is estimated at 134 Mkm².

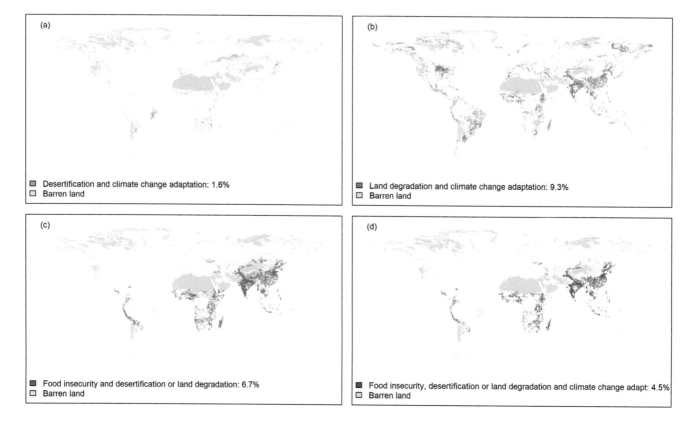

Figure 6.3 | Example of overlap between land challenges. (a) Overlap between the desertification (from land use) challenge and the climate change adaptation (strong dissimilarity in seasonal cycles) challenge. **(b)** Overlap between the land degradation (soil erosion proxy) challenge and the climate change adaptation challenge. **(c)** Overlap between the desertification or land degradation challenges and the food insecurity (chronic undernourishment) challenge. **(d)** Overlap between challenges shown in C and the climate change adaptation challenge. For challenges definitions, see text; references as in Figure 6.2.

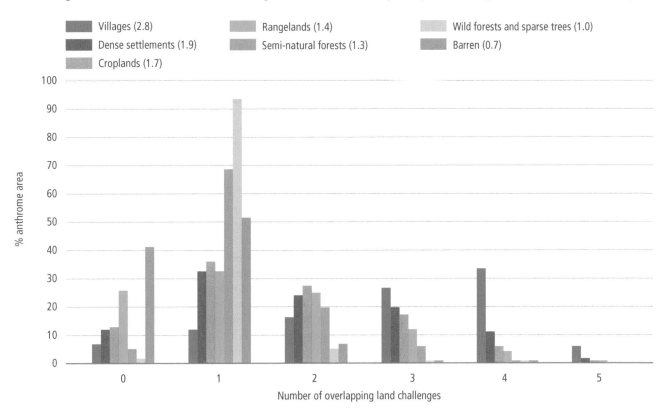

Figure 6.4 | Percentage distribution of land-use type (or anthrome) area by number of overlapping land challenges for the villages, dense settlements, croplands, rangelands, semi-natural forests, wild forests and sparse trees and barren land-use types. Values in brackets show the mean number of land challenges per land-use type. Land challenges include desertification (from land use), land degradation (soil erosion proxy), climate change adaptation (seasonal dissimilarity proxy), food security (chronic undernourishment), biodiversity (threatened hot spots), groundwater stress (over abstraction) and water quality (critical nitrogen and phosphorus loads).

Since land-based challenges overlap, part of the ice-free land area is exposed to combinations of two or more challenges. For instance, land degradation (severe soil erosion) or desertification from land use and food insecurity (chronic undernourishment) are combined with a strong climate change adaptation challenge (dissimilarity in seasonal cycles) in 4.5% of the ice-free land area (Figure 6.3).

The global distribution of land area by the number of overlapping land challenges (Figure 6.4) shows: the least exposure to land challenges in barren lands; less frequent exposure to two or more challenges in wild forests than in semi-natural forests; more frequent exposure to two or more challenges in agricultural anthromes (croplands and

rangelands) and dense settlements than in forests; most frequent exposure to three or more challenges in villages compared to other land-use types. Therefore, land-use types intensively used by humans are, on average, exposed to a larger number of challenges than land-use types (or anthromes) least exposed to human use.

Case studies located in different world regions are presented for each anthrome, in order to provide historical context on the interlinkages between multiple challenges and responses (Box 6.1). Taken together, these case studies illustrate the large contrast across anthromes in land-based interventions, and show the way these interventions respond to combinations of challenges.

Box 6.1 | Case studies by anthrome type showing historical interlinkages between land-based challenges and the development of local responses

A. Croplands. Land degradation, groundwater stress and food insecurity: Soil and water conservation measures in the Tigray region of Ethiopia

In northern Ethiopia, the Tigray Region is a drought-prone area that has been subjected to severe land degradation (Frankl et al. 2013) and to recurrent drought and famine during 1888–1892, 1973–1974 and 1984–1985 (Gebremeskel et al. 2018). The prevalence of stunting and being underweight among children under five years is high (Busse et al. 2017) and the region was again exposed to a severe drought during the strong El Niño event of 2015–2016. Croplands are the dominant land-use type, with approximately 90% of the households relying on small-scale plough-based cultivation. Gullies affect nearly all slopes and frequently exceed 2 m in depth and 5 m in top width. Landsat imagery shows that cropland area peaked in 1984–1986, and increased erosion rates in the 1980s and 1990s caused the drainage density and volume to peak in 1994 (Frankl et al. 2013). Since around 2000, the large-scale implementation of soil and water conservation (SWC) measures, integrated catchment management,

Box 6.1 (continued)

conservation agriculture and indigenous tree regeneration has started to yield positive effects on the vegetation cover and led to the stabilisation of about 25% of the gullies by 2010 (Frankl et al. 2013). Since 1991, farmers have provided labour for SWC in January as a free service for 20 consecutive working days, followed by food for work for the remaining days of the dry season. Most of the degraded landscapes have been restored, with positive impacts over the last two decades on soil fertility, water availability and crop productivity. However, misuse of fertilisers, low survival of tree seedlings and lack of income from exclosures may affect the sustainability of these land restoration measures (Gebremeskel et al. 2018).

B. Rangelands. Biodiversity hotspot, land degradation and climate change:
Pasture intensification in the Cerrados of Brazil

Cerrados are a tropical savannah ecoregion in Brazil corresponding to a biodiversity hot spot with less than 2% of its region protected in national parks and conservation areas (Cava et al. 2018). Extensive cattle ranching (limited mechanisation, low use of fertiliser and seed inputs) has led to pasture expansion, including clearing forests to secure properties rights, occurring mainly over 1950–1975 (Martha et al. 2012). Despite observed productivity gains made over the last three decades (Martha et al. 2012), more than half of the pasture area is degraded to some extent, and challenges remain to reverse grassland degradation while accommodating growing demand and simultaneously avoiding the conversion of natural habitats (de Oliveira Silva et al. 2018). The largest share of production is on unfertilised pastures, often sown with perennial forage grasses of African origin, mainly *Brachiaria spp.* (Cardoso et al. 2016). This initial intensification era was partly at the expense of significant uncontrolled deforestation, and average animal stocking rates remained well below the potential carrying capacity (Strassburg et al. 2014). Changes in land use are difficult to reverse since pasture abandonment does not lead to the spontaneous restoration of old-growth savannah (Cava et al. 2018); moreover, pasture to crop conversion is frequent, supporting close to half of cropland expansion in Mato Grosso state over 2000–2013 (Cohn et al. 2016). Pasture intensification through liming, fertilisation and controlled grazing could increase soil organic carbon and reduce net GHG emission intensity per unit meat product, but only at increased investment cost per unit of area (de Oliveira Silva et al. 2017). Scenarios projecting a decoupling between deforestation and increased pasture intensification, provide the basis for a Nationally Determined Contribution (NDC) of Brazil that is potentially consistent with accommodating an upward trend in livestock production to meet increasing demand (de Oliveira Silva et al. 2018). Deforestation in Brazil has declined significantly between 2004 and 2014 in the national inventory, but recent data and analyses suggest that the decrease in deforestation and the resulting GHG emissions reductions have slowed down or even stopped (UNEP 2017).

C. Semi-natural forests. Biodiversity hotspot, land degradation, climate change
and food insecurity: Restoration and resilience of tropical forests in Indonesia

During the last two decades, forest cover in Indonesia declined by 150,000 km^2 in the period 1990–2000 (Stibig et al. 2014) and approximately 158,000 km^2 in the period 2000–2012 (Hansen et al. 2013), most of which was converted to agricultural lands (e.g., oil palm, pulpwood plantations). According to recent estimates, deforestation in Indonesia mainly concerns primary forests, including intact and degraded forests, thus leading to biodiversity loss and reduced carbon sequestration potentials (e.g., Margono et al. 2014). For example, Graham et al. (2017) estimated that the following strategies to reduce deforestation and forest degradation may cost-effectively increase carbon sequestration and reduce carbon emissions in 30 years: reforestation (3.54 GtCO₂), limiting the expansion of oil palm and timber plantations into forest (3.07 GtCO₂ and 3.05 GtCO₂, respectively), reducing illegal logging (2.34 GtCO₂), and halting illegal forest loss in protected areas (1.52 GtCO₂) at a total cost of 15.7 USD tC⁻¹. The importance of forest mitigation in Indonesia is indicated by the NDC, where between half and two-thirds of the 2030 emission target relative to a business-as-usual scenario is from reducing deforestation, forest degradation, peatland drainage and fires (Grassi et al. 2017). Avoiding deforestation and reforestation could have multiple co-benefits by improving biodiversity conservation and employment opportunities, while reducing illegal logging in protected areas. However, these options could also have adverse side effects if they deprive local communities of access to natural resources (Graham et al. 2017). The adoption of the Roundtable on Sustainable Palm Oil certification in oil palm plantations reduced deforestation rates by approximately 33% in the period 2001–2015 (co-benefits with mitigation), and fire rates much more than for non-certified plantations (Carlson et al. 2018). However, given that large-scale oil palm plantations are one of the largest drivers of deforestation in Indonesia, objective information on the baseline trajectory for land clearance for oil palm is needed to further assess commitments, regulations and transparency in plantation development (Gaveau et al. 2016). For adaptation options, the community forestry scheme *Hutan Desa* (Village Forest) in Sumatra and Kalimantan helped to avoid deforestation (co-benefits with mitigation) by between 0.6 and 0.9 ha km^{-2} in Sumatra and 0.6 and 0.8 ha km^{-2} in Kalimantan in the period 2012–2016; Santika et al. 2017), improve local livelihood options, and restore degraded ecosystems (positive side effects for NCP provision) (e.g., Pohnan et al. 2015). Finally, the establishment of Ecosystem Restoration Concessions in Indonesia (covering more than 5,500 km^2 of forests now, and 16,000 km^2 allocated for the future) facilitates the planting of commercial timber species (co-benefits with mitigation), while assisting natural regeneration, preserving important habitats and species, and improving local well-being and incomes (positive side effects for Nature's Contributions to People provision), at relatively lower costs compared with timber concessions (Silalahi et al. 2017).

Box 6.1 (continued)

D. Villages. Land degradation, groundwater overuse, climate change and food insecurity: Climate smart villages in India

Indian agriculture, which includes both monsoon-dependent rainfed (58%) and irrigated agriculture, is exposed to climate variability and change. Over the past years, the frequency of droughts, cyclones, and hailstorms has increased, with severe droughts in eight of 15 years between 2002 and 2017 (Srinivasa Rao et al. 2016; Mujumdar et al. 2017). Such droughts result in large yield declines for major crops like wheat in the Indo-Gangetic Plain (Zhang et al. 2017). The development of a submersible pump technology in the 1990s, combined with public policies that provide farmers with free electricity for groundwater irrigation, resulted in a dramatic increase in irrigated agriculture (Shah et al. 2012). This shift has led to increased dependence on irrigation from groundwater and induced a groundwater crisis, with large impacts on socio-ecosystems. An increasing number of farmers report bore-well failures, either due to excessive pumping of an existing well or a lack of water in new wells. The decrease in the groundwater table level has suppressed the recharge of river beds, turning permanent rivers into ephemeral streams (Srinivasan et al. 2015). Wells have recently been drilled in upland areas, where groundwater irrigation is also increasing (Robert et al. 2017). Additional challenges include declining soil organic matter and fertility under monocultures and rice/wheat systems. Unoccupied land is scarce, meaning that the potential for expanding the area farmed is very limited (Aggarwal et al. 2018). In rural areas, diets are deficient in protein, dietary fibre and iron, and mainly comprised of cereals and pulses grown and/or procured through welfare programmes (Vatsala et al. 2017). Cultivators are often indebted, and suicide rates are much higher than the national average, especially for those strongly indebted (Merriott 2016). Widespread use of diesel pumps for irrigation, especially for paddies, high use of inorganic fertilisers and crop residue burning lead to high GHG emissions (Aggarwal et al. 2018). The Climate-Smart Village (CSV) approach aims at increasing farm yield, income, input use efficiency (water, nutrients, and energy) and reducing GHG emissions (Aggarwal et al. 2018). Climate-smart agriculture interventions are considered in a broad sense by including practices, technologies, climate information services, insurance, institutions, policies, and finance. Options differ based on the CSV site, its agro-ecological characteristics, level of development, and the capacity and interest of farmers and the local government (Aggarwal et al. 2018). Selected interventions included crop diversification, conservation agriculture (minimum tillage, residue retention, laser levelling), improved varieties, weather-based insurance, agro-advisory services, precision agriculture and agroforestry. Farmers' cooperatives were established to hire farm machinery, secure government credit for inputs, and share experiences and knowledge. Tillage practices and residue incorporation increased rice–wheat yields by 5–37%, increased income by 28–40%, reduced GHG emissions by 16–25%, and increased water-use efficiency by 30% (Jat et al. 2015). The resulting portfolio of options proposed by the CSV approach has been integrated with the agricultural development strategy of some states like Haryana.

E. Dense settlements. Climate change and food: Green infrastructures

Extreme heat events have led to particularly high rates of mortality and morbidity in cities, as urban populations are pushed beyond their adaptive capacities, leading to an increase in mortality rates of 30–130% in major cities in developed countries (Norton et al. 2015). Increased mortality and morbidity from extreme heat events are exacerbated in urban populations by the urban heat island effect (Gabriel and Endlicher 2011; Schatz and Kucharik 2015), which can be limited by developing green infrastructure in cities. Urban green infrastructure includes public and private green spaces – such as remnant native vegetation, parks, private gardens, golf courses, street trees, urban farming – and more engineered options, such as green roofs, green walls, biofilters and raingardens (Norton et al. 2015). Increasing the amount of vegetation, or green infrastructure, in a city is one way to help reduce urban air temperature maxima and variation. Increasing vegetation by 10% in Melbourne, Australia was estimated to reduce daytime urban surface temperatures by approximately 1°C during extreme heat events (Coutts and Harris 2013). Urban farming (a type of urban green infrastructure) is largely driven by the desire to reconnect food production and consumption (Whittinghill and Rowe 2012) (Chapter 5). Even though urban farming can only meet a very small share of the overall urban food demand, it provides fresh and local food, especially perishable fruits and crops that are usually shipped from far and sold at high prices (Thomaier et al. 2015). Food-producing urban gardens and farms are often started by grassroots initiatives (Ercilla-Montserrat et al. 2019) that occupy vacant urban spaces. In recent years, a growing number of urban farming projects (termed Zero-Acreage farming, or Z-farming, Thomaier et al. 2015) were established in and on existing buildings, using rooftop spaces or abandoned buildings through contracts between food businesses and building owners. Almost all Z-farms are located in cities with more than 150,000 inhabitants, with a majority in North American cities such as New York City, Chicago and Toronto (Thomaier et al. 2015). They depend on the availability of vacant buildings and roof tops, thereby competing with other uses, such as roof-based solar systems. Urban farming, however, has potentially high levels of soil pollution and air pollutants, which may lead to crop contamination and health risks. These adverse effects could be reduced on rooftops (Harada et al. 2019) or in controlled environments.

6

6.1.4   Challenges represented in future scenarios

In this section, the evolution of several challenges (climate change, mitigation, adaptation, desertification, land degradation, food insecurity, biodiversity and water) in the future are assessed, focusing on global analyses. The effect of response options on these land challenges in the future is discussed in Section 6.4.4. Where possible, studies quantifying these challenges in the Shared Socio-economic Pathways (SSPs) (O'Neill et al. 2014) (Chapter 1, Cross-Chapter Box 1, and Cross-Chapter Box 9 in this chapter), should be used to assess which future scenarios could experience multiple challenges in the future.

Climate change: Without any additional efforts to mitigate, global mean temperature rise is expected to increase by anywhere from 2°C to 7.8°C in 2100 relative to the 1850–1900 reference period (Clarke et al. 2014; Chapter 2). The level of warming varies, depending on the climate model (Collins et al. 2013), uncertainties in the Earth system (Clarke et al. 2014), and socio-economic/technological assumptions (Clarke et al. 2014; Riahi et al. 2017). Warming over land is 1.2 to 1.4 times higher than global mean temperature rise; warming in the Arctic region is 2.4 to 2.6 times higher than warming in the tropics (Collins et al. 2013). Increases in global mean temperature are accompanied by increases in global precipitation; however, the effect varies across regions, with some regions projected to see increases in precipitation and others to see decreases (Collins et al. 2013) (Chapter 2). Additionally, climate change also has implications for extreme events (e.g., drought, heat waves, etc.), freshwater availability, and other aspects of the terrestrial system (Chapter 2).

Mitigation: Challenges to mitigation depend on the underlying emissions and 'mitigative capacity', including technology availability, policy institutions, and financial resources (O'Neill et al. 2014). Challenges to mitigation are high in SSP3 and SSP5, medium in SSP2, and low in SSP1 and SSP4 (O'Neill et al. 2014, 2017; Riahi et al. 2017).

Adaptation: Challenges to adaptation depend on climate risk and adaptive capacity, including technology availability, effectiveness of institutions, and financial resources (O'Neill et al. 2014). Challenges to adaptation are high in SSP3 and SSP4, medium in SSP2, and low in SSP1 and SSP5 (O'Neill et al. 2014, 2017; Riahi et al. 2017).

Desertification: The combination of climate and land-use changes can lead to decreases in soil cover in drylands (Chapter 3). Population living in drylands is expected to increase by 43% in the SSP2-Baseline, due to both population increases and an expansion of dryland area (UNCCD 2017).

Land degradation: Future changes in land use and climate have implications for land degradation, including impacts on soil erosion, vegetation, fire, and coastal erosion (Chapter 4; IPBES 2018). For example, soil organic carbon is expected to decline by 99 GtCO$_2$e in 2050 in an SSP2-Baseline scenario, due to both land management and expansion in agricultural area (Ten Brink et al. 2018).

Food insecurity: Food insecurity in future scenarios varies significantly, depending on socio-economic development and study. For example, the population at risk of hunger ranges from 0 to 800 million in 2050 (Hasegawa et al. 2015a; Ringler et al. 2016; Fujimori et al. 2018; Hasegawa et al. 2018; Fujimori et al. 2019; Baldos and Hertel 2015) and 0–600 million in 2100 (Hasegawa et al. 2015a). Food prices in 2100 in non-mitigation scenarios range from 0.9 to about two times their 2005 values (Hasegawa et al. 2015a; Calvin et al. 2014; Popp et al. 2017). Food insecurity depends on both income and food prices (Fujimori et al. 2018). Higher income (e.g., SSP1, SSP5), higher yields (e.g., SSP1, SSP5), and less meat intensive diets (e.g., SSP1) tend to result in reduced food insecurity (Hasegawa et al. 2018; Fujimori et al. 2018).

Biodiversity: Future species extinction rates vary from modest declines to 100-fold increases from 20th century rates, depending on the species (e.g., plants, vertebrates, invertebrates, birds, fish, corals), the degree of land-use change, the level of climate change, and assumptions about migration (Pereira et al. 2010). Mean species abundance (MSA) is also estimated to decline in the future by 10–20% in 2050 (Van Vuuren et al. 2015; Pereira et al. 2010). Scenarios with greater cropland expansion lead to larger declines in MSA (UNCCD 2017) and species richness (Newbold et al. 2015).

Water stress: Changes in water supply (due to climate change) and water demand (due to socio-economic development) in the future have implications for water stress. Water withdrawals for irrigation increase from about 2500 km^3 yr^{-1} in 2005 to between 2900 and 9000 km^3 yr^{-1} at the end of the century (Chaturvedi et al. 2013; Wada and Bierkens 2014; Hejazi et al. 2014a; Kim et al. 2016; Graham et al. 2018; Bonsch et al. 2015); total water withdrawals at the end of the century range from 5000 to 13,000 km^3 yr^{-1} (Wada and Bierkens 2014; Hejazi et al. 2014a; Kim et al. 2016; Graham et al. 2018). The magnitude of change in both irrigation and total water withdrawals depend on population, income, and technology (Hejazi et al. 2014a; Graham et al. 2018). The combined effect of changes in water supply and water demand will lead to an increase of between 1 billion and 6 billion people living in water-stressed areas (Schlosser et al. 2014; Hanasaki et al. 2013; Hejazi et al. 2014b). Changes in water quality are not assessed here but could be important (Liu et al. 2017).

Scenarios with multiple challenges: Table 6.2 summarises the challenges across the five SSP Baseline scenarios.

Table 6.2 | Assessment of future challenges to climate change, mitigation, adaptation, desertification, land degradation, food insecurity, water stress, and biodiversity in the SSP Baseline scenarios.

SSP	Summary of challenges
SSP1	SSP1 (Van Vuuren et al. 2017b) has low challenges to mitigation and adaptation. The resulting Baseline scenario includes: – continued, but moderate, *climate change*: global mean temperature increases by 3 to 3.5°C in 2100 (Huppmann et al. 2018; Riahi et al. 2017) – low levels of *food insecurity*: malnourishment is eliminated by 2050 (Hasegawa et al. 2015a) – declines in *biodiversity*: biodiversity loss increases from 34% in 2010 to 38% in 2100 (UNCCD 2017) – high *water stress*: global water withdrawals decline slightly from the baseline in 2071–2100, but about 2.6 billion people live in water stressed areas (Hanasaki et al. 2013). Additionally, this scenario is likely to have lower challenges related to desertification, land degradation, and biodiversity loss than SSP2 as it has lower population, lower land-use change and lower climate change (Riahi et al. 2017).
SSP2	SSP2 (Fricko et al. 2017) is a scenario with medium challenges to mitigation and medium challenges to adaptation. The resulting Baseline scenario includes: – continued *climate change*: global mean temperature increases by 3.8°C to 4.3°C in 2100 (Fricko et al. 2017; Huppmann et al. 2018; Riahi et al. 2017) – increased challenges related to *desertification*: the population living in drylands is expected to increase by 43% in 2050 (UNCCD 2017) – increased *land degradation*: soil organic carbon is expected to decline by 99 GtCO2e in 2050 (Ten Brink et al. 2018) – low levels of *food insecurity*: malnourishment is eliminated by 2100 (Hasegawa et al. 2015a) – declines in *biodiversity*: biodiversity loss increases from 34% in 2010 to 43% in 2100 (UNCCD 2017) – high *water stress*: global water withdrawals nearly doubles from the baseline in 2071–2100, with about 4 billion people living in water stressed areas (Hanasaki et al. 2013).
SSP3	SSP3 (Fujimori et al. 2017) is a scenario with high challenges to mitigation and high challenges to adaptation. The resulting Baseline scenario includes: – continued *climate change*: global mean temperature increases by 4°C to 4.8°C in 2100 (Huppmann et al. 2018; Riahi et al. 2017) – high levels of *food insecurity*: about 600 million malnourished in 2100 (Hasegawa et al. 2015a) – declines in *biodiversity*: biodiversity loss increases from 34% in 2010 to 46% in 2100 (UNCCD 2017) – high *water stress*: global water withdrawals more than double from the baseline in 2071–2100, with about 5.5 billion people living in water stressed areas (Hanasaki et al. 2013). Additionally, this scenario is likely to have higher challenges to desertification, land degradation, and biodiversity loss than SSP2 as it has higher population, higher land-use change and higher climate change (Riahi et al. 2017).
SSP4	SSP4 (Calvin et al. 2017) has high challenges to adaptation but low challenges to mitigation. The resulting Baseline scenario includes: – continued *climate change*: global mean temperature increases by 3.4°C to 3.8°C in 2100 (Calvin et al. 2017; Huppmann et al. 2018; Riahi et al. 2017) – high levels of *food insecurity*: about 400 million malnourished in 2100 (Hasegawa et al. 2015a) – high *water stress*: about 3.5 billion people live in water stressed areas in 2100 (Hanasaki et al. 2013). Additionally, this scenario is likely to have similar effects on biodiversity loss as SSP2 as it has similar land-use change and similar climate change (Riahi et al. 2017).
SSP5	SSP5 (Kriegler et al. 2017) has high challenges to mitigation but low challenges to adaptation. The resulting Baseline scenario includes: – continued *climate change*: global mean temperature increases by 4.6°C to 5.4°C in 2100 (Kriegler et al. 2017; Huppmann et al. 2018; Riahi et al. 2017) – low levels of *food insecurity*: malnourishment is eliminated by 2050 (Hasegawa et al. 2015a) – increased water use and water scarcity: global water withdrawals increase by about 80% in 2071–2100, with nearly 50% of the population living in water stressed areas (Hanasaki et al. 2013). Additionally, this scenario is likely to have higher effects on biodiversity loss as SSP2 as it has similar land-use change and higher climate change (Riahi et al. 2017).

6.2 Response options, co-benefits and adverse side effects across the land challenges

This section describes the integrated response options available to address the land challenges of climate change mitigation, climate change adaptation, desertification, land degradation and food security. These can be categorised into options that rely on (i) land management, (ii) value chain management, and (iii) risk management (Figure 6.5). The land management integrated response options can be grouped according to those that are applied in agriculture, in forests, on soils, in other/all ecosystems and those that are applied specifically for carbon dioxide removal (CDR). The value chain management integrated response options can be categorised as those based demand management and supply management. The risk management options are grouped together (Figure 6.5).

Note that the integrated response options are not mutually exclusive – for example, cropland management might also increase soil organic matter stocks – and a number of the integrated response options are comprised of a number of practices – for example, improved cropland management is a collection of practices consisting of:

1. management of the crop, including high-input carbon practices, for example, improved crop varieties, crop rotation, use of cover crops, perennial cropping systems, agricultural biotechnology
2. nutrient management: including optimised fertiliser application rate, fertiliser type [organic and mineral], timing, precision application, inhibitors
3. reduced tillage intensity and residue retention
4. improved water management: including drainage of waterlogged mineral soils and irrigation of crops in arid/semi-arid conditions
5. improved rice management, including water management such as mid-season drainage and improved fertilisation and residue management in paddy rice systems.

In this section, we deal only with integrated response options, not the policies that are currently or could be implemented to enable their application; that is the subject of Chapter 7. Also note that enabling conditions such as indigenous and local knowledge, gender issues, governance and so on are not categorised as integrated response options (Section 6.1.2). Some suggested methods to address land challenges are better described as *overarching frameworks* than as integrated response options. For example, *climate smart agriculture* is a collection of integrated

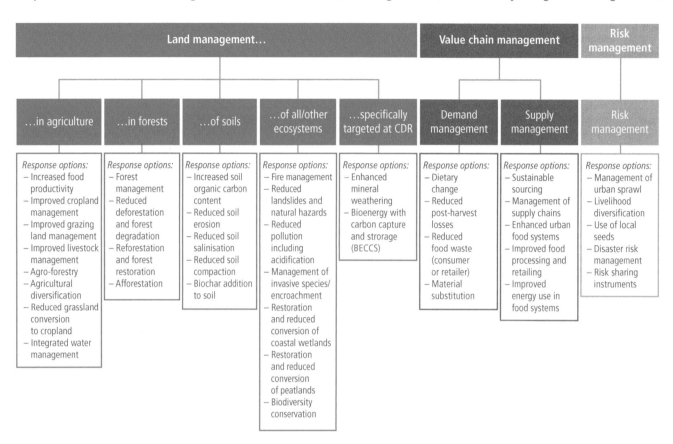

Figure 6.5 | Broad categorisation of response options categorised into three main classes and eight sub-classes.

response options aimed at delivering mitigation and adaptation in agriculture, including improved cropland management, grazing land management and livestock management. Table 6.3 shows how a number of overarching frameworks are comprised of a range of integrated response options.

Similarly, policy goals, such as *land degradation neutrality* (discussed further in Chapter 7), are not considered as integrated response options. For this reason, *land degradation neutrality*, and overarching frameworks, such as those described in Table 6.3 do not appear as response options in the following sections, but the component integrated response options that contribute to these policy goals or overarching frameworks are addressed in detail.

Table 6.3 | Examples of overarching frameworks that consist of a range of response options.

Framework (definition used)	Nature-based solutions (IUCN)	Agro-ecology (FAO)	Climate smart agriculture (FAO)	Ecosystem-based adaptation (CBD)	Conservation agriculture (FAO)	Community-based adaptation (IIED)	Integrated landscape management including integrated coastal zone management (FAO)	Precision agriculture (FAO)	Sustainable forest management (UN)	Sustainable intensification (Cross-Chapter Box 5 in Chapter 5)	Organic agriculture (FAO)
Response options based on land management											
Increased food productivity			●		●		●	●		●	
Improved cropland management		●	●		●	●	●	●		●	●
Improved grazing land management		●	●	●		●	●			●	●
Improved livestock management		●	●			●	●			●	

Framework (definition used)	Nature-based solutions (IUCN)	Agro-ecology (FAO)	Climate smart agriculture (FAO)	Ecosystem-based adaptation (CBD)	Conservation agriculture (FAO)	Community-based adaptation (IIED)	Integrated landscape management including integrated coastal zone management (FAO)	Precision agriculture (FAO)	Sustainable forest management (UN)	Sustainable intensification (Cross-Chapter Box 5 in Chapter 5)	Organic agriculture (FAO)
Agroforestry		●	●	●		●	●			●	●
Agricultural diversification		●	●			●				●	●
Reduced grassland conversion to cropland		●		●		●	●				
Integrated water management	●	●	●	●	●	●	●	●		●	●
Forest management	●			●		●	●		●		
Reduced deforestation and forest degradation		●		●		●	●				
Reforestation and forest restoration	●	●		●		●			●		
Afforestation				●		●	●				
Increased soil organic carbon content		●	●	●	●		●			●	●
Reduced soil erosion		●	●	●	●		●			●	●
Reduced soil salinisation		●	●	●	●		●	●		●	●
Reduced soil compaction		●	●	●	●		●			●	●
Biochar addition to soil		●	●								
Fire management		●	●	●		●	●		●		
Reduced landslides and natural hazards		●	●	●		●	●				
Reduced pollution including acidification							●	●		●	●
Management of invasive species/encroachment	●	●		●		●	●		●		●
Restoration and reduced conversion of coastal wetlands		●		●		●	●				
Restoration and reduced conversion of peatlands		●	●	●		●	●				
Biodiversity conservation	●	●	●	●	●	●	●		●	●	
Enhanced weathering of minerals											
Bioenergy and BECCS							●				
Response options based on value chain management											
Dietary change		●									●
Reduced post-harvest losses		●	●			●		●			●
Reduced food waste (consumer or retailer)		●									
Material substitution											
Sustainable sourcing		●	●			●					●
Management of supply chains		●	●								
Enhanced urban food systems		●	●			●	●	●		●	●
Improved food processing and retailing		●									
Improved energy use in food systems		●	●		●			●		●	
Response options based on risk management											
Management of urban sprawl				●		●	●				
Livelihood diversification		●	●	●		●	●	●			
Use of local seeds	●	●	●	●		●	●				
Disaster risk management	●			●			●				●
Risk sharing instruments										●	

Table 6.4 | Mapping of response options considered in this report (SRCCL) and SR15.

SRCCL Response option/options	SR15 Response option/options
Afforestation	Afforestation
Reforestation and forest restoration	Reforestation and reduced land degradation and forest restoration
Agricultural diversification	Mixed crop-livestock systems
Agroforestry	Agroforestry and silviculture
Biochar addition to soil	Biochar
Biodiversity conservation	Biodiversity conservation
Bioenergy and bioenergy with carbon capture and storage (BECCS)	BECCS (through combustion, gasification, or fermentation)
Dietary change	Dietary changes, reducing meat consumption
Disaster risk management	Climate services
	Community-based adaptation
Enhanced urban food systems	Urban and peri-urban agriculture and forestry
Enhanced weathering of minerals	Mineralisation of atmospheric carbon dioxide (CO_2) through enhanced weathering of rocks
Fire management	Fire management and (ecological) pest control
Forest management	Forest management
Improved cropland management	Methane reductions in rice paddies
Improved cropland management	Nitrogen pollution reductions, e.g., by fertiliser reduction, increasing nitrogen fertiliser efficiency, sustainable fertilisers
	Precision agriculture
	Conservation agriculture
Improved food processing and retailing	
Improved grazing land management	Livestock and grazing management, e.g., methane and ammonia reductions in ruminants through feeding management or feed additives, or manure management for local biogas production to replace traditional biomass use
Improved livestock management	Manure management
Increased energy efficiency in food systems	
Increased food productivity	Increasing agricultural productivity
Increased soil organic carbon content	Changing agricultural practices enhancing soil carbon
	Soil carbon enhancement, enhancing carbon sequestration in biota and soils, e.g., with plants with high carbon sequestration potential – also agriculture, forestry and other land-use (AFOLU) measure.
Integrated water management	Irrigation efficiency
Livelihood diversification	
Management of invasive species/encroachment	
Management of supply chains	
Management of urban sprawl	Urban ecosystem services
	Climate resilient land use
Material substitution	Material substitution of fossil CO_2 with bio-CO_2 in industrial application (e.g., the beverage industry)
	Carbon capture and usage (CCU); bioplastics (bio-based materials replacing fossil fuel uses as feedstock in the production of chemicals and polymers), carbon fibre
Reduced soil erosion	
Reduced soil compaction	
Reduced deforestation	Reduced deforestation, forest protection, avoided forest conversion
Reduced food waste (consumer or retailer)	Reduction of food waste (incl. reuse of food processing waste for fodder)
Reduced grassland conversion to cropland	
Reduced landslides and natural hazards	
Reduced pollution including acidification	Reduced air pollution
Reduced post-harvest losses	
Reduced soil salinisation	
Restoration and reduced conversion of coastal wetlands	Managing coastal stress
	Restoration of wetlands (e.g., coastal and peat-land restoration, blue carbon) and wetlands management
Restoration and reduced conversion of peatlands	
Risk sharing instruments	Risk sharing
Sustainable sourcing	
Use of local seeds	

SR15 considered a range of response options (from a mitigation/adaptation perspective only). Table 6.4 shows how the SR15 options map on to the response options considered in this report (SRCCL). Note that this report excludes most of the energy-related options from SR15, as well as green infrastructure and sustainable aquaculture.

Before providing the quantitative assessment of the impacts of each response option in addressing mitigation, adaptation, desertification, land degradation and food security in Section 6.3, the integrated response options are descried in Section 6.2.1 and any context specificities in the effects are noted.

6.2.1 Integrated response options based on land management

6.2.1.1 Integrated response options based on land management in agriculture

Integrated response options based on land management in agriculture are described in Table 6.5, which also notes any context specificities, and provides the evidence base for the effects of the response options.

6.2.1.2 Integrated response options based on land management in forests

Integrated response options based on land management in forests are described in Table 6.6, which also notes any context specificities, and provides the evidence base for the effects of the response options.

6.2.1.3 Integrated response options based on land management of soils

Integrated response options based on land management of soils are described in Table 6.7, which also notes any context specificities, and provides the evidence base for the effects of the response options.

6.2.1.4 Integrated response options based on land management of all/other ecosystems

Integrated response options based on land management in all/other ecosystems are described in Table 6.8, which also notes any context specificities, and provides the evidence base for the effects of the response options.

6.2.1.5 Integrated response options based on land management specifically for carbon dioxide removal (CDR)

Integrated response options based on land management specifically for CDR are described in Table 6.9, which also notes any context specificities, and provides the evidence base for the effects of the response options.

Table 6.5 | Integrated response options based on land management in agriculture.

Integrated response option	Description	Context and caveats	Supporting evidence
Increased food productivity	Increased food productivity arises when the output of food commodities increases per unit of input, e.g., per unit of land or water. It can be realised through many other interventions such as improved cropland, grazing land and livestock management.	Many interventions to increase food production, particularly those predicated on very large inputs of agro-chemicals, have a wide range of negative externalities leading to the proposal of sustainable intensification as a mechanism to deliver future increases in productivity that avoid these adverse outcomes. Intensification through additional input of nitrogen fertiliser, for example, would result in negative impacts on climate, soil, water and air pollution. Similarly, if implemented in a way that over-exploits the land, significant negative impacts would occur, but if achieved through sustainable intensification, and used to spare land, it could reduce the pressure on land.	Cross-Chapter Box 6 in Chapter 5; Chapter 3 Balmford et al. 2018; Burney et al. 2010; Foley et al. 2011; Garnett et al. 2013; Godfray et al. 2010; IPBES 2018; Lal 2016; Lamb et al. 2016; Lobell et al. 2008; Shcherbak et al. 2014; Smith et al. 2013; Tilman et al. 2011
Improved cropland management	Improved cropland management is a collection of practices consisting of a) *management of the crop*: including high input carbon practices, for example, improved crop varieties, crop rotation, use of cover crops, perennial cropping systems, integrated production systems, crop diversification, agricultural biotechnology, b) *nutrient management*: including optimised fertiliser application rate, fertiliser type (organic manures, compost and mineral), timing, precision application, nitrification inhibitors, c) *reduced tillage intensity and residue retention*, d) *improved water management*: including drainage of waterlogged mineral soils and irrigation of crops in arid/semi-arid conditions, e) *improved rice management*: including water management such as mid-season drainage and improved fertilisation and residue management in paddy rice systems, and f) *biochar application*.	Improved cropland management can reduce GHG emissions and create soil carbon sinks, though if poorly implemented, it could increase nitrous oxide and methane emissions from nitrogen fertilisers, crop residues and organic amendments. It can improve resilience of food crop production systems to climate change, and can be used to tackle desertification and land degradation by improving sustainable land management. It can also contribute to food security by closing crop yield gaps to increase food productivity.	Chapter 4; Chapter 3; Chapter 2; Chapter 5 Bryan et al. 2009; Chen et al. 2010; Labrière et al. 2015; Lal 2011; Poeplau and Don 2015; Porter et al. 2014; Smith 2008b; Smith et al. 2014; Tilman et al. 2011

Integrated response option	Description	Context and caveats	Supporting evidence
Improved grazing land management	Improved grazing land management is a collection of practices consisting of a) *management of vegetation*: including improved grass varieties/sward composition, deep rooting grasses, increased productivity, and nutrient management, b) *animal management*: including appropriate stocking densities fit to carrying capacity, fodder banks, and fodder diversification, and c) *fire management*: improved use of fire for sustainable grassland management, including fire prevention and improved prescribed burning (see also fire management as a separate response option) (Table 6.8).	Improved grazing land management can increase soil carbon sinks, reduce GHG emissions, improve the resilience of grazing lands to future climate change, help reduce desertification and land degradation by optimising stocking density and reducing overgrazing, and can enhance food security through improved productivity.	Chapter 2; Chapter 3; Chapter 4; Chapter 5; Section 6.4 Archer et al. 2011; Briske et al. 2015; Conant et al. 2017; Herrero et al. 2016; Porter et al. 2014; Schwilch et al. 2014; Smith et al. 2014; Tighe et al. 2012
Improved livestock management	Improved livestock management is a collection of practices consisting of a) *improved feed and dietary additives* (e.g., bioactive compounds, fats), used to increase productivity and reduce emissions from enteric fermentation; b) *breeding* (e.g., breeds with higher productivity or reduced emissions from enteric fermentation), c) *herd management,* including decreasing neo-natal mortality, improving sanitary conditions, animal health and herd renewal, and diversifying animal species, d) *emerging technologies* (of which some are not legally authorised in several countries) such as propionate enhancers, nitrate and sulphate supplements, archaea inhibitors and archaeal vaccines, methanotrophs, acetogens, defaunation of the rumen, bacteriophages and probiotics, ionophores/antibiotics; and e) *improved manure management,* including manipulation of bedding and storage conditions, anaerobic digesters; biofilters; dietary change and additives, soil-applied and animal-fed nitrification inhibitors, urease inhibitors, fertiliser type, rate and timing, manipulation of manure application practices, and grazing management.	Improved livestock management can reduce GHG emissions, particularly from enteric methane and manure management. It can improve the resilience of livestock production systems to climate change by breeding better adapted livestock. It can help with desertification and land degradation, e.g., through use of more efficient and adapted breeds to allow reduced stocking densities. Improved livestock sector productivity can also increase food production.	Chapter 2; Chapter 3; Chapter 4; Chapter 5 Archer et al. 2011; Herrero et al. 2016; Miao et al. 2015; Porter et al. 2014; Rojas-Downing et al. 2017; Smith et al. 2008, 2014; Squires and Karami 2005; Tighe et al. 2012
Agroforestry	Agroforestry involves the deliberate planting of trees in croplands and silvo-pastoral systems.	Agroforestry sequesters carbon in vegetation and soils. The use of leguminous trees can enhance biological nitrogen fixation and resilience to climate change. Soil improvement and the provision of perennial vegetation can help to address desertification and land degradation. Agroforestry can increase agricultural productivity, with benefits for food security. Additionally, agroforestry can enable payments to farmers for ecosystem services and reduce vulnerability to climate shocks.	Antwi-Agyei et al. 2014; Benjamin et al. 2018; Guo et al. 2018; den Herder et al. 2017; Mbow et al. 2014a; Mosquera-Losada et al. 2018; Mutuo et al. 2005; Nair and Nair 2014; Ram et al. 2017; Rosenstock et al. 2014; Sain et al. 2017; Santiago-Freijanes et al. 2018; Sida et al. 2018; Vignola et al. 2015; Yirdaw et al. 2017
Agricultural diversification	Agricultural diversification includes a set of agricultural practices and products obtained in the field that aim to improve the resilience of farmers to climate variability and climate change and to economic risks posed by fluctuating market forces. In general, the agricultural system is shifted from one based on low-value agricultural commodities to one that is more diverse, composed of a basket of higher value-added products.	Agricultural diversification is targeted at adaptation but could also deliver a small carbon sink, depending on how it is implemented. It could reduce pressure on land, benefitting desertification, land degradation, food security and household income. However, the potential to achieve household food security is influenced by the market orientation of a household, livestock ownership, non-agricultural employment opportunities, and available land resources.	Birthal et al. 2015; Campbell et al. 2014; Cohn et al. 2017; Lambin and Meyfroidt 2011; Lipper et al. 2014; Massawe et al. 2016; Pellegrini and Tasciotti 2014; Waha et al. 2018
Reduced grassland conversion to cropland	Grasslands can be converted to croplands by ploughing of grassland and seeding with crops. Since croplands have a lower soil carbon content than grasslands and are also more prone to erosion than grasslands, reducing conversion of grassland to croplands will prevent soil carbon losses by oxidation and soil loss through erosion. These processes can be reduced if the rate of grassland conversion to cropland is reduced.	Stabilising soils by retaining grass cover also improves resilience, benefitting adaptation, desertification and land degradation. Since conversion of grassland to cropland usually occurs to remedy food security challenges, food security could be adversely affected, since more land is required to produce human food from livestock products on grassland than from crops on cropland.	Chapter 3; Chapter 4; Chapter 5 Clark and Tilman 2017; Lal 2001; de Ruiter et al. 2017; Poore and Nemecek 2018

Integrated response option	Description	Context and caveats	Supporting evidence
Integrated water management	Integrated water management is the process of creating holistic strategies to promote integrated, efficient, equitable and sustainable use of water for agroecosystems. It includes a collection of practices including water-use efficiency and irrigation in arid/semi-arid areas, improvement of soil health through increases in soil organic matter content, and improved cropland management, agroforestry and conservation agriculture. Increasing water availability, and reliability of water for agricultural production, can be achieved by using different techniques of water harvesting, storage, and its judicious utilisation through farm ponds, dams, and community tanks in rainfed agriculture areas can benefit adaptation.	These practices can reduce aquifer and surface water depletion, and prevent over-extraction, and the management of climate risks. Many technical innovations, e.g., precision water management, can have benefits for both adaptation and mitigation, although trade-offs are possible. Maintaining the same level of yield through use of site-specific water management-based approach could have benefits for both food security and mitigation.	Chapter 3; Chapter 4; Chapter 5 Brindha and Pavelic 2016; Jat et al. 2016; Jiang 2015; Keesstra et al. 2018; Liu et al. 2017; Nejad 2013; Rao et al. 2017b; Shaw et al. 2014; Sapkota et al. 2017; Scott et al. 2011; Waldron et al. 2017

Table 6.6 | Integrated response options based on land management in forests.

Integrated response option	Description	Context and caveats	Supporting evidence
Forest management	Forest management refers to management interventions in forests for the purpose of climate change mitigation. It includes a wide variety of practices affecting the growth of trees and the biomass removed, including improved regeneration (natural or artificial) and a better schedule, intensity and execution of operations (thinning, selective logging, final cut, reduced impact logging, etc.). Sustainable forest management is the stewardship and use of forests and forest lands in a way, and at a rate, that maintains their biodiversity, productivity, regeneration capacity, vitality and their potential to fulfil, now and in the future, relevant ecological, economic and social functions, at local, national, and global levels, and that does not cause damage to other ecosystems.	Sustainable forest management can enhance the carbon stock in biomass, dead organic matter, and soil – while providing wood-based products to reduce emissions in other sectors through material and energy substitution. A trade-off exists between different management strategies: higher harvest decreases the carbon in the forest biomass in the short term but increases the carbon in wood products and the potential for substitution effects. Sustainable forest management, also through close-to-nature silvicultural techniques, can potentially offer many co-benefits in terms of climate change mitigation, adaptation, biodiversity conservation, microclimatic regulation, soil erosion protection, coastal area protection and water and flood regulation. Forest management strategies aimed at increasing the biomass stock levels may have adverse side effects, such as decreasing the stand-level structural complexity, biodiversity and resilience to natural disasters. Forest management also affects albedo and evapotranspiration.	Chapter 2; Chapter 4 D'Amato et al. 2011; Dooley and Kartha 2018; Ellison et al. 2017; Erb et al. 2017; Grassi et al. 2018; Griscom et al. 2017; Jantz et al. 2014; Kurz et al. 2016; Locatelli 2011; Luyssaert et al. 2018; Nabuurs et al. 2017; Naudts et al. 2016; Pingoud et al. 2018; Putz et al. 2012; Seidl et al. 2014; Smith et al. 2014; Smyth et al. 2014; Stanturf et al. 2015
Reduced deforestation and forest degradation	Reduced deforestation and forest degradation includes conservation of existing carbon pools in forest vegetation and soil by controlling the drivers of deforestation (i.e., commercial and subsistence agriculture, mining, urban expansion) and forest degradation (i.e., overharvesting including fuelwood collection, poor harvesting practices, overgrazing, pest outbreaks, and extreme wildfires), also through establishing protected areas, improving law enforcement, forest governance and land tenure, supporting community forest management and introducing forest certification.	Reducing deforestation and forest degradation is a major strategy to reduce global GHG emissions. The combination of reduced GHG emissions and biophysical effects results in a large climate mitigation effect, with benefits also at local level. Reduced deforestation preserves biodiversity and ecosystem services more efficiently and at lower costs than afforestation/reforestation. Efforts to reduce deforestation and forest degradation may have potential adverse side effects, for example, reducing availability of land for farming, restricting the rights and access of local people to forest resources (e.g., firewood), or increasing the dependence of local people to insecure external funding.	Chapter 2 Alkama and Cescatti 2016; Baccini et al. 2017; Barlow et al. 2016; Bayrak et al. 2016; Caplow et al. 2011; Curtis et al. 2018; Dooley and Kartha 2018; Griscom et al. 2017; Hansen et al. 2013; Hosonuma et al. 2012; Houghton et al. 2015; Lewis et al. 2015; Pelletier et al. 2016; Rey Benayas et al. 2009
Reforestation and forest restoration	Reforestation is the conversion to forest of land that has previously contained forests but that has been converted to some other use. Forest restoration refers to practices aimed at regaining ecological integrity in a deforested or degraded forest landscape. As such, it could fall under reforestation if it were re-establishing trees where they have been lost, or under forest management if it were restoring forests where not all trees have been lost. For practical reasons, here forest restoration is treated together with reforestation.	Reforestation is similar to afforestation with respect to the co-benefits and adverse side effects among climate change mitigation, adaptation, desertification, land degradation and food security (see row on Afforestation below). Forest restoration can increase terrestrial carbon stocks in deforested or degraded forest landscapes and can offer many co-benefits in terms of increased resilience of forests to climate change, enhanced connectivity between forest areas and conservation of biodiversity hotspots. Forest restoration may threaten livelihoods and local access to land if subsistence agriculture is targeted.	Chapter 2 Dooley and Kartha 2018; Ellison et al. 2017; Locatelli 2011; Locatelli et al. 2015b; Smith et al. 2014; Stanturf et al. 2015

6

Integrated response option	Description	Context and caveats	Supporting evidence
Afforestation	Afforestation is the conversion to forest of land that historically have not contained forests (see also 'reforestation').	Afforestation increases terrestrial carbon stocks but can also change the physical properties of land surfaces, such as surface albedo and evapotranspiration with implications for local and global climate. In the tropics, enhanced evapotranspiration cools surface temperatures, reinforcing the climate benefits of CO_2 sequestration in trees. At high latitudes and in areas affected by seasonal snow cover, the decrease in surface albedo after afforestation becomes dominant and causes an annual average warming that counteracts carbon benefits. Net biophysical effects on regional climate from afforestation is seasonal and can reduce the frequency of climate extremes, such as heat waves, improving adaptation to climate change and reducing the vulnerability of people and ecosystems. Afforestation helps to address land degradation and desertification, as forests tend to maintain water quality by reducing runoff, trapping sediments and nutrients, and improving groundwater recharge. However, food security could be hampered since an increase in global forest area can increase food prices through land competition. Other adverse side effects occur when afforestation is based on non-native species, especially with the risks related to the spread of exotic fast-growing tree species. For example, exotic species can upset the balance of evapotranspiration regimes, with negative impacts on water availability, particularly in dry regions.	Chapter 2; Chapter 3; Chapter 4; Chapter 5 Alkama and Cescatti 2016; Arora and Montenegro 2011; Bonan 2008; Boysen et al. 2017a; Brundu and Richardson 2016; Cherubini et al. 2017; Ciais et al. 2013; Ellison et al. 2017; Findell et al. 2017; Medugu et al. 2010; Kongsager et al. 2016; Kreidenweis et al. 2016; Lejeune et al. 2018; Li et al. 2015; Locatelli et al. 2015b; Perugini et al. 2017; Salvati et al. 2014; Smith et al. 2013, 2014; Trabucco et al. 2008

Table 6.7 | Integrated response options based on land management of soils.

Integrated response option	Description	Context and caveats	Supporting evidence
Increased soil organic carbon content	Practices that increase soil organic matter content include a) *land-use change* to an ecosystem with higher equilibrium soil carbon levels (e.g., from cropland to forest), b) *management of the vegetation*: including high input carbon practices, for example, improved varieties, rotations and cover crops, perennial cropping systems, biotechnology to increase inputs and recalcitrance of below ground carbon, c) *nutrient management and organic material input* to increase carbon returns to the soil, including: optimised fertiliser and organic material application rate, type, timing and precision application, d) *reduced tillage intensity and residue retention*, and e) *improved water management*: including irrigation in arid/semi-arid conditions.	Increasing soil carbon stocks removes CO_2 from the atmosphere and increases the water-holding capacity of the soil, thereby conferring resilience to climate change and enhancing adaptation capacity. It is a key strategy for addressing both desertification and land degradation. There is some evidence that crop yields and yield stability increase by increased organic matter content, though some studies show equivocal impacts. Some practices to increase soil organic matter stocks vary in their efficacy. For example, the impact of no-till farming and conservation agriculture on soil carbon stocks is often positive, but can be neutral or even negative, depending on the amount of crop residues returned to the soil. If soil organic carbon stocks were increased by increasing fertiliser inputs to increase productivity, emissions of nitrous oxide from fertiliser use could offset any climate benefits arising from carbon sinks. Similarly, if any yield penalty is incurred from practices aimed at increasing soil organic carbon stocks (e.g., through extensification), emissions could be increased through indirect land-use change, and there could also be adverse side effects on food security.	Bestelmeyer and Briske 2012; Cheesman et al. 2016; Frank et al. 2017; Gao et al. 2018; Hijbeek et al. 2017b; Keesstra et al. 2016; Lal 2016; Lambin and Meyfroidt 2011; de Moraes Sá et al. 2017; Palm et al. 2014; Pan et al. 2009; Paustian et al. 2016; Powlson et al. 2014, 2016; Schjønning et al. 2018; Smith et al. 2013, 2014, 2016c; Soussana et al. 2019; Steinbach and Alvarez 2006; VandenBygaart 2016
Reduced soil erosion	Soil erosion is the removal of soil from the land surface by water, wind or tillage, which occurs worldwide but it is particularly severe in Asia, Latin America and the Caribbean, and the Near East and North Africa. Soil erosion management includes conservation practices (e.g., the use of minimum tillage or zero tillage, crop rotations and cover crops, rational grazing systems), engineering-like practices (e.g., construction of terraces and contour cropping for controlling water erosion), or forest barriers and strip cultivation for controlling wind erosion. In eroded soils, the advance of erosion gullies and sand dunes can be limited by increasing plant cover, among other practices.	The fate of eroded soil carbon is uncertain, with some studies indicating a net source of CO_2 to the atmosphere and others suggesting a net sink. Reduced soil erosion has benefits for adaptation as it reduces vulnerability of soils to loss under climate extremes, increasing resilience to climate change. Some management practices implemented to control erosion, such as increasing ground cover, can reduce the vulnerability of soils to degradation/landslides, and prevention of soil erosion is a key measure used to tackle desertification. Because it protects the capacity of land to produce food, it also contributes positively to food security.	Chapter 3 Chen 2017; Derpsch et al. 2010; FAO and ITPS 2015; FAO 2015; Garbrecht et al. 2015; Jacinthe and Lal 2001; Lal and Moldenhauer 1987; Lal 2001; Lugato et al. 2016; de Moraes Sá et al. 2017; Poeplau and Don 2015; Smith et al. 2001, 2005; Stallard 1998; Van Oost et al. 2007

6

Integrated response option	Description	Context and caveats	Supporting evidence
Reduced soil salinisation	Soil salinisation is a major process of land degradation that decreases soil fertility and affects agricultural production, aquaculture and forestry. It is a significant component of desertification processes in drylands. Practices to reduce soil salinisation include improvement of water management (e.g., water-use efficiency and irrigation/drainage technology in arid/semi-arid areas, surface and groundwater management), improvement of soil health (through increase in soil organic matter content) and improved cropland, grazing land and livestock management, agroforestry and conservation agriculture.	Techniques to prevent and reverse soil salinisation may have small benefits for mitigation by enhancing carbon sinks. These techniques may benefit adaptation and food security by maintaining existing crop systems and closing yield gaps for rainfed crops. These techniques are central to reducing desertification and land degradation, since soil salinisation is a primary driver of both.	Section 3.6; Chapter 4; Chapter 5 Baumhardt et al. 2015; Dagar et al. 2016; Datta et al. 2000; DERM 2011; Evans and Sadler 2008; He et al. 2015; D'Odorico et al. 2013; Kijne et al. 1988; Qadir et al. 2013; Rengasamy 2006; Singh 2009; UNCTAD 2011; Wong et al. 2010
Reduced soil compaction	Reduced soil compaction mainly includes agricultural techniques (e.g., crop rotations, control of livestock density) and control of agricultural traffic.	Techniques to reduce soil compaction have variable impacts on GHG emissions but may benefit adaptation by improving soil climatic resilience. Since soil compaction is a driver of both desertification and land degradation, a reduction of soil compaction could benefit both. It could also help close yield gaps in rainfed crops.	Chamen et al. 2015; Epron et al. 2016; FAO and ITPS 2015; Hamza and Anderson 2005; Soane and Van Ouwerkerk 1994; Tullberg et al. 2018
Biochar addition to soil	The use of biochar, a solid product of the pyrolysis process, as a soil amendment increases the water-holding capacity of soil. It may therefore provide better access to water and nutrients for crops and other vegetation types (so can form part of cropland, grazing land and forest management).	The use of biochar increases carbon stocks in the soil. It can enhance yields in the tropics (but less so in temperate regions), thereby benefitting both adaptation and food security. Since it can improve soil water-holding capacity and nutrient-use efficiency, and can ameliorate heavy metal pollution and other impacts, it can benefit desertification and land degradation. The positive impacts could be tempered by additional pressure on land if large quantities of biomass are required as feedstock for biochar production.	Chapter 2; Chapter 3; Chapter 4; Chapter 5 Jeffery et al. 2017; Smith 2016; Sohi 2012; Woolf et al. 2010

Table 6.8 | Integrated response options based on land management of all/other ecosystems.

Integrated response option	Description	Context and caveats	Supporting evidence
Fire management	Fire management is a land management option aimed at safeguarding life, property and resources through the prevention, detection, control, restriction and suppression of fire in forest and other vegetation. It includes the improved use of fire for sustainable forestry management, including wildfire prevention and prescribed burning. Prescribed burning is used to reduce the risk of large, uncontrollable fires in forest areas, and controlled burning is among the most effective and economic methods of reducing fire danger and stimulating natural reforestation under the forest canopy and after clear felling.	The frequency and severity of large wildfires have increased around the globe in recent decades, which has impacted on forest carbon budgets. Fire can cause various GHG emissions such as carbon dioxide (CO_2), methane (CH_4), and nitrous oxide (N_2O), and others such as carbon monoxide (CO), volatile organic carbon, and smoke aerosols. Fire management can reduce GHG emissions and can reduce haze pollution, which has significant health and economic impacts. Fire management helps to prevent soil erosion and land degradation and is used in rangelands to conserve biodiversity and to enhance forage quality.	Chapter 2; Cross-Chapter Box 3 in Chapter 2 Esteves et al. 2012; FAO 2006; Lin et al. 2017; O'Mara 2012; Rulli et al. 2006; Scasta et al. 2016; Seidl et al. 2014; Smith et al. 2014; Tacconi 2016; Valendik et al. 2011; Westerling et al. 2006; Whitehead et al. 2008; Yong and Peh 2016
Reduced landslides and natural hazards	Landslides are mainly triggered by human activity (e.g., legal and illegal mining, fire, deforestation) in combination with climate. Management of landslides and natural hazards (e.g., floods, storm surges, droughts) is based on vegetation management (e.g., afforestation) and engineering works (e.g., dams, terraces, stabilisation and filling of erosion gullies).	Management of landslides and natural hazards is important for adaptation and is a crucial intervention for managing land degradation, since landslides and natural hazards are among the most severe degradation processes. In countries where mountain slopes are planted with food crops, reduced landslides will help deliver benefits for food security. Most deaths caused due to different disasters have occurred in developing countries, where poverty, poor education and health facilities and other aspects of development, increase exposure, vulnerability and risk.	Noble et al. 2014; Arnáez J et al. 2015; Campbell 2015; FAO and ITPS 2015; Gariano and Guzzetti 2016; Mal et al. 2018
Reduced pollution including acidification	Management of air pollution is connected to climate change by emission sources of air-polluting materials and their impacts on climate, human health and ecosystems, including agriculture. Acid deposition is one of the many consequences of air pollution, harming trees and other vegetation, as well as being a significant driver of land degradation. Practices that reduce acid deposition include prevention of emissions of nitrogen oxides (NOx) and sulphur dioxide (SO_2), which also reduce GHG emissions and other short-lived climate pollutants (SLCPs). Reductions of SLCPs reduce warming in the near term and the overall rate of warming, which can be crucial for plants that are sensitive to even small increases in temperature.	There are a few potential adverse side effects of reduction in air pollution to carbon sequestration in terrestrial ecosystems, because some forms of air pollutants can enhance crop productivity by increasing diffuse sunlight, compared to direct sunlight. Reactive nitrogen deposition could also enhance CO_2 uptake in boreal forests and increase soil carbon pools to some extent. Air pollutants have different impacts on climate depending primarily on the composition, with some aerosols (and clouds seeded by them) increasing the reflection of solar radiation to space leading to net cooling, while others (e.g., black carbon and tropospheric ozone) having a net warming effect. Therefore, control of these different pollutants will have both positive and negative impacts on climate mitigation.	Chapter 2 Anderson et al. 2017; Chum et al. 2011; Carter et al. 2015; Coakley 2005; Maaroufi et al. 2015; Markandya et al. 2018; Melamed and Schmale 2016; Mostofa et al. 2016; Nemet et al. 2010; Ramanathan et al. 2001; Seinfeld and Pandis; Smith et al. 2015; UNEP 2017; UNEP and WMO 2011; Wild et al. 2012; Xu et al. 2013; Xu and Ramanathan 2017

6

Integrated response option	Description	Context and caveats	Supporting evidence
Reduced pollution including acidification *continued*	Management of harmful air pollutants such as fine particulate matter (PM$_{2.5}$) and ozone (O$_3$) also mitigate the impacts of incomplete fossil fuel combustion and GHG emissions. In addition, management of pollutants such as tropospheric O$_3$ has beneficial impacts on food production, since O$_3$ decreases crop production. Control of urban and industrial air pollution would also mitigate the harmful effects of pollution and provide adaptation co-benefits *via* improved human health. Management of pollution contributes to aquatic ecosystem conservation since controlling air pollution, rising atmospheric CO$_2$ concentrations, acid deposition, and industrial waste will reduce acidification of marine and freshwater ecosystems.		
Management of invasive species/ encroachment	Agriculture and forests can be diverse, but often much of the diversity is non-native. Invasive species in different biomes have been introduced intentionally or unintentionally through export of ornamental plants or animals, and through the promotion of modern agriculture and forestry. Non-native species tend to be more numerous in larger than in smaller human-modified landscapes (e.g., over 50% of species in an urbanised area or extensive agricultural fields can be non-native). Invasive alien species in the USA cause major environmental damage amounting to almost 120 billion USD yr^{-1}. There are approximately 50,000 foreign species and the number is increasing. About 42% of the species on the Threatened or Endangered species lists are at risk primarily because of alien-invasive species. Invasive species can be managed through manual clearance of invasive species, while in some areas, natural enemies of the invasive species are introduced to control them.	Exotic species are used in forestry where local indigenous forests cannot produce the type, quantity and quality of forest products required. Planted forests of exotic tree species make significant contributions to the economy and provide multiple products and Nature's Contributions to People. In general, exotic species are selected to have higher growth rates than native species and produce more wood per unit of area and time. In 2015, the total area of planted forest with non-native tree species was estimated to be around 0.5 Mkm^2. Introduced species were dominant in South America, Oceania and Eastern and Southern Africa, where industrial forestry is dominant. The use of exotic tree species has played an important role in the production of roundwood, fibre, firewood and other forest products. The challenge is to manage existing and future plantation forests of alien trees to maximise current benefits, while minimising present and future risks and negative impacts, and without compromising future benefits. In many countries or regions, non-native trees planted for production or other purposes often lead to sharp conflicts of interest when they become invasive, and to negative impacts on Nature's Contributions to People and nature conservation.	Brundu and Richardson 2016; Cossalter and Pye-Smith 2003; Dresner et al. 2015; Payn et al. 2015; Pimentel et al. 2005; Vilà et al. 2011
Restoration and reduced conversion of coastal wetlands	Coastal wetland restoration involves restoring degraded/ damaged coastal wetlands, including mangroves, salt marshes and seagrass ecosystems.	Coastal wetland restoration and avoided coastal wetland impacts have the capacity to increase carbon sinks and can provide benefits by regulating water flow and preventing downstream flooding. Coastal wetlands provide a natural defence against coastal flooding and storm surges by dissipating wave energy, reducing erosion and by helping to stabilise shore sediments. Since large areas of global coastal wetlands are degraded, restoration could provide benefits land degradation. Since some areas of coastal wetlands are used for food production, restoration could displace food production and damage local food supply (Section 6.3.4), though some forms (e.g., mangrove restoration) can improve local fisheries.	Griscom et al. 2017; Lotze et al. 2006; Munang et al. 2014; Naylor et al. 2000
Restoration and reduced conversion of peatlands	Peatland restoration involves restoring degraded/ damaged peatlands, which both increases carbon sinks, but also avoids ongoing CO$_2$ emissions from degraded peatlands. So, as well as protecting biodiversity, it both prevents future emissions and creates a sink.	Avoided peat impacts and peatland restoration can provide significant mitigation, though restoration can lead to an increase in methane emissions, particularly in nutrient rich fens. There may also be benefits for climate adaptation by regulating water flow and preventing downstream flooding. Considering that large areas of global peatlands are degraded, peatland restoration is a key tool in addressing land degradation. Since large areas of tropical peatlands and some northern peatlands have been drained and cleared for food production, their restoration could displace food production and damage local food supply, potentially leading to adverse impacts on food security locally, though the global impact would be limited due to the relatively small areas affected.	Griscom et al. 2017; Jauhiainen et al. 2008; Limpens et al. 2008; Munang et al. 2014

6

Integrated response option	Description	Context and caveats	Supporting evidence
Biodiversity conservation	Biodiversity conservation refers to practices aimed at maintaining components of biological diversity. It includes conservation of ecosystems and natural habitats, maintenance and recovery of viable populations of species in their natural surroundings (*in-situ* conservation) and, in the case of domesticated or cultivated species, in the surroundings where they have developed their distinctive properties outside their natural habitats (*ex-situ* conservation). Examples of biodiversity conservation measures are establishment of protected areas to achieve specific conservation objectives, preservation of biodiversity hotspots, land management to recover natural habitats, interventions to expand or control selective plant or animal species in productive lands or rangelands (e.g., rewilding).	Biodiversity conservation measures interact with the climate system through many complex processes, which can have either positive or negative impacts. For example, establishment of protected areas can increase carbon storage in vegetation and soil, and tree planting to promote species richness and natural habitats can enhance carbon uptake capacity of ecosystems. Management of wild animals can influence climate *via* emissions of GHGs (from anaerobic fermentation of plant materials in the rumen), impacts on vegetation (*via* foraging), changes in fire frequency (as grazers lower grass and vegetation densities as potential fuels), and nutrient cycling and transport (by adding nutrients to soils). Conserving and restoring megafauna in northern regions also prevents thawing of permafrost and reduces woody encroachment, thus avoiding methane emissions and increases in albedo. Defaunation affects carbon storage in tropical forests and savannahs. In the tropics, the loss of mega-faunal frugivores is estimated be responsible for up to 10% reduction in carbon storage of global tropical forests. Frugivore rewilding programmes in the tropics are seen as carbon sequestration options that can be equally effective as tree planting schemes. Biodiversity conservation measures generally favour adaptation, but can interact with food security, land degradation or desertification. Protected areas for biodiversity reduce the land available for food production, and abundancies of some species (such as large animals) can influence land degradation processes by grazing, trampling and compacting soil surfaces, thereby altering surface temperatures and chemical reactions affecting sediment and carbon retention.	Bello et al. 2015; Campbell et al. 2008; Cromsigt et al. 2018; Kapos et al. 2008; Osuri et al. 2016; Schmitz et al. 2018; Secretariat of the Convention on Biological Diversity 2008

Table 6.9 | Integrated response options based on land management specifically for carbon dioxide removal (CDR).

Integrated response option	Description	Context and caveats	Supporting evidence
Enhanced weathering of minerals	The enhanced weathering of minerals that naturally absorb CO_2 from the atmosphere has been proposed as a CDR technology with a large mitigation potential. The rocks are ground to increase the surface area and the ground minerals are then applied to the land where they absorb atmospheric CO_2.	Enhanced mineral weathering can remove atmospheric carbon dioxide (CO_2). Since ground minerals can increase pH, there could be some benefits for efforts to prevent or reverse land degradation where acidification is the driver of degradation. Since increasing soil pH in acidified soils can increase productivity, the same effect could provide some benefit for food security. Minerals used for enhanced weathering need to be mined, and mining has large impacts locally, though the total area mined is likely to be small on the global scale.	Beerling et al. 2018; Lenton 2010; Schuiling and Krijgsman 2006; Smith et al. 2016a; Taylor et al. 2016
Bioenergy and bioenergy with carbon capture and storage (BECCS)	Bioenergy production can mitigate climate change by delivering an energy service, therefore avoiding combustion of fossil energy. It is the most common renewable energy source used in the world today and has a large potential for future deployment (see Cross Chapter Box 7 in this chapter). BECCS entails the use of bioenergy technologies (e.g., bioelectricity or biofuels) in combination with CO_2 capture and storage (see also Glossary). BECCS simultaneously provides energy and can reduce atmospheric CO_2 concentrations (Chapter 2; Cross-Chapter Box 7 in this chapter) for a discussion of potentials and atmospheric effects); thus, BECCS is considered a CDR technology. While several BECCS demonstration projects exist, it has yet to be deployed at scale. Bioenergy and BECCS are widely-used in many future scenarios as a climate change mitigation option in the energy and transport sector, especially those scenarios aimed at a stabilisation of global climate at 2°C or less above pre-industrial levels.	Bioenergy and BECCS can compete for land and water with other uses. Increased use of bioenergy and BECCS can result in large expansion of cropland area, significant deforestation, and increased irrigation water use and water scarcity. Large-scale use of bioenergy can result in increased food prices and can lead to an increase in the population at risk of hunger. As a result of these effects, large-scale bioenergy and BECCS can have negative impacts for food security. Interlinkages of bioenergy and BECCS with climate change adaptation, land degradation, desertification, and biodiversity are highly dependent on local factors such as the type of energy crop, management practice, and previous land use. For example, intensive agricultural practices aiming to achieve high crop yields, as is the case for some bioenergy systems, may have significant effects on soil health, including depletion of soil organic matter, resulting in negative impacts on land degradation and desertification. However, with low inputs of fossil fuels and chemicals, limited irrigation, heat/drought tolerant species, using marginal land, biofuel programmes can be beneficial to future adaptation of ecosystems.	Cross-Chapter Box 7 in Chapter 6 IPCC SR15 (IPCC 2018); Chapter 2; Chapter 4; Section 6.4; Chapter 7 Baker et al. 2019; Calvin et al. 2014; Chaturvedi et al. 2013; Chum et al. 2011; Clarke et al. 2014; Correa et al. 2017; Creutzig et al. 2015; Dasgupta et al. 2014; Don et al. 2012; Edelenbosch et al. 2017; IPCC 2012; Favero and Mendelsohn 2014; FAO 2011a; Fujimori et al. 2019; Fuss et al. 2016, 2018; Hejazi et al. 2015; Kemper 2015; Kline et al. 2017; Lal 2014; Lotze-Campen et al. 2013; Mello et al. 2014; Muratori et al. 2016; Noble et al. 2014; Obersteiner et al. 2016;

6

Integrated response option	Description	Context and caveats	Supporting evidence
Bioenergy and bioenergy with carbon capture and storage (BECCS) *continued*		Planting bioenergy crops, like perennial grasses, on degraded land can increase soil carbon and ecosystem quality (including biodiversity), thereby helping to preserve soil quality, reverse land degradation, prevent desertification processes, and reduce food insecurity. These effects depend on the scale of deployment, the feedstock, the prior land use, and which other response options are included (see Section 6.4.4.2). Large-scale production of bioenergy can require significant amounts of land, increasing potential pressures for land conversion and land degradation. Low levels of bioenergy deployment require less land, leading to smaller effects on forest cover and food prices; however, these land requirements could still be substantial. In terms of feedstocks, in some regions, they may not need irrigation, and thus would not compete for water with food crops. Additionally, the use of residues or microalgae could limit competition for land and biodiversity loss; however, residues could result in land degradation or decreased soil organic carbon. Whether woody bioenergy results in increased competition for land or not is disputed in the literature, with some studies suggesting reduced competition and others suggesting enhanced competition. One study noted that this effect changes over time, with complementarity between woody bioenergy and forest carbon sequestration in the near-term, but increased competition for land with afforestation/reforestation in the long term. Additionally, woody bioenergy could also result in land degradation.	Popp et al. 2011b, 2014, 2017; Riahi et al. 2017; Robertson et al. 2017a; Sánchez et al. 2017; Searchinger et al. 2018; Sims et al. 2014; Slade et al. 2014; Smith et al. 2016c; Tian et al. 2018; Torvanger 2018; Van Vuuren et al. 2011, 2015, 2016; Wise et al. 2015

6.2.2   Integrated response options based on value chain management

6.2.2.1   Integrated response options based on value chain management through demand management

Integrated response options based on value chain management through demand management are described in Table 6.10, which also notes any context specificities, and provides the evidence base for the effects of the response options.

6.2.2.2   Integrated response options based on value chain management through supply management

Integrated response options based on value chain management through supply management are described in Table 6.11, which also notes any context specificities, and provides the evidence base in for effects of the response options.

6.2.3   Integrated response options based on risk management

6.2.3.1   Risk management options

Integrated response options based on risk management are described in Table 6.12, which also notes any context specificities, and provides the evidence base for the effects of the response options.

6

Table 6.10 | Integrated response options based on value chain management through demand management.

Integrated response option	Description	Context and caveats	Supporting evidence
Dietary change	Sustainable healthy diets represent a range of dietary changes to improve human diets, to make them healthy in terms of the nutrition delivered, and also (economically, environmentally and socially) sustainable. A 'contract and converge' model of transition to sustainable healthy diets would involve a reduction in over-consumption (particularly of livestock products) in over-consuming populations, with increased consumption of some food groups in populations where minimum nutritional needs are not met. Such a conversion could result in a decline in undernourishment, as well as reduction in the risk of morbidity and mortality due to over-consumption.	A dietary shift away from meat can reduce GHG emissions, reduce cropland and pasture requirements, enhance biodiversity protection, and reduce mitigation costs. Additionally, dietary change can both increase potential for other land-based response options and reduce the need for them by freeing land. By decreasing pressure on land, demand reduction through dietary change could also allow for decreased production intensity, which could reduce soil erosion and provide benefits to a range of other environmental indicators such as deforestation and decreased use of fertiliser (nitrogen and phosphorus), pesticides, water and energy, leading to potential benefits for adaptation, desertification, and land degradation.	Chapter 5; Section 6.4.4.2 Aleksandrowicz et al. 2016; Bajželj et al. 2014a; Bonsch et al. 2016; Erb et al. 2016; Godfray et al. 2010; Haberl et al. 2011; Havlík et al. 2014; Muller et al. 2017; Smith et al. 2013; Springmann et al. 2018; Stehfest et al. 2009; Tilman and Clark 2014; Wu et al. 2019
Reduced post-harvest losses	Approximately one-third of the food produced for human consumption is wasted in post-production operations. Most of these losses are due to poor storage management. Post-harvest food losses underlie the food system's failure to equitably enable accessible and affordable food in all countries. Reduced post-harvest food losses can improve food security in developing countries (while food loss in developed countries mostly occurs at the retail/consumer stage). The key drivers for post-harvest waste in developing countries are structural and infrastructure deficiencies. Thus, reducing food waste at the post-harvest stage requires responses that process, preserve and, where appropriate, redistribute food to where it can be consumed immediately.	Differences exist between farm food waste reduction technologies between small-scale agricultural systems and large-scale agricultural systems. A suite of options includes farm-level storage facilities, trade or exchange processing technologies including food drying, on-site farm processing for value addition, and improved seed systems. For large-scale agri-food systems, options include cold chains for preservation, processing for value addition and linkages to value chains that absorb the harvests almost instantly into the supply chain. In addition to the specific options to reduce food loss and waste, there are more systemic possibilities related to food systems. Improving and expanding the 'dry chain' can significantly reduce food losses at the household level. Dry chains are analogous to the cold chain and refers to the 'initial dehydration of durable commodities to levels preventing fungal growth' followed by storage in moisture-proof containers. Regional and local food systems are now being promoted to enable production, distribution, access and affordability of food. Reducing post-harvest losses has the potential to reduce emissions and could simultaneously reduce food costs and increase availability. The perishability and safety of fresh foods are highly susceptible to temperature increase.	Chapter 5 Ansah et al. 2017; Bajželj et al. 2014b; Billen et al. 2018; Bradford et al. 2018; Chaboud and Daviron 2017; Göbel et al. 2015; Gustavsson et al. 2011; Hengsdijk and de Boer 2017; Hodges et al. 2011; Ingram et al. 2016; Kissinger et al. 2018; Kumar and Kalita 2017; Ritzema et al. 2017; Sheahan and Barrett 2017a; Wilhelm et al. 2016)
Reduced food waste (consumer or retailer)	Since approximately 9–30% of all food is wasted, reducing food waste can reduce pressure on land (see also reducing post-harvest losses).	Reducing food waste could lead to a reduction in cropland area and GHG emissions, resulting in benefits for mitigation. By decreasing pressure on land, food waste reduction could allow for decreased production intensity, which could reduce soil erosion and provide benefits to a range of other environmental indicators such as deforestation and decreases in use of fertiliser (N and P), pesticides, water and energy, leading to potential benefits for adaptation, desertification, and land degradation.	Alexander et al. 2016; Bajželj et al. 2014b; Gustavsson et al. 2011; Kummu et al. 2012; Muller et al. 2017; Smith et al. 2013; Vermeulen et al. 2012b
Material substitution	Material substitution involves the use of wood or agricultural biomass (e.g., straw bales) instead of fossil fuel-based materials (e.g., concrete, iron, steel, aluminium) for building, textiles or other applications.	Material substitution reduces carbon emissions – both because the biomass sequesters carbon in materials while re-growth of forests can lead to continued sequestration, and because it reduces the demand for fossil fuels, delivering a benefit for mitigation. However, a potential trade-off exists between conserving carbon stocks and using forests for wood products. If the use of material for substitution was large enough to result in increased forest area, then the adverse side effects for adaptation and food security would be similar to that of reforestation and afforestation. In addition, some studies indicate that wooden buildings, if properly constructed, could reduce fire risk compared to steel, creating a co-benefit for adaptation. The effects of material substitution on land degradation depend on management practice; some forms of logging can lead to increased land degradation. Long-term forest management with carbon storage in long-lived products also results in atmospheric CO_2 removal.	Chapter 4 Dugan et al. 2018; Eriksson et al. 2012; Gustavsson et al. 2006; Iordan et al. 2018; Kauppi et al. 2018; Kurz et al. 2016; Leskinen et al. 2018; McLaren 2012; Miner 2010; Oliver and Morecroft 2014; Ramage et al. 2017; Sathre and O'Connor 2010; Smyth et al. 2014

6

Table 6.11 | Integrated response options based on value chain management through supply management.

Integrated response option	Description	Context and caveats	Supporting evidence
Sustainable sourcing	Sustainable sourcing includes approaches to ensure that the production of goods is done in a sustainable way, such as through low-impact agriculture, zero-deforestation supply chains, or sustainably harvested forest products. Currently around 8% of global forest area has been certified in some manner, and 25% of global industrial roundwood comes from certified forests. Sustainable sourcing also aims to enable producers to increase their percentage of the final value of commodities. Adding value to products requires improved innovation, coordination and efficiency in the food supply chain, as well as labelling to meet consumer demands. As such, sustainable sourcing is an approach that combines both supply- and demand-side management. Promoting sustainable and value-added products can reduce the need for compensatory extensification of agricultural areas and is a specific commitment of some sourcing programmes (such as forest certification programmes). Table 7.3 (Chapter 7) provides examples of the many sustainable sourcing programmes now available globally.	Sustainable sourcing is expanding but accounts for only a small fraction of overall food and material production; many staple food crops do not have strong sustainability standards. Sustainable sourcing provides potential benefits for both climate mitigation and adaptation by reducing drivers of unsustainable land management, and by diversifying and increasing flexibility in the food system to climate stressors and shocks. Sustainable sourcing can lower expenditure for food processors and retailers by reducing losses. Adding value to products can extend a producer's marketing season and provide unique opportunities to capture niche markets, thereby increasing their adaptive capacity to climate change. Sustainable sourcing can also provide significant benefits for food security, while simultaneously creating economic alternatives for the poor. Sustainable sourcing programmes often also have positive impacts on the overall efficiency of the food supply chain and can create closer and more direct links between producers and consumers. In some cases, processing of value-added products could lead to higher emissions or demand for resources in the food system, potentially leading to small adverse impacts on land degradation and desertification challenges.	Chapter 2; Chapter 3; Chapter 5; Section 6.4 Accorsi et al. 2017; Bajželj et al. 2014a; Bustamante et al. 2014; Clark and Tilman 2017; Garnett 2011; Godfray et al. 2010; Hertel 2015; Ingram et al. 2016; James and James 2010; Muller et al. 2017; Springer et al. 2015; Tayleur et al. 2017; Tilman and Clark 2014
Management of supply chains	Management of supply chains include a set of polycentric governance processes focused on improving efficiency and sustainability across the supply chain for each product, to reduce climate risk and profitably reduce emissions. Trade-driven food supply chains are becoming increasingly complex and are contributing to emissions. Improved management of supply chains can include 1) better food transport and increasing the economic value or reduce risks of commodities through production processes (e.g., packaging, processing, cooling, drying, extracting) and 2) improved policies for stability of food supply, as globalised food systems and commodity markets are vulnerable to food price volatility. The 2007–2008 food price shocks negatively affected food security for millions, most severely in Sub-Saharan Africa. Increasing the stability of food supply chains is a key goal to increase food security, given that climate change threatens to lead to more production shocks in the future.	Successful implementation of supply chain management practices is dependent on organisational capacity, the agility and flexibility of business strategies, the strengthening of public-private policies and effectiveness of supply-chain governance. Existing practices include a) greening supply chains (e.g., utilising products and services with a reduced impact on the environment and human health), b) adoption of specific sustainability instruments among agri-food companies (e.g., eco-innovation practices), c) adopting emission accounting tools (e.g., carbon and water foot-printing), and d) implementing 'demand forecasting' strategies (e.g., changes in consumer preference for 'green' products). In terms of food supply, measures to improve stability in traded markets can include i) financial and trade policies, such as reductions on food taxes and import tariffs, (ii) shortening food supply chains (SFSCs), (iii) increasing food production, (iv) designing alternative distribution networks, (v) increasing food market transparency and reducing speculation in futures markets, (vi) increasing storage options, and (vii) increasing subsidies and food-based safety nets.	Chapter 5 Barthel and Isendahl 2013; Haggblade et al. 2017; Lewis and Witham 2012; Michelini et al. 2018; Minot 2014; Mundler and Rumpus 2012; Tadasse et al. 2016; Wheeler and von Braun 2013; Wilhelm et al. 2016; Wodon and Zaman 2010; World Bank 2011
Enhanced urban food systems	Urban areas are becoming the principal territories for intervention in improving food access through innovative strategies that aim to reduce hunger and improve livelihoods. Interventions include urban and peri-urban agriculture and forestry and local food policy and planning initiatives such as Food Policy Councils and city-region-wide regional food strategies. Such systems have demonstrated inter-linkages of the city and its citizens with surrounding rural areas to create sustainable, and more nutritious food supplies for the city, while improving the health status of urban dwellers, reducing pollution levels, adapting to and mitigating climate change, and stimulating economic development. Options include support for urban and peri-urban agriculture, green infrastructure (e.g., green roofs), local markets, enhanced social (food) safety nets and development of alternative food sources and technologies, such as vertical farming.	Urban territorial areas have a potential to reduce GHG emissions through improved food systems to reduce vehicle miles of food transportation, localised carbon capture and food waste reduction. The benefits of urban food forests that are intentionally planted woody perennial food-producing species, are also cited for their carbon sequestration potentials. However, new urban food systems may have diverse and unexpected adverse side effects with climate systems, such as lower efficiencies in food supply and higher costs than modern large-scale agriculture. Diversifying markets, considering value-added products in the food supply system may help to improve food security by increasing its economic performance and revenues to local farmers.	Akhtar et al. 2016; Benis and Ferrão 2017; Brinkley et al. 2013; Chappell et al. 2016; Dubbeling 2014; Goldstein et al. 2016; Kowalski and Conway 2018; Lee-Smith 2010; Barthel and Isendahl 2013; Lwasa et al. 2014, 2015; Revi et al. 2014; Specht et al. 2014; Tao et al. 2015

Integrated response option	Description	Context and caveats	Supporting evidence
Improved food processing and retailing	Improved food processing and retailing involves several practices related to a) greening supply chains (e.g., utilising products and services with a reduced impact on the environment and human health), b) adoption of specific sustainability instruments among agri-food companies (e.g., eco-innovation practices), c) adopting emission accounting tools (e.g., carbon and water foot-printing), d) implementing 'demand forecasting' strategies (e.g., changes in consumer preference for 'green' products) and, e) supporting polycentric supply-chain governance processes.	Improved food processing and retailing can provide benefits for climate mitigation since GHG-friendly foods can reduce agri-food GHG emissions from transportation, waste and energy use. In cases where climate extremes and natural disasters disrupt supply chain networks, improved food processing and retailing can benefit climate adaptation by buffering the impacts of changing temperature and rainfall patterns on upstream agricultural production. It can provide benefits for food security by supporting healthier diets and reducing food loss and waste. Successful implementation is dependent on organisational capacity, the agility and flexibility of business strategies, the strengthening of public-private policies and effectiveness of supply-chain governance.	Chapter 2; Chapter 5

Avetisyan et al. 2014; Garnett et al. 2013; Godfray et al. 2010; Mohammadi et al. 2014; Porter et al. 2016; Ridoutt et al. 2016; Song et al. 2017 |
| Improved energy use in food systems | Agriculture's energy efficiency can be improved to reduce the dependency on non-renewable energy sources. This can be realised either by decreased energy inputs, or through increased outputs per unit of input. In some countries, managerial inefficiency (rather than a technology gap) is the main source for energy-efficiency loss. Heterogenous patterns of energy efficiency exist at the national scale and promoting energy-efficient technologies along with managerial capacity development can reduce the gap and provide large benefits for climate adaptation. Improvements in carbon monitoring and calculation techniques such as the foot-printing of agricultural products can enhance energy-efficiency transition management and uptake in agricultural enterprises. | Transformation to low-carbon technologies such as renewable energy and energy efficiency can offer opportunities for significant climate change mitigation, for example, by providing a substitute to transport fuel that could benefit marginal agricultural resources, while simultaneously contributing to long-term economic growth. In poorer nations, increased energy efficiency in agricultural value-added production, in particular, can provide large mitigation benefits. Under certain scenarios, the efficiency of agricultural systems can stagnate and could exert pressure on grasslands and rangelands, thereby impacting on land degradation and desertification. Rebound effects can also occur, with adverse impacts on emissions. | Al-Mansour F and Jejcic V 2017; Baptista et al. 2013; Begum et al. 2015; Gunatilake et al. 2014; Jebli and Youssef 2017; Van Vuuren et al. 2017b |

Table 6.12 | Integrated response options based on risk management.

Integrated response option	Description	Context and caveats	Supporting evidence
Management of urban sprawl	Unplanned urbanisation leading to sprawl and extensification of cities along the rural-urban fringe has been identified as a driver of forest and agricultural land loss and a threat to food production around cities. It has been estimated that urban expansion will result in a 1.8–2.4% loss of global croplands by 2030. This rapid urban expansion is especially strong in new emerging towns and cities in Asia and Africa. Policies to prevent such urbanisation have included integrated land-use planning, agricultural zoning ordinances and agricultural districts, urban redevelopment, arable land reclamation, and transfer/purchase of development rights or easements.	The prevention of uncontrolled urban sprawl may provide adaptation co-benefits, but adverse side effects for adaptation might arise due to restricted ability of people to move in response to climate change.	Barbero-Sierra et al. 2013; Bren d'Amour et al. 2016; Cai et al. 2013; Chen 2007; Francis et al. 2012; Gibson et al. 2015; Lee et al. 2015; Qian et al. 2015; Shen et al. 2017; Tan et al. 2009
Livelihood diversification	When households' livelihoods depend on a small number of sources of income without much diversification, and when those income sources are in fields that are highly climate dependent, like agriculture and fishing, this dependence can put food security and livelihoods at risk. Livelihood diversification (drawing from a portfolio of dissimilar sources of livelihood as a tool to spread risk) has been identified as one option to increase incomes and reduce poverty, increase food security, and promote climate resilience and risk reduction.	Livelihood diversification offers benefits for desertification and land degradation, particularly through non-traditional crops or trees in agroforestry systems which improve soil. Livelihood diversification may increase on-farm biodiversity due to these investments in more ecosystem-mimicking production systems, like agroforestry and polycultures. Diversification into non-agricultural fields, such as wage labour or trading, is increasingly favoured by farmers as a low-cost strategy, particularly to respond to increasing climate risks.	Adger 1999; Ahmed and Stepp 2016; Antwi-Agyei et al. 2014; Barrett et al. 2001; Berman et al. 2012; Bryceson 1999; DiGiano and Racelis 2012; Ellis 1998, 2008; Little et al. 2001; Ngigi et al. 2017; Rakodi 1999; Thornton and Herrero 2014
Use of local seeds	Using local seeds (also called seed sovereignty) refers to use of non-improved, non-commercial seeds varieties. These can be used and stored by local farmers as low-cost inputs and can often help contribute to the conservation of local varieties and landraces, increasing local biodiversity. Many local seeds also require no pesticide or fertiliser use, leading to less land degradation in their use.	Use of local seeds is important in the many parts of the developing world that do not rely on commercial seed inputs. Promotion of local seed-saving initiatives can include seed networks, banks and exchanges, and non-commercial open source plant breeding. These locally developed seeds can help protect local agrobiodiversity and can often be more climate resilient than generic commercial varieties, although the impacts on food security and overall land degradation are inconclusive.	Bowman 2015; Campbell and Veteto 2015; Coomes et al. 2015; Kloppenberg 2010; Luby et al. 2015; Van Niekerk and Wynberg 2017; Patnaik et al. 2017; Reisman 2017; Vasconcelos et al. 2013; Wattnem 2016

6

Integrated response option	Description	Context and caveats	Supporting evidence
Disaster risk management	Disaster risk management encompasses many approaches to try to reduce the consequences of climate- and weather-related disasters and events on socio-economic systems. The Hyogo Framework for Action is a UN framework for nations to build resilience to disasters through effective integration of disaster risk considerations into sustainable development policies. For example, in Vietnam a national strategy on disasters based on Hyogo has introduced the concept of a 'four-on-the-spot' approach for disaster risk management of: proactive prevention, timely response, quick and effective recovery, and sustainable development. Other widespread approaches to disaster risk management include using early warning systems that can encompass 1) education systems, 2) hazard and risk maps, 3) hydrological and meteorological monitoring (such as flood forecasting or extreme weather warnings), and 4) communications systems to pass on information to enable action. These approaches have long been considered to reduce the risk of household asset damage during one-off climate events and are increasingly being combined with climate adaptation policies.	Community-based disaster risk management has been pointed to as one of the most successful ways to ensure that information reaches the people who need to be participants in risk reduction. Effective disaster risk management approaches must be 'end-to-end,' reaching communities at risk and supporting and empowering vulnerable communities to take appropriate action. The most effective early warning systems are not simply technical systems of information dissemination, but utilise and develop community capacities, create local ownership of the system, and are based on a shared understanding of needs and purpose. Tapping into existing traditional or local knowledge has also been recommended for disaster risk management approaches to reducing vulnerability.	Ajibade and McBean 2014; Alessa et al. 2016; Bouwer et al. 2014; Carreño et al. 2007; Cools et al. 2016; Djalante et al. 2012; Garschagen 2016; Maskrey 2011; Mercer 2010; Schipper and Pelling 2006; Sternberg and Batbuyan 2013; Thomalla et al. 2006; Vogel and O'Brien 2006
Risk-sharing instruments	Risk-sharing instruments can encompass a variety of approaches. Intra-household risk pooling is a common strategy in rural communities, such as through extended family financial transfers; one study found that 65% of poor households in Jamaica report receiving transfers, and such transfers can account for up to 75% of household income or more after crisis events. Community rotating savings and credit associations (ROSCAs) have long been used for general risk pooling and can be a source of financing to cope with climate variability as well. Credit services have been shown to be important for adaptation actions and risk reduction. Insurance of various kinds is also a form of risk pooling. Commercial crop insurance is one of the most widely used risk-hedging financial vehicles, and can involve both traditional indemnity-based insurance that reimburses clients for estimated financial losses from shortfalls, or index insurance that pays out the value of an index (such as weather events) rather than actual losses; the former is more common for large farms in the developed world and the latter for smaller non-commercial farms in developing countries.	Locally developed risk-pooling measures show general positive impacts on household livelihoods. However, more commercial approaches have mixed effects. Commercial crop insurance is highly subsidised in much of the developed world. Index insurance programmes have often failed to attract sufficient buyers or have remained financially unfeasible for commercial insurance sellers. The overall impact of index insurance on food production supply and access has also not been assessed. Traditional crop insurance has generally been seen as positive for food security as it leads to expansion of agricultural production areas and increased food supply. However, insurance may also 'mask' truly risky agriculture and prevent farmers from seeking less risky production strategies. Insurance can also provide perverse incentives for farmers to bring additional lands into crop production, leading to greater risk of degradation.	Akter et al. 2016; Annan and Schlenker 2015; Claassen et al. 2011a; Fenton et al. 2017; Giné et al. 2008; Goodwin and Smith 2003; Hammill et al. 2008; Havemenn and Muccione 2011; Jaworski 2016; Meze-Hausken et al. 2009; Morduch and Sharma 2002; Bhattamishra and Barrett 2010; Peterson 2012; Sanderson et al. 2013; Skees and Collier 2012; Smith and Glauber 2012

Cross-Chapter Box 7 | Bioenergy and bioenergy with carbon capture and storage (BECCS) in mitigation scenarios

Katherine Calvin (The United States of America), Almut Arneth (Germany), Luis Barioni (Brazil), Francesco Cherubini (Norway/Italy), Annette Cowie (Australia), Joanna House (United Kingdom), Francis X. Johnson (Sweden), Alexander Popp (Germany), Joana Portugal Pereira (Portugal/United Kingdom), Mark Rounsevell (United Kingdom), Raphael Slade (United Kingdom), Pete Smith (United Kingdom)

Bioenergy and BECCS potential
Using biomass to produce heat, electricity and transport fuels (bioenergy) instead of coal, oil, and natural gas can reduce GHG emissions. Combining biomass conversion technologies with systems that capture CO_2 and inject it into geological formations, BECCS can deliver net negative emissions. The net climate effects of bioenergy and BECCS depend on the magnitude of bioenergy supply chain emissions and land/climate interactions, described further below.

Cross-Chapter Box 7 (continued)

Biomass in 2013 contributed about 60 EJ (10%) to global primary energy[4] (WBA 2016). In 2011, the IPCC Special Report on Renewable Energy Sources concluded that biomass supply for energy could reach 100–300 EJ yr^{-1} by 2050 with the caveat that the technical potential[5] cannot be determined precisely while societal preferences are unclear; that deployment depends on 'factors that are inherently uncertain'; and that biomass use could evolve in a 'sustainable' or 'unsustainable' way, depending on the governance context (IPCC 2012). The IPCC WGIII AR5 report noted, in addition, that high deployment levels would require extensive use of technologies able to convert lignocellulosic biomass such as forest wood, agricultural residues, and lignocellulosic crops. The IPCC Special Report on Global Warming of 1.5°C (SR15) noted that high levels of bioenergy deployment may result in adverse side effects for food security, ecosystems, biodiversity, water use, and nutrients (de Coninck et al. 2018).

Although estimates of potential are uncertain, there is *high confidence* that the most important factors determining future biomass supply are land availability and land productivity. These factors are, in turn, determined by competing uses of land and a myriad of environmental and economic considerations (Dornburg et al. 2010; Batidzirai et al. 2012; Erb et al. 2012; Slade 2014, Searle and Malins 2014). Overlaying estimates of technical potential with such considerations invariably results in a smaller estimate. Recent studies that have attempted to do this estimate that 50–244 EJ biomass could be produced on 0.1–13 Mkm^2 (Fuss et al. 2018; Schueler et al. 2016; Searle and Malins 2014; IPCC 2018; Wu et al. 2019; Heck et al. 2018; de Coninck et al. 2018). While preferences concerning economic, social and environmental objectives vary geographically and over time, studies commonly estimate 'sustainable' potentials by introducing restrictions intended to protect environmental values and avoid negative effects on poor and vulnerable segments in societies.

Estimates of global geological CO_2 storage capacity are large – ranging from 1680 GtCO_2 to 24,000 GtCO_2 (McCollum et al. 2014) – however, the potential of BECCS may be significantly constrained by socio-political and technical and geographical considerations, including limits to knowledge and experience (Chapters 6 and 7).

Bioenergy and BECCS use in mitigation scenarios

Most mitigation scenarios include substantial deployment of bioenergy technologies (Clarke et al. 2014; Fuss et al. 2014; IPCC 2018). Across all scenarios, the amount of bioenergy and BECCS ranges from 0 EJ yr^{-1} to 561 EJ yr^{-1} in 2100 (Figure 1 in this box, left panel). Notably, all 1.5°C pathways include bioenergy, requiring as much as 7 Mkm^2 to be dedicated to the production of energy crops in 2050 (Rogelj et al. 2018a). If BECCS is excluded as a mitigation option, studies indicate that more biomass may be required in order to substitute for a greater proportion of fossil fuels (Muratori et al. 2016; Rose et al. 2014).

Different Integrated Assessment Models (IAMs) use alternative approaches to land allocation when determining where and how much biomass is used, with some relying on economic approaches and some relying on rule-based approaches (Popp et al. 2014). Despite these differences, a consistent finding across models is that increasing biomass supply to the extent necessary to support deep decarbonisation is likely to involve substantial land-use change (Popp et al. 2017) (Cross-Chapter Box 9 in this chapter). In model runs, bioenergy deployment and the consequent demand for biomass and land, is influenced by assumptions around the price of bioenergy, the yield of bioenergy crops, the cost of production (including the costs of fertiliser and irrigation if used), the demand for land for other uses, and the inclusion of policies (e.g., subsidies, taxes, constraints) that may alter land-use or bioenergy demand. In general, higher carbon prices result in greater bioenergy deployment (Cross-Chapter Box 7, Figure 1, right panel) and a larger percentage of BECCS. Other factors can also strongly influence bioenergy use, including the cost and availability of fossil fuels (Calvin et al. 2016a), socio-economics (Popp et al. 2017), and policy (Calvin et al. 2014; Reilly et al. 2012).

Co-benefits, adverse side effect, and risks associated with bioenergy

The production and use of biomass for bioenergy can have co-benefits, adverse side effects, and risks for land degradation, food insecurity, GHG emissions, and other environmental goals. These impacts are context specific and depend on the scale of deployment, initial land use, land type, bioenergy feedstock, initial carbon stocks, climatic region and management regime (Qin et al. 2016; Del Grosso et al. 2014; Alexander et al. 2015; Popp et al. 2017; Davis et al. 2013; Mello et al. 2014; Hudiburg et al. 2015; Carvalho et al. 2016; Silva-Olaya et al. 2017; Whitaker et al. 2018; Robledo-Abad et al. 2017; Jans et al. 2018).

---

[4]   Of this, more than half was traditional biomass, predominately used for cooking and heating in developing regions, bioelectricity accounted for about 1.7 EJ, and transport biofuels for 3.19 EJ. (Cross-Chapter Box 12 in Chapter 7).

[5]   The future availability of biomass is usually discussed in terms of a hierarchy of potentials: theoretical>technical>economic. Caution is required, however, as these terms are not always defined consistently and estimates depend on the specific definitions and calculation methodologies.

6

Cross-Chapter Box 7 (continued)

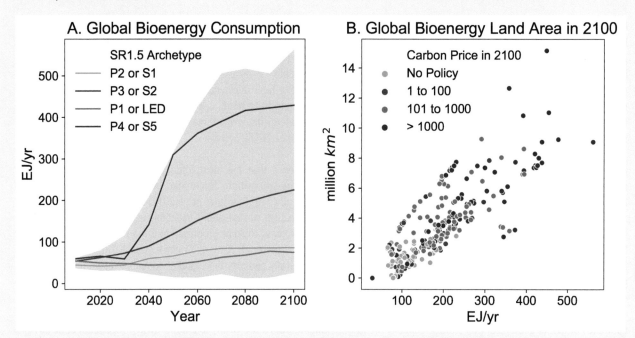

Cross-Chapter Box 7, Figure 1 | Global bioenergy consumption in IAM scenarios. Data is from an update of the Integrated Assessment Modelling Consortium (IAMC) Scenario Explorer developed for the SR15 (Huppmann et al. 2018; Rogelj et al. 2018a). The left panel **A.** shows bioenergy deployment over time for the entire scenario database (grey areas) and the four illustrative pathways from SR15 (Rogelj et al. 2018a). The right panel **B.** shows global land area for energy crops in 2100 versus total global bioenergy consumption in 2100; colours indicate the carbon price in 2100 (in 2010 USD per tCO₂). Note that this figure includes 409 scenarios, many of which exceed 1.5°C.

Synergistic outcomes with bioenergy are possible, for example, strategic integration of perennial bioenergy crops with conventional crops can provide multiple production and environmental benefits, including management of dryland salinity, enhanced biocontrol and biodiversity, and reduced eutrophication (Davis et al. 2013; Larsen et al. 2017; Cacho et al. 2018; Odgaard et al. 2019). Additionally, planting perennial bioenergy crops on low-carbon soil could enhance soil carbon sequestration (Bárcena et al. 2014; Schröder et al. 2018; Walter et al. 2015; Robertson et al. 2017a; Rowe et al. 2016; Chadwick et al. 2014; Immerzeel et al. 2014; Del Grosso et al. 2014; Mello et al. 2014; Whitaker et al. 2018). However, large-scale expansion of bioenergy may also result in increased competition for land (DeCicco 2013; Humpenöder et al. 2018; Bonsch et al. 2016; Harris et al. 2015; Richards et al. 2017; Ahlgren et al. 2017; Bárcena et al. 2014), increased GHG emissions from land-use change and land management, loss in biodiversity, and nutrient leakage (Harris et al. 2018; Harper et al. 2018; Popp et al. 2011b; Wiloso et al. 2016; Behrman et al. 2015; Valdez et al. 2017; Hof et al. 2018). If biomass crops are planted on land with a high carbon stock, the carbon loss due to land conversion may take decades to over a century to be compensated by either fossil fuel substitution or CCS (Harper et al. 2018). Competition for land may be experienced locally or regionally and is one of the determinants of food prices, food security (Popp et al. 2014; Bailey 2013; Pahl-Wostl et al. 2018; Rulli et al. 2016; Yamagata et al. 2018; Franz et al. 2017; Kline et al. 2017; Schröder et al. 2018) and water availability (Rulli et al. 2016; Bonsch et al. 2015; Pahl-Wostl et al. 2018; Bailey 2013; Chang et al. 2016; Bárcena et al. 2014).

Experience in countries at quite different levels of economic development (Brazil, Malawi and Sweden) has shown that persistent efforts over several decades to combine improved technical standards and management approaches with strong governance and coherent policies, can facilitate long-term investment in more sustainable production and sourcing of liquid biofuels (Johnson and Silveira 2014). For woody biomass, combining effective governance with active forest management over long time periods can enhance substitution-sequestration co-benefits, such as in Sweden where bioenergy has tripled during the last 40 years (currently providing about 25% of total energy supply) while forest carbon stocks have continued to grow (Lundmark et al. 2014). A variety of approaches are available at landscape level and in national and regional policies to better reconcile food security, bioenergy and ecosystem services, although more empirical evidence is needed (Mudombi et al. 2018; Manning et al. 2015; Kline et al. 2017; Maltsoglou et al. 2014; Lamers et al. 2016).

Thus, while there is *high confidence* that the technical potential for bioenergy and BECCS is large, there is also *very high confidence* that this potential is reduced when environmental, social and economic constraints are considered. The effects of bioenergy production on land degradation, water scarcity, biodiversity loss, and food insecurity are scale and context specific (*high confidence*). Large areas of monoculture bioenergy crops that displace other land uses can exacerbate these challenges, while integration into sustainably managed agricultural landscapes can ameliorate them (*medium confidence*).

Inventory reporting for BECCS and bioenergy

One of the complications in assessing the total GHG flux associated with bioenergy under United Nations Framework Convention on Climate Change (UNFCCC) reporting protocols is that fluxes from different aspects of bioenergy lifecycle are reported in different sectors and are not linked. In the energy sector, bioenergy is treated as carbon neutral at the point of biomass combustion because all change in land carbon stocks due to biomass harvest or land-use change related to bioenergy are reported under agriculture, forestry and other land-use (AFOLU) sector. Use of fertilisers is captured in the agriculture sector, while fluxes related to transport/conversion and removals due to CCS are reported in the energy sector. IAMs follow a similar reporting convention. Thus, the whole lifecycle GHG effects of bioenergy systems are not readily observed in national GHG inventories or modelled emissions estimates (see also IPCC 2006; SR15 Chapter 2 Technical Annex; Chapter 2).

Bioenergy in this report

Bioenergy and BECCS are discussed throughout this special report. Chapter 1 provides an introduction to bioenergy and BECCS and its links to land and climate. Chapter 2 discusses mitigation potential, land requirements and biophysical climate implications. Chapter 4 includes a discussion of the threats and opportunities with respect to land degradation. Chapter 5 discusses linkages between bioenergy and BECCS and food security. Chapter 6 synthesises the co-benefits and adverse side effects for mitigation, adaptation, desertification, land degradation, and food security, as well as barriers to implementation (e.g., cost, technological readiness, etc.). Chapter 7 includes a discussion of risk, policy, governance, and decision-making with respect to bioenergy and BECCS.

6.3   Potentials for addressing the land challenges

In this section, we assess how each of the integrated response options described in Section 6.2 address the land challenges of climate change mitigation (Section 6.3.1), climate change adaptation (Section 6.3.2), desertification (Section 6.3.3), land degradation (Section 6.3.4), and food security (Section 6.3.5). The quantified potentials across all of mitigation, adaptation, desertification, land degradation and food security are summarised and categorised for comparison in Section 6.3.6.

6.3.1   Potential of the integrated response options for delivering mitigation

In this section, the impacts of integrated response options on climate change mitigation are assessed.

6.3.1.1   Integrated response options based on land management

In this section, the impacts on climate change mitigation of integrated response options based on land management are assessed. Some of the caveats of these potential mitigation studies are discussed in Chapter 2 and Section 6.2.1.

Integrated response options based on land management in agriculture

Increasing the productivity of land used for food production can deliver significant mitigation by avoiding emissions that would occur if increased food demand were met through expansion of the agricultural land area (Burney et al. 2010). If pursued through increased agrochemical inputs, numerous adverse impacts on GHG emissions (and other environmental sustainability) can occur (Table 6.5), but, if pursued through sustainable intensification, increased food productivity could provide high levels of mitigation. For example, yield improvement has been estimated to have contributed to emissions savings of >13 $GtCO_2$ yr^{-1} since 1961 (Burney et al. 2010) (Table 6.13). This can also reduce the GHG intensity of products (Bennetzen et al. 2016a,b) which means a smaller environmental footprint of production, since demand can be met using less land and/or with fewer animals.

Improved cropland management could provide moderate levels of mitigation (1.4–2.3 $GtCO_2e$ yr^{-1}) (Smith et al. 2008, 2014; Pradhan et al. 2013) (Table 6.13). The lower estimate of potential is from Pradhan et al. (2013) for decreasing emissions intensity, and the upper end of technical potential is estimated by adding technical potentials for cropland management (about 1.4 $GtCO_2e$ yr^{-1}), rice management (about 0.2 $GtCO_2e$ yr^{-1}) and restoration of degraded land (about 0.7 $GtCO_2e$ yr^{-1}) from Smith et al. (2008) and Smith et al. (2014). Note that much of this potential arises from soil carbon sequestration so there is an overlap with that response option. (Section 6.3.1.1).

Grazing lands can store large stocks of carbon in soil and root biomass compartments (Conant and Paustian 2002; O'Mara 2012; Zhou et al. 2017). The global mitigation potential is moderate (1.4–1.8 $GtCO_2$ yr^{-1}), with the lower value in the range for technical potential taken from Smith et al. (2008) which includes only grassland management measures, and the upper value in the range from Herrero et al. (2016), which includes also indirect effects and some components of livestock management, and soil carbon sequestration, so there is overlap with these response options (Section 6.3.1.1).

Table 6.13 | Mitigation effects of response options based on land management in agriculture.

Integrated response option	Potential	Confidence	Citation
Increased food productivity	>13 GtCO₂e yr⁻¹	Low confidence	Chapter 5 Burney et al. 2010
Improved cropland management[a]	1.4–2.3 GtCO₂e yr⁻¹	Medium confidence	Chapter 2; Chapter 5 Pradhan et al. 2013; Smith et al. 2008, 2014
Improved grazing land management[a]	1.4–1.8 GtCO₂e yr⁻¹	Medium confidence	Chapter 2; Chapter 5 Conant et al. 2017; Herrero et al. 2016; Smith et al. 2008, 2014
Improved livestock management[a]	0.2–2.4 GtCO₂e yr⁻¹	Medium confidence	Chapter 2; Chapter 5 Herrero et al. 2016; Smith et al. 2008, 2014
Agroforestry	0.1–5.7 GtCO₂e yr⁻¹	Medium confidence	Chapter 2 Albrecht and Kandji 2003; Dickie et al. 2014; Griscom et al. 2017; Hawken 2017; Zomer et al. 2016
Agricultural diversification	>0	Low confidence	Campbell et al. 2014; Cohn et al. 2017
Reduced grassland conversion to cropland	0.03–0.7 GtCO₂e yr⁻¹	Low confidence	Note high value not shown in Chapter 2; calculated from values in Griscom et al. 2017; Krause et al. 2017; Poeplau et al. 2011
Integrated water management	0.1–0.72 GtCO₂ yr⁻¹	Low confidence	IPCC 2014; Howell et al. 2015; Li et al. 2006; Rahman and Bulbul 2015; Smith et al. 2008, 2014

[a] Note that Chapter 2 reports mitigation potential for subcategories within this response option and not the combined total reported here.

Conant et al. (2005) caution that increases in soil carbon stocks could be offset by increases in N_2O fluxes.

The mitigation potential of improved livestock management is also moderate (0.2–1.8 GtCO₂e yr⁻¹; Smith et al. (2008) including only direct livestock measures; Herrero et al. (2016) include also indirect effects, and some components of grazing land management and soil carbon sequestration) to high (6.13 GtCO₂e yr⁻¹) (Pradhan et al. 2013) (Table 6.13). There is an overlap with other response options (Section 6.3.1.1).

Zomer et al. (2016) reported that the trees agroforestry landscapes have increased carbon stock by 7.33 GtCO₂ between 2000–2010, which is equivalent to 0.7 GtCO₂ yr⁻¹. Estimates of global potential range from 0.1 GtCO₂ yr⁻¹ to 5.7 GtCO₂ yr⁻¹ (from an optimum implantation scenario of Hawken (2017), based on an assessment of all values in Griscom et al. (2017), Hawken (2017), Zomer et al. (2016) and Dickie et al. (2014) (Table 6.13).

Agricultural diversification mainly aims at increasing climate resilience, but it may have a small (but globally unquantified) mitigation potential as a function of type of crop, fertiliser management, tillage system, and soil type (Campbell et al. 2014; Cohn et al. 2017).

Reducing conversion of grassland to cropland could provide significant climate mitigation by retaining soil carbon stocks that might otherwise be lost. When grasslands are converted to croplands, they lose about 36% of their soil organic carbon stocks after 20 years (Poeplau et al. 2011). Assuming an average starting soil organic carbon stock of grasslands of 115 tC ha⁻¹ (Poeplau et al. 2011), this is equivalent to a loss of 41.5 tC ha⁻¹ on conversion to cropland. Mean annual global cropland conversion rates (1961–2003) have been around 47,000 km² yr⁻¹ (Krause et al. 2017), or 940000 km² over a 20-year period. The equivalent loss of soil organic carbon over 20 years would therefore be 14 GtCO₂e = 0.7 GtCO₂ yr⁻¹. Griscom et al. (2017) estimate a cost-effective mitigation potential of 0.03 GtCO₂ yr⁻¹ (Table 6.13).

Integrated water management provides moderate benefits for climate mitigation due to interactions with other land management strategies. For example, promoting soil carbon conservation (e.g., reduced tillage) can improve the water retention capacity of soils. Jat et al. (2015) found that improved tillage practices and residue incorporation increased water-use efficiency by 30%, rice–wheat yields by 5–37%, income by 28–40% and reduced GHG emission by 16–25%. While irrigated agriculture accounts for only 20% of the total cultivated land, the energy consumption from groundwater irrigation is significant. However, current estimates of mitigation potential are limited to reductions in GHG emissions mainly in cropland and rice cultivation (Smith et al. 2008, 2014) (Chapter 2 and Table 6.13). Li et al. (2006) estimated a 0.52–0.72 GtCO₂ yr⁻¹ reduction using the alternate wetting and drying technique. Current estimates of N_2O release from terrestrial soils and wetlands accounts for 10–15% of anthropogenically fixed nitrogen on the Earth System (Wang et al. 2017).

Table 6.13 summarises the mitigation potentials for agricultural response options, with confidence estimates based on the thresholds outlined in Table 6.53 in Section 6.4.6, and indicative (not exhaustive) references upon which the evidence in based.

Integrated response options based on land management in forests

Forest management could potentially contribute to moderate mitigation benefits globally, up to about 2 GtCO₂e yr⁻¹ (Chapter 2, Table 6.14). For managed forests, the most effective forest carbon mitigation strategy is the one that, through increasing biomass productivity, optimises the carbon stocks (in forests and in long-lived products) as well as the wood substitution effects for a given time frame (Smyth et al. 2014; Grassi et al. 2018; Nabuurs et al. 2007; Lewis et al. 2019; Kurz et al. 2016; Erb et al. 2017). Estimates of the mitigation potential vary also depending on the counterfactual, such as business-as-usual management (e.g., Grassi et al. 2018) or other scenarios. Climate change will affect the mitigation potential of forest management due to an increase in extreme events like fires, insects and pathogens (Seidl et al. 2017). More detailed estimates are available at regional or biome level. For instance, according to Nabuurs et al. (2017), the implementation of Climate-Smart Forestry (a combination of forest management, expansion of forest areas, energy substitution, establishment of forest reserves, etc.) in the European Union has the potential to contribute to an additional 0.4 $GtCO_2$ yr^{-1} mitigation by 2050. Sustainable forest management is often associated with a number of co-benefits for adaptation, ecosystem services, biodiversity conservation, microclimatic regulation, soil erosion protection, coastal area protection and water and flood regulation (Locatelli 2011). Forest management mitigation measures are more likely to be long-lasting if integrated into adaptation measures for communities and ecosystems, for example, through landscape management (Locatelli et al. 2011). Adoption of reduced-impact logging and wood processing technologies along with financial incentives can reduce forest fires, forest degradation, maintain timber production, and retain carbon stocks (Sasaki et al. 2016). Forest certification may support sustainable forest management, helping to prevent forest degradation and over-logging (Rametsteiner and Simula 2003). Community forest management has proven a viable model for sustainable forestry, including for carbon sequestration (Chhatre and Agrawal 2009) (Chapter 7, Section 7.7.4).

Reducing deforestation and forest degradation rates represents one of the most effective and robust options for climate change mitigation, with large mitigation benefits globally (Chapters 2 and 4, and Table 6.14). Because of the combined climate impacts of GHGs and biophysical effects, reducing deforestation in the tropics has a major climate mitigation effect, with benefits at local levels too (Alkama and Cescatti 2016) (Chapter 2). Reduced deforestation and forest degradation typically lead to large co-benefits for other ecosystem services (Table 6.14).

A large range of estimates exist in the scientific literature for the mitigation potential of reforestation and forest restoration, and they sometimes overlap with estimates for afforestation. At global level, the overall potential for these options is large, reaching about 10 $GtCO_2$ yr^{-1} (Chapter 2 and Table 6.14). The greatest potential for these options is in tropical and subtropical climate (Houghton and Nassikas 2018). Furthermore, climate change mitigation benefits of afforestation, reforestation and forest restoration are reduced at high latitudes owing to the surface albedo feedback (Chapter 2).

Table 6.14 summarises the mitigation potentials for forest response options, with confidence estimates based on the thresholds outlined in Table 6.53 in Section 6.3.6, and indicative (not exhaustive) references upon which the evidence in based.

Table 6.14 | Mitigation effects of response options based on land management in forests.

Integrated response option	Potential	Confidence	Citation
Forest management	0.4–2.1 $GtCO_2e$ yr^{-1}	Medium confidence	Chapter 2 Griscom et al. 2017; Sasaki et al. 2016
Reduced deforestation and forest degradation	0.4–5.8 $GtCO_2e$ yr^{-1}	High confidence	Chapter 2 Baccini et al. 2017; Griscom et al. 2017; Hawken 2017; Houghton et al. 2015; Houghton and Nassikas 2018; Smith et al. 2014
Reforestation and forest restoration	1.5–10.1 $GtCO_2e$ yr^{-1}	Medium confidence	Chapter 2 Dooley and Kartha 2018; Griscom et al. 2017; Hawken 2017; Houghton and Nassikas 2017 Estimates partially overlapping with Afforestation
Afforestation	0.5–8.9 $GtCO_2e$ yr^{-1}	Medium confidence	Chapter 2 Fuss et al. 2018; Hawken 2017; Kreidenweis et al. 2016; Lenton 2010. Estimates partially overlapping with Reforestation

Integrated response options based on land management of soils

The global mitigation potential for increasing soil organic matter stocks in mineral soils is estimated to be in the range of 1.3–5.1 $GtCO_2e$ yr^{-1}, though the full literature range is wider (Fuss et al. 2018; Lal 2004; de Coninck et al. 2018; Sanderman et al. 2017; Smith et al. 2008; Smith 2016) (Table 6.15).

The management and control of erosion may prevent losses of organic carbon in water- or wind- transported sediments, but since the final fate of eroded material is still debated, ranging from a source of 1.36–3.67 $GtCO_2$ yr^{-1} (Jacinthe and Lal 2001; Lal 2004) to a sink of 0.44–3.67 $GtCO_2$ yr^{-1} (Smith et al. 2001; Stallard 1998; Van Oost et al. 2007) (Table 6.15), the overall impact of erosion control on mitigation is context-specific and uncertain at the global level (Hoffmann et al. 2013).

Salt-affected soils are highly constrained environments that require permanent prevention of salinisation. Their mitigation potential is likely to be small (Wong et al. 2010; UNCTAD 2011; Dagar et al. 2016).

Soil compaction prevention could reduce N_2O emissions by minimising anoxic conditions favourable for denitrification (Mbow et al. 2010), but its carbon sequestration potential depends on crop management, and the global mitigation potential, though globally unquantified, is likely to be small (Chamen et al. 2015; Epron et al. 2016; Tullberg et al. 2018) (Table 6.15).

For biochar, a global analysis of technical potential, in which biomass supply constraints were applied to protect against food insecurity, loss of habitat and land degradation, estimated technical potential abatement of 3.7–6.6 $GtCO_2e$ yr^{-1} (including 2.6–4.6 $GtCO_2e$ yr^{-1} carbon stabilisation). Considering all published estimates by Woolf et al. (2010), Smith (2016), Fuss et al. (2018), Griscom et al. (2017), Hawken (2017), Paustian et al. (2016), Powell and Lenton (2012),

Table 6.15 | Mitigation effects of response options based on land management of soils.

Integrated response option	Potential	Confidence	Citation
Increased soil organic carbon content	0.4–8.6 GtCO₂e yr⁻¹	High confidence	Chapter 2 Conant et al. 2017; Dickie et al. 2014; Frank et al. 2017; Fuss et al. 2018; Griscom et al. 2017; Hawken 2017; Henderson et al. 2015; Herrero et al. 2016; Paustian et al. 2016; Powlson et al. 2014; Sanderman et al. 2017; Smith 2016; Smith et al. 2016b; Sommer and Bossio 2014; Zomer et al. 2016
Reduced soil erosion	Source of 1.36–3.67 to sink of 0.44–3.67 GtCO₂e yr⁻¹	Low confidence	Chapter 2 Jacinthe and Lal 2001; Lal 2004; Smith et al. 2001, 2005; Stallard 1998; Van Oost et al. 2007
Reduced soil salinisation	>0	Low confidence	Dagar et al. 2016; UNCTAD 2011; Wong et al. 2010
Reduced soil compaction	>0	Low confidence	Chamen et al. 2015; Epron et al. 2016; Tullberg et al. 2018
Biochar addition to soil	0.03–6.6 GtCO₂e yr⁻¹	Medium confidence	Chapter 2 Dickie et al. 2014; Fuss et al. 2018; Griscom et al. 2017; Hawken 2017; IPCC 2018; Lenton 2010, 2014; Paustian et al. 2016; Powell and Lenton 2012; Pratt and Moran 2010; Roberts et al. 2009; Smith 2016; Woolf et al. 2010

Dickie et al. (2014), Lenton (2010), Lenton (2014), Roberts et al. (2009), Pratt and Moran (2010) and IPCC (2018), the low value for the range of potentials of 0.03 GtCO₂e yr⁻¹ is for the 'plausible' scenario of Hawken, (2017) (Table 6.15). Fuss et al. (2018) propose a range of 0.5–2 GtCO₂e yr⁻¹ as the sustainable potential for negative emissions through biochar, similar to the range proposed by Smith (2016) and IPCC (2018).

Table 6.15 summarises the mitigation potentials for soil-based response options, with confidence estimates based on the thresholds outlined in Table 6.53 in Section 6.3.6, and indicative (not exhaustive) references upon which the evidence in based.

Integrated response options based on land management in all/other ecosystems

For fire management, total emissions from fires have been in the order of 8.1 GtCO₂e yr⁻¹ for the period 1997–2016 (Chapter 2 and Cross-Chapter Box 3) and there are important synergies between air pollution and climate change control policies. Reduction in fire CO₂ emissions due to fire suppression and landscape fragmentation associated with increases in population density is calculated to enhance land carbon uptake by 0.48 GtCO₂e yr⁻¹ for the 1960–2009 period (Arora and Melton 2018) (Table 6.16).

Management of landslides and natural hazards is a key climate adaptation option but, due to limited global areas vulnerable to landslides and natural hazards, its mitigation potential is likely to be modest (Noble et al. 2014).

In terms of management of pollution, including acidification, United Nations Environment Programme (UNEP) and World Meterological Organization (WMO) (2011) and Shindell et al. (2012) identified measures targeting reduction in short-lived climate pollutant (SLCP) emissions that reduce projected global mean warming about 0.5°C by 2050. Bala et al. (2013) reported that a recent coupled modelling study showed nitrogen deposition and elevated CO₂ could have a synergistic effect, which could explain 47% of terrestrial carbon uptake in the 1990s. Estimates of global terrestrial carbon uptake due to current nitrogen deposition ranges between 0.55 and 1.28 GtCO₂ yr⁻¹ (De Vries et al. 2006, 2009; Bala et al. 2013; Zaehle and Dalmonech 2011) (Table 6.16).

There are no global data on the impacts of management of invasive species/encroachment on mitigation.

Coastal wetland restoration could provide high levels of climate mitigation, with avoided coastal wetland impacts and coastal wetland restoration estimated to deliver 0.3–3.1 GtCO₂e yr⁻¹ in total when considering all global estimates from Griscom et al. (2017), Hawken (2017), Pendleton et al. (2012), Howard et al. (2017) and Donato et al. (2011) (Table 6.16).

Peatland restoration could provide moderate levels of climate mitigation, with avoided peat impacts and peat restoration estimated to deliver 0.6–2 GtCO₂e yr⁻¹ from all global estimates published in Griscom et al. (2017), Hawken (2017), Hooijer et al. (2010), Couwenberg et al. (2010) and Joosten and Couwenberg (2008), though there could be an increase in methane emissions after restoration (Jauhiainen et al. 2008) (Table 6.16).

Mitigation potential from biodiversity conservation varies depending on the type of intervention and specific context. Protected areas are estimated to store over 300 Gt carbon, roughly corresponding to 15% of terrestrial carbon stocks (Campbell et al. 2008; Kapos et al. 2008). At global level, the potential mitigation resulting from protection of these areas for the period 2005–2095 is, on average, about 0.9 GtCO₂-eq yr⁻¹ relative to a reference scenario (Calvin et al. 2014). The potential effects on the carbon cycle of management of wild animal species are context dependent. For example, moose browsing in boreal forests can decrease the carbon uptake of ecosystems by up to 75% (Schmitz et al. 2018), and reducing moose density through active population management in Canada is estimated to be a carbon sink equivalent to about 0.37 GtCO₂e yr⁻¹ (Schmitz et al. 2014).

Table 6.16 summarises the mitigation potentials for land management response options in all/other ecosystems, with confidence estimates based on the thresholds outlined in Table 6.53 in Section 6.3.6, and indicative (not exhaustive) references upon which the evidence in based.

6

Table 6.16 | Mitigation effects of response options based on land management in all/other ecosystems.

Integrated response option	Potential	Confidence	Citation
Fire management	0.48–8.1 GtCO₂e yr⁻¹	Medium confidence	Chapter 2, Cross-Chapter Box 3 in Chapter 2 Arora and Melton 2018; Tacconi 2016
Reduced landslides and natural hazards	>0	Low confidence	
Reduced pollution including acidification	(i) Reduce projected warming ~0.5°C by 2050; (ii) Reduce terrestrial carbon uptake 0.55–1.28 GtCO₂e yr⁻¹	(i) and (ii) Medium confidence	(i) Shindell et al. 2012; UNEP and WMO 2011 (ii) Bala et al. 2013
Management of invasive species/encroachment	No global estimates	No evidence	
Restoration and reduced conversion of coastal wetlands	0.3–3.1 GtCO₂e yr⁻¹	Medium confidence	Chapter 2 Donato et al. 2011; Duarte et al. 2013; Hawken 2017; Howard et al. 2017; Pendleton et al. 2012
Restoration and reduced conversion of peatlands	0.6–2 GtCO₂e yr⁻¹	Medium confidence	Chapter 2 Couwenberg et al. 2010; Griscom et al. 2017; Hawken 2017; Hooijer et al. 2010; Joosten and Couwenberg 2008
Biodiversity conservation	~0.9 GtCO₂e yr⁻¹	Low confidence	Chapter 2 Calvin et al. 2014; Schmitz et al. 2014

Integrated response options based on land management specifically for carbon dioxide removal (CDR)

Enhanced mineral weathering provides substantial climate mitigation, with a global mitigation potential in the region of about 0.5–4 GtCO₂e yr⁻¹ (Beerling et al. 2018; Lenton 2010; Smith et al. 2016a; Taylor et al. 2016) (Table 6.17).

The mitigation potential for bioenergy and BECCS derived from bottom-up models is large (IPCC 2018) (Chapter 2 and Cross-Chapter Box 7 in this chapter), with technical potential estimated at 100–300 EJ yr⁻¹ (Chum et al. 2011; Cross-Chapter Box 7 in

Chapter 6) or up to about 11 GtCO₂ yr⁻¹ (Chapter 2). These estimates, however, exclude N₂O associated with fertiliser application and land-use change emissions. Those effects are included in the modelled scenarios using bioenergy and BECCS, with the sign and magnitude depending on where the bioenergy is grown (Wise et al. 2015), at what scale, and whether nitrogen fertiliser is used.

Table 6.17 summarises the mitigation potentials for land management options specifically for CDR, with confidence estimates based on the thresholds outlined in Table 6.53 in Section 6.3.6, and indicative (not exhaustive) references upon which the evidence in based.

Table 6.17 | Mitigation effects of response options based on land management specifically for CDR.

Integrated response option	Potential	Confidence	Citation
Enhanced weathering of minerals	0.5–4 GtCO₂ yr⁻¹	Medium confidence	Chapter 2 Beerling et al. 2018; Lenton 2010; Smith et al. 2016a; Taylor et al. 2016
Bioenergy and BECCS	0.4–11.3 GtCO₂ yr⁻¹	Medium confidence	Chapter 2 IPCC 2018; Fuss et al. 2018; McLaren 2012; Lenton 2010, 2014; Powell and Lenton 2012

6.3.1.2 Integrated response options based on value chain management

In this section, the impacts on climate change mitigation of integrated response options based on value chain management are assessed.

Integrated response options based on value chain management through demand management

Dietary change and waste reduction can provide large benefits for mitigation, with potentials of 0.7–8 GtCO₂ yr⁻¹ for both (Aleksandrowicz et al. 2016; Bajželj et al. 2014b; Dickie et al. 2014; Hawken 2017; Hedenus et al. 2014; Herrero et al. 2016; Popp et al. 2010; Smith et al. 2013; Springmann et al. 2016; Stehfest et al. 2009;

Tilman and Clark 2014). Estimates for food waste reduction (Bajželj et al. 2014b; Dickie et al. 2014; Hiç et al. 2016; Hawken 2017) include both consumer/retailed waste and post-harvest losses (Table 6.18).

Some studies indicate that material substitution has the potential for significant mitigation, with one study estimating a 14–31% reduction in global CO₂ emissions (Oliver et al. 2014); other studies suggest more modest potential (Gustavsson et al. 2006) (Table 6.18).

Table 6.18 summarises the mitigation potentials for demand management options, with confidence estimates based on the thresholds outlined in Table 6.53 in Section 6.3.6, and indicative (not exhaustive) references upon which the evidence in based.

Table 6.18 | Mitigation effects of response options based on demand management.

Integrated response option	Potential	Confidence	Citation
Dietary change	0.7–8 GtCO$_2$ yr^{-1}	High confidence	Chapter 2; Chapter 5 Aleksandrowicz et al. 2016; Bajželj et al. 2014b; Dickie et al. 2014; Hawken 2017; Hedenus et al. 2014; Herrero et al. 2016; Popp et al. 2010; Smith et al. 2013; Springmann et al. 2016; Stehfest et al. 2009; Tilman and Clark 2014
Reduced post-harvest losses	4.5 GtCO$_2$ yr^{-1}	High confidence	Chapter 5 Bajželj et al. 2014b
Reduced food waste (consumer or retailer)	0.8–4.5 GtCO$_2$ yr^{-1}	High confidence	Chapter 5 Bajželj et al. 2014b; Dickie et al. 2014; Hiç et al. 2016; Hawken 2017
Material substitution	0.25–1 GtCO$_2$ yr^{-1}	Medium confidence	Chapter 2 Dugan et al. 2018; Gustavsson et al. 2006; Kauppi et al. 2001; Leskinen et al. 2018; McLaren 2012; Miner 2010; Sathre and O'Connor 2010; Smyth et al. 2017

Integrated response options based on value chain management through supply management

While sustainable sourcing presumably delivers a mitigation benefit, there are no global estimates of potential. Palm oil production alone is estimated to contribute 0.038 to 0.045 GtC yr^{-1}, and Indonesian palm oil expansion contributed up to 9% of tropical land-use change carbon emissions in the 2000s (Carlson and Curran 2013), however, the mitigation benefit of sustainable sourcing of palm oil has not been quantified. There are no estimates of the mitigation potential for urban food systems.

Efficient use of energy and resources in food transport and distribution contribute to a reduction in GHG emissions, estimated to be 1% of global CO$_2$ emissions (James and James 2010; Vermeulen et al. 2012b). Given that global CO$_2$ emissions in 2017 were 37 GtCO$_2$, this equates to 0.37 GtCO$_2$ yr^{-1} (covering food transport and distribution, improved efficiency of food processing and retailing, and improved energy efficiency) (Table 6.19).

Table 6.19 summarises the mitigation potentials for supply management options, with confidence estimates based on the thresholds outlined in Table 6.53 in Section 6.3.6, and indicative (not exhaustive) references upon which the evidence in based.

Table 6.19 | Mitigation effects of response options based on supply management.

Integrated response option	Potential	Confidence	Citation
Sustainable sourcing	No global estimates	No evidence	
Management of supply chains	No global estimates	No evidence	
Enhanced urban food systems	No global estimates	No evidence	
Improved food processing and retailing	See improved energy efficiency		
Improved energy use in food systems	0.37 GtCO$_2$ yr^{-1}	Low confidence	James and James 2010; Vermeulen et al. 2012b

6.3.1.3   Integrated response options based on risk management

In this section, the impacts on climate change mitigation of integrated response options based on risk management are assessed. In general, because these options are focused on adaptation and other benefits, the mitigation benefits are modest, and mostly unquantified.

Extensive and less dense urban development tends to have higher energy usage, particularly from transport (Liu et al. 2015), such that a 10% reduction of very low-density urban fabrics is correlated with 9% fewer emissions per capita in Europe (Baur et al. 2015). However, the exact contribution to mitigation from the prevention of land conversion in particular has not been well quantified (Thornbush et al. 2013). Suggestions from select studies in the USA are that biomass decreases by half in cases of conversion from forest to urban land uses (Briber et al. 2015), and a study in Bangkok found a decline by half in carbon sinks in the urban area in the past 30 years (Ali et al. 2018).

There is no literature specifically on linkages between livelihood diversification and climate mitigation benefits, although some

forms of diversification that include agroforestry would likely result in increased carbon sinks (Altieri et al. 2015; Descheemaeker et al. 2016). There is no literature exploring linkages between local seeds and GHG emission reductions, although use of local seeds likely reduces emissions associated with transport for commercial seeds, though the impact has not been quantified.

While disaster risk management can presumably have mitigation co-benefits, as it can help reduce food loss on-farm (e.g., crops destroyed before harvest or avoided animal deaths during droughts and floods meaning reduced production losses and wasted emissions), there is no quantified global estimate for this potential.

Risk-sharing instruments could have some mitigation co-benefits if they buffer household losses and reduce the need to expand agricultural lands after experiencing risks. However, the overall impacts of these are unknown. Further, commercial insurance may induce producers to bring additional land into crop production, particularly marginal or land with other risks that may be more environmentally sensitive (Claassen et al. 2011a). Policies to deny crop insurance to farmers who have converted grasslands in the USA resulted in a 9% drop in conversion, which likely has positive mitigation impacts (Claassen et al. 2011a). Estimates of emissions from cropland conversion in the USA in 2016 were 23.8 MtCO₂e, only some of which could be attributed to insurance as a driver.

Table 6.20 summarises the mitigation potentials for risk management options, with confidence estimates based on the thresholds outlined in Table 6.53 in Section 6.3.6, and indicative (not exhaustive) references upon which the evidence is based.

Table 6.20 | Mitigation effects of response options based on risk management.

Integrated response option	Potential	Confidence	Citation
Management of urban sprawl	No global estimates	No evidence	
Livelihood diversification	No global estimates	No evidence	
Use of local seeds	No global estimates	No evidence	
Disaster risk management	No global estimates	No evidence	
Risk-sharing instruments	>0.024 GtCO₂e yr⁻¹ for crop insurance; likely some benefits for other risk-sharing instruments	Low confidence	Claassen et al. 2011b; EPA 2018

6.3.2   Potential of the integrated response options for delivering adaptation

In this section, the impacts of integrated response options on climate change adaptation are assessed.

6.3.2.1   Integrated response options based on land management

In this section, the impacts on climate change adaptation of integrated response options based on land management are assessed.

Integrated response options based on land management in agriculture

Increasing food productivity by practices such as sustainable intensification improves farm incomes and allows households to build assets for use in times of stress, thereby improving resilience (Campbell et al. 2014). By reducing pressure on land and increasing food production, increased food productivity could be beneficial for adaptation (Campbell et al. 2014) (Chapter 2 and Section 6.3). Pretty et al. (2018) report that 163 million farms occupying 4.53 Mkm² have passed a redesign threshold for application of sustainable intensification, suggesting the minimum number of people benefitting from increased productivity and adaptation benefits under sustainable intensification is >163 million, with the total likely to be far higher (Table 6.21).

Improved cropland management is a key climate adaptation option, potentially affecting more than 25 million people, including a wide range of technological decisions by farmers. Actions towards adaptation fall into two broad overlapping areas: (i) accelerated adaptation to progressive climate change over decadal time scales, for example integrated packages of technology, agronomy and policy options for farmers and food systems, including changing planting dates and zones, tillage systems, crop types and varieties, and (ii) better management of agricultural risks associated with increasing climate variability and extreme events, for example, improved climate information services and safety nets (Vermeulen et al. 2012b; Challinor et al. 2014; Lipper et al. 2014; Lobell 2014). In the same way, improved livestock management is another

technological adaptation option potentially benefitting between 1 million and 25 million people. Crop and animal diversification are considered the most promising adaptation measures (Porter et al. 2014; Rojas-Downing et al. 2017). In grasslands and rangelands, regulation of stocking rates, grazing field dimensions, establishment of exclosures and locations of drinking fountains and feeders are strategic decisions by farmers to improve grazing management (Taboada et al. 2011; Mekuria and Aynekulu 2013; Porter et al. 2014).

Around 30% of the world's rural population use trees across 46% of all agricultural landscapes (Lasco et al. 2014), meaning that up to 2.3 billion people benefit from agroforestry globally (Table 6.21).

Agricultural diversification is key to achieving climatic resilience (Campbell et al. 2014; Cohn et al. 2017). Crop diversification is one important adaptation option to progressive climate change (Vermeulen et al. 2012a) and it can improve resilience by engendering a greater ability to suppress pest outbreaks and dampen pathogen transmission, as well as by buffering crop production from the effects of greater climate variability and extreme events (Lin 2011).

Reduced conversion of grassland to cropland may lead to adaptation benefits by stabilising soils in the face of extreme climatic events (Lal 2001), thereby increasing resilience, but since it would likely have a negative impact on food production/security (since croplands produce more food per unit area than grasslands), the wider adaptation impacts would likely be negative. However, there is no literature quantifying the global impact of avoidance of conversion of grassland to cropland on adaptation.

Integrated water management provides large co-benefits for adaptation (Dillon and Arshad 2016) by improving the resilience of food crop production systems to future climate change (Porter et al. 2014) (Chapter 2 and Table 6.7). Improving irrigation systems and integrated water resource management, such as enhancing urban and rural water supplies and reducing water evaporation losses (Dillon and Arshad 2016), are significant options for enhancing climate adaptation. Many technical innovations (e.g., precision water management) can lead to beneficial adaptation outcomes by increasing water availability and the reliability of agricultural production, using different techniques of water harvesting, storage, and its judicious utilisation through

6

farm ponds, dams and community tanks in rainfed agriculture areas. Integrated water management response options that use freshwater would be expected to have few adverse side effects in regions where water is plentiful, but large adverse side effects in regions where water is scarce (Grey and Sadoff 2007; Liu et al. 2017; Scott et al. 2011).

Table 6.21 summarises the potentials for adaptation for agricultural response options, with confidence estimates based on the thresholds outlined in Table 6.53 in Section 6.3.6, and indicative (not exhaustive) references upon which the evidence in based.

Table 6.21 | Adaptation effects of response options based on land management in agriculture.

Integrated response option	Potential	Confidence	Citation
Increased food productivity	>163 million people	Medium confidence	Pretty et al. 2018
Improved cropland management	>25 million people	Low confidence	Challinor et al. 2014; Lipper et al. 2014; Lobell 2014; Vermeulen et al. 2012b
Improved grazing land management	1–25 million people	Low confidence	Porter et al. 2014
Improved livestock management	1–25 million people	Low confidence	Porter et al. 2014; Rojas-Downing et al. 2017
Agroforestry	2300 million people	Medium confidence	Lasco et al. 2014
Agricultural diversification	>25 million people	Low confidence	Campbell et al. 2014; Cohn et al. 2017; Vermeulen et al. 2012b
Reduced grassland conversion to cropland	No global estimates	No evidence	
Integrated water management	250 million people	Low confidence	Dillon and Arshad 2016; Liu et al. 2017

Integrated response options based on land management in forestry

Forest management positively impacts on adaptation through limiting the negative effects associated with pollution (of air and fresh water), infections and other diseases, exposure to extreme weather events and natural disasters, and poverty (e.g., Smith et al. 2014). There is high agreement on the fact that reduced deforestation and forest degradation positively impact on adaptation and resilience of coupled human-natural systems. Based on the number of people affected by natural disasters (CRED 2015), the number of people depending to varying degrees on forests for their livelihoods (World Bank et al. 2009) and the current deforestation rate (Keenan et al. 2015), the estimated global potential effect for adaptation is largely positive for forest management, and moderately positive for reduced deforestation when cumulated until the end of the century (Table 6.22). The uncertainty of these global estimates is high, for example, the impact of reduced deforestation may be higher when the large biophysical impacts on the water cycle (and thus drought) from deforestation (e.g., Alkama and Cescatti 2016) are taken into account (Chapter 2).

More robust qualitative, and some quantitative, estimates are available at local and regional level. According to Karjalainen et al. (2009), reducing deforestation and habitat alteration contributes to limiting infectious diseases such as malaria in Africa, Asia and Latin America, thus lowering the expenses associated with healthcare treatments. Bhattacharjee and Behera (2017) found that human lives lost due to floods increase with reducing forest cover and increasing deforestation rates in India. In addition, maintaining forest cover in urban contexts reduces air pollution and therefore avoids mortality of about one person per year per city in US, and up to 7.6 people per year in New York City (Nowak et al. 2014). There is also evidence

that reducing deforestation and forest degradation in mangrove plantations potentially improves soil stabilisation, and attenuates the impact of tropical cyclones and typhoons along the coastal areas in South and Southeast Asia (Chow 2018). At local scale, co-benefits between REDD+ and adaptation of local communities can potentially be substantial (Long 2013; Morita and Matsumoto 2018), even if often difficult to quantify, and not explicitly acknowledged (McElwee et al. 2017b).

Forest restoration may facilitate the adaptation and resilience of forests to climate change by enhancing connectivity between forest areas and conserving biodiversity hotspots (Locatelli et al. 2011, 2015b; Ellison et al. 2017; Dooley and Kartha 2018). Furthermore, forest restoration may improve ecosystem functionality and services, provide microclimatic regulation for people and crops, wood and fodder as safety nets, soil erosion protection and soil fertility enhancement for agricultural resilience, coastal area protection, water and flood regulation (Locatelli et al. 2015b).

Afforestation and reforestation are important climate change adaptation response options (Reyer et al. 2009; Ellison et al. 2017; Locatelli et al. 2015b), and can potentially help a large proportion of the global population to adapt to climate change and to associated natural disasters (Table 6.22). For example, trees generally mitigate summer mean warming and temperature extremes (Findell et al. 2017; Sonntag et al. 2016).

Table 6.22 summarises the potentials for adaptation for forest response options, with confidence estimates based on the thresholds outlined in Table 6.53 in Section 6.3.6, and indicative (not exhaustive) references upon which the evidence in based.

Table 6.22 | Adaptation effects of response options based on land management in forests.

Integrated response option	Potential	Confidence	Citation
Forest management	>25 million people	Low confidence	CRED 2015; World Bank et al. 2009
Reduced deforestation and forest degradation	1–25 million people	Low confidence	CRED 2015; Keenan et al. 2015; World Bank et al. 2009. The estimates consider a cumulated effect until the end of the century.
Reforestation and forest restoration	See afforestation		
Afforestation	>25 million people	Medium confidence	CRED 2015; Reyer et al. 2009; Smith et al. 2014; Sonntag et al. 2016; World Bank et al. 2009. The estimates consider a cumulated effect until the end of the century.

Integrated response options based on land management of soils

Soil organic carbon increase is promoted as an action for climate change adaptation. Since increasing soil organic matter content is a measure to address land degradation (see Section 6.2.1), and restoring degraded land helps to improve resilience to climate change, soil carbon increase is an important option for climate change adaptation. With around 120,000 km^2 lost to degradation every year, and over 3.2 billion people negatively impacted by land degradation globally (IPBES 2018), practices designed to increase soil organic carbon have a large potential to address adaptation challenges (Table 6.23).

Since soil erosion control prevents land degradation and desertification, it improves the resilience of agriculture to climate change and increases food production (Lal 1998; IPBES 2018), though the global number of people benefitting from improved resilience to climate change has not been reported in the literature. Using figures from (FAO and ITPS 2015), IPBES (2018) estimates that land losses due to erosion are equivalent to 1.5 Mkm^2 of land used for crop production to 2050, or 45,000 km^2 yr^{-1} (Foley et al. 2011). Control of soil erosion (water and wind) could benefit 11 Mkm^2 of degraded land (Lal 2014), and improve the resilience of at least some of the 3.2 billion people affected by land degradation (IPBES 2018), suggesting positive impacts on adaptation. Management of erosion is an important climate change adaptation measure, since it reduces the vulnerability of soils to loss under climate extremes, thereby increasing resilience to climate change (Garbrecht et al. 2015).

Prevention and/or reversion of topsoil salinisation may require a combined management of groundwater, irrigation techniques, drainage, mulching and vegetation, with all of these considered relevant for adaptation (Qadir et al. 2013; UNCTAD 2011; Dagar et al. 2016). Taking into account the widespread diffusion of salinity problems, many people can benefit from its implementation by farmers. The relation between compaction prevention and/or reversion and climate adaption is less evident, and can be related to better hydrological soil functioning (Chamen et al. 2015; Epron et al. 2016; Tullberg et al. 2018).

Biochar has the potential to benefit climate adaptation by improving the resilience of food crop production systems to future climate change by increasing yield in some regions and improving water holding capacity (Woolf et al. 2010; Sohi 2012) (Chapter 2 and Section 6.4). By increasing yield by 25% in the tropics (Jeffery et al. 2017), this could increase food production for 3.2 billion people affected by land degradation (IPBES 2018), thereby potentially improving their resilience to climate change shocks (Table 6.23). A requirement for large areas of land to provide feedstock for biochar could adversely impact on adaptation, though this has not been quantified globally.

Table 6.23 summarises the potentials for adaptation for soil-based response options, with confidence estimates based on the thresholds outlined in Table 6.53 in Section 6.3.6, and indicative (not exhaustive) references upon which the evidence in based.

Table 6.23 | Adaptation effects of response options based on land management of soils.

Integrated response option	Potential	Confidence	Citation
Increased soil organic carbon content	Up to 3200 million people	Low confidence	IPBES 2018
Reduced soil erosion	Up to 3200 million people	Low confidence	IPBES 2018
Reduced soil salinisation	1–25 million people	Low confidence	Dagar et al. 2016; Qadir et al. 2013; UNCTAD 2011
Reduced soil compaction	<1 million people	Low confidence	Chamen et al. 2015; Epron et al. 2016; Tullberg et al. 2018
Biochar addition to soil	Up to 3200 million people; but potential negative (unquantified) impacts from land required from feedstocks	Low confidence	Jeffery et al. 2017

Integrated response options based on land management across all/other ecosystems

For fire management, Doerr et al. (2016) showed that the number of people killed by wildfire was 1940, and the total number of people affected was 5.8 million from 1984 to 2013, globally. Johnston et al. (2012) showed that the average mortality attributable to landscape fire smoke exposure was 339,000 deaths annually. The regions most affected were sub-Saharan Africa (157,000) and Southeast Asia (110,000). Estimated annual mortality during La Niña was 262,000, compared with around 100,000 excess deaths across Indonesia, Malaysia and Singapore (Table 6.24).

Management of landslides and natural hazards are usually listed among planned adaptation options in mountainous and sloped hilly areas, where uncontrolled runoff and avalanches may cause climatic disasters, affecting millions of people from both urban and rural areas. Landslide control requires increasing plant cover and engineering practices (see Table 6.8).

For management of pollution, including acidification, Anenberg et al. (2012) estimated that, for particulate matter ($PM_{2.5}$) and ozone, respectively, fully implementing reduction measures could reduce global population-weighted average surface concentrations by 23–34% and 7–17% and avoid 0.6–4.4 and 0.04–0.52 million annual premature deaths globally in 2030. UNEP and WMO (2011) considered emission control measures to reduce ozone and black carbon (BC) and estimated that 2.4 million annual premature deaths (with a range of 0.7 million to 4.6 million) from outdoor air pollution could be avoided. West et al. (2013) estimated global GHG mitigation brings co-benefits for air quality and would avoid 0.5 ± 0.2, 1.3 ± 0.5, and 2.2 ± 0.8 million premature deaths in 2030, 2050, and 2100, respectively.

There are no global data on the impacts of management of invasive species/encroachment on adaptation.

6

Coastal wetlands provide a natural defence against coastal flooding and storm surges by dissipating wave energy, reducing erosion, and by helping to stabilise shore sediments, so restoration may provide significant benefits for adaptation. The Ramsar Convention on Wetlands covers 1.5 Mkm^2 across 1674 sites (Keddy et al. 2009). Coastal floods currently affect 93–310 million people (in 2010) globally, and this could rise to 600 million people in 2100 with sea level rise, unless adaptation measures are taken (Hinkel et al. 2014). The proportion of the flood-prone population that could avoid these impacts through restoration of coastal wetlands has not been quantified, but this sets an upper limit.

Avoided peat impacts and peatland restoration can help to regulate water flow and prevent downstream flooding (Munang et al. 2014), but the global potential (in terms of number of people who could avoid flooding through peatland restoration) has not been quantified.

There are no global estimates about the potential of biodiversity conservation to improve the adaptation and resilience of local communities to climate change, in terms of reducing the number of people affected by natural disasters. Nevertheless, it is widely recognised that biodiversity, ecosystem health and resilience improves the adaptation potential (Jones et al. 2012). For example, tree species mixture improves the resistance of stands to natural disturbances, such as drought, fires, and windstorms (Jactel et al. 2017), as well as stability against landslides (Kobayashi and Mori 2017). Moreover, protected areas play a key role for improving adaptation (Watson et al. 2014; Lopoukhine et al. 2012), through reducing water flow, stabilising rock movements, creating physical barriers to coastal erosion, improving resistance to fires, and buffering storm damages (Dudley et al. 2010). Of the largest urban areas worldwide, 33 out of 105 rely on protected areas for some, or all, of their drinking water (Secretariat of the Convention on Biological Diversity 2008), indicating that many millions are likely to benefit from conservation practices.

Table 6.24 summarises the potentials for adaptation for soil-based response options, with confidence estimates based on the thresholds outlined in Table 6.53 in Section 6.3.6, and indicative (not exhaustive) references upon which the evidence in based.

Table 6.24 | Adaptation effects of response options based on land management of soils.

Integrated response option	Potential	Confidence	Citation
Fire management	>5.8 million people affected by wildfire; max. 0.5 million deaths per year by smoke	Medium confidence	Doerr and Santín 2016; Johnston et al. 2012; Koplitz et al. 2016
Reduced landslides and natural hazards	>25 million people	Low confidence	Arnáez J et al. 2015; Gariano and Guzzetti 2016
Reduced pollution including acidification	Prevent 0.5–4.6 million annual premature deaths globally	Medium confidence	Anenberg et al. 2012; Shindell et al. 2012; West et al. 2013; UNEP and WMO 2011
Management of invasive species/encroachment	No global estimates	No evidence	
Restoration and reduced conversion of coastal wetlands	Up to 93–310 million people	Low confidence	Hinkel et al. 2014
Restoration and reduced conversion of peatlands	No global estimates	No evidence	
Biodiversity conservation	Likely many millions	Low confidence	Secretariat of the Convention on Biological Diversity 2008

Integrated response options based on land management specifically for CDR

Enhanced weathering of minerals has been proposed as a mechanism for improving soil health and food security (Beerling et al. 2018), but there is no literature estimating the global adaptation benefits.

Large-scale bioenergy and BECCS can require substantial amounts of cropland (Popp et al. 2017; Calvin et al. 2014; Smith et al. 2016a), forestland (Baker et al. 2019; Favero and Mendelsohn 2017), and water (Chaturvedi et al. 2013; Hejazi et al. 2015; Popp et al. 2011a; Smith et al. 2016a; Fuss et al. 2018); suggesting that bioenergy and BECCS could have adverse side effects for adaptation. In some contexts – for example, low inputs of fossil fuels and chemicals, limited irrigation, heat/drought tolerant species, and using marginal land – bioenergy can have co-benefits for adaptation (Dasgupta et al. 2014; Noble et al. 2014). However, no studies were found that quantify the magnitude of the effect.

Table 6.25 summarises the impacts on adaptation of land management response options specifically for CDR, with confidence estimates based on the thresholds outlined in Table 6.53 in Section 6.3.6, and indicative (not exhaustive) references upon which the evidence in based.

Table 6.25 | Adaptation effects of response options based on land management specifically for CDR.

Integrated response option	Potential	Confidence	Citation
Enhanced weathering of minerals	No global estimates	No evidence	
Bioenergy and BECCS	Potentially large negative consequences	Low confidence	Fuss et al. 2018; Muller et al. 2017; Smith et al. 2016a

6.3.2.2   Integrated response options based on value chain management

In this section, the impacts on climate change adaptation of integrated response options based on value chain management are assessed.

Integrated response options based on value chain management through demand management

Decreases in pressure on land and decreases in production intensity associated with sustainable healthy diets or reduced food waste could also benefit adaptation; however, the size of this effect is not well quantified (Muller et al. 2017).

Reducing food waste losses can relieve pressure on the global freshwater resource, thereby aiding adaptation. Food losses account for 215 km^3 yr^{-1} of freshwater resources, which Kummu et al. (2012) report to be about 12–15% of the global consumptive water use.

Given that 35% of the global population is living under high water stress or shortage (Kummu et al. 2010), reducing food waste could benefit 320–400 million people (12–15% of the 2681 million people affected by water stress/shortage).

While no studies report quantitative estimates of the effect of material substitution on adaptation, the effects are expected to be similar to reforestation and afforestation if the amount of material substitution leads to an increase in forest area. Additionally, some studies indicate that wooden buildings, if properly constructed, could reduce fire risk, compared to steel, which softens when burned (Gustavsson et al. 2006; Ramage et al. 2017).

Table 6.26 summarises the impacts on adaptation of demand management options, with confidence estimates based on the thresholds outlined in Table 6.53 in Section 6.3.6, and indicative (not exhaustive) references upon which the evidence in based.

Table 6.26 | Adaptation effects of response options based on demand management.

Integrated response option	Potential	Confidence	Citation
Dietary change	No global estimates	No evidence	Muller et al. 2017
Reduced post-harvest losses	320–400 million people	Medium confidence	Kummu et al. 2012
Reduced food waste (consumer or retailer)	No global estimates	No evidence	Muller et al. 2017
Material substitution	No global estimates	No evidence	

Integrated response options based on value chain management through supply management

It is estimated that 500 million smallholder farmers depend on agricultural businesses in developing countries (IFAD 2013), meaning that better promotion of value-added products and improved efficiency and sustainability of food processing and retailing could potentially help up to 500 million people to adapt to climate change. However, figures on how sustainable sourcing in general could help farmers and forest management is mostly unquantified. More than 1 million farmers have currently been certified through various schemes (Tayleur et al. 2017), but how much this has helped them prepare for adaptation is unknown.

Management of supply chains has the potential to reduce vulnerability to price volatility. Consumers in lower-income countries are most affected by price volatility, with sub-Saharan Africa and South Asia at highest risk (Regmi and Meade 2013; Fujimori et al. 2019). However, understanding of the stability of food supply is one of the weakest links in global food system research (Wheeler and von Braun 2013) as instability is driven by a confluence of factors (Headey and Fan 2008). Food price spikes in 2007 increased the number of people below the poverty line by between 100 million people (Ivanic and Martin 2008) and 450 million people (Brinkman et al. 2009), and caused welfare losses of 3% or more for poor households in many countries (Zezza et al. 2009). Food

price stabilisation by China, India and Indonesia alone in 2007/2008 led to reduced staple food price for 2 billion people (Timmer 2009). Presumably, spending less on food frees up money for other activities, including adaptation, but it is unknown how much (Zezza et al. 2009; Ziervogel and Ericksen 2010). In one example, reduction in staple food price costs to consumers in Bangladesh from food stability policies saved rural households 887 million USD2003 total (Torlesse et al. 2003). Food supply stability through improved supply chains also potentially reduces conflicts (by avoiding food price riots, which occurred in countries with over 100 million total in population in 2007/2008), and thus increases adaptation capacity (Raleigh et al. 2015).

There are no global estimates of the contribution of improved food transport and distribution, or of urban food systems, in contributing to adaptation, but since the urban population in 2018 was 4.2 billion people, this sets the upper limit on those who could benefit.

Given that 65% (760 million) of working adults in poverty make a living through agriculture, increased energy efficiency in agriculture could benefit these 760 million people.

Table 6.27 summarises the impacts on adaptation of supply management options, with confidence estimates based on the thresholds outlined in Table 6.53 in Section 6.3.6, and indicative (not exhaustive) references upon which the evidence in based.

6

Table 6.27 | Adaptation effects of response options based on demand management.

Integrated response option	Potential	Confidence	Citation
Sustainable sourcing	>1 million	Low confidence	Tayleur et al. 2017
Management of supply chains	>100 million	Medium confidence	Campbell et al. 2016; Ivanic and Martin 2008; Timmer 2009; Vermeulen et al. 2012b
Enhanced urban food systems	No global estimates	No evidence	
Improved food processing and retailing	500 million people	Low confidence	IFAD 2013; World Bank 2017
Improved energy use in food systems	760 million	Low confidence	IFAD 2013; World Bank 2017

6.3.2.3 Integrated response options based on risk management

In this section, the impacts on climate change adaptation of integrated response options based on risk management are assessed.

Reducing urban sprawl is likely to provide adaptation co-benefits *via* improved human health (Frumkin 2002; Anderson 2017), as sprawl contributes to reduced physical activity, worse air pollution, and exacerbation of urban heat island effects and extreme heat waves (Stone et al. 2010). The most sprawling cities in the US have experienced extreme heat waves, more than double those of denser cities, and 'urban albedo and vegetation enhancement strategies have significant potential to reduce heat-related health impacts' (Stone et al. 2010). Other adaption co-benefits are less well understood. There are likely to be cost savings from managing planning growth (one study found 2% savings in metropolitan budgets, which can then be spent on adaptation planning) (Deal and Schunk 2004).

Diversification is a major adaptation strategy and form of risk management, as it can help households smooth out income fluctuations and provide a broader range of options for the future (Osbahr et al. 2008; Adger et al. 2011; Thornton and Herrero 2014). Surveys of farmers in climate variable areas find that livelihood diversification is increasingly favoured as an adaptation option (Bryan et al. 2013), although it is not always successful, since it can increase exposure to climate variability (Adger et al. 2011). There are more than 570 million small farms in the world (Lowder et al. 2016), and many millions of smallholder agriculturalists already practice livelihood diversification by engaging in multiple forms of off-farm income (Rigg 2006). It is not clear, however, how many farmers have not yet practiced diversification and thus how many would be helped by supporting this response option.

Currently, millions of farmers still rely to some degree on local seeds. Use of local seeds can facilitate adaptation for many smallholders, as moving to use of commercial seeds can increase costs for farmers (Howard 2015). Seed networks and banks protect local agrobiodiversity and landraces, which are important to facilitate adaptation, as local landraces may be resilient to some forms of climate change (Coomes et al. 2015; Van Niekerk and Wynberg 2017; Vasconcelos et al. 2013).

Disaster risk management is an essential part of adaptation strategies. The Famine Early Warning Systems Network funded by the US Agency for International Development (USAID) has operated across three continents since the 1980s, and many millions of people across 34 countries have access to early information on drought. Such information can assist communities and households in adapting to onset conditions (Hillbruner and Moloney 2012). However, concerns have been raised as to how many people are actually reached by disaster risk management and early warning systems; for example, less than 50% of respondents in Bangladesh had heard a cyclone warning before it hit, even though an early warning system existed (Mahmud and Prowse 2012). Further, there are concerns that current early warning systems 'tend to focus on response and recovery rather than on addressing livelihood issues as part of the process of reducing underlying risk factors,' (Birkmann et al. 2015), leading to less adaptation potential being realised.

Local risk-sharing instruments like rotating credit or loan groups can help buffer farmers against climate impacts and help facilitate adaptation. Both index and commercial crop insurance offers some potential for adaptation, as it provides a means of buffering and transferring weather risk, saving farmers the cost of crop losses (Meze-Hausken et al. 2009; Patt et al. 2010). However, overly subsidised insurance can undermine the market's role in pricing risks and thus depress more rapid adaptation strategies (Skees and Collier 2012; Jaworski 2016) and increase the riskiness of decision-making (McLeman and Smit 2006). For example, availability of crop insurance was observed to reduce farm-level diversification in the US, a factor cited as increasing adaptive capacity (Sanderson et al. 2013) and crop insurance-holding soybean farmers in the USA have been less likely to adapt to extreme weather events than those not holding insurance (Annan and Schlenker 2015). It is unclear how many people worldwide use insurance as an adaptation strategy; Platteau et al. (2017) suggest that less than 30% of smallholders take out any form of insurance, but it is likely in the millions.

Table 6.28 summarises the impacts on adaptation of risk management options, with confidence estimates based on the thresholds outlined in Table 6.53 in Section 6.3.6, and indicative (not exhaustive) references upon which the evidence in based.

Table 6.28 | Adaptation effects of response options based on risk management.

Integrated response option	Potential	Confidence	Citation
Management of urban sprawl	Unquantified but likely to be many millions	Low confidence	Stone et al. 2010
Livelihood diversification	>100 million likely	Low confidence	Morton 2007; Rigg 2006
Use of local seeds	Unquantified but likely to be many millions	Low confidence	Louwaars 2002; Santilli 2012
Disaster risk management	>100 million	High confidence	Hillbruner and Moloney 2012
Risk sharing instruments	Unquantified but likely to be several million	Low confidence	Platteau et al. 2017

6.3.3 Potential of the integrated response options for addressing desertification

In this section, the impacts of integrated response options on desertification are assessed.

6.3.3.1 Integrated response options based on land management

In this section, the impacts on desertification of integrated response options based on land management are assessed.

Integrated response options based on land management in agriculture

Burney et al. (2010) estimated that an additional global cropland area of 11.11–15.14 Mkm^2 would have been needed if productivity had not increased between 1961 and 2000. Given that agricultural expansion is a main driver of desertification (FAO and ITPS 2015), increased food productivity could have prevented up to 11.11–15.14 Mkm^2 from exploitation and desertification (Table 6.10).

Improved cropland, livestock and grazing land management are strategic options aimed at prevention of desertification, and may include crop and animal selection, optimised stocking rates, changed tillage and/or cover crops, to land-use shifting from cropland to rangeland, in general targeting increases in ground cover by vegetation, and protection against wind erosion (Schwilch et al. 2014; Bestelmeyer et al. 2015). Considering the widespread distribution of deserts and desertified lands globally, more than 10 Mkm^2 could benefit from improved management techniques.

Agroforestry can help stabilise soils to prevent desertification (Section 6.3.2.1), so given that there is around 10 Mkm^2 of land with more than 10% tree cover (Garrity 2012), agroforestry could benefit up to 10 Mkm^2 of land.

Agricultural diversification to prevent desertification may include the use of crops with manures, legumes, fodder legumes and cover crops combined with conservation tillage systems (Schwilch et al. 2014). These practices can be considered to be part of improved crop management options (see above) and aim at increasing ground coverage by vegetation and controlling wind erosion losses.

Since shifting from grassland to the annual cultivation of crops increases erosion and soil loss, there are significant benefits for desertification control, by stabilising soils in arid areas (Chapter 3). Cropland expansion during 1985 to 2005 was 359,000 km^2, or 17,400 km^2 yr^{-1} (Foley et al. 2011). Not all of this expansion will be from grasslands or in desertified areas, but this value sets the maximum contribution of prevention of conversion of grasslands to croplands, a small global benefit for desertification control (Table 6.10).

Integrated water management strategies such as water-use efficiency and irrigation, improve soil health through increase in soil organic matter content, thereby delivering benefits for prevention or reversal of desertification (Baumhardt et al. 2015; Datta et al. 2000; Evans and Sadler 2008; He et al. 2015) (Chapter 3). Climate change will amplify existing stress on water availability and on agricultural systems, particularly in semi-arid environments (IPCC AR5 2014) (Chapter 3). In 2011, semi-arid ecosystems in the southern hemisphere contributed 51% of the global net carbon sink (Poulter et al. 2014). These results suggest that arid ecosystems could be an important global carbon sink, depending on soil water availability.

Table 6.29 summarises the impacts on desertification of agricultural options, with confidence estimates based on the thresholds outlined in Table 6.53 in Section 6.3.6, and indicative (not exhaustive) references upon which the evidence in based.

Table 6.29 | Effects on desertification of response options in agriculture.

Integrated response option	Potential	Confidence	Citation
Increased food productivity	11.1–15.1 Mkm^2	Low confidence	Burney et al. 2010
Improved cropland management	10 Mkm^2	Low confidence	Schwilch et al. 2014
Improved grazing land management	0.5–3 Mkm^2	Low confidence	Schwilch et al. 2014
Improved livestock management	0.5–3 Mkm^2	Low confidence	Miao et al. 2015; Squires and Karami 2005
Agroforestry	10 Mkm^2 (with >10% tree cover)	Medium confidence	Garrity 2012
Agricultural diversification	0.5–3 Mkm^2	Low confidence	Lambin and Meyfroidt 2011; Schwilch et al. 2014
Reduced grassland conversion to cropland	Up to 17,400 km^2 yr^{-1}	Low confidence	Foley et al. 2011
Integrated water management	10,000 km^2	Low confidence	Pierzynski et al. 2017; UNCCD 2012

Integrated response options based on land management in forestry

Forests are important to help to stabilise land and regulate water and microclimate (Locatelli et al. 2015b). Based on the extent of dry forest at risk of desertification (Núñez et al. 2010; Bastin et al. 2017), the estimated global potential effect for avoided desertification is large for both forest management and for reduced deforestation and forest degradation when cumulated for at least

20 years (Table 6.30). The uncertainty of these global estimates is high. More robust qualitative and some quantitative estimates are available at regional level. For example, it has been simulated that human activity (i.e., land management) contributed to 26% of the total land reverted from desertification in Northern China between 1981 and 2010 (Xu et al. 2018). In Thailand, it was found that the desertification risk is reduced when the land use is changed from bare lands to agricultural lands and forests, and

595

from non-forests to forests; conversely, the desertification risk increases when converting forests and denuded forests to bare lands (Wijitkosum 2016).

Afforestation, reforestation and forest restoration are land management response options that are used to prevent desertification. Forests tend to maintain water and soil quality by reducing runoff and trapping sediments and nutrients (Medugu et al. 2010; Salvati et al. 2014), but planting of non-native species in semi-arid regions can deplete soil water resources if they have high evapotranspiration rates (Zeng et al. 2016; Yang et al. 2014). Afforestation and reforestation programmes can be deployed over large areas of the Earth, so can create synergies in areas prone to desertification. Global estimates of land potentially available for afforestation are up to 25.8 Mkm^2 by the end of the century, depending on a variety of assumptions on socio-economic developments and climate policies (Griscom et al. 2017;

Kreidenweis et al. 2016; Popp et al. 2017). The higher end of this range is achieved under the assumption of a globally uniform reward for carbon uptake in the terrestrial biosphere, and it is halved by considering tropical and subtropical areas only to minimise albedo feedbacks (Kreidenweis et al. 2016). When safeguards are introduced (e.g., excluding existing cropland for food security, boreal areas, etc.), the area available declines to about 6.8 Mkm^2 (95% confidence interval of 2.3 and 11.25 Mkm^2), of which about 4.72 Mkm^2 is in the tropics and 2.06 Mkm^2 is in temperate regions (Griscom et al. 2017) (Table 6.30).

Table 6.30 summarises the impacts on desertification of forestry options, with confidence estimates based on the thresholds outlined in Table 6.53 in Section 6.3.6, and indicative (not exhaustive) references upon which the evidence in based.

Table 6.30 | Effects on desertification of response options in forests.

Integrated response option	Potential	Confidence	Citation
Forest management	>3 Mkm^2	Low confidence	Bastin et al. 2017; Núñez et al. 2010
Reduced deforestation and forest degradation	>3 Mkm^2 (effects cumulated for at least 20 years)	Low confidence	Bastin et al. 2017; Keenan et al. 2015; Núñez et al. 2010
Reforestation and forest restoration	See afforestation		
Afforestation	2–25.8 Mkm^2 by the end of the century	Medium confidence	Griscom et al. 2017; Kreidenweis et al. 2016; Popp et al. 2017

Integrated response options based on land management of soils

With more than 2.7 billion people affected globally by desertification (IPBES 2018), practices to increase soil organic carbon content are proposed as actions to address desertification, and could be applied to an estimated 11.37 Mkm^2 of desertified soils (Lal 2001) (Table 6.31).

Control of soil erosion could have large benefits for desertification control. Using figures from FAO et al. (2015), IPBES (2018) estimated that land losses due to erosion to 2050 are equivalent to 1.5 Mkm^2 of land from crop production, or 45,000 km^2 yr^{-1} (Foley et al. 2011) so soil erosion control could benefit up to 1.50 Mkm^2 of land in the coming decades. Lal (2001) estimated that desertification control (using soil erosion control as one intervention) could benefit 11.37 Mkm^2 of desertified land globally (Table 6.10).

Oldeman et al. (1991) estimated that the global extent soil affected by salinisation is 0.77 Mkm^2 yr^{-1}, which sets the upper limit on the area that could benefit from measures to address soil salinisation (Table 6.31).

In degraded arid grasslands, shrublands and rangelands, desertification can be reversed by alleviation of soil compaction through installation of enclosures and removal of domestic livestock (Allington et al. 2010), but there are no global estimates of potential (Table 6.31).

Biochar could potentially deliver benefits in efforts to address desertification though improving water-holding capacity (Woolf et al. 2010; Sohi 2012), but the global effect is not quantified.

Table 6.31 summarises the impacts on desertification of soil-based options, with confidence estimates based on the thresholds outlined in Table 6.53 in Section 6.3.6, and indicative (not exhaustive) references upon which the evidence in based.

Table 6.31 | Effects on desertification of land management of soils.

Integrated response option	Potential	Confidence	Citation
Increased soil organic carbon content	Up to 11.37 Mkm^2	Medium confidence	Lal 2001
Reduced soil erosion	Up to 11.37 Mkm^2	Medium confidence	Lal 2001
Reduced soil salinisation	0.77 Mkm^2 yr^{-1}	Medium confidence	Oldeman et al. 1991
Reduced soil compaction	No global estimates	No evidence	FAO and ITPS 2015; Hamza and Anderson 2005
Biochar addition to soil	No global estimates	No evidence	

6

Integrated response options based on land management across all/other ecosystems

For fire management, Arora and Melton (2018) estimated, using models and GFED4.1s0 data, that burned area over the 1997–2014 period was 4.834–4.855 $Mkm^2 yr^{-1}$. Randerson et al. (2012) estimated small fires increased total burned area globally by 35% from 3.45 to 4.64 $Mkm^2 yr^{-1}$ during the period 2001–2010. Tansey et al. (2004) estimated that over 3.5 $Mkm^2 yr^{-1}$ of burned areas were detected in the year 2000 (Table 6.32).

Although slope and slope aspect are predictive factors of desertification occurrence, the factors with the greatest influence are land cover factors, such as normalised difference vegetation index (NDVI) and rangeland classes (Djeddaoui et al. 2017). Therefore, prevention of landslides and natural hazards exert indirect influence on the occurrence of desertification.

The global extent of chemical soil degradation (salinisation, pollution and acidification) is about 1.03 $Mkm^2 yr^{-1}$ (Oldeman et al. 1991), giving the maximum extent of land that could benefit from the management of pollution and acidification.

There are no global data on the impacts of management of invasive species/encroachment on desertification, though the impact is presumed to be positive. There are no studies examining the potential role of restoration and avoided conversion of coastal wetlands on desertification.

There are no impacts of peatland restoration for prevention of desertification, as peatlands occur in wet areas and deserts in arid areas, so they are not connected.

For management of pollution, including acidification, Oldeman et al. (1991) estimated the global extent of chemical soil degradation, with 0.77 $Mkm^2 yr^{-1}$ affected by salinisation, 0.21 $Mkm^2 yr^{-1}$ affected by pollution, and 0.06 $Mkm^2 yr^{-1}$ affected by acidification (total: 1.03 $Mkm^2 yr^{-1}$), so this is the area that could potentially benefit from pollution management measures.

Biodiversity conservation measures can interact with desertification, but the literature contains no global estimates of potential.

Table 6.32 summarises the impacts on desertification of options on all/other ecosystems, with confidence estimates based on the thresholds outlined in Table 6.53 in Section 6.3.6, and indicative (not exhaustive) references upon which the evidence in based.

Table 6.32 | Effects on desertification of response options on all/other ecosystems.

Integrated response option	Potential	Confidence	Citation
Fire management	Up to 3.5–4.9 $Mkm^2 yr^{-1}$	Medium confidence	Arora and Melton 2018; Randerson et al. 2012; Tansey et al. 2004
Reduced landslides and natural hazards	>0	Low confidence	Djeddaoui et al. 2017; Noble et al. 2014
Reduced pollution including acidification	1.03 $Mkm^2 yr^{-1}$	Low confidence	Oldeman et al. 1991
Management of invasive species/encroachment	No global estimates	No evidence	
Restoration and reduced conversion of coastal wetlands	No global estimates	No evidence	
Restoration and reduced conversion of peatlands	No impact		
Biodiversity conservation	No global estimates	No evidence	

Integrated response options based on land management specifically for carbon dioxide removal (CDR)

While spreading of crushed minerals onto land as part of enhanced weathering may provide soil/plant nutrients in nutrient-depleted soils (Beerling et al. 2018), there is no literature reporting on the potential global impacts of this in addressing desertification.

Large-scale production of bioenergy can require significant amounts of land (Smith et al. 2016a; Clarke et al. 2014; Popp et al. 2017), with as much as 15 Mkm^2 in 2100 in 2°C scenarios (Popp et al. 2017), increasing pressures for desertification (Table 6.33).

Table 6.33 summarises the impacts on desertification of options specifically for CDR, with confidence estimates based on the thresholds outlined in Table 6.53 in Section 6.3.6, and indicative (not exhaustive) references upon which the evidence in based.

Table 6.33 | Effects on desertification of response options specifically for CDR.

Integrated response option	Potential	Confidence	Citation
Enhanced weathering of minerals	No global estimates	No evidence	
Bioenergy and BECCS	Negative impact on up to 15 Mkm^2	Low confidence	Clarke et al. 2014; Popp et al. 2017; Smith et al. 2016a

6

6.3.3.2 Integrated response options based on value chain management

In this section, the impacts on desertification of integrated response options based on value chain management are assessed.

Integrated response options based on value chain management through demand management

Dietary change and waste reduction both result in decreased cropland and pasture extent (Bajželj et al. 2014a; Stehfest et al. 2009; Tilman and Clark 2014), reducing the pressure for desertification (Table 6.34).

Reduced post-harvest losses could spare 1.98 Mkm^2 of cropland globally (Kummu et al. 2012). Not all of this land could be subject to desertification pressure, so this represents the maximum area that could be relieved from desertification pressure by reduction of post-harvest losses. No studies were found linking material substitution to desertification.

Table 6.34 summarises the impacts on desertification of demand management options, with confidence estimates based on the thresholds outlined in Table 6.53 in Section 6.3.6, and indicative (not exhaustive) references upon which the evidence in based.

Table 6.34 | Effects on desertification of response options based on demand management.

Integrated response option	Potential	Confidence	Citation
Dietary change	0.80–5 Mkm^2	Low confidence	Alexander et al. 2016; Bajželj et al. 2014b; Stehfest et al. 2009; Tilman and Clark 2014
Reduced post-harvest losses	<1.98 Mkm^2	Low confidence	Kummu et al. 2012
Reduced food waste (consumer or retailer)	1.4 Mkm^2	Low confidence	Bajželj et al. 2014b
Material substitution	No global estimates	No evidence	

Integrated response options based on value chain management through supply management

There are no global estimates of the impact on desertification of sustainable sourcing, management of supply chains, enhanced urban food systems, improved food processing, or improved energy use in agriculture.

Table 6.35 summarises the impacts on desertification of supply management options, with confidence estimates based on the thresholds outlined in Table 6.53 in Section 6.3.6, and indicative (not exhaustive) references upon which the evidence in based.

Table 6.35 | Effects on desertification of response options based on supply management.

Integrated response option	Potential	Confidence	Citation
Sustainable sourcing	No global estimates	No evidence	
Management of supply chains	No global estimates	No evidence	
Enhanced urban food systems	No global estimates	No evidence	
Improved food processing and retailing	No global estimates	No evidence	
Improved energy use in food systems	No global estimates	No evidence	

6.3.3.3 Integrated response options based on risk management

In this section, the impacts on desertification of integrated response options based on risk management are assessed.

There are regional case studies of urban sprawl contributing to desertification in Mediterranean climates in particular (Barbero-Sierra et al. 2013; Stellmes et al. 2013), but no global figures.

Diversification may deliver some benefits for addressing desertification when it involves greater use of tree crops that may reduce the need for tillage (Antwi-Agyei et al. 2014). Many anti-desertification programmes call for diversification (Stringer et al. 2009), but there is little evidence on how many households had done so (Herrmann and Hutchinson 2005). There are no numbers for global impacts.

The literature is unclear on whether the use of local seeds has any relationship to desertification, although some local seeds are more likely to adapt to arid climates and less likely to degrade land than commercially introduced varieties (Mousseau 2015). Some anti-desertification programmes have also shown more success using local seed varieties (Bassoum and Ghiggi 2010; Nunes et al. 2016).

Some disaster risk management approaches can have impacts on reducing desertification, like the Global Drought Early Warning System (GDEWS) (currently in development), which will monitor precipitation, soil moisture, evapotranspiration, river flows, groundwater, agricultural productivity and natural ecosystem health. It may have some potential co-benefits to reduce desertification (Pozzi et al. 2013). However, there are no figures yet for how much land area will be covered by such early warning systems.

Risk-sharing instruments, such as pooling labour or credit, could help communities invest in anti-desertification actions, but evidence is missing. Commercial crop insurance is likely to deliver no co-benefits for prevention and reversal of desertification, as evidence suggests that subsidised insurance, in particular, can increase crop production in marginal lands. Crop insurance could have been responsible for shifting up to 0.9% of rangelands to cropland in the Upper Midwest of the USA (Claassen et al. 2011a).

Table 6.36 summarises the impact on desertification for options based on risk management, with confidence estimates based on the thresholds outlined in Table 6.53 in Section 6.3.6, and indicative (not exhaustive) references upon which the evidence in based.

Table 6.36 | Effects on desertification of response options based on risk management.

Integrated response option	Potential	Confidence	Citation
Management of urban sprawl	>5000 km^2	Low confidence	Barbero-Sierra et al. 2013
Livelihood diversification	No global estimates	Low confidence	Herrmann and Hutchinson 2005
Use of local seeds	No global estimates	No evidence	
Disaster risk management	No global estimates	No evidence	Pozzi et al. 2013
Risk-sharing instruments	Likely negative impacts but not quantified	Low confidence	Claassen et al. 2011a

6.3.4 Potential of the integrated response options for addressing land degradation

In this section, the impacts of integrated response options on land degradation are assessed.

6.3.4.1 Integrated response options based on land management

In this section, the impacts on land degradation of integrated response options based on land management are assessed.

Integrated response options based on land management in agriculture

Burney et al. (2010) estimated that an additional global cropland area of 11.11–15.14 Mkm^2 would have been needed if productivity had not increased between 1961 and 2000. As for desertification, given that agricultural expansion is a main driver of land degradation (FAO and ITPS 2015), increased food productivity has prevented up to 11.11–15.14 Mkm^2 from exploitation and land degradation (Table 6.37).

Land degradation can be addressed by the implementation of improved cropland, livestock and grazing land management practices, such as those outlined in the recently published Voluntary Guidelines for Sustainable Soil Management (FAO 2017b). Each one could potentially affect extensive surfaces, not less than 10 Mkm^2. The guidelines include a list of practices aimed at minimising soil erosion, enhancing soil organic matter content, fostering soil nutrient balance and cycles, preventing, minimising and mitigating soil salinisation and alkalinisation, soil contamination, soil acidification, soil sealing, soil compaction, and improving soil water management. Land cover and land cover change are key factors and indicators of land degradation. In many drylands, land cover is threatened by overgrazing, so management of stocking rate and grazing can help to prevent the advance of land degradation (Smith et al. 2016a).

Agroforestry can help stabilise soils to prevent land degradation; so, given that there is around 10 Mkm^2 of land with more than 10% tree cover (Garrity 2012), agroforestry could benefit up to 10 Mkm^2 of land.

Agricultural diversification usually aims at increasing climate and food security resilience, such as under 'climate smart agriculture' approaches (Lipper et al. 2014). Both objectives are closely related to land degradation prevention, potentially affecting 1–5 Mkm^2.

Shifting from grassland to tilled crops increases erosion and soil loss, so there are significant benefits for addressing land degradation, by stabilising degraded soils (Chapter 3). Since cropland expansion during 1985 to 2005 was 17,400 km^2 yr^{-1} (Foley et al. 2011) – and not all of this expansion will be from grasslands or degraded land – the maximum contribution of prevention of conversion of grasslands to croplands is 17,400 km^2 yr^{-1}, a small global benefit for control of land degradation (Table 6.37).

Most land degradation processes that are sensitive to climate change pressures (e.g., erosion, decline in soil organic matter, salinisation, waterlogging, drying of wet ecosystems) can benefit from integrated water management. Integrated water management options include management to reduce aquifer and surface water depletion, and to prevent over-extraction, and provide direct co-benefits for prevention of land degradation. Land management practices implemented for climate change mitigation may also affect water resources. Globally, water erosion is estimated to result in the loss of 23–42 MtN and 14.6–26.4 MtP annually (Pierzynski et al. 2017). Forests influence the storage and flow of water in watersheds (Eisenbies et al. 2007) and are therefore important for regulating how climate change will impact on landscapes.

Table 6.37 summarises the impact on land degradation of options in agriculture, with confidence estimates based on the thresholds outlined in Table 6.53 in Section 6.3.6, and indicative (not exhaustive) references upon which the evidence in based.

6

Table 6.37 | Effects on land degradation of response options in agriculture.

Integrated response option	Potential	Confidence	Citation
Increased food productivity	11.11–15.14 Mkm^2	Medium confidence	Burney et al. 2010
Improved cropland management	10 Mkm^2	Low confidence	Lal 2015; Smith et al. 2016a
Improved grazing land management	10 Mkm^2	Low confidence	Smith et al. 2016a
Improved livestock management	10 Mkm^2	Low confidence	Lal 2015; Smith et al. 2016a
Agroforestry	10 Mkm^2 (with >10% tree cover)	Medium confidence	Garrity 2012
Agricultural diversification	1–5 Mkm^2	Medium confidence	Lambin and Meyfroidt 2011
Reduced grassland conversion to cropland	Up to 17,400 km^2 yr^{-1}	Low confidence	Foley et al. 2011
Integrated water management	0.01 Mkm^2	Medium confidence	Pierzynski et al. 2017; UNCCD 2012

Integrated response options based on land management in forestry

Based on the extent of forest exposed to degradation (Gibbs and Salmon 2015), the estimated global potential effect for reducing land degradation, for example, through reduced soil erosion (Borrelli et al. 2017), is large for both forest management and for reduced deforestation and forest degradation when cumulated for at least 20 years (Table 6.38). The uncertainty of these global estimates is high. More robust qualitative, and some quantitative, estimates are available at regional level. For example, in Indonesia, Santika et al. (2017) demonstrated that reduced deforestation (Sumatra and Kalimantan islands) contributed to significantly reduced land degradation.

Forest restoration is a key option to achieve the overarching frameworks to reduce land degradation at global scale, such as, for example, Zero Net Land Degradation (ZNLD; UNCCD 2012) and Land Degradation Neutrality (LDN), not only in drylands (Safriel 2017). Indeed, it has been estimated that more than 20 Mkm^2 are suitable for forest and landscape restoration, of which 15 Mkm^2 may be

devoted to mixed plant mosaic restoration (UNCCD 2012). Moreover, the Bonn Challenge[6] aims to restore 1.5 Mkm^2 of deforested and degraded land by 2020, and 3.5 Mkm^2 by 2030. Under a restoration and protection scenario (implementing restoration targets), Wolff et al. (2018) simulated that there will be a global increase in net tree cover of about 4 Mkm^2 by 2050 (Table 6.38). At local level, Brazil's Atlantic Restoration Pact aims to restore 0.15 Mkm^2 of forest areas in 40 years (Melo et al. 2013). The Y Ikatu Xingu campaign (launched in 2004) aims to contain deforestation and forest degradation processes by reversing the liability of 3000 km^2 in the Xingu Basin, Brazil (Durigan et al. 2013).

Afforestation and reforestation are land management options frequently used to address land degradation (see Section 6.3.3.1 for details, and Table 6.38).

Table 6.38 summarises the impact on land degradation of options in forestry, with confidence estimates based on the thresholds outlined in Table 6.53 in Section 6.3.6, and indicative (not exhaustive) references upon which the evidence in based.

Table 6.38 | Effects on land degradation of response options in forestry.

Integrated response option	Potential	Confidence	Citation
Forest management	>3 Mkm^2	Low confidence	Gibbs and Salmon 2015
Reduced deforestation and forest degradation	>3 Mkm^2 (effects cumulated for at least 20 years)	Low confidence	Gibbs and Salmon 2015; Keenan et al. 2015
Reforestation and forest restoration	20 Mkm^2 suitable for restoration >3 Mkm^2 by 2050 (net increase in tree cover for forest restoration)	Medium confidence	UNCCD 2012; Wolff et al. 2018
Afforestation	2–25.8 Mkm^2 by the end of the century	Low confidence	Griscom et al. 2017; Kreidenweis et al. 2016; Popp et al. 2017

Integrated response options based on land management of soils

Increasing soil organic matter content is a measure to address land degradation. With around 120,000 km^2 lost to degradation every year, and over 3.2 billion people negatively impacted on by land degradation globally (IPBES 2018), practices designed to increase soil organic carbon have a large potential to address land degradation, estimated to affect more than 11 Mkm^2 globally (Lal 2004) (Table 6.39).

Control of soil erosion could have large benefits for addressing land degradation. Soil erosion control could benefit up to 1.50 Mkm^2 of land to 2050 (IPBES 2018). Lal (2004) suggested that interventions to

prevent wind and water erosion (two of the four main interventions proposed to address land degradation), could restore 11 Mkm^2 of degraded and desertified soils globally (Table 6.39).

Oldeman et al. (1991) estimated that the global extent soil affected by salinisation is 0.77 Mkm^2 yr^{-1}, which sets the upper limit on the area that could benefit from measures to address soil salinisation (Table 6.39). The global extent of chemical soil degradation (salinisation, pollution and acidification) is about 1.03 Mkm^2 (Oldeman et al. 1991) giving the maximum extent of land that could benefit from the management of pollution and acidification.

---

[6]   www.bonnchallenge.org/content/challenge.

Biochar could provide moderate benefits for the prevention or reversal of land degradation, by improving water-holding capacity and nutrient-use efficiency, managing heavy metal pollution, and other co-benefits (Sohi 2012), though the global effects are not quantified.

Table 6.39 summarises the impact on land degradation of soil-based options, with confidence estimates based on the thresholds outlined in Table 6.53 in Section 6.3.6, and indicative (not exhaustive) references upon which the evidence in based.

Table 6.39 | Effects on land degradation of soil-based response options.

Integrated response option	Potential	Confidence	Citation
Increased soil organic carbon content	11 Mkm^2	Medium confidence	Lal 2004
Reduced soil erosion	11 Mkm^2	Medium confidence	Lal 2004
Reduced soil salinisation	0.77 Mkm^2 yr^{-1}	Medium confidence	FAO 2018a; Qadir et al. 2013
Reduced soil compaction	10 Mkm^2	Low confidence	FAO and ITPS 2015; Hamza and Anderson 2005
Biochar addition to soil	Positive but not quantified globally	Low confidence	Chapter 4

Integrated response options based on land management across all/other ecosystems

For fire management, details of estimates of the impact of wildfires (and thereby the potential impact of their suppression) are given in Section 6.3.3.1 (Table 6.40).

Management of landslides and natural hazards aims at controlling a severe land degradation process affecting sloped and hilly areas, many of them with poor rural inhabitants (FAO and ITPS 2015; Gariano and Guzzetti 2016), but the global potential has not been quantified.

There are no global data on the impacts of management of invasive species/encroachment on land degradation, though the impact is presumed to be positive.

Since large areas of coastal wetlands are degraded, restoration could potentially deliver moderate benefits for addressing land degradation, with 0.29 Mkm^2 globally considered feasible for restoration (Griscom et al. 2017) (Table 6.40).

Considering that large areas (0.46 Mkm^2) of global peatlands are degraded and considered suitable for restoration (Griscom et al. 2017), peatland restoration could deliver moderate benefits for addressing land degradation (Table 6.40).

There are no global estimates of the effects of biodiversity conservation on reducing degraded lands. However, at local scale, biodiversity conservation programmes have been demonstrated to stimulate gain of forest cover across large areas over the last three decades (e.g., in China; Zhang et al. 2013). Management of wild animals can influence land degradation processes by grazing, trampling and compacting soil surfaces, thereby altering surface temperatures and chemical reactions affecting sediment and carbon retention (Cromsigt et al. 2018).

Table 6.40 summarises the impact on land degradation of options in all/other ecosystems, with confidence estimates based on the thresholds outlined in Table 6.53 in Section 6.3.6, and indicative (not exhaustive) references upon which the evidence in based.

Table 6.40 | Effects on land degradation of response options in all/other ecosystems.

Integrated response option	Potential	Confidence	Citation
Fire management	Up to 3.5–4.9 Mkm^2 yr^{-1}	Medium confidence	Arora and Melton 2018; Randerson et al. 2012; Tansey et al. 2004
Reduced landslides and natural hazards	1–5 Mkm^2	Low confidence	FAO and ITPS 2015; Gariano and Guzzetti 2016
Reduced pollution including acidification	~1.03 Mkm^2	Low confidence	Oldeman et al. 1991
Management of invasive species/encroachment	No global estimates	No evidence	
Restoration and reduced conversion of coastal wetlands	0.29 Mkm^2	Medium confidence	Griscom et al. 2017
Restoration and reduced conversion of peatlands	0.46 Mkm^2	Medium confidence	Griscom et al. 2017
Biodiversity conservation	No global estimates	No evidence	

Integrated response options based on land management specifically for carbon dioxide removal (CDR)

While spreading of crushed minerals onto land as part of enhanced weathering can provide soil/plant nutrients in nutrient-depleted soils, increase soil organic carbon stocks and help to replenish eroded soil (Beerling et al. 2018), there is no literature on the global potential for addressing land degradation.

Large-scale production of bioenergy can require significant amounts of land (Smith et al. 2016a; Clarke et al. 2014; Popp et al. 2017) – as much as 15 Mkm^2 in 2°C scenarios (Popp et al. 2017) – therefore increasing pressures for land conversion and land degradation (Table 6.13). However, bioenergy production can either increase (Robertson et al. 2017b; Mello et al. 2014) or decrease (FAO 2011b; Lal 2014) soil organic matter, depending on where it is produced and how it is managed. These effects are not included in the quantification in Table 6.41.

Table 6.41 summarises the impact on land degradation of options specifically for CDR, with confidence estimates based on the thresholds outlined in Table 6.53 in Section 6.3.6, and indicative (not exhaustive) references upon which the evidence in based.

Table 6.41 | Effects on land degradation of response options specifically for CDR.

Integrated response option	Potential	Confidence	Citation
Enhanced weathering of minerals	Positive but not quantified	Low confidence	Beerling et al. 2018
Bioenergy and BECCS	Negative impact on up to 15 Mkm^2	Low confidence	Clarke et al. 2014; Popp et al. 2017; Smith et al. 2016a

6.3.4.2   Integrated response options based on value chain management

In this section, the impacts on land degradation of integrated response options based on value chain management are assessed.

Integrated response options based on value chain management through demand management

Dietary change and waste reduction both result in decreased cropland and pasture extent (Bajželj et al. 2014a; Stehfest et al. 2009; Tilman and Clark 2014), reducing the pressure for land degradation (Table 6.15). Reduced post-harvest losses could spare 1.98 Mkm^2 of cropland globally (Kummu et al. 2012) meaning that land degradation pressure could be relieved from this land area through reduction of post-harvest losses. The effects of material substitution on land degradation depend on management practice; some forms of logging can lead to increased land degradation (Chapter 4).

Table 6.42 summarises the impact on land degradation of demand management options, with confidence estimates based on the thresholds outlined in Table 6.53 in Section 6.3.6, and indicative (not exhaustive) references upon which the evidence in based.

Table 6.42 | Effects on land degradation of response options based on demand management.

Integrated response option	Potential	Confidence	Citation
Dietary change	4–28 Mkm^2	High confidence	Alexander et al. 2016; Bajželj et al. 2014b; Stehfest et al. 2009; Tilman and Clark 2014
Reduced post-harvest losses	1.98 Mkm^2	Medium confidence	Kummu et al. 2012
Reduced food waste (consumer or retailer)	7 Mkm^2	Medium confidence	Bajželj et al. 2014b
Material substitution	No global estimates	No evidence	

Integrated response options based on value chain management through supply management

There are no global estimates of the impact on land degradation of enhanced urban food systems, improved food processing, retailing, or improved energy use in food systems.

There is evidence that sustainable sourcing could reduce land degradation, as the explicit goal of sustainable certification programmes is often to reduce deforestation or other unsustainable land uses. Over 4 Mkm^2 of forests are certified for sustainable harvesting (PEFC and FSC 2018), although it is not clear if all these lands would be at risk of degradation without certification. While the food price instability of 2007/2008 increased financial investment in crop expansion (especially through so-called land grabbing), and thus better management of supply chains might have reduced this amount, no quantification of the total amount of land acquired, nor the possible impact of this crop expansion on degradation, has been recorded (McMichael and Schneider 2011; McMichael 2012).

Table 6.43 summarises the impact on land degradation of supply management options, with confidence estimates based on the thresholds outlined in Table 6.53 in Section 6.3.6, and indicative (not exhaustive) references upon which the evidence in based.

Table 6.43 | Effects on land degradation of response options based on supply management.

Integrated response option	Potential	Confidence	Citation
Sustainable sourcing	>4 Mkm^2	Low confidence	Auld et al. 2008
Management of supply chains	No global estimates	No evidence	
Enhanced urban food systems	No global estimates	No evidence	
Improved food processing and retailing	No global estimates	No evidence	
Improved energy use in food systems	No global estimates	No evidence	

6.3.4.3   Integrated response options based on risk management

In this section, the impacts on land degradation of integrated response options based on risk management are assessed.

Urban expansion has been identified as a major culprit in soil degradation in some countries; for example, urban expansion in China has now affected 0.2 Mkm^2, or almost one-sixth of the cultivated land total, causing an annual grain yield loss of up to 10 Mt, or around 5–6% of cropland production. Cropland production losses of 8–10% by 2030 are expected under model scenarios of urban expansion (Bren d'Amour et al. 2016). Pollution from urban development has included water and soil pollution from industry, and wastes and sewage, as well as acid deposition from increasing energy use in cities (Chen 2007), all resulting in major losses to Nature's Contributions to People from urban conversion (Song and Deng 2015). Soil sealing from urban expansion is a major loss of soil productivity across many areas. The World Bank has estimated that new city dwellers in developing countries will require 160–500 m^2 per capita, converted from non-urban to urban land (Barbero-Sierra et al. 2013; Angel et al. 2005).

Degradation can be a driver leading to livelihood diversification (Batterbury 2001; Lestrelin and Giordano 2007). Diversification has the potential to deliver some reversal of land degradation, if diversification involves adding non-traditional crops or trees that may reduce the need for tillage (Antwi-Agyei et al. 2014). China's Sloping Land Conversion Program has had livelihood diversification benefits and is said to have prevented degradation of 93,000 km^2 of land (Liu et al. 2015). However, Warren (2002) provides conflicting evidence that more diverse-income households had increased degradation on their lands in Niger. Palacios et al. (2013) associate landscape fragmentation with increased livelihood diversification in Mexico.

Use of local seeds may play a role in addressing land degradation due to the likelihood of local seeds being less dependent on inputs such as chemical fertilisers or mechanical tillage; for example, in India, local legumes are retained in seed networks while commercial crops like sorghum and rice dominate food markets (Reisman 2017). However, there are no global figures.

Disaster Risk Management systems can have some positive impacts on prevention and reversal of land degradation, such as the Global Drought Early Warning System (Pozzi et al. 2013) (Section 6.3.3.3).

Risk-sharing instruments could have benefits for reduced degradation, but there are no global estimates. Commercial crop insurance is likely to deliver no co-benefits for prevention and reversal of degradation. One study found a 1% increase in farm receipts generated from subsidised farm programmes (including crop insurance and others) increased soil erosion by 0.3 t ha^{-1} (Goodwin and Smith 2003). Wright and Wimberly (2013) found a 5310 km^2 decline in grasslands in the Upper Midwest of the USA during 2006–2010, due to crop conversion driven by higher prices and access to insurance.

Table 6.44 summarises the impact on land degradation of risk management options, with confidence estimates based on the thresholds outlined in Table 6.53 in Section 6.3.6, and indicative (not exhaustive) references upon which the evidence in based.

Table 6.44 | Effects on land degradation of response options based on risk management.

Integrated response option	Potential	Confidence	Citation
Management of urban sprawl	>0.2 Mkm^2	Medium confidence	Chen 2007; Zhang et al. 2000
Livelihood diversification	>0.1 Mkm^2	Low confidence	Liu and Lan 2015
Use of local seeds	No global estimates	No evidence	
Disaster risk management	No global estimates	No evidence	Pozzi et al. 2013
Risk-sharing instruments	Variable, but negative impact on >5000 km^2 in Upper Midwest USA	Low confidence	Goodwin and Smith 2003; Wright and Wimberly 2013

6.3.5   Potential of the integrated response options for addressing food security

In this section, the impacts of integrated response options on food security are assessed.

6.3.5.1   Integrated response options based on land management

In this section, the impacts on food security of integrated response options based on land management are assessed.

Integrated response options based on land management in agriculture

Increased food productivity has fed many millions of people who would otherwise not have been fed. Erisman et al. (2008) estimated that more than 3 billion people worldwide could not have been fed without increased food productivity arising from nitrogen fertilisation (Table 6.45).

Improved cropland management to achieve food security aims at closing yield gaps by increasing use efficiency of essential inputs such as water and nutrients. Large production increases (45–70% for most crops) are possible from closing yield gaps to 100% of attainable yield, by increasing fertiliser use and irrigation, but overuse of nutrients could cause adverse environmental impacts (Mueller et al. 2012). This improvement can impact on 1000 million people.

Improved grazing land management includes grasslands, rangelands and shrublands, and all sites on which pastoralism is practiced. In general terms, continuous grazing may cause severe damage to topsoil quality, for example, through compaction. This damage may be reversed by short grazing-exclusion periods under rotational grazing systems (Greenwood and McKenzie 2001; Drewry 2006; Taboada et al. 2011). Due to the widespread diffusion of pastoralism, improved grassland management may potentially affect more than 1000 million people, many of them under subsistence agricultural systems.

Meat, milk, eggs and other animal products, including fish and other seafoods, will play an important role in achieving food security (Reynolds et al. 2015). Improved livestock management with different animal types and feeds may also impact on one million people (Herrero et al. 2016). Ruminants are efficient converters of grass into human-edible energy, and protein and grassland-based food production can produce food with a comparable carbon footprint to mixed systems (O'Mara 2012). However, in the future, livestock production will increasingly be affected by competition for natural resources, particularly land and water, competition between food and feed, and by the need to operate in a carbon-constrained economy (Thornton et al. 2009).

Currently, more than 1.3 billion people are on degrading agricultural land, and the combined impacts of climate change and land degradation could reduce global food production by 10% by 2050. Since agroforestry could help to address land degradation, up to 1.3 billion people could benefit in terms of food security through agroforestry.

Agricultural diversification is not always economically viable; technological, biophysical, educational and cultural barriers may emerge that limit the adoption of more diverse farming systems by farmers (Section 6.4.1). Nevertheless, diversification could benefit 1000 million people, many of them under subsistence agricultural systems (Birthal et al. 2015; Massawe et al. 2016; Waha et al. 2018).

Cropland expansion during 1985 to 2005 was 17,000 km^2 yr^{-1} (Foley et al. 2005). Given that cropland productivity (global average of 250 kg protein ha^{-1} yr^{-1} for wheat; Clark and Tilman 2017) is greater than that of grassland (global average of about 10 kg protein ha^{-1} yr^{-1} for beef/mutton; Clark and Tilman 2017), prevention of this conversion to cropland would have led to a loss of about 0.4 Mt protein yr^{-1} globally. Given an average protein consumption in developing countries of 25.5 kg protein yr^{-1} (equivalent to 70 g person^{-1} day^{-1}; FAO 2018b; OECD and FAO 2018), this is equivalent to the protein consumption of 16.4 million people each year (Table 6.45).

Integrated water management provides direct benefits to food security by improving agricultural productivity (Chapter 5; Godfray and Garnett 2014; Tilman et al. 2011), thereby potentially impacting on the livelihood and well-being of more than 1000 million people (Campbell et al. 2016) affected by hunger and highly impacted on by climate change. Increasing water availability and reliable supply of water for agricultural production using different techniques of water harvesting, storage, and its judicious utilisation through farm ponds, dams and community tanks in rainfed agriculture areas have been presented by Rao et al. (2017a) and Rivera-Ferre et al. (2016).

Table 6.45 summarises the impact on food security of options in agriculture, with confidence estimates based on the thresholds outlined in Table 6.53 in Section 6.3.6, and indicative (not exhaustive) references upon which the evidence in based.

Table 6.45 | Effects on food security of response options in agriculture.

Integrated response option	Potential	Confidence	Citation
Increased food productivity	3000 million people	High confidence	Erisman et al. 2008
Improved cropland management	>1000 million people	Low confidence	Campbell et al. 2014; Lipper et al. 2014
Improved grazing land management	>1000 million people	Low confidence	Herrero et al. 2016
Improved livestock management	>1000 million people	Low confidence	Herrero et al. 2016
Agroforestry	Up to 1300 million people	Low confidence	Sascha et al. 2017
Agricultural diversification	>1000 million people	Low confidence	Birthal et al. 2015; Massawe et al. 2016; Waha et al. 2018
Reduced grassland conversion to cropland	Negative impact on 16.4 million people	Low confidence	Clark and Tilman 2017; FAO 2018b
Integrated water management	>1000 million people	High confidence	Campbell et al. 2016

Integrated response options based on land management in forestry

Forests play a major role in providing food to local communities (non-timber forest products, mushrooms, fodder, fruits, berries, etc.), and diversify daily diets directly or indirectly through improving productivity, hunting, diversifying tree-cropland-livestock systems, and grazing in forests. Based on the extent of forest contributing to food supply, considering the people undernourished (FAO et al. 2013; Rowland et al. 2017), and the annual deforestation rate (Keenan et al. 2015), the global potential to enhance food security is moderate for forest management and small for reduced deforestation (Table 6.46). The uncertainty of these global estimates is high. More robust qualitative, and some quantitative, estimates are available at regional level. For example, managed natural forests, shifting cultivation and agroforestry systems are demonstrated to be crucial to food security and nutrition for hundreds of millions of people in rural landscapes worldwide (Sunderland et al. 2013; Vira et al. 2015). According to Erb et al. (2016), deforestation would not be needed to feed the global population by 2050, in terms of quantity and quality of food. At local level, Cerri et al. (2018) suggested that reduced deforestation, along with integrated cropland-livestock management, would positively impact on more than 120 million people in the Cerrado, Brazil. In Sub-Saharan Africa, where population and food demand are projected to continue to rise substantially, reduced deforestation may have strong positive effects on food security (Doelman et al. 2018).

Afforestation and reforestation negatively impact on food security (Boysen et al. 2017a; Frank et al. 2017; Kreidenweis et al. 2016). It is estimated that large-scale afforestation plans could cause increases in food prices of 80% by 2050 (Kreidenweis et al. 2016), and more general mitigation measures in the agriculture, forestry and other land-use (AFOLU) sector can translate into a rise in undernourishment of 80–300 million people (Frank et al. 2017) (Table 6.16). For reforestation, the potential adverse side effects with food security are smaller than afforestation, because forest regrows on recently deforested areas, and its impact would be felt mainly through impeding possible expansion of agricultural areas. On a smaller scale, forested land also offers benefits in terms of food supply, especially when forest is established on degraded land, mangroves and other land that cannot be used for agriculture. For example, food from forests represents a safety net during times of food and income insecurity (Wunder et al. 2014) and wild harvested meat and freshwater fish provides 30–80% of protein intake for many rural communities (McIntyre et al. 2016; Nasi et al. 2011).

Table 6.46 summarises the impact on food security of options in forestry, with confidence estimates based on the thresholds outlined in Table 6.53 in Section 6.3.6, and indicative (not exhaustive) references upon which the evidence in based.

Table 6.46 | Effects on food security of response options in forestry.

Integrated response option	Potential	Confidence	Citation
Forest management	Positive impact on <100 million people	Low confidence	FAO et al. 2013; Rowland et al. 2017
Reduced deforestation and forest degradation	Positive impact on <1 million people	Low confidence	FAO et al. 2013; Keenan et al. 2015; Rowland et al. 2017
Reforestation and forest restoration	See Afforestation		
Afforestation	Negative impact on >100 million people	Medium confidence	Boysen et al. 2017a; Frank et al. 2017; Kreidenweis et al. 2016

Integrated response options based on land management of soils

Increasing soil organic matter stocks can increase yield and improve yield stability (Lal 2006; Pan et al. 2009; Soussana et al. 2019), though this is not universally seen (Hijbeek et al. 2017), Lal (2006) concludes that crop yields can be increased by 20–70 kg ha^{-1}, 10–50 kg ha^{-1} and 30–300 kg ha^{-1} for wheat, rice and maize, respectively, for every 1 tC ha^{-1} increase in soil organic carbon in the root zone. Increasing soil organic carbon by 1 tC ha^{-1} could increase food grain production in developing countries by 32 Mt yr^{-1} (Lal 2006). Frank et al. (2017) estimate that soil carbon sequestration could reduce calorie loss associated with agricultural mitigation measures by 65%, saving 60–225 million people from undernourishment compared to a baseline without soil carbon sequestration (Table 6.47).

Lal (1998) estimated the risks of global annual loss of food production due to accelerated erosion to be as high as 190 Mt yr^{-1} of cereals, 6 Mt yr^{-1} of soybean, 3 Mt yr^{-1} of pulses and 73 Mt yr^{-1} of roots and tubers. Considering only cereals, if we estimate per-capita annual grain consumption in developing countries to be 300 kg yr^{-1} (based on data included in FAO 2018b; FAO et al. 2018; Pradhan et al. 2013; World Bank 2018a), the loss of 190 Mt yr^{-1} of cereals is equivalent to that consumed by 633 million people, annually (Table 6.47).

Though there are biophysical barriers, such as access to appropriate water sources and limited productivity of salt-tolerant crops, prevention/reversal of soil salinisation could benefit 1–100 million people (Qadir et al. 2013). Soil compaction affects crop yields, so prevention of compaction could also benefit an estimated 1–100 million people globally (Anderson and Peters 2016).

Biochar on balance, could provide moderate benefits for food security by improving yields by 25% in the tropics, but with more limited impacts in temperate regions (Jeffery et al. 2017), or through improved water-holding capacity and nutrient-use efficiency (Sohi 2012) (Chapter 5). These benefits could, however, be tempered by additional pressure on land if large quantities of biomass are required as feedstock for biochar production, thereby causing potential conflicts with food security (Smith 2016). Smith (2016) estimated that 0.4–2.6 Mkm^2 of land would be required for biomass feedstock to deliver 2.57 GtCO$_2$e yr^{-1} of CO$_2$ removal. If biomass production occupied 2.6 Mkm^2 of cropland, equivalent to around 20% of the global cropland area, this could potentially have a large effect on food security, although Woolf et al. (2010) argue that abandoned cropland could be used to supply biomass for biochar, thus avoiding competition with food production. Similarly, Woods et al. (2015) estimate that 5–9 Mkm^2 of land is available for biomass production without compromising food security and biodiversity, considering marginal and degraded land and land released by pasture intensification (Table 6.47).

Table 6.47 summarises the impact on food security of soil-based options, with confidence estimates based on the thresholds outlined in Table 6.53 in Section 6.3.6, and indicative (not exhaustive) references upon which the evidence in based.

6

Table 6.47 | Effects on food security of soil-based response options.

Integrated response option	Potential	Confidence	Citation
Increased soil organic carbon content	60–225 million people	Low confidence	Frank et al. 2017
Reduced soil erosion	633 million people yr^{-1}	Low confidence	FAO 2018b; FAO et al. 2018; Lal 1998; Pradhan et al. 2013; World Bank 2018a
Reduced soil salinisation	1–100 million people	Low confidence	Qadir et al. 2013
Reduced soil compaction	1–100 million people	Low confidence	Anderson and Peters 2016
Biochar addition to soil	Range from positive impact in the tropics from biochar addition to soil to a maximum potential negative impact on >100 million people by worst-case conversion of 20% of global cropland	Low confidence	Jeffery et al. 2017; worse-case negative impacts calculated from area values in Smith 2016

Integrated response options based on land management across all/other ecosystems

FAO (2015) calculated that damage from forest fires between 2003 and 2013 impacted on a total of 49,000 km^2 of crops, with the vast majority in Latin America. Based on the world cereal yield in 2013 reported by Word Bank (2018b) (3.8 t ha^{-1}), the loss of 49,000 km^2 of crops is equivalent to 18.6 Mt yr^{-1} of cereals lost. Assuming annual grain consumption per capita to be 300 kg yr^{-1} (estimated, based on data included in FAO 2018b; FAO et al. 2018; Pradhan et al. 2013; World Bank 2018a), the loss of 18.6 Mt yr^{-1} would remove cereal crops equivalent to that consumed by 62 million people (Table 6.48).

Landslides and other natural hazards affect 1–100 million people globally, so preventing them could provide food security benefits to these people.

In terms of measures to tackle pollution, including acidification, Shindell et al. (2012) considered about 400 emission control measures to reduce ozone and black carbon (BC). This strategy increases annual crop yields by 30–135 Mt due to ozone reductions in 2030 and beyond. If annual grain consumption per capita is assumed as 300 kg yr^{-1} (estimated based on data included in FAO 2018b; FAO et al. 2018; Pradhan et al. 2013; World Bank 2018a), increase in annual crop yields by 30–135 Mt would feed 100–450 million people.

There are no global data on the impacts of management of invasive species/encroachment on food security.

Since large areas of converted coastal wetlands are used for food production (e.g., mangroves converted for aquaculture; Naylor et al. 2000), restoration of coastal wetlands could displace food production and damage local food supply, potentially leading to adverse impacts on food security. However, these effects are likely to be very small,

given that only 0.3% of human food comes from the oceans and other aquatic ecosystems (Pimentel 2006), and that the impacts could be offset by careful management, such as the careful siting of ponds within mangroves (Naylor et al. 2000) (Table 6.46).

Around 14–20% (0.56–0.80 Mkm^2) of the global 4 Mkm^2 of peatlands are used for agriculture, mostly for meadows and pasture, meaning that, if all of these peatlands were removed from production, 0.56–0.80 Mkm^2 of agricultural land would be lost. Assuming livestock production on this land (since it is mostly meadow and pasture) with a mean productivity of 9.8 kg protein ha^{-1} yr^{-1} (calculated from land footprint of beef/mutton (Clark and Tilman 2017), and average protein consumption in developing countries of 25.5 kg protein yr^{-1} (equivalent to 70 g per person per day; (FAO 2018b; OECD and FAO 2018)), this would be equivalent to 21–31 million people no longer fed from this land (Table 6.46)).

There are no global estimates on how biodiversity conservation improves nutrition (i.e., the number of nourished people). Biodiversity, and its management, is crucial for improving sustainable and diversified diets (Global Panel on Agriculture and Food Systems for Nutrition 2016). Indirectly, the loss of pollinators (due to combined causes, including the loss of habitats and flowering species) would contribute to 1.42 million additional deaths per year from non-communicable and malnutrition-related diseases, and 27.0 million lost disability-adjusted life years (DALYs) per year (Smith et al. 2015). However, at the same time, some options to preserve biodiversity, like protected areas, may potentially conflict with food production by local communities (Molotoks et al. 2017).

Table 6.48 summarises the impact on food security of response options in all/other ecosystems, with confidence estimates based on the thresholds outlined in Table 6.53 in Section 6.3.6, and indicative (not exhaustive) references upon which the evidence is based.

Table 6.48 | Effects on food security of response options in all/other ecosystems.

Integrated response option	Potential	Confidence	Citation
Fire management	~62 million people	Low confidence	FAO 2015, 2018b; FAO et al. 2018; Pradhan et al. 2013; World Bank 2018a,b
Reduced landslides and natural hazards	1–100 million people	Low confidence	Campbell 2015
Reduced pollution including acidification	Increase annual crop yields 30–135 Mt globally; feeds 100–450 million people	Low confidence	FAO 2018b; FAO et al. 2018; Pradhan et al. 2013; World Bank 2018a
Management of invasive species/encroachment	No global estimates	No evidence	
Restoration and reduced conversion of coastal wetlands	Very small negative impact but not quantified	Low confidence	
Restoration and reduced conversion of peatlands	Potential negative impact on 21–31 million people	Low confidence	Clark and Tilman 2017; FAO 2018b
Biodiversity conservation	No global estimates	No evidence	

Integrated response options based on land management specifically for CDR

The spreading of crushed minerals on land as part of enhanced weathering on nutrient-depleted soils can potentially increase crop yield by replenishing plant available silicon, potassium and other plant nutrients (Beerling et al. 2018), but there are no estimates in the literature reporting the potential magnitude of this effect on global food production.

Competition for land between bioenergy and food crops can lead to adverse side effects for food security. Many studies indicate that bioenergy could increase food prices (Calvin et al. 2014; Popp et al. 2017; Wise et al. 2009). Only three studies were found linking bioenergy to the population at risk of hunger; they estimate an increase in the population at risk of hunger of between 2 million and 150 million people (Table 6.49).

Table 6.49 summarises the impact on food security of response options specifically for CDR, with confidence estimates based on the thresholds outlined in Table 6.53 in Section 6.3.6, and indicative (not exhaustive) references upon which the evidence in based.

Table 6.49 | Effects on food security of response options specifically for CDR.

Integrated response option	Potential	Confidence	Citation
Enhanced weathering of minerals	No global estimates	No evidence	
Bioenergy and BECCS	Negative impact on up to 150 million people	Low confidence	Chapter 7; Chapter 7 SM Baldos and Hertel 2014; Fujimori et al. 2018

6.3.5.2 Integrated response options based on value chain management

In this section, the impacts on food security of integrated response options based on value chain management are assessed.

Integrated response options based on value chain management through demand management

Dietary change can free up agricultural land for additional production (Bajželj et al. 2014a; Stehfest et al. 2009; Tilman and Clark 2014) and reduce the risk of some diseases (Tilman and Clark 2014; Aleksandrowicz et al. 2016), with large positive impacts on food security (Table 6.50).

Kummu et al. (2012) estimate that an additional billion people could be fed if food waste was halved globally. This includes both post-harvest losses and retail and consumer waste. Measures such as improved food transport and distribution could also contribute to this waste reduction (Table 6.50).

While no studies quantified the effect of material substitution on food security, the effects are expected to be similar to reforestation and afforestation if the amount of material substitution leads to an increase in forest area.

Table 6.50 summarises the impact on food security of demand management options, with confidence estimates based on the thresholds outlined in Table 6.53 in Section 6.3.6, and indicative (not exhaustive) references upon which the evidence in based.

Table 6.50 | Effects on food security of demand management options.

Integrated response option	Potential	Confidence	Citation
Dietary change	821 million people	High confidence	Aleksandrowicz et al. 2016; Tilman and Clark 2014
Reduced post-harvest losses	1000 million people	Medium confidence	Kummu et al. 2012
Reduced food waste (consumer or retailer)	700–1000 million people	Medium confidence	FAO 2018b; Kummu et al. 2012
Material substitution	No global estimates	No evidence	

Integrated response options based on value chain management through supply management

Since 810 million people are undernourished (FAO 2018b), this sets the maximum number of those who could potentially benefit from sustainable sourcing or better management of supply chains. Currently, however, only 1 million people are estimated to benefit from sustainable sourcing (Tayleur et al. 2017). For the others, food price spikes affect food security and health; there are clearly documented effects of stunting among young children as a result of the 2007/2008 food supply crisis (de Brauw 2011; Arndt et al. 2016; Brinkman et al. 2009; Darnton-Hill and Cogill 2010) with a 10% increase in wasting attributed to the crisis in South Asia

(Vellakkal et al. 2015). There is conflicting evidence on the impacts of different food price stability options for supply chains, and little quantification (Byerlee et al. 2006; del Ninno et al. 2007; Alderman 2010; Braun et al. 2014). Reduction in staple food prices due to price stabilisation resulted in more expenditure on other foods and increased nutrition (e.g., oils, animal products), leading to a 10% reduction in malnutrition among children in one study (Torlesse et al. 2003). Comparison of two African countries shows that protectionist policies (food price controls) and safety nets to reduce price instability resulted in a 20% decrease in risk of malnutrition (Nandy et al. 2016). Models using policies for food aid and domestic food reserves to achieve food supply and price stability showed the most effectiveness of all options in achieving climate mitigation and food security goals (e.g., more effective than carbon taxes) as they did not exacerbate food insecurity and did not reduce ambitions for achieving temperature goals (Fujimori et al. 2019).

For urban food systems, increased food production in cities, combined with governance systems for distribution and access can improve food security, with a potential to produce 30% of food consumed in cities. The urban population in 2018 was 4.2 billion people, so 30% represents 1230 million people who could benefit in terms of food security from improved urban food systems (Table 6.51).

It is estimated that 500 million smallholder farmers depend on agricultural businesses in developing countries (World Bank 2017),

which sets the maximum number of people who could benefit from improved efficiency and sustainability of food processing, retail and agri-food industries.

Up to 2500 million people could benefit from increased energy efficiency in agriculture, based on the estimated number of people worldwide lacking access to clean energy and instead relying on biomass fuels for their household energy needs (IEA 2014).

Table 6.51 summarises the impact on food security of supply management options, with confidence estimates based on the thresholds outlined in Table 6.53 in Section 6.3.6, and indicative (not exhaustive) references upon which the evidence in based.

Table 6.51 | Effects on food security of supply management options.

Integrated response option	Potential	Confidence	Citation
Sustainable sourcing	>1 million people	Low confidence	Tayleur et al. 2017
Management of supply chains	>1 million people	Low confidence	FAO 2018b; Kummu et al. 2012
Enhanced urban food systems	Up to 1260 million people	Low confidence	Benis and Ferrão 2017; Padgham et al. 2014; Specht et al. 2014; Zeeuw and Drechsel 2015
Improved food processing and retailing	500 million people	Low confidence	World Bank 2017
Improved energy use in food systems	Up to 2500 million people	Low confidence	IEA 2014

6.3.5.3   Integrated response options based on risk management

In this section, the impacts on food security of integrated response options based on risk management are assessed.

Evidence in the USA indicates ambiguous trends between sprawl and food security: on the one hand, most urban expansion in the USA has primarily been on lands of low and moderate soil productivity with only 6% of total urban land on highly productive soil; on the other hand, highly productive soils have experienced the highest rate of conversion of any soil type (Nizeyimana et al. 2001). Specific types of agriculture are often practiced in urban-influenced fringes, such as fruits, vegetables, and poultry and eggs in the USA, the loss of which can have an impact on the types of nutritious foods available in urban areas (Francis et al. 2012). China is also concerned with food security implications of urban sprawl, and a loss of 30 Mt of grain production from 1998 to 2003 in eastern China was attributed to urbanisation (Chen 2007). However, overall global quantification has not been attempted.

Diversification is associated with increased welfare and incomes and decreased levels of poverty in several country studies (Arslan et al. 2018; Asfaw et al. 2018). These are likely to have large food security benefits (Barrett et al. 2001; Niehof 2004), but there is little global quantification.

Local seed use can provide considerable benefits for food security because of the increased ability of farmers to revive and strengthen local food systems (McMichael and Schneider 2011); studies have reported more diverse and healthy food in areas with strong food sovereignty networks (Coomes et al. 2015; Bisht et al. 2018). Women, in particular, may benefit from seed banks for low-value but nutritious crops (Patnaik et al. 2017). Many hundreds of millions

of smallholders still rely on local seeds and they provide for many hundreds of millions of consumers (Altieri et al. 2012; McGuire and Sperling 2016). Therefore, keeping their ability to do so through seed sovereignty is important. However, there may be lower food yields from local and unimproved seeds, so the overall impact of local seed use on food security is ambiguous (McGuire and Sperling 2016).

Disaster risk management approaches can have important impacts on reducing food insecurity, and current warning systems for drought and storms currently reach over 100 million people. When these early warning systems can help farmers harvest crops in advance of impending weather events, or otherwise make agricultural decisions to prepare for adverse events, there are likely to be positive impacts on food security (Fakhruddin et al. 2015). Surveys with farmers reporting food insecurity from climate impacts have indicated their strong interest in having such early warning systems (Shisanya and Mafongoya 2016). Additionally, famine early warning systems have been successful in Sahelian Africa to alert authorities of impending food shortages so that food acquisition and transportation from outside the region can begin, potentially helping millions of people (Genesio et al. 2011; Hillbruner and Moloney 2012).

Risk-sharing instruments are often aimed at sharing food supplies and reducing risk, and thus are likely to have important, but unquantified, benefits for food security. Crop insurance, in particular, has generally led to (modest) expansions in cultivated land area and increased food production (Claassen et al. 2011a; Goodwin et al. 2004).

Table 6.52 summarises the impact on food security of risk management options, with confidence estimates based on the thresholds outlined in Table 6.53 in Section 6.3.6, and indicative (not exhaustive) references upon which the evidence in based.

Table 6.52 | Effects on food security of risk management options.

Integrated response option	Potential	Confidence	Citation
Management of urban sprawl	>1 million likely	Low confidence	Bren d'Amour et al. 2016; Chen 2017
Livelihood diversification	>100 million	Low confidence	Morton 2007
Use of local seeds	>100 million	Low confidence	Altieri et al. 2012
Disaster risk management	>100 million	Medium confidence	Genesio et al. 2011; Hillbruner and Moloney 2012
Risk-sharing instruments	>1 million likely	Low confidence	Claassen et al. 2011a; Goodwin et al. 2004

6.3.6   Summarising the potential of the integrated response options across mitigation, adaptation, desertification land degradation and food security

Using the quantification provided in Tables 6.13 to 6.52, the impacts are categorised as either positive or negative, and are designated as large, moderate and small, according to the criteria given in Table 6.53.[7]

Table 6.53 | Key for criteria used to define the magnitude of the impact of each integrated response option.

	Mitigation	Adaptation	Desertification	Land degradation	Food
Large positive	More than 3 GtCO$_2$-eq yr^{-1}	Positively impacts more than around 25 million people	Positively impacts more than around 3 million km^2	Positively impacts more than around 3 million km^2	Positively impacts more than around 100 million people
Moderate positive	0.3–3 GtCO$_2$-eq	1 million to 25 million	0.5–3 million km^2	0.5–3 million km^2	1 million to 100 million
Small positive	>0	Under 1 million	>0	>0	Under 1 million
Negligible	0	No effect	No effect	No effect	No effect
Small negative	<0	Under 1 million	<0	<0	Under 1 million
Moderate negative	−0.3 to −3 GtCO$_2$-eq	1 million to 25 million	0.5 to 3 million km^2	0.5 to 3 million km^2	1 million to 100 million
Large negative	More than −3 GtCO$_2$-eq yr^{-1}	Negatively impacts more than around 25 million people	Negatively impacts more than around 3 million km^2	Negatively impacts more than around 3 million km^2	Negatively impacts more than around 100 million people

Note: All numbers are for global scale; all values are for technical potential. For mitigation, the target is set at around the level of large single mitigation measure (about 1 GtC yr^{-1} = 3.67 GtCO$_2$-eq yr^{-1}) (Pacala and Socolow 2004), with a combined target to meet 100 GtCO$_2$ in 2100, to go from baseline to 2°C (Clarke et al. 2014). For adaptation, numbers are set relative to the about 5 million lives lost per year attributable to climate change and a carbon-based economy, with 0.4 million per year attributable directly to climate change. This amounts to 100 million lives predicted to be lost between 2010 and 2030 due to climate change and a carbon-based economy (DARA 2012), with the largest category representing 25% of this total. For desertification and land degradation, categories are set relative to the 10–60 million km^2 of currently degraded land (Gibbs and Salmon 2015) with the largest category representing 30% of the lower estimate. For food security, categories are set relative to the roughly 800 million people currently undernourished (HLPE 2017) with the largest category representing around 12.5% of this total.

Tables 6.54 to 6.61 summarise the potentials of the integrated response options across mitigation, adaptation, desertification, land degradation and food security. Cell colours correspond to the large, moderate and small impact categories shown in Table 6.53.

As seen in Tables 6.54 to 6.61, three response options across the 14 for which there are data for every land challenge: *increased food productivity, agroforestry* and *increased soil organic carbon content*, deliver large benefits across all five land challenges.

A further six response options: *improved cropland management, improved grazing land management, improved livestock management, agroforestry, fire management* and *reduced post-harvest losses*, deliver either large or moderate benefits for all land challenges.

Three additional response options: *dietary change, reduced food waste* and *reduced soil salinisation*, each missing data to assess global potential for just one of the land challenges, deliver large or moderate benefits to the four challenges for which there are global data.

---

[7]   Note: 1) The response options often overlap, so are not additive. For example, increasing food productivity will involve changes to cropland, grazing land and livestock management, which in turn may include increasing soil carbon stocks. Therefore, the response options cannot be summed or regarded as entirely mutually exclusive interventions. 2) The efficacy of a response option for addressing the primary challenge for which it is implemented needs to be weighed against any co-benefits and adverse side effects for the other challenges. For example, if a response option has a major impact in addressing one challenge but results in relatively minor and manageable adverse side effects for another challenge, it may remain a powerful response option despite the adverse side effects, particularly if they can be minimised or managed. 3) Though the impacts of integrated response options have been quantified as far as possible in Section 6.3, there is no equivalence implied in terms of benefits or adverse side effects, either in number or in magnitude of the impact – that is, one benefit *does not equal* one adverse side effect. As a consequence (i) large benefits for one challenge might outweigh relatively minor adverse side effects in addressing another challenge, and (ii) some response options may deliver mostly benefits with few adverse side effects, but the benefits might be small in magnitude, that is, the response options do no harm, but present only minor co-benefits. A number of benefits and adverse side effects are context specific; the context specificity has been discussed in Section 6.2 and is further examined Section 6.4.5.1.

Eight response options: *increased food productivity, reforestation and forest restoration, afforestation, increased soil organic carbon content, enhanced mineral weathering, dietary change, reduced post-harvest losses*, and *reduced food waste*, have large mitigation potential (>3 GtCO₂e yr⁻¹) without adverse impacts on other challenges.

Sixteen response options: *increased food productivity, improved cropland management, agroforestry, agricultural diversification, forest management, increased soil organic carbon content, reduced landslides and natural hazards, restoration and reduced conversion of coastal wetlands, reduced post-harvest losses, sustainable sourcing, management of supply chains, improved food processing and retailing, improved energy use in food systems, livelihood diversification, use of local seeds,* and *disaster risk management*, have large adaptation potential at global scale (positively affecting more than 25 million people) without adverse side effects for other challenges.

Thirty-three of the 40 response options can be applied without requiring land-use change and limiting available land. A large number of response options do not require dedicated land, including several land management options, all value chain options, and all risk management options. Four options, in particular, could greatly increase competition for land if applied at scale: *afforestation, reforestation,* and land used to provide feedstock for *bioenergy (with or without BECCS)* and *biochar*, with three further options: *reduced grassland conversion to croplands, restoration and reduced conversion of peatlands,* and *restoration and reduced conversion of coastal wetlands* having smaller or variable impacts on competition for land. Other options such as *reduced deforestation and forest degradation,* restrict land conversion for other options and uses.

Some response options can be more effective when applied together – for example, *dietary change* and *waste reduction* expand the potential to apply other options by freeing as much as 25 Mkm² (4–25 Mkm² for dietary change; Alexander et al. 2016; Bajželj et al. 2014b; Stehfest et al. 2009; Tilman and Clark 2014 and 7 Mkm² for reduced food waste; Bajželj et al. 2014b).

In terms of the categories of response options, most agricultural land management response options (all except for reduced grassland conversion to cropland which potentially adversely affects food security), deliver benefits across the five land challenges (Table 6.54). Among the forest land management options, afforestation and reforestation have the potential to deliver large co-benefits across all land challenges except for food security, where these options provide a threat due to competition for land (Table 6.55). Among the soil-based response options, some global data are missing, but none except biochar shows any potential for negative impacts, with that potential negative impact arising from additional pressure on land if large quantities of biomass feedstock are required for biochar production (Table 6.56). Where global data exists, most response options in other/all ecosystems deliver benefits, except for a potential moderate negative impact on food security by restoring peatlands currently used for agriculture (Table 6.57). Of the two response options specifically targeted at CDR, there are missing data for enhanced weathering of minerals for three of the challenges, but large-scale bioenergy and BECCS show a potential large benefit for mitigation, but small to large adverse impacts on the other four land challenges (Table 6.58), mainly driven by increased pressure on land due to feedstock demand.

While data allow the impact of material substitution to be assessed only for mitigation, the three other demand-side response options: *dietary change, reduced post-harvest losses,* and *reduced food waste* provide large or moderate benefits across all challenges for which data exist (Table 6.59). Data is not available for any of the supply-side response options to assess the impact on more than three of the land challenges, but there are large to moderate benefits for all those for which data are available (Table 6.60). Data are not available to assess the impact of risk-management-based response options on all of the challenges, but there are small to large benefits for all of those for which data are available (Table 6.61).

Table 6.54 | Summary of direction and size of impact of land management options in agriculture on mitigation, adaptation, desertification, land degradation and food security.

Integrated response option	Mitigation	Adaptation	Desertification	Land degradation	Food security	Context and evidence base for magnitude of effect
Increased food productivity						These estimates assume that increased food production is implemented sustainably (e.g., through sustainable intensification: Garnett et al. 2013; Pretty et al. 2018) rather than through increasing external inputs, which can have a range of negative impacts. **Mitigation:** *Large benefits* (Table 6.13). **Adaptation:** *Large benefits* (Campbell et al. 2014) (Chapter 2 and Table 6.21). **Desertification:** *Large benefits* (Dai 20100 (Chapter 3 and Table 6.29). **Land degradation:** *Large benefits* (Clay et al. 1995) (Chapter 4 and Table 6.37). **Food security:** *Large benefits* (Godfray et al. 2010; Godfray and Garnett 2014; Tilman et al. 2011) (Chapter 5 and Table 6.45).
Improved cropland management						**Mitigation:** *Moderate benefits* by reducing GHG emissions and creating soil carbon sinks (Smith et al. 2008, 2014) (Chapter 2 and Table 6.13). **Adaptation:** *Large benefits* by improving the resilience of food crop production systems to future climate change (Porter et al. 2014) (Chapter 2 and Table 6.21). **Desertification:** *Large benefits* by improving sustainable use of land in dry areas (Bryan et al. 2009; Chen et al. 2010) (Chapter 3 and Table 6.29). **Land degradation:** *Large benefits* by forming a major component of sustainable land management (Labrière et al. 2015) (Chapter 4 and Table 6.37). **Food security:** *Large benefits* by improving agricultural productivity for food production (Porter et al. 2014) (Chapter 5 and Table 6.45).
Improved grazing land management						**Mitigation:** *Moderate benefits* by increasing soil carbon sinks and reducing GHG emissions (Herrero et al. 2016) (Chapter 2 and Table 6.13). **Adaptation:** *Moderate benefits* by improving the resilience of grazing lands to future climate change (Porter et al. 2014) (Chapter 2 and Table 6.21). **Desertification:** *Moderate benefits* by tackling overgrazing in dry areas to reduce desertification (Archer et al. 2011) (Chapter 3 and Table 6.29). **Land degradation:** *Large benefits* by optimising stocking density to reduce land degradation (Tighe et al. 2012) (Chapter 4, Table 6.37 and Table 6.45). **Food security:** *Large benefits* by improving livestock sector productivity to increase food production (Herrero et al. 2016) (Chapter 5 and Table 6.45).
Improved livestock management						**Mitigation:** *Moderate benefits* by reducing GHG emissions, particularly from enteric methane and manure management (Smith et al. 2008, 2014) (Chapter 2 and Table 6.13). **Adaptation:** *Moderate benefits* by improving resilience of livestock production systems to climate change (Porter et al. 2014) (Chapter 2 and Table 6.21). **Desertification:** *Moderate benefits* by tackling overgrazing in dry areas (Archer et al. 2011) (Chapter 3 and Table 6.29). **Land degradation:** *Large benefits* by reducing overstocking which can reduce land degradation (Tighe et al. 2012) (Chapter 4, Table 6.37 and Table 6.45). **Food security:** *Large benefits* by improving livestock sector productivity to increase food production (Herrero et al. 2016) (Chapter 5 and Table 6.45).
Agroforestry						**Mitigation:** *Large benefits* by increasing carbon sinks in vegetation and soils (Delgado 2010; Mbow et al. 2014a; Griscom et al. 2017) (Chapter 2 and Table 6.13). **Adaptation:** *Large benefits* by improving the resilience of agricultural lands to climate change (Mbow et al. 2014a) (Chapter 2 and Table 6.21). **Desertification:** *Large benefits* through, for example, providing perennial vegetation in dry areas (Nair et al. 2010; Lal 2001) (Chapter 3 and Table 6.29). **Land degradation:** *Large benefits* by stabilising soils through perennial vegetation (Narain et al. 1997; Lal 2001) (Chapter 4 and Table 6.37). **Food production:** *Large benefits* since well-planned agroforestry can enhance productivity (Bustamante et al. 2014; Sascha et al. 2017) (Chapter 5 and Table 6.45).
Agricultural diversification						Agricultural diversification is a collection of practices aimed at deriving more crops or products per unit of area (e.g., intercropping) or unit of time (e.g., double cropping, ratoon crops, etc.). **Mitigation:** *Limited benefits* (Table 6.13). **Adaptation:** *Large benefits* through improved household income (Pellegrini and Tasciotti 2014) (Table 6.21). **Desertification:** *Moderate benefits*, limited by global dryland cropped area (Table 6.29). **Land degradation:** *Large benefits* by reducing pressure on land (Lambin and Meyfroidt 2011) (Table 6.37). **Food security:** *Large benefits* for food security by provision of more diverse foods (Birthal et al. 2015; Massawe et al. 2016; Waha et al. 2018) (Chapter 5 and Table 6.45).
Reduced grassland conversion to cropland		ND				**Mitigation:** *Moderate benefits* by retaining soil carbon stocks that might otherwise be lost. Historical losses of soil carbon have been in the order of 500 GtCO₂ (Sanderman et al. 2017) (Table 6.13). Mean annual global cropland conversion rates (1961–2003) have been 0.36% per year (Krause et al. 2017), that is, around 47,000 km² yr⁻¹ – so preventing conversion could potentially save moderate emissions of CO₂. **Adaptation:** No literature (Table 6.21). **Desertification:** *Limited benefits* by shifting from annual crops to permanent vegetation cover under grass in dry areas (Table 6.29) (Chapter 3). **Land degradation:** *Limited benefits* by shifting from annual crops to permanent vegetation cover under grass (Chapter 4 and Table 6.37). **Food security:** *Moderate negative impacts*, since more land is required to produce human food from livestock products on grassland than from crops on cropland, meaning that a shift to grassland could reduce total productivity and threaten food security (Clark and Tilman 2017) (Chapter 5 and Table 6.45).
Integrated water management						**Mitigation:** *Moderate benefits* by reducing GHG emissions mainly in cropland and rice cultivation (Smith et al. 2008, 2014) (Chapter 2 and Table 6.13). **Adaptation:** *Large benefits* by improving the resilience of food crop production systems to future climate change (Porter et al. 2014) (Chapter 2 and Table 6.21). **Desertification:** *Limited benefits* by improving sustainable use of land in dry areas (Chapter 3 and Table 6.29). **Land degradation:** *Limited benefits* by forming a major component of sustainable land and water management (Chapter 4 and Table 6.37). **Food security:** *Large benefits* by improving agricultural productivity for food production (Godfray and Garnett 2014; Tilman et al. 2011) (Chapter 5 and Table 6.45).

Legend:
- ▮ Large positive
- ▮ Small positive
- ▮ Small negative
- ▮ Large negative
- ▮ Moderate positive
- ▮ Negligible/no effect
- ▮ Moderate negative
- ▮ Variable

Note: Cell colours correspond to the large, moderate and small categories shown in Table 6.53. ND = no data.

6

Table 6.55 | Summary of direction and size of impact of land management options in forests on mitigation, adaptation, desertification, land degradation and food security.

Integrated response option	Mitigation	Adaptation	Desertification	Land degradation	Food security	Context and evidence base for magnitude of effect
Forest management						**Mitigation:** *Moderate benefits* by conserving and enhancing carbon stocks in forests and long-lived products, through, for example, selective logging (Smith et al. 2014) (Table 6.14). **Adaptation:** *Large benefits*, including through improving ecosystem functionality and services, with mostly qualitative evidence at global scale and more robust estimates at regional level and local scale (Locatelli et al. 2015b) (Table 6.22). **Desertification and land degradation:** *Large benefits* by helping to stabilise land and regulate water and microclimate (Locatelli et al. 2015b) (Chapters 3 and 4, and Tables 6.30 and 6.38). **Food security:** *Moderate benefits* with mostly qualitative estimate at global level, by providing food to local communities, and diversify daily diets (Chapter 5 and Table 6.46).
Reduced deforestation and forest degradation						**Mitigation:** *Large benefits* by maintaining carbon stocks in forest ecosystems (Chapter 2 and Table 6.14). **Adaptation:** *Moderate benefits* at global scale when effect is cumulated until the end of the century; local scale, co-benefits between REDD+ and adaptation of local communities can be more substantial (Long 2013; Morita and Matsumoto 2018), even if often difficult to quantify and not explicitly acknowledged (McElwee et al. 2017a) (Table 6.22). **Desertification and land degradation:** *Large benefits* at global scale when effects are cumulated for at least 20 years, for example, through reduced soil erosion (Borrelli et al. 2017) (Tables 6.30 and 6.38). The uncertainty of these global estimates is high, while more robust qualitative and some quantitative estimates are available at regional level. **Food security:** *Small benefits*; difficult to quantify at global level (Chapter 5 and Table 6.46).
Reforestation and forest restoration						**Mitigation:** *Large benefits* by rebuilding the carbon stocks in forest ecosystems, although decreases in surface albedo can reduce the net climate benefits, particularly in areas affected by seasonal snow cover (Sonntag et al. 2016; Mahmood et al. 2014) (Chapter 2 and Table 6.14). **Adaptation:** *Large benefits* by provision of Nature's Contributions to People, including improving ecosystem functionality and services, providing microclimatic regulation for people and crops, wood and fodder as safety nets, soil erosion protection and soil fertility enhancement for agricultural resilience, coastal area protection, water and flood regulation (Locatelli et al. 2015b) (Table 6.22). **Desertification:** *Large benefits* through restoring forest ecosystems in dryland areas (Medugu et al. 2010; Salvati et al. 2014) (Chapter 3 and Table 6.30). **Land degradation:** *Large benefits* by re-establishment of perennial vegetation (Ellison et al. 2017) (Chapter 4 and Table 6.38). **Food security:** *Moderate negative impacts* due to potential competition for land for food production (Frank et al. 2017) (Chapter 5 and Table 6.46).
Afforestation						**Mitigation:** *Large benefits* for mitigation (Chapter 2 and Table 6.14), especially if it occurs in the tropics and in areas that are not significantly affected by seasonal snow cover. **Adaptation:** *Large benefits* on adaptation (Kongsager et al. 2016; Reyer et al. 2009) (Chapter 2 and Table 6.22). **Desertification:** *Large benefits* by providing perennial vegetation in dry areas to help control desertification (Medugu et al. 2010; Salvati et al. 2014) (Chapter 3 and Table 6.30). **Land degradation:** *Large benefits* by stabilising soils through perennial vegetation (Lal 2001) (Chapter 4 and Table 6.38). **Food security:** *Large negative impacts* due to competition for land for food production (Kreidenweis et al. 2016; Smith et al. 2013) (Chapter 5 and Table 6.46).

▮ Large positive   ▮ Small positive   ▮ Large negative

▮ Moderate positive   ▮ Moderate negative

Note: Cell colours correspond to the large, moderate and small categories shown in Table 6.53. ND = no data.

Table 6.56 | Summary of direction and size of impact of soil-based land management options on mitigation, adaptation, desertification, land degradation and food security.

Integrated response option	Mitigation	Adaptation	Desertification	Land degradation	Food security	Context and evidence base for magnitude of effect
Increased soil organic carbon content						**Mitigation:** *Large benefits* by creating soil carbon sinks (Table 6.15). **Adaptation:** *Large benefits* by improving resilience of food crop production systems to climate change (IPBES 2018) (Chapter 2 and Table 6.24). **Desertification:** *Large benefits* by improving soil health and sustainable use of land in dry areas (D'Odorico et al. 2013) (Chapter 3 and Table 6.31). **Land degradation:** *Large benefits* since it forms a major component of recommended practices for sustainable land management (Altieri and Nicholls 2017) (Chapter 4 and Table 6.39). **Food security:** *Large benefits* since it can increase yield and yield stability to enhance food production, though this is not always the case (Pan et al. 2009; Soussana et al. 2019; Hijbeek et al. 2017b; Schjønning et al. 2018) (Chapter 5 and Table 6.47).
Reduced soil erosion						**Mitigation:** *Large benefits or large negative impacts*, since the final fate of eroded material is still debated – for example, at the global level, it is debated whether it is a large source or a large sink (Hoffmann et al. 2013) (Chapter 2 and Table 6.15). **Adaptation:** *Large benefits* since soil erosion control prevents **desertification** (*large benefits*) and **land degradation** (*large benefits*), thereby improving the resilience of agriculture to climate change (Lal 1998; FAO and ITPS 2015) (Chapters 2, 3 and 4, and Tables 6.23, 6.30 and 6.39). **Food security:** *Large benefits* mainly through the preservation of crop productivity (Lal 1998) (Chapter 5 and Table 6.47).
Reduced soil salinisation	ND					Techniques to prevent and reverse soil salinisation include groundwater management by drainage systems and/or crop rotation and use of amendments to alleviate soil sodicity. **Mitigation:** There are no studies to quantify the global impacts (Table 6.15). **Adaptation:** *Moderate benefits* by allowing existing crop systems to be maintained, reducing the need to abandon land (Dagar et al. 2016; UNCTAD 2011) (Table 6.23). **Desertification and land degradation:** *Moderate benefits* since soil salinisation is a main driver of both desertification and land degradation (Rengasamy 2006; Dagar et al. 2016) (Chapters 3 and 4, and Tables 6.31 and 6.39). **Food security:** *Moderate benefits* by maintaining existing cropping systems and helping to close yield gaps in rainfed crops (Table 6.47).
Reduced soil compaction	ND		ND			Techniques to prevent and reverse soil compaction are based on the combination of suitable crop rotations, tillage and regulation of agricultural traffic (Hamza and Anderson 2005). **Mitigation:** The global mitigation potential has not been quantified (Chamen et al. 2015; Epron et al. 2016; Tullberg et al. 2018) (Table 6.15). **Adaptation:** *Limited benefits* by improving productivity but on relatively small global areas (Table 6.22). **Desertification:** no global data (Table 6.31). **Land degradation:** *Large benefits* since soil compaction is a main driver of land degradation (FAO and ITPS 2015) (Table 6.39). **Food security:** *Moderate benefits* by helping to close yield gaps where compaction is a limiting factor (Anderson and Peters 2016) (Table 6.47).
Biochar addition to soil		ND	ND			**Mitigation:** *Large benefits* by increasing recalcitrant carbon stocks in the soil (Smith 2016; Fuss et al. 2018; IPCC 2018) (Chapter 2 and Table 6.15). **Adaptation:** There are no global estimates of the impact of biochar on climate adaptation (Table 6.23). **Desertification:** There are no global estimates of the impact of biochar on desertification (Table 6.31). **Land degradation:** *Limited benefits* by improving the soil water-holding capacity, nutrient-use efficiency, and potentially ameliorating heavy metal pollution (Sohi 2012) (Table 6.39). **Food security:** *Limited benefits* by increasing crop yields in the tropics – though not in temperate regions (Jeffery et al. 2017) – but potentially *Large negative impacts* by creating additional pressure on land if large quantities of biomass feedstock are required for biochar production (Table 6.47).

Legend:
- Large positive
- Moderate positive
- Small positive
- Large negative
- Variable

Note: Cell colours correspond to the large, moderate and small categories shown in Table 6.53. ND = no data.

Table 6.57 | Summary of direction and size of impact of land management in all/other ecosystems on mitigation, adaptation, desertification, land degradation and food security.

Integrated response option	Mitigation	Adaptation	Desertification	Land degradation	Food security	Context and evidence base for magnitude of effect
Fire management						**Mitigation:** *Large benefits* by reduced size, severity and frequency of wildfires, thereby preventing emissions and preserving carbon stocks (Arora and Melton 2018) (Table 6.16, Chapter 2, and Cross-Chapter Box 3 in Chapter 2). **Adaptation:** *Moderate benefits* by reducing mortality attributable to landscape fire smoke exposure, fire management provides adaptation benefits (Doerr and Santín 2016; Johnston et al. 2012; Koplitz et al. 2016) (Table 6.24). **Desertification:** *Large benefits* since control of wildfires and long-term maintenance of tree stock density protects against soil erosion (Neary et al. 2009; Arora and Melton 2018) (Table 6.32). **Land degradation:** *Large benefits* by stabilising forest ecosystems (Neary et al. 2009; Arora and Melton 2018) (Table 6.40). **Food security:** *Moderate benefits* by maintaining forest food product availability and preventing fire expansion to agricultural land (FAO 2015; Keenan et al. 2015; FAO et al. 2018; Pradhan et al. 2013; World Bank 2018a, b) (Table 6.48).
Reduced landslides and natural hazards						**Mitigation:** The prevention of landslides and natural hazards benefits mitigation, but because of the limited impact on GHG emissions and eventual preservation of topsoil carbon stores, the impact is estimated to be small globally (IPCC AR5 WG2, Chapter 14) (Table 6.16). **Adaptation:** Provides structural/physical adaptations to climate change (IPCC AR5 WG2, Chapter 14) (Table 6.24). **Desertification:** Due to the small global areas affected within global drylands, the benefits for desertification control are limited (Chapter 3 and Table 6.32). **Land degradation:** Since landslides and natural hazards are among the most severe degradation processes, prevention will have a large positive impact on land degradation (FAO and ITPS 2015) (Chapter 4 and Table 6.40). **Food security:** In countries where mountain slopes are cropped for food, such as in the Pacific Islands (Campbell 2015), the management and prevention of landslides can deliver benefits for food security, though the global areas are limited (Table 6.48).
Reduced pollution including acidification						**Mitigation:** *Large benefits* since measures to reduce emissions of short-lived climate pollutants (SLCPs) can slow projected global mean warming (UNEP and WMO 2011), with early intervention providing 0.5°C cooling by 2050 (UNEP and WMO 2011) (Table 6.16). But *moderate negative impacts* are also possible since reduced reactive nitrogen deposition could decrease terrestrial carbon uptake (Table 6.16). **Adaptation:** *Moderate benefits* since controlling particulate matter (PM$_{2.5}$) and ozone improves human health (Anenberg et al. 2012) (Table 6.24). **Desertification:** *Moderate benefits* since salinisation, pollution and acidification are stressors for desertification (Oldeman et al. 1991) (Table 6.32). **Land degradation:** *Moderate benefits* since acid deposition is a significant driver of land degradation (Oldeman et al. 1991; Smith et al. 2015) (Table 6.40). **Food security:** *Large benefits* since ozone is harmful to crops, so measures to reduce air pollution would be expected to increase crop production (FAO 2018b; FAO et al. 2018; Shindell et al. 2012; World Bank 2018a) (Table 6.48).
Management of invasive species/ encroachment	ND	ND	ND	ND	ND	There is no literature that assesses the global potential of management of invasive species on **mitigation, adaptation, desertification, land degradation** or on **food security** (Tables 6.16, 6.24, 6.33, 6.40 and 6.48).
Restoration and reduced conversion of coastal wetlands						**Mitigation:** *Large benefits* since coastal wetland restoration and avoided coastal wetland impacts deliver moderate carbon sinks by 2030 (Griscom et al. 2017) (Table 6.16). **Adaptation:** *Large benefits* by providing a natural defence against coastal flooding and storm surges by dissipating wave energy, reducing erosion and by helping to stabilise shore sediments (Table 6.24). **Desertification:** There is likely *negligible impact* of coastal wetland restoration for prevention of desertification (Table 6.32). **Land degradation:** *Limited benefits* since large areas of global coastal wetlands are degraded (Lotze et al. 2006; Griscom et al. 2017) (Table 6.40). **Food security:** *Small benefits* to *small adverse impacts* since large areas of converted coastal wetlands are used for food production (e.g., mangroves converted for aquaculture), restoration could displace food production and damage local food supply, though mangrove restoration can also restore local fisheries (Naylor et al. 2000) (Table 6.48).
Restoration and reduced conversion of peatlands		ND				**Mitigation:** *Moderate benefits* since avoided peat impacts and peat restoration deliver moderate carbon sinks by 2030 (Griscom et al. 2017) (Table 6.16), though there can be increases in methane emissions after restoration (Jauhiainen et al. 2008). **Adaptation:** Likely to be benefits by regulating water flow and preventing downstream flooding (Munang et al. 2014) (Table 6.24), but the global potential has not been quantified. **Desertification:** No impact since peatlands occur in wet areas and deserts in dry areas. **Land degradation:** *Moderate benefits* since large areas of global peatlands are degraded (Griscom et al. 2017) (Table 6.40). **Food security:** *Moderate adverse impacts* since restoration of large areas of tropical peatlands and some northern peatlands that have been drained and cleared for food production, could displace food production and damage local food supply (Table 6.48).
Biodiversity conservation			ND	ND	ND	**Mitigation:** *Moderate benefits* from carbon sequestration in protected areas (Calvin et al. 2014) (Table 6.16). **Adaptation:** *Moderate benefits* – likely many millions benefit from the adaptation and resilience of local communities to climate change (Secretariat of the Convention on Biological Diversity 2008) (Table 6.24), though global potential is poorly quantified. **Desertification:** No global data (Table 6.32). **Land degradation:** No global data (Table 6.40). **Food security:** No global data (Table 6.48).

▮ Large positive	▮ Small positive	▮ Variable
▮ Moderate positive	▮ Moderate negative	▮ Negligible/no effect

Note: Cell colours correspond to the large, moderate and small categories shown in Table 6.53. ND = no data.

Table 6.58 | Summary of direction and size of impact of land management options specifically for CDR on mitigation, adaptation, desertification, land degradation and food security.

Integrated response option	Mitigation	Adaptation	Desertification	Land degradation	Food security	Context and evidence base for magnitude of effect
Enhanced weathering of minerals		ND	ND		ND	**Mitigation:** *Moderate to large benefits* by removing atmospheric CO_2 (Table 6.17; Lenton 2010; Smith et al. 2016a; Taylor et al. 2016). **Adaptation:** There is no literature to assess the global impacts of enhanced mineral weathering on adaptation (Table 6.25) nor on **desertification** (Table 6.33). **Land degradation:** *Limited benefits* expected since ground minerals can increase pH where acidification is the driver of degradation (Table 6.41; Taylor et al. 2016). **Food security:** Though there may be co-benefits for food production (Beerling et al. 2018), these have not been quantified globally (Table 6.49).
Bioenergy and BECCS						**Mitigation:** *Large benefits* of large-scale bioenergy and BECCS by potential to remove large quantities of CO_2 from the atmosphere (Table 6.17). **Adaptation:** *Limited adverse impacts* of large-scale bioenergy and BECCS by increasing pressure on land (Table 6.25). **Desertification:** Up to 15 million km^2 of additional land is required in 2100 in 2°C scenarios, which will increase pressure for desertification and land degradation (Sections 6.3.3.1 and 6.3.4.1). This defines the maximum area potentially impacted, though the actual area affected by this additional pressure is not easily quantified. **Land degradation:** Up to 15 million km^2 of additional land is required in 2100 in 2°C scenarios, which will increase pressure for desertification and land degradation (Sections 6.3.3.1; 6.3.4.1). This defines the maximum area potentially impacted, though the actual area affected by this additional pressure is not easily quantified. **Food security:** *Large adverse impacts* of large-scale bioenergy and BECCS through increased competition for land for food (Table 6.49). These potentials and effects assume large areas of bioenergy crops, resulting in large mitigation potentials (i.e., >3 $GtCO_2$ yr^{-1}). The sign and magnitude of the effects of bioenergy and BECCS depends on the scale of deployment, the type of bioenergy feedstock, which other response options are included, and where bioenergy is grown (including prior land use and indirect land-use change emissions). For example, limiting bioenergy production to marginal lands or abandoned cropland would have negligible effects on biodiversity, food security, and potentially small co-benefits for land degradation; however, the benefits for mitigation would also be smaller (Cross-Chapter Box 7 in this chapter, and Table 6.13).

■ Large positive   ■ Small positive   ■ Small negative   ■ Large negative

Note: Cell colours correspond to the large, moderate and small categories shown in Table 6.53. ND = no data.

Table 6.59 | Summary of direction and size of impact of demand management options on mitigation, adaptation, desertification, land degradation and food security.

Integrated response option	Mitigation	Adaptation	Desertification	Land degradation	Food security	Context and evidence base for magnitude of effect
Dietary change		ND				**Mitigation:** *Large benefits* for mitigation by greatly reducing GHG emissions (Chapter 5 and Table 6.18). **Adaptation:** While it would be expected to help with adaptation by reducing agricultural land area, there are no studies providing global quantifications (Table 6.26). **Desertification:** Potential *moderate benefits* by decreasing pressure on land – restricted by relatively limited global area (Table 6.34). **Land degradation:** *Large benefits* by decreasing pressure on land (Table 6.42). **Food security:** Large benefits by decreasing competition for land, allowing more food to be produced from less land (Table 6.50).
Reduced post-harvest losses						**Mitigation:** *Large benefits* by reducing food sector GHG emissions and reducing the area required to produce the same quantity of food (Table 6.18), though increased use of refrigeration could increase emissions from energy use. **Adaptation:** *Large benefits* by reducing pressure on land (Table 6.26). **Desertification and land degradation:** *Moderate benefits* for both by reducing pressure on land (Table 6.34 and Table 6.42). **Food security:** *Large benefits* since most of the food wasted in developing countries arises from post-harvest losses (Ritzema et al. 2017) (Chapter 5 and Table 6.50).
Reduced food waste (consumer or retailer)		ND				**Mitigation:** *Large benefits* by reducing food sector GHG emissions and reducing the area required to produce the same quantity of food (Table 6.18). **Adaptation:** While it would be expected to help with adaptation by reducing agricultural land area, there are no studies quantifying global adaptation impacts (Table 6.26). **Desertification:** *Moderate benefits* by reducing pressure on land (Table 6.34). **Land degradation:** *Large benefits* by reducing pressure on land (Table 6.42). **Food security:** *Large benefits* since 30% of all food produced globally is wasted (Kummu et al. 2012) (Table 6.50).
Material substitution		ND	ND	ND	ND	**Mitigation:** *Moderate benefits* through long-lived carbon storage, and by substitution of materials with higher embedded GHG emissions (Table 6.18). No global studies available to assess the quantitative impact on **adaptation, desertification, land degradation** or **food security** (Tables 6.26, 6.34, 6.42 and 6.50).

■ Large positive   ■ Moderate positive

Note: Cell colours correspond to the large, moderate and small categories shown in Table 6.53. ND = no data.

Table 6.60 | Summary of direction and size of impact of supply management options on mitigation, adaptation, desertification, land degradation and food security.

Integrated response option	Mitigation	Adaptation	Desertification	Land degradation	Food security	Context and evidence base for magnitude of effect
Sustainable sourcing	ND		ND			**Mitigation:** No studies available to assess the global impact (Table 6.19). **Adaptation:** *Moderate benefits* by diversifying and increasing flexibility in the food system to climate stressors and shocks while simultaneously creating economic alternatives for the poor (thereby strengthening adaptive capacity) and lowering expenditures of food processors and retailers by reducing losses (Muller et al. 2017) (Chapter 5 and Table 6.27). **Desertification:** No studies available to assess the global impact (Table 6.35 and Table 6.43). **Land degradation:** Potentially *large benefits*, as over 4 Mkm² is currently certified for sustainable forest production, which could increase in future (Table 6.44). **Food security:** *Moderate benefits* by diversifying markets and developing value-added products in the food supply system, by increasing its economic performance and revenues to local farmers (Reidsma et al. 2010), by strengthening the capacity of food production chains to adapt to future markets and to improve income of smallholder farmers (Murthy and Madhava Naidu 2012) (Chapter 5 and Table 6.51). It may also provide more direct links between producers and consumers.
Management of supply chains	ND		ND	ND		**Mitigation:** There are no studies assessing the mitigation potential globally (Table 6.19). **Adaptation:** *Large benefits* by improving resilience to price increases or reducing volatility of production (Fafchamps et al. 1998; Haggblade et al. 2017) (Table 6.27). **Desertification** and **land degradation:** No studies assessing global potential (Tables 6.35 and 6.43). **Food security:** *Moderate benefits* through helping to manage food price increases and volatility (Vellakkal et al. 2015; Arndt et al. 2016) (Table 6.51).
Enhanced urban food systems	ND	ND	ND	ND		There are no studies that assess the global potential to contribute to **mitigation, adaptation, desertification** or **land degradation** (Tables 6.19, 6.27, 6.35 and 6.43). **Food security:** *Large benefits* by increasing food access to urban dwellers and shortening of supply chains (Chappell et al. 2016) (Chapter 5 and Table 6.51).
Improved food processing and retailing			ND	ND		**Mitigation:** *Moderate benefits* through reduced energy consumption, climate-friendly foods and reduced GHG emissions from transportation (Avetisyan et al. 2014), waste (Porter et al. 2016), and energy use (Mohammadi et al. 2014; Song et al. 2017) (Table 6.19). **Adaptation:** Large benefits among poor farmers through reduced costs and improved resilience (Table 6.27). **Desertification** and **land degradation:** There are no studies assessing global potential (Tables 6.35 and Table 6.43). **Food security:** *Large benefits* by supporting healthier diets and reducing food loss and waste (Garnett 2011) (Chapter 5 and Table 6.51).
Improved energy use in food systems			ND	ND		**Mitigation:** *Moderate benefits* by reducing GHG emissions through decreasing use of fossil fuels and energy-intensive products, though the emission reduction is not accounted for in the agriculture, forestry and other land-use (AFOLU) sector (Smith et al. 2014; IPCC AR5 WG3 Chapter 11) (Table 6.19). **Adaptation:** *Large benefits* for small farmers by reducing costs and increasing their resilience to climate change (Table 6.27). **Desertification** and **land degradation:** There are no studies assessing global potential (Tables 6.35 and 6.43). **Food security:** *Large benefits*, largely by improving efficiency for 2.5 million people still using traditional biomass for energy (Chapter 5 and Table 6.51).

▮ Large positive   ▮ Moderate positive

Note: Cell colours correspond to the large, moderate and small categories shown in Table 6.53. ND = no data.

Table 6.61 | Summary of direction and size of impact of risk management options on mitigation, adaptation, desertification, land degradation and food security.

Integrated response option	Mitigation	Adaptation	Desertification	Land degradation	Food security	Context and evidence base for magnitude of effect
Management of urban sprawl	ND					**Mitigation:** There are no studies assessing the global potential (Table 6.20). **Adaptation:** *Moderate benefits* – though poorly quantified globally, likely to affect many millions of people (Table 6.28). **Desertification:** *Limited benefits* – though poorly quantified globally, 5000 km² is at risk from urban sprawl in Spain alone (Table 6.36). **Land degradation:** *Limited benefits* – though poorly quantified globally, urban sprawl effects millions of ha of land (Table 6.44). **Food security:** *Moderate benefits* estimated from impacts on food supply in models (Bren d'Amour et al. 2016) (Table 6.52).
Livelihood diversification	ND					**Mitigation:** There are no studies assessing the global potential (Table 6.20). **Adaptation:** *Large benefits* through helping households to buffer income fluctuations and providing a broader range of options for the future (Table 6.28; Ahmed and Stepp 2016; Thornton and Herrero 2014). **Desertification:** There are no studies assessing the global potential, although there are anecdotal reports of *limited benefits* from improved land management resulting from diversification (Batterbury 2001; Herrmann and Hutchinson 2005; Stringer et al. 2009) (Table 6.36). **Land degradation:** *Limited benefits*, for example, improved land-use mosaics (Palacios et al. 2013), larger-scale adoption in China's Sloping Land Conversion Program to diversify income and reduce degradation has impacted on 0.1 Mkm² (Liu and Lan 2015) (Table 6.44). **Food security:** *Large benefits* since many of the world's 700 million smallholders practice diversification, helping to provide economic access to food (Morton 2007) (Table 6.52).
Use of local seeds	ND		ND	ND		**Mitigation:** There are no studies assessing the global potential (Table 6.19). **Adaptation:** *Large benefits* given that 60 to 100% of seeds used in various countries of the global South are likely local farmer-bred (non-commercial) seed, and moving to the use of commercial seed would increase costs considerably for these farmers. Seed networks and banks protect local agrobiodiversity and landraces, which are important to facilitate adaptation, and can provide crucial lifelines when crop harvests fail (Louwaars 2002; Howard 2015; Coomes et al. 2015; Van Niekerk and Wynberg 2017; Vasconcelos et al. 2013; Reisman 2017) (Table 6.28). **Desertification and land degradation:** There are no studies assessing global potential (Tables 6.36 and Table 6.44). **Food security:** *Large benefits* since local seeds increase the ability of farmers to revive and strengthen local food systems; several studies have reported more diverse and healthy food in areas with strong food sovereignty networks (Coomes et al. 2015; Bisht et al. 2018) (Table 6.52).
Disaster risk management	ND		ND	ND		**Mitigation:** There are no studies to assess the global mitigation potential of different Disaster Risk Management (DRM) approaches (Table 6.19). **Adaptation:** *Large benefits* due to widespread use of early warning systems that reach hundreds of millions (Hillbruner and Moloney 2012; Mahmud and Prowse 2012; Birkmann et al. 2015) (Table 6.28). **Desertification and land degradation.** There are no studies assessing the global potential (Tables 6.36 and Table 6.44). **Food security:** *Moderate benefits* by helping farmers to harvest crops in advance of impending weather events, or otherwise to make agricultural decisions to prepare for adverse events (Fakhruddin et al. 2015; Genesio et al. 2011; Hillbruner and Moloney 2012) (Table 6.52).
Risk-sharing instruments			ND			**Mitigation:** *Variable impacts* – poor global coverage in the literature, though studies from the US suggest a small increase in emissions from crop insurance and likely benefits from other risk-sharing instruments (Table 6.20). **Adaptation:** *Moderate benefits* by buffering and transferring weather risk, saving farmers the cost of crop losses. However, overly subsidised insurance can undermine the market's role in pricing risks and thus depress more rapid adaptation strategies (Meze-Hausken et al. 2009; Skees and Collier 2012; Jaworski 2016) (Table 6.28). **Desertification:** The impacts of risk-sharing globally have not been quantified (Table 6.36). **Land degradation:** *Variable impacts* as evidence suggests that subsidised insurance in particular can increase crop production in marginal lands, and reforming this would lead to benefits (Table 6.44). **Food security:** *Small to moderate benefits* for food security, as risk-sharing often promotes food-supply sharing (Table 6.52).

■ Large positive   ■ Moderate positive   ▦ Small positive   ■ Variable

Note: Cell colours correspond to the large, moderate and small categories shown in Table 6.53. ND = no data.

6.4   Managing interactions and interlinkages

Having assessed the potential of each response option for contributing to addressing mitigation, adaptation, desertification, land degradation and food security in Section 6.3, this section assesses the feasibility of each response option with respect to cost, barriers, and issues of saturation and reversibility (Section 6.4.1), before assessing the sensitivity of the response options to future climate change (Section 6.4.2) and examining the contribution of each response option to ecosystem services (classified according to Nature's Contribution to People (IPBES 2018), and to sustainable development (assessed against the UN SDGs) (6.4.3). Section 6.4.4 examines opportunities for implementation of integrated response options, paving the way to potential policies examined in Chapter 7, before the consequences of delayed action are assessed in Section 6.4.5.

6.4.1   Feasibility of the integrated response options with respect to costs, barriers, saturation and reversibility

For each of the response options, Tables 6.62–6.69 summarise the feasibility with respect to saturation and reversibility and cost, technological, institutional, socio-cultural and environmental and geophysical barriers (the same barrier categories used in SR15).

Many land management options face issues of saturation and reversibility; however, these are not of concern for the value chain and risk management options. Reversibility is an issue for all options that increase terrestrial carbon stock, either through increased soil carbon or changes in land cover (e.g., reforestation, afforestation), since future changes in climate or land cover could result in reduced carbon storage (Smith 2013). In addition, the benefits of options that improve land management (e.g., improved cropland management, improved grazing management) will cease if the practice is halted, reversing any potential benefits.

The cost of the response options varies substantially, with some options having relatively low costs (e.g., the cost of agroforestry is less than 10 USD tCO_2e^{-1}) while others have much higher costs (e.g., the cost of BECCS could be as much as 250 USD tCO_2e^{-1}). In addition to cost, other economic barriers may prevent implementation; for example, agroforestry is a low-cost option (Smith et al. 2014), but lack of reliable financial support could be a barrier (Hernandez-Morcillo et al. 2018). Additionally, there are a number of reasons why even no-cost options are not adopted, including risk aversion, lack of information, market structure, externalities, and policies (Jaffe 2019).

Some of the response options have technological barriers that may limit their wide-scale application in the near term. For example, BECCS has only been implemented at small-scale demonstration facilities (Kemper 2015); challenges exist with upscaling these options to the levels discussed in this chapter.

Many response options have institutional and socio-cultural barriers. Institutional barriers include governance, financial incentives and financial resources. For example, the management of supply chains includes challenges related to political will within trade regimes, economic laissez-faire policies that discourage interventions in markets, and the difficulties of coordination across economic sectors (Poulton et al. 2006; Cohen et al. 2009; Gilbert 2012). Implementation of other options, for example, BECCS, is limited by the absence of financial incentives.

Options like dietary change face socio-cultural barriers; while diets have changed in the past, they are deeply culturally embedded and behaviour change is extremely difficult to effect, even when health benefits are well known (Macdiarmid et al. 2018). For some options, the specific barrier is dependent on the region. For example, barriers to reducing food waste in industrialised countries include inconvenience, lack of financial incentives, lack of public awareness, and low prioritisation (Kummu et al. 2012; Graham-Rowe et al. 2014). Barriers in developing countries include reliability of transportation networks, market reliability, education, technology, capacity, and infrastructure (Kummu et al. 2012).

Table 6.62 | Feasibility of land management response options in agriculture, considering cost, technological, institutional, socio-cultural and environmental and geophysical barriers and saturation and reversibility. See also Appendix.

Response option	Saturation	Reversibility	Cost	Technological	Institutional	Socio-cultural	Environmental and geophysical	Context and sources
Increased food productivity								**Biophysical:** Only if limited by climatic and environmental factors. (Barnes and Thomson 2014; Martin et al. 2015a; Olesen and Bindi 2002; Pretty and Bharucha 2014; Schut et al. 2016.)
Improved cropland management								**Institutional:** Only in some regions (e.g., poor sustainability frameworks). (Bryan et al. 2009; Bustamante et al. 2014; Madlener et al. 2006; Reichardt et al. 2009; Roesch-McNally et al. 2017; Singh and Verma 2007; Smith et al. 2008, 2014.)
Improved grazing land management								**Institutional:** Only in some regions (e.g., need for extension services). (Herrero et al. 2016; McKinsey and Company 2009; Ndoro et al. 2014; Singh and Verma 2007; Smith et al. 2008, 2015.)
Improved livestock management								**Economic:** Improved productivity is cost negative, but others (e.g., dietary additives) are expensive. **Institutional:** Only in some regions (e.g., need for extension services). (Beauchemin et al. 2008; Herrero et al. 2016; McKinsey and Company 2009; Rojas-Downing et al. 2017; Smith et al. 2008; Thornton et al. 2009; Ndoro et al. 2014.)
Agroforestry								**Economic:** Low cost but may lack reliable financial support. **Institutional:** only in some regions (e.g., seed availability). (Lillesø et al. 2011; Meijer et al. 2015; Sileshi et al. 2008; Smith et al. 2007, 2014.)
Agricultural diversification								More support from extension services, access to inputs and markets, economic incentives for producing a certain crop or livestock product, research and investments focused on adapted varieties and climatic resilient systems, a combination of agricultural and non-agricultural activities (e.g., off-farm jobs) are all important interventions aimed at overcoming barriers to agricultural diversification. (Ahmed and Stepp 2016; Barnes et al. 2015; Barnett and Palutikof 2015; Martin and Lorenzen 2016; Roesch-McNally et al. 2016; Waha et al. 2018.)
Reduced grassland conversion to cropland								**Economic:** Avoiding conversion is low cost, but there may be significant opportunity costs associated with foregone production of crops. **Institutional:** only in some regions (e.g., poor governance to prevent conversion.)
Integrated water management								**Institutional:** Effective implementation is dependent on the adoption of a combination of 'hard', infrastructural, and 'soft' institutional measures. **Socio-cultural:** Education can be a barrier and some strategies (e.g., site-specific water management, drip irrigation) can be expensive. Cultural/behavioural barriers are likely to be small. (Dresner et al. 2015; Erwin 2009; Lotze et al. 2006; Thornton et al. 2009.)

Saturation and reversibility

▮ Not important

▦ A concern

Cost

▮ Low cost (<10 USD tCO$_2$e^{-1} or <20 USD ha^{-1})

▦ Medium cost (10–100 USD tCO$_2$e^{-1} or <20-200 USD ha^{-1})

▦ High cost (>100 USD tCO$_2$e^{-1} or 200 USD ha^{-1})

Technological, institutional, socio-cultural and environmental and geophysical barriers

▮ High current feasibility (no barriers)

▦ Medium current feasibility (moderate barriers)

▦ Low current feasibility (large barriers)

▮ Variable barriers

Note: The cost thresholds in USD tCO$_2$e^{-1} are from Griscom et al. (2017); thresholds in USD ha^{-1} are chosen to be comparable, but precise conversions will depend on the response option.

6

Table 6.63 | Feasibility of land management response options in forests, considering cost, technological, institutional, socio-cultural and environmental and geophysical barriers and saturation and reversibility. See also Appendix.

Response option	Saturation	Reversibility	Cost	Technological	Institutional	Socio-cultural	Environmental and geophysical	Context and sources
Forest management								Seidl et al. 2014
Reduced deforestation and forest degradation								**Economic:** Requires transaction and administration costs Busch and Engelmann 2017; Kindermann et al. 2008; Overmars et al. 2014
Reforestation and forest restoration								Strengers et al. 2008
Afforestation								Medugu et al. 2010; Kreidenweis et al. 2016

Note: See note for Table 6.62.

Table 6.64 | Feasibility of land management response options for soils, considering cost, technological, institutional, socio-cultural and environmental and geophysical barriers and saturation and reversibility. See also Appendix.

Response option	Saturation	Reversibility	Cost	Technological	Institutional	Socio-cultural	Environmental and geophysical	Context and sources
Increased soil organic carbon content								**Institutional:** Only in some regions (e.g., lack of institutional capacity). (Smith et al. 2008; McKinsey and Company 2009; Baveye et al. 2018; Bustamante et al. 2014; Reichardt et al. 2009; Smith 2004; Smith et al. 2007; Wollenberg et al. 2016).
Reduced soil erosion								Haregeweyn et al. 2015
Reduced soil salinisation								Barriers depend on how salinisation and sodification are implemented. (Bhattacharyya et al. 2015; CGIAR 2016; Dagar et al. 2016; Evans and Sadler 2008; Greene et al. 2016; Machado and Serralheiro 2017.)
Reduced soil compaction								Antille et al. 2016; Chamen et al. 2015
Biochar addition to soil								Saturation and reversibility issues lower than for soil organic carbon. **Economic:** In general, biochar has high costs. However, a small amount of biochar potential could be available at negative cost, and some at low cost, depending on markets for the biochar as a soil amendment. **Institutional:** Only in some regions (e.g., lack of quality standards). (Dickinson et al. 2014; Guo et al. 2016; Meyer et al. 2011; Shackley et al. 2011; Woolf et al. 2010) (Chapter 4.)

Saturation and reversibility
- Not important
- A concern

Cost
- Low cost (<10 USD tCO$_2$e^{-1} or <20 USD ha^{-1})
- Medium cost (10–100 USD tCO$_2$e^{-1} or <20-200 USD ha^{-1})
- High cost (>100 USD tCO$_2$e^{-1} or 200 USD ha^{-1})

Technological, institutional, socio-cultural and environmental and geophysical barriers
- High current feasibility (no barriers)
- Medium current feasibility (moderate barriers)
- Low current feasibility (large barriers)
- Variable barriers

Note: See note for Table 6.62.

Table 6.65 | Feasibility of land management response options in any/other ecosystems, considering cost, technological, institutional, socio-cultural and environmental and geophysical barriers and saturation and reversibility. See also Appendix.

Response option	Saturation	Reversibility	Cost	Technological	Institutional	Socio-cultural	Environmental and geophysical	Context and sources
Fire management								**Economic:** The cost of its implementation is moderate, since it requires constant maintenance, and can be excessive for some local communities. (Freeman et al. 2017; Hurteau et al. 2014; North et al. 2015.)
Reduced landslides and natural hazards								Gill and Malamud 2017; Maes et al. 2017; Noble et al. 2014
Reduced pollution including acidification								Begum et al. 2011; Shah et al. 2018; Yamineva and Romppanen 2017; WMO 2015
Management of invasive species/ encroachment								**Technological:** In the case of natural enemies. **Socio-cultural:** Education can be a barrier, where populations are unaware of the damage caused by the invasive species, but cultural/ behavioural barriers are likely to be small. **Institutional:** Where agricultural extension and advice services are poorly developed. **Source:** Dresner et al. 2015
Restoration and reduced conversion of coastal wetlands								**Economic:** Can be cost-effective at scale. **Institutional:** Only in some regions (e.g., poor governance of wetland use). **Socio-cultural:** Educational barriers (e.g., lack of knowledge of impact of wetland conversion), though cultural/behavioural barriers are likely to be small. (Erwin 2009; Lotze et al. 2006.)
Restoration and reduced conversion of peatlands								**Institutional:** Only in some regions (e.g., lack of inputs). (Bonn et al. 2014; Worrall et al. 2009.)
Biodiversity conservation								**Economic:** While protected areas and other forms of biodiversity conservation can be cost-effective, they are often underfunded relative to needs. **Institutional:** There have been challenges in getting systematic conservation planning to happen, due to institutional fragmentation and overlapping mandates. **Socio-cultural:** Despite the fact that biodiversity conservation may provide co-benefits, such as water or carbon protection, local populations often have had social and cultural conflicts with protected areas and other forms of exclusionary biodiversity conservation that are imposed in a top-down fashion or which restrict livelihood options. (Emerton et al. 2006; Hill et al. 2015; Langford et al. 2011; Larsen et al. 2012; Schleicher 2018; Wei et al. 2018; Wilkie et al. 2001.)

Note: See note for Table 6.62.

Table 6.66 | Feasibility of land management response options specifically for carbon dioxide removal (CDR), considering cost, technological, institutional, socio-cultural and environmental and geophysical barriers and saturation and reversibility. See also Appendix.

Response option	Saturation	Reversibility	Cost	Technological	Institutional	Socio-cultural	Environmental and geophysical	Context and sources
Enhanced weathering of minerals								Permanence not an issue on the decadal timescales. **Institutional:** Only in some regions (e.g., lack of infrastructure for this new technology). **Socio-cultural:** Could occur in some regions, for example, due to minerals lying under undisturbed natural areas where mining might generate public acceptance issues. (Renforth et al. 2012; Smith et al. 2016a; Taylor et al. 2016.)
Bioenergy and BECCS								**Economic:** While most estimates indicate the cost of BECCS as less than 200 USD tCO$_2^{-1}$, there is significant uncertainty. **Technological:** While there are a few small BECCS demonstration facilities, BECCS has not been implemented at scale. (IPCC 2018; Chapter 7; Kemper 2015; Sanchez and Kammen 2016; Vaughan and Gough 2016.)

Note: See note for Table 6.62.

6

Table 6.67 | Feasibility of demand management response options, considering economic, technological, institutional, socio-cultural and environmental and geophysical barriers and saturation and reversibility. See also Appendix.

Response option	Saturation	Reversibility	Cost	Technological	Institutional	Socio-cultural	Environmental and geophysical	Context and sources
Dietary change								**Institutional:** Only in some regions (e.g., poorly developed dietary health advice). (Hearn et al. 1998; Lock et al. 2005; Macdiarmid et al. 2018; Wardle et al. 2000).
Reduced post-harvest losses								
Reduced food waste (consumer or retailer)								Specific barriers differ between developed and developing countries. (Diaz-Ruiz et al. 2018; Graham-Rowe et al. 2014; Kummu et al. 2012.)
Material substitution								Gustavsson et al. 2006; Ramage et al. 2017

Note: See note for Table 6.62.

Table 6.68 | Feasibility of supply management response options, considering cost, technological, institutional, socio-cultural and environmental and geophysical barriers and saturation and reversibility. See also Appendix.

Response option	Saturation	Reversibility	Cost	Technological	Institutional	Socio-cultural	Environmental and geophysical	Context and sources
Sustainable sourcing								**Economic:** The cost of certification and sustainable sourcing can lead to higher production costs. **Institutional:** There are some barriers to adopting sustainable sourcing in terms of getting governments on board with market-based policies. **Socio-cultural:** Barriers include consumers unfamiliar with sustainably sourced goods. (Capone et al. 2014; Ingram et al. 2016.)
Management of supply chains								**Economic:** Supply chain management and management of price volatility faces challenges from businesses in terms of economic costs of change. **Technological:** Barriers like supply chain tracking. **Institutional:** Barriers like political will against government action in markets. (Cohen et al. 2009; Gilbert 2012; Poulton et al. 2006.)
Enhanced urban food systems								
Improved food processing and retailing								**Economic:** The implementation of strategies to improve the efficiency and sustainability of retail and agri-food industries can be expensive. **Institutional:** Successful implementation is dependent on organisational capacity, the agility and flexibility of business strategies, the strengthening of public-private policies and effectiveness of supply-chain governance.
Improved energy use in food systems								Baudron et al. 2015; Vlontzos et al. 2014

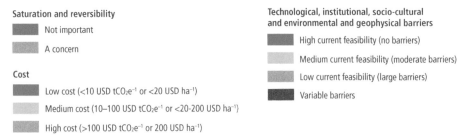

Saturation and reversibility
- Not important
- A concern

Cost
- Low cost (<10 USD tCO₂e⁻¹ or <20 USD ha⁻¹)
- Medium cost (10–100 USD tCO₂e⁻¹ or <20-200 USD ha⁻¹)
- High cost (>100 USD tCO₂e⁻¹ or 200 USD ha⁻¹)

Technological, institutional, socio-cultural and environmental and geophysical barriers
- High current feasibility (no barriers)
- Medium current feasibility (moderate barriers)
- Low current feasibility (large barriers)
- Variable barriers

Note: See note for Table 6.62.

Table 6.69 | Feasibility of risk management response options, considering cost, technological, institutional, socio-cultural and environmental and geophysical barriers and saturation and reversibility. See also Appendix.

Response option	Saturation	Reversibility	Cost	Technological	Institutional	Socio-cultural	Environmental and geophysical	Context and sources
Management of urban sprawl								There are economic and political forces that benefit from less-regulated urban development. (Tan et al. 2009.)
Livelihood diversification								**Economic:** Expanded diversification can cost additional financial resources. **Socio-cultural:** Problems with adoption of new or unfamiliar crops and livelihoods. (Ahmed and Stepp 2016; Berman et al. 2012; Ngigi et al. 2017.)
Use of local seeds								**Economic:** Local seeds are highly cost effective, and do not require new technology. **Institutional:** Barriers from agronomy departments and businesses promoting commercial seeds. **Socio-cultural:** Preferences for some non-local seed sourced crops. (Reisman 2017; Timmermann and Robaey 2016.)
Disaster risk management								**Economic:** Disaster risk management (DRM) systems can be initially costly, but usually pay for themselves over time. **Institutional:** Some barriers in terms of getting initial support behind new systems. (Birkmann et al. 2015; Hallegatte 2012.)
Risk-sharing instruments								There are few barriers to risk-sharing instruments, as they are often low cost and low technology. **Socio-cultural:** Some barriers to instruments like crop insurance, which some farmers in developing countries are not familiar with. (Goodwin and Smith 2013.)

Note: See note for Table 6.62.

6.4.2 Sensitivity of the integrated response options to climate change impacts

With continued increases in warming, there are risks to the efficacy of some of the response options due to future climate change impacts, such as increased climate variability and extreme events. While many of the response options can help increase capacity to deliver adaptation benefits (Section 6.3.2), beyond certain thresholds of climate impacts they may be less effective or increasingly risky options. This requires that some response options need to anticipate these climate impacts in their implementation. We outline some of these impacts below.

Agriculture response options: Increased food productivity as a response option is highly sensitive to climate change impacts. Chapter 5 (Section 5.2.3.1) notes that global mean yields of some crops (maize and soybean) decrease with warming, while others (rice and wheat) increase with warming, up to a threshold of 3°C. Similarly, improved cropland management response options that rely on crop diversification or improved varieties may face challenges in efficacy from production declines. Improved grazing land management may continue to be feasible as a response option in the future under climate change in northern regions, but will likely become more difficult in tropical regions and Australia as temperature rises will reduce the carrying capacity of lands (Nardone et al. 2010) (Section 5.2.3.2). Improved livestock management also faces numerous challenges, particularly related to stresses on animals from temperatures, water, and diseases; overall, livestock numbers are projected to decline 7.5–9.6% by 2050 (Rivera-Ferre et al. 2016; Boone et al. 2018) (Section 5.2.3.2). Pastoralists may also be less likely to implement improved measures due to other risks and vulnerabilities under climate change (Thornton et al. 2009).

The impact of climate change on agroforestry is more difficult to model than single crops in process-based crop models, as agroforestry systems are far more complex (Luedeling et al. 2014); thus, it is unknown how the efficacy of this response option might be impacted. Agricultural diversification has been promoted as an adaptive strategy to climate impacts, given that diversity is known to increase resiliency of agricultural and natural systems, such as in resistance to increased pests or diseases; it can also provide diversified income portfolios when some crops may become sensitive to climate events (Bradshaw et al. 2004; Lin 2011). Diversified farms are expected to increase in Africa by 2060 as specialised farms with single crops face challenges under climate change (Seo 2010). However, it is not known if these options and advantages of diversification have a temperature threshold beyond which they are less effective.

Reduced grassland conversion is not likely to be affected as a response option *per se* since it is directed at conserving natural grassland areas, but these areas may face increased pressures for conversion if farmers experience crop failures under climate change and need to expand the cultivated area holdings to make up for losses. Lobell et al. (2013) have estimated the impacts of investment decisions to adapt to the effects of climate change on crop yields to 2050 and find that cropland will expand over 23% more land area (over 3 Mkm^2), mostly in Latin America and Sub-Saharan Africa.

Integrated water management to improve water availability and reliability of water for agricultural production is likely to become more challenging in future scenarios of water declines, which are likely to be regionally uneven (Sections 2.6 and 6.4.4).

Forest response options: The availability of forest management as a response option can be impacted on by climate-induced changes, including increased diseases, pests and fires (Dale et al. 2001; Logan et al. 2003) (Section 4.5.1.2). These impacts will affect reforestation and afforestation response options as well. Locatelli et al. (2015a) note

6

623

that climate change will influence seedling establishment, tree growth and mortality, and the presence of invasive species and/or pests; these can be buffered with modified silvicultural practices, including species selection (Pawson et al. 2013). Climate change can also alter the sink capacity for vegetation carbon sequestration, reducing the potential for reducing emissions from deforestation and forest degradation (REDD), reforestation and afforestation (Bonan 2008; Malhi et al. 2002).

Soil management: Climate change can alter the sink capacity for soil carbon sequestration, reducing the potential for increased soil organic carbon as an option. Projected climate change can reduce soil resilience to extreme weather, pests and biological invasion, environmental pollutants and other pressures, making reduced soil erosion and reduced soil compaction as response options harder to achieve (Smith et al. 2015). Climate change will likely increase demand for irrigation in dryland areas, which can increase risks of salinisation, diminishing the effectiveness of this response (Smith et al. 2015). Biochar additions to soil may be affected by future climatic changes, such as rising soil temperatures, but little is known, given that most research on the subject is from laboratory and not *in situ* field experiments. There are also wide estimates of the stability and residence times of biochar from this literature (Gurwick et al. 2013).

Other ecosystem management: Fire management is likely to become more challenging in a changing climate; some studies suggest an 50% increase in fire occurrence by the end of the century in circumboreal forests (Flannigan et al. 2009). Landslide risks are related to climate through total rainfall, rainfall intensity, air temperature and the general weather system (Gariano and Guzzetti 2016); thus reduced landslides and natural hazards as a response option will be made more difficult by increasing storms and seasonality of rainfall events projected for many areas of the world. Reduced pollution is likely less affected by climate change and can continue to be an option, despite increasing temperatures.

Conversely, some invasive species may thrive under climate change, such as moving to new areas or being less susceptible to control protocols (Hellmann et al. 2008). Conversion of coastal wetlands will be more difficult to halt if loss of productive land elsewhere encourages development on these lands, but coastal wetlands will likely adapt to increased CO_2 and higher sea levels through sediment accretion, which will also enhance their capacity to act as carbon sinks (Duarte et al. 2013). While subarctic peatlands are at risk due to warming, these are not the main peatlands that are at risk from agricultural conversion (Tarnocai 2006). Peatlands, such as those in the tropics, may be more vulnerable in hotter scenarios to water table alterations and fire risk (Gorham 1991). Biodiversity conservation, such as through protected areas or corridors, may be threatened by increased land expansion under agriculture in climate change scenarios, including the newly available land in northern climates that may become agriculturally suited (Gimona et al. 2012), lessening the effectiveness of this response option.

Carbon dioxide removal (CDR): The efficacy of enhanced weathering is not likely to be affected by future climate changes. On the other hand, climate change will affect the productivity of bioenergy crops (Cronin et al. 2018), influencing the mitigation potential of bioenergy

and BECCS (Calvin et al. 2013; Kyle et al. 2014). There is uncertainty in the sign and magnitude of the effect of climate change on bioenergy crop yields. As a result, there is uncertainty in whether climate change will increase or decrease the potential of bioenergy and BECCS.

Demand management of value chains: For most response options in demand-side management, the tools are generally not made more difficult by future climate changes. For example, dietary change is not likely to be affected by climate change; in fact, the opposite is more likely, that diets will shift in response to climate change impacts as reflected in high prices for some staple grains and meats, the productivity of which may be reduced (Tigchelaar et al. 2018). However, there is some indication that fruit and vegetable production will also be reduced in future scenarios, making healthier diets potentially harder to achieve in some regions (Springmann et al. 2016). Reduced post-harvest losses and reduced food waste may become an even more important option if water or heat stresses under climate change reduce overall harvests. Material substitution does have risks related to the availability of products if there are declines in the growth of forest and other biomass in certain future scenarios over time, although some evidence indicates that biomass may increase in the short term with limited warming (Boisvenue and Running 2006).

Supply management of value chains: Sustainable sourcing relies on being able to produce consumer goods sustainably (palm oil, timber, cocoa, etc.), and these may be at risk; for example, areas suitable for oil palm production are estimated to decrease by 75% by 2100 (Paterson et al. 2017). Improved management of supply chains is likely to increase in importance as a tool to manage food security, given that climate change threatens to lead to more production shocks in the future (Baldos and Hertel 2015). For enhanced urban food systems, climate stresses like heat island effects or increased water scarcity in urban areas may reduce the viability of food production in certain urban systems (da Silva et al. 2012). Improved food processing and retailing and improved energy use in agriculture are not likely to be impacted on by climate change.

Risk management options: Most risk management response options are not affected by climate impacts *per se*, although the increased risks that people may face will increase the need for funding and support to deploy these options. For example, disaster risk management will likely increase in importance in helping people adapt to longer-term climate changes (Begum et al. 2014); it is also likely to cost more as increased impacts of climate change, such as intensification or frequency of storm events, may increase. Management of urban sprawl may also be challenged by increased migration driven by climate change, as people displaced by climate change may move to unregulated urban areas (Adamo 2010). Livelihood diversification can assist in adapting to climate changes and is not likely to be constrained as a response option, as climate-sensitive livelihoods may be replaced by others that are less so. Use of local seeds as an effective response option may depend on the specific types of seeds and crops used, as some may not be good choices under increased heat and water stress (Gross et al. 2017). Risk-sharing instruments are unlikely to be affected by climate change, with the exception of index and crop insurance, which may become unaffordable if too many climate shocks result in insurance claims, decreasing the ability of the industry to provide this tool (Mills 2005).

Cross-Chapter Box 8 | Ecosystem services and Nature's Contributions to People, and their relation to the land–climate system

Pamela McElwee (The United States of America), Jagdish Krishnaswamy (India), Lindsay Stringer (United Kingdom)

This Cross-Chapter Box describes the concepts of *ecosystem services* (ES) and *Nature's Contributions to People* (NCP), and their importance to land–climate interactions. ES have become a useful concept to describe the benefits that humans obtain from ecosystems and have strong relevance to sustainable land management (SLM) decisions and their outcomes, while NCP is a new approach championed by the Intergovernmental Science-Policy Platform on Biodiversity and Ecosystem Services (IPBES) (explained below). It is timely that this SRCCL report includes attention to ES/NCP, as the previous Special Report on land-use, land-use change and forestry (LULUCF) did not make use of these concepts and focused mostly on carbon fluxes in land–climate interactions (IPCC 2000). The broader mandate of SRCCL is to address climate, but also land degradation, desertification and food security issues, all of which are closely linked to the provisioning of various ES/NCP, and the decision and outline for SRCCL explicitly request an examination of how desertification and degradation 'impacts on ecosystem services (e.g., water, soil and soil carbon and biodiversity that underpins them)'. Attention to ES/NCP is particularly important in discussing co-benefits, trade-offs and adverse side effects of potential climate change mitigation, land management, or food security response options, as many actions may have positive impacts on climate mitigation or food production, but may also come with a decline in ES provisioning, or adversely impact on biodiversity (Section 6.4.3). This box considers the importance of the ES/NCP concepts, how definitions have changed over time, continuing debates over operationalisation and use of these ideas. It concludes by looking at how ES/NCP are treated in various chapters in this report.

While the first uses of the term 'ecosystem services' appeared in the 1980s (Lele et al. 2013; Mooney and Ehrlich 1997), the roots of interest in ES extend back to the late 1960s and the extinction crisis, with concern that species decline might cause loss of valuable benefits to humankind (King 1966; Helliwell 1969; Westman 1977). While concern over extinction was explicitly linked to biodiversity loss, later ideas beyond biodiversity have animated interest in ES, including the multi-functional nature of ecosystems. A seminal paper by Costanza et al. (1997) attempted to put an economic value on the stocks of global ES and natural capital on which humanity relied. Attention to ES expanded rapidly after the Millennium Ecosystem Assessment (MA, 2005), and the linkages between ES and economic valuation of these functions were addressed by the Economics of Ecosystems and Biodiversity study (TEEB 2009). The ES approach has increasingly been used in global and national environmental assessments, including the UK National Ecosystem Assessment (Watson et al. 2011), and recent and ongoing regional and global assessments organised by the Intergovernmental Science-Policy Platform on Biodiversity and Ecosystem Services (IPBES) (Díaz et al. 2015). IPBES has recently completed an assessment on land degradation and restoration that addresses a range of ES issues of relevance to the SRCCL report (IPBES 2018).

The MA defined ES as 'the benefits that ecosystems provide to people,' and identified four broad groupings of ES: *provisioning services* such as food, water, or timber; *regulating services* that have impacts on climate, diseases or water quality, among others; *cultural services* that provide recreational, aesthetic, and spiritual benefits; and *supporting services* such as soil formation, photosynthesis, and nutrient cycling (MA 2005). The MA emphasised that people are components of ecosystems engaged in dynamic interactions, and particularly assessed how changes in ES might impact human well-being, such as access to basic materials for living (shelter, clothing, energy); health (clean air and water); social relations (including community cohesion); security (freedom from natural disasters); and freedom of choice (the opportunity to achieve) (MA 2005). Upon publication of the MA, incorporation of ES into land-use change assessments increased dramatically, including studies on how to maximise provisioning of ES alongside human well-being (Carpenter et al. 2009); how intensive food production to feed growing populations required trading off a number of important ES (Foley et al. 2005); and how including ES in general circulation models indicated increasing vulnerability to ES change or loss in future climate scenarios (Schröter et al. 2005).

Starting in 2015, IPBES introduced a new related concept to ES, that of *Nature's Contributions to People* (NCP), which are defined as 'all the contributions, both positive and negative, of living nature (i.e., diversity of organisms, ecosystems and their associated ecological and evolutionary processes) to the quality of life of people' (Díaz et al. 2018). NCP are divided into regulating NCP, non-material NCP, and material NCP, a different approach than used by the MA (see Figure 1). However, IPBES has stressed that NCP are a particular *way to think* of ES, rather than a replacement for ES. The concept of NCP is proposed to be a broader umbrella to engage a wider range of scholarship – particularly from the social sciences and humanities – and a wider range of values, from intrinsic to instrumental to relational – particularly those held by indigenous peoples and local communities (Redford and Adams 2009; Schröter et al. 2014; Pascual et al. 2017; Díaz et al. 2018). The differences between the MA and IPBES approaches can be seen in Table 1.

While there are many similarities between ES and NCP, as seen above, the IPBES's decision to use the NCP concept has been controversial, with some people arguing that an additional term is superfluous; that it incorrectly associates ES with economic

6

Cross-Chapter Box 8 (continued)

Cross-Chapter Box 8, Table 1 | Comparison of MA and IPBES categories and types of ecosystem services (ES) and Nature's Contributions to People (NCP).

MA category	MA: ES	IPBES category	IPBES: NCP
Supporting services	Soil formation		
	Nutrient cycling		
	Primary production		
Regulating services		**Regulating contributions**	Habitat creation and maintenance
	Pollination		Pollination and dispersal of seeds and other propagules
	Air-quality regulation		Regulation of air quality
	Climate regulation		Regulation of climate
	Water regulation		Regulation of ocean acidification
	See above		Regulation of freshwater quantity, flow and timing
	Water purification and waste treatment		Regulation of freshwater and coastal water quality
	Erosion regulation		Formation, protection and decontamination of soils and sediments
	Natural hazard regulation		Regulation of hazards and extreme events
	Pest regulation and disease regulation		Regulation of organisms detrimental to humans
Provisioning services	Fresh water	**Material contributions**	Energy
	Food		Food and feed
	Fibre		Materials and assistance
	Medicinal and biochemical and genetic		Medicinal, biochemical and genetic resources
Cultural services	Aesthetic values	**Non-material contributions**	Learning and inspiration
	Recreation and ecotourism		Physical and psychological experiences
	Spiritual and religious values		Supporting identities
			Maintenance of options

Sources: MA 2005; Díaz et al. 2018.

valuation; and that the NCP concept is not useful for policy uptake (Braat 2018; Peterson et al. 2018). Others have argued that the MA's approach is outdated, does not explicitly address biodiversity, and confuses different concepts, like economic goods, ecosystem functions, and general benefits (Boyd and Banzhaf 2007). Moreover, for both ES and NCP approaches, it has been difficult to make complex ecological processes and functions amenable to assessments that can be used and compared across wider landscapes, different policy actors, and multiple stakeholders (de Groot et al. 2002; Naeem et al. 2015; Seppelt et al. 2011). There remain competing categorisation schemes for ES, as well as competing metrics on how most ES might be measured (Wallace 2007; Potschin and Haines-Young 2011; Danley and Widmark 2016; Nahlik et al. 2012). The implications of these discussions for this SRCCL report is that many areas of uncertainty remain with regard to much ES/NCP measurement and valuation, which will have ramifications for choosing response options and policies.

This report addresses ES/NCP in multiple ways. Individual chapters have used the term 'ES' in most cases, especially since the preponderance of existing literature uses the ES terminology. For example, Chapter 2 discusses CO_2 fluxes, nutrients and water budgets as important ES deriving from land–climate interactions. Chapters 3 and 4 discuss issues such as biomass production, soil erosion, biodiversity loss, and other ES affected by land-use change. Chapter 5 discusses both ES and NCP issues surrounding food system provisioning and trade-offs.

In Chapter 6, the concept of NCP is used. For example, Tables 6.70 to 6.72, possible response options to respond to climate change, to address land degradation or desertification, and to ensure food security, are cross-referenced against the 18 NCPs identified by Díaz et al. (2018) to see where there are co-benefits and adverse side effects. For instance, while BECCS may deliver on climate mitigation, it results in a number of adverse side effects that are significant with regard to water provisioning, food and feed availability, and loss of supporting identities if BECCS competes against local land uses of cultural importance. Chapter 7 has Section 7.2.2.2, explicitly covering risks due to loss of biodiversity and ES, and Table 7.1 which includes policy responses to various land–climate–society hazards, some of which are likely to enhance risk of loss of biodiversity and ES. A case study on the impact of renewable energy on biodiversity and ES is also included. Chapter 7 also notes that, because there is no Sustainable Development Goal covering freshwater biodiversity and aquatic ecosystems, this policy gap may have adverse consequences for the future of rivers and associated ES.

6.4.3 Impacts of integrated response options on Nature's Contributions to People (NCP) and the UN Sustainable Development Goals (SDGs)

In addition to evaluating the importance of our response options for climate mitigation, adaptation, land degradation, desertification and food security, it is also necessary to pay attention to other co-benefits and trade-offs that may be associated with these responses. How the different options impact progress toward the Sustainable Development Goals (SDGs) can be a useful shorthand for looking at the social impacts of these response options. Similarly, looking at how these response options increase or decrease the supply of ecosystem services/Nature's Contributions to People (NCP) (see Cross-Chapter Box 8 in Chapter 6) can be a useful shorthand for a more comprehensive environmental impact beyond climate and land. Such evaluations are important as response options may lead to unexpected trade-offs with social goals (or potential co-benefits) and impacts on important environmental indicators such as water or biodiversity. Similarly, there may be important synergies and co-benefits associated with some response options that may increase their cost-effectiveness or attractiveness. As we note in Section 6.4.4, many of these synergies are not automatic, and are dependent on well-implemented and coordinated activities in appropriate environmental contexts (Section 6.4.4.1), often requiring institutional and enabling conditions for success and participation of multiple stakeholders (Section 6.4.4.3).

In the following sections and tables, we evaluate each response option against 17 SDGs and 18 NCPs. Some of the SDG categories appear similar to each other, such as SDG 13 on 'climate action' and an NCP titled 'climate regulation'. However, SDG 13 includes targets for both mitigation and adaptation, so options were weighed by whether they were useful for one or both. On the other hand, the NCP 'regulation of climate' does not include an adaptation component, and refers specifically to 'positive or negative effects on emissions of GHGs and positive or negative effects on biophysical feedbacks from vegetation cover to atmosphere, such as those involving albedo, surface roughness, long-wave radiation, evapotranspiration (including moisture-recycling) and cloud formation or direct and indirect processes involving biogenic volatile organic compounds (BVOC), and regulation of aerosols and aerosol precursors by terrestrial plants and phytoplankton' (Díaz et al. 2018).

In all tables, colours represent the direction of impact: positive (blue) or negative (brown), and the scale of the impact (dark colours for large impact and/or strong evidence to light colours for small impact and/or less certain evidence). Supplementary tables show the values and references used to define the colour coding used in all tables. In cases where there is no evidence of an interaction, or at least no literature on such interactions, the cell is left blank. In cases where there are both positive and negative interactions and the literature is uncertain about the overall impact, a note appears in the box. In all cases, many of these interactions are contextual, or the literature only refers to certain

co-benefits in specific regions or ecosystems, so readers are urged to consult the supplementary tables for the specific caveats that may apply.

6.4.3.1 Impacts of integrated response options on NCP

Tables 6.70–6.72 summarise the impacts of the response options on NCP supply. Examples of synergies between response options and NCP include positive impacts on habitat maintenance (NCP 1) from activities like invasive species management and agricultural diversification. For the evaluation process, we considered that NCP are about ecosystems, therefore options which may have overall positive effects, but which are *not* ecosystem-based are not included; for example, improved food transport and distribution could reduce ground-level ozone and thus improve air quality, but this is not an ecosystem-based NCP. Similarly, energy-efficiency measures would increase energy availability, but the 'energy' NCP refers specifically to biomass-based fuel provisioning. This necessarily means that the land management options have more direct NCP effects than the value chain or governance options, which are less ecosystem focused.

In evaluating NCP, we have also tried to avoid 'indirect' effects – that is, a response option might increase household income which could then be invested in habitat-saving actions, or dietary change would lead to conservation of natural areas, which would then lead to increased water quality. Similarly, material substitution would increase wood demand, which in turn might lead to deforestation, which might have water regulation effects. These can all be considered *indirect* impacts on NCP, which were not evaluated.[8] Instead, the assessment focuses as much as possible on *direct* effects only: for example, local seeds policies preserve local landraces, which *directly* contribute to 'maintenance of genetic options' for the future. Therefore, this NCP table is a conservative estimation of NCP effects; there are likely many more secondary effects, but they are too difficult to assess, or the literature is not yet complete or conclusive.

Further, many NCPs trade-off with one another (Rodríguez et al. 2006), so supply of one might lead to less availability of another – for example, use of ecosystems to produce bioenergy will likely lead to decreases in water availability if mono-cropped high-intensity plantations are used (Gasparatos et al. 2011).

Overall, several response options stand out as having co-benefits across 10 or more NCP with no adverse impacts, including: improved cropland management, agroforestry, forest management and forest restoration, increased soil organic content, fire management, restoration and avoided conversion of coastal wetlands, and use of local seeds. Other response options may have strengths in some NCP but require trade-offs with others. For example, reforestation and afforestation bring many positive benefits for climate and water quality but may trade-off with food production (Table 6.70). Several response options, including increased food productivity, bioenergy and BECCS, and some risk-sharing instruments, like crop insurance, have significant negative consequences across multiple NCPs.

---

[8] The exception is NCP 6, regulation of ocean acidification, which is by itself an indirect impact. Any option that sequesters CO_2 would lower the atmospheric CO_2 concentration, which then indirectly increases the seawater pH. Therefore, any action that directly increases the amount of sequestered carbon is noted in this column, but not any action that avoids land-use change and, therefore, indirectly avoids CO_2 emissions.

Table 6.70 | Impacts on Nature's Contributions to People (NCP) of integrated response options based on land management.

Integrated response options based on land management	Habitat creation and maintenance	Pollination and dispersal of seeds and other propagules	Regulation of air quality	Regulation of climate	Regulation of ocean acidification	Regulation of freshwater quantity, flow and timing	Regulation of freshwater and coastal water quality	Formation, protection and decontamination of soils and sediments	Regulation of hazards and extreme events	Regulation of organisms detrimental to humans	Energy	Food and feed	Materials and assistance	Medicinal, biochemical and genetic resources	Learning and inspiration	Physical and psychological experiences	Supporting identities	Maintenance of options
Increased food productivity																		
Improved cropland management																		
Improved grazing land management																		
Improved livestock management																		
Agroforestry																		
Agricultural diversification																		
Avoidance of conversion of grassland to cropland																		
Integrated water management													+ or −					
Forest management and forest restoration									+ or −				+ or −					
Reduced deforestation and forest degradation																		
Reforestation									+ or −									
Afforestation									+ or −	+ or −								
Increased soil organic carbon content																		
Reduced soil erosion																		
Reduced soil salinisation																		
Reduced soil compaction																		
Biochar addition to soil																		
Fire management																		
Reduced landslides and natural hazards																		
Reduced pollution including acidification																		
Management of invasive species/encroachment																		
Restoration and avoided conversion of coastal wetlands													+ or −					
Restoration and avoided conversion of peatlands																		
Biodiversity conservation													+ or −					
Enhanced weathering of minerals																		
Bioenergy and BECCS[9]																		

Legend:
- Large positive impacts, strong evidence
- Medium positive impacts, some evidence
- Small positive impacts or low evidence
- Low negative impacts or low evidence
- Medium negative impacts, medium evidence
- Large negative impacts, high evidence

[9] Note that this refers to large areas of bioenergy crops capable of producing large mitigation benefits (>3 GtCO$_2$ yr^{-1}). The effect of bioenergy and BECCS on NCPs is scale and context dependent (see Cross-Chapter Box 7 in Chapter 6 and Section 6.3).

6

Table 6.71 | Impacts on NCP of integrated response options based on value chain management.

Integrated response options based on value chain management	Habitat creation and maintenance	Pollination and dispersal of seeds and other propagules	Regulation of air quality	Regulation of climate	Regulation of ocean acidification	Regulation of freshwater quantity, flow and timing	Regulation of freshwater and coastal water quality	Formation, protection and decontamination of soils and sediments	Regulation of hazards and extreme events	Regulation of organisms detrimental to humans	Energy	Food and feed	Materials and assistance	Medicinal, biochemical and genetic resources	Learning and inspiration	Physical and psychological experiences	Supporting identities	Maintenance of options
Dietary change																		
Reduced post-harvest losses																		
Reduced food waste (consumer or retailer)																		
Material substitution																		
Sustainable sourcing																		
Management of supply chains																		
Enhanced urban food systems																		
Improved food processing and retail																		
Improved energy use in food systems																		

Legend:
- Large positive impacts, strong evidence
- Medium positive impacts, some evidence
- Small positive impacts or low evidence
- Low negative impacts or low evidence

Table 6.72 | Impacts on NCP of integrated response options based on risk management.

Integrated response options based on risk management	Habitat creation and maintenance	Pollination and dispersal of seeds and other propagules	Regulation of air quality	Regulation of climate	Regulation of ocean acidification	Regulation of freshwater quantity, flow and timing	Regulation of freshwater and coastal water quality	Formation, protection and decontamination of soils and sediments	Regulation of hazards and extreme events	Regulation of organisms detrimental to humans	Energy	Food and feed	Materials and assistance	Medicinal, biochemical and genetic resources	Learning and inspiration	Physical and psychological experiences	Supporting identities	Maintenance of options
Management of urban sprawl																		
Livelihood diversification																		
Use of local seeds																		
Disaster risk management																		
Risk-sharing instruments																		

Legend:
- Large positive impacts, strong evidence
- Medium positive impacts, some evidence
- Small positive impacts or low evidence
- Low negative impacts or low evidence
- Medium negative impacts, medium evidence

6

6.4.3.2   Impacts of integrated response options on the UN SDGs

Tables 6.73–6.75 summarise the impact of the integrated response options on the UN SDGs. Some of the synergies between response options and SDGs in the literature include positive poverty eradication impacts (SDG 1) from activities like improved water management or improved management of supply chains, or positive gender impacts (SDG 5) from livelihood diversification or use of local seeds. Because many land management options only produce indirect or unclear effects on SDGs, we did not include these where there was no literature. Therefore, the value chain and governance options appear to offer more direct benefits for SDG.

However, it is noted that some SDG are internally difficult to assess because they contain many targets, not all of which could be evaluated (e.g., SDG 17 is about partnerships, but has targets ranging from foreign aid to debt restructuring, technology transfer to trade openness). Additionally, it is noted that some SDG contradict one another – for example, SDG 9 to increase industrialisation and infrastructure and SDG 15 to improve life on land. More industrialisation is likely to lead to increased resource demands with negative effects on habitats. Therefore, a positive association on one SDG measure might be directly correlated with a negative measure on another, and the table needs to be read with caution for that reason. The specific caveats on each of these interactions can be found in the supplementary material tables in the Chapter 6 Appendix.

Overall, several response options have co-benefits across 10 or more SDGs with no adverse side effects on any SDG: increased food production, improved grazing land management, agroforestry, integrated water management, reduced post-harvest losses, sustainable sourcing, livelihood diversification and disaster risk management. Other response options may have strengths in some SDGs but require trade-offs with others. For example, use of local seeds brings many positive benefits for poverty and hunger reduction, but may reduce international trade (SDG 17). Other response options like enhanced urban food systems, management of urban sprawl, or management of supply chains are generally positive for many SDGs but may trade-off with one, like clean water (SDG 6) or decent work (SDG 8), as they may increase water use or slow economic growth. Several response options, including avoidance of grassland conversion, reduced deforestation and forest degradation, reforestation and afforestation, biochar, restoration and avoided conversion of peatlands and coastlands, have trade-offs across multiple SDGs, primarily as they prioritise land health over food production and poverty eradication. Several response options such as bioenergy and BECCS and some risk-sharing instruments, such as crop insurance, trade-off over multiple SDG with potentially significant adverse consequences.

Overall, across categories of SDG and NCPs; 17 of 40 options deliver co-benefits or no adverse side effects for the full range of NCPs and SDGs. This includes most agriculture- and soil-based land management options, many ecosystem-based land management options, forest management, reduced post-harvest losses, sustainable sourcing, improved energy use in food systems, and livelihood diversification. Only three options (afforestation, bioenergy and BECCS and some types of risk-sharing instruments, such as crop insurance) have potentially adverse side effects for five or more NCPs or SDGs.

Table 6.73 | Impacts of integrated response options based on land management on the UN SDGs.

Integrated response options based on land management	GOAL 1: No poverty	GOAL 2: Zero hunger	GOAL 3: Good health and well-being	GOAL 4: Quality education	GOAL 5: Gender equality	GOAL 6: Clean water and sanitation	GOAL 7: Affordable and clean energy	GOAL 8: Decent work and economic growth	GOAL 9: Industry, innovation and infrastructure	GOAL 10: Reduced inequality	GOAL 11: Sustainable cities and communities	GOAL 12: Responsible consumption and production	GOAL 13: Climate action	GOAL 14: Life below water	GOAL 15: Life on land	GOAL 16: Peace, justice and strong institutions	GOAL 17: Partnerships to achieve the goals
Increased food productivity																	
Improved cropland management																	
Improved grazing land management																	
Improved livestock management																	
Agroforestry																	
Agricultural diversification										+ or −							
Avoidance of conversion of grassland to cropland																	
Integrated water management																	
Forest management and forest restoration																	
Reduced deforestation and forest degradation	+ or −																
Reforestation	+ or −																
Afforestation																	
Increased soil organic carbon content																	
Reduced soil erosion																	
Reduced soil salinisation																	
Reduced soil compaction																	
Biochar addition to soil																	
Fire management																	
Reduced landslides and natural hazards																	
Reduced pollution, including acidification																	
Management of invasive species/encroachment																	
Restoration and avoided conversion of coastal wetlands	+ or −	+ or −															
Restoration and avoided conversion of peatlands																	
Biodiversity conservation																	
Enhanced weathering of minerals																	
Bioenergy and BECCS[10]	+ or −		+ or −														

Legend:
- Large positive impacts, strong evidence
- Medium positive impacts, some evidence
- Small positive impacts or low evidence
- Low negative impacts or low evidence
- Medium negative impacts, medium evidence
- Large negative impacts, high evidence

[10] Note that this refers to large areas of bioenergy crops capable of producing large mitigation benefits (>3 GtCO$_2$ yr^{-1}). The effect of bioenergy and BECCS on SDGs is scale and context dependent (see Cross-Chapter Box 7 in Chapter 6 and Section 6.3).

6

Table 6.74 | Impacts of integrated response options based on value chain interventions on the UN SDGs.

Integrated response options based on value chain management	GOAL 1: No poverty	GOAL 2: Zero hunger	GOAL 3: Good health and well-being	GOAL 4: Quality education	GOAL 5: Gender equality	GOAL 6: Clean water and sanitation	GOAL 7: Affordable and clean energy	GOAL 8: Decent work and economic growth	GOAL 9: Industry, innovation and infrastructure	GOAL 10: Reduced inequality	GOAL 11: Sustainable cities and communities	GOAL 12: Responsible consumption and production	GOAL 13: Climate action	GOAL 14: Life below water	GOAL 15: Life on land	GOAL 16: Peace, justice and strong institutions	GOAL 17: Partnerships to achieve the goals
Dietary change																	
Reduced post-harvest losses																	
Reduced food waste (consumer or retailer)																	
Material substitution																	
Sustainable sourcing																	
Management of supply chains																	
Enhanced urban food systems																	
Improved food processing and retail																	
Improved energy use in food systems																	

- ■ Large positive impacts, strong evidence
- ■ Medium positive impacts, some evidence
- ■ Small positive impacts or low evidence
- ■ Low negative impacts or low evidence
- ■ Medium negative impacts, medium evidence

Table 6.75 | Impacts of integrated response options based on risk management on the UN SDGs.

Integrated response options based on risk management	GOAL 1: No poverty	GOAL 2: Zero hunger	GOAL 3: Good health and well-being	GOAL 4: Quality education	GOAL 5: Gender equality	GOAL 6: Clean water and sanitation	GOAL 7: Affordable and clean energy	GOAL 8: Decent work and economic growth	GOAL 9: Industry, innovation and infrastructure	GOAL 10: Reduced inequality	GOAL 11: Sustainable cities and communities	GOAL 12: Responsible consumption and production	GOAL 13: Climate action	GOAL 14: Life below water	GOAL 15: Life on land	GOAL 16: Peace, justice and strong institutions	GOAL 17: Partnerships to achieve the goals
Management of urban sprawl																	
Livelihood diversification																	
Use of local seeds																	
Disaster risk management																	
Risk-sharing instruments												+ or –					

- ■ Large positive impacts, strong evidence
- ■ Medium positive impacts, some evidence
- ■ Small positive impacts or low evidence
- ■ Low negative impacts or low evidence

6

6.4.4 Opportunities for implementing integrated response options

6.4.4.1 Where can the response options be applied?

As shown in Section 6.1.3, a large part of the land area is exposed to overlapping land challenges, especially in villages, croplands and rangelands. The deployment of land management responses may vary with local exposure to land challenges. For instance, with croplands exposed to a combination of land degradation, food insecurity and climate change adaptation challenges, maximising the co-benefits of land management responses would require selecting responses having only co-benefits for these three overlapping challenges, as well as for climate change mitigation, which is a global challenge. Based on these criteria, Figure 6.6 shows the potential deployment area of land management responses across land-use types (or anthromes).

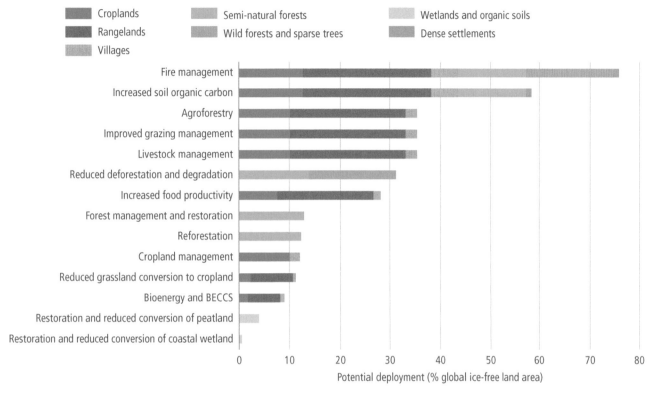

Figure 6.6 | **Potential deployment area of land management responses (see Table 6.1) across land-use types (or anthromes, see Section 6.3), when selecting responses having only co-benefits for local challenges and for climate change mitigation and no large adverse side effects on global food security.** See Figure 6.2 for the criteria used to map challenges considered (desertification, land degradation, climate change adaptation, chronic undernourishment, biodiversity, groundwater stress and water quality). No response option was identified for barren lands.

Land management responses having co-benefits across the range of challenges, including climate change mitigation, could be deployed between one land-use type (coastal wetlands, peatlands, forest management and restoration, reforestation) and five (increased soil organic carbon) or six (fire management) land-use types (Figure 6.6). Fire management and increased soil organic carbon have a large potential since they could be deployed with mostly co-benefits and few adverse effects over 76% and 58% of the ice-free land area. In contrast, other responses have a limited area-based potential due to biophysical constraints (e.g., limited extent of organic soils and of coastal wetlands for conservation and restoration responses), or due to the occurrence of adverse effects. Despite strong co-benefits for climate change mitigation, the deployment of bioenergy and BECCS would have co-benefits on only 9% of the ice-free land area (Figure 6.6), given adverse effects of this response option for food security, land degradation, climate change adaptation and desertification (Tables 6.62–6.69).

Without including the global climate change mitigation challenge, there are up to five overlapping challenges on lands that are not barren (Figure 6.7A, calculated from the overlay of individual challenges shown in Figure 6.2) and up to nine land management response options having only co-benefits for these challenges and for climate change mitigation (Figure 6.7B). Across countries, the mean number of land management response options with mostly co-benefits declines ($p<0.001$, Spearman rank order correlation) with the mean number of land challenges. Hence, the higher the number of land challenges per country, the fewer the land management response options having only co-benefits for the challenges encountered.

Enabling conditions (see Section 6.1.2.2) for the implementation of land management responses partly depend on human development (economics, health and education) as estimated by a country scale composite index, the Human Development Index (HDI) (UNDP 2018) (Figure 6.7C). Across countries, HDI is negatively correlated ($p<0.001$, Spearman rank order correlation) with the mean number of land challenges. Therefore, on a global average, the higher the number of local challenges faced, the fewer the land management responses having only co-benefits, and the lower the human development (Figure 6.7) that could favour the implementation of these responses.

6

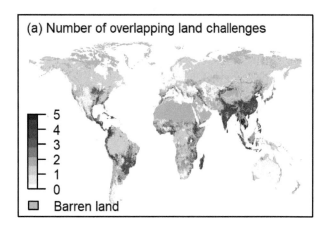

(a) Number of overlapping land challenges

5
4
3
2
1
0

☐ Barren land

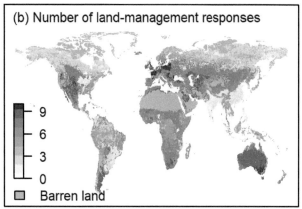

(b) Number of land-management responses

9

6

3

0

☐ Barren land

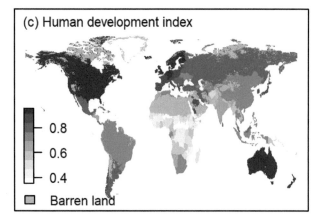

(c) Human development index

0.8

0.6

0.4

☐ Barren land

Figure 6.7 | Global distributions of: **(a)** number of overlapping land challenges (desertification, land degradation, climate change adaptation, chronic undernourishment, biodiversity, groundwater stress and water quality (Figure 6.2); **(b)** number of land management responses providing medium-to-large co-benefits and no adverse side effects (see Figure 6.6) across challenges; **(c)** Human Development Index (HDI) by country. The HDI (UNDP 2018) is a country-based composite statistical index measuring average achievement in three basic dimensions of human development: a long and healthy life (estimated from life expectancy at birth), knowledge (estimated from years of schooling), and a decent standard of living (estimated from gross national income per capita).

6.4.4.2   Interlinkages and response options in future scenarios

This section assesses more than 80 articles quantifying the effect of various response options in the future, covering a variety of response options and land-based challenges. These studies cover spatial scales ranging from global (Popp et al. 2017; Fujimori et al. 2019) to regional (Calvin et al. 2016a; Frank et al. 2015) to country level (Gao and Bryan 2017; Pedercini et al. 2018). This section focuses on models that can quantify interlinkages between response options, including agricultural economic models, land system models, and Integrated Assessment Models (IAMs). The IAM and non-IAM literature, however, is also categorised separately to elucidate what is and is not included in global mitigation scenarios, like those included in the SR15. Results from bottom-up studies and models (e.g., Griscom et al. 2017) are assessed in Sections 6.2–6.3.

Response options in future scenarios

More than half of the 40 land-based response options discussed in this chapter are represented in global IAMs models used to develop and analyse future scenarios, either implicitly or explicitly (Table 6.76). For example, all IAMs include improved cropland management, either explicitly through technologies that improve nitrogen use efficiency (Humpenöder et al. 2018) or implicitly through marginal abatement cost curves that link reductions in nitrous oxide emissions from crop production to carbon prices (most other models).

However, the literature discussing the effect of these response options on land-based challenges is more limited (Table 6.76). There are 57 studies (43 IAM studies) that articulate the effect of response options on mitigation, with most including bioenergy and BECCS or a combination of reduced deforestation, reforestation, and afforestation; 37 studies (21 IAM studies) discuss the implications of response options on food security, usually using food price as a metric. While a small number of non-IAM studies examine the effects of response options on desertification (three studies) and land degradation (five studies), no IAM studies were identified. However, some studies quantify these challenges indirectly using IAMs, either via climate outputs from the representative concentration pathways (RCPs) (Huang et al. 2016) or by linking IAMs to other land and ecosystem models (Ten Brink et al. 2018; UNCCD 2017).

For many of the scenarios in the literature, land-based response options are included as part of a suite of mitigation options (Popp et al. 2017; Van Vuuren et al. 2015). As a result, it is difficult to isolate the effect of an individual option on land-related challenges. A few studies focus on specific response options (Calvin et al. 2014; Popp et al. 2014; Kreidenweis et al. 2016; Humpenöder et al. 2018), quantifying the effect of including an individual option on a variety of sustainability targets.

6

Table 6.76 | Number of IAM and non-IAM studies including specific response options (rows) and quantifying particular land challenges (columns). The third column shows how many IAM models include the individual response option. The remaining columns show challenges related to climate change (C), mitigation (M), adaptation (A), desertification (D), land degradation (L), food security (F), and biodiversity/ecosystem services/sustainable development (B). Additionally, counts of total (left value) and IAM-only (right value) studies are included. Some IAMs include agricultural economic models, which can also be run separately; these models are not counted as IAM literature when used on their own. Studies using a combination of IAMs and non-IAMs are included in the total only. A complete list of studies is included in the Appendix.

Category	Response option	IAMs[a]	Studies [Total/IAM]						
			C	M	A	D	L[b]	F[c]	B
Land management	Increased food productivity		1/1	18/14	5/1	2/0	3/0	18/9	12/6
	Improved cropland management		0/0	15/11	7/2	0/0	0/0	13/6	7/4
	Improved grazing land management		0/0	1/0	1/0	0/0	0/0	1/0	0/0
	Improved livestock management		0/0	10/6	1/0	2/0	2/0	7/3	5/2
	Agroforestry		0/0	0/0	0/0	0/0	0/0	0/0	0/0
	Agricultural diversification		0/0	0/0	0/0	0/0	0/0	0/0	0/0
	Reduced grassland conversion to cropland		0/0	2/2	0/0	0/0	0/0	1/1	1/1
	Integrated water management		1/0	17/12	5/2	0/0	2/0	13/7	20/13
	Forest management		0/0	2/0	0/0	1/0	1/0	2/0	2/0
	Reduced deforestation and forest degradation		2/2	24/20	1/0	1/0	1/0	14/9	14/8
	Reforestation and forest restoration		3/3	19/18	1/1	1/0	2/0	9/8	9/6
	Afforestation		3/3	24/21	2/1	0/0	0/0	10/9	8/7
	Increased soil organic carbon content		0/0	3/1	0/0	0/0	0/0	1/1	0/0
	Reduced soil erosion		0/0	0/0	0/0	0/0	0/0	0/0	0/0
	Reduced soil salinisation		0/0	0/0	0/0	0/0	0/0	0/0	0/0
	Reduced soil compaction		0/0	0/0	0/0	0/0	0/0	0/0	0/0
	Biochar addition to soil		0/0	0/0	0/0	0/0	0/0	0/0	0/0
	Fire management		0/0	1/1	0/0	0/0	0/0	0/0	0/0
	Reduced landslides and natural hazards		0/0	0/0	0/0	0/0	0/0	0/0	0/0
	Reduced pollution, including acidification		2/2	18/16	2/1	0/0	0/0	10/7	6/6
	Management of invasive species/encroachment		0/0	0/0	0/0	0/0	0/0	0/0	0/0
	Restoration and reduced conversion of coastal wetlands		0/0	0/0	0/0	1/0	1/0	0/0	1/0
	Restoration and reduced conversion of peatlands		0/0	0/0	0/0	0/0	0/0	0/0	0/0
	Biodiversity conservation		1/0	7/3	0/0	1/0	3/0	4/2	8/1
	Enhanced weathering of minerals		0/0	0/0	0/0	0/0	0/0	0/0	0/0
	Bioenergy and BECCS		5/4	50/40	7/4	0/0	2/0	25/18	21/13
Value chain management	Dietary change		0/0	15/12	1/0	2/0	2/0	13/9	10/7
	Reduced post-harvest losses		0/0	5/4	0/0	0/0	0/0	2/2	2/1
	Reduced food waste (consumer or retailer)		0/0	6/4	0/0	0/0	0/0	4/2	3/1
	Material substitution		0/0	0/0	0/0	0/0	0/0	0/0	0/0
	Sustainable sourcing		0/0	0/0	0/0	0/0	0/0	0/0	0/0
	Management of supply chains		1/1	11/9	8/1	2/0	3/0	17/9	7/3
	Enhanced urban food systems		0/0	0/0	0/0	0/0	0/0	0/0	0/0
	Improved food processing and retailing		0/0	0/0	0/0	0/0	0/0	0/0	0/0
	Improved energy use in food systems		0/0	0/0	0/0	0/0	0/0	0/0	0/0
Risk management	Management of urban sprawl		0/0	0/0	0/0	1/0	1/0	0/0	1/0
	Livelihood diversification		0/0	0/0	0/0	0/0	0/0	0/0	0/0
	Use of local seeds		0/0	0/0	0/0	0/0	0/0	0/0	0/0
	Disaster risk management		0/0	0/0	0/0	0/0	0/0	0/0	0/0
	Risk sharing instruments		0/0	0/0	0/0	0/0	0/0	0/0	0/0

IAMs: How many models include the individual response option

- ▇ All models
- ▇ More than half
- ▇ Less than half
- ☐ No models

Columns C, M, A, D, L, F and B: Number of total studies

- ☐ 0
- ▇ 1–5
- ▇ 6–10
- ▇ 11–15
- ▇ 16+

[a] Only IAMs that are used in the papers assessed are included in this column.
[b] There are many indicators for land degradation (Chapter 4). In this table, studies are categorised as quantifying land degradation if they explicitly discuss land degradation.
[c] Studies are categorised is quantifying food security if they report food prices or the population at risk of hunger.

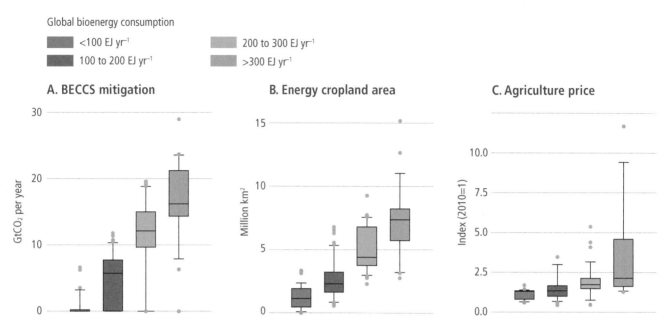

Figure 6.8 | Correlation between bioenergy use and other indicators. Panel A shows global CO$_2$ sequestration by BECCS in 2100. **Panel B** shows global energy cropland area in 2100. **Panel C** shows agricultural prices in 2100 indexed to 2010. Data are based on the amount of bioenergy used globally in 2100. All scenario data that include bioenergy consumption and the variable of interest are included in the figure; the resulting number of scenarios varies per panel, with 352 in panel A, 262 in panel B, and 172 in panel C. The boxes represent the interquartile range (i.e., the middle 50% of all scenarios). The line in the middle of the box represents the median, and the 'whiskers' represent the 5 to 95% range of scenarios. Data is from an update of the Integrated Assessment Modelling Consortium (IAMC) Scenario Explorer developed for the SR15 (Huppmann et al. 2018; Rogelj et al. 2018b).

Interactions and interlinkages between response options

The effect of response options on desertification, land degradation, food security, biodiversity, and other SDGs depends strongly on which options are included, and the extent to which they are deployed. For example, Sections 2.6 and 6.3.6, and Cross-Chapter Box 7 note that bioenergy and BECCS has a large mitigation potential, but could potentially have adverse side effects for land degradation, food security, and other SDGs. Global modelling studies demonstrate that these effects are dependent on scale. Increased use of bioenergy can result in increased mitigation (Figure 6.8, panel A) and reduced climate change, but can also lead to increased energy cropland expansion (Figure 6.8, panel B), and increased competition for land, resulting in increased food prices (Figure 6.8, panel C). However, the exact relationship between bioenergy deployment and each sustainability target depends on a number of other factors, including the feedstock used, the underlying socio-economic scenario, assumptions about technology and resource base, the inclusion of other response options, and the specific model used (Calvin et al. 2014; Clarke et al. 2014; Popp et al. 2014, 2017; Kriegler et al. 2014).

The previous sections have examined the effects of individual land-response options on multiple challenges. A number of studies using global modelling and analyses have examined interlinkages and interaction effects among land response options by incrementally adding or isolating the effects of individual options. Most of these studies focus on interactions with bioenergy and BECCS (Table 6.77). Adding response options that require land (e.g., reforestation, afforestation, reduced deforestation, avoided grassland conversion, or biodiversity conservation) results in increased food prices (Calvin et al. 2014; Humpenöder et al. 2014; Obersteiner et al.

2016; Reilly et al. 2012) and potentially increased temperature through biophysical climate effects (Jones et al. 2013). However, this combination can result in reduced water consumption (Hejazi et al. 2014b), reduced cropland expansion (Calvin et al. 2014; Humpenöder et al. 2018), increased forest cover (Calvin et al. 2014; Humpenöder et al. 2018; Wise et al. 2009) and reduced biodiversity loss (Pereira et al. 2010), compared to scenarios with bioenergy and BECCS alone. While these options increase total mitigation, they reduce mitigation from bioenergy and BECCS as they compete for the same land (Wu et al. 2019; Baker et al. 2019; Calvin et al. 2014; Humpenöder et al. 2014).

The inclusion of land-sparing options (e.g., dietary change, increased food productivity, reduced food waste, management of supply chains) in addition to bioenergy and BECCS results in reduced food prices, reduced agricultural land expansion, reduced deforestation, reduced mitigation costs, reduced water use, and reduced biodiversity loss (Bertram et al. 2018; Wu et al. 2019; Obersteiner et al. 2016; Stehfest et al. 2009; Van Vuuren et al. 2018). These options can increase bioenergy potential, resulting in increased mitigation than from bioenergy and BECCS alone (Wu et al. 2019; Stehfest et al. 2009; Favero and Massetti 2014).

Other combinations of land response options create synergies, alleviating land pressures. The inclusion of increased food productivity and dietary change can increase mitigation, reduce cropland use, reduce water consumption, reduce fertiliser application, and reduce biodiversity loss (Springmann et al. 2018; Obersteiner et al. 2016). Similarly, improved livestock management, combined with increased food productivity, can reduce agricultural land expansion (Weindl et al. 2017). Reducing disturbances (e.g., fire management) in combination with afforestation can increase the terrestrial carbon sink, resulting in increased mitigation potential and reduced mitigation cost (Le Page et al. 2013).

Studies including multiple land response options often find that the combined mitigation potential is not equal to the sum of individual mitigation potential as these options often share the same land. For example, including both afforestation and bioenergy and BECCS results in a cumulative reduction in GHG emissions of 1200 GtCO$_2$ between 2005 and 2100, which is much lower than the sum of the contributions of bioenergy (800 GtCO$_2$) and afforestation (900 GtCO$_2$) individually (Humpenöder et al. 2014). More specifically, Baker et al. (2019) find that woody bioenergy and afforestation are complementary in the near term, but become substitutes in the long term, as they begin to compete for the same land. Similarly, the combined effect of increased food productivity, dietary change and reduced waste on GHG emissions is less than the sum of the individual effects (Springmann et al. 2018).

Table 6.77 | Interlinkages between bioenergy and BECCS and other response options. Table indicates the combined effects of multiple land-response options on climate change (C), mitigation (M), adaptation (A), desertification (D), land degradation (L), food security (F), and biodiversity/ecosystem services/sustainable development (O). Each cell indicates the implications of adding the option specified in the row in addition to bioenergy and BECCS. Blue colours indicate positive interactions (e.g., including the option in the second column increases mitigation, reduces cropland area, or reduces food prices relative to bioenergy and BECCS alone). Yellow indicates negative interactions; grey indicates mixed interactions (some positive, some negative). Note that only response option combinations found in the assessed literature are included in the interest of space.

	C[a]	M[b]	A	D	L[c]	F	O[d]	Context and sources
Increased food productivity		▉			▉			Humpenöder et al. 2018; Obersteiner et al. 2016
Increased food productivity; improved livestock management		▉						Van Vuuren et al. 2018
Improved cropland management							▉	Humpenöder et al. 2018
Integrated water management		▉					▉	O: Reduces water use, but increases fertiliser use. (Humpenöder et al. 2018)
Reduced deforestation		▉			▉			Calvin et al. 2014; Humpenöder et al. 2018
Reduced deforestation, avoided grassland conversion		▉			▉		▉	O: Reduces biodiversity loss and fertiliser, but increases water use. (Calvin et al. 2014; Obersteiner et al. 2016)
Reforestation		▉						Reilly et al. 2012
Reforestation, afforestation, avoided grassland conversion	▒	▉						Calvin et al. 2014; Hejazi et al. 2014a; Jones et al. 2013
Afforestation		▉						Humpenöder et al. 2014
Biodiversity conservation		▒			▉		▒	M: Reduces emissions but also reduces bioenergy potential. O: Reduces biodiversity loss but increases water use. (Obersteiner et al. 2016; Wu et al. 2019)
Reduced pollution		▉						Van Vuuren et al. 2018
Dietary change		▉			▉			Bertram et al. 2018; Stehfest et al. 2009; Wu et al. 2019
Reduced food waste; dietary change					▉			Van Vuuren et al. 2018
Management of supply chains		▉						Favero and Massetti 2014
Management of supply chains; increased productivity		▉						Wu et al. 2019
Reduced deforestation; improved cropland management; improved food productivity; integrated water management					▉		▉	Humpenöder et al. 2018
Reduced deforestation; management of supply chains; integrated water management; improved cropland management; increased food productivity					▉			Bertram et al. 2018
Reduced deforestation; management of supply chains; integrated water management; improved cropland management; increased food productivity; dietary change					▉			Bertram et al. 2018

▉ Positive interactions   ▒ Negative interactions   ▒ Mixed interactions

[a] Includes changes in biophysical effects on climate (e.g., albedo).
[b] Either through reduced emissions, increased mitigation, reduced mitigation cost, or increased bioenergy potential. For increased mitigation, a positive indicator in this column only indicates that total mitigation increases and not that the total is greater than the sum of the individual options.
[c] Use changes in cropland or forest as an indicator (reduced cropland expansion or reduced deforestation are considered positive).
[d] Includes changes in water use or scarcity, fertiliser use, or biodiversity.

Land-related response options can also interact with response options in other sectors. For example, limiting deployment of a mitigation response option will either result in increased climate change or additional mitigation in other sectors. A number of studies have examined limiting bioenergy and BECCS. Some such studies show increased emissions (Reilly et al. 2012). Other studies meet the same climate goal, but reduce emissions elsewhere *via* reduced energy demand (Grubler et al. 2018; Van Vuuren et al. 2018), increased fossil carbon capture and storage (CCS), nuclear energy, energy efficiency and/or renewable energy (Van Vuuren et al. 2018; Rose et al. 2014; Calvin et al. 2014; Van Vuuren et al. 2017b), dietary change (Van Vuuren et al. 2018), reduced non-CO$_2$ emissions (Van Vuuren et al. 2018), or lower population (Van Vuuren et al. 2018). The co-benefits and adverse side effects of non-land mitigation options are discussed in SR15, Chapter 5. Limitations on bioenergy and BECCS can result in increases in the cost of mitigation (Kriegler et al. 2014; Edmonds et al. 2013). Studies have also examined limiting CDR, including reforestation, afforestation, and bioenergy and BECCS (Kriegler et al. 2018a,b). These studies find that limiting CDR can increase mitigation costs, increase food prices, and even preclude limiting warming to less than 1.5°C above pre-industrial levels (Kriegler et al. 2018a,b; Muratori et al. 2016).

In some cases, the land challenges themselves may interact with land-response options. For example, climate change could affect the production of bioenergy and BECCS. A few studies examine these effects, quantifying differences in bioenergy production (Calvin et al. 2013; Kyle et al. 2014) or carbon price (Calvin et al. 2013) as a result of climate change. Kyle et al. (2014) find increase in bioenergy production due to increases in bioenergy yields, while Calvin et al. (2013) find declines in bioenergy production and increases in carbon price due to the negative effects of climate on crop yield.

Gaps in the literature

Not all of the response options discussed in this chapter are included in the assessed literature, and many response options are excluded from the IAM models. The included options (e.g., bioenergy and BECCS; reforestation) are some of the largest in terms of mitigation potential (see Section 6.3). However, some of the options excluded also have large mitigation potential. For example, biochar, agroforestry, restoration/avoided conversion of coastal wetlands, and restoration/avoided conversion of peatland all have mitigation potential of about 1 $GtCO_2$ yr^{-1} (Griscom et al. 2017). Additionally, quantifications of and response options targeting land degradation and desertification are largely excluded from the modelled studies, with a few notable exceptions (Wolff et al. 2018; Gao and Bryan 2017; Ten Brink et al. 2018; UNCCD 2017). Finally, while a large number of papers have examined interactions between bioenergy and BECCS and other response options, the literature examining other combinations of response options is more limited.

6.4.4.3 Resolving challenges in response option implementation

The 40 response options assessed in this chapter face a variety of barriers to implementation that require action across multiple actors to overcome (Section 6.4.1). Studies have noted that, while adoption of response options by individuals may depend on individual assets and motivation, larger structural and institutional factors are almost always equally important if not more so (Adimassu et al. 2016; Djenontin et al. 2018), though harder to capture in research variables (Schwilch et al. 2014). These institutional and governance factors can create an enabling environment for sustainable land management (SLM) practices, or challenges to their adoption (Adimassu et al. 2013). Governance factors include the institutions that manage rules and policies, the social norms and collective actions of participants (including civil society actors and the private sector), and the interactions between them (Ostrom 1990; Huntjens et al. 2012; Davies 2016). Many of Ostrom's design principles for successful governance can be applied to response options for SLM; these principles are:

(i) clearly defined boundaries, (ii) understanding of both benefits and costs, (iii) collective choice arrangements, (iv) monitoring, (v) graduated sanctions, (vi) conflict-resolution mechanisms, (vii) recognition of rights, and (viii) nested (multi-scale) approaches. Unfortunately, studies of many natural resources and land management policy systems – in particular, in developing countries – often show the opposite: a lack of flexibility, strong hierarchical tendencies, and a lack of local participation in institutional frameworks (Ampaire et al. 2017). Analysis of government effectiveness (GE) – defined as quality of public services, policy formulation and implementation, civil service and the degree of its independence from political pressures, as well as credibility of the government's commitment to its policies (Kaufmann et al. 2010) – has been shown to play a key role in land management. GE mediates land-user actions on land management and investment, and government policies and laws can help land users adopt sustainable land management practices (Nkonya et al. 2016) (Figure 6.9).

It is simply not a matter of putting the 'right' institutions or policies in place, however, as governance can be undermined by inattention to power dynamics (Fabinyi et al. 2014). Power shapes how actors gain access and control over resources, and negotiate, transform and adopt certain response options or not. These variable dynamics of power between different levels and stakeholders have an impact on the ability to implement different response options. The inability of many national governments to address social exclusion in general will have an effect on the implementation of many response options. Further, response options themselves can become avenues for actors to exert power claims over others (Nightingale 2017). For example, there have been many concerns that reduced deforestation and forest degradation projects run the risk of reversing trends towards decentralisation in forest management and creating new power disparities between the state and local actors (Phelps et al. 2010). Below we assess how two important factors – the involvement of stakeholders, and the coordination of action across scales – will help in moving from response options to policy implementation, a theme Chapter 7 takes up in further detail.

Involvement of stakeholders

A wide range of stakeholders are necessary for successful land, agricultural and environmental policy, and implementing response options requires that a range of actors, including businesses, consumers, land managers, indigenous peoples and local communities, scientists, and policymakers work together for success. Diverse stakeholders have a particularly important role to play in defining problems, assessing knowledge and proposing solutions (Stokes et al. 2006; Phillipson et al. 2012). Lack of connection between science knowledge and on-the-ground practice has hampered adoption of many response options in the past; simply presenting 'scientifically' derived response options is not enough (Marques et al. 2016). For example, the importance of recognising and incorporating local knowledge and indigenous knowledge is increasingly emphasised in successful policy implementation (see Cross-Chapter Box 13 in Chapter 7), as local practices of water management, soil fertility management, improved grazing, restoration and sustainable management of forests are often well-aligned with response options assessed by scientists (Marques et al. 2016).

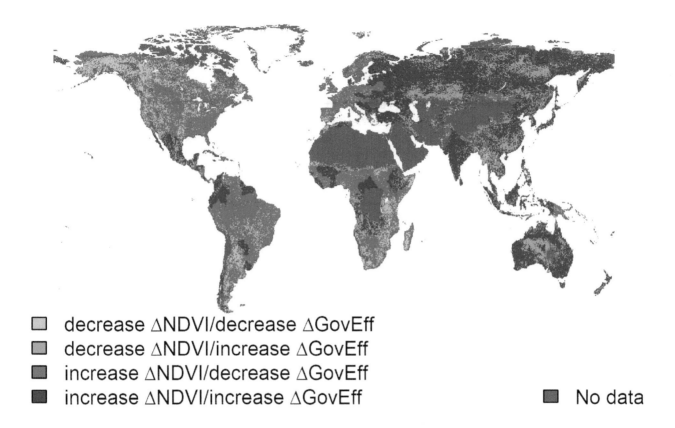

decrease △NDVI/decrease △GovEff

decrease △NDVI/increase △GovEff

increase △NDVI/decrease △GovEff

increase △NDVI/increase △GovEff   No data

Figure 6.9 | Relationship between changes in government effectiveness (GE) and changes in land management. Notes: △NDVI = Change in Normalized Difference Vegetation Index (baseline year 2001, Endline year 2010). Source of NDVI data: MODIS △GovEff = Change in GE (baseline year 2001, endline year 2010). (World Bank; Nkonya et al. 2016).

Stakeholder engagement is an important approach for successful environmental and climate policy and planning. Tools such as stakeholder mapping, in which affected and interested parties are identified and described in terms of their interrelationships and current or future objectives and aspirations, and scenario-based stakeholder engagement, which combines stakeholder analysis with climate scenarios, are increasingly being applied to facilitate better planning outcomes (Tompkins et al. 2008; Pomeroy and Douvere 2008; Star et al. 2016). Facilitated dialogues early in design processes have shown good success in bringing multiple and sometimes conflicting stakeholders to the table to discuss synergies and trade-offs around policy implementation (Gopnik et al. 2012). Knowledge exchange, social learning, and other concepts are also increasingly being incorporated into understanding how to facilitate sustainable land management (Djenontin et al. 2018), as evidence suggests that negotiating the complexity of socio-ecological systems (SES) requires flexible learning arrangements, in particular for multiple stakeholders (Gerlak and Heikkila 2011; Armitage et al. 2018; Heikkila and Gerlak 2018). Social learning has been defined as 'a change in understanding and skills that becomes situated in groups of actors/communities of practice through social interactions,' (Albert et al. 2012), and social learning is often linked with attempts to increase levels of participation in decision-making, from consultation to more serious community control (Collins and Ison 2009; McCrum et al. 2009). Learning also facilitates responses to emerging problems and helps actors in SESs grapple with complexity. One outcome of

learning can be adaptive risk management (ARM), in which 'one takes action based on available information, monitors what happens, learns from the experience and adjusts future actions based on what has been learnt' (Bidwell et al. 2013). Suggestions to facilitate social learning, ARM, and decision-making include extending science-policy networks and using local bridging organisations, such as extension services, for knowledge co-production (Bidwell et al. 2013; Böcher and Krott 2014; Howarth and Monasterolo 2017) (see further discussion in Chapter 7, Section 7.5).

Ensuring that women are included as key stakeholders in response option implementation is also important, as gender norms and roles affect vulnerability and access to resources, and gender inequality limits the possible range of responses for adoption by women (Lambrou and Piana 2006). For example, environmental change may increase women's workload as their access to natural resources may decline, or they may have to take up low-wage labour if agriculture becomes unsuitable in their local areas under climate change (Nelson et al. 2002). Every response option considered in this chapter potentially has a gender dimension to it that needs to be taken into consideration (Tables 6.73–6.75 note how response options intersect with SDG 5 Gender Equality); for example, to address food security through sustainable intensification will clearly have to address female farmers in Africa (Kondylis et al. 2016; Garcia and Wanner 2017) (for further information, see Cross-Chapter Box 11 in Chapter 7).

6

Challenges of coordination

Coordinated action to implement the response options will be required across a range of actors, including business, consumers, land managers, indigenous peoples and local communities and policymakers to create enabling conditions. Conjoining response options to maximise social, climatic and environmental benefits will require framing of such actions as strong pathways to sustainable development (Ayers and Dodman 2010). As the chapter has pointed out, there are many potential options for synergies, especially among several response options that might be applied together and in coordination with one another (such as dietary change and improved land management measures). This coordination will help ensure that synergies are met and trade-offs minimised, but this will require deliberate coordination across multiple scales, actors and sectors. For example, there are a variety of response options available at different scales that could form portfolios of measures applied by different stakeholders from farm to international scales. Agricultural diversification and use of local seeds by smallholders can be particularly useful poverty eradication and biodiversity conservation measures, but are only successful when higher scales, such as national and international markets and supply chains, also value these goods in trade regimes, and consumers see the benefits of purchasing these goods. However, the land and food sectors face particular challenges of institutional fragmentation, and often suffer from a lack of engagement between stakeholders at different scales (Biermann et al. 2009; Deininger et al. 2014) (see Chapter 7, Section 7.6.2).

Many of the response options listed in this chapter could be potentially implemented as 'community-based' actions, including community-based reforestation, community-based insurance, or community-based disaster risk management. Grounding response options in community approaches aims to identify, assist and implement activities 'that strengthen the capacity of local people to adapt to living in a riskier and less predictable climate' (Ayers and Forsyth 2009). Research shows that people willingly come together to provide mutual aid and protection against risk, to manage natural resources, and to work cooperatively to find solutions to environmental provisioning problems. Some activities that fall under this type of collective action include the creation of institutions or rules, working cooperatively to manage a resource by restricting some activities and encouraging others, sharing information to improve public goods, or mobilising resources (such as capital) to fix a collective problem (Ostrom 2000; Poteete and Ostrom 2004), or engagement in participatory land-use planning (Bourgoin 2012; Evers and Hofmeister 2011). These participatory processes 'are likely to lead to more beneficial environmental outcomes through better informed, sustainable decisions, and win-win solutions regarding economic and conservation objectives' (Vente et al. 2016), and evaluations of community-based response options have been generally positive (Karim and Thiel 2017; Tompkins and Adger 2004).

Agrawal (2001) has identified more than 30 different indicators that have been important in understanding who undertakes collective action for the environment, including: the size of the group undertaking action; the type and distribution of the benefits from the action; the heterogeneity of the group; the dependence of the group on these benefits; the presence of leadership; presence of social capital and trust; and autonomy and independence to make and enforce rules. Alternatively, when households expect the government to undertake response actions, they have less incentive to join in collective action, as the state role has 'crowded out' local cooperation (Adger 2009). High levels of social trust and capital can increase willingness of farmers to engage in response options, such as improved soil management or carbon forestry (Stringer et al. 2012; Lee 2017), and social capital helps with connectivity across levels of SESs (Brondizio et al. 2009). Dietz et al. (2013) lay out important policy directions for more successful facilitation of collective action across scales and stakeholders. These include: providing information; dealing with conflict; inducing rule compliance; providing physical, technical or institutional infrastructure; and being prepared for change. The adoption of participatory protocols and structured processes to select response options together with stakeholders will likely lead to greater success in coordination and participation (Bautista et al. 2017; Franks 2010; Schwilch et al. 2012a).

However, wider adoption of community-based approaches is potentially hampered by several factors, including the fact that most are small-scale (Forsyth 2013; Ensor et al. 2014) and it is often unclear how to assess criteria of success (Forsyth 2013). Others also caution that community-based approaches often are not able to adequately address the key drivers of vulnerability such as inequality and uneven power relations (Nagoda and Nightingale 2017).

Moving from response options to policies

Chapter 7 discusses in further depth the risks and challenges involved in formulating policy responses that meet the demands for sustainable land management and development outcomes, such as food security, community adaptation and poverty alleviation. Table 7.1 in Chapter 7 maps how specific response options might be turned into policies; for example, to implement a response option aimed at agricultural diversification, a range of policies from elimination of agricultural subsidies (which might favour single crops) to environmental farm programmes and agro-environmental payments (to encourage alternative crops). Oftentimes, any particular response option might have a variety of potential policy pathways that might address different scales or stakeholders or take on different aspects of coordination and integration (Section 7.6.1). Given the unique challenges of decision-making under uncertainty in future climate scenarios, Chapter 7 particularly discusses the need for flexible, iterative, and adaptive processes to turn response options into policy frameworks.

Cross-Chapter Box 9 | Climate and land pathways

Katherine Calvin (The United States of America), Edouard Davin (France/Switzerland), Margot Hurlbert (Canada), Jagdish Krishnaswamy (India), Alexander Popp (Germany), Prajal Pradhan (Nepal/Germany)

Future development of socio-economic factors and policies influence the evolution of the land–climate system, among others, in terms of the land used for agriculture and forestry. Climate mitigation policies can also have a major impact on land use, especially in scenarios consistent with the climate targets of the Paris Agreement. This includes the use of bio-energy or CDR, such as bioenergy with carbon capture and storage (BECCS) and afforestation. Land-based mitigation options have implications for GHG fluxes, desertification, land degradation, food insecurity, ecosystem services and other aspects of sustainable development.

Shared Socio-economic Pathways

The five pathways are based on the Shared Socio-economic Pathways (SSPs) (O'Neill et al. 2014; Popp et al. 2017; Riahi et al. 2017; Rogelj et al. 2018b) (Cross-Chapter Box 1 in Chapter 1). SSP1 is a scenario with a broad focus on sustainability, including human development, technological development, nature conservation, globalised economy, economic convergence and early international cooperation (including moderate levels of trade). The scenario includes a peak and decline in population, relatively high agricultural yields and a move towards food produced in low-GHG emission systems (Van Vuuren et al. 2017b). Dietary change and reductions in food waste reduce agricultural demands, and effective land-use regulation enables reforestation and/or afforestation. SSP2 is a scenario in which production and consumption patterns, as well as technological development, follows historical patterns (Fricko et al. 2017). Land-based CDR is achieved through bioenergy and BECCS and, to a lesser degree, by afforestation and reforestation. SSP3 is a scenario with slow rates of technological change and limited land-use regulation. Agricultural demands are high due to material-intensive consumption and production, and barriers to trade lead to reduced flows for agricultural goods. In SSP3, forest mitigation activities and abatement of agricultural GHG emissions are limited due to major implementation barriers such as low institutional capacities in developing countries and delays as a consequence of low international cooperation (Fujimori et al. 2017). Emissions reductions are achieved primarily through the energy sector, including the use of bioenergy and BECCS.

Policies in the Pathways

SSPs are complemented by a set of shared policy assumptions (Kriegler et al. 2014), indicating the types of policies that may be implemented in each future world. Integrated Assessment Models (IAMs) represent the effect of these policies on the economy, energy system, land use and climate with the caveat that they are assumed to be effective or, in some cases, the policy goals (e.g., dietary change) are imposed rather than explicitly modelled. In the real world, there are various barriers that can make policy implementation more difficult (Section 7.4.9). These barriers will be generally higher in SSP3 than SSP1.

SSP1: A number of policies could support SSP1 in future, including: effective carbon pricing, emission trading schemes (including net CO_2 emissions from agriculture), carbon taxes, regulations limiting GHG emissions and air pollution, forest conservation (mix of land sharing and land sparing) through participation, incentives for ecosystem services and secure tenure, and protecting the environment, microfinance, crop and livelihood insurance, agriculture extension services, agricultural production subsidies, low export tax and import tariff rates on agricultural goods, dietary awareness campaigns, taxes on and regulations to reduce food waste, improved shelf life, sugar/fat taxes, and instruments supporting sustainable land management, including payment for ecosystem services, land-use zoning, REDD+, standards and certification for sustainable biomass production practices, legal reforms on land ownership and access, legal aid, legal education, including reframing these policies as entitlements for women and small agricultural producers (rather than sustainability) (Van Vuuren et al. 2017b; O'Neill et al. 2017) (Section 7.4).

SSP2: The same policies that support SSP1 could support SSP2 but may be less effective and only moderately successful. Policies may be challenged by adaptation limits (Section 7.4.9), inconsistency in formal and informal institutions in decision-making (Section 7.5.1) or result in maladaptation (Section 7.4.7). Moderately successful sustainable land management policies result in some land competition. Land degradation neutrality is moderately successful. Successful policies include those supporting bioenergy and BECCS (Rao et al. 2017b; Fricko et al. 2017; Riahi et al. 2017) (Section 7.4.6).

SSP3: Policies that exist in SSP1 may or may not exist in SSP3, and are ineffective (O'Neill et al. 2014). There are challenges to implementing these policies, as in SSP2. In addition, ineffective sustainable land management policies result in competition for land between agriculture and mitigation. Land degradation neutrality is not achieved (Riahi et al. 2017). Successful policies include those supporting bioenergy and BECCS (Kriegler et al. 2017; Fujimori et al. 2017; Rao et al. 2017b) (Section 7.4.6). Demand-side food policies are absent and supply-side policies predominate. There is no success in advancing land ownership and access policies for agricultural producer livelihood (Section 7.6.5).

6

Cross-Chapter Box 9 (continued)

Land-use and land-cover change

In SSP1, sustainability in land management, agricultural intensification, production and consumption patterns result in reduced need for agricultural land, despite increases in per capita food consumption. This land can instead be used for reforestation, afforestation and bioenergy. In contrast, SSP3 has high population and strongly declining rates of crop yield growth over time, resulting in increased agricultural land area. SSP2 falls somewhere in between, with societal as well as technological development following historical patterns. Increased demand for land mitigation options such as bioenergy, reduced deforestation or afforestation decreases availability of agricultural land for food, feed and fibre. In the climate policy scenarios consistent with the Paris Agreement, bioenergy/BECCS and reforestation/afforestation play an important role in SSP1 and SSP2. The use of these options, and the impact on land, is larger in scenarios that limit radiative forcing in 2100 to 1.9 W m^{-2} than in the 4.5 W m^{-2} scenarios. In SSP3, the expansion of land for agricultural production implies that the use of land-related mitigation options is very limited, and the scenario is characterised by continued deforestation.

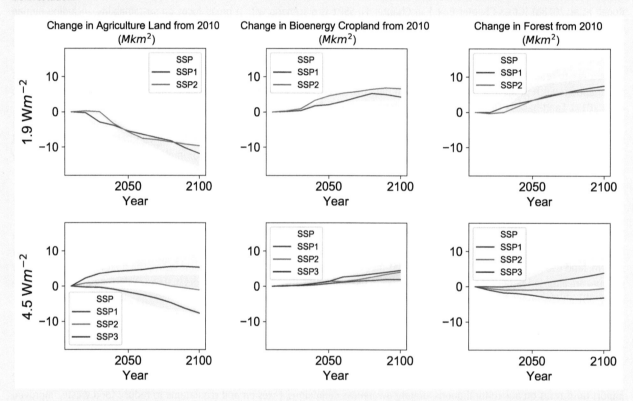

Cross-Chapter Box 9, Figure 1 | Changes in agriculture land (left), bioenergy cropland (middle) and forest (right) under three different SSPs (colours) and two different warming levels (rows). Agricultural land includes both pasture and cropland. Colours indicate SSPs, with SSP1 shown in green, SSP2 in yellow, and SSP3 in red. For each pathway, the shaded areas show the range across all IAMs; the line indicates the median across models. There is no SSP3 in the top row, as 1.9 W m^{-2} is infeasible in this world. Data is from an update of the Integrated Assessment Modelling Consortium (IAMC) Scenario Explorer developed for the SR15 (Huppmann et al. 2018; Rogelj et al. 2018a).

Implications for mitigation and other land challenges

The combination of baseline emissions development, technology options, and policy support makes it much easier to reach the climate targets in the SSP1 scenario than in the SSP3 scenario. As a result, carbon prices are much higher in SSP3 than in SSP1. In fact, the 1.9 W m^{-2} target was found to be infeasible in the SSP3 world (Table 1 in Cross-Chapter Box 9). Energy system CO$_2$ emissions reductions are greater in SSP3 than in SSP1 to compensate for the higher land-based CO$_2$ emissions.

Accounting for mitigation and socio-economics alone, food prices (an indicator of food insecurity) are higher in SSP3 than in SSP1 and higher in the 1.9 W m^{-2} target than in the 4.5 W m^{-2} target (Table 1 in Cross-Chapter Box 9). Forest cover is higher in SSP1 than SSP3 and higher in the 1.9 W m^{-2} target than in the 4.5 W m^{-2} target. Water withdrawals and water scarcity are, in general, higher in SSP3 than SSP1 (Hanasaki et al. 2013; Graham et al. 2018) and higher in scenarios with more bioenergy (Hejazi et al. 2014b); however, these indicators have not been quantified for the specific SSP-representative concentration pathways (RCP) combinations discussed here.

Cross-Chapter Box 9 (continued)

Cross-Chapter Box 9, Table 1 | Quantitative indicators for the pathways. Each cell shows the mean, minimum, and maximum value across IAM models for each indicator and each pathway in 2050 and 2100. All IAMs that provided results for a particular pathway are included here. Note that these indicators exclude the implications of climate change. Data is from an update of the IAMC Scenario Explorer developed for the SR15 (Huppmann et al. 2018; Rogelj et al. 2018b).

		SSP1		SSP2		SSP3	
		1.9 W m^{-2} mean (max., min.)	4.5 W m^{-2} mean (max., min.)	1.9 W m^{-2} mean (max., min.)	4.5 W m^{-2} mean (max., min.)	1.9 W m^{-2} mean (max., min.)	4.5 W m^{-2} mean (max., min.)
Population (billion)	2050	8.5 (8.5, 8.5)	8.5 (8.5, 8.5)	9.2 (9.2, 9.2)	9.2 (9.2, 9.2)	N/A	10.0 (10.0, 10.0)
	2100	6.9 (7.0, 6.9)	6.9 (7.0, 6.9)	9.0 (9.0, 9.0)	9.0 (9.1, 9.0)	N/A	12.7 (12.8, 12.6)
Change in GDP per capita (% rel to 2010)	2050	170.3 (380.1, 130.9)	175.3 (386.2, 166.2)	104.3 (223.4, 98.7)	110.1 (233.8, 103.6)	N/A	55.1 (116.1, 46.7)
	2100	528.0 (1358.4, 408.2)	538.6 (1371.7, 504.7)	344.4 (827.4, 335.8)	356.6 (882.2, 323.3)	N/A	71.2 (159.7, 49.6)
Change in forest cover (Mkm^2)	2050	3.4 (9.4, -0.1)	0.6 (4.2, −0.7)	3.4 (7.0, −0.9)	−0.9 (2.9, −2.5)	N/A	−2.4 (−1.0, −4.0)
	2100	7.5 (15.8, 0.4)	3.9 (8.8, 0.2)	6.4 (9.5, −0.8)	−0.5 (5.9, −3.1)	N/A	−3.1 (−0.3, −5.5)
Change in cropland (Mkm^2)	2050	−1.2 (−0.3, −4.6)	0.1 (1.5, −3.2)	−1.2 (0.3, −2.0)	1.2 (2.7, −0.9)	N/A	2.3 (3.0, 1.2)
	2100	−5.2 (−1.8, −7.6)	−2.3 (−1.6, −6.4)	−2.9 (0.1, −4.0)	0.7 (3.1, −2.6)	N/A	3.4 (4.5, 1.9)
Change in energy cropland (Mkm^2)	2050	2.1 (5.0, 0.9)	0.8 (1.3, 0.5)	4.5 (7.0, 2.1)	1.5 (2.1, 0.1)	N/A	1.3 (2.0, 1.3)
	2100	4.3 (7.2, 1.5)	1.9 (3.7, 1.4)	6.6 (11.0, 3.6)	4.1 (6.3, 0.4)	N/A	4.6 (7.1, 1.5)
Change in pasture (Mkm^2)	2050	−4.1 (−2.5, −5.6)	−2.4 (−0.9, −3.3)	−4.8 (−0.4, −6.2)	−0.1 (1.6, −2.5)	N/A	2.1 (3.8, −0.1)
	2100	−6.5 (−4.8, −12.2)	−4.6 (−2.7, −7.3)	−7.6 (−1.3, −11.7)	−2.8 (1.9, −5.3)	N/A	2.0 (4.4, −2.5)
Change in other natural land (Mkm^2)	2050	0.5 (1.0, −4.9)	0.5 (1.7, −1.0)	−2.2 (0.6, −7.0)	−2.2 (0.7, −2.2)	N/A	−3.4 (−2.0, −4.4)
	2100	0.0 (7.1, −7.3)	1.8 (6.0, −1.7)	−2.3 (2.7, −9.6)	−3.4 (1.5, −4.7)	N/A	−6.2 (−5.4, −6.8)
Carbon price (2010 USD per tCO$_2$)[a]	2050	510.4 (4304.0, 150.9)	9.1 (35.2, 1.2)	756.4 (1079.9, 279.9)	37.5 (73.4, 13.6)	N/A	67.2 (75.1, 60.6)
	2100	2164.0 (350, 37.7, 262.7)	64.9 (286.7, 42.9)	4353.6 (10149.7, 2993.4)	172.3 (597.9, 112.1)	N/A	589.6 (727.2, 320.4)
Food price (Index 2010=1)	2050	1.2 (1.8, 0.8)	0.9 (1.1, 0.7)	1.6 (2.0, 1.4)	1.1 (1.2, 1.0)	N/A	1.2 (1.7, 1.1)
	2100	1.9 (7.0, 0.4)	0.8 (1.2, 0.4)	6.5 (13.1, 1.8)	1.1 (2.5, 0.9)	N/A	1.7 (3.4, 1.3)
Increase in Warming above pre-industrial (°C)	2050	1.5 (1.7, 1.5)	1.9 (2.1, 1.8)	1.6 (1.7, 1.5)	2.0 (2.0, 1.9)	N/A	2.0 (2.1, 2.0)
	2100	1.3 (1.3, 1.3)	2.6 (2.7, 2.4)	1.3 (1.3, 1.3)	2.6 (2.7, 2.4)	N/A	2.6 (2.6, 2.6)
Change in per capita demand for food, crops (% rel to 2010)[b]	2050	6.0 (10.0, 4.5)	9.1 (12.4, 4.5)	4.6 (6.7, −0.9)	7.9 (8.0, 5.2)	N/A	2.4 (5.0, 2.3)
	2100	10.1 (19.9, 4.8)	15.1 (23.9, 4.8)	11.6 (19.2, −10.8)	11.7 (19.2, 4.1)	N/A	2.0 (3.4, −1.0)
Change in per capita demand for food, animal products (% rel to 2010)[b,c]	2050	6.9 (45.0, −20.5)	17.9 (45.0, −20.1)	7.1 (36.0, 1.9)	10.3 (36.0, −4.2)	N/A	3.1 (5.9, 1.9)
	2100	−3.0 (19.8, −27.3)	21.4 (44.1, −26.9)	17.0 (39.6, −24.1)	20.8 (39.6, −5.3)	N/A	−7.4 (−0.7, −7.9)
Agriculture, forestry and other land-use (AFOLU) CH$_4$ Emissions (% relative to 2010)	2050	−39.0 (−3.8, −68.9)	−2.9 (22.4, −23.9)	−11.7 (31.4, −59.4)	7.5 (43.0, −15.5)	N/A	15.0 (20.1, 3.1)
	2100	−60.5 (−41.7, −77.4)	−47.6 (−24.4, −54.1)	−40.3 (33.1, −58.4)	−13.0 (63.7, −45.0)	N/A	8.0 (37.6, −9.1)
AFOLU N$_2$O Emissions (% relative to 2010)	2050	−13.1 (−4.1, −26.3)	0.1 (34.6, −14.5)	8.8 (38.4, −14.5)	25.4 (37.4, 5.5)	N/A	34.0 (50.8, 29.3)
	2100	−42.0 (4.3, −49.4)	−25.6 (−3.4, −51.2)	−1.7 (46.8, −37.8)	19.5 (66.7, −21.4)	N/A	53.9 (65.8, 30.8)
Cumulative Energy CO$_2$ Emissions until 2100 (GtCO$_2$)		428.2 (1009.9, 307.6)	2787.6 (3213.3, 2594.0)	380.8 (552.8, −9.4)	2642.3 (2928.3, 2515.8)	N/A	2294.5 (2447.4, 2084.6)
Cumulative AFOLU CO$_2$ Emissions until 2100 (GtCO$_2$)		−127.3 (5.9, −683.0)	−54.9 (52.1, −545.2)	−126.8 (153.0, −400.7)	40.8 (277.0, −372.9)	N/A	188.8 (426.6, 77.9)

[a] SSP2–19 is infeasible in two models. One of these models sets the maximum carbon price in SSP1–19; the carbon price range is smaller for SSP2–19 as this model is excluded there. Carbon prices are higher in SSP2–19 than SSP1–19 for every model that provided both simulations.

[b] Food demand estimates include waste.

[c] Animal product demand includes meat and dairy.

Climate change results in higher impacts and risks in the 4.5 W m⁻² world than in the 1.9 W m⁻² world for a given SSP and these risks are exacerbated in SSP3 compared to SSP1 and SSP2 due to the population's higher exposure and vulnerability. For example, the risk of fire is higher in warmer worlds; in the 4.5 W m⁻² world, the population living in fire prone regions is higher in SSP3 (646 million) than in SSP2 (560 million) (Knorr et al. 2016). Global exposure to multi-sector risk quadruples between 1.5°C[11] and 3°C and is a factor of six higher in SSP3-3°C than in SSP1–1.5°C (Byers et al. 2018). Future risks resulting from desertification, land degradation and food insecurity are lower in SSP1 compared to SSP3 at the same level of warming. For example, the transition moderate-to-high risk of food insecurity occurs between 1.3 and 1.7°C for SSP3, but not until 2.5 to 3.5°C in SSP1 (Section 7.2).

Summary

Future pathways for climate and land use include portfolios of response and policy options. Depending on the response options included, policy portfolios implemented, and other underlying socio-economic drivers, these pathways result in different land-use consequences and their contribution to climate change mitigation. Agricultural area declines by more than 5 Mkm² in one SSP but increases by as much as 5 Mkm² in another. The amount of energy cropland ranges from nearly zero to 11 Mkm², depending on the SSP and the warming target. Forest area declines in SSP3 but increases substantially in SSP1. Subsequently, these pathways have different implications for risks related to desertification, land degradation, food insecurity, and terrestrial GHG fluxes, as well as ecosystem services, biodiversity, and other aspects of sustainable development.

6.4.5   Potential consequences of delayed action

Delayed action, in terms of overall GHG mitigation across both land and energy sectors, as well as delayed action in implementing the specific response options outlined in this chapter, will exacerbate the existing land challenges due to the continued impacts of climate change and socio-economic and other pressures. It can decrease the potential of response options and increase the costs of deployment, and will deprive communities of immediate co-benefits, among other pressures. The major consequences of delayed action are outlined below.

Delayed action exposes vulnerable people to continued and increasing climate impacts: Slower or delayed action in implementing response options exacerbates existing inequalities and impacts. This will increase the number of people vulnerable to climate change, due to population increases and increasing climate impacts (IPCC 2018; AR 5). Future climate change will lead to exacerbation of the existing land challenges, increased pressure on agricultural livelihoods, potential for rapid land degradation, and millions more people exposed to food insecurity (Schmidhuber and Tubiello 2007) (Chapters 3, 4 and 5). Delay can also bring political risks and significant social impacts, including risks to human settlements (particularly in coastal areas), large-scale migration, and conflict (Barnett and Adger 2007; Hsiang et al. 2013). Early action reducing vulnerability and exposure can create an opportunity for a virtual circle of benefits: increased resilient livelihoods, reduced degradation of land, and improved food security (Bohle et al. 1994).

Delayed action increases requirements for adaptation: Failure to mitigate climate change will increase requirements for adaptation. For example, it is likely that by 2100 with no mitigation or adaptation, 31–69 million people world-wide could be exposed to flooding

(Rasmussen et al. 2017; IPCC SR15) (Chapter 3); such outcomes could be prevented with investments in both mitigation and adaptation now. Some specific response options (e.g., reduced deforestation and forest degradation, reduced peatland and wetland conversion) prevent further detrimental effects to the land surface; delaying these options could lead to increased deforestation, conversion, or degradation, serving as increased sources of GHGs and having concomitant negative impacts on biodiversity and ecosystem services (Section 6.2). Response options aimed at land restoration and rehabilitation can serve as adaptation mechanisms for communities facing climatic stresses like precipitation variability and changes in land quality, as well as provide benefits in terms of mitigation.

Delayed action increases response costs and reduces economic growth: Early action on reducing emissions through mitigation is estimated to result in smaller temperature increases as well as lower mitigation costs than delayed action (Sanderson et al. 2016; Luderer et al. 2013, 2018; Rose et al. 2017; Van Soest et al. 2017; Fujimori et al. 2017). The cost of inaction to address mitigation, adaptation, and sustainable land use exceeds the cost of immediate action in most countries, depending on how damage functions and the social cost of carbon are calculated (Dell et al. 2012; Moore and Diaz 2015). Costs of acting now would be one to two orders of magnitude lower than economic damages from delayed action, including damage to assets from climate impacts, as well as potentially reduced economic growth, particularly in developing countries (Luderer et al. 2016; Moore and Diaz 2015; Luderer et al. 2013). Increased health costs and costs of energy (e.g., to run air-conditioners to combat increased heat waves) in the US by the end of the century alone are estimated to range from 10–58% of US GDP in 2010 (Deschênes and Greenstone 2011).

---

[11]   Pathways that limit radiative forcing in 2100 to 1.9 W m⁻² result in median warming in 2100 to 1.5°C in 2100 (Rogelj et al. 2018b). Pathways limiting radiative forcing in 2100 to 4.5 W m⁻² result in median warming in 2100 above 2.5°C (IPCC 2014).

Delay also increases the costs of both mitigation and adaptation actions at later dates. In models of climate-economic interactions, deferral of emissions reduction now requires trade-offs leading to higher costs of several orders of magnitude and risks of higher temperatures in the longer term (Luderer et al. 2013). Further, the cost of action is likely to increase over time due to the increased severity of challenges in future scenarios.

Conversely, timely implementation of response options brings economic benefits. Carbon pricing is one economic component to encourage adoption of response options (Jakob et al. 2016), but carbon pricing alone can induce higher risk in comparison to other scenarios and pathways that include additional targeted sustainability measures, such as promotion of less material- and energy-intensive lifestyles and healthier diets, as noted in our response options (Bertram et al. 2018). While the short-term costs of deployment of actions may increase, better attainment of a broad set of sustainability targets can be achieved through these combined measures (Bertram et al. 2018).

There are also investments now that can lead to immediate savings in terms of avoided damages; for example, for each dollar spent on disaster risk management, countries accrue avoided disaster-related economic losses of 4 USD or more (Mechler 2016). While they can require upfront investment, the economic benefits of actions to ensure sustainable land management, such as increased soil organic carbon, can more than double the economic value of rangelands and improve crop yields (Chapter 4 and Section 6.2).

Delayed action reduces future policy space and decreases efficacy of some response options: The potential for some response options decreases as climate change increases; for example, climate alters the sink capacity for soil and vegetation carbon sequestration, reducing the potential for increased soil organic carbon, afforestation and reforestation (Section 6.4.2). Additionally, climate change affects the productivity of bioenergy crops, influencing the potential mitigation of bioenergy and BECCS (Section 6.4.4).

For response options in the supply chain, demand-side management, and risk management, while the consequences of delayed action are apparent in terms of continued GHG emissions from drivers, the tools for response options are not made more difficult by delay and could be deployed at any time. Additionally, given increasing pressures on land as a consequence of delay, some policy response options may become more cost-effective while others become costlier. For example, over time, land-based mitigation measures like forest and ecosystem protection are likely to increase land scarcity, leading to higher food prices; while demand-side measures, like reduced-impact diets and reducing waste, are less likely to raise food prices in economic models (Stevanović et al. 2017).

For risk management, some response options provide timely and rapidly deployable solutions for preventing further problems, such as disaster risk management and risk-sharing instruments. For example, early warning systems serve multiple roles in protecting lives and property and helping people adapt to longer-term climate changes, and can be used immediately.

Delaying action can also result in problems of irreversibility of biophysical impacts and tipping points: Early action provides a potential way to avoid irreversibility – such as degradation of ecosystems that cannot be restored to their original baseline – and tipping points, whereby ecological or climate systems abruptly shift to a new state. Ecosystems, such as peatlands, are particularly vulnerable to irreversibility because of the difficulties of rewetting to original states (Section 6.2), and dryland grazing systems are vulnerable to tipping points when ground cover falls below 50%, after which productivity falls, infiltration declines, and erosion increases (Chapters 3 and 4). Further, tipping points can be especially challenging for human populations to adapt to, given the lack of prior experience with such system shifts (Kates et al. 2012; Nuttall 2012).

Policy responses require lead time for implementation; delay makes this worse: For all the response options, particularly those that need to be deployed through policy implementation, there are unavoidable lags in this cycle. 'Policy lags', by which implementation is delayed by the slowness of the policy implementation cycle, are significant across many land-based, response options (Brown et al. 2019). Further, the behavioural change necessary to achieve some demand-side and risk management response options often takes a long time, and delay only lengthens this process (Stern 1992; Steg and Vlek 2009). For example, actively promoting the need for healthier and more sustainable diets through individual dietary decisions is an important underpinning and enabling step for future changes, but is likely to be a slow-moving process, and delay in beginning will only exacerbate this.

Delay can lead to lock-in: Delay in implementation can cause 'lock-in' as decisions made today can constrain future development and pathways. For example, decisions made now on where to build infrastructure, make investments and deploy technologies, will have longer-term (decades-long) ramifications due to the inertia of capital stocks (Van Soest et al. 2017). In tandem, the vulnerability of the poor is likely to be exacerbated by climate change, creating a vicious circle of lock in whereby an increasing share of the dwindling carbon budget may be needed to assist with improved energy use for the poorest (Lamb and Rao 2015).

Delay can increase the need for widespread deployment of land-based mitigation (afforestation, BECCS) (IPCC 2018; Strefler et al. 2018): Further delays in mitigation could result in an increased need for carbon dioxide removal (CDR) options later; for example, delayed mitigation requires a 10% increase in cumulative CDR over the century (IPCC 2018). Similarly, strengthening near-term mitigation effort can reduce the CDR requirements in 2100 by a factor of 2 to 8 (Strefler et al. 2018). Conversely, scenarios with limited CDR require earlier emissions reductions (Van Vuuren et al. 2017b) and may make more stringent mitigation scenarios, like the 1.5°C, infeasible (Kriegler et al. 2018a,b).

Frequently Asked Questions

FAQ 6.1 | What types of land-based options can help mitigate and adapt to climate change?

Land-based options that help mitigate climate change are various and differ greatly in their potential. The options with moderate-to-large mitigation potential, and no adverse side effects, include options that decrease pressure on land (e.g., by reducing the land needed for food production) and those that help to maintain or increase carbon stores both above-ground (e.g., forest measures, agroforestry, fire management) and below-ground (e.g., increased soil organic matter or reduced losses, cropland and grazing land management, urban land management, reduced deforestation and forest degradation). These options also have co-benefits for adaptation by improving health, increasing yields, flood attenuation and reducing urban heat island effects. Another group of practices aim at reducing greenhouse gas (GHG) emission sources, such as livestock management or nitrogen fertilisation management. Land-based options delivering climate change adaptation may be structural (e.g., irrigation and drainage systems, flood and landslide control), technological (e.g., new adapted crop varieties, changing planting zones and dates, using climate forecasts), or socio-economic and institutional (e.g., regulation of land use, associativity between farmers). Some adaptation options (e.g., new planting zones, irrigation) may have adverse side effects for biodiversity and water. Adaptation options may be planned, such as those implemented at regional, national or municipal level (top-down approaches), or autonomous, such as many technological decisions taken by farmers and local inhabitants. In any case, their effectiveness depends greatly on the achievement of resilience against extreme events (e.g., floods, droughts, heat waves, etc.).

FAQ 6.2 | Which land-based mitigation measures could affect desertification, land degradation or food security?

Some options for mitigating climate change are based on increasing carbon stores, both above-ground and below-ground, so mitigation is usually related to increases in soil organic matter content and increased land cover by perennial vegetation. There is a direct relationship, with very few or no adverse side effects for prevention or reversal of desertification and land degradation and the achievement of food security. This is because desertification and land degradation are closely associated with soil organic matter losses and the presence of bare ground surfaces. Food security depends on the achievement of healthy crops and high and stable yields over time, which is difficult to achieve in poor soils that are low in organic matter.

FAQ 6.3 | What is the role of bioenergy in climate change mitigation, and what are its challenges?

Plants absorb carbon as they grow. If plant-based material (biomass) is used for energy, the carbon it absorbed from the atmosphere is released back. Traditional use of bioenergy for cooking and heating is still widespread throughout the world. Modern conversion to electricity, heat, gas and liquid fuels can reduce the need to burn fossil fuels and this can reduce GHG emissions, helping to mitigate climate change. However, the total amount of emissions avoided depends on the type of biomass, where it is grown, how it is converted to energy, and what type of energy source it displaces. Some types of bioenergy require dedicated land (e.g., canola for biodiesel, perennial grasses, short rotation woody crops), while others can be co-produced or use agricultural or industrial residues (e.g., residues from sugar and starch crops for ethanol, and manure for biogas). Depending on where, how, and the amount of bioenergy crops that are grown, the use of dedicated land for bioenergy could compete with food crops or other mitigation options. It could also result in land degradation, deforestation or biodiversity loss. In some circumstances, however, bioenergy can be beneficial for land, for example, by increasing soil organic carbon. The use of co-products and residues for bioenergy limits the competition for land with food but could result in land degradation if carbon and nutrient-rich material is removed that would otherwise be left on the land. On the other hand, the by-products of some bioenergy conversion processes can be returned to the land as a fertiliser and may have other co-benefits (e.g., reducing pollution associated with manure slurry).

6

References

Accorsi, R., A. Gallo, and R. Manzini, 2017: A climate driven decision-support model for the distribution of perishable products. *J. Clean. Prod.*, **165**, 917–929, doi:10.1016/j.jclepro.2017.07.170.

Adamo, S.B., 2010: Environmental migration and cities in the context of global environmental change. *Curr. Opin. Environ. Sustain.*, **2**, 161–165, doi:10.1016/J.COSUST.2010.06.005.

Adger, W.N., 1999: Social vulnerability to climate change and extremes in coastal Vietnam. *World Dev.*, **27**, 249–269, doi:10.1016/S0305-750X(98)00136-3.

Adger, W.N., 2009: Social capital, collective action, and adaptation to climate change. *Econ. Geogr.*, **79**, 387–404, doi:10.1111/j.1944-8287.2003.tb00220.x.

Adger, W.N., N. Brooks, G. Bentham, and M. Agnew, 2004: *New Indicators of Vulnerability and Adaptive Capacity*. Tyndall Centre for Climate Change Research, Norwich, UK, 128 pp.

Adger, W.N. et al., 2011: Resilience implications of policy responses to climate change. *Wiley Interdiscip. Rev. Clim. Chang.*, **2**, 757–766, doi:10.1002/wcc.133.

Adimassu, Z., A. Kessler, and L. Stroosnijder, 2013: Exploring co-investments in sustainable land management in the Central Rift Valley of Ethiopia. *Int. J. Sustain. Dev. World Ecol.*, **20**, 32–44, doi:10.1080/13504509.2012.740690.

Adimassu, Z., S. Langan, and R. Johnston, 2016: Understanding determinants of farmers' investments in sustainable land management practices in Ethiopia: Review and synthesis. *Environ. Dev. Sustain.*, **18**, 1005–1023, doi:10.1007/s10668-015-9683-5.

Aggarwal, P.K. et al., 2018: The climate-smart village approach: Framework of an integrative strategy for scaling up adaptation options in agriculture. *Ecol. Soc.*, **23**, art14, doi:10.5751/ES-09844-230114.

Agrawal, A., 2001: Common property institutions and sustainable governance of resources. *World Dev.*, **29**, 1649–1672, doi:10.1016/S0305-750X(01)00063-8.

Ahlgren, E.O., M. Börjesson Hagberg, and M. Grahn, 2017: Transport biofuels in global energy–economy modelling – A review of comprehensive energy systems assessment approaches. *GCB Bioenergy*, **9**, 1168–1180, doi:10.1111/gcbb.12431.

Ahmed, S., and J.R. Stepp, 2016: Beyond yields: Climate change effects on specialty crop quality and agroecological management. *Elem. Sci. Anthr.*, **4**, 92, doi:10.12952/journal.elementa.000092.

Ajibade, I., and G. McBean, 2014: Climate extremes and housing rights: a political ecology of impacts, early warning and adaptation constraints in Lagos slum communities. *Geoforum*, **55**, 76–86, doi:10.1016/j.geoforum.2014.05.005.

Akhtar, P., Y. Tse, Z. Khan, and R. Rao-Nicholson, 2016: Data-driven and adaptive leadership contributing to sustainability: Global agri-food supply chains connected with emerging markets. *Int. J. Prod. Econ.*, **181**, 392–401.

Akter, S., T.J. Krupnik, F. Rossi, and F. Khanam, 2016: The influence of gender and product design on farmers' preferences for weather-indexed crop insurance. *Glob. Environ. Chang.*, **38**, 217–229, doi:10.1016/j.gloenvcha.2016.03.010.

Al-Mansour F., and Jejcic V, 2017: A model calculation of the carbon footprint of agricultural products: The case of Slovenia. *Energy*, **136**, 7–15, doi:10.1016/j.energy.2016.10.099.

Albert, C., T. Zimmermann, J. Knieling, and C. von Haaren, 2012: Social learning can benefit decision-making in landscape planning: Gartow case study on climate change adaptation, Elbe valley biosphere reserve. *Landsc. Urban Plan.*, **105**, 347–360, doi:10.1016/j.landurbplan.2011.12.024.

Albrecht, A., and S.T. Kandji, 2003: Carbon sequestration in tropical agroforestry systems. *Agric. Ecosyst. Environ.*, **99**, 15–27, doi:10.1016/S0167-8809(03)00138-5.

Alderman, H., 2010: Safety nets can help address the risks to nutrition from increasing climate variability. *J. Nutr.*, **140**, 148S–152S, doi:10.3945/jn.109.110825.

Aleksandrowicz, L., R. Green, E.J.M. Joy, P. Smith, and A. Haines, 2016: The impacts of dietary change on greenhouse gas emissions, land use, water use, and health: A systematic review. *PLoS One*, **11**, e0165797, doi:10.1371/journal.pone.0165797.

Alessa, L. et al., 2016: The role of indigenous science and local knowledge in integrated observing systems: Moving toward adaptive capacity indices and early warning systems. *Sustain. Sci.*, **11**, 91–102, doi:10.1007/s11625-015-0295-7.

Alexander, P. et al., 2015: Drivers for global agricultural land use change: The nexus of diet, population, yield and bioenergy. *Glob. Environ. Chang.*, **35**, 138–147, doi:10.1016/j.gloenvcha.2015.08.011.

Alexander, P., C. Brown, A. Arneth, J. Finnigan, and M.D.A. Rounsevell, 2016: Human appropriation of land for food: The role of diet. *Glob. Environ. Chang.*, **41**, 88–98, doi:10.1016/J.GLOENVCHA.2016.09.005.

Ali, G., N. Pumijumnong, and S. Cui, 2018: Valuation and validation of carbon sources and sinks through land cover/use change analysis: The case of Bangkok metropolitan area. *Land use policy*, **70**, 471–478, doi:10.1016/J.LANDUSEPOL.2017.11.003.

Alkama, R., and A. Cescatti, 2016: Biophysical climate impacts of recent changes in global forest cover. *Science.*, **351**, 600–604, doi:10.1126/science.aac8083.

Allington, G., and T. Valone, 2010: Reversal of desertification: the role of physical and chemical soil properties. *J. Arid Environ.*, **74**, 973–977.

Altieri, M.A., and C.I. Nicholls, 2017: The adaptation and mitigation potential of traditional agriculture in a changing climate. *Clim. Change*, **140**, 33–45, doi:10.1007/s10584-013-0909-y.

Altieri, M.A., F.R. Funes-Monzote, and P. Petersen, 2012: Agroecologically efficient agricultural systems for smallholder farmers: Contributions to food sovereignty. *Agron. Sustain. Dev.*, **32**, 1–13, doi:10.1007/s13593-011-0065-6.

Altieri, M.A., C.I. Nicholls, A. Henao, and M.A. Lana, 2015: Agroecology and the design of climate change-resilient farming systems. *Agron. Sustain. Dev.*, **35**, 869–890, doi:10.1007/s13593-015-0285-2.

Ampaire, E.L. et al., 2017: Institutional challenges to climate change adaptation: A case study on policy action gaps in Uganda. *Environ. Sci. Policy*, **75**, 81–90, doi:10.1016/j.envsci.2017.05.013.

Anderson, C.M., C.B. Field, and K.J. Mach, 2017: Forest offsets partner climate-change mitigation with conservation. *Front. Ecol. Environ.*, **15**, 359–365, doi:10.1002/fee.1515.

Anderson, H.R., 2017: Implications for the science of air pollution and health. *Lancet Respir. Med.*, **5**, 916–918, doi:10.1016/S2213-2600(17)30396-X.

Anderson, K., and G. Peters, 2016: The trouble with negative emissions. *Science.*, **354**, 182–183, doi:10.1126/science.aah4567.

Anenberg, S. et al., 2012: Global air quality and health co-benefits of mitigating near-term climate change through methane and black carbon emission controls. *Environ. Health Perspect.*, **120**, 831–839. doi:10.1289/ehp.1104301.

Angel, S. et al., 2005: *The Dynamics of Global Urban Expansion*. World Bank, Transport and Urban Development Department, Washington, DC, USA, 205 pp.

Annan, F., and W. Schlenker, 2015: Federal crop insurance and the disincentive to adapt to extreme heat. *Am. Econ. Rev.*, **105**, 262–266, doi:10.1257/aer.p20151031.

Ansah, I.G.K., B.K.D. Tetteh, and S.A. Donkoh, 2017: Determinants and income effect of yam postharvest loss management: Evidence from the Zabzugu District of Northern Ghana. *Food Secur.*, **9**, 611–620, doi:10.1007/s12571-017-0675-1.

Antille, D.L., J.M. Bennett, and T.A. Jensen, 2016: Soil compaction and controlled traffic considerations in Australian cotton-farming systems. *Crop Pasture Sci.*, **67**, 1, doi:10.1071/CP15097.

Antwi-Agyei, P., L.C. Stringer, and A.J. Dougill, 2014: Livelihood adaptations to climate variability: Insights from farming households in Ghana. *Reg. Environ. Chang.*, **14**, 1615–1626, doi:10.1007/s10113-014-0597-9.

Archer, S. et al., 2011: Brush management as a rangeland conservation strategy: A critical evaluation. In: *Conservation benefits of rangeland practices: assessment, recommendations, and knowledge gaps.* [Briske, D.D. (ed.)]. USDA-NRCS, Washington DC, USA, pp 105–170.

Armitage, D. et al., 2018: An approach to assess learning conditions, effects and outcomes in environmental governance. *Environ. Policy Gov.*, **28**, 3–14, doi:10.1002/eet.1781.

Arnáez J, Lana-Renault N, Lasanta T, Ruiz-Flaño P, and Castroviejo J, 2015: Effects of farming terraces on hydrological and geomorphological processes: A review. *Catena*, **128**, 122–134.

Arndt, C., M. Hussain, V. Salvucci, and L. Østerdal, 2016: Effects of food price shocks on child malnutrition: The Mozambican experience 2008/2009. *Econ. Hum. Biol.*, **22**, 1–13.

Arora, V.K., and A. Montenegro, 2011: Small temperature benefits provided by realistic afforestation efforts. *Nat. Geosci.*, **4**, 514–518, doi:10.1038/ngeo1182.

Arora, V.K., and J.R. Melton, 2018: Reduction in global area burned and wildfire emissions since 1930s enhances carbon uptake by land. *Nat. Commun.*, **9**, 1326, doi:10.1038/s41467-018-03838-0.

Arslan, A. et al., 2018: Diversification under climate variability as part of a CSA strategy in rural Zambia. *J. Dev. Stud.*, **54**, 457–480, doi:10.1080/00 220388.2017.1293813.

Asfaw, S., G. Pallante, and A. Palma, 2018: Diversification strategies and adaptation deficit: Evidence from rural communities in Niger. *World Dev.*, **101**, 219–234, doi:10.1016/J.WORLDDEV.2017.09.004.

Auld, G., L.H. Gulbrandsen, and C.L. McDermott, 2008: Certification schemes and the impacts on forests and forestry. *Annu. Rev. Environ. Resour.*, **33**, 187–211, doi:10.1146/annurev.environ.33.013007.103754.

Avetisyan, M., T. Hertel, and G. Sampson, 2014: Is local food more environmentally friendly? The GHG emissions impacts of consuming imported versus domestically produced food. *Environ. Resour. Econ.*, **58**, 415–462, doi:10.1007/s10640-013-9706-3.

Ayers, J., and T. Forsyth, 2009: Community-based adaptation to climate change. *Environ. Sci. Policy Sustain. Dev.*, **51**, 22–31, doi:10.3200/ENV.51.4.22-31.

Ayers, J., and D. Dodman, 2010: Climate change adaptation and development I: The state of the debate. *Prog. Dev. Stud.*, **10**, 161–168, doi:10.1177/146 499340901000205.

Baccini, A. et al., 2017: Tropical forests are a net carbon source based on aboveground measurements of gain and loss. *Science*, **358**, 230–234, doi:10.1126/science.aam5962.

Bailey, R., 2013: The "food versus fuel" nexus. *Handb. Glob. Energy Policy*, **8**, 265–281, doi:10.1002/9781118326275.ch16.

Bajželj, B., K. Richards, B. Bajželj, and K.S. Richards, 2014a: The positive feedback loop between the impacts of climate change and agricultural expansion and relocation. *Land*, **3**, 898–916, doi:10.3390/land3030898.

Bajželj, B. et al., 2014b: Importance of food-demand management for climate mitigation. *Nat. Clim. Chang.*, **4**, 924–929, doi:10.1038/nclimate2353.

Baker, J.S., C.M. Wade, B.L. Sohngen, S. Ohrel, and A.A. Fawcett, 2019: Potential complementarity between forest carbon sequestration incentives and biomass energy expansion. *Energy Policy*, **126**, 391–401, doi:10.1016/J.ENPOL.2018.10.009.

Bala, G., N. Devaraju, R.K. Chaturvedi, K. Caldeira, and R. Nemani, 2013: Nitrogen deposition: How important is it for global terrestrial carbon uptake? *Biogeosciences*, **10**, 7147–7160, doi:10.5194/bg-10-7147-2013.

Baldos, U.L.C., and T.W. Hertel, 2014: Global food security in 2050: The role of agricultural productivity and climate change. *Aust. J. Agric. Resour. Econ.*, **58**, 554–570, doi:10.1111/1467-8489.12048.

Baldos, U.L.C., and T.W. Hertel, 2015: The role of international trade in managing food security risks from climate change. *Food Secur.*, **7**, 275–290, doi:10.1007/s12571-015-0435-z.

Balmford, A. et al., 2018: The environmental costs and benefits of high-yield farming. *Nat. Sustain.*, **1**, 477, doi:10.1038/s41893-018-0138-5.

Baptista, F. et al., 2013: *Energy efficiency in agriculture.* Conference Paper; 5th International Congress on Energy and Environment Engineering and Management.

Barbero-Sierra, C., M.J. Marques, and M. Ruíz-Pérez, 2013: The case of urban sprawl in Spain as an active and irreversible driving force for desertification. *J. Arid Environ.*, **90**, 95–102, doi:10.1016/j.jaridenv.2012.10.014.

Bárcena, T.G. et al., 2014: Soil carbon stock change following afforestation in Northern Europe: A meta-analysis. *Glob. Chang. Biol.*, **20**, 2393–2405, doi:10.1111/gcb.12576.

Barlow, J. et al., 2016: Anthropogenic disturbance in tropical forests can double biodiversity loss from deforestation. *Nature*, **535**, 144–147, doi:10.1038/nature18326.

Barnes, A.P., and S.G. Thomson, 2014: Measuring progress towards sustainable intensification: How far can secondary data go? *Ecol. Indic.*, **36**, 213–220, doi:10.1016/j.ecolind.2013.07.001.

Barnes, A.P., H. Hansson, G. Manevska-Tasevska, S.S. Shrestha, and S.G. Thomson, 2015: The influence of diversification on long-term viability of the agricultural sector. *Land use policy*, **49**, 404–412, doi:10.1016/j.landusepol.2015.08.023.

Barnett, J., and W.N. Adger, 2007: Climate change, human security and violent conflict. *Polit. Geogr.*, **26**, 639–655, doi:10.1016/j.polgeo.2007.03.003.

Barnett, J., and J. Palutikof, 2015: 26 The limits to adaptation. A comparative analysis. In: *Applied Studies in Climate Adaptation*, John Wiley & Sons, Ltd., UK, 231 pp.

Barrett, C., T. Reardon, and P. Webb, 2001: Nonfarm income diversification and household livelihood strategies in rural Africa: Concepts, dynamics, and policy implications. *Food Policy*, **26**, 315–331, doi:10.1016/S0306-9192(01)00014-8.

Barthel, S., and C. Isendahl, 2013: Urban gardens, agriculture, and water management: Sources of resilience for long-term food security in cities. *Ecol. Econ.*, **86**, 224–234, doi:10.1016/J.ECOLECON.2012.06.018.

Bassoum, S., and D. Ghiggi, 2010: Sahel vert: A project of Centre Mampuya, Senegal. In: *International Symposium on Urban and Peri-Urban Horticulture in the Century of Cities: Lessons, Challenges, Opportunities* 1021, 367–372.

Bastin, J.-F. et al., 2017: The extent of forest in dryland biomes. *Science*, **356**, 635–638, doi:10.1126/science.aam6527.

Batidzirai, B., E.M.W. Smeets, and A.P.C. Faaij, 2012: Harmonising bioenergy resource potentials: Methodological lessons from review of state of the art bioenergy potential assessments. *Renew. Sustain. Energy Rev.*, **16**, 6598–6630, doi:10.1016/j.rser.2012.09.002.

Batterbury, S., 2001: Landscapes of diversity: A local political ecology of livelihood diversification in South-Western Niger. *Ecumene*, **8**, 437–464, doi:10.1177/096746080100800404.

Baudron, F. et al., 2015: Re-examining appropriate mechanization in Eastern and Southern Africa: Two-wheel tractors, conservation agriculture, and private sector involvement. *Food Secur.*, **7**, 889–904, doi:10.1007/s12571-015-0476-3.

Baumhardt, R.L., B.A. Stewart, and U.M. Sainju, 2015: North American soil degradation: Processes, practices, and mitigating strategies. *Sustain.*, **7**, 2936–2960, doi:10.3390/su7032936.

Baur, A.H., M. Förster, and B. Kleinschmit, 2015: The spatial dimension of urban greenhouse gas emissions: Analyzing the influence of spatial structures and LULC patterns in European cities. *Landsc. Ecol.*, **30**, 1195–1205, doi:10.1007/s10980-015-0169-5.

Bautista, S. et al., 2017: Integrating knowledge exchange and the assessment of dryland management alternatives: A learning-centered participatory approach. *J. Environ. Manage.*, **195**, 35–45, doi:10.1016/j.jenvman.2016.11.050.

Baveye, P.C., J. Berthelin, D. Tessier, and G. Lemaire, 2018: The "4 per 1000" initiative: A credibility issue for the soil science community? *Geoderma*, **309**, 118–123, doi:10.1016/j.geoderma.2017.05.005.

6

Bayrak, M., L. Marafa, M.M. Bayrak, and L.M. Marafa, 2016: Ten years of REDD+: A critical review of the impact of REDD+ on forest-dependent communities. *Sustainability*, **8**, 620, doi:10.3390/su8070620.

Beauchemin, K.A., M. Kreuzer, F. O'Mara, and T.A. McAllister, 2008: Nutritional management for enteric methane abatement: A review. *Aust. J. Exp. Agric.*, **48**, 21–27.

Beerling, D.J. et al., 2018: Farming with crops and rocks to address global climate, food and soil security. *Nat. Plants*, **4**, 138.

Begum, B.A. et al., 2011: Long–range transport of soil dust and smoke pollution in the South Asian region. *Atmos. Pollut. Res.*, **2**, 151–157, doi:10.5094/APR.2011.020.

Begum, R., K. Sohag, S. Abdullah, and M. Jaafar, 2015: CO2 emissions, energy consumption, economic and population growth in Malaysia. *Renew. Sustain. Energy Rev.*, **41**, 594–601, doi:10.1016/j.rser.2014.07.205.

Begum, R.A., M.S.K. Sarkar, A.H. Jaafar, and J.J. Pereira, 2014: Toward conceptual frameworks for linking disaster risk reduction and climate change adaptation. *Int. J. Disaster Risk Reduct.*, **10**, 362–373, doi:10.1016/J.IJDRR.2014.10.011.

Behrman, K.D., T.E. Juenger, J.R. Kiniry, and T.H. Keitt, 2015: Spatial land use trade-offs for maintenance of biodiversity, biofuel, and agriculture. *Landsc. Ecol.*, **30**, 1987–1999, doi:10.1007/s10980-015-0225-1.

Bello, C. et al., 2015: Defaunation affects carbon storage in tropical forests. *Sci. Adv.*, **1**, e1501105, doi:10.1126/sciadv.1501105.

Benis, K., and P. Ferrão, 2017: Potential mitigation of the environmental impacts of food systems through urban and peri-urban agriculture (UPA): A life cycle assessment approach. *J. Clean. Prod.*, **140**, 784–795, doi:10.1016/j.jclepro.2016.05.176.

Benjamin, E.O., O. Ola, and G. Buchenrieder, 2018: Does an agroforestry scheme with payment for ecosystem services (PES) economically empower women in sub-Saharan Africa? *Ecosyst. Serv.*, **31**, 1–11, doi:10.1016/j.ecoser.2018.03.004.

Bennetzen, E., P. Smith, and J. Porter, 2016a: Agricultural production and greenhouse gas emissions from world regions: The major trends over 40 years. *Glob. Environ. Chang.*, **37**, 43–55, doi:https://doi.org/10.1016/j.gloenvcha.2015.12.004.

Bennetzen, E.H., P. Smith, and J.R. Porter, 2016b: Decoupling of greenhouse gas emissions from global agricultural production: 1970-2050. *Glob. Chang. Biol.*, **22**, 763–781, doi:10.1111/gcb.13120.

Berman, R., C. Quinn, and J. Paavola, 2012: The role of institutions in the transformation of coping capacity to sustainable adaptive capacity. *Environ. Dev.*, **2**, 86–100, doi:10.1016/j.envdev.2012.03.017.

Bertram, C. et al., 2018: Targeted policies can compensate most of the increased sustainability risks in 1.5°C mitigation scenarios. *Environ. Res. Lett.*, **13**, 64038, doi:10.1088/1748-9326/aac3ec.

Bestelmeyer, B., and D. Briske, 2012: Grand challenges for resilience-based management of rangelands. *Rangel. Ecol. Manag.*, **65**, 654–663, doi:10.2111/REM-D-12-00072.1.

Bestelmeyer, B.T. et al., 2015: Desertification, land use, and the transformation of global drylands. *Front. Ecol. Environ.*, **13**, 28–36, doi:10.1890/140162.

Bhattacharjee, K., and B. Behera, 2017: Forest cover change and flood hazards in India. *Land use policy*, **67**, 436–448, doi:10.1016/j.landusepol.2017.06.013.

Bhattacharyya, R. et al., 2015: Soil degradation in India: Challenges and potential solutions. *Sustainability*, **7**, 3528–3570, doi:10.3390/su7043528.

Bhattamishra, R., and C.B. Barrett, 2010: Community-based risk management arrangements: a review. *World Dev.*, **38**, 923–932, doi:10.1016/j.worlddev.2009.12.017.

Bidwell, D., T. Dietz, and D. Scavia, 2013: Fostering knowledge networks for climate adaptation. *Nat. Clim. Chang.*, **3**, 610–611, doi:10.1038/nclimate1931.

Biermann, F., P. Pattberg, H. Van Asselt, and F. Zelli, 2009: The fragmentation of global governance architectures: A framework for analysis. *Glob. Environ. Polit.*, **9**, 14–40.

Billen, G. et al., 2018: Opening to distant markets or local reconnection of agro-food systems? Environmental consequences at regional and global scales. *Agroecosystem Diversity*, Elsevier, 391–413.

Birkmann, J. et al., 2015: Scenarios for vulnerability: Opportunities and constraints in the context of climate change and disaster risk. *Clim. Change*, **133**, 53–68, doi:10.1007/s10584-013-0913-2.

Birthal, P.S., D. Roy, and D.S. Negi, 2015: Assessing the impact of crop diversification on farm poverty in India. *World Dev.*, **72**, 70–92, doi:10.1016/J.WORLDDEV.2015.02.015.

Bisht, I.S. et al., 2018: Farmers' rights, local food systems, and sustainable household dietary diversification: A case of Uttarakhand Himalaya in north-western India. *Agroecol. Sustain. Food Syst.*, **42**, 77–113, doi:10.1080/21683565.2017.1363118.

Böcher, M., and M. Krott, 2014: The RIU model as an analytical framework for scientific knowledge transfer: The case of the "decision support system forest and climate change." *Biodivers. Conserv.*, **23**, 3641–3656, doi:10.1007/s10531-014-0820-5.

Bockstael, E., and F. Berkes, 2017: Using the capability approach to analyze contemporary environmental governance challenges in coastal Brazil. *Int. J. Commons*, **11**, 799–822, doi:10.18352/ijc.756.

Bohle, H.G., T.E. Downing, and M.J. Watts, 1994: Climate change and social vulnerability: Toward a sociology and geography of food insecurity. *Glob. Environ. Chang.*, **4**, 37–48.

Boisvenue, C., and S.W. Running, 2006: Impacts of climate change on natural forest productivity – Evidence since the middle of the 20th century. *Glob. Chang. Biol.*, **12**, 862–882, doi:10.1111/j.1365-2486.2006.01134.x.

Bonan, G.B., 2008: Forests and climate change: Forcings, feedbacks, and the climate benefits of forests. *Science*, **320**, 1444–1449, doi:10.1126/science.1155121.

Bonn, A., M. Reed, C. Evans, H. Joosten, and C.B. Services, 2014: Investing in nature: Developing ecosystem service markets for peatland restoration. *Ecosystem*, **9**, 54–65, doi:10.1016/j.ecoser.2014.06.011.

Bonsch, M. et al., 2015: Environmental flow provision: Implications for agricultural water and land-use at the global scale. *Glob. Environ. Chang.*, **30**, 113–132, doi:10.1016/j.gloenvcha.2014.10.015.

Bonsch, M. et al., 2016: Trade-offs between land and water requirements for large-scale bioenergy production. *GCB Bioenergy*, **8**, 11–24, doi:10.1111/gcbb.12226.

Boone, R.B., R.T. Conant, J. Sircely, P.K. Thornton, and M. Herrero, 2018: Climate change impacts on selected global rangeland ecosystem services. *Glob. Chang. Biol.*, **24**, 1382–1393, doi:10.1111/gcb.13995.

Borrelli, P. et al., 2017: An assessment of the global impact of 21st century land use change on soil erosion. *Nat. Commun.*, **8**, 2013, doi:10.1038/s41467-017-02142-7.

Bourgoin, J., 2012: Sharpening the understanding of socio-ecological landscapes in participatory land-use planning. A case study in Lao PDR. *Appl. Geogr.*, **34**, 99–110, doi:10.1016/j.apgeog.2011.11.003.

Bouwer, L.M., E. Papyrakis, J. Poussin, C. Pfurtscheller, and A.H. Thieken, 2014: The costing of measures for natural hazard mitigation in Europe. *Nat. Hazards Rev.*, **15**, 4014010, doi:10.1061/(ASCE)NH.1527-6996.0000133.

Bowman, A., 2015: Sovereignty, risk and biotechnology: Zambia's 2002 GM controversy in retrospect. *Dev. Change*, **46**, 1369–1391, doi:10.1111/dech.12196.

Boyd, J., and S. Banzhaf, 2007: What are ecosystem services? The need for standardized environmental accounting units. *Ecol. Econ.*, **63**, 616–626, doi:10.1016/J.ECOLECON.2007.01.002.

Boysen, L.R., W. Lucht, and D. Gerten, 2017: Trade-offs for food production, nature conservation and climate limit the terrestrial carbon dioxide removal potential. *Glob. Chang. Biol.*, **23**, 4303–4317, doi:10.1111/gcb.13745.

Braat, L., 2018: Five reasons why the Science publication "Assessing nature's contributions to people" (Diaz et al. 2018) would not have been accepted in Ecosystem Services. *Ecosyst. Serv.*, **30**, A1–A2, doi:10.1016/j.ecoser.2018.02.002.

6

Bradford, K.J. et al., 2018: The dry chain: Reducing postharvest losses and improving food safety in humid climates. *Trends Food Sci. Technol.*, **71**, 84–93, doi:10.1016/j.tifs.2017.11.002.

Bradshaw, B., H. Dolan, and B. Smit, 2004: Farm-level adaptation to climatic variability and change: Crop diversification in the Canadian prairies. *Clim. Change*, **67**, 119–141, doi:10.1007/s10584-004-0710-z.

Braun, J. von, B. Algieri, M. Kalkuhl, 2014: World food system disruptions in the early 2000s: causes, impacts and cures. *World Food Policy*, **1**, 34–55, doi:10.18278/wfp.1.1.3.

de Brauw, A., 2011: Migration and child development during the food price crisis in El Salvador. *Food Policy*, **36**, 28–40, doi:10.1016/j.foodpol.2010.11.002.

Bren d'Amour, C. et al., 2016: Future urban land expansion and implications for global croplands. *Proc. Natl. Acad. Sci.*, **114**, 8939–8944, doi:10.1073/pnas.1606036114.

Brenkert, A.L., and E.L. Malone, 2005: Modeling vulnerability and resilience to climate change: A case study of India and Indian states. *Clim. Change*, **72**, 57–102, doi:10.1007/s10584-005-5930-3.

Briber, B.M. et al., 2015: Tree productivity enhanced with conversion from forest to urban land covers. *PLoS One*, **10**, e0136237, doi:10.1371/journal.pone.0136237.

Brindha, K., and P. Pavelic, 2016: Identifying priority watersheds to mitigate flood and drought impacts by novel conjunctive water use management. *Mitig. Adapt. Strateg. Glob. Chang.*, **75**, 399, doi:10.1007/s12665-015-4989-z.

Ten Brink, B.J.E. et al., 2018: Chapter 7: Scenarios of IPBES, land degradation and restoration. In: *Land: The IPBES Assessment Report on Degradation and Restoration*, [L. Montanarella, R. Scholes, and A. Brainich (eds.)], Intergovernmental Ecosystem, Platform on Biodiversity and Services, Bonn, Germany, 531–589 pp.

Brinkley, C., E. Birch, and A. Keating, 2013: Feeding cities: Charting a research and practice agenda towards food security. *J. Agric. food Syst. community Dev.*, **3**, 81–87, doi:10.5304/jafscd.2013.034.008.

Brinkman, H., S. De Pee, I. Sanogo, L. Subran, and M. Bloem, 2009: High food prices and the global financial crisis have reduced access to nutritious food and worsened nutritional status and health. *J. Nutr.*, **140**, 153S–161S, doi:10.3945/jn.109.110767.

Briske, D.D. et al., 2015: Climate-change adaptation on rangelands: Linking regional exposure with diverse adaptive capacity. *Front. Ecol. Environ.*, **13**, 249–256, doi:10.1890/140266.

Brondizio, E.S., E. Ostrom, and O.R. Young, 2009: Connectivity and the governance of multilevel social-ecological systems: The role of social capital. *Annu. Rev. Environ. Resour.*, **34**, 253–278, doi:10.1146/annurev.environ.020708.100707.

Brooks, N., W.N. Adger, and P.M. Kelly, 2005: The determinants of vulnerability and adaptive capacity at the national level and the implications for adaptation. *Glob. Environ. Chang.*, **15**, 151–163, doi:10.1016/j.gloenvcha.2004.12.006.

Brown, C, P. Alexander, A. Arneth, I. Holman, and M. Rounsevell, 2019: Achievement of Paris climate goals unlikely due to time lags in the land system. *Nat. Clim. Chang.*, 1.

Brundu, G., and D.M. Richardson, 2016: Planted forests and invasive alien trees in Europe: A code for managing existing and future plantings to mitigate the risk of negative impacts from invasions. *NeoBiota*, **30**, 5–47, doi:10.3897/neobiota.30.7015.

Bryan, E., T.T. Deressa, G.A. Gbetibouo, and C. Ringler, 2009: Adaptation to climate change in Ethiopia and South Africa: Options and constraints. *Environ. Sci. Policy*, **12**, 413–426, doi:10.1016/j.envsci.2008.11.002.

Bryan, E. et al., 2013: Adapting agriculture to climate change in Kenya: Household strategies and determinants. *J. Environ. Manage.*, **114**, 26–35, doi:10.1016/j.jenvman.2012.10.036.

Bryceson, D.F., 1999: African rural labour, income diversification & livelihood approaches: A long-term development perspective. *Rev. Afr. Polit. Econ.*, **26**, 171–189, doi:10.1080/03056249908704377.

Burney, J., S.J. Davis, and D.B. Lobell, 2010: Greenhouse gas mitigation by agricultural intensification. *Proc. Natl. Acad. Sci.*, **107**, 12052–12057, doi:10.1073/pnas.0914216107.

Busch, J., and J. Engelmann, 2017: Cost-effectiveness of reducing emissions from tropical deforestation, 2016–2050. *Environ. Res. Lett.*, **13**, 15001, doi:10.1088/1748-9326/aa907c.

Busse, H., W. Jogo, G. Leverson, F. Asfaw, and H. Tesfay, 2017: Prevalence and predictors of stunting and underweight among children under 5 years in Tigray, Ethiopia: Implications for nutrition-sensitive agricultural interventions. *J. Hunger Environ. Nutr.*, 1–20, doi:10.1080/19320248.2017.1393364.

Bustamante, M. et al., 2014: Co-benefits, trade-offs, barriers and policies for greenhouse gas mitigation in the agriculture, forestry and other land use (AFOLU) sector. *Glob. Chang. Biol.*, **20**, 3270–3290, doi:10.1111/gcb.12591.

Byerlee, D., T. Jayne, and R. Myers, 2006: Managing food price risks and instability in a liberalizing market environment: Overview and policy options. *Food Policy*, **31**, 275–287, doi:10.1016/j.foodpol.2006.02.002.

Byers, E., M. Gidden, and D. Lecl, 2018: Global exposure and vulnerability to multi-sector development and climate change hotspots. *Environ. Res. Lett.*, **13**, 55012, doi:10.1088/1748-9326/aabf45.

Cacho, J.F., M.C. Negri, C.R. Zumpf, and P. Campbell, 2018: Introducing perennial biomass crops into agricultural landscapes to address water quality challenges and provide other environmental services. *Wiley Interdiscip. Rev. Energy Environ.*, **7**, e275, doi:10.1002/wene.275.

Cai, H., X. Yang, and X. Xu, 2013: Spatiotemporal patterns of urban encroachment on cropland and its impacts on potential agricultural productivity in China. *Remote Sens.*, **5**, 6443–6460, doi:10.3390/rs5126443.

Calvin, K. et al., 2013: Implications of simultaneously mitigating and adapting to climate change: Initial experiments using GCAM. *Clim. Change*, **117**, 545–560, doi:10.1007/s10584-012-0650-y.

Calvin, K. et al., 2014: Trade-offs of different land and bioenergy policies on the path to achieving climate targets. *Clim. Change*, **123**, 691–704, doi:10.1007/s10584-013-0897-y.

Calvin, K. et al., 2016: Implications of uncertain future fossil energy resources on bioenergy use and terrestrial carbon emissions. *Clim. Change*, **136**, 57–68, doi:10.1007/s10584-013-0923-0.

Calvin, K. et al., 2017: The SSP4: A world of deepening inequality. *Glob. Environ. Chang.*, **42**, 284–296, doi:10.1016/j.gloenvcha.2016.06.010.

Campbell, B.C., and J.R. Veteto, 2015: Free seeds and food sovereignty: Anthropology and grassroots agrobiodiversity conservation strategies in the US South. *J. Polit. Ecol.*, **22**, 445–465, doi:10.2458/v22i1.21118.

Campbell, B.M., P. Thornton, R. Zougmoré, P. Van Asten, and L. Lipper, 2014: Sustainable intensification: What is its role in climate smart agriculture? *Curr. Opin. Environ. Sustain.*, **8**, 39–43, doi:10.1016/J.COSUST.2014.07.002.

Campbell, B.M. et al., 2016: Reducing risks to food security from climate change. *Glob. Food Sec.*, **11**, 34–43, doi:10.1016/j.gfs.2016.06.002.

Campbell, J.E., D.B. Lobell, R.C. Genova, and C.B. Field, 2008: The global potential of bioenergy on abandoned agriculture lands. *Environ. Sci. Technol.*, **42**, 5791–5794, doi:10.1021/es800052w.

Campbell, J.R., 2015: Development, global change and traditional food security in Pacific Island countries. *Reg. Environ. Chang.*, **15**, 1313–1324, doi:10.1007/s10113-014-0697-6.

Caplow, S., P. Jagger, K. Lawlor, and E. Sills, 2011: Evaluating land use and livelihood impacts of early forest carbon projects: Lessons for learning about REDD+. *Environ. Sci. Policy*, **14**, 152–167, doi:10.1016/j.envsci.2010.10.003.

Capone, R., H. El Bilali, P. Debs, G. Cardone, and N. Driouech, 2014: Food system sustainability and food security: Connecting the dots. *J. Food Secur.*, **2**, 13–22, doi:10.12691/JFS-2-1-2.

Cardoso, A.S. et al., 2016: Impact of the intensification of beef production in Brazil on greenhouse gas emissions and land use. *Agric. Syst.*, **143**, 86–96, doi:10.1016/J.AGSY.2015.12.007.

Carlson, K.M., and L.M. Curran, 2013: Refined carbon accounting for oil palm agriculture: Disentangling potential contributions of indirect emissions and smallholder farmers. *Carbon Manag.*, **4**, 347–349, doi:10.4155/cmt.13.39.

6

Carlson, K.M. et al., 2018: Effect of oil palm sustainability certification on deforestation and fire in Indonesia. *Proc. Natl. Acad. Sci. U.S.A.*, **115**, 121–126, doi:10.1073/pnas.1704728114.

Carmenta, R., A. Zabala, W. Daeli, and J. Phelps, 2017: Perceptions across scales of governance and the Indonesian peatland fires. *Glob. Environ. Chang.*, **46**, 50–59, doi:10.1016/j.gloenvcha.2017.08.001.

Carpenter, S.R. et al., 2009: Science for managing ecosystem services: Beyond the Millennium Ecosystem Assessment. *Proc. Natl. Acad. Sci. U.S.A.*, **106**, 1305–1312, doi:10.1073/pnas.0808772106.

Carreño, M.L., O.D. Cardona, and A.H. Barbat, 2007: A disaster risk management performance index. *Nat. Hazards*, **41**, 1–20, doi:10.1007/s11069-006-9008-y.

Carter, D.R., R.T. Fahey, K. Dreisilker, M.B. Bialecki, and M.L. Bowles, 2015: Assessing patterns of oak regeneration and C storage in relation to restoration-focused management, historical land use, and potential trade-offs. *For. Ecol. Manage.*, **343**, 53–62, doi:10.1016/j.foreco.2015.01.027.

Carvalho, J.L.N., T.W. Hudiburg, H.C.J. Franco, and E.H. Delucia, 2016: Contribution of above- and belowground bioenergy crop residues to soil carbon. *GCB Bioenergy*, **9**, 1333–1343, doi:10.1111/gcbb.12411.

Cava, M.G.B., N.A.L. Pilon, M.C. Ribeiro, and G. Durigan, 2018: Abandoned pastures cannot spontaneously recover the attributes of old-growth savannas. *J. Appl. Ecol.*, **55**, 1164–1172, doi:10.1111/1365-2664.13046.

Cavanagh, C., A. Chemarum, P. Vedeld, and J. Petursson, 2017: Old wine, new bottles? Investigating the differential adoption of "climate-smart" agricultural practices in western Kenya. *J. Rural Stud.*, **56**, 114–123, doi:10.1016/j.jrurstud.2017.09.010.

Cerri, C.E.P. et al., 2018: Reducing Amazon deforestation through agricultural intensification in the cerrado for advancing food security and mitigating climate change. *Sustainability*, **10**, 989, doi:10.3390/su10040989.

CGIAR, 2016: *The drought crisis in the Central Highlands of Vietnam*. CGIAR Research Program on Climate Change, Agriculture and Food Security (CCAFS), Vietnam.

Chaboud, G., and B. Daviron, 2017: Food losses and waste: Navigating the inconsistencies. *Glob. Food Sec.*, **12**, 1–7, doi:10.1016/j.gfs.2016.11.004.

Chadwick, D.R. et al., 2014: Optimizing chamber methods for measuring nitrous oxide emissions from plot-based agricultural experiments. *Eur. J. Soil Sci.*, **65**, 295–307, doi:10.1111/ejss.12117.

Challinor, A.J. et al., 2014: A meta-analysis of crop yield under climate change and adaptation. *Nat. Clim. Chang.*, **4**, 287–291, doi:10.1038/NCLIMATE2153.

Chamen, W., A.P. Moxey, W. Towers, B. Balana, and P.D. Hallett, 2015: Mitigating arable soil compaction: A review and analysis of available cost and benefit data. *Soil Tillage Res.*, **146**, 10–25, doi:10.1016/J.STILL.2014.09.011.

Chang, Y., G. Li, Y. Yao, L. Zhang, and C. Yu, 2016: Quantifying the water-energy-food nexus: Current status and trends. *Energies*, MDPI AG., 17, doi:10.3390/en9020065.

Chappell, M.J., J.R. Moore, and A.A. Heckelman, 2016: Participation in a city food security program may be linked to higher ant alpha- and beta-diversity: An exploratory case from Belo Horizonte, Brazil. *Agroecol. Sustain. Food Syst.*, **40**, 804–829, doi:10.1080/21683565.2016.1160020.

Chaturvedi, V. et al., 2013: Climate mitigation policy implications for global irrigation water demand. *Mitig. Adapt. Strateg. Glob. Chang.*, **20**, 389–407, doi:10.1007/s11027-013-9497-4.

Cheesman, S., C. Thierfelder, N.S. Eash, G.T. Kassie, and E. Frossard, 2016: Soil carbon stocks in conservation agriculture systems of Southern Africa. *Soil Tillage Res.*, **156**, 99–109, doi:10.1016/J.STILL.2015.09.018.

Chen, J., 2007: Rapid urbanization in China: A real challenge to soil protection and food security. *Catena*, **69**, 1–15, doi:10.1016/J.CATENA.2006.04.019.

Chen, L., J. Wang, W. Wei, B. Fu, and D. Wu, 2010: Effects of landscape restoration on soil water storage and water use in the Loess Plateau Region, China. *For. Ecol. Manage.*, **259**, 1291–1298, doi:10.1016/J.FORECO.2009.10.025.

Chen, W., 2017: Environmental externalities of urban river pollution and restoration: A hedonic analysis in Guangzhou (China). *Landsc. Urban Plan.*, **157**, 170–179, doi:10.1016/j.landurbplan.2016.06.010.

Cherubini, F., S. Vezhapparambu, W. Bogren, R. Astrup, and A.H. Strømman, 2017: Spatial, seasonal, and topographical patterns of surface albedo in Norwegian forests and cropland. *Int. J. Remote Sens.*, **38**, 4565–4586, doi:10.1080/01431161.2017.1320442.

Chhatre, A., and A. Agrawal, 2009: Trade-offs and synergies between carbon storage and livelihood benefits from forest commons. *Proc. Natl. Acad. Sci.*, **106**, 17667–17670, doi:10.1073/pnas.0905308106.

Chow, J., 2018: Mangrove management for climate change adaptation and sustainable development in coastal zones. *J. Sustain. For.*, **37**, 139–156, doi:10.1080/10549811.2017.1339615.

Chum, H., A. Faaij, J. Moreira, G. Berndes, P. Dhamija, H. Dong, B. Gabrielle, A. Goss Eng, W. Lucht, M. Mapako, O. Masera Cerutti, T. McIntyre, T. Minowa, and K. Pingoud, 2011: Bioenergy. In: IPCC Special Report on Renewable Energy Sources and Climate Change Mitigation, [Edenhofer, O., R. Pichs-Madruga, Y. Sokona, K. Seyboth, P. Matschoss, S. Kadner, T. Zwickel, P. Eickemeier, G. Hansen, S. Schlömer, C. von Stechow (eds.)]. Cambridge University Press, Cambridge, United Kingdom and New York, NY, USA, United Kingdom and New York, NY, USA, pp. 209–332.

Ciais, P. Ciais, P., C. Sabine, G. Bala, L. Bopp, V. Brovkin, J. Canadell, A. Chhabra, R. DeFries, J. Galloway, M. Heimann, C. Jones, C. Le Quéré, R.B. Myneni, S. Piao, and P. Thornton, 2013a: Carbon and Other Biogeochemical Cycles. In: Climate Change 2013: The Physical Science Basis. Contribution of Working Group I to the Fifth Assessment Report of the Intergovernmental Panel on Climate Change [Stocker, T.F., D. Qin, G.-K. Plattner, M. Tignor, S.K. Allen, J. Boschung, A. Nauels, Y. Xia, V. Bex and P.M. Midgley (eds.)]. Cambridge University Press, Cambridge, United Kingdom and New York, NY, USA, pp. 465–570.

Claassen, R., F. Carriazo, J. Cooper, D. Hellerstein, and K. Ueda, 2011a: *Grassland to Cropland Conversion in the Northern Plains. The Role of Crop Insurance, Commodity, and Disaster Programs*. United States Department of Agriculture, Washington, DC, USA, 85 pp.

Claassen, R., J.C. Cooper, and F. Carriazo, 2011b: Crop insurance, disaster payments, and land use change: the effect of sodsaver on incentives for grassland conversion. *J. Agric. Appl. Econ.*, **43**, 195–211, doi:10.1017/S1074070800004168.

Clark, M., and D. Tilman, 2017: Comparative analysis of environmental impacts of agricultural production systems, agricultural input efficiency, and food choice. *Environ. Res. Lett.*, **12**, 64016, doi:10.1088/1748-9326/aa6cd5.

Clarke L., K. Jiang, K. Akimoto, M. Babiker, G. Blanford, K. Fisher-Vanden, J.-C. Hourcade, V. Krey, E. Kriegler, A. Löschel, D. McCollum, S. Paltsev, S. Rose, P.R. Shukla, M. Tavoni, B.C.C. van der Zwaan, and D.P. van Vuuren, 2014: Assessing Transformation Pathways. In: Climate Change 2014: Mitigation of Climate Change. Contribution of Working Group III to the Fifth Assessment Report of the Intergovernmental Panel on Climate Change [Edenhofer, O., R. Pichs-Madruga, Y. Sokona, E. Farahani, S. Kadner, K. Seyboth, A. Adler, I. Baum, S. Brunner, P. Eickemeier, B. Kriemann, J. Savolainen, S. Schlömer, C. von Stechow, T. Zwickel and J.C. Minx (eds.)]. Cambridge University Press, Cambridge, United Kingdom and New York, NY, USA.

Clay, D.C. et al., 1995: *Promoting Food Security in Rwanda through Sustainable Agricultural Productivity: Meeting the Challenges of Population Pressure, Land Degradation, and Poverty*. Michigan State University, Department of Agricultural, Food, and Resource Economics, Michigan, USA., 136 pp.

Coakley, J., 2005: Atmospheric physics: reflections on aerosol cooling. *Nature*, **438**, 1091–1092, doi:10.1038/4381091a.

Cohen, M.J., J. Clapp, and Centre for International Governance Innovation., 2009: *The Global Food Crisis: Governance Challenges and Opportunities*. Wilfrid Laurier University Press, Ontario, Canada, 267 pp.

Cohn, A.S., J. Gil, T. Berger, H. Pellegrina, and C. Toledo, 2016: Patterns and processes of pasture to crop conversion in Brazil: Evidence from Mato Grosso State. *Land use policy*, **55**, 108–120, doi:10.1016/J.LANDUSEPOL.2016.03.005.

Cohn, A.S. et al., 2017: Smallholder agriculture and climate change. *Annu. Rev. Environ. Resour.*, **42**, 347–375, doi:10.1146/annurev-environ-102016-060946.

6

Collins, K., and R. Ison, 2009: Jumping off Arnstein's ladder: Social learning as a new policy paradigm for climate change adaptation. *Environ. Policy Gov.*, **19**, 358–373, doi:10.1002/eet.523.

Collins, M., R. Knutti, J. Arblaster, J.-L. Dufresne, T. Fichefet, P. Friedlingstein, X. Gao, W.J. Gutowski, T. Johns, G. Krinner, M. Shongwe, C. Tebaldi, A.J. Weaver and M. Wehner, 2013: Long-term Climate Change: Projections, Commitments and Irreversibility. In: Climate Change 2013: The Physical Science Basis. Contribution of Working Group I to the Fifth Assessment Report of the Intergovernmental Panel on Climate Change [Stocker, T.F., D. Qin, G.-K. Plattner, M. Tignor, S.K. Allen, J. Boschung, A. Nauels, Y. Xia, V. Bex and P.M. Midgley (eds.)]. Cambridge University Press, Cambridge, United Kingdom and New York, NY, USA.

Conant, R.T., and K. Paustian, 2002: Potential soil carbon sequestration in overgrazed grassland ecosystems. *Global Biogeochem. Cycles*, **16**, 90-1-90–99, doi:10.1029/2001GB001661.

Conant, R.T., and K. Paustian, S.J. Del Grosso, and W.J. Parton, 2005: Nitrogen pools and fluxes in grassland soils sequestering carbon. *Nutr. Cycl. Agroecosystems*, **71**, 239–248, doi:10.1007/s10705-004-5085-z.

Conant, R.T., C.E.P. Cerri, B.B. Osborne, and K. Paustian, 2017: Grassland management impacts on soil carbon stocks: a new synthesis. *Ecol. Appl.*, **27**, 662–668, doi:10.1002/eap.1473.

de Coninck, H., A. Revi, M. Babiker, P. Bertoldi, M. Buckeridge, A. Cartwright, W. Dong, J. Ford, S. Fuss, J.-C. Hourcade, D. Ley, R. Mechler, P. Newman, A. Revokatova, S. Schultz, L. Steg, and T. Sugiyama, 2018: Strengthening and Implementing the Global Response. In: Global Warming of 1.5°C: an IPCC special report on the impacts of global warming of 1.5°C above pre-industrial levels and related global greenhouse gas emission pathways, in the context of strengthening the global response to the threat of climate change, [Masson-Delmotte, V., P. Zhai, H.-O. Pörtner, D. Roberts, J. Skea, P.R. Shukla, A. Pirani, W. Moufouma-Okia, C. Péan, R. Pidcock, S. Connors, J.B.R. Matthews, Y. Chen, X. Zhou, M.I. Gomis, E. Lonnoy, T. Maycock, M. Tignor, and T. Waterfield (eds.)]. World Meteorological Organization, Geneva, Switzerland, pp. 313–444.

Cools, J., D. Innocenti, and S. O'Brien, 2016: Lessons from flood early warning systems. *Environ. Sci. Policy*, **58**, 117–122, doi:10.1016/J.ENVSCI.2016.01.006.

Coomes, O.T. et al., 2015: Farmer seed networks make a limited contribution to agriculture? Four common misconceptions. *Food Policy*, **56**, 41–50, doi:10.1016/J.FOODPOL.2015.07.008.

Correa, D.F., H.L. Beyer, H.P. Possingham, S.R. Thomas-Hall, and P.M. Schenk, 2017: Biodiversity impacts of bioenergy production: microalgae vs. first generation biofuels. *Renew. Sustain. Energy Rev.*, **74**, 1131–1146, doi:10.1016/J.RSER.2017.02.068.

Cossalter, C., and C. Pye-Smith, 2003: *Fast-wood Forestry: Myths and Realities*. The Center for International Forestry Research (CIFOR), Bogor Barat, Indonesia.

Costanza, R. et al., 1997: The value of the world's ecosystem services and natural capital. *Nature*, **387**, 253–260, doi:10.1038/387253a0.

Coutts, A., and R. Harris, 2013: *A Multi-Scale Assessment of Urban Heating in Melbourne During an Extreme Heat Event and Policy Approaches for Adaptation*. Victorian Centre for Climate Change Adaption Research, Victoria, Australia, 64 pp.

Couwenberg, J., R. Dommain, and H. Joosten, 2010: Greenhouse gas fluxes from tropical peatlands in South-East Asia. *Glob. Chang. Biol.*, **16**, 1715–1732, doi:10.1111/j.1365-2486.2009.02016.x.

CRED, 2015: *The human cost of natural disasters 2015: A Global Perspective*. CRED, Brussels, Belgium, 55 pp.

Creutzig, F. et al., 2015: Bioenergy and climate change mitigation: An assessment. *GCB Bioenergy*, **7**, 916–944, doi:10.1111/gcbb.12205.

Cromsigt, J.P.G.M. et al., 2018: Trophic rewilding as a climate change mitigation strategy? *Philos. Trans. R. Soc. B Biol. Sci.*, **373**, 20170440, doi:10.1098/rstb.2017.0440.

Cronin, J., G. Anandarajah, and O. Dessens, 2018: Climate change impacts on the energy system: A review of trends and gaps. *Clim. Change*, **151**, 79–93, doi:10.1007/s10584-018-2265-4.

Curtis, P.G., C.M. Slay, N.L. Harris, A. Tyukavina, and M.C. Hansen, 2018: Classifying drivers of global forest loss. *Science (80–.)*., **361**, 1108–1111, doi:10.1126/science.aau3445.

D'Amato, A.W., J.B. Bradford, S. Fraver, and B.J. Palik, 2011: Forest management for mitigation and adaptation to climate change: Insights from long-term silviculture experiments. *For. Ecol. Manage.*, **262**, 803–816, doi:10.1016/j.foreco.2011.05.014.

D'Odorico, P. et al., 2013: Vegetation-microclimate feedbacks in woodland-grassland ecotones. *Glob. Ecol. Biogeogr.*, **22**, 364–379, doi:10.1111/geb.12000.

Dagar, J., P. Sharma, D. Sharma, and A. Singh, 2016: *Innovative Saline Agriculture*. Springer India, New Delhi, 1–519 pp.

Dai, Z., 2010: Intensive agropastoralism: Dryland degradation, the grain-to-green program and islands of sustainability in the Mu Us sandy land of China. *Agric. Ecosyst. Environ.*, **138**, 249–256.

Dale, V.H. et al., 2001: Climate change and forest disturbances: Climate change can affect forests by altering the frequency, intensity, duration, and timing of fire, drought, introduced species, insect and pathogen outbreaks, hurricanes, windstorms, ice storms, or landslides. *Bioscience*, **51**, 723–734, doi:10.1641/0006-3568(2001)051[0723:ccafd]2.0.co;2.

Dallimer, M., L.C. et al., 2018: Who uses sustainable land management practices and what are the costs and benefits? Insights from Kenya. *L. Degrad. Dev.*, **29**, 2822–2835, doi:10.1002/ldr.3001.

Danley, B., and C. Widmark, 2016: Evaluating conceptual definitions of ecosystem services and their implications. *Ecol. Econ.*, **126**, 132–138, doi:10.1016/J.ECOLECON.2016.04.003.

DARA, 2012: *Climate Vulnerability Monitor*. DARA, Madrid, Spain, 331 pp.

Darnton-Hill, I., and B. Cogill, 2010: Maternal and young child nutrition adversely affected by external shocks such as increasing global food prices. *J. Nutr.*, **140**, 162S–169S, doi:10.3945/jn.109.111682.

Dasgupta, P., J.F. Morton, D. Dodman, B. Karapinar, F. Meza, M.G. Rivera-Ferre, A. Toure Sarr, and K.E. Vincent, 2014: Rural areas. In: Climate Change 2014: Impacts, Adaptation, and Vulnerability. Part A: Global and Sectoral Aspects. Contribution of Working Group II to the Fifth Assessment Report of the Intergovernmental Panel on Climate Change [Field, C.B., V.R. Barros, D.J. Dokken, K.J. Mach, M.D. Mastrandrea, T.E. Bilir, M. Chatterjee, K.L. Ebi, Y.O. Estrada, R.C. Genova, B. Girma, E.S. Kissel, A.N. Levy, S. MacCracken, P.R. Mastrandrea, and L.L.White (eds.)]. Cambridge University Press, Cambridge, United Kingdom and New York, NY, USA, pp. 613–657.

Datta, K.K., C. De Jong, and O.P. Singh, 2000: Reclaiming salt-affected land through drainage in Haryana, India: A financial analysis. *Agric. Water Manag.*, **46**, 55–71, doi:10.1016/S0378-3774(00)00077-9.

Davies, J., 2016: Enabling governance for sustainable land management. In: *Land Restoration*, Academic Press, 67–76, doi:10.1016/B978-0-12-801231-4.00006-9.

Davis, S.C. et al., 2013: Management swing potential for bioenergy crops. *GCB Bioenergy*, **5**, 623–638, doi:10.1111/gcbb.12042.

Deal, B., and D. Schunk, 2004: Spatial dynamic modeling and urban land use transformation: A simulation approach to assessing the costs of urban sprawl. *Ecol. Econ.*, **51**, 79–95, doi:10.1016/j.ecolecon.2004.04.008.

DeCicco, J.M., 2013: Biofuel's carbon balance: Doubts, certainties and implications. *Clim. Change*, **121**, 801–814, doi:10.1007/s10584-013-0927-9.

Deininger, K., T. Hilhorst, and V. Songwe, 2014: Identifying and addressing land governance constraints to support intensification and land market operation: Evidence from 10 African countries. *Food Policy*, **48**, 76–87.

Delgado, C.L., 2010: Sources of growth in smallholder agriculture integration of smallholders with processors in sub-saharan Africa: The role of vertical and marketers of high value-added items. *Agrekon*, **38**, 165-198, doi:10.1080/03031853.1999.9524913.

Dell, M., B.F. Jones, and B.A. Olken, 2012: Temperature shocks and economic growth: Evidence from the last half century. *Am. Econ. J. Macroecon.*, **4**, 66–95.

DERM, 2011: *Salinity Management Handbook. Second Edition*. Brisbane, Australia: Department of Environment and Resource Management, 188 pp.

Derpsch, R., T. Friedrich, A. Kassam, and H. Li, 2010: Current status of adoption of no-till farming in the world and some of its main benefits. *Int. J. Agric. Biol. Eng.*, **3**, 1–25, doi:10.25165/IJABE.V3I1.223.

Descheemaeker, K. et al., 2016: Climate change adaptation and mitigation in smallholder crop–livestock systems in sub-Saharan Africa: A call for integrated impact assessments. *Reg. Environ. Chang.*, **16**, 2331–2343, doi:10.1007/s10113-016-0957-8.

Deschênes, O., and M. Greenstone, 2011: Climate change, mortality, and adaptation: Evidence from annual fluctuations in weather in the US. *Am. Econ. J. Appl. Econ.*, **3**, 152–185.

Diaz-Ruiz, R., M. Costa-Font, and J.M. Gil, 2018: Moving ahead from food-related behaviours: An alternative approach to understand household food waste generation. *J. Clean. Prod.*, **172**, 1140–1151.

Díaz, S. et al., 2015: The IPBES conceptual framework – Connecting nature and people. *Curr. Opin. Environ. Sustain.*, **14**, 1–16, doi:10.1016/J. COSUST.2014.11.002.

Díaz, S. et al., 2018: Assessing nature's contributions to people. *Science.*, **359**, 270–272, doi:10.1126/science.aap8826.

Dickie, I.A. et al., 2014: Conflicting values: Ecosystem services and invasive tree management. *Biol. Invasions*, **16**, 705–719, doi:10.1007/s10530-013-0609-6.

Dickinson, D. et al., 2014: Cost-benefit analysis of using biochar to improve cereals agriculture. *GCB Bioenergy*, **7**, 850–864, doi:10.1111/gcbb.12180.

Dietz, T., E. Ostrom, and P. Stern, 2013: The struggle to govern the commons. *Science.*, **302**, 1907–1912, doi: 10.1126/science.1091015.

DiGiano, M.L., and A.E. Racelis, 2012: Robustness, adaptation and innovation: Forest communities in the wake of Hurricane Dean. *Appl. Geogr.*, **33**, 151–158, doi:10.1016/j.apgeog.2011.10.004.

Dillon, P., and M. Arshad, 2016: Managed aquifer recharge in integrated water resource management. In: *Integrated Groundwater Management* [A.J. Jakeman, O. Barreteau, R.J. Hunt, J.-D. Rinaudo, and A. Ross (eds.).]. Springer, Cham, Switzerland, 435–452 pp.

Djalante, R., F. Thomalla, M.S. Sinapoy, and M. Carnegie, 2012: Building resilience to natural hazards in Indonesia: Progress and challenges in implementing the Hyogo Framework for Action. *Nat. Hazards*, **62**, 779–803, doi:10.1007/s11069-012-0106-8.

Djeddaoui, F., M. Chadli, and R. Gloaguen, 2017: Desertification susceptibility mapping using logistic regression analysis in the Djelfa area, Algeria. *Remote Sens.*, **9**, 1031, doi:10.3390/rs9101031.

Djenontin, I., S. Foli, and L. Zulu, 2018: Revisiting the factors shaping outcomes for forest and landscape restoration in sub-Saharan Africa: A way forward for policy, practice and research. *Sustainability*, **10**, 906.

Doelman, J.C. et al., 2018: Exploring SSP land-use dynamics using the IMAGE model: Regional and gridded scenarios of land-use change and land-based climate change mitigation. *Glob. Environ. Chang.*, **48**, 119–135, doi:10.1016/J.GLOENVCHA.2017.11.014.

Doerr, S., and C. Santín, 2016: Global trends in wildfire and its impacts: Perceptions versus realities in a changing world. *Philos. Trans. R. Soc. B Biol. Sci.*, **371**, 20150345, doi:10.1098/rstb.2015.0345.

Don, A. et al., 2012: Land-use change to bioenergy production in Europe: Implications for the greenhouse gas balance and soil carbon. *Glob. Chang. Biol. Bioenergy*, **4**, 372–391, doi:10.1111/j.1757-1707.2011.01116.x.

Donato, D. et al., 2011: Mangroves among the most carbon-rich forests in the tropics. *Nat. Geosci.*, **4**, 293.

Dooley, K., and S. Kartha, 2018: Land-based negative emissions: Risks for climate mitigation and impacts on sustainable development. *Int. Environ. Agreements Polit. Law Econ.*, **18**, 79–98, doi:10.1007/s10784-017-9382-9.

Dornburg, V. et al., 2010: Bioenergy revisited: Key factors in global potentials of bioenergy. *Energy Environ. Sci.*, **3**, 258–267, doi:10.1039/B922422J.

Dresner, M., C. Handelman, S. Braun, and G. Rollwagen-Bollens, 2015: Environmental identity, pro-environmental behaviors, and civic engagement of volunteer stewards in Portland area parks. *Environ. Educ. Res.*, **21**, 991–1010, doi:10.1080/13504622.2014.964188.

Drewry, J., 2006: Natural recovery of soil physical properties from treading damage of pastoral soils in New Zealand and Australia: A review. *Agric. Ecosyst. Environ.*, **114**, 159–169.

Duarte, C.M., I.J. Losada, I.E. Hendriks, I. Mazarrasa, and N. Marbà, 2013: The role of coastal plant communities for climate change mitigation and adaptation. *Nat. Clim. Chang.*, **3**, 961–968, doi:10.1038/nclimate1970.

Dubbeling, M., 2014: *Integrating Urban and Peri-Urban Agriculture and Forestry (UPAF) in City Climate Change Strategies*. RUAF Foundation, Leusden, The Netherlands, 18 pp.

Dudley, N. et al., 2010: *Natural Solutions: Protected Areas Helping People Cope with Climate Change*. Gland, Switzerland, Washington DC and New York, USA, 130 pp.

Dugan, A.J. et al., 2018: A systems approach to assess climate change mitigation options in landscapes of the United States forest sector. *Carbon Balance Manag.*, **13**, 13, doi:10.1186/s13021-018-0100-x.

Durigan, G., N. Guerin, and J.N.M.N. da Costa, 2013: Ecological restoration of Xingu Basin headwaters: Motivations, engagement, challenges and perspectives. *Philos. Trans. R. Soc. B Biol. Sci.*, **368**, 20120165–20120165, doi:10.1098/rstb.2012.0165.

Edelenbosch, O.Y. et al., 2017: Decomposing passenger transport futures: Comparing results of global integrated assessment models. *Transp. Res. Part D Transp. Environ.*, **55**, 281–293, doi:10.1016/j.trd.2016.07.003.

Edmonds, J. et al., 2013: Can radiative forcing be limited to 2.6 Wm-2 without negative emissions from bioenergy and CO_2 capture and storage? *Clim. Change*, **118**, 29-43, doi:10.1007/s10584-012-0678-z.

Eisenbies, M.H., W.M. Aust, J.A. Burger, and M.B. Adams, 2007: Forest operations, extreme flooding events, and considerations for hydrologic modeling in the Appalachians — A review. *For. Ecol. Manage.*, **242**, 77–98.

Ellis, E.C., and N. Ramankutty, 2008: Putting people in the map: Anthropogenic biomes of the world. *Front. Ecol. Environ.*, **6**, 439–447, doi:10.1890/070062.

Ellis, F., 1998: Household strategies and rural livelihood diversification. *J. Dev. Stud.*, **35**, 1–38, doi:10.1080/00220389808422553.

Ellis, F., 2008: The determinants of rural livelihood diversification in developing countries. *J. Agric. Econ.*, **51**, 289–302, doi:10.1111/j.1477-9552.2000. tb01229.x.

Ellison, D. et al., 2017: Trees, forests and water: Cool insights for a hot world. *Glob. Environ. Chang.*, **43**, 51–61, doi:10.1016/j.gloenvcha.2017.01.002.

Emerton, L., J. Bishop, and L. Thomas, 2006: *Sustainable Financing of Protected Areas: A Global Review of Challenges and Options*. IUCN – The World Conservation Union, Gland, Switzerland, 109 pp.

Ensor, J., R. Berger, and S. Huq, 2014: *Community-Based Adaptation to Climate Change: Emerging Lessons*. Practical Action Publishing, Rugby, UK, pp. 183–197.

EPA, 2018: *Risk Management Plan RPM*esubmit User's Manual*. Washington DC: United States Environmental Protection Agency, 160 pp.

Epron, D. et al., 2016: Effects of compaction by heavy machine traffic on soil fluxes of methane and carbon dioxide in a temperate broadleaved forest. *For. Ecol. Manage.*, **382**, 1–9, doi:10.1016/J.FORECO.2016.09.037.

Erb, K.H., H. Haberl, and C. Plutzar, 2012: Dependency of global primary bioenergy crop potentials in 2050 on food systems, yields, biodiversity conservation and political stability. *Energy Policy*, **47**, 260–269, doi:https://doi.org/10.1016/j.enpol.2012.04.066.

Erb, K.H. et al., 2016: Exploring the biophysical option space for feeding the world without deforestation. *Nat. Commun.*, **7**, 11382.

Erb, K.H. et al., 2017: Unexpectedly large impact of forest management and grazing on global vegetation biomass. *Nature*, **553**, 73–76, doi:10.1038/nature25138.

6

Ercilla-Montserrat, M. et al., 2019: Analysis of the consumer's perception of urban food products from a soilless system in rooftop greenhouses: A case study from the Mediterranean area of Barcelona (Spain). *Agric. Human Values*, 1–19, doi:10.1007/s10460-019-09920-7.

Eriksson, L.O. et al., 2012: Climate change mitigation through increased wood use in the European construction sector – Towards an integrated modelling framework. *Eur. J. For. Res.*, **131**, 131–144, doi:10.1007/s10342-010-0463-3.

Erisman, J.W., M.A. Sutton, J. Galloway, Z. Klimont, and W. Winiwarter, 2008: How a century of ammonia synthesis changed the world. *Nat. Geosci.*, **1**, 636–639, doi:10.1038/ngeo325.

Erwin, K.L., 2009: Wetlands and global climate change: The role of wetland restoration in a changing world. *Wetl. Ecol. Manag.*, **17**, 71–84, doi:10.1007/s11273-008-9119-1.

Esteves, T. et al., 2012: Mitigating land degradation caused by wildfire: Application of the PESERA model to fire-affected sites in central Portugal. *Geoderma*, **191**, 40–50.

Evans, R.G., and E.J. Sadler, 2008: Methods and technologies to improve efficiency of water use. *Water Resour. Res.*, **44**, 1–15, doi:10.1029/2007WR006200.

Evers, M., and S. Hofmeister, 2011: Gender mainstreaming and participative planning for sustainable land management. *J. Environ. Plan. Manag.*, **54**, 1315–1329, doi:10.1080/09640568.2011.573978.

Fabinyi, M., L. Evans, and S.J. Foale, 2014: Social-ecological systems, social diversity, and power: Insights from anthropology and political ecology. *Ecol. Soc.*, **19**(4): 28, doi:10.5751/ES-07029-190428.

Fafchamps, M., C. Udry, and K. Czukas, 1998: Drought and saving in West Africa: Are livestock a buffer stock? *J. Dev. Econ.*, **55**, 273–305, doi:10.1016/S0304-3878(98)00037-6.

Fakhruddin, S.H.M., A. Kawasaki, and M.S. Babel, 2015: Community responses to flood early warning system: Case study in Kaijuri Union, Bangladesh. *Int. J. Disaster Risk Reduct.*, **14**, 323–331, doi:10.1016/J.IJDRR.2015.08.004.

FAO, 2006: *Fire Management: Voluntary Guidelines. Principles and Strategic Actions*. Food and Agriculture Organization of the United Nations, Rome, Italy, 2 pp.

FAO, 2011a: *The Global Bioenergy Partnership Sustainability Indicators for Bioenergy*. Food and Agriculture Organization of the United Nations, Rome, Italy, 223 pp.

FAO, 2011b: *The State of the World's Land and Water Resources for Food and Agriculture. Managing Systems at Risk*. Food and Agriculture Organization of the United Nations, Rome, Italy, and Earthscan, London, UK, 308 pp.

FAO, 2015: *The Impact of Disasters on Agriculture and Food Security*. Food and Agriculture Organization of the United Nations, Rome, Italy, 54 pp.

FAO, 2017a: *The Future of Food and Agriculture – Trends and Challenges*. Food and Agriculture Organization of the United Nations, Rome, Italy 180 pp.

FAO, 2017b: *Voluntary Guidelines for Sustainable Soil Management*. Food and Agriculture Organization of the United Nations, Rome, Italy, 26 pp.

FAO, 2018a: *Handbook for Saline Soil Management*. Food and Agriculture Organization of the United Nations, Rome, Italy, 144 pp.

FAO, 2018b: *The Future of Food and Agriculture. Alternative pathways to 2050*. Food and Agriculture Organization of the United Nations, Rome, Italy, 64 pp.

FAO, 2018c: *The State of the World's Forests*. Food and Agriculture Organization of the United Nations, Rome, Italy, 180 pp.

FAO and ITPS, 2015: *Status of the World's Soil Resources (SWSR) – Main Report*. Food and Agriculture Organization of the United Nations, Rome, Italy, 650 pp.

FAO, IFAD and WFP, 2013: *The State of Food Insecurity in the World*. Food and Agriculture Organization of the United Nations, Rome, Italy, 56 pp.

FAO, IFAD, UNICEF, WFP and WHO, 2018: *The State of Food Security and Nutrition in the World 2018*. Food and Agriculture Organization of the United Nations, Rome, Italy, 202 pp.

Favero, A., and E. Massetti, 2014: Trade of woody biomass for electricity generation under climate mitigation policy. *Resour. Energy Econ.*, **36**, 166–190, doi:10.1016/J.RESENEECO.2013.11.005.

Favero, A., and R. Mendelsohn, 2014: Using markets for woody biomass energy to sequester carbon in forests. *J. Assoc. Environ. Resour. Econ.*, **1**, 75–95, doi:10.1086/676033.

Favero, A., and R. Mendelsohn, 2017: The land-use consequences of woody biomass with more stringent climate mitigation scenarios. *J. Environ. Prot. (Irvine,. Calif.)*, **8**, 61–73, doi:10.4236/jep.2017.81006.

Fenton, A., J. Paavola, and A. Tallontire, 2017: The role of microfinance in household livelihood adaptation in Satkhira District, Southwest Bangladesh. *World Dev.*, **92**, 192–202, doi:10.1016/j.worlddev.2016.12.004.

Findell, K.L. et al., 2017: The impact of anthropogenic land use and land cover change on regional climate extremes. *Nat. Commun.*, **8**, 989, doi:10.1038/s41467-017-01038-w.

Flannigan, M., B. Stocks, M. Turetsky, and M. Wotton, 2009: Impacts of climate change on fire activity and fire management in the circumboreal forest. *Glob. Chang. Biol.*, **15**, 549–560, doi:10.1111/j.1365-2486.2008.01660.x.

Foley, J.A. et al., 2005: Global consequences of land use. *Science.*, **309**, 570–574, doi:10.1126/SCIENCE.1111772.

Foley, J.A. et al., 2011: Solutions for a cultivated planet. *Nature*, **478**, 337–342, doi:10.1038/nature10452.

Forsyth, T., 2013: Community-based adaptation: A review of past and future challenges. *Wiley Interdiscip. Rev. Clim. Chang.*, **4**, 439–446, doi:10.1002/wcc.231.

Francis, C.A. et al. 2012: Farmland conversion to non-agricultural uses in the US and Canada: Current impacts and concerns for the future. *Int. J. Agric. Sustain.*, **10**, 8–24, doi:10.1080/14735903.2012.649588.

Frank, S. et al., 2015: The dynamic soil organic carbon mitigation potential of European cropland. *Glob. Environ. Chang.*, **35**, 269–278, doi:10.1016/j.gloenvcha.2015.08.004.

Frank, S. et al., 2017: Reducing greenhouse gas emissions in agriculture without compromising food security? *Environ. Res. Lett.*, **12**, 105004, doi:10.1088/1748-9326/aa8c83.

Frankl, A., J. Poesen, M. Haile, J. Deckers, and J. Nyssen, 2013: Quantifying long-term changes in gully networks and volumes in dryland environments: The case of Northern Ethiopia. *Geomorphology*, **201**, 254–263, doi:10.1016/J.GEOMORPH.2013.06.025.

Franks, J., 2010: Boundary organizations for sustainable land management: The example of Dutch Environmental Co-operatives. *Ecol. Econ.*, **70**, 283–295, doi:10.1016/j.ecolecon.2010.08.011.

Franz, M., N. Schlitz, and K.P. Schumacher, 2017: Globalization and the water-energy-food nexus – Using the global production networks approach to analyze society-environment relations. *Environ. Sci. Policy*, **90**, 201–212, doi:10.1016/j.envsci.2017.12.004.

Freeman, J., L. Kobziar, E.W. Rose, and W. Cropper, 2017: A critique of the historical-fire-regime concept in conservation. *Conserv. Biol.*, **31**, 976–985, doi:10.1111/cobi.12942.

Fricko, O. et al., 2017: The marker quantification of the Shared Socioeconomic Pathway 2: A middle-of-the-road scenario for the 21st century. *Glob. Environ. Chang.*, **42**, 251–267, doi:10.1016/j.gloenvcha.2016.06.004.

Fridahl, M., and M. Lehtveer, 2018: Bioenergy with carbon capture and storage (BECCS): Global potential, investment preferences, and deployment barriers. *Energy Research & Social Science*, Vol. 42, August 2018, pp. 155–165. doi:10.1016/j.erss.2018.03.019.

Frumkin, H., 2002: Urban sprawl and public health. *Public Health Rep.*, **117**, 201–217, doi:10.1093/phr/117.3.201.

Fujimori, S. et al., 2017: SSP3: AIM implementation of shared socioeconomic pathways. *Glob. Environ. Chang.*, **42**, 268–283, doi:10.1016/j.gloenvcha.2016.06.009.

Fujimori, S. et al., 2018: Inclusive climate change mitigation and food security policy under 1.5°C climate goal. *Environ. Res. Lett.*, **13**, 74033, doi:10.1021/es5051748.

Fujimori, S. et al., 2019: A multi-model assessment of food security implications of well below 2°C scenario vans. *Nat. Commun.*, **2**, 386–396.

Fuss, S. et al., 2014: Betting on negative emissions. *Nat. Clim. Chang.*, **4**, 850, doi:10.1038/nclimate2392.

Fuss, S. et al., 2016: Research priorities for negative emissions. *Environ. Res. Lett.*, **11**, 115007, doi:10.1088/1748-9326/11/11/115007.

Fuss, S. et al., 2018: Negative emissions — part 2: Costs, potentials and side effects. *Environ. Res. Lett.*, **13**, 63002, doi:10.1088/1748-9326/aabf9f.

Gabriel, K.M.A., and W.R. Endlicher, 2011: Urban and rural mortality rates during heat waves in Berlin and Brandenburg, Germany. *Environ. Pollut.*, **159**, 2044–2050, doi:10.1016/J.ENVPOL.2011.01.016.

Gao, B. et al., 2018: Chinese cropping systems are a net source of greenhouse gases despite soil carbon sequestration. *Glob. Chang. Biol.*, doi:10.1111/gcb.14425.

Gao, L., and B.A. Bryan, 2017: Finding pathways to national-scale land-sector sustainability. *Nature*, **544**, 217–222, doi:10.1038/nature21694.

Garbrecht, J., M. Nearing, J. Steiner, X. Zhang, and M. Nichols, 2015: Can conservation trump impacts of climate change on soil erosion? An assessment from winter wheat cropland in the Southern Great Plains of the United States. *Weather Clim. Extrem.*, **10**, 32–39, doi:10.1016/j.wace.2015.06.002.

Garcia, A.S., and T. Wanner, 2017: Gender inequality and food security: Lessons from the gender-responsive work of the International Food Policy Research Institute and the Bill and Melinda Gates Foundation. *Food Secur.*, **9**, 1091–1103, doi:10.1007/s12571-017-0718-7.

Gariano, S., and F. Guzzetti, 2016: Landslides in a changing climate. *Earth-Science Rev.*, **162**, 227–252.

Garnett, T., 2011: Where are the best opportunities for reducing greenhouse gas emissions in the food system (including the food chain)? *Food Policy*, **36**, S23–S32.

Garnett, T. et al., 2013: Sustainable intensification in agriculture: Premises and policies. *Science.*, **341**, 33–34, doi:10.1126/science.1234485.

Garrity, D., and Nair, P.K. (eds.) 2012: *Agroforestry – The Future of Global Land Use*. Heidelberg, Germany: Springer Netherlands, 21–27.

Garschagen, M., 2016: Decentralizing urban disaster risk management in a centralized system? Agendas, actors and contentions in Vietnam. *Habitat Int.*, **52**, 43–49, doi:10.1016/j.habitatint.2015.08.030.

Gasparatos, A., P. Stromberg, and K. Takeuchi, 2011: Biofuels, ecosystem services and human wellbeing: Putting biofuels in the ecosystem services narrative. *Agric. Ecosyst. Environ.*, **142**, 111–128, doi:10.1016/j.agee.2011.04.020.

Gassert, F., M. Luck, M. Landis, P. Reig, T. Shiao, 2014: *Aqueduct Global Maps 2.1 Indicators: Constructing Decision-Relevant Global Water Risk Indicators*. World Resources Institute, Washington DC, USA.

Gaveau, D.L.A. et al., 2016: Rapid conversions and avoided deforestation: Examining four decades of industrial plantation expansion in Borneo. *Sci. Rep.*, **6**, 32017, doi:10.1038/srep32017.

Gebremeskel, G., T.G. Gebremicael, and A. Girmay, 2018: Economic and environmental rehabilitation through soil and water conservation, the case of Tigray in northern Ethiopia. *J. Arid Environ.*, **151**, 113–124, doi:10.1016/J.JARIDENV.2017.12.002.

Genesio, L. et al., 2011: Early warning systems for food security in West Africa: Evolution, achievements and challenges. *Atmos. Sci. Lett.*, **12**, 142–148, doi:10.1002/asl.332.

Gerlak, A.K., and T. Heikkila, 2011: Building a theory of learning in collaboratives: Evidence from the everglades restoration program. *J. Public Adm. Res. Theory*, **21**, 619–644, doi:10.1093/jopart/muq089.

Gibbs, H.K., and J.M. Salmon, 2015: Mapping the world's degraded lands. *Appl. Geogr.*, **57**, 12–21, doi:10.1016/J.APGEOG.2014.11.024.

Gibson, J., G. Boe-Gibson, and G. Stichbury, 2015: Urban land expansion in India 1992–2012. *Food Policy*, **56**, 100–113, doi:10.1016/J.FOODPOL.2015.08.002.

Gilbert, C.L., 2012: International agreements to manage food price volatility. *Glob. Food Sec.*, **1**, 134–142, doi:10.1016/J.GFS.2012.10.001.

Gill, A.M., and S.L. Stephens, 2009: Scientific and social challenges for the management of fire-prone wildland–urban interfaces. *Environ. Res. Lett.*, **4**, 34014, doi:10.1088/1748-9326/4/3/034014.

Gill, J.C., and B.D. Malamud, 2017: Anthropogenic processes, natural hazards, and interactions in a multi-hazard framework. *Earth-Science Rev.*, **166**, 246–269, doi:10.1016/j.earscirev.2017.01.002.

Gimona, A., L. Poggio, I. Brown, and M. Castellazzi, 2012: Woodland networks in a changing climate: Threats from land use change. *Biol. Conserv.*, **149**, 93–102, doi:10.1016/J.BIOCON.2012.01.060.

Giné, X., R. Townsend, and J. Vickery, 2008: *Patterns of Rainfall Insurance Participation in Rural India*. World Bank, Washington, DC, USA, 47 pp.

Global Panel on Agriculture and Food Systems for Nutrition, 2016: *Food Systems and Diets:Facing the Challenges of the 21st Century*. University of London Institutional Repository, London, UK, 132 pp.

Göbel, C., N. Langen, A. Blumenthal, P. Teitscheid, and G. Ritter, 2015: Cutting food waste through cooperation along the food supply chain. *Sustainability*, **7**, 1429–1445.

Godfray, H.C.J., and T. Garnett, 2014: Food security and sustainable intensification. *Philos. Trans. R. Soc. B Biol. Sci.*, **369**, 20120273–20120273, doi:10.1098/rstb.2012.0273.

Godfray, H.C.J. et al., 2010: Food security: The challenge of feeding 9 billion people. *Science.*, **327**, 812–818, doi:10.1126/science.1185383.

Goldstein, B., M. Hauschild, J. Fernandez, and M. Birkved, 2016: Testing the environmental performance of urban agriculture as a food supply in northern climates. *J. Clean. Prod.*, **135**, 984–994.

Goodwin, B.K., and V.H. Smith, 2003: An ex post evaluation of the conservation reserve, federal crop insurance, and other government programs: Program participation and soil erosion. *J. Agric. Resour. Econ.*, **28**, 201–216, doi:10.2307/40987182.

Goodwin, B.K., and V.H. Smith, 2013: What harm is done by subsidizing crop insurance? *Am. J. Agric. Econ.*, **95**, 489–497, doi:10.1093/ajae/aas092.

Goodwin, M.L. Vandeveer, and J.L. Deal, 2004: An empirical analysis of acreage effects of participation in the federal crop insurance program. *Am. J. Agric. Econ.*, **86**, 1058–1077, doi:10.1111/j.0002-9092.2004.00653.x.

Gopnik, M. et al., 2012: Coming to the table: Early stakeholder engagement in marine spatial planning. *Mar. Policy*, **36**, 1139–1149.

Gorham, E., 1991: Northern peatlands: Role in the carbon cycle and probable responses to climatic warming. *Ecol. Appl.*, **1**, 182–195, doi:10.2307/1941811.

Graham-Rowe, E., D.C. Jessop, and P. Sparks, 2014: Identifying motivations and barriers to minimising household food waste. *Resour. Conserv. Recycl.*, **84**, 15–23, doi:10.1016/j.resconrec.2013.12.005.

Graham, N.T. et al., 2018: Water sector assumptions for the shared socioeconomic pathways in an integrated modeling framework. *Water Resour. Res.*, **0**, doi:10.1029/2018WR023452.

Graham, V., S.G. Laurance, A. Grech, and O. Venter, 2017: Spatially explicit estimates of forest carbon emissions, mitigation costs and REDD+ opportunities in Indonesia. *Environ. Res. Lett.*, **12**, 44017, doi:10.1088/1748-9326/aa6656.

Grassi, G., J. et al., 2017: The key role of forests in meeting climate targets requires science for credible mitigation. *Nat. Clim. Chang.*, **7**, 220, doi:10.1038/nclimate3227.

Grassi, G., R. Pilli, J. House, S. Federici, and W.A. Kurz, 2018: Science-based approach for credible accounting of mitigation in managed forests. *Carbon Balance Manag.*, **13**, 8, doi:10.1186/s13021-018-0096-2.

Greene, R., W. Timms, P. Rengasamy, M. Arshad, and R. Cresswell, 2016: Soil and aquifer salinization: Toward an integrated approach for salinity management of groundwater. In: *Integrated Groundwater Management*, [A.J. Jakeman, O. Barreteau, R.J. Hunt, J.-D. Rinaudo, and A. Ross (eds.)]. Springer, Cham, Switzerland, 377–412.

Greenwood, K., and B. McKenzie, 2001: Grazing effects on soil physical properties and the consequences for pastures: A review. *Aust. J. Exp. Agric.*, **41**, 1231–1250.

6

Grey, D., and C.W. Sadoff, 2007: Sink or swim? Water security for growth and development. *Water Policy*, **9**, 545–571, doi:10.2166/wp.2007.021.

Griscom, B.W. et al., 2017: Natural climate solutions. *Proc. Natl. Acad. Sci.*, **114**, 11645–11650, doi:10.1073/pnas.1710465114.

de Groot, R.S., M.A. Wilson, and R.M. Boumans, 2002: A typology for the classification, description and valuation of ecosystem functions, goods and services. *Ecol. Econ.*, **41**, 393–408, doi:10.1016/S0921-8009(02)00089-7.

Gross, C.L., M. Fatemi, and I.H. Simpson, 2017: Seed provenance for changing climates: Early growth traits of nonlocal seed are better adapted to future climatic scenarios, but not to current field conditions. *Restor. Ecol.*, **25**, 577–586, doi:10.1111/rec.12474.

Del Grosso, S., P. Smith, M. Galdos, A. Hastings, and W. Parton, 2014: Sustainable energy crop production. *Curr. Opin. Environ. Sustain.*, doi:10.1016/j.cosust.2014.07.007.

Grubler, A. et al., 2018: A low energy demand scenario for meeting the 1.5°C target and sustainable development goals without negative emission technologies. *Nat. Energy*, **3**, 515–527, doi:10.1038/s41560-018-0172-6.

Gunatilake, H., D. Roland-Holst, and G. Sugiyarto, 2014: Energy security for India: Biofuels, energy efficiency and food productivity. *Energy Policy*, **65**, 761–767.

Guo, J., B. Wang, G. Wang, Y. Wu, and F. Cao, 2018: Vertial and seasonal variations of soil carbon pools in ginkgo agroforestry systems in eastern China. *Catena*, **171**, 450–459, doi:10.1016/j.catena.2018.07.032.

Guo, X. et al., 2016: Application of goethite modified biochar for tylosin removal from aqueous solution. *Colloids Surfaces A Physicochem. Eng. Asp.*, **502**, 81–88, doi:10.1016/J.COLSURFA.2016.05.015.

Gurwick, N.P., L.A. Moore, C. Kelly, and P. Elias, 2013: A systematic review of biochar research, with a focus on its stability in situ and its promise as a climate mitigation strategy. *PLoS One*, **8**, e75932, doi:10.1371/journal.pone.0075932.

Gustavsson, J., C. Cederberg, U. Sonesson, R. van Otterdijk, and A. Meybeck, 2011: *Global Food Losses and Food Waste – Extent, Causes and Prevention.* Food and Agriculture Organization of the United Nations, Rome, Italy, 37 pp.

Gustavsson, L. et al., 2006: The role of wood material for greenhouse gas mitigation. *Mitig. Adapt. Strateg. Glob. Chang.*, **11**, 1097–1127, doi:10.1007/s11027-006-9035-8.

Haberl, H. et al., 2011: Global bioenergy potentials from agricultural land in 2050: Sensitivity to climate change, diets and yields. *Biomass and Bioenergy*, **35**, 4753-4769, doi:10.1016/j.biombioe.2011.04.035.

Haggblade, S., N.M. Me-Nsope, and J.M. Staatz, 2017: Food security implications of staple food substitution in Sahelian West Africa. *Food Policy*, **71**, 27–38, doi:10.1016/J.FOODPOL.2017.06.003.

Hallegatte, S., 2012: *A Cost Effective Solution to Reduce Disaster Losses in Developing Countries: Hydro-Meteorological Services, Early Warning, and Evacuation.* World Bank Policy Research Working Paper No. 6058, World Bank, Washington, DC, USA.

Hallegatte, S. et al., 2015: *Shock Waves: Managing the Impacts of Climate Change on Poverty.* World Bank, Washington, DC, USA.

Hammill, A., R. Matthew, and E. McCarter, 2008: Microfinance and climate change adaptation. *IDS Bull.*, **39**, 113–122.

Hamza, M.A., and W.K. Anderson, 2005: Soil compaction in cropping systems: A review of the nature, causes and possible solutions. *Soil Tillage Res.*, **82**, 121–145, doi:10.1016/J.STILL.2004.08.009.

Hanasaki, N. et al., 2013: A global water scarcity assessment under shared socio-economic pathways – part 2: Water availability and scarcity. *Hydrol. Earth Syst. Sci.*, **17**, 2393–2413, doi:10.5194/hess-17-2393-2013.

Hansen, M.C. et al., 2013: High-resolution global maps of 21st-century forest cover change. *Science.*, **342**, 850–853, doi:10.1126/science.1244693.

Harada, Y. et al., 2019: The heavy metal budget of an urban rooftop farm. *Sci. Total Environ.*, **660**, 115–125, doi:10.1016/J.SCITOTENV.2018.12.463.

Haregeweyn, N. et al., 2015: Soil erosion and conservation in Ethiopia. *Prog. Phys. Geogr.*, **39**, 750–774, doi:10.1177/0309133315598725.

Harper, A.B. et al., 2018: Land-use emissions play a critical role in land-based mitigation for Paris climate targets. *Nat. Commun.*, **9**, 2938, doi:10.1038/s41467-018-05340-z.

Harris, E., T. Ladreiter-Knauss, K. Butterbach-Bahl, B. Wolf, and M. Bahn, 2018: Land-use and abandonment alters methane and nitrous oxide fluxes in mountain grasslands. *Sci. Total Environ.*, **628**, 997–1008, doi:10.1016/j.scitotenv.2018.02.119.

Harris, Z.M., R. Spake, and G. Taylor, 2015: Land use change to bioenergy: A meta-analysis of soil carbon and GHG emissions. *Biomass and Bioenergy*, **82**, 27–39, doi:10.1016/j.biombioe.2015.05.008.

Hasegawa, T. et al., 2015: Consequence of climate mitigation on the risk of hunger. *Environ. Sci. Technol.*, **49**, 7245–7253, doi:10.1021/es5051748.

Hasegawa, T. et al., 2018: Risk of increased food insecurity under stringent global climate change mitigation policy. *Nat. Clim. Chang.*, **8**, 699–703, doi:10.1038/s41558-018-0230-x.

Havemenn, T., and V. Muccione, 2011: Mechanisms for Agricultural Climate Change Mitigation Incentives for Smallholders. *CCAFS Report no. 6. CGIAR Research Program on Climate Change, Agriculture and Food Security (CCAFS).* Copenhagen, Denmark.

Havlík, P. et al., 2014: Climate change mitigation through livestock system transitions. *Proc. Natl. Acad. Sci.*, **111**, 3709–3714, doi:10.1073/pnas.1308044111.

Hawken, P., 2017: *Drawdown: The Most Comprehensive Plan Ever Proposed to Reverse Global Warming*, Penguin, London, UK, 265 pp.

He, B., Y. Cai, W. Ran, X. Zhao, and H. Jiang, 2015: Spatiotemporal heterogeneity of soil salinity after the establishment of vegetation on a coastal saline field. *Catena*, **127**, 129–134, doi:10.1016/j.catena.2014.12.028.

Headey, D., and S. Fan, 2008: Anatomy of a crisis: The causes and consequences of surging food prices. *Agric. Econ.*, **39**, 375–391, doi:10.1111/j.1574-0862.2008.00345.x.

Hearn, M.D. et al., 1998: Environmental influences on dietary behavior among children: Availability and accessibility of fruits and vegetables enable consumption. *J. Heal. Educ.*, **29**, 26–32, doi:10.1080/10556699.1998.10603294.

Heck, V., D. Gerten, W. Lucht, and A. Popp, 2018: Biomass-based negative emissions difficult to reconcile with planetary boundaries. *Nat. Clim. Chang.*, **8**, 151–155, doi:10.1038/s41558-017-0064-y.

Hedenus, F., S. Wirsenius, and D.J.A. Johansson, 2014: The importance of reduced meat and dairy consumption for meeting stringent climate change targets. *Clim. Change*, **124**, 79–91.

Heikkila, T., and A.K. Gerlak, 2018: Working on learning: How the institutional rules of environmental governance matter. *J. Environ. Plan. Manag.*, 1–18, doi:10.1080/09640568.2018.1473244.

Hejazi, M. et al., 2014a: Long-term global water projections using six socioeconomic scenarios in an integrated assessment modeling framework. *Technol. Forecast. Soc. Change*, **81**, 205–226, doi:10.1016/j.techfore.2013.05.006.

Hejazi, M.I. et al., 2014b: Integrated assessment of global water scarcity over the 21st century under multiple climate change mitigation policies. *Hydrol. Earth Syst. Sci.*, **18**, 2859–2883, doi:10.5194/hess-18-2859-2014.

Hejazi, M.I. et al., 2015: 21st century United States emissions mitigation could increase water stress more than the climate change it is mitigating. *Proc. Natl. Acad. Sci.*, **112**, 10635–10640, doi:10.1073/pnas.1421675112.

Helliwell, D.R., 1969: Valuation of wildlife resources. *Reg. Stud.*, **3**, 41–47, doi:10.1080/09595236900185051.

Hellmann, J.J., J.E. Byers, B.G. Bierwagen, and J.S. Dukes, 2008: Five potential consequences of climate change for invasive species. *Conserv. Biol.*, **22**, 534–543, doi:10.1111/j.1523-1739.2008.00951.x.

Henderson, B.B. et al., 2015: Greenhouse gas mitigation potential of the world's grazing lands: Modeling soil carbon and nitrogen fluxes of mitigation practices. *Agric. Ecosyst. Environ.*, **207**, 91–100, doi:10.1016/j.agee.2015.03.029.

6

Hengsdijk, H., and W.J. de Boer, 2017: Post-harvest management and post-harvest losses of cereals in Ethiopia. *Food Secur.*, **9**, 945–958, doi:10.1007/s12571-017-0714-y.

Den Herder, M. et al., 2017: Current extent and stratification of agroforestry in the European Union. *Agric. Ecosyst. Environ.*, **241**, 121–132, doi:10.1016/j.agee.2017.03.005.

Hernandez-Morcillo, M., P. Burgess, J. Mirck, A. Pantera, and T. Plieninger, 2018: Scanning agroforestry-based solutions for climate change mitigation and adaptation in Europe. *Environ. Sci. Policy*, **80**, 44–52.

Herrero, M. et al., 2016: Greenhouse gas mitigation potentials in the livestock sector. *Nat. Clim. Chang.*, **6**, 452–461, doi:10.1038/nclimate2925.

Herrmann, S., and C. Hutchinson, 2005: The changing contexts of the desertification debate. *J. Arid Environ.*, **63**, 538–555.

Hertel, T.W., 2015: The challenges of sustainably feeding a growing planet. *Food Secur.*, **7**, 185–198, doi:10.1007/s12571-015-0440-2.

Hiç, C., P. Pradhan, D. Rybski, and J.P. Kropp, 2016: Food surplus and its climate burdens. *Environ. Sci. Technol.*, **50**, 4269–4277.

Hijbeek, R., M.K. Van Ittersum, H.F.M. Ten Berge, G. Gort, H. Spiegel, and A.P. Whitmore, 2017: Do a organic inputs matter – A meta-analysis of additional yield effects for arable crops in Europe. *Plant Soil*, **411**, 293–303.

Hill, R. et al., 2015: Collaboration mobilises institutions with scale-dependent comparative advantage in landscape-scale biodiversity conservation. *Environ. Sci. Policy*, **51**, 267–277, doi:10.1016/J.ENVSCI.2015.04.014.

Hillbruner, C., and G. Moloney, 2012: When early warning is not enough – Lessons learned from the 2011 Somalia famine. *Glob. Food Sec.*, **1**, 20–28, doi:10.1016/J.GFS.2012.08.001.

Hinkel, J. et al., 2014: Coastal flood damage and adaptation costs under 21st century sea-level rise. *Proc. Natl. Acad. Sci.*, **111**, 3292–3297.

HLPE, 2017: *Nutrition and Food Systems. A Report by the High Level Panel of Experts on Food Security and Nutrition of the Committee on World Food Security.* Food and Agriculture Organization for the United Nations, Rome, Italy.

Hodges, R.J., J.C. Buzby, and B. Bennett, 2011: Postharvest losses and waste in developed and less developed countries: Opportunities to improve resource use. *J. Agric. Sci.*, doi:10.1017/S0021859610000936.

Hof, C. et al., 2018: Bioenergy cropland expansion may offset positive effects of climate change mitigation for global vertebrate diversity. *Proc. Natl. Acad. Sci.*, **115**, 13294 LP-13299, doi:10.1073/pnas.1807745115.

Hoffmann, T. et al., 2013: Humans and the missing C-sink: Erosion and burial of soil carbon through time. *Earth Surf. Dyn.*, **1**, 45–52.

Hooijer, A. et al., 2010: Current and future CO_2 emissions from drained peatlands in Southeast Asia. *Biogeosciences*, **7**, 1505–1514, doi:10.5194/bg-7-1505-2010.

Hosonuma, N. et al., 2012: An assessment of deforestation and forest degradation drivers in developing countries. *Environ. Res. Lett.*, **7**, 44009, doi:10.1088/1748-9326/7/4/044009.

Houghton, R.A., and A.A. Nassikas, 2017: Global and regional fluxes of carbon from land use and land cover change 1850-2015. *Global Biogeochem. Cycles*, **31**, 456–472, doi:10.1002/2016GB005546.

Houghton, R.A., and A.A. Nassikas, 2018: Negative emissions from stopping deforestation and forest degradation, globally. *Glob. Chang. Biol.*, **24**, 350–359, doi:10.1111/gcb.13876.

Houghton, R.A., B. Byers, and A.A. Nassikas, 2015: A role for tropical forests in stabilizing atmospheric CO_2. *Nat. Clim. Chang.*, **5**, 1022–1023, doi:10.1038/nclimate2869.

Howard, J. et al., 2017: Clarifying the role of coastal and marine systems in climate mitigation. *Front. Ecol. Environ.*, **15**, 42–50, doi:10.1002/fee.1451.

Howard, P.H., 2015: Intellectual property and consolidation in the seed industry. *Crop Sci.*, **55**, 2489, doi:10.2135/cropsci2014.09.0669.

Howarth, C., and I. Monasterolo, 2017: Opportunities for knowledge co-production across the energy-food-water nexus: Making interdisciplinary approaches work for better climate decision making. *Environ. Sci. Policy*, **75**, 103–110, doi:10.1016/j.envsci.2017.05.019.

Howell, T.A., S.R. Evett, J.A. Tolk, K.S. Copeland, and T.H. Marek, 2015: Evapotranspiration, water productivity and crop coefficients for irrigated sunflower in the U.S. Southern High Plains. *Agric. Water Manag.*, **162**, 33–46, doi:10.1016/J.AGWAT.2015.08.008.

Hsiang, S.M., M. Burke, and E. Miguel, 2013: Quantifying the influence of climate on human conflict. *Science*, **341**, 1235367, doi: 10.1126/science.1235367.

Huang, J., H. Yu, X. Guan, G. Wang, and R. Guo, 2016: Accelerated dryland expansion under climate change. *Nat. Clim. Chang.*, **6**, 166–171, doi:10.1038/nclimate2837.

Hudiburg, T.W., S.C. Davis, W. Parton, and E.H. Delucia, 2015: Bioenergy crop greenhouse gas mitigation potential under a range of management practices. *GCB Bioenergy*, doi:10.1111/gcbb.12152.

Humpenöder, F. et al., 2014: Investigating afforestation and bioenergy CCS as climate change mitigation strategies. *Environ. Res. Lett.*, **9**, 64029, doi:10.1088/1748-9326/9/6/064029.

Humpenöder, F. et al., 2018: Large-scale bioenergy production: How to resolve sustainability trade-offs? *Environ. Res. Lett.*, **13**, 24011, doi:10.1088/1748-9326/aa9e3b.

Huntjens, P. et al., 2012: Institutional design propositions for the governance of adaptation to climate change in the water sector. *Glob. Environ. Chang.*, **22**, 67–81, doi:10.1016/j.gloenvcha.2011.09.015.

Huppmann, D. et al., 2018: IAMC 1.5°C scenario explorer and data hosted by IIASA. In: *Integrated Assessment Modeling Consortium & International Institute for Applied Systems Analysis, doi: 10.5281/zenodo.3363345.*

Hurteau, M.D., J.B. Bradford, P.Z. Fulé, A.H. Taylor, and K.L. Martin, 2014: Climate change, fire management, and ecological services in the southwestern US. *For. Ecol. Manage.*, **327**, 280–289, doi:10.1016/J.FORECO.2013.08.007.

IEA, 2014: *Key World Energy Statistics*. International Engery Agency (IEA), Paris, France, 82 pp.

IFAD, 2013: *Smallholders, Food Security, and the Environment*. International Fund for Agricultural Development, Rome, Italy, 54 pp.

Immerzeel, D.J., P.A. Verweij, F. Van der Hilst, and A.P.C. Faaij, 2014: Biodiversity impacts of bioenergy crop production: A state-of-the-art review. *GCB Bioenergy*, **6**, 183–209, doi:10.1111/gcbb.12067.

Ingram, J. et al., 2016: Food security, food systems, and environmental change. *Solutions Journal*, **7**, 63–73.

Iordan, C.-M., X. Hu, A. Arvesen, P. Kauppi, and F. Cherubini, 2018: Contribution of forest wood products to negative emissions: Historical comparative analysis from 1960 to 2015 in Norway, Sweden and Finland. *Carbon Balance Manag.*, **13**, 12, doi:10.1186/s13021-018-0101-9.

IPBES, 2018: *Summary for Policymakers of the Thematic Assessment Report on Land Degradation and Restoration of the Intergovernmental Science-Policy Platform on Biodiversity and Ecosystem Services* [R. Scholes et al. (eds.)]. IPBES Secretariat, Bonn, Germany, 1–31 pp.

IPCC, 2000: *Land Use, Land-use Change and Forestry.* [R.T. Watson, I.R. Noble, B. Bolin, N.H. Ravindranath, D.J. Verardo, and D.J. Dokken, (eds.)] Cambridge University Press, Cambridge, UK.

IPCC, 2006: 2006 IPCC Guidelines for National Greenhouse Gas Inventories – A Primer. [Eggleston H.S., K. Miwa, N. Srivastava, and K. Tanabe (eds.)]. Institute for Global Environmental Strategies (IGES) for the Intergovernmental Panel on Climate Change. IGES, Japan, 20 pp.

IPCC, 2012: Renewable Energy Sources and Climate Change Mitigation: Special Report of the Intergovernmental Panel on Climate Change, [Edenhofer, O., R. Pichs-Madruga, Y. Sokona, K. Seyboth, P. Matschoss, S. Kadner, T. Zwickel, P. Eickemeier, G. Hansen, S. Schlömer, C. von Stechow (eds.)]. Cambridge University Press, Cambridge, United Kingdom and New York, NY, USA, United Kingdom and New York, NY, USA, 1088 pp.

6

IPCC, 2014: Climate change 2014: mitigation of climate change. Contribution of Working Group III to the Fifth Assessment Report of the Intergovernmental Panel on Climate Change. [Edenhofer, O., R. Pichs-Madruga, Y. Sokona, E. Farahani, S. Kadner, K. Seyboth, A. Adler, I. Baum, S. Brunner, P. Eickemeier, B. Kriemann, J. Savolainen, S. Schlömer, C. von Stechow, T. Zwickel and J.C. Minx (eds.)].Cambridge University Press, Cambridge, United Kingdom and New York, NY, USA.

IPCC, 2018: Global Warming of 1.5°C An IPCC special report on the impacts of global warming of 1.5°C above pre-industrial levels and related global greenhouse gas emission pathways, in the context of strengthening the global response to the threat of climate change, [V. Masson-Delmotte, P. Zhai, H.-O. Pörtner, D. Roberts, J. Skea, P.R. Shukla, A. Pirani, W. Moufouma-Okia, C. Péan, R. Pidcock, S. Connors, J.B.R. Matthews, Y. Chen, X. Zhou, M.I. Gomis, E. Lonnoy, T. Maycock, M. Tignor, and T. Waterfield (eds.)]. World Meteorological Organization, Geneva, Switzerland, 1552 pp.

Ivanic, M., and W. Martin, 2008: Implications of higher global food prices for poverty in low-income countries. *Agric. Econ.*, **39**, 405–416, doi:10.1111/j.1574-0862.2008.00347.x.

Jacinthe, P.A., and R. Lal, 2001: A mass balance approach to assess carbon dioxide evolution during erosional events. *L. Degrad. Dev.*, **12**, 329–339, doi:10.1002/ldr.454.

Jactel, H. et al., 2017: Tree diversity drives forest stand resistance to natural disturbances. *Curr. For. Reports*, **3**, 223–243, doi:10.1007/s40725-017-0064-1.

Jaffe, A., 2019: *Barriers to Adoption of No-Cost Options for Mitigation of Agricultural Emissions: A Typology*. Motu, New Zealand, 9 pp.

Jakob, M. et al., 2016: Carbon pricing revenues could close infrastructure access gaps. *World Dev.*, **84**, 254–265.

James, S.J., and C. James, 2010: The food cold-chain and climate change. *Food Res. Int.*, **43**, 1944–1956, doi:10.1016/j.foodres.2010.02.001.

Jans, Y., G. Berndes, J. Heinke, W. Lucht, and D. Gerten, 2018: Biomass production in plantations: land constraints increase dependency on irrigation water. *GCB Bioenergy*, **10**, 628–644, doi:10.1111/gcbb.12530.

Jantz, P., S. Goetz, and N. Laporte, 2014: Carbon stock corridors to mitigate climate change and promote biodiversity in the tropics. *Nat. Clim. Chang.*, **4**, 138–142, doi:10.1038/nclimate2105.

Jat, H.S. et al., 2015: Management influence on maize-wheat system performance, water productivity and soil biology. *Soil Use Manag.*, **31**, 534–543, doi:10.1111/sum.12208.

Jat, M. et al., 2016: Climate change and agriculture: Adaptation strategies and mitigation opportunities for food security in South Asia and Latin America. *Adv. Agron. (Vol. 137)*, 127–235, doi:10.1016/bs.agron.2015.12.005.

Jauhiainen, J., S. Limin, H. Silvennoinen, and H. Vasander, 2008: Carbon dioxide and methane fluxes in drained tropical peat before and after hydrological restoration. *Ecology*, **89**, 3503–3514, doi:10.1890/07-2038.1.

Jaworski, A., 2016: Encouraging climate adaptation through reform of federal crop insurance subsidies. *New York Univ. Law Rev.*, **91**, 1684.

Jebli, M., and S. Youssef, 2017: The role of renewable energy and agriculture in reducing CO_2 emissions: Evidence for North Africa countries. *Ecol. Indic.*, **74**, 295–301.

Jeffery, S. et al., 2017: Biochar boosts tropical but not temperate crop yields. *Environ. Res. Lett.*, **12**, 53001, doi:10.1088/1748-9326/aa67bd.

Jiang, Y., 2015: China's water security: Current status, emerging challenges and future prospects. *Environ. Sci. Policy*, **54**, 106–125, doi:10.1016/j.envsci.2015.06.006.

Johnson, F.X., and S. Silveira, 2014: Pioneer countries in the transition to alternative transport fuels: Comparison of ethanol programmes and policies in Brazil, Malawi and Sweden. *Environ. Innov. Soc. Transitions*, **11**, 1–24, doi:10.1016/j.eist.2013.08.001.

Johnston, F.H. et al., 2012: Estimated global mortality attributable to smoke from landscape fires. *Environ. Health Perspect.*, doi:10.1289/ehp.1104422.

Jones, A.D. et al., 2013: Greenhouse gas policy influences climate via direct effects of land-use change. *J. Clim.*, **26**, 3657–3670, doi:10.1175/JCLI-D-12-00377.1.

Jones, H.P., D.G. Hole, and E.S. Zavaleta, 2012: Harnessing nature to help people adapt to climate change. *Nat. Clim. Chang.*, **2**, 504–509, doi:10.1038/nclimate1463.

Joosten, H., and J. Couwenberg, 2008: Peatlands and carbon. In. *Assessment on Peatlands, Biodiversity and Climate Change. Main Report*. Global Envionment Centre, Kulala Lumpur, Malaysia, and Wetlands International, Wageningen, Netherlands, 99–117 pp.

Kapos, V. et al., 2008: *Carbon and Biodiversity: A Demonstration Atlas*. Cambridge, UK, 32 pp.

Karim, M.R., and A. Thiel, 2017: Role of community based local institution for climate change adaptation in the Teesta riverine area of Bangladesh. *Clim. Risk Manag.*, **17**, 92–103, doi:10.1016/j.crm.2017.06.002.

Karjalainen, E., T. Sarjala, and H. Raitio, 2009: Promoting human health through forests: Overview and major challenges. *Environ. Health Prev. Med.*, **15**, 1, doi:10.1007/s12199-008-0069-2.

Kates, R.W., W.R. Travis, and T.J. Wilbanks, 2012: Transformational adaptation when incremental adaptations to climate change are insufficient. *Proc. Natl. Acad. Sci.*, **109**, 7156–7161, doi:10.1073/pnas.1115521109.

Kaufmann, D., A. Kraay, and M. Mastruzzi, 2010: *The Worldwide Governance Indicators: Methodology and Analytical Issues*. Global Economy and Development at Brookings, Washington DC, USA, 29 pp.

Kauppi, P. et al., 2001: Technological and economic potential of options to enhance, maintain, and manage biogeological carbon reservoirs and geo-engineering. In: *Climate Change 2001: Mitigation*. Contribution of Working Group III to the Third Assessment Report of the Intergovernmental Panel on Climate Change, [B. Metz, O. Davidson, R. Swart, and J. Pan, (eds.)], Vol. 3, Cambridge University Press, Cambridge, UK, 301–344.

Kauppi, P.E., V. Sandström, and A. Lipponen, 2018: Forest resources of nations in relation to human well-being. *PLoS One*, **13**, e0196248, doi:10.1371/journal.pone.0196248.

Keddy, P., L. Fraser, A. Solomeshch, and W. Junk, 2009: Wet and wonderful: The world's largest wetlands are conservation priorities. *Bioscience*, **59**, 39–51.

Keenan, R.J. et al., 2015: Dynamics of global forest area: Results from the FAO Global Forest Resources Assessment 2015. *For. Ecol. Manage.*, **352**, 9–20, doi:10.1016/J.FORECO.2015.06.014.

Keesstra, S. et al., 2018: The superior effect of nature based solutions in land management for enhancing ecosystem services. *Sci. Total Environ.*, **610–611**, 997–1009.

Keesstra, S.D. et al., 2016: The significance of soils and soil science towards realization of the United Nations Sustainable Development Goals. *SOIL*, **2**, 111–128, doi:10.5194/soil-2-111-2016.

Kemper, J., 2015: Biomass and carbon dioxide capture and storage: A review. *Int. J. Greenh. Gas Control*, **40**, 401–430, doi:10.1016/j.ijggc.2015.06.012.

Kijne, J.W., S.A. Prathapar, M.C.S. Wopereis, and K.L. Sahrawat, 1988: *How to Manage Salinity in Irrigated Lands: A Selective Review with Particular Reference to Irrigation in Developing Countries*. International Irrigation Management Institute (IIMI), Colombo, Sri Lanka, 33 pp.

Kim, S.H. et al., 2016: Balancing global water availability and use at basin scale in an integrated assessment model. *Clim. Change*, doi:10.1007/s10584-016-1604-6.

Kindermann, G. et al., 2008: Global cost estimates of reducing carbon emissions through avoided deforestation. *Proc. Natl. Acad. Sci.*, **105**, 10302–10307, doi:10.1073/pnas.0710616105.

King, R.T., 1966: Wildlife and man. *New York Conservationist*, **20**, 8–11.

Kissinger, M., C. Sussmann, and C. Dorward, 2018: Local or global: A biophysical analysis of a regional food system. *Renew. Agric. Food Syst.*, 1-11, doi: 10.1017/S1742170518000078

Kline, K.L. et al., 2017: Reconciling food security and bioenergy: Priorities for action. *GCB Bioenergy*, **9**, 557-576, doi:10.1111/gcbb.12366.

Kloppenberg, J., 2010: Impeding dispossession, enabling repossession: Biological open source and the recovery of seed sovereignty. *J. Agrar. Chang.*, **10**, 367–388, doi:10.1111/j.1471-0366.2010.00275.x.

Knorr, W., A. Arneth, and L. Jiang, 2016: Demographic controls of future global fire risk. *Nat. Clim. Chang.*, **6**, 781–785, doi:10.1038/nclimate2999.

Kobayashi, Y., and A.S. Mori, 2017: The potential role of tree diversity in reducing shallow landslide risk. *Environ. Manage.*, **59**, 807–815, doi:10.1007/s00267-017-0820-9.

Kondylis, F., V. Mueller, G. Sheriff, and S. Zhu, 2016: Do female instructors reduce gender bias in diffusion of sustainable land management techniques? Experimental evidence from Mozambique. *World Dev.*, **78**, 436–449.

Kongsager, R., B. Locatelli, and F. Chazarin, 2016: Addressing climate change mitigation and adaptation together: A global assessment of agriculture and forestry projects. *Environ. Manage.*, **57**, 271–282, doi:10.1007/s00267-015-0605-y.

Koplitz, S.N. et al., 2016: Public health impacts of the severe haze in Equatorial Asia in September–October 2015: Demonstration of a new framework for informing fire management strategies to reduce downwind smoke exposure. *Environ. Res. Lett.*, **11**, 94023.

Kowalski, J., and T. Conway, 2018: Branching out: The inclusion of urban food trees in Canadian urban forest management plans. *Urban For. Urban Green.*

Krause, A. et al., 2017: Global consequences of afforestation and bioenergy cultivation on ecosystem service indicators. *Biogeosciences*, **14**, 4829–4850 doi:10.5194/bg-14-4829-2017.

Kreidenweis, U. et al., 2016: Afforestation to mitigate climate change: Impacts on food prices under consideration of albedo effects. *Environ. Res. Lett.*, **11**, 85001, doi:10.1088/1748-9326/11/8/085001.

Kriegler, E. et al., 2014: The role of technology for achieving climate policy objectives: Overview of the EMF 27 study on global technology and climate policy strategies. *Clim. Change*, **123**, 353–367, doi:10.1007/s10584-013-0953-7.

Kriegler, E. et al., 2017: Fossil-fueled development (SSP5): An energy and resource intensive scenario for the 21st century. *Glob. Environ. Chang.*, **42**, doi:10.1016/j.gloenvcha.2016.05.015.

Kriegler, E. et al., 2018a: Short term policies to keep the door open for Paris climate goals. *Environ. Res. Lett.*, **13**, 74022, doi:10.1088/1748-9326/aac4f1.

Kriegler, E. et al., 2018b: Pathways limiting warming to 1.5°C: A tale of turning around in no time? *Philos. Trans. R. Soc. A Math. Phys. Eng. Sci.*, **376**, 20160457, doi:10.1098/rsta.2016.0457.

Kumar, D., and P. Kalita, 2017: Reducing postharvest losses during storage of grain crops to strengthen food security in developing countries. *Foods*, **6**, 8, doi:10.3390/foods6010008.

Kummu, M., P.J. Ward, H. de Moel, and O. Varis, 2010: Is physical water scarcity a new phenomenon? Global assessment of water shortage over the last two millennia. *Environ. Res. Lett.*, **5**, 34006, doi:10.1088/1748-9326/5/3/034006.

Kummu, M. et al., 2012: Lost-food, wasted resources: Global food supply chain losses and their impacts on freshwater, cropland, and fertiliser use. *Sci. Total Environ.*, **438**, 477–489, doi:10.1016/J.SCITOTENV.2012.08.092.

Kurz, W., C. Smyth, and T. Lemprière, 2016: Climate change mitigation through forest sector activities: principles, potential and priorities. *Unasylva*, **67**, 61–67.

Kyle, P., C. Müller, K. Calvin, and A. Thomson, 2014: Meeting the radiative forcing targets of the representative concentration pathways in a world with agricultural climate impacts. *Earth's Futur.*, **2**, 83–98, doi:10.1002/2013EF000199.

Labrière, N., B. Locatelli, Y. Laumonier, V. Freycon, and M. Bernoux, 2015: Soil erosion in the humid tropics: A systematic quantitative review. *Agric. Ecosyst. Environ.*, **203**, 127–139, doi:10.1016/j.agee.2015.01.027.

Lal, R., 1998: Soil erosion impact on agronomic productivity and environment quality. *CRC. Crit. Rev. Plant Sci.*, **17**, 319–464, doi:10.1080/07352689891304249.

Lal, R., 2001: Soil degradation by erosion. *L. Degrad. Dev.*, **12**, 519–539, doi:10.1002/ldr.472.

Lal, R., 2004: Soil carbon sequestration impacts on global climate change and food security. *Science.*, **304**, 1623–1627.

Lal, R., 2006: Enhancing crop yields in the developing countries through restoration of the soil organic carbon pool in agricultural lands. *L. Degrad. Dev.*, **17**, 197–209, doi:10.1002/ldr.696.

Lal, R., 2011: Sequestering carbon in soils of agro-ecosystems. *Food Policy*, **36**, S33–S39.

Lal, R., 2014: Soil carbon management and climate change. In: *Soil Carbon* [A.E. Hartemink and K. McSweeney (eds.)]. Springer, Switzerland, 439–462.

Lal, R., 2015: Restoring soil quality to mitigate soil degradation. *Sustainability*, **7**, 5875–5895, doi:10.3390/su7055875.

Lal, R., 2016: Soil health and carbon management. *Food Energy Secur.*, **5**, 212–222, doi:10.1002/fes3.96.

Lal, R., and W.C. Moldenhauer, 1987: Effects of soil erosion on crop productivity. *CRC. Crit. Rev. Plant Sci.*, **5**, 303–367, doi:10.1080/07352688709382244.

Lamb, A. et al., 2016: The potential for land sparing to offset greenhouse gas emissions from agriculture. *Nat. Clim. Chang.*, **6**, 488–492, doi:10.1038/nclimate2910.

Lamb, W.F., and N.D. Rao, 2015: Human development in a climate-constrained world: what the past says about the future. *Glob. Environ. Chang.*, **33**, 14–22.

Lambin, E.F., and P. Meyfroidt, 2011: Global land use change, economic globalization, and the looming land scarcity. *Proc. Natl. Acad. Sci. U.S.A.*, **108**, 3465–3472, doi:10.1073/pnas.1100480108.

Lambrou, Y., and G. Piana, 2006: *Gender: The missing component of the response to climate change*. Food and Agriculture Organization of the United Nations (FAO), Rome, Italy.

Lamers, P., E. Searcy, J.R. Hess, and H. Stichnothe, 2016: *Developing the Global Bioeconomy: Technical, Market, and Environmental Lessons from Bioenergy*. Academic Press, Elsevier, London, UK and San Diego, CA, USA and Camridge, MA, USA and Oxford, UK, 220 pp.

Langford, W.T. et al., 2011: Raising the bar for systematic conservation planning. *Trends Ecol. Evol.*, **26**, 634–640, doi:10.1016/J.TREE.2011.08.001.

Larsen, F.W., W.R. Turner, and T.M. Brooks, 2012: Conserving critical sites for biodiversity provides disproportionate benefits to people. *PLoS One*, **7**, e36971, doi:10.1371/journal.pone.0036971.

Larsen, S. et al., 2017: Possibilities for near-term bioenergy production and GHG-mitigation through sustainable intensification of agriculture and forestry in Denmark. *Environ. Res. Lett.*, **12**, 114032, doi:10.1088/1748-9326/aa9001.

Lasco, R.D., R.J.P. Delfino, D.C. Catacutan, E.S. Simelton, and D.M. Wilson, 2014: Climate risk adaptation by smallholder farmers: the roles of trees and agroforestry. *Curr. Opin. Environ. Sustain.*, doi:10.1016/j.cosust.2013.11.013.

Lee-Smith, D., 2010: Cities feeding people: an update on urban agriculture in equatorial Africa. *Environ. Urban.*, **22**, 483–499, doi:10.1177/0956247810377383.

Lee, J., 2017: Farmer participation in a climate-smart future: Evidence from the Kenya Agricultural Carbon Project. *Land use policy*, **68**, 72–79, doi:10.1016/j.landusepol.2017.07.020.

Lee, Y., J. Ahern, and C. Yeh, 2015: Ecosystem services in peri-urban landscapes: The effects of agricultural landscape change on ecosystem services in Taiwan's western coastal plain. *Landsc. Urban Plan.*, **139**, 137–148, doi:10.1016/j.landurbplan.2015.02.023.

Lejeune, Q., E. Davin, L. Gudmundsson, J. Winckler, and S. Seneviratne, 2018: Historical deforestation locally increased the intensity of hot days in northern mid-latitudes. *Nat. Clim. Chang.*, **8**, 386–390, doi:10.1038/s41558-018-0131-z.

Lele, S., O. Springate-Baginski, R. Lakerveld, D. Deb, and P. Dash, 2013: Ecosystem services: Origins, contributions, pitfalls, and alternatives. *Conserv. Soc.*, **11**, 343, doi:10.4103/0972-4923.125752.

Lenton, T.M., 2010: The potential for land-based biological CO_2 removal to lower future atmospheric CO_2 concentration. *Carbon Manag.*, **1**, 145–160, doi:10.4155/cmt.10.12.

Lenton, T.M., 2014: The global potential for carbon dioxide removal. In: *Geoengin. Clim. Syst.*, The Royal Society of Chemistry, Cambridge, UK, 52–79.

6

Leskinen, P. et al., 2018: *Substitution Effects of Wood-Based Products in Climate Change Mitigation*. From Science to Policy 7, European Forest Institute, 28 pp.

Lestrelin, G., and M. Giordano, 2007: Upland development policy, livelihood change and land degradation: Interactions from a Laotian village. *L. Degrad. Dev.*, **18**, 55–76, doi:10.1002/ldr.756.

Lewis, K., and C. Witham, 2012: Agricultural commodities and climate change. *Clim. Policy*, **12**, S53–S61, doi:10.1080/14693062.2012.728790.

Lewis, S.L., D.P. Edwards, and D. Galbraith, 2015: Increasing human dominance of tropical forests. *Science.*, **349**, 827–832, doi:10.1126/science.aaa9932.

Lewis, S.L., C.E. Wheeler, E.T.A. Mitchard, and A. Koch, 2019: Restoring natural forests is the best way to remove atmospheric carbon. *Nature*, **568**, 25–28, doi:10.1038/d41586-019-01026-8.

Li, J., R. Xu, D. Tiwari, and G. Ji, 2006: Effect of low-molecular-weight organic acids on the distribution of mobilized Al between soil solution and solid phase. *Appl. Geochemistry*, **21**, 1750–1759, doi:10.1016/J.APGEOCHEM.2006.06.013.

Li, Y. et al., 2015: Local cooling and warming effects of forests based on satellite observations. *Nat. Commun.*, **6**, 6603, doi:10.1038/ncomms7603.

Lillesø, J.B.L. et al., 2011: Innovation in input supply systems in smallholder agroforestry: Seed sources, supply chains and support systems. *Agrofor. Syst.*, **83**, 347–359, doi:10.1007/s10457-011-9412-5.

Limpens, J. et al., 2008: Peatlands and the carbon cycle: From local processes to global implications – A synthesis. *Biogeosciences*, **5**, 1475–1491.

Lin, B.B., 2011: Resilience in agriculture through crop diversification: Adaptive management for environmental change. *Bioscience*, **61**, 183–193, doi:10.1525/bio.2011.61.3.4.

Lin, Y., L.S. Wijedasa, and R.A. Chisholm, 2017: Singapore's willingness to pay for mitigation of transboundary forest-fire haze from Indonesia. *Environ. Res. Lett.*, **12**, 24017, doi:10.1088/1748-9326/aa5cf6.

Lipper, L. et al., 2014: Climate-smart agriculture for food security. *Nat. Clim. Chang.*, **4**, 1068–1072, doi:10.1038/nclimate2437.

Little, P.D., K. Smith, B.A. Cellarius, D.L. Coppock, and C. Barrett, 2001: Avoiding disaster: Diversification and risk management among East African herders. *Dev. Change*, **32**, 401–433, doi:10.1111/1467-7660.00211.

Liu, J. et al., 2017: Water scarcity assessments in the past, present, and future. *Earth's Futur.*, **5**, 545–559, doi:10.1002/2016EF000518.

Liu, Y., Y. Zhou, and W. Wu, 2015: Assessing the impact of population, income and technology on energy consumption and industrial pollutant emissions in China. *Appl. Energy*, **155**, 904–917, doi:10.1016/J.APENERGY.2015.06.051.

Liu, Z., and J. Lan, 2015: The sloping land conversion program in China: Effect on the livelihood diversification of rural households. *World Dev.*, **70**, 147–161, doi:10.1016/j.worlddev.2015.01.004.

Lobell, D. et al., 2008: Prioritizing climate change adaptation needs for food security in 2030. *Science.*, **319**, 607–610, doi:10.1126/science.1152339.

Lobell, D.B., 2014: Climate change adaptation in crop production: Beware of illusions. *Glob. Food Sec.*, **3**, 72–76, doi:10.1016/j.gfs.2014.05.002.

Lobell, D.B., U.L.C. Baldos, and T.W. Hertel, 2013: Climate adaptation as mitigation: The case of agricultural investments. *Environ. Res. Lett.*, **8**, 15012, doi:10.1088/1748-9326/8/1/015012.

Locatelli, B., 2011: *Synergies Between Adaptation and Mitigation in a Nutshell*. Centre for International Forestry Research, Bogor, Indonesia, 4 pp.

Locatelli, B., V. Evans, A. Wardell, A. Andrade, and R. Vignola, 2011: Forests and climate change in Latin America: Linking adaptation and mitigation. *Forests*, **2**, 431–450, doi:10.3390/f2010431.

Locatelli, B. et al., 2015a: Tropical reforestation and climate change: Beyond carbon. *Restor. Ecol.*, **23**, 337–343, doi:10.1111/rec.12209.

Locatelli, B., C. Pavageau, E. Pramova, and M. Di Gregorio, 2015b: Integrating climate change mitigation and adaptation in agriculture and forestry: Opportunities and trade-offs. *Wiley Interdiscip. Rev. Clim. Chang.*, **6**, 585–598, doi:10.1002/wcc.357.

Lock, K., J. Pomerleau, L. Causer, D.R. Altmann, and M. McKee, 2005: The global burden of disease attributable to low consumption of fruit and vegetables: Implications for the global strategy on diet. *Bull. World Health Organ.*, **83**, 100–108, doi:10.1590/S0042-96862005000200010.

Logan, J.A., J. Régnière, and J.A. Powell, 2003: Assessing the impacts of global warming on forest pest dynamics. *Front. Ecol. Environ.*, **1**, 130–137, doi:10.1890/1540-9295(2003)001[0130:ATIOGW]2.0.CO;2.

Long, A., 2013: REDD+, adaptation, and sustainable forest management: Toward effective polycentric global forest governance. *Trop. Conserv. Sci.*, **6**, 384–408.

Lopoukhine, N. et al., 2012: Protected areas: Providing natural solutions to 21st Century challenges. *SAPIENS*, **5**, 117–131.

Lotze-Campen, H. et al., 2013: Impacts of increased bioenergy demand on global food markets: An AgMIP economic model intercomparison. *Agric. Econ.*, **45**, 103–116, doi:10.1111/agec.12092.

Lotze, H. et al., 2006: Depletion, degradation, and recovery potential of estuaries and coastal seas. *Science*, **312**, 1806–1809, doi:10.1126/science.1128035.

Louwaars, N.P., 2002: Seed policy, legislation and law. *J. New Seeds*, **4**, 1–14, doi:10.1300/J153v04n01_01.

Lowder, S.K., J. Skoet, and T. Raney, 2016: The number, size, and distribution of farms, smallholder farms, and family farms worldwide. *World Dev.*, **87**, 16–29, doi:10.1016/j.worlddev.2015.10.041.

Luby, C.H., J. Kloppenburg, T.E. Michaels, and I.L. Goldman, 2015: Enhancing freedom to operate for plant breeders and farmers through open source plant breeding. *Crop Sci.*, **55**, 2481, doi:10.2135/cropsci2014.10.0708.

Luderer, G., R.C. Pietzcker, C. Bertram, E. Kriegler, M. Meinshausen, and O. Edenhofer, 2013: Economic mitigation challenges: how further delay closes the door for achieving climate targets. *Environ. Res. Lett.*, **8**, 34033.

Luderer, C. Bertram, K. Calvin, E. De Cian, and E. Kriegler, 2016: Implications of weak near-term climate policies on long-term mitigation pathways. *Clim. Change*, **136**, 127–140.

Luderer et al., 2018: Residual fossil CO_2 emissions in 1.5–2°C pathways. *Nat. Clim. Chang.*, **8**, 626.

Luedeling, E., R. Kindt, N.I. Huth, and K. Koenig, 2014: Agroforestry systems in a changing climate – Challenges in projecting future performance. *Curr. Opin. Environ. Sustain.*, **6**, 1–7, doi:10.1016/j.cosust.2013.07.013.

Lugato, E., K. Paustian, P. Panagos, A. Jones, and P. Borrelli, 2016: Quantifying the erosion effect on current carbon budget of European agricultural soils at high spatial resolution. *Glob. Chang. Biol.*, **22**, 1976–1984, doi:10.1111/gcb.13198.

Lundmark, T. et al., 2014: Potential roles of Swedish forestry in the context of climate change mitigation. *For. 2014, Vol. 5*, **5**, 557–578, doi:10.3390/F5040557.

Luyssaert, S. et al., 2018: Trade-offs in using European forests to meet climate objectives. *Nature*, **562**, 259–262, doi:10.1038/s41586-018-0577-1.

Lwasa, S., F. Mugagga, B. Wahab, D. Simon, and J.C. Climate, 2014: Urban and peri-urban agriculture and forestry: Transcending poverty alleviation to climate change mitigation and adaptation. *Urban Clim.*, **7**, 92–106, doi:10.1016/j.uclim.2013.10.007.

Lwasa, S., F. et al., 2015: A meta-analysis of urban and peri-urban agriculture and forestry in mediating climate change. *Curr. Opin. Environ. Sustain.*, **13**, 68–73, doi:10.1016/j.cosust.2015.02.003.

Maaroufi, N.I. et al., 2015: Anthropogenic nitrogen deposition enhances carbon sequestration in boreal soils. *Glob. Chang. Biol.*, **21**, 3169–3180, doi:10.1111/gcb.12904.

Macdiarmid, J.I., H. Clark, S. Whybrow, H. de Ruiter, and G. McNeill, 2018: Assessing national nutrition security: The UK reliance on imports to meet population energy and nutrient recommendations. *PLoS One*, **13**, e0192649, doi:10.1371/journal.pone.0192649.

Machado, R., and R. Serralheiro, 2017: Soil salinity: Effect on vegetable crop growth. Management practices to prevent and mitigate soil salinization. *Horticulturae*, **3**, 30, doi:10.3390/horticulturae3020030.

Madlener, R., C. Robledo, B. Muys, and J.T.B. Freja, 2006: A sustainability framework for enhancing the long-term success of LULUCF projects. *Clim. Change*, **75**, 241–271, doi:10.1007/s10584-005-9023-0.

Maes, J. et al., 2017: Landslide risk reduction measures: A review of practices and challenges for the tropics. *Prog. Phys. Geogr.*, **41**, 191–221, doi:10.1177/0309133316689344.

Mahmood, R. et al., 2014: Land cover changes and their biogeophysical effects on climate. *Int. J. Climatol.*, **34**, 929–953, doi:10.1002/joc.3736.

Mahmud, T., and M. Prowse, 2012: Corruption in cyclone preparedness and relief efforts in coastal Bangladesh: Lessons for climate adaptation? *Glob. Environ. Chang.*, **22**, 933–943, doi:10.1016/J.GLOENVCHA.2012.07.003.

Mal, S., R.B. Singh, C. Huggel, and A. Grover, 2018: Introducing linkages between climate change, extreme events, and disaster risk reduction. Springer, Cham, Switzerland, 1–14 pp.

Malhi, Y., P. Meir, and S. Brown, 2002: Forests, carbon and global climate. *Philos. Trans. R. Soc. London. Ser. A Math. Phys. Eng. Sci.*, **360**, 1567–1591, doi:10.1098/rsta.2002.1020.

Maltsoglou, I. et al., 2014: Combining bioenergy and food security: An approach and rapid appraisal to guide bioenergy policy formulation. *Biomass and Bioenergy*, **79**, 80–95, doi:10.1016/j.biombioe.2015.02.007.

Manning, P., G. Taylor, and M.E. Hanley, 2015: Bioenergy, food production and biodiversity – An unlikely alliance? *GCB Bioenergy*, **7**, 570–576, doi:10.1111/gcbb.12173.

Margono, B.A., P.V. Potapov, S. Turubanova, F. Stolle, and M.C. Hansen, 2014: Primary forest cover loss in Indonesia over 2000–2012. *Nat. Clim. Chang.*, **4**, 730–735, doi:10.1038/nclimate2277.

Markandya, A. et al., 2018: Health co-benefits from air pollution and mitigation costs of the Paris Agreement: A modelling study. *Lancet Planet. Heal.*, **2**, e126–e133, doi:10.1016/S2542-5196(18)30029-9.

Marques, M. et al., 2016: Multifaceted impacts of sustainable land management in drylands: A review. *Sustainability*, **8**, 177, doi:10.3390/su8020177.

Martha, G.B., E. Alves, and E. Contini, 2012: Land-saving approaches and beef production growth in Brazil. *Agric. Syst.*, **110**, 173–177, doi:10.1016/J.AGSY.2012.03.001.

Martin, A., N. Gross-Camp, and A. Akol, 2015: Towards an explicit justice framing of the social impacts of conservation. *Conserv. Soc.*, **13**, 166, doi:10.4103/0972-4923.164200.

Martin, S.M., and K. Lorenzen, 2016: Livelihood diversification in rural Laos. *World Dev.*, **83**, 231–243, doi:10.1016/J.WORLDDEV.2016.01.018.

Maskrey, A., 2011: Revisiting community-based disaster risk management. *Environ. Hazards*, **10**, 42–52, doi:10.3763/ehaz.2011.0005.

Massawe, F., S. Mayes, and A. Cheng, 2016: Crop diversity: An unexploited treasure trove for food security. *Trends Plant Sci.*, **21**, 365–368, doi:10.1016/J.TPLANTS.2016.02.006.

Mbow, C., M. Van Noordwijk, and P.A. Minang, 2014a: Agroforestry solutions to address food security and climate change challenges in Africa. *Curr. Opin. Environ. Sustain.*, **6**, 61–67, doi:10.1016/J.COSUST.2013.10.014.

Mbow, C., P. Smith, D. Skole, L. Duguma, and M. Bustamante, 2014b: Achieving mitigation and adaptation to climate change through sustainable agroforestry practices in Africa. *Curr. Opin. Environ. Sustain.*, **6**, 8–14, doi:10.1016/j.cosust.2013.09.002.

McCollum, D., N. Bauer, K. Calvin, A. Kitous, and K. Riahi, 2014: Fossil resource and energy security dynamics in conventional and carbon-constrained worlds. *Clim. Change*, **123**, doi:10.1007/s10584-013-0939-5.

McCrum, G., K. et al., 2009: Adapting to climate change in land management: The role of deliberative workshops in enhancing social learning. *Environ. Policy Gov.*, **19**, 413–426, doi:10.1002/eet.525.

McElwee, P., T. Nghiem, H. Le, and H. Vu, 2017a: Flood vulnerability among rural households in the Red River Delta of Vietnam: Implications for future climate change risk and adaptation. *Nat. Hazards*, **86**, 465–492, doi:10.1007/s11069-016-2701-6.

McElwee, P. et al., 2017b: Using REDD+ policy to facilitate climate adaptation at the local level: Synergies and challenges in Vietnam. *Forests*, **8**, 1–25, doi:10.3390/f8010011.

McGuire, S., and L. Sperling, 2016: Seed systems smallholder farmers use. *Food Secur.*, **8**, 179–195, doi:10.1007/s12571-015-0528-8.

McIntyre, P.B., C.A.R. Liermann, and C. Revenga, 2016: Linking freshwater fishery management to global food security and biodiversity conservation. *Proc. Natl. Acad. Sci.*, **113**, 12880–12885.

McKinsey and Company, 2009: *Pathways to a Low-Carbon Economy: Version 2 of the Global Greenhouse Gas Abatement Cost Curve*. McKinsey and Company, Stockholme, 192 pp.

McLaren, D., 2012: A comparative global assessment of potential negative emissions technologies. *Process Saf. Environ. Prot.*, **90**, 489–500, doi:10.1016/J.PSEP.2012.10.005.

McLeman, R., and B. Smit, 2006: Migration as an adaptation to climate change. *Clim. Change*, **76**, 31–53.

McMichael, P., 2012: The land grab and corporate food regime restructuring. **39**, 681–701, doi:10.1080/03066150.2012.661369.

McMichael, P., and M. Schneider, 2011: Food security politics and the millennium development goals. *Third World Q.*, **32**, 119–139, doi:10.1080/01436597.2011.543818.

Mechler, R., 2016: Reviewing estimates of the economic efficiency of disaster risk management: opportunities and limitations of using risk-based cost–benefit analysis. *Nat. Hazards*, **81**, 2121–2147, doi:10.1007/s11069-016-2170-y.

Medugu, N.I., M.R. Majid, F. Johar, and I.D. Choji, 2010: The role of afforestation programme in combating desertification in Nigeria. *Int. J. Clim. Chang. Strateg. Manag.*, **2**, 35–47, doi:10.1108/17568691011020247.

Meijer, S.S., D. Catacutan, O.C. Ajayi, G.W. Sileshi, and M. Nieuwenhuis, 2015: The role of knowledge, attitudes and perceptions in the uptake of agricultural and agroforestry innovations among smallholder farmers in sub-Saharan Africa. *Int. J. Agric. Sustain.*, **13**, 40–54, doi:10.1080/14735903.2014.912493.

Mekuria, W., and E. Aynekulu, 2013: Exclosure land management for restoration of the soils in degraded communal grazing lands in northern Ethiopia. *L. Degrad. Dev.*, **24**, 528–538, doi:10.1002/ldr.1146.

Melamed, M., and J. Schmale, 2016: Sustainable policy – Key considerations for air quality and climate change. *Curr. Opin. Environ. Sustain.*, **23**, 85–91.

Mello, F.F.C. et al., 2014: Payback time for soil carbon and sugar-cane ethanol. *Nat. Clim. Chang.*, **4**, 605–609, doi:10.1038/nclimate2239.

Melo, F.P.L. et al., 2013: Priority setting for scaling-up tropical forest restoration projects: Early lessons from the Atlantic Forest Restoration Pact. *Environ. Sci. Policy*, **33**, 395–404, doi:10.1016/j.envsci.2013.07.013.

Mercer, J., 2010: Policy arena disaster risk reduction or climate change adaptation: Are we reinventing the wheel? *J. Int. Dev.*, **22**, 247–264, doi:10.1002/jid.

Merriott, D., 2016: Factors associated with the farmer suicide crisis in India. *J. Epidemiol. Glob. Health*, **6**, 217–227, doi:10.1016/J.JEGH.2016.03.003.

Meyer, S., B. Glaser, and P. Quicker, 2011: Technical, economical, and climate-related aspects of biochar production technologies: A literature review. *Environ. Sci. Technol.*, **45**, 9473–9483, doi:10.1021/es201792c.

Meze-Hausken, E., A. Patt, and S. Fritz, 2009: Reducing climate risk for micro-insurance providers in Africa: A case study of Ethiopia. *Glob. Environ. Chang.*, **19**, 66–73, doi:10.1016/j.gloenvcha.2008.09.001.

Miao, L. et al., 2015: Footprint of research in desertification management in China. *L. Degrad. Dev.*, **26**, 450–457, doi:10.1002/ldr.2399.

Michelini, L., L. Principato, and G. Iasevoli, 2018: Understanding food sharing models to tackle sustainability challenges. *Ecol. Econ.*, **145**, 205–217, doi:10.1016/J.ECOLECON.2017.09.009.

Millenium Ecosystem Assessment (MA), 2005: *Ecosystems and Human Well-Being: Synthesis*. Island Press, Washington, DC., 137 pp.

Mills, E., 2005: Insurance in a climate of change. *Science*, **309**, 1040–1044, doi:10.1126/science.1112121.

6

Miner, R., 2010: *Impact of the Global Forest Industry on Atmospheric Greenhouse Gases*. Food and Agriculture Organization of the United Nations (FAO), Rome, Italy, 71 pp.

Minot, N., 2014: Food price volatility in sub-Saharan Africa: Has it really increased? *Food Policy*, **45**, 45–56, doi:10.1016/J.FOODPOL.2013.12.008.

Mittermeier, R.A., W.R. Turner, F.W. Larsen, T.M. Brooks, and C. Gascon, 2011: Global biodiversity conservation: The critical role of hotspots. In:*Biodiversity Hotspots* [F. Zachos and J. Habel (eds.)]. Springer Berlin Heidelberg, Berlin, Heidelberg, 3–22 pp.

Mohammadi, A. et al., 2014: Energy use efficiency and greenhouse gas emissions of farming systems in north Iran. *Renew. Sustain. Energy Rev.*, **30**, 724–733, doi:10.1016/J.RSER.2013.11.012.

Molotoks, A., M. Kuhnert, T. Dawson, and P. Smith, 2017: Global hotspots of conflict risk between food security and biodiversity conservation. *Land*, **6**, 67, doi:10.3390/land6040067.

Mooney, H.A., and P.R. Ehrlich, 1997: Ecosystem services: A fragmentary history. In: *Nature's Services: Societal Dependence on Natural Ecosystems*. Island Press, Washington, DC and Covelo, California, USA, 392 pp.

Moore, F.C., and D.B. Diaz, 2015: Temperature impacts on economic growth warrant stringent mitigation policy. *Nat. Clim. Chang.*, **5**, 127.

de Moraes Sá, J.C. et al., 2017: Low-carbon agriculture in South America to mitigate global climate change and advance food security. *Environ. Int.*, **98**, 102–112, doi:10.1016/J.ENVINT.2016.10.020.

Morduch, J., and M. Sharma, 2002: Strengthening public safety nets from the bottom up. *Dev. Policy Rev.*, **20**, 569–588, doi:10.1111/1467-7679.00190.

Morita, K., and K. Matsumoto, 2018: Synergies among climate change and biodiversity conservation measures and policies in the forest sector: A case study of Southeast Asian countries. *For. Policy Econ.*, **87**, 59–69.

Morton, J.F., 2007: The impact of climate change on smallholder and subsistence agriculture. *Proc. Natl. Acad. Sci. U.S.A.*, **104**, 19680–19685, doi:10.1073/pnas.0701855104.

Mosquera-Losada, M.R. et al., 2018: Agroforestry in the European common agricultural policy. *Agrofor. Syst.*, **92**, 1117, doi:10.1007/s10457-018-0251-5.

Moss, R.H., A.L. Brenkert, and E.L. Malone, 2001: *Vulnerability to climate change. A quantitative approach*. Pacific Northwest National Laboratory (PNNL-SA-33642). Prepared for the US Department of Energy, US Department of Commerce, Springfield, VA, USA, 88 pp.

Mostofa, K. et al., 2016: Reviews and syntheses: Ocean acidification and its potential impacts on marine ecosystems. *Biogeosciences*, **13**, 1767–1786, doi:10.5194/bg-13-1767-2016.

Mousseau, F., 2015: The untold success story of agroecology in Africa. *Development*, **58**, 341–345, doi:10.1057/s41301-016-0026-0.

Mudombi, S. et al., 2018: Multi-dimensional poverty effects around operational biofuel projects in Malawi, Mozambique and Swaziland. *Biomass and bioenergy*, **114**, 41–54, doi:10.1016/j.biombioe.2016.09.003.

Mueller, N.D. et al., 2012: Closing yield gaps through nutrient and water management. *Nature*, **490**, 254–257, doi:10.1038/nature11420.

Mujumdar, M. et al., 2017: Anomalous convective activity over sub-tropical east Pacific during 2015 and associated boreal summer monsoon teleconnections. *Clim. Dyn.*, **48**, 4081–4091, doi:doi.org/10.1007/s00382-016-3321-2.

Muller, A. et al., 2017: Strategies for feeding the world more sustainably with organic agriculture. *Nat. Commun.*, **8**, 1290, doi:10.1038/s41467-017-01410-w.

Munang, R., J. Andrews, K. Alverson, and D. Mebratu, 2014: Harnessing ecosystem-based adaptation to address the social dimensions of climate change. *Environ. Sci. Policy Sustain. Dev.*, **56**, 18–24, doi:10.1080/00139157.2014.861676.

Mundler, P., and L. Rumpus, 2012: The energy efficiency of local food systems: A comparison between different modes of distribution. *Food Policy*, **37**, 609–615, doi:10.1016/J.FOODPOL.2012.07.006.

Muratori, M., K. Calvin, M. Wise, P. Kyle, and J. Edmonds, 2016: Global economic consequences of deploying bioenergy with carbon capture and storage (BECCS). *Environ. Res. Lett.*, **11**, 95004, doi:10.1088/1748-9326/11/9/095004.

Murthy, P.S., and M. Madhava Naidu, 2012: Sustainable management of coffee industry by-products and value addition – A review. *Resour. Conserv. Recycl.*, **66**, 45–58, doi:10.1016/j.resconrec.2012.06.005.

Mutoko, M., C. Shisanya, and L. Hein, 2014: Fostering technological transition to sustainable land management through stakeholder collaboration in the western highlands of Kenya. *Land use policy*, **41**, 110–120.

Mutuo, P.K., G. Cadisch, A. Albrecht, C.A. Palm, and L. Verchot, 2005: Potential of agroforestry for carbon sequestration and mitigation of greenhouse gas emissions from soils in the tropics. *Nutr. Cycl. Agroecosystems*, **71**, 43–54, doi:10.1007/s10705-004-5285-6.

Nabuurs, G.J., A. Pussinen, J. Van Brusselen, and M.J. Schelhaas, 2007: Future harvesting pressure on European forests. *Eur. J. For. Res.*, **126**, 391–400, doi:10.1007/s10342-006-0158-y.

Nabuurs, G.J. et al., 2017: By 2050 the mitigation effects of EU forests could nearly double through climate smart forestry. *Forests*, **8**, 1–14, doi:10.3390/f8120484.

Naeem, S. et al., 2015: Get the science right when paying for nature's services. *Science.*, **347**, 1206–1207, doi:10.1126/science.aaa1403.

Nagoda, S., and A.J. Nightingale, 2017: Participation and power in climate change adaptation policies: Vulnerability in food security programs in Nepal. *World Dev.*, doi:10.1016/j.worlddev.2017.07.022.

Nahlik, A.M., M.E. Kentula, M.S. Fennessy, and D.H. Landers, 2012: Where is the consensus? A proposed foundation for moving ecosystem service concepts into practice. *Ecol. Econ.*, **77**, 27–35, doi:10.1016/J.ECOLECON.2012.01.001.

Nair, P., and V. Nair, 2014: "Solid–fluid–gas": The state of knowledge on carbon-sequestration potential of agroforestry systems in Africa. *Curr. Opin. Environ. Sustain.*, **6**, 22–27, doi:10.1016/j.cosust.2013.07.014.

Nair, P., V.D. Nair, B.M. Kumar, and J.M. Showalter, 2010: Carbon sequestration in agroforestry systems. *Advances in Agronomy*, **108**, 237.

Nandy, S., A. Daoud, and D. Gordon, 2016: Examining the changing profile of undernutrition in the context of food price rises and greater inequality. *Soc. Sci. Med.*, **149**, 153–163, doi:10.1016/j.socscimed.2015.11.036.

Narain, P., R.K. Singh, N.S. Sindhwal, and P. Joshie, 1997: Agroforestry for soil and water conservation in the western Himalayan Valley Region of India 2. Crop and tree production. *Agrofor. Syst.*, **39**, 191–203, doi:10.1023/A:1005900229886.

Nardone, A., B. Ronchi, N. Lacetera, M.S. Ranieri, and U. Bernabucci, 2010: Effects of climate changes on animal production and sustainability of livestock systems. *Livest. Sci.*, **130**, 57–69, doi:10.1016/J.LIVSCI.2010.02.011.

Nasi, R., A. Taber, and N. Van Vliet, 2011: Empty forests, empty stomachs? Bushmeat and livelihoods in the Congo and Amazon Basins. *Int. For. Rev.*, **13**, 355–368.

Naudts, K., Y. Chen, M.J. McGrath, J. Ryder, A. Valade, J. Otto, and S. Luyssaert, 2016: Europe's forest management did not mitigate climate warming. *Science.*, **351**, 597–601, doi:10.1126/science.aad7270.

Naylor, R.L. et al., 2000: Effect of aquaculture on world fish supplies. *Nature*, **405**, 1017–1024, doi:10.1038/35016500.

Ndoro, J.T., M. Mudhara, and M. Chimonyo, 2014: Cattle commercialization in rural South Africa: Livelihood drivers and implications for livestock marketing extension. *J. Hum. Ecol.*, **45**, 207–221.

Neary, D.G., G.G. Ice, and C.R. Jackson, 2009: Linkages between forest soils and water quality and quantity. *For. Ecol. Manage.*, **258**, 2269-2281, doi:10.1016/j.foreco.2009.05.027.

Nejad, A.N., 2013: Soil and water conservation for desertification control in Iran. In: *Combating Desertification in Asia, Africa and the Middle East: Proven practices* [G. Heshmati and V. Squires (eds.)]. Springer, Dordrecht, 377–400 pp.

Nelson, V., K. Meadows, T. Cannon, J. Morton, and A. Martin, 2002: Uncertain predictions, invisible impacts, and the need to mainstream gender in climate change adaptations. *Gend. Dev.*, **10**, 51–59, doi:10.1080/13552070215911.

Nemet, G.F., T. Holloway, and P. Meier, 2010: Implications of incorporating air-quality co-benefits into climate change policymaking. *Environ. Res. Lett.*, **5**, 14007, doi:10.1088/1748-9326/5/1/014007.

Netzel, P., and T. Stepinski, 2018: Climate similarity search: GeoWeb tool for exploring climate variability. *Bull. Am. Meteorol. Soc.*, **99**, 475–477, doi:10.1175/BAMS-D-16-0334.1.

Newbold, T. et al., 2015: Global effects of land use on local terrestrial biodiversity. *Nature*, **520**, 45–50, doi:10.1038/nature14324.

Ngigi, M.W., U. Mueller, and R. Birner, 2017: Gender differences in climate change adaptation strategies and participation in group-based approaches: An intra-household analysis from rural Kenya. *Ecol. Econ.*, **138**, 99–108, doi:10.1016/j.ecolecon.2017.03.019.

Niehof, A., 2004: The significance of diversification for rural livelihood systems. *Food Policy*, **29**, 321–338, doi:10.1016/j.foodpol.2004.07.009.

Van Niekerk, J., and R. Wynberg, 2017: Traditional seed and exchange systems cement social relations and provide a safety net: a case arotudym KwaZulu-Natal, South Africa. *Agroecol. Sustain. Food Syst.*, **41**, 1–25, doi:10.1080/21683565.2017.1359738.

Nightingale, A.J., 2017: Power and politics in climate change adaptation efforts: Struggles over authority and recognition in the context of political instability. *Geoforum*, **84**, 11–20, doi:10.1016/j.geoforum.2017.05.011.

Nigussie, Z. et al., 2017: Factors influencing small-scale farmers' adoption of sustainable land management technologies in north-western Ethiopia. *Land use policy*, **67**, 57–64.

del Ninno, C., P.A. Dorosh, and K. Subbarao, 2007: Food aid, domestic policy and food security: Contrasting experiences from South Asia and sub-Saharan Africa. *Food Policy*, **32**, 413–435, doi:10.1016/j.foodpol.2006.11.007.

Nizeyimana, E.L. et al., 2001: Assessing the impact of land conversion to urban use on soils with different productivity levels in the USA. *Soil Sci. Soc. Am. J.*, **65**, 391, doi:10.2136/sssaj2001.652391x.

Nkonya, E., A. Mirzabaev, and J. von Braun, 2016: *Economics of Land Degradation and Improvement – A Global Assessment for Sustainable Development*. Springer, Heidelberg, New York, Dordrecht, London, 695 pp.

Noble, I.R. et al., 2014: Adaptation Needs and Options. In: Climate Change 2014: Impacts, Adaptation, and Vulnerability. Part A: Global and Sectoral Aspects. Contribution of Working Group II to the Fifth Assessment Report of the Intergovernmental Panel on Climate Change, [Field, C.B., V.R. Barros, D.J. Dokken, K.J. Mach, M.D. Mastrandrea, T.E. Bilir, M. Chatterjee, K.L. Ebi, Y.O. Estrada, R.C. Genova, B. Girma, E.S. Kissel, A.N. Levy, S. MacCracken, P.R. Mastrandrea, and L.L.White (eds.)]. Cambridge University Press, Cambridge, United Kingdom and New York, NY, USA, 833–868.

North, M.P. et al., 2015: Reform forest fire management. *Science.*, **349**, 1280–1281, doi:10.1126/science.aab2356.

Norton, B.A. et al., 2015: Planning for cooler cities: A framework to prioritise green infrastructure to mitigate high temperatures in urban landscapes. *Landsc. Urban Plan.*, **134**, 127–138, doi:10.1016/J.LANDURBPLAN.2014.10.018.

Nowak, D.J., S. Hirabayashi, A. Bodine, and E. Greenfield, 2014: Tree and forest effects on air quality and human health in the United States. *Environ. Pollut.*, **193**, 119–129, doi:10.1016/j.envpol.2014.05.028.

Nunes, A. et al., 2016: Ecological restoration across the Mediterranean Basin as viewed by practitioners. *Sci. Total Environ.*, **566**, 722–732.

Núñez, M., B. et al., 2010: Assessing potential desertification environmental impact in life cycle assessment. *Int. J. Life Cycle Assess.*, **15**, 67–78, doi:10.1007/s11367-009-0126-0.

Nussbaum, M., and A. Sen, 1993: *The Quality of Life*. Clarendon Press, Oxford, UK, 468 pp.

Nuttall, M., 2012: Tipping points and the human world: Living with change and thinking about the future. *Ambio*, **41**, 96–105, doi:10.1007/s13280-011-0228-3.

O'Mara, F., 2012: The role of grasslands in food security and climate change. *Ann. Bot.*, **110**, 1263–1270, doi:10.1093/aob/mcs209.

O'Neill, B.C. et al., 2014: A new scenario framework for climate change research: The concept of shared socioeconomic pathways. *Clim. Change*, **122**, 387–400, doi:10.1007/s10584-013-0905-2.

O'Neill, B.C. et al., 2017: The roads ahead: Narratives for shared socioeconomic pathways describing world futures in the 21st century. *Glob. Environ. Chang.*, **42**, 169–180, doi:10.1016/J.GLOENVCHA.2015.01.004.

Obersteiner, M. et al., 2016: Assessing the land resource-food price nexus of the Sustainable Development Goals. *Sci. Adv.*, **2**, e1501499, doi:10.1126/sciadv.1501499.

Odgaard, M. V, M.T. Knudsen, J.E. Hermansen, and T. Dalgaard, 2019: Targeted grassland production – A Danish case study on multiple benefits from converting cereal to grasslands for green biorefinery. *J. Clean. Prod.*, **223**, 917–927, doi:10.1016/j.jclepro.2019.03.072.

OECD and FAO, 2018: *OECD-FAO Agricultural Outlook 2018 – 2027. Meat.* Food and Agriculture Organization of the United Nations, Rome, Italy, 29 pp.

Oldeman, L., R. Hakkeling, and W. Sombroek, 1991: *World Map of the Status of Human-Induced Soil Degradation: An Explanatory Note.* Global Assessment of Soil Degradation (GLASOD). CIP-Gegevens Koninklijke Bibliotheek, The Hauge, Netherlands, 41 pp.

Olesen, J., and M. Bindi, 2002: Consequences of climate change for European agricultural productivity, land use and policy. *Eur. J. Agron.*, **16**, 239–262.

de Oliveira Silva, R. et al., 2017: Sustainable intensification of Brazilian livestock production through optimized pasture restoration. *Agric. Syst.*, **153**, 201–211, doi:10.1016/J.AGSY.2017.02.001.

de Oliveira Silva, R., L.G. Barioni, J.A., G. Queiroz Pellegrino, and D. Moran, 2018: The role of agricultural intensification in Brazil's nationally determined contribution on emissions mitigation. *Agric. Syst.*, **161**, 102–112, doi:10.1016/J.AGSY.2018.01.003.

Oliver, C.D., N.T. Nassar, B.R. Lippke, and J.B. McCarter, 2014: Carbon, fossil fuel, and biodiversity mitigation with wood and forests. *J. Sustain. For.*, **33**, 248–275, doi:10.1080/10549811.2013.839386.

Oliver, T.H., and M.D. Morecroft, 2014: Interactions between climate change and land use change on biodiversity: Attribution problems, risks, and opportunities. *Wiley Interdiscip. Rev. Clim. Chang.*, **5**, 317–335, doi:10.1002/wcc.271.

Oloo, J.O., and P. Omondi, 2017: Strengthening local institutions as avenues for climate change resilience. *Int. J. Disaster Resil. Built Environ.*, **8**, 573–588, doi:10.1108/IJDRBE-12-2013-0047.

Van Oost, K. et al., 2007: The impact of agricultural soil erosion on the global carbon cycle. *Science*, **318**, 626–629, doi:10.1126/science.1145724.

Osbahr, H., C. Twyman, W. Neil Adger, and D.S.G. Thomas, 2008: Effective livelihood adaptation to climate change disturbance: Scale dimensions of practice in Mozambique. *Geoforum*, **39**, 1951–1964, doi:10.1016/j.geoforum.2008.07.010.

Ostrom, E., 1990: An institutional approach to the study of self-organization and self-governance in CPR situations. In: *Governing the Commons: The evolution of institutions for collective action*, Cambridge Univesity Press, Cambridge, UK, 29–57, doi:10.1017/CBO9780511807763.

Ostrom, E., 2000: Collective action and the evolution of social norms. *J. Econ. Perspect.*, **14**, 137–158, doi:10.1257/jep.14.3.137.

Osuri, A.M. et al., 2016: Contrasting effects of defaunation on aboveground carbon storage across the global tropics. *Nat. Commun.*, **7**, 11351, doi:10.1038/ncomms11351.

Overmars, K.P. et al., 2014: Estimating the opportunity costs of reducing carbon dioxide emissions via avoided deforestation, using integrated assessment modelling. *Land use policy*, **41**, 45–60, doi:10.1016/J.LANDUSEPOL.2014.04.015.

Pacala, S., and R. Socolow, 2004: Stabilization wedges: Solving the climate problem for the next 50 years with current technologies. *Science.*, **305**, 968 LP-972.

Padgham, J., J. Jabbour, and K. Dietrich, 2014: Managing change and building resilience: A multi-stressor analysis of urban and peri-urban agriculture in Africa and Asia. *Urban Clim.*, **1**, 183–204, doi:10.1016/j.uclim.2015.04.003.

Le Page, Y. et al., 2013: Sensitivity of climate mitigation strategies to natural disturbances. *Environ. Res. Lett.*, **8**, 15018, doi:10.1088/1748-9326/8/1/015018.

6

Pahl-Wostl, C., A. Bhaduri, and A. Bruns, 2018: Editorial special issue: The nexus of water, energy and food – An environmental governance perspective. *Environ. Sci. Policy*, **90**, 161–163, doi:10.1016/j.envsci.2018.06.021.

Palacios, M., E. et al., 2013: Landscape diversity in a rural territory: Emerging land use mosaics coupled to livelihood diversification. *Land use policy*, **30**, 814–824, doi:10.1016/j.landusepol.2012.06.007.

Palm, C., H. Blanco-Canqui, F. DeClerck, L. Gatere, and P. Grace, 2014: Conservation agriculture and ecosystem services: An overview. *Agric. Ecosyst. Environ.*, **187**, 87–105, doi:10.1016/J.AGEE.2013.10.010.

Pan, G., P. Smith, and W. Pan, 2009: The role of soil organic matter in maintaining the productivity and yield stability of cereals in China. *Agric. Ecosyst. Environ.*, **129**, 344–348, doi:10.1016/J.AGEE.2008.10.008.

Pascual, U. et al., 2017: Valuing nature's contributions to people: the IPBES approach. *Curr. Opin. Environ. Sustain.*, **26–27**, 7–16, doi:10.1016/J.COSUST.2016.12.006.

Paterson, R.R.M., L. Kumar, F. Shabani, and N. Lima, 2017: World climate suitability projections to 2050 and 2100 for growing oil palm. *J. Agric. Sci.*, **155**, 689–702, doi:10.1017/S0021859616000605.

Patnaik, A., J. Jongerden, and G. Ruivenkamp, 2017: Repossession through sharing of and access to seeds: Different cases and practices. *Int. Rev. Sociol.*, **27**, 179–201, doi:10.1080/03906701.2016.1235213.

Patt, A., P. Suarez, and U. Hess, 2010: How do small-holder farmers understand insurance, and how much do they want it? Evidence from Africa. *Glob. Environ. Chang.*, **20**, 153–161, doi:10.1016/j.gloenvcha.2009.10.007.

Paustian, K. et al., 2016: Climate-smart soils. *Nature*, **532**, 49–57, doi:10.1038/nature17174.

Pawson, S.M. et al., 2013: Plantation forests, climate change and biodiversity. *Biodivers. Conserv.*, **22**, 1203–1227, doi:10.1007/s10531-013-0458-8.

Payn, T. et al., 2015: Changes in planted forests and future global implications. *For. Ecol. Manage.*, **352**, 57–67, doi:10.1016/j.foreco.2015.06.021.

Pedercini, M., G. Zuellich, K. Dianati, and S. Arquitt, 2018: Toward achieving Sustainable Development Goals in Ivory Coast: Simulating pathways to sustainable development. *Sustain. Dev.*, **26**, 588-595, doi:10.1002/sd.1721.

PEFC, and FSC, 2018: *Double Certification FSC and PEFC – Estimations for Mid 2018*. Geneva, Switzerland,.

Pellegrini, L., and L. Tasciotti, 2014: Crop diversification, dietary diversity and agricultural income: Empirical evidence from eight developing countries. *Can. J. Dev. Stud. / Rev. Can. d'études du développement/ Rev. Can. d'études du développement*, **35**, 211–227, doi:10.1080/02255189.2014.898580.

Pelletier, J. et al., 2016: The place of community forest management in the REDD+ landscape. *Forests*, **7**, 170, doi:10.3390/f7080170.

Pendleton, L. et al., 2012: Estimating global "blue carbon" emissions from conversion and degradation of vegetated coastal ecosystems. *PLoS One*, **7**, e43542, doi: 10.1371/journal.pone.0043542.

Pereira, H.M. et al., 2010: Scenarios for global biodiversity in the 21st century. *Science.*, **330**, 1496–1501, doi:10.1126/science.1196624.

Perugini, L. et al., 2017: Biophysical effects on temperature and precipitation due to land cover change. *Environ. Res. Lett.*, **12**, 053002, doi:10.1088/1748-9326/aa6b3f.

Peterson, G.D. et al., 2018: Welcoming different perspectives in IPBES: "Nature's contributions to people" and "Ecosystem services." *Ecol. Soc.*, **23**, art39, doi:10.5751/ES-10134-230139.

Peterson, N.D., 2012: Developing climate adaptation: The intersection of climate research and development programmes in index insurance. *Dev. Change*, **43**, 557–584, doi:10.1111/j.1467-7660.2012.01767.x.

Phelps, J., E. Webb, and A. Agrawal, 2010: Does REDD+ threaten to recentralize forest governance? *Science.*, **328**, 312–313, doi:10.1126/science.1187774.

Phillipson, J., P. Lowe, A. Proctor, and E. Ruto, 2012: Stakeholder engagement and knowledge exchange in environmental research. *J. Environ. Manage.*, **95**, 56–65.

Pierzynski, G., C.L. Brajendra, and R. Vargas, 2017: *Threats to Soils: Global Trends and Perspectives. Global Land Outlook Working Paper 28*. Secretariat of the United Nations Convention to Combat Desertification, Bonn, Germany, 340 pp.

Pimentel, D., 2006: Soil erosion: A food and environmental threat. *Environ. Dev. Sustain.*, **8**, 119–137, doi:10.1007/s10668-005-1262-8.

Pimentel, D., R. Zuniga, and D. Morrison, 2005: Update on the environmental and economic costs associated with alien-invasive species in the United States. *Ecol. Econ.*, **52**, 273–288.

Pingoud, K., T. Ekholm, R. Sievänen, S. Huuskonen, and J. Hynynen, 2018: Trade-offs between forest carbon stocks and harvests in a steady state – A multi-criteria analysis. *J. Environ. Manage.*, **210**, 96–103, doi:10.1016/j.jenvman.2017.12.076.

Platteau, J.-P., O. De Bock, and W. Gelade, 2017: The demand for microinsurance: A literature review. *World Dev.*, **94**, 139–156, doi:10.1016/j.worlddev.2017.01.010.

Poeplau, C., and A. Don, 2015: Carbon sequestration in agricultural soils via cultivation of cover crops – A meta-analysis. *Agric. Ecosyst. Environ.*, **200**, 33–41, doi:10.1016/J.AGEE.2014.10.024.

Poeplau, C. et al., 2011: Temporal dynamics of soil organic carbon after land-use change in the temperate zone – Carbon response functions as a model approach. *Glob. Chang. Biol.*, **17**, 2415–2427, doi:10.1111/j.1365-2486.2011.02408.x.

Pohnan, E., H. Ompusunggu, and C. Webb, 2015: Does tree planting change minds? Assessing the use of community participation in reforestation to address illegal logging in West Kalimantan. *Trop. Conserv. Sci.*, **8**, 45–57, doi:10.1177/194008291500800107.

Pomeroy, R., and F. Douvere, 2008: The engagement of stakeholders in the marine spatial planning process. *Mar. Policy*, **32**, 816–822.

Poore, J., and T. Nemecek, 2018: Reducing food's environmental impacts through producers and consumers. *Science.*, **360**, 987–992, doi:10.1126/science.aaq0216.

Popp, A., H. Lotze-Campen, and B. Bodirsky, 2010: Food consumption, diet shifts and associated non-CO_2 greenhouse gases from agricultural production. *Glob. Environ. Chang.*, **20**, 451–462.

Popp, A. et al., 2011a: The economic potential of bioenergy for climate change mitigation with special attention given to implications for the land system. *Environ. Res. Lett.*, **6**, 34017, doi:10.1088/1748-9326/6/3/034017.

Popp, A. et al., 2011b: On sustainability of bioenergy production: Integrating co-emissions from agricultural intensification. *Biomass and Bioenergy*, **35**, 4770–4780, doi:10.1016/j.biombioe.2010.06.014.

Popp, A. et al., 2014: Land-use transition for bioenergy and climate stabilization: Model comparison of drivers, impacts and interactions with other land use based mitigation options. *Clim. Change*, **123**, 495–509, doi:10.1007/s10584-013-0926-x.

Popp, A. et al., 2017: Land-use futures in the shared socio-economic pathways. *Glob. Environ. Chang.*, **42**, 331–345, doi:10.1016/j.gloenvcha.2016.10.002.

Porter, J.R., L. Xie, A.J. Challinor, K. Cochrane, S.M. Howden, M.M. Iqbal, D.B. Lobell, and M.I. Travasso, 2014: Food Security and Food Production Systems. In: Climate Change 2014: Impacts, Adaptation, and Vulnerability. Part A: Global and Sectoral Aspects. Contribution of Working Group II to the Fifth Assessment Report of the Intergovernmental Panel on Climate Change, [Field, C.B., V.R. Barros, D.J. Dokken, K.J. Mach, M.D. Mastrandrea, T.E. Bilir, M. Chatterjee, K.L. Ebi, Y.O. Estrada, R.C. Genova, B. Girma, E.S. Kissel, A.N. Levy, S. MacCracken, P.R. Mastrandrea, and L.L. White (eds.)]. Cambridge University Press, Cambridge, United Kingdom and New York, NY, USA, pp. 485–533.

Porter, S.D., D.S. Reay, P. Higgins, and E. Bomberg, 2016: A half-century of production-phase greenhouse gas emissions from food loss & waste in the global food supply chain. *Sci. Total Environ.*, **571**, 721–729, doi:10.1016/j.scitotenv.2016.07.041.

Poteete, A.R., and E. Ostrom, 2004: In pursuit of comparable concepts and data about collective action. *Agric. Syst.*, **82**, 215–232, doi:10.1016/j.agsy.2004.07.002.

Potschin, M.B., and R.H. Haines-Young, 2011: Ecosystem services. *Prog. Phys. Geogr. Earth Environ.*, **35**, 575–594, doi:10.1177/0309133311423172.

Poulter, B. et al., 2014: Contribution of semi-arid ecosystems to interannual variability of the global carbon cycle. *Nature*, **509**, 600–603, doi:10.1038/nature13376.

Poulton, C., J. Kydd, S. Wiggins, and A. Dorward, 2006: State intervention for food price stabilisation in Africa: Can it work? *Food Policy*, **31**, 342–356, doi:10.1016/J.FOODPOL.2006.02.004.

Powell, T.W.R., and T.M. Lenton, 2012: Future carbon dioxide removal via biomass energy constrained by agricultural efficiency and dietary trends. *Energy Environ. Sci.*, **5**, 8116–8133.

Powlson, D.S. et al., 2014: Limited potential of no-till agriculture for climate change mitigation. *Nat. Clim. Chang.*, **4**, 678–683, doi:10.1038/nclimate2292.

Powlson, D.S., C.M. Stirling, C. Thierfelder, R.P. White, and M.L. Jat, 2016: Does conservation agriculture deliver climate change mitigation through soil carbon sequestration in tropical agro-ecosystems? *Agric. Ecosyst. Environ.*, **220**, 164–174, doi:10.1016/j.agee.2016.01.005.

Pozzi, W. et al., 2013: Toward global drought early warning capability: Expanding international cooperation for the development of a framework for monitoring and forecasting. *Bull. Am. Meteorol. Soc.*, **94**, 776–785, doi:10.1175/BAMS-D-11-00176.1.

Pradhan, P., D.E. Reusser, and J.P. Kropp, 2013: Embodied greenhouse gas emissions in diets. *PLoS One*, **8**, e62228.

Pratt, K., and D. Moran, 2010: Evaluating the cost-effectiveness of global biochar mitigation potential. *Biomass and bioenergy*, **34**, 1149–1158.

Pretty, J., and Z.P.Z. Bharucha, 2014: Sustainable intensification in agricultural systems. *Ann. Bot.*, **114**, 1571–1596, doi:10.1093/aob/mcu205.

Pretty, J. et al., 2018: Global assessment of agricultural system redesign for sustainable intensification. *Nat. Sustain.*, **1**, 441, doi:10.1038/s41893-018-0114-0.

Putz, F.E. et al., 2012: Sustaining conservation values in selectively logged tropical forests: The attained and the attainable. *Conserv. Lett.*, **5**, 296–303, doi:10.1111/j.1755-263X.2012.00242.x.

Qadir, M., A.D. Noble, and C. Chartres, 2013: Adapting to climate change by improving water productivity of soils in dry areas. *L. Degrad. Dev.*, **24**, 12–21, doi:10.1002/ldr.1091.

Qian, J., Y. Peng, C. Luo, C. Wu, and Q. Du, 2015: Urban land expansion and sustainable land use policy in Shenzhen: A case study of China's rapid urbanization. *Sustainability*, **8**, 16, doi:10.3390/su8010016.

Qin, Z., J.B. Dunn, H. Kwon, S. Mueller, and M.M. Wander, 2016: Soil carbon sequestration and land use change associated with biofuel production: Empirical evidence. *GCB Bioenergy*, **8**, 66–80, doi:10.1111/gcbb.12237.

Rahman, M.R., and S.H. Bulbul, 2015: Adoption of water saving irrigation techniques for sustainable rice production in Bangladesh. *Environ. Ecol. Res.*, **3**, 1–8, doi:10.13189/EER.2015.030101.

Rakodi, C., 1999: A capital assets framework for analysing household livelihood strategies: Implications for policy. *Dev. Policy Rev.*, **17**, 315–342, doi:10.1111/1467-7679.00090.

Raleigh, C., H.J. Choi, and D. Kniveton, 2015: The devil is in the details: An investigation of the relationships between conflict, food price and climate across Africa. *Glob. Environ. Chang.*, **32**, 187–199, doi:10.1016/j.gloenvcha.2015.03.005.

Ram, A. et al., 2017: Reactive nitrogen in agroforestry systems of India. In: *The Indian Nitrogen Assessment*. Woodhead Publishing, Duxford, UK and Cambridge, MA, USA and Kidlington, UK, 207–218.

Ramage, M.H. et al., 2017: The wood from the trees: The use of timber in construction. *Renew. Sustain. Energy Rev.*, **68**, 333–359, doi:10.1016/J.RSER.2016.09.107.

Ramanathan, V., P.J. Crutzen, J.T. Kiehl, and D. Rosenfeld, 2001: Aerosols, climate, and the hydrological cycle. *Science*, **294**, 2119–2124, doi:10.1126/science.250.4988.1669.

Rametsteiner, E., and M. Simula, 2003: Forest certification – An instrument to promote sustainable forest management? *J. Environ. Manage.*, **67**, 87–98, doi:10.1016/S0301-4797(02)00191-3.

Randerson, J.T., Y. Chen, G.R. Van der Werf, B.M. Rogers, and D.C. Morton, 2012: Global burned area and biomass burning emissions from small fires. *J. Geophys. Res. Biogeosciences*, **117**, doi:10.1029/2012JG002128.

Rao, C.S. et al., 2017a: Farm ponds for climate-resilient rainfed agriculture. *Curr. Sci.*, **112**, 471.

Rao, S. et al., 2017b: Future air pollution in the Shared Socio-economic Pathways. *Glob. Environ. Chang.*, **42**, 346–358, doi:10.1016/j.gloenvcha.2016.05.012.

Rasmussen, D.J. et al., 2017: Coastal flood implications of 1.5°C, 2.0°C, and 2.5°C temperature stabilization targets in the 21st and 22nd century. *Environmental Research Letters*, **13**, 034040, doi:10.1088/1748-9326/aaac87.

Rawlins, A., and J. Morris, 2010: Social and economic aspects of peatland management in Northern Europe, with particular reference to the English case. *Geoderma*, **154**, 242–251, doi:10.1016/j.geoderma.2009.02.022.

Redford, K.H., and W.M. Adams, 2009: Payment for ecosystem services and the challenge of saving nature. *Conserv. Biol.*, **23**, 785–787, doi:10.1111/j.1523-1739.2009.01271.x.

Regmi, A., and B. Meade, 2013: Demand side drivers of global food security. *Glob. Food Sec.*, **2**, 166–171, doi:10.1016/j.gfs.2013.08.001.

Reichardt, M., C. Jürgens, U. Klöble, J. Hüter, and K. Moser, 2009: Dissemination of precision farming in Germany: Acceptance, adoption, obstacles, knowledge transfer and training activities. *Precis. Agric.*, **10**, 525–545, doi:10.1007/s11119-009-9112-6.

Reidsma, P., F. Ewert, A.O. Lansink, and R. Leemans, 2010: Adaptation to climate change and climate variability in European agriculture: The importance of farm level responses. *Eur. J. Agron.*, **32**, 91–102, doi:10.1016/j.eja.2009.06.003.

Reilly, J. et al., 2012: Using land to mitigate climate change: Hitting the target, recognizing the trade-offs. *Environ. Sci. Technol.*, **46**, 5672–5679, doi:10.1021/es2034729.

Reisman, E., 2017: Troubling tradition, community, and self-reliance: Reframing expectations for village seed banks. *World Dev.*, **98**, 160–168, doi:10.1016/J.WORLDDEV.2017.04.024.

Renforth, P. et al., 2012: Contaminant mobility and carbon sequestration downstream of the Ajka (Hungary) red mud spill: The effects of gypsum dosing. *Sci. Total Environ.*, **421–422**, 253–259, doi:10.1016/J.SCITOTENV.2012.01.046.

Rengasamy, P., 2006: World salinization with emphasis on Australia. *J. Exp. Bot.*, **57**, 1017–1023, doi:10.1093/jxb/erj108.

Revi, A. et al., 2014: Urban Areas. In: Climate Change 2014: Impacts, Adaptation, and Vulnerability. Part A: Global and Sectoral Aspects. Contribution of Working Group II to the Fifth Assessment Report of the Intergovernmental Panel on Climate Change, C. [Field, C.B., V.R. Barros, D.J. Dokken, K.J. Mach, M.D. Mastrandrea, T.E. Bilir, M. Chatterjee, K.L. Ebi, Y.O. Estrada, R.C. Genova, B. Girma, E.S. Kissel, A.N. Levy, S. MacCracken, P.R. Mastrandrea, and L.L.White (eds.)]. Cambridge University Press, Cambridge, United Kingdom and New York, NY, USA, 535–612.

Rey Benayas, J.M., A.C. Newton, A. Diaz, and J.M. Bullock, 2009: Enhancement of biodiversity and ecosystem services by ecological restoration: A meta-analysis. *Science*, **325**, 1121–1124, doi:10.1126/science.1172460.

Reyer, C., M. Guericke, and P.L. Ibisch, 2009: Climate change mitigation via afforestation, reforestation and deforestation avoidance: And what about adaptation to environmental change? *New For.*, **38**, 15–34, doi:10.1007/s11056-008-9129-0.

6

Reynolds, L.P., M.C. Wulster-Radcliffe, D.K. Aaron, and T.A. Davis, 2015: Importance of animals in agricultural sustainability and food security. *J. Nutr.*, **145**, 1377–1379.

Riahi, K. et al., 2017: The Shared Socioeconomic Pathways and their energy, land use, and greenhouse gas emissions implications: An overview. *Glob. Environ. Chang.*, **42**, 153–168, doi:10.1016/j.gloenvcha.2016.05.009.

Richards, M. et al., 2017: High resolution spatial modelling of greenhouse gas emissions from landuse change to energy crops in the United Kingdom. *GCB Bioenergy*, doi:10.1111/gcbb.12360.

Ridoutt, B. et al., 2016: Climate change adaptation strategy in the food industry – Insights from product carbon and water footprints. *Climate*, **4**, 26, doi:10.3390/cli4020026.

Rigg, J., 2006: Land, farming, livelihoods, and poverty: Rethinking the links in the Rural South. *World Dev.*, **34**, 180–202, doi:10.1016/j.worlddev.2005.07.015.

Ringler, C. et al., 2016: Global linkages among energy, food and water: an economic assessment. *J. Environ. Stud. Sci.*, **6**, 161–171, doi:10.1007/s13412-016-0386-5.

Ritzema, R.S. et al., 2017: Is production intensification likely to make farm households food-adequate? A simple food availability analysis across smallholder farming systems from East and West Africa. *Food Secur.*, **9**, 115–131, doi:10.1007/s12571-016-0638-y.

Rivera-Ferre, M.G. et al., 2016: Re-framing the climate change debate in the livestock sector: Mitigation and adaptation options. *Wiley Interdiscip. Rev. Clim. Chang.*, **7**, 869–892, doi:10.1002/wcc.421.

Robert, M. et al., 2017: Farm typology in the Berambadi Watershed (India): Farming systems are determined by farm size and access to groundwater. *Water*, **9**, 51, doi:10.3390/w9010051.

Roberts, K.G., B.A. Gloy, S. Joseph, N.R. Scott, and J. Lehmann, 2009: Life cycle assessment of biochar systems: Estimating the energetic, economic, and climate change potential. *Environ. Sci. Technol.*, **44**, 827–833.

Robertson, A.D. et al., 2017a: Climate change impacts on yields and soil carbon in row crop dryland agriculture. *J. Environ. Qual.*, **47**, 684–694, doi:10.2134/jeq2017.08.0309.

Robertson, G.P. et al., 2017b: Cellulosic biofuel contributions to a sustainable energy future: Choices and outcomes. *Science.*, **356**, eaal2324, doi:10.1126/science.aal2324.

Robledo-Abad, C. et al., 2017: Bioenergy production and sustainable development: Science base for policymaking remains limited. *GCB Bioenergy*, **9**, 541–556, doi:10.1111/gcbb.12338.

Rodríguez, J. et al., 2006: Trade-offs across space, time, and ecosystem services. *Ecol. Soc.*, **11**, 28.

Roesch-McNally, G.E., S. Rabotyagov, J.C. Tyndall, G. Ettl, and S.F. Tóth, 2016: Auctioning the forest: A qualitative approach to exploring stakeholder responses to bidding on forest ecosystem services. *Small-scale For.*, **15**, 321–333, doi:10.1007/s11842-016-9327-0.

Roesch-McNally, G.E. et al., 2017: The trouble with cover crops: Farmers' experiences with overcoming barriers to adoption. *Renew. Agric. Food Syst.*, **33**, 322–333, doi:10.1017/S1742170517000096.

Rogelj, J., D. Shindell, K. Jiang, S. Fifita, P. Forster, V. Ginzburg, C. Handa, H. Kheshgi, S. Kobayashi, E. Kriegler, L. Mundaca, R. Séférian, and M.V.Vilariño, 2018a: Mitigation Pathways Compatible with 1.5°C in the Context of Sustainable Development. In: Global Warming of 1.5°C an IPCC special report on the impacts of global warming of 1.5°C above pre-industrial levels and related global greenhouse gas emission pathways, in the context of strengthening the global response to the threat of climate change, sustainable development, and efforts to eradicate poverty [Masson-Delmotte, V., P. Zhai, H.-O. Pörtner, D. Roberts, J. Skea, P.R. Shukla, A. Pirani, W. Moufouma-Okia, C. Péan, R. Pidcock, S. Connors, J.B.R. Matthews, Y. Chen, X. Zhou, M.I.Gomis, E. Lonnoy, T.Maycock, M.Tignor, and T. Waterfield (eds.)]. World Meteorological Organization, Geneva, Switzerland, pp. 93–174.

Rogelj, J. et al., 2018b: Scenarios towards limiting global mean temperature increase below 1.5°C. *Nat. Clim. Chang.*, **8**, 325–332, doi:10.1038/s41558-018-0091-3.

Rojas-Downing, M.M., A.P. Nejadhashemi, T. Harrigan, and S.A. Woznicki, 2017: Climate change and livestock: Impacts, adaptation, and mitigation. *Clim. Risk Manag.*, **16**, 145–163, doi:10.1016/j.crm.2017.02.001.

Rose, S.K. et al., 2014: Bioenergy in energy transformation and climate management. *Clim. Change*, **123**, doi:10.1007/s10584-013-0965-3.

Rose, S.K., R. Richels, G. Blanford, and T. Rutherford, 2017: The Paris Agreement and next steps in limiting global warming. *Clim. Change*, **142**, 255–270.

Rosenstock, T.S. et al., 2014: Agroforestry with N2-fixing trees: Sustainable development's friend or foe? *Curr. Opin. Environ. Sustain.*, doi:10.1016/j.cosust.2013.09.001.

Rowe, H. et al., 2016: Integrating legacy soil phosphorus into sustainable nutrient management strategies for future food, bioenergy and water security. *Nutr. Cycl. Agroecosystems*, doi:10.1007/s10705-015-9726-1.

Rowland, D., A. Ickowitz, B. Powell, R. Nasi, and T. Sunderland, 2017: Forest foods and healthy diets: Quantifying the contributions. *Environ. Conserv.*, **44**, 102–114, doi:10.1017/S0376892916000151.

de Ruiter, H. et al., 2017: Total global agricultural land footprint associated with UK food supply 1986-2011. *Global Environmental Change*, **43**, 72–81.

Rulli, M., S. Bozzi, M. Spada, D. Bocchiola, and R. Rosso, 2006: Rainfall simulations on a fire disturbed Mediterranean area. *J. Hydrol.*, **327**, 323–338.

Rulli, M.C., D. Bellomi, A. Cazzoli, G. De Carolis, and P. D'Odorico, 2016: The water-land-food nexus of first-generation biofuels. *Sci. Rep.*, **6**, 22521, doi:10.1038/srep22521.

Safriel, U., 2017: Land Degradation Neutrality (LDN) in drylands and beyond – Where has it come from and where does it go. *Silva Fenn.*, **51**, 1650, doi:10.14214/sf.1650.

Sain, G. et al., 2017: Costs and benefits of climate-smart agriculture: The case of the Dry Corridor in Guatemala. *Agric. Syst.*, **151**, 163–173, doi:10.1016/J.AGSY.2016.05.004.

Salvati, L., A. Sabbi, D. Smiraglia, and M. Zitti, 2014: Does forest expansion mitigate the risk of desertification? Exploring soil degradation and land-use changes in a Mediterranean country. *Int. For. Rev.*, **16**, 485–496, doi:10.1505/146554814813484149.

Sanchez, D.L., and D.M. Kammen, 2016: A commercialization strategy for carbon-negative energy. *Nat. Energy*, **1**, 15002, doi:10.1038/nenergy.2015.2.

Sánchez, J., M.D. Curt, and J. Fernández, 2017: Approach to the potential production of giant reed in surplus saline lands of Spain. *GCB Bioenergy*, **9**, 105-118, doi:10.1111/gcbb.12329.

Sanderman, J., T. Hengl, and G.J. Fiske, 2017: Soil carbon debt of 12,000 years of human land use. *Proc. Natl. Acad. Sci. U.S.A.*, **114**, 9575–9580, doi:10.1073/pnas.1706103114.

Sanderson, B.M., B.C. O'Neill, and C. Tebaldi, 2016: What would it take to achieve the Paris temperature targets? *Geophys. Res. Lett.*, **43**, doi:10.1002/2016GL069563.

Sanderson, M. et al., 2013: Diversification and ecosystem services for conservation agriculture: Outcomes from pastures and integrated crop-livestock systems. *Renew. Agric. Food Syst.*, **28**, 194, doi:10.1017/s1742170513000124.

Santiago-Freijanes, J.J. et al., 2018: Understanding agroforestry practices in Europe through landscape features policy promotion. *Agrofor. Syst.*, **92**, 1105–1115, doi:10.1007/s10457-018-0212-z.

Santika, T. et al., 2017: Community forest management in Indonesia: Avoided deforestation in the context of anthropogenic and climate complexities. *Glob. Environ. Chang.*, **46**, 60–71, doi:10.1016/J.GLOENVCHA.2017.08.002.

Santilli, J., 2012: *Agrobiodiversity and the Law: Regulating Genetic Resources, Food Security and Cultural Diversity*. Earthscan, New York, USA, 288 pp.

Sapkota, T. et al., 2017: Reducing global warming potential through sustainable intensification of basmati rice-wheat systems in India. *Sustainability*, **9**, 1044, doi:10.3390/su9061044.

6

Sasaki, N. et al., 2016: Sustainable management of tropical forests can reduce carbon emissions and stabilize timber production. *Front. Environ. Sci.*, **4**, 50.

Sascha, A., G. Cheppudira, U. Shaanker, and J. Chris, 2017: Evaluating realized seed dispersal across fragmented tropical landscapes: A two fold approach using parentage analysis and the neighbourhood model. *New Phytol.*, **214**, 1307-1316, doi: 10.1111/nph.14427.

Sathre, R., and J. O'Connor, 2010: Meta-analysis of greenhouse gas displacement factors of wood product substitution. *Environ. Sci. Policy*, **13**, 104–114.

Scasta, J. et al., 2016: Patch-burn grazing (PBG) as a livestock management alternative for fire-prone ecosystems of North America. *Renew. Agric. Food Syst.*, **31**, 550–567.

Schatz, J., and C.J. Kucharik, 2015: Urban climate effects on extreme temperatures in Madison, Wisconsin, USA. *Environ. Res. Lett.*, **10**, 94024, doi:10.1088/1748-9326/10/9/094024.

Schipper, L., and M. Pelling, 2006: Disaster risk, climate change and international development: Scope for, and challenges to, integration. *Disasters*, **30**, 19–38, doi:10.1111/j.1467-9523.2006.00304.x.

Schjønning, P. et al., 2018: The role of soil organic matter for maintaining crop yields: Evidence for a renewed conceptual basis. *Advances in Agronomy*, **150**, 35–79.

Schleicher, J., 2018: The environmental and social impacts of protected areas and conservation concessions in South America. *Curr. Opin. Environ. Sustain.*, **32**, 1–8, doi:10.1016/J.COSUST.2018.01.001.

Schlosser, C., K. Strzepek, and X. Gao, 2014: The future of global water stress: An integrated assessment. *Earth's Futur.*, **2**, 341–361, doi:10.1002/2014EF000238.Received.

Schmidhuber, J., and F.N. Tubiello, 2007: Global food security under climate change. *Proc. Natl. Acad. Sci.*, **104**, 19703–19708, doi:10.1073/pnas.0701976104.

Schmitz, O.J. et al., 2014: Animating the carbon cycle. *Ecosystems*, **17**, 344–359, doi:10.1007/s10021-013-9715-7.

Schmitz, O.J. et al., 2018: Animals and the zoogeochemistry of the carbon cycle. *Science*, **362**, eaar3213, doi:10.1126/SCIENCE.AAR3213.

Schröder, P. et al., 2018: Intensify production, transform biomass to energy and novel goods and protect soils in Europe – A vision how to mobilize marginal lands. *Sci. Total Environ.*, **616–617**, 1101–1123, doi:10.1016/j.scitotenv.2017.10.209.

Schröter, D. et al., 2005: Ecosystem service supply and vulnerability to global change in Europe. *Science*, **310**, 1333–1337, doi:10.1126/science.1115233.

Schröter, M. et al., 2014: Ecosystem services as a contested concept: A synthesis of critique and counter-arguments. *Conserv. Lett.*, **7**, 514–523, doi:10.1111/conl.12091.

Schueler, V., S. Fuss, J.C. Steckel, U. Weddige, and T. Beringer, 2016: Productivity ranges of sustainable biomass potentials from non-agricultural land. *Environ. Res. Lett.*, doi:10.1088/1748-9326/11/7/074026.

Schuiling, R.D., and P. Krijgsman, 2006: Enhanced weathering: An effective and cheap tool to sequester CO_2. *Clim. Change*, **74**, 349–354, doi:10.1007/s10584-005-3485-y.

Schut, M. et al., 2016: Sustainable intensification of agricultural systems in the Central African Highlands: The need for institutional innovation. *Agric. Syst.*, **1**, 165–176, doi:10.1016/j.agsy.2016.03.005.

Schwilch, G. et al., 2012a: A structured multi-stakeholder learning process for Sustainable Land Management. *J. Environ. Manage.*, **107**, 52–63, doi:10.1016/j.jenvman.2012.04.023.

Schwilch, G., F. Bachmann, and J. de Graaff, 2012b: Decision support for selecting SLM technologies with stakeholders. *Appl. Geogr.*, **34**, 86–98, doi:10.1016/j.apgeog.2011.11.002.

Schwilch, G., H.P. Liniger, and H. Hurni, 2014: Sustainable Land Management (SLM) practices in drylands: How do they address desertification threats? *Environ. Manage.*, **54**, 983–1004, doi:10.1007/s00267-013-0071-3.

Scott, C.A. et al., 2011: Policy and institutional dimensions of the water–energy nexus. *Energy Policy*, **39**, 6622–6630, doi:10.1016/J.ENPOL.2011.08.013.

Searchinger, T.D. et al., 2018: Europe's renewable energy directive poised to harm global forests. *Nat. Commun.*, **9**, 10–13, doi:10.1038/s41467-018-06175-4.

Searle, S., and C. Malins, 2014: A reassessment of global bioenergy potential in 2050. *GCB Bioenergy*, **7**, 328–336, doi:10.1111/gcbb.12141.

Secretariat of the Convention on Biological Diversity, 2008: *Protected Areas in Today's World: Their Values and Benefits for the Welfare of the Planet*. 96 pp.

Seidl, R., M.J. Schelhaas, W. Rammer, and P.J. Verkerk, 2014: Increasing forest disturbances in Europe and their impact on carbon storage. *Nat. Clim. Chang.*, **4**, 806–810, doi:10.1038/nclimate2318.

Seidl, R. et al., 2017: Forest disturbances under climate change. *Nat. Clim. Chang.*, **7**, 395.

Seinfeld, J.H., and S.N. Pandis, *Atmospheric Chemistry and Physics: From Air pollution to Climate Change*. 3rd Edition, Wiley, New Jersey, USA.

Sen, A., 1992: *Inequality Reexamined*. Clarendon Press, Oxford, UK, 206 pp.

Seo, S.N., 2010: Is an integrated farm more resilient against climate change? A micro-econometric analysis of portfolio diversification in African agriculture. *Food Policy*, **35**, 32–40, doi:10.1016/J.FOODPOL.2009.06.004.

Seppelt, R., C.F. Dormann, F.V. Eppink, S. Lautenbach, and S. Schmidt, 2011: A quantitative review of ecosystem service studies: Approaches, shortcomings and the road ahead. *J. Appl. Ecol.*, **48**, 630–636, doi:10.1111/j.1365-2664.2010.01952.x.

Shackley, S., J. Hammond, J. Gaunt, and R. Ibarrola, 2011: The feasibility and costs of biochar deployment in the UK. *Carbon Manag.*, **2**, 335–356, doi:10.4155/cmt.11.22.

Shah, P., A. Bansal, and R.K. Singh, 2018: Life cycle assessment of organic, BCI and conventional cotton: A comparative study of cotton cultivation practices in India. In: *Designing Sustainable Technologies, Products and Policies.* Springer International Publishing AG, Switzerland, 67–77.

Shah, T., M. Giordano, and A. Mukherji, 2012: Political economy of the energy-groundwater nexus in India: exploring issues and assessing policy options. *Hydrogeol. J.*, **20**, 995–1006, doi:10.1007/s10040-011-0816-0.

Shaw, C., S. Hales, P. Howden-Chapman, and R. Edwards, 2014: Health co-benefits of climate change mitigation policies in the transport sector. *Nat. Clim. Chang.*, **4**, 427–433, doi:10.1038/nclimate2247.

Shcherbak, I., N. Millar, and G.P. Robertson, 2014: Global metaanalysis of the nonlinear response of soil nitrous oxide (N_2O) emissions to fertilizer nitrogen. *Proc. Natl. Acad. Sci. U.S.A.*, **111**, 9199–9204, doi:10.1073/pnas.1322434111.

Sheahan, M., and C.B. Barrett, 2017: Ten striking facts about agricultural input use in sub-Saharan Africa. *Food Policy*, **67**, 12–25, doi:10.1016/J.FOODPOL.2016.09.010.

Shen, X. et al., 2017: Local interests or centralized targets? How China's local government implements the farmland policy of Requisition–Compensation Balance. *Land use policy*, **67**, 716–724, doi:10.1016/J.LANDUSEPOL.2017.06.012.

Shindell, D. et al., 2012: Simultaneously mitigating near-term climate change and improving human health and food security. *Science*, **335**, 183–189, doi:10.1126/science.1210026.

Shisanya, S., and P. Mafongoya, 2016: Adaptation to climate change and the impacts on household food security among rural farmers in uMzinyathi District of Kwazulu-Natal, South Africa. *Food Secur.*, **8**, 597–608, doi:10.1007/s12571-016-0569-7.

Sida, T.S., F. Baudron, K. Hadgu, A. Derero, and K.E. Giller, 2018: Crop vs. tree: Can agronomic management reduce trade-offs in tree-crop interactions? *Agric. Ecosyst. Environ.*, **260**, 36–46, doi:10.1016/J.AGEE.2018.03.011.

Silalahi, M. et al., 2017: Indonesia's ecosystem restoration concessions. *Unasylva*, **68**, 63–70.

Sileshi, G., E. Kuntashula, P. Matakala, and P.N., 2008: Farmers' perceptions of tree mortality, pests and pest management practices in agroforestry in Malawi, Mozambique and Zambia, *Agroforestry Systems*, **72** (2), pp. 87–101.

Silva-Olaya, A.M. et al., 2017: Modelling SOC response to land use change and management practices in sugarcane cultivation in South-Central Brazil. *Plant Soil*, **410**, 483-498, doi:10.1007/s11104-016-3030-y.

da Silva, J., S. Kernaghan, and A. Luque, 2012: A systems approach to meeting the challenges of urban climate change. *Int. J. Urban Sustain. Dev.*, **4**, 125–145, doi:10.1080/19463138.2012.718279.

Sims, R., R. Schaeffer, F. Creutzig, X. Cruz-Núñez, M. D'Agosto, D. Dimitriu, M.J. Figueroa Meza, L. Fulton, S. Kobayashi, O. Lah, A. McKinnon, P. Newman, M. Ouyang, J.J. Schauer, D. Sperling, and G. Tiwari, 2014: Transport. In: Climate Change 2014: Mitigation of Climate Change. Contribution of Working Group III to the Fifth Assessment Report of the Intergovernmental Panel on Climate Change. [Edenhofer, O., R. Pichs-Madruga, Y. Sokona, E. Farahani, S. Kadner, K. Seyboth, A. Adler, I. Baum, S. Brunner, P. Eickemeier, B. Kriemann, J. Savolainen, S. Schlömer, C. von Stechow, T. Zwickel and J.C. Minx (eds.)]. Cambridge University Press, Cambridge, United Kingdom and New York, NY, USA, pp. 599–670.

Singh, G., 2009: Salinity-related desertification and management strategies: Indian experience. *L. Degrad. Dev.*, **20**, 367–385, doi:10.1002/ldr.933.

Singh, S.N., and A. Verma, 2007: Environmental review: The potential of nitrification inhibitors to manage the pollution effect of nitrogen fertilizers in agricultural and other soils: A review. *Environ. Pract.*, **9**, 266–279, doi:10.1017/S1466046607070482.

Skees, J.R., and B. Collier, 2012: The roles of weather insurance and the carbon market. In: *Greening the Financial Sector*. Springer Berlin Heidelberg, Berlin, Heidelberg, 111–164 pp.

Slade, R., A. Bauen, and R. Gross, 2014: Global bioenergy resources. *Nat. Clim. Chang.*, **4**, 99–105, doi:10.1038/nclimate2097.

Smit, B., Pilifosova, O., Burton, I., Challenger, B., Huq, S., Klein, R.J.T., and Yohe, G., 2001: Adaptation to climate change in the context of sustainable development and equity. In: *Climate Change 2001: Impacts, Adaptation and Vulnerability. Contribution of Working Group II to the Third Assessment Report of the Intergovernmental Panel on Climate Change*. [McCarthy, J., Canziani, O., Leary, N., Dokken, D., and White, K. (eds.)] pp 877–912

Smith, P., 2004: Monitoring and verification of soil carbon changes under Article 3.4 of the Kyoto Protocol. *Soil Use Manag.*, **20**, 264–270, doi:10.1111/j.1475-2743.2004.tb00367.x.

Smith, P., 2008: Land use change and soil organic carbon dynamics. *Nutr. Cycl. Agroecosystems*, **81**, 169–178, doi:10.1007/s10705-007-9138-y.

Smith, P., 2013: Delivering food security without increasing pressure on land. *Glob. Food Sec.*, **2**, 18–23, doi:10.1016/j.gfs.2012.11.008.

Smith, P., 2016: Soil carbon sequestration and biochar as negative emission technologies. *Glob. Chang. Biol.*, **22**, 1315–1324, doi:10.1111/gcb.13178.

Smith, P., D. Martino, Z. Cai, D. Gwary, H. Janzen, P. Kumar, B. McCarl, S. Ogle, F. O'Mara, C. Rice, B. Scholes, O. Sirotenko, 2007: Agriculture. In: Climate Change 2007: Mitigation. Contribution of Working Group III to the Fourth Assessment Report of the Intergovernmental Panel on Climate Change, [B. Metz, O. Davidson, P. Bosch, R. Dave, and L. Meyer, (eds.)], Cambridge University Press, Cambridge, United Kingdom and New York, NY, USA, pp 497–540.

Smith, P. et al., 2008: Greenhouse gas mitigation in agriculture. *Philos. Trans. R. Soc. Lond. B. Biol. Sci.*, **363**, 789–813, doi:10.1098/rstb.2007.2184.

Smith, P. et al., 2013: How much land-based greenhouse gas mitigation can be achieved without compromising food security and environmental goals? *Glob. Chang. Biol.*, **19**, 2285–2302, doi:10.1111/gcb.12160.

Smith, P., M. Bustamante, H. Ahammad, H. Clark, H. Dong, E.A. Elsiddig, H. Haberl, R. Harper, J. House, M. Jafari, O. Masera, C. Mbow, N.H. Ravindranath, C.W. Rice, C. Robledo Abad, A. Romanovskaya, F. Sperling, and F. Tubiello, 2014: Agriculture, Forestry and Other Land Use (AFOLU). Climate Change 2014: Mitigation of Climate Change. Contribution of Working Group III to the Fifth Assessment Report of the Intergovernmental Panel on Climate Change, [Edenhofer, O., R. Pichs-Madruga, Y. Sokona, E. Farahani, S. Kadner, K. Seyboth, A. Adler, I. Baum, S. Brunner, P. Eickemeier, B. Kriemann, J. Savolainen, S. Schlömer, C. von Stechow, T. Zwickel and J.C. Minx (eds.)].Cambridge University Press, Cambridge, United Kingdom and New York, NY, USA, pp. 811–922.

Smith, P. et al., 2015: Biogeochemical cycles and biodiversity as key drivers of ecosystem services provided by soils. **1**, 665–685, doi:10.5194/soil-1-665-2015.

Smith, P. et al., 2016a: Biophysical and economic limits to negative CO_2 emissions. *Nat. Clim. Chang.*, **6**, 42–50, doi:10.1038/NCLIMATE2870.

Smith, P. et al., 2016b: Global change pressures on soils from land use and management. *Glob. Chang. Biol.*, **22**, 1008–1028, doi:10.1111/gcb.13068.

Smith, P., R.S. Haszeldine, and S.M. Smith, 2016c: Preliminary assessment of the potential for, and limitations to, terrestrial negative emission technologies in the UK. *Environ. Sci. Process. Impacts*, **18**, 1400–1405, doi:10.1039/C6EM00386A.

Smith, S.V., W.H. Renwick, R.W. Buddemeier, and C.J. Crossland, 2001: Budgets of soil erosion and deposition for sediments and sedimentary organic carbon across the conterminous United States. *Global Biogeochem. Cycles*, **15**, 697–707, doi:10.1029/2000GB001341.

Smith, S.V., R.O. Sleezer, W.H. Renwick, and R.W. Buddemeier, 2005: Fates of eroded soil organic carbon: Mississippi Basin case study. *Ecol. Appl.*, **15**, 1929–1940, doi:10.1890/05-0073.

Smith, V.H., and J.W. Glauber, 2012: Agricultural insurance in developed countries: Where have we been and where are we going? *Appl. Econ. Perspect. Policy*, **34**, 363–390, doi:10.1093/aepp/pps029.

Smyth, C., G. Rampley, T.C. Lemprière, O. Schwab, and W.A. Kurz, 2017: Estimating product and energy substitution benefits in national-scale mitigation analyses for Canada. *GCB Bioenergy*, **9**, 1071–1084, doi:10.1111/gcbb.12389.

Smyth, C.E. et al., 2014: Quantifying the biophysical climate change mitigation potential of Canada's forest sector. *Biogeosciences*, **11**, 3515–3529, doi:10.5194/bg-11-3515-2014.

Soane, B.D., and C. Van Ouwerkerk, 1994: Soil compaction problems in world agriculture. *Dev. Agric. Eng.*, **11**, 1–21, doi:10.1016/B978-0-444-88286-8.50009-X.

Van Soest, H.L. et al., 2017: Early action on Paris Agreement allows for more time to change energy systems. *Clim. Change*, **144**, 165–179.

Sohi, S., 2012: Carbon storage with benefits. *Science*, **338**, 1034–1035, doi:10.1126/science.1227620.

Sommer, R., and D. Bossio, 2014: Dynamics and climate change mitigation potential of soil organic carbon sequestration. *J. Environ. Manage.*, **144**, 83–87.

Song, G., M. Li, P. Fullana-i-Palmer, D. Williamson, and Y. Wang, 2017: Dietary changes to mitigate climate change and benefit public health in China. *Sci. Total Environ.*, **577**, 289–298, doi:10.1016/j.scitotenv.2016.10.184.

Song, W., and X. Deng, 2015: Effects of urbanization-induced cultivated land loss on ecosystem services in the North China Plain. *Energies*, **8**, 5678–5693, doi:10.3390/en8065678.

Sonntag, S., J. Pongratz, C.H. Reick, and H. Schmidt, 2016: Reforestation in a high-CO_2 world – Higher mitigation potential than expected, lower adaptation potential than hoped for. *Geophys. Res. Lett.*, **43**, 6546–6553, doi:10.1002/2016GL068824.

Soussana, J. et al., 2019: Matching policy and science: Rationale for the "4 per 1000-soils for food security and climate" initiative. *Soil Tillage Res.*, **188**, 3–15.

Specht, K. et al., 2014: Urban agriculture of the future: An overview of sustainability aspects of food production in and on buildings. *Agric. Human Values*, **31**, 33–51, doi:10.1007/s10460-013-9448-4.

Springer, N.P. et al., 2015: Sustainable sourcing of global agricultural raw materials: Assessing gaps in key impact and vulnerability issues and indicators. *PLoS One*, **10**, e0128752, doi:10.1371/journal.pone.0128752.

Springmann, M. et al., 2016: Global and regional health effects of future food production under climate change: A modelling study. *Lancet*, **387**, 1937–1946, doi:10.1016/S0140-6736(15)01156-3.

Springmann, M. et al., 2018: Options for keeping the food system within environmental limits. *Nature*, **562**, 519–525, doi:10.1038/s41586-018-0594-0.

Squires, V., and E. Karami, 2005: Livestock management in the Arid zone: Coping strategies. *J. Rangel. Sci.*, **5**, 336–346.

6

Srinivasa Rao, C., K.A. Gopinath, J.V.N.S. Prasad, Prasannakumar, and A.K. Singh, 2016: Climate resilient villages for sustainable food security in tropical India: Concept, process, technologies, institutions, and impacts. *Adv. Agron.*, **140**, 101–214, doi:10.1016/BS.AGRON.2016.06.003.

Srinivasan, V. et al., 2015: Why is the Arkavathy River drying? A multiple-hypothesis approach in a data-scarce region. *Hydrol. Earth Syst. Sci*, **19**, 1950–2015, doi:10.5194/hess-19-1905-2015.

Stallard, R.F., 1998: Terrestrial sedimentation and the carbon cycle: Coupling weathering and erosion to carbon burial. *Global Biogeochem. Cycles*, **12**, 231–257, doi:10.1029/98GB00741.

Stanturf, J.A. et al., 2015: *Forest Landscape Restoration as a Key Component of Climate Change Mitigation and Adaptation*. International Union of Forest Research Organizations (IUFRO), Vienna, Austria, 76 pp.

Star, J. et al., 2016: Supporting adaptation decisions through scenario planning: Enabling the effective use of multiple methods. *Clim. Risk Manag.*, **13**, 88–94.

Steg, L., and C. Vlek, 2009: Encouraging pro-environmental behaviour: An integrative review and research agenda. *J. Environ. Psychol.*, **29**, 309–317.

Stehfest, E. et al., 2009: Climate benefits of changing diet. *Clim. Change*, **95**, 83–102, doi:10.1007/s10584-008-9534-6.

Steinbach, H.S., and R. Alvarez, 2006: Changes in soil organic carbon contents and nitrous oxide emissions after introduction of no-till in Pampean agroecosystems. *J. Environ. Qual.*, **35**, 3–13, doi:doi:10.2134/jeq2005.0050.

Stellmes, M., A. Röder, T. Udelhoven, and J. Hill, 2013: Mapping syndromes of land change in Spain with remote sensing time series, demographic and climatic data. *Land use policy*, **30**, 685–702, doi:10.1016/j.landusepol.2012.05.007.

Stern, P.C., 1992: Psychological dimensions of global environmental change. *Annu. Rev. Psychol.*, **43**, 269–302.

Sternberg, T., and B. Batbuyan, 2013: Integrating the Hyogo framework into Mongolia's disaster risk reduction (DRR) policy and management. *Int. J. Disaster Risk Reduct.*, **5**, 1-9, doi:10.1016/j.ijdrr.2013.05.003.

Stevanović, M. et al., 2017: Mitigation strategies for greenhouse gas emissions from agriculture and land-use change: Consequences for food prices. *Environ. Sci. Technol.*

Stibig, H.-J., F. Achard, S. Carboni, R. Raši, and J. Miettinen, 2014: Change in tropical forest cover of Southeast Asia from 1990 to 2010. *Biogeosciences*, **11**, 247–258, doi:10.5194/bg-11-247-2014.

Stokes, K.E., W.I. Montgomery, J.T.A. Dick, C.A. Maggs, and R.A. McDonald, 2006: The importance of stakeholder engagement in invasive species management: A cross-jurisdictional perspective in Ireland. *Biodivers. Conserv.*, **15**, 2829–2852.

Stone, B., J.J. Hess, and H. Frumkin, 2010: Urban form and extreme heat events: Are sprawling cities more vulnerable to climate change than compact cities? *Environ. Health Perspect.*, **118**, 1425–1428, doi:10.1289/ehp.0901879.

Strassburg, B.B.N. et al., 2014: When enough should be enough: Improving the use of current agricultural lands could meet production demands and spare natural habitats in Brazil. *Glob. Environ. Chang.*, **28**, 84–97, doi:10.1016/J.GLOENVCHA.2014.06.001.

Strefler, J., T. Amann, N. Bauer, E. Kriegler, and J. Hartmann, 2018: Potential and costs of carbon dioxide removal by enhanced weathering of rocks. *Environ. Res. Lett.*, **13**, 34010.

Strengers, B.J., J.G. Van Minnen, and B. Eickhout, 2008: The role of carbon plantations in mitigating climate change: Potentials and costs. *Clim. Change*, **88**, 343–366, doi:10.1007/s10584-007-9334-4.

Stringer, L.C. et al., 2009: Adaptations to climate change, drought and desertification: Local insights to enhance policy in southern Africa. *Environ. Sci. Policy*, **12**, 748–765.

Stringer, L.C. et al., 2012: Challenges and opportunities for carbon management in Malawi and Zambia. *Carbon Manag.*, **3**, 159–173, doi:10.4155/cmt.12.14.

Sunderland, T.C.H. et al., 2013: *Food security and nutrition: The role of forests*. Center for International Forestry Research (CIFOR), Bogor Barat, Indonesia.

Taboada, M., G. Rubio, and E. Chaneton, 2011: Grazing impacts on soil physical, chemical, and ecological properties in forage production systems. In: *Soil Management: Building a Stable Base for Agriculture* [J. Hatfield and T. Sauer (eds.)]. Soil Science Society of America, Wisconsin, USA, pp 301–320.

Tacconi, L., 2016: Preventing fires and haze in Southeast Asia. *Nat. Clim. Chang.*, **6**, 640–643, doi:10.1038/nclimate3008.

Tadasse, G., B. Algieri, M. Kalkuhl, and J. von Braun, 2016: Drivers and triggers of international food price spikes and volatility. In: *Food Price Volatility and Its Implications for Food Security and Policy* [M. Kalkuh, J. Von Braun, and M. Torero, (eds.)]. Springer International Publishing, Cham, Switzerland, 59–82 pp.

Tan, R., V. Beckmann, L. Van den Berg, and F. Qu, 2009: Governing farmland conversion: Comparing China with the Netherlands and Germany. *Land use policy*, **26**, 961–974, doi:10.1016/J.LANDUSEPOL.2008.11.009.

Tansey, K. et al., 2004: Vegetation burning in the year 2000: Global burned area estimates from SPOT VEGETATION data. *J. Geophys. Res. Atmos.*, **109**, D14, doi:10.1029/2003JD003598.

Tao, Y. et al., 2015: Variation in ecosystem services across an urbanization gradient: A study of terrestrial carbon stocks from Changzhou, China. *Ecol. Appl.*, **318**, 210-216, doi:10.1016/j.ecolmodel.2015.04.027.

Tarnocai, C., 2006: The effect of climate change on carbon in Canadian peatlands. *Glob. Planet. Change*, **53**, 222–232, doi:10.1016/J.GLOPLACHA.2006.03.012.

Tayleur, C. et al., 2017: Global coverage of agricultural sustainability standards, and their role in conserving biodiversity. *Conserv. Lett.*, **10**, 610–618, doi:10.1111/conl.12314.

Taylor, L.L. et al., 2016: Enhanced weathering strategies for stabilizing climate and averting ocean acidification. *Nat. Clim. Chang.*, **6**, 402–406, doi:10.1038/nclimate2882.

TEEB, 2009: *TEEB for Policy Makers: Responding to the Value of Nature*. The Economics of Ecosystems and Biodiversity Project. [Online] Available at: http://www.teebweb.org/our-publications/.

Thomaier, S. et al., 2015: Farming in and on urban buildings: Present practice and specific novelties of Zero-Acreage Farming (ZFarming). *Renew. Agric. Food Syst.*, **30**, 43–54, doi:10.1017/S1742170514000143.

Thomalla, F., T. Downing, E. Spanger-Siegfried, G. Han, and J. Rockström, 2006: Reducing hazard vulnerability: Towards a common approach between disaster risk reduction and climate adaptation. *Disasters*, **30**, 39–48, doi:10.1111/j.1467-9523.2006.00305.x.

Thornbush, M., O. Golubchikov, and S. Bouzarovski, 2013: Sustainable cities targeted by combined mitigation–adaptation efforts for future-proofing. *Sustain. Cities Soc.*, **9**, 1–9, doi:10.1016/j.scs.2013.01.003.

Thornton, P.K., and M. Herrero, 2014: Climate change adaptation in mixed crop–livestock systems in developing countries. *Glob. Food Sec.*, **3**, 99–107, doi:10.1016/J.GFS.2014.02.002.

Thornton, P.K., J. Van de Steeg, A. Notenbaert, and M. Herrero, 2009: The impacts of climate change on livestock and livestock systems in developing countries: A review of what we know and what we need to know. *Agric. Syst.*, **101**, 113–127, doi:10.1016/J.AGSY.2009.05.002.

Tian, X., B. Sohngen, J. Baker, S. Ohrel, and A.A. Fawcett, 2018: Will US forests continue to be a carbon sink? *Land Econ.*, **94**, 97–113.

Tigchelaar, M., D.S. Battisti, R.L. Naylor, and D.K. Ray, 2018: Future warming increases probability of globally synchronized maize production shocks. *Proc. Natl. Acad. Sci. U.S.A.*, **115**, 6644–6649, doi:10.1073/pnas.1718031115.

Tighe, M., R.E. Haling, R.J. Flavel, and I.M. Young, 2012: Ecological succession, hydrology and carbon acquisition of biological soil crusts measured at the micro-scale. *PLoS One*, **7**, e48565, doi:10.1371/journal.pone.0048565.

Tilman, D., and M. Clark, 2014: Global diets link environmental sustainability and human health. *Nature*, **515**, 518–522, doi:10.1038/nature13959.

Tilman, D., C. Balzer, J. Hill, and B.L. Befort, 2011: Global food demand and the sustainable intensification of agriculture. *Proc. Natl. Acad. Sci. U.S.A.*, **108**, 20260–20264, doi:10.1073/pnas.1116437108.

Timmer, C., 2009: Preventing food crises using a food policy approach. *J. Nutr.*, **140**, 224S–228S.

Timmermann, C., and Z. Robaey, 2016: Agrobiodiversity under different property regimes. *J. Agric. Environ. Ethics*, **29**, 285–303, doi:10.1007/s10806-016-9602-2.

Tompkins, E., and W.N. Adger, 2004: Does adaptive management of natural resources enhance resilience to climate change? *Ecol. Soc.*, **9**, 10.

Tompkins, E.L., R. Few, and K. Brown, 2008: Scenario-based stakeholder engagement: Incorporating stakeholders preferences into coastal planning for climate change. *J. Environ. Manage.*, **88**, 1580–1592.

Torlesse, H., L. Kiess, and M.W. Bloem, 2003: Association of household rice expenditure with child nutritional status indicates a role for macroeconomic food policy in combating malnutrition. *J. Nutr.*, **133**, 1320–1325, doi:10.1093/jn/133.5.1320.

Torvanger, A., 2018: Governance of bioenergy with carbon capture and storage (BECCS): Accounting, rewarding, and the Paris agreement. *Clim. Policy*, **0**, 1–13, doi:10.1080/14693062.2018.1509044.

Trabucco, A., R.J. Zomer, D.A. Bossio, O. Van Straaten, and L.V. Verchot, 2008: Climate change mitigation through afforestation/reforestation: A global analysis of hydrologic impacts with four case studies. *Agric. Ecosyst. Environ.*, **126**, 81–97, doi:10.1016/j.agee.2008.01.015.

Tschakert, P., B. van Oort, A.L. St. Clair, and A. LaMadrid, 2013: Inequality and transformation analyses: A complementary lens for addressing vulnerability to climate change. *Clim. Dev.*, **5**, 340–350, doi:10.1080/17565529.2013.828583.

Tullberg, J., D.L. Antille, C. Bluett, J. Eberhard, and C. Scheer, 2018: Controlled traffic farming effects on soil emissions of nitrous oxide and methane. *Soil Tillage Res.*, **176**, 18–25, doi:10.1016/J.STILL.2017.09.014.

UNCCD, 2012: *Zero Net Land Degradation. A Sustainable Development Goal for Rio+20. To Secure the Contribution of our Planet's Land and Soil to Sustainable Development, including Food Security and Poverty Eradication.* UNCCD, Bonn, Germany, 32 pp.

UNCCD, 2017: *Global Land Outlook*. UNCCD, Bonn, Germany, 340 pp.

UNCTAD, 2011: *Water for food – Innovative Water Management Technologies for Food Security and Poverty Alleviation*. New York, USA, and Geneva, Switzerland, 32 pp.

UNDP, 2018: *Human Development Indicies and Indicators*. Communications Development Incorporated, Washington DC, USA, 123 pp.

UNEP, 2017: *The Emissions Gap Report 2017*. United Nations Environment Programme (UNEP), Nairobi, Kenya, 116 pp.

UNEP and WMO, 2011: *Integrated Assessment of Black Carbon and Tropospheric Ozone: Summary for Decision Makers*. United Nations Environment Programme (UNEP), Nairobi, Kenya, 36 pp.

Valdez, Z.P., W.C. Hockaday, C.A. Masiello, M.E. Gallagher, and G. Philip Robertson, 2017: Soil carbon and nitrogen responses to nitrogen fertilizer and harvesting rates in switchgrass cropping systems. *Bioenergy Res.*, **10**, 456–464, doi:10.1007/s12155-016-9810-7.

Valendik, E.N. et al., 2011: *Tekhnologii Kontroliruyemykh Vyzhiganiy v Lesakh Sibiri: Kollektivnaya Monografiya [Technologies of Controlled Burning in Forests of Siberia]* [E.S. Petrenko, (ed.)]. Siberian Federal University, Krasnoyarsk, Russia, 60 pp.

VandenBygaart, A.J., 2016: The myth that no-till can mitigate global climate change. *Agric. Ecosyst. Environ.*, **216**, 98–99, doi:10.1016/J.AGEE.2015.09.013.

Vasconcelos, A.C.F. et al., 2013: Landraces as an adaptation strategy to climate change for smallholders in Santa Catarina, Southern Brazil. *Land use policy*, **34**, 250–254, doi:10.1016/J.LANDUSEPOL.2013.03.017.

Vatsala, L., J. Prakash, and S. Prabhavathi, 2017: Food security and nutritional status of women selected from a rural area in South India. *J. Food, Nutr. Popul. Heal.*, **1**, 10.

Vaughan, N.E., and C. Gough, 2016: Expert assessment concludes negative emissions scenarios may not deliver. *Environ. Res. Lett.*, **11**, 95003, doi:10.1088/1748-9326/11/9/095003.

Vellakkal, S. et al, 2015: Food price spikes are associated with increased malnutrition among children in Andhra Pradesh, India. *J. Nutr.*, **145**, 1942–1949, doi:10.3945/jn.115.211250.

Vente, J. de, M. Reed, L. Stringer, S. Valente, and J. Newig, 2016: How does the context and design of participatory decision making processes affect their outcomes? Evidence from sustainable land management in global drylands. *Ecol. Soc.*, **21**, doi:10.5751/ES-08053-210224.

Vermeulen, S.J. et al., 2012a: Options for support to agriculture and food security under climate change. *Environ. Sci. Policy*, **15**, 136–144, doi:10.1016/J.ENVSCI.2011.09.003.

Vermeulen, S.J., B.M. Campbell, and J.S.I. Ingram, 2012b: Climate change and food systems. *Annu. Rev. Environ. Resour.*, **37**, 195–222, doi:10.1146/annurev-environ-020411-130608.

Vignola, R. et al., 2015: Ecosystem-based Adaptation for smallholder farmers: Definitions, opportunities and constraints. *Agric. Ecosyst. Environ.*, **211**, 126–132, doi:10.1016/J.AGEE.2015.05.013.

Vilà, M. et al., 2011: Ecological impacts of invasive alien plants: A meta-analysis of their effects on species, communities and ecosystems. *Ecol. Lett.*, **14**, 702–708, doi:10.1111/j.1461-0248.2011.01628.x.

Vira, B., C. Wildburger, and S. Mansourian, (eds.) 2015: *Forests and Food: Addressing Hunger and Nutrition Across Sustainable Landscapes*. Open Book Publishers, Cambridge, UK, 280 pp.

Vlontzos, G., S. Niavis, and B. Manos, 2014: A DEA approach for estimating the agricultural energy and environmental efficiency of EU countries. *Renew. Sustain. Energy Rev.*, **40**, 91–96, doi:10.1016/J.RSER.2014.07.153.

Vogel, C., and K. O'Brien, 2006: Who can eat information? Examining the effectiveness of seasonal climate forecasts and regional climate-risk management strategies. *Clim. Res.*, **33**, 111–122, doi:10.3354/cr033111.

De Vries, W., G.J. Reinds, P. Gundersen, and H. Sterba, 2006: The impact of nitrogen deposition on carbon sequestration in European forests and forest soils. *Glob. Chang. Biol.*, **12**, 1151–1173, doi:10.1111/j.1365-2486.2006.01151.x.

De Vries, W. et al., 2009: The impact of nitrogen deposition on carbon sequestration by European forests and heathlands. *For. Ecol. Manage.*, **258**, 1814–1823, doi:10.1016/j.foreco.2009.02.034.

Van Vuuren, D.P. et al., 2011: RCP2.6: Exploring the possibility to keep global mean temperature increase below 2 degrees C. *Clim. Change*, **109**, 95, doi:10.1007/s10584-011-0152-3.

Van Vuuren, D.P. et al., 2015: Pathways to achieve a set of ambitious global sustainability objectives by 2050: Explorations using the IMAGE integrated assessment model. *Technol. Forecast. Soc. Change*, **98**, 303–323, doi:10.1016/j.techfore.2015.03.005.

Van Vuuren, D.P. et al., 2016: Carbon budgets and energy transition pathways. *Environ. Res. Lett.*, **11**, 75002, doi:10.1088/1748-9326/11/7/075002.

Van Vuuren, D.P., A.F. Hof, M.A.E. Van Sluisveld, and K. Riahi, 2017: Open discussion of negative emissions is urgently needed. *Nat. Energy*, **2**, 902–904, doi:10.1038/s41560-017-0055-2.

Van Vuuren, D.P. et al., 2018: Alternative pathways to the 1.5°C target reduce the need for negative emission technologies. *Nat. Clim. Chang.*, **8**, 391–397, doi:10.1038/s41558-018-0119-8.

Wada, Y., and M.F.P. Bierkens, 2014: Sustainability of global water use: past reconstruction and future projections. *Environ. Res. Lett.*, **9**, 104003, doi:10.1088/1748-9326/9/10/104003.

Waha, K. et al. 2018: Agricultural diversification as an important strategy for achieving food security in Africa. *Glob. Chang. Biol.*, **24**, 3390–3400, doi:10.1111/gcb.14158.

Waldron, A. et al., 2017: Agroforestry can enhance food security while meeting other Sustainable Development Goals. *Trop. Conserv. Sci.*, **10**, 194008291772066, doi:10.1177/1940082917720667.

Wallace, K.J., 2007: Classification of ecosystem services: Problems and solutions. *Biol. Conserv.*, **139**, 235–246, doi:10.1016/J.BIOCON.2007.07.015.

Walter, K., A. Don, and H. Flessa, 2015: No general soil carbon sequestration under Central European short rotation coppices. *GCB Bioenergy*, **7**, 727–740, doi:10.1111/gcbb.12177.

Wang, M. et al., 2017: On the long-term hydroclimatic sustainability of perennial bioenergy crop expansion over the United States. *J. Clim.*, **30**, doi:10.1175/JCLI-D-16-0610.1.

Wardle, J., K. Parmenter, and J. Waller, 2000: Nutrition knowledge and food intake. *Appetite*, **34**, 269–275, doi:10.1006/APPE.1999.0311.

Warren, A., 2002: Land degradation is contextual. *L. Degrad. Dev.*, **13**, 449–459, doi:10.1002/ldr.532.

Watson, J.E.M., N. Dudley, D.B. Segan, and M. Hockings, 2014: The performance and potential of protected areas. *Nature*, **515**, 67–73, doi:10.1038/nature13947.

Watson, R. et al., 2011: *UK National Ecosystem Assessment: Understanding Nature's Value to Society. Synthesis of Key Findings*. UNEP-WCMC, LWEC, UK, 1466 pp.

Wattnem, T., 2016: Seed laws, certification and standardization: Outlawing informal seed systems in the Global South. *J. Peasant Stud.*, **43**, 850–867, doi:10.1080/03066150.2015.1130702.

WBA, 2016: *WBA Global Bioenergy Statistics 2016*. World Bioenergy Association, Stockholm, Sweden, 1–80 pp.

Wei, F., S. Wang, L. Zhang, C. Fu, and E.M. Kanga, 2018: Balancing community livelihoods and biodiversity conservation of protected areas in East Africa. *Curr. Opin. Environ. Sustain.*, **33**, 26–33, doi:10.1016/J.COSUST.2018.03.013.

Weindl, I. et al., 2017: Livestock and human use of land: Productivity trends and dietary choices as drivers of future land and carbon dynamics. *Glob. Planet. Change*, **159**, 1–10, doi:10.1016/j.gloplacha.2017.10.002.

West, J.J. et al., 2013: Co-benefits of mitigating global greenhouse gas emissions for future air quality and human health. *Nat. Clim. Chang.*, **3**, 885–889, doi:10.1038/nclimate2009.

Westerling, A., H. Hidalgo, D. Cayan, and T. Swetnam, 2006: Warming and earlier spring increase Western U.S. forest wildfire activity. *Science*, **313**, 940–943, doi:10.1126/science.262.5135.885.

Westman, W., 1977: How much are nature's services worth? *Science.*, **197**, 960–964.

Wheeler, T., and J. von Braun, 2013: Climate change impacts on global food security. *Science.*, **341**, 508–513, doi:10.1126/science.1239402.

Whitaker, J. et al., 2018: Consensus, uncertainties and challenges for perennial bioenergy crops and land use. *GCB Bioenergy*, **10**, 150–164, doi:10.1111/gcbb.12488.

Whitehead, P.J., P. Purdon, J. Russell-Smith, P.M. Cooke, and S. Sutton, 2008: The management of climate change through prescribed Savanna burning: Emerging contributions of indigenous people in Northern Australia. *Public Adm. Dev.*, **28**, 374–385, doi:10.1002/pad.512.

Whittinghill, L.J., and D.B. Rowe, 2012: The role of green roof technology in urban agriculture. *Renew. Agric. Food Syst.*, **27**, 314–322, doi:10.1017/S174217051100038X.

Wijitkosum, S., 2016: The impact of land use and spatial changes on desertification risk in degraded areas in Thailand. *Sustain. Environ. Res.*, **26**, 84–92, doi:https://doi.org/10.1016/j.serj.2015.11.004.

Wild, M., A. Roesch, and C. Ammann, 2012: Global dimming and brightening – Evidence and agricultural implications. *CAB Rev.*, **7**, doi:10.1079/PAVSNNR20127003.

Wilhelm, M., C. Blome, V. Bhakoo, and A. Paulraj, 2016: Sustainability in multi-tier supply chains: Understanding the double agency role of the first-tier supplier. *J. Oper. Manag.*, **41**, 42–60, doi:10.1016/j.jom.2015.11.001.

Wilkie, D.S., J.F. Carpenter, and Q. Zhang, 2001: The under-financing of protected areas in the Congo Basin: So many parks and so little willingness-to-pay. *Biodivers. Conserv.*, **10**, 691–709, doi:10.1023/A:1016662027017.

Wiloso, E.I., R. Heijungs, G. Huppes, and K. Fang, 2016: Effect of biogenic carbon inventory on the life cycle assessment of bioenergy: Challenges to the neutrality assumption. *J. Clean. Prod.*, doi:10.1016/j.jclepro.2016.03.096.

Wise, M. et al., 2009: Implications of limiting CO_2 concentrations for land use and energy. *Science*, **324**, 1183–1186, doi:10.1126/science.1168475.

Wise, M. et al., 2015: An approach to computing marginal land use change carbon intensities for bioenergy in policy applications. *Energy Econ.*, **50**, 337–347, doi:10.1016/J.ENECO.2015.05.009.

WMO, 2015: Monitoring ocean carbon and ocean acidification. *WMO Greenhouse Gas Bulletin No. 10.*, Vol 64(1)–2015.

Wodon, Q., and H. Zaman, 2010: Higher food prices in sub-Saharan Africa: Poverty impact and policy responses. *World Bank Res. Obs.*, **25**, 157–176, doi:10.1093/wbro/lkp018.

Wolff, S., E.A. Schrammeijer, C.J.E. Schulp, and P.H. Verburg, 2018: Meeting global land restoration and protection targets: What would the world look like in 2050? *Glob. Environ. Chang.*, **52**, 259–272, doi:10.1016/j.gloenvcha.2018.08.002.

Wollenberg, E. et al., 2016: Reducing emissions from agriculture to meet the 2°C target. *Glob. Chang. Biol.*, **22**, 3859–3864, doi:10.1111/gcb.13340.

Wong, V.N.L., R.S.B. Greene, R.C. Dalal, and B.W. Murphy, 2010: Soil carbon dynamics in saline and sodic soils: a review. *Soil Use Manag.*, **26**, 2–11, doi:10.1111/j.1475-2743.2009.00251.x.

Woods, J. et al., 2015: Land and bioenergy. In: *Bioenergy & Sustainability: Bridging the Gaps*. [G.M. Souza, R.L. Victoria, C.A. Joly, and L.M. Verdade (eds.)]. SCOPE, Paris, France, 258–300 pp.

Woolf, D., J.E. Amonette, F.A. Street-Perrott, J. Lehmann, and S. Joseph, 2010: Sustainable biochar to mitigate global climate change. *Nat. Commun.*, **1**, 1–9, doi:10.1038/ncomms1053.

World Bank, 2011: *Rising Global Interest in Farmland: Can it Yield Sustainable and Equitable Benefits?* World Bank, Washington DC, USA.

World Bank, 2017: *Future of Food: Shaping the Food System to Deliver Jobs*. World Bank, Washington DC, USA, 36 pp.

World Bank, 2018a: *Commodity Markets Outlook. Oil Exporters: Policies and Challenges*. World Bank, Washington DC, USA, 82 pp.

World Bank, 2018b: *Strengthening Forest Fire Management in India*. World Bank, Washington DC, USA, 234 pp.

World Bank, FAO, and IFAD, 2009: Module 15: Gender and forestry. In: *Gender in Agriculture Sourcebook*. The International Bank for Reconstruction and Development/World Bank, Washington DC, USA, 643–674 pp.

Worrall, F. et al., 2009: Can carbon offsetting pay for upland ecological restoration? *Sci. Total Environ.*, **408**, 26–36, doi:10.1016/J.SCITOTENV.2009.09.022.

Wright, C.K., and M.C. Wimberly, 2013: Recent land use change in the Western Corn Belt threatens grasslands and wetlands. *Proc. Natl. Acad. Sci.*, **110**, 4134–4139, doi:10.1073/pnas.1215404110.

Wu, W. et al., 2019: Global advanced bioenergy potential under environmental protection policies and societal transformation measures. *GCB Bioenergy*, gcbb.12614, doi:10.1111/gcbb.12614.

Wunder, S., A. Angelsen, and B. Belcher, 2014: Forests, livelihoods, and conservation: Broadening the empirical base. *World Dev.*, **64**, S1–S11, doi:10.1016/j.worlddev.2014.03.007.

Xie, H., and C. Ringler, 2017: Agricultural nutrient loadings to the freshwater environment: The role of climate change and socioeconomic change. *Environ. Res. Lett.*, **12**, 104008, doi:10.1088/1748-9326/aa8148.

Xu, D., A. Song, D. Li, X. Ding, and Z. Wang, 2018: Assessing the relative role of climate change and human activities in desertification of North China from 1981 to 2010. *Front. Earth Sci.*, **13**, 43–54, doi:10.1007/s11707-018-0706-z.

Xu, Y., and V. Ramanathan, 2017: Well below 2°C: mitigation strategies for avoiding dangerous to catastrophic climate changes. *Proc. Natl. Acad. Sci.*, **114**, 10315–10323.

Xu, Y., D. Zaelke, G.J.M. Velders, and V. Ramanathan, 2013: The role of HFCs in mitigating 21st century climate change. *Atmos. Chem. Phys.*, **13**, 6083–6089, doi:10.5194/acp-13-6083-2013.

6

Yamagata, Y., N. Hanasaki, A. Ito, T. Kinoshita, D. Murakami, and Q. Zhou, 2018: Estimating water–food–ecosystem trade-offs for the global negative emission scenario (IPCC-RCP2.6). *Sustain. Sci.*, **13**, 301–313, doi:10.1007/s11625-017-0522-5.

Yamineva, Y., and S. Romppanen, 2017: Is law failing to address air pollution? Reflections on international and EU developments. *Rev. Eur. Comp. Int. Environ. law*, **26**, 189–200.

Yang, L., L. Chen, W. Wei, Y. Yu, and H. Zhang, 2014: Comparison of deep soil moisture in two re-vegetation watersheds in semi-arid regions. *J. Hydrol.*, **513**, 314–321, doi:doi.org/10.1016/j.jhydrol.2014.03.049.

Yirdaw, E., M. Tigabu, and A. Monge, 2017: Rehabilitation of degraded dryland ecosystems – review. *Silva Fenn.*, **51**, 1673, doi:10.14214/sf.1673.

Yong, D., and K. Peh, 2016: South-east Asia's forest fires: Blazing the policy trail. *Oryx*, **50**, 207–212.

Zaehle, S., and D. Dalmonech, 2011: Carbon–nitrogen interactions on land at global scales: Current understanding in modelling climate biosphere feedbacks. *Curr. Opin. Environ. Sustain.*, **3**, 311–320.

Zeeuw, H. de, and P. Drechsel, 2015: Cities and agriculture: developing resilient urban food systems. Routledge Taylor & Francis Group, London, UK and New York, USA, 431 pp.

Zeng, Y. et al., 2016: Revegetation in China's Loess Plateau is approaching sustainable water resource limits. *Nat. Clim. Chang.*, **6**, 1019–1022, doi:10.1038/nclimate3092.

Zezza, A., G. Carletto, B. Davis, K. Stamoulis, and P. Winters, 2009: Rural income generating activities: Whatever happened to the institutional vacuum? Evidence from Ghana, Guatemala, Nicaragua and Vietnam. *World Dev.*, **37**, 1297–1306, doi:10.1016/j.worlddev.2008.11.004.

Zhang, K. et al. 2013: Sustainability of social–ecological systems under conservation projects: Lessons from a biodiversity hotspot in western China. *Biol. Conserv.*, **158**, 205–213, doi:10.1016/j.biocon.2012.08.021.

Zhang, P. et al., 2000: China's forest policy for the 21st century. *Science.*, **288**, 2135–2136, doi:10.1126/science.288.5474.2135.

Zhang, X., R. Obringer, C. Wei, N. Chen, and D. Niyogi, 2017: Droughts in India from 1981 to 2013 and implications to wheat production. *Sci. Rep.*, **7**, 44552, doi:10.1038/srep44552.

Zhou, G. et al., 2017: Grazing intensity significantly affects belowground carbon and nitrogen cycling in grassland ecosystems: A meta-analysis. *Glob. Chang. Biol.*, **23**, 1167–1179, doi:10.1111/gcb.13431.

Ziervogel, G., and P.J. Ericksen, 2010: Adapting to climate change to sustain food security. *Wiley Interdiscip. Rev. Clim. Chang.*, **1**, 525–540, doi:10.1002/wcc.56.

Zomer, R. et al., 2016: Global tree cover and biomass carbon on agricultural land: The contribution of agroforestry to global and national carbon budgets. *Sci. Rep.*, **6**, 29987.

7

Risk management and decision-making in relation to sustainable development

Coordinating Lead Authors:
Margot Hurlbert (Canada), Jagdish Krishnaswamy (India)

Lead Authors:
Edouard Davin (France/Switzerland), Francis X. Johnson (Sweden), Carlos Fernando Mena (Ecuador), John Morton (United Kingdom), Soojeong Myeong (The Republic of Korea), David Viner (United Kingdom), Koko Warner (The United States of America), Anita Wreford (New Zealand), Sumaya Zakieldeen (Sudan), Zinta Zommers (Latvia)

Contributing Authors:
Rob Bailis (The United States of America), Brigitte Baptiste (Colombia), Kerry Bowman (Canada), Edward Byers (Austria/Brazil), Katherine Calvin (The United States of America), Rocio Diaz-Chavez (Mexico), Jason Evans (Australia), Amber Fletcher (Canada), James Ford (United Kingdom), Sean Patrick Grant (The United States of America), Darshini Mahadevia (India), Yousef Manialawy (Canada), Pamela McElwee (The United States of America), Minal Pathak (India), Julian Quan (United Kingdom), Balaji Rajagopalan (The United States of America), Alan Renwick (New Zealand), Jorge E. Rodríguez-Morales (Peru), Charlotte Streck (Germany), Wim Thiery (Belgium), Alan Warner (Barbados)

Review Editors:
Regina Rodrigues (Brazil), B.L. Turner II (The United States of America)

Chapter Scientist:
Thobekile Zikhali (Zimbabwe)

This chapter should be cited as:
Hurlbert, M., J. Krishnaswamy, E. Davin, F.X. Johnson, C.F. Mena, J. Morton, S. Myeong, D. Viner, K. Warner, A. Wreford, S. Zakieldeen, Z. Zommers, 2019: Risk Management and Decision making in Relation to Sustainable Development. In: Climate Change and Land: an IPCC special report on climate change, desertification, land degradation, sustainable land management, food security, and greenhouse gas fluxes in terrestrial ecosystems [P.R. Shukla, J. Skea, E. Calvo Buendia, V. Masson-Delmotte, H.-O. Pörtner, D.C. Roberts, P. Zhai, R. Slade, S. Connors, R. van Diemen, M. Ferrat, E. Haughey, S. Luz, S. Neogi, M. Pathak, J. Petzold, J. Portugal Pereira, P. Vyas, E. Huntley, K. Kissick, M. Belkacemi, J. Malley, (eds.)]. https://doi.org/10.1017/9781009157988.009

Table of contents

Executive summary 675

7.1 Introduction and relation to other chapters 677

   7.1.1 Findings of previous IPCC assessments
       and reports 677

   Box 7.1: Relevant findings of recent IPCC reports 678

   7.1.2 Treatment of key terms in the chapter 679

   7.1.3 Roadmap to the chapter 679

7.2 Climate-related risks for land-based
    human systems and ecosystems 680

   7.2.1 Assessing risk 680

   7.2.2 Risks to land systems arising from climate change 680

   7.2.3 Risks arising from responses to climate change ... 686

   7.2.4 Risks arising from hazard, exposure
       and vulnerability 688

7.3 Consequences of climate – land change for human
    well-being and sustainable development 690

   7.3.1 What is at stake for food security? 690

   7.3.2 Risks to where and how people live: Livelihood
       systems and migration 690

   7.3.3 Risks to humans from disrupted
       ecosystems and species 691

   7.3.4 Risks to communities and infrastructure 691

   Cross-Chapter Box 10 | Economic dimensions
   of climate change and land 692

7.4 Policy instruments for land and climate 695

   7.4.1 Multi-level policy instruments 695

   7.4.2 Policies for food security and social protection 696

   7.4.3 Policies responding to climate-related extremes 699

   7.4.4 Policies responding to greenhouse gas (GHG) fluxes 701

   Case study: Including agriculture in the
   New Zealand Emissions Trading Scheme (ETS) 703

   7.4.5 Policies responding to desertification and
       degradation – Land Degradation Neutrality (LDN) 705

   7.4.6 Policies responding to land degradation 706

   Case study: Forest conservation instruments:
   REDD+ in the Amazon and India 709

   7.4.7 Economic and financial instruments
       for adaptation, mitigation, and land 711

   7.4.8 Enabling effective policy instruments –
       policy portfolio coherence 713

   7.4.9 Barriers to implementing policy responses 714

Cross-Chapter Box 11 | Gender in inclusive
approaches to climate change, land
and sustainable development 717

7.5 Decision-making for climate change and land 719

   7.5.1 Formal and informal decision-making 720

   7.5.2 Decision-making, timing, risk, and uncertainty ... 720

   7.5.3 Best practices of decision-making toward
       sustainable land management (SLM) 723

   7.5.4 Adaptive management 723

   7.5.5 Performance indicators 725

   7.5.6 Maximising synergies and
       minimising trade-offs 725

Cross-Chapter Box 9 | Climate and land pathways ... 727

Case study: Green energy: Biodiversity
conservation vs global environment targets? 735

7.6 Governance: Governing the land–climate interface 736

   7.6.1 Institutions building adaptive
       and mitigative capacity 736

   7.6.2 Integration – Levels, modes and scale
       of governance for sustainable development 737

Case study: Governance: Biofuels and bioenergy ... 738

Cross-Chapter Box 12 | Traditional biomass use:
Land, climate and development implications 740

   7.6.3 Adaptive climate governance
       responding to uncertainty 742

Box 7.2: Adaptive governance and interlinkages
of food, fibre, water, energy and land 743

   7.6.4 Participation 745

Cross-Chapter Box 13 | Indigenous and local
knowledge (ILK) 746

   7.6.5 Land tenure 749

   7.6.6 Institutional dimensions
       of adaptive governance 753

   7.6.7 Inclusive governance for
       sustainable development 754

7.7 Key uncertainties and knowledge gaps 754

Frequently Asked Questions 755

   FAQ 7.1: How can indigenous knowledge and
   local knowledge inform land-based mitigation
   and adaptation options? 755

   FAQ 7.2: What are the main barriers
   to and opportunities for land-based
   responses to climate change? 756

References 757

Executive summary

Increases in global mean surface temperature are projected to result in continued permafrost degradation and coastal degradation (*high confidence*), increased wildfire, decreased crop yields in low latitudes, decreased food stability, decreased water availability, vegetation loss (*medium confidence*), decreased access to food and increased soil erosion (*low confidence*). There is *high agreement* and *high evidence* that increases in global mean temperature will result in continued increase in global vegetation loss, coastal degradation, as well as decreased crop yields in low latitudes, decreased food stability, decreased access to food and nutrition, and *medium confidence* in continued permafrost degradation and water scarcity in drylands. Impacts are already observed across all components (*high confidence*). Some processes may experience irreversible impacts at lower levels of warming than others. There are high risks from permafrost degradation, and wildfire, coastal degradation, stability of food systems at 1.5°C while high risks from soil erosion, vegetation loss and changes in nutrition only occur at higher temperature thresholds due to increased possibility for adaptation (*medium confidence*). {7.2.2.1, 7.2.2.2, 7.2.2.3; 7.2.2.4; 7.2.2.5; 7.2.2.6; 7.2.2.7; Figure 7.1}

These changes result in compound risks to food systems, human and ecosystem health, livelihoods, the viability of infrastructure, and the value of land (*high confidence*). The experience and dynamics of risk change over time as a result of both human and natural processes (*high confidence*). There is *high confidence* that climate and land changes pose increased risks at certain periods of life (i.e., to the very young and ageing populations) as well as sustained risk to those living in poverty. Response options may also increase risks. For example, domestic efforts to insulate populations from food price spikes associated with climatic stressors in the mid-2000s inadequately prevented food insecurity and poverty, and worsened poverty globally. {7.2.1, 7.2.2, 7.3, Table 7.1}

There is significant regional heterogeneity in risks: tropical regions, including Sub-Saharan Africa, Southeast Asia and Central and South America are particularly vulnerable to decreases in crop yield (*high confidence*). Yield of crops in higher latitudes may initially benefit from warming as well as from higher carbon dioxide (CO$_2$) concentrations. But temperate zones, including the Mediterranean, North Africa, the Gobi desert, Korea and western United States are susceptible to disruptions from increased drought frequency and intensity, dust storms and fires (*high confidence*). {7.2.2}

Risks related to land degradation, desertification and food security increase with temperature and can reverse development gains in some socio-economic development pathways (*high confidence*). SSP1 reduces the vulnerability and exposure of human and natural systems and thus limits risks resulting from desertification, land degradation and food insecurity compared to SSP3 (*high confidence*). SSP1 is characterised by low population growth, reduced inequalities, land-use regulation, low meat consumption, increased trade and few barriers to adaptation or mitigation. SSP3 has the opposite characteristics. Under SSP1, only a small fraction of the dryland population (around 3% at 3°C for the year 2050) will be exposed and vulnerable to water stress. However under SSP3, around 20% of dryland populations (for the year 2050) will be exposed and vulnerable to water stress by 1.5°C and 24% by 3°C. Similarly under SSP1, at 1.5°C, 2 million people are expected to be exposed and vulnerable to crop yield change. Over 20 million are exposed and vulnerable to crop yield change in SSP3, increasing to 854 million people at 3°C (*low confidence*). Livelihoods deteriorate as a result of these impacts, livelihood migration is accelerated, and strife and conflict is worsened (*medium confidence*). {Cross-Chapter Box 9 in Chapters 6 and 7, 7.2.2, 7.3.2, Table 7.1, Figure 7.2}

Land-based adaptation and mitigation responses pose risks associated with the effectiveness and potential adverse side-effects of measures chosen (*medium confidence*). Adverse side-effects on food security, ecosystem services and water security increase with the scale of bioenergy and bioenergy with carbon capture and storage (BECCS) deployment. In a SSP1 future, bioenergy and BECCS deployment up to 4 million km^2 is compatible with sustainability constraints, whereas risks are already high in a SSP3 future for this scale of deployment. {7.2.3}

There is *high confidence* that policies addressing vicious cycles of poverty, land degradation and greenhouse gas (GHG) emissions implemented in a holistic manner can achieve climate-resilient sustainable development. Choice and implementation of policy instruments determine future climate and land pathways (*medium confidence*). Sustainable development pathways (described in SSP1) supported by effective regulation of land use to reduce environmental trade-offs, reduced reliance on traditional biomass, low growth in consumption and limited meat diets, moderate international trade with connected regional markets, and effective GHG mitigation instruments) can result in lower food prices, fewer people affected by floods and other climatic disruptions, and increases in forested land (*high agreement, limited evidence*) (SSP1). A policy pathway with limited regulation of land use, low technology development, resource intensive consumption, constrained trade, and ineffective GHG mitigation instruments can result in food price increases, and significant loss of forest (*high agreement, limited evidence*) (SSP3). {3.7.5, 7.2.2, 7.3.4, 7.5.5, 7.5.6, Table 7.1, Cross-Chapter Box 9 in Chapters 6 and 7, Cross-Chapter Box 12 in Chapter 7}

Delaying deep mitigation in other sectors and shifting the burden to the land sector, increases the risk associated with adverse effects on food security and ecosystem services (*high confidence*). The consequences are an increased pressure on land with higher risk of mitigation failure and of temperature overshoot and a transfer of the burden of mitigation and unabated climate change to future generations. Prioritising early decarbonisation with minimal reliance on carbon dioxide removal (CDR) decreases the risk of mitigation failure (*high confidence*). {2.5, 6.2, 6.4, 7.2.1, 7.2.2, 7.2.3, 7.5.6, 7.5.7, Cross-Chapter Box 9 in Chapters 6 and 7}

Trade-offs can occur between using land for climate mitigation or Sustainable Development Goal (SDG) 7 (affordable clean energy) with biodiversity, food, groundwater and riverine ecosystem services (*medium confidence*). There is *medium confidence* that trade-offs currently do not figure into climate policies and decision making. Small hydro power installations (especially in clusters) can impact downstream river ecological connectivity for fish (*high agreement, medium evidence*). Large scale solar farms and wind turbine installations can impact endangered species and disrupt habitat connectivity (*medium agreement, medium evidence*). Conversion of rivers for transportation can disrupt fisheries and endangered species (through dredging and traffic) (*medium agreement, low evidence*). {7.5.6}

The full mitigation potential assessed in this report will only be realised if agricultural emissions are included in mainstream climate policy (*high agreement, high evidence*). Carbon markets are theoretically more cost-effective than taxation but challenging to implement in the land-sector (*high confidence*) Carbon pricing (through carbon markets or carbon taxes) has the potential to be an effective mechanism to reduce GHG emissions, although it remains relatively untested in agriculture and food systems. Equity considerations can be balanced by a mix of both market and non-market mechanisms (*medium evidence, medium agreement*). Emissions leakage could be reduced by multi-lateral action (*high agreement, medium evidence*). {7.4.6, 7.5.5, 7.5.6, Cross-Chapter Box 9 in Chapters 6 and 7}

A suite of coherent climate and land policies advances the goal of the Paris Agreement and the land-related SDG targets on poverty, hunger, health, sustainable cities and communities, responsible consumption and production, and life on land. There is *high confidence* that acting early will avert or minimise risks, reduce losses and generate returns on investment. The economic costs of action on sustainable land management (SLM), mitigation, and adaptation are less than the consequences of inaction for humans and ecosystems (*medium confidence*). Policy portfolios that make ecological restoration more attractive, people more resilient – expanding financial inclusion, flexible carbon credits, disaster risk and health insurance, social protection and adaptive safety nets, contingent finance and reserve funds, and universal access to early warning systems – could save 100 billion USD a year, if implemented globally. {7.3.1, 7.4.7, 7.4.8, 7.5.6, Cross-Chapter Box 10 in Chapter 7}

Coordination of policy instruments across scales, levels, and sectors advances co-benefits, manages land and climate risks, advances food security, and addresses equity concerns (*medium confidence*). Flood resilience policies are mutually reinforcing and include flood zone mapping, financial incentives to move, and building restrictions, and insurance. Sustainability certification, technology transfer, land-use standards and secure land tenure schemes, integrated with early action and preparedness, advance response options. SLM improves with investment in agricultural research, environmental farm practices, agri-environmental payments, financial support for sustainable agricultural water infrastructure (including dugouts), agriculture emission trading, and elimination of agricultural subsidies (*medium confidence*). Drought resilience policies (including drought preparedness planning, early warning and

monitoring, improving water use efficiency), synergistically improve agricultural producer livelihoods and foster SLM. {3.7.5, Cross-Chapter Box 5 in Chapter 3, 7.4.3, 7.4.6, 7.5.6, 7.4.8, , 7.5.6, 7.6.3}

Technology transfer in land-use sectors offers new opportunities for adaptation, mitigation, international cooperation, R&D collaboration, and local engagement (*medium confidence*). International cooperation to modernise the traditional biomass sector will free up both land and labour for more productive uses. Technology transfer can assist the measurement and accounting of emission reductions by developing countries. {7.4.4, 7.4.6, Cross-Chapter Box 12 in Chapter 7}

Measuring progress towards goals is important in decision-making and adaptive governance to create common understanding and advance policy effectiveness (*high agreement, medium evidence*). Measurable indicators, selected with the participation of people and supporting data collection, are useful for climate policy development and decision-making. Indicators include the SDGs, nationally determined contributions (NDCs), land degradation neutrality (LDN) core indicators, carbon stock measurement, measurement and monitoring for REDD+, metrics for measuring biodiversity and ecosystem services, and governance capacity. {7.5.5, 7.5.7, 7.6.4, 7.6.6}

The complex spatial, cultural and temporal dynamics of risk and uncertainty in relation to land and climate interactions and food security, require a flexible, adaptive, iterative approach to assessing risks, revising decisions and policy instruments (*high confidence*). Adaptive, iterative decision making moves beyond standard economic appraisal techniques to new methods such as dynamic adaptation pathways with risks identified by trigger points through indicators. Scenarios can provide valuable information at all planning stages in relation to land, climate and food; adaptive management addresses uncertainty in scenario planning with pathway choices made and reassessed to respond to new information and data as it becomes available. {3.7.5, 7.4.4, 7.5.2, 7.5.3, 7.5.4, 7.5.7, 7.6.1, 7.6.3}

Indigenous and local knowledge (ILK) can play a key role in understanding climate processes and impacts, adaptation to climate change, sustainable land management (SLM) across different ecosystems, and enhancement of food security (*high confidence*). ILK is context-specific, collective, informally transmitted, and multi-functional, and can encompass factual information about the environment and guidance on management of resources and related rights and social behaviour. ILK can be used in decision-making at various scales and levels, and exchange of experiences with adaptation and mitigation that include ILK is both a requirement and an entry strategy for participatory climate communication and action. Opportunities exist for integration of ILK with scientific knowledge. {7.4.1, 7.4.5, 7.4.6, 7.6.4, Cross-Chapter Box 13 in Chapter 7}

Participation of people in land and climate decision making and policy formation allows for transparent effective solutions and the implementation of response options that advance

synergies, reduce trade-offs in SLM (*medium confidence*), and overcomes barriers to adaptation and mitigation (*high confidence*). Improvements to SLM are achieved by: (i) engaging people in citizen science by mediating and facilitating landscape conservation planning, policy choice, and early warning systems (*medium confidence*); (ii) involving people in identifying problems (including species decline, habitat loss, land-use change in agriculture, food production and forestry), selection of indicators, collection of climate data, land modelling, agricultural innovation opportunities. When social learning is combined with collective action, transformative change can occur addressing tenure issues and changing land-use practices (*medium confidence*). Meaningful participation overcomes barriers by opening up policy and science surrounding climate and land decisions to inclusive discussion that promotes alternatives. {3.7.5, 7.4.1, 7.4.9; 7.5.1, 7.5.4, 7.5.5, 7.5.7, 7.6.4, 7.6.6}

Empowering women can bolster synergies among household food security and SLM (*high confidence*). This can be achieved with policy instruments that account for gender differences. The overwhelming presence of women in many land based activities including agriculture provides opportunities to mainstream gender policies, overcome gender barriers, enhance gender equality, and increase SLM and food security (*high confidence*). Policies that address barriers include gender qualifying criteria and gender appropriate delivery, including access to financing, information, technology, government transfers, training, and extension may be built into existing women's programmes, structures (civil society groups) including collective micro enterprise (*medium confidence*). {Cross-Chapter Box 11 in Chapter 7}

The significant social and political changes required for sustainable land use, reductions in demand and land-based mitigation efforts associated with climate stabilisation require a wide range of governance mechanisms. The expansion and diversification of land use and biomass systems and markets requires hybrid governance: public-private partnerships, transnational, polycentric, and state governance to insure opportunities are maximised, trade-offs are managed equitably and negative impacts are minimised (*medium confidence*). {7.4.6, 7.6.2, 7.6.3, Cross-Chapter Box 7 in Chapter 6}

Land tenure systems have implications for both adaptation and mitigation, which need to be understood within specific socio-economic and legal contexts, and may themselves be impacted by climate change and climate action (*limited evidence*, *high agreement*). Land policy (in a diversity of forms beyond focus on freehold title) can provide routes to land security and facilitate or constrain climate action, across cropping, rangeland, forest, freshwater ecosystems and other systems. Large-scale land acquisitions are an important context for the relations between tenure security and climate change, but their scale, nature and implications are imperfectly understood. There is *medium confidence* that land titling and recognition programmes, particularly those that authorize and respect indigenous and communal tenure, can lead to improved management of forests, including for carbon storage. Strong public coordination (government and public administration)

can integrate land policy with national policies on adaptation and reduce sensitivities to climate change. {7.6.2; 7.6.3; 7.6.4, 7.6.5}

Significant gaps in knowledge exist when it comes to understanding the effectiveness of policy instruments and institutions related to land-use management, forestry, agriculture and bioenergy. Interdisciplinary research is needed on the impacts of policies and measures in land sectors. Knowledge gaps are due in part to the highly contextual and local nature of land and climate measures and the long time periods needed to evaluate land-use change in its socio-economic frame, as compared to technological investments in energy or industry that are somewhat more comparable. Significant investment is needed in monitoring, evaluation and assessment of policy impacts across different sectors and levels. {7.7}

7.1 Introduction and relation to other chapters

Land is integral to human habitation and livelihoods, providing food and resources, and also serves as a source of identity and cultural meaning. However, the combined impacts of climate change, desertification, land degradation and food insecurity pose obstacles to resilient development and the achievement of the Sustainable Development Goals (SDGs). This chapter reviews and assesses literature on risk and uncertainty surrounding land and climate change, policy instruments and decision-making that seek to address those risks and uncertainties, and governance practices that advance the response options with co-benefits identified in Chapter 6, lessen the socio-economic impacts of climate change and reduce trade-offs, and advance SLM.

7.1.1 Findings of previous IPCC assessments and reports

This chapter builds on earlier assessments contained in several chapters of the IPCC Fifth Assessment Report (the contributions of both Working Groups II and III), the IPCC Special Report on Managing the Risks of Extreme Events and Disasters to Advance Climate Change Adaptation (SREX) (IPCC 2012), and the IPCC Special Report on Global Warming of 1.5°C (SR15) (IPCC 2018a). The findings most relevant to decision-making on and governance of responses to land-climate challenges are set out in Box 7.1.

Box 7.1 | Relevant findings of recent IPCC reports

Climate change and sustainable development pathways
"Climate change poses a moderate threat to current sustainable development and a severe threat to future sustainable development" (Denton et al. 2014; Fleurbaey et al. 2014).

Significant transformations may be required for climate-resilient pathways (Denton et al. 2014; Jones et al. 2014).

The design of climate policy is influenced by (i) differing ways that individuals and organisations perceive risks and uncertainties, and (ii) the consideration of a diverse array of risks and uncertainties – as well as human and social responses – which may be difficult to measure, are of low probability but which would have a significant impact if they occurred (Kunreuther et al. 2014; Fleurbaey et al. 2014; Kolstad et al. 2014).

Building climate-resilient pathways requires iterative, continually evolving and complementary processes at all levels of government (Denton et al. 2014; Kunreuther et al. 2014; Kolstad et al. 2014; Somanthan et al. 2014; Lavell et al. 2012).

Important aspects of climate-resilient policies include local level institutions, decentralisation, participatory governance, iterative learning, integration of local knowledge, and reduction of inequality (Dasgupta et al. 2014; Lavell et al. 2012; Cutter et al. 2012b; O'Brien et al. 2012; Roy et al. 2018).

Climate action and sustainable development are linked: adaptation has co-benefits for sustainable development, while "sustainable development supports, and often enables, the fundamental societal and systems transitions and transformations that help limit global warming" (IPCC 2018a). Redistributive policies that shield the poor and vulnerable can resolve trade-offs between mitigation objectives and the hunger, poverty and energy access SDGs.

Land and rural livelihoods
Policies and institutions relating to land, including land tenure, can contribute to the vulnerability of rural people, and constrain adaptation. Climate policies, such as encouraging cultivation of biofuels, or payments under REDD+, will have significant secondary impacts, both positive and negative, in some rural areas (Dasgupta et al. 2014).

"Sustainable land management is an effective disaster risk reduction tool" (Cutter et al. 2012a).

Risk and risk management
A variety of emergent risks not previously assessed or recognised, can be identified by taking into account: (i) the "interactions of climate change impacts on one sector with changes in exposure and vulnerability, as well as adaptation and mitigation actions", and (ii) "indirect, trans-boundary, and long-distance impacts of climate change" including price spikes, migration, conflict and the unforeseen impacts of mitigation measures (Oppenheimer et al. 2014).

"Under any plausible scenario for mitigation and adaptation, some degree of risk from residual damages is unavoidable" (Oppenheimer et al. 2014).

Decision-making
"Risk management provides a useful framework for most climate change decision-making. Iterative risk management is most suitable in situations characterised by large uncertainties, long time frames, the potential for learning over time, and the influence of both climate as well as other socio-economic and biophysical changes" (Jones et al. 2014).

"Decision support is situated at the intersection of data provision, expert knowledge, and human decision making at a range of scales from the individual to the organisation and institution" (Jones et al. 2014).

"Scenarios are a key tool for addressing uncertainty", either through problem exploration or solution exploration (Jones et al. 2014).

Box 7.1 (continued)

Governance

There is no single approach to adaptation planning and both top-down and bottom-up approaches are widely recognised. "Institutional dimensions in adaptation governance play a key role in promoting the transition from planning to implementation of adaptation" (Mimura et al. 2014). Adaptation is also essential at all scales, including adaptation by local governments, businesses, communities and individuals (Denton et al. 2014).

"Strengthened multi-level governance, institutional capacity, policy instruments, technological innovation and transfer and mobilisation of finance, and changes in human behaviour and lifestyles are enabling conditions that enhance the feasibility of mitigation and adaptation options for 1.5°C-consistent systems transitions" (IPCC 2018b).

Governance is key for vulnerability and exposure represented by institutionalised rule systems and habitualised behaviour and norms that govern society and guide actors, and "it is essential to improve knowledge on how to promote adaptive governance within the framework of risk assessment and risk management" (Cardona 2012).

7.1.2   Treatment of key terms in the chapter

While the term **risk** continues to be subject to a growing number of definitions in different disciplines and sectors, this chapter takes as a starting point the definition used in the IPCC Special Report on Global Warming of 1.5°C (SR15) (IPCC 2018a), which reflects definitions used by both Working Group II and Working Group III in the Fifth Assessment Report (AR5): "The potential for adverse consequences where something of value is at stake and where the occurrence and degree of an outcome is uncertain" (Allwood et al. 2014; Oppenheimer et al. 2014). The SR15 definition further specifies: "In the context of the assessment of climate impacts, the term risk is often used to refer to the potential for adverse consequences of a climate-related hazard, or of adaptation or mitigation responses to such a hazard, on lives, livelihoods, health and well-being, ecosystems and species, economic, social and cultural assets, services (including ecosystem services), and infrastructure." In SR15, as in the IPCC SREX and AR5 WGII, risk is conceptualised as resulting from the interaction of vulnerability (of the affected system), its exposure over time (to a hazard), as well as the (climate-related) impact and the likelihood of its occurrence (AR5 2014; IPCC 2018a, 2012). In the context of SRCCL, risk must also be seen as including risks to the implementation of responses to land–climate challenges from economic, political and governance factors. Climate and land risks must be seen in relation to human values and objectives (Denton et al. 2014). Risk is closely associated with concepts of vulnerability and resilience, which are themselves subject to differing definitions across different knowledge communities.

Risks examined in this chapter arise from more than one of the major land–climate–society challenges (desertification, land degradation, and food insecurity), or partly stem from mitigation or adaptation actions, or cascade across different sectors or geographical locations. They could thus be seen as examples of **emergent risks**: "aris[ing] from the interaction of phenomena in a complex system" (Oppenheimer et al. 2014, p.1052). Stranded assets in the coal sector due to proliferation of renewable energy and government response could be examples of emergent risks (Saluja and Singh 2018; Marcacci 2018). Additionally, the absence of an explicit goal for conserving freshwater ecosystems and ecosystem services in SDGs (in contrast to a goal – 'life below water' – exclusively for marine biodiversity) is related to its trade-offs with energy and irrigation goals, thus posing a substantive risk (Nilsson et al. 2016b; Vörösmarty et al. 2010).

Governance is not previously well defined in IPCC reports, but is used here to include all of the processes, structures, rules and traditions that govern, which may be undertaken by actors including governments, markets, organisations, or families (Bevir 2011), with particular reference to the multitude of actors operating in respect of land–climate interactions. Such definitions of governance allow for it to be decoupled from the more familiar concept of government and studied in the context of complex human–environment relations and environmental and resource regimes (Young 2017a). Governance involves the interactions among formal and informal institutions through which people articulate their interests, exercise their legal rights, meet their legal obligations, and mediate their differences (UNDP 1997).

7.1.3   Roadmap to the chapter

This chapter firstly discusses risks and their drivers, at various scales, in relation to land-climate challenges, including risks associated with responses to climate change (Section 7.2). The consequences of the principal risks in economic and human terms, and associated concepts such as tipping points and windows of opportunity for response are then described (Section 7.3). Policy responses at different scales to different land-climate risks, and barriers to implementation, are described in Section 7.4, followed by an assessment of approaches to decision-making on land-climate challenges (Section 7.5), and questions of the governance of the land-climate interface (Section 7.6). Key uncertainties and knowledge gaps are identified in Section 7.7.

7

7.2   Climate-related risks for land-based human systems and ecosystems

This section examines risks that climate change poses to selected land-based human systems and ecosystems, and then further explores how social and economic choices, as well as responses to climate change, will exacerbate or lessen risks. 'Risk' is defined as *the potential for adverse consequences for human or ecological systems, recognising the diversity of values and objectives associated with such systems*. The interacting processes of climate change, land change, and unprecedented social and technological change, pose significant risk to climate-resilient sustainable development. The pace, intensity, and scale of these sizeable risks affect the central issues in sustainable development: access to ecosystem services (ES) and resources essential to sustain people in given locations; how and where people live and work; and the means to safeguard human well-being against disruptions (Warner et al. 2019). In the context of climate change, adverse consequences can arise from the potential *impacts o*f climate change as well as human *responses to* climate change. Relevant adverse consequences include those on lives, livelihoods, health and well-being, economic, social and cultural assets and investments, infrastructure, services (including ES), ecosystems and species (see Glossary). Risks result from dynamic interactions between climate-related hazards with the exposure and vulnerability of the affected human or ecological system to the hazards. Hazards, exposure and vulnerability may change over time and space as a result of socio-economic changes and human decision-making ('risk management'). Numerous uncertainties exist in the scientific understanding of risk (Section 1.2.2).

7.2.1   Assessing risk

This chapter applies and further improves methods used in previous IPCC reports including AR5 and the Special Report on Global Warming of 1.5°C (SR15) to assess risks. Evidence is drawn from published studies, which include observations of impacts from human-induced climate change and model projections for future climate change. Such projections are based on Integrated Assessment Models (IAMs), Earth System Models (ESMs), regional climate models and global or regional impact models examining the impact of climate change on various indicators (Cross-Chapter Box 1 in Chapter 1). Results of laboratory and field experiments that examine impacts of specific changes were also included in the review. Risks under different future socio-economic conditions were assessed using recent publications based on Shared Socio-economic Pathways (SSPs). SSPs provide storylines about future socio-economic development and can be combined with Representative Concentration Pathways RCPs (Riahi et al. 2017) (Cross-Chapter Box 9 in Chapters 6 and 7). Risk arising from land-based mitigation and adaptation choices is assessed using studies examining the adverse side effects of such responses (Section 7.2.3).

Burning embers figures introduced in the IPCC Third Assessment Report through to the Fifth Assessment Report, and the SR15, were developed for this report to illustrate risks at different temperature thresholds. Key components involved in desertification, land degradation and food security were identified, based on discussions with authors in Chapters 3, 4 and 5. The final list of burning embers in Figure 7.1 is not intended to be fully comprehensive, but represents processes for which sufficient literature exists to make expert judgements. Literature used in the burning embers assessment is summarised in tables in Supplementary Material. Following an approach articulated in O'Neill, B.C. et al., (2017), expert judgements were made to assess thresholds of risk (O'Neill, B.C. et al., 2017). To further strengthen replicability of the method, a predefined protocol based on a modified Delphi process was followed (Mukherjee et al. 2015). This included two separate anonymous rating rounds, feedback in between rounds and a group discussion to achieve consensus.

Burning embers provide ranges of a given variable (typically global mean near-surface air temperature) for which risks transitions within four categories: undetectable, moderate, high and very high. **Moderate risk** indicates that impacts are detectable and attributable to climate-related factors. **High risk** indicates widespread impacts on larger numbers or proportion of population/area, but with the potential to adapt or recover. **Very high risk** indicates severe and possibly irreversible impacts with limited ability of societies and ecosystems to adapt to them. Transitions between risk categories were assigned confidence levels based on the amount, and quality, of academic literature supporting judgements: L = low, M = medium, and H = high. Further details of the procedure are provided in Supplementary Material.

7.2.2   Risks to land systems arising from climate change

At current levels of global mean surface temperature (GMST) increase, impacts are already detectable across numerous land-related systems (*high confidence*) (Chapters 2, 3, 4 and 6). There is *high confidence* that unabated future climate change will result in continued changes to processes involved in desertification, land degradation and food security, including: water scarcity in drylands; soil erosion; coastal degradation; vegetation loss; fire; permafrost thaw; and access, stability, utilisation and physical availability of food (Figure 7.1). These changes will increase risks to food systems, the health of humans and ecosystems, livelihoods, the value of land, infrastructure and communities (Section 7.3). Details of the risks, and their transitions, are described in the following subsections.

7.2.2.1   Crop yield in low latitudes

There is *high confidence* that climate change has resulted in decreases in yield (of wheat, rice, maize, soy) and reduced food availability in low-latitude regions (IPCC, 2018) (Section 5.2.2). Countries in low-latitude regions are particularly vulnerable because the livelihoods of high proportions of the population are dependent on agricultural production. Even moderate temperature increases (1°C to 2°C) have negative yield impacts for major cereals, because the climate of many tropical agricultural regions is already quite close to the high-temperature thresholds for suitable production of these cereals (Rosenzweig et al. 2014). Thus, by 1.5°C global mean temperature GMT, or between approximately 1.6°C and approximately 2.6°C of

Risks to humans and ecosystems from changes in land-based processes as a result of climate change

Increases in global mean surface temperature (GMST), relative to pre-industrial levels, affect processes involved in **desertification** (water scarcity), **land degradation** (soil erosion, vegetation loss, wildfire, permafrost thaw) and **food security** (crop yield and food supply instabilities). Changes in these processes drive risks to food systems, livelihoods, infrastructure, the value of land, and human and ecosystem health. Changes in one process (e.g., wildfire or water scarcity) may result in compound risks. Risks are location-specific and differ by region.

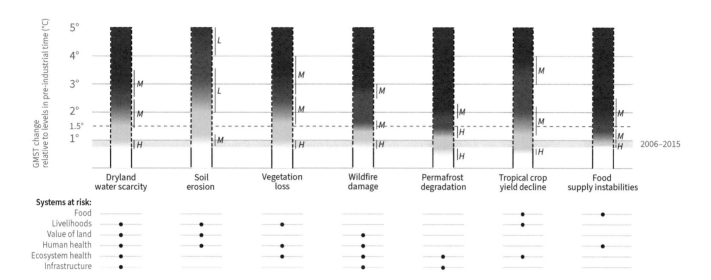

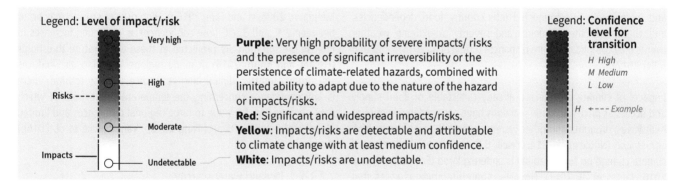

Figure 7.1 | Risks to selected land system elements as a function of global mean surface temperature increase since pre-industrial times. Impacts on human and ecological systems include: 1) economic loss and declines in livelihoods and ecosystem services from water scarcity in drylands, 2) economic loss and declines in livelihoods and ecosystem services from reduced land productivity due to soil erosion, 3) vegetation loss and shifts in vegetation structure, 4) damage to infrastructure, altered land cover, accelerated erosion and increased air pollution from fires, 5) damage to natural and built environment from permafrost thaw related ground instability, 6) changes to crop yield and food availability in low-latitude regions and 7) increased disruption of food supply stability. Risks are global (2, 3, 4, 7) and specific to certain regions (1, 5, 6). Selected components are illustrative and not intended to be fully comprehensive of factors influencing food security, land degradation and desertification. The supporting literature and methods are provided in Supplementary Material. Risk levels are estimated assuming medium exposure and vulnerability driven by moderate trends in socioeconomic conditions broadly consistent with an SSP2 pathway.

local warming, risks to yields may already transition to *high* in West Africa, Southeast Asia and Central and South America (Faye et al. 2018) (*medium confidence*). For further information see Section 5.3.2.1. By contrast, higher latitudes may initially benefit from warming as well as well higher CO$_2$ concentrations (IPCC 2018a). Wheat yield losses are expected to be lower for the USA (−5.5 ± 4.4% per degree Celsius) and France (−6.0 ± 4.2% per degree Celsius) compared to India (−9.1 ± 5.4% per degree Celsius) (Zhao et al. 2017). Very high risks to low-latitude yields may occur between 3°C and 4°C (*medium confidence*). At these temperatures, catastrophic reductions in crop

yields may occur, of up to 60% in low latitudes (Rosenzweig et al. 2014) (Sections 5.2.2 and 5.2.3). Some studies report significant population displacement from the tropics related to systemic livelihood disruption in agriculture systems (Tittonell 2014; Montaña et al. 2016; Huber-Sannwald et al. 2012; Wise et al. 2016; Tanner et al. 2015; Mohapatra 2013). However, at higher temperatures of warming, all regions of the world face risks of declining yields as a result of extreme weather events and reduced heat tolerance of maize, rice, wheat and soy (Zhao et al. 2017; IPCC 2018a).

7

7.2.2.2   Food supply instability

Stability of food supply is expected to decrease as the magnitude and frequency of extreme events increase, disrupting food chains in all areas of the world (*medium evidence, high agreement*) (Wheeler and Von Braun 2013; Coates 2013; Puma et al. 2015; Deryng et al. 2014; Harvey et al. 2014b; Iizumi et al. 2013; Seaman et al. 2014) (Sections 5.3.2, 5.3.3, 5.6.2 and 5.7.1). While international trade in food is assumed to be a key response for alleviating hunger, historical data and economic models suggest that international trade does not adequately redistribute food globally to offset yield declines or other food shortages when weather extremes reduce crop yields (*medium confidence*) (Schmitz et al. 2012; Chatzopoulos et al. 2019; Marchand et al. 2016; Gilbert 2010; Wellesley et al. 2017). When droughts, heat waves, floods or other extremes destroy crops, evidence has shown that exports are constrained in key producing countries contributing to price spikes and social tension in importing countries which reduce access to food (*medium evidence, medium agreement*) (von Uexkull et al. 2016; Gleick 2014; Maystadt and Ecker 2014; Kelley et al. 2015; Church et al. 2017; Götz et al. 2013; Puma et al. 2015; Willenbockel 2012; Headey 2011; Distefano et al. 2018; Brooks 2014). There is little understanding of how food system shocks cascade through a modern interconnected economy. Reliance on global markets may reduce some risks, but the ongoing globalisation of food trade networks exposes the world food system to new impacts that have not been seen in the past (Sections 5.1.2, 5.2.1, 5.5.2.5, 5.6.5 and 5.7.1). The global food system is vulnerable to systemic disruptions and increasingly interconnected inter-country food dependencies, and changes in the frequency and severity of extreme weather events may complicate future responses (Puma et al. 2015; Jones and Hiller 2017).

Impacts of climate change are already detectable on food supply and access as price and trade reactions have occurred in response to heatwaves, droughts and other extreme events (*high evidence, high agreement*) (Noble et al. 2014; O'Neill, B.C. et al., 2017). The impact of climate change on food stability is underexplored (Schleussner et al. 2016; James et al. 2017). However, some literature assesses that by about 2035, daily maximum temperatures will exceed the 90th percentile of historical (1961–1990) temperatures on 25–30% of days (O'Neill, B.C. et al., 2017, Figures 11–17) with negative shocks to food stability and world food prices. O'Neill, B.C. et al., (2017) remark that in the future, return periods for precipitation events globally (land only) will reduce from one-in-20-year (historical) to about one-in-14-year or less by 2046–2065 in many areas of the world. Domestic efforts to insulate populations from food price spikes associated with climatic stressors in the mid-2000s have been shown to inadequately shield from poverty, and worsen poverty globally (Diffenbaugh et al. 2012; Meyfroidt et al. 2013; Hertel et al. 2010). The transition to *high risk* is estimated to occur around 1.4°C, possibly by 2035, due to changes in temperature and heavy precipitation events (*medium confidence*) (O'Neill, B.C. et al.. 2017; Fritsche et al. 2017; Harvey et al. 2014b). *Very high risk* may occur between 1.5°C and 2.5°C (*medium confidence*) and 4°C of warming is considered catastrophic (IPCC 2018c; Noble et al. 2014) for food stability and access because a combination of extreme events, compounding political and social

factors, and shocks to crop yields can heavily constrain options to ensure food security in import-reliant countries.

7.2.2.3   Soil erosion

Soil erosion increases risks of economic loss and declines in livelihoods due to reduced land productivity. In the EU, on-site costs of soil erosion by wind has been reported at an average of 55 USD per hectare annually, but up to 450 USD per hectare for sugar beet and oilseed rape (Middleton et al. 2017). Farmers in the Dapo watershed in Ethiopia lose about 220 USD per hectare of maize due to loss of nitrogen through soil erosion (Erkossa et al. 2015). Soil erosion not only increases crop loss but has been shown to have reduced household food supply with older farmers most vulnerable to losses from erosion (Ighodaro et al. 2016). Erosion also results in increased risks to human health, through air pollution from aerosols (Middleton et al. 2017), and brings risks of reduced ES including supporting services related to soil formation.

At current levels of warming, changes in erosion are already detected in many regions. Attribution to climate change is challenging as there are other powerful drivers of erosion (e.g., land use), limited global-scale studies (Li and Fang 2016a; Vanmaercke et al. 2016a) and the absence of formal detection and attribution studies (Section 4.2.3). However, studies have found an increase in short-duration and high-intensity precipitation, due to anthropogenic climate change, which is a causative factor for soil erosion (Lenderink and van Meijgaard 2008; Li and Fang 2016b). High risks of erosion may occur between 2°C and 3.5°C (*low confidence*) as continued increases in intense precipitation are projected at these temperature thresholds (Fischer and Knutti 2015) in many regions. Warming also reduces soil organic matter, diminishing resistance against erosion. There is *low confidence* concerning the temperature threshold at which risks become *very high* due to large regional differences and limited global-scale studies (Li and Fang 2016b; Vanmaercke et al. 2016b) (Section 4.4).

7.2.2.4   Dryland water scarcity

Water scarcity in drylands contributes to changes in desertification and hazards such as dust storms, increasing risks of economic loss, declines in livelihoods of communities and negative health effects (*high confidence*) (Section 3.1.3). Further information specific to costs and impacts of water scarcity and droughts is detailed in Cross-Chapter Box 5 in Chapter 3.

The IPCC AR5 report and the SR15 concluded that there is *low confidence* in the direction of drought trends since 1950 at the global scale. While these reports did not assess water scarcity with a specific focus on drylands, they indicated that there is *high confidence* in observed drought increases in some regions of the world, including in the Mediterranean and West Africa (IPCC AR5) and that there is *medium confidence* that anthropogenic climate change has contributed to increased drying in the Mediterranean region (including southern Europe, northern Africa and the western Asia and the Middle east) and that this tendency will continue to increase under higher levels of global warming (IPCC 2018d). Some parts of the drylands have experienced decreasing precipitation over recent decades (IPCC AR5) (Chapter 3 and Section 3.2), consistent with the fact that climate change is implicated in desertification trends in some regions (Section 3.2.2). Dust storms, linked to changes in precipitation and vegetation, appear to be occurring with greater frequency in some deserts and their margins (Goudie 2014) (Section 3.3.1). There is therefore *high confidence* that the transition from *undetectable* to *moderate risk* associated with water scarcity in drylands occurred in recent decades in the range 0.7°C to 1°C (Figure 7.1).

Between 1.5°C and 2.5°C, the risk level is expected to increase from *moderate* to *high* (*medium confidence*). Globally, at 2°C an additional 8% of the world population (of population in 2000) will be exposed to new forms of or aggravated water scarcity (IPCC 2018d). However, at 2°C, the annual warming over drylands will reach 3.2°C–4.0°C, implying about 44% more warming over drylands than humid lands (Huang et al. 2017), thus potentially aggravating water scarcity issues through increased evaporative demand. Byers et al. (2018a) estimate that 3–22% of the drylands population (range depending on socio-economic conditions) will be exposed and vulnerable to water stress. The Mediterranean, North Africa and the Eastern Mediterranean will be particularly vulnerable to water shortages, and expansion of desert terrain and vegetation is predicted to occur in the Mediterranean biome, an unparalleled change in the last 10,000 years (*medium confidence*) (IPCC 2018d). At 2.5°C–3.5°C risks are expected to become *very high* with migration from some drylands resulting as the only adaptation option (*medium confidence*). Scarcity of water for irrigation is expected to increase, in particular in Mediterranean regions, with limited possibilities for adaptation (Haddeland et al. 2014).

7.2.2.5   Vegetation degradation

There are clear links between climate change and vegetation cover changes, tree mortality, forest diseases, insect outbreaks, forest fires, forest productivity and net ecosystem biome production (Allen et al. 2010; Bentz et al. 2010; Anderegg et al. 2013; Hember et al. 2017; Song et al. 2018; Sturrock et al. 2011). Forest dieback, often a result of drought and temperature changes, not only produces risks to forest ecosystems but also to people with livelihoods dependent on forests. A 50-year study of temperate forest, dominated by beech (*Fagus sylvatica* L.), documented a 33% decline in basal area and a 70% decline in juvenile tree species, possibly as a result of interacting pressures of drought, overgrazing and pathogens (Martin et al. 2015). There is *high confidence* that such dieback impacts ecosystem properties and services including soil microbial community structure (Gazol et al. 2018). Forest managers and users have reported negative emotional impacts from forest dieback such as pessimism about losses, hopelessness and fear (Oakes et al. 2016). Practices and policies such as forest classification systems, projection of growth, yield and models for timber supply are already being affected by climate change (Sturrock et al. 2011).

While risks to ecosystems and livelihoods from vegetation degradation are already detectable at current levels of GMT increase, risks are expected to reach *high* levels between 1.6°C

and 2.6°C (*medium confidence*). Significant uncertainty exists due to countervailing factors: CO_2 fertilisation encourages forest expansion but increased drought, insect outbreaks, and fires result in dieback (Bonan 2008; Lindner et al. 2010). The combined effects of temperature and precipitation change, with CO_2 fertilisation, make future risks to forests very location specific. It is challenging therefore to make global estimates. However, even locally specific studies make clear that *very high risks* occur between 2.6°C and 4°C (*medium confidence*). Australian tropical rainforests experience significant loss of biodiversity with 3.5°C increase. At this level of increase there are no areas with greater than 30 species, and all endemics disappear from low- and mid-elevation regions (Williams et al. 2003). Mountain ecosystems are particularly vulnerable (Loarie et al. 2009).

7.2.2.6   Fire damage

Increasing fires result in heightened risks to infrastructure, accelerated erosion, altered hydrology, increased air pollution, and negative mental health impacts. Fire not only destroys property but induces changes in underlying site conditions (ground cover, soil water repellency, aggregate stability and surface roughness) which amplifies runoff and erosion, increasing future risks to property and human lives during extreme rainfall events (Pierson and Williams 2016). Dust and ash from fires can impact air quality in a wide area. For example, a dust plume from a fire in Idaho, USA, in September 2010 was visible in MODIS satellite imagery and extended at least 100 km downwind of the source area (Wagenbrenner et al. 2013). Individuals can suffer from property damage or direct injury, psychological trauma, depression, and post traumatic stress disorder, and have reported negative impacts to well-being from loss of connection to landscape (Paveglio et al. 2016; Sharples et al. 2016a). Costs of large wildfires in the USA can exceed 20 million USD per day (Pierson et al. 2011) and has been estimated at 8.5 billion USD per year in Australia (Sharples et al. 2016b). Globally, human exposure to fire will increase due to projected population growth in fire-prone regions (Knorr et al. 2016a).

It is not clear how quickly, or even if, systems can recover from fires. Longevity of effects may differ depending on cover recruitment rate and soil conditions, recovering in one to two seasons or over 10 growing seasons (Pierson et al. 2011). In Russia, one-third of forest area affected by fires turned into unproductive areas where natural reforestation is not possible within 2–3 lifecycles of major forest forming species (i.e., 300–600 years) (Shvidenko et al. 2012).

Risks under current warming levels are already *moderate* as anthropogenic climate change has caused significant increases in fire area (*high confidence*) due to availability of detection and attribution studies (Cross-Chapter Box 3 in Chapter 2). This has been detected and attributed regionally, notably in the western USA (Abatzoglou and Williams 2016; Westerling et al. 2006; Dennison et al. 2014), Indonesia (Fernandes et al. 2017) and other regions (Jolly et al. 2015). Regional increases have been observed despite a global-average declining trend induced by human fire-suppression strategies, especially in savannahs (Yang et al. 2014a; Andela et al. 2017).

High risks of fire may occur between 1.3°C and 1.7°C (*medium confidence*). Studies note heightened risks above 1.5°C as fire, weather, and land prone to fire increase (Abatzoglou et al. 2019a), with *medium confidence* in this transition, due to complex interplay between (i) global warming, (ii) CO$_2$-fertilisation, and (iii) human/economic factors affecting fire risk. Canada, the USA and the Mediterranean may be particularly vulnerable as the combination of increased fuel due to CO$_2$ fertilisation, and weather conditions conducive to fire increase risks to people and property. Some studies show substantial effects at 3°C (Knorr et al. 2016b; Abatzoglou et al. 2019b), indicating a transition to *very high risks* (*medium confidence*). At high warming levels, climate change may become the primary driver of fire risk in the extratropics (Knorr et al. 2016b; Abatzoglou et al. 2019b; Yang et al. 2014b). Pyroconvection activity may increase, in areas such as southeast Australia (Dowdy and Pepler 2018), posing major challenges to adaptation.

7.2.2.7　Permafrost

There is a risk of damage to the natural and built environment from permafrost thaw-related ground instability. Residential, transportation, and industrial infrastructure in the pan-Arctic permafrost area are particularly at risk (Hjort et al. 2018). *High risks* already exist at low temperatures (*high confidence*). Approximately, 21–37% of Arctic permafrost is projected to thaw under a 1.5°C of warming (Hoegh-Guldberg et al. 2018). This increases to *very high risk* around 2°C (between 1.8°C and 2.3°C) of temperature increase since pre-industrial times (*medium confidence*) with 35–47% of the Arctic permafrost thawing (Hoegh-Guldberg et al. 2018). If climate stabilised at 2°C, still approximately 40% of permafrost area would be lost (Chadburn et al. 2017), leading to nearly four million people and 70% of current infrastructure in the pan-Arctic permafrost area exposed to permafrost thaw and high hazard (Hjort et al. 2018). Indeed between 2°C and 3°C a collapse of permafrost may occur with a drastic biome shift from tundra to boreal forest (Drijfhout et al. 2015; SR15). There is mixed evidence of a tipping point in permafrost collapse, leading to enhanced greenhouse gas (GHG) emission – particularly methane – between 2°C and 3°C (Hoegh-Guldberg et al. 2018).

7.2.2.8　Risks of desertification, land degradation and food insecurity under different Future Development Pathways

Socio-economic developments and policy choices that govern land–climate interactions are an important driver of risk, along with climate change (*very high confidence*). Risks under two different Shared Socio-economic Pathways (SSPs) were assessed using emerging literature. SSP1 is characterised by low population growth, reduced inequalities, land-use regulation, low meat consumption, and moderate trade (Riahi et al. 2017; Popp et al. 2017a). SSP3 is characterised by high population growth, higher inequalities, limited land-use regulation, resource-intensive consumption including meat-intensive diets, and constrained trade (for further details see Chapter 1 and Cross-Chapter Box 9 in Chapters 6 and 7). These two SSPs, among the set of five SSPs, were selected because they illustrate contrasting futures, ranging from low (SSP1) to high (SSP3) challenges to mitigation and adaptation. Figure 7.2 shows that for

a given global mean temperature (GMT) change, risks are different under SSP1 compared to SSP3. In SSP1, global temperature change does not increase above 3°C even in the baseline case (i.e., with no additional mitigation measures) because in this pathway the combination of low population and autonomous improvements, for example, in terms of carbon intensity and/or energy intensity, effectively act as mitigation measures (Riahi et al. 2017). Thus Figure 7.2 does not indicate risks beyond this point in either SSP1 and SSP3. Literature based on such socio-economic and climate models is still emerging and there is a need for greater research on impacts of different pathways. There are few SSP studies exploring aspects of desertification and land degradation, but a greater number of SSP studies on food security (Supplementary Material). SSP1 reduces the vulnerability and exposure of human and natural systems and thus limits risks resulting from desertification, land degradation and food insecurity compared to SSP3 (*high confidence*).

Changes to the water cycle due to global warming are an essential driver of desertification and of the risks to livelihood, food production and vegetation in dryland regions. Changes in water scarcity due to climate change have already been detected in some dryland regions (Section 7.2.2.4) and therefore the transition to *moderate risk* occurred in recent decades (*high confidence*). IPCC (2018d) noted that in the case of risks to water resources, socio-economic drivers are expected to have a greater influence than the changes in climate (*medium confidence*). Indeed, in SSP1 there is only moderate risk even at 3°C of warming, due to the lower exposure and vulnerability of human population (Hanasaki et al. 2013a; Arnell and Lloyd-Hughes 2014; Byers et al. 2018b). Considering drylands only, Byers et al. (2018b) estimate, using a time-sampling approach for climate change and the 2050 population, that at 1.5°C, 2°C and 3°C, the dryland population exposed and vulnerable to water stress in SSP1 will be 2%, 3% and 3% respectively, thus indicating relatively stable *moderate risks*. In SSP3, the transition from *moderate* to *high risk* occurs in the range 1.2°C to 1.5°C (*medium confidence*) and the transition from *high* to *very high risk* is in the range 1.5°C to 2.8°C (*medium confidence*). Hanasaki et al. (2013b) found a consistent increase in water stress at higher warming levels due in large part to growth in population and demand for energy and agricultural commodities, and to a lesser extent due to hydrological changes induced by global warming. In SSP3, Byers et al. (2018b) estimate that at 1.5°C, 2°C and 3°C, the population exposed and vulnerable to water stress in drylands will steadily increase from 20% to 22% and 24% respectively, thus indicating overall much higher risks compared to SSP1 for the same global warming levels.

SSP studies relevant to land degradation assess risks such as: number of people exposed to fire; the costs of floods and coastal flooding; and loss of ES including the ability of land to sequester carbon. The risks related to permafrost melting (Section 7.2.2.7) are not considered here due to the lack of SSP studies addressing this topic. Climate change impacts on various components of land degradation have already been detected (Sections 7.2.2.3, 7.2.2.5 and 7.2.2.6) and therefore the transition from *undetectable* to *moderate risk* is in the range 0.7°C to 1°C (*high confidence*). Less than 100 million people are exposed to habitat degradation at 1.5°C under SSP1 in non-dryland regions, increasing to 257 million at 2°C (Byers

Different socioeconomic pathways affect levels of climate related risks

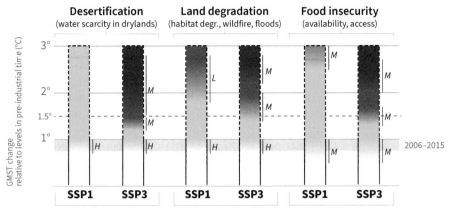

<table>
<tr><td></td><td></td><td></td></tr>
</table>

Socio-economic choices can reduce or exacerbate climate related risks as well as influence the rate of temperature increase. The **SSP1** pathway illustrates a world with low population growth, high income and reduced inequalities, food produced in low GHG emission systems, effective land use regulation and high adaptive capacity. The **SSP3** pathway has the opposite trends. Risks are lower in SSP1 compared with SSP3 given the same level of GMST increase.

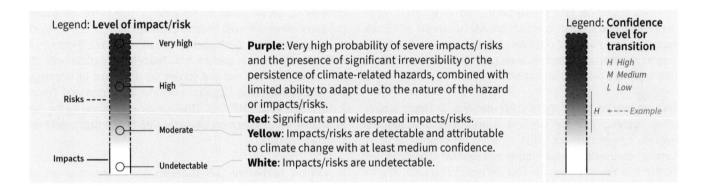

Figure 7.2 | Risks associated with desertification, land degradation and food security due to climate change and patterns of socio-economic development. Increasing risks associated with desertification include population exposed and vulnerable to water scarcity in drylands. Risks related to land degradation include increased habitat degradation, population exposed to wildfire and floods and costs of floods. Risks to food security include availability and access to food, including population at risk of hunger, food price increases and increases in disability adjusted life years attributable due to childhood underweight. The risks are assessed for two contrasted socio-economic futures (SSP1 and SSP3) under unmitigated climate change {3.6, 4.3.1.2, 5.2.2, 5.2.3, 5.2.4, 5.2.5, 6.2.4, 7.3}. Risks are not indicated beyond 3°C because SSP1 does not exceed this level of temperature change.

et al. 2018). This suggests a gradual transition to *high risk* in the range 1.8°C to 2.8°C, but a *low confidence* is attributed due to the very limited evidence to constrain this transition.

By contrast in SSP3, there are already 107 million people exposed to habitat degradation at 1.5°C, increasing to 1156 million people at 3°C (Byers et al. 2018b). Furthermore, Knorr et al. (2016b) estimate that 646 million people will be exposed to fire at 2°C warming, the main risk driver being the high population growth in SSP3 rather than increased burned area due to climate change. Exposure to extreme rainfall, a causative factor for soil erosion and flooding, also differs under SSPs. Under SSP1 up to 14% of the land and population experience five-day extreme precipitation events. Similar levels of exposure occur at lower temperatures in SSP3 (Zhang et al. 2018b). Population exposed to coastal flooding is lowest under SSP1 and higher under SSP3 with a limited effect of enhanced protection in SSP3 already after 2°C warming (Hinkel et al. 2014). The transition from *high* to *very high risk* will occur at 2.2°C to 2.8°C in SSP3 (*medium confidence*), whereas this level of risk is not expected to be reached in SSP1.

The greatest number of SSP studies explore climate change impacts relevant to food security, including population at risk of hunger, food price increases, increases in disability adjusted life years (Hasegawa et al. 2018a; Wiebe et al. 2015a; van Meijl et al. 2018a; Byers et al. 2018b). Changes in crop yields and food supply stability have already been attributed to climate change (Sections 7.2.2.1 and 7.2.2.2) and the transition from *undetectable* to *moderate risk* is placed at 0.5°C to 1°C (*medium confidence*). At 1.5°C, about two million people are exposed and vulnerable to crop yield change in SSP1 (Hasegawa et al. 2018b; Byers et al. 2018b), implying *moderate risk*. A transition from *moderate* to *high risk* is expected above 2.5°C (*medium confidence*) with population at risk of hunger of the order of 100 million (Byers et al. 2018b). Under SSP3, *high risks* already exist at 1.5°C (*medium confidence*), with 20 million people exposed and vulnerable to crop yield change. By 2°C, 178 million are vulnerable and 854 million people are vulnerable at 3°C (Byers et al. 2018b). This is supported by the higher food prices increase of up to 20% in 2050 in an RCP6.0 scenario (i.e., slightly below 2°C) in SSP3 compared to up to 5% in SSP1 (van Meijl et al. 2018). Furthermore in SSP3, restricted trade increase this price effect (Wiebe et al. 2015). In SSP3, the transition from *high* to *very high risk* is in the range 2°C to 2.7°C (*medium confidence*) while this transition is never reached in SSP1. This overall confirms that socio-economic development, by affecting exposure and vulnerability, has an even larger effect than climate change for future trends in the population at risk of hunger (O'Neill, B.C. et al., 2017, p.32). Changes can also threaten development gains (*medium confidence*). Disability adjusted life years due to childhood underweight decline in both SSP1 and SSP3 by 2030 (by 36.4 million disability adjusted life years in SSP1 and 16.2 million in SSP3). However by 2050, disability adjusted life years increase by 43.7 million in SSP3 (Ishida et al. 2014).

7.2.3   Risks arising from responses to climate change

7.2.3.1   Risk associated with land-based adaptation

Land-based adaptation relates to a particular category of adaptation measures relying on land management (Sanz et al. 2017). While most land-based adaptation options provide co-benefits for climate mitigation and other land challenges (Chapter 6 and Section 6.4.1), in some contexts adaptation measures can have adverse side effects, thus implying a risk to socio-ecological systems.

One example of risk is the possible decrease in farmer income when applying adaptive cropland management measures. For instance, conservation agriculture including the principle of no-till farming, contributes to soil erosion management (Chapter 6 and Section 6.2). Yet, no-till management can reduce crop yields in some regions, and although this effect is minimised when no-till farming is complemented by the other two principles of conservation agriculture (residue retention and crop rotation), this could induce a risk to livelihood in vulnerable smallholder farming systems (Pittelkow et al. 2015).

Another example is the use of irrigation against water scarcity and drought. During the long lasting drought from 2007–2009 in California, USA, farmers adapted by relying on groundwater withdrawal and caused groundwater depletion at unsustainable levels (Christian-Smith et al. 2015). The long-term effects of irrigation from groundwater may cause groundwater depletion, land subsidence, aquifer overdraft, and saltwater intrusion (Tularam and Krishna 2009). Therefore, it is expected to increase the vulnerability of coastal aquifers to climate change due to groundwater usage (Ferguson and Gleeson 2012). The long-term practice of irrigation from groundwater may cause a severe combination of potential side effects and consequently irreversible results.

7.2.3.2   Risk associated with land-based mitigation

While historically land-use activities have been a net source of GHG emissions, in future decades the land sector will not only need to reduce its emissions, but also to deliver negative emissions through carbon dioxide removal (CDR) to reach the objective of limiting global warming to 2°C or below (Section 2.5). Although land-based mitigation in itself is a risk-reduction strategy aiming at abating climate change, it also entails risks to humans and ecosystems, depending on the type of measures and the scale of deployment. These risks fall broadly into two categories: risk of mitigation failure – due to uncertainties about mitigation potential, potential for sink reversal and moral hazard; and risks arising from adverse side effects – due to increased competition for land and water resources. This section focuses specifically on bioenergy and bioenergy with carbon capture and storage (BECCS) since it is one of the most prominent land-based mitigation strategies in future mitigation scenarios (along with large-scale forest expansion, which is discussed in Cross-Chapter Box 1 in Chapter 1). Bioenergy and BECCS is assessed in Chapter 6 as being, at large scales, the only response option with adverse side effects across all dimensions (adaptation, food security, land degradation and desertification) (Section 6.4.1).

Risk of mitigation failure. The mitigation potential from bioenergy and BECCS is highly uncertain, with estimates ranging from 0.4 to 11.3 $GtCO_2e$ yr^{-1} for the technical potential, while consideration of sustainability constraints suggest an upper end around 5 $GtCO_2e$ yr^{-1} (Chapter 2, Section 2.6). In comparison, IAM-based mitigation pathways compatible with limiting global warming at 1.5°C project bioenergy and BECCS deployment exceeding this range (Figure 2.24 in Chapter 2). There is *medium confidence* that IAMs currently do not reflect the lower end and exceed the upper end of bioenergy and BECCS mitigation potential estimates (Anderson and Peters 2016; Krause et al. 2018; IPCC 2018c), with implications for the risk associated with reliance on bioenergy and BECCS deployment for climate mitigation.

In addition, land-based CDR strategies are subject to a risk of carbon sink reversal. This implies a fundamental asymmetry between mitigation achieved through fossil fuel emissions reduction compared to CDR. While carbon in fossil fuel reserves – in the case of avoided fossil fuel emissions – is locked permanently (at least over a time scale of several thousand years), carbon sequestered into the terrestrial biosphere – to compensate fossil fuel emissions – is subject to various disturbances, in particular from climate change and associated extreme events (Fuss et al. 2018; Dooley and Kartha 2018). The probability of sink reversal therefore increases with climate change, implying that the effectiveness of land-based mitigation depends on emission reductions in other sectors and can be sensitive to temperature overshoot (*high confidence*). In the case of bioenergy associated with CCS (BECCS), the issue of the long-term stability of the carbon storage is linked to technical and geological constraints, independent of climate change but presenting risks due to limited knowledge and experience (Chapter 6 and Cross-Chapter Box 7 in Chapter 6).

Another factor in the risk of mitigation failure, is the moral hazard associated with CDR technologies. There is *medium evidence* and *medium agreement* that the promise of future CDR deployment – bioenergy and BECCS in particular – can deter or delay ambitious emission reductions in other sectors (Anderson and Peters 2016; Markusson et al. 2018a; Shue 2018a). The consequences are an increased pressure on land with higher risk of mitigation failure and of temperature overshoot, and a transfer of the burden of mitigation and unabated climate change to future generations. Overall, there is therefore *medium evidence* and *high agreement* that prioritising early decarbonisation with minimal reliance on CDR decreases the risk of mitigation failure and increases intergenerational equity (Geden et al. 2019; Larkin et al. 2018; Markusson et al. 2018b; Shue 2018b).

Risk from adverse side-effects. At large scales, bioenergy (with or without CCS) is expected to increase competition for land, water resources and nutrients, thus exacerbating the risks of food insecurity, loss of ES and water scarcity (Chapter 6 and Cross-Chapter Box 7 in Chapter 6). Figure 7.3 shows the risk level (from *undetectable* to *very high*, aggregating risks of food insecurity, loss of ES and water scarcity) as a function of the global amount of land (million km^2) used for bioenergy, considering second generation bioenergy. Two illustrative future Socio-economic Pathways (SSP1 and SSP3; see Section 7.2.2 for more details) are depicted: in SSP3 the competition for land is exacerbated compared to SSP1 due to higher food demand resulting from larger population growth and higher consumption of meat-based products. The literature used in this assessment is based on IAM and non-IAM-based studies examining the impact of bioenergy crop deployment on various indicators, including food security (food prices or population at risk of hunger with explicit consideration of exposure and vulnerability), SDGs, ecosystem losses, transgression of various planetary boundaries and water consumption (see Supplementary Material). Since most of the assessed literature is centred around 2050, prevailing demographic and economic conditions for this year are used for the risk estimate. An aggregated risk metric including risks of food insecurity, loss of ES and water

scarcity is used because there is no unique relationship between bioenergy deployment and the risk outcome for a single system. For instance, bioenergy deployment can be implemented in such a way that food security is prioritised at the expense of natural ecosystems, while the same scale of bioenergy deployment implemented with ecosystem safeguards would lead to a fundamentally different outcome in terms of food security (Boysen et al. 2017a). Considered as a combined risk, however, the possibility of a negative outcome on either food security, ecosystems or both can be assessed with less ambiguity and independently of possible implementation choices.

In SSP1, there is *medium confidence* that 1 to 4 million km^2 can be dedicated to bioenergy production without significant risks to food security, ES and water scarcity. At these scales of deployment, bioenergy and BECCS could have co-benefits for instance by contributing to restoration of degraded land and soils (Cross-Chapter Box 7 in Chapter 6). Although currently degraded soils (up to 20 million km^2) represent a large amount of potentially available land (Boysen et al. 2017a), trade-offs would occur already at smaller scale due to fertiliser and water use (Hejazi et al. 2014; Humpenöder et al. 2017; Heck et al. 2018a; Boysen et al. 2017b). There is *low confidence* that the transition from moderate to high risk is in the range 6–8.7 million km^2. In SSP1, (Humpenöder et al. 2017) found no important impacts on sustainability indicators at a level of 6.7 million km^2, while (Heck et al. 2018b) note that several planetary boundaries (biosphere integrity; land-system change; biogeochemical

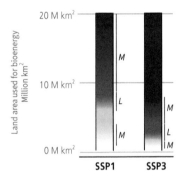

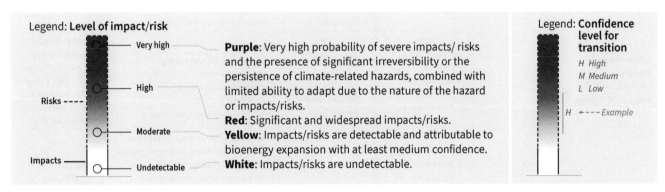

Figure 7.3 | Risks associated with bioenergy crop deployment as a land-based mitigation strategy under two SSPs (SSP1 and SSP3). The assessement is based on literature investigating the consequences of bioenergy expansion for food security, ecosystem loss and water scarcity. These risk indicators were aggregated as a single risk metric in the figure. In this context, very high risk indicates that important adverse consequences are expected for all these indicators (more than 100 million people at risk of hunger, major ecosystem losses and severe water scarcity issues). The climate scenario considered is a mitigation scenario consistent with limiting global warming at 2°C (RCP2.6), however some studies considering other scenarios (e.g., no climate change) were considered in the expert judgement as well as results from other SSPs (e.g., SSP2). The literature supporting the assessment is provided in Table SM7.3.

flows; freshwater use) would be exceeded above 8.7 million km². There is *very high confidence* that all the risk transitions occur at lower bioenergy levels in SSP3, implying higher risks associated with bioenergy deployment, due to the higher competition for land in this pathway. In SSP3, land-based mitigation is therefore strongly limited by sustainability constraints such that moderate risk occur already between 0.5 and 1.5 million km² (*medium confidence*). There is *medium confidence* that a bioenergy footprint beyond 4 to 8 million km² would entail very high risk with transgression of most planetary boundaries (Heck et al. 2018b), strong decline in sustainability indicators (Humpenöder et al. 2017) and increase in the population at risk of hunger well above 100 million (Fujimori et al. 2018a; Hasegawa et al. 2018b).

7.2.4   Risks arising from hazard, exposure and vulnerability

Table 7.1 shows hazards from land-climate-society interactions identified in previous chapters, or in other IPCC reports (with supplementary hazards appearing in the Appendix); the regions that are exposed or will be exposed to these hazards; components of the land-climate systems and societies that are vulnerable to the hazard; the risk associated with these impacts and the available indicative policy responses. The last column shows representative supporting literature.

Included are forest dieback, extreme events in multiple economic and agricultural regimes (also see Sections 7.2.2.1 and 7.2.2.2), disruption in flow regimes in river systems, climate change mitigation impacts (Section 7.2.3.2), competition for land (plastic substitution by cellulose, charcoal production), land degradation and desertification (Section 7.2.2.8), loss of carbon sinks, permafrost destabilisation (Section 7.2.2.7), and stranded assets (Section 7.3.4). Other hazards such as from failure of carbon storage, renewable energy impacts on land use, wild-fire in forest-urban transition context, extreme events effects on cultural heritage and urban air pollution from surrounding land use are covered in Table 7.1 extension in the appendix as well in Section 7.5.6.

Table 7.1 | Characterising land–climate risk and indicative policy responses. Table shows hazards from land–climate–society interactions identified in previous chapters or in *other* IPCC reports; the regions that are exposed or will be exposed to these hazards; components of the land-climate systems and societies that are vulnerable to the hazard; the risk associated with these impacts and the available policy responses and response options from Chapter 6. The last column shows representative supporting literature.

Land–climate–society interaction hazard	Exposure	Vulnerability	Risk	Policy response (indicative)	References
Forest dieback	Widespread across biomes and regions	Marginalised population with insecure land tenure	– Loss of forest-based livelihoods – Loss of identity	– Land rights – Community-based conservation – Enhanced political enfranchisement – Manager–scientist partnerships for adaptation silviculture	Allen et al. 2010; McDowell and Allen 2015; Sunderlin et al. 2017; Belcher et al. 2005; Soizic et al. 2013; Nagel et al. 2017
		Endangered species and ecosystems	– Extinction – Loss of ecosystem services (ES) – Cultural loss	– Effective enforcement of protected areas and curbs on illegal trade – Ecosystem restoration – Protection of indigenous people	Bailis et al. 2015; Cameron et al. 2016
Extreme events in multiple economic and agricultural regimes	Global	– Food-importing countries – Low-income indebtedness – Net food buyer	– Conflict – Migration – Food inflation – Loss of life – Disease, malnutrition – Farmer distress	– Insurance – Social protection encouraging diversity of sources – Climate smart agriculture – Land rights and tenure – Adaptive public distribution systems	Fraser et al. 2005; Schmidhuber and Tubiello 2007; Lipper et al. 2014a; Lunt et al. 2016; Tigchelaar et al. 2018; Casellas Connors and Janetos 2016
Disruption of flow regimes in river systems	– 1.5 billion people, Regional (e.g., South Asia, Australia) – Aral sea and others	– Water-intensive agriculture – Freshwater, estuarine and near coastal ecosystems – Fishers – Endangered species and ecosystems	– Loss of livelihoods and identity – Migration – Indebtedness	– Build alternative scenarios for economies and livelihoods based on non-consumptive use (e.g., wild capture fisheries) – Define and maintain ecological flows in rivers for target species and ES – Experiment with alternative, less water-consuming crops and water management strategies – Redefine SDGs to include freshwater ecosystems or adopt alternative metrics of sustainability Based on Nature's Contributions to People (NCP)	Craig 2010; Di Baldassarre et al. 2013; Verma et al. 2009; Ghosh et al. 2016; Higgins et al. 2018; Hall et al. 2013; Youn et al. 2014

Land–climate–society interaction hazard	Exposure	Vulnerability	Risk	Policy response (indicative)	References
Depletion/exhaustion of groundwater	– Widespread across semi-arid and humid biomes – India, China and the USA – Small Islands	– Farmers, drinking water supply – Irrigation – See forest note above – Agricultural production – Urban sustainability (Phoenix, US) – Reduction in dry-season river flows – Sea level rise	– Food insecurity – Water insecurity – Distress migration – Conflict – Disease – Inundation of coastal regions, estuaries and deltas	– Monitoring of emerging groundwater-climate linkages – Adaptation strategies that reduce dependence on deep groundwater – Regulation of groundwater use – Shift to less water-intensive rainfed crops and pasture – Conjunctive use of surface and groundwater	Wada et al. 2010; Rodell et al. 2009; Taylor et al. 2013; Aeschbach-Hertig and Gleeson 2012
Climate change mitigation impacts	Across various biomes, especially semi-arid and aquatic, where renewable energy projects (solar, biomass, wind and small hydro) are sited	– Fishers and pastoralists – Farmers – Endangered range restricted species and ecosystems	– Extinction of species – Downstream loss of ES – Loss of livelihoods and identity of fisher/pastoralist communities – Loss of regional food security	– Avoidance and informed siting in priority basins – Mitigation of impacts – Certification	Zomer et al. 2008; Nyong et al. 2007; Pielke et al. 2002; Schmidhuber and Tubiello 2007; Jumani et al. 2017; Eldridge et al. 2011; Bryan et al. 2010; Scarlat and Dallemand 2011
Competition for land e.g., plastic substitution by cellulose, charcoal production	Peri-urban and rural areas in developing countries	– Rural landscapes; farmers; charcoal suppliers; small businesses	– Land degradation; loss of ES; GHG emissions; lower adaptive capacity	– Sustainability certification; producer permits; subsidies for efficient kilns	Woollen et al. 2016; Kiruki et al. 2017a
Land degradation and desertification	Arid, semi-arid and sub-humid regions	– Farmers – Pastoralists – Biodiversity	– Food insecurity – Drought – Migration – Loss of agro and wild biodiversity	– Restoration of ecosystems and management of invasive species – Climate smart agriculture and livestock management – Managing economic impacts of global and local drivers – Changes in relief and rehabilitation policies – Land degradation neutrality	Fleskens, Luuk, Stringer 2014; Lambin et al. 2001; Cowie et al. 2018a; Few and Tebboth 2018; Sandstrom and Juhola 2017
Loss of carbon sinks	Widespread across biomes and regions	– Tropical forests – Boreal soils	– Feedback to global and regional climate change	– Conservation prioritisation of tropical forests – Afforestation	Barnett et al. 2005; Tribbia and Moser 2008
Permafrost destabilisation	Arctic and Sub-Arctic regions	– Soils – Indigenous communities – Biodiversity	– Enhanced GHG emissions	– Enhanced carbon uptake from novel ecosystem after thaw – Adapt to emerging wetlands	Schuur et al. 2015
Stranded assets	– Economies transitioning to low-carbon pathways – Oil economies – Coastal regions facing inundation	– Coal-based power – Oilrefineries – Plastic industry – Large dams – Coastal infrastructure	– Disruption of regional economies and conflict – Unemployment – Pushback against renewable energy – Migration	– Insurance and tax cuts – Long-term power purchase agreements – Economic and technical support for transitioning economies – transforming oil wealth into renewable energy leadership – Redevelopment using adaptation – OPEC investment in information sharing for transition	Farfan and Breyer 2017; Ansar et al. 2013; Van de Graaf 2017; Trieb et al. 2011

7.3 Consequences of climate – land change for human well-being and sustainable development

To further explore what is at stake for human systems, this section assesses literature about potential consequences of climate and land change for human well-being and ecosystems upon which humans depend. Risks described in Section 7.2 have significant social, spiritual, and economic ramifications for societies across the world and this section explores potential implications of the risks outlined above to food security, livelihood systems, migration, ecosystems, species, infectious disease, and communities and infrastructure. Because food and livelihood systems are deeply tied to one another, combinations of climate and land change could pose higher present risks to humans and ecosystems than examination of individual elements alone might suggest.

7.3.1 What is at stake for food security?

This section examines risks to food security when access to food is jeopardised by yield shortfall and instability related to climate stressors. Past assessments of climate change impacts have sometimes assumed that, when grain and food yields in one area of the world are lower than expected, world trade can redistribute food adequately to ensure food security. There is *medium confidence* that severe and spatially extensive climatic stressors pose *high risk* to stability of and access to food for large numbers of people across the world.

The 2007–2008, and 2010–2011 droughts in several regions of the world resulted in crop yield decline that in turn led some governments to protect their domestic grain supplies rather than engaging in free trade to offset food shortfalls in other areas of the world. These responses cascaded and strongly affected regional and global food prices. Simultaneous crop yield impacts combined with trade impacts have proven to play a larger and more pervasive role in global food crises than previously thought (Sternberg 2012, 2017; Bellemare 2015; Chatzopoulos et al. 2019). There is *high confidence* that regional climate extremes already have significant negative domestic and international economic impacts (Chatzopoulos et al. 2019).

7.3.2 Risks to where and how people live: Livelihood systems and migration

There is *high confidence* that climate and land change interact with social, economic, political, and demographic factors that affect how well and where people live (Sudmeier-Rieux et al. 2017; Government Office for Science 2011; Laczko and Piguet 2014; Bohra-Mishra and Massey 2011; Raleigh et al. 2015; Warner and Afifi 2011; Hugo 2011; Warner et al. 2012). There is *high evidence and high agreement* that people move to manage risks and seek opportunities for their safety and livelihoods, recognising that people respond to climatic change and land-related factors in tandem with other variables (Hendrix and Salehyan 2012; Lashley and Warner 2015; van der Geest and Warner 2014; Roudier et al. 2014; Warner and Afifi 2014; McLeman 2013;

Kaenzig and Piguet 2014; Internal Displacement Monitoring Centre 2017; Warner 2018; Cohen and Bradley 2010; Thomas and Benjamin 2017). People move towards areas offering safety and livelihoods such as in rapidly growing settlements in coastal zones (Black et al. 2013; Challinor et al. 2017; Adger et al. 2013); burgeoning urban areas also face changing exposure to combinations of storm surges and sea level rise, coastal erosion and soil and water salinisation, and land subsidence (Geisler and Currens 2017; Maldonado et al. 2014; Bronen and Chapin 2013).

There is *medium confidence* that livelihood-related migration can accelerate in the short-to-medium term when weather-dependent livelihood systems deteriorate in relation to changes in precipitation, changes in ecosystems, and land degradation and desertification (Abid et al. 2016; Scheffran et al. 2012; Fussell et al. 2014; Bettini and Gioli 2016; Reyer et al. 2017; Warner and Afifi 2014; Handmer et al. 2012; Nawrotzki and Bakhtsiyarava 2017; Nawrotzki et al. 2016; Steffen et al. 2015; Black et al. 2013). Slow onset climate impacts and risks can exacerbate or otherwise interact with social conflict corresponding with movement at larger scales (see Section 7.2.3.2). Long-term deterioration in habitability of regions could trigger spatial population shifts (Denton et al. 2014).

There is *medium evidence* and *medium agreement* that climatic stressors can worsen the complex negative impacts of strife and conflict (Schleussner et al. 2016; Barnett and Palutikof 2014; Scheffran et al. 2012). Climate change and human mobility could be a factor that heightens tensions over scarce strategic resources, a further destabilising influence in fragile states experiencing socio-economic and political unrest (Carleton and Hsiang 2016a). Conflict and changes in weather patterns can worsen conditions for people working in rainfed agriculture or subsistence farming, interrupting production systems, degrading land and vegetation further (Papaioannou 2016; Adano and Daudi 2012). In recent decades, droughts and other climatic stressors have compounded livelihood pressures in areas already torn by strife (Tessler et al. 2015; Raleigh et al. 2015), such as in the Horn of Africa. Seizing of agricultural land by competing factions, preventing food distribution in times of shortage have, in this region and others, contributed to a triad of food insecurity, humanitarian need, and large movements of people (Theisen et al. 2011; Mohmmed et al. 2018; Ayeb-Karlsson et al. 2016; von Uexkull et al. 2016; Gleick 2014; Maystadt and Ecker 2014). People fleeing complex situations may return if peaceful conditions can be established. Climate change and development responses induced by climate change in countries and regions are likely to exacerbate tensions over water and land, and its impact on agriculture, fisheries, livestock and drinking water downstream. Shared pastoral landscapes used by disadvantaged or otherwise vulnerable communities are particularly impacted on by conflicts that are likely to become more severe under future climate change (Salehyan and Hendrix 2014; Hendrix and Salehyan 2012). Extreme events could considerably enhance these risks, in particular long-term drying trends (Kelley et al. 2015; Cutter et al. 2012a). There is *medium evidence* and *medium agreement* that governance is key in magnifying or moderating climate change impact and conflict (Bonatti et al. 2016).

There is *low evidence* and *medium agreement* that longer-term deterioration in the habitability of regions could trigger spatial population shifts (Seto 2011). Heat waves, rising sea levels that salinise and inundate coastal and low-lying aquifers and soils, desertification, loss of geologic sources of water such as glaciers and freshwater aquifers could affect many regions of the world and put life-sustaining ecosystems under pressure to support human populations (Flahaux and De Haas 2016; Chambwera et al. 2015; Tierney et al. 2015; Lilleør and Van den Broeck 2011).

7.3.3 Risks to humans from disrupted ecosystems and species

Risks of loss of biodiversity and ecosystem services (ES)

Climate change poses significant threat to species survival, and to maintaining biodiversity and ES. Climate change reduces the functionality, stability, and adaptability of ecosystems (Pecl et al. 2017). For example, drought affects cropland and forest productivity and reduces associated harvests (provisioning services). In additional, extreme changes in precipitation may reduce the capacity of forests to provide stability for groundwater (regulation and maintenance services). Prolonged periods of high temperature may cause widespread death of trees in tropical mountains, boreal and tundra forests, impacting on diverse ES, including aesthetic and cultural services (Verbyla 2011; Chapin et al. 2010; Krishnaswamy et al. 2014). According to the Millennium Ecosystem Assessment (2005), climate change is likely to become one of the most significant drivers of biodiversity loss by the end of the century.

There is *high confidence* that climate change already poses a moderate risk to biodiversity, and is projected to become a progressively widespread and high risk in the coming decades; loss of Arctic sea ice threatens biodiversity across an entire biome and beyond; the related pressure of ocean acidification, resulting from higher concentrations of carbon dioxide in the atmosphere, is also already being observed (UNEP 2009). There is ample evidence that climate change and land change negatively affects biodiversity across wide spatial scales. Although there is relatively *limited evidence* of current extinctions caused by climate change, studies suggest that climate change could surpass habitat destruction as the greatest global threat to biodiversity over the next several decades (Pereira et al. 2010). However, the multiplicity of approaches and the resulting variability in projections make it difficult to get a clear picture of the future of biodiversity under different scenarios of global climatic change (Pereira et al. 2010). Biodiversity is also severely impacted on by climate change induced land degradation and ecosystem transformation (Pecl et al. 2017). This may affect humans directly and indirectly through cascading impacts on ecosystem function and services (Millennium Assessment 2005). Climate change related human migration is likely to impact on biodiversity as people move into and contribute to land stress in biodiversity hotspots now and in the future; and as humans concurrently move into areas where biodiversity is also migrating to adapt to climate change (Oglethorpe et al. 2007).

Climate and land change increases risk to respiratory and infectious disease

In addition to risks related to nutrition articulated in Figure 7.1, human health can be affected by climate change through extreme heat and cold, changes in infectious diseases, extreme events, and land cover and land use (Hasegawa et al. 2016; Ryan et al. 2015; Terrazas et al. 2015; Kweka et al. 2016; Yamana et al. 2016). Evidence indicates that action to prevent the health impacts of climate change could provide substantial economic benefits (Martinez et al. 2015; Watts et al. 2015).

Climate change exacerbates air pollution with increasing UV and ozone concentration. It has negative impacts on human health and increases the mortality rate, especially in urban region (Silva et al. 2016, 2013; Lelieveld et al. 2013; Whitmee et al. 2015; Anenberg et al. 2010). In the Amazon, research shows that deforestation (both net loss and fragmentation) increases malaria, where vectors are expected to increase their home range (Alimi et al. 2015; Ren et al. 2016), confounded with multiple factors, such as social-economic conditions and immunity (Tucker Lima et al. 2017; Barros and Honório 2015). Deforestation has been shown to enhance the survival and development of major malaria vectors (Wang et al. 2016). The World Health Organization estimates 60,091 additional deaths for climate change induced malaria for the year 2030 and 32,695 for 2050 (World Health Organization 2014).

Human encroachment on animal habitat, in combination with the bushmeat trade in Central African countries, has contributed to the increased incidence of zoonotic (i.e., animal-derived) diseases in human populations, including the Ebola virus epidemic (Alexander et al. 2015a; Nkengasong and Onyebujoh 2018). The composition and density of zoonotic reservoir populations, such as rodents, is also influenced by land use and climate change (*high confidence*) (Young et al. 2017a). The bushmeat trade in many regions of central and west African forests (particularly in relation to chimpanzee and gorilla populations) elevates the risk of Ebola by increasing human–animal contact (Harrod 2015).

7.3.4 Risks to communities and infrastructure

There is *high confidence* that policies and institutions which accentuate vicious cycles of poverty and ill-health, land degradation and GHG emissions undermine stability and are barriers to achieving climate-resilient sustainable development. There is *high confidence* that change in climate and land pose high periodic and sustained risk to the very young, those living in poverty, and ageing populations. Older people are particularly exposed, due to more restricted access to resources, changes in physiology, and the decreased mobility resulting from age, which may limit adaptive capacity of individuals and populations as a whole (Filiberto et al. 2010).

Combinations of food insecurity, livelihood loss related to degrading soils and ecosystem change, or other factors that diminish the habitability of where people live, disrupt social fabric and are currently detected in most regions of the world (Carleton and

Hsiang 2016b) There is *high confidence* that coastal flooding and degradation already poses widespread and rising future risk to infrastructure value and stranded infrastructure, as well as livelihoods made possible by urban infrastructure (Radhakrishnan et al. 2017; Pathirana et al. 2018; Pathirana et al. 2018; Radhakrishnan et al. 2018; EEA 2016; Pelling and Wisner 2012; Oke et al. 2017; Parnell and Walawege 2011; Uzun and Cete 2004; Melvin et al. 2017).

There is *high evidence* and *high agreement* that climate and land change pose a high risk to communities. Interdependent infrastructure systems, including electric power and transportation, are highly vulnerable and interdependent (Below et al. 2012; Adger et al. 2013; Pathirana et al. 2018; Conway and Schipper 2011; Caney 2014; Chung Tiam Fook 2017). These systems are exposed to disruption from severe climate events such as weather-related power interruptions

lasting for hours to days (Panteli and Mancarella 2015). Increased magnitude and frequency of high winds, ice storms, hurricanes and heat waves have caused widespread damage to power infrastructure and also severe outages, affecting significant numbers of customers in urban and rural areas (Abi-Samra and Malcolm 2011).

Increasing populations, enhanced per capita water use, climate change, and allocations for water conservation are potential threats to adequate water availability. As climate change produces variations in rainfall, these challenges will intensify, evidenced by severe water shortages in recent years in Cape Town, Los Angeles, and Rio de Janeiro, among other places (Watts et al. 2018; Majumder 2015; Ashoori et al. 2015; Mini et al. 2015; Otto et al. 2015; Ranatunga et al. 2014; Ray and Shaw 2016; Gopakumar 2014) (Cross-Chapter Box 5 in Chapter 3).

Cross-Chapter Box 10 | Economic dimensions of climate change and land

Koko Warner (The United States of America), Aziz Elbehri (Morocco), Marta Guadalupe Rivera Ferre (Spain), Alisher Mirzabaev (Germany/Uzbekistan), Lindsay Stringer (United Kingdom), Anita Wreford (New Zealand)

Sustainable land management (SLM) makes strong social and economic sense. Early action in implementing SLM for climate change adaptation and mitigation provides distinct societal advantages. Understanding the full scope of what is at stake from climate change presents challenges because of inadequate accounting of the degree and scale at which climate change and land interactions impact society, and the importance society places on those impacts (Santos et al. 2016) (Sections 7.2.2, 5.3.1, 5.3.2 and 4.1). The consequences of inaction and delay bring significant risks, including irreversible change and loss in land ecosystem services (ES) – including food security – with potentially substantial economic damage to many countries in many regions of the world (high confidence).

This cross-chapter box brings together the salient economic concepts underpinning the assessments of SLM and mitigation options presented in this report. Four critical concepts are required to help assess the social and economic implications of land-based climate action:

i.   Value to society
ii. Damages from climate and land-induced interventions on land ecosystems
iii. Costs of action and inaction
iv. Decision-making under uncertainty

i.   Value to society
Healthy functioning land and ecosystems are essential for human health, food and livelihood security. Land derives its value to humans from being a finite resource and vital for life, providing important ES from water recycling, food, feed, fuel, biodiversity and carbon storage and sequestration.

Many of these ES may be difficult to estimate in monetary terms, including when they hold high symbolic value, linked to ancestral history, or traditional and indigenous knowledge systems (Boillat and Berkes 2013). Such incommensurable values of land are core to social cohesion – social norms and institutions, trust that enables all interactions, and sense of community.

ii. Damages from climate and land-induced interventions on land ecosystems
Values of many land-based ES and their potential loss under land–climate change interaction can be considerable: in 2011, the global value of ES was 125 trillion USD per year and the annual loss due to land-use change was between 4.3 and 20.2 trillion USD per year from 2007 (Costanza et al. 2014; Rockström et al. 2009). The annual costs of land degradation are estimated to be about 231 billion USD per year or about 0.41% of the global GDP of 56.49 trillion USD in 2007 (Nkonya et al. 2016) (Sections 4.4.1 and 4.4.2).

Cross-Chapter Box 10 (continued)

Studies show increasingly negative effects on GDP from damage and loss to land-based values and service as global mean temperatures increase, although the impact varies across regions (Kompas et al. 2018).

iii. Costs of action and inaction

Evidence suggests that the cost of inaction in mitigation and adaptation, and land use, exceeds the cost of interventions in both individual countries, regions, and worldwide (Nkonya et al. 2016). Continued inaction reduces the future policy option space, dampens economic growth and increases the challenges of mitigation as well as adaptation (Moore and Diaz 2015; Luderer et al. 2013). The cost of reducing emissions is estimated to be considerably less than the costs of the damages at all levels (Kainuma et al. 2013; Moran 2011; Sánchez and Maseda 2016).

The costs of adapting to climate impacts are also projected to be substantial, although evidence is limited (summarised in Chambwera et al. 2014a). Estimates range from 9 to 166 billion USD per year at various scales and types of adaptation, from capacity building to specific projects (Fankhauser 2017). There is insufficient literature about the costs of adaptation in the agriculture or land-based sectors (Wreford and Renwick 2012) due to lack of baselines, uncertainty around biological relationships and inherent uncertainty about anticipated avoided damage estimates, but economic appraisal of actions to maintain the functions of the natural environment and land sector generate positive net present values (Adaptation Sub-committee 2013).

Preventing land degradation from occurring is considered more cost-effective in the long term compared to the magnitude of resources required to restore already degraded land (Cowie et al. 2018a) (Section 3.6.1). Evidence from drylands shows that each US dollar invested in land restoration provides between 3 and 6 USD in social returns over a 30-year period, using a discount rate between 2.5 and 10% (Nkonya et al. 2016). SLM practices reverse or minimise economic losses of land degradation, estimated at between 6.3 and 10.6 trillion USD annually, (ELD Initiative 2015) more than five times the entire value of agriculture in the market economy (Costanza et al. 2014; Fischer et al. 2017; Sandifer et al. 2015; Dasgupta et al. 2013) (Section 3.7.5).

Across other areas such as food security, disaster mitigation and risk reduction, humanitarian response, and healthy diet (to address malnutrition as well as disease), early action generates economic benefits greater than costs (*high evidence, high agreement*) (Fankhauser 2017; Wilkinson et al. 2018; Venton 2018; Venton et al. 2012; Clarvis et al. 2015; Nugent et al. 2018; Watts et al. 2018; Bertram et al. 2018) (Sections 6.3 and 6.4).

iv. Decision-making under uncertainty

Given that significant uncertainty exists regarding the future impacts of climate change, effective decisions must be made under unavoidable uncertainty (Jones et al., 2014).

Approaches that allow for decision-making under uncertainty are continually evolving (Section 7.5). An emerging trend is towards new frameworks that will enable multiple decision-makers with multiple objectives to explore the trade-offs between potentially conflicting preferences to identify strategies that are robust to deep uncertainties (Singh et al. 2015; Driscoll et al. 2016; Araujo Enciso et al. 2016; Herman et al. 2014; Pérez et al. 2016; Girard et al. 2015; Haasnoot et al. 2018; Roelich and Giesekam 2019).

Valuation of benefits and damages and costing interventions: Measurement issues

Cost appraisal tools for climate adaptation are many and their suitability depends on the context (Section 7.5.2.2). Cost-benefit analysis (CBA) and cost-effectiveness analysis (CEA) are commonly applied, especially for current climate variability situations. However, these tools are not without criticism and their limitations have been observed in the literature (see Rogelj et al. 2018). In general, measuring costs and providing valuations are influenced by four conditions: measurement and valuation; the time dimension; externalities; and aggregate versus marginal costs.

Measurement and value issues

ES not traded in the market fall outside the formal or market-based valuation and so their value is either not accounted for or underestimated in both private and public decisions (Atkinson et al. 2018). Environmental valuation literature uses a range of techniques to assign monetary values to environmental outcomes where no market exists (Atkinson et al. 2018; Dallimer et al. 2018), but some values remain inestimable. For some indigenous cultures and peoples, land is not considered something that can be sold and bought, so economic valuations are not meaningful even as proxy approaches (Boillat and Berkes 2013; Kumpula et al. 2011; Pert et al. 2015; Xu et al. 2005).

While a rigorous CBA is broader than a purely financial tool and can capture non-market values where they exist, it can prioritise certain values over others (such as profit maximisation for owners, efficiency from the perspective of supply chain processes, and judgements about which parties bear the costs). Careful consideration must be given to whose perspectives are considered when undertaking a CBA and also to the limitations of these methods for policy interventions.

Time dimension (short versus long term) and the issue of discount rates

Economics uses a mechanism to convert future values to present day values known as discounting, or the pure rate of time preference. Discount rates are increasingly being chosen to reflect concerns about intergenerational equity, and some countries (e.g., the UK and France) apply a declining discount rate for long-term public projects. The choice of discount rate has important implications for policy evaluation (Anthoff, Tol, and Yohe, 2010; Arrow et al., 2014; Baral, Keenan, Sharma, Stork, and Kasel, 2014; Dasgupta et al., 2013; Lontzek, Cai, Judd, and Lenton, 2015; Sorokin et al., 2015; van den Bergh and Botzen, 2014) (*high evidence, high agreement*). Stern (2007), for example, used a much lower discount rate (giving almost equal weight to future generations) than the mainstream authors (e.g., Nordhaus (1941) and obtained much higher estimates of the damage of climate change).

Positive and negative externalities (consequences and impacts not accounted for in market economy),

All land use generates externalities (unaccounted for side effects of an activity). Examples include loss of ES (e.g., reduced pollinators; soil erosion, increased water pollution, nitrification, etc.). Positive externalities include sequestration of carbon dioxide (CO_2) and improved soil water filtration from afforestation. Externalities can also be social (e.g., displacement and migration) and economic (e.g., loss of productive land). In the context of climate change and land, the major externality is the agriculture, forestry and other land-use (AFOLU) sourced greenhouse gas (GHG) emissions. Examples of mechanisms to internalise externalities are discussed in 7.5.

Aggregate versus marginal costs

Costs of climate change are often referred to through the marginal measure of the social cost of carbon (SCC), which evaluate the total net damages of an extra metric tonne of CO_2 emissions due to the associated climate change (Nordhaus 2014). The SCC can be used to determine a carbon price, but SCC depends on discount rate assumptions and may neglect processes, including large losses of biodiversity, political instability, violent conflicts, large-scale migration flows, and the effects of climate change on the development of economies (Stern 2013; Pezzey 2019).

At the sectoral level, marginal abatement cost (MAC) curves are widely used for the assessment of costs related to CO_2 or GHG emissions reduction. MAC measures the cost of reducing one more GHG unit and MAC curves are either expert-based or model-derived and offer a range of approaches and assumptions on discount rates or available abatement technologies (Moran 2011).

7.3.4.1   Windows of opportunity

Windows of opportunity are important learning moments wherein an event or disturbance in relation to land, climate, and food security triggers responsive social, political, policy change (*medium agreement*). Policies play an important role in windows of opportunity and are important in relation to managing risks of desertification, soil degradation, food insecurity, and supporting response options for SLM (*high agreement*) (Kivimaa and Kern 2016; Gupta et al. 2013b; Cosens et al. 2017; Darnhofer 2014; Duru et al. 2015) (Chapter 6).

A wide range of events or disturbances may initiate windows of opportunity – ranging from climatic events and disasters, recognition of a state of land degradation, an ecological social or political crisis, and a triggered regulatory burden or opportunity. Recognition of a degraded system such as land degradation and desertification (Chapters 3 and 4) and associated ecosystem feedbacks, allows for strategies, response options and policies to address the degraded

state (Nyström et al. 2012). Climate related disasters (flood, droughts, etc.) and crisis may trigger latent local adaptive capacities leading to systemic equitable improvement (McSweeney and Coomes 2011), or novel and innovative recombining of sources of experience and knowledge, allowing navigation to transformative social ecological transitions (Folke et al. 2010). The occurrence of a series of punctuated crises such as floods or droughts, qualify as windows of opportunity when they enhance society's capacity to adapt over the long term (Pahl-Wostl et al. 2013). A disturbance from an ecological, social, or political crisis may be sufficient to trigger the emergence of new approaches to governance wherein there is a change in the rules of the social world such as informal agreements surrounding human activities or formal rules of public policies (Olsson et al. 2006; Biggs et al. 2017) (Section 7.6). A combination of socio-ecological changes may provide windows of opportunity for a socio-technical niche to be adopted on a greater scale, transforming practices towards SLM such as biodiversity-based agriculture (Darnhofer 2014; Duru et al. 2015).

Policy may also create windows of opportunity. A disturbance may cause inconvenience, including high costs of compliance with environmental regulations, thereby initiating a change of behaviour (Cosens et al. 2017). In a similar vein, multiple regulatory requirements existing at the time of a disturbance may result in emergent processes and novel solutions in order to correct for piecemeal regulatory compliance (Cosens et al. 2017). Lastly, windows of opportunity can be created by a policy mix or portfolio that provides for creative destruction of old social processes and thereby encourages new innovative solutions (Kivimaa et al. 2017b) (Section 7.4.8).

7.4   Policy instruments for land and climate

This section outlines policy responses to risk. It describes multi-level policy instruments (Section 7.4.1), policy instruments for social protection (Section 7.4.2), policies responding to hazard (Section 7.4.3), GHG fluxes (Section 7.4.4), desertification (Section 7.4.5), land degradation (Section 7.4.6), economic instruments (Section 7.4.7), enabling effective policy instruments through policy mixes (Section 7.4.8), and barriers to SLM and overcoming these barriers (Section 7.4.9).

Policy instruments are used to influence behaviour and effect a response – to do, not do, or continue to do certain things (Anderson 2010) – and they can be invoked at multiple levels (international, national, regional, and local) by multiple actors (Table 7.2). For efficiency, equity and effectiveness considerations, the appropriate choice of instrument for the context is critical and, across the topics addressed in this report, the instruments will vary considerably. A key consideration is whether the benefits of the action will generate private or public social net benefits. Pannell (2008) provides a widely-used framework for identifying the appropriate type of instrument depending on whether the actions encouraged by the instrument are private or public, and positive or negative. Positive incentives (such as financial or regulatory instruments) are appropriate where the public net benefits are highly positive and the private net benefits are close to zero. This is likely to be the case for GHG mitigation measures such as carbon pricing. Many other GHG mitigation measures (more effective water or fertiliser use, better agricultural practices, less food waste, agroforestry systems, better forest management) discussed in previous chapters may have substantial private as well as public benefit. Extension (knowledge provision) is recommended when public net benefits are *highly positive*, and private net benefits are *slightly positive* – again for some GHG mitigation measures, and for many adaptations, food security and SLM measures. Where the private net benefits are *slightly positive* but the public net benefits *highly negative*, negative incentives (such as regulations and prohibitions) are appropriate, (e.g., over-application of fertiliser).

While Pannell's (2008) framework is useful, it does not address considerations relating to the timescale of actions and their consequences, particularly in the long time-horizons involved under climate change: private benefits may accrue in the short term but become negative over time (Outka 2012) and some of the changes

necessary will require transformation of existing systems (Park et al. 2012; Hadarits et al. 2017) necessitating a more comprehensive suite of instruments. Furthermore, the framework applies to private land ownership, so where land is in different ownership structures, different mechanisms will be required. Indeed, land tenure is recognised as a factor in barriers to sustainable land management and an important governance consideration (Sections 7.4.9 and 7.6.4). A thorough analysis of the implications of policy instruments temporally, spatially and across other sectors and goals (e.g., climate versus development) is essential before implementation to avoid unintended consequences and achieve policy coherence (Section 7.4.8).

7.4.1   Multi-level policy instruments

Policy responses and planning in relation to land and climate interactions occur at and across multiple levels, involve multiple actors, and utilise multiple planning mechanisms (Urwin and Jordan 2008). Climate change is occurring on a global scale while the impacts of climate change vary from region to region and even within a region. Therefore, in addressing local climate impacts, local governments and communities are key players. Advancing governance of climate change across all levels of government and relevant stakeholders is crucial to avoid policy gaps between local action plans and national/sub-national policy frameworks (Corfee-Morlot et al. 2009).

This section of the chapter identifies policies by level that respond to land and climate problems and risks. As risk management in relation to land and climate occurs at multiple levels by multiple actors, and across multiple sectors in relation to hazards (as listed on Table 7.2), risk governance, or the consideration of the landscapes of risk arising from Chapters 2 to 6 is addressed in Sections 7.5 and 7.6. Categories of instruments include regulatory instruments (command and control measures), economic and market instruments (creating a market, sending price signals, or employing a market strategy), voluntary of persuasive instruments (persuading people to internalise behaviour), and managerial (arrangements including multiple actors in cooperatively administering a resource or overseeing an issue) (Gupta et al. 2013a; Hurlbert 2018b).

Given the complex spatial and temporal dynamics of risk, a comprehensive, portfolio of instruments and responses is required to comprehensively manage risk. Operationalising a portfolio response can mean layering, sequencing or integrating approaches. Layering means that, within a geographical area, households are able to benefit from multiple interventions simultaneously (e.g., those for family planning and those for livelihoods development). A sequencing approach starts with those interventions that address the initial binding constraints, and then adding further interventions later (e.g., the poorest households first receive grant-based support before then gaining access to appropriate microfinance or market-oriented initiatives). Integrated approaches involve cross-sectoral support within the framework of one programme (Scott et al. 2016; Tengberg and Valencia 2018) (Sections 7.4.8, 7.5.6 and 7.6.3).

7

Climate-related risk could be categorised by climate impacts such as flood, drought, cyclone, and so on (Christenson et al. 2014). Table 7.2 outlines instruments relating to impacts responding to the risk of climate change, food insecurity, land degradation and desertification, and hazards (flood, drought, forest fire), and GHG fluxes (climate mitigation).

7.4.2   Policies for food security and social protection

There is *medium evidence* and *high agreement* that a combination of structural and non-structural policies are required in averting and minimising as well as responding to land and climate change risk, including food and livelihood security. If disruptions to elements of food security are long-lasting, policies are needed to change practices.

If disruptions to food and livelihood systems are temporary, then policies aimed at stemming worsening human well-being and stabilising short-term income fluctuations in communities (such as

Table 7.2 | Policies/instruments that address multiple land-climate risks at different jurisdictional levels.

Scale	Policy/instrument	Food security	Land degradation and desertification	Sustainable land management (SLM)	Climate related extremes	GHG flux/climate change mitigation
Global/cross-border	Finance mechanisms (also national)	●	●	●	●	●
	Certification (also national)		●	●		●
	Standards (including risk standards) (also national)		●	●	●	●
	Market-based systems (also national)			●		●
	Payments for ecosystem services (also national)		●	●	●	●
	Disaster assistance (also national)				●	
National	Taxes	●		●		●
	Subsidies	●	●	●		●
	Direct income payments (with cross-compliance)	●	●	●		
	Border adjustments (e.g., tariffs)	●				●
	Grants	●	●	●	●	
	Bonds	●	●	●		●
	Forecast-based finance, targeted microfinance	●	●	●		●
	Insurance (various forms)	●			●	
	Hazard information and communication (also sub-national and local)	●			●	
	Drought preparedness plans (also sub-national and local)	●			●	
	Fire policy (suppression or prescribed fire management)			●	●	●
	Regulations	●	●	●	●	●
	Land ownership laws (reform of, if necessary, for secure land title, or access/control)	●	●	●		
	Protected area designation and management		●	●		
	Extension – including skill and community development for livelihood diversification (also sub-national and local)	●	●	●	●	●
Sub-national	Spatial and land-use planning	●	●		●	
	Watershed management	●	●			
Local	Land-use zoning, spatial planning and integrated land-use planning	●		●	●	
	Community-based awareness programmes	●	●	●	●	●

This table highlights policy and instruments addressing key themes identified in this chapter; a " ● " indicates the relevance of the policy or instrument to the corresponding theme.

increasing rural credit or providing social safety-net programmes) may be appropriate (Ward 2016).

7.4.2.1 Policies to ensure availability, access, utilisation and stability of food

Food security is affected by interactions between climatic factors (rising temperatures, changes in weather variability and extremes), changes in land use and land degradation, and Socio-economic Pathways and policy choices related to food systems (see Figures 7.1 and 7.2). As outlined in Chapter 5, key aspects of food security are food availability, access to food, utilisation of food, and stability of food systems.

While comprehensive reviews of policy are rare and additional data is needed (Adu et al. 2018), evidence indicates that the results of food security interventions vary widely due to differing values underlying the design of instruments. A large portfolio of measures is available to shape outcomes in these areas from the use of tariffs or subsidies, to payments for production practices (OECD 2018). In the past, efforts to increase food production through significant investment in agricultural research, including crop improvement, have benefited farmers by increasing yields and reducing losses, and have helped consumers by lowering food prices (Pingali 2012, 2015; Alston and Pardey 2014; Popp et al. 2013). Public spending on agriculture research and development (R&D) has been more effective at raising sustainable agriculture productivity than irrigation or fertiliser subsidies (OECD 2018). Yet, on average, between 2015 and 2017, governments spent only around 14% of total agricultural support on services, including physical and knowledge infrastructure, transport and information and communications technology.

In terms of increasing food availability and supply, producer support, including policies mandating subsidies or payments, have been used to boost production of certain commodities or protect ES. Incentives can distort markets and farm business decisions in both negative and positive ways. For example, the European Union promotes meat and dairy production through voluntary coupled direct payments. These do not yet internalise external damage to climate, health, and groundwater (Velthof et al. 2014; Bryngelsson et al. 2016). In most countries, producer support has been declining since the mid-1990s (OECD 2018). Yet new evidence indicates that a government policy supporting producer subsidy could encourage farmers to adopt new technologies and reduce GHG emissions in agriculture (*medium evidence, high agreement*). However, this will require large capital (Henderson 2018). Since a 1995 reform in its forest law, Costa Rica has effectively used a combination of fuel tax, water tax, loans and agreements with companies, to pay landowners for agroforestry, reforestation and sustainable forest management (Porras and Asquith 2018).

Inland capture fisheries and aquaculture are an integral part of nutrition security and livelihoods for large numbers of people globally (Welcomme et al. 2010; Hall et al. 2013; Tidwell and Allan 2001; Youn et al. 2014) and are increasingly vulnerable to climate change and competing land and water use (Allison et al. 2009; Youn et al. 2014). Future production may increase in some high-latitude regions (*low confidence*) but production is likely to decline in low-latitude regions under future warming (*high confidence*) (Brander and Keith 2015; Brander 2007). However over-exploitation and degradation of rivers has resulted in a decreasing trend in the contribution of capture fisheries to protein security in comparison to managed aquaculture (Welcomme et al. 2010). Aquaculture, however, competes for land and water resources with many negative ecological and environmental impacts (Verdegem and Bosma 2009; Tidwell and Allan 2001). Inland capture fisheries are undervalued in national and regional food security, ES and economy, are data deficient and are neglected in terms of supportive policies at national levels, and absent in SDGs (Cooke et al. 2016; Hall et al. 2013; Lynch et al. 2016). Revival of sustainable capture fisheries and converting aquaculture to environmentally less-damaging management regimes, is likely to succeed with the following measures: investment in recognition of their importance, improved valuation and assessment, secure tenure and adoption of social, ecological and technological guidelines, upstream-downstream river basin cooperation, and maintenance of ecological flow regimes in rivers (Youn et al. 2014; Mostert et al. 2007; Ziv et al. 2012; Hurlbert and Gupta 2016; Poff et al. 2003; Thomas 1996; FAO 2015a).

Extension services, and policies supporting agricultural extension systems, are also critical. Smallholder farmer-dominated agriculture is currently the backbone of global food security in the developing world. Without education and incentives to manage land and forest resources in a manner that allows regeneration of both the soils and wood stocks, smallholder farmers tend to generate income through inappropriate land management practices, engage in agricultural production on unsuitable land and use fertile soils, timber and firewood for brick production and construction. Also, they engage in charcoal production (deforestation) as a coping mechanism (increasing income) against food deficiency (Munthali and Murayama 2013). Through extension services, governments can play a proactive role in providing information on climate and market risks, animal and plant health. Farmers with greater access to extension training retain more crop residues for mulch on their fields (Jaleta et al. 2015, 2013; Baudron et al. 2014).

Food security cannot be achieved by increasing food availability alone. Policy instruments, which increase access to food at the household level, include safety-net programming and universal basic income. The graduation approach, developed and tested over the past decade using randomised control trials in six countries, has lasting positive impacts on income, as well as food and nutrition security (Banerjee et al. 2015; Raza and Poel 2016) (*robust evidence, high agreement*). The graduation approach layers and integrates a series of interventions designed to help the poorest: consumption support in the form of cash or food assistance, transfer of an income-generating asset (such as a livestock) and training on how to maintain the asset, assistance with savings and coaching or mentoring over a period of time to reinforce learning and provide support. Due to its success, the graduation approach is now being scaled up, and is now used in more than 38 countries and included by an increasing number of governments in social safety-net programmes (Hashemi and de Montesquiou 2011).

At the national and global levels, food prices and trade policies impact on access to food. Fiscal policies, such as taxation, subsidies, or tariffs, can be used to regulate production and consumption of certain foods and can affect environmental outcomes. In Denmark, a tax on saturated fat content of food adopted to encourage healthy eating habits accounted for 0.14% of total tax revenues between 2011 and 2012 (Sassi et al. 2018). A global tax on GHG emissions, for example, has large mitigation potential and will generate tax revenues, but may also result in large reductions in agricultural production (Henderson 2018). Consumer-level taxes on GHG-intensive food may be applied to address competitiveness issues between different countries, if some countries use taxes while others do not. However, increases in prices might impose disproportionate financial burdens on low-income households, and may not be publicly acceptable. A study examining the relationship between food prices and social unrest found that, between 1990 and 2011, whereas food price stability has not been associated with increases in social unrest (Bellemare 2015).

Interventions that allow people to maximise their productive potential while protecting the ES may not ensure food security in all contexts. Some household land holdings are so small that self-sufficiency is not possible (Venton 2018). Value chain development has, in the past, increased farm income but delivered fewer benefits to vulnerable consumers (Bodnár et al. 2011). Ultimately, a mix of production activities and consumption support is needed. Consumption support can be used to help achieve the second important element of food security – access to food.

Agricultural technology transfer can help optimise food and nutrition security (Section 7.4.4.3). Policies that affect agricultural innovation span sectors and include 'macro-economic policy-settings; institutional governance; environmental standards; investment, land, labor and education policies; and incentives for investment, such as a predictable regulatory environment and robust intellectual property rights'.

The scientific community can partner across sectors and industries for better data sharing, integration, and improved modelling and analytical capacities (Janetos et al. 2017; Lunt et al. 2016). To better predict, respond to, and prepare for concurrent agricultural failures, and gain a more systematic assessment of exposure to agricultural climate risk, large data gaps need to be filled, as well as gaps in empirical foundation and analytical capabilities (Janetos et al. 2017; Lunt et al. 2016). Data required include global historical datasets, many of which are unreliable, inaccessible, or not available (Maynard 2015; Lunt et al. 2016). Participation in co-design for scenario planning can build social and human capital while improving understanding of food system risks and creating innovative ways for collectively planning for a more equitable and resilient food system (Himanen et al. 2016; Meijer et al. 2015; Van Rijn et al. 2012).

Demand management for food, including promoting healthy diets, reducing food loss and waste, is covered in Chapter 5. There is a gap in knowledge regarding what policies and instruments support demand management. There is *robust evidence* and *robust agreement* that changes in household wealth and parents' education can drive changes in diet and improvements in nutrition (Headey et al. 2017). Bangladesh has managed to sustain a rapid reduction in the rate of child undernutrition for at least two decades. Rapid wealth accumulation and large gains in parental education are the two largest drivers of change (Headey et al. 2017). Educating consumers, and providing affordable alternatives, will be critical to changing unsustainable food-use habits relevant to climate change.

7.4.2.2   Policies to secure social protection

There is *medium evidence* and *high agreement* from all regions of the world that safety nets and social protection schemes can provide stability which prevents and reduces abject poverty (Barrientos 2011; Hossain 2018; Cook and Pincus 2015; Huang and Yang 2017; Slater 2011; Sparrow et al. 2013; Rodriguez-Takeuchi and Imai 2013; Bamberg et al. 2018) in the face of climatic stressors and land change (Davies et al. 2013; Cutter et al. 2012b; Pelling 2011; Ensor 2011).

The World Bank estimates that, globally, social safety net transfers have reduced the absolute poverty gap by 45% and the relative poverty gap by 16% (World Bank 2018). Adaptive social protection builds household capacity to deal with shocks as well as the capacity of social safety nets to respond to shocks. For low-income communities reliant on land and climate for their livelihoods and well-being, social protection provides a way for vulnerable groups to manage weather and climatic variability and deteriorating land conditions to household income and assets (*robust evidence, high agreement*) (Baulch et al. 2006; Barrientos 2011; Harris 2013; Fiszbein et al. 2014; Kiendrebeogo et al. 2017; Kabeer et al. 2010; FAO 2015b; Warner et al. 2018; World Bank 2018).

A lifecycle approach to social protection is one approach, which some countries (such as Bangladesh) are using when developing national social protection policies. These policies acknowledge that households face risks across the lifecycle that they need to be protected from. If shocks are persistent, or occur numerous times, then policies can address concerns of a more structural nature (Glauben et al. 2012). Barrett (2005), for example, distinguishes between the role of safety nets (which include programmes such as emergency feeding programmes, crop or unemployment insurance, disaster assistance, etc.) and cargo nets (which include land reforms, targeted microfinance, targeted school food programmes, etc.). While the former prevents non-poor and transient poor from becoming chronically poor, the latter is meant to lift people out of poverty by changing societal or institutional structures. The graduation approach has adopted such systematic thinking with successful results (Banerjee et al. 2015).

Social protection systems can provide buffers against shocks through vertical or horizontal expansion, 'piggybacking' on pre-established programmes, aligning social protection and humanitarian systems or refocusing existing resources (Wilkinson et al. 2018; O'Brien et al. 2018; Jones and Presler-Marshall 2015). There is increasing evidence that forecast-based financing, linked to a social protection, can be used to enable anticipatory actions based on forecast triggers, and guarantee funding ahead of a shock (Jjemba et al. 2018). Accordingly, scaling up social protection based on an early warning could enhance timeliness, predictability and adequacy of social protection benefits (Kuriakose et al. 2012; Costella et al. 2017a; Wilkinson et al. 2018; O'Brien et al. 2018).

Countries at high risk of natural disasters often have lower safety-net coverage percent (World Bank 2018), and there is *medium evidence* and *medium agreement* that those countries with few financial and other buffers have lower economic and social performance (Cutter et al. 2012b; Outreville 2011a). Social protection systems have also been seen as an unaffordable commitment of public budget in many developing and low-income countries (Harris 2013). National systems may be disjointed and piecemeal, and subject to cultural acceptance and competing political ideologies (Niño-Zarazúa et al. 2012). For example, Liberia and Madagascar each have five different public works programmes, each with different donor organisations and different implementing agencies (Monchuk 2014). These implementation shortcomings mean that positive effects of social protection systems might not be robust enough to shield recipients completely against the impacts of severe shocks or from long-term losses and damages from climate change (*limited evidence*, *high agreement*) (Davies et al. 2009; Umukoro 2013; Béné et al. 2012; Ellis et al. 2009).

There is increasing support for establishment of public-private safety nets to address climate-related shocks, which are augmented by proactive preventative (adaptation) measures and related risk transfer instruments that are affordable to the poor (Kousky et al. 2018b). Studies suggest that the adaptive capacity of communities has improved with regard to climate variability, like drought, when ex-ante tools, including insurance, have been employed holistically; providing insurance in combination with early warning and institutional and policy approaches reduces livelihood and food insecurity as well as strengthens social structures (Shiferaw et al. 2014; Lotze-Campen and Popp 2012). Bundling insurance with early warning and seasonal forecasting can reduce the cost of insurance premiums (Daron and Stainforth 2014). The regional risk insurance scheme, African Risk Capacity, has the potential to significantly reduce the cost of insurance premiums (Siebert 2016) while bolstering contingency planning against food insecurity.

Work-for-insurance programmes applied in the context of social protection have been shown to improve livelihood and food security in Ethiopia (Berhane 2014; Mohmmed et al. 2018) and Pakistan. The R4 Rural Resilience Initiative in Ethiopia is a widely cited example of a programme that serves the most vulnerable and includes aspects of resource management, and access by the poor to financial services, including insurance and savings (Linnerooth-Bayer et al. 2018). Weather index insurance (such as index-based crop insurance) is being presented to low-income farmers and pastoralists in developing countries (e.g., Ethiopia, India, Kazakhstan, South Asia) to complement informal risk sharing, reducing the risk of lost revenue associated with variations in crop yield, and provide an alternative to classic insurance (Bogale 2015a; Conradt et al. 2015; Dercon et al. 2014; Greatrex et al. 2015; McIntosh et al. 2013). The ability of insurance to contribute to adaptive capacity depends on the overall risk management and livelihood context of households – studies find that agriculturalists and foresters working on rainfed farms/land with

more years of education and credit but limited off-farm income are more willing to pay for insurance than households who have access to remittances (such as from family members who have migrated) (Bogale 2015a; Gan et al. 2014; Hewitt et al. 2017; Nischalke 2015). In Europe, modelling suggests that insurance incentives, such as vouchers, would be less expensive than total incentivised damage reduction and may reduce residential flood risk in Germany by 12% in 2016 and 24% by 2040 (Hudson et al. 2016).

7.4.3 Policies responding to climate-related extremes

7.4.3.1 Risk management instruments

Risk management addressing climate change has broadened to include mitigation, adaptation and disaster preparedness in a process using instruments facilitating contingency and cross-sectoral planning (Hurlimann and March 2012; Oels 2013), social community planning, and strategic, long-term planning (Serrao-Neumann et al. 2015a). A comprehensive consideration integrates principles from informal support mechanisms to enhance formal social protection programming (Mobarak and Rosenzweig 2013; Stavropoulou et al. 2017) such that the social safety net, disaster risk management, and climate change adaptation are all considered to enhance livelihoods of the chronic poor (see char dwellers and recurrent floods in Jamuna and Brahmaputra basins of Bangladesh Awal 2013) (Section 7.4.7). Iterative risk management is an ongoing process of assessment, action, reassessment and response (Mochizuki et al. 2015) (Sections 7.5.2 and 7.4.7.2).

Important elements of risk planning include education, and creation of hazard and risk maps. Important elements of predicting include hydrological and meteorological monitoring to forecast weather, seasonal climate forecasts, aridity, flood and extreme weather. Effective responding requires robust communication systems that pass on information to enable response (Cools et al. 2016).

Gauging the effectiveness of policy instruments is challenging. Timescales may influence outcomes. To evaluate effectiveness researchers, programme managers and communities strive to develop consistency, comparability, comprehensiveness and coherence in their tracking. In other words, practitioners utilise a consistent and operational conceptualisation of adaptation; focus on comparable units of analysis; develop comprehensive datasets on adaptation action; and are coherent with an understanding of what constitutes real adaptation (Ford and Berrang-Ford 2016). Increasing the use of systematic reviews or randomised evaluations may also be helpful (Alverson and Zommers 2018).

Many risk management policy instruments are referred to by the International Organization of Standardization which lists risk management principles, guidelines, and frameworks for explaining the elements of an effective risk management programme (ISO 2009). The standard provides practical risk management instruments and makes a business case for risk management investments (McClean et al. 2010). Insurance addresses impacts associated with extreme weather events (storms, floods, droughts, temperature extremes), but

7

it can provide disincentives for reducing disaster risk at the local level through the transfer of risk spatially to other places or temporally to the future (Cutter et al. 2012b) and uptake is unequally distributed across regions and hazards (Lal et al. 2012). Insurance instruments (Sections 7.4.2 and 7.4.6) can take many forms (traditional indemnity based, market-based crop insurance, property insurance), and some are linked to livelihoods sensitive to weather as well as food security (linked to social safety-net programmes) and ecosystems (coral reefs and mangroves). Insurance instruments can also provide a framework for risk signals to adaptation planning and implementation and facilitate financial buffering when climate impacts exceed current capabilities delivered through both public and private finance (Bogale 2015b; Greatrex et al. 2015; Surminski et al. 2016). A holistic consideration of all instruments responding to extreme impacts of climate change (drought, flood, etc.) is required when assessing if policy instruments are promoting livelihood capitals and contributing to the resilience of people and communities (Hurlbert 2018b). This holistic consideration of policy instruments leads to a consideration of risk governance (Section 7.6).

Early warning systems are critical policy instruments for protecting lives and property, adapting to climate change, and effecting adaptive climate risk management (*high confidence*) (Selvaraju 2011; Cools et al. 2016; Travis 2013; Henriksen et al. 2018; Seng 2013; Kanta Kafle 2017; Garcia and Fearnley 2012). Early warning systems exist at different levels and for different purposes, including the Food and Agriculture Organization of the United Nations' Global Information and Early Warning System on Food and Agriculture (GIEWS), United States Agency for International Development (USAID) Famine Early Warning System Network (FEWS-NET), national and local extreme weather, species extinction, community-based flood and landslide, and informal pastoral drought early warning systems (Kanta Kafle 2017). Medium-term warning systems can identify areas of concern, hotspots of vulnerabilities and sensitivities, or critical zones of land degradation (areas of concern) (see Chapter 6) critical to reduce risks over five to 10 years (Selvaraju 2012). Early warning systems for dangerous climate shifts are emerging, with considerations of rate of onset, intensity, spatial distribution and predictability. Growing research in the area is considering positive and negative lessons learned from existing hazard early warning systems, including lead time and warning response (Travis 2013).

For effectiveness, communication methods are best adapted to local circumstances, religious and cultural-based structures and norms, information technology, and local institutional capacity (Cools et al. 2016; Seng 2013). Considerations of governance or the actors and architecture within the socio-ecological system, is an important feature of successful early warning system development (Seng 2013). Effective early warning systems consider the critical links between hazard monitoring, risk assessment, forecasting tools, warning and dissemination (Garcia and Fearnley 2012). These effective systems incorporate local context by defining accountability, responsibility, acknowledging the importance of risk perceptions and trust for an effective response to warnings. Although increasing levels and standardisation nationally and globally is important, revising these systems through participatory approaches cognisant of the tension

with technocratic approaches improves success (Cools et al. 2016; Henriksen et al. 2018; Garcia and Fearnley 2012).

7.4.3.2   Drought-related risk minimising instruments

A more detailed review of drought instruments, and three broad policy approaches for responding to drought, is provided in Cross-Chapter Box 5 in Chapter 3. Three broad approaches include: (i) early warning systems and response to the disaster of drought (through instruments such as disaster assistance or crop insurance); (ii) disaster response ex-ante preparation (through drought preparedness plans); and (iii) drought risk mitigation (proactive polices to improve water-use efficiency, make adjustments to water allocation, funds or loans to build technology such as dugouts or improved soil management practices).

Drought plans are still predominantly reactive crisis management plans rather than proactive risk management and reduction plans. Reactive crisis management plans treat only the symptoms and are inefficient drought management practices. More efficient drought preparedness instruments are those that address the underlying vulnerability associated with the impacts of drought, thereby building agricultural producer adaptive capacity and resilience (*high confidence*) (Cross-Chapter Box 5 in Chapter 3).

7.4.3.3   Fire-related risk minimising instruments

There is *robust evidence* and *high agreement* that fire strategies need to be tailored to site-specific conditions in an adaptive application that is assessed and reassessed over time (Dellasala et al. 2004; Rocca et al. 2014). Strategies for fire management include fire suppression, prescribed fire and mechanical treatments (such as thinning the canopy), and allowing wildfire with little or no active management (Rocca et al. 2014). Fire suppression can degrade the effectiveness of forest fire management in the long run (Collins et al. 2013).

Different forest types have different fire regimes and require different fire management policies (Dellasala et al. 2004). For instance, Cerrado, a fire dependent savannah, utilises a different fire management policy and fire suppression policy (Durigan and Ratter 2016). The choice of strategy depends on local considerations, including land ownership patterns, dynamics of local meteorology, budgets, logistics, federal and local policies, tolerance for risk and landscape contexts. In addition, there are trade-offs among the management alternatives and often no single management strategy will simultaneously optimise ES, including water quality and quantity, carbon sequestration, or run-off erosion prevention (Rocca et al. 2014).

7.4.3.4   Flood-related risk minimising instruments

Flood risk management consists of command and control measures, including spatial planning and engineered flood defences (Filatova 2014), financial incentive instruments issued by regional or national governments to facilitate cooperative approaches through local planning, enhancing community understanding and political support for safe development patterns and building standards, and regulations requiring local government participation and support for local flood planning (Burby and May 2009). However, Filatova (2014) found that if autonomous adaptation is downplayed, people are more likely to make land-use choices that collectively lead to increased flood risks and leave costs to governments. Taxes and subsidies that do not encourage (and even counter) perverse behaviour (such as rebuilding in flood zones) are important instruments mitigating this cost to government. Flood insurance has been found to be maladaptive as it encourages rebuilding in flood zones (O'Hare et al. 2016) and government flood disaster assistance negatively impacts on average insurance coverage the following year (Kousky et al. 2018a). Modifications to flood insurance can counter perverse behaviour. One example is the provision of discounts on flood insurance for localities that undertake one of 18 flood mitigation activities, including structural mitigation (constructing dykes, dams, flood control reservoirs), and non-structural initiatives such as point source control and watershed management efforts, education and maintenance of flood-related databases (Zahran et al. 2010). Flood insurance that provides incentives for flood mitigation, marketable permits and transferable development rights (see Case study: Flood and food security in Section 7.6) instruments can provide price signals to stimulate autonomous adaptation, countering barriers of path dependency, and the time lag between private investment decisions and consequences (Filatova 2014). To build adaptive capacity, consideration needs to be made of policy instruments responding to flood, including flood zone mapping, land-use planning, flood zone building restrictions, business and crop insurance, disaster assistance payments, preventative instruments, (including environmental farm planning, e.g., soil and water management (see Chapter 6)), farm infrastructure projects, and recovery from debilitating flood losses ultimately through bankruptcy (Hurlbert 2018a). Non-structural measures have been found to advance sustainable development as they are more reversible, commonly acceptable and environmentally friendly (Kundzewicz 2002).

7.4.4   Policies responding to greenhouse gas (GHG) fluxes

7.4.4.1   GHG fluxes and climate change mitigation

Pathways reflecting current nationally stated mitigation ambitions as submitted under the Paris Agreement would not limit global warming to 1.5°C with no or limited overshoot, but instead result in a global warming of about 3°C by 2100 with warming continuing afterward (IPCC 2018d). Reversing warming after an overshoot of 0.2°C or higher during this century would require deployment of CDR at rates and volumes that might not be achievable given considerable implementation challenges (IPCC 2018d). This gap (Höhne et al. 2017; Rogelj et al. 2016) creates a significant risk of global warming impacting on land degradation, desertification, and food security (IPCC 2018d) (Section 7.2). Action can be taken by 2030 adopting already known cost-effective technology (United Nations Environment Programme 2017), improving the finance, capacity building, and technology transfer mechanisms of the United Nations Framework Convention on Climate Change (UNFCCC), improving food security (listed by 73 nations in their nationally determined contributions (NDCs)) and nutritional security (listed by 25 nations)

(Richards et al. 2015). UNFCCC Decision 1. CP21 reaffirmed the UNFCCC target that 'developed country parties provide USD 100 billion annually by 2020 for climate action in developing countries' (Rajamani 2011) and a new collective quantified goal above this floor is to be set, taking into account the needs and priorities of developing countries (Fridahl and Linnér 2016).

Mitigation policy instruments to address this shortfall include financing mechanisms, carbon pricing, cap and trade or emissions trading, and technology transfer. While climate change is a global commons problem containing free-riding issues cost-effective international policies that ensure that countries get the most environmental benefit out of mitigation investments promote an international climate policy regime (Nordhaus 1999; Aldy and Stavins 2012). Carbon pricing instruments may provide an entry point for inclusion of appropriate agricultural carbon instruments. Models of cost-efficient distribution of mitigation across regions and sectors typically employ a global uniform carbon price, but such treatment in the agricultural sector may impact on food security (Section 7.4.4.4).

One policy initiative to advance climate mitigation policy coherence in this section is the phase out of subsidies for fossil fuel production (see also Section 7.4.8). The G20 agreed in 2009, and the G7 agreed in 2016, to phase out these subsidies by 2025. Subsidies include lower tax rates or exemptions and rebates of taxes on fuels used by particular consumers (diesel fuel used by farming, fishing, etc.), types of fuel, or how fuels are used. The OECD estimates the overall value of these subsides to be 160–200 billion USD annually between 2010 and 2014 (OECD 2015). The phase-out of fossil fuel subsidies has important economic, environmental and social benefits. Coady et al. (2017) estimate the economic and environmental benefits of reforming fossil fuel subsidies could be valued worldwide at 4.9 trillion USD in 2013, and 5.3 trillion USD in 2015. Eliminating subsidies could have reduced emissions by 21%, raised 4% of global GDP as revenue (in 2013), and improved social welfare (Coady et al. 2017).

Legal instruments addressing perceived deficiencies in climate change mitigation include human rights and liability. Developments in attribution science are improving the ability to detect human influence on extreme weather. Marjanac et al. (2017) argue that this broadens the legal duty of government, business and others to manage foreseeable harms, and may lead to more climate change litigation (Marjanac et al. 2017). Peel and Osofsky (2017) argue that courts are becoming increasingly receptive to employ human rights claims in climate change lawsuits (Peel and Osofsky 2017); citizen suits in domestic courts are not a universal phenomenon and, even if unsuccessful, Estrin (2016) concludes they are important in underlining the high level of public concern.

7.4.4.2   Mitigation instruments

Similar instruments for mitigation could be applied to the land sector as in other sectors, including: market-based measures such as taxes and cap and trade systems; standards and regulations; subsidies and tax credits; information instruments and management tools; R&D investment; and voluntary compliance programmes. However, few regions have implemented agricultural mitigation instruments

(Cooper et al. 2013). Existing regimes focus on subsidies, grants and incentives, and voluntary offset programmes.

7.4.4.3   Market-based instruments

Although carbon pricing is recognised to be an important cost-effective instrument in a portfolio of climate policies (*high evidence, high agreement*) (Aldy et al. 2010), as yet, no country is exposing their agricultural sector emissions to carbon pricing in any comprehensive way. A carbon tax, fuel tax, and carbon markets (cap and trade system or Emissions Trading System (ETS), or baseline and credit schemes, and voluntary markets) are predominant policy instruments that implement carbon pricing. The advantage of carbon pricing is environmental effectiveness at relatively low cost (*high evidence, high agreement*) (Baranzini et al. 2017; Fawcett et al. 2014). Furthermore, carbon pricing could be used to raise revenue to reinvest in public spending, either to help certain sectors transition to lower carbon systems, or to invest in public spending unrelated to climate change. Both of these options may make climate policies more attractive and enhance overall welfare (Siegmeier et al. 2018), but there is, as yet, no evidence of the effectiveness of emissions pricing in agriculture (Grosjean et al. 2018). There is, however, a clear need for progress in this area as, without effective carbon pricing, the mitigation potential identified in chapters 5 and 6 of this report will not be realised (*high evidence, high agreement*) (Boyce 2018).

The price may be set at the social cost of carbon (the incremental impact of emitting an additional tonne of CO_2, or the benefit of slightly reducing emissions), but estimates of the SCC vary widely and are contested (*high evidence, high agreement*) (Pezzey 2019). An alternative to the SCC includes a pathways approach that sets an emissions target and estimates the carbon prices required to achieve this at the lowest possible cost (Pezzey 2019). Theoretically, higher costs throughout the entire economy result in reduction of carbon intensity, as consumers and producers adjust their decisions in relation to prices corrected to reflect the climate externality (Baranzini et al. 2017).

Both carbon taxes and cap and trade systems can reduce emissions, but cap and trade systems are generally more cost effective (*medium evidence, high agreement*) (Haites 2018a). In both cases, the design of the system is critical to its effectiveness at reducing emissions (*high evidence, high agreement*) (Bruvoll and Larsen 2004; (Lin and Li 2011). The trading system allows the achievement of emission reductions in the most cost-effective manner possible and results in a market and price on emissions that create incentives for the reduction of carbon pollution. The way allowances are allocated in a cap and trade system is critical to its effectiveness and equity. Free allocations can be provided to trade-exposed sectors, such as agriculture, either through historic or output-based allocations, the choice of which has important implications (Quirion 2009). Output-based allocations may be most suitable for agriculture, also minimising leakage risk (see below in this section) (Grosjean et al. 2018; Quirion 2009). There is *medium evidence* and *high agreement* that properly designed, a cap and trade system can be a powerful policy instrument (Wagner 2013) and may collect more rents than a variable carbon tax (Siegmeier et al. 2018; Schmalensee and Stavins 2017).

In the land sector, carbon markets are challenging to implement. Although several countries and regions have an ETS in place (for example, the EU, Switzerland, the Republic of Korea, Quebec in Canada, California in the USA (Narassimhan et al. 2018)), none have included non-CO_2 (methane and nitrous oxide) emissions from agriculture. New Zealand is the only country currently considering ways to incorporate agriculture into its ETS (see Case study: Including agriculture in the New Zealand Emissions Trading Scheme).

Three main reasons explain the lack of implementation to date:

1. The large number of heterogeneous buyers and sellers, combined with the difficulties of monitoring, reporting and verification (MRV) of emissions from biological systems introduce potentially high levels of complexity (and transaction costs). Effective policies therefore depend on advanced MRV systems which are lacking in many (particularly developing) countries (Wilkes et al. 2017). This is discussed in more detail in the case study on the New Zealand Emissions Trading Scheme.

2. Adverse distributional consequences (Grosjean et al. 2018) (*medium evidence, high agreement*). Distributional issues depend, in part, on the extent that policy costs can be passed on to consumers, and there is *medium evidence* and *medium agreement* that social equity can be increased through a combination of non-market and market-based instruments (Haites 2018b).

3. Regulation, market-based or otherwise, adopted in only one jurisdiction and not elsewhere may result in 'leakage' or reduced effectiveness – where production relocates to weaker regulated regions, potentially reducing the overall environmental benefit. Although modelling studies indicate the possibility of leakage following unilateral agricultural mitigation policy implementation (e.g., Fellmann et al. 2018), there is no empirical evidence from the agricultural sector yet available. Analysis from other sectors shows an overestimation of the extent of carbon leakage in modelling studies conducted before policy implementation compared to evidence after the policy was implemented (Branger and Quirion 2014). Options to avoid leakage include: border adjustments (emissions in non-regulated imports are taxed at the border, and payments made on products exported to non-regulated countries are rebated); differential pricing for trade-exposed products; and output-based allocation (which effectively works as a subsidy for trade-exposed products). Modelling shows that border adjustments are the most effective at reducing leakage, but may exacerbate regional inequality (Böhringer et al. 2012) and through their trade-distorting nature may contravene World Trade Organization rules. The opportunity for leakage would be significantly reduced, ideally through multilateral commitments (Fellmann et al. 2018) (*medium evidence, high agreement*) but could also be reduced through regional or bi-lateral commitments within trade agreements.

Case study | Including agriculture in the New Zealand Emissions Trading Scheme (ETS)

New Zealand has a high proportion of agricultural emissions at 49% (Ministry of the Environment 2018) – the next-highest developed country agricultural emitter is Ireland at around 32% (EPA 2018) – and is considering incorporating agricultural non-CO_2 gases into the existing national ETS. In the original design of the ETS in 2008, agriculture was intended to be included from 2013, but successive governments deferred the inclusion (Kerr and Sweet 2008) due to concerns about competitiveness, lack of mitigation options and the level of opposition from those potentially affected (Cooper and Rosin 2014). Now though, as the country's agricultural emissions are 12% above 1990 levels, and the country's total gross emissions have increased 19.6% above 1990 levels (New Zealand Ministry for the Environment 2018), there is a recognition that, without any targeted policy for agriculture, only 52% of the country's emissions face any substantive incentive to mitigate (Narassimhan et al. 2018). Including agriculture in the ETS is one option to provide incentives for emissions reductions in that sector. Other options are discussed in Section 7.4.4. Although some producer groups raise concern that including agriculture will place New Zealand producers at a disadvantage compared with their international competitors who do not face similar mechanisms (New Zealand Productivity Commission 2018), there is generally greater acceptance of the need for climate policies for agriculture.

The inclusion of non-CO_2 emissions from agriculture within an ETS is potentially complex, however, due to the large number of buyers and sellers if obligations are placed at farm level, and different choices of how to estimate emissions from biological systems in cost-effective ways. New Zealand is currently investigating practical and equitable approaches to include agriculture through advice being provided by the Interim Climate Change Committee (ICCC 2018). Main questions centre around the point of obligation for buying and selling credits, where trade-offs have to be made between providing incentives for behaviour change at farm level and the cost and complexity of administering the scheme (Agriculture Technical Advisory Group 2009; Kerr and Sweet 2008). The two potential points of obligation are at the processor level or at the individual farm level. Setting the point of obligation at the processor level means that farmers would face limited incentive to change their management practices, unless the processors themselves rewarded farmers for lowered emissions. Setting it at the individual farm level would provide a direct incentive for farmers to adopt mitigation practices, however, the reality of having thousands of individual points of obligation would be administratively complex and could result in high transaction costs (Beca Ltd 2018).

Monitoring, reporting and verification (MRV) of agricultural emissions presents another challenge, especially if emissions have to be estimated at farm level. Again, trade-offs have to be made between accuracy and detail of estimation method and the complexity, cost and audit of verification (Agriculture Technical Advisory Group 2009).

The ICCC is also exploring alternatives to an ETS to provide efficient abatement incentives (ICCC 2018).

Some discussion in New Zealand also focuses on a differential treatment of methane compared to nitrous oxide. Methane is a short-lived gas with a perturbation lifetime of 12 years in the atmosphere; nitrous oxide on the other hand is a long-lived gas and remains in the atmosphere for 114 years (Allen et al. 2016). Long-lived gases have a cumulative and essentially irreversible effect on the climate (IPCC 2014b) so their emissions need to reduce to net-zero in order to avoid climate change. Short-lived gases, however, could potentially be reduced to a certain level and then stabilised, and would not contribute further to warming, leading to suggestions of treating these two gases separately in the ETS or alternative policy instruments, possibly setting different budgets and targets for each (New Zealand Productivity Commission 2018). Reisinger et al. (2013) demonstrate that different metrics can have important implications globally and potentially at national and regional scales on the costs and levels of abatement.

While the details are still being agreed on in New Zealand, almost 80% of nationally determined contributions committed to action on mitigation in agriculture (FAO 2016), so countries will be looking for successful examples.

Australia's Emissions Reduction Fund, and the preceding Carbon Farming Initiative, are examples of baseline-and-credit schemes, which creates credits for activities that generate emissions below a baseline – effectively a subsidy (Freebairn 2016). It is a voluntary scheme, and has the potential to create real and additional emission reductions through projects reducing emissions and sequestering carbon (Verschuuren 2017) (*low evidence, low agreement*). Key success factors in the design of such an instrument are policy-certainty for at least 10 to 20 years, regulation that focuses on projects and not uniform rules, automated systems for all phases of the projects, and a wider focus of the carbon farming initiative on adaptation, food security, sustainable farm business, and creating jobs (Verschuuren 2017). A recent review highlighted the issue of permanence and reversal, and recommended that projects detail how they will maintain carbon in their projects, and deal with the risk of fire.

7.4.4.4   Technology transfer and land-use sectors

Technology transfer has been part of the UNFCCC process since its inception and is a key element of international climate mitigation and adaptation efforts under the Paris Agreement. The IPCC definition of 'technology transfer' includes transfer of knowledge and technological cooperation (see Glossary) and can include modifications to suit local conditions and/or integration with indigenous technologies (Metz et al. 2000). This definition suggests greater heterogeneity in the applications for climate mitigation and adaptation, especially in land-use sectors where indigenous knowledge may be important for long-term climate resilience (Nyong et al. 2007). For land-use sectors, the typical reliance on trade and patent data for empirical analyses is generally not feasible as the 'technology' in question is often related to resource management and is neither patentable nor tradable (Glachant and Dechezleprêtre 2017) and ill-suited to provide socially beneficially innovation for poorer farmers in developing countries (Lybbert and Sumner 2012; Baker et al. 2017).

Technology transfer has contributed to emissions reductions (*medium confidence*). A detailed study for nearly 4000 Clean Development Mechanism (CDM) projects showed that 39% of projects had a stated and actual technology transfer component, accounting for 59% of emissions reductions; however, the more land-intensive projects (e.g., afforestation, bioenergy) showed lower percentages (Murphy et al. 2015). Bioenergy projects that rely on agricultural residues offer substantially more development benefits than those based on industrial residues from forests (Lee and Lazarus 2013). Energy projects tended to have a greater degree of technology transfer under the CDM compared to non-energy projects (Gandenberger et al. 2016). However, longer-term cooperation and collaborative R&D approaches to technology transfer will be more important in land-use sectors (compared to energy or industry) due to the time needed for improved resource management and interaction between researchers, practitioners and policymakers. These approaches offer longer-term technology transfer that is more difficult to measure compared to specific cooperation projects; empirical research on the effects of R&D collaboration could help to avoid the 'one-policy-fits-all' approach (Ockwell et al. 2015).

There is increasing recognition of the role of technology transfer in climate adaptation, but in the land-use sector there are inherent adoption challenges specific to adaptation, due to uncertainties arising from changing climatic conditions, agricultural prices, and suitability under future conditions (Biagini et al. 2014). Engaging the private sector is important, as adoption of new technologies can only be replicated with significant private sector involvement (Biagini and Miller 2013).

7.4.4.5   International cooperation under the Paris Agreement

New cooperative mechanisms under the Paris Agreement illustrate the shift away from the Kyoto Protocol's emphasis on obligations of developed country Parties to pursue investments and technology transfer, to a more pragmatic, decentralised and collaborative approach (Savaresi 2016; Jiang et al. 2017). These approaches can effectively include any combination of measures or instruments related to adaptation, mitigation, finance, technology transfer and capacity building, which could be of particular interest in land-use sectors where such aspects are more intertwined than in energy or industry sectors. Article 6 sets out several options for international cooperation (Gupta and Dube 2018).

The close relationship between emission reductions, adaptive capacity, food security and other sustainability and governance objectives in the land sectors means that Article 6 could bring co-benefits that increase its attractiveness and the availability of finance, while also bringing risks that need to be monitored and mitigated against, such as uncertainties in measurements and the risk of non-permanence (Thamo and Pannell 2016; Olsson et al. 2016; Schwartz et al. 2017). There has been progress in accounting for land-based emissions, mainly forestry and agriculture (*medium evidence, low agreement*), but various challenges remain (Macintosh 2012; Pistorius et al. 2017; Krug 2018).

Like the CDM and other existing carbon trading mechanisms, participation in Article 6.2 and 6.4 of the Paris Agreement requires certain institutional and data management capacities in the land sector to effectively benefit from the cooperation opportunities (Totin et al. 2018). While the rules for the implementation of the new mechanisms are still under development, lessons from REDD+ (reducing emissions from deforestation and forest degradation) may be useful, which is perceived as more democratic and participative than the CDM (Maraseni and Cadman 2015). Experience with REDD+ programmes emphasise the necessity to invest in 'readiness' programmes that assist countries to engage in strategic planning and build management and data collection systems to develop the capacity and infrastructure to participate in REDD+ (Minang et al. 2014). The overwhelming majority of countries (93%) cite weak forest sector governance and institutions in their applications for REDD+ readiness funding (Kissinger et al. 2012). Technology transfer for advanced remote sensing technologies that help to reduce uncertainty in monitoring forests helps to achieve REDD+ 'readiness' (Goetz et al. 2015).

As well as new opportunities for finance and support, the Paris cooperation mechanisms and the associated roles for technology transfer bring new challenges, particularly in reporting, verifying and accounting in land-use sectors. Since developing countries must now achieve, measure and communicate emission reductions, they now have value for both developing and developed countries in achieving their NDCs, but reductions cannot be double-counted (i.e., towards multiple NDCs). All countries have to prepare and communicate NDCs, and many countries have included in their NDCs either economy-wide targets that include the land-use sectors, or specific targets for the land-use sectors. The Katowice climate package clarifies that all Parties have to submit 'Biennial Transparency Reports' from 2024 onwards, using common reporting formats, following most recent IPCC Guidelines (use of the 2013 Supplement on Wetlands is encouraged), identifying key categories of emissions, ensuring time-series consistency, and providing completeness and uncertainty assessments as well as quality control (UNFCCC 2018a; Schneider and La Hoz Theuer 2019). In total, the ambiguity in how countries incorporate land-use sectors into their NDC is estimated to lead to an uncertainty of more than 2 GtCO$_2$ in 2030 (Fyson and Jeffery 2018). Uncertainty is lower if the analysis is limited to countries that have provided separate land-use sector targets in their NDCs (Benveniste et al. 2018).

7.4.5 Policies responding to desertification and degradation – Land Degradation Neutrality (LDN)

Land Degradation Neutrality (LDN) (SDG Target 15.3), evolved from the concept of Net Zero Land Degradation, which was introduced by the United Nations Convention to Combat Desertification (UNCCD) to promote SLM (Kust et al. 2017; Stavi and Lal 2015; Chasek et al. 2015). Neutrality here implies no net loss of the land-based natural resource and ES relative to a baseline or a reference state (UNCCD 2015; Kust et al. 2017; Easdale 2016; Cowie et al. 2018a; Stavi and Lal 2015; Grainger 2015; Chasek et al. 2015). LDN can be achieved by reducing the rate of land degradation (and concomitant loss of ES) and increasing the rate of restoration and rehabilitation of degraded or desertified land. Therefore, the rate of global land degradation is not to exceed that of land restoration in order to achieve LDN goals (adopted as national platform for actions by more than 100 countries) (Stavi and Lal 2015; Grainger 2015; Chasek et al. 2015; Cowie et al. 2018a; Montanarella 2015). Achieving LDN would decrease the environmental footprint of agriculture, while supporting food security and sustaining human well-being (UNCCD 2015; Safriel 2017; Stavi and Lal 2015; Kust et al. 2017).

Response hierarchy – avoiding, reducing and reversing land degradation – is the main policy response (Chasek et al. 2019, Wonder and Bodle 2019, Cowie et al. 2018, Orr et al. 2017). The LDN response hierarchy encourages through regulation, planning and management instruments, the adoption of diverse measures to avoid, reduce and reverse land degradation in order to achieve LDN (Cowie et al. 2018b; Orr et al. 2017).

Chapter 3 categorised policy responses into two categories; (i) avoiding, reducing and reversing it through SLM; and (ii) providing alternative livelihoods with economic diversification. LDN could be achieved through planned effective actions, particularly by motivated stakeholders – those who play an essential role in a land-based climate change adaptation (Easdale 2016; Qasim et al. 2011; Cowie et al. 2018a; Salvati and Carlucci 2014). Human activities impacting the sustainability of drylands is a key consideration in adequately reversing degradation through restoration or rehabilitation of degraded land (Easdale 2016; Qasim et al. 2011; Cowie et al. 2018a; Salvati and Carlucci 2014).

LDN actions and activities play an essential role for a land-based approach to climate change adaptation (UNCCD 2015). Policies responding to degradation and desertification include improving market access, gender empowerment, expanding access to rural advisory services, strengthening land tenure security, payments for ES, decentralised natural resource management, investing in R&D, modern renewable energy sources and monitoring of desertification and desert storms, developing modern renewable energy sources, and developing and strengthening climate services. Policy supporting economic diversification includes investing in irrigation, expanding agricultural commercialisation, and facilitating structural transformations in rural economies (Chapter 3). Policies and actions also include promoting indigenous and local knowledge (ILK), soil conservation, agroforestry, crop-livestock interactions as an approach to manage land degradation, and forest-based activities such as afforestation, reforestation, and changing forest

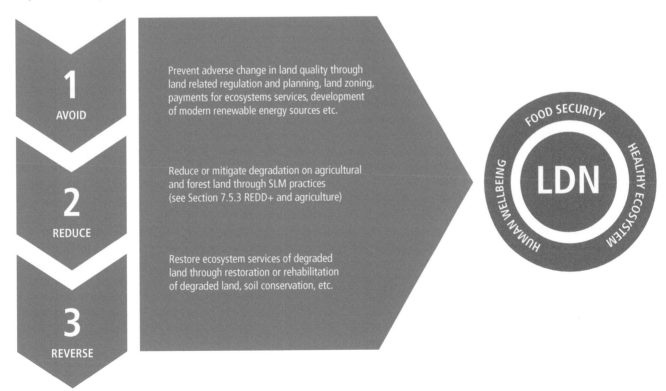

Figure 7.4 | LDN response hierarchy. Source: Adapted from (Liniger et al. 2019; UNCCD/Science-Policy-Interface 2016).

management (Chapter 4). Measures identified for achievement of LDN include effective financial mechanisms (for implementation of land restoration measures and the long-term monitoring of progress), parameters for assessing land degradation, detailed plans with quantified objectives and timelines (Kust et al. 2017; Sietz et al. 2017; Cowie et al. 2018a; Montanarella 2015; Stavi and Lal 2015).

Implementing the international LDN target into national policies has been a challenge (Cowie et al. 2018a; Grainger 2015) as baseline land degradation or desertification information is not always available (Grainger 2015) and challenges exist in monitoring LDN as it is a dynamic process (Sietz et al. 2017; Grainger 2015; Cowie et al. 2018a). Wunder and Bodle (2019) propose that LDN be implemented and monitored through indicators at the national level. Effective implementation of global LDN will be supported by integrating lessons learned from existing programmes designed for other environmental objectives and closely coordinate LDN activities with actions for climate change adaptation and mitigation at both global and national levels (*high confidence*) (Stavi and Lal 2015; Grainger 2015).

7.4.6   Policies responding to land degradation

7.4.6.1   Land-use zoning

Land-use zoning divides a territory (including local, sub-regional or national) into zones with different rules and regulations for land use (mining, agriculture, urban development, etc.), management practices and land-cover change (Metternicht 2018). While the policy instrument is zoning ordinances, the process of determining these regulations is covered in integrated land-use planning (Section 7.6.2). Urban zoning can guide new growth in urban communities outside forecasted hazard areas, assist relocating existing dwellings to safer sites and manage post-event redevelopment in ways to reduce future vulnerability (Berke and Stevens 2016). Holistic integration of climate mitigation and adaptation are interdependent and can be implemented by restoring urban forests, and improving parks (Brown 2010; Berke and Stevens 2016). Zoning ordinances can contribute to SLM through protection of natural capital by preventing or limiting vegetation clearing, avoiding degradation of planning for rehabilitation of degraded land or contaminated sites, promoting conservation and enhancement of ecosystems and ecological corridors (Metternicht 2018; Jepson and Haines 2014). Zoning ordinances can also encourage higher density development, mixed use, local food production, encourage transportation alternatives (bike paths and transit-oriented development), preserve a sense of place, and increase housing diversity and affordability (Jepson and Haines 2014). Conservation planning varies by context and may include one or several adaptation approaches, including protecting current patterns of biodiversity, large intact natural landscapes, and geophysical settings. Conservation planning may also maintain and restore ecological connectivity, identify and manage areas that provide future climate space for species expected to be displaced by climate change, and identify and protect climate refugia (Stevanovic et al. 2016; Schmitz et al. 2015).

Anguelovski et al. (2016) studied land-use interventions in eight cities in the global north and south, and concluded that historic trends of socio-economic vulnerability can be reinforced. They also found that vulnerability could be avoided with a consideration of the distribution of adaptation benefits and prioritising beneficial outcomes for disadvantaged and vulnerable groups when making future adaptation plans. Concentration of adaptation resources within wealthy business districts creating ecological enclaves exacerbated climate risks elsewhere and building of climate adaptive infrastructure such as sea walls or temporary flood barriers occurred at the expense of underserved neighbourhoods (Anguelovski et al. 2016a).

7.4.6.2   Conserving biodiversity and ecosystem services (ES)

There is *limited evidence* but *high agreement* that ecosystem-based adaptation (biodiversity, ecosystem services (ES), and Nature's Contribution to People (see Chapter 6)) and incentives for ES – including payment for ecosystem services (PES) – play a critical part of an overall strategy to help people adapt to the adverse effects of climate change on land (UNEP 2009; Bonan 2008; Millar et al. 2007; Thompson et al. 2009).

Ecosystem-based adaptation can promote socio-ecological resilience by enabling people to adapt to the impacts of climate change on land and reduce their vulnerability (Ojea 2015). Ecosystem-based adaptation can promote nature conservation while alleviating poverty and even provide co-benefits by removing GHGs (Scarano 2017) and protecting livelihoods (Munang et al. 2013). For example, mangroves provide diverse ES such as carbon storage, fisheries, non-timber forest products, erosion protection, water purification, shore-line stabilisation, and also regulate storm surge and flooding damages, thus enhancing resilience and reducing climate risk from extreme events such as cyclones (Rahman et al. 2014; Donato et al. 2011; Das and Vincent 2009; Ghosh et al. 2015; Ewel et al. 1998).

There has been considerable increase in the last decade of PES, or programmes that exchange value for land management practices intended to ensure ES (Salzman et al. 2018; Yang and Lu 2018; Barbier 2011). However, there is a deficiency in comprehensive and reliable data concerning the impact of PES on ecosystems, human well-being, their efficiency, and effectiveness (Pynegar et al. 2018; Reed et al. 2014; Salzman et al. 2018; Barbier 2011; Yang and Lu 2018). While some studies assess ecological effectiveness and social equity, fewer assess economic efficiency (Yang and Lu 2018). Part of the challenge surrounds the fact that the majority of ES are not marketed, so determining how changes in ecosystems structures, functions and processes influence the quantity and quality of ES flows to people is challenging (Barbier 2011). PES include agri-environmental targeted outcome-based payments, but challenges exist in relation to scientific uncertainty, pricing, timing of payments, increasing risk to land managers, World Trade Organization compliance, and barriers of land management and scale (Reed et al. 2014).

PES is contested (Wang and Fu 2013; Czembrowski and Kronenberg 2016; Perry 2015) for four reasons: (i) understanding and resolving trade-offs between conflicting groups of stakeholders (Wam et al.

2016; Matthies et al. 2015); (ii) knowledge and technology capacity (Menz et al. 2013); (iii) challenges integrating PES with economic and other policy instruments (Ring and Schröter-Schlaack 2011; Tallis et al. 2008; Elmqvist et al. 2003; Albert et al. 2014); and (iv) top-down climate change mitigation initiatives which are still largely carbon-centric, with limited opportunities for decentralised ecological restoration at local and regional scales (Vijge and Gupta 2014).

These challenges and contestations can be resolved with the participation of people in establishing PES, thereby addressing trust issues, negative attitudes, and resolving trade-offs between issues (such as retaining forests that consume water versus the provision of run-off, or balancing payments to providers versus cost to society) (Sorice et al. 2018; Matthies et al. 2015). Similarly, a 'co-constructive' approach is used involving a diversity of stakeholders generating policy-relevant knowledge for sustainable management of biodiversity and ES at all relevant spatial scales, by the current Intergovernmental Science-Policy Platform on Biodiversity and Ecosystem Services (IPBES) initiative (Díaz et al. 2015). Invasive species are also best identified and managed with the participation of people through collective decisions, coordinated programmes, and extensive research and outreach to address their complex social-ecological impacts (Wittmann et al. 2016; Epanchin-Niell et al. 2010).

Ecosystem restoration with co-benefits for diverse ES can be achieved through passive restoration, passive restoration with protection, and active restoration with planting (Birch et al. 2010; Cantarello et al. 2010). Taking into account the costs of restoration and co-benefits from bundles of ES (carbon, tourism, timber), the benefit-cost ratio (BCR) of active restoration and passive restoration with protection was always less than 1, suggesting that financial incentives would be required. Passive restoration was the most cost-effective with a BCR generally between 1 and 100 for forest, grassland and shrubland restoration (TEEB 2009; Cantarello et al. 2010). Passive restoration is generally more cost-effective, but there is a danger that it could be confused with abandoned land in the absence of secure tenure and a long time period (Zahawi et al. 2014). Net social benefits of degraded land restoration in dry regions range from about 200–700 USD per hectare (Cantarello et al. 2010). Investments in active restoration could benefit from analyses of past land use, the natural resilience of the ecosystem, and the specific objectives of each project (Meli et al. 2017). One successful example is the Working for Water Programme in South Africa that linked restoration through removal of invasive species and enhanced water security (Milton et al. 2003).

Forest, water and energy cycle interactions and teleconnections such as contribution to rainfall potentially (Aragão 2012; Ellison et al. 2017; Paul et al. 2018; Spracklen et al. 2012) provide a foundation for achieving forest-based adaptation and mitigation goals. They are, however, poorly integrated in policy and decision-making, including PES (Section 2.5.4).

7.4.6.3 Standards and certification for sustainability of biomass and land-use sectors

During the past two decades, standards and certification have emerged as important sustainability and conservation instruments for agriculture, forestry, bioenergy, land-use management and bio-based products (Lambin et al. 2014; Englund and Berndes 2015; Milder et al. 2015; Giessen et al. 2016a; Endres et al. 2015; Byerlee et al. 2015; van Dam et al. 2010). Standards are normally voluntary, but can also become obligatory through legislation. A standard provides specifications or guidelines to ensure that materials, products, processes and services are fit for purpose, whereas certification is the procedure through which an accredited party confirms that a product, process or service is in conformity with certain standards. Standards and certification are normally carried out by separate organisations for legitimacy and accountability (Section 7.6.6). The International Organization for Standardization is a key source for global environmental standards. Those with special relevance for land and climate include a recent standard on combating land degradation and desertification (ISO 2017) and an earlier standard on sustainable bioenergy and biomass use (ISO 2015; Walter et al. 2018). Both aim to support the long-term transition to a climate-resilient bioeconomy; there is *medium evidence* on the sustainability implications of different bioeconomy pathways, but *low agreement* as to which pathways are socially and environmentally desirable (Priefer et al. 2017; Johnson 2017; Bennich et al. 2017a).

Table 7.3 provides a summary of selected standards and certification schemes with a focus on land use and climate: the tickmark shows inclusion of different sustainability elements, with all recognising the inherent linkages between the biophysical and social aspects of land use. Some certification schemes and best practice guidelines are specific to a particular agriculture crop (e.g., soya, sugarcane) or a tree (e.g., oil palm) while others are general. International organisations promote sustainable land and biomass use through good practice guidelines, voluntary standards and jurisdictional approaches (Scarlat and Dallemand 2011; Stattman et al. 2018a). Other frameworks, such as the Global Bioenergy Partnership (GBEP) focus on monitoring land and biomass use through a set of indicators that are applied across partner countries, thereby also promoting technology/knowledge transfer (GBEP 2017). The Economics of Land Degradation (ELD) Initiative provides common guidelines for economic assessments of land degradation (Nkonya et al. 2013).

Whereas current standards and certification focus primarily on land, climate and biomass impacts where they occur, more recent analysis considers trade-related land-use change by tracing supply chain impacts from producer to consumer, leading to the notion of 'imported deforestation' that occurs from increasing demand and trade in unsustainable forest and agriculture products, which is estimated to account for 26% of all tropical deforestation (Pendrill et al. 2019). Research and implementation efforts aim to improve supply chain transparency and promote commitments to 'zero deforestation' (Gardner et al. 2018a; Garrett et al. 2019; Newton et al. 2018; Godar and Gardner 2019; Godar et al. 2015, 2016). France has developed specific policies on imported deforestation

Table 7.3 | Selected standards and certification schemes and their components or coverage.

Acronym	Scheme, programme or standard	Commodity/process, relation to others	Type of mechanism	Environmental						Socio-economic		
				GHG emissions	Biodiversity	Carbon stock	Soil	Air	Water	Land-use management[a]	Land rights	Food security[b]
ISCC	International Sustainability and Carbon Certification	All feedstocks, all supply chains	Certification	√	√	√	√	√	√	√	√	√
Bonsucro	Bonsucro EU	Sugar cane and derived products	Certification	√	√	√	√	√	√	√	√	
RTRS	Roundtable on Responsible Soy EU	Soy-based products	Certification	√	√	√	√	√	√	√	√	
RSB	Roundtable on Sustainable Biomaterials EU	Biomass for biofuels and biomaterials	Certification	√	√	√	√	√	√	√	√	√
SAN	Sustainable Agriculture	Various agricultural crops and commodities; linked to Rain Forest Alliance	Technical Network		√	√	√	√	√	√		
RSPO RED	Roundtable on Sustainable Palm Oil RED	Palm oil products	Certification	√	√	√	√	√	√	√	√	√
PEFC	Programme for Endorsement of Forest Certification	Forest management	Certification		√	√	√	√	√	√	√	c
FSC	Forest Stewardship Council	Forest management	Certification		√	√	√	√	√	√	√	
SBP	Sustainable Biomass Programme	Woody biomass (e.g., wood pellets, wood chips); linked to PEFC and FSC	Certification	√	√	√	√	√	√	√	√	
WOCAT	World Overview of Conservation Approaches and Technologies	Global network on sustainable land management	Best Practice Network		√	√	√	√	√	√		
ISO 13065: 2015	Bioenergy	Biomass and bioenergy, including conversion processes	Standard	√	√	√	√	√	√	√	√	√[d]
ISO 14055-1: 2017	Land Degradation and Desertification	Land-use management, including restoration of degraded land	Standard	√			√	√	√	√	√	

Source: Modified from (European Commission 2012; Díaz-Chavez 2015).

√ indicates that the issue is addressed in the standard or scheme

[a] includes restoration of degraded land in some cases (especially ISO 14055–1)

[b] where specifically indicated

[c] reference to the RSB certification/standard

[d] where specifically noted

that are expected to eventually include a 'zero deforestation' label (Government of France 2019).

The sustainability of biofuels and bioenergy has been in particular focus during the past decade or so due to biofuel mandates and renewable energy policies in the USA, EU and elsewhere (van Dam et al. 2010; Scarlat and Dallemand 2011). The European Union Renewable Energy Directive (EU-RED) established sustainability criteria in relation to EU renewable energy targets in the transport sector (European Commission 2012), which subsequently had impacts on land use and trade with third-party countries (Johnson et al. 2012). In particular, the EU-RED marked a departure in the context of Kyoto/UNFCCC guidelines by extending responsibility for emissions beyond the borders of final use, and requiring developing countries wishing to sell into the EU market to meet the sustainability criteria (Johnson 2011b). The recently revised EU-RED provides sustainability criteria that include management of land and forestry as well as socio-economic aspects (European Union 2018; Faaij 2018; Stattman et al. 2018b). Standards and certification aim to address potential conflicts between different uses of biomass, and most schemes also consider co-benefits and synergies (see Cross-Chapter Box 7 in Chapter 6). Bioenergy may offer additional income and livelihoods to farmers as well as improvements in technical productivity and multi-functional landscapes (Rosillo Callé and Johnson 2010a; Kline et al. 2017; Araujo Enciso et al. 2016). Results depend on the commodities involved, and also differ between rural and urban areas.

Analyses on the implementation of standards and certification for land and biomass use have focused on their stringency, effectiveness and geographical scope as well as socio-economic impacts such as land tenure, gender and land rights (Diaz-Chavez 2011; German and Schoneveld 2012; Meyer and Priess 2014). The level of stringency and enforcement varies with local environmental conditions, governance approaches and the nature of the feedstock produced (Endres et al. 2015; Lambin et al. 2014; Giessen et al. 2016b; Stattman et al. 2018b). There is *low evidence* and *low agreement* on how the application and use of standards and certification has actually improved sustainability beyond the local farm, factory or plantation level; the lack of harmonisation and consistency across countries that has been observed, even within a common market or economic region such as the EU, presents a barrier to wider market impacts (Endres et al. 2015; Stattman et al. 2018b; ISEAL Alliance 2018). In the

forest sector, there is evidence that certification programmes such as the Forest Stewardship Council (FSC) have reduced deforestation in the aggregate, as well as reducing air pollution (Miteva et al. 2015; Mcdermott et al. 2015). Certification and standards cannot address global systemic concerns such as impacts on food prices or other market-wide effects, but rather are aimed primarily at insuring best practices in the local context. More general approaches to certification such as the Gold Standard are designed to accelerate progress toward the SDGs as well as the Paris Climate Agreement by certifying investment projects while also emphasising support to governments (Gold Standard).

7.4.6.4   Energy access and biomass use

Access to modern energy services is a key component of SDG 7, with an estimated 1.1 billion people lacking access to electricity, while nearly 3 billion people rely on traditional biomass (fuelwood, agriculture residues, animal dung, charcoal) for household energy needs (IEA 2017). Lack of access to modern energy services is significant in the context of land-climate systems because heavy reliance on traditional biomass can contribute to land degradation, household air pollution and GHG emissions (see Cross-Chapter Box 12 in Chapter 7). A variety of policy instruments and programmes have been aimed at improving energy access and thereby reducing the heavy reliance on traditional biomass (Table 7.2); there is *high evidence* and *high agreement* that programmes and policies that reduce dependence on traditional biomass will have benefits for health and household productivity, as well as reducing land degradation (Section 4.5.4) and GHG emissions (Bailis et al. 2015; Cutz et al. 2017a; Masera et al. 2015; Goldemberg et al. 2018a; Sola et al. 2016a; Rao and Pachauri 2017; Denton et al. 2014). There can be trade-offs across different options, especially between health and climate benefits, since more efficient wood stoves might have only limited effect, whereas gaseous and liquid fuels (e.g., biogas, LPG, bioethanol) will have highly positive health benefits and climate benefits that vary depending on specific circumstances of the substitution (Cameron et al. 2016; Goldemberg et al. 2018b). Unlike traditional biomass, modern bioenergy offers high-quality energy services, although, for household cookstoves, even the cleanest options using wood may not perform as well in terms of health and/or climate benefits (Fuso Nerini et al. 2017; Goldemberg et al. 2018b).

Case study | Forest conservation instruments: REDD+ in the Amazon and India

More than 50 countries have developed national REDD+ strategies, which have key conditions for addressing deforestation and forest degradation (improved monitoring capacities, understanding of drivers, increased stakeholder involvement, and providing a platform to secure indigenous and community land rights). However, to achieve its original objectives and to be effective under current conditions, forest-based mitigation actions need to be incorporated in national development plans and official climate strategies, and mainstreamed across sectors and levels of government (Angelsen et al. 2018a).

The Amazon region can illustrate the complexity of the implementation of REDD+, in the most biodiverse place on the planet, with millions of inhabitants and hundreds of ethnic groups, under the jurisdiction of eight countries. While different experiences can be drawn at different spatial scales, at the regional-level, for example, Amazon Fund (van der Hoff et al. 2018), at the subnational level (Furtado 2018), and at the local level (Alvarez et al. 2016; Simonet et al. 2019), there is *medium evidence* and high agreement that

Case Study (continued)

REDD+ has stimulated sustainable land-use investments but is also competing with other land uses (e.g., agroindustry) and scarce international funding (both public and private) (Bastos Lima et al. 2017b; Angelsen et al. 2018b).

In the Amazon, at the local level, a critical issue has been the incorporation of indigenous people in the planning and distribution of benefits of REDD+ projects. While REDD+, in some cases, has enhanced participation of community members in the policy-planning process, fund management, and carbon baseline establishment, increasing project reliability and equity (West 2016), it is clear that, in this region, insecure and overlapping land rights, as well as unclear and contradictory institutional responsibilities, are probably the major problems for REDD+ implementation (Loaiza et al. 2017). Despite legal and rhetoric recognition of indigenous land rights, effective recognition is still lacking (Aguilar-Støen 2017). The key to the success of REDD+ in the Amazon, has been the application of both incentives and disincentives on key safeguard indicators, including land security, participation, and well-being (Duchelle et al. 2017).

On the other hand, at the subnational level, REDD+ has been unable to shape land-use dynamics or landscape governance, in areas suffering strong exogenous factors, such as extractive industries, and in the absence of effective regional regulation for sustainable land use (Rodriguez-Ward et al. 2018; Bastos Lima et al. 2017b). Moreover, projects with weak financial incentives, engage households with high off-farm income, which are already better off than the poorest families (Loaiza et al. 2015). Beyond operational issues, clashing interpretations of results might create conflict between implementing countries or organisations and donor countries, which have revealed concerns over the performance of projects (van der Hoff et al. 2018) REDD+ Amazonian projects often face methodological issues, including how to assess the opportunity cost among landholders, and informing REDD+ implementation (Kweka et al. 2016). REDD+ based projects depend on consistent environmental monitoring methodologies for measuring, reporting and verification and, in the Amazon, land-cover estimates are crucial for environmental monitoring efforts (Chávez Michaelsen et al. 2017).

In India, forests and wildlife concerns are on the concurrent list of the Constitution since an amendment in 1976, thus giving the central or federal government a strong role in matters related to governance of forests. High rates of deforestation due to development projects led to the Forest (Conservation) Act (1980) which requires central government approval for diversion of forest land in any state or union territory.

Before 2006, forest diversion for development projects leading to deforestation needed clearance from the Central Government under the provisions of the Forest (Conservation Act) 1980. In order to regulate forest diversion, and as payment for ES, a net present value (NPV) frame-work was introduced by the Supreme Court of India, informed by the Kanchan Chopra committee (Chopra 2017). The Forest (Conservation) Act of 1980 requires compensatory afforestation in lieu of forest diversion, and the Supreme Court established the Compensatory Afforestation Fund Management and Planning Authority (CAMPA) which collects funds for compensatory afforestation and on account of NPV from project developers.

As of February 2018, 6825 million USD had accumulated in CAMPA funds in lieu of NPV paid by developers diverting forest land throughout India for non-forest use. Funds are released by the central government to state governments for afforestation and conservation-related activities to 'compensate' for diversion of forests. This is now governed by legislation called the CAMPA Act, passed by the Parliament of India in July 2016. The CAMPA mechanism has, however, invited criticism on various counts in terms of undervaluation of forest, inequality, lack of participation and environmental justice (Temper and Martinez-Alier 2013).

The other significant development related to forest land was the landmark legislation called the Scheduled Tribes and Other Traditional Forest Dwellers (Recognition of Forest Rights) Act, 2006 or Forest Rights Act (FRA) passed by the Parliament of India in 2007. This is the largest forest tenure legal instrument in the world and attempted to undo historical injustice to forest dwellers and forest-dependent communities whose traditional rights and access were legally denied under forest and wildlife conservation laws. The FRA recognises the right to individual land titles on land already cleared, as well as community forest rights such as collection of forest produce. A total of 64,328 community forest rights and a total of 17,040,343 individual land titles had been approved and granted up to the end of 2017. Current concerns on policy and implementation gaps are about strengths and pitfalls of decentralisation, identifying genuine right holders, verification of land rights using technology and best practices, and curbing illegal claims (Sarap et al. 2013; Reddy et al. 2011; Aggarwal 2011; Ramnath 2008; Ministry of Environment and Forests and Ministry and Tribal Affairs, Government of India 2010).

Case Study (continued)

As per the FRA, the forest rights shall be conferred free of all encumbrances and procedural requirements. Furthermore, without the FRA's provision for getting the informed consent of local communities for both diversion of community forest land and for reforestation, there would be legal and administrative hurdles in using existing forest land for implementation of India's ambitious Green India Mission that aims to respond to climate change by a combination of adaptation and mitigation measures in the forestry sector. It aims to increase forest/tree cover to the extent of 5 million hectares (Mha) and improve quality of forest/tree cover on another 5 Mha of forest/non-forest lands and support forest-based livelihoods of 3 million families and generate co-benefits through ES (Government of India 2010).

Thus, the community forest land recognised under FRA can be used for the purpose of compensatory afforestation or restoration under REDD+ only with informed consent of the communities and a decentralised mechanism for using CAMPA funds. India's forest and forest restoration can potentially move away from a top-down carbon centric model with the effective participation of local communities (Vijge and Gupta 2014; Murthy et al. 2018a).

India has also experimented with the world's first national inter-governmental ecological fiscal transfer (EFT) from central to local and state government to reward them for retaining forest cover. In 2014, India's 14th Finance Commission added forest cover to the formula that determines the amount of tax revenue the central government distributes annually to each of India's 29 states. It is estimated that, in four years, it would have distributed 6.9–12 billion USD per year to states in proportion to their 2013 forest cover, amounting to around 174–303 USD per hectare of forest per year (Busch and Mukherjee 2017). State governments in India now have a sizeable fiscal incentive based on extent of forest cover at the time of policy implementation, contributing to the achievement of India's climate mitigation and forest conservation goals. India's tax revenue distribution reform has created the world's first EFTs for forest conservation, and a potential model for other countries. However, it is to be noted that EFT is calculated based on a one-time estimate of forest cover prior to policy implementation, hence does not incentivise ongoing protection and this is a policy gap. It's still too early but its impact on trends in forest cover in the future and its ability to conserve forests without other investments and policy instruments is promising but untested (Busch and Mukherjee 2017; Busch 2018).

In order to build on the new promising policy developments on forest rights and fiscal incentives for forest conservation in India, incentivising ongoing protection, further investments in monitoring (Busch 2018), decentralisation (Somanathan et al. 2009) and promoting diverse non-agricultural forest and range of land-based livelihoods (e.g., sustainable non-timber forest product extraction, regulated pastures, carbon credits for forest regeneration on marginal agriculture land and ecotourism revenues) as part of individual and community forest tenure and rights are ongoing concerns. Decentralised sharing of CAMPA funds between government and local communities for forest restoration as originally suggested and filling in implementation gaps could help reconcile climate change mitigation through forest conservation, REDD+ and environmental justice (Vijge and Gupta 2014; Temper and Martinez-Alier 2013; Badola et al. 2013; Sun and Chaturvedi 2016; Murthy et al. 2018b; Chopra 2017; Ministry of Environment, Forest and Climate Change, and Ministry of Tribal Affairs, Government of India 2010).

7.4.7   Economic and financial instruments for adaptation, mitigation, and land

There is an urgent need to increase the volume of climate financing and bridge the gap between global adaptation needs and available funds (*medium confidence*) (Masson-Delmotte et al. 2018; Kissinger et al. 2019; Chambwera and Heal 2014), especially in relation to agriculture (FAO 2010). The land sector offers the potential to balance the synergies between mitigation and adaptation (Locatelli et al. 2016) – although context and unavailability of data sets makes cost comparisons between mitigation and adaptation difficult (UNFCCC 2018b). Estimates of adaptation costs range from 140 to 300 billion USD by 2030, and between 280 and 500 billion USD by 2050; (UNEP 2016). These figures vary according to methodologies and approaches (de Bruin et al. 2009; IPCC 2014 2014; OECD 2008; Nordhaus 1999; UNFCCC 2007; Plambeck et al. 1997).

7.4.7.1   Financing mechanisms for land mitigation and adaptation

There is a startling array of diverse and fragmented climate finance sources: more than 50 international public funds, 60 carbon markets, 6000 private equity funds, 99 multilateral and bilateral climate funds (Samuwai and Hills 2018). Most public finance for developing countries flows through bilateral and multilateral institutions such as the World Bank, the International Monetary Fund, International Finance Corporation, regional development banks, as well as specialised multilateral institutions such as the Global Environmental Fund, and the EU Solidarity Fund. Some governments have established state investment banks (SIBs) to close the financing gap, including the UK (Green Investment Bank), Australia (Clean Energy Finance Corporation) and in Germany (Kreditanstalt für Wiederaufbau) the Development Bank has been involved in supporting low-carbon finance (Geddes et al. 2018). The Green Climate Fund (GCF) now offers additional finance, but is still a new institution with policy gaps, a lengthy and cumbersome process related to approval (Brechin and Espinoza 2017; Khan and Roberts 2013; Mathy and Blanchard 2016), and challenges with adequate and sustained funding (Schalatek and Nakhooda 2013). Private adaptation finance exists, but is difficult to define, track, and coordinate (Nakhooda et al. 2016).

The amount of funding dedicated to agriculture, land degradation or desertification is very small compared to total climate finance (FAO 2010). Funding for agriculture (rather than mitigation) is accessed through the smaller adaptation funds (Lobell et al. 2013). Focusing on synergies, between mitigation, adaptation, and increased productivity, such as through climate-smart agriculture (CSA) (Lipper et al. 2014b) (Section 7.5.6), may leverage greater financial resources (Suckall et al. 2015; Locatelli et al. 2016). Payments for ecosystem services (Section 7.4.6) are another emerging area to encourage environmentally desirable practices, although they need to be carefully designed to be effective (Engel and Muller 2016).

The UNCCD established the Land Degradation Neutrality Fund (LDN Fund) to mobilise finance and scale-up land restoration and sustainable business models on restored land to achieve the target of a land degradation neutral world (SDG target 15.3) by 2030. The LDN Fund generates revenues from sustainable use of natural resources, creating green job opportunities, sequestering CO_2, and increasing food and water security (Cowie et al. 2018a; Akhtar-Schuster et al. 2017). The fund leverages public money to raise private capital for SLM and land restoration projects (Quatrini and Crossman 2018; Stavi and Lal 2015). Many small-scale projects are demonstrating that sustainable landscape management (Section 7.6.3) is key to achieving LDN, and it is also more financially viable in the long term than the unsustainable alternative (Tóth et al. 2018; Kust et al. 2017).

7.4.7.2 Instruments to manage the financial impacts of climate and land change disruption

Comprehensive risk management (Section 7.4.3.1) designs a portfolio of instruments which are used across a continuum of preemptive, planning and assessment, and contingency measures in order to bolster resilience (Cummins and Weiss 2016) and address limitations of any one instrument (Surminski 2016; Surminski et al. 2016; Linnerooth-bayer et al. 2019). Instruments designed and applied in isolation have shown short-term results, rather than sustained intended impacts (Vincent et al. 2018). Risk assessments limited to events and impacts on particular asset classes or sectors can misinform policy and drive misallocation of funding (Gallina et al. 2016; Jongman et al. 2014).

Comprehensive risk assessment combined with risk layering approaches that assign different instruments to different magnitude and frequency of events, have better potential to provide stability to societies facing disruption (Mechler et al. 2014; Surminski et al. 2016). Governments and citizens define limits of what they consider acceptable risks, risks for which market or other solutions can be developed and catastrophic risks that require additional public protection and intervention. Different financial tools may be used

for these different categories of risk or phases of the risk cycle (preparedness, relief, recovery, reconstruction).

In order to protect lives and livelihoods early action is critical, including a coordinated plan for action agreed in advance, a fast, evidence-based decision-making process, and contingency financing to ensure that the plan can be implemented (Clarke and Dercon 2016a). Forecast-based finance mechanisms incorporate these principles, using climate or other indicators to trigger funding and action prior to a shock (Wilkinson 2018). Forecast-based mechanisms can be linked with social protection systems by providing contingent scaled-up finance quickly to vulnerable populations following disasters, enhancing scalability, timeliness, predictability and adequacy of social protection benefits (Wilkinson 2018; Costella et al. 2017b; World Food Programme 2018).

Measures in advance of risks set aside resources before negative impacts related to adverse weather, climatic stressors, and land changes occur. These tools are frequently applied in extreme event, rapid onset contexts. These measures are the main instruments for reducing fatalities and limiting damage from extreme climate and land change events (Surminski et al. 2016). Finance tools in advance of risk include insurance (macro, meso, micro), green bonds, and forecast-based finance (Hunzai et al. 2018).

There is *high confidence* that insurance approaches that are designed to effectively reduce and communicate risks to the public and beneficiaries, designed to reduce risk and foster appropriate adaptive responses, and provide value in risk transfer, improve economic stability and social outcomes in both higher – and lower-income contexts (Kunreuther and Lyster 2016; Outreville 2011b; Surminski et al. 2016; Kousky et al. 2018b), bolster food security, help keep children in school, and help safeguard the ability of low-income households to pay for essentials like medicines (Shiferaw et al. 2014; Hallegatte et al. 2017).

Low-income households show demand for affordable risk transfer tools, but demand is constrained by liquidity, lack of assets, financial and insurance literacy, or proof of identity required by institutions in the formal sector (Eling et al. 2014; Cole 2015; Cole et al. 2013; Ismail et al. 2017). Microinsurance participation takes many forms, including through mobile banking (Eastern Africa, Bangladesh), linked with social protection or other social stabilisation programmes (Ethiopia, Pakistan, India), through flood or drought protection schemes (Indonesia, the Philippines, the Caribbean, and Latin America), often in the form of weather index insurance. The insurance industry faces challenges due to low public awareness of how insurance works. Other challenges include risk, low capacity in financial systems to administer insurance, data deficits, and market imperfections (Mechler et al. 2014; Feyen et al. 2011; Gallagher 2014; Kleindorfer et al. 2012; Lazo et al.; Meyer and Priess 2014; Millo 2016).

Countries also request grant assistance, and contingency debt finance that includes dedicated funds, set aside for unpredictable climate-related disasters, household savings, and loans with 'catastrophe risk deferred drawdown option' (which allows countries to divert loans from development objectives such as health, education, and infrastructure to make immediate disbursement of funds in the event of a disaster) (Kousky and Cooke 2012; Clarke and Dercon 2016b). Contingency finance is suited to manage frequently occurring, low-impact events (Campillo et al. 2017; Mahul and Ghesquiere 2010; Roberts 2017) and may be linked with social protection systems. These instruments are limited by uncertainty surrounding the size of contingency fund reserves, given unpredictable climate disasters (Roberts 2017) and lack of borrowing capacity of a country (such as small island states) (Mahul and Ghesquiere 2010).

In part because of its link with debt burden, contingency, or post-event finance can disrupt development and is not suitable for higher consequence events and processes such as weather extremes or structural changes associated with climate and land change. Post-event finance of negative impacts such as sea level rise, soil salinisation, depletion of groundwater, and widespread land degradation, is likely to become infeasible for multiple, high-cost events and processes. There is *high confidence* that post-extreme event assistance may face more severe limitations, given the impacts of climate change (Linnerooth-bayer et al. 2019; Surminski et al. 2016; Deryugina 2013; Dillon et al. 2014; Clarke 2016; Shreve and Kelman 2014; Von Peter et al. 2012).

In a catastrophe risk pool, multiple countries in a region pool risks in a diversified portfolio. Examples include African Risk Capacity (ARC), the Caribbean Catastrophe Risk Insurance Facility (CCRIF), and the Pacific Catastrophe Risk Assessment and Financing Initiative (PCRAFI) (Bresch et al. 2017; Iyahen and Syroka 2018). ARC payouts have been used to assist over 2.1 million food insecure people and provide more than 900,000 cattle with subsidised feed in the affected countries (Iyahen and Syroka 2018). ARC has also developed the Extreme Climate Facility, which is designed to complement existing bilateral, multilateral and private sources of finance to enable proactive adaptation (Vincent et al. 2018). It provides beneficiaries the opportunity to increase their benefit by reducing exposure to risk through adaptation and risk reduction measures, thus side-stepping 'moral hazard' problems sometimes associated with traditional insurance.

Governments pay coupon interest when purchasing catastrophe (CAT) bonds from private or corporate investors. In the case of the predefined catastrophe, the requirement to pay the coupon interest or repay the principal may be deferred or forgiven (Nguyen and Lindenmeier 2014). CAT bonds are typically short term instruments (three to five years) and the payout is triggered once a particular threshold of disaster/damage is passed (Härdle and Cabrera 2010; Campillo et al. 2017; Estrin and Tan 2016; Hermann et al. 2016; Michel-Kerjan 2011; Roberts 2017). The primary advantage of CAT bonds is their ability to quickly disburse money in the event of a catastrophe (Estrin and Tan 2016). Green bonds, social impact bonds, and resilience bonds are other instruments that can be used to fund land-based interventions. However, there are significant barriers for developing country governments to enter into the bond market: lack of familiarity with the instruments; lack of capacity and resources to deal with complex legal arrangements; limited or non-existent data and modelling of disaster exposure; and other political disincentives linked to insurance. For these reasons, the utility and application

of bonds is currently largely limited to higher-income developing countries (Campillo et al. 2017; Le Quesne 2017).

7.4.7.3 Innovative financing approaches for transition to low-carbon economies

Traditional financing mechanisms have not been sufficient and thereby leave a gap in facilitating a rapid transition to a low-carbon economy or building resilience (Geddes et al. 2018). More recently there have been developments in more innovative mechanisms, including crowdfunding (Lam and Law 2016), often supported by national governments (in the UK through regulatory and tax support) (Owen et al. 2018). Crowdfunding has no financial intermediaries and thus low transaction costs, and the projects have a greater degree of independence than bank or institution funding (Miller et al. 2018). Other examples of innovative mechanisms are community shares for local projects, such as renewable energy (Holstenkamp and Kahla 2016), or Corporate Power Purchase Agreements (PPAs) used by companies such as Google and Apple to purchase renewable energy directly or virtually from developers (Miller et al. 2018). Investing companies benefit from avoiding unpredictable price fluctuations as well as increasing their environmental credentials. A second example is auctioned price floors, or subsidies that offer a guaranteed price for future emission reductions, currently being trialled in developing countries, by the World Bank Group, known as the Pilot Auction Facility for Methane and Climate Change Mitigation (PAF) (Bodnar et al. 2018). Price floors can maximise the climate impact per public dollar while incentivising private investment in low-carbon technologies, and ideally would be implemented in conjunction with complementary policies such as carbon pricing.

In order for climate finance to be as effective and efficient as possible, cooperation between private, public and third sectors (e.g., non-governmental organisations (NGOs), cooperatives, and community groups) is more likely to create an enabling environment for innovation (Owen et al. 2018). While innovative private sector approaches are making significant progress, the existence of a stable policy environment that provides certainty and incentives for long-term private investment is critical.

7.4.8 Enabling effective policy instruments – policy portfolio coherence

An enabling environment for policy effectiveness includes: (i) the development of comprehensive policies, strategies and programmes (Section 7.4); (ii) human and financial resources to ensure that policies, programmes and legislation are translated into action; (iii) decision-making that draws on evidence generated from functional information systems that make it possible to monitor trends, track and map actions, and assess impact in a manner that is timely and comprehensive (Section 7.5); (iv) governance coordination mechanisms and partnerships; and (v) a long-term perspective in terms of response options, monitoring, and maintenance (FAO 2017a) (Section 7.6).

7

A comprehensive consideration of policy portfolios achieves sustainable land and climate management (*medium confidence*) (Mobarak and Rosenzweig 2013; Stavropoulou et al. 2017; Jeffrey et al. 2017; Howlett and Rayner 2013; Aalto et al. 2017; Brander and Keith 2015; Williams and Abatzoglou 2016; Linnerooth-Bayer and Hochrainer-Stigler 2015; FAO 2017b; Bierbaum and Cowie 2018). Supporting the study of enabling environments, the study of policy mixes has emerged in the last decade in regards to the mix or set of instruments that interact together and are aimed at achieving policy objectives in a dynamic setting (Reichardt et al. 2015). This includes studying the ultimate objectives of a policy mix – such as biodiversity (Ring and Schröter-Schlaack 2011) – the interaction of policy instruments within the mix (including climate change mitigation and energy (del Río and Cerdá 2017)) (see Trade-offs and synergies, Section 7.5.6), and the dynamic nature of the policy mix (Kern and Howlett 2009).

Studying policy mixes allows for a consideration of policy coherence that is broader than the study of discrete policy instruments in rigidly defined sectors, but entails studying policy in relation to the links and dependencies among problems and issues (FAO 2017b). Consideration of policy coherence is a new approach, rejecting simplistic solutions, but acknowledging inherently complex processes involving collective consideration of public and private actors in relation to policy analysis (FAO 2017b). A coherent, consistent mix of policy instruments can solve complex policy problems (Howlett and Rayner 2013) as it involves lateral, integrative, and holistic thinking in defining and solving problems (FAO 2017b). Such a consideration of policy coherence is required to achieve sustainable development (FAO 2017b; Bierbaum and Cowie 2018). Consideration of policy coherence potentially addresses three sets of challenges: challenges that exist with assessing multiple hazards and sectors (Aalto et al. 2017; Brander and Keith 2015; Williams and Abatzoglou 2016); challenges in mainstreaming adaptation and risk management into ongoing development planning and decision-making (Linnerooth-Bayer and Hochrainer-Stigler 2015); and challenges in scaling-up community and ecosystem-based initiatives in countries overly focused on sectors, instead of sustainable use of biodiversity and ES (Reid 2016). There is a gap in integrated consideration of adaptation, mitigation, climate change policy and development. A study in Indonesia found that, while internal policy coherence between mitigation and adaptation is increasing, external policy coherence between climate change policy and development objectives is still required (Di Gregorio et al. 2017).

There is *medium evidence* and *high agreement* that a suite of agricultural business risk programmes (which would include crop insurance and income stability programmes) increase farm financial performance, reduce risk, and also reinforce incentives to adopt stewardship practices (beneficial management practices) improving the environment (Jeffrey et al. 2017). Consideration of the portfolio of instruments responding to climate change and its associated risks, and the interaction of policy instruments, improve agricultural producer livelihoods (Hurlbert 2018b). In relation to hazards, or climate-related extremes (Section 7.4.3), the policy mix has been found to be a key determinant of the adaptive capacity of agricultural producers. In relation to drought, the mix of policy instruments including crop insurance, SLM practices, bankruptcy and insolvency,

co-management of community in water and disaster planning, and water infrastructure programmes are effective at responding to drought (Hurlbert 2018b; Hurlbert and Mussetta 2016; Hurlbert and Pittman 2014; Hurlbert and Montana 2015; Hurlbert 2015a; Hurlbert and Gupta 2018). Similarly, in relation to flood, the mix of policy instruments including flood zone mapping, land-use planning, flood zone building restrictions, business and crop insurance, disaster assistance payments, preventative instruments, such as environmental farm planning (including soil and water management (Chapter 6)) and farm infrastructure projects, and recovery from debilitating flood losses, ultimately through bankruptcy, are effective at responding to flood (Hurlbert 2018a) (see Case study: Flood and flood security in Section 7.6.3).

In respect of land conservation and management goals, consideration of differing strengths and weakness of instruments is necessary. While direct regulation may secure effective minimum standards of biodiversity conservation and critical ES provision, economic instruments may achieve reduced compliance costs as costs are borne by policy addressees (Rogge and Reichardt 2016). In relation to GHG emissions and climate mitigation, a comprehensive mix of instruments targeted at emissions reductions, learning, and R&D is effective (*high confidence*) (Fischer and Newell 2008). The policy coherence between climate policy and public finance is critical in ensuring the efficiency, effectiveness and equity of mitigation policy, and ultimately to make stringent mitigation policy more feasible (Siegmeier et al. 2018). Recycling carbon tax revenue to support clean energy technologies can decrease losses from unilateral carbon mitigation targets, with complementary technology polices (Corradini et al. 2018).

When evaluating a new policy instrument, its design in relation to achieving an environmental goal or solving a land and climate change issue, includes consideration of how the new instrument will interact with existing instruments operating at multiple levels (international, regional, national, sub-national, and local) (Ring and Schröter-Schlaack 2011) (Section 7.4.1).

7.4.9   Barriers to implementing policy responses

There are barriers to implementing the policy instruments that arise in response to the risks from climate-land interactions. Such barriers to climate action help determine the degree to which society can achieve its sustainable development objectives (Dow et al. 2013; Langholtz et al. 2014; Klein et al. 2015). However, some policies can also be seen as being designed specifically to overcome barriers, while some cases may actually create or strengthen barriers to climate action (Foudi and Erdlenbruch 2012; Linnerooth-Bayer and Hochrainer-Stigler 2015). The concept of barriers to climate action is used here in a sense close to that of 'soft limits' to adaptation (Klein, et al. 2014). 'Hard limits' by contrast are seen as primarily biophysical. Predicted changes in the key factors of crop growth and productivity – temperature, water, and soil quality – are expected to pose limits to adaptation in ways that affect the world population's ability to get enough food in the future (Altieri et al. 2015; Altieri and Nicholls 2017).

This section assesses research on barriers specific to policy implementation in adaptation and mitigation respectively, then addresses the cross-cutting issue of inequality as a barrier to climate action, including the particular cases of corruption and elite capture, before assessing how policies on climate and land can be used to overcome barriers.

7.4.9.1 Barriers to adaptation

There are human, social, economic, and institutional barriers to adaptation to land-climate challenges as described in Table 7.4 (*medium evidence, high agreement*). Considerable literature exists around changing behaviours through response options targeting social and cultural barriers (Rosin 2013; Eakin 2016; Marshall et al. 2012) (Chapter 6).

Since the publication of the IPCC's Fifth Assessment Report (AR5) (IPCC 2014), research is emerging, examining the role of governance, institutions and (in particular) policy instruments, in creating or overcoming barriers to adaptation to land and climate change in the land-use sector (Foudi and Erdlenbruch 2012; Linnerooth-Bayer and Hochrainer-Stigler 2015). Evidence shows that understanding the local context and targeted approaches are generally most successful (Rauken et al. 2014). Understanding the nature of constraints to adaptation is critical in determining how barriers may be overcome. Formal institutions (rules, laws, policies) and informal institutions (social and cultural norms and shared understandings) can be barriers and enablers of climate adaptation (Jantarasami et al. 2010).

Governments play a key role in intervening and confronting existing barriers by changing legislation, adopting policy instruments, providing additional resources, and building institutions and knowledge exchange (Ford and Pearce 2010; Measham et al. 2011; Mozumder et al. 2011; Storbjörk 2010). Understanding institutional barriers is important in addressing barriers (*high confidence*). Institutional barriers may exist due to the path-dependent nature of institutions governing natural resources and public good, bureaucratic structures that undermine horizontal and vertical integration (Section 7.6.2), and lack of policy coherence (Section 7.4.8).

7.4.9.2 Barriers to land-based climate mitigation

Barriers to land-based mitigation relate to full understanding of the permanence of carbon sequestration in soils or terrestrial biomass, the additionality of this storage, its impact on production and production shifts to other regions, measurement and monitoring systems and costs (Smith et al. 2007). Agricultural producers are more willing to expand mitigation measures already employed (including efficient and effective management of fertiliser, including manure and slurry) and less favourable to those not employed, such as using dietary additives, adopting genetically improved animals, or covering slurry tanks and lagoons (Feliciano et al. 2014). Barriers identified in land-based mitigation include physical environmental constraints such as lack of information, education, and suitability for size and location of farm. For instance, precision agriculture is not viewed as efficient in small-scale farming (Feliciano et al. 2014).

Property rights may be a barrier when there is no clear single-party land ownership to implement and manage changes (Smith et al. 2007). In forestry, tenure arrangements may not distribute obligations and incentives for carbon sequestration effectively between public management agencies and private agents with forest licences. Including carbon in tenure and expanding the duration of tenure may provide stronger incentive for tenure holders to manage carbon as well as timber values (Williamson and Nelson 2017). Effective policy will require answers as to the current status of agriculture in regard to GHG emissions, the degree that emissions are to change, the best pathway to achieve the change, and an ability to know when the target level of change is achieved (Smith et al. 2007). Forest governance may not have the structure to advance mitigation and adaptation. Currently top-down traditional modes do not have the flexibility or responsiveness to deal with the complex, dynamic, spatially diverse, and uncertain features of climate change (Timberlake and Schultz 2017; Williamson and Nelson 2017).

In respect of forest mitigation, two main institutional barriers have been found to predominate. First, forest management institutions do not consider climate change to the degree necessary for enabling effective climate response, and do not link adaptation and mitigation. Second, institutional barriers exist if institutions are not forward looking, do not enable collaborative adaptive management, do not

Table 7.4 | Soft barriers and limits to adaptation.

Category	Description	References
Human	– Cognitive and behavioural obstacles – Lack of knowledge and information	Hornsey et al. 2016; Prokopy et al. 2015; Wreford et al. 2017
Social	– Undermined participation in decision-making and social equity	Burton et al. 2008; Laube et al. 2012
Economic	– Market failures and missing markets: transaction costs and political economy; ethical and distributional issues – Perverse incentives – Lack of domestic funds; inability to access international funds	Chambwera et al. 2014b; Wreford et al. 2017; Rochecouste et al. 2015; Baumgart-Getz et al. 2012
Institutional	– Mal-coordination of policies and response options; unclear responsibility of actors and leadership; misuse of power; all reducing social learning – Government failures – Path-dependent institutions	Oberlack 2017; Sánchez et al. 2016; Greiner and Gregg 2011
Technological	– Systems of mixed crop and livestock – Polycultures	Nalau and Handmer 2015

7

promote flexible approaches that are reversible as new information becomes available, do not promote learning and allow for diversity of approaches that can be tailored to different local circumstances (Williamson and Nelson 2017).

Land-based climate mitigation through expansions and enhancements in agriculture, forestry and bioenergy has great potential but also poses great risks; its success will therefore require improved land-use planning, strong governance frameworks and coherent and consistent policies. 'Progressive developments in governance of land and modernisation of agriculture and livestock and effective sustainability frameworks can help realise large parts of the technical bioenergy potential with low associated GHG emissions' (Smith et al. 2014b, p. 97).

7.4.9.3   Inequality

There is *medium evidence* and *high agreement* that one of the greatest challenges for land-based adaptation and SLM is posed by inequalities that influence vulnerability and coping and adaptive capacity – including age, gender, wealth, knowledge, access to resources and power (Kunreuther et al. 2014; IPCC 2012; Olsson et al. 2014). Gender is the dimension of inequality that has been the focus of most research, while research demonstrating differential impacts, vulnerability and adaptive capacity based on age, ethnicity and indigeneity is less well developed (Olsson et al. 2015a). Cross-Chapter Box 11 in Chapter 7 sets out both the contribution of gender relations to differential vulnerability and available policy instruments for greater gender inclusivity.

One response to the vulnerability of poor people and other categories differentially affected is effective and reliable social safety nets (Jones and Hiller 2017). Social protection coverage is low across the world and informal support systems continue to be the key means of protection for a majority of the rural poor and vulnerable (Stavropoulou et al. 2017) (Section 7.4.2). However, there is a gap in knowledge in understanding both positive and negative synergies between formal and informal systems of social protection and how local support institutions might be used to implement more formal forms of social protection (Stavropoulou et al. 2017).

7.4.9.4   Corruption and elite capture

Inequalities of wealth and power can allow processes of corruption and elite capture (where public resources are used for the benefit of a few individuals in detriment to the larger populations) which can affect both adaptation and mitigation actions, at levels from the local to the global that, in turn, risk creating inequitable or unjust outcomes (Sovacool 2018) (*limited evidence, medium agreement*). This includes risks of corruption in REDD+ processes (Sheng et al. 2016; Williams and Dupuy 2018) and of corruption or elite capture in broader forest governance (Sundström 2016; Persha and Andersson 2014), as well as elite capture of benefits from planned adaptation at a local level (Sovacool 2018).

Peer-reviewed empirical studies that focus on corruption in climate finance and interventions, particularly at a local level, are rare, due in part to the obvious difficulties of researching illegal and clandestine activity (Fadairo et al. 2017). At the country level, historical levels of corruption are shown to affect current climate polices and global cooperation (Fredriksson and Neumayer 2016). Brown (2010) sees three likely inlets of corruption into REDD+: in the setting of forest baselines, the reconciliation of project and natural credits, and the implementation of control of illegal logging. The transnational and north-south dimensions of corruption are highlighted by debates on which US legislative instruments (e.g., the Lacey Act, the Foreign Corrupt Practices Act) could be used to prosecute the northern corporations that are involved in illegal logging (Gordon 2016; Waite 2011).

Fadairo et al. (2017) carried out a structured survey of perceptions of households in forest-edge communities served by REDD+, as well as those of local officials, in south eastern Nigeria. They report high rates of agreement that allocation of carbon rights is opaque and uncertain, distribution of benefits is untimely, uncertain and unpredictable, and the REDD+ decision-making process is vulnerable to political interference that benefits powerful individuals. Only 35% of respondents had an overall perception of transparency in REDD+ process as 'good'. Of eight institutional processes or facilities previously identified by the government of Nigeria and international agencies as indicators of commitment to transparent and equitable governance, only three were evident in the local REDD+ office as 'very functional' or 'fairly functional'.

At the local level, the risks of corruption and elite capture of the benefits of climate action are high in decentralised regimes (Persha and Andersson 2014). Rahman (2018) discusses elicitation of bribes (by local-level government staff) and extortion (by criminals) to allow poor rural people to gather forest products. The results are a general undermining of households' adaptive capacity and perverse incentives to over-exploit forests once bribes have been paid, leading to over-extraction and biodiversity loss. Where there are pre-existing inequalities and conflict, participation processes need careful management and firm external agency to achieve genuine transformation and avoid elite capture (Rigon 2014). An illustration of the range of types of elite capture is given by Sovacool (2018) for adaptation initiatives including coastal afforestation, combining document review and key informant interviews in Bangladesh, with an analytical approach from political ecology. Four processes are discussed: enclosure, including land grabbing and preventing the poor establishing new land rights; exclusion of the poor from decision-making over adaptation; encroachment on the resources of the poor by new adaptation infrastructure; and entrenchment of community disempowerment through patronage. The article notes that observing these processes does not imply they are always present, nor that adaptation efforts should be abandoned.

7.4.9.5   Overcoming barriers

Policy instruments that strengthen agricultural producer assets or capital reduce vulnerability and overcome barriers to adaptation (Hurlbert 2018b, 2015b). Additional factors like formal education and knowledge of traditional farming systems, secure tenure rights, access to electricity and social institutions in rice-farming areas of Bangladesh have played a positive role in reducing adaptation barriers (Alam 2015). A review of more than 168 publications over 15 years about adaptation of water resources for irrigation in Europe found the highest potential for action is in improving adaptive capacity and responding to changes in water demands, in conjunction with alterations in current water policy, farm extension training, and viable financial instruments (Iglesias and Garrote 2015). Research on the Great Barrier Reef, the Olifants River in Southern Africa, and fisheries in Europe, North America, and the Antarctic Ocean, suggests that the leading factor in harnessing the adaptive capacity of ecosystems is to reduce human stressors by enabling actors to collaborate across diverse interests, institutional settings, and sectors

(Biggs et al. 2017; Schultz et al. 2015; Johnson and Becker 2015). Fostering equity and participation are correlated with the efficacy of local adaptation to secure food and livelihood security (Laube et al. 2012). In this chapter, we examine the literature surrounding appropriate policy instruments, decision-making, and governance practices to overcome limits and barriers to adaptation.

Incremental adaptation consists of actions where the central aim is to maintain the essence and integrity of a system or process at a given site, whereas transformational adaptation changes the fundamental attributes of a system in response to climate and its effects; the former is characterised as doing different things and the latter, doing things differently (Noble et al. 2014). Transformational adaptation is necessary in situations where there are hard limits to adaptation or it is desirable to address deficiencies in sustainability, adaptation, inclusive development and social equity (Kates et al. 2012; Mapfumo et al. 2016). In other situations, incremental changes may be sufficient (Hadarits et al. 2017).

Cross-Chapter Box 11 | Gender in inclusive approaches to climate change, land and sustainable development

Margot Hurlbert (Canada), Brigitte Baptiste (Colombia), Amber Fletcher (Canada), Marta Guadalupe Rivera Ferre (Spain), Darshini Mahadevia (India), Katharine Vincent (United Kingdom)

Gender is a key axis of social inequality that intersects with other systems of power and marginalisation – including race, culture, class/socio-economic status, location, sexuality, and age – to cause unequal experiences of climate change vulnerability and adaptive capacity. However, 'policy frameworks and strong institutions that align development, equity objectives, and climate have the potential to deliver "triple-wins"' (Roy et al. 2018), including enhanced gender equality. Gender in relation to this report is introduced in Chapter 1, referred to as a leverage point in women's participation in decisions relating to land desertification (Section 3.6.3), land degradation (Section 4.1.6), food security (Section 5.2.5.1), and enabling land and climate response options (Section 6.1.2.2).

Focusing on 'gender' as a relational and contextual construct can help avoid homogenising women as a uniformly and consistently vulnerable category (Arora-Jonsson 2011; Mersha and Van Laerhoven 2016; Ravera et al. 2016). There is high agreement that using a framework of intersectionality to integrate gender into climate change research helps to recognise overlapping and interconnected systems of power (Djoudi et al. 2016; Fletcher 2018; Kaijser and Kronsell 2014; Moosa and Tuana 2014; Thompson-Hall et al. 2016), which create particular inequitable experiences of climate change vulnerability and adaptation. Through this framework, both commonalities and differences may be found between the experiences of rural and urban women, or between women in high-income and low-income countries, for example.

In rural areas, women generally experience greater vulnerability than men, albeit through different pathways (Djoudi et al., 2016; Goh, 2012; Jost et al., 2016; Kakota, Nyariki, Mkwambisi, & Kogi-Makau, 2011). In masculinised agricultural settings of Australia and Canada, for example, climate adaptation can increase women's work on- and off-farm, but without increasing recognition for women's undervalued contributions (Alston et al. 2018a; Fletcher and Knuttila 2016). A study in rural Ethiopia found that male-headed households had access to a wider set of adaptation measures than female-headed households (Mersha and Van Laerhoven 2016).

Due to engrained patriarchal social structures and gendered ideologies, women may face multiple barriers to participation and decision-making in land-based adaptation and mitigation actions in response to climate change (high confidence) (Alkire et al. 2013a; Quisumbing et al. 2014). These barriers include: (i) disproportionate responsibility for unpaid domestic work, including care-giving activities (Beuchelt and Badstue 2013) and provision of water and firewood (UNEP, 2016); (ii) risk of violence in both public and private spheres, which restricts women's mobility for capacity-building activities and productive work outside the home (Day et al., 2005; Jost et al., 2016; UNEP, 2016); (iii) less access to credit and financing (Jost et al. 2016); (iv) lack of organisational social capital, which may help in accessing credit (Carroll et al. 2012); (v) lack of ownership of productive assets and resources (Kristjanson et al., 2014;

Cross-Chapter Box 11 (continued)

Meinzen-Dick et al., 2010), including land. Constraints to land access include not only state policies, but also customary laws (Bayisenge 2018) based on customary norms and religion that determine women's rights (Namubiru-Mwaura 2014a).

Differential vulnerability to climate change is related to inequality in rights-based resource access, established through formal and informal tenure systems. In only 37% of 161 developing and developed countries do men and women have equal rights to use and control land, and in 59% customary, traditional, and religious practices discriminate against women (OECD 2014), even if the law formally grants equal rights. Women play a significant role in agriculture, food security and rural economies globally, forming 43% of the agricultural labour force in developing countries (FAO, IFAD, UNICEF, & WHO, 2018, p. 102), ranging from 25% in Latin America (FAO, 2017, p. 89) to nearly 50% in Eastern Asia and Central and South Europe (FAO, 2017, p. 88) and 47% in Sub-Saharan Africa (FAO, 2017, pp. 88). Further, the share of women in agricultural employment has been growing in all developing regions except East Asia and Southeast Asia (FAO, 2017, p. 88). At the same time, women constitute less than 5% of landholders (with legal rights and/or use-rights (Doss et al. 2018a) in North Africa and West Asia, about 15% in Sub-Saharan Africa, 12% in Southern and Southeastern Asia, 18% in Latin America and Caribbean (FAO 2011b, p. 25), 10% in Bangladesh, 4% in Nigeria (FAO 2015c). Patriarchal structures and gender roles can also affect women's control over land in developed countries (Carter 2017; Alston et al. 2018b). Thus, longstanding gender inequality in land rights, security of tenure, and decision-making may constrict women's adaptation options (Smucker and Wangui 2016).

Adaptation options related to land and climate (see Chapter 6) may produce environment and development trade-offs as well as social conflicts (Hunsberger et al. 2017) and changes with gendered implications. Women's strong presence in agriculture provides an opportunity to bring gender dimensions into climate change adaptation, particularly regarding food security (Glemarec 2017; Jost et al. 2016; Doss et al. 2018b). Some studies point to a potentially emancipatory role played by adaptation interventions and strategies, albeit with some limitations depending on context. For example, in developing contexts, male out-migration may cause women in socially disadvantaged groups to engage in new livelihood activities, thus challenging gendered roles (Djoudi and Brockhaus 2011; Alston 2006). Collective action and agency of women in farming households, including widows, have led to prevention of crop failure, reduced workload, increased nutritional intake, increased sustainable water management, diversified and increased income and improved strategic planning (Andersson and Gabrielsson 2012). Women's waged labour can help stabilise income from more land- and climate-dependent activities such as agriculture, hunting, or fishing (Alston et al., 2018; Ford and Goldhar, 2012). However, in developed contexts like Australia, women's participation in off-farm employment may exacerbate existing masculinisation of agriculture (Clarke and Alston 2017).

Literature suggests that land-based mitigation measures may lead to land alienation, either through market or appropriation (acquisition) by the government, may interfere with traditional livelihoods in rural areas, and lead to decline in women's livelihoods (Hunsberger et al. 2017). If land alienation is not prevented, existing inequities and social exclusions may be reinforced (medium agreement) (Mustalahti and Rakotonarivo 2014; Chomba et al. 2016; Poudyal et al. 2016). These activities also can lead to land grabs, which remain a focal point for research and local activism (Borras Jr. et al. 2011; White et al. 2012; Lahiff 2015). Cumulative effects of land-based mitigation measures may put families at risk of poverty. In certain contexts, they lead to increased conflicts. In conflict situations, women are at risk of personal violence, including sexual violence (UNEP, 2016).

Policy instruments for gender-inclusive approaches to climate change, land and sustainable development

Integrating, or mainstreaming, gender into land and climate change policy requires assessments of gender-differentiated needs and priorities, selection of appropriate policy instruments to address barriers to women's sustainable land management (SLM), and selection of gender indicators for monitoring and assessment of policy (medium confidence) (Huyer et al. 2015a; Alston 2014). Important sex-disaggregated data can be obtained at multiple levels, including the intra-household level (Seager 2014; Doss et al. 2018b), village- and plot-level information (Theriault et al. 2017a), and through national surveys (Agarwal 2018a; Doss et al. 2015a). Gender-disaggregated data provides a basis for selecting, monitoring and reassessing policy instruments that account for gender-differentiated land and climate change needs (medium confidence) (Rao 2017a; Arora-Jonsson 2014; Theriault et al. 2017b; Doss et al. 2018b). While macro-level data can reveal ongoing gender trends in SLM, contextual data are important for revealing intersectional aspects, such as the difference made by family relations, socio-economic status, or cultural practices about land use and control (Rao 2017a; Arora-Jonsson 2014; Theriault et al. 2017b), as well as on security of land holding (Doss et al. 2018b). Indices such as the Women's Empowerment in Agriculture Index (Alkire et al. 2013b) may provide useful guidelines for quantitative data collection on gender and SLM, while qualitative studies can reveal the nature of agency and whether policies are likely to be accepted, or not, in the context of local structures, meanings, and social relations (Rao 2017b).

Cross-Chapter Box 11 (continued)

Women's economic empowerment, decision-making power and voice is a necessity in SLM decisions (Mello and Schmink 2017a; Theriault et al. 2017b). Policies that address barriers include: gender considerations as qualifying criteria for funding programmes or access to financing for initiatives; government transfers to women under the auspices of anti-poverty programmes; spending on health and education; and subsidised credit for women (medium confidence) (Jagger and Pender 2006; Van Koppen et al. 2013a; Theriault et al. 2017b; Agarwal 2018b). Training and extension for women to facilitate sustainable practices is also important (Mello and Schmink 2017b; Theriault et al. 2017b). Such training could be built into existing programmes or structures, such as collective microenterprise (Mello and Schmink 2017b). Huyer et al. (2015) suggest that information provision (e.g., information about SLM) could be effectively dispersed through women's community-based organisations, although not in such a way that it overwhelms these organisations or supersedes their existing missions. SLM programmes could also benefit from intentionally engaging men in gender-equality training and efforts (Fletcher 2017), thus recognising the relationality of gender. Recognition of the household level, including men's roles and power relations, can help avoid the decontextualised and individualistic portrayal of women as purely instrumental actors (Rao 2017b).

Technology, policy, and programmes that exacerbate women's workloads or reinforce gender stereotypes (MacGregor 2010; Huyer et al. 2015b), or which fail to recognise and value the contributions women already make (Doss et al. 2018b), may further marginalise women. Accordingly, some studies have described technological and labour interventions that can enhance sustainability while also decreasing women's workloads; for example, Vent et al. (2017) described the system of rice intensification as one such intervention. REDD+ initiatives need to be aligned with the Sustainable Development Goals (SDGs) to achieve complementary synergies with gender dimensions.

Secure land title and/or land access and control for women increases SLM by increasing women's conservation efforts, increasing their productive and environmentally beneficial agricultural investments, such as willingness to engage in tree planting and sustainable soil management (high confidence) as well as improving cash incomes (Higgins et al. 2018; Agarwal 2010; Namubiru-Mwaura 2014b; Doss et al. 2015b; Van Koppen et al. 2013b; Theriault et al. 2017b; Jagger and Pender 2006). According FAO (2011b, p. 5), if women had the same access to productive resources as men, the number of hungry people in the world could be reduced by 12–17%. Policies promoting secure land title include legal reforms at multiple levels, including national laws on land ownership, legal education, and legal aid for women on land ownership and access (Argawal 2018). Policies to increase women's access to land could occur through three main avenues of land acquisition: inheritance/family (Theriault et al. 2017b), state policy, and the market (Agarwal 2018). Rao (2017) recommends framing land rights as entitlements rather than as instrumental means to sustainability. This reframing may address persistent, pervasive gender inequalities (FAO 2015d).

7.5   Decision-making for climate change and land

The risks posed by climate change generate considerable uncertainty and complexity for decision-makers responsible for land-use decisions (*robust evidence, high agreement*). Decision-makers balance climate ambitions, encapsulated in the NDCs, with other SDGs, which will differ considerably across different regions, sociocultural conditions and economic levels (Griggs et al. 2014). The interactions across SDGs also factor into decision-making processes (Nilsson et al. 2016b). The challenge is particularly acute in least developed countries where a large share of the population is vulnerable to climate change. Matching the structure of decision-making processes to local needs while connecting to national strategies and international regimes is challenging (Nilsson and Persson 2012). This section explores methods of decision-making to address the risks and inter-linkages outlined in the above sections. As a result, this section outlines policy inter-linkages with SDGs and NDCs, trade-offs and synergies in specific measures, possible challenges as well as opportunities going forward.

Even in cases where uncertainty exists, there is *medium evidence* and *high agreement* in the literature that it need not present a barrier to taking action, and there are growing methodological developments and empirical applications to support decision-making. Progress has been made in identifying key sources of uncertainty and addressing them (Farber 2015; Lawrence et al. 2018; Bloemen et al. 2018). Many of these approaches involve principles of robustness, diversity, flexibility, learning, or choice editing (Section 7.5.2).

Since the IPCC's Fifth Assessment Report (*Foundations for Decision Making*) chapter on Contexts for Decision-making (Jones et al. 2014) considerable advances have been made in decision-making under uncertainty, both conceptually and in economics (Section 7.5.2), and in the social/qualitative research areas (Sections 7.5.3 and 7.5.4). In the land sector, the degree of uncertainty varies and is particularly challenging for climate change adaptation decisions (Hallegatte 2009; Wilby and Dessai 2010). Some types of agricultural production decisions can be made in short timeframes as changes are observed, and will provide benefits in the current time period (Dittrich et al. 2017).

7.5.1   Formal and informal decision-making

Informal decision-making facilitated by open platforms can solve problems in land and resource management by allowing evolution and adaptation, and incorporation of local knowledge (*medium confidence*) (Malogdos and Yujuico 2015a; Vandersypen et al. 2007). Formal centres of decision-making are those that follow fixed procedures (written down in statutes or moulded in an organisation backed by the legal system) and structures (Onibon et al. 1999). Informal centres of decision-making are those following customary norms and habits based on conventions (Onibon et al. 1999) where problems are ill-structured and complex (Waddock 2013).

7.5.1.1   Formal Decision Making

Formal decision-making processes can occur at all levels, including the global, regional, national and sub-national levels (Section 7.4.1). Formal decision-making support tools can be used, for example, by farmers, to answer 'what-if' questions as to how to respond to the effects of changing climate on soils, rainfall and other conditions (Wenkel et al. 2013).

Optimal formal decision-making is based on realistic behaviour of actors, important in land–climate systems, assessed through participatory approaches, stakeholder consultations and by incorporating results from empirical analyses. Mathematical simulations and games (Lamarque et al. 2013), behavioural models in land-based sectors (Brown et al. 2017), agent-based models and micro-simulations are examples useful to decision-makers (Bishop et al. 2013). These decision-making tools are expanded on in Section 7.5.2.

There are different ways to incorporate local knowledge, informal institutions and other contextual characteristics that capture non-deterministic elements, as well as social and cultural beliefs and systems more generally, into formal decision-making (*medium evidence, medium agreement*) (Section 7.6.4). Classic scientific methodologies now include participatory and interdisciplinary methods and approaches (Jones et al. 2014). Consequently, this broader range of approaches may capture informal and indigenous knowledge, improving the participation of indigenous peoples in decision-making processes, and thereby promote their rights to self-determination (Malogdos and Yujuico 2015b) (Cross-Chapter Box 13 in Chapter 7).

7.5.1.2   Informal decision-making

Informal institutions have contributed to sustainable resources management (common pool resources) through creating a suitable environment for decision-making. The role of informal institutions in decision-making can be particularly relevant for land-use decisions and practices in rural areas in the global south and north (Huisheng 2015). Understanding informal institutions is crucial for adapting to climate change, advancing technological adaptation measures, achieving comprehensive disaster management and advancing collective decision-making (Karim and Thiel 2017). Informal institutions have been found to be a crucial entry point in dealing with vulnerability of communities and exclusionary tendencies impacting on marginalised and vulnerable people (Mubaya and Mafongoya 2017).

Many studies underline the role of local/informal traditional institutions in the management of natural resources in different parts of the world (Yami et al. 2009; Zoogah et al. 2015; Bratton 2007; Mowo et al. 2013; Grzymala-Busse 2010). Traditional systems include: traditional silvopastoral management (Iran), management of rangeland resources (South Africa), natural resource management (Ethiopia, Tanzania, Bangladesh) communal grazing land management (Ethiopia) and management of conflict over natural resources (Siddig et al. 2007; Yami et al. 2011; Valipour et al. 2014; Bennett 2013; Mowo et al. 2013).

Formal–informal institutional interaction could take different shapes such as: complementary, accommodating, competing, and substitutive. There are many examples when formal institutions might obstruct, change, and hinder informal institutions (Rahman et al. 2014; Helmke and Levitsky 2004; Bennett 2013; Osei-Tutu et al. 2014). Similarly, informal institutions can replace, undermine, and reinforce formal institutions (Grzymala-Busse 2010). In the absence of formal institutions, informal institutions gain importance, requiring focus in relation to natural resources management and rights protection (Estrin and Prevezer 2011; Helmke and Levitsky 2004; Kangalawe et al. 2014; Sauerwald and Peng 2013; Zoogah et al. 2015).

Community forestry comprises 22% of forests in tropical countries in contrast to large-scale industrial forestry (Hajjar et al. 2013) and is managed with informal institutions, ensuring a sustainable flow of forest products and income, utilising traditional ecological knowledge to determine access to resources (Singh et al. 2018). Policies that create an open platform for local debates and allow actors their own active formulation of rules strengthen informal institutions. Case studies in Zambia, Mali, Indonesia and Bolivia confirm that enabling factors for advancing the local ownership of resources and crafting durability of informal rules require recognition in laws, regulations and policies of the state (Haller et al. 2016).

7.5.2   Decision-making, timing, risk, and uncertainty

This section assesses decision-making literature, concluding that advances in methods have been made in the face of conceptual risk literature and, together with a synthesis of empirical evidence, near-term decisions have significant impact on costs.

7.5.2.1   Problem structuring

Structured decision-making occurs when there is scientific knowledge about cause and effect, little uncertainty, and agreement exists on values and norms relating to an issue (Hurlbert and Gupta 2016). This decision space is situated within the 'known' space where cause and effect is understood and predictable (although uncertainty is not quite zero) (French 2015). Figure 7.5 displays the structured problem area in the bottom left-hand corner corresponding with the 'known' decision-making space. Decision-making surrounding quantified risk assessment and risk management (Section 7.4.3.1) occurs within this decision-making space. Examples in the land and climate area include cost-benefit analysis surrounding implementation of irrigation projects (Batie 2008) or adopting soil erosion practices by agricultural producers based on anticipated profit (Hurlbert 2018b). Comprehensive risk management also occupies this decision space (Papathoma-Köhle et al. 2016), encompassing risk assessment, reduction, transfer, retention, emergency preparedness and response, and disaster recovery by combining quantified proactive and reactive approaches (Fra.Paleo 2015) (Section 7.4.3).

A moderately structured decision space is characterised as one where there is either some disagreement on norms, principles, ends and goals in defining a future state, or there is some uncertainty surrounding land and climate including land use, observations of land-use changes, early warning and decision support systems, model structures, parameterisations, inputs, or from unknown futures informing integrated assessment models and scenarios (see Chapter 1, Section 1.2.2 and Cross-Chapter Box 1 in Chapter 1). Environmental decision-making often takes place in this space where there is limited information and ability to process it, and individual stakeholders make different decisions on the best future course of action (*medium confidence*) (Waas et al. 2014; Hurlbert and Gupta 2016, 2015; Hurlbert 2018b). Figure 7.5 displays the moderately structured problem space characterised by disagreement surrounding norms on the top left-hand side. This corresponds with the complex decision-making space, the realm of social sciences and qualitative knowledge, where cause and effect is difficult to relate with any confidence (French 2013).

The moderately structured decision space characterised by uncertainty surrounding land and climate on the bottom right-hand side of Figure 7.5 corresponds to the knowable decision-making space, where the realm of scientific inquiry investigates cause and effects. Here there is sufficient understanding to build models, but not enough understanding to define all parameters (French 2015).

The top right-hand corner of Figure 7.5 corresponds to the 'unstructured' problem or chaotic space where patterns and relationships are difficult to discern and unknown unknowns reside (French 2013). It is in the complex but knowable space, the structured and moderately structured space, that decision-making under uncertainty occurs.

7.5.2.2   Decision-making tools

Decisions can be made despite uncertainty (*medium confidence*), and a wide range of possible approaches are emerging to support decision-making under uncertainty (Jones et al. 2014), applied both to adaptation and mitigation decisions.

Traditional approaches for economic appraisal, including cost-benefit analysis and cost-effectiveness analysis referred to in Section 7.5.2.1 do not handle or address uncertainty well (Hallegatte 2009; Farber 2015) and favour decisions with short-term benefits (see Cross-Chapter Box 10 in this chapter). Alternative economic decision-making approaches aim to better incorporate uncertainty while delivering adaptation goals, by selecting projects that meet their purpose across a variety of plausible futures (Hallegatte et al.

2012) – so-called 'robust' decision-making approaches. These are designed to be less sensitive to uncertainty about the future (Lempert and Schlesinger 2000).

Much of the research for adaptation to climate change has focused around three main economic approaches: real options analysis, portfolio analysis, and robust decision-making. Real options analysis develops flexible strategies that can be adjusted when additional climate information becomes available. It is most appropriate for large irreversible investment decisions. Applications to climate adaptation are growing quickly, with most studies addressing flood risk and sea-level rise (Gersonius et al. 2013; Woodward et al. 2014; Dan 2016), but studies in land-use decisions are also emerging, including identifying the optimal time to switch land use in a changing climate (Sanderson et al. 2016) and water storage (Sturm et al. 2017; Kim et al. 2017). Portfolio analysis aims to reduce risk by diversification, by planting multiple species rather than only one, for example, in forestry (Knoke et al. 2017) or crops (Ben-Ari and Makowski 2016), or in multiple locations. There may be a trade-off between robustness to variability and optimality (Yousefpour and Hanewinkel 2016; Ben-Ari and Makowski 2016); but this type of analysis can help identify and quantify trade-offs. Robust decision-making identifies how different strategies perform under many climate outcomes, also potentially trading off optimality for resilience (Lempert 2013).

Multi-criteria decision-making continues to be an important tool in the land-use sector, with the capacity to simultaneously consider multiple goals across different domains (e.g., economic, environmental, social) (Bausch et al. 2014; Alrø et al. 2016), and so is useful as a mitigation as well as an adaptation tool. Lifecycle assessment can also be used to evaluate emissions across a system – for example, in livestock production (McClelland et al. 2018) – and to identify areas to prioritise for reductions. Bottom-up marginal abatement cost curves calculate the most cost effective cumulative potential for mitigation across different options (Eory et al. 2018).

In the climate adaptation literature, these tools may be used in adaptive management (Section 7.5.4), using a monitoring, research, evaluation and learning process (cycle) to improve future management strategies (Tompkins and Adger 2004). More recently these techniques have been advanced with iterative risk management (IPCC 2014a) (Sections 7.4.1 and 7.4.7), adaptation pathways (Downing 2012), and dynamic adaptation pathways (Haasnoot et al. 2013) (Section 7.6.3). Decision-making tools can be selected and adapted to fit the specific land and climate problem and decision-making space. For instance, dynamic adaptation pathways processes (Haasnoot et al. 2013; Wise et al. 2014) identify and sequence potential actions based on alternative potential futures and are situated within the complex, unstructured space (see Figure 7.5). Decisions are made based on trigger points, linked to indicators and scenarios, or changing performance over time (Kwakkel et al. 2016). A key characteristic of these pathways is that, rather than making irreversible decisions now, decisions evolve over time, accounting for learning (Section 7.6.4), knowledge, and values. In New Zealand, combining dynamic adaptive pathways and a form of real options analysis with multiple-criteria decision analysis has enabled risk that changes over time to be included in the assessment of adaptation options through a participatory learning process (Lawrence et al. 2019).

Scenario analysis is also situated within the complex, unstructured space (although, unlike adaptation pathways, it does not allow for changes in pathway over time) and is important for identifying technology and policy instruments to ensure spatial-temporal coherence of land-use allocation simulations with scenario storylines (Brown and Castellazzi 2014) and identifying technology and policy instruments for mitigation of land degradation (Fleskens et al. 2014).

While economics is usually based on the idea of a self-interested, rational agent, more recently insights from psychology are being used to understand and explain human behaviour in the field of behavioural economics (Shogren and Taylor 2008; Kesternich et al. 2017), illustrating how a range of cognitive factors and biases can affect choices (Valatin et al. 2016). These insights can be critical in supporting decision-making that will lead to more desirable outcomes relating to land and climate change. One example of this is 'policy nudges' (Thaler and Sunstein 2008) which can 'shift choices in socially desirable directions' (Valatin et al. 2016). Tools can include framing tools, binding pre-commitments, default settings, channel factors, or broad choice bracketing (Wilson et al. 2016). Although relatively few empirical examples exist in the land sector, there is evidence that nudges could be applied successfully, for example, in woodland creation (Valatin et al. 2016) and agri-environmental schemes (Kuhfuss et al. 2016) (*medium certainty, low evidence*).

Consumers can be 'nudged' to consume less meat (Rozin et al. 2011) or to waste less food (Kallbekken and Sælen 2013).

Programmes supporting and facilitating desired practices can have success at changing behaviour, particularly if they are co-designed by the end-users (farmers, foresters, land users) (*medium evidence, high agreement*). Programmes that focus on demonstration or trials of different adaptation and mitigation measures, and facilitate interaction between farmers and industry specialists are perceived as being successful (Wreford et al. 2017; Hurlbert 2015b) but systematic evaluations of their success at changing behaviour are limited (Knook et al. 2018).

Different approaches to decision-making are appropriate in different contexts. Dittrich et al. (2017) provide a guide to the appropriate application in different contexts for adaptation in the livestock sector in developed countries. While considerable advances have been made in theoretical approaches, a number of challenges arise when applying these in practice, and partly relate to the necessity of assigning probabilities to climate projects, and the complexity of the approaches being a prohibitive factor beyond academic exercises. Formalised expert judgement can improve how uncertainty is characterised (Kunreuther et al. 2014) and these methods have been improved utilising Bayesian belief networks to synthesise expert judgements and include fault trees and reliability block diagrams to overcome standard reliability techniques (Sigurdsson et al. 2001) as well as mechanisms incorporating transparency (Ashcroft et al. 2016).

Categorisation of climate change decision tools against the decision-making process

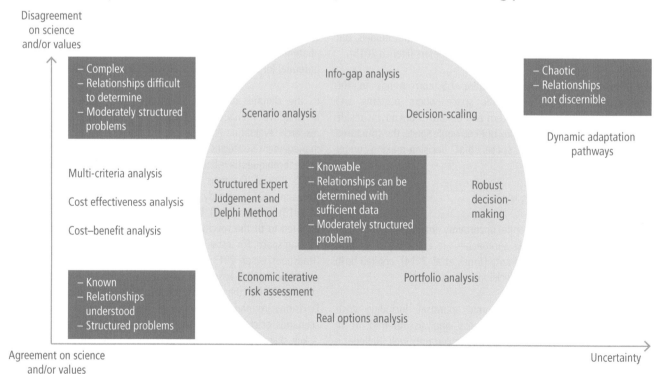

Figure 7.5 | Structural and uncertain decision making.

It may also be beneficial to combine decision-making approaches with the precautionary principle, or the idea that lack of scientific certainty is not to postpone action when faced with serious threats or irreversible damage to the environment (Farber 2015). The precautionary principle requires cost-effective measures to address serious but uncertain risks (Farber 2015). It supports a rights-based policy instrument choice as consideration is whether actions or inactions harm others moving beyond traditional risk-management policy considerations that surround net benefits (Etkin et al. 2012). Farber, (2015) concludes that the principle has been successfully applied in relation to endangered species and situations where climate change is a serious enough problem to justify some response. There is *medium confidence* that combining the precautionary principle with integrated assessment models, risk management, and cost-benefit analysis in an integrated, holistic manner, would be a good combination of decision-making tools supporting sustainable development (Farber 2015; Etkin et al. 2012).

7.5.2.3 Cost and timing of action

The Cross-Chapter Box 10 on Economic dimensions of climate change and land deals with the costs and timing of action. In terms of policies, not only is timing important, but the type of intervention itself can influence returns (*high evidence, high agreement*). Policy packages that make people more resilient – expanding financial inclusion, disaster risk and health insurance, social protection and adaptive safety nets, contingent finance and reserve funds, and universal access to early warning systems (Sections 7.4.1 and 7.6.3) – could save 100 billion USD a year, if implemented globally (Hallegatte et al. 2017). In Ethiopia, Kenya and Somalia, every 1 USD spent on safety-net/resilience programming results in net benefits of between 2.3 and 3.3 USD (Venton 2018). Investing in resilience-building activities, which increase household income by 365 to 450 USD per year in these countries, is more cost effective than providing ongoing humanitarian assistance.

There is a need to further examine returns on investment for land-based adaptation measures, both in the short and long term. Other outstanding questions include identifying specific triggers for early response. Food insecurity, for example, can occur due to a mixture of market and environmental factors (changes in food prices, animal or crop prices, rainfall patterns) (Venton 2018). The efficacy of different triggers, intervention times and modes of funding are currently being evaluated (see, for example, forecast-based finance study; Alverson and Zommers 2018). To reduce losses and maximise returns on investment, this information can be used to develop: 1) coordinated, agreed plans for action; 2) a clear, evidence-based decision-making process, and; 3) financing models to ensure that the plans for early action can be implemented (Clarke and Dercon 2016a).

7.5.3 Best practices of decision-making toward sustainable land management (SLM)

Sustainable land management (SLM) is a strategy and also an outcome (Waas et al. 2014) and decision-making practices are fundamental in achieving it as an outcome (*medium evidence,*

medium agreement). SLM decision-making is improved (*medium evidence* and *high agreement*) with ecological service mapping with three characteristics: robustness (robust modelling, measurement, and stakeholder-based methods for quantification of ES supply, demand and/or flow, as well as measures of uncertainty and heterogeneity across spatial and temporal scales and resolution); transparency (to contribute to clear information-sharing and the creation of linkages with decision support processes); and relevancy to stakeholders (people-centric in which stakeholders are engaged at different stages) (Willemen et al. 2015; Ashcroft et al. 2016). Practices that advance SLM include remediation practices, as well as critical interventions that are reshaping norms and standards, joint implementation, experimentation, and integration of rural actors' agency in analysis and approaches in decision-making (Hou and Al-Tabbaa 2014). Best practices are identified in the literature after their implementation demonstrates effectiveness at improving water quality, the environment, or reducing pollution (Rudolph et al. 2015; Lam et al. 2011).

There is *medium evidence* and *medium agreement* about what factors consistently determine the adoption of agricultural best management practices (Herendeen and Glazier 2009) and these positively correlate to education levels, income, farm size, capital, diversity, access to information, and social networks. Attending workshops for information and trust in crop consultants are also important factors in adoption of best management practices (Ulrich-Schad et al. 2017; Baumgart-Getz et al. 2012). More research is needed on the sustained adoption of these factors over time (Prokopy et al. 2008).

There is *medium evidence* and *high agreement* that SLM practices and incentives require mainstreaming into relevant policy; appropriate market-based approaches, including payment for ES and public-private partnerships, need better integration into payment schemes (Tengberg et al. 2016). There is *medium evidence* and *high agreement* that many of the best SLM decisions are made with the participation of stakeholders and social learning (Section 7.6.4) (Stringer and Dougill 2013). As stakeholders may not be in agreement, either practices of mediating agreement, or modelling that depicts and mediates the effects of stakeholder perceptions in decision-making may be applicable (Hou 2016; Wiggering and Steinhardt 2015).

7.5.4 Adaptive management

Adaptive management is an evolving approach to natural resource management founded on decision-making approaches in other fields (such as business, experimental science, and industrial ecology) (Allen et al. 2011; Williams 2011) and decision-making that overcomes management paralysis and mediates multiple stakeholder interests through use of simple steps. Adaptive governance considers a broader socio-ecological system that includes the social context that facilitates adaptive management (Chaffin et al. 2014). Adaptive management steps include evaluating a problem and integrating planning, analysis and management into a transparent process to build a road map focused on achieving fundamental objectives. Requirements of success are clearly articulated objectives, the explicit acknowledgment of uncertainty, and a transparent response to all stakeholder interests in the decision-making process (Allen et al. 2011). Adaptive management builds on this foundation by incorporating a formal iterative process, acknowledging uncertainty and achieving management objectives through a structured feedback process that includes stakeholder participation (Foxon et al. 2009) (Section 7.6.4). In the adaptive management process, the problem and desired goals are identified, evaluation criteria formulated, the system boundaries and context are ascertained, trade-offs evaluated, decisions are made regarding responses and policy instruments, which are implemented, and monitored, evaluated and adjusted (Allen et al. 2011). The implementation of policy strategies and monitoring of results occurs in a continuous management cycle of monitoring, assessment and revision (Hurlbert 2015b; Newig et al. 2010; Pahl-Wostl et al. 2007), as illustrated in Figure 7.6.

A key focus on adaptive management is the identification and reduction of uncertainty (as described in Chapter 1, Section 1.2.2 and Cross-Chapter Box 1 on Scenarios) and partial controllability, whereby policies used to implement an action are only indirectly responsible (for example, setting a harvest rate) (Williams 2011). There is *medium evidence* and *high agreement* that adaptive management is an ideal method to resolve uncertainty when uncertainty and controllability (resources will respond to management) are both high (Allen et al. 2011). Where uncertainty is high, but controllability is low, developing and analysing scenarios may be more appropriate (Allen et al. 2011). Anticipatory governance has developed combining scenarios and forecasting in order to creatively design strategy to address 'complex, fuzzy and wicked challenges' (Ramos 2014; Quay 2010) (Section 7.5). Even where there is low controllability, such as in the case of climate change, adaptive management can help mitigate impacts, including changes in water availability and shifting distributions of plants and animals (Allen et al. 2011).

There is *medium evidence* and *high agreement* that adaptive management can help reduce anthropogenic impacts of changes of land and climate, including: species decline and habitat loss (participative identification, monitoring, and review of species at risk as well as decision-making surrounding protective measures) (Fontaine 2011; Smith 2011) including quantity and timing of harvest of animals (Johnson 2011a), human participation in natural resource-based recreational activities, including selection fish harvest quotas and fishing seasons from year to year (Martin and Pope 2011), managing competing interests of land-use planners and conservationists in public lands (Moore et al. 2011), managing endangered species and minimising fire risk through land-cover management (Breininger et al. 2014), land-use change in hardwood forestry through mediation of hardwood plantation forestry companies and other stakeholders, including those interested in water, environment or farming (Leys and Vanclay 2011), and SLM protecting biodiversity, increasing carbon storage, and improving livelihoods (Cowie et al. 2011). There is *medium evidence* and *medium agreement* that, despite abundant literature and theoretical explanation, there has remained imperfect realisation of adaptive management because of several challenges: lack of clarity in definition and approach, few success stories on which to build an experiential base practitioner knowledge of adaptive management, paradigms surrounding management, policy and funding that favour reactive approaches instead of the proactive adaptive management approach, shifting objectives that do not allow for the application of the approach, and failure to acknowledge social uncertainty (Allen et al. 2011). Adaptive management includes participation (Section 7.6.4), the use of indicators (Section 7.5.5), in order to avoid maladaptation and trade-offs while maximising synergies (Section 7.5.6).

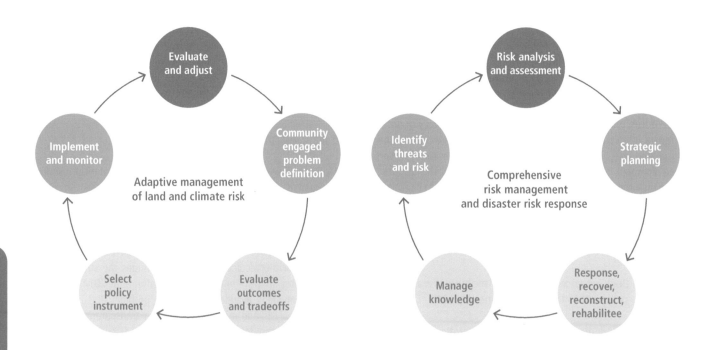

Figure 7.6 | Adaptive governance, management and comprehensive iterative risk management. Source: Adapted from Ammann 2013; Allen et al. 2011.

7.5.5   Performance indicators

Measuring performance is important in adaptive management decision-making, policy instrument implementation and governance, and can help evaluate policy effectiveness (*medium evidence, high agreement*) (Wheaton and Kulshreshtha 2017; Bennett and Dearden 2014; Oliveira Júnior et al. 2016; Kaufmann 2009). Indicators can relate to specific policy problems (climate mitigation, land degradation), sectors (agriculture, transportation, etc.), and policy goals (SDGs, food security).

It is necessary to monitor and evaluate the effectiveness and efficiency of performing climate actions to ensure the long-term success of climate initiatives or plans. Measurable indicators are useful for climate policy development and decision-making processes since they can provide quantifiable information regarding the progress of climate actions. The Paris Agreement (UNFCCC 2015) focused on reporting the progress of implementing countries' pledges – that is, NDCs and national adaptation needs in order to examine the aggregated results of mitigation actions that have already been implemented. For the case of measuring progress toward achieving LDN, it was suggested to use land-based indicators – that is, trends in land cover and land productivity or functioning of the land, and trends in carbon stock above and below ground (Cowie et al. 2018a). There is *medium evidence* and *high agreement* that indicators for measuring biodiversity and ES in response to governance at local to international scales meet the criteria of parsimony and scale specificity, are linked to some broad social, scientific and political consensus on desirable states of ecosystems and biodiversity, and include normative aspects such as environmental justice or socially just conservation (Layke 2009; Van Oudenhoven et al. 2012; Turnhout et al. 2014; Häyhä and Franzese 2014; Guerry et al. 2015; Díaz et al. 2015).

Important in making choices of metrics and indicators is understanding that the science, linkages and dynamics in systems are complex, not amenable to be addressed by simple economic instruments, and are often unrelated to short-term management or governance scales (Naeem et al. 2015; Muradian and Rival 2012). Thus, ideally, stakeholders participate in the selection and use of indicators for biodiversity and ES and monitoring impacts of governance and management regimes on land–climate interfaces. The adoption of non-economic approaches that are part of the emerging concept of Nature's Contributions to People (NCP) could potentially elicit support for conservation from diverse sections of civil society (Pascual et al. 2017).

Recent studies increasingly incorporate the role of stakeholders and decision-makers in the selection of indicators for land systems (Verburg et al. 2015) including sustainable agriculture (Kanter et al. 2016), bioenergy sustainability (Dale et al. 2015), desertification (Liniger et al. 2019), and vulnerability (Debortoli et al. 2018). Kanter et al. (2016) propose a four-step 'cradle-to-grave' approach for agriculture trade-off analysis, which involves co-evaluation of indicators and trade-offs with both stakeholders and decision-makers.

7.5.6   Maximising synergies and minimising trade-offs

Synergies and trade-offs to address land and climate-related measures are identified and discussed in Chapter 6. Here we outline policies supporting Chapter 6 response options (see Table 7.5), and discuss synergies and trade-offs in policy choices and interactions among policies. Trade-offs will exist between broad policy approaches. For example, while legislative and regulatory approaches may be effective at achieving environmental goals, they may be costly and ideologically unattractive in some countries. Market-driven approaches such as carbon pricing are cost-effective ways to reduce emissions, but may not be favoured politically and economically (Section 7.4.4). Information provision involves little political risk or ideological constraints, but behavioural barriers may limit their effectiveness (Henstra 2016). This level of trade-off is often determined by the prevailing political system.

Synergies and trade-offs also result from interaction between policies (policy interplay; Urwin and Jordan 2008) at different levels of policy (vertical) and across different policies (horizontal) (Section 7.4.8). If policy mixes are designed appropriately, acknowledging and incorporating trade-offs and synergies, they are better placed to deliver an outcome such as transitioning to sustainability (Howlett and Rayner 2013; Huttunen et al. 2014) (*medium evidence* and *medium agreement*). However, there is *limited evidence* and *medium agreement* that evaluating policies for coherence in responding to climate change and its impacts is not occurring, and policies are instead reviewed in a fragmented manner (Hurlbert and Gupta 2016).

Table 7.5 | Selection of policies/programmes/instruments that support response options.

Category	Integrated response option	Policy instrument supporting response option
Land management in agriculture	Increased food productivity	Investment in agricultural research for crop and livestock improvement, agricultural technology transfer, inland capture fisheries and aquaculture {7.4.7} agricultural policy reform and trade liberalisation
	Improved cropland, grazing, and livestock management	Environmental farm programmes/agri-environment schemes, water-efficiency requirements and water transfer {3.7.5}, extension services
	Agroforestry	Payment for ecosystem services (ES) {7.4.6}
	Agricultural diversification	Elimination of agriculture subsidies {5.7.1}, environmental farm programmes, agri-environmental payments {7.4.6}, rural development programmes
	Reduced grassland conversion to cropland	Elimination of agriculture subsidies, remove insurance incentives, ecological restoration {7.4.6}
	Integrated water management	Integrated governance {7.6.2}, multi-level instruments {7.4.1}
Land management in forests	Forest management, reduced deforestation and degradation, reforestation and forest restoration, afforestation	REDD+, forest conservation regulations, payments for ES, recognition of forest rights and land tenure {7.4.6}, adaptive management of forests {7.5.4}, land-use moratoriums, reforestation programmes and investment {4.9.1}
Land management of soils	Increased soil organic carbon content, reduced soil erosion, reduced soil salinisation, reduced soil compaction, biochar addition to soil	Land degradation neutrality (LDN) {7.4.5}, drought plans, flood plans, flood zone mapping {7.4.3}, technology transfer {7.4.4}, land-use zoning {7.4.6}, ecological service mapping and stakeholder-based quantification {7.5.3}, environmental farm programmes/agri-environment schemes, water-efficiency requirements and water transfer {3.7.5}
Land management in all other ecosystems	Fire management	Fire suppression, prescribed fire management, mechanical treatments {7.4.3}
	Reduced landslides and natural hazards	Land-use zoning {7.4.6}
	Reduced pollution – acidification	Environmental regulations, climate mitigation (carbon pricing) {7.4.4}
	Management of invasive species/encroachment	Invasive species regulations, trade regulations {5.7.2, 7.4.6}
	Restoration and reduced conversion of coastal wetlands	Flood zone mapping {7.4.3}, land-use zoning {7.4.6}
	Restoration and reduced conversion of peatlands	Payment for ES {7.4.6; 7.5.3}, standards and certification programmes {7.4.6}, land-use moratoriums
	Biodiversity conservation	Conservation regulations, protected areas policies
Carbon dioxide removal (CDR) land management	Enhanced weathering of minerals	No data
	Bioenergy and bioenergy with carbon capture and storage (BECCS)	Standards and certification for sustainability of biomass and land use {7.4.6}
Demand management	Dietary change	Awareness campaigns/education, changing food choices through nudges, synergies with health insurance and policy {5.7.2}
	Reduced post-harvest losses Reduced food waste (consumer or retailer), material substitution	Agricultural business risk programmes {7.4.8}; regulations to reduce and taxes on food waste, improved shelf life, circularising the economy to produce substitute goods, carbon pricing, sugar/fat taxes {5.7.2}
Supply management	Sustainable sourcing	Food labelling, innovation to switch to food with lower environmental footprint, public procurement policies {5.7.2}, standards and certification programmes {7.4.6}
	Management of supply chains	Liberalised international trade {5.7.2}, food purchasing and storage policies of governments, standards and certification programmes {7.4.6}, regulations on speculation in food systems
	Enhanced urban food systems	Buy local policies; land-use zoning to encourage urban agriculture, nature-based solutions and green infrastructure in cities; incentives for technologies like vertical farming
	Improved food processing and retailing, improved energy use in food systems	Agriculture emission trading {7.4.4}; investment in R&D for new technologies; certification
Risk management	Management of urban sprawl	Land-use zoning {7.4.6}
	Livelihood diversification	Climate-smart agriculture policies, adaptation policies, extension services {7.5.6}
	Disaster risk management	Disaster risk reduction {7.5.4; 7.4.3}, adaptation planning
	Risk-sharing instruments	Insurance, iterative risk management, CAT bonds, risk layering, contingency funds {7.4.3}, agriculture business risk portfolios {7.4.8}

Cross-Chapter Box 9 | Climate and land pathways
This is a duplicate of Cross-Chapter Box 9 in Chapter 6.

Katherine Calvin (The United States of America), Edouard Davin (France/Switzerland), Margot Hurlbert (Canada), Jagdish Krishnaswamy (India), Alexander Popp (Germany), Prajal Pradhan (Nepal/Germany)

Future development of socio-economic factors and policies influence the evolution of the land–climate system, among others, in terms of the land used for agriculture and forestry. Climate mitigation policies can also have a major impact on land use, especially in scenarios consistent with the climate targets of the Paris Agreement. This includes the use of bio-energy or CDR, such as bioenergy with carbon capture and storage (BECCS) and afforestation. Land-based mitigation options have implications for GHG fluxes, desertification, land degradation, food insecurity, ecosystem services and other aspects of sustainable development.

Shared Socio-economic Pathways
The five pathways are based on the Shared Socio-economic Pathways (SSPs) (O'Neill et al. 2014; Popp et al. 2017; Riahi et al. 2017; Rogelj et al. 2018b) (Cross-Chapter Box 1 in Chapter 1). SSP1 is a scenario with a broad focus on sustainability, including human development, technological development, nature conservation, globalised economy, economic convergence and early international cooperation (including moderate levels of trade). The scenario includes a peak and decline in population, relatively high agricultural yields and a move towards food produced in low-GHG emission systems (Van Vuuren et al. 2017b). Dietary change and reductions in food waste reduce agricultural demands, and effective land-use regulation enables reforestation and/or afforestation. SSP2 is a scenario in which production and consumption patterns, as well as technological development, follows historical patterns (Fricko et al. 2017). Land-based CDR is achieved through bioenergy and BECCS and, to a lesser degree, by afforestation and reforestation. SSP3 is a scenario with slow rates of technological change and limited land-use regulation. Agricultural demands are high due to material-intensive consumption and production, and barriers to trade lead to reduced flows for agricultural goods. In SSP3, forest mitigation activities and abatement of agricultural GHG emissions are limited due to major implementation barriers such as low institutional capacities in developing countries and delays as a consequence of low international cooperation (Fujimori et al. 2017). Emissions reductions are achieved primarily through the energy sector, including the use of bioenergy and BECCS.

Policies in the Pathways
SSPs are complemented by a set of shared policy assumptions (Kriegler et al. 2014), indicating the types of policies that may be implemented in each future world. Integrated Assessment Models (IAMs) represent the effect of these policies on the economy, energy system, land use and climate with the caveat that they are assumed to be effective or, in some cases, the policy goals (e.g., dietary change) are imposed rather than explicitly modelled. In the real world, there are various barriers that can make policy implementation more difficult (Section 7.4.9). These barriers will be generally higher in SSP3 than SSP1.

SSP1: A number of policies could support SSP1 in future, including: effective carbon pricing, emission trading schemes (including net CO_2 emissions from agriculture), carbon taxes, regulations limiting GHG emissions and air pollution, forest conservation (mix of land sharing and land sparing) through participation, incentives for ecosystem services and secure tenure, and protecting the environment, microfinance, crop and livelihood insurance, agriculture extension services, agricultural production subsidies, low export tax and import tariff rates on agricultural goods, dietary awareness campaigns, taxes on and regulations to reduce food waste, improved shelf life, sugar/fat taxes, and instruments supporting sustainable land management, including payment for ecosystem services, land-use zoning, REDD+, standards and certification for sustainable biomass production practices, legal reforms on land ownership and access, legal aid, legal education, including reframing these policies as entitlements for women and small agricultural producers (rather than sustainability) (Van Vuuren et al. 2017b; O'Neill, B.C. et al., 2017) (Section 7.4).

SSP2: The same policies that support SSP1 could support SSP2 but may be less effective and only moderately successful. Policies may be challenged by adaptation limits (Section 7.4.9), inconsistency in formal and informal institutions in decision-making (Section 7.5.1) or result in maladaptation (Section 7.4.7). Moderately successful sustainable land management policies result in some land competition. Land degradation neutrality is moderately successful. Successful policies include those supporting bioenergy and BECCS (Rao et al. 2017b; Fricko et al. 2017; Riahi et al. 2017) (Section 7.4.6).

SSP3: Policies that exist in SSP1 may or may not exist in SSP3, and are ineffective (O'Neill et al. 2014). There are challenges to implementing these policies, as in SSP2. In addition, ineffective sustainable land management policies result in competition for land between agriculture and mitigation. Land degradation neutrality is not achieved (Riahi et al. 2017). Successful policies include those supporting bioenergy and BECCS (Kriegler et al. 2017; Fujimori et al. 2017; Rao et al. 2017b) (Section 7.4.6). Demand-side food policies are absent and supply-side policies predominate. There is no success in advancing land ownership and access policies for agricultural producer livelihood (Section 7.6.5).

Cross-Chapter Box 9 (continued)

Land-use and land-cover change

In SSP1, sustainability in land management, agricultural intensification, production and consumption patterns result in reduced need for agricultural land, despite increases in per capita food consumption. This land can instead be used for reforestation, afforestation and bioenergy. In contrast, SSP3 has high population and strongly declining rates of crop yield growth over time, resulting in increased agricultural land area. SSP2 falls somewhere in between, with societal as well as technological development following historical patterns. Increased demand for land mitigation options such as bioenergy, reduced deforestation or afforestation decreases availability of agricultural land for food, feed and fibre. In the climate policy scenarios consistent with the Paris Agreement, bioenergy/BECCS and reforestation/afforestation play an important role in SSP1 and SSP2. The use of these options, and the impact on land, is larger in scenarios that limit radiative forcing in 2100 to 1.9 W m^{-2} than in the 4.5 W m^{-2} scenarios. In SSP3, the expansion of land for agricultural production implies that the use of land-related mitigation options is very limited, and the scenario is characterised by continued deforestation.

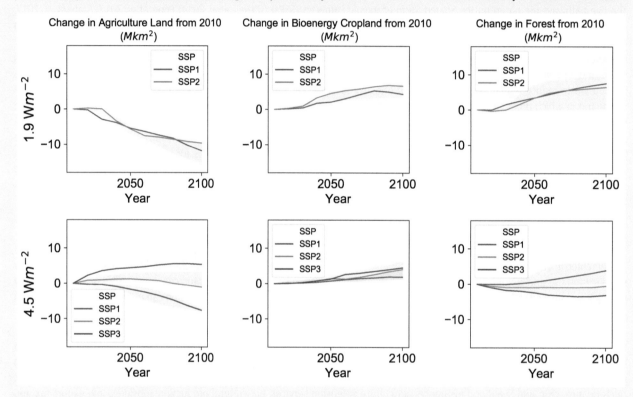

Cross-Chapter Box 9, Figure 1 | Changes in agriculture land (left), bioenergy cropland (middle) and forest (right) under three different SSPs (colours) and two different warming levels (rows). Agricultural land includes both pasture and cropland. Colours indicate SSPs, with SSP1 shown in green, SSP2 in yellow, and SSP3 in red. For each pathway, the shaded areas show the range across all IAMs; the line indicates the median across models. There is no SSP3 in the top row, as 1.9 W m^{-2} is infeasible in this world. Data is from an update of the Integrated Assessment Modelling Consortium (IAMC) Scenario Explorer developed for the SR15 (Huppmann et al. 2018; Rogelj et al. 2018a).

Implications for mitigation and other land challenges

The combination of baseline emissions development, technology options, and policy support makes it much easier to reach the climate targets in the SSP1 scenario than in the SSP3 scenario. As a result, carbon prices are much higher in SSP3 than in SSP1. In fact, the 1.9 W m^{-2} target was found to be infeasible in the SSP3 world (Table 1 in Cross-Chapter Box 9). Energy system CO_2 emissions reductions are greater in SSP3 than in SSP1 to compensate for the higher land-based CO_2 emissions.

Accounting for mitigation and socio-economics alone, food prices (an indicator of food insecurity) are higher in SSP3 than in SSP1 and higher in the 1.9 W m^{-2} target than in the 4.5 W m^{-2} target (Table 1 in Cross-Chapter Box 9). Forest cover is higher in SSP1 than SSP3 and higher in the 1.9 W m^{-2} target than in the 4.5 W m^{-2} target. Water withdrawals and water scarcity are, in general, higher in SSP3 than SSP1 (Hanasaki et al. 2013; Graham et al. 2018) and higher in scenarios with more bioenergy (Hejazi et al. 2014b); however, these indicators have not been quantified for the specific SSP-representative concentration pathways (RCP) combinations discussed here.

Cross-Chapter Box 9 (continued)

Cross-Chapter Box 9, Table 1 | Quantitative indicators for the pathways. Each cell shows the mean, minimum, and maximum value across IAM models for each indicator and each pathway in 2050 and 2100. All IAMs that provided results for a particular pathway are included here. Note that these indicators exclude the implications of climate change. Data is from an update of the IAMC Scenario Explorer developed for the SR15 (Huppmann et al. 2018; Rogelj et al. 2018b).

		SSP1		SSP2		SSP3	
		1.9 W m^{-2} mean (max., min.)	4.5 W m^{-2} mean (max., min.)	1.9 W m^{-2} mean (max., min.)	4.5 W m^{-2} mean (max., min.)	1.9 W m^{-2} mean (max., min.)	4.5 W m^{-2} mean (max., min.)
Population (billion)	2050	8.5 (8.5, 8.5)	8.5 (8.5, 8.5)	9.2 (9.2, 9.2)	9.2 (9.2, 9.2)	N/A	10.0 (10.0, 10.0)
	2100	6.9 (7.0, 6.9)	6.9 (7.0, 6.9)	9.0 (9.0, 9.0)	9.0 (9.1, 9.0)	N/A	12.7 (12.8, 12.6)
Change in GDP per capita (% rel to 2010)	2050	170.3 (380.1, 130.9)	175.3 (386.2, 166.2)	104.3 (223.4, 98.7)	110.1 (233.8, 103.6)	N/A	55.1 (116.1, 46.7)
	2100	528.0 (1358.4, 408.2)	538.6 (1371.7, 504.7)	344.4 (827.4, 335.8)	356.6 (882.2, 323.3)	N/A	71.2 (159.7, 49.6)
Change in forest cover (Mkm^2)	2050	3.4 (9.4, -0.1)	0.6 (4.2, −0.7)	3.4 (7.0, −0.9)	−0.9 (2.9, −2.5)	N/A	−2.4 (−1.0, −4.0)
	2100	7.5 (15.8, 0.4)	3.9 (8.8, 0.2)	6.4 (9.5, −0.8)	−0.5 (5.9, −3.1)	N/A	−3.1 (−0.3, −5.5)
Change in cropland (Mkm^2)	2050	−1.2 (−0.3, −4.6)	0.1 (1.5, −3.2)	−1.2 (0.3, −2.0)	1.2 (2.7, −0.9)	N/A	2.3 (3.0, 1.2)
	2100	−5.2 (−1.8, −7.6)	−2.3 (−1.6, −6.4)	−2.9 (0.1, −4.0)	0.7 (3.1, −2.6)	N/A	3.4 (4.5, 1.9)
Change in energy cropland (Mkm^2)	2050	2.1 (5.0, 0.9)	0.8 (1.3, 0.5)	4.5 (7.0, 2.1)	1.5 (2.1, 0.1)	N/A	1.3 (2.0, 1.3)
	2100	4.3 (7.2, 1.5)	1.9 (3.7, 1.4)	6.6 (11.0, 3.6)	4.1 (6.3, 0.4)	N/A	4.6 (7.1, 1.5)
Change in pasture (Mkm^2)	2050	−4.1 (−2.5, −5.6)	−2.4 (−0.9, −3.3)	−4.8 (−0.4, −6.2)	−0.1 (1.6, −2.5)	N/A	2.1 (3.8, −0.1)
	2100	−6.5 (−4.8, −12.2)	−4.6 (−2.7, −7.3)	−7.6 (−1.3, −11.7)	−2.8 (1.9, −5.3)	N/A	2.0 (4.4, −2.5)
Change in other natural land (Mkm^2)	2050	0.5 (1.0, −4.9)	0.5 (1.7, −1.0)	−2.2 (0.6, −7.0)	−2.2 (0.7, −2.2)	N/A	−3.4 (−2.0, −4.4)
	2100	0.0 (7.1, −7.3)	1.8 (6.0, −1.7)	−2.3 (2.7, −9.6)	−3.4 (1.5, −4.7)	N/A	−6.2 (−5.4, −6.8)
Carbon price (2010 USD per tCO$_2$)[a]	2050	510.4 (4304.0, 150.9)	9.1 (35.2, 1.2)	756.4 (1079.9, 279.9)	37.5 (73.4, 13.6)	N/A	67.2 (75.1, 60.6)
	2100	2164.0 (350, 37.7, 262.7)	64.9 (286.7, 42.9)	4353.6 (10149.7, 2993.4)	172.3 (597.9, 112.1)	N/A	589.6 (727.2, 320.4)
Food price (Index 2010=1)	2050	1.2 (1.8, 0.8)	0.9 (1.1, 0.7)	1.6 (2.0, 1.4)	1.1 (1.2, 1.0)	N/A	1.2 (1.7, 1.1)
	2100	1.9 (7.0, 0.4)	0.8 (1.2, 0.4)	6.5 (13.1, 1.8)	1.1 (2.5, 0.9)	N/A	1.7 (3.4, 1.3)
Increase in Warming above pre-industrial (°C)	2050	1.5 (1.7, 1.5)	1.9 (2.1, 1.8)	1.6 (1.7, 1.5)	2.0 (2.0, 1.9)	N/A	2.0 (2.1, 2.0)
	2100	1.3 (1.3, 1.3)	2.6 (2.7, 2.4)	1.3 (1.3, 1.3)	2.6 (2.7, 2.4)	N/A	2.6 (2.6, 2.6)
Change in per capita demand for food, crops (% rel to 2010)[b]	2050	6.0 (10.0, 4.5)	9.1 (12.4, 4.5)	4.6 (6.7, −0.9)	7.9 (8.0, 5.2)	N/A	2.4 (5.0, 2.3)
	2100	10.1 (19.9, 4.8)	15.1 (23.9, 4.8)	11.6 (19.2, −10.8)	11.7 (19.2, 4.1)	N/A	2.0 (3.4, −1.0)
Change in per capita demand for food, animal products (% rel to 2010)[b,c]	2050	6.9 (45.0, −20.5)	17.9 (45.0, −20.1)	7.1 (36.0, 1.9)	10.3 (36.0, −4.2)	N/A	3.1 (5.9, 1.9)
	2100	−3.0 (19.8, −27.3)	21.4 (44.1, −26.9)	17.0 (39.6, −24.1)	20.8 (39.6, −5.3)	N/A	−7.4 (−0.7, −7.9)
Agriculture, forestry and other land-use (AFOLU) CH$_4$ Emissions (% relative to 2010)	2050	−39.0 (−3.8, −68.9)	−2.9 (22.4, −23.9)	−11.7 (31.4, −59.4)	7.5 (43.0, −15.5)	N/A	15.0 (20.1, 3.1)
	2100	−60.5 (−41.7, −77.4)	−47.6 (−24.4, −54.1)	−40.3 (33.1, −58.4)	−13.0 (63.7, −45.0)	N/A	8.0 (37.6, −9.1)
AFOLU N$_2$O Emissions (% relative to 2010)	2050	−13.1 (−4.1, −26.3)	0.1 (34.6, −14.5)	8.8 (38.4, −14.5)	25.4 (37.4, 5.5)	N/A	34.0 (50.8, 29.3)
	2100	−42.0 (4.3, −49.4)	−25.6 (−3.4, −51.2)	−1.7 (46.8, −37.8)	19.5 (66.7, −21.4)	N/A	53.9 (65.8, 30.8)
Cumulative Energy CO$_2$ Emissions until 2100 (GtCO$_2$)		428.2 (1009.9, 307.6)	2787.6 (3213.3, 2594.0)	380.8 (552.8, −9.4)	2642.3 (2928.3, 2515.8)	N/A	2294.5 (2447.4, 2084.6)
Cumulative AFOLU CO$_2$ Emissions until 2100 (GtCO$_2$)		−127.3 (5.9, −683.0)	−54.9 (52.1, −545.2)	−126.8 (153.0, −400.7)	40.8 (277.0, −372.9)	N/A	188.8 (426.6, 77.9)

[a] SSP2–19 is infeasible in two models. One of these models sets the maximum carbon price in SSP1–19; the carbon price range is smaller for SSP2–19 as this model is excluded there. Carbon prices are higher in SSP2–19 than SSP1–19 for every model that provided both simulations.

[b] Food demand estimates include waste.

[c] Animal product demand includes meat and dairy.

Cross-Chapter Box 9 (continued)

Climate change results in higher impacts and risks in the 4.5 W m^{-2} world than in the 1.9 W m^{-2} world for a given SSP and these risks are exacerbated in SSP3 compared to SSP1 and SSP2 due to the population's higher exposure and vulnerability. For example, the risk of fire is higher in warmer worlds; in the 4.5 W m^{-2} world, the population living in fire prone regions is higher in SSP3 (646 million) than in SSP2 (560 million) (Knorr et al. 2016). Global exposure to multi-sector risk quadruples between 1.5°C[1] and 3°C and is a factor of six higher in SSP3-3°C than in SSP1-1.5°C (Byers et al. 2018). Future risks resulting from desertification, land degradation and food insecurity are lower in SSP1 compared to SSP3 at the same level of warming. For example, the transition moderate-to-high risk of food insecurity occurs between 1.3 and 1.7°C for SSP3, but not until 2.5 to 3.5°C in SSP1 (Section 7.2).

Summary

Future pathways for climate and land use include portfolios of response and policy options. Depending on the response options included, policy portfolios implemented, and other underlying socio-economic drivers, these pathways result in different land-use consequences and their contribution to climate change mitigation. Agricultural area declines by more than 5 Mkm^2 in one SSP but increases by as much as 5 Mkm^2 in another. The amount of energy cropland ranges from nearly zero to 11 Mkm^2, depending on the SSP and the warming target. Forest area declines in SSP3 but increases substantially in SSP1. Subsequently, these pathways have different implications for risks related to desertification, land degradation, food insecurity, and terrestrial GHG fluxes, as well as ecosystem services, biodiversity, and other aspects of sustainable development.

7.5.6.1 Trade-offs and synergies between ecosystem services (ES)

Unplanned or unintentional trade-offs and synergies between policy driven response options related to ecosystem services (ES) can happen over space (e.g., upstream-downstream, integrated watershed management, Section 3.7.5.2) or intensify over time (reduced water in future dry-season due to growing tree plantations, Section 6.4.1). Trade-offs can occur between two or more ES (land for climate mitigation vs food; Sections 6.2, 6.3, 6.4, Cross-Chapter Box 8 in Chapter 6; Cross-Chapter Box 9 in Chapters 6 and 7), and between scales, such as forest biomass-based livelihoods versus global ES carbon storage (Chhatre and Agrawal 2009) (*medium evidence, medium agreement*). Trade-offs can be reversible or irreversible (Rodríguez et al. 2006; Elmqvist et al. 2013) (for example, a soil carbon sink is reversible) (Section 6.4.1.1).

Although there is *robust evidence* and *high agreement* that ES are important for human well-being, the relationship between poverty alleviation and ES can be surprisingly complex, understudied and dependent on the political economic context; current evidence is largely about provisioning services and often ignores multiple dimensions of poverty (Suich et al. 2015; Vira et al. 2012). Spatially explicit mapping and quantification of stakeholder choices in relation to distribution of various ES can help enhance synergies and reduce trade-offs (Turkelboom et al. 2018; Locatelli et al. 2014) (Section 7.5.5).

7.5.6.2 Sustainable Development Goals (SDGs): Synergies and trade-offs

The Sustainable Development Goals (SDGs) are an international persuasive policy instrument that apply to all countries, and measure sustainable and socially just development of human societies at all scales of governance (Griggs et al. 2013). The UN SDGs rest on the premise that the goals are mutually reinforcing and there are inherent linkages, synergies and trade-offs (to a greater or lesser extent) between and within the sub-goals (Fuso Nerini et al. 2018; Nilsson et al. 2016b; Le Blanc 2015). There is *high confidence* that opportunities, trade-offs and co-benefits are context – and region-specific and depend on a variety of political, national and socio-economic factors (Nilsson et al. 2016b) depending on perceived importance by decision-makers and policymakers (Figure 7.7 and Table 7.6). Aggregation of targets and indicators at the national level can mask severe biophysical and socio-economic trade-offs at local and regional scales (Wada et al. 2016).

There is *medium evidence* and *high agreement* that SDGs must not be pursued independently, but in a manner that recognises trade-offs and synergies with each other, consistent with a goal of 'policy coherence'. Policy coherence also refers to spatial trade-offs and geopolitical implications within and between regions and countries implementing SDGs. For instance, supply-side food security initiatives of land-based agriculture are impacting on marine fisheries globally through creation of dead-zones due to agricultural run-off (Diaz and Rosenberg 2008).

SDGs 6 (clean water and sanitation), 7 (affordable and clean energy) and 9 (industry, innovation and infrastructure) are important SDGs related to mitigation with adaptation co-benefits, but they have local

---

[1] Pathways that limit radiative forcing in 2100 to 1.9 W m^{-2} result in median warming in 2100 to 1.5°C in 2100 (Rogelj et al. 2018b). Pathways limiting radiative forcing in 2100 to 4.5 W m^{-2} result in median warming in 2100 above 2.5°C (IPCC 2014).

trade-offs with biodiversity and competing uses of land and rivers (see Case study: Green energy: Biodiversity conservation vs global environment targets) (*medium evidence*, *high agreement*) (Bogardi et al. 2012; Nilsson and Berggren 2000; Hoeinghaus et al. 2009; Winemiller et al. 2016). This has occurred despite emerging knowledge about the role that rivers and riverine ecosystems play in human development and in generating global, regional and local ES (Nilsson and Berggren 2000; Hoeinghaus et al. 2009). The transformation of river ecosystems for irrigation, hydropower and water requirements of societies worldwide is the biggest threat to freshwater and estuarine biodiversity and ecosystems services (Nilsson and Berggren 2000; Vörösmarty et al. 2010). These projects address important energy and water-related demands, but their economic benefits are often overestimated in relation to trade-offs with respect to food (river capture fisheries), biodiversity and downstream ES (Winemiller et al. 2016). Some trade-offs and synergies related to SDG7 impact on aspirations of greater welfare and well-being, as well as physical and social infrastructure for sustainable development (Fuso Nerini et al. 2018) (Section 7.5.6.1, where trade-offs exist between climate mitigation and food).

There are also spatial trade-offs related to large river diversion projects and export of 'virtual water' through water-intensive crops produced in one region and exported to another, with implications for food security, water security and downstream ES of the exporting region (Hanasaki et al. 2010; Verma et al. 2009). Synergies include cropping adaptations that increase food system production and eliminate hunger (SDG2) (Rockström et al. 2017; Lipper et al. 2014a; Neufeldt et al. 2013). Well-adapted agricultural systems are shown to have synergies, positive returns on investment and contribute to safe drinking water, health, biodiversity and equity goals (DeClerck 2016). Assessing the water footprint of different sectors at the river basin scale can provide insights for interventions and decision-making (Zeng et al. 2012).

Sometimes the trade-offs in SDGs can arise in the articulation and nested hierarchy of 17 goals and the targets under them. In terms of aquatic life and ecosystems, there is an explicit SDG for sustainable management of marine life (SDG 14, Life below water). There is no equivalent goal exclusively for freshwater ecosystems, but hidden under SDG 6 (Clean water and sanitation) out of six listed targets, the sixth target is about protecting and restoring water-related ecosystems, which suggests a lower order of global priority compared to being listed as a goal in itself (e.g., SDG 14).

There is *limited evidence* and *limited agreement* that binary evaluations of individual SDGs and synergies and trade-offs that categorise interactions as either 'beneficial' or 'adverse' may be subjective and challenged further by the fact that feedbacks can often not be assigned as unambiguously positive or negative (Blanc et al. 2017). The IPCC Special Report on Global Warming of 1.5°C (SR15) notes: 'A reductive focus on specific SDGs in isolation may undermine the long-term achievement of sustainable climate change mitigation' (Holden et al. 2017). Greater work is needed to tease out these relationships; studies have started that include quantitative modelling (see Karnib 2017) and nuanced scoring scales (ICSU 2017) of these relationships.

A nexus approach is increasingly being adopted to explore synergies and trade-offs between a select subset of goals and targets (such as the interaction between water, energy and food – see for example, Yumkella and Yillia 2015; Conway et al. 2015; Ringler et al. 2015). However, even this approach ignores systemic properties and interactions across the system as a whole (Weitz et al. 2017a). Pursuit of certain targets in one area can generate rippling effects across the system, and these in turn can have secondary impacts on yet other targets. Weitz et al. (2017a) found that SDG target 13.2 (climate change policy/planning) is influenced by actions in six other targets. SDG 13.1 (climate change adaption) and also SDG 2.4 (food production) receive the most positive influence from progression in other targets.

There is *medium evidence* and *high agreement* that, to be effective, truly sustainable, and to reduce or mitigate emerging risks, SDGs need knowledge dissemination and policy initiatives that recognise and assimilate concepts of co-production of ES in socio-ecological systems, cross-scale linkages, uncertainty, spatial and temporal trade-offs between SDGs and ES that acknowledge biophysical, social and political constraints and understand how social change occurs at various scales (Rodríguez et al. 2006; Norström et al. 2014; Palomo et al. 2016). Several methods and tools are proposed in literature to address and understand SDG interactions. Nilsson et al. (2016a) suggest going beyond a simplistic framing of synergies and trade-offs to understanding the various relationship dimensions, and proposing a seven-point scale to understand these interactions.

This approach, and the identification of clusters of synergy, can help indicate that government ministries work together or establish collaborations to reach their specific goals. Finally, context-specific analysis is needed. Synergies and trade-offs will depend on the natural resource base (such as land or water availability), governance arrangements, available technologies, and political ideas in a given location (Nilsson et al. 2016b). Figure 7.7 shows that, at the global scale, there is less uncertainty in the evidence surrounding SDGs, but also less agreement on norms, priorities and values for SDG implementation. Although there is *some agreement* on the regional and local scale surrounding SDGs, there is *higher certainty* on the science surrounding ES.

Figure 7.7 and Table 7.6 | Risks at various scales, levels of uncertainty and agreement in relation to trade-offs among SDGs and other goals.

	Land-climate-society hazard	SDGs impacted by or involved in mutual trade-offs	Selected literature
a	Decline of freshwater and riverine ecosystems	2, 3, 6, 7, 8, 12, 16, 18	Falkenmark 2001; Zarfl et al. 2014; Canonico et al. 2005
b	Forest browning	3, 8, 13, 15	Verbyla 2011; Krishnaswamy et al. 2014; McDowell and Allen 2015b; Anderegg et al. 2013; Samanta et al. 2010
c	Exhaustion of groundwater	1, 3, 6, 8, 11, 12, 13, 18	Barnett and O'Neill 2010; Wada et al. 2010; Harootunian 2018; Dalin et al. 2017; Rockström, Johan Steffen et al. 2009; Falkenmark 2001
d	Loss of biodiversity	6, 7, 12, 15, 18	Pereira et al. 2010; Pascual et al. 2017; Pecl et al. 2017; Jumani et al. 2017, 2018
e	Extreme events in cities and towns	3, 6, 11, 13	Douglas et al. 2008; Stone et al. 2010; Chang et al. 2007; Hanson et al. 2011
f	Stranded assets	8, 9, 11, 12, 13	Ansar et al. 2013; Chasek et al. 2015; Melvin et al. 2017; Surminski 2013; Hallegatte et al. 2013; Larsen et al. 2008; Nicholls and Cazenave 2010
g	Expansion of the agricultural frontier into tropical forests	15, 13	Celentano et al. 2017; Nepstad et al. 2008; Bogaerts et al. 2017; Fearnside 2015; Beuchle et al. 2015; Grecchi et al. 2014
h	Food and nutrition security	2, 1, 3, 10, 11	Hasegawa et al. 2018a; Frank et al. 2017; Fujimori et al. 2018b; Zhao et al. 2017
i	Emergence of infectious diseases	3, 1, 6, 10, 11, 12, 13	Wu et al. 2016; Patz et al. 2004; McMichael et al. 2006; Young et al. 2017b; Smith et al. 2014a; Tjaden et al. 2017; Naicker 2011
j	Decrease in agricultural productivity	2, 1, 3, 10, 11, 13	Porter et al. 2014; Müller et al. 2013; Rosenzweig et al. 2014
k	Expansion of farm and fish ponds	1, 2, 3, 6, 8, 10, 13, 14	Kale 2017; Boonstra and Hanh 2015

Sustainable Development Goals

1. No poverty
2. Zero hunger
3. Good health and well-being
4. Quality education
5. Gender equality
6. Clean water and sanitation
7. Affordable and clean energy
8. Decent work and economic growth
9. Industry, innovation and infrastructure

10. Reduced inequality
11. Sustainable cities and communities
12. Responsible consumption and production
13. Climate action
14. Life below water
15. Life on Land
16. Peace and justice strong institutions
17. Partnerships to achieve the goals

7.5.6.3   Forests and agriculture

Retaining existing forests, restoring degraded forest and afforestation are response options for climate change mitigation with adaptation benefits (Section 6.4.1). Policies at various levels of governance that foster ownership, autonomy, and provide incentives for forest cover can reduce trade-offs between carbon sinks in forests and local livelihoods (especially when the size of forest commons is sufficiently large) (Chhatre and Agrawal 2009; Locatelli et al. 2014) (see Table 7.6 this section; Case study: Forest conservation instruments: REDD+ in the Amazon and India, Section 7.4.6).

Forest restoration for mitigation through carbon sequestration and other ES or co-benefits (e.g., hydrologic, non-timber forest products, timber and tourism) can be passive or active (although both types largely exclude livestock). Passive restoration is more economically viable in relation to restoration costs as well as co-benefits in other ES, calculated on a net present value basis, especially under flexible carbon credits (Cantarello et al. 2010). Restoration can be more cost effective with positive socio-economic and biodiversity conservation outcomes, if costly and simplistic planting schemes are avoided (Menz et al. 2013). Passive restoration takes longer to demonstrate co-benefits and net economic gains. It can be confused with land abandonment in some regions and countries, and therefore secure land-tenure at individual or community scales is important for its success (Zahawi et al. 2014). Potential approaches include improved markets and payment schemes for ES (Tengberg et al. 2016) (Section 7.4.6).

Proper targeting of incentive schemes and reducing poverty through access to ES requires knowledge regarding the distribution of beneficiaries, information about those whose livelihoods are likely to be impacted, and in what manner (Nayak et al. 2014; Loaiza et al. 2015; Vira et al. 2012). Institutional arrangements to govern ecosystems are believed to synergistically influence maintenance of carbon storage and forest-based livelihoods, especially when they incorporate local knowledge and decentralised decision-making (Chhatre and Agrawal 2009). Earning carbon credits from reforestation with native trees involves the higher cost of certification and validation processes, increasing the temptation to choose fast-growing (perhaps non-native) species with consequences for native biodiversity. Strategies and policies that aggregate landowners or forest dwellers are needed to reduce the cost to individuals and payment for ecosystem services (PES) schemes can generate synergies (Bommarco et al. 2013; Chhatre and Agrawal 2009). Bundling several PES schemes that address more than one ES can increase income generated by forest restoration (Brancalion et al. 2012).

In the forestry sector, there is evidence that adaptation and mitigation can be fostered in concert. A recent assessment of the California Forestry Offset Project shows that, by compensating individuals and industries for forest conservation, such programmes can deliver mitigation and sustainability co-benefits (Anderson et al. 2017). Adaptive forest management focusing on reintroducing native tree species can provide both mitigation and adaptation benefit by reducing fire risk and increasing carbon storage (Astrup et al. 2018).

In the agricultural sector, there has been little published empirical work on interactions between adaptation and mitigation policies. Smith and Oleson (2010) describe potential relationships, focusing particularly on the arable sector, predominantly on mitigation efforts, and more on measures than policies. The considerable potential of the agro-forestry sector for synergies and contributing to increasing resilience of tropical farming systems is discussed in Verchot et al. (2007) with examples from Africa.

Climate-smart agriculture (CSA) has emerged in recent years as an approach to integrate food security and climate challenges. The three pillars of CSA are to: (1) adapt and build resilience to climate change; (2) reduce GHG emissions, and; (3) sustainably increase agricultural productivity, ultimately delivering 'triple-wins' (Lipper et al. 2014c). While the idea is conceptually appealing, a range of criticisms, contradictions and challenges exist in using CSA as the route to resilience in global agriculture, notably around the political economy (Newell and Taylor 2017), the vagueness of the definition, and consequent assimilation by the mainstream agricultural sector, as well as issues around monitoring, reporting and evaluation (Arakelyan et al. 2017).

Land-based mitigation is facing important trade-offs with food production, biodiversity and local biogeophysical effects (Humpenöder et al. 2017; Krause et al. 2017; Robledo-Abad et al. 2017; Boysen et al. 2016, 2017a,b). Synergies between bioenergy and food security could be achieved by investing in a combination of instruments, including technology and innovations, infrastructure, pricing, flex crops, and improved communication and stakeholder engagement (Kline et al. 2017). Managing these trade-offs might also require demand-side interventions, including dietary change incentives (Section 5.7.1).

Synergies and trade-offs also result from interaction between policies (Urwin and Jordan 2008) at different levels of policy (vertical) and across different policies (horizontal) – see also Section 7.4.8. If policy mixes are designed appropriately, acknowledging and incorporating trade-offs and synergies, they are more apt to deliver an outcome such as transitioning to sustainability (Howlett and Rayner 2013; Huttunen et al. 2014) (*medium evidence* and *medium agreement*). However, there is *medium evidence* and *medium agreement* that evaluating policies for coherence in responding to climate change and its impacts is not occurring, and policies are instead reviewed in a fragmented manner (Hurlbert and Gupta 2016).

7.5.6.4   Water, food and aquatic ecosystem services (ES)

Trade-offs between some types of water use (e.g., irrigation for food security) and other ecosystem services (ES) are expected to intensify under climate change (Hanjra and Ejaz Qureshi 2010). There is an urgency to develop approaches to understand and communicate this to policymakers and decision-makers (Zheng et al. 2016). Reducing water use in agriculture (Mekonnen and Hoekstra 2016) through policies on both the supply and demand side, such as a shift to less water-intensive crops (Richter et al. 2017; Fishman et al. 2015), and a shift in diets (Springmann et al. 2016) has the potential to reduce trade-offs between food security and freshwater aquatic ES (*medium evidence*, *high agreement*). There is strong evidence that improved efficiency in irrigation can actually increase overall water use in agriculture, and therefore its contribution to improved flows in rivers is questionable (Ward and Pulido-Velazquez 2008).

There are now powerful new analytical approaches, high-resolution data and decision-making tools that help to predict cumulative impacts of dams, assess trade-offs between engineering and environmental goals, and can help funders and decision-makers compare alternative sites or designs for dam-building as well as to manage flows in regulated rivers based on experimental releases and adaptive learning. This could minimise ecological costs and maximise synergies with other development goals under climate change (Poff et al. 2003; Winemiller et al. 2016). Furthermore, the adoption of metrics based on the emerging concept of Nature's Contributions to People (NCP) under the IPBES framework brings in non-economic instruments and values that, in combination with conventional valuation of ES approaches, could elicit greater support for non-consumptive water use of rivers for achieving SDG goals (De Groot et al. 2010; Pascual et al. 2017).

7.5.6.5 Considering synergies and trade-offs to avoid maladaptation

Coherent policies that consider synergies and trade-offs can also reduce the likelihood of maladaptation, which is the opposite of sustainable adaptation (Magnan et al. 2016). Sustainable adaptation 'contributes to socially and environmentally sustainable development pathways including both social justice and environmental integrity' (Eriksen et al. 2011). In IPCC's Fifth Assessment Report (AR5) there was *medium evidence* and *high agreement* that maladaptation is 'a cause of increasing concern to adaptation planners, where intervention in one location or sector could increase the vulnerability of another location or sector, or increase the vulnerability of a group to future climate change' (Noble et al. 2014). AR5 recognised that maladaptation arises not only from inadvertent, badly planned adaptation actions, but also from deliberate decisions where wider considerations place greater emphasis on short-term outcomes ahead of longer-term threats, or that discount, or fail to consider, the full range of interactions arising from planned actions (Noble et al. 2014).

Some maladaptations are only beginning to be recognised as we become aware of unintended consequences of decisions. An example prevalent across many countries is irrigation as an adaptation to water scarcity. During a drought from 2007–2009 in California, farmers adapted by using more groundwater, thereby depleting groundwater elevation by 15 metres. This volume of groundwater depletion is unsustainable environmentally and also emits GHG emissions during the pumping (Christian-Smith et al. 2015). Despite the three years of drought, the agricultural sector performed financially well, due to the groundwater use and crop insurance payments. Drought compensation programmes through crop insurance policies may reduce the incentive to shift to lower water-use crops, thereby perpetuating the maladaptive situation. Another example of maladaptation that may appear as adaptation to drought is pumping out groundwater and storing in surface

farm ponds, with consequences for water justice, inequity and sustainability (Kale 2017). These examples highlight the potential for maladaptation from farmers' adaptation decisions as well as the unintended consequences of policy choices; the examples illustrate the findings of Barnett and O'Neill (2010) that maladaptation can include: high opportunity costs (including economic, environmental, and social); reduced incentives to adapt (adaptation measures that reduce incentives to adapt by not addressing underlying causes); and path dependency or trajectories that are difficult to change.

In practice, maladaptation is a specific instance of policy incoherence, and it may be useful to develop a framework in designing policy to avoid this type of trade-off. This would specify the type, aim and target audience of an adaptation action, decision, project, plan, or policy designed initially for adaptation, but actually at high risk of inducing adverse effects, either on the system in which it was developed, or another connected system, or both. The assessment requires identifying system boundaries, including temporal and geographical scales at which the outcomes are assessed (Magnan 2014; Juhola et al. 2016). National-level institutions that cover the spectrum of sectors affected, or enhanced collaboration between relevant institutions, is expected to increase the effectiveness of policy instruments, as are joint programmes and funds (Morita and Matsumoto 2018).

As new knowledge about trade-offs and synergies amongst land-climate processes emerges regionally and globally, concerns over emerging risks and the need for planning policy responses grow. There is *medium evidence* and *medium agreement* that trade-offs currently do not figure into existing climate policies including NDCs and SDGs being vigorously pursued by some countries (Woolf et al. 2018). For instance, the biogeophysical co-benefits of reduced deforestation and re/afforestation measures (Chapter 6) are usually not accounted for in current climate policies or in the NDCs, but there is increasing scientific evidence to include them as part of the policy design (Findell et al. 2017; Hirsch et al. 2018; Bright et al. 2017).

Case study | Green energy: Biodiversity conservation vs global environment targets?

Green and renewable energy and transportation are emerging as important parts of climate change mitigation globally (*medium evidence, high agreement*) (McKinnon 2010; Zarfl et al. 2015; Creutzig et al. 2017). Evidence is, however, emerging across many biomes (from coastal to semi-arid and humid) about how green energy may have significant trade-offs with biodiversity and ecosystem services, thus demonstrating the need for closer environmental scrutiny and safeguards (Gibson et al. 2017; Hernandez et al. 2015). In most cases, the accumulated impact of pressures from decades of land use and habitat loss set the context within which the potential impacts of renewable energy generation need to be considered.

Until recently, small hydropower projects (SHPs) were considered environmentally benign compared to large dams. SHPs are poorly understood, especially since the impacts of clusters of small dams are just becoming evident (Mantel et al. 2010; Fencl et al. 2015; Kibler and Tullos 2013). SHPs (<25/30 MW) are labelled 'green' and are often exempt from environmental scrutiny (Abbasi and Abbasi 2011; Pinho et al. 2007; Premalatha et al. 2014b; Era Consultancy 2006). Being promoted in mountainous global biodiversity hotspots, SHPs have changed the hydrology, water quality and ecology of headwater streams and neighbouring forests significantly. SHPs have created dewatered stretches of stream immediately downstream and introduced sub-daily to sub-weekly hydro-pulses that have transformed the natural dry-season flow regime. Hydrologic and ecological connectivity have been impacted, especially for endemic fish communities and forests in some sites of significant biodiversity values (*medium evidence, medium agreement*) (Jumani et al. 2017, 2018; Chhatre and Lakhanpal 2018; Anderson et al. 2006; Grumbine and Pandit 2013). In some sites, local communities have opposed SHPs due to concerns about their impact on local culture and livelihoods (Jumani et al. 2017, 2018; Chhatre and Lakhanpal 2018).

Semi-arid and arid regions are often found suitable for wind and solar farms which may impact endemic biodiversity and endangered species (Collar et al. 2015, Thaker, M, Zambre, A. Bhosale 2018). The loss of habitat for these species over the decades has been largely due to agricultural intensification driven by irrigation and bad management in designated reserves (Collar et al. 2015; Ledec, George C.; Rapp, Kennan W.; Aiello 2011) but intrusion of power lines is a major worry for highly endangered species such as the Great Indian Bustard (Great Indian Bustard (Ardeotis nigriceps) and conservation and mitigation efforts are being planned to address such concerns (Government of India 2012). In many regions around the world, wind-turbines and solar farms pose a threat to many other species especially predatory birds and insectivorous bats (*medium evidence, medium agreement*) (Thaker, M, Zambre, A. Bhosale 2018) and disrupt habitat connectivity (Northrup and Wittemyer 2013).

Additionally, conversion of rivers into waterways has emerged as a fuel-efficient (low carbon emitting) and environment-friendly alternative to surface land transport (IWAI 2016; Dharmadhikary, S., and Sandbhor 2017). India's National Waterways seeks to cut transportation time and costs and reduce carbon emissions from road transport (Admin 2017). There is some evidence that dredging and under-water noise could impact the water quality, human health and habitat of fish species (Junior et al. 2012; Martins et al. 2012), disrupt artisanal fisheries and potentially impact species that rely on echo-location (*low evidence, medium agreement*) (Dey Mayukh 2018). Off-shore renewable energy projects in coastal zones have been known to have similar impacts on marine fauna (Gill 2005). The Government of India has decided to support studies of the impact of waterways on the endangered Gangetic dolphin in order in order to plan mitigation measures.

Responses to mitigating and reducing the negative impacts of small dams include changes in SHP operations and policies to enable the conservation of river fish diversity. These include mandatory environmental impact assessments, conserving remaining undammed headwater streams in regulated basins, maintaining adequate environmental flows, and implementing other adaptation measures based on experiments with active management of fish communities in impacted zones (Jumani et al. 2018). Location of large solar farms needs to be carefully scrutinised (Sindhu et al. 2017). For mitigating negative impacts of power lines associated with solar and wind farms in bustard habitats, suggested measures include diversion structures to prevent collision, underground cables and avoidance in core wildlife habitat, as well as incentives for maintaining low-intensity rainfed agriculture and pasture around existing reserves, and curtailing harmful infrastructure in priority areas (Collar et al. 2015). Mitigation for minimising the ecological impact of inland waterways on biodiversity and fisheries is more complicated, but may involve improved boat technology to reduce underwater noise, maintaining ecological flows and thus reduced dredging, and avoidance in key habitats (Dey Mayukh 2018).

The management of ecological trade-offs of green energy and green infrastructure and transportation projects may be crucial for long-term sustainability and acceptance of emerging low-carbon economies.

7.6   Governance: Governing the land–climate interface

Building on the definition in Section 7.1.2, governance situates decision-making and selection or calibration of policy instruments within the reality of the multitude of actors operating in respect of land and climate interactions. Governance includes all of the processes, structures, rules and traditions that govern; governance processes may be undertaken by actors including a government, market, organisation, or family (Bevir 2011). Governance processes determine how people in societies make decisions (Patterson et al. 2017) and involve the interactions among formal and informal institutions (Section 7.4.1) through which people articulate their interests, exercise their legal rights, meet their legal obligations, and mediate their differences (Plummer and Baird 2013).

The act of governance 'is a social function centred on steering collective behaviour toward desired outcomes [sustainable climate-resilient development] and away from undesirable outcomes' (Young 2017a). This definition of governance allows for it to be decoupled from the more familiar concept of government and studied in the context of complex human–environment relations and environmental and resource regimes (Young 2017a) and used to address the interconnected challenges facing food and agriculture (FAO 2017b). These challenges include assessing, combining, and implementing policy instruments at different governance levels in a mutually reinforcing way, managing trade-offs while capitalising on synergies (Section 7.5.6), and employing experimentalist approaches for improved and effective governance (FAO 2017b), for example, adaptive climate governance (Section 7.6.3). Emphasising governance also represents a shift of traditional resource management (focused on hierarchical state control) towards recognition that political and decision-making authority can be exercised through interlinked groups of diverse actors (Kuzdas et al. 2015).

This section will start by describing institutions and institutional arrangements – the core of a governance system (Young 2017) – that build adaptive and mitigative capacity. The section then outlines modes, levels and scales of governance for sustainable climate-resilient development. It does on to describe adaptive climate governance that responds to uncertainty, and explore institutional dimensions of adaptive governance that create an enabling environment for strong institutional capital. We then discuss land tenure (an important institutional context for effective and appropriate selection of policy instruments), and end with the participation of people in decision-making through inclusive governance.

7.6.1   Institutions building adaptive and mitigative capacity

Institutions are rules and norms held in common by social actors that guide, constrain, and shape human interaction. Institutions can be formal – such as laws, policies, and structured decision-making processes (Section 7.5.1.1) – or informal – such as norms, conventions, and decision-making following customary norms and habits (Section 7.5.1.2). Organisations – such as parliaments,

regulatory agencies, private firms, and community bodies – as well as people, develop and act in response to institutional frameworks and the incentives they frame. 'Institutions can guide, constrain, and shape human interaction through direct control, through incentives, and through processes of socialization' (IPCC 2014d, p. 1768). Nations with 'well developed institutional systems are considered to have greater adaptive capacity', and better institutional capacity to help deal with risks associated with future climate change (IPCC, 2001, p. 896). Institutions may also prevent the development of adaptive capacity when they are 'sticky' or characterised by strong path dependence (Mahoney 2000; North 1991) and prevent changes that are important to address climate change (Section 7.4.9).

Formal and informal governance structures are composed of these institutionalised rule systems that determine vulnerability as they influence power relations, risk perceptions and establish the context wherein risk reduction, adaptation and vulnerability are managed (Cardona 2012). Governance institutions determine the management of a community's assets, the community members' relationships with one another, and with natural resources (Hurlbert and Diaz 2013). Traditional or locally evolved institutions, backed by cultural norms, can contribute to resilience and adaptive capacity. Anderson et al. (2010) suggest that these are a particular feature of dry land societies that are highly prone to environmental risk and uncertainty. Concepts of resilience, and specifically the resilience of socio-ecological systems, have advanced analysis of adaptive institutions and adaptive governance in relation to climate change and land (Boyd and Folke 2011a). In their characterisation, 'resilience is the ability to reorganise following crisis, continuing to learn, evolving with the same identity and function, and also innovating and sowing the seeds for transformation. It is a central concept of adaptive governance' (Boyd and Folke 2012). In the context of complex and multi-scale socio-ecological systems, important features of adaptive institutions that contribute to resilience include the characteristics of an adaptive governance system (Section 7.6.6).

There is *high confidence* that adaptive institutions have a strong learning dimension and include:

1. Institutions advancing the capacity to learn through availability, access to, accumulation of, and interpretation of information (such as drought projections, costing of alternatives land, food, and water strategies). Government-supported networks, learning platforms, and facilitated interchange between actors with boundary and bridging organisations, creating the necessary self-organisation to prepare for the unknown. Through transparent, flexible networks, whole sets of complex problems of land, food and climate can be tackled to develop shared visions and critique land and food management systems assessing gaps and generating solutions.

2. Institutions advancing learning by experimentation (in interpretation of information, new ways of governing, and treating policy as an ongoing experiment) through many interrelated decisions, but especially those that connect the social to the ecological and entail anticipatory planning by considering a longer-term time frame. Mechanisms to do so include ecological stewardship, and rituals and beliefs of indigenous societies that sustain ES.

3. Institutions that decide on pathways to realise system change through cultural, inter and intra organisational collaboration, with a flexible regulatory framework allowing for new cognitive frames of 'sustainable' land management and 'safe' water supply that open alternative pathways (Karpouzoglou et al. 2016; Bettini et al. 2015; Boyd et al. 2015; Boyd and Folke 2011b, and 2012).

Shortcomings of resilience theory include limits in relation to its conceptualisation of social change (Cote and Nightingale 2012), its potential to be used as a normative concept, implying politically prescriptive policy solutions (Thorén and Olsson 2017; Weichselgartner and Kelman 2015; Milkoreit et al. 2015), its applicability to local needs and experiences (Forsyth 2018), and its potential to hinder evaluation of policy effectiveness (Newton 2016; Olsson et al. 2015b). Regardless, concepts of adaptive institutions building adaptive capacity in complex socio-ecological systems governance have progressed (Karpouzoglou et al. 2016; Dwyer and Hodge 2016) in relation to adaptive governance (Koontz et al. 2015).

The study of institutions of governance, levels, modes, and scale of governance, in a multi-level and polycentric fashion is important because of the multi-scale nature of the challenges to resilience, dissemination of ideas, networking and learning.

7.6.2   Integration – Levels, modes and scale of governance for sustainable development

Different types of governance can be distinguished according to intended levels (e.g., local, regional, global), domains (national, international, transnational), modes (market, network, hierarchy), and scales (global regimes to local community groups) (Jordan et al. 2015b). Implementation of climate change adaptation and mitigation has been impeded by institutional barriers, including multi-level governance and policy integration issues (Biesbroek et al. 2010). To overcome these barriers, climate governance has evolved significantly beyond the national and multilateral domains that tended to dominate climate efforts and initiatives during the early years of the UNFCCC. The climate challenge has been placed in an Earth System context, showing the existence of complex interactions and governance requirements across different levels, and calling for a radical transformation in governance, rather than minor adjustments (Biermann et al. 2012). Climate governance literature has expanded since AR5 in relation to the sub-national and transnational levels, but all levels and their interconnection is important. Expert thinking has evolved from implementing good governance at high levels (with governments) to a decentred problem-solving approach consistent with adaptive governance. This approach involves iterative bottom-up and experimental mechanisms that might entail addressing tenure of land or forest management through a territorial approach to development, thereby supporting multi-sectoral governance in local, municipal and regional contexts (FAO 2017b).

Local action in relation to mitigation and adaptation continues to be important by complementing and advancing global climate policy (Ostrom 2012). Sub-national governance efforts for climate policy, especially at the level of cities and communities, have become

significant during the past decades (*medium evidence, medium agreement*) (Castán Broto 2017; Floater et al. 2014; Albers et al. 2015; Archer et al. 2014). A transformation of sorts has been underway through deepening engagement from the private sector and NGOs as well as government involvement at multiple levels. It is now recognised that business organisations, civil society groups, citizens, and formal governance all have important roles in governance for sustainable development (Kemp et al. 2005).

Transnational governance efforts have increased in number, with applications across different economic sectors, geographical regions, civil society groups and NGOs. When it comes to climate mitigation, transnational mechanisms generally focus on networking and may not necessarily be effective in terms of promoting real emissions reductions (Michaelowa and Michaelowa 2017). However, acceleration in national mitigation measures has been determined to coincide with landmark international events such as the lead up to the Copenhagen Climate Change Conference 2009 (Iacobuta et al. 2018). There is a tendency for transnational governance mechanisms to lack monitoring and evaluation procedures (Jordan et al. 2015a).

To address shortcomings of transnational governance, polycentric governance considers the interaction between actors at different levels of governance (local, regional, national, and global) for a more nuanced understanding of the variation in diverse governance outcomes in the management of common-pool resources (such as forests) based on the needs and interests of citizens (Nagendra and Ostrom 2012). A more 'polycentric climate governance' system has emerged that incorporates bottom-up initiatives that can support and synergise with national efforts and international regimes (Ostrom 2010). Although it is clear that many more actors and networks are involved, the effectiveness of a more polycentric system remains unclear (Jordan et al. 2015a).

There is *high confidence* that a hybrid form of governance, combining the advantages of centralised governance (with coordination, stability, compliance) with those of more horizontal structures (that allow flexibility, autonomy for local decision-making, multi-stakeholder engagement, co-management) is required for effective mainstreaming of mitigation and adaptation in sustainable land and forest management (Keenan 2015; Gupta 2014; Williamson and Nelson 2017; Liniger et al. 2019). Polycentric institutions self-organise, developing collective solutions to local problems as they arise (Koontz et al. 2015). The public sector (governments and administrative systems) are still important in climate change initiatives as these actors retain the political will to implement and make initiatives work (Biesbroek et al. 2018).

Sustainable development hinges on the holistic integration of interconnected land and climate issues, sectors, levels of government, and policy instruments (Section 7.4.8) that address the increasing volatility in oscillating systems and weather patterns (Young 2017b; Kemp et al. 2005). Climate adaptation and mitigation goals must be integrated or mainstreamed into existing governance mechanisms around key land-use sectors such as forestry and agriculture. In the EU, mitigation has generally been well-mainstreamed in regional policies but not adaptation (Hanger et al. 2015). Climate change adaptation has been impeded by institutional barriers, including the inherent challenges of multi-level governance and policy integration (Biesbroek et al. 2010).

Integrative polycentric approaches to land use and climate interactions take different forms and operate with different institutions and governance mechanisms. Integrative approaches can provide coordination and linkages to improve effectiveness and efficiency and minimise conflicts (*high confidence*). Different types of integration with special relevance for the land–climate interface can be characterised as follows:

1. **Cross-level integration:** local and national level efforts must be coordinated with national and regional policies and also be capable of drawing direction and financing from global regimes, thus requiring multi-level governance. Integration of SLM to prevent, reduce and restore degraded land is advanced with national and subnational policy, including passing the necessary laws to establish frameworks and provide financial incentives. Examples include: integrated territorial planning addressing specific land-use decisions; local landscape participatory planning with farmer associations, microenterprises, and local institutions identifying hot spot areas, identifying land-use pressures and scaling out SLM response options (Liniger et al. 2019).

2. **Cross-sectoral integration:** rather than approach each application or sector (e.g., energy, agriculture, forestry) separately, there is a conscious effort at co-management and coordination in policies and institutions, such as with the energy–water–food nexus (Biggs et al. 2015).

3. **End-use/market integration:** often involves exploiting economies of scope across products, supply chains, and infrastructure (Nuhoff-Isakhanyan et al. 2016; Ashkenazy et al. 2017). For instance, land-use transport models consider land use, transportation, city planning, and climate mitigation (Ford et al. 2018).

4. **Landscape integration:** rather than physical separation of activities (e.g., agriculture, forestry, grazing), uses are spatially integrated by exploiting natural variations while incorporating local and regional economies (Harvey et al. 2014a). In an assessment of 166 initiatives in 16 countries, integrated landscape initiatives were found to address the drivers of agriculture, ecosystem conservation, livelihood preservation and institutional coordination. However, such initiatives struggled to move from planning to implementation due to lack of government and financial support, and powerful stakeholders sidelining the agenda (Zanzanaini et al. 2017). Special care helps ensure that initiatives don't exacerbate socio-spatial inequalities across diverse developmental and environmental conditions (Anguelovski et al. 2016b). Integrated land-use planning, coordinated through multiple government levels, balances property rights, wildlife and forest conservation, encroachment of settlements and agricultural areas and can reduce conflict (*high confidence*) (Metternicht 2018). Land-use planning can also enhance management of areas prone to natural disasters, such as floods, and resolve issues of competing land uses and land tenure conflicts (Metternicht 2018).

Another way to analyse or characterise governance approaches or mechanisms might be according to a temporal scale with respect to relevant events – for example, those that may occur gradually versus abruptly (Cash et al. 2006). Desertification and land degradation are drawn-out processes that occur over many years, whereas extreme events are abrupt and require immediate attention. Similarly, the frequency of events might be of special interest – for example, events that occur periodically versus those that occur infrequently and/or irregularly. In the case of food security, abrupt and protracted events of food insecurity might occur. There is a distinction between 'hunger months' and longer-term food insecurity. Some indigenous practices already incorporate hunger months whereas structural food deficits have to be addressed differently (Bacon et al. 2014). Governance mechanisms that facilitate rapid response to crises are quite different from those aimed at monitoring slower changes and responding with longer-term measures.

Case study | Governance: Biofuels and bioenergy

New policies and initiatives during the past decade or so have increased support for bioenergy as a non-intermittent (stored) renewable with wide geographic availability that is cost-effective in a range of applications. Significant upscaling of bioenergy requires dedicated (normally land-based) sources in addition to use of wastes and residues. As a result, a disadvantageous high land-use intensity compared to other renewables (Fritsche et al. 2017) that, in turn, place greater demands on governance. Bioenergy, especially traditional fuels, currently provides the largest share of renewable energy globally and has a significant role in nearly all climate stabilisation scenarios, although estimates of its potential vary widely (see Cross-Chapter Box 7 in Chapter 6). Policies and governance for bioenergy systems and markets must address diverse applications and sectors across levels from local to global; here we briefly review the literature in relation to governance for modern bioenergy and biofuels with respect to land and climate impacts, whereas traditional biomass use (see Glossary) (> 50% of energy used today with greater land use and GHG emissions impacts in low- and medium-income countries (Bailis et al. 2015; Masera et al. 2015; Bailis et al. 2017a; Kiruki et al. 2017b)) is addressed elsewhere (Sections 4.5.4 and 7.4.6.4 and Cross-Chapter Box 12 in Chapter 7). The bioenergy lifecycle is relevant in accounting for – and attributing – land impacts and GHG emissions (Section 2.5.1.5). Integrated responses across different sectors can help to reduce negative impacts and promote sustainable development opportunities (Table 6.9, Table 6.58, Chapter 6).

Case study (continued)

It is very likely that bioenergy expansion at a scale that contributes significantly to global climate mitigation efforts (see Cross-Chapter Box 7 in Chapter 6) will result in substantial land-use change (Berndes et al. 2015; Popp et al. 2014a; Wilson et al. 2014; Behrman et al. 2015; Richards et al. 2017; Harris et al. 2015; Chen et al. 2017a). There is *medium evidence* and *high agreement* that land-use change at such scale presents a variety of positive and negative socio-economic and environmental impacts that lead to risks and trade-offs that must be managed or governed across different levels (Pahl-Wostl et al. 2018a; Kurian 2017; Franz et al. 2017; Chang et al. 2016; Larcom and van Gevelt 2017; Lubis et al. 2018; Alexander et al. 2015b; Rasul 2014; Bonsch et al. 2016; Karabulut et al. 2018; Mayor et al. 2015). There is *medium evidence* and *high agreement* that impacts vary considerably according to factors such as initial land-use type, choice of crops, initial carbon stocks, climatic region, soil types and the management regime and adopted technologies (Qin et al. 2016; Del Grosso et al. 2014; Popp et al. 2017; Davis et al. 2013; Mello et al. 2014; Hudiburg et al. 2015; Carvalho et al. 2016; Silva-Olaya et al. 2017; Whitaker et al. 2018; Alexander et al. 2015b).

There is *medium evidence* and *high agreement* that significant socio-economic impacts requiring additional policy responses can occur when agricultural lands and/or food crops are used for bioenergy, due to competition between food and fuel (Harvey and Pilgrim 2011; Rosillo Callé and Johnson 2010b), including impacts on food prices (Martin Persson 2015; Roberts and Schlenker 2013; Borychowski and Czyżewski 2015; Koizumi 2014; Muratori et al. 2016; Popp et al. 2014b; Araujo Enciso et al. 2016) and impacts on food security (Popp et al. 2014b; Bailey 2013; Pahl-Wostl et al. 2018b; Rulli et al. 2016; Yamagata et al. 2018; Kline et al. 2017; Schröder et al. 2018; Franz et al. 2017; Mohr et al. 2016). Additionally, crops such as sugarcane, which are water-intensive when used for ethanol production, have a trade-off with water and downstream ES and other crops more important for food security (Rulli et al. 2016; Gheewala et al. 2011). Alongside negative impacts that might fall on urban consumers (who purchase both food and energy), there is *medium evidence* and *medium agreement* that rural producers or farmers can increase income or strengthen livelihoods by diversifying into biofuel crops that have an established market (Maltsoglou et al. 2014; Mudombi et al. 2018a; Gasparatos et al. 2018a,b,c; von Maltitz et al. 2018; Kline et al. 2017; Rodríguez Morales and Rodríguez López 2017; Dale et al. 2015; Lee and Lazarus 2013; Rodríguez-Morales 2018). A key governance mechanism that has emerged in response to such concerns, (especially during the past decade) are standards and certification systems that include food security and land rights in addition to general criteria or indicators related to sustainable use of land and biomass (Section 7.4.6.3). There is *medium evidence* and *medium agreement* that policies promoting use of wastes and residues, use of non-edible crops and/or reliance on degraded and marginal lands for bioenergy could reduce land competition and associated risk for food security (Manning et al. 2015; Maltsoglou et al. 2014; Zhang et al. 2018a; Gu and Wylie 2017; Kline et al. 2017; Schröder et al. 2018; Suckall et al. 2015; Popp et al. 2014a; Lal 2013).

There is *medium evidence* and *high agreement* that good governance, including policy coherence and coordination across the different sectors involved (agriculture, forestry, livestock, energy, transport) (Section 7.6.2) can help to reduce the risks and increase the co-benefits of bioenergy expansion (Makkonen et al. 2015; Di Gregorio et al. 2017; Schut et al. 2013; Mukhtarov et al.; Torvanger 2019a; Müller et al. 2015; Nkonya et al. 2015; Johnson and Silveira 2014; Lundmark et al. 2014; Schultz et al. 2015; Silveira and Johnson 2016; Giessen et al. 2016b; Stattman et al. 2018b; Bennich et al. 2017b). There is *medium evidence* and *high agreement* that the nexus approach can help to address interconnected biomass resource management challenges and entrenched economic interests, and leverage synergies in the systemic governance of risk. (Bizikova et al. 2013; Rouillard et al. 2017; Pahl-Wostl 2017a; Lele et al. 2013; Rodríguez Morales and Rodríguez López 2017; Larcom and van Gevelt 2017; Pahl-Wostl et al. 2018a; Rulli et al. 2016; Rasul and Sharma 2016; Weitz et al. 2017b; Karlberg et al. 2015).

A key issue for governance of biofuels and bioenergy, as well as land-use governance more generally, during the past decade is the need for new governance mechanisms across different levels as land-use policies and bioenergy investments are scaled up and result in wider impacts (Section 7.6). There is *low evidence* and *medium agreement* that hybrid governance mechanisms can promote sustainable bioenergy investments and land-use pathways. This hybrid governance can include multi-level, transnational governance, and private-led or partnership-style (polycentric) governance, complementing national-level, strong public coordination (government and public administration) (Section 7.6.2) (Pahl-Wostl 2017a; Pacheco et al. 2016; Winickoff and Mondou 2017; Nagendra and Ostrom 2012; Jordan et al. 2015a; Djalante et al. 2013; Purkus, A, Gawel, E. and Thrän, D. 2012; Purkus et al. 2018; Stattman et al.; Rietig 2018; Cavicchi et al. 2017; Stupak et al. 2016; Stupak and Raulund-Rasmussen 2016; Westberg and Johnson 2013; Giessen et al. 2016b; Johnson and Silveira 2014; Stattman et al. 2018b; Mukhtarov et al.; Torvanger 2019b).

7

Cross-Chapter Box 12 | Traditional biomass use: Land, climate and development implications

Francis X. Johnson (Sweden), Fahmuddin Agus (Indonesia), Rob Bailis (The United States of America), Suruchi Bhadwal (India), Annette Cowie (Australia), Tek Sapkota (Nepal)

Introduction and significance

Most biomass used for energy today is in traditional forms (fuelwood, charcoal, agricultural residues) for cooking and heating by some 3 billion people worldwide (IEA 2017). Traditional biomass has high land and climate impacts, with significant harvesting losses, greenhouse gas (GHG) emissions, soil impacts and high conversion losses (Cutz et al. 2017b; Masera et al. 2015; Ghilardi et al. 2016a; Bailis et al. 2015; Fritsche et al. 2017; Mudombi et al. 2018b). In addition to these impacts, indoor air pollution from household cooking is a leading cause of mortality in low- and medium-income countries and especially affects women and children (Smith et al. 2014a; HEI/IHME 2018; Goldemberg et al. 2018b). In rural areas, the significant time needed for gathering fuelwood imposes further costs on women and children (Njenga and Mendum 2018; Gurung and Oh 2013a; Behera et al. 2015a).

Both agricultural and woody biomass can be upgraded and used sustainably through improved resource management and modern conversion technologies, providing much greater energy output per unit of biomass (Cutz et al. 2017b; Hoffmann et al. 2015a; Gurung and Oh 2013b). More relevant than technical efficiency is the improved quality of energy services: with increasing income levels and/or access to technologies, households transition over time from agricultural residues and fuelwood to charcoal and then to gaseous or liquid fuels and electricity (Leach 1992; Pachauri and Jiang 2008; Goldemberg and Teixeira Coelho 2004; Smeets et al. 2012a). However, most households use multiple stoves and/or fuels at the same time, known as 'fuel stacking' for economic flexibility and also for socio-cultural reasons (Ruiz-Mercado and Masera 2015a; Cheng and Urpelainen 2014; Takama et al. 2012).

Urban and rural use of traditional biomass

In rural areas, fuelwood is often gathered at no cost to the user, and burned directly whereas, in urban areas, traditional biomass use may often involve semi-processed fuels, particularly in Sub-Saharan Africa where charcoal is the primary urban cooking fuel. Rapid urbanisation and/or commercialisation drives a shift from fuelwood to charcoal, which results in significantly higher wood use (*very high confidence*) due to losses in charcoal supply chains and the tendency to use whole trees for charcoal production (Santos et al. 2017; World Bank. 2009a; Hojas-Gascon et al. 2016a; Smeets et al. 2012b). One study in Myanmar found that charcoal required 23 times the land area of fuelwood (Win et al. 2018). In areas of woody biomass scarcity, animal dung and agricultural residues, as well as lower-quality wood, are often used (Kumar Nath et al. 2013a; Go et al. 2019a; Jagger and Kittner 2017; Behera et al. 2015b). The fraction of woody biomass harvested that is not 'demonstrably renewable' is the fraction of non-renewable biomass (fNRB) under UNFCCC accounting; default values for fNRB for least-developed countries and small island developing states ranged from 40–100% (CDM Executive Board 2012). Uncertainties in woodfuel data, complexities in spatiotemporal woodfuel modelling and rapid forest regrowth in some tropical regions present sources of variation in such estimates, and some fNRB values are *likely* to have been overestimated (McNicol et al. 2018a; Ghilardi et al. 2016b; Bailis et al. 2017b).

GHG emissions and traditional biomass

Due to over-harvesting, incomplete combustion and the effects of short-lived climate pollutants, traditional woodfuels (fuelwood and charcoal) contribute 1.9–2.3% of global GHG emissions; non-renewable biomass is concentrated especially in 'hotspot' regions of East Africa and South Asia (Bailis et al. 2015). The estimate only includes woody biomass and does not account for possible losses in soil carbon or the effects of nutrient losses from use of animal dung, which can be significant in some cases (Duguma et al. 2014a; Achat et al. 2015a; Sánchez et al. 2016). Reducing emissions of black carbon alongside GHG reductions offers immediate health co-benefits (Shindell et al. 2012; Pandey et al. 2017; Weyant et al. 2019a; Sparrevik et al. 2015). Significant GHG emissions reductions, depending on baseline or reference use, can be obtained through fuel-switching to gaseous and liquid fuels, sustainable harvesting of woodfuels, upgrading to efficient stoves, and adopting high-quality processed fuels such as wood pellets (*medium evidence, high agreement*) (Wathore et al. 2017; Jagger and Das 2018; Quinn et al. 2018; Cutz et al. 2017b; Carter et al. 2018; Bailis et al. 2015; Ghilardi et al. 2018; Weyant et al. 2019b; Hoffmann et al. 2015b).

Land and forest degradation

Land degradation is itself a significant source of GHG emissions and biodiversity loss, with over-harvesting of woodfuel as a major cause in some regions and especially in Sub-Saharan Africa (Pearson et al. 2017; Joana Specht et al. 2015a; Kiruki et al. 2017b; Ndegwa et al. 2016; McNicol et al. 2018b). Reliance on traditional biomass is quite land-intensive: supplying one household sustainably for a year can require more than half a hectare of land, which, in dryland countries such as Kenya, can result in substantial percentage of total tree cover (Fuso Nerini et al. 2017). In Sub-Saharan Africa and in some other regions, land degradation is widely associated with charcoal production (*high confidence*), often in combination with timber harvesting or clearing land for agriculture

Cross-Chapter Box 12 (continued)

(Kiruki et al. 2017a; Ndegwa et al. 2016; Hojas-Gascon et al. 2016b). Yet charcoal makes a significant contribution to livelihoods in many areas and thus, in spite of the ecological damage, halting charcoal production is difficult due to the lack of alternative livelihoods and/or the affordability of other fuels (Smith et al. 2015; Zulu and Richardson 2013a; Jones et al. 2016a; World Bank 2009b).

Use of agricultural residues and animal dung for bioenergy

Although agricultural wastes and residues from almost any crop can be used in many cases for bioenergy, excessive removal or reduction of forest (or agricultural) biomass can contribute to a loss of soil carbon, which can also, in turn, contribute to land degradation (James et al. 2016; Blanco-Canqui and Lal 2009a; Carvalho et al. 2016; Achat et al. 2015b; Stavi and Lal 2015). Removals are limited to levels at which problems of soil erosion, depletion of soil organic matter, soil nutrient depletion and decline in crop yield are effectively mitigated (Ayamga et al. 2015a; Baudron et al. 2014; Blanco-Canqui and Lal 2009b). Application or recycling of residues may, in some cases, be more valuable for soil improvement (*medium confidence*). Tao et al. (2017) used leftover oil palm fruit bunches and demonstrated that application of 30 to 90 t ha^{-1} empty fruit bunches maintains high palm oil yield with low temporal variability. A wide variety of wastes from palm oil harvesting can be used for bioenergy, including annual crop residues (Go et al. 2019b; Ayamga et al. 2015b; Gardner et al. 2018b).

Animal dung is a low-quality fuel used where woody biomass is scarce, such as in South Asia and some areas of eastern Africa (Duguma et al. 2014b; Behera et al. 2015b; Kumar Nath et al. 2013b). Carbon and nutrient losses can be significant when animal dung is dried and burned as cake, whereas using dung in a biodigester provides high-quality fuel and preserves nutrients in the by-product slurry (Clemens et al. 2018; Gurung and Oh 2013b; Quinn et al. 2018).

Production and use of biochar

Converting agricultural residues into biochar can also help to reverse trends of soil degradation (Section 4.10.7). The positive effects of using biochar have been demonstrated in terms of soil aggregate improvement, increase of exchangeable cations, cation exchange capacity, available phosphorus, soil pH and carbon sequestration as well as increased crop yields (Huang et al. 2018; El-Naggar et al. 2018; Wang et al. 2018; Oladele et al. 2019; Blanco-Canqui and Lal 2009b). The level of biochar effectiveness varies depending on the kind of feedstock, soil properties and rate of application (Shaaban et al. 2018; Pokharel and Chang 2019). In addition to adding value to an energy product, the use of biochar offers a climate-smart approach to addressing agricultural productivity (Solomon and Lehmann 2017).

Relationship to food security and other Sustainable Development Goals (SDGs)

The population that is food insecure also intersects significantly with those relying heavily on traditional biomass such that poor and vulnerable populations often expend considerable time (gathering fuel) or use a significant share of household income for low-quality energy services (Fuso Nerini et al. 2017; McCollum et al. 2018; Rao and Pachauri 2017; Pachauri et al. 2018; Muller and Yan 2018; Takama et al. 2012). Improvements in energy access and reduction or elimination of traditional biomass use thus have benefits across multiple SDGs (*medium evidence, high agreement*) (Masera et al. 2015; Rao and Pachauri 2017; Pachauri et al. 2018; Hoffmann et al. 2017; Jeuland et al. 2015; Takama et al. 2012; Gitau et al. 2019; Quinn et al. 2018; Ruiz-Mercado and Masera 2015b; Duguma et al. 2014b; Sola et al. 2016b). Improved energy access contributes to adaptive capacity, although charcoal production itself can also serve as a diversification or adaptation strategy (Perera et al. 2015; Ochieng et al. 2014; Sumiya 2016; Suckall et al. 2015; Jones et al. 2016b).

Socio-economic choices and shifts

When confronted with the limitations of higher-priced household energy alternatives, climate mitigation policies can result in trade-offs with health, energy access and other SDGs (Cameron et al. 2016; Fuso Nerini et al. 2018). The poorest households have no margin to pay for higher-cost efficient stoves; a focus on product-specific characteristics, user needs and/or making clean options more available would improve the market take-up (*medium confidence*) (Takama et al. 2012; Mudombi et al. 2018c; Khandelwal et al. 2017; Rosenthal et al. 2017; Cundale et al. 2017; Jürisoo et al. 2018). Subsidies for more efficient end-use technologies, in combination with promotion of sustainable harvesting techniques, would provide the highest emissions reductions while improving energy services (Cutz et al. 2017a).

Knowledge gaps

Unlike analyses on modern energy sources, scientific assessments on traditional biomass use are complicated by its informal nature and the difficulty of tracing data and impacts; more systematic analytical efforts are needed to address this research gap

(Cerutti et al. 2015). In general, traditional biomass use is associated with poverty. Therefore, efforts to reduce the dependence on fuelwood use are to be conducted in coherence with poverty alleviation (McCollum et al. 2018; Joana Specht et al. 2015b; Zulu and Richardson 2013b). The substantial potential co-benefits suggest that the traditional biomass sector remains under-researched and under-exploited in terms of cost-effective emissions reductions, as well as for synergies between climate stabilisation goals and other SDGs.

7.6.3   Adaptive climate governance responding to uncertainty

In the 1990s, adaptive governance emerged from adaptive management (Holling 1978, 1986), combining resilience and complexity theory, and reflecting the trend of moving from government to governance (Hurlbert 2018b). Adaptive governance builds on multi-level and polycentric governance. Adaptive governance is 'a process of resolving trade-offs and charting a course for sustainability' (Boyle et al. 2001, p. 28) through a range of 'political, social, economic and administrative systems that develop, manage and distribute a resource in a manner promoting resilience through collaborative, flexible and learning-based issue management across different scales' (Hurlbert 2018, p. 25). There is *medium evidence* and *medium agreement* that few alternative governance theories handle processes of change characterised by nonlinear dynamics, threshold effects, cascades and limited predictability; however, the majority of literature relates to the USA or Canada (Karpouzoglou et al. 2016). Combining adaptive governance with other theories has allowed good evaluation of important governance features such as power and politics, inclusion and equity, short-term and long-term change, and the relationship between public policy and adaptive governance (Karpouzoglou et al. 2016).

There is *robust evidence* and *high agreement* that resource and disaster crises are crises of governance (Pahl-Wostl 2017b; Villagra and Quintana 2017; Gupta et al. 2013b). Adaptive governance of risk has emerged in response to these crises and involves four critical pillars (Fra.Paleo 2015):

1. Sustainability as a response to environmental degradation, resource depletion and ES deterioration
2. Recognition that governance is required as government is unable to resolve key societal and environmental problems, including climate change and complex problems
3. Mitigation as a means to reduce vulnerability and avoid exposure
4. Adaptation responds to changes in environmental conditions.

Closely related to (and arguably components of) adaptive governance are adaptive management (Section 7.5.4) (a regulatory environment that manages ecological system boundaries through hypothesis testing, monitoring, and re-evaluation (Mostert et al. 2007)), adaptive co-management (flexible community-based resource management (Plummer and Baird 2013)), and anticipatory governance (flexible decision-making through the use of scenario planning and reiterative policy review (Boyd et al. 2015)). Adaptive governance can be conceptualised as including multilevel governance with a balance between top-down and bottom-up decision-making that is performed by many actors (including citizens) in both formal and informal networks, allowing policy measures and governance arrangements to be tailored to local context and matched at the appropriate scale of the problem, allowing for opportunities for experimentation and learning by individuals and social groups (Rouillard et al. 2013; Hurlbert 2018b).

There is *high confidence* that anticipation is a key component of adaptive climate governance wherein steering mechanisms in the present are developed to adapt to and/or shape uncertain futures (Vervoort and Gupta 2018; Wiebe et al. 2018; Fuerth 2009). Effecting this anticipatory governance involves simultaneously making short-term decisions in the context of longer-term policy visioning, anticipating future climate change models and scenarios in order to realise a more sustainable future (Bates and Saint-Pierre 2018; Serrao-Neumann et al. 2013; Boyd et al. 2015). Utilising the decision-making tools and practices in Section 7.5, policymakers operationalise anticipatory governance through a foresight system considering future scenarios and models, a networked system for integrating this knowledge into the policy process, a feedback system using indicators (Section 7.5.5) to gauge performance, an open-minded institutional culture allowing for hybrid and polycentric governance (Fuerth and Faber 2013; Fuerth 2009).

There is *high confidence* that, in order to manage uncertainty, natural resource governance systems need to allow agencies and stakeholders to learn and change over time, responding to ecosystem changes and new information with different management strategies and practices that involve experimentation (Camacho 2009; Young 2017b). There is emerging literature on experimentation in governance surrounding climate change and land use (Kivimaa et al. 2017a) including policies such as REDD+ (Kaisa et al. 2017). Governance experiment literature could be in relation to scaling up policies from the local level for greater application, or downscaling policies addressing broad complex issues such as climate change, or addressing necessary change in social processes across sectors (such as water energy and food) (Laakso et al. 2017). Successful development of new policy instruments occurred in a governance experiment relating to coastal policy adapting to rising sea levels and extreme weather events through planned retreat (Rocle and Salles 2018). Experiments in emissions trading between 1968 and 2000 in the USA helped to realise specific models of governance and material practices through mutually supportive lab experiments and field applications that advanced collective knowledge (Voß and Simons 2018).

There is *high confidence* that an SLM plan is dynamic and adaptive over time to (unforeseen) future conditions by monitoring indicators as early warnings or signals of tipping points, initiating a process of change in policy pathway before a harmful threshold is reached (Stephens et al. 2018, 2017; Haasnoot et al. 2013; Bloemen et al. 2018) (Section 7.5.2.2). This process has been applied in relation to coastal sea level rise, starting with low-risk, low-cost measures and working up to measures requiring greater investment after review and reevaluation (Barnett et al. 2014). A first measure was stringent controls of new development, graduating to managed relocation of low-lying critical infrastructure, and eventually movement of habitable dwellings to more elevated parts of town, as flooding and inundation triggers are experienced (Haasnoot et al. 2018; Lawrence et al. 2018; Barnett et al. 2014; Stephens et al. 2018). Nanda et al. (2018) apply the concept to a wetland in Australia to identify a mix of short- and long-term decisions, and Prober et al. (2017) develop adaptation pathways for agricultural landscapes, also in Australia. Both studies identify that longer-term decisions may involve a considerable change to institutional arrangements at different scales. Viewing climate mitigation as a series of connected decisions over a long time period and not an isolated decision, reduces the fragmentation and uncertainty endemic of models and effectiveness of policy measures (Roelich and Giesekam 2019).

There is *medium evidence* and *high agreement* that participatory processes in adaptive governance within and across policy regimes overcome limitations of polycentric governance, allowing priorities to be set in sustainable development through rural land management and integrated water resource management (Rouillard et al. 2013). Adaptive governance addresses large uncertainties and their social amplification through differing perceptions of risk (Kasperson 2012; Fra.Paleo 2015) offering an approach to co-evolve with risk by implementing policy mixes and assessing effectiveness in an ongoing process, making mid-point corrections when necessary (Fra.Paleo 2015). In respect of climate adaptation to coastal and riverine land erosion due to extreme weather events impacting on communities, adaptive governance offers the capacity to monitor local socio-economic processes and implement dynamic locally informed institutional responses. In Alaska, adaptive governance responded to the dynamic risk of extreme weather events and issue of climate migration by providing a continuum of policy from protection in place to community relocation, integrating across levels and actors in a more effective and less costly response option than other governance systems (Bronen and Chapin 2013). In comparison to other governance initiatives of ecosystem management aimed at conservation and sustainable use of natural capital, adaptive governance has visible effects on natural capital by monitoring, communicating and responding to ecosystem-wide changes at the landscape level (Schultz et al. 2015). Adaptive governance can be applied to manage drought assistance as a common property resource. Adaptive governance can manage complex, interacting goals to create innovative policy options, facilitated through nested and polycentric systems of governance, effected by watershed or catchment management groups in areas of natural resource management (Nelson et al. 2008).

There is *medium evidence* and *high agreement* that transformational change is a necessary societal response option to manage climate risks which is uniquely characterised by the depth of change needed to reframe problems and change dominant mindsets, the scope of change needed (that is larger than just a few people) and the speed of change required to reduce emissions (O'Brien et al. 2012; Termeer et al. 2017). Transformation of governance occurs with changes in values to reflect an understanding that the environmental crisis occurs in the context of our relation with the earth (Hordijk et al. 2014; Pelling 2010). Transformation can happen by intervention strategies that enable small in-depth wins, amplify these small wins through integration into existing practices, and unblock stagnations (locked in structures) preventing transformation by confronting social and cognitive fixations with counterintuitive interventions (Termeer et al. 2017). Iterative consideration of issues and reformulation of policy instruments and response options facilitates transformation by allowing experimentation (Monkelbaan 2019).

Box 7.2 | Adaptive governance and interlinkages of food, fibre, water, energy and land

Emerging literature and case studies recognise the connectedness of the environment and human activities, and the interrelationships of multiple resource use practices in an attempt to understand synergies and trade-offs (Albrecht et al. 2018). Sustainable adaptation – or actions contributing to environmentally and socially sustainable development pathways (Eriksen et al. 2011) – requires consideration of the interlinkage of different sectors (Rasul and Sharma 2016). Integrating considerations can address sustainability (Hoff 2011) showing promise (Allan et al. 2015) for effective adaptation to climate impacts in many drylands (Rasul and Sharma 2016).

Case studies of integrated water resources management (IWRM), landscape- and ecosystem-based approaches illustrate important dimensions of institutions, institutional coordination, resource coupling and local and global connections (Scott et al. 2011). Integrated governance, policy coherence, and use of multi-functional systems are required to advance synergies across land, water, energy and food sectors (Liu et al. 2017).

Case study: Flood and food security
Between 2003 and 2013, floods were the natural disaster that most impacted on crop production (FAO 2015b) (albeit in certain contexts, such as riverine ecosystems and flood plain communities, floods can be beneficial).

7

Box 7.2 (continued)

In developing countries, flood jeopardises primary access to food and impacts on livelihoods. In Bangladesh, the 2007 flood reduced average consumption by 103Kcal/cap/day (worsening the existing 19.4% calories deficit), and in Pakistan, the 2010 flood resulted in a loss of 205 Kcal/cap/day (or 8.5% of the Pakistan average food supply). The 2010 flood affected more than 4.5 million workers, two-thirds employed in agriculture; and 79% of farms lost greater than one-half of their expected income (Pacetti et al. 2017).

Policy instruments and responses react to the sequential and cascading impacts of flood. In a Malawi study, flood impacts cascaded through labour, trade and transfer systems. First a harvest failure occurred, followed by the decline of employment opportunities and reduction in real wages, followed by a market failure or decline in trade, ultimately followed by a failure in informal safety nets (Devereux 2007). Planned policy responses include those that address the sequential nature of the cascading impacts, starting with 'productivity-enhancing safety nets' addressing harvest failure, then public works programmes addressing the decline in employment opportunities, followed by food price subsidies to address the market failure, and finally food aid to address the failure of informal safety nets (Devereux 2007). In another example in East Africa's range lands, flood halted livestock sales, food prices fell, and grain production ceased. Local food shortages couldn't be supplemented with imports due to destruction of transport links, and pastoral incomes were inadequate to purchase food. Livestock diseases became rampant and eventually food shortages led to escalating prices. Due to the contextual nature and timing of events, policy responses initially addressed mobility and resource access, and eventually longer-term issues such as livestock disease (Little et al. 2001).

In North America, floods are often described in terms of costs. For instance, the 1997 Red River Basin flood cost Manitoba, Canada 1 billion USD and the USA 4 billion USD in terms of impact on agriculture and food production (Adaptation to Climate Change Team 2013). In Canada, floods accounted for 82% of disaster financial assistance spent from 2005–2014 (Public Safety Canada 2017) and this cost may increase in the future. Future climate change may result in a 2 meter in sea level by 2100, costing from 507 to 882 billion USD, affecting 300 American cities (losing one-half of their homes) and the wholesale loss of 36 cities (Lemann 2018).

Policy measures are important as an increasingly warming world may make post-disaster assistance and insurance increasingly unaffordable (Surminski et al. 2016). Historic legal mechanisms for retreating from low-lying and coastal areas have failed to encourage relocation of people out of flood plains and areas of high risk (Stoa 2015). In some places, cheap flood insurance and massive aid programmes have encouraged the populating of low-lying flood-prone and coastal areas (Lemann 2018). Although the state makes disaster assistance payments, it is local governments that determine vulnerability through flood zone mapping, restrictions from building in flood zones, building requirements (Stoa 2015), and integrated planning for flood. A comprehensive policy mix (Section 7.4.8) (implemented through adaptive management as illustrated in Figure 7.6) reduces vulnerability (Hurlbert 2018a,b). Policy mixes that allow people to respond to disasters include bankruptcy, insolvency rules, house protected from creditors, income minimums, and basic agricultural implement protection laws. The portfolio of policies allows people to recover and, if necessary, migrate to other areas and occupations (Hurlbert 2018b).

At the international level, reactionary disaster response has evolved to proactive risk management that combines adaptation and mitigation responses to ensure effective risk response, build resilient systems and solve issues of structural social inequality (Innocenti and Albrito 2011). Advanced measures of preparedness are the main instruments to reduce fatalities and limit damage, as illustrated in Figure 7.8. The Sendai Declaration (Sendai Framework for Disaster Risk Reduction 2015–2030), is an action plan to reduce mortality, the number of affected people and economic losses, using four priorities: understanding disaster risk; strengthening its governance to enhance the ability to manage disaster risk; investing in resilience; and enhancing disaster preparedness. There is *medium evidence* and *high agreement* that the Sendai Declaration significantly refers to adaptive governance and could be a window of opportunity to transform disaster risk reduction to address the causes of vulnerability (Munene et al. 2018). Addressing disasters increasingly requires individual, household, community and national planning and commitment to a new path of resilience and shared responsibility through whole community engagement and linking private and public infrastructure interests (Rouillard et al. 2013). It is recommended that a vision and overarching framework of governance be adopted to allow participation and coordination by government, NGOs, researchers and the private sector, individuals in the neighbourhood community. Disaster risk response is enhanced with complementary structural and non-structural measures, implemented together with measurable scorecard indicators (Chen 2011).

Box 7.2 (continued)

Adaptive Governance

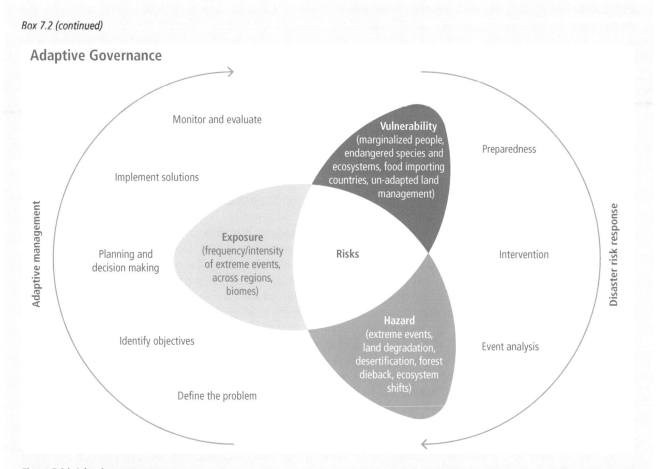

Figure 7.8 | Adaptive governance.

Adaptive management identifies and responds to exposure and vulnerability to land and climate change impacts by identifying problems and objectives, making decisions in relation to response options, and instruments advancing response options in the context of uncertainty. These decisions are continuously monitored, evaluated and adjusted to changing conditions. Similarly disaster risk management responds to hazards through preparation, prevention, response, analysis, and reconstruction in an iterative process.

7.6.4   Participation

It is recognised that more benefits are derived when citizens actively participate in land and climate decision-making, conservation, and policy formation (*high confidence*) (Jansujwicz et al. 2013; Coenen and Coenen 2009; Hurlbert and Gupta 2015). Local leaders supported by strong laws, institutions, and collaborative platforms, are able to draw on local knowledge, challenge external scientists, and find transparent and effective solutions for climate and land conflicts (Couvet and Prevot 2015; Johnson et al. 2017). Meaningful participation is more than providing technical/scientific information to citizens in order to accept decisions already made – rather, it allows citizens to deliberate about climate change impacts to determine shared responsibilities, creating genuine opportunity to construct, discuss and promote alternatives (*high confidence*) (Lee et al. 2013; Armeni 2016; Pieraccini 2015; Serrao-Neumann et al. 2015b; Armeni 2016). Participation is an emerging quality of collective action and social learning processes (Castella et al. 2014) when

barriers for meaningful participation are surpassed (Clemens et al. 2015). The absence of systematic leadership, the lack of consensus on the place of direct citizen participation, and the limited scope and powers of participatory innovations, limits the utility of participation (Fung 2015).

Multiple methods of participation exist, including multi-stakeholder forums, participatory scenario analyses, public forums and citizen juries (Coenen and Coenen 2009). No one method is superior, but each method must be tailored for local context (*high confidence*) (Blue and Medlock 2014; Voß and Amelung 2016). Strategic innovation in developing policy initiatives requires a strategic adaptation framework involving pluralistic and adaptive processes and use of boundary organisations (Head 2014).

The framing of a land and climate issue can influence the manner of public engagement (Hurlbert and Gupta 2015) and studies have found that local frames of climate change are particularly important (Hornsey et al. 2016; Spence et al. 2012), emphasising diversity of perceptions to adaptation and mitigation options (Capstick et al. 2015) – although Singh and Swanson (2017) found little evidence that framing impacted on the perceived importance of climate change.

Recognition and use of indigenous and local knowledge (ILK) is an important element of participatory approaches of various kinds. ILK can be used in decision-making on climate change adaptation, SLM and food security at various scales and levels, and is important for long-term sustainability (*high confidence*). Cross-Chapter Box 13 discusses definitional issues associated with ILK, evidence of its usefulness in responses to land-climate challenges, constraints on its use, and possibilities for its incorporation in decision-making.

Cross-Chapter Box 13 | Indigenous and local knowledge (ILK)

John Morton (United Kingdom), Fatima Denton (The Gambia), James Ford (United Kingdom), Joyce Kimutai (Kenya), Pamela McElwee (The United States of America), Marta Rivera Ferre (Spain), Lindsay Stringer (United Kingdom).

Indigenous and local knowledge (ILK) can play a key role in climate change adaptation (*high confidence*) (Mapfumo et al. 2017; Nyong et al. 2007; Green and Raygorodetsky 2010; Speranza et al. 2010; Alexander et al. 2011; Leonard et al. 2013; Nakashima et al. 2013; Tschakert 2007). The Summary for Policymakers of the Contribution of Working Group II to the Fifth Assessment Report of the Intergovernmental Panel on Climate Change (IPCC 2014b, p. 26) states that 'Indigenous, local, and traditional knowledge systems and practices, including indigenous peoples' holistic view of community and environment, are a major resource for adapting to climate change, but these have not been used consistently in existing adaptation efforts. Integrating such forms of knowledge with existing practices increases the effectiveness of adaptation' (see also Ford et al. 2016). The IPCC's Special Report on Global Warming of 1.5°C (SR15) (IPCC 2018a; de Coninck et al. 2018) confirms the effectiveness and potential feasibility of adaptation options based on ILK, but also raises concerns that such knowledge systems are being threatened by multiple socio-economic and environmental drivers (*high confidence*). The Intergovernmental Science-Policy Platform on Biodiversity and Ecosystem Services (IPBES) Land Degradation and Restoration Assessment (IPBES 2018) finds the same – that ILK can support adaptation to land degradation, but is threatened.

A variety of terminology has been used to describe ILK: indigenous knowledge, local knowledge, traditional knowledge, traditional ecological knowledge, and other terms are used in overlapping and often inconsistent ways (Naess 2013). SR15 (IPCC 2018a) reserves 'indigenous knowledge' for culturally distinctive ways of knowing associated with 'societies with long histories of interaction with their natural surroundings', while using 'local knowledge' for 'understandings and skills developed by individuals and populations, specific to the places where they live', but not all research studies observe this distinction. This Special Report generally uses ILK as a combined term for these forms of knowledge, but in some sections the terminology used follows that from the research literature assessed.

In contrast to scientific knowledge, ILK is context-specific, collective, transmitted informally, and is multi-functional (Mistry and Berardi 2016; Naess 2013; Janif et al. 2016). Persson et al. (2018) characterise ILK as 'practical experience', as locally held knowledges are acquired through processes of experience and interaction with the surrounding physical world. ILK is embedded in local institutions (Naess 2013) and in cultural aspects of landscape and food systems (Fuller and Qingwen 2013; Koohafkan and Altieri 2011). ILK can encompass such diverse content as factual information about the environment, guidance on rights and management, value statements about interactions with others, and cosmologies and worldviews that influence how information is perceived and acted on, among other topics (Spoon 2014; Usher 2000).

This cross-chapter box assesses evidence for the positive role of ILK in understanding climate change and other environmental processes, and in managing land sustainably in the face of climate change, desertification, land degradation and food insecurity. It also assesses constraints on and threats to the use of ILK in these challenges, and processes by which ILK can be incorporated in decision-making and governance processes.

ILK in understanding and responding to climate change impacts

ILK can play a role in understanding climate change and other environmental processes, particularly where formal data collection is sparse (Alexander et al. 2011; Schick et al. 2018), and can contribute to accurate predictions of impending environmental change (Green and Raygorodetsky 2010; Orlove et al. 2010) (*medium confidence*). At both global level (Alexander et al. 2011; Green and Raygorodetsky 2010), and local level (Speranza et al. 2010; Ayanlade et al. 2017), strong correlations between local perceptions of climate change and meteorological data have been shown, as calendars, almanacs, and other seasonal and interannual systems knowledge embedded in ILK hold information about environmental baselines (Orlove et al. 2010; Cochran et al. 2016).

Cross-Chapter Box 13 (continued)

ILK is strongly associated with sustainable management of natural resources, (including land), and with autonomous adaptation to climate variability and change, while also serving as a resource for externally-facilitated adaptation (Stringer et al. 2009). For example, women's traditional knowledge adds value to a society's knowledge base and supports climate change adaptation practices (Lane and McNaught 2009). In dryland environments, populations have historically demonstrated remarkable resilience and innovation to cope with high climatic variability, manage dynamic interactions between local communities and ecosystems, and sustain livelihoods (Safriel and Adeel 2008; Davies 2017). There is *high confidence* that pastoralists have created formal and informal institutions based on ILK for regulating grazing, collection and cutting of herbs and wood, and use of forests across the Middle East and North Africa (Louhaichi and Tastad 2010; Domínguez 2014; Auclair et al. 2011), Mongolia (Fernandez-Gimenez 2000), the Horn of Africa (Oba 2013) and the Sahel (Krätli and Schareika 2010). Herders in both the Horn of Africa and the Sahel have developed complex livestock breeding and selection systems for their dryland environment (Krätli 2008; Fre 2018). Numerous traditional water harvesting techniques are used across the drylands to adapt to climate variability: planting pits (*zai, ngoro*) and micro-basins and contouring hill slopes and terracing (Biazin et al. 2012), alongside the traditional *ndiva* water harvesting system in Tanzania to capture runoff in community-managed micro-dams for small-scale irrigation (Enfors and Gordon 2008).

Across diverse agro-ecological systems, ILK is the basis for traditional practices to manage the landscape and sustain food production, while delivering co-benefits in the form of biodiversity and ecosystem resilience at a landscape scale (*high confidence*). Flexibility and adaptiveness are hallmarks of such systems (Richards 1985a; Biggs et al. 2013), and documented examples include: traditional integrated watershed management in the Philippines (Camacho et al. 2016); widespread use of terracing, with benefits, in cases of both intensifying and decreasing rainfall (Arnáez et al. 2015; Chen et al. 2017b) and management of water harvesting and local irrigation systems in the Indo-Gangetic Plains (Rivera-Ferre et al. 2016). Rice cultivation in East Borneo is sustained by traditional forms of shifting cultivation, often involving intercropping of rice with bananas, cassava and other food crops (Siahaya et al. 2016), although the use of fire in land clearance implies trade-offs for climate change mitigation which have been sparsely assessed. Indigenous practices for enhanced soil fertility have been documented among South Asian farmers (Chandra et al. 2011; Dey and Sarkar 2011) and among Mayan farmers, where management of carbon has positive impacts on mitigation (Falkowski et al. 2016). Korean traditional groves or 'bibosoop' have been shown to reduce wind speed and evaporation in agricultural landscapes (Koh et al. 2010). Particularly in the context of changing climates, agriculture based on ILK that focuses on biodiversification, soil management, and sustainable water harvesting holds promise for long-term resilience (Altieri and Nicholls 2017) and rehabilitation of degraded land (Maikhuri et al. 1997). ILK is also important in other forms of ecosystem management, such as forests and wetlands, which may be conserved by efforts such as sacred sites (Ens et al. 2015; Pungetti et al. 2012). ILK can also play an important role in ecological restoration efforts, including for carbon sinks, through knowledge surrounding species selection and understanding of ecosystem processes, like fire (Kimmerer 2000).

Constraints on the use of ILK

Use of ILK as a resource in responding to climate change can be constrained in at least three ways (*high confidence*). First, the rate of climate change and the scale of its impacts may render incremental adaptation based on the ILK of smallholders and others, less relevant and less effective (Lane and McNaught 2009; Orlowsky and Seneviratne 2012; Huang et al. 2016; Morton 2017). Second, maintenance and transmission of ILK across generations may be disrupted, for example, by formal education, missionary activity, livelihood diversification away from agriculture, and a general perception that ILK is outdated and unfavourably contrasted with scientific knowledge (Speranza et al. 2010), and by HIV-related mortality (White and Morton 2005). Urbanisation can erode ILK, although ILK is constantly evolving, and becoming integrated into urban environments (Júnior et al. 2016; Oteros-Rozas et al. 2013; van Andel and Carvalheiro 2013). Third, ILK holders are experiencing difficulty in using ILK due to loss of access to resources, such as through large-scale land acquisition (Siahaya et al. 2016; Speranza et al. 2010; de Coninck et al. 2018). The increasing globalisation of food systems and integration into global market economy also threatens to erode ILK (Gómez-Baggethun et al. 2010; Oteros-Rozas et al. 2013; McCarter et al. 2014). The potential role that ILK can play in adaptation at the local level depends on the configuration of a policy–institutions–knowledge nexus (Stringer et al. 2018), which includes power relations across levels and interactions with government strategies (Alexander et al. 2011; Naess 2013).

Incorporation of ILK in decision-making

ILK can be used in decision-making on climate change adaptation, sustainable land management (SLM) and food security at various scales and levels, and is important for long-term sustainability (*high confidence*). Respect for ILK is both a requirement and an entry strategy for participatory climate action planning and effective communication of climate action strategies (Nyong et al. 2007). The nature, source, and mode of knowledge generation are critical to ensure that sustainable solutions are community-owned and fully integrated within the local context (Mistry and Berardi 2016). Integrating ILK with scientific information is a prerequisite for such community-owned solutions. Scientists can engage farmers as experts in processes of knowledge co-production (Oliver et al. 2012), helping to introduce, implement, adapt and promote locally appropriate responses (Schwilch et al. 2011). Specific approaches to decision-making that aim to integrate indigenous and local knowledge include some versions of decision support systems (Jones et al. 2014) as well as citizen science and participatory modelling (Tengö et al. 2014).

ILK can be deployed in the practice of climate governance, especially at the local level where actions are informed by the principles of decentralisation and autonomy (Chanza and de Wit 2016; Harmsworth and Awatere 2013). International environmental agreements are also increasingly including attention to ILK and diverse cultural perspectives, for reasons of social justice and inclusive decision-making (Brondizio and Tourneau 2016). However, the context-specific, and dynamic nature of ILK and its embeddedness in local institutions and power relations needs consideration (Naess 2013). It is also important to take a gendered approach so as not to further marginalise certain knowledge, as men and women hold different knowledge, expertise and transmission patterns (Díaz-Reviriego et al. 2017).

Citizen science

Citizen science is a democratic approach to science involving citizens in collecting, classifying, and interpreting data to influence policy and assist decision processes, including issues relevant to the environment (Kullenberg and Kasperowski 2016). It has flourished in recent years due to easily available technical tools for collecting and disseminating information (e.g., cell phone-based apps, cloud-based services, ground sensors, drone imagery, and others), recognition of its free source of labour, and requirements of funding agencies for project-related outreach (Silvertown 2009). There is significant potential for combining citizen science and participatory modelling to obtain favourable outcomes and improve environmental decision-making (*medium confidence*) (Gray et al. 2017). Citizen participation in land-use simulation integrates stakeholders' preferences through the generation of parameters in analytical and discursive approaches (Hewitt et al. 2014), and thereby supports the translation of narrative scenarios to quantitative outputs (Mallampalli et al. 2016), supports the development of digital tools to be used in co-designing decision-making participatory structures (Bommel et al. 2014), and supports the use of games to understand the preferences of local decision-making when exploring various balanced policies about risks (Adam et al. 2016).

There is *medium confidence* that citizen science improves SLM through mediating and facilitating landscape conservation decision-making and planning, as well as boosting environmental awareness and advocacy (Lange and Hehl-Lange 2011; Bonsu et al. 2017; Graham et al. 2015; Bonsu et al. 2017; Lange and Hehl-Lange 2011; Sayer et al. 2015; McKinley et al. 2017; Johnson et al. 2017, 2014; Gray et al. 2017). One study found limited evidence of direct conservation impact (Ballard et al. 2017) and most of the cases derive from rich industrialised countries (Loos et al. 2015). There are many practical challenges to the concept of citizen science at the local level. These include differing methods and the lack of universal implementation framework (Conrad and Hilchey 2011; Jalbert and Kinchy 2016; Stone et al. 2014). Uncertainty related to citizen science needs to be recognised and managed (Swanson et al. 2016; Bird et al. 2014; Lin et al. 2015) and citizen science projects around the world need better coordination to understand significant issues, such as climate change (Bonney et al. 2014).

Participation, collective action, and social learning

As land and climate issues cannot be solved by one individual, a diverse collective action issue exists for land-use policies and planning practices (Moroni 2018) at local, national, and regional levels. Collective action involves individuals and communities in land-planning processes in order to determine successful climate adaptation and mitigation (Nkoana et al. 2017; Liu and Ravenscroft 2017; Nieto-Romero et al. 2016; Nikolakis et al. 2016), or as Sarzynski (2015) finds, a community 'pulling together' to solve common adaptation and land-planning issues.

Collective action offers solutions for emerging land and climate change risks, including strategies that target maintenance or change of land-use practices, increase livelihood security, share risk through pooling, and sometimes also aim to promote social and economic goals such as reducing poverty (Samaddar et al. 2015; Andersson and Gabrielsson 2012). Collective action has resulted in the successful implementation of national-level land transfer policies (Liu and Ravenscroft 2017), rural development and land sparing (Jelsma et al. 2017), and the development of tools to identify shared objectives, trade-offs and barriers to land management (Nieto-Romero et al. 2016; Nikolakis et al. 2016). Collective action can also produce mutually binding agreements, government regulation, privatisation, and incentive systems (IPCC 2014c).

Successful collective action requires understanding and implementation of factors that determine successful participation in climate adaptation and mitigation (Nkoana et al. 2017). These include ownership, empowerment or self-reliance, time effectiveness, economic and behavioural interests, livelihood security, and the requirement for plan implementation (Samaddar et al. 2015; Djurfeldt et al. 2018; Sánchez and Maseda 2016). In a UK study, dynamic trust relations among members around specific issues, determined the potential of agri-environmental schemes to offer landscape-scale environmental protection (Riley et al. 2018). Collective action is context specific and rarely scaled up or replicated in other places (Samaddar et al. 2015).

Collective action in land-use policy has been shown to be more effective when implemented as bundles of actions rather than as single-issue actions. For example, land tenure, food security, and market access can mutually reinforce each other when they are interconnected (Corsi et al. 2017). For example, Liu and Ravenscroft (2017) found that financial incentives embedded in collective forest reforms in China have increased forest land and labour inputs in forestry.

A product of participation, equally important in practical terms, is social learning (*high confidence*) (Reed et al. 2010; Dryzek and Pickering 2017; Gupta 2014), which is learning in and with social groups through interaction (Argyris 1999) including collaboration and organisation which occurs in networks of interdependent stakeholders (Mostert et al. 2007). Social learning is defined as a change in understanding measured by a change in behaviour, and perhaps worldview, by individuals and wider social units, communities of practice and social networks (Reed et al. 2010; Gupta 2014). Social learning is an important factor contributing to long-term climate adaptation whereby individuals and organisations engage in a multi-step social process, managing different framings of issues while raising awareness of climate and land risks and opportunities, exploring policy options and institutionalising new rights, responsibilities, feedback and learning processes (Tàbara et al. 2010). It is important for engaging with uncertainty (Newig et al. 2010) and addressing the increasing unequal geography of food security (Sonnino et al. 2014).

Social learning is achieved through reflexivity or the ability of a social structure, process, or set of ideas to reconfigure itself after reflection on performance through open-minded people interacting iteratively to produce reasonable and well-informed opinions (Dryzek and Pickering 2017). These processes develop through skilled facilitation attending to social differences and power, resulting in a shared view of how change might happen (Harvey et al. 2012; Ensor and Harvey 2015). When combined with collective action, social learning can make transformative change, measured by a change in worldviews (beliefs about the world and reality) and understanding of power dynamics (Gupta 2014; Bamberg et al. 2015).

7.6.5 Land tenure

Land tenure, defined as 'the terms under which land and natural resources are held by individuals, households or social groups', is a key dimension in any discussion of land–climate interactions, including the prospects for both adaptation and land-based mitigation, and possible impacts on tenure and thus land security of both climate change and climate action (Quan and Dyer 2008) (*medium evidence, high agreement*).

Discussion of land tenure in the context of land–climate interactions in developing countries needs to consider the prevalence of informal, customary and modified customary systems of land tenure: estimates range widely, but perhaps as much as 65% of the world's total land area is managed under some form of these local, customary or communal tenure systems, and only a small fraction of this (around 15%) is formally recognised by governments (Rights and Resources Initiative 2015a). These customary land rights can extend across many categories of land, but are difficult to assess properly

due to poor reporting, lack of legal recognition, and lack of access to reporting systems by indigenous and rural peoples (Rights and Resources Initiative 2018a). Around 521 million ha of forest land is estimated to be legally owned, recognised, or designated for use by indigenous and local communities as of 2017 (Rights and Resources Initiative 2018b), predominantly in Latin America, followed by Asia. However, in India approximately 40 million ha of forest land is managed under customary rights not recognised by the government (Rights and Resources Initiative 2015b). In 2005 only 1% of land in Africa was legally registered (Easterly 2008a).

Much of the world's carbon is stored in the biomass and soil on the territories of customary landowners, including indigenous peoples (Walker et al. 2014; Garnett et al. 2018), making securing of these land tenure regimes vital in land and climate protection. These lands are estimated to hold at least 293 GtC of carbon, of which around one-third (72 GtC) is located in areas where indigenous peoples and local communities lack formal recognition of their tenure rights (Frechette et al. 2018).

Understanding the interactions between land tenure and climate change has to be based on underlying understanding of land tenure and land policy and how they relate to sustainable development, especially in low- and middle-income countries: such understandings have changed considerably over the last three decades, and now show that informal or customary systems can provide secure tenure (Toulmin and Quan 2000). For smallholder systems, Bruce and Migot-Adholla (1994) (among other authors) established that African customary tenure can provide the necessary security for long-term investments in farm fertility such as tree-planting. For pastoral systems, Behnke (1994), Lane and Moorehead (1995) and other authors showed the rationality of communal tenure in situations of environmental variability and herd mobility. However, where customary systems are unrecognised or weakened by governments, or the rights from them are undocumented or unenforced, tenure insecurity may result (Lane 1998; Toulmin and Quan 2000). There is strong empirical evidence of the links between secure communal tenure and lower deforestation rates, particularly for intact forests (Nepstad et al. 2006; Persha et al. 2011; Vergara-Asenjo and Potvin 2014). Securing and recognising tenure for indigenous communities (such as through revisions to legal or policy frameworks) has been shown to be highly cost effective in reducing deforestation and improving land management in certain contexts, and is therefore also apt to help improve indigenous communities' ability to adapt to climate changes (Suzuki 2012; Balooni et al. 2008; Ceddia et al. 2015; Pacheco et al. 2012; Holland et al. 2017).

Rights to water for agriculture or livestock are linked to land tenure in complex ways still little understood and neglected by policymakers and planners (Cotula 2006a). Provision of water infrastructure tends to increase land values, but irrigation schemes often entail reallocation of land rights (Cotula 2006b) and new inequalities based on water availability such as the creation of a category of tailenders (farmers at the downstream end of distribution channels) in large-scale irrigation (Chambers 1988) and disruption of pastoral grazing patterns through use of riverine land (Behnke and Kerven 2013).

Understanding land tenure under climate change also has to take account of the growth in large-scale land acquisitions (LSLAs), also referred to as land-grabbing, in developing countries. These LSLAs are defined by acquisition of more than 200 ha per deal (Messerli et al. 2014a). Klaus Deininger (2011) links the growth in demand for land to the 2007–2008 food price spike, and demonstrates that high levels of demand for land at the country level are statistically associated with weak recognition of land rights. Land grabs, where LSLAs occur despite local use of lands, are often driven by direct collaboration of politicians, government officials and land agencies (Koechlin et al. 2016), involving corruption of governmental land agencies, failures to register community land claims and illegal lands uses, and lack of the rule of law and enforcement in resource extraction frontiers (Borras Jr et al. 2011). Though data is poor, overall, small- and medium-scale domestic investment has in fact been more important than foreign investment (Deininger 2011; Cotula et al. 2014). There are variations in estimates of the scale of LSLAs: Nolte et al. (2016) concluded that deals totalled 42.2 million ha worldwide. Cotula et al. (2014) using cross-checked data for completed lease agreements in Ethiopia, Ghana and Tanzania conclude that they cover 1.9%, 1.9% and 1.1% respectively of each country's total land suitable for agriculture. The literature expresses different views on whether these acquisitions concern marginal lands or lands already in use, thereby displacing existing users (Messerli et al. 2014b). Land-grabbing is associated with, and may be motivated by, the acquisition of rights to water, and erosion of those rights for other users such as those downstream (Mehta et al. 2012). Quantification of the acquisition of water rights resulting from LSLAs raises major issues of definition, data availability, and measurement. One estimate of the total acquisition of gross irrigation water associated with land-grabbing across the 24 countries most affected is 280 billion m^3 (Rulli et al. 2013).

While some authors see LSLAs as investments that can contribute to more efficient food production at larger scales (World Bank 2011; Deininger and Byerlee 2012), others have warned that local food security may be threatened by them (Daniel 2011; Golay and Biglino 2013; Lavers 2012). Reports suggest that recent land-grabbing has affected 12 million people globally in terms of declines in welfare (Adnan 2013; Davis et al. 2014). De Schutter (2011) argues that large-scale land acquisitions will: a) result in types of farming less liable to reduce poverty than smallholder systems, b) increase local vulnerability to food price shocks by favouring export agriculture and c) accelerate the development of a market for land, with detrimental impacts on smallholders and those depending on common property resources. Land-grabbing can threaten not only agricultural lands of farmers, but also protected ecosystems, like forests and wetlands (Hunsberger et al. 2017; Carter et al. 2017; Ehara et al. 2018).

The primary mechanisms for combating LSLAs have included restrictions on the size of land sales (Fairbairn 2015), pressure on agribusiness companies to agree to Voluntary Guidelines on the Responsible Governance of Tenure of Land, Fisheries and Forests in the Context of National Food Security, known as VGGT, or similar principles (Collins 2014; Goetz 2013), attempts to repeal biofuels standards (Palmer 2014), strengthening of existing land law and land registration systems (Bebbington et al. 2018), use of community

monitoring systems (Sheil et al. 2015), and direct protests against land acquisitions (Hall et al. 2015; Fameree 2016).

Table 7.7 sets out, in highly summarised form, some key findings on the multi-directional inter-relations between land tenure and climate change, with particular reference to developing countries. The rows represent different categories of landscape or resource systems. For each system the second column summarises current understandings on land tenure and sustainable development, in many cases predating concerns over climate change. The third column summarises the most important implications of land tenure systems, policy about land tenure, and the implementation of that policy, for vulnerability and adaptation to climate change, and the fourth column gives a similar summary for mitigation of climate change. The fifth column summarises key findings on how climate change and climate action (both adaptation and mitigation) will impact land tenure, and the final column, findings on implications of climate change for evolving land policy.

Table 7.7 | Major findings on the interactions between land tenure and climate change.

Landscape or natural resource system	State of understanding of land tenure, land policy and sustainable development	Implications of land tenure for vulnerability and adaptation to climate change	Implications of land tenure for mitigation of climate change	Impacts of climate change and climate action on land tenure	Implications of climate change and climate action for land policy
Smallholder cropland	In South Asia and Latin America, the poor suffer from limited access, including insecure tenancies, though this has been partially alleviated by land reform.[1] In Africa informal/customary systems may provide considerable land tenure security and enable long-term investment in land management, but are increasingly weakened by demographic pressures on available land resources increase. However, creation of freehold rights through conventional land titling is not a necessary condition for tenure security and may be cost-ineffective or counterproductive.[2,3,4,5] Alternative approaches utilising low-cost technologies and participatory methods are available.[6] Secure and defendable land tenure, including modified customary tenure, has been positively correlated with food production increases.[7,8,9]	Insecure land rights are one factor deterring adaptation and accentuating vulnerability.[10,11] Specific dimensions of inequity in customary systems may act as constraints on adaptation in different contexts.[12] Large-scale land acquisitions (LSLAs) may be associated with monoculture and other unsustainable land-use practices, have negative consequences for soil degradation[13] and disincentivise more sustainable forms of agriculture.[14]	Secure land rights, including through customary systems, can incentivise farmers to adopt long-term climate-smart practices,[15] e.g., planting trees in mixed cropland/forest systems.[16]	Increased frequency and intensity of extreme weather can lead to displacement and effective loss of land rights.[17] REDD+ (reducing emissions from deforestation and forest degradation) programmes tend slightly to increase land tenure insecurity on agricultural forest frontier lands – but not in forests.[18]	Landscape governance and resource tenure reforms at farm and community levels can facilitate and incentivise planning for landscape management and enable the integration of adaptation and mitigation strategies.[11]
Rangelands	Communal management of rangelands in pastoral systems is a rational and internally sustainable response to climate variability and the need for mobility. Policies favouring individual or small group land-tenure may have negative impacts on both ecosystems and livelihoods.[19,20,21]	Many pastoralists in lands at risk from desertification do not have secure land tenure, and erosion of traditional communal rangeland tenure has been identified as a determinant of increasing vulnerability to drought and climate change and as a driver of dryland degradation.[22,23,24,25,26]	Where pastoralists' traditional land use does not have legal recognition, or where pastoralists are unable to exclude others from land use, this presents significant challenges for carbon sequestration initiatives.[27,28]	Increasing conflict on rangelands is a possible result of climate change and environmental pressures, but depends on local institutions.[29] Where land-use rights for pastoralists are absent or unenforced, demonstrated potential for carbon sequestration may assist advocacy.[28]	Carbon sequestration initiatives on rangelands may require clarification and maintenance of land rights.[27,28]

Landscape or natural resource system	State of understanding of land tenure, land policy and sustainable development	Implications of land tenure for vulnerability and adaptation to climate change	Implications of land tenure for mitigation of climate change	Impacts of climate change and climate action on land tenure	Implications of climate change and climate action for land policy
Forests	Poor management of state and open-access forests has been combated in recent years by a move towards forest decentralisation and community co-management.[30,31,32,33,34,35] Land tenure systems have complex interactions with deforestation processes. Land tenure security is generally associated with less deforestation, regardless of whether the tenure form is private, customary or communal.[33,36,37,38] Historical injustices towards forest dwellers can be ameliorated with appropriate policy, e.g., 2006 Forest Rights Act in India.[39]	Land tenure security can lead to improved adaptation outcomes[40,41,42,43] but land tenure policy for forests that focuses narrowly on cultivation has limited ability to reduce ecological vulnerability or enhance adaptation.[39] Secure rights to land and forest resources can facilitate efforts to stabilise shifting cultivation and promote more sustainable resource use if appropriate technical and market support are available.[44]	Land tenure insecurity has been identified as a key driver of deforestation and land degradation, leading to loss of sinks and creating sources of GHGs.[45,46,47,48,49] While land tenure systems interact with land-based mitigation actions in complex ways,[36] forest decentralisation and community co-management has shown considerable success in slowing forest loss and contributing to carbon mitigation.[30,31,32,33,34,35] Communal tenure systems may lower transaction costs for REDD+ schemes, though with risk of elite capture of payments.[16]	Findings on both direction of change in tenure security and extent to which this has been influenced by REDD+ are very diverse.[18] The implications of land-based mitigation – e.g., bioenergy with carbon capture and storage (BECCS) – on land tenure systems is currently understudied, but evidence from biofuels expansion shows negative impacts on local livelihoods and loss of forest sinks where LSLAs override local land tenure.[50,51]	Forest tenure policies under climate change need to accommodate and enable evolving and shifting boundaries linked to changing forest livelihoods.[10] REDD+ programmes need to be integrated with national-level forest tenure reform.[18]
Poor and informal urban settlements	Residents of poor and informal urban settlements enjoy varying degrees of tenure security from different forms of tenure. Security will be increased by building on de facto rights rather than through abrupt changes in tenure systems.[52]	Public land on the outskirts of urban areas can be used to adapt to increasing flood risks by protecting natural assets.[53] Secure land titles in hazardous locations may make occupants reluctant to move and raise the costs of compensation and resettlement.[17]	Urban land-use strategies such as tree planting, establishing public parks, can save energy usage by moderating urban temperature and protect human settlement from natural disaster such as flooding or heatwaves.[54]	Without proper planning, climate hazards can undermine efforts to recognise and strengthen informal tenure rights without proper planning.[55,56]	Climate risks increase the requirements for land-use planning and settlement that increases tenure security, with direct involvement of residents, improved use of public land, and innovative collaboration with private and traditional land owners.[56,57]
Riverscapes and riparian fringes	Well-defined but spatially flexible community tenure can support regulated and sustainable artisanal capture fisheries and biodiversity.[58,59,60,61,62,63,64]	Unequal land rights and absence of land management arrangements in floodplains increases vulnerability and constrains adaptation.[65] Marginalised or landless fisherfolk will be empowered by tenurial rights and associated identity to respond more effectively to ecological changes in riverscapes, including riparian zones.[66,67,68,69]	Mitigation measures such as protection of riparian forests and grasslands can potentially play a major role, provided rights to land and trees are sufficiently clear.[70,71]		Secured but spatially flexible tenure will enable climate change mitigation in riverscapes to be synergised with local livelihoods and ecological security.[67,72]

Sources: 1) Binswanger et al. 1995; 2) Schlager and Ostrom 1992; 3) Toulmin and Quan 2000; 4) Bruce and Migot-Adholla 1994; 5) Easterly 2008; 6) McCall and Dunn 2012; 7) Maxwell and Wiebe 1999; 8) Holden and Ghebru 2016; 9) Corsi et al. 2017; 10) Quan et al. 2017; 11) Harvey et al. 2014; 12) Antwi-Agyei et al. 2015; 13) Balehegn 2015; 14) Friis and Nielsen, 2016; 15) Scherr et al. 2012; 16) Barbier and Tesfaw 2012; 17) Mitchell 2010; 18) Sunderlin et al. 2018; 19) Behnke 1994; 20) Lane and Moorehead 1995; 21) Davies et al. 2015; 22) Morton 2007; 23) López-i-Gelats et al. 2016; 24) Oba 1994; 25) Fraser et al. 2011; 26) Dougill et al. 2011; 27) Roncoli et al. 2007; 28) Tennigkeit and Wilkes 2008; 29) Adano et al. 2012; 30) Agrawal et al. 2008; 31) Chhatre and Agrawal, 2009; 32) Gabay and Alam, 2017; 33) Holland et al. 2017; 34) Larson and Pulhin, 2012; 35) Pagdee et al. 2006; 36) Robinson et al. 2014; 37) Blackman et al. 2017; 38) Robinson et al. 2001; 39) Ramnath 2008; 40) Suzuki 2012; 41) Balooni et al. 2008; 42) Ceddia et al. 2015; 43) Pacheco et al. 2012; 44) Garnett et al. 2018; 45) Clover and Eriksen, 2009; 46) Damnyag et al. 2012; 47) Finley-Brook 2007; 48) Robinson et al. 2014; 49) Stickler et al. 2017; 50) Romijn 2011; 51) Aha and Ayitey 2017; 52) Payne 2001; 53) Barbedo et al. 2015; 54) Zhao et al. 2018; 55) Mitchell et al. 2015; 56) Mitchell et al. 2015; 57) Satterthwaite 2007; 58) Thomas 1996; 59) Welcomme et al. 2010; 60) Silvano and Valbo-Jørgensen 2008; 61) Biermann et al. 2012; 62) Abbott et al. 2007; 63) Béné et al. 2011; 64) McGrath et al. 1993; 65) Barkat et al. 2001; 66) FAO 2015; 67) Hall et al. 2013; 68) Berkes 2001; 69) ISO 2017; 70) Rocheleau and Edmunds 1997; 71) Baird and Dearden 2003; 72) Béné et al. 2010.

In drylands, weak land tenure security, either for households disadvantaged within a customary tenure system or more widely as such a system is eroded, can be associated with increased vulnerability and decreased adaptive capacity (*limited evidence, high agreement*). There is *medium evidence* and *medium agreement* that land titling and recognition programmes, particularly those that authorise and respect indigenous and communal tenure, can lead to improved management of forests, including for carbon storage (Suzuki 2012; Balooni et al. 2008; Ceddia et al. 2015; Pacheco et al. 2012), primarily by providing legally secure mechanisms for exclusion of others (Nelson et al. 2001; Blackman et al. 2017). However, these titling programmes are highly context-dependent and there is also evidence that titling can exclude community and common management, leading to more confusion over land rights, not less, where poorly implemented (Broegaard et al. 2017). For all the systems, an important finding is that land policies can provide both security and flexibility in the face of climate change, but through a diversity of forms and approaches (recognition of customary tenure, community mapping, redistribution, decentralisation, co-management, regulation of rental markets, strengthening the negotiating position of the poor) rather than sole focus on freehold title (*medium evidence, high agreement*) (Quan and Dyer, 2008; Deininger and Feder 2009; St. Martin 2009). Land policy can be climate-proofed and integrated with national policies such as National Adaptation Programme of Action NAPAs (Quan and Dyer 2008). Land administration systems have a vital role in providing land tenure security, especially for the poor, especially when linked to an expanded range of information relevant to mitigation and adaptation (Quan and Dyer 2008; van der Molen and Mitchell 2016). Challenges to such a role include outdated and overlapping national land and forest tenure laws, which often fail to recognise community property rights and corruption in land administration (Monterrosso et al. 2017), as well as lack of political will and the costs of improving land administration programmes (Deininger and Feder 2009).

7.6.6   Institutional dimensions of adaptive governance

Institutional systems that demonstrate the institutional dimensions, or indicators (Table 7.8) enhance the adaptive capacity of the socio-ecological system to a greater degree than institutional systems that do not demonstrate these dimensions (*high confidence*) (Gupta et al. 2010; Mollenkamp and Kasten 2009). Governance processes and policy instruments supporting these characteristics are context specific (*medium evidence, high agreement*) (Biermann 2007; Gunderson and Holling 2001; Hurlbert and Gupta 2017; Bastos Lima et al. 2017a; Gupta et al. 2013a; Mollenkamp and Kasten 2009; Nelson et al. 2010; Olsson et al. 2006; Ostrom 2011; Pahl-Wostl 2009; Verweij et al. 2006; Weick and Sutcliffe 2001).

Consideration of these indicators is important when implementing climate change mitigation instruments. For example, a 'variety,' redundancy, or duplication of climate mitigation policy instruments is an important consideration for meeting Paris Agreement commitments. Given that 58% of EU emissions are outside of the EU Emissions Trading System, implementation of a 'redundant' carbon tax may add co-benefits (Baranzini et al. 2017). Further, a carbon tax phased in over time through a schedule of increases

allows for 'learning.' The tax revenues could be earmarked to finance additional climate change mitigation and/or redistributed to achieve the indicator of 'fair governance – equity'. It is recommended that carbon pricing measures be implemented using information-sharing and communication devices to enable public acceptance, openness, provide measurement and accountability (Baranzini et al. 2017; Siegmeier et al. 2018).

The impact of flood on a socio-ecological system is reduced with the governance indicator of both leadership and resources (Emerson and Gerlak 2014). 'Leadership' pertains to a broad set of stakeholders that facilitate adaptation (and might include scientists and leaders in NGOs) and those that respond to flood in an open, inclusive, and fair manner identifying the most pressing issues and actions needed. Resources are required to support this leadership and includes upfront financial investment in human capital, technology, and infrastructure (Emerson and Gerlak 2014).

Policy instruments advancing the indicator of 'participation' in community forest management include favourable loans, tax measures, and financial support to catalyse entrepreneurial leadership, and build in rewards for supportive and innovative elites to reduce elite capture and ensure more inclusive participation (Duguma et al. 2018) (Section 7.6.4).

Table 7.8 | Institutional dimensions or indicators of adaptive governance. This table represents a summation of characteristics, evaluative criteria, elements, indicators or institutional design principles that advance adaptive governance.

Indicators/Institutional dimensions	Description	References
Variety	Room for a variety of problem frames reflecting different opinions and problem definitions	Biermann 2007 Gunderson and Holling 2001 Hurlbert and Gupta 2017 Bastos Lima et al. 2017a Gupta, J., van der Grijp, N., Kuik 2013 Mollenkamp and Kasten 2009 Nelson et al. 2010 Olsson et al. 2006 Ostrom 2011 Pahl-Wostl 2009 Verweij et al. 2006 Weick and Sutcliffe 2001
Variety	Participation. Involving different actors at different levels, sectors, and dimensions	
Variety	Availability of a wide range or diversity of policy options to address a particular problem	
Variety	Redundancy or duplication of measures, back-up systems	
Learning	Trust	
Learning	Single loop learning or ability to improve routines based on past experience	
Learning	Double loop learning or changed underlying assumptions of institutional patterns	
Learning	Discussion of doubts (openness to uncertainties, monitoring and evaluation of policy experiences)	
Learning	Institutional memory (monitoring and evaluation of policy experiences over time)	
Room for autonomous change	Continuous access to information (data institutional memory and early warning systems)	
Room for autonomous change	Acting according to plan (especially in relation to disasters)	
Room for autonomous change	Capacity to improvise (in relation to self-organisation and fostering social capital)	
Leadership	Visionary (long-term and reformist)	
Leadership	Entrepreneurial; leads by example	
Leadership	Collaborative	
Resources	Authority resources or legitimate forms of power	
Resources	Human resources of expertise, knowledge and labour	
Resources	Financial resources	
Fair governance	Legitimacy or public support	
Fair governance	Equity in relation to institutional fair rules	
Fair governance	Responsiveness to society	
Fair governance	Accountability in relation to procedures	

7.6.7   Inclusive governance for sustainable development

Many sustainable development efforts fail because of lack of attention to societal issues, including inequality, discrimination, social exclusion and marginalisation (see Cross-Chapter Box 11 in this chapter) (Arts 2017a). However, the human-rights-based approach of the 2030 Agenda and Sustainable Development Goals commits to leaving no one behind (Arts 2017b). Inclusive governance focuses attention in issues of equity and the human-rights-based approach for development as it includes social, ecological and relational components used for assessing access to, as well as the allocations of rights, responsibilities and risks with respect to social and ecological resources (*medium agreement*) (Gupta and Pouw 2017).

Governance processes that are inclusive of all people in decision-making and management of land, are better able to make decisions addressing trade-offs of sustainable development (Gupta et al. 2015) and achieve SDGs focusing on social and ecological inclusiveness (Gupta and Vegelin 2016). Citizen engagement is important in enhancing natural resource service delivery by citizen inclusion in management and governance decisions (Section 7.5.5). In governing natural resources, focus is now not only on rights of citizens in relation to natural resources, but also on citizen obligations, responsibilities (Karar and Jacobs-Mata 2016; Chaney and Fevre 2001), feedback and learning processes (Tàbara et al. 2010). In this respect, citizen engagement is also an imperative, particularly for

analysing and addressing aggregated informal coping strategies of local residents in developing countries, which are important drivers of natural resource depletions (but often overlooked in conventional policy development processes in natural resource management) (Ehara et al. 2018).

Inclusive adaptive governance makes important contributions to the management of risk. Inclusive governance concerning risk integrates people's knowledge and values by involving them in decision-making processes where they are able to contribute their respective knowledge and values to make effective, efficient, fair, and morally acceptable decisions (Renn and Schweizer 2009). Representation in decision-making would include major actors – government, economic sectors, the scientific community and representatives of civil society (Renn and Schweizer 2009). Inclusive governance focuses attention on the well-being and meaningful participation in decision-making of the poorest (in income), vulnerable (in terms of age, gender, and location), and the most marginalised, and is inclusive of all knowledges (Gupta et al. 2015).

7.7   Key uncertainties and knowledge gaps

Uncertainties in land, society and climate change processes are outlined in Section 7.2 and Chapter 1. This chapter has reviewed literature on risks arising from GHG fluxes, climate change, land degradation, desertification and food security, policy instruments responding to these risks, as well as decision-making and adaptive climate and land governance, in the face of uncertainty.

More research is required to understand the complex interconnections of land, climate, water, society, ES and food, including:

- new models that allow incorporation of considerations of justice, inequality and human agency in socio-environmental systems
- understanding how policy instruments and response options interact and augment or reduce risks in relation to acute shocks and slow-onset climate events
- understanding how response options, policy and instrument portfolios can reduce or augment the cascading impacts of land, climate and food security and ES interactions through different domains such as health, livelihoods and infrastructure, especially in relation to non-linear and tipping-point changes in natural and human systems
- consideration of trade-offs and synergies in climate, land, water, ES and food policies
- the impacts of increasing use of land due to climate mitigation measures such as BECCS, carbon-centric afforestation/REDD+ and their impacts on human conflict, livelihoods and displacement
- understanding how different land tenure systems, both formal and informal, and the land policies and administration systems that support them, can constrain or facilitate climate adaptation and mitigation, and on how forms of climate action can enhance or undermine land tenure security and land justice
- expanding understanding of barriers to implementation of land-based climate policies at all levels from the local to the global, including methods for monitoring and documenting corruption, misappropriation and elite capture in climate action
- identifying characteristics and attributes signalling impending socio-ecological tipping points and collapse

- understanding the full cost of climate change in the context of disagreement on accounting for climate change interactions and their impact on society, as well as issues of valuation, and attribution uncertainties across generations
- new models and Earth observation to understand the complex interactions described in this section
- the impacts, monitoring, effectiveness, and appropriate selection of certification and standards for sustainability (Section 7.4.6.3) (Stattman et al. 2018) and the effectiveness of its implementation through the landscape governance approach (Pacheco et al. 2016) (Section 7.6.3).

Actions to mitigate climate change are rarely evaluated in relation to impact on adaptation, SDGs, and trade-offs with food security. For instance, there is a gap in knowledge in the optimal carbon pricing or emission trading scheme together with monitoring, reporting and verification system for agricultural emissions that will advance GHG reductions, food security, and SLM. Better understanding is needed of the triggers and leveraging actions that build sustainable development and SLM, as well as the effective organisation of the science and society interaction jointly shaping policies in the future. What societal interaction in the future will form inclusive and equitable governance processes and achieve inclusive governance institutions, especially including land tenure?

As there is a significant gap in NDCs and achieving commitments to keep global warming well below 2°C (Section 7.4.4.1), governments might consider evaluating national, regional, and local gaps in knowledge surrounding response options, policy instruments portfolios, and SLM supporting the achievement of NDCs in the face of land and climate change.

Frequently Asked Questions

FAQ 7.1 | How can indigenous knowledge and local knowledge inform land-based mitigation and adaptation options?

Indigenous knowledge (IK) refers to the understandings, skills and philosophies developed by societies with long histories of interaction with their natural surroundings. Local knowledge (LK) refers to the understandings and skills developed by individuals and populations, specific to the place where they live. These forms of knowledge, jointly referred to as Indigenous and Local Knowledge or ILK, are often highly context specific and embedded in local institutions, providing biological and ecosystem knowledge with landscape information. For example, they can contribute to effective land management, predictions of natural disasters, and identification of longer-term climate changes, and ILK can be particularly useful where formal data collection on environmental conditions may be sparse. ILK is often dynamic, with knowledge holders often experimenting with mixes of local and scientific approaches. Water management, soil fertility practices, grazing systems, restoration and sustainable harvesting of forests, and ecosystem-based adaptation are many of the land management practices often informed by ILK. ILK can also be used as an entry point for climate adaptation by balancing past experiences with new ways to cope. To be effective, initiatives need to take into account the differences in power between the holders of different types of knowledge. For example, including indigenous and/ or local people in programmes related to environmental conservation, formal education, land management planning and security tenure rights is key to facilitate climate change adaptation. Formal education is necessary to enhance adaptive capacity of ILK, since some researchers have suggested that these knowledge systems may become less relevant in certain areas where the rate of environmental change is rapid and the transmission of ILK between generations is becoming weaker.

FAQ 7.2 | **What are the main barriers to and opportunities for land-based responses to climate change?**

Land-based responses to climate change can be mitigation (e.g., renewable energy, vegetation or crops for biofuels, afforestation) or adaptation (e.g., change in cropping pattern, less water-intensive crops in response to moisture stress), or adaptation with mitigation co-benefits (e.g., dietary shifts, new uses for invasive tree species, siting solar farms on highly degraded land). Productive land is an increasingly scarce resource under climate change. In the absence of adequate deep mitigation in the less land-intensive energy sector, competition for land and water for mitigation and for other sectors such as food security, ecosystem services (ES) and biodiversity conservation could become a source of conflict and a barrier to land-based responses.

Barriers to land-based mitigation include opposition due to real and perceived trade-offs between land for mitigation and food security and ES. These can arise due to absence of or uncertain land and water rights. Significant upscaling of mitigation requires dedicated (normally land-based) sources in addition to use of wastes and residues. This requires high land-use intensity compared to other mitigation options that, in turn, place greater demands on governance. A key governance mechanism that has emerged in response to such concerns, especially during the past decade are standards and certification systems that include food security, and land and water rights, in addition to general criteria or indicators related to sustainable use of land and biomass, with an emphasis on participatory approaches. Other governance responses include linking land-based mitigation (e.g., forestry) to secure tenure and support for local livelihoods. A barrier to land-based mitigation is our choice of development pathway. Our window of opportunity – whether or not we face barriers or opportunities to land-based mitigation – depends on socio-economic decisions or pathways. If we have high population growth and resource intensive consumption (i.e., SSP3) we will have more barriers. High population and low land-use regulation results in less available space for land-based mitigation. But if we have the opposite trends (SSP1), we can have more opportunities.

Other barriers can arise when, in the short term, adaptation to a climate stress (e.g., increased dependence on groundwater during droughts) can become unsustainable in the longer term, and become a maladaptation. Policies and approaches that lead to land management that synergises multiple ES and reduce trade-offs could find greater acceptance and enjoy more success.

Opportunities to obtain benefits or synergies from land-based mitigation and adaptation arise from their relation to the land availability and the demand for such measures in rural areas that may otherwise lack incentives for investment in infrastructure, livelihoods and institutional capacity. After decades of urbanisation around the world, facilitated by significant investment in urban infrastructure and centralised energy and agricultural systems, rural areas have been somewhat neglected; this is even as farmers in these areas provide critical food and materials needed for urban areas. As land and biomass becomes more valuable, there will be benefits for farmers, forest owners and associated service providers as they diversify and feed into economic activities supporting bioenergy, value-added products, preservation of biodiversity and carbon sequestration (storage).

A related opportunity for benefits is the potentially positive transformation in rural and peri-urban landscapes that could be facilitated by investments that prioritise more effective management of ES and conservation of water, energy, nutrients and other resources that have been priced too low in relation to their environmental or ecological value. Multifunctional landscapes supplying food, feed, fibre and fuel to both local and urban communities, in combination with reduced waste and healthier diets, could restore the role of rural producers as stewards of resources rather than providing food at the lowest possible price. Some of these landscape transformations will function as both mitigation and adaptation responses by increasing resilience, even as they provide value-added bio-based products.

Governments can introduce a variety of regulations and economic instruments (taxes, incentives) to encourage citizens, communities and societies to adopt sustainable land management practices, with further benefits in addition to mitigation. Windows of opportunity for redesigning and implementing mitigation and adaptation can arise in the aftermath of a major disaster or extreme climate event. They can also arise when collective action and citizen science motivate voluntary shifts in lifestyles supported by supportive top-down policies.

References

Aalto, J., M. Kämäräinen, M. Shodmonov, N. Rajabov, and A. Venäläinen, 2017: Features of Tajikistan's past and future climate. *Int. J. Climatol.*, **37**, 4949–4961, doi:10.1002/joc.5135.

Abatzoglou, J.T., and A.P. Williams, 2016: Impact of anthropogenic climate change on wildfire across western US forests. *Proc. Natl. Acad. Sci.*, **113** (42), 11770–11775, doi:10.1073/pnas.1607171113.

Abatzoglou, J.T., A. Park Williams, and R. Barbero, 2019a: Global emergence of anthropogenic climate change in fire weather indices. *Geophys. Res. Lett.*, **46**, 326–336, doi:10.1029/2018GL080959.

Abbasi, T., and S.A. Abbasi, 2011: Small hydro and the environmental implications of its extensive utilization. *Renew. Sustain. energy Rev.*, **15**, 2134–2143, doi:10.1016/j.rser.2010.11.050.

Abbott, J. et al., 2007: Rivers as resources, rivers as borders: Community and transboundary management of fisheries in the Upper Zambezi River floodplains. *Can. Geogr. / Le Géographe Can.*, **51**, 280–302, doi:10.1111/j.1541-0064.2007.00179.x.

Abi-Samra, N.C., and W.P. Malcolm, 2011: *Extreme Weather Effects on Power Systems*. IEEE Power and Energy Society General Meeting, IEEE, Michigan, USA, 1–5, doi:10.1109/PES.2011.6039594.

Abid, M., U.A. Schneider, and J. Scheffran, 2016: Adaptation to climate change and its impacts on food productivity and crop income: Perspectives of farmers in rural Pakistan. *J. Rural Stud.*, **47**, 254–266, doi:10.1016/j.jrurstud.2016.08.005.

Achat, D.L., M. Fortin, G. Landmann, B. Ringeval, and L. Augusto, 2015a: Forest soil carbon is threatened by intensive biomass harvesting. *Sci. Rep.*, **5**, 15991, doi:10.1038/srep15991.

Adam, C., F. Taillandier, E. Delay, O. Plattard, and M. Toumi, 2016: *SPRITE – Participatory Simulation for Raising Awareness About Coastal Flood Risk on the Oleron Island*. Conference paper, SpringerInternational Publishing, Cham, Switzerland, 33–46 pp.

Adano, W., and F. Daudi, 2012: *Link Between Climate change, Conflict and Governance in Africa. Institute for Security Studies*, 234, Pretoria, South Africa.

Adano, W.R., T. Dietz, K. Witsenburg, and F. Zaal, 2012: Climate change, violent conflict and local institutions in Kenya's drylands. *J. Peace Res.*, **49**, 65–80, doi:10.1177/0022343311427344.

Adaptation Sub-committee, 2013: *Managing the Land in a Changing Climate Chapter 5: Regulating Services – Coastal Habitats*. Committee on Climate Change, London, UK, pp. 92–107.

Adaptation to Climate Change Team, 2013: *Summary for Decision Makers. Climate Change Adaptation and Canada's Crop and Food Supply*, 1–33. SFU Vancouver, British Columbia, Canada.

Adger, W.N., T. Quinn, I. Lorenzoni, C. Murphy, and J. Sweeney, 2013: Changing social contracts in climate-change adaptation. *Nat. Clim. Chang.*, **3**, 330–333, doi:10.1038/nclimate1751.

Inland Waterways Authority of India, 2016. Consolidated Environmental Impact Assessment Report of National Waterways-1: Volume 3. http://documents. worldbank.org/curated/en/190981468255890798/pdf/SFG2231-V1-REVISED-EA-P148775-Box396336B-PUBLIC-Disclosed-12-5-2016.pdf.

Adnan, S., 2013: Land grabs and primitive accumulation in deltaic Bangladesh: Interactions between neoliberal globalization, state interventions, power relations and peasant resistance. *J. Peasant Stud.*, **40**, 87–128, doi:10.1080/03066150.2012.753058.

Adu, M., D. Yawson, F. Armah, E. Abano, and R. Quansah, 2018: Systematic review of the effects of agricultural interventions on food security in northern Ghana. *PLoS One*, **13**, doi:10.1371/journal.pone.0203605.

Aeschbach-Hertig, W., and T. Gleeson, 2012: Regional strategies for the accelerating global problem of groundwater depletion. *Nat. Geosci.*, **5**, 853–861, https://doi.org/10.0.4.14/ngeo1617.

Agarwal, B., 2010: *Gender and Green Governance: The Political Economy of Women's Presence Within and Beyond Community Forestry.* Oxford University Press, Oxford, UK, 488 pages. DOI: 10.1177/2321023013482799.

Agarwal, B., 2018a: Gender equality, food security and the sustainable development goals. *Curr. Opin. Environ. Sustain.*, **34**, 26–32, doi:10.1016/J.cosust.2018.07.002.

Agarwal, B., 2018b: Gender equality, food security and the sustainable development goals. *Curr. Opin. Environ. Sustain.*, **34**, 26–32, doi:10.1016/j.cosust.2018.07.002.

Aggarwal, A., 2011: Implementation of Forest Rights Act, changing forest landscape, and 'politics of REDD+' in India. *J. Resour. Energy Dev.*, **8**, 131–148. https://pdfs.semanticscholar.org/60b4/5b258310186884f99d3333072943dba63e6f.pdf.

Agrawal, A., A. Chhatre, and R. Hardin, 2008: Changing governance of the world's forests. *Science*, **320**, 1460–1462.

Agriculture Technical Advisory Group, 2009: *Point of Obligation Designs and Allocation Methodologies for Agriculture and the New Zealand Emissions Trading Scheme*. Ministry of Agriculture and Forestry Pastoral House, Wellington, New Zealand, www.parliament.nz/resource/0000077853.

Aguilar-Støen, M., 2017: Better safe than sorry? Indigenous peoples, carbon cowboys and the governance of REDD in the Amazon. *Forum Dev. Stud.*, **44**, 91–108, doi:10.1080/08039410.2016.1276098.

Aha, B., and J.Z. Ayitey, 2017: Biofuels and the hazards of land grabbing: Tenure (in)security and indigenous farmers' investment decisions in Ghana. *Land Use Policy*, **60**, 48–59, doi:10.1016/J.LANDUSEPOL.2016.10.012.

Akhtar-Schuster, M. et al., 2017: Unpacking the concept of land degradation neutrality and addressing its operation through the Rio Conventions. *J. Environ. Manage.*, **195**, 4–15, doi:10.1016/j.jenvman.2016.09.044.

Alam, K., 2015: Farmers' adaptation to water scarcity in drought-prone environments: A case study of Rajshahi District, Bangladesh. *Agric. Water Manag.*, **148**, 196–206, doi:10.1016/j.agwat.2014.10.011.

Albers, R.A.W. et al., 2015: Overview of challenges and achievements in the climate adaptation of cities and in the Climate Proof Cities program. *Building and Environment*, **83**, 1–10, doi:10.1016/j.buildenv.2014.09.006.

Albert, C., J. Hauck, N. Buhr, and C. von Haaren, 2014: What ecosystem services information do users want? Investigating interests and requirements among landscape and regional planners in Germany. *Landsc. Ecol.*, **29**, 1301–1313, doi:10.1007/s10980-014-9990-5.

Albrecht, T.R., A. Crootof, and C.A. Scott, 2018: The water-energy-food nexus: A comprehensive review of nexus-specific methods. *Environ. Res. Lett.*, **13** (4), 043002, doi:10.1088/1748-9326/aaa9c6.

Aldy, J., A. Krupnick, R. Newell, I. Parry, and W.A. Pizer, 2010: Designing climate mitigation policy. *J. Econ. Lit.*, **48**, 903–934, doi:10.3386/w15022.

Aldy, J.E., and R.N. Stavins, 2012: The promise and problems of pricing carbon. *J. Environ. Dev.*, **21**, 152–180, doi:10.1177/1070496512442508.

Alexander, C. et al., 2011: Linking indigenous and scientific knowledge of climate change. *Bioscience*, **61**, 477–484, doi:10.1525/bio.2011.61.6.10.

Alexander, K.A. et al., 2015a: What factors might have led to the emergence of Ebola in West Africa? *PLoS Negl. Trop. Dis.*, **9** (6): e0003652, doi:10.1371/journal.pntd.0003652.

Alexander, P. et al., 2015b: Drivers for global agricultural land use change: The nexus of diet, population, yield and bioenergy. *Glob. Environ. Chang.*, **35**, 138–147, doi:10.1016/j.gloenvcha.2015.08.011.

Alimi, T.O. et al., 2015: Predicting potential ranges of primary malaria vectors and malaria in northern South America based on projected changes in climate, land cover and human population. *Parasit. Vectors*, **8**, 431, doi:10.1186/s13071-015-1033-9.

Alkire, S. et al., 2013a: The women's empowerment in agriculture index. *World Dev.*, **52**, 71–91, doi:10.1016/j.worlddev.2013.06.007.

Alkire, S. et al., 2013b: The women's empowerment in agriculture index. *World Dev.*, **52**, 71–91, doi:10.1016/j.worlddev.2013.06.007.

Allan, T., M. Keulertz, and E. Woertz, 2015: The water–food–energy nexus: An introduction to nexus concepts and some conceptual and operational problems. *International Journal of Water Resources Development*, **31**, 301–311, doi:10.1080/07900627.2015.1029118.

Allen, C.D. et al., 2010: A global overview of drought and heat-induced tree mortality reveals emerging climate change risks for forests. *For. Ecol. Manage.*, **259**, 660–684, doi:10.1016/j.foreco.2009.09.001.

Allen, C.R., J.J. Fontaine, K.L. Pope, and A.S. Garmestani, 2011: Adaptive management for a turbulent future. *J. Environ. Manage.*, **92**, 1339–1345, doi:10.1016/j.jenvman.2010.11.019.

Allen, M.R. et al., 2016: New use of global warming potentials to compare cumulative and short-lived climate pollutants. *Nat. Clim. Chang.*, **6**, 773. doi:10.1038/nclimate2998.

Allison, E.H. et al., 2009: Vulnerability of national economies to the impacts of climate change on fisheries. *Fish Fish.*, **10**, 173–196, doi:10.1111/j.1467-2979.2008.00310.x.

Allwood, J.M., V. Bosetti, N.K. Dubash, L. Gómez-Echeverri, and C. von Stechow, 2014: Glossary. In: Climate Change 2014: Mitigation of Climate Change. Contribution of Working Group III to the Fifth Assessment Report of the Intergovernmental Panel on Climate Change [Edenhofer, O., R. Pichs-Madruga, Y. Sokona, E. Farahani, S. Kadner, K. Seyboth, A. Adler, I. Baum, S. Brunner, P. Eickemeier, B. Kriemann, J. Savolainen, S. Schlömer, C. von Stechow, T. Zwickel and J.C. Minx (eds.)]. Cambridge University Press, Cambridge and New York, New York, USA.

Alrø, H.F., H. Moller, J. Læssøe, and E. Noe, 2016: Opportunities and challenges for multicriteria assessment of food system sustainability. *Ecol. Soc.*, **21**, 38, doi:10.5751/ES-08394-210138.

Alston, J.M., and P.G. Pardey, 2014: Agriculture in the global economy. *J. Econ. Perspect.*, **28**, 121–46, doi:10.1257/jep.28.1.121.

Alston, M., 2006: The gendered impact of drought. In: *Rural Gender Relations: Issues And Case Studies* [Bock, B.B., and S. Shortall, (eds.)]. CABI Publishing, Cambridge, Oxfordshire, UK, pp. 165–180.

Alston, M., 2014: Gender mainstreaming and climate change. *Women's Studies International Forum*, **47**, 287–294, doi:10.1016/j.wsif.2013.01.016.

Alston, M., J. Clarke, and K. Whittenbury, 2018a: Contemporary feminist analysis of Australian farm women in the context of climate changes. *Soc. Sci.*, **7**, 16, doi:10.3390/socsci7020016.

Altieri, M.A., and C.I. Nicholls, 2017: The adaptation and mitigation potential of traditional agriculture in a changing climate. *Clim. Change*, **140**, 33–45, doi:10.1007/s10584-013-0909-y.

Altieri, M.A., C.I. Nicholls, A. Henao, and M.A. Lana, 2015: Agroecology and the design of climate change-resilient farming systems. *Agron. Sustain. Dev.*, **35** (3), 869–890, doi:10.1007/s13593-015-0285-2.

Alvarez, G., M. Elfving, and C. Andrade, 2016: REDD+ governance and indigenous peoples in Latin America: the case of Suru Carbon Project in the Brazilian Amazon Forest. *Lat. Am. J. Manag. Sustain. Dev.*, **3**, 133, doi:10.1504/LAJMSD.2016.083705.

Alverson, K., and Z. Zommers, eds., 2018: *Resilience The Science of Adaptation to Climate Change*. Elsevier Science BV, 360 pp, doi:https://doi.org/10.1016/C2016-0-02121-6.

Ammann, W.J., 2013: Disaster risk reduction. *Encyclopedia of Earth Sciences Series*.

van Andel, T., and L.G. Carvalheiro, 2013: Why urban citizens in developing countries use traditional medicines: the case of suriname. *Evid. Based. Complement. Alternat. Med.*, **2013**, 687197, doi:10.1155/2013/687197.

Andela, N. et al., 2017: A human-driven decline in global burned area. *Science*, **356**, 1356–1362, doi:10.1126/science.aal4108.

Anderegg, W.R. L., J.M. Kane, and L.D.L. Anderegg, 2013: Consequences of widespread tree mortality triggered by drought and temperature stress. *Nat. Clim. Chang.*, **3**, 30–36, doi:10.1038/nclimate1635.

Anderson, C.M., C.B. Field, and K.J. Mach, 2017: Forest offsets partner climate-change mitigation with conservation. *Front. Ecol. Environ.*, **15**, 359–365, doi:10.1002/fee.1515.

Anderson, E.P., M.C. Freeman, and C.M. Pringle, 2006: Ecological consequences of hydropower development in Central America: Impacts of small dams and water diversion on neotropical stream fish assemblages. *River Res. Appl.*, **22**, 397–411, doi:10.1002/rra.899.

Anderson, J.E. (ed.), 2010: *Public Policymaking: An Introduction*. Cengage Learning, Massachusetts, USA, 352 pp.

Anderson, K., and G. Peters, 2016: The trouble with negative emissions. *Science*, **354**, 182–183, doi:10.1126/science.aah4567.

Anderson, S., J. Morton, and C. Toulmin, 2010: Climate change for agrarian societies in drylands: Implications and future pathways. In: *Social Dimensions of Climate Change: Equity and vulnerability in a warming world* [Mearns, R. and A. Norton (eds.)]. The World Bank, Washington, DC, USA, pp. 199–230.

Andersson, E., and S. Gabrielsson, 2012: 'Because of poverty, we had to come together': Collective action for improved food security in rural Kenya and Uganda. *Int. J. Agric. Sustain.*, **10**, 245–262, doi:10.1080/14735903.2012.666029.

Anenberg, S.C., L.W. Horowitz, D.Q. Tong, and J.J. West, 2010: An estimate of the global burden of anthropogenic ozone and fine particulate matter on premature human mortality using atmospheric modeling. *Environ. Health Perspect.*, **118** (9), 1189–95, doi:10.1289/ehp.0901220.

Angelsen, A. et al., 2018a: Conclusions: Lessons for the path to a transformational REDD+. In: *Transforming REDD+: Lessons and new directions* [Angelsen, A., C. Martius, V. De Sy, A. Duchelle, A. Larson, and T. Pham (eds.)]. CIFOR, Bogor, Indonesia, pp. 203–214.

Anguelovski, I. et al., 2016a: Equity impacts of urban land use planning for climate adaptation: Critical perspectives from the Global North and South. *J. Plan. Educ. Res.*, **36**, 333–348, doi:10.1177/0739456X16645166.

Ansar, A., B.L. Caldecott, and J. Tilbury, 2013: *Stranded assets and the fossil fuel divestment campaign: What does divestment mean for the valuation of fossil fuel assets?* Smith School of Enterprise and the Environment, University of Oxford, Oxford, UK, pp. 1–81.

Anthoff, D., R.S.J. Tol, and G. Yohe, 2010: Discounting for climate change. *Economics: The Open-Access, Open-Assessment E-Journal*, **3**, 2009–24, doi:10.5018/economics-ejournal.ja.2009-24.

Antwi-Agyei, P., A.J. Dougill, and L.C. Stringer, 2015: Impacts of land tenure arrangements on the adaptive capacity of marginalized groups: The case of Ghana's Ejura Sekyedumase and Bongo districts. *Land Use Policy*, **49**, 203–212, doi:10.1016/j.landusepol.2015.08.007.

Aragão, L.E.O.C., 2012: Environmental science: The rainforest's water pump. *Nature*, **489**, 217–218. doi:10.1038/nature11485.

Arakelyan, I., D. Moran, and A. Wreford, 2017: Climate smart agriculture: A critical review. In: *Making Climate Compatible Development Happen* [Nunan, F. (ed.)]. Routledge, London, UK, pp. 262.

Araujo Enciso, S.R., T. Fellmann, I. Pérez Dominguez, and F. Santini, 2016: Abolishing biofuel policies: Possible impacts on agricultural price levels, price variability and global food security. *Food Policy*, **61**, 9–26, doi:10.1016/J.FOODPOL.2016.01.007.

Archer, D. et al., 2014: Moving towards inclusive urban adaptation: Approaches to integrating community-based adaptation to climate change at city and national scale. *Clim. Dev.*, **6**, 345–356, doi:10.1080/17565529.2014.918868.

Argyris, C. (ed.), 1999: *On Organizational Learning*. Wiley-Blackwell, 480 pp. ISBN: 978-0-631-21309-3. www.wiley.com/en-us/On+Organizational+Learning%2C+2nd+Edition-p-9780631213093.

Armeni, C., 2016: Participation in environmental decision-making: Reflecting on planning and community benefits for major wind farms. *J. Environ. Law*, **28**, 415–441, doi:10.1093/jel/eqw021.

Arnáez, J., N. Lana-Renault, T. Lasanta, P. Ruiz-Flaño, and J. Castroviejo, 2015: Effects of farming terraces on hydrological and geomorphological processes. A review. *CATENA*, **128**, 122–134, doi:10.1016/J.CATENA.2015.01.021.

Arnell, N.W., and B. Lloyd-Hughes, 2014: The global-scale impacts of climate change on water resources and flooding under new climate and socio-economic scenarios. *Clim. Change*, **122**, 127–140, doi:10.1007/s10584-013-0948-4.

Arora-Jonsson, S., 2011: Virtue and vulnerability: Discourses on women, gender and climate change. *Glob. Environ. Chang.*, **21**, 744–751, doi:10.1016/j.gloenvcha.2011.01.005.

Arora-Jonsson, S., 2014: Forty years of gender research and environmental policy: Where do we stand? *Womens. Stud. Int. Forum*, **47**, 295–308, doi:10.1016/J.WSIF.2014.02.009.

Arrow, K.J. et al., 2014: Should governments use a declining discount rate in project analysis? *Rev. Environ. Econ. Policy*, doi:10.1093/reep/reu008.

Arts, K., 2017a: Inclusive sustainable development: A human rights perspective. *Curr. Opin. Environ. Sustain.*, 24, 58–62, doi:10.1016/j.cosust.2017.02.001.

Ashcroft, M. et al., 2016: Expert judgement. *Br. Actuar. J.*, **21**, 314–363, doi:10.1017/S1357321715000239.

Ashkenazy, A. et al., 2017: Operationalising resilience in farms and rural regions – Findings from fourteen case studies. *J. Rural Stud.*, **59**, 211–221, doi:10.1016/J.JRURSTUD.2017.07.008.

Ashoori, N., D.A. Dzombak, and M.J. Small, 2015: Sustainability Review of Water-Supply Options in the Los Angeles Region. *J. Water Resour. Plan. Manag.*, **141** (12): A4015005, doi:10.1061/ (ASCE)WR.1943-5452.0000541.

Astrup, R., R.M. Bright, P.Y. Bernier, H. Genet, and D.A. Lutz, 2018: A sensible climate solution for the boreal forest. *Nat. Clim. Chang.*, **8**, 11–12, doi:10.1038/s41558-017-0043-3.

Atkinson, G., B. Groom, N. Hanley, and S. Mourato, 2018: Environmental valuation and benefit-cost analysis in UK policy. *J. Benefit-Cost Anal.*, **9**, 1–23, doi:10.1017/bca.2018.6.

Auclair, L., P. Baudot, D. Genin, B. Romagny, and R. Simenel, 2011: Patrimony for resilience: Evidence from the forest Agdal in the Moroccan High Atlas Mountains. *Ecol. Soc.*, **16**, art24, doi:10.5751/ES-04429-160424.

Awal, M.A., 2013: Social safety net, disaster risk management and climate change adaptation: Examining their integration potential in Bangladesh. *Int. J. Sociol. Study*, **1**, 62–72. https://pdfs.semanticscholar.org/d353/9315f1f8162618e98958e662a23210fb4d6e.pdf.

Ayamga, E.A., F. Kemausuor, and A. Addo, 2015a: Technical analysis of crop residue biomass energy in an agricultural region of Ghana. *Resour. Conserv. Recycl.*, **96**, 51–60, doi:10.1016/j.resconrec.2015.01.007.

Ayanlade, A., M. Radeny, and J. Morton, 2017: Comparing smallholder farmers' perception of climate change with meteorological data: A case study from south-western Nigeria. *Weather Clim. Extrem.*, **15**, 24–33, https://doi.org/10.1016/j.wace.2016.12.001.

Ayeb-Karlsson, S., K. van der Geest, I. Ahmed, S. Huq, and K. Warner, 2016: A people-centred perspective on climate change, environmental stress, and livelihood resilience in Bangladesh. *Sustain. Sci.*, **11**, 679–694, doi:10.1007/s11625-016-0379-z.

Bacon, C.M. et al., 2014: Explaining the 'hungry farmer paradox': Smallholders and fair trade cooperatives navigate seasonality and change in Nicaragua's corn and coffee markets. *Glob. Environ. Chang.*, **25**, 133–149, doi:10.1016/j.gloenvcha.2014.02.005.

Badola, R., S.C. Barthwal, and S.A. Hussain, 2013: Payment for ecosystem services for balancing conservation and development in the rangelands of the Indian Himalayan region. In: Ning W., G.S Rawat, S. Joshi, M. Ismail, E. Sharma, 2013. *High-Altitude Rangelands their Interfaces Hindu Kush Himalayas*, International Centre for Integrated Mountain Development, Kathmandu, Nepal, 175 pp.

Bailey, R., 2013: The 'Food versus fuel' nexus. In: *The Handbook of Global Energy Policy* [Goldthau, A. (ed.)]. Wiley-Blackwell, New Jersey, USA, doi:10.1002/9781118326275.

Bailis, R., R. Drigo, A. Ghilardi, and O. Masera, 2015: The carbon footprint of traditional woodfuels. *Nat. Clim. Chang.*, **5**, 266–272, doi:10.1038/nclimate2491.

Bailis, R., Y. Wang, R. Drigo, A. Ghilardi, and O. Masera, 2017a: Getting the numbers right: Revisiting woodfuel sustainability in the developing world. *Environ. Res. Lett.*, **12**, 115002, doi:10.1088/1748-9326/aa83ed.

Baird, I.G., and P. Dearden, 2003: Biodiversity conservation and resource tenure regimes: A case study from Northeast Cambodia. *Environ. Manage.*, **32**, 541–550, doi:10.1007/s00267-003-2995-5.

Baker, D., A. Jayadev, and J. Stiglitz, 2017: *Innovation, Intellectual Property, and Development: A Better Set of Approaches for the 21st Century*. Access IBSA, Center for Economic and Policy Research (CEPR), Washington DC, USA. http://ip-unit.org/wp-content/uploads/2017/07/IP-for-21st-Century-EN.pdf.

Di Baldassarre, G., M. Kooy, J.S. Kemerink, and L. Brandimarte, 2013: Towards understanding the dynamic behaviour of floodplains as human-water systems. *Hydrol. Earth Syst. Sci.*, **17**, 3235–3244, doi:10.5194/hess-17-3235-2013.

Balehegn, M., 2015: Unintended consequences: The ecological repercussions of land grabbing in Sub-Saharan Africa. *Environment*, **57**, 4–21, doi:10.1080/00139157.2015.1001687.

Ballard, H.L., C.G.H. Dixon, and E.M. Harris, 2017: Youth-focused citizen science: Examining the role of environmental science learning and agency for conservation. *Biol. Conserv.*, **208**, 65–75, doi:10.1016/j.biocon.2016.05.024.

Balooni, K., J.M. Pulhin, and M. Inoue, 2008: The effectiveness of decentralisation reforms in the Philippines's forestry sector. *Geoforum*, **39**, 2122–2131, doi:10.1016/j.geoforum.2008.07.003.

Bamberg, S., J. Rees, and S. Seebauer, 2015: Collective climate action: Determinants of participation intention in community-based pro-environmental initiatives. *J. Environ. Psychol.*, **43**, 155–165, doi:10.1016/J.JENVP.2015.06.006.

Bamberg, S., J.H. Rees, and M. Schulte, 2018: Environmental protection through societal change: What psychology knows about collective climate action– And what it needs to find out. In: *Psychology and Climate Change* [Clayton, S. and C. Manning (eds.)]. Academic Press, Elsevier, Massachusetts, USA, 312pp., doi:10.1016/C2016-0-04326-7.

Banerjee, A. et al., 2015: A multifaceted program causes lasting progress for the very poor: Evidence from six countries. *Science*, **348** (6236), 1260799, doi:10.1126/science.1260799.

Baral, H., R.J. Keenan, S.K. Sharma, N.E. Stork, and S. Kasel, 2014: Economic evaluation of ecosystem goods and services under different landscape management scenarios. *Land Use Policy*, **39**, 54–64, doi:10.1016/j.landusepol.2014.03.008.

Baranzini, A. et al., 2017: Carbon pricing in climate policy: Seven reasons, complementary instruments, and political economy considerations. *Wiley Interdiscip. Rev. Clim. Chang.*, **8:e462**, doi:10.1002/wcc.462.

Barbedo, J., M. Miguez, D. van der Horst, P. Carneiro, P. Amis, and A. Ioris, 2015: Policy dimensions of land use change in peri-urban floodplains: The case of Paraty. *Ecol. Soc.*, **20**, doi:10.5751/ES-07126-200105.

Barbier, E.B., 2011: Pricing nature. *Annual Review of Resource Economics*, **3**, 337–353, doi:10.1146/annurev-resource-083110-120115.

Barbier, E.B., and A.T. Tesfaw, 2012: Can REDD+ save the forest? The role of payments and tenure. *Forests*, **3**, 881–895, doi:10.3390/f3040881.

Barkat, A., S. uz Zaman, S. Raihan, M. Rahman Chowdhury, and E. Director, 2001: *Political economy of Khas land in Bangladesh*. Association for Land Reform and Development, Dhaka, 270 pp. www.hdrc-bd.com/admin_panel/images/notice/1389588575.3. cover 11 text.pdf.

Barnett, J., and S. O'Neill, 2010: Maladaptation. *Global Environmental Change*, **20**, 211–213, doi:10.1016/j.gloenvcha.2009.11.004.

Barnett, J., and J.P. Palutikof, 2014: The limits to adaptation: A comparative analysis. In: *Applied Studies in Climate Adaptation* [Palutikof, J.P., S.L. Boulter, J. Barnett, D. Rissik (eds.)]. John Wiley & Sons, West Sussex, UK, pp. 231–240, doi:10.1002/9781118845028.ch26.

Barnett, J. et al., 2014: A local coastal adaptation pathway. *Nat. Clim. Chang.*, **4**, 1103–1108, doi:10.1038/nclimate2383.

Barnett, T.P., J.C. Adam, and D.P. Lettenmaier, 2005: Potential impacts of a warming climate on water availability in snow-dominated regions. *Nature*, **438**, 303–309, doi:10.1038/nature04141.

Barrett, C.B., 2005: Rural poverty dynamics: Development policy implications. *Agric. Econ.*, **32**, 45–60, doi:10.1111/j.0169-5150.2004.00013.x.

Barrientos, A., 2011: Social protection and poverty. *Int. J. Soc. Welf.*, **20**, 240–249, doi:10.1111/j.1468-2397.2011.00783.x.

Barros, F.S.M., and N.A. Honório, 2015: Deforestation and malaria on the Amazon frontier: Larval clustering of Anopheles darlingi (Diptera: Culicidae) determines focal distribution of malaria. *Am. J. Trop. Med. Hyg.*, **93**, 939–953, doi:10.4269/ajtmh.15-0042.

Bastos Lima, M.G. et al., 2017a: The sustainable development goals and REDD+: Assessing institutional interactions and the pursuit of synergies. *Int. Environ. Agreements Polit. Law Econ.*, **17**, 589–606, doi:10.1007/s10784-017-9366-9.

Bastos Lima, M.G., I.J. Visseren-Hamakers, J. Braña-Varela, and A. Gupta, 2017b: A reality check on the landscape approach to REDD+: Lessons from Latin America. *For. Policy Econ.*, **78**, 10–20, doi:10.1016/J.FORPOL.2016.12.013.

Bates, S., and P. Saint-Pierre, 2018: Adaptive policy framework through the lens of the viability theory: A theoretical contribution to sustainability in the Anthropocene Era. *Ecol. Econ.*, **145**, 244–262, doi:10.1016/j.ecolecon.2017.09.007.

Batie, S.S., 2008: Wicked problems and applied economics. *Am. J. Agric. Econ.*, **90**, 1176–1191, doi:10.1111/j.1467-8276.2008.01202.x.

Baudron, F., M. Jaleta, O. Okitoi, and A. Tegegn, 2014: Conservation agriculture in African mixed crop-livestock systems: Expanding the niche. *Agric. Ecosyst. Environ.*, **187**, 171–182, doi:10.1016/j.agee.2013.08.020.

Baulch, B., J. Wood, and A. Weber, 2006: Developing a social protection index for Asia. *Dev. Policy Rev.*, **24**, 5–29, doi:10.1111/j.1467-7679.2006.00311.x.

Baumgart-Getz, A., L.S. Prokopy, and K. Floress, 2012: Why farmers adopt best management practice in the United States: A meta-analysis of the adoption literature. *J. Environ. Manage.*, **96**, 17–25, doi:10.1016/j.jenvman.2011.10.006.

Bausch, J., L. Bojo'rquez-Tapia, and H. Eakin, 2014: Agroenvironmental sustainability assessment using multi-criteria decision analysis and system analysis. *Sustain. Sci.*, **9**, 303–319, doi:https://doi.org/10.1007/s11625-014-0243-y.

Bayisenge, J., 2018: From male to joint land ownership: Women's experiences of the land tenure reform programme in Rwanda. *J. Agrar. Chang.*, **18**, 588–605, doi:10.1111/joac.12257.

Bebbington, A.J. et al., 2018: Resource extraction and infrastructure threaten forest cover and community rights. *Proc. Natl. Acad. Sci.*, **115**, 13164–13173, doi:10.1073/PNAS.1812505115.

Beca Ltd, 2018: *Assessment of the Administration Costs and Barriers of Scenarios to Mitigate Biological emissions from Agriculture*. Beca Limited. New Zealand. www.mpi.govt.nz/dmsdocument/32146/direct.

Behera, B., B. Rahut, A. Jeetendra, and A. Ali, 2015a: Household collection and use of biomass energy sources in South Asia. *Energy*, **85**, 468–480, doi:10.1016/j.energy.2015.03.059.

Behnke, R., 1994: Natural resource management in pastoral Africa. *Dev. Policy Rev.*, **12**, 5–28, doi:10.1111/j.1467-7679.1994.tb00053.x.

Behnke, R., and C. Kerven, 2013: Counting the costs: Replacing pastoralism with irrigated agriculture in the Awash valley, north-eastern Ethiopia. In: *Pastoralism and Development in Africa: Dynamic Changes at the Margins*, [Catley, A., J. Lind, and I. Scoones (eds.)]. Routledge, London, UK, pp. 49.

Behrman, K.D., T.E. Juenger, J.R. Kiniry, and T.H. Keitt, 2015: Spatial land use trade-offs for maintenance of biodiversity, biofuel, and agriculture. *Landsc. Ecol.*, **30**, 1987–1999, doi:10.1007/s10980-015-0225-1.

Belcher, B., M. Ruíz-Pérez, and R. Achdiawan, 2005: Global patterns and trends in the use and management of commercial NTFPs: Implications for livelihoods and conservation. *World Dev.*, **33**, 1435–1452, doi:10.1016/j.worlddev.2004.10.007.

Bellemare, M.F., 2015: Rising food prices, food price volatility, and social unrest. *Am. J. Agric. Econ.*, **97**, 1–21, doi:10.1093/ajae/aau038.

Below, T.B. et al., 2012: Can farmers' adaptation to climate change be explained by socio-economic household-level variables? *Glob. Environ. Chang.*, **22**, 223–235, doi:10.1016/j.gloenvcha.2011.11.012.

Ben-Ari, T., and D. Makowski, 2016: Analysis of the trade-off between high crop yield and low yield instability at the global scale. *Environ. Res. Lett.*, **11**, 104005 doi:10.1088/1748-9326/11/10/104005.

Béné, C., B. Hersoug, and E.H. Allison, 2010: Not by rent alone: Analysing the pro-poor functions of small-scale fisheries in developing countries. *Dev. Policy Rev.*, **28**, 325–358, doi:10.1111/j.1467-7679.2010.00486.x.

Béné, C., et al., 2011: Testing resilience thinking in a poverty context: Experience from the Niger River Basin. *Glob. Environ. Chang.*, **21**, 1173–1184, doi:10.1016/j.gloenvcha.2011.07.002.

Béné, C., S. Devereux, and R. Sabates-Wheeler, 2012: Shocks and social protection in the Horn of Africa: Analysis from the Productive Safety Net programme in Ethiopia. *IDS Working Paper*, **2012**, 1–120, doi:10.1111/j.2040-0209.2012.00395.x.

Bennett, J.E., 2013: Institutions and governance of communal rangelands in South Africa. *African J. Range Forage Sci.*, **30**, 77–83, doi:10.2989/10220119.2013.776634.

Bennett, N.J., and P. Dearden, 2014: From measuring outcomes to providing inputs: Governance, management, and local development for more effective marine protected areas. *Mar. Policy*, **50**, 96–110, doi:10.1016/j.marpol.2014.05.005.

Bennich, T., S. Belyazid, T. Bennich, and S. Belyazid, 2017a: The route to sustainability – Prospects and challenges of the bio-based economy. *Sustainability*, **9**, 887, doi:10.3390/su9060887.

Bentz, B.J. et al., 2010: Climate change and bark beetles of the western United States and Canada: Direct and indirect effects. *Bioscience*, **60**, 602–613, doi:10.1525/bio.2010.60.8.6.

Benveniste, H., O. Boucher, C. Guivarch, H. Le Treut, and P. Criqui, 2018: Impacts of nationally determined contributions on 2030 global greenhouse gas emissions: Uncertainty analysis and distribution of emissions. *Environ. Res. Lett.*, **13**, 014022, doi:10.1088/1748-9326/aaa0b9.

van den Bergh, J.C.J.M., and W.J.W. Botzen, 2014: A lower bound to the social cost of CO_2 emissions. *Nat. Clim. Chang.*, **4**, 253–258, doi:10.1038/nclimate2135.

Berhane, G., 2014: Can social protection work in Africa? The impact of Ethiopia's productive safety net programme. *Econ. Dev. Cult. Change*, **63**, 1–26, doi:10.1086/677753.

Berke, P.R., and M.R. Stevens, 2016: Land use planning for climate adaptation. *J. Plan. Educ. Res.*, **36**, 283–289, doi:10.1177/0739456X16660714.

Berkes, F., Mahon R., McConney P., Pollnac R., and Pomeroy R. 2001: *Managing Small-Scale Fisheries: Alternative Directions and Methods*. IDRC. Canada. https://idl-bnc-idrc.dspacedirect.org/handle/10625/31968.

Berndes, G., S. Ahlgren, P. Böorjesson, and A.L. Cowie, 2015: Bioenergy and land use change-state of the art. In: *Advances in Bioenergy: The Sustainability Challenge* [Lund, P.D., J. Byrne, G. Berndes, and I. Vasalos (eds.)], Wiley, 560 pp., doi:10.1002/9781118957844.

Bertram, M.Y. et al., 2018: Investing in non-communicable diseases: An estimation of the return on investment for prevention and treatment services. *The Lancet*, **391**, 2071–2078, doi:10.1016/S0140-6736(18)30665-2.

Bettini, G., and G. Gioli, 2016: Waltz with development: Insights on the developmentalization of climate-induced migration. *Migr. Dev.*, **5**, 171–189, doi:10.1080/21632324.2015.1096143.

Bettini, Y., R.R. Brown, and F.J. de Haan, 2015: Exploring institutional adaptive capacity in practice: Examining water governance adaptation in Australia. *Ecol. Soc.*, **20**, Art. 47, doi:10.5751/ES-07291-200147.

Beuchelt, T.D., and L. Badstue, 2013: Gender, nutrition- and climate-smart food production: Opportunities and trade-offs. *Food Secur.*, **5**, 709–721, doi:10.1007/s12571-013-0290-8.

Beuchle, R. et al., 2015: Land cover changes in the Brazilian Cerrado and Caatinga biomes from 1990 to 2010 based on a systematic remote sensing sampling approach. *Appl. Geogr.*, **58**, 116–127, doi:10.1016/J.APGEOG.2015.01.017.

Bevir, M., 2011: *The SAGE handbook of governance*. Sage Publishing, pp 592. California, USA.

Biagini, B., and A. Miller, 2013: Engaging the private sector in adaptation to climate change in developing countries: Importance, status, and challenges. *Clim. Dev.*, **5**, 242–252, doi:10.1080/17565529.2013.821053.

Biagini, B., L. Kuhl, K.S. Gallagher, and C. Ortiz, 2014: Technology transfer for adaptation. *Nat. Clim. Chang.*, **4**, 828–834, doi:10.1038/NCLIMATE2305.

Biazin, B., G. Sterk, M. Temesgen, A. Abdulkedir, and L. Stroosnijder, 2012: Rainwater harvesting and management in rainfed agricultural systems in Sub-Saharan Africa – A review. *Phys. Chem. Earth, Parts A/B/C*, **47–48**, 139–151, doi:10.1016/J.PCE.2011.08.015.

Bierbaum, R., and A. Cowie, 2018: *Integration: To Solve Complex Environmental Problems. Scientific and Technical Advisory Panel to the Global Environment Facility*. Washington, DC, USA, www.stapgef.org.

Biermann, F., 2007: 'Earth System governance' as a crosscutting theme of global change research. *Glob. Environ. Chang.*, **17**, 326–337, doi:10.1016/j.gloenvcha.2006.11.010.

Biermann, F. et al., 2012: Science and government. Navigating the anthropocene: Improving Earth System governance. *Science*, **335**, 1306–1307, doi:10.1126/science.1217255.

Biesbroek, G.R. et al., 2010: Europe adapts to climate change: Comparing National Adaptation Strategies. *Glob. Environ. Chang.*, **20**, 440–450, doi:10.1016/j.gloenvcha.2010.03.005.

Biesbroek, R., B.G. Peters, and J. Tosun, 2018: Public bureaucracy and climate change adaptation. *Rev. Policy Res.*, doi:10.1111/ropr.12316.

Biggs, E.M., E.L. Tompkins, J. Allen, C. Moon, and R. Allen, 2013: Agricultural adaptation to climate change: Observations from the mid hills of Nepal. *Clim. Dev.*, **5**, 165–173, doi:10.1080/17565529.2013.789791.

Biggs, E.M. et al., 2015: Sustainable development and the water-energy-food nexus: A perspective on livelihoods. *Environ. Sci. Policy*, **54**, 389–397, doi:10.1016/j.envsci.2015.08.002.

Biggs, H.C., J.K. Clifford-Holmes, S. Freitag, F.J. Venter, and J. Venter, 2017: Cross-scale governance and ecosystem service delivery: A case narrative from the Olifants River in north-eastern South Africa. *Ecosyst. Serv.*, doi:10.1016/j.ecoser.2017.03.008.

Binswanger, H.P., K. Deininger, and F. Gershon, 1995: Power, distortions, revolt, and reform in agricultural land relations. *Handb. Dev. Econ.*, **3**, 2659–2772, doi:https://doi.org/10.1016/S1573-4471 (95)30019-8.

Birch, J.C. et al., 2010: Cost-effectiveness of dryland forest restoration evaluated by spatial analysis of ecosystem services. *Proc. Natl. Acad. Sci.*, **107**, 21925–21930.

Bird, T.J. et al., 2014: Statistical solutions for error and bias in global citizen science datasets. *Biol. Conserv.*, **173**, 144–154, doi:10.1016/J.BIOCON.2013.07.037.

Bishop, I.D., C.J. Pettit, F. Sheth, and S. Sharma, 2013: Evaluation of data visualisation options for land use policy and decision-making in response to climate change. *Environ. Plan. B Plan. Des.*, **40**, 213–233, doi:10.1068/b38159.

Bizikova, L., D. Roy, D. Swanson, H.D. Venema, M. McCandless, 2013: *The Water-Energy-Food Security Nexus: Towards a Practical Planning and Decision-support Framework for Landscape Investment and Risk Management*. International Institute for Sustainable Development, Manitoba, Canada. www.iisd.org/sites/default/files/publications/wef_nexus_2013.pdf.

Black, R., N.W. Arnell, W.N. Adger, D. Thomas, and A. Geddes, 2013: Migration, immobility and displacement outcomes following extreme events. *Environ. Sci. Policy*, **27**, S32-S43, doi:10.1016/j.envsci.2012.09.001.

Blackman, A., L. Corral, E.S. Lima, and G.P. Asner, 2017: Titling indigenous communities protects forests in the Peruvian Amazon. *Proc. Natl. Acad. Sci.*, **114**, 4123–4128, doi:10.1073/pnas.1603290114.

Le Blanc, D., 2015: Towards integration at last? The sustainable development goals as a network of targets. *Sustain. Dev.*, **23**, 176–187, doi: https://doi.org/10.1002/sd.1582.

Le Blanc, D., C. Freire, and M. Vierro, 2017: *Mapping the linkages between oceans and other Sustainable Development Goals: A preliminary exploration*. DESA Working Paper No. 149, New York, USA, 34 pp., doi:10.18356/3adc8369-en.

Blanco-Canqui, H., and R. Lal, 2009a: Crop residue removal impacts on soil productivity and environmental quality. *CRC. Crit. Rev. Plant Sci.*, **28**, 139–163, doi:10.1080/07352680902776507.

Bloemen, P., M. Van Der Steen, and Z. Van Der Wal, 2018: Designing a century ahead: Climate change adaptation in the Dutch Delta. *Policy and Society*, **38**, 58–76, doi:10.1080/14494035.2018.1513731.

Blue, G., and J. Medlock, 2014: Public engagement with climate change as scientific citizenship: A case study of worldwide views on global warming. *Sci. Cult. (Lond).*, **23**, 560–579, doi:10.1080/09505431.2014.917620.

Bodnár, F., B. de Steenhuijsen Piters, and J. Kranen, 2011: Improving Food Security: A systematic review of the impact of interventions in agricultural production, value chains, market regulation and land security. Ministry of Foreign Affairs of the Netherlands, The Hague, Netherlands. https://europa.eu/capacity4dev/hunger-foodsecurity-nutrition/document/improving-food-security-systematic-review-impact-interventions-agricultural-production-valu.

Bodnar, P. et al., 2018: Underwriting 1.5°C: Competitive approaches to financing accelerated climate change mitigation. *Clim. Policy*, **18**, 368–382, doi:10.1080/14693062.2017.1389687.

Bogaerts, M. et al., 2017: Climate change mitigation through intensified pasture management: Estimating greenhouse gas emissions on cattle farms in the Brazilian Amazon. *J. Clean. Prod.*, **162**, 1539–1550, doi:10.1016/J.JCLEPRO.2017.06.130.

Bogale, A., 2015a: Weather-indexed insurance: An elusive or achievable adaptation strategy to climate variability and change for smallholder farmers in Ethiopia. *Clim. Dev.*, **7**, 246–256, doi:10.1080/17565529.2014.934769.

Bogardi, J.J. et al., 2012: Water security for a planet under pressure: Interconnected challenges of a changing world call for sustainable solutions. *Curr. Opin. Environ. Sustain.*, **4**, 35–43, doi:10.1016/j.cosust.2011.12.002.

Bohra-Mishra, P., and D.S. Massey, 2011: Environmental degradation and out-migration: New evidence from Nepal. In: *Migration and Climate Change* [Piguet, E., A. Pécoud and P. de Guchteneire (eds.)]. Cambridge University Press, Cambridge, United Kingdom and New York, NY, USA.

Böhringer, C., J.C. Carbone, and T.F. Rutherford, 2012: Unilateral climate policy design: Efficiency and equity implications of alternative instruments to reduce carbon leakage. *Energy Econ.*, **34**, S208–S217, doi:10.1016/j.eneco.2012.09.011.

Boillat, S., and F. Berkes, 2013: Perception and interpretation of climate change among Quechua farmers of Bolivia: Indigenous knowledge as a resource for adaptive capacity. *Ecol. Soc.*, **18**, Art. 21, doi:10.5751/ES-05894-180421.

Bommarco, R., D. Kleijn, and S.G. Potts, 2013: Ecological intensification: Harnessing ecosystem services for food security. *Trends Ecol. Evol.*, **28**, 230–238, doi:10.1016/j.tree.2012.10.012.

Bommel, P. et al., 2014: A further step towards participatory modelling. Fostering stakeholder involvement in designing models by using executable UML. *J. Artif. Soc. Soc. Simul.*, **17**, 1–9, doi:10.18564/jasss.2381.

Bonan, G.B., 2008: Forests and climate change: Forcings, feedbacks, and the climate benefits of forests. *Science*, **320**, 1444–1449, doi:10.1126/science.1155121.

Bonatti, M. et al., 2016: Climate vulnerability and contrasting climate perceptions as an element for the development of community adaptation

7

strategies: Case studies in southern Brazil. *Land Use Policy*, **58**, 114–122, doi:10.1016/j.landusepol.2016.06.033.

Bonney, R. et al., 2014: Citizen science. Next steps for citizen science. *Science*, **343**, 1436–1437, doi:10.1126/science.1251554.

Bonsch, M. et al., 2016: Trade-offs between land and water requirements for large-scale bioenergy production. *GCB Bioenergy*, **8**, 11–24, doi:10.1111/gcbb.12226.

Bonsu, N.O., Á.N. Dhubháin, and D. O'Connor, 2017: Evaluating the use of an integrated forest land-use planning approach in addressing forest ecosystem services conflicting demands: Expereince within an Irish forest landscape. *Futures*, **86**, 1–17, doi:10.1016/j.futures.2016.08.004.

Boonstra, W.J., and T.T.H. Hanh, 2015: Adaptation to climate change as social-ecological trap: A case study of fishing and aquaculture in the Tam Giang Lagoon, Vietnam. *Environ. Dev. Sustain.*, **17**, 1527–1544.

Borras Jr., S.M., R. Hall, I. Scoones, B. White, and W. Wolford, 2011: Towards a better understanding of global land grabbing: An editorial introduction. *J. Peasant Stud.*, **38**, 209–216, doi:10.1080/03066150.2011.559005.

Borychowski, M., and A. Czyżewski, 2015: Determinants of prices increase of agricultural commodities in a global context. *Management*, **19**, 152–167, doi:10.1515/manment-2015-0020.

Boyce, J.K., 2018: Carbon pricing: Effectiveness and equity. *Ecol. Econ.*, **150**, 52–61, doi:10.1016/j.ecolecon.2018.03.030.

Boyd, E., and C. Folke (eds.), 2011a: *Adapting Institutions: Governance, Complexity and Social-Ecological Resilience*. Cambridge University Press, Cambridge, UK, 290 pp.

Boyd, E., and C. Folke, 2012: Adapting institutions, adaptive governance and complexity: An introduction. In: *Adapting Institutions: Governance, Complexity and Social-Ecological Resilience* [Boyd, E. and C. Folke, (eds.)]. Cambridge University Press, Cambridge, United Kingdom and New York, NY, USA, pp. 1–8.

Boyd, E., B. Nykvist, S. Borgström, and I.A. Stacewicz, 2015: Anticipatory governance for social-ecological resilience. *Ambio*, **44**, 149–161, doi:10.1007/s13280-014-0604-x.

Boyle, M., Kay, J.J.; Pond, B., 2001: Monitoring in support of policy: An adaptive ecosystem approach. *Encyclopedia Glob. Environ. Chang.*, **4**, 116–137. http://documentacion.ideam.gov.co/openbiblio/bvirtual/017931/DocumentosIndicadores/Temasvarios/Docum8.pdf.

Boysen, L.R., W. Lucht, D. Gerten, and V. Heck, 2016: Impacts devalue the potential of large-scale terrestrial CO_2 removal through biomass plantations. *Environ. Res. Lett.*, **11**, 095010, doi:10.1088/1748-9326/11/9/095010.

Boysen, L.R., W. Lucht, and D. Gerten, 2017a: Trade-offs for food production, nature conservation and climate limit the terrestrial carbon dioxide removal potential. *Glob. Chang. Biol.*, **23**, 4303–4317, doi:10.1111/gcb.13745.

Boysen, L.R. et al., 2017b: The limits to global-warming mitigation by terrestrial carbon removal. *Earth's Future*, **5**, 463–474, doi:10.1002/2016EF000469.

Brancalion, P.H.S., R.A.G. Viani, B.B.N. Strassburg, and R.R. Rodrigues, 2012: Finding the money for tropical forest restoration. *Unasylva*, **239** (63), 41–50. www.fao.org/3/i2890e/i2890e07.pdf.

Brander, K., 2015: Improving the reliability of fishery predictions under climate change. *Curr. Clim. Chang. Reports*, **1**, 40–48, doi:10.1007/s40641-015-0005-7.

Brander, K.M., 2007: Global fish production and climate change. *Proc. Natl. Acad. Sci.*, **104**, 19709–19714, doi:10.1073/pnas.0702059104.

Branger, F., and P. Quirion, 2014: Climate policy and the 'carbon haven' effect. *Wiley Interdiscip. Rev. Clim. Chang.*, **5**, 53–71, doi:10.1002/wcc.245.

Bratton, M., 2007: Formal versus informal institutions in Africa. *J. Democr.*, **18**, 96–110, doi:10.1353/jod.2007.0041.

Brechin, S.R., and M.I. Espinoza, 2017: A case for further refinement of the green climate fund's 50:50 ratio climate change mitigation and adaptation allocation framework: Toward a more targeted approach. *Clim. Change*, **142**, 311–320, doi:10.1007/s10584-017-1938-8.

Breininger, D., B. Duncan, M. Eaton, F. Johnson, and J. Nichols, 2014: Integrating land cover modeling and adaptive management to conserve endangered species and reduce catastrophic fire risk. *Land*, **3**, 874–897, doi:10.3390/land3030874.

Brenkert-Smith, H., P.A. Champ, and N. Flores, 2006: Insights into wildfire mitigation decisions among wildland-urban interface residents. *Soc. Nat. Resour.*, **19**, 759–768, doi:10.1080/08941920600801207.

Bresch, D.N. et al., 2017: *Sovereign Climate and Disaster Risk Pooling*. World Bank Technical Contribution to the G20. World Bank. Washington DC, USA, 76 pp. http://documents.worldbank.org/curated/en/837001502870999632/pdf/118676-WP-v2-PUBLIC.pdf.

Bright, R.M. et al., 2017: Local temperature response to land cover and management change driven by non-radiative processes. *Nat. Clim. Chang.*, **7**, 296–302, doi:10.1038/nclimate3250.

Broegaard, R.B., T. Vongvisouk, and O. Mertz, 2017: Contradictory land use plans and policies in Laos: Tenure security and the threat of exclusion. *World Dev.*, **89**, 170–183, doi:10.1016/J.WORLDDEV.2016.08.008.

Brondizio, E.S., and F.-M. L. Tourneau, 2016: Environmental governance for all. *Science*, **352**, 1272–1273, doi:10.1126/science.aaf5122.

Bronen, R., and F.S. Chapin, 2013: Adaptive governance and institutional strategies for climate-induced community relocations in Alaska. *Proc. Natl. Acad. Sci.*, **110**, 9320–9325, doi:10.1073/pnas.1210508110.

Brooks, J., 2014: Policy coherence and food security: The effects of OECD countries' agricultural policies. *Food Policy*, 44, 88–94, doi:10.1016/j.foodpol.2013.10.006.

Brown, C., P. Alexander, S. Holzhauer, and M.D.A. Rounsevell, 2017: Behavioral models of climate change adaptation and mitigation in land-based sectors. *Wiley Interdiscip. Rev. Clim. Chang.*, **8**, e448, doi:10.1002/wcc.448.

Brown, I., and M. Castellazzi, 2014: Scenario analysis for regional decision-making on sustainable multifunctional land uses. *Reg. Environ. Chang.*, **14**, 1357–1371, doi:10.1007/s10113-013-0579-3.

Brown, M.L., 2010: Limiting corrupt incentives in a global REDD regime. *Ecol. Law Q.*, **37**, 237–267, doi:10.15779/Z38HC41.

Bruce, J.W., and S.E. Migot-Adholla, 1994: Introduction: Are indigenous African tenure systems insecure? In: *Searching for Land Tenure Security in Africa* [Bruce, J.W., and S.E. Migot-Adholla (ed.)]. World Bank, Washington, DC, USA, pp. 282.

de Bruin, K., R. Dellink, S. Agrawala, and R. Dellink, 2009: Economic aspects of adaptation to climate change: Integrated assessment modelling of adaptation costs and benefits. *OECD Environ. Work. Pap.*, **22**, 36–38, doi:10.1787/225282538105.

Bruvoll, A., and B.M. Larsen, 2004: Greenhouse gas emissions in Norway: Do carbon taxes work? *Energy Policy*, **32**, 493–505, doi:10.1016/S0301-4215 (03)00151-4.

Bryan, B.A., D. King, and E. Wang, 2010: Biofuels agriculture: Landscape-scale trade-offs between fuel, economics, carbon, energy, food, and fiber. *Gcb Bioenergy*, **2**, 330–345, doi:https://doi.org/10.1111/j.1757-1707.2010.01056.x.

Bryngelsson, D., S. Wirsenius, F. Hedenus, and U. Sonesson, 2016: How can the EU climate targets be met? A combined analysis of technological and demand-side changes in food and agriculture. *Food Policy*, **59**, 152–164, doi:10.1016/j.foodpol.2015.12.012.

Burby, R.J., and P.J. May, 2009: Command or cooperate? Rethinking traditional central governments' hazard mitigation policies. In: NATO Science for Peace and Security Series – E: Human and Societal Dynamics [Fra Paleo, U. (ed.)]. IOS Press Ebooks, Amsterdam, Netherlands, pp. 21–33. doi:10.3233/978-1-60750-046-9-21.

Burton, R.J.F., C. Kuczera, and G. Schwarz, 2008: Exploring farmers' cultural resistance to voluntary agrienvironmental schemes. *Sociol. Ruralis*, **48**, 16–37, doi:10.1111/j.1467-9523.2008.00452.x.

Busch, J., 2018: Monitoring and evaluating the payment-for-performance premise of REDD+: The case of India's ecological fiscal transfers. *Ecosyst. Heal. Sustain.*, **4**, 169–175, doi:10.1080/20964129.2018.1492335.

7

Busch, J., and A. Mukherjee, 2017: Encouraging state governments to protect and restore forests using ecological fiscal transfers: India's tax revenue distribution reform. *Conserv. Lett.*, **00**, 1–10, doi:10.1111/conl.12416.

Byerlee, D., D. Byerlee, and X. Rueda, 2015: From public to private standards for tropical commodities: A century of global discourse on land governance on the forest frontier. *Forests*, **6**, 1301–1324, doi:10.3390/f6041301.

Byers, E. et al., 2018a: Global exposure and vulnerability to multi-sector development and climate change hotspots. *Environ. Res. Lett.*, **13**, 055012, doi:10.1088/1748-9326/aabf45.

Byers, E., et al., 2018b: Global exposure and vulnerability to multi-sector development and climate change hotspots. *Environ. Res. Lett.*, **13**, 055012, doi:10.1088/1748-9326/aabf45.

Camacho, A.E., 2009: Adapting governance to climate change: Managing uncertainty through a learning infrastructure. *Emory Law J.*, **59**, 1–77, doi:10.2139/ssrn.1352693.

Camacho, L.D., D.T. Gevaña, Antonio P. Carandang, and S.C. Camacho, 2016: Indigenous knowledge and practices for the sustainable management of Ifugao forests in Cordillera, Philippines. *Int. J. Biodivers. Sci. Ecosyst. Serv. Manag.*, **12**, 5–13, doi:10.1080/21513732.2015.1124453.

Cameron, C. et al., 2016: Policy trade-offs between climate mitigation and clean cook-stove access in South Asia. *Nat. Energy*, **1**, 15010, doi:10.1038/nenergy.2015.10.

Campillo, G., M. Mullan, and L. Vallejo, 2017: *Climate Change Adaptation and Financial Protection*. OECD Environment Working Papers, No. 120, OECD Publishing, Paris, France, pp 59. doi:10.1787/0b3dc22a-en.

Caney, S., 2014: Climate change, intergenerational equity and the social discount rate. *Polit. Philos. Econ.*, **13** (4), 320–342, doi:10.1177/1470594X14542566.

Canonico, G.C., A. Arthington, J.K. McCrary, and M.L. Thieme, 2005: The effects of introduced tilapias on native biodiversity. *Aquat. Conserv. Mar. Freshw. Ecosyst.*, **15**, 463–483, doi:10.1002/aqc.699.

Cantarello, E. et al., 2010: Cost-effectiveness of dryland forest restoration evaluated by spatial analysis of ecosystem services. *Proc. Natl. Acad. Sci.*, **107**, 21925–21930, doi:10.1073/pnas.1003369107.

Capstick, S., L. Whitmarsh, W. Poortinga, N. Pidgeon, and P. Upham, 2015: International trends in public perceptions of climate change over the past quarter century. *Wiley Interdiscip. Rev. Clim. Chang.*, **6**, 35–61, doi:10.1002/wcc.321.

Cardona, O., and M.K. van Aalst, 2012: Determinants of Risk: Exposure and Vulnerability. In: Managing the Risks of Extreme Events and Disasters to Advance Climate Change Adaptation [Field, C.B., V. Barros, T.F. Stocker, D. Qin, D.J. Dokken, K.L. Ebi, M.D. Mastrandrea, K.J. Mach, G.-K. Plattner, S.K. Allen, M. Tignor, and P.M. Midgley (eds.)], Cambridge University Press, Cambridge, UK, and New York, NY, USA, 582 pp.

Carleton, T.A., and S.M. Hsiang, 2016a: Social and economic impacts of climate. *Science*, **353**, aad9837, doi:10.1126/science.aad9837.

Carroll, T. et al., 2012: *Catalyzing Smallholder Agricultural Finance*. Dalberg Global Development Advisors. https://oneacrefund.org/documents/101/Dalberg_Skoll_Citi_Catalyzing_Smallholder_Agricultural_Finance_Farm_Finance.pdf.

Carter, A., 2017: Placeholders and changemakers: Women farmland owners navigating gendered expectations. *Rural Sociol.*, **82**, 499–523, doi:10.1111/ruso.12131.

Carter, E. et al., 2018: Development of renewable, densified biomass for household energy in China. *Energy Sustain. Dev.*, **46**, 42–52, doi:10.1016/j.esd.2018.06.004.

Carter, S. et al., 2017: Large scale land acquisitions and REDD plus: A synthesis of conflicts and opportunities. *Environ. Res. Lett.*, **12** (3): 035010, doi:10.1088/1748-9326/aa6056.

Carvalho, J.L.N., T.W. Hudiburg, H.C.J. Franco, and E.H. Delucia, 2016: Contribution of above- and belowground bioenergy crop residues to soil carbon. *GCB Bioenergy*, **9**, 1333–1343, doi:10.1111/gcbb.12411.

Casellas Connors, J.P., and A. Janetos, 2016: *Assessing the Impacts of Multiple Breadbasket Failures*. AGU Fall Meeting Abstracts, American Geophysical Union, Washington, DC, USA.2016AGUFMNH21B..07C.

Cash, D.W. et al., 2006: Scale and cross-scale dynamics: Governance and information in a multilevel world. *Ecol. Soc.*, **11**, art8, doi:10.5751/ES-01759-110208.

Castán Broto, V., 2017: Urban governance and the politics of climate change. *World Dev.*, **93**, 1–15, doi:10.1016/j.worlddev.2016.12.031.

Castella, J.-C., J. Bourgoin, G. Lestrelin, and B. Bouahom, 2014: A model of the science-practice-policy interface in participatory land use planning: Lessons from Laos. *Landsc. Ecol.*, **29**, 1095–1107, doi:10.1007/s10980-014-0043-x.

Cavicchi, B. et al., 2017: The influence of local governance: Effects on the sustainability of bioenergy innovation. *Sustainability*, **9**, 406, doi:10.3390/su9030406.

CDM Executive Board, 2012: *Default Values of Fraction of Non-Renewable Biomass for Least Developed Countries and Small Island Developing States*. UNFCCC/CCNUCC, 13 pp. https://cdm.unfccc.int/Reference/Notes/meth/meth_note12.pdf.

Ceddia, M., U. Gunter, and A. Corriveau-Bourque, 2015: Land tenure and agricultural expansion in Latin America: The role of indigenous peoples' and local communities' forest rights. *Glob. Environ. Chang.*, **35**, 316–322, doi:10.1016/j.gloenvcha.2015.09.010.

Celentano, D. et al., 2017: Degradation of Riparian Forest affects soil properties and ecosystem services provision in eastern Amazon of Brazil. *L. Degrad. Dev.*, **28**, 482–493, doi:10.1002/ldr.2547.

Cerutti, P.O. et al., 2015: The socio-economic and environmental impacts of wood energy value chains in Sub-Saharan Africa: A systematic map protocol. *Environ. Evid.*, **4**, 12, doi:10.1186/s13750-015-0038-3.

Chadburn, S.E., (2017). An observation-based constraint on permafrost loss as a function of global warming. *Nature Climate Change*, **7**, 340–344, doi:10.1038/nclimate3262.

Chaffin, B.C., H. Gosnell, and B.A. Cosens, 2014: A decade of adaptive governance scholarship: Synthesis and future directions. *Ecol. Soc.*, **19**, Art. 56, doi:10.5751/ES-06824-190356.

Challinor, A.J., W.N. Adger, and T.G. Benton, 2017: Climate risks across borders and scales. *Nat. Clim. Chang.*, **7**, 621–623, doi:10.1038/nclimate3380.

Chambers, R. (ed.), 1988: *Managing Canal Irrigation: Practical Analysis from South Asia*. Cambridge University Press, Cambridge, UK, 279 pp.

Chambwera, M., and G. Heal, 2014: Economics of Adaptation. In: Climate Change 2014 Impacts, Adaptation and Vulnerability: Part A: Global and Sectoral Aspects [Field, C.B., V.R. Barros, D.J. Dokken, K.J. Mach, M.D. Mastrandrea, T.E. Bilir, M. Chatterjee, K.L. Ebi, Y.O. Estrada, R.C. Genova, B. Girma, E.S. Kissel, A.N. Levy, S. MacCracken, P.R. Mastrandrea, and L.L. White (eds.)], Cambridge University Press, Cambridge, United Kingdom and New York, NY, USA, pp. 945–977.

Chandra, A., P.P. Saradhi, R.K. Maikhuri, K.G. Saxena, and K.S. Rao, 2011: Traditional agrodiversity management: A case study of central Himalayan village ecosystem. *J. Mt. Sci.*, **8**, 62–74, doi:10.1007/s11629-011-1081-3.

Chaney, P., and R. Fevre, 2001: Inclusive governance and 'minority' groups: The role of the third sector in Wales. *Voluntas*, **12**, 131–156, doi:10.1023/A:1011286602556.

Chang, S.E., T.L. McDaniels, J. Mikawoz, and K. Peterson, 2007: Infrastructure failure interdependencies in extreme events: Power outage consequences in the 1998 ice storm. *Nat. Hazards*, **41**, 337–358, doi:https://doi.org/10.1007/s11069-006-9039-4.

Chang, Y., G. Li, Y. Yao, L. Zhang, and C. Yu, 2016: Quantifying the water-energy-food nexus: Current status and trends. *Energies*, **9**, 65, doi:10.3390/en9020065.

Chanza, N., and A. de Wit, 2016: Enhancing climate governance through indigenous knowledge: Case in sustainability science. *S. Afr. J. Sci.*, **112**, 1–7, doi:10.17159/sajs.2016/20140286.

Chapin, F.S. et al., 2010: Resilience of Alaska's boreal forest to climatic change. *Can. J. For. Res.*, **40**, 1360–1370, doi:10.1139/X10-074.

Chasek, P., U. Safriel, S. Shikongo, and V.F. Fuhrman, 2015: Operationalizing zero net land degradation: The next stage in international efforts to combat desertification? *J. Arid Environ.*, **112**, 5–13, doi:10.1016/j.jaridenv.2014.05.020.

Chatzopoulos, T., I. Pérez Domínguez, M. Zampieri, and A. Toreti, 2019: Climate extremes and agricultural commodity markets: A global economic analysis of regionally simulated events. *Weather Clim. Extrem.*, doi:10.1016/j.wace.2019.100193. In press.

Chávez Michaelsen, A. et al., 2017: Effects of drought on deforestation estimates from different classification methodologies: Implications for REDD+ and other payments for environmental services programs. *Remote Sens. Appl. Soc. Environ.*, **5**, 36–44, doi:10.1016/J.RSASE.2017.01.003.

Chen, D., W. Wei, and L. Chen, 2017a: Effects of terracing practices on water erosion control in China: A meta-analysis. *Earth-Science Rev.*, **173**, 109–121, doi:10.1016/J.EARSCIREV.2017.08.007.

Chen, J., 2011: Modern disaster theory: Evaluating disaster law as a portfolio of legal rules. *Emory Int. Law Rev.*, **25**. University of Louisville School of Law Legal Studies Research Paper No. 2011–05. Available at SSRN: https://ssrn.com/abstract=1910669.

Chen, Y., Y. Tan, and Y. Luo, 2017b: Post-disaster resettlement and livelihood vulnerability in rural China. *Disaster Prev. Manag.*, **26**, 65–78, doi:10.1108/DPM-07-2016-0130.

Cheng, C.-Y., and J. Urpelainen, 2014: Fuel stacking in India: Changes in the cooking and lighting mix, 1987-2010. *Energy*, **76**, 306–317, doi:10.1016/j.energy.2014.08.023.

Chhatre, A., and A. Agrawal, 2009: Trade-offs and synergies between carbon storage and livelihood benefits from forest commons. *Proc. Natl. Acad. Sci.*, **106**, 17667–17670, doi:10.1073/pnas.0905308106.

Chhatre, A., and S. Lakhanpal, 2018: For the environment, against conservation: Conflict between renewable energy and biodiversity protection in India. In: *Conservation and Development in India* [Bhagwat, S. (ed.)]. Routledge, London, UK, pp. 52–72.

Chomba, S., J. Kariuki, J.F. Lund, and F. Sinclair, 2016: Roots of inequity: How the implementation of REDD+ reinforces past injustices. *Land Use Policy*, **50**, 202–213, doi:10.1016/j.landusepol.2015.09.021.

Chopra, K., 2017: Land and forest policy: Resources for development or our natural resources? In: *Development and Environmental Policy in India* [Chopra, K. (ed.)]. Springer, Singapore, pp. 13–25.

Christenson, E., M. Elliott, O. Banerjee, L. Hamrick, and J. Bartram, 2014: Climate-related hazards: A method for global assessment of urban and rural population exposure to cyclones, droughts, and floods. *Int. J. Environ. Res. Public Health*, **11**, 2169–2192, doi:10.3390/ijerph110202169.

Christian-Smith, J., M.C. Levy, and P.H. Gleick, 2015: Maladaptation to drought: A case report from California, USA. *Sustain. Sci.*, **10**, 491–501, doi:10.1007/s11625-014-0269-1.

Chung Tiam Fook, T., 2017: Transformational processes for community-focused adaptation and social change: A synthesis. *Clim. Dev.*, **9**, 5–21, doi:10.1080/17565529.2015.1086294.

Church, S.P. et al., 2017: Agricultural trade publications and the 2012 Midwestern US drought: A missed opportunity for climate risk communication. *Clim. Risk Manag.*, **15**, 45–60, doi:10.1016/j.crm.2016.10.006.

Clarke, D., and S. Dercon, 2016a: *Dull Disasters? How Planning Ahead Will Make a Difference.* pp 154, Oxford University Press, Oxford. http://documents.worldbank.org/curated/en/962821468836117709/Dull-disasters-How-planning-ahead-will-make-a-difference.

Clarke, D.J., 2016: A theory of rational demand for index insurance. *Am. Econ. J. Microeconomics*, **8**, 283–306, doi:10.1257/mic.20140103.

Clarke, J., and M. Alston, 2017: Understanding the 'local' and 'global': Intersections engendering change for women in family farming in Australia. In: *Women in Agriculture Worldwide: Key Issues and Practical Approaches* [Fletcher, A. and W. Kubik (eds.)]. Routledge, Abingdon, UK, and New York, USA, pp. 13–22.

Clarvis, M.H., E. Bohensky, and M. Yarime, 2015: Can resilience thinking inform resilience investments? Learning from resilience principles for disaster risk reduction. *Sustain.*, **7**, 9048–9066, doi:10.3390/su7079048.

Clemens, H., R. Bailis, A. Nyambane, and V. Ndung'u, 2018: Africa Biogas Partnership Program: A review of clean cooking implementation through market development in East Africa. *Energy Sustain. Dev.*, **46**, 23–31, doi:10.1016/j.esd.2018.05.012.

Clemens, M., J. Rijke, A. Pathirana, J. Evers, and N. Hong Quan, 2015: Social learning for adaptation to climate change in developing countries: Insights from Vietnam. *J. Water Clim. Chang.*, **7**, 365–378, doi:10.2166/wcc.2015.004.

Clover, J., and S. Eriksen, 2009: The effects of land tenure change on sustainability: Human security and environmental change in southern African savannas. *Environ. Sci. Policy*, **12**, 53–70, doi:10.1016/j.envsci.2008.10.012.

Coady, D., I. Parry, L. Sears, and B. Shang, 2017: How large are global fossil fuel subsidies? *World Dev.*, 91, 11–27, doi:10.1016/j.worlddev.2016.10.004.

Coates, J., 2013: Build it back better: Deconstructing food security for improved measurement and action. *Glob. Food Sec.*, 2, 188–194, doi:10.1016/j.gfs.2013.05.002.

Cochran, F.V. et al., 2016: Indigenous ecological calendars define scales for climate change and sustainability assessments. *Sustain. Sci.*, **11**, 69–89, doi:10.1007/s11625-015-0303-y.

Coenen, F., and F.H.J.M. Coenen (eds.), 2009: *Public Participation and Better Environmental Decisions.* Springer Netherlands, Dordrecht, Netherlands, 183–209 pp.

Cohen, R., and M. Bradley, 2010: Disasters and displacement: Gaps in protection. *J. Int. Humanit. Leg. Stud.*, **1**, 1–35, doi:10.1163/18781521 0X12766020139884.

Cole, S., 2015: Overcoming barriers to microinsurance adoption: Evidence from the field. *Geneva Pap. Risk Insur. – Issues Pract.*, **40**, 720–740.

Cole, S. et al., 2013: Barriers to household risk management: Evidence from India. *Am. Econ. J. Appl. Econ.*, **5**, 104–135, doi:10.1257/app.5.1.104.

Collar, N.J., P. Patil, and G.S. Bhardwaj, 2015: What can save the Great Indian Bustard *Ardeotis nigriceps*? *Bird. ASIA*, **23**, 15–24.

Collins, A.M., 2014: Governing the global land grab: What role for gender in the voluntary guidelines and the principles for responsible investment? *Globalizations*, **11**, 189–203, doi:10.1080/14747731.2014.887388.

Collins, R.D., R. de Neufville, J. Claro, T. Oliveira, and A.P. Pacheco, 2013: Forest fire management to avoid unintended consequences: A case study of Portugal using system dynamics. *J. Environ. Manage.*, **130**, 1–9, doi:10.1016/j.jenvman.2013.08.033.

de Coninck, H. et al., 2018: Strengthening and Implementing the Global Response. In: Global Warming of 1.5°C: An IPCC Special Report on the Impacts of Global Warming of 1.5C Above Pre-Industrial Levels and Related Global Greenhouse Gas Emission Pathways, in the Context of Strengthening the Global Response to the Threat of Climate Change [Masson-Delmotte, V., P. Zhai, H.-O. Pörtner, D. Roberts, J. Skea, P.R. Shukla, A. Pirani, W. Moufouma-Okia, C. Péan, R. Pidcock, S. Connors, J.B.R. Matthews, Y. Chen, X. Zhou, M.I. Gomis, E. Lonnoy, T. Maycock, M. Tignor, and T. Waterfield (eds.)]. Cambridge University Press, Cambridge, United Kingdom and New York, NY, USA.

Conrad, C.C., and K.G. Hilchey, 2011: A review of citizen science and community-based environmental monitoring: Issues and opportunities. *Environ. Monit. Assess.*, **176**, 273–291, doi:10.1007/s10661-010-1582-5.

Conradt, S., R. Finger, and M. Spörri, 2015: Flexible weather index-based insurance design. *Clim. Risk Manag.*, **10**, 106–117, doi:10.1016/j.crm.2015.06.003.

Conway, D., and E.L. F. Schipper, 2011: Adaptation to climate change in Africa: Challenges and opportunities identified from Ethiopia. *Glob. Environ. Chang.*, **21**, 227–237, doi:10.1016/j.gloenvcha.2010.07.013.

Conway, D., et al., 2015: Climate and southern Africa's water-energy-food nexus. *Nat. Clim. Chang.*, **5**, 837, doi:10.1038/nclimate2735.

Cook, S., and J. Pincus, 2015: Poverty, inequality and social protection in Southeast Asia: An Introduction. *Southeast Asian Econ.*, **31**, 1–17, doi:10.1355/ae31-1a.

Cooke, S.J., et al., 2016: On the sustainability of inland fisheries: Finding a future for the forgotten. *Ambio*, **45**, 753–764, doi:10.1007/s13280-016-0787-4.

Cools, J., D. Innocenti, and S. O'Brien, 2016: Lessons from flood early warning systems. *Environ. Sci. Policy*, **58**, 117–122, doi:10.1016/J.ENVSCI.2016.01.006.

Cooper, M.H., and C. Rosin, 2014: Absolving the sins of emission: The politics of regulating agricultural greenhouse gas emissions in New Zealand. *J. Rural Stud.*, **36**, 391–400, doi:10.1016/j.jrurstud.2014.06.008.

Cooper, M.H., J. Boston, and J. Bright, 2013: Policy challenges for livestock emissions abatement: Lessons from New Zealand. *Clim. Policy*, **13**, 110–133, doi:10.1080/14693062.2012.699786.

Corfee-Morlot, J. et al., 2009: *Cities, Climate Change and Multilevel Governance*. OECD Environmental Working Papers N° 14, 2009, OECD publishing, Paris, France, pp. 1–125.

Corradini, M., V. Costantini, A. Markandya, E. Paglialunga, and G. Sforna, 2018: A dynamic assessment of instrument interaction and timing alternatives in the EU low-carbon policy mix design. *Energy Policy*, 120, 73–84, doi:10.1016/j.enpol.2018.04.068.

Corsi, S., L.V. Marchisio, and L. Orsi, 2017: Connecting smallholder farmers to local markets: Drivers of collective action, land tenure and food security in East Chad. *Land Use Policy*, **68**, 39–47, doi:10.1016/J.LANDUSEPOL.2017.07.025.

Cosens, B., et al., 2017: The role of law in adaptive governance. *Ecol. Soc.*, **22**, Art. 30, doi:10.5751/ES-08731-220130.

Costanza, R. et al., 2014: Changes in the global value of ecosystem services. *Glob. Environ. Chang.*, **26**, 152–158, doi:10.1016/j.gloenvcha.2014.04.002.

Costella, C. et al., 2017a: Scalable and sustainable: How to build anticipatory capacity into social protection systems. *IDS Bull.*, **48**, 31–46, doi:10.19088/1968-2017.151.

Cote, M., and A.J. Nightingale, 2012: Resilience thinking meets social theory: Situating social change in socio-ecological systems (SES) research. *Prog. Hum. Geogr.*, 36, 475–489, doi:10.1177/0309132511425708.

Cotula, L. (ed.), 2006a: *Land and Water Rights in the Sahel: Tenure Challenges of Improving Access to Water for Agriculture*. International Institute for Environment and Development, Drylands Programme, London, UK, 92 pp.

Cotula, L. (ed.), 2006b: *Land and Water Rights in the Sahel: Tenure Challenges of Improving Access to Water for Agriculture*. International Institute for Environment and Development, Drylands Programme, London, UK, 92 pp.

Cotula, L. et al., 2014: Testing claims about large land deals in Africa: Findings from a multi-country study. *J. Dev. Stud.*, **50**, 903–925, doi:10.1080/00220388.2014.901501.

Couvet, D., and A.C. Prevot, 2015: Citizen-science programs: Towards transformative biodiversity governance. *Environ. Dev.*, **13**, 39–45, doi:10.1016/j.envdev.2014.11.003.

Cowie, A.L. et al., 2011: Towards sustainable land management in the drylands: Scientific connections in monitoring and assessing dryland degradation, climate change and biodiversity. *L. Degrad. Dev.*, **22**, 248–260, doi:10.1002/ldr.1086.

Cowie, A.L. et al., 2018a: Land in balance: The scientific conceptual framework for land degradation neutrality. *Environ. Sci. Policy*, **79**, 25–35, doi:10.1016/j.envsci.2017.10.011.

Craig, R.K., 2010: 'Stationary is dead' – Long live transformation: Five principles for climate change adaptation law. *Harvard Environ. Law Rev.*, **34**, 9–73, doi:10.2139/ssrn.1357766.

Creutzig, F. et al., 2017: The underestimated potential of solar energy to mitigate climate change. *Nat. Energy*, **2**, 17140, doi:https://doi.org/10.1038/nenergy.2017.140.

Cummins, J.D., and M.A. Weiss, 2016: Equity capital, internal capital markets, and optimal capital structure in the US property-casualty insurance industry. *Annu. Rev. Financ. Econ.*, **8**, 121–153, doi:10.1146/annurev-financial-121415-032815.

Cundale, K., R. Thomas, J.K. Malava, D. Havens, K. Mortimer, and L. Conteh, 2017: A health intervention or a kitchen appliance? Household costs and benefits of a cleaner burning biomass-fuelled cookstove in Malawi. *Soc. Sci. Med.*, **183**, 1–10, doi:10.1016/j.socscimed.2017.04.017.

Cutter, S., Osman-Elasha, B., Campbell, J., Cheong, S.M., McCormick, S., Pulwarty, R., Supratid, S., Ziervogel, G., Calvo, E., Mutabazi, K., Arnall, A., Arnold, M., Bayer, J.L., Bohle, H.G., Emrich, C., Hallegatte, S., Koelle, B., Oettle, N., Polack, E., Ranger, N., 2012a: Managing the Risks from Climate Extremes at the Local Level. In: Managing the Risks of Extreme Events and Disasters to Advance Climate Change Adaptation [Field, C.B., V. Barros, T.F. Stocker, D. Qin, D.J. Dokken, K.L. Ebi, M.D. Mastrandrea, K.J. Mach, G.-K. Plattner, S.K. Allen, M. Tignor, and P.M. Midgley (eds.)]. Cambridge University Press, Cambridge, UK, 582 pp.

Cutter, S., B. Osman-Elasha, J. Campbell, S.-M. Cheong, S. McCormick, R. Pulwarty, S. Supratid, and G. Ziervogel, 2012b: Managing the Risks from Climate Extremes at the Local Level. In: Managing the Risks of Extreme Events and Disasters to Advance Climate Change Adaptation: Special Report of the Intergovernmental Panel on Climate Change [Field, C.B., V. Barros, T.F. Stocker, D. Qin, D.J. Dokken, K.L. Ebi, M.D. Mastrandrea, K.J. Mach, G.-K. Plattner, S.K. Allen, M. Tignor, and P.M. Midgley (eds.)]. Cambridge University Press, Cambridge, UK, and New York, NY, USA, 291–338 pp.

Cutz, L., O. Masera, D. Santana, and A.P. C. Faaij, 2017a: Switching to efficient technologies in traditional biomass intensive countries: The resultant change in emissions. *Energy*, **126**, 513–526, doi:10.1016/J.ENERGY.2017.03.025.

Czembrowski, P., and J. Kronenberg, 2016: Hedonic pricing and different urban green space types and sizes: Insights into the discussion on valuing ecosystem services. *Landsc. Urban Plan.*, **146**, 11–19, doi:10.1016/j.landurbplan.2015.10.005.

Dale, V.H., R.A. Efroymson, K.L. Kline, and M.S. Davitt, 2015: A framework for selecting indicators of bioenergy sustainability. *Biofuels, Bioprod. Biorefining*, **9**, 435–446, doi:10.1002/bbb.1562.

Dalin, C., Y. Wada, T. Kastner, and M.J. Puma, 2017: Groundwater depletion embedded in international food trade. *Nature*, **543**, 700.

Dallimer, M. et al., 2018: Who uses sustainable land management practices and what are the costs and benefits? Insights from Kenya. *L. Degrad. Dev.*, **29**, 2822–2835, doi:10.1002/ldr.3001.

van Dam, J., M. Junginger, and A.P. C. Faaij, 2010: From the global efforts on certification of bioenergy towards an integrated approach based on sustainable land use planning. *Renew. Sustain. Energy Rev.*, **14**, 2445–2472, doi:10.1016/J.RSER.2010.07.010.

Damnyag, L., O. Saastamoinen, M. Appiah, and A. Pappinen, 2012: Role of tenure insecurity in deforestation in Ghana's high forest zone. *For. Policy Econ.*, **14**, 90–98, doi:10.1016/j.forpol.2011.08.006.

Dan, R., 2016: Optimal adaptation to extreme rainfalls in current and future climate. *Water Resour. Res.*, **53**, 535–543, doi:10.1002/2016WR019718.

Daniel, S., 2011: Land grabbing and potential implications for world food security. In: *Sustainable Agricultural Development* [Behnassi, M., S.A. Shahid, J. D'Silva (eds.)]. Springer Netherlands, Dordrecht, Netherlands, pp. 25–42.

Darnhofer, I., 2014: Socio-technical transitions in farming: Key concepts. In: *Transition Pathways Towards Sustainability in Agriculture: Case Studies from Europe* [Sutherland, L.-A., L. Zagata (eds.)]. CABI, Oxfordshire, UK, pp. 246.

Daron, J.D., and D.A. Stainforth, 2014: Assessing pricing assumptions for weather index insurance in a changing climate. *Clim. Risk Manag.*, **1**, 76–91, doi:10.1016/j.crm.2014.01.001.

Das, S., and J.R. Vincent, 2009: Mangroves protected villages and reduced death toll during Indian super cyclone. *Proc. Natl. Acad. Sci.*, **106**, 7357–7360, doi:10.1073/pnas.0810440106.

Dasgupta, P., A.P. Kinzig, and C. Perrings, 2013: The value of biodiversity. In: *Encyclopedia of Biodiversity: Second Edition* [Levin, S. (ed.)]. Academic Press, Elsevier, Massachusetts, USA, pp. 5504.

Dasgupta, P., J.F. Morton, D. Dodman, B. Karapinar, F. Meza, M.G. Rivera-Ferre, A. Toure Sarr, and K.E. Vincent, 2014: Rural Areas. In: Climate Change 2014: Impacts, Adaptation, and Vulnerability. Part A: Global and Sectoral Aspects. Contribution of Working Group II to the Fifth Assessment Report of the Intergovernmental Panel on Climate Change [Field, C.B., V.R. Barros, D.J. Dokken, K.J. Mach, M.D. Mastrandrea, T.E. Bilir, M. Chatterjee, K.L. Ebi, Y.O. Estrada, R.C. Genova, B. Girma, E.S. Kissel, A.N. Levy, S. MacCracken, P.R. Mastrandrea, and L.L. White (eds.)]. Cambridge University Press, Cambridge, United Kingdom and New York, NY, USA, pp. 613–657.

Davies, J. (ed.), 2017: *The Land in Drylands: Thriving in Uncertainty Through Diversity*. Global Land Outlook Working Paper, UNCCD, Bonn, Germany, 18 pp.

Davies, J., C. Ogali, P. Laban, and G. Metternicht, 2015: *Homing in on the Range: Enabling Investments for Sustainable Land Management*. IUCN, Global Drylands Initiative. **vii**, 23 pp. https://portals.iucn.org/library/node/47775.

Davies, M., B. Guenther, J. Leavy, T. Mitchell, and T. Tanner, 2009: *Climate Change Adaptation, Disaster Risk Reduction, and Social Protection: Complementary Roles in Agriculture and Rural Growth?*Institute of Development Studies Working Papers, University of Sussex, Brighton, United Kingdom. 1–37 pp, doi:10.1111/j.2040-0209.2009.00320_2.x.

Davies, M., C. Béné, A. Arnall, T. Tanner, A. Newsham, and C. Coirolo, 2013: Promoting resilient livelihoods through adaptive social protection: Lessons from 124 programmes in South Asia. *Dev. Policy Rev.*, 31, 27–58, doi:10.1111/j.1467-7679.2013.00600.x.

Davis, K.F., P. D'Odorico, and M.C. Rulli, 2014: Land grabbing: A preliminary quantification of economic impacts on rurallivelihoods. *Popul. Environ.*, **36**, 180–192, doi:10.1007/s11111-014-0215-2.

Davis, S.C. et al., 2013: Management swing potential for bioenergy crops. *GCB Bioenergy*, 5, 623–638, doi:10.1111/gcbb.12042.

Debortoli, N.S., J.S. Sayles, D.G. Clark, and J.D. Ford, 2018: A systems network approach for climate change vulnerability assessment. *Environ. Res. Lett.*, **13**, 104019, doi:10.1088/1748-9326/aae24a.

DeClerck, F., 2016: IPBES: Biodiversity central to food security. *Nature*, **531**, 305, doi:https://doi.org/10.1038/531305e.

Deininger, K., 2011: Challenges posed by the new wave of farmland investment. *J. Peasant Stud.*, **38**, 217–247, doi:10.1080/03066150.2011.559007.

Deininger, K., and O. Feder, 2009: Land registration, governance, and development: Evidence and implications for policy. *World Bank Res. Obs.*, **24**, 233–266.http://documents.worldbank.org/curated/en/869031468150595587/Land-registration-governance-and-development-evidence-and-implications-for-policy.

Deininger, K., and D. Byerlee, 2011: The rise of large farms in land abundant countries: Do they have a future?World Development, 40, 701–714, doi:10.1016/j.worlddev.2011.04.030.

Dellasala, D.A., J.E. Williams, C.D. Williams, and J.F. Franklin, 2004: Beyond smoke and mirrors: A synthesis of fire policy and science. *Conserv. Biol.*, **18**, 976–986, doi:10.1111/j.1523-1739.2004.00529.x.

Dennison, P.E., S.C. Brewer, J.D. Arnold, and M.A. Moritz, 2014: Large wildfire trends in the western United States, 1984–2011. *Geophys. Res. Lett.*, **41**, 2928–2933, doi:10.1002/2014GL059576.

Denton, F., T.J. Wilbanks, A.C. Abeysinghe, I. Burton, Q. Gao, M.C. Lemos, T. Masui, K.L. O'Brien, and K. Warner, 2014: Climate-Resilient Pathways: Adaptation, Mitigation, and Sustainable Development. In: Climate Change 2014 Impacts, Adaptation and Vulnerability: Part A: Global and Sectoral Aspects [Field, C.B., V.R. Barros, D.J. Dokken, K.J. Mach, M.D. Mastrandrea,

T.E. Bilir, M. Chatterjee, K.L. Ebi, Y.O. Estrada, R.C. Genova, B. Girma, E.S. Kissel, A.N. Levy, S. MacCracken, P.R. Mastrandrea, and L.L. White (eds.)]. Cambridge University Press, Cambridge, United Kingdom and New York, NY, USA, 1101–1131.

Dercon, S., R.V. Hill, D. Clarke, I. Outes-Leon, and A. Seyoum Taffesse, 2014: Offering rainfall insurance to informal insurance groups: Evidence from a field experiment in Ethiopia. *J. Dev. Econ.*, **106**, 132–143, doi:10.1016/j.jdeveco.2013.09.006.

Deryng, D., D. Conway, N. Ramankutty, J. Price, and R. Warren, 2014: Global crop yield response to extreme heat stress under multiple climate change futures. *Environ. Res. Lett.*, **9**, 041001, doi:10.1088/1748-9326/9/3/034011.

Deryugina, T., 2013: Reducing the cost of ex post bailouts with ex ante regulation: Evidence from building codes. *SSRN Electron. J.*, 2009, 1–37, doi:10.2139/ssrn.2314665.

Devereux, S., 2007: The impact of droughts and floods on food security and policy options to alleviate negative effects. *Agricultural Economics*, **37**, 47–58, doi:10.1111/j.1574-0862.2007.00234.x.

Dey, Mayukh 2018: Conserving river dolphins in a changing soundscape: acoustic and behavioural responses of Ganges river dolphins to anthropogenic noise in the Ganges River, India. MSc thesis, xii+121p, Tata Institute of Fundamental Research, India.

Dey, P., and A. Sarkar, 2011: Revisiting indigenous farming knowledge of Jharkhand (India) for conservation of natural resources and combating climate change. *Indian J. Tradit. Knowl.*, **10**, 71–79. www.fao.org/fsnforum/sites/default/files/discussions/contributions/Indian_J_Traditional_Knowledge.pdf.

Dharmadhikary, S., and Sandbhor, J., 2017: *National Inland Waterways in India: A Strategic Status Report*. Manthan Adhyayan Kendra and SRUTI, Manthan, India. 67 pp.

Diaz-Chavez, R., 2015: Assessing sustainability for biomass energy production and use., in Rosillo-Calle F, de Groot P, S Hemstock and Woods J., *The Biomass assessment Handbook. Energy for a sustainable environment*, Second Edition, Routledge, Editors: pp.181–209.

Diaz-Chavez, R.A., 2011: Assessing biofuels: Aiming for sustainable development or complying with the market? *Energy Policy*, **39**, 5763–5769, doi:10.1016/J.ENPOL.2011.03.054.

Díaz-Reviriego, I., Á. Fernández-Llamazares, P.L. Howard, J.L. Molina, and V. Reyes-García, 2017: Fishing in the Amazonian forest: A gendered social network puzzle. *Soc. Nat. Resour.*, **30**, 690–706, doi:10.1080/08941920.2016.1257079.

Diaz, R.J., and R. Rosenberg, 2008: Spreading dead zones and consequences for marine ecosystems. *Science*, **321**, 926–929, doi:10.1126/science.1156401.

Díaz, S. et al., 2015: The IPBES Conceptual Framework– Connecting nature and people. *Curr. Opin. Environ. Sustain.*, **14**, 1–16, doi:10.1016/j.cosust.2014.11.002.

Diffenbaugh, N.S., T.W. Hertel, M. Scherer, and M. Verma, 2012: Response of corn markets to climate volatility under alternative energy futures. *Nat. Clim. Chang.*, 2, 514–518, doi:10.1038/nclimate1491.

Dillon, R.L., C.H. Tinsley, and W.J. Burns, 2014: Near-misses and future disaster preparedness. *Risk Anal.*, **34**, 1907–1922, doi:10.1111/risa.12209.

Distefano, T., F. Laio, L. Ridolfi, and S. Schiavo, 2018: Shock transmission in the international food trade network. *PLoS One*, 13, e0200639, doi:10.1371/journal.pone.0200639.

Dittrich, R., A. Wreford, C.F. E. Topp, V. Eory, and D. Moran, 2017: A guide towards climate change adaptation in the livestock sector: Adaptation options and the role of robust decision-making tools for their economic appraisal. *Reg. Environ. Chang.*, **17**, doi:10.1007/s10113-017-1134-4.

Djalante, R., C. Holley, F. Thomalla, and M. Carnegie, 2013: Pathways for adaptive and integrated disaster resilience. *Nat. Hazards*, **69**, 2105–2135, doi:10.1007/s11069-013-0797-5.

Djoudi, H., and M. Brockhaus, 2011: Is adaptation to climate change gender neutral? Lessons from communities dependent on livestock and forests in northern Mali. *Int. For. Rev.*, **13**, 123–135, doi:10.1505/146554811797406606.

Djoudi, H., B. Locatelli, C. Vaast, K. Asher, M. Brockhaus, and B. Basnett Sijapati, 2016: Beyond dichotomies: Gender and intersecting inequalities in climate change studies. *Ambio*, **45**, 248–262, doi:10.1007/s13280-016-0825-2.

Domínguez, P., 2014: Current situation and future patrimonializing perspectives for the governance of agropastoral resources in the Ait Ikis transhumants of the High Atlas (Morocco), pp. 148–166. In Herrera, P M., Davies, J., and Baena, P M. First Edition. Routledge. London. United Kingdom. https://doi.org/10.4324/9781315768014pp 320.

Donato, D.C., J.B. Kauffman, D. Murdiyarso, S. Kurnianto, M. Stidham, and M. Kanninen, 2011: Mangroves among the most carbon-rich forests in the tropics. *Nat. Geosci.*, **4**, 293, doi:10.1038/ngeo1123.

Dooley, K., and S. Kartha, 2018: Land-based negative emissions: Risks for climate mitigation and impacts on sustainable development. *Int. Environ. Agreements Polit. Law Econ.*, **18**, 79–98, doi:10.1007/s10784-017-9382-9.

Doss, C., C. Kovarik, A. Peterman, A. Quisumbing, and M. van den Bold, 2015a: Gender inequalities in ownership and control of land in Africa: Myth and reality. *Agric. Econ.*, **46**, 403–434, doi:10.1111/agec.12171.

Doss, C., R. Meinzen-Dick, A. Quisumbing, and S. Theis, 2018a: Women in agriculture: Four myths. *Glob. Food Sec.*, **16**, 69–74, doi:10.1016/J. GFS.2017.10.001.

Dougill, A.J., E.D. G. Fraser, and M.S. Reed, 2011: Anticipating vulnerability to climate change in dryland pastoral systems: Using dynamic systems models for the Kalahari. *Ecol. Soc.*, **15**, ART. 17, www.ecologyandsociety. org/vol15/iss2/art17/.

Douglas, I., K. Alam, M. Maghenda, Y. Mcdonnell, L. Mclean, and J. Campbell, 2008: Unjust waters: Climate change, flooding and the urban poor in Africa. *Environ. Urban.*, **20**, 187–205, doi:10.1177/0956247808089156.

Dow, K., F. Berkhout, and B.L. Preston, 2013: Limits to adaptation to climate change: A risk approach. *Curr. Opin. Environ. Sustain.*, **5**, 384–391, doi:10.1016/j.cosust.2013.07.005.

Dowdy, A.J., and A. Pepler, 2018: Pyroconvection risk in Australia: Climatological changes in atmospheric stability and surface fire weather conditions. *Geophys. Res. Lett.*, **45**, 2005–2013, doi:10.1002/2017GL076654.

Downing, T., 2012: Views of the frontiers in climate change adaptation economics. *Wiley Interdiscip. Rev. Clim. Chang.*, **3**, 161–170, doi:10.1002/ wcc.157.

Driscoll, D.A., M. Bode, R.A. Bradstock, D.A. Keith, T.D. Penman, and O.F. Price, 2016: Resolving future fire management conflicts using multicriteria decision-making. *Conserv. Biol.*, **30**, 196–205, doi:10.1111/cobi.12580.

Dryzek, J.S., and J. Pickering, 2017: Deliberation as a catalyst for reflexive environmental governance. *Ecol. Econ.*, **131**, 353–360, doi:10.1016/j. ecolecon.2016.09.011.

Duchelle, A.E. et al., 2017: Balancing carrots and sticks in REDD+: Implications for social safeguards. *Ecol. Soc.*, **22**, art2, doi:10.5751/ES-09334-220302.

Duguma, L.A., P.A. Minang, O.E. Freeman, and H. Hager, 2014a: System-wide impacts of fuel usage patterns in the Ethiopian highlands: Potentials for breaking the negative reinforcing feedback cycles. *Energy for Sustainable Development*, 20, 77–85, doi:10.1016/j.esd.2014.03.004.

Duguma, L.A., P.A. Minang, D. Foundjem-Tita, P. Makui, and S.M. Piabuo, 2018: Prioritizing enablers for effective community forestry in Cameroon. *Ecol. Soc.*, **23**, Art. 1, doi:10.5751/ES-10242-230301.

Durigan, G., and J.A. Ratter, 2016: The need for a consistent fire policy for Cerrado conservation. *J. Appl. Ecol.*, **53**, 11–15, doi:10.1111/1365-2664.12559.

Duru, M., O. Therond, and M. Fares, 2015: Designing agroecological transitions; A review. *Agron. Sustain. Dev.*, **35**, 1237–1257, doi:10.1007/ s13593-015-0318-x.

Dwyer, J., and I. Hodge, 2016: Governance structures for social-ecological systems: Assessing institutional options against a social residual claimant. *Environ. Sci. Policy*, **66**, 1–10, doi:10.1016/j.envsci.2016.07.017.

Eakin, H.C., 2016: Cognitive and institutional influences on farmers' adaptive capacity: Insights into barriers and opportunities for transformative change in central Arizona. *Regional Environmental Change*, **16**, 801–814, doi:https://doi.org/10.1007/s10113-015-0789-y.

Easdale, M.H., 2016: Zero net livelihood degradation – The quest for a multidimensional protocol to combat desertification. *SOIL*, **2**, 129–134, doi:10.5194/soil-2-129-2016.

Easterly, W., 2008a: Institutions: Top down or bottom up? *Am. Econ. Rev.*, **98**, 95–99, doi:10.1257/aer.98.2.95.

Easterly, W., 2008b: Institutions: Top down or bottom up? *Am. Econ. Rev.*, **98**, 95–99, doi:10.1257/aer.98.2.95.

EEA, 2016: *Urban Adaptation to Climate Change in Europe: Transforming Cities in a Changing Climate*. EEA Report No 12/2016, Copenhagen, Denmark, 135 pp.

Ehara, M. et al., 2018: Addressing maladaptive coping strategies of local communities to changes in ecosystem service provisions using the DPSIR Framework. *Ecol. Econ.*, **149**, 226–238 doi:10.1016/j. ecolecon.2018.03.008.

El-Naggar, A. et al., 2018: Influence of soil properties and feedstocks on biochar potential for carbon mineralization and improvement of infertile soils. *Geoderma*, **332**, 100–108, doi:10.1016/j.geoderma.2018.06.017.

ELD Initiative, 2015: *The Value of Land: Prosperous Lands and Positive Rewards Through Sustainable Land Management*. ELD Secretariat, Bonn, Germany. ELD Initiative (2015). https://reliefweb.int/sites/reliefweb.int/ files/resources/ELD-main-report_05_web_72dpi.pdf.

Eldridge, D.J. et al., 2011: Impacts of shrub encroachment on ecosystem structure and functioning: Towards a global synthesis. *Ecol. Lett.*, **14**, 709–722, doi:10.1111/j.1461-0248.2011.01630.x.

Eling, M., S. Pradhan, and J.T. Schmit, 2014: The determinants of microinsurance demand. *Geneva Pap. Risk Insur. – Issues Pract.*, **39**, 224–263, doi:10.1057/ gpp.2014.5.

Ellis, F., S. Devereux, and P. White, 2009: Social Protection in Africa. *Enterp. Dev. Microfinance*, **20**, 158–160, doi:10.3362/1755-1986.2009.015.

Ellison, D. et al., 2017: Trees, forests and water: Cool insights for a hot world. *Glob. Environ. Chang.*, **43**, 51–61, doi:10.1016/j.gloenvcha.2017.01.002.

Elmqvist, T. et al., 2003: Response diversity, ecosystem change, and resilience. *Front. Ecol. Environ.*, **1**, 488–494, doi:10.1890/1540-9295 (2003)001[0488:rdecar]2.0.co; 2.

Elmqvist, T., M. Tuvendal, J. Krishnaswamy, and K. Hylander, 2013: Managing trade-offs in ecosystem services. In: *Values, Payments Institutions Ecosystem Management* [Kumar, P., and I. Thiaw (eds.)]. Edward Elgar Publishing Ltd, Cheltenham, UK, pp. 70–89.

Emerson, K., and A.K. Gerlak, 2014: Adaptation in collaborative governance regimes. *Environ. Manage.*, **54**, 768–781, doi:10.1007/s00267-014-0334-7.

Endres, J. et al., 2015: Sustainability certification. In: *Bioenergy & Sustainability: Bridging the Gaps* [Souza, G., R. Victoria, C.A. Joly, L.M. Verdade, (eds.)]. pp 660–680. SCOPE, Paris, France.

Enfors, E.I., and L.J. Gordon, 2008: Dealing with drought: The challenge of using water system technologies to break dryland poverty traps. *Glob. Environ. Chang.*, **18**, 607–616, doi:10.1016/J.GLOENVCHA.2008.07.006.

Engel, S., and A. Muller, 2016: Payments for environmental services to promote 'climate-smart agriculture'? Potential and challenges. *Agric. Econ.*, **47**, 173–184, doi:10.1111/agec.12307.

Englund, O., and G. Berndes, 2015: How do sustainability standards consider biodiversity? *Wiley Interdiscip. Rev. Energy Environ.*, **4**, 26–50, doi:10.1002/ wene.118.

Ens, E.J. et al., 2015: Indigenous biocultural knowledge in ecosystem science and management: Review and insight from Australia. *Biol. Conserv.*, **181**, 133–149, doi:10.1016/J.BIOCON.2014.11.008.

Ensor, J., 2011: *Uncertain Futures: Adapting Development to a Changing Climate*. Practical Action Publishing, London, UK, 108 pp.

Ensor, J., and B. Harvey, 2015: Social learning and climate change adaptation: Evidence for international development practice. *Wiley Interdiscip. Rev. Clim. Chang.*, **6**, 509–522, doi:10.1002/wcc.348.

Eory, V., C.F.E. Topp, A. Butler, and D. Moran, 2018: Addressing uncertainty in efficient mitigation of agricultural greenhouse gas emissions. *J. Agric. Econ.*, **69**, 627–645, doi:10.1111/1477-9552.12269.

EPA, 2018: *Ireland's Final Greenhouse Gas Emissions: 1990–2016.* Environmental Protection Agency, Dublin, Ireland, 12 pp.

Epanchin-Niell, R.S. et al., 2010: Controlling invasive species in complex social landscapes. *Front. Ecol. Environ.*, **8**, 210–216, doi:10.1890/090029.

Era Consultancy, 2006: *The Environment Impact Assessment (EIA) Notification, 2006.* The Ministry of Environment, Forest and Climate Change, New Delhi, India, 15 pp.

Eriksen, S. et al., 2011: When not every response to climate change is a good one: Identifying principles for sustainable adaptation. *Clim. Dev.*, **3**, 7–20, doi:10.3763/cdev.2010.0060.

Erkossa, T., A. Wudneh, B. Desalegn, and G. Taye, 2015: Linking soil erosion to on-site financial cost: Lessons from watersheds in the Blue Nile basin. *Solid Earth*, **6**, 765–774, doi:10.5194/se-6-765-2015.

Estrin, D., 2016: *Limiting Dangerous Climate Change the Critical Role of Citizen Suits and Domestic Courts – Despite the Paris Agreement.* CIGI Papers No. 101, Centre for International Governance Innovation, Ontario, Canada, 36 pp.

Estrin, D., and S.V. Tan, 2016: *Thinking Outside the Boat about Climate Change Loss and Damage: Innovative Insurance, Financial and Institutional Mechanisms to Address Climate Harm Beyond the Limits of Adaptation.* International Workshop Report, Washington, DC, USA, 24 pp.

Estrin, S., and M. Prevezer, 2011: The role of informal institutions in corporate governance: Brazil, Russia, India, and China compared. *Asia Pacific J. Manag.*, **28**, 41–67, doi:10.1007/s10490-010-9229-1.

Etkin, D., J. Medalye, and K. Higuchi, 2012: Climate warming and natural disaster management: An exploration of the issues. *Clim. Change*, **112**, 585–599, doi:10.1007/s10584-011-0259-6.

European Commission, 2012: *Renewable Energy Progress and Biofuels Sustainability.* ECOFYS BV, Utrecht. Netherlands, 410 pp.

European Union, 2018: *Directives Directive (EU) 2018/2001 of the European Parliament and of the Council of 11 December 2018 on the Promotion of the Use of Energy from Renewable Sources.* Official Journal of the European Union , Cardiff, UK, 128 pp. https://eur-lex.europa.eu/legal-content/EN/TXT/PDF/?uri=CELEX:32018L2001&from=EN.

Ewel, K., R. Twilley, and J.I.N. Ong, 1998: Different kinds of mangrove forests provide different goods and services. *Glob. Ecol. Biogeogr. Lett.*, **7**, 83–94, doi:10.2307/2997700.

Faaij, A.P., 2018: Securing Sustainable Resource Availability of Biomass for Energy Applications in Europe; Review of Recent Literature. The Role of Biomass for Energy and Materials for GHG Mitigation from a Global and European Perspective. University of Groningen. The Netherlands, 26 pp. https://pdfs.semanticscholar.org/48c6/62527d3a7a7ea491d531472dc63a1ae76efb.pdf.

Fadairo, O.S., R. Calland, Y. Mulugetta, and J. Olawoye, 2017: A corruption risk assessment for reducing emissions from deforestation and forest degradation in Nigeria. *Int. J. Clim. Chang. Impacts Responses*, **10**, 1–21, doi:10.18848/1835-7156/CGP/v10i01/1-21.

Fairbairn, M., 2015: Foreignization, financialization and land grab regulation. *J. Agrar. Chang.*, **15**, 581–591, doi:10.1111/joac.12112.

Falkenmark, M., 2001: The greatest water problem: The inability to link environmental security, water security and food security. *Int. J. Water Resour. Dev.*, **17**, 539–554, doi:10.1080/07900620120094073.

Falkowski, T.B., S.A. W. Diemont, A. Chankin, and D. Douterlungne, 2016: Lacandon Maya traditional ecological knowledge and rainforest restoration: Soil fertility beneath six agroforestry system trees. *Ecol. Eng.*, **92**, 210–217, doi:10.1016/J.ECOLENG.2016.03.002.

Fameree, C., 2016: Political contestations around land deals: Insights from Peru. *Can. J. Dev. Stud.. Revue canadienne d'études du développement*, **37**, 541–559, doi:10.1080/02255189.2016.1175340.

Fankhauser, S., 2017: Adaptation to Climate Change. *Annual Review of Resource Economics*, **9**, 209–230, doi:10.1146/annurev-resource-100516-033554.

FAO, 2010: *Climate-Smart Agriculture: Policies, Practices and Financing for Food Security, Adaptation and Mitigation.* Food and Agriculture Organization of the United Nations, Rome, Italy, 49 pp.

FAO, 2011a: *State of Food and Agriculture 2010–2011.* Food and Agriculture Organization of the United Nations, Rome, Italy, 147 pp.

FAO, 2011b: *The State of Food and Agriculture Women in Agriculture 2010–11. Closing the Gender Gap for Development.* Food and Agriculture Organization of the United Nations, Rome, Italy, 160 pp.

FAO, 2015a: Voluntary Guidelines for Securing Sustainable Small-Scale Fisheries in the Context of Food Security and Poverty Eradication. Food and Agriculture Organization of the United Nations, Rome, Italy, 34 pp.

FAO, 2015b: *The Impact of Disasters on Agriculture and Food Security.* Food and Agriculture Organization of the United Nations, Rome, Italy, 54 pp.

FAO, 2015c: *Gender and Land Statistics Recent developments in FAO's Gender and Land Rights Database.* Food and Agriculture Organization of the United Nations, Rome, Italy, 35 pp.

FAO, 2016: *The Agriculture Sectors in the Intended Nationally Determined Contributions: Analysis.* Food and Agriculture Organization of the United Nations, Rome, Italy, 92 pp.

FAO, 2017a: *The Future of Food and Agriculture: Trends and Challenges.* Food and Agriculture Organization of the United Nations, Rome, Italy, 180 pp.

FAO, 2017b: *FAO Cereal Supply and Demand Brief.* Food and Agriculture Organization of the United Nations, Rome, Italy.

Farber, D.A., 2015: Coping with uncertainty: Cost-benefit analysis, the precautionary principle, and climate change. *Washingt. Law Rev.*, **54**, 23–46, doi:10.1525/sp.2007.54.1.23.

Farfan, J., and C. Breyer, 2017: Structural changes of global power generation capacity towards sustainability and the risk of stranded investments supported by a sustainability indicator. *J. Clean. Prod.*, **141**, 370–384, doi:10.1016/j.jclepro.2016.09.068.

Fawcett, A., L. Clarke, S. Rausch, and J.P. Weyant, 2014: Overview of EMF 24 policy scenarios. *Energy J.*, **35**, 33–60, doi:10.5547/01956574.35.SI1.3.

Faye, B. et al., 2018: Impacts of 1.5 versus 2.0°c on cereal yields in the West African Sudan Savanna. *Environ. Res. Lett.*, **13**034014, doi:10.1088/1748-9326/aaab40.

Fearnside, P.M., 2015: Deforestation soars in the Amazon. *Nature*, **521**, 423–423, doi:10.1038/521423b.

Feliciano, D., C. Hunter, B. Slee, and P. Smith, 2014: Climate change mitigation options in the rural land use sector: Stakeholders' perspectives on barriers, enablers and the role of policy in North East Scotland. *Environ. Sci. Policy*, **44**, 26–38, doi:10.1016/j.envsci.2014.07.010.

Fellmann, T. et al., 2018: Major challenges of integrating agriculture into climate change mitigation policy frameworks. *Mitigation and Adaptation Strategies for Global Change*, **23**, 451–468, doi:10.1007/s11027-017-9743-2.

Fencl, J.S., M.E. Mather, K.H. Costigan, and M.D. Daniels, 2015: How big of an effect do small dams have? Using geomorphological footprints to quantify spatial impact of low-head dams and identify patterns of across-dam variation. *PLoS One*, **10**, e0141210, doi:10.1371/journal.pone.0141210.

Ferguson, G., and T. Gleeson, 2012: Vulnerability of coastal aquifers to groundwater use and climate change. *Nat. Clim. Chang.*, **2**, 342–345, doi:10.1038/nclimate1413.

Fernandes, K. et al., 2017: Heightened fire probability in Indonesia in non-drought conditions: The effect of increasing temperatures. *Environ. Res. Lett.*, **12**, 054002, doi:10.1088/1748-9326/aa6884.

Fernandez-Gimenez, M.E., 2000: The role of Mongolian nomadic pastoralists' ecological knowledge in rangeland management. *Ecol. Appl.*, **10**, 1318–1326, doi:10.1890/1051-0761 (2000)010[1318:TROMNP]2.0.CO; 2.

Few, R., and M.G.L. Tebboth, 2018: Recognising the dynamics that surround drought impacts. *J. Arid Environ.*, **157**, 113–115, doi:10.1016/j.jaridenv.2018.06.001.

Feyen, E., R. Lester, and R. Rocha, 2011: What Drives the Development of the Insurance Sector? An Empirical Analysis based on a Panel of Developed and Developing Countries. Policy Research Working Paper Series 5572, The World Bank, Washington, DC, USA, 46 pp.

Filatova, T., 2014: Market-based instruments for flood risk management: A review of theory, practice and perspectives for climate adaptation policy. *Environ. Sci. Policy*, **37**, 227–242, doi:10.1016/j.envsci.2013.09.005.

Filiberto, B.D., E. Wethington, and K. Pillemer, 2010: Older people and climate change: Vulnerability and health effects. *Generations*, **33**, 19–25, www.ingentaconnect.com/content/asag/gen/2009/00000033/00000004/art00004#expand/collapse.

Findell, K.L. et al., 2017: The impact of anthropogenic land use and land cover change on regional climate extremes. *Nat. Commun.*, **8**, 989, doi:10.1038/s41467-017-01038-w.

Finley-Brook, M., 2007: Indigenous land tenure insecurity fosters illegal logging in Nicaragua. *Int. For. Rev.*, **9**, 850–864, doi:10.1505/ifor.9.4.850.

Fischer, C., and R.G. Newell, 2008: Environmental and technology policies for climate mitigation. *J. Environ. Econ. Manage.*, **55**, 142–162, doi:10.1016/j.jeem.2007.11.001.

Fischer, E.M., and R. Knutti, 2015: Anthropogenic contribution to global occurrenceof heavy-precipitation andhigh-temperature extremes. *Nat. Clim. Chang.*, **5**, 560–564, doi:10.1038/nclimate2617.

Fischer, J. et al., 2017: Reframing the food-biodiversity challenge. *Trends Ecol. Evol.*, **32**, 335–345, doi:10.1016/j.tree.2017.02.009.

Fishman, R., N. Devineni, and S. Raman, 2015: Can improved agricultural water use efficiency save India's groundwater? *Environ. Res. Lett.*, **10**, 084022, doi:10.1088/1748-9326/10/8/084022.

Fiszbein, A., R. Kanbur, and R. Yemtsov, 2014: Social protection and poverty reduction: Global patterns and some targets. *World Dev.*, **61**, 167–177, doi:10.1016/j.worlddev.2014.04.010.

Flahaux, M.-L., and H. De Haas, 2016: African migration: Trends, patterns, drivers. *Comp. Migr. Stud.*, **4**, 1–25, doi:10.1186/s40878-015-0015-6.

Fleskens, L., L.C. Stringer, 2014: Land management and policy responses to mitigate desertification and land degradation. *L. Degrad. Dev.*, **25**, 1–4, doi:10.1002/ldr.2272.

Fleskens, L., D. Nainggolan, and L.C. Stringer, 2014: An exploration of scenarios to support sustainable land management using integrated environmental socio-economic models. *Environ. Manage.*, **54**, 1005–1021, doi:10.1007/s00267-013-0202-x.

Fletcher, A.J., 2017: 'Maybe tomorrow will be better': Gender and farm work in a changing climate. In: *Climate Change and Gender in Rich Countries: Work, Public Policy and Action* [Cohen, M. (ed.)]. Routledge, Abingdon, UK, and New York, USA, pp. 185–198.

Fletcher, A.J., 2018: What works for women in agriculture? In: *Women in Agriculture Worldwide: Key Issues and Practical Approaches* [Fletcher, A.J. and W. Kubik, (eds.)]. Routledge, Abingdon, UK, and New York, USA, pp. 257–268.

Fletcher, A.J., and E. Knuttila, 2016: Gendering change: Canadian farm women respond to drought. In: *Vulnerability and Adaptation to Drought: The Canadian Prairies and South America* [Diaz, H., M. Hurlbert, and J. Warren (eds.)]. University of Calgary Press, Calgary, Canada, pp. 159–177.

Fleurbaey M., S. Kartha, S. Bolwig, Y.L. Chee, Y. Chen, E. Corbera, F. Lecocq, W. Lutz, M.S. Muylaert, R.B. Norgaard, C. Oker-eke, and A.D. Sagar, 2014: Sustainable Development and Equity. In: Climate Change 2014: Mitigation of Climate Change. Contribution of Working Group III to the Fifth Assessment Report of the Intergovernmental Panelon Climate Change [Edenhofer, O., R. Pichs-Madruga, Y. Sokona, E. Farahani, S. Kadner, K. Seyboth, A. Adler, I. Baum, S. Brunner, P. Eickemeier, B. Kriemann, J. Savolainen, S. Schlömer, C. von Stechow, T. Zwickel and J.C. Minx (eds.)]. Cambridge University Press, Cambridge, United Kingdom and New York, NY, USA, pp. 283–350.

Floater, G., P. Rode, B. Friedel, and A. Robert, 2014: *Steering Urban Growth: Governance, Policy and Finance*. New Climate Economy Cities Paper 02, LSE Cities, London School of Economics and Political Science, London, UK, 49 pp.

Folke, C. et al., 2010: Resilience thinking: Integrating resilience, adaptability and transformability. *Ecol. Soc.*, **15**, ART. 20, doi:10.5751/ES-03610-150420.

Fontaine, J.J., 2011: Improving our legacy: Incorporation of adaptive management into state wildlife action plans. *J. Environ. Manage.*, **92**, 1403–1408, doi:10.1016/j.jenvman.2010.10.015.

Ford, A., R. Dawson, P. Blythe, and S. Barr, 2018: Land use transport models for climate change mitigation and adaptation planning. *J. Transp. Land Use*, **11**, 83–101, doi:10.5198/jtlu.2018.1209.

Ford, J.D., and T. Pearce, 2010: What we know, do not know, and need to know about climate change vulnerability in the western Canadian Arctic: A systematic literature review. *Environ. Res. Lett.*, **5**, 014008, doi:10.1088/1748-9326/5/1/014008.

Ford, J.D., and C. Goldhar, 2012: Climate change vulnerability and adaptation in resource dependent communities: A case study from West Greenland. *Clim. Res.*, **54**, 181–196, doi:10.3354/cr01118.

Ford, J.D., and L. Berrang-Ford, 2016: The 4Cs of adaptation tracking: Consistency, comparability, comprehensiveness, coherency. *Mitig. Adapt. Strateg. Glob. Chang.*, **21**, 839–859, doi:10.1007/s11027-014-9627-7.

Ford, J.D., L. Cameron, J. Rubis, M. Maillet, D. Nakashima, A.C. Willox, and T. Pearce, 2016: Including indigenous knowledge and experience in IPCC assessment reports. *Nat. Clim. Chang.*, **6**, 349–353, doi:10.1038/nclimate2954.

Forsyth, T., 2018: Is resilience to climate change socially inclusive? Investigating theories of change processes in Myanmar. *World Dev.*, **111**, 13–26, doi:10.1016/j.worlddev.2018.06.023.

Foudi, S., and K. Erdlenbruch, 2012: The role of irrigation in farmers' risk management strategies in France. *Eur. Rev. Agric. Econ.*, **39**, 439–457, doi:10.1093/erae/jbr024.

Foxon, T.J., M.S. Reed, and L.C. Stringer, 2009: Governing long-term social–Ecological change: What can the adaptive management and transition management approaches learn from each other? *Change*, **20**, 3–20, doi:10.1002/eet.

FraPaleo, U. (ed.), 2015: Risk Governance: The Articulation of Hazard, Politics and Ecology. Springer Netherlands, Dordrecht, Netherlands, 515 pp.

Frank, S. et al., 2017: Reducing greenhouse gas emissions in agriculture without compromising food security? *Environ. Res. Lett.*, **12**, 105004, doi:10.1088/1748-9326/aa8c83.

Franz, M., N. Schlitz, and K.P. Schumacher, 2017: Globalization and the water-energy-food nexus – Using the global production networks approach to analyze society-environment relations. *Environmental Science and Policy*, **90**, 201–212, doi:10.1016/j.envsci.2017.12.004.

Fraser, E.D.G., W. Mabee, and F. Figge, 2005: A framework for assessing the vulnerability of food systems to future shocks. *Futures*, **37**, 465–479, doi:10.1016/J.FUTURES.2004.10.011.

Fraser, E.D.G. et al., 2011: Assessing vulnerability to climate change in dryland livelihood systems: Conceptual challenges and interdisciplinary solutions. *Ecol. Soc.*, **16**, art3, doi:10.5751/ES-03402-160303.

Fre, Z., 2018: *Knowledge Sovereignty Among African Cattle Herders*. UCL Press, London, UK, 216 pp.

Frechette, A., C. Ginsburg, W. Walker, S. Gorelik, S. Keene, C. Meyer, K. Reytar, and P. Veit, 2018: *A Global Baseline of Carbon Storage in Collective Lands*. Washington, DC, USA, 12 pp.

Fredriksson, P.G., and E. Neumayer, 2016: Corruption and climate change policies: Do the bad old days matter? *Environ. Resour. Econ.*, **63**, 451–469, doi:10.1007/s10640-014-9869-6.

Freebairn, J., 2016: A comparison of policy instruments to reduce greenhouse gas emissions. *Econ. Pap.*, **35**, 204–215, doi:10.1111/1759-3441.12141.

French, S., 2013: Cynefin, statistics and decision analysis. *J. Oper. Res. Soc.*, **64**, 547–561, doi:10.1057/jors.2012.23.

French, S., 2015: Cynefin: Uncertainty, small worlds and scenarios. *J. Oper. Res. Soc.*, **66**, 1635–1645, doi:10.1057/jors.2015.21.

Fridahl, M., and B.O. Linnér, 2016: Perspectives on the Green Climate Fund: Possible compromises on capitalization and balanced allocation. *Clim. Dev.*, **8**, 105–109, doi:10.1080/17565529.2015.1040368.

Friis, C., and J.Ø. Nielsen, 2016: Small-scale land acquisitions, large-scale implications: Exploring the case of Chinese banana investments in Northern Laos. *Land Use Policy*, **57**, 117–129, doi:10.1016/j.landusepol.2016.05.028.

Fritsche, U. et al., 2017: *Energy and Land Use: Global Land Outlook Working Paper.* United Nations Convention to Combat Desertification (UNCCD). Bonn, Germany, 60 pp. doi:10.13140/RG.2.2.24905.44648.

Fuerth, L.S., 2009: Operationalizing anticipatory governance. *Prism*, **4**, 31–46, https://cco.ndu.edu/Portals/96/Documents/prism/prism_2-4/Prism_31-46_Fuerth.pdf.

Fuerth, L.S., and E.M. H. Faber, 2013: Anticipatory governance: Winning the future. *Futurist*, **47**, 42–49. www.dropbox.com/s/4ax1mpkt27rohq0/Futurist.pdf?dl=0.

Fujimori, S. et al., 2018a: Inclusive climate change mitigation and food security policy under 1.5°C climate goal. *Environ. Res. Lett.*, **13**, 074033, doi:10.1088/1748-9326/aad0f7.

Fuller, T., and M. Qingwen, 2013: Understanding agricultural heritage sites as complex adaptive systems: The challenge of complexity. 4, 195–201, doi:10.5814/J.ISSN.1674-764X.2013.03.002.

Fung, A., 2015: Putting the public back into governance: The challenges of citizen participation and its future. *Public Adm. Rev.*, **75**, 513–522, doi:10.1111/puar.12361.

Furtado, F., 2018: A construção da natureza e a natureza da construção: Políticas de incentivo aos serviços ambientais no Acre e no Mato Grosso. *Estud. Soc. e Agric*, **26**, 123–147, https://revistaesa.com/ojs/index.php/esa/article/view/1152/558.

Fuso Nerini, F., C. Ray, and Y. Boulkaid, 2017: The cost of cooking a meal. The case of Nyeri County, Kenya. *Environ. Res. Lett.*, **12**, 065007, doi:10.1088/1748-9326/aa6fd0.

Fuso Nerini, F. et al., 2018: Mapping synergies and trade-offs between energy and the Sustainable Development Goals. *Nat. Energy*, **3**, 10–15, doi:10.1038/s41560-017-0036-5.

Fuss, S. et al., 2014: Betting on negative emissions. *Nat. Clim. Chang.*, **4**, 850–853, doi:10.1038/nclimate2392.

Fuss, S.et al., 2018: Negative emissions – Part 2: Costs, potentials and side effects. *Environ. Res. Lett.*, **13**, 063002, doi:10.1088/1748-9326/aabf9f.

Fussell, E., L.M. Hunter, and C.L. Gray, 2014: Measuring the environmental dimensions of human migration: The demographer's toolkit. *Glob. Environ. Chang.*, **28**, 182–191, doi:10.1016/j.gloenvcha.2014.07.001.

Fyson, C., and L. Jeffery, 2018: Examining treatment of the LULUCF sector in the NDCs. In: *20th EGU Gen. Assem. EGU2018, Proc. from Conf. held 4–13 April. 2018 Vienna, Austria*, **20**, 16542, https://meetingorganizer.copernicus.org/EGU2018/EGU2018-16542.pdf.

Gabay, M., and M. Alam, 2017: Community forestry and its mitigation potential in the Anthropocene: The importance of land tenure governance and the threat of privatization. *For. Policy Econ.*, **79**, 26–35, doi:10.1016/j.forpol.2017.01.011.

Gallagher, J., 2014: Learning about an infrequent event: Evidence from flood insurance take-up in the United States. *Am. Econ. J. Appl. Econ.*, **6**, 206–233, doi:10.1257/app.6.3.206.

Gallina, V., S. Torresan, A. Critto, A. Sperotto, T. Glade, and A. Marcomini, 2016: A review of multi-risk methodologies for natural hazards: Consequences and challenges for a climate change impact assessment. *J. Environ. Manage.*, **168**, 123–132, doi:10.1016/j.jenvman.2015.11.011.

Gan, J., A. Jarrett, and C.J. Gaither, 2014: Wildfire risk adaptation: Propensity of forestland owners to purchase wildfire insurance in the southern United States. *Can. J. For. Res.*, **44**, 1376–1382, doi:10.1139/cjfr-2014-0301.

Gandenberger, C., M. Bodenheimer, J. Schleich, R. Orzanna, and L. Macht, 2016: Factors driving international technology transfer: Empirical insights from a CDM project survey. *Clim. Policy*, **16**, 1065–1084, doi:10.1080/14693062.2015.1069176.

Garcia, C., and C.J. Fearnley, 2012: Evaluating critical links in early warning systems for natural hazards. *Environmental Hazards*, **11**, 123–137, doi:10.1080/17477891.2011.609877.

Gardner, T.A. et al., 2018a: Transparency and sustainability in global commodity supply chains. *World Development*, **121**, 163–177, doi:10.1016/j.worlddev.2018.05.025.

Garnett, S.T. et al., 2018: A spatial overview of the global importance of indigenous lands for conservation. *Nat. Sustain.*, **1**, 369–374, doi:10.1038/s41893-018-0100-6.

Garnett, T. et al., 2013: Sustainable Intensification in Agriculture: Premises and Policies. *Science*, **341**, 33–34, doi:10.1126/science.1234485.

Garrett, R.D. et al., 2019: Criteria for effective zero-deforestation commitments. *Global Environmental Change*, 54, 135–147, doi:10.1016/j.gloenvcha.2018.11.003.

Gasparatos, A. et al., 2018a: Mechanisms and indicators for assessing the impact of biofuel feedstock production on ecosystem services. *Biomass and Bioenergy*, **114**, 157–173, doi:10.1016/j.biombioe.2018.01.024.

Gasparatos, A.et al., 2018b: Survey of local impacts of biofuel crop production and adoption of ethanol stoves in southern Africa. *Sci. Data*, **5**, 180186, doi:10.1038/sdata.2018.186.

Gasparatos, A. et al., 2018c: Using an ecosystem services perspective to assess biofuel sustainability. *Biomass and Bioenergy*, **114**, 1–7, doi:10.1016/j.biombioe.2018.01.025.

Gazol, A. et al., 2018: Beneath the canopy: Linking drought-induced forest die off and changes in soil properties. *For. Ecol. Manage.*, 422, 294–302, doi:10.1016/j.foreco.2018.04.028.

GBEP, 2017: *The Global Bioenergy Partnership: A Global Commitment to Bioenergy*. Food and Agriculture Organization of the United Nations, Rome, Italy, 4 pp.

Geddes, A., T.S. Schmidt, and B. Steffen, 2018: The multiple roles of state investment banks in low-carbon energy finance: An analysis of Australia, the UK and Germany. *Energy Policy*, **115**, 158–170, doi:10.1016/j.enpol.2018.01.009.

Geden, O., G.P. Peters, and V. Scott, 2019: Targeting carbon dioxide removal in the European Union. *Clim. Policy*, **19**, 487–494, doi:10.1080/14693062.2018.1536600.

van der Geest, K., and K. Warner, 2014: Vulnerability, coping and loss and damage from climate events. In: *Hazards, Risks and, Disasters in Society* [Shroder, J., A. Collins Jones, S. Bernard Manyena, J. Jayawickrama (eds.)]. Academic Press, Elsevier, Massachusetts, USA, pp. 424, doi:10.1016/b978-0-12-396451-9.00008-1.

Geisler, C., and B. Currens, 2017: Impediments to inland resettlement under conditions of accelerated sea level rise. *Land Use Policy*, **66**, 322–330, doi:10.1016/j.landusepol.2017.03.029.

German, L., and G. Schoneveld, 2012: A review of social sustainability considerations among EU-approved voluntary schemes for biofuels, with implications for rural livelihoods. *Energy Policy*, **51**, 765–778, doi:10.1016/J.ENPOL.2012.09.022.

Gersonius, B., R. Ashley, A. Pathirana, and C. Zevenbergen, 2013: Climate change uncertainty: Building flexibility into water and flood risk infrastructure. *Clim. Change*, **116**, 411–423, doi:10.1007/s10584-012-0494-5.

Gheewala, S.H., G. Berndes, and G. Jewitt, 2011: The bioenergy and water nexus. *Biofuels, Bioprod. Biorefining*, **5**, 353–360, doi:10.1002/bbb.316.

Ghilardi, A. et al., 2016a: Spatiotemporal modeling of fuelwood environmental impacts: Towards improved accounting for non-renewable biomass. *Environ. Model. Softw.*, **82**, 241–254, doi:10.1016/j.envsoft.2016.04.023.

Ghilardi, A., A. Tarter, and R. Bailis, 2018: Potential environmental benefits from woodfuel transitions in Haiti: Geospatial scenarios to 2027. *Environ. Res. Lett.*, **13**, 035007, doi:10.1088/1748-9326/aaa846.

Ghosh, A., S. Schmidt, T. Fickert, and M. Nüsser, 2015: The Indian Sundarban mangrove forests: History, utilization, conservation strategies and local perception. *Diversity*, **7**, 149–169, doi:10.3390/d7020149.

Ghosh, S. et al., 2016: Indian Summer Monsoon Rainfall: Implications of contrasting trends in the spatial variability of means and extremes. *PLoS One*, **11**, e0158670, doi:10.1371/journal.pone.0158670.

Gibson, L., E.N. Wilman, and W.F. Laurance, 2017: How Green is 'Green' energy? *Trends Ecol. Evol.*, **32**, 922–935, doi:10.1016/j.tree.2017.09.007.

Giessen, L., S. Burns, M.A. K. Sahide, and A. Wibowo, 2016a: From governance to government: The strengthened role of state bureaucracies in forest and agricultural certification. *Policy Soc.*, **35**, 71–89, doi:10.1016/j.polsoc.2016.02.001.

Gilbert, C.L., 2010: How to understand high food prices. *J. Agric. Econ.*, **61**, 398–425, doi:10.1111/j.1477-9552.2010.00248.x.

Gill, A.B., 2005: Offshore renewable energy: Ecological implications of generating electricity in the coastal zone. *J. Appl. Ecol.*, **42**, 605–615, doi:10.1111/j.1365-2664.2005.01060.x.

Girard, C., M. Pulido-Velazquez, J.-D. Rinaudo, C. Pagé, and Y. Caballero, 2015: Integrating top-down and bottom-up approaches to design global change adaptation at the river basin scale. *Glob. Environ. Chang.*, **34**, 132–146, doi:10.1016/j.gloenvcha.2015.07.002.

Girma, H.M., R.M. Hassan, and G. Hertzler, 2012: Forest conservation versus conversion under uncertain market and environmental forest benefits in Ethiopia: The case of Sheka forest. *For. Policy Econ.*, **21**, 101–107, doi:10.1016/j.forpol.2012.01.001.

Gitau, J.K. et al., 2019: Implications on livelihoods and the environment of uptake of Gasifier cook stoves among Kenya's rural households. *Appl. Sci.*, **9**, 1205, doi:10.3390/app9061205.

Glachant, M., and A. Dechezleprêtre, 2017: What role for climate negotiations on technology transfer? *Clim. Policy*, **17**, 962–981, doi:10.1080/14693062.2016.1222257.

Glauben, T., T. Herzfeld, S. Rozelle, and X. Wang, 2012: Persistent poverty in rural China: Where, why, and how to escape? *World Dev.*, **40**, 784–795, doi:10.1016/j.worlddev.2011.09.023.

Gleick, P.H., 2014: Water, drought, climate change, and conflict in Syria. *Weather. Clim. Soc.*, **6**, 331–340, doi:10.1175/WCAS-D-13-00059.1.

Glemarec, Y., 2017: Addressing the gender differentiated investment risks to climate-smart agriculture. *AIMS Agric. Food*, **2**, 56–74, doi:10.3934/agrfood.2017.1.56.

Go, A.W., A.T. Conag, R.M. B. Igdon, A.S. Toledo, and J.S. Malila, 2019a: Potentials of agricultural and agroindustrial crop residues for the displacement of fossil fuels: A Philippine context. *Energy Strateg. Rev.*, **23**, 100–113, doi:10.1016/j.esr.2018.12.010.

Godar, J., and T. Gardner, 2019: Trade and land use telecouplings. In: *Telecoupling* [Friis, C., J.Ø. Nielsen (eds.)]. Springer International Publishing, Cham, Switzerland, pp. 149–175.

Godar, J., U.M. Persson, E.J. Tizado, and P. Meyfroidt, 2015: Methodological and ideological options towards more accurate and policy relevant footprint analyses: Tracing fine-scale socio-environmental impacts of production to consumption. *Ecol. Econ.*, **112**, 25–35, doi:10.1016/j.ecolecon.2015.02.003.

Godar, J., C. Suavet, T.A. Gardner, E. Dawkins, and P. Meyfroidt, 2016: Balancing detail and scale in assessing transparency to improve the governance of agricultural commodity supply chains. *Environ. Res. Lett.*, **11**, 035015, doi:10.1088/1748-9326/11/3/035015.

Goetz, A., 2013: Private Governance and Land Grabbing: The Equator Principles and the Roundtable on Sustainable Biofuels, *Globalizations*, **10**, 199–204, doi:10.1080/14747731.2013.760949.

Goetz, S.J., M. Hansen, R.A. Houghton, W. Walker, N. Laporte, and J. Busch, 2015: Measurement and monitoring needs, capabilities and potential for addressing reduced emissions from deforestation and forest degradation under REDD+. *Environ. Res. Lett.*, **10**, 123001, doi:10.1088/1748-9326/10/12/123001.

Goh, A.H. X., 2012: A literature review of the gender-differentiated impacts of climate change on women's and men's assets and well-being in developing countries. CAPRi Working Paper No. 106, International Food Policy Research Institute, Washington, DC, USA, 44 pp, doi:10.2499/CAPRiWP106.

Golay, C., and I. Biglino, 2013: Human rights responses to land grabbing: A right to food perspective. *Third World Q.*, **34**, 1630–1650, doi:10.1080/01436597.2013.843853.

Gold Standard: 2018: Gold standard for the global goals. The Gold Standard Foundation. Geneva, Switzerland, www.goldstandard.org/our-work/what-we-do.

Goldemberg, J., and S. Teixeira Coelho, 2004: Renewable energy-traditional biomass vs. modern biomass. *Energy Policy*, **32**, 711–714, doi:10.1016/S0301-4215 (02)00340-3.

Goldemberg, J., J. Martinez-Gomez, A. Sagar, and K.R. Smith, 2018a: Household air pollution, health, and climate change: Cleaning the air. *Environ. Res. Lett.*, **13**, 030201, doi:10.1088/1748-9326/aaa49d.

Gómez-Baggethun, E., S. Mingorría, V. Reyes-García, L. Calvet, and C. Montes, 2010: Traditional ecological knowledge trends in the transition to a market economy: Empirical study in the Doñana natural areas. *Conserv. Biol.*, **24**, 721–729, doi:10.1111/j.1523-1739.2009.01401.x.

Gopakumar, G., 2014: *Transforming Urban Water Supplies in India*: The Role of Reform and Partnerships in Globalization, 1st Edition. Routledge, Abingdon, UK, and New York, USA, 168 pp.

Gordon, S.M., 2016: The foreign corrupt practices act: Prosecute corruption and end transnational illegal logging. *Bost. Coll. Environ. Aff. Law Rev.*, **43**111, https://lawdigitalcommons.bc.edu/ealr/vol43/iss1/5.

Götz, L., T. Glauben, and B. Brümmer, 2013: Wheat export restrictions and domestic market effects in Russia and Ukraine during the food crisis. *Food Policy*, **38**, 214–226, doi:10.1016/j.foodpol.2012.12.001.

Goudie, A.S., 2014: Desert dust and human health disorders. *Environ. Int.*, **63**, 101–113, doi:10.1016/J.ENVINT.2013.10.011.

Goulder, L.H., and R.C. Williams, 2012: *The Choice of Discount Rate for Climate Change Policy Evaluation*. NBER Working Paper No. 18301, Climate Change Economics (CCE), World Scientific Publishing Co. Pte. Ltd., pp. 1250024-1-1, doi:10.3386/w18301.

Government of France, 2019: Ending deforestation caused by importing unsustainable products Gouvernement.fr., Paris, France, www.gouvernement.fr/en/ending-deforestation-caused-by-importing-unsustainable-products.

Government of India, *National Mission for a Green India. Under the National Action Plan on Climate Change*. New Delhi, India, 37 pp., www.indiaenvironmentportal.org.in/files/green-india-mission.pdf.

Government of India, 2012: *Guidelines for Preparation of State Action Plan for Bustards' Recovery Programme*. Ministry of Environment and Forests, New Delhi, India, 29 pp, www.indiaenvironmentportal.org.in/files/file/Bustards%E2%80%99%20Recovery%20Programme.pdf.

Government Office for Science, 2011: Migration and global environmental change: Future challenges and opportunities. Foresight: Migration and Global Environmental Change. The Final Project Report. London, UK, 234 pp. https://eprints.soas.ac.uk/22475/1/11-1116-migration-and-global-environmental-change.pdf.

Van de Graaf, T., 2017: Is OPEC dead? Oil exporters, the Paris Agreement and the transition to a post-carbon world. *Energy Res. Soc. Sci.*, **23**, 182–188, doi:10.1016/j.erss.2016.10.005.

Graham, L.J., R.H. Haines-Young, and R. Field, 2015: Using citizen science data for conservation planning: Methods for quality control and downscaling for use in stochastic patch occupancy modelling. *Biol. Conserv.*, **192**, 65–73, doi:10.1016/j.biocon.2015.09.002.

Grainger, A., 2015: Is land degradation neutrality feasible in dry areas? *J. Arid Environ.*, **112**, 14–24, doi:10.1016/j.jaridenv.2014.05.014.

Gray, S. et al., 2017: Combining participatory modelling and citizen science to support volunteer conservation action. *Biol. Conserv.*, **208**, 76–86, doi:10.1016/J.BIOCON.2016.07.037.

Greatrex, H. et al., 2015: Scaling up index insurance for smallholder farmers: Recent evidence and insights. *CCAFS Rep.*, **14**, 1–32, doi:1904-9005.

Grecchi, R.C., Q.H.J. Gwyn, G.B. Bénié, A.R. Formaggio, and F.C. Fahl, 2014: Land use and land cover changes in the Brazilian Cerrado: A multidisciplinary approach to assess the impacts of agricultural expansion. *Appl. Geogr.*, **55**, 300–312, doi:10.1016/J.APGEOG.2014.09.014.

Green, D., and G. Raygorodetsky, 2010: Indigenous knowledge of a changing climate. *Clim. Change*, **100**, 239–242, doi:10.1007/s10584-010-9804-y.

Di Gregorio, M. et al., 2017: Climate policy integration in the land use sector: Mitigation, adaptation and sustainable development linkages. *Environ. Sci. Policy*, **67**, 35–43, doi:10.1016/j.envsci.2016.11.004.

Greiner, R., and D. Gregg, 2011: Farmers' intrinsic motivations, barriers to the adoption of conservation practices and effectiveness of policy instruments: Empirical evidence from northern Australia. *Land Use Policy*, **28**, 257–265, doi:10.1016/j.landusepol.2010.06.006.

Griggs, D. et al., 2013: Sustainable development goals for people and planet. *Nature*, **495**, 305. doi:10.1038/495305a.

Griggs, D.et al., 2014: An integrated framework for sustainable development goals. *Ecol. Soc.*, **19**, art49-art49, doi:10.5751/ES-07082-190449.

De Groot, R.S., R. Alkemade, L. Braat, L. Hein, and L. Willemen, 2010: Challenges in integrating the concept of ecosystem services and values in landscape planning, management and decision-making. *Ecol. Complex.*, **7**, 260–272, doi:10.1016/j.ecocom.2009.10.006.

Grosjean, G. et al., 2018: Options to overcome the barriers to pricing European agricultural emissions. *Clim. Policy*, **18**, 151–169, doi:10.1080/14693062.2016.1258630.

Del Grosso, S., P. Smith, M. Galdos, A. Hastings, and W. Parton, 2014: Sustainable energy crop production. *Curr. Opin. Environ. Sustain.*, **9–10**, 20–25, doi:10.1016/j.cosust.2014.07.007.

Grumbine, R.E., and M.K. Pandit, 2013: Threats from India's Himalaya dams. *Science*, **339**, 36–37, doi:10.1126/science.1227211.

Grzymala-Busse, A., 2010: The best laid plans: The impact of informal rules on formal institutions in transitional regimes. *Stud. Comp. Int. Dev.*, **45**, 311–333, doi:10.1007/s12116-010-9071-y.

Gu, Y., and B.K. Wylie, 2017: Mapping marginal croplands suitable for cellulosic feedstock crops in the Great Plains, United States. *GCB Bioenergy*, **9**, 836–844, doi:10.1111/gcbb.12388.

Guerry, A.D. et al., 2015: Natural capital and ecosystem services informing decisions: From promise to practice. *Proc. Natl. Acad. Sci.*, **112**, 7348–7355, doi:10.1073/pnas.1503751112.

Gunderson, L.H., and C. Holling (eds.), 2001: *Panarchy: Understanding Transformations in Human and Natural Systems*. Island Press, Washington, DC, USA, 507 pp.

Gupta, H., and L.C. Dube, 2018: Addressing biodiversity in climate change discourse: Paris mechanisms hold more promise. *Int. For. Rev.*, **20**, 104–114, doi:10.1505/146554818822824282.

Gupta, J. (ed.), 2014: *The History of Global Climate Governance*. Cambridge University Press, Cambridge, UK, and New York, NY, USA, 1–244 pp.

Gupta, J., and C. Vegelin, 2016: Sustainable development goals and inclusive development. *Int. Environ. Agreements Polit. Law Econ.*, **16**, 433–448, doi:10.1007/s10784-016-9323-z.

Gupta, J., and N. Pouw, 2017: Towards a transdisciplinary conceptualization of inclusive development. *Curr. Opin. Environ. Sustain.*, **24**, 96–103, doi:10.1016/j.cosust.2017.03.004.

Gupta, J., C. Termeer, J. Klostermann, S. Meijerink, M. van den Brink, P. Jong, S. Nooteboom, and E. Bergsma, 2010: The adaptive capacity wheel: A method to assess the inherent characteristics of institutions to enable the adaptive capacity of society. *Environ. Sci. Policy*, **13**, 459–471, doi:10.1016/j.envsci.2010.05.006.

Gupta, J., N. van der Grijp, and O. Kuik, 2013a: *Climate Change, Forests, and REDD Lessons for Institutional Design*. Routledge, Abingdon, UK, and New York, USA, 288 pp.

Gupta, J., C. Pahl-Wostl, and R. Zondervan, 2013b: 'Glocal' water governance: A multi-level challenge in the anthropocene. *Curr. Opin. Environ. Sustain.*, **5**, 573–580, doi:10.1016/j.cosust.2013.09.003.

Gupta, J., N.R. M. Pouw, and M.A. F. Ros-Tonen, 2015: Towards an elaborated theory of inclusive development. *Eur. J. Dev. Res.*, **27**, 541–55, doi:10.1057/ejdr.2015.30.

Gurung, A., and S.E. Oh, 2013a: Conversion of traditional biomass into modern bioenergy systems: A review in context to improve the energy situation in Nepal. *Renew. Energy*, 50, 206–213, doi:10.1016/j.renene.2012.06.021.

Haasnoot, M., J.H. Kwakkel, W.E. Walker, and J. ter Maat, 2013: Dynamic adaptive policy pathways: A method for crafting robust decisions for a deeply uncertain world. *Glob. Environ. Chang.*, **23**, 485–498, doi:10.1016/j.gloenvcha.2012.12.006.

Haasnoot, M., S. van 't Klooster, and J. van Alphen, 2018: Designing a monitoring system to detect signals to adapt to uncertain climate change. *Glob. Environ. Chang.*, **52**, 273–285, doi:10.1016/j.gloenvcha.2018.08.003.

Hadarits, M., J. Pittman, D. Corkal, H. Hill, K. Bruce, and A. Howard, 2017: The interplay between incremental, transitional, and transformational adaptation: A case study of Canadian agriculture. *Reg. Environ. Chang.*, **17**, 1515–1525, doi:10.1007/s10113-017-1111-y.

Haddeland, I. et al., 2014: Global water resources affected by human interventions and climate change. *Proc. Natl. Acad. Sci. U.S.A.*, **111**, 3251–3256, doi:10.1073/pnas.1222475110.

Haites, E., 2018a: Carbon taxes and greenhouse gas emissions trading systems: What have we learned? *Clim. Policy*, **18**, 955–966, doi:10.1080/14693062.2018.1492897.

Hajjar, R., R.A. Kozak, H. El-Lakany, and J.L. Innes, 2013: Community forests for forest communities: Integrating community-defined goals and practices in the design of forestry initiatives. *Land Use Policy*, **34**, 158–167, doi:10.1016/j.landusepol.2013.03.002.

Hall, R. et al., 2015: Resistance, acquiescence or incorporation? An introduction to landgrabbing and political reactions 'from below'. *J. Peasant Stud.*, **42**, 467–488, doi:10.1080/03066150.2015.1036746.

Hall, S.J., R. Hilborn, N.L. Andrew, and E.H. Allison, 2013: Innovations in capture fisheries are an imperative for nutrition security in the developing world. *Proc. Natl. Acad. Sci.*, **110**, 8393–8398, doi:10.1073/pnas.1208067110.

Hallegatte, S., 2009: Strategies to adapt to an uncertain climate. *Glob. Environ. Chang.*, **19**, 240–247, doi:10.1016/j.gloenvcha.2008.12.003.

Hallegatte, S., A. Shah, R.J. Lempert, C. Brown, and S. Gill, 2012: *Investment Decision-Making Under Deep Uncertainty – Application to Climate Change*. Policy Research Working Paper; No. 6193. World Bank, Washington, DC, USA, 41 pp https://openknowledge.worldbank.org/bitstream/handle/10986/12028/wps6193.pdf?sequence=1&isAllowed=y License: CC BY 3.0 IGO.

Hallegatte, S., C. Green, R.J. Nicholls, and J. Corfee-Morlot, 2013: Future flood losses in major coastal cities. *Nat. Clim. Chang.*, **3**, 802, doi:10.1038/nclimate1979.

Hallegatte, S., A. Vogt-Schilb, M. Bangalore, and J. Rozenberg, 2017: *Unbreakable: Building the Resilience of the Poor in the Face of Natural Disasters*. Climate Change and Development Series. World Bank, Washington, DC, USA, 201 pp.

Haller, T., G. Acciaioli, and S. Rist, 2016: Constitutionality: Conditions for crafting local ownership of institution-building processes. *Soc. Nat. Resour.*, 29, 68–87, doi:10.1080/08941920.2015.1041661.

Hanasaki, N., T. Inuzuka, S. Kanae, and T. Oki, 2010: An estimation of global virtual water flow and sources of water withdrawal for major crops and livestock products using a global hydrological model. *J. Hydrol.*, **384**, 232–244, doi:10.1016/j.jhydrol.2009.09.028.

Hanasaki, N. et al., 2013a: A global water scarcity assessment under shared socio-economic pathways – Part 2: Water availability and scarcity. *Hydrol. Earth Syst. Sci.*, **17**, 2393–2413, doi:10.5194/hess-17-2393-2013.

Handmer, J., Y. Honda, Z.W. Kundzewicz, N. Arnell, G. Benito, J. Hatfield, I.F. Mohamed, P. Peduzzi, S. Wu, B. Sherstyukov, K. Takahashi, and Z. Yan, 2012: Changes in Impacts of Climate Extremes: Human Systems and Ecosystems. In: Managing the Risks of Extreme Events and Disasters to Advance Climate Change Adaptation: Special Report of the Intergovernmental Panel on Climate Change [Field, C.B., V. Barros, T.F. Stocker, D. Qin, D.J. Dokken, K.L. Ebi, M.D. Mastrandrea, K.J. Mach, G.-K. Plattner, S.K. Allen, M. Tignor, and P.M. Midgley (eds.)]. Cambridge University Press, Cambridge, UK, and New York, NY, USA, 582 pp.

Hanger, S., C. Haug, T. Lung, and L.M. Bouwer, 2015: Mainstreaming climate change in regional development policy in Europe: Five insights from the 2007–2013 programming period. *Reg. Environ. Chang.*, **15**, 973–985, doi:10.1007/s10113-013-0549-9.

Hanjra, M.A., and M. Ejaz Qureshi, 2010: Global water crisis and future food security in an era of climate change. *Food Policy*, **35**, 365–377, doi:10.1016/j.foodpol.2010.05.006.

Hanson, S., R. Nicholls, N. Ranger, S. Hallegatte, J. Corfee-Morlot, C. Herweijer, and J. Chateau, 2011: A global ranking of port cities with high exposure to climate extremes. *Clim. Change*, **104**, 89–111, doi:10.1007/s10584-010-9977-4.

Härdle, W.K., and B.L. Cabrera, 2010: Calibrating CAT bonds for Mexican earthquakes. *J. Risk Insur.*, **77**, 625–650, doi:10.1111/j.1539-6975.2010.01355.x.

Harmsworth, G., and S. Awatere, 2013: 2013. Indigenous Māori knowledge and perspectives of ecosystems. In: *Ecosystem services in New Zealand – Conditions and trends* [J. Dymond (ed.)]. Manaaki Whenua Press, Lincoln, New Zealand., pp. 274–286.

Harootunian, G., 2018: California: It's Complicated: Drought, drinking water and drylands. *Resilience: The Science of Adaptation to Climate Change* [Alverson, K., and Z. Zommers (eds.)]. Elsevier, London, UK, 127–142, https://doi.org/10.1016/B978-0-12-811891-7.00010-4.

Harris, E., 2013: Financing social protection floors: Considerations of fiscal space. *Int. Soc. Secur. Rev.*, **66**, 111–143, doi:10.1111/issr.12021.

Harris, Z.M., R. Spake, and G. Taylor, 2015: Land use change to bioenergy: A meta-analysis of soil carbon and GHG emissions. *Biomass and Bioenergy*, 82, 27–39, doi:10.1016/j.biombioe.2015.05.008.

Harrod, K.S., 2015: Ebola: History, treatment, and lessons from a new emerging pathogen. *Am. J. Physiol. – Lung Cell. Mol. Physiol.*, **308**, L307–L313, doi:10.1152/ajplung.00354.2014.

Harvey, B., J. Ensor, L. Carlile, B. Garside, and Z. Patterson, 2012: *Climate Change Communication and Social Learning – Review and Strategy Development for CCAFS*. CCAFS Working Paper No. 22. CGIAR Research Program on Climate Change, Agriculture and Food Security (CCAFS), Copenhagen, Denmark, 53 pp.

Harvey, C.A. et al., 2014a: Climate-smart landscapes: Opportunities and challenges for integrating adaptation and mitigation in tropical agriculture. *Conserv. Lett.*, **7**, 77–90, doi:10.1111/conl.12066.

Harvey, C.A. et al., 2014b: Extreme vulnerability of smallholder farmers to agricultural risks and climate change in Madagascar. *Philos. Trans. R. Soc. B Biol. Sci.*, **369**, 20130089, doi:10.1098/rstb.2013.0089.

Harvey, M., and S. Pilgrim, 2011: The new competition for land: Food, energy, and climate change. *Food Policy*, **36**, S40-S51, doi:10.1016/j.foodpol.2010.11.009.

Hasegawa, T., S. Fujimori, K. Takahashi, T. Yokohata, and T. Masui, 2016: Economic implications of climate change impacts on human health through undernourishment. *Clim. Change*, **136**, 189–202, doi:10.1007/s10584-016-1606-4.

Hasegawa, T.et al., 2018a: Risk of increased food insecurity under stringent global climate change mitigation policy. *Nat. Clim. Chang.*, **8**, 699–703, doi:10.1038/s41558-018-0230-x.

Hashemi, S.M. and de Montesquiou, A. (eds.), 2011: *Reaching the Poorest: Lessons from the Graduation Model*. Focus Note 69, Washington, DC, USA, 16 pp.

Häyhä, T., and P.P. Franzese, 2014: Ecosystem services assessment: A review under an ecological-economic and systems perspective. *Ecol. Modell.*, **289**, 124–132, doi:10.1016/j.ecolmodel.2014.07.002.

Head, B.W., 2014: Evidence, uncertainty, and wicked problems in climate change decision-making in Australia. *Environ. Plan. C Gov. Policy*, **32**, 663–679, doi:10.1068/c1240.

Headey, D., 2011: Rethinking the global food crisis: The role of trade shocks. *Food Policy*, 36, 136–146, doi:10.1016/j.foodpol.2010.10.003.

Headey, D., J. Hoddinott, and S. Park, 2017: Accounting for nutritional changes in six success stories: A regression-decomposition approach. *Glob. Food Sec.*, **13**, 12–20, doi:10.1016/j.gfs.2017.02.003.

Heck, V., D. Gerten, W. Lucht, and A. Popp, 2018a: Biomass-based negative emissions difficult to reconcile with planetary boundaries. *Nat. Clim. Chang.*, **8**, 151–155, doi:10.1038/s41558-017-0064-y.

HEI/IHME, 2018: *A Special Report on Global Exposure To Air Pollution and Its Disease Burden*. Health Effects Institute, Boston, USA, 24 pp.

Hejazi, M.I. et al., 2014: Integrated assessment of global water scarcity over the 21st century under multiple climate change mitigation policies. *Hydrol. Earth Syst. Sci.*, **18**, 2859–2883, doi:10.5194/hess-18-2859-2014.

Helmke, G., and S. Levitsky, 2004: Informal institutions and comparative politics: A research agenda. *Perspect. Polit.*, **2**, 725–740, doi:10.1017/S1537592704040472.

Hember, R.A., W.A. Kurz, and N.C. Coops, 2017: Relationships between individual-tree mortality and water-balance variables indicate positive trends in water stress-induced tree mortality across North America. *Glob. Chang. Biol.*, **23**, 1691–1710, doi:10.1111/gcb.13428.

Henderson, B., 2018: *A Global Economic Evaluation of GHG Mitigation Policies for Agriculture*. Joint Working Party on Agriculture and the Environment. Organisation for Economic Co-operation and Development, Paris, France, 38 pp. www.oecd.org/officialdocuments/publicdisplaydocumentpdf/?cote=COM/TAD/CA/ENV/EPOC (2018)7/FINAL&docLanguage=En.

Hendrix, C.S., and I. Salehyan, 2012: Climate change, rainfall, and social conflict in Africa. *J. Peace Res.*, **49**, 35–50, doi:10.1177/0022343311426165.

Henriksen, H.J., M.J. Roberts, P. van der Keur, A. Harjanne, D. Egilson, and L. Alfonso, 2018: Participatory early warning and monitoring systems: A Nordic framework for web-based flood risk management. *Int. J. Disaster Risk Reduct.*, doi:10.1016/j.ijdrr.2018.01.038.

Henstra, D., 2016: The tools of climate adaptation policy: Analysing instruments and instrument selection. *Clim. Policy*, **16**, 496–521, doi:10.1080/14693062.2015.1015946.

Herendeen, N., and N. Glazier, 2009: Agricultural best management practices for Conesus Lake: The role of extension and soil/water conservation districts. *J. Great Lakes Res.*, **35**, 15–22, doi:10.1016/j.jglr.2008.08.005.

Herman, J.D., H.B. Zeff, P.M. Reed, and G.W. Characklis, 2014: Beyond optimality: Multistakeholder robustness tradeoffs for regional water portfolio planning under deep uncertainty. *Water Resour. Res.*, **50**, 7692–7713, doi:10.1002/2014WR015338.

Hermann, A., Koferl, P., Mairhofer, J.P., 2016: *Climate Risk Insurance: New Approaches and Schemes*. Economic Research Working Paper. Germany, 22 pp. www.allianz.com/content/dam/onemarketing/azcom/Allianz_com/migration/media/economic_research/publications/working_papers/en/ClimateRisk.pdf.

Hernandez, R.R., M.K. Hoffacker, M.L. Murphy-Mariscal, G.C. Wu, and M.F. Allen, 2015: Solar energy development impacts on land cover change and protected areas. *Proc. Natl. Acad. Sci.*, **112**, 13579–13584.

Hertel, T.W., M.B. Burke, and D.B. Lobell, 2010: The poverty implications of climate-induced crop yield changes by 2030. *Glob. Environ. Chang.*, **20**, 577–585, doi:10.1016/j.gloenvcha.2010.07.001.

Hewitt, K. et al., 2017: Identifying emerging issues in disaster risk reduction, migration, climate change and sustainable development. *Identifying*

Emerging Issues in Disaster Risk Reduction, Migration, Climate Change and Sustainable Development. Springer International Publishing, Cham, Switzerland, doi:10.1007/978-3-319-33880-4, 281 pp.

Hewitt, R., H. van Delden, and F. Escobar, 2014: Participatory land use modelling, pathways to an integrated approach. *Environ. Model. Softw.*, **52**, 149–165, doi:10.1016/J.ENVSOFT.2013.10.019.

Higgins, S.A., I. Overeem, K.G. Rogers, and E.A. Kalina, 2018: River linking in India: Downstream impacts on water discharge and suspended sediment transport to deltas. *Elem Sci Anth*, **6**, 20, doi:10.1525/elementa.269.

Himanen, S.J., P. Rikkonen, and H. Kahiluoto, 2016: Codesigning a resilient food system. *Ecol. Soc.*, **21**, Art. 41, doi:10.5751/ES-08878-210441.

Hinkel, J. et al., 2014: Coastal flood damage and adaptation costs under 21st century sea-level rise. *Proc. Natl. Acad. Sci.*, 111, 3292–3297, doi:10.1073/pnas.1222469111.

Hirsch, A.L. et al., 2018: Biogeophysical impacts of land use change on climate extremes in low-emission scenarios: Results from HAPPI-Land. *Earth's Futur.*, **6**, 396–409, doi:10.1002/2017EF000744.

Hjort, J., Karjalainen, O., Aalto, J., Westermann, S., Romanovsky, V.E., Nelson, F.E., Luoto, M. (2018). Degrading permafrost puts Arctic infrastructure at risk by mid-century. *Nature Communications*, **9** (1), 5147, doi:10.1038/s41467-018-07557-4.

Hoegh-Guldberg, O. et al., 2018: Impacts of 1.5°C Global Warming on Natural and Human Systems. In: Global Warming of 1.5°C. An IPCC Special Report on the Impacts of Global Warming of 1.5°C Above Pre-Industrial Levels and Related Global Greenhouse Gas Emission Pathways, in the Context of Strengthening the Global Response to the Threat of Climate Change [Masson-Delmotte, V., P. Zhai, H.-O. Pörtner, D. Roberts, J. Skea, P.R. Shukla, A. Pirani, W. Moufouma-Okia, C. Péan, R. Pidcock, S. Connors, J.B.R. Matthews, Y. Chen, X. Zhou, M.I. Gomis, E. Lonnoy, T. Maycock, M. Tignor, and T. Waterfield (eds.)]. Cambridge University Press, Cambridge, UK, and New York, NY, USA, 630 pp.

Hoeinghaus, D.J. et al., 2009: Effects of river impoundment on ecosystem services of large tropical rivers: Embodied energy and market value of artisanal fisheries. *Conserv. Biol.*, **23**, 1222–1231, doi:10.1111/j.1523-1739.2009.01248.x.

Hoff, H., 2011: *Understanding the Nexus*. Background Paper for the Bonn2011 Conference: The Water, Energy and Food Security Nexus, Stockholm Environment Institute, Stockholm, 1–52 pp.

van der Hoff, R., R. Rajão, and P. Leroy, 2018: Clashing interpretations of REDD+ 'results' in the Amazon Fund. *Clim. Change*, 150, 433–445, doi:10.1007/s10584-018-2288-x.

Hoffmann, H., G. Uckert, C. Reif, K. Müller, and S. Sieber, 2015a: Traditional biomass energy consumption and the potential introduction of firewood efficient stoves: Insights from western Tanzania. *Reg. Environ. Chang.*, **15**, 1191–1201, doi:10.1007/s10113-014-0738-1.

Hoffmann, H.K., K. Sander, M. Brüntrup, and S. Sieber, 2017: Applying the water-energy-food nexus to the charcoal value chain. *Front. Environ. Sci.*, **5**, 84, doi:10.3389/fenvs.2017.00084.

Höhne, N. et al., 2017: The Paris Agreement: Resolving the inconsistency between global goals and national contributions. *Clim. Policy*, **17**, 16–32, doi:10.1080/14693062.2016.1218320.

Hojas-Gascon, L., H.D. Eva, D. Ehrlich, M. Pesaresi, F. Achard, and J. Garcia, 2016a: Urbanization and forest degradation in east Africa – A case study around Dar es Salaam, Tanzania. 2016 IEEE International Geoscience and Remote Sensing Symposium (IGARSS), IEEE. Institute Of Electrical And Electronics Engineers. Beijing, China, 7293–7295.

Holden, E., K. Linnerud, and D. Banister, 2017: The imperatives of sustainable development. *Sustain. Dev.*, **25**, 213–226, doi:10.1002/sd.1647.

Holden, S.T., and H. Ghebru, 2016: Land tenure reforms, tenure security and food security in poor agrarian economies: Causal linkages and research gaps. *Glob. Food Sec.*, **10**, 21–28, doi:10.1016/j.gfs.2016.07.002.

Holland, M.B., K.W. Jones, L. Naughton-Treves, J.L. Freire, M. Morales, and L. Suárez, 2017: Titling land to conserve forests: The case of Cuyabeno Reserve in Ecuador. *Glob. Environ. Chang.*, **44**, 27–38, doi:10.1016/j.gloenvcha.2017.02.004.

Holling, C.S. (ed.), 1978: *Adaptive Environmental Assessment and Management*. John Wiley & Sons, Chichester, UK, 402 pp.

Holling, C.S., 1986: Adaptive environmental management. *Environment: Science and Policy for Sustainable Development*, **28**, 39, doi:10.1080/00139157.1986.9928829.

Holstenkamp, L., and F. Kahla, 2016: What are community energy companies trying to accomplish? An empirical investigation of investment motives in the German case. *Energy Policy*, **97**, 112–122, doi:10.1016/j.enpol.2016.07.010.

Hordijk, M., L.M. Sara, and C. Sutherland, 2014: Resilience, transition or transformation? A comparative analysis of changing water governance systems in four southern cities. *Environ. Urban.*, **26**, 130–146, doi:10.1177/0956247813519044.

Hornsey, M.J., E.A. Harris, P.G. Bain, and K.S. Fielding, 2016: Meta-analyses of the determinants and outcomes of belief in climate change. *Nat. Clim. Chang.*, **6**, 622–626, doi:10.1038/nclimate2943.

Hossain, M., 2018: Introduction: Pathways to a sustainable economy. In: *Pathways to a Sustainable Economy*. Springer International Publishing, Cham, Switzerland, pp. 1–1.

Hou, D., 2016: Divergence in stakeholder perception of sustainable remediation. *Sustain. Sci.*, **11**, 215–230, doi:10.1007/s11625-015-0346-0.

Hou, D., and A. Al-Tabbaa, 2014: Sustainability: A new imperative in contaminated land remediation. *Environ. Sci. Policy*, **39**, 25–34, doi:10.1016/j.envsci.2014.02.003.

Howlett, M., and J. Rayner, 2013: Patching vs packaging in policy formulation: Assessing policy portfolio design. *Polit. Gov.*, **1**, 170, doi:10.17645/pag.v1i2.95.

Huang, J., and G. Yang, 2017: Understanding recent challenges and new food policy in China. *Glob. Food Sec.*, 12, 119–126, doi:10.1016/j.gfs.2016.10.002.

Huang, J., H. Yu, X. Guan, G. Wang, and R. Guo, 2016: Accelerated dryland expansion under climate change. *Nat. Clim. Chang.*, **6**, 166–171, doi:10.1038/nclimate2837.

Huang, J., H. Yu, A. Dai, Y. Wei, and L. Kang, 2017: Drylands face potential threat under 2°C global warming target. *Nat. Clim. Chang.*, **7**, 417–422, doi:10.1038/nclimate3275.

Huang, R., D. Tian, J. Liu, S. Lv, X. He, and M. Gao, 2018: Responses of soil carbon pool and soil aggregates associated organic carbon to straw and straw-derived biochar addition in a dryland cropping mesocosm system. *Agric. Ecosyst. Environ.*, **265**, 576–586, doi:10.1016/J.AGEE.2018.07.013.

Huber-Sannwald, E. et al., 2012: Navigating challenges and opportunities of land degradation and sustainable livelihood development in dryland social-ecological systems: A case study from Mexico. *Philos. Trans. R. Soc. B Biol. Sci.*, **367**, 3158–77. doi:10.1098/rstb.2011.0349.

Hudiburg, T.W., S.C. Davis, W. Parton, and E.H. Delucia, 2015: Bioenergy crop greenhouse gas mitigation potential under a range of management practices. *GCB Bioenergy*, 7, 366–374, doi:10.1111/gcbb.12152.

Hudson, P., W.J. W. Botzen, L. Feyen, and J.C. J.H. Aerts, 2016: Incentivising flood risk adaptation through risk based insurance premiums: Trade-offs between affordability and risk reduction. *Ecol. Econ.*, **125**, 1–13, doi:10.1016/J.ECOLECON.2016.01.015.

Hugo, G.J., 2011: Lessons from past forced resettlement for climate change migration. In: E. Piguet, A. Pécoud and P. de Guchteneire (eds.), *Migration and Climate Change*, UNESCO Publishing/Cambridge University Press, pp. 260–288.

Huisheng, S., 2015: Between the formal and informal: Institutions and village governance in rural China. *An Int. J.*, **13**, 24–44. https://muse.jhu.edu/article/589970.

Humpenöder, F. et al., 2017: Large-scale bioenergy production: How to resolve sustainability trade-offs? *Environ. Res. Lett.*, **13**, 1–15, doi:10.1088/1748-9326/aa9e3b.

Hunsberger, C. et al., 2017: Climate change mitigation, land grabbing and conflict: Towards a landscape-based and collaborative action research agenda. *Can. J. Dev. Stud.*, **38**, 305–324, doi:10.1080/02255189.2016.1 250617.

Hunzai, K., T. Chagas, L. Gilde, T. Hunzai, and N. Krämer, 2018: *Finance Options and Instruments for Ecosystem-Based Adaptation. Overview and Compilation of Ten Examples.* Deutsche Gesellschaft für Internationale Zusammenarbeit (GIZ) GmbH, Bonn, Germany, 76 pp.

Hurlbert, M., 2015a: Climate justice: A call for leadership. *Environ. Justice*, **8**, 51–55, doi:10.1089/env.2014.0035.

Hurlbert, M., 2015b: Learning, participation, and adaptation: Exploring agri-environmental programmes. *J. Environ. Plan. Manag.*, **58**, 113–134, doi:10.1080/09640568.2013.847823.

Hurlbert, M., 2018a: The challenge of integrated flood risk governance: Case studies in Alberta and Saskatchewan, Canada. *Int. J. River Basin Manag.*, **16**, 287–297, doi:10.1080/15715124.2018.1439495.

Hurlbert, M., and J. Pittman, 2014: Exploring adaptive management in environmental farm programs in Saskatchewan, Canada. *J. Nat. Resour. Policy Res.*, **6**, 195–212, doi:10.1080/19390459.2014.915131.

Hurlbert, M., and J. Gupta, 2015: The split ladder of participation: A diagnostic, strategic, and evaluation tool to assess when participation is necessary. *Environ. Sci. Policy*, **50**, 100–113, doi:10.1016/j.envsci.2015.01.011.

Hurlbert, M., and J. Gupta, 2016: Adaptive governance, uncertainty, and risk: Policy framing and responses to climate change, drought, and flood. *Risk Anal.*, **36**, 339–356, doi:10.1111/risa.12510.

Hurlbert, M., and P. Mussetta, 2016: Creating resilient water governance for irrigated producers in Mendoza, Argentina. *Environ. Sci. Policy*, **58**, 83–94, doi:10.1016/j.envsci.2016.01.004.

Hurlbert, M., and J. Gupta, 2017: The adaptive capacity of institutions in Canada, Argentina, and Chile to droughts and floods. *Reg. Environ. Chang.*, **17**, 865–877, doi:10.1007/s10113-016-1078-0.

Hurlbert, M.A., 2018b: *Adaptive Governance of Disaster: Drought and Flood in Rural Areas.* Springer, Cham, Switzerland, 258 pp, DOI: 10.1007/978-3-319-57801-9.

Hurlbert, M.A., and H. Diaz, 2013: Water governance in Chile and Canada: A comparison of adaptive characteristics. *Ecol. Soc.*, **18**, 61, doi:10.5751/ES-06148-180461.

Hurlbert, M., and E. Montana, 2015: Dimensions of adaptive water governance and drought in Argentina and Canada. *J. Sustain. Dev.*, **8**, 120–137, doi:10.5539/jsd.v8n1p120.

Hurlbert, M.A., and J. Gupta, 2018: An institutional analysis method for identifying policy instruments facilitating the adaptive governance of drought. *Environ. Sci. Policy*, **93**, 221–231, doi:10.1016/j. envsci.2018.09.017.

Hurlimann, A.C., and A.P. March, 2012: The role of spatial planning in adapting to climate change. *Wiley Interdiscip. Rev. Clim. Chang.*, **3**, 477–488, doi:10.1002/wcc.183.

Huttunen, S., P. Kivimaa, and V. Virkamäki, 2014: The need for policy coherence to trigger a transition to biogas production. *Environ. Innov. Soc. Transitions*, **12**, 14–30, doi:10.1016/j.eist.2014.04.002.

Huyer, S., J. Twyman, M. Koningstein, J. Ashby, and S. Vermeulen, 2015a: Supporting women farmers in a changing climate: Five policy lessons. CCAFS Policy Brief 10, CGIAR Research Program on Climate Change, Agriculture and Food Security (CCAFS), 8 pp.

Iacobuta, G., N.K. Dubash, P. Upadhyaya, M. Deribe, and N. Höhne, 2018: National climate change mitigation legislation, strategy and targets: A global update. *Clim. Policy*, 18, 1114–1132, doi:10.1080/14693062.20 18.1489772.

ICCC, 2018: *Interim Climate Change Committee Terms of Reference and Appointment.* Ministry for the Environment, Wellington, New Zealand, 7 pp.

ICSU, 2017: *A Guide to SDG Interactions: From Science to Implentation.* International Science Council, Paris, France, 239 pp.

IEA, 2017: *World Energy Outlook 2017.* International Energy Agency, Paris, France, 753 pp.

Ighodaro, I.D., F.S. Lategan, and W. Mupindu, 2016: The impact of soil erosion as a food security and rural livelihoods risk in South Africa. *J. Agric. Sci.*, **8**, 1, doi:10.5539/jas.v8n8p1.

Iglesias, A., and L. Garrote, 2015: Adaptation strategies for agricultural water management under climate change in Europe. *Agric. Water Manag.*, 155, 113–124, doi:10.1016/j.agwat.2015.03.014.

Iizumi, T. et al., 2013: Prediction of seasonal climate-induced variations in global food production. *Nat. Clim. Chang.*, 3, 904–908, doi:10.1038/nclimate1945.

Innocenti, D., and P. Albrito, 2011: Reducing the risks posed by natural hazards and climate change: The need for a participatory dialogue between the scientific community and policy makers. *Environ. Sci. Policy*, 14, 730–733, doi:10.1016/j.envsci.2010.12.010.

Internal Displacement Monitoring Center, 2017: Global Disaster Displacement Risk – A Baseline for Future Work. Internal Displacement Monitoring Centre (IDMC), Geneva, Switzerland, 40 pp.

IPBES, 2018: *The Assessment Report on Land Degradation and Restoration.* IPBES Secretariat, Bonn, Germany, 744 pp.

IPCC, 2000: *Methodological and Technological Issues in Technology Transfer* [Metz, B., O. Davidson, J.-W. Martens, S. Van Rooijen, and L. Van Wie Mcgrory (eds.)]. Cambridge University Press, Cambridge, UK, 466 pp.

IPCC, 2001: Climate Change 2001: Impacts, Adaptation and Vulnerability. In: Contribution of Working Group II to the Third Assessment Report of the Intergovernmental Panel on Climate Change [McCarthy, J.J., O.F. Canziani, N.A. Leary, D.J. Dokken and K.S. White (eds)]. Cambridge University Press, Cambridge, UK, and New York, USA, pp. 1032.

IPCC, 2012: Managing the Risks of Extreme Events and Disasters to Advance Climate Change Adaptation. A Special Report of Working Groups I and II of the Intergovernmental Panel on Climate Change [Field, C.B., V. Barros, T.F. Stocker, D. Qin, D.J. Dokken, K.L. Ebi, M.D. Mastrandrea, K.J. Mach, G.-K. Plattner, S.K. Allen, M. Tignor, and P.M. Midgley (eds.)]. Cambridge University Press, Cambridge, UK, and New York, USA, 594 pp.

IPCC, 2014a: Climate Change 2014: Impacts, Adaptation, and Vulnerability. Part A: Global and Sectoral Aspects. Contribution of Working Group II to the Fifth Assessment Report of the Intergovernmental Panel on Climate Change [Field, C.B., V.R. Barros, D.J. Dokken, K.J. Mach, M.D. Mastrandrea, T.E. Bilir, M. Chatterjee, K.L. Ebi, Y.O. Estrada, R.C. Genova, B. Girma, E.S. Kissel, A.N. Levy, S. MacCracken, P.R. Mastrandrea, and L.L. White (eds.)]. Cambridge University Press, Cambridge, United Kingdom and New York, NY, USA, 1132 pp.

IPCC, 2014b: Summary for Policymakers. In: Climate Change 2014: Impacts, Adaptation, and Vulnerability. Part A: Global and Sectoral Aspects. Contribution of Working Group II to the Fifth Assessment Report of the Intergovernmental Panel on Climate Change [Field, C.B., V.R. Barros, D.J. Dokken, K.J. Mach, M.D. Mastrandrea, T.E. Bilir, M. Chatterjee, K.L. Ebi, Y.O. Estrada, R.C. Genova, B. Girma, E.S. Kissel, A.N. Levy, S. MacCracken, P.R. Mastrandrea, and L.L. White Field, C.B., V.R. Barros, D.J. Dokken, K. (eds.)]. Cambridge University Press, Cambridge, United Kingdom and New York, NY, USA.

IPCC, 2014c: Climate Change 2014: Synthesis Report. Contribution of Working Groups I, II and III to the Fifth Assessment Report of the Intergovernmental Panel on Climate Change [Pachauri, R.K., and L.A. Meyer (eds.)]. IPCC, Geneva, Switzerland, 151 pp.

IPCC, 2014d: Annex II: Glossary [Agard, J., E.L.F. Schipper, J. Birkmann, M. Campos, C. Dubeux, Y. Nojiri, L. Olsson, B. Osman-Elasha, M. Pelling, M.J. Prather, M.G. Rivera-Ferre, O.C. Ruppel, A. Sallenger, K.R. Smith, A.L. St. Clair, K.J. Mach, M.D. Mastrandrea, and T.E. Bilir (eds.)]. In: Climate Change 2014: Impacts, Adaptation, and Vulnerability. Part B: Regional Aspects. Contribution of Working Group II to the Fifth Assessment Report of the Intergovernmental Panel on Climate Change [Barros, V.R., C.B. Field, D.J. Dokken, M.D. Mastrandrea, K.J. Mach, T.E. Bilir, M. Chatterjee, K.L. Ebi,

7

775

Y.O. Estrada, R.C. Genova, B. Girma, E.S. Kissel, A.N. Levy, S. MacCracken, P.R. Mastrandrea, and L.L. White (eds.)]. Cambridge University Press, Cambridge, United Kingdom and New York, NY, USA, pp. 1757–1776.

IPCC, 2018a: Summary for Policymakers. In: Global Warming of 1.5°C. An IPCC Special Report on the impacts of global warming of 1.5°C above pre-industrial levels and related global greenhouse gas emission pathways, in the context of strengthening the global response to the threat of climate change, sustainable development, and efforts to eradicate poverty [Masson-Delmotte, V., P. Zhai, H.-O. Pörtner, D. Roberts, J. Skea, P.R. Shukla, A. Pirani, W. Moufouma-Okia, C. Péan, R. Pidcock, S. Connors, J.B.R. Matthews, Y. Chen, X. Zhou, M.I. Gomis, E. Lonnoy, T. Maycock, M. Tignor, and T. Waterfield (eds.)]. World Meteorological Organization, Geneva, Switzerland, 32 pp.

IPCC, 2018b: Global Warming of 1.5°C. An IPCC Special Report on the Impacts of Global Warming of 1.5°C Above Pre-Industrial Levels and Related Global Greenhouse Gas Emission Pathways, in the Context of Strengthening the Global Response to the Threat of Climate Change, Sustainable Development, and Efforts to Eradicate Poverty [Masson-Delmotte, V., P. Zhai, H.-O. Pörtner, D. Roberts, J. Skea, P.R. Shukla, A. Pirani, W. Moufouma-Okia, C. Péan, R. Pidcock, S. Connors, J.B. R. Matthews, Y. Chen, X. Zhou, M.I. Gomis, E. Lonnoy, T. Maycock, M. Tignor, and T. Waterfield (eds.)]. In Press.

ISEAL Alliance, 2018: *Private Sustainability Standards and the EU Renewable Energy Directive*. ISEAL Alliance, London, UK, www.isealalliance.org/impacts-and-benefits/case-studies/private-sustainability-standards-and-eu-renewable-energy.

Ishida, H. et al., 2014: Global-scale projection and its sensitivity analysis of the health burden attributable to childhood undernutrition under the latest scenario framework for climate change research. *Environ. Res. Lett.*, **9**, 064014, doi:10.1088/1748-9326/9/6/064014.

Ismail, F. et al., 2017: *Market Trends in Family and General Takaful*. MILLIMAN, Washington, DC, USA.

ISO, 2009: *Australia and New Zealand Risk Management Standards 31000:2009*. International Organization for Standardization, ISO Central Secretariat, Geneva, Switzerland.

ISO, 2015: *ISO 13065:2015 – Sustainability Criteria for Bioenergy*. International Organization for Standardization, ISO Central Secretariat, Geneva, Switzerland, 57 pp.

ISO, 2017: *Environmental Management – Guidelines for Establishing Good Practices for Combatting Land Degradation and Desertification – Part 1: Good Practices Framework*. International Organization for Standardization, ISO Central Secretariat, Geneva, Switzerland, 31 pp.

IWAI, 2016: *Consolidated Environmental Impact Assessment Report of National Waterways-1 : Volume – 3*. Environmental Impact Assessment Reports, World Bank, Washington, DC, USA.

Iyahen, E., and J. Syroka, 2018: Managing risks from climate change on the African continent: The African risk capacity (arc) as an innovative risk financing mechanism. In: *Resilience: The Science of Adaptation to Climate Change* [Zommers, Z., and K. Alverson (eds.)]. Elsevier.

Jagger, P., and J. Pender, 2006: Influences of programs and organizations on the adoption of sustainable land management technologies in Uganda. In: *Strategies for Sustainable Land Management in the East African Highlands* [Pender, J., F. Place, S. Ehui, (eds.)]. International Food Policy Research Institute, Washington, DC, USA, pp. 277–306.

Jagger, P., and N. Kittner, 2017: Deforestation and biomass fuel dynamics in Uganda. *Biomass and Bioenergy*, **105**, 1–9, doi:10.1016/j.biombioe.2017.06.005.

Jagger, P., and I. Das, 2018: Implementation and scale-up of a biomass pellet and improved cookstove enterprise in Rwanda. *Energy Sustain. Dev.*, **46**, 32–41, doi:10.1016/j.esd.2018.06.005.

Jalbert, K., and A.J. Kinchy, 2016: Sense and influence: Environmental monitoring tools and the power of citizen science. *J. Environ. Policy Plan.*, **18**, 379–397, doi:10.1080/1523908X.2015.1100985.

Jaleta, M., M. Kassie, and B. Shiferaw, 2013: Tradeoffs in crop residue utilization in mixed crop-livestock systems and implications for conservation agriculture. *Agric. Syst.*, **121**, 96–105, doi:10.1016/j.agsy.2013.05.006.

Jaleta, M., M. Kassie, and O. Erenstein, 2015: Determinants of maize stover utilization as feed, fuel and soil amendment in mixed crop-livestock systems, Ethiopia. *Agric. Syst.*, **134**, 17–23, doi:10.1016/j.agsy.2014.08.010.

James, J., R. Harrison, J. James, and R. Harrison, 2016: the effect of harvest on forest soil carbon: A meta-analysis. *Forests*, **7**, 308, doi:10.3390/f7120308.

James, R., R. Washington, C.F. Schleussner, J. Rogelj, and D. Conway, 2017: Characterizing half-a-degree difference: A review of methods for identifying regional climate responses to global warming targets. *Wiley Interdiscip. Rev. Clim. Chang.*, **8**, e457, doi:10.1002/wcc.457.

Janetos, A., C. Justice, M. Jahn, M. Obersteiner, J. Glauber, and W. Mulhern, 2017: *The Risks of Multiple Breadbasket Failures in the 21st Century: A Science Research Agenda*. The Frederick S. Pardee Center for the Study of the Longer-Range Future, Massachusetts, USA, 24 pp.

Janif, S.Z., P.D. Nunn, P. Geraghty, W. Aalbersberg, F.R. Thomas, and M. Camailakeba, 2016: Value of traditional oral narratives in building climate-change resilience: Insights from rural communities in Fiji. *Ecol. Soc.*, **21**, art7, doi:10.5751/ES-08100-210207.

Jansujwicz, J.S., A.J. K. Calhoun, and R.J. Lilieholm, 2013: The Maine Vernal Pool Mapping and Assessment Program: Engaging municipal officials and private landowners in community-based citizen science. *Environ. Manage.*, **52**, 1369–1385, doi:10.1007/s00267-013-0168-8.

Jantarasami, L.C., J.J. Lawler, and C.W. Thomas, 2010: Institutional barriers to climate change adaptation in US National parks and forests. *Ecol. Soc.*, **15**, 33, doi:10.5751/ES-03715-150433.

Jeffrey, S.R., D.E. Trautman, and J.R. Unterschultz, 2017: Canadian agricultural business risk management programs: Implications for farm wealth and environmental stewardship. *Can. J. Agric. Econ. Can. d'agroeconomie*, **65**, 543–565, doi:10.1111/cjag.12145.

Jelsma, I., M. Slingerland, K. Giller, J.B.-J. of R. Studies, 2017: Collective action in a smallholder oil palm production system in Indonesia: The key to sustainable and inclusive smallholder palm oil? *Journal of Rural Studies*, **54**, 198–210, doi:10.1016/j.jrurstud.2017.06.005.

Jepson, E.J., and A.L. Haines, 2014: Zoning for sustainability: A review and analysis of the zoning ordinances of 32 cities in the United States. *J. Am. Plan. Assoc.*, **80**, 239–252, doi:10.1080/01944363.2014.981200.

Jeuland, M., S.K. Pattanayak, and R. Bluffstone, 2015: The economics of household air pollution. *Annu. Rev. Resour. Econ.*, **7**, 81–108, doi:10.1146/annurev-resource-100814-125048.

Jiang, J., W. Wang, C. Wang, and Y. Liu, 2017: Combating climate change calls for a global technological cooperation system built on the concept of ecological civilization. *Chinese J. Popul. Resour. Environ.*, **15**, 21–31, doi:10.1080/10042857.2017.1286145.

Jjemba, E.W., B.K. Mwebaze, J. Arrighi, E. Coughlan de Perez, and M. Bailey, 2018: Forecast-based financing and climate change adaptation: Uganda makes history using science to prepare for floods. In: *Resilience: The Science of Adaptation to Climate Change* [Alverson, K. and Z. Zommers (eds.)]. Elsevier, Oxford, UK, pp. 237–243.

Joana Specht, M. et al., 2015a: Burning biodiversity: Fuelwood harvesting causes forest degradation in human-dominated tropical landscapes. *Glob. Ecol. Conserv.*, **3**, 200–209, doi:10.1016/j.gecco.2014.12.002.

Johnson, B.B., and M.L. Becker, 2015: Social-ecological resilience and adaptive capacity in a transboundary ecosystem. *Soc. Nat. Resour.*, **28**, 766–780, doi:10.1080/08941920.2015.1037035.

Johnson, F.A., 2011a: Learning and adaptation in the management of waterfowl harvests. *J. Environ. Manage.*, **92**, 1385–1394, doi:10.1016/j.jenvman.2010.10.064.

Johnson, F.X., 2011b: Regional-global linkages in the energy-climate-development policy nexus: The case of biofuels in the EU Renewable energy directive. *Renew. Energy Law Policy Rev.*, **2**, 91–106, doi:10.2307/24324724.

Johnson, F.X., 2017: Biofuels, bioenergy and the bioeconomy in North and South. *Ind. Biotechnol.*, **13**, 289–291, doi:10.1089/ind.2017.29106.fxj.

Johnson, F.X., and S. Silveira, 2014: Pioneer countries in the transition to alternative transport fuels: Comparison of ethanol programmes and policies in Brazil, Malawi and Sweden. *Environ. Innov. Soc. Transitions*, **11**, 1–24, doi:10.1016/j.eist.2013.08.001.

Johnson, F.X., H. Pacini, and E. Smeets, 2012: Transformations In EU biofuels markets under the Renewable Energy Directive and the implications for land use, trade and forests. CIFOR, Bogor, Indonesia.

Johnson, M.F. et al., 2014: Network environmentalism: Citizen scientists as agents for environmental advocacy. *Glob. Environ. Chang.*, **29**, 235–245, doi:10.1016/J.GLOENVCHA.2014.10.006.

Gray, S.et al., 2017: Combining participatory modelling and citizen science to support volunteer conservation action. *Biol. Conserv.*, **208**, 76–86, doi:10.1016/j.biocon.2016.07.037.

Jolly, W.M., M.A. Cochrane, P.H. Freeborn, Z.A. Holden, T.J. Brown, G.J. Williamson, and D.M. J.S. Bowman, 2015: Climate-induced variations in global wildfire danger from 1979 to 2013. *Nat. Commun.*, **6**, 7537, doi:10.1038/ncomms8537.

Jones, A., and B. Hiller, 2017: Exploring the dynamics of responses to food production shocks. *Sustainability*, **9**, 960, doi:10.3390/su9060960.

Jones, D., C.M. Ryan, and J. Fisher, 2016a: Charcoal as a diversification strategy: The flexible role of charcoal production in the livelihoods of smallholders in central Mozambique. *Energy for Sustainable Development*, **32**, 14–21, doi:10.1016/j.esd.2016.02.009.

Jones, N., and E. Presler-Marshall, 2015: Cash transfers. In: *International Encyclopedia of the Social & Behavioral Sciences: Second Edition*. Elsevier.

Jones, R.N. A. Patwardhan, S.J. Cohen, S. Dessai, A. Lammel, R.J. Lempert, M.M.Q. Mirza, and H. von Storch, 2014: Foundations for Decision-Making. In: Climate Change 2014: Impacts, Adaptation, and Vulnerability. Part A: Global and Sectoral Aspects. Contribution of Working Group II to the Fifth Assessment Report of the Intergovernmental Panel on Climate Change [Field, C.B., V.R. Barros, D.J. Dokken, K.J. Mach, M.D. Mastrandrea, T.E. Bilir, M. Chatterjee, K.L. Ebi, Y.O. Estrada, R.C. Genova, B. Girma, E.S. Kissel, A.N. Levy, S. MacCracken, P.R. Mastrandrea, and L.L. White (eds.)]. Cambridge University Press, Cambridge, United Kingdom and New York, NY, USA, pp. 195–228.

Jongman, B. et al., 2014: Increasing stress on disaster-risk finance due to large floods. *Nat. Clim. Chang.*, **4**, 264–268, doi:10.1038/nclimate2124.

Jordan, A.J. et al., 2015a: Emergence of polycentric climate governance and its future prospects. *Nat. Clim. Chang.*, **5**, 977–982, doi:10.1038/nclimate2725.

Jordan, R., A. Crall, S. Gray, T. Phillips, and D. Mellor, 2015b: Citizen science as a distinct field of inquiry. *Bioscience*, **65**, 208–211, doi:10.1093/biosci/biu217.

Jost, C. et al., 2016: Understanding gender dimensions of agriculture and climate change in smallholder farming communities. *Clim. Dev.*, **8**, 133–144, doi:10.1080/17565529.2015.1050978.

Juhola, S., E. Glaas, B.O. Linnér, and T.S. Neset, 2016: Redefining maladaptation. *Environ. Sci. Policy*, **55**, 135–140, doi:10.1016/j.envsci.2015.09.014.

Jumani, S., S. Rao, S. Machado, and A. Prakash, 2017: Big concerns with small projects: Evaluating the socio-ecological impacts of small hydropower projects in India. *Ambio*, **46**, 500–511, doi:10.1007/s13280-016-0855-9.

Jumani, S. et al., 2018: Fish community responses to stream flow alterations and habitat modifications by small hydropower projects in the Western Ghats biodiversity hotspot, India. *Aquat. Conserv. Mar. Freshw. Ecosyst.*, **28**, 979–993.

Junior, S., S.R. Santos, M. Travassos, and M. Vianna, 2012: Impact on a fish assemblage of the maintenance dredging of a navigation channel in a tropical coastal ecosystem. *Brazilian J. Oceanogr.*, **60**, 25–32, doi:10.1590/S1679-87592012000100003.

Júnior, W.S. F., F.R. Santoro, I. Vandebroek, and U.P. Albuquerque, 2016: Urbanization, modernization, and nature knowledge. In: *Introduction to Ethnobiology* [Albuquerque, U.P., R.R. N.Alves (eds.)]. Springer International Publishing, Cham, Switzerland, pp. 251–256, doi:10.1007/978-3-319-28155-1.

Jürisoo, M., F. Lambe, and M. Osborne, 2018: Beyond buying: The application of service design methodology to understand adoption of clean cookstoves in Kenya and Zambia. *Energy Res. Soc. Sci.*, **39**, 164–176, doi:10.1016/j.erss.2017.11.023.

Kabeer, N., K. Mumtaz, and A. Sayeed, 2010: Beyond risk management: Vulnerability, social protection and citizenship in Pakistan. *J. Int. Dev.*, **22**, 1–19, doi:10.1002/jid.1538.

Kaenzig, R., and E. Piguet, 2014: Migration and climate change in Latin America and the Caribbean. In: *People on the Move in a Changing Climate. The Regional Impact of Environmental Change on Migration* [Piguet, E., F. Laczko (eds.)]. Springer Netherlands, Dordrecht, Netherlands, pp. 253.

Kaijser, A., and A. Kronsell, 2014: Climate change through the lens of intersectionality. *Env. Polit.*, **23**, 417–433, doi:10.1080/09644016.2013.835203.

Kainuma, M., K. Miwa, T. Ehara, O. Akashi, and Y. Asayama, 2013: A low-carbon society: Global visions, pathways, and challenges. *Clim. Policy*, **13**, 5–21, doi:10.1080/14693062.2012.738016.

Kaisa, K.K. et al., 2017: Analyzing REDD+ as an experiment of transformative climate governance: Insights from Indonesia. *Environ. Sci. Policy*, **73**, 61–70, doi:10.1016/j.envsci.2017.03.014.

Kakota, T., D. Nyariki, D. Mkwambisi, and W. Kogi-Makau, 2011: Gender vulnerability to climate variability and household food insecurity. *Clim. Dev.*, **3**, 298–309, doi:10.1080/17565529.2011.627419.

Kale, E., 2017: Problematic uses and practices of farm ponds in Maharashtra. *Econ. Polit. Wkly.*, **52**, 20–22.

Kallbekken, S., and H. Sælen, 2013: 'Nudging' hotel guests to reduce food waste as a win-win environmental measure. *Econ. Lett.*, **119**, 325–327, doi:10.1016/j.econlet.2013.03.019.

Kangalawe, R.Y.M, Noe. C, Tungaraza. F.S.K, G. Naimani, M. Mlele, 2014: Understanding of traditional knowledge and indigenous institutions on sustainable land management in Kilimanjaro region, Tanzania. *Open J. Soil Sci.*, **4**, 469–493, doi:10.4236/ojss.2014.413046.

Kanta Kafle, S., 2017: Disaster early warning systems in Nepal: Institutional and operational frameworks. *J. Geogr. Nat. Disasters*, doi:10.4172/2167-0587.1000196.

Kanter, D.R. et al., 2016: Evaluating agricultural trade-offs in the age of sustainable development. *Agric. Syst.*, **163**, 73–88, doi:10.1016/J.AGSY.2016.09.010.

Karabulut, A.A., E. Crenna, S. Sala, and A. Udias, 2018: A proposal for integration of the ecosystem-water-food-land-energy (EWFLE) nexus concept into life cycle assessment: A synthesis matrix system for food security. *J. Clean. Prod.*, **172**, 3874–388, doi:10.1016/j.jclepro.2017.05.092.

Karar, E., and I. Jacobs-Mata, 2016: Inclusive governance: The role of knowledge in fulfilling the obligations of citizens. *Aquat. Procedia*, **6**, 15–22, doi:10.1016/j.aqpro.2016.06.003.

Karim, M.R., and A. Thiel, 2017: Role of community based local institution for climate change adaptation in the Teesta riverine area of Bangladesh. *Clim. Risk Manag.*, **17**, 92–103 doi:10.1016/j.crm.2017.06.002.

Karlberg, L. et al., 2015: Tackling Complexity: Understanding the Food-Energy-Environment Nexus in Ethiopia's Lake Tana Sub-basin. *Water Altern.*, **8**, 710–734.

Karnib, A., 2017: A quantitative nexus approach to analyse the interlinkages across the sustainable development goals. *J. Sustain. Dev.*, **10**, 173–180, doi:10.5539/jsd.v10n5p173.

Karpouzoglou, T., A. Dewulf, and J. Clark, 2016: Advancing adaptive governance of social-ecological systems through theoretical multiplicity. *Environ. Sci. Policy*, **57**, 1–9, doi:10.1016/j.envsci.2015.11.011.

Kasperson, R.E., 2012: Coping with deep uncertainty: Challenges for environmental assessment and decision-making. In: *Uncertainty and Risk: Multidisciplinary Perspectives* [Bammer, G., and M. Smithson (ed.)].

Earthscan Risk in Society Series, London, UK, pp. 382, doi:10.1111/j.1468-5973.2009.00565.x.

Kates, R.W., W.R. Travis, and T.J. Wilbanks, 2012: Transformational adaptation when incremental adaptations to climate change are insufficient. *Proc. Natl Acad. Sci. Usa*, **109**, 7156–7161.

Kaufmann, D., A. Kraay, M. Mastruzzi, 2009: Governance Matters VIII Aggregate and Individual Governance Indicators 1996–2008 (English). Policy Research Working Paper No. WPS 4978. World Bank, Washington, DC, USA, doi:10.1080/713701075.

Kaval, P., J. Loomis, and A. Seidl, 2007: Willingness-to-pay for prescribed fire in the Colorado (USA) wildland urban interface. *For. Policy Econ.*, **9**, 928–937.

Keenan, R.J., 2015: Climate change impacts and adaptation in forest management: A review. *Ann. For. Sci.*, **72**, 145–167, doi:10.1007/s13595-014-0446-5.

Kelkar, N., 2016: Digging our rivers' graves?*Dams, Rivers, People Newsl.*, **14**, 1–6.

Kelley, C.P., S. Mohtadi, M.A. Cane, R. Seager, and Y. Kushnir, 2015: Climate change in the Fertile Crescent and implications of the recent Syrian drought. *Proc. Natl. Acad. Sci.*, **112**, 3241–3246, doi:10.1073/pnas.1421533112.

Kemp, R., S. Parto, and R. Gibson, 2005: Governance for sustainable development: Moving from theory to practice. *International J. Sustain. Dev.*, **8**, doi:10.1504/IJSD.2005.007372.

Kern, F., and M. Howlett, 2009: Implementing transition management as policy reforms: A case study of the Dutch energy sector. *Policy Sci.*, **42**, 391–408, doi:10.1007/s11077-009-9099-x.

Kerr, S., and A. Sweet, 2008: Inclusion of agriculture into a domestic emissions trading scheme: New Zealand's experience to date. *Farm Policy J.*, **5**.

Kesternich, M., C. Reif, and D. Rübbelke, 2017: Recent trends in behavioral environmental economics. *Environ. Resour. Econ.*, **67**, 403–411, doi:10.1007/s10640-017-0162-3.

Khan, M.R., and J.T. Roberts, 2013: Adaptation and international climate policy. *Wiley Interdiscip. Rev. Clim. Chang.*, **4**, 171–189, doi:10.1002/wcc.212.

Khandelwal, M. et al., 2017: Why have improved cook-stove initiatives in india failed? *World Dev.*, **92**, 13–27, doi:10.1016/j.worlddev.2016.11.006.

Kibler, K.M., and D.D. Tullos, 2013: Cumulative biophysical impact of small and large hydropower development in Nu River, China. *Water Resour. Res.*, **49**, 3104–3118.

Kiendrebeogo, Y., K. Assimaidou, and A. Tall, 2017: Social protection for poverty reduction in times of crisis. *J. Policy Model.*, **39**, 1163–1183, doi:10.1016/j.jpolmod.2017.09.003.

Kim, K., T. Park, S. Bang, and H. Kim, 2017: Real Options-based framework for hydropower plant adaptation to climate change. *J. Manag. Eng.*, **33**, 04016049, doi:10.1061/ (ASCE)ME.1943-5479.0000496.

Kimmerer, R.W., 2000: Native knowledge for native ecosystems. *J. For.*, **98**, 4–9, doi:10.1093/jof/98.8.4.

Kiruki, H.M., E.H. van der Zanden, Ž. Malek, and P.H. Verburg, 2017a: Land cover change and woodland degradation in a charcoal producing semi-arid area in Kenya. *L. Degrad. Dev.*, **28**, 472–481, doi:10.1002/ldr.2545.

Kissinger, G., M. Herold, and V. De Sy, 2012: *Drivers of Deforestation and Forest Degradation: A Synthesis Report for REDD + Policymakers*. Lexeme Consulting, Vancouver, Canada, 48 pp.

Kissinger, G., A. Gupta, I. Mulder, and N. Unterstell, 2019: Climate financing needs in the land sector under the Paris Agreement: An assessment of developing country perspectives. *Land Use Policy*, **83**, 256–269, doi:10.1016/j.landusepol.2019.02.007.

Kivimaa, P., and F. Kern, 2016: Creative destruction or mere niche support? Innovation policy mixes for sustainability transitions. *Res. Policy*, **45**, 205–217, doi:10.1016/j.respol.2015.09.008.

Kivimaa, P., M. Hildén, D. Huitema, A. Jordan, and J. Newig, 2017a: Experiments in climate governance – A systematic review of research on energy and built environment transitions. *J. Clean. Prod.*, **169**, 17–29, doi:10.1016/j.jclepro.2017.01.027.

Kivimaa, P., H.L. Kangas, and D. Lazarevic, 2017b: Client-oriented evaluation of 'creative destruction' in policy mixes: Finnish policies on building energy efficiency transition. *Energy Research and Social Science*, **33**, 115–127, doi:10.1016/j.erss.2017.09.002.

Klein, R.J.T., G.F. Midgley, B.L. Preston, M. Alam, F.G.H. Berkhout, K.D., and M. Shaw, 2014: Adaptation Opportunities, Constraints, and Limits. In: Climate Change 2014: Impacts, Adaptation, and Vulnerability. Part A: Global and Sectoral Aspects. Contribution of Working Group II to the Fifth Assessment Report of the Intergovernmental Panel on Climate Change [Field, C.B., V.R. Barros, D.J. Dokken, K.J. Mach, M.D. Mastrandrea, T.E. Bilir, M. Chatterjee, K.L. Ebi, Y.O. Estrada, R.C. Genova, B. Girma, E.S. Kissel, A.N. Levy, S. MacCracken, P. Mastreanda, and L. White (eds.)]. Cambridge University Press, Cambridge, UK, and New York, NY, USA, 899–943.

Kleindorfer, P.R., H. Kunreuther, and C. Ou-Yang, 2012: Single-year and multi-year insurance policies in a competitive market. *J. Risk Uncertain.*, **45**, 51–78, doi:10.1007/s11166-012-9148-2.

Kline, K.L. et al., 2017: Reconciling food security and bioenergy: Priorities for action. *GCB Bioenergy*, **9**, 557–576, doi:10.1111/gcbb.12366.

Knoke, T., K. Messerer, and C. Paul, 2017: The role of economic diversification in forest ecosystem management. *Curr. For. Reports*, **3**, 93–106, doi:10.1007/s40725-017-0054-3.

Knook, J., V. Eory, M. Brander, and D. Moran, 2018: Evaluation of farmer participatory extension programmes. *J. Agric. Educ. Ext.*, **24**, 309–325, doi:10.1080/1389224X.2018.1466717.

Knorr, W., A. Arneth, and L. Jiang, 2016a: Demographic controls of future global fire risk. *Nat. Clim. Chang.*, **6**, 781–785, doi:10.1038/nclimate2999.

Koechlin, L., J. Quan, and H. Mulukutla, 2016: *Tackling corruption in land governance*. A LEGEND Analytical paper.

Koh, I., S. Kim, and D. Lee, 2010: Effects of bibosoop plantation on wind speed, humidity, and evaporation in a traditional agricultural landscape of Korea: Field measurements and modeling. *Agric. Ecosyst. Environ.*, **135**, 294–303, doi:10.1016/J.AGEE.2009.10.008.

Koizumi, T., 2014: Biofuels and food security. SpringerBriefs in Applied Sciences and Technology.

Kolstad, C., K. Urama, J. Broome, A. Bruvoll, M. Cariño Olvera, D. Fullerton, C. Gollier, W.M. Hanemann, R. Hassan, F. Jotzo, M.R. Khan, L. Meyer, and L. Mundaca, 2014: Social, Economic and Ethical Concepts and Methods. In: Climate Change 2014: Mitigation of Climate Change. Contribution of Working Group III to the Fifth Assessment Report of the Intergovernmental Panel on Climate Change [Edenhofer, O., R. Pichs-Madruga, Y. Sokona, E. Farahani, S. Kadner, K. Seyboth, A. Adler, I. Baum, S. Brunner, P. Eickemeier, B. Kriemann, J. Savolainen, S. Schlömer, C. von Stechow, T. Zwickel and J.C. Minx (eds.)]. Cambridge University Press, Cambridge, United Kingdom and New York, NY, USA.

Kompas, T., V.H. Pham, and T.N. Che, 2018: The effects of climate change on GDP by country and the global economic gains from complying with the Paris climate accord. *Earth's Futur.*, **6**, 1153–1173, doi:10.1029/2018EF000922.

Koohafkan, P., and M.A. Altieri, 2011: *Globally Important Agricultural Heritage Systems A Legacy for the Future GIAHS Globally Important Agricultural Heritage Systems*. Food and Agriculture Organization of the United Nations, Rome, Italy.

Koontz, T.M., D. Gupta, P. Mudliar, and P. Ranjan, 2015: Adaptive institutions in social-ecological systems governance: A synthesis framework. *Environ. Sci. Policy*, 53, 139–151, doi:10.1016/j.envsci.2015.01.003.

Van Koppen, B., L. Hope, and W. Colenbrander, 2013a: *Gender Aspects of Small-scale Private Irrigation in Africa, IWMI Working Paper*. International Water Management Institute, Colombo, Sri Lanka.

Kousky, C., and R. Cooke, 2012: Explaining the failure to insure catastrophic risks. *Geneva Pap. Risk Insur. – Issues Pract.*, **37**, 206–227, doi:10.1057/gpp.2012.14.

Kousky, C., E.O. Michel-Kerjan, and P.A. Raschky, 2018a: Does federal disaster assistance crowd out.

Krätli, S., 2008: Cattle breeding, complexity and mobility in a structurally unpredictable environment: The WoDaaBe herders of Niger. *Nomad. Peoples*, **12**, 11–41, doi:10.3167/np.2008.120102.

Krätli, S., and N. Schareika, 2010: Living off uncertainty: The intelligent animal production of dryland pastoralists. *Eur. J. Dev. Res.*, **22**, 605–622, doi:10.1057/ejdr.2010.41.

Krause, A. et al., 2017: Global consequences of afforestation and bioenergy cultivation on ecosystem service indicators. *Biogeosciences*, **14**, 4829–4850, doi:10.5194/bg-14-4829-2017.

Krause, A. et al., 2018: Large uncertainty in carbon uptake potential of land-based climatechange mitigation efforts. *Glob. Chang. Biol.*, **24**, 3025–3038, doi:10.1111/gcb.14144.

Krishnaswamy, J., R. John, and S. Joseph, 2014: Consistent response of vegetation dynamics to recent climate change in tropical mountain regions. *Glob. Chang. Biol.*, **20**, 203–215, doi:10.1111/gcb.12362.

Kristjanson, P., A. Waters-Bayer, N. Johnson, A. Tipilda, J. Njuki, I. Baltenweck, D. Grace, and S. MacMillan, 2014: Livestock and women's livelihoods. In: *Gender in Agriculture: Closing the Knowledge Gap* [Quisumbing, A.R., R. Meinzen-Dick, T.L. Raney, A. Croppenstedt, J.A. Behrman, and A. Peterman (eds.)]. International Food Policy Research Institute, Washington, DC, USA, 209–234 pp.

Krug, J.H. A., 2018: Accounting of GHG emissions and removals from forest management: A long road from Kyoto to Paris. *Carbon Balance Manag.*, **13**, 1, doi:10.1186/s13021-017-0089-6.

Kuhfuss, L., R. Préget, S. Thoyer, N. Hanley, P. Le Coent, and M. Désolé, 2016: Nudges, social norms, and permanence in agri-environmental schemes. *Land Econ.*, **92**, 641–655, doi:10.3368/le.92.4.641.

Kullenberg, C., and D. Kasperowski, 2016: What is citizen science? A scientometric meta-analysis. *PLoS One*, **11**, e0147152, doi:10.1371/journal.pone.0147152.

Kumar Nath, T., T. Kumar Baul, M.M. Rahman, M.T. Islam, and M. Harun-Or-Rashid, 2013a: *Traditional Biomass Fuel Consumption by Rural Households in Degraded Sal (Shorea Robusta) Forest Areas of Bangladesh*. International Journal of Emerging Technology and Advanced Engineering, **3**, 537–544 pp.

Kumpula, T., A. Pajunen, E. Kaarlejärvi, B.C. Forbes, and F. Stammler, 2011: Land use and land cover change in Arctic Russia: Ecological and social implications of industrial development. *Glob. Environ. Chang.*, **21**, 550–562, doi:10.1016/j.gloenvcha.2010.12.010.

Kundzewicz, Z.W., 2002: Non-structural flood protection and sustainability. *Water Int.*, **27**, 3–13, doi:10.1080/02508060208686972.

Kunreuther, H., and R. Lyster, 2016: The role of public and private insurance in reducing losses from extreme weather events and disasters. *Asia Pacific J. Environ. Law*, **19**, 29–54.

Kunreuther, H., S. Gupta, V. Bosetti, R. Cooke, V. Dutt, M. Ha-Duong, H. Held, J. Llanes-Regueiro, A. Patt, E. Shittu, and E. Weber, 2014: Integrated Risk and Uncertainty Assessment of Climate Change Response Policies. In: Climate Change 2014: Mitigation of Climate Change. Contribution of Working Group III to the Fifth Assessment Report of the Intergovernmental Panel on Climate Change [Edenhofer, O., R. Pichs-Madruga, Y. Sokona, E. Farahani, S. Kadner, K. Seyboth, A. Adler, I. Baum, S. Brunner, P. Eickemeier, B. Kriemann, J. Savolainen, S. Schlömer, C. von Stechow, T. Zwickel and J.C. Minx (eds.)]. Cambridge University Press, Cambridge, United Kingdom and New York, NY, USA.

Kuriakose, A.T., R. Heltberg, W. Wiseman, C. Costella, R. Cipryk, and S. Cornelius, 2012: Climate-Responsive Social Protection Climate - responsive Social Protection. Social Protection and Labor Strategy No.1210, World Bank, Washington, DC, USA.

Kurian, M., 2017: The water-energy-food nexus: Trade-offs, thresholds and transdisciplinary approaches to sustainable development. *Environ. Sci. Policy*, **68**, 97–106, doi:10.1016/j.envsci.2016.11.006.

Kust, G., O. Andreeva, and A. Cowie, 2017: Land degradation neutrality: Concept development, practical applications and assessment. *J. Environ. Manage.*, **195**, 16–24, doi:10.1016/j.jenvman.2016.10.043.

Kuzdas, C., A. Wiek, B. Warner, R. Vignola, and R. Morataya, 2015: Integrated and participatory analysis of water governance regimes: The case of the Costa Rican dry tropics. *World Dev.*, **66**, 254–266, doi:10.1016/j.worlddev.2014.08.018.

Kwakkel, J.H., M. Haasnoot, and W.E. Walker, 2016: Comparing robust decision-making and dynamic adaptive policy pathways for model-based decision support under deep uncertainty. *Environ. Model. Softw.*, **86**, 168–183, doi:10.1016/j.envsoft.2016.09.017.

Kweka, E.J., E.E. Kimaro, and S. Munga, 2016: Effect of deforestation and land use changes on mosquito productivity and development in Western Kenya highlands: Implication for malaria risk. *Front. public Heal.*, **4**, 238, doi:10.3389/fpubh.2016.00238.

Laakso, S., A. Berg, and M. Annala, 2017: Dynamics of experimental governance: A meta-study of functions and uses of climate governance experiments. *J. Clean. Prod.*, **169**, 8–16, doi:10.1016/j.jclepro.2017.04.140.

Laczko, F., and E. Piguet, 2014: Regional perspectives on migration, the environment and climate change. In: *People on the Move in an Changing Climate: The Regional Impact of Environmental Change on Migration*. Springer Netherlands, Dordrecht, Netherlands, pp. 253.

Lahiff, E., 2015: The great African land grab? Agricultural investments and the global food system. *J. Peasant Stud.*, **42**, 239–242, doi:10.1080/03066150.2014.978141.

Lal, P.N. et al., 2012: National Systems for Managing the Risks from Climate Extremes and Disasters. In: Managing the Risks of Extreme Events and Disasters to Advance Climate Change Adaptation. A Special Report of Working Groups I and II of the Intergovernmental Panel on Climate Change [Field, C.B., Barros, T.F. Stocker, D. Qin, D.J. Dokken, K.L. Ebi, M.D. Mastrandrea, K.J. Mach, G.-K. Plattner, S.K. Allen, M. Tignor, and P.M. Midgley (eds.)]. Cambridge University Press, Cambridge, UK, and New York, NY, USA, 339–392.

Lal, R., 2013: Food security in a changing climate. *Ecohydrol. Hydrobiol.*, **13**, 8–21, doi:10.1016/j.ecohyd.2013.03.006.

Lam, P.T. I., and A.O. K. Law, 2016: Crowdfunding for renewable and sustainable energy projects: An exploratory case study approach. *Renew. Sustain. Energy Rev.*, **60**, 11–20, doi:10.1016/j.rser.2016.01.046.

Lam, Q.D., B. Schmalz, and N. Fohrer, 2011: The impact of agricultural Best Management Practices on water quality in a North German lowland catchment. *Environ. Monit. Assess.*, **183**, 351–379, doi:10.1007/s10661-011-1926-9.

Lamarque, P., A. Artaux, C. Barnaud, L. Dobremez, B. Nettier, and S. Lavorel, 2013: Taking into account farmers' decision-making to map fine-scale land management adaptation to climate and socio-economic scenarios. *Landsc. Urban Plan.*, **119**, 147–157, doi:10.1016/j.landurbplan.2013.07.012.

Lambin, E.F. et al., 2001: The causes of land-use and land-cover change: Moving beyond the myths. *Glob. Environ. Chang.*, **11**, 261–269, doi:10.1016/S0959-3780 (01)00007-3.

Lambin, E.F. et al., 2014: Effectiveness and synergies of policy instruments for land use governance in tropical regions. *Glob. Environ. Chang.*, **28**, 129–140, doi:10.1016/J.GLOENVCHA.2014.06.007.

Lane, C., and R. Moorehead, 1995: New directions in rangeland and resource tenure and policy. In: *Living with Uncertainty: New Directions in Pastoral Development in Africa* [Scoones, I. (ed.)]. Practical Action Publishing, Warwickshire, UK, pp. 116–133.

Lane, C.R., 1998: *Custodians of the Commons: Pastoral Land Tenure in East and West Africa*. Earthscan, London, UK, 238 pp.

Lane, R., and R. McNaught, 2009: Building gendered approaches to adaptation in the Pacific. *Gend. Dev.*, **17**, 67–80, doi:10.1080/13552070802696920.

Lange, E., and S. Hehl-Lange, 2011: Citizen participation in the conservation and use of rural landscapes in Britain: The Alport Valley case study. *Landsc. Ecol. Eng.*, **7**, 223–230, doi:10.1007/s11355-010-0115-2.

7

Langholtz, M. et al., 2014: Climate risk management for the US cellulosic biofuels supply chain. *Clim. Risk Manag.*, **3**, 96–115, doi:10.1016/j.crm.2014.05.001.

Larcom, S., and T. van Gevelt, 2017: Regulating the water-energy-food nexus: Interdependencies, transaction costs and procedural justice. *Environ. Sci. Policy*, **72**, 55–64, doi:10.1016/j.envsci.2017.03.003.

Larkin, A., J. Kuriakose, M. Sharmina, and K. Anderson, 2018: What if negative emission technologies fail at scale? Implications of the Paris Agreement for big emitting nations. *Clim. Policy*, **18**, 690–714, doi:10.1080/14693062.2017.1346498.

Larsen, P.H. et al., 2008: Estimating future costs for Alaska public infrastructure at risk from climate change. *Glob. Environ. Chang.*, **18**, 442–457, doi:10.1016/j.gloenvcha.2008.03.005.

Larson, A., and J. Pulhin, 2012: Enhancing forest tenure reforms through more responsive regulations. *Conserv. Soc.*, **10**, 103, doi:10.4103/0972-4923.97482.

Lashley, J.G., and K. Warner, 2015: Evidence of demand for microinsurance for coping and adaptation to weather extremes in the Caribbean. *Clim. Change*, **133**, 101–112, doi:10.1007/s10584-013-0922-1.

Laube, W., B. Schraven, and M. Awo, 2012: Smallholder adaptation to climate change: Dynamics and limits in Northern Ghana. *Clim. Change*, **111**, 753–774, doi:10.1007/s10584-011-0199-1.

Lavell, A., M. Oppenheimer, C. Diop, J. Hess, R. Lempert, J. Li, R. Muir-Wood, and S. Myeong, 2012: Climate Change: New Dimensions in Disaster Risk, Exposure, Vulnerability, and Resilience. In: Managing the Risks of Extreme Events and Disasters to Advance Climate Change Adaptation: Special Report of the Intergovernmental Panel on Climate Change [Field, C.B., V. Barros, T.F. Stocker, D. Qin, D.J. Dokken, K.L. Ebi, M.D. Mastrandrea, K.J. Mach, G.-K. Plattner, S.K. Allen, M. Tignor, and P.M. Midgley (eds.)]. A Special Report of Working Groups I and II of the Intergovernmental Panel on Climate Change (IPCC). Cambridge University Press, Cambridge, UK, and New York, NY, USA, pp. 25–64.

Lavers, T., 2012: 'Land grab' as development strategy? The political economy of agricultural investment in Ethiopia. *J. Peasant Stud.*, **39**, 105–132, doi:10.1080/03066150.2011.652091.

Lawrence, J., R. Bell, P. Blackett, S. Stephens, and S. Allan, 2018: National guidance for adapting to coastal hazards and sea-level rise: Anticipating change, when and how to change pathway. *Environ. Sci. Policy*, **82**, 100–107, doi:10.1016/j.envsci.2018.01.012.

Lawrence, J., R. Bell, and A. Stroombergen, 2019: A hybrid process to address uncertainty and changing climate risk in coastal areas using dynamic adaptive pathways planning, multi-criteria decision analysis andreal options analysis: A New Zealand application. *Sustain.*, **11**, 1–18, doi:10.3390/su11020406.

Layke, C., 2009: *Measuring Nature's Benefits: A Preliminary Roadmap for Improving Ecosystem Service Indicators*. World Resources Institute, Washington, DC, USA, 36 pp.

Lazo, J.K., A. Bostrom, R. Morss, J. Demuth, and H. Lazrus, 2014: *Communicating hurricane warnings: Factors affecting protective behavior*. Conference on Risk, Perceptions, and Response, Harvard University, Massachusetts, USA, 33 pp.

Leach, G., 1992: The energy transition. *Energy Policy*, **20**, 116–123, doi:10.1016/0301-4215 (92)90105-B.

Ledec, George C., Rapp, W. Kennan, R.G.Aiello, 2011: *Greening the wind: Environmental and Social Considerations For Wind Power Development*. World Bank, Washington, DC, USA, 172 pp.

Lee, C.M., and M. Lazarus, 2013: Bioenergy projects and sustainable development: Which project types offer the greatest benefits? *Clim. Dev.*, **5**, 305–317, doi:10.1080/17565529.2013.812951.

Lee, M. et al., 2013: Public participation and climate change infrastructure. *J. Environ. Law*, **25**, 33–62, doi:10.1093/jel/eqs027.

Lele, U., M. Klousia-Marquis, and S. Goswami, 2013: Good Governance for Food, Water and Energy Security. *Aquat. Procedia*, 1, 44–63, doi:10.1016/j.aqpro.2013.07.005.

Lelieveld, J., C. Barlas, D. Giannadaki, and A. Pozzer, 2013: Model calculated global, regional and megacity premature mortality due to air pollution. *Atmos. Chem. Phys.*, **13**, 7023–7037, doi:10.5194/acp-13-7023-2013.

Lemann, A.B., 2018: Stronger than the storm: Disaster law in a defiant age. *Louisiana Law Review*, **78**, 437–497.

Lempert, R., 2013: Scenarios that illuminate vulnerabilities and robust responses. *Clim. Change*, **117**, 627–646, doi:10.1007/s10584-012-0574-6.

Lempert, R.J., and M.E. Schlesinger, 2000: Robust strategies for abating climate change. *Clim. Change*, **45**, 387–401, doi:10.1023/A:1005698407365.

Lenderink, G., and E. van Meijgaard, 2008: Increase in hourly precipitation extremes beyond expectations from temperaturechanges. *Nat. Geosci.*, **1**, 511–514, doi:10.1038/ngeo262.

Leonard, S., M. Parsons, K. Olawsky, and F. Kofod, 2013: The role of culture and traditional knowledge in climate change adaptation: Insights from East Kimberley, Australia. *Glob. Environ. Chang.*, **23**, 623–632, doi:10.1016/J.GLOENVCHA.2013.02.012.

Leys, A.J., and J.K. Vanclay, 2011: Social learning: A knowledge and capacity building approach for adaptive co-management of contested landscapes. *Land Use Policy*, **28**, 574–584, doi:10.1016/j.landusepol.2010.11.006.

Li, Z., and H. Fang, 2016a: Impacts of climate change on water erosion: A review. *Earth-Science Rev.*, **163**, 94–117, doi:10.1016/J.EARSCIREV.2016.10.004.

Lilleør, H.B., and K. Van den Broeck, 2011: Economic drivers of migration and climate change in LDCs. *Glob. Environ. Chang.*, **21**, S70–S81, doi:10.1016/j.gloenvcha.2011.09.002.

Lin, B., and X. Li, 2011: The effect of carbon tax on per capita CO_2 emissions. *Energy Policy*, **39**, 5137–5146, doi:10.1016/j.enpol.2011.05.050.

Lin, Y.-P., D. Deng, W.-C. Lin, R. Lemmens, N.D. Crossman, K. Henle, and D.S. Schmeller, 2015: Uncertainty analysis of crowd-sourced and professionally collected field data used in species distribution models of Taiwanese moths. *Biol. Conserv.*, **181**, 102–110, doi:10.1016/J.BIOCON.2014.11.012.

Lindner, M. et al., 2010: Climate change impacts, adaptive capacity, and vulnerability of European forest ecosystems. *For. Ecol. Manage.*, **259**, 698–709, doi:10.1016/J.FORECO.2009.09.023.

Liniger, H., N. Harari, G. van Lynden, R. Fleiner, J. de Leeuw, Z. Bai, and W. Critchley, 2019: Achieving land degradation neutrality: The role of SLM knowledge in evidence-based decision-making. *Environ. Sci. Policy*, **94**, 123–134, doi:10.1016/j.envsci.2019.01.001.

Linnerooth-bayer, J., S. Surminski, L.M. Bouwer, I. Noy, and R. Mechler, 2018: Insurance as a Response to Loss and Damage? In: *Loss and Damage from Climate Change: Concepts, Methods and Policy Options* [Mechler, R., L.M. Bouwer, T. Schinko, S. Surminski, and J. Linnerooth-bayer (eds.)]. SpringerInternational Publishing, Cham, Switzerland, pp. 483–512.

Linnerooth-Bayer, J., and S. Hochrainer-Stigler, 2015: Financial instruments for disaster risk management and climate change adaptation. *Clim. Change*, **133**, 85–100, doi:10.1007/s10584-013-1035-6.

Lipper, L. et al., 2014a: Climate-smart agriculture for food security. *Nat. Clim. Chang.*, **4**, 1068–1072, doi:10.1038/nclimate2437.

Little, P.D., H. Mahmoud, and D.L. Coppock, 2001: When deserts flood: Risk management and climatic processes among East African pastoralists. *Clim. Res.*, **19**, 149–159, doi:10.3354/cr019149.

Liu, J. et al., 2017: Challenges in operationalizing the water-energy-food nexus. *Hydrol. Sci. J.*, **62**, 1714–1720, doi:10.1080/02626667.2017.1353695.

Liu, P., and N. Ravenscroft, 2017: Collective action in implementing top-down land policy: The case of Chengdu, China. *Land Use Policy*, **65**, 45–52, doi:10.1016/J.LANDUSEPOL.2017.03.031.

Loaiza, T., U. Nehren, and G. Gerold, 2015: REDD+ and incentives: An analysis of income generation in forest-dependent communities of the Yasuní Biosphere Reserve, Ecuador. *Appl. Geogr.*, **62**, 225–236, doi:10.1016/J.APGEOG.2015.04.020.

Loaiza, T, M.O. Borja, U. Nehren, and G. Gerold, 2017: Analysis of land management and legal arrangements in the Ecuadorian Northeastern Amazon as preconditions for REDD+ implementation. *For. Policy Econ.*, **83**, 19–28, doi:10.1016/J.FORPOL.2017.05.005.

Loarie, S.R., P.B. Duffy, H. Hamilton, G.P. Asner, C.B. Field, and D.D. Ackerly, 2009: The velocity of climate change. *Nature*, **462**, 1052–1055, doi:10.1038/nature08649.

Lobell, D.B., U.L. C. Baldos, and T.W. Hertel, 2013: Climate adaptation as mitigation: The case of agricultural investments. *Environ. Res. Lett.*, **8**, 1–12, doi:10.1088/1748-9326/8/1/015012.

Locatelli, B., P. Imbach, and S. Wunder, 2014: Synergies and trade-offs between ecosystem services in Costa Rica. *Environ. Conserv.*, **41**, 27–36, doi:10.1017/S0376892913000234 .

Locatelli, B., G. Fedele, V. Fayolle, and A. Baglee, 2016: Synergies between adaptation and mitigation in climate change finance. *Int. J. Clim. Chang. Strateg. Manag.*, **8**, 112–128, doi:10.1108/IJCCSM-07-2014-0088.

Lontzek, T.S., Y. Cai, K.L. Judd, and T.M. Lenton, 2015: Stochastic integrated assessment of climate tipping points indicates the need for strict climate policy. *Nat. Clim. Chang.*, **5**, 441–444, doi:10.1038/nclimate2570.

Loos, J., A.I. Horcea-Milcu, P. Kirkland, T. Hartel, M. Osváth-Ferencz, and J. Fischer, 2015: Challenges for biodiversity monitoring using citizen science in transitioning social-ecological systems. *J. Nat. Conserv.*, **26**, 45–48, doi:10.1016/j.jnc.2015.05.001.

López-i-Gelats, F., E.D. G. Fraser, J.F. Morton, and M.G. Rivera-Ferre, 2016: What drives the vulnerability of pastoralists to global environmental change? A qualitative meta-analysis. *Glob. Environ. Chang.*, **39**, 258–274, doi:10.1016/J.GLOENVCHA.2016.05.011.

Lotze-Campen, H., and A. Popp, 2012: Agricultural adaptation options: Production technology, insurance, trade. In: *Climate Change, Justice and Sustainability* [Edenhofer, O., J. Wallacher, H. Lotze-Campen, M. Reder, B. Knopf (eds.)]. Springer Netherlands, Dordrecht, Netherlands, pp. 171–178.

Louhaichi, M., and A. Tastad, 2010: The Syrian Steppe: Past trends, current status, and future priorities. *Rangelands*, **32**, 2–7, doi:10.2307/40588043.

Lubis, R.F., R. Delinom, S. Martosuparno, and H. Bakti, 2018: Water-food nexus in Citarum Watershed, Indonesia. *Earth Environ. Sci.*, **118**, 012023, IOP Conference Series: Earth and Environmental Science, doi:10.1088/1755-1315/118/1/012023.

Luderer, G., R.C. Pietzcker, C. Bertram, E. Kriegler, M. Meinshausen, and O. Edenhofer, 2013: Economic mitigation challenges: How further delay closes the door for achieving climate targets. *Environmental Research Letters*, **8**, 3, doi:10.1088/1748-9326/8/3/034033.

Lundmark, T. et al., 2014: Potential roles of Swedish forestry in the context of climate change mitigation. *Forests*, **5**, 557–578, doi:10.3390/f5040557.

Lunt, T., A.W. Jones, W.S. Mulhern, D.P. M. Lezaks, and M.M. Jahn, 2016: Vulnerabilities to agricultural production shocks: An extreme, plausible scenario for assessment of risk for the insurance sector. *Clim. Risk Manag.*, **13**, 1–9, doi:10.1016/j.crm.2016.05.001.

Lybbert, T.J., and D.A. Sumner, 2012: Agricultural technologies for climate change in developing countries: Policy options for innovation and technology diffusion. *Food Policy*, **37**, 114–123, doi:10.1016/j. foodpol.2011.11.001.

Lynch, A.J. et al., 2016: The social, economic, and environmental importance of inland fish and fisheries. *Environ. Rev.*, **24**, 115–121, doi:10.1139/er-2015-0064 .

MacGregor, S., 2010: 'Gender and climate change': From impacts to discourses. *J. Indian Ocean Reg.*, **6**, 223–238, doi:10.1080/19480881.2010.536669.

Macintosh, A.K., 2012: LULUCF in the post-2012 regime: Fixing the problems of the past? *Clim. Policy*, **12**, 341–355, doi:10.1080/14693062.2011.605711.

Magnan, A., 2014: Avoiding maladaptation to climate change: Towards guiding principles. *S.A.P.I.E.N.S.*, **7**, 1–11.

Magnan, A.K. et al., 2016: Addressing the risk of maladaptation to climate change. *Wiley Interdiscip. Rev. Clim. Chang.*, **7**, 646–665, doi:10.1002/wcc.409.

Mahul, O., and F. Ghesquiere, 2010: Financial protection of the state against natural disasters: A primer. Policy Research working paper No. WPS 5429, World Bank, Washington, DC, USA, 26 pp, doi:10.1596/1813-9450-5429.

Maikhuri, R.K., R.L. Senwal, K.S. Rao, and K.G. Saxena, 1997: Rehabilitation of degraded community lands for sustainable development in Himalaya: A case study in Garhwal Himalaya, India. *Int. J. Sustain. Dev. World Ecol.*, **4**, 192–203, doi:10.1080/13504509709469954.

Majumder, M., 2015: *Impact of Urbanization on Water Shortage in Face of Climatic Aberrations*. Springer Singapore, Singapore, 98 pp.

Makkonen, M., S. Huttunen, E. Primmer, A. Repo, and M. Hildén, 2015: Policy coherence in climate change mitigation: An ecosystem service approach to forests as carbon sinks and bioenergy sources. *For. Policy Econ.*, **50**, 153–162, doi:10.1016/j.forpol.2014.09.003.

Maldonado, J.K., C. Shearer, R. Bronen, K. Peterson, and H. Lazrus, 2014: The impact of climate change on tribal communities in the US: Displacement, relocation, and human rights. In: *Climate Change and Indigenous Peoples in the United States: Impacts, Experiences and Actions* [Maldonado, J.K., C. Benedict, R. Pandya (eds.)]. Springer International Publishing, Cham, Switzerland, 174pp.

Mallampalli, V.R. et al., 2016: Methods for translating narrative scenarios into quantitative assessments of land use change. *Environ. Model. Softw.*, **82**, 7–20, doi:10.1016/J.ENVSOFT.2016.04.011.

Malogdos, F.K., and E. Yujuico, 2015a: Reconciling formal and informal decision-making on ecotourist infrastructure in Sagada, Philippines. *J. Sustain. Tour.*, doi:10.1080/09669582.2015.1049608.

von Maltitz, G.P. et al., 2018: Institutional arrangements of outgrower sugarcane production in southern Africa. *Dev. South. Afr.*, **36**, 175–197, doi:10.1080/0376835X.2018.1527215.

Maltsoglou, I. et al., 2014: Combining bioenergy and food security: An approach and rapid appraisal to guide bioenergy policy formulation. *Biomass and Bioenergy*, **79**, 80–95, doi:10.1016/j.biombioe.2015.02.007.

Manning, P., G. Taylor, and M.E. Hanley, 2015: Bioenergy, food production and biodiversity – An unlikely alliance? *GCB Bioenergy*, **7**, 570–576, doi:10.1111/gcbb.12173.

Mantel, S.K., D.A. Hughes, and N.W. J. Muller, 2010: Ecological impacts of small dams on South African rivers part 1: Drivers of change-water quantity and quality. *Water Sa*, **36**, 351–360.

Mapfumo, P., F. Mtambanengwe, and R. Chikowo, 2016: Building on indigenous knowledge to strengthen the capacity of smallholder farming communities to adapt to climate change and variability in southern Africa. *Clim. Dev.*, **8**, 72–82, doi:10.1080/17565529.2014.998604.

Mapfumo, P. et al., 2017: Pathways to transformational change in the face of climate impacts: An analytical framework. *Clim. Dev.*, **9**, 439–451, doi:10.1 080/17565529.2015.1040365.

Maraseni, T.N., and T. Cadman, 2015: A comparative analysis of global stakeholders' perceptions of the governance quality of the clean development mechanism (CDM) and reducing emissions from deforestation and forest degradation (REDD+). *Int. J. Environ. Stud.*, **72**, 288–304, doi:10.1080/00207233.2014.993569.

Marcacci, S., 2018: *India Coal Power is About to Crash: 65% of Existing Coal Costs More Than New Wind and Solar*. Forbes Energy Innovation, www.forbes.com/sites/energyinnovation/2018/01/30/india-coal-power-is-about-to-crash-65-of-existing-coal-costs-more-than-new-wind-and-solar/#68419e4c0fab.

Marchand, P. et al., 2016: Reserves and trade jointly determine exposure to food supply shocks. *Environ. Res. Lett.*, **11**, 1–11, doi:10.1088/1748-9326/11/9/095009.

Marjanac, S., L. Patton, and J. Thornton, 2017: Acts of god, human infuence and litigation. *Nat. Geosci.*, **10**, 616–619, doi:10.1038/ngeo3019.

Markusson, N., D. McLaren, and D. Tyfield, 2018a: Towards a cultural political economy of mitigation deterrence by negative emissions technologies (NETs). *Glob. Sustain.*, **1**, e10, doi:10.1017/sus.2018.10.

Marshall, N., S. Park, W.N. Adger, K. Brown, and S. Howden, 2012: Transformational capacity and the influence of place and identity. *Environ. Res. Lett.*, **7**, 1–9, doi:10.1088/1748-9326/7/3/034022.

Martin, D.R., and K.L. Pope, 2011: Luring anglers to enhance fisheries. *J. Environ. Manage.*, **92**, 1409–1413, doi:10.1016/j.jenvman.2010.10.002.

St. Martin, K., 2009: Toward a cartography of the commons: Constituting the political and economic possibilities of place. *Prof. Geogr.*, **61**, 493–507, doi:10.1080/00330120903143482.

Martin Persson, U., 2015: The impact of biofuel demand on agricultural commodity prices: A systematic review. *Wires Energy and Environment*, **4**, 410–428, doi:10.1002/wene.155.

Martinez, G., E. Williams, and S. Yu, 2015: The economics of health damage and adaptation to climate change in Europe: A review of the conventional and grey literature. *Climate*, **3**, 522–541, doi:10.3390/cli3030522.

Martins, M. et al., 2012: Impact of remobilized contaminants in Mytilus edulis during dredging operations in a harbour area: Bioaccumulation and biomarker responses. *Ecotoxicol. Environ. Saf.*, **85**, 96–103, doi:10.1016/j.ecoenv.2012.08.008.

Masera, O.R., R. Bailis, R. Drigo, A. Ghilardi, and I. Ruiz-Mercado, 2015: Environmental burden of traditional bioenergy use. *Annu. Rev. Environ. Resour.*, **40**, 121–150, doi:10.1146/annurev-environ-102014-021318.

Mathy, S., and O. Blanchard, 2016: Proposal for a poverty-adaptation-mitigation window within the Green Climate Fund. *Clim. Policy*, **16**, 752–767, doi:10.1080/14693062.2015.1050348.

Matthies, B.D., T. Kalliokoski, T. Ekholm, H.F. Hoen, and L.T. Valsta, 2015: Risk, reward, and payments for ecosystem services: A portfolio approach to ecosystem services and forestland investment. *Ecosyst. Serv.*, **16**, 1–12, doi:10.1016/j.ecoser.2015.08.006.

Maxwell, D., and K. Wiebe, 1999: Land tenure and food security: Exploring dynamic linkages. *Dev. Change*, **30**, 825–849, doi:10.1111/1467-7660.00139.

Maynard, T., 2015: *Food System Shock: The Insurance Impacts of Acute Disruption to Global Food Supply.* Lloyd's Emerging Risk Report. Lloyd's, London, UK, 27 pp.

Mayor, B., E. López-Gunn, F.I. Villarroya, and E. Montero, 2015: Application of a water-energy-food nexus framework for the Duero river basin in Spain. *Water Int.*, **40**, 791–808, doi:10.1080/02508060.2015.1071512.

Maystadt, J.F., and O. Ecker, 2014: Extreme weather and civil war: Does drought fuel conflict in Somalia through livestock price shocks? *Am. J. Agric. Econ.*, **96**, 1157–1182, doi:10.1093/ajae/aau010.

McCall, M.K., and C.E. Dunn, 2012: Geo-information tools for participatory spatial planning: Fulfilling the criteria for 'good' governance? *Geoforum*, **43**, 81–94, doi:10.1016/j.geoforum.2011.07.007.

McCarter, J., M.C. Gavin, S. Baereleo, and M. Love, 2014: The challenges of maintaining indigenous ecological knowledge. *Ecol. Soc.*, **19**, art39, doi:10.5751/ES-06741-190339.

McClean, C., R. Whiteley, and N.M. Hayes, 2010: *ISO 31000 — The New, Streamlined Risk Management Standard.* Forrester Research Inc, Cambridge, USA, 1–4 pp.

McClelland, S.C., C. Arndt, D.R. Gordon, and G. Thoma, 2018: Type and number of environmental impact categories used in livestock life cycle assessment: A systematic review. *Livest. Sci.*, **209**, 39–45, doi:10.1016/j.livsci.2018.01.008.

McCollum, D.L. et al., 2018: Connecting the sustainable development goals by their energy inter-linkages. *Environ. Res. Lett.*, **13**, 033006, doi:10.1088/1748-9326/aaafe3.

Mcdermott, C.L., L.C. Irland, and P. Pacheco, 2015: Forest certification and legality initiatives in the Brazilian Amazon: Lessons for effective and equitable forest governance. *For. Policy Econ.*, **50**, 134–142, doi:10.1016/j.forpol.2014.05.011.

McDowell, N.G., and C.D. Allen, 2015a: Darcy's law predicts widespread forest mortality under climate warming. *Nat. Clim. Chang.*, **5**, 669–672, doi:10.1038/nclimate2641.

McGrath, D.G., F. de Castro, C. Futemma, B.D. de Amaral, and J. Calabria, 1993: Fisheries and the evolution of resource management on the lower Amazon floodplain. *Hum. Ecol.*, **21**, 167–195, doi:10.1007/BF00889358.

McIntosh, C., A. Sarris, and F. Papadopoulos, 2013: Productivity, credit, risk, and the demand for weather index insurance in smallholder agriculture in Ethiopia. *Agric. Econ. (United Kingdom)*, **44**, 399–417, doi:10.1111/agec.12024.

McKinley, D.C. et al., 2017: Citizen science can improve conservation science, natural resource management, and environmental protection. *Biol. Conserv.*, **208**, 15–28, doi:10.1016/j.biocon.2016.05.015.

McKinnon, A., 2010: Green logistics: The carbon agenda. *Electron. Sci. J. Logist.*, **6**, 1–9.

McLeman, R.A. (ed.), 2013: *Climate and Human Migration: Past Experiences, Future Challenges.* Cambridge University Press, Cambridge, UK, and New York, NY, USA, doi:10.1017/CBO9781139136938.

McMichael, A.J., R.E. Woodruff, and S. Hales, 2006: Climate change and human health: Present and future risks. *Lancet*, **367**, 859–869, doi:10.1016/S0140-6736 (06)68079-3.

McNicol, I.M., C.M. Ryan, and E.T. A. Mitchard, 2018a: Carbon losses from deforestation and widespread degradation offset by extensive growth in African woodlands. *Nat. Commun.*, **9**, 3045, doi:10.1038/s41467-018-05386-z.

McSweeney, K., and O.T. Coomes, 2011: Climate-related disaster opens a window of opportunity for rural poor in north-eastern Honduras. *Proc. Natl. Acad. Sci.*, **108**, 5203–5208, doi:10.1073/pnas.1014123108.

Measham, T.G., 2011: Adapting to climate change through local municipal planning: Barriers and challenges. *Mitig. Adapt. Strateg. Glob. Chang.*, **16**, 889–909, doi:10.1007/s11027-011-9301-2.

Mechler, R. et al., 2014: Managing unnatural disaster risk from climate extremes. *Nat. Clim. Chang.*, **4**, 235–237, doi:10.1038/nclimate2137.

Mehta, L., G.J. Veldwisch, and J. Franco, 2012: Introduction to the special issue: Water grabbing? Focus on the (re)appropriation of finite water resources. *Water Altern.*, **5**, 193–207.

Meijer, S.S., D. Catacutan, O.C. Ajayi, G.W. Sileshi, and M. Nieuwenhuis, 2015: The role of knowledge, attitudes and perceptions in the uptake of agricultural and agroforestry innovations among smallholder farmers in Sub-Saharan Africa. *Int. J. Agric. Sustain.*, doi:10.1080/14735903.2014.912493.

van Meijl, H. et al., 2018a: Comparing impacts of climate change and mitigation on global agriculture by 2050. *Environ. Res. Lett.*, **13**, 064021, doi:10.1088/1748-9326/aabdc4.

Meinzen-dick, R. et al., 2010: Engendering Agricultural Research. *IFPRI Disscussion Pap. 973*, **72**, International Food Policy Research Institute, 63 pp.

Mekonnen, M.M., and A.Y. Hoekstra, 2016: Sustainability: Four billion people facing severe water scarcity. *Sci. Adv.*, **2**, e1500323, doi:10.1126/sciadv.1500323.

Meli, P. et al., 2017: A global review of past land use, climate, and active vs. passive restoration effects on forest recovery. *PLoS One*, **12**, e0171368, doi:10.1371/journal.pone.0171368.

Mello, D., and M. Schmink, 2017a: Amazon entrepreneurs: Women's economic empowerment and the potential for more sustainable land use practices. *Womens. Stud. Int. Forum*, **65**, 28–36, doi:10.1016/J.WSIF.2016.11.008.

Mello, F.F. C. et al., 2014: Payback time for soil carbon and sugar-cane ethanol. *Nat. Clim. Chang.*, **4**, 605–609, doi:10.1038/nclimate2239.

Melvin, A.M. et al., 2017: Climate change damages to Alaska public infrastructure and the economics of proactive adaptation. *Proc. Natl. Acad. Sci.*, **114**, E122-E131, doi:10.1073/pnas.1611056113.

Menz, M.H. M., K.W. Dixon, and R.J. Hobbs, 2013: Hurdles and opportunities for landscape-scale restoration. *Science*, **339**, 526–527, doi:10.1126/science.1228334.

Mersha, A.A., and F. Van Laerhoven, 2016: A gender approach to understanding the differentiated impact of barriers to adaptation:

Responses to climate change in rural Ethiopia. *Reg. Environ. Chang.*, **16**, 1701–1713, doi:10.1007/s10113-015-0921-z.

Messerli, P., M. Giger, M.B. Dwyer, T. Breu, and S. Eckert, 2014a: The geography of large-scale land acquisitions: Analysing socio-ecological patterns of target contexts in the Global South. *Appl. Geogr.*, **53**, 449–459, doi:10.1016/J.APGEOG.2014.07.005.

Metternicht, G. (ed.), 2018: *Contributions of Land Use Planning to Sustainable Land Use and Management*. SpringerInternational Publishing, Cham, Switzerland, 35–51 pp.

Meyer, M.A., and J.A. Priess, 2014: Indicators of bioenergy-related certification schemes – An analysis of the quality and comprehensiveness for assessing local/regional environmental impacts. *Biomass and Bioenergy*, **65**, 151–169, doi:10.1016/J.BIOMBIOE.2014.03.041.

Meyfroidt, P., E.F. Lambin, K.H. Erb, and T.W. Hertel, 2013: Globalization of land use: Distant drivers of land change and geographic displacement of land use. *Curr. Opin. Environ. Sustain.*, **5**, 438–444, doi:10.1016/j.cosust.2013.04.003.

Michaelowa, K., and A. Michaelowa, 2017: Transnational climate governance initiatives: Designed for effective climate change mitigation? *Int. Interact.*, **43**, 129–155, doi:10.1080/03050629.2017.1256110.

Michel-Kerjan, E., 2011: *Catastrophe Financing for Governments: Learning from the 2009–2012 MultiCat Program in Mexico*. Press release, World Bank, Washington, DC, USA, www.worldbank.org/en/news/press-release/2012/10/12/mexico-launches-second-catastrophe-bond-to-provide-coverage-against-earthquakes-and-hurricanes.

Middleton, N., U. Kang, N. Middleton, and U. Kang, 2017: Sand and dust storms: Impact mitigation. *Sustainability*, **9**, 1053, doi:10.3390/su9061053.

Milder, J.C. et al., 2015: An agenda for assessing and improving conservation impacts of sustainability standards in tropical agriculture. *Conserv. Biol.*, **29**, 309–320, doi:10.1111/cobi.12411.

Milkoreit, M., M.L. Moore, M. Schoon, and C.L. Meek, 2015: Resilience scientists as change-makers – Growing the middle ground between science and advocacy? *Environ. Sci. Policy*, **53**, 87–95, doi:10.1016/j.envsci.2014.08.003.

Millar, C.I., N.L. Stephenson, and S.L. Stephens, 2007: Climate change and forests of the future: Managing in the face of uncertainty. *Ecol. Appl.*, **17**, 2145–2151, doi:10.1890/06-1715.1.

Miller, L., R. Carriveau, and S. Harper, 2018: Innovative financing for renewable energy project development – Recent case studies in North America. *Int. J. Environ. Stud.*, **75**, 121–134, doi:10.1080/00207233.2017.1403758.

Millo, G., 2016: The Income Elasticity of Nonlife Insurance: A Reassessment. *J. Risk Insur.*, **83**, 335–362, doi:10.1111/jori.12051.

Milton, S.J., W.R. J. Dean, and D.M. Richardson, 2003: Economic incentives for restoring natural capital in southern African rangelands. *Front. Ecol. Environ.*, **1**, 247–254, doi:10.1890/1540-9295 (2003)001[0247:EIFRNC]2.0.CO; 2.

Mimura, N., R.S. Pulwarty, D.M. Duc, I. Elshinnawy, M.H. Redsteer, H.Q. Huang, J.N. Nkem, and R.A. Sanchez, Rodriguez, 2014: Adaptation Planning and Implementation. In: Climate Change 2014: Impacts, Adaptation, and Vulnerability. Part A: Global and Sectoral Aspects. Contribution of Working Group II to the Fifth Assessment Report of the Intergovernmental Panel on Climate Change [Field, C.B., V.R. Barros, D.J. Dokken, K.J. Mach, M.D. Mastrandrea, T.E. Bilir, M. Chatterjee, K.L. Ebi, Y.O. Estrada, R.C. Genova, B. Girma, E.S. Kissel, A.N. Levy, S. MacCracken, P.R. Mastrandrea, and L.L. White (eds.)]. Cambridge University Press, Cambridge, United Kingdom and New York, NY, USA, pp. 869–898.

Minang, P.A. et al., 2014: REDD+ readiness progress across countries: Time for reconsideration. *Clim. Policy*, **14**, 685–708, doi:10.1080/14693062.2014.905822.

Mini, C., T.S. Hogue, and S. Pincetl, 2015: The effectiveness of water conservation measures on summer residential water use in Los Angeles, California. *Resour. Conserv. Recycl.*, **94**, 136–145, doi:10.1016/j.resconrec.2014.10.005.

Ministry of Environment and Forests and Ministry, and N.D. Tribal Affairs, Government of India, 2010: *Manthan: Report of the National Committee on Forest Rights Act*. New Delhi, India, 284 pp.

Mistry, J., and A. Berardi, 2016: Bridging indigenous and scientific knowledge. *Science*, 352, 1274–1275, doi:10.1126/science.aaf1160.

Mitchell, D., 2010: Land tenure and disaster risk management. *L. Tenure J.*, **1**, 121–141.

Mitchell, D., S. Enemark, and P. van der Molen, 2015: Climate resilient urban development: Why responsible land governance is important. *Land Use Policy*, **48**, 190–198, doi:10.1016/J.LANDUSEPOL.2015.05.026.

Miteva, D.A., C.J. Loucks, and S.K. Pattanayak, 2015: Social and environmental impacts of forest management certification in Indonesia. *PLoS One*, **10**, e0129675, doi:10.1371/journal.pone.0129675.

Mobarak, A.M., and M.R. Rosenzweig, 2013: Informal risk sharing, index insurance, and risk taking in developing countries. *American Economic Review*, 103, 375–380, doi:10.1257/aer.103.3.375.

Mochizuki, J., S. Vitoontus, B. Wickramarachchi, S. Hochrainer-Stigler, K. Williges, R. Mechler, and R. Sovann, 2015: Operationalizing iterative risk management under limited information: Fiscal and economic risks due to natural disasters in Cambodia. *Int. J. Disaster Risk Sci.*, **6**, 321–334, doi:10.1007/s13753-015-0069-y.

Mohapatra, S., 2013: Displacement due to climate change and international law. *Int. J. Manag. Soc. Sci. Res.* **2**, 1–8.

Mohmmed, A. et al., 2018: Assessing drought vulnerability and adaptation among farmers in Gadaref region, Eastern Sudan. *Land Use Policy*, **70**, 402–413, doi:10.1016/j.landusepol.2017.11.027.

Mohr, A., T. Beuchelt, R. El Schneider, and D. Virchow, 2016: Food security criteria for voluntary biomass sustainability standards and certifications. *Biomass and Bioenergy*, **89**, 133–145, doi:10.1016/j.biombioe.2016.02.019.

van der Molen, P., and D. Mitchell, 2016: Climate change, land use and land surveyors. *Surv. Rev.*, **48**, 148–155, doi:10.1179/1752270615Y.0000000029.

Mollenkamp, S., and B. Kasten, 2009: Institutional Adaptation to Climate Change: The Current Status and Future Strategies in the Elbe Basin, Germany. In: *Climate Change Adaptation in the Water Sector* [Ludwig, F., P. Kabat, H. Van Schaik, M. Michael Van Der Valk (eds.)]. Earthscan, London, UK, pp. 227–249.

Monchuk, V., 2014: Reducing Poverty and Investing in People: The New Role of Safety Nets in Africa. World Bank, Washington, DC, USA, 20 pp.

Monkelbaan, J., 2019: *Governance for the Sustainable Development Goals: Exploring an Integrative Framework of Theories, Tools, and Competencies*. Springer Singapore, XXI, 214 pp.

Montaña, E., H.P. Diaz, and M. Hurlbert, 2016: Development, local livelihoods, and vulnerabilities to global environmental change in the South American Dry Andes. *Reg. Environ. Chang.*, 16, 2215–2228, doi:10.1007/s10113-015-0888-9.

Montanarella, L., 2015: The importance of land restoration for achieving a land degradation-neutral world. In: *Land Restoration: Reclaiming Landscapes for a Sustainable Future* [Chabay, I., M. Frick, and J. Helgeson (eds.)]. Academic Press, Elsevier, Massachusetts, USA, pp. 249–258.

Monterrosso, I., P. Cronkleton, D. Pinedo, and A. Larson, 2017: *Reclaiming Collective Rights: Land and Forest Tenure Reforms in Peru (1960–2016)*. CIFOR Working Paper no. 224, Center for International Forestry Research (CIFOR), Bogor, Indonesia, 31 pp.

Moore, C.T., E.V. Lonsdorf, M.G. Knutson, H.P. Laskowski, and S.K. Lor, 2011: Adaptive management in the US National Wildlife Refuge System: Science-management partnerships for conservation delivery. *J. Environ. Manage.*, **92**, 1395–1402, doi:10.1016/j.jenvman.2010.10.065.

Moore, F.C., and D.B. Diaz, 2015: Temperature impacts on economic growth warrant stringent mitigation policy. 5, 127–132, doi:10.1038/NCLIMATE2481.

7

Moosa, C.S., and N. Tuana, 2014: Mapping a research agenda concerning gender and climate change: A review of the literature. *Hypatia*, **29**, 677–694, doi:10.1111/hypa.12085.

Moran, D. et al., 2010: Marginal abatement cost curves for UK agricultural greenhouse gas emissions. *J. Agric. Econ.*, **62**, 93–118, doi:10.1111/j.1477-9552.2010.00268.x .

Morita, K., and K. Matsumoto, 2018: Synergies among climate change and biodiversity conservation measures and policies in the forest sector: A case study of Southeast Asian countries. *For. Policy Econ.*, **87**, 59–69, doi:10.1016/j.forpol.2017.10.013.

Moroni, S., 2018: Property as a human right and property as a special title. Rediscussing private ownership of land. *Land Use Policy*, **70**, 273–280, doi:10.1016/J.LANDUSEPOL.2017.10.037.

Morton, J.F., 2017: Climate change and African agriculture: Unlocking the potential of research and advisory services. *Making Climate Compatible Development Happen* [Nunan, F., (ed.)]. Routledge, London, UK, pp. 87–113.

Morton, J.F., 2007: The impact of climate change on smallholder and subsistence agriculture. *Proc. Natl. Acad. Sci. U.S.A.*, **104**, 19680–19685, doi:10.1073/pnas.0701855104.

Mostert, E., C. Pahl-Wostl, Y. Rees, B. Searle, D. Tàbara, and J. Tippett, 2007: Social learning in European river-basin management: Barriers and fostering mechanisms from 10 river basins. *Ecol. Soc.*, **12**, ART. 19, doi:10.5751/ES-01960-120119.

Mowo, J., Z. Adimassu, D. Catacutan, J. Tanui, K. Masuki, and C. Lyamchai, 2013: The importance of local traditional institutions in the management of natural resources in the highlands of East Africa. *Hum. Organ.*, **72**, 154–163, doi:10.17730/humo.72.2.e1x3101741127x35.

Mozumder, P., R. Helton, and R.P. Berrens, 2009: Provision of a wildfire risk map: Informing residents in the wildland urban interface. *Risk Anal. An Int. J.*, **29**, 1588–1600, doi:10.1111/j.1539-6924.2009.01289.x.

Mozumder, P., E. Flugman, and T. Randhir, 2011: Adaptation behavior in the face of global climate change: Survey responses from experts and decision makers serving the Florida Keys. *Ocean Coast. Manag.*, 54, 37–44, doi:10.1016/j.ocecoaman.2010.10.008.

Mubaya, C.P., and P. Mafongoya, 2017: The role of institutions in managing local level climate change adaptation in semi-arid Zimbabwe. *Clim. Risk Manag.*, 16, 93–105, doi:10.1016/j.crm.2017.03.003.

Mudombi, S. et al., 2018a: Multi-dimensional poverty effects around operational biofuel projects in Malawi, Mozambique and Swaziland. *Biomass and Bioenergy*, **114**, 41–54, doi:10.1016/j.biombioe.2016.09.003.

Mudombi, S et al., 2018c: User perceptions about the adoption and use of ethanol fuel and cookstoves in Maputo, Mozambique. *Energy Sustain. Dev.*, **44**, 97–108, doi:10.1016/j.esd.2018.03.004.

Mukherjee, N. et al., 2015: The Delphi technique in ecology and biological conservation: Applications and guidelines. *Methods Ecol. Evol.*, **6**, 1097–1109, doi:10.1111/2041-210X.12387.

Mukhtarov, F., P. Osseweijer, and R. Pierce, Global governance of biofuels: A case for public-private governance? *Bio-based and Applied Economics*, **3**, 285–294, doi:10.13128/BAE-14767.

Müller, A. et al., 2015: *IASS Working Paper The Role of Biomass in the Sustainable Development Goals: A Reality Check and Governance Implications*. Institute for Advanced Sustainability Studies (IASS), Potsdam, Germany, 35 pp.

Muller, C., and H. Yan, 2018: Household fuel use in developing countries: Review of theory and evidence. *Energy Econ.*, **70**, 429–439, doi:10.1016/j.eneco.2018.01.024.

Müller, C. et al., 2013: Assessing agricultural risks of climate change in the 21st century in a global gridded crop model intercomparison. *Proc. Natl. Acad. Sci.*, **9**, 3268–3273, doi:10.1073/pnas.1222463110.

Munang, R., I. Thiaw, K. Alverson, M. Mumba, J. Liu, and M. Rivington, 2013: Climate change and ecosystem-based adaptation: A new pragmatic approach to buffering climate change impacts. *Curr. Opin. Environ. Sustain.*, **5**, 67–71, doi:10.1016/j.cosust.2012.12.001.

Munene, M.B., Å.G. Swartling, and F. Thomalla, 2018: Adaptive governance as a catalyst for transforming the relationship between development and disaster risk through the Sendai Framework? *Int. J. Disaster Risk Reduct.*, **28**, 653–663, doi:10.1016/j.ijdrr.2018.01.021.

Munthali, K., and Y. Murayama, 2013: Interdependences between smallholder farming and environmental management in rural Malawi: A case of agriculture-induced environmental degradation in Malingunde Extension Planning Area (EPA). *Land*, 2, 158–175, doi:10.3390/land2020158.

Muradian, R., and L. Rival, 2012: Between markets and hierarchies: The challenge of governing ecosystem services. *Ecosyst. Serv.*, **1**, 93–100, doi:10.1016/j.ecoser.2012.07.009.

Muratori, M., K. Calvin, M. Wise, P. Kyle, and J. Edmonds, 2016: Global economic consequences of deploying bioenergy with carbon capture and storage (BECCS). *Environ. Res. Lett.*, **11**, 1–9, doi:10.1088/1748-9326/11/9/095004.

Murphy, K., G.A. Kirkman, S. Seres, and E. Haites, 2015: Technology transfer in the CDM: An updated analysis. *Clim. Policy*, **15**, 127–145, doi:10.1080/14693062.2013.812719.

Murthy, I.K., V. Varghese, P. Kumar, and S. Sridhar, 2018a: Experience of participatory forest management in India: Lessons for governance and institutional arrangements under REDD+. In: *Global Forest Governance and Climate Change* [Nuesiri, E.O. (ed.)]. Palgrave Macmillan, Cham, Switzerland, 175–201, doi:10.1007/978-3-319-71946-7.

Mustalahti, I., and O.S. Rakotonarivo, 2014: REDD+ and empowered deliberative democracy: Learning from Tanzania. *World Dev.*, **59**, 199–211, doi:10.1016/j.worlddev.2014.01.022.

Naeem, S. et al., 2015: Get the science right when paying for nature's services. *Science*, **347**, 1206–1207, doi:10.1126/science.aaa1403.

Naess, L.O., 2013: The role of local knowledge in adaptation to climate change. *Wiley Interdiscip. Rev. Chang.*, 4, 99–106, doi:10.1002/Wcc.204.

Nagel, L.M. et al., 2017: Adaptive silviculture for climate change: A national experiment in manager-scientist partnerships to apply an adaptation framework. *J. For.*, **115**, 167–178, doi:10.5849/jof.16-039.

Nagendra, H., and E. Ostrom, 2012: Polycentric governance of multifunctional forested landscapes. *Int. J. Commons*, **6**, 104–133, doi:10.18352/ijc.321.

Naicker, P., 2011: The impact of climate change and other factors on zoonotic diseases. *Arch. Clin. Microbiol.*, **2**, 1–6, doi:10:3823/226.

Nakashima, D., K.G. McLean, H.D. Thulstrup, A.R. Castillo, and J.T. Rubis, 2013: *Weathering Uncertainty: Traditional Knowledge for Climate Change Assessment and Adaptation*. UNESCO, Paris, France, and UNU, Darwin, Australia, 120 pp.

Nakhooda, S., C. Watson, and L. Schalatek, 2016: The Global Climate Finance Architecture. *Clim. Financ. Fundam.*, **5**, Heinrich Boll Stiftung North America and Overseas Development Institute, Washington DC, USA and London, UK, 5 pp.

Nalau, J., and J. Handmer, 2015: When is transformation a viable policy alternative? *Environ. Sci. Policy*, 54, 349–356, doi:10.1016/j.envsci.2015.07.022.

Namubiru-Mwaura, E., 2014a: *Land Tenure and Gender: Approaches and Challenges for Strengthening Rural Women's Land Rights*. Women's Voice, Agency, and Participation Research Series No. 6., World Bank Group, Washington, DC, USA, 32 pp.

Nanda, A.V., J. Rijke, L. Beesley, B. Gersonius, M.R. Hipsey, and A. Ghadouani, 2018: Matching ecosystem functions with adaptive ecosystem management : Decision pathways to overcome institutional barriers. *Water*, **10**, 672, doi:10.3390/w10060672.

Narassimhan, E. et al., 2018: Carbon pricing in practice: A review of existing emissions trading systems. *Climate Policy*, **18**, 967–9913062, doi:10.1080/14693062.2018.1467827.

Nawrotzki, R.J., and M. Bakhtsiyarava, 2017: International climate migration: Evidence for the Climate Inhibitor Mechanism and the agricultural pathway. *Popul. Space Place*, **23**, e2033, doi:10.1002/psp.2033.

Nawrotzki, R.J., A.M. Schlak, and T.A. Kugler, 2016: Climate, migration, and the local food security context: Introducing Terra Populus. *Popul. Environ.*, **38**, 164–184, doi:10.1007/s11111-016-0260-0.

Nayak, R.R., S. Vaidyanathan, and J. Krishnaswamy, 2014: Fire and grazing modify grass community response to environmental determinants in savannas: Implications for sustainable use. *Agric. Ecosyst. Environ.*, **185**, 197–207, doi:10.1016/j.agee.2014.01.002.

Ndegwa, G.M., U. Nehren, F. Grüninger, M. Iiyama, and D. Anhuf, 2016: Charcoal production through selective logging leads to degradation of dry woodlands: A case study from Mutomo District, Kenya. *J. Arid Land*, **8**, 618–631, doi:10.1007/s40333-016-0124-6.

Nelson, G.C., V. Harris, and S.W. Stone, 2001: Deforestation, land use, and property rights: Empirical evidence from Darien, Panama. *Land Econ.*, **77**, 187, doi:10.2307/3147089.

Nelson, R., M. Howden, and M.S. Smith, 2008: Using adaptive governance to rethink the way science supports Australian drought policy. *Environ. Sci. Policy*, **11**, 588–601, doi:10.1016/j.envsci.2008.06.005.

Nelson, R. et al., 2010: The vulnerability of Australian rural communities to climate variability and change: Part II – Integrating impacts with adaptive capacity. *Environ. Sci. Policy*, **13**, 18–27, doi:10.1016/j.envsci.2009.09.007.

Nepstad, D. et al., 2006: Inhibition of Amazon deforestation and fire by parks and indigenous lands. *Conserv. Biol.*, **20**, 65–73, doi:10.1111/j.1523-1739.2006.00351.x.

Nepstad, D.C., C.M. Stickler, B. Soares-Filho, and F. Merry, 2008: Interactions among Amazon land use, forests and climate: Prospects for a near-term forest tipping point. *Philos. Trans. R. Soc. London B. Biol. Sci.*, **363**, 1737–1746, doi:10.1098/rstb.2007.0036.

Neufeldt, H. et al., 2013: Beyond climate-smart agriculture: Toward safe operating spaces for global food systems. *Agric. Food Secur.*, **2**, 1–6, doi:10.1186/2048-7010-2-12.

New Zealand Ministry for the Environment, 2018: *New Zealand's Greenhouse Gas Inventory 1990–2016*. New Zealand Ministry for the Environment, Wellington, New Zealand, 497 pp.

New Zealand Productivity Commission, 2018: *Low-Emissions Economy: Final Report*. New Zealand Productivity Commission, Wellington, New Zealand, 588 pp.

Newell, P., and O. Taylor, 2017: Contested landscapes: The global political economy of climate-smart agriculture. *J. Peasant Stud.*, **45**, 108–129, doi:10.1080/03066150.2017.1324426.

Newig, J., D. Gunther, and C. Pahl-Wostl, 2010: Synapses in the network: Learning in governance networks in the context of environmental management. *Ecol. Soc.*, **15**, 24, 1–16.

Newton, A.C., 2016: Biodiversity risks of adopting resilience as a policy goal. *Conserv. Lett.*, **9**, 369–376, doi:10.1111/conl.12227.

Newton, P. et al., 2018: The role of zero-deforestation commitments in protecting and enhancing rural livelihoods. *Curr. Opin. Environ. Sustain.*, **32**, 126–133, doi:10.1016/j.cosust.2018.05.023.

Nguyen, T., and J. Lindenmeier, 2014: Catastrophe risks, cat bonds and innovation resistance. *Qual. Res. Financ. Mark.*, **6**, 75–92, doi:10.1108/QRFM-06-2012-0020.

Nicholls, R.J., and A. Cazenave, 2010: Sea-level rise and its impact on coastal zones. *Science*, **328**, 1517–1520, doi:10.1126/science.1185782.

Nieto-Romero, M., A. Milcu, J. Leventon, F. Mikulcak, and J. Fischer, 2016: The role of scenarios in fostering collective action for sustainable development: Lessons from central Romania. *Land Use Policy*, **50**, 156–168, doi:10.1016/J.LANDUSEPOL.2015.09.013.

Nikolakis, W., S. Akter, and H. Nelson, 2016: The effect of communication on individual preferences for common property resources: A case study of two Canadian First Nations. *Land Use Policy*, **58**, 70–82, doi:10.1016/J.LANDUSEPOL.2016.07.007.

Nilsson, C., and K. Berggren, 2000: Alterations of Riparian Ecosystems caused by river regulation: Dam operations have caused global-scale ecological changes in riparian ecosystems. How to protect river environments and human needs of rivers remains one of the most important questions of our time. *AIBS Bull.*, **50**, 783–792, doi:10.1641/0006-3568 (2000)050[0783:AORECB]2.0.CO; 2.

Nilsson, M., and Å. Persson, 2012: Can Earth System interactions be governed? Governance functions for linking climate change mitigation with land use, freshwater and biodiversity protection. *Ecol. Econ.*, **75**, 61–71, doi:10.1016/J.ECOLECON.2011.12.015.

Nilsson, M., D. Griggs, and M. Visback, 2016a: Map the interactions between sustainable development Goa. *Nature*, **534**, 320–322, doi:10.1038/534320a.

Nilsson, M., D. Griggs, and M. Visbeck, 2016b: Map the interactions between sustainable development goals. *Nature*, **534**, 320–323, doi:10.1038/534320a.

Niño-Zarazúa, M., A. Barrientos, S. Hickey, and D. Hulme, 2012: Social protection in Sub-Saharan Africa: Getting the politics right. *World Dev.*, **40**, 163–176, doi:10.1016/j.worlddev.2011.04.004.

Nischalke, S.M., 2015: Adaptation options adaptation options to improve food security in a changing climate in the Hindu Kush-Himalayan region. *Handbook of Climate Change Adaptation*, Springer Berlin, Berlin, Germany, 1423–1442.

Njenga, M., and R. Mendum, 2018: *Recovering Bioenergy in Sub-Saharan Africa: Gender Dimensions, Lessons and Challenges*. International Water Management Institute (IWMI), CGIAR Research Program on Water, Land and Ecosystems (WLE), Colombo, Sri Lanka, 96 pp, doi:10.5337/2018.226.

Nkengasong, J.N., and P. Onyebujoh, 2018: Response to the Ebola virus disease outbreak in the Democratic Republic of the Congo. *Lancet*, **391**, 2395–2398, doi:10.1016/S0140-6736 (18)31326-6.

Nkoana, E.M., T. Waas, A. Verbruggen, C.J. Burman, and J. Hugé, 2017: Analytic framework for assessing participation processes and outcomes of climate change adaptation tools. *Environ. Dev. Sustain.*, **19**, 1731–1760, doi:10.1007/s10668-016-9825-4.

Nkonya, E., J. von Braun, A. Mirzabaev, Q.B. Le, H.Y. Kwon, and O. Kirui, 2013: Economics of Land Degradation Initiative: Methods and Approach for Global and National Assessments. ZEF – Discussion Papers on Development Policy No. 183, Bonn, Germany, 41 pp, doi:10.2139/ssrn.2343636.

Nkonya, E., T. Johnson, H.Y. Kwon, and E. Kato, 2015: Economics of land degradation in Sub-Saharan Africa. In: *Economics of Land Degradation and Improvement – A Global Assessment for Sustainable Development* [Nkonya, E., A. Mirzabaev, J. von Braun (eds.)]. Springer International Publishing, Cham, Switzerland, pp. 215–259, doi:10.1007/978-3-319-19168-3.

Nkonya, E. et al., 2016: Global cost of land degradation. In: *Economics of Land Degradation and Improvement – A Global Assessment for Sustainable Development* [Nkonya, E., A. Mirzabaev, and J. Von Braun (eds.)]. Springer International Publishing, Cham, Switzerland, pp. 117–165, doi:10.1007/978-3-319-19168-3_6.

Nolte, K., W. Chamberlain, and M. Giger, 2016: *International Land Deals for Agriculture. Fresh Insights from the Land Matrix: Analytical Report II*. Centre for Development and Environment, University of Bern, Centre de coopération internationale en recherche agronomique pour le développement, German Institute of Global and Area Studies, University of Pretoria, Bern Open Publishing, Germany, doi:10.7892/boris.85304, 56 pp.

Nordhaus, W., 2014: Estimates of the social cost of carbon: Concepts and results from the DICE-2013R model and alternative approaches. *J. Assoc. Environ. Resour. Econ.*, **1**, 273–312, doi:10.1086/676035.

Nordhaus, W.D., 1999: Roll the DICE Again: The economics of global warming. *Draft Version*, **28**, 1999, 79 pp.

Norström, A. et al., 2014: Three necessary conditions for establishing effective Sustainable Development Goals in the Anthropocene. *Ecol. Soc.*, **19**, Art. 8, doi:10.5751/ES-06602-190308.

North, D., 1991: Institutions. *J. Econ. Perspect.*, **5**, 97–112, doi:10.1257/jep.5.1.97.

7

Northrup, J.M., and G. Wittemyer, 2013: Characterising the impacts of emerging energy development on wildlife, with an eye towards mitigation. *Ecol. Lett.*, **16**, 112–125, doi:10.1111/ele.12009.

Nugent, R. et al., 2018: Investing in non-communicable disease prevention and management to advance the Sustainable Development Goals. *The Lancet*, **391**, 2029–2035, doi:10.1016/S0140-6736 (18)30667-6.

Nuhoff-Isakhanyan, G., E. Wubben, and S.W. F. Omta, 2016: Sustainability benefits and challenges of inter-organizational collaboration in bio-based business: A systematic literature review. *Sustainability*, **8**, 307, doi:10.3390/su8040307.

Nyong, A., F. Adesina, and B. Osman Elasha, 2007: The value of indigenous knowledge in climate change mitigation and adaptation strategies in the African Sahel. *Mitig. Adapt. Strateg. Glob. Chang.*, **12**, 787–797, doi:10.1007/s11027-007-9099-0.

Nyström, M. et al., 2012: Confronting feedbacks of degraded marine ecosystems. *Ecosystems*, **15**, 695–710, doi:10.1007/s10021-012-9530-6.

IPCC, 2012: Toward a sustainable and resilient future. *Managing the Risks of Extreme Events and Disasters to Advance Climate Change Adaptation* [Field, C.B., V. Barros, T.F. Stocker, D. Qin, D.J. Dokken, K.L. Ebi, M.D. Mastrandrea, K.J. Mach, G.-K. Plattner, S.K. Allen, M. Tignor, and P.M. Midgley (eds.)]., Cambridge University Press, Cambridge, UK, and New York, NY, USA, 437–486 pp.

O'Brien, C.O. et al., 2018: *Shock-Responsive Social Protection Systems Research Synthesis Report*. Oxford Policy Management, Oxford, UK, 89 pp.

O'Hare, P., I. White, and A. Connelly, 2016: Insurance as maladaptation: Resilience and the 'business as usual' paradox. *Environ. Plan. C Gov. Policy*, **34**, 1175–1193, doi:10.1177/0263774X15602022.

O'Neill, B.C. et al., 2017a: IPCC reasons for concern regarding climate change risks. *Nat. Clim. Chang.*, **7**, 28–37, doi:10.1038/nclimate3179.

Oakes, L.E., N.M. Ardoin, and E.F. Lambin, 2016: 'I know, therefore I adapt?' Complexities of individual adaptation to climate-induced forest dieback in Alaska. *Ecol. Soc.*, **21**, art40, doi:10.5751/ES-08464-210240.

Oba, G., 1994: The importance of pastoralists' indigenous coping strategies for planning drought management in the arid zone of Kenya. *Nomad. People.*, **5**, 89–119, doi:10.2307/43123620.

Oba, G., 2013: The sustainability of pastoral production in Africa. In: Pastoralism and Development in Africa: Dynamic Change at the Margins [Andy Catley, Jeremy Lind, Ian Scoones (Eds.)]. Routledge, London, UK and New York, USA, 54–61, doi:10.4324/9780203105979.

Oberlack, C., 2017: Diagnosing institutional barriers and opportunities for adaptation to climate change. *Mitig. Adapt. Strateg. Glob. Chang.*, **22**, 805–838, doi:10.1007/s11027-015-9699-z.

Ochieng, C., S. Juhola, and F.X. Johnson, 2014: The societal role of charcoal production in climate change adaptation of the arid and semi-arid lands (ASALs) of Kenya. *Climate Change Adaptation and Development: Transforming Paradigms and Practices* [Inderberg, T.H., S.H. Eriksen, K.L. O'Brien, and L. Sygna (eds.)]. Routledge, London, UK.

Ockwell, D., A. Sagar, and H. de Coninck, 2015: Collaborative research and development (R&D) for climate technology transfer and uptake in developing countries: Towards a needs driven approach. *Clim. Change*, **131**, 401–415, doi:10.1007/s10584-014-1123-2.

OECD, 2014: *Social Institutions and Gender Index*. OECD Development Centre, Paris, France, 68 pp.

OECD, 2015: Climate Finance in 2013–14 and the USD 100 billion goal. World Economic Forum, Cologny, Switzerland, doi:10.1787/9789264249424-en, 64 pp.

OECD, 2018: *Joint Working Party on Agriculture and the Environment: A Global Economic Evaluation Of GHG Mitigation Policies For Agriculture*. Paris, France, 38 pp.

Oels, A., 2013: Rendering climate change governable by risk: From probability to contingency. *Geoforum*, **45**, 17–29, doi:10.1016/j.geoforum.2011.09.007.

Oglethorpe, J., J. Ericson, R. Bilsborrow, and J. Edmond, 2007: *People on the Move: Reducing the Impact of Human Migration on Biodiversity*. World Wildlife Fund and Conservation International Foundation, Washington, DC, USA, doi:10.13140/2.1.2987.0083, 92 pp.

Ojea, E., 2015: Challenges for mainstreaming ecosystem-based adaptation into the international climate agenda. *Curr. Opin. Environ. Sustain.*, **14**, 41–48, doi:10.1016/j.cosust.2015.03.006.

Oke, T.R., G. Mills, A. Christen, and J.A. Voogt, 2017: *Urban climates*. Cambridge University Press, Cambridge, UK, and New York, NY, USA, doi:10.1017/9781139016476, 526 pp.

Oladele, S.O., A.J. Adeyemo, and M.A. Awodun, 2019: Influence of rice husk biochar and inorganic fertilizer on soil nutrients availability and rain-fed rice yield in two contrasting soils. *Geoderma*, **336**, 1–11, doi:10.1016/J. GEODERMA.2018.08.025.

Oliveira Júnior, J.G. C., R.J. Ladle, R. Correia, and V.S. Batista, 2016: Measuring what matters – Identifying indicators of success for Brazilian marine protected areas. *Mar. Policy*, **74**, 91–98, doi:10.1016/j.marpol.2016.09.018.

Oliver, D.M., R.D. Fish, M. Winter, C.J. Hodgson, A.L. Heathwaite, and D.R. Chadwick, 2012: Valuing local knowledge as a source of expert data: Farmer engagement and the design of decision support systems. *Environ. Model. Softw.*, **36**, 76–85, doi:10.1016/J.ENVSOFT.2011.09.013.

Olsson, A., S. Grönkvist, M. Lind, and J. Yan, 2016: The elephant in the room – A comparative study of uncertainties in carbon offsets. *Environmental Science & Policy*, **56**, 32–38, doi:10.1016/j.envsci.2015.11.004.

Olsson, L., M. Opondo, P. Tschakert, A. Agrawal, S.H. Eriksen, S. Ma, L.N. Perch, and S.A. Zakieldeen, 2014: Livelihoods and Poverty. In: Climate Change 2014: Impacts, Adaptation, and Vulnerability. Part A: Global and Sectoral Aspects. Contribution of Working Group II to the Fifth Assessment Report of the Intergovernmental Panel on Climate Change [Field, C.B., V.R. Barros, D.J. Dokken, K.J. Mach, M.D. Mastrandrea, T.E. Bilir, M. Chatterjee, K.L. Ebi, Y.O. Estrada, R.C. Genova, B. Girma, E.S. Kissel, A.N. Levy, S. MacCracken, P.R. Mastrandrea, and L.L. White (eds.)]. Cambridge University Press, Cambridge, UK, and New York, NY, USA, pp. 793–832.

Olsson, L., A. Jerneck, H. Thoren, J. Persson, and D. O'Byrne, 2015b: Why resilience is unappealing to social science: Theoretical and empirical investigations of the scientific use of resilience. *Sci. Adv.*, **1**, 1–12, doi:10.1126/sciadv.1400217.

Olsson, P., L.H. Gunderson, S.R. Carpenter, P. Ryan, L. Lebel, C. Folke, and C.S. Holling, 2006: Shooting the rapids: Navigating transitions to adaptive governance of social-ecological systems. *Ecol. Soc.*, **11**, ART. 18, 1–18

Onibon, A., B. Dabiré, and L. Ferroukhi, 1999: Local practices and the decentralization and devolution of natural resource management in French-speaking West Africa. *Unasylva*, **50**, no. 4.

Oppenheimer, M., M. Campos, R. Warren, J. Birkmann, G. Luber, B. O'Neill, and K. Takahashi, 2014: Emergent Risks and Key Vulnerabilities. Climate Change 2014 Impacts, Adaptation and Vulnerability: Part A: Global and Sectoral Aspects. *Contribution of Working Group II to the Fifth Assessment Report of the Intergovernmental Panel on Climate Change* [Field, C.B., V.R. Barros, D.J. Dokken, K.J. Mach, M.D. Mastrandrea, T.E. Bilir, M. Chatterjee, K.L. Ebi, Y.O. Estrada, R.C. Genova, B. Girma, E.S. Kissel, A.N. Levy, S. MacCracken, P.R. Mastrandrea, and L.L. White (eds.)]. Cambridge University Press, Cambridge, United Kingdom and New York, NY, USA, pp. 1039–1100.

OECD, 2008: *Economic Aspects of Adaptation to Climate Change: Costs, Benefits and Policy Instruments*. OECD Development Centre, Paris, France, 133 pp.

Orlove, B., C. Roncoli, M. Kabugo, and A. Majugu, 2010: Indigenous climate knowledge in southern Uganda: The multiple components of a dynamic regional system. *Clim. Change*, **100**, 243–265, doi:10.1007/s10584-009-9586-2.

Orlowsky, B., and S.I. Seneviratne, 2012: Global changes in extreme events: Regional and seasonal dimension. *Clim. Change*, **110**, 669–696, doi:10.1007/s10584-011-0122-9.

Orr, A.L. et al., 2017: *Scientific Conceptual Framework for Land Degradation Neutrality. A Report of the Science-Policy Interface*. United Nations Convention to Combat Desertification (UNCCD), Bonn, Germany, 128 pp.

Osei-Tutu, P., M. Pregernig, and B. Pokorny, 2014: Legitimacy of informal institutions in contemporary local forest management: Insights from Ghana. *Biodivers. Conserv.*, **23**, 3587–3605, doi:10.1007/s10531-014-0801-8.

Ostrom, E., 2010: Beyond markets and states: Polycentric governance of complex economic systems. *Am. Econ. Rev.*, **100**, 641–672, doi:10.1257/aer.100.3.641.

Ostrom, E., 2011: Background on the institutional analysis and development framework. *Policy Stud. J.*, **39**, 7–27, doi:10.1111/j.1541-0072.2010.00394.x.

Ostrom, E., 2012: Nested externalities and polycentric institutions: Must we wait for global solutions to climate change before taking actions at other scales? *Econ. Theory*, **49**, 353–369, doi:10.1007/s00199-010-0558-6.

Oteros-Rozas, E. et al., 2013: Traditional ecological knowledge among transhumant pastoralists in Mediterranean Spain. *Ecol. Soc.*, **18**, art33, doi:10.5751/ES-05597-180333.

Otto, F.E. L. et al., 2015: Explaining extreme events of 2014 from a climate perspective: Factors other than climate change, main drivers of 2014/2015 water shortage in Southeast Brazil. *Bull. Am. Meteorol. Soc.*, **96**, S35–S40, doi:10.1175/BAMS-D-15-00120.1.

Van Oudenhoven, A.P. E., K. Petz, R. Alkemade, L. Hein, and R.S. de Groot, 2012: Framework for systematic indicator selection to assess effects of land management on ecosystem services. *Ecol. Indic.*, **21**, 110–122, doi:10.1016/j.ecolind.2012.01.012.

Outka, U., 2012: Environmental law and fossil fuels: Barriers to renewable energy. *Vanderbilt Law Rev.*, **65**, 1679–1721.

Outreville, J.F., 2011a: The relationship between insurance growth and economic development – 80 empirical papers for a review of the literature. ICER Working Papers 12-2011, ICER – International Centre for Economic Research, Torino, Italy, 51 pp.

Owen, R., G. Brennan, and F. Lyon, 2018: Enabling investment for the transition to a low carbon economy: Government policy to finance early stage green innovation. *Curr. Opin. Environ. Sustain.*, **31**, 137–145, doi:10.1016/j.cosust.2018.03.004.

Pacetti, T., E. Caporali, and M.C. Rulli, 2017: Floods and food security: A method to estimate the effect of inundation on crops availability. *Adv. Water Resour.*, 110, 494–504, doi:10.1016/j.advwatres.2017.06.019.

Pachauri, S., and L. Jiang, 2008: The household energy transition in India and China. *Energy Policy*, **36**, 4022–4035, doi:10.1016/j.enpol.2008.06.016.

Pachauri, S., N.D. Rao, and C. Cameron, 2018: Outlook for modern cooking energy access in Central America. *PLoS One*, **13**, e0197974, doi:10.1371/journal.pone.0197974.

Pacheco, P., D. Barry, P. Cronkleton, and A. Larson, 2012: The recognition of forest rights in Latin America: Progress and shortcomings of forest tenure reforms. *Soc. Nat. Resour.*, **25**, 556–571, doi:10.1080/08941920.2011.574314.

Pacheco, P., R. Poccard-Chapuis, I. Garcia Drigo, M.-G. Piketty, and M. Thales, 2016: *Linking Sustainable Production and Enhanced Landscape Governance in the Amazon: Towards Territorial Certification (Terracert)*. CIRAD, Montpellier, France.

Pagdee, A., Y.S. Kim, and P.J. Daugherty, 2006: What makes community forest management successful: A meta-study from community forests throughout the world. *Soc. Nat. Resour.*, **19**, 33–52, doi:10.1080/08941920500323260.

Pahl-Wostl, C., 2009: A conceptual framework for analysing adaptive capacity and multi-level learning processes in resource governance regimes. *Glob. Environ. Chang.*, **19**, 354–365, doi:10.1016/j.gloenvcha.2009.06.001.

Pahl-Wostl, C., 2017a: Governance of the water-energy-food security nexus: A multi-level coordination challenge. *Environmental Science and Policy*, **92**, 356–367, doi:10.1016/j.envsci.2017.07.017.

Pahl-Wostl, C., 2017b: An evolutionary perspective on water governance: From understanding to transformation. *Water Resour. Manag.*, **31**, 2917–2932, doi:10.1007/s11269-017-1727-1.

Pahl-Wostl, C. et al., 2007: Managing change toward adaptive water management through social learning. *Ecol. Soc.*, **12**, 1–18. doi:30.

Pahl-Wostl, C. et al., 2013: Towards a sustainable water future: Shaping the next decade of global water research. *Curr. Opin. Environ. Sustain.*, **5**, 708–714, doi:10.1016/j.cosust.2013.10.012.

Pahl-Wostl, C., A. Bhaduri, and A. Bruns, 2018a: Editorial special issue: The nexus of water, energy and food – An environmental governance perspective. *Environ. Sci. Policy*, **90**, 161–163, doi:10.1016/j.envsci.2018.06.021.

Palmer, J.R., 2014: Biofuels and the politics of land use change: Tracing the interactions of discourse and place in European policy making. *Environ. Plan. A*, **46**, 337–352, doi:10.1068/a4684.

Palomo, I., M.R. Felipe-Lucia, E.M. Bennett, B. Martín-López, and U. Pascual, 2016: Disentangling the pathways and effects of ecosystem service co-production. *Advances in Ecological Research*, 54, 245–283, doi:10.1016/bs.aecr.2015.09.003.

Pandey, A. et al., 2017: Aerosol emissions factors from traditional biomass cookstoves in India: Insights from field measurements. *Atmos. Chem. Phys.*, **17**, 13721–13729, doi:10.5194/acp-17-13721-2017.

Pannell, D., 2008: Public benefits, private benefits, and policy mechanism choice for land use change for environmental benefits. *Land Econ.*, **84**, 225–240, doi:10.3368/le.84.2.225.

Panteli, M., and P. Mancarella, 2015: Influence of extreme weather and climate change on the resilience of power systems: Impacts and possible mitigation strategies. *Electr. Power Syst. Res.*, **127**, 259–270, doi:10.1016/j.epsr.2015.06.012.

Papaioannou, K.J., 2016: Climate shocks and conflict: Evidence from colonial Nigeria. *Polit. Geogr.*, **50**, 33–47, doi:10.1016/j.polgeo.2015.07.001.

Papathoma-Köhle, M., C. Promper, and T. Glade, 2016: A common methodology for risk assessment and mapping of climate change related hazards – Implications for climate change adaptation policies. *Climate*, 4, 8, doi:10.3390/cli4010008.

Park, S.E., N. Marshall, E. Jakku, A. Dowd, S. Howden, E. Mendham, and A. Fleming, 2012: Informing adaptation responses through theories of transformation. *Glob. Environ. Chang.*, **22**, 115–126, doi:10.1016/j.gloenvcha.2011.10.003.

Parnell, S., and R. Walawege, 2011: Sub-Saharan African urbanisation and global environmental change. *Glob. Environ. Chang.*, **21**, S12–S20, doi:10.1016/j.gloenvcha.2011.09.014.

Pascual, U. et al., 2017: Valuing nature's contributions to people: The IPBES approach. *Curr. Opin. Environ. Sustain.*, **26–27**, 7–16, doi:10.1016/j.cosust.2016.12.006.

Pathirana, A., Radhakrishnan, M., Ashley, R. et al, 2018: Managing urban water systems with significant adaptation deficits– Unified framework for secondary cities: Part II– The practice. *Clim. Change*, **149**, 57–74. doi:10.1007/s10584-017-2059-0.

Patterson, J. et al., 2017: Exploring the governance and politics of transformations towards sustainability. *Environ. Innov. Soc. Transitions*, **24**, 1–16, doi:10.1016/j.eist.2016.09.001.

Patz, J.A. et al., 2004: Unhealthy landscapes: Policy recommendations on land use change and infectious disease emergence. *Environ. Health Perspect.*, **112**, 1092–1098, doi:10.1289/EHP.6877.

Paul, S., S. Ghosh, K. Rajendran, and R. Murtugudde, 2018: Moisture supply from the Western Ghats forests to water deficit east coast of India. *Geophys. Res. Lett.*, **45**, 4337–4344, doi:10.1029/2018GL078198.

Paveglio, T.B., C. Kooistra, T. Hall, and M. Pickering, 2016: Understanding the effect of large wildfires on residents' well-being: What factors influence wildfire impact? *Forest Science*, **62**, 59–69, doi:10.5849/forsci.15-021.

Payne, G., 2001: Urban land tenure policy options: Titles or rights? *Habitat Int.*, **3**, 415–429, doi:10.1016/S0197-3975 (01)00014-5 .

Pearson, T.R.H., S. Brown, L. Murray, and G. Sidman, 2017: Greenhouse gas emissions from tropical forest degradation: An underestimated source. *Carbon Balance Manag.*, **12**, 3, doi:10.1186/s13021-017-0072-2.

Pecl, G.T. et al., 2017: Biodiversity redistribution under climate change: Impacts on ecosystems and human well-being. *Science*, **355**, eaai9214, doi:10.1126/science.aai9214.

Peel, J., and H.M. Osofsky, 2017: A Rights Turn in Climate Change Litigation? *Transnational Environmental Law*, **7**, 37–67, doi:10.1017/S2047102517000292.

Pelling, M., 2010: *Adaptation to Climate Change: From Resilience to Transformation: From Resilience to Transformation*. Routledge, Abingdon, UK, and New York, USA, 224 pp.

Pelling, M., and B. Wisner, 2012: African cities of hope and risk. In: *Disaster Risk Reduction: Cases from Urban Africa* [Pelling, M., B. Wisner (eds.)]. Routledge, London, UK, pp. 17–42.

Pendrill, F., M. Persson, J. Godar, and T. Kastner, 2019: Deforestation displaced: Trade in forest-risk commodities and the prospects for a global forest transition. *Environ. Res. Lett.*, **14**, 5, doi:10.1088/1748-9326/ab0d41.

Pereira, H.M. et al., 2010: Scenarios for global biodiversity in the 21st century. *Science*, **330**, 1496–1501, doi:10.1126/science.1196624.

Perera, N., E. Boyd Gill Wilkins, and R. Phillips Itty, 2015: *Literature Review on Energy Access and Adaptation to Climate Change*. UK Department for International Development, London, UK, 89 pp.

Pérez, I., M.A. Janssen, and J.M. Anderies, 2016: Food security in the face of climate change: Adaptive capacity of small-scale social-ecological systems to environmental variability. *Glob. Environ. Chang.*, **40**, 82–91, doi:10.1016/j.gloenvcha.2016.07.005.

Perry, J., 2015: Climate change adaptation in the world's best places: A wicked problem in need of immediate attention. *Landsc. Urban Plan.*, **133**, 1–11, doi:10.1016/j.landurbplan.2014.08.013.

Persha, L., and K. Andersson, 2014: Elite capture risk and mitigation in decentralized forest governance regimes. *Glob. Environ. Chang.*, **24**, 265–276, doi:10.1016/J.GLOENVCHA.2013.12.005.

Persha, L., A. Agrawal, and A. Chhatre, 2011: Social and ecological synergy: Local rulemaking, forest livelihoods, and biodiversity conservation. *Science*, **331**, 1606–1608, doi:10.1126/science.1199343.

Persson, J., E.L. Johansson, and L. Olsson, 2018: Harnessing local knowledge for scientific knowledge production: Challenges and pitfalls within evidence-based sustainability studies. *Ecol. Soc.*, **23**, art38, doi:10.5751/ES-10608-230438.

Pert, P.L. et al., 2015: Mapping cultural ecosystem services with rainforest aboriginal peoples: Integrating biocultural diversity, governance and social variation. *Ecosyst. Serv.*, **13**, 41–56, doi:10.1016/j.ecoser.2014.10.012.

Von Peter, G., S. Von Dahlen, and S. Saxena, 2012: *Unmitigated disasters? New evidence on the macroeconomic cost of natural catastrophes*. BIS Working Papers No. 394, BIS, Basel, Switzerland, www.bis.org.

Pezzey, J.C.V., 2019: Why the social cost of carbon will always be disputed. *Wiley Interdiscip. Rev. Clim. Chang.*, **10**, 1–12, doi:10.1002/wcc.558.

Pielke, R.A. et al., 2002: The influence of land use change and landscape dynamics on the climate system: Relevance to climate-change policy beyond the radiative effect of greenhouse gases. *Philos. Trans. R. Soc. A Math. Phys. Eng. Sci.*, **360**, 1705–1719, doi:10.1098/rsta.2002.1027.

Pieraccini, M., 2015: Rethinking participation in environmental decision-making: Epistemologies of marine conservation in Southeast England. *J. Environ. Law*, **27**, 45–67, doi:10.1093/jel/equ035.

Pierson, F.B., and C.J. Williams, 2016: *Ecohydrologic Impacts of Rangeland Fire on Runoff and Erosion: A Literature Synthesis*. General Technical Report (GTR), Department of Agriculture, Forest Service, Rocky Mountain Research Station, Fort Collins, USA, 110 pp.

Pierson, F.B. et al., 2011: Fire, plant invasions, and erosion events on Western Rangelands. *Rangel. Ecol. Manag.*, **64**, 439–449, doi:10.2111/REM-D-09-00147.1.

Pingali, P., 2015: Agricultural policy and nutrition outcomes – Getting beyond the preoccupation with staple grains. *Food Secur.*, **7**, 583–591, doi:10.1007/s12571-015-0461-x.

Pingali, P.L., 2012: Green Revolution: Impacts, limits, and the path ahead. *Proc. Natl. Acad. Sci.*, 31, 12302–12308, doi:10.1073/pnas.0912953109.

Pinho, P., R. Maia, and A. Monterroso, 2007: The quality of Portuguese Environmental Impact Studies: The case of small hydropower projects. *Environ. Impact Assess. Rev.*, **27**, 189–205, doi:10.1016/j.eiar.2006.10.005.

Pistorius, T., S. Reinecke, and A. Carrapatoso, 2017: A historical institutionalist view on merging LULUCF and REDD+ in a post-2020 climate agreement. *Int. Environ. Agreements Polit. Law Econ.*, **17**, 623–638, doi:10.1007/s10784-016-9330-0.

Pittelkow, C.M. et al., 2015: Productivity limits and potentials of the principles of conservation agriculture. *Nature*, **517**, 365–368, doi:10.1038/nature13809.

Plambeck, E.L., C. Hope, and J. Anderson, 1997: The Page95 model: Integrating the science and economics of global warming. *Energy Econ.*, **19**, 77–101, doi:10.1016/S0140-9883 (96)01008-0.

Plummer, R., and J. Baird, 2013: Adaptive co-management for climate change adaptation: Considerations for the barents region. *Sustain.*, **5**, 629–642, doi:10.3390/su5020629.

Poff, N.L. et al., 2003: River flows and water wars: Emerging science for environmental decision-making. *Front. Ecol. Environ.*, **1**, 298–306, doi:10.1890/1540-9295 (2003)001[0298:RFAWWE]2.0.CO; 2.

Pokharel, P., and S.X. Chang, 2019: Manure pellet, woodchip and their biochars differently affect wheat yield and carbon dioxide emission from bulk and rhizosphere soils. *Sci. Total Environ.*, **659**, 463–472, doi:10.1016/J.SCITOTENV.2018.12.380.

Popp, A. et al., 2014a: Land use transition for bioenergy and climate stabilization: Model comparison of drivers, impacts and interactions with other land use based mitigation options. *Clim. Change*, **123**, 495–509, doi:10.1007/s10584-013-0926-x.

Popp, A. et al., 2017: Land use futures in the shared socio-economic pathways. *Glob. Environ. Chang.*, **42**, 331–345, doi:10.1016/J.GLOENVCHA.2016.10.002.

Popp, J., K. Peto, and J. Nagy, 2013: Pesticide productivity and food security. A review. *Agron. Sustain. Dev.*, **33**, 243–255, doi:10.1007/s13593-012-0105-x.

Popp, J., Z. Lakner, M. Harangi-Rákos, and M. Fári, 2014b: The effect of bioenergy expansion: Food, energy, and environment. *Renew. Sustain. Energy Rev.*, **32**, 559–578, doi:10.1016/j.rser.2014.01.056.

Porras, I., and N. Asquith, 2018: *Ecosystems, Poverty Alleviation and Conditional Transfers Guidance for Practitioners*. IIED, London, UK, 59 pp.

Porter, J.R., L. Xie, A.J. Challinor, K. Cochrane, S.M. Howden, M.M. Iqbal, D.B. Lobell, and M.I. Travasso, 2014: Food security and food production systems. Climate Change 2014: Impacts, Adaptation, and Vulnerability. Part A: Global and Sectoral Aspects. Contribution of Working Group II to the Fifth Assessment Report of the Intergovernmental Panel on Climate Change [Field, C.B., V.R. Barros, D.J. Dokken, K.J. Mach, M.D. Mastrandrea, T.E. Bilir, M. Chatterjee, K.L. Ebi, Y.O. Estrada, R.C. Genova, B. Girma, E.S. Kissel, A.N. Levy, S. MacCracken, P.R. Mastrandrea, and L.L. White (eds.)]. Cambridge University Press, Cambridge, United Kingdom and New York, NY, USA, pp. 485–533.

Poudyal, M. et al., 2016: Can REDD+ social safeguards reach the 'right' people? Lessons from Madagascar. *Glob. Environ. Chang.*, **37**, 31–42, doi:10.1016/j.gloenvcha.2016.01.004.

Premalatha, M., T. Abbasi, and S.A. Abbasi, 2014a: Wind energy: Increasing deployment, rising environmental concerns. *Renew. Sustain. Energy Rev.*, **31**, 270–288, doi:10.1016/j.rser.2013.11.019.

Premalatha, M., T. Abbasi, and S.A. Abbasi, 2014b: A critical view on the eco-friendliness of small hydroelectric installations. *Sci. Total Environ.*, **481**, 638–643, doi:10.1016/j.scitotenv.2013.11.047.

Priefer, C., J. Jörissen, and O. Frör, 2017: Pathways to shape the bioeconomy. *Resources*, **6**, 10, doi:10.3390/resources6010010.

Prober, S.M. et al., 2017: Informing climate adaptation pathways in multi-use woodland landscapes using the values-rules-knowledge framework. *Agric. Ecosyst. Environ.*, **241**, 39–53, doi:10.1016/j.agee.2017.02.021.

Prokopy, L.S., K. Floress, D. Klotthor-Weinkauf, and A. Baumgart-Getz, 2008: Determinants of agricultural best management practice adoption: Evidence from the literature. *J. Soil Water Conserv.*, **63**, 300–311, doi:10.2489/jswc.63.5.300.

Prokopy, L.S. et al., 2015: Farmers and climate change: A cross-national comparison of beliefs and risk perceptions in high-income countries. *Environ. Manage.*, **56**, 492–504, doi:10.1007/s00267-015-0504-2.

Public Safety Canada, 2017: *2016–2017 Evaluation of the Disaster Financial Assistance Arrangements.* Public Safety Canada, Ottawa, Canada, www.publicsafety.gc.ca/cnt/rsrcs/pblctns/vltn-dsstr-fnncl-ssstnc-2016-17/index-en.aspx, 20 pp.

Puma, M.J., S. Bose, S.Y. Chon, and B.I. Cook, 2015: Assessing the evolving fragility of the global food system. *Environ. Res. Lett.*, **10**, 1–15, doi:10.1088/1748-9326/10/2/024007.

Pungetti, G., G. Oviedo, and D. Hooke, 2012: *Sacred Species and Sites: Advances in Biocultural Conservation.* Cambridge University Press, Cambridge, UK, and New York, NY, USA, 472 pp.

Purkus, Alexandra; Gawel, Erik; Thrän, D., 2012: *Bioenergy Governance Between Market and Government Failures: A New Institutional Economics Perspective.* UFZ Discussion Papers 13/2012, Helmholtz Centre for Environmental Research (UFZ), Division of Social Sciences (ÖKUS), Leipzig, Germany, 27 pp.

Purkus, A., E. Gawel, and D. Thrän, 2018: Addressing uncertainty in decarbonisation policy mixes – Lessons learned from German and European bioenergy policy. *Energy Res. Soc. Sci.*, **33**, 82–94, doi:10.1016/j.erss.2017.09.020.

Pynegar, E.L., J.P.G. Jones, J.M. Gibbons, and N.M. Asquith, 2018: The effectiveness of payments for ecosystem services at delivering improvements in water quality: Lessons for experiments at the landscape scale. *PeerJ*, **6**, e5753, doi:10.7717/peerj.5753.

Qasim, S., R.P. Shrestha, G.P. Shivakoti, and N.K. Tripathi, 2011: Socio-economic determinants of land degradation in Pishin sub-basin, Pakistan. *Int. J. Sustain. Dev. World Ecol.*, **18**, 48–54, doi:10.1080/13504509.2011.543844.

Qin, Z., J.B. Dunn, H. Kwon, S. Mueller, and M.M. Wander, 2016: Soil carbon sequestration and land use change associated with biofuel production: Empirical evidence. *GCB Bioenergy*, **8**, 66–80, doi:10.1111/gcbb.12237.

Quan, J., and N. Dyer, 2008: Climate Change and Land Tenure: The Implications of Climate Change for Land Tenure and Land Policy. Land Tenure Working Paper 2, Food and Agriculture Organization of the United Nations, Rome, Italy, 62 pp.

Quan, J., L.O. Naess, A. Newsham, A. Sitoe, and M.C. Fernandez, 2017: The political economy of REDD+ in Mozambique: Implications for climate compatible development. In: Making Climate Compatible Devevelopment Happen [Nunan, F. (ed)]. Routledge, London, UK, pp. 151–181, doi:10.4324/9781315621579.

Quatrini, S., and N.D. Crossman, 2018: Most finance to halt desertification also benefits multiple ecosystem services: A key to unlock investments in land degradation neutrality? *Ecosyst. Serv.*, 31, 265–277, doi:10.1016/j.ecoser.2018.04.003.

Quay, R., 2010: Anticipatory Governance. *J. Am. Plan. Assoc.*, **76**, 496–511, doi:10.1080/01944363.2010.508428.

Le Quesne, F., 2017: *The Role of Insurance in Integrated Disaster and Climate Risk Management: Evidence and Lessons Learned.* UNU-EHS, Bonn, Germany, 64 pp.

Quinn, A.K. et al., 2018: An analysis of efforts to scale up clean household energy for cooking around the world. *Energy Sustain. Dev.*, **46**, 1–10, doi:10.1016/j.esd.2018.06.011.

Quirion, P., 2009: Historic versus output-based allocation of GHG tradable allowances: A comparison. *Clim. Policy*, **9**, 575–592, doi:10.3763/cpol.2008.0618.

Quisumbing, A.R. et al., 2014: Closing the knowledge gap on gender in agriculture. In: *Gender in Agriculture* [Quisumbing, A.R., R. Meinzen-Dick, T.L. Raney, A. Croppenstedt, J.A. Behrman, A. Peterman (eds.)]. Springer Netherlands, Dordrecht, Netherlands, pp. 3–27.

Radeloff, V.C. et al., 2018: Rapid growth of the US wildland-urban interface raises wildfire risk. *Proc. Natl. Acad. Sci.*, **115**, 3314–3319.

Radhakrishnan, M., Nguyen, H., Gersonius, B. et al., 2018: Coping capacities for improving adaptation pathways for flood protection in Can Tho, Vietnam. *Clim. Change*, **149**, 29–41, doi:10.1007/s10584-017-1999-8.

Radhakrishnan, M., A. Pathirana, R. Ashley, and C. Zevenbergen, 2017: Structuring climate adaptation through multiple perspectives: Framework and case study on flood risk management. *Water*, **9**, 129, doi:10.3390/w9020129.

Rahman, M.M., M.N.I. Khan, A.K.F. Hoque, I. Ahmed, 2014: Carbon stock in the Sundarbans mangrove forest: Spatial variations in vegetation types and salinity zones. *Wetl. Ecol. Manag.*, **23**, 269–283, doi:10.1007/s11273-014-9379-x.

Rahman, H.M.T., S.K. Sarker, G.M. Hickey, M. Mohasinul Haque, and N. Das, 2014: Informal institutional responses to government interventions: Lessons from Madhupur National Park, Bangladesh. *Environ. Manage.*, **54**, 1175–1189, doi:10.1007/s00267-014-0325-8.

Rahman, M.A., 2018: Governance matters: Climate change, corruption, and livelihoods in Bangladesh. *Clim. Change*, **147**, 313–326, doi:10.1007/s10584-018-2139-9.

Rajamani, L., 2011: The cancun climate agreements: Reading the text, subtext and tea leaves. *Int. Comp. Law Q.*, **60**, 499–519, doi:10.1017/S0020589311000078.

Raju, K.V, A. Aziz, S.S.M. Sundaram, M. Sekher, S.P. Wani, and T.K. Sreedevi, 2008: *Guidelines for Planning and Implementation of Watershed Development Program in India: A Review. Global Theme on Agroecosystems Report no. 48.* International Crops Research Institute for the Semi-Arid Tropics, Patancheru, Hyderabad, India, http://oar.icrisat.org/2353/, 92 pp.

Raleigh, C., H.J. Choi, and D. Kniveton, 2015: The devil is in the details: An investigation of the relationships between conflict, food price and climate across Africa. *Glob. Environ. Chang.*, **32**, 187–199, doi:10.1016/j.gloenvcha.2015.03.005.

Ramnath, M., 2008: Surviving the Forest Rights Act: Between Scylla and Charybdis. *Econ. Polit. Wkly.*, **43**, 37–42.

Ramos, J.M., 2014: Anticipatory governance: Traditions and trajectories for strategic design. *J. Futur. Stud.*, **19**, 35–52.

Ranatunga, T., S.T.Y. Tong, Y. Sun, and Y.J. Yang, 2014: A total water management analysis of the Las Vegas Wash watershed, Nevada. *Phys. Geogr.*, **35**, 220–244, doi:10.1080/02723646.2014.908763.

Singh, R.K. et al., 2018. Classification and management of community forests in Indian Eastern Himalayas: Implications on ecosystem services, conservation and livelihoods. *Ecological Processes*, **7**, 27, 1–15 doi:10.1186/s13717-018-0137-5.

Singh, S.P., and M. Swanson, 2017: How issue frames shape beliefs about the importance of climate change policy across ideological and partisan groups. *PLoS One*, **12**, 1–14, doi:10.1371/journal.pone.0181401.

Rao, N., 2017a: Assets, agency and legitimacy: Towards a relational understanding of gender equality policy and practice. *World Dev.*, **95**, 43–54, doi:10.1016/j.worlddev.2017.02.018.

Rao, N.D., and S. Pachauri, 2017: Energy access and living standards: Some observations on recent trends. *Environ. Res. Lett.*, **12**, 025011, doi:10.1088/1748-9326/aa5b0d.

Rasul, G., 2014: Food, water, and energy security in South Asia: A nexus perspective from the Hindu Kush Himalayan region. *Environ. Sci. Policy*, **39**, 35–48, doi:10.1016/j.envsci.2014.01.010.

Rasul, G., and B. Sharma, 2016: The nexus approach to water-energy-food security: An option for adaptation to climate change. *Clim. Policy*, **16**, 682–702, doi:10.1080/14693062.2015.1029865.

Rauken, T., P.K. Mydske, and M. Winsvold, 2014: Mainstreaming climate change adaptation at the local level. *Local Environ.*, **20**, 408–423, doi:10.1 080/13549839.2014.880412.

Ravera, F., I. Iniesta-Arandia, B. Martín-López, U. Pascual, and P. Bose, 2016: Gender perspectives in resilience, vulnerability and adaptation to global environmental change. *Ambio*, **45**, 235–247, doi:10.1007/ s13280-016-0842-1.

Ray, B., and R. Shaw, 2016: Water stress in the megacity of kolkata, india, and its implications for urban resilience. In: *Urban Disasters and Resilience in Asia* [Shaw, R., Atta-ur-Rahman, A. Surjan, G. Ara Parvin (eds.)]. Elsevier, Oxford, UK, pp. 317–336.

Raza, W., and E. Poel, 2016: Impact and spill-over effects of an asset transfer program on malnutrition: Evidence from a randomized control trial in Bangladesh. *J. Health Econ.*, **62**, 105–120, doi:10.1016/j. jhealeco.2018.09.011.

Reddy, M.G., K.A. Kumar, P.T. Rao, and O. Springate-Baginski, 2011: Issues related to Implementation of the Forest rights Act in Andhra Pradesh. *Econ. Polit. Wkly.*, **46**, 73–81.

Reed, M. et al., 2010: What is Social Learning? *Ecol. Soc.*, **15**, r1.

Reed, M.S. et al., 2014: Improving the link between payments and the provision of ecosystem services in agri-environment schemes. *Ecosyst. Serv.*, **9**, 44–53, doi:10.1016/j.ecoser.2014.06.008.

Reichardt, K., K.S. Rogge, and S. Negro, 2015: Unpacking the policy processes for addressing systemic problems: The case of the technological innovation system of offshore wind in Germany. *Renewable and Sustainable Energy Reviews*, **80**, 1217–1226, doi:10.1016/j.rser.2017.05.280.

Reid, H., 2016: Ecosystem- and community-based adaptation: Learning from community-based natural resource management management. *Clim. Dev.*, **8**, 4–9, doi:10.1080/17565529.2015.1034233.

Reisinger, A., P. Havlik, K. Riahi, O. van Vliet, M. Obersteiner, and M. Herrero, 2013: Implications of alternative metrics for global mitigation costs and greenhouse gas emissions from agriculture. *Clim. Change*, **117**, 677–690, doi:10.1007/s10584-012-0593-3.

Ren, Z. et al., 2016: Predicting malaria vector distribution under climate change scenarios in China: Challenges for malaria elimination. *Sci. Rep.*, **6**, 20604, doi:10.1038/srep20604.

Renn, O., and P. Schweizer, 2009: Inclusive Risk Governance: Concepts andapplication to environmental policy making. *Environ. Policy Gov.*, **19**, 174–185, doi:10.1002/eet.507.

Reyer, C.P.O. et al., 2017: Turn down the heat: Regional climate change impacts on development. *Regional Environmental Change*, **17**, 1563–1568, doi:10.1007/s10113-017-1187-4.

Riahi, K. et al., 2017: The shared socio-economic pathways and their energy, land use, and greenhouse gas emissions implications: An overview. *Glob. Environ. Chang.*, **42**, 153–168, doi:10.1016/J.GLOENVCHA.2016.05.009.

Richards, M., T.B. Bruun, B.M. Campbell, L.E. Gregersen, S. Huyer, et al., 2015: *How Countries Plan to Address Agricultural Adaptation and Mitigation: An Analysis of Intended Nationally Determined Contributions.* CGIAR Research Program on Climate Change, Agriculture and Food Security (CCAFS), Copenhagen, Denmark, 1–8 pp.

Richards, M. et al., 2017: High-resolution spatial modelling of greenhouse gas emissions from land use change to energy crops in the United Kingdom. *GCB Bioenergy*, **9**, 627–644, doi:10.1111/gcbb.12360.

Richards, P., 1985a: Indigenous agricultural revolution: Ecology and food production in West Africa. *American Anthropology*, **89**, 240–241, doi:10.1525/aa.1987.89.1.02a01040.

Richards, P. (ed.), 1985b: *Indigenous Agricultural Revolution: Ecology and Food Production in West Africa.* Westview Press, Colorado, USA, 192 pp.

Richter, B.D. et al., 2017: Opportunities for saving and reallocating agricultural water to alleviate water scarcity. *Water Policy*, **19**, 886–907, doi:10.2166/ wp.2017.143.

Rietig, K., 2018: The links among contested knowledge, beliefs, and learning in European climate governance: From consensus to conflict in reforming biofuels policy. *Policy Stud. J.*, **46**, 137–159, doi:10.1111/psj.12169.

Rights and Resources Initiative, 2015a: *Who Owns the World's Land? A Global Baseline of Formally Recognized Indigenous and Community Land Rights.* Rights and Resources Initiative, Washington DC, USA, 44 pp.

Rights and Resources Initiative, 2018a: At a crossroads: Consequential trends in recognition of community-based forest tenure from 2002–2017. Rights and Resources Initiative, Washington DC, USA.

Rights and Resources Initiative, 2018b: At a Crossroads: Consequential Trends in Recognition of Community-based Forest Tenure From 2002–2017. *Rights Resour. Initiat.*,

Rigon, A., 2014: Building local governance: Participation and Elite capture in slum-upgrading in Kenya. *Dev. Change*, **45**, 257–283, doi:10.1111/ dech.12078.

Van Rijn, F., E. Bulte, and A. Adekunle, 2012: Social capital and agricultural innovation in Sub-Saharan Africa. *Agric. Syst.*, **108**, 112–122, doi:10.1016/j. agsy.2011.12.003.

Riley, M., H. Sangster, H. Smith, R. Chiverrell, and J. Boyle, 2018: Will farmers work together for conservation? The potential limits of farmers' cooperation in agri-environment measures. *Land Use Policy*, **70**, 635–646, doi:10.1016/J.LANDUSEPOL.2017.10.049.

Ring, I., and C. Schröter-Schlaack, 2011: *Instruments Mixes for Biodiversity Policies.* Helmholtz Centre for Environmental Research – UFZ, Leipzig, Germany, 119–144 pp.

Ringler, E., A. Pašukonis, W.T. Fitch, L. Huber, W. Hödl, and M. Ringler, 2015: Flexible compensation of uniparental care: Female poison frogs take over when males disappear. *Behav. Ecol.*, **26**, 1219–1225, doi:10.1093/ beheco/arv069.

del Río, P., and E. Cerdá, 2017: The missing link: The influence of instruments and design features on the interactions between climate and renewable electricity policies. *Energy Res. Soc. Sci.*, **33**, 49–58, doi:10.1016/j. erss.2017.09.010.

Rivera-Ferre, M.G. et al., 2016: Local agriculture traditional knowledge to ensure food availability in a changing climate: Revisiting water management practices in the Indo-Gangetic Plains. *Agroecol. Sustain. Food Syst.*, **40**, 965–987, doi:10.1080/21683565.2016.1215368.

Roberts, J.T. et al., 2017: How will we pay for loss and damage? *Ethics, Policy Environ.*, **20**, 208–226, doi:10.1080/21550085.2017.1342963.

Roberts, M.J., and W. Schlenker, 2013: Identifying supply and demand elasticities of agricultural commodities: Implications for the US ethanol mandate. *Am. Econ. Rev.*, 103, 2265–95, doi:10.1257/aer.103.6.2265.

Robinson, B.E., M.B. Holland, and L. Naughton-Treves, 2014: Does secure land tenure save forests? A meta-analysis of the relationship between land tenure and tropical deforestation. *Glob. Environ. Chang.*, **29**, 281–293, doi:10.1016/J.GLOENVCHA.2013.05.012.

Robledo-Abad, C. et al., 2017: Bioenergy production and sustainable development: Science base for policymaking remains limited. *GCB Bioenergy*, **9**, 541–556, doi:10.1111/gcbb.12338.

Rocca, M.E., P.M. Brown, L.H. MacDonald, and C.M. Carrico, 2014: Climate change impacts on fire regimes and key ecosystem services in Rocky Mountain forests. *For. Ecol. Manage.*, **327**, 290–305, doi:10.1016/j. foreco.2014.04.005.

Rochecouste, J.-F., P. Dargusch, D. Cameron, and C. Smith, 2015: An analysis of the socio-economic factors influencing the adoption of conservation agriculture as a climate change mitigation activity in Australian dryland grain production. *Agric. Syst.*, **135**, 20–30, doi:10.1016/j.agsy.2014.12.002.

Rocheleau, D., and D. Edmunds, 1997: Women, men and trees: Gender, power and property in forest and agrarian landscapes. *World Dev.*, **25**, 1351–1371, doi:10.1016/S0305-750X (97)00036-3.

Rockström, Johan Steffen, W. et al., 2009: A safe operating space for humanity. *Nature*, doi:10.1038/461472a.

Rockström, J. et al., 2009: A safe operating space for humanity. *Nature*, **461**, 472–475, doi:10.1038/461472a.

Rockström, J. et al., 2017: Sustainable intensification of agriculture for human prosperity and global sustainability. *Ambio*, **46**, 4–17, doi:10.1007/s13280-016-0793-6.

Rocle, N., and D. Salles, 2018: 'Pioneers but not guinea pigs': Experimenting with climate change adaptation in French coastal areas. *Policy Sci.*, **51**, 231–247, doi:10.1007/s11077-017-9279-z.

Rodell, M., I. Velicogna, and J.S. Famiglietti, 2009: Satellite-based estimates of groundwater depletion in India. *Nature*, **460**, 999–1002, doi:10.1038/nature08238.

Rodríguez-Morales, J.E., 2018: Convergence, conflict and the historical transition of bioenergy for transport in Brazil: The political economy of governance and institutional change. *Energy Res. Soc. Sci.*, **44**, 324–335, doi:10.1016/j.erss.2018.05.031.

Rodriguez-Takeuchi, L., and K.S. Imai, 2013: Food price surges and poverty in urban colombia: New evidence from household survey data. *Food Policy*, **43**, 227–236, doi:10.1016/j.foodpol.2013.09.017.

Rodriguez-Ward, D., A.M. Larson, and H.G. Ruesta, 2018: Top-down, bottom-up and sideways: The multilayered complexities of multi-level actors shaping forest governance and REDD+ arrangements in Madre de Dios, Peru. *Environ. Manage.*, **62**, 98–116, doi:10.1007/s00267-017-0982-5.

Rodríguez, J., T.D. Beard Jr., E. Bennett, G. Cumming, S. Cork, J. Agard, A. Dobson, and G. Peterson, 2006: Trade-offs across space, time, and ecosystem services. *Ecol. Soc.*, **11**, ART. 28.

Rodríguez Morales, J.E., and F. Rodríguez López, 2017: The political economy of bioenergy in the United States: A historical perspective based on scenarios of conflict and convergence. *Energy Res. Soc. Sci.*, 27, 141–150, doi:10.1016/j.erss.2017.03.002.

Roelich, K., and J. Giesekam, 2019: Decision-making under uncertainty in climate change mitigation: Introducing multiple actor motivations, agency and influence. *Clim. Policy*, **19**, 175–188, doi:10.1080/14693062.2018.1479238.

Rogelj, J. et al., 2016: Paris Agreement climate proposals need a boost to keep warming well below 2°C. *Nature*, **534**, 631–639, doi:10.1038/nature18307.

Rogelj, J., D. Shindell, K. Jiang, S. Fifita, P. Forster, V. Ginzburg, C. Handa, H. Kheshgi, S. Kobayashi, E. Kriegler, L. Mundaca, R. Séférian, and M.V.Vilariño, 2018: Mitigation Pathways Compatible with 1.5°C in the Context of Sustainable Development. In: Global Warming of 1.5°C: An IPCC special report on the impacts of global warming of 1.5°C above pre-industrial levels and related global greenhouse gas emission pathways, in the context of strengthening the global response to the threat of climate change [Masson-Delmotte, V., P. Zhai, H.-O. Pörtner, D. Roberts, J. Skea, P.R. Shukla, A. Pirani, W. Moufouma-Okia, C. Péan, R. Pidcock, S. Connors, J.B.R. Matthews, Y. Chen, X. Zhou, M.I. Gomis, E. Lonnoy, T. Maycock, M. Tignor, and T. Waterfield (eds.)]. Cambridge University Press, Cambridge, UK, and New York, NY, USA, 93–174.

Rogge, K.S., and K. Reichardt, 2016: Policy mixes for sustainability transitions: An extended concept and framework for analysis. *Res. Policy*, 45, 1620–1635, doi:10.1016/j.respol.2016.04.004.

Romijn, H.A., 2011: Land clearing and greenhouse gas emissions from Jatropha biofuels on African Miombo Woodlands. *Energy Policy*, **39**, 5751–5762, doi:10.1016/J.ENPOL.2010.07.041.

Roncoli, C., C. Jost, C. Perez, K. Moore, A. Ballo, S. Cissé, and K. Ouattara, 2007: Carbon sequestration from common property resources: Lessons from community-based sustainable pasture management in north-central Mali. *Agric. Syst.*, **94**, 97–109, doi:10.1016/j.agsy.2005.10.010.

Rosenthal, J., A. Quinn, A.P. Grieshop, A. Pillarisetti, and R.I. Glass, 2017: Clean cooking and the SDGs: Integrated analytical approaches to guide energy interventions for health and environment goals. *Energy for Sustainable Development*, **42**, 152–159, doi:10.1016/j.esd.2017.11.003.

Rosenzweig, C. et al., 2014: Assessing agricultural risks of climate change in the 21st century in a global gridded crop model intercomparison. *Proc. Natl. Acad. Sci.*, **111**, 3268–3273, doi:10.1073/pnas.1222463110.

Rosillo Callé, F., and F.X. Johnson (eds.), 2010a: *Food versus fuel: An Informed Introduction to Biofuels.* Zed Books, London, UK, 217 pp.

Rosin, C., 2013: Food security and the justification of productivism in New Zealand. *J. Rural Stud.*, **29**, 50–58, doi:10.1016/j.jrurstud.2012.01.015.

Roudier, P., B. Muller, P. Aquino, C. Roncoli, M.A. Soumaré, L. Batté, and B. Sultan, 2014: The role of climate forecasts in smallholder agriculture: Lessons from participatory research in two communities in Senegal. *Clim. Risk Manag.*, 2, 42–55, doi:10.1016/j.crm.2014.02.001.

Rouillard, J., D. Benson, A.K. Gain, and C. Giupponi, 2017: Governing for the nexus: Empirical, theoretical and normative Dimensions. In: *Water-Energy-Food Nexus: Principles and Practices* [Salam, P.A., S. Shrestha , V.P. Pandey, A.K. Anal (eds.)]. John Wiley & Sons Inc., New Jersey, USA.

Rouillard, J.J., K.V. Heal, T. Ball, and A.D. Reeves, 2013: Policy integration for adaptive water governance: Learning from Scotland's experience. *Environ. Sci. Policy*, **33**, 378–387, doi:10.1016/j.envsci.2013.07.003.

Roy, J., P. Tschakert, H. Waisman, P. Abdul Halim, P. Antwi-Agyei, P. Dasgupta, B. Hayward, M. Kanninen, D. Liverman, C. Okereke, P.F. Pinho, K. Riahi, and A.G. Suarez Rodriguez, 2018: Sustainable Development , Poverty Eradication and Reducing Inequalities. Global Warming of 1.5°C an IPCC special report on the impacts of global warming of 1.5°C above pre-industrial levels and related global greenhouse gas emission pathways, in the context of strengthening the global response to the threat of climate change [Masson-Delmotte, V., P. Zhai, H.-O. Pörtner, D. Roberts, J. Skea, P.R. Shukla, A. Pirani, W. Moufouma-Okia, C. Péan, R. Pidcock, S. Connors, J.B.R. Matthews, Y. Chen, X. Zhou, M.I. Gomis, E. Lonnoy, T. Maycock, M. Tignor, and T. Waterfield (eds.)]. Cambridge University Press, Cambridge, UK, and New York, NY, USA, 445–538.

Rozin, P., S. Scott, M. Dingley, J.K. Urbanek, H. Jiang, and M. Kaltenbach, 2011: Nudge to nobesity I: Minor changes in accessibility decrease food intake. *Judgm. Decis. Mak.*, **6**, 323–332.

Rudolph, D.L., J.F. Devlin, and L. Bekeris, 2015: Challenges and a strategy for agricultural BMP monitoring and remediation of nitrate contamination in unconsolidated aquifers. *Groundw. Monit. Remediat.*, **35**, 97–109, doi:10.1111/gwmr.12103.

Ruiz-Mercado, I., and O. Masera, 2015a: Patterns of stove use in the context of fuel-device stacking: rationale and implications. *Ecohealth*, **12**, 42–56, doi:10.1007/s10393-015-1009-4.

Rulli, M.C., A. Saviori, and P. D'Odorico, 2013: Global land and water grabbing. *Proc. Natl. Acad. Sci.*, **110**, 892–897, doi:10.1073/PNAS.1213163110.

Rulli, M.C., D. Bellomi, A. Cazzoli, G. De Carolis, and P. D'Odorico, 2016: The water-land-food nexus of first-generation biofuels. *Sci. Rep.*, **6**, 22521, doi:10.1038/srep22521.

Ryan, S.J., A. McNally, L.R. Johnson, E.A. Mordecai, T. Ben-Horin, K. Paaijmans, and K.D. Lafferty, 2015: Mapping physiological suitability limits for malaria in africa under climate change. *Vector-Borne Zoonotic Dis.*, **15**, 718–725, doi:10.1089/vbz.2015.1822.

Safriel, U., 2017: Land degradation neutrality (LDN) in drylands and beyond – Where has it come from and where does it go. *Silva Fenn.*, **51**, 1650, doi:10.14214/sf.1650.

Safriel, U., and Z. Adeel, 2008: Development paths of drylands: Thresholds and sustainability. *Sustain. Sci.*, **3**, 117–123, doi:10.1007/s11625-007-0038-5.

Salehyan, I., and C.S. Hendrix, 2014: Climate shocks and political violence. *Glob. Environ. Chang.*, **28**, 239–250, doi:10.1016/j.gloenvcha.2014.07.007.

Saluja, N and Singh, S., 2018: Coal-fired power plants set to get renewed push. *Economic Times*, New Delhi, India, https://economictimes.indiatimes.com/industry/energy/power/coal-fired-power-plants-set-to-get-renewed-push/articleshow/64769464.cms.

Salvati, L., and M. Carlucci, 2014: Zero Net Land Degradation in Italy: The role of socio-economic and agroforest factors. *J. Environ. Manage.*, **145**, 299–306, doi:10.1016/j.jenvman.2014.07.006.

7

Salzman, J., G. Bennett, N. Carroll, A. Goldstein, and M. Jenkins, 2018: The global status and trends of payments for ecosystem services. *Nat. Sustain.*, **1**, 136–144, doi:10.1038/s41893-018-0033-0.

Samaddar, S. et al., 2015: Evaluating effective public participation in disaster management and climate change adaptation: Insights from Northern Ghana through a user-based approach. *Risk, Hazards Cris. Public Policy*, **6**, 117–143, doi:10.1002/rhc3.12075.

Samanta, A., S. Ganguly, H. Hashimoto, S. Devadiga, E. Vermote, Y. Knyazikhin, R.R. Nemani, and R.B. Myneni, 2010: Amazon forests did not green-up during the 2005 drought. *Geophys. Res. Lett.*, **37**, 1–5, doi:10.1029/2009GL042154.

Samuwai, J., and J. Hills, 2018: Assessing climate finance readiness in the Asia-Pacific Region. *Sustainability*, **10**, 1–18, doi:10.3390/su10041192.

Sánchez, B. et al., 2016: Management of agricultural soils for greenhouse gas mitigation: Learning from a case study in NE Spain. *J. Environ. Manage.*, **170**, 37–49, doi:10.1016/j.jenvman.2016.01.003.

Sánchez, J.M.T., and R.C. Maseda, 2016: Forcing and avoiding change. Exploring change and continuity in local land use planning in Galicia (Northwest of Spain) and The Netherlands. *Land Use Policy*, **50**, 74–82, doi:10.1016/J.LANDUSEPOL.2015.09.006.

Sanderson, T., G. Hertzler, T. Capon, and P. Hayman, 2016: A real options analysis of Australian wheat production under climate change. *Aust. J. Agric. Resour. Econ.*, **60**, 79–96, doi:10.1111/1467-8489.12104.

Sandifer, P.A., A.E. Sutton-Grier, and B.P. Ward, 2015: Exploring connections among nature, biodiversity, ecosystem services, and human health and well-being: Opportunities to enhance health and biodiversity conservation. *Ecosyst. Serv.*, **12**, 1–15, doi:10.1016/j.ecoser.2014.12.007.

Sandstrom, S., and S. Juhola, 2017: Continue to blame it on the rain? Conceptualization of drought and failure of food systems in the Greater Horn of Africa. *Environ. Hazards*, **16**, 71–91, doi:10.1080/17477891.2016.1229656.

Santos, M.J., S.C. Dekker, V. Daioglou, M.C. Braakhekke, and D.P. van Vuuren, 2017: Modeling the Effects of Future Growing Demand for Charcoal in the Tropics. *Front. Environ. Sci.*, **5**, 1–12, doi:10.3389/fenvs.2017.00028.

Sanz, M.J. et al., 2017: *Sustainable Land Management Contribution to Successful Land-Based Climate Change Adaptation and Mitigation. A Report of the Science-Policy Interface*. A Report of the Science-Policy Interface. United Nations Convention to Combat Desertification (UNCCD), Bonn, Germany, 170 pp.

Sarap, K., T.K. Sarangi, and J. Naik, 2013: Implementation of Forest Rights Act 2006 in Odisha: Process, constraints and outcome. *Econ. Polit. Wkly.*, **48**, 61–67.

Sarzynski, A., 2015: Public participation, civic capacity, and climate change adaptation in cities. *Urban Clim.*, **14**, 52–67, doi:10.1016/J.UCLIM.2015.08.002.

Sassi, F. et al., 2018: Equity impacts of price policies to promote healthy behaviours. *The Lancet*, **391**, 2059–2070, doi:10.1016/S0140-6736(18)30531-2.

Satterthwaite, D. (ed.), 2007: *Climate Change and Urbanization: Effects and Implications for Urban Governance*. UNDESA, United Nations Expert Group Meeting on Population Distribution, Urbanization, Internal Migration and Development, New York, USA, 29 pp.

Satterthwaite, D., D. Archer, S. Colenbrander, D. Dodman, J. Hardoy, and S. Patel, 2018: *Responding to climate change in cities and in their informal settlements and economies*. IIED and IIED-América Latina, London, UK, 61 pp.

Sauerwald, S., and M.W. Peng, 2013: Informal institutions, shareholder coalitions, and principal-principal conflicts. *Asia Pacific J. Manag.*, **30**, 853–870, doi:10.1007/s10490-012-9312-x.

Savaresi, A., 2016: The Paris Agreement: A new beginning? *J. Energy Nat. Resour. Law*, **34**, 16–26, doi:10.1080/02646811.2016.1133983.

Sayer, J., C. Margules, I.C. Bohnet, A.K. Boedhihartono, 2015: The role of citizen science in landscape and seascape approaches to integrating conservation and development. *Land*, **4**, 1200–1212, doi:10.3390/land4041200.

Scarano, F.R., 2017: Ecosystem-based adaptation to climate change: Concept, scalability and a role for conservation science. *Perspect. Ecol. Conserv.*, **15**, 65–73, doi:10.1016/j.pecon.2017.05.003.

Scarlat, N., and J.-F. Dallemand, 2011: Recent developments of biofuels/bioenergy sustainability certification: A global overview. *Energy Policy*, **39**, 1630–1646, doi:10.1016/J.ENPOL.2010.12.039.

Schalatek, L., and S. Nakhooda, 2013: The Green Climate Fund. *Clim. Financ. Fundam.*, **11**, Heinrich Boll Stiftung North America and Overseas Development Institute, Washington DC, USA and London, UK, pp. 1–4.

Scheffran, J., E. Marmer, and P. Sow, 2012: Migration as a contribution to resilience and innovation in climate adaptation: Social networks and co-development in Northwest Africa. *Appl. Geogr.*, **33**, 119–127, doi:10.1016/j.apgeog.2011.10.002.

Scherr, S.J., S. Shames, and R. Friedman, 2012: From climate-smart agriculture to climate-smart landscapes. *Agric. Food Secur.*, **1**, 12, doi:10.1186/2048-7010-1-12.

Schick, A. et al., 2018: People-centered and ecosystem-based knowledge co-production to promote proactive biodiversity conservation and sustainable development in Namibia. *Environ. Manage.*, **62**, 858–876, doi:10.1007/s00267-018-1093-7.

Schlager, E., and E. Ostrom, 1992: Property-rights regimes and natural resources: A conceptual analysis. *Land Econ.*, **68**, 249, doi:10.2307/3146375.

Schleussner, C.F. et al., 2016: Differential climate impacts for policy-relevant limits to global warming: The case of 1.5°C and 2°C. *Earth Syst. Dyn.*, **7**, 327–351, doi:10.5194/esd-7-327-2016.

Schmalensee, R., and R.N. Stavins, 2017: Lessons learned from three decades of experience with cap and trade. *Rev. Environ. Econ. Policy*, **11**, 59–79, doi:10.1093/reep/rew017.

Schmidhuber, J., and F.N. Tubiello, 2007: Global food security under climate change. *Proc. Natl. Acad. Sci. U.S.A.*, **104**, 19703–19708, doi:10.1073/pnas.0701976104.

Schmitz, C. et al., 2012: Trading more food: Implications for land use, greenhouse gas emissions, and the food system. *Glob. Environ. Chang.*, **22**, 189–209, doi:10.1016/j.gloenvcha.2011.09.013.

Schmitz, O.J. et al., 2015: Conserving biodiversity: Practical guidance about climate change adaptation approaches in support of land-use planning. *Source Nat. Areas J.*, **35**, 190–203, doi:10.3375/043.035.0120.

Schneider, L., and S. La Hoz Theuer, 2019: Environmental integrity of international carbon market mechanisms under the Paris Agreement. *Clim. Policy*, **19**, 386–400, doi:10.1080/14693062.2018.1521332.

Schröder, P. et al., 2018: Intensify production, transform biomass to energy and novel goods and protect soils in Europe – A vision how to mobilize marginal lands. *Sci. Total Environ.*, **616–617**, 1101–1123, doi:10.1016/j.scitotenv.2017.10.209.

Schultz, L., C. Folke, H. Österblom, and P. Olsson, 2015: Adaptive governance, ecosystem management, and natural capital. *Proc. Natl. Acad. Sci.*, **112**, 7369–7374, doi:10.1073/pnas.1406493112.

Schut, M., N.C. Soares, G. Van De Ven, and M. Slingerland, 2013: Multi-actor governance of sustainable biofuels in developing countries: The case of Mozambique. *Energy Policy*, **65**, 631–643, doi:10.1016/j.enpol.2013.09.007.

De Schutter, O., 2011: How not to think of land-grabbing: Three critiques of large-scale investments in farmland. *J. Peasant Stud.*, **38**, 249–279, doi:10.1080/03066150.2011.559008.

Schuur, E.A.G. et al., 2015: Climate change and the permafrost carbon feedback. *Nature*, **520**, 171–179, doi:10.1038/nature14338.

Schwartz, N.B., M. Uriarte, R. DeFries, V.H. Gutierrez-Velez, and M.A. Pinedo-Vasquez, 2017: Land use dynamics influence estimates of carbon sequestration potential in tropical second-growth forest. *Environ. Res. Lett.*, **12**, 074023, doi:10.1088/1748-9326/aa708b.

Schwilch, G. et al., 2011: Experiences in monitoring and assessment of sustainable land management. *L. Degrad. Dev.*, **22**, 214–225, doi:10.1002/ldr.1040.

Scott, C.A., S.A. Pierce, M.J. Pasqualetti, A.L. Jones, B.E. Montz, and J.H. Hoover, 2011: Policy and institutional dimensions of the water-energy nexus. *Energy Policy*, **39**, 6622–6630, doi:10.1016/j.enpol.2011.08.013.

Scott, D., C.M. Hall, and S. Gössling, 2016: A report on the Paris Climate Change Agreement and its implications for tourism: Why we will always have Paris. *J. Sustain. Tour.*, **24**, 933–948, doi:10.1080/09669582.2016.1187623.

Seager, J., 2014: Disasters are gendered: What's new? In: *Reducing Disaster: Early Warning Systems for Climate Change* [Singh, A. and Z. Zommers (eds.)]. Springer Netherlands, Dordrecht, Netherlands, pp. 265–281.

Seaman, J.A., G.E. Sawdon, J. Acidri, and C. Petty, 2014: The household economy approach. Managing the impact of climate change on poverty and food security in developing countries. *Clim. Risk Manag.*, **4–5**, 59–68, doi:10.1016/j.crm.2014.10.001.

Selvaraju, R., 2011: Climate risk assessment and management in agriculture. In: *Building Resilience for Adaptation to Climate Change in the Agriculture Sector* [Meybeck, A., J. Lankoski, S. Redfern, N. Azzu, V. Gitz (eds.)]. Proceedings of a Joint FAO/OECD Workshop, Food and Agriculture Organization of the United Nations, Rome, Italy, pp. 71–89.

Seng, D.C., 2012: Improving the governance context and framework conditions of natural hazard early warning systems. *J. Integr. Disaster Risk Manag.*, **2**, 1–25, doi:10.5595/idrim.2012.0020.

Serrao-Neumann, S., B.P. Harman, and D. Low Choy, 2013: The role of anticipatory governance in local climate adaptation: Observations from Australia. *Plan. Pract. Res.*, **28**, 440–463, doi:10.1080/02697459.2013.795788.

Serrao-Neumann, S., F. Crick, B. Harman, G. Schuch, and D.L. Choy, 2015a: Maximising synergies between disaster risk reduction and climate change adaptation: Potential enablers for improved planning outcomes. *Environ. Sci. Policy*, **50**, 46–61, doi:10.1016/j.envsci.2015.01.017.

Serrao-Neumann, S., B. Harman, A. Leitch, and D. Low Choy, 2015b: Public engagement and climate adaptation: insights from three local governments in Australia. *J. Environ. Plan. Manag.*, **58**, 1196–1216, doi:10.1080/09640568.2014.920306.

Seto, K.C., 2011: Exploring the dynamics of migration to mega-delta cities in Asia and Africa: Contemporary drivers and future scenarios. *Glob. Environ. Chang.*, **21**, S94-S107, doi:10.1016/j.gloenvcha.2011.08.005.

Shaaban, M. et al., 2018: A concise review of biochar application to agricultural soils to improve soil conditions and fight pollution. *J. Environ. Manage.*, **228**, 429–440, doi:10.1016/J.JENVMAN.2018.09.006.

Sharples, J.J. et al., 2016a: Natural hazards in Australia: Extreme bushfire. *Clim. Change*, **139**, 85–99, doi:10.1007/s10584-016-1811-1.

Sheil, D., M. Boissière, and G. Beaudoin, 2015: Unseen sentinels: Local monitoring and control in conservation's blind spots. *Ecol. Soc.*, **20**, art39, doi:10.5751/ES-07625-200239.

Sheng, J., X. Han, H. Zhou, and Z. Miao, 2016: Effects of corruption on performance: Evidence from the UN-REDD Programme. *Land Use Policy*, **59**, 344–350, doi:10.1016/j.landusepol.2016.09.014.

Shiferaw, B. et al., 2014: Managing vulnerability to drought and enhancing livelihood resilience in Sub-Saharan Africa: Technological, institutional and policy options. *Weather Clim. Extrem.*, **3**, 67–79, doi:10.1016/j.wace.2014.04.004.

Shindell, D. et al., 2012: Simultaneously mitigating near-term climate change and improving human health and food security. *Science*, **335**, 183–189, doi:10.1126/science.1210026.

Shogren, J.F., and L.O. Taylor, 2008: On behavioural-environmental economics. *Rev. Environ. Econ. Policy*, **2**, 26–44, doi:10.1093/reep/rem027.

Shreve, C.M., and I. Kelman, 2014: Does mitigation save? Reviewing cost-benefit analyses of disaster risk reduction. *International Journal of Disaster Risk Reduction*, **10**, 213–235, doi:10.1016/j.ijdrr.2014.08.004.

Shue, H., 2018a: Mitigation gambles: Uncertainty, urgency and the last gamble possible. *Philos. Trans. R. Soc. A Math. Eng. Sci.*, **376**, 20170105, doi:10.1098/rsta.2017.0105.

Shvidenko, A.Z., D.G. Shchepashchenko, E.A. Vaganov, A.I. Sukhinin, S.S. Maksyutov, I. McCallum, and I.P. Lakyda, 2012: Impact of wildfire in Russia between 1998–2010 on ecosystems and the global carbon budget. *Dokl. Earth Sci.*, **441**, 1678–1682, doi:10.1134/s1028334x11120075.

Siahaya, M.E., T.R. Hutauruk, H.S.E.S. Aponno, J.W. Hatulesila, and A.B. Mardhanie, 2016: Traditional ecological knowledge on shifting cultivation and forest management in East Borneo, Indonesia. *Int. J. Biodivers. Sci. Ecosyst. Serv. Manag.*, **12**, 14–23, doi:10.1080/21513732.2016.1169559.

Siddig, E.F.A., K. El-Harizi, and B. Prato, 2007: *Managing conflict over natural resources in greater Kordofan, Sudan: Some recurrent patterns and governance implications*. IFPRI Discussion Paper 00711, International Food Policy Research Institute, Washington DC, USA, 98 pp.

Siebert, A., 2016: Analysis of the future potential of index insurance in the West African Sahel using CMIP5 GCM results. *Clim. Change*, **134**, 15–28, doi:10.1007/s10584-015-1508-x.

Siegmeier, J. et al., 2018: The fiscal benefits of stringent climate change mitigation: An overview. **3062**, Climate Policy, **18**, 352–367, doi:10.1080/14693062.2017.1400943.

Sietz, D., L. Fleskens, and L.C. Stringer, 2017: Learning from non-linear ecosystem dynamics is vital for achieving land degradation neutrality. *L. Degrad. Dev.*, **28**, 2308–2314, doi:10.1002/ldr.2732.

Sigurdsson, J.H., L.A. Walls, and J.L. Quigley, 2001: Bayesian belief nets for managing expert judgement and modelling reliability. *Qual. Reliab. Eng. Int.*, **17**, 181–190, doi:10.1002/qre.410.

Silva-Olaya, A.M. et al., 2017: Modelling SOC response to land use change and management practices in sugarcane cultivation in South-Central Brazil. *Plant Soil*, **410**, 483–498, doi:10.1007/s11104-016-3030-y.

Silva, R.A. et al., 2013: Global premature mortality due to anthropogenic outdoor air pollution and the contribution of past climate change. *Environ. Res. Lett.*, **8**, 031002, doi:10.1088/1748-9326/8/3/034005.

Silva, R. A et al., 2016: The effect of future ambient air pollution on human premature mortality to 2100 using output from the ACCMIP model ensemble. *Atmos. Chem. Phys.*, **16**, 9847–9862, doi:10.5194/acp-16-9847-2016.

Silvano, R.A.M., and J. Valbo-Jørgensen, 2008: Beyond fishermen's tales: Contributions of fishers' local ecological knowledge to fish ecology and fisheries management. *Environ. Dev. Sustain.*, **10**, 657–675, doi:10.1007/s10668-008-9149-0.

Silveira, S., and F.X. Johnson, 2016: Navigating the transition to sustainable bioenergy in Sweden and Brazil: Lessons learned in a European and International context. *Energy Res. Soc. Sci.*, **13**, 180–193, doi:10.1016/j.erss.2015.12.021.

Silvertown, J., 2009: A new dawn for citizen science. *Trends in Ecology & Evolution*, **24**, 467–471, doi:10.1016/j.tree.2009.03.017.

Simonet, G., J. Subervie, D. Ezzine-de-Blas, M. Cromberg, and A.E. Duchelle, 2019: Effectiveness of a REDD+ project in reducing deforestation in the Brazilian Amazon. *Am. J. Agric. Econ.*, **101**, 211–229, doi:10.1093/ajae/aay028.

Sindhu, S., V. Nehra, and S. Luthra, 2017: Investigation of feasibility study of solar farms deployment using hybrid AHP-TOPSIS analysis: Case study of India. *Renew. Sustain. Energy Rev.*, **73**, 496–511, doi:10.1016/j.rser.2017.01.135.

Singh, R., P.M. Reed, and K. Keller, 2015: Many-objective robust decision-making for managing an ecosystem with a deeply uncertain threshold response. *Ecol. Soc.*, **20**, 1–32, doi:10.5751/ES-07687-200312.

Slater, R., 2011: Cash transfers, social protection and poverty reduction. *Int. J. Soc. Welf.*, **20**, 250–259, doi:10.1111/j.1468-2397.2011.00801.x.

Smeets, E., F.X. Johnson, and G. Ballard-Tremeer, 2012a: Traditional and improved use of biomass for energy in Africa. In: *Bioenergy for Sustainable*

Development in Africa [Janssen, R., D. Rutz (eds.)]. Springer Netherlands, Dordrecht, Netherlands, pp. 3–12.

Smith, C.B., 2011: Adaptive management on the central Platte River – Science, engineering, and decision analysis to assist in the recovery of four species. *J Env. Manag.*, **92**, 1414–1419, doi:10.1016/j.jenvman.2010.10.013.

Smith, H.E., F. Eigenbrod, D. Kafumbata, M.D. Hudson, and K. Schreckenberg, 2015: Criminals by necessity: The risky life of charcoal transporters in Malawi. *For. Trees Livelihoods*, **24**, 259–274, doi:10.1080/14728028.2015.1062808.

Smith, K.R. et al., 2014a: millions Dead: How do we know and what does it mean? Methods used in the comparative risk assessment of household air pollution. *Annu. Rev. Public Health*, **35**, 185–206, doi:10.1146/annurev-publhealth-032013-182356.

Smith, P., and J.E. Olesen, 2010: Synergies between the mitigation of, and adaptation to, climate change in agriculture. *J. Agric. Sci.*, **148**, 543–552, doi:10.1017/S0021859610000341.

Smith, P. et al., 2007: Policy and technological constraints to implementation of greenhouse gas mitigation options in agriculture. *Agric. Ecosyst. Environ.*, **118**, 6–28, doi:10.1016/j.agee.2006.06.006.

Smith, P., M. Bustamante, H. Ahammad, H. Clark, H. Dong, E.A. Elsiddig, H. Haberl, R. Harper, J. House, M. Jafari, O. Masera, C. Mbow, N.H. Ravindranath, C.W. Rice, C. Robledo Abad, A. Romanovskaya, F. Sperling, and F. Tubiello, 2014b: Agriculture, Forestry and Other Land Use (AFOLU). In: Climate Change 2014: Mitigation of Climate Change. Contribution of Working Group III to the Fifth Assessment Report of the Intergovernmental Panel on Climate Change [Edenhofer, O., R. Pichs-Madruga, Y. Sokona, E. Farahani, S. Kadner, K. Seyboth, A. Adler, I. Baum, S. Brunner, P. Eickemeier, B. Kriemann, J. Savolainen, S. Schlömer, C. von Stechow, T. Zwickel and J.C. Minx (eds.)]. Cambridge University Press, Cambridge, United Kingdom and New York, NY, USA, 811–922.

Smith, P. et al., 2016: Biophysical and economic limits to negative CO_2 emissions. *Nat. Clim. Chang.*, **6**, 42–50, doi:10.1038/nclimate2870.

Smucker, T.A., and E.E. Wangui, 2016: Gendered knowledge and adaptive practices: Differentiation and change in Mwanga District, Tanzania. *Ambio*, **45**, 276–286, doi:10.1007/s13280-016-0828-z.

Le Saout, S. et al., 2013: Protected areas and effective biodiversity conservation. *Science*, **342**, 803–805, doi:10.1126/science.1239268.

Sola, P., C. Ochieng, J. Yila, and M. Iiyama, 2016a: Links between energy access and food security in Sub-Saharan Africa: An exploratory review. *Food Secur.*, **8**, 635–642, doi:10.1007/s12571-016-0570-1.

Solomon et al., 2015: *Socio-Economic Scenarios of Low Hanging Fruits for Developing Climate-Smart Biochar Systems in Ethiopia: Biomass Resource Availability to Sustainably Improve Soil Fertility, Agricultural Productivity and Food and Nutrition Security.* School of Integrative Plant Science Soil and Crop Sciences Section, Cornell University, New York, USA, 89 pp.

Somanathan, E., R. Prabhakar, and B.S. Mehta, 2009: Decentralization for cost-effective conservation. *Proc. Natl. Acad. Sci.*, **106**, 4143–4147, doi:10.1073/pnas.0810049106.

Somanthan, E., T. Sterner, T. Sugiyama, D. Chimanikire, N.K. Dubash, J. Essandoh-Yeddu, S. Fifita, L. Goulder, A. Jaffe, X. Labandeira, S. Managi, C. Mitchell, J.P. Montero, F. Teng, and T. Zylicz, 2014: *15.* National and Sub-National Policies and Institutions. In: Climate Change 2014: Mitigation of Climate Change. Contribution of Working Group III to the Fifth Assessment Report of the Intergovernmental Panel on Climate Change [Edenhofer, O., R. Pichs-Madruga, Y. Sokona, E. Farahani, S. Kadner, K. Seyboth, A. Adler, I. Baum, S. Brunner, P. Eickemeier, B. Kriemann, J. Savolainen, S. Schlömer, C. von Stechow, T. Zwickel and J.C. Minx (eds.)]. Cambridge University Press, Cambridge, United Kingdom and New York, NY, USA, pp. 1141–1206.

Song, X.-P. et al., 2018: Global land change from 1982 to 2016. *Nature*, **560**, 639–643, doi:10.1038/s41586-018-0411-9.

Sonnino, R., C. Lozano Torres, and S. Schneider, 2014: Reflexive governance for food security: The example of school feeding in Brazil. *J. Rural Stud.*, **36**, 1–12, doi:10.1016/j.jrurstud.2014.06.003.

Sorice, M.G., C. Josh Donlan, K.J. Boyle, W. Xu, and S. Gelcich, 2018: Scaling participation in payments for ecosystem services programs. *PLoS One*, **13**, e0192211, doi:10.1371/journal.pone.0192211.

Sorokin, A. et al., 2015: The economics of land degradation in Russia. In: *Economics of Land Degradation and Improvement – A Global Assessment for Sustainable Development* [Nkonya, E., A. Mirzabaev, J. von Braun (eds.)]. Springer International Publishing, Cham, Switzerland, pp. 541–576.

Sovacool, B.K., 2018: Bamboo beating bandits: Conflict, inequality, and vulnerability in the political ecology of climate change adaptation in Bangladesh. *World Dev.*, **102**, 183–194, doi:10.1016/J.WORLDDEV.2017.10.014.

Sparrevik, M., C. Adam, V. Martinsen, and G. Cornelissen, 2015: Emissions of gases and particles from charcoal/biochar production in rural areas using medium-sized traditional and improved 'retort' kilns. Biomass and Bioenergy, **72**, 65–73, doi:10.1016/j.biombioe.2014.11.016.

Sparrow, R., A. Suryahadi, and W. Widyanti, 2013: Social health insurance for the poor: Targeting and impact of Indonesia's Askeskin programme. *Soc. Sci. Med.*, **96**, 264–271, doi:10.1016/j.socscimed.2012.09.043.

Spence, A., W. Poortinga, and N. Pidgeon, 2012: The psychological distance of climate change. *Risk Anal.*, **32**, 957–972, doi:10.1111/j.1539-6924.2011.01695.x.

Speranza, C.I., B. Kiteme, P. Ambenje, U. Wiesmann, and S. Makali, 2010: Indigenous knowledge related to climate variability and change: Insights from droughts in semi-arid areas of former Makueni District, Kenya. *Clim. Change*, **100**, 295–315, doi:10.1007/s10584-009-9713-0.

Spoon, J., 2014: Quantitative, qualitative, and collaborative methods: Approaching indigenous ecological knowledge heterogeneity. *Ecol. Soc.*, **19**, art33, doi:10.5751/ES-06549-190333.

Spracklen, D. V, S.R. Arnold, and C.M. Taylor, 2012: Observations of increased tropical rainfall preceded by air passage over forests. *Nature*, **489**, 282, doi:10.1038/nature11390.

Springmann, M., H.C.J. Godfray, M. Rayner, and P. Scarborough, 2016: Analysis and valuation of the health and climate change cobenefits of dietary change. *Proc. Natl. Acad. Sci.*, **113**, 4146–4151, doi:10.1073/pnas.1523119113.

Stattman, S. et al., 2018a: Toward sustainable biofuels in the European Union? Lessons from a decade of hybrid biofuel governance. *Sustainability*, **10**, 4111, doi:10.3390/su10114111.

Stattman, S.L., A. Gupta, and L. Partzsch, 2016: Biofuels in the European Union: Can Hybrid Governance Promote Sustainability? In: ECPR General Conference, Prague, Czech Republic, pp. 1–17.

Stavi, I., and R. Lal, 2015: Achieving zero net land degradation: Challenges and opportunities. *J. Arid Environ.*, **112**, 44–51, doi:10.1016/j.jaridenv.2014.01.016.

Stavropoulou, M., R. Holmes, and N. Jones, 2017: Harnessing informal institutions to strengthen social protection for the rural poor. *Glob. Food Sec.*, **12**, 73–79, doi:10.1016/j.gfs.2016.08.005.

Steffen, W. et al., 2015: Planetary boundaries: Guiding human development on a changing planet. *Science*, **347**, 1259855, doi:10.1126/science.1259855.

Stephens, S., R. Bell, and J. Lawrence, 2017: Applying principles of uncertainty within coastal hazard assessments to better support coastal adaptation. *J. Mar. Sci. Eng.*, **5**, 40, doi:10.3390/jmse5030040.

Stephens, S.A., R.G. Bell, and J. Lawrence, 2018: Developing signals to trigger adaptation to sea-level rise. *Environ. Res. Lett.*, **13**, 1–12, doi:10.1088/1748-9326/aadf96.

Stern, N., 2007: *The Economics of Climate Change.* Cambridge University Press, Cambridge, UK, and New York, USA, 692 pp.

Stern, N., 2013: The structure of economic modeling of the potential impacts of climate change: Grafting gross underestimation of risk onto already narrow science models. *J. Econ. Lit.*, **51**, 838–859, doi:10.257/jel.51.3.838.

Sternberg, T., 2012: Chinese drought, bread and the Arab Spring. *Appl. Geogr.*, **34**, 519–524, doi:10.1016/j.apgeog.2012.02.004.

Sternberg, T., 2017: Climate hazards in Asian drylands. *Climate Hazard Crises in Asian Societies and Environments* [Sternberg, T. (ed.)]. Routledge, Abingdon, UK, and New York, USA.

Stevanovic, M. et al., 2016: The impact of high-end climate change on agricultural welfare. *Sci. Adv.*, **2**, e1501452–e1501452, doi:10.1126/sciadv.1501452.

Stickler, M.M., H. Huntington, A. Haflett, S. Petrova, and I. Bouvier, 2017: Does de facto forest tenure affect forest condition? Community perceptions from Zambia. *For. Policy Econ.*, **85**, 32–45, doi:10.1016/j.forpol.2017.08.014.

Stoa, R.B., 2015: Droughts, floods, and wildfires: Paleo perspectives on diaster law in the Anthropocene. *Georg. Int. Environ. Law Rev.*, **27**, 393–446.

Stone, B., J.J. Hess, and H. Frumkin, 2010: Urban form and extreme heat events: Are sprawling cities more vulnerable to climate change than compact cities? *Environ. Health Perspect.*, **118**, 1425, doi:10.1289/ehp.0901879.

Stone, J. et al., 2014: Risk reduction through community-based monitoring: The vigías of Tungurahua, Ecuador. *J. Appl. Volcanol.*, **3**, 11, doi:10.1186/s13617-014-0011-9.

Storbjörk, S., 2010: 'It takes more to get a ship to change course': Barriers for organizational learning and local climate adaptation in Sweden. *J. Environ. Policy Plan.*, **12**, 235–254, doi:10.1080/1523908X.2010.505414.

Stringer, L.C., and A.J. Dougill, 2013: Channelling science into policy: Enabling best practices from research on land degradation and sustainable land management in dryland Africa. *J. Environ. Manage.*, **114**, 328–335, doi:10.1016/j.jenvman.2012.10.025.

Stringer, L.C. et al., 2009: Adaptations to climate change, drought and desertification: Local insights to enhance policy in southern Africa. *Environ. Sci. Policy*, **12**, 748–765, doi:10.1016/J.ENVSCI.2009.04.002.

Stringer, L.C. et al., 2018: A new framework to enable equitable outcomes: Resilience and nexus approaches combined. *Earth's Futur.*, **6**, 902–918, doi:10.1029/2017EF000694.

Stupak, I., and K. Raulund-Rasmussen, 2016: Historical, ecological, and governance aspects of intensive forest biomass harvesting in Denmark. *Wiley Interdiscip. Rev. Energy Environ.*, **5**, 588–610, doi:10.1002/wene.206.

Stupak, I. et al., 2016: A global survey of stakeholder views and experiences for systems needed to effectively and efficiently govern sustainability of bioenergy. *Wiley Interdiscip. Rev. Energy Environ.*, **5**, 89–118, doi:10.1002/wene.166.

Sturm, M., M.A. Goldstein, H.P. Huntington, and T.A. Douglas, 2017: Using an option pricing approach to evaluate strategic decisions in a rapidly changing climate: Black-Scholes and climate change. *Clim. Change*, **140**, 437–449, doi:10.1007/s10584-016-1860-5.

Sturrock, R.N. et al., 2011: Climate change and forest diseases. *Plant Pathol.*, **60**, 133–149, doi:10.1111/j.1365-3059.2010.02406.x.

Suckall, N., L.C. Stringer, and E.L. Tompkins, 2015: Presenting triple-wins? Assessing projects that deliver adaptation, mitigation and development co-benefits in rural Sub-Saharan Africa. *Ambio*, **44**, 34–41, doi:10.1007/s13280-014-0520-0.

Sudmeier-Rieux, K., M. Fernández, J.C. Gaillard, L. Guadagno, and M. Jaboyedoff, 2017: Exploring linkages between disaster risk reduction, climate change adaptation, migration and sustainable development. In: *Identifying Emerging Issues in Disaster Risk Reduction, Migration, Climate Change and Sustainable Development* [Sudmeier-Rieux, K., M. Fernández, I.M. Penna, M. Jaboyedoff, J.C. Gaillard (eds.)]. Springer International Publishing, Cham, Switzerland, pp. 1–11.

Suich, H., C. Howe, and G. Mace, 2015: Ecosystem services and poverty alleviation: A review of the empirical links. *Ecosyst. Serv.*, **12**, 137–147, doi:10.1016/j.ecoser.2015.02.005.

Sumiya, B., 2016: Energy poverty in context of climate change: What are the possible impacts of improved modern energy access on adaptation capacity of communities? *Int. J. Environ. Sci. Dev.*, **7**, 7, doi:10.7763/IJESD.2016.V7.744.

Sun, K., and S.S. Chaturvedi, 2016: Forest conservation and climate change mitigation potential through REDD+ mechanism in Meghalaya, north-eastern India: A review. *Int. J. Sci. Environ. Technol.*, **5**, 3643–3650.

Sunderlin, W., C. de Sassi, A. Ekaputri, M. Light, and C. Pratama, 2017: REDD+ contribution to well-being and income is marginal: The perspective of local stakeholders. *Forests*, **8**, 125, doi:10.3390/f8040125.

Sunderlin, W.D. et al., 2018: Creating an appropriate tenure foundation for REDD+: The record to date and prospects for the future. *World Dev.*, **106**, 376–392, doi:10.1016/J.WORLDDEV.2018.01.010.

Sundström, A., 2016: Understanding illegality and corruption in forest governance. *J. Environ. Manage.*, **181**, 779–790, doi:10.1016/j.jenvman.2016.07.020.

Surminski, S., 2013: Private-sector adaptation to climate risk. *Nat. Clim. Chang.*, **3**, 943–945, doi:10.1038/nclimate2040.

Surminski, S. et al., 2016: *Submission to the UNFCCC Warsaw International Mechanism by the Loss and Damage Network*, 8 pp.

Surminski, S., L.M. Bouwer, and J. Linnerooth-Bayer, 2016: How insurance can support climate resilience. *Nat. Clim. Chang.*, **6**, 333–334, doi:10.1038/nclimate2979.

Suzuki, R., 2012: *Linking Adaptation and Mitigation through Community Forestry: Case Studies from Asia. RECOFTC – The Center for People and Forests*. RECOFTC, The Center for People and Forests, Bangkok, Thailand, 80 pp.

Swanson, A., M. Kosmala, C. Lintott, and C. Packer, 2016: A generalized approach for producing, quantifying, and validating citizen science data from wildlife images. *Conserv. Biol.*, **30**, 520–531, doi:10.1111/cobi.12695.

Tàbara, J.D. et al., 2010: The climate learning ladder. A pragmatic procedure to support climate adaptation. *Environ. Policy Gov.*, **20**, 1–11, doi:10.1002/eet.530.

Takama, T., S. Tsephel, and F.X. Johnson, 2012: Evaluating the relative strength of product-specific factors in fuel switching and stove choice decisions in Ethiopia. A discrete choice model of household preferences for clean cooking alternatives. *Energy Econ.*, **34**, 1763–1773, doi:10.1016/J.ENECO.2012.07.001.

Tallis, H., P. Kareiva, M. Marvier, and A. Chang, 2008: An ecosystem services framework to support both practical conservation and economic development. *Proc. Natl. Acad. Sci.*, **105**, 9457–9464, 10.1073/pnas.0705797105.

Tanner, T. et al., 2015: Livelihood resilience in the face of climate change. *Nat. Clim. Chang.*, **5**, 23–26, doi:10.1038/nclimate2431.

Tao, H.-H. et al., 2017: Long-term crop residue application maintains oil palm yield and temporal stability of production. *Agron. Sustain. Dev.*, **37**, 33, doi:10.1007/s13593-017-0439-5.

Taylor, R.G. et al., 2013: Ground water and climate change. *Nat. Clim. Chang.*, **3**, 322–329, doi:10.1038/nclimate1744.

Teeb, T., 2009: The economics of ecosystems and biodiversity for national and international policy makers – Summary: Responding to the value of nature 2009. TEEB – The Economics of Ecosystems and Biodiversity for National and International Policy Makers, Geneva, Switzerland, 39 pp.

Temper, L., and J. Martinez-Alier, 2013: The god of the mountain and Godavarman: Net Present Value, indigenous territorial rights and sacredness in a bauxite mining conflict in India. *Ecol. Econ.*, **96**, 79–87, doi:10.1016/j.ecolecon.2013.09.011.

Tengberg, A., and S. Valencia, 2018: Integrated approaches to natural resources management-theory and practice. *L. Degrad. Dev.*, **29**, 1845–1857, doi:10.1002/ldr.2946.

Tengberg, A., F. Radstake, K. Zhang, and B. Dunn, 2016: Scaling up of sustainable land management in the western People's Republic of China: Evaluation of a 10-Year partnership. *L. Degrad. Dev.*, **27**, 134–144, doi:10.1002/ldr.2270.

Tengö, M., E.S. Brondizio, T. Elmqvist, P. Malmer, and M. Spierenburg, 2014: Connecting diverse knowledge systems for enhanced ecosystem governance: The multiple evidence base approach. *Ambio*, **43**, 579–591, doi:10.1007/s13280-014-0501-3.

Tennigkeit, T., and W. Andreas, 2008: *Working Paper: An Assessment of the Potential for Carbon Finance in Rangelands*. ICRAF Working Paper No.68, World Agroforestry Centre, Beijing, China, 31 pp.

Termeer, C.J.A.M., A. Dewulf, and G.R. Biesbroek, 2017: Transformational change: Governance interventions for climate change adaptation from a continuous change perspective. *J. Environ. Plan. Manag.*, **60**, 558–576, doi:10.1080/09640568.2016.1168288.

Terrazas, W.C.M. et al., 2015: Deforestation, drainage network, indigenous status, and geographical differences of malaria in the state of Amazonas. *Malar. J.*, **14**, 379, doi:10.1186/s12936-015-0859-0.

Tessler, Z.D. et al., 2015: Profiling risk and sustainability in coastal deltas of the world. *Science*, **349**, 638–643, doi:10.1126/science.aab3574.

Thaker, M, Zambre, A. Bhosale, H., 2018: Wind farms have cascading impacts on ecosystems across trophic levels. *Nat. Ecol. Evol.*, **2**, 1854–1858, doi:10.1038/s41559-018-0707-z.

Thaler, R.H., and C.R. Sunstein (eds.), 2008: *Nudge: Improving decisions about health, wealth, and happiness*. Penguin, New York, USA, 1–293 pp.

Thamo, T., and D.J. Pannell, 2016: Challenges in developing effective policy for soil carbon sequestration: Perspectives on additionality, leakage, and permanence. *Clim. Policy*, **16**, 973–992, doi:10.1080/14693062.2015.1075372.

Theisen, O.M., H. Holtermann, and H. Buhaug, 2011: Climate wars? Assessing the claim that drought breeds conflict. *Int. Secur.*, **36**, 79–106, doi:10.1162/isec_a_00065.

Theriault, V., M. Smale, and H. Haider, 2017a: How does gender affect sustainable intensification of cereal production in the West African Sahel? Evidence from Burkina Faso. *World Dev.*, **92**, 177–191, doi:10.1016/J.WORLDDEV.2016.12.003.

Thomas, A., and L. Benjamin, 2017: Policies and mechanisms to address climate-induced migration and displacement in Pacific and Caribbean small island developing states. *Int. J. Clim. Chang. Strateg. Manag.*, **10**, 86–104, doi:10.1108/IJCCSM-03-2017-0055.

Thomas, D.H.L., 1996: Fisheries tenure in an African floodplain village and the implications for management. *Hum. Ecol.*, **24**, 287–313, doi:10.1007/BF02169392.

Thompson-Hall, M., E.R. Carr, and U. Pascual, 2016: Enhancing and expanding intersectional research for climate change adaptation in agrarian settings. *Ambio*, **45**, 373–382, doi:10.1007/s13280-016-0827-0.

Thompson, I., B. Mackey, S. McNulty, and A. Mosseler, 2009: *Forest Resilience, Biodiversity, and Climate Change: A Synthesis of the Biodiversity/Resilience/Stability Relationship in Forest Ecosystems*. Secretariat of the Convention on Biological Diversity, Montreal, Canada, 67 pp.

Thorén, H., and L. Olsson, 2017: Is resilience a normative concept? *Resilience*, **6**, 112–128, doi:10.1080/21693293.2017.1406842.

Tidwell, J.H., and G.L. Allan, 2001: Fish as food: Aquaculture's contribution: Ecological and economic impacts and contributions of fish farming and capture fisheries. *EMBO Rep.*, **2**, 958–963, doi:10.1093/embo-reports/kve236.

Tierney, J.E., C.C. Ummenhofer, and P.B. DeMenocal, 2015: Past and future rainfall in the Horn of Africa. *Sci. Adv.*, **1**, e1500682, doi:10.1126/sciadv.1500682.

Tigchelaar, M., D. Battisti, R.. Naylor, and D.. Ray, 2018: Future warming increases probability of globally synchronized maize production shocks. *Proc. Natl. Acad. Sci.*, **115**, 6644–6649, doi:10.1073/pnas.1718031115.

Timberlake, T.J., and C.A. Schultz, 2017: Policy, practice, and partnerships for climate change adaptation on US national forests. *Clim. Change*, **144**, 257–269, doi:10.1007/s10584-017-2031-z.

Tittonell, P., 2014: Livelihood strategies, resilience and transformability in African agroecosystems. *Agric. Syst.*, **126**, 3–14, doi:10.1016/j.agsy.2013.10.010.

Tjaden, N.B. et al., 2017: Modelling the effects of global climate change on Chikungunya transmission in the 21st century. *Sci. Rep.*, **7**, 3813, doi:10.1038/s41598-017-03566-3.

Tompkins, E.L., and W.N. Adger, 2004: Does adaptive management of natural resources enhance resilience to climate change? *Ecol. Soc.*, **9**, 10.

Torvanger, A., 2019a: Governance of bioenergy with carbon capture and storage (BECCS): Accounting, rewarding, and the Paris Agreement. *Clim. Policy*, **19**, 329–341, doi:10.1080/14693062.2018.1509044.

Tóth, G., T. Hermann, M.R. da Silva, and L. Montanarella, 2018: Monitoring soil for sustainable development and land degradation neutrality. *Environ. Monit. Assess.*, **57**, 190, doi:10.1007/s10661-017-6415-3.

Totin, E. et al., 2018: Institutional perspectives of climate-smart agriculture: A systematic literature review. *Sustainability*, **10**, 1990, doi:10.3390/su10061990.

Toulmin, C., and J. Quan, 2000: *Evolving Land Rights, Policy and Tenure in Africa*. IIED and Natural Resources Institute, London, UK, 324 pp.

Travis, W.R., 2013: Design of a severe climate change early warning system. *Weather Clim. Extrem.*, **2**, 31–38, doi:10.1016/j.wace.2013.10.006.

Tribbia, J., and S.C. Moser, 2008: More than information: What coastal managers need to plan for climate change. *Environ. Sci. Policy*, **11**, 315–328, doi:10.1016/J.ENVSCI.2008.01.003.

Trieb, F., H. Müller-Steinhagen, and J. Kern, 2011: Financing concentrating solar power in the Middle East and North Africa – Subsidy or investment? *Energy Policy*, **39**, 307–317, doi:10.1016/j.enpol.2010.09.045.

Tschakert, P., 2007: Views from the vulnerable: Understanding climatic and other stressors in the Sahel. *Glob. Environ. Chang.*, **17**, 381–396, doi:10.1016/j.gloenvcha.2006.11.008.

Tucker Lima, J.M., A. Vittor, S. Rifai, and D. Valle, 2017: Does deforestation promote or inhibit malaria transmission in the Amazon? A systematic literature review and critical appraisal of current evidence. *Philos. Trans. R. Soc. Lond. B. Biol. Sci.*, **372**, 20160125, doi:10.1098/rstb.2016.0125.

Tularam, G., and M. Krishna, 2009: Long-term consequences of groundwater pumping in Australia: A review of impacts around the globe. *J. Appl. Sci. Environ. Sanit.*, **4**, 151–166.

Turkelboom, F. et al., 2018: When we cannot have it all: Ecosystem services trade-offs in the context of spatial planning. *Ecosyst. Serv.*, **29**, 566–578, doi:10.1016/j.ecoser.2017.10.011.

Turnhout, E., K. Neves, and E. de Lijster, 2014:'Measurementality'in biodiversity governance: Knowledge, transparency, and the Intergovernmental Science-Policy Platform on Biodiversity and Ecosystem Services (IPBES). *Environ. Plan. A*, **46**, 581–597, doi:10.1068/a4629.

von Uexkull, N., M. Croicu, H. Fjelde, and H. Buhaug, 2016: Civil conflict sensitivity to growing-season drought. *Proc. Natl. Acad. Sci.*, **113**, 12391–12396, doi:10.1073/pnas.1607542113.

Ulrich-Schad, J.D., S. Garcia de Jalon, N. Babin, A. Pape, L.S. Prokopy, 2017: Measuring and understanding agricultural producers' adoption of nutrient best management practices. *J. Soil Water Conserv.*, **72**, 506–518, doi:10.2489/jswc.72.5.506.

Umukoro, N., 2013: Poverty and social protection in Nigeria. *J. Dev. Soc.*, **29**, 305–322, doi:10.1177/0169796X13494281.

Cowie, A.L., 2016: Land in balance: The scientific conceptual framework for land degradation neutrality. *Sci. Br.*, **79**, 25–35, doi:10.1016/j.envsci.2017.10.011.

UNCCD, 2015: *Land Degradation Neutrality: The Target Setting Programme*. Global Mechanism of the UNCCD, Bonn, Germany, 22 pp.

UNDP, 2014: *Governance for Sustainable Human Development*. United Nations Development Programme, New York, USA, pp. 2–3.

UNEP, 2009: *Statement by Ahmed Djoghlaf Executive Secretary at the Meeting of Steering Committee Global Form on Oceans, Coasts and Islands*. Secretariat of the Convention on Biological Diversity, United Nations, Montreal, Canada, 3 pp.

UNEP, 2016: *The Adaptation Finance Gap Report 2016*. United Nations Environment Programme, Nairobi, Kenya, 84 pp.

UNFCCC, 2007: Climate Change: Impacts, Vulnerabilities and Adaptation in Developing Countries. Climate Change Secretariat (UNFCCC), Bonn, Germany, 64 pp.

UNFCCC, 2018a: *Paris Rulebook: Proposal by the President, Informal Compilation of L-documents*. UNFCCC, Katowice, Poland, 133 pp.

UNFCCC, 2016: Paris Agreement. *Paris Agreement– Pre 2020 Action*. Paris, France, 25 pp.

United Nations Environment Programme, 2017: *The Emissions Gap Report 2017: A UN Environment Synthesis Report*. The Emissions Gap Report 2017, United Nations Environment Programme (UNEP), Nairobi, Kenya, 1–86 pp.

Urwin, K., and A. Jordan, 2008: Does public policy support or undermine climate change adaptation? Exploring policy interplay across different scales of governance. *Glob. Environ. Chang.*, **18**, 180–191, doi:10.1016/j.gloenvcha.2007.08.002.

Usher, P.J., 2000: Traditional ecological knowledge in environmental assessment and management. *ARCTIC*, **53**, 183–193 pp.

Uzun, B., and M. Cete, 2004: A Model for Solving Informal Settlement Issues in Developing Countries. *Planning, Valuat. Environ.* FIG Working Week, Athens, Greece, 7 pp.

Valatin, G., D. Moseley, and N. Dandy, 2016: Insights from behavioural economics for forest economics and environmental policy: Potential nudges to encourage woodland creation for climate change mitigation and adaptation? *For. Policy Econ.*, **72**, 27–36, doi:10.1016/j.forpol.2016.06.012.

IPCC, 2018: Summary for Policymakers. In: Global Warming of 1.5°C. An IPCC Special Report on the impacts of global warming of 1.5°C above pre-industrial levels and related global greenhouse gas emission pathways, in the context of strengthening the global response to the threat of climate change, sustainable development, and efforts to eradicate poverty [Masson-Delmotte, V., P. Zhai, H.-O. Pörtner, D. Roberts, J. Skea, P.R. Shukla, A. Pirani, W. Moufouma-Okia, C. Péan, R. Pidcock, S. Connors, J.B.R. Matthews, Y. Chen, X. Zhou, M.I. Gomis, E. Lonnoy, T. Maycock, M. Tignor, and T. Waterfield (eds.)]. Cambridge University Press, Cambridge, United Kingdom and New York, NY, USA, 24 pp.

Valipour, A., T. Plieninger, Z. Shakeri, H. Ghazanfari, M. Namiranian, and M.J. Lexer, 2014: Traditional silvopastoral management and its effects on forest stand structure in Northern Zagros, Iran. *For. Ecol. Manage.*, **327**, 221–230, doi:10.1016/j.foreco.2014.05.004.

Vandersypen, K., A.C.T. Keita, Y. Coulibaly, D. Raes, and J.Y. Jamin, 2007: Formal and informal decision-making on water management at the village level: A case study from the Office du Niger irrigation scheme (Mali). *Water Resour. Res.*, **43**, 1–10, doi:10.1029/2006WR005132.

Vanmaercke, M. et al., 2016a: How fast do gully headcuts retreat? *Earth-Science Rev.*, **154**, 336–355, doi:10.1016/J.EARSCIREV.2016.01.009.

Velthof, G.L. et al., 2014: The impact of the Nitrates Directive on nitrogen emissions from agriculture in the EU-27 during 2000–2008. *Sci. Total Environ.*, **468–469**, 1225–1233, doi:10.1016/j.scitotenv.2013.04.058.

Vent, O., Sabarmatee, and N. Uphoff, 2017: The system of rice intensification and its impacts on women: Reducing pain, discomfort, and labor in rice farming while enhancing households' food security. In: *Women in Agriculture Worldwide: Key issues and practical approaches* [Fletcher, A., and W. Kubik (eds.)]. Routledge, London, UK and New York, USA, pp. 55–76.

Venton, C.C., 2018: *The Economics of Resilience to Drought*. USAID Centre for Resilience, 130 pp.

Venton, C.C.C., C. Fitzgibbon, T. Shitarek, L. Coulter, and O. Dooley, 2012: *The Economics of Early Response and Disaster Resilience: Lessons from Kenya and Ethiopia*. Economics of Resilience Final Report, UK Department of International Development, UK, 1–84 pp.

Verburg, P.H. et al., 2015: Land system science and sustainable development of the Earth System: A global land project perspective. *Anthropocene*, **12**, 29–41, doi:10.1016/j.ancene.2015.09.004.

Verbyla, D., 2011: Browning boreal forests of western North America. *Environ. Res. Lett.*, **6**, 41003, doi:10.1088/1748-9326/6/4/041003.

Verchot, L.V. et al., 2007: Climate change: Linking adaptation and mitigation through agroforestry. *Mitig. Adapt. Strateg. Glob. Chang.*, **12**, 901–918, doi:10.1007/s11027-007-9105-6.

Verdegem, M.C.J., and R.H. Bosma, 2009: Water withdrawal for brackish and inland aquaculture, and options to produce more fish in ponds with present water use. *Water Policy*, **11**, 52–68, doi:10.2166/wp.2009.003.

Vergara-Asenjo, G., and C. Potvin, 2014: Forest protection and tenure status: The key role of indigenous peoples and protected areas in Panama. *Glob. Environ. Chang.*, **28**, 205–215, doi:10.1016/J.GLOENVCHA.2014.07.002.

Verma, S., D.A. Kampman, P. van der Zaag, and A.Y. Hoekstra, 2009: Going against the flow: A critical analysis of inter-state virtual water trade in the context of India's National River Linking Program. *Phys. Chem. Earth, Parts A/B/C*, **34**, 261–269, doi:10.1016/j.pce.2008.05.002.

Verschuuren, J., 2017: Towards a regulatory design for reducing emissions from agriculture: Lessons from Australia's carbon farming initiative. *Clim. Law*, **7**, 1–51, doi:10.1163/18786561-00701001.

Vervoort, J., and A. Gupta, 2018: Anticipating climate futures in a 1.5°C era: The link between foresight and governance. *Curr. Opin. Environ. Sustain.*, **31**, 104–111, doi:10.1016/j.cosust.2018.01.004.

Verweij, M. et al., 2006: Clumsy solutions for a complex world: The case of climate change. *Public Adm.*, **84**, 817–843, doi:10.1111/j.1467-9299.2006.00614.x.

Vijge, M.J., and A. Gupta, 2014: Framing REDD+ in India: Carbonizing and centralizing Indian forest governance? *Environ. Sci. Policy*, **38**, 17–27, doi:10.1016/j.envsci.2013.10.012.

Villagra, P., and C. Quintana, 2017: Disaster governance for community resilience in coastal towns: Chilean case studies. *Int. J. Environ. Res. Public Health*, **14**, 1063, doi:10.3390/ijerph14091063.

Vincent, K., S. Besson, T. Cull, and C. Menzel, 2018: Sovereign insurance to incentivize the shift from disaster response to adaptation to climate change – African Risk Capacity's Extreme Climate Facility. *Clim. Dev.*, **10**, 385–388, doi:10.1080/17565529.2018.1442791.

Vira, B., B. Adams, C. Agarwal, S. Badiger, R. a Hope, J. Krishnaswamy, and C. Kumar, 2012: Negotiating trade-offs: Choices about ecosystem services for poverty alleviation. *Econ. Polit. Wkly.*, **47**, 67.

Vörösmarty, C.J. et al., 2010: Global threats to human water security and river biodiversity. *Nature*, **467**, 555–561, doi:10.1038/nature09440.

Voß, J.-P., and N. Amelung, 2016: Innovating public participation methods: Technoscientization and reflexive engagement. *Soc. Stud. Sci.*, **46**, 749–772, doi:10.1177/0306312716641350.

Voß, J.P., and A. Simons, 2018: A novel understanding of experimentation in governance: Co-producing innovations between 'lab' and 'field'. *Policy Sci.*, **51**, 213–229, doi:10.1007/s11077-018-9313-9.

Waas, T. et al., 2014: Sustainability assessment and indicators: Tools in a decision-making strategy for sustainable development. *Sustain.*, **6**, 5512–5534, doi:10.3390/su6095512.

Wada, Y. et al., 2010: Global depletion of groundwater resources. *Geophys. Res. Lett.*, **37**, 1–5, doi:10.1029/2010GL044571.

Wada, Y., A.K. Gain, and C. Giupponi, 2016: Measuring global water security towards sustainable development goals. *Environ. Res. Lett.*, **11**, 2–13, doi:10.1088/1748-9326/11/12/124015.

Waddock, S., 2013: The wicked problems of global sustainability need wicked (good) leaders and wicked (good) collaborative solutions. *J. Manag. Glob. Sustain.*, **1**, 91–111, doi:10.13185/JM2013.01106.

Wagenbrenner, N.S., M.J. Germino, B.K. Lamb, P.R. Robichaud, and R.B. Foltz, 2013: Wind erosion from a sagebrush steppe burned by wildfire: Measurements of PM10 and total horizontal sediment flux. *Aeolian Res.*, **10**, 25–36, doi:10.1016/j.aeolia.2012.10.003.

Wagner, G., 2013: Carbon Cap and Trade. *Encycl. Energy, Nat. Resour. Environ. Econ.*, **1–3**, 1–5, doi:10.1016/B978-0-12-375067-9.00071-1.

Waite, S.H., 2011: Blood forests: Post Lacey Act, why cohesive global governance is essential to extinguish the market for illegally harvested timber. *Seattle J. Environ. Law*, **2**, 317–342.

Walker, W. et al., 2014: Forest carbon in Amazonia: The unrecognized contribution of indigenous territories and protected natural areas. *Carbon Manag.*, **5**, 479–485, doi:10.1080/17583004.2014.990680.

Walter, A., J.E.A. Seabra, P.G. Machado, B. de Barros Correia, and C.O.F. de Oliveira, 2018: Sustainability of biomass. In: *Biomass and Green Chemistry*, Springer International Publishing, Cham, Switzerland, pp. 191–219.

Wam, H.K., N. Bunnefeld, N. Clarke, and O. Hofstad, 2016: Conflicting interests of ecosystem services: Multi-criteria modelling and indirect evaluation of trade-offs between monetary and non-monetary measures. *Ecosyst. Serv.*, **22**, 280–288, doi:10.1016/j.ecoser.2016.10.003.

Wang, C. et al., 2018: Effects of biochar amendment on net greenhouse gas emissions and soil fertility in a double rice cropping system: A 4-year field experiment. *Agric. Ecosyst. Environ.*, **262**, 83–96, doi:10.1016/J.AGEE.2018.04.017.

Wang, S., and B. Fu, 2013: Trade-offs between forest ecosystem services. *For. Policy Econ.*, **26**, 145–146, doi:10.1016/j.forpol.2012.07.014.

Wang, X. et al., 2016: Life-table studies revealed significant effects of deforestation on the development and survivorship of Anopheles minimus larvae. *Parasit. Vectors*, **9**, 323, doi:10.1186/s13071-016-1611-5.

Ward, F.A., and M. Pulido-Velazquez, 2008: Water conservation in irrigation can increase water use. *Proc. Natl. Acad. Sci.*, **105**, 18215–18220, doi:10.1073pnas.0805554105.

Ward, P.S., 2016: Transient poverty, poverty dynamics, and vulnerability to poverty: An empirical analysis using a balanced panel from rural China. *World Dev.*, **78**, 541–553, doi:10.1016/j.worlddev.2015.10.022.

Warner, K., 2018: Coordinated approaches to large-scale movements of people: Contributions of the Paris Agreement and the global compacts for migration and on refugees. *Popul. Environ.*, **39**, 384–401, doi:10.1007/s11111-018-0299-1.

Warner, K., and T. Afifi, 2011: Environmentally induced migration in the context of social vulnerability. *Int. Migr.*, **49**, 242 pp, doi:10.1111/j.1468-2435.2011.00697.x.

Warner, K., and T. Afifi, 2014: Where the rain falls: Evidence from 8 countries on how vulnerable households use migration to manage the risk of rainfall variability and food insecurity. *Clim. Dev.*, **6**, 1–17, doi:10.1080/17565529.2013.835707.

Warner, K. et al., 2012: *Evidence from the Frontlines of Climate Change: Loss and Damage to Communities Despite Coping and Adaptation*. UNU-EHS, Bonn, Germany, 85 pp.

Warner, K. et al., 2018: Characteristics of transformational adaptation in land-society-climate interactions. *Sustainability*, **11**, 356, doi:10.3390/su11020356.

Wathore, R., K. Mortimer, and A.P. Grieshop, 2017: In-use emissions and estimated impacts of traditional, natural- and forced-draft cookstoves in rural Malawi. *Environ. Sci. Technol.*, **51**, 1929–1938, doi:10.1021/acs.est.6b05557.

Watts, N. et al., 2015: Health and climate change: Policy responses to protect public health. *Lancet*, **386**, 1861–1914, doi:10.1016/S0140-6736(15)60854-6.

Watts, N. et al., 2018: The 2018 report of the Lancet Countdown on health and climate change: shaping the health of nations for centuries to come. *Lancet*, **392**, 2479–2514, doi:10.1016/S0140-6736(18)32594-7.

Weichselgartner, J., and I. Kelman, 2015: Geographies of resilience: Challenges and opportunities of a descriptive concept. *Prog. Hum. Geogr.*, **39 (3)**, 249–267, doi:10.1177/0309132513518834.

Weick, K.E., and K.M. Sutcliffe (eds.), 2001: *Managing the Unexpected. Resilient Performance in a Time of Change*. Jossey-Bass, California, USA, 200 pp.

Weitz, N., H. Carlsen, M. Nilsson, and K. Skånberg, 2017a: Towards systemic and contextual priority setting for implementing the 2030 Agenda. *Sustainability Science*, **13**, 531–548, doi:10.1007/s11625-017-0470-0.

Weitz, N., C. Strambo, E. Kemp-Benedict, and M. Nilsson, 2017b: Closing the governance gaps in the water–energy–food nexus: Insights from integrative governance. *Glob. Environ. Chang.*, **45**, 165–173, doi:10.1016/j.gloenvcha.2017.06.006.

Welcomme, R.L. et al., 2010: Inland capture fisheries. *Philos. Trans. R. Soc. London B Biol. Sci.*, **365**, 2881–2896, doi:10.1098/rstb.2010.0168.

Wellesley, L., F. Preston, J. Lehne, and R. Bailey, 2017: Chokepoints in global food trade: Assessing the risk. *Res. Transp. Bus. Manag.*, **25**, 15–28, doi:10.1016/j.rtbm.2017.07.007.

Wenkel, K.-O. et al., 2013: LandCaRe DSS – An interactive decision support system for climate change impact assessment and the analysis of potential agricultural land use adaptation strategies. *J. Environ. Manage.*, **127**, S168–S183, doi:10.1016/J.JENVMAN.2013.02.051.

West, T.A.P., 2016: Indigenous community benefits from a de-centralized approach to REDD+ in Brazil. *Clim. Policy*, **16**, 924–939, doi:10.1080/14693062.2015.1058238.

Westberg, C.J., and F.X. Johnson, 2013: *The Path Not Yet Taken: Bilateral Agreements to Promote Sustainable Biofuels under the EU Renewable Energy Directive Stockholm Environment Institute, Working Paper 2013–02*. SEI Working Paper No. 2013–02, Stockholm Environment Institute, Stockholm, Sweden, 41 pp.

Westerling, A.L., H.G. Hidalgo, D.R. Cayan, and T.W. Swetnam, 2006: Warming and earlier spring increase Western US forest wildfire activity. *Science*, **313**, 940–943, doi:10.1126/SCIENCE.1128834.

Weyant, C.L. et al., 2019a: Emission measurements from traditional biomass cookstoves in South Asia and Tibet. *Environ. Sci. Technol.*, **53**, 3306–3314, doi:10.1021/acs.est.8b05199.

Weyant, C.L. et al., 2019b: Emission measurements from traditional biomass cookstoves in South Asia and Tibet. *Environ. Sci. Technol.*, **53**, 3306–3314, doi:10.1021/acs.est.8b05199.

Wheaton, E., and S. Kulshreshtha, 2017: Environmental sustainability of agriculture stressed by changing extremes of drought and excess moisture: A conceptual review. *Sustain.*, **9**, 970, doi:10.3390/su9060970.

Wheeler, T., and J. Von Braun, 2013: Climate change impacts on global food security. *Science*, **341**, 508–513, doi:10.1126/science.1239402.

Whitaker, J. et al., 2018: Consensus, uncertainties and challenges for perennial bioenergy crops and land use. *GCB Bioenergy*, **10**, 150–164, doi:10.1111/gcbb.12488.

White, B., S.M. Borras, R. Hall, I. Scoones, and W. Wolford, 2012: The new enclosures: Critical perspectives on corporate land deals. *J. Peasant Stud.*, **39**, 619–647, doi:10.1080/03066150.2012.691879.

White, J., and J. Morton, 2005: Mitigating impacts of HIV/AIDS on rural livelihoods: NGO experiences in Sub-Saharan Africa. *Dev. Pract.*, **15**, 186–199, doi:10.1080/09614520500041757.

Whitmee, S. et al., 2015: Safeguarding human health in the Anthropocene epoch: Report of the Rockefeller Foundation-Lancet Commission on planetary health. *Lancet*, **386**, 1973–2028, doi:10.1016/S0140-6736(15)60901-1.

Wiebe, K. et al., 2015a: Climate change impacts on agriculture in 2050 under a range of plausible socio-economic and emissions scenarios. *Environ. Res. Lett.*, **10**, 085010, doi:10.1088/1748-9326/10/8/085010.

Wiebe, K. et al., 2015b: Climate change impacts on agriculture in 2050 under a range of plausible socio-economic and emissions scenarios. *Environ. Res. Lett.*, **10**, 085010, doi:10.1088/1748-9326/10/8/085010.

Wiebe, K. et al., 2018: Scenario development and foresight analysis: Exploring options to inform choices. *Annual Review of Environment and Resources*, **43**, 545–570, doi:10.1146/annurev-environ-102017-030109.

Wiggering, H., and U. Steinhardt, 2015: A conceptual model for site-specific agricultural land use. *Ecol. Modell.*, **295**, 42–46, doi:10.1016/j.ecolmodel.2014.08.011.

Wilby, R.L., and S. Dessai, 2010: Robust adaptation to climate change. *Weather*, **65**, 180–185, doi:10.1002/wea.543.

Wilkes, A., A. Reisinger, E. Wollenberg, and S. Van Dijk, 2017: *Measurement, Reporting and Verification of Livestock GHG Emissions by Developing*

Countries in the UNFCCC: Current Practices and Opportunities for Improvement. CCAFS Rep. No. 17, Wageningen, Netherlands, 114 pp.

Wilkinson, E. et al., 2018: *Forecasting Hazards, Averting Disasters – Implementing Forecast-Based Early Action at Scale*. Overseas Development Institute, London, UK, 38 pp.

Willemen, L., B. Burkhard, N. Crossman, E.G. Drakou, and I. Palomo, 2015: Editorial: Best practices for mapping ecosystem services. *Ecosystem Services*, **13**, 1–5, doi:10.1016/j.ecoser.2015.05.008.

Willenbockel, D., 2012: *Extreme weather events and crop price spikes in a changing climate. Illustrative global simulation scenarios*. Oxfam Research Reports, Oxford, UK, 59 pp.

Williams, A.P., and J.T. Abatzoglou, 2016: Recent advances and remaining uncertainties in resolving past and future climate effects on global fire activity. *Curr. Clim. Chang. Reports*, **2**, 1–14, doi:10.1007/s40641-016-0031-0.

Williams, B.K., 2011: Adaptive management of natural resources-framework and issues. *J. Environ. Manage.*, **92**, 1346–1353, doi:10.1016/j.jenvman.2010.10.041.

Williams, D.A., and K.E. Dupuy, 2018: Will REDD+ Safeguards Mitigate Corruption? Qualitative evidence from Southeast Asia. *J. Dev. Stud.*, **55**, 2129–2144, doi:10.1080/00220388.2018.1510118.

Williams, S.E., E.E. Bolitho, and S. Fox, 2003: Climate change in Australian tropical rainforests: An impending environmental catastrophe. *Proc. R. Soc. London. Ser. B Biol. Sci.*, **270**, 1887–1892, doi:10.1098/rspb.2003.2464.

Williamson, T.B., and H.W. Nelson, 2017: Barriers to enhanced and integrated climate change adaptation and mitigation in Canadian forest management. *Can. J. For. Res.*, **47**, 1567–1576, doi:10.1139/cjfr-2017-0252.

Wilson, G.L., B.J. Dalzell, D.J. Mulla, T. Dogwiler, and P.M. Porter, 2014: Estimating water quality effects of conservation practices and grazing land use scenarios. *J. Soil Water Conserv.*, **69**, 330–342, doi:10.2489/jswc.69.4.330.

Wilson, R.S. et al., 2016: A typology of time-scale mismatches and behavioral interventions to diagnose and solve conservation problems. *Conserv. Biol.*, **30**, 42–49, doi:10.1111/cobi.12632.

Win, Z.C. et al., 2018: Differences in consumption rates and patterns between firewood and charcoal: A case study in a rural area of Yedashe Township, Myanmar. *Biomass and Bioenergy*, **109**, 39–46, doi:10.1016/j.biombioe.2017.12.011.

Winemiller, K.O. et al., 2016: DEVELOPMENT AND ENVIRONMENT. Balancing hydropower and biodiversity in the Amazon, Congo, and Mekong. *Science*, **351**, 128–129, doi:10.1126/science.aac7082.

Winickoff, D.E., and M. Mondou, 2017: The problem of epistemic jurisdiction in global governance: The case of sustainability standards for biofuels. *Soc. Stud. Sci.*, **47**, 7–32, doi:10.1177/0306312716667855.

Wise, R.M. et al., 2014: Reconceptualising adaptation to climate change as part of pathways of change and response. *Glob. Environ. Chang.*, **28**, 325–336, doi:10.1016/j.gloenvcha.2013.12.002.

Wise, R.M. et al., 2016: How climate compatible are livelihood adaptation strategies and development programs in rural Indonesia? *Clim. Risk Manag.*, **12**, 100–114, doi:10.1016/j.crm.2015.11.001.

Wittmann, M., S. Chandra, K. Boyd, and C. Jerde, 2016: Implementing invasive species control: A case study of multi-jurisdictional coordination at Lake Tahoe, USA. *Manag. Biol. Invasions*, **6**, 319–328, doi:10.3391/mbi.2015.6.4.01.

Wodon, Q., and H. Zaman, 2010: Higher food prices in Sub-Saharan Africa: Poverty impact and policy responses. *World Bank Res. Obs.*, **25**, 157–176, doi:10.1093/wbro/lkp018.

Woodward, M., Z. Kapelan, and B. Gouldby, 2013: Adaptive flood risk management under climate change uncertainty using real options and optimisation. *Journ. Risk Anal.*, **34**, 75–92, doi:10.1111/risa.12088.

Woolf, D., D. Solomon, and J. Lehmann, 2018: Land restoration in food security programmes: Synergies with climate change mitigation. *Clim. Policy*, **18**, 1–11, doi:10.1080/14693062.2018.1427537.

Woollen, E. et al., 2016: Charcoal production in the Mopane woodlands of Mozambique: What are the trade-offs with other ecosystem services? *Philos. Trans. R. Soc. B Biol. Sci.*, **371**, 20150315, doi:10.1098/rstb.2015.0315.

World Bank, 2009a: *Environmental crisis or sustainable development opportunity?* World Bank, Washington, DC, USA.

World Bank, 2009b: *Environmental Crisis or Sustainable Development Opportunity?*

World Bank, 2018: *The State of Social Safety Nets 2018*. Washington, DC, USA, 165 pp.

World Food Programme, 2018: *Food Security Climate Resilience Facility (FoodSECuRE)*, Rome, Italy, 2 pp.

World Health Organization, 2014: *Quantitative Risk Assessment of the Effects of Climate Change on Selected Causes of Death, 2030s and 2050s*. World Health Organization, Geneva, Switzerland, 115 pp.

Wreford, A., and A. Renwick, 2012: Estimating the costs of climate change adaptation in the agricultural sector. *CAB Rev. Perspect. Agric. Vet. Sci. Nutr. Nat. Resour.*, **7**, 1–10, doi:10.1079/PAVSNNR20127040.

Wreford, A., A. Ignaciuk, and G. Gruère, 2017: Overcoming barriers to the adoption of climate-friendly practices in agriculture. *OECD Food, Agric. Fish. Pap.*, **101**, 1–40, doi:10.1787/97767de8-en.

Wu, X., Y. Lu, S. Zhou, L. Chen, and B. Xu, 2016: Impact of climate change on human infectious diseases: Empirical evidence and human adaptation. *Environ. Int.*, **86**, 14–23, doi:10.1016/J.ENVINT.2015.09.007.

Wunder, S., and R. Bodle, 2019: Achieving land degradation neutrality in Germany: Implementation process and design of a land use change based indicator. *Environ. Sci. Policy*, **92**, 46–55, doi:10.1016/J.ENVSCI.2018.09.022.

Xu, J. et al., 2005: Integrating sacred knowledge for conservation: Cultures and landscapes in Southwest China. *Ecol. Soc.*, **10**, ART. 7, doi:10.5751/ES-01413-100207.

Yamagata, Y., N. Hanasaki, A. Ito, T. Kinoshita, D. Murakami, and Q. Zhou, 2018: Estimating water-food-ecosystem trade-offs for the global negative emission scenario (IPCC-RCP2.6). *Sustain. Sci.*, **13**, 301–313, doi:10.1007/s11625-017-0522-5.

Yamana, T.K., A. Bomblies, and E.A.B. Eltahir, 2016: Climate change unlikely to increase malaria burden in West Africa. *Nat. Clim. Chang.*, **6**, 1009–1013, doi:10.1038/nclimate3085.

Yami, M., C. Vogl, and M. Hauser, 2009: Comparing the effectiveness of informal and formal institutions in sustainable common pool resources management in Sub-Saharan Africa. *Conserv. Soc.*, **7**, 153, doi:10.4103/0972-4923.64731.

Yami, M., C. Vogl, and M. Hauser, 2011: Informal institutions as mechanisms to address challenges in communal grazing land management in Tigray, Ethiopia. *Int. J. Sustain. Dev. World Ecol.*, **18**, 78–87, doi:10.1080/13504509.2010.530124.

Yang, J. et al., 2014a: Spatial and temporal patterns of global burned area in response to anthropogenic and environmental factors: Reconstructing global fire history for the 20th and early 21st centuries. *J. Geophys. Res. Biogeosciences*, **119**, 249–263, doi:10.1002/2013JG002532.

Yang, W., and Q. Lu, 2018: Integrated evaluation of payments for ecosystem services programs in China: A systematic review. *Ecosyst. Heal. Sustain.*, **4**, 73–84, doi:10.1080/20964129.2018.1459867.

Youn, S.-J. et al., 2014: Inland capture fishery contributions to global food security and threats to their future. *Glob. Food Sec.*, **3**, 142–148, doi:10.1016/j.gfs.2014.09.005.

Young, H.S. et al., 2017a: Interacting effects of land use and climate on rodent-borne pathogens in central Kenya. *Philos. Trans. R. Soc. B Biol. Sci.*, **372**, 20160116, doi:10.1098/rstb.2016.0116.

Young, O.R., 2017a: *Governing Complex Systems. Social Capital for the Anthropocene*. Massachusetts Institute of Technology, Massachusetts, USA, 296 pp.

Young, O.R., 2017b: Beyond regulation: Innovative strategies for governing large complex systems. *Sustain.*, **9**, 938, doi:10.3390/su9060938.

7

Yousefpour, R., and M. Hanewinkel, 2016: Climate change and decision-making under uncertainty. *Curr. For. Reports*, **2**, 143–149, doi:10.1007/s40725-016-0035-y.

Yumkella, K.K., and P.T. Yillia, 2015: Framing the water-energy-nexus for the Post-2015 Development Agenda. *Aquat. Procedia*, **5**, 8–12, doi:10.1016/j.aqpro.2015.10.003.

Zahawi, R.A., J.L. Reid, and K.D. Holl, 2014: Hidden costs of passive restoration. *Restor. Ecol.*, **22**, 284–287, doi:10.1111/rec.12098.

Zahran, S., S.D. Brody, W.E. Highfield, and A. Vedlitz, 2010: Non-linear incentives, plan design, and flood mitigation: The case of the Federal Emergency Management Agency's community rating system. *J. Environ. Plan. Manag.*, **53**, 219–239, doi:10.1080/09640560903529410.

Zanzanaini, C. et al., 2017: Integrated landscape initiatives for agriculture, livelihoods and ecosystem conservation: An assessment of experiences from South and Southeast Asia. *Landsc. Urban Plan.*, 165, 11–21, doi:10.1016/j.landurbplan.2017.03.010.

Zarfl, C., A.E. Lumsdon, J. Berlekamp, L. Tydecks, and K. Tockner, 2015: A global boom in hydropower dam construction. *Aquat. Sci.*, **77**, 161–170, doi:10.1007/s00027-014-0377-0.

Zeng, Z., J. Liu, P.H. Koeneman, E. Zarate, and A.Y. Hoekstra, 2012: Assessing water footprint at river basin level: A case study for the Heihe River Basin in Northwest China. *Hydrol. Earth Syst. Sci.*, **16**, 2771–2781, doi:10.5194/hess-16-2771-2012.

Zhang, J., C. He, L. Chen, and S. Cao, 2018a: Improving food security in China by taking advantage of marginal and degraded lands. *J. Clean. Prod.*, **171**, 1020–1030, doi:10.1016/j.jclepro.2017.10.110.

Zhang, W., T. Zhou, L. Zou, L. Zhang, and X. Chen, 2018b: Reduced exposure to extreme precipitation from 0.5°C less warming in global land monsoon regions. *Nat. Commun.*, **9**, 3153, doi:10.1038/s41467-018-05633-3.

Zhao, C. et al., 2017: Temperature increase reduces global yields of major crops in four independent estimates. *Proc. Natl. Acad. Sci.*, **114**, 9326–9331, doi:10.1073/pnas.1701762114.

Zhao, L. et al., 2018: Interactions between urban heat islands and heat waves. *Environ. Res. Lett.*, **13**, 1–11, doi:10.1088/1748-9326/aa9f73.

Zheng, H. et al., 2016: Using ecosystem service trade-offs to inform water conservation policies and management practices. *Front. Ecol. Environ.*, **14**, 527–532, doi:10.1002/fee.1432.

Ziv, G., E. Baran, S. Nam, I. Rodríguez-Iturbe, and S.A. Levin, 2012: Trading-off fish biodiversity, food security, and hydropower in the Mekong River Basin. *Proc. Natl. Acad. Sci.*, **109**, 5609–5614, doi:10.1073/pnas.1201423109.

Zomer, R.J., A. Trabucco, D.A. Bossio, and L.V. Verchot, 2008: Climate change mitigation: A spatial analysis of global land suitability for clean development mechanism afforestation and reforestation. *Agric. Ecosyst. Environ.*, **126**, 67–80, doi:10.1016/j.agee.2008.01.014.

Zoogah, D.B., M.W. Peng, and H. Woldu, 2015: Institutions, resources, and organizational effectiveness in Africa. *Acad. Manag. Perspect.*, **29**, 7–31, doi:10.5465/amp.2012.0033.

Zulu, L.C., and R.B. Richardson, 2013a: Charcoal, livelihoods, and poverty reduction: Evidence from Sub-Saharan Africa. *Energy Sustain. Dev.*, **17**, 127–137, doi:10.1016/j.esd.2012.07.007.

Senyolo, M.P., T.B. Long, V. Blok, O. Omta, 2018: How the characteristics of innovations impact their adoption: An exploration of climate-smart agricultural innovations in South Africa. *Journal of Cleaner Production*, **172**, 3825–3840, doi:1016/j.jclepro.2017.06.019.

Annexes

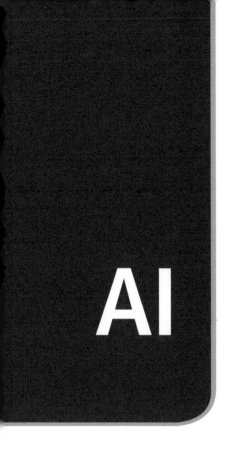

Annex I: Glossary

Coordinating Editor:
Renée van Diemen (The Netherlands/United Kingdom)

Editorial Team:
Tim Benton (United Kingdom), Eduardo Calvo (Peru), Annette Cowie (Australia), Valérie Masson-Delmotte (France), Aziz Elbehri (Morocco), Karlheinz Erb (Italy), Giacomo Grassi (Italy/European Union), J.B. Robin Matthews (United Kingdom), Hans-Otto Pörtner (Germany), Andy Reisinger (New Zealand), Debra Roberts (South Africa), Priyadarshi Shukla (India), Andrey Sirin (The Russian Federation), Jim Skea (United Kingdom), Murat Türkeş (Turkey), Nora M. Weyer (Germany), Sumaya Zakieldeen (Sudan), Panmao Zhai (China)

Notes:
Note that subterms are in italics beneath main terms.

This glossary defines some specific terms as the Lead Authors intend them to be interpreted in the context of this report. Blue, italicised words indicate that the term is defined in the Glossary.

This annex should be cited as:
IPCC, 2019: Annex I: Glossary [van Diemen, R. (ed.)]. In: *Climate Change and Land: an IPCC special report on climate change, desertification, land degradation, sustainable land management, food security, and greenhouse gas fluxes in terrestrial ecosystems* [P.R. Shukla, J. Skea, E. Calvo Buendia, V. Masson-Delmotte, H.-O. Pörtner, D. C. Roberts, P. Zhai, R. Slade, S. Connors, R. van Diemen, M. Ferrat, E. Haughey, S. Luz, S. Neogi, M. Pathak, J. Petzold, J. Portugal Pereira, P. Vyas, E. Huntley, K. Kissick, M. Belkacemi, J. Malley, (eds.)]. https://doi.org/10.1017/9781009157988.010

AI

1.5°C pathway See *Pathways*.

2030 Agenda for Sustainable Development A UN resolution in September 2015 adopting a plan of action for people, planet and prosperity in a new global development framework anchored in 17 *Sustainable Development Goals* (UN, 2015).

Acceptability of policy or system change The extent to which a policy or system change is evaluated unfavourably or favourably, or rejected or supported, by members of the general public (public acceptability) or politicians or governments (political acceptability). Acceptability may vary from totally unacceptable/fully rejected to totally acceptable/fully supported; individuals may differ in how acceptable policies or system changes are believed to be.

Acclimatisation A change in functional or morphological traits occurring once or repeatedly (e.g., seasonally) during the lifetime of an individual organism in its natural environment. Through acclimatisation the individual maintains performance across a range of environmental conditions. For a clear differentiation between findings in laboratory and field studies, the term acclimation is used in ecophysiology for the respective phenomena when observed in well-defined experimental settings. The term (adaptive) plasticity characterises the generally limited scope of changes in phenotype that an individual can reach through the process of acclimatisation.

Activity A practice or ensemble of practices that take place on a delineated area over a given period of time.

Activity data Data on the magnitude of a human *activity* resulting in emissions or removals taking place during a given period of time. In the *Agriculture, Forestry and Other Land Use (AFOLU)* sector, data on area of different *land uses*, management systems, animal numbers, lime and fertiliser use are examples of activity data.

Adaptability See *Adaptive capacity*.

Adaptation In *human systems,* the process of adjustment to actual or expected *climate* and its effects, in order to moderate harm or exploit beneficial opportunities. In natural systems, the process of adjustment to actual climate and its effects; human intervention may facilitate adjustment to expected climate and its effects.

Incremental adaptation
Adaptation that maintains the essence and integrity of a system or process at a given scale (Park et al., 2012).

Transformational adaptation
Adaptation that changes the fundamental attributes of a *social-ecological system* in anticipation of climate change and its impacts.

Adaptation limits
The point at which an actor's objectives (or system needs) cannot be secured from intolerable risks through adaptive actions.

- Hard adaptation limit: No adaptive actions are possible to avoid intolerable risks.

- Soft adaptation limit: Options are currently not available to avoid intolerable risks through adaptive action.

See also *Adaptation options*, *Adaptive capacity* and *Maladaptive actions (Maladaptation)*.

Adaptation behaviour See *Human behaviour*.

Adaptation limits See *Adaptation*.

Adaptation options The array of strategies and measures that are available and appropriate for addressing *adaptation*. They include a wide range of actions that can be categorised as structural, institutional, ecological or behavioural. See also *Adaptive capacity*, and *Maladaptive actions (Maladaptation)*.

Adaptation pathways See *Pathways*

Adaptive capacity The ability of systems, *institutions*, humans and other organisms to adjust to potential damage, to take advantage of opportunities, or to respond to consequences (IPCC, 2014; MA, 2005). See also *Adaptation, Adaptation options,* and *Maladaptive actions (Maladaptation)*.

Adaptive governance See *Governance*.

Adverse side-effect The negative effects that a policy or measure aimed at one objective might have on other objectives, without yet evaluating the net effect on overall social welfare. Adverse side-effects are often subject to uncertainty and depend on, among others, local circumstances and implementation practices. See also *Co-benefits* and *Risk*.

Aerosol A suspension of airborne solid or liquid particles, with a typical size between a few nanometres and 10 µm that reside in the *atmosphere* for at least several hours. The term aerosol, which includes both the particles and the suspending gas, is often used in this report in its plural form to mean aerosol particles. Aerosols may be of either natural or *anthropogenic* origin. Aerosols may influence *climate* in several ways: through both interactions that scatter and/or absorb radiation and through interactions with cloud microphysics and other cloud properties, or upon deposition on snow or ice covered surfaces thereby altering their *albedo* and contributing to *climate feedback*. Atmospheric aerosols, whether natural or anthropogenic, originate from two different pathways: emissions of primary particulate matter (PM), and formation of secondary PM from gaseous *precursors*. The bulk of aerosols are of natural origin. Some scientists use group labels that refer to the chemical composition, namely: sea salt, organic carbon, *black carbon* (BC), mineral species (mainly desert dust), sulphate, nitrate, and ammonium. These labels are, however, imperfect as aerosols combine particles to create complex mixtures. See also *Short-lived climate forcers (SLCF)*.

Afforestation Conversion to *forest* of land that historically has not contained forests. [Note: For a discussion of the term forest and related terms such as afforestation, *reforestation* and *deforestation,* in the context of reporting and accounting Article 3.3 and 3.4 activities under the *Kyoto Protocol,* see 2013 Revised Supplementary Methods and Good Practice Guidance Arising from the Kyoto Protocol.] See also *Reforestation, Deforestation, Forest* and *Reducing Emissions from Deforestation and Forest Degradation (REDD+)*.

Agreement In this report, the degree of agreement within the scientific body of knowledge on a particular finding is assessed based on multiple lines of *evidence* (e.g., mechanistic understanding, theory, data, models, expert judgement) and expressed qualitatively (Mastrandrea et al., 2010). See also *Confidence, Likelihood,* and *Uncertainty*.

Agriculture, Forestry and Other Land Use (AFOLU) In the context of national *greenhouse gas (GHG)* inventories under the *United Nations Convention on Climate Change (UNFCCC)*, AFOLU is the sum of the GHG inventory sectors Agriculture and *Land Use, Land-Use Change and Forestry (LULUCF)*; see the 2006 IPCC Guidelines for National GHG Inventories for details. Given the difference in estimating the 'anthropogenic' *carbon dioxide* (CO_2) removals between countries and the global modelling community, the land-related net GHG emissions from global models included in this report are not necessarily directly comparable with LULUCF estimates in national GHG Inventories.

FOLU (Forestry and Other Land Use) – also referred to as LULUCF The subset of AFOLU emissions and removals of greenhouse gases (GHGs) resulting from direct human-induced *land use, land-use change,* and forestry activities excluding agricultural emissions.

See also *Land-Use Change (LUC)* and *Land Use, Land-Use Change and Forestry (LULUCF)*.

Agrobiodiversity 'The variety and variability of animals, plants and micro-organisms that are used directly or indirectly for food and agriculture, including crops, livestock, forestry and fisheries. It comprises the diversity of genetic resources (varieties, breeds) and species used for food, fodder, fibre, fuel and pharmaceuticals. It also includes the diversity of non-harvested species that support production (soil micro-organisms, predators, pollinators), and those in the wider environment that support agro-ecosystems (agricultural, pastoral, forest and aquatic) as well as the diversity of the agro-ecosystems' (FAO, 2005).

Agroecology 'The science and practice of applying ecological concepts, principles and knowledge (i.e., the interactions of, and explanations for, the diversity, abundance and activities of organisms) to the study, design and management of sustainable agroecosystems. It includes the roles of human beings as a central organism in agroecology by way of social and economic processes in farming systems. Agroecology examines the roles and interactions among all relevant biophysical, technical and socioeconomic components of farming systems and their surrounding landscapes' (IPBES, 2019).

Agroforestry Collective name for land-use systems and technologies where woody perennials (trees, shrubs, palms, bamboos, etc.) are deliberately used on the same land-management units as agricultural crops and/or animals, in some form of spatial arrangement or temporal sequence. In agroforestry systems there are both ecological and economical interactions between the different components. Agroforestry can also be defined as a dynamic, ecologically based, natural resource management system that, through the integration of trees on farms and in the agricultural landscape, diversifies and sustains production for increased social, economic and environmental benefits for land users at all levels (FAO, 2015a).

Air pollution Degradation of air quality with negative effects on human health, the natural or built environment, due to the introduction by natural processes or human activity in the *atmosphere* of substances (gases, *aerosols*) which have a direct (primary pollutants) or indirect (secondary pollutants) harmful effect. See also *Short-lived climate forcers (SLCF)*.

Albedo The proportion of sunlight (solar radiation) reflected by a surface or object, often expressed as a percentage. Clouds, snow and ice usually have high albedo; soil surfaces cover the albedo range from high to low; vegetation in the dry season and/or in arid zones can have high albedo, whereas photosynthetically active vegetation and the ocean have low albedo. The Earth's planetary albedo changes mainly through varying cloudiness, snow, ice, leaf area and land cover changes.

Ambient persuasive technology Technological systems and environments that are designed to change human cognitive processing, attitudes and behaviours without the need for the user's conscious attention.

Anomaly The deviation of a variable from its value averaged over a *reference period*. See also *Reference period*.

Anthromes '*Human systems*, with natural *ecosystems* embedded within them' (Ellis and Ramankutty 2008). The anthrome classification system is based on human population density and *land use*, and comprises the following classes: dense settlements, villages, croplands, rangeland, forested (then broadened to seminatural) and wildlands (Ellis et al. 2010).

Anthropocene A proposed new geological epoch resulting from significant human-driven changes to the structure and functioning of the Earth System, including the *climate system*. Originally proposed in the Earth System science community in 2000, the proposed new epoch is undergoing a formalisation process within the geological community based on the stratigraphic *evidence* that human activities have changed the Earth System to the extent of forming geological deposits with a signature that is distinct from those of the *Holocene*, and which will remain in the geological record. Both the stratigraphic and Earth System approaches to defining the Anthropocene consider the mid-20th century to be the most appropriate starting date, although others have been proposed and continue to be discussed. The Anthropocene concept has been taken up by a diversity of disciplines and the public to denote the substantive influence humans have had on the state, dynamics and future of the Earth System. See also *Holocene*.

Anthropogenic Resulting from or produced by human activities. See also *Anthropogenic emissions,* and *Anthropogenic removals*.

Anthropogenic emissions Emissions of *greenhouse gases (GHGs), precursors* of GHGs and *aerosols* caused by human activities. These activities include the burning of *fossil fuels, deforestation, land use and land-use changes (LUC)*, livestock production, fertilisation, waste management, and industrial processes. See also *Anthropogenic,* and *Anthropogenic removals*.

Anthropogenic removals The withdrawal of *greenhouse gases (GHGs)* from the atmosphere as a result of deliberate human activities. These include enhancing biological sinks of CO_2 and using chemical engineering to achieve long term removal and storage. *Carbon capture and storage (CCS)* from industrial and energy-related sources, which alone does not remove CO_2 from the atmosphere, can help reduce atmospheric CO_2 if it is combined with *bioenergy production (BECCS)*. [Note: In the 2006 IPCC Guidelines for National GHG Inventories, which are used in reporting of emissions to the UNFCCC, 'anthropogenic' land-related GHG *fluxes* are defined as all

AI

those occurring on *'managed land'*, i.e. 'where human interventions and practices have been applied to perform production, ecological or social functions'. However, some removals (e.g. removals associated with CO_2 fertilisation and N deposition) are not considered as 'anthropogenic', or are referred to as 'indirect' anthropogenic effects, in some of the scientific literature assessed in this report. As a consequence, the land-related net GHG emission estimates from global models included in this report are not necessarily directly comparable with LULUCF estimates in national GHG Inventories. See also *Anthropogenic emissions, Bioenergy with carbon dioxide capture and storage (BECCS), Carbon dioxide capture and storage (CCS)* and *Land use, Land-use change, and Forestry (LULUCF).*

Aridity   The state of a long-term climatic feature characterised by low average precipitation or available water in a region. Aridity generally arises from widespread persistent *atmospheric* subsidence or anticyclonic conditions, and from more localised subsidence in the lee side of mountains (adapted from Gbeckor-Kove, 1989; Türkeş, 1999).

Atmosphere   The gaseous envelope surrounding the Earth, divided into five layers — the *troposphere* which contains half of the Earth's atmosphere, the *stratosphere*, the mesosphere, the thermosphere, and the exosphere, which is the outer limit of the atmosphere. The dry atmosphere consists almost entirely of nitrogen (78.1% volume mixing ratio) and oxygen (20.9% volume mixing ratio), together with a number of trace gases, such as argon (0.93 % volume mixing ratio), helium and radiatively active *greenhouse gases (GHGs)* such as *carbon dioxide (CO_2)* (0.04% volume mixing ratio) and *ozone (O_3)*. In addition, the atmosphere contains the GHG water vapour (H_2O), whose amounts are highly variable but typically around 1% volume mixing ratio. The atmosphere also contains clouds and aerosols. See also *Carbon dioxide (CO_2), Ozone (O_3), Troposphere, Stratosphere, Greenhouse gas (GHG),* and *Hydrological cycle.*

Atmosphere-ocean general circulation model (AOGCM)   See *Climate model.*

Atmospheric boundary layer   The atmospheric layer adjacent to the Earth's surface that is affected by friction against that boundary surface, and possibly by transport of heat and other variables across that surface (AMS, 2000). The lowest 100 m of the boundary layer (about 10% of the boundary layer thickness), where mechanical generation of turbulence is dominant, is called the surface boundary layer or surface layer.

Attribution   See *Detection and attribution.*

Baseline scenario   In much of the literature the term is also synonymous with the term business-as-usual (BAU) *scenario*, although the term BAU has fallen out of favour because the idea of business as usual in century-long socio-economic *projections* is hard to fathom. In the context of *transformation pathways*, the term baseline scenarios refers to scenarios that are based on the assumption that no mitigation policies or measures will be implemented beyond those that are already in force and/or are legislated or planned to be adopted. Baseline scenarios are not intended to be predictions of the future, but rather counterfactual constructions that can serve to highlight the level of emissions that would occur without further policy effort. Typically, baseline scenarios are then compared to

mitigation scenarios that are constructed to meet different goals for *greenhouse gas (GHG)* emissions, atmospheric concentrations or temperature change. The term baseline scenario is often used interchangeably with reference scenario and no policy scenario. See also *Emission scenario,* and *Mitigation scenario.*

Biochar   Relatively stable, carbon-rich material produced by heating *biomass* in an oxygen-limited environment. Biochar is distinguished from charcoal by its application: biochar is used as a soil amendment with the intention to improve soil functions and to reduce *greenhouse gas (GHG)* emissions from biomass that would otherwise decompose rapidly (IBI, 2018).

Biodiversity   Biodiversity or biological diversity means the variability among living organisms from all sources including, among other things, terrestrial, marine and other aquatic *ecosystems* and the ecological complexes of which they are part; this includes diversity within species, between species and of ecosystems (UN, 1992). See also *Ecosystem,* and *Ecosystem service.*

Bioenergy   Energy derived from any form of biomass or its metabolic by-products. See also *Biomass* and *Biofuel.*

Bioenergy with carbon dioxide capture and storage (BECCS)   *Carbon dioxide capture and storage (CCS)* technology applied to a *bioenergy* facility. Note that depending on the total emissions of the BECCS supply chain, *carbon dioxide (CO_2)* can be removed from the *atmosphere*. See also *Bioenergy,* and *Carbon dioxide capture and storage (CCS).*

Biofuel   A fuel, generally in liquid form, produced from *biomass*. Biofuels include bioethanol from sugarcane, sugar beet or maize, and biodiesel from canola or soybeans. See also *Biomass,* and *Bioenergy.*

Biogeochemical effects   Processes through which land affects *climate*, excluding *biophysical effects*. These processes include changes in net emissions of *carbon dioxide (CO_2)* towards the *atmosphere*, net emissions of *aerosols* (mineral and organic), ozone deposition on *ecosystems*, and net emissions of biogenic volatile organic compounds (BVOCs) and their subsequent changes in atmospheric chemistry. See also *Biophysical effects.*

Biomass   Organic material excluding the material that is fossilised or embedded in geological formations. Biomass may refer to the mass of organic matter in a specific area (ISO, 2014). See also *Bioenergy,* and *Biofuel.*

Traditional biomass
The combustion of wood, charcoal, agricultural residues and/or animal dung for cooking or heating in open fires or in inefficient stoves as is common in low-income countries.

Biome   'Global-scale zones, generally defined by the type of plant life that they support in response to average rainfall and temperature patterns. For example, tundra, coral reefs or savannas' (IPBES, 2019).

Biophysical effects   The range of physical processes through which *land* affects *climate*. These processes include changes in hydrology (e.g. water vapor fluxes at the land/atmosphere interface), heat exchanges via convective *fluxes* (latent and sensible), radiation (solar and infra-red, absorbed and emitted), and momentum (e.g. affecting wind speed).

Black carbon (BC)   A relatively pure form of carbon, also known as soot, arising from the incomplete combustion of *fossil fuels, biofuel,* and *biomass*. It stays in the *atmosphere* only for days or weeks. Black carbon is a climate forcing agent with strong warming effect, both in the atmosphere and when deposited on snow or ice. See also *Atmosphere,* and *Aerosol*.

Blue carbon   All biologically-driven carbon *fluxes* and storage in marine systems that are amenable to management can be considered as blue carbon. Coastal blue carbon focuses on rooted vegetation in the coastal zone, such as tidal marshes, mangroves and seagrasses. These *ecosystems* have high carbon burial rates on a per unit area basis and accumulate carbon in their soils and sediments. They provide many non-climatic benefits and can contribute to ecosystem-based adaptation. If degraded or lost, coastal blue carbon ecosystems are likely to release most of their carbon back to the *atmosphere*. There is current debate regarding the application of the blue carbon concept to other coastal and non-coastal processes and ecosystems, including the open ocean. See also *Ecosystem services,* and *Carbon sequestration*.

Business as usual (BAU)   See *Baseline scenario*.

Carbon budget   Refers to three concepts in the literature: (1) an assessment of *carbon cycle sources* and *sinks* on a global level, through the synthesis of *evidence* for *fossil-fuel* and cement emissions, *land-use change* emissions, ocean and land CO_2 sinks, and the resulting atmospheric *carbon dioxide* (CO_2) growth rate. This is referred to as the global carbon budget; (2) the estimated cumulative amount of global carbon dioxide emissions that that is estimated to limit global surface temperature to a given level above a *reference period*, taking into account global surface temperature contributions of other *greenhouse gases* (GHGs) and climate forcers; (3) the distribution of the carbon budget defined under (2) to the regional, national, or sub-national level based on considerations of *equity*, costs or efficiency. See also *Remaining carbon budget*.

Carbon cycle   The flow of carbon (in various forms, e.g., as *carbon dioxide* (CO_2), carbon in *biomass*, and carbon dissolved in the ocean as carbonate and bicarbonate) through the *atmosphere*, hydrosphere, terrestrial and marine biosphere and lithosphere. In this report, the reference unit for the global carbon cycle is $GtCO_2$ or GtC (one Gigatonne = 1 Gt = 10^{15} grams; 1 GtC corresponds to 3.667 $GtCO_2$).

Carbon dioxide (CO_2)   A naturally occurring gas, CO_2 is also a by-product of burning *fossil fuels* (such as oil, gas and coal), of burning *biomass*, of *land-use changes (LUC)* and of industrial processes (e.g., cement production). It is the principal *anthropogenic* greenhouse gas (GHG) that affects the Earth's radiative balance. It is the reference gas against which other GHGs are measured and therefore has a Global Warming Potential (GWP) of 1. See also *Greenhouse gas (GHG), Land use,* and *Land-use change*.

Carbon dioxide capture and storage (CCS)   A process in which a relatively pure stream of *carbon dioxide (CO_2)* from industrial and energy-related sources is separated (captured), conditioned, compressed and transported to a storage location for long-term isolation from the atmosphere. Sometimes referred to as Carbon Capture and Storage. See also *Carbon dioxide capture and utilisation (CCU), Bioenergy with carbon dioxide capture and storage (BECCS),* and *Sequestration*.

Carbon dioxide capture and utilisation (CCU)   A process in which *carbon dioxide (CO_2)* is captured and then used to produce a new product. If the CO_2 is stored in a product for a *climate*-relevant time horizon, this is referred to as carbon dioxide capture, utilisation and storage (CCUS). Only then, and only combined with CO_2 recently removed from the *atmosphere*, can CCUS lead to carbon dioxide removal. CCU is sometimes referred to as Carbon dioxide capture and use. See also *Carbon dioxide capture and storage (CCS)*.

Carbon dioxide capture, utilisation and storage (CCUS)   See *Carbon dioxide capture and utilisation (CCU)*.

Carbon dioxide removal (CDR)   *Anthropogenic* activities removing *carbon dioxide (CO_2)* from the *atmosphere* and durably storing it in geological, terrestrial, or ocean reservoirs, or in products. It includes existing and potential anthropogenic enhancement of biological or geochemical sinks and direct air capture and storage, but excludes natural CO_2 *uptake* not directly caused by human activities. See also *Mitigation (of climate change), Greenhouse gas removal (GGR), Negative emission technologies,* and *Sink*.

Carbon intensity   The amount of emissions of *carbon dioxide (CO_2)* released per unit of another variable such as *gross domestic product (GDP)*, output energy use or transport.

Carbon price   The price for avoided or released *carbon dioxide (CO_2)* or CO_2-*equivalent emissions*. This may refer to the rate of a carbon tax, or the price of emission permits. In many models that are used to assess the economic costs of *mitigation*, carbon prices are used as a proxy to represent the level of effort in mitigation policies. See also *Mitigation*.

Carbon sequestration   The process of storing carbon in a carbon *pool*. See also *Blue carbon, Carbon dioxide capture and storage (CCS), Uptake,* and *Sink*.

Carbon sink   See *Sink*.

Carbon stock   The quantity of carbon in a carbon *pool*. See also *Pool, carbon and nitrogen*.

Citizen science   A voluntary participation of the public in the collection and/or processing of data as part of a scientific study (Silvertown, 2009).

Clean Development Mechanism (CDM)   A mechanism defined under Article 12 of the *Kyoto Protocol* through which investors (governments or companies) from developed (Annex B) countries may finance *greenhouse gas (GHG)* emission reduction or removal projects in developing countries (Non-Annex B), and receive Certified Emission Reduction Units (CERs) for doing so. The CERs can be credited towards the commitments of the respective developed countries. The CDM is intended to facilitate the two objectives of promoting *sustainable development* (SD) in developing countries and of helping *industrialised countries* to reach their emissions commitments in a cost-effective way.

Climate   Climate in a narrow sense is usually defined as the average weather, or more rigorously, as the statistical description in terms of the mean and variability of relevant quantities over a period

AI

of time ranging from months to thousands or millions of years. The classical period for averaging these variables is 30 years, as defined by the World Meteorological Organization. The relevant quantities are most often surface variables such as temperature, precipitation and wind. Climate in a wider sense is the state, including a statistical description, of the *climate system*.

Climate change   A change in the state of the *climate* that can be identified (e.g., by using statistical tests) by changes in the mean and/or the variability of its properties and that persists for an extended period, typically decades or longer. Climate change may be due to natural internal processes or external *forcings* such as modulations of the solar cycles, volcanic eruptions and persistent *anthropogenic* changes in the composition of the *atmosphere* or in *land use*. Note that the *United Nations Framework Convention on Climate Change (UNFCCC),* in its Article 1, defines climate change as: 'a change of climate which is attributed directly or indirectly to human activity that alters the composition of the global atmosphere and which is in addition to natural climate variability observed over comparable time periods'. The UNFCCC thus makes a distinction between climate change attributable to human activities altering the atmospheric composition and climate variability attributable to natural causes. See also *Climate variability, Global warming, Ocean acidification,* and *Detection and attribution*.

Climate extreme (extreme weather or climate event)   The occurrence of a value of a weather or *climate* variable above (or below) a threshold value near the upper (or lower) ends of the range of observed values of the variable. For simplicity, both extreme weather events and extreme climate events are referred to collectively as 'climate extremes.' See also *Extreme weather event*.

Climate feedback   An interaction in which a perturbation in one *climate* quantity causes a change in a second and the change in the second quantity ultimately leads to an additional change in the first. A negative feedback is one in which the initial perturbation is weakened by the changes it causes; a positive feedback is one in which the initial perturbation is enhanced. The initial perturbation can either be externally forced or arise as part of internal variability.

Climate governance   See *Governance*.

Climate model   A numerical representation of the *climate system* based on the physical, chemical and biological properties of its components, their interactions and *feedback* processes and accounting for some of its known properties. The climate system can be represented by models of varying complexity; that is, for any one component or combination of components a spectrum or hierarchy of models can be identified, differing in such aspects as the number of spatial dimensions, the extent to which physical, chemical or biological processes are explicitly represented, or the level at which empirical parametrizations are involved. There is an evolution towards more complex models with interactive chemistry and biology. *Climate models* are applied as a research tool to study and simulate the *climate* and for operational purposes, including monthly, seasonal and interannual climate predictions. See also *Earth system model (ESM)*.

Climate projection   Simulated response of the *climate system* to a *scenario* of future emissions or concentrations of *greenhouse gases (GHGs)* and *aerosols*, and changes in land use, generally derived using *climate models*. Climate projections are distinguished from climate predictions by their dependence on the emission/concentration/*radiative forcing* scenario used, which is in turn based on assumptions concerning, for example, future socioeconomic and technological developments that may or may not be realised.

Climate-resilient development pathways (CRDPs)   Trajectories that strengthen *sustainable development* and efforts to eradicate *poverty* and reduce inequalities while promoting *fair* and cross-scalar *adaptation* to and *resilience* in a changing *climate*. They raise the *ethics, equity,* and *feasibility* aspects of the deep *societal transformation* needed to drastically reduce emissions to limit *global warming* (e.g., to 2°C) and achieve desirable and liveable futures and *well-being* for all.

Climate-resilient pathways   Iterative processes for managing change within complex systems in order to reduce disruptions and enhance opportunities associated with *climate change*. See also *Climate-resilient development pathways (CRDPs), Development pathways, Pathways,* and *Transformation pathways*.

Climate sensitivity   The change in the annual *global mean surface temperature* in response to a change in the atmospheric *carbon dioxide (CO$_2$)* concentration or other radiative forcing.

Equilibrium climate sensitivity
An estimate of the *global mean surface temperature* response to a doubling of the atmospheric *carbon dioxide (CO$_2$)* concentration that is evaluated from model output or observations for evolving non-equilibrium conditions. It is a measure of the strengths of the *climate feedbacks* at a particular time and may vary with *forcing* history and *climate* state, and therefore may differ from *equilibrium climate sensitivity*.

Transient climate response
The change in the *global mean surface temperature*, averaged over a 20-year period, centred at the time of atmospheric *carbon dioxide (CO$_2$)* doubling, in a *climate model* simulation in which CO$_2$ increases at 1% yr^{-1} from *pre-industrial*. It is a measure of the strength of *climate feedbacks* and the timescale of ocean heat uptake.

See also *Climate model*, and *Global mean surface temperature (GMST)*.

Climate services   Information and products that enhance users' knowledge and understanding about the *impacts* of *climate change* and/or *climate variability* so as to aid decision-making of individuals and organisations and enable preparedness and early climate change action. Such services involve high-quality data from national and international databases on temperature, rainfall, wind, soil moisture and ocean conditions, as well as maps, *risk* and *vulnerability* analyses, assessments, and long-term *projections* and *scenarios*. Depending on the user's needs, these data and information products may be combined with non-meteorological data, such as agricultural production, health trends, population distributions in high-risk areas, road and infrastructure maps for the delivery of goods, and other socio-economic variables (WMO, 2019).

Climate-smart agriculture (CSA)   An approach to agriculture that aims to transform and reorient agricultural systems to effectively support development and ensure *food security* in a changing climate by: sustainably increasing agricultural productivity and incomes; adapting and building resilience to climate change; and reducing and/or removing *greenhouse gas* emissions, where possible (FAO, 2018).

Climate system   The system consisting of five major components: the *atmosphere*, the hydrosphere, the cryosphere, the lithosphere and the biosphere and the interactions between them. The climate system evolves in time under the influence of its own internal dynamics and because of external *forcings* such as volcanic eruptions, solar variations, orbital forcing, and *anthropogenic* forcings such as the changing composition of the atmosphere and *land-use change*.

Climate target   A temperature limit, concentration level, or emissions reduction goal used towards the aim of avoiding dangerous *anthropogenic* interference with the *climate system*. For example, national climate targets may aim to reduce *greenhouse gas* emissions by a certain amount over a given time horizon, for example those under the *Kyoto Protocol*.

Climate variability   Variations in the mean state and other statistics (such as standard deviations, the occurrence of extremes, etc.) of the *climate* on all spatial and temporal scales beyond that of individual weather events. Variability may be due to natural internal processes within the *climate system* (internal variability), or to variations in natural or *anthropogenic* external *forcing* (external variability). See also *Climate change*.

CO₂ equivalent (CO₂-eq) emission   The amount of *carbon dioxide (CO₂)* emission that would cause the same integrated *radiative forcing* or temperature change, over a given time horizon, as an emitted amount of a *greenhouse gas (GHG)* or a mixture of GHGs. There are a number of ways to compute such equivalent emissions and choose appropriate time horizons. Most typically, the CO₂-equivalent emission is obtained by multiplying the emission of a GHG by its Global Warming Potential (GWP) for a 100 year time horizon. For a mix of GHGs it is obtained by summing the CO₂-equivalent emissions of each gas. CO₂-equivalent emission is a common scale for comparing emissions of different GHGs but does not imply equivalence of the corresponding climate change responses. There is generally no connection between CO₂-equivalent emissions and resulting CO₂-equivalent concentrations.

CO₂ fertilisation   The enhancement of plant growth as a result of increased atmospheric *carbon dioxide (CO₂)* concentration. The magnitude of CO₂ fertilisation depends on nutrients and water availability.

Co-benefits   The positive effects that a policy or measure aimed at one objective might have on other objectives, thereby increasing the total benefits for society or the environment. Co-benefits are often subject to uncertainty and depend on local circumstances and implementation practices, among other factors. Co-benefits are also referred to as ancillary benefits. See also *Adverse side-effects*, and *Risk*.

Collective action   A number of people working together voluntarily to achieve some common objective (Meinzen-Dick and Di Gregorio, 2004).

Conference of the Parties (COP)   The supreme body of UN conventions, such as the *United Nations Framework Convention on Climate Change (UNFCCC)*, comprising parties with a right to vote that have ratified or acceded to the convention. See also *United Nations Convention to Combat Desertification (UNCCD)*.

Confidence   The robustness of a finding based on the type, amount, quality and consistency of *evidence* (e.g., mechanistic understanding, theory, data, models, expert judgment) and on the degree of *agreement* across multiple lines of evidence. In this report, confidence is expressed qualitatively (Mastrandrea et al., 2010). See also *Likelihood*, and *Uncertainty*.

Convection   Vertical motion driven by buoyancy forces arising from static instability, usually caused by near-surface cooling or increases in salinity in the case of the ocean and near-surface warming or cloud-top radiative cooling in the case of the *atmosphere*. In the atmosphere, convection gives rise to cumulus clouds and precipitation and is effective at both scavenging and vertically transporting chemical species. In the ocean, convection can carry surface waters to deep within the ocean.

Coping capacity   The ability of people, *institutions*, organisations, and systems, using available skills, values, beliefs, resources, and opportunities, to address, manage, and overcome adverse conditions in the short to medium term. (UNISDR, 2009; IPCC, 2012a). See also *Resilience*.

Cost-benefit analysis   Monetary assessment of all negative and positive impacts associated with a given action. Cost-benefit analysis enables comparison of different interventions, investments or strategies and reveal how a given investment or policy effort pays off for a particular person, company or country. Cost-benefit analyses representing society's point of view are important for *climate change* decision making, but there are difficulties in aggregating costs and benefits across different actors and across timescales. See also *Discounting*.

Cost-effectiveness   A measure of the cost at which a policy goal or outcome is achieved. The lower the cost the greater the cost effectiveness.

Coupled Model Intercomparison Project (CMIP)   A climate modelling activity from the World Climate Research Programme (WCRP) which coordinates and archives *climate model* simulations based on shared model inputs by modelling groups from around the world. The CMIP3 multi-model data set includes *projections* using SRES scenarios. The CMIP5 data set includes projections using the *Representative Concentration Pathways (RCPs)*. The CMIP6 phase involves a suite of common model experiments as well as an ensemble of CMIP-endorsed model intercomparison projects (MIPs).

Cumulative emissions   The total amount of emissions released over a specified period of time. See also *Carbon budget*.

Decarbonisation   Process by which countries, individuals or other entities aim to achieve zero fossil carbon existence. Typically refers to a reduction of the carbon emissions associated with electricity, industry and transport.

Decoupling   Decoupling (in relation to *climate change*) is where economic growth is no longer strongly associated with consumption of *fossil fuels*. Relative decoupling is where both grow but at different rates. Absolute decoupling is where economic growth happens but fossil fuels decline.

Deforestation   Conversion of *forest* to non-forest. [Note: For a discussion of the term forest and related terms such as *afforestation*, *reforestation* and *deforestation* in the context of reporting and accounting Article 3.3 and 3.4 activities under the Kyoto Protocol, see 2013 Revised Supplementary Methods and Good Practice Guidance Arising from the Kyoto Protocol.] See also *Reducing Emissions from Deforestation and Forest Degradation (REDD+)*.

Deliberative governance   See *Governance*.

Demand and supply-side measures

Demand-side measures
Policies and programmes for influencing the demand for goods and/ or services. In the energy sector, demand-side management aims at reducing the demand for electricity and other forms of energy required to deliver energy services.

Supply-side measures
Policies and programmes for influencing how a certain demand for goods and/or services is met. In the energy sector, for example, supply-side mitigation measures aim at reducing the amount of greenhouse gas (GHG) emissions emitted per unit of energy produced.

See also *Mitigation measures*.

Demand-side measures   See *Demand and supply-side measures*.

Desertification   *Land degradation* in arid, semi-arid, and dry sub-humid areas resulting from many factors, including climatic variations and human activities (UNCCD, 1994).

Detection   See *Detection and attribution*.

Detection and attribution   Detection of change is defined as the process of demonstrating that climate or a system affected by climate has changed in some defined statistical sense, without providing a reason for that change. An identified change is detected in observations if its *likelihood* of occurrence by chance due to internal variability alone is determined to be small, for example, <10%. Attribution is defined as the process of evaluating the relative contributions of multiple causal factors to a change or event with a formal assessment of *confidence*.

Development pathways   See *Pathways*.

Diet   'The kinds of food that follow a particular pattern that a person or community eats' (FAO, 2014). See also *Dietary patterns*.

Dietary patterns   The quantities, proportions, variety or combinations of different foods and beverages in diets, and the frequency with which they are habitually consumed (Dietary Guidelines Advisory Committee, 2015). See also *Diet*.

Dietary and nutrition transitions   Modernisation, urbanisation, economic development, and increased wealth lead to predictable shifts in diet, referred to as 'nutrition transitions' (Misra and Khurana, 2008; Popkin, 2006). Over historical time there have been a number of dietary *transitions* but in recent decades the prime transition has been associated with changes from subsistence towards eating diets rich in calories and relatively poor in nutrition (the 'westernised diet') that are obesogenic. From a public health perspective, a new dietary transition is in focus, from the obesogenic diet to one promoting health.

Disaster   A 'serious disruption of the functioning of a community or a society at any scale due to hazardous events interacting with conditions of exposure, vulnerability and capacity, leading to one or more of the following: human, material, economic and environmental losses and impacts' (UNISDR, 2017). See also *Exposure*, *Risk*, *Vulnerability* and *Hazard*.

Disaster risk management (DRM)   Processes for designing, implementing, and evaluating strategies, policies, and measures to improve the understanding of current and future *disaster* risk, foster disaster risk reduction and transfer, and promote continuous improvement in disaster preparedness, prevention and protection, response, and recovery practices, with the explicit purpose of increasing human security, *well-being*, quality of life, and *sustainable development* (UNISDR, 2017).

Discounting   A mathematical operation that aims to make monetary (or other) amounts received or expended at different times (years) comparable across time. The discounter uses a fixed or possibly time-varying discount rate from year to year that makes future value worth less today (if the discount rate is positive). The choice of discount rate(s) is debated as it is a judgement based on hidden and/or explicit values.

Discount rate   See *Discounting*.

(Internal) Displacement   The forced movement of people within the country they live in. Internally displaced persons (IDPs) are 'Persons or groups of persons who have been forced or obliged to flee or to leave their homes or places of habitual residence, in particular as a result of or in order to avoid the effects of armed conflict, situations of generalized violence, violations of human rights or natural or human-made *disasters*, and who have not crossed an internationally recognized State border' (UN, 1998). See also *Migration*.

Displacement   In land system science, displacement denotes the increasing spatial separation between the location of agricultural and forestry production and the place of consumption of these products, as it occurs with trade. Displacement disconnects spatially environmental impacts from their socioeconomic drivers.

Downscaling   Method that derives local- to regional-scale (up to 100 km) information from larger-scale models or data analyses. Two main methods exist: dynamical downscaling and empirical/statistical downscaling. The dynamical method uses the output of regional *climate models*, global models with variable spatial resolution, or high-resolution global models. The empirical/statistical methods [are based on observations and] develop statistical relationships that link the large-scale atmospheric variables with local/ regional climate variables. In all cases, the quality of the driving model remains an important limitation on quality of the downscaled information. The two methods can be combined, e.g., applying empirical/statistical downscaling to the output of a regional climate model, consisting of a dynamical downscaling of a global climate model.

Drainage   'Artificial lowering of the soil water table' (IPCC, 2013). See also *Rewetting*.

Drought A period of abnormally dry weather long enough to cause a serious hydrological imbalance. Drought is a relative term, therefore any discussion in terms of precipitation deficit must refer to the particular precipitation-related activity that is under discussion. For example, shortage of precipitation during the growing season impinges on crop production or *ecosystem* function in general (due to *soil moisture* drought, also termed agricultural drought), and during the *runoff* and percolation season primarily affects water supplies (hydrological drought). Storage changes in soil moisture and groundwater are also affected by increases in actual *evapotranspiration* in addition to reductions in precipitation. A period with an abnormal precipitation deficit is defined as a meteorological drought.

Megadrought
A very lengthy and pervasive drought, lasting much longer than normal, usually a decade or more.

Early warning systems (EWS) The set of technical, financial and *institutional capacities* needed to generate and disseminate timely and meaningful warning information to enable individuals, communities and organisations threatened by a *hazard* to prepare to act promptly and appropriately to reduce the possibility of harm or loss. Dependent upon context, EWS may draw upon scientific and/ or *Indigenous knowledge*. EWS are also considered for ecological applications e.g., conservation, where the organisation itself is not threatened by hazard but the *ecosystem* under conservation is (an example is coral bleaching alerts), in agriculture (for example, warnings of ground frost, hailstorms) and in fisheries (storm and tsunami warnings) (UNISDR, 2009; IPCC, 2012a).

Earth system feedbacks See *Climate feedback*.

Earth system model (ESM) A coupled atmosphere–ocean general circulation model in which a representation of the *carbon cycle* is included, allowing for interactive calculation of atmospheric *carbon dioxide (CO$_2$)* or compatible emissions. Additional components (e.g., atmospheric chemistry, *ice sheets*, dynamic vegetation, nitrogen cycle, but also urban or crop models) may be included. See also *Climate model*.

Ecological cascade A series of secondary extinctions as a result of the extinction of a key species within an *ecosystem* (Soulé, 2010).

Ecosystem A functional unit consisting of living organisms, their non-living environment and the interactions within and between them. The components included in a given ecosystem and its spatial boundaries depend on the purpose for which the ecosystem is defined: in some cases they are relatively sharp, while in others they are diffuse. Ecosystem boundaries can change over time. Ecosystems are nested within other ecosystems and their scale can range from very small to the entire biosphere. In the current era, most ecosystems either contain people as key organisms, or are influenced by the effects of human activities in their environment. See also *Ecosystem services*.

Ecosystem services Ecological processes or functions having monetary or non-monetary value to individuals or society at large. These are frequently classified as (1) supporting services such as productivity or *biodiversity* maintenance, (2) provisioning services such as food or fibre, (3) regulating services such as climate regulation or *carbon sequestration*, and (4) cultural services such as tourism or spiritual and aesthetic appreciation. See also *Ecosystem*.

Effective climate sensitivity See *Climate sensitivity*.

Effective radiative forcing See *Radiative forcing*.

El Niño-Southern Oscillation (ENSO) The term El Niño was initially used to describe a warm-water current that periodically flows along the coast of Ecuador and Peru, disrupting the local fishery. It has since become identified with warming of the tropical Pacific Ocean east of the dateline. This oceanic event is associated with a fluctuation of a global-scale tropical and subtropical surface pressure pattern called the Southern Oscillation. This coupled atmosphere–ocean phenomenon, with preferred time scales of two to about seven years, is known as the El Niño-Southern Oscillation (ENSO). It is often measured by the surface pressure anomaly difference between Tahiti and Darwin and/or the *sea surface temperatures* in the central and eastern equatorial Pacific. During an ENSO event, the prevailing trade winds weaken, reducing upwelling and altering ocean currents such that the sea surface temperatures warm, further weakening the trade winds. This phenomenon has a great impact on the wind, sea surface temperature and precipitation patterns in the tropical Pacific. It has climatic effects throughout the Pacific region and in many other parts of the world, through global *teleconnections*. The cold phase of ENSO is called La Niña.

Embodied (embedded) [emissions, water, land] The total emissions [water use, *land use*] generated [used] in the production of goods and services regardless of the location and timing of those emissions [water use, land use] in the production process. This includes emissions [water use, land use] within the country used to produce goods or services for the country's own use, but also includes the emissions [water use, land use] related to the production of such goods or services in other countries that are then consumed in another country through imports. Such emissions [water, land] are termed 'embodied' or 'embedded' emissions, or in some cases (particularly with water) as 'virtual water use' (David and Caldeira, 2010; Allan, 2005; MacDonald et al., 2015).

Emission scenario A plausible representation of the future development of emissions of substances that are radiatively active (e.g., *greenhouse gases (GHGs), aerosols*) based on a coherent and internally consistent set of assumptions about driving forces (such as demographic and socio-economic development, technological change, energy and *land use*) and their key relationships. Concentration *scenarios*, derived from emission scenarios, are often used as input to a *climate model* to compute *climate projections*. See also *Baseline scenario, Mitigation scenario, Representative Concentration Pathways (RCPs)* (under *Pathways*), *Shared socio-economic pathways (SSPs)* (under *Pathways*), *Scenario, Socio-economic scenario*, and *Transformation pathway*.

Emission trajectories A *projected* development in time of the emission of a *greenhouse gas (GHG)* or group of GHGs, *aerosols*, and GHG *precursors*. See also *Pathways*.

Energy access Access to clean, reliable and affordable energy services for cooking and heating, lighting, communications, and productive uses (with special reference to *Sustainable Development Goal 7*) (AGECC, 2010). See also *Traditional biomass*.

Enabling conditions (for adaptation and mitigation options) Conditions that affect the *feasibility* of *adaptation* and *mitigation* options, and can accelerate and scale-up systemic transitions that would limit temperature increase and enhance capacities of systems and societies to adapt to the associated *climate change*, while achieving *sustainable development*, eradicating *poverty* and reducing *inequalities*. Enabling conditions include finance, technological innovation, strengthening policy instruments, *institutional capacity*, *multi-level governance,* and changes in *human behaviour* and lifestyles. They also include inclusive processes, attention to power asymmetries and unequal opportunities for development and reconsideration of values.

Energy efficiency The ratio of output or useful energy or energy services or other useful physical outputs obtained from a system, conversion process, transmission or storage activity to the input of energy (measured as kWh kWh^{-1}, tonnes kWh^{-1} or any other physical measure of useful output like tonne-km transported). Energy efficiency is often described by energy intensity. In economics, energy intensity describes the ratio of economic output to energy input. Most commonly energy efficiency is measured as input energy over a physical or economic unit, i.e. kWh USD^{-1} (energy intensity), kWh tonne^{-1}. For buildings, it is often measured as kWh m^{-2}, and for vehicles as km liter^{-1} or liter km^{-1}. Very often in policy 'energy efficiency' is intended as the measures to reduce energy demand through technological options such as insulating buildings, more efficient appliances, efficient lighting, efficient vehicles, etc.

Energy security The goal of a given country, or the global community as a whole, to maintain an adequate, stable and predictable energy supply. Measures encompass safeguarding the sufficiency of energy resources to meet national energy demand at competitive and stable prices and the resilience of the energy supply; enabling development and deployment of technologies; building sufficient infrastructure to generate, store and transmit energy supplies and ensuring enforceable contracts of delivery.

Enhanced weathering Enhancing the removal of *carbon dioxide (CO$_2$)* from the *atmosphere* through dissolution of silicate and carbonate rocks by grinding these minerals to small particles and actively applying them to soils, coasts or oceans.

(Model) Ensemble A group of parallel model simulations characterising historical climate conditions, *climate* predictions, or *climate projections*. Variation of the results across the ensemble members may give an estimate of modelling-based *uncertainty*. Ensembles made with the same model but different initial conditions only characterise the uncertainty associated with internal *climate variability*, whereas multi-model ensembles including simulations by several models also include the impact of model differences. Perturbed parameter ensembles, in which model parameters are varied in a systematic manner, aim to assess the uncertainty resulting from internal model specifications within a single model. Remaining sources of uncertainty unaddressed with model ensembles are related to systematic model errors or biases, which may be assessed from systematic comparisons of model simulations with observations wherever available. See also *Climate projection*.

Equality A principle that ascribes equal worth to all human beings, including equal opportunities, rights, and obligations, irrespective of origins.

Inequality
Uneven opportunities and social positions, and processes of discrimination within a group or society, based on gender, class, ethnicity, age, and (dis)ability, often produced by uneven development. Income inequality refers to gaps between highest and lowest income earners within a country and between countries. See also *Equity and Fairness*.

Equilibrium climate sensitivity See *Climate sensitivity*.

Equity The principle of being fair and impartial, and a basis for understanding how the *impacts* and responses to *climate change*, including costs and benefits, are distributed in and by society in more or less equal ways. It is often aligned with ideas of *equality*, fairness and justice and applied with respect to *equity* in the responsibility for, and distribution of, climate impacts and policies across society, generations, and gender, and in the sense of who participates and controls the processes of decision making.

Distributive equity
Equity in the consequences, outcomes, costs and benefits of actions or policies. In the case of *climate change* or climate policies for different people, places and countries, including equity aspects of sharing burdens and benefits for *mitigation* and *adaptation*.

Gender equity
Equity between women and men with regard to their rights, resources and opportunities. In the case of *climate change*, gender equity recognises that women are often more vulnerable to the *impacts* of climate change and may be disadvantaged in the process and outcomes of climate policy.

Inter-generational equity
Equity between generations. In the context of *climate change*, inter-generational equity acknowledges that the effects of past and present emissions, *vulnerabilities* and policies impose costs and benefits for people in the future and of different age groups.

Procedural equity
Equity in the process of decision making including recognition and inclusiveness in participation, equal representation, bargaining power, voice and equitable access to knowledge and resources to participate.

See also *Equality and Fairness*.

Evaporation The physical process by which a liquid (e.g., water) becomes a gas (e.g., water vapour).

Evapotranspiration The combined processes through which water is transferred to the atmosphere from open water and ice surfaces, bare soil, and vegetation that make up the Earth's surface.

Potential Evapotranspiration
The potential rate of water loss without any limits imposed by the water supply. See also *Evaporation*.

Evidence Data and information used in the scientific process to establish findings. In this report, the degree of evidence reflects the amount, quality, and consistency of scientific/technical information on which the Lead Authors base their findings. See also *Agreement, Confidence, Likelihood,* and *Uncertainty*.

Exposure   The presence of people; *livelihoods*; species or *ecosystems*; environmental functions, services, and resources; infrastructure; or economic, social, or cultural assets in places and settings that could be adversely affected. See also *Hazard, Risk,* and *Vulnerability*.

Extratropical Cyclone   Any cyclonic-scale storm that is not a *tropical cyclone*. Usually refers to a middle- or high-latitude migratory storm system formed in regions of large horizontal temperature variations. Sometimes called extratropical storm or extratropical low.   See also *Tropical cyclone*.

Extreme weather or climate event   See *Climate extreme (extreme weather or climate event)*.

Extreme weather event   An event that is rare at a particular place and time of year. Definitions of 'rare' vary, but an extreme weather event would normally be as rare as or rarer than the 10th or 90th percentile of a probability density function estimated from observations. By definition, the characteristics of what is called extreme weather may vary from place to place in an absolute sense. When a pattern of extreme weather persists for some time, such as a season, it may be classed as an extreme climate event, especially if it yields an average or total that is itself extreme (e.g., *drought* or heavy rainfall over a season). See also *Heat wave,* and *Climate extreme (extreme weather or climate event)*.

Fairness   Impartial and just treatment without favouritism or discrimination in which each person is considered of equal worth with equal opportunity. See also *Equity,* and *Equality*.

Feasibility   The degree to which climate goals and response options are considered possible and/or desirable. Feasibility depends on geophysical, ecological, technological, economic, social and *institutional* conditions for change. Conditions underpinning feasibility are dynamic, spatially variable, and may vary between different groups. See also *Enabling conditions*.

Feedback   See *Climate feedback*.

Flexible governance   See *Governance*.

Flood   The overflowing of the normal confines of a stream or other body of water, or the accumulation of water over areas that are not normally submerged. Floods include river (fluvial) floods, flash floods, urban floods, rain (pluvial) floods, sewer floods, coastal floods, and glacial lake outburst floods.

Flux   A movement (a flow) of matter (e.g., water vapor, particles), heat or energy from one place to another, or from one medium (e.g., land surface) to another (e.g., atmosphere).

Food loss and waste   'The decrease in quantity or quality of food'. Food waste is part of food loss and refers to discarding or alternative (non-food) use of food that is safe and nutritious for human consumption along the entire food supply chain, from primary production to end household consumer level. Food waste is recognised as a distinct part of food loss because the drivers that generate it and the solutions to it are different from those of food losses (FAO, 2015b).

Food security   A situation that exists when all people, at all times, have physical, social and economic access to sufficient, safe and nutritious food that meets their dietary needs and food preferences for an active and healthy life (FAO, 2001). [Note: Whilst the term 'food security' explicitly includes nutrition within it 'dietary needs … for an active and healthy life', in the past the term has sometimes privileged the supply of calories (energy), especially to the hungry. Thus, the term 'food and nutrition security' is often used (with the same definition as food security) to emphasise that the term food covers both energy and nutrition (FAO, 2009).]

Food system   All the elements (environment, people, inputs, processes, infrastructures, institutions, etc.) and activities that relate to the production, processing, distribution, preparation and consumption of food, and the output of these activities, including socio-economic and environmental outcomes (HLPE, 2017). [Note: Whilst there is a global food system (encompassing the totality of global production and consumption), each location's food system is unique, being defined by that place's mix of food produced locally, nationally, regionally or globally.]

Forcing   See *Radiative forcing*.

Forest   A vegetation type dominated by trees. Many definitions of the term forest are in use throughout the world, reflecting wide differences in biogeophysical conditions, social structure and economics. [Note: For a discussion of the term forest in the context of National GHG inventories, see the 2006 IPCC Guidelines for National GHG Inventories and information provided by the United Nations Framework Convention on Climate Change (UNFCCC, 2019).] See also *Afforestation, Deforestation,* and *Reforestation*.

Fossil fuels   Carbon-based fuels from fossil hydrocarbon deposits, including coal, oil, and natural gas.

Framework Convention on Climate Change   See *United Nations Framework Convention on Climate Change (UNFCCC)*.

Gender equity   See *Equity*.

Glacier   A perennial mass of ice, and possibly firn and snow, originating on the *land* surface by the recrystallisation of snow and showing evidence of past or present flow. A glacier typically gains mass by accumulation of snow, and loses mass by melting and ice discharge into the sea or a lake if the glacier terminates in a body of water. Land ice masses of continental size (>50 000 km²) are referred to as *ice sheets*.

Global climate model (also referred to as general circulation model, both abbreviated as GCM)   See *Climate model*.

Global mean surface temperature (GMST)   Estimated global average of near-surface air temperatures over land and sea-ice, and *sea surface temperatures* over ice-free ocean regions, with changes normally expressed as departures from a value over a specified *reference period*. When estimating changes in GMST, near-surface air temperature over both land and oceans are also used. See also *Global mean surface air temperature (GSAT), Land surface air temperature,* and *Sea surface temperature (SST)*.

Global mean surface air temperature (GSAT) Global average of near-surface air temperatures over land and oceans. Changes in GSAT are often used as a measure of global temperature change in *climate models* but are not observed directly. See also *Global mean surface temperature (GMST)*, and *Land surface air temperature*.

Global warming An increase in *global mean surface temperature (GMST) averaged* over a 30-year period, or the 30-year period centred on a particular year or decade, expressed relative to *pre-industrial* levels unless otherwise specified. For 30-year periods that span past and future years, the current multi-decadal warming trend is assumed to continue. See also *Climate change*, and *Climate variability*.

Governance A comprehensive and inclusive concept of the full range of means for deciding, managing, implementing and monitoring policies and measures. Whereas government is defined strictly in terms of the nation-state, the more inclusive concept of governance recognises the contributions of various levels of government (global, international, regional, sub-national and local) and the contributing roles of the private sector, of nongovernmental actors, and of civil society to addressing the many types of issues facing the global community, and the local context where the effectiveness of policies and measures are determined.

Adaptive governance
An emerging term in the literature for the evolution of formal and informal institutions of governance that prioritise planning, implementation and evaluation of policy through iterative *social learning*; in the context of *climate change*, governance facilitating social learning to steer the use and protection of natural resources, and *ecosystem services*, particularly in situations of complexity and uncertainty.

Climate governance
Purposeful mechanisms and measures aimed at steering social systems towards preventing, mitigating, or adapting to the risks posed by climate change (Jagers and Stripple, 2003).

Deliberative governance
Involves decision making through inclusive public conversation which allows opportunity for developing policy options through public discussion rather than collating individual preferences through voting or referenda (although the latter governance mechanisms can also be proceeded and legitimated by public deliberation processes).

Flexible governance
Strategies of governance at various levels, which prioritise the use of *social learning* and rapid feedback mechanisms in planning and policy making, often through incremental, experimental and iterative management processes.

Governance capacity
The ability of governance *institutions*, leaders, and non-state and civil society to plan, co-ordinate, fund, implement, evaluate and adjust policies and measures over the short, medium and long term, adjusting for *uncertainty*, rapid change and wide ranging impacts and multiple actors and demands.

Multi-level governance
Negotiated, non-hierarchical exchanges between institutions at the transnational, national, regional and local levels.

Participatory governance
A governance system that enables direct public engagement in decision-making using a variety of techniques for example, referenda, community deliberation, citizen juries or participatory budgeting. The approach can be applied in formal and informal institutional contexts from national to local, but is usually associated with devolved decision making (Fung and Wright, 2003; Sarmiento and Tilly, 2018).

Governance capacity See *Governance*.

Grazing land The sum of rangelands and *pastures* not considered as cropland, and subject to livestock grazing or hay production. It includes a wide range of *ecosystems*, e.g. systems with vegetation that fall below the threshold used in the *forest* land category, silvo-pastoral systems, as well as natural, managed grasslands and semideserts.

Green infrastructure The interconnected set of natural and constructed ecological systems, green spaces and other landscape features. It includes planted and indigenous trees, wetlands, parks, green open spaces and original grassland and woodlands, as well as possible building and street level design interventions that incorporate vegetation. Green infrastructure provides services and functions in the same way as conventional infrastructure (Culwick and Bobbins, 2016).

Greenhouse gas (GHG) Gaseous constituents of the *atmosphere*, both natural and *anthropogenic*, that absorb and emit radiation at specific wavelengths within the spectrum of terrestrial radiation emitted by the Earth's surface, the atmosphere itself, and by clouds. This property causes the greenhouse effect. Water vapour (H_2O), *carbon dioxide* (CO_2), *nitrous oxide* (N_2O), *methane* (CH_4) and *ozone* (O_3) are the primary GHGs in the Earth's atmosphere. Moreover, there are a number of entirely human-made GHGs in the atmosphere, such as the *halocarbons* and other chlorine- and bromine-containing substances, dealt with under the Montreal Protocol. Beside CO_2, N_2O and CH_4, the *Kyoto Protocol* deals with the GHGs sulphur hexafluoride (SF_6), hydrofluorocarbons (HFCs) and perfluorocarbons (PFCs).

Greenhouse gas removal (GGR) Withdrawal of a *greenhouse gas (GHG)* and/or a *precursor* from the *atmosphere* by a sink. See also *Carbon dioxide removal (CDR)*, and *Negative emissions*.

Gross domestic product (GDP) The sum of gross value added, at purchasers' prices, by all resident and non-resident producers in the economy, plus any taxes and minus any subsidies not included in the value of the products in a country or a geographic region for a given period, normally one year. GDP is calculated without deducting for depreciation of fabricated assets or depletion and degradation of natural resources.

Halocarbons A collective term for the group of partially halogenated organic species, which includes the chlorofluorocarbons (CFCs), hydrochlorofluorocarbons (HCFCs), hydrofluorocarbons (HFCs), halons, methyl chloride and methyl bromide. Many of the halocarbons have large Global Warming Potentials. The chlorine and bromine-containing halocarbons are also involved in the depletion of the ozone layer.

Hazard The potential occurrence of a natural or human-induced physical event or trend that may cause loss of life, injury, or other

AI

health impacts, as well as damage and loss to property, infrastructure, *livelihoods*, service provision, *ecosystems* and environmental resources. See also *Disaster, Exposure, Risk*, and *Vulnerability*.

Heatwave A period of abnormally hot weather. Heatwaves and warm spells have various and in some cases overlapping definitions. See also *Extreme weather event*.

Holocene The current interglacial geological epoch, the second of two epochs within the Quaternary period, the preceding being the Pleistocene. The International Commission on Stratigraphy defines the start of the Holocene at 11,700 years before 2000 (ICS, 2019).See also *Anthropocene*.

Human behaviour The way in which a person acts in response to a particular situation or stimulus. Human actions are relevant at different levels, from international, national, and *sub-national actors,* to NGO, firm-level actors, and communities, households, and individual actions.

Adaptation behaviour
Human actions that directly or indirectly affect the risks of climate change *impacts*.

Mitigation behaviour
Human actions that directly or indirectly influence *mitigation*.

Human behavioural change A transformation or modification of human actions. Behaviour change efforts can be planned in ways that mitigate *climate change* and/or reduce negative consequences of climate change *impacts*.

Human rights Rights that are inherent to all human beings, universal, inalienable, and indivisible, typically expressed and guaranteed by law. They include the right to life, economic, social, and cultural rights, and the right to development and self-determination (based upon the definition by the UN Office of the High Commissioner for Human Rights).

Procedural rights
Rights to a legal procedure to enforce substantive rights.

Substantive rights
Basic human rights, including the right to the substance of being human such as life itself, liberty and happiness.

Human security A condition that is met when the vital core of human lives is protected, and when people have the freedom and capacity to live with dignity. In the context of *climate change,* the vital core of human lives includes the universal and culturally specific, material and non-material elements necessary for people to act on behalf of their interests and to live with dignity.

Human system Any system in which human organisations and *institutions* play a major role. Often, but not always, the term is synonymous with society or social system. Systems such as agricultural systems, urban systems, political systems, technological systems, and economic systems are all human systems in the sense applied in this report.

Hydrological cycle The cycle in which water evaporates from the oceans and the land surface, is carried over the Earth in atmospheric circulation as water vapour, condenses to form clouds, precipitates as rain or snow, which on land can be intercepted by trees and vegetation, potentially accumulating as snow or ice, provides *runoff* on the land surface, infiltrates into soils, recharges groundwater, discharges into streams, and ultimately, flows out into the oceans as rivers, polar *glaciers* and *ice sheets*, from which it will eventually evaporate again. The various systems involved in the hydrological cycle are usually referred to as hydrological systems.

Ice sheet An ice body originating on land that covers an area of continental size, generally defined as covering >50,000 km². An ice sheet flows outward from a high central ice plateau with a small average surface slope. The margins usually slope more steeply, and most ice is discharged through fast flowing ice streams or outlet *glaciers*, often into the sea or into ice shelves floating on the sea. There are only two ice sheets in the modern world, one on Greenland and one on Antarctica. The latter is divided into the East Antarctic Ice Sheet (EAIS), the West Antarctic Ice Sheet (WAIS) and the Antarctic Peninsula ice sheet. During glacial periods there were other ice sheets. See also *Glacier*.

Impacts (consequences, outcomes) The consequences of realised risks on natural and *human systems*, where risks result from the interactions of climate-related *hazards* (including *extreme weather and climate events*), *exposure*, and *vulnerability*. Impacts generally refer to effects on lives, *livelihoods*, health and *well-being, ecosystems* and species, economic, social and cultural assets, services (including *ecosystem services*), and infrastructure. Impacts may be referred to as consequences or outcomes, and can be adverse or beneficial. See also *Adaptation, Exposure, Hazard, Loss and Damage, and losses and damages,* and *Vulnerability*.

(climate change) Impact assessment The practice of identifying and evaluating, in monetary and/or non-monetary terms, the effects of climate change on natural and human systems.

Incremental adaptation See *Adaptation*.

Indigenous knowledge The understandings, skills and philosophies developed by societies with long histories of interaction with their natural surroundings. For many Indigenous peoples, Indigenous knowledge informs decision-making about fundamental aspects of life, from day-to-day activities to longer term actions. This knowledge is integral to cultural complexes, which also encompass language, systems of classification, resource use practices, social interactions, values, ritual and spirituality. These distinctive ways of knowing are important facets of the world's cultural diversity. (UNESCO, 2018). See also *Local knowledge*.

Indirect land-use change See *Land-use change*.

Industrial revolution A period of rapid industrial growth with far-reaching social and economic consequences, beginning in Britain during the second half of the 18th century and spreading to Europe and later to other countries including the United States. The invention of the steam engine was an important trigger of this development. The industrial revolution marks the beginning of a strong increase in the use of *fossil fuels,* initially coal, and hence emission of *carbon dioxide (CO₂)*. See also *Pre-industrial*.

Industrialised/developed/developing countries There are a diversity of approaches for categorising countries on the basis of their level of economic development, and for defining terms such as industrialised, developed, or developing. Several categorisations are used in this report. (1) In the United Nations system, there is no established convention for designating of developed and developing countries or areas. (2) The United Nations Statistics Division specifies developed and developing regions based on common practice. In addition, specific countries are designated as Least Developed Countries (LDC), landlocked developing countries, *small island developing states (SIDS)*, and transition economies. Many countries appear in more than one of these categories. (3) The World Bank uses income as the main criterion for classifying countries as low, lower middle, upper middle, and high income. (4) The United Nations Development Programme (UNDP) aggregates indicators for life expectancy, educational attainment, and income into a single composite Human Development Index (HDI) to classify countries as low, medium, high, or very high human development.

Inequality   See *Equality*.

Institution   Rules, norms and conventions held in common by social actors that guide, constrain and shape human interaction. Institutions can be formal, such as laws and policies, or informal, such as norms and conventions. Organisations - such as parliaments, regulatory agencies, private firms, and community bodies - develop and act in response to institutional frameworks and the incentives they frame. Institutions can guide, constrain and shape human interaction through direct control, through incentives, and through processes of socialisation. See also *Institutional capacity*.

Institutional capacity   Building and strengthening individual organisations and providing technical and management training to support integrated planning and decision-making processes between organisations and people, as well as empowerment, social capital, and an enabling environment, including the culture, values and power relations (Willems and Baumert, 2003).

Integrated assessment   A method of analysis that combines results and models from the physical, biological, economic and social sciences and the interactions among these components in a consistent framework to evaluate the status and the consequences of environmental change and the policy responses to it. See also *Integrated assessment model (IAM)*.

Integrated assessment model (IAM)   Models that integrate knowledge from two or more domains into a single framework. They are one of the main tools for undertaking *integrated assessments*. One class of IAM used in respect of climate change *mitigation* may include representations of: multiple sectors of the economy, such as energy, *land use* and *land use change*; interactions between sectors; the economy as a whole; associated *greenhouse gas (GHG)* emissions and *sinks*; and reduced representations of the *climate system*. This class of model is used to assess linkages between economic, social and technological development and the evolution of the climate system. Another class of IAM additionally includes representations of the costs associated with climate change *impacts*, but includes less detailed representations of economic systems. These can be used to assess impacts and mitigation in a *cost-benefit* framework and have been used to estimate the *social cost of carbon*.

Integrated response options   In this report, integrated response options are those options that simultaneously address

more than one *land challenge*. These can be categorised into options that rely on a) *land management,* b) value chain management, and c) *risk management*. Integrated response options are not mutually exclusive. See also *Land challenge*.

Integrated water resources management (IWRM)   A process which promotes the coordinated development and management of water, land and related resources in order to maximise economic and social welfare in an equitable manner without compromising the sustainability of vital *ecosystems*.

Inter-generational equity   See *Equity*.

Internal variability   See *Climate variability*.

Irreversibility   A perturbed state of a dynamical system is defined as irreversible on a given timescale if the recovery timescale from this state due to natural processes is substantially longer than the time it takes for the system to reach this perturbed state. See also *Tipping point*.

Kyoto Protocol   The Kyoto Protocol to the *United Nations Framework Convention on Climate Change (UNFCCC)* is an international treaty adopted in December 1997 in Kyoto, Japan, at the Third Session of the *Conference of the Parties* (COP3) to the UNFCCC. It contains legally binding commitments, in addition to those included in the UNFCCC. Countries included in Annex B of the Protocol (mostly OECD countries and countries with economies in transition) agreed to reduce their anthropogenic *greenhouse gas (GHG)* emissions (carbon dioxide (CO_2), methane (CH_4), nitrous oxide (N_2O), hydrofluorocarbons (HFCs), perfluorocarbons (PFCs), and sulphur hexafluoride (SF_6) by at least 5% below 1990 levels in the first commitment period (2008–2012). The Kyoto Protocol entered into force on 16 February 2005 and as of May 2018 had 192 Parties (191 States and the European Union). A second commitment period was agreed in December 2012 at COP18, known as the Doha Amendment to the Kyoto Protocol, in which a new set of Parties committed to reduce GHG emissions by at least 18% below 1990 levels in the period from 2013 to 2020. However, as of May 2018, the Doha Amendment had not received sufficient ratifications to enter into force. See also *Paris Agreement*.

Land   The terrestrial portion of the biosphere that comprises the natural resources (soil, near surface air, vegetation and other biota, and water), the ecological processes, topography, and human settlements and infrastructure that operate within that system (FAO, 2007; UNCCD, 1994).

Land challenges   In this report, land challenges refers to land-based *mitigation* and *adaptation, desertification, land degradation* and *food security*.

Land cover   The biophysical coverage of *land* (e.g., bare soil, rocks, forests, buildings and roads or lakes). Land cover is often categorised in broad land-cover classes (e.g., deciduous forest, coniferous forest, mixed forest, grassland, bare ground). [Note: In some literature assessed in this report, land cover and *land use* are used interchangeably, but the two represent distinct classification systems. For example, the land cover class woodland can be under various land uses such as livestock grazing, recreation, conservation, or wood harvest.] See also *Land cover change*, and *Land-use change*.

AI

Land cover change Change from one *land cover* class to another, due to change in land use or change in natural conditions (Pongratz et al., 2018). See also *Land-use change*, and *Land management change*.

Land degradation A negative trend in land condition, caused by direct or indirect human-induced processes including *anthropogenic* climate change, expressed as long-term reduction or loss of at least one of the following: biological productivity, ecological integrity or value to humans. [Note: This definition applies to *forest* and non-forest land. Changes in land condition resulting solely from natural processes (such as volcanic eruptions) are not considered to be land degradation. Reduction of biological productivity or ecological integrity or value to humans *can* constitute degradation, but any one of these changes need not necessarily be considered degradation.]

Land degradation neutrality A state whereby the amount and quality of land resources necessary to support *ecosystem* functions and services and enhance *food security* remain stable or increase within specified temporal and spatial scales and ecosystems (UNCCD, 2019).

Land management Sum of *land-use* practices (e.g., sowing, fertilizing, weeding, harvesting, thinning, clear-cutting) that take place within broader land-use categories. (Pongratz et al., 2018)

Land management change
A change in land management that occurs within a *land-use* category. See also *Land-use change*.

Land potential The inherent, long-term potential of the *land* to sustainably generate *ecosystem services*, which reflects the capacity and resilience of the land-based natural capital, in the face of ongoing environmental change (UNEP, 2016).

Land rehabilitation Direct or indirect actions undertaken with the aim of reinstating a level of *ecosystem* functionality, where the goal is provision of goods and services rather than ecological restoration (McDonald, et al., 2016).

Land restoration The process of assisting the recovery of *land* from a degraded state (McDonald et al., 2016; IPBES, 2018).

Land surface air temperature (LSAT) The near-surface air temperature over land, typically measured at 1.25–2 m above the ground using standard meteorological equipment.

Land use The total of arrangements, activities and inputs applied to a parcel of land. The term land use is also used in the sense of the social and economic purposes for which land is managed (e.g., grazing, timber extraction, conservation and city dwelling). In national GHG inventories, land use is classified according to the IPCC land use categories of *forest* land, cropland, grassland, *wetlands*, settlements, other lands (see the 2006 IPCC Guidelines for National GHG Inventories for details). See also *Land-use change* and *Land management*.

Land-use change (LUC) The change from one *land use* category to another. [Note: In some of the scientific literature assessed in this report, land-use change encompasses changes in land-use categories as well as changes in *land management*.]

Indirect land-use change (iLUC)
Land use change outside the area of focus, that occurs as a consequence of change in use or management of land within the area of focus, such as through market or policy drivers. For example, if agricultural land is diverted to *biofuel* production, *forest* clearance may occur elsewhere to replace the former agricultural production. See also *Afforestation, Agriculture, Forestry and Other Land Use (AFOLU), Deforestation, Land use, land-use change and forestry (LULUCF), Reforestation, the IPCC Special Report on Land Use, Land-Use Change, and Forestry (IPCC, 2000),* and the *2006 IPCC Guidelines for National GHG Inventories (IPCC, 2006)*.

Land use, land-use change and forestry (LULUCF) In the context of national *greenhouse gas (GHG)* inventories under the *United Nations Framework Convention on Climate Change (UNFCCC, 2019)*, LULUCF is a GHG inventory sector that covers anthropogenic emissions and removals of GHG in managed lands, excluding non-CO_2 agricultural emissions. Following the 2006 IPCC Guidelines for National GHG Inventories, 'anthropogenic' land-related GHG fluxes are defined as all those occurring on 'managed land', i.e., 'where human interventions and practices have been applied to perform production, ecological or social functions'. Since managed land may include *carbon dioxide (CO_2)* removals not considered as 'anthropogenic' in some of the scientific literature assessed in this report (e.g., removals associated with CO_2 fertilisation and N deposition), the land-related net GHG emission estimates from global models included in this report are not necessarily directly comparable with LULUCF estimates in National GHG Inventories. See also *Land-use change (LUC)*.

Latent heat flux The turbulent flux of heat from the Earth's surface to the atmosphere that is associated with *evaporation* or condensation of water vapour at the surface; a component of the surface energy budget. See also *Atmosphere*, and *Flux*.

Leakage The effects of policies that result in a displacement of the environmental impact, thereby counteracting the intended effects of the initial policies.

Lifecycle assessment (LCA) Compilation and evaluation of the inputs, outputs and the potential environmental impacts of a product or service throughout its life cycle (ISO, 2018).

Likelihood The chance of a specific outcome occurring, where this might be estimated probabilistically. Likelihood is expressed in this report using a standard terminology (Mastrandrea et al., 2010). See also *Agreement, Evidence, Confidence,* and *Uncertainty*.

Livelihood The resources used and the activities undertaken in order to live. Livelihoods are usually determined by the entitlements and assets to which people have access. Such assets can be categorised as human, social, natural, physical, or financial.

Local knowledge The understandings and skills developed by individuals and populations, specific to the places where they live. Local knowledge informs decision-making about fundamental aspects of life, from day-to-day activities to longer term actions. This knowledge is a key element of the social and cultural systems which influence observations of, and responses to *climate change*; it also informs *governance* decisions (UNESCO, 2018). See also *Indigenous knowledge*.

AI

Lock-in A situation in which the future development of a system, including infrastructure, technologies, investments, *institutions*, and behavioural norms, is determined or constrained ('locked in') by historic developments.

Long-lived climate forcers (LLCF) A set of well-mixed greenhouse gases with long atmospheric lifetimes. This set of compounds includes c*arbon dioxide (CO_2)* and *nitrous oxide (N_2O)*, together with some fluorinated gases. They have a warming effect on climate. These compounds accumulate in the atmosphere at decadal to centennial timescales, and their effect on *climate* hence persists for decades to centuries after their emission. On timescales of decades to a century already emitted emissions of long-lived climate forcers can only be abated by *greenhouse gas removal (GGR)*. See also *Short-lived climate forcers (SLCF)*.

Loss and Damage, and losses and damages Research has taken Loss and Damage (capitalised letters) to refer to political debate under the *United Nations Framework Convention on Climate Change (UNFCCC)* following the establishment of the Warsaw Mechanism on Loss and Damage in 2013, which is to 'address loss and damage associated with impacts of climate change, including extreme events and slow onset events, in developing countries that are particularly vulnerable to the adverse effects of climate change.' Lowercase letters (losses and damages) have been taken to refer broadly to harm from (observed) *impacts* and (projected) *risks* (Mechler et al., 2018).

Maladaptive actions (Maladaptation) Actions that may lead to increased *risk* of adverse climate-related outcomes, including via increased *greenhouse gas (GHG)* emissions, increased *vulnerability* to *climate change,* or diminished welfare, now or in the future. Maladaptation is usually an unintended consequence.

Malnutrition Deficiencies, excesses, or imbalances in a person's intake of energy and/or nutrients. The term malnutrition addresses three broad groups of conditions: undernutrition, which includes wasting (low weight-for-height), stunting (low height-for-age) and underweight (low weight-for-age); micronutrient-related malnutrition, which includes micronutrient deficiencies (a lack of important vitamins and minerals) or micronutrient excess; and overweight, obesity and diet-related noncommunicable diseases (such as heart disease, stroke, diabetes and some cancers) (WHO, 2018). Micronutrient deficiencies are sometimes termed 'hidden hunger' to emphasise that people can be malnourished in the sense of deficient without being deficient in calories. Hidden hunger can apply even where people are obese.

Managed forest *Forests* subject to human interventions (notably silvicultural management such as planting, pruning, thinning), timber and fuelwood harvest, protection (fire suppression, insect supression) and management for amenity values or conservation, with defined geographical boundaries (Ogle et al., 2018). [Note: For a discussion of the term 'forest' in the context of National GHG inventories, see the 2006 IPCC Guidelines for National GHG Inventories.] See also *Managed land.*

Managed grassland Grasslands on which human interventions are carried out, such as grazing domestic livestock or hay removal.

Managed land In the context of national *greenhouse gas (GHG)* inventories under the *United Nations Framework Convention on Climate Change (UNFCCC),* the 2006 IPCC Guidelines for National GHG Inventories (IPCC, 2006) defines managed land 'where human interventions and practices have been applied to perform production, ecological or social functions'. The IPCC (2006) defines *anthropogenic* GHG emissions and removals in the LULUCF sector as all those occurring on 'managed land'. The key rationale for this approach is that the preponderance of anthropogenic effects occurs on managed lands. [Note: More details can be found in 2006 IPCC Guidelines for National GHG Inventories, Volume 4, Chapter 1.]

Market failure When private decisions are based on market prices that do not reflect the real scarcity of goods and services but rather reflect market distortions, they do not generate an efficient allocation of resources but cause welfare losses. A market distortion is any event in which a market reaches a market clearing price that is substantially different from the price that a market would achieve while operating under conditions of perfect competition and state enforcement of legal contracts and the ownership of private property. Examples of factors causing market prices to deviate from real economic scarcity are environmental externalities, public goods, monopoly power, information asymmetry, transaction costs, and non-rational behaviour.

Measurement, reporting and verification (MRV)

Measurement

'The process of data collection over time, providing basic datasets, including associated accuracy and precision, for the range of relevant variables. Possible data sources are field measurements, field observations, detection through remote sensing and interviews' (UN REDD, 2009).

Reporting

'The process of formal reporting of assessment results to the UNFCCC, according to predetermined formats and according to established standards, especially the Intergovernmental Panel on Climate Change (IPCC) Guidelines and GPG (Good Practice Guidance)' (UN REDD, 2009).

Verification

'The process of formal verification of reports, for example, the established approach to verify national communications and national inventory reports to the UNFCCC' (UN REDD, 2009).

Megadrought See *Drought.*

Methane (CH_4) One of the six *greenhouse gases (GHGs)* to be mitigated under the *Kyoto Protocol.* Methane is the major component of natural gas and associated with all hydrocarbon fuels. Significant *anthropogenic* emissions also occur as a result of animal husbandry and paddy rice production. Methane is also produced naturally where organic matter decays under anaerobic conditions, such as in *wetlands*.

Migrant See *Migration.*

Migration 'The movement of a person or a group of persons, either across an international border, or within a State. It is a population movement, encompassing any kind of movement of people, whatever its length, composition and causes; it includes migration of refugees, displaced persons, economic migrants, and persons moving for other purposes, including family reunification' (IOM, 2018).

Migrant

'Any person who is moving or has moved across an international border or within a State away from his/her habitual place of residence, regardless of (1) the person's legal status; (2) whether the movement is voluntary or involuntary; (3) what the causes for the movement are; or (4) what the length of the stay is' (IOM, 2018).

See also *(Internal) Displacement.*

Millennium Development Goals (MDGs) A set of eight time-bound and measurable goals for combating *poverty,* hunger, disease, illiteracy, discrimination against women and environmental degradation. These goals were agreed at the UN Millennium Summit in 2000 together with an action plan to reach the goals by 2015.

Mitigation (of climate change) A human intervention to reduce emissions or enhance the *sinks* of *greenhouse gases.*

Mitigation behaviour See *Human behaviour.*

Mitigation measures In climate policy, *mitigation* measures are technologies, processes or practices that contribute to mitigation, for example renewable energy technologies, waste minimisation processes, public transport commuting practices.

Mitigation option A technology or practice that reduces *greenhouse gas (GHG)* emissions or enhances sinks.

Mitigation pathways See *Pathways.*

Mitigation scenario A plausible description of the future that describes how the (studied) system responds to the implementation of *mitigation* policies and measures. See also *Emission scenario, Pathways, Socio-economic scenarios,* and *Stabilisation (of GHG or CO₂-equivalent concentration).*

Monitoring and evaluation (M&E) Mechanisms put in place at national to local scales to respectively monitor and evaluate efforts to reduce *greenhouse gas* emissions and/or adapt to the impacts of *climate change* with the aim of systematically identifying, characterising and assessing progress over time.

Motivation (of an individual) An individual's reason or reasons for acting in a particular way; individuals may consider various consequences of actions, including financial, social, affective, and environmental consequences. Motivation can arise from factors external or internal to the individual.

Multi-level governance See *Governance.*

Narratives (in the context of scenarios) Qualitative descriptions of plausible future world evolutions, describing the characteristics, general logic and developments underlying a particular quantitative set of *scenarios.* Narratives are also referred to in the literature as 'storylines'. See also *Scenario, Scenario storyline,* and *Pathways.*

Nationally Determined Contributions (NDCs) A term used under the *United Nations Framework Convention on Climate Change (UNFCCC)* whereby a country that has joined the *Paris Agreement* outlines its plans for reducing its emissions. Some countries NDCs

also address how they will adapt to climate change impacts, and what support they need from, or will provide to, other countries to adopt low-carbon pathways and to build climate resilience. According to Article 4 paragraph 2 of the *Paris Agreement,* each Party shall prepare, communicate and maintain successive NDCs that it intends to achieve. In the lead up to the 21st *Conference of the Parties* in Paris in 2015, countries submitted Intended Nationally Determined Contributions (INDCs). As countries join the Paris Agreement, unless they decide otherwise, this INDC becomes their first Nationally Determined Contribution (NDC).

Negative emissions Removal of *greenhouse gases (GHGs)* from the *atmosphere* by deliberate human activities, i.e., in addition to the removal that would occur via natural *carbon cycle* processes. See also *Net negative emissions, Net-zero emissions, Carbon dioxide removal (CDR),* and *Greenhouse gas removal (GGR).*

Negative emissions technologies An activity or mechanism that results in *negative emissions.*

Net negative emissions A situation of net negative emissions is achieved when, as result of human activities, more *greenhouse gases (GHG)* are removed from the *atmosphere* than are emitted into it. Where multiple greenhouse gases are involved, the quantification of *negative emissions* depends on the climate metric chosen to compare emissions of different gases (such as global warming potential, global temperature change potential, and others, as well as the chosen time horizon). See also *Negative emissions, Net-zero emissions* and *Net-zero CO₂ emissions.*

Net-zero CO₂ emissions Conditions in which any remaining *anthropogenic* carbon dioxide (CO₂) emissions are balanced by anthropogenic CO₂ removals over a specified period. See also *Net-zero emissions,* and *Net negative emissions.*

Net-zero emissions Net-zero emissions are achieved when emissions of *greenhouse gases (GHGs)* to the *atmosphere* are balanced by *anthropogenic removals.* Where multiple greenhouse gases are involved, the quantification of net-zero emissions depends on the climate metric chosen to compare emissions of different gases (such as global warming potential, global temperature change potential, and others, as well as the chosen time horizon). See also *Net-zero CO₂ emissions, Negative emissions,* and *Net negative emissions.*

Nitrous oxide (N₂O) One of the six *greenhouse gases (GHGs)* to be mitigated under the *Kyoto Protocol.* The main anthropogenic source of N₂O is agriculture (soil and animal manure management), but important contributions also come from sewage treatment, *fossil fuel* combustion, and chemical industrial processes. N₂O is also produced naturally from a wide variety of biological sources in soil and water, particularly microbial action in wet tropical *forests.*

Non-overshoot pathways See *Pathways.*

Nutrition transition A predictable change in *dietary patterns* associated with a country's economic development whereby 'problems of under- and overnutrition often coexist, reflecting the trends in which an increasing proportion of people consume the types of diets associated with a number of chronic diseases' (Popkin, 1994).

Ocean acidification (OA)   A reduction in the *pH* of the ocean, accompanied by other chemical changes, over an extended period, typically decades or longer, which is caused primarily by *uptake of carbon dioxide (CO₂)* from the *atmosphere*, but can also be caused by other chemical additions or subtractions from the ocean. *Anthropogenic* ocean acidification refers to the component of pH reduction that is caused by human activity (IPCC, 2011, p. 37). See also *Climate change*.

Ocean fertilisation   Deliberate increase of nutrient supply to the near-surface ocean in order to enhance biological production through which additional *carbon dioxide (CO₂)* from the *atmosphere* is sequestered. This can be achieved by the addition of micro-nutrients or macro-nutrients. Ocean fertilisation is regulated by the London Protocol

Overshoot   See *Temperature overshoot*.

Overshoot pathways   See *Pathways*.

Ozone (O₃)   The triatomic form of oxygen (O₃). In the *troposphere*, it is created both naturally and by photochemical reactions involving gases resulting from human activities (smog). Tropospheric ozone acts as a *greenhouse gas (GHG)*. In the *stratosphere*, it is created by the interaction between solar ultraviolet radiation and molecular oxygen (O₂). Stratospheric ozone plays a dominant role in the stratospheric radiative balance. Its concentration is highest in the ozone layer.

Paris Agreement   The Paris Agreement under the *United Nations Framework Convention on Climate Change (UNFCCC)* was adopted on December 2015 in Paris, France, at the 21st session of the *Conference of the Parties (COP)* to the UNFCCC. The agreement, adopted by 196 Parties to the UNFCCC, entered into force on 4 November 2016 and as of May 2018 had 195 Signatories and was ratified by 177 Parties. One of the goals of the Paris Agreement is 'Holding the increase in the global average temperature to well below 2°C above pre-industrial levels and pursuing efforts to limit the temperature increase to 1.5°C above pre-industrial levels', recognising that this would significantly reduce the risks and impacts of climate change. Additionally, the Agreement aims to strengthen the ability of countries to deal with the impacts of climate change. The Paris Agreement is intended to become fully effective in 2020. See also *Kyoto Protocol,* and *Nationally Determined Contributions (NDCs)*.

Participatory governance   See *Governance*.

Pasture   Area covered with grass or other plants used or suitable for grazing of livestock; grassland.

Pathways   The temporal evolution of natural and/or *human systems* towards a future state. Pathway concepts range from sets of quantitative and qualitative *scenarios* or *narratives* of potential futures to solution-oriented decision-making processes to achieve desirable societal goals. Pathway approaches typically focus on biophysical, techno-economic, and/or socio-behavioural trajectories and involve various dynamics, goals, and actors across different scales.

1.5°C pathway
A pathway of emissions of *greenhouse gases* and other climate forcers that provides an approximately one-in-two to two-in-three chance, given current knowledge of the climate response, of *global*

warming either remaining below 1.5°C or returning to 1.5°C by around 2100 following an overshoot.
See also *Temperature overshoot*.

Adaptation pathways
A series of *adaptation* choices involving trade-offs between short-term and long-term goals and values. These are processes of deliberation to identify solutions that are meaningful to people in the context of their daily lives and to avoid potential maladaptation.

Development pathways
Development pathways are trajectories based on an array of social, economic, cultural, technological, institutional, and biophysical features that characterise the interactions between human and natural systems and outline visions for the future, at a particular scale.

Mitigation pathways
A mitigation pathway is a temporal evolution of a set of *mitigation scenario* features, such as *greenhouse gas (GHG)* emissions and socio-economic development.

Overshoot pathways
Pathways that exceed the stabilisation level (concentration, *forcing*, or temperature) before the end of a time horizon of interest (e.g., before 2100) and then decline towards that level by that time. Once the target level is exceeded, removal by *sinks* of *greenhouse gases (GHGs)* is required. See also *Temperature overshoot*.

Non-overshoot pathways
Pathways that stay below the stabilisation level (concentration, forcing, or temperature) during the time horizon of interest (e.g., until 2100).

Representative Concentration Pathways (RCPs)
Scenarios that include time series of emissions and concentrations of the full suite of *greenhouse gases (GHGs)* and *aerosols* and chemically active gases, as well as *land use/land cover* (Moss et al., 2008). The word representative signifies that each RCP provides only one of many possible scenarios that would lead to the specific *radiative forcing* characteristics. The term pathway emphasises the fact that not only the long-term concentration levels, but also the trajectory taken over time to reach that outcome are of interest (Moss et al., 2010). RCPs were used to develop *climate projections* in CMIP5. See also *Coupled Model Intercomparison Project (CMIP)*, and *Shared Socio-economic Pathways (SSPs)*.

- RCP2.6: One pathway where radiative forcing peaks at approximately 3 W m⁻² and then declines to be limited at 2.6 W m⁻² in 2100 (the corresponding Extended Concentration Pathway, or ECP, has constant emissions after 2100).

- RCP4.5 and RCP6.0: Two intermediate stabilisation pathways in which radiative forcing is limited at approximately 4.5 W m⁻² and 6.0 W m⁻² in 2100 (the corresponding ECPs have constant concentrations after 2150).

- RCP8.5: One high pathway which leads to >8.5 W m⁻² in 2100 (the corresponding ECP has constant emissions after 2100 until 2150 and constant concentrations after 2250).

Shared Socio-economic Pathways (SSPs)
Shared Socio-economic Pathways (SSPs) were developed to complement the Representative Concentration Pathways (RCPs) with varying socio-economic challenges to *adaptation* and *mitigation*

(O'Neill et al., 2014). Based on five *narratives*, the SSPs describe alternative socio-economic futures in the absence of climate policy intervention, comprising sustainable development (SSP1), regional rivalry (SSP3), inequality (SSP4), fossil–fueled development (SSP5), and a middle-of-the-road development (SSP2) (O'Neill et al., 2017; Riahi et al., 2017). The combination of SSP-based socio-economic scenarios and RCP-based *climate projections* provides an integrative frame for climate *impact* and policy analysis.

Transformation pathways
Trajectories describing consistent sets of possible futures of *greenhouse gas (GHG)* emissions, atmospheric concentrations, or *global mean surface temperatures* implied from *mitigation* and *adaptation* actions associated with a set of broad and irreversible economic, technological, societal, and behavioural changes. This can encompass changes in the way energy and infrastructure are used and produced, natural resources are managed and *institutions* are set up and in the pace and direction of technological change (TC).

See also *Scenario, Scenario storyline, Emission scenario, Mitigation scenario, Baseline scenario, Stabilisation (of GHG or CO2-equivalent concentration),* and *Narratives*.

Peat   Soft, porous or compressed, sedentary deposit of which a substantial portion is partly decomposed plant material with high water content in the natural state (up to about 90 percent) (IPCC, 2013). See also *Peatlands*.

Peatlands   Peatland is a land where soils are dominated by *peat*. See also *Reservoir*, and *Sink*.

Peri-urban areas   Parts of a city that appear to be quite rural but are in reality strongly linked functionally to the city in its daily activities.

Permafrost   Ground (soil or rock and included ice and organic material) that remains at or below 0°C for at least two consecutive years.

pH   A dimensionless measure of the acidity of a solution given by its concentration of hydrogen ions ($[H^+]$). pH is measured on a logarithmic scale where pH = $-\log_{10}[H^+]$. Thus, a pH decrease of 1 unit corresponds to a 10-fold increase in the concentration of H^+, or acidity. See also *Ocean acidification*.

Phenology   The relationship between biological phenomena that recur periodically (e.g., development stages, migration) and *climate* and seasonal changes.

Planetary health   The Rockefeller-Lancet Commission defines planetary health as 'the achievement of the highest attainable standard of health, well-being, and equity worldwide through judicious attention to the human systems — political, economic, and social — that shape the future of humanity *and* the Earth's natural systems that define the safe environmental limits within which humanity can flourish. Put simply, planetary health is the health of human civilisation and the state of the natural systems on which it depends' (Whitmee et al., 2015).

Political economy   The set of interlinked relationships between people, the state, society and markets as defined by law, politics, economics, customs and power that determine the outcome of trade and transactions and the distribution of wealth in a country or economy.

Pool, carbon and nitrogen   A *reservoir* in the earth system where elements, such as carbon and nitrogen, reside in various chemical forms for a period of time.

Poverty   A complex concept with several definitions stemming from different schools of thought. It can refer to material circumstances (such as need, pattern of deprivation or limited resources), economic conditions (such as standard of living, *inequality* or economic position) and/or social relationships (such as social class, dependency, exclusion, lack of basic security or lack of entitlement). See also *Poverty eradication*.

Poverty eradication   A set of measures to end *poverty* in all its forms everywhere. See also *Sustainable Development Goals (SDGs)*.

Precursors   Atmospheric compounds that are not *greenhouse gases (GHGs)* or *aerosols*, but that have an effect on GHG or aerosol concentrations by taking part in physical or chemical processes regulating their production or destruction rates. See also *Aerosol,* and *Greenhouse gas (GHG)*.

Pre-industrial   The multi-century period prior to the onset of large-scale industrial activity around 1750. The *reference period* 1850–1900 is used to approximate pre-industrial *global mean surface temperature (GMST)*. See also *Industrial revolution*.

Primary production   The synthesis of organic compounds by plants and microbes, on land or in the ocean, primarily by photosynthesis using light and *carbon dioxide (CO_2)* as sources of energy and carbon respectively. It can also occur through chemosynthesis, using chemical energy, e.g., in deep sea vents.

Gross Primary Production (GPP)
The total amount of carbon fixed by photosynthesis over a specific time period.

Net primary production (NPP)
The amount of carbon accumulated through photosynthesis minus the amount lost by plant respiration over a specified time period that would prevail in the absence of land use.

Procedural equity   See *Equity*

Procedural rights   See *Human rights*.

Projection   A potential future evolution of a quantity or set of quantities, often computed with the aid of a model. Unlike predictions, projections are conditional on assumptions concerning, for example, future socio-economic and technological developments that may or may not be realised. See also *Climate projection, Scenario,* and *Pathways*.

Radiative forcing   The change in the net, downward minus upward, radiative flux (expressed in W m^{-2}) at the tropopause or top of *atmosphere* due to a change in a driver of *climate change*, such as a change in the concentration of *carbon dioxide (CO_2)*, the concentration of volcanic aerosols or the output of the Sun. The traditional radiative forcing is computed with all tropospheric properties held fixed at their unperturbed values, and after allowing for stratospheric temperatures, if perturbed, to readjust to radiative-dynamical equilibrium. Radiative forcing is called instantaneous if no change in stratospheric temperature is accounted for. The radiative forcing once rapid adjustments are accounted for is termed the effective radiative forcing. Radiative forcing is not to be confused with cloud radiative forcing, which describes an unrelated measure of the impact of clouds on the radiative flux at the top of the atmosphere.

Reasons for concern (RFCs) Elements of a classification framework, first developed in the IPCC Third Assessment Report, which aims to facilitate judgments about what level of *climate change* may be dangerous (in the language of Article 2 of the *UNFCCC*) by aggregating risks from various sectors, considering hazards, exposures, vulnerabilities, capacities to adapt, and the resulting impacts.

Reducing Emissions from Deforestation and Forest Degradation (REDD+) REDD+ refers to reducing emissions from *deforestation*; reducing emissions from forest degradation; conservation of forest carbon stocks; sustainable management of forests; and enhancement of forest carbon stocks (see UNFCCC decision 1/CP.16, para. 70).

Reference period The period relative to which *anomalies* are computed. See also *Anomalies*.

Reference scenario See *Baseline scenario*.

Reforestation Conversion to *forest* of *land* that has previously contained forests but that has been converted to some other use. [Note: For a discussion of the term forest and related terms such as *afforestation, reforestation* and *deforestation* in the context of reporting and accounting Article 3.3 and 3.4 activities under the *Kyoto Protocol*, see 2013 Revised Supplementary Methods and Good Practice Guidance Arising from the Kyoto Protocol.] See also *Afforestation, Deforestation, and Reducing Emissions from Deforestation* and *Forest Degradation (REDD+)*.

Region A relatively large-scale land or ocean area characterised by specific geographical and climatological features. The *climate* of a land-based region is affected by regional and local scale features like topography, *land use* characteristics and large water bodies, as well as remote influences from other regions, in addition to global climate conditions. The IPCC defines a set of standard regions for analyses of observed climate trends and climate model projections (see IPCC 2018, Fig. 3.2; AR5, SREX).

Remaining carbon budget Cumulative global *carbon dioxide (CO$_2$)* emissions from the start of 2018 to the time that CO$_2$ emissions reach net-zero that would result, at some probability, in limiting *global warming* to a given level, accounting for the impact of other *anthropogenic emissions*. See also *Carbon budget*.

Representative concentration pathways (RCPs) See *Pathways*.

Reservoir A component or components of the climate system where a *greenhouse gas (GHG)* or a *precursor* of a greenhouse gas is stored (UNFCCC Article 1.7).

Resilience The capacity of interconnected social, economic and ecological systems to cope with a hazardous event, trend or disturbance, responding or reorganising in ways that maintain their essential function, identity and structure. Resilience is a positive attribute when it maintains capacity for *adaptation*, learning and/or *transformation* (adapted from the Arctic Council, 2013). See also *Hazard, Risk*, and *Vulnerability*.

Respiration The process whereby living organisms convert organic matter to *carbon dioxide (CO$_2$)*, releasing energy and consuming molecular oxygen.

Rewetting 'The deliberate action of changing a drained soil into a wet soil, e.g. by blocking *drainage* ditches, disabling pumping facilities or breaching obstructions' (IPCC, 2013). See also *Drainage*.

Risk The potential for adverse consequences for human or ecological systems, recognising the diversity of values and objectives associated with such systems. In the context of *climate change*, risks can arise from potential *impacts* of climate change as well as human responses to climate change. Relevant adverse consequences include those on lives, *livelihoods*, health and *well-being*, economic, social and cultural assets and investments, infrastructure, services (including *ecosystem services*), *ecosystems* and species.

In the context of climate change impacts, risks result from dynamic interactions between climate-related *hazards* with the *exposure* and *vulnerability* of the affected human or ecological system to the hazards. Hazards, exposure and vulnerability may each be subject to *uncertainty* in terms of magnitude and likelihood of occurrence, and each may change over time and space due to socio-economic changes and human decision-making (see also *risk management, adaptation,* and *mitigation*).

In the context of climate change responses, risks result from the potential for such responses not achieving the intended objective(s), or from potential trade-offs with, or negative side-effects on, other societal objectives, such as the *Sustainable Development Goals* (see also *risk trade-off*). Risks can arise for example from uncertainty in implementation, effectiveness or outcomes of climate policy, climate-related investments, technology development or adoption, and system *transitions*.

Risk assessment The qualitative and/or quantitative scientific estimation of *risks*. See also *Risk management,* and *Risk perception*.

Risk management Plans, actions, strategies or policies to reduce the *likelihood* and/or magnitude of adverse potential consequences, based on assessed or perceived *risks*. See also *Risk assessment,* and *Risk perception*.

Risk perception The subjective judgment that people make about the characteristics and severity of a *risk*. See also *Risk assessment,* and *Risk management*.

Risk trade-off The change in portfolio of *risks* that occurs when a countervailing risk is generated (knowingly or inadvertently) by an intervention to reduce the target risk (Wiener and Graham, 2009). See also *Adverse side-effect*, and *Co-benefits*.

Runoff The flow of water over the surface or through the subsurface, which typically originates from the part of liquid precipitation and/or snow/ice melt that does not evaporate or refreeze, and is not transpired. See also *Hydrological cycle*.

Saline soils Soils with levels of soluble salts (commonly sulphates and chlorides of calcium and magnesium) in the saturation extract high enough to negatively affect plant growth. Saline soils are usually flocculated and have good water permeability (Well and Brady, 2016). See also *Soil salinity* and *Sodic soils*.

Scenario   A plausible description of how the future may develop based on a coherent and internally consistent set of assumptions about key driving forces (e.g., rate of technological change (TC), prices) and relationships. Note that scenarios are neither predictions nor forecasts, but are used to provide a view of the implications of developments and actions. See also *Baseline scenario, Emission scenario, Mitigation scenario* and *Pathways*.

Scenario storyline   A *narrative* description of a *scenario* (or family of scenarios), highlighting the main scenario characteristics, relationships between key driving forces and the dynamics of their evolution. Also referred to as 'narratives' in the scenario literature.

Sea ice   Ice found at the sea surface that has originated from the freezing of seawater. Sea ice may be discontinuous pieces (ice floes) moved on the ocean surface by wind and currents (pack ice), or a motionless sheet attached to the coast (land-fast ice). Sea ice concentration is the fraction of the ocean covered by ice. Sea ice less than one year old is called first-year ice. Perennial ice is sea ice that survives at least one summer. It may be subdivided into second-year ice and multi-year ice, where multi-year ice has survived at least two summers.

Sea level change (sea level rise/sea level fall)   Change to the height of sea level, both globally and locally (relative sea level change) due to (1) a change in ocean volume as a result of a change in the mass of water in the ocean, (2) changes in ocean volume as a result of changes in ocean water density, (3) changes in the shape of the ocean basins and changes in the Earth's gravitational and rotational fields, and (4) local subsidence or uplift of the land. Global mean sea level change resulting from change in the mass of the ocean is called barystatic. The amount of barystatic sea level change due to the addition or removal of a mass of water is called its sea level equivalent (SLE). Sea level changes, both globally and locally, resulting from changes in water density are called steric. Density changes induced by temperature changes only are called thermosteric, while density changes induced by salinity changes are called halosteric. Barystatic and steric sea level changes do not include the effect of changes in the shape of ocean basins induced by the change in the ocean mass and its distribution.

Sea surface temperature (SST)   The subsurface bulk temperature in the top few meters of the ocean, measured by ships, buoys, and drifters. From ships, measurements of water samples in buckets were mostly switched in the 1940s to samples from engine intake water. Satellite measurements of skin temperature (uppermost layer; a fraction of a millimetre thick) in the infrared or the top centimetre or so in the microwave are also used, but must be adjusted to be compatible with the bulk temperature.

Sendai Framework for Disaster Risk Reduction   The Sendai Framework for Disaster Risk Reduction 2015–2030 outlines seven clear targets and four priorities for action to prevent new, and to reduce existing disaster risks. The voluntary, non-binding agreement recognises that the State has the primary role to reduce disaster risk but that responsibility should be shared with other stakeholders

including local government, the private sector and other stakeholders, with the aim for the substantial reduction of disaster risk and losses in lives, livelihoods and health and in the economic, physical, social, cultural and environmental assets of persons, businesses, communities and countries.

Sequestration   See *Uptake* and *Carbon sequestration*.

Shared socio-economic pathways (SSPs)   See *Pathways*.

Short-lived climate forcers (SLCF)   A set of compounds that are primarily composed of those with short lifetimes in the atmosphere compared to well-mixed *greenhouse gases (GHGs)*, and are also referred to as near-term climate forcers. This set of compounds includes *methane (CH$_4$)*, which is also a well-mixed greenhouse gas, as well as *ozone (O$_3$)* and *aerosols*, or their *precursors*, and some halogenated species that are not well-mixed greenhouse gases. These compounds do not accumulate in the atmosphere at decadal to centennial timescales, and so their effect on *climate* is predominantly in the first decade after their emission, although their changes can still induce long-term climate effects such as sea level change. Their effect can be cooling or warming. A subset of exclusively warming short-lived climate forcers is referred to as short-lived climate pollutants. See also *Long-lived climate forcers (LLCF)*.

Short-lived climate pollutants (SLCP)   See *Short-lived climate forcers (SLCF)*.

Sink   Any process, activity or mechanism which removes a *greenhouse gas*, an *aerosol* or a *precursor* of a greenhouse gas from the *atmosphere* (UNFCCC Article 1.8). See also *Sequestration, Source,* and *Uptake*.

Small Island Developing States (SIDS)   Small Island Developing States (SIDS), as recognised by the United Nations OHRLLS (Office of the High Representative for the Least Developed Countries, Landlocked Developing Countries and Small Island Developing States), are a distinct group of developing countries facing specific social, economic and environmental vulnerabilities (UN-OHRLLS, 2011). They were recognised as a special case both for their environment and development at the Rio Earth Summit in Brazil in 1992. Fifty eight countries and territories are presently classified as SIDS by the UN OHRLLS, with 38 being UN member states and 20 being Non-UN Members or Associate Members of the Regional Commissions (UN-OHRLLS, 2018).

Social costs   The full costs of an action in terms of social welfare losses, including external costs associated with the impacts of this action on the environment, the economy (GDP, employment) and on the society as a whole.

Social cost of carbon (SCC)   The net present value of aggregate climate damages (with overall harmful damages expressed as a number with positive sign) from one more tonne of carbon in the form of *carbon dioxide (CO$_2$)*, conditional on a global emissions trajectory over time.

Social-ecological system   An integrated system that includes human societies and *ecosystems*, in which humans are part of nature. The functions of such a system arise from the interactions and interdependence of the social and ecological subsystems. The system's structure is characterised by reciprocal feedbacks, emphasising that humans must be seen as a part of, not apart from, nature (Arctic Council, 2016; Berkes and Folke, 1998).

Social inclusion A process of improving the terms of participation in society, particularly for people who are disadvantaged, through enhancing opportunities, access to resources, and respect for rights (UN DESA 2016).

Social learning A process of social interaction through which people learn new behaviours, capacities, values, and attitudes.

Societal (social) transformation See *Transformation*.

Socio-economic scenario A *scenario* that describes a possible future in terms of population, *gross domestic product (GDP)*, and other socio-economic factors relevant to understanding the implications of *climate change*. See also *Baseline scenario, Emission scenario, Mitigation scenario,* and *Pathways*.

Socio-technical transitions Where technological change is associated with social systems and the two are inextricably linked.

Sodic soils Soils with disproportionately high concentration of sodium (Na^+) in relation to calcium (Ca^{2+}) and magnesium (Mg^{2+}) adsorbed at the cation exchange site on the surface of soil particles. Sodic soils are characterised by a poor soil structure and poor aeration (NDSU, 2014). See also *Soil salinity*.

Soil carbon sequestration (SCS) Land management changes which increase the *soil organic carbon* content, resulting in a net removal of *carbon dioxide (CO_2)* from the *atmosphere*.

Soil conservation The maintenance of soil fertility through controlling erosion, preserving soil organic matter, ensuring favourable soil physical properties, and retaining nutrients (Young, 1989).

Soil erosion The displacement of the soil by the action of water or wind. Soil erosion is a major process of *land degradation*.

Soil organic carbon Carbon contained in *soil organic matter*.

Soil organic matter The organic component of soil, comprising plant and animal residue at various stages of decomposition, and soil organisms.

Soil moisture Water stored in the soil in liquid or frozen form. Root-zone soil moisture is of most relevance for plant activity.

Soil salinity The concentration of soluble salts in the water extracted from a saturated soil (saturation extract), comprising chlorides and sulphates of Sodium (Na^+), calcium (Ca^{2+}) and magnesium (Mg^{2+}) as well as carbonate salts (adapted from FAO, 1985). See also *Saline soils,* and *Sodic soils*.

Source Any process or activity which releases a *greenhouse gas*, an *aerosol* or a *precursor* of a greenhouse gas into the atmosphere (UNFCCC Article 1.9). See also *Sink*.

Stabilisation (of GHG or CO₂-equivalent concentration) A state in which the *atmospheric* concentrations of one *greenhouse gas (GHG)* (e.g., carbon dioxide) or of a *CO₂-equivalent* basket of GHGs (or a combination of GHGs and *aerosols*) remains constant over time.

Stranded assets Assets exposed to devaluations or conversion to 'liabilities' because of unanticipated changes in their initially expected revenues due to innovations and/or evolutions of the business context, including changes in public regulations at the domestic and international levels.

Stratosphere The highly stratified region of the *atmosphere* above the *troposphere* extending from about 10 km (ranging from 9 km at high latitudes to 16 km in the tropics on average) to about 50 km altitude. See also *Atmosphere,* and *Troposphere*.

Subnational actors State/provincial, regional, metropolitan and local/municipal governments as well as non-party stakeholders, such as civil society, the private sector, cities and other subnational authorities, local communities and indigenous peoples.

Substantive rights See *Human rights*.

Supply-side measures See *Demand and supply-side measures*.

Surface temperature See *Global mean surface temperature (GMST), Land surface air temperature,* and *Sea surface temperature (SST)*.

Sustainability A dynamic process that guarantees the persistence of natural and human systems in an equitable manner.

Sustainable development (SD) Development that meets the needs of the present without compromising the ability of future generations to meet their own needs (WCED, 1987) and balances social, economic and environmental concerns. See also *Sustainable Development Goals (SDGs),* and *Development pathways* (under *Pathways*).

Sustainable Development Goals (SDGs) The 17 global goals for development for all countries established by the United Nations through a participatory process and elaborated in the *2030 Agenda for Sustainable Development*, including ending *poverty* and hunger; ensuring health and *well-being*, education, gender *equality*, clean water and energy, and decent work; building and ensuring resilient and sustainable infrastructure, cities and consumption; reducing *inequalities*; protecting land and water ecosystems; promoting peace, justice and partnerships; and taking urgent action on *climate change*. See also *Sustainable development (SD)*.

Sustainable forest management The stewardship and use of *forests* and forest lands in a way, and at a rate, that maintains their *biodiversity*, productivity, regeneration capacity, vitality and their potential to fulfil, now and in the future, relevant ecological, economic and social functions, at local, national, and global levels, and that does not cause damage to other *ecosystems* (Forest Europe, 1993).

Sustainable intensification (of agriculture) Increasing yields from the same area of land while decreasing negative environmental impacts of agricultural production and increasing the provision of environmental services (CGIAR, 2019). [Note: this definition is based on the concept of meeting demand from a finite land area, but it is scale-dependent. Sustainable intensification at a given scale (e.g., global or national) may require a decrease in production intensity at smaller scales and in particular places (often associated with previous, unsustainable, intensification) to achieve *sustainability* (Garnett et al., 2013).]

AI

Sustainable land management The stewardship and use of *land* resources, including soils, water, animals and plants, to meet changing human needs, while simultaneously ensuring the long-term productive potential of these resources and the maintenance of their environmental functions (Adapted from WOCAT, undated).

Technology transfer The exchange of knowledge, hardware and associated software, money and goods among stakeholders, which leads to the spread of technology for *adaptation* or *mitigation*. The term encompasses both diffusion of technologies and technological cooperation across and within countries.

Teleconnections A statistical association between climate variables at widely separated, geographically-fixed spatial locations. Teleconnections are caused by large spatial structures such as basin-wide coupled modes of ocean-atmosphere variability, Rossby wave-trains, mid-latitude jets and storm tracks, etc.

Temperature overshoot The temporary exceedance of a specified level of *global warming*, such as 1.5°C. Overshoot implies a peak followed by a decline in global warming, achieved through *anthropogenic removal of carbon dioxide (CO_2)* exceeding remaining CO_2 emissions globally. See also *Pathways (Subterms: Overshoot pathways, Non-overshoot Pathways)*.

Tier In the context of the IPCC Guidelines for National Greenhouse Gas Inventories, a tier represents a level of methodological complexity. Usually three tiers are provided. Tier 1 is the basic method, Tier 2 intermediate and Tier 3 most demanding in terms of complexity and data requirements. Tiers 2 and 3 are sometimes referred to as higher tier methods and are generally considered to be more accurate (IPCC, 2019).

Tipping point A level of change in system properties beyond which a system reorganises, often abruptly, and does not return to the initial state even if the drivers of the change are abated. For the *climate system*, it refers to a critical threshold beyond which global or regional climate changes from one stable state to another stable state. Tipping points are also used when referring to *impact*: the term can imply that an impact tipping point is (about to be) reached in a natural or *human system*. See also *Irreversibility*.

Transformation A change in the fundamental attributes of natural and human systems.

Societal (social) transformation
A profound and often deliberate shift initiated by communities toward sustainability, facilitated by changes in individual and collective values and behaviours, and a fairer balance of political, cultural, and *institutional* power in society.

Transformation pathways See *Pathways*.

Transformational adaptation See *Adaptation*.

Transformative change A system wide change that alters the fundamental attributes of the system.

Transient climate response to cumulative CO_2 emissions (TCRE) The transient global average surface temperature change per unit cumulative *carbon dioxide (CO_2)* emissions, usually 1000 GtC. TCRE combines both information on the airborne fraction of cumulative CO_2 emissions (the fraction of the total CO_2 emitted that remains in the atmosphere, which is determined by *carbon cycle* processes) and on the transient climate response (TCR). See also *Transient climate response (TCR) (under Climate sensitivity)*.

Transit-oriented development (TOD) An approach to urban development that maximises the amount of residential, business and leisure space within walking distance of efficient public transport, so as to enhance mobility of citizens, the viability of public transport and the value of urban land in mutually supporting ways.

Transition The process of changing from one state or condition to another in a given period of time. Transition can occur in individuals, firms, cities, regions and nations, and can be based on incremental or transformative change.

Tropical cyclone The general term for a strong, cyclonic-scale disturbance that originates over tropical oceans. Distinguished from weaker systems (often named tropical disturbances or depressions) by exceeding a threshold wind speed. A tropical storm is a tropical cyclone with one-minute average surface winds between 18 and 32 m s^{-1}. Beyond 32 m s^{-1}, a tropical cyclone is called a hurricane, typhoon, or cyclone, depending on geographic location. See also *Extratropical cyclone*.

Troposphere The lowest part of the *atmosphere*, from the surface to about 10 km in altitude at mid-latitudes (ranging from 9 km at high latitudes to 16 km in the tropics on average), where clouds and weather phenomena occur. In the troposphere, temperatures generally decrease with height. See also *Atmosphere*, and *Stratosphere*.

Uncertainty A state of incomplete knowledge that can result from a lack of information or from disagreement about what is known or even knowable. It may have many types of sources, from imprecision in the data to ambiguously defined concepts or terminology, incomplete understanding of critical processes, or uncertain *projections* of *human behaviour*. Uncertainty can therefore be represented by quantitative measures (e.g., a probability density function) or by qualitative statements (e.g., reflecting the judgment of a team of experts) (see IPCC, 2004; Mastrandrea et al., 2010; Moss and Schneider, 2000). See also *Confidence*, and *Likelihood*.

United Nations Convention to Combat Desertification (UNCCD) A legally binding international agreement linking environment and development to sustainable land management, established in 1994. The Convention's objective is 'to combat desertification and mitigate the effects of drought in countries experiencing drought and/or desertification'. The Convention specifically addresses the arid, semi-arid and dry sub-humid areas, known as the drylands, and has a particular focus on Africa. As of October 2018, the UNCCD had 197 Parties.

United Nations Framework Convention on Climate Change (UNFCCC) The UNFCCC was adopted in May 1992 and opened for signature at the 1992 Earth Summit in Rio de Janeiro. It entered into force in March 1994 and as of May 2018 had 197 Parties (196 States and the European Union). The Convention's ultimate objective is the 'stabilisation of greenhouse gas concentrations in the atmosphere at a level that would prevent dangerous anthropogenic interference with the climate system'. The provisions of the Convention are pursued and implemented by two treaties: the *Kyoto Protocol* and the *Paris Agreement*.

AI

AI

Urban green infrastructure   Public and private green spaces, including remnant native vegetation, parks, private gardens, golf courses, street trees, urban farming and engineered options such as green roofs, green walls, biofilters and raingardens (Norton et al., 2015).

Urban and Peri-urban agriculture   'The cultivation of crops and rearing of animals for food and other uses within and surrounding the boundaries of cities, including fisheries and forestry' (EPRS, 2014).

Uptake   The addition of a substance of concern to a *reservoir*. See also *Carbon sequestration*, and *Sink*.

Vegetation browning   A decrease in photosynthetically active plant *biomass* which is inferred from satellite observations.

Vegetation greening   An increase in photosynthetically active plant *biomass* which is inferred from satellite observations.

Vulnerability   The propensity or predisposition to be adversely affected. Vulnerability encompasses a variety of concepts and elements including sensitivity or susceptibility to harm and lack of capacity to cope and adapt. See also *Exposure, Hazard,* and *Risk*.

Water cycle   See *Hydrological cycle.*

Well-being   A state of existence that fulfils various human needs, including material living conditions and quality of life, as well as the ability to pursue one's goals, to thrive, and feel satisfied with one's life. Ecosystem well-being refers to the ability of *ecosystems* to maintain their diversity and quality.

Wetland   Land that is covered or saturated by water for all or part of the year (e.g., *peatland*).

References

Allan, J.A., 2005: Virtual water: A strategic resource global solutions to regional deficits. *Groundwater*, **36**(4), 545–546, doi: 10.1111/j.1745-6584.1998.tb02825.x.

AMS, 2000: AMS Glossary of Meteorology, 2nd ed. American Meteorological Society, Boston, MA, Retrieved from: http://amsglossary.allenpress.com/glossary/browse.

Arctic Council, 2013: Glossary of terms. In: *Arctic Resilience Interim Report 2013*. Stockholm Environment Institute and Stockholm Resilience Centre, Stockholm, Sweden, pp. viii.

Arctic Council, 2016: *Arctic Resilience Report 2016* [M. Carson and G. Peterson (eds.)]. Stockholm Environment Institute and Stockholm Resilience Centre, Stockholm, Sweden.

Berkes, F. and C. Folke, 1998: *Linking Social and Ecological Systems: Management Practices and Social Mechanisms for Building Resilience*. Cambridge University Press, Cambridge, United Kingdom and New York, NY, USA, 459 pp.

CGIAR, 2019: Sustainable intensification of agriculture: oxymoron or real deal? The Consultative Group on International Agricultural Research (CGIAR). Retrieved from: https://wle.cgiar.org/thrive/big-questions/sustainable-intensification-agriculture-oxymoron-or-real-deal/sustainable-1 .

Culwick, C. and K. Bobbins, 2016: *A Framework for a Green Infrastructure Planning Approach in the Gauteng City–Region*. GCRO Research Report No. 04, Gauteng City–Region Observatory (GRCO), Johannesburg, South Africa, 127 pp.

Davis, S.J. and K. Caldeira, 2010: Consumption-based accounting of CO_2 emissions. *Proceedings of the National Academy of Sciences of the United States of America*, **107**(12), 5687–5692, doi:10.1073/pnas.0906974107.

Dietary Guidelines Advisory Committee, 2015: *Scientific Report of the 2015 Dietary Guidelines Advisory Committee*. US Department of Agriculture, Agricultural Research Service, Washington DC, United States of America.

Ellis, E.C. and N. Ramankutty, 2008: Putting people in the map: anthropogenic biomes of the world. *Front. Ecol. Environ.*, **6**(8), 439–447, doi: 10.1890/070062.

Ellis, E.C., K.K. Goldewijk, S. Siebert, D. Lightman and N. Ramankutty, 2010: Anthropogenic transformation of the biomes, 1700 to 2000. *Global Ecology and Biogeography*, **19**(5), 589–606, doi:10.1111/j.1466-8238.2010.00540.x.

EPRS, 2014: Urban and Peri-Urban Agriculture. European Parliamentary Research Service. Retrieved from: https://epthinktank.eu/2014/06/18/urban-and-peri-urban-agriculture/

FAO, 1985: Irrigation water management: *Training manual no.1 – Introduction to irrigation*. eds, C. Brouwer, A. Goffeau, and M. Heibloem. Food and Agriculture Organization of the United Nations (FAO), Rome, Italy.

FAO, 2000: *Land cover classification system (LCCS): Classification concepts and user manual*. [A. Di Gregorio and L.J.M. Jansen (eds.)]. Food and Agriculture Organization of the United Nations (FAO), Rome, Italy, 179 pp.

FAO, 2001: Glossary. In: *The State of Food Insecurity in the World 2001*. Food and Agriculture Organization of the United Nations (FAO), Rome, Italy, pp. 49–50.

FAO, 2005: *Building on gender, agrobiodiversity and local knowledge: A training manual*. Food and Agriculture Organization of the United Nations (FAO), Rome, Italy.

FAO, 2007: *Land evaluation: Towards a revised framework. Land and water discussion paper*. Food and Agriculture Organization of the United Nations (FAO), Rome, Italy.

FAO, 2009: Declaration of the World Summit on Food Security. WSFS 2009/2, Food and Agriculture Organization of the United Nations (FAO), Rome, Italy.

FAO, 2013: *Food wastage footprint: Impacts on natural resources. Summary report*. Food and Agriculture Organization of the United Nations (FAO), Rome, Italy, 63 pp.

FAO, 2015a: Agroforestry. Food and Agriculture Organization of the United Nations (FAO). Retrieved from: http://www.fao.org/forestry/agroforestry/80338/en/.

FAO, 2015b: Food waste. Food and Agriculture Organization of the United Nations (FAO). Retrieved from: http://www.fao.org/platform-food-loss-waste/food-waste/definition/en/.

FAO, 2018: Climate–Smart Agriculture. Food and Agriculture Organization of the United Nations (FAO). Retrieved from: www.fao.org/climate–smart–agriculture.

Forest Europe, 1993: Resolution H1 General Guidelines for the Sustainable Management of Forests in Europe. Second Ministerial Conference on the Protection of Forests in Europe, Helsinki, 16–17 June 1993. Retrieved from: https://www.foresteurope.org/docs/MC/MC_helsinki_resolutionH1.pdf.

Fung, A. and E.O. Wright (eds.), 2003: *Deepening Democracy: Institutional Innovations in Empowered Participatory Governance*. Verso, London, UK, 312 pp.

Garnett, T. et al, 2013: Sustainable intensification in Agriculture: Premises and Policies. *Science*, **341**(6141), 33. doi:10.1126/science.1234485.

Gbeckor-Kove, N. 1989: Lectures on drought, desertification, in drought and desertification. WMO, TDNo. 286, World Meteorological Organization, Geneva, Switzerland.

HLPE, 2017: *Nutrition and food systems*. The High Level Panel of Experts on Food Security and Nutrition of the Committee on World Food Security. Rome, Italy, 152 pp.

IBI, 2018: Frequently Asked Questions About Biochar: What is biochar? International Biochar Initiative (IBI). Retrieved from: https://biochar–international.org/faqs.

ICS, 2019: Formal subdivision of the Holocene Series/Epoch. International Commission on Stratigraphy (ICS). Retrieved from http://www.stratigraphy.org/index.php/ics-news-and-meetings/125-formal-subdivision-of-the-holocene-series-epoch.

IOM, 2018: Key Migration Terms. International Organization for Migration (IOM). Retrieved from: www.iom.int/key–migration–terms.

IPBES, 2018: The IPBES assessment report on land degradation and restoration. [Montanarella, L., Scholes, R., and Brainich, A. (eds.)]. Secretariat of the Intergovernmental Science-Policy Platform on Biodiversity and Ecosystem services, Bonn, Germany, 744 pp.

IPBES, 2019: Glossary. Secretariat of the Intergovernmental Science-Policy Platform on Biodiversity and Ecosystem services, Bonn, Germany. Retrieved from: https://www.ipbes.net/glossary.

IPCC, 2000: Land Use, Land–Use Change, and Forestry: A Special Report of the IPCC. [Watson, R.T., I.R. Noble, B. Bolin, N.H. Ravindranath, D.J. Verardo, and D.J. Dokken (eds.)]. Cambridge University Press, Cambridge, UK, 375 pp.

IPCC, 2003: Definitions and Methodological Options to Inventory Emissions from Direct Human–induced Degradation of Forests and Devegetation of Other Vegetation Types. [Penman, J., M. Gytarsky, T. Hiraishi, T. Krug, D. Kruger, R. Pipatti, L. Buendia, K. Miwa, T. Ngara, K. Tanabe, and F. Wagner (eds.)]. Institute for Global Environmental Strategies (IGES), Hayama, Kanagawa, Japan, 32 pp.

IPCC, 2004: IPCC Workshop on Describing Scientific Uncertainties in Climate Change to Support Analysis of Risk of Options. Workshop Report. Intergovernmental Panel on Climate Change (IPCC), Geneva, Switzerland, 138 pp.

IPCC, 2006: 2006 IPCC Guidelines for National Greenhouse Gas Inventories. [H.S. Eggleston, L. Buendia, K. Miwa, T. Ngara, K. Tanabe (eds)]. Institute for Global Environmental Strategies (IGES), Hayama, Kanagawa, Japan, 20 pp.

IPCC, 2011: Workshop Report of the Intergovernmental Panel on Climate Change Workshop on Impacts of Ocean Acidification on Marine Biology and Ecosystems. [Field, C.B., V. Barros, T.F. Stocker, D. Qin, K.J. Mach, G.–K. Plattner, M.D. Mastrandrea, M. Tignor, and K.L. Ebi (eds.)]. IPCC Working

Group II Technical Support Unit, Carnegie Institution, Stanford, California, United States of America, 164 pp.

IPCC, 2012a: Managing the Risks of Extreme Events and Disasters to Advance Climate Change Adaptation. A Special Report of Working Groups I and II of the Intergovernmental Panel on Climate Change (IPCC). [Field, C.B., V. Barros, T.F. Stocker, D. Qin, D.J. Dokken, K.L. Ebi, M.D. Mastrandrea, K.J. Mach, G.-K. Plattner, S.K. Allen, M. Tignor, and P.M. Midgley (eds.)]. Cambridge University Press, Cambridge, UK and New York, NY, USA, 582 pp.

IPCC, 2012b: Meeting Report of the Intergovernmental Panel on Climate Change Expert Meeting on Geoengineering. IPCC Working Group III Technical Support Unit, Potsdam Institute for Climate Impact Research, Potsdam, Germany, 99 pp.

IPCC, 2013: 2013 Supplement to the 2006 IPCC Guidelines for National Greenhouse Gas Inventories: Wetlands. [Hiraishi, T., T. Krug, K. Tanabe, N. Srivastava, J. Baasansuren, M. Fukuda, and T.G. Troxler (eds.)]. IPCC, Switzerland.

ISO, 2014: ISO 16559:2014(en) Solid biofuels – Terminology, definitions and descriptions. International Standards Organisation (ISO). Retrieved from: https://www.iso.org/obp/ui/#iso:std:iso:16559:ed-1:v1:en.

ISO, 2018: ISO 14044:2006. Environmental management – Life cycle assessment – Requirements and guidelines. International Standards Organisation (ISO). Retrieved from: www.iso.org/standard/38498.html.

Jagers, S.C. and J. Stripple, 2003: Climate Governance Beyond the State. Global Governance, 9(3), 385–399, www.jstor.org/stable/27800489.

MacDonald, G.K., K.A. Brauman, S. Sun, K.M. Carlson, E.S. Cassidy, J.S. Gerber, and P.C. West, 2015: Rethinking agricultural trade relationships in an era of globalization. BioScience, 65(3), 275–289, doi:10.1093/biosci/biu225.

Mastrandrea, M.D. et al., 2010: Guidance Note for Lead Authors of the IPCC Fifth Assessment Report on Consistent Treatment of Uncertainties. Intergovernmental Panel on Climate Change (IPCC), Geneva, Switzerland, 6 pp.

McDonald, T., J. Jonson, and K.W. Dixon, 2016: National standards for the practice of ecological restoration in Australia. Restoration Ecology, 24(S1) S4-S32, doi:10.1111/rec.12359.

MA, 2005: Appendix D: Glossary. In: Ecosystems and Human Well–being: Current States and Trends. Findings of the Condition and Trends Working Group [Hassan, R., R. Scholes, and N. Ash (eds.)]. Millennium Ecosystem Assessment (MA). Island Press, Washington DC, USA, pp. 893–900.

Meinzen-Dick, R. and M. Di Gregorio, 2004: Collective Action and Property Rights for Sustainable Deveopment. International Food and Policy Research Institute. Washington DC, USA.

Mitchell, T. and S. Maxwell, 2010: Defining climate compatible development. CDKN ODI Policy Brief November 2010/A, Climate & Development Knowledge Network (CDKN), 6 pp.

Misra, A., L. Khurana, 2008: Obesity and the metabolic syndrome in developing countries. The Journal of Clinical Endocrinology & Metabolism, 93(11), 3–30, doi: 10.1210/jc.2008-1595.

Moss, R.H. and S.H. Schneider, 2000: Uncertainties in the IPCC TAR: Recommendations to Lead Authors for More Consistent Assessment and Reporting. In: Guidance Papers on the Cross Cutting Issues of the Third Assessment Report of the IPCC [Pachauri, R., T. Taniguchi, and K. Tanaka (eds.)]. Intergovernmental Panel on Climate Change (IPCC), Geneva, Switzerland, pp. 33–51

Moss, R.H. et al., 2008: Towards New Scenarios for Analysis of Emissions, Climate Change, Impacts, and Response Strategies. Technical Summary. Intergovernmental Panel on Climate Change (IPCC), Geneva, Switzerland, 25 pp.

Moss, R.H. et al., 2010: The next generation of scenarios for climate change research and assessment. Nature, 463(7282), 747–756, doi:10.1038/nature08823.

MRFCJ, 2018: Principles of Climate Justice. Mary Robinson Foundation For Climate Justice (MRFCJ). Retrieved from: www.mrfcj.org/principles–of–climate–justice.

NDSU, 2014: Saline and Sodic Soils. North Dakota State University (NDSU). Retrieved from: https://www.ndsu.edu/soilhealth/wp-content/uploads/2014/07/Saline-and-Sodic-Soils-2-2.pdf.

Norton, B.A., A.M. Coutts, S.J. Livesley, R.J. Harris, A.M. Hunter, and N.S.G. Williams, 2015: Planning for cooler cities: A framework to prioritise green infrastructure to mitigate high temperatures in urban landscapes. Landscape and Urban Planning, 134, 127–138, doi:10.1016/j.landurbplan.2014.10.018.

O'Neill, B.C. et al., 2014: A new scenario framework for climate change research: the concept of shared socioeconomic pathways. Climatic Change, 122(3), 387–400, doi:10.1007/s10584–013–0905–2.

O'Neill, B.C. et al., 2017: The roads ahead: Narratives for shared socioeconomic pathways describing world futures in the 21st century. Global Environmental Change, 42, 169–180, doi:10.1016j.gloenvcha.2015.01.004.

Park, S. E. et al., 2012: Informing adaptation responses to climate change through theories of transformation. Global Environmental Change, 22(1), 115–126. doi:10.1016/j.gloenvcha.2011.10.003.

Peters, B.G. and J. Pierre, 2001: Developments in intergovernmental relations: towards multi–level governance. Policy & Politics, 29(2), 131–135, doi:10.1332/0305573012501251.

Popkin, B.M., 1994: The nutrition transition in low-income countries: an emerging crisis. Nutr. Rev. 52(9), 285–298, doi: https://doi.org/10.1111/j.1753-4887.1994.tb01460.x.

Popkin, B.M., 2006: Global nutrition dynamics: the world is shifting rapidly toward a diet linked with noncommunicable diseases. The American Journal of Clinical Nutrition, 84(2), 289–298, doi: https://doi.org/10.1093/ajcn/84.1.289.

Pongratz, J. et al., 2018: Models meet data: Challenges and opportunities in implementing land management in Earth system models. Global Change Biology, 24(4) 1470–1487, doi: 10.1111/gcb.13988.

Riahi, K. et al., 2017: The Shared Socioeconomic Pathways and their energy, land use, and greenhouse gas emissions implications: An overview. Global Environmental Change, 42, 153–168, doi:10.1016/j.gloenvcha.2016.05.009.

Rulli, M.C. and P. D'Odorico, 2014: Food appropriation through large scale land acquisitions. Environmental Research Letters, 9(6), doi: 10.1088/1748-9326/9/6/064030.

Sarmiento, H. and C. Tilly, 2018: Governance Lessons from Urban Informality. Politics and Governance, 6(1), 199–202, doi:10.17645/pag.v6i1.1169.

Silvertown, J. (2009). A new dawn for citizen science. Trends in Ecology & Evolution, 24(9), 467–471, doi: 10.1016/J.TREE.2009.03.017.

Soulé, M.E., 2010: Conservation relevance of ecological cascades. In: Trophic Cascades: Predators, Prey, and the Changing Dynamics of Nature [J. Terborgh and J.A. Estes (eds.)]. Island Press, 337–352.

Tabara, J.D., J. Jager, D. Mangalagiu, and M. Grasso, 2018: Defining transformative climate science to address high–end climate change. Regional Environmental Change, 1–12, doi:10.1007/s10113–018–1288–8.

Termeer, C.J.A.M., A. Dewulf, and G.R. Biesbroek, 2017: Transformational change: governance interventions for climate change adaptation from a continuous change perspective. Journal of Environmental Planning and Management, 60(4), 558–576, doi:10.1080/09640568.2016.1168288.

Türkeş M, 1999: Vulnerability of Turkey to desertification with respect to precipitation and aridity conditions. Turkish J. Eng. Environ. Sci., 23, 363–380.

Türkeş, M., 2017: General Climatology: Fundamentals of Atmosphere, Weather and Climate. Revised Second Edition, Kriter Publisher Physical Geography Series No: 4, ISBN: 978-605-9336-28-4, xxiv + 520 pp. Kriter Publisher, Berdan Matbaası: İstanbul. (In Turkish).

United Nations Secretary General's Advisory Group on Energy and Climate (AGECC), 2010: Energy for a Sustainable Future. New York, NY, USA.

UN, 1992: Article 2: Use of Terms. In: Convention on Biological Diversity. United Nations (UN), pp. 3–4.

UN, 1998: *Guiding Principles on Internal Displacement*. E/CN.4/1998/53/Add.2,United Nations (UN) Economic and Social Council, 14 pp.

UN, 2015: *Transforming Our World: The 2030 Agenda for Sustainable Development*. A/RES/70/1, United Nations General Assembly (UNGA), New York, NY, USA, 35 pp.

UNCCD, 1994: *United Nations Convention to Combat Desertification in countries experiencing serious drought and/or desertification, particularly in Africa*. A/AC.241/27, United Nations General Assembly (UNGA), New York, NY, USA, 58 pp.

UNCCD, 2019: Achieving Land Degradation Neutrality. United Nations Convention to Combat Desertification (UNCCD). Retrieved from: https://www.unccd.int/actions/achieving-land-degradation-neutrality.

UN DESA, 2016: Identifying social inclusion and exclusion. In: *Leaving no one behind: the imperative of inclusive development*. Report on the World Social Situation 2016. ST/ESA/362, United Nations Department of Economic and Social Affairs (UN DESA), New York, NY, USA, pp. 17–31.

UNESCO, 2018: Local and Indigenous Knowledge Systems. United Nations Educational,Scientific and Cultural Organization (UNESCO). Retrieved from: www.unesco.org/new/en/natural–sciences/priority–areas/links/related–information/what–is–local–and–indigenous–knowledge.

UNEP, 2016: Unlocking the sustainable potential of land resources: evaluating systems, strategies and tools. Factsheet. United Nations Environment Programme (UNEP). Retrieved from: http://hdl.handle.net/20.500.11822/7711.

UNFCCC, 1992: *United Nations Framework Convention on Climate Change*. United Nations, New York, United States. Retrieved from: https://unfccc.int/files/essential_background/background_publications_htmlpdf/application/pdf/conveng.pdf.

UNFCCC, 2019: *Land Use, Land-Use Change and Forestry (LULUCF)*. United Nations Framework Convention on Climatic Change (UNFCCC), Bonn, Germany. Retrieved from: https://unfccc.int/topics/land-use/workstreams/land-use--land-use-change-and-forestry-lulucf.

UNISDR, 2009: *2009 UNISDR Terminology on Disaster Risk Reduction*. United Nations International Strategy for Disaster Reduction (UNISDR), Geneva, Switzerland, 30 pp.

UNOHCHR, 2018:What are Human rights? UN Office of the High Commissioner for Human Rights (UNOHCHR). Retrieved from: www.ohchr.org/EN/Issues/Pages/whatarehumanrights.aspx.

UN–OHRLLS, 2011: Small Island Developing States: Small Islands Big(ger) Stakes. Office for the High Representative for the Least Developed Countries, Landlocked Developing Countries and Small Island Developing States (UN–OHRLLS), New York, NY, USA, 32 pp.

UN–OHRLLS, 2018: Small Island Developing States: Country profiles. Office for the High Representative for the Least Developed Countries, Landlocked Developing Countries and Small Island Developing States (UN–OHRLLS). Retrieved from: http://unohrlls.org/about–sids/country–profiles.

UN–REDD, 2009: *Measurement, Assessment, Reporting and Verification (MARV): Issues and Options for REDD*. Draft Discussion Paper, United Nations Collaborative Programme on Reducing Emissions from Deforestation and Forest Degradation in Developing Countries (UN–REDD), Geneva, Switzerland, 12 pp.

Vandergeten, E., H. Azadi, D. Teklemariam, J. Nyssen, F. Witlox and E. Vanhaute, 2016: Agricultural Outsourcing or Land Grabbing: A Meta-Analysis. *Landscape Ecology*, **31**(7), 1395–1417, doi: 10.1007/s10980-016-0365-y.

WCED, 1987: *Our Common Future*. World Commission on Environment and Development (WCED), Geneva, Switzerland, 400 pp., doi:10.2307/2621529.

Well, R.R. and N.C. Brady, 2016: *Nature and property of soils, 15th Edition*. Pearson Education Limited, Harlow, England, 420–430 pp.

Wiener, J.B. and J.D. Graham, 2009: *Risk vs. risk: Tradeoffs in protecting health and the environment*. Harvard University Press, Cambridge, MA, USA.

Willems, S. and K. Baumert, 2003: *Institutional Capacity and Climate Actions*. COM/ENV/EPOC/IEA/SLT(2003)5, Organisation for Economic Co–operation and Development (OECD) International Energy Agency (IEA), Paris, France, 50 pp.

Whitmee, S. et al., 2015: Safeguarding human health in the Anthropocene epoch: Report of the Rockefeller Foundation-Lancet Commission on planetary health. *The Lancet*, **366**(10007), 1973–2028.

WHO, 2018: Malnutrition. World Health Organization (WHO). Retrieved from: http://www.who.int/topics/malnutrition/en/.

WMO, 2019: What are weather/climate services? Global Framework for Climate Services, Geneva, Switzerland. Retrieved from: https://www.wmo.int/gfcs/what_are_climate_weather_services.

WOCAT, undated: Glossary. World Overview of Conservation Approaches and Technologies (WOCAT). Retrieved from: https://www.wocat.net/en/glossary.

Young, A. 1989: *Agroforestry for soil conservation*. International Council for Research in Agroforestry, Nairobi, Kenya, 318p.

AII

Annex II: Acronyms

This annex should be cited as:
IPCC, 2019: Annex II: Acronyms. In: *Climate Change and Land: an IPCC special report on climate change, desertification, land degradation, sustainable land management, food security, and greenhouse gas fluxes in terrestrial ecosystems* [P.R. Shukla, J. Skea, E. Calvo Buendia, V. Masson-Delmotte, H.-O. Pörtner, D. C. Roberts, P. Zhai, R. Slade, S. Connors, R. van Diemen, M. Ferrat, E. Haughey, S. Luz, S. Neogi, M. Pathak, J. Petzold, J. Portugal Pereira, P. Vyas, E. Huntley, K. Kissick, M. Belkacemi, J. Malley, (eds.)]. https://doi.org/10.1017/9781009157988.011

3-NOP	3-Nitrooxypropanol	**CAFO**	Concentrated Animal Feeding Operations
A/F	Afforestation/Reforestation	**CAMPA**	Compensatory Afforestation Management and Planning Authority
ABC	Low Carbon Agriculture Plan		
ABM	Agent-Based Model	**CATDDO**	Catastrophe Risk Deferred Drawdown Option
ADB	Asian Development Bank	**CBA**	Community-based Adaptation/Cost Benefit Analysis
AerChemMIP	Aerosol Chemistry Model Intercomparison Project		
		CBD	United Nations Convention on Biodiversity
AFOLU	Agriculture, Forestry and Other Land Use	**CC**	Carbon Capture
AgMIP	Agricultural Model Intercomparison and Improvement Project	**CCS**	Carbon dioxide Capture and Storage
		CCA	Climate Change Adaptation
AGRA	Alliance for a Green Revolution in Africa	**CCAFS**	CGIAR Research Program on Climate Change, Agriculture and Food Security
AI	Aridity Index		
AMO	Atlantic Multi-decadal Oscillation	**CCN**	Cloud Condensation Nuclei
AR	Afforestation and Reforestation	**CCRIF**	Caribbean Catastrophe Risk Insurance Facility
AR5	Fifth Assessment Report	**CCS**	Carbon Capture and Storage
ARC	Africa Risk Capacity	**CCU**	Carbon Capture and Usage
ARM	Adaptive Risk Management	**CDM**	Clean Development Mechanism
ASF	Animal Sourced Foods	**CDP**	Carbon Disclosure Project
AVHRR	Advanced High Resolution Radiometer	**CDR**	Carbon Dioxide Removal
AWD	Alternate Wetting and Drying	**CEA**	Cost Effectiveness Analysis
BAPEN	British Association for Parenteral and Enteral Nutrition	**CEC**	Commission of the European Communities
		ÇEMGM	Conservation Reserve Program
BAU	Business as Usual	**CEN-SAD**	The Community of Sahel-Saharan States
BC	Biochar	**CFS**	Committee on World Food Security
BC	Black Carbon	**CGIAR-**	Consultative Group on International Agricultural Research
BCR	Benefit Cost Ratio		
BE	Bioenergy	**CGRA**	Coordinated Global and Regional Assessment
BECCS	Bioenergy with Carbon Capture and Storage	**CH₄**	Methane
BEFM	Bio-Economic Farm Model	**CHP**	Combined Heat and Power
BEST	Berkeley Earth Surface Temperature	**CHIRPS**	Climate Hazards Group Infrared Precipitation with Station
BPE	Biomass Production Efficiency		
BrC	Brown Carbon	**CHMI**	Czech Hydrometeorological Institute
BTV	Blue-Tongue Virus	**CHP**	Combined Heat and Power
BUR	Biannual Update Report	**CI**	Confidence Interval
BVOC	Biogenic Volatile Organic Compounds	**CIAT**	International Centre for Tropical Agriculture
C	Carbon	**CIPHET**	Central Institute of Post-Harvest Engineering and Technology
CA	Conservation Agriculture		
CABI	Centre for Agriculture and Bioscience International	**CIREN**	Centro de Información de Recursos Naturales
		CM	Cultured Meat
CACILM	Central Asian Countries Initiative for Land Management	**CMIE**	Centre of Monitoring the Indian Economy
		CMIP5	Coupled Model Intercomparison Project Phase 5

AII

CMIP6	Coupled Model Intercomparison Project Phase 6		**ENSO**	El Niño/Southern Oscillation
CO	Carbon monoxide		**EPA**	UN Environmental Protection Agency
CO$_2$	Carbon dioxide		**ES**	Ecosystem Service
CO$_2$e	Carbon dioxide equivalent		**ESM**	Earth System Models
CO$_2$eq	Carbon dioxide equivalent		**ESL**	Extreme Sea Level
COP	Conference of the Parties		**ETS**	Emissions Trading Scheme
CRCM	Canadian Regional Climate Model		**EU**	European Union
CRISPR	Clustered Regularly Interspaced Short Palindromic Repeats		**EU-RED**	European Union Renewable Energy Directive
CRP	Conservation Reserve Program		**EVI**	Enhanced Vegetation Index
CRU	Climatic Research Unit of the University of East Anglia		**EW/EA**	Early Warning and Early Action
			EWS	Early Warning Systems
CRUTEM	Dataset of the Climatic Research Unit of the University of East Anglia		**FACE**	Free-Air Carbon dioxide Enrichment
			FAO	United Nations Food and Agricultural Organisation
CRV	Climate Resilient Villages		**FAO-FRA**	FAO Global Forest Resources Assessment
CSA	Climate Smart Agriculture		**FAOSTAT**	FAO Database
CSO	Civil Society Organisation		**FAQ**	Frequently Asked Questions
CSV	Climate Smart Village		**Fe**	Iron
DALY	Disability-Adjusted Life-Year		**FF**	Fossil Fuel
DERM	Department of Environment and Resource Management, India		**FIES**	Food Insecurity Experience Scale
			fNRB	fraction of Non-Renewable Biomass
DESIRE	Desertification Mitigation and Remediation of Land		**FRA**	Forest Rights Act (India)
			FSC	Forest Stewardship Council
DGVM	Dynamic Global Vegetation Model		**FWL**	Food Waste and Loss
DJF	December-January-February		**GATT**	General Agreement on Tariffs and Trade
DMP	Dust Mass Path		**GBEP**	Global Bioenergy Partnership
DRM	Disaster Risk Management		**GCF**	Green Climate Fund
DOD	Dust Optical Depth		**GCM**	Global Climate Model or General Circulation Model
DRR	Disaster Risk Reduction			
EBA	Ecosystem Based Adaptation		**GCP**	Global Carbon Project
EC	Elemental Carbon or European Commission		**GDEWS**	Global Drought Early Warning System
ECA	European Court of Auditors		**GDP**	Gross Domestic Product
EDGAR	Emissions Database for Global Atmospheric Research		**GE**	Government Effectiveness or General Equilibrium
EEA	European Environment Agency		**GEF**	Global Environmental Facility
EF	Emission Factor		**GFCS**	Global Framework for Climate Services
EFSA	European Food Safety Authority		**GFED**	Global Fire Emissions Database
EFT	Ecological Fiscal Transfer		**GFGP**	Conversion of Cropland to Forests and Grasslands Program (Grain for Green Programme)
ELD	Economics of Land Degradation or Economics of Land Degradation Initiative			
EM-DAT	International Disaster Database		**GGCM**	Global Gridded Crop Models

All

Gha	Gigahectares		**IAM**	Integrated Assessment Model
GHCN	Global Historical Climatology Network		**ICARDA**	International Center for Agriculture Research in the Dry Areas
GHG	Greenhouse Gas		**ICCC**	International Conference on Climate Change
GHGE	Greenhouse Gas Emissions		**ICOS**	Integrated Carbon Observation System
GHGI	Greenhouse Gas Inventory		**ICPAC**	IGAD Climate Prediction and Applications Centre
GIAHS	Globally Important Agricultural Heritage Site		**ICSU**	International Council for Science
GIEWS	FAO Global Information and Early Warning System		**ICT**	Information and Communication Technology
GIMMS3g	Global Inventory Modelling and Mapping Studies		**ICTP**	International Centre for Theoretical Physics
GISTEMP	Goddard's Global Surface Temperature Analysis		**IEA**	International Energy Agency
GIZ	German Society for International Cooperation		**IFAD**	International Fund for Agricultural Development
GJ	Gigajoules		**IFPRI**	International Food Policy Research Institute
GLASOD	Global Assessment of Human-Induced Soil Degradation		**IGAD**	Intergovernmental Authority on Development
GLASS	Global Land Atmosphere System Study		**IHME**	Institute for Health Metrics and Evaluation
GLOBIOM	GLObal BIOsphere Management model		**IK**	Indigenous Knowledge
GLOF	Glacial Lake Outburst Floods		**ILK**	Indigenous and Local Knowledge
GLOMAP	Global Model of Aerosol Processes		**iLUC**	indirect Land Use Change
GM	Genetically Modified		**IMAGE**	Integrated Model to Assess Global Environment
GMO	Genetically Modified Organism		**INDC**	Intended Nationally Determined Contribution
GMSL	Global Mean Sea Level		**INPE**	Instituto Nacional de Pesquisas Espaciais (National Institute for Space Research)
GMST	Global Mean Surface Temperature		**IOD**	Indian Ocean Dipole
GMT	Global Mean Temperature		**IPBES**	Intergovernmental Science-Policy Platform on Biodiversity and Ecosystem Services
GNR	Global Nutrition Report		**IPCC**	Intergovernmental Panel on Climate Change
GPP	Gross Primary Productivity/Gross Primary Production		**IPO**	Inter-decadal Pacific Oscillation
GWP	Global Warming Potential		**IQR**	Interquartile Range
Ha	Hectares		**IRD**	Integrated Rural Development
H$_2$	Hydrogen		**ISEAL**	ISEAL Alliance (International Social and Environmental Accreditation and Labelling Alliance)
HadCM3	Hadley Centre Coupled Model, version 3		**ISO**	International Organization for Standardization
HANPP	Human Appropriation of Net Primary Production		**ITPS**	Intergovernmental Technical Panel on Soils
HAPPI	Half a degree Additional warming, Projections, Prognosis and Impacts		**ITRDB**	International Tree Ring Data Bank
HDI	Human Development Index		**IUCN**	International Union for Conservation of Nature
HEI	Healthcare Environment Inspectorate		**IWAI**	Inland Waterways Authority of India
HHP	High Hydrostatic Pressure		**IWM**	Integrated Watershed Management
HKH	Hindu-Kush Himalayan		**JJA**	June-July-August
HLPE	High Level Panel of Experts			
HYDE	History Database of the Global Environment			

K	Potassium
km	Kilometres
kt	Kilotonnes
kWh	Kilowatt hours
L	Litres
LADA	Land Degradation Assessment in Drylands
LAI	Leaf Area Index
LAPA	Local Adaptation Plan of Action
LCA	Lifecycle Analysis/Life-Cycle Assessment
LCC	Land-Cover Conversions
LCCS	Land Cover Classification System
LDC	Least Developed Countries
LDN	Land Degradation Neutrality
LED	Low Energy Demand
LGP	Length of Growing Period
LiDAR	Light Detection and Ranging
LIMCOM	Limpopo Watercourse Commission
LK	Local Knowledge
LM	Land Management
LMIC	Low- and Middle-Income Country
LPG	Liquified Petroleum Gas
LSAT	Land Surface Air Temperature
LSLA	Large-Scale Land Acquisition
LTKA	Local and Traditional Knowledge in Agriculture
LUC	Land Use Change
LUH2	Harmonised Land Use Change Data
LULCC	Land Use Land Cover Change
LULUCF	Land Use, Land Use Change and Forestry
MA	Millennium Ecosystem Assessment
MAC	Marginal Abatement Cost
MACC	Marginal Abatement Cost Curve
MAgPIE	Model of Agricultural Production and its Impact on the Environment
MAM	March-April-May
MAR	Managed Aquifer Recharge
MCF	Methyl Chloroform
MCS	Mesoscale Convective Systems
MEA	Millennium Ecosystem Assessment
MED	Middle Eastern Dust Storms

MERIS	Medium Resolution Imaging Spectrometer
MESSAGE	Model for Energy Supply Systems And their General Environmental impact
MFS	Microfinance Services
Mha	Megahectare
MIRWH	Mechanized Micro Rainwater Harvesting
MJ	Megajoules
MND	Micro Nutrient Deficiency
MODIS	Moderate Resolution Imaging Spectroradiometer
MOFPI	Ministry of Food Processing Industries
MRV	Monitoring, Reporting and Verification
MSA	Mean Species Abundance
Mt	Megatonnes
MTOE	Million Tonnes of Oil Equivalent
N	Nitrogen
N_2O	Nitrous Oxide
NAMA	Nationally Appropriate Mitigation Action
NAP	National Adaptation Plan
NAPA	National Adaptation Program of Action
NARC	Nepal Agricultural Research Council
NASA/GISS	US National Aeronautics and Space Administration
NBS	Nature Based Solutions
NCE	New Climate Economy
NCD	Non-Communicable Diseases
NCP	Nature's Contributions to People
NDC	Nationally Determined Contribution
NDVI	Normalised Difference Vegetation Index
NENA	Near East and North Africa
NEON	National Science Foundation's National Ecological Observatory Network
NEP	National Energy Programme
NET	Negative Emission Technology
NFPP	National Forest Protection Program
NGO	Non-Governmental Organisation
NH_3	Ammonia
NIAB	National Institute of Agricultural Botany
NOAA	US National Oceanic and Atmospheric Administration

All

All

NOAA AVHRR	US National Oceanic and Atmospheric Administration Advanced Very High Resolution Radiometer	**PPCDAm**	Interministerial Plan of Action for Prevention and Control of Deforestation of the Legal Amazon
NOAA ESRL	US National Oceanic and Atmospheric Administration Earth System Research Laboratory	**PPI**	Plant Phenology Index
		PRIMAP	Potsdam Real-time Integrated Model for the probabilistic Assessment of emission Paths
NPP	Net Primary Production/Net Primary Productivity	**RAP**	Representative Agricultural Pathways
NPV	Net Present Value	**RCM**	Regional Climate Models
NRC	National Research Council	**RCP**	Representative Concentration Pathway
NRDC	Natural Resources Defence Council	**REDD**	Reducing Emissions from Deforestation and Degradation
NTFP	Non-Timber Forest Product	**REDD+**	Reducing Emissions from Deforestation and Degradation and the role of conservation, sustainable management of forests and enhancement of forest carbon stocks in developing countries
NUS	Neglected and Underutilised Species		
O_2	Oxygen		
O_3	Ozone		
OC	Organic Carbon		
OECD	Organisation for Economic Co-operation and Development	**RegCM**	Regional Climate Model
OH	Hydroxyl Radical	**REMIND**	REgional Model of INvestments and Development
ORCHIDEE	Organising Carbon and Hydrology In Dynamic Ecosystems	**REN21**	Renewable Energy Policy Network for the 21st Century
OSS	Sahara and Sahel Observatory	**RF**	Radiative Forcing
P	Precipitation or Phosphorus	**RESTREND**	Residual Trends
PAF	Pilot Auction Facility	**RIL**	Reduced Impact Logging
PAGGW	Pan-African Agency of the Great Green Wall	**ROSCA**	Rotating Saving and Credit Association
PAH	Polycyclic Aromatic Hydrocarbon	**RPDS**	Research Program on Dryland Systems
PAYT	Pay As You Throw	**RUE**	Rain Use Efficiency
PBAP	Primary Biological Aerosol Particles	**RUSLE**	Revised Universal Soil Loss Equation
PCP	Precipitation	**RVI**	Reconstructed Vegetation Index
PCRAFI	Pacific Catastrophe Risk Assessment and Financing Initiative	**RWH**	Rainwater Harvesting
		RX1day	Annual Maximum 1-day Precipitation
PEFC	Programme for the Endorsement of Forest Certification	**SAH**	Sahara
		SAS	South Asia
PES	Payment for Environmental Services/Payment for Ecosystem Services	**SAT**	Surface Air Temperature
		SCC	Social Cost of Carbon
PET	Potential Evapotranspiration	**SD**	Standard Deviation or Sustainable Development
PHL	Post Harvest Losses		
PICS	Purdue Improved Crop Storage	**SDG**	Sustainable Development Goal
PM	Particulate Matter	**SDS-WAS**	Sand and Dust Storm Warning Advisory and Assessment System
PM2.5	Particulate matter with size less than 2.5 μm		
PoU	Prevalence of Undernourishment	**SERFOR**	National Forest Service and Wildlife Authority
PPA	Power Purchase Agreements	**SES**	Socio-Ecological Systems
		SFM	Sustainable Forest Management

SFSC	Shortening Food Supply Chain		**TSS**	Time Series Segmentation
SI	Sustainable Intensification		**UGI**	Urban Green Infrastructure
SIB	State Investment Bank		**UHI**	Urban Heat Island
SICAR	National Land Registry System		**UNCBD**	United Nations Convention on Biodiversity
SIDS	Small Island Developing States		**UNCCD**	United Nations Convention to Combat Desertification
SLCF	Short-Lived Climate Forcer		**UNCED**	United Nations Conference on Environment and Development
SLCP	Short Lived Climate Pollutant			
SLM	Sustainable Land Management		**UNCTAD**	United Nations Conference on Trade and Development
SM	Supplementary Materials			
SO₂	Sulfur dioxide		**UNDP**	United Nations Development Programme
SOA	Secondary Organic Aerosols		**UN-EMG**	UN Environment Management Group
SOC	Soil Organic Carbon		**UNEP**	United Nations Environment Programme
SOFI	The State of Food Security and Nutrition in the World		**UNEP-GEF**	UN Environment Global Environment Facility
SOM	Soil Organic Matter		**UNFCCC**	United Nations Framework Convention on Climate Change
SON	September-October-November		**UNICEF**	United Nations Children's Fund
SPOT VGT	Satellite Pour l'Observation de la Terre Vegetation (Satellite for the Observation of the Earth Vegetation)		**UNISDR**	UN Sendai Framework for Disaster Risk Reduction
SR15	IPCC Special Report on Global Warming of 1.5°C		**UN-REDD**	United Nations Collaborative Programme on Reducing Emissions fom Deforestation and Forest Degradation in Developing Countries
SRCCL	IPCC Special Report on Climate Change and Land		**UPAF**	Urban and Peri-urban Agriculture and Forestry
SRES	IPCC Special Report on Emission Scenarios		**USAID**	United States Agency for International Development
SREX	IPCC Special Report on Managing the Risks of Extreme Events and Disasters to Advance Climate Change Adaptation		**USD**	United States Dollars
			USDA	United States Department of Agriculture
SR-LULUCF	IPCC Special Report on Land Use, Land-Use Change and Forestry		**USEPA**	United States Environmental Protection Agency
SROCC	IPCC Special Report on the Ocean Cryosphere and Climate Change		**UTFI**	Underground Taming of Floods for Irrigation
			UV	Ultraviolet
SSP	Shared Socio-economic Pathways		**VGGT**	Voluntary Guidelines on the Responsible Governance of Tenure of Land, Fisheries and Forests in the Context of National Food Security
SSSA	Soil Science Society of America			
SST	Sea Surface Temperature			
SWAT	Soil and Water Assessment Tool		**VOC**	Volatile Organic Compound
SWC	Soil and Water Conservation		**VOD**	Vegetation Optical Depth
SYR	IPCC Synthesis Report		**w/**	With
t	Tonnes		**w/o**	Without
TCR	Transient Climate Response		**W**	Watts
TEEB	The Economics of Ecosystems and Biodiversity		**WAF**	West Africa
Tg	Teragrams		**WaTEM/SEDEM**	Spatially distributed sediment delivery model combining the WaTEM and SEDEM models
TS	Technical Summary			

All

WBA	World Bioenergy Association
WBCSD	World Business Council for Sustainable Development
WEC	World Energy Council
WEF	World Economic Forum
WEO	World Energy Outlook
WEPP	Water Erosion Prediction Project
WET	Wetland Extent Trends
WFP	World Food Programme
WGI	IPCC Working Group I
WGII	IPCC Working Group II
WGIII	IPCC Working Group III
WHO	World Health Organisation
WMO	World Meteorological Organisation
WOCAT	World Overview of Conservation Approaches and Technologies
WRI	World Resources Institute
WSOA	Water Soluble Organic Compounds
WTO	World Trade Organisation
WUE	Water Use Efficiency
yr	Year
ZNLD	Zero Net Land Degradation

Annex III: Contributors to the IPCC Special Report on Climate Change and Land

This annex should be cited as:

IPCC, 2019: Annex III: Contributors to the IPCC Special Report on Climate Change and Land. In: *Climate Change and Land: an IPCC special report on climate change, desertification, land degradation, sustainable land management, food security, and greenhouse gas fluxes in terrestrial ecosystems* [P.R. Shukla, J. Skea, E. Calvo Buendia, V. Masson-Delmotte, H.-O. Pörtner, D. C. Roberts, P. Zhai, R. Slade, S. Connors, R. van Diemen, M. Ferrat, E. Haughey, S. Luz, S. Neogi, M. Pathak, J. Petzold, J. Portugal Pereira, P. Vyas, E. Huntley, K. Kissick, M. Belkacemi, J. Malley, (eds.)]. https://doi.org/10.1017/9781009157988.012

ABDULLA, Amjad
Vice-Chair IPCC WGIII
Maldives

AGUS, Fahmuddin
Indonesian Soil Research Institute
Indonesia

AKHTAR-SCHUSTER, Mariam
DLR Project Management Agency, German
Aerospace Center
Germany

ALDRIAN, Edvin
Vice-Chair IPCC WGI
Indonesia

ALEXANDER, Peter
School of Geosciences, University of
Edinburgh
United Kingdom

ALMAZROUI, Mansour
King Abdulaziz University
Saudi Arabia

ALOUI, Hamda
Ministry of Local Affairs and Environment
Tunisia

ANDEREGG, William
University of Utah
The United States of America

ARIAS-NAVARRO, Cristina
French National Institute for Agricultural
Research
Spain

ARMSTRONG, Edward
University of Bristol
United Kingdom

ARNETH, Almut
Karlsruhe Institute of Technology
Germany

ARTAXO, Paulo
University of São Paulo
Brazil

AWAN, Abdul Rasul
Nuclear Institute for Agriculture and Biology
Pakistan

BAI, Yuping
Institute of Geographic Sciences and Natural
Resources Research, Chinese Academy of
Sciences
China

BAILIS, Rob
Stockholm Environment Institute
The United States of America

BAPTISTE, Brigitte
Alexander von Humboldt Biological
Resources Research Institute
Colombia

BARBOSA, Humberto
Federal University of Alagoas
Brazil

BARIONI, Luis
Embrapa Agricultural Informatics
Brazil

BASTOS, Ana
Ludwig Maximilian University of Munich
Portugal/Germany

BELLARD, Céline
University of Paris Sud
France

BENKEBLIA, Noureddine
The University of the West Indies
Jamaica

BENTON, Tim
University of Leeds
United Kingdom

BERNIER, Pierre
Natural Resources Canada, Canadian Forest
Service
Canada

BERNSTEN, Terje Koren
University of Oslo
Norway

BHADWAL, Suruchi
The Energy and Resources Institute
India

BLANCHARD, Julia
University of Tasmania
Canada/Australia

BOWMAN, Kerry
University of Toronto
Canada

BURRELL, Arden
University of Leicester
Australia

BYERS, Edward
International Institute for Applied Systems
Analysis
Austria/Brazil

CAI, Peng
Xinjiang Institute of Ecology and Geography,
Chinese Academy of Science
China

CALVIN, Katherine
Pacific Northwest National Laboratory
The United States of America

CALVO, Eduardo
Vice-Chair IPCC TFI
Peru

CAMPBELL, Donovan
University of the West Indies
Jamaica

CAVALETT, Otavio
The Norwegian University of Science and
Technology
Brazil

CHALLINOR, Andrew
University of Leeds
United Kingdom

CHERUBINI, Francesco
The Norwegian University of Science and
Technology
Italy

CONNORS, Sarah
TSU IPCC WGI, University of Paris-Saclay
France/United Kingdom

CONTRERAS, Erik Mencos
Columbia University
The United States of America/Mexico

COWIE, Annette
University of New England
Australia

CREWS, Timothy
The Land Institute, Kansas
The United States of America

DALLIMER, Martin
University of Leeds
United Kingdom

DAVIN, Edouard
ETH Zurich
France/Switzerland

DE KLEIN, Cecile
AgResearch
New Zealand

DE NOBLET-DUCOUDRÉ, Nathalie
Laboratoire des Sciences du Climat et de
l'Environnement
France

DEBONNE, Niels
Vrije Universiteit Amsterdam
The Netherlands

DELUSCA, Kenel
Ministry of Environment
Haiti

DENTON, Fatima
United Nations University
The Gambia

DIAZ-CHAVEZ, Rocio
Stockholm Environment Institute
Mexico

DRIOUECH, Fatima
Vice-Chair IPCC WGI, Mohammed VI
Polytechnic University
Morocco

EEKHOUT, Joris
Spanish National Research Council
The Netherlands

EL-ASKARY, Hesham
Chapman University
Egypt

ELBEHRI, Aziz
Food and Agriculture Organization of the
United Nations
Morocco

ERB, Karlheinz
Institute of Social Ecology Vienna
Italy

ESPINOZA, Jhan Carlo
Instituto Geofísico del Perú
Peru

EVANS, Jason
University of New South Wales
Australia

FANZO, Jessica
Johns Hopkins
The United States of America

FERRAT, Marion
TSU IPCC WGIII, Imperial College London
France

FLETCHER, Amber
University of Regina
Canada

FLORES-RENTERÍA, Dulce
Centre for Research and Advanced Studies
of the National Polytechnic Institute
Mexico

FORD, James
Priestley International Centre for Climate,
University of Leeds
United Kingdom

FRANK, Stefan
International Institute for Applied Systems
Analysis
Austria

FUGLESTVEDT, Jan
Vice-Chair IPCC WGI, Centre for
International Climate and Environmental
Research
Norway

GARCÍA-OLIVA, Felipe
Universidad Nacional Autónoma de México
Mexico

GRANT, Sean Patrick
Indiana University, Richard M. Fairbanks
School of Public Health
The United States of America

GRASSI, Giacomo
European Commission Joint Research Centre
Italy/European Union

HAMDI, Rafiq
Royal Meteorological Institute of Belgium
Belgium

HAROLD, Jordan
University of East Anglia
United Kingdom

HAUGHEY, Eamon
TSU IPCC WGIII, Trinity College Dublin
Ireland

HERMANS, Kathleen
Helmholtz Centre for Environmental
Research
Germany

HERRERO, Mario
Commonwealth Scientific and Industrial
Research Organisation
Costa Rica

HETEM, Robyn
University of the Witwatersrand
South Africa

HILLERBRAND, Rafaela
Karlsruhe Institute of Technology
Germany

HOUGHTON, Richard
Woods Hole Research Center
The United States of America

HOUSE, Joanna I.
University of Bristol
United Kingdom

HOWDEN, Mark
Vice-Chair IPCC WGII, Australian National
University
Australia

HUMPENÖDER, Florian
Potsdam Institute for Climate Impact
Research
Germany

AIII

HUPPMANN, Daniel
International Institute for Applied Systems
Analysis
Austria

HURLBERT, Margot
Johnson Shoyama Graduate School of Public
Policy, University of Regina
Canada

HUSSEIN, Ismail Abdel Galil
Ministry of Agriculture and Land
Reclamation
Egypt

IQBAL, Muhammad Mohsin
Global Change Impact Studies Centre
Pakistan

JANZ, Baldur
Karlsruhe Institute of Technology
Germany

JIA, Gensuo
Institute of Atmospheric Physics, Chinese
Academy of Sciences
China

JOBBAGY, Esteban
Universidad Nacional de San Luis
Argentina

JOHANSEN, Tom Gabriel
InfoDesignLab
Norway

JOHNSON, Francis X.
Stockholm Environment Institute
Sweden

KANTER, David
New York University
The United States of America

KASTNER, Thomas
Senckenberg Biodiversity and Climate
Research Centre
Austria

KHAN, Amanullah
Department of Agronomy, University of
Agriculture Peshawar
Pakistan

KITAJIMA, Kaoru
Graduate School of Agriculture, Kyoto
University
Japan

KNOWLES, Tony
Cirrus Group
South Africa

KOROTKOV, Vladimir
Institute of Global Climate and Ecology
The Russian Federation

KRIEWALD, Steffen
Potsdam Institute for Climate Impact
Research
Germany

KRINNER, Gerhard
Institut des Géosciences de l'Environnement
France

KRISHNAPILLAI, Murukesan V.
College of Micronesia-FSM
India

KRISHNASWAMY, Jagdish
Ashoka Trust for Research in Ecology and
the Environment
India

KUBIK, Zaneta
University of Bonn
Poland

KURZ, Werner
Canadian Forest Service,
Natural Resources Canada
Canada

KUST, German
Institute of Geography and Moscow State
University
The Russian Federation

LANIGAN, Gary
Teagasc
Ireland

LE HOANG, Anh
Ministry of Agriculture and Rural
Development of Vietnam
Viet Nam

LEE, Woo-kyun
Korea University, Seoul
The Republic of Korea

LENNARD, Christopher
University of Cape Town
South Africa

LI, Diqiang
Chinese Academy of Forestry
China

LIWENGA, Emma
Institute of Resource Assessment, University
of Dar Es Salaam
The United Republic of Tanzania

LONGVA, Ylva
School of Geosciences,
University of Edinburgh
United Kingdom

LÓPEZ, Daniel
Fundación Entretantos
Spain

LÜDELING, Eike
University of Bonn
Germany

LWASA, Shuaib
Makerere University
Uganda

MAHADEVIA, Darshini
School of Arts and Sciences, Ahmedabad
University
India

MAHMOUD, Nagmeldin
Vice-Chair IPCC WGIII, Higher Council for
Environment and Natural Resources
Sudan

MANIALAWY, Yousef
University of Toronto
Canada

MASON-D'CROZ, Daniel
Commonwealth Scientific and Industrial
Research Organisation
The United States of America/Colombia

MASSON-DELMOTTE, Valérie
Co-Chair IPCC WGI, Laboratoire des Sciences
du Climat et de l'Environnement
France

MBOW, Cheikh
START-International
Senegal

AIII

MCCARL, Bruce
Texas A&M University
The United States of America

MCDERMID, Sonali
New York University
India/The United States of America

MCELWEE, Pamela
Rutgers University
The United States of America

MEIJER, Johan
Netherlands Environmental Assessment
Agency
The Netherlands

MENA, Carlos Fernando
Universidad San Francisco de Quito
Ecuador

MEYFROIDT, Patrick
Earth and Life Institute, Université
Catholique de Louvain
Belgium

MEZA, Francisco
Pontificia Universidad Catolica de Chile
Chile

MICHAELIDES, Katerina
University of Bristol
Cyprus/United Kingdom

MIRZABAEV, Alisher
Center for Development Research, University
of Bonn
Germany/Uzbekistan

MOHAMMED, Ali
Desert Research Center-Cairo
Egypt

MORELLI, Angela
InfoDesignLab
Norway/Italy

MORENO, José Manuel
Department of Environmental Sciences,
University of Castilla-La Mancha
Spain

MORTON, John
Natural Resources Institute, University of
Greenwich
United Kingdom

MOUFOUMA OKIA, Wilfran
TSU IPCC WGI
France

MYEONG, Soojeong
Korea Environment Institute
The Republic of Korea

NARAYANAPPA, Devaraju
Laboratoire des Sciences du Climat et de
l'Environnement
India/France

NEDJRAOUI, Dalila
University of Science and Technology Houari
Boumediene
Algeria

NEOFOTIS, Peter
Michigan State University
The United States of America

NEOGI, Suvadip
TSU IPCC WGIII, Ahmedabad University
India

NKEM, Johnson
African Climate Policy Centre
Cameroon

NKONYA, Ephraim
International Food Policy Research Institute
The United Republic of Tanzania

NOBLE, Ian
Australian National University
Australia

OLSSON, Lennart
Lund University
Sweden

OSMAN ELASHA, Balgis
African Development Bank
Sudan

O'SULLIVAN, Michael
University of Exeter
United Kingdom

PANT, Laxmi
University of Guelph
Canada

PATHAK, Minal
TSU IPCC WGIII, Ahmedabad University
India

PEÑUELAS, Josep
National Research Council of Spain
Spain

PETZOLD, Jan
TSU IPCC WGII, Alfred-Wegener-Institut
Germany

PICHS-MADRUGA, Ramón
Vice-Chair IPCC WGIII, Centre for World
Economy Studies
Cuba

POLOCZANSKA, Elvira
TSU IPCC WGII, Alfred-Wegener-Institut
United Kingdom/Australia

POPP, Alexander
Potsdam Institute for Climate Impact
Research
Germany

PORTER, John R.
University of Copenhagen
United Kingdom/Denmark

PÖRTNER, Hans-Otto
Co-Chair IPCC WGII, Alfred-Wegener-Institut
Germany

PORTUGAL PEREIRA, Joana
TSU IPCC WGIII, Imperial College London
United Kingdom

PRADHAN, Prajal
Potsdam Institute for Climate Impact
Research
Nepal/Germany

QUAN, Julian
Natural Resources Institute,
University of Greenwich
United Kingdom

QUESADA, Benjamin
Universidad del Rosario
Colombia

RAHIMI, Mohammad
University of Semnan
Iran

RAJAGOPALAN, Balaji
University of Colorado
The United States of America

AIII

REISINGER, Andy
Vice-Chair IPCC WG III, New Zealand
Agricultural Greenhouse Gas Research
Centre
New Zealand

RENOU-WILSON, Florence
University Collge Dublin
Ireland

RENWICK, Alan
Lincoln University
New Zealand

RIVERA-FERRE, Marta Guadalupe
University of Vic - Central University of
Catalonia
Spain

ROBERTS, Debra C.
Co-Chair IPCC WGII, EThekwini Municipality
South Africa

RODRIGUES, Regina
Brazilian Agriculture Research Corporation
Brazil

RODRIGUES, Renato
Brazilian Agriculture Research Corporation
Brazil

RODRÍGUEZ-MORALES, Jorge E.
Pontifical Catholic University of Peru
Peru

ROE, Stephanie
University of Virginia
The Philippines/The United States of America

ROHDE, Robert A.
Berkeley Earth
The United States of America

ROSENZWEIG, Cynthia
NASA Goddard Institute for Space Studies
The United States of America

ROUNSEVELL, Mark
Karlsruhe Institute of Technology
United Kingdom

RUANE, Alexander
NASA Goddard Institute for Space Studies
The United States of America

SAIGUSA, Nobuko
Center for Global Environmental Research,
National Institute for Environmental Studies
Japan

SANKARAN, Mahesh
National Centre for Biological Sciences, Tata
Institute of Fundamental Research
India

SANZ SÁNCHEZ, María José
Basque Centre for Climate Change
Spain

SAPKOTA, Tek
International Maize and Wheat
Improvement Center
Nepal

SEMENOV, Sergey
Vice-Chair IPCC WGII, Institute of Global
Climate and Ecology
The Russian Federation

SHEVLIAKOVA, Elena
National Oceanic and Atmospheric
Administration
The United States of America

SHUKLA, Priyadarshi
Co-Chair IPCC WGIII, Ahmedabad University
India

SIRIN, Andrey
Institute of Forest Science, Russian Academy
of Sciences
The Russian Federation

SKEA, Jim
Co-Chair IPCC WGIII, Imperial College
London
United Kingdom

SLADE, Raphael
TSU IPCC WGIII, Imperial College London
United Kingdom

SLOT, Martijn
Smithsonian Tropical Research Institute
Panama

SMITH, Pete
University of Aberdeen
United Kingdom

SOKONA, Youba
Vice-Chair IPCC, South Centre
Mali

SOMMER, Rolf
International Center for Tropical Agriculture
Germany

SONWA, Denis Jean
Centre for International Forestry Research
Cameroon

SOUSSANA, Jean-Francois
French National Institute for Agricultural,
Environment and Food Research
France

SPENCE, Adrian
International Centre for Environmental and
Nuclear Sciences
Jamaica

SPORRE, Moa
Lund University
Norway

STRECK, Charlotte
Climate Focus
Germany

STRINGER, Lindsay
University of Leeds
United Kingdom

STROHMEIER, Stefan Martin
International Center for Agricultural
Research in the Dry Areas (ICARDA)
Austria

SUKUMAR, Raman
Indian Institute of Science
India

SULMAN, Benjamin
Oak Ridge National Laboratory
The United States of America

SYKES, Alasdair
Scotland's Rural College
United Kingdom

TABOADA, Miguel Angel
National Agricultural Technology Institute
Argentina

AIII

TENA, Fasil
Ethiopia Pilot Program for Country
Partnership
Ethiopia

TENGBERG, Anna
Lund University
Sweeden

THIERY, Wim
Vrije Universiteit Brussel
Belgium

TUBIELLO, Francesco
Food and Agriculture Organization of the
United Nations
The United States of America/Italy

TÜRKEŞ, Murat
Center for Climate Change andPolicy
Studies, Boğaziçi University
Turkey

TURNER II, Billie
Arizona State University
The United States of America

VALENTINI, Riccardo
University of Tuscia
Italy

VAN DER ESCH, Stefan
Netherlands Environmental Assessment
Agency
The Netherlands

VAN DIEMEN, Renée
TSU IPCC WGIII, Imperial College London
The Netherlands/United Kingdom

VÁZQUEZ MONTENEGRO, Ranses José
Instituto de Meteorología
Cuba

VERA, Carolina
Vice-Chair IPCC WGI
Argentina

VERCHOT, Louis
International Center for Tropical Agriculture
Colombia/The United States of America

VILLAMOR, Grace
Scion – New Zealand
The Philippines

VINCENT, Katharine
Kulima Integrated Development Solutions
(Pty) Ltd
South Africa

VINER, David
Mott MacDonald
United Kingdom

VIZZARRI, Matteo
European Commission Joint Research Centre
Italy

WAHA, Katharina
Commonwealth Scientific and Industrial
Research Organisation
Germany/Australia

WARNER, Alan
University of Toronto
Barbados

WARNER, Koko
UNFCCC Adaptation Programme,
Subprogramme Climate Impacts,
Vulnerability, and Impacts
The United States of America

WELTZ, Mark
United States Department of Agriculture,
Agricultural Research Service
The United States of America

WEYER, Nora. M
TSU IPCC WGII, Alfred-Wegener-Institut
Germany

WILLIAMSON, Phil
The University of East Anglia
United Kingdom

WREFORD, Anita
New Zealand Forest Research Institute
New Zealand

WU, Jianguo
Chinese Research Academy
of Environmental Sciences
China

XU, Yinlong
Institute of Environment and Sustainable
Development in Agriculture, Chinese
Academy of Agricultural Sciences
China

YAMAGATA, Yoshiki
National Institute for Environmental Studies
Japan

YASSAA, Noureddine
Vice-Chair IPCC WGI, University of Science
and Technology Houari Boumediene
Algeria

ZAKIELDEEN, Sumaya
University of Khartoum
Sudan

ZATARI, Taha
Vice-Chair IPCC WGII
Saudi Arabia

ZHAI, Panmao
Co-Chair IPCC WGI
China

ZHOU, Yuyu
Iowa State University
China

ZOMMERS, Zinta
Food and Agriculture Organization of the
United Nations
Latvia

AIII

Annex IV: Reviewers of the IPCC Special Report on Climate Change and Land

This annex should be cited as:

IPCC, 2019: Annex IV: Reviewers of the IPCC Special Report on Climate Change and Land. In: *Climate Change and Land: an IPCC special report on climate change, desertification, land degradation, sustainable land management, food security, and greenhouse gas fluxes in terrestrial ecosystems* [P.R. Shukla, J. Skea, E. Calvo Buendia, V. Masson-Delmotte, H.-O. Pörtner, D. C. Roberts, P. Zhai, R. Slade, S. Connors, R. van Diemen, M. Ferrat, E. Haughey, S. Luz, S. Neogi, M. Pathak, J. Petzold, J. Portugal Pereira, P. Vyas, E. Huntley, K. Kissick, M. Belkacemi, J. Malley, (eds.)]. https://doi.org/10.1017/9781009157988.013

ABABNEH, Linah
The Swedish University of Agricultural
Sciences and Cornell University
Sweden

ABAYAZID, Hala
National Water Research Center
Egypt

ABELDAÑO, Roberto
Universidad de la Sierra Sur
Mexico

ACÁCIO, Vanda
Instituto Superior de Agronomia, University
of Lisbon
Portugal

ACEVEDO, Alberto
Instituto Nacional de Tecnología
Agropecuaria
Argentina

ACHARYA, Prasannalakshmi (Lakshmi)
Massey University
New Zealand

ACIKGOZ, Nazimi
Ege University
Turkey

ADEAGA, Olusegun
University of Lagos
Nigeria

ADEGOKE, ABIODUN
Obafemi Awolowo University
Nigeria

ADLAN, Asia
University of Khartoum
Sudan

ADOJOH, Onema
Missouri University of Science and
Technology
The United States of America

AGRAWAL, Mahak
International Society of City and Regional
Planners
India

AGUIRRE-VILLEGAS, Horacio
University of Wisconsin-Madison
The United States of America

AHMED, Essam Hassan Mohamed
NSfCS Egypt
The United States of America

AHO, Hanna
Fern
Belgium

AKCA, Erhan
Adiyaman University
Turkey

AKHTAR-SCHUSTER, Mariam
DLR Project Management Agency, German
Aerospace Center
Germany

AL MADHOUN, Wesam
Universiti Teknologi Petronas
Malaysia

ALABI, Ayotomiwa
Centre for Petroleum Energy Economics and
Law, University of Ibadan
Nigeria

ALCANTARA CERVANTES, Viridiana
Federal Office for Agriculture and Food
Germany

ALEXANDROV, Georgii
A.M. Obukhov Institute of Atmospheric
Physics of Russian Academy of Sciences
The Russian Federation

AL-HASANI, Alaa
Ministry of Agriculture
Iraq

ALMAGRO BONMATÍ, María
Basque Centre for Climate Change
Spain

AMADOU, Mahamadou Laouali
AGRHYMET
The Niger

AMANI, Abou
United Nations Educational, Scientific and
Cultural Organization
France

AMBINAKUDIGE, Shrinidhi
Mississippi State University
The United States of America

AMPARO MARTINEZ ARROYO, Maria
Directora General del Instituto Nacional de
Ecologia y Cambio Climatico
Mexico

ANDRESEN, Louise
University of Gothenburg
Sweden

ANNOR, Thompson
Kwame Nkrumah University of Science and
Technology, Kumasi
Ghana

ANORUO, Chukwuma
Imo State University Owerri and University
of Nigeria
Nigeria

ARIBO, Lawrence
Uganda National Meteorological Authority
Uganda

ASIZUA, Denis
National Agricultural Research Organisation
Uganda

ASLAM, Hasnat
Flood Emergency Resilience and
Reconstruction Project, University
of the Punjab
Pakistan

ATIENO, Lucy
Sustainable Travel and Tourism Agenda
Kenya

AUBINET, Marc
University of Liege
Belgium

AUSSEIL, Anne-Gaelle
Manaaki Whenua Landcare Research
New Zealand

AUSTRHEIM, Gunnar
Norwegian University of Science and
Technology
Norway

AYANLADE, Sina
Obafemi Awolowo University, Ile-Ife
Nigeria

BAA, Ojong.E nee Enokenwa
Rhodes University
South Africa

BABIKER, Mustafa
Saudi Aramco
Saudi Arabia

BADI, Wafae
Moroccan Met Service
Morocco

BAEK, Aram
Korea Meteorological Administration
The Republic of Korea

BAJZELJ, Bojana
The Waste and Resources Action Programme
United Kingdom

BAKHTIARI, Fatemeh
Consultant
Denmark

BALKANSKI, Yves
Laboratory of Climate Science and
Environment
France

BARBIERI, Lindsay
University of Vermont, Gund Institute for
the Environment and CGIAR program on
Climate Change Agriculture and Food
Security
The United States of America

BATTIPAGLIA, Giovanna
University of Campania Luigi Vanvitelli
Italy

BAUMEL, Alex
Aix-Marseille University
France

BELLASSEN, Valentin
The Institut National de la Recherche
Agronomique
France

BELLÙ, Lorenzo Giovanni
Food and Agriculture Organization of the
United Nations
Italy

BENALI, Abdel-Hai
El-oued University
Algeria

BENAVIDES-MENDOZA, Adalberto
Autonomous Agrarian
University Antonio Narro
Mexico

BENJAMINSEN, Tor A.
Norwegian University of Life Sciences
Norway

BENKEBLIA, Noureddine
The University of the West Indies
Jamaica

BERK, Marcel
Ministry of Economic Affairs and Climate
Policy
The Netherlands

BERLINER, Derek
University of Cape Town
South Africa

BERNIER, Pierre
Canadian Forest Service
Canada

BERNOUX, Martial
Food and Agriculture Organization of the
United Nations
Italy

BERNTSEN, Terje
University of Oslo
Norway

BERTRAND, Guillaume
Université Franche Comté
France

BHATT, J.R
Ministry of Environment, Forest
and Climate Change
India

BJERKE, Jarle W.
Norwegian Institute for Nature Research
Norway

BLANCO-SEPULVEDA, Rafael
University of Malaga
Spain

BLARD, Pierre-Henri
Université de Lorraine and Université Libre
de Bruxelles
France

BOO, Kyung-On
Korea Meteorological Administration
The Republic of Korea

BOOTH, Mary
Partnership for Policy Integrity
The United States of America

BOREVITZ, Justin
Australian National University
Australia

BOWMAN, Kevin
Jet Propulsion Laboratory
The United States of America

BRAGHIERE, Renato
The Institut National de la Recherche
Agronomique
France

BRAHMA, Biplab
Assam University
India

BRENNA, Stefano
Regional Agency for Agriculture and Forests
of Lombardy
Italy

BROCCA, Luca
National Research Council of Italy
Italy

BRÜMMER, Bernhard
University of Göttingen
Germany

BRUN, Eric
Ministère de la Transition écologique
et solidaire
France

BRYN, Anders
Natural History Museum and
University of Oslo
Norway

BURBA, George
University of Nebraska and LI-COR
Biosciences
The United States of America

CAI, Rongshuo
Third Institute of Oceanography,
Ministry of Natural Resoureces
China

AIV

CAMPBELL, Kristin
Institute for Governance and Sustainable
Development
The United States of America

CARRIL, Andrea Fabiana
Centro de Investigaciones del Mar
y de la Atmósfera – CONICET UBA
Argentina

CASCONE, Carmela
Italian National Institute for Environmental
Protection and Research
Italy

CASTRUCCI, Luca
Pacific Northwest National Laboratory
The United States of America

CAYUELA, Maria Luz
Spanish National Research Council
Spain

CESARIO, Manuel
Academia Magdalena
Brazil

CESCHIA, Eric
Centre d'Etudes Spatiales de la BIOsphère
France

CHANG, Ching-Cheng
Institute of Economics, Academia Sinica
China

CHANG'A, Ladislaus
Tanzania Meteorological Agency
The United Republic of Tanzania

CHARLERY, Debra
Ministry of Education, Innovation, Gender
Relations and Sustainable Development
Saint Lucia

CHEN, Dexiang
Research Institute of Tropical Forestry,
Chinese Academy of Forestry
China

CHENG, Kun
Nanjing Agricultural University
China

CHERCHI, Annalisa
Euro-Mediterranean Center for Climate
Change – National Institute of Geophysics
and Volcanology
Italy

CHERYL, Jeffers
Department of Environment, Ministry of
Agriculture, Marine Resources, Cooperatives,
Environment and Human Settlements
Saint Kitts and Nevis

CHIARELLI, Paulo José
Ministry of External Relations of Brazil
Brazil

CHINTALA, Rajesh
Innovation Center for United States Dairy
The United States of America

CHOTTE, Jean-Luc
French National Research Institute for
Sustainable Development
France

CHRISTENSEN, Tina
Danish Meteorological Institute
Denmark

CHRISTENSEN, Knud
Fremsyn
Denmark

CHRISTOPHERSEN, Oyvind
Norwegian Pollution Control Authority
Norway

CLAEYS, Florian
French Ministry of Agriculture and Food
France

CLAYTON, Susan
The College of Wooster
The United States of America

COLLINS, William
University of Reading
United Kingdom

CONNORS, Sarah
TSU IPCC WGI, University of Paris Saclay
France

COOK, Jolene
Department for Business, Energy and
Industrial Strategy
United Kingdom

COOPER, David
Centre on Biological Diversity
Canada

CORTÉS, Amparo
University of Barcelona
Spain

COSENTINO, Vanina Rosa Noemí
National Institute of Agricultural Technology
Argentina

COURAULT, Romain
Sorbonne Université Faculté des Lettres
France

COWIE, Annette
New South Wales Department
of Primary Industries
Australia

CRAIG, Marlies
TSU IPCC WGII
South Africa

CRUZ DE CARVALHO, Maria Helena
Paris-Est Créteil University
France

CUDENNEC, Christophe
Agrocampus Ouest
France

CUI, Xuefeng
Beijing Normal University
China

DAIOGLOU, Vassilis
PBL Netherlands Environmental Assessment
Agency
The Netherlands

DAKA, Julius
Zambia Institute of Environmental
Management
Zambia

DALE, Daniel Danano
Food and Agriculture Organization of the
United Nations
Italy

DALEI, Narendra
University of Petroleum and Energy Studies
India

DAMEN, Beau
Food and Agriculture Organization of the
United Nations
Thailand

DAMSKI, Juhani
Finnish Meteorological Institute
Finland

DARWISH, Talal
National Center for Remote Sensing
Lebanon

DAVIES, Elizabeth Penelope
Ford Foundation
The United States of America

DAVIN, Edouard
ETH Zurich
Switzerland

DDAMULIRA, Robert
University of Delaware
The United States of America

DE GUZMAN, Emmanuel M.
Climate Change Commission
The Philippines

DE KLEIN, Cecile
AgResearch
New Zealand

DE LA VEGA NAVARRO, Angel
National Autonomous University of Mexico
Mexico

DE VENTE, Joris
Spanish National Research Council
Spain

DEISSENBERG, Christophe
Aix-Marseille University
Luxembourg

DELLASALA, Dominick
Consultant
The United States of America

DENG, Xiangzheng
Chinese Academy of Sciences
China

DERMAUX, Valerie
Ministry of Agriculture
France

DERYNG, Delphine
Climate Analytics
Germany

DETCHON, Reid
United Nations Foundation
The United States of America

DEVANEY, John
Trinity College Dublin
Ireland

DI GREGORIO, Monica
University of Leeds
United Kingdom

DI VITTORIO, Alan
Lawrence Berkeley National Laboratory
The United States of America

DIEGO, Elizabeth
Water Resources Authority
Kenya

DIGA, Girma
Ethiopian Institute of Agricultural Research
Ethiopia

DIRMEYER, Paul
George Mason University
The United States of America

DIXON, John
Australian National University
Australia

DJEMOUAI, Kamal
Consultant
Algeria

DOMÍNGUEZ-NÚÑEZ, José Alfonso
Technical University of Madrid
Spain

DOMKE, Grant
United States Department of Agriculture
Forest Service
The United States of America

DOYLE, Moira
University of Buenos Aires
Argentina

DUBE, Lokesh Chandra
Ministry of Environment, Forest
and Climate Change
India

DUGAN, Andrew
Drax Group plc
United Kingdom

DUMANSKI, Julian
Consultant
Canada

DUMBLE, Paul
Paul's Environment Ltd
United Kingdom

DUPAR, Mairi
Overseas Development Institute
United Kingdom

DUVEILLER, Gregory
European Commission Joint
Research Centre
Italy

EEKHOUT, Joris
Superior Council of Scientific Investigations
Spain

EHARA, Makoto
Forestry and Forest Products Research
Institute
Japan

EHRENSPERGER, Albrecht
University of Bern
Switzerland

EL RAEY, Mohamed
Institute of Graduate Studies and Research,
Alexandria University
Egypt

ELBEHRI, Aziz
Food and Agriculture Organization
of the United Nations
Italy

ELHAG, Mustafa
National Center for Research
Sudan

ELJADID, Ali Geath
University of Tripoli
Libya

ELOUISSI, Abdelkader
University of Mascara
Algeria

AIV

ENGELBRECHT, Francois
Global Change Institute, University
of the Witwatersrand
South Africa

ENGSTRÖM, Erik
Swedish Meteorological and Hydrological
Institute
Sweden

ERB, Karlheinz
University of Natural Resources and Life
Sciences, Institue of Social Ecology
Austria

ERLEWEIN, Alexander
German Society for International
Cooperation
Germany

ERŞAHIN, Sabit
Department of Forest Engineering, Çankırı
Karatekin University
Turkey

ESPINOZA, Jhan Carlo
Geographical Institute of Peru
France/Peru

ESSAM, Heggy
University of Southern California
The United States of America

ETKIN, David
York University
Canada

ETZION, Dror
McGill University
Canada

FALGE, Eva
German Weather Service
Germany

FAN, Jingli
China University of Mining and Technology,
Beijing
China

FARHAN, Kiran
The Urban Unit
Pakistan

FARRELL, Aidan
The University of the West Indies
Trinidad and Tobago

FEDERICI, Sandro
Food and Agriculture Organization
of the United Nations
Italy

FELICIA, Wu
Michigan State University
The United States of America

FELICIEN, Ana
Venezuelan Institute for Scientific Research
Venezuela

FENSHOLT, Rasmus
University of Copenhagen
Denmark

FERRARA, Vincenza
Azienda Agricola DORA di Vincenza Ferrara
Italy

FILECCIA, Turi
Consultant
Italy

FINLAY, Kerri
University of Regina
Canada

FISCHLIN, Andreas
Vice-Chair IPCC WGII and ETH Zurich
Switzerland

FLUCK, Hannah
Historic England
United Kingdom

FORSYTH, Timothy
London School of Economics and Political
Science
United Kingdom

FRA.PALEO, Urbano
University of Extremadura
Spain

FRANCAVIGLIA, Rosa
Agriculture and Environment Research
Center
Italy

FRANCESCO, Nocera
University of Catania
Italy

FUGLESTVEDT, Jan
Vice-Chair IPCC WGI and Centre
for International Climate Research
Norway

GALOS, Borbala
University of Sopron
Hungary

GAYO, Eugenia
Center for Climate and the Resilience
Research (CR)2
Chile

GEDEN, Oliver
German Institute for International
and Security Affairs
Germany

GEHL, Georges
Ministry of Sustainable Development
and Infrastructure, Department of the
Environment
Luxembourg

GEORGE, Burba
LI-COR Biosciences and University
of Nebraska
The United States of America

GHAHREMAN, Nozar
University of Tehran
Iran

GIBEK, Jakub
Ministry of Environment, Department
of Air Protection and Climate
Poland

GIGER, Markus
University of Bern
Switzerland

GIRKIN, Nicholas
Teagasc
Ireland

GLASER, Paul
University of Minnesota
The United States of America

GOHEER, Muhammad
Global Change Impact Studies Centre
Pakistan

GONZALEZ, Patrick
University of California, Berkeley
The United States of America

GRAF, Alexander
Forschungszentrum Jülich, Institute for Bio-
and Geosciences, Agrosphere (IBG 3)
Germany

GRIFFITHS, Thomas
Forest Peoples Programme
United Kingdom

GRILLAKIS, Manolis
Technical University of Crete
Greece

GROVER, Samantha
RMIT University
Australia

GRUJIC, Gordana
Oasis
Serbia

GUINEY, Itchell
Department of Environmental Affairs
South Africa

HABERL, Helmut
University of Natural Resources
and Life Sciences
Austria

HAINES, Anna
University of Wisconsin
The United States of America

HALL, Natasha
Middlesex University
United Kingdom

HALOFSKY, Jessica
University of Washington and United States
Department of Agriculture Forest Service
The United States of America

HAMADAIN, Nagla
Director of Department in Forests
Management
Sudan

HAMDI, Rafiq
Royal Meteorological Institute of Belgium
Belgium

HAMZAWI, Nancy
Environment and Climate Change Canada
Canada

HASHIMOTO, Shoji
Forestry and Forest Products Research
Institute
Japan

HASSANI, Samir
Renewable Energy Development Center
Algeria

HAUGEN, Jon Magnar
Norwegian Ministry of Agriculture and Food
Norway

HAUGHEY, Eamon
TSU IPCC WGIII, Trinity College Dublin
Ireland

HAVALIGI, Neeraja
Oregon State University and Greater
Portland Sustainability Education Network
The United States of America

HAVERD, Vanessa
Commonwealth Scientific and Industrial
Research Organisation
Australia

HE, Chunyang
Beijing Normal University
China

HEGERL, Gabriele
University of Edinburgh
United Kingdom

HEINRICH, Viola
University of Bristol
United Kingdom

HERBERT, Annika
University of Witwatersrand
South Africa

HIEDERER, Roland
European Commission Joint Research Centre
Italy

HIJBEEK, Renske
Wageningen University
The Netherlands

HILMI, Nathalie Jeanne Marie
Monaco Scientific Center
France

HIRATSUKA, Motoshi
Waseda University
Japan

HIRCHE, Azziz
University of Science and Technology –
Houari Boumediene
Algeria

HOSSAIN, Md Moazzem
Griffith University
Australia

HOWDEN, Mark
Vice-Chair IPCC WGII and Australian
National University
Australia

HUBERTY, Brian
United States Federal Government
The United States of America

HUDSON, Paul
University of Potsdam
Germany

HUMPHREYS, Stephen
London School of Economics
and Political Science
United Kingdom

HUNT, Maya
Ministry of Primary Industries
New Zealand

HUNTER, Nina
University of KwaZulu-Natal
South Africa

HUQIANG, Zhang
Australian Bureau of Meteorology
Australia

HURLBERT, Margot
Johnson-Shoyama Graduate School
of Public Policy
Canada

HURTADO ALBIR, Francisco Javier
European Patent Office
Germany

AIV

HUSZÁR, András
Ministry of Innovation and Technology,
Department for Climate Policy
Hungary

IBRAHIM, Tarig
The National Centre for Research
Sudan

IESE, Viliamu
The University of the South Pacific
Fiji

IGBINE, Lizzy
Nigerian Women Agro Allied Farmers
Association
Nigeria

IIZUMI, Toshichika
National Agriculture and Food Research
Organization
Japan

ILORI, Christopher
Simon Fraser University
Canada

IMAM, Ahmed
Desert Research Center
Egypt

INDRAWAN, Mochamad
University of Indonesia
Indonesia

IQBAL, Muhammad Mohsin
Global Change Impact Studies Center
Pakistan

ITO, Akihiko
National Institute for Environmental Studies
Japan

IVERSON, Louis
United States Department of Agriculture
Forest Service
The United States of America

JADRIJEVIC GIRARDI, Martiza
Ministry of Environment
Chile

JAFARI, Mostafa
Research Institute of Forests and
Rangelands, Agricultural Research,
Education and Extension Organization
Iran

JANOT, Noémie
University of Lorraine
France

JAYANARAYANAN, Sanjay
Indian Institute of Tropical Meteorology
India

JIANJUN, Huai
Northwest Agriculture and Forestry
University
China

JOHANSSON, Maria Ulrika
Stockholm University, Department of
Ecology, Environment and Plant Sciences
Sweden

JONARD, Mathieu
Université catholique de Louvain
Belgium

JONCKHEERE, Inge
Food and Agriculture Organization of the
United Nations
Italy

JONES, Chris
Met Office Hadley Centre
United Kingdom

JOSEPH, Shijo
Kerala Forest Research Institute
India

JOSHI, Kirti Kusum
Tribhuvan University
Nepal

JU, Hui
Chinese Academy of Agricultural Sciences
The Comoros

KABIDI, khadija
National Meteorological Direction
Morocco

KAHURI, Serah
Kenya Forest Service
Kenya

KAIMOWITZ, David
Ford Foundation
Nicaragua

KALLIOKOSKI, Tuomo
University of Helsinki
Finland

KANG, Xiaoming
Institute of Wetland Research, Chinese
Academy of Forestry
China

KANKAL, Bhushan
TSU IPCC WGIII, Ahmedabad University
India

KANNINEN, Markku
University of Helsinki
Finland

KAPOS, Valerie
UN Environment World Conservation
Monitoring Centre
United Kingdom

KARUNANAYAKE, A. K.
Department of Meteorology
Sri Lanka

KASWAMILA, Abiud
Dodoma University
The United Republic of Tanzania

KATE, Dooley
University of Melbourne
Australia

KENTARCHOS, Anastasios
European Commission
Belgium

KERNS, Becky
United States Department of Agriculture
Forest Service
The United States of America

KETO, Aila
Griffith University and Australian Rainforest
Conservation Society
Australia

KHALILZADEH, Ameneh
Atomic Energy Organization of Iran
Iran

KHAMAN, Azadeh
Department of the Environment
Iran

KHAN, Amanullah
The University of Agriculture, Peshawar
Pakistan

KHATTABI, Abdellatif
Ecole Nationale Forestiere d'ingenieurs
Morocco

KHESHGI, Haroon
ExxonMobil Research and Engineering
Company
The United States of America

KHODKE, Aditi
Institute for Global Environmental Strategies
Japan

KIM, Raehyun
National Institute of Forest Science
The Republic of Korea

KIM, Junhwan
Rural Development Administration, National
Institute of Crop Science
The Republic of Korea

KISHWAN, Jagdish
Network for Certification and Conservation
of Forests
India

KITAJIMA, Kaoru
Kyoto University
Japan

KOLKA, Randall
United States Department of Agriculture
Forest Service
The United States of America

KONDO, Hiroaki
National Institute of Advanced Industrial
Science and Technology, Japan Weather
Association
Japan

KOSONEN, Kaisa
Greenpeace International
Finland

KOUADIO, Boyossoro Hélène
Université Félix Houphouet Boigny
Côte d'Ivoire

KOUTROULIS, Aristeidis
Technical University of Crete
Greece

KRISNAWATI, Haruni
Forestry and Environment Research,
Development and Innovation Agency
Indonesia

KRUG, Thelma
Vice-Chair IPCC and National Institute for
Space Research
Brazil

KUMAR, Suresh
Central Arid Zone Research Institute
India

KUNA, Birgit
German Aerospace Center, Department of
Environment and Sustainability
Germany

KUSCH-BRANDT, Sigrid
University of Padua
Germany

KVALEVAG, Maria
Norwegian Environment Agency
Norway

LA SCALA JR., Newton
Faculty of Agricultural and Veterinary
Sciences
Brazil

LABINTAN, Constant
Laboratory of Population Dynamics,
Benin Center for Scientific Research and
Innovation
Benin

LAHIRI, Souparna
Global Forest Coalition
India

LAHOZ, William
Norwegian Institute for Air Research
Norway

LAJTHA, Kate
Oregon State University
The United States of America

LAL, Rattan
The Ohio State University
The United States of America

LARSEN, Morten Andreas Dahl
Technical University of Denmark
Denmark

LAWRENCE, Judy
New Zealand Climate Change Research
Institute, Victoria University of Wellington
New Zealand

LE HOANG, Anh
Ministry of Agriculture and Rural
Development of Vietnam
Viet Nam

LEE, Leonie
Ministry of Environment
and Water Resources
Singapore

LEE, Janice Ser Huay
Nanyang Technological University
of Singapore
Singapore

LEFFERTSTRA, Harold
Consultant
Norway

LEHTONEN, Heikki
Natural Resources Institute Finland
Finland

LEITE, Edson
Brazilian Agricultural Research
Organization, Ministry of Agriculture,
Livestock and Food Supply
Brazil

LEJEUNE, Quentin
Climate Analytics
Germany

LEO, Meyer
ClimateContact-Consultancy
The Netherlands

LEONARD, Stephen
Center for International Forestry Research
Indonesia

LEONARD, Sunday
Scientific and Technical Advisory Panel
of the Global Environment Facility, UN
Environment
The United States of America

LEY, Debora
Latinoamérica Renovable
Guatemala

AIV

LI, Xin-Rong
Cold and Arid Regions Environmental and
Engineering Research Institute, Chinese
Academy of Sciences
China

LI, Wei
Laboratory of Climate Science and
Environment
France

LI, Changxiao
Southwest University
China

LIBONATI, Renata
Federal University of Rio de Janeiro
Brazil

LICKEL, Sara
Secours catholique – Caritas France
France

LIM, Lee-Sim
School of Distance Education, Universiti
Sains Malaysia
Malaysia

LINDER, Sofie
Swedish National Heritage Board
Sweden

LIPKA, Oksana
World Wildlife Foundation Russia
The Russian Federation

LIU, Yaming
China Meteorological Administration
China

LIU, Yuanbo
Nanjing Institute of Geography and
Limnology, Chinese Academy of Sciences
China

LIU, Junguo
Southern University of Science and
Technology
China

LLANSOLA-PORTOLES, Manuel
The French National Center for Scientific
Research
France

LOCKWOOD, Dale
Colorado State University
The United States of America

LOHILA, Annalea
Finnish Meteorological Institute
Finland

LORENZ, William
University of Southern Queensland
Australia

LORITE, Ignacio
Institute of Agricultural and Fisheries
Research and Training
Spain

LOUAPRE, Philippe
Université Bourgogne Franche-Comté
France

LOVERA-BILDERBEEK, Simone
Global Forest Coalition
Paraguay

LU, Fei
Research Center for Eco-Environmental
Sciences, Chinese Academy of Sciences
China

LUCON, Oswaldo
University of São Paulo
Brazil

LUENING, Sebastian
Institute for Hydrography, Geoecology and
Climate Sciences
Portugal

LUGENDO, Prudence
Platform for Agricultural Policy Analysis and
Coordination
The United Republic of Tanzania

LUISE, Anna
Institute for Environmental Protection and
Research
Italy

LUND, Marianne Tronstad
Center for International Climate Research
Norway

LUYSSAERT, Sebastiaan
Vrije University Amsterdam
Belgium

LYAMBAI, Martin
Pan African University, Institute
for Energy and Water Sciences
including Climate Change
Zambia

MABDAL, Prafulla Kumar
Government of West Bengal
India

MACDONALD, Gordon
University of Leicester
Canada

MADARI, Beata Emoke
Embrapa
Brazil

MADHOUN, Wesam
Universiti Teknologi PETRONAS
Malaysia

MAGOSAKI, Kaoru
Ministry of Foreign Affairs
Japan

MAHADEVIA, Darshini
CEPT University
India

MAHMUD, Mastura
Universiti Kebangsaan Malaysia
Malaysia

MAILUMO, Daniel
College of Advanced and Professional
Studies
Nigeria

MAINALY, Jony
Kathmandu School of Law
Nepal

MANDYCH, Anatoliy
Institute of Geography, Russian
Academy of Sciences
The Russian Federation

MANI, Francis Sundresh
The University of the South Pacific
Fiji

MANSOUR, Amany
Desert Research Center
Egypt

MARBAIX, Philippe
Université catholique de Louvain
Belgium

MARGARITA, Roldán
Polytechnic University of Madrid
Spain

MARSH, Anne
United States Department of Agriculture
Forest Service
The United States of America

MARTICORENA, Beatrice
French National Centre for Scientific
Research
France

MARUNYE, Joalane
National University of Lesotho
Lesotho

MASOODIAN, Seyed Abolfazl
University of Isfahan
Iran

MATIAS FIGUEROA, Carlos
National Institute of Ecology and Climate
Change
Mexico

MATSUMOTO, Ken'ichi
Nagasaki University
Japan

MAY, Wilhelm
Lund University
Denmark

MELTON, Joe
Environment and Climate Change Canada
Canada

MENDEZ GAONA, Fernando
Faculty of Exact and Natural Sciences,
National University of Asuncion
Paraguay

MERDAS, Saifi
Centre of Scientific and Technical Research
of Arid Regions
Algeria

METTERNICHT, Graciela
The University of New South Wales
Australia

METZKER, Thiago
Institute of Environmental Good
Brazil

MIDDLETON, Beth
United States Geological Survey
The United States of America

MIDDLETON, Nicholas
Universty of Oxford
United Kingdom

MIDGLEY, Pauline
Consultant
Germany

MIGONGO-BAKE, Elizabeth
Consultant
Kenya

MILLSTONE, Carina
Feedback Global Ltd
United Kingdom

MINNEROP, Petra
School of Social Science, University of
Dundee
United Kingdom

MINTENBECK, Katja
TSU IPCC WGII, Alfred-Wegener-Institut
Germany

MIRZABAEV, Alisher
Center for Development Research,
University of Bonn
Germany/Uzbekistan

MISHRA, Santosh Kumar
Shreemati Nathibai Damodar Thackersey
Women's University
India

MIYAMOTO, Asako
Forestry and Forest Products Research
Institute
Japan

MOGHIM, Sanaz
Sharif University of Technology
Iran

MOGHIMI, Ebrahim
University of Tehran
Iran

MOHAMED ABULEIF, Khalid
Ministry of Petroleum and Mineral
Resources
Saudi Arabia

MOLINA, Tomas
University of Barcelona
Spain

MORECROFT, Mike
Natural England
United Kingdom

MORENO RODRIGUEZ, Jose Manuel
University of Castilla-La Mancha
Spain

MORHART, Christopher
University of Freiburg
Germany

MORITA, Kanako
Forestry and Forest Products Research
Institute
Japan

MORRIS, Paul
University of Leeds
United Kingdom

MORTON, John
University of Greenwich
United Kingdom

MOSTAFAVI DARANI, Sayed Masoud
Islamic Republic of Iran Meteorological
Organization
Iran

MOTTET, Anne
Food and Agriculture Organization
of the United Nations
Italy

MOUFOUMA OKIA, Wilfran
TSU IPCC WGI
France

MSONGALELI, Barnabas
University of Dodoma
The United Republic of Tanzania

MUJURU, Lizzie
Bindura University of Science Education
Zimbabwe

AIV

MUKHERJI, Aditi
International Centre for Integrated
Mountain Development
Nepal

MURI, Helene
Norwegian University of Science and
Technology
Norway

MUSAH SURUGU, Justice Issah
United Nations Frameword Convention
on Climate Change and United Nations
University
Germany

MUSTAFA, Sawsan
Ministry of Animal Resources
Sudan

MUTEMI, Joseph
University of Nairobi
Kenya

NABEL, Julia
Max Planck Institute for Meteorology
Germany

NABUURS, Gert-Jan
Wageningen University
The Netherlands

NADELHOFFER, Knute
University of Michigan
The United States of America

NARAYANAPPA, Devaraju
The Laboratory of Climate and
Environmental Sciences
France

NARESH KUMAR, Soora
Indian Agricultural Research Institute
India

NAVARRA, Antonio
Euro-Mediterranean Center for Climate
Change and National Institute of Geophysics
and Volcanology
Italy

NDIONE, Jacques Andre
Ecological Monitoring Center
Senegal

NEMITZ, Dirk
United Nations Frameword Convention on
Climate Change
Germany

NEOFOTIS, Peter
Michigan State University
The United States of America

NEOGI, Suvadip
TSU IPCC WGIII, Ahmedabad University
India

NHEMACHENA, Charles
International Water Management Institute
South Africa

NISHIOKA, Shuzo
Institute for Global Environmental Strategies
Japan

NORDÉN, Jenni
Norwegian Institute for Nature Research
Norway

NORSE, David
Institute of Sustainable Resources, University
College London
United Kingdom

NORTH, Michelle
University of KwaZulu-Natal
South Africa

O'HEHIR, Colin
Department of Communications, Climate
Action and Environment
Ireland

ODHIAMBO, Franklin
United Nations Environment
Kenya

ODIGIE-EMMANUEL, Omoyemen Lucia
Nigerian Law School and Centre for Human
Rights and Climate Change Research
Nigeria

OHREL, Sara
United States Environmental Protection
Agency
The United States of America

OJEDA, Gerardo
Universidad Nacional Abierta y a Distancia
Colombia

OKEM, Andrew
TSU IPCC WGII
South Africa

**OLIVEIRA DE ALMEIDA MACHADO,
Pedro Luiz**
Embrapa
Brazil

OLSEN, Siri Lie
Norwegian Institute for Nature Research
Norway

OLUOWO, Elohor Freeman
Norhed International Program, Nha Trang
University
Nigeria

ONDIMU, Kennedy
Ministry of Environment and Forestry
Kenya

ONYENEKE, Robert
Alex Ekwueme Federal University
Nigeria

ORANYE, Nkechinyelu
Centre for Petroleum, Energy Economics and
Law, University of Ibadan
Nigeria

ORDÓÑEZ DÍAZ, José Antonio Benjamín
Instituto Tecnológico y de Estudios
Superiores de Monterrey
Mexico

ORR, Barron Joseph
United Nations Convention to Combat
Desertification
Germany

OTT, Cordula
Centre for Development and Environment,
University of Bern
Switzerland

OTTO, Ilona M.
Potsdam Institute for Climate Impact
Research
Germany

OUILLON, Sylvain
Research Institute for Development
France

OULD-DADA, Zitouni
Food and Agriculture Organization
of the United Nations
Italy

OWIDHI, Mark
University of Nairobi
Kenya

OWOEYE, Idowu
University of Ibadan
Nigeria

PACALA, Stephen
Princeton University
The United States of America

PANMAO, Zhai
Co-Chair IPCC WGI and Chinese
Academy of Meteorological Sciences
China

PANT, Harshit
Govind Ballabh Pant National Institute
of Himalayan Environment and
Sustainable Development
India

PANT, Laxmi
University of Guelph
Canada

PARHIZKAR, Davood
Islamic Republic of Iran
Meteorological Organization
Iran

PARK, Taehyun
Greenpeace East Asia
The Republic of Korea

PARTANEN, Antti-Ilari
Finnish Meteorological Institute
Finland

PATHAK, Minal
TSU IPCC WGIII, Ahmedabad University
India

PATRA, Prabir
Japan Agency for Marine-Earth
Science and Technology
Japan

PATRICK, Gonzalez
University of California, Berkeley
The United States of America

PATRIZIO, Piera
International Institute for Applied Systems
Analysis
Austria

PAULOSE, Hanna
United Nations Entity for Gender Equality
and the Empowerment of Women
The United States of America

PAZMINO, Daniel
Ecoimpacto
Ecuador

PEDACE, Roque
Climate Action Network – Latin America
Argentina

PENG, Jian
University of Oxford
United Kingdom

PENNOCK, Daniel
University of Saskatchewan
Canada

PENNY, Davies
Ford Foundation
The United States of America

PENUELAS, Josep
Supreme Council for Scientific Research
and Ecological and Forestry Applications
Research Centre
Spain

PEREIRA, Christopher
Secretariat of the Convention
on Biological Diversity
Canada

PERLMAN, Kelsey
Consultant
France

PETRIE, Matthew
University of Nevada Las Vegas
The United States of America

PETZOLD, Jan
TSU IPCC WGII, Alfred-Wegener-Institut
Germany

PHAM, Quangha
Institute for Agricultural Environment
Viet Nam

PIAO, Shilong
Peking University
China

PINGOUD, Kim
Kim Pingoud Consulting
Finland

PINO MAESO, Don Alfonso
Ministry of the Ecological Transition
Spain

PITMAN, Andrew
The University of New South Wales
Australia

PLASENCIA GONZALEZ, Felix
Ministry of People's Power
for Foreign Affairs
Venezuela

PODWOJEWSKI, Pascal
Research Institute for Development
France

POLACK, Sharelle
Global Alliance for Improved Nutrition
Switzerland

POLOCZANSKA, Elvira
TSU IPCC WGII, Alfred-Wegener-Institut
United Kingdom/Australia

PÖRTNER, Hans-Otto
Co-Chair IPCC WGII, Alfred-Wegener-Institut
Germany

PORTUGAL PEREIRA, Joana
TSU IPCC WGIII, Imperial College London
United Kingdom

PRESTELE, Reinhard
Ludwig Maximilian University of Munich
Germany

PRIETO AMPARAN, Jesus Alejandro
Autonomous University of Chihuahua
Mexico

PRINCE, Stephen
University of Maryland
The United States of America

QUAIFE, Tristan
National Centre for Earth Observation,
University of Reading
United Kingdom

AIV

QUESADA, Benjamin
Karlsruhe Institute of Technology
Germany

QUILLEROU, Emmanuelle
University of Brest
France

RADUNSKY, Klaus
Umweltbundesamt
Austria

RAGHAVAN, Krishnan
Indian Institute of Tropical Meteorology
India

RAHAL, Farid
University of Sciences and Technology
of Oran – Mohamed Boudiaf
Algeria

RAHMAN, Muhammad Ashfaqur
International Center for Theoretical Physics
Italy

RAIS AKHTAR, Rais
International Institute of Health
Management and Research
India

RAKONCZAY, Zoltán
European Commission
Belgium

RAWE, Tonya
CARE International
The United States of America

REAY, Dave
University of Edinburgh
United Kingdom

REIDSMA, Pytrik
Wageningen University
The Netherlands

REISINGER, Andy
Vice-Chair IPCC WGIII and New Zealand
Agricultural Greenhouse Gas Research
Centre
New Zealand

RENWICK, Douglas
Nottingham Business School
United Kingdom

REUTER, Thomas
University of Melbourne
Australia

RIVERA-FERRE, Marta guadalupe
University of Vic – Central University of
Catalonia
Spain

RIYAZ, Mahmood
Maldivian Coral Reef Society
Maldives

ROBERTS, Debra C.
Co-Chair IPCC WGII and EThekwini
Municipality
South Africa

RODERICK, Michael
Australian National University
Australia

ROLLS, Will
University of Leeds
United Kingdom

ROMANENKOV, Vladimir
Lomonosov Moscow State University
The Russian Federation

ROMERO, José
Swiss Federal Office for the Environment
Switzerland

RONCHAIL, Josyane
University Paris Diderot
France

ROSALES BENITES DE FRANCO, Marina
Federico Villarreal National University
Peru

RUDGE RAMOS RIBEIRO, Rodrigo
Pisando Verde
Brazil

RUMMUKAINEN , Markku
Swedish Meteorological and Hydrological
Institute
Sweden

RUMPEL, Cornelia
French National Center for Scientific
Research
France

RÜTTING, Tobias
University of Gothenburg
Sweden

SABLÉ, Anne-Laure
Consultant
France

SALAZAR VARGAS, Maria del Pilar
National Institute of Ecology and Climate
Change
Mexico

SAMADI, Vidya
University of South Carolina
The United States of America

SANDKER, Marieke
Food and Agriculture Organization
of the United Nations
Italy

SANE, Youssouph
National Agency of Civil Aviation
and Meteorology
Senegal

SCHEYVENS, Henry
Institute for Global Environmental Strategies
Japan

SCHNEIDER, Laura
The State University of New Jersey
The United States of America

SCHULZ, Astrid
German Advisory Council on Global Change
Germany

SCHUT, Antonius
Wageningen University
The Netherlands

SCHWINGSHACKL, Clemens
ETH Zurich
Switzerland

SEARCHINGER, Timothy
Princeton University
The United States of America

SEGNON, Alcade
University of Abomey-Calavi
Benin

SEMENOV, Sergey
Vice-Chair IPCC WGI and Institute of Global
Climate and Ecology
The Russian Federation

SENEVIRATNE, Sonia
ETH Zurich
Switzerland

ŞENYAZ, Ahmet
Ministry of Agriculture and Forestry
Turkey

SEYMOUR, Frances
World Resources Institute
The United States of America

SHAHIN, Md
National University
Bangladesh

SHAHSAVANI, Abbas
Shahid Beheshti University
of Medical Sciences
Iran

SHEPHERD, Anita
University of Aberdeen
United Kingdom

SHILOMBOLENI, Helena
University of Waterloo
Canada

SHIVAKOTI, Binaya Raj
Institute for Global Environmental Strategies
Japan

SIMS, Ralph
Massey University
New Zealand

SINGH, Chandni
Indian Institute for Human Settlements
Myanmar

SINGH, Nayanika
Ministry of Environment, Forest
and Climate Change
India

SIORAK, Nicolas
Business Alliance for Climate Resilience
France

SKEA, Jim
Co-Chair IPCC WGIII and Imperial
College London
United Kingdom

SKEIE, Ragnhild Bieltvedt
Center for International Climate Research
Norway

SLADE, Raphael
TSU IPCC WGIII, Imperial College London
United Kingdom

SLOT, Martijn
Smit
The Netherlands

SMITH, Donald
McGill University
Canada

SMITH, Aaron
The Norwegian Institute of Bioeconomy
Research
Norway

SMITH, Tanya
Tsleil-Waututh First Nation
Canada

SMOLKER, Rachel
Biofuelwatch
The United States of America

SOHNGEN, Brent
Ohio State University
The United States of America

SOLAYMANI OSBOOEI, HAMIDREZA
Forest, Range and Watershed
Management Organization
Iran

SOMMER, Rolf
International Center for Tropical Agriculture
Kenya

SOMOGYI, Zoltán
NARIC Forest Research Institute
Hungary

SOORA, Naresh Kumar
Indian Agricultural Research Institute
India

SÖRENSSON, Anna
Center for Sea and Atmospheric Research
Argentina

SOUZA, José João
Federal University of Rio Grande do Norte
Brazil

SOW, Samba
National Institute of Pedology
Senegal

SPAGNUOLO, Francesca
Scuola Superiore Sant'Anna
Italy

SPORRE, Moa
Lund University
Sweden

SRINIDHI, Arjuna
Watershed Organisation Trust
India

STABINSKY, Doreen
College of the Atlantic
The United States of America

STEFANESCU, Mihaela
Ministry of Waters and Forests
Romania

STEHFEST, Elke
PBL Netherlands Environmental
Assessment Agency
The Netherlands

STEPHAN GRUBER, Stephan
Carleton University
Canada

STJERN, Camilla
Center for International Climate Research
Norway

STONE, Kelly
ActionAid USA
The United States of America

STRANDBERG, Gustav
Swedish Meteorological and Hydrological
Institute
Sweden

AIV

STURGISS, Rob
Department of Environment,
Australian Government
Australia

SULISTIAWATI, Sulistiawati
Medical Faculty Airlangga University
Indonesia

SULMAN, Benjamin
Oak Ridge National Laboratory
The United States of America

SULTAN, Benjamin
Oceans, Climate and Resources Department,
Research Institute for Development
France

SUMMER, Heike
Office of Environment
Liechtenstein

SUN, Jianqi
Institute of Atmsopheric Physics, Chinese
Academy of Sciences
China

SUSANTO, RADEN
University of Maryland
The United States of America

SUSCA, Tiziana
Edinburgh Napier University
United Kingdom

SYMEONAKIS, Elias
Manchester Metropolitan University
United Kingdom

SZLAFSZTEIN, Claudio
Federal University of Pará
Brazil

SZOPA, Sophie
Laboratory of Climate Science
and Environment
France

TABAJ, Kristi
Consultant
The United States of America

TACHIIRI, Kaoru
Japan Agency for Marine-Earth Science
and Technology
Japan

TAIMAR, Ala
Estonian Meteorological
and Hydrological Institute
Estonia

TAJBAKHSH MOSALMAN, Sahar
Islamic Republic of Iran
Meteorological Organization
Iran

TAKATA, Yusuke
National Agriculture and Food
Research Organization
Japan

TALLEY, Trigg
United States Department of State
The United States of America

TAN, Xianchun
Institute of Science and Development,
Chinese Academy of Sciences
China

TANAKA, Kenzo
Forestry and Forest Products
Research Institute
Japan

TAYLOR, Betsy
Consultant
The United States of America

TAYLOR, David
National University of Singapore
Singapore

TCHOUAFFE, Norbert François
Pan African Institute for development
Cameroon

TEASLEY, Rebecca
University of Minnesota Duluth
The United States of America

TELLRO WAI, Nadji
Ministry of Environment, Water and Fishery
Chad

TEN VEEN, Rianne
Green Creation
The Netherlands

TENGBERG, Anna
Stockholm International Water Institute
Sweden

TIBIG, Lourdes
Climate Change Commission
The Philippines

TITILOLA, Bolanle Mutiat
University of Ibadan
Nigeria

TIWARI, Pushp Raj
University of Hertfordshire
United Kingdom

TÖLLE, Merja
Institute of Geography, Climatology,
Climate Dynamics and Climate Change
Germany

TONOSAKI, Kochi
Organizatoion for Landscape and Urban
Green Infrastructure
Japan

TORBERT, Henry Allen
United States Department of Agriculture
The United States of America

TOURAY, Lamin Mai
Department of Water Resources
The Gambia

TOUSSAINT, Renaud
University of Strasbourg and
University of Oslo
France

TREUHAFT, Robert
California Institute of Technology
The United States of America

TUBIELLO, Francesco
Food and Agriculture Organization
of the United Nations
Italy

TURNER II, Billie
Arizona State University
The United States of America

TYNER, Wallace
Purdue University
The United States of America

VALKAMA, Elena
Natural Resources Institute Finland
Finland

VAN DIEMEN, Renée
TSU IPCC WGIII, Imperial College London
The Netherlands/United Kingdom

VAN YPERSELE, Jean-Pascal
Université catholique de Louvain
Belgium

VAN ZWIETEN, Lukas
New South Wales Department
of Primary Industries
Australia

VANDERSTRAETEN, Martine
Belgian Federal Science Policy
Belgium

VANUYTRECHT, Eline
Katholieke Universiteit Leuven
Belgium

VARGA, György
Research Centre for Astronomy
and Earth Sciences
Hungary

VELASCO MUNGUIRA, Aida
Ministry of Agriculture and Fisheries,
Food and Environment
Spain

VERA, Carolina
Vice-Chair IPCC WGI, University of Buenos
Aires and National Scientific and Technical
Research Council
Argentina

VERMEULEN, Sonja
World Wide Fund for Nature
United Kingdom

VICENTE VICENTE, Jose Luis
Mercator Research Institute on Global
Commons and Climate Change
Germany

VIGLIZZO, Ernesto
National Scientific and Technical
Research Council
Argentina

VINCKE, Caroline
Université catholique de Louvain
Belgium

VIZZARRI, Matteo
European Commission Joint Research Centre
Italy

VON MALTITZ, Graham
Council for Scientific and Industrial Research
South Africa

VON UEXKULL, Nina
Uppsala University
Sweden

WAFULA, James
Africa Climate Leadership Program
Kenya

WAHA, Katharina
Commonwealth Scientific and Industrial
Research Organisation
Australia

WALDTEUFEL, Philippe
Scientific Research National Center
France

WANG, Kaicun
Beijing Normal University
China

WANG, Jing
China Agricultural University
China

WANG, Guosheng
China National Academy of Forestry
Inventory and Planning
China

WANG, Changke
National Climate Center
China

WARNER, Koko
United Nations Framework Convention on
Climate Change
Germany

WEBB, Nicholas
United States Department of Agriculture
The United States of America

WEBER, Bettina
Max Planck Institute for Chemistry
Germany

WEI, Chao
National Climate Center
China

WEYER, Nora M.
TSU IPCC WGII, Alfred-Wegener-Institut
Germany

WIBIG, Joanna
University of Lodz
Poland

WIEDMANN, Tommy
University of New South Wales
Australia

WIGGINS, Meredith
Historic England
United Kingdom

WILKE, Nicole
Federal Ministry for the Environment,
Nature Conservation and Nuclear Safety
International Climate Policy
Germany

WITHANACHCHI, Sisira
University of Kassel
Germany

WOODFINE, Anne
Consultant
United Kingdom

WU, Jianguo
Chinese Research Academy
of Environmental Sciences
China

WU, Bo
Institute of Desertification Studies,
Chinese Academy of Forestry
China

XIA, Chaozong
China National Academy of Forestry
Inventory and Planning
China

XU, Duanyang
Chinese Academy of Sciences
China

XU, Xiyan
Institute of Atmospheric Physics
China

AIV

XU, Weimu
Trinity College Dublin
Ireland

YABI, Ibouraïma
University of Abomey-Calavi
Benin

YAHYA, Mohammed Yahya Said
Associate Institute of Climate
Change and Adaptation
Kenya

YANG, Fulin
China Meteorological Administration
China

YASSAA, Noureddine
Vice-Chair IPCC WGI and Centre for
Renewable Energy Development
Algeria

YU, Qiang
Northwest A&F University
China

ZAELKE, Durwood
Institute for Governance
and Sustainable Development
The United States of America

ZAITON IBRAHIM, Zelina Binti
University Putra Malaysia
Malaysia

ZARGARLELLAHI, Hanieh
Geological Survey of Iran
Iran

ZARIN, Daniel
Climate and Land Use Alliance
The United States of America

ZHANG, Chengyi
China Metoerological Administration
China

ZHANG, Qiang
Consultant
China

ZHANG, Guobin
National Forestry and Grassland
Administration
China

ZHOU, Guangsheng
Chinese Academy of Meteorological
Sciences
China

ZHOU, Yuyu
Iowa State University
China

ZHOU, Jizhong
University of Oklahoma
The United States of America

ZIAYAN, Sadegh
Iran Meteorological Organization
Iran

ZOPATTI, Alvaro Gabriel
National Directorate of Climate Change
Argentina

ZWARTZ, Dan
Ministry for the Environment
New Zealand

AIV

Index

This index should be cited as:
IPCC, 2019: Index. In: *Climate Change and Land: an IPCC special report on climate change, desertification, land degradation, sustainable land management, food security, and greenhouse gas fluxes in terrestrial ecosystems* [P.R. Shukla, J. Skea, E. Calvo Buendia, V. Masson-Delmotte, H.-O. Pörtner, D. C. Roberts, P. Zhai, R. Slade, S. Connors, R. van Diemen, M. Ferrat, E. Haughey, S. Luz, S. Neogi, M. Pathak, J. Petzold, J. Portugal Pereira, P. Vyas, E. Huntley, K. Kissick, M. Belkacemi, J. Malley, (eds.)]. https://doi.org/10.1017/9781009157988.014

Note: * indicates the term also appears in the Glossary and n indicates a footnote. Italicised page numbers denote tables, figures, associated captions and boxed material. Supplementary Material is listed by section number, for example 5.SM.5.1, 6.SM.6.4.1

1.5°C pathway* 22–23, 32*n*, 195–199, 200–201, 373, *581*, 686
1.5°C warming
  compared to 2°C *83*, 137, 146, 256, 279, 295, 362, 449, 683
  crop productivity 454, 680–681
  limiting to 22–23, 49, 55, *83*, 138, 348, 373, 449
    benefits to coastal regions 372
    and land-use change 136, 138, 373, 449
  risks 15–16, 67, 277, 449, *644*, 675, 683, 684, *730*
  SSP scenarios 50, 67, 278, 675, 684–685, *730*
2°C warming
  compared to 1.5°C *83*, 137, 146, 256, 279, 295, 362, 449, 683
  crop productivity 454
  limiting to 22, 49, 55, 348
    benefits to coastal regions 372
    and land-use change 136, 373, 449
  pathways 22, 195, 197–199, 200–201
  risks 15–16, 683, 684
  SSP scenarios 50, 278, 684–686
3°C warming
  pathways 22, 701
  risks 15–16, 373, *644*, 683, 684, *730*
  SSP scenarios 50, 67, 278, 675, 685, *730*
4 per 1000 initiative 387
2030 Agenda for Sustainable Development* 388

A

acceptability (of policy or system change)* 490, 510, 698, *754*
acclimation 201–202
acclimatisation* *see* acclimation
acidification *573–574*
  ocean acidification 627*n*, 691
  soils 355, *357*, 376, 399, *575*
    see also reduced pollution including acidification
activity data* 160, *164*
adaptation* 79, 80, 102–103, 138, 389, 558
  agriculture 280, 383, 512
  autonomous 466, 512, 701
  barriers to *448*, 470, 475, 513, 715, *715*, 716, 717, 738
  challenge 558, *559–561*, 561, 564, *565*
  co-benefits 102–103, *392*
  community-based 474–475, 518, *566–567*
  consequences of delay 644–645
  costs *693*, 711, 723

decision making approaches 721–723
  incorporating ILK 512, *747–748*
demand-side adaptation 439, 472–473
dietary changes 472–473
ecosystem-based 19, 282, 381, 468, 470, *566–567*, 706–707
FAQs 646
financing mechanisms 474, 711–712
food system 439, 440, *441*, 470–473, 513
future scenarios *13*, 564, *565*
gender and *447–448*
governance 737–738, 5.SM.5.5
incremental 466, *466*, *467*, 717, *747*
indigenous and local knowledge 512, *746–748*
institutional measures 473–475
knowledge gaps 513
  and land tenure *751–752*
mitigation co-benefits and synergies 22, 102–103, 383, 391, *392*
  with food security *448*, 492, *493*, 500, 507, 513–514
near-term actions 33–34
planning 473, 474, 475
policies 27–29, 105–106, *509*
risk associated with 686
  in shared socio-economic pathways (SSPs) *13*, 564, *565*
social learning and 749
supply-side adaptation 470–472
synergies with mitigation *448*, 492, *493*, 499–502, 507, 513–514
technologies 475
transformational 360, 466–467, *466*, *467*, 717
urban areas *188*, 391–393, *392*, 505, 507, 706–707
adaptation limits* 714–717
  desertification 20, 252–253, 291
  land degradation 21, 348, 388
adaptation options* 18–22, *24–26*, 81, 589–594, 721–722
  adverse side effects 686, *718*
  agriculture 460, *469*, 470–471, 499–502, 589–590
  agroecology 468–470, *469*, 499–500
  agroforestry 382, 383–384
  barriers 79
  biophysical 470–471
  community involvement 403, 474–475, *562*
  demand-side measures 472–473, 593
  diversification 468–470, *469*, 589
  early warning systems 475
  FAQs 107, 646, 755
  financial instruments 474, 475
  flexible livelihoods 471
  food system 464–475, *467*, *509*, 513–514
  gender and *718*
  global potential *24–26*
  indigenous and local knowledge *469*, 470, 512
  institutional 473–475
  migration 285, 380, 466, 683
  reducing meat consumption 472–473
  region-specific 107, *469*, *561–563*, 591, 592
  risk management 467–468, 594

soil management 470–471, 591–592
  supply-side measures 470–472, 593, *594*
  sustainable food systems *465–466*
  sustainable integrated agricultural systems 499–502
  sustainable land management (SLM) 388, *465*
  synergies and trade-offs 492, *493*, 499–502, 686, *718*
  transport and trade 471–472
  urban green infrastructure (UGI) 391–393, *392*, *563*
  water management 471
adaptation pathways* 104, 721–722, 743
adaptation potential 589–594, 609–610, *609*, *611–617*, 7.SM.7.1
  agricultural response options 589–590, *590*, *611*
  demand management options 593, *593*, *615*
  forest response options 590, *590*, *612*
  land management options 589–592, *611–615*
    all/other ecosystems 591–592, *592*, *614*
    specifically for CDR 592, *592*, *615*
  risk management options 594, *594*, *617*
  soil-based response options 591, *591*, *613*
  supply management options 593, *594*, *616*
adaptive capacity* 16, 557, 717, 736–737, 753, *754*
  by continent 5.SM.5.2
  corruption and 716
  culture and beliefs 470
  dryland areas 16, 753
  enhancing 22, 28, 104, 107, 701
  forested areas 103
  indigenous people 470, 755
  inequality and 716
  insurance and 594, 699
  knowledge and 104, 755
  land tenure and 27
  mitigation and 103
  oasis populations 301–302
  pastoralists 22, 276, *448*
    in shared socio-economic pathways *13*, *14*, 92–93
  smallholders 22
  strengthening 286–288
  sustainable sourcing and *578*, *616*
  to floods 701
  transformational 466–467
  vulnerable groups 104, 518, 691
  women 353, *448*, 716, *717*
adaptive governance* 723, 737, 742–743, *743–745*
  inclusive 754
  indicators/institutional dimensions 753, *754*
  social-ecological systems 391
adaptive institutions 736–737
adaptive management 68, 351, 721, 723–725, *724*, *745*
adaptive risk management (ARM) 639
adverse side effects* 19, 609, *611–615*, 625
  adaptation options 686, *718*
  afforestation 374, 605, *612*
  biochar 399, *613*

bioenergy and BECCS 373–374, *581–582*, 592, *615*, *687*

coastal protection measures 402–403

forest area expansion 97, *99*, *100*

of mitigation 138

on NCPs or SDGs 630

peatland restoration/reduced conversion *614*

reducing deforestation and forest degradation *562*

reducing grassland conversion to cropland *611*

reforestation and forest restoration 605, *612*

risk from 687–688

AerChemMIP (Aerosol Chemistry Model Intercomparison Project) 169

aerosols* *139*, 166–170, 268–269, *269*, 271, 293

carbonaceous aerosols *149*, 167–169, *573*

deposition on snow 166, 269

fire emissions *149*, *573*, *683*

net cooling effect of dust emissions 377

secondary organic aerosols (SOA) 166, 167, 169, 170

transport 269, 271

afforestation* 19, 385–386, *567*, *572*

adaptation potential 590, *590*

adverse side effects 374, 605, *612*

best practice *25*

CO_2 emissions 45, 155

combined with bioenergy and BECCS *637*

compensatory afforestation *710*

defined *98*

feasibility *620*, 6.SM.6.4.1

global potential *25*

green walls/dams 294–296, *297*

impact of delayed action *645*

impact on desertification 596, *596*

impact on food security 605, *605*

impact on land degradation 374–375, 385–386, 600, *600*

impact on NCP *628*, 6.SM.6.4.3

impact on SDGs *631*, 6.SM.6.4.3

increasing 385–386

interlinkages 636–637

Karapìnar wind erosion area 293, *293*

mitigation potential 191, 196, 585, *585*, *637*

policy instruments *726*

potential across land challenges 610, *612*

risk of land degradation 374

sensitivity to climate change impacts 623–624

short-term static abatement costs 102

side effects and trade-offs 97, *99*

water balance *98*

afforestation/reforestation 8–9, 191, 492

future scenarios 198–199, 373

land type used 374–375

mitigation potential *48*, *49*

AFOLU (agriculture, forestry and other land use)* 8–11, 138, 151

CH_4 emissions/removals 8, 11, *151*, 159–160, *160*

CO_2 emissions/removals 8–9, 133–134, *151*, 152–155, *152*, *153*, *154*, *156*, 157

emissions 151, *151*, *152*, 157

food system emissions *10–11*, 475–476

GHG emissions 8–11, *10–11*, *82–83*, 133–134, 151, *154*, *156*

gross CO_2 flux 134, *152*, 157

mitigation 199, 480

N_2O emissions/removals 8, 11, 133, 134, *151*, 160–162, *161*, *163*

net anthropogenic emissions 8–11, *10–11*, 133–134, *151*

net CO_2 flux 133–134, 152–153, *152*

regional differences in emissions 155, *156*

total net GHG emissions 11

Africa

agricultural emissions 159

charcoal production 375, *740–741*

conflict 380

conservation agriculture 501

crop production 300–301, 452–453, 454, 682

deforestation *185*

desertification 263, 305

drought 258–259, 276, *290–291*, 682–683

dryland areas *255*, 682–683

dryland population 256–258, *257*

dust emissions 166, 167, 268–269

floods *744*

food loss and waste 100–101, 682

food security 450, *465–466*, 472

Great Green Wall of the Sahara and the Sahel 296, *297*

Green Dam project in Algeria 295–296, *296*

invasive plants 262, 298

irrigation 180, 288–289

land degradation 263, 375, 380

land tenure 287, *750*, *751*

Limpopo River basin 263, 305

oases 300–302, *300*, *302*

pastoralists 276

poverty in dryland areas 257

rainfall erosivity in Niger Basin 370

rainfall patterns *176*, 180, 186, 258, 305, 450, *451*

river basin degradation 263, 305, 370

sustainable food systems *465–466*

traditional biomass 375

urbanisation 285

vegetation greening 263

water scarcity 301, 682–683

agreement* *4n*, 26

agricultural commercialism 289

agricultural diversification *567*, *570*

adaptation potential 589, *590*

feasibility *619*, 6.SM.6.4.1

impact on desertification 595, *595*

impact on food security 604, *604*

impact on land degradation 599, *600*

impact on NCP *628*, 6.SM.6.4.3

impact on SDGs *631*, 6.SM.6.4.3

mitigation potential 584, *584*

policy instruments *726*

potential across land challenges 610, *611*

sensitivity to climate change impacts 623

smallholders 640

agricultural intensification 195, 197, *502*, *562*, 583

adverse effects 252–253, 276, 291–292, *735*

future pathways 30, 195, *642*

global 5.SM.5.5

sustainable *481–482*, 501–502, *502–505*, *566–567*, 583, 589

agricultural land

BVOC emissions 170

CO_2 emissions 376

degradation *352*, 373, 376, 402

global trends in land use *444*

intensive management 373

land use/cover change *642*

see also croplands; pasture

agricultural productivity 379, *379*, 603–604

impact of desertification and climate change 273, 276, 279

livestock 454–458, *455*

see also crop productivity; crop yields

agricultural response options 100, 189, *569–571*, 610

adaptation potential 589–590, *590*, *611*

feasibility *619*, 6.SM.6.4.1

impact on desertification 595, *595*, *611*

impact on food security 603–604, *604*, *611*

impact on land degradation 599, *600*, *611*

impact on NCP *628*, 6.SM.6.4.3

impact on SDGs *631*, 6.SM.6.4.3

mitigation potential 189, 583–584, *584*, *611*

potential across land challenges *611*

sensitivity to climate change impacts 623

synergies and trade-offs 733

agricultural services 286–287

agriculture

adaptation and mitigation 103, 470–471, 733

adaptation policies 473

agricultural expansion *481–482*

agronomic practices 381, 382–383

best practice 723

climate-smart agriculture 474, 500, *563*, 565–566, *566–567*, 733, *751*, 5.SM.5.5

CO_2 land-atmosphere exchange 376

conservation agriculture 100, 192, 281, 470, 471, 500–501

controlled traffic farming *503*

dependency 5.SM.5.2

desertification and 273, 276, 279, 279–283

diversification 468, *469*, *504*, 589

dryland areas 16, 257, 259

emissions pricing 702, *703*

energy crops 374

energy efficiency *579*

extensification 511

financing mechanisms 712

flooding 147–148

GHG emissions 159, 160, *160*, *161*, 376, 475–478, 511, 702, *703*

croplands and soils 159–160, 16*1*, 162, 16*3*, 476, 477

enteric fermentation *160*, 189, 477

global trends 444, 445, 496

livestock 159, 16*0*, *161*, 162, 476, 477–478, *478*

rice cultivation 159, 16*0*, 477

see also AFOLU

GHG fluxes 376
GHG mitigation 190, 480–486
global status and trends 85–88, *87*
Hindu-Kush Himalayan Region *469*
impacts of climate change 373, 451–460, *461*
impacts of precipitation extremes 147–148
improved efficiency *503*
improved market access 286
indigenous and local knowledge (ILK) 283–284,
        381, 384, *747*
institutional adaptation options 473
integrated agricultural systems 499–502, *504*
intensification
        see agricultural intensification
invasive plants impacts 298, 299
land degradation 372, 376
land use *82–83*, 85–86, *87*
large-scale land acquisition (LSLA) 91, 750, *751*
maladaptation 734
mitigation barriers 715, 716
mixed farming 384, 500
mountain agriculture 301, *469*
nitrogen fertilisation 159
no-till farming 292, 376, 383, 471, 686
oasis agriculture 300–301
Pacific island communities *517*, 518
pastoralism 257, 276, 384
perennial grains and SOC *392*, 393–395
policies 286–287, 473, *482*, 508, 697, 701–702,
        703, 714
precision agriculture 100, *503, 566–567*
re-vegetation of saline land 283
research and development 697
resilience 591
response options, mitigation potential 189
rice cultivation 384
risk management 102
smallholder farming systems 459–460,
        499–500, 593, 594, 697
smallholder plantations 397–398
standards and certification schemes
        707–709, *708*
sustainable farming systems 381, 384, *465–466*
sustainable land management 100
synergies 731
urban and peri-urban agriculture *188*, 505, 507
vulnerability 5.SM.5.2
water use 7.SM.7.1
        see also AFOLU; agricultural response
                options; agroforestry; irrigation
agriculture, forestry and other land use
        see AFOLU
agrobiodiversity* 468
agroecology* 381, 499–500, *566–567*
        food systems and 468–470, *469*
agroforestry* *280*, 382, 383–384, *567, 570*
        adaptation potential 589, *590*
        carbon sequestration potential *485*
        co-benefits *504*
        feasibility 618, *619*, 6.SM.6.4.1
        and food systems 470, 485, *504*
        GHG mitigation 485, *485*
        impact on desertification 595, *595*

impact on food security 604, *604*
impact on land degradation 599, *600*
impact on NCP *628*, 6.SM.6.4.3
impact on SDGs *631*, 6.SM.6.4.3
mitigation potential 189, 584, *584*
policy instruments *726*
potential across land challenges 609, 610, *611*
potential deployment area *633*
sensitivity to climate change impacts 623
agronomic response measures 381,
        382–383, *382*
agropastoralists 257, 439, 455–456, 5.SM.5.1
air pollution 160, *187–188*, 590
        from fire 683
        indoor 288, 709, *740*
        management *573–574*
        short-lived climate pollutants* 451, 586, *740*
        urban *187–188*, 603, 691
                see also reduced pollution
                        including acidification
Alaska 743
albedo* *139*
        aerosol deposition and 166
        albedo-induced surface temperature
                changes 172
        croplands 181–182, *181*
        deforestation and 177
        forest management 191–192
        forest vs. non-forest *98*
        impact of afforestation 374
        impact of biochar 399
        land cover changes 12, *172*, 374
        land degradation and surface albedo
                change 377
        radiative forcing from changes in 138
        seasonal vegetation change *139*
        snow-albedo feedback 178, *179*, 183–184
        surface albedo change and feedbacks to climate
                182–184, 269–270
Algeria 263, 301
        Green Dam project 295–296, *296*
Amazon
        biodiversity 352
        BVOC emissions 169
        deforestation 106, *149*, 175, *185*, 481–482
        deforestation and malaria 691
        drought 146
        drought induced fires *149*, 155
        global warming and local climate feedbacks 45,
                183, *183*
        land rights 106, 378
        peatlands 397–398
        REDD+ *709–710*
Amazon biome *481*
ammonia (NH₃) 376, 497, 5.SM.5.3
animal feed 276, 473, 485, 5.SM.5.3
anthromes* 279–280, *280*, 558, *559*
        area exposed to land challenges *560*
        defined 86
        local response to land challenges *561–563*
        overlapping land challenges 558, *560, 561, 561,*
                561–563
anthropogenic* defining 155

anthropogenic drivers
        of coastal degradation 354, 402–403
        of desertification 251, 259–260, 264, 268
        interaction with climate change 259–260, *382*
        of land degradation 349, 354–355
anthropogenic emissions* 8–9, *10–11*, 11, *41,*
        44, 45–46, 84, *151,* 152–155, 199, 349
        aerosols 166, 167, 168–169, 170
        carbon dioxide (CO₂) 152–155, *152, 154,*
                156, 157
        estimating 153–155, *154, 163–164*
        gross emissions 157
        methane (CH₄) 159–160, *160*
        nitrous oxide (N₂O) 160–162, *161*
        rapid reduction 34, 79
        regional trends 155, *156*
        separating from non-anthropogenic 151, 199
                see also greenhouse gas emissions
anthropogenic removals* 8, *46,* 152–155,
        157, 188
        estimating 153–155, *154*
        negative emissions technologies 348, 398, 399,
                441, 492
                see also carbon dioxide removal (CDR)
anthropogenic warming 133, 147, 175
anticipatory governance 724, 742
aquaculture 697
        adaptation options 471
        GHG emissions 162, 478
        GHG mitigation 486–487
        impacts of climate change 459
AR5 *see* IPCC Fifth Assessment Report
Arabian Peninsula 258, 274
        oases 300–302, *300*
Aral Sea 264, 293, 294
Arctic region
        permafrost thaw 684, *689*, 7.SM.7.1
        sea ice 179, 691
        soils 204
        vegetation increase 377, 456
        warming 168, 172, 362, 377, 564
Argentina 265, 452, *481*
arid ecosystems 252, 271, 595
aridity* 142
        aridity index (AI) 254, *254*, 260
        future projections 276–277
Asia
        agricultural emissions 159
        black carbon emissions 168
        crop production and food security 452
        deforestation and rainfall patterns *185*
        desertification and land degradation 263–264
        dryland areas *255*
        dryland populations 257, *257*
        floods 472, *744*
        greening trend 263
        Hindu-Kush Himalayan region 452
        invasive plants in Pakistan 299–300
        land tenure *751*
        monsoon rainfall *176*
        pastoral systems 456
        peatland degradation 397–398
        peatland fires 397

Index

reforestation in South Korea 395–396
river basin degradation 263
soil erosion in Central Asia 293–294
Sundarbans mangroves 400
traditional biomass use and land
        degradation 375
    see also China
atmosphere* 185–186
atmospheric CO₂ 79, 84–85, *140*, *171*, 172,
    184, 254
changes in 7.SM.7.2
desertification feedbacks to climate 268, *269*
effect of increasing levels 88, 144, *165*
        on crops 451–452, *453*, 454, 458, 463–464
        on food quality 463–464
        on livestock 454–455, *455*, *456*
        on soil organic carbon 134, 204
        on vegetation 79, 134, 144, *165*, 202–203,
            251, 297, 362, *457*, 463
forestation and 179
impact on food security 5.SM.5.2
increase due to land cover change 172–173,
        174, *174*
potential impact of mitigation 157
regional warming due to increase 135,
        172–173, *173*, 174, *174*
removal 133–134, 135–136, 157, 492, 494
    see also CO₂ fertilisation
atmospheric inversions *164*
attribution*
desertification 265–268
land degradation 360, 362
soil erosion 682
of vegetation changes to human activity 266
Australia
climate change and crop production 452
desertification 264
dryland areas *255*
dryland population *257*
mesoscale convective systems (MCS) 370
monsoon rainfall *176*
autonomous adaptation* 466, 512, 701

B

Bangkok flood 472
Bangladesh 698, *744*
barren lands *560*, 561, *561*
barriers 28, 34, 42
economic 42, 618, *619–623*, *715*, 6.SM.6.4.1
environmental *619–623*, 6.SM.6.4.1
geophysical *619–623*, 6.SM.6.4.1
inequality as 716
institutional 618, *619–623*, 715–716, *715*, 737,
        738, 6.SM.6.4.1
multiple 62
overcoming 34, 70, 103, 513, 717
region specific 292
socio-cultural 42, 618, *619–623*, *715*,
        6.SM.6.4.1
technological barriers 62–63, 618, *619–623*,
        715, 6.SM.6.4.1

to adaptation *448*, 470, 475, 513, 715, *715*,
        716, 717, 738
to addressing desertification 292
to addressing land degradation 55
to community-based adaptation 475
to early warning systems 475
to implementing policy response 28, 43,
        714–717
to implementing SLM 28, 284, 389–391
to integrated response options 618, *619–623*,
        6.SM.6.4.1
to mitigation 79, *188*, 292, 513, 715–716
to participation and decision making *717–718*
to urban agriculture *188*
baseline scenario* 195–196, 197, 564, *565*, 684
baseline values 260, 350, 365
baseline-and-credit schemes *703*
**BECCS (bioenergy with carbon dioxide capture
    and storage)*** 196, 198–199, 373, 492, 494,
    513–514
large land-area need 19, 97
mitigation potential 49, 193, 201, 494
risks of 686–688, *687*
        see also bioenergy and BECCS
behavioural change *95*, 291, 390, 645
best practice *25*, 391, 707, 723
Biennial Transparency Reports 704
bio-economic farm model (BEFM) 460
bioaerosols 168
biochar* 100, *392*, 398–400, 492, *573*
adaptation potential *493*, 591, *591*
best practice *25*
combined with other response options 374, *567*
demand for land 19, 610
feasibility *620*, 6.SM.6.4.1
global potential *25*
impact on desertification 596, *596*
impact on food security 605, *606*
impact on land degradation 374, 399–400,
        601, *601*
impact on NCP *628*, 6.SM.6.4.3
impact on SDGs *631*, 6.SM.6.4.3
mitigation potential 192–193, 399, *493*,
        585–586, *586*
mitigation, role in 398–399
negative effects 399, 610
policy instruments *726*
potential across land challenges 610, *613*
production 192, 398, 400, 605, *741*
sensitivity to climate change impacts 624
bioclimates *141*
biodiversity* 79
agroecosystems *504*
drylands 271–272, 278–279
forest *98–99*, 352
future scenarios 564, *565*
green energy and *735*
impact of bioenergy 97
impact of climate change 404
impact of desertification 263, 271–272,
        278–279
impact of forest area expansion *98–99*
impact of grazing and fire regimes 281–282

impact of invasive plant species 297–300
loss 19–20, 88, 263, 683
risk to 691
SLM practices and 306
threatened hotspots 558, *559*, *560*, *562*
trade-offs 730–731, *735*
biodiversity conservation 567, *575*, 706–707
adaptation potential 592, *592*
combined with bioenergy and BECCS *637*
feasibility *621*, 6.SM.6.4.1
impact on desertification 597
impact on food security 606, *606*
impact on land degradation 601, *601*
impact on NCP *628*, 6.SM.6.4.3
impact on SDGs *631*, 6.SM.6.4.3
mitigation potential 586, *587*
policy instruments *726*
potential across land challenges *614*
sensitivity to climate change impacts 624
bioenergy* 19, 288, 492, 494, 646
agricultural and food waste streams 486, *741*
alternative land use to livestock production 511
biomass supply 97, 193–194, 373–374, 375,
        386, *581*
competition for land 42, 53–54, 62, *99*, 607
cropland *30–32*, *31–32*, 62, 194, 196, 199,
        642–644, 646
crops 97, 193–194, 373–374, 492, *576*, 646
energy access 709
GHG emissions 49, 193–194, 196, *583*
global consumption *582*
governance *738–739*
impacts on land degradation 373–374
land area required 19, 97, 687–688, *687*, *739*,
        7.SM.7.1, 7.SM.7.3
        modelled pathways 22–23, *30–32*, 49, 97,
            373, 449
mitigation potential 193–194, 201
potential scale 373
reducing/reversing land degradation 374–375
risks due to 373–374, 686–688, *687*
risks under different SSPs 7.SM.7.1, 7.SM.7.3
short term net emissions 193–194
socio-economic impacts *739*
sustainability standards and certification 707,
        708, 709
synergistic outcomes *582*
technology transfer 704
trade-offs with SDGs 7.SM.7.1
traditional biomass 20, 288, 375, 709, *740–742*
bioenergy and BECCS 19, 193–194, *567*,
575–576, *580–583*
adaptation potential 592, *592*
adverse side effects *581–582*, 592, 615
best practice *25*, 707
co-benefits *581–582*, 592, 615
feasibility 618, *621*, 6.SM.6.4.1
global potential *25–26*
impact of delayed action 645
impact on desertification 597, *597*
impact on food security 607, *607*
impact on land degradation 373, 601, *602*
impact on NCP *628*, 6.SM.6.4.3

impact on SDGs *631*, 6.SM.6.4.3
interlinkages and interactions with other
       response options 636–638, *636, 637*
inventory reporting *583*
limiting 637–638
mitigation potential 193, *580–583*, 587, *587*, 637
modelled pathways 22–23, *72–74*, 97, 373, 494
   mitigation scenarios 196–199, *580–583*
policy instruments *726*
potential across land challenges 610, *615*
potential deployment area 633, *633*
risks due to 686–688, 7.SM.7.1, 7.SM.7.3
scale of deployment 62, 63, 67
sensitivity to climate change impacts 624
biofuel* 288, 486, *580–583*
crops 193–194, 283
governance *738–739*
sustainability *708*, 709
biogenic volatile organic compounds (BVOCs)
   169–170
contribution to climate change 170
decrease in emissions 192
future trends 170
oxidation 168, 169, 170
and tropospheric ozone 170
biogeochemical effects* 134–135, 136, *139, 140,*
173, 174, 175
aerosols deposition and 166
changes in anthropogenic land cover 243,
     243–247
cooling 179
deforestation/forestation *98*, 176–177, 178–180
dynamics of soil organic carbon 203
forest response options 191–192
global warming 172, 174–175, 177
regional warming 172–173
warming 135, 176–177, 179
biogeochemical models 158
biological soil crusts 356, *358*
biomass* *580–583*
for bioenergy 19, 193–194
Biomass Production Efficiency (BPE) 368
burning 162, 168, 169, 7.SM.7.1
burning emissions 162, 168, 169
feedstock 399, 605, 610
field measurements *163*
fuelwood 288
harvested 86, 351, 352
potentials *581*
resource management *739*
sustainability standards and certification
     707–709, *708*
traditional biomass 20, 288, 375, 709, *740–742*
water content 262
biomes* *141*, 279–280, *280*
Amazon biome *481*
biome shifts 140, 371, 684, 7.SM.7.1
biophysical effects* 135, *139, 173, 174, 175*
bioenergy deployment 194
changes in anthropogenic land cover 135, 243,
     243–247
cooling 172–173, 174–175, 177, 178, 179
deforestation/forestation *98*, 176–180

forest response options 191–192
global 172, 174–175, 177, 178, 179
regional 172–173, 174, 175–176, 177–178, 179
seasonal 178–179
warming 172, 174, 175, 177–178, 179, 197
biophysical models 262, 364, 366
biotic degradation processes 355–356, 371–372
black carbon (BC)* 167, 168, 451, 591, 606, *740*
blockchains 513
Bolivia 452
bookkeeping/accounting models 9, 152–155,
154, **163**
boreal forest 15, 179–180, 191
climate related risks 7.SM.7.1
shift to woodland/shrubland 7.SM.7.1
boreal regions 12
area burned by fires 156
BVOC emissions 169
climate feedbacks 182–183, *182, 183*
deforestation *177*, 179
evolution of natural vegetation 172
fires 148–*149*, *150*
forest management 191–192
forestation *177*, 179, 191
peatlands 397, 398
water use efficiency *165*
Botswana 287
Brazil *482, 562, 582*
Bromus tectorum (Cheatgrass) 299
Brossentia papyrifera (Paper Mulberry) 299
brown carbon (BrC) 167, 168
buffelgrass 299
Burkina Faso 263
burning embers diagrams *14*, 680, *681, 685, 687,*
7.SM.7.1–3
bush encroachment *see* woody encroachment
business as usual (BAU)* 199, *200*

C

Canada
fire *149*, 684
floods *744*
canopy cover 367, 368, *369*
cap and trade systems 702
capabilities 557–558
capital markets 511
carbohydrate dilution 463
carbon balance 191, 193, 201–202
carbon budget* 485, *573*
global carbon budget 157, 385
carbon capture and storage (CCS)* *99*, 373
carbon cycle* 84
future terrestrial carbon source/sink 137
impact of desertification 268
impact of extreme rainfall 148
impact of fire *149*
impact of heat extremes and drought 146
impact of land degradation 376
peatlands, wetlands and coastal habitats 193
rebound effect 157
wild animal management and 586

carbon dioxide (CO$_2$)* 8–9, *10–11*, 152–157
4p1000 initiative 387
biogeochemical effects of land use change
     171–173, *173*, 174–175, *174, 175,*
     176–177, *177*, 243, *243–245*
CO$_2$ equivalent emissions *151*
cumulative emissions 243, *243–245*
emission reduction and removal 195–197,
     196, 485
emissions 79, *82–83*, 89, 133–134, 137, *151*
   due to deforestation 153, 176–177, *177*, 476
   due to land degradation 153, 376–377
   due to land use change 195–196, 476
   fire emissions *149*, 586
   food system emissions 475–476, *477*, 478
   fossil fuel emissions 153, *153*
   peatland emissions 159, 397, 476, 477
   soil emissions 134, 203
   transport emissions 478
estimating emissions 134, 153, 155
fluxes *87*, 88, *154*
   AFOLU fluxes 8–9, 133–134, *151*, 152–155,
     152, *153, 154, 156*, 157
   anthropogenic land flux 8–9, 133–134, *151,*
     154, 156, 163–164
   forest fluxes 154
   LULUCF fluxes 199–200, *200*
   total net land-atmosphere flux 8, 133–134,
     152–157, *152*
gross emissions/removals 134, *152*, 157
land cover change and 8–9, *10–11*, 133–134,
     151, 172, 243, *243–245*
land sink processes *153*, 155–156, 157
negative emissions 135–136, 198–199
net anthropogenic flux due to land cover change
     8–9, *10–11*, 133–134, *151, 156*
net emissions *6–7*, 8, *154, 171*
net FOLU emissions *6–7*
net negative emissions 198
non-AFOLU emissions *10–11*, *151, 153*, 155
permafrost release of 134, 184
release from deep soil 203
sequestration through forest area expansion *99*
sink 84, *87*, 172, 180, *182*, 397
   see also atmospheric CO$_2$; carbon dioxide
     removal (CDR); CO$_2$ fertilisation
carbon dioxide removal (CDR)* 135–136,
492, 494
future pathways 22–23
increased need 645
land area needed for 373
land management response options *97*
mitigation pathways 196–197, *196*, 198–199
Paris Agreement 449
potential scale 373
reducing/reversing land degradation 374–375
risk of land degradation 373–374
sustainable forest management (SFM)
     and 386–387
synergies and trade-offs 492, *493*, 494
Carbon Disclosure Project 511
carbon footprint 479, 491, 505, 511
carbon intensity* 702

carbon monoxide (CO) *149*

carbon pools 84, 191, 368
    conservation of 191, *571*
    and erosion 376
    permafrost 184

carbon price* 645, *694*, 701, 702, 753

carbon rights 715, 716

carbon sequestration* 18, 84, 281, 715
    agroforestry 485, *485*
    aquaculture 486–487
    coastal wetlands 193
    compensation by albedo changes 377
    cropland soils 192, 483
    decrease 147
    dryland areas 271
    forests 191, 385–386
    grasslands and rangelands 483
    grazing lands 192
    impact of fire *149*
    impact of heat extremes and drought 146, 147
    land degradation 376
    nitrogen deposition and 203
    projected 278

carbon sink* 351, *352*, 368
    enhancing 388
    forests 21, 156, 180, 386
    future trends 137
    impacts of heat extremes and drought 146
    importance of arid ecosystems 271
    land sink process *153*, 155–157
    loss of *689*
    reversal 686

carbon stocks* 351, *352*
    desertification and 270, 271
    forest carbon stocks 191, 351, *352*, 367, 368–369, 385–387
    modelling 201
    peatlands 397, 398
    vulnerability to extreme events 147

carbon tax 68, 498, 510, 702, 714, 753
    policy in SSPs *727*

carbonaceous aerosols *149*, 167–169, *573*

cascading impacts 354, 376, 682, 690, 691, *744*, 755

cascading risks 15, 679

case studies
    avoiding coastal maladaptation *392*, 402–403
    biochar *392*, 398–400
    climate change and soil erosion 292–294
    climate smart villages in India *563*
    conservation agriculture 561–562
    degradation and management of peat soils *392*, 397–398
    desertification 292–305
    flood and food security *743–745*
    governance of biofuels and bioenergy *738–739*
    green energy trade-offs with biodiversity and ES *735*
    green walls/dams 294–296, *297*
    integrated watershed management (IWM) 302–305, *303*, *304*
    interlinkages between land challenges 561–563
    invasive plant species 297–300, *298*

New Zealand Emissions Trading Scheme (ETS) *703*
    oases in hyper-arid areas 300–302, *300*, *302*
    overlapping land challenges *561–563*
    pasture intensification *562*
    perennial grains and SOC *392*, 393–395
    REDD+ in Amazon and India *709–711*
    reforestation *392*, 395–397
    saltwater intrusion *392*, 401–402
    soil and water conservation *561–562*
    tropical cyclone damage *392*, 400–401
    tropical forests restoration and resilience *562*
    urban green infrastructure (UGI) 391–393, *392*, *563*

catastrophe (CAT) bonds 713

catastrophe risk pool 713

cellular agriculture 487

Cenchrus ciliaris L. (Buffelgrass) 299

Central America 265, 460, 518

Central Asia 264, 293–294

Cerrados, Brazil *562*

certification schemes 602, 707–709, *708*

CFS (Committee on World Food Security) 5.SM.5.5

charcoal 375, *740–741*

cheatgrass (*Bromus tectorum*) 299

childhood stunting 445, *445*, 446, 607

childhood wasting 445, *445*, 446, 607

Chile 265, 292

China
    afforestation programmes 98, 294–295
    crop production 452
    desertification and land degradation 263, 294–295, 396, 603
    dust storms 294–295
    land degradation control policies 396–397
    reforestation 396–397
    rice cultivation 452
    sand movement and railways 275
    Sloping Land Conversion Program 603

citizen engagement 754

citizen science* 512, 748

civil society organisations (CSOs) 5.SM.5.5

Clean Development Mechanism (CDM)* 704

climate*
    climate zone shifts 8, 15, 133, 140, 143, 205
    desertification feedbacks to 268–270, *269*
    dust and 166–167
    effect on land sink 155–156
    future pathways *641–644*
    global impacts of land cover change 171–172, *171*, *172*, *173*, 174–175, *175*, *182*
    impact of deforestation/forestation 176–180
    impact of land condition changes 12, 134–135, 171–186
    local effects 180, 377
    non-local and downwind effects 135, 180, 184–186, *185*
    novel unprecedented climates 143
    projections* *140*, 176, 184, 277
    regional *see* regional climate
    seasonal 173, 178, 179, *572*
    see also climate system

climate change* 7–8, 79–80, 133, *171*, 756
    amplification 172, 178, 377
    biogeochemical warming 172–173, *173*, *174*
    biophysical cooling 172–173, *173*, 174–175, *174*, *175*
    BVOC contribution to 170
    capacity to respond 80
    desertification and 251–252, 258–260
    dust emissions and 167
    equity 446–447, *447–448*
    financial impacts and instruments for managing 712–713
    fire and *148–150*
    food security and 439–440
    future scenarios 564, *565*
    gender and 104–105, 446–447, *447–448*
    impacts *see* climate change impacts
    increasing risks and impacts *14–15*, 15–17
    and indigenous food systems *469*
    influence on land use 90–91
    land challenges and response options 553–555
    land degradation in the context of 347–348, 353–365
    land tenure and 749–750
    observed change *82–83*
    reducing negative impacts 79
    risk management and decision making for sustainable development 675–677, *678*
    risk transitions *14*, 680–688, 7.SM.7.1–3
    role of ILK in understanding *746–747*
    socio-economic challenges 81
    sustainable development pathways *678*
    upper atmosphere effects 185

Climate Change Adaptation (CCA) 467
    see also adaptation

climate change impacts 84–85, 89–90, *90*, 623–624
    agriculture 373, 451–460, *461*, 623
        crop production 8, 380, 451–454, 458–459, 5.SM.5.2
        livestock 276, 454–458, *455*, 5.SM.5.2
        pastoral systems 276, 455–458, 5.SM.5.1
    analysis methods 460
    aquaculture 459
    aridity 142
    biodiversity 404, 624
    climate change impact-land management interactions 351, *352*
    conflict 380–381
    desertification 7–8, 258–260
    direct effects on plant and animal biology 463
    dust emissions and dust storms 167, 277
    ecosystem services 404
    energy infrastructure 275
    feedbacks to climate 12, 171–180, 182–185
    food prices 460–462, *461*, 494–497, *495*, 685–686
    food security 7–8, 15, 142–143, 442, *443*, 450–464, 519, 690
    food systems 89–90, 442, *443*, 450–464
    forests 367–368, 371–372
    gender and 274, 5.SM.5.1
    human health 274, 691, 5.SM.5.1

invasive plants 297–298
on land 140–148
land degradation 351, 360–363, 367–368, 369–373
land ecosystems 84–85, 143–144
on land use 462
local and regional impacts of land cover change 135, *182*
migration 380–381
oases 300–302
peatland degradation 397–398
pests and diseases 458
pollinators 458–459
poverty 259, 272–273, 279
on rangelands 372, 454–456
regionally distributed 143
sensitivity of integrated response options 623–624
short lived climate forcers 167, 169
soil erosion 360, 361–362, 363, 624
terrestrial biogeochemical cycles 157
urban areas *186*, 447, *752*
water resources 205, 274

climate extreme (extreme weather or climate event)* 16, 133, 144–148, *688*
and conflict 518, 690
drylands 259
financial impacts and instruments for managing 712–713
and fire *149*
food system and security 142–143, 450–451, *451*, 464, 500, 514–515, *515*, *516*, 5.SM.5.2
frequency and intensity 133, *145*, 147
historical land cover change impacts 174
and migration 516, *517*, 518, 690–691
policies responses 699–701, 714
precipitation extremes 12, 15, 147–148, 302, 361, 7.SM.7.1
resilience to 28, 285, 500, 513
soil moisture and 135, 184
spatial and temporal scales *145*
temperature extremes 12, 145–147, 174, 176, 179, 362, *516*, *563*
urban areas *186–187*
vulnerability and exposure to 133, 138
see also extreme weather events

climate feedbacks* 138, *140*
boreal regions 182–183, *183*
BVOCs and 169, 170
from desertification 12, 268–270, *269*, *382*
from high latitude land surface changes 183–184
from land degradation 375–377
permafrost carbon feedbacks 183, 184
surface albedo changes 182–184
vegetation greening 172
wind and solar energy installations 288

climate finance 34, 387, 711–713, 716
climate governance* 90–91, 104, 737–738, *748*
climate impact models 513
climate models* 147, *173*, 174–176, *174*, *176*, 276–278, 370
see also CMIP; Earth system models

climate pathways *641–644*
climate policies 27, 68, 639, *678*
and corruption 716
measurable indicators 725
policy integration 103
scenarios consistent with Paris Agreement *642*
climate-resilient pathways* *678*
climate services* 52, 288, *493*, 513
climate shocks 143, 379, 513, 514–515, *515*
climate-smart agriculture (CSA)* 474, 500, *563*, 565–566, *566–567*, 733, *751*, 5.SM.5.5
climate-smart forestry 585
climate-smart villages (CSV) *563*
climate system* 5, *6–7*
aerosols and 166–170
assessing land processes in 91–92
consequences of land-based adaptation and mitigation 47–49, 189–195
future scenarios 92–93, *93–96*, 195–201
land and *83*, 84–85, *90*, 137–138
land forcing and feedbacks 46–47, *139–140*, 171–186
see also land-climate interactions
climate targets* 49, 195–201, *641–644*
climate variability* *140*, 278
CO_2 land sink and 155
fire upsurges *149*
and food security 450–451, 464
impacts on land 140–148, 205
impacts on livelihoods 516, *517*, 518
migration and 516–518, *517*
clouds 166, 168, 169, 177, 377
CMIP (Coupled Model Intercomparison Project)*
carbon and nitrogen cycle feedbacks 157
CMIP6 global emissions pathways 168, 170
desertification projections 277–278
dust emissions 167
CO_2 equivalent emissions* 151
CO_2 fertilisation 134, 155, **165**, 202–203, 266, 362
crops 451–452, 454, 463–464
in drylands 251, 262, *267*
and fire risk 683, 684
greening trends 144, 266, *267*
increased CO_2 removals 8, 155, *165*, 202
and nutritional quality 455, 463–464
and rangeland productivity 455, *455*, *457*
co-benefits* 18–19, 80, *392*, 609, *625*
agroforestry 383, 485, *504*
of biochar 399
bioenergy 374, 492, *581–582*, 592, *739*
carbon dioxide removal (CDR) 374, 492
dietary change 22, 510
disaster risk management 588
ecosystem-based adaptation 706
forest area expansion *99*, *100*
integrated response options 627, 633
integrated response options and SDGs 630
integrated water management 589–590
land management 633
mitigation 138
mitigation and adaptation 19–20, 21
near-term action 33–34
policy design 28

re-vegetation of saline land with halophytes 283
REDD+ and adaptation 590
reducing deforestation and forest degradation *562*
reducing reliance on traditional biomass 375
reducing urban sprawl 594
responses to land degradation 381
risk-sharing instruments 588–589
of Sustainable Development Goals 730–731
sustainable forest management (SFM) 351–352
sustainable intensification 501
sustainable land management 21, 403
urban agriculture 505, 507
coastal communities, risks to 372–373, 400, 7.SM.7.2
coastal degradation 402, 7.SM.7.1, 7.SM.7.2
impact of tropical cyclones 400
saltwater intrusion 401–402
coastal erosion 8, 354, *356*, 370, 372–373
climate related risks 7.SM.7.1
exceeding limits to adaptation 21
result of sea walls 402–403
under different SSPs 7.SM.7.1
wetlands and 372
coastal flooding 402–403, 592, 692, 7.SM.7.1, 7.SM.7.2
coastal maladaptation *392*, 402–403
coastal wetlands
see restoration and reduced conversion of coastal wetlands
coffee crop 372, 383, 460
collective action* 284, 640, 745, 748–749
Committee on World Food Security (CFS) 5.SM.5.5
commodity markets 515, *515*, *516*
commodity-based systems *465*
community-based adaptation (CBA) 474–475, 518, *566–567*
community-based disaster risk management *580*
community forest land *710–711*, *752*
community forestry 385, 720
compensatory afforestation *710*
competition for land 90–91, *100*, 373, 610, *689*
afforestation 610
bioenergy 42, 53–54, 62, *99*
bioenergy and BECCS *581–582*, 607
food systems 449, *502*
land-based CDR 492, 494, 687–688
land-based response options 18–19, *24–26*, 97
compound events 144, 146
comprehensive risk management 712, 721, *724*
conditional probabilistic futures *94*
Conferences of the Parties (COPs)* 473
confidence* 4n, *24–26*, 91, *92*
conflict 89, 150, 275, 380–381, 445, 518, 690
Congo Basin 353, 397
congress weed 298
conservation agriculture 100, 192, 281, 470, 471, 500–501, *566–567*
case study *561–562*
risk associated with 686
conservation planning 706

consumption 106
   grain 605
contingency finance 712–713
controlled traffic farming *503*
convection* *139*, 180, 205
cooperation mechanisms 704–705
coping capacity* 388
corporate social responsibility 106
corruption 716, 750
cost-benefit analysis (CBA)* 96, *693*, *694*, 721
   reforestation 396
   sustainable land management (SLM) 381–382
cost-effectiveness* 102, *693*, 707, 721, 723
costs *692–694*, 711–713
   of action *693*, 723
   of delayed action 102, 348, 644–645
   of drought *290*
   of flooding *744*, 7.SM.7.1
   of inaction 102, 298, 644–645, *693*
   of integrated response options *24–26*, 618,
      619–623, 6.SM.6.4.1
   of land degradation *692–693*
   of mitigation 102
   of SLM technologies 285
   social cost of carbon (SCC) 102, *694*, 702
   of soil erosion 682
   of wildfires 683
Coupled Model Intercomparison Project*
   see CMIP
cover crops 181–182, 192, 376, 471
crisis management *290*, 700
CRISPR systems 513
crop insurance *580*, 588–589, 594, 599, 603,
   608, 699
   see also risk sharing instruments
crop-livestock integration 384, 485–486, *504*
crop models 380, 453–454, *453*
crop production
   adaptation options 470–471
   crop suitability 454
   economic mitigation potential 486
   fruit and vegetables 454
   GHG mitigation 483, 486
   global trends *444*
   impacts of climate change 8, 380, 451–454,
      458–459, 5.SM.5.2
   improved crop management *493*
   land area suitable for 454
   projected impacts 453–454
   sustainable intensification *481–482*
   technical mitigation potential 483
   see also crops; rice cultivation
crop productivity 273, *366*, 373, 518
   assessing climate change impacts 380
   changes for 1.5°C and 2.0°C 279, 454, 680–681
   saline lands 283
   temperature and crop suitability 300–301
   see also crop yields
crop yields 362–363, 379–380, 500
   closing yield gaps *466*, 501, 603
   global trends *444*, 451–452
   impact of climate change 8, 5.SM.5.2

impact of extreme weather and climate 143,
   464, 690
   increasing 605, 606, 607
   low altitudes 680–681, *681*
   projected *453*
   risks from climate change 680–681, *681*, 685,
      7.SM.7.1
   under different SSPs 7.SM.7.1
   warming temperatures and 5.SM.5.2
croplands 22, 79, *561–562*, *633*
   acidification 355
   albedo 181–182, *181*
   biochar biomass production 605
   bioenergy cropland *31–32*, 62, 194, 196, 199,
      642–644, 646
   current land use 79, 85–86, *85*
   dryland areas 254, *256*
   expansion *562*, 595, 602, 603, 604
   GHG emissions 476, *477*
      N_2O emissions 134, 162, 181–182, 476, *477*
   GHG mitigation 22, 483
   global trends 86, *87*
   impact of urban expansion 603
   integrated crop-soil-water management
      280–281
   land tenure 287
   mitigation potential 189
   nitrogen addition to soils 134, 162, *163*
   overlapping land challenges *560*, 561, *561*, 633
   projected land use change *30–32*, *461*, 462
   reduction in 197
   regional and local temperature change 194
   smallholders *751*
   soil carbon sequestration 192, 483
   soil erosion 293, 294, 596
   soil organic carbon 393–395
      see also improved cropland
         management; reduced grassland
         conversion to cropland
crops 79, 84–85
   agronomic practices 382–383
   bioenergy 97, 193–194, 373–374, 492, *576*, 646
   biofuel 193–194, 283, *739*
   cover crops 181–182, 192, 376, 471
   diversification 468, *469*, 589
   genetics 513
   indigenous *469*
   loss 606, *606*
   nutrient quality 463
   oasis areas 300–301
   perennial 194, 383, *392*, 393–395, 485
   perennial grains *392*, 393–395
   pests and diseases 458
   pollinators 458–459
   reduced nutritional value 7.SM.7.2
   suitability 372, 454
   viability under climate change 301
   see also local seeds
cross-level integration 738
cross-sectorial integration 738
cultural policy instruments 106
cultural values 470
cultured meat (CM) 199, 487

customary norms 106, 720
cyclones *see* tropical cyclones
Czech Republic 453

D

dairy systems 483
dams and dam-building 734, *735*
data sources 91–92
date palms 300–301
decarbonisation* 97, 675
decentralised governance 287
decision making 68, 70, 638–639, *678*,
   719–725, *726*
   adaptive management 723–725, *724*
   cost-benefit analysis 96
   economic approaches 721
   effectiveness 28–29
   FAQs 755–756
   formal/informal 720
   futures analysis *93*, *96*
   in global models *96*
   human-environment interactions *360*
   indigenous and local knowledge (ILK) *747–748*
   knowledge gaps 755
   participation 754
   performance indicators 725
   precautionary approach 96
   problem structuring 720–721
   response to key challenges 103
   synergies and trade-offs 725, *726*, 730–734
   tools 721–723, *722*, 734
   under uncertainty *96*, *693*, 719, 721–723, *722*
deforestation* 79, 368, *562*
   and agricultural expansion *481*
   agricultural expansion *481–482*
   albedo impacts 377
   boreal regions *177*, 179
   BVOCs emissions 170
   CO_2 emissions 8–9, 153, 155, 176–177,
      177, 476
   community-managed forests 385
   definition 155
   drivers 367
   emissions estimates 45, 385–386
   emissions reduction 388
   and fire *149*
   historical land cover change impacts 170, 174
   impact on climate 12, 176–180
   imported deforestation 707, 709
   and land tenure 749, *752*
   mangroves 402
   mitigation potential 189–191
   and net forest area increase *98*
   non-local and downwind rainfall effects
      185–186, *185*
   physical effects 377
   REDD+ 385–386, *709–710*
   reducing/halting 100, 385–386, 388,
      481–482, *562*
   seasonal impacts 178, 179, 179–180
   simultaneous cooling and warming response
      176–177

soil N$_2$O emissions 162
and spread of malaria 691
temperate regions *177*, 178–179
tropical regions 177–178, *177*, 385–386, *562*
  see also reduced deforestation
    and forest degradation
delayed action 34, 67, 102, 348, 554, 644–645
deliberative governance* 745
deltas 372–373, 401–402
demand management response options
  101–102, 195, *577*, 610
  adaptation potential *493*, 593, *593*
  delayed action 645
  feasibility 618, *622*, 6.SM.6.4.1
  impact on desertification 598, *598*
  impact on food security 607, *607*
  impact on land degradation 602, *602*
  impact on NCP *629*, 6.SM.6.4.3
  impact on SDGs *632*, 6.SM.6.4.3
  mitigation potential *190*, 191, 195, *493*,
    587, *588*
  policy instruments 698, *726*
  potential across land challenges *615*
  sensitivity to climate change impacts 624
  uncertainties in potentials 5.SM.5.3
demand-side adaptation 439, 472–473
demand-side management 101–102, 195, *493*, 698
demand-side mitigation options 487–491
dense settlements *560*, 561, *561*, *563*, *633*
  see also urban areas
desertification* 5, *6–7*, 7, 50–52, 89, 249–343,
  558, *689*
  adaptation limits 20, 252–253, 291
  addressing 19–20, 255–256, 279–305
    barriers to 292
    costs of 285
    potential for 24–*26*, 595–599, 609–610,
      609, 611–617
  afforestation/reforestation programmes
    294–296, *297*
  anthropogenic drivers 251, 259–260, 264, 268
  assessing 260–265
  attribution methods 265–268
  biodiversity and 263, 271–272, 278–279
  case studies 266–268, 292–305
  challenge *559*, *560*, 561, 564, *565*
  climate related risks 7.SM.7.1, 7.SM.7.3
  defined 4*n*, 107, 254
  desertification syndrome vs. drylands
    development 260
  detection and assessment methods 255,
    260–262
  difference from land degradation 107, 254
  drivers 251, 258–260, 264, 268, *382*
  ecosystem services and 270–271, 278–279
  FAQs 107, 306, 646
  feedbacks to climate 12, 268–270, *269*, *382*
  financing mechanisms 712
  fire and *149*, 259, 597, *597*
  future projections 276–279
  future scenarios 564, *565*, 634, *635*
  global scale 255, 260–262, *599*
  hotspots 292

impact of climate change 7–8, 258–260
impact of integrated response options 19–20,
  595–599, *611–617*
impact of risk management 598–599, *599*, *617*
impacts of 270–276, 278–279
indicators of 255
knowledge gaps 305–306
local case studies 266–268
location-specific trends 263–265
maladaptation 291–292
migration and 295
near-term action 33–34
on-the-ground actions 279–283, *280*
policy responses 285–289, *290–291*, *696*,
  705–706
previous IPCC and related reports 256
processes 258–259
regional scale 263–265, 277
research and development investment 287–288
risks from climate change *681*, 682–683
  under different SSPs 684, 685, 7.SM.7.1,
    7.SM.7.3
  SLM practices 255–256, 279–283
    adoption of 283–284, 285–288
  socio-economic impacts of 272–276, 279
  socio-economic response 283–285, 288–289
  soil erosion and 292–294, 596
  technologies 285, 287–288
  urban response options *188*
detection and attribution* 91, 5.SM.5.2
developing countries 372, 390, 512, 618, 701
  agriculture 384, 697, 699
  citizen engagement 754
  deforestation 368, 369, 385
  dryland populations 257–258
  early warning systems 91, 253, *290*
  emissions *186*, 704, 709
  energy sources 288, 375, 494
  finance 387, 701, 711–712, 713
  floods *744*
  food loss and waste 100–101, 440, 490, *577*
  food security 465, 468, 472, 697
  food systems 593, 604, 605
  GHGI reporting *164*
  land tenure 749–750, *751–752*
  large-scale land acquisitions 91, 750
  livelihoods 378–379, 608
  NDCs 199, 704
  poverty 53, 257–258, *290*, 378–379
  risk sharing 699
  social protection systems 699
  technology transfer 68, 704
  urbanisation *186*, 603
  vulnerability 449, 452, *573*
  women 285, *718*
    see also REDD+
development pathways* 16, 756
DGVM
  see Dynamic Global Vegetation Models (DGVMs)
diet* 79–80, 86, 101, 5.SM.5.4
  addressing climate change 22, 519
  changing *see* dietary change
  cultured meat (CM) 487

dietary diversity 468, *469*
dietary patterns 101
  GHG emissions for different diets 479–480
  and health 497–499, *498*
  indigenous communities 106
  insect-based diets 490
  local produce 491
  low GHG emission diets 497–499
  mitigation potential 195, 487–489, *488*
  near-term action 34
  and poverty 442
  reducing meat consumption 489, 498–499,
    5.SM.5.3, 5.SM.5.4
  rural diets 605
  sensitivity to climate change impacts 624
  sustainable 497–499
  traditional *469*
  urban diets 505
dietary change 101, 195, 196, *469*, 490, 497–499,
  567, *577*
  adaptation potential 593, *593*
  co-benefits 510
  combined with bioenergy and BECCS 637
  feasibility *622*, 6.SM.6.4.1
  impact on desertification 598, *598*
  impact on food security 607, *607*
  impact on land degradation 602, *602*
  impact on NCP *629*, 6.SM.6.4.3
  impact on SDGs *632*, 6.SM.6.4.3
  interlinkages 636
  mitigation potential 487–489, *488*, 587, *588*
  policy instruments *726*
  potential across land challenges 609, 610, *615*
  sensitivity to climate change impacts 624
  socio-cultural barriers 618
  technical mitigation potential uncertainties
    5.SM.5.3
direct aerosol effect 169, 170, 192
disaster risk management (DRM) *567*, 580,
  744–745
  adaptation potential 594, *594*
  feasibility *623*, 6.SM.6.4.1
  impact on desertification 598, *599*
  impact on food security 608, *609*
  impact on land degradation 603, *603*
  impact on NCP *629*, 6.SM.6.4.3
  impact on SDGs *632*, 6.SM.6.4.3
  mitigation potential 588
  policy instruments *726*
  potential across land challenges 610, *617*
  sensitivity to climate change impacts 624
disaster risk reduction (DRR) 33, 396, 467,
  474–475, *744*
discount rates* *694*
diseases *see* human health; pests and diseases
disruptive technology 511
diurnal temperature 178, 179, 180, *186–187*
diversification
  agricultural production systems *504*
  crop diversification 468, *469*, 589
  dietary diversity 468
  economic 285, 288–289
  energy supply 20

food system 22, 468–470, *469*
    intensively managed systems *504*
    livestock production systems 589
        see also agricultural diversification;
           livelihood diversification
downwind effects 135, 184–186, *185*
drainage* 89, 355
    peatlands 160, 397–398
    wetlands 193
drought* 7, *82–83*, 145–147, 276
    adaptation measures 686
    climate related risks 7.SM.7.1
    conflict and 516, 518
    costs *290*
    defined 254
    and desertification 265
    drivers of 258–259, 266
    Dust Bowl-type 514–515
    early warning systems (EWS) 475, 594, 598, 608
    fires *149*
    food security and 450, *451*, 464, 516, 690
    frequency and intensity 15
    impact on food security 5.SM.5.2
    impact on land 146–147
    inter-annual variability 145–146
    livestock and 457–458
    maladaptation 734
    migration and 276, 285, 518
    policy responses to *290–291*, 714
    projections 144, 277–278
    risks and risk management *290–291*,
        682–683, 700
    soil moisture and 184
    trends 145–146
    vegetation response 202
drought risk mitigation *290–291*
dryland areas 5, *6–7*, 7, 16, 89
    adaptive capacity 16, 753
    addressing desertification 279–284, 286–288
    agriculture 16, 257, 259
    biodiversity 271–272, 278–279, *735*
    classification 254, *255*
    climate-driven changes in aridity 142
    delineating in increasing CO_2 environment 254
    drivers of vegetation change 265–268, *267*
    economic diversification 285, 288–289
    ecosystem services 270–271
    expansion 278
    food demand 259
    food security 8
    future projections 277–279
    geographic distribution 254–255, *254, 255*
    impacts of climate change 142, 278–279
    indigenous and local knowledge (ILK) 284, *747*
    land degradation 262
    land tenure 753
    land use/cover 254–255, *256*, 257
    livelihoods 257–258, 304
    near-term actions 33–34
    poverty 257, 259, 260
    precipitation 258, 262, 265, 278
    renewable energy *735*
    soil erosion 258, 259, 278

tipping points 645
vulnerability and risk of desertification
    277–278, 645
water scarcity *681*, 682–683, 684, *685*, 7.SM.7.1
dryland populations 16, 89, 255–256, *257*
    anthropogenic drivers of desertification
        259–260
    vulnerability and resilience 256–258
dust 166–167, 268–269, *269*, 377, 683
dust mass path (DMP) 167
dust storms 7, 268, 271, 275, 682
    combating 20, 283, 294–295
    health effects 274
    impacts of climate change 277
dynamic adaptation pathways 721–722
Dynamic Global Vegetation Models (DGVMs)
    153, *154*, 155, 156–157, *156*, **163**
    temperature response to land cover change
        173, 175

E

early warning systems (EWS)* 91–92, 102, 475,
  598, 700
    effectiveness 594
    food security related 91–92, 594, 608
    near-term actions 33
    sand and dust storms 288
Earth system models (ESMs)* *95, 140*, **163**
    soil and plant processes 201–204
    sources of uncertainty 144, 156–157, 201–202
    temperature response to land cover change 175
    underestimating emissions 168
East Asia 168, *176*, 395–396
    see also China
ecological cascades* 251
ecological fiscal transfer (EFT) *711*
economic barriers 618, *619–623*, 715, 6.SM.6.4.1
economic decision making approaches 721
economic diversification 285, 288–289
economic growth 644–645
economic health costs 510
economic mitigation potential 483, 486, 489
economic policy instruments 105–106
economics *692–694*
ecosystem-based adaptation (EbA) 19, 282, 381,
  468, 470, *566–567*, 706–707
ecosystem services (ES)* *79, 625–626*
    aquatic 733–734
    cultural 353
    drylands 270–271
    forest area expansion *99*
    green energy and *735*
    human-environment interactions *360*
    impact of afforestation and reforestation 97
    impact of bioenergy crop deployment 97,
        687–688, *687*
    impact of climate change 404
    impact of invasive plant species 297–300
    market-based policy instruments 105
    near-term actions and 33–34
    non-material 81

PES schemes 105, 287, 706–707, 733
risk to 691
SLM practices and 306
synergies and trade-offs 730, 731, *731, 735*
valuing 5, 79, 81, 350, *692–694*
ecosystems*
    changes in distribution 172, *182*
    climate-driven changes 143–144, 355
    climate-related risks 680–688, *688–689*
    health and resilience 592
    impact on climate 84
    impacts of bioenergy crop deployment
        687–688, *687*
    impacts of climate change 7–8, 84–85, 270–272
    impacts of desertification 270–272
    impacts of land degradation 88–89, 354–356
    implications of forest area increase *98–99*
    land ecosystems 79, 84–85
    land management response options 97
    managed/unmanaged *139–140*, 172, 191, 203,
        270–272, 356
    management *747*
    models *95*, 266
    plant composition changes 355
    resilience 146, 147, 265
    restoration 707
    risk of delayed action 34
    vulnerability to irreversibility 645
EDGAR
    see Emissions Database for Global
        Atmospheric Research
education 286, 512
Egypt 301, *302*
El Niño-Southern Oscillation (ENSO)* 146, *149*,
  305, 361, 397, 450
elemental carbon (EC) 167–168
elite capture 716
emergent risks *678*, 679
emissions
    aerosols 166, 168–169
    emissions pricing 702, *703*
    estimating 9, 152–155, *154, 156, 163–164*
    fugitive emissions 159
    knowledge gaps 513
    land sector net emissions 8–11, *10–11*
    net negative 198, *580*
    net reductions needed to limit global warming to
        2°C or 1.5°C 197–199
    peatland 159–160
    policies 701–705
    regional differences 155, *156*
    short-lived climate forcers (SLCF) 166–167, 168,
        169, 170
    soil carbon and microbial processes 201
    urban areas *186*
    see also AFOLU; anthropogenic emissions;
        greenhouse gas emissions
Emissions Database for Global Atmospheric
  Research (EDGAR) 160, *160, 161*
Emissions Trading System (ETS) 702, *703*, 753
enabling conditions* 27–29, 103–106, 558,
  638–640
    coordinated action 554, 640

food security related adaptation *467*
food system policies 27–28, 440, 507–512
land management responses *633*
policy effectiveness 713–714
for REDD+ *385*
end-use/market integration *738*
energy access 20, 709, *741*
energy demand *731*
energy infrastructure *275*
energy services *740*
energy use
see improved energy use in food systems
enhanced urban food systems 567, *578*
adaptation potential *593*
feasibility *622*, 6.SM.6.4.1
impact on desertification *598*
impact on food security 608, *608*
impact on land degradation *602*
impact on NCP *629*, 6.SM.6.4.3
impact on SDGs *632*, 6.SM.6.4.3
policy instruments *726*
potential across land challenges *616*
sensitivity to climate change impacts *624*
Enhanced Vegetation Index (EVI) *363*
enhanced weathering* of minerals 194, *567*, *575*
adaptation potential *592*
feasibility *621*, 6.SM.6.4.1
impact on desertification *597*
impact on food security 607, *607*
impact on land degradation 601, *602*
impact on NCP *628*, 6.SM.6.4.3
impact on SDGs *631*, 6.SM.6.4.3
mitigation potential 587, *587*
potential across land challenges 610, *615*
sensitivity to climate change impacts *624*
ensemble (models)* *150*, 262, *267*, *278*
enteric contamination 462–463
enteric fermentation *160*, 189, 477, *484*
environmental barriers *619–623*, 6.SM.6.4.1
environmental risk 275–276
equality*
in decision making *754*
enhancing gender equality 70
impacts on gender equality *274*
inequality as barrier to climate action 716
see also gender equality; inequality
equity* 42–43, 68, *717–719*
climate change and food system 58, 446–447,
447–448
in decision making 638–639
see also gender equity
erosion 258, 259, 263, 264, 265
gully erosion 302, 303, *304*, 359, 7.SM.7.1
and precipitation 361–362, 370–371, 682
see also coastal erosion; soil erosion;
water erosion; wind erosion
ES *see* ecosystem services
Ethiopia
drought 450, *451*
invasive plants 298
Tigray region croplands *561–562*
work-for-insurance programmes 699
Eucalyptus camaldulensis 299

Eurasia 172–173, *174*
Europe
agricultural emissions 159
black carbon emissions 168
crop production and climate change 453
cropland albedo *181*
desertification 264
dryland areas *255*, 264
dryland population *257*
food safety 463
forestation 179
European Union (EU)
emissions 753
mitigation and adaptation 737–738
Renewable Energy Directive (EU-RED) 709
eutrophication *358*
evaporation* *362*
evapotranspiration* *98*, 137, 177, *178*
and soil moisture content 362
tropical regions 183
EVI (Enhanced Vegetation Index) *363*
expanded policies 508–510, *509*, 5.SM.5.5
expert judgement 680, 722, 7.SM.7.1
explicit nitrogen stress *453*, 454
exploratory scenario analysis *93–94*
exposure* 680, 683, 684–686, 688, *688–689*, *745*
communities and infrastructure 691, 692
in different SSPs 684–686, *730*
to fire 7.SM.7.1
to flood 7.SM.7.1
extension 695, 697, *719*
externalities *694*
extreme weather events* 85, 144
adaptive governance and 742–743
and crop yields 102, 143, 459–460, *516*,
681–682
early warning systems 475
food systems and security 459–460, 462, 465
food prices and markets 514–515, *515*, *516*
food supply stability 15, 147–148, 464,
514–515, 682
infrastructure vulnerability 692
insurance 594, 699–700, 712
and land tenure *751*
and poverty 259, 272, 306
and smallholders 459–460
see also climate extreme

F

Fakara region, Niger 263
famine early warning systems 594, 608, 700
FAOSTAT emissions data *151*, 153, *154*, *156*, 157,
160, *160*, *161*, **164**, *200*
FAQ *see* frequently asked questions
farmer-led innovations 284
feasibility* *618*, *619–623*, 6.SM.6.4.1
feedstock 19, 605, 610
fertiliser *6–7*, 79, 86, 500
CO_2 fertilisation effect 202–203, 262, 362
N_2O fluxes 160, *161*, 162, *163*
nitrogen fertilisation 134, 159, 203
overfertilisation *357*

synthetic nitrogen 160, 162, 476, *477*
FIES *see* Food Insecurity Experience Scale
financial policy instruments 105–106, 474, 698,
710–711
risk management 467–468, 700–701
financing mechanisms 34, 387, 475, 711–713
fire 133, *148–150*
area burnt *148–149*, *150*, 156, 168, 683,
7.SM.7.1
biomass burning 162, 168, 169
climate change and fire regimes *149–150*,
683–684
damage 606, *681*, 683–684
desertification and *149*, 259
emissions *149*, 162, 168, 169, 270, 376, 397,
573, 7.SM.7.1
fire weather season length *14*, 45, *149–150*,
7.SM.7.1
forests *149*, 372, 700
frequency 382, 7.SM.7.1
future projections *149–150*
impact of invasive plant species 297, 299
increased burning 359
land degradation *149*, 376
management *see* fire management
peatlands *149*, 397, 398
rangeland management 281–282
regimes *149–150*, 376, 683–684, 700
risk management 700
risks *149–150*, 593, *681*, 683–684, 7.SM.7.1
societal impacts 397, 683
fire management *280*, 281–282, *567*, *573*
adaptation potential 591, *592*
feasibility *621*, 6.SM.6.4.1
forests 700
impact on desertification 597, *597*
impact on food security 606, *606*
impact on land degradation 601, *601*
impact on NCP *628*, 6.SM.6.4.3
impact on SDGs *631*, 6.SM.6.4.3
interlinkages 636–637
mitigation potential 586, *587*
policy instruments *726*
potential across land challenges 609, *614*
potential deployment area 633, *633*
sensitivity to climate change impacts *624*
fisheries 471, 478, 5.SM.5.2
flexitarian diet 488, 489, 498, 510
floodplains *752*
floods* 147–148, *358*
Bangkok flood 472
coastal flooding 402–403, 592, 692, 7.SM.7.1,
7.SM.7.2
costs *744*, 7.SM.7.1
fisheries management 471
food safety and human health 462–463
governance indicators 753
impact on food security *743–745*, 5.SM.5.2
impact of peatlands 592
increased likelihood 371
insurance 701, *744*
land degradation and GHGs 377
Limpopo River basin 305

policies 714, *744*
recurring episodes 355
risk management 700–701
risks *685*
under different SSPs 7.SM.7.1
food access 15, *443*, 513, 514, 697–698
gender and *447*
impact of climate change *443*, 460–462, *461*, 690
impact of climate drivers 5.SM.5.2
and poverty 446
risks to *685*, 7.SM.7.1
under different SSPs 7.SM.7.1
Food and Agriculture Organization of the UN
(FAO) *83*, 473–474
food aid 607
food availability *443*, 472, 513–514
crop production 451–454
gender and *447*
global trends *444*, 445
impact of climate change 15, *443*, 450,
451–460
impact of climate drivers 450, 5.SM.5.2
livestock production 454–458, *455*
policies 697
risks to *685*, 7.SM.7.1
seasonal supply shocks 467
under different SSPs 7.SM.7.1
food demand *6*, *82–83*, 259
projected increase 88, *502–505*
food-energy-water nexus *466*, 514
food insecurity 508
assessing 442, 446
defining 442
future scenarios 564, *565*
migration and 516–518, *517*
spatial distribution 446
status and trends 445–446, *445*
see also food security*
Food Insecurity Experience Scale (FIES) 442, 446
food loss and waste* 5, 100–102, 195, 473,
490–491
demand-side 101–102, 195
education 512
GHG emissions 476, 490–491
policy response 510
reduction 22, 498, 507, 512
reduction and use of 485–486
SDGs and 507
supply-side 100–101
see also reduced food waste (consumer or
retailer); reduced post-harvest losses
food prices 15, 510, 605, 7.SM.7.1
controls 607
impact of climate change 460–462, *461*,
685–686
impact of climate drivers 5.SM.5.2
impact of large-based CDR 492, 494
shocks 508
spikes 514–515, *515*, *516*, 593, 682
stability 607
taxes 698
food processing 478–479
see also improved food processing and retailing

food production 606, 697
and bioenergy deployment 7.SM.7.1
building resilience into *466*
climate drivers relevant to 450–451
risks to in dryland areas 684, *685*
transformative change *465–466*
see also increased food productivity
food quality 463–464
food safety 462–463, 472
food security* 20, 56–59, 79–80, 81, 89–90,
437–550, *441*, *443*, 558
adaptation *443*, 472, 475
adaptation options *467*
assessing 442, 446
case studies *561*, *563*
challenge 88, *559*, *560*, 561
climate drivers relevant to 450–451, 5.SM.5.2
climate related risks 7.SM.7.1, 7.SM.7.3
co-benefits with agroforestry GHG mitigation 485
community-based adaptation (CBA) 474–475
conflict and 518
defined *4n*, 89, 442
detection and attribution methods 5.SM.5.2
dietary diversity 468
early warning systems 475
ecosystem-based adaptation (EbA) 468, 470
enabling conditions for adaptation *467*
equity and 446–447, *447–448*
extreme climate events 464–465
FAQs 646
financial resources 475
floods and *743–745*
food aid 464
food loss and waste 490–491, 507
food prices 446, 447, 460–462, *461*, 494–497, *495*
spikes 515–516, *515*, *516*
food safety and quality 462–464
four pillars
see food access; food availability;
food stability; food utilisation
framing and context 441–450
future challenges 514–518
future scenarios 450, 634, *635*
gendered approach to climate change impacts
446, *447–448*, 456, 5.SM.5.1
global initiatives 473–474
governance 5.SM.5.5
health and 447, 450, 462–463
hidden hunger 442, 445–446
Hindu-Kush Himalayan Region *469*
hunger, risk of 460–462, *461*, *495*, 685–686,
7.SM.7.1
impact of climate change 7–8, 15, 142–143,
442, *443*, 450–464, 519, 690
impact of climate drivers 5.SM.5.2
impact of desertification 272, 273
impact of integrated response options *24–26*,
603–608, *609*, 610, *611–617*
impact of land-based CDR and bioenergy 492,
494, 607, *607*, *615*, 687–688
impact of land degradation 379–380
impact of land grabbing 91, 750
impact of land-use change *461*, 462

impact of risk management options 608, *609*
incorporating ILK in decision making *747–748*
incremental adaptation *467*
integrated approaches 492
knowledge gaps and key research areas
513–514
links to SDGs *506*, 507
migration and 516–518, *517*
mitigation *443*
mobilising knowledge 512
near-term action 33–34
as outcome of food system 442
policies 474, 696–698, *696*
pollinators, role of 458–459
potential for addressing *24–26*, 603–608, *609*,
610, *611–617*
prevalence of undernourishment (PoU) 464
protein availability 463
risk management 467–468, 475, 608
risks from bioenergy and BECCS 687–688
risks from climate change 680–682, *681*
under different SSPs 685–686, 685,
7.SM.7.1, 7.SM.7.3
spatial distribution of food insecurity 446
status and trends 445–446, *445*
sustainable intensification 501–502, 507
synergies with adaptation and mitigation *448*
synergies and trade-offs 29, 492, 494, 730, 731
temporal scales 738
trade policies 508
and traditional biomass use *741*
transformational change in food systems
466–467, *467*
urban areas *188*, 449, 505, 507, 607
see also food insecurity
food sovereignty 508, 608
food stability *443*, 479, 513, 514, 593, 607
gender and *447*
impact of climate change *443*, 464, 682, 690
impact of climate drivers 5.SM.5.2
risks to 7.SM.7.1
food supply 5, 697
instability *681*, 682
risks to 7.SM.7.1
stability 15, 685
food system* 15–16, 56–59, *60*, 89–90, *90*, *443*
adaptation challenges 464–465
adaptation options 21–22, *443*, 449, 464–475,
513–514
agroecology 468–470, *469*
capacity building and education 512
climate change response options 448–449
climate drivers important to 450–451
culture and beliefs 470
defined *8n*
demand-side adaptation 472–473, 513–514
demand-side mitigation 487–491, 513–514
detection and attribution methods of climate
change impacts 5.SM.5.2
dietary preferences and consumer choice 489
diversification 22, 468–470, *469*
economic health costs 510
enabling conditions 27–28, 440, 507–512

expanded policies 508–510, *509*, 5.SM.5.5
food security as an outcome of 442
food supply and required food *444*
future projections 479–480, 5.SM.5.2
gender and equity 58, 446–447, *447–448*
GHG emissions 8, *10–11*, 11, 475–480, *476*,
    477, *478*, 513, 519
global system 8, *10–11*, 11
global trends *444*, 445
governance 507, 5.SM.5.5
impact on climate change 475–480
impact of climate change 450–464
impact of climate extremes 465, 514–515,
    515, *516*
indigenous food systems *469*
institutional measures 473–475, 512
integrated agricultural systems 499–502, *504*
interlinkages *441*
investment and insurance 511
Just Transitions to sustainability 511
knowledge gaps 511, 513–514
land competition 449, *502*
local system 608
markets and trade 80, 472, 508, 511
mitigation 21–22, *443*, 449, 480–491, 513
mitigation potential 449
mobilising knowledge 512
Paris Agreement and 449
policy responses 27–28, 474, 507–510, *509*
    acceptable to the public 490, 510, 698
    agriculture and trade 508
    health related 510
previous reports 448–449, 450
processing 472, 508
production
    see crop production; livestock
        production systems
projected emissions 480
response options related to 21–22, 492, *493*
risk management 467–468
scenario analysis *93–94*
in SSPs *13*
supply chains 491, 513
supply-side adaptation 470–472, 513–514
supply-side mitigation 480–487, *484*, 513–514
sustainable *465–466*, *502–505*
synergies and trade-offs 492, *493*, 494,
    513–514, 733–734, 5.SM.5.5
transformational change 449
transport and storage 471–472
urban areas 505, 507
value chain management 100–102
vulnerability to climate change 680–682, *681*
    see also enhanced urban food systems
food utilisation *443*, 513–514, 7.SM.7.2
    gender and *447*
    impact of climate change *443*, 462–464
    impact of climate drivers 5.SM.5.2
    risks to 7.SM.7.1
forest carbon density 395, 396
forest carbon sink 21, 156, 180, 386
forest carbon stocks 21, 351, *352*, 367, 368–369,
    385–387

forest certification schemes 585, 602
forest conservation instruments *709–711*
forest degradation 367–369
    albedo impacts 377
    charcoal production 375
    defined 350
    emissions 385–386
    local land users 353
    reducing 189–191, 385–386, 388, *562*
    traditional biomass use *740–741*
        see also reduced deforestation
            and forest degradation
forest dieback 371–372, 683, *688*
forest governance 715–716
forest management 368–369, *567*, *571*
    adaptation potential 590, *590*
    feasibility *620*, 6.SM.6.4.1
    illegal logging 716
    impact on desertification 595–596, *596*
    impact on food security 604, *605*
    impact on land degradation 600, *600*
    impact on NCP *628*, 6.SM.6.4.3
    impact on SDGs *631*, 6.SM.6.4.3
    mitigation potential 189–192, 584–585, *585*
    policy instruments *726*
    potential across land challenges 610, *612*
    potential deployment area *633*
    rice cultivation 384
    water balance 371
forest mitigation 715–716, 733
forest productivity 351–352, *352*, 353
    climate related risks 7.SM.7.1
forest response options 100, 189–192, *571–572*
    adaptation potential 590, *590*, *612*
    feasibility *620*, 6.SM.6.4.1
    impact on desertification 595–596, *596*, *612*
    impact on food security 604–605, *605*, *612*
    impact on land degradation 600, *600*, *612*
    impact on NCP *628*, 6.SM.6.4.3
    impact on SDGs *631*, 6.SM.6.4.3
    mitigation potential 584–585, *585*, *612*
    policy instruments *726*
    potential across land challenges 610, *612*
    sensitivity to climate change impacts 623–624
    synergies and trade-offs 733
forestation
    climate impacts 176–180
    combined forestation and irrigation 185, *185*
    seasonal impacts 178–179, *179–180*
        see also afforestation; reforestation
forests* *6*, *85*
    adaptive capacity 103
    albedo 191–192
    biodiversity *98–99*, 352
    biogeochemical and biophysical processes *98*
    BVOC emissions 169, 192
    carbon stocks 351, *352*
    certification schemes 352–353, 707–709, *708*
    community-managed forests 385
    cover change and climate feedbacks 12,
        176–180, *185*
    current land use 85–86, *85*
    dryland areas 254, *256*

effect on temperature variation 174
emissions contributions 368
fire *149*, 372, 700
fire management 700
food from 605
global change and mitigation scenarios *197*
global trends in tree-cover 86–88, *87*
growth rates 386
harvest 386–387
hydrological cycle 371
impacts of climate change 367–368, 371–372
impacts of flooding 148
impacts of heat extremes and drought 146
land degradation 89
land management response options *97*
land productivity trends *366*, 367
land sink and 156
land tenure 749, *752*
large-scale conversion non-forest to forest land
    98–100
managed forest for bioenergy 194
managed forest CO_2 emissions 9, 155
net area increase *98*
plant water transport 202
productive capacity 368–369
projected land use change *30–32*, *461*, *642*
regrowth 352, 353
restoration *98*, 191, 733
secondary organic aerosols (SOA) 169
sequestration potential *99*
water balance *98*
wildfires 372
    see also reforestation and forest restoration;
        sustainable forest management
formal institutions 720, *747*
fossil fuels*
    emissions 153, *153*, 168
    reducing use of 486
    subsidies 701
framing and context 40–43, 77–129
    dealing with uncertainties 91–93, 96
    enabling response to key challenges 103–106
    interdisciplinary nature of SRCCL 106
    key challenges related to land use change
        88–96
    of land and climate issues 745–746
    objectives and scope 81–84
    previous reports *83*
    response options 96–103, *97*
frequently asked questions (FAQs)
    1.1 What are the approaches to study
        the interactions between land
        and climate? 107
    1.2 How region-specific are the impacts
        of different land-based adaptation
        and mitigation options? 107
    1.3 What is the difference between
        desertification and land degradation
        and where are they happening? 107
    2.1 How does climate change affect land use
        and land cover? 205
    2.2 How do the land and land use contribute to
        climate change? 205

2.3 How does climate change affect water resources? 205

3.1 How does climate change affect desertification? 306

3.2 How can climate change induced desertification be avoided, reduced or reversed? 306

3.3 How do sustainable land management practices affect ecosystem services and biodiversity? 306

4.1 How do climate change and land degradation interact with land use? 404

4.2 How does climate change affect land-related ecosystem services and biodiversity? 404

5.1 How does climate change affect food security? 519

5.2 How can changing diets help address climate change? 519

6.1 What types of land-based options can help mitigate and adapt to climate change? 646

6.2 Which land-based mitigation measures could affect desertification land degradation or food security? 646

6.2 What is the role of bioenergy in climate change mitigation, and what are its challenges? 646

7.1 How can ILK inform land-based mitigation and adaptation options? 755

7.2 What are the main barriers to and opportunities for land-based responses to climate change? 756

futures analysis 80, *93–96*

G

GCM

see general circulation models; global climate models

GDEWS (Global Drought Early Warning System) 598

GEF (Global Environmental Facility) 387, 388, 390

gender equality 274, *631–632, 717*

land use and land management 286

and response option implementation 639

women's empowerment 29, 70, 286, *448, 488,* 639, *718–719*

gender equity* 80, 104–105

conservation agriculture 501

dryland areas 257–258

food security and climate change 446, *447–448*

women's empowerment policies 286

gender inclusive approaches 446, *447–448,* 456, *717–719,* 5.SM.5.1

gender inequality 353, 639, 716, *717–718*

general circulation models (GCMs) 147, 370

geophysical barriers *619–623,* 6.SM.6.4.1

GFED4s (Global Fire Emissions Database v.4) *148, 149*

GFGP (Grain for Green Program) 396

GHGI *see* greenhouse gas inventories

gilir balik cultivation 384

glaciers* 294

GLASOD (Global Assessment of Human-induced Soil Degradation) 261

global climate models (GCMs)* *173,* 174–175, 176, 276–278

Global Commission on Adaptation 474

Global Drought Early Warning System (GDEWS) 598

global emissions pathways 168, 170

Global Environmental Facility (GEF) 387, 388, 390

Global Fire Emissions Database v.4 (GFED4s) *148, 149*

global land cover map *297*

global land system *82–83, 87*

current patterns in land use/cover 85–86, *85*

future trends 88, *93–96*

past and ongoing trends 86–88

status and dynamics 79, 84–88

global mean surface air temperature* *see* GSAT

global mean surface temperature* *see* GMST

global warming* 7–8, 140–142, 362

afforestation/reforestation and 49, 191

consequences of 44–45, 133, 140, 205, 277

climate feedbacks 136, 182–184, *183*

climate variability changes 140

climate/weather extremes 144–148, 361–362, 464

for crop yields 451–454

for ecosystems 143–144, 251, 683

for fire regimes 1*49–150,* 683–684

for food systems and security 142–143, 451–460, 463

hydrological changes 684

increased emissions 376–377

land cover and productivity changes 182–183, *183*

for land degradation 347, 360–362

for livestock systems 454, 456, 458

for soil 184, 258

deforestation/forestation and 177–178, 179

delayed action 34

dryland water scarcity and 682–683

land cover change and 12, 47, 135, 171–172, 174–175, 205

net emissions reductions needed to limit to 2°C or 1.5°C 197–199, 686

Paris Agreement 81, 449, 480, 492, 701

regional climate change feedbacks 136, 182, *183*

reversing after temperature overshoot 701

risk and *14–15,* 15–17, 138, 251, 682, 683–686, 7.SM.7.1–3

socio-economic pathways and 684–686

vulnerable populations 278, 459–460, 464, 682–686, 691–692

see also 1.5°C warming; 2°C warming; 3°C warming; climate change

Global Warming Potential (GWP) 11*n,* 151*n*

GMST (global mean surface temperature)* *6,* 7, 133, 140–142, *141, 142*

impacts at different temperatures 680–686, *681, 685*

impacts of 7.SM.7.1–3

projected *13,* 675

risk as a function of 680, *681*

governance* 28–29, 638, *679,* 736–754, 5.SM.5.5

adaptation 471, 474

of biochar 400

of biofuels and bioenergy *738–739*

capacity 743

climate change 5.SM.5.5

climate policy integration 103

combating desertification and dust storms 295

coordination 80

decentralised 287

definitions 679, 736

disaster risk response *744–745*

enabling response to key challenges 103–104

experimentation 736, 742

FAQs 755–756

food systems and security 507, 5.SM.5.5

forest governance *709–711,* 715–716

global experimentalist 5.SM.5.5

hybrid forms 737, *739*

implementing sustainable land management 353

inclusive 80, 754

indicators of adaptive governance 753, *754*

institutions 736–737, 5.SM.5.5

integrated 737–738, 5.SM.5.5

integrated watershed management (IWM) 304, 305

land governance 90–91, 374–375

land tenure 749–753, *751–752*

levels and modes 737–738

market-based policies 105

modes 104

participation 745–746, 748–749, *748,* 753

participatory governance 391, 743, 745–746

policy instruments 105–106

polycentric approach 104, *578,* 737, 738, 5.SM.5.5

for sustainable development 737–738, 754

temporal scales 738

uncertainty, responding to 742–743

windows of opportunity 694–695, 756

see also policies

government effectiveness (GE) 638, *639*

graduation approach 697, 698

Grain for Green Program (GFGP) 396

grasslands 281–283, 494

dryland areas 254–255, *256*

impact of flooding 148

increased plant diversity *504*

multi-species *504*

N₂O emissions 162

productivity trends *366*

projected land use change *461*

soil compaction 596

see also reduced grassland conversion to cropland

grazing land*

current land use 85–86, *85*

dryland areas 254–255, 257

N₂O emissions 162

past and ongoing trends 86, *87*

see also improved grazing land management

Index

grazing practices *280*, 281–282, 355, 376, 5.SM.5.3

Greece 453

green infrastructure*
see urban green infrastructure

Green Revolution *469*

green walls/dams 294–296, *297*

greenhouse gas(GHG)* 81, 137, 205
impact of desertification 270
impact of land degradation 376–377
see also carbon dioxide (CO₂); methane (CH₄); nitrous oxide (N₂O); ozone (O₃)

greenhouse gas emissions *6–7*, 137, *139, 140*, 376
agricultural 159, 160, *160, 161*, 376, 475–478, 511, 702, *703*
croplands and soils 159–160, 16*1*, 162, 16*3*, 476, 477
enteric fermentation 16*0*, 189, 477
global trends *444*, 445, 496
livestock 159, 16*0*, 16*1*, 162, 476, 477–478, *478*, 5.SM.5.3
rice cultivation 159, 16*0*, 477
see also AFOLU
anthropogenic *see* anthropogenic emissions
aquaculture and fisheries 478
bioenergy emissions 194, 196
by food type 513
calculating estimates 152–153
CO₂ equivalent (CO₂-eq) emissions *151*
deforestation 176–177, 385–386, 388
desertification and 270
diet type and 479–480, 497–499, 513, 5.SM.5.4
fire emissions *149*, 162, 270, 376, 397, *573*, 7.SM.7.1
food system 475–480, *476, 477, 478*, 490–491, 519
forest degradation 385–386, 388
global meat consumption 5.SM.5.4
hotspots 397
impact of delayed action 34
indirect land use change (iLUC) 194, 199
land degradation processes 376–377
mitigation pathways 195–196, *196*, 197–199, *197*
non-AFOLU *10–11, 151*
over-consumption of food 490
peatland drainage and management 397, 398
permafrost thawing 134, 137, 184
rapid reduction 34, 79
rebound effect 477–478
reducing with biochar 398–399
reducing by dietary change 497–499
reducing with novel technologies 485–486
regional differences 155, *156*, 158
soil emissions 134, 137, 184, 398–399, 476
spatial distribution *477, 478*
traditional biomass use 375, *740*
transport emissions 478–479

greenhouse gas fluxes 133–134
anthropogenic land CO₂ flux 8–9, *154*
between land and atmosphere 151–165, *163–165*
CH₄ 157–160, 15*8*, 16*0*

CO₂ 8–9, 152–157
gross emissions/removals 15*2*, 157
N₂O 160–162, 16*3*
total net flux of CO₂ 8, 152
bioenergy and BECCS *583*
and climate change mitigation 701–702
desertification and 270
estimation methods and approaches 9, 134, 152–155, *154, 156, 163–164*
forest CO₂ fluxes *154*
future trends *200*
GHGI reporting 164
GHGIs vs. global model estimates 134, 153, *154*, 155
impact of extreme rainfall on carbon fluxes 148
land use effects 151, *151*, 152–153, 159–160
LULUCF CO₂ fluxes 199–200, *200*
managed/unmanaged lands 133–134, *139*, 152, *152, 154*, 155, *164*
plant processes 201
policies *696*, 701–705

greenhouse gas inventories (GHGIs) 9, 153–155, *156*, 513
country reporting **164**
vs. global model estimates 134, 153, *154*, 155

greenhouse gas removal (GGR)* *188*

gross primary production (GPP) 146

groundwater
depletion/exhaustion 271, *689*
irrigation 584, 686, 734
oasis areas 301
over-extraction 271, 289
saltwater intrusion 401–402
vegetation and 268

groundwater stress 558
anthrome area exposed to *560*
case studies *561–562, 563*
global distribution *559*

growing season 144, 182

growth dilution 463

GSAT (global mean surface air temperature)*
land-to-climate feedbacks 182–184, *182, 183*
response to land cover change 171–173, *172, 173*, 174–175, *175*
large-scale deforestation/forestation 177–180, *177*, 243, *246–247*

gully erosion 302, 303, *304*, 359, 7.SM.7.1

H

habitat degradation 685, 7.SM.7.1

Hadley circulation 277

halophytes 283

HANPP (Human Appropriation of Net Primary Production) *87*

hazards* 688, *732, 745*
moral hazard 686–687
non-climatic 7.SM.7.1
policy response 688–689, 714, *726*
see also reduced landslides and natural hazards

health *see* human health

heatwaves* 7, 15, 133, 145, 146, 362

impact on food system and security *516*
soil moisture and 135, 184

hidden hunger 442, 445–446, *469*

high latitude regions
aquaculture 697
crop yields 680–681, 7.SM.7.1
land surface changes and climate feedbacks 183–184
see also boreal regions

Hindu-Kush Himalayan Region 452, *469*

hotspots
desertification 292
GHG emissions 397
land degradation 365
threatened biodiversity 558, *559, 560, 562*

human activity
and fire *148*
interaction with land system *360*
land degradation and 349, 367
vegetation changes attributed to 266

Human Appropriation of Net Primary Production (HANPP) *87*

human barriers to adaptation *715*

human behaviour* *722*
as consumers 101–102, 105
and decision making 720
diet and lifestyle change 101, 618
modelling *96*

human behavioural change* 105, 618, 722

Human Development Index (HDI) 633, *634*, 5.SM.5.2

human footprint 367

human health
and air pollution 288, 451, 691
childhood stunting 445, *445*, 446, 607
childhood wasting 445, *445*, 446, 607
diet and 497–499, *498*, 5.SM.5.4
economic costs 510
extreme temperatures *187*, 447, *563*
food safety 462–463
gender and climate change impacts 5.SM.5.1
impact of climate change 274, 462–463, 691
impact of dust storms 274
impact of food aid 464
impact of lack of clean water 402
infectious disease 590, 691
invasive plants 298, 299
malnutrition 442, 445–446, *445*
micronutrient deficiency 442, 445–446
nutrition-related risks to 7.SM.7.2
obesity/overweight *444*, 445, *445*, 446
plant allergy 298, 299
policies 28, 510
respiratory disease 298, 299, 691
traditional biomass use 375, *740*
undernourishment 442
and urban sprawl 594
urbanisation *187–188*

human systems*
climate-related risks 680–688, *688–689*
communities and infrastructure 691–692
consequences of climate-land change 690–695

Hungary 453

Index

hunger 460–462, *461*, *495*, 685–686
 hidden hunger 442, 445–446
 population at risk of 7.SM.7.1
hurricanes 7.SM.7.1
hydrological cycle* 205
 BVOC emissions and 169
 feedback to climate 138
 intensification 147, 360–362, 370–371
 response to deforestation 175–176
 role of forests 371
hydrological systems 355, 377
hydroxyl radical (OH) 158–159, 169, 170
hyper-arid areas 254, *255*, *256*, *257*
 infrastructure 275
 oases 300–302

I

ice-free land area 5, *6–7*, 558, *560*, 633
 at risk of land degradation 365, 367
 overlapping challenges *560*, 561
ILK *see* indigenous and local knowledge
impacts*
 at different global mean surface temperatures
  680–686, *681*, *685*, 7.SM.7.1–3
 of BVOCs on climate 169, 170
 of climate change *see* climate change impacts
 of climate variability
  on land 140–148, 205
  on livelihoods 516, *517*, 518
 of compound events 144, 146
 of deforestation/forestation on climate 176–180
 of delayed action 34
 of desertification 270–276, 278–279
 of extreme weather and climate 143,
  144–148, *145*
  precipitation extremes 147–148
  temperature extremes 145–147
 of flooding 147–148
 of food systems on climate change 89–90, *90*,
  475–480
 of heat extremes and drought
  on food system and security 51*6*
  on land 146–147
 of historical anthropogenic land cover change
  171–174, *172*, *173*, *174*
 of increased atmospheric CO_2 144
 knowledge gaps 513
 of land-based CDR and bioenergy 492, 494
 of land cover change on climate 135, 171–182,
  243, *243–247*
 of land degradation on climate change
  375–377, *382*
 of land use and land cover 86–88, *87*
 regional 84, *94*
 of short-lived climate forcers 166, 167–168,
  169, 170
imported deforestation 707, 709
improved cropland management 565, *566*, *569*
 adaptation potential 589, *590*
 combined with bioenergy and BECCS *637*
 feasibility *619*, 6.SM.6.4.1
 future scenarios 634, *635*

impact on desertification 595, *595*
 impact on food security 603, *604*
 impact on land degradation 599, *600*
 impact on NCP *628*, 6.SM.6.4.3
 impact on SDGs *631*, 6.SM.6.4.3
 mitigation potential 583, *584*
 policy instruments *726*
 potential across land challenges 609, 610, *611*
 potential deployment area *633*
improved energy use in food systems 567, *579*
 adaptation potential 593, *594*
 feasibility *622*, 6.SM.6.4.1
 impact on desertification 598
 impact on food security 608, *608*
 impact on land degradation 602
 impact on NCP *629*, 6.SM.6.4.3
 impact on SDGs *632*, 6.SM.6.4.3
 mitigation potential 588, *588*
 policy instruments *726*
 potential across land challenges 610, *616*
 sensitivity to climate change impacts 624
improved food processing and retailing *188*,
 491, 567, *579*
 adaptation potential 593, *594*
 feasibility *622*, 6.SM.6.4.1
 impact on desertification 598
 impact on food security 608, *608*
 impact on land degradation 602
 impact on NCP *629*, 6.SM.6.4.3
 impact on SDGs *632*, 6.SM.6.4.3
 mitigation potential 588, *588*
 policy instruments *726*
 potential across land challenges 610, *616*
 sensitivity to climate change impacts 624
improved grazing land management 566, *570*
 adaptation potential 589, *590*
 feasibility *619*, 6.SM.6.4.1
 impact on desertification 595, *595*
 impact on food security 604, *604*
 impact on land degradation 599, *600*
 impact on NCP *628*, 6.SM.6.4.3
 impact on SDGs *631*, 6.SM.6.4.3
 mitigation potential 583–584, *584*
 potential across land challenges 609, *611*
 potential deployment area *633*
 sensitivity to climate change impacts 623
improved livestock management *493*, 566, *570*,
 5.SM.5.3
 adaptation potential 589, *590*
 combined with bioenergy and BECCS *637*
 costs 102, *619*
 feasibility *619*, 6.SM.6.4.1
 impact on desertification 595, *595*
 impact on food security 604, *604*
 impact on land degradation 599, *600*
 impact on NCP *628*, 6.SM.6.4.3
 impact on SDGs *631*, 6.SM.6.4.3
 interlinkages 636
 mitigation potential 584, *584*
 potential across land challenges 609, *611*
 potential deployment area *633*
 sensitivity to climate change impacts 623
inclusive governance 80

increased food productivity *566*, *569*, 697
 adaptation potential 589, *590*
 combined with bioenergy and BECCS *637*
 feasibility *619*, 6.SM.6.4.1
 impact on desertification 595, *595*
 impact on food security 603, *604*
 impact on land degradation 599, *600*
 impact on NCP *628*, 6.SM.6.4.3
 impact on SDGs *631*, 6.SM.6.4.3
 interlinkages 636
 mitigation potential 583, *584*
 policy instruments *726*
 potential across land challenges 609, 610, *611*
 potential deployment area *633*
 sensitivity to climate change impacts 623
increased soil organic carbon content 382–383,
 567, *572*
 adaptation potential 591, *591*
 feasibility *620*, 6.SM.6.4.1
 impact on desertification 596, *596*
 impact on food security 605, *606*
 impact on land degradation 600, *601*
 impact on NCP *628*, 6.SM.6.4.3
 impact on SDGs *631*, 6.SM.6.4.3
 mitigation potential 585, *586*
 policy instruments *726*
 potential across land challenges 609, 610, *613*
 potential deployment area 633, *633*
 sensitivity to climate change impacts 624
incremental adaptation* 466, *466*, *467*, 717, *747*
India
 climate smart villages (CSV) *563*
 crop production 452
 integrated watershed management (IWM)
  304–305
 irrigation 180, 185–186
 monsoon *176*, 180
 net forest area increase *98*
 REDD+ *710–711*
indicators 753, *754*
indigenous* and local knowledge* (ILK) 104,
 746, *746–748*
 adaptation decision making 474
 addressing desertification 283–284
 addressing land degradation 384–385
 agricultural practices 29
 decision making 720
 dryland areas 258, 284
 food storage 472
 food systems and security 512
 informing mitigation and adaptations options 755
 mitigation and adaptation strategies 470
 response option implementation 638
 supply side adaptation 471, 472
 urbanisation 289
indigenous food systems 469, 738
indigenous peoples 27, 28–29, 43, 62, 80,
 746–748
 adaptive capacity 470, 755
 land rights 91, 106, 749–750
 stakeholder involvement and decision making
  104, 638, 640, *710*, 720
indirect aerosol effect 169, 170, 192

Index

indirect land-use change (iLUC) 54, 194, 199

Indonesia 385, 474, *562*, 588
    mangrove deforestation 402, 487
    peatlands 397–398

Indus delta 401–402

inequality*
    barrier to climate action 716
    gender 353, 639, 716, *717–718*
    and land degradation 347
    in land use and management 353
    and maladaptation 474
    reduced *631–632*
    social 481, *744*, 754
    in SSPs *13*, 92–93
    of water management benefits 304, 305
    wealth and power 716

informal institutions 720, *747*

infrastructure
    areas *85*, 86
    coastal 7.SM.7.2
    green infrastructure *735*
    risks to 275, 684, 691–692

insect-based diets 490

institutional barriers 618, *619–623*, 715–716, *715*, 737, 738, 6.SM.6.4.1

institutional capacity* 33

institutional dimensions 753, *754*

institutions* 507, 640, 5.SM.5.5
    adaptation measures 473–475
    adaptive 736–737
    based on ILK *747*
    building adaptive and mitigative capacity 736–737
    decision making 720
    local 287

insurance 102, 467, 511, 700, 712
    crop insurance *580*, 588–589, 594, 599, 603, 608, 699
    flood insurance 701, *744*
    social protection policies 699
    see also risk sharing instruments

integrated agricultural systems 499–502, *504*

integrated assessment models (IAMs)* *95*, *163*, 195, *641*, *643*
    bioenergy and BECCS *30*, *581*, *582*
    integrated response options 634–638, *635*, *636*, 6.SM.6.5.4
    socioeconomic development, mitigation responses and land *30*, *32*

integrated catchment management 561–562

integrated coastal zone management *566–567*

integrated crop-soil-water management 280–281, *280*

integrated landscape management *566–567*

integrated response options* 18–19, 61–63, *63–66*, 551–672, 565–569, *566*
    adverse side effects 627, *628–629*, 630, *631–632*, 633
    barriers to implementation 618, *619–623*, 638, 6.SM.6.4.1
    co-benefits 627, *628–629*, 630, *631–632*, 633
    delayed action 554, 644–645
    enabling conditions 554, 558, 633

FAQs 646
    feasibility 618, *619–623*
    framing within social-ecological systems (SES) 556–558, *557*
    future scenarios 634, *635*
    IAMs and non-IAMs studies 634–638, *635*, *636*, 6.SM.6.5.4
    impact on NCP 627, *628–629*, 6.SM.6.4.3
    impact on SDGs 627, 630, *631–632*, 6.SM.6.4.3
    implementation 18, 633–645
        challenges 638–640
        coordination 640
        stakeholder involvement 638–639
    interactions and interlinkages 636–638, *636*, *637*
    knowledge gaps 638
    land management based
        see land management response options
    mapped from SR15 *568*, 569
    overarching frameworks 565–566, *566–567*
    overlapping challenges 609–610, 633
    potential 583–610, *609*, *611–617*
        for addressing desertification 595–599
        for addressing food security 603–608, *609*
        for addressing land degradation 599–603
        for delivering adaptation 589–594, 609–610
        for delivering mitigation 583–589, 609–610
        global contribution *24–26*
    potential deployment area 633, *633*, *634*
    risk management based
        see risk management response options
    sensitivity to climate change impacts 623–624
    size of negative/positive impact 609–610, *609*, *611–617*
    synergies and trade-offs 627, 636–637
    value chain management based
        see demand management response options; supply management response options

integrated soil fertility management 383

integrated water management *567*, *571*
    adaptation potential 589–590, *590*
    combined with bioenergy and BECCS *637*
    feasibility *619*, 6.SM.6.4.1
    impact on desertification 595, *595*
    impact on food security 604, *604*
    impact on land degradation 599, *600*
    impact on NCP *628*, 6.SM.6.4.3
    impact on SDGs *631*, 6.SM.6.4.3
    mitigation potential 584, *584*
    policy instruments *726*
    potential across land challenges *611*
    sensitivity to climate change impacts 623

integrated watershed management (IWM) 302–305, *303*, *304*

intercropping 280–281, 384, *504*

interdisciplinary work 106, 268, 514

interlinkages 61–66
    adaptive governance *743–745*
    between integrated response options 636–638, *636*, *637*
    between land challenges 551–672
    between land challenges and local response *561–563*

desertification, land degradation, food security and GHG fluxes 551–672
    land, water, energy and food sectors *743–745*
    migration, conflict and climate change 380–381
    poverty, land degradation and climate change 378–379, *379*
    of SDGs *506*, 507
    see also integrated response options

intra-linkages of SDGs *506*, 507

invasive species *358*, 707
    invertebrates 355
    plants 259, 262, 297–300, *298*, 355

invasive species/encroachment management *567*, *574*
    adaptation potential 591
    feasibility *621*, 6.SM.6.4.1
    impact on desertification 597
    impact on food security 606
    impact on land degradation 601
    impact on NCP *628*, 6.SM.6.4.3
    impact on SDGs *631*, 6.SM.6.4.3
    mitigation potential 586
    policy instruments *726*
    potential across land challenges *614*
    sensitivity to climate change impacts 624

investment 33–34, 645
    crop expansion 602
    food system 511
    irrigation 288–289
    land degradation responses 381–382
    land restoration programmes 396
    research and development 287–288
    sustainable land management (SLM) 285, 387, 390, 391

IPBES Land Degradation and Restoration Assessment report 256, 259, 388, *746*

IPCC Fifth Assessment Report (AR5)
    desertification 256
    ecosystem change 143, 371
    food system 448–449, 450
    indigenous and local knowledge *746*
    land aerosols emission 166
    land degradation 350
    land-climate interactions 137–138
    maladaptation 734
    precipitation 361
    risk 679

IPCC Special Report on Climate Change and Land (SRCCL)
    interdisciplinary nature 106
    objectives and scope 81–84, *83*
    overview *84*

IPCC Special Report on Emission Scenarios (SRES) 175

IPCC Special Report on Extreme Events (SREX) *83*, 351

IPCC Special Report on Global Warming of 1.5°C (SR15) *83*, 362
    bioenergy and CDR 373
    crop productivity changes 454
    desertification 256
    food security 449
    indigenous and local knowledge *746*

land-climate interactions 137
precipitation 361
risk 679
IPCC Special Report on Land Use, Land-Use Change and Forestry (SR-LULUCF) 350–351
Iran 264, 278, 402, 460, 470
irreversibility* *69*, 645, *681*, *685*, *687*
irrigation 86, 254, 595, 731
combined forestation and irrigation 185, *185*
development and investment 288–289
downwind and non-local climate effects 185–186, *185*
excessive water use and soil erosion 293, 294
groundwater extraction 401, 402, *563*, 686, 734
impacts on climate 180, *181*
integrated crop-soil-water management 280
land and water rights 749, 750
oasis agriculture 301
risks of 686
isoprene epoxydiol-derived SOA (IEPOX-SOA) 169
IWM *see* integrated watershed management

J

Jordan, Badia region 302–304
Just Transitions 511

K

Karapinar, Turkey 292–293, *293*
Kenya 263, 380, 384, *448*, 460
agricultural commercialisation 289
pastoral systems 287, 457
knowledge gaps
desertification 305–306
dust storms 277
emissions 513
food system and security 511, 513–514
integrated response options 638
interlinkages 755
land degradation 403–404
risk management and decision making for sustainable development 755
traditional biomass *741–742*
knowledge transfer 33, 512

L

La Niña 146, *149*, 305, 591
land* 40–43
area required for bioenergy 687–688, *687*, *739*, 7.SM.7.1, 7.SM.7.3
climate drivers of form and function 140
and climate policies *678*
climate-related extremes 133, 138
CO_2 source and sink 8
competing demand 373, 449, 470, 492, 494, *502*
contribution to climate change 205
defined 4*n*, 349
demand for 750, 756
grazing value 272

human use 5, *6–7*
impact of delayed action 34
land-climate interactions 107
natural response to human induced environmental change 8–9, *10–11*, *151*, *154*
warming *6*, 7–8, 140–142
land abandonment 518
land-atmosphere exchanges, and fire *149*
land-atmosphere feedback loops 268
land-based mitigation *752*, 756
barriers to 715–716
land challenges* 558–565, *561–563*, *565*, *642–644*
consequences of delayed action 554, 644–645
future scenarios 564, *565*
historic and local response to *561–563*
impact of bioenergy and BECCS *580–583*
interaction with response options 638
local 633
overarching frameworks for addressing 565–566, *566–567*, 600
overlapping 558, *560*, 561, *561*, *561–563*, 633, *633*, *634*
policy responses 27–29
response options
see integrated response options
spatial distribution 558, *559*
land-climate interactions 5, 44–49, 131–248, *625–626*
albedo and land use change 138, 177
biogeochemical interactions *140*, 176–177, *177*
biogeochemical warming 172–173, *173*, 174–175, *174*, *175*, 176–177
GHG fluxes 151–162, 16*3–165*
regional 172
biophysical interactions *139*, 177, *177*
biophysical cooling 172–173, *173*, 174–175, *174*, *175*
biophysical warming 174, *175*
regional 172–173, *174*, 175
climate variability/change and impacts on land 1, 133, 140–148, 205
changes in aridity 142
changes in terrestrial ecosystems 143–144
climate extremes 138, 143, 144–148, *145*
food security 142–143
global land surface air temperature 140–142, *141*, *142*
heavy precipitation 137, 147–148
soil carbon response to warming 203–204
temperature extremes, heatwaves and drought 145–147
thermal response of plant and ecosystem production 201–202
water transport 202
cooling response 177, 194
ES/NCP concepts *625–626*
FAQs 107, 205
future terrestrial carbon source/sink 137
hydrological feedback to climate 138
land area precipitation change 137
land-based climate change adaptation and mitigation 138

land-based GHGs 137
land-based water cycle changes 137, 202
land-climate feedbacks *149*, 182–184, *182*, *183*
land cover change and impacts on climate 171–176, 205
amplifying/dampening climate changes 182–184, 18*2*, *183*
deforestation/forestation 176–180
global climate 171–172, *171*, *172*, *173*, 174–175, *175*
regional climate *171*, *172*, 172–174, *174*, 175–176, *176*
urbanisation 18*6–188*
land management impacts 180–182, *181*
land processes underlying *139–140*
plant and soil processes 201–204
previous IPCC and other reports 137–138
regional variations 191–192
response options effect on climate 188–201
mitigation pathways 195–199, *196*, *197*, *198*
SLCFs emissions and impacts 166–170
warming response 194
land cover* 79, 84–88, *297*
biophysical climate interactions 135, *139*
climate feedbacks 135, 182–184, *182*, *183*
dryland areas 254–255, *256*
impact of climate change 205
impact of precipitation extremes 147–148
remote sensing 367, 368
land cover change* 12, 85–88, *171*, *369*
and albedo 12, 181–182, 183–184, 374
and BVOC emission 170
due to global warming 182–183
impact on climate 135, 171–176, 243, *243–247*
deforestation/forestation 176–180
land management changes 180–182
net anthropogenic CO_2 flux *151*
non-local and downwind effects 135, 184–186, *185*
pathways *30–32*, 195–199, *197*, *642–644*
land degradation* *6–7*, 53–56, 84, 88–89, 345–436, 558, *689*
adaptation limits 388
addressing 21, 381–403
barriers to SLM implementation 389–391
local responses *561–563*
near-term action 33–34
on-the-ground actions 381–384
potential for 24–*26*, 599–603, 609–610, *609*, *611–617*
agricultural land 373, 376
assessment approaches 363–367
attribution 360, 362, 363, 404
case studies 391–403, *392*, *561–563*
challenge *559*, *560*, 561, 564, *565*
and climate change 5, 7–8, 353–365, *356–359*
climate-change-related drivers 359
complex linkages 360, 362–363
direct impacts 360–362, 367–368, 369–373
indirect impacts 362–363, 369, 373
land management interactions 351, *352*
climate-induced vegetation change 361, 367–368, 371–372

coastal areas 372–373, *392*, 400–403, *401*
costs *692–693*
defined *4n*, 88, 107, 254, 349–350
difference from desertification 107, 254
diverging and conflicting views 349, 350, 365, 404
drivers 354, 359–360, *360*, 361, *382*
dust emissions 377
FAQs 107, 646
feedbacks on climate *356–359*, 375–377, *382*
financing mechanisms 712
and fire *149*
future scenarios 564, *565*, 634, *635*
global status and trends 365–369, *559*
hotspots 365
human activity and 349, 367
impact on food security 379–380, 604
impact on greenhouse gases (GHGs) 376–377
impact on migration and conflict 380–381
impact on poverty and livelihoods 377–379
impacts of bioenergy and CDR technologies 373–375, 601–602, *602*, *615*
impacts of integrated response options 599–603, 609–610, *609*, *611–617*
indigenous and local knowledge (ILK) 384–385
intensified land use and *502, 505*
knowledge gaps 403–404
land degradation neutrality (LDN) 20, 27, 52, 387–388, 705–706, *705*
and land tenure *752*
local communities and 353
mapping with biophysical models 262
physical effects 377
policy responses 27, 387–388, *696*, 705–711, *705*
previous reports 350–351, 388
processes 354–356, *356–359*, 369–373
projections 369–373
recovery rates 359
resilience 388–389
risks from bioenergy and land-based CDR 373–374
risks from climate change 369–373, *681*, 682, 683–684, 7.SM.7.1, 7.SM.7.3
under different SSPs 684–685, *685*, 7.SM.7.1, 7.SM.7.3
seminatural ecosystems 376
socio-economic drivers 359–360
socio-economic impacts of 377–381
soil carbon loss 381
surface albedo change 377
thresholds 389
traditional biomass use 375, *740–741*
uncertainty in risk assessment 369
urban areas *186*, 188
water erosion risk 370–371
see also desertification; forest degradation
land degradation neutrality (LDN)* 387–388, 566, *705*, 705–706
achieving targets 294, 296, 305
concept 350
framework elements 286
policies promoting 20, 27, 52
in SSPs *727*

Land Degradation Neutrality Fund 712

land governance 90–91, 374–375
land grabbing 91, 750, *751*
land management*
ecosystems and climate change 84
futures analysis and decision making 80
gender inequality 353
GHG emissions 79, 81
impacts on climate 180–182
informed by ILK 755
interactions with climate change impacts 351, *352*
non-local and downwind effects 184–186, *185*
protected areas 100
soil erosion and projected rainfall 371
land management response options *97*, 100, 189–193, *611–615*
agriculture *see* agricultural response options
all/other ecosystems 573–575, *614*
adaptation potential 591–592, *592*, *614*
feasibility *621*, 6.SM.6.4.1
impact on desertification 597, *597*, *614*
impact on food security 606, *606*, *614*
impact on land degradation 601, *601*, *614*
impact on NCP *628*, 6.SM.6.4.3
impact on SDGs *631*, 6.SM.6.4.3
mitigation potential 586, *587*, *614*
policy instruments *726*
potential across land challenges *614*
sensitivity to climate change impacts *624*
co-benefits 633
for CO$_2$ removal *97*, 575–576, *580–583*, 610
adaptation potential 592, *592*
feasibility *621*, 6.SM.6.4.1
impact on desertification 597, *597*
impact on food security 607, *607*
impact on land degradation 601–602, *602*
impact on NCP 6.SM.6.4.3
impact on SDGs *631*, 6.SM.6.4.3
mitigation potential 587, *587*
policy instruments *726*
potential across land challenges *615*
risks of 686–688, *687*
sensitivity to climate change impacts *624*
demand for land 18–19, *24–26*, 97
feasibility 618, *619–621*, 6.SM.6.4.1
forests *see* forest response options
global potential *24–26*
impact on NCP 627, *628*, 6.SM.6.4.3
impact on SDGs *631*, 6.SM.6.4.3
implementation 633
mitigation potential *190*, 586, 587, *587*
policy instruments *726*
potential across land challenges *611–615*
potential deployment area 633, *633*
sensitivity to climate change impacts *624*
soils *see* soil-based response options
land ownership *see* land tenure
land pathways *30–32*, *641–644*
land productivity 182–183, *183*, 365, *366*, 367
land rehabilitation* 351
land resources 79, 80, 104–105
land restoration* 33–34, 294–295, 351, 396
financing mechanisms 712
Karapınar wind erosion area 293, *293*

returns from 285
see also restoration and reduced conversion of coastal wetlands; restoration and reduced conversion of peatlands
land rights 29, 353, *709–711*
barrier to climate action 715
land sharing *502–505*
land sink 84, *154*, 165
process *153*, 155–157
land sparing 488, 489, *502–505*, 636
land surface air temperature* *see* LSAT
land surface models 202
land system models 634–638, *635*, *636*
land systems
global 79, 84–88
human interactions with *360*
risks from climate change 133, 680–686, *681*
land tenure 27, 695, 749–753, *751–752*
barrier to climate action 715
forest rights *710–711*
indigenous and community land rights 106, *709–711*
land grabbing 91, 750, *751*
security 287, 383, *719*
land use* 5, *6–7*, 79
anthromes and land challenges 558, *559–560*, 561, *561–563*
BVOC emissions 135, 170
carbon dioxide (CO$_2$) emissions 133–134, 135–136, 151–157
and climate change 84–85, 133, 134–135, 205, 404, 462
and desertification 89
dietary habits and 79–80, 472–473, 487–489
dryland areas 254–255, *256*, 257
FAQs 205, 404
and food loss and waste 490–491
food system 79–80, *90*, 472–473, 487–489
future pathways 22–23, *30–32*, 88, *93–96*, 195, 373, *641–644*
GHG fluxes 8–9, 79, 81, 84, 133–134, 135–136, 151–162, *163–165*
global patterns *6*, 79, *82*, 84–88, *85*, *87*, *559–560*
integrated landscape initiatives 738
intensity *82–83*, 86, *502–505*
and land degradation 88–89, 404
methane (CH$_4$) emissions 84, 133–134, 136, 151, 159–160, *160*
mitigation response options 79, 97, *98–100*, 100, 135–136, 373–374, 487–489, *641–644*
nitrous oxide (N$_2$O) emissions 133–134, 136, 151, 160–162, *161*, *163*
observations 91–92
policies and governance 737–738, *738–739*, 748–749, *751–752*
pressures and impacts *87*
rotational 157
suitability for CDR 374
transition from livestock production 511
uncertainties related to 91–93, 96

land-use change (LUC)* 5, 6
  assessing 91
  biodiversity losses 88
  bioenergy expansion 739
  BVOC emissions 170
  carbonaceous aerosols emissions 168
  climate drivers 93
  CO_2 emissions 10–11, 195–196, 198
  competition for land 90–91
  drivers 79, 93, 373
  impact on climate 138
  impact on food systems and security 461, 462, 514
  key challenges 88–96
  large-scale CDR 19, 492, 494
  large-scale conversion non-forest to forest land
    98–100
  measuring and monitoring 33
  projected 30–32, 196–197, 197, 461, 462, 642
  regional variation 462
  response options and 18–19, 22–23, 25–26, 188
  socio-economic drivers 79, 93
  soil N_2O emissions 162
  urbanisation 186
land use land cover change (LULCC) 174, 176
land use, land-use change and forestry
  (LULUCF)* 154, 199–200, 200
land-use zoning 706
landscape approaches 505
landscape integration 738
landscape transformations 756
landslides 370, 7.SM.7.1
  see also reduced landslides and natural hazards
large-scale land acquisition (LSLA) 91, 750, 751
Latin America and Caribbean 159, 751
leakage* 702
learning 736–737
legal instruments 105, 701
lifecycle analysis/assessment (LCA)* 490, 721
likelihood* 4n, 92, 363
Limpopo River basin 263, 305
livelihood diversification 285, 471, 567, 579
  adaptation potential 594, 594
  feasibility 623, 6.SM.6.4.1
  impact on desertification 598, 599
  impact on food security 608, 609
  impact on land degradation 603, 603
  impact on NCP 629, 6.SM.6.4.3
  impact on SDGs 632, 6.SM.6.4.3
  mitigation potential 588
  policy instruments 726
  potential across land challenges 610, 617
  sensitivity to climate change impacts 624
livelihoods* 5, 79
  and agricultural productivity 604
  climate stressors 690–691
  coastal 7.SM.7.2
  dryland areas 257–258, 304, 684, 685
  food systems and security 90, 5.SM.5.2
  gender and climate change impacts 5.SM.5.1
  impact of climate variability/extremes 516,
    517, 518
  impact of climate-related land degradation
    377–379, 380, 388

  impact of peatland degradation 397–398
  impact of vegetation degradation 683
  and migration 690–691
  near-term actions 33–34
  oasis populations 301
  Pacific island communities 517
  pastoral 457–458
  risks to 683, 684, 685
  safety nets 380
  synergies and trade-offs 5.SM.5.3
  women's 717–718
    see also livelihood diversification
livestock
  enteric fermentation 160, 189, 477, 484
  feed 276, 444, 456–457, 473, 485, 5.SM.5.3
  genetics 513
  GHG emissions 22, 477–478, 478, 5.SM.5.3
    reduction methods 486, 5.SM.5.3
  higher temperatures and 455, 455, 456
  impact of climate change 276, 454–458, 455,
    5.SM.5.2
  manure and N_2O emissions 11, 162, 376, 477,
    478, 5.SM.5.3
  meat production and dietary consumption 473,
    479–480, 487–490, 604, 5.SM.5.4
  methane (CH_4) emissions 79, 134, 159,
    477–478, 486
  pests and diseases 456, 457, 458
    see also improved livestock management
livestock production systems
  adaptation options 22, 471
  crop-livestock integration 384, 485–486, 504
  current land use 85–86
  dairy systems 483
  diversification 589
  economic mitigation potential 483, 486
  GHG emissions 159, 160, 161, 162, 476,
    477–478, 478
  grazing and fire management 281–283
  impacts of climate change 454–458, 455
  intensive and localised herding 302
  migratory livestock systems 257, 293
  mitigation strategies 22, 483, 484, 485–486,
    5.SM.5.3
  past and ongoing trends 86, 87
  peatlands 606
  productivity 454–458, 455
  smallholder systems 459–460
  sustainable intensification 481–482
  technical mitigation potential 483, 484,
    485–486
  transition away from 511
    see also livestock; pastoralists
Local Adaptation Plan of Action (LAPA) 474
local institutions 287
local knowledge*
  see indigenous and local knowledge
local seeds 468, 567, 579
  adaptation potential 594, 594
  feasibility 623, 6.SM.6.4.1
  impact on desertification 598, 599
  impact on food security 608, 609
  impact on land degradation 603, 603

  impact on NCP 629, 6.SM.6.4.3
  impact on SDGs 632, 6.SM.6.4.3
  mitigation potential 588
  potential across land challenges 610, 617
  sensitivity to climate change impacts 624
  smallholders 640
lock-in* 645
logging practices 351–352, 353, 376
  illegal 716
  NFPP policy in China 396
long-term projections 94
low-carbon economies 713
low latitude regions 67, 372
  aquaculture 697
  crop yields 143, 680–681, 681
  drought trends 146
low-regret measures 83
LSAT (land surface air temperature)* 7, 133,
  140–142, 141, 142
  changes in annual mean 177, 186–187
  and desertification 251
  extremes 174, 176
  and land sink 155
  land-to-climate feedbacks 182–184, 182, 183
  observed mean 6, 7, 44
  regional temperature changes 172–174, 174,
    175, 177–180, 177
  response to land cover change 135, 171,
    172–174, 184–185
    large-scale deforestation/forestation
      177–180, 177
  response to land management change 180
    cropland albedo 181–182, 181
    irrigation 180, 181
  response to urbanisation 186–187
  seasonal changes 172–174, 174, 182–183
  and soil moisture 145, 184
LSLA see large-scale land acquisition
LUC see land-use change
LULCC see land use land cover change
LULUCF see land use, land-use change and forestry

M

maladaptation/maladapative actions* 103, 734
  avoiding coastal maladaptation 392, 402–403
  flood insurance 701
  inequalities and 474
  to desertification 20, 291–292
Malawi 289, 582, 744
malnutrition* 607
  climate change and food quality 463–464
  climate related risks to nutrition 7.SM.7.1
  defining 442
  food prices and 495, 497
  gendered approach to health and nutrition
    5.SM.5.1
  global trends 445–446, 445
  Hindu-Kush Himalayan Region 469
  protein deficiency 463–464
  risk of hunger 685–686
managed forest* 155, 164, 191, 194, 562

managed land* 133–134, *139*, 152, *152*, *154*, 155, *164*, 194

management of supply chains *567*, *578*
adaptation potential 593, *594*
feasibility *622*, 6.SM.6.4.1
impact on desertification 598
impact on food security 607–608, *608*
impact on land degradation 602
impact on NCP *629*, 6.SM.6.4.3
impact on SDGs *632*, 6.SM.6.4.3
institutional barriers 618, 6.SM.6.4.1
mitigation potential 588
policy instruments *726*
potential across land challenges 610, *616*
sensitivity to climate change impacts 624

management of urban sprawl *567*, *579*
adaptation potential 594, *594*
feasibility *623*, 6.SM.6.4.1
impact on desertification 598, *599*
impact on food security 608, *609*
impact on land degradation 603, *603*
impact on NCP *629*, 6.SM.6.4.3
impact on SDGs *632*, 6.SM.6.4.3
mitigation potential 588
policy instruments *726*
potential across land challenges *617*
sensitivity to climate change impacts 624

mangroves 193, 400, 401, *401*
climate related risks 7.SM.7.1
deforestation 402, 487, 590

manure *160*, *161*, 162, 189
GHG emissions 477, 478
green manure cover crops 181–182, 192, 376, 471
management 483, *484*, 5.SM.5.3

marginal abatement cost curves (MACCs) 102

market access 286

market-based instruments 105, 702, *703*, 725

market failure* 508

material substitution *567*, *577*
adaptation potential 593
feasibility *622*, 6.SM.6.4.1
impact on desertification 598
impact on food security 607, *607*
impact on land degradation 602, *602*
impact on NCP *629*, 6.SM.6.4.3
impact on SDGs *632*, 6.SM.6.4.3
mitigation potential 587, *588*
policy instruments *726*
potential across land challenges *615*
sensitivity to climate change impacts 624

mean global annual temperature 177–178

measurement, reporting and verification (MRV)* 29

meat consumption 489, 498–499, 5.SM.5.3, 5.SM.5.4

Mechanized Micro Rainwater Harvesting (MIRWH) 303–304, *303*

'Mediterranean' diet *488*, 489, 497, *498*

Mediterranean region
crop yields 56
desertification 264, 682
drought 15, 45, 146, 682
drylands 264, 277, 278–279

erosion and projected rainfall 302, 362, 370
fire risk *69*, 684
water scarcity 682, 683

megadroughts* 145, *145*

mesoscale convective systems (MCS) 370

mesquite 298, 299

metal toxicity *357*

methane (CH$_4$)* 157–160, *158*, *160*, 366–367, 451, *703*
AFOLU emissions 8–11, *10–11*
agricultural emissions *6–7*, 84, *160*, 195, 196, 476
enteric fermentation *160*, 189, 477
livestock 79, 134, 159, 477–478, 486
rice cultivation 11, 79, 134, 159, 160, 476, 477
anthropogenic emissions 134, *151*
atmospheric lifetime 169, 170, 192
atmospheric trends 157–159
emissions from flooded soil 398
emissions from peatlands 159–160, 476, *477*, 586
emissions from permafrost thawing 137, 184
emissions from wetlands 157–158, 159
fire emissions *149*
globally averaged atmospheric concentration 11, 157
land degradation and 376, 377
land use effects 159–160, 195, 196
mitigation options 486
non-AFOLU emissions *10–11*, *151*
projected emissions 195–196, *196*, 198, 199

Mexico
agriculture 402, 460, 518
drylands 265, 402
invasive plants 299
livelihoods 447, 603
migration 285, 518
salinisation of agricultural land 402

microfinance services (MFS) 475

microinsurance 712

micrometeorological flux measurements *164*

micronutrient deficiency 442, 445–446, 463–464

mid latitudes
bioenergy crops 194
crop yields 7.SM.7.1
forests 368
impact of deforestation 174, 179
precipitation 137
snow cover 184

Middle East 264, 292–293, *293*, 302–304

migration* 16, 285, 690–691
adaptation strategy 285, 380, 388, 683
climate change and 259
desertification and 380–381
dryland areas 259
due to long-term deterioration in habitability 690, 691
environmental risk response 275–276
food security and 516–518, *517*
impact on biodiversity 691
of labour 289
land degradation and 380–381
Pacific region *517*, 518

Millennium Ecosystem Assessment (MA) *625–626*

mineral dust *see* dust

mitigation (of climate change)* 80, 138, *642–644*, 646
adaptation co-benefits and synergies 22, 102–103, 383, 391, *392*
with food security *448*, 492, *493*, 500, 507, 513–514
agriculture 181–182, 281, 282
barriers for *188*, 292, 513
co-benefits 102–103
costs 102
decision making approaches 721–723
demand-side mitigation 487–491, 513–514
dietary changes 472–473
financing mechanisms 711–712
food system 21–22, 101, *443*, 449, 480–491, 513–514
food system interlinkages *441*
future scenarios *13*, 564, *565*
governance 737–738, 5.SM.5.5
green walls/dams programmes 294–296
and greenhouse gas fluxes 157, 701–702
impact on food prices 494–497, *495*
integrated pathways 195–199
knowledge gaps 513
and land tenure *751–752*
market-led response 105
near-term actions 33–34
policies 27–28, *509*, *696*, 701–702, 714
renewable/green energy and transport *735*
risk associated with 686–688, *689*
supply-side mitigation 101, 480–487, *484*, 513–514
synergies with adaptation and food security *448*

mitigation failure 686–687

mitigation measures* 19, 695
barriers to 715–716
gender inclusive approach 43, *717–719*

mitigation options* 18–23, *24–26*, 81, 135–136, 188–199, 583–589
adverse side effects 687–688, *718*
agriculture 480–486, 583–584
agroforestry 383, 485, *485*
in aquaculture 486–487
barriers 79
biochar 100, 398–400
cellular agriculture and cultured meat 487
demand management 101–102, 587
dietary change 487–490, *488*, 497–499
estimating costs 102
FAQs 107, 646, 755
food system 480–491, *509*, 513–514
forest-based *99–100*, 584–585
global potential *24–26*
impact on land degradation 373–375
land management 586–587
land use 79, 97, *98–100*, 100, 373–374, 487–489, *641–644*
land-demanding 34, 79, 97, *98–100*, 196–197
region-specific impacts 107
renewable energy 288, 486
risk management options 102, 588–589
shortening supply chains 491

soil management 100, 382–383, 483, 585–586
supply management 100–101, 588
sustainable intensification *481–482*, 501–502, *502–505*
urban green infrastructure (UGI) *188*, 392–393
urban and per-urban agriculture* 505, 507
    see also REDD+
mitigation pathways* 34, 195–199, *196*, *197*, *198*
archetypes 198–199, *198*
assumptions 195
economics of 102
from the food system 480
GHG mitigation in croplands and soils 483
linking socioeconomic development and land use *30–32*
mitigation potential 189, *190*, 193–195, 201, 583–589, 609–610, *609*, 610, *611–617*
agricultural response options 189, *190*, 483, 583–584, *584*, *611*
combined land response options 636–637
demand management options 191, 195, 587, *588*, *615*
dietary mitigation potential 487–490, *488*, 497–499
economic potential 189, 483, 486, 489
of food systems 449, 487–491, 497–499
forest response options 189–192, 584–585, *585*, *612*
land management options 583–587, *611–615*
    all/other ecosystems 193, 586, *587*, *614*
    specifically for CDR 193–194, 587, *587*, *615*
risk management 588–589, *589*, *617*
soil-based response options 192–193, 483, 585–586, *586*, *613*
supply management options 490–491, 588, *588*, *616*
uncertainties 489–490, 5.SM.5.3
mitigation scenarios* *95*, 634, *635*
bioenergy and BECCS in *580–583*
forest area expansion *99–100*
with large land requirements 19, 97
mitigation strategies 197–199
integrating cultural beliefs and ILK 470
Karapìnar wind erosion area 293
livestock mitigation strategies 5.SM.5.3
novel technologies 485–486
mixed farming 384, 500
models
evaluation and testing *95–96*
futures analysis 80, *93–96*
uncertainties 92–93
    see also specific types of model
Moderate Resolution Imaging Spectroradiometer (MODIS) 363
monitoring, reporting and verification (MRV) 702, *703*
monsoons 176, *176*, 277
moral hazard 686–687
Morocco 263, 284
drought and conflict 518
oases 300, *300*, 301
mountain regions *185*, 294, 301, 452, *469*, 591
multi-criteria decision making 721

multi-level governance* 391, 737–738, 742
multi-level policy instruments 695–696, *696*
multi-species grasslands *504*
multiple policy pathways 510
mycotoxins 462, 463

N

National Adaptation Plans (NAPs) 387, 473
National Adaptation Programs of Action (NAPAs) 473
National Forest Protection Program (NFPP) 396
Nationally Determined Contributions (NDCs)* 81, 199–201, *200*, 387, 562
agricultural adaptation and mitigation 449
CO_2 mitigation potential 199–200, *200*
food systems 449
land-use sectors 704–705
LULUCF contribution 200, *200*
traditional biomass use 375
natural capital accounting 511
natural disasters 590, 592, 699, *743–745*
and food security 518, *743–744*
and migration 276, 518
planning 738, *744*
    see also landslides; tropical cyclones
natural grasslands *85*, 86
natural land *30–32*
natural resource management 287
Nature-Based Solutions 381, 391, 403, *566–567*
NCP (Nature's Contributions to People) *500–501*, *625–626*
impacts of integrated response options 627, *628–629*, 6.SM.6.4.3
urban areas 603
NDVI (Normalised Difference Vegetation Index) 261–262, *261*, 363, 365
as proxy for NPP 363
near-term action 33–34, 554–555, 644–645
costs vs. benefits 102
pathways 198–199
negative emissions* 135–136, 374, 386, 387, 686
    see also carbon dioxide removal (CDR)
negative emissions technologies* 348, 398, 399, 441, 492
neglected and underutilised species (NUS) *469*
Nepal *469*, 474
net negative emissions* 198, *580*
net primary production/productivity *see* NPP
net-zero emissions* *703*
nexus approach 103, 514, 731, *739*
NFPP (National Forest Protection Program) 396
NGOs (non-governmental organisations) 737
Niger Basin 370
Nigeria 452–453
nitrogen addition to soils 134, 159, 162, *163*, 203
emissions 162, 476, *477*
nitrogen cycle 202–203
nitrous oxide (N₂O)* 160–162, *161*, *163*, 366–367, *703*
AFOLU emissions 8–11, *10–11*

agricultural emissions *6*, 79, 84, 134, 195, 476, 478, 5.SM.5.3
anthropogenic emissions 134, *151*
aquaculture emissions 478
atmospheric trends 160–162, *161*
cropland emissions 134, 181–182
data sources 160–161
fertilizer application and 134, 181–182, 476, *477*
fire emissions *149*
food system emissions 478, *479*
grazing land emissions 134
land degradation and 376, *377*
land use effects 162, *163*
livestock system emissions 134, 478, 5.SM.5.3
non-AFOLU emissions *10–11*, *151*
projected emissions 195, 196, *196*, 198, 199
soil emissions 134, 192, 193, 398
sources 160–161
no-till farming 292, 376, 383, 471, 686
non-governmental organisations (NGOs) 737
non-local effects 184–186, *185*
Normalised Difference Vegetation Index *see* NDVI
normative scenarios *94*
North America
agricultural emissions 159
BVOC emissions 169
desertification 265
dryland areas *255*, *257*, 265
dust emissions 167
floods *744*
historical land cover change effects 172–173, *174*
invasive plants 259, 299
monsoon rainfall *176*
rangelands 372
Northern Ireland 370–371
novel technologies 485–486, 513–514
NOx concentrations 169, 170
NPP (net primary production/productivity) 5, 86, *87*, *140*, 363–364
livestock production systems 456
NDVI as proxy for 363
of rangelands 456, *457*
relationship with rainfall 363–364
nutrient depletion 355
nutrition 5.SM.5.1, 7.SM.7.1
nutrition transition* 480, 489, 505

O

oases 300–302, *300*, *302*
obesity/overweight 5, *6–7*, *444*, 445, *445*, 446, 510
global trends *444*, 445, 446
ocean acidification* 627n, 691
off-site feedbacks 269
Okavango Basin 263, 397
Oman *300*, 301
on-the-ground actions
addressing desertification 279–283, *280*
addressing land degradation 381–384
organic agriculture *566–567*

organic carbon (OC) 167–168
out-migration 259, 276, 285, 518
over-consumption (of food) 490
over-exploitation of land resources 79
over-extraction of groundwater 271, 289
overarching frameworks 565–566, *566–567*, 600
    food security 5.SM.5.5
Owena River Basin 263
ozone (O₃)* 135, 451, *573–574*, 691
    BVOC emissions and 169, 170, 192
    pollution management *573–574*, 591, 606

P

Pacific region *255*, *517*, 518
Pakistan 299–300, 452, *744*
palm oil 398, *562*, 588, 624, *741*
Paris Agreement* 81, 199–201
    cooperation mechanisms 704–705
    food systems in NDCs 449
    global initiatives for food security 473, 474
    performance indicators 725
Parthenium hysterophorus (Congress weed) 298
participatory governance* 391, 743, 745–746
particulate matter 591
pastoralists/pastoral systems 8, 257, 384
    adaptive capacity 22, 276, *448*
    agropastoralists 257, 439, 455–456, 5.SM.5.1
    bush encroachment 282–283
    community-based natural resource
        management 287
    drought and 276, 457–458
    fire management 281–282
    gender and climate change impacts 5.SM.5.1
    grazing management 281–283, 604, 623
    impacts of climate change 276, 455–458,
        5.SM.5.1
    impacts of desertification 276
    institutions based on ILK *747*
    land tenure 749, *751*
    risks to 276
    traditional practices 284
    vulnerability 257, 439, 454–456, 458
pasture*
    conversion for energy crops 373
    current global extent *85*
    dryland areas 254–255
    intensification *562*
    intensive *6–7*
    N₂O emissions 134, 162
    past and ongoing trends 86, *87*
    projected land use change *30–32*
    reduction in 196–197, 494
pathways* 16–17, *94*, 737
    1.5°C pathway 22–23, 32*n*, 195–199, 200–201,
        373, *581*, 686
    2°C pathway 22, 195, 197–199, 200–201
    3°C pathway 22, 701
    adaptation pathways 104, 721–722, 743
    archetypes 198–199, *198*
    climate pathways *641–644*
    climate-resilient pathways *678*

defined *93*
future pathways for climate and land use
    641–644
global emissions 168, 170
integrated pathways for climate change
    mitigation 195–199
land pathways *30–32*, *641–644*
least cost pathway 195
linking socioeconomic development, mitigation
    responses and land *30–32*
multiple policy pathways 510
near-term action 198–199
policy pathways 640, 742–743
representative agricultural pathways 460
sustainable development pathways *678*
    see also mitigation pathways; RCPs; SSPs
pathways analysis *94*, *95*
payment for ecosystem services *see* PES
payment for environmental services *see* PES
peatlands* 397–398
    carbon balance 193
    carbon stored in 355, 397, 398
    climate change and 193
    degradation 89, 355, 397–398, 476
    emissions 153, 159–160, 397, 476, *477*
    fire *149*, 153, 397, 398
    management 159–160, *392*, 398
    methane fluxes 159–160
    sustainable management 100
        see also restoration and reduced conversion
            of peatlands
perennial crops 194, 383, *392*, *393–395*, 485
perennial grains 383, *392*, *393–395*
performance indicators 725, 742, 743
peri-urban areas* 135, *188*, 505, 507
permafrost* 184, *357*, *382*
    climate feedbacks 183, 184
    release of CO₂ 134, 184
    release of methane (CH₄) 134, 137, 184
    risks due to degradation *681*, 684, *689*,
        7.SM.7.1
    soil carbon loss 8, 134, 184
    thaw-related ground instability 21, 684
PES (payment for ecosystem/environmental
    services) 105, 287, 706–707, 733
pests and diseases 8, *358*, 457, 458
phenomics-assisted breeding 513
phosphorus cycle 202–203
PICS (Purdue Improved Crop Storage) 472
planetary boundaries transgression 7.SM.7.1
plant diversity 201, *504*
policies 27–29, 80, 640, 695–717, *726*
    acceptability 490, 510, 698, 754
    agricultural 286–287, 473, *482*, 508, 697,
        701–702, *703*, 714
    barriers to implementation 714–717
    for biofuels and bioenergy *738–739*
    for climate extremes 699–701, 714
    climate policies 27, 68, 103, 639, *642*, *678*, 716
    coherence 713–714, 730, 734, *739*
    coordination 28, 390, *739*
    cost and timing of action 723
    cross-sector policies 104

customary norms 106
for decentralised resource management 287
delayed action and 645
development 742
drivers of desertification 259–260
for economic diversification 288–289
economic and financial instruments 105–106,
    474, 698, *710–711*, 711–713
effectiveness 699
enabling environment 508–510, 713–714
expanded policies 508–510, *509*, 5.SM.5.5
for food security 474, 696–698, *696*, 719
food system policies 27–28, 474, 507–510, *509*,
    5.SM.5.5
forest conservation instruments *709–711*
gender-inclusive approaches *718–719*
health related 28, 510
for improvement of IWM 305
incorporating ILK 258
integration 737–738
for investment in R&D 287–288
land tenure 287, *709–711*, 719, *751–752*, 753
legal and regulatory instruments 105
maladaptation 734
market-based instruments 702, *703*
mitigation policies 27–28, *509*, *696*, 701–702, 714
multi-level policy instruments 695–696, *696*
multiple policy pathways 510
policy and planning scenarios *94*
response to climate-related extremes
    699–701, 714
response to desertification 258, 285–288
response to drought *290–291*, 714
response to floods 714, *744*
response to GHG fluxes 701–705, 714
response to land degradation 27, 387–388,
    396–397, *696*, 705–711, *705*
response to land-climate-society interaction
    hazards *688–689*
rights-based instruments 106
risk management 467–468, 700–701
in shared socio-economic pathways *641–644*
social and cultural instruments 106
for social protection 696–697, 698–699, 716
standards and certification 707–709, *708*, *739*
for sustainable development 288–289
for sustainable land management 27–29, 258,
    285–289, *696*, 723
synergies and trade-offs 725, *726*, 730–734
timescales 695
windows of opportunity 694–695
policy goals 566
policy lags 645
policy nudges 722
policy pathways 640, 742–743
pollinators 458–459, 606
pollution 357
    see also air pollution
polycentric governance 104, *578*, 737, 738,
    5.SM.5.5
population distribution 558, *560*
population growth 5, 88, 301, *517*, 692
    in SSPs *13*, *14*, 16

urban areas *186*
pore volume loss 354–355
portfolio analysis 721
portfolio response 695
potential evapotranspiration (PET) 276–277
poverty*
    agroforestry adoption 384
    climate change and 259, 279, 691–692
    climate change and food security 446–447
    climate change-poverty linkages 378–379, *379*
    and diet 442, 446
    in dryland areas 257, 259, 260, 279
    ecosystem services and 730
    impacts of climate-related land degradation
        377–379
    impacts of desertification 259, 272–273, 279
    land degradation-poverty linkages 378–379, *379*
    projected 279
    reduction 289
    social protection policies 698–699
    and traditional biomass use *741–742*
poverty eradication* 19, 20, 27, 33, 640
power dynamics 638
precautionary principle 723
precipitation
    adaptation technologies 475
    anomalies *451*
    and carbon sequestration 271
    change in rainfall patterns 137, 147, 369–370
        food and livelihood security 516, 518
    dryland areas 258, 262, 265, 271
    enhanced by irrigation 180, *181*
    and erosion 361–362, 370–371, 682
    extreme rainfall events 12, 15, 137, 147–148,
        302, 361, 369–371, 7.SM.7.1
    extreme snowfall 147
    impact on food security 5.SM.5.2
    impacts of precipitation extremes on different
        land cover types 147–148
    increased frequency and intensity 7, 137,
        147–148, 360–362, 369–371
    intensity 7, 147, 360–362, 369–371
    land cover changes and 174, 175–176, *176*
    land cover induced changes *176*
    mesoscale convective systems (MCS) 370
    monsoon areas 176, *176*
    and net primary production (NPP) 363–364
    non-local and downwind rainfall effects
        185–186, *185*
    projected 278, 302, 305, 360–362, 370–371
        monsoon rainfall 175–176, *176*
    rainfall patterns and deforestation/forestation
        178, 179
    rainfall and SST anomalies 258–259, 266
    Sahel rainfall 258, 377, 450, *451*
    snow 147, 178, 179, 183–184, 361
    tropical regions 183
    under different SSPs 7.SM.7.1
    urbanisation *187*
    variation and effect on livestock *455*
    and vegetation variability 183, 265–266, 361, *451*
precision agriculture 100, *503*, *566–567*
precursors* *139*, 166, 167–168, 169, 170

prevalence of undernourishment (PoU) 464
primary biological aerosol particles (PBAP)
    see bioaerosols
primary production* 201–202
private net benefits 695
***Prosopis juliflora* (Mesquite)** 298, 299
protected areas 100
public net benefits 695
public policy organisations 105–106
Purdue Improved Crop Storage (PICS) 472

R

radiation
    aerosol interaction 166–167, 168
    aerosol scattering 166, 167–168, 169, 170
    interactions with dust 166–167
radiation absorption, aerosols 166, 167, 168
radiative forcing*
    BVOC emissions and 170
    carbonaceous aerosols and 168
    changes in albedo induced by land
        use change 138
    direct aerosol effect 169, 170, 192
    dryland areas 270
    dust emissions and 167
    indirect aerosol effect 169, 170, 192
    net positive response to historic land cover
        change 172
rain use efficiency (RUE) 265, 266, 363–364
rainfall *see* precipitation
rainwater harvesting (RWH) *280*, 281
range expansion 143, 462
rangelands 494, *562*, *633*
    Badia region, Jordan 302–304
    composition 456
    degradation rate 276
    fire management 281, 282
    floods *744*
    grazing practices 281–282, 376
    impact of climate change 372, 454–456
    integrated watershed management (IWM)
        302–304
    land tenure-climate change interactions *751*
    N_2O emissions 162
    overlapping land challenges *560*, 561, *561*, 633
    productivity trends *366*
    selective grazing 355
    soil compaction reduction 596
    soil erosion 293
    sustainable management 281–282, 284
RAPs (representative agricultural pathways) 460
RCPs (representative concentration pathways)*
    92, 93, 680
    combined with SSPs *13*, *30–32*
    desertification projections 276–277
    forestation impacts 179
    GHG emissions and removals 195–197, *196*
    land cover change scenarios *30–32*, *173*,
        174–176, *175*, *176*
    mitigation scenarios 195–199
    RCP1.9 mitigation pathway 198–199, *198*
    RCP2.6 mitigation pathway 198, *198*

SLCF emissions 169
real options analysis 721
recycling 512
**REDD+ (Reducing Emissions from Deforestation
    and Forest Degradation*)** 385–386, 388, 704,
    709–711
    corruption risk 716
    and land tenure *751*, *752*
reduced deforestation and forest degradation
    385–386, *567*, *571*
    adaptation potential 590, *590*
    combined with bioenergy and BECCS *637*
    feasibility *620*, 6.SM.6.4.1
    impact on desertification 595–596, *596*
    impact on food security 604, *605*
    impact on land degradation 600, *600*
    impact on NCP *628*, 6.SM.6.4.3
    impact on SDGs *631*, 6.SM.6.4.3
    mitigation potential 21, 585, *585*
    policy instruments *726*
    potential across land challenges 610, *612*
    potential deployment area *633*
    sensitivity to climate change impacts 624
reduced food waste (consumer or retailer)
    100–102, *567*, *577*
    adaptation potential 593, *593*
    combined with bioenergy and BECCS *637*
    feasibility *622*, 6.SM.6.4.1
    impact on desertification 598, *598*
    impact on food security 607, *607*
    impact on land degradation 602, *602*
    impact on NCP *629*, 6.SM.6.4.3
    impact on SDGs *632*, 6.SM.6.4.3
    mitigation potential 587, *588*
    policy instruments *726*
    potential across land challenges 609, 610, *615*
    sensitivity to climate change impacts 624
reduced grassland conversion to cropland *567*,
    570, 588–589
    adaptation potential 589
    combined with bioenergy and BECCS *637*
    feasibility *619*, 6.SM.6.4.1
    impact on desertification 595, *595*
    impact on food security 604, *604*
    impact on land degradation 599, *600*
    impact on NCP *628*, 6.SM.6.4.3
    impact on SDGs *631*, 6.SM.6.4.3
    mitigation potential 584, *584*
    policy instruments *726*
    potential across land challenges 610, *611*
    potential deployment area *633*
    sensitivity to climate change impacts 623
reduced landslides and natural hazards *567*,
    573, *726*
    adaptation potential 591, *592*
    feasibility *621*, 6.SM.6.4.1
    impact on desertification 597, *597*
    impact on food security 606, *606*
    impact on land degradation 601, *601*
    impact on NCP *628*, 6.SM.6.4.3
    impact on SDGs *631*, 6.SM.6.4.3
    mitigation potential 586, *587*
    policy instruments *726*

potential across land challenges 610, *614*
sensitivity to climate change impacts 624
reduced pollution including acidification *567,*
573–574, 586
adaptation potential 591, *592*
combined with bioenergy and BECCS *637*
feasibility *621,* 6.SM.6.4.1
impact on desertification 597, *597*
impact on food security 606, *606*
impact on land degradation 600, *601*
impact on NCP *628,* 6.SM.6.4.3
impact on SDGs *631,* 6.SM.6.4.3
mitigation potential 586, *587*
policy instruments *726*
potential across land challenges *614*
sensitivity to climate change impacts 624
reduced post-harvest losses 100–101, *567, 577*
adaptation potential 593, *593*
feasibility *622,* 6.SM.6.4.1
impact on desertification 598, *598*
impact on food security 607, *607*
impact on land degradation 602, *602*
impact on NCP *629,* 6.SM.6.4.3
impact on SDGs *632,* 6.SM.6.4.3
mitigation potential 587, *588*
policy instruments *726*
potential across land challenges 609, 610, *615*
sensitivity to climate change impacts 624
reduced soil compaction *567, 573,* 596
adaptation potential 591, *591*
feasibility *620,* 6.SM.6.4.1
impact on desertification 596, *596*
impact on food security 605, *606*
impact on land degradation *601*
impact on NCP *628,* 6.SM.6.4.3
impact on SDGs *631,* 6.SM.6.4.3
mitigation potential 585, *586*
policy instruments *726*
potential across land challenges *613*
sensitivity to climate change impacts 624
reduced soil erosion 280–281, 292–294, *567, 572*
adaptation potential 591, *591*
feasibility *620,* 6.SM.6.4.1
impact on desertification 596, *596*
impact on food security 605, *606*
impact on land degradation 600, *601*
impact on NCP *628,* 6.SM.6.4.3
impact on SDGs *631,* 6.SM.6.4.3
mitigation potential 585, *586*
policy instruments *726*
potential across land challenges *613*
sensitivity to climate change impacts 624
reduced soil salinisation 283, *567, 573*
adaptation potential 591, *591*
feasibility *620,* 6.SM.6.4.1
impact on desertification 596, *596*
impact on food security 605, *606*
impact on land degradation 600, *601*
impact on NCP *628,* 6.SM.6.4.3
impact on SDGs *631,* 6.SM.6.4.3
mitigation potential 585, *586*
policy instruments *726*

potential across land challenges 609, *613*
sensitivity to climate change impacts 624
Reducing Emissions from Deforestation and
Forest Degradation* *see* REDD+
reforestation*
case studies *392,* 395–397
defined *98*
green walls/dams 294–296, *297*
implications *98–99*
mitigation potential 191
mitigation scenarios *99–100*
reducing/reversing land degradation 374–375,
395–397
risk of land degradation 374
side effects and trade-offs 97, *99, 100*
water balance *98*
reforestation and forest restoration *567, 571*
adaptation potential 590, *590*
adverse side effects 605, *612*
best practice 25
combined with bioenergy and BECCS *637*
feasibility *620,* 6.SM.6.4.1
global potential 25
impact on desertification 596, *596*
impact on food security 605, *605*
impact on land degradation 600, *600*
impact on NCP *628,* 6.SM.6.4.3
impact on SDGs *631,* 6.SM.6.4.3
mitigation potential 585, *585*
policy instruments *726*
potential across land challenges 610, *612*
potential deployment area *633*
sensitivity to climate change impacts
623–624, *624*
regional climate 12, 47, 53, *94,* 107, *139–140,*
166, 182–186, *182, 183,* 205
impacts of aerosols 166–167, 168, 268, 377
impacts of bioenergy deployment 194
impacts of land cover change *171,* 172–174,
174, 175–176, *176,* 179, *572*
impacts of land degradation 377
regional climate models (RCMs) 147, 276–277, 278
regulatory policy instruments 105
remote sensing 33, 52, 56
assessing desertification 253, 261–262,
263–264
assessing land carbon fluxes 153, 155,
163–164, 165
assessing land degradation 348, 363–364,
365, 367
crop production 452
dust emissions 167, 683
early warning systems 475, 513
estimating CO_2 emissions 153, *163*
forest loss 86, 153, 155, 367–368
greening and browning 7, 143–144, 265–266
limitations *7n,* 91, 262
mangroves 402
monitoring risks to food security 475, 513
vegetation indices 143–144, *165,* 265–266, 558
renewable energy 287–288, 289, 709, *735*
see also bioenergy
representative agricultural pathways (RAPs) 460

representative concentration pathways *see* RCPs
residual risks 291
residual trends *see* RESTREND method
resilience* 103, 105–106, 388–389
of agriculture 280, 591
aquaculture 471
building via agroecology 499–500
of crop production systems 591
dryland populations 256–258, 284, 285
ecosystem resilience 265
food systems 468, *469,* 591
investing in 723
mountain communities *469*
pastoral communities 276
socio-ecological 106, 706, 736–737
sustainable food systems *465–466*
to climate-related land degradation 378
to extreme climate events 28, 500, 513
to land degradation 388–389
respiration* 201–202, 204
response options 18–23, *24–26,* 96–103, *97*
addressing desertification 279–292, 595–599
addressing food security 603–608, *609*
addressing land degradation 381–388, 599–603
barriers to 556, 756, 6.SM.6.4.1
climate consequences 135–136, 188–201
community approaches 640
competition for land 18–19, *24–26,* 97
demand management 101–102, *190,* 191, 195
enabling 103–106, 554, 558, 633
food system 21–22, 449, 492, *493*
increased demand for land conversion 19
and land-use change 18–19, 22–23, *25–26,* 188
with large land-area need 19, 97
locally appropriate 384–385
mitigation potential 189–193, *190,* 194–195
opportunities 756
Paris Agreement and 199–201
reduced demand for land conversion 18–19
regional variation in mitigation benefits 194
risks arising from 686–688
role of ILK *746–747*
socio-economic co-benefits 33–34
and sustainable development 18, 627, 630,
631–632, 6.SM.6.4.3
upscaling 33, 554–555, *738,* 756
urban *188*
see also integrated response options; land
management response options; policies;
risk management response options;
value chain management
restoration and reduced conversion of coastal
wetlands 400–401, *401, 567, 574*
adaptation potential 592, *592*
feasibility *621,* 6.SM.6.4.1
impact on desertification 597
impact on food security 606, *606*
impact on land degradation 601, *601*
impact on NCP *628,* 6.SM.6.4.3
impact on SDGs *631,* 6.SM.6.4.3
mitigation potential 193, 586, *587*
policy instruments *726*
potential across land challenges 610, *614*

potential deployment area *633*

sensitivity to climate change impacts 624

restoration and reduced conversion of peatlands 159–160, 398, *557, 567, 574*

adaptation potential 592, *592*

feasibility *621*, 6.SM.6.4.1

impact on desertification 597

impact on food security 606, *606*

impact on land degradation 601, *601*

impact on NCP *628*, 6.SM.6.4.3

impact on SDGs *631*, 6.SM.6.4.3

mitigation potential 193, 586, *587*

policy instruments *726*

potential across land challenges 610, *614*

potential deployment area *633*

sensitivity to climate change impacts 624

RESTREND method 262, 265, 363–364

reversibility

impact of delayed action 34, 645

integrated response options 618, *619–623*, 6.SM.6.4.1

of trade-offs 730

Revised Universal Soil Loss Equation (RUSLE) 364

rewetting* 193, 377

rice cultivation 384, 452, 476, *477*, 584

methane (CH$_4$) emissions 159, *160*, 476, *477*

reducing emissions 189

rights based policy instruments 106

Rio Conventions 387

risk* *14*, 15–17, 673–692, *678*, 7.SM.7.1–3

adaptive governance of 742

categories 680

climate-related 696

current levels 15

definition 91, 679, 680

drought risk mitigation *290–291*

fire *148–150*

from adverse side-effects 687–688

from changes in land processes *14*

from climate change responses 374, 686–688

from disrupted ecosystems and species 691

from land-climate-society hazards 688, *688–689, 732*

future risk of desertification 277–278

of hunger 685–686

increasing with warming *14–15*, 15

of land degradation from bioenergy and CDR 374

of land degradation under climate change 363

layering approaches 712

of mitigation failure 686–687

monitoring 475

of natural disasters 699

near-term action and 34

policy responses 695–717

reduction initiatives 471

related to bioenergy crop deployment 19, *581–582*, 687–688, *687*, 7.SM.7.1, 7.SM.7.3

related to drylands water scarcity 16, 684, *685*, 7.SM.7.1, 7.SM.7.3

related to food security 16–17, 685–686, *685*, 7.SM.7.1, 7.SM.7.3

related to land degradation 684–685, *685*, 7.SM.7.1, 7.SM.7.3

SSPs and 16–17, 7.SM.7.1

SSPs and level of climate related risk 684–686, *685*

to biodiversity and ecosystem services 691

to communities and infrastructure 691–692

to crop yields 362–363

to humans 691

to land systems from climate change 680–686, *681*

to where and how people live 690–692

transitions *14*, 680–688, 7.SM.7.1–3

see also decision making

risk assessment* 369, 471, 680–688

risk management* 67–74, *97*, 102, *678*, 695

adaptation options 467–468

comprehensive 712, 721, *724*

food systems 467–468

land management in terms of 33

policy instruments 105–106, 699–701

proactive *744*

risk management response options 19, *97*, 102, *579–580*

adaptation potential 594, *594*

delayed action 645

feasibility 618, *623*, 6.SM.6.4.1

global potential *24*

impact on desertification 598–599, *599*

impact on food security 608, *609*

impact on land degradation 603, *603*

impact on NCP *629*, 6.SM.6.4.3

impact on SDGs *632*, 6.SM.6.4.3

mitigation potential 588–589, *589*

policy instruments *726*

potential across land challenges 610, *617*

sensitivity to climate change impacts 624

risk perception* 515

risk sharing instruments 567, 580, 699, 712

adaptation potential 594, *594*

feasibility *623*, 6.SM.6.4.1

impact on desertification 599, *599*

impact on food security 608, *609*

impact on land degradation 603, *603*

impact on NCP *629*, 6.SM.6.4.3

impact on SDGs *632*, 6.SM.6.4.3

mitigation potential 588–589, *589*

policy instruments *726*

potential across land challenges *617*

sensitivity to climate change impacts 624

river basin degradation 263, 264

rivers 697, 731

climate related risks 7.SM.7.1

disruption of flow regimes *688*

for transport *735*

riverscapes and riparian fringes *752*

robust decision making 721

RUE *see* rain use efficiency (RUE)

ruminants 473, 479–480, 487–490, 604

GHG emissions 473, 477, *478*, 479–480, 5.SM.5.3

methane (CH$_4$) emissions 159

transition of land use for 511

see also livestock

rural areas

autonomous adaptation 466

benefits of land-based response options 756

climate change and food security 446–447

drought 518

sustainability of rural communities 502

traditional biomass use *740*

vulnerability 449

RUSLE (Revised Universal Soil Loss Equation) 364

Russia 264

S

safety nets 459, 510, 697, 698–699, *744*

access to UGI 392

sagebrush ecosystems 299

Sahara 288, 296, *297*

Sahel region 263, 268, 276, 452

Great Green Wall initiative 296, *297*

rainfall 180, 258, 377, 450, *451*

saline soils* 258, 283

salinisation 355, *357*

of oasis areas 301

of river basins 263, 264

saltwater intrusion *392*, 401–402

saltwater lakes 402

sand aerosols 268–269, *269*

sand dunes 265, 277

impact on infrastructure 275

preventing movement 293, *293*

stabilisation 283

sand storms 268, 283

satellite observations *see* remote sensing

saturation of integrated response options 618, *619–623*

Saudi Arabia 264, 275, 300, *300*

savannah 86, 265, 270, *562*

burning 133, 162, 168

grazing and fire management 281–282, 700

woody encroachment 270, 282–283, 355, 456

SCC *see* social cost of carbon

scenario analysis 722

scenario storyline* *93, 94*

scenarios* 34, 88, *93–96*

alternative diets in 487–489, *488*

baseline scenario 195–196, 197, 564, *565*, 684

exploratory scenario analysis *93–94*

futures analysis 80, 564

integrated response options in 634, *635*

land challenges in 564, *565*

land cover changes in 174–176

land-use change *93–96*

limitations *95–96*

methods and applications *93–95*

regional scale *94*

uncertainties from unknown futures 92–93

see also mitigation scenarios

SDG *see* Sustainable Development Goals

sea ice* extent *174*, 179

sea level rise* 8, 372

adaptive governance and 743

climate related risks 7.SM.7.1, 7.SM.7.2
climate-change-induced 372
and coastal flooding 592
and migration *517*
saltwater intrusion 401–402
socio-economic effects 372
sea surface temperature (SST)* 142, *174*
anomalies and rainfall 258–259, 266
changes in 186
climate change and 174, *174*, 258–259
sea walls 402–403
seasonal variations *139, 140*
secondary organic aerosols (SOA) 166, 167, 169, 170
seed sovereignty *see* local seeds
self-regulation 106
semi-arid ecosystems 271, 595
semi-natural forests *560, 561, 561, 562, 633*
Sendai Framework for Disaster Risk Reduction* *744*
sequestration* *see* carbon sequestration
shared socio-economic pathways *see* SSPs
shock scenarios *94*
short-lived climate forcers (SLCF)* *99,* 166–170
short-lived climate pollutants (SLCP)* 451, 586, *740*
silicate minerals 374
silvopasture systems *504*
sink* 84
of atmospheric CH₄ 159
atmospheric hydroxyl radical (OH) sink 158–159
capacity 624
forest carbon sink 21, 156, 180, 386
increasing 157, *165*
land sink 8–9, 84, *153, 154,* 155–157
non-anthropogenic land sink 157
ocean sink *153,* 157
see also carbon sink
Siwa oasis 301, *302*
SLCF *see* short-lived climate forcers
SLCP *see* short-lived climate pollutants
small hydropower projects (SHPs) *735*
small islands
coastal degradation 400, 403
food security *517,* 518
Small Island Developing States (SIDS) 193, 400, 403, 473
smallholders 593, 608, 697
adaptive capacity 22
agroecology 499–500
climate change impacts 459–460
conservation agriculture 500
land tenure 749, 750, *751*
livelihood diversification 594
poverty eradication 640
risk management 594
snow accumulation 361
snow-albedo feedback 178, 179, 183–184
snow melt 361
social barriers to adaptation *715*
social capital 284, *390*
social cost of carbon (SCC)* 102, *694,* 702

social-ecological systems (SES) 104, 556–558, *557*
resilience 106
social learning 639
social learning* 639, 745, 749
social policy instruments 106
social protection policies 696–697, 698–699, 716
societal transformation* 512
socio-cultural barriers 618, *619–623,* 6.SM.6.4.1
socio-ecological resilience 706
socio-economic drivers
desertification 259–260, 684
land degradation 359–360, 684–685
land-use change 79, *93*
socio-economic pathways *see* SSPs
socio-economic responses 283–284, 285
socio-economic systems 33–34, 272–276, 279
sodic soils* 258
sodification *357*
soil
biological soil crusts 356, *358*
carbon 381, 382–383, 398, 584
carbon management 189, 382–383, 584
carbon uptake 278
chemical degradation processes 355, *357*
CO₂ release from deep soil 203
compaction/hardening *357*
degradation 89, 350, 393–395, 456
increased water scarcity 274
and urban sprawl 603
direct temperature effects 362
dryland areas 271
enhancing carbon storage 100
GHG emissions 476
GHG mitigation 483
global extent of chemical degradation 597, 600
impact of flooding 148
indicators of land degradation 364
integrated crop-soil-water management 280–281
land management response options *97*
methane (CH₄) uptake 159
microbial and mesofaunal composition changes 355–356, *358*
microbial processes 201, 202–203, 204
N₂O emissions 11, 162
nutrient depletion *357*
nutrient dynamics 202–203, 204
pore volume loss 354–355
processes 201–204
productivity 608
quality 148
regional variation 204
response to warming 203–204
rewetting 162
sustainable land management 100
temperature 181
see also reduced soil compaction; reduced soil salinisation
soil-based response options 100, 192–193, 381, 382–383, *572–573,* 610
adaptation potential 591, *591, 613*
feasibility *620,* 6.SM.6.4.1
impact on desertification 596, *596, 613*

impact on food security 605, *606, 613*
impact on land degradation 600–601, *601, 613*
impact on NCP *628,* 6.SM.6.4.3
impact on SDGs *631,* 6.SM.6.4.3
mitigation potential 192–193, 585–586, *586, 613*
policy instruments *726*
potential across land challenges *613*
sensitivity to climate change impacts 624
soil carbon sequestration (SCS)* 100, 483, *484,* 583–584, 605, 624
agroforestry 485, *485*
impact of desertification 278
measures to combat desertification 20
soil conservation* 287, 292–294, 382–383
soil erosion* 5, 354, *356,* 367
adaptation limits 21
case studies 292–294
caused by human activity 293
in Central Asia 293–294
costs 682
cropping methods and 280–281
dryland areas 258, 259
erosivity of rainfall 370–371
field-based data 366
hotspots of desertification 292–294, 295
impact of climate change 258, 292–294, 360, 361–362, 363, 393
impact on GHG 376
irrigation and excessive water use 293, 294
land sink and 156
management and mitigation potential 192
observed erosion rates 361–362, *361*
projected 278
rainfall intensity and 258, 361
reduction methods 292–294
risk due to precipitation changes 370–371
risks from climate change *681,* 682, 7.SM.7.1
RUSLE model 366
vegetation cover and 362
see also reduced soil erosion; wind erosion
soil management 280–281, 382–383, 398–400, 500–501
adaptation options 470–471
precision agriculture *503*
short-term static abatement costs 102
see also reduced soil compaction; reduced soil salinisation
soil moisture* 146, 303, *303,* 362
climate feedbacks 184
soil carbon and 204
soil organic carbon (SOC)* 201, 281
climate change and 134, 258
conservation agriculture 500–501, 584
emissions 201
land degradation 351, *352,* 366
land degradation response measures 382–383, 393–395, 397, 398
loss 381, 584
organic matter inputs by plants 204
peatlands 397, 398
perennial grains and 383, *392,* 393–395
permafrost storage 134, 184
response to warming 203–204

sequestration 192, 199
urbanisation *187*
vertical distribution 203
   see also increased soil organic carbon content
soil organic matter (SOM)* 393
  decline *357*
  indicators of land degradation/improvement 364
  pool depletion 355
  soil microbial processes and 203, 204
soil salinity*
  climate change and 258
  combating 283
  sea water intrusion 402
    see also reduced soil salinisation
soil and water conservation (SWC) 561–562
  Ethiopian Tigray region croplands *561–562*
solar power 275, 377, *735*
solar radiation 177
Somalia 518
South America
  biome shifts 371
  Cerrados pasture intensification *562*
  crop production 452
  desertification 265
  dryland areas *255*
  dryland population *257*
  monsoon rainfall *176*
  soil erosion and no-till farming 292
  sustainable agricultural intensification *481–482*
South Asia 257, 264, *751*
South Korea 395–396
Southeast Asia 168, *185*, 397–398, 472
species
  compositional shifts *358*
  extinction rates 564
  impact of climate change 8, 7.SM.7.1
  loss *358*
  range expansion 143
    see also invasive species
SR-LULUCF
  see IPCC Special Report on Land Use, Land-Use
    Change and Forestry
SR15
  see IPCC Special Report on Global Warming
    of 1.5°C
SRCCL
  see IPCC Special Report on Climate Change
    and Land
SRES *see* IPCC Special Report on Emission Scenarios
SREX *see* IPCC Special Report on Extreme Events
SSPs (shared socio-economic pathways)* *13*,
  92–93, 195, *196*, 278, *641–644*, 680
  land challenges in 564, *565*
  land use, prices and risk of hunger 460–462, *461*
  land use/cover change *30–32*
  mitigation and climate impacts on food security
    495–496
  mitigation responses and land *30–32*
  risks related to bioenergy crop deployment 19,
    687–688, *687*, 7.SM.7.1, 7.SM.7.3
  risks related to drylands water scarcity 16, 684,
    685, 7.SM.7.1, 7.SM.7.3

risks related to food security 16–17, 685–686,
  685, 7.SM.7.1, 7.SM.7.3
risks related to land degradation 684–685, *685*,
  7.SM.7.1, 7.SM.7.3
stakeholder engagement 28, 62, 639, 640, 723
stakeholder involvement 293, 638–639, *709–710*
  decision making 29, 96, 721, 723, 725
  knowledge sharing 288
  participatory planning *43*, *94*, 640, 720
  valuing ecosystem services 350, 725
standards 707–709, *708*
storage, food system 472
stranded assets* *689*
structural transformations 289
stylised scenarios *93–94*
subnational governance 737, 5.SM.5.5
subsidence 354–355, *357*, 372, 684
subsidies 697, 701, *741*
Sudan 263, 275, 380
Sundarbans mangroves 400
supply chains *493*, 513
  shortening 195, 491
  sustainability 707–709, *708*
    see also management of supply chains
supply management response options 100–102,
  578–579
  adaptation potential *493*, 593, *594*
  combined with bioenergy and BECCS *637*
  delayed action 645
  feasibility 618, *622*, 6.SM.6.4.1
  impact on desertification 598
  impact on food security 607–608, *608*
  impact on land degradation 602, *602*
  impact on NCP *629*, 6.SM.6.4.3
  impact on SDGs *632*, 6.SM.6.4.3
  mitigation potential *493*, 588, *588*
  policy instruments *726*
  potential across land challenges 610, *616*
  sensitivity to climate change impacts 624
supply shocks 514–515, *515*
supply-side adaptation 470–472
supply-side issues of land degradation 379–380
supply-side mitigation 480–487
surface roughness *139*, 177, 377
surface runoff 371, 391, 392
sustainability*
  community-owned solutions 104
  education 512
  food supply 79–80
  gender agency 104–105
  response options to key challenges 96
  standards and certification 707–709, *708*
sustainable adaptation *743–745*
sustainable agriculture 381–384, *465–466*, 507
sustainable certification programmes 602,
  707–709
  palm oil 398, *562*, 624, *708*
sustainable development (SD)*
  co-benefits of combating desertification 19–20
  consequences of climate-land change 690–695
  contribution of response options 18, 627, 630,
    631–632, 6.SM.6.4.3
  gender-inclusive approaches *717–719*

governance for 737–738, 742–743, 754
knowledge gaps 755
land tenure-climate change interactions
  751–752
near-term action 34
risk management and decision making 673–800
Sustainable Development Goals (SDGs)* 21, 79,
  388, *506*, 507
  and bioenergy deployment 7.SM.7.1
  climate change mitigation 97
  desertification and 272
  global scale 731, *732*
  governance 5.SM.5.5
  impacts of integrated response options 627,
    630, *631–632*, 6.SM.6.4.3
  Land Degradation Neutrality (LDN) 705–706, *705*
  local and regional scale 730–731, *732*
  synergies and trade-offs *506*, 507, 730–731, *732*
  and traditional biomass use *741*
sustainable development pathways *678*
sustainable diets 497–499
sustainable farming *see* sustainable agriculture
sustainable food systems *465–466*
sustainable forest management (SFM)* 21, 100,
  351–353, 369, 385–387, *566–567*, 571, 585
  CO$_2$ removal (CDR) technologies 386–387
  defined 21*n*, 351
  REDD+ 385–386
sustainable intensification (of agriculture)*
  481–482, 501–502, *502–505*, *566–567*, 583, 589
sustainable land management (SLM)* 21, 100,
  138, 306, 351–353, 381, 404, *625–626*
  adaptive governance 743
  addressing desertification 255–256, 279–283
  addressing land degradation 381–384
  adoption of 283–284, 285, 286, 387–388, *390*
  barriers to implementation 28, 389–391
  best practice 391, 723
  cross-level integration 738
  decision making 723
  defined 21*n*, 351
  economic assessment *692–694*
  farming systems 381, *465*
  financing mechanisms 712
  gender-inclusive approach 80, 104–105
  indigenous and local knowledge 381, *747–748*
  investment 285, 387, 390, 391
  migration and 259
  near-term actions 33–34
  policies 27–29, 258, 285–289, *696*, 723
  resilience considerations 388–389
  soil erosion reduction 294
  women and *717–719*
sustainable soil management 500–501
sustainable sourcing *567*, *578*
  adaptation potential 593, *594*
  feasibility *622*, 6.SM.6.4.1
  impact on desertification 598
  impact on food security 607, *608*
  impact on land degradation 602, *602*
  impact on NCP *629*, 6.SM.6.4.3
  impact on SDGs *632*, 6.SM.6.4.3
  mitigation potential 588, *588*

Index

policy instruments *726*
potential across land challenges 610, *616*
sensitivity to climate change impacts 624
Sweden *582*
synergies 33, 388, *506*, 507
adaptation, mitigation and food security *448*
agricultural sector 733
between food security and bioenergy 733
between integrated response options and
SDGs 630
between LDN and NDCs 388
climate-smart agriculture 500
ecosystem services (ES) 730, 731, *731*, 735
empowering women 29, *448*
food system 490–491, 492, *493*, 513–514
forestry sector 733
integrated response options 627, 636–637
mitigation and adaptation 22, *448*, 499–502,
507, 756
mitigation strategies 5.SM.5.3
policy choices 725
policy interactions 733
Sustainable Development Goals (SDGs)
730–731, *732*
synthetic aperture radar (SAR) 364
Syria 264, 275, 518

T

taxation 105, 698, 701, *727*
carbon tax 68, 498, 510, 702, 714, 753
**TCRE (transient climate response to cumulative
CO₂ emissions)** 243, *243–247*
technical mitigation potential
crop production 21, 483
cropland soil carbon sequestration 483
dietary change 21, 487–489, *488*
food system 21
livestock sector 21, 483, *484*, 485–486
uncertainties 5.SM.5.3
**technical potential of integrated response
options** *609*
technological barriers 618, *619–623*, *715*,
6.SM.6.4.1
technology, adopting 33, 389–391
technology transfer* 33, 698, 704
teleconnections* 184–186, 373, 379
telecoupling 88, 514
temperate forest *149*, 179–180, 192, 596
temperate regions 12, 45, *150*, *504*
biochar 605
peatlands 397
projected impacts 456
seasonal climate 173, 174
soil erosion 362
water use efficiency 144, *165*
temperature
albedo-induced surface temperature changes 172
biogeochemical cooling 179
biogeochemical warming 135, 176–177, 179
biophysical cooling 172–173, 174–175, 177,
178, 179

biophysical warming 172, 174, 175, 177–178,
179, 197
changes due to deforestation/forestation
176–180
diurnal 178, 179, 180, *186–187*
effect on soils 203–204, 362
extremes 145–147, 174, 176, *186–187*
increase and BVOC emissions 170
increase and crop yields 680–681
increase and crops 143, 300–301, 453, 454
increase and desertification 276–278
increase and food security 5.SM.5.2
increase and livestock production 454, *455*
increase and soil erosion 682
interannual growing-season variability 467
irrigation effect on 180, *181*
land cover change impacts 174, 176
and livestock 455, *455*, 456
local change due to bioenergy crops 194
local surface temperature 179–180, 181–182
mean global annual surface air temperature
171–173, *172*, *173*, *175*
mean surface air temperature *174*
plant and ecosystem production thermal
response 201–202
projected 276–278, 362, 363, 373
regional changes 84–85
sea surface temperature (SST)* 142, 174, *174*,
186, 258–259, 266
since pre-industrial period *6, 7*
urban areas *186–187*
see also global warming; GMST; GSAT; LSAT
temperature overshoot* 675, 686, 701
terracing 383
tier* methods 160, *164*
timber yield 352
Time Series Segmentation-RESTREND
see TSS-RESTREND
tipping points* 389, 645, 679, 743
desertification 265
peatlands 62
permafrost collapse 684
sagebrush ecosystems 299
socio-ecological 755
Tracking Adaptation 474
trade 80, 86, *87*, 101, 472, 690
trade policies 508
trade-offs 80, 281, *506*, 507, *625*
acceptable levels 403
agricultural sector 733
barriers to land-based mitigation 756
between adaptation and mitigation 103
between ecosystem services 730
between integrated response options
and SDGs 630
CDR and bioenergy 492, 494, *739*
conservation agriculture 501
conventional and cultured meat 487
food security 492, 494
food system 513–514
forest management 191, 352, 368, 733
green energy with biodiversity and ES *735*
integrated response options 627

land use intensity and long-term sustainability *504*
land use/management decisions 350, 353
land-based mitigation 733
mitigation strategies 5.SM.5.3
policy choices 28–29, 725
policy interactions 733
renewable energy *735*
socio-economic 97
in Sustainable Development Goals (SDGs)
730–731, *732*
traditional biomass 20, 288, 375, 709, *740–742*
transformation*
in governance 737, 743
societal 512
transformational adaptation* 360, 466–467,
466, 467, 717
transformational change 21, 385, 390, 449, 743
transformative change* *465–466*, 749
transitions* 390–391, 511
risk *14*, 680–688, 7.SM.7.1–3
transnational governance 737
transport 471–472
GHG emissions 478–479
infrastructure 275, 379–380, 472
waterways *735*
tree mortality 202, 371–372, 7.SM.7.1
treeline migration 172, 182
tropical cyclones* 372, *392*, 400–401, 518
climate related risks 7.SM.7.1
early warning systems (EWS) 594
impact on food security 5.SM.5.2
tropical regions 15
BVOC emissions 169
crop yields 680–681, *681*, 7.SM.7.1
deforestation/forestation *149*, 177–178, *177*,
179–180, *185*, 191
forest restoration and resilience *562*
land cover change and climate feedbacks 12,
175, 177–178, *177*, 179–180
peatlands 397, 398
vegetation greening/browning 183
troposphere*, ozone in 170
TSS-RESTREND 266, *267*
Tunisia 300, *300*, 301
Turkey 264, 292–293, *293*
Tuvalu 403, *517*

U

UGI *see* urban green infrastructure
UHI *see* urban heat island
uncertainty* 91
adaptive governance 742–743
adaptive management and 724
assessing desertification 255
assessing risks of land degradation 369
from bioenergy and CDR 374
contributing factors 89
costs of mitigation 102
dealing with 91–93, 96
in decision making 96
decision making under *693*, 719, 721–723, *722*

demand-side mitigation potential 489–490, 5.SM.5.3

drivers of land use 88

Earth system models (ESMs) 201–202

futures analysis *93, 94*

knowledge gaps and 305–306, 403–404

model parameters *94*

in models 80, 92–93

in observations 91–92

plant and soil processes 201, 202

projecting land-climate interactions 201–203

unknown futures 92–93

undernourishment 5, 442, 605

risk of under different SSPs *495*, 7.SM.7.1

underweight, global trends *444, 445*, 446

UNEP Emissions Gap Report 201

United Nations Convention to Combat Desertification (UNCCD)*

definition of land degradation 350

report *83*

United Nations Framework Convention on Climate Change (UNFCCC)* 473, 701, 704

United States of America 265

biofuel modelling studies 194

dust emissions 167

invasive plants 299

mesoscale convective systems (MCS) 370

United States Environmental Protection Agency (USEPA) 160

unprecedented climatic conditions 15

unused land *85*, 86

uptake* 79, 278, 386

CH_4 in upland soils 159

enhanced by CO_2 fertilisation *165*

global terrestrial carbon uptake 586, 596

urban agriculture*

see urban and peri-urban agriculture*

urban areas 17, 86, *186–188, 563*

adaptation 706–707

aerosols 166, 168

carbonaceous aerosols 168

climate change and food security 447

diets 505

food forests *578*

food security *188*, 449, 505, 507, 607

increased heat *186–187*, 505, *563*

infrastructure *188*, 391–393, *392, 563*

land tenure-climate change interactions *752*

maintaining forest cover 590

mitigation strategies 505, 706

pollution *187–188*, 603, 691

soil degradation 603

surface runoff 391, 392

traditional biomass use *740*

urban planning *186*

urban sprawl 505, 507

urbanisation 86, 88, 391

and climate change *186–188*, 285

economic transformations 289

vulnerability 706

zoning 706

see also enhanced urban food systems; management of urban sprawl

urban green infrastructure (UGI)* *188*, 391–393, *392, 563*

urban heat island (UHI) *186–187*, 505, *563*

urban and peri-urban agriculture* 505, 507, *563*, 608

V

Vallerani system 303, *303*

value chain management 19, *97*, 100–102, *566, 567, 577–579*

adaptation effects 593, *594*

barriers 618, 6.SM.6.4.1

delayed action 645

demand-side

see demand management response options

feasibility 618, *622*, 6.SM.6.4.1

global potential *24*

impact on desertification 598

impact on food security 607–608

impact on land degradation 602

impact on NCP *629*, 6.SM.6.4.3

impact on SDGs *632*, 6.SM.6.4.3

mitigation potential 587–588

policy instruments *726*

potential across land challenges 610, *615–616*

sensitivity to climate change impacts 624

supply-side

see supply management response options

value to society *692*

Vanuatu *517*

vegan diet 487, *488*

vegetation

acclimation 201–202

assessing changes in 261–262

bioaerosol emissions 168

biophysical climate interactions *139*

clearing processes 355

cover change 7.SM.7.1

albedo impacts 377

changing rainfall regimes 361

climate feedbacks 270, 377

climate induced 258, 259, 369–370, 371–372

drivers of 265–268, *267*

due to land use 84

degradation *681*, 683, 7.SM.7.1

drought response 202

drylands 281–282

GHG flux 201, 270

impacts on monsoon rains 183

increasing CO_2 and 463

inherent interannual variability 363

invasive plants 259, 270

Karapìnar wind erosion area 293, *293*

photosynthetic activity 143–144

plant biodiversity 271–272

protection against erosion 362

re-vegetation of saline land 283

restoration 162, 293

risks to in dryland areas 684, *685*

Sahel vegetation dynamics 258

SOM inputs 204

spatial mosaic 371

stressors 363

thermal response of plant respiration 201–202

trends 365

variability with rainfall 265–266, 361, *451*

vegetation browning* 7, 133, 143, 144, 183

vegetation greening* 7, 143–144, 261–262, 363, 365

Africa 263

Asia 263

Australia 264

boreal regions 172

climate feedbacks 172

global trends 133

tropical regions 183

vegetation optical depth (VOD) 262, 265, 364

villages

climate smart villages (CSV) *563*

deployment of response options 633, *633*

overlapping land challenges *560*, 561, *561, 563*

visions *94, 95, 96*

VOD *see* vegetation optical depth (VOD)

Volta River basin 263

voluntary agreements 106

voluntary carbon market 385, 386, 388

vulnerability* 16, 103, 557, 688, *688–689, 745*

access to land-based resources 104

in agriculture 5.SM.5.2

livestock production systems 454–455

pastoral systems 439, 454–456, 458

production/crop yields 464

smallholder farmers 459–460

in aquaculture and fisheries 459, 5.SM.5.2

and delayed action 644–645

differential *718*

dryland populations 16, 256–258, 272, 277–278, 452

of ecosystems to irreversibility 645

gender and 446, *447–448*, 717

global food system 682

and land tenure *751–752*

of oases 301–302

prediction and assessment 288

reduction via microfinance 475

rural areas 449

of soils 591

to climate-related extremes 138

to climate-related land degradation 16, 378

to desertification 16, 277–278

to drought 464

to flood *744*

to price volatility 593

urban areas 706

vulnerable groups adaptive capacity 104, 518, 691

W

waste burning *160*

water

for bioenergy 7.SM.7.1

biomass water content 262

climate change impacts on resources 205
conservation 383
contaminated 462–463
demand *98*, 301, 304, 731
drinking water 301, 402
for food production 450
forest area increase and water balance *98*
future scenarios of water stress 564, *565*
groundwater irrigation 734
harvesting systems 284, 303, 304
impacts of desertification on water use 274
integrated crop-soil-water management 280–281
land-based water cycle changes 137
oasis agriculture and 301
transport through soil-plant-atmosphere continuum 202
water use trade-offs 733–734
water erosion 295, *356*
climate related risks 7.SM.7.1
direct measurements of 361–362
precipitation changes and 370–371
water management *291*, 294, 471
see also integrated water management
water policy 717
water quality 558, *559*, *560*
water rights 749, 750
water scarcity *681*, 682–683, 684, *685*, 692
adaptation measures 686
climate related risks 7.SM.7.1
impacts of desertification 274
under different SSPs 16, 7.SM.7.1

water security 731
water stress 564, *565*, 595, 7.SM.7.1
water use efficiency (WUE) *165*, 595
water vapour 185–186, 202
water-soluble organic compounds (WSOC) 168
waterlogging 355, *358*
well-being* 5, 79, 81, *625*
consequences of climate-land change 690–695
negative impacts of fire 683
negative impacts of forest dieback 683
Western North Pacific, monsoon rainfall *176*
wetlands* 86, *358*, *633*
carbon balance 193
coastal wetland management 400–401
drainage for agriculture 89
dryland areas *256*
GHG release 376–377
loss of 354
methane (CH_4) emissions 157–158, 159
mitigation potential of protection and restoration 193
restoration 401, *401*
and sea level rise 372
wild forests *633*
overlapping land challenges *560*, 561, *561*
wildfires 133, *148–150*, 259, 299, 372, *685*, 7.SM.7.1
damage 606, *681*, 683–684
see also fire; fire management
wind energy 275, *735*
wind erosion 271, 294, *356*, 362, 595, 600
Central Asian drylands 293
climate change and 277
climate related risks 7.SM.7.1
Green Dam regions, Algeria 295
prevention and mitigation in Karapìnar, Turkey 292–293, *293*

windows of opportunity 694–695, 756
women *447–448*, *717–719*
access to resources 104–105
adaptive capacity 353, *448*, 716, *717*
empowering 29, 70, 286, *448*, *488*, 639, *718–719*, *719*
inequality in land use and management 353
role in response option implementation 639
traditional knowledge *747*
vulnerability 456, *717*
woody biomass *582*, 637
woody encroachment 270, 355, *358*, 456
clearance *280*, 282–283
wildlife biodiversity 272
see also invasive species/encroachment management
World Atlas of Desertification 256, 365

Z

zero tillage *see* no-till farming
Zimbabwe, smallholder farmers 459
zoning ordinances 706